W9-AEN-993

WITHDRAWN

Library
Brevard Junior College
Cocoa, Florida

The Focal

Encyclopedia of Film & Television

Techniques

The Focal **Encyclopedia of Film & Television** Techniques

HASTINGS HOUSE, PUBLISHERS
New York

© 1969 by FOCAL PRESS LIMITED

First American Edition 1969

Hastings House, Publishers, Inc.,
New York, N.Y. 10016.

SBN 8038-2268-5

All Rights Reserved. No part of this publication may be reproduced, stored in a retrieval system, or transmitted, in any form or by any means, electronic, mechanical, photocopying, recording or otherwise, without the prior permission of the Copyright owner.

Printed and bound in Great Britain by
Staples Printers Limited at their Rochester, Kent, establishment

Editorial Board

RAYMOND SPOTTISWOODE
General Editor

BERNARD HAPPÉ
Editor: Film

ERIC VAST
Editor: Television

DAPHNE P. BUCKMASTER
Administrative Editor

P. C. POYNTER
Art Editor

A. KRASZNA-KRAUSZ
Chairman of the Editorial Board

Preface

1

The *Focal Encyclopedia of Film and Television* grew out of an attempt to produce a comprehensive encyclopedia of the cinema embracing its aesthetics and history, its methods of production and distribution and its impact on society. Focal Press began this project in 1964, when it set up an Editorial Board to consider a one-volume work on the lines of the first edition of its *Focal Encyclopedia of Photography*. It was soon apparent that to compress all aspects of the cinema into a single book would require so much selection and omission that it could not rank as a major work of reference. On the other hand, an expansion into several volumes on the cinema alone ran the risk of stretching the framework a great deal beyond what its readership was likely to demand.

2

It was then that the fact of increasing convergence between film and its near neighbour, television, became particularly evident.

The older medium had proved the foundation on which television, a thriving entertainments and electronics industry, was being built. The cinema, shorn of its mass audiences, was beginning to explore new horizons of possibility. The techniques of both were closely interwoven, but in growing apart from one another, gaps in knowledge had been created which an encyclopedia linking the two media might help to bridge. Such a work could fill an acute and growing need, command a much wider readership and

rightly call for a wider framework than one based on film alone.
Equally, it would need to solve many new problems.
The history of film can be studied at first hand through commercial revivals, film societies and archives. Television lives in the instant; what is once past is accessible only through the private libraries of the producing organisations.
On the technical plane, film is grounded on photography, of which the basic concepts are familiar to everyone who owns a camera. Television is a province of advanced electronics, in which fewer readers feel at home.

3
The need to map, delimit and suitably co-ordinate the subject matter thus became imperative. It was decided to tackle first the area of technology, where there is much in common between film and television. It is this subject matter to which the present volume is devoted.
A good deal would need to be said about underlying branches of science: optics, chemistry, the photographic image, modulation systems, acoustics, recording methods and the like. Elementary knowledge within the reach of the educated reader must, however, be taken for granted.
As the work grew, these selective principles were extended in two directions. A historical framework was provided, with condensed technical histories and chronologies of the media, and short biographical sketches of the pioneers.
A survey of the roles of principal creative technicians was added in the form of more subjective entries, the work of leading film and television practitioners in each field, designed to illustrate how tools and techniques are applied in daily use and how different skills compete and co-operate; this without any attempt to exhaust a subject whose proper place is outside the concept of the present volume.

4
In a book intended for the widest English-speaking audience, the problem arose of reconciling British and North American terminology.
Wherever there was a significant risk of misunderstanding, the American equivalent of a British term is given on the first occasion of its use in an entry, e.g. aerial (antenna), telerecording (kinescoping), telecine (film scanner). Where standards differ, as in television transmission, equal coverage of the principal systems has been aimed at, to suit British and American readers alike.

5
A more delicate editorial problem concerned the inclusion of trade names and the mention of commercial equipment. To list these at length would open the Editors to a charge of bias for their omissions, and would soon date the work as specific models passed into obsolescence.
To omit them all would give an air of unreality to large areas of the book.
We have sought a solution by providing definition entries of only those processes and pieces of equipment which serve a unique function or are household terms in the media.

6
The plan of the book necessarily became more complex during the editing: to the principal techniques of film and television had to be added the increasing number of inter-relationships which have developed between them.
Longer articles were devoted to more than a hundred main subjects, and these were buttressed by some fifteen hundred shorter entries, ranging from a few lines to a page.
Close on a thousand line drawings are the work of the publisher's Art Department, making it possible to establish a uniform style of illustration throughout the book.
Pictorial symbols, ⊕ and □, designate the entries mainly concerned with film or television, or equally with both.
Cross-references direct the reader to fuller treatments or to descriptions of closely-related subjects. Guides to books and technical literature are confined to readily accessible sources, where the reader will often find more extensive bibliographies to aid in further research.
The Encyclopedia is completed by an Index of some 10,000 references to give full access to material which would otherwise remain embedded in long articles.

7
Because of the fragmentation of the coverage inevitable in a work arranged alphabetically, we have supplemented the entries with a synoptic survey, considerably longer than any single article, embracing the whole field of technology covered by the Encyclopedia.
This survey forms a Basic Anatomy: it relates the many sections of the subject to one another, and gives the reader a perspective on the processes by which films and television programmes advance from studio or actuality towards their final shape on home or cinema screen. It is interspersed with an analytical breakdown as a further guide to the appropriate entries in the Encyclopedia.

8

Throughout the work we have faced the difficulty of combining two different worlds of knowledge, the photographic and the electronic. The reader's background predisposes him strongly towards one or the other; and yet his own enquiry, more often than not, is likely to be directed towards "the other man's job".

Whenever treatment in depth is essential it has been hard to find a common denominator acceptable to every reader without reducing the level of information too low to be of genuine value. When dealing with more sophisticated subject areas, this contingency was met by providing the less initiated reader with a summary introduction, followed by a more thorough treatment within the body of the article for those sufficiently familiar with the discipline in question.

9

Focal Press set up an Editorial Board under the chairmanship of the Publisher, consisting of a General Editor, a Film Editor and a Television Editor, assisted by an Administrative Editor and an Art Editor.

Contributions by nearly a hundred experts were scrutinized by the three principal editors in the first place, and re-writing was undertaken where necessary to equalize the level of approach across the whole subject within the limits already mentioned. Comparison of entries revealed gaps and overlaps, and while these were being filled, outside experts were consulted wherever the complexity of the subject matter demanded it.

All entries were again considered by the three principal editors during the proof stages of the book, and every attempt was made to update the whole to the spring of 1969.

10

The full title of the present volume is the *Focal Encyclopedia of Film and Television Techniques*. In concentrating on the tools of film and television making it is both comprehensive and self-contained.

A second independent volume now in preparation, the *Focal Encyclopedia of Film and Television Production*, will be complementary to the first and will cover in far greater detail the applications of film and television techniques to production methods, examine creative skills, and survey the organization and economics of the two industries.

Authors and their Principal Articles

Articles	Author
Sound Mixer Sound Recording (Film)	**ALDRED, JOHN BRIAN** Sound Supervisor, Paramount Pictures (UK) Ltd., London
Narrow Gauge Production Techniques	**BADDELEY, W. HUGH, F.R.P.S.** Managing Director, Gateway Film Productions Ltd. and Gateway Educational Films Ltd.
Large Screen TV Station Timing Synchronizing Pulse Generators	**BARRETT, J. H., B.Sc.** Formerly Supervisory Design Engineer, Rediffusion Television, London
Biographies (TV) History (TV)	**BELL, ROBERT WALTON** Historian, Marconi Company Ltd, Chelmsford
Microwave Links	**BELL, THOMAS ANTHONY KERR** Supervisory Engineer, Master Control and Telecine, Thames Television Ltd., London; formerly Supervisory Communications Engineer, Rediffusion Television Ltd., London
Lighting Equipment (TV)	**BENTHAM, FREDERICK P., F.I.E.S.** Director of Research and Development, Strand Electric & Engineering Co. Ltd., London; Founder Member and Vice-Chairman, Association of British Theatre Technicians; Master of the Art Workers Guild
Image Formation	**BERG, W. F., D.Sc., Ph.D., F.R.P.S., F.Inst.P.** Professor and Director of Photographisches Institut der Eidgenössischen Technischen Hochschule, Zürich, Switzerland; General Editor, Focal Library
Studio Complex (TV) *Brief entries in the fields of:* Transfer Characteristics; Telecine; Studio Complex; Picture Monitors; Video Mixing Equipment	**BERKELEY, P. R., M.I.E.E.** Head of Engineering Projects Group, Thames Television Ltd., London
Duplicating Processes (Black and White)	**BOMBACK, RICHARD HENRY, B.Sc., F.R.P.S., A.R.I.C., M.B.K.S.T.S.** Formerly Technical Manager, Pathé Laboratories Ltd.

Closed Circuit Systems

BRACE, JOHN E. H., B.Sc.
Manager, Electro-Optical Systems Division, The Marconi Company Ltd., Chelmsford

Colour Principles (Film and TV)

BROCKLEBANK, RALPH WILFRED, B.A.(Cantab) (Natural Sciences)
Lecturer, Goethean Science Foundations, Stourbridge; Chairman, Midland Section, The Colour Groups of Great Britain.

Standards (USA)
Index

BUCKMASTER, DAPHNE P.
Administrative Editor (Encyclopedias), Focal Press Ltd., London

Vision Mixer

BULTITUDE, F. B.
Chief Engineer, Southern Television Ltd., Southampton

Medical Cinematography & TV

CARDEW, PETER N., M.R.C.S., L.R.C.P., F.R.P.S., F.I.I.P.
Director of Department of Audio Communication, St. Mary's Hospital, London

Art Director (Film)

CARTER, MAURICE G., G.F.A.D.
Supervising Art Director, Pinewood Film Studios, Iver Heath

Outside Broadcast Units

CARTWRIGHT, A. E., C.Eng., M.I.E.E.
Engineer, Studio Planning and Installation Department, BBC, London

Biographies (Film)
Chronology of Principal Inventions (Film)
History (Film)

COE, BRIAN W., M.B.K.S.
Curator, Kodak Museum, Harrow

Colour Cinematography
Integral Tripack Systems

COOTE, JACK HOWARD, F.R.P.S., F.B.K.S.
Manager, Technical Service Division, Ilford Ltd., Ilford

Processing
Silver Recovery

CORBETT, DENNIS JOHN, C.Eng., M.I.E.R.E., M.B.K.S., A.R.P.S.
Film Technical and Training Manager, Film Services Department, BBC Television, London; Author of *Motion Picture and Television Film: Image Control and Processing Techniques.*

Coating of Lenses
Lens Performance
Lens Types
Stabilization (Image)

COX, ARTHUR
Consultant; formerly Vice-President, Bell & Howell Company, Chicago; author of *Photographic Optics*

Film Storage

CRICKS, R. HOWARD, Hon. F.B.K.S., F.R.P.S. (deceased)
Kinematograph Engineer, London; Author of *The Complete Projectionist* and *Photographic Illumination*

Brief entries in the fields of:
Electronic Special Effects;
Transmitters and Aerials; Receiver

CROSSLEY, W. J., C.Eng., M.I.E.E.
Head of Engineering Operations, London Weekend Television Ltd., London

Viscous Processing

DAVIES, B. J., B.Sc., A.Inst.P., F.B.K.S.T.S.
Technical Service, Kodak Ltd., London

Lighting Equipment (Film)

EARLE-KNIGHT, T., M.A.S.E.E., M.B.K.S.T.S.
Chief Engineer, Film Production Division, Rank Organization, Iver Heath

Editor

FALKENBERG, PAUL V.
Film Producer and Lecturer; Contributor to *Film Culture, Cinema Journal, Film Comment*

Special Effects (Film)

FIELDING, RAYMOND, Ph.D.
Department of Radio-TV-Film, School of Communications, Temple University, Philadelphia, Pa.; President of the University Film Association; Vice-President of the International Congress of Schools of Cinema and Television; Past-President of the Industry Film Producers Association

Article	Author
Sensitometry and Densitometry	**FOAKES, M. C., B.Sc.** Senior Equipment Technician, Film Testing Department, Kodak Ltd., Harrow
Director (Film)	**FREND, CHARLES** Film Director, London
Chemistry of the Photographic Process	**GLOYNS, F. P., M.Sc., Ph.D., F.R.I.C., F.R.P.S.** Technological Manager, Rank Film Processing Ltd., Denham
Library (Film)	**GRENFELL, DAVID** Chief Librarian, Radio Telefis Eireann, Dublin; formerly Chief Film Librarian, Visnews Ltd., London
Film Manufacture	**GRIMSHAW, HERBERT EDMUND BRIAN** Product Planning Manager, Consumer Markets Division, Kodak Ltd., London. Formerly Chairman, British Kinematograph Sound and Television Society
Picture Monitors	**HALL, RONALD C., C.Eng., M.I.E.E.** Joint Development Manager, Television Group, Electrical & Musical Industries Ltd., Hayes
Animation Techniques	**HALAS, JOHN, F.S.I.A.** Director of Halas & Batchelor Cartoon Films Ltd.; Director-General of the International Animated Film Association; Film Producer and Director; Author of *Technique of Film Animation*, etc.
Library (TV)	**HANFORD, ANNE** Film Librarian, BBC Television Film Studios, London
Exposure Control Film Dimensions and Physical Characteristics Imbibition Printing Laboratory Organization Sensitometry of Colour Films Tone Reproduction Wide Screen Processes	**HAPPÉ, L. BERNARD, B.Sc., F.R.P.S., F.B.K.S.T.S.** Technical Manager, Technicolor Ltd., London; Chairman, Technical Committee, National Film Archive; Member of British Standards Institution committees on motion picture and TV practice
Special Effects (TV)	**HEIGHTMAN, ANTHONY NORMAN, C.Eng., M.I.E.R.E.** Studio Engineering Manager, Broadcasting Division, The Marconi Company Ltd., Chelmsford
Cassette Loading Continuous Projection Daylight Projection	**HIND, H. S., C.Eng., M.I.E.E., F.R.P.S., Hon.F.B.K.S.T.S., F.R.S.A.** Managing Director, Sound-Services Ltd., London
Communications	**HOLT, L.** Television Consultant; formerly Deputy Head of Project and Planning Department, Rediffusion Television Ltd., London
Producer (TV)	**HOOD, STUART** Freelance Writer and Producer, London
Visual Principles	**HOWE, J. A. M., M.A., Ph.D.** Deputy Chairman and Lecturer, Department of Machine Intelligence and Perception, University of Edinburgh
Producer (Film)	**HYMAN, K.** Executive Vice-President, World Wide Productions, Warner Brothers, Seven Arts, Inc., London
Transmitters & Aerials	**JAMES, ALAN, B.Sc., F.I.E.E.** Head of Engineering Information Service, Independent Television Authority, London

Authors and their Principal Articles

Articles	Author
Camera (Colour) (TV)	**JAMES, I. J. P., B.Sc., C.Eng.** General Manager, Television Group, EMI Electronics Ltd., Hayes
Animated Captions	**JAMES, KENNETH R., A.R.C.A.** Consultant, Visual Information Design; formerly with Rediffusion Television, Ltd.
Cold Climate Cinematography Tropical Cinematography	**JOHN, D. H. O., B.Sc., A.R.C.Sc., F.R.I.C., F.R.P.S.** Chief Librarian and Information Officer, Ilford Ltd., Ilford; Editor, *Photographic Abstracts*
Cameraman (TV)	**JONES, PETER** Freelance Television Director/Producer
Picture Locking Techniques	**KITCHIN, H. D., M.I.E.R.E.** Chief of Electronic Equipment Group, Broadcasting Division, The Marconi Co. Ltd., Chelmsford
Cinema Theatre Drive-In Cinema Theatre Rear Projection Screen Luminance Screens	**KNOPP, LESLIE, M.B.E., Ph.D., M.Sc., F.R.P.S., Hon. F.B.K.S.T.S., F.I.E.S.** Technical Consultant to Royal Naval Film Corporation and Cinematograph Exhibitors' Association, London
Cameraman (Film)	**LASSALLY, WALTER** Director of Photography, Lassally Productions Ltd., London
Kalvar Process	**LINDEMEYER, ROBERT B.** Formerly Director of Operations, Metro/Kalvar Inc., Darion, Conn.
Colour Principles (TV) Colour Systems NTSC Colour System PAL Colour System SECAM Colour System	**LORD, ARTHUR VALENTINE, M.Sc., F.I.E.E.** Head of Signal Origination Group, BBC Research Department, Kingswood Warren
High-Speed Cinematography	**LUNN, GEORGE H., A.Inst.P., A.R.P.S., F.I.I.P.** Head of Photography, U.K. Atomic Energy Authority, Aldermaston; British National Delegate for High-Speed Photography
Thermoplastic Recording	**MacCOUN, T. F.** Market Development Manager, General Electric Company, New York
Acoustics; Microphones Sound Sound Recording Systems	**MACKENZIE, G. W., M.I.E.R.E.** Lecturer and writer on Acoustics and Audio Engineering for Sound Broadcasting
Videotape Editing	**MAKEPIECE, A. P. W., B.Sc., M.B.K.S.T.S.** Director, Audio-Visual Aids Unit, University of Bristol; formerly Videotape Editor, BBC Television Centre, London
Animation Techniques	**MANVELL, R.** Formerly Director, British Film Academy; Visiting Fellow, University of Sussex; Head of Department of Film History, London School of Film Technique; Author of *Technique of Film Animation*, etc.
Satellite Communications	**MATHEWS, L. F., C.Eng., F.I.E.R.E., F.B.I.S.** Director and General Manager, Midlands ATV Network Ltd., Birmingham

Article	Author
Special Events	**METCALFE, MIKE, I.E.R.E., M.R.T.S.** Chief Engineer, Electronic Filming Development, Mole-Richardson (England) Ltd.; formerly Programme Liaison Engineer, Rediffusion Television Ltd., London
Master Control	**MIRZWINSKI, HENRYK, B.Sc.(Eng.), C.Eng., M.I.E.E.** Leader of Systems Engineering Group, Broadcasting Division, Marconi Company Ltd., Chelmsford
Sound Reproduction in the Cinema	**MOIR, JAMES, C.(Eng.), F.I.E.E.** Acoustical Consultant, James Moir and Associates, Chipperfield
Cinemacrography Cinemicrography Time Lapse Cinematography	**MOREMAN, KENNETH** Director of Photography, Chester Beatty Research Institute, London
Videotape Recording	**MORRISON, ERIC FRASER** Project Engineer, Video Engineering Department, Ampex Corporation, Redwood City, California. Formerly Head of Video Engineering, Ampex Electronics Ltd., Reading.
Lawrence Tube Receiver Scanning	**MUDD, L. T., C.(Eng.), M.I.E.E.** President, Society of Relay Engineers, London
Cineradiography	**NEWMAN, LEONARD A., A.Inst.E., F.I.X-ray T.** Product Engineer, Medical Division, Philips Electrical Ltd., London
Animation Equipment	**NIELSON, PETER** Director, Oxberry Division, Berkey Technical (UK) Ltd., Thetford
Transfer Characteristics	**PACKHAM, DENNIS GEORGE** Technical Controller and Director of Tyne Tees Television, Ltd., Newcastle upon Tyne
Camera (TV)	**PARKER-SMITH, N. N., B.Sc., F.I.E.E.** Technical Manager, Broadcasting Division, The Marconi Company Ltd., Chelmsford
Studio Complex (Film)	**PEERS, VICTOR A.** Formerly Director and General Manager, Granada Television Ltd., London; Production Manager, Gaumont British Picture Corporation, London
Director (TV)	**PRICE, ERIC FRANCIS** Film and Television Director; Production Adviser, Jamaica Broadcasting Corporation
Automation in the Cinema	**PULMAN, R. R. E., F.B.K.S.** Projection Engineer, Operating Services Control (Engineering), Rank Organization, Whyteleafe
Standards Conversion	**RAINGER, PETER, B.Sc.(Eng.), C.Eng., F.I.E.E.** Head of Studio Group, Designs Department, BBC, London
Standards (British) Standards (International)	**RAYMONT, WILLIAM J., M.B.K.S.** Cinematograph Engineer, London
Brief entries in the fields of: Electronics; TV Waveform	**REIS, C. W. B., B.Sc.** Technical Consultant, William Reis Associates
Videotape Recording (Colour)	**ROBINSON, DAVID P., M.A., M.I.E.E., C.Eng.** Chief Engineer, Dolby Laboratories, London; formerly of BBC Designs Department, London

Article	Author
Chronology of Principal Inventions (TV)	**ROUSE, D. G.** Chief Assistant, Patent Office, Marconi Company, Ltd., Chelmsford
Underwater Cinematography	**SAYER, STANLEY** Cameraman, Technicolor Ltd.
Transmission Networks	**SEWTER, JOHN BARON, C.Eng., M.I.E.E.** Assistant Staff Engineer, Post Office Telecommunications Headquarters, London
Brief entries in the fields of: TV Camera	**SOUTHGATE, JOHN W., C.Eng., M.I.E.R.E., B.K.S.T.S.** Assistant General Manager, Mole-Richardson (England) Ltd., London; formerly Head of Films, Rediffusion Television Ltd., London
Brief entries in the fields of: Station Timing; Master Control and Presentation; TV Sound; Colour Systems	**SPONG, LESLIE** Supervisory Engineer, ATV Network Ltd., Boreham Wood
Telerecording	**SPOONER, ARCHER MICHAEL, Ph.D., B.Sc.(Eng.), C.Eng., F.I.E.E.** Engineering Consultant, Redifon Ltd., Aylesbury
Camera (Film) Computer Animation Film Storage Stereoscopic Cinematography	**SPOTTISWOODE, RAYMOND, M.A.(Oxon), M.B.K.S.T.S.** Fellow of the Society of Motion Picture and Television Engineers; Author of *Film and its Techniques*
Holography	**SPOTTISWOODE, S. J. P.**
News Programmes	**SWEENY, W. H. O., F.B.K.S.T.S.** Formerly Chief Engineer, Independent Television News, London
Sound Recording (TV)	**TASKER, J. E., M.B.K.S.T.S.** Head of Sound, Thames Television Ltd., Teddington
Telecine (colour)	**TOWNSEND, G. BORIS, Ph.D., C.Eng., A.M.B.I.M., F.Inst.P., F.I.E.E., A.K.C.** Fellow of the Royal Television Society; Head of Engineering Research, Thames Television Ltd., Teddington
Colour Printers (additive) Projector	**TOWNSLEY, MALCOLM G.** Consulting Engineer; formerly Vice-President, Engineering, Bell & Howell Company, Chicago, U.S.A.
Television Principles Transmission & Modulation	**TUCKER, JOHN D., C.Eng., A.M.I.E.R.E., M.I.E.E.** Manager, Broadcast Equipment Division, EMI Electronics Ltd., Hayes
Communications Power Supplies Phosphors for Television Tubes Transfer Characteristics Transients	**VAST, ERIC C., B.Sc., D.I.C., M.I.E.E.** Assistant Technical Controller, Thames Television Ltd., London
Narrow Gauge Film & Equipment	**WEST, PETER A., M.B.K.S.T.S.** News Film Cameraman, Independent Television News, London
Power Supplies	**WHITTLE, D., M.A., A.M.I.E.E.** Chief Engineer, Southern Television Ltd., Southampton
Magnetic Recording Materials	**WOOLER, J.** Director & General Manager, EMI Tape Ltd., Hayes
Designer	**YATES, MICHAEL, M.S.I.A.** Head of Design, London Weekend Television Ltd.

Key to Signatures

A.C.	Arthur Cox
A.E.C.	A. E. Cartwright
A.H.	Anne Hanford
A.J.	Alan James
A.M.S.	A. M. Spooner
A.N.H.	A. N. Heightman
A.P.W.M.	A. P. W. Makepiece
A.V.L.	A. V. Lord
B.J.D.	B. J. Davies
B.W.C.	Brian W. Coe
C.F.	Charles Frend
C.W.	Catherine Ware
C.W.B.R.	C. W. B. Reis
D.G.	David Grenfell
D.G.R.	D. G. Rouse
D.H.O.J.	D. H. O. John
D.J.C.	Dennis J. Corbett
D.P.	D. Packham
D.P.B.	Daphne P. Buckmaster
D.P.R.	David P. Robinson
D.W.	D. Whittle
E.C.V.	Eric C. Vast
E.F.M.	E. Fraser Morrison
E.F.P.	Eric F. Price
F.B.B.	F. B. Bultitude
F.P.B.	Frederick P. Bentham
F.P.G.	F. P. Gloyns
G.B.T.	G. Boris Townsend
G.H.L.	George H. Lunn
G.W.M.	G. W. Mackenzie
H.D.K.	H. D. Kitchin
H.E.B.G.	H. E. B. Grimshaw
H.M.	Henryk Mirzwinski
H.S.H.	H. S. Hind
I.J.P.J.	I. J. P. James
J.A.M.H.	J. A. M. Howe
J.B.A.	John B. Aldred
J.B.S.	J. B. Sewter
J.D.T.	John D. Tucker
J.E.H.B.	J. E. H. Brace
J.E.T.	J. E. Tasker
J.H.	John Halas
J.H.B.	J. H. Barrett
J.H.C.	Jack H. Coote
J.M.	James Moir
J.W.	J. Wooler
J.W.S.	John W. Southgate
K.H.	K. Hyman
K.M.	Kenneth Moreman
K.R.J.	Kenneth R. James
L.A.N.	Leonard A. Newman
L.B.H.	L. Bernard Happé
L.F.M.	L. F. Mathews
L.H.	L. Holt

L.K.	Leslie Knopp	R.H.C.	R. Howard Cricks
L.S.	L. Spong	R.J.S.	Raymond J. Spottiswoode
L.T.M.	L. T. Mudd	R.M.	Roger Manvell
M.C.F.	M. C. Foakes	R.R.E.P.	R. R. E. Pulman
M.G.C.	Maurice G. Carter	R.W.B.	Robert W. Bell
M.G.T.	Malcolm G. Townsley	R.W.Br.	Ralph W. Brocklebank
M.M.	Mike Metcalfe	S.H.	Stuart Hood
M.Y.	Michael Yates	S.J.P.S.	S. J. P. Spottiswoode
N.N.P-S.	N. N. Parker-Smith	S.S.	Stanley Sayer
P.A.W.	Peter A. West	T.A.K.B.	T. A. K. Bell
P.J.	Peter Jones	T.E-K.	T. Earle-Knight
P.N.	Peter Nielson	T.F.M.	T. F. MacCoun
P.N.C.	Peter N. Cardew	V.A.P.	Victor A. Peers
P.R.	Peter Rainger	W.F.B.	W. F. Berg
P.R.B.	P. R. Berkeley	W.H.B.	W. Hugh Baddeley
P.V.F.	Paul V. Falkenberg	W.H.O.S.	W. H. O. Sweeny
R.B.L.	R. B. Lindmeyer	W.J.C.	W. J. Crossley
R.C.H.	Ronald C. Hall	W.J.R.	William J. Raymont
R.F.	Raymond Fielding	W.L.	Walter Lassally
R.H.B.	R. H. Bomback		

Abbreviations

Å Angstrom
a.c. alternating current
AFNOR Association Française de Normalisation
AGC automatic gain control
a.m. amplitude modulation
AR aspect ratio
ART Added Reference Transmission
ASA American Standards Association
ASC American Society of Cinematographers
ASLIB Association of Specialist Libraries and Information Bureaux
B & H Bell & Howell
b & w black and white
BBC British Broadcasting Corporation
bit binary digit
BREMA British Radio Equipment Manufacturers' Association
BS British Standard
BSI Bristish Standards Institution
BSS British Standards Specification
C Celsius (centigrade)
CATV central aerial television
cc cubic centimetre
CCIR Comité Consultatif Internationale des Radiocommunications
CCITT Comité Consultatif Internationale Télégraphique et Téléphonique
CCTV closed circuit television
CCU camera control unit
cd/m² candela per square metre
CIE Commission Internationale d'Éclairage
cm centimetre
COMSAT Communications Satellite Corporation
cpm cycles per minute
cps cycles per second
CPS cathode potential stabilized
CRT cathode ray tube
CS Cinemascope (perforation)
CT colour temperature (lamp)
CTV community television
cu.ft. cubic foot
c.w. continuous wave
D density
dB decibel
d.c. direct current
dBm decibels above (below) one milliwatt
DIN Deutsche Industrie-Norm (German Standard)
DP double play (tape)
dyn/cm² dyne per square centimetre
E voltage
EBU European Broadcasting Union
EHT extra high tension

EMI	Electrical & Musical Industries (Britain)
ERP	effective radiated power
EVR	electronic video-recording
F	Fahrenheit
f	frequency or focal length
FA	full aperture (of film frame)
fc	foot candle
FCC	Federal Communications Commission (US)
f.m.	frequency modulation
fps	feet per second or frames per second
ft.	foot
FX	effects (film)
γ	gamma
G	granularity
GEC	General Electric Corporation
GHz	gigaHertz (10^9 Hz)
gm	gram
h.f.	high frequency
HI	high intensity
HT	high tension
Hz	Hertz (cycles per second)
I	illuminance or current
IEC	International Electrotechnical Commission
i.f.	intermediate frequency
in.	inch
ips	inches per second
ISO	International Standardization Organization
ITA	Independent Television Authority (Britain)
K	Degrees Kelvin
KHz	kiloHertz (10^3 Hz)
KS	Kodak Standard (perforation)
kV	kilovolt
kVA	kilovolt ampere
kW	kilowatt
LC	inductance/capacity
LCT	low colour temperature
l.f.	low frequency
lm	lumen
log	logarithm
LP	long play (tape)
mA	milliampere
MCR	mobile control room
MEV	million electron volts
MHz	megaHertz (10^6 Hz)
μsec	microsecond
mil	milli-inch (0.001 in.)
mm	millimetre
msec	millisecond
MTF	modulation transfer function
mV	millivolt
NA	numerical aperture
NBC	National Broadcasting Corporation (US)
neg	negative
NHK	Nippon Hoso Kyokai (Japan Broadcasting Company)
nm	nanometre (10^{-9}metres)
non-flam	non-inflammable (film base)
NPN	type of transistor
NRDC	National Research Development Corporation (Britain)
NSC	network switching centre
nsec	nanosecond (10^{-9} sec)
NTSC	National Television System Committee (US)
OB	outside broadcast
OTF	optical transfer function
PAL	Phase Alternation Line
pF	picofarad (10^{-12} farad)
pH	measure of alkalinity or acidity
PLUGE	picture line-up generating equipment
PNP	type of transistor
pos	positive
ppm	pictures per minute
PPM	peak programme meter
pps	pictures per second
PSC	programme switching centre
psi	pounds per square inch
PTT	Posts, Telephones and Telegraphs
QP	quadruple play (tape)
RA	reduced aperture (of film frame)
RCA	Radio Corporation of America
r.f.	radio frequency
RGB	red, green, blue
RH	relative humidity
RIAA	Record Industry Association of America
RMA	Radio Manufacturers Association (US)
rms	root mean square
rpm	revolutions per minute
RT	reverberation time
RTF	Radio Television Française
RX	receiver
SCR	silicon-controlled rectifier
sec	second
SECAM	Séquentiel Couleur à Mémoire

SMPE	Society of Motion Picture Engineers (until 1949)
SMPTE	Society of Motion Picture and Television Engineers (US)
SP	standard play (tape)
SPL	sound pressure level
Syncom	Synchronous Communication (satellite)
T	periodic time
T-C	telecine (film scanner)
3-D	three-dimensional (film)
T-number	transmission number (lens)
torr	torricelli (unit of vacuum)
TP	triple play (tape)
T-stop	stop using transmission number (lens)
TX	transmitter
UHF	ultra-high frequency
USA	USA Standards Institute
u,v,	uniform chromaticity (colour triangle)
v	volt
VHF	very high frequency
VITS	vertical interval test signal
VTR	videotape recording
VU	volume unit
w	watt

Acknowledgements

We are indebted to the following sources for the diagrams on the pages shown:

Acoustical Design of Broadcasting Studios by H. L. Kirke and A. B. Howe, *Journal of the Institution of Electrical Engineers*, Vol. 78, No. 472 (page 5 bottom); *Acoustics, Noise and Buildings* by P. H. Parkin and H. R. Humphrey (Faber and Faber) (page 6); Associated British Picture Corporation (page 399 (b)); British Broadcasting Corporation (page 5 top); British Space Development Co. Ltd. (page 640); Canadian Broadcasting Corporation (page 822 left).

Colour in Business Science and Industry, *2nd Edn* by D. B. Judd and G. Wyszecki (Wiley) (page 186 top); Dawe Instruments Ltd. (page 681); Electrical and Musical Industries Ltd., (pages 821 left, 824 top, 826 bottom); English Electric Valve Company Ltd., (pages 79, 80 bottom); *Experimental Psychology* by R. S. Woodworth and H. Schlosberg (Methuen) (page 974 top); *Eye and Brain* by R. L. Gregory (Weidenfeld & Nicholson) (page 973); Fernseh, GmbH. (page 819 top right).

Infra-red Characteristics of Various Colour Release Prints and Their Effects on Color Television Reproduction by H. N. Kozanowski, Radio Corporation of America (page 842 bottom); Marconi Company Ltd. (pages 154, 155, 819 top left, 837).

Measurement of Colour, *3rd Edn* by W. D. Wright (Hilger & Watts) (pages 183 left, 187 top); National Film Archive, London (page 399 (c)).

A New Approach to Clinical Film Recording With Special Reference to Endoscopy by Dr. G. Berci, *British Medical and Biological Illustration*, Vol. 16 (page 457 top); Philips Electrical Ltd. (page 153).

Radio, Vols. 1, 2, 3, by J. D. Tucker and D. F. Wilkinson, (English Universities Press) (pages 888, 889); Rank Cintel Television (pages 825, 826 top, 827 (all), 828 (both), 829, 838 (both)); Rank, Taylor, Hobson (page 81 right).

Sound Absorbing Materials by E. J. Evans and E. N. Bazley (Her Majesty's Stationery Office) (pages 7, 8 bottom).

Television Recording by W. D. Kemp, *Proceedings of the Institute of Electrical Engineers*, Vol. 99, No. 17 (page 853); Thorne-AEI Radio Valves and Tubes Ltd. (page 670); Visnews Ltd. (page 399 (a) and (d)).

These symbols, denoting film (*left*) and television (*right*), are prefixed to all entries to indicate their principal areas of reference.

For a comprehensive view of film and television techniques, the reader is referred to the survey

FILM & TELEVISION: A BASIC ANATOMY

on pp. 993–1070, which also contains a classified analysis of the entries in the main body of the Encyclopedia.

INDEX on page 1071.

⊕A and B Printing. Method of printing, now virtually universal for narrow-gauge original materials, to incorporate optical effects and provide splice-free screen presentation, in which the negative shots are not joined in sequence in a single roll, but distributed between two rolls, the A and B rolls, which are made up to equal length with appro-

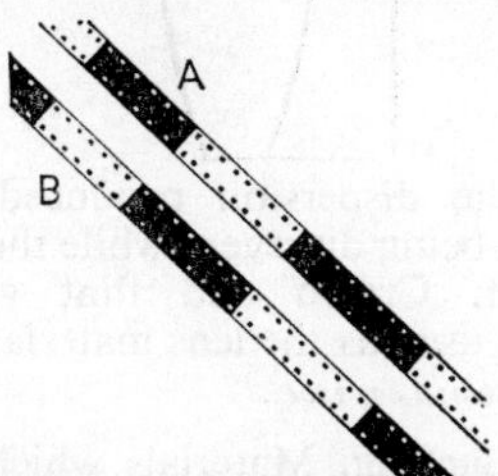

priate sections of black leader. In particular, where dissolves are to be made, the overlapping portions are laid in on the two rolls. The rolls are printed in succession on to the same length of film. On the A roll, say, a 3 ft. fade-out is timed to occur at a certain footage; when the B roll is run through, a 3 ft. fade-in starts at the same footage, so that the final effect is a complete dissolve. This method has the advantage that dissolves and fades require no intermediate stages of duplication, as they do in single-roll printing, so that their picture quality is as good as that of ordinary shots with cuts at both ends.

For more complicated optical effects, A, B and C printing, with three synchronized rolls, is sometimes employed.

Since narrow-gauge projector masks do not fully conceal the width of splices, advantage can be taken of the A and B method to alternate all shots, making use of the fact that frame-line joins overlap in one direction only. Each join is therefore overlapped towards the black leader which parallels the following shot on the other roll, thus rendering the join invisible in the resulting print. This is called checkerboard cutting.

See: *Checkerboard printing; Narrow gauge production techniques.*

⊕A and B Winding. Two forms of winding used for rolls of film in which only one edge is perforated. With the emulsion surface

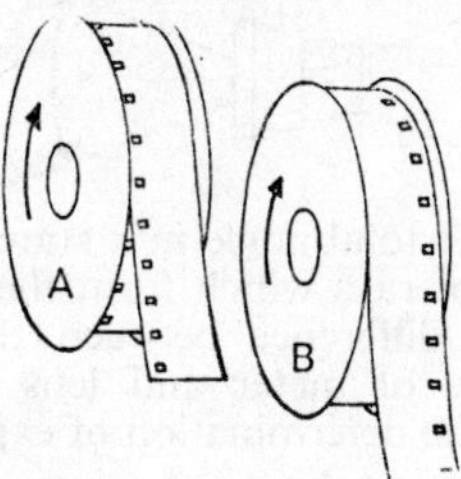

inwards and the film unwinding clockwise, the perforated edge is toward the operator for A winding, and away from him for B winding.

See: *Narrow gauge production techniques.*

⊕**Aberration.** In optics, the inherent deficiencies of a lens or optical system which cause imperfect formation of the image.

See: *Astigmatism; Barrel distortion; Coma; Curvature of field; Lens performance; Optical principles; Pincushion distortion; Spherical aberration.*

⊕**Abrasion.** Unwanted mark or scratch on the surface of film caused by rubbing action. Film emulsion surface is extremely delicate, and can be abraded by rubbing contact between adjacent layers, or by the friction of particles of hardened emulsion, with consequent loss of picture quality.

See: *Cinch marks; Lubrication; Release print examination and maintenance.*

⊕**Academy Aperture.** Standard aperture size used in 35mm cameras and projectors since the introduction of sound-on-film motion picture processes. So-named from its standardization by the (American) Academy of Motion Picture Arts and Sciences.

See: *Full aperture; Reduced aperture; Silent aperture.*

⊕**Academy Standards.** Standards established by the (American) Academy of Motion Picture Arts and Sciences for standard practices in film technology. The Academy's influence is reflected in the terms Academy Aperture, the standard 35mm camera mask size, and Academy Leader, the length of film carrying identification and projection operation information which is attached to the beginning of reels of release prints.

⊕□**Acceptance Angle.** Applied to an exposure meter, acceptance angle is the angle of the cone of rays which the optical design allows to strike the cathode of the photo-cell. Applied to a camera lens, the acceptance

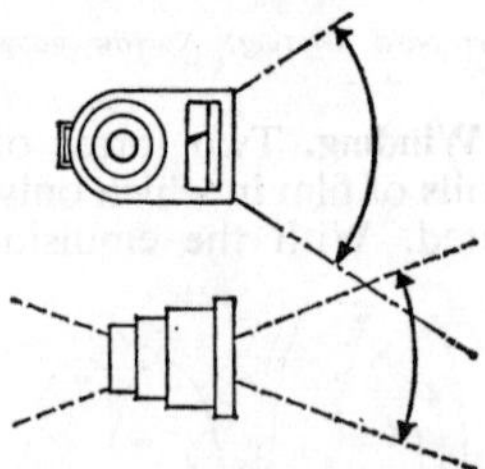

angle is the total angle in a stated plane of the cone of rays which form the image on film. The difference between the acceptance angle of meter and lens can cause errors in the determination of exposure.

See: *Exposure control.*

□**Acceptance Test Method.** Method for calculating the exact amplitude-frequency and phase-frequency responses of a television circuit by measurement of the output waveform when a sine-squared pulse is applied, the half-amplitude duration of which is $T = 1/2fc$ where fc is the designed cutoff frequency of the system. The waveform is photographed from an oscilloscope, and measurements from the photograph are made with a travelling microscope. The results are compared with the relevant specifications and are in a form which is a permanent record of the circuit's performance.

See: *Transmission networks.*

⊕**Acetate.** Safety film of low inflammability and slow burning characteristics was originally manufactured by coating the emulsion on a support of cellulose acetate: the term acetate is thus sometimes used (incorrectly) for all safety film.

See: *Nitrate base; Non-flam film.*

⊕□**Achromatic Lens.** Lens corrected for chromatic aberration. In its simplest form consists of a pair of lenses, or doublet, made from different materials and designed so that the dispersion produced by one element

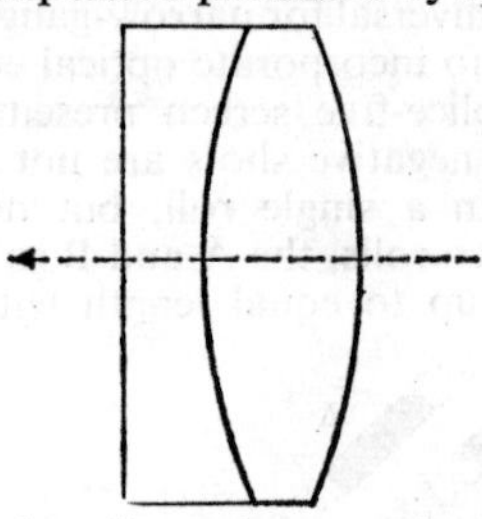

corrects the dispersion produced by the other, one being divergent while the other is convergent. Crown and flint glass are frequently used as the lens materials.

See: *Aberration; Lens types.*

⊕□**Acoustic Backing.** Materials which absorb sound energy used to control the acoustics of studios and theatres; they usually consist of porous structures either in blanket form or faced with perforated or slotted surfaces. Open texture plaster mixtures may also be applied or sprayed where required.

See: *Acoustics; Sound reproduction in the cinema.*

⊕□**Acoustic Feedback.** A positive feedback loop, usually involving a loudspeaker and microphone in which sound is fed back and amplified until the system overloads and a loud howl is set up. Also called howl round.

See: *Communications.*

⊕ACOUSTICS

□

In its strict sense, acoustics is the science of sound. In a broader sense, it refers to all those conditions of a given environment which affect the character of sound recorded, produced or reproduced in it.

The acoustics of the open air pose the recordist with several severe practical problems. The amount of energy radiated by most sources of sound is very small, and consequently a serious problem is to maintain a satisfactory level of wanted sound against the background of noise due to unwanted sounds. Another problem is that the effect of the wind striking the microphone can be very strong compared with that from the wanted source. Unless adequate precautions are taken with microphones, the resulting output may be quite useless.

Recording indoors seems an obvious solution; by reflection from the walls, ceiling and floor the source of sound can be reinforced so that an adequate level is easily obtained and the effects of noise and wind can be eliminated.

But other problems then arise, the principal one being that the size, shape and type of surface on the walls, ceiling and floor have a profound effect on the quality of sound. Care has to be taken with the design and construction of the room if the acoustics are to be acceptable for both direct listening and recording. For recording, the ease or difficulty of microphone placement is determined by the room acoustics. The effect of the room on reproduced music can be most marked, more so with stereophonic reproduction than with monophonic; for example, the positioning of the loudspeakers and the effect of the room surfaces have a considerable bearing on the spatial relationships on the sound stage between the loudspeakers.

ROOM ACOUSTICS

When a source of sound is placed a little way from the microphone near the centre of a room, the first sound energy to reach the microphone is that which travelled by the direct path. The next is that from first reflections, i.e. the sound reflected once by the walls. There is some loss of sound energy on reflection and from travelling further in the air, so that the energy arriving by this path is less than that which came direct. But the total energy received is greater than if there were no walls.

Next to arrive is the energy which has been reflected twice; these second reflections, having struck two of the walls, have suffered more loss than the first. The energy at the microphone thus increases again.

EFFECT OF REFLECTED SOUND. Thus, as time goes on from the moment of switching on, the energy at the microphone increases as more energy comes in by the reflected paths. But at the same time the contributions from the reflections decrease; since the later reflections have struck the surfaces more often, they have travelled considerable distances in the air, and accordingly have suffered more loss. Eventually, contributions from later reflections are so small that they can be ignored and the received energy has reached its maximum value.

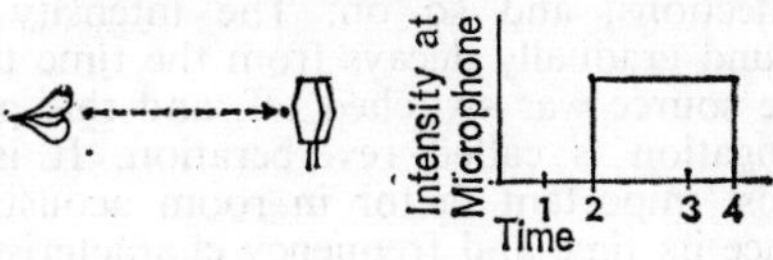

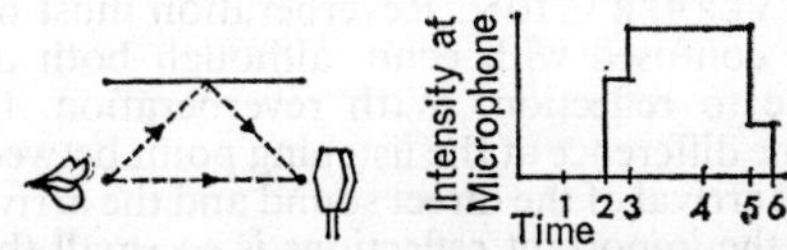

DIRECT AND REFLECTED SOUND. (*Top*) Direct sound switched on at time 1 reaches microphone at time 2 and intensity rises to maximum. At time 3, sound is switched off, and intensity drops to zero at time 4. (*Bottom*) Reflected sound is added, energy by this path arriving at time 3 and thus increasing total intensity at microphone. The longer path also delays drop of intensity to time 6.

As there are many reflections in a room, the energy at the microphone does not increase in discrete steps. It tends to be a continuous rise, steep at first and then gradually levelling out. Since the source is supplying energy at a certain rate into the room and the sound waves lose energy on being reflected, there comes a time when the energy is being lost at the same rate at which it is being supplied; at this point the intensity of sound in the room is constant and is called the equilibrium intensity.

The effect of the reflections from the walls is therefore to increase the intensity of sound available from the source. In practice, this is important in getting adequate volume in a given room, without strain on the part of the speaker or musician.

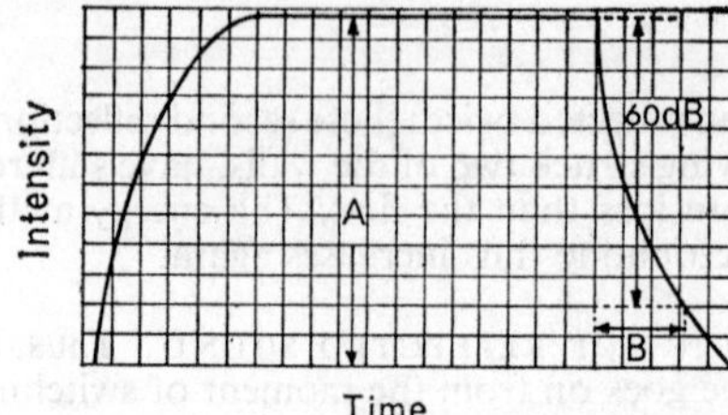

TIME AND INTENSITY VARIATION. An idealized curve. In practice, intensity fluctuates about this curve due to interference between component sound waves. A. Equilibrium intensity. B. Reverberation time.

When the sound is switched off, the energy contributed by the direct sound is the first to disappear, since it has come by the shortest path.

Next the contributions due to the first reflections go, then those due to the second reflections, and so on. The intensity of sound gradually decays from the time that the source was switched off, and this prolongation is called reverberation. It is a most important factor in room acoustics, since its time and frequency characteristics control the sound quality of a room.

REVERBERATION. Reverberation must not be confused with echo, although both are due to reflection. With reverberation, the time difference at the listening point between the arrival of the direct sound and the arrival of the important reflections is so small that it is impossible to distinguish one from the other.

With an echo, the time difference between the direct sound and a reflection is long enough for the listener easily to recognize the arrival of the reflection, if it is loud enough. The smallest time difference detectable by the ear is of the order of 100 milliseconds.

Reverberation is a desirable, indeed essential, quality of room acoustics; an echo is a serious acoustic fault.

The effect of the reverberant energy depends on the length of time for which it is present. In a room with highly absorbing surfaces, the energy contributed by reflection is low and there is little reverberation. If the surfaces are made less absorbent, the reverberant energy is greater and the sound in the room continues to be audible for a longer time. The difference in sound quality is immediately obvious.

Thus the reverberation time can be controlled by the absorption qualities of the room surfaces.

REVERBERATION TIME. The reverberation time, i.e., the period during which the reverberation dies away, is defined as the time taken for the sound intensity to fall by 60 dB from the equilibrium intensity.

The absorption of a material depends on the area of the material which is in contact with the sound waves and on its absorption coefficient. The latter is a measure of how efficient the material is as an absorber; a coefficient of 1·0 represents a material which absorbs all the sound energy striking it; a coefficient of 0 represents a material which reflects all the energy and absorbs none. The product of the surface area and the absorption coefficient gives the absorption in absorption units.

Another factor which affects the reverberation time is the volume of the room, and the three factors are related by Sabine's formula:

$$\text{R.T.} = 0{\cdot}05\,\frac{\text{volume}}{\text{total absorption}}\ \text{seconds}$$

The derivation of this simple relationship makes certain assumptions which, although acceptable when the volume is large, lead to significant errors when the volume is small and the absorption high.

To meet this objection, Eyring modified Sabine's formula as follows:

$$\text{R.T.} = \frac{0{\cdot}05\ \text{volume}}{-\text{S}\log_e(1-\alpha)}\ \text{seconds}$$

where S = total surface area

$$\alpha = \frac{\text{total absorption in absorption units}}{\text{total surface area}}$$

This relationship shows that, for a given volume, the reverberation time is determined by the total absorption present. The question arises as to what value of reverberation time is required. The effect of reverberation, providing extremes are avoided, depends on the subjective judgment of an observer and it is consequently impossible to lay down hard-and-fast rules which will yield universally acceptable reverberation time. However, much empirical data has been accumulated over the years and from this an acoustic consultant can obtain an optimum reverberation time for the volume of the room or hall he is designing. Then the required absorption can be calculated. This gives a guide to the types of wall and ceiling treatment to be applied. Adjustment is necessary during the final stages of construction to obtain results approaching the optimum.

TYPICAL REVERBERATION TIMES

Location	*Reverberation Time (secs)*
Domestic living room	0·35
Broadcast talks studio	0·3
Theatre	1·0
Medium-sized broadcast music studio	1·2
Large broadcast music studio ...	1·8
Large concert hall	1·8

These figures emphasize the fact that a fairly short reverberation time is necessary for speech to be easily heard. This is because speech is made up of syllables and, if the delay time is prolonged, each syllable will interfere with the succeeding one until the sound becomes confused and the meaning difficult to follow.

REQUIREMENTS FOR MUSIC. For music, fullness of tone is required both by the audience and the performers. It is essential that the players hear themselves and each other easily. If they cannot, they tend to force the tone to obtain adequate volume and ensemble playing suffers. These effects are apparent to the audience and, with the absence of reflections farther down the hall, the final result is said to be dry; the music lacks resonance. Therefore, to obtain the degree of reflection needed, auditoria for music should have a fairly long reverberation time.

Another desirable quality for music is good definition, or clarity, and, as with speech, this is obtained with a short reverberation time. An acceptable compromise has therefore to be secured by the designer to reconcile these two conflicting requirements—a fairly long reverberation time for fullness of tone, and a short one for good definition.

REVERBERATION AND FREQUENCY. Information on the behaviour of the reverberation at different frequencies is also required. This is best seen in a graph showing the reverberation time at all frequencies in the audio band. The question arises as to the ideal shape for this graph. There is no universally accepted shape, various authorities differing in their ideas, but in general a fairly flat graph is the aim.

Excessive rise in the reverberation time at one frequency, or over a band of frequencies, is usually regarded as a bad defect in the acoustic performance of a room or hall. For example, if the low-frequency reverberation time in a concert hall is high

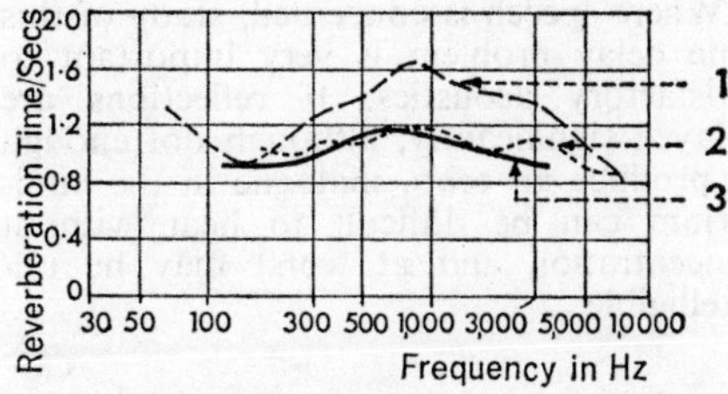

STUDIO REVERBERATION. Time/frequency curves for small orchestral studio. 1. Empty. 2. With orchestra. 3. Calculated curve with orchestra.

with respect to the other frequencies, there is a tendency to boominess owing to the bass instruments receiving considerable reinforcement from the acoustics of the hall. In small rooms, such as those used for speech studios, there can be difficulty with the reverberation time rising considerably at particular frequencies. This is an example of a defect usually called coloration.

In designing for certain reverberation time in a concert hall, allowance must also be made for the absorption effect of the audience. With orchestral broadcasting studios, the effect of the orchestral players must be noted; there is a definite relationship between the size of a studio and the maximum number of players it should accommodate. Graphs show the effect of

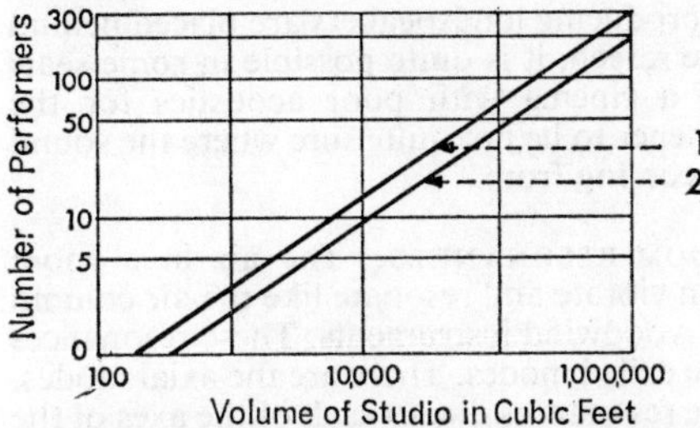

STUDIO CAPACITY. Relation between number of performers and studio volume. 1. Maximum number of performers. 2. Optimum number of performers.

these factors on the reverberation time, and how, for example, it differs according to whether the hall is empty or the orchestra is present. To minimize the effect of variations in the size of an audience, seating is often specially designed so that unoccupied seats provide some absorption.

TIME DELAY. A distinction has already been drawn between reverberation and echo, and the point made that the difference lay in the time delay which the reflections had undergone.

Where speech is concerned, study of this time delay problem is very important to satisfactory acoustics. If reflections are delayed significantly, although not enough to produce an echo, dialogue in the auditorium can be difficult to hear without concentration and at worst may be unintelligible.

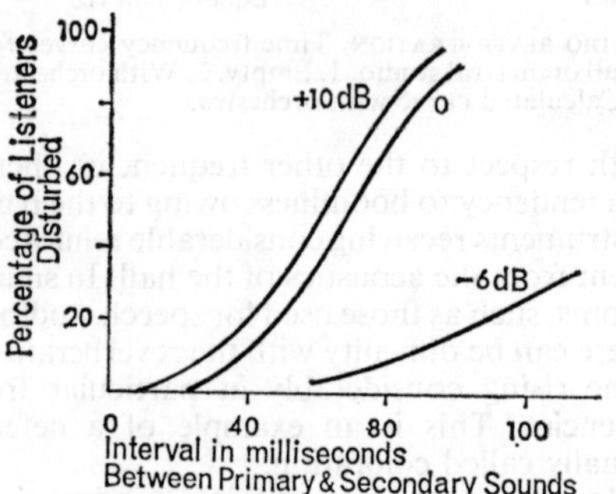

DISTURBANCE EFFECTS. When sound reaches listeners by two paths, the delay must be small if the secondary sound is greater than the primary one. Otherwise, intelligibility is likely to be seriously disturbed.

In a cinema, for example, it is essential that the audience hear the dialogue clearly with no difficulty and that the sound appears to come from the screen. Although the reproducing loudspeakers are placed behind the screen, it is quite possible in some seats of a cinema with poor acoustics for the listener to be not quite sure where the sound is coming from.

ROOM RESONANCES. The air in a room can vibrate and resonate like the air column in woodwind instruments. These resonances are called modes. There are the axial modes, the resonances due to each of the axes of the room – length, width and height. Then there are modes which use two of the axes, for example the length and the width; these are called the tangential modes. Finally, there are those which use all three dimensions, the so-called oblique modes. All these types can, like the air column, support a fundamental and its harmonics, the lowest fundamental being due to the longest axial dimension.

The resonances of a small room lie within the audible range, and this makes the design of small studios difficult as the effect of the small size shows as a defect in the frequency response.

With a large concert hall, theatre, or cinema, the large size can cause a defect in the time response which leads to the production of echoes. While small changes in shape have a considerable effect on the time delays in large halls, they have little effect on the modes of small rooms.

The axial modes are the most important. They contain most of the reverberant energy and hence determine the reverberation time. The effect of a particular mode depends on its bandwidth and how far it is away from the adjacent modes. The bandwidth determines the chances of the mode being set into resonance. If it does resonate, and is fairly isolated, then there is considerable reinforcement of the original sound by the room at the resonant frequency, and strong coloration results. It is therefore highly desirable to make the axial modes equally spaced throughout the important low frequency range, up to 300 Hz (= hertz = cycles per second) approximately, so that no mode is isolated. Since the mode frequencies depend on the room dimensions, control of their spacing can be achieved by a suitable choice of length, width and height.

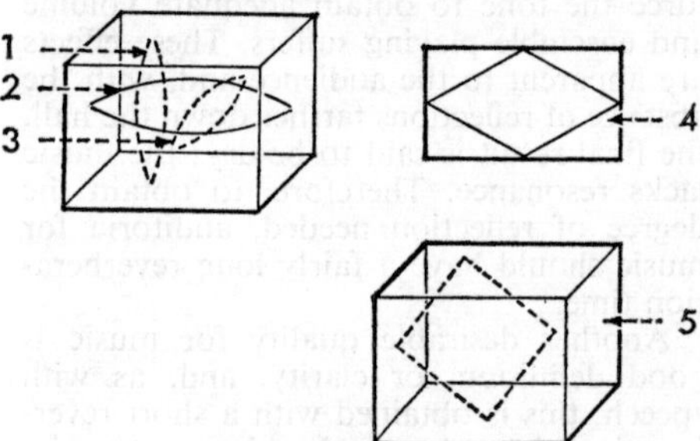

RESONANCE MODES. Axial modes due to room resonating along: 1. the height axis; 2. the length axis; 3. the width axis. 4. Tangential modes when the room resonates along two of the axes, e.g. length and width. 5. Oblique modes involving all three axes.

Another factor controlling room resonances is the distribution of the absorption materials. These are most effective if they are broken up into relatively small patches and randomly distributed.

SOUND ABSORPTION

There are now many sound-absorbing materials and methods used in practice by architects, broadcasting organizations, etc., but they can be roughly divided into two main categories: porous absorbers and resonant absorbers.

POROUS ABSORBERS. Porous materials such as felt, glass wool and acoustic plaster, contain air in a network of interconnected

ABSORPTION CO-EFFICIENTS

Material	Thickness	Frequency in Hz						
		125	250	500	1,000	2,000	4,000	8,000
Brickwork (plain or painted)		0·05	0·04	0·02	0·04	0·05	0·05	
Coke Breeze Blocks	3 in.	0·2	0·45	0·6	0·4	0·45	0·4	
Acoustic Plaster	1 in.	0·05	0·1	0·15	0·15	0·2	0·25	
Sprayed Asbestos	1 in.	0·2	0·35	0·6	0·75	0·75	0·75	
Glass-wool	2 in.	0·2	0·45	0·65	0·75	0·8	0·8	
Glass-wool	4 in.	0·45	0·75	0·8	0·85	0·9	0·85	
Glass-wool—Resin Bonded	1 in.	0·1	0·35	0·55	0·65	0·75	0·85	0·75
Glass-wool—Bitumen Bonded	1 in.	0·1	0·35	0·5	0·55	0·7	0·7	0·75
Mineral Wool Tiles 12 in. sq.	¾ in.	0·1	0·2	0·5	0·85	0·85	0·85	
Flexible Polyurethene Foam	2 in.	0·25	0·5	0·85	0·95	0·9	0·9	
Rigid Polyurethene Foam	2 in.	0·2	0·4	0·65	0·55	0·7	0·7	
Expanded Polystyrene (on 2-in. thick battens)	1 in.	0·1	0·25	0·55	0·2	0·1	0·15	
Perforated Fibre-board (Tiles on 2-in. thick battens)	2·4 cm.	0·3	0·45	0·5	0·55	0·65	0·8	
Rockwool (covered by Hardboard, ⅛-in. thick open 5% area)	2 in.	0·35	0·75	0·85	0·7	0·45	0·25	
Axminster Carpet	0·3–0·35 in.	—	0·05	0·15	0·3	0·45	0·55	
Turkey Carpet	0·6 in.	—	0·1	0·25	0·5	0·65	0·7	
Axminster Carpet (on needleloom underfelt)...	0·5–0·6 in.	—	0·15	0·4	0·6	0·75	0·75	
Cork Tiles	$\frac{9}{16}$ in.	—	0·05	0·15	0·25	0·25	—	
Sorbo-Rubber (with thin layer of hard-wearing surface)	$\frac{5}{16}$ in.	—	0·05	0·15	0·1	0·05	—	
Plate Glass Windows	¼ in.	0·1	—	0·04	—	0·02	—	
Audience-absorption units per person when seated in fully-upholstered seats...		2·0	4·3	5·0	5·0	5·5	5·0	

Absorption co-efficients obtained by reverberation method.

cavities. When the sound waves set up air vibrations, heat is generated owing to friction within the material and the sound wave loses energy.

Porous materials have absorption coefficients which vary widely with frequency. Taking two examples, the coefficient for a carpet with a felt underlay laid on the floor varies from 0·15 at 250 Hz to 0·75 at 2,000 Hz, and that for glass wool varies from 0·25 to 0·7 over the same frequency range. The absorption is generally much greater at the higher frequencies.

The coefficient also depends on the thickness of the material used. The figures quoted above for glass wool were for one-inch-thick layer with a rigid backing. If the thickness is doubled, the coefficient becomes 0·45 at 250 Hz and 0·8 at 2,000 Hz.

The reason why thickness affects absorption efficiency can be explained in terms of the behaviour of standing waves. When a wave is reflected from a rigid surface, the air particles have zero velocity (node) at the surface, and the first position of maximum velocity (antinode) is a quarter of a wavelength away from the surface. The absorption of sound energy in a porous material depends mainly on the particle velocity of air enclosed inside the pores. At low frequencies – long wavelengths – the distance of the antinode from the reflecting wall is large, and the particle velocity within, say, a one-inch-thick layer is low. At higher

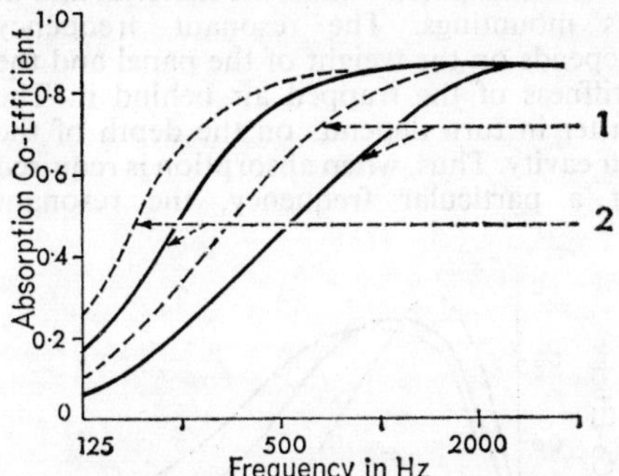

ABSORPTION COEFFICIENT. Porous material: variation with rigid backing (*solid line*) and 1-in. air space (*dotted line*) and with different thicknesses. 1. 1 in. 2. 2 in.

frequencies, i.e., short wavelengths, however, the particle velocity within the absorber is high. Increasing the thickness increases the particle velocity within the material for the longer wavelengths and thus increases the absorption at the lower frequencies.

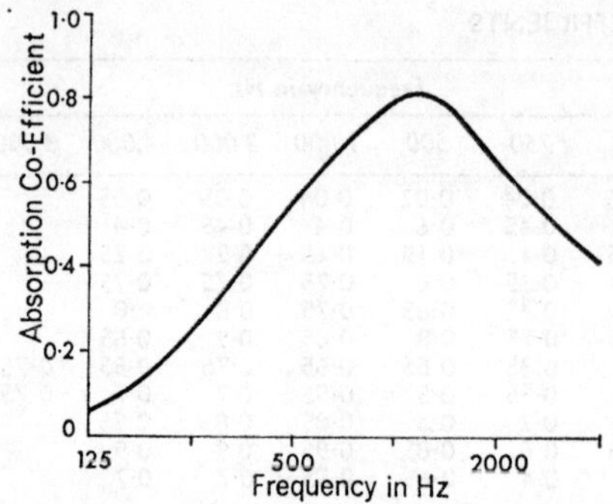

ABSORPTION COEFFICIENT. Typical covering for porous materials: ⅞ in. hardboard with $\frac{3}{16}$ in. holes and 10 per cent open area over absorbent 1 in. thick. Absorption coefficient falls above 1000 Hz.

Another way to increase the absorption at the low frequencies is to mount the material away from the wall, leaving an air space. This is equivalent to increasing the thickness.

Because of the thickness requirements, porous materials are impracticable for low-frequency absorption. Below approximately 300 Hz, therefore, some other, more effective means of absorption is required.

RESONANT ABSORBERS. Resonant methods are used for low-frequency absorption. In the first type, non-porous panels are caused to vibrate by the sound wave and heat is dissipated within the material and at its mountings. The resonant frequency depends on the weight of the panel and the stiffness of the trapped air behind it. The latter in turn depends on the depth of the air cavity. Thus, when absorption is required at a particular frequency, the resonant frequency of the absorber can be adjusted by a suitable choice of the panel weight and air cavity depth. The absorption is often increased by using a layer of porous material behind the panel. Wood, hardboard and roofing felt are examples of materials used for the panels.

The second form of resonant absorber is the Helmholtz resonator. Essentially, this consists of a rigid box with a narrow neck. The volume of the neck is much smaller than that of the box and resonance occurs due to the mass of air in the neck and the compliance of the air trapped in the box. The resonant frequency is thus controlled by the volume of the box and of the neck.

To make the device act as an absorber, some porous material such as gauze is placed across or inside the neck. When sound waves strike the neck, the resonance action causes the air in the neck to vibrate and, in passing through the gauze, heat is produced and energy is thus absorbed from the sound wave.

Resonant methods are most often used to provide general low-frequency absorption in large orchestral studios and in concert halls.

They are also used in small studios to control the room resonances.

All resonant methods depend on a body being set into vibration. But a vibrating body can also act as a source of sound, and it is essential to ensure that the energy is dissipated quickly, otherwise the so-called absorber may continue to vibrate after the original sound has died away, and radiate back to the room. The absorber has what is called a decay time of its own, and this should be less than the reverberation time of the studio.

The absorption of the resonant types is generally quoted in the same manner as that of porous materials, the absorption coefficient being given at various frequencies.

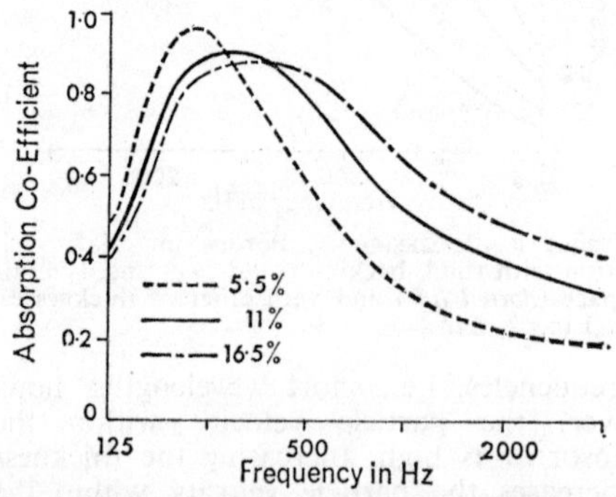

RESONANT ABSORPTION. Amount of absorption obtained by a panel of ½ in. plywood perforated with $\frac{3}{16}$ in. holes, over absorbent 2⅜ in. thick. Percentages shown are the amount of open area.

AIR ABSORPTION. It is often necessary to include a factor to show how the absorption varies with air conditions. This absorption is only important above 4,000 Hz; it increases with frequency and also depends on humidity and temperature.

SOUND INSULATION

The sound from the speaker or artist plus the reverberation should be the only sounds to reach the microphone. The studio must be insulated against all other possible sources of unwanted sound.

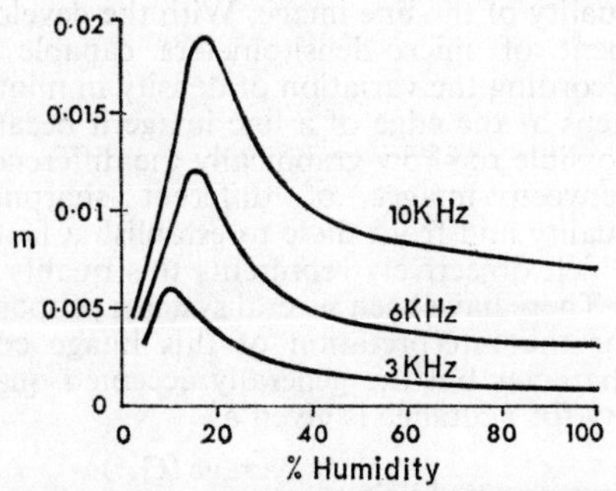

AIR ABSORPTION. Relation between factor m, humidity and frequency at 20°C. Air absorption may be expressed as $4mV$, where V=volume in cu. ft.

There has been over the years a gradual rise in the noise level in everyday surroundings. Traffic along roads and streets has become heavier, and the landing and take-off of modern aircraft causes considerable noise.

MAXIMUM PERMISSIBLE NOISE LEVELS (DB ABOVE 0·0002 DYN/CM²)

Octave Band Hz	*Criteria* A	B	C	D
37–75	53	54	57	60
75–150	38	43	47	51
150–300	28	35	39	43
300–600	18	28	32	37
600–1,200	12	23	28	32
1,200–2,400	11	20	25	30
2,400–4,800	10	17	22	28
4,800–9,600	22	22	22	27

Criterion A
Concert-halls with best possible conditions.

Criterion B
Concert-halls where A is impossible; opera houses; theatres (more than 500 seats); broadcasting studios.

Criterion C
Theatres (up to 500 seats); assembly halls; conference halls (for 50); classrooms; music rooms.

Criterion D
Cinemas; churches; courtrooms; conference rooms (for 20).

SOUND TRAVEL. There are two ways sound can travel in a building: airborne and by impact.

Sound by airborne paths can be direct, for example, through an open window or past a badly fitting door. Indirect transmission occurs when the sound waves from outside cause ceilings, windows, walls, etc., to vibrate and radiate into the room, or when the sound actually travels through the walls, ceiling, floor, etc.

Impact sounds include footsteps on an uncovered floor, machinery vibrations and slamming doors. The source causes vibrations in the structure itself, and these vibrations can easily travel considerable distances through steel, concrete and brick.

NOISE PREVENTION. If possible, of course, means should be found to stop the noise from being generated. The impact noise due to footsteps along a corridor can easily be reduced by covering the floor with carpet or linoleum; doors can be fitted with slow-closing mechanisms to prevent slamming; proper design of water fittings can reduce considerably the noises set up in a plumbing system. In ventilation systems, the trunking can be an excellent generator and transmission path for both airborne and impact noise and here careful design and layout are necessary. Since the steel core of a building transmits impact noises, it has been omitted in some broadcasting studio centres. Instead, the building core consists of a massive structure to retain strength but impede the transmission of sounds from floor to floor.

Although a certain amount can be done to prevent noise being generated, by far the biggest problem is to keep out the noises over which there is no control. Here the best form of insulation is isolation. By careful planning of buildings, the areas requiring quiet conditions can be isolated from sources of noise. The first step in the design of broadcasting centres, film studios, recording studios and concert halls is to arrange that the areas which can tolerate somewhat noisy conditions shield those areas in which quiet conditions are imperative.

A source of noise must not be placed next to a quiet area. In one case, a gymnasium was placed immediately above a lecture theatre with disastrous effects on the lecturer and his audience. Some relief in such situations can be provided, but the best answer is proper layout. G. W. M.

See: *Loudspeaker; Microphones; Sound and sound terms; Sound recording; Sound recording systems.*

Books: *Acoustics*, by G. W. Mackenzie (London), 1964; *Acoustics, Noise and Buildings*, by P. H. Parkin and H. R. Humphreys (London), 1958; *Acoustical Techniques and Transducers*, by M. L. Gayford (London), 1961; *Music, Acoustics and Architecture*, by L. Beranek (New York), 1962.

⊕**Acres, Birt, 1854–1918.** Born in USA of English parents, educated in Paris, and joined the photographic works of Elliott and Company in Barnet, where he developed his interest in photography. He became interested in the recording and reproduction of movement in 1893 and began experiments in this field. He patented his Kinetic Lantern apparatus for taking and showing moving pictures in 1895, and entered into an association with R. W. Paul to develop workable apparatus. Also in 1895 he successfully filmed the Derby and the opening of the Kiel Canal.

Conflicting personalities soon led to a break with Paul, and Acres continued to work on his own, becoming in January 1896 the first Englishman to give successful shows of projected animated pictures to photographic groups. His belief in this new medium as a device for serious historical record-making and his objection to its frivolous commercial exploitation led him to delay public presentation until after others had already entered the field in the early months of 1896. Acres continued to develop and improve his system; in 1898 he marketed the Birtac camera-projector, one of the first systems for home movie-making, using 17·5 mm film.

⊕**Action.** Term sometimes used to designate picture film in contrast to sound film. "Action!" is the order given by the director of a film when the picture camera and sound recorder are running to speed, indicating that the action of the shot is to begin.

⊕**Acutance.** Acutance or acuity is a quantity ☐used to provide an objective measurement of that character of a photographic image which is subjectively assessed as its sharpness. It is based on the measurement of the spatial rate of change of density across the boundary of the image by the use of a recording micro-densitometer.

For many years, following classical work in astronomical optics, it was considered that the performance of a photographic lens would be satisfactorily represented by the measurement of its resolution, or resolving power, and this was extended to the evaluation of photographic emulsions in the same terms. However, it was found that the assessment of the ability of a system to resolve a fine line pattern was by no means a satisfactory method of rating its capacity to produce sharp pictures, and the concept arose that for this the number of lines resolved was less important than the quality of the line image. With the development of micro-densitometers capable of recording the variation of density in minute steps at the edge of a line image it became possible to show graphically the differences between images of different sharpness quality and from these to establish a factor which objectively represents this quality.

There have been several systems proposed for the interpretation of this image edge character but the generally accepted quantity for acutance is given as

$$\text{acutance} = \frac{\text{average}\,(G_x^2)}{DS}$$

where average (G_x^2) represents the mean of the squares of the density gradients $\Delta D/\Delta x$ across the boundary between a light and a dark area in the image having a density difference DS.

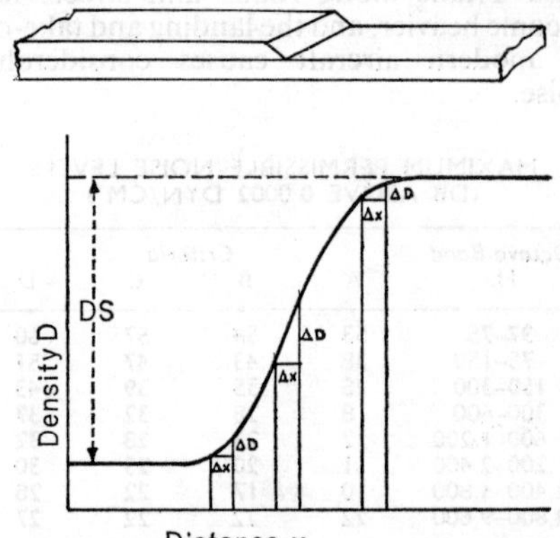

DENSITY GRADATION. In a photographic image from a knife-edge exposure the boundary between the light and dark areas has a gradation of density which can be shown as a curved trace by a micro-densitometer. Gradient G at any point X is $\Delta D/\Delta x$.

Initially, acutance measurements were made on photographic emulsions in which the image boundary was produced by the exposure of a knife-edge object in contact with the film. Later the same basis of assessment was used for the evaluation of lenses.

It was found that the focus position of a lens for maximum resolution in the image was not necessarily the same as its position for the image of greatest sharpness in terms of acutance; in the case of a point object, the image of maximum resolution was the smallest circle but was surrounded by zones of varying density rings, whereas the image of maximum sharpness was larger in diameter but had extremely clearly defined edges with minimum density variation outside the central image.

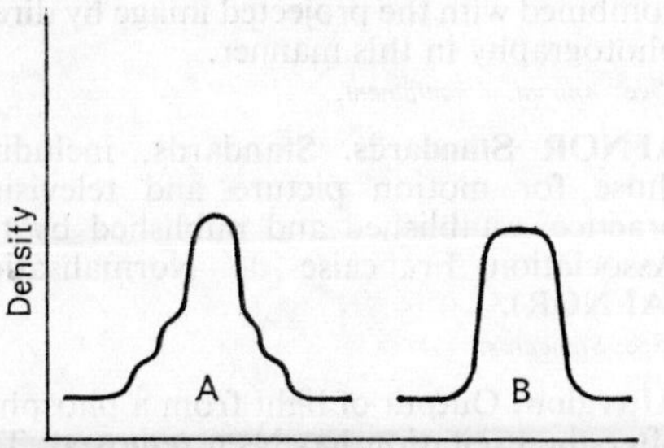

SHARPNESS AND RESOLUTION. Micro-densitometer traces across the photographic images of a point source formed by a lens focused (A) for maximum resolution and (B) for maximum sharpness.

It was also established by subjective viewing tests that acutance ratings of image forming systems correlated generally very satisfactorily with the viewer's judgment of sharpness, where with pure resolution ratings the correlation was by no means uniform and might even be anomalous.

Within recent years the basic concept of acutance rating based on a single image edge profile has been extended and the performance of photographic image forming systems is more and more based on the assessment of spatial frequency analysis with contrast reproduction in what is termed the modulation transfer function.

Where it is desired to specify the subjective sharpness resulting from the operation of a sequence of processes such as camera – film – projector – viewer, the modulation transfer functions of each step may be combined to provide a quantity known as the SMT acutance or system modulation transfer acutance. L.B.H.

See: *Definition; Image formation; Lens performance; Resolution; SMT acutance*

⊕□**Add-A-Vision.** A combined film and television camera system based on the Mitchell BNC but of British design. A 10 : 1 zoom lens forms the film image, but a light-splitting prism in the shape of an optical cube mounted in front of the shutter bleeds off a proportion of the incoming light to form a second image on the faceplate of a Plumbicon tube set at right-angles to the camera axis. Two light-splitting ratios are available, 3 : 1 and 9 : 1, both favouring the film channel, the latter to suit colour emulsions.

The Plumbicon signal after processing feeds an electronic viewfinder, together with floor monitors accessible to the film's director and lighting cameraman. A videotape recorder enables the programme to be stored for future transmission or used as an editing guide. Multi-camera techniques may be employed with centralized programme control.

A variant of Add-A-Vision known as EFS (Electronic Filming System) is basically similar, but employs the Mitchell Mark II camera, also equipped with a 10 : 1 zoom lens. Focus, zoom and aperture controls are by servo to make possible one-man operation of the camera.

Monitoring signals from two or more EFS cameras may be fed to a control desk where the director, normally with two assistants, selects the channel to be used and places an identifying mark on a control track as a subsequent indication to the film editor.

As with other multi-camera arrangements, the need for continuous variation of set lighting as the action proceeds requires a lighting control console equipped with a memory, so that a series of set-ups can be established during rehearsals and played off in pre-set sequence while filming.

See: *Camera; Electronicam; Gemini; System 35.*

⊕□**Additive Colour Process.** System of colour reproduction based on the addition of light of the three primary colours, red, green and blue, on a single surface, rather than by the use of coloured dyes or pigments within a three-layer recording medium. In motion picture work, additive colour processes are obsolete, but colour television systems are fundamentally additive.

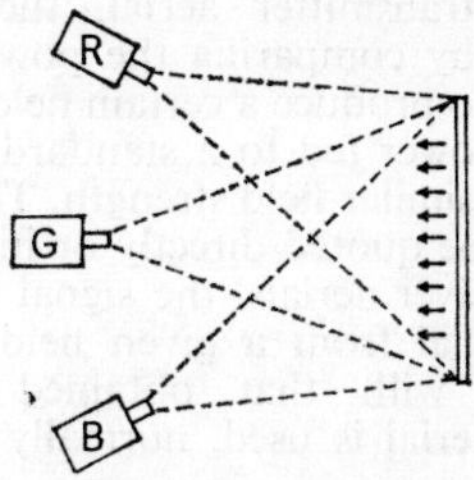

See: *Colour principles; Colour systems; Subtractive processes; Trinoscope*

⊕**Additive Printing.** Process for the printing of colour positive film in which the intensity and colour of the exposing light is modulated by separate variation of its red, green and blue components, which are subsequently combined (added) at the point of which the film is exposed.

See: *Colour printers (additive).*

⊕**Advance (Sound and Picture).** When film is projected, the picture gate and sound head are not adjacent to one another, but are staggered along the length of the film. Thus the printed relationship of picture and

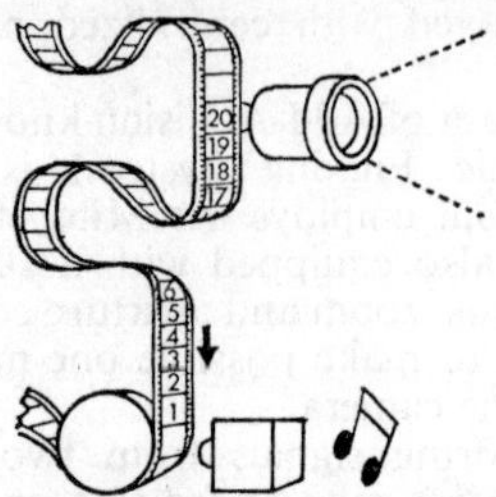

corresponding sound cannot be a parallel one, and, in fact, the sound is generally ahead of the picture when the two are correctly synchronized. The sound advance is 21 frames for 35mm optical and 26 frames for 16mm optical.

See: *Film dimensions and physical characteristics; Penthouse head.*

□**Aerial Gain.** The power gain of a given aerial (antenna) compared with an isotropic radiator, i.e., a radiator which will radiate uniformly in all directions. Thus, a measure of the directivity of the aerial compared to a standard aerial. An isotropic radiator is impossible to achieve, and is purely theoretical, so it is more usual to compare the gain to a simple dipole. A dipole itself has a gain of 1·64 compared with the theoretical omnidirectional aerial.

For a transmitter aerial, the gain is measured by comparing the power fed to the aerial to produce a certain field strength with the power fed to a standard aerial to produce a similar field strength. This power gain may be quoted directly or in decibels.

For receiver aerials, the signal produced by the aerial from a given field strength compared with that obtained from a standard aerial is used, normally stated in decibels.

See: *Dipole; Receiver; Transmitters and aerials.*

⊕**Aerial Image Photography.** Method used for special effects cinematography in which the

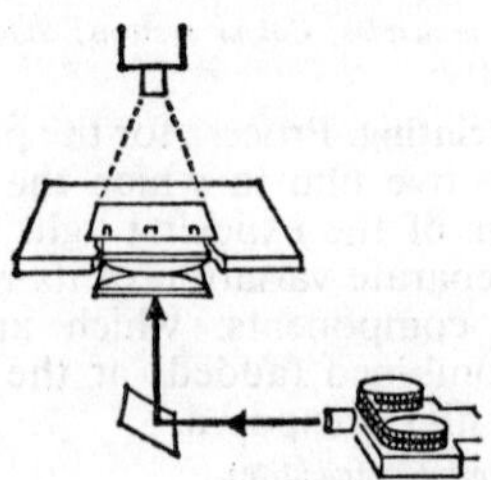

subject photographed is an optical image in space rather than one formed on a screen. Titles, models or other film material may be combined with the projected image by direct photography in this manner.

See: *Animation equipment.*

⊕**AFNOR Standards.** Standards, including
□those for motion picture and television practice, established and published by the Association Française de Normalisation (AFNOR).

See: *Standards.*

□**Afterglow.** Output of light from a phosphor after the excitation has been removed. The rate of decay, or disappearance of the light emission, is the afterglow characteristic.

See: *Phosphors for* TV *tubes; Telecine; Telerecording.*

□**AGC Detector.** Device which provides an output proportional to the signal level for application to variable gain devices earlier in the chain which can thus be made to maintain a constant output level. Commonly employed in radio and television receivers.

⊕**Agfacolor.** Negative/positive integral tripack colour film process first introduced for professional 35mm motion picture production in Germany about 1940. Now in use as Agfa-Gevaert.

See: *Colour cinematography.*

⊕**Agitation.** In continuous film processing, a means of carrying away oxidation products from the neighbourhood of the emulsion surface, where pockets of exhausted developer may give rise to unwanted streaks.

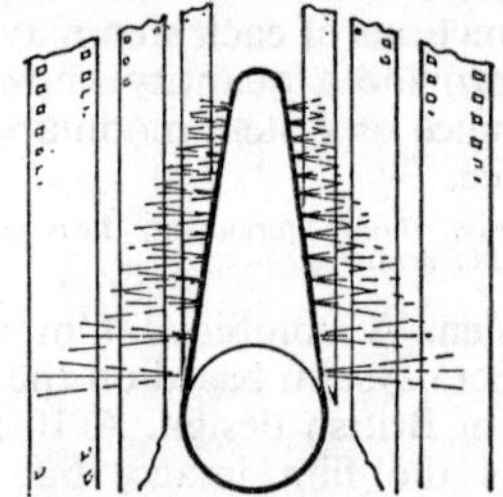

Agitation, also called turbulation, is often effected by compressed air or nitrogen jets, by mechanical stirring, and by spraying developer on to the emulsion surface from submerged jets.

See: *Chemistry of the photographic process; Processing; Spray processing; Turbulation.*

⊕**Air Knife.** Compressed air jets which wipe excess solution from films in continuous processing machines, so that each solution

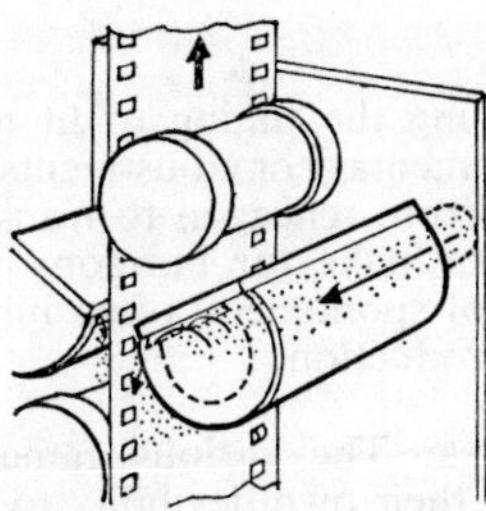

is not unduly contaminated by the preceding one. The superficial liquid is wiped back into the tank by the jet. The terms air-squeegee or blow-off are used for the same operation. At high processing speeds the quantity of liquid carried on the surface of the film may be considerable and a series, or bank, of air knives in succession may be necessary.

See: *Processing; Squeegee.*

⊕□**Ambient Lighting.** In a cinema theatre, the continuous general lighting of the auditorium which must be maintained even when the picture is being projected, for safety precautions and the convenience of the audience. To avoid reduction of the projected picture quality it is important to limit the general ambient light falling on the screen to the minimum.

In home television viewing the ambient lighting is the general room illumination, and television picture reproduction characteristics are designed with this factor in mind.

See: *Visual principles.*

⊕□**Amplifier.** Device for increasing the power associated with a phenomenon with the least possible alteration to its quality, and with the introduction of the least possible noise. Amplifiers perform their function by making the input control a larger amount of power which is supplied by a local source to the amplifier output.

⊕**Anaglyphic Process.** Method of printing the two views of a stereoscopic camera in complementary colours on a single length of film. The superimposed images on the screen are sorted out by equipping the audience with spectacles carrying filters of the same two colours, mounted so that each eye sees only the picture destined for it.

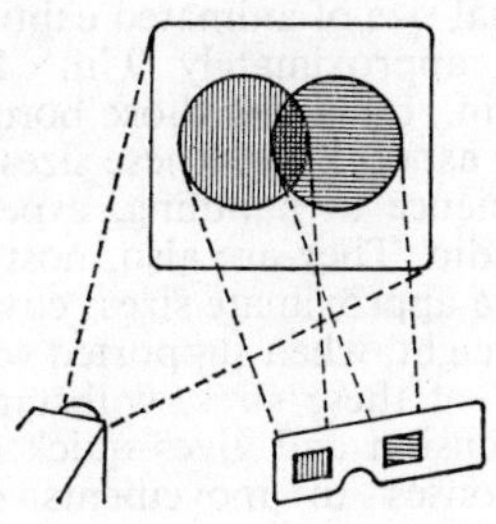

See: *Stereoscopic cinematography.*

⊕**Anamorphic Processes.** Techniques by which the image on the film is distorted by the use in the camera or printer of optical systems having different magnifications in the vertical and horizontal dimensions. The most frequently used is one in which the

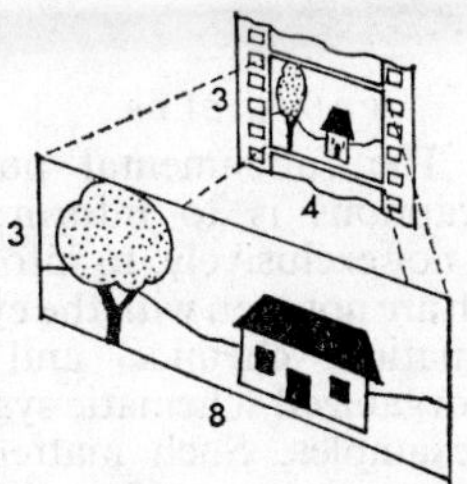

picture on the print is compressed or squeezed horizontally, the effect being corrected by projection on to a wide screen with an anamorphic lens giving a corresponding lateral expansion.

See: *Wide screen processes.*

⊕□**Angle, Camera.** Field of view of a camera when it is set up to shoot. The qualifying terms "high", "low", "wide", are based on an imaginary norm which more or less corresponds to a 35mm camera with a two-inch lens pointed at a scene from shoulder height.

See: *Cameraman.*

⊕**Animascope.** Trade name of an American animation process which enables complex movements of human characters and models to be reduced to animated form, in place of the synthesis of such movements from drawings which adds much to the cost and complexity of conventional animation techniques.

Basically the process is a photographic one, in which actors (in animal dress if the story calls for animal characters) are recorded on a special type of film which generates outlines and suppresses internal detail, as well as flattening the colour to give a cartoon effect.

Backgrounds are drawn and painted, and combined with the live action strip by special effects techniques such as the blue screen process.

See: *Animation equipment; Animation techniques.*

□ANIMATED CAPTIONS

Animated captions for television differ from film animations in method and purpose. Live television animation techniques should not be confused with those of film animation where movement is built up frame by frame. The TV technique is best described as live action animation, movement being hand-operated by simple mechanical means. The action takes place in front of the television camera and within the aspect ratio of the receiver, the caption being supported with a hardboard backing on a vertical easel for transmission.

PRINCIPLES

PURPOSE. The fundamental purpose of animated captions is to inform; in particular, but not exclusively, to inform about things which are not seen with the eye. Statistical information, chemical and physical structures, organized schematic systems and maps are examples. Such matters, which comprise the largest area of need for visual information, are readily interpreted in animated caption form.

In the preparation of captions, the design (intention) must be clearly defined before the basic techniques of animation can be usefully employed. In other words, the technique must be made to interpret the idea. The technique of making and handling being basically simple, subject matter is necessarily designed to eliminate all non-essentials.

The contributing factors of line, tone, shape, proportion and symbolic design, two-dimensional and diagrammatic in character, all combine to aid the viewers' perception. For children's programmes and educational television, it is essential to associate the diagram with some actual experience (e.g., the see-saw and the lever principle). The use of animated captions in television assists this process. With more sophisticated viewers and subject matter, the abstraction of visual symbolism from such examples as molecular structures can be clearly illustrated by means of animated captions.

ADVANTAGES. The advantages of designing and making animated captions for use with live television cameras are twofold. First, when information becomes available at short notice, it can be interpreted and made ready for transmission without additional processing; secondly, and more important, the visual can be animated and held in vision, varying the timing to fit modifications in commentary or adjustments in speed of delivery from rehearsal to transmission. These points make for harmony with the conditions of spontaneity often inherent in television production.

TECHNIQUES. The various camera techniques lend their own flexibility to those of live television animation. These include superimposing or inlaying one picture on another, dissolving or mixing from one caption to another, and the tracking or panning effects of the movements of the camera itself.

Interpretation from verbal information or abstract idea into visual form disciplines the method and style of the animation: the subject matter usually determines the way in which animated captions are constructed. For this reason, any general description of techniques must be confined to basic aspects of construction, and a few specific applications.

The concept of live television animated captions is founded on the idea of concealing and revealing items of information at will: to achieve this effect, six basic directions of movement are used: horizontal, vertical, diagonal, circular, level and scissor.

These movements may be used individually for simple animation effects, or in combination one with another, giving remarkable flexibility of animation and design techniques considering the limitations of the simple materials used to make captions. The basic movements provide the background or vehicle for design, combination giving scope for inventive animated illustration.

SIZE. The aspect ratio of television pictures is three vertical units to four horizontal units. These proportions may have to be modified to suit the special requirements of animation, e.g. where camera movements over the caption area are necessary.

The actual size of animated captions may vary from approximately 9 in. × 12 in. to 18 in. × 24 in., excluding those border areas outside the aspect ratio. These sizes are best for convenience of handling, expecially in the TV studio. They are also most efficient within these approximate sizes: cover paper of heavy weight, when supported within the framework of these sizes, withstands considerable tension and gives quick and efficient responses to movements such as pulling, without folding, rippling or binding. Of course, smaller and larger sizes may be used when necessary.

MATERIALS AND TOOLS

The electronics of TV requires the use of special materials. The following selection has been found adequate for most purposes, but no doubt it could be extended by intending designers of animated captions.

PAPER AND BOARD. Essentials are, first, heavy duty cover papers (5-sheet or thereabouts) in black, off-white and two intermediate greys. Cover paper of this quality is self-coloured throughout, eliminating any tonal changes in cut edges. It is easily cut with a sharp knife, which makes for ease and speed in carrying out the work.

Black provides the basic colour; cut-out shapes within areas of black are provided electronically by rendering such shapes invisible on the television tube unless or until it becomes necessary to reveal them.

The next essential is a good quality, though not very thick, mounting board. This is used for making runners or tracks between which the moving or animated parts of the caption are to ride. Cut into strips, and used in conjunction with double-sided adhesive tape, tracks can be made and applied to the caption with speed and efficiency, enabling movements to be perfectly controlled during transmission.

Printed textured papers, e.g., with dots and straight or wavy lines, simplify the work of defining areas or abstract qualities such as air, sea and land. Textured papers are useful in helping to obscure cut-outs where they are not intended to be visible; for example, an area may be divided without the separate parts being visible in advance.

OTHER MATERIALS. Acetate or celluloid, both clear and frosted, used together with the foregoing materials, provide a very wide range of animation effects. These are obtainable in varying gauges of thickness. The thinnest gauge consistent with stability in use is about 1/5000 in., but this must vary with the purpose, e.g., a wider area will require a thicker gauge.

Various elastics, such as shirring elastic, facilitate the handling of certain movements, and may themselves provide lines of direction and movement which cannot be made in other ways.

TOOLS AND FIXATIVES. Metal and leather punches of varying size and shape are essential, and may be used for punching out anything from a simple dot or dash line to an abstract arrangement of a complex molecular chain or structure.

Movements controlled by pivots and axes require the use of office-type split pins or paper clips. Press studs, and metal lace-hole pieces which can be punched into cover paper or card, are useful additional aids.

Double-sided adhesive tape is needed. In recent years, the development of easily-handled tape of this kind has immensely simplified the making of animated captions. The tape may be obtained in various widths, ½ in. and 1 in. being the most useful. Cellulose tape is a necessary aid but should be used only behind the scenes in parts of the animation where it will not appear in vision.

In addition to these basic materials, a steel cutting edge such as a 2 ft. steel rule and a cutting knife with replaceable blades, drawing instruments and a good range of art materials are necessary, especially white pencils and graphic designers' colours.

SIMPLE ANIMATION METHODS

STRAIGHT LINE. The fundamental principle of animated caption making can be shown by describing how to animate a simple straight line. First, a line of dots is

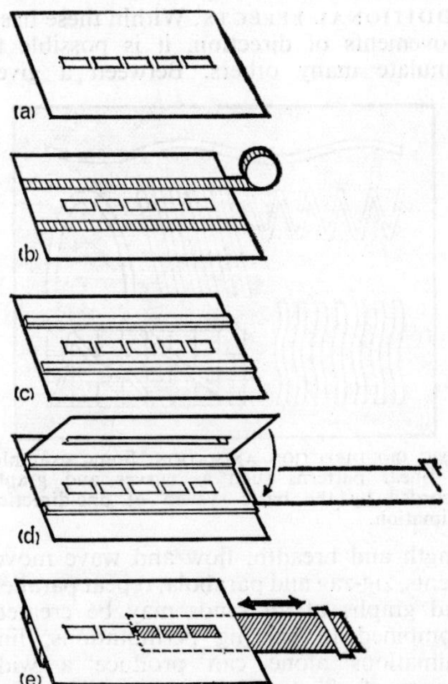

BASIC ONE-DIRECTION ANIMATION. (a) Line of cut-out dots. (b) Application of double-sided adhesive tape. (c) Runners. (d) Backing and slide. (e) Movement.

cut out or punched out in the black cover paper. This is followed by laying on the double-sided adhesive tape in parallel lines (in this instance at points equidistant from the original cut-out line).

The parallels may be at any angle in relation to the original cut-out; the angle is determined by the relationship of the cut-out to any other lines or shapes in an animation.

Strips of mounting board are held in place by adhesive tape. These strips are to act as runners between which the concealing strip of cover paper rides.

The backing paper, showing white line to coincide with cut-outs, is placed in position after further layers of double-sided adhesive have been affixed to the runners. A slide, also made of cover paper, is cut to the width between the runners so that it will ride freely. A crossing piece of paper fixed to the end of the runner controls the length of the runner in relation to any cut-out line or shape.

When the model is completed, the action of withdrawing a slide will apparently make a line consisting of a series of white dots move as required when properly arranged in front of a camera. If a white line is painted on the slide itself, the impression created is of a diminishing line.

ADDITIONAL EFFECTS. Within these basic movements of direction, it is possible to simulate many others. Between a given

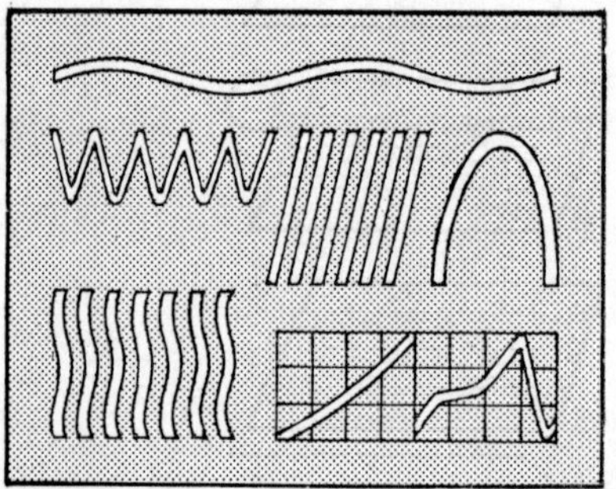

BASIC ONE-DIRECTION ANIMATION. Some examples of linear patterns such as curves and graphs recorded by the basic system of one-direction animation.

length and breadth, flow and wave movements, zig-zag and parabola, repeat parallels and graphs of all kinds may be created. Combined in varying permutations, line animations alone can produce a wide variety of effects. If these cut-out designs are combined with slides on which other lines of direction are painted the effects can be varied still further.

Holding to the basic one-direction animation, it is possible to vary tonal arrangements as well as linear movements. Slides may be constructed with various tones and arrangements of tones.

Experiments with a simple one-piece, one-direction animation, combining various linear and tonal arrangements in juxtaposition, will reveal almost limitless possibilities for animation.

LEVER AND CIRCLE. Simple lever animation is held within the aspect ratio of the frame in order to control the movement in two dimensions. The fulcrum of the lever

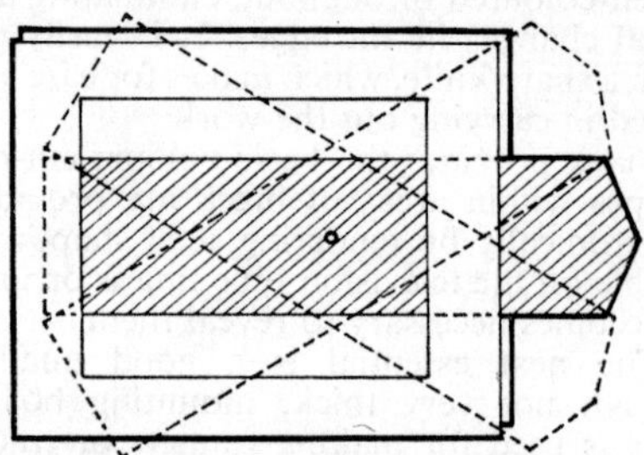

LEVER ANIMATION. The frame is used to limit the movement of the lever to the required extent by blocking the free end of the lever piece.

may be anywhere within the frame, depending upon the purpose of the illustration. The pivoting point is made with the smallest size of office paper fastener of the split pin type. The card is reinforced on the back with cellulose tape to reduce wear.

A circle model is similarly held in the frame, and the centre of movement may also be placed at any convenient position. One aspect of circular motion is simulated in animation by cutting or punching out on a surface layer of caption card, and by having the circular movement operating behind it.

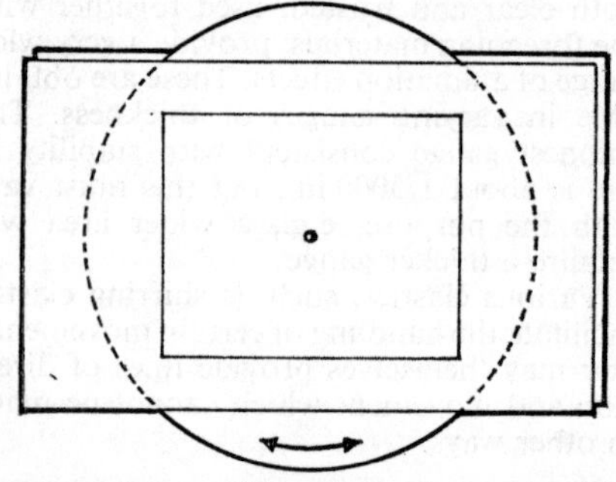

CIRCLE MOVEMENT. Circumference is supported by upper surface of caption and circle centre is fixed to backing piece. Rectangle shows camera area.

This can show the effect of a satellite moving around the earth. A white spot on a circle which turns behind a circle of punched-out dots in a black surface card effectively illustrates this method.

DOUBLE DEPTH ANIMATIONS

COMBINED MOVEMENTS. There are many possibilities for combining different directions in double depth. A horizontal sliding movement may be followed by a vertical sliding movement. If the upper layer (on which the design or pattern is drawn) is transparent or translucent, then the two directions can be operated simultaneously and give scope for fascinating developments in animated caption design.

Sliding movement may be combined with lever movement as an alternative to combining the sliding movements only. The

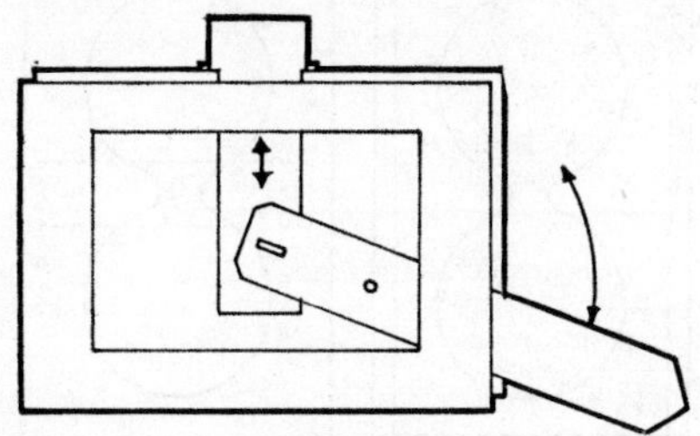

COMBINED MOVEMENT. Pivoted lever actuates vertical sliding action; slot in end of lever allows for compound movement.

pivot point of the lever is affixed to the upper layer of cover paper and the sliding piece is set one level below it. A slot where the lever is attached to the slide allows for the increasing and decreasing diameter as the mechanism moves up and down.

Further combined movements, such as the piston and circular ones, will readily suggest themselves.

It will now be evident that further layers of animation can be added to the two depths mentioned. Three and four layers of movement, one over the other, will work before the camera very satisfactorily. Care should be taken in design to ensure that openings are sufficiently generous not to permit cast shadows to intrude or obscure the illustration in any particular. At any level, any of the basic movements may be used to convey the intentions of the designer.

CUTTING TECHNIQUES. Cutting to reveal areas and the slides at different angles gives differing simulations of movement. A vertical line may be revealed by means of a horizontal pull simply by cutting the sliding piece at an angle other than a right-angle. The designer can relate the angle at which the slide is cut to the desired speed of reveal.

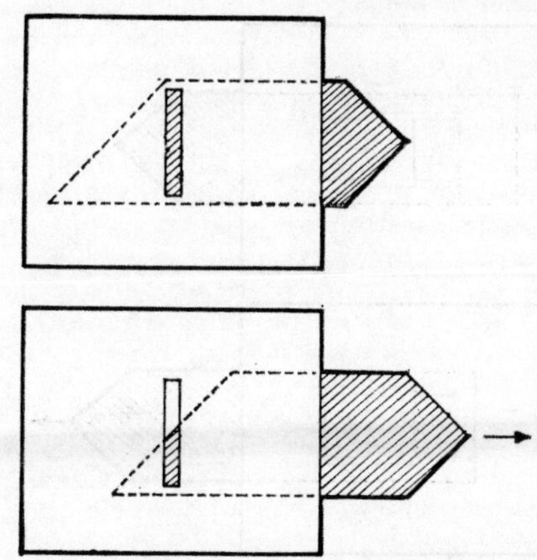

CUT-OUT TECHNIQUE. A line or column may be animated vertically by a horizontal movement of the operating slide, if the end of the slide is cut at an angle to the vertical. Slope of this angle controls rate of vertical movement.

Three different directions may be revealed in sequence by means of one horizontal movement. In this case, a white line is painted on the sliding piece itself in order to delay the appearance of the lower two directions.

Two pairs of similar rectangles, each pair cut one in a surface layer and the second in a sliding piece, may be placed together in order to make one larger rectangle. As the sliding opening moves from left to right, the left-hand surface opening is being closed,

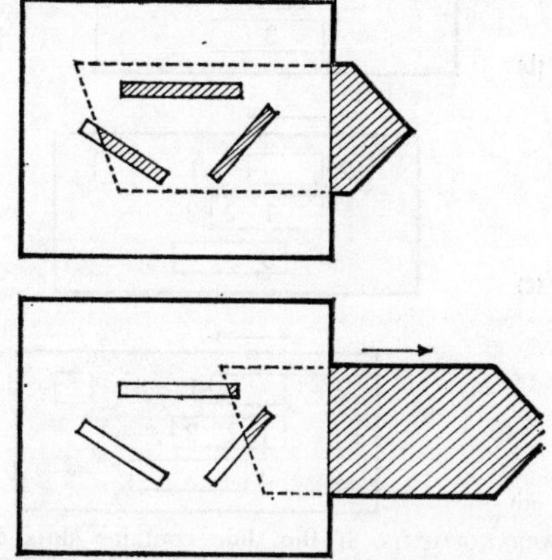

CUT-OUT TECHNIQUE. Different angling of cut-out slots produces different directions and speeds of movement with a single horizontal movement of the operating slide.

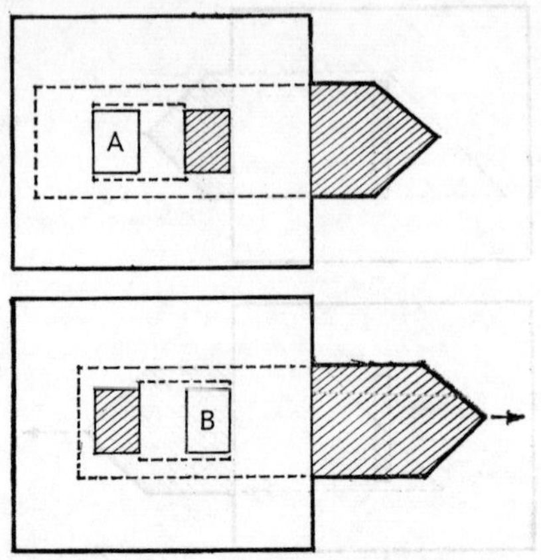

CUT-OUT TECHNIQUE. Identical techniques may be used for appearances and disappearances, or replacement of one symbol by another, as A by B.

and the right-hand opening is being revealed. Pushing the slide back will, of course, reverse the opening–closing procedure. This technique is very useful where a visual comparison is desired.

This method may be extended as follows: the numbers 1, 2 and 3 are randomly distributed within the caption area and are

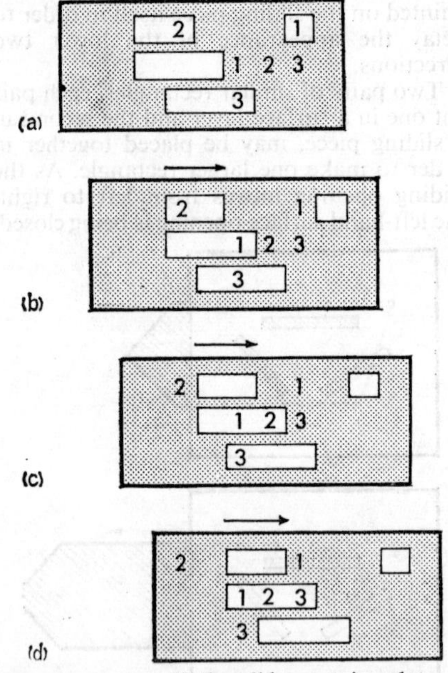

REARRANGEMENTS. If the slide contains slots of calculated size and position, it may be used to rearrange numbers or other symbols so that, starting from a jumbled order in (a) and going through intermediate stages in (b) and (c), a correct ordering results in (d).

visible through three openings of progressive size in a moving frame; a further opening is cut to the left and centrally in this frame equal to the area of all three numbers. As it moves from left to right, first the visible number 1 is concealed and the covered number 1 is revealed. When finally the numbers 3 are concealed and revealed (with a quick movement), it will appear that the random order has re-assembled itself in progressive order in the centre of the screen. In order to ensure that the action stops precisely where required, a crossing strip of paper should be affixed (with double-sided adhesive tape) to the concealed end of the slide, so placed that it will hold the movement at any predetermined position.

Sometimes it is necessary to control a

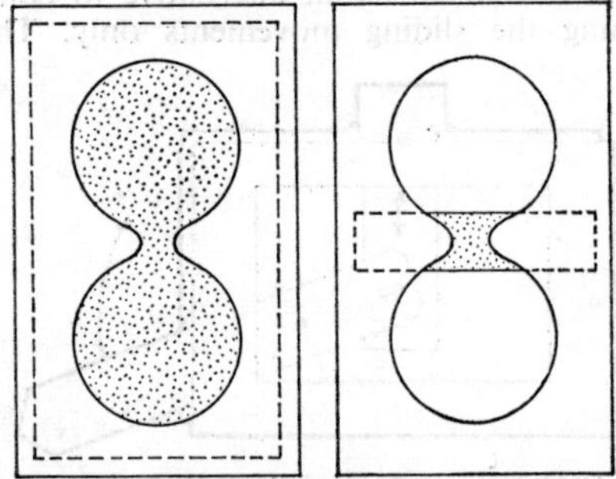

MECHANICAL SUPPORTS. Weak areas liable to curl or break are suitably strengthened by transparent bridging pieces.

movement by stopping the action in two directions. An opening cut in the backing of the slide will reveal the moving piece within.

A crossing strip attached to it will control the movement to any degree that the opening allows. The amount cut out of the backing is determined by the design requirements on the face of the caption.

LINKING SHAPES. Some shapes need support within an animated caption. The central shapes may tend to curl, thus changing the shape of the design, unless tied together in some way. Clear acetate used as a narrow strip across the centre, and secured with double-sided adhesive tape, or as a sheet covering the whole area, overcomes this problem. Where acetate is used to support and cover a whole area, it may further be used to carry some part of the design.

Sometimes an odd-shaped end of a slide may need to be held down. In such a case, acetate over the surface opening will prevent curling and distortion of the picture.

BAR GRAPHS. Another elementary application of technique is that of a simple bar graph. The increase in TV sets over a twenty-year period could be shown by bringing up on the screen a caption with title and years revealed as a basis on which to build. Each unit would represent *x* TV sets. As each year is separately revealed, the visual impact of growth is forcefully shown by the basic horizontal slide animation.

FURTHER APPLICATIONS

As with mechanics, there are only a few primary principles on which all applications are based. The rectangles described demonstrate some ways and means. All manner of shapes and forms are cut out at various levels of any animated caption. Most of the examples given convey factual information. However, a succession of animations and illustrations used in conjunction with the great flexibility of the television cameras themselves could well illustrate a story.

Every subject requires its own solution. With the foregoing basic movements and constructions on which to build, the graphic designer may resolve further problems of communication in this field. K.R.J.

See: *Special effects* (TV).

⊕ANIMATION EQUIPMENT

The purpose of an animation stand is to photograph art work precisely positioned on pegs and lying beneath the lens of a camera pointing vertically downwards. The camera moves on a vertical axis; this calls for the utmost rigidity in the stand, and determines its basic design. The art work attached to the peg tracks (there are usually two or more of these) can be moved east/west, i.e., to left and right, or north/south, i.e., away from or towards the operator. The double-layered sliding device which effects this movement is called a compound.

Modern animation practice calls for a variety of sophisticated equipment to provide the complicated, fast-moving effects expected of animated film today. This equipment must be precision-built, rugged and reliable with automatic operation where feasible. It was very different in the early days.

EVOLUTION

Most early animation stands were either made by the animator or the local engineering works. The animator generally used wood and had a crude slide for the camera which was helpful in changing field size but seldom any good for tracking. The engineer used lathe beds, or copied their principles to provide a smooth track when controlled by a lead screw.

In the early days, the animator used any camera that happened to be available, adapting it to his own needs where possible.

One good idea was to mount a camera with a zoom lens in the ceiling and position it over a table in the centre of the room. It was not completely successful, because few zoom lenses allow tracking into a small field and their definition is not usually as high as that of a normal lens.

Some early stands looked rather like four-poster beds; the camera was mounted vertically in the roof and the animation table with the lights moved up and down beneath it. However, most animation stand designers favoured twin columns which were fixed to the wall while a heavy lead screw raised and lowered the camera bracket. Some modern types have twin columns but these are now part of the base, making the machine independent of the wall, and the lead screw has generally been replaced by chain.

Many stands are now produced with a single column. In one make, this is offset to one side, so that, when combined with a rotating compound as well as a table-top rotation, it allows long north/south or any angle tracks to be made.

Fades, dissolves and follow focus have only recently become automatic; there are many stands still in use where the animator must operate the zoom hand wheel and focus the lens between each frame.

MODERN EQUIPMENT

A modern animation stand comprises the following units: Basic stand; zoom counter; zoom motor; camera, with auto fade and dissolves, etc.; stop motion motor; automatic focus; cam lift; table top and peg tracks; floating pegs; platen; table rotation; compound; compound rotation; pantograph; shadow board; aerial image projector.

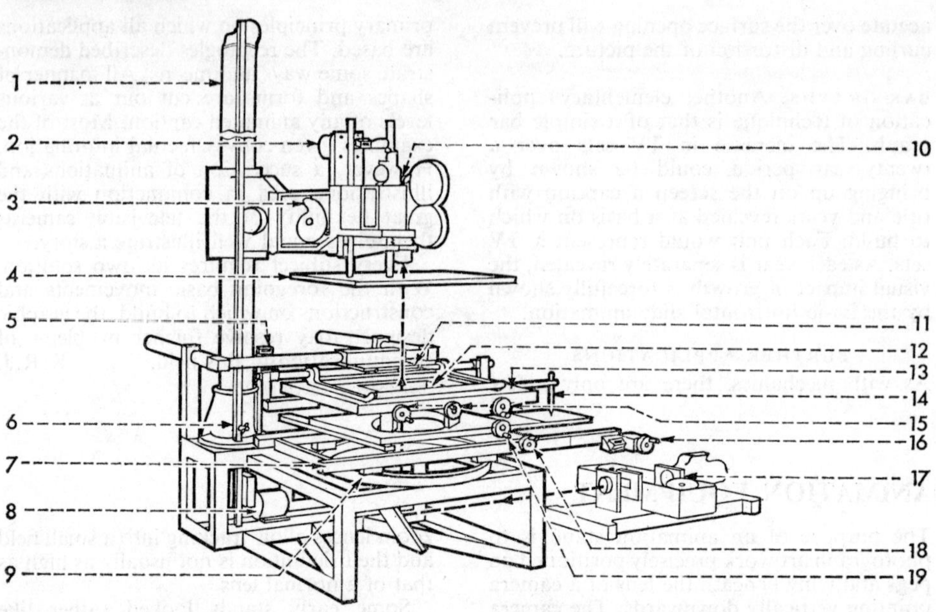

ANIMATION STAND. Complete camera equipment required for production of cartoon films and technical animation. 1. Column. 2. Stop motion motor. 3. Follow focus cam. 4. Shadow board. 5. Zoom counter. 6. Cam lift. 7. Compound. 8. Zoom motor. 9. Compound rotation. 10. Camera. 11. Floating pegs. 12. Platen. 13. Table top. 14. Pantograph pointer. 15. Peg track hand wheels. 16. Table east-west handle. 17. Aerial image projector. 18. Table north-south handle. 19. Table rotation handle.

BASIC STAND. Of very rigid construction to allow the camera to move up and down without vibration, the stand must also maintain the camera in exactly the same position horizontally during vertical travel; this means there cannot be any play in the bearings.

In the camera this is usually accomplished by having 16 adjustable ball races running on two $4\frac{1}{2}$ in. centreless steel ground tubes which are hard chromed. Some manufacturers use cotton-reel shaped rollers; others square key steel bars to run their ball races on.

The height of the stand can vary between 8 ft. and 13 ft., although there is at least one of 17 ft. Most, however, are around 12 ft. high.

The movement up and down the stand of the camera is called zooming or tracking, and on a 12 ft. stand using a 50mm lens, the field size (size of picture measured across horizontally) ranges between 2 in. and 30 in.

On a basic stand, this movement is operated by a hand wheel, but nearly all stands are now motorized.

ZOOM COUNTER. The position of the camera on the stand is always registered by some form of scale; usually this is a counter reading in 1/100th parts of an inch. The accurate positioning of the camera is very necessary in planning a zoom, or when it is necessary to repeat either for double exposure or for reshooting.

ZOOM MOTOR. This can be a simple variable a.c. motor giving a speed range of 5 : 1. Most stands, however, use a d.c. motor with a speed range of 30 : 1. If an electronic control is incorporated, the range can be extended to 100 : 1 as well as being interlocked with a motorized compound for traversing the art work.

ZOOM JOG SWITCH. A trailing switch with two push buttons, one for up and the other for down, enables the operator to look through the viewfinder and zoom the camera without using the control panel.

CAMERA. For animation, the camera has to be something more than just a studio camera, which relies on the film editor and the optical printer to achieve its special effects.

A typical process camera designed specifically for animation will give utmost precision of film registration with great reliability by using a shuttle-type gate with a fixed register pin. A long series of double exposures, running both forward and reverse, can be made with perfect registration.

This precision is possible because the film register pins are fixed. The film to be advanced is lifted clear of the register pins and on to the transport pins. After being advanced it is deposited back on the register pins. The stripper and pressure plates which transfer the film between register and transport pins are deeply recessed and superfinished to prevent film scratch and emulsion pick-up.

A 170-degree rotating shutter with a manually controlled variable shutter mechanism for fades and dissolves is standard equipment.

When the automatic fade/dissolve mechanism is fitted, a cam controls the variable shutter to carry out the fade or dissolve in the number of frames selected, which may be any number from 8 to 128. As the exposure curve required for fades and dissolves is different, the mechanism contains two cams, one giving a logarithmic fade curve and the other a sinusoidal dissolve curve.

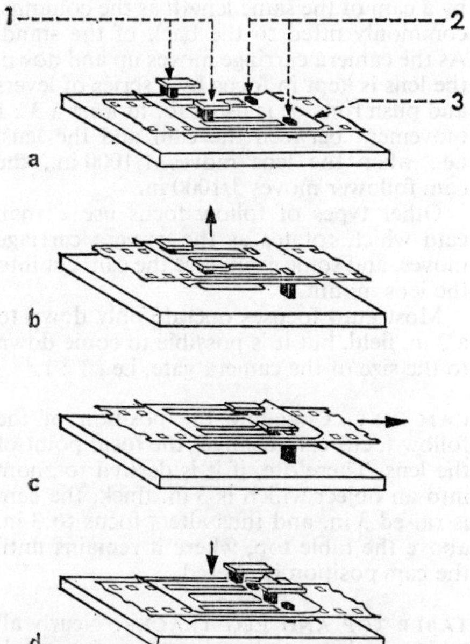

FILM INTERMITTENT MOVEMENT. (a) Film in exposure position. 1. Transport pins. 2. Fixed register pins. 3. Stripper and pressure plate. (b) Film being lifted on to transport pins. (c) Film being transported. (d) Film being lowered on to fixed register pins.

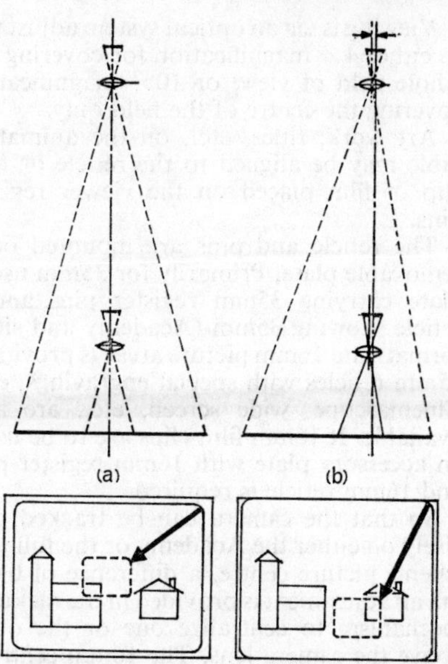

OFF-CENTRE ZOOM. Means of de-centering zoom motion without altering camera track or displacing cels. (a) Lens zooms in on-centre of scene. (b) Displaced lens tracks in off-centre of scene.

The lens focusing movement is linear (i.e., non-rotating) on precision ball bearings. The overall movement is $2\frac{1}{2}$ in., providing focus down to 1 : 1 with a 50mm lens.

In some cameras, the lens is itself fixed to a north/south and an east/west sliding device, or compound, so that it may be shifted slightly in any direction with respect to the film. When the lens has been offset in this way and the camera is zoomed, it moves in towards a point on the table displaced from the optical axis of the lens.

The stripper pressure plate assembly is spring-loaded so that, without modification, up to three films (or two films and a splice) can be accommodated for bi-pack work. The back pressure plate is removable to permit viewing through the camera gate. The 35mm front aperture plate is also readily removable, so that alternative apertures, e.g., Cinemascope, wide screen, apertures for split screen, etc., may be fitted.

Conversion between 35mm and 16mm is by substituting one shuttle movement and sprocket assembly for the other. Since no realignment of the camera is necessary, the whole operation takes only a few seconds.

For viewing, the camera body is racked across, bringing into position above the camera lens an engraved ground glass reticle and register pins to match the camera aperture and register pins.

Viewing is via an optical system adjustable to either 4× magnification for covering the whole field of view, or 10× magnification covering the centre of the field only.

Art work, titles, etc., on the animation table may be aligned to the reticle or to a clip of film placed on the viewer register pins.

The reticle and pins are mounted on a removable plate. Primarily for 35mm use, a plate carrying 35mm register pins and a reticle showing 35mm (Academy and silent format) and 16mm picture areas is provided; 35mm reticles with special engravings, e.g., Cinemascope, wide screen, etc., are also available. If 16mm film clips are to be used, an accessory plate with 16mm register pins and 16mm reticle is required.

So that the camera can be tracked precisely on either the Academy or the full gate (silent) picture centre, a difference of 0·056 in., an adjustment is provided in the rackover mechanism to centralize one or the other above the camera lens. The 16mm centre is the same as the full gate centre.

CAMERA ACCESSORIES. To project through the rackover viewfinder, an accessory 100-watt projector lamp is mounted on the viewer eyepiece. Either the reticle engravings or a film clip may then be projected on to the table top.

The back pressure plate and the sprocket assembly can be removed and a viewing scope of 10× magnification fitted to allow viewing through the back of the camera. This method of viewing is recommended where the greatest possible precision of alignment is required, e.g., for matte work.

To project through the camera gate, a prism is mounted on the gate and the 100-watt projection lamp fitted at the side of the camera so that light is reflected down through the gate to project the whole field of view.

Film take-up is by electric torque motors. The take-up tension is controllable to accommodate the different requirements of 16mm, 35mm, bi-pack, etc. The standard equipment is two motors; for bi-pack work, two additional motors are added.

Magazines for 16mm and 35mm film are available in 400 ft., 1,000 ft. and 1,200 ft. sizes. The bi-pack magazine comprises a normal 400 ft. magazine as the top half which fits to a special 400 ft. bottom half. This allows the camera to be used as a contact printer.

An electrically operated shutter fits between the camera and the lens for automatic cel cycling, double frame exposure, and running the camera without exposing film. The automatic cel cycling feature provides automatic exposure of every second, third, fourth, etc., to twenty-fourth frame.

The camera is racked over electrically and the projector lamp (with blower) is automatically switched on.

OTHER FEATURES

STOP MOTION MOTOR. The simplest design is of the slipping clutch type. A felt pad is spring loaded between two discs, one of which is connected to the motor which is turning continuously. The other, connected to the camera drive, has a cut-out which is engaged by a plunger operated by an electrical solenoid. For a single frame or picture, the solenoid is energized for a split second, the plunger is pulled out of the cut-out and the clutch rotates the camera drive. As the plunger is spring loaded, it returns into the cut-out. For continuous running, the solenoid is energized all the time. A better type of stop motion motor has several speeds and uses magnetic clutches which are operated by contacts on the solenoid plunger.

AUTOMATIC FOCUS. Focus is maintained by a cam of the same length as the columns, commonly fitted to the back of the stand. As the camera carriage moves up and down, the lens is kept in focus by a series of levers and push rods. It is usual for at least a 3 : 1 movement between the cam and the lens, i.e., when the lens moves 1/1000 in., the cam follower moves 3/1000 in.

Other types of follow focus use a snail cam which rotates as the camera carriage moves, and some even have the cam cut into the lens mount.

Most auto focuses operate only down to a 3 in. field, but it is possible to come down to the size of the camera gate, i.e., 1 : 1.

CAM LIFT. Changing the position of the follow focus cam changes the focal point of the lens. Therefore, if it is desired to zoom into an object which is 3 in. thick, the cam is raised 3 in. and thus alters focus to 3 in. above the table top, where it remains until the cam position is altered.

TABLE TOP AND PEG TRACKS. Nearly all table tops have a glass aperture of approximately 12 in. × 9 in. which is used for photographing transparencies, bi-pack, mattes or for back projection. They can have up to six peg tracks; usually there are three, two at 12 in. field and one at 18 in. field. There are two main standards of pegs: Acme, which has two outside flat pegs of $\frac{1}{8}$ in. × $\frac{3}{8}$ in. and a $\frac{1}{4}$ in. diameter centre peg, and Oxberry, which has the two outside pegs $\frac{1}{4}$ in. × $\frac{3}{8}$ in. and a smaller centre peg which allows a ruler to be put across the two outside pegs, e.g., for aligning titles, etc.

The peg tracks can be moved across the table by either a gear rack and wheel or a lead screw. Both types are controlled by hand wheels with counters reading in 1/100th in.

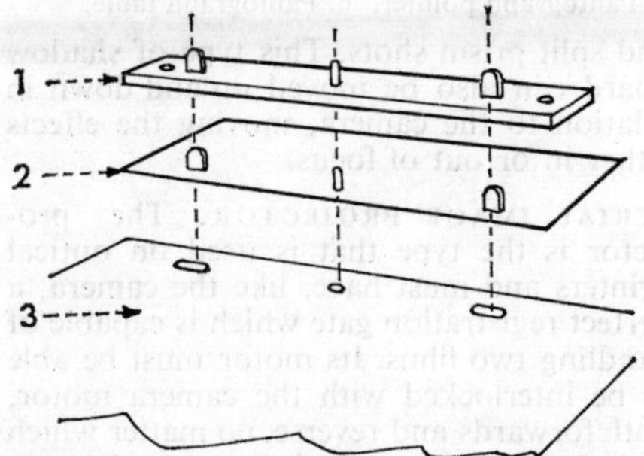

PEG BARS AND CELS. Method of positioning cels in identical relation to stand and camera so as to secure perfect registration of movement and position. 1. Peg bar. 2. Tape-on peg. 3. Cel.

When more than two peg tracks are used, it is necessary to remove the pegs from the inner track and tape the cel to the peg bar.

FLOATING PEGS. An additional set of pegs with a north/south and east/west movement completely free of the table top, fixed either to the base or to the columns. These floating pegs are extremely useful when a cel has to move independently of the cels fixed to the peg tracks, or if a cel is required to remain stationary while the table top moves.

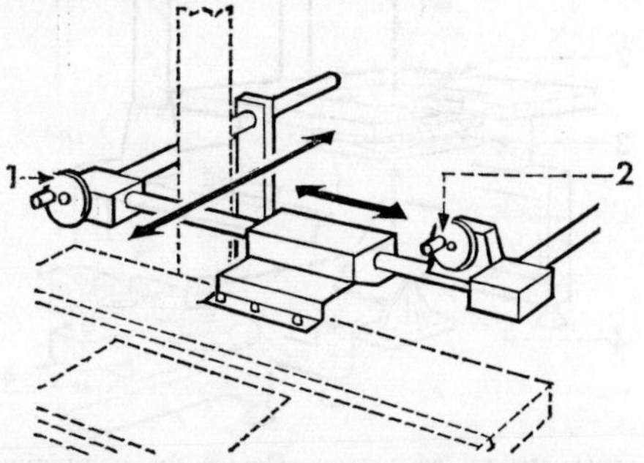

FLOATING PEGS. Device for providing additional cel movement by means of an auxiliary compound independent of the table top. 1. North-south handwheel. 2. East-west handwheel.

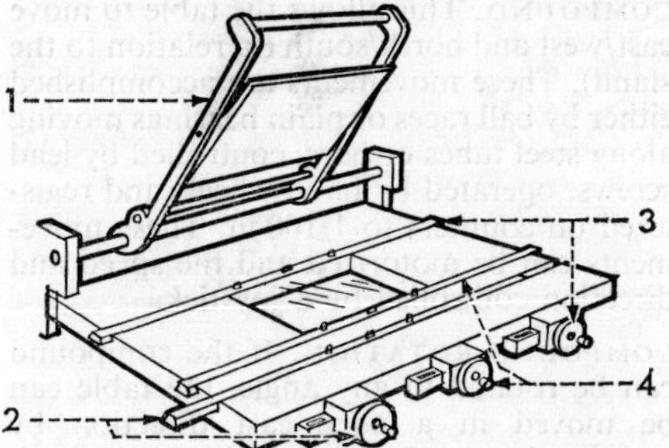

PLATEN AND PEG TRACKS. Means of sliding the peg tracks to provide accurately controllable movements under the camera. Platen secures cel flatness. Aperture beneath may admit aerial image from special projector. 1. Platen. 2, 3, 4. Peg tracks and operating handwheels.

PLATEN. The platen is a flat sheet of glass held in a frame which can be lifted either by hand or by a foot switch. Its main use is to hold cels flat during exposure, and it is normally fixed to the table top. Some makers provide additional fixing either to the base or to the columns.

TABLE ROTATION. Because table tops have a glass insert, the rotation unit cannot have a centre pivot and must take the form of a large ring. As this has to be turned by a hand wheel, the ring is cut into a gear wheel

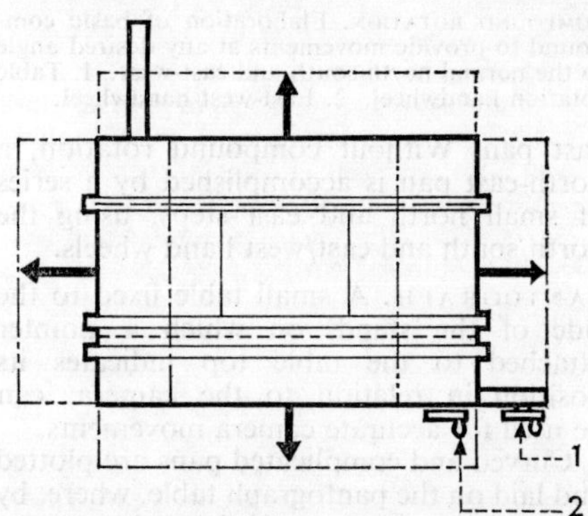

COMPOUND. Fundamental device for providing fully calibrated movements of table top and peg tracks beneath the camera. 1. East-west handwheel for compound. 2. North-south handwheel.

as well as a bearing. As with other movements, a counter registers the position of rotation to 1/10th of a degree.

So that fast spins can easily be photographed, it is usual to have a means of disengagement of the gear wheel, allowing the table top to be rotated freely without the use of the hand wheel.

COMPOUND. This allows the table to move east/west and north/south (in relation to the stand). These movements are accomplished either by ball races or plain bearings moving along steel tubes or bars, controlled by lead screws, operated by hand wheels and registered on counters to 1/100 in. These movements can be motorized and the speed and direction controlled by a joystick.

COMPOUND ROTATION. If the compound can be rotated to any angle, the table can be moved in a north-east direction by rotating the compound until the north/south runners are pointing north-east. Turning one hand wheel then produces a smooth north-east pan. Without compound rotation, a north-east pan is accomplished by a series of small north and east steps, using the north/south and east/west hand wheels.

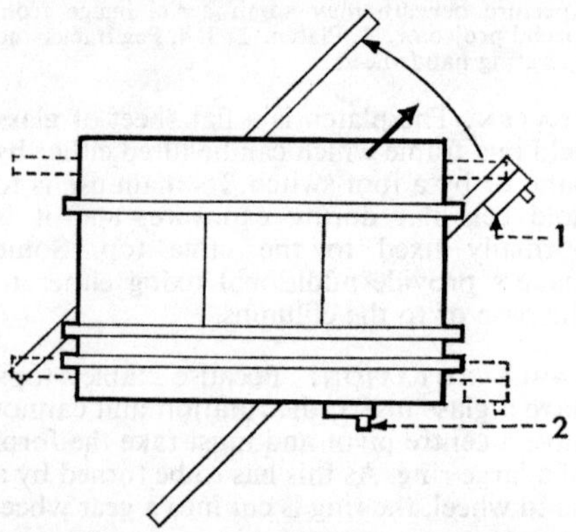

COMPOUND ROTATION. Elaboration of basic compound to provide movements at any desired angle to the normal north-south and east-west. 1. Table rotation handwheel. 2. East-west handwheel.

PANTOGRAPH. A small table fixed to the side of the stand, on which a pointer attached to the table top indicates its position in relation to the camera, can be used for accurate camera movements.

Curved and complicated pans are plotted and laid on the pantograph table, where, by using the pointer to follow the course plotted, these movements can be effected without having to look through the viewfinder. It is even possible to control the movement entirely with a joystick.

SHADOW BOARD. Owing to the undesirable reflecting properties of the glass platen, a shadow board is placed in front of the camera with a hole big enough for the lens.

A special effects shadow board is made to take an effects attachment which can accomplish wipes, ripple dissolves, distortion and split prism shots. This type of shadow board can also be moved up and down in relation to the camera, moving the effects either in or out of focus.

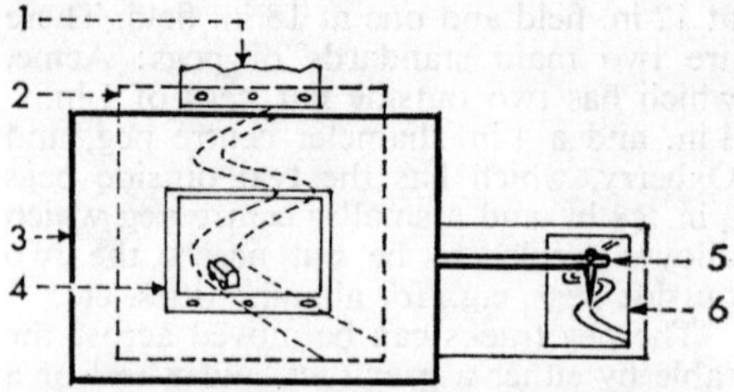

PANTOGRAPH. Device for guiding the operator in the transfer of a movement from a sketch to the cels mounted on peg-tracks and compound. 1. Floating pegs. 2. Cel. 3. Table top. 4. Cel on peg track. 5. Pantograph pointer. 6. Pantograph table.

AERIAL IMAGE PROJECTOR. The projector is the type that is used on optical printers and must have, like the camera, a perfect registration gate which is capable of handling two films. Its motor must be able to be interlocked with the camera motor, both forwards and reverse, no matter which way the camera motor is running. It must also be capable of stop motion in and out of sync.

Because of the efficiency of the aerial

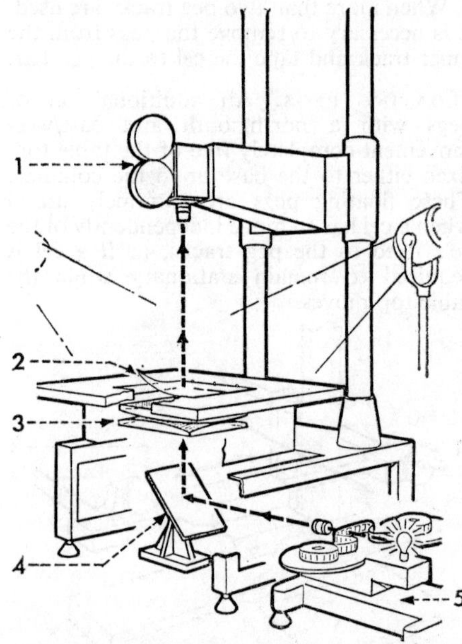

AERIAL IMAGE PROJECTOR. Device for combining filmed live action with animation by means of an aerial image projected into the plane of the cel beneath the rostrum camera. 1. Camera. 2. Cel. 3. Condenser lenses. 4. Mirror. 5. Projector.

image, the light needed is far less than for back projection, because all the light from the projector is focused into the camera lens.

As no screen is used in aerial image photography, the problems of screen grain and top lighting are eliminated; dual condensers refract the light into the camera lens.

A cel with either a cartoon figure or title placed on the table top creates its own matte and, in a single run, the projected picture and cel can be rephotographed.

The aerial image projector can be used very much like an optical printer, and such effects as fades, dissolves, push-offs, skip frame and freeze frame can easily be shot. Because the field size is almost 12 in., some painted matte work is possible.

THE FUTURE

Animation stands of the future will be automated as much as possible. Already computerized stands have been built to work from punched tape which operates all the camera controls as well as the table movements. It is only a matter of time before automatic cel changing becomes possible. P.N.

See: *Animation techniques.*

Books: *Technique of Film Animation*, by J. Halas and R. Manvell (London), 1968; *Animation Techniques and Commercial Film Productions*, by E. L. Levitan (New York), 1962.

⊕ANIMATION TECHNIQUES

□

The animated film is a special branch of the cinema in which the cine-photography of live action is replaced by cine-photography either of drawings and paintings or of three-dimensional objects and puppets, basically by using the stop-motion or stop-frame process. This means that the illusion of movement can be introduced into the graphic image, as well as into puppetry.

RANGE AND HISTORY

The range of stop-frame animation includes the cartoon film, the puppet film, object animation (frequently seen in television commercials), the silhouette film, as well as many kinds of experimental film-making in which various devices are used that in some instances replace, or partially replace, the stop-frame process. In many notable films, cine-photography has been altogether omitted by drawing the image direct on the celluloid strip.

INSTRUCTION AND PUBLICITY. Whatever the final technical process used, animation has developed from the entertainment cartoon into fields that include technical instruction, advertising and public relations. The particular contribution of animation to education has been in the mobile diagram and drawing, which have a unique value in the demonstration of complex actions and processes.

Animation in all its forms has reached a high artistic and technical maturity in permanent, specialized studios existing in countries all over the world. But in its fullest technical form it has always been a costly branch of film-making. Since World War II it has increasingly moved away from the production of short and feature entertainment films into advertising and instruction. The cartoon film has for the most part proved uneconomic in the post-war cinema, and represents now a very small percentage of cinema programmes. New systems of production are being devised in order to bring the cost of animation within the limits set by the television budget.

In the state-sponsored industries of the communist countries, however, elaborate forms of animation are still used for the projection of legendary and fairy tales, while the more advanced cartoon animation in America, Britain, France, Italy, Poland, Czechoslovakia and Yugoslavia has attempted satire and off-beat humour in various highly stylized graphic forms. Contemporary animation offers unique opportunities to the serious graphic artist, who conceives the technical realization of his work in terms of time and motion as well as composition in line and colour. It offers opportunities, too, in experimentation with graphics and typography (for example, in television commercials, industrial films, and in feature film titles). Of all branches of the cinema, it is by far the most flexible, and represents the widest range of contemporary experiment active in motion pictures.

ORIGINS. The traditional form of cartoon animation was developed in America before World War I, following the pioneer cartoons with white matchstick figures on black backgrounds made in France by Emile Cohl around 1908. Cartoons reached the cinemas in popular form with such series as Pat Sullivan's Felix the Cat and Max Fleischer's Koko the Clown during the period 1913–17.

These films were simply drawn, but they made use of cel animation. With the coming of sound, the tradition of American cartoon-making was dominated by the work of Walt Disney, whose finest cartoons were completed during the period 1928–40, during which he produced the Mickey Mouse and Donald Duck series, the Silly Symphonies, and his earlier features, from *Snow White and the Seven Dwarfs* (1938) to *Fantasia* and *Pinocchio* (1940). Max Fleischer's series featuring the cartoon character Popeye were also very popular.

ADVANCED STYLES. Less conventional work began to develop early in Europe with the abstract films of Hans Richter and Oscar Fischinger, followed by the experiments in direct drawing and painting on celluloid by Len Lye and Norman McLaren, whose long career experimenting with the medium has been carried out under the sponsorship of the National Film Board of Canada. More advanced graphic styles were shown in the work completed before World War II in France by Hector Hoppin and Anthony Gross, and during the war in Britain by the Halas and Batchelor unit and in the United States by the UPA group. In the USSR, cartoon and puppet animation developed on more traditional lines.

Its increasing appearance on television has given it a vast new outlet and the impetus to experiment further with quicker and less expensive methods of production.

BASIC PRINCIPLES

Animation might begin with the substitution of a drawn figure or a puppet which exactly reproduces the appearance and movement of a living character. This would be a wasteful, costly and laborious effort, although animators have been known to take the motion picture record of a living person and trace his movements frame by frame in order to reproduce some at least of their characteristics by means of a drawn figure. But the charm and humour of cartoon films have always been their artificiality and their open defiance of the natural laws that govern the physical behaviour of living things. The drawing of Felix the Cat or of Mickey Mouse always remained a drawing, a series of outlines in motion which behaved as such, as in Paul Klee's remark about "taking a line for a gentle walk".

EXAGGERATION AND DEFIANCE OF NATURAL LAWS. In the natural world, a cat or a mouse, like other living things, must respond to the forces of gravity and friction which under normal conditions are absolute; their bodies, like ours, cannot remain unaffected by the influence of the temperature, the winds, and the actions of earth, fire and water. Newton's fundamental principles of motion apply to all objects in the natural world: a body which is still tends to remain still, while a body in motion tends to remain in motion; its state of stillness or of motion can be changed only by the direct application of an external force, while every such action causes an equal reaction in the opposite direction.

The animator begins by both making visible and exaggerating the effect of these laws. A door is slammed violently: as it comes to rest in the jamb it visibly bends and then straightens. A soft ball at rest is given a sudden kick: the kicking boot enters its side and distends it with an exaggerated reaction. A figure skating sees the approaching hole in the ice and elongates into a few straight lines with hair on end in the attempt to pull back from danger. A character falls from a rooftop and momentarily squashes flat on impact with the pavement. This is known as "squash" in animation, and represents the essential resilience of the cartoon figures to natural forces. Almost all cartoon films emphasize for comic effect the natural laws of motion and of stress.

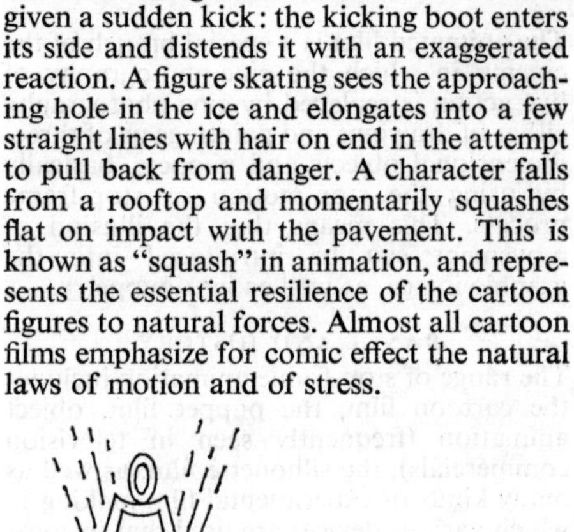

SHAPE DISTORTION. Method of conveying impact by distortion of shape. 1. Weight distorts in direction of fall. 2. Flattens on impact. 3. Recovers to normal shape.

The next stage comes with the defiance of these laws in the form of graphic wishful thinking. For example, the cartoon figure defies gravity by unconsciously walking on air across a precipice, though gravity may suddenly assert itself and, with a flurry of limbs, he may double-take and then suddenly fall. In cartoon, speed, time and space mean nothing; a character can fly over the horizon or expand or contract in bulk according to the needs of the situation.

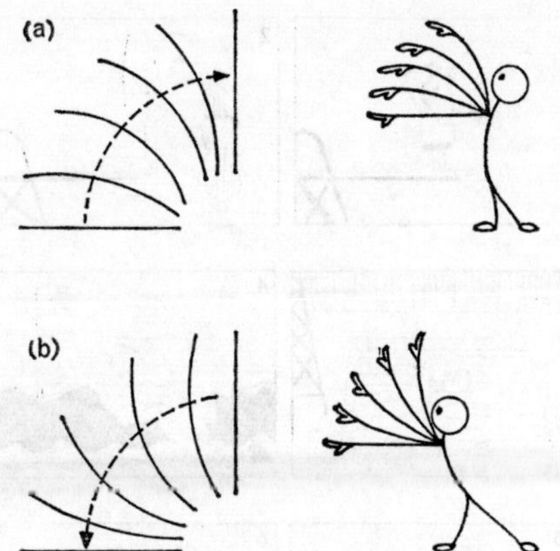

DEPICTION OF MOVEMENT. Curvature of arm to give a flowing and sympathetic movement as it travels upwards (a) and downwards (b), illustrating uses of distortion.

EVOLUTION OF CARTOON CHARACTERS. Simple geometrical shapes retain identity under severe distortions to which they are often subjected in depicting rapid movement.

CHARACTERIZATION. The exaggerations, distortions and liberties taken with the forces of nature are, to a considerable extent, the very essence of animation. Cartoon figures are designed to carry out these extraordinary activities and at the same time to exaggerate certain recognizable aspects of human and animal characters in appearance and behaviour. Their bodies, and particularly their heads, are mostly distorted for this purpose. Mister Magoo's head represents one-third of his total height, and is as large as his torso. Popeye's muscular forearms are as large as his thighs. This essential, easily recognizable, characteristic of a cartoon figure, must emerge from the key outlines that represent his face, and the range of his changing expressions must remain within close limits. Even the most subtle of cartoon characters are strictly limited from any human, psychological point of view.

Most of these factors in animation result from the need for the characters to retain their simplified identity while in motion. Provided this need is fulfilled, there seems to be no limit to the graphic stylization animation may employ as it moves away from the representational. Cartoon figures may be slightly moulded to suggest a three-dimensional body as in many of Disney's cartoons, or kept essentially flat within their containing outlines, like Mister Magoo and the characters in most of the UPA films. Or they may be composed of squares or oblongs with the bare suggestions of faces, like certain of the Yugoslav designs. The whole range of contemporary graphics is open to the inventive cartoonist, and John Hubley, a notable American artist, has designed a number of animated films in which no outlines are used; the compositions are achieved entirely through the mobile relationships of colour.

AUDIENCE RESPONSE. A factor in animation which is very little discussed is the nature of perception to be expected in audiences. Any departure from more or less realistic drawing relies upon the varying degrees of response in recognition (or visual "reading") by each individual member of any audience. A type character—a scientist, for example, conceived in a few deft lines – must be instantaneously recognized for what he is, and changes in contemporary graphic styles must be reflected alike in the animator's work and the audience's response to it. The art lies in reducing what are relatively complex implications of character and function to a simplified graphic form which immediately suggests these characteristics in contemporary terms. Some fantasy figures and lightning caricatures in television commercials, for example, must be recognized and responded to instantaneously if their significance and their humour is to be fully appreciated during the few seconds of their appearance on the screen.

Similarly, with the use of symbols and diagrams, audiences must be able to recognize quickly whatever is presented. Twenty or more years ago, Otto Neurath's symbolic men, seen as silhouette diagrams, could be immediately presented in multiples showing, for example, some aspect or other in human statistics; their introduction was indeed advanced at this period. Nowadays, their use has lost its original stimulus. Contemporary audiences accept mobile symbols and diagrams with a much greater complexity of content and a greater elaboration of graphic style and colour.

The public has become familiar with charts, symbolic maps, diagrams and various uses of symbolic human figures through their constant employment in the press, in advertising and in other forms of popular presentation. The animator can draw on this popular acceptance and extend it in mobile forms. New symbols are constantly being created—among them, road signs, letterheads and commercial trade symbols – and this graphic proliferation has become an accepted part of our contemporary method of communication.

PRODUCTION TECHNIQUES

The technical realization of the animated cartoon has passed through a number of evolutionary stages. First, the succession of outline drawings on paper were photographed one by one in series. As the static background began to assume some importance as a setting for the action, areas of it were kept clear so that it could be used repeatedly with only the drawings of the moving figures changed as each photographic frame was exposed. From this followed the all-important technique of cel animation, the superimposition of the moving parts of the picture on to the static background by drawing or painting them on to transparent celluloid sheets. Cel animation remains to this day the basic principle of the cartoon film, whether for comic, serious or technical subjects.

Cel animation allows for the logical breakdown of the technical work, and so for greater elaboration. Several cels can be superimposed on the single, static background, and the artists involved can specialize in the development of individual characters whose movements are confined to a single series of cels, each of which makes its particular contribution to the final, composite picture under the camera.

TREATMENT AND STORYBOARD. The production of an animated cartoon normally takes the following course. A written treatment is needed to set down the outline continuity of the story or action. This determines the main dramatic development of the film and the part the various characters will play in it; at the same time, the directing animator makes his preliminary sketches

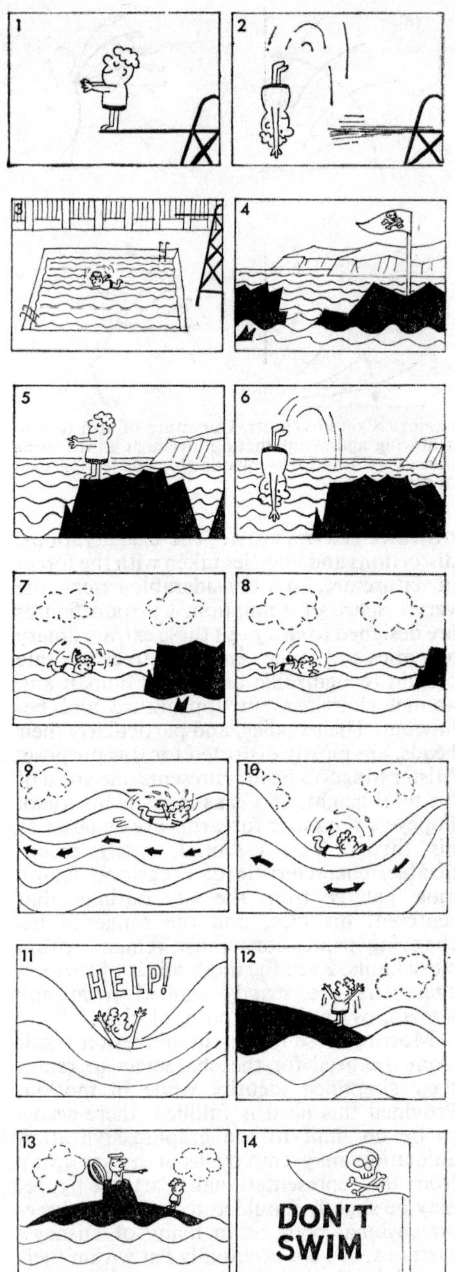

STORYBOARD LAYOUT. Key elements explaining plot of cartoon story for initial discussions with sponsor or distributor, and to determine balance of individual sequences in longer film. Visual style is established, while motion can be indicated by arrows, as in 9 and 10. The storyboard thus lays down the choreography of the animated film.

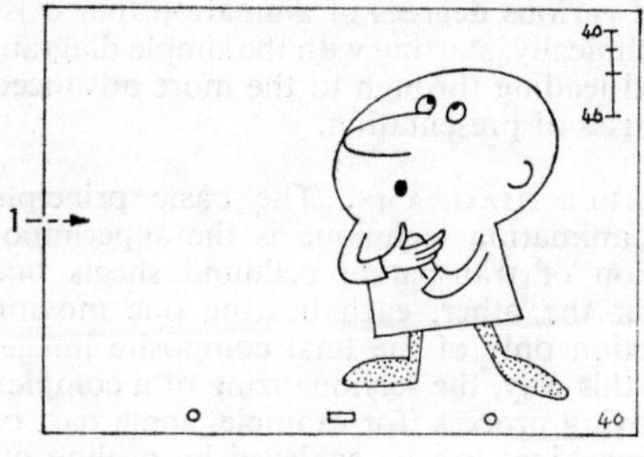

PLANNING OF CARTOON ANIMATION. 1. Extremes of movement depicted by animator to provide key drawings and set style and character of sequence. 2. Representative in-between animation drawn by assistant animators. Scale at upper right indicates span of frames within which action has to be carried out and spacing of movement.

which begin to disclose the visual nature of these characters and the general graphic design and appearance the film will assume. Where colour is to be used, as it frequently is, the colour design as well as the graphic style has to be tried out and decided.

While the dramatic and aesthetic ideas take shape, the script can be elaborated until the action in each shot is predetermined, together with any dialogue, narration, commentary or key sound effects that are needed. From the script a storyboard is created. This shows the action in the form of a series of still sketches, much like a lengthy comic strip, revealing scene by scene the flow of the action and the general graphic manner in which it will be presented. It may take many sketches to show enough detail for one minute's action. The storyboard is, in a sense, the first demonstration of the choreography of the film once it assumes motion, though naturally at this stage the nature of the movement can only be indicated by means of verbal description or superimposed arrows.

PLANNING THE MOVEMENT. Parallel with the storyboard come the research drawings for the characters and the backgrounds. The essential composition for each character must be simplified in such a way that a chain of artists of varying seniority can contribute to his realization through thousands of repeat drawings. These sketches must all be made with a view to the ultimate movement of which the finished drawings will be components. The sketches must suggest the lively character and the movement to come.

Once the visual essentials – representing not only basic expressions seen from varying viewpoints, but the scale of the various parts of the face, head and body – have been agreed, then the key or principal animators can begin to work out the actual movement which will bring the characters to life. For this they may well need some form of prerecording of voice and music in order to achieve a witty or apt synchronization of their drawn movements with the sound.

Many, though not all, cartoon films have their tracks pre-recorded so that the animators may hear them constantly played back for appropriate matching when they begin the laborious task of rendering in successive phases each record of the action, including lip-synchronization with the dialogue. This, for a sound film, means twenty-four phase-drawings for each second on the screen, though in practice there are many short cuts in the process of realizing movement in these terms. Movements and points of characterization may be rehearsed by means of photographed pencil tests, that is, animated sketches projected on the screen to check the effect of details in the action.

THE FINAL ANIMATION. For full-scale animation, an instructional work-book is drawn up by the director from the script and storyboard, giving, shot-by-shot and frame-by-frame, every detail of timing, camera movement (if any), and shot punctuation. This work is essential alike to the composer and the animator, especially when the music, voices and effects are to be composed and pre-recorded before the final animation begins. The music and dialogue may need special time-charts to guide the composer, while the cameraman, who is to photograph each batch of backgrounds and cels, needs a camera-exposure chart (or dope sheet) to give him his instructions, which are derived by the director from the work-book.

The key animators, while observing their technical instructions closely, are primarily responsible for the creative vitality of the work. They draw the key phases of movement, the ones that determine the life and expression of a character, while their assistants complete the in-between phases. This work may be done direct on the cels with coloured chinagraph pencils, or by the traditional system of copying the animators' work, which is on drawing paper, on to the cels and colouring them with opaque paints.

When the photography is completed and processed, editing follows, though in the case of animation the editing is almost entirely predetermined at work-book stage.

PUPPETRY. Other forms of animation, using puppets, for example, follow a similar pattern. The dolls have to be characterized just as exactly as the drawn characters, but in their case the final animation is achieved through frame-by-frame adjustment with their gestures and expression changed; they perform within miniature sets which are designed and constructed like studio sets for live-action film, in order to contain the action and give it place and atmosphere. Exact synchronization between sound and visual action is just as important in puppet and object animation as it is in the cartoon.

GRAPHIC TECHNIQUES AND PRESENTATION

Since a great part of the use of modern animation is in the educational, scientific and industrial field, it is necessary to analyse the various degrees of animation that exist technically, starting with the simple diagram and leading through to the more advanced phases of presentation.

SIMPLE DIAGRAMS. The basic principle of animation technique is the superimposition of transparent celluloid sheets one over the other, each bearing one moving section only of the final composite image. In this way, the sectionalizing of a complex moving process (for example, some part of a machine) can be achieved by peeling off layer after layer of the total image, so that the whole may clearly be seen as the sum of the parts.

The technique employed here is to produce a single drawing that contains all the necessary visual information. The animator can achieve a continuous, smooth movement by scratching off, or eliminating, the drawing in stages during the process of filming. When the film is reversed and run forwards, the illusion is created of the diagram being assembled in all the stages up to its fullest form.

It is also common practice to devise a mobile design in segments which can be moved by hand – for instance, a clock-face with rotating hands. The hands can be adjusted manually between camera exposures.

Camera dissolves and superimposition can also be introduced to elaborate the presentation of diagrams in various phases of their development on the screen. Such methods are of great importance in clarifying a process or operation stage by stage or phase by phase.

For example, one particular element can be isolated and emphasized to the exclusion of the rest, and followed through, while other elements are kept in abeyance until their turn comes.

Familiar and basic symbols help to clarify an idea and to save a considerable amount of screen time: a flag represents a nation, a clock the time, an arrow a direction.

ELABORATE DIAGRAMS. Whereas most of the movements in simple diagram animation can be achieved manually or mechanically under the camera, elaborate diagrams need substantial pre-planning. That is, the movements should be planned and timed in storyboard form. Animation can, more flexibly and significantly than live action, slow down or accelerate the presentation of movement in any process which is being demonstrated. It is also very economical, because important elements in a process may be presented in emphatic slow time, while the unimportant can be speeded up or merely flashed by. The transformation of decayed animal organisms into oil, which took nature several million years, can be clearly shown in a few minutes.

COMPOSITE DIAGRAMS. Animation can superimpose upon the living image the moving diagram which simplifies, and at the same time analyses, the working principles within.

Much of the plant, machinery or laboratory equipment used today is highly complex, and considerable scientific or technological understanding is needed to appreciate the processes and principles involved in its operation. According to the level of knowledge required in particular cases for particular audiences, animation can be graded from the simplest, basic form of demonstration to the fullest and most elaborate. The animator adapts his particular skill in graphic demonstration to the requirements of the expert who is advising him on the film.

A point of some importance here (and great importance in the film of technical instruction) is that carefully-designed, animated films have proved easier to memorize than live-action films aiming to give comparable information.

ANALYTICAL DIAGRAMS. The animated film can incorporate live-action photography, and then cut away to mobile diagrams when identification and added clarification become necessary.

The live-action picture has the obvious advantage of immediate authenticity. Here is a shot of an atomic reactor; here is a railway signal box; here is a coalmine. But the complex surface of actuality has no real meaning unless the principles that make it function are demonstrated.

To move over from the actuality of the live-action picture into the purely analytical presentation of the animated drawing or diagram is to achieve the best of both worlds. For example, the effects of magnetism and gravity can be made visible; the action of physical laws can be demonstrated; or radioactivity, the invisible forces within atomic energy, can be clearly shown.

GRAPHIC STYLE. An advantage of animation is that it can use every variety of graphic style and presentation the animator may devise to illuminate the principles involved, ranging from the wholly imaginative or symbolic image to the simplest diagrammatic representation, according to what is most appropriate. Aesthetic qualities can therefore be added to purely intellectual concepts. This lifts animation from the level of a purely representational picture or diagram in motion to a figurative level, where the power of suggestion begins to operate, and the imagination, as distinct from the craftsmanship, of the animator is brought more fully into play. This can be particularly useful in the industrial film intended for more general audiences, where demonstration of the nature of what is happening is of greater importance than a detailed and accurate reconstruction of processes.

Take, for instance, the drawing of a simplified germ representing disease. If it is transformed by animation into a graphic monster it can acquire a heightened significance. The animator can further develop the shape of the monster into some destructive force which wipes out a whole population. Such a sequence, designed in a powerful graphic style, can combine aesthetic quality with memorable dramatic content.

COLOUR SYMBOLISM. Colour can be used to emphasize forms of motion designed to make a point clear, saving perhaps much verbal explanation. Colour differentiations can convey instantaneously what might well have to be laboriously pointed out in words. Colour, too, can be used symbolically to convey psychological implications, or merely to create atmosphere and feeling. Colour has seldom been used symbolically in the live-action film. In animation it is consistently used non-representationally to assist an argument or create a mood.

SYNCHRONIZED SOUND AND VISION. Sound can be so exactly timed in the animated film that comment (whether by word, by significant natural or artificial sound, or by musical counterpoint) can be synchronized, if need be, as closely to the action as one twenty-fourth part of a second. This is achieved by pre-recording the sound track, and so having it available for exact timing, which may prove very important when dealing with a subject where significant time or rhythmic processes are involved. The emphatic use of naturalistic or artificial sound can help to reinforce the image.

This is also true of musical effects which have been specially composed.

THE HUMAN FIGURE. In cartoons, the more naturalistic the human figures are the less natural they seem, for cartoon is essentially a medium that thrives on stylization. If the need for a naturalistic figure is genuine, live action, not animation, would be more appropriate.

Human figures in cartoon can range from simplified drawings, both designed and played relatively straight, to grotesques made up of a few simple lines. Some element of caricature can in fact humanize a subject far more readily than a stereotyped performance by a live actor.

TIMING. The smoothness and the mood of all moving objects in animation depend on timing, and the expertise with which it is judged. The nature and quality of the movement depend on whether the image is held still under the camera for one or more frames.

Single-frame animation, with changes introduced on the basis of 24 phases per second of action, creates an illusion of completely smooth movement. Double-frame animation, with changes introduced in every alternate frame, or 12 changes per second of action, creates comparatively smooth movement, since the eye still fails to detect the displacement on the screen. Three-frame animation, with changes introduced at every third frame, is possible, but must be devised with special skill in order to avoid any unacceptable jerkiness of movement; some jerkiness is inevitable.

With four-frame animation the observable jerkiness of movement becomes a positive technical factor to be exploited – a form of slip-frame technique which has become fashionable in commercials and achieves certain kinds of comic graphic effect.

Six-frame animation, and so on up to a hold lasting a complete second (24-frame animation), is also fashionable (particularly in television titling) in sequences using still photographs pixilated into a fantastic form of jerky action. This technique produces the illusion that a photographed human character has been released from the laws of gravity.

Timing is a complex art and technique of its own, and many contributory factors combine to achieve the general effect, including the relative size of the figures to the background and their tonal relationships.

STUDIO PERSONNEL

The working method of an animation studio is determined by the type of animation that the studio specializes in. Many artists engaged in diagrammatic animation like to carry out the total operation from story-planning to photography. Others, with the obligation of higher output, subdivide the work into specialized departments. The experimentalists and avant-gardists of the medium usually insist on working on their own, especially those who paint their visuals directly on to the film itself.

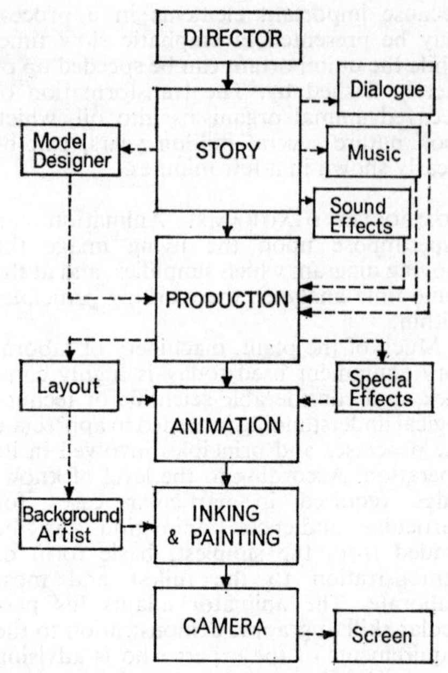

ANIMATION PRODUCTION. Flow chart showing typical organization of main processes (central column) and collateral activities in making cartoon films and technical animation.

All studios cater for two basic functions: creative and technical. The creative functions consist of conceiving the story, visualizing it in terms of animation, designing the characters and backgrounds and inventing animation ideas. The technical side consists of preparing the work in terms of stop motion photography and making charts for the cameraman showing detailed arrangements of the drawings and backgrounds. Preparing the timing of each individual scene and deciding for how many frames each individual drawing should be held is the work of the animation director or the animator and demands a thorough understanding of animation techniques. Studios maintaining a high output usually subdivide such technical functions among several departments.

WRITER. The writer's function is the conception of the film script. This may be supplied by free-lance writers, but most studios find, especially in the production of television serials, that only visually-minded script writers, understanding the limitations of animated film, can effectively contribute. So the tendency is to train special writers who work with the animation director and the storyboard artist inside the studio.

STORYBOARD ARTIST. The storyboard artist may be capable of taking the place of the writer in cases where no dialogue is required. The writing of cartoon dialogue is at least as difficult as writing live action dialogue. The storyboard artist must imagine the written text in pictures. The quality of the drawings matters less than visual ideas and the flow of action from one situation to another. Story continuity must be seen also in terms of the smooth flow of subsequent stages of production, timing of the action and animation. Approximately 50 rough sketches, size 3 in. × 4 in., are sufficient for one minute running film length.

DIRECTOR. The animation director times the action of the film and supervises the creative and technical aspects of the work in the various departments. His talent and technical knowledge usually is the prime factor in the value of an animated film. His knowledge and experience have to extend into every stage as the film develops and takes its final shape. His sense of timing is the most important factor in the whole process.

DESIGNER. The designer is the counterpart of the art director in live action film. He determines the style of a film, its visual character, the colour theme and colour continuity, the tones of backgrounds and figures and textures. He may also create the actual characters in a film which he usually draws jointly with the director. The designer's capability must suit the special requirements of cartoon techniques, especially in the case of characters' shape and design. Cartoon characters are drawn in several thousand stages and angles by the other members of the studio in many different forms and sizes so they must be conceived as simple shapes with the utmost expressiveness. The designer also prepares visual proportion charts of the characters to safeguard uniformity during the production, especially if many animation units are working on the same character.

LAYOUT ARTIST. In smaller animation units, providing the layouts may be the work of the director, designer or key animator, but in high production it is the work of a specialist. Up to this point, the plans for a film have been roughed out by the storyboard artist and the action timing estimated by the director. The sound-track is recorded already and its length is known. It is now essential to give a definition of what should happen in every scene and this is what the layout artist does under the director's instructions. The layout provides the outlines of what shapes a scene should contain stylistically and also in terms of the graphic organization of the compositions of shapes and forms. Such composition is related to the choreography of animated figures, establishes a smooth scene-to-scene continuity and provides the field sizes for each shot. The layouts contain, therefore, all the essential visual information for the animation, background and camera departments.

ANIMATORS. Though the medium itself has been called animation, this is, in fact, usually a small part of the overall activities of a studio, but an important one. The function of animators is to create the movement of figures or objects in a film. This function varies a great deal in different films. Some films depend entirely on animation, some replace animation with camera mobility.

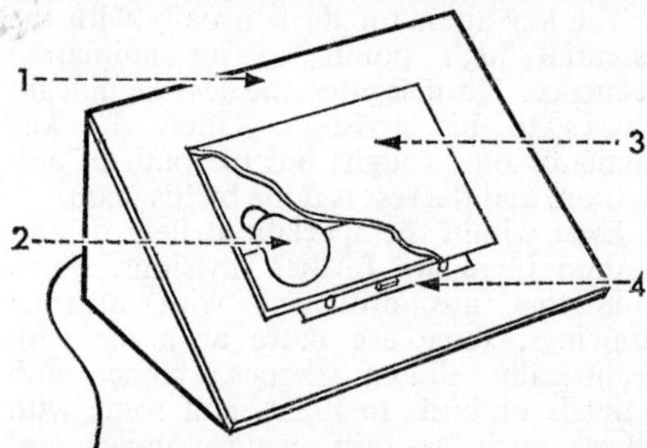

ANIMATION DESK. Basic equipment of the animation artist. 1. Wooden top. 2. Light bulb. 3. Ground glass. 4. Peg bar for registering cels.

During the 1960s, the art of animation has almost been forgotten owing to the influence of quick television productions. In the 1970s it is expected to come back once again as a unique art. The best animation is produced when the animator can forget the outline of the character and concentrate mainly on its movement. A phase of animation may consist of 12 to 96 individual drawings, according to the speed and length of a scene. When these drawings are in firm consecutive

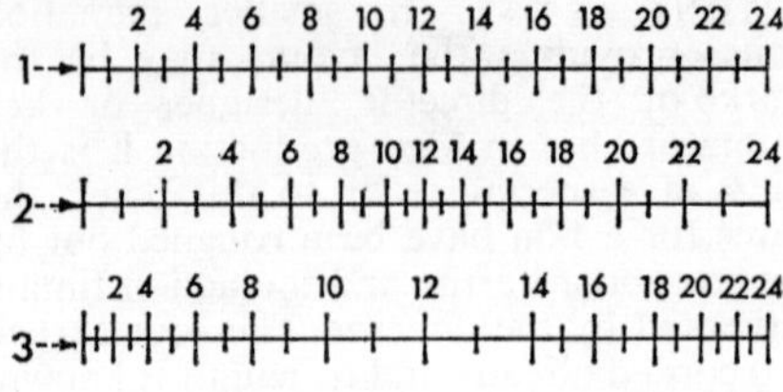

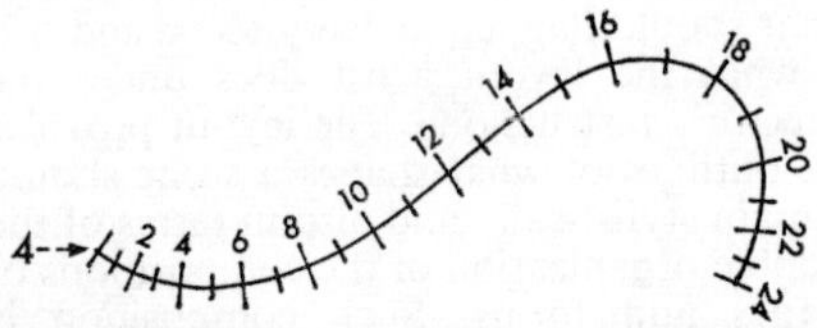

ANIMATING MOVEMENT. Numbers on scales represent position of object under camera, frame by frame (24 frames=1 second). 1. Equal spacing to give constant speed. 2. Unequal spacing to give fast, slow, fast movement of object. 3. Slow, fast, slow movement. 4. Perspective view of object going round bend and animated at different speeds so as to appear at a constant speed. Same principles apply to phasing of cartoon animation.

order, conveying the full characteristic expression of the figure, good animation emerges. The drawing skill, though important, is less essential than an ability to understand the mechanics of movement and to give technical information to the cameraman about the order in which the individual animation phases should be photographed.

The key animator deals usually with the essential high points in an animation sequence. He delegates the less significant phases to his assistant. Often the key animator only roughs out the path of animation, and the rest is done by his team.

Even within the specialized field of animation there are further divisions. Some animators are proficient with abstract drawings, some are more at home with realistically drawn shapes, some with animals or birds in flight, and some with effects such as rain, water ripples and explosions. In *Animal Farm* each animator was cast as a character part, as in a live action feature film.

Today's style of abstract animated figures needs more humour than animation. Up-to-date animated films run at twice the speed of films of ten years ago, so many less drawings are needed. Single frame animation has given place to one drawing for every two, or even three, frames. Good animation, nevertheless, requires such basic qualities as craftsmanship, sense of characterization, feeling for movement, understanding of timing and knowledge of film techniques.

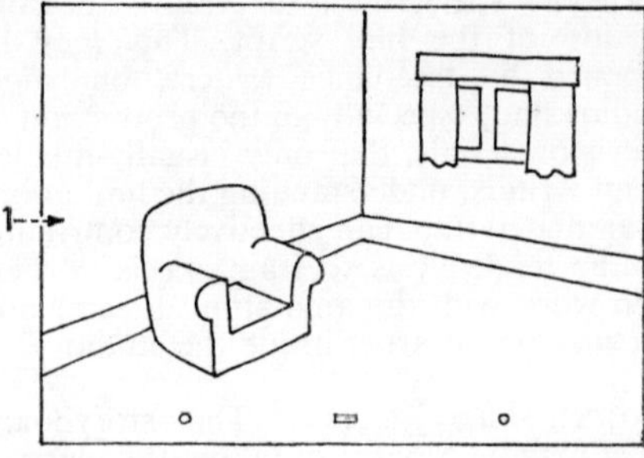

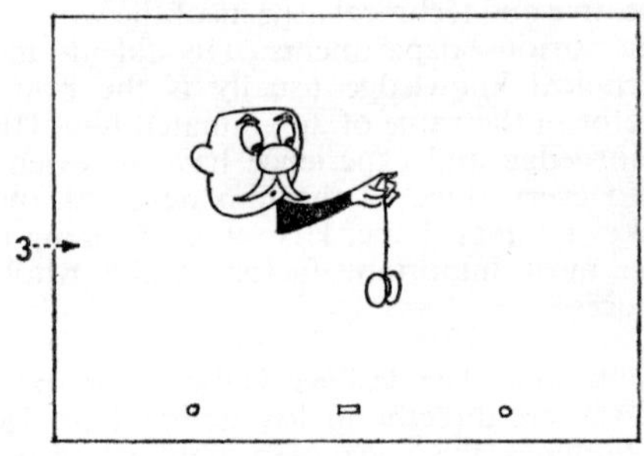

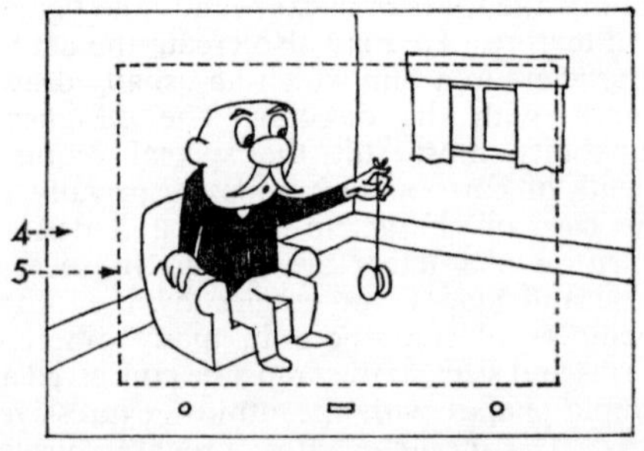

ELEMENTS OF CARTOON SCENE. From bottom to top of peg board, 1. Background opaque. 2. First cel of stationary animation registered to chair. 3. Second cel, phase of moving animation. 4. Complete set-up showing camera field, 5.

TRACER. The up-to-date tendency is to machine copy the animated drawings on to celluloid sheets, but the finer animation is still hand-traced. The tracer's responsibility is to copy the animated drawings in perfect registration on to celluloid sheets, sometimes with coloured inks, sometimes in black Indian ink, sometimes with coloured wax pencils. This job appears to be repetitious, but it can be quite creative and in every case it must be highly skilful.

COLOURIST. Most animated cartoons are produced with transparent celluloids, so that the background paintings are fully visible, and the animated figures and objects are coloured with opaque paint, usually on the back of the traced figures. The new graphic techniques have provided new approaches for colouring the moving characters. The traced characters are painted on both sides of the celluloids and much more freely than was the practice some time ago. Most of the flat colours used for the characters have given place to texturized tones or even translucent surfaces with the object of integrating the moving figures with the tones of the background. The total impression is a fresher and more creative appearance similar to the work of modern graphic artists.

BACKGROUND ARTIST. This artist provides the finished backgrounds for the individual scenes of a film. Painting ability, a good sense of tones, colour and textures, and precision work are essential. It is also essential for a background artist to know what happens when a background travels at a given speed under the camera and what is the effect of a painted overlay in the foreground in relationship to the mobile background. At a given speed, for instance, certain vertically painted objects in the foreground have a stroboscopic effect causing acute eyestrain.

CAMERAMAN. The animation cameraman photographs his objects from a fixed position with fixed lighting, while photographing live is recording a performance with constant movement of the camera and adjustments of lighting. The latter demands a creative approach; the former a technical one and a somewhat different temperament. Nevertheless, modern animation photography does require a complete understanding of the range of optical superimpositions, a knowledge of the characteristics of a great many negative films in black-and-white and colour, and familiarity with the mobility of modern rostrum camera stands.

EDITOR. All animation shots are pre-timed by the director; the function of the editor is to assemble the individual shots in continuity and to fit them to the various soundtracks. There are, however, occasions when, with the utmost care in timing, the visuals still do not fit the dialogue, and this must be dealt with by the editor. The provision of the sound effects track is also the editor's responsibility. So is the analysis of the characteristics of a music track or dialogue track and the provision of information to the animators in terms of counted film frames. The editor is also in charge of dubbing the separate sound track and contacting the laboratories for the final showprints.

CAMERA TECHNIQUES

The work of the animation cameraman differs substantially from that of the live-action cameraman. It involves frame-by-frame recording of the painted background and celluloids as a series of static combinations. The combinations are provided for him by the animator on a camera exposure chart showing the correct sequence and combination of each drawing, celluloid and background, with instructions regarding movements such as panning and tracking.

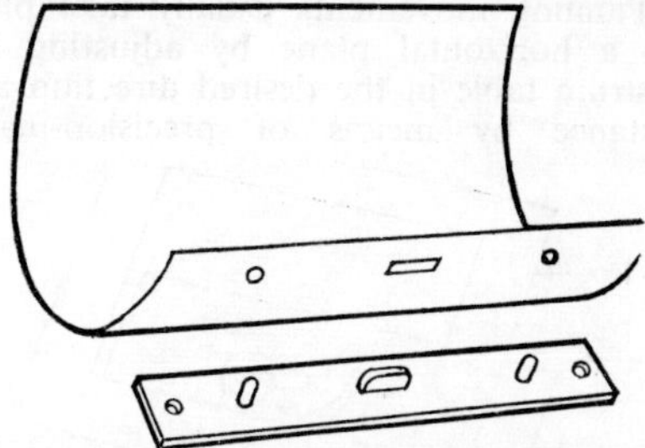

PEG BAR. Cel or animation paper is perforated to give precise positioning on pegs. Ensures correct registration of images, both fixed and moving, under animation camera.

SETTING UP. The cameraman, following the camera instruction chart, positions the background on the rostrum table below the camera, and places the appropriate set of celluloids through which the animation is created over the background on to the registration pin bars which hold them firmly in position. He presses a glass sheet over them in order to keep them meticulously in place while he exposes the film for one or more frames according to the camera chart.

Then he prepares for the next change; this might involve the adjustment of the cellu-

PANNING AND TRACKING. (*Top*) Background moved in opposite direction to animation action or cel to produce effect of car passing scenery. (*Bottom*) Camera tracks in to produce larger image of portion of scene.

loids and the registration pin bars in order to achieve a panning movement in the backgrounds, or adjustment of the camera position in relation to the rostrum table for a tracking movement. Panning means moving the art work from west to east or from north to south, or a combination of these directions. Tracking means moving the camera away from or towards the object to make it appear smaller or bigger. Naturally, the lens is adjusted during these moves to keep the object constantly in focus. In modern animation cameras, focusing is automatic.

Panning movements usually take place on a horizontal plane by adjusting the rostrum table in the desired direction and distance by means of precision-made,

FIELD MASKING. 1. Mask cut to precise size of desired camera field and registered on pegs. Field masks are also used to assess free space round titles, so as to gauge their suitability to different aspect ratios.

screw-driven rods. The tracking movement is usually achieved by moving the rostrum table vertically towards the camera or vice-versa. Once again, either the camera or the camera combined with the rostrum table is driven by the aid of precision screw rods to achieve smooth movement.

CAMERA EFFECTS. In spite of the fact that the rostrum cameraman's work is substantially predetermined compared with that of the live-action cameraman, his contribution can still be an important factor in the production of animated films. The skilful use of lens techniques can achieve effects which would be practically impossible with hand-drawn animation, so saving precious hours of labour. These camera effects include the superimposition of images and variations of forms to make objects appear soft and transparent.

With the animation camera, it is comparatively easy to expose the negative in

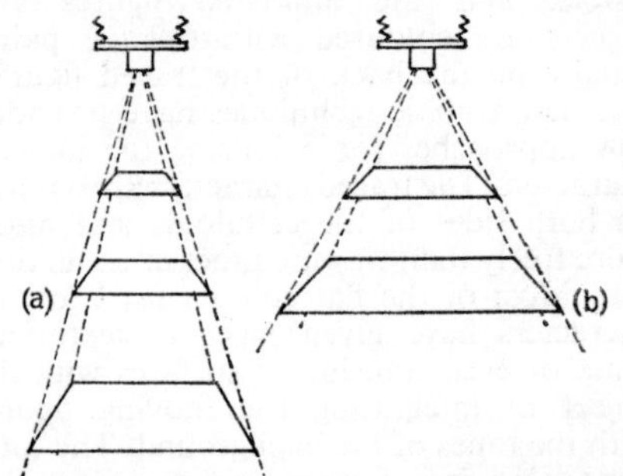

CAMERA FIELD OF VIEW. At equal distances above table, normal lens (a) gives smaller field than wide-angle lens (b). Exact focal length and height above table must be known in establishing field sizes.

several stages, and this method can be useful if sections of the visual require part-exposure only.

This device can be especially useful when a static caption is to be superimposed on a moving background, or similarly if the caption has to move over a still picture. By providing for separate exposures on the same negative film, the cameraman can photograph captions and the background at different distances and, if necessary, diffuse these elements or put them out of focus.

When parts of a scene require to be photographed with reduced density, the whole scene is photographed twice (in exact register) with percentage exposures, the reduced parts being removed during one exposure. In this way, for example, shadows, fire and smoke can be given a certain degree of transparency. The background or other parts of the picture show through, because they have had full exposure. Mist or fog effects can also be obtained by giving a very low extra exposure on a light grey card, either moving or still as desired. The mist can be faded in or out entirely independently of the remainder of the picture.

THE CAMERAMAN'S ROLE. The cameraman must realize the potentiality of his rostrum camera, the characteristics of his negative film stock, and the different approaches to the various types and forms of animation. For instance, in simple diagrammatic animation, the demands on the cameraman's skill are greater than in full animation, where all the necessary movement has already been incorporated on the celluloids. In the latter case, his function is more or less confined to changing the celluloids, adjusting the backgrounds and exposing the film – a routine operation.

In diagram animation, especially if the movement is to be achieved by animated cut-out objects, he may have to effect the animation himself frame-by-frame under the camera. This requires additional skills and an understanding of some of the basic principles of animation dynamics. Also needed is an understanding of available colour systems, especially as most animation is photographed today on colour negative.

Modern camera techniques have been gradually developing towards a wholly electronically-controlled operation. One of the latest animation rostrums is operated electronically by programmed punch cards. All camera movements are automated and the cameraman need only supervise the operation and change the celluloids. Thus the production is speeded up and the camera moves can be precisely calculated.

RELATIVE COSTS

Animation costs, like those of any manufactured article, depend on labour, time and material, with the difference that the process of manufacture requires a high degree of both aesthetic consideration and technical knowledge. For this reason, a good animator is rare and commands a higher salary than other kinds of film technician.

Generally speaking, the simpler the animation, the higher is the skill required. Only those who know animation well and, in the case of technical films, have acquired a sufficient grasp of the given subject under the guidance of an expert, can appreciate the innumerable different animation approaches and decide which is the one best suited to a given film or film sequence.

The longer the time spent during the stages of planning, the greater the chances of solving the problems of animation economically. For this reason, it may be advisable to invest more money at this stage in order to spend less on production.

There are overlaps in every type of animation, but the four main categories may be summarized as diagram animation, multiple techniques (with live action), object animation in three dimensions, and cartoon.

DIAGRAM ANIMATION. If the aim is the simple presentation of a plain fact, such as how a machine moves, a sequence can be built up and manually operated under the camera at far less cost than if the machine were cut away and photographed with a live camera. Only flat movement in two dimensions is achieved by this technique. Such limited animation has been cleverly utilized in many small-budget television programmes for subjects concerning science and economics. To compare animation costs, this most basic type of animation may be rated at one unit per minute.

MULTIPLE TECHNIQUES. This category offers a wide range of manipulation. It combines animation with live-action photography by means of superimposition, split screen arrangement and back projection as the case may demand. Some of these effects can be achieved either under the animation camera, by photographing the animation on to the live-action negative, or by the laboratory printing the various negatives through an optical printer.

This technique can be expensive but it has several advantages. It can combine pure reality by showing the real object, for instance a jet engine, photographed by a live-action camera, with animation representing the direction and the speed of the airflow inside the engine, which cannot be seen by the naked eye.

Scientific theories, mathematical formulae, complex internal processes, mechanical developments in medicine and in industrial research – all these aspects which have defied presentation through live cinematography can be brought to life.

It is essential that the artist who plans the animation should also understand the potentiality of the stop-motion camera, and exercise control from design to finished film. Since the variations of such techniques are extremely diverse, it is very difficult to foretell the cost of such an operation. But the average cost is likely to fall between one and a half and three units per minute of film.

OBJECT ANIMATION IN THREE DIMENSIONS. The different methods and styles of animating drawings can be used with objects, too.

This technique is used more in films for entertainment than for instruction, and the results are generally rather static. Yet it could be very useful in demonstrating certain processes and operations where the spatial nature of the models is of prime importance.

Molecular construction, for instance, can be handled by this technique. The substances from which the model is made may range from carved wood to various pliable plastics, or, even better, wire constructions. The form of animation may demand an entirely new puppet or model for several single frames, or the puppet or model may be constructed in such a way that a single one may be operated throughout the whole sequence.

The cost varies accordingly, and the construction of the background also influences the cost. In the case of three-dimensional molecules, for instance, the background may be only an inexpensive flat back-cloth. A minute of film using this technique may cost anything from one to five units.

The objects to be animated remain motionless during the exposure of each frame. After the first phase has been recorded on the first frame of the film, the aspect of the objects is modified by the animation team while the shutter is closed. This second aspect (or phase) is then recorded on the second frame, and so on. The frame-by-frame changes in the aspect of the objects can be made in the following ways:

1. Modification of lighting and exposure.
2. Shifting the position of the camera or the objects.
3. Movement of articulated objects.
4. Substitution of objects. (This last is the most difficult, the costliest, and the most versatile device of the four.)

In each case, strict rigidity of the objects during exposure and high precision in their movement is necessary. The phases of movement are calculated and numbered in advance on what is known as a score. Test negatives are made and screened for rhythm control, and, if necessary, the score is modified before the final shooting, which takes much less time than its preparation.

This process requires a team of from three to seven persons, and is costlier than the animated cartoon. It can be used for black-and-white or colour animation. It excels in industrial and advertising films because of the lyrical effects it can produce. It outclasses the animated cartoon in the movements specific to the cinema: zooms, the rotation of real volumes in space, and lighting effects. However, it lends itself less readily than the animated cartoons to the demonstration of processes and to the projection of literary or comic action.

CARTOON ANIMATION. The essence of cartoon animation is animating figures or personalities, a technique which may require a separate drawing for every film frame and a staff of artists to transfer the animated drawings on to celluloids. Background artists are needed to prepare the layouts and sets, and other technical staff are necessary to control the whole production from the storyboard of a film to the creation of personalities, and to undertake the various stages of animation, photography, editing, and the synchronization of the sound. The cost can reach ten units per minute of film, though it may be considerably less.

There is room for economy in cartoon animation, however. Character animation can be exposed in double frames or complex figures can be portrayed in silhouettes and shadows. Scenes can be planned so that some part of the action takes place out of frame; animation can be conceived in cycles, and the camera itself can become mobile and replace the need for animation. Even with these devices, cartoon animation can still cost more than diagram animation, but it may be the only way to present some types of information entertainingly and memorably. Animated cartoon films, being basically visual, have the advantage of being intelligible all over the world.

RECENT DEVELOPMENTS

Animation is in the process of undergoing substantial changes. The laborious techniques involved in what might be termed classical cartoon animation are giving ground to forms of automation. Many of these were already in use by the middle 1960s, while in puppet animation various electronic controls had been introduced experimentally.

ANIMOGRAPHE. A notable centre for developments in cartoon work is the Research Studio of Radio-Diffusion-Française under Jean Dejoux, who has experimented with a number of prototypes for his Animographe, which is designed primarily to produce rapid animation for low-budget television programmes.

The animators work directly on twin strips of 70mm film, drawing small images lengthwise along the strips, successive frames being drawn alternately on each strip. The strips have a frosted surface capable of taking impressions from every kind of marking device from pencil to pigment. The Animographe produces an immediate playback of the work done, running either forwards or backwards, and satisfactory animation can be achieved with projector speeds as low as 11 down to 4 frames per second. The speed can be instantly adjusted down to zero to suit the nature of the action.

When the work is done, this apparatus can produce an image which may either be televised direct or recorded on black-and-white or colour film. It is intended to control the timing of the final Animographe projection by electronic means which will memorize the best of the rehearsal playbacks.

The saving in animation time in comparison with the classical method is very great, while the possibilities for developing mobile diagrams and drawings in educational film are obvious.

TECHNAMATION. The Animographe is not alone in the field. During the New York World's Fair, 1964–65, many of the stands using films for demonstration introduced new systems; one of them in particular, Technamation, is capable of further development.

Technamation involves the complex use of polarized light. In a light-polarizing film or filter, the molecules are oriented in a given direction or axis. When the light, thus polarized, enters a second filter whose polarizing axis is parallel to that of the original filter, it is transmitted freely by the second filter. However, if the axis of the second filter is at right-angles to that of the first, the axes are crossed and practically no light will pass through the second filter.

The amount of light which can pass through the two filters, then, is dependent on the angular relation of their polarizing axes. Each motion area is associated with a light-polarizing area having several different directions or axes of polarization. By varying the polarizing axis of an incident beam of light, that is, by constantly turning a light-polarizing filter in front of a light source, the relation of its axis to those of the polarizing areas continually and progressively changes. One area after another appears to change progressively in density and the illusion is created.

COMPUTER SYSTEMS. The Bell Telephone Laboratories developed a technique of film animation by computer which proved most suitable for films dealing with mathematics but might become flexible enough to deal with other kinds of subject.

The future of animation, with its present wide range of technical and artistic experimentation, lies on the one hand in its service to industry, education and public relations, and on the other in forms of innovation which may well have a profound effect on the cinema as a whole. With new and cheaper methods of production, the entertainment cartoon may return to the cinemas, and its use on television be extended to programmes other than commercials. J.H.&R.M.

See: *Animation cameras; Computer animation; Multiplane photography; Rostrum photography.*

Books: *The Technique of Film Animation*, by J. Halas and R. Manvell (London), 1968; *How to Cartoon*, by J. Halas and B. Privett (London), 1951.

⊕**Anschütz, Ottomar, 1846–1907.** German photographer specializing in photography of movement. Devised a multiple camera system in 1885 for recording sequences of photographs of objects in movement, the cameras using a focal plane shutter system developed by Anschütz in 1882. Sets of his pictures were sold as a modified Zoetrope device, called the Tachyscope, demonstrated in 1887. In that year he constructed the Electrical Tachyscope, using a large disc carrying pictures printed from his multiple camera negatives, intermittently illuminated by a flashing Geissler tube. A coin-fed version of this machine was introduced for public display in 1892. Later a revised version of the apparatus used a long band of pictures intermittently illuminated.

⊕**Anscochrome.** Integral tripack reversal colour film process used for motion picture work in both 35mm and 16mm. Anscochrome originals are suitable for direct projection and may be printed on to a corresponding reversal stock.

See: *Colour cinematography.*

⊕**Anscocolor.** Integral tripack colour film process using reversal material for original photography introduced for motion picture production in the United States in 1941 as 16mm and subsequently as 35mm. Now obsolete.

See: *Colour cinematography.*

⊕**Answer Print.** First print of a picture to reach the standard of customer acceptance and which the laboratory charges to the customer. The answer print is the standard by which subsequent prints are judged. Sometimes called a grading print.

See: *Grading; Laboratory organization.*

⊕**Anti-Halation Coating.** Layer applied to cinematograph film, usually during manufacture, to absorb light passing through the emulsion and prevent it being reflected from the base and causing halation around bright areas. In reversal materials and some colour stocks, this layer is often applied between the emulsion and the support and may contain colloidal silver which is bleached out during processing. Negative materials are usually coated on a grey base which may, in addition, have a jet-black pigment layer on the surface opposite the emulsion which is mechanically wiped off during processing.

See: *Film manufacture; Halation; Processing.*

⊕□**Anti-reflection Coating.** Thin coating of transparent substance, deposited on the surface of glass (e.g., a lens), to reduce the loss of light by reflection at the surface. The transparent substance has a refractive index less than that of the glass, and is deposited to a thickness of one-quarter of a wavelength of light. Light falling normally on the front surface of the coating is reflected partly at this front surface and partly at the rear surface in contact with the glass. The path difference between these reflected rays when they return to the air is twice the coating thickness, i.e., a half wavelength of light. If the refractive index of the coating is so chosen that the intensity of the two reflections is equal, there is destructive interference between the two reflected rays and no energy is lost by reflection. As the wavelength of light varies with colour, the full effect of the coating is obtained only at the colour for which it is designed.

See: *Coating of lenses; Light; Optical principles.*

□**Aperture Correction.** Method of compensating for loss of higher picture frequencies caused by the scanning spot in a camera tube having a finite size, and thus failing to respond sharply to sudden vertical boundaries between dark and light areas.

Consider a square spot of finite size scanning a sharp black-to-white transition. The resulting signal output changes level with a linear slope. In practical electron devices the spot tends to be circular or nearly so, and to have a Gaussian distribution of energy, so that the signal changes with a more rounded transition. This effectively reduces the high frequency content of the signal, and compensation must be made by increasing the gain in the high frequencies, taking care not to exceed the bandwidth of the channel or unduly increase noise or introduce phase distortion.

See: *Camera* (TV)*; Gaussian distribution; Telecine.*

⊕**Aperture Lens.** Lens placed close to the negative aperture in the projection head of an optical printer using a relay system to produce an image of the lamp filament in the plane of the copy lens diaphragm. Also, a lens used at a projector gate for field flattening.

See: *Copy lens; Field flattener; Relay lens.*

⊕□**Aperture, Lens.** Adjustable orifice controlling the amount of light transmitted by a

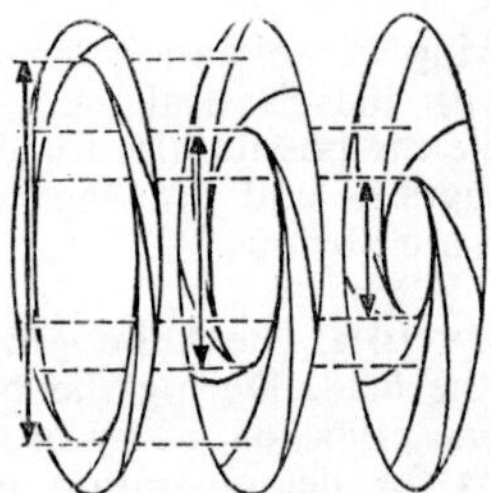

lens. The maximum diameter of the aperture in relation to the focal length of the lens determines its theoretical speed. Its effective speed depends also on the transmission of the glass elements of the lens.

See: *Focal length; Lens performance; Lens types; Optical principles; Speed.*

⊕**Aperture Plate.** Plate in a film camera, printer or projector through which the photographic image is exposed or projected.

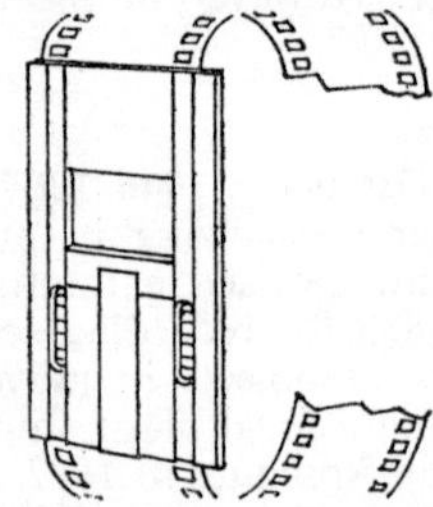

For each gauge of film, the projector aperture is slightly smaller than the printer aperture in order to mask imperfections at the edges of the image. Thus the aperture plate frames the image, and is usually removable so that it can be cleaned or exchanged to give a different film format.

See: *Aperture sizes; Aspect ratio; Film dimensions and physical characteristics.*

⊕**Aperture Sizes.** Dimensions of the apertures used in motion picture cameras to expose cinematograph film and in projectors for its presentation on the screen of the cinema theatre have been internationally standarddized for all gauges of film now in commercial use.

It is general practice for the camera aperture to be slightly larger all round than the mask used in the projector, so as to avoid the image of an unexposed area of negative being projected, but camera masks or reduced apertures smaller than the projector aperture height are sometimes used in photography of wide-screen formats. The printed area on the positive and release print is always made larger than the projector aperture and usually corresponds at least to the area exposed on the camera negative.

The generally accepted dimensions for camera and projector apertures for various gauges of film and formats of presentation are shown in the following table. L. B. H.

See: *Aspect ratio; Film dimensions and physical characteristics; Wide screen processes.*

APERTURE SIZES

		Inches		Millimetres	
		Width A	Height B	Width A	Height B
65 mm.	Camera	2·072	·906	52·60	23·00
70 mm.	Projector	1·913	·866	48·60	22·00
35 mm.	Full aperture: Camera	·945	·709	24·00	18·00
35 mm.	Full aperture: Projector	906	·679	23·00	17·25
35 mm.	Academy aperture: Camera	·868	·631	22·05	16·03
35 mm.	Academy aperture: Projector (AR 1.33:1)	·825	·600	20·96	15·25
35 mm.	Wide Screen Projector (AR 1.66:1) ...	·825	·497	20·96	12·62
35 mm.	,, ,, ,, (AR 1.75:1) ...	·825	·472	20·96	11·99
35 mm.	,, ,, ,, (AR 1.85:1) ...	·825	·446	20·96	11·33
35 mm.	Half-frame Camera (Techniscope, Ultra semi-scope, etc.)	·868	·373	22·05	9·47
35 mm.	Anamorphic systems: Camera	·870	·735	22·10	18·67
35 mm.	Anamorphic systems: Projector (AR 2.35:1)	·839	·715	21·31	18·16
16 mm.	Camera	·404	·295	10·26	7·49
16 mm.	Projector	·380	·284	9·65	7·21
9·5 mm.	Camera	·335	·256	8·50	6·50
9·5 mm.	Projector	·315	·242	8·00	6·15
Super-8	Camera	·245	·166	6·24	4·22
Super-8	Projector	·215	·158	5·46	4·01
8 mm.	Camera	·192	·145	4·88	3·68
8 mm.	Projector	·172	·129	4·37	3·28

AR=Aspect Ratio

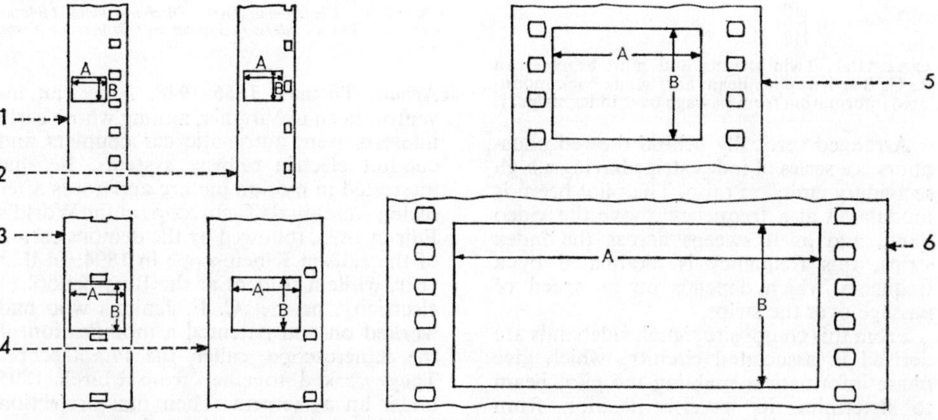

APERTURE SIZES. 1. 8mm film. 2. Super-8 film. 3. 9.5mm film. 4. 16mm film. 5. 35mm film. 6. 70mm film. (All to a uniform scale.)

⊕**Apex System.** System of photographic exposure rating, short for additive photographic exposure. The lens aperture, exposure time, lighting level and film sensitivity (speed) are all expressed in units on a logarithmic scale to base 2 and may therefore be conveniently added to arrive at the required exposure conditions. On this scale, the lens aperture is rated as the log to the base 2 of the square of the *f*-number.

See: *Aperture, lens; ASA speed rating; DIN speed rating; Exposure control; Speed.*

⊕**Apostilb.** Metric unit of diffuse luminance. □Brightness of a uniform diffusing surface whose total emission is one lumen per square metre.

See: *Light units.*

□**Apple Tube.** Colour display tube with spaced vertical red, green and blue phosphor strips, the spacing varying at the top and bottom of the tube, so that the face somewhat resembles an apple. An indexing device to locate the beam precisely as it sweeps across the strip works in conjunction with an indexing electron beam or pilot beam. This beam and the writing beam (which alone is modulated by the picture signal) are very close together and fall on the same small area of the tube face as they scan it.

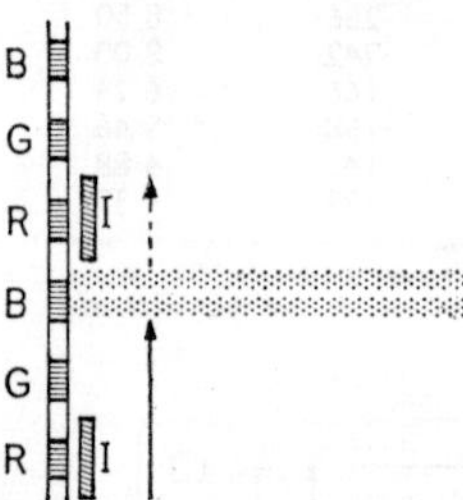

APPLE TUBE. Twin writing and pilot beams scan R, G, B strips; pilot beam also sends back modulated information from passage over index strips, I.

Arranged vertically behind the red phosphors is a series of index strips having a high secondary emission ratio. The pilot beam is modulated at a frequency above the video band, and as it sweeps across the index strips, this frequency is modulated by a frequency which depends on its speed of passage over the strips.

From this composite signal, sidebands are derived in associated circuitry which give phase information enabling the pilot beam to determine its precise position from moment to moment. By alteration of the phase, the current in the writing beam can be increased as the beam passes over a red, green or blue phosphor strip, so that the colour reproduced is fully determined.

See: *Colour display tubes; Phosphors for TV tubes; Writing beam.*

⊕**Arc Lamp.** Type of lamp in which light is produced by a bridge of incandescent particles which carries electric current from one electrode to another. The carbon arc is still employed for high-power film studio and location lighting, as well as for projection. The electrodes are of carbon and require frequent adjustment, whether manual or automatic. Each arc may pass 300 amperes or more and a d.c. supply is needed.

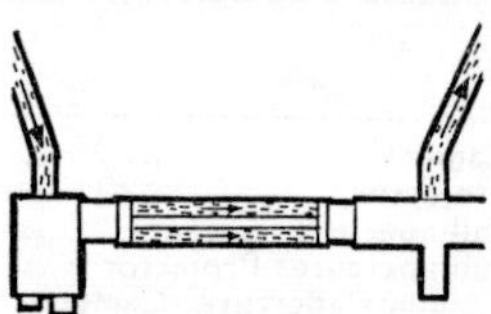

XENON ARC. Water jacket carries away heat from compact high-temperature source.

In xenon and mercury arcs, the electrodes are mounted in a small quartz tube filled with gas, often at pressures of 100 atmospheres; the tube is usually water-cooled. These arcs need no adjustment, and are used as intense light sources in compact film and slide projectors. Their light output has a discontinuous spectrum, but the rendering of colour film can be improved by suitable choice of gas and pressure.

The colour temperature of carbon arcs is close to that of sunlight, which makes them most suitable as boosters for location colour shooting.

See: *Brute; Cameraman* (FILM)*; Discharge lamp; Lighting equipment* (FILM)*; Mercury vapour arc; Projector; Xenon lamp.*

⊕**Armat, Thomas, 1866–1948.** American inventor, born in Virginia, among whose early interests were automatic car couplers and conduit electric railway systems. Became interested in motion picture apparatus after seeing Anschütz's Tachyscope at the World's Fair in 1893, followed by the demonstration of the Edison Kinetoscope in 1894. In that year, while a student at the Bliss School of electricity, he met C. F. Jenkins who had worked on and patented a modification of the Kinetoscope, called the Phantascope. They worked together from March 1895 under an agreement. Their first projection machine failed; Armat worked on a new version using a mutilated gear intermittent system on which he filed a patent on 28th August 1895 (issued 20th July 1897).

Heavy wear produced by this design made it impractical and Armat next tried a modification of the beater movement in a more practical machine demonstrated in September 1895 at the Cotton States Exposition, Atlanta, Georgia. He filed a patent for this device on 19th February 1896. The Vitascope projector was demonstrated to Thomas Edison, with whom manufacture of the machine was arranged. Edison's agents arranged a public show using the Vitascope in Koster and Bial's Music Hall in New York, 23rd April 1896. Armat himself operated the projector, showing among others R. W. Paul's film of breaking waves at Dover. A patent for an improved intermittent mechanism, using a Maltese cross device, was filed in September 1896, issued 21st March 1897, and was used in the second version of the Vitascope. After some trouble with both the Edison Company and the Biograph Company over patents, Armat later joined with them to form the Motion Pictures Patents Company, later to run foul of the Sherman Antitrust legislation.

⊕ART DIRECTOR

The art director is responsible for the background against which the actors appear in every phase of a film. He designs and prepares all settings both indoors and out and provides every property required by the script.

Colour combinations in the film are his responsibility as well as the full gamut of effects from snow and rain to fog and fire. He is part artist, part organizer, part dreamer, but he must be able to assess money values in relation to what the audience will see on the screen. He needs a working knowledge of camera lenses, film stock, laboratory processes, trick shots, sound requirements and constructional problems, and must also be aware of the rules of editing film in relation to camera angles and direction of travel. He should be able to advise or know where to find out about many specialist techniques, including casting in plaster or fibreglass, painting a backing and constructing a window frame or staircase.

FUNCTIONS

There is some confusion nowadays between the functions of the art director and the production designer in feature films. Originally the production designer was someone whose scope and talents were much wider than those of the art director – his task was not only to give sketched ideas of film sets but to concern himself with every artistic aspect of the film from each individual camera set-up through to the design of the costumes and credit titles. Today the credit Production Designer is often used to cover the functions and responsibilities undertaken by the art director.

TRAINING. The training of an art director usually includes a course at an art or architectural college, then a first job as draughtsman in an interior decorator's or architect's office before going on to a film studio drawing office. From this basic beginning to the moment when the art director is likely to have enough knowledge and experience to attempt his first film can take from five to ten years.

As a fully-fledged art director, he will be in charge of a varying number of draughtsmen according to the size and scope of the film production. He will also have under his control a staff consisting of an assistant art director, set dresser, construction manager, property buyer and scenic artist. This team, including the art director himself, usually works on a freelance basis and is engaged on merit from film to film.

It is through this group that the art director disseminates his ideas and orders amongst the studio construction workers to culminate in the completion of the studio set or outdoor location ready for the director, cameraman, actors and other technicians to move in and film the particular scene. There are hundreds, sometimes thousands, of decisions to be taken, and details to be settled before each setting is completed. And, during the time it is actually being used for filming, the set has to be serviced and maintained in exactly the same condition.

The art director is normally engaged for a period of pre-production work and retained throughout the shooting period. The pre-production period varies in length. It can extend from four or five weeks to four or five months, depending on the complexity and the size of the production. Sometimes intense pre-planning is impossible owing to the lack of a completely finished script during the pre-production and even the early shooting periods.

WORKING PROCEDURE. At the very start of his assignment, the art director discusses the visual interpretation of the script with the producer and director. Usually at this stage he is admonished on the need for economy and ingenuity.

The art director then gets down to the drawing board and begins to produce set sketches. These are drawn to show the main shooting angle of the set and coloured to give the lighting, atmosphere and character suggested to him by the director's interpretation of the scene to be filmed. From this sketch is made a rough quarter-inch-to-the-foot plan of the proposed set showing distances between doors, windows and furniture. These distances affect the placing of actors and camera movements. Both sketch and plans are checked with the director.

Meanwhile the art director picks his art department team and gets them installed ready to take over the sketches and plans and convert them into the detailed drawings that will give the workshops precise instructions for every item to be made.

BUDGETING. Early in the pre-production period the budget must be carefully considered. The sets are among the items of the film's budget least able to be accurately costed. In the struggle to make the budget fit the pockets of the film's financiers, it is often the amounts to be spent on sets which are most heavily slashed. As it is virtually impossible during the preparation period to produce all the sketches and working drawings that a quantity surveyor would require before estimating the cost of building a house, the art director is forced to make an inspired guess and from then on must bear the responsibility of keeping to the budgeted figure.

As a result, some studios may employ set estimators who, when all drawings for a set are complete, make estimates based on exact known costing of materials and

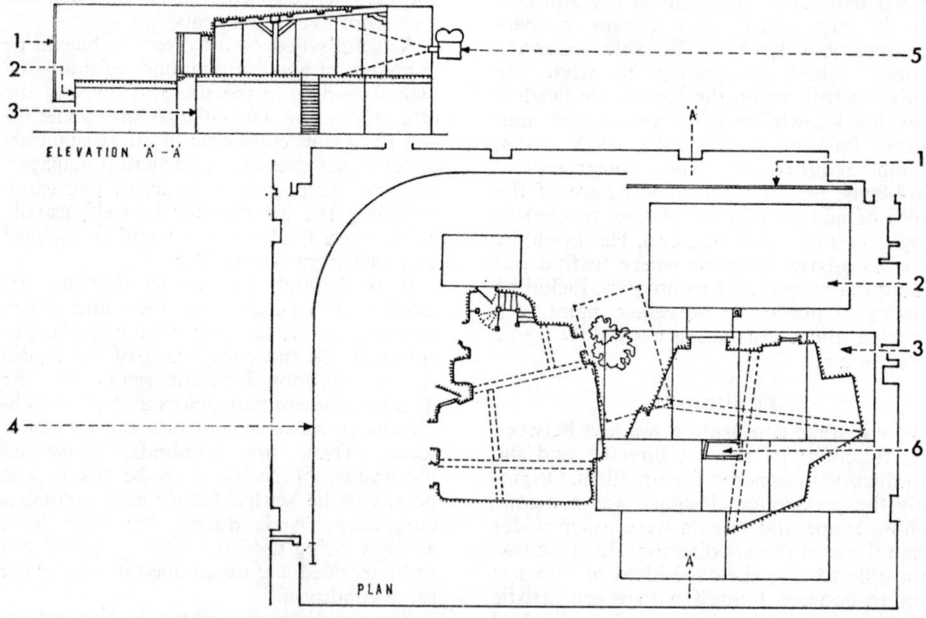

TYPICAL STUDIO SET LAYOUT. Lack of space forces sets to overlap each other and pieces of first set to be shot must be removed overnight so that second set can be completed for shooting on following day. Two types of vista backings are shown: 4, Painted cycloramic cloth backing for static vista; 1, back lighted blue backing for travelling matte to provide moving picture vista for windows and doors leading on to terrace levels 2 and 3 seen from camera position, 5. Rostrums in area 6 provide various levels for set with terraces and stairways.

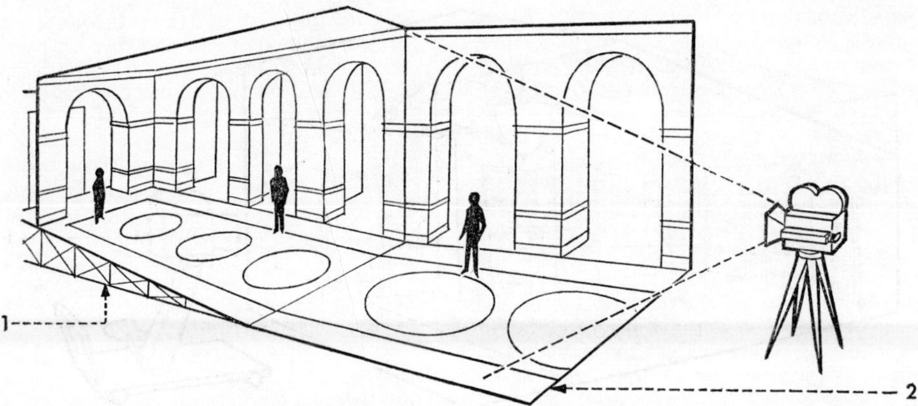

PERSPECTIVE BUILT SET. Gives impression of much greater length than is available to build in as a normal set. Foreground or action area is built to normal scale, 2. At a predetermined distance, design is continued but requires ramp, 1, and all lateral lines above the camera lens centre decline pro rata to the foreshortening.

labour, but these are often less accurate than the predictions of art directors.

Another of the art director's anxieties in the pre-production stage is the need to state the exact stage space required for the progressive erection of sets. Since at this point in time he has only vague ideas of the size and even the number of sets that will be needed, his task is not a happy one. However, the film's budget cannot be completed without this heavy cost being included in as exact terms as possible. This space requirement can be accurately costed only when the production manager has given the art director a complete schedule of all the shooting.

This the art director uses to produce plans which are then worked out into a kind of jig-saw puzzle, fitting construction into precise studio areas so that every foot of space is used and no money wasted.

SET DESIGN AND CONSTRUCTION

BASIC DESIGN PRINCIPLES. The main artistic aim of the art director is to fill the frame shape with a series of compositions of the maximum interest in shape and colour to fit the action and intention of the story. While a sketch of the main shooting angle will serve as a general image of the set, it is essential to remember that the camera is an eye that travels in all dimensions. It is therefore wise to build a model of the set so that it can be viewed from all the possible angles that can be reached by the camera.

It must also be possible to light the set for all angles of shooting. This can be checked by the use of camera angles on scale drawings to discover the maximum limits of ceilings required and the minimum height for a wall to fill the camera frame and yet allow adequate open areas for the placing of lights.

Other essential factors in the design are that any sections of the set not seen within the frame must be removable (able to float) so that the technicians have room to manoeuvre the camera, lights and sound equipment into their required positions. These technical requirements often fight the design of the set and it is the task of the art director and his construction manager to find ingenious ways of creating a set that is good to look at and at the same time easily and quickly workable for the shooting unit.

PROPERTIES. By the time a set has reached the building stage, it is possible for the art director to collaborate with the set dresser to produce a list of properties required to dress it.

Together with the buyer they visit the shops and stores that are willing to hire out their stock for the period it will be needed for filming. Special attention is paid to action props. These vary from guns that fire blanks to non-intoxicating whisky for the actors to drink. When these objects are breakable, care must be taken to have exact duplicate replacements in the studio so that no filming time is lost. Many properties are specially made from designs prepared by the art department, including breakaway furniture for fights, portraits of actors as

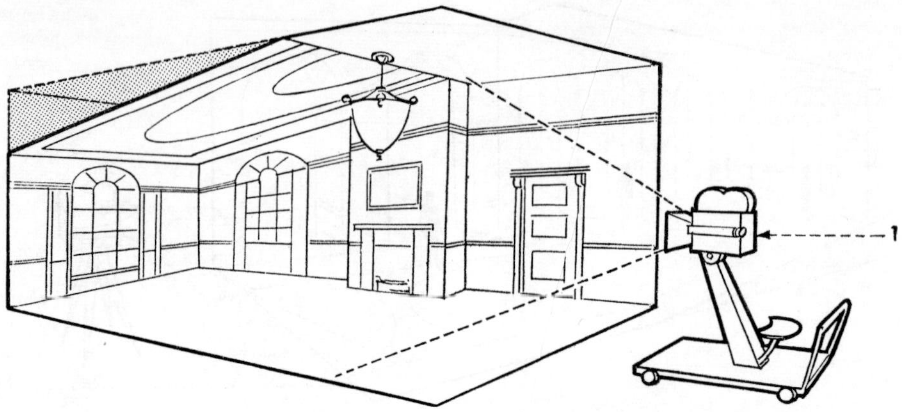

FORCED PERSPECTIVE CONSTRUCTION. Accentuates force of perspective line in a set and adds to dramatic power. Camera, 1, is relatively free to move but action is generally kept within normal scale foreground of set.

characters in the film, laser beam guns and all the million and one strange objects conceived by script writers.

Even an outdoor location can consume much of the art director's time. If, for instance, the film requires a large period battle scene, he will have to produce battering rams that work, arrow-firing machines that can fire real flaming arrows, swords and spears that are safe in the hands of extras, and period saddles, spurs, bridles, saddle cloths and banners. It is a formidable task to produce all the bits and pieces for an army of perhaps 3,000 men when each man may have as many as ten different pieces of equipment for the scene.

COLOUR PROBLEMS. Colour plays an important part in both the design and mood of a set.

As all photographers know, there is a great discrepancy between the colour the eye sees and that reproduced in a colour print of the same subject. The art director needs to know what colours to present to the camera to compensate for the errors and limitations of film stock.

There has to be close collaboration between the costume designer and art director over colour combinations used for artistes' costumes and the settings. Otherwise the dress chosen for the star may be identical to the wall colour or in uneasy contrast to the chair in which she is to sit. In practice there may be many costume colours in a set, and it will be difficult to choose a wall colour that not only harmonizes with all the costume shades but also enhances the mood of the scene as a whole. The colour of the walls and set decoration is a matter of choice, but the furniture and other properties, which are normally hired for the occasion, present a much less controllable problem.

Many types of furniture and other *objets d'art* are rare in their design and the range of colour restricted. Very warm colours in woodwork and veneers tend to become hot reds due to the colour film stock deviations from true colour reproduction.

CONSTRUCTION METHODS. Once the detailed scale drawings have gone to the construction shops, the construction manager marks out the portion of the sound stage that the set will occupy. He does this with the aid of the stage plan prepared by the art director, showing the siting of each set and its sequence of building. This stage plan has to be related to the shooting schedule which is arranged to the budget's best possible advantage – usually so that each actor does all his scenes in the shortest consecutive period of time.

The set starts to rise in its appointed position. First the flats – framed plywood wall pieces used many times over – are set up to form the basic wall areas. Apertures and spaces are left to accommodate the doors, windows and any plaster decorations. Where there are windows, backings will have to be painted showing suitable scenery to represent the view from the interior.

The art director finds an actual setting for the scenic artist's subject matter, be it a jungle, a garden or a street. This is photographed by a stills photographer whose pictures give the scenic artist an exact knowledge of what is to be reproduced on the backings.

Sometimes, to find these settings, the art director has to climb on roofs and up mountains to get as close as he can to the exact angle to relate to his interior set. Often these backing views have to match what has already been filmed on location as the exterior view of the interior being built at the studios. The scenic artist may also be required to decorate the walls of the set itself.

PERSPECTIVE BUILDING. Where studio space is restricted for one reason or another it is possible to resort to perspective building. This system involves the drawing to scale of, say, the whole length of a corridor. The art director knows from his script that only the foreground will be used by the actors, who are to walk a short way along the corridor before exiting through a door to left or right. Once this distance is established, the remainder of the set in background can be built with a squeezed-up foreshortening which will reduce the space needed for a 150-ft. corridor to a length of perhaps 60 ft.

Although it saves stage space, perspective building is an expensive form of set construction in other ways since normal flats and stock pieces cannot be used. In perspective building, every piece, including mouldings and doors, has to diminish and has to be specially built for one particular set.

The floor must be ramped up to bring the eye level to the height that it would appear in the camera viewfinder if the whole length of set had been built. On the very rare occasions when people are seen moving in the perspective areas of these sets, dwarfs or children dressed as adults are used.

Another application of perspective building is to make the shape of a set more dramatic, for instance by putting the upper part of the set above the eyeline into a forced ramp down. This gives a similar effect to placing a camera at a very low level in a room.

Sets seldom look exciting if they are filmed at camera levels that incline the camera downwards, but directors are often reluctant to use low angles as the actors tend to cover each other in the frame. Also, cameramen take much longer to light the set for a low-angle shot because ceilings have to be placed, precluding the use of lamps on rails above the set. This drawback can be overcome if the art director designs the set so that it is deeply indented and puts in bulky furniture.

Lamps can be hidden in the indentations and behind the furniture.

LOCATION DESIGN. The art director's work is not limited to building studio sets. Almost all films have open-air shots and it is usually his duty, together with the director, production manager and possibly location manager, to find the best and most accessible places where these can be shot.

Finding exterior locations can take an art director to any country in the world. Only very seldom is a location found which con-

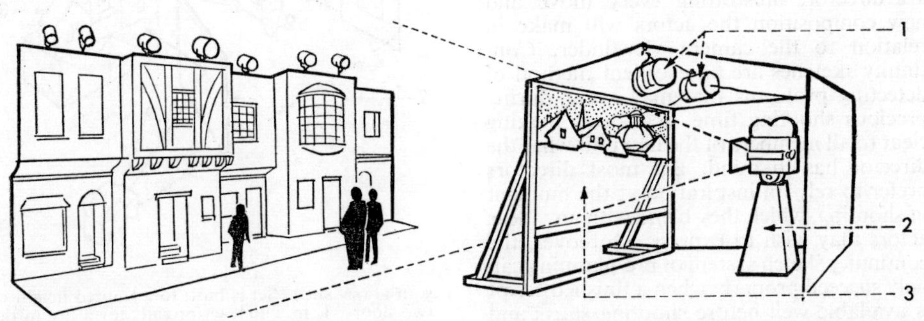

PAINTED MATTE OR GLASS SHOT. Sheet of optical glass is erected in a stout frame, 3, far enough from the camera for focus to be held on the glass surface and the set beyond. Camera is mounted on solid base, 2, to avoid vibration. On the glass, the top line of the set is drawn; artist paints in architecture above the line. Glass above roof tops may be left clear and film of red sky afterwards inserted in composite. The whole assembly is enclosed by black velvet. The painting is illuminated to a suitable balance with the built set by lights, 1, hung clear of the reflective angle of the glass with the camera.

forms to the script or entirely satisfies the director.

Doors may not be in the right places, television aerials may be visible on ancient palaces to be used in period stories, and trees have nearly always been planted in the wrong place for film cameramen.

Often large pieces of architecture, up to the size of whole towns, have to be built in the most difficult places, sometimes with local labour in areas where the inhabitants have never seen a film in their lives. The art director must adapt himself to all these circumstances and be prepared to rough it before the comforts and amenities the unit will require for the filming period have been brought into existence.

LIAISON. There should be the closest possible liaison between director, cameraman and art director so that sets can be built which will allow plenty of continuous movement to the camera. Sequences can be made interesting and exciting with the use of the camera crane – changing levels and passing through from room to room and from exterior to interior. However, the art director must be able to design for these movements right from the start, which means that the director must commit himself to exact moves even before the set is built.

As a rule, only very experienced directors are willing to fix shooting angles in this way.

CONTINUITY SKETCHES. The question of pre-planning of angles and camera movements can often be resolved with the aid of storyboards or continuity sketches. These form a series of action sketches made by the art director, illustrating every move and new composition the actors will make in relation to the camera viewfinder. Continuity sketches are an excellent method of detecting problems that may occur during precious shooting time, as well as making clear to all members of the film unit what the director has in mind. But most directors prefer to rely on inspiration at the moment of shooting, which they blend with any ideas actors may wish to express. Moreover, the continuity sketch system of pre-planning can only succeed properly when a finished script is available well before shooting starts and when the director is a highly-experienced man with a good visual mind.

SPECIAL EFFECTS

The art director must have a good knowledge of all special effects (trick) work, as it is impossible to design a set-up for the camera unless he understands the method by which it can be obtained. For travelling mattes it is imperative that each angle be drawn out and pre-set to match the background plate.

This requirement also applies with back projection, as it is essential to calculate the length of the projection throw and the size of the projected image required. These calculations should be put into diagram form by the art department and displayed on the sound stage or back projection tunnel to make sure that all personnel concerned with the shooting know the limits imposed.

MODELS. The art director is normally responsible for the design of model shots but can draw on the assistance of the special effects department, which has the specialist knowledge of how to achieve the shot once it has been sketched out. In the same way, the art director supervises all the various effects that may be required on the film such as fire, smoke, steam, snow, wind, fog and rain.

Where the art director is concerned with non-static sets, such as ships' decks, aeroplanes and other structures requiring violent movements, he calls in companies who

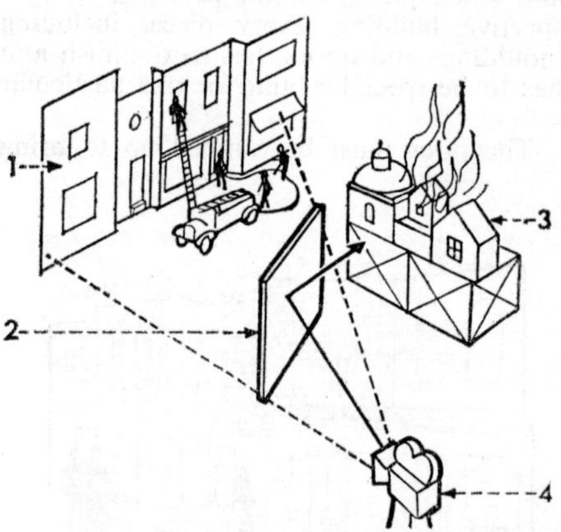

SCHUFFTAN SHOT. Set is built to a limited height of two floors, 1, in which script calls for a fire in the upper storeys, built in model form, 3, on rostrum draped in black velvet. Interposed at 45° to the faces of set and model is a sheet of optical mirror, 2. Viewing the whole through the camera lens, 4, silver from the lower back of the mirror is scraped away until through the clear glass so formed the set is seen to join reflected image of model in top half, where silver has been left intact.

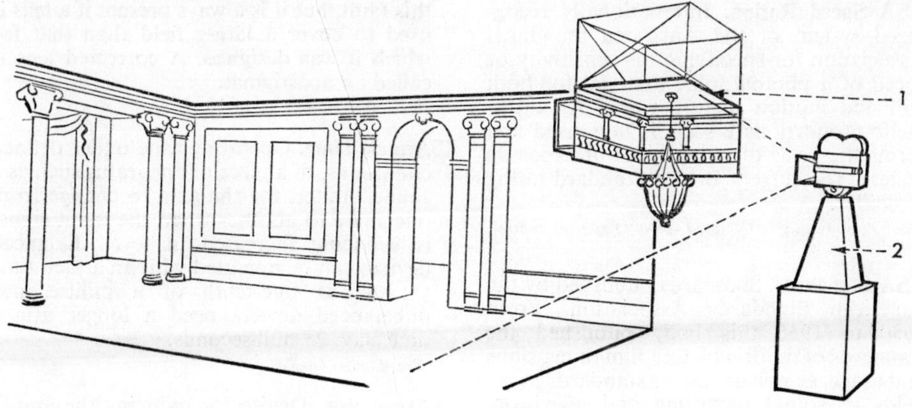

HANGING MODEL MATTE SHOT. To give greater height in the set when lack of studio height limits use of a normal fully-built set. Model of upper portion of set, 1, is hung from above and positioned so that bottom edge joins accurately to top line of built set when seen through the lens of the camera mounted on rigid base, 2.

specialize in hydraulic mountings and who are able to calculate the movements and the stresses, weights and safety factors required.

Many trick processes have been devised to avoid the building of the costly upper portions of sets. In recent years these have suffered a setback as they often inhibit the free movement of the camera and need care and time to light. Sometimes a set is built up to the minimum height necessary to contain the head of a standing man in the immediate foreground. A model of the remaining top portion of the set is hung above this line in front of the camera so that the bottom edge of the model exactly matches to the top edge of the set itself. The lighting of the model and set are exactly balanced, giving a composite shot of a lofty interior or of tall buildings in a street. However, the camera must maintain a static position and can only pan on a "nodal head" – a camera attachment which gives the camera a turning movement in the lateral and vertical planes exactly on the centre or nodal point of the lens.

With a normal camera mounting, the model portion would be seen to separate from the built set when the camera turned from side to side.

MOCK-UPS. Mock-ups of the interiors of cars, aeroplanes and trains are drawn up by the art department, usually from photographs, and measurements are taken from the actual vehicle which has been filmed from the exterior. These mock-ups are used with a travelling matte screen or back projection screen behind them to give the effect of moving backgrounds, and are built on springs or hydraulic rockers to simulate the movements of travel. They are built in small sections which can easily be taken apart to allow camera and lights to change their position as one side or the other is favoured.

All trick processes requiring sets involve the most precise drawings from the art department and constant checking by the art director for deviations in alignment.

In set design much can be done to avoid building too much height to the set by, for instance, having a foreground ceiling low enough to mask much of what would be seen at a higher level beyond. Since most studio stages have been built without reference to camera angles – that is, the outside limits that the camera can see with its widest lens – they are too low to allow for the building of a set that would cover the full height seen unless some design device is used to cut down the height vision. The most common of these devices is, of course, roof beams.

But these, in their turn, produce problems in lighting and clearance for the sound boom, so they have to be constructed to float away, that is, to be lifted off. M. G. C.

See: *Cameraman* (FILM); *Director* (FILM); *Special effects* (FILM).

Books: *Technique of Special Effects Cinematography*, by R. Fielding (London), 1965; *Designing for Moving Pictures*, by E. Carrick (London), 1941.

⊕**ASA Speed Rating.** Internationally recognized system of the American Standards Association for specifying the sensitivity or speed of a photographic material for both still and motion picture work. An arithmetic proportionate scale is now used and forms the basis of many types of exposure meters. The current British Standard rating is similar.

See: *Apex system; DIN speed rating; Exposure control; Speed.*

⊕□**ASA Standards.** Standards published by the American Standards Association. From 1930 to 1966 this body published the standards of motion picture film dimensions and usage, as well as similar standards in the fields of sound recording and television. Since 1966 the organization has been the USA Standards Institute.

See: *Standards.*

⊕□**Aspect Ratio.** Ratio of width to height of the picture image projected on the screen or printed on the film, the height being taken as unity. The long-established film aspect ratio, still retained for narrow-gauge film, is 4 : 3 (1·33 : 1).

Wide-screen cinema presentation systems may have aspect ratios from 1·65 : 1 up to 2·55 : 1.

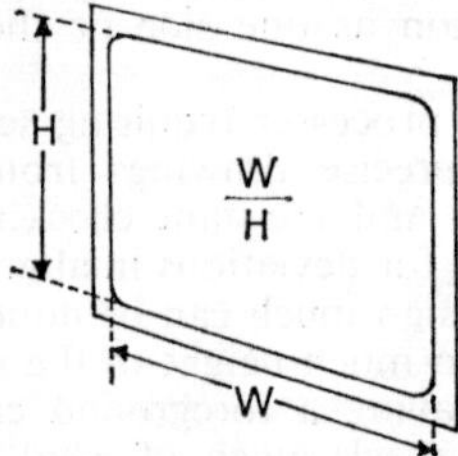

The television aspect ratio is 4 : 3. Early systems were 5 : 4 to provide the maximum rectangle obtainable on the circular picture tubes then in use.

See: *Screens; Wide screen processes.*

⊕□**Aspheric Corrector Plate.** Lens, one surface of which is specially shaped and is not part of the surface of a sphere as are the surfaces of most lenses. Used in some large-screen TV projectors and some wide-range zoom lenses.

See: *Large-screen TV.*

⊕□**Astigmatism.** Lens aberration by which a point object not lying on the lens axis is imaged as two short lines in different positions in space. Lens design can correct this fault, but it is always present if a lens is used to cover a larger field than that for which it was designed. A corrected lens is called an anastigmat.

See: *Aberration.*

□**Attack Time.** Operating time under defined conditions, of a piece of apparatus such as a sound limiter, for the gain to change from one value to another when the input signal is suddenly increased in level. Balanced devices can be operated with an attack time as low as one-tenth of a millisecond; unbalanced devices need a longer attack time, say, 25 milliseconds.

See: *Limiter (sound).*

⊕□**Attenuator.** Device for reducing the amplitude of an electrical signal without introducing appreciable distortion.

⊕□**Audio Spectrum.** Band of sound frequencies to which the normal human ear responds. It is often considered as ranging from 20 to 20,000 Hz, but these are the extreme limits, and 30 to 15,000 Hz comprises all the useful audio frequencies. This is therefore the minimum pass-band of a high-fidelity audio system, but for purely technical reasons the actual pass-band is often made much greater.

See: *Sound and sound terms.*

⊕**Auditorium.** In motion picture theatres, the whole of the space which may be occupied by the audience viewing the screen.

See: *Cinema theatre.*

⊕**Auto-Masking.** In colour photography, a system whereby the masking required to correct for the unwanted colour absorptions of the component colour images of an integral tripack material is formed within the emulsion layers during the process of development. Coloured couplers are used in the layers which are changed where the developed image is formed to produce dye images of another colour. In the best-known colour negative material using this system, the coupler used to form the cyan (blue-green) image is itself orange in colour, while the coupler used to form the magenta image is yellow. The unchanged coupler in each layer absorbs coloured light to substantially the same extent as the unwanted absorption of the dye image. Thus the unwanted absorption of blue light by the magenta image is balanced by the blue absorption of the yellow coupler. Similarly, the unwanted blue and green absorption of the cyan image is balanced by the absorption of the orange coupler wherever a cyan image has not been formed.

See: *Chemistry of the photographic process; Integral tripack; Masking (colour).*

☐**Automatic Frequency Control.** Automatic control of the frequency of an oscillator, e.g., in a TV transmitter or receiver, in order to reduce undesirable changes to a minimum. A change in output frequency is used to adjust the circuit in such a manner that the oscillator frequency returns toward its original value.

See: *Receiver; Transmitters and aerials.*

☐**Automatic Light Control.** Device on a telecine for altering the intensity of light falling on the film. Variations in the average density of film or slides may cause the signal level from a telecine channel to fall outside acceptable limits. To reduce such variations, the effective light level from the projector may be adjusted. Automatic light control accomplishes this by a servo mechanism, altering the voltage of the projector lamp (subject to disadvantages of time lag and change of colour temperature), or by interposing a variable neutral density filter or shutter in the light path.

See: *Telecine.*

☐**Automatic Voltage Regulator.** Device for automatically keeping output voltage at a predetermined level, irrespective of supply or load variations.

See: *Power supplies.*

⊕AUTOMATION IN CINEMA THEATRES

To improve operational efficiency and ensure high standards of picture and sound presentation, many cinemas, especially in Europe, have adopted some degree of automation. The mechanization of routine tasks allows the skilled projectionist to concentrate on the more discriminating features of good presentation.

The initial application of automation was to the recurrent changeovers when both picture and sound reproduction have to be switched from projector to projector without the change being visible or audible to the audience. The changeover is a moment of potential hazard involving the accurate timing of light source energizing, projector starting, simultaneous opening and closing of light dowsers and the selection of sound tracks. The regular, repetitive nature of all these functions made automation possible and in many cinemas the changeover has been automatic for several years.

METHODS. There are several methods of automatic changeover but the usual way is for the outgoing reel of film to originate, at the correct time, a pulse which starts a motor driving a spindle on which a series of cams has been mounted. Each cam operates a microswitch at one point in its rotation and each cam can be adjusted on the spindle to fit in with the timing sequence of a changeover and the run-up time of the particular projector.

The pulse is normally obtained from the film by means of a small piece of adhesive silver foil which either makes physical electrical contact in the form of a bridge, or reflects light on to a photo-transistor. In both cases, the electrical impulse which results triggers the pulse.

At the start of a film showing, it is generally necessary to lower the auditorium and screen lighting, open the screen curtains, change from the sound channel carrying disc or tape non-synchronous music to the film sound track, whether optical, single magnetic or multi-track magnetic, and open the light dowser (the projector having already run up to full speed). The correct masking aspect ratio, lens and aperture plate have to be pre-selected and the light source energized. When the film finishes, the reverse procedure is generally required.

These operations can be automated and, if necessary, a timing device can be introduced to effect the operations at a predetermined time.

A typical method is to use a circular drum perforated with lines of holes so arranged that a pulse from any source rotates the drum just one step or line of holes. A row of microswitches is mounted so that each vertical line of holes passes a microswitch at some time in the drum's rotation. Pins can then be inserted in any of the holes and the head of the pin activates the microswitch on a particular step of the drum rotation. These microswitches can be arranged to act as parallel switch operations to the manual controls of the lighting dimmers, curtain motors, sound channel selection, projector starting, light source energizing, opening or closing of light dowsers, etc., and it therefore becomes possible to automate the whole sequence of operations required at the start or conclusion of a film presentation.

More sophisticated equipments are now available and the latest consists of a wall-mounting cabinet incorporating a programme matrix, which has replaced the rotating drum. By inserting pins into the appropriate sockets all the above functions can be controlled accurately and easily.

Most of the operations involved in a changeover lend themselves to automation because it is only necessary to provide an electronic "finger" to press the operating buttons that already exist, i.e., the closing of a microswitch in parallel.

LIGHT SOURCE. The most difficult function is the energizing of the light source. With Xenon lamps or pulsed discharge lamps it is simple, but the carbon-arc lamp has to be physically struck by touching the two electrodes together when they are alive. The answer lies in the introduction of a metal pellet which is placed between the electrodes. This melts when current flows through it and, in dropping away, strikes the arc. Another method is to introduce a solenoid which, when energized, moves the negative electrode toward the positive and, as they touch, cuts out of circuit, leaving the negative electrode to return to its normal position under spring control.

While carbon-arc lamps are more difficult to automate than discharge lamps, there is at least one carbon-arc lamp that has been designed for automation and incorporates the three main requirements, solenoid-operated striking, a solenoid-operated light dowser, and an electronic control system that maintains the arc gap at its right distance. Carbon electrodes have to be fed forward as they burn away and such a control is necessary if the carbon-arc lamp is to need the minimum of attention, which is an essential part of automation.

FILM LENGTH. But even with the most reliable of automation systems, the changeover is still a potential hazard and the aim should be to decrease the number of changeovers. For many years, the standard maximum length of film reel for projection has been 2,000 ft., and the usual projected length is between 1,600 and 1,800 ft. With the advent of automation, the reel lengths run on each projector have been increased to about 6,000 ft. and may be as high as 12,000 ft. in one continuous run. The Xenon lamp or pulsed discharge lamp which burns steadily for long periods of time is thus advantageous, although special carbon arc lamps with longer feed travel have been produced to suit the increased reel length.

ASPECT RATIO. For many years the cinema has had to deal with a series of differing picture shapes. These have now resolved themselves into three major aspect ratios. To accommodate them, it is necessary to provide motorized variable picture masking, a series of lenses and a series of specially shaped aperture plates.

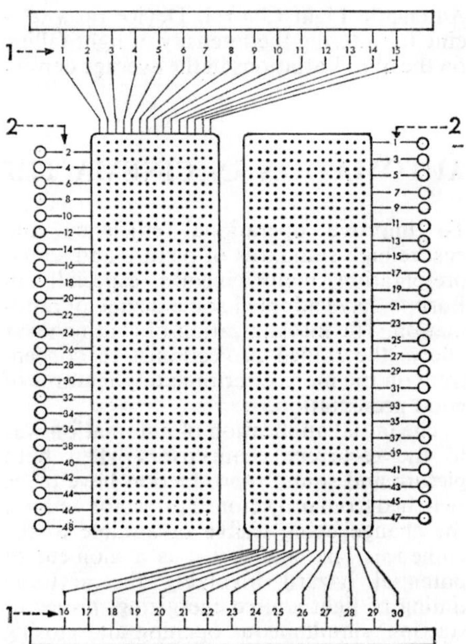

PROGRAMMING MULTIPLE OPERATIONS IN CINEMA. Insertion of a diode pin into any hole in the multilayer matrices causes any function, 1, to be switched on or off as the particular horizontal line, 2, is scanned in sequence at the arrival of each signal pulse from the motion-picture projector.

A motorized lens turret on the projector coupled with spring-loaded operation of a sliding multi-aperture plate allows the complete automation of the sequence. A pulse from the film can be made to operate the whole sequence of a change in picture aspect ratio without any manual act whatever and without even stopping the projector.

ADDITIONAL FUNCTIONS. The more sophisticated equipments allow for the film itself to originate pulses which operate appropriate warning signals in the manager's office at a preselected period before a mass audience movement, or in the refrigerator rooms prior to a sales interval, and which illuminate category boards, warning signals and notices.

A companion wall-mounting cabinet to the one just described and carrying another programme matrix can be used to automate the control of all-day and every-day heating, ventilation, lighting (interior and exterior), canopy lighting and signs, car park lighting, cleaners' lighting and a host of other functions.

In this instrument, a pulse generator scans a line of matrix holes every half-hour. Two types of pin are used, each including a diode transistor: a green pin to initiate a function and a red pin to switch it off. Thus it becomes possible to programme the engineering functions of the theatre throughout the 24 hours, before the projectionist arrives and after he has departed.

Manual override is provided by three-position switches for each function: automatic control, a blocking position, and an override "on" without disturbing the programme matrix.

Any automatic system devised for the presentation of picture and sound in cinemas must be reliable and be provided with full manual overrides. In addition, there are usually local safety regulations to be observed.

Reliable automation equipment is, in effect, an assistant projectionist working 24 hours each day, never ill or on holiday, always on time, accurate and possessing a potential that is only just being realized. It is thus a means of improving the operational efficiency while maintaining the standards of picture and sound presentation in a highly competitive business. R. R. E. P.

See: *Changeover; Cinema theatres; Projector.*

Literature: *Evolution of Projection Practices in the UK*, by R. R. E. Pulman (*Jnl. SMPTE*, Oct. 1967).

⊕**Auto-Optical.** A method of incorporating optical effects (fades, dissolves and superimpositions) in imbibition prints without going through an intermediate duplicating stage with its consequent loss of image quality.

This is achieved by A & B printing of the matrices.

See: *Imbibition printing; Technicolor.*

⊕**Autotransformer.** Electrical transformer in ☐which the secondary (output) circuit shares

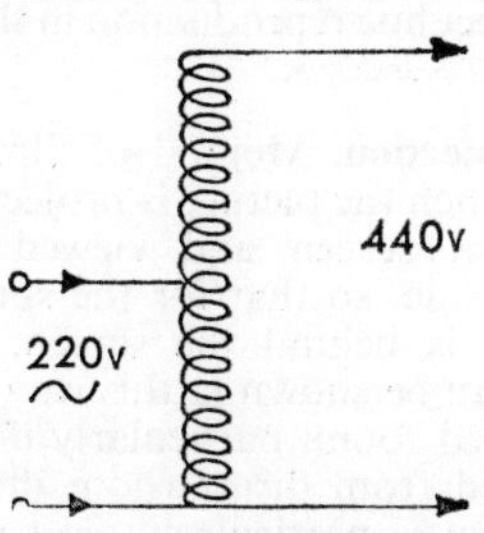

part of the winding with the primary (input) circuit.

A simple means of raising or lowering the voltage in a.c. mains circuits when the common winding is no disadvantage.

See: *Power supplies.*

☐**Ayrton, William Edward, 1847–1908.** English electrical engineer and physicist, born in London and educated at University College, London. Devoted himself to the study of electrical physics and spent some years in the public service of India and in teaching science in Japan, where he collaborated with Professor John Perry in scientific experimental work. Their proposal for "seeing by electricity" – one of the earliest to be put forward – was described in a letter to *Nature* in 1880.

⊕**Azimuth Error.** Optical film sound tracks are normally recorded and reproduced by passing light through a narrow slit set at right-angles to the direction of movement of the film. If, because of faulty adjustment or wear, this angle is other than a right-angle, there is said to be an azimuth error. It causes waveform distortion and loss of signal. Its avoidance is equally important in magnetic recording and reproduction.

B

⊕**Backing Removal.** Removal, during processing, of the anti-halation backing coated on the base side of many types of cinematograph film, including colour materials. Rotating pads with a suitable alkaline solution are often used on continuous processing machines.

See: *Integral tripack; Processing.*

⊕**Back Light.** Lighting so arranged as to □illuminate objects in the scene from behind. Such lighting would not alter the appearance of objects as seen by the camera, were it not that back light spills over and illuminates their sides and edges to produce a rim of

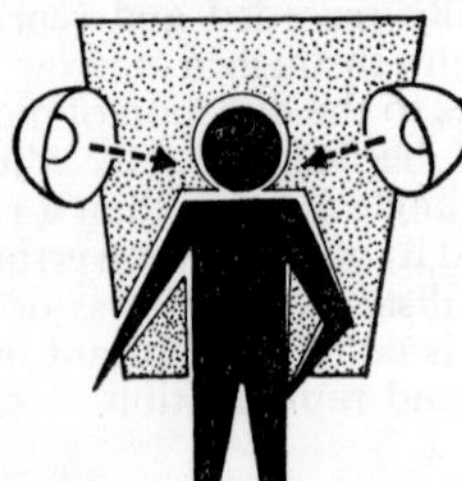

light which helps to make them stand out from a darker background. Back lighting requires careful positioning of sources and calculation of exposure.

See: *Cameraman* (FILM).

□**Back Porch.** Period of time, longer than that of the front porch, which follows the line-sync pulse within the line-blanking interval in the standard TV waveform. It enables black-level clamp circuits to operate, and in the NTSC and PAL colour systems provides space for the colour burst, a few cycles of the sub-carrier frequency transmitted once each line to provide a phase reference and so ensure correct hue reproduction in the receiver.

See: *Television principles.*

⊕**Back Projection.** Method of film presentation in which the picture is projected on to a translucent screen and viewed from the opposite side, so that for the spectator the projector is behind the screen. Since the picture can be shown in this way to viewers in a lighted room, particularly if the screen is shielded from direct room illumination, this system is particularly used where it is impossible or undesirable to darken the area for normal projection.

In special-effects cinematography, back-projection photography is one of the most important methods for showing actors against a background, either still or motion picture, photographed separately on a previous occasion. The foreground action must be carefully lighted to avoid illumination falling on the translucent screen and the projector and camera must be accurately lined up so that the background scene appears uniformly illuminated on the screen. Back projection of small areas, such as the view through a porthole or car window, presents little difficulty, but where the background to be shown is large in comparison with the foreground action, problems of inadequate picture brightness and uniformity are considerable. Directional screens concentrating the light transmitted towards the camera position have been devised and, for very large screens, two or three separate projectors superimposing their projected pictures have been used; background scenes are sometimes printed and projected using large area film, such as 70mm, for the same reason.

From the production point of view, back projection has many advantages, since the result is obtained at a single stage in the camera and the matching and synchronization of foreground and background action is achieved by direct viewing, but for extremely large background scenes, especially in wide-screen processes, travelling matte methods can provide better quality results with less complication in the studio.

See: *Rear projection; Screens; Special effects* (FILM).

⊕□**Baffle.** Louvred shutter mounted on the front of a studio lamp to direct light where

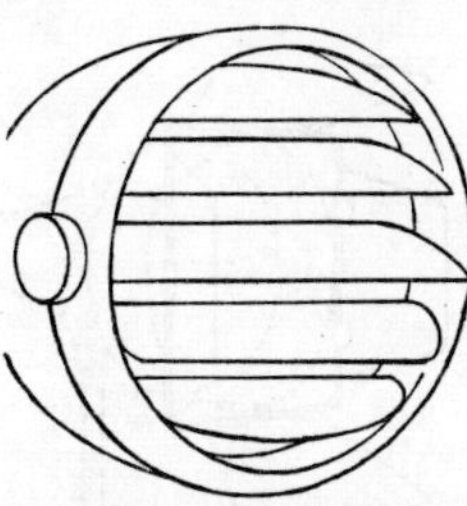

it is wanted, and control its intensity without alteration of colour temperature.

□**Bain, Alexander, 1810–1877.** Scottish inventor of the chemical telegraph and several other important telegraphic devices. Born at Wattem, Caithness-shire, served as an apprentice to a clockmaker at Wick and came to London in 1837 as a journeyman. Attended lectures and became interested in applying electricity to clocks. Was one of the first to devise a method of working a number of clocks off a standard timekeeper (an invention also claimed by Wheatstone). He was also early in the field with the printing telegraph. The chemical telegraph was the most important of his inventions and Bain described it in a paper read by him before the Society of Arts in 1866. The most valuable part of the invention was the strips of perforated paper which Wheatstone subsequently used. Electric fire-alarms and sounding apparatus were also among his inventions.

□**Baird, John Logie, 1888–1946.** One of the leading British television pioneers. Born at Helensburgh, Scotland, and studied engineering at Larchfield Academy and Glasgow University. His mechanical system provided the first public television service in the world. His versatile inventive powers started him on exploiting first, in Glasgow, a chemically treated damp-proof undersock and then, for health reasons, jam manufacture in Trinidad. This latter enterprise failed badly and he returned to England. He began working on television in 1923 while convalescing at Hastings and it was there that his first practical results were achieved. Moving to London, Baird set up a laboratory in Frith Street, Soho, and later at 133 Long Acre. In 1926, at Frith Street, he demonstrated a mechanical scanning system before members of the Royal Institution, and the following year formed the Baird Television Development Company. He was now working on Noctavision, colour television and stereoscopic television, and in 1928 successfully demonstrated television transmission from London to New York and to the liner *Berengaria* in mid-Atlantic. In 1929, the British Broadcasting Corporation (BBC) opened a limited television service using the Baird system and this was continued until 1934. Meanwhile, Baird put his Televisor receiver on the market. Large-screen television was his next interest and this was given a public showing in 1930 at London's Coliseum. In 1931, he televised the Derby for the first time and in 1935 demonstrated at the Dominion Theatre, London, large-screen colour television.

When the London Television Station opened in 1936 the Baird system shared transmissions with the Marconi-EMI all-electronic system. After a period during which the two systems were evaluated, the Baird system was abandoned in favour of the all-electronic one. Baird retired, to continue working on his large-screen colour project. In 1945 he formed his own small company, John Logie Baird Limited, with the object of developing this and producing large home television sets. His death the following year came after years of hard work and ill-health.

□**Balanced Pair.** Cable containing two wire circuits which are symmetrical with respect to earth.

See: *Transmission networks.*

⊕**Balancing Stripe.** Stripe of the same thickness as the magnetic stripe on cinematograph film for sound recording purposes, applied to the opposite edge to enable the film to be wound in a flat smooth roll.

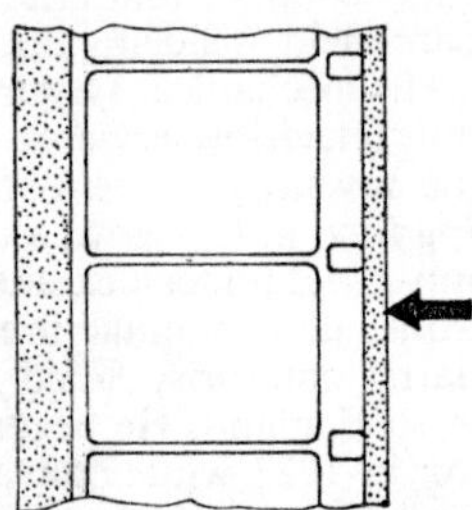

Balancing stripes are not required on 35mm and 70mm prints made for multi-channel stereophonic sound reproduction, because striped zones on both edges of the film are already used.

See: *Film dimensions and physical characteristics.*

□**Banana Tube.** A colour display device so called from the shape of the tube which, however, does not itself form the picture, as in the shadow-mask tube and most other types.

A single electron gun scans the side wall of a cylindrical tube on which a single phosphor strip is coated in the black-and-white version, and three colour-emitting strips in the colour version. The spot is

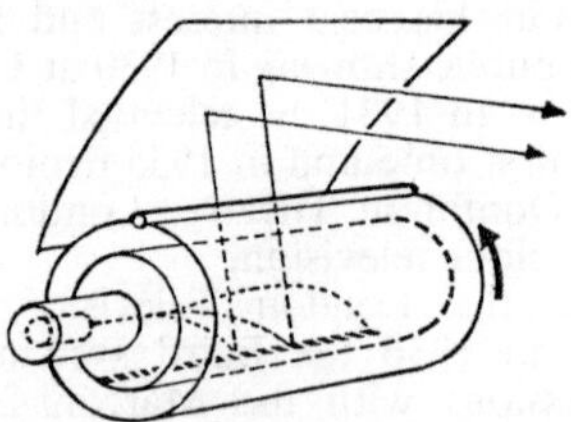

deflected along the tube for the horizontal scan, and a transverse deflection raises or lowers it so that the red, green and blue strips can be scanned as the colour signal demands.

Vertical scanning is achieved mechanically by means of a series of cylindrical lenses which rotate around the banana tube. This device is not at present in commercial use.

See: *Colour display tubes.*

□**Band.** Range of frequencies. The VHF and UHF frequency spectrum is divided for television purposes into five groups or bands of frequencies by Article 5 of the International Radio Regulations of the International Telecommunication Union. This is a world-wide agreement covering virtually all countries. The frequencies covered by each band are:

Band	Limits (MHz)
I	41 – 68
II	87·5–100
III	162 –230
IV	470 –558
V	582 –860

In certain countries, sections of these bands are used for purposes other than television (e.g., FM broadcasting and amateur transmissions).

See: *Channel; Transmission and modulation.*

□**Bank.** Row of push-button switches, one for each source, corresponding to a particular function of a switching equipment or vision mixer, such as cutting, preview, or A and B banks of an A/B mixer.

See: *Vision mixer.*

⊕**Barndoor.** Pairs of hinged doors mounted □on the rim of a studio lamp or round the lens of a spot light which may be opened or

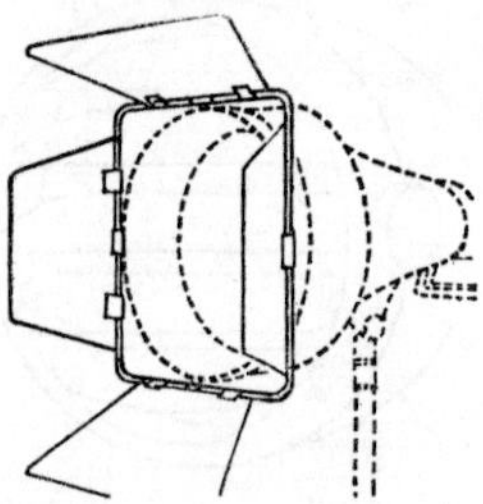

closed to control the shape of the beam and avoid unwanted spill light.

⊕**Barn-Door Wipe.** Special printing effect providing the transition from one scene to another in such a way that the new scene appears to open out on the screen from the middle of the previous image.

See: *Optical effects.*

⊕**Barney.** Flexible sound-damping cover thrown over a film camera and usually zipped into place to provide some reduction of camera noise for sound recording when a more elaborate rigid blimp is not available or is not convenient to use.

⊕**Barrel Distortion.** Defect in a lens or cathode ☐ray deflection system which causes a square

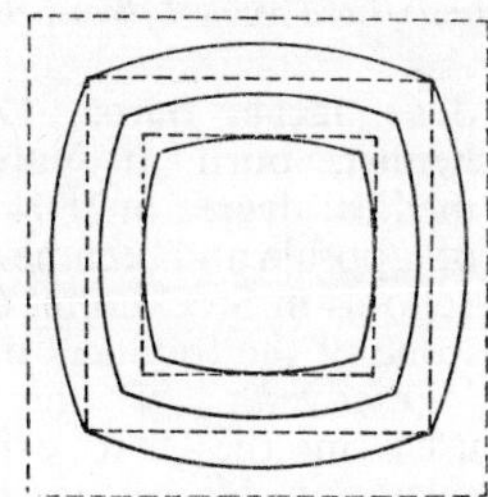

to appear barrel-shaped, compressed toward the corners.

See: *Aberration.*

⊕**Base.** Flexible transparent support on which is coated the sensitive emulsion of photographic materials and the oxide layer of magnetic recording tapes. Film base was originally a form of cellulose nitrate but the use of other cellulose derivatives of low inflammability is now universal.

See: *Film manufacture.*

☐**Base Band.** Narrow band of frequencies at the low-frequency end of a broad band channel. Loose term for the lowest frequencies of a broad band channel.

See: *Transmission and modulation.*

⊕**Basher.** Small studio lamp which can be held in the hand or fixed to the front of a blimp. The light source may be a 500-watt tungsten bulb, or a 250-watt tungsten-halogen type.

Simple means may be provided to narrow or widen the beam for use as a spot or a flood.

See: *Lighting equipment* (FILM).

⊕**Batch.** Emulsion mix or chemical solution made up at a single time. Batch replenishment is used in small film processing machines. Emulsion batches have distinct characteristics which differentiate the rolls of film made from them in respect of speed, colour response, etc. While commercial tolerances hold these differences to a minimum, they are often of importance to professional users.

See: *Processing.*

⊕**Bath.** Generic term for a container in which film is immersed during the several stages of processing, e.g., developing, fixing, bleaching, washing.

See: *Processing.*

⊕**Beaded Screen.** Projection screen with spherical-section reflectors embedded in its surface, causing it to return a high proportion of light towards the source, namely the projector. Beaded screens therefore have a high light gain for viewers near the projection centreline, and are of great value for narrow-gauge film projection when the light source is of low power and the audience

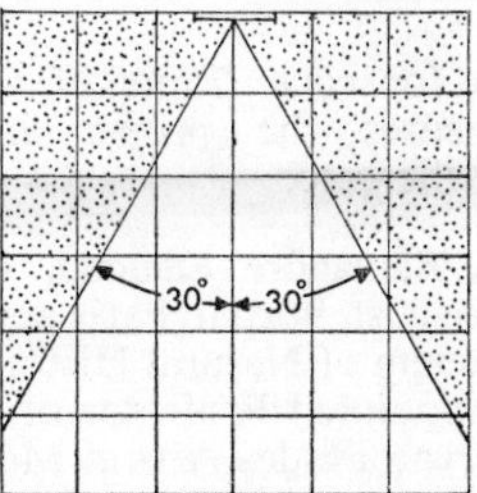

can be concentrated into the central area. Either side of this area, the screen brightness falls off rapidly; even more serious, there is more or less severe brightness difference from left to right of the screen.

See: *Screen luminance; Screens.*

☐**Beam Current.** The current of a scanning beam in camera or cathode ray tubes.

☐**Beam Landing Errors.** Errors which can occur when the electron beam does not strike the target correctly, owing to distortions of the magnetic fields. This may happen in cameras and cathode ray tubes in areas where two fields interact, such as line and frame deflecting fields.

See: *Camera* (TV).

⊕**Beam-Splitting Systems.** Devices for splitting ☐a light beam to form two or more separate images from a single lens. Often used in colour TV cameras to form the three primary colour images. Beam-splitting can be accomplished by prisms, semi-reflecting surfaces and dichroic mirrors.

See: *Camera (colour).*

☐**Beam Tilt.** Tilt above or below the horizontal axis of the pattern of radiation from a directional aerial array.

The angle between the horizontal and the direction of radiation pattern is known as the angle of beam tilt.

See: *Transmitters and aerials.*

⊕**Beater Mechanism.** Pull-down mechanism used only in simple narrow-gauge projectors. A single continuous sprocket effects the film feed and take-up, while the eccentrically mounted beater roller, making one

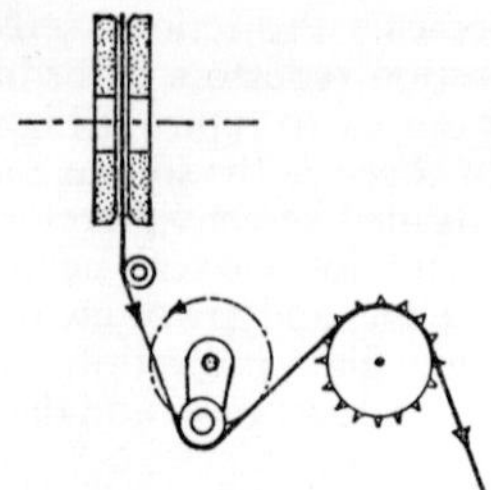

revolution for each frame of film pulled down, ensures the proper intermittent motion at the gate.

□**Becquerel, Alexandre Edmond, 1820–1891.** French physicist, born in Paris and educated at the Museum of Natural History in Paris. In 1853, appointed Professor of Physics at the Conservatoire des Arts et Métiers. Was especially interested in the study of light and investigated the photo-chemical effects and spectroscopic character of solar radiation and electric light, and the phenomena of phosphorescence. His discovery in 1839 that certain chemicals when charged with electricity give off light was not followed by any suggestion as to its practical use. Alexandre Becquerel was the son of another French physicist, Antoine Cesar (1788–1878), and his own son Antoine Henri Becquerel (1852–1908) continued the family tradition and was the discoverer of radioactivity in 1896.

⊕**Bell-Howell Mechanism.** Film camera intermittent movement designed by A. S. Howell more than 50 years ago, and still unexcelled for the steadiness with which it registers the film.

It is characterized by having fixed pilot pins, on to which the film is placed by a movable shuttle as it comes to rest.

See: *Camera* (FILM).

⊕**Bell-Howell Perforation.** Form of sprocket hole used on 35mm motion picture film; has rounded ends and is now normally used in the United States and Western Europe for

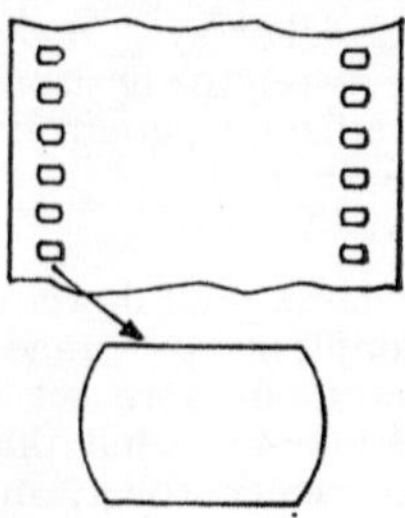

35mm negative and duplicating materials only, not for release prints.

See: *Cinemascope perforation; Dubray-Howell perforation; Film dimensions and physical characteristics; Kodak perforation.*

□**Berzelius, Jöns Jacob, Baron, 1779–1848.** Swedish chemist, born at Vafversunda, received a medical degree in 1804 from the University of Uppsala and became a lecturer at medical schools in Stockholm. Generally regarded as one of the founders of modern chemistry. Discoverer of the element selenium used in the first experiments leading to television. He became interested in the exact determination of atomic and molecular weights and over a period of ten years ascertained this for some two thousand simple and compound bodies. One of his beliefs was that the whole of chemistry revolved around oxygen and he therefore used oxygen as a basis of reference for these calculations. He continued the efforts of other great chemists to establish a convenient system of chemical nomenclature and may be said to have introduced the modern system of chemical symbols. Besides selenium, Berzelius also discovered the element thorium and was the first to isolate titanium, silicon and zirconium.

⊕□**Bias.** Steady voltage applied to an element in an electrical circuit, causing it to take on a potential different from normal. The classical use of bias is to set the mean potential of the control element of a valve (tube) or transistor to the point where it operates near the centre of the straight part of the characteristic curve.

□**Bidwell, Shelford, 1848–1909.** British pioneer of phototelegraphy. Born at Thetford, Norfolk, and took his degrees at Cambridge. For a time he practised law but then turned to scientific study, specializing in electricity, magnetism and physiological optics. About 1880, began investigations into the photoelectric properties of selenium and in 1881 gave details, in a lecture to the Royal Institution, of an apparatus he had devised for picture transmission over wires (preserved today at the Science Museum, London). Bidwell, whose studies included also meteorology, was the author of many scientific papers, one of his better-known works being *Curiosities of Light and Vision* (1899). He was elected a Fellow of the Royal Society (1886) and President of the Physical Society (1897–99).

⊕**Bilateral Sound Track.** Type of symmetrical, variable-area, optical sound track on film, recognizable by the fact that in the print the central axis and surrounding modulated area are transparent, whereas in the equally common duplex track they are opaque.

See: *Sound recording* (FILM).

⊕**Bin.** Container for loose film, often made of fibre, placed in cutting rooms and other

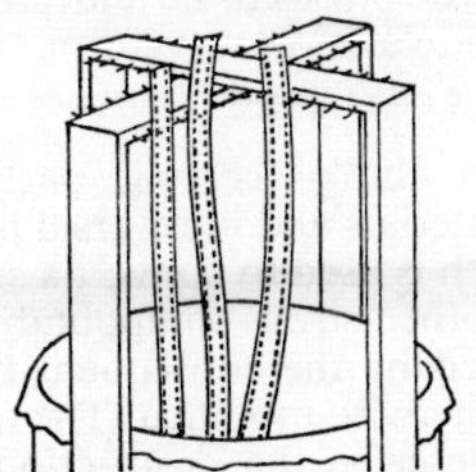

rooms where film is handled. Scrap film is often dumped in a waste bin.

See: *Editor*.

⊕**Binaural Recording and Reproduction.** Sound systems which attempt, with more or less fidelity, to reproduce human methods of locating the source of sound. The fact of having two ears separated by a solid body, the head, is of considerable help in locating the source of a sound. The two ears provide separate information about the frequency, intensity and phase of incoming trains of waves, which the brain uses to fix the source. The contribution of these factors is still a matter of research, but there is no doubt

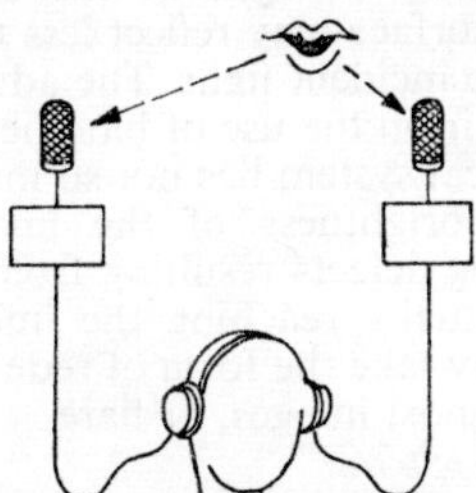

TRUE BINAURAL SYSTEM. Complete channel separation from original sound to hearer's ears.

that a binaural transmission system gives more pleasing and realistic results than a monaural one. However, except in large-scale cinema installations, this improvement seldom justifies the cost and complication of providing an additional sound channel all the way from microphone to loudspeakers.

The terms binaural and stereophonic are often used synonymously, though binaural systems more correctly are those with two channels only, and complete aural separation by headphones at the receiving end. Stereophonic systems often employ three or more channels, and sound distribution is by groups of loudspeakers.

See: *Sound recording* (FILM); *Sound reproduction in the cinema.*

⊕**Bipack.** As a general term, the running of two films simultaneously through a camera or printer aperture to expose both together; often used in special-effects techniques. When applied to a colour process, bipack denotes the location of two colour-sensitive emulsions on separate films which are exposed simultaneously. Now obsolete.

See: *Colour cinematography.*

⊕**Bipost Lamp.** Studio spotlamp in which the
□connections are brought out to two posts at the base of the bulb. In the medium base bulb (500 watts), these posts are 0·875 in. between centres, and in the large base (mogul) bulb (1–5 kW) 1·500 in.

See: *Lighting equipment* (FILM).

□**Black Bar.** Number of lines of black used as a reference or test signal.

See: *Picture line-up generating equipment.*

□**Black Crushing.** Compression of low values of signal (i.e., black) resulting in loss of significant detail in the darker picture areas.

See: *Pedestal; Transfer characteristics.*

□**Blacker-than-Black.** Excursion of the television video waveform signal downwards below the nominal black level, e.g., the excursion of the synchronizing pulses to zero signal.

□**Black Level.** The potential at any given point in the video signal path corresponding to regions of zero illumination or blackness in the scene being transmitted. Its value is entirely arbitrary and may be positive, negative or zero with respect to earth.

At certain points in the signal path it is important that the d.c. component be preserved or restored, e.g., the input to a cathode ray display tube (kinescope). At points such as this, black level is a fixed quantity. At other points, for instance where a.c. coupling has been used between amplifier stages, the d.c. component is not present, and in these circumstances the potential at black level depends on the mean value of the picture signal content.

See: *Picture monitors; Receiver; Television principles.*

□**Black Level Clamp.** Circuit which establishes the signal level corresponding to black at a finite level. Necessary after a.c. coupling to restore the d.c. component in the TV signal, and to eliminate low-frequency distortion and hum. The signal is fed through a capacitor and shorted to a fixed direct voltage during line-blanking intervals. Used also in line clamp amplifiers.

See: *Picture monitors; Receiver; Television principles.*

□**Black Stretch.** Non-linearity applied to the television signal so that the part toward black is increased in amplitude relative to the rest of the signal. The effect is to make detail in the black more visible, and correct the crushing caused by the non-linear toe of the typical transfer curve. As in film technique, a slightly higher contrast than is theoretically desirable often improves the picture.

See: *Transfer characteristics.*

⊕**Blank.** Clear processed film on to which colour images are transferred in the imbibition process; the blank usually carries a sound track image in black and white.

See: *Imbibition process.*

□**Blanking.** The operation of removing part of a signal to delete unwanted sections and to produce a waveform of the required shape and duration, as with synchronizing pulses.

See: *Scanning; Television principles.*

□**Blanking Pulse.** Addition to a television camera output to suppress spurious signals. The scanning beam in a television camera traces out the pattern of a raster across an image of the studio scene but it is important that information is only transmitted during the forward trace and not during flyback. Nevertheless, spurious signals are generated during flyback which must be suppressed. To do this, a negative pulse, known as a line-blanking pulse, is added separately to each camera output during the periods between the end of one line and the commencement of the next. Similarly, a field-blanking pulse of several lines duration is also added during the periods between the end of one field and the commencement of the next. Both these sets of pulses are then limited at black level, giving rise to intervals of zero picture content referred to respectively as line-blanking and field-blanking. It is within these blanking intervals that the synchronizing pulses and equalizing pulses are subsequently added as appropriate.

The word suppression is sometimes used as an alternative to blanking.

See: *Scanning; Television principles.*

⊕**Blanking Shutter.** In motion picture printing equipment, an opaque shutter which can be introduced into the path of the illumination so as to cut off practically instantaneously the exposure of the film being printed.

See: *A and B printing.*

□**Blanking Signal.** Suppresses noise and retrace signals inserted into the picture signal. It also provides a suitable amplitude level to allow sync pulses to be added to form the composite television waveform.

See: *Blanking pulse; Scanning; Television principles.*

⊕**Bleach.** In film processing, particularly of colour materials and in reversal processes, a bleach bath is used to convert a silver image to some other silver compound which can be removed by the subsequent fixing solution or otherwise modified. The term is also used to describe the destruction of a dye image by chemical means.

See: *Chemistry of the photographic process; Integral tripack.*

⊕**Blimp.** Soundproof housing which surrounds a film camera when it is used to record scenes involving dialogue, and which prevents the noise of the camera from being superimposed on the recorded sound. Cameras are sometimes called self-blimped when the normal camera housing silences the noise of the mechanism without the addition of an external blimp.

See: *Camera* (FILM).

⊕□**Bloom.** Thin coating of transparent substance applied to a glass surface to reduce the reflection of light at the surface. A bloomed surface may reflect less than 1 per cent of the incident light. The advantage to be gained from the use of bloomed surfaces in an optical system lies not so much in the increased brightness of the image as in diminishing defects resulting from reflected light eventually reaching the image. Such defects may take the form of reduced image contrast, ghost images, or flare.

See: *Coating of lenses.*

⊕**Bloop.** Noise made by the passage of a splice through an optical film reproducer.

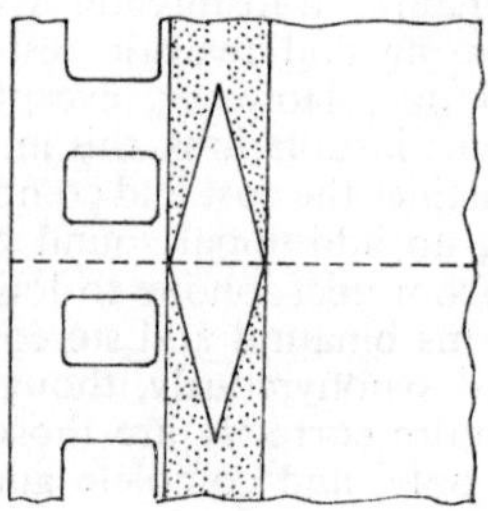

This noise is caused by sudden modulation of the light beam by a transparent line resulting from scraping the emulsion at the splice over too wide an area. Normally, however, the term bloop means the patch, fogging mark or stencil by which the unwanted noise is rendered inaudible.

⊕**Blooping Machine.** Machine designed to punch out a blooping hole in a negative optical sound track, or to apply a blooping patch on a positive track to silence the passage of splices through an optical film reproducing machine.

⊕**Blow-Up.** Film enlargement. A single frame of film may be said to be blown up to a still picture of increased size. Similarly, 16mm film can be blown up to 35mm in an optical or projection printer.

See: *Printers.*

⊕**Blue.** Colloquial term for a certain duplicating positive film of medium fine grain coated on a bluish-grey coloured support. Also called lavender.

See: *Duplicating processes (black-and-white).*

⊕**Blue-Back Shot.** Scene in which the action is photographed in front of a blue backing to provide the foreground component of a travelling-matte trick shot.

See: *Special effects* (FILM).

⊕**Blue-Screen Process.** System of travelling matte preparation in which the foreground action is photographed against a uniformly illuminated blue background using colour negative film in the camera. Light from the backing thus exposes only the blue-sensitive layer of the film while the foreground objects are recorded normally. By printing from this negative with colour filters and combining the resultant images, the required silhouette masks, or mattes, are prepared which allow a dupe negative to be printed in which the foreground action is combined with a normal background scene.

See: *Special effects* (FILM).

□**Blumlein, Alan Dower, 1904–1942.** Prolific British inventor in the electronics field, educated at Highgate School and the City & Guilds College, London. Graduated in 1925 and joined the International Western Electric Corporation where his work was concerned with line communications. In 1929 moved to the research department of the Columbia Graphophone Company. Here his work on sound recording resulted in a method of stereophonic recording which he patented in 1931. When the firm of Electric and Musical Industries Limited was formed, by a merger between the Columbia and HMV companies, Blumlein began to take a leading part in the major programme of television development embarked upon by the new company, and it was in this field that he was to make his greatest contributions. When the world's first public high-definition service opened from Alexandra Palace in 1936 using the Marconi-EMI system, almost every part of the EMI equipment owed something to Blumlein. He was mainly responsible for establishing the type of waveform which has been used by British television ever since and among his many television circuit devices were black spotting to make interference less conspicuous, a method of sync pulse separation, and spot wobble. In wartime projects his inventions included a type of aircraft altimeter and various radar devices. He died at the age of 38 in a wartime air accident.

Altogether, in only seventeen years, he registered 132 patents.

⊕**Boom.** Jib of a mobile camera mount, □which may be raised above a set and projected out over it.

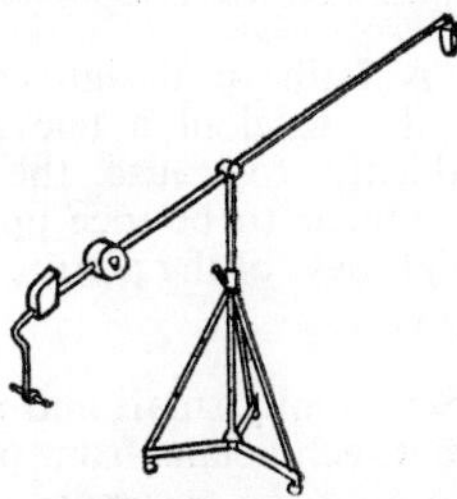

Also, long arm carrying a microphone, which may likewise be projected over the set and rotated until it is in the best position to pick up dialogue free from background sounds.

⊕**Boom Operator.** Sound technician, a mem- □ber of the film unit or television crew responsible for manipulating the microphone and boom to the requirements of the sound recordist (mixer). During shooting, he must handle the boom and microphone so as to compensate for any irregularities on the part of the artistes from the pattern set during rehearsal, and thus maintain the desired sound standard. In doing this, he must avoid casting shadows of microphone or boom on artistes or part of the set being photographed.

See: *Sound recording.*

⊕**Booster.** Auto-transformer for boosting the □voltage on photoflood lamps so as greatly to increase the light output for a given wattage at the expense of lamp life. The over-voltage is applied by a switch during shooting only, normal voltage being used for setting up the lighting.

See: *Lighting equipment.*

⊕**Booster (Chemical).** Popular term for a replenisher solution in film processing, usually of greater concentration than the circulating bath to which it is added.

See: *Processing.*

⊕**Booster Light.** Lamp designed to provide □filler light and boost shadow illumination on exteriors.

□**Bootstrap Circuit.** Circuit in which the output is directly connected to the input as where the output load is connected between the cathode and negative HT and where the input is applied between cathode and grid.

□**Bothway Broad Band Channel.** Channel capable of carrying broad band signals, i.e., a frequency band exceeding the audio range, in both directions at the same time.

See: *Transmission and modulation.*

□**Bouncing.** A fault or design condition in which the d.c. level of a television signal varies suddenly to cause the waveform display to appear to bounce up and down and the brightness of the picture monitor to vary sharply.

⊕**Box Set.** Set usually small and so enclosed □as to prevent technicians from photographing or recording the necessary action until one or more walls are temporarily removed.

□**Braun, Karl Ferdinand, 1850–1918.** German physicist, born at Fulda, Hesse, Germany, and educated at Marburg and Berlin. Made notable contributions to electronics with his development of the cathode ray tube, carrying further the work of Crookes, Hittorf and others. With his cathode ray oscillograph (1897) he was able to guide and deflect the previously uncontrolled cathode rays and make the stream of electrons trace patterns on a fluorescent screen. The Braun tube had all the elements of the kinescope or television tube except the means of controlling the stream of electrons, which flowed at a constant intensity. This improvement was later introduced by von Rosing. Braun became Professor of Physics at Marburg and afterwards at Karlsruhe. Later he was Director of the Physical Institute at Strasbourg. Apart from his work with cathode rays, Braun was active in the field of wireless telegraphy and his inventions were incorporated in the Telefunken system. In 1909 he shared the Nobel Prize for physics with Marconi.

⊕**Bray, John R., b. 1879.** American cartoonist who developed in 1914 the system of drawing cartoons on cels in collaboration with Earl Hurd.

⊕**Breakdown.** Breaking down of a script according to actors, locations, etc. In editing, breakdown is the process of reducing a roll of film to its component shots. Usually applied to rushes (dailies), when a single roll of picture or sound may contain 30 or more separate scenes and takes.

See: *Editor.*

□**Break Point.** An abrupt change of shape in a gamma correction circuit. While such a circuit properly requires a smooth curve, the gain is more commonly varied in three discrete steps. If a sawtooth signal is passed through such a circuit, and then displayed on a waveform monitor, it is seen to consist of three straight lines, each at an angle relative to the others. The break point is the point at which the change in gain occurs (*arrowed*).

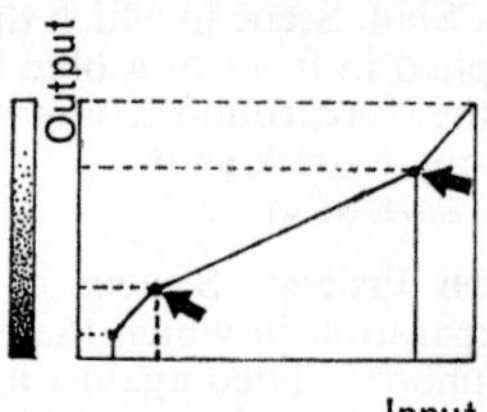

BREAK POINT. Upper point initiates white stretch, lower point black stretch.

See: *Transfer characteristics.*

⊕**Breathing.** Fairly common camera fault, caused by fore-and-aft movement in the gate, in which the image moves in and out of focus, sometimes with an unpleasant distorting effect. Even small amounts of breathing are noticeable on the screen to the critical eye and are sometimes difficult to cure in the camera.

See: *Camera* (FILM).

⊕**Brewster, Percy D., 1886–1952.** American inventor, graduate of Cornell University in 1906. Developed a number of colour processes, with patents for bipack colour systems in 1915 and 1917. In 1935 he introduced a camera using a rotating mirror system to produce three separation negatives simultaneously. These were used to make three-colour subtractive prints, with cyan and magenta images on the opposite sides of double-coated film and a yellow image produced by imbibition printing in register. Held 360 patents on a variety of processes.

⊕**Brewster, Sir David, 1781–1868.** Doctor and inventor. Introduced the refracting stereoscope in 1844, and a two-lens stereo camera in 1849 – the first devices for stereoscopic photography.

⊕□**Brightness.** Property of a surface to emit or reflect light in the direction from which it is viewed, also called luminance. The brightness of a small element of a source is defined as:

$$\frac{\text{luminous intensity of the element}}{\text{projected area of the element perpendicular to the direction of view}}$$

The unit of brightness is the candela per unit area: the nit is one candela per square metre, and the stilb is one candela per square centimetre.

See: *Light units; Screen luminance.*

□**Brightness Control.** Component of a television receiver or monitor governing the bias applied to the display cathode ray tube and consequently operating on all signal levels. In practice it is used to set correctly the darker tones of the picture in relation to the ambient lighting; the lighter tones are then set to the desired level of brightness by adjustment of the contrast control.

Unless the receiver contains a means of correctly establishing the d.c. component of the signal at the point of application to the cathode ray tube, the setting of these two controls is to some extent interdependent. In this event, moreover, the levels of reproduced brightness corresponding to the various levels of applied signal amplitude also depend on the mean level of picture signal being transmitted.

See: *Receiver.*

⊕**Brightness, Screen.** Measure of the light reflected from a screen when a film projector is directed on to it, with the motor running but no film in the gate. It thus takes into account the brightness of the light source, light losses in lens system and shutter, and screen reflectance.

Screen brightness is measured by a photometer in foot-lamberts or candelas per unit area, usually nits (1 nit = 1 candela/m²). The photometer is often applied to different parts of the screen surface to measure the fall-off in brightness at sides and corners. National standards are commonly set up for 35mm film projection, e.g., in Britain a recommended 8 to 16 foot-lamberts.

See: *Light units; Screen luminance.*

□**Brightness Separation.** The commonest method of obtaining a signal suitable for use as an electronic switch for overlay or inlay special effects is to rely on the differing grey-scale values of the composite picture. For overlay purposes, the lighting has to be most carefully adjusted to give the maximum separation between the subject to be keyed-in and its background.

See: *Special effects* (TV).

□**Broad Band.** Band of frequencies exceeding the audio range and usually of several MHz.

□**Broad Pulses.** Field-synchronizing pulses in the standard TV waveform, so called because they are broader (i.e., of longer duration) than line-synchronizing pulses. Their frequency, however, is higher, so that the receiver can distinguish between field- and line-sync pulses. In the American 525-line system there are six broad pulses, and in the CCIR 625-line system five broad pulses after each field.

□**Browne, Cecil Oswald, 1905–1942.** British television pioneer of the 1930s, working for the Graphophone Company Limited and later EMI. Was killed with A. D. Blumlein in an air accident in World War II.

⊕**Brute.** Trade name often applied generically. High-intensity arc spotlamp frequently used in lighting film sets for colour, and is the

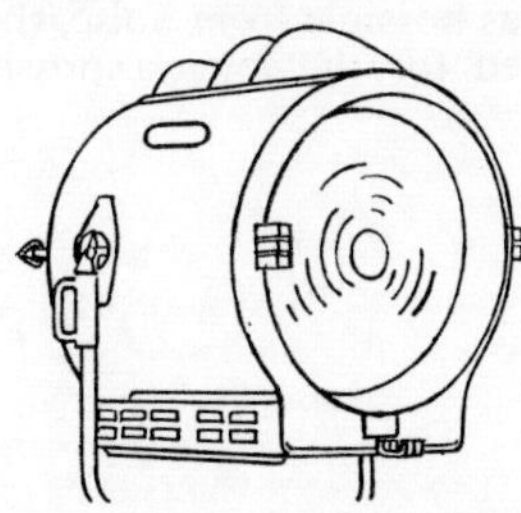

largest standard studio lamp. Motor-driven carbon feed with rotating positive carbon to maintain symmetrical crater. Twenty-four-inch diameter lens, current rating 225 amps.

See: *Lighting equipment* (FILM).

⊕**BSI Standards.** Standards and recom-
□mended practices including those for the motion picture and television industries, established and published in Great Britain by the British Standards Institute.
See: *Standards.*

⊕**Buckle Switch.** Trip switch incorporated in studio film cameras which breaks the camera current should the film begin to pile up inside on account of breakage, torn perforations or incorrect threading.

⊕**Build-Up.** Sections of leader (blank film) which are used to replace missing pieces of picture or sound track during the film editing process, to maintain synchronism or to keep a shot at its correct length for subsequent negative cutting.

□**Burn Out.** Television images are said to suffer from burnt out whites, or bleached whites, when they appear on a television screen to lack tonal gradation in the white and near-white portions. The four most common causes of this picture degradation are: (1) Over-exposure of the studio camera tube. (2) Unsuitable film densities or gamma, effectively giving overall distortion. (3) Amplitude non-linearity in the vision signal chain, causing the higher amplitude portions to be compressed or, in the most severe cases, clipped completely. (4) Incorrect setting of receiver brightness and contrast controls.

□**Burst Gating.** Process of separating the colour burst from the complete colour signal. The encoded colour television signal, as transmitted, must include a phase reference for the chrominance detector in the receiver. This is done by transmitting a short burst of the colour sub-carrier during the line-blanking interval that follows the synchronizing signal. To keep the signal as immune as possible from noise, the burst is maintained for the longest possible time. This usually allows about 10 cycles of the sub-carrier to occur during the burst interval.

To avoid phase errors between the burst and the encoded picture signal, the burst is applied to the video waveform in the encoder or colorplexer. This is done by matrixing the burst gating pulse from the studio sync pulse generator, into the encoder modulators. The sub-carrier is thus produced by these modulators for the allotted period.

In a receiver-type signal decoder, the burst is gated into the phase reference circuits by a pulse derived from the flyback of the horizontal scan. In other types of decoder, the gating pulse is derived from the signal synchronizing waveform.
See: *Colour systems; Colorplexer; Receiver.*

□**Burst Keying Pulse.** Pulse for correctly timing the colour burst in a TV colour system, usually by a gating circuit.

□**Burst-Locked Oscillator.** Oscillator, e.g. in a colour receiver, locked to the colour burst for subsequent application to later circuits.
See: *Colorplexer; Colour principles* (TV).

□**Burst Phase.** Phase of subcarrier signal forming the colour burst with which reference the modulated subcarrier signal is compared.
See: *Colorplexer; Colour principles* (TV).

□**Burst Signal.** In a colour television transmission, the burst of high frequency associated with the colour information.
See: *Colorplexer; Colour principles* (TV).

⊕**Buzz Track.** Sound track, free of speech or other pronounced modulations, which carries a low background sound to match that of the dialogue tracks with which it is intercut. The use of completely silent track between snatches of dialogue produces an unnatural effect.

C

□**Cable Compensation Circuits.** Circuits designed to compensate for losses in television signal links and not designed for a wide frequency range. Some units can be adjusted over a wide range of gain and frequency with reasonably constant phase equalisation.

⊕**Callier Effect.** Effect of light scatter in optical systems, first studied by André Callier (1877–1938) in 1909. In an optical

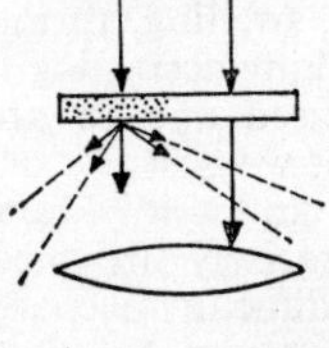

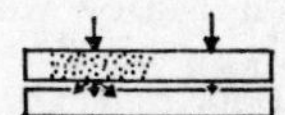

CALLIER EFFECT. (*Above*) Scattered light does not enter copying lens in optical printing. (*Below*) All light falls on printed film in contact printing.

film printer where the negative is specularly illuminated, a certain amount of the light is scattered by the silver grains which make up the image and, as a result, does not pass through the copying lens on to the positive. This effect is greatest where the negative is most dense and least where its density is least, so that in the positive print the contrast between light and dark areas is increased in comparison with a print made with the negative and positive in contact and illuminated with diffuse light.

Optical printing is therefore always regarded as producing a print of greater tonal contrast than contact printing, but the extent of the difference depends greatly on the characteristics of the whole optical system and on the structure of the negative image; in colour negatives, where the image consists of transparent dyes, the effect is much less than in a black-and-white negative.

The ratio of the densities of the film when measured using specular and diffuse illumination is known as the Callier quotient or coefficient, denoted by Q.

⊕**Cam.** A revolving device often used inside the shuttle framework of a camera or pro-

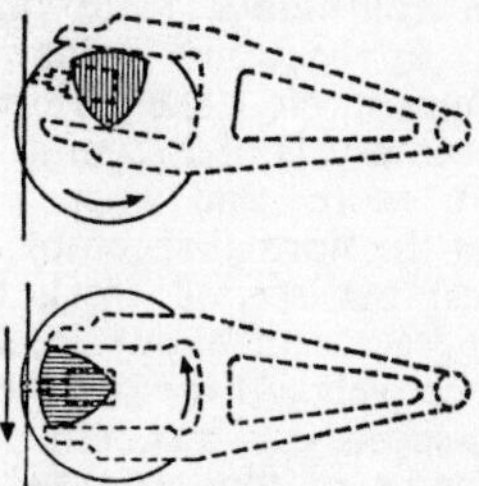

jector to actuate the claws in an intermittent movement.

⊕CAMERA

In film making, the camera is the first important tool to be used. Historically, the camera (and the projector, which closely resembled it) was the first instrument to be developed. For nearly a century this primacy has given it a key role in the evolution of film techniques. Even in television, where the wholly different electronic camera takes first place, the film camera has by no means been ousted as a recorder, and indeed TV demands have played an important role in the past decade in reshaping its design.

DEVELOPMENT

In the earliest period, it was sufficient if the film camera worked at all. It was a large and ungainly box, and inventors concentrated their efforts on steadying the jerky and flickering pictures which marred the first generation of movies. In its middle period, the camera, by now a highly efficient machine, was loaded with all manner of refinements and accessories to suit the elaborate epics of the day. These cameras, developed 30–40 years ago, are still standard for large studio productions.

The latest period shows a complete reversal of trend. It was not the avant-garde or documentary cinema, but the demands of wartime reporting, which led to the lightening and simplification of the camera, even at the expense of precision, so that the tripod could be discarded and the film camera become an integral part of its operator, whether he were running, jumping or standing still. This trend, further accentuated by TV, helped to popularize 16mm films for professional use, and led to the development of today's highly automated cameras, which include advanced 8mm designs.

RANGE OF APPLICATIONS. The motion picture camera, in its conventional form, is in fact a device for recording a series of static pictures in more or less rapid succession, on a continuous band of film. In specialized applications (plant growth, traffic movement, etc.), the pictures may be recorded at intervals of a second, a minute, an hour, or more; and when the film is projected at the normal speed of 24 frames (or pictures) per second (fps), this gives time compressions of about 25, 1,500 and 80,000 respectively. At the other end of the scale, high-speed cameras, still with continuous lengths of film, can be driven at speeds of 20,000 fps, giving time magnifications up to about 850. Speeds up to many millions of frames per second are attainable, but these require different methods of image analysis. The normal film camera operates in the narrow range between 16 and 25 fps.

HISTORY. The earliest attempts to make photographic records of movement were hampered by the lack of a flexible support for the image, which could conveniently be stored in rolled-up form. Up to 1890, nothing better than a glass plate was available, and it was the French astronomer P. J. C. Janssen (of Norwegian birth) who hit upon the idea of a circular plate turned in jerks to expose a sector at a time to light passing through a lens. This photographic revolver (1873) inspired the celebrated French physiologist E. J. Marey (1830–1904) to build his photographic rifle in 1882, based precisely on Janssen's principles but attaining a speed of 10 fps. With his rifle, Marey was able to record and analyse the motion of birds in flight.

But glass plates were clumsy, and the individual pictures or frames were little bigger than a postage stamp. The next step was to record the image on the new material, celluloid, which George Eastman began to commercialize in the USA in 1887, as did other inventors in Britain, France and Germany about the same time. Marey soon found that glass plates, whatever their other inconveniences, gave him pictures which were precisely positioned, so that any movements he recorded were those of his subject alone. But early celluloid was a highly unstable material, swelling, shrinking, curling and even cracking according to the way it was manufactured and to atmospheric conditions when it was used.

To move the film rapidly forward, and then hold it precisely in place during exposure, posed difficult mechanical problems which Marey was far from solving successfully in his first camera, developed between 1888 and 1891. Many inventors, among them the French chemist and artist L. A. A. LePrince and the English photographer W. Friese-Greene, pursued the same aim independently, some using simple unperforated film clamped between rollers, others devising crude perforations, and others again, like Edison, oscillating between the two methods.

However disputed the claims to the invention of the motion picture camera may be, it was the American inventor Thomas Edison in 1890–91 who fixed the dimensions

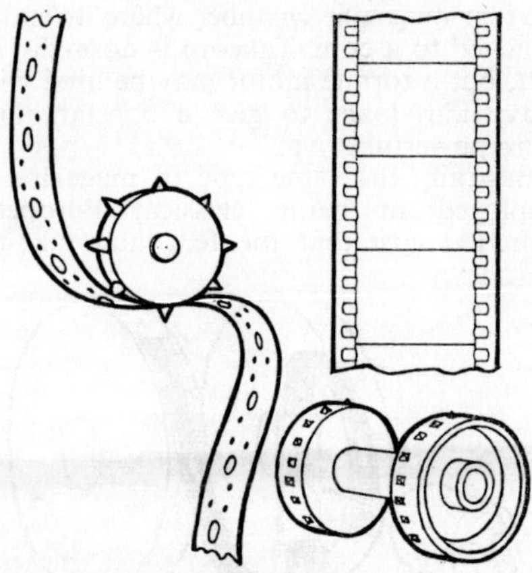

PRINCIPLE OF PERFORATIONS. Thomas Edison made use of Wheatstone's device in the automatic telegraph (before 1870) of perforating a paper tape (*left*) to ensure a uniform motion, a principle also employed in steam organs and other automatic musical instruments of the period. His 35mm film (*right*) with four perforations each side of each picture, remains substantially unchanged today.

of standard film at an overall width of 1⅜ in. (so conveniently close to 35mm), and a width of exactly 1 in. between perforations and thus available for the picture. These perforations were the most vital feature of Edison's film. Standardized at four each side of each frame or picture ¾ in. high, thus giving a pitch or spacing of $\frac{3}{16}$ in., they enabled the film to be driven regularly by toothed sprocket wheels or pulled down intermittently by a claw type of mechanism.

By a historical quirk, it was not Edison who developed the standard camera mechanism. He seems to have been hypnotized by the possibilities of continuously-moving film, on which the movement was arrested by a rapidly rotating disc with a narrow slit-like opening. His early cameras gave an exposure of only 1/50 sec. to each frame, though he was taking them at a rate of no more than 10 per second.

It was left to other inventors to devise the modern intermittent movement: R. W. Paul in England in 1895 with his Maltese cross, the Lumière brothers in France a few weeks earlier with their shuttle and claw. Neither was strictly original, the Maltese cross movement having been adapted from a watchmaker's device of the seventeenth century, and the idea of the shuttle borrowed from the contemporary sewing machine.

As in the modern camera, the film was drawn out of a box, or magazine, by a sprocket wheel and then passed down through a gate, the front of which was perforated by an aperture which framed on the film behind it a picture of exactly the required size.

In front of the aperture was placed the lens, and between them a black revolving disc with a cut-out sector obscured the film while it was being pulled down, and allowed light to reach it through the aperture when it was stationary. If the cut-out section of the shutter was a half-circle, the exposure at 10 fps was 1/20 of a second, 2½ times as long as with the Edison camera. After exposure, the film was wound up into a second magazine.

All these basic components were equally essential to a projector, and in the earliest days the camera and projector were one and the same instrument, conversion requiring only the addition of a source of light. But at the end of the century, when the cinema revealed itself as more than a nine days' wonder, more specialized instruments began to emerge.

Camera requirements soon established themselves, and the most important of these became known as registration, or the identical positioning of every successive frame of film in relation to the fixed camera body. A high order of registration was and is needed because the inch-wide picture on the negative might be enlarged several hundred times on to the cinema screen. The problem was a multiple one: film shrinkage must be brought under strict control, perforation standards made extremely exact, and intermittent movements developed which would lock the film

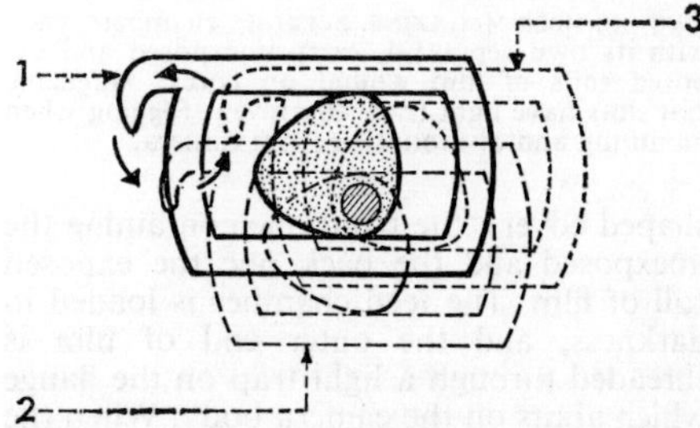

LIMIÈRE'S INTERMITTENT MOVEMENT (1894). Louis Lumière, working with his brother Auguste, hit on the idea of adapting the shuttle action of the sewing machine to the problem of alternately moving and arresting film in his camera-projector, the Cinématographe. 1. Claw engages with film perforation at start of movement. 2. Shuttle, shown at bottom of stroke. 3. Shuttle and claw withdrawn, and returning upwards to start new cycle.

in position after each pull-down movement, so as to produce a rock-steady image on the large screen.

THE MODERN CAMERA

There are now four film gauges in general camera use: 65, 35, 16 and Super-8, together with the obsolescent 8mm. (In the Soviet Union, 70mm film is used in cameras, but in the West only for cinema release prints.) Cameras have many common features irrespective of gauge, and these may conveniently be described together, along with the accessories normally available for the different types of camera.

Cameras comprise five main groups of components: film chambers with feed and take-up arrangements; a pull-down movement; lenses with mounts and accessories; viewfinding devices; enclosing structure and drive.

FILM FEED. The classic type of film chamber or magazine was pioneered by American constructors. It is mounted externally to the camera, usually on top of it, and consists of two separate chambers, closed by disc-

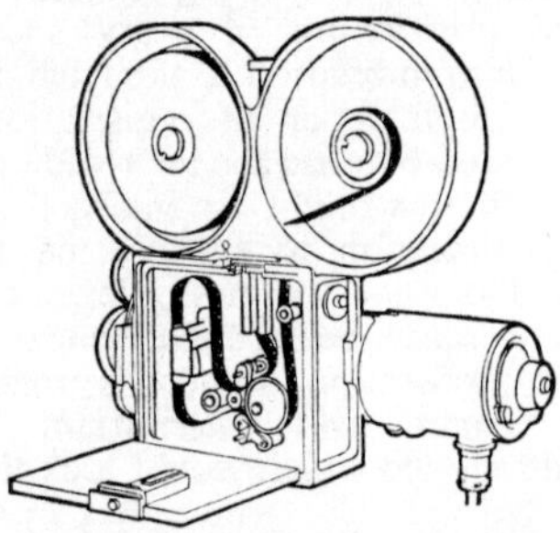

TWO-CHAMBER MAGAZINE. Separate chambers, each with its own screw lid, carry unexposed and exposed rolls of film wound on cores. Magazine exit slots have light traps to prevent fogging when mounting and dismounting from camera.

shaped covers, the front one containing the unexposed and the back one the exposed roll of film. The feed chamber is loaded in darkness, and the outer end of film is threaded through a light trap on the flange which abuts on the camera body. When the camera is threaded, film is led to the left round the feed sprocket and through the picture gate, with suitable loops top and bottom to allow for the change from continuous to intermittent movement and vice versa. The film then passes round the right-hand side of the feed sprocket, which thus acts as a take-up sprocket also, and enters the rear magazine chamber where its end is attached to a core. Take-up is normally by belt, but a torque motor may be used with heavy film loads to give a constant and more powerful drive.

Basically the same type of magazine is employed in many classical European cameras, save that the feed and take-up

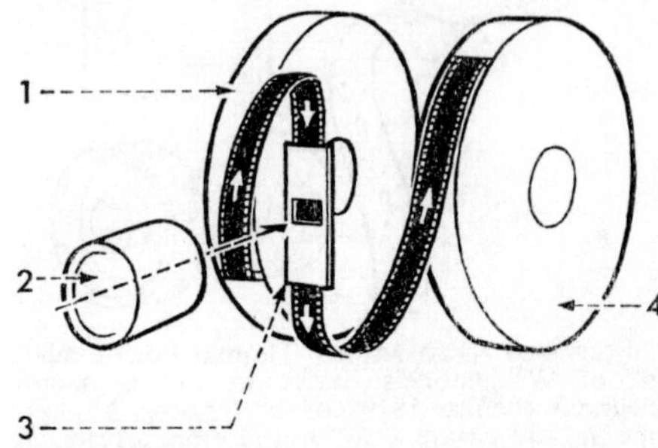

COAXIAL CHAMBER MAGAZINE. Two separate film chambers may be mounted coaxially, making for a very compact layout and simple drive, but with a more complicated film path including loops of ram's horn shape above and below the aperture. If the camera is blimped, these magazines are buried in the camera body. 1. Feed chamber. 2. Lens. 3. Aperture plate. 4. Take-up chamber.

chambers are completely separate, and can in fact be mounted on the same axis, side by side. This leads to the use of ram's-horn loops to displace the film into a different plane without warping it.

A compact German camera introduced in 1937 (Arriflex) popularized many innovations, among them a more complex type of magazine. This is a container with a single chamber carrying both rolls of film. Space

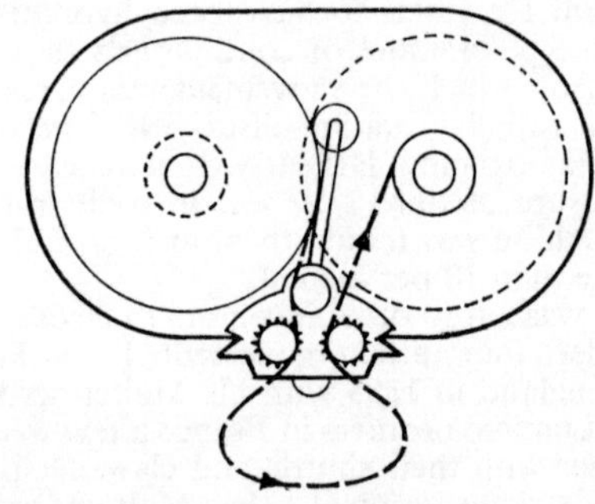

SINGLE-CHAMBER MAGAZINE. Space can be saved in a single-chamber magazine because the exposed roll, when taken-up on the right, can overlap the space it previously occupied on the left before exposure. A roller, bearing on the surface of the roll, indicates its position on the outside of the magazine and thus gives a simple measure of unexposed footage.

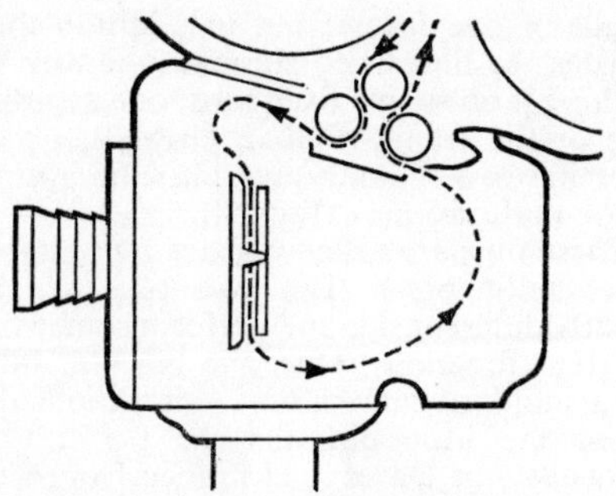

ARRIFLEX THREADING PATH. The magazine contains sprockets which engage with the camera mechanism when it is properly seated. These provide drive and take-up, so that no separate drive sprocket is necessary; threading is correspondingly simplified. If the loop has been formed to the proper size when loading it is only necessary to insert the film in the aperture and make sure that the claw engages properly with it.

is saved because, as one roll shrinks during running, the other grows, so that the roll spaces can be allowed to overlap. More important, the feed and take-up sprocket is incorporated in the magazine itself, so that all the necessary threading is effected in advance in the darkroom or changing-bag. The magazine, with its loop formed outside, is simply clamped on to the camera, and a gear on the magazine thereupon engaged with one on the camera body. The loop is slipped through the picture gate which is then shut, and the camera is ready to run.

This convenient device was still further

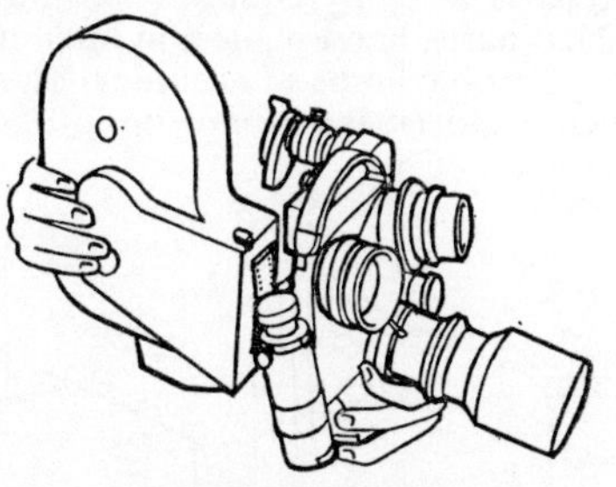

CAMEFLEX THREADING PATH. Loops formed inside magazine provide ultimate simplification of film threading. Claws are of the ratchet type and engage automatically with film perforations.

simplified on a well-known French camera introduced shortly after World War II (Cameflex). Here the loops are formed inside the magazine, which fits to the back of the camera, so that the film, which is flat against the base of the magazine, is in the proper vertical plane. When the magazine is clipped to the camera body, it abuts directly on the picture gate, and the film is at once pulled down by ratchet-type claws. Thus no threading of the camera is needed, and in fact the magazines can be mounted when the motor is actually running, without damage to the film. When time is short, useful seconds can be saved with this type of magazine.

A wholly different trend is represented by the daylight loading spool, much used in amateur cameras but pioneered in the '30s

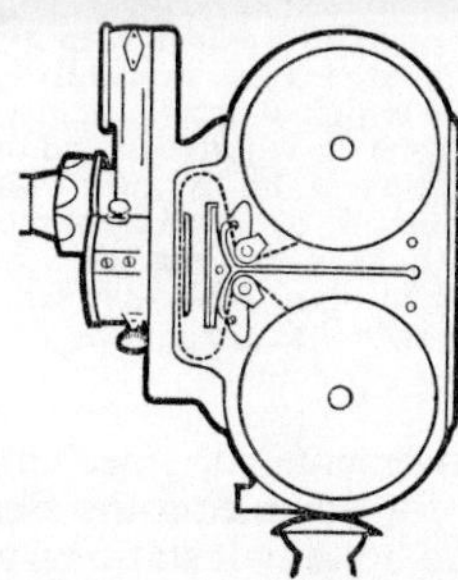

DAYLIGHT LOADING SPOOLS. Film spools are purchased already loaded with film; the end is protected from fogging by a length of black leader which must be run off before shooting begins. Take-up is on a similar spool which afterwards is sent to the laboratory. Daylight spools are limited to 100 ft. of 35mm film.

by an American manufacturer of compact 35mm and 16mm cameras. Daylight loading spools are wound with raw stock (usually up to 100 ft.) by the manufacturer. Solid sides prevent the ingress of light, and several turns of blank film form a protective leader at both ends. When the camera has been threaded in subdued daylight, the film end is attached to the core of a similar empty spool, both spools being enclosed in the camera body.

Film cartridges or cassettes, now popular on 8mm cameras, have combined feed and take-up chambers, and are loaded with raw stock and supplied fully threaded by the manufacturers. The aperture is often formed in the face of the cartridge, and a fixed projection may be moulded on one of the outer surfaces, which engages with a sensor arm on the camera body. When the cartridge is inserted in the camera, the projection and sensor together indicate the photographic speed of the film, and the sensor is linked to an automatic exposure device.

PULL-DOWN MOVEMENTS AND SHUTTERS. Pull-down movements, often called inter-

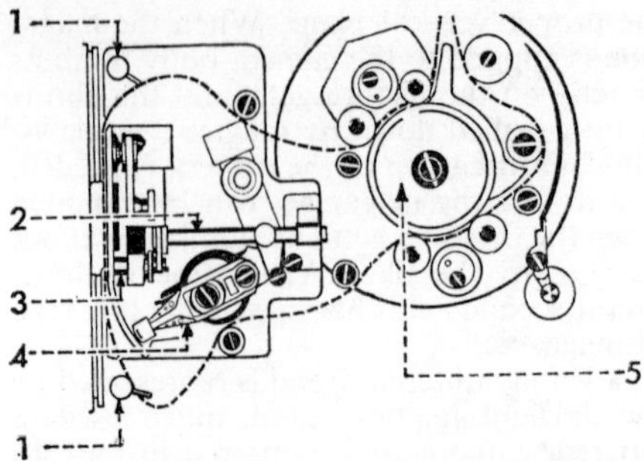

MITCHELL INTERMITTENT MOVEMENT. High-precision 35mm intermittent movement with register pins. Film enters and leaves mechanism on main sprocket, 5. It is pulled intermittently by double claws, 4, here shown withdrawn and moving back to upper position. Film is meanwhile fixed by twin register pins, 3, operated by a piston motion of the shaft, 2. Movable and fixed guide rollers complete the film path. Levers, 1, serve to detach the baseplate with its mechanism for inspection and cleaning.

mittent movements or mechanisms, are responsible for advancing the film a frame at a time and holding it stationary while the lens records an image on it through the aperture. The film intermittent, the gate, aperture and pressure plate, are essentially parts of one unit. Their combined function is to ensure that, after each of the 24 pull-down operations which occur within a second, the film is arrested and stabilized in the same position, with a permissible error of only a few ten-thousandths of an inch.

Many of the intermittent movements of highest precision, still employed where the greatest accuracy of image registration is imperative, were devised between 40 and 50 years ago. Like most other movements, they employ claws to pull the film down by the height of one frame; but in addition they register the film when stationary, either by pushing it on to two fixed pins, one at either side of the frame (Bell & Howell), or by inserting two moving pins into the appropriate perforations (Mitchell).

These pins are called register pins (sometimes pilot pins). The two pins are of slightly different size and perform somewhat different functions. One, the big pin, fully fits a standard camera film perforation; the other, the little pin, fits the perforation vertically but leaves a clearance from side to side. In this way the film is held with the minimum number of constraints, and slight lateral dimensional changes are taken up without producing buckling.

The flatness of the film in the focal plane of the lens is as important as vertical and horizontal register. It is ensured by applying a uniform back pressure to the film by means of a spring-loaded pressure plate. As all the rollers and sprockets in a camera are recessed so that the delicate emulsion in the picture area cannot touch them, the pressure plate, which must apply pressure over the whole image area, is the point of greatest danger in inflicting abrasions and scratches. Sometimes, therefore, the pressure plate contains a set of freely-revolving horizontal rollers, sometimes runnels which help to eliminate emulsion build-up, the most frequent cause of scratches. In some older designs, the pressure plate moves back and forth intermittently, and touches the film only when it is at rest.

An intermittent movement of the register-pin type is a highly complex assembly of precision parts, hand-made and fitted to the closest possible limits of accuracy. The cost of such a movement cannot be justified in

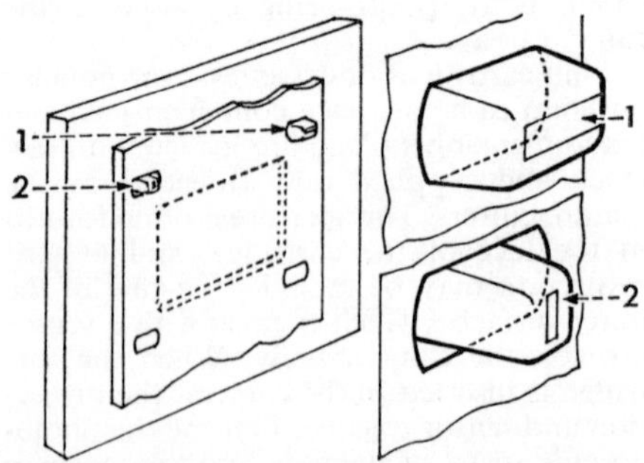

PILOT PIN REGISTRATION. 1. Big pin fully fits Bell & Howell 35mm negative perforation on all sides. 2. Little pin fits vertically, but horizontal clearance (exaggerated in diagram) allows minute expansion and shrinkage of film without buckling.

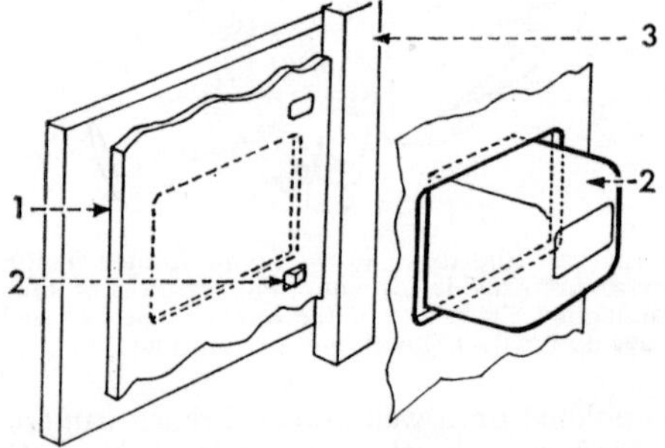

EDGE AND PIN REGISTRATION. Pressure on edge of film, 1, locates it against guided edge, 3. Pin, 2, fully fitting vertically, provides up-and-down positioning, whilst side-to-side clearances allow for dimensional changes in film.

many series-produced cameras, and simpler solutions have been found. Indeed, these solutions are so simple that it is surprising what a high degree of image steadiness is possible with cameras lacking positive registration.

Most of these employ a type of intermittent movement in which the claws, after pulling down the film, withdraw horizontally over a comparatively long period. This dwelling time arrests all trace of film movement, and positions the film just before exposure takes place. Spring pressure applied to one edge of the film causes it to be guided by the other edge, which runs along a fixed rail, thus providing additional steadiness.

It may be said that cameras without register pins have adequate picture steadiness for all but the most critical applications, such as back projection and optical processes. On the other hand, they are more easily deranged by lack of correct maintenance, and when faulty are harder to put right with certainty.

Since narrow-gauge production is less critical in its applications, and the smaller film is more likely to move integrally, registration by pin is much less often employed than in 35mm practice. However, it is noteworthy that one 16mm German camera employs a register pin, though its 35mm counterpart does not do so.

LENSES AND ACCESSORIES. There are two methods of mounting lenses on a camera: the single-screw or bayonet mount, and the turret. The turret is a rotating disc which carries three or more lens positions, and can be turned to bring the lens required in front of the picture aperture. The early single mount, which involved the removal of one lens and its replacement by another, soon gave way to the turret for all except special applications such as gun cameras and certain sound cameras in which transmission of mechanical noise through the large area of the turret presented problems.

One handicap of the turret – that long-focus lenses, which are also physically long, interfere with the field of view of neighbouring wide-angle lenses – was overcome by the divergent-axis turret in which the pivot was so angled that each lens assumed the correct straight-ahead alignment when moved into the filming position.

However, the increasing popularity of the zoom lens, replacing a whole battery of fixed-focus lenses, has made the turret less necessary, while the weight of the zoom lens makes it advisable to build it into the camera

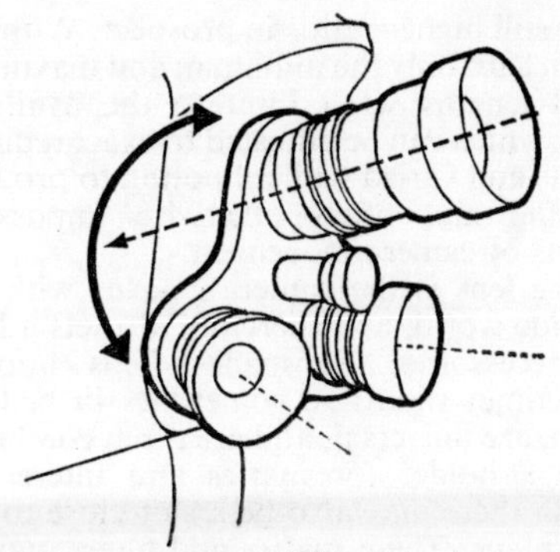

DIVERGENT-AXIS TURRET. Divergence helps to prevent wide-angle lenses seeing long-focus lens mounted on same turret. Angled pivot ensures that when lens is in shooting position, its axis is perpendicular to film plane.

body. Virtually all 8mm cameras now dispense with turrets, and some very recent designs of professional 16mm camera are based on the sole use of zoom lenses, However, the majority of 35mm and 16mm cameras still employ a lens turret.

Traditionally, 35mm cameras are equipped with a battery of separate lenses, ranging from 18·5mm (ultra wide-angle) to 150mm (medium long-focus). On the Academy aperture, this range gives horizontal angles of about 60° to 8½°, and on full aperture about 65° to 9½°. To give the same angles of view 16mm cameras need lenses of only about half these focal lengths, while 2 : 1 anamorphic lenses, whether 35 or 16mm,

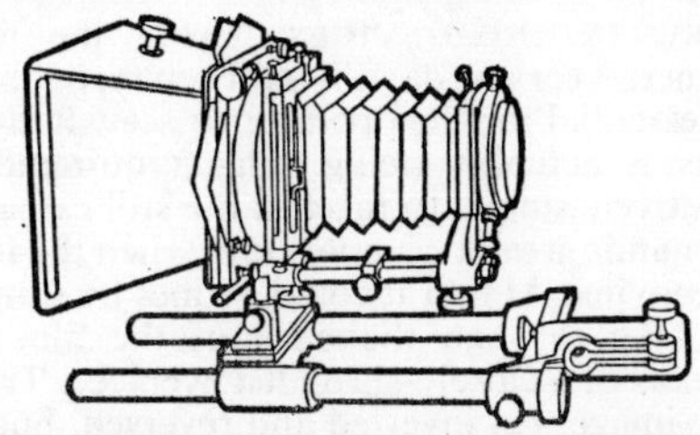

MATTE-BOX ASSEMBLY. Common in film cameras of older design, this type of assembly comprises a flexible bellows, a rigid sunshade and a variety of slots for mattes and filters, with provision for accurately locating these and moving them across the lens field as desired. The assembly slides on rails rigidly attached to the camera.

approximately double the horizontal field angle, while leaving the vertical angle unchanged. (The actual angles depend on the precise width of the aperture and the degree of masking in printing and projection.)

Today, however, the zoom lens provides a stepless increment of focal length over a range of 3 : 1 up to 10 : 1 and even 20 : 1, with still higher ratios in prospect. With the zoom lens, only the minimum and maximum focal lengths set a limit to the available field, which can be adjusted to exact requirements and varied while shooting to produce striking and often otherwise impossible effects of camera movement.

The lens is the camera's nexus with the outside world, and therefore attracts a host of accessories. Sunshades (lens hoods), sometimes rigid and sometimes in bellows form, are universal, and are often combined with a holder for mattes and filters, for which there may also be a slot close to the film plane. These mattes and filters may be of many forms: mattes for different special effects and filters for colour temperature conversion and correction, as well as black-and-white compensating filters and neutral density filters. In the more advanced types of camera, focus and iris adjustments on the lenses may be coupled to controls which operate them remotely at the side or back of the camera.

In many 8mm cameras, a photocell which receives light from the scene is coupled to the lens diaphragm to provide automatic exposure control. Within a few years, cameras of this type may be fitted with automatic focus control as well, in which light from the central part of the scene (where the chief object of interest usually lies) is assessed and electronically maximized by running the lens to and fro through the plane of sharpest focus.

VIEWFINDING DEVICES. From the film camera's earliest days, it has been recognized as essential for the operator to see what the lens is actually seeing. This requirement, relatively simple to meet in the still camera, demands greater complication when the film is moving. At first it could be met by simply looking through the back of the film by means of a closely-shrouded eyepiece. True, the image was inverted and reversed, but at least the operator could tell if his picture were properly framed and, though not very accurately, whether it was in sharp focus. But with the introduction of anti-halation backings to film all translucency was lost, and this method had to be abandoned.

Another simple solution was to move the film sideways by rather more than its width, and put in its place a ground glass on which the operator could examine the image. However, this image remained inverted and reversed, and the system made it impossible to view the film picture when the camera was running.

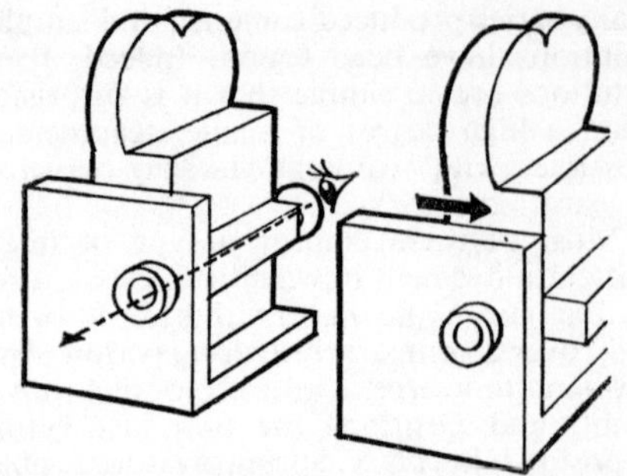

RACK-OVER VIEWFINDING. The lens mount is attached to a heavy casting, on the base of which the camera body can be moved sideways by a handle. In the viewfinding position (*left*), the focusing telescope is aligned behind the lens. In the shooting position (*right*), the camera is racked over to a precise position where the film aperture replaces the telescope. Great accuracy of machining and fitting is required by this arrangement.

A better scheme was then developed, though it still suffered the last-mentioned disadvantage. This was rack-over viewfinding, in use today in many cameras of older design or manufacture. The lens mounting or turret is attached rigidly to a baseplate which in turn is secured to the tripod. On this baseplate, as on a lathe bed, the camera body with magazine and viewfinder tube can be moved from side to side, or racked over, by means of a convenient handle at the back. At one limit of its travel, the film aperture is aligned behind the lens and at the other, the viewfinder tube or focusing microscope, as it is more properly called. The image is bright and has correct geometry; the only drawback is that it is lost to view when the camera is racked over for filming. The operator must then look through a side finder, often called a monitor, which can be corrected for parallax to avoid framing errors on near subjects.

For the highly mobile camera, something still better was needed, and the reflex finder

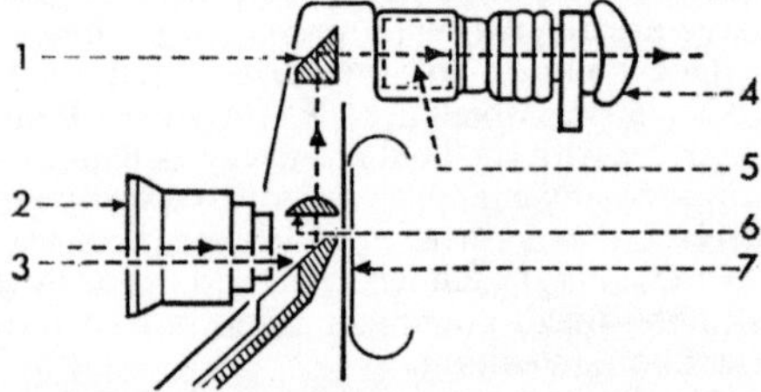

MIRROR SHUTTER (REFLEX) VIEWFINDING. Light normally forming image through lens, 2, on film, 7, is reflected from mirror shutter, 3, during pull-down, and instead forms an image on the ground glass, 6, which, through a prism system, 1 and 5, is seen by the cameraman through an eyepiece, 4.

was introduced in Germany in the 1930s, and has since been adopted by an increasing number of manufacturers. The reflex finder is based on the idea that the ordinary camera shutter wastes the light which falls on it when each frame of film is being pulled down. If the opaque shutter blades were made to reflect, the light falling on them could be made to form an image which the operator would see during the pull-down period. Thus, the film and the operator's eye would alternately see the lens image, and would do so for approximately the same time, since intermittent movements require about 180° for pull-down.

Mechanically, this aim was achieved by setting the shutter blades at 45° to the film plane and front-silvering them. The reflected light formed an image on a ground glass which was routed through prisms to a conveniently placed eyepiece at the back of the camera. The mirror shutter, though it suffered from various minor drawbacks such as the flickering view it gave the operator, was such a large step forward that it was adopted for virtually all new designs of hand-held camera, and recently for studio cameras as well.

Lately, especially in the 16mm field, the mirror reflex has met competition from an even simpler device, a semi-reflecting mirror mounted behind the lens but in front of the shutter, which is sometimes of the oscillating rather than revolving type to economize space. This mirror feeds a flickerless image to the operator's eye; though there is some loss of light to the film, it amounts to only half a stop or less. Zoom lenses now often incorporate this type of reflex finder, so that they may be used on cameras not equipped with mirror shutters.

ENCLOSING STRUCTURE AND DRIVE. Studio cameras, in which weight is of little moment, and which have heavy magazines and rack-over viewfinding, must be built with the utmost rigidity and tend to be of box-like construction. Field and hand-held cameras, by contrast, are encased in light-alloy castings shaped to the mechanism within and adapted to the operator's stance when filming. Some cameras are intended for shoulder-holding, and these may have the controls set into hand-grips for stability while focus and diaphragm adjustments are made or the motor switched on and off.

The smallest 8mm cameras are now often electric-motor driven, but 16mm cameras, many of them designed in an earlier epoch when light motors and batteries were not available, often retain spring drive. A few makes of 35mm spring-driven camera remain in use, one of them unique in being able to drive 200 ft. of film at a single wind. Nowadays, rechargeable batteries can be carried by the operator in a bandolier or back-pack, so that he has virtually unlimited motive power to drive his film.

Studio cameras have wholly different provisions for the drive. Heavy motors are needed for 1,000 ft. magazines, and interchangeable types are provided for single-phase and three-phase a.c., three-phase interlock, and variable-speed (wild) running, usually in the range of 8–24 fps, for both a.c. and battery supplies.

With these powerful motors, a built-in safety device is essential in the event of film running off the sprocket inside the camera, or failing to take up. A camera jam does not merely waste valuable film and studio time; much more serious, it may distort the alignment of precision parts and put the camera out of action for days or weeks. Studio cameras therefore incorporate a trip or buckle switch which interrupts the motor supply current as soon as film starts to build up inside the camera. Simpler cameras sometimes have a slipping clutch or shear pin which breaks the drive if the load is excessive.

CAMERA AUXILIARIES AND ACCESSORIES

No camera would function properly without some of the following items, but none is likely to be equipped with all of them.

FOOTAGE COUNTERS. An auxiliary drive is usually taken from a main shaft to a Veeder-type counter like a car mileage recorder which accurately totals the footage exposed and must be reset to zero when a new magazine is started. A simpler device is employed in compact single-chamber magazines; a roller rests on the outer edge of the unexposed film roll within and rides down with it as it is used up. A simple pointer connected to the roller indicates approximate footage on a scale outside the magazine.

FRAME COUNTERS. Some cameras with variable shutters have accurate counters recording the passage of individual frames, so that in making fades, dissolves or trick effects, an exact number of frames can be counted either forwards or backwards.

SHUTTER CONTROL. With a variable-aperture shutter, it is useful to be able to read the setting at a glance. Often control and indicator are combined into a small replica of the shutter mounted behind the camera.

INCHING KNOB. A knob which, when pushed in, engages with one of the camera drive shafts. By turning it, the operator can advance the mechanism a little at a time to make sure that the film is running correctly after he has threaded it. With a mirror shutter, he may need to inch it on until the mirror is in the right position for viewing the image.

TACHOMETER. With variable-speed (wild) motors, it is important to know the exact speed at which the camera is running so that the rheostat may be reset if necessary. This is the function of the tachometer, graduated in fps, which works on the principle of the car speedometer.

MONITOR VIEWFINDER. An auxiliary finder is needed if the camera has no reflex shutter. This usually provides a large picture which may be viewed with both eyes at a

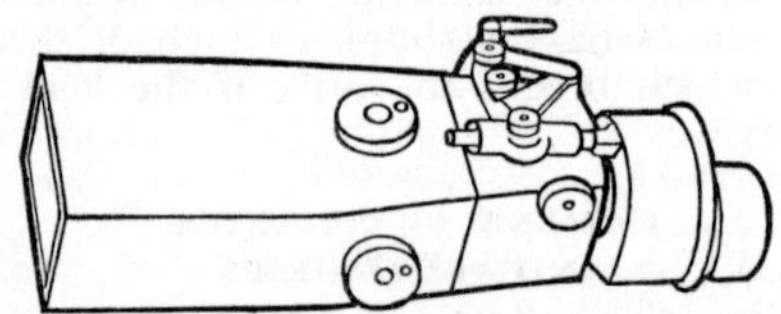

AUXILIARY FINDER. Large finder, with rectangular viewing aperture, mounted on outside of camera. Handwheels operate movable viewing mattes for different lens focal length and aspect ratios. Cam action sometimes provided to compensate for parallax errors on close-ups.

distance; the field may be equipped with variable matting lines to suit different lenses and aspect ratios. Automatic parallax compensation is often provided.

An entirely different type of monitor finder makes use of a reflex image which is scanned by a TV camera tube and thus transmitted to one or more television screens, which may be within the operator's view, at the back of the studio, and in distant offices and viewing rooms.

SERVO CONTROLS. Increasing use is being made of servo mechanisms attached to lens focus and diaphragm rings. Very compact motor units on the camera provide the necessary torque, and at the operating end, which may be as remote as desired, both position and speed of movement can be closely controlled. Zoom lenses can be coupled to servos in the same way, and preset zoom speeds brought into action at the touch of a button. Simple motorized zooms, without remote control, are available on some 8mm cameras.

SYNC GENERATOR. A pulse generator may be attached to one of the drive shafts of the camera to produce a series of electric pulses at a frequency which is usually 48 Hz for 24 fps and 32 Hz for 16 fps. This pulse is recorded on ¼-in. magnetic tape which also carries the dialogue. The camera and tape recorder, driven by separate non-synchronous motors, may not be in precise synchronization. The sync pulse is therefore used to control the speed of a re-recorder, bringing it into synchronism with the camera which generated the sync pulses.

SPECIALIZED CAMERAS

The basic camera so far described is essentially a silent camera, a term which, though universal, is somewhat misleading because the instrument is far from silent in operation. It is therefore not adapted to filming dialogue, since its mechanical noise would be audible to the microphone. Its counterpart, the sound camera, is in fact silent and therefore suitable for studio use.

SOUND CAMERAS. Silent cameras may be adapted for sound shooting by putting them in a blimp or soundproof enclosure, through which the viewfinder and the necessary controls project. Sound cameras of recent design are, however, self-blimped; their

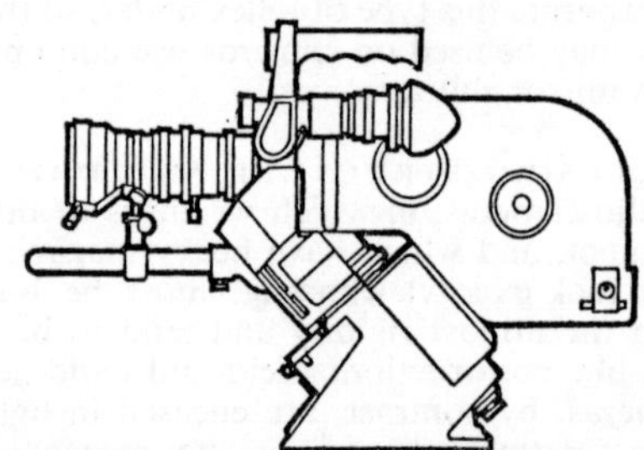

SELF-BLIMPED CAMERA. Modern 16mm camera in which external silencing is cut to a minimum by careful design of intermittent motion, loops, and film path.

housings, magazines and doors are padded or acoustically treated so that mechanical noise is reduced to an absolute minimum. Much thought has been put into the design of quieter intermittent movements, for it is here that most of the noise is set up. Modern directional microphones severely attenuate sounds coming from behind them, and thus help to produce a sound track with an acceptably quiet background.

ANIMATION CAMERAS. These are cameras incorporating standard high-precision register-pin movements, vertically mounted and moving up and down in relation to the artwork, which is laid out flat on a table furnished with calibrated x- and y-axis controls. Automatic focusing is usually incorporated, providing for accurate movement in all three dimensions, though the camera can move only in one dimension.

STOP-MOTION CAMERAS. Many cameras of normal design can be equipped with a control, sometimes called an intervalometer, arranged to advance the film one frame at a time at predetermined intervals. This device, employed for puppet animation as well as scientific study, thus gives a variable time compression.

HIGH-SPEED CAMERAS. The type of intermittent movement already described sets an upper speed limit of about 300 fps for 35mm film and 600 fps for 16mm film. High-speed cameras take over above this range. Much higher speeds are possible if the film runs continuously, and the image may be arrested by interposing a rotating prism between lens and film. For a short period, the image is moved downward by refraction at the same rate as the film is moving downward, so that there is no relative movement between the two, and a blur-free picture results. Frame rates as high as 20,000 fps may be attained with this device, sometimes known as an optical compensator.

For the highest speeds of all, the film remains stationary, and a limited number of pictures is distributed over its surface by rotating lenses or mirrors. Normally the pictures are intended for static analysis, but they can be made into motion pictures by optical transfer to standard film. By these and other means, frame rates of 30 million fps have been attained, but the total taking time is limited to a few microseconds.

TRENDS IN CAMERA DESIGN

The many wide-screen processes of the '50s threw up a host of different camera designs, some employing triple films, others a standard 35mm film travelling horizontally and thus producing a larger-than-standard frame, eight perforations long. Though some cameras of these types remain in use, they are no longer in production, and cameras for the biggest wide-screen processes now universally employ 65mm film, except in Russia where the 70mm gauge prevails. These 65mm cameras are now well developed, and hand-held models with zoom lenses are available.

In the 35mm field, stability has also been reached with a long-established range of studio and lightweight cameras. Most of these can be equipped with anamorphic lenses for various wide-screen processes, and can be adapted to two-perforation pulldown for the Techniscope half-size frame.

It is in the narrower gauges, where the impact of television practices is greatest, that the biggest advances are likely. Automatic exposure control is making its way into the professional field, and one recent design of 16mm hand-held camera is equipped with three miniature servo motors, one for diaphragm setting, one for focus, and one for zooming. The diaphragm control works from a photoresistance which receives its light through the lens (sometimes called the TTL method) and is in turn coupled to a differential amplifier which reverses and sets the movement of the servo motor. The sensitivity range of the system is such that it will correctly expose films of 12 to 1600 ASA speed rating at camera speeds from 12 to 50 fps.

Another interesting new 16mm camera is of the continuous-motion film type, normally used only for high-speed cameras. In this design, the moving film is engaged by a register pin which draws with it a rocking mirror, the function of which is to displace the image at precisely the same speed as that of the film itself, thus producing an unblurred image, as if the film were stationary. The exposure at 24 fps is a normal 1/48 second. This camera has a built-in single-system magnetic recording head; the camera film is pre-striped. Weight with zoom lens and 400 ft. of film is only 15 lb.

These cameras are representative of the movement towards complete automation and built-in solid-state recording equipment, both of which aid the capture of picture and sound under conditions which, even a few years ago, would have made effective camera work impossible. R.J.S.

See: *Animation equipment; Cameraman; Cineradiography; Cold climate cinematography; High-speed cinematography; Lens performance; Lens types; Medical cinematography; Narrow gauge film and equipment; Stereocinematography; Time lapse cinematography; Tropical cinematography.*

Book: *Technique of the Motion Picture Camera*, by H. M. R. Souto (London), 1969.

□CAMERA

The camera is that part of the television system which converts the scene to be televised into the form of an electrical signal. Although the term is sometimes employed when the image to be transmitted is on film, equipment designed specially for that purpose is more usually known as a telecine or film scanner, the term camera being reserved for channels which directly televise live scenes.

ELEMENTS OF THE CAMERA CHANNEL

The camera normally comprises two items, the camera and the camera control unit (CCU), though these are occasionally combined when mobility is important.

TUBE AND LENSES. The nucleus of the camera is the tube which receives the optical image and produces a corresponding electrical output. The image is formed on the face-plate of the camera tube by a lens, basically similar to a normal photographic lens, and until the mid-1960s broadcast cameras were invariably provided with a lens turret, normally designed to mount four lenses. Today, improvement in the zoom lens has enabled designers to meet the TV producer's needs with a single variable focal length lens. An essential part of the camera is the electronic viewfinder which enables the camera operator to frame precisely the scene being transmitted.

$4\frac{1}{2}$-IN. IMAGE ORTHICON CAMERA. A modern zoom-only camera. The tilting viewfinder, 1, greatly facilitates operation at extreme angles of tilt. The zoom control, 2, for the zoom lens, 3, is mounted on the panning handle.

CAMERA CONTROL UNIT. To keep to a minimum the amount of electronics housed in the actual camera, only those functions are performed there which it would be impracticable to carry out remotely, and the majority of electrical functions are therefore performed in the CCU.

Early camera channels employing the thermionic valve (tube) as the active circuit element normally required constant monitoring from the CCU position and for this reason the CCU was mounted in a console-type desk, each camera channel being associated with its own picture and waveform monitor which enabled the electrical output of the channel to be both viewed as a picture and monitored as a time-varying voltage. As circuit techniques improved, and particularly with the introduction of the transistor as the active circuit element, the stability achieved made it possible to dispense with the individual CCU operators and it became more usual to mount the CCU in a standard equipment rack, several channels being associated with a single picture and waveform monitor which is predominantly used for setting up. With this type of installation it is common practice to relay the essential controls required for operational use to a remote control panel in the studio control area.

CABLE AND ANCILLARY EQUIPMENT. An important part of the channel is the camera cable which interconnects the camera and CCU, and which may require some 30 to 40 ways, including several co-axial cables to carry high-frequency signals. Modern camera cables are designed for maximum flexibility to reduce the drag on the camera during operation, a total length of up to 2,000 ft. often being used.

Ancillary camera equipment includes pedestals or tripods fitted with pan and tilt heads capable of bearing the combined weight of the camera and lens which normally lies in the region of 100 to 200 lb. Most cameras are also fitted to take a variety of devices such as cue holders, prompters and auxiliary spotlights.

Apart from mains power, the camera channel must be provided with standard synchronizing signals, normally derived from a central source in the studio area, or, in the case of outside broadcast vehicles, from a local synchronizing pulse generator.

THE CAMERA TUBE

The camera tube is the principal component of the camera and has largely dictated the lines along which cameras have developed. In its basic form, the modern camera tube comprises a photo-cathode upon which is formed an optical image of the scene to be televised, and an electron-beam scanning system by which this photo-cathode is directly or indirectly interrogated to produce an output signal the instantaneous value of which follows the changes in light intensity across the image. Camera tube development has followed two major avenues. First are the photo-emissive tubes having a photo-cathode which emits electrons (usually referred to as photo-electrons) upon exposure to light. Second are the photo-conductive tubes having a photo-cathode which changes in conductivity when exposed to light.

Of historic importance only are the image dissector invented by Farnsworth and the iconoscope invented by Zworykin in America and also developed in Great Britain during the early 1930s where it was known as the Emitron.

IMAGE DISSECTOR. In the image dissector, photo-electrons emitted by the photo-emissive light-sensitive surface at one end of the tube are focused to form an electron image on a metal plate having a small central

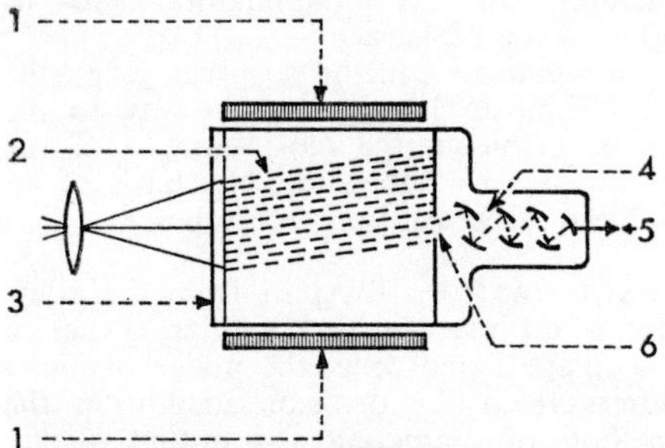

IMAGE DISSECTOR. An early camera tube of low sensitivity. 1. Focus and deflection coils. 2. Photo-electrons. 3. Photo-cathode. 4. Electron multiplier. 5. Output signal. 6. Aperture.

aperture. A set of deflection coils is provided, by means of which this image is deflected both horizontally and vertically so that each point on the image passes in turn across the aperture. The electrons passing through the aperture are amplified by means of an electron multiplier to provide a satisfactory output level for feeding to an external amplifier. The disadvantage of the image dissector is the absence of storage, the input to the electron multiplier at any instant being only those photo-electrons emitted during the short time equivalent to a single picture element.

ICONOSCOPE. In the iconoscope, the light image is formed on a light-sensitive mosaic consisting of small particles, the number of which is large compared to the number of picture elements which the television system is capable of resolving. In operation, electrons are liberated from the mosaic in proportion to the light intensity on any particular area. The liberated electrons are removed by means of a collector so that a charge pattern of the image is built up on the mosaic. This pattern is erased by the electron scanning beam which passes over each picture element once on each complete scan, that is, every 1/25 sec. on systems employing 25 pps (pictures per second). The output of the tube is derived from the flow of current in the external circuit during this discharge process.

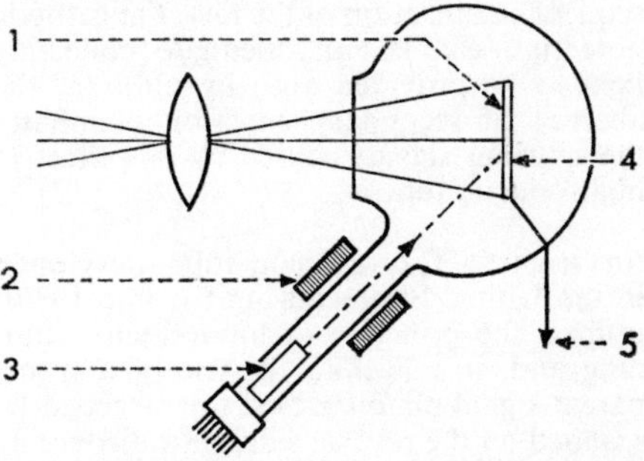

ICONOSCOPE. The first type of camera tube to employ full storage. 1. Mosaic. 2. Deflection coils. 3. Electron gun. 4. Signal plate. 5. Output signal.

As compared with the image dissector, the storage characteristic of the iconoscope offers a theoretical gain equal to the number of picture elements in the picture. This figure is partly offset by the absence of an electron multiplying system, and to obtain an output signal which is large compared to the noise generated in the coupling resistor and first stage of the external amplifier, a scene illumination of at least 300 fc (foot-candles) is required, using a lens of aperture f4, a setting which offers a very restricted depth of focus with the large image size employed with this type of tube.

IMAGE ICONOSCOPE. In an effort to improve the sensitivity of the iconoscope, the image iconoscope (or Super-Emitron) was developed. The basic difference is the provision of a separate photo-cathode designed for maximum emission of photo-electrons which are focused to form an electron image on a mosaic or target of secondary-emitting material. The sensitivity of the image iconoscope is a function of the efficiency of the photo-emissive photo-cathode and the secondary emission ratio of the target. Although a theoretical improvement of 10 to 12 times can be obtained by the use of an image section, the smaller image size employed with this tube necessitates the use of a smaller lens for a given angle of view. A real improvement of only 5 : 1 is achieved in practice.

Tubes of the iconoscope type employ a high-velocity scanning beam, the resulting secondary emission stabilizing the target at a potential close to that of the final anode, and a relatively weak field thus exists between the two. The collection efficiency of electrons emitted as the result of incident light rays is therefore low. The use of a low-velocity scanning beam, and the consequent stabilization of the target at cathode potential, enables an adequate collecting field to be provided and, in addition, the absence of secondary emission eliminates the spurious signals caused by this effect in high-velocity tubes.

ORTHICON. The orthicon tube, developed in the United States during the late 1930s, utilized the principle of low-velocity scanning and, in addition, the use of a transparent signal plate enabled the target to be scanned on the reverse side, thus dispensing with the oblique scanning process and much simplifying the construction of the tube, all the basic elements being on the same axis or in line, as the name implies.

In the absence of light, the target assumes cathode potential so that the electrons, arriving with zero velocity, fail to land. With the incidence of light, photo-electrons are emitted, to be collected by the anode, thus establishing a positive charge pattern. The removal of this charge pattern by the landing of electrons from the scanning beam, with consequent restabilization to cathode potential, provides the output signal. Great care must be exercised in the design of the scanning system to ensure that the low-velocity beam arrives at the target at normal incidence, and in focus, over the entire scanned area, as any failure in this respect introduces serious shading signals.

EMITRON. Although the orthicon may be considered the forerunner of the modern low-velocity camera tube, the sensitivity was only slightly greater than that of the iconoscope, and it was not until the introduction of the CPS (cathode potential stabilized) Emitron that the first high-sensitivity camera tube became available. The improvement in

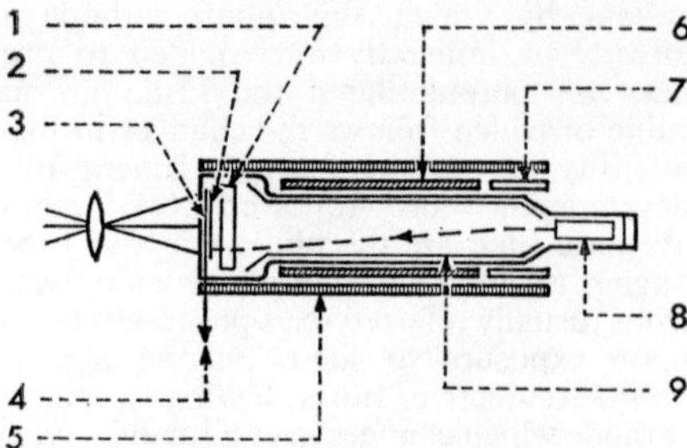

CPS EMITRON. The first high-sensitivity camera tube of the orthicon type using low-velocity scanning. 1. Decelerator. 2. Mosaic. 3. Signal plate. 4. Output signal. 5. Focus coil. 6. Deflection coils. 7. Alignment coils. 8. Electron gun. 9. Wall anode.

sensitivity is not due to any fundamental change in principle, for the CPS Emitron is a member of the orthicon family and operates in basically the same way. General improvement in techniques and materials gives a much enhanced target sensitivity, however, and the tube, which employs an image diagonal of 2·2 in., can produce good pictures with a scene illumination as low as 30 fc with a *f* 2 lens.

In regular use in the post-war years, the CPS Emitron has now given way to the image orthicon, the most modern of the photo-emissive tubes, and the basis of all modern cameras using such tubes.

IMAGE ORTHICON. Apart from the addition of an image section, similar to that of the image iconoscope, the image orthicon differs from the orthicon mainly in the method of extracting the output signal. Instead of obtaining the output from a

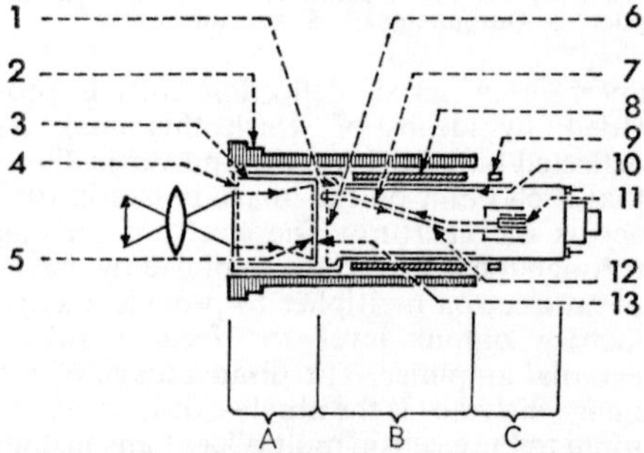

IMAGE ORTHICON. The latest type of photo-emissive camera tube. 1. Decelerator. 2. Focus coil. 3. Target cup. 4. Photo-cathode. 5. Target mesh. 6. Field mesh. 7. Scanning beam. 8. Deflection coils. 9. Alignment coils. 10. Wall coating. 11. Electron gun. 12. Return beam. 13. Target. A. Image section. B. Scanning section. C. Gun/multiplier section.

signal plate, use is made of the return scanning beam, which comprises those electrons which failed to strike the target and which is therefore greatest when it is passing over unilluminated areas. The great advantage of this method of signal extraction is that the return beam can be amplified by an electron multiplier, gains of up to 2,000 being attainable, the output level thus being raised well above the noise level of the first stage of the external amplifier.

An important feature of the image orthicon is the inclusion of a target mesh placed very close to and in front of the target, spacings of 0·0001 in. to 0·001 in. being common. In operation, the photo-electrons emitted by the photo-cathode penetrate the target mesh to form a focused electron image on the target, where they give rise to secondary emission forming a positive charge pattern. The secondary electrons are collected by the target mesh, which is held at a few volts above cathode potential, i.e., the potential to which the target is stabilized.

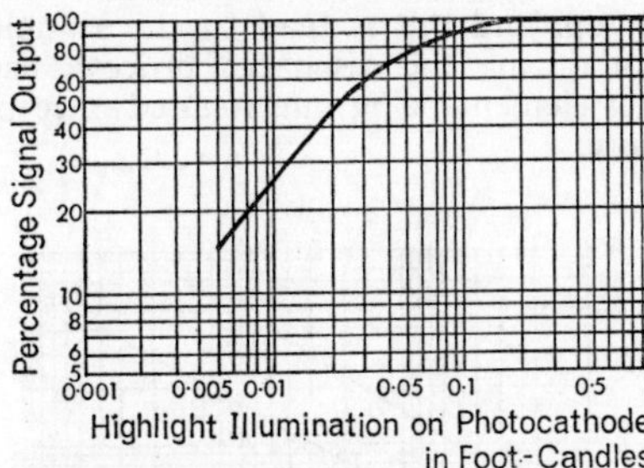

IMAGE ORTHICON CHARACTERISTIC. Operating curve for a typical 4½ in. close-spaced tube showing the knee.

The collection efficiency is very high until the point is reached where the flow of photo-electrons is sufficient to raise the potential of the target to that of the mesh. At this stage, the secondary electrons are no longer collected by the mesh and tend to fall back on the target, thus limiting the intensity of the charge pattern. It is this feature which provides the well-known knee in the operating characteristic of the image orthicon and enables the tube to handle scenes of very high contrast without overloading or unduly compressing the contrast scale in the darker regions of the picture.

A further feature of this behaviour of the secondary electrons when the target potential reaches that of the mesh is that large area highlights tend to be depressed with respect to small area highlights. This is due to the fact that, in the latter case, electrons emitted by the target do not all return to the same spot but tend to be redistributed over adjacent areas which are consequently depressed below their proper value. The subjective effect of this redistribution tends to enhance areas of fine detail so that it is normally considered as a desirable form of distortion.

Since the output level is high compared with the noise level in the external amplifier, the signal-to-noise ratio is determined by the noise level in the return scanning beam. For optimum performance, the beam must therefore be adjusted to a level which is just sufficient to discharge the target, the noise in the picture rapidly increasing when this value is exceeded. The noise in the beam is, however, proportional to the square root of the beam current, and the best performance therefore results when the tube is designed to require a high minimum beam current. In practice, this condition can be realized by decreasing the target/mesh spacing and thereby increasing the target capacitance. While close-spaced tubes are capable of better performance in this respect, the higher capacitance also demands an increased flow of photo-electrons to produce the same voltage excursion of the target relative to the mesh, and the tube consequently requires more light.

In studio conditions, where adequate lighting levels are available, it is therefore customary to employ close-spaced tubes. Where sensitivity is important, it is usual to employ wide-spaced tubes. While the exact degree to which an image orthicon tube should be exposed, with respect to the knee on the taking characteristic, has been the subject of much discussion, a commonly-accepted operating point is between half a stop and one stop above the commencement of the knee. In these conditions, typical close-spaced and wide-spaced tubes would require scene illuminations of approximately 40 and 20 fc respectively, both figures relating to a lens aperture of *f* 4 and assuming a maximum scene reflectivity of 60 per cent.

The image orthicon was first introduced by RCA in the United States as a 3 in. tube with a 1·6 in. image diagonal, and in this form it has been extensively developed in many countries, a large number of variants using different photo-cathode and target materials now being available. Development of a larger tube, 4½ in. in diameter, was also started in the United States, but it was not at first brought into commercial use because of optical problems resulting from the larger image size.

Later, it was realized that the desire for improved resolution which had prompted this move might equally be achieved by employing a larger target area only, maintaining the existing image diagonal of 1·6 in. on the photo-cathode, the fine texture of which causes no material loss of resolution. The 4½ in. image orthicon operating on this principle was developed in Britain and first introduced into service in the mid-1950s. An essential feature of this tube is the magnification which occurs in the image section, the scanned area on the target being three times that occupied by the image on the photo-cathode.

PHOTO-CONDUCTIVE TUBES. Although small in size and simple to operate, the photo-conductive tube has only recently gained a significant foothold in the broadcasting field for live black-and-white television cameras. Before the mid-1960s, several tubes, which appeared under a variety of names according to the source of origin but which are most generally referred to as vidicons, were developed for telecine and closed-circuit applications. The limitations which restrict the adoption of the simple vidicon for live cameras are lag and the presence of spurious signals, neither of which can be reduced to an acceptable level for broadcast purposes at normal studio lighting levels.

Vidicon cameras, however, are widely used in closed circuit applications, where standards are often less exacting, and in telecines, where lighting levels are adequate.

In the early 1960s, Philips introduced the lead-oxide vidicon, or Plumbicon, which had been under development in Holland for many years. Both the lag and the level of spurious signals have been reduced to acceptable proportions in the Plumbicon, and it is making a significant challenge to the image orthicon in the broadcast field. A major factor in the introduction of the Plumbicon has been the difficulty of producing the tube in adequate numbers, a considerably tighter manufacturing control being required than for the vidicon.

VIDICON. The standard vidicon is 1 in. in diameter and employs an image diagonal of 0·625 in. The rear surface of the faceplate is coated with a transparent signal plate on which is deposited a photo-conductive layer. In operation, the layer is stabilized at cathode potential by the action of the low-velocity scanning beam, and the signal plate is maintained at a positive potential normally referred to as the target voltage. Owing to the potential gradient thus established across the photo-conductive layer, a current flows whose magnitude is a function of the light intensity at any particular point on the layer. A positive charge pattern is thus built up, and this is read by the scanning beam during the discharge process, each target element being interrogated once per picture.

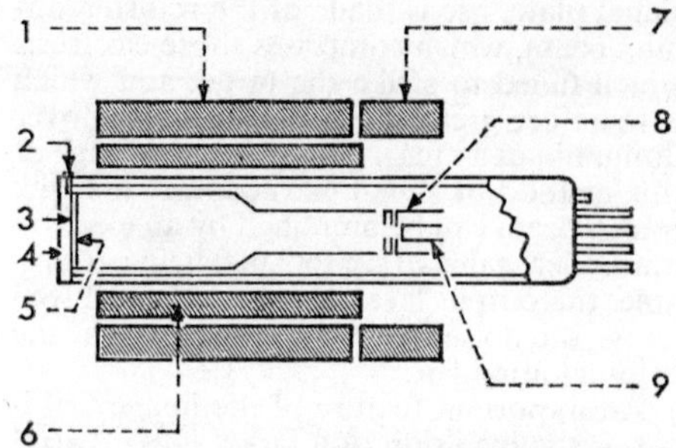

VIDICON. A modern type of photo-conductive camera tube. 1. Focus coil. 2. Target connection. 3. Target. 4. Glass faceplate. 5. Mesh. 6. Deflection coils. 7. Alignment coils. 8. Grid. 9. Cathode.

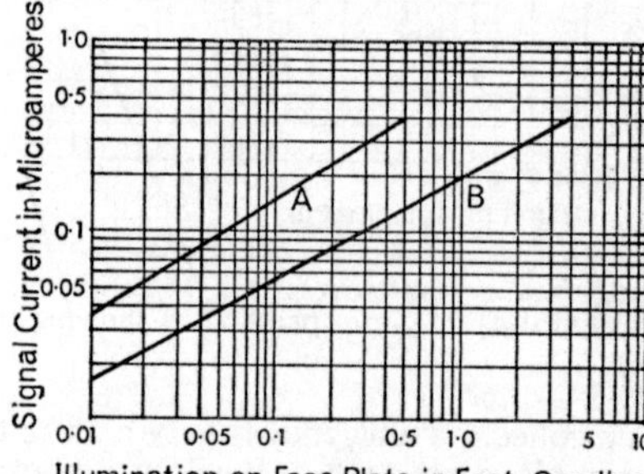

VIDICON OPERATING CHARACTERISTIC. Curves for a typical 1-in. tube showing how an increased sensitivity may be obtained at higher settings of target voltage, but at the expense of increased dark current. A. Dark current 0·2μ amps. B. Dark current 0·02μ amps.

The picture lag, which makes the tube unsuitable for moving scenes except at very high light levels, arises through two causes. The first type of lag is in the photo-conductive material and arises from the failure of the conductivity to change instantly with changes in illumination. While photo-conductive lag is less significant at higher light levels, the performance is first limited by the illumination in the darker areas of the picture, so that smearing is a particular problem in scenes of high contrast. The second type of lag is due to the time constant formed by the target capacitance and the source impedance of the scanning beam which leads to the failure of the beam to erase completely the charge pattern in a single sweep.

The second limitation of the vidicon is the significant magnitude of the signal current which flows in the absence of light. This unwanted component, known as the dark current, is very dependent on both temperature and target voltage, so that the maintenance of a stable black-level in the picture is very difficult. In earlier tubes, where the photo-conductive layer was evaporated on to the faceplate under vacuum, and which are characterized by the side-pip which was used in this process, the thickness of the layer was not uniform, and this caused shading errors owing to the dependence of dark current on the layer thickness. Modern techniques now allow the layer to be deposited before the tube is evacuated, and, although the dark current is still significant, the greater uniformity achievable by this method of construction greatly reduces the shading errors.

PLUMBICON. This tube has a diameter of approximately 1·25 in. and employs an image diagonal of 0·8 in. While operational experience is as yet limited, it can be at once said that neither lag nor dark current are likely to limit the use of the Plumbicon in live cameras. The inherent resolution is lower than that obtainable with the image orthicon but this disadvantage can be largely overcome by careful design of the input stage to the amplifier, thus achieving the best possible signal-to-noise ratio and enabling a considerably higher degree of high frequency or aperture correction to be applied than is possible with the image orthicon. A more significant disadvantage of the original tube was its poor spectral response, there being little or no response at the far red end of the spectrum. There is every indication, however, that this limitation will be overcome.

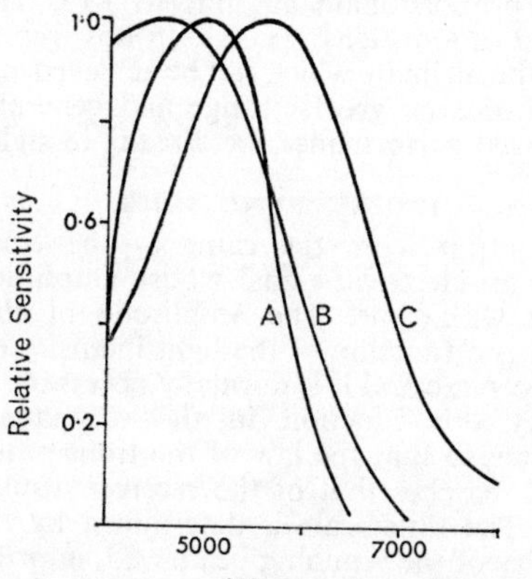

SPECTRAL RESPONSE. The response of typical camera tubes. A. Plumbicon. B. Image orthicon. C. Vidicon.

A basic feature of the Plumbicon, which may limit its acceptance for outdoor use where lighting cannot be controlled, is its linear taking characteristic and the absence of a knee. It is therefore necessary to set the beam current at a level capable of discharging the brightest parts of the scene which may be encountered from time to time, from specular reflections for instance. However, this limitation has not prevented the widespread adoption of the Plumbicon for colour television cameras where the advantages of smaller size and simplicity more than offset the disadvantage of a lower contrast handling ability.

CAMERA LENSES

The television camera lens is basically similar to its photographic equivalent and, since the resolving power obtainable on a photographic plate is greatly superior to that which can be obtained using a broadcast television system, it is often supposed that the performance of the lens is of an order better than is required for television, and that it does not therefore significantly limit the overall performance of the system. This supposition is quite incorrect, and it is most important that lenses for television be designed specifically to suit the parameters of the television system. A photographic lens gives the best subjective performance when it is designed to portray the finest detail possible even though the extent to which such detail is resolved may be very

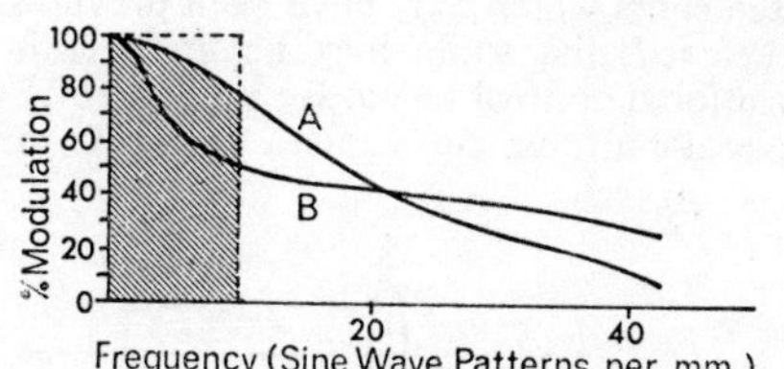

TELEVISION LENSES. Comparison between a photographic lens, B, and one specially corrected for television, A.

slight. If used for television, such a lens would portray detail which could not be transmitted by the overall system because of the limited bandwidth available for the transmission of the video signal. The improved performance which can be obtained by using a specially computed television lens is achieved by ensuring that the lens transmits the finest detail which the system can transmit with the maximum possible depth of modulation, and no attention is paid to the transmission of detail that is finer than this.

VARIABLE FOCAL LENGTH LENSES. The variable focal length or zoom lens first found extensive use in television in the 1950s. The range of these early zooms was limited to 3 or 5 : 1 and it was not until the introduction of the modern 10 : 1 lens in the early 1960s that serious thought was given to the possibility of zoom-only working. Marconi introduced such a camera in Britain in 1965.

When zoom lenses were first used for television, it was largely for their ability to simulate the effect of a rapidly-approaching object or scene, from which effect, of course, they derive their name. In modern practice, however, this effect is rarely used, and the variable focal length lens may more sensibly be considered as a continuously variable lens turret. To facilitate the repetition of fixed shots which may have been previously rehearsed, the zoom may be fitted with a positional control servo. By pressing one of several buttons, the camera operator may

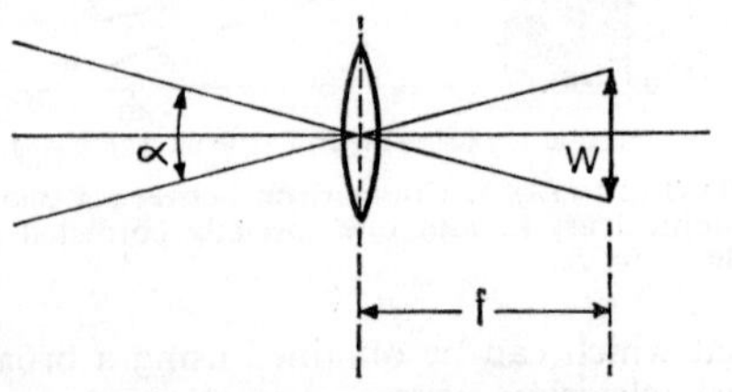

ANGLE OF VIEW. The angle of view α of a camera lens is related to the focal length f and the image width w by the equation $\tan \alpha/2 = w/2f$.

select one of a number of viewing angles, provision normally being made by which the particular angle of view associated with each button can be preselected. A further refinement is the ability to select the speed at which the change of focal length occurs, a range of almost-imperceptible to almost-instantaneous usually being available.

Alternatively, in cases where rehearsal is not appropriate, such as the televising of sporting events, the cameraman may select a velocity control servo, the direction and speed of the zoom being controlled from a single hand-grip. Although the basic cost of such a full-facility zoom lens may be relatively high, the actual camera is made simpler by the omission of the lens turret and there is also a potential saving in the number of cameras required, since it is no longer essential to make lens changes only when off-air.

The performance of the modern zoom is similar to that of an equivalent series of fixed lenses. A parameter relating only to zoom lenses is the accuracy with which correct focus is maintained over the zoom, that is, the ability of the lens to retain in focus an object at a fixed distance while zooming. To ensure proper functioning in this respect, it is important that the rear conjugate is properly set by using the normal camera focusing movement.

For normal use with image orthicon cameras, lenses are available having a range typically of 1·6–16 in. focal length, and 4°–44° horizontal angle of view, at a maximum aperture of f4. It is relatively simple to change the image size on such lenses so that the above lens can be adapted for use with Plumbicons at a maximum aperture of f2, the focal lengths being scaled down in the ratio of the image sizes in order to preserve the same range of viewing angles. This ability to change the image size may also be exploited by the provision of range extenders, a ×3 extender raising the range of focal lengths of the image orthicon lens to 4·8–48 in., although the aperture is reduced by a proportionate amount to f12. The present zoom lens by no means represents the ultimate which can be achieved and lenses of even greater range and generally improved performance are already in sight.

CIRCUIT FUNCTIONS

The output from the camera tube comprises an electrical signal whose amplitude varies with time. The amplitude of this signal is a function of the light intensity on the faceplate and it is normally necessary to modify this function in the subsequent circuitry so that the law of the transmitted signal matches that of the receiver display tube. The time scale is determined by the velocity of the scanning beam and, in order to preserve the proper geometry of the picture, it is important that the scanning velocity be maintained constant over the active scanning period.

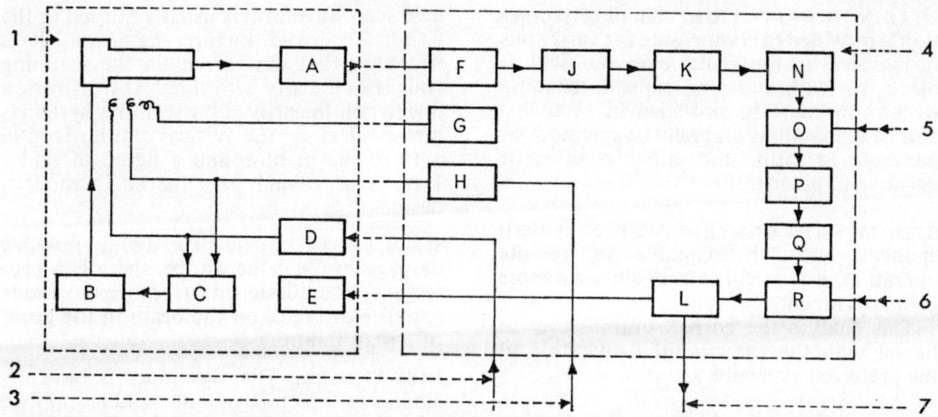

IMAGE ORTHICON CAMERA CHANNEL. The basic circuit elements of a modern camera channel. 1. Camera. 2. System line drive imput. 3. System field drive input. 4. Camera control unit. 5. System blanking input. 6. System sycnhronizing pulses input. 7. Output. A. Head amplifier. B. Tube supplies. C. Scan protection. D. Line deflection. E. View-finder. F. Cable corrector. G. Focus current regulator. H. Field deflection. J. Aperture corrector. K. Shading corrector. L. Output stage. N. Remote gain stage. O. Blanking stage. P. Clippers. Q. Gamma corrector. R. Sync. inserter.

HEAD AMPLIFIER. The function of this unit is to raise the video signal to a sufficient level for transmission to the CCU without introducing significant crosstalk in the camera cable, a level of 0·5–1 volt normally being employed. The output signal obtainable from an image orthicon is usually 10–40 μA at a signal-to-noise ratio of 38 dB, and it is not difficult to design an amplifier which will reduce this signal-to-noise ratio by no more than 0·5 dB. The maximum output of photo-conductive tubes is normally 0·3 μA, and every precaution must be taken to keep down the noise level at the input to the amplifier.

A good design should achieve a signal-to-noise ratio in the region of 50 dB.

CABLE CORRECTION. Since cameras may be used with varying amounts of camera cable, a variable amount of cable correction must be provided. A typical loss figure for co-axial camera cables is 10 dB at 10 MHz for a 1,000 ft. run. It is important that the correcting network be properly matched to the attenuation characteristic of the cable, the loss being proportional to the square root of the frequency, as streaking effects will otherwise occur on the picture.

APERTURE CORRECTION. The limiting resolution obtainable from a camera tube is restricted by the diameter of the scanning spot, which as a rule is large compared to a picture element. The output signal generated by scanning across an abrupt black-to-white transition is therefore not an infinitely short transient but a form of ramp function whose duration is related to the diameter of the spot. Since this diameter is constant, it is possible to compensate for the effect by modifying the amplitude versus frequency response of the video amplifier. Simple high-frequency boost is unsatisfactory owing to the introduction of unwanted phase change.

In practice, a form of phaseless boost is employed by passing the signal through a low-pass filter, to the output of which is added a usually variable amount of signal which has been obtained from a high-pass filter. From 6–10 dB of the high-frequency component can be added before high-frequency noise becomes objectionable.

Because of the bandwidth limitation imposed by the system, it is common practice to adjust the amount of horizontal aperture correction so that a test pattern of resolution bars at a pitch equivalent to 400 lines per picture height is resolved at 100 per cent depth of modulation. While improvement of the horizontal definition in this way is relatively simple, an equivalent correction in the vertical direction is more difficult owing to the necessity for line-period delay lines. The availability of ultrasonic delay lines now makes vertical aperture correction practicable, however, and its use is becoming fairly widespread particularly with Plumbicon camera tubes, the inherent resolution of which is lower than that of the image orthicon.

SHADING CORRECTION. Shading correction is provided to compensate for variations in the level of the signal over the field at black, or with the lens capped. Shading waveforms must be provided in both line and field directions and may be sawtooth or parabola in form, and either positive or negative in polarity.

REMOTE GAIN CONTROL. Since modern channels must be capable of remote operation, it is useful to provide for remote gain adjustment.

This enables the correct output level to be set with the camera tube operating at the preferred exposure.

BLANKING STAGE. Although it is usual to provide blanking at the camera, the beam being cut off during the retrace period to prevent unwanted retrace patterns appearing on the picture, a significant amount of random noise is usually present on the output signal during the period between active picture lines.

This is removed and the exact active line length established by the introduction of system blanking.

CLIPPERS. Both black and white clippers are normally provided to prevent unwanted signal excursions above nominal peak-white or below black.

GAMMA CORRECTOR. A typical receiver display tube has a power law relating the light output to the signal input of between $\gamma = 2$ and $\gamma = 3$. It is therefore necessary to provide complementary gamma correction to the transmitted signal. In practice, an adequate degree of performance can be obtained using a characteristic with three linear portions each of different gain, both the gain of each section and the level of the change-over or break points being adjustable. The sense of the correction is to stretch the black parts of the picture, and the degree of correction which can be tolerated is often governed by the exaggeration of the noise in the dark areas. It is normal to provide for a gain in the lowest portion of the characteristic of up to 16 dB above that of the average slope.

DEFLECTION CIRCUITS. Modern cameras invariably employ magnetic deflection, and since a current of several amperes is required for the line scan, this circuit uses low impedance coils, and the circuitry is contained in the camera. The lower frequency field scan waveform is usually housed in the CCU. To avoid picture distortion, it is necessary that the current in the scanning coils rises linearly with time. Any distortion due to non-linearity of the scanning system is measured along the vertical and horizontal axes of the picture, and a figure of under 1 per cent should be achieved in modern designs.

SCAN PROTECTION. To avoid possible damage to the camera tube, there is a protection circuit designed to remove the anode supplies or blank off the beam in the event of a scan failure.

VIEWFINDER. The viewfinder is basically similar to a picture monitor and is supplied with a fully processed signal, an extra conductor being used in the camera cable to provide for this. This arrangement also makes it possible for split-field pictures to be displayed to assist the camera operator in setting up special effects.

A picture brightness of up to 200 foot-lamberts should be available on modern cameras.

INTERCOMMUNICATIONS. Although an essential part of the camera channel, the talk-back facilities need no special mention as the techniques are quite standard.

TEMPERATURE CONTROL. Optimum performance is obtained from image orthicons by closely controlling the temperature of the tube to about 40°C, and the combination of a heater in the deflection yoke and a cooling fan may be used to maintain this condition automatically.

OPERATIONAL CONTROL PANEL. To enable the camera channel to be controlled from a remote operational control position, provision is normally made to relay the iris and black-level controls and also several auxiliary controls such as gain, electronic cap, overscan for use during stand-by periods, and mains on/off. An excellent way of adjusting the iris and black level takes the form of a quadrant iris control, with a rotary black-level control forming the knob of the quadrant. A further refinement of this technique allows the operator to switch the output of the associated channel to a single picture monitor by pressure on the knob.

N. N. P-S.

See: *Cameraman* (TV); *Camera* (*colour*); *Lens performance; Lens types; Shot box; Sticking; Telecine; Telerecording.*

Books: *Television Engineering Handbook*, by D. G. Fink (New York), 1957; *Television Engineering, Vol.* 1, by S. W. Amos and D. C. Birkinshaw (London), 1957.

□CAMERA (COLOUR)

Colour television cameras are going through a transitional phase since many important components are in a state of rapid development, for example, wide-range zoom lenses, dichroic beam-splitters, very sensitive pick-up tubes and solid-state signal processing circuits, particularly of the integrated circuit type. For some time to come widely differing types of colour camera are likely to compete for acceptance, and the design that will ultimately establish itself cannot be predicted with certainty.

BASIC PRINCIPLES

FLYING-SPOT ANALYSER. The simplest device for originating a colour television signal is the flying-spot analyser, which may be applied to a stationary transparent object such as a slide or a frame of film in a telecine (film scanner). In this device a high-voltage electron beam in a cathode ray tube produces an unmodulated raster on the tube face – that is, it scans the phosphor surface, without modulation, at the appropriate line and field rates. The white spot of light on the tube face is then focused by a lens on to the coloured slide or frame, which is therefore itself scanned (hence the term flying-spot). As the focused spot of light traverses the photographic film, it is appropriately coloured by the components of redness, greenness and blueness in every element of the film. The three colour components are then separated by means of beam-splitting mirrors (usually dichroic); these are arranged

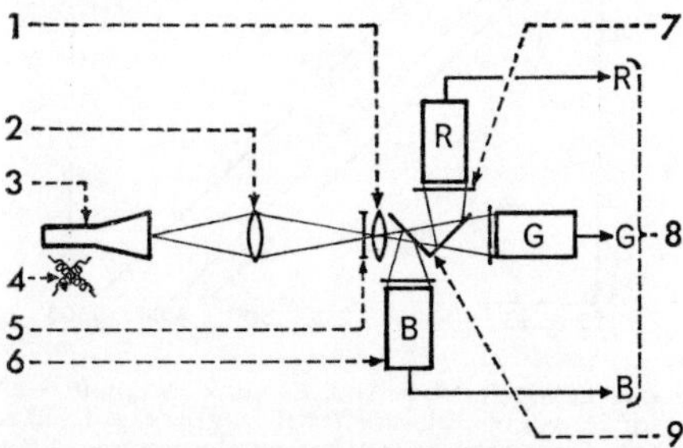

FLYING-SPOT ANALYSER. Raster formed by scanning coils, 4, on face of flying-spot tube, 3, is focused by lens, 2, on to slide, 5. Light traversing the slide is separated into its three colour components by dichroic mirrors, 9. These components are then directed into photocells, 6, which produce the three signals, R, G, and B. Each photocell is associated with a suitable trimming filter, 7. Immediately behind the slide is a field lens, 1, which improves the light collection. The three signals, R, G and B, are passed to the channels, 8.

in such a way that the three colour components are directed into three photosensitive cells.

The outputs of these cells are fed to the three signal channels. Since the spot of light on the cathode ray tube screen is focused by the lens on only one point of the film at any instant, the three signals corresponding to the colour of this point are produced simultaneously and therefore register in time exactly.

BANDWIDTH. At first sight it would appear that these three signal channels would require a bandwidth three times that of a black-and-white channel.

It is here that NTSC principles, which govern the design of modern cameras, and underlie all modern colour television, come into effect. The eye does not perceive detail in colour, and a satisfactory picture can be obtained if all fine details (corresponding to the high signal frequencies) are transmitted in monochrome only, provided that the relative luminosities are not altered. A "detail" signal can be derived from the three colour signals by adding their high frequencies; this is known as the mixed highs principle.

The colour information can then be transmitted by a narrow band of frequencies, approximately 1 MHz in the USA system, compared to a total channel width of 4 MHz.

Further developments led to the NTSC constant luminance proposals in which the red, green and blue signals are added (matrixed) to form a monochrome signal of normal bandwidth (4 MHz), and two narrow-band colour difference signals are formed in suitable matrices. Thus, by a purely signal-processing technique, it is possible to separate the signals into two categories: the brightness, or monochrome information, and the colour information. Both signals are transmitted in one channel of 4 MHz bandwidth by adding to the monochrome information a subcarrier wave (3·58 MHz) modulated by the narrow band colour information.

CAMERA TYPES. When the time came to develop direct-view colour cameras for transmitting live action, the same signal processing was adopted, but the problems were greatly increased because the flying-spot principle was no longer applicable. The first camera employed three independent

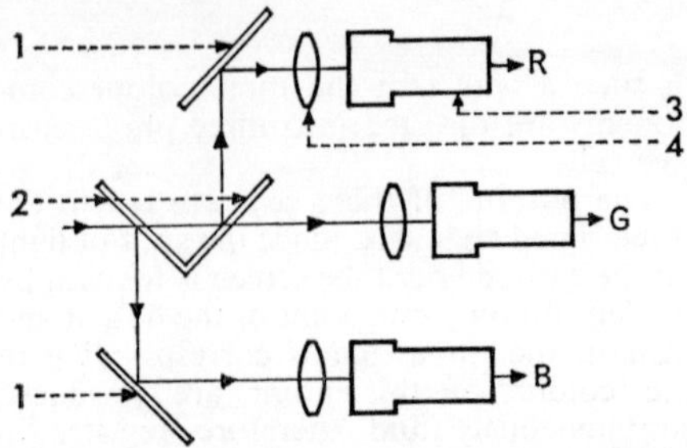

SIMPLE 3-TUBE CAMERA. The simplest form of 3-tube camera has an objective lens, 4, arranged in front of each pick-up tube, 3. Incoming light is split by dichroic mirrors, 2, into its three components, red, green and blue. Two front-surface mirrors, 1, reflect light into the blue and red tubes, so that the three tubes are parallel, thus ensuring that the earth's magnetic field acts in the same manner on them all, which facilitates registration.

pick-up tubes and remains an important type of colour camera. Two difficulties had to be faced: the need for precision optical beam-splitting arrangements, and the need for scanning with great accuracy the three electron beams so that completely registered output signals were obtained and maintained.

To produce a 4 MHz monochrome signal of adequate resolution by matrixing the signals, an accuracy of registration of 1 part in 3,000 all over the three scanned rasters is required.

To mitigate these difficulties, a second important class of camera was developed in which a separate tube was used for the monochrome signal (just as in a black-and-white camera), the colour information being derived from one or more other tubes the characteristics of which could be less critical.

The four-tube type of camera employs three tubes to produce the red, green and blue signals. In one form of two-tube camera, a single colouring tube provides the colour-difference signals by means of a stripe filter consisting of a recurring series of very narrow red, blue, green and black stripes with the aid of which a standard NTSC signal is derived, by suitable circuitry, from the outputs of these two tubes.

Thus colour camera problems fall into five categories: optical systems, colorimetry, pick-up tubes, signal processing, and circuitry.

OPTICS

ANGLES OF VIEW. In considering any type of camera, it is helpful to know the horizontal angle of view of the scene to be taken. This angle $\alpha°$ is given by the equation $\cot \alpha/2 = 2F/W$, where F is the focal length of the lens and W is the width of the photosensitive surface. For example, an image orthicon tube has W equal to 32mm, and for a camera lens of 50mm focal length the angle of view is 35°.

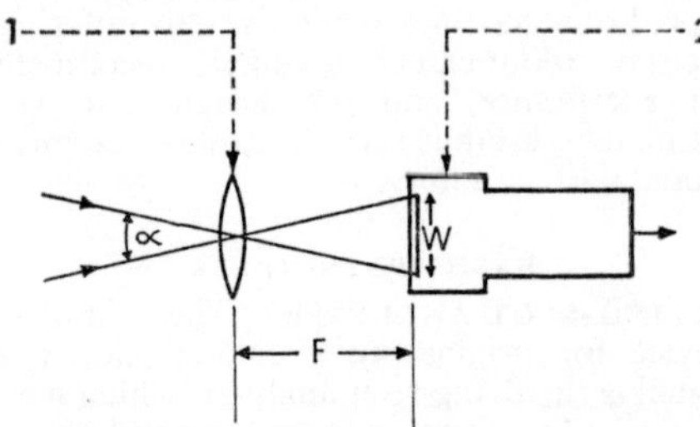

FOCAL LENGTH, IMAGE WIDTH AND ANGLE OF VIEW. Distance between lens, 1, and image produced on photo surface W of pick-up tube, 2, is equal to F, the focal length, assuming object at infinity. Horizontal angle of view, α, can be calculated from equation $\cot \alpha/2 = 2F/W$.

It is usual with studio cameras to cover a range of angles of 4° to 40°, and this is commonly achieved with a four-lens turret, or a single zoom lens.

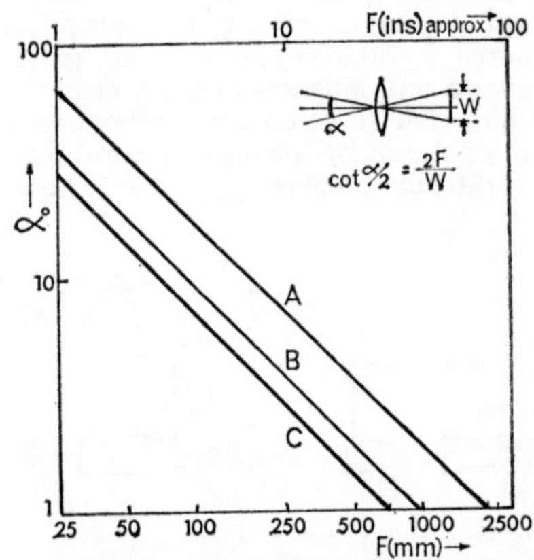

FOCAL LENGTH AND ANGLE OF VIEW. Angle of view α for lenses of different focal length, expressed in inches (*top scale*) or millimetres (*bottom scale*), for typical pick-up tubes A. Image orthicon (W = 1·27 in.). B. Plumbicon (W = 0·635 in.). C. 1 in. vidicon (W = 0·50 in.).

RELAY LENS. In a three-tube colour camera, it is impossible to arrange the tubes (which in the case of image orthicons are 80mm in diameter at the photosensitive ends) directly behind a lens having a back-

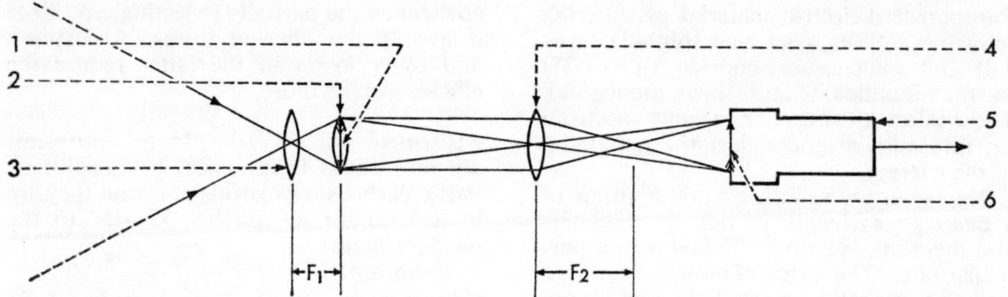

PRINCIPLE OF RELAY LENS. Objective lens, 3, produces an image at a distance F_1, its focal length. This image, 1, is relayed by means of lens, 4, to the photosensitive surface, 6, of the tube, 5. A field lens, 2, is placed in the plane of the image, 1, produced by objective lens, 3. A 1 : 1 relay system produces an image on the pick-up tube the same size as the image produced by the objective lens, and the spacing of field lens, 2, and pick-up tube face, 6, is $4 \times F_2$.

focal length of, for example, 35mm. The solution is to provide a relay lens system.

The relay lens may be chosen, for example, to have a focal length of 12 cm, and if unit magnification is desired between the real image produced by the objective lens and the image focused on the pick-up tube photosensitive surface, the distance between these images is approximately four times the focal length, that is, 48 cm, which easily permits the light to be split by suitable mirrors between the three pick-up tubes. To cover a range of viewing angles, a turret of objective lenses can be arranged to produce real images in the same plane.

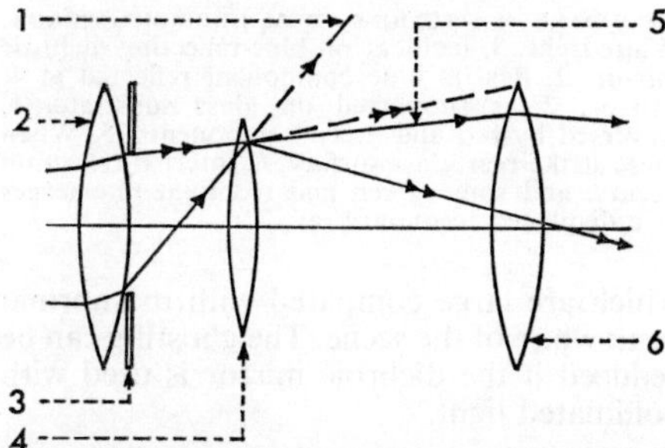

FUNCTION OF FIELD LENS IN RELAY SYSTEM. Field lens, 4, is used with relay lens, 6, to converge rays so that extreme rays, 1, are brought within entrance pupil of relay lens. Objective lens, 2, produces a real image in plane of field lens, which is relayed by relay lens. Extreme rays as shown by exit pupil, 3, of objective lens are brought within entrance pupil, 5, of relay system.

To improve light collection by the relay system, a field lens can be situated in the real image plane; this reduces vignetting and makes the illumination more uniform over the field.

DICHROIC MIRRORS. The light energy passing through the relay lens is split between the red, green and blue tubes by means of dichroic mirrors. A typical dichroic mirror consists of a glass substrate on which is deposited alternate layers of

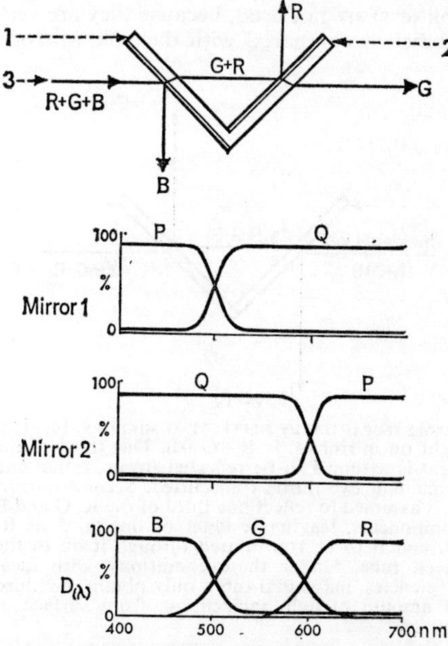

ACTION OF DICHROIC MIRRORS. (*Top*) Incident white light, 3, has its blue component reflected by blue-reflecting dichroic mirror, 1. Green and red components pass through it unaffected and strike red-reflecting mirror, 2, which reflects red component and transmits green. (*Below*) Characteristics of mirrors 1 and 2. P. Reflectance. Q. Transmittance. Blue transmittance of mirror 2 is immaterial, since blue component has already been removed by reflection from mirror 1. (*Bottom*) Effective passbands of the three colour components are shown by curves R, G and B.

transparent dielectric material of differing refractive indices, e.g., zinc sulphide (μ = 2·4) and magnesium fluoride (μ = 1·37) (or titanium dioxide and silicon monoxide). These select a band of the visible spectrum for transmission and reflect the remainder of the energy.

The layers are usually an odd multiple of a quarter wavelength of light in thickness and there may be up to 20 layers in a particular filter. The virtue of a dichroic mirror is that practically all the light that is not reflected is transmitted through the mirror, there being very little absorption.

By overlapping the characteristics of two dichroic mirrors, it is possible to separate out three energy regions of the visible spectrum corresponding to red, green and blue light.

The first mirror reflects the energy in the blue region and transmits the longer wavelength green and red energy. The second mirror is chosen to reflect the red energy and to transmit the green.

Partially reflecting mirrors (neutral beam-splitters) are not used, because they are very inefficient compared with dichroic mirrors.

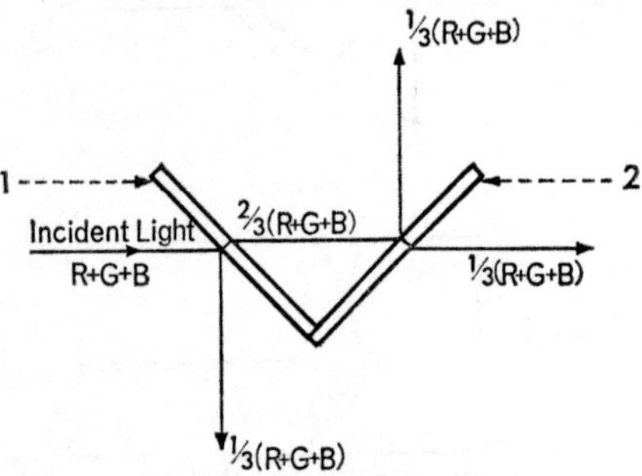

LOSSES IN PARTIALLY REFLECTING MIRRORS. Incident light on mirror, 1, is R+G+B. One third of the light is assumed to be reflected towards blue and remaining two thirds transmitted. Second mirror, 2, is assumed to reflect one third of the R, G and B components, leaving residual of one third of R, G. and B to be transmitted through it on to the green tube. Under these conditions, with ideal efficiencies, individual tubes only obtain one third of amount of light incident on front surface of mirror.

Assume partially reflecting mirrors which, for simplicity, are 100 per cent efficient, i.e., have no absorption and reflection losses. To obtain equal amounts of light on each pick-up tube, the first mirror reflects one-third of the light and transmits two-thirds, and the second mirror is arranged to reflect half and to transmit half of the incident light. The dichroic mirror is therefore three times as efficient as the partially reflecting type since it avoids the filtering losses. Absorption and other losses in the latter reduce the efficiency even more.

DICHROIC PROBLEMS. Many problems are introduced by practical beam-splitters, particularly as for reasons of layout they are in general set at roughly 35°–45° to the incident light.

These are:

1. Astigmatism and coma introduced by the thickness of the oblique glass substrate (this may be from about ⅛ in. to ¼ in. in practice).

2. Secondary images (ghosts) caused by reflections from the rear surface of the substrate. These can be troublesome when the scene contains bright sources of light,

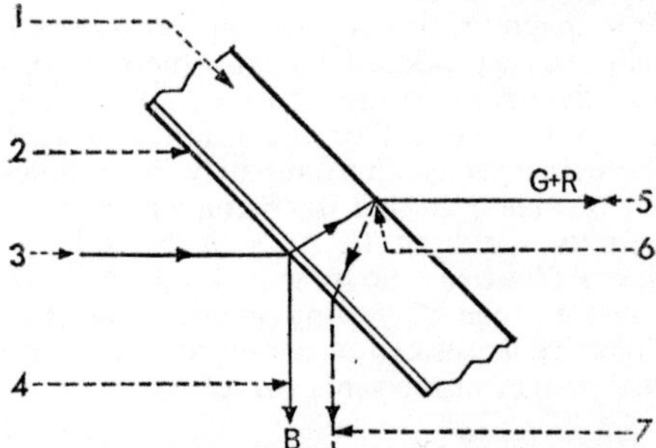

SECONDARY REFLECTIONS FROM DICHROIC MIRROR. White light, 3, incident on blue-reflecting dichroic mirror, 2, has its blue component reflected at 4. Mirror, 2, is supported on glass substrate, 1, traversed by red and green components, 5. When these strike rear glass surface, 6, internal reflection occurs, and some green and red light re-emerges as a displaced secondary ray, 7.

which are large compared with the normal peak white of the scene. The ghosting can be reduced if the dichroic mirror is used with collimated light.

3. The dichroic spectral characteristic is a function of the angle of incidence of the light rays. A dichroic mirror functions because of interference between the rays reflected from the top layer(s) and those reflected from lower layers, and since the path difference for a single layer is $2 \mu t \cos r$, where μ = refractive index of layer, t is thickness of layer and r is angle of refraction in the layer, it is apparent that, as the angle of incidence increases, the steep part of the curve shifts to lower wavelengths. Rays of light falling on the dichroic mirror vary in angle of incidence if they are coming from

different points in the scene, and if from a single point in the scene they pass through different parts of the lens. In the first case, the variation in angle is a function of the angle of view of the objective lens as given by equation $\cot \alpha/2 = 2F/W$, and in the second case is a function of the relative aperture of the lens ($\sin \mu/2 = 1/2 \times f_{NO}$). Thus the wider the aperture the greater the variation in angle.

These problems are reduced in the prism type of beam-splitter in which the dichroic layers may be bonded to the glass surfaces.

COLORIMETRY

TAKING CHARACTERISTICS. The spectral characteristics for the three analysing tubes are calculated with respect to the specified chromaticity coefficients of the reproducing primary colours. In the NTSC specification, the primaries are as given in the table below.

SPECIFIED CHROMATICITY COEFFICIENTS FOR NTSC RECEIVER PRIMARIES

Colour	x	y	z	Y (relative)
Red	0·670	0·330	0·000	0·30
Green	0·210	0·710	0·080	0·59
Blue	0·140	0·080	0·780	0·11
White	0·310	0·316	0·374	1·00

NOTE: $x+y+z=1$

The equations determining the red, green and blue taking characteristics are respectively:

$$\bar{r} = K_1[(y_g z_b - z_g y_b)\bar{x} + (z_g x_b - x_g z_b)\bar{y} + (x_g y_b - y_g x_b)\bar{z}]$$

$$\bar{g} = K_2[(z_r y_b - y_r z_b)\bar{x} + (x_r z_b - z_r x_b)\bar{y} + (y_r x_b - x_r y_b)\bar{z}]$$

$$\bar{b} = K_3[(y_r z_g - z_r y_g)\bar{x} + (z_r x_g - x_r z_g)\bar{y} + (x_r y_g - y_r x_g)\bar{z}]$$

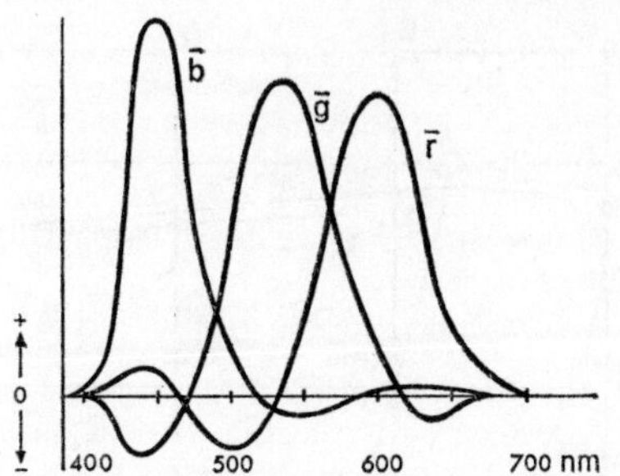

FCC SPECIFIED COLOUR PRIMARIES. Theoretical taking characteristics for red, green and blue tubes in accordance with FCC specified primaries. Note negative lobes, indicated by minus values.

where $\bar{x}$, $\bar{y}$ and $\bar{z}$ are the normalized distribution coefficients for the standard observer. (Recently shadow-mask colour tubes have been placed on the market with sulphide phosphors as well as rare-earth red phosphors having a chromaticity specification differing from the NTSC. If these primaries are used, colorimetry will not be correct.)

The curves produced when the values for $\bar{r}$, $\bar{g}$ and $\bar{b}$ are inserted in the equations have minor positive and negative lobes, the latter extending below zero. Although it is theoretically possible to obtain these negative lobes by means of extra pick-up tubes, the signals of which are subtracted from the signals generated by tubes providing the positive portions, in practice this is unnecessary and various compromise curves using only positive lobes have been suggested, which give adequate colour analyses. One suggested set of values $\bar{R}$, $\bar{G}$ and $\bar{B}$ is given below, and the errors obtained are shown in the u, v (uniform chromaticity) colour triangle.

The length and direction of the arrows indicate the errors. Improvement can be obtained by matrixing the signals in

SUGGESTED PRACTICAL VALUES FOR $\bar{R}$, $\bar{G}$ AND $\bar{B}$ USING POSITIVE VALUES ONLY

λ	Proposed Values			λ	Proposed Values			λ	Proposed Values		
nm	$\bar{R}$	$\bar{G}$	$\bar{B}$	nm	$\bar{R}$	$\bar{G}$	$\bar{B}$	nm	$\bar{R}$	$\bar{G}$	$\bar{B}$
400	—	—	0·038	500	—	0·390	0·115	600	1·000	0·135	—
410	—	—	0·177	510	—	0·613	0	610	0·972	0·012	—
420	—	—	0·365	520	—	0·835	—	620	0·843	0	—
430	—	—	0·785	530	—	0·960	—	630	0·640	—	—
440	—	—	0·990	540	—	1·000	—	640	0·450	—	—
450	—	—	1·000	550	0	0·960	—	650	0·286	—	—
460	—	—	0·940	560	0·275	0·865	—	660	0·170	—	—
470	—	—	0·720	570	0·560	0·710	—	670	0·090	—	—
480	—	—	0·448	580	0·760	0·520	—	680	0·050	—	—
490	—	0·180	0·246	590	0·920	0·310	—	690	0·023	—	—
500	—	0·390	0·115	600	1·000	0·135	—	700	0·012	—	—

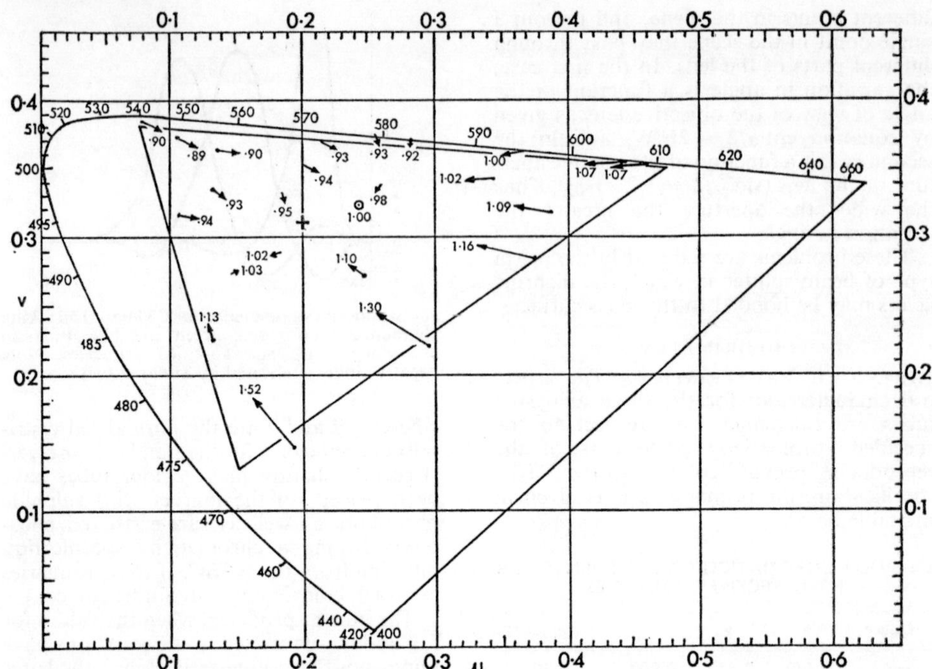

ERRORS RESULTING FROM TAKING POSITIVE LOBES ONLY. Original colour is represented by tail of arrow, reproduced colour by position of arrowhead. Number near arrowhead represents luminance value of reproduced colour relative to unity value for original colour.

suitable proportions. For example, by matrixing the signals to produce a modified green signal

$$G = -0{\cdot}06R + 1{\cdot}24G - 0{\cdot}18B$$

the errors are reduced, as shown in the next u,v colour triangle. The signal-to-noise ratios of modern cameras are so good that some loss of signal strength can be tolerated. Hence, 3×3 matrices for colour correction are employed in most cameras so that all three colour signals can be modified.

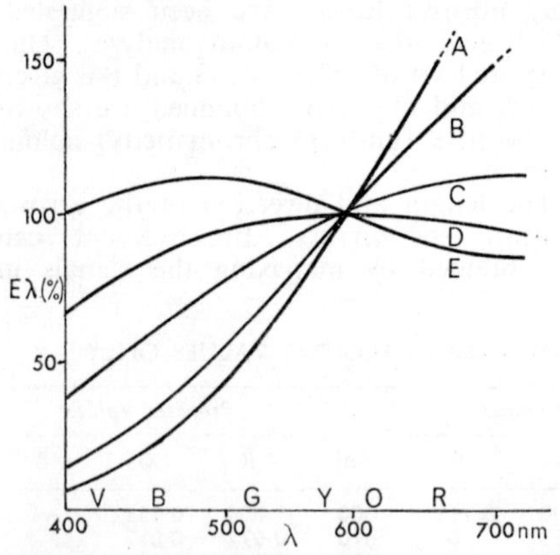

ENERGY DISTRIBUTION OF TYPICAL LIGHT SOURCES. Relation of energy, E_λ, to wavelength for various colour temperatures over the visible spectrum from 400-700 nm. A. 2,500°K (vacuum lamp). B. 3,000°K (studio lighting). C. 4,000°K (carbon arc). D. 5,000°K (sunlight). E. 6,000°K (daylight). All curves shown relative to arbitrary 100 per cent at wavelength of 590 nm. E_λ represents the energy radiated by a black body source.

ILLUMINATION. The curves $\bar{r}$, $\bar{g}$ and $\bar{b}$ specified assume that the illumination is equal-energy white, that is, the curve of light energy versus wavelength is flat over the visible spectrum. However, in normal studios, the colour temperature of the illumination is designed to be between 3,000°K and 3,200°K. It is usual to correct for the spectral characteristic of illumination $E(\lambda)$ by the following equations: $\bar{r}$ corrected $= \bar{r}/E(\lambda)$; $\bar{g}$ corrected $= \bar{g}/E(\lambda)$; and $\bar{b}$ corrected $= \bar{b}/E(\lambda)$.

SPECTRAL CHARACTERISTICS OF TUBES. A further factor involved in the colour analysis is the spectral characteristic $T(\lambda)$ of the pick-up tube. The principal characteristics of interest are the S10 and S11 characteristics of the photo-emissive tubes such as the image orthicon, the Sb_2S_3 vidicon characteristic and the lead oxide vidicon (Plumbicon) characteristic which has only recently been extended beyond 640 nm.

TRIMMING FILTERS. The dichroic mirror characteristics are designed to be as efficient as possible in splitting the light energy between the three pick-up tubes. A correction filter $F(\lambda)$ is included in the light path to each pick-up tube to give additional filtering or trimming.

Taking into account all factors gives three similar equations, one for each path, of which the red one is

$$E(\lambda) \times D_R(\lambda) \times T_R(\lambda) = \bar{R}(\lambda)$$

$D_R(\lambda)$ is the effective spectral characteristic of the dichroic mirrors to light being directed into the red channel. The spectral responses of the three pick-up tubes may be the same or, as indicated, may be different for the three channels.

PICK-UP TUBES

There are at the moment four types of pick-up tube available which have application to colour television cameras, namely, the 4½ in. image orthicon, the 3 in. image orthicon, the vidicon and the lead oxide vidicon (Plumbicon). Several other types are in the process of development with a view to obtaining greater sensitivity, improved contrast characteristics and spectral responses, and examples are the selenicon – a selenium-layer vidicon – and the bombardment-induced-conductivity type, such as the ebicon and the SEC vidicon. However, these are still in the research stage, as are solid-state matrix-type pick-up devices, which may revolutionize future camera design.

The 4½ in. and 3 in. image orthicon tubes are already in standard use with monochrome cameras. The 4½ in. tube gives a slightly better signal-to-noise ratio output than the 3 in. tube, at the expense of rather more light input, and for this reason is to be preferred in a separate luminance camera to the 3 in. tube; alternatively for a trichromatic camera using image orthicons, the 3 in. tubes are preferable, not only on

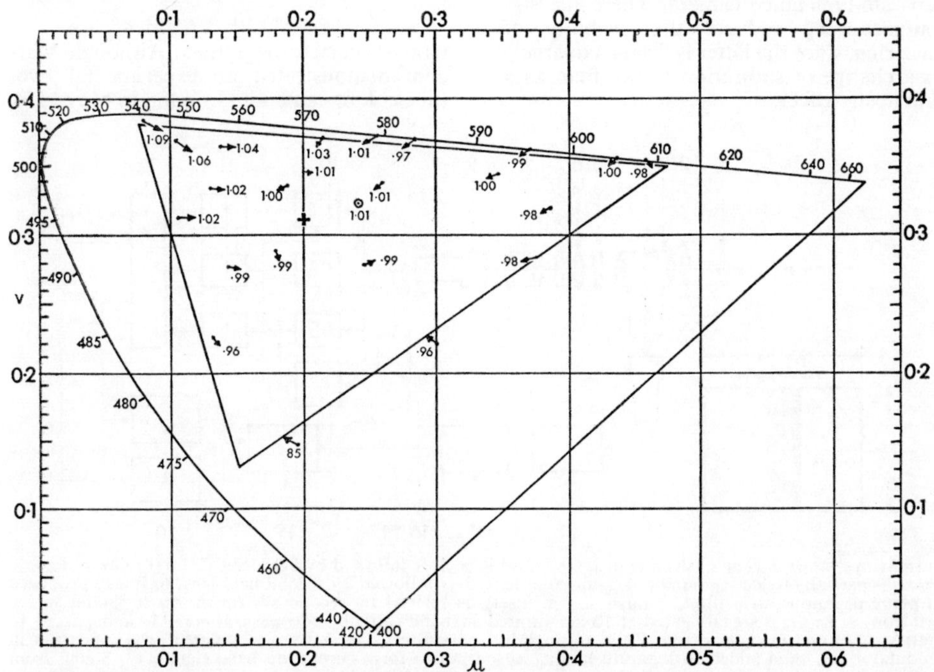

IMPROVEMENT BY MODIFYING GREEN MATRIX. Green signal made by matrix G = −0·06R+1·24G−0·18B. Shorter arrows indicate reduced errors. Coefficients add up to unity so that on grey scale no effect is obtained.

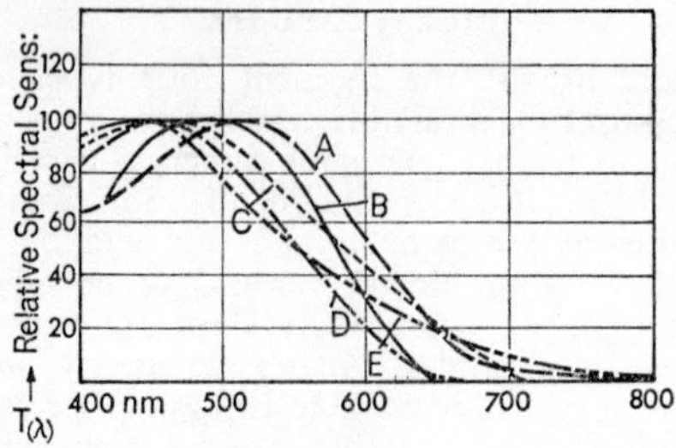

RELATIVE SPECTRAL SENSITIVITY, T(λ), OF PICK-UP TUBES. A. Extended-red Plumbicon. B. Plumbicon. C. Image orthicon with S-10 characteristic. D. Image orthicon with S-11 characteristic. E. Antimony trisulphide vidicon.

account of the size, but also because the matrixing operation for the monochrome signal produces about 3 dB improvement on the green tube output, thus compensating for the lower individual tube signal-to-noise ratio.

The standard antimony trisulphide vidicon, while giving pictures of excellent resolution, gamma and signal-to-noise ratio, suffers from smearing with movement, and for broadcast purposes is not suitable, except possibly as a colour tube in a separate-luminance camera. There the signal-processing reduces the visibility of smearing, since the latter is displayed largely as a change of saturation rather than as a luminosity effect.

The best tube available at the moment for colour cameras is the lead-oxide vidicon, since it has the advantages of the standard vidicon in simplicity, but is reasonably free of smearing and has a greater sensitivity, as well as a low dark current. With care in amplifier design it can give an excellent signal-to-noise ratio.

The spectral characteristics of the original production Plumbicons had a restricted red response above about 640 nm. However, Plumbicons with extended red characteristics are being introduced and these should give improved red colorimetry and increased overall camera sensitivity.

CAMERA DESIGN

It is interesting to compare the differences in approach adopted by different camera designers over the past fifteen years, conditioned by the techniques and components available, for example, pick-up tubes, zoom lenses, dichroic filters and beam splitters. The table indicates how several manufacturers have tackled the design of two-, three- and four-tube cameras respectively, and the following notes indicate in more detail some of the significant aspects of their design.

TWO-TUBE CAMERAS

IMAGE ORTHICON TUBES. Although Marconi demonstrated an experimental two-tube colour camera in London in May 1954,

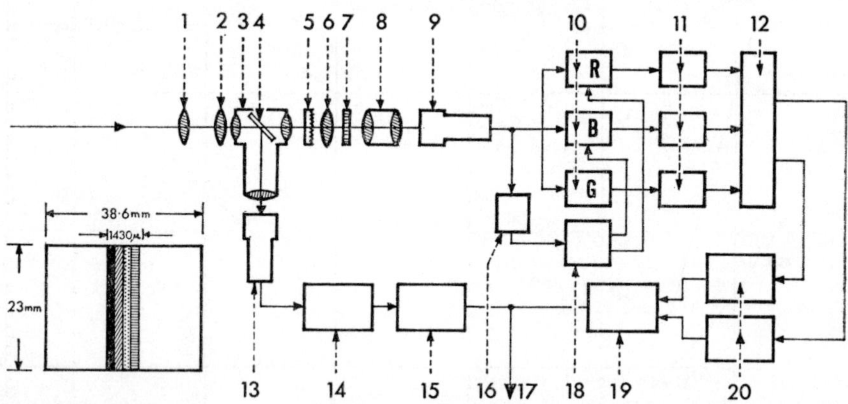

OPTICAL SYSTEM OF 2-TUBE CAMERA (NHK). Objective lens, 1, is followed by field lens, 2. First relay system, 3, contains partially reflecting mirror, 4. Lenticular lens, 5, is followed by second field lens, 6. Image produced in plane of colour strip filter, 7 (also shown inset), is relayed by second system, 8, on to colour image orthicon, 9. R, G, B signals gated at 10 are limited in bandwidth by low-pass filters, 11, and passed to matrix, 12. I and Q signals from 12 go through low-pass filters, 20, and are modulated on a subcarrier in modulator, 19, then added to separate luminance output to form composite NTSC signal, 17. Signal from separate luminance image orthicon, 13, passes through time delay, 14, and processing amplifier, 15. Signals derived from colour tube, 9, are fed to limiting amplifier, 16, to actuate gate-pulse generator, 18, controlling gates, 10.

the first broadcast colour signals from a two-tube camera were transmitted from Japan during the Olympic Games in October 1964. The camera was developed by Toshiba with the assistance of NHK, which provided the prototype. The luminance tube was a 4½ in. image orthicon and the colouring tube was a 3 in. image orthicon. After the Olympic Games the former was modified to a 3 in. tube for better sensitivity. The sensitivity during the opening ceremony was about f6·3 to f8 at 150 foot/candles.

The head of this two-tube camera is 78 cm long × 46 cm wide and 56 cm high and weighs about 120 kilograms. The turret lenses are imaged on to a field lens which is followed by a first relay system employing two separate lenses spaced apart so that the light between them is collimated. A semi-reflecting mirror is placed at 45° to the central rays between the two component lenses to direct the light into another lens (forming part of the first relay system) to focus the image on to the photocathode of the luminance tube.

The image produced by the straight portion of the first relay system is focused on a second field lens associated with a tri-colour stripe filter. This image is then relayed to a colouring image orthicon tube.

A lenticular lens is inserted in front of the second field lens to limit the bandwidth of the incident light so as to eliminate spurious beat patterns with the stripe filter. The stripe filter consists of a recurrent series of red, blue and green stripes interspersed with a black stripe, the total width of any four stripes being 0·43mm; there are 75 to 80 such quadruplets in the filter width.

The output from the colouring tube is therefore a dot sequential RBG colour signal interspersed with black keying pulses. The keying pulses are limited and delayed to form three gating pulses, which are used to gate the dot sequential colour signals to form three simultaneous signals which are band limited and suitably processed before feeding into the NTSC type encoder to be subsequently mixed with the separate luminance signal.

PLUMBICON TUBES. During 1968, Ampex in conjunction with the American Broadcasting Company developed a small two-tube camera particularly for mobile applications. Two Plumbicons are used, one giving a standard luminance signal on 525 lines 60 fields/second, and the second a blue/red field sequential output, i.e., one field of blue and one field of red alternately, taken through a rotating filter disc. To provide simultaneous signals it is necessary to store these latter signals for 1/60 sec. each and then to replay them. In practice the camera provides three output signals, Y, R-Y and B-Y, and since the latter two signals are only narrow-band signals, they are field-stored relatively easily by means of a rotating disc field store.

THREE-TUBE CAMERAS

IMAGE ORTHICON TUBES. The first practical camera designed for the NTSC system in the USA (and still in use in large numbers today) has three image orthicon tubes in a relay lens configuration (the RCA TK41). The relay lens is formed from two lenses each having a focal length of 9½ in.; to improve the shape of the camera the pick-up tubes are arranged to be practically parallel and two front-surface mirrors direct the light from the dichroic mirrors into the red and blue tubes.

The objective lenses are mounted on a rotating turret to give a choice of angles of view, as in normal monochrome practice. The widest angle lens has a focal length of about 28mm. Associated with each objective lens is a field lens which redirects all the light from the image plane into the relay lens system.

The two dichroic mirrors are arranged at 76° to each other and the various glass surfaces are suitably bloomed to increase their efficiency and to reduce internal reflections.

The effective aperture of the lens system is adjusted by means of a remotely controlled iris stop located between the components of the relay lenses.

To operate the tubes correctly over their dynamic characteristics neutral density filters are placed in front of the red and green tubes. The glass substrate of the dichroic mirrors is about ¼ in. thick and, because the rays of light passing through the mirrors are displaced differently in the horizontal and vertical directions, astigmatism is introduced. To correct for this, two optically ground plates are mounted ahead of the relay lens to introduce negative astigmatism. A single-plate astigmatism corrector is also mounted in the blue channel, so that the light rays in the red, green and blue channels pass through the same total thickness of plate.

The studio lighting is generally about 300/400 fc with a lens aperture between f5·6 and f8.

Later versions of the TK41 camera use a

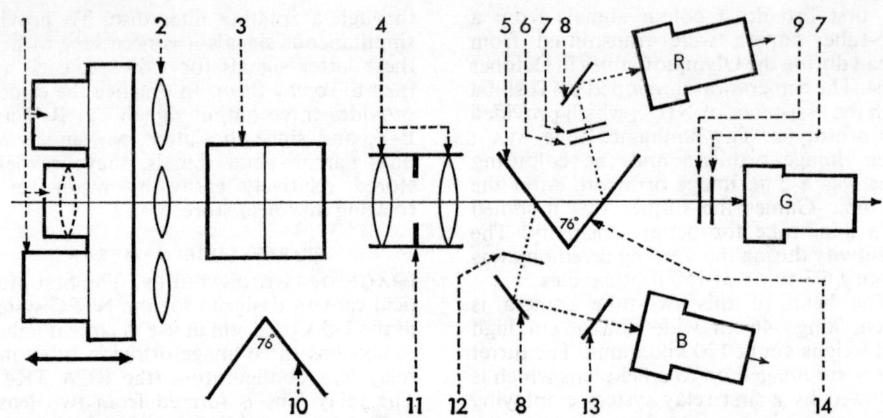

OPTICAL SYSTEM OF 3-IMAGE ORTHICON (RCA) CAMERA. Objective lenses, 1, mounted on focusing turret produce real images in plane of condenser lenses, 2. Real image is relayed by lens, 4, to pick-up tubes R, G, B, via dichroic mirrors, 5 and 14. Amount of light into pick-up tubes is controlled by iris stop, 11. Front surface mirrors, 8, fold optical paths to red and blue tubes. Trimming filters 6, 9 and 12 are inserted in the three light paths. Red and green tubes have neutral density filters, 7. Pieces of plate glass, 3 (shown in side view, 10), form vertical astigmatism corrector. Another plate glass, 13, corrects horizontal astigmatism in optical path to blue tube.

prism type beam-splitter to improve efficiency and reduce astigmatic effects.

PLUMBICON TUBES AND DIRECT IMAGING. An interesting colour camera using three Plumbicons is that developed by Philips: it employs a prism beam-splitter designed to reduce polarization and other colour defects of dichroic mirros.

The object of the prism design is to reduce the angle of incidence to a low value in order to separate the red from green light as efficiently as possible.

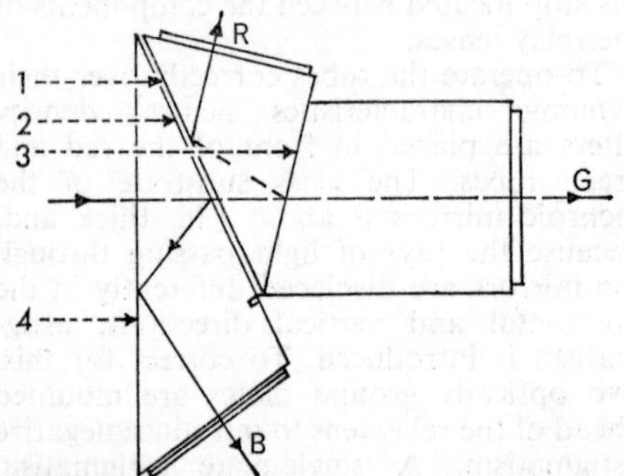

PRISM BEAM SPLITTER (PHILIPS). Light incident on first prism, 4, is internally reflected by blue reflecting surface, 2. Blue light is then totally internally reflected from front surface and directed via trimming filter to blue tube. Red and green light passes via small air gap through front surface, 1, of second prism. Red light reflected by red reflecting layer, 3, is then totally internally reflected and sent to red tube. A third prism equalizes path length of green light going to green tube.

The beam-splitter is composed of three separate prisms, the first of which has a blue reflecting surface, which is separated by a small air gap from the second prism which has a red-reflecting dichroic surface cemented to the third prism. After reflection, the blue component of the incoming light from a zoom lens undergoes total reflection and leaves the system via a trimming filter. The red component is reflected by a red-reflecting dichroic mirror and after total reflection leaves the system.

The green component passes right through the three prisms and the green trimming filter.

The central ray is incident at 25·5° to the blue-reflecting surface and is incident at 13° to the red-reflecting surface. The angles are respectively equivalent in air to 41° and 20°. The reduction of colour errors in the separation of red from green is more important than of green from blue, particularly when reproducing skin tones, and this explains the difference in choice of angle.

The prism system is suitable for use with an $f\,2$ objective lens and produces a target image of 9 × 12mm. The prism assembly behaves as a block of glass 63mm long and a 10 : 1 zoom lens has been designed by Angénieux with suitable corrections and covers a focal range of 18mm to 180mm at $f\,2{\cdot}2$.

Pictures of good signal-to-noise ratio are obtained with 150–200 fc studio illumination at $f\,4$.

Another feature introduced into this camera is the technique known as contours-out-of-green which was evolved to reduce registration errors in the three-tube camera.

FOUR-TUBE CAMERAS

IMAGE ORTHICON AND VIDICON TUBES. RCA introduced a developmental four-tube camera in 1962 which formed the basis of the TK42 camera; this camera used a 4½ in. image orthicon for the luminance channel and three vidicons for the colouring information. The optical system incorporated a Rank-Taylor-Hobson Varotal III lens having a total focal length range of 1·6 in. to 40 in. when extended by an auxiliary

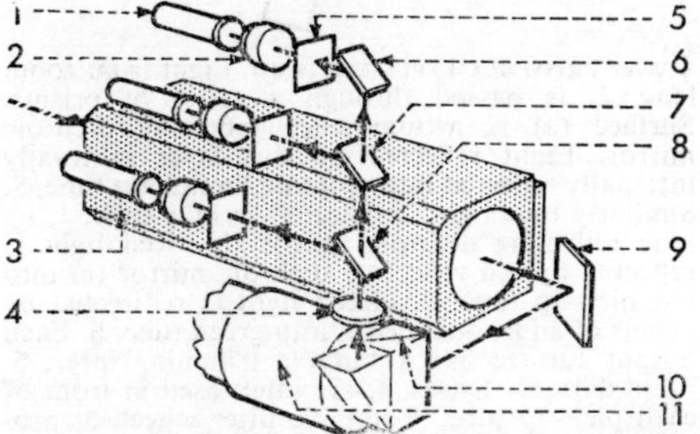

OPTICAL SYSTEM OF 4-TUBE CAMERA (RCA). Light from zoom lens, 3, a Varotal III, is directed via front surface mirror, 9, on to prism structure, 10. This reflects 20 per cent of the incident light into the 4½ in. separate luminance image orthicon, 11, and 80 per cent via field lens, 4, to three vidicon chrominance tubes, 1, each with relay lens, 2, and trimming filter, 5. 6. Front surface mirror. 7. Red dichroic. 8. Blue dichroic.

wide-angle adaptor. The back-focal distance of about 11 in. made it possible to locate all beam-splitting and dichroic elements in this space.

A prism beam-splitter allowed 20 per cent of the available light from the lens to illuminate the 4½ in. tube target and the remaining 80 per cent to be transferred via relay lens systems to each of the vidicons.

The sensitivity of this camera was about the same as the original three-tube image orthicon camera.

PLUMBICON TUBES WITH DIRECT AND RELAY OPTICS. A four-tube camera with Plumbicons developed by General Electric (type PE350) uses a combination of direct and relay optics. The zoom lens images on to the separate luminance tube via a right-angled prism and the remainder of the light for the colouring tubes is relayed by dichroic mirrors and relay lenses.

PLUMBICON TUBES WITH RELAY OPTICS. A colour camera which uses a relay optical system for both the luminance tube and the colouring tubes is the Marconi Mark VII. The object is to adapt the camera for any standard television zoom lens designed for an image-orthicon format. The camera also takes a single fixed focus lens. The optical system is mounted on a fixed plate enclosed by a dust-proof cover and movement of the whole optical bedplate permits focusing adjustments to be made for fixed lens and zoom back focus tolerances. A light-splitting prism is mounted behind the field lens and has a surface which reflects a portion of the light into the luminance tube of the camera and transmits the remainder to the colour tubes. The surface is designed so that it reflects only that amount of light which is necessary to provide the correct signal in the luminance tube, thus the maximum available light is transmitted to the colour tubes. Prism type dichroic surfaces are used to split the light into its red, green and blue components which then pass through colour trimming filters and relay lenses into the three colouring camera tubes.

For black-and-white operation the light-splitting surface is replaced by a fully reflecting surface to pass all the incoming light to the luminance tube.

Considerable use is made in the camera and camera control unit of small circuit modules in which the circuit is formed by the thin-film deposition technique on a glass substrate. This process yields highly stable resistors having a temperature coefficient of 10 parts per million per degree C. Over sixty of these modules measuring 4·1×2·8×1·4 cm. are used in the colour camera channel.

PLUMBICON TUBES AND DIRECT IMAGING OPTICS. In 1967 EMI introduced a professional broadcast colour camera (type 2001) using four Plumbicon tubes. This camera depended on the development of a series of zoom lenses in which the back focal distance of each lens was sufficient to be used in conjunction with a prism beam splitter assembly for four different tube outputs – that is, one output for the separate luminance Plumbicon tube, the other three outputs for the red, green and blue tubes. This arrangement gives the advantage of direct imaging optics, and the elimination of relay systems results in a considerable saving of weight. The zoom lens unit is detachable from the camera for easy transportation and to permit the use of alternative zoom lenses having different characteristics. The basic studio lens is the Angénieux 10×18mm (i.e., 18–180mm) lens having a maximum aperture of f2·2 and a minimum focusing distance of 3 ft., giving angles of view from 3° to 50°. Alternative zoom lenses are the Angénieux 18×27½mm and a Rank-Taylor-Hobson lens which covers a 16 : 1 range.

In the prism optical system used in this camera, internal reflection reduces the angles of incidence, effectively minimizing the amount of polarization introduced by the dichroic surfaces. The luminance-reflecting dichroic surface is designed to reduce the amount of blue light going into the luminance tube so that the shape of the overall separate luminance characteristic is substantially photopic. Approximately 40 per cent of the incident light goes into the luminance tube and the residual 60 per cent is used for the colouring tubes.

A novel feature of this camera is the way in which the prisms are used to splay out the light into the four pick-up tubes. These are arranged approximately at right-angles to the optical axis and in addition radiate diagonally into the four corners of the camera case. This capstan arrangement reduces the length of the camera, and the zoom unit is contained within the camera case, which is only a little longer than a standard type camera. The viewfinder tube is arranged to fit between the two top diagonal pick-up tubes.

A manually operated filter wheel is provided for insertion of colour correction filters and neutral-density filters between the zoom unit and the prism assembly. The zoom pre-set buttons are located in a shot-box built into the rear of the camera near the focus handle. In this separate luminance camera, aperture correction is included in the luminance tube output to give contours-out-of-luminance.

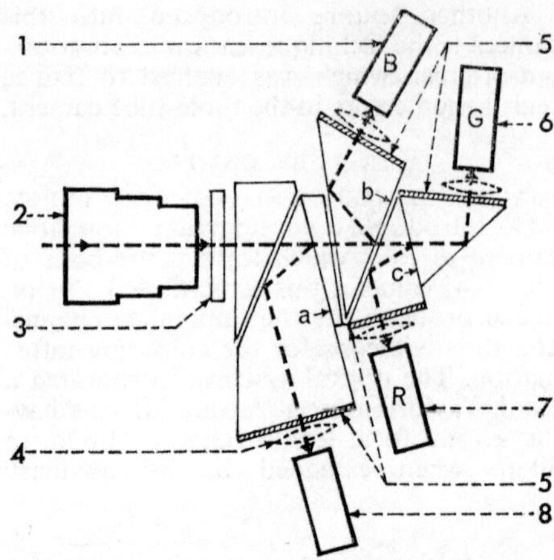

4-WAY PRISM BEAM SPLITTER (EMI). Light from zoom lens, 2, is passed through a series of prisms. Surface (a) is a luminance reflecting dichroic mirror. Light reflected by this layer is totally internally reflected into separate luminance tube, 8. Similarly blue light is reflected to blue tube, 1, by blue reflecting dichroic mirror (b). Red light is reflected by red reflecting dichroic mirror (c) into red pick-up tube, 7. Green light is redirected by means of additional prism into green tube, 6. Each output surface has a suitable trimming filter, 5. Field-flattener lenses, 4, are interposed in front of each pick-up tube. Rotatable filter wheel, 3, provides colour correction, neutral density, or capping filters.

The sensitivity of this camera is such that excellent pictures are obtained at f4 with 100 to 150 fc incident illumination. The

TYPES OF COLOUR TV CAMERA

Number of Tubes	Manufacturer	Type Number	Configuration	Pickup Tubes	Remarks
2	NHK		YRGB	2 Image Orthicons	I tube Y, I tube RGB by grating (relay optics)
	Ampex	BC100	RYB	2 Plumbicons	I tube Y, I tube RB by rotating filter and field store
3	RCA	TK41	RGB	3 Image Orthicons	Relay optics
	Packard Bell	PC100	RYB	3 Vidicons	Relay optics
	Philips	PC80	RGB	3 Plumbicons	Zoom + prism for RGB
4	RCA	TK42	YRGB	Image Orthicon + 3 Vidicons	Zoom + prism for Y + relay for RGB
	General Electric	PE350	YRGB	4 Plumbicons	Zoom + prism for Y + relay for RGB
	Marconi	Mk. VII	YRGB	4 Plumbicons	Zoom + relay optics for YRGB
	EMI	2001	YRGB	4 Plumbicons	Zoom+prism for YRGB. ΔL-correction

Y: Luminance; R,G,B: Red, green, blue

linear characteristic of the Plumbicon tube demands accurate limiting circuits, in order to avoid overloading on highlights. Three separate circuits operate at suitable intervals in descending order, thus preventing objectionable smearing and streaking effects in the camera output.

The four-signal output is processed to three signals as described below.

The reference signal for colour balance adjustments when setting up the camera is the separate luminance signal, whereas for scan registration the signal used is the green signal, because green register is the most critical.

In other words, the red, blue and Y signals are adjusted in turn to register with the green signal.

The scan circuits have been designed without linearity controls. This avoids having to readjust linearity every time the amplitude control is adjusted.

One of the remote operational control panels enables the lens iris adjustment to be set for the correct light level in the camera. The other, a colour balance control panel, has two joy-sticks, one for adjusting the black levels of the three colouring tubes in varying proportions in order to match the darker colours of another camera, and the second for controlling the gains of the three colour amplifiers, so as to adjust the effective colour temperature. This is useful on outside broadcasts where the colour temperature of the sky is continuously varying.

SIGNAL PROCESSING

DELTA-L SYSTEM. The separate luminance camera has raised a number of difficulties concerning signal processing to suit the NTSC specification. The NTSC system departs from a truly constant luminance system in order to simplify receiver decoding. Thus the monochrome signal transmitted is $E'_Y = 0{\cdot}30E_R{}^{1/\gamma} + 0{\cdot}59E_G{}^{1/\gamma} + 0{\cdot}11E_B{}^{1/\gamma}$, which is not the same as a gamma-corrected true luminance signal $E_Y{}^{1/\gamma} = (0{\cdot}30E_R + 0{\cdot}59E_G + 0{\cdot}11E_B)^{1/\gamma}$ such as is obtained from a separate luminance tube. The signal E'_Y cannot be obtained from a single pick-up tube.

In American four-tube camera television broadcasts the practice has been to ignore this distinction and to be satisfied with the colour errors introduced by using $E_Y{}^{1/\gamma}$ instead of E'_Y demanded by the signal specification. (A certain amount of correction can be obtained by suitable choice of spectral characteristics.)

While these errors are admittedly of a minor nature it is desirable to correct them, as may be done by a technique called ΔL-correction applied to the four-tube camera. A luminance-difference signal ΔL is formed from the RGB signals (from the colouring tubes) by matrixing techniques; thus $\Delta L = E_y{}^{1/\gamma} - E'_y$, where the suffix y indicates that these signals are derived from the relatively low resolution colour tubes. The signal ΔL is then subtracted from the separate luminance signal $E_Y{}^{1/\gamma}$ to make a monochrome signal of the form required for the NTSC specification.

The luminance-difference signal ΔL can be generated from the E_R, E_G and E_B signals by using normal gamma-correction circuits or by a technique of electronic approximation known as multi-linear matrixing. This technique makes use of stable amplifiers and stable limiting circuits and once set up does not need adjustment.

FOUR-SIGNAL TO THREE-SIGNAL PROCESSING. As four signals are derived from the four different pick-up tubes in the camera, the philosophy has been to convert the four signals to three signals, so that the picture can be monitored on a standard RGB monitor.

This technique is not only useful in broadcasting but also for closed circuit industrial applications. For example, the four-signal to three-signal matrix produces three signals R′G′B′, which can be applied directly to RGB monitors and standard colour encoders. The matrix ensures that colour fidelity is not degraded by the use of a separate luminance tube and in effect the signals derived from the camera employ the Delta-L process.

The narrow band, R, G, and B signals are matrixed to produce a signal $y = 0{\cdot}3R + 0{\cdot}59G + 0{\cdot}11B$, which is then gamma-corrected to form $y^{1/\gamma}$. The gamma-corrected separate luminance signal $Y^{1/\gamma}$ and the gamma-corrected $R^{1/\gamma}$ are then matrixed with $y^{1/\gamma}$ to form the red output signal $R' = Y^{1/\gamma} - y^{1/\gamma} + R^{1/\gamma}$. Since at low frequencies $Y^{1/\gamma}$ and $y^{1/\gamma}$ are equal, $R' = R^{1/\gamma}$ plus the high frequency components of $Y^{1/\gamma}$. The green and blue signals G′ and B′ are formed in similar matrixes.

CONTOURING (EDGE ENHANCEMENT). The resolution of an accurately focused camera is determined by a number of factors, such as the resolution of the optical lens system, the resolution of the target structure of the pick-up tube, the latter's electron beam resolution and the bandwidth and flatness of response of the amplifiers, etc. The overall horizontal resolution of a channel (excluding the display monitor) can be assessed by measuring the percentage modulation of the channel output corresponding to an input test pattern of vertical stripes but of varying spacings. The stripes are normally black and white, but to obtain the modulation transfer index (MTF) sine wave stripes are preferred.

The horizontal resolution can be improved by providing the amplifier with a network having a rising frequency characteristic reciprocally related to the modulation transfer index. For example, if the MTF is 50 per cent at a frequency of 4 MHz, then the amplifier gain should be increased by 2 (6 dB) at this frequency, giving an effective resolution of 100 per cent. An additional property of this network is that it should have a time delay which is constant with varying frequency, i.e., a linear phase delay.

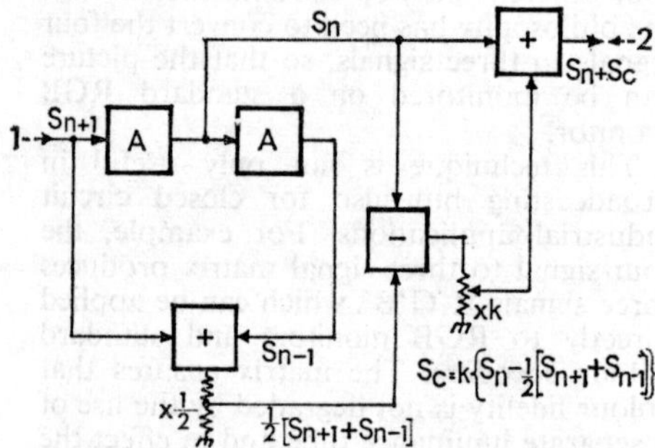

BLOCK DIAGRAM OF APERTURE CORRECTOR. Input signal at 1 is S_{n+1} which passes into a time-delay unit A, producing delayed signal S_n which is passed into further time-delay unit A, the output signal of which is S_{n-1}. Original signal and twice delayed signal are added and then reduced in amplitude by a factor of 2, giving a signal $\frac{1}{2}[S_{n+1}+S_{n-1}]$. The latter signal is subtracted from the once-delayed signal S_n to form correction signal S_c. Correction signal is added to signal S_n to produce aperture corrected signal S_n+S_c at 2. A potential divider is provided for controlling the amount of correction. The term, k, represents fractional amount of correction signal.

A network for achieving this so-called aperture correction employs two time-delay lines, the delay of each being approximately a picture element duration (say 0·1 microsecond). By delaying the main output by one unit of time and subtracting from it two signals of reduced amplitude, one undelayed and one delayed 2 units, it is possible to

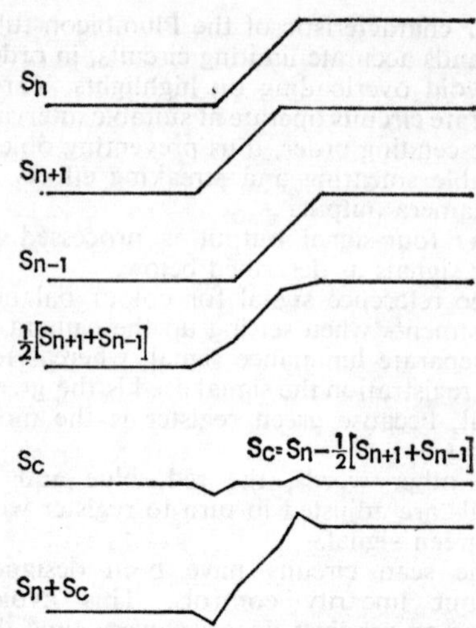

APERTURE CORRECTOR WAVEFORMS. Stage-by-stage modification of waveform. It is assumed that input signal S_{n+1} is a ramp transition from black to white; output signal, S_n+S_c, represents enhanced version of transition.

produce an output signal of which the resolution is enhanced. In practice a single delay network is used.

An analogous form of aperture correction to improve the vertical resolution of the channel was devised by Gibson and Schroeder but uses unit time delays equal to the line scanning period (64 microseconds for 625 and 63·5 microseconds for 525 lines). By this means the increase (say) in brightness of one horizontal line of the picture is effectively increased visually by decreasing the brightness of the preceding and following lines. In practice, interlacing of the scanning lines causes the enhancing effects to operate over the next-but-one line, and this is equivalent to having the peak correction corresponding to 312½ lines rather than to 625 lines.

By combining the above two networks, both horizontal and vertical aperture correction (or contouring) is obtained and a considerable enhancement in the picture sharpness is evident, particularly with pick-up tubes having relatively low resolution, such as the Plumbicon. In the case of a separate luminance type colour camera, the natural place to connect such a contouring unit is in the separate luminance signal amplifier: there is no point in enhancing the colouring signals.

In the case of a three-tube camera a technique known as contours-out-of-green has become popular, since it not only increases the resolution of the green signals, but is used to enhance the overall picture by generating grey scale transients which improves the allowable tolerance in misregistration in forming the luminance signal. This effect is achieved by taking the green contour signal and adding it to the red and blue signals in addition to the green signal.

FLARE CORRECTION. All lens and optical systems suffer from flare, which represents a worsening of the picture contrast caused by light scattered from the various surfaces, e.g., lens surfaces, dichroic surfaces and filter surfaces. Light from the lighter parts of the scene tends to scatter and this adds to the darker areas, thus giving an effect of haze over the picture. Great pains are taken by optical manufacturers to reduce these effects by carefully coating the surface with anti-reflection layers.

An additional source of flare is found in the Plumbicon tube, resulting from the light scattered by the target layer itself. This effect is reduced to a minimum by making the glass front surface as thick as practicable by the addition of a glass stud, increasing the normal thickness from about 2mm to about 8mm. However, it is found that the total amount of flare is still excessive, showing itself as a need to reduce the black (set-up) controls, when the average light in a picture increases, going from a low key picture to a high key picture.

An automatic form of flare correction consists in outline of an integrating network, which measures the average picture content over a field period or so and then generates a pulse, the amplitude of which is approximately proportional to the mean level. The pulse is added to the pick-up tube signal during the line clamping period, so that the black level is automatically reduced as the flare increases. Since the flare in the red Plumbicon tube output is greater than in the blue tube output, because the absorption of red energy by the layer is less for red than blue light, it is usual to incorporate flare correction for each colour and to provide more correction for the green and red tubes than for the blue tube.

COMPARISON OF THREE-TUBE AND FOUR-TUBE CAMERAS. Because a three-tube camera has one less tube and fewer associated circuit components than a four-tube camera its capital cost is less, fewer wires are required in the camera cable and a slightly smaller and lighter camera is possible. The advantages derived from the extra tube and components are: better resolution, increased tolerances to misregistration and reduced lag, leading to improved performance and lower operating costs. The misregistration permissible in the three-tube camera is approximately $\pm\frac{1}{4}$ of a picture element, whereas in a four-tube camera it is about $\pm\frac{1}{2}$ a picture element for the colouring tubes and approximately ± 1 picture element for the luminance tube to the green tube.

CIRCUITRY

SOLID-STATE COMPONENTS. All colour cameras employ transistors, semiconductor diodes and other solid-state components, especially in monolithic and thin film forms. These provide a much better performance than valves, and the solid-state design gives greater stability, less power consumption, reduction in size and improved reliability. Liberal employment of negative feedback in the amplifier chains not only stabilizes the gain, but ensures excellent linearity.

In gamma-correction circuits, semiconductor diodes lend themselves to accurate law matching and similarly, transistors facilitate accurate d.c. clamping, and limiting circuits; these are particularly necessary with pick-up tubes such as the Plumbicon, which have linear output characteristics and therefore tend to produce high output signals on specular reflections. Semiconductor diodes and also opto-electronic devices are extremely useful in gain-control circuits.

Shading correction is required with vidicons and image orthicons, but is unnecessary with the Plumbicon, which has negligible dark current.

CABLES. Special camera cables developed for four-tube cameras contain approximately 100 wires. Two types of camera cable are in use, one of light construction in relatively short lengths for studios, and the other a heavier type for running in ducts and for longer distances. As transistor circuits require lower voltages and higher currents than valve circuits, excessive voltage drop may occur, and in long cables may be as much or more than the terminal voltage. To counteract this difficulty the terminal voltage of the camera has to be measured via a separate lead in the cable, so that the voltage output from the power supply stabilizer compensates for the voltage drop in the cable.

Solid-state circuits also contribute to short warm-up time and stability of registration in the scanning circuits, and thermistors are used for temperature compensation.

Pulse supply arrangements are undergoing rapid development; there is a trend towards a phase-lock loop for timing the scanning relative to the blanking pulses, thereby automatically correcting the delay of different lengths of cable. Encoder performance is improved also by being transistorized and hot-electron diodes have advantages in modulators because of their reduced hole-storage hangover.

FUTURE DEVELOPMENTS

Future developments depend on the introduction of smaller pick-up tubes such as the 1 in. diameter and half-size Plumbicon tubes, with electrostatic focus and magnetic deflection. The elimination of the focus coil saves considerable size and weight, and permits greater flexibility in design of optical systems. An even more useful tube in these respects would be the fully electrostatic tube with electrostatic focus and deflection, since no external coils would be required.

To take full advantage of smaller tubes, lens dimensions will have to be reduced, with a consequent reduction in effective aperture. To maintain an equivalent sensitivity, it will be necessary to develop even more sensitive pick-up tubes.

Another area for development is in camera cable utilization. Ideally a single co-axial cable is adequate to operate a camera, if suitable multiplexing techniques are adopted for signal and control transfer.

I.J.P.J.

See: *Camera* (TV); *Colour principles.; Lens performance; Lens types.*

Books: *Principles of Colour Television*, Eds. K. McIlwain and C. E. Dean (New York), 1956; *Colour Television Engineering*, by J. W. Wentworth (New York), 1955; *Colour Television Explained*, by W. A. Holm (London), 1963.

□**Camera Cable.** Connecting cable between the television camera and the camera control unit to provide the various potentials to the camera tube, and to return the vision from the tube to the camera control unit for processing.

The camera cable has to be able to stand sharp curves, and stranded conductors are preferred to give improved flexibility. Very thin insulation is seldom successful under practical conditions, because, in a broadcast studio, the cable must withstand being run over by camera dollies, and for outside use it must be waterproof. It is customary for the conductors to be covered by a tough sheath of rubber, sometimes a double sheath.

A typical cable consists of 18 control cores, 3 coaxial cores, 3 screened quads (groups of four cables), and 6 conductivity wires, with screened braiding and oversheath. For broadcasting, lengths up to 1,000 ft. or 2,000 ft. are used, depending upon the type of camera, and even these figures are exceeded on the latest cameras.

Difficulties arise with such long lengths of cable: the voltage to the camera tube heaters drops, and there are h.f. losses in the video signal. Compensation is provided within the camera control unit to overcome these losses and high voltage transformation techniques are now used to avoid the heater voltage drop.

At the present time, a typical cable able to withstand rough usage has a diameter of approximately $\frac{7}{8}$ in., and is available in lengths up to 2,000 ft. in multiples of 100 ft. Typical signals carried by individual conductors within a modern broadcast camera are as follows: mains line; mains neutral; utility mains line; focus; high tension; decelerator; beam focus, focus mod. alignment; EHT; HT negative; talk back and line scan reverse; iris motor; align X; align Y; video out and screen; viewfinder video and screen; line drive; target blanking; line drive to CCU; image focus; camera field scan; beam control; viewfinder field scan; multifocus; programme sound; camera reverse talk back and cues.

Cables are normally provided with a protective screw-on cap to prevent damage to pins during transit.

See: *Camera* (TV); *Camera (colour).*

□**Camera Control Operator.** Operator of the main variable controls of a camera channel, which are brought out to panels in the control room. Formerly the control operators had a wide choice of adjustments and usually were well occupied as lighting levels varied greatly. Modern techniques of setting up a camera channel and leaving it at its optimum working point reduce the necessary controls to iris, channel gain, and lift. Accordingly, there are fewer control operators and nowadays they are more concerned with touching up the picture and picture matching.

See: *Cameraman* (TV).

□**Camera Control Unit (CCU).** Equipment in a television camera channel which contains the major part of the electronic circuitry necessary to operate the camera channel. It processes the synchronizing signals and vision signals into a form suitable for use within the studio or control room, and provides operating signals for the camera tube and viewfinder. In the case of broadcast channels, it is normally associated with a separate power supply unit, and remote camera control panel, but for industrial and closed-circuit use, the whole is usually contained within a single unit.

Camera control units may be mounted in control rooms or, in a large studio complex, may be gathered together into a central apparatus room.

See: *Camera* (TV).

⊕CAMERAMAN

The work of the cameraman consists of transforming ideas in the mind and words on paper into images on film. This process has three pre-requisites: technical knowledge, artistic sensibility and the ability to collaborate with other people in the creative process, both in accepting suggestions from others without rancour and freely giving one's own ideas for the collective end result.

The head of a production team or unit is the director, and all really good films bear his imprint first and foremost. But the cameraman is the vital link in getting the director's ideas on to the screen and can make or mar the final result.

PREPARATORY WORK

In the preparatory stages of a film, discussions usually take place between the director, the art director, the production manager and some other members of the unit to determine the visual style of the picture and to make practical preparations to enable the film to come in on schedule and budget.

The choice, for a feature film, of colour or black-and-white, and of wide screen or Cinemascope, are normally made by the production company on purely commercial grounds. The cameraman may be consulted, however, as to the merits of one wide screen system over another. Shooting on 65mm, for example, would mean much heavier and bulkier camera equipment, leading to an additional expenditure of time and money. While the cameraman is usually free to choose the types of film stock and the processing laboratory, contractual arrangements are sometimes made between the production company on the one hand and the laboratory and/or raw stock manufacturer on the other which cover all the company's output and preclude any choice on the part of the cameraman.

MOOD. The overall aim of the cameraman is to create a mood and an atmosphere appropriate to the subject. In a thriller, for example, with people prowling about and hiding in corners, the suspense is lost if all is as bright as day; on the other hand, if the director wants a light romantic feeling in a comedy, it shouldn't look like Goya's etchings.

Visual styles have varied a great deal in the short history of the cinema, and are to some extent dependent on the types of film stock and the sorts of lighting units in current use, as well as on the fashion of the period. Thus the use of strong diffusion was common in the '30s, while sharp, contrasty photography, sometimes with deep focus, was in vogue in the post-war period.

STYLE. Photographic style is controlled by the character of the lighting, the granularity of the film stock, the skilful use of composition, and, if desired, by the use of specialized processing techniques.

The lighting of a film can be characterized, in part at least, by the terms high key and low key, the former signifying a high general light level, with shadows filled-in to light or medium grey and a low overall contrast. In low key work, the contrast is relatively high, shadows are allowed to go black or very dark and highlights are maintained at full brilliance.

A further variation of style is obtained by the use of reflected, rather than direct, light or by the use of very fast film stocks, enabling the cameraman to use available light with only the minimum supplementary photographic lighting units. This, combined with natural interiors rather than sets, gives rise to a distinctive style similar in atmosphere to newsreel or TV reportage. At the other end of the stylistic scale lies a glossy, glamorous technique, making use of spotlights under studio conditions, as seen in many television commercials.

COMPOSITION. Composition is largely a matter of individual taste, but governed by the classic rules of composition in painting modified by the additional factor of movement. It demands a fixed frame within which to compose, and the original proportions of 3 : 4 were close to the artist's conception of the ideal rectangle. With the introduction of wide screen and Cinemascope, this ideal proportion and often the fixed frame itself have disappeared, as compatible composition – a picture which can be screened on wide screen as well as later on television – is often demanded. An extreme compromise such as this rules out any effective composition and leaves the cameraman with a so-called safe area in which action is sure not to be cut off.

EQUIPMENT. Just before shooting begins, the cameraman tours the proposed locations, stating the numbers and types of lighting units he needs for each. Economizing here can cost dearly, for though lamp hire is expensive, being short of light at the time of shooting is a disaster. Also, just before shooting proper starts, tests are often made of artistes and of make-up and costumes to be sure that everything photographs as expected. Shortly before shooting any particular sequence in a studio, the cameraman is called on to the set being built and asked to determine the placing of the lighting units and rails as well as to prepare for any special camera or lighting rostrums that are needed.

MAKING THE FILM

Once shooting begins, the cameraman is the key man on the floor after the director. Not only the final look of a sequence, but also the time taken to shoot it, depends largely on him. He selects the camera position, or set-up, together with the director and his own second-in-command, the camera operator, and then lights the scene. The camera operator works out the details of the set-up and of any camera movement that has been decided upon, and operates the camera during the shot. The third member of the team is known colloquially as the focus puller, for his main function is to keep the shot continually in focus. He also changes the film magazines and is in charge of all the camera equipment. The fourth member is the clapper/loader who loads the film into the magazines, keeps the camera notes or sheets, and announces the scene number before clapping, i.e., producing a synchronized audible and visible signal with the clapper board. The last member is the grip, who manoeuvres the dolly during the tracking shots and does most of the heavy work in getting the camera into position.

SETTING UP. Usually, the shooting script is written not in terms of shots but of master scenes which must be broken down on the floor into actual set-ups for shooting. This work, sometimes called the découpage, is done between the director, the cameraman and the camera operator. The director indicates how he wants the action to be staged, and the characters to move, and the other two make suggestions as to the most effective and economical way of covering this in set-ups and camera movement. It is at this stage that the scene as it finally appears on the screen is created.

Any given scene can be staged in many ways for the camera – in long shot, medium shot, over-shoulder shots and close-ups, for example, or in one continuous tracking and panning shot, to name just two possibilities. Here the director's as well as the cameraman's style shows itself, while the final pattern may well derive from the operator's suggestions.

In any situation where there is much work to be done on the lighting, the cameraman is preoccupied with planning this and often leaves the details of a set-up to the director and the operator, keeping an ear tuned to their deliberations, however, to make sure the total effect blends well and is feasible from the lighting point of view. Choice of lens, however, is the cameraman's prerogative, and this is a very important factor in any set-up, governing as it does the perspective of a shot, as well as the relative sharpness of the background.

Decisions taken at the preparatory stage about the visual style of the film now have to be carried out in detail, making sure that the problems of detailed execution do not put out of mind the original overall design. Every single scene or sequence presents so many varied problems of staging and lighting that it is only too easy to forget what follows and what precedes it. But it is vital to maintain continuity of lighting mood and style with scenes that may have been shot weeks before in a totally different place, or that have not yet been shot at all. "Every frame a Rembrandt" is therefore not a satisfactory ideal to aspire to in motion picture shooting.

INTERIOR LIGHTING. The problems of interior lighting can be grouped roughly into two sections: lighting the set, and lighting the artistes. The latter is as much a question of psychology as of photography and considerable diplomacy has often to be exercised to maintain good relations on the floor, as some of the stars can be very demanding. In studio work, the normal procedure is to rough in the lighting of the set first, if it is a large one, then to rehearse the first sequence with the actors. During this time the principal positions of the actors are marked on the floor, after which the stand-ins take over; this enables the cameraman to light figures and faces without tiring the artistes.

Source lighting, i.e., recreating the light that would normally fall on a scene, say from windows or lighting fixtures, is the basic pattern employed by many cameramen for the first stage of set lighting. In the second stage, the stand-ins are used to light all the principal positions of the artistes in the scene, and the special lighting for these positions has to be blended into the general set lighting, so that no-one appears to step into a spotlight, as it were, on cue to speak his lines.

The whole effect must also harmonize with the decor, the atmosphere, the time of day and type of scene. Furthermore, to avoid waste of time later, lamps must be so placed for the long shots that at least some of them will be able to function for mid- and close-shots with little or no adjustment.

Thus it is important for the cameraman to know how the entire sequence is to be staged before starting to light the first shot – otherwise he may find he has worked himself into an impasse by placing, for example, an important key light in a part of the set that will be seen immediately afterwards. All the time the lighting is being carried out, the cameraman goes back and forth to look through the camera, by now positioned for the shot by the operator, to make sure that the pattern he has been building up is really effective from that viewpoint.

CLOSE-UPS. When lighting close-ups, the face of the player becomes, so to speak, a landscape which needs very subtle control of the height, strength and angle of the lamps to bring it out to best advantage. Hours can easily be spent lighting a face, starting with the stand-in for the roughing-in process, then continuing the more subtle adjustments with the actual artiste, as no two faces are sufficiently alike for this delicate process. Diffusion filters or gauzes are normally employed on close-ups of women and sometimes on men also. Particularly difficult are close-ups that come as the climax of a track-in which starts as a general shot, for then the most advantageous lighting positions can often not be used as the units themselves would be seen in the earlier part of the shot. So for best results on extreme close-ups it is necessary to make these a separate set-up. Good composition, too, is very important in close-ups, and by no means achieved by just carefully centering the subject, as is so often advocated. But the compatible composition mentioned earlier makes effective close-up composition even more difficult.

NATURAL INTERIORS. In a studio, lamps can normally be placed anywhere the cameraman wishes – the set is ringed with lamps of his choice, and it is often possible to place lamps on the floor and hide small ones behind pieces of furniture and even behind actors.

In natural interiors, however, placement of lamps can become a great problem, particularly if the room being used is small and low-ceilinged. In these circumstances, it is essential not to try to turn the place into a miniature studio, but to adapt to, and make good use of, local conditions. By using the natural view out of the windows and taking advantage of the fact that walls cannot normally be knocked down for unusual camera angles, such as behind bed-heads or sets of shelves, a greater feeling of realism is often obtainable from such locations, which tend to impose a style of their own.

It is important that a fast film stock be employed, so that lamps can be as small as possible. Camera movement being more restricted than in a studio, the understanding of the director is essential, and cameras, dollies and tracks should be as light and small as possible. Windows are normally covered with filter gelatines or tinted perspex, and must be changed according to the light levels outside.

STUDIO EXTERIORS. When exteriors have to be simulated in a studio, problems of quite a different nature arise. Here the aim is for the artificial garden, square or street to look real, and this is almost entirely up to the cameraman. However well these sets have been built, there is always a feeling of unreality about them as first seen under the working lights of a stage.

Source lighting is the principle once again employed here, and sun and skyshine are simulated as well as possible. The appearance of multiple shadows, never present in real exteriors, is particularly difficult to avoid altogether, and the reproduction of an overcast day is much more difficult than that of sunshine, since all lighting units throw shadows of some sort, and a great deal of light is often needed to cover the area of these exterior-interiors. Where a garden or yard is attached to an interior on a set, the cameraman must create the feeling that inside is inside and outside is outside, and do so within the acceptance range of the film stock, which is considerably smaller than that of the human eye.

DAY EXTERIORS. Good day exteriors depend quite a lot on the correct choice of the time of day for shooting. This is often difficult, owing to other schedule requirements. During the preparatory stages of a film, when the cameraman tours the proposed locations, he indicates at what time of day he considers them best lit. This is more critical in sunshine than in cloudy weather. Use of arc lamps, which are heavy and bulky, is still favoured for fill light in day exteriors, although banks of photofloods or tungsten halogen lamps can usefully serve in this capacity too.

Once exterior shooting has begun, the cameraman is consulted each day as to which sequence is best to work on, according to the weather and continuity requirements, and he is often expected to be a bit of a weather prophet. He is also consulted as to the order of the shots to be taken for best use of available time and light. Since the use of lamps is more limited on exteriors, the cameraman has more time to work through the camera itself, and controls his effects mainly through composition and the use of filters.

Lighting continuity is a great problem in Britain, where there is no climate but only samples of weather. Too much importance has been given to this factor in the past, however, as it is usually possible to disguise changes of light, if not too drastic, by devices of cutting and of camera set-up; the only thing which really has to be avoided is the continual intercutting of shots of greatly differing contrast. Sun and dull light, however, can be intercut, providing that a little care is taken in the planning of the sequence and that violent light changes are bridged by close-ups or cut-aways. It is very important for the cameraman to know how many shots a sequence will comprise, and what precedes and follows it, so that he may bridge effectively from day to night, say, and match interior and exterior conditions.

Dusk and dawn sequences are particularly difficult, but often very effective, and impossible to fake at other times of day with any degree of verisimilitude. Again, the sequence must be carefully planned in advance, bearing in mind that the suitable light usually only lasts from 30–45 minutes, less in more southerly latitudes. Balancing lights have to be placed in or near their positions well in advance, and diffusers or gauzes kept handy by each lamp to vary its strength as the daylight falls or rises. As in day-for-night shooting, actors in mid- and close-shot are lit to enable the laboratory to print the background sufficiently dark without losing important detail in the faces.

EXTERIOR LIGHTING. Night exteriors can usually be lit in the same way as natural interiors, except that larger units further away from the subject have to be employed. In countryside locations, such as fields and woods, far from natural sources of light other than moonlight, the lighting has to be particularly unobtrusive as scenes of this kind, when too obviously lit, tend to look artificial. On the other hand, sufficient light must be present for the audience to follow the action without difficulty.

Filming this type of scene, "day-for-night" (i.e., using daylight to produce a night effect) is often therefore the best solution. It is also usually considerably cheaper, bearing in mind that true night shooting involves overtime and the hiring of large lamps. In black-and-white, day-for-night shooting is best accomplished in sunlight using the Wratten 72B filter which adequately darkens any blue sky included. Where no sunlight is available, it is best to exclude the sky, underexpose shots by up to two stops, and notify the laboratory that a night effect is wanted. Actors in medium-close shot and close-up usually have to be lit for best effect in these conditions, in order to separate them properly from the background. Reflectors or arcs are normally employed for this purpose, the latter being preferable as it is very hard for an actor not to squint when a reflector is directed fully at him.

SHOOTING IN COLOUR. Colour films call for the greatest possible degree of collaboration between director, cameraman and art director to achieve a pleasing aesthetic colour style. Natural colour, that is as near to life as possible, is difficult to achieve as the raw stocks themselves exaggerate colour so that dark colours render darker and light ones lighter than in life. This gives an effect of heightened contrast, which is further increased by the presence of a black surround to the colour picture as seen in the cinema, and tends to give a certain gaudiness to the colours. Pastel effects are therefore particularly difficult to achieve. The best results are usually seen in studio scenes, where the colour of the sets as well as of the costumes can be carefully controlled and verified by tests before shooting starts.

Outdoor scenes, on the other hand, present great difficulties, if natural or pastel colour is desired, since grass and trees always photograph greener and sky bluer than in nature. Shooting in overcast lighting conditions softens this effect considerably, but these conditions cannot be produced at will. So to soften the colour it is often necessary to employ diffusion filters or gauzes, which cut down the contrast, as it is not possible to modify this in the laboratory, as can be done with black-and-white. In additive printing systems it is possible to vary, in the grading stage, any one of the primary colours without affecting the others. If, however, the colour grading is done by means of filters overall, no colour can be corrected without materially affecting also its opposite. Low contrast print films have now been developed for television use, but these do not yet provide a really satisfactory solution.

Colour film is somewhat slower than black-and-white and needs more light for comparable scenes. Modern colour films are, nevertheless, reasonably fast and do not require the players to be roasted under closely packed arc lights, as in the early days of colour.

It is normal to keep the contrast ratio somewhat lower than for black-and-white filming, a maximum ratio of key to fill light of 4 : 1 being advocated. These rules must be interpreted imaginatively, however; many effect shots in colour employ a contrast ratio similar to black-and-white. Colour negative allows more latitude here than colour reversal film.

In some ways, use of colour makes the cameraman's job easier, as two of the prime pre-requisites for black-and-white photography – good modelling and separation – are virtually taken care of by the colour itself.

In colour day-for-night shooting, the convention is to use back light, underexpose and instruct the laboratory to print the scene with a blue tone. This derives from a stage convention for moonlight and has little to do with filmic realism. No satisfactory solution has yet been found for this problem, as the monochromatic effect closest to realistic night conditions is not obtainable by the use of existing filter combinations; but further individual tests could be useful in providing an acceptable solution in any particular instance.

Dramatic and aesthetic use of colour in films is still in its infancy, and the cameraman can do much to bring this medium to more effective application if he is consulted in good time before the start of a production.

DOCUMENTARY FILMING. This now covers a very wide area indeed, and cameramen trained in this field usually have a great range of experience to call on if they move over to feature films. On the other hand, a cameraman used exclusively to the luxury of the full facilities of a studio would often find himself in great difficulty on a documentary location, where objects have a habit of being large and immovable and no easy overhead arrangements are available for lighting. Common sense and the ability to adapt to difficult and varying conditions are therefore especially necessary for documentary work, which can range from scenes similar to those in a feature film, employing sets and actors, to highly specialized scientific and medical subjects calling for continuous close collaboration with an expert in these fields.

Where people are not principally involved as subjects, the difficulties usually centre around the placing of the film equipment. There must be no undue interference with the process of production or study being filmed, yet the film must give real insight into the heart of the process itself. On documentaries of this type, the cameraman is normally his own operator and works with one assistant. The whole team consists of four or five people.

Wherever possible, equipment is small and light so as to speed up preparations as much as possible. This is particularly important where people are being filmed at work, so that some degree of naturalness remains after the preparations are complete and filming begins. By means of small hand-held 16mm synchronous cameras, it is possible to film many subjects with the minimum of fuss and thus to avoid the greatest problem in documentary filming of this sort, namely that the subject tends to get tired, bored or even vanish altogether before the cameraman is ready to press the button.

Where this technique is used for TV reportage, the cameraman virtually becomes his own director as well, as there is often no time for the latter to check through the camera. The technique of using hand-held 16mm cameras, fast film and available light has made possible a degree of intimacy and immediacy in synchronous sound documentary shooting such as has never before been achieved.

The possibilities are even applicable to features, although this field is as yet little explored.

POST-SHOOTING DUTIES

After the picture is shot and finally edited, the cameraman is usually called in to the laboratory to supervise the grading of the shots in the first married print. Having already supplied a guide sheet to the grader, he then sits with him and views the first print, making suggestions for modifications as he goes along. In general, only adjustments of density (brightness) can be made at this stage, as the contrast is fixed by standard laboratory practice and can be altered only in the reel as a whole. If higher or lower contrast is desired for a particular sequence, this must be achieved in the shooting and negative developing stages. Colour grading is further complicated, of course, by the problems of matching colour tones from scene to scene, and here it is best to aim for overall balance, rather than trying to match the colour of individual objects (such as grass or the sky) from shot to shot.

Finally, it would be ideal if the cameraman were able to make spot checks on release prints to make sure that quality is being maintained once the film has passed into the hands of the distributor, but this is only rarely possible. By this time he is often already away on a new assignment, and even the grading itself sometimes has to be supervised by the director or editor in his absence.

Although the camera itself continues to improve, it is the vision of the man who uses it which makes the difference between a mechanical record and an imaginative interpretation of the subject.

W. L.

See: *Camera; Cameraman* (TV).

Book: *Technique of the Motion Picture Camera*, by H. M R. Souto (London) 1969.

□CAMERAMAN

The television cameraman is a specialist in a specialized industry: the techniques of his craft are geared to the particular problems of television. A comparison between the work of the television cameraman and the film cameraman, however, is useful – if only to illustrate the differences between them.

FUNCTIONS AND QUALITIES

In motion picture work, the lighting cameraman is responsible for lighting each scene, and ultimately for the overall quality of the pictures taken with the camera. The camera operator in films performs the mechanics of camerawork (framing, panning, tilting, etc.), and is usually provided with a focus puller who adjusts the focus on each shot where necessary.

Television cameramen, on the other hand, have nothing to do with lighting a scene – this is the responsibility of the lighting supervisor, or whatever title he is given from studio to studio. The set is lit for the cameraman, and he has little control over picture quality, which is the job of the vision control supervisor who works in close liaison with the lighting supervisor. The cameraman's contribution to picture quality is confined to his duty to place his camera for each shot within the range of the existing lighting arrangements.

The television cameraman's work is similar, in some ways, to that of the camera operator in films. But there are major differences. The television camera is a large, box-like piece of electronic equipment having multiple lenses and an electronic viewfinder. It contains no film, which is a disadvantage in one sense since the stock in a film camera can be changed to meet varying light conditions.

Because the television camera requires a comparatively high range of minimum light levels to produce an acceptable picture, the cameraman must often operate with his lens irises set to between $f5{\cdot}6$ and $f8$, where there is a fairly narrow depth of field, and accurate focusing is essential. He must make all adjustments of focusing himself – he has no focus puller – and he consequently carries more responsibility for good camerawork than his counterpart in films.

Another important point is that the television cameraman has no second chances when his programme is on the air. A shot or a camera movement which is not to his or the director's liking cannot be retaken time after time until it is perfect. Instead, he must develop his technique to the point where he can perform the intricacies of television camerawork to perfection.

During a television play, for example, one cameraman alone may have to frame as many as 200 to 250 shots, each one exactly as rehearsed, in a categorized order, and from different positions on an overcrowded studio floor. Further, the television programme is a continuous affair, and these movements and changes of shot must be made in the space of perhaps fifty minutes' air time, and integrated with the work of other cameramen (each with a similar number of shots to present), sound operators, stage hands, artistes, and so on.

PLACE IN THE TEAM. Because the television cameraman is denied the comparative luxury of time, he cannot line up each shot carefully, taking and retaking until it is satisfactory. Instead he must work at great speed, framing and focusing instinctively with a perfection that leaves little room for human error.

For these reasons it is often thought that the television cameraman is the most important technician on the operational side of television. This is also probably due to the fact that the cameraman operates what appears to be the most important piece of television equipment – and certainly the camera *is* television. But it is a mistake to credit the cameraman with such importance.

He is no more and no less essential to a production than those other technicians whose function it is to support the work of the cameraman – vision control operators, maintenance engineers, trackers, and so on. Further, every artisan on the studio floor is vital to a television programme, and the duties and skills of each section must be co-ordinated to produce a composite, professional result.

The cameraman, then, must adjust his technique to provide for the requirements of the boom operator, lighting supervisor, designer, director, etc. They, in turn, are restricted by the limitations imposed on them by the cameraman.

Television is a constant battle to equate the possible with the impracticable, and because of this all sections have evolved individual techniques which allow them to present a high standard of work within these limitations. However, owing to the very nature of his occupation, there can be no doubt that the cameraman carries the greatest responsibility for combining highly professional skill with cast-iron team work.

DUTIES AND RESPONSIBILITIES

To say that the cameraman's first duty is to be technically proficient might sound a little obvious, but it is a fact nevertheless. Television camerawork is highly specialized, and a technically incompetent operator, apart from lowering the standard of camerawork, becomes a constant embarrassment to the efforts of other sections.

TECHNICAL KNOWLEDGE. The cameraman's knowledge must extend over a wide field. He should have a grasp of basic optics, for his work is dependent on the principles of the behaviour of light waves.

He must have an understanding of lenses. Each lens has a characteristic of its own – an individual function for which it was designed. There is a right and wrong lens for every shot the cameraman takes, and he must select the right one instantly and infallibly.

He must be conversant with the depth of field, the perspective qualities, and the degree of distortion inherent in every lens he might use. He must know, instinctively, the angle of view accepted by each of the focal lengths of the lenses on his camera. He will often be called upon to change from a wide shot to a close-up in seconds, and he must select the right lens without delay and reframe the next shot with a minimum of effort.

Although some technical knowledge is essential, the cameraman does not need to be familiar with the intricate details of the components of his camera. He must know what happens when he operates a certain control, but he need not know why. The television camera is a complicated piece of equipment, and the cameraman's knowledge need only extend to the flexibility and limitations it offers him.

The basic function of the camera is to convert light waves into electronic impulses. The electronic equipment necessary to achieve this is intricate, and its design and maintenance is a specialized science. For this reason alone it is advisable that the work be left to experts – it is more important that the cameraman concentrates on becoming an expert in his own field.

The television cameraman must learn, too, how to prevent damage to his camera (physically and electronically), and must co-operate with the vision control operator to maintain an electronic output of maximum quality.

Cameras are usually on some form of movable mounting. These range from the simple basic tripod with castor wheels, through the various types of pedestals and motorized dollies to the highly versatile cranes. The cameraman must be thoroughly familiar with every type of mounting he is called upon to operate. He must know just what forms of camera movement he can perform with each, and the restrictions imposed on him by the peculiarities of individual models. He is responsible for ensuring that his equipment is maintained in good condition, for without smooth-working equipment competent camerawork is impossible.

The electronic impulses are usually fed through a cable from the camera to a camera control unit, where they can be monitored and electronically adjusted to optimum quality by the vision control supervisor.

The camera cable, necessary though it is, can be a hindrance to the cameraman, but it must be taken into account when planning and executing an involved camera movement.

It is common on a fast moving programme, which utilizes perhaps three or four cameras engaged in a large number of camera movements, to delegate two or three assistants to the all-important task of clearing cables from the path of these moving cameras.

PRACTICAL CAMERA WORK. The art of good camerawork embraces more than technical knowledge and skill. Familiarity with equipment and the ability to manipulate it to a high degree of performance merely provide the cameraman with the means to express himself. And in expressing himself he must aim for the ideals which represent his major duty. His work must be artistically pleasing and unobtrusive.

However, because the television cameraman must compromise with many of his shots, there is the ever-present possibility that he might allow his standards to fall below professionally acceptable levels. This is something the cameraman must guard against.

If his work is to be artistic and unobtrusive he must be acquainted with the principles of composition, and the niceties of framing pictures in studio conditions. Any reasonably intelligent person can be taught the principles of composition. But in applying them, the cameraman must have a sympathy for the mood of the production and an instinct for interpreting the director's intentions.

He must, then, apply his knowledge of basic principles together with his skills in other departments (camera movement, lenses, etc.) to help the director realize the dream which is the programme script.

CREATIVE RESPONSIBILITY. Those engaged in television are continually striving to create a feeling of reality in artificial surroundings. The viewer's attention is vital to the success of a production, and anything which breaks his concentration is bad technique.

If the director has planned a sequence which terminates in a dramatic close-up of an artiste's face, and that close-up is out of focus, or shot from the wrong angle so that the artiste is in shadow, the impact is lacking. The viewer who might have been held by the dramatic pull of the sequence is subconsciously disturbed. This has been brought about by a fault of technique, and it should never be allowed to happen.

A cameraman who cannot hold his camera steady while moving it slowly towards an artiste, or who hunts for focus at a vital moment, is guilty of making his camerawork obtrusive. He must be a master of good technique, therefore, if he is to justify the responsibility imposed on him.

But there is a limit to what can be taught to a cameraman. Unless he is blessed with inherent artistic feelings and manual coordination, he can never be anything but a merely useful cameraman – and there are many of those. But the first-class cameraman contributes something extra to a production, albeit in a subtle manner. This involves a feeling for the mood of a sequence and thorough understanding of the director's intentions.

Some cameramen are born with the ability to absorb these feelings instinctively. But those not so fortunate acquire them to some extent by studying the motives behind every shot they are required to take. Why does the director want this shot from this angle? What is the overall mood of the sequence? If this actor is trying to create a feeling of loneliness, would a high angle shot contribute to the effect the director is seeking? These are the sort of questions which run continually through a good cameraman's mind.

He has a duty and a responsibility beyond the framing of a shot according to a set of rules. The cameraman must be the unobtrusive, artistic eye that mirrors the mind of the director.

WORKING WITH OTHERS

Because the camera is so important in television, there is the danger that the cameraman may come to regard his work as being of greater consequence than any other. But it is not.

Pictures without sound can be meaningless; without the benefit of carefully placed lights, they are flat and uninteresting. In the studio, the picture does not exist at all without sets and/or artists.

COMPROMISE. Because of the continuous nature of a television programme, all departments must compromise with one another to some extent. The sound balancer cannot insist on the ideal conditions which exist in a recording studio, for example. And in the same way the cameraman finds that he cannot always frame his shots according to the rules of good composition.

The headroom which a cameraman likes to allow for a given shot may have to be reduced to allow the boom operator to place his microphone near an artiste. An exciting camera movement may have to be changed so that the lighting supervisor can accommodate it within his lighting arrangements.

These are two simple examples of restrictions imposed on cameramen by other sections. The cameraman who attempts to resist such restrictions as these has no place on the studio floor.

But in co-operating with other sections the cameraman must be careful never to compromise to the point where his technique falls below standard. He should refuse to shoot close-ups of artists with a wide-angle lens, for example, because of the facial distortion which will result. Another means of taking the shot with the correct lens must be found, and it then becomes the responsibility of other sections to compromise.

On occasions such as these, much will depend on the cameraman's tact and approach. But if he is to maintain a harmonious relationship with other sections he must learn something of their problems. The cameraman cannot expect co-operation from a boom operator who has been asked to do something which is unprofessional – or even impossible.

UNDERSTANDING OTHER TECHNIQUES. Many cameramen utilize some free time in the studio by acquainting themselves with the operating techniques of other departments. This must be done under supervision, of course. But until a cameraman has attempted to operate a boom under typical studio conditions he cannot hope to appreciate the special problems of the boom operator.

These opportunities are not always available, but cameramen learn much by discussing television with other operators. Many find that, if they are not operating a camera on a particular programme, a visit to the vision control room can be most rewarding – and revealing. It is surprising how careless or incompetent camerawork increases the problems of the vision control operator, and directly affects the service given to the cameraman.

He must learn something of television lighting if his pictures are to do justice to the efforts of the lighting supervisor. He needs to be sympathetic to the special problems of the artists in the programme, and do nothing to detract from their performance.

This overall realization of what is happening during a television production is essential if the cameraman is to integrate with other sections – indeed, it is necessary for his individual success as a cameraman.

In addition he must be consistent in his technique. If he allows a certain amount of headroom on each shot during rehearsals, he must allow the same amount during transmission in fairness to the boom operator. If he moves from one set to another in twelve seconds when on the air – instead of eight seconds as rehearsed – he can set in motion a series of errors which may jeopardize the success of the programme.

Many directors are more concerned with the success of the visual side of their programmes than other aspects. It is the cameraman's responsibility to provide the right visual content without interfering with the equally professional efforts of his colleagues in other departments.

USE AND CARE OF EQUIPMENT

The cameraman must have good tools if he is to be responsible for first-class camerawork and produce artistic pictures. Yet he has surprisingly little control over much of his equipment.

Except in the smaller studios, cameras are maintained by qualified electronic engineers, and camera mountings and dollies are the responsibility of mechanical engineers. The quality of the pictures emanating from television cameras is determined largely by the abilities of the vision control operator and the lighting supervisor.

The cameraman depends on these other sections to provide him with a service; without their ministrations he is helpless. But he must be aware of the optimum conditions he is entitled to, and insist on them.

These conditions have become fairly well

defined over the years, and although cameramen will always argue as to what is the ideal camera or dolly, a majority vote would probably lay down the following requirements.

CAMERA. The camera should be light and easily movable. It must retain some bulk, however, since television cameramen seem to enjoy the solid feel of the camera. This is more than just a whim for they often have to move the pedestal by pushing against the camera. If the television camera weighed no more than the average 8mm cine camera, it would be impossible to maintain a steady shot during the movement.

LENSES. Lenses are usually the cameraman's responsibility and need handling with care. It is his job to ensure that they are always free from grease and dirt.

At least three lenses are desirable if the camera is to be a flexible piece of equipment, though four lenses are preferable. These lenses are mounted so that any one can be easily and quickly selected. Sound engineers would add that any good system of lens selection is also silent.

Lenses are normally mounted on a turret at the front of the camera, and the required lens is brought into the taking position by turning a handle which rotates the turret. Some cameras have been fitted with electric lens changes, but they have not been entirely successful. Cameramen generally prefer some form of manual change over which they can exercise full control.

The method employed in mounting the lenses on the turret is important. The mounting must be firm, yet so designed that lenses can be removed and replaced in a few seconds. It is also necessary to provide for supplementary and zoom lenses to be accommodated without modifications being made to the camera or turret.

PANNING HEAD. A camera is invariably mounted on a panning head, which in turn is fitted to the top of the dolly (pedestal, crane, etc.). The panning head has to be smooth in operation, having no flat spots which interfere with the cameraman's attempts at smooth panning and tilting. Once again, panning heads need provision for receiving the camera with a minimum of mechanical contrivance, and for holding the camera firm once it has been mounted.

The ideal head is robust, needing little maintenance, and is provided with individual pan and tilt controls to suit the varying requirements of cameramen.

The camera is panned and tilted by means of a panning handle attached to the panning head. It is necessary for this to be clamped firmly to the head with no play at the fitting, yet to be easily removable and adjustable by the cameraman without the need for tools. The ideal handle is telescopic, with a simple adjustment to suit the individual needs of cameramen.

It is most important that there is a means of balancing the camera accurately on the panning head. This is particularly vital when extra equipment (cueing devices, long focal length lenses, lighting attachments, etc.) is mounted on the camera, affecting its weight distribution.

Good camerawork is impossible if the panning heads and camera mountings are out of adjustment, or are not smooth-working. Although major maintenance and adjustments are best attended to by qualified engineers, cameramen can do much to keep their equipment in perfect working order.

DOLLIES. Pedestal dollies should be heavy enough to be unaffected by slight irregularities on the studio floor, yet light enough to be moved easily by the cameraman – especially when his camera is on the air and he is attempting to track or crab.

The central column of the pedestal is best designed so that it can be raised or lowered smoothly without the cameraman having to release his hold on the camera. And once the cameraman has raised or lowered the column so that his camera is at the required height for the shot in hand, the column should remain at that height without the need for any locking device.

Most pedestals have three wheels or castors, with provision for setting these wheels in such a condition that they turn in unison. This is known as the crab position – the normal working state. And it is desirable that a simple control is available to the cameraman with which he can alter the state of the dolly so that one wheel or castor only is driven by the steering device. This condition enables the cameraman to swivel the base of the dolly into a more convenient working position.

The steering device itself must be accessible to the cameraman wherever he stands in relation to his dolly. For this reason it usually takes the form of a wheel running around the outside of the central column, and just below the base of the camera.

Trackable dollies – i.e., those moved around the studio floor by a member of the camera crew called a tracker – can be either manually tracked or motor driven. The cameraman normally sits with his camera at the front end of a pivoted arm which can be raised or lowered either by himself or by the tracker, depending on the type of dolly.

But whether manually tracked or motor driven, the dolly must be easily manoeuvrable and smooth running. The pivoted arm, too, should be simple to raise and lower, and gentle enough in operation to enable the camera height to be adjusted unobtrusively when the camera is on the air. If it is motor driven, the controls are best given to the cameraman.

Some means of tonguing or gibbing the pivoted arm from side to side is necessary, or alternatively a method of adjusting the wheels of the dolly easily to the crab position, so that the camera can be moved to an off-centre position without the need for the dolly itself to be repositioned.

CONTROLS. The cameraman wants as few electronic controls as possible to operate so that he can concentrate on the art of camerawork without the need to adjust electronic levels or other settings at the same time. Iris rings, filter selectors and other devices can be remotely controlled by the vision control operator.

VIEWFINDER. Without a basically good picture to start with, the cameraman cannot be expected to be responsible for good camerawork. But it is just as important that the viewfinder also provides him with a first-class picture. It does not matter that those in the vision control and production control rooms can see a crisp picture. If the cameraman's viewfinder is faulty, the cameraman can hardly be blamed if his shots are rarely correctly framed and focused.

Most cameramen prefer an electronic viewfinder to an optical viewfinder. An electronic viewfinder is virtually a small television set mounted on the top of the camera, showing a picture which corresponds in every way to the adjusted television picture emanating from that camera. With this type of viewfinder the cameraman can see the television quality of each shot he frames; an optical viewfinder merely indicates the area of the scene contained in his shot.

The viewfinder should be provided with a hood – which is a box-like projection attached to the viewfinder screen, and having an open end through which the cameraman can see the viewfinder picture. This hood is necessary to prevent extraneous light falling on the viewfinder picture and debasing its quality. Since the cameraman must often work with his face pressed close to the hood, it is desirable for the edges around the opening to be padded for comfort.

The cameraman does not take all his shots from one height only, and if the viewfinder is mounted so that it is always parallel to the top of the camera body, the cameraman has great difficulty in seeing his viewfinder picture when working with the camera set at very high or low positions. For this reason, the viewfinder itself should be centrally mounted in a cradle, enabling it to be tilted up or down through a fairly wide angle.

If the cameraman then raises his camera on a pedestal to the limit of its travel and points the camera downwards to present a high-angle shot of an artist, he can tilt his viewfinder until he can see the picture from a comfortable standing position. Without this provision he has to stand on something which is high enough to enable him to look along the line of the downward-pointing camera. Moving into and out of this position is a laborious business, and he certainly cannot perform any camera movement under these circumstances. A fixed viewfinder with a movable hood is a poor substitute for a fully swivelling viewfinder.

EARPHONES. To receive instructions from the director, the cameraman must be provided with a pair of earphones. As he works for long periods under the hot studio lights, it is essential for the earphones to be light and comfortable, and, of course, fully adjustable to meet individual needs. They may be easily attached to the back of the camera by a simple plug which is pushed into a socket, and they can be provided with a lightweight microphone to enable the cameraman to communicate with the director and the vision control operator.

Ideally, there is provision on the camera for the cameraman to adjust the level of the sound in his earphones to compensate for the varying levels produced on the studio floor.

OUTSIDE BROADCASTS

The needs of the outside broadcast cameraman are highly specialized. It is most important that his cameras are light enough to carry from the outside broadcast vehicles to their mounting positions. This often entails taking them up ladders to high and remote platforms, or across a mile or so of rough countryside to cover part of a horse-race or motor-cycle scramble.

It follows, then, that the cameras must be particularly robust, and provided with effective waterproof jackets. The same might be said of the cameramen!

The cameraman often needs long focal length lenses on actuality outside broadcasts (such as race meetings, processions, football matches) because of the long distances sometimes involved. And since he is often tied to a fixed camera position during an outside broadcast, the zoom lens is a particularly vital piece of equipment, for it compensates, to some extent, for lack of camera mobility.

Except on rare occasions, the outside broadcast cameraman does not need the sophisticated pedestals and dollies used by his colleagues in the studio. A simple, sturdy tripod mounted on castors or freely-rotating wheels is his maid-of-all-work. Trackable dollies need pneumatic tyres to allow for the indifferent floor surfaces over which they are moved.

For static positions, camera mountings must be easily transportable yet rigid once placed in position. They should either be provided with prongs which can be driven into earth, wooden platforms, etc., or have a heavy solid base to guarantee rigidity.

A great asset to the outside broadcast cameraman is the mobile camera, often called the creepie peepie, which functions without cables. This enables the cameraman to operate in places which would be denied him were he using a conventional camera. Exciting pictures taken from helicopters and lifeboats, for example, are typical of the opportunities available to the creepie peepie cameraman.

Finally, a word about colour television. The television cameraman is in the sad position of being the only technician unable to enjoy the beauty of the colour medium, for his viewfinder picture is monochrome. There are sound technical reasons for this, but the colour television cameraman must be waiting for the day when these problems are overcoming. P.J.

See: *Camera* (TV); *Cameraman* (FILM).

Book: *Technique of the TV Cameraman*, by P. Jones (London), 1965.

⊕□**Camera Operator.** Technician in charge of the film camera, responsible for manipulating it so that each picture on the negative contains all the director requires to be seen and excludes all he wishes to be excluded. Throughout each shot he must pay proper regard to picture composition and his camera manipulation must be smooth and accurate. In television, he is responsible for moving the camera and base in any direction in response to instructions he receives over headphones from the director, as well as movements of the camera itself. He is able to watch the picture covered by the camera lens in a small picture monitor attached to the camera.

See: *Cameraman; Focus puller*.

□**Campbell Swinton, Alan Archibald, 1863–1930.** British inventor born in Edinburgh and educated there and abroad. Worked from 1882 to 1887 at the Elswick works of Sir W. G. Armstrong, settling then in London as a consulting electrical engineer. He was the first to propose, in 1908, an all-electronic system of television – or distant electric vision as he termed it – the details of which were surprisingly akin to those of the high-definition all-electronic system of today. In 1896, was instrumental in launching the young Marconi on his successful career by giving him a letter of introduction to Sir William Preece of the Post Office. He took a keen interest in the development of electrical physics, and after Roentgen had discovered X-rays in 1895 was one of the first to experiment with them and use them for taking photographs. Campbell Swinton's famous prophesy in 1908 regarding television was made in a letter to *Nature*, his suggestions being that the problem might be solved by using two cathode ray tubes, one at the transmitter and one at the receiver. He put his detailed proposals in a lecture to the Roentgen Society in 1911 and, in 1926, in a further letter to *Nature*, he described some experiments he had been doing. These had not been very successful owing to the lack of the necessary techniques at the time, but he continued to advocate his proposed system, pointing out the limitations of mechanical methods and urging that the work should be taken up by some large research laboratory. A member of the Institutions of Civil, Electrical and Mechanical Engineers, he was also a Fellow of the Royal Society, elected President of the Roentgen Society of Great Britain (1913), and Chairman of Council of the Royal Society of Arts (1917–19).

⊕**Can.** Circular metal cans of various sizes are normally used for the handling and storage of film. For the storage of magnetic film, cans have the additional advantage of providing some security against demagnetization by stray magnetic fields.

See: *Film storage*.

⊕**Candela.** Unit of light intensity, defined as ☐the luminous intensity equal to one-sixtieth of the luminous intensity per square centimetre of a black-body radiator maintained at the temperature of melting platinum (2042°K). The brightness or luminance of a source is now commonly expressed in candelas per square metre.

See: *Light units*.

⊕**Capstan.** Enlarged shaft or cylinder, some-☐times an extension of the driving motor

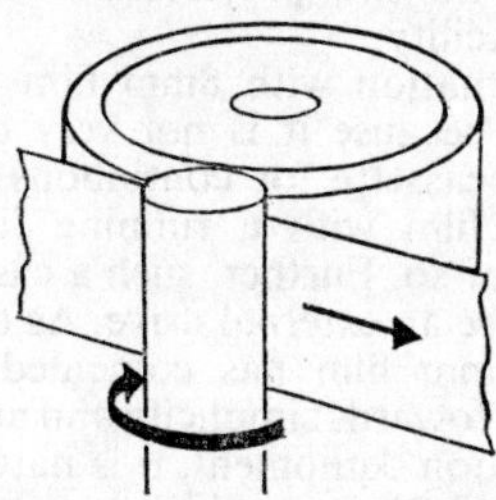

spindle, against which magnetic tape is pressed by a larger pinch roller to impart a steady movement to the tape as it passes record and reproduce heads.

See: *Sound recording*.

⊕**Caption.** Graphic material prepared for ☐television presentation, generally containing lettering. The standard dimension for the caption card is 12 in. × 10 in., with the transmitted area 11¼ in. × 8½ in. The British Standard for slides and opaques for television, BS 2948, specifies dimensions in detail. The equivalent American standard is PH 22.94-1954, revised 1960.

Also used to denote lettering superimposed on motion picture film for the translation of foreign dialogue.

See: *Foreign release: Master control*.

⊕**Caption Roller.** Device for presenting a ☐continuous length of caption material to a camera. It generally takes the form of a pair of rollers, between which the caption (12 in. wide) is wound at a controlled speed.

⊕**Caption Stand.** Device for holding caption ☐material at a level above the floor convenient for viewing by a camera.

⊕**Cartridge.** Container in which a roll of film may be loaded in advance so as to facilitate

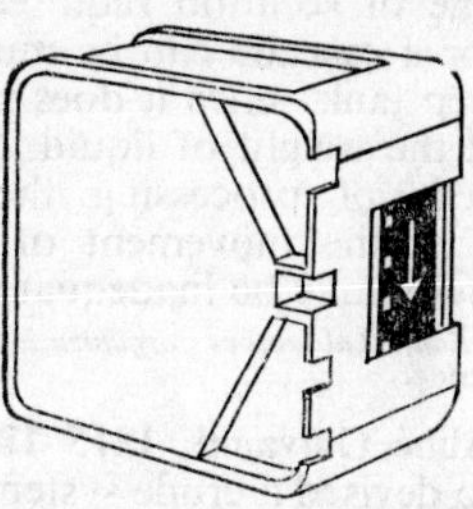

CARTRIDGE. Provides automatic drive coupling and film threading. External notches engage camera mechanism to indicate film speed, colour temperature, etc.

its use in a motion picture camera or projector. Cartridge loading cameras were developed for use in difficult operating conditions, such as aerial combat photography, and to simplify equipment used by the amateur. The cartridge normally contains feed and take-up reels for the film which can be engaged with drives from the camera mechanism and often includes the spring-loaded pressure plate which forms the rear part of the gate assembly of the intermittent movement.

Recent developments of cartridges for use in amateur movie cameras include coded notches on the cartridge body which indicate the type of film loaded inside and its photographic speed and which can provide the automatic exposure system of the camera with this information.

See: *Camera* (FILM); *Cassette loading for projectors; Narrow gauge film and equipment*.

⊕**Cascade.** Processing method by which chemical solutions and wash water, par-

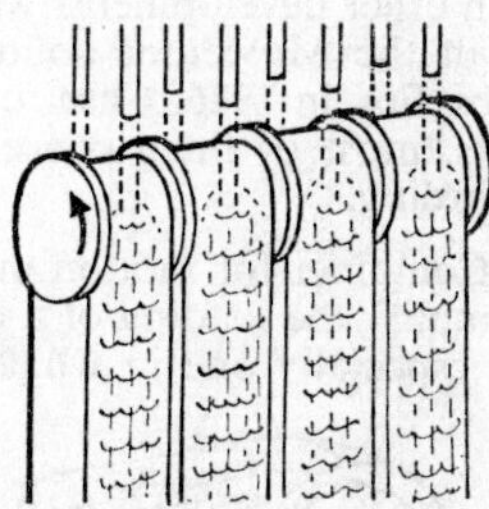

ticularly the latter, are applied to the film in a continuous processing machine in the form of jets of liquid which run down the film strands from top to bottom under gravity, rather than by the immersion of the film under the surface of the solution in a deep tank. The passage of the falling liquid over the film surface provides a degree of agitation and the system is economical in the volume of solution required. The enclosure for a cascade can be much simpler than a deep tank, since it does not need to withstand the weight of liquid, but at very high speeds of processing the agitation provided by the movement of the liquid under gravity may be inadequate.

See: *Agitation; Laboratory organization; Processing; Spray processing.*

□**Caselli, Abbé Giovanni, 1815–1891.** Italian priest who devised a crude system of phototelegraphy, which made possible the first long-distance transmissions of images by electricity. His work and experiments were aided by Napoleon III. Caselli was reported to have been ten years in working out his system, which was in actual practice between Amiens and Paris (1865–69). It appeared to be a modification of the chemical recorder, an iron stylus being made to mark on paper sensitized with cyanide of potash.

⊕**Case, Theodore W., 1888–1944.** American inventor who graduated at Yale University, where his first experiments on sound recording began. In 1914, he started a laboratory at Auburn, New York. The first outcome of his work was the thallium sulphide photocell – the Thalofide cell – used in World War I in an infra-red communications system.

After the war, he discovered the principle of the barium photoelectric cell. By 1923, he had designed a crude sound-recording camera, and, with Lee de Forest, developed the Phonofilm system, but then discovered that an electrical discharge through an argon-filled tube could be modulated. This led to the Aeo-light system, which with other developments was used as the basis of the Movietone sound system, acquired by Fox in 1926. Case took out 62 US patents, most of them concerned with sound recording.

⊕**Cassette.** Container for motion picture film used to simplify the loading of a camera or projector, especially one in which the film is used in the form of a continuous endless loop.

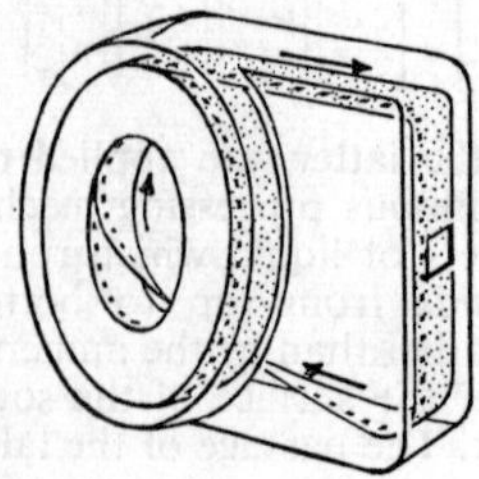

See: *Cartridge; Charger*

⊕**Cassette Loading for Projectors.** Method of loading which uses magazines, or cassettes, containing a continuous loop of film which does not have to be laced in the projector; this is effected automatically by the insertion of the cassette.

The use of cassettes for 16mm projectors has not, except in isolated cases, been generally adopted. In the first place, the bulk of a cassette would be considerable to house a film of up to 800 ft. Secondly, the cassette would have to handle the film in a loop form, and this demands some form of drive to the cassette. The average 16mm projector does not provide any simple method whereby this could be achieved and certainly no two projectors would have a similar facility.

The situation with 8mm film is entirely different because it is not very difficult to design a cassette for continuous operation of 8mm film with a running time of 20 minutes or so. Further, such a cassette does not require an external drive. As the growth of the 8mm film has coincided with the tendency towards simplicity and automation in projection equipment, it is natural that a number of 8mm machines have been designed solely for use with cassettes. The first and most notable of these is the silent 8mm single-concept film projector, which handles up to 45 ft. of 8mm film in its cassette. Projectors are available to accept these loop cassettes with either standard 8mm film or type S (Super-8), but there are no projectors which can accept both types.

In the field of 8mm sound projectors employing cassettes, there are a few units of reasonable size incorporating their own rear projection screens. These units have, up to the present, been designed to reproduce sound from a magnetic stripe, but there is now available in the US an 8mm cassette projector handling optical recorded sound.

See: *Daylight projection unit; Projector; Rear projection.*

□**Cathode Follower Stage.** Type of circuit in which the output signal is taken from the cathode of a valve (electron tube) rather than the anode. The output voltage follows the input voltage, as it is of the same polarity, and the gain is always less than unity. The input impedance is very high, and the output impedance very low, so that this stage can be used as an impedance converter, or transformer. Also called a grounded-plate amplifier. The equivalent transistor circuit is known as an emitter-follower, common-collector or grounded-collector amplifier.

□**Cavity Resonator.** Type of tuned circuit used at ultra-high frequencies (Bands IV and V). Instead of the more conventional inductor and capacitor, a hole or cavity inside a piece of metal can be made to resonate with a very high efficiency.

See: *Transmission networks; Transmitters and aerials.*

⊕**Cel.** Transparent layer of celluloid containing two or more holes or slots which mount on pins under an animation camera to secure perfect registration. Painted on the cels are phases of action which build up under the animation camera into continuous motion.

See: *Animation equipment; Animation techniques.*

⊕**Celluloid.** Cellulose derivatives used as the base for coating photographic materials such as motion picture film. Also refers to the transparent sheets on which the component drawings of an animated film are prepared.

See: *Film manufacture.*

⊕**Cement, Film.** Adhesive for securing splices in film. In many film joiners, splices are made by overlapping the ends of two pieces of film from the surface of one of which the emulsion has been scraped. Film cement is then used to dissolve the base locally, so that pressure contact will unite the two ends into what is effectively a single piece of film.

See: *Editor; Narrow gauge production techniques.*

⊕**Censor Title.** Title which in some countries appears on the screen in advance of the main titles to indicate that a film has been passed for exhibition either to general or to restricted audiences.

See: *Release print make-up.*

⊕**Changing Bag.** Light-tight black bag with close-fitting sleeves through which a film cameraman or his assistant puts his hands when loading unexposed film into magazines, or unloading exposed film.

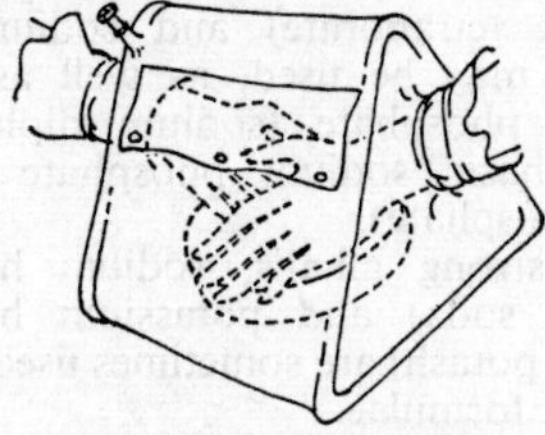

Often used on location when darkrooms are not available.

⊕**Change-Over.** In film projection, the switch from one film spool to the next to be shown, which is held ready on a second machine. The change-over occurs every 20, 30, 40 or 60 minutes, depending on legal restrictions and the type of machine used. The traditional magazine size is 2,000 ft. of 35mm film (about 20 minutes running time).

Change-overs must be very accurately timed if the audience is not to notice them, and the approaching end of the reel is signalled by cue marks at the upper right-hand corner of the screen. Sound must also be changed over. Automatic change-over systems are now in use which double the time before the operator has to mount a new film magazine on the projector.

See: *Automation in the cinema; Release print make-up.*

□**Channel.** Band of frequencies allocated to a specific use, e.g., a single television transmitter. Television bands are subdivided into numbered channels, and the standard channel width therefore determines the number of channels the band can accommodate.

TELEVISION CHANNELS (BRITAIN)

Channels	Band Limits MHz	Channel Width MHz	Band Number & Designation
1–5	41–68	5	I, VHF
6–13	174–216	5	III, VHF
21–34	470–582	8	IV, UHF
39–68	614–854	8	V, UHF

TELEVISION CHANNELS (US)

Channels	Band Limits MHz	Channel Width MHz	Designation
2–4	54–72	6	VHF
5–6	76–88	6	VHF
7–13	174–216	6	VHF
14–83	470–890	6	UHF

See: *Transmission and modulation.*

□**Channel Gain Stability.** Constancy of gain of a channel in the presence of changes due to ageing, supply voltage, temperature, etc.

⊕**Characteristic Curve.** Curve relating exposure of a photographic material to the density of the image produced under stated conditions of development.

When density is plotted against the logarithm of exposure, the long central part of the curve is substantially straight, i.e., given logarithmic increases of exposure produce proportionate increases of density. A tonal transmission system confined to

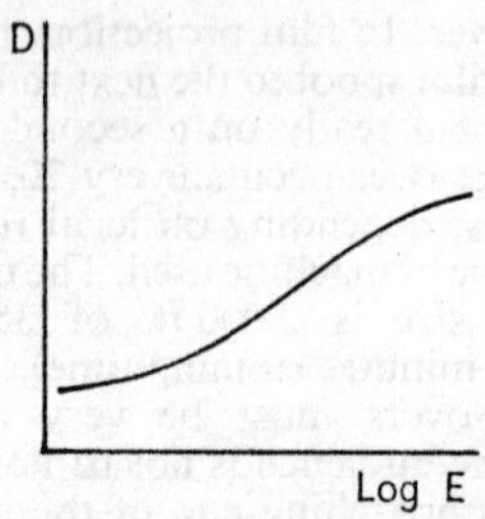

CHARACTERISTIC CURVE. Plot of density against log exposure gives a fairly straight central portion with a curved shoulder and toe.

this part of the curve is therefore linear.

At its extremities, the curve reveals an S form. At the upper end, the shoulder, increases of exposure produce less than proportionate increases of density. In this region, for negative materials, the highlights of the subject are compressed and do not reproduce in their full tonal range.

At the lower end, the toe, exposure density sinks into the fog level, or general background density of the unexposed part of the film. The toe of the negative material causes tonal compression of shadows.

The slope of the straight line portion of the curve, measured by the tangent of the angle it makes with the exposure (log E) axis, is known as gamma (γ) and is an important measure of the inherent contrast of the emulsion. When the slope of the curve is steep, a relatively small increase of exposure produces a large density change, and the material is said to be high-contrast, and vice versa with a gently sloping low-contrast curve.

Photographic materials used as negatives have normally a fairly low contrast (gamma values of the order of 0·6 to 0·7) with as long a straight-line portion of the characteristic curve as possible, and extended toe and shoulder. Positive stocks and sound recording films are of higher contrast, with gamma values of 2·5 to 3·0, and show a limited toe with practically no shoulder within the range of densities normally used.

The characteristic curve is also called a D log E curve or H & D curve after Hurter and Driffield, pioneers of sensitometry.

See: *Laboratory organization; Sensitometry and densitometry; Sensitometry of colour films.*

⊕**Charger.** Simple form of film-loading cartridge used in 9·5mm motion picture cameras.

⊕**Checkerboard Cutting.** When negative is prepared for A & B printing (or A, B & C), it must be laid out checkerboard fashion so that the different shots are correctly distributed between the different rolls. The spaces between the end of one shot and the start of the next are filled with leader.

See: *A and B printing.*

⊕**Check Print.** Preliminary print of a motion picture film prepared by the laboratory for submission to the customer for approval before proceeding with further work.

⊕**Chemical Mixer.** Film laboratory operator who carries out chemical handling and weighing, and mixing the solutions used on film processing machines. Particularly where automatic replcnishment systems are in use, his duties may extend to operating supply valves on circulating systems, recording temperatures, flow rates, etc.

⊕**Chemicals.** The range of chemicals generally employed in the processing of motion picture cinematograph film is limited in comparison with the very large number which are used in photography as a whole, and may be conveniently summarized under the following headings.

DEVELOPING AGENTS. Almost all black-and white developer formulae are based on combinations of metol (monomethyl para-amino-phenol sulphate) or phenidone (1-phenyl-3-pyrazolidone) with hydroquinone, although ethylene-diamine may sometimes be encountered for special applications.

In colour developers the active agents are usually proprietary compounds, such as C.D.2 (2-amino-5-diethylaminotoluene monohydrochloride) and C.D.3 (4-amino-N-ethyl-N (β-methane sulphonamide ethyl) m-toluidene sesquisulphate), both of which may be known by other trade names. Diethyl-paraphenylene-diamine and hydroxylamine sulphate are also used.

Benzyl alcohol, although not itself a developing agent, plays an important part as an activator in some colour developers.

ALKALIS IN DEVELOPERS. All developers contain alkalis and one of the most widely employed is sodium carbonate (soda, soda ash), but in fully buffered developers borax (sodium tetraborate) and sodium metaborate) may be used, as well as dibasic sodium phosphate (sodium diphosphate) and tribasic sodium phosphate (sodium orthophosphate).

The strong alkalis, sodium hydroxide (caustic soda) and potassium hydroxide (caustic potash) are sometimes used in high-contrast formulae.

PRESERVATIVES IN DEVELOPERS. The purpose of the preservative in a developer is to limit the effect of oxidation, and for this purpose sodium sulphite is almost universal. Other preservative chemicals are sodium bisulphite and sodium or potassium metabisulphite. These may also be used as components in acid fixing and stop baths.

RESTRAINERS IN DEVELOPERS. The most widely used restrainer is potassium bromide, but sodium bromide may be used occasionally. Both these chemicals are also used as the source of the halide in bleach baths.

FIXING SOLUTIONS. As in general photographic practice, the usual fixing agent is hypo (sodium thiosulphate), but for special purposes ammonium thiocyanate or potassium thiocyanate (sulphocyanide) may be employed. Both the latter chemicals may be used as silver halide solvents in certain reversal developer formulae.

BLEACHING SOLUTIONS. The bleaching operation of converting a developed silver image to silver halide forms an important part of many reversal and colour processing systems and the active agents are generally potassium or sodium bichromate and potassium or sodium ferricyanide.

HARDENERS. Chemicals used for hardening the gelatin of the film in the course of processing are usually alum (potash alum – potassium aluminium sulphate), chrome alum (potassium chromium sulphate) or formalin (formaldehyde).

Although not strictly a hardening agent, sodium sulphate is often used to control the swelling of the emulsion at some stages.

ACIDS. The most frequently used acids are acetic acid, boric acid (boracic acid) and citric acid, while dilute sulphuric acid is often used for the control of pH of processing solutions.

MISCELLANEOUS. Sodium chloride (common salt) may occasionally be used as a halide source in bleaches.

Sodium hexametaphosphate (trade name Calgon) has important uses as a water softener in preventing the formation of calcium scale deposits.

Sodium hydrosulphite is used as a chemical blackener in some forms of reversal process.

Sodium pentachlorphenate is sometimes used as an algicide to prevent the growth of moulds and fungus in developing solutions.

Sulphamic acid may be employed in cleaning film processing equipment, particularly for the removal of calcium deposits caused by hard water.

Wetting agents are often added to processing solutions and wash water to assist the uniformity of wetting the surface of the film. They are usually proprietary substances such as Photo-Flo, Teepol, etc. L. B. H.

See: *Chemistry of the photographic process; Laboratory organization; Processing.*

□**Chemical Telegraph.** Early technique in which telegraph signals were displayed at a distant station by marks made by chemically decomposing suitable chemicals.

See: *History* (TV).

⊕CHEMISTRY OF THE PHOTOGRAPHIC PROCESS

The photographic emulsion consists essentially of a dispersion of very small crystals of silver halide (silver chloride, bromide or iodide) in a gelatin gel.

Silver halide crystals are pale yellow and metallic silver is black. When such crystals are exposed to light, very small specks of silver (containing only a few atoms) are generated on their surface, i.e., a latent image is formed.

These silver specks can catalyse the conversion of the silver halide into silver by certain chemical reducing agents; that is to say, the conversion of silver halide crystals into silver goes more quickly in the case of crystals which bear metallic silver particles than in the case of those which do not.

BASIC PRINCIPLES

Chemical reducing agents which convert silver halide into silver at a convenient rate in aqueous solutions at room temperature, and whose reaction is powerfully catalysed by metallic silver, are called developing agents.

After treatment with a developing agent the emulsion contains a developed image which consists of black silver particles mixed with unconverted silver halide crystals. To stabilize this image and to make it permanent, the unconverted silver halide crystals must be removed. This is done by treating the developed emulsion with a solution which readily dissolves silver halide but which has no effect on metallic silver. Such a solution is called a fixer.

Since development and fixation are chemical reactions, both are accelerated by increase of temperature.

The development or fixation of a silver halide crystal takes place within a gelatin layer. This layer contains only a small volume of treating solution which is exhausted during the reaction unless more chemicals diffuse into the layer from outside and the products of reaction are removed by diffusing out into the solution. It is therefore important that the solution at the emulsion surface is well agitated so that it approximates to the same composition as the bulk of the treating solution.

PHOTOGRAPHIC DEVELOPING AGENTS. Considering that the history of photographic developing goes back for nearly a century, it is surprising how small is the number of developing agents in practical use for film processing. Compounds with suitable properties are mainly dihydroxy-, hydroxyamino-, or diamino-benzene derivatives. Among the commonest are 1:4 dihydroxybenzene (or hydroquinone), 1-hydroxy-4-methylaminobenzene (metol and various trade names), and 1-phenyl-3-pyrazolidone (Phenidone). These are widely used in black-and-white developers.

Colour developing agents are generally derivatives of 1:4 diaminobenzene, common examples of such bases being shown below. They are used in the form of their salts.

DEVELOPERS

COMPOSITION. A practical photographic developing solution contains a number of constituents, each with its particular function. They are:

The solvent. Almost invariably water. Serves to dissolve the other constituents and to swell the gelatin of the emulsion.

The developing agent. Usually one or more of those already described, mixtures of metol and hydroquinone or of phenidone and hydroquinone being common in film processing. Its function is to convert the exposed silver halide crystals into black metallic silver.

The preservative. Most commonly sodium sulphite. Takes part in the developing reaction and ensures that the developing agent which is used up forms a soluble product. Also serves to protect the developing agent from the action of the oxygen in the air.

The alkali. Controls the pH of the developer. The higher the pH of the developer the greater its activity, other factors being equal. It is important that the alkali is selected to give a well-buffered solution, that is, one which is as resistant as possible to altering its pH. In film processes, the most commonly used alkalis are sodium borates and carbonates.

The restrainer. Reduces the rate of development, particularly of the silver halide crystals which have not been exposed to light. It is said to reduce chemical fog. Sodium or potassium bromide are almost always used.

FORMULATION. Developers are always formulated for a particular use and vary widely in composition. Developers for negative films are characterized by having very low concentrations of soluble bromide, low pH values and fairly high concentrations of sulphites. They develop slowly, since negative emulsions are not usually fully developed but are stopped at predetermined contrasts lower than their maximum. Their low activity minimizes the graininess of the image, and the low concentration of bromide gives to the emulsion the highest attainable sensitivity to light.

Positive emulsions, on the other hand, are more fully developed to a high contrast and it is important that the chemical fog developed should be low. Developers for positive films have, therefore, a high pH and a relatively high concentration of bromide.

A wide range of developer composition is possible for black-and-white films, and the precise formula used at any time in a film processing laboratory depends on the machine on which the film is being processed as well as the type of film. It is common practice to convert the used negative developer into positive developer by addition of extra chemicals.

DEVELOPMENT REACTIONS. During the process of development, the silver halide crystals are converted into metallic silver (are reduced) while the developing agent itself is converted into another material (oxidized developing agent). The oxidized developing agent combines with some of the sulphite present and becomes an inactive soluble compound. The developing agent in solution may also combine with the oxygen of the air. The end product is the same as if the oxidation had been brought about with silver halide. The two reactions may be represented chemically:

Oxidation by silver halide:

$2AgBr + C_6H_4(OH)_2 + Na_2SO_3 \longrightarrow 2Ag$
Silver + hydroquinone + sodium ⟶ silver
sulphite

$+ C_6H_3(OH)_2SO_3Na + NaBr + HBr$
+ sodium hydroquinone monosulphonate + sodium bromide + hydrobromic acid

Oxidation by aerial oxygen:

$2C_6H_4(OH)_2 + O_2 + 2Na_2SO_3 \longrightarrow$
Hydroquinone + oxygen + sodium sulphite ⟶

$2C_6H_3(OH)_2SO_3Na + NaOH$
sodium hydroquinone monosulphonate + sodium hydroxide

Under all practical conditions, both reactions occur simultaneously and lead to loss of developing agent with consequent lowering of photographic activity. Oxidation by silver halide leads to an increase in the concentration of soluble bromide in the developer and thus to loss of activity. Acid is also produced and leads to a lowering of pH and a reduction of activity. Oxidation by aerial oxygen produces sodium hydroxide, which causes pH to rise and thus increases activity.

In an actual practical situation, it is sometimes quite difficult to forecast how developer activity will alter with time, as the relative importance of the two reactions above may vary widely on different occasions. In film laboratory practice, variations in the photographic activity of developers cannot be tolerated and solutions are, therefore, continuously replenished.

REPLENISHER. No practical method has yet been discovered for removing soluble bromide from developers. Its concentration in a developer can be reduced only by rejecting part of the solution and replacing it with fresh developing solution, without bromide. Such a bromide-free solution is called a replenisher.

The rate at which replenisher must be added depends on the rate at which bromide is being added to the system. This, in turn, depends on the area of film being processed per unit of time and on the average density to which it is being developed.

As an example, it is found that 35mm black-and-white film during processing liberates about 8–10 gm of potassium bromide for each 1,000 ft. The composition of the replenisher must be such that, when it is added at a rate which holds the concentration of bromide constant in the developer, the concentrations of all other photographically active materials also remain constant.

The typical black-and-white film replenisher solution therefore contains no bromide, a greater concentration of developing agent and of sulphite than the solution which it is to maintain, and a pH either above or below that of the developer, depending on the prevailing conditions.

DEVELOPING AGENT COMBINATIONS. It is usual in film laboratory practice to use in a developer a combination of developing agents rather than a single one. The commonest combination is metol and hydroquinone. In a metol-hydroquinone (MQ) developer, the reduction of the silver halide crystals is almost exclusively brought about by the metol, with formation of oxidized metol.

This oxidized metol either reacts with some of the sulphite in the solution or with some of the hydroquinone. In the latter case, metol is regenerated and oxidized hydroquinone is produced which reacts with sulphite to give inactive hydroquinone monosulphonate. The net effect of oxidizing an MQ developer, therefore, is to leave the concentration of metol substantially unaltered while that of the hydroquinone decreases. Changes in developer activity are much smaller than in a developer containing only metol.

Metol may be replaced by Phenidone in phenidone/hydroquinone (PQ) developers. The mechanism of regeneration of oxidized Phenidone by hydroquinone is analogous to that of metol but the concentration of Phenidone is even less affected.

In commercial practice in film processing laboratories it is impossible to standardize exactly the conditions of use of developers; the relative importance of aerial oxidation depends on the amount of work being processed. Replenishment can be accurately worked out for only one set of operating conditions, and so commercial developers are always liable to small changes in composition. This difficulty is dealt with by making regular chemical analyses of the developers to determine developing agents, sulphite, alkalinity, bromide and pH.

COLOUR DEVELOPMENT

Colour development is the name given to a development process in which a coloured insoluble dyestuff is formed in association with the silver. There are certain developing agents, mainly derivatives of diaminobenzene, whose oxidation products can readily be converted, by combination with chemicals referred to as colour couplers, into insoluble dyes. The colour developing agents currently used are salts of various bases, which themselves are substituted p-phenylene diamines. Examples have been mentioned above.

These developing agents act in essentially the same way as the black-and-white developing agents already described. The developing agent reduces the silver halide crystal to silver and is itself oxidized. This oxidized developing agent is capable of reacting with another chemical present in the system, the colour coupler, to give a coloured insoluble dye. The oxidized developing agent can also react with sodium sulphite, and this material is therefore only present in colour developing solutions at low concentration. Colour development gives two images, silver and dye, mixed together.

COLOUR COUPLERS. By altering the nature of the colour coupler, the same oxidized developing agent can give dyes of different hues; couplers can be selected in fact which give, with the same developing agent, yellow, magenta and cyan colours. For example, yellow couplers are commonly substituted acetoacetanilides, magenta couplers are substituted nitriles or pyrazolones, and cyan couplers substituted naphthols.

A typical cyan-forming coupling reaction may be represented:

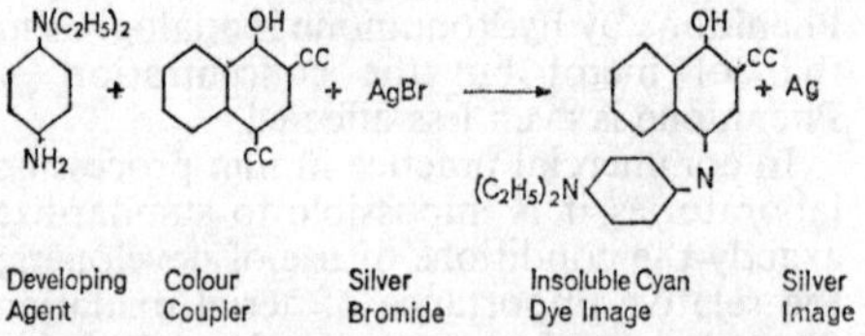

The colour coupler must be present in adequate concentration at the silver halide crystal being developed. The coupler, if soluble, may be dissolved in the developer solution. In this case, the developer is formulated to give one colour and, if three colours are needed, three different developing solutions must be used. This happens with Kodachrome and similar processes.

The coupler required in the dye-forming reaction need not be put in the solution but may already be present in the gelatin layer as coated. Development of such a coupler-containing emulsion layer with a colour developer naturally gives rise to both silver and insoluble dye. In integral tripack films, three emulsion layers, sensitive to blue, green and red light, are coated one on top of the other on to the same piece of base, and each layer contains a colour coupler which on development gives rise to a dye of colour complementary to the colour to which the layer is sensitive.

It is thus possible, with a single development, to produce yellow, magenta and cyan dyes in the three layers.

It is vital that each colour coupler should stay in its own layer and there must be no diffusion of coupler either during coating or on subsequent storage. Methods of making couplers fast to diffusion in gelatin layers fall into two main groups. In the first, the coupler is modified by inclusion in its molecule of long carbon chains, which give the coupler affinity for gelatin rather as some dyes are fixed by textile fibres. Typical groupings such as stearylamino are used.

In the second type, the coupler, by inclusion of suitable groupings, is made highly insoluble in water. It is dissolved in a high-boiling-point organic solvent and the solvent is then emulsified in a gelatin solution and added to the photographic emulsion. The coupler is therefore present in the gelatin of the emulsion layer in the minute drops of solvent, rather as droplets of fat are present in milk. The droplets are very small compared with the silver halide crystals. During development, oxidized developing agent is formed, and this penetrates the solvent droplets and couples there, forming dye.

Where the coupler is included in the emulsion layer, in the areas of slight exposure where only a weak image is formed, only a small proportion of the coupler present is used up. The remainder is left in the layer.

In certain cases, this residual coupler can react slowly with the dye formed in the layer and destroy it. To prevent this happening, the film may be treated with a solution of formalin which combines with residual coupler which is then no longer available to cause fading of the dye image.

AUTO-MASKING. It is a difficulty of colour photography that the dyes which can be formed by colour development do not have the transmission characteristics desirable in a colour negative. What is required is that each dye should absorb strongly in one colour band and transmit completely in the other two. This does not happen and there is generally some undesired absorption in the other colour bands.

For example, a magenta, or green-absorbing, dye has some absorption in the blue and red bands. The bad effects of this can be overcome by an auto-masking technique which depends on certain properties of colour couplers.

The cyan-forming coupling reaction given above has been chosen as having a special interest. The residue of the developing agent couples with the naphthol in the -4- position, displacing the chloro- group which occupied the position. Many different groups may occupy the -4- position in the coupler and, in each case, the group is displaced and the same dye is formed. If the benzene-azo-group, for example, is introduced into the coupler, it gives it a yellow colour, i.e., it absorbs blue light. When it reacts with oxidized developing agent, cyan dye is formed and yellow coupler is used up. The absorption in the blue band rises because of the undesired blue absorption of the cyan dye formed, and falls because of the destruction of the blue absorbing coupler.

By choosing the correct concentration of coloured colour coupler, the two effects may be made to cancel each other. The result is as if there were a cyan dye in which changes in red absorption did not cause undesired changes in blue absorption or as if it were a more perfect dye. Naturally, the colour of the coloured colour coupler residual in the unexposed regions must be compensated by filtration during printing.

COMPETITIVE COUPLING. In some processes of colour development, as for example in the colour development stage of certain reversal films, the colour image produced has too high a contrast. This can be overcome by making use of competitive coupling. In the developing solution, there is included a suitable soluble colour coupler which couples as usual with oxidized developing agent to give a soluble dye; the favoured coupler is citrazinic acid. When development occurs, some of the oxidized developing agent combines with the coupler anchored in the emulsion to give an insoluble dye, and some combines with the citrazinic acid to give a soluble blue dye, which subsequently washes out of the film. The effect is to reduce the contrast of the developed dye image.

FIXATION

The function of a photographic fixer is to dissolve residual silver halide from the emulsion after development.

COMPOSITION. The normal constituents of a fixer and their functions are:

The solvent. Water is invariably used.

The silver solvent. Sodium thiosulphate (hypo) is generally used or occasionally ammonium thiosulphate. Hypo operates by forming a soluble complex ion (AgS_2O_3).

The acid. Invariably some alkaline developer is carried through the wash into the fixer. This can cause precipitation of silver sludge and therefore the fixer must be acidified to prevent this. Since sodium thiosulphate is decomposed by strong acids, a weak acid, almost invariably acetic, is used.

The preservative. In the presence even of a weak acid there is a tendency for hypo to precipitate sulphur. This is prevented by adding sodium sulphite in which sulphur dissolves to form sodium thiosulphate.

The hardener. Hardens the gelatin of the film, thus improving its mechanical properties. In fixers, potassium alum is nearly always the hardening agent.

The normal range of pH values for acid fixers is within the range 4·0–4·5.

ACTIVITY. The activity of a fixer, and therefore the time which it takes to fix a film, depends on a number of factors, chief among them being:

1. Temperature of the solution; increase of temperature reduces fixing time.
2. Agitation of the solution at the film surface; increase of agitation reduces fixing time.
3. Concentration of hypo; increase of this reduces fixing time.
4. Concentration of dissolved silver; increase of this increases fixing time.
5. Concentration of restrainers from the film, particularly iodide; increase of this increases fixing time.

LIFE. In a film-processing laboratory, fixers can be used for very long periods by removing dissolved silver from them. Electrolytic silver recovery is used; the silver, which is an objectionable constituent in fixers, is a valuable by-product of film processing. The fixer is circulated through tanks where it is very vigorously agitated and is electrolysed with carbon anodes and stainless steel cathodes.

The concentration of hypo, sulphite, alum and acetic acid must be maintained by the dissolution of the solids in the fixer. Fixers become diluted because the film enters them saturated with water and leaves saturated with fixer. Additionally, alkali from the developer is carried in and neutralizes some of the acid.

The use of ammonium thiosulphate rather than hypo in fixers considerably reduces fixation times but is much more costly.

REVERSAL PROCESSING

A reversal process may be employed with both black-and-white and colour films and is used to produce a positive image instead of a negative.

BLACK-AND-WHITE. In an exposed film, a negative image is first produced by an active black-and-white developer. This silver image is then dissolved away and the residual silver halide is exposed to light and developed with a second developer. The solution used to dissolve away (or bleach) the negative silver usually contains potassium dichromate and sulphuric acid. Its action is to oxidize the silver metal to silver ion which accumulates in the solution.

The highlights in the positive image arise from the densest parts of the negative image. As the positive highlights should be of very low density, it is most important that no exposed silver halide crystals in these areas should escape negative development and later be developed in the second developer. To avoid this difficulty, first developers for reversal processes usually contain some silver halide solvent, potassium thiocyanate being most commonly used. They have as developing agents metol-hydroquinone or Phenidone-hydroquinone and are usually of high pH to permit full development in a reasonably short time. They also contain high concentrations of bromide for negative developers, in order effectively to minimize the production of chemical fog, which leads to poor blacks in the final positive.

In the black-and-white reversal process, the dichromate bleach is usually followed by a solution of sodium sulphite, which reduces any small residues of dichromate. The second developer which follows is a conventional active formula without silver solvent. Finally, a fixer removes any silver halide crystals which have escaped reduction in either developer.

COLOUR. Multilayer colour films may be processed by reversal. The first developer is an active black-and-white developer substantially as described above, which produces a silver image in each layer. The film after development is washed, exposed to white light, and then developed in a high activity colour developer, as previously described. This developer converts the exposed residual silver halide into a dye plus silver image, a different colour in each layer if the film is of the coupler-in-the-layer type.

The film has, therefore, a silver negative image, a silver positive image and a dye positive image. It only remains to remove all the silver, and this is done by first treating the film with a bleaching solution containing ferricyanide and a bromide to convert the silver into silver bromide, and then dissolving the silver bromide in a solution of hypo.

If the multilayer film does not contain a non-diffusing coupler in each layer, then exposure to white light followed by a single colour development stage must be replaced by selective re-exposure of the layers one by one, each layer being developed separately in a coupler containing colour-forming developer.

In certain modern types of film the reversal exposure has been eliminated and a chemical reversing agent is included in the colour developer.

WASHING

Any processing sequence contains a number of washes – relatively short washes between steps in the process and a long wash at the end. The intermediate washes prevent contamination of each solution by the preceding one; the final wash reduces to a low value the concentration of soluble salts in the film. Inadequate final washing is likely to cause the image to fade on keeping.

DURATION. The time for which it is necessary to wash a film to get a sufficiently low concentration of hypo in the gelatin layer to ensure good stability depends on several factors. Among the more important are:

1. Temperature of the water.
2. State of agitation of the water at the film surface.
3. Concentration of hypo in the wash water.
4. Thickness of the gelatin layer.

If sufficient time is allowed for attainment of equilibrium, the concentration of hypo within the layer is equal to that in the wash water. In black-and-white films, archival fastness is attained with about 0·5 gm/l. of hypo in the layer. It is very important to allow sufficient washing time; the same effect cannot be achieved by using a very rapid flow of water.

MECHANICAL WASHING. In a processing machine, water is used most efficiently in a cascade system. This consists of a number of wash tanks connected in series through which the water flows, entering at the point remote from the hypo. The hypo-laden film enters first the most contaminated wash water and proceeds into cleaner and cleaner water as it becomes washed.

DRYING MARKS. After washing on a film processing machine, there is always some form of squeegee to remove surplus water from the film before it is dried. It is difficult to remove every trace, and on the base side of the film the residual water film draws up into drops; as water is usually hard and contains dissolved salts, these leave marks after evaporation (so-called drying marks). This difficulty may be avoided by providing, just before the squeegee, a solution of a surface active agent (wetting agent), which causes the water film on the base after squeegeeing to spread out and not to draw up into droplets.

STOP BATHS

These are solutions whose function is to stop the action of the developer sharply by suddenly acidifying the emulsion layer. They are usually used after a very short water rinse. The stop bath must be very well buffered and of not too low pH. Decomposition of sodium carbonate from the developer must be avoided, as this can cause the formation of minute bubbles of carbon dioxide gas within the gelatin layer. The agitation at the film surface should be good to avoid local exhaustion. A typical stop bath may contain acetic acid and sodium acetate, some sodium sulphite and a hardener such as potassium alum. F.P.G.

See: *Chemicals; Duplicating processes; Integral tripack; Processing.*

Books: *Photographic Processing Chemistry*, by L. F. A. Mason (London), 1966; *Photographic Emulsion Chemistry*, by G. F. Duffin (London), 1966.

⊕**Chopping.** Limiting the amplitude of a ☐voltage waveform so that it cannot rise above a controlled level. A sine wave so limited appears as if the peaks had been sliced off, and if repeatedly chopped approximates to a square-wave of low amplitude. Chopping may occur in audio or in power supply circuits.

⊕**Chrétien, Henri, 1879–1956.** French scientist who in 1927 developed and demonstrated the Hypergonar lens, the first anamorphic lens system suitable for motion pictures. It was used by Claude Autant-Lara in 1929 for several short documentary films. A similar lens was employed in the Full-vue process, demonstrated in England in 1930 but not adopted at the time by film makers. Chrétien employed his system at the Paris Exposition in 1937, projecting a picture 10×60 metres in size, using two projectors fitted with anamorphic lenses. Chrétien's system was revived by Twentieth Century Fox in 1953 in the Cinemascope process. Anamorphic lenses are now employed in a number of wide-screen processes.

⊕**Chroma.** Indication of degree of colour ☐saturation. Two parameters together define any colour in the visible spectrum. One of these describes the dominant hue. This is the pure spectral colour, or wavelength, which when mixed with white light produces a colour equivalent to the sample colour. The second parameter indicates the amount of white light that has to be added to obtain this colour match.

A colour that requires no white to be added is a pure spectral colour and is said to be saturated. As white light is added to a spectral colour, it becomes paler and desaturated.

See: *Colour principles.*

⊕**Chromatic Aberration.** Defect of a lens ☐appearing as coloured fringes around the image, resulting from the material of the lens having different refractive indices for different colours of light, so that the blue rays come to a focus at a different point from the red rays. This defect is reduced in achromatic lenses.

See: *Aberration.*

⊕**Chromaticity.** Objective term for the defi-☐nition of the characteristic of a colour in colorimetry, corresponding to the subjective qualities of hue and saturation.

⊕**Chromaticity Coefficient.** Measure of the ☐purity of a colour. Often arbitrary scales are quoted for practical work.

⊕**Chromaticity Diagram.** System of representing colours in problems of colorimetry as points on a two-dimensional diagram whose co-ordinates may be employed in calculation, generally in the form known as the colour triangle.

□**Chromatron Tube.** A colour display tube of relatively simple design and potentially high efficiency but which has not yet emerged from the laboratory stage of development.

See: *Colour display tubes.*

□**Chrominance.** The part of the colour information which defines hue as opposed to brightness. The analysis of a colour, in a colorimetric system, can be separated into two parts. One concerns the intensity of the colour and its luminosity. The other concerns the dominant hue and the degree of saturation of the colour. This latter part is called the chrominance information.

In a colour television system, the chrominance information is derived in the encoder or colorplexer. This is then used to modulate a subcarrier. The modulated signal, called the chrominance signal, is added to the luminance signal to form the composite video signal for transmission to the colour receiver. On this colour receiver, the loss of the chrominance signal results in the picture becoming monochrome, but there is no change in the luminosity of any of the picture detail.

See: *Colour principles* (TV); *Colorplexer.*

□**Chrominance Channel.** That part of the frequency spectrum of a colour transmission containing chrominance information.

□**Chrominance Vector.** In a colour television signal, the finite mathematical vector whose angle represents the hue and whose length represents saturation. The reference frequency burst gives the necessary phase information.

⊕CHRONOLOGY OF PRINCIPAL INVENTIONS

1725 Schulze, J. H.: Experiments on light sensitivity of silver salts; contact images (from stencils) on liquid mixtures of chalk and silver nitrate in a bottle; no fixing.

1777 Scheele, C. W.: Blackening of silver chloride in the violet and the blue of the spectrum quicker than by other colours.

1816 Niépce, J. Nicéphore: Camera photographs on paper sensitized with silver chloride. Partially fixed.

1819 Herschel, Sir John F. W.: Discovery of thiosulphates and of the solution of silver halides by "hypo".

1824 Persistence of vision investigated.

1826 Paris, J. A.: Thaumotrope introduced.

1832 Plateau, J. A.: Phenakistiscope introduced.

1835-37 Daguerre, L. J. M.: Daguerreotype (direct photography on silvered copper plates with a silver iodide surface); development of the latent image by mercury vapour. 1837, fixing with salt solution.

1835 Talbot, W. H. Fox: "Photogenic drawings", copied on paper sensitized with silver chloride; fixed with potassium iodide or by prolonged washing in salt water. Also tiny camera photographs.

1855 Maxwell, Clerk: Theory of colour photography.

1856 Patents for stereoscopic Phenakistiscopes.

1858 d'Almeida, J. C.: Anaglyphic process.

1861 Du Mont, T. H.: Patent for photographs in rapid succession; Maxwell, Clerk: Colour photograph demonstrated.

1869 Projection Phenakistiscopes.

1872 Muybridge, E.: Chronophotography began.

1874 Janssen, P.: Photographic revolver.

1876 Donisthorpe, W.: Kinesigraph patented.

1877 Reynaud, E.: Praxinoscope patented; Edison, T.: Phonograph introduced.

1882 Marey, E. J.: Photographic gun.

1884 Choreutoscope patented.

1887 Marey, E. J.: Film camera demonstrated.

1889 Eastman, G.: Celluloid roll film introduced.

1891 Edison, T.: Kinetograph camera and Kinetoscope viewer patented.

1892 Reynaud, E.: Théâtre Optique opened.

1893 Kinetoscope marketed.

1895 Lumière, A. & L.: Cinématographe patented and publicly demonstrated; Acres, B.: Kinetic Lantern patented and privately demonstrated; Kinetoscopes synchronized with Phonographs publicly demonstrated.

1896 Paul, R. W.: Animatographe publicly demonstrated. Widespread use of film apparatus; hand colouring used in early films; Demeny, G.: 60mm film apparatus; Baron, A.: Patent for 360° panoramic projection.

1897 Grimoin Samson, R.: Cinéorama process of 360° panoramic projection; Grivolas, C.: Patent for stereoscopic films; Biograph 68mm film.

1898 Lumiere, A. & L.: 75mm film; Wilson, G. R.: Patent for anaglyph and polarizing stereo projection.

1899 Several disc and film systems demonstrated; Lee, F. M., and Turner, E. R.: Patent for three-colour additive process; Birtac and Biokam apparatus marketed for amateur use.

1904 Fleming, A.: Diode valve.

1905 Pathécolor stencil process.

1906 Lauste, E. A.: Sound-on-film patent; Smith, G. A.: Two-colour additive process patented.

1907 de Forest, Lee: Triode valve patented.

1908 Cellulose acetate safety film introduced.

1909 Standardization of 35mm format; Kinemacolor demonstrated.

1910 Hepworth, C.: Vivaphone commercially exploited.

1911 Lauste, E. A.: Sound-on-film processes demonstrated; Gaumont Chronophone system demonstrated; Kinemacolor commercially introduced.

1912 Aeroscope compressed air camera; Pathé Kok projector (28mm); Edison, T.: Home Kinetoscope.

1913 Panchromatic film commercially introduced; Gaumont Chronochrome three-colour additive process demonstrated.

1914 Patents for cels for cartoon films.

1915 Kodachrome two-colour subtractive system demonstrated.

1919 de Forest, Lee: Photion modulator developed; Tri-Ergon company formed; Prizma two-colour subtractive process demonstrated.

1921 de Forest, Lee: Sound-on-film process demonstrated.

1922 Wente, E. C.: Light valve modulator demonstrated; Polychromide two-colour subtractive process demonstrated.

1923 Phonofilm process demonstrated; first two-colour subtractive Technicolor process used; 16mm and 9·5mm systems introduced for amateur use.

1924 Petersen, A. C. G., and Poulsen, A.: Sound-on-film process patented.

1925 Anaglyph stereoscopic process exploited commercially.

1926 Vitaphone sound-on-disc process used for short films; Magnascope variable screen size process.

1927 First Vitaphone feature film *The Jazz Singer*; Chrétien, H.: Anamorphic lens patented; Gance, A.: Three-film projection system demonstrated.

1928 RCA Photophone, Fox Movietone and GPP sound-on-film processes introduced; two-colour subtractive Technicolor imbibition printing introduced; two-colour additive Raycol process demonstrated; first amateur colour process 16mm Kodacolor (three-colour additive lenticular process).

1929 Blattner, L.: Magnetic recording system demonstrated; 56mm Magnafilm, 65mm Vitascope, 70mm Grandeur, wide-film processes.

1930 Fulvue anamorphic lens process demonstrated; 16mm sound projectors introduced.

1931 Two-colour subtractive Multicolor process; three-colour additive mosaic process Dufaycolor demonstrated.

1932 Three-colour imbibition Technicolor prints; Polaroid polarizing material introduced; 8mm system introduced.

1933 Ives, F. E. and H. E.: Parallax stereogram demonstrated.

1934 Gasparcolor three-colour subtractive process introduced; three-strip Technicolor cameras used; three-colour additive Opticolor demonstrated; 16mm Dufaycolor film introduced.

1935 First tripack process Kodachrome demonstrated; first three-colour Technicolor feature film; Audioscopics–anaglyph stereo process demonstrated; 16mm and 8mm Kodachrome films introduced.

1937 Agfacolor three-colour reversal film introduced; Chrétien, H.: two-film anamorphic process demonstrated; 16mm Agfacolor films introduced.

1939 Stereoscopic projection using polarizers.

1941 Stereophonic sound used in *Fantasia*; Agfacolor negative positive process developed; parallax stereogram process introduced in USSR.

1944 Magnetic tape recording process developed.

1948 Two-colour Trucolor subtractive process introduced.
1949 Magnetic recording (non-synchronized) in use; several negative-positive subtractive processes based on Agfacolor introduced.
1950 Magnetic tape and film (synchronized) used for most original recording.
1951 35mm magnetic film used for multi-track stereophonic sound at Festival of Britain; Eastman colour films using integral colour masking introduced; stereoscopic films with polarized projection at Festival of Britain.
1952 Multitrack magnetic sound used in Cinerama process; Cinerama three-film process.
1953 Magnetic stripe tracks for stereophonic Cinemascope prints; Cinemascope introduced; 3-D films reintroduced.
1954 VistaVision introduced; Cinemascope 55 process used; magnetic striping on 16mm film.
1955 Five-track magnetic sound used on 70mm Todd-AO; Dynamic Frame process demonstrated; Disney 11-film Circarama process; Todd-AO 65/70mm process.
1956 Technirama introduced.
1958 Cinemiracle three-film process.
1959 Russian Kinopanorama 22-film process.
1964 Techniscope process introduced.
1965 Super-8mm format introduced.

B.W.C.

See: *History* (FILM).

□CHRONOLOGY OF PRINCIPAL INVENTIONS

The following is a table of the major introductions of technique and apparatus which advanced the television art to its present standard.

Wherever possible reference has been made to Letters Patent granted in respect of the inventions. The specifications of these Letters Patent may be inspected and copies obtained from the Patent Office Library of the reader's country. It should be noted that the quotation of a patent number does not necessarily indicate that the invention was first patented in that country.

1839 Becquerel, E. (France): Observation of photo-electric effect.
1842 Bain, A. (Great Britain): Facsimile apparatus. At the transmitter, the picture was set up by type faces over which an electrical contactor moved. At the receiver, a contactor moved over chemically prepared paper to form an impression of the original picture by discoloration. It was proposed that the contactors should be synchronized by electrically controlled swinging pendulums. British Patent 9745: 1843.
1847 Bakewell, F. (Great Britain): Chemical telegraph. BP 12352: 1848.
1855 Casselli, Abbé G. (France): Early work on facsimile transmission. BP 2532.
1859 Plucker, Julius: Early work on cathode rays.
1861 Caselli, Abbé G. (France): Phototelegraphy system. BP 2395.
1873 May, J., and Smith, Willoughby (Eire): Photo-electric effect in selenium observed and investigated.
1878 Crookes, Sir William: Cathode ray tube.
1878 Senlacq, M. (of Ardres): Telectrascope described in *English Mechanic*.
1880 Carey, George (of Boston): Proposed in *Design and Work* (26th June 1880) a television apparatus using a plurality of selenium cells to analyse the picture, each cell having a separate transmission wire.
1880 Ayrton, W. E., and Perry, J.: Proposal described in *Nature*. Generally similar to Carey's except for the receiver, for which two possible approaches were suggested.
1881 Bidwell, Shelford: Demonstrated his pinhole box camera containing a selenium cell. Scanning was effected by moving the box.
1884 Nipkow, P. (Germany): The Electrical Telescope. This was his rotating disc scanner. German Patent 30105.
1887 Hertz, H. R.: Work on properties of alkaline metals when electrified under the influence of light.
1889 Weiller, Lazare (Germany): Rotating mirror scanning system using rotating drum provided with small mirrors which reflected the image on to a selenium screen.
1895 Berglund, R. (Great Britain): Photo-telegraphic system employing scanning at the transmitter and receiver.

1897 Braun, K. F. (Germany): Cathode ray oscilloscope tube.

1897 Sczcepanik, J., and Kleinburg: Scanning in a zigzag fashion. Vibrating mirror scanning system using two mirrors, one of which vibrated much faster than the other. BP 5031.

1904 Fleming, J. A. (Great Britain): Thermionic vacuum diode.

1904 Wehnelt, A. R. B.: Hot cathode in Braun tube.

1905 Elster, Julius, and Geitel, Hans (Germany): Developed photo-electric cell.

1907 Rosing, Boris (Russia): Cathode ray oscilloscope tube used at the receiver and synchronization between transmitter scanning and receiver scanning employed. Saw-tooth deflection and synchronizing voltages for the cathode ray tube at the receiver were generated by several rheostats attached to the rotating mirrors at the transmitter. BP 27570 (1907), 5486, 5959 (1911).

1908 Campbell Swinton, A. A.: Use of a cathode ray tube at the transmitter as well as at the receiver proposed in a letter to *Nature* (18th June 1908).

1908 de Forest, Lee (USA): Thermionic triode amplifier. BP 1427.

1909 Andersen, A. C. and L. S. (Denmark): Colour television system proposed. Mechanical scanning system using a selenium cell and a dispersive prism. BP 3018.

1910 Gaede molecular air pump invented in Germany.

1911 Sinding-Larsen, A.: Transmission of television signals by wireless proposed. Three wireless channels proposed, one for picture signals, the other two for the synchronizing signals of the mechanical scanning arrangements. BP 14503.

1918 Siemens and Halske; Levy, T.; and Armstrong, E. H.: Superheterodyne receiver principle developed independently. BPs 135177, 133306, 137271.

1919 Slepian, J. (USA): Vacuum tube electron multiplier. US Patent 1450265.

1923 Stephenson, W. S., and Walton, G. W.: Interlace scanning proposed, in a mechanical scanning television system. BP 218766.

1923 Westinghouse (USA): Picture signals and synchronizing information combined on one carrier. A photoelectric cell and a cathode ray tube operated as a flying-spot scanner at the transmitter. A single wireless carrier wave carried four modulations – picture signals, two synchronization signals, and sound. BP 225553.

1923 Zworykin, V. K. (USA): Iconoscope camera tube. Blanking and synchronizing signal transmission during flyback. US Patent 2141059.

1929 Walton, G. W.: Electron multiplier in photoelectric vacuum tube. BP 338548.

1930 Farnsworth, P. T. (USA): Image dissector tube and associated circuitry. BPs 368309, 368721.

1930 Telefunken (Germany): Interlace scanning applied to electronic television receiver. BP 380602.

1931 Telefunken (Germany): Silver-activated zinc cadmium sulphide used for cathode ray tube screen. BP 410414.

1932 Telefunken (Germany): Proposed synchronizing signals blacker-than-black or whiter-than-white for separation purposes. BP 413561.

1934 Lubszynski, H. G., and Rodda, S. (Great Britain): Image Iconoscope. BP 442666.

1934 Bedford, A. V. (USA) Line and frame synchronizing signals locked in a fixed relationship with the mains a.c. BP 458161.

1937 Blumlein, A. D. (Great Britain): Spot wobbling. BP 503555.

1938 Rose, A. (USA): Orthicon camera tube. BP 530409.

1940 Rose, A. (USA): Image orthicon camera tube. BP 613003.

1949 Rose, A. (USA): RCA shadow mask colour tube. BP 674114.

1950 Vidicon camera tube described in *Electronics* for May 1950.

1951 Hazeltine Corporation (USA): PAL colour television system. BP 702182; thereafter developed in Germany under Dr. W. Bruch.

1951 IRE-RTMA Radio Fall Meetings, Toronto, October 1951: NTSC colour television system. (First public disclosure *Electronics*, February 1952.)

1953 Georges, Henri, de France (France): SECAM colour television system. BPs 792863, 821246.

1953 Philips (Holland): Plumbicon camera tube. BP 763745. D.G.R.

See: *History* (TV).

⊕ **CIE System.** The Commission Internationale ☐ d'Éclairage have established several recognized international standards in colorimetry, in particular, the specification of three standard illuminants of defined spectral energy distribution, the definition of the colour-matching characteristics of a Standard Observer, and from these the establishment of three colour stimuli known as XYZ on which colorimetric specifications and calculations for colour matching are based.

See: *Colour principles.*

⊕ **Cinch Marks.** Abrasions or scratches running lengthwise on a piece of film; usually caused by trying to tighten a loosely wound roll of film by pulling the end of the film so that the layers rub against one another.

See: *Winding.*

⊕ **Cinecolor.** Process for the production of motion pictures in colour, widely used between 1935 and 1955.

Originally Cinecolor was a two-colour process in which prints from a bipack original negative were made on duplitized stock which was processed by dye-toning to provide blue-green and orange-red images in register on opposite surfaces of the film.

In its later form it was a three-colour system, the yellow and magenta images being produced successively on one surface and the cyan on the opposite side.

The process became obsolete after the introduction of integral tripack colour positive material.

See: *Colour cinematography.*

⊕ CINEMACROGRAPHY

Cinemacrography is cinematography of small objects without using the compound microscope. It is complementary to cinemicrography and in any particular situation the decision as to which technique to employ depends on the nature of the object, any manipulative techniques to be used during filming, and the purpose for which the film is required. Magnification likewise depends both on the size of the object and the freedom of movement which is needed to express both normal movement and behaviour.

Cinemacrography aims at producing an image of the object on film but this image does not necessarily have to be larger than life. In some circumstances, the object may need magnifying before it can be appreciated but this is not always the case.

Television advertising, for instance, makes much use of cinemacrography in the so-called "pack shots" but these are normally filmed at a scale of less than 1 : 1. Their visual impact rests as much on the manner in which they are presented to the viewer as on the filming technique. The advertising matter is shot either with the standard motion picture camera lens or by using lenses specially computed for working at close range, the picture being composed to allow for the superimposition of titles. There must be absolutely no float of the film image, otherwise on projection after superimposition, the titles appear to move against the background. Apart from the filming of the advertising material, which could be a carton of soap powder or a cutout of a tin of cat food, the rest of the film technique is carried out on the animation rostrum. Pack shots and biological teaching films are two examples of the kind of material which is to be found in the world of cinemacrography.

EQUIPMENT

The traditional methods of obtaining film images of comparatively small objects are either by the use of supplementary lenses placed in front of the normal lens or by increasing the distance between the lens and film plane by means of extension tubes. Advances in lens design have suggested yet another method of taking close-up pictures, i.e., by using an endoscope coupled to the motion picture camera.

There are, however, other ways of achieving a similar result, one being to use two similar lenses in tandem, back to back, a second the substitution of a microscope objective for the camera lens, and a third by using a specially computed close-up attachment with a focusing zoom lens. It is also possible to use the standard camera lens if it is turned round, i.e., with the back of the lens towards the object. Most of these methods are means of minimizing the spherical aberration which may occur when using long extension tubes with a normal camera lens.

SUPPLEMENTARY LENSES. A converging positive supplementary lens is a simple thin lens which is fitted immediately in front of the camera lens with which it is to be used.

Its magnifying power is usually expressed in diopters (the diopter being the reciprocal of the focal length in metres, e.g., 1 metre F.L. = 1 diopter, and ½ metre F.L. = 2 diopters). When one of these supplementary lenses is used with a camera lens focused on infinity, the resulting optical combination brings objects that are situated at the focal length of the supplementary lens into sharp focus on the film plane. If the camera lens is focused on a distance closer than infinity, the plane of sharp focus with the supplementary lens in position is thereby brought closer, thus yielding a bigger image. The distance of sharp focus is then:

$$\frac{\text{Distance camera lens is focused on} \times \text{F.L. supplementary lens}}{\text{Distance camera lens is focused on} + \text{F.L. supplementary lens}}$$

When using supplementary lenses, the *f*-number of the lens combination is not changed if the camera lens is focused on infinity, otherwise the effective value of the lens aperture is as follows:

$$\text{Effective } f\text{-number} = \frac{\text{marked } f\text{-number} \times D}{D+F}$$

where D = focal length of supplementary lens
F = focal length of camera lens.

EXTENSION TUBES. There is a theoretical limit to the magnification which can be obtained by using extension tubes but, in practice, before this point is reached, other factors intervene. The first is the problem of keeping the set-up sufficiently rigid to prevent unwanted movement, the second the rapid increase in *f*-number which results with increase in extension tube length, Lastly, there is the problem of lighting the object adequately without harming it.

An extension tube consists of a metal tube of similar diameter to the lens mount, threaded at both ends. In use, the lens is removed from the motion picture camera and attached to the extension tube which is then fitted to the camera in the place normally occupied by the lens.

Similarly, a microscope objective can be used in place of the camera lens. In this case, the tube length should approximate the mechanical tube length for which the objective was computed. Modern objectives usually work best with a tube length of 160mm but this can vary considerably depending on the manufacture and age of the microscope objective, and whether it is used with a covered or uncovered object.

Provided great care is taken to support and mount the extension tubes, it is practical to have a total tube length of up to 300mm (12 in.).

MACRO LENSES. In effect these are camera lenses, the only difference being that they are so computed that the conjugate distances are reversed, i.e., the larger conjugate is at the back of the lens and the shorter in front. Their maximum aperture is about *f* 4·5, regardless of focal length. They are fitted with an iris diaphragm for reducing the aperture and increasing the depth of field. If used without an intervening microscope body tube, a camera bellows or extension tube should be employed.

ULTRA CLOSE-UP ATTACHMENT FOR ZOOM LENS. The design of zoom lenses, which depends on a fixed distance between lens mount and camera face, precludes the use of extension tubes for macro work, and the supplementary lenses normally available have a minimum working distance of 60 cm. More powerful supplementary lenses cannot be used as they introduce a marked degradation of image quality. There is, however, an ultra close-up attachment made by Optec of London for use with Angénieux and Som Berthiot zoom lenses which permits focusing down to just under 22 cm, giving a minimum field size of 22×30mm.

ENDOSCOPES. In some circumstances, an endoscope, which is an optical instrument used for the visual examination of the interior of a body cavity or hollow viscus, can be coupled to the motion picture camera. A special form of this instrument, the Modelscope (Optec of London), allows cinemacrography of inaccessible small objects. A wide-angle lens of extremely short focal length is fitted to the end of a slender repeater-type telescope. It has a depth of field from about 5mm to infinity, the entire instrument being 5mm in diameter and 300mm (12 in.) long. A unique feature of this instrument is the undistorted perspective it provides of both the outside and inside of scale models.

ADDITIONAL EQUIPMENT. Whereas in cinemicrography there is a wealth of commercially made equipment readily available, e.g., microscopes and light sources, the same is not true of cinemacrography. The scientist or cameraman may have to improvize and adapt existing apparatus to suit his particular purpose. Light sources can be

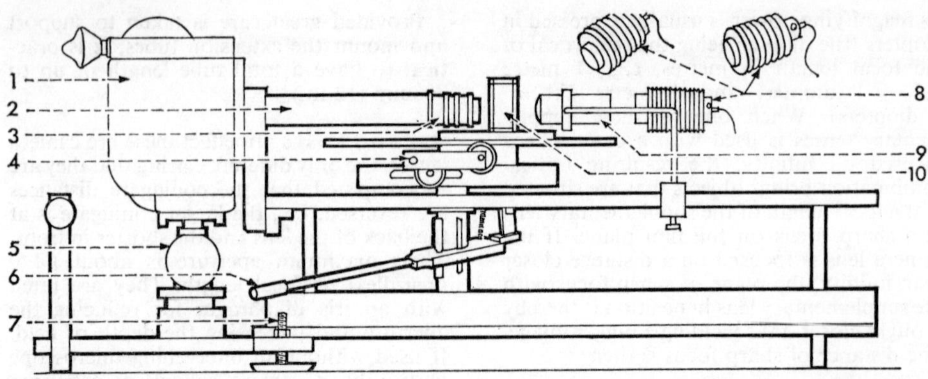

SPECIALIZED EQUIPMENT FOR CINEMACROGRAPHY. Camera, 1, extension tube, 2, and specimen holder, 10, are all mounted on same rigid base, so will tend to move as a whole in event of accidental displacement. 3. Camera lens. 4. Fore-and-aft movement of specimen holder for making initial set-up. 5. Pre-set camera adjustment. 6. Control handle for vertical movement via a plunger with arresting device. 7. Levelling screw (1 of 2) for baseplate. 8. Lamphouses with ball-and-socket adjustment. 9. Dark-ground condenser.

constructed from a photoflood bulb with a 3-litre round-bottomed flask full of distilled water acting as a combined bull's eye condenser and heat filter. An extra pair of hands can be provided by using a motor car dip switch as a foot control for the camera. Disused lathe beds can be brought into use as tripod stands able to provide movement in the three planes of space and be sufficiently solid to dampen unwanted vibration. Even the standard tripod can be modified so that split plates are used to assemble the camera instead of the conventional screw fitting, thus saving time in field work.

TECHNICAL PROBLEMS

ILLUMINATION. One of the major problems is to provide sufficient light for filming and yet not overheat or damage the object. Most methods used in cinemicrography are also applicable to cinemacrography. An additional technique is to pipe light on to the specimen by means of a quartz rod or bundle of fibres. The light has no heating effect on the specimen and takes up very little room. It can be "sprayed" on to the object from a ¼ in. rod giving about a ½ in. spread of light. With transmitted light, the special macro lenses already mentioned can be used in conjunction with special large field condensers.

The normal substage condenser of a compound microscope should not be used in cinemacrography as it does not provide a large enough field of illumination to cover the objects usually encountered in this type of work. The selection of suitable condenser lenses depends upon both the size of the object and the focal length of the camera lens. They are usually plano-convex, their diameter being somewhat greater than the largest dimension of the object to ensure adequate and efficient illumination over the entire object area.

The focal length of the condenser should not be less than that of the camera lens and

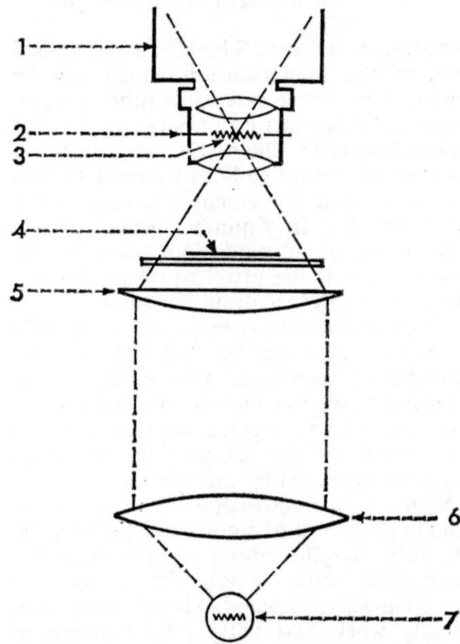

ILLUMINATION BY TRANSMITTED LIGHT. 1. Cine camera. 2. Lens diaphragm. 3. Image of lamp filament. 4. Object. 5. Plano-convex condenser. 6. Lamp condenser. 7. Light source.

preferably somewhat longer. The object is placed between the flat side of the condenser and the camera lens and as close to the former as is practical.

Köhler's illumination is rarely used for cinemacrography as it requires optical equipment of a high standard, and equally high picture quality can be obtained by using a simpler system of illumination by transmitted light.

The specimen is placed in position close to the plane side of the condenser lens. The position of the film camera and lens is then adjusted until the image of the specimen, as seen through the watching eyepiece, is at the required magnification.

Next, the light source is centred with respect to the condenser and camera, the camera lens diaphragm is closed and the specimen removed. The position of the lamp house and the focus of the lamp condenser are adjusted until an image of the lamp filament is formed on the camera lens diaphragm; this image should be slightly larger than the diameter of the lens. The specimen is replaced and a check made to see that it is adequately illuminated.

If the condenser is moved closer to or further from the specimen, the illumination beam becomes larger or smaller respectively. Whenever the condenser is moved, the lamp must be readjusted to refocus the filament image on the camera lens diaphragm.

The lens diaphragm is opened to an aperture rather smaller than its maximum (e.g. an f 4·5 lens should be reduced to f 5.6 or f 8). The lamp diaphragm can be left fully open since it is not used as a field stop as in cinemicrography. The focus of the image and the evenness of illumination through the watching eyepiece should be given a final check.

As some specimens may be transparent and practically colourless, dark-ground illumination can be used with advantage to show up their internal structure. The same principle applies in cinemacrography as in cinemicrography when employing dark-ground illumination, i.e., to prevent any direct light from the lamp house from reaching the camera lens, so that the object is seen only when light scattered by it is collected by the camera lens.

EFFECT OF DISTANCE ON f-NUMBER. As long as the motion picture camera is used to film an object at a distance of more than 10 times the focal length of the lens, the f-numbers marked on the lens barrel are operative. When the object is at a distance of less than this, the effective f-number is:

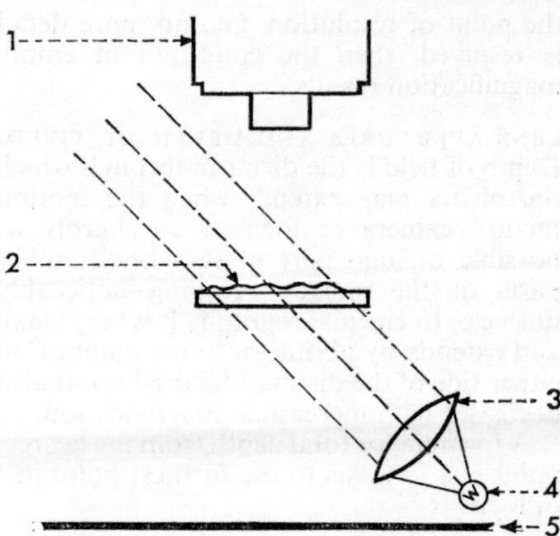

DARK-GROUND ILLUMINATION. Direct light from lamp screened from lens, so object seen only when light scattered by it reaches lens. 1. Cine camera and lens. 2. Object. 3. Lamp condenser. 4. Light source. 5. Black back drop.

$$\frac{\text{nominal } f\text{-number} \times (\text{distance of lens from object} + 1)}{\text{distance of lens from object}}$$

Also, when filming with extension tubes, the nominal f-numbers are no longer valid when the scale of reproduction is larger than 1/10. The increased exposure is given by multiplying the nominal f-number by the scale of reproduction+1.

e.g., Reproduction of object $\frac{1}{2}$ size nominal f-number = 4

∴.Effective f-number is $4 \times (\frac{1}{2}+1) = 6$.

Reproduction of object same size nominal f-number = 4

∴.Effective f-number is $4 \times (1+1) = 8$.

EXPOSURE INCREASE WITH MAGNIFICATION

	Magnification				
	x2	x4	x8	x12	x15
Nominal f/No.	4	4	4	4	4
Effective f/No.	12	20	36	52	64

RESOLVING POWER AND EMPTY MAGNIFICATION. The resolving power of a lens is its ability to render fine detail. Two lines extremely close together are said to be resolved when they can be distinguished as two lines. If an image of these lines is formed but does not show any separation between them, then the lines are not resolved. If magnification is carried beyond the point of resolution, i.e., no more detail is resolved, then the condition of empty magnification results.

LENS APERTURES AND DEPTH OF FIELD. Depth of field is the distance through which the object may extend, when the motion picture camera is focused as sharply as possible on one part of it, without other parts of the image becoming noticeably unsharp. In cinemacrography it is very small and extends by about the same amount on either side of the distance focused on. It also decreases with increase in magnification.

A formula for total depth from the nearest point just in focus to the furthest point just in focus is

$$\text{Total depth} = \frac{2\,cf(m+1)}{m^2}$$

where f = *f* number
c = permissible diameter of the circle of confusion in the negative (usually 1/1000 of the focal length of the lens)
m = magnification.

The smallest camera lens aperture can be used to obtain maximum depth of field but this may result in an image that is less sharp, because of diffraction effects, especially when using macro lenses. Focusing should therefore be carried out at the aperture to be used in filming.

EXPOSURE DETERMINATION. Methods of exposure determination are similar to those used in cinemicrography.

USE OF FILTERS. Particular attention should be paid when using reversal colour film to any "colour cast" which may occur when using large plano-convex condensers in conjunction with macro lenses. It will probably be greenish in colour and can be offset by the use of an appropriate magenta-coloured compensating filter.

VIEWFINDER SYSTEMS AND PARALLAX. Some form of direct or reflex viewing is necessary in cinemacrography. Most modern motion picture cameras have this facility, and zoom lenses are usually supplied with a reflex finder.

The main fault with all separate viewfinders is the introduction of a parallax error. The vantage point from which the observer sees the scene through the viewfinder is not the same point as the forward nodal point of the lens. When the object is at infinity the difference is negligible, but when the camera is used for cinemacrography the difference is appreciable. The exception to this rule is when filming at comparatively long range. Then it may be possible to use a tilting device to frame the specimen, knowing by trial and error that anything within the frame will be recorded.

K.M.

See: *Cinemicrography; Medical cinematography; Time lapse cinematography.*

Books: *Macrophoto and Cine Methods*, by A. and I. Tölke (London), 1969; *Photomicrography*, by C. P. Shillaber (New York), 1963; *Photography for the Scientist*, by C. E. Engel (New York), 1968.

⊕**Cinemascope.** Trade name of the most widely used method of anamorphic wide-screen presentation; camera lenses producing images on 35mm film with a 2 : 1 lateral compression are used, with compensating horizontal expansion on projection.

See: *Wide screen processes.*

⊕**Cinemascope Perforation.** Sprocket holes on stock used for positive prints of Cinemascope films with four magnetic sound tracks. Also known colloquially as foxholes, from Twentieth Century-Fox, the company which introduced the process.

See: *Film dimensions and physical characteristics.*

⊕CINEMA THEATRE

HISTORY

It was at the end of the first decade of the present century that buildings began to be erected especially for the exhibition of motion picture films. Before this time, films were shown in any kind of permanent or temporary premises that could accommodate a seated audience. In towns and cities, concert halls, lecture halls and the like were hired by exhibitors often on a weekly basis and not infrequently as a one-night stand. In the country and urban districts, enterprising showmen would travel from fairground to fairground, giving their performances in marquees in competition with the sideshows and circus.

The exhibition of films during interludes became popular in music halls and variety theatres, so much so that the time devoted to them increased as more and longer films became available until, eventually, they occupied the whole programme.

MUSIC HALL INFLUENCE. Early cinemas built around the 1910 era closely followed the design and decor of the Edwardian music hall. In Britain and in America, only the boxes and the bars were missing – the former because they occupied a position that was unsatisfactory for viewing the screen, and the latter because of the difficulties of the licensing laws when children were on the premises. But the cinemas were complete with proscenium, stage and orchestra pit, with stalls and circle and, in some cases, with a gallery. In continental Europe the boxes and bars remained. The boxes, on two or three levels, extended along the side walls of the auditorium.

The music hall concept in the design of cinema auditoria remained during the boom period between World Wars I and II. Proscenium, stage and orchestra pit prevailed; and even when the orchestra became too expensive to maintain, its place was occupied by the decorated and illuminated organ console. Such was the cycle of events, that music hall acts and tableaux provided the interlude between the film programmes.

EMERGENCE OF THE CINEMA. Cinemas increased both in number and size. The smallest town could boast two or three cinemas, and in the cities, halls accommodating some 3,000 or more people were not uncommon. In Western Europe there was one cinema seat for every fifteen members of the population, and in industrial towns and cities there was one seat for every ten. In 1935 there were in Britain 4,300 cinemas, providing one seat for every 12 members of the population.

Commercial considerations were paramount; upon a given site, an auditorium of greatest possible seating capacity had to be created. Auditoria were rectangular in plan to give maximum seating density. Preferably the auditorium had to be not too wide to permit the economical erection of a large balcony, cantilevered so that supporting columns were unnecessary, thus giving an unobstructed view from below.

The introduction of sound films in 1928/29 had little effect upon auditorium design. If on completion an auditorium was unsatisfactory acoustically, sound absorbent materials were placed upon walls and ceiling until more or less acceptable results were achieved. Although the pioneer work of W. C. Sabine some fifty years previously had laid firm foundations for the practice of architectural acoustics, and there was a substantial literature on the subject, the acoustics of most cinemas were usually corrected after building.

CINEMA DESIGN. The decor of cinemas built in the early thirties changed from the crimson, cream and gilt of the music hall to more grandiose and exotic motifs, perhaps reflective of the romantic adventure films that were popular at that time. Schemes of decoration, often flamboyant and garishly illuminated by multi-colour systems of lighting, became the vogue. The café or restaurant in close proximity to the balcony foyer became a necessary adjunct to the exhibition of films. But in the late thirties a few architects who had established themselves in other fields of activities introduced more modest styles. Perhaps because the film was beginning to be recognized as an independent art form, a combination of culture and entertainment, these architects were more conventional or traditional in their approach to cinema design. They set up a high standard of dignity and restraint, with the result that there are in Britain and elsewhere several cinemas of unique architectural quality.

World War II caused an almost complete cessation in cinema construction, and it was not until the 1950s that restrictions began to be lifted and economic conditions permitted the construction of new buildings. At this time, the design of cinemas began to be influenced strongly by many new factors. The mode of life of the civilized world had changed. Standards of education had improved substantially, people were enjoying higher living standards, and more time was available for leisure and recreation. Radio, and particularly television, had become the principal media for mass entertainment. In this era of sophistication and, perhaps, of disillusionment, films of the character so popular before the war were no longer acceptable. Cinema attendances declined.

DEMANDS OF THE WIDE SCREEN. In 1952 a considerable technical advance in motion pictures was presented to the public. This technique, known as Cinerama, had for its object the creation of an impressive visual and aural impact upon the audience. This was achieved by projecting the picture upon a large, semi-circular screen and reproducing the sound stereophonically.

Three separate photographic images were projected on to the large screen to provide the complete picture, and consequently three projectors, electrically interlocked to provide accurate synchronism, were required; these were spaced approximately 47° apart and equidistant from the screen. Five speaker systems were placed behind the screen and several ambient speakers in the auditorium.

The technique could not be widely used because not many auditoria could accommodate the large screen satisfactorily or meet the exacting projection requirements which the system demanded. Additionally, because of their high production costs, very few films were produced. Thus, the original system was adopted only in large cities with a population that would support the showing of one film for a long period.

During the war period and the subsequent decade, there had been substantial advances in the techniques of cinematography. The quality of photographic emulsions and particularly of colour systems had greatly improved. There had also been substantial advances in the design and manufacture of lenses both for camera and projector use. These lenses were of much shorter focal length than those previously available and of much larger aperture, thereby transmitting more light than hitherto. Thus, the area of the screen in a cinema could be increased four-fold without impairing the acceptability of picture quality, in terms both of the resolution of the picture image and of its brightness.

Taking advantage of these improvements, Twentieth Century-Fox in 1953 introduced the Cinemascope system. It was recognized that if larger screens were to be fitted in existing cinemas, restrictions upon the height of the screen would appear sooner than on the width. Very frequently the height of the screen was governed by the line of sight of members of the audience sitting in the rear stalls, the maximum vertical angle of view being limited by the soffit of the balcony; and in all cases, the increase in screen height would involve removing seats from the front stalls because of unsatisfactory viewing angles.

WIDE SCREEN TECHNIQUE. The Cinemascope system involved the use of a cylindrical (anamorphic) auxiliary lens in front of the normal camera lens. The cylindrical lens with its axis disposed vertically, had the effect of compressing the image laterally without disturbing it vertically. The anamorphic lens chosen for Cinemascope had a lateral compression ratio of 2 : 1. The compressed photographic image on the negative was transferred to the positive release print, and the fitting of an anamorphic lens of similar ratio in front of the projector lens enabled the image to be expanded so that an undistorted picture was projected on the screen.

Although the ratio of compression and expansion was 2 : 1, the width of the screen picture was not in fact doubled, because at this time Twentieth Century-Fox introduced a stereophonic sound system which required four magnetic tracks to be placed upon release prints, for which a part of the picture area had to be yielded up. Unfortunately, cinema owners in general would not accept the stereophonic sound system because of the high cost of the necessary additional equipment. Twentieth Century-Fox was compelled to add an optical sound track, which absorbed still more of the picture area, and thus the aspect ratio (the ratio of the height to the width) of the screen pictures was reduced from a possible 1 : 2·75 to 1 : 2·35.

Because the screen picture produced from a normal release print appeared unacceptably small when compared with the Cinemascope picture, cinema owners adopted the practice of projecting the

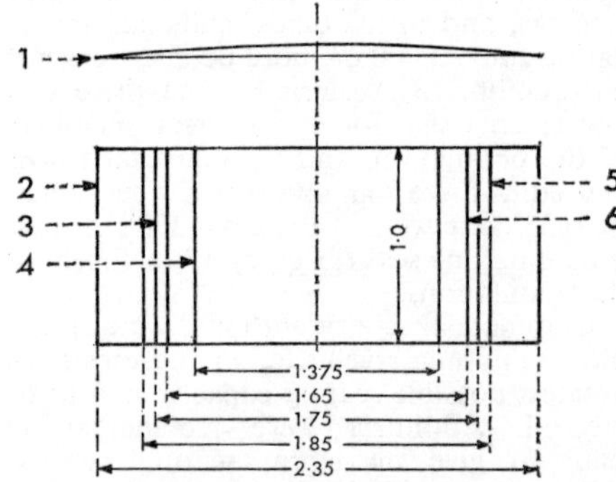

ASPECT RATIOS FOR 35MM PROJECTION. 1. Screen may be flat or curved. If curved, radius of curvature is usually distance from projector lens to screen. 2. Cinemascope picture. 3. International preferred ratio for wide screen projection. 4. Basic international format. 5. International maximum ratio for wide screen projection. 6. International minimum ratio for wide screen projection.

normal release print at a greater magnification than hitherto. This was accomplished by fitting projector lenses of shorter focal length. But because of the previously mentioned restrictions upon increasing the height of the picture, it was necessary to fit plates to the film gate of the projector with apertures of less than standard height. This departure from the then standard practice was loosely termed the wide-screen technique.

The size of the screen and its height-to-width ratio installed in any particular cinema invariably depended not only on the sight lines of the audience, but also upon structural restrictions. In many cases, the width of the proscenium opening was a determining factor, and in others the presence of exit doorways on each side of the front wall of the auditorium restricted the width of the screen. There was, however, a general tendency to install screens as wide as possible, with the result that aspect ratios of 1 : 1·9 and greater were not uncommon. The projection of films intended to be shown at a ratio of 1 : 1·375 but cut down 30 per cent. in height, often produced ludicrous results – where in close-ups artists appeared semi-decapitated or important action was cropped off and unseen by the audience.

This position has, however, been regularized. Several countries have adopted national standards for wide-screen pictures and there is currently a proposed international standard, that requires the picture on the negative to be composed for an aspect ratio of 1 : 1·75, but capable of being projected at any ratio between 1 : 1·375 and 1 : 1·85 (1 : 1·75 being the preferred ratio).

Hitherto the standard film used in cinemas was 35mm wide, but the desire for very large screen pictures culminated in the introduction in 1957 of 70mm release prints, first adopted by the late Michael Todd in collaboration with the American Optical Corporation. By reason of the smaller magnification required, the photographic quality of the screen picture was much enhanced, while the picture could be adequately illuminated because of the larger area of the aperture in the projector gate.

The success of this format led to further developments. The Cinerama Corporation adopted this format and, by the use of slight anamorphic compression of the film image and of specially corrected lenses for the projectors, found that a very satisfactory picture could be projected on to the deeply curved Cinerama screen if restricted to an included arc of 120°. This eliminated the mismatch of the three separate screen pictures which was a distracting element of the three-projector system.

Other similar systems based upon the use of 70mm film, deeply curved screens and the use of specially-corrected projector lenses are known as Dimension 150 and Vistarama – in the case of the latter, the screen embraces an arc of 70°.

SMALLER CINEMAS. Despite these new techniques and improvements in the art of cinematography, cinema attendances continued to fall throughout the period 1950 – 1968. In America, Britain and many other western countries more than 55 per cent of cinemas closed and only in those countries where living and educational standards were lower did they continue to flourish.

In this era of sophistication and discernment, films of a character and type so popular before the war were no longer acceptable. Producers turned to other subject matter or to different treatments for their films. Although the mammoth epic pictures became more spectacular and glamorous, demanding large audiences for their support, the new cinemas built were smaller than their predecessors. In many cases large cinemas were divided and reconstructed to form two smaller ones.

The diminution in size was undoubtedly the direct result of the reduced audiences, but it enabled the architect to design more intimate and more comfortable auditoria; and it enabled the art and techniques of cinematography to be presented with greater perfection.

While in Britain the closure of cinemas has greatly exceeded the number of new cinemas constructed, in European continental countries there is a resurgence of cinema attendances, and in France, Germany and Spain the number of cinemas in operation in 1968 is substantially greater than in 1957. In this ten-year period the number of cinemas in Spain and in Japan has doubled and in the USA a substantial number of new cinemas, particularly drive-ins, has been built during the last two years.

AUDITORIUM DESIGN

In the design of a cinema, the architect must accept as the primary requirement of his brief that his structure is required solely to accommodate this technique; and that the audience shall be presented with an unobstructed view of the screen picture,

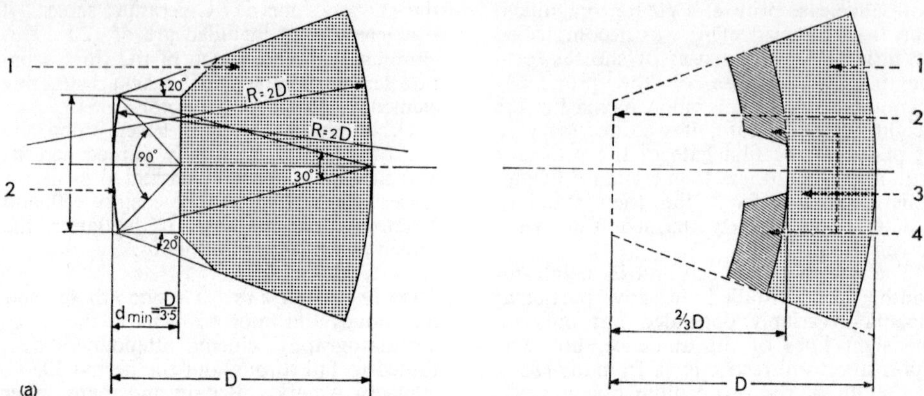

FRENCH STANDARD FOR SEATING AREAS OF AUDITORIA WITH CINEMASCOPE SCREENS. (a) Stalls. (b) Balcony. D. Distance from screen to rearmost row of seats. 1. Seating area. 2. Position of screen. 3. Projection room in balcony void. 4. Permissible additional balcony seating.

which must be of adequate size, of satisfactory brightness and free from objectionable distortion. The audience must also be presented with sound that is clear, and free from coloration and other acoustic defects, and speech must be readily intelligible.

FRENCH STANDARD. The first determinant is the relationship between the size of the screen and the auditorium. Only one country has determined this relationship by means of a national standard. In this specification (AFNOR S27.001) the maximum length of the auditorium is determined by the width of the screen picture when a Cinemascope film of the ratio 1 : 2·35 is being projected. The length of the auditorium is twice the picture width. In order to provide the most favourable viewing conditions, the horizontal angle from the edges of the screen subtending the eyes of persons sitting in the rearmost row of seats should be 30°, while the subtended angle of view of persons sitting in the front row should not exceed 90°. It is also required that these latter seats shall not be placed nearer the screen than D/3·5, where D= the distance from the screen to the rearmost row of seats. The specification provides a seating area conforming to the following:

$$S = 0.95D^2$$

$$D = \sqrt{\frac{S}{0{\cdot}95}}$$

where D = length of auditorium in feet from screen to rearmost row of seats, and S = seating capacity of the stalls.

The French specification also provides for a balcony and, to avoid trapezoidal distortion of the screen picture, recommends that the projection room be contained within the balcony structure. The distance from the front of the balcony to the screen is therefore limited by the focal length of the lenses available to give satisfactory resolution of the picture and the optical system to give adequate distribution of screen luminance. Thus the specification determines that the front of the projection room should not be nearer to the screen than 2/3D, whence –

$$S_b = 0{\cdot}034D^2 \text{ approx.}$$

$$\text{or } D = \sqrt{\frac{S_b}{0{\cdot}034}} \text{ approx.}$$

where S_b=seating capacity of the balcony.

The seating capacity given by this formula may be increased up to about 20 per cent by extending the balcony forward on each side of the light beam of the projectors.

The French specification does not deal with the relative sizes of screen and seating area for auditoria designed for the projection of 70mm film, but adaptations of the Cinemascope standard have been recommended for projection upon deeply-curved screens having included angles of 120° and 180°. In this recommendation, the seating areas are reoriented to be normal to 60° segments of the screens, and those portions of the seating areas which overlap are considered to be suitable for viewing. The seating capacity may be approximated by –

$S = 0{\cdot}045D^2$ for 120° screen.

$S = 0{\cdot}02D^2$ for 180° screen.

Professor E. Goldovsky of Moscow University has given much consideration to auditorium design for many years, and his recommendations have been adopted in the design of a large number of Russian cinemas. The same general shape of auditorium is adopted for both 70mm and 35mm projection, and the recommended screen sizes in relation to auditorium length are slightly larger than the French.

CINERAMA LAYOUT. The recommended layout of an auditorium for the presentation of Cinerama films is based upon an auditorium with a balcony and accommodating

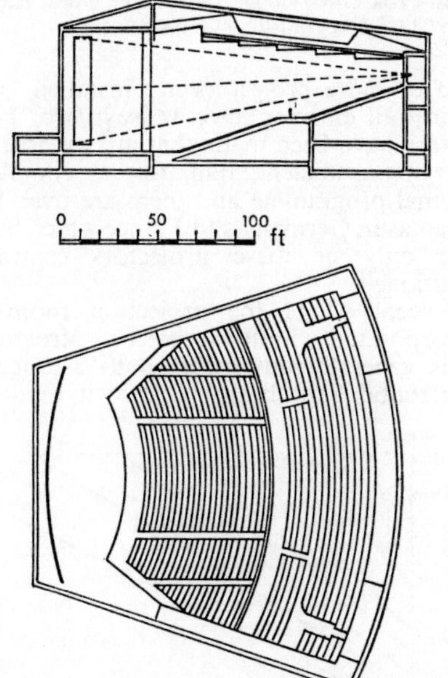

RUSSIAN LARGE CINEMA DESIGN. Recommendation for seating areas of large cinemas projecting 35 and 70mm films, the latter of Todd-AO type.

an audience of 1,200 persons. Here the seating capacity is –

$S = 0{\cdot}93D^2$ (stalls)

$S = {\cdot}043D^2$ (balcony)

The arrangement provides for a curved screen embracing an arc of 146° with three projectors angularly spaced at 46°52′. The auditorium would, of course, be suitable for the Cinerama single projector system using 70mm film, but the included angle of the screen would be reduced to 120°.

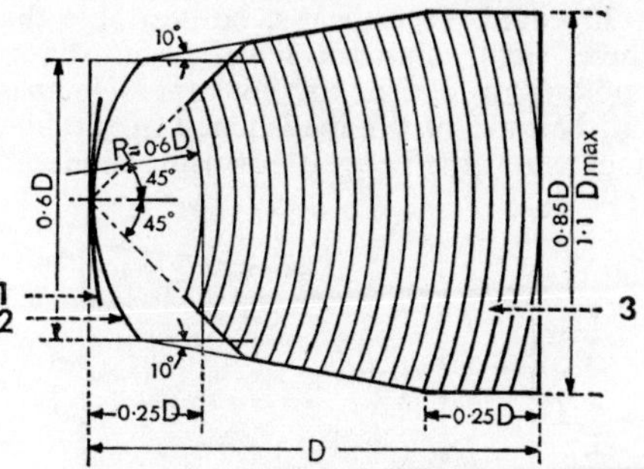

RUSSIAN 35MM CINEMA. Recommended layout for seating area. Same general shape of auditorium adopted for 70mm projection. 1. Screen for Cinemascope. 2. Screen for Todd-AO, Super Panavision, Technirama. 3. Seating area.

ADAPTING TO 16MM. Because of the limitations on the degree of magnification that can be made of the small photographic image, and on the quantity of light and its accompanying heat that can pass through the aperture of the projector gate without damage to the film, it is not possible to present 16mm film to give the same visual impact as with the larger formats; nor indeed would this generally be desirable. It is true that in many countries there is a large number of cinemas showing 16mm exclusively, but to small audiences, and under less exacting conditions.

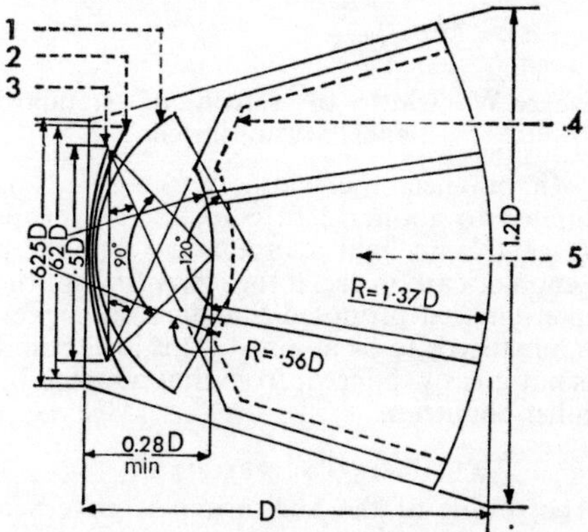

SEATING AREAS OF MODERN BRITISH CINEMAS. Accommodation up to 750 persons. D. Length of auditorium. 1. 120-degree curved screen for Cinerama and Dimension-150. 2. Screen for Todd-AO, Super Panavision, Technirama. 3. Screen for Cinemascope. 4. Limit of seating area for Cinerama and Dimension-150. 5. Seating area for Todd-AO, Super Panavision, Technirama, Cinemascope.

It would, nevertheless, be desirable that these small cinemas have a fan-shaped auditorium similar to the larger cinemas, but because of the inadequacy of the light source and the necessity of using high gain

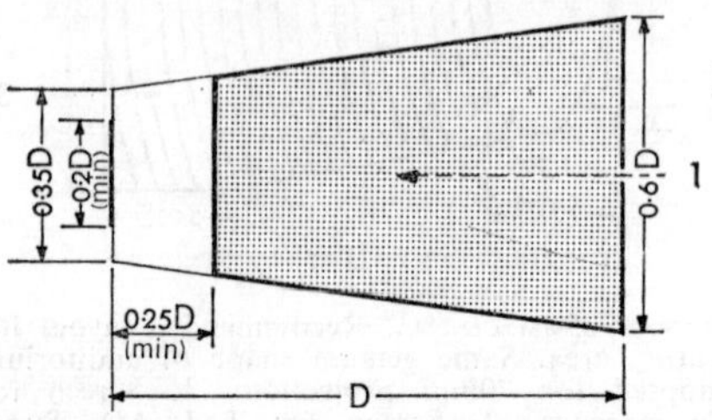

SEATING AREA FOR 16MM PROJECTION. D. Distance from screen to rear row of seats. 1. Recommended seating area.

screens, that is to say, screens having a partially specular reflection characteristic, the audience must be grouped centrally along the length of the auditorium. Thus the auditorium becomes more nearly rectangular, with a length some five or six times the width of the screen picture, with the width of the seated area two-and-a-half to three times the picture width. A typical arrangement is based upon the screen being of the standard aspect ratio of 1 : 1·375. In halls of this type, the seating capacity may be determined by:

$$S = 1.5W^2$$

$$W = \sqrt{\frac{S}{1.5}}$$

where W = width of screen of standard aspect ratio, in feet.

In practice, the width of the screen is limited to about 12 ft. 6 in. for an incandescent lamp light source and 18 ft. for a xenon or carbon arc, if the lower limit of the international proposed Standard for screen luminance is to be attained. This lower limit is not greatly different from that adopted by other countries.

AUDITORIUM PROFILES

The profile of the auditorium is governed by two major factors: the provision of an unobstructed view of the screen by every member of the audience, and the position of the projector light beam.

PROJECTOR LOCATION. Dealing with the latter first, it is obvious that the plane of the aperture of the projector should be parallel with that of the screen, (or the chord of the screen if it is curved), and lines passing through the centre of both planes and normal to them, should be coincident. This cannot, of course, be achieved where

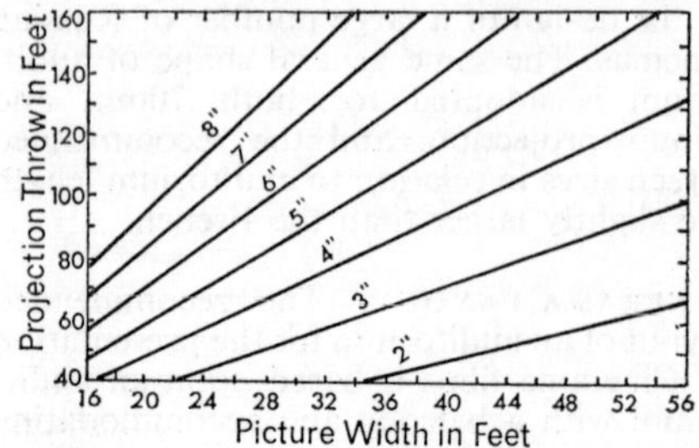

LENSES FOR CINEMASCOPE PROJECTION. Focal length of lens related to throw and picture width.

two or more projectors are installed, and nearly all cinemas have at least two. Projectors have been made that are capable of carrying sufficient film for a complete normal programme and there are over 100 cinemas in Germany and France which have one only of these projectors centrally positioned.

Except where the projection room is incorporated within the balcony structure, it is generally not practical to arrange a horizontal light beam because it involves

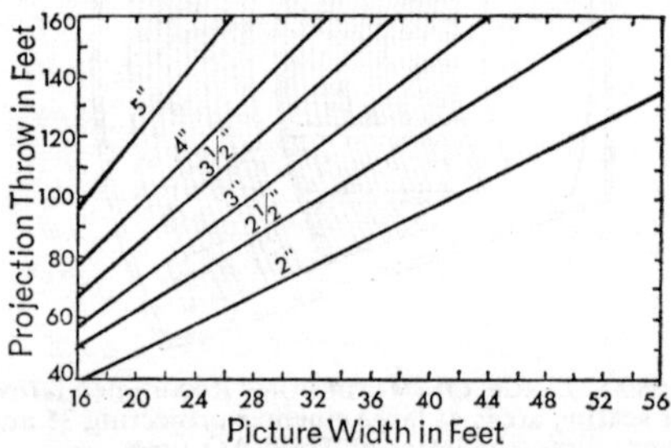

LENSES FOR STANDARD 35MM PROJECTION. Focal length of lens related to throw and picture width.

elevating the screen and thereby reducing the seating area. With cinemas of the stadium type or with one stalls level only, the effectiveness of a screen having directional reflection characteristics (which are commonly installed) would be reduced.

The French specification previously referred to, in dealing with the location of the projectors, stipulates that the projector beam, with respect to the horizontal, shall not result in a trapezoidal distortion of the picture greater than 5 per cent when projecting 35mm film. The recommended limit is 3 per cent. While this specification is useful in determining the desirable results to be achieved, it contains within itself so many variables as to be of little assistance to architects or others planning an auditorium.

When the focal length of the projector lens is in excess of about 5 in., the maximum angle of the light beam to the horizontal is about 10°, and with a 2 in. lens, about 6°. In neither case would the screen be vertical, but would be tilted backwards to minimize the distortion and to reflect the light from the screen to the centre of the seating area.

With 70mm projection systems, including single lens Cinerama, the maximum rake

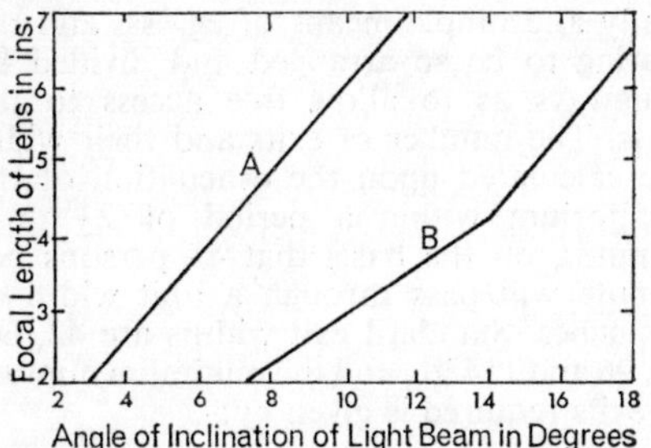

LIGHT BEAM ANGLES. Recommended limiting angle, A, and maximum angle, B, based on focal length of projection lens.

of the light beam is 4°, although here again, as the focal length of the projector lens is reduced, the less should be the angular departure from the horizontal.

The minimum height of the light beam from any part of the auditorium floor (except close to the screen) should be 6 ft. 6 in., otherwise the shadow of persons who are standing is likely to fall upon the screen.

This dimension invariably determines the level of the floor at the rear of the auditorium in stalls-only or stadium-type cinemas.

AUDIENCE SIGHT LINES. The requirement that every member of the audience must have an unobstructed view of the screen is difficult to attain, although it has been achieved in many modern cinemas.

The first consideration is the sight lines of the audience to the lower edge of the screen.

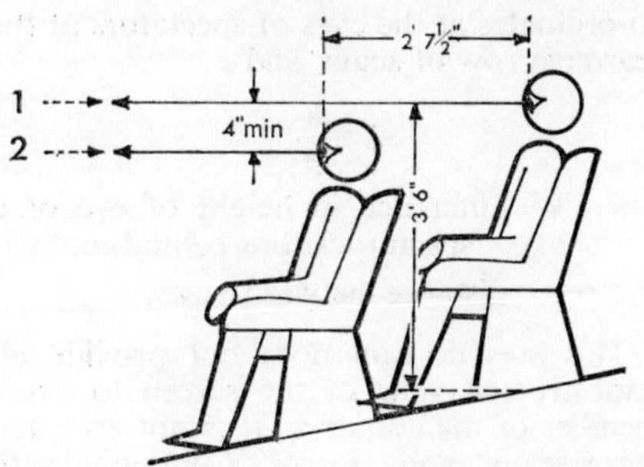

RAKE OF CINEMA FLOOR. French standard. Sight lines 1 and 2 converge on to lower edge of screen. When plotted over several rows of seats, a curved line of floor is produced.

Some European specifications require that the profile of the stalls floor shall be established upon the basis of a clearance of heads by a minimum of 4 in. (10 cm) and preferably 4¾ in. (12 cm) when the seats are spaced 2 ft. 7½ in. (80 cm) apart.

If the sight line of each member of the audience converges on the lower edge of the screen picture, this requirement results in a curved, saucer-shaped floor for the auditorium; the equation for the curve is:

$$y = x\left(\frac{y_o}{x_o} + \lambda \log \frac{x}{x_o}\right).$$

Where x and y are the abscissa and ordinate of the profile at the eyes of spectators at lowest level and x_o and y_o

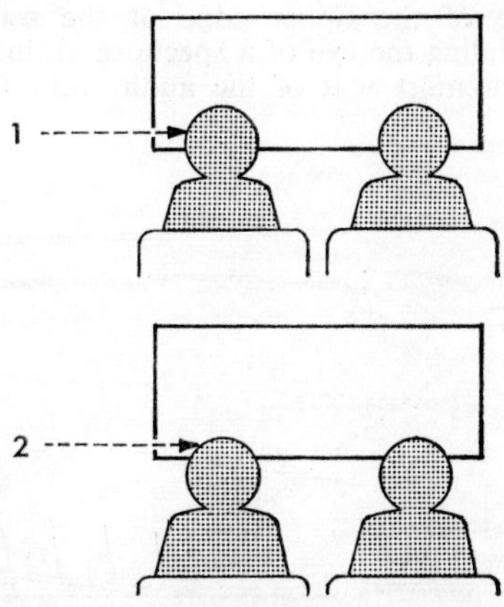

CLEAR VIEW OF SCREEN. Obstruction of picture from rear row of seats at stalls level (French Cinemascope layout) by heads of persons sitting in row immediately in front. Head clearance at 3¼ in., 1, and 4¼ in. 2. Ideal clearance should be at least 5 in.

co-ordinates at the eyes of spectators at the rearmost row of seats, and

$$\lambda=\frac{k}{p}$$

where k = difference in height of eye of a spectator sitting behind another.

p = distance between rows.

This specification does not provide an unobstructed view of the screen to every member of the audience. It is not accepted in Britain and many Commonwealth countries, because the licensing or security authorities impose a maximum rake of floor of 10°. Thus with a standard seat spacing of 2 ft. 6 in., the distance k is 3 in., and with a seat spacing of 3 ft. 0 in. which is coming into use, k is only 3½ in.

It is therefore becoming the general practice to step each row of seats at least over the rearmost two-thirds of the seating area. This should provide a difference of eye level at each row of 5 in. or of 10 in. (which allows 5 in. steppings in the gangways) or a combination of both.

Steps with 5 in. risers have been found from experience to be most conducive to the safety of the public walking up or down gangways in the conditions of a darkened auditorium.

A limiting requirement, first determined by a committee on Eyestrain in Cinemas, set up by the Illuminating Engineering Society at the request of licensing authorities, and now imposed by authorities in many European countries, prescribes that the angle between the horizontal and the centre of the upper edge of the screen subtending the eye of a spectator sitting in the foremost seat of the auditorium shall not exceed 35°. This determines the position of the front row of seating.

Balconies are arranged as low as practicable so as to avoid unnecessary stair-climbing by the public, but high enough not to interfere with the sight lines of the audience sitting in the rear stalls.

Balcony seating is invariably arranged on steppings of 10 in. height (5 in. risers in gangways) with a seat spacing of 3 ft. 0 in. to 3 ft. 6 in. back to back. Balcony seating does not, as a rule, offer difficulties with sight lines, except perhaps in the front row where security authorities require a rester rail at a height of 3 ft. 6 in. above the floor.

SECURITY ARRANGEMENTS. The regulations of security authorities usually require auditoria to be provided with an adequate number of exits to afford the audience ready and ample means of egress, and the seating to be so arranged and divided by gangways as to allow free access to the exits. The number of exits and their width are calculated upon the evacuation of the auditorium within a period of 2½ to 3 minutes, on the basis that 45 persons per minute will pass through a unit width of 21 inches. Standard exit widths are 42, 60, 78, 96 and 114 in., and the minimum number of exits required is given by:–

EXIT REQUIREMENTS

Number of persons accommodated	*Number of exits*
1– 600	2
601–1,000	3
1,001–1,400	4
1,401–1,700	5
1,701–2,000	6
2,001–2,500	7

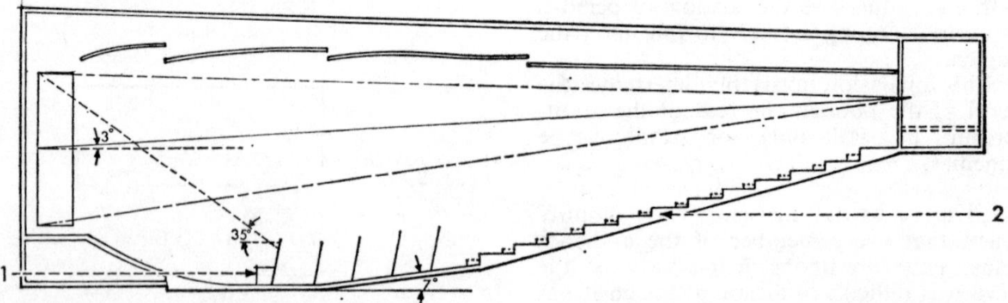

DESIGN OF MODERN CINEMA. Profile of modern cinema seating 685 persons and providing good conditions for viewing picture. Position of front row of seats, 1, is determined by angle subtending observer's eye between horizontal and top edge of picture. Height of steppings for seats, 2, is 10 in., with intermediate steps of 5 in. each in gangway. Spacing of seats, 36 in.

The security authorities usually prescribe the width of the space between rows of seats and the distance from any seat to the nearest gangway. Where the seatway – the space between rows of seats, measured between perpendiculars between the back of one seat and the nearest part of the seat behind (when the seat is up) – is 12 in., the distance to the nearest gangway must not exceed 10 to 12 feet (depending on the authority concerned). A 12 in. seatway usually provides a seat spacing of 2 ft. 6 in. back to back, and the current tendency is to increase this to 3 ft. or more to give greater comfort. In such cases the distance to the gangway may be increased to 15 ft. Thus with chairs of the usual 20 in. width, it is the practice to arrange between 13 and 18 seats between gangways.

The travel distance from any seat in the auditorium to an exit should not exceed 100 ft. in the case of premises wholly fire-resistant, and 80 ft. when partially fire-resistant (one-hour fire-resistant), provided there is no stage risk.

Longitudinal gangways are usually required to be 3 ft. 6 in. wide (minimum) and the width of cross gangways is dependent upon the number of longitudinal gangways.

ACOUSTICS. Although the fan-shaped auditorium is determined for the optimum viewing of the screen, it is also eminently suitable for the reproduction of sound. Nevertheless, in order to secure optimum reverberation time and satisfactory sound-energy decay characteristics, acoustic treatment is invariably necessary to rear walls and the rearmost areas of the side walls. It was the practice to fit acoustic tiles in these areas, but unfortunately the tiles do not give a pleasing decorative effect and their absorption qualities are destroyed when re-decoration takes place. It has become a widespread practice to apply rock-wool or fibreglass blankets or other acoustic materials to these areas, and to cover the whole of the walls of the auditorium with acoustically inert and flameproof curtain material, which as a rule merges into or harmonizes with the screen curtains. With appropriate lighting effects, the decor can be most attractive. Specially-treated canvas or hessian has been used instead of curtaining.

Gangways and seatways should be heavily carpeted with felt or rubber underlay, and seats should be well upholstered, not only for the comfort of the audience, but for acoustic reasons.

When deciding upon a scheme of decoration, care must be taken to ensure that surface finishes will not convey specular reflections from the screen or other light sources that will impinge upon the field of vision of the audience. Walls and carpets should have surface finishes or colours such that their reflection characteristics differ from those of the upholstery, seat standards and other objects which may impede members of the audience in their movements about the auditorium during the performance under prevailing low levels of illumination.

SERVICES

LIGHTING. The lighting of the auditorium usually comprises three separate systems:

1. *Decorative lighting* is in use when the performance has stopped. As its name implies, it should enhance the decorative effect and decor of the auditorium. The system is sometimes subdivided into two or more colour-lighting systems, but it is always controlled by dimmers so that the lighting is not instantaneously switched on or off. Although a part of the decorative lighting system, the lighting of the screen curtains, usually by three colours, is separately controlled. This lighting is by floods, spots or battens and it is desirable that they be concealed from the field of vision of the audience. They are therefore recessed in the ceiling and in the stage where this is provided.
2. *Safety lighting* is provided to all parts of the premises to which the public has access and must, in the absence of adequate daylight, be kept on during the whole time the public is present. Thus the safety lighting in the auditorium, which must be of sufficient intensity to enable the public to see its way out, is kept on during the performance. The illumination provided by this system should be between 0·001 and 0·0025 lumens/sq. ft., and should be substantially uniform over the entire area of the auditorium.

   The safety lighting must be supplied from a source separate from that of the general lighting, i.e. decorative and management.
3. *Management lighting* is that lighting, other than the safety lighting, which in the absence of adequate daylight is on during the whole time the public is present. Not only must the management lighting be sufficient to enable the public to see the way out, but it is provided also to encourage orderly conduct in all parts of the audit-

orium and to contribute to the easy movement of the audience to and from their seats. The level of illumination in the auditorium provided by the management lighting is about 0·005 lumens/sq. ft.

It is important that the sources of both the safety and management lighting should originate behind and above spectators seated in the auditorium and looking towards the screen. But care must be taken to avoid loss of contrast in the screen picture by stray light falling upon it, and the distraction likely to result from light sources or light reflected from fittings placed within the field of view of spectators when watching the screen.

HEATING AND VENTILATION. It is a general minimum standard that every member of the audience shall be provided with 1,000 cubic feet of fresh air per hour; the present day practice is to increase this quantity to 1,250/1,300 cubic feet.

The air is taken from atmosphere at high level to prevent so far as is possible contamination by traffic etc., and is discharged by means of a fan (usually of a centrifugal type) into the auditorium preferably at low level, below the screen.

Before entering the auditorium, the air is cleaned either by water sprays or filters and then passed through a heater battery to raise its temperature in winter to an appropriate level. In some luxury cinemas, provision is made for cooling the air during hot weather, and also providing humidity control.

The air is extracted from the auditorium usually at high level at the rear of the auditorium and preferably well removed from the projector light beam. The capacity of the extract fan is generally between two-thirds and three-quarters of that of the inlet fan. This deficiency in extract fan capacity tends to promote a positive air pressure in the auditorium which is relieved by the escape of air through exit and other door openings in the auditorium, thus avoiding incoming cold draughts.

Heat is generally provided by one or more hot-water boilers; steam boilers are seldom used because of the risks and complications involved. The current trend is the adoption of oil-fired boilers which are fully automatic and demand little attention by the staff. The greatest demand on the heating system is the heater battery for the auditorium. Foyers, vestibules, toilets etc., are usually heated by radiators, and the demands are comparatively small.

Radiators or coils are frequently provided in the auditorium for background heating, i.e. to warm the structure before the entry of the public and to prevent draughts caused by the intake of warm air becoming chilled on striking the cold wall and ceiling surfaces, and being deflected down on to the audience.

The heating and ventilation system is thermostatically controlled and is designed to maintain a temperature of 22°C in the auditorium and 18°C in other parts of the premises when the external temperature is 0°C. When cooling plants are installed, it is usual to provide an auditorium temperature of 13 – 16°C with an external temperature of 25°C.

Ventilation arrangements separate from that of the auditorium must be provided for other parts of the premises. Toilets, for instance, must have their own air-extract systems, the capacity of which must exceed that of any inlet system that may be fitted.

Space heating is required only during cold weather, but domestic hot water is required throughout the year; it is therefore a common practice to instal a small boiler for domestic hot water services; this, preferably, should be on a closed system, i.e. the domestic water should be indirectly heated via a calorifier. In small premises the hot water may economically be provided by local gas or electric heaters.

PROJECTION ROOMS

Projection rooms require to be substantially constructed and large enough to enable the operators to work freely at the

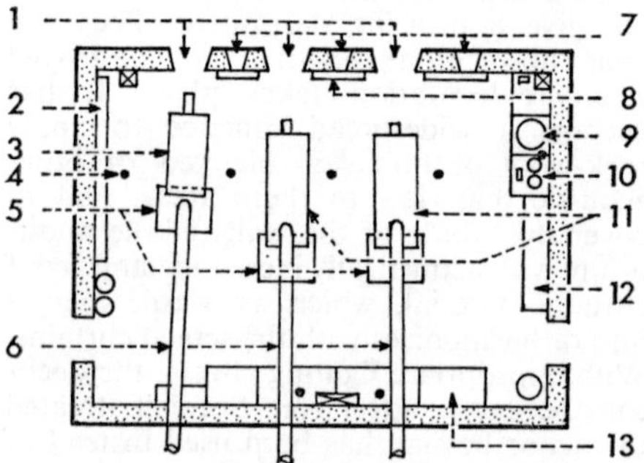

LAYOUT OF MODERN PROJECTION ROOM. 1. Projection ports. 2. Switchboards. 3. Effects lantern. 4. Lights. 5. Rectifiers. 6. Arc exhaust pipes. 7. Observation ports. 8. Curtain and masking controls. 9. Turntable. 10. Tape deck. 11. Film projectors. 12. Rewind bench. 13. Main amplifiers.

projectors and other equipment. All the equipment and controls should be so placed that they are readily accessible with the least physical effort or movement.

The position of the projectors is determined by the relationship of the projection room and the screen. It is desirable that each arc rectifier be placed immediately behind the projector which it supplies. The rewind bench with film bins on each side or below, should be placed at the rear wall, centrally behind the projectors, so that the travel distance from each projector is the same. Where both 70mm and 35mm films are used, a rewind machine for each is recommended, the former being power-driven.

Where safety film only is used, film rewinding and storage may take place in the projection room: nitrate cellulose film which is highly inflammable has now passed out of use.

CONTROLS AND SWITCHGEAR. Controls for curtains, variable masking, change-over of projectors and sound fader should be in duplicate, mounted on the front wall and on the working side of each projector. The dimmer controls for the decorative lighting of auditorium and screen curtains may be similarly duplicated and mounted, but if only one set is provided the working side of projector No. 2 is preferred. A transcription turntable and tape playback machine for non-synchronous music is also near the working side of this projector. These machines should be stereophonic when a multi-channel sound system is installed.

It is desirable to arrange within the projection room or in a room immediately adjacent, a switchboard for all the electrical circuits which come within the primary control of the operator. This will include switchgear for management, safety and decorative lighting, ventilating fans and small power, in addition to the supplies to the rectifiers and sound equipment. Sometimes, control for the boiler equipment, water circulating pumps, etc. is included. In such cases, and where there is remote control for power-consuming equipment, e.g. ventilating fans, suitable ammeters should be placed in the circuits and mounted on the switchboard to indicate that the equipment is working satisfactorily.

It is not uncommon to provide a multi-purpose effects lantern in the projection room. This lantern should be provided with the same kind of light source as the projectors so that there will be no discernible disparity in colour quality and screen luminance. The lanterns are provided with a lens of suitable focal length, slide carrier for the projection of transparencies, and optical effects attachments. These usually comprise a motor-driven turntable carrying a variety of colour filters for flood- or spot-lighting the screen curtains, and for giving animated effects such as moving clouds, dissolving colours, etc.

Projection rooms require a supply of fresh air delivered at low level, and an extract at high level. Most regulations require a free area of not less than 27 sq. in. each for inlet and extract per projector. Additionally, each projector, if the light source is a carbon or xenon arc, requires an extract duct from the arc lantern to open air; the capacity of the duct varies with the current consumption of the arc, and lies between 30 cu. ft. per minute for a 60 amp carbon arc, and 100 cu. ft. per minute for 130 amp.

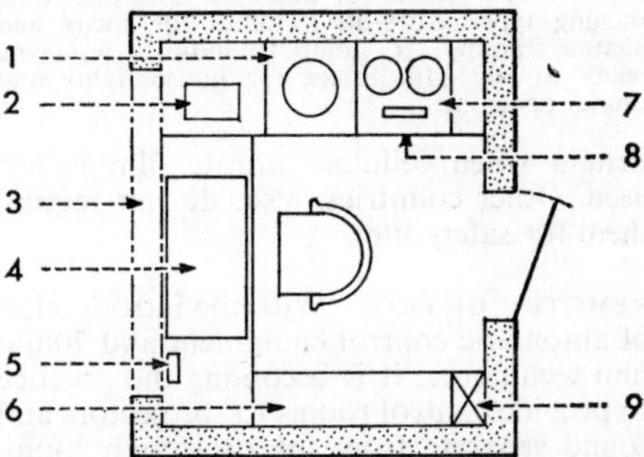

LAYOUT OF PROJECTOR REMOTE CONTROL ROOM. 1. Turntable. 2. Sound control switches. 3. Observation port opening on to auditorium. 4. Main control for projectors, including framing, focusing, sound level control, house light dimmers. 5. Internal phone. 6. Workbench. 7. Tape deck. 8. Bench with tape and disc storage. 9. Ventilation trunking.

The openings on the front wall of the projection room for the light beam and for observing the screen picture should be covered with good quality, 1/8 in. plate glass.

They should be easily removable for cleaning both surfaces, but they must fit their frames tightly to prevent sounds passing into the auditorium. The observation ports should be of such size and fitted at such height above the floor as will enable the operator comfortably to see the entire screen.

The metal shutters formerly required to cover the projection and observation ports in case of fire are not now required in

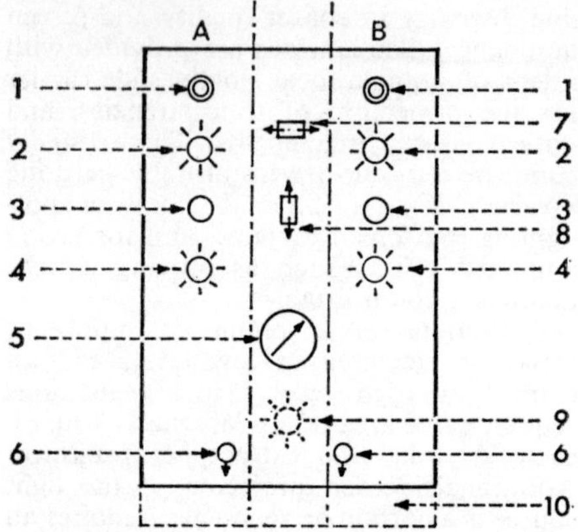

PROJECTOR REMOTE CONTROL BOARD. British Standard layout for 35mm projection with metal rectifiers or equivalent. A. No. 1 projector. B. No. 2 projector. 1. Switch to stop projector. 2. Amber lamp indicating projector ready to run. 3. Push-button to start projector. 4. Green lamp indicating projector running. 5. Sound volume control. 6. Switches for non-sync controls. 7, 8. Inching type centre-off switches for focus and picture framing. 9. Lamp to indicate non-sync ready to run. 10. Space for house lights and similar controls.

Britain when cellulose nitrate film is not used. Other countries, also, do not require them for safety film.

REMOTE CONTROL. With the introduction of automatic control equipment and 70mm film techniques, it is becoming the practice to provide control rooms for projectors and sound separate from the projection room. The front of the control room is open to the auditorium so that the operator may hear the sound being reproduced therein and can clearly see the screen without obstruction. Provision is made on the control board not only for controlling the sound volume and starting the projectors, but also for starting non-synchronous equipment, for focusing and framing the picture, and for controlling curtains, variable masking, and decorative lighting. Indeed, except for renewing carbons, and rewinding and rethreading film, the entire performance may be remotely controlled from this room. The layout of the control board has recently become the subject of a British Standard.

Automatic control of projection equipment with an overall remote control has been particularly successful in those cinemas on the European continent equipped with single projectors. A programme is initiated, the picture focused and framed and the sound level is adjusted from a remote panel, upon which the entire day's exhibition is controlled automatically, including the rewinding of the film after each programme is shown. During the process of rewinding, which takes about 15 minutes, the house lights are brought up for a sales interval, subsequently lowered and the exhibition recommenced, all operations being effected automatically. L. K.

See: *Automation in the cinema; Drive-in cinema theatre; Projector; Rear projection; Screen luminance; Screens; Sound reproduction in the cinema.*

Literature: *Cinema Buildings*, by L. Knopp and A. Wylson (*Architects' Jnl*, Vol. 145, No. 9, 1967); *Cinema Buildings*, by A. Wylson (*Architects' Jnl*, Vol. 147, No. 25, 1968); *Cinema Theatre Design*, by G. G. Graham and others (*SMPTE Jnl*, Vol. 75, No. 3, 1966).

⊕CINEMICROGRAPHY

Cinemicrography is the technique of combining the motion picture camera and lens with the compound microscope. It is this use of the compound microscope that distinguishes cinemicrography from cinemacrography, not the final magnification on the film.

Historically it is not known when the first motion picture record was made of moving objects seen through the microscope, but even before the turn of the century E. J. Marey, in his book *Le Mouvement* (1894), had devoted a whole chapter to a description of his "chronophotographic microscopy". It was not very many years later (1909) before another Frenchman, J. Comandon, succeeded in filming the movement of bacteria. Comandon could be regarded as the first person to realize the enormous scientific potential generated by linking the motion picture camera with the microscope.

EQUIPMENT

For successful filming through a microscope, several criteria have to be established. First, the microscope must be provided with good optics. Second, the cameraman needs a sound knowledge of the principles of microscopy and considerable experience of photography. Third, a wide variety of equipment must be available to cover any contingency, and lastly, the film maker must have a flair for technique, and great patience.

The function of the cine camera is to

record the magnified image of the object as seen through the eyepiece of the microscope, but it plays no part in its formation. The camera lens merely acts as a transfer lens, as the eye does when looking down a microscope, and is focused at infinity to receive the parallel rays of light coming from the eyepiece.

THE COMPOUND MICROSCOPE. The optical components of the compound microscope consist of two lens systems. The objective forms a real magnified image of the specimen in the draw tube of the microscope. The eyepiece further magnifies the image formed by the objective to give a virtual image suitable for observation and photography.

Besides the objective and eyepiece, there are two other essential items. The substage condenser, located below the microscope stage, collects light from the source and concentrates it, in the form of a cone, on to the object and hence into the objective. The field lens is usually an integral part of the lamp housing.

MOUNTING OF MICROSCOPE AND CAMERA. The area devoted to cinemicrography should be situated as far from vibrations as is practicable in order to reduce unsharp pictures to a minimum. Ideally, the elimination of vibration should take place at source but this is largely impractical. Nevertheless, the movement of people near the equipment can be stopped and direct contact of the camera with the microscope avoided, the former by housing the apparatus in a small room barely large enough for two people to work in, and the latter by interposing a light trap between microscope and camera. The microscope should also be mounted on a rigid support which is both large and heavy compared with the instrument itself.

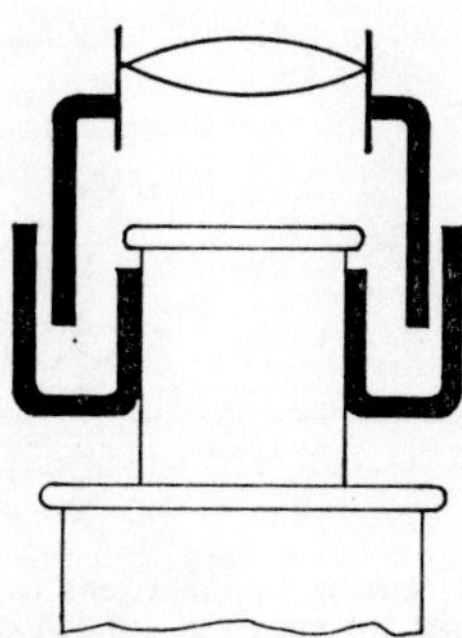

MICROSCOPE LIGHT TRAP. Camera (*above*) and microscope (*below*) separately mounted. Interleaving light trap prevents stray light interfering with image.

If, despite all these precautions, mechanical disturbance is still a problem, then microscope and camera must be insulated against it. A foam rubber underlay fitted to the bench carrying the equipment is often of value in damping vibration, or mountings composed of natural or synthetic rubber can be used. Alternatively, microscope and camera can be locked together mechanically. In this way, sudden movements are transmitted through the whole system, so that camera, microscope and specimen vibrate together and the resulting film may then not show any undue movement. This technique is especially useful if the equipment has to be used in the field or must be readily transportable, and consequently not too heavy.

WATCHING EYEPIECE. It is customary to interpose a prism beam-splitter between the eyepiece of the microscope and the motion picture camera lens to enable the cinemicrographer to study the field of view while filming. This is normally essential to ensure that the phenomenon to be filmed is actually taking place, and that the specimen is still in focus, adequately illuminated and in frame.

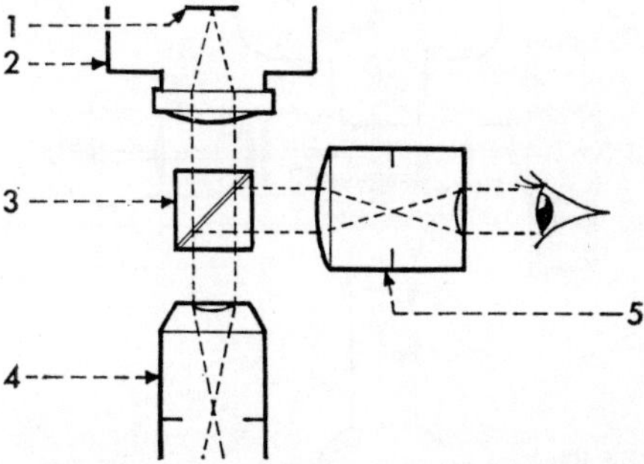

WATCHING EYEPIECE. Convenient method of studying field of view while filming. Cine camera, 2, with film plane, 1. 3. Prism beam-splitter. 4. Microscope eyepiece. 5. Watching eyepiece.

The motion picture or TV camera viewfinder in most cases is a poor substitute for a watching eyepiece. It may be sited in the wrong place, making viewing uncomfortable, difficult or impossible. If the camera does not possess continuous reflex viewing, the viewfinder will certainly not be usable; even if it does, the coarseness of the ground glass viewing screen makes critical focusing of the microscope image difficult. A reflex finder does, nevertheless, enable image composition to be carried out before filming starts. Watching eyepieces which allow 90 per cent of the light leaving the microscope to reach the film are the most suitable, the remaining 10 per cent being used to monitor the event. The ratio 90 : 10 is particularly advantageous in high-speed cinematography or when illumination brightness is low.

A more sophisticated version has a built-in photoelectric cell which, on moving into the light path, needs connection only to a microammeter to record the light incident upon it.

MICRO-SLATING. One of the problems in cinemicrography is how to add information to the film while the experiment is in progress. One method is to couple a projection

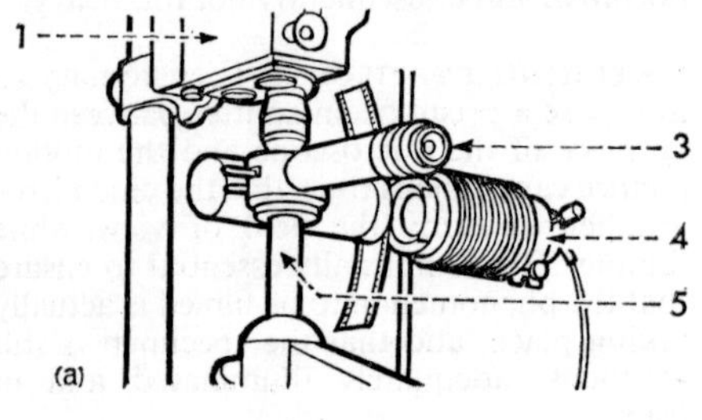

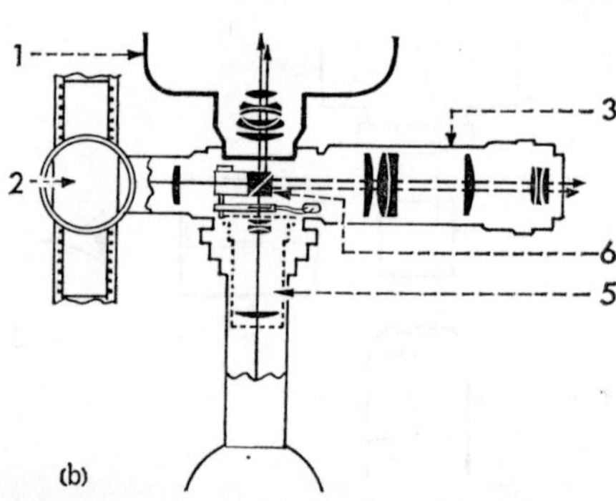

MICRO-SLATING DEVICE. Method of adding time information, arrows, titles, etc., to image of specimen recorded on film. (a) External view. (b) Optical system. 1. Cine camera. 2. Slating information on frame of film. 3. Watching eyepiece. 4. Source of illumination for film frame. 5. Microscope eyepiece. 6. Prism beam splitter combining slating information and microscope image for transmission to cine film and watching eyepiece.

tube to the watching eyepiece. The image of such things as arrows, timepieces and short titles can then be projected into the same plane as the image of the specimen and recorded there, thus making subsequent evaluation of the footage taken more valuable.

MEASUREMENT OF MAGNIFICATION. In cinemicrography, it is not as easy to measure the linear magnification of an object as when using a camera fitted with a ground glass screen.

The best method is to substitute a stage micrometer for the specimen, then film two or three frames of the magnified image of this scale as seen through the watching eyepiece. After the film has been processed, the photographic image of the scale can be measured and compared with the same distance on the stage micrometer. Once this has been done for a given combination of objective and eyepiece, it need not be repeated unless one of the optical components in the system is changed.

ILLUMINATION

The fundamental principle of any microscope lighting system involves the illumination of the specimen with sufficient evenness and with light of the appropriate quality and brightness to ensure proper exposure of the cine film with adequate contrast, brilliance and uniformity frame by frame. The exposure must not be unduly prolonged, especially when filming living material, and the whole illuminating system, from the lamp house to the substage condenser, must be sufficiently flexible to provide any adjustments required.

KÖHLER'S ILLUMINATION. There are various methods of illuminating the specimen, but the one recommended is that advocated by August Köhler. Basically this consists of using the field lens to focus an image of the light source on the substage

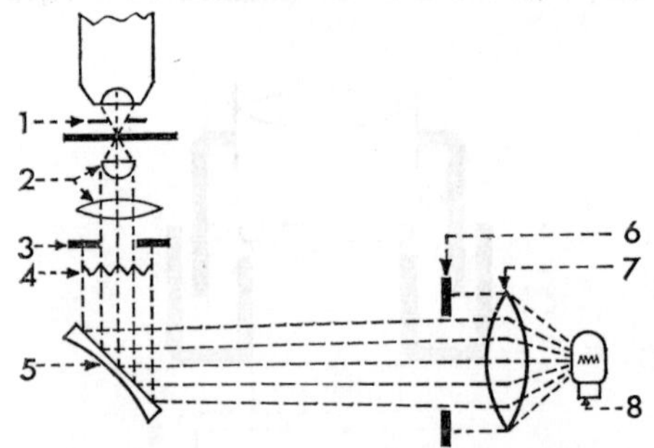

KÖHLER'S ILLUMINATION. Field lens focuses image of light source on substage condenser, which in turn focuses image of field stop in plane of specimen. 1. Image of field stop. 2. Substage condenser. 3. Iris diaphragm. 4. Image of lamp filament. 5. Front surface mirror. 6. Field stop. 7. Field lens. 8. Light source.

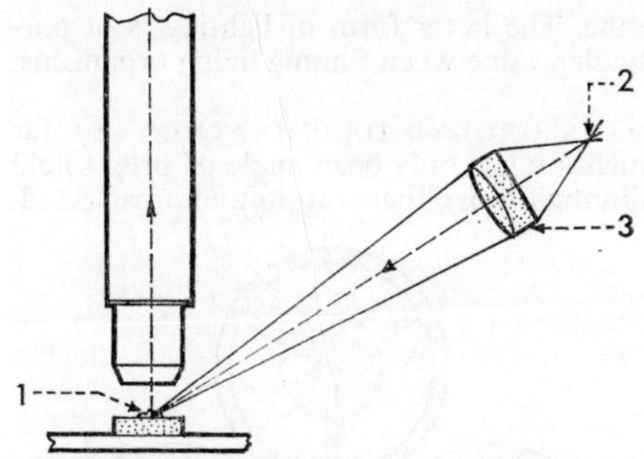

OBLIQUE OVERSTAGE ILLUMINATION. Method of lighting opaque object. 1. Object. 2. Light source. 3. Lens system focusing light on object.

condenser, which in turn focuses an image of the field stop in the plane of the specimen.

When the illuminating system is properly aligned, a uniformly illuminated field is provided with practically no restriction as to the structure of the light source, e.g., coil filament, high intensity arc of small area or other non-uniform source.

REFLECTED LIGHT. Transmitted light is the common form of illumination used in cinemicrography, but there are occasions when a record of an opaque object is needed, e.g., a metallurgical specimen or a coloured crystal. There are two ways of lighting these. The simplest and least expensive is by oblique overstage illumination.

A refinement is to place a small parabolic mirror close to the objective and slightly overstage. Specimens having unpolished or rough surfaces are suitable subjects for this technique, but this and the previous method can be used only with objectives of comparatively long focal lengths.

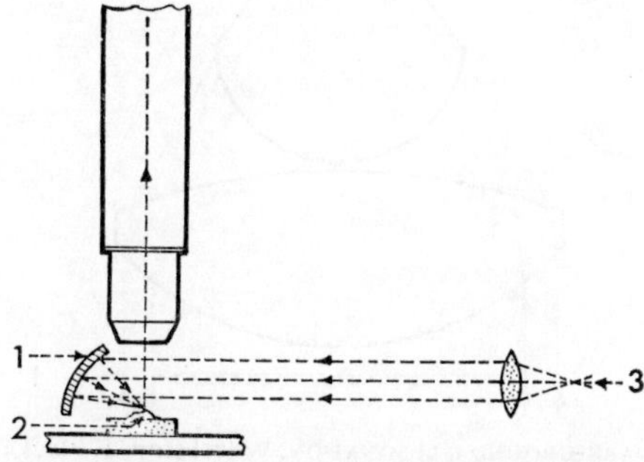

IMPROVED OVERSTAGE ILLUMINATION. Lighting of specimen, 2, with aid of small parabolic mirror, 1, receiving collimated light from source, 3. Suitable for specimens having rough or unpolished surfaces.

Oblique lighting can cause unwanted hard shadows, depending on the specimen, but these may be reduced by using a white card as a reflector opposite the lamp and close to the specimen.

A Lieberkühn illuminator makes it possible to illuminate an opaque object without creating the harsh shadows of unidirectional oblique lighting. This special concave mirror fits over the objective, the front lens of which protrudes through a central hole. The focal length of the mirror must match that of the objective to ensure that the specimen is both in focus and correctly illuminated. Parallel light reflected from the plane mirror

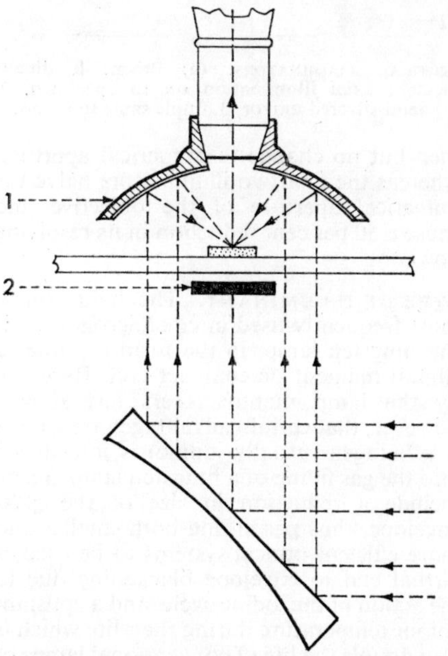

LIEBERKÜHN ILLUMINATOR. Concave mirror, 1, of focal length matched to that of microscope objective. 2. Wheel stop to block central rays from entering objective. Avoids the harsh shadows of uni-directional oblique lighting.

of the microscope is focused by the Lieberkühn illuminator on to the object. This type of illumination is suitable for both low and medium power objectives and when filming polished metal surfaces, fibres or any other small opaque object. Vertical illuminators employing incident axial illumination with either a prism or a mirror to direct light on to the specimen are not entirely satisfactory. If possible, the third type, with a plane-parallel slip of glass, should be used.

However, this leads to some reduction in the quantity of light reflected from the speci-

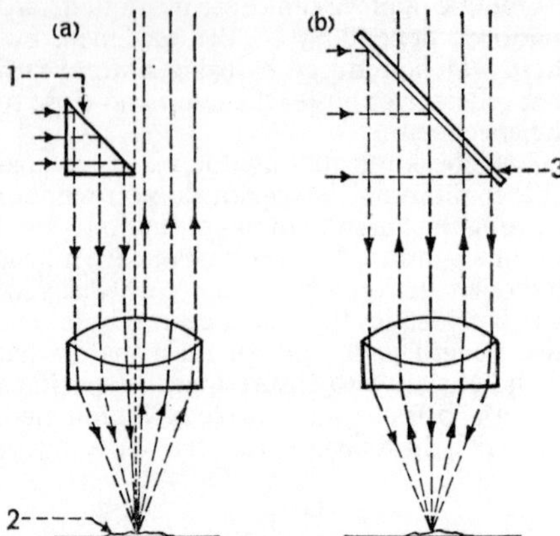

VERTICAL ILLUMINATORS. (a) Prism, 1, directs incident axial illumination on to specimen, 2. (b) Semi-silvered mirror, 3, fulfils same function.

men but no change in numerical aperture, whereas the first two illuminators halve the numerical aperture of the objective and cause a 50 per cent reduction in its resolving power.

TYPE OF ILLUMINANT. The light source most frequently used in cinemicrography is the tungsten lamp, in the form of either a ribbon filament or compact coil. Research by the lamp manufacturers has shown, however, that certain advantages are gained if a halogen (usually iodine) is introduced into the gas filling of a tungsten lamp. These include a reduction in size of the glass envelope, thus permitting both smaller and more efficient optical systems to be used, a virtual end to envelope blackening due to the action of the iodine cycle, and a constant colour temperature during their life which is also double the life of conventional lamps of the same colour temperature. Modern microscope lamps, including the tungsten halogen variety, are usually designed for low voltage use and have a colour temperature suitable for use with colour film balanced to 3,200°K or 3,400°K. It is sometimes necessary to interpose a heat filter, or a water cell, between the light source and the microscope to absorb unwanted infra-red radiation which might damage or kill the specimen. There are other light sources which may be needed for special purposes, amongst which is the zirconium arc lamp of value in infra-red cinemicrography, the xenon high-pressure lamp with a colour temperature of 5,200–5,800°K useful for polarized light work, the mercury high-pressure lamp for fluorescence studies and the xenon flash tube. The latter form of lighting is of particular value when filming living organisms.

DARK GROUND ILLUMINATION. So far mention has only been made of bright field illumination, either transmitted or reflected.

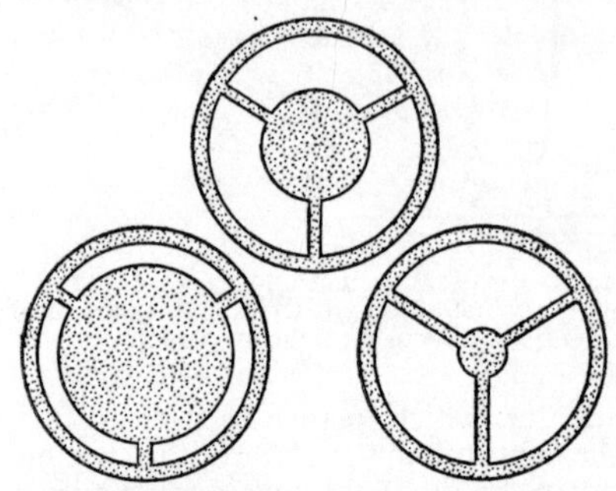

WHEEL STOPS. Opaque stops used in dark-ground illumination to stop axial rays from light source from reaching microscope objective. They are mounted just below the substage condenser.

There are occasions, however, when dark ground illumination is required, e.g., when filming large particles such as fibres and crystals, medium size pigment particles and marine plankton, or living bacteria and particles which are difficult to see by bright field methods because of their minute size. In dark ground illumination, the cone of light that normally illuminates the subject must not be allowed to enter the microscope objective. Only light that is scattered or reflected by the specimen is received by the objective, and the microscopist sees the

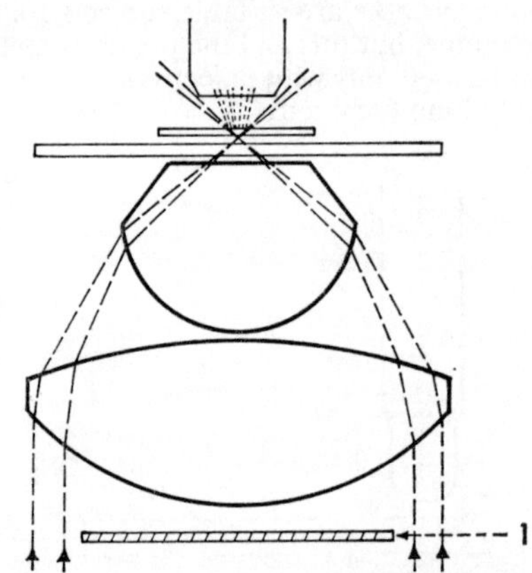

DARK-GROUND ILLUMINATION. Wheel stop, 1, blocks axial rays, while substage condenser above forms hollow cone of light with apex in plane of specimen. In absence of specimen, microscope field is dark; when present, it appears as bright object against dark background.

object bright against a black background.

There are two methods of achieving dark ground illumination. The simpler is to insert an opaque stop in the light path just below the Abbe substage condenser to stop the axial rays from the light source from reaching the objective. These opaque stops, sometimes called wheel stops, can be either made or purchased. This method is most efficient when used with the lower power objectives.

The second method is to use a special dark ground condenser, either paraboloid or cardioid, depending on whether low/medium or high power pictures are needed.

In either method of dark ground illumination, the essential principle is the formation of a hollow cone of light whose apex occurs in the plane of the specimen. Thus, when no object is present, the microscope field is dark, and when a specimen does appear it is seen as a bright object visible against a dark background.

EXPOSURE

In cinemicrography, the only really satisfactory method of ensuring correctly exposed film is by trial and error. Nevertheless, a knowledge of some of the factors affecting exposure enables an intelligent estimate to be made, thus cutting down waste of time and film.

SHUTTER SPEED. In cinemicrography, unlike photomicrography, exposure cannot be modified by a change in shutter speed. This is a function of the rate of filming and is normally unchangeable, except with those motion picture cameras which have a variable aperture shutter. The choice of the correct camera speed is ultimately governed by the desired rate of movement of the object on projection. Other factors to be taken into consideration are magnification in the microscope and the natural rate of movement of the specimen.

MAGNIFICATION. The illuminating system of the microscope concentrates on the object a definite intensity of light per unit area. This light is employed to form the image, and its intensity varies inversely as the square of the magnification.

NUMERICAL APERTURE. Every objective has a figure engraved on its mount, usually next to the designation of focal length. This is its numerical aperture (NA), which is important in determining its capacity to resolve fine detail and its light-gathering power. The larger the NA of an objective, the better is its resolving power and light-gathering property, e.g., an objective with an NA of 0·65 has less resolving power than one with an NA of 1·00. The NA also governs the depth of field (i.e., vertical resolution) which, with a 3mm objective of NA 1·30, is of the order of 1/50,000 of an inch (0·5μ).

The resolution of detail increases directly with the increase in NA but the light-gathering power increases as the square of the NA. This means that by using an objective of double the NA of the previous one the light intensity is increased four times for the same magnification. This partially offsets the reduction in light intensity consequent upon an increase in magnification.

Magnification should, however, be kept to the minimum commensurate with the production of a screen image in which all the detail resolved by the microscope is clearly distinguishable. Under good projection conditions it is possible for a person 20 ft. away from the screen to distinguish detail made up of lines or dots separated by about 5mm.

SPECIMEN DENSITY. Specimens for filming can be broadly classified into groups: e.g., fixed and stained histological sections, erratically moving pigmented or slow moving transparent creatures, imperceptibly moving tissue cultures or cells in division, the growth of crystals, and so on. The cinemicrographer will soon become familiar with one type of specimen and in the light of experience can then compare the relative density of one type of specimen against another.

These are some of the special factors which govern exposure. There are others, photographic in character, which must also be borne in mind.

FILTERS. There are different types of filter for colour and black-and-white motion picture film. A colour film must be of the right type, either daylight or artificial light, for use with the microscope illumination system. Even so, the colour quality of the light may still differ significantly from that for which the film is balanced, and the resulting film may be unacceptable because of its lack of colour balance.

In such a case, the light source must be compensated either by changing the voltage supplied to the lamp or by using filters of the type known as colour temperature or light balancing. These are either bluish or yellowish in colour. The blue series raise the colour temperature of the light source while the yellow series lower it. Most of these filters possess a filter factor necessitating ⅓ stop to ⅔ stop increase in exposure.

Sometimes off-colour results are obtained despite the fact that full correction has been made for any colour temperature variations. These may be due to emulsion manufacturing tolerances, reciprocity effects, the colour of the glass slides, mounting medium or coverslips associated with the microscope preparation, the colour transmitted by the heat absorbing glass or liquid filter, or discoloration due to the microscope optics and the mounting medium used in cementing the objectives.

These off-colour effects are not usually very pronounced and can be corrected by using a series of colour compensating filters. They also may call for a ⅓ stop to 1⅓ stop increase in exposure which must be included in the final exposure calculation.

With black-and-white motion picture films, the main reason for employing filters is to control image contrast and detail. All filters used for this purpose have filter factors, some quite large, which must be included in any exposure estimate together with that of any heat absorbing filter placed in the optical path.

LIGHT SOURCE. Mains voltage variation alters both the colour temperature of a tungsten lamp and its intensity. It is therefore advisable to incorporate into the lamp circuit a voltage stabilizer or at least a voltage regulator. Then, any correction in colour temperature by the use of filters is based on the knowledge that it will not be nullified by voltage variations.

As a tungsten lamp ages, the glass envelope becomes blackened with a deposit of tungsten from the lamp filament. This decreases the light ouput and alters its colour temperature. It is worth while to log the number of hours the lamp has been burning and to discard it after three-quarters of its useful life. This ageing is one of the reasons why the tungsten halogen lamp is fast becoming so popular among cinemicrographers.

DETERMINING THE BASIC EXPOSURE. This should be established under the conditions which are likely to occur during the actual filming. A short sequence is shot varying the light intensity by means of neutral density filters. The exposed film is either sent to a commercial firm for processing, choosing the same one for the actual run, or else processed in the laboratory using a comparable developer to that used commercially. Having established a basic exposure for a certain combination of optical components and photographic materials, it is then comparatively simple to adjust the basic exposure if one or more of these are subsequently changed.

Once the basic exposure has been determined a photoelectric exposure meter can be used for further work. A note should therefore be made of the exposure meter reading of the light intensity used to establish the basic exposure. It can be most easily obtained by placing the photo cell against the watching eyepiece lens. This avoids disturbing the camera set-up and also enables monitoring of the light intensity to be undertaken at any moment while filming, without obtaining blank or otherwise useless frames in the sequence.

CHARACTERISTICS. The film chosen must be of high enough contrast to record adequately specimen detail, of sufficient speed to give a correctly exposed image at the appropriate filming speed and of fine enough grain to give on projection a satisfactory picture.

FILMS

FORMAT. The sizes currently available are 8mm, Super-8, 16mm and 35mm. Until recently, commercial film production has always used a 35mm motion picture camera, and it is only natural that the early workers employed this gauge for their classic studies in the biological sciences. Motion picture cameras made in this gauge are solid, dependable and extremely well made, possessing a remarkably steady film transport which is complete with film locating pins, thus making irregular film movement almost impossible.

The high standard of engineering associated with 35mm motion picture cameras applies also to the wide range of accessory equipment which, like the cameras, is readily available either by purchase or by hire through the film trade. The picture size 18×24mm compares very favourably with that of the second choice (16mm), which is only $7{\cdot}5 \times 10{\cdot}5$mm. Four major disadvantages must, however, be mentioned. The first is the high cost of both apparatus and film stock and second the weight of the camera and the necessity for an extremely rigid construction to support the equipment in apposition to the microscope. The third reason concerns the lack of projection facilities for the screening of 35mm film outside of the commercial cinema, while the last major objection rests on the difficulty of transporting such heavy equipment if much field work is to be undertaken.

The 16mm camera can be likened to a research tool rather than a professional cameraman's motion picture camera. It was once the amateur's cine camera but has long since been promoted from that humble position. Today the 16mm size is probably used more frequently than any other gauge to record the behaviour of industrial machinery, the finer points of a new surgical technique, the slow growth of a human tissue culture, or anything else needed by man for research, teaching or demonstration purposes. The choice of cameras and ancillary equipment is reasonably wide, and the range of sensitive emulsions sufficient to cover all foreseeable needs. When finance, portability or expediency dictates, then 16mm is the format to choose.

Super 8 is the logical replacement for the old 8mm, but the smallness of the picture, even though it has a picture area 50 per cent larger than standard 8mm, and the limited facilities, including photographic emulsions, which this gauge has to offer, as yet present no challenge to the 16 or 35mm format.

FUTURE DEVELOPMENTS

The motion picture camera is already a precision scientific tool with the ability to record accurately whatever is placed before it. Camera lenses are available in a wide range of focal lengths and the minimum working distance of some of them can be measured in inches rather than feet. It is, perhaps, in the realm of the photographic emulsion that some developments may be expected. Speed may be increased without grain size becoming too objectionable, and colour films will certainly stand some improvement.

Nevertheless, there will probably be fewer changes in the accepted pattern of motion picture recording than in the construction of the microscopes and their ancillary equipment. Zoom lenses are already available for dissecting microscopes, and this type of objective will probably be fitted as standard to all research microscopes, including the high dry and oil immersion series. Interference objectives should be available in a wider range of magnifications, and continuous exposure monitoring coupled with automatic control of focus may be commercially available in the not too distant future.

However, it is in the applications of cinemicrography and cinemacrography that the greatest advances will probably take place. When the scientist, electronic engineer and cameraman work together as a team and pool their combined learning and resources, discoveries will be made and a fresh insight gained into the anatomy and physiology of minute living organisms, or the behaviour of materials under stress as shown by the high-speed camera. Medical applications of cinemacrography will include high-speed motion pictures of the human larynx in colour complete with synchronized sound. The telemicroscope, now in its infancy, will enable magnified views to be obtained of an object at any distance from 5 cm (2 in.) upwards. Magnifications will range from $\times 4$ to $\times 250$. This instrument will span the gap between the microscope and the telescope.

Under-water cinemacrography using a periscope instead of the conventional under-water gear will enable extensive documentation to take place in clear shallow waters. The technique has already been proven and gives freedom of access to the camera at all times, providing a more realistic picture of life below the water's surface than when using a standard under-water motion picture camera.

Looking further ahead, perhaps the day will dawn when the photographic emulsion will take second place to the videotape. The object will be seen through the conventional optical systems but will be recorded in colour or monochrome, via the television camera, on to magnetic or thermoplastic tape. Image converters and the amplification factor inherent in television electronics will be of inestimable value in developments along these lines. The retrieval system will be such that film or single pictures can be viewed on monitor screens, and then conventional photographic methods used to prepare visual material for other purposes, e.g., publication, film loops, etc. The electron microscopist will stop filming the fluorescent screen in favour of videotape when it comes.

Even videotape could, with advantage, be replaced by other media which would record an ephemeral or transient image on a re-usable medium. The reason a change in recording techniques is desirable is because the sequential scanning of the television picture and videotape playback puts a limit on television electronics. When this is removed, the range of framing rates at present available only to the motion picture cameraman will be for all to use. One possible technique would be to record a series of images as electron charges on a non-photographic film or drum (a similar process to xerography) which can be wiped clean and re-used any number of times. The image could then be observed by television scanning techniques at the standard framing rate (25 fps) without destroying the electron charge image. The word "tecography" (i.e., dying away writing) has been suggested to describe this technique. K. M.

See: *Cinemacrography; Medical cinematography; Time lapse cinematography.*

Books: *Photomicrography*, by C. P. Shillaber (New York), 1963; *Photography for the Scientist*, by C. E. Engel (New York), 1968.

⊕**Cinemiracle.** Form of multiple-film wide-screen presentation using three projectors to show three matching panels on a deeply curved screen.

See: *Wide-screen processes.*

⊕**Cinéorama.** Early form of multiple-film wide-screen presentation (circa 1896) using ten projectors to cover 360° viewing angle on the inner surface of a cylindrical screen.

See: *History* (FILM).

⊕CINERADIOGRAPHY

In December 1895 Professor Röntgen announced his discovery of X-rays and a copy of his first paper with some original radiographs passed to Dr. John Macintyre from Lord Kelvin. In 1896 Dr. Macintyre demonstrated Röntgen X-rays or the New Photography before students and faculty of the University of Glasgow and the Philosophical Society of Glasgow. In 1897 Dr. Macintyre made the first animated X-ray film showing the movement of the leg of a frog. This film was produced from a series of radiographs put together and photographed in sequence on to cine film.

HISTORY

Dr. Russell Reynolds started, in 1921, experiments filming the fluorescent X-ray screen with a normal cine camera and was the first person to synchronize the X-ray output with the camera shutter phase. This was the forerunner of pulsed cineradiography and, in 1925, he produced his first medical film on a 16mm camera at 1½ frames per second for a child's chest and 4 fps for adult bone structure and joints. There were at this time serious limitations due to the brightness of fluorescent screens and a lack of wide-aperture optics.

By 1930, Dr. Jankers of Bonn had produced some 35mm films (24 × 16mm format) at 22 fps and even some cardiac films at 100 fps. Improvements in equipment continued until, by 1933, a Zeiss lens had been developed with an aperture of *f*0·85.

Dr. Russell Reynolds used this lens in a 16mm camera with a specially developed fluorescent screen to produce the most successful installation to date. The 16mm cine camera operated between 3⅛ and 50 fps through a variable gearbox, and the X-ray source was interrupted during the film pull-down period.

Apparatus produced in Rochester by 1948 used a 35mm camera with 185° shutter and a large synchronous lead shutter to protect the patient from radiation during the film pull-down period. The *f*0·85 lens, which produced an image of 15mm diameter on 18 × 24mm format, was still used, and at this time Eastman Kodak produced a lens with an aperture of *f*0·7 specifically for this apparatus. About this time, a mirror lens was designed by Taylor, Taylor and Hobson, especially for cineradiography, which allowed routine work to be undertaken in hospital departments without subjecting patients to excessive radiation.

During the 1940s, experiments in the production of an evacuated X-ray image intensifier had been taking place at Philips of Eindhoven, and by 1950 a commercial intensifier with a 5 in. field was available, having a minimum light gain of 1,000 times over a standard fluorescent screen. It was an obvious step to couple such an intensifier to a cine camera and, following this, production intensifier installations became obtainable complete with viewing system and cine camera.

EQUIPMENT

IMAGE INTENSIFIER. The modern image intensifier covers a 9 in. field with provision for electronically doubling the size of the centre of the image, and a photo-multiplier which samples the illumination at the centre of the image and is arranged so as to ensure

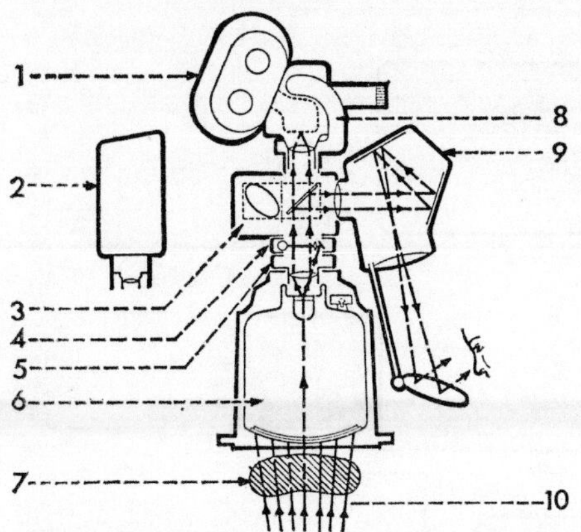

X-RAY IMAGE INTENSIFIER AND OPTICS. 1. Cine camera. 2. Television camera. 3. Image distributor to which up to three optical systems can be connected simultaneously (3rd, behind plane of drawing). 4. Photo pick-up for photo-timing signal or brightness measurement in cinefluorography. 5. Lens holder with lens producing parallel light beam for further distribution. 6. Image intensifier. 7. Test object. 8. Support for cine camera. 9. Viewer for fluoroscopy. 10. X-ray beam.

constant film blackening.

The image intensifier consists of an evacuated glass cylinder about 10 in. in diameter with a fluorescent screen deposited on one end of the inside surface. This screen mainly consists of a metallic salt of zinc, cadmium or a combination of both which will fluoresce in the presence of X radiation. The efficiency of the screen depends on the amount of absorbed radiation which is proportional to the layer thickness which is, unfortunately, inversely proportional to resolution.

In contact with this fluorescent screen is a photocathode prepared from an alkali metallic oxide (e.g., caesium). This is vaporized or sputtered on to the surface of the fluorescent screen and has the property of emitting electrons from one face in proportion to light falling on the other. Thus an electron pattern is formed on the surface of the photocathode in proportion to the light and shade of the image on the fluorescent screen. These electrons are accelerated to a cylindrical anode by 25 kV and focused electronically on to an output phosphor about 1 in. in diameter. In front of this output phosphor is mounted the basic or prime objective, which is collimated to give a parallel beam of light; thus any optical system coupling into this is focused to infinity.

One image distributor with beam-splitting device incorporates three semi-transparent mirrors on a motorized turret. This can be automatically rotated to allow 90 per cent of the light to be transmitted to a cine camera and the remaining 10 per cent to a television camera for monitoring purposes. Alternatively, the whole of the light can be reflected to the television camera for fluoroscopic viewing. A total of six combinations is possible.

LIGHT AMPLIFIER. An alternative system to the X-ray image intensifier uses a specially constructed fluorescent screen focused on to the face of a $4\frac{1}{2}$ in. image orthicon TV tube, the light amplification in this instance being achieved by the gain of the video amplifier. The visual display is on a conventional TV monitor and all cineradiographic films are taken from a similar monitor located elsewhere. A variation of this system has a light amplifier before the image orthicon, which means that the television chain does not have to work at such a high gain and a better signal-to-noise ratio is obtained.

Mirror optics in these systems focus the fluorescent screen on to the image orthicon or light amplifier and an aperture of about $f0{\cdot}7$ is obtained with very little distortion or aberration at the edge of the image.

APPLICATIONS

Cineradiography is used in a variety of investigations, the most usual being examination of the digestive tract, arteriography and cardiac investigations – the latter requiring the more complex equipment and greater operational skill for satisfactory results.

GENERAL PROCEDURE. For normal cineradiographic studies the patient is required to swallow a quantity of X-ray-opaque material, usually barium, and a film is taken during its natural journey through the body, the organs to be examined being clearly outlined by the opaque material.

In the case of cardiac examinations, a catheter is inserted into an artery, pushed to the site of the examination and, if a large vessel is required to be filled quickly, about 40 cc. of X-ray-opaque medium is injected in about $1-1\frac{1}{2}$ sec. at a pressure of approximately 100 lb. per sq. in.

Alternatively, very small quantities of opaque medium can be fed to the entrance of a particular vessel or valve and dribbled from the catheter so that the flow characteristics can be observed.

Most fluorescent screens and, in particular, X-ray intensifying devices present a low-contrast image, and the X-ray factors,

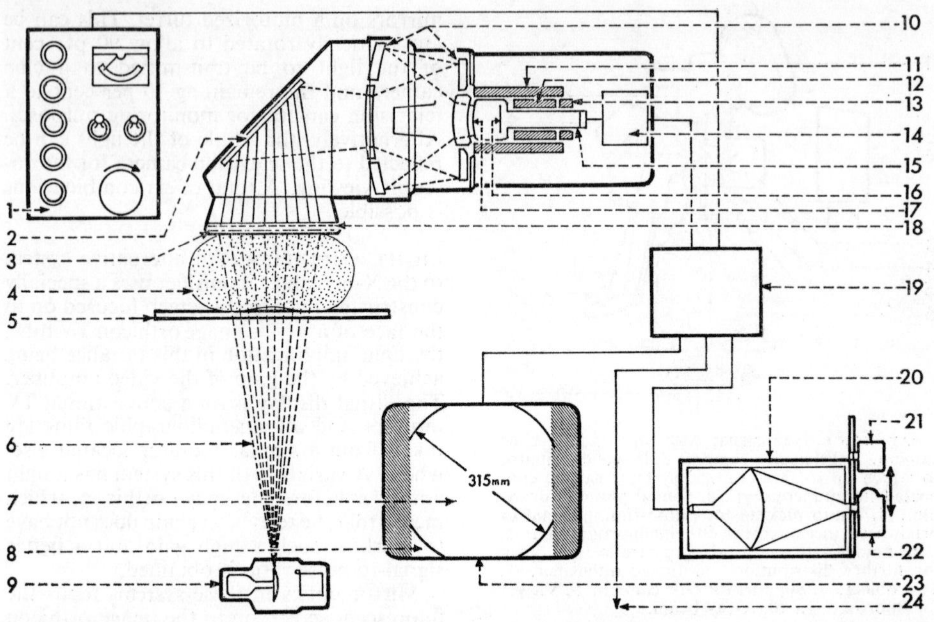

FLUORESCENT SCREEN AND TV IMAGE. **1. Control panel for fluoroscopic current and camera. 2. Mirror. 3. Anti-scatter grid. 4. Patient. 5. Table top. 6. X-ray beam. 7. 315mm image. 8. Raster. 9. X-ray source. 10. Optical system. 11. Focusing coil. 12. Deflecting coil. 13. Alignment coil. 14. Low-noise head amplifier. 15. Image orthicon camera tube. 16. Target. 17. Photocathode. 18. 315mm fluorescent screen. 19. Electronic rack. 20. Remote recording unit. 21. 100mm spot film camera. 22. 16mm cine camera. 23. Viewing monitor. 24. Output to additional monitors.**

the film and the processing techniques have to be carefully chosen to produce a diagnostic result. As a generalization, it may be said that the voltage on the X-ray tube varies the contrast, while the current through the tube varies the brightness – the higher the voltage the more penetrating the radiation and therefore the less contrasty the image. However, unless a minimum current is reached, the picture assumes a scintillating appearance, on account of individual X-ray quanta becoming visible.

FILMS AND PROCESSING. The choice of film is always a compromise between the radiation dose to the patient and the diagnostic quality of the final product, but as a general rule fast films with a rating between 200 and 400 ASA are preferred to minimize the radiation dose – the slower films being used where it is necessary to record fine detail. Processing is also a compromise, but an image of maximum contrast requires X-ray developers, or the film is fed through an automatic X-ray processor – either system producing a gamma of about 1·3. Commercial film processors have not proved popular in hospital departments because it is often necessary to process as little as 10–15 ft. of film and the majority of commercial processors are constructed to handle lengths in excess of 100 ft.

SPECIALIZED TECHNIQUES. The technique of switching or pulsing the X-ray output in synchronism with the camera shutter is already an established technique which has the advantage of reducing the X-ray dose to the patient and restricting the load on the X-ray generator.

A further development in this field is to use pulses of very short duration so that each frame may have a sharp image and be examined individually. Pulse lengths of 1–10 msec. are usual in this type of installation.

Bi-plane cineradiography is a technique developed in the last few years for cardiac work. Two X-ray cine films are taken along axes at right-angles to one another. The X-ray beam is pulsed in synchronism with the camera shutters and the two cameras are run 180° out of phase, so that only one plane is exposed at any one instant. This prevents the scattered radiation from one plane causing unwanted blackening on the film in the other plane. Camera speeds up to 200 fps have been successfully achieved.

In cinecardiography, paper chart recordings are normally made of blood pressure, oxygen and CO_2 content of the blood and the electrical activity of the heart (ECG) and it is often necessary to relate one particular cine frame with any part of these recordings. Many cameras are therefore fitted with event markers in the form of a miniature light source, fogging the corner of the film through a pinhole in the gate. The frames can then be marked sequentially from an external time base, or single frames can be marked to coincide with any part of the examination procedure.

Conversely, a commutator can be fitted to the camera shutter shaft, either directly or through reduction gearing, to record a pulse at regular intervals on the ECG or other recording instrument in step with the camera.

Another recent development is the production of stereo cineradiographic films. Two X-ray tubes are mounted beneath the examination table with the output windows about 4 in. apart. The tubes are pulsed alternately at about 60 fps each, and the X-ray images recorded on a 35mm cine camera running at 120 fps. Thus every two adjacent frames make a stereoscopic pair. If they are projected on to a screen as a cine film and a synchronous shutter held in front of the eyes, such that the left eye is uncovered when the corresponding frame is projected, and similarly for the right eye, a satisfactory stereo X-ray film results.

MONITOR FILMING. Cineradiographic films from a television monitor are, generally speaking, taken at a much lower X-ray dose rate than films from the output phosphor of an image intensifier. This is because the gain required to increase the image brilliance is provided by the amplifiers in the TV installation. Approximately one-third of the available information fed into the TV camera is lost when it passes through the electronic circuitry and the lens-film combination of the cine camera.

Filming the TV monitor has also special problems because the image consists of a spot which scans the monitor tube horizontally from top to bottom, twice in 1/25 sec., to make one complete image frame; each frame therefore consists of two fields each having a duration of 1/50 sec.

Any cine camera that is used in this type of equipment should be synchronized with the mains frequency and polarized to make the shutter open when the scanning spot is just at the top of the monitor tube. The shutter must then remain open for 1/50 sec. or a multiple of 1/50 sec. The film pull-down must take place during a similar time.

FILMING SPEEDS AND SHUTTER PHASE FOR TV FILMING

Filming speed (fps)	*Shutter phase (degrees)*	*Information recorded per cine frame (TV fields)*	*Information lost per pull-down (TV fields)*
25	180	1	1
12½	180	2	2
8⅓	180	3	3
6¼	180	4	4
16⅔	240	2	1
8⅓	240	4	2

Film blackening and contrast range are preset by the contrast and brightness controls on the monitor and kept constant by the automatic gain control in the video amplifier. Although some distortion of the grey scale is inevitable under certain signal conditions, it does not detract from the diagnostic quality of the end result.

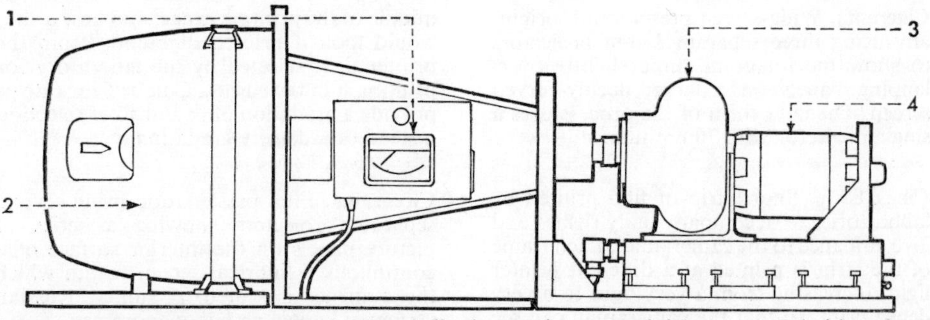

CAMERA FILMING TV MONITOR IMAGE. 1. Meter indicating monitor screen brightness. 2. Television picture monitor. 3. Cine camera. 4. Camera drive motor.

TYPICAL RESOLVING POWER OF VARIOUS SYSTEMS

System	Line pairs/cm.
Conventional fluoroscopy	11
X-ray intensifier	15
X-ray intensifier (magnified image) ...	22
X-ray intensifier with TV (625 lines) ...	9
X-ray intensifier with TV (819 lines) ...	12
Conventional radiography with intensifying screens	28
X-ray intensifier/35 mm. fast film ...	11
X-ray intensifier/35 mm. slow film ...	13
X-ray intensifier/16 mm. fast film ...	8
X-ray intensifier/16 mm. slow film ...	12
X-ray intensifier/TV monitor and 16 mm. film	8
X-ray intensifier/70 mm.	14
Fluorescent screen/image orthicon TV with 16 mm. film	9
Fluorescent screen/image orthicon TV (1,025 lines)	11

A small efficient camera has been recently developed in Britain to run at 16⅔ fps. It has a synchronized motor, and the 240° shutter is designed so that the motor need not be polarized.

VIDEO RECORDINGS. Television pictures can now be recorded and stored successfully by means of a videotape recorder, modern equipment being portable, a little over twice the size of a domestic tape recorder, and weighing about 90 lb. A stationary image of any part of the film can be played back and the tape can be stored indefinitely provided care is taken to keep it away from magnetic fields and conditions of high humidity and temperature.

Videotape recording has the great advantage of immediate play-back so that the success of the examination can be ascertained before the patient is removed from the examination room.

FUTURE TRENDS

There are several current developments which, if brought to fruition, will revolutionize cineradiography. The most promising of these is the coupling of TV and cine cameras to image intensifiers by fibre optics which will give increased light transmission and information. Experiments are proceeding with a fibre optic block built into the neck of the image intensifier, one end having the output phosphor deposited upon it and the other end having the input screen of the TV tube deposited directly on to it. This will reduce the optical aberrations and increase the light transmission.

Another application for fibre optics is to couple the optical block directly to the output phosphor of the intensifier and form the image on the other face of the block. The cine film could then be moved across the face of this image with no intermediate optics.

The development of solid-state image intensifiers with high gain and minimum decay characteristics will reduce the bulk and complexity of installations, giving at the same time much larger viewing areas.

The development of high-gain, high-resolution TV systems with low-noise, wide-bandwidth video amplifiers can confidently be expected in the near future, and these will materially improve the quality of pictures on the TV monitor and the resultant films. These developments, together with improved electronic techniques for stabilizing exposure and contrast, may well help to reduce many of the variables which are such a hazard to the production of satisfactory cineradiographic films. L. A. N.

See: *Medical cinematography and television.*

Book: *X-Ray Physics and Equipment*, by F. Jaundrell-Thompson and W. J. Ashworth (Oxford), 1965.

⊕**Cinerama.** Wide-screen presentation originally using three separate 35mm projectors to show the image as three slightly overlapping panels on a large, deeply-curved screen. The later form of the process uses a single projector with 70mm film.

See: *Wide screen processes.*

⊕**Cinex Strip.** Short strip of film printed by laboratories to accompany daily rushes and give guidance to the cameraman. Each frame of the strip is printed at a different printer light, increasing from a very light to a very dense print, so that the cameraman can see at a glance whether his exposure is near the middle of the printing range, and how a shot would look if printed differently from the printer light selected by the laboratory for printing it in the rushes. Colour Cinex strips provide a gradation of colour filter selection instead of a density gradation.

See: *Laboratory organization.*

⊕**Circarama.** Film presentation using several separate projectors showing a series of picture panels on the interior surface of a continuous cylindrical screen within which the audience is seated. A similar Russian system is known as Kinopanorama.

See: *Wide screen processes.*

⊕Circle of Least Confusion. Spot of light of ☐finite size which constitutes the optical image of an object point when formed by a practical lens.

Owing to spherical aberration in a lens system there can be no perfect focus, that is, at no position can a point image be produced to correspond to a point object. In any plane perpendicular to the optical axis, a point object of light is produced as a circular patch of light. The best focus is where this patch of light is at its minimum size, and the patch of light in this position is known as the circle of least confusion.

See: *Optical principles.*

☐**Circular Waveguide.** Circular metal tube along the inside of which radio waves travel, reflected from the sides. The diameter of the tube must be at least 0·584 times the wavelength of the waves, for one mode.

⊕**Circulating System.** System of pumps and pipes by which the solutions in a continuous film developing machine are taken from and returned to storage tanks where the active ingredients undergo replenishment adjusted to keep their strength constant. Solution storage tanks are temperature controlled.

See: *Processing.*

☐**Clamp Amplifier.** Video amplifier which re-establishes the d.c. level of the signal with respect to black on a line-by-line basis. Normally the back porch which lies between the trailing edge of the sync pulse and the trailing edge of the corresponding blanking pulse is used as a reference since its level does not vary with picture content and it is not affected by disturbance of the sync pulse amplitude.

See: *Picture monitors; Television principles.*

☐**Clamping Diode.** Diode used in a circuit to clamp the signal level. Also used in pairs or bridge formation and rendered operative by a clamping pulse. Used to maintain black level in some receivers.

☐**Clamp Pulse.** Accurately timed pulse which renders a clamp circuit operative, usually causing it to become conductive.

⊕**Clapper Board.** Pair of hinged boards clapped together in film dialogue shooting before or after each take, when the picture and sound cameras are running at synchronous speed. The picture frame on which the clappers close is synchronized in the cutting room with the modulations resulting from the bang, thus establishing synchronism between sound and picture tracks. Clapper boards have been dispensed with in many modern types of sound recording systems.

See: *Editor; Pulse sync.*

⊕**Clappers.** Junior member of the camera crew who operates the clappers on the number board at the start of each shot. He also loads and unloads film magazines.

⊕**Class "A" Amplifier.** A method of operation ☐in which a valve or transistor acts as a linear device. Any variation of signal on the control electrode, negative or positive, causes a similar change in the output current.

See: *Transmitters and aerials.*

⊕**Class "B" Amplifier.** A method of operation ☐of an amplifying valve or transistor in which bias is applied to the point where only positive excursions of the signal fed to the control electrode cause output current to flow.

See: *Transmitters and aerials.*

☐**Class "C" Amplifier.** A method of operation used in radio frequency amplifiers in which valves are biased in such a way that only the positive peaks of the sinusoidal signals fed to the control grids cause the valves to conduct. Tuned circuits are connected in the anode circuits, and the resulting pulses of current at the anode cause excitation of the tuned circuit, resulting in a complete sine wave being produced at the output.

See: *Transmitters and aerials.*

⊕**Claw.** Device used in cameras and narrow-gauge projectors for providing intermittent

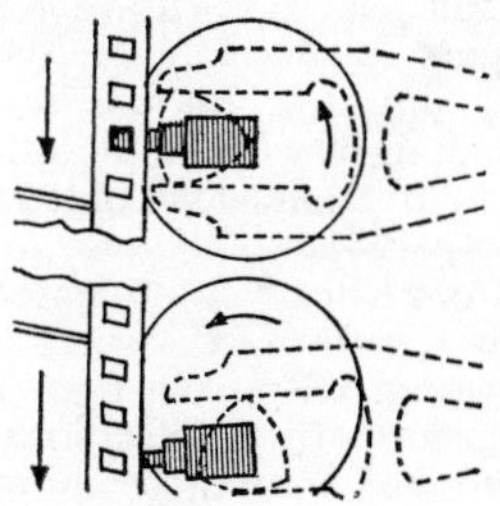

movement of the film. The claw engages a perforation, and thus pulls down the film a distance equal to the height of one frame; it then withdraws to go back to the initial position.

Double claws are often used, one above the other, so that if a perforation is broken, the next one is engaged. The claw is one example of a pull-down or intermittent mechanism.

□**Clean Effects.** Background sound from a broadcast without added commentary.

See: *Outside broadcast units.*

⊕**Cleaner.** Film laboratory operator who cleans rolls of assembled picture and sound track negative in preparation work and during their use for general release printing. At one time, these operations were done manually, but with the greater use of high-speed cleaning machines, the operator is now often concerned with the functioning of a group of machines rather than cleaning the film himself.

See: *Laboratory organization; Ultrasonic cleaning.*

⊕**Clip.** In editing, a short section removed from a picture shot. More often called a cut or trim.

□**Clippers.** Electrical circuits used to clip off or remove parts of a waveform which exceed the desired limits.

□CLOSED CIRCUIT SYSTEMS

Though there is no precise and generally accepted definition of closed-circuit television, the most usual criteria are point-to-point signal transmission by cable or directional radiation, and audience limitation by physical control or non-standard transmission.

There are no essential technical differences between closed-circuit and broadcast television and, apart from very specialized applications, similar equipment is used for both.

That television is a highly effective visual communications medium is obvious from its enormous social impact in the realms of entertainment. Its very success in this field, however, undoubtedly tended to inhibit its development for other purposes until quite recently, most of the best talent in the industry being devoted to the development of equipment for entertainment broadcasting. Paradoxically, military organizations throughout the world were generally slow to appreciate the potential of television for their purposes, and it is only during the past few years that they have spent substantially on it.

After some fifteen years of development, closed-circuit television still represents only a small part of the total activity of the electronics industry. Its use is, however, now expanding very fast, and in time can be expected to overhaul and outstrip its still flourishing parent.

Apart from its capacity to entertain, television can perform a variety of useful functions more effectively than any other existing means. Its applications may be loosely grouped into those in which television performs a warning function, those in which it constitutes an information medium, and those in which it facilitates the acquisition of knowledge and experience.

In all of these, the specific virtues of television are the ability to enlarge, brighten and disseminate material widely; to overcome distance, eliminate hazard, enter confined spaces, enhance contrast, record and repeat.

EQUIPMENT

A closed-circuit television system essentially comprises a pick-up device, a signal distribution system and one or more displays.

The pick-up device is most usually a conventional television camera employing a vidicon camera tube. The image orthicon is also frequently used for applications in which its size and relatively high cost are not a problem, and the lead-oxide photoconductive tube (Plumbicon) is coming into ever wider use, particularly in colour systems. Flying-spot scanners are frequently used in applications involving the transmission of static documentary information or film. The modern trend is towards very simple cameras with the minimum or even a complete absence of operational controls.

Signal distribution in closed-circuit systems is usually by cable. Over short distances, transmission at video frequencies through coaxial cable is generally most satisfactory. Balanced telephone pairs can be used as an alternative to coaxial cable, but normally limit performance in respect of bandwidth and are highly susceptible to electrical interference. Carrier systems are employed where distances are long or the system is complex.

The use of standard broadcast VHF channel frequencies permits the employment of domestic receivers as displays but imposes severe distance limitations. A technically preferable solution is the HF system, normally using a carrier frequency in the region 5 MHz to 10 MHz such as is extensively used for the distribution of public broadcast television by wire in urban communities.

Image display in a closed-circuit system is usually on a conventional direct-view monitor or domestic receiver with a screen size ranging from 6 in. to 27 in. diagonal.

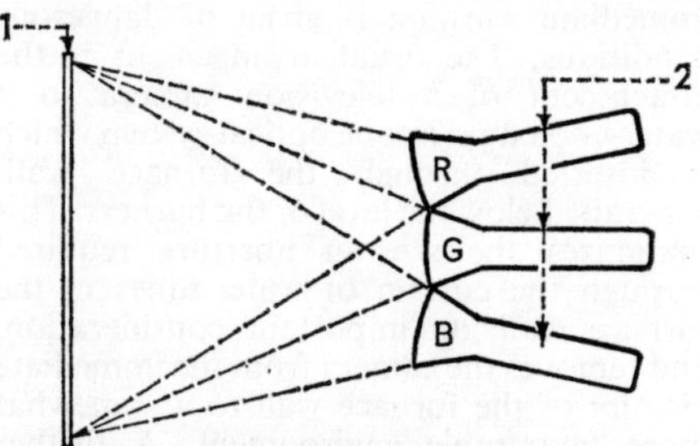

LARGE SCREEN TELEVISION PROJECTOR. Direct system using high-voltage projection tubes. 1. Screen. 2. Three 5 in. cathode ray tubes in 16 in. Schmidt optical systems for colour or black-and-white.

Projection displays, with either cathode ray tubes and Schmidt optical systems or oil-film light-valve techniques (Eidophors) are used for both black-and-white and colour systems requiring a large-screen display. Screen sizes up to full cinema dimensions are available.

In addition to the basic system elements, the whole range of available television techniques, such as switching, mixing, special effects, picture inversion, polarity reversal, crispening, and so on, are common in closed-circuit applications. But because the transmission bandwidth restrictions associated with television broadcast rarely apply to closed-circuit systems, there is a trend towards non-standard high-definition scanning systems and other technical niceties which are impracticable in broadcasting.

ENTERTAINMENT SYSTEMS

Closed-circuit television is frequently encountered as an entertainment medium as a supplement to, a substitute for, or in competition with, the more ubiquitous broadcast product.

HOTELS. In many hotels around the world, a locally generated closed-circuit programme supplements the local broadcast programmes with material of specific interest to hotel guests. A typical system incorporates live cameras in a small presentation studio and telecine equipment and/or videotape for continuous transmission of recorded material. Simple vision mixing arrangements facilitate the assembly of a varied programme which is fed on a vacant channel into the hotel distribution system.

PASSENGER SHIPS. Similar systems are quite commonplace in large ocean-going ships to provide passenger entertainment when the ship is out of range of broadcast transmissions.

In addition to simple live pick-up and dual telecine facilities, these installations are of particular interest in respect of their off-air reception arrangements. They make it possible to receive off-air programmes on any normal broadcast standard and distribute them to the receivers in the public rooms and cabins, notwithstanding the use of entirely conventional USA standard domestic receivers throughout. This is achieved by channel conversion from CCIR to FCC standards for 625-line reception and by scan conversion using vidicon cameras for 405-line and 819-line reception.

AEROPLANES. A recent development is the provision of television entertainment in long-distance passenger aeroplanes, so far mainly in the USA. In such an installation, the programme is derived from a small helical-scan videotape recorder installed in the aircraft's galley or baggage hold and displayed on a number of 8-in. and 11-in. video monitors in the passenger compartments.

One arrangement is to mount the monitors in pods immediately beneath the overhead racks, slightly angled towards the centre aisle and suitably spaced to allow an adequate view to each passenger. The programme sound is distributed by a separate circuit and available through earphones to each passenger, thus permitting individual passengers to read or sleep without interference from the television system.

As a rival to the conventional projection of films in aeroplanes, the television system has the advantage that curtaining of the aircraft windows is unnecessary owing to the much higher brightness of the television picture as compared with the cinema screen.

THEATRE TELEVISION. The use of projection television in theatres and other auditoria for the closed-circuit transmission of special events is very long-established. It has probably been used most widely for the presentation of major boxing contests, and on such occasions even communications satellites have been used for transatlantic transmission. Other sporting events presented on theatre television include horse-racing and motor-racing.

In the USA, experiments have been made in the use of theatre television to present simultaneous transmission of live theatre performances from New York theatres to provincial towns and it seems probable that this will become an established form of entertainment in the future.

PAY-TELEVISION. This is another example of the use of closed-circuit television for entertainment. In most pay-TV systems, signal distribution is by wire, and special equipment at the transmission point detects the operation of individual receivers for billing purposes. An alternative and less sophisticated arrangement involves inserting a coin or token into a slot-meter attached to the receiver.

It is likely that the use of closed-circuit television as a private and public entertainment medium will continue to expand. The increasing availability of communications satellite and other long-distance wide-band circuits will facilitate intercontinental closed-circuit transmissions, and the presentation of major cultural and sporting events in several continents simultaneously can be expected to become common within the next decade.

WARNING SYSTEMS

As a warning system, television has the unique attribute of conveying an actual image of the subject, thereby eliminating the possibility of false information being transmitted owing to system malfunction. It is, in fact, inherently a safe system.

POWER STATIONS. Closed-circuit television is used extensively as a warning system in modern thermal power-stations. The most common application is the transmission of an image of the water-level gauge on the boiler drum to the boiler control panel. Plant safety requirements dictate that this gauge shall be continuously and instantaneously visible to operating staff, but for simple engineering reasons it has to be located on the boiler drum at a height of up to 150 ft. above the station operating floor. Failure of the water supply to a modern water-tube boiler can result in catastrophic damage within minutes, and damage to turbines can result from the water rising above a certain level in the drum.

By reason of its safety characteristics, television provides a uniquely satisfactory solution to the remote monitoring of this gauge and has been adopted as the standard technique in new power-stations in Britain.

Continuous monitoring of furnace flame conditions is also achieved most effectively with television. This is particularly valuable during the critical lighting-up process of a large modern power-station furnace, as immediate warning is given of dangerous conditions. The usual arrangement is the attachment of a television camera to a water-cooled periscope optical system which is inserted through the furnace wall, generally below the level of the burners. This minimizes the size of aperture required through the curtain of water-tubes in the furnace wall, an important consideration, and removes the camera from the immediate vicinity of the furnace wall to a somewhat more favourable environment. A further refinement is the provision of a pneumatically operated carriage which, in conjunction with a thermal overload switch, automatically withdraws the periscope from the furnace in the event of coolant failure or when the television system is not in use.

Television has also been successfully used in power-stations for the observation of smoke-stacks, to warn of black smoke emission, and to assist in the control of complex coal conveyor systems, drawing immediate attention to spillage or mechanical breakdown at critical points in the system.

STEEL INDUSTRY. Another industry in which television has been extensively employed as a warning device is the steel industry. Most steel-making processes involve hazards to equipment which can be minimized by providing process operators with a clear view of conditions at critical points in the process. A typical process is the heating of large steel slabs in reheat furnaces prior to rolling. Slabs are fed in a transverse position into the reheat furnace at one end by means of a hydraulic ram, each new slab entering the furnace causing another slab to drop out at the far end, this procedure being under the control of a single operator situated at the input end.

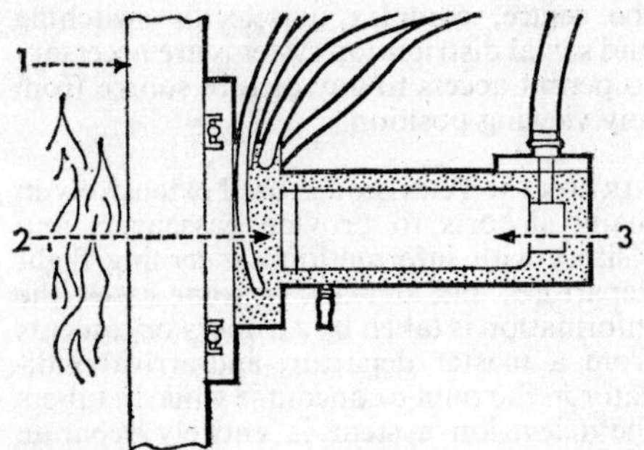

FURNACE MONITOR. Vidicon channel for monitoring interior of steel slab reheat furnace with viewing aperture (not shown). 1. Furnace wall. 2. Protective housing. 3. Vidicon camera.

It occasionally happens that two adjacent slabs become welded together or that one slab rides up on to the one in front of it, whereupon any subsequent endeavour to introduce another slab may result in serious structural damage to the furnace. It is now common practice to install television cameras to provide views of the furnace interior at the output end, and of the drop-out table, to give the operator immediate warning of trouble. To provide the interior view, a camera has to be mounted close to the furnace wall in an area of high ambient temperature, and for this purpose special enclosures have been developed which also provide protection against the intense heat radiated from the interior of the furnace through the viewing aperture.

Television has also provided an excellent solution to an operating problem in large slabbing mills. It is customary to situate the control pulpit on the input side of the mill, the other side being thus obscured from the operator's view by the mill itself. During the process of rolling, "fishtailing" of the slabs (that is, splitting and spreading of one end) or misalignment may occur on the unseen side which would damage the mill if a normal return pass were to be attempted. Again, the view provided by the television camera warns the operator and enables him to take corrective action.

SECURITY. The use of television as a means of improving the security of premises is quite commonplace. In many industrial organizations, unattended entrances are made secure by the use of television in conjunction with remotely operated gates, and specific areas can be kept under continuous observation from a distance to reduce unauthorized ingress, pilferage, and so on. Successful experiments have also been made with closed-circuit television to supplement traditional security measures in prisons.

ROCKET SITES. Very extensive use is made of closed-circuit television at space-rocket launching sites to provide maximum safety for personnel and equipment during the launching process. On a large vehicle, up to fifty separate television cameras may be employed to provide a view in the launch control centre of conditions at critical points on the launch pad and the surrounding area normally situated at a considerable distance from the control centre. Critical junctions in fuelling lines, meters, mechanical actuators and other important system elements are continuously observed.

SHIPS. The modern trend in cargo ship design, for both wet and dry cargoes, is to place all machinery and accommodation aft. It is, therefore, advantageous also to situate the navigation bridge aft, but this introduces serious safety problems in confined or congested waters, as the view immediately ahead is obscured.

Television provides a very satisfactory solution to this problem, and such ships are quite commonly fitted with a television camera in the forepeak with a display in the navigation bridge. Remote pan and tilt facilities are generally provided, together with a means of changing the field of view, such as a lens turret, a zoom lens, and an optical attachment to the camera to assist in range assessment.

INFORMATION SYSTEMS

Television constitutes a highly effective medium where complex information has to be speedily transferred. In such applications, it can be likened to a visual extension of the telephone and might reasonably, therefore, be regarded as having an important future in industrial communications.

AIR TRAFFIC CONTROL. The increasing density and speed of commercial air traffic throughout the world is making traditional methods of control obsolete, and every appropriate new technique is carefully evaluated in the quest for greater speed and efficiency.

For example, at the Southern Air Traffic Control Centre at London Airport, television was selected as the best available way of speeding up communications between individual controllers. With the aid of surveillance radar in the airways, the radar controllers follow, on their displays, aircraft movements throughout the control area and assist the procedural controllers directly responsible for the aircraft by monitoring the progress of aircraft in each sector. At times of high traffic density or emergency, the radar controller may take over complete control, through his procedural controller, of a group of aircraft.

Hitherto, details of each aircraft in the sector were prepared and updated by an assistant, seated alongside each radar controller, receiving the information by radio direct from the aircraft and recording it with chinagraph pencil on an edge-lit Perspex progress board. Appropriate information was exchanged between individual assistants via an intercom system. During periods of intense activity, however, danger arose because several minutes could elapse between the receipt of important information by the first assistant and his passing it to another.

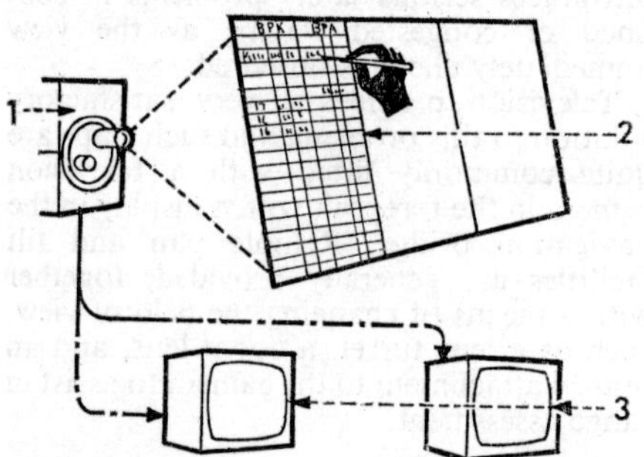

AIR TRAFFIC CONTROL. Control assistant updates progress board, 2, scanned from behind by vidicon camera, 1. Controllers on monitors, 3, can select picture from outputs of 16 progress boards and cameras.

The control centre was therefore rearranged with all the control assistants situated together in a group and physically separated from the radar controllers; a particular control assistant can thus see information on aircraft in adjacent airways merely by looking at the progress boards next to his. Each of the edge-lit Perspex progress boards is viewed from behind by a television camera, a total of 16 boards and cameras being employed. Next to each radar controller is a small television monitor and a bank of push-buttons by means of which the controller can select on his monitor the picture from any of the 16 cameras. As there are some fifty viewing positions throughout the centre, complex uniselector switching and signal distribution systems are necessary to permit access to any picture source from any viewing position.

AIRPORTS. Television is used extensively in many airports to provide passengers and visitors with information concerning flight departures and arrivals. In some cases, the information is taken by a camera or cameras from a master departure-and-arrival indicator in the public concourse while in others the television system is entirely separate from the master indicator.

TRAFFIC CONTROL. Increasing use is being made of television as an aid to traffic control in busy urban areas or other high traffic density situations. In several European cities, for example Munich and Frankfurt, complex television systems provide centralized observation of traffic conditions at numerous points, and similar systems are under consideration or actually in the planning stage in many countries.

Working installations exist, for example, in the Lincoln Tunnel, New York, the Mersey Tunnel in Liverpool, and many motorways and expressways around the world, such as the John Lodge motorway, Detroit, and the overhead section of the M4 motorway in London. The television cameras are usually mounted on high buildings or specially erected gantries, fitted with remotely operated pan and tilt heads and zoom lenses. Signals from the cameras are carried to the receiving point by microwave link or by cable and control of the pan and tilt heads and zoom lenses over the long distances normally involved can be achieved economically by modulated audio tone signals.

DOCUMENT TRANSMISSION. An application of closed-circuit television which is already of great importance and will grow fast in the future is its use for transmitting documentary or tabulated information between widely separated points. For example, a television installation enables drawing prints to be checked against masters in the print store from several places in a large plant, obviating the need for frequent time-consuming visits to the print store by many people. In one such system, the camera in the print store is mounted above a viewing table and equipped with pan and tilt facilities and a zoom lens. The remote viewer, having requested on the associated intercom system the required master drawing, which is placed on the viewing table by the print-store assistant, can control the camera orientation and the field of view from the viewing position. This enables him to select the general area of interest and then to increase the magnification substantially until the particular drawing detail of interest is clearly legible.

Similar applications are commonplace in banks, where television is, for example, used to transmit customer account statements from records departments to managers' offices and business-handling areas, eliminating time delays in sending messengers between departments. It is also frequently employed as a rapid means of signature verification before paying out on cheques.

INSTRUCTIONAL SYSTEMS

Perhaps the most significant class of closed-circuit television applications is in the field of training and education. There can be little doubt that the advent of television in these fields has caused a major impact, the effect of which may not be fully appreciated for many years.

EDUCATIONAL TELEVISION. The use of television in educational establishments in Britain, following the lead given by the USA, increased dramatically during the mid-1960s, and systems in use range from the simplest single camera and monitor arrangement as an enlarging device within a single room to full-scale production centres involving multi-camera studios, telecine, video recorders and all the other complexities of a broadcasting system including in certain cases the transmitter.

Several colleges, in addition to comprehensive studio facilities, also operate mobile

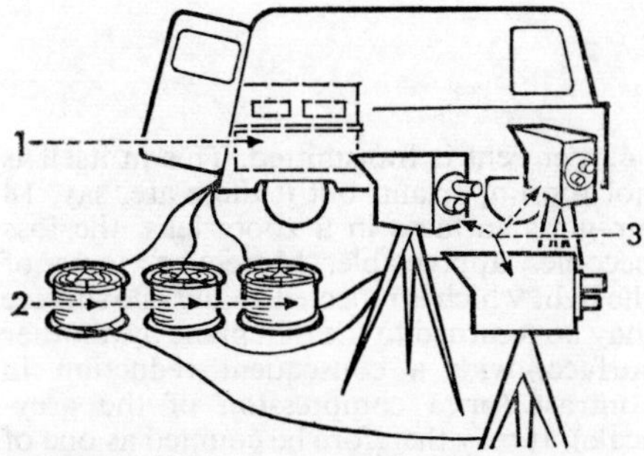

MOBILE RECORDING UNIT. Type suitable for educational and industrial use. 1. Video mixing console with VTR facilities. 2. Cable drums. 3. Vidicon cameras.

recording units which contain all the facilities necessary to record outside events on video tape for subsequent replay in the classroom.

This concept is used extensively by the United States Continental Army Command (CONARC), which operates one of the largest training television systems in the world. The mobile recording units can be taken to any installation or into the field to record field exercises or special material required for training courses at any of the playback installations situated at 17 CONARC establishments across the USA. Each mobile unit contains a video recorder, film projector, slide projector and multiplexer (to permit incorporation of filmed or slide material into recordings), three vidicon cameras, monitoring switching and test equipment. They are fully air conditioned.

Television is becoming an indispensable facility in teacher-training colleges to permit group observation of teaching situations without classroom interference. In Britain, a two-camera system at Oastler College, Huddersfield, is installed in a specially designed studio-classroom, one camera being placed in one of two alternative positions either in front of the classroom and to one side, giving a view of the children, or at the rear of the classroom to provide a view mainly of the teacher. The second camera is fitted to an overhead railway system which runs the length of the room and can thus be positioned anywhere along this horizontal axis by the manipulation of a pulley system operated from the control room. The person in charge of the proceedings is situated in the control room, overlooking the studio-classroom, and can talk to the students in the viewing room which seats about eighty people. The students in the viewing room can, by means of a two-way talkback system, put questions to the controller via the lecturer in the viewing room.

MEDICAL TELEVISION. Medical education is another field in which the use of television is increasing. Many teaching hospitals around the world are equipped with television cameras in operating theatres, providing a close-up view of surgical operations to observers outside the theatre. In the most sophisticated of such installations, colour cameras and large-screen projection systems are employed, giving vastly magnified images in full colour to audiences of several hundred people.

Television can also be used most effectively to provide students with an excellent view of clinical procedures, psychiatric procedures, autopsies and so on where the presence of an audience is undesirable or impracticable or where a satisfactory view cannot be obtained by a large group. The attachment of a television camera to an optical microscope is a good example of the latter situation.

TRAINING SIMULATORS. The increasing use of simulators for complex training procedures offers considerable scope for the use of television. An example of this is the simulation of visual landing and take-off exercises.

In a typical system, a fully detailed terrain model of an airport and its surroundings is built to a scale of, say, 2,000 : 1. This model is sometimes built in the form of an endless belt mounted between two rollers.

A television camera, normally a colour camera, represents the aircraft and is mounted above the model in such a manner that it can be moved relative to the model to represent all the normal degrees of freedom of an aircraft in flight. In practice, some of these movements are made in the special optical system fitted to the camera and, where the model is on an endless belt, forward movement is represented by movement of the model past the camera rather than vice versa. The optical system must have sufficient depth of field that detail in the picture from within half an inch, when the "aircraft" is on the runway, to "infinity" is in focus simultaneously. This results in very small apertures (as small as *f* 60) and hence the need for very high levels of illumination on the model, particularly for colour.

The image is normally displayed on a projection screen in front of the simulator fuselage, the pilot's window being partially obscured so that only the projection screen is in view. The projector is mounted immediately above the fuselage and in systems in which actual pitching and yawing motion is applied to the simulator fuselage the complete projector assembly and screen moves with it. Full colour systems capable of motion are operating at the training establishments of many of the world's major airlines including TWA in Kansas City, American Airlines in Fort Worth, Texas, and Qantas in Sydney, Australia. J. E. H. B.

See: *Camera* (TV); *Large screen television; Medical cinematography.*

Books: *TV in Science and Industry*, by C. K. Zworykin, E. G. Ramberg and L. E. Flory (New York), 1958; *Closed Circuit TV Handbook*, by L. A. Wortman (Slough), 1965.

□**Closed Loop Drive System.** Tape drive system which includes a feedback signal in the loop. Used for accurate positional control of quadruplex heads in transverse-scan videotape recorders.

See: *Videotape recording.*

⊕**Close-Up.** Shot taken at what appears to be □a short distance from its subject (often the human face), so as to give a close view of it. Also called a close shot.

⊕**C-Mount.** Standard form of interchangeable screw-threaded mount used for the lenses of 16mm motion picture cameras. The C-type has a major diameter of 1·00 in. with 23 threads to the inch and the focal plane is 0·690 in. behind the seating face.

See: *Narrow gauge film and equipment.*

⊕**Coating.** Application of emulsion or mag- □netic material to base film.

See: *Film manufacture.*

⊕COATING OF LENSES
□

When light passes from one medium to another of different refractive index, a small part of it is reflected and the greater part is transmitted. The relative proportions of reflected and transmitted light depend on the difference in refractive index, and on the angle that a particular light ray makes with the surface.

REFLECTION LOSSES. For glass of the type normally used in a high-performance lens, about 5½ per cent is reflected and 94½ per cent is transmitted. This in itself is not too important, but if there are, say, 18 air–glass surfaces in a zoom lens, the loss becomes appreciable. Moreover, some of the light which is reflected at any one surface may be returned to the film plane by another surface, with a consequent reduction in contrast (or a compression of the grey-scale). It may therefore be counted as one of the most important developments of recent times that a means has been found to reduce this reflection light loss.

The technique of effecting this reduction is based upon the mechanism of light reflection. When a ray of light encounters an air–glass interface, it gives rise to a series of rays which spread out in all directions. This happens for all the rays in an incident beam of light encountering the interface. Along each light ray there is, however, a periodically varying pattern of electric and magnetic

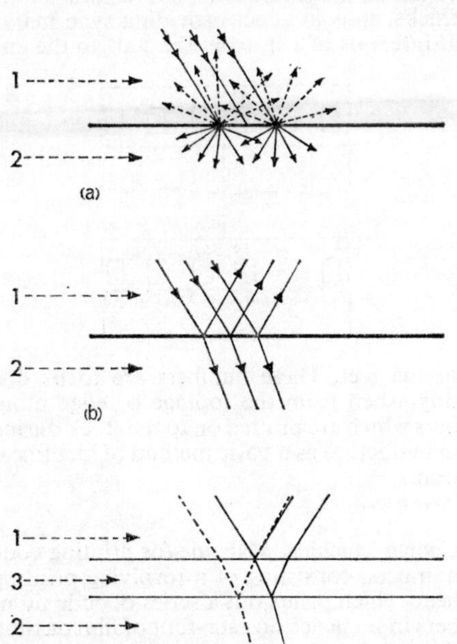

REDUCING SURFACE REFLECTIONS. (a) When light rays in air, 1, encounter surface of glass, 2, secondary rays are sent out in all directions. (b) As a result of interference, only a few of these survive. (c) By using same principle for rays encountering surfaces of interposed thin coating, 3, light reflections may be minimized or eliminated.

forces. Under suitable conditions, the forces arising from two or more light rays which go through a point in space will reinforce one another, so that optical effects may be observed at that point. Under other conditions, the forces arising from two or more rays may neutralize one another at a point in space, and no radiant energy appears at that point. This phenomenon is known as the interference of light.

Because of this interference phenomenon, only a few of the myriad rays, created where the incident rays meet the interface, are able to survive, and these surviving rays are those which are normally considered as reflected and refracted rays.

COATING PRINCIPLES. A glass surface coated with an extremely thin film of low index transparent material results in surviving reflected and refracted rays from two surfaces. The first of these is the surface of the film material, the interface between air and the film material. The second is the interface between the film material and glass.

If the film material is properly chosen, and if its thickness is properly controlled to be one-quarter of the wavelength of light of a particular colour, there is interference between the surviving reflected rays from both surfaces; they neutralize one another and there is no resultant reflected ray. In other words, reflection has been inhibited.

The reduction of reflection can be optimized for only one wavelength of light. Other wavelengths are reflected, but with a much reduced intensity. As a result, the coated surface appears to have a coloured bloom.

When the reflection is minimized for greenish-yellow light, the surfaces have a blue colour. When reflection is minimized for the blue end of the spectrum, the surfaces have a straw colour.

Surface coating or blooming is now an everyday process in lens manufacture, with magnesium fluoride as the coating material. The coating is tough and has good resistance to abrasion.

MULTIPLE COATING. In recent times, a further development has become important, particularly in zoom lenses with their numerous lens surfaces. Coatings of different thicknesses, giving the surfaces different-coloured appearances, are used on the various lens element surfaces. This means that there is no residual tint in the light which is transmitted through the complete lens.

Alternatively, a number of thin films, of high and low refractive index, may be coated on the glass surface, one over another, in place of the single-layer coating. This gives a more efficient suppression of reflection for light throughout the visible range of the spectrum. Such coatings often have a pink or green tinge.

Multiple-layer coatings of the same kind may be used to make colour filters, or to make dichroic reflectors, i.e., surfaces which will reflect light of one colour and transmit light of a complementary colour. A. C.

See: *Dichroic reflector; Lens performance; Lens types.*

Book: *Photographic Optics*, by A. Cox (London), 1966.

□**Coaxial Cable.** Type of feeder cable widely used between aerial and receiver. So called because the inner conductor (single or stranded) is coaxial with the outer conductor, a braided sheath. The two are separated by

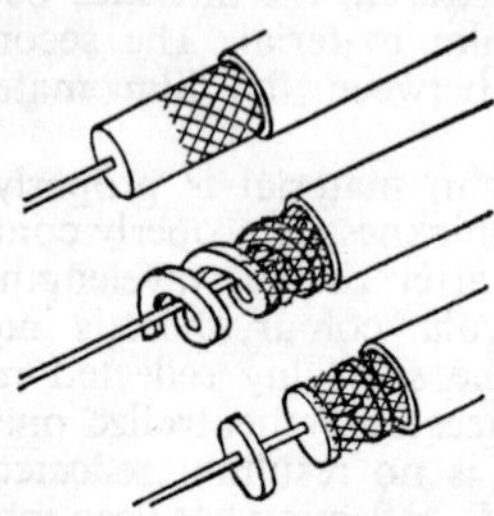

a dielectric layer, usually solid polythene, but in low-loss UHF cables a spiral structure is sometimes adopted, or an air dielectric with polythene disc spacers to provide mechanical support. The braided sheath is insulated and weather protected by a stout outer covering. In the US, twin wire feeders are more common.

Losses in coaxial cable increase as the square root of the frequency transmitted. They decrease with increasing diameter, so UHF cable is thicker than VHF cable.

See: *Transmission and modulation.*

□**Co-Channel Interference.** Interference between TV stations operating on the same channel. To economize channel allocations, and at the same time to provide an adequate service in mountainous areas or other pockets of bad reception, it is usual to allocate several low-power repeater stations to the same channel. These stations are geographically separated as widely as possible, but atmospheric conditions in the troposphere (the region of cloud formation) can cause freak transmission of signals over distances far greater than low-power ground-wave signals can reach, amounting sometimes to several hundred miles.

Hence co-channel interference can be severe, especially in hot summer weather, and other methods besides geographical separation must be taken to reduce it. Thus co-channel stations often use different polarizations, since a horizontally polarized receiving aerial (antenna) is insensitive to a vertically polarized signal, and vice versa. Such a signal, however, always contains horizontally polarized components, so separation is by no means complete.

A further technique concentrates on reducing the visual unpleasantness of the co-channel interference by slightly offsetting the carrier frequency of the potentially interfering station. This causes the interference pattern of broad vertical stripes to break down into a finer and less disturbing pattern.

See: *Transmitters and aerials.*

⊕**Code Numbers.** Identical numbers often printed during the editing process along the edges of synchronized picture and sound tracks, thus in effect providing sync marks at intervals of 1 ft. from the start to the end

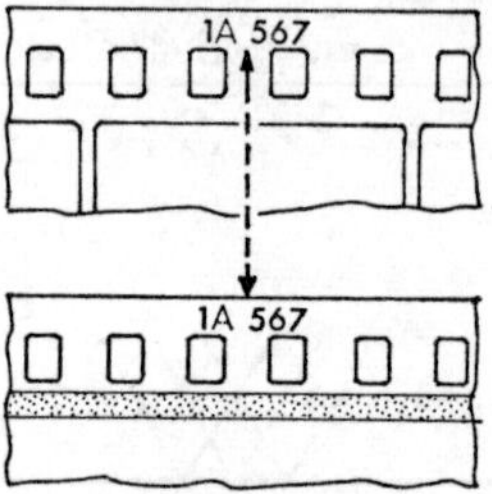

of the reel. These numbers are to be distinguished from the footage or edge numbers which are printed on to the stock during manufacture as a basic method of identification.

See: *Editor.*

⊕**Coding Machine.** Machine for printing code numbers, consisting of a revolving printing head which prints out a series of code numbers in sequence on each foot of film passing through it.

See: *Code numbers.*

⊕□**Coercivity.** In a ferro-magnetic material, the magnetizing force which must be applied to that material to reduce the magnetic flux density to zero.

□**Cogging.** Break-up of vertical edges caused by a displacement of the raster from one field to the next. A common receiver fault.

⊕**Cohl, Emile, 1857–1938.** French cartoonist and photographer, and pioneer of the animated cartoon. For 20 years an illustrator of comic journals, Cohl resented that an idea of his was borrowed by Gaumont, but at their request joined them to turn it into a trick film. Soon he was devising new animation processes based on the American invention of one turn, one picture. His first film, *Fantasmagorie*, almost two minutes long, was shown in August 1908. In 1909, following Stuart Blackton, he combined animation with real life in *Joyeux Microbes.*

⊕COLD CLIMATE CINEMATOGRAPHY

Problems of cinematography at low temperatures fall into two categories. First, the equipment itself is difficult to handle, as the familiar techniques are awkward or laborious to perform; second, the operator tends to be more sluggish in his movements, and less alert mentally. Appropriate precautions can go far to eliminate most of the first group of difficulties, but the personal shortcomings offer an ever-present challenge which only lengthy training and determined effort can adequately overcome.

CAMERA CHOICE. Reliability must be considered first when choosing a camera for use at low temperatures. It is well to seek the advice of some manufacturer of repute, if the stay in cold areas is likely to be prolonged, since his experience could prove of great value.

Clockwork motors have a tendency to become sluggish when cold, and at very low temperatures the springs may snap; on the other hand, the output of dry cells and accumulators greatly decreases in sub-zero conditions, so that some slowing down of the mechanism may follow unless precautions are taken.

It is often suggested that in order to prevent the camera from becoming cold, it should be kept inside the outer clothing, thus making use of the natural heat of the body; with small 8mm models this advice may be of value, but most cine cameras are too bulky to be thus protected.

A more satisfactory method is to use an electric heating pad built into a well-lagged camera case and fed from dry cells or an accumulator. The latter can be kept in a pocket and hence should remain reasonably warm. Careful practical trials in cold conditions are necessary, not only to determine the wattage required to maintain the camera temperature around 40°F, under severe climatic conditions, but also to discover the effective life of each dry cell used. The current drain is much greater than that normally demanded from a dry cell, so the

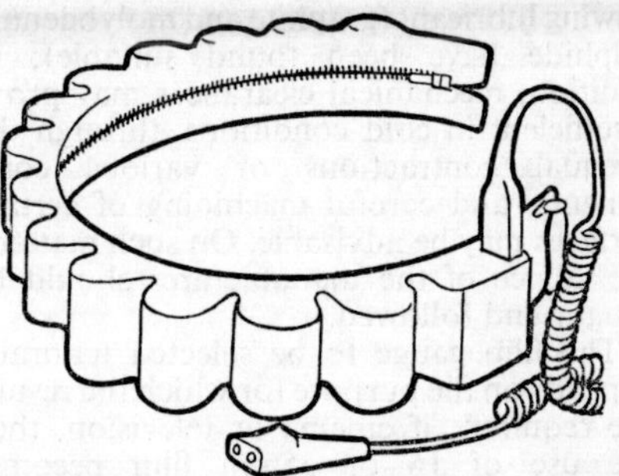

BATTERY BELT. Device to hold small rechargeable cells inside clothing to keep temperature and thus voltage reasonably high.

life will be reduced accordingly; moreover, if the cells are permitted to become cold, their output will also suffer. Portable accumulators can help to solve these problems, but then the charging frequency will have to be determined.

If it is possible to keep the instrument reasonably warm, the various hazards discussed here may prove irrelevant. Otherwise, it should be recalled that film base tends to become brittle when cold, and is

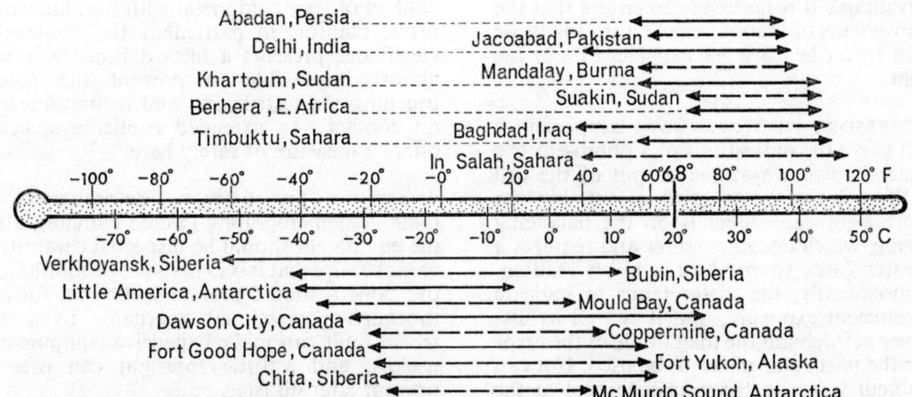

TEMPERATURE EXTREMES. Range of temperatures in typical hot and cold regions shows that difference from normal 20°C is much greater in cold than in hot regions.

then liable to snap if sharply bent. Since it necessarily follows a jerky, tortuous path inside the camera, a real hazard can exist, as, even if the film itself does not tear, there is a possibility that the perforations will. The use of high framing rates is consequently inadvisable, unless the operator is confident that the risk is minimal. The smoother the film path through the camera the better, but normally this path is intricate.

If sub-zero temperatures are to be encountered, it is wise to have the instrument cleaned and re-oiled with a special free-flowing lubricant (graphite and molybdenum sulphide have been found suitable); in addition, mechanical clearances may prove insufficient in cold conditions, through the unequal contractions of various components, and careful machining of certain surfaces may be advisable. On such matters, the advice of the manufacturer should be sought and followed.

The film gauge to be selected naturally depends on the purpose for which the results are required; if cinema or television, then the use of 16 or 35mm film becomes obligatory, and the framing rate should be 24 fps, to enable a sound track to be added later. The economy and lightness of the 8mm camera are advantages which can be enjoyed if no widespread public showings of the film are intended.

For any serious work, the use of lenses of differing focal lengths is necessary; the grandeur of a mountain range ten miles away is lost on the screen, unless a long focus lens is employed. The choice between the increasingly popular zoom lens and the well-established turret head must be left to the operator, but where portability is of great importance the zoom lens has a real advantage. It is necessary to ensure that the movements of a zoom mechanism (or those of a turret head) work satisfactorily in the cold.

ACCESSORY CHOICE. Low temperatures can affect the indications of a photo-electric meter in two ways: the output of the cell itself is decreased to some extent, but the more significant effect is on the balancing spring, which becomes stiffer and requires a greater force to produce a given reading. Consequently, the meter tends to indicate insufficient exposure, and it is well to discover beforehand the magnitude of the error for the particular meter to be used. If it can be kept warm in the pocket, except at the moment of use, the error will probably be insignificant.

In polar regions, when the sun is low in the sky, the light is lacking in blue and hence the colour balance on daylight film shows marked red bias. This may be acceptable, and even pleasing, for future audiences if it is not too pronounced or too persistent; however, it is normally preferable to impose some measure of suppression on the red by employing colour temperature correction filters, but only experience can show the extent of correction necessary.

Up mountains, conversely, ultra-violet light abounds, leading to a blue bias on the film; an efficient UV filter is therefore required, of a deeper absorbence than that normally used at ground level.

Opinions vary on the value of a polarizing filter to reduce the glare from sunlit snow; a few experiments at home in carefully controlled conditions, at different lighting angles, will soon reveal to the operator the limitations and the value of this particular technique.

A sturdy tripod is essential; the operator will probably be far too cold to expect even moderate success in any attempt to hand-hold the camera. To prevent the narrow legs from slipping through the snow, a useful device can be made from an equilateral triangle of stout canvas (with sides about 2 ft.), in each apex of which is sewn a metal ring; if the legs of the tripod are placed in these rings, there should be no risk of slipping, even in snow-drifts.

The use of a deep lens hood is recommended; not only is glare from sun and snow likely to prove troublesome but, in a blizzard, an efficient hood can offer valuable physical protection to the lens. It is wise to cover all exposed metal parts of the camera with some insulating material, since the contact of very cold metal with the skin can prove painful; in particular, the eye-level viewfinder presents a hazard because it is almost impossible to prevent the face touching the instrument, and it may freeze on contact. An extended rubber eyepiece offers a measure of safety here.

CHECKING EQUIPMENT. Before undertaking cinematography in cold climates, all the equipment should be inspected carefully to make sure that it is in first-class condition; the camera should preferably be sent for a thorough professional overhaul. Even a trivial fault can make expensive equipment useless, and a little foresight can often prevent later disaster.

If sub-zero temperatures are to be met, the behaviour of the equipment in such conditions should be determined. For instance, attention should be paid to the convenience of the controls, such as the focusing ring and the starter button; these may prove satisfactory in normal conditions but difficult to operate with cold, heavily gloved hands. The manipulation of the focusing ring can be eased by screwing in radially a threaded rod, as already applied to some zoom lenses.

It may not prove possible to keep the camera warm, as suggested above, and it then becomes important to know just how it will behave at sub-zero temperatures; one satisfactory way of doing this is to leave it in a cold meat or fish store overnight (the temperature of a domestic refrigerator is unlikely to be low enough). If the cold camera is then used to expose film, examination of the results should reveal the extent of the faults likely to be met later.

The air in the camera must be dry before using the cold store, and the camera should therefore be kept in a sealed box for two days with a supply of silica gel desiccant; neglect of this precaution will probably allow the moisture in the air to freeze, and ice on the shutter can clog it badly.

Even if it is intended to keep the camera warm, knowledge of its ability to work at low temperatures is useful, as the heating equipment may fail. Further, it is worth discovering whether the reloading operations can be performed effectively with fingers numb from cold, or whether it will be necessary to retire to warmer surroundings for this task.

USING EQUIPMENT. It is wise to waste a few frames of film before starting the day's work by making a brief exposure in relatively warm surroundings; the tendency of the cold film to stick at a bend in the camera, and then to snap, may thereby be overcome. Cold film base also assumes an edge of razor-like sharpness and should be handled with great care.

When a camera is brought into a comparatively warm place at the end of a trip in the cold, moisture readily condenses and freezes on it; the deposit can easily jam the mechanism, and it takes a surprisingly long time to evaporate. This condensation may best be avoided by surrounding the cold equipment with a plastic bag before entering the warm room; moisture will form on the outside of the bag, where it will do no harm, and after a couple of hours the equipment can be handled normally.

The effect of lowered temperature on film speed may be ignored at normal low temperatures, but at −50°F to −70°F, one stop may be lost, i.e., the speed is halved. On the other hand, if, unknown to the operator, the camera mechanism slows down, the exposure time is increased and overexposure may result. The extent of this trouble naturally depends on the mechanism employed, its lubrication efficiency, and its specific temperature; the extent of exposure errors shown in a film exposed during the cold store test provides valuable information.

Exposure determination itself may prove difficult, as the operator's normal experience can be misleading in unusual surroundings,

COLD CLIMATE PROCESSING

Temperature:	*40° F.* *4½° C.*	*30° F.* *−1° C.*	*20° F.* *−7° C.*
DEVELOPMENT: use Kodak developer D.8, and ethylene glycol:			
Solvent	Water	Water	25% glycol
Approximate time ... ... (mins.)	10	18	70
RINSING: use dilute acetic acid:			
Acetic acid	3%	3%	1%, in 25% glycol
Approximate time ... ... (mins.)	1	1	2
FIXING: use commercial ammonium thiosulphate fixer, diluted:			
Conc. fixer ... ... (volume)	2	2	1
Water ... ... (volume)	3	3	3
Ethylene glycol ... ... (volume)	0	0	1
Approximate time ... ... (mins.)	12	18	30
WASHING: use several changes of water, at least four:			
Time per change ... ... (mins.)	5	5	10
Solvent	Water	Water	25% glycol
DRYING: in air, using heat and/or circulation if possible; preliminary bath in alcohol decreases drying time:			
Alcohol bath ... ... (mins.)	1½	1½	5

which may be of surprisingly low contrast if the weather is at all dull. Brilliant sunshine in the clear air up mountains, however, can lead to unexpectedly high contrast, and much reliance has to be placed on any experience of this kind the operator may have had.

Periodic inspection of the equipment used is very desirable; brittle film at low temperatures can cause dust inside the camera, particularly around the gate, and a systematic clean after every 400 ft. of film exposed is recommended.

PROCESSING. It is usually unnecessary to process cine films at low temperatures, as it is normally possible to heat the solutions. However, if development in the cold is essential (possibly to check on camera exposure levels), a special technique must be adopted. The table shows an example.

Some preliminary experiments with this technique are clearly advisable before attempting to employ it in adverse climatic conditions.

STORAGE. There are few problems associated with the low temperature storage of either unprocessed or processed materials, as the action of harmful influences is greatly decreased in these conditions. Condensation of moisture may occur if films are sealed when relatively warm and moist, and then cooled; they should be sealed in conditions as near as possible to those of storage, and if there is a danger of condensation, the inclusion of a small sachet of silica gel as a drying agent is a wise precaution. After storage, materials should be allowed to regain the surrounding temperature before being projected, and several hours are sometimes needed for this warming up. The increase in temperature normally associated with projection can produce patches of moisture on the image, giving an extraordinary effect on the screen. D.H.O.J.

See: *Camera* (FILM)*; Film storage; Narrow gauge film and equipment; Processing; Tropical cinematography.*

Book: *Photography on Expeditions*, by D. H. O. John (London), 1965.

⊕□**Cold-Mirror Reflector.** Type of studio lamp in which the standard internal aluminized reflector is replaced by a special dichroic coating which reflects the visible light with

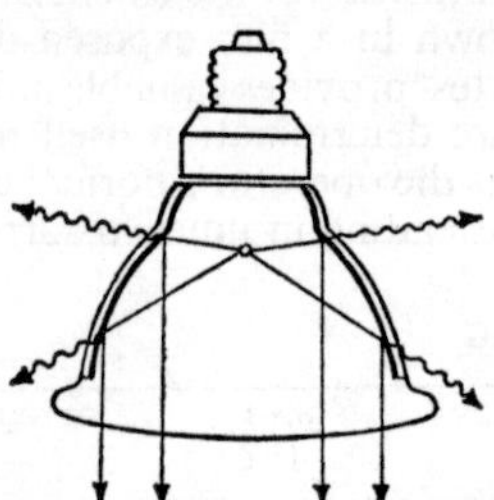

little loss, but transmits infra-red wavelengths (heat radiation). These pass through the side wall of the lamp, and thus are not focused along with the visible light on to the subject. In this way, some two-thirds of the heat content is removed from the light beam.

Discomfort to actors is much reduced, with little loss of light efficiency.

See: *Lighting equipment.*

□**Collector.** Anode which collects the secondary electrons emitted from the mosaic of an iconoscope or similar device.

⊕□**Collimation.** Process of setting up a lens system to produce a parallel beam from a source emitting rays which diverge or converge. Collimators, which produce these parallel beams, play a part in many pieces of optical equipment such as relay systems in TV colour cameras. Collimators are also used for checking the calibration and performance of lenses.

□**Colorplexer.** Device which accepts the red, green and blue separation signals and produces a complete encoded video signal. Composite sync. signals are also added.

The red, green and blue separation signals are added together in a matrix in proportions which are equal to the relative luminance of their three respective primary colours. The resultant signal is proportional to the luminance of the scanned picture element, and can be used by a monochrome receiver to reproduce a black-and-white picture. It is called the E_Y signal.

The red, green and blue signals are adjusted in level and polarity and added together in another matrix. The proportions and polarity of the three signals are such that when all three primary signals are equal, as during the scanning of colourless detail, the resultant output is zero. The output of this matrix is called E_I. A similar process, but with different polarity and level adjustments, is carried out in yet another matrix. The output of this one is called E_Q.

The two signals, E_I and E_Q, are used to amplitude modulate two subcarriers. When either of the two modulating signals is zero, the output from their respective modulators is zero. The same carrier frequency is used for both modulators, but they differ in phase by 90°. When the two modulator outputs are combined, therefore, the phase of the resultant signal depends on the proportions of the three primary separation signals. As the scanned colour desaturates, and tends towards white, the amplitude of the resultant subcarrier tends towards zero.

The two signals, E_I and E_Q, are band-limited by appropriate filters and then added to the E_Y signal. The reference burst is also added at this point with a phase angle of +57° relative to E_I. The bandwidth of the E_Q channel is limited to approximately 0·8 MHz, while that of the E_I channel is approximately 1·5 MHz. This allows a wide single-sideband channel that operates for colour detail where the eye has high resolution, and a low bandwidth double-sideband channel that operates for colours which are not critical in resolution. The phase angle of the reference burst is 180° relative to the E_B–E_Y component of the encoded signal for convenience of decoding in the receiver.

In the PAL system, E_U and E_V signals (corresponding to E_B–E_Y and E_R–E_Y components) are matrixed from the separation signals instead of E_I and E_Q. Both E_U and E_V are limited in bandwidth to 1·2 MHz and are then applied to the subcarrier modulators in a similar manner to the NTSC E_I and E_Q signals. The final encoded waveform therefore consists of E_U and E_V chrominance components added to the E_Y signal together with composite sync. The burst waveform instead of being always on the $-E_B - E_Y$ axis as in the NTSC system, alternates line by line plus and minus 45° from that axis.

See: *Colour principles* (TV); *NTSC colour system; PAL colour system; SECAM colour system.*

⊕**Colour Balance.** Appearance of a colour print which results from the relative exposures which have been given to the three components, yellow, cyan and magenta, forming the image. Sometimes also termed the "ratio" of a colour print. If the printing exposures have been correctly balanced, taking into account the characteristics of the projection light, the print is termed "neutral balance" or "on ratio". Excess of yellow printing or deficiency of cyan and magenta together results in a yellow overall appearance, while excess of yellow and cyan gives a green balance, excess of yellow and magenta a red one, and so on.

See: *Sensitometry of colour films.*

□**Colour Bar.** Part of a colour test signal, comprising six vertical bars which have colours that occur at points spread throughout the reproducible spectrum. Peak white and black level bars are also provided. The red, green and blue bars correspond to the system primary colours. The other three are yellow (red+green), magenta (red+blue) and cyan (green+blue).

Colours that are 100 per cent saturated as well as having 100 per cent luminance are very abnormal. To avoid overload conditions at the transmitter, therefore, it is normal to transmit the colour bars with reduced luminance.

The bars are generated by multivibrators triggered by signals from the studio synchronizing generator. The outputs from these multivibrators are equivalent to the separation signals from a picture source. These are encoded in a colorplexer in the normal manner.

See: *Colorplexer.*

□**Colour Burst.** Feature of the NTSC and PAL television colour signal consisting of a short burst or pulse of subcarrier frequency occupying the back porch of the line sync pulse, and therefore transmitted once per line. In the NTSC and PAL systems, the phase of the subcarrier determines the hue of the

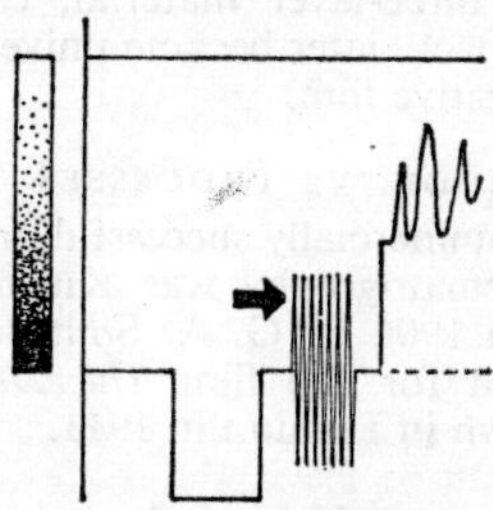

colour reproduced by the receiver; to ensure that this hue corresponds with that of the original scene, a phase reference must be provided. This is the function of the colour burst which initiates each line with a reference signal of known hue.

In the PAL system, the phase of the colour burst oscillates ±45° line by line, and thus controls the (R−Y) and (B−Y) colour difference signals which are transmitted on alternate lines. In the SECAM system, no colour burst is needed.

See: *Colour systems.*

⊕COLOUR CINEMATOGRAPHY

Colour cinematography may conveniently be reviewed through its history. In the past sixty years, virtually every possible colour process – and many an impossible one – has been put on record in the world's patent offices.

Colour film systems are of two basic kinds, additive and subtractive.

Additive systems attracted the early inventors because they made it possible to record and project colour with black-and-white emulsions, the only ones then known. In the camera, this entailed filming through primary colour filters on to alternate frames, or on to film embossed with thousands of cylindrical lenses (lenticular film), or employing film incorporating a mosaic of minute three-colour filters. Projection was by a blending of light coloured by passing through the same or similar filters.

Though research on additive films was only finally abandoned after World War II, subtractive processes had long before taken the lead. To begin with, additive analysis was often employed in the camera, separate two- or three-colour records being made through filters on to black-and-white emulsions.

But for projection of positive prints, synthesis was effected by three pigmented layers in superimposition in the complementary colours of magenta (minus-green), cyan (minus-red) and yellow (minus-blue). A similar three-layer material, called an integral tripack, later became universal as a camera negative film.

ADDITIVE PROCESSES

The first commercially successful process of colour cinematography was Kinemacolor; patented in 1906 by G. A. Smith and still remembered for the film *The Durbar of Delhi*, shown in London in 1911.

TWO-COLOUR. Kinemacolor was a two-colour additive process which depended upon recording alternating orange-red and blue-green images. The speed of taking and projecting was 32 frames a second, and colour mixture resulted from persistence of vision. But at such critically low speeds of alternation – only just above the flicker limit – viewers were prone to eye strain and there was the inevitable risk of time-parallax occurring whenever fast-moving objects crossed the field of view.

It was because of the inherent limitations of additive processes that E. J. Wall wrote, in 1906, "What we want is a length of cinematograph film each picture in which shall be a record of the movement at the instant of exposure and at the same time in itself a complete colour record."

In the 1920s work continued on two other additive methods – the lenticular system, based mainly on the ideas of R. Berthon, a Frenchman, and Keller-Dorian, and the mosaic screen system typified by Dufaycolor. Both these processes continued to receive attention up to the start of World War II, and even later Eastman Kodak continued to work on lenticular film with a view to making it acceptable for motion picture purposes.

LENTICULAR. The earliest patent proposing a lenticular system was granted in 1909 to R. Berthon. The idea was to place a three-colour filter comprising three horizontal bands on the lens of a camera and

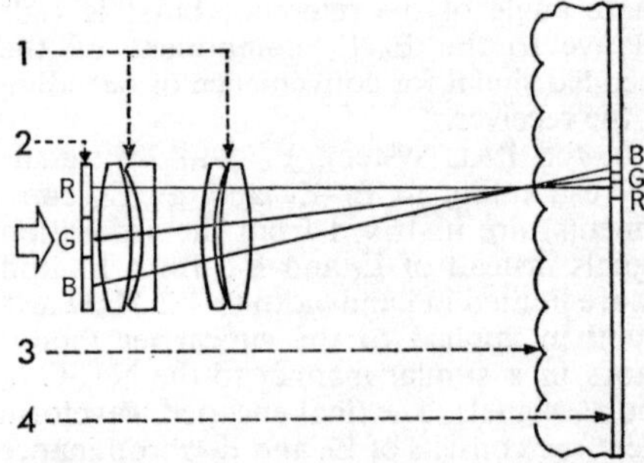

LENTICULAR PROCESS. 1. Wide aperture camera lens focused on film comprising panchromatic emulsion layer, 4, coated on embossed film base arranged so that lenticulations, 3, face lens. In use, a banded tricolour filter, 2, placed in front of the lens is imaged by each tiny embossed lens to form a corresponding pattern of minute filter images on the emulsion layer, thereby separating the recorded image of the subject into tricolour components.

then to record a series of minute images of the object on a sheet of negative material bearing a panchromatic emulsion layer on one side and a series of transparent embossed semi-cylindrical lines on the other. The embossed surface was to face the lens, and the direction of the lines was to run parallel to the bands of the filter in the lens.

With this arrangement, light passing through the camera lens and its banded filter is also transmitted by the tiny embossed cylindrical lenses, each one of which images the bands on the film. The film was processed by a straightforward black-and-white reversal procedure and, after processing, the same film could be projected through a matched optical system with a banded filter on the projection lens.

It is easy to understand the attraction of an additive lenticular process such as this, because the cost of manufacturing the film and processing it would be almost as low as for a black-and-white production.

In 1925 Eastman Kodak acquired the rights to the Berthon Keller-Dorian patents and in 1928 announced their 16mm Kodacolor process for amateur cinematography.

In Germany, Siemens Halske took over Berthon's patents and Perutz coated lenticular embossed film was made in Germany in 1936 and a documentary was shown at the Paris Exhibition in 1937.

Capstaff of Eastman Kodak continued research on the lenticular process after all other activity in this direction had ceased. His work was done in conjunction with 20th Century-Fox, but no films were made for general release.

MOSAIC. Dufaycolor used yet another system. A tri-colour mosaic screen was formed between the film base and its emulsion layer, so that by exposing through the film support, a very large number of minute separation records were made simultaneously behind a myriad of tiny filter elements.

The Dufay process, named after the Frenchman, Louis Dufay, originated in France. The first commercially available material was in the form of Dioptochrom plates, made by the Guilleminot Company of Paris in 1910.

The filter elements in the Dufay reseau were minute. Nearly one million filters covered the area of a single 35mm motion picture frame. Each individual element was no more than ·025mm wide.

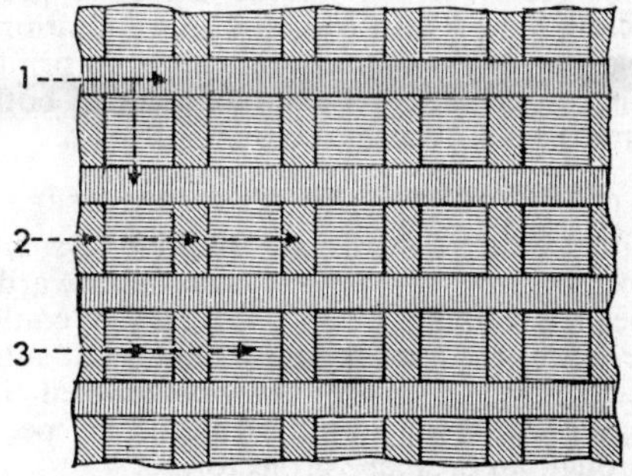

DUFAYCOLOR RESEAU. 1. Red filter elements. 2. Green filter elements. 3. Blue filter elements.

Dufaycolor roll-film had a speed of about 10 ASA and was first sold in 1930. In 1934 a 16mm version was made available for amateur cinematography. Both these films were processed by reversal and were viewed or projected in the normal way, except that considerably more light was required than for a black-and-white transparency.

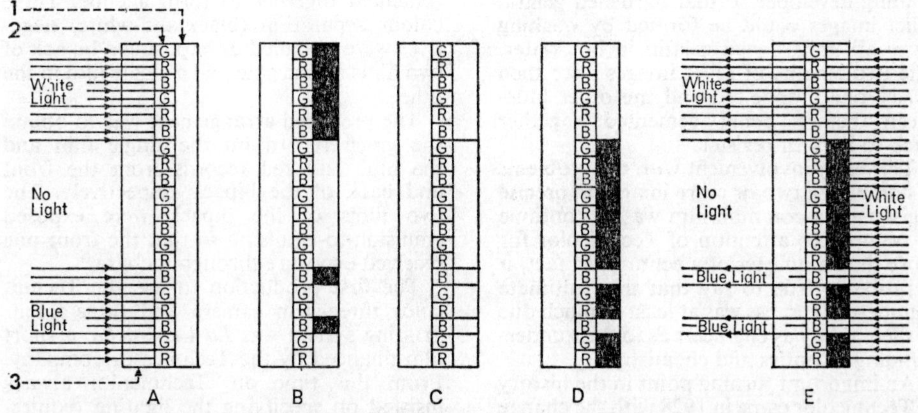

DUFAYCOLOR SYSTEM. (A) Panchromatic emulsion layer, 1, is coated on a reseau pattern, 2, of extremely small red, green and blue filter elements supported on a transparent film base, 3, through which exposure is made. (B) Film after latent image development in black-and-white developer. (C) Result of removing negative silver image by bleaching. (D) After remaining silver halide has been fogged and developed to form a positive image, this image may be viewed by transmitted light at (E), reproducing original scene.

NEGATIVE-POSITIVE. In 1936 it was considered practical for the first time to make positives from Dufaycolor camera negatives. This was in fact the first commercial use of a negative-positive colour process. Several short but important films were made; but the one problem that could not be solved with Dufaycolor any more than with any other additive process was the serious loss of light resulting from projection through a pattern of red, green and blue filters. At least three-quarters of the available projection light is thus lost, and this limitation, together with unsolved difficulties in printing, caused the final abandonment of both lenticular and mosaic colour processes.

TECHNICOLOR AND DYE-TRANSFER

After attempts with an additive system, Technicolor's efforts were directed towards the production of positive images that could be dyed in appropriately coloured subtractive dyes before being combined in register ready for projection at normal speed through an ordinary projector.

TWO-COLOUR PROCESS. In its earliest form, a Technicolor two-colour subtractive print was made from two separation negatives exposed in a beam-splitter camera, each record being in the form of a black-and-white silver image. A skip-printer was used to separate the two colour components from the single length of camera film on which they were recorded, and two continuous positive prints were produced on separate specially thin-based films. The positive images were then developed in a tanning developer so that hardened gelatin relief images could be formed by washing away all unhardened gelatin in hot water. The two lengths of relief images were then dyed – one orange-red and the other blue-green – before being cemented together back-to-back in register.

This early involvement with the problems of combining two or more images in precise register on a common film was to continue to occupy the attention of Technicolor for more than a quarter of a century. In fact, it is probably true to say that their ultimate commercial success was at least as much due to their ability as engineers as to their understanding of optics and chemistry.

An important turning point in the history of Technicolor came in 1928 with the change to an imbibition dye-transfer procedure. Initially the process still used only two colours, but the transfer of both dye images to a common receptor or blank film avoided the troubles that had been experienced with cupping or distortion of the older cemented films during projection. Transferring the dye images also made it possible to obtain many prints from the same set of gelatin relief images or matrices, simply by re-dyeing them between each successive transfer.

THREE-STRIP CAMERA. Another important advance in the development of the Technicolor system came in 1932 with the introduction of the three-colour beam-splitter camera that was to dominate professional colour cinematography for so many years.

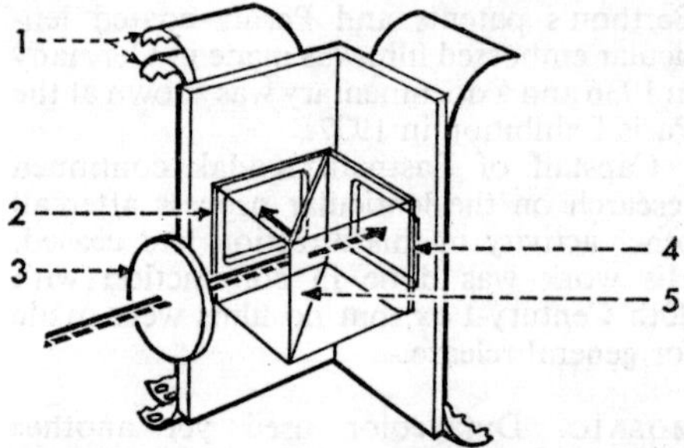

TECHNICOLOR THREE-STRIP CAMERA. Light from subject passes through lens, 3, and enters prism block where part of it is reflected by interface, 5, to gate, 2, containing bipack of two films, 1, the front film blue-sensitive and the rear film red-sensitive. The remaining light is transmitted to aperture, 4, and exposes the green-sensitive film. Separate R, G and B records are thus obtained.

The heart of the Technicolor camera was a prism block comprising two 45° prisms cemented together to form a cube. Three colour separation (black-and-white) negatives were obtained by exposing a bipack of two films in one gate and a single film in the other.

The preferred arrangement was to obtain the green record on the single film and the blue and red records from the front and back of the bipack respectively. The two films of the bipack were exposed emulsion-to-emulsion so that the front one received exposure through its base.

The first production to use the Technicolor three-strip camera and three-colour printing system was *La Cucuracha*, a short film financed by the Technicolor company. From this time on, Technicolor always insisted on specifying the lighting requirements on the set or on location, and they maintained a very high standard of negative quality by carrying out the whole of the processing under their own control.

IMBIBITION TRANSFER PRINTING. The principles of the dye-transfer or imbibition colour printing process date back to the beginning of this century, when two Frenchmen, Charles Cros and L. Didier, independently developed methods of producing hardened gelatin images that could be differentially dyed with aqueous solutions of dye so that, after washing to clear the highlights, a dye image could be transferred to a wet sheet of gelatin-coated paper.

All these early attempts to transfer the dye contained in a matrix to another support or blank, depended upon planographic surfaces – in other words there was no washing or etching away of the gelatin layer to form a relief image.

Sanger Shepherd in 1920 had the idea of forming a gelatin relief matrix that could be dyed and then used to transfer the dye image to another support. At that time, the procedure was to coat a sheet of celluloid with gelatin and then to sensitize it with a solution of bichromate before exposure. But while gelatin relief matrices had to be made from bichromate sensitized layers, the value of the process was very limited, since only contact printing was possible with a material of such low sensitivity. It was Warnerke who first made relief matrices from silver sensitized gelatin layers. He used pyro-ammonia as the developer and omitted the usual sulphite. In the absence of a preservative, a pyro developer tans gelatin in the area of reduced silver.

MATRIX MAKING. After Technicolor began to utilize the tanning development procedure to produce their gelatin matrices, they did a great deal of work on the chemistry and the techniques of matrix making. Both E. M. Wall and L. T. Troland worked on these problems while they were with Technicolor.

No matter whether the original film is shot with a three-strip camera or with a multi-layer film, the imbibition transfer process must start with three colour separation negatives – red, green and blue records of the original scene. Each of the separation negatives is used to print a corresponding geletin relief image or matrix.

It is always necessary to expose through the support of a matrix film because this is the only way in which lightly exposed highlight areas of an image can be retained on the film when unexposed, and therefore unhardened, parts of the gelatin layer come to be washed away in hot water. If the emulsion layer were to be exposed with its outer surface towards the negative in the ordinary way, then all the highlight and low density areas of the image would be undermined with unhardened gelatin which would dissolve in hot water.

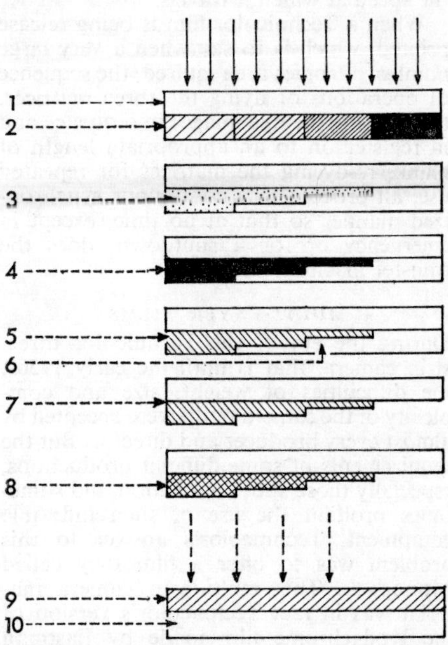

STEPS IN FORMING AND USING DYE TRANSFER MATRIX. Film emulsion exposed through base, 1, to one of three silver separation negative images, 2. Exposed matrix film, 3, developed in tanning developer to form positive silver image, 4. At this stage, silver is embedded in hardened gelatin, 5. Unhardened gelatin, 6, is removed by washing film in hot water, resulting in formation of gelatin relief image, 7, able to absorb dye in proportion to amount of gelatin present, 8. When dyed matrix is brought into contact with plain gelatin layer, 9, coated on film support, 10, dye is transferred to new surface.

After exposure, the matrix films are developed on a continuous processing machine in a tanning developer of the pyro or pyrocatechin type which hardens all gelatin in the region of the developed silver image. After hot water treatment, the resulting gelatin relief images are dyed in appropriate solutions of yellow, magenta and cyan dyes.

TRANSFERRING THE DYE. Substantially complete transfer of the dye from a matrix to a receptor surface requires that the matrix and the blank remain in intimate undisturbed contact for an appreciable length of time – of the order of a few minutes. To satisfy this requirement within the context of continuous processing, Technicolor uses an endless pin-belt on which a matrix film and a blank film can be held in register for whatever time is required, depending only on the length of the belt and the speed at which it travels.

When a Technicolor film is being release printed, which is to say when a very large number of copies are required, the sequence of operations of dying the three matrices, transferring the dye images in sequence and in register on to an appropriate length of blank, re-dyeing the matrices for repeated use, all proceed in a continuous synchronized manner so that at no time, except in emergency or for a shutdown, does the transfer machine ever stop.

MULTI-LAYER FILMS

During the era of the Technicolor three-strip camera, that is until the early 1950s, the difficulties of weight, size and complexity of the camera outfit were accepted by almost every producer and director. But the requirements of some difficult productions, especially those shot on location, did sometimes prohibit the use of such inflexible equipment. Technicolor's answer to this problem was to offer a film they called Monopack. This multi-layer camera material was in fact Technicolor's version of the Kodachrome film made by Eastman Kodak, but processed by Technicolor.

Dr. Mees once reported that when the Kodachrome process was first worked out, it was thought that Kodachrome film might be used both for taking and for release printing, but it was found that better results were obtained by making separation negatives from the Kodachrome camera original and using them to produce gelatin matrices for printing by Technicolor's dye-transfer process.

ORIGINS. With very few exceptions, typified by the Technicolor imbibition transfer system, almost all modern processes of colour photography are based on the principle of using a multi-layer colour film. Even before film became available as a support for photographic images, Louis Ducos du Hauron had, in 1891, suggested using three layers of emulsion in superimposition on the same plate, and his name for such a tripack was Polyfolium Chromodialytique. For many reasons, not least among them the fact that there were no colour sensitizers available in those days, the prophetic ideas of du Hauron could not then be realized.

Several subsequent workers understood the potentialities of a three-layer monopack, but none succeeded in developing the idea very far until about 1911, when Rudolf Fischer proposed to incorporate substances in emulsion layers that would, when the layer had been exposed, react during the course of development in such a way as to produce coloured images in situ with the usual images of reduced silver.

The principal difficulty Fischer encountered at that time was with the colour-forming substances that he proposed to include in his emulsion layers. They were soluble in aqueous solutions and therefore able to diffuse quite easily between the three emulsion layers of the tripack, which of course resulted in unsharp images and unsatisfactory colour reproduction.

ANCHORING THE DYES. Between 1913 and 1932 nothing significant was done with the Fischer idea. But in 1932 Wilhelm Schneider of Agfa at Wolfen patented a way of anchoring dyes in gelatin layers so that they would stay put and not diffuse. This can now be seen as the breakthrough that led to the modern Agfacolor process. But a few more years elapsed before Schneider realized in 1935 that his non-diffusing components might be used to render Fischer's colour-forming substances immobile in a photographic emulsion layer.

Even then, the first attempts to combine the Fischer and Schneider ideas were unsuccessful because although colour formers could be anchored in their intended emulsion layers, they were now rendered insoluble so that little or no reaction resulted during the course of colour development. This difficulty was overcome by adding acid salts to the long-chain hydrocarbon residue that had been attached to the molecule of each of the required colour formers.

AGFACOLOR FILMS. The first version of the new Agfacolor process – there had been an earlier additive system with the same name – was a reversal material. Between 1936 and 1939 the film was available for use in 35mm still cameras and 16mm amateur movie cameras.

The first commercial use of Agfacolor as a negative-positive process for the production of cinema release prints came in 1941, although this was not the first time that the negative-positive principle of colour printing had been employed. A number of feature films were produced on Agfacolor in Germany during the course of World War II, but after that period Agfa Filmfabrik in Wolfen was taken over by the Russians. This meant that Agfa's other photographic manufacturing centre at Leverkusen in West Germany had to expand its activities to take in the production of Agfacolor negative and positive films as well as the Agfacolor paper they had been making since 1935.

EARLY KODACHROME PROCESSING. In its early form, the Kodachrome process depended upon successive stages of controlled diffusion of colour developing and bleaching solutions. The composite colour image was built, layer by layer, from the base upwards. First of all, after a normal type of black-and-white negative development, the three residual positive images were all subjected to cyan colour development. Then the cyan images that were not required in the top and middle layers of the tripack had to be destroyed. This was done by controlling the diffusion of a bleach so that its action did not penetrate any lower than the middle layer.

With the unwanted cyan images destroyed, a magenta colour developer was used to form magenta images in the top and middle layers. The magenta image in the top layer then had to be removed by another stage of controlled bleaching before being reformed as the yellow positive image. In all there were some thirty steps involved in this early form of Kodachrome processing, and it was soon replaced by a simpler system based on the selective re-exposure of each of the three image layers.

SIMPLIFIED PROCESS. The revised procedure – virtually unchanged today – is first to develop the exposed film in a normal type of black-and-white developer so as to produce negative silver images in all three layers. After this first development stage, the film is washed but not fixed and the back of the tripack is exposed to red light, through the film support. The use of coloured light ensures that only the bottom, red-sensitive layer of the tripack is affected. The film is then immersed in a colour coupling developer containing a cyan dye former, the result of which is that a positive cyan dye image is formed together with a silver image in the bottom layer.

After further washing, the front of the film is exposed to blue light so that only the outer or uppermost layer of the tripack is affected. The middle layer, although still sensitive to blue and green light, is protected from the blue exposure by the yellow

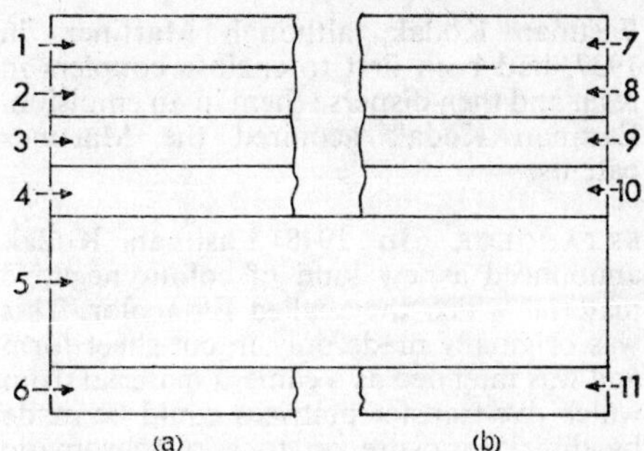

FILM OF KODACHROME TYPE. (a) Film before processing: 1. Blue-sensitive emulsion layer. 2. Yellow filter layer. 3. Green- and blue-sensitive emulsion layer. 4. Red- and blue-sensitive emulsion layer. 5. Base on which emulsions are coated. 6. Dyed anti-halation layer. (b) After processing: image formed in layer 7 is yellow, filter layer, 8, is now colourless, image formed in layer 9 is magenta and in layer 10 cyan. Anti-halation dye in 11 is removed in processing.

filter layer that is used beneath the blue-sensitive layer. The film is then immersed in yellow-forming developer to produce a combined yellow dye and black silver image in the outer layer.

Now only the middle, green-record layer requires treatment and the unreduced silver halide in this layer is either fogged with white light or by the application of a chemical foggant before being immersed in a magenta colour-forming developer.

Finally, all the reduced silver comprising both negative and positive images and any colloidal silver used in filter or anti-halation layers is removed. A composite dye image remains to form the colour picture. All of these stages of development, washing, re-exposure and bleaching are carried out on continuous-running machines, often at speeds of over 100 ft. a minute.

KODACOLOR. The first Eastman Kodak coupler-containing film was Kodacolor – a colour negative material for the amateur. This was launched in 1941 and by 1942 large-scale production of the film and an associated colour paper had started, and in the last quarter of a century has grown to become one of the most important processes for amateur photography.

The colour couplers used in Kodacolor, instead of being anchored in their emulsion layers by reason of their molecular size and shape, are embedded or encapsulated in resin-like materials that are insoluble yet permeable in water. Much of the work on this method of producing diffusion-fast couplers was done by Jelly and Vittum of

Eastman Kodak, although Martinez, in 1937, had been first to enclose couplers in resin and then disperse them in an emulsion. Eastman Kodak acquired the Martinez patents.

EKTACOLOR. In 1948 Eastman Kodak announced a new kind of colour negative material which they called Ektacolor. This was originally made only in cut-sheet form and was intended as a camera material from which dye-transfer matrices could be made by direct exposure on to a panchromatic matrix material known as Pan-Matrix.

Ektacolor was the first colour negative material to incorporate means of forming its own colour correction masking images during development – the masking images compensating for deficiencies of the dyes that comprise the colour negative image.

Instead of absorbing only in the primary colour bands red, green or blue, the best available dyes formed from colour couplers also absorb light of one or both the colours they should freely transmit. Generally speaking, magentas and cyans are the least perfect, while yellows are usually fairly satisfactory. Most magentas absorb green light as they should, but also absorb some of the blue light they should transmit; while cyan dyes usually absorb some blue and green as well as the red they should absorb.

In Ektacolor film, the undesirable blue absorption of the magenta dye is compensated by the provision of a correction mask which is coloured yellow, while the unwanted blue and green absorptions of the cyan dye are compensated by means of a mask which is coloured orange-red. These yellow and orange-red masking images are formed within the appropriate image layers of the film in an entirely automatic manner during the normal process of colour development.

AUTOMATIC COLOUR-MASKING. Like any other multi-layer negative material, Ektacolor comprises emulsion layers sensitive to red, green and blue, and incorporated in these layers are cyan, magenta and yellow couplers, the first two of which are themselves coloured at the time of incorporation in the emulsions. The cyan coupler is orange-red while the magenta coupler is yellow. After exposure and colour development, three silver images are formed in the film, as well as three corresponding dye images. After removal of the reduced silver and remaining halides, the negative dye images remain, together with the unused couplers in the three layers. In Ektacolor, the original colours of the cyan and magenta couplers are not changed or destroyed when the developed colour images are formed, and the remaining, unused couplers comprise positive images which are utilized to form the required colour correction masks.

COLOURED COUPLERS. Wherever there is a magenta image density, there is also an unwanted yellow density accompanying it by reason of the absorption of the magenta dye itself. Where there is a maximum density of magenta, there is also a maximum density of unwanted yellow, and no additional yellow due to coloured coupler. The density of the yellow-coloured coupler layer is so calculated that the superfluous yellow density is just matched by the yellow due to the coloured coupler in areas where no silver is reduced. Where there is less than a maximum density of magenta, there the whole of the yellow-coloured coupler will not have been used and the residue will add just sufficient yellow density to make the same overall total.

This uniform yellow density, which is continuous throughout the magenta image layer, cancels the effect of the unwanted yellow component of the magenta image, and only density due to the true magenta component of the image modulates light transmitted by the layer. The same principle is employed in masking the cyan image, but an orange-red coloured coupler is used because the cyan dye is responsible for unwanted absorption of both blue and green light.

EASTMAN COLOR NEGATIVE. With the experience of Kodacolor and Ektacolor films behind them, it was not long before Eastman Kodak announced a motion picture version of a coloured coupler negative film, which they decided to call Eastman Color Negative. The first motion picture production to use the new material was shot in 1951, and the release prints were made on Eastman Color Print film.

The speed of Eastman Color Negative was originally 16 ASA, and this compared favourably with the effective speed of Technicolor's three-strip camera. Consequently it was not very long before the number of producers who were prepared to use a special camera in preference to Eastman Color Negative film in any ordinary camera had dwindled to such a low level that it was no longer economic for Technicolor to offer the services of their cameras and crews. By

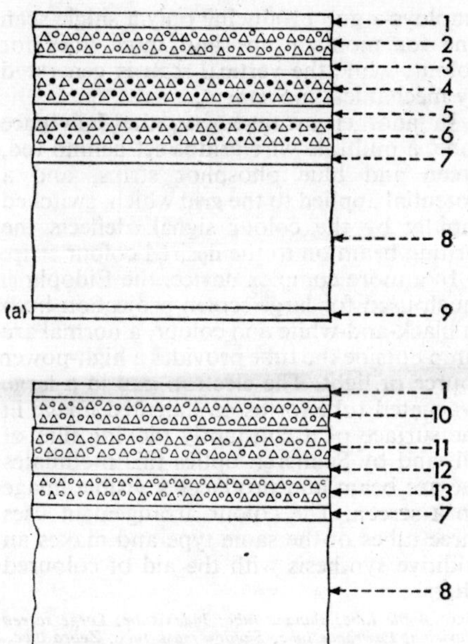

SCHEMATIC CROSS-SECTION OF EASTMAN COLOR FILM. (a) Negative film: 1. Gelatin supercoating. 2. Blue-sensitive emulsion with uncoloured yellow dye coupler. 3. Yellow filter layer. 4. Blue- and green-sensitive emulsion with yellow-coloured magenta dye coupler. 5. Gelatin interlayer. 6. Blue- and red-sensitive emulsion containing reddish-orange-coloured cyan dye coupler. 7. Substratum layer. 8. Film base. 9. Anti-halation layer. (b) Print film: 1. Gelatin supercoat protecting green-sensitive emulsion layer, 10, containing uncoloured magenta dye coupler. 11. Red-sensitive emulsion layer incorporating uncoloured cyan dye coupler. 12. Gelatin interlayer. 13. Blue-sensitive emulsion layer incorporating uncoloured yellow dye coupler. Substratum, 7, binds all layers on to film base, 8. 14. Removable anti-halation backing layer.

the end of the 1950s almost all colour film production was being shot with colour negative film in normal cameras.

PROFESSIONAL 16MM FILMS. The first professional 16mm camera film was Kodachrome Commercial Film, which yielded a low contrast positive that could be printed on to Kodachrome Duplicating Film. In 1955 the advent of Eastman Reversal Color Print Film helped to improve print quality and in 1958 Eastman Ektachrome Commercial Film, a low contrast camera material, began to replace Kodachrome because of superior image characteristics and more general availability of processing. More recently, two fast Ektachrome films were introduced together with new processing procedures that cut the time of wet treatment to about 17 minutes. These two new materials, one of them balanced to daylight and the other to tungsten illumination, can, with suitable adjustment of first development time, be exposed at effective speeds between 125 and 500 ASA.

Improvements in colour negative continued with a speed of 50 ASA and better grain and colour rendering.

While the advantages of the integral colour masking used in Eastman Color Negative has placed this material in a dominating position internationally for professional motion picture photography, the corresponding positive printing stock is not restricted to Eastman Color Positive, and several other positives, such as Gevacolor, Ferraniacolor and Fujicolor, are now manufactured and widely used in some places.

COLOUR FILMS IN TELEVISION. There is now a rapidly growing use of colour film for television which is having an influence on the materials employed. While, in many cases, original photography is carried out on Eastman Color Negative, including the 16mm negative now available, it has been suggested that television transmission requires a different characteristic of positive stock from that used in the cinema and special low contrast colour positive materials have been produced for this purpose.

It appears that generally 16mm film will be widely used and in some applications the reversal system seems to be preferred to the negative-positive method. J.H.C.

See: *Chemistry of the photographic process; Colour principles; Imbibition printing; Integral tripack; Sensitometry of colour films; Technicolor.*

Books: *History of Colour Photography*, by J. S. Friedman (London), 1968; *Colour Cinematography*, by A. Cornwell-Clyne (London), 1951.

⊕**Colour Difference Process.** An improved laboratory method of handling the components of a travelling matte scene photographed by the blue-screen process, allowing the better rendering of foreground objects.

See: *Special effects* (FILM).

□**Colour Display Tubes.** Picture tubes employed in colour TV receivers to translate a standard colour signal, suitably decoded, into a picture. The tube most commonly employed is the shadow mask or tricolour tube, in which three electron guns are con-

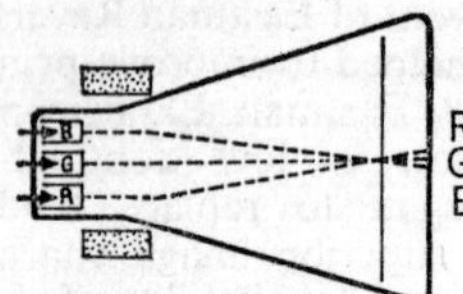

SHADOW MASK TUBE. Perforated mask mounted behind phosphor screen enables each of three guns to "see" only one colour dot in each triad.

verged on a plate perforated with a very large number of small holes mounted about half an inch behind the phosphor screen. This screen is coated with a repeated pattern of three phosphor dots, emitting red, green and blue light respectively, each set of which is called a triad.

The tube's electron optics are so arranged that the three beams converge on the holes and, diverging on the far side, fall precisely each on one spot of the triad. By modulating the beams with the appropriate colour signals, any combination of the primary colours can be reproduced.

Other types of colour display tube, still experimental, employ a simpler gun structure with only one picture-writing beam, but more associated circuitry, e.g., apple tube, zebra tube. Another type, the banana tube, employs a gun producing only a single scan line for monochrome and three lines for colour, while the vertical scan is generated by mechanical movement.

In another arrangement, the Lawrence tube, a multiple wire grid is set behind red, green and blue phosphor strips, and a potential applied to the grid which, switched rapidly by the colour signal, deflects the writing beam on to the desired colour strip.

In a more complex device, the Eidophor, much used for large-screen projection both in black-and-white and colour, a normal arc lamp outside the tube provides a high-power source of light. The electron gun in a large evacuated tube forms an indented image in the surface of a constantly moving film of oil, and by Schlieren optics this modulates the arc beam and produces a visible image on a screen. The colour arrangement uses three tubes of the same type and makes an additive synthesis with the aid of coloured filters.

See: *Apple tube; Banana tube; Index tube; Large screen television; Lawrence tube; Shadow mask tube; Zebra tube.*

□**Colour Killer.** Device in a colour television receiver which makes the chrominance channel inoperative when a monochrome signal is received.

⊕COLOUR PRINCIPLES
□

Colour is a property of the perceived world as it is experienced through the sense of vision. Other properties – size, shape, distance, texture, movement – may be detected by touch or other senses, and in so far as they appear to the visual sense it is through differences between one patch of colour and another in space or in time. Thus colour can be defined as that quality by which we can distinguish visually between one uniform area and another, apart from all variations in extent or duration.

FUNCTIONS OF COLOUR SENSE

Interest in colour may be directed either towards the subjective quality of the visual experience or towards the correlation of this experience with other features of the objective world. The actual appearance of colours is the common starting point for both directions of interest.

On the subjective side, there is an immediate interaction between colours and feelings, through association and memory. Different combinations of colours may evoke responses of warmth, quiet, cheerfulness and so on; on the other hand, the mood or attitude of the subject, as well as his cultural background and training, affects his response and may, for example, determine whether he perceives a particular colour as dull or as subtle and interesting, or it may even change the appearance of the colour.

There is the possibility that colour influences behaviour, and though there may be few proven facts to rely upon, colour is certainly used in some fields (e.g., interior decoration, marketing, the entertainment industry) to induce people to do what they might otherwise avoid.

On the objective side, colours give information that helps us to distinguish and identify objects, and recognize what state they are in – whether fruit is ripe, meat fresh, or a child healthy. This is probably the natural biological significance of colour vision: the meanings of colours in different contexts have to be learned through experience.

Apart from natural colours, meanings may also be assigned deliberately, as in colour coding, signal colours, and brand identification colours; but if natural colours are to give consistent information about the nature of an object, there must be some systematic link between its perceived colour and its other properties. Therefore the objective study of colour must start with a systematic description of perceived colours and the range of variation that colours display.

VARIABLES OF PERCEIVED COLOUR

The appearance of colours can be described and ordered in terms of three independent variables: hue, saturation and brightness. Three variables are necessary and sufficient, so the range of perceived colours may be represented by a three-dimensional spatial co-ordinate system.

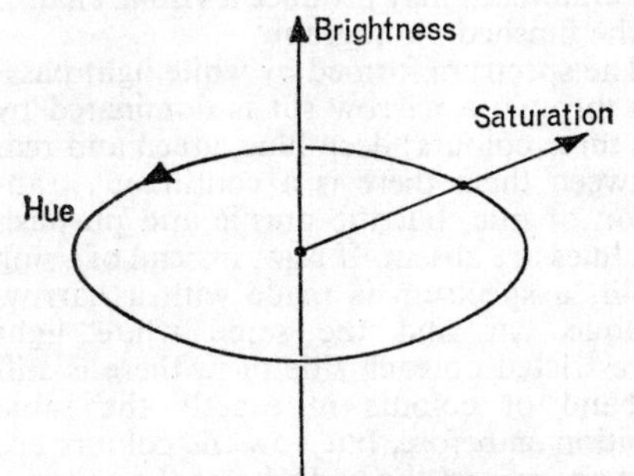

DIMENSIONS OF PERCEIVED COLOUR. Appearance of colours described in terms of three independent variables: hue, saturation, brightness.

HUE. Hue is a variable describing the similarity of a colour to one in the series ranging circularly through red, yellow, green, blue, purple and back to red. Colours either have a hue that can be fitted somewhere into this series (e.g., flesh pink a yellowish red hue, tobacco brown a reddish yellow hue, olive drab a greenish yellow hue, etc.), and are called chromatic colours; or they have no hue at all (black, white and grey) and are called achromatic colours.

For any point in the hue sequence there is another that is least like it, and opposite to it in character; such pairs of opposite hues are loosely called complementary colours. Two pairs of opposite hues (red and green, blue and yellow), suffice to define any hue, much as directions are defined by the cardinal points of the compass. These four are called unique hues.

SATURATION. This is the variable that describes the position of a chromatic colour in a radial series in terms of its distance from the nearest achromatic colour, in other words, its difference from grey. Neutral or truly achromatic whites, greys and blacks have zero saturation; yellowish whites, greenish greys and bluish blacks, etc., have very low saturation. Dull, greyish and pale colours have moderately low saturation whereas vivid and brilliant colours have high saturation.

CHROMA. The distance from grey, taken as a measure of the amount of colouredness of the colour, is called chroma. Sometimes the word saturation is restricted to mean the proportion of colouredness to uncolouredness in the colour; in these terms, when an object is moved into shade, the saturation of its colour remains the same but the chroma is reduced, because the object now appears less colourful.

BRIGHTNESS. The position of any colour in a vertical series on a scale from darker to lighter is called brightness. It can be found by comparing the colour with the nearest one of a series of achromatic colours. Consideration of this variable leads at once to the concept of light as something that shines through space and makes objects visible – more light makes everything look brighter.

The total amount of light that appears to be coming from the direction of a colour patch is called the luminosity of the colour, but the proportion of total light which appears to be reflected by the area of colour seen is called the lightness of the colour.

Luminosity ranges from total darkness up to dazzling or blinding light, and applies to any perceived patch, including a primary source of light. Lightness ranges between extremes of absolute black (seeming to reflect no light) and absolute white (seeming to reflect all the light); it applies only to illuminated objects, or secondary sources, and can be assessed only where colours can be related together with some standard.

The moon, for instance, though a secondary source, is unrelated to other objects and does not look at all grey; in comparison with the sun it merely appears to be rather dim, with a low luminosity, and it is not directly perceptible that this is partly because it has a rather low reflection factor.

VISUAL ADAPTATION. Where ordinary objects appear in a scene, the lightness, or greyness, or degree of blackness of the colours is assessed as a property of the objects, independently of the level of illumination, and seems to remain about constant even when the total amount of light changes a great deal. This shows that what is perceived is a proportion or relationship, rather than an absolute quantity. In other words, in vision there is adaptation to different conditions rather than a fixed scale of reference.

Adaptation is important for chromatic differences as well as brightness differences, and when the illuminant changes in colour, for instance, from direct sunlight to light from a clear blue sky, the colours of objects seen in context do not appear to change to the extent that they do if the two illuminants can be made to shine side by side. A wall may seem perfectly white in either the yellowish or the bluish illuminant, or in a mixture of the two, but if the two shine from different sides and something casts two shadows, the difference in colour in the shadows is quite marked.

Observations of this kind show that perceived colours cannot be specified in terms of object properties alone, nor yet in conjunction with the illuminant properties; the nature of the light reflected from the object (or transmitted or emitted by it) has to be assessed in relation to the whole complex of light affecting the eye. Thus it is more important that any attempt to reproduce a coloured scene should get the internal relationships properly balanced than that individual colours be exactly matched. Indeed, it is quite possible for a picture where every colour is different from the original to be so well balanced that it still yields all the same perceptions of hue, saturation and brightness as the original scene.

THE SPECTRUM

The features of the objective world, as deduced from experience, that correlate with colour are: the nature of radiant energy; the atomic and molecular structure of matter and its ability to emit or absorb energy; the function of the eye, which enables it to detect selected samples of radiant energy; and the system of nerves leading from the eye into the brain, which carries codified signals corresponding to the pattern of sensation.

RADIANT ENERGY. Radiant energy can be analysed in terms of waves into a scale with more or less power at each wavelength. Only energy with wavelengths between about 380 and 780 nm (nanometre = 10^{-9} metre) is detected by the eye, and this alone can be called light. If a narrow beam of light in dark surroundings is dispersed by a prism or diffraction grating, the range of visible wavelengths is spread out in a sequence called the spectrum.

The eye sees the spectrum as a range of brilliant colours, from violet at the short-wave end, through blue, green and yellow to red at the long-wave end. There may be energy beyond each end, but the eye cannot detect it and it remains invisible. Ultraviolet energy, with wavelengths below 380nm, may be indirectly important: first because fluorescent materials may absorb UV energy and re-emit visible energy, so if there is UV energy in the illuminant the material will change appearance; secondly because most photographic emulsions respond to UV energy, so invisible energy in the illuminant may produce a visible change in the finished photograph.

The spectrum formed by white light passing through a narrow slit is dominated by the three colours: deep blue, green and red. Between them there is a continuous transition of hue, but the purple and purplish red hues are absent. If now, instead of using a slit, a spectrum is made with a narrow opaque bar and the same white light unrestricted on each side of it, there is still a band of colours in exactly the same position as before, but now the colours are in every respect the opposite of the narrow-

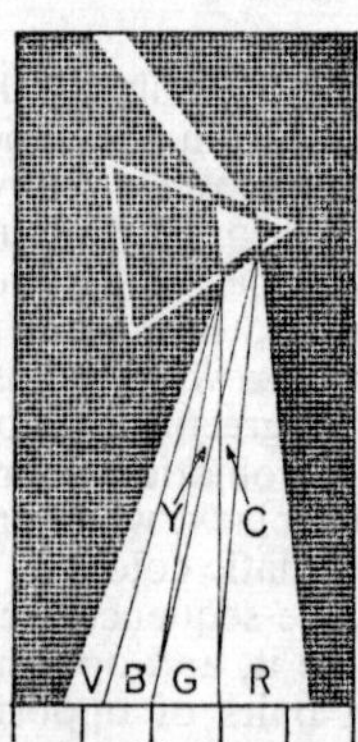

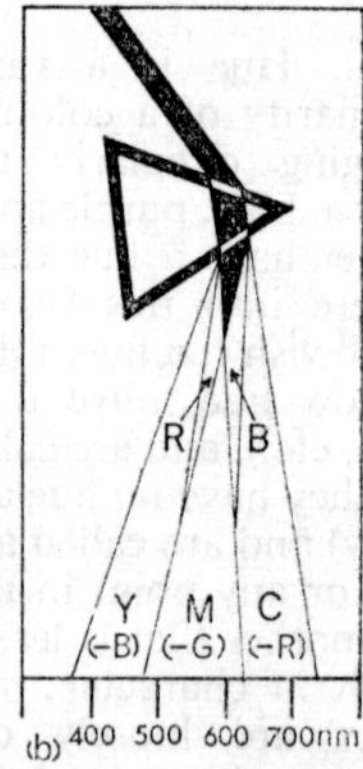

SPECTRUM FORMED BY DISPERSION OF WHITE LIGHT. (a) Light through narrow slit, in darkness, gives spectrum dominated by deep blue, green, red. (b) When slit is replaced by opaque strip, dark shadow in the light gives complementary-colour spectrum, dominated by yellow, magenta, cyan blue. Wavelength scales denote (a) portions of energy transmitted, and (b) portions of energy subtracted from total light.

slit spectrum. At the short-wave end the white light darkens gradually into yellow; this passes rapidly through orange-pink into a broad band of magenta-red, which in turn passes through purplish pink and pale blue into a strong turquoise or cyan blue, which then fades out into white light again at the long-wave end of the spectrum. Every colour of the narrow-slit spectrum is replaced by its complementary: yellow instead of deep blue, magenta-red instead of green, and cyan-blue instead of red. This new spectrum has no greens in it, but it does have the purples and purplish reds which are missing in the other spectrum. It is the spectrum of darkening, where each wavelength of energy in turn is removed from the total light.

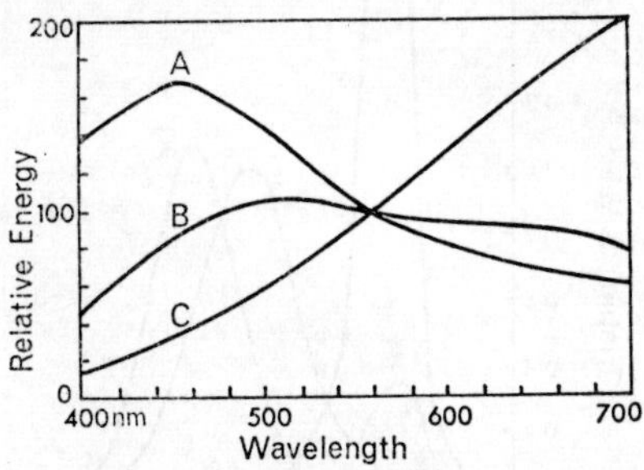

ENERGY DISTRIBUTION CURVES. Three common light sources normalized at 560 nm. A. North sky. B. Sunlight. C. Gas-filled tungsten lamp.

Whatever source of light is taken to make a spectrum, the pattern comes in the same place geometrically, but the relative intensity of light in the different wavelength positions may vary considerably. In general, a yellowish light, such as from an incandescent tungsten lamp, has less energy at the short-wave end, while a bluish light, such as from a clear sky, has less energy at the long-wave end. Some special light sources, for instance, gas discharge tubes, have packets of energy concentrated at particular wavelengths; with a narrow slit spectrum, such a light shows a number of distinct images of the slit, each at its appropriate place and in the appropriate colour, with dark spaces between, and for this reason is called a line spectrum. Measures of the intensity of energy at each wavelength, when plotted in a graph or tabulated, give the spectral power distribution of the light source.

SPECTRAL REFLECTANCE. When light passes through or into an object, some of the energy may be absorbed, and this absorption may be stronger at some wavelengths than at others, depending on the atomic or molecular structure of the material. Thus the spectral distribution of light reflected or transmitted by the object (except for the direct reflection from the outer surface) is different from that of the incident light. For instance, a yellow dye absorbs short-wave light strongly, but not middle- or long-wave light, so that if white light shines through the dye, the transmitted light is deficient in short-wave energy relative to the incident light, and looks yellow, like the short-wave end of the spectrum of darkening.

Thus the colour of a secondary source (and for vision this means every object not emitting light) is determined by the combination of the spectral power distribution of the illuminant and the ability of the object to alter the balance of that distribution. This latter is generally given by the relative spectral reflectance (or transmittance) of the object, that is, the proportion of incident light reflected (or transmitted) at each wavelength.

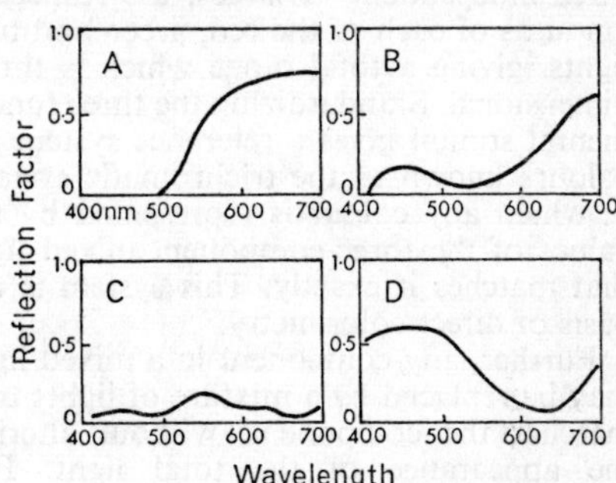

SPECTRAL REFLECTANCE CURVES. Typical coloured objects. A. Yellow daffodil. B. Pink rose. C. Green grass. D. Cyan-blue book.

Spectral reflection curves of the simpler sort give a rough idea of what colour an object has: a level curve is a neutral colour, high for white, low for black, midway for grey. A curve low at the short-wave end and high at the long-wave end is a warm colour, while sloping the other way gives a cool colour; high in the middle and low at the ends is greenish, while low in the middle and high at the ends is pinkish or purplish.

If the curve has one or two steep parts with extremes of high or low in other parts, the colour is brilliant or highly saturated, while gentle undulating curves generally indicate soft gentle colours. But in the latter case it is very hard to tell from the curve what the actual colour is, without taking into account the details of the eye's response to different wavelengths.

EYE FUNCTION. The eye does not make a complete wavelength analysis of light: it does not perceive the colours of the spectrum in the undispersed white light. Yet colour differences seen by the eye do correlate with variations in spectral distribution. The key to this correlation is found in the laws of the mixing of coloured lights.

Light from the yellow part of the narrow-slit spectrum can be exactly matched with a mixture of red and green spectrum lights; the mixed light looks yellow although there are no "yellow wavelengths" present. By changing the proportion of red to green in the mixture, any of the other colours between the two (e.g., orange) can be matched.

By adding light of a third wavelength, from the blue end of the spectrum, the mixed light can be made to match almost any other colour, including white.

The mixed light is then determined by three independent variables, the respective amounts of each of the red, green and blue lights, giving a total range which is three-dimensional. Standardizing the three fundamental stimuli gives a reference system for colours known as the trichromatic system, in which any colour is represented by the values of the three-component mixed light that matches it exactly. This system is the basis of direct colorimetry.

Further, any component in a mixed light may be replaced by a mixture of lights that matches the component, without altering the appearance of the total light. For instance, a blue and a yellow light may be mixed, giving a light which is white in colour; the yellow light may then be replaced by a mixture of red and green lights that matches the yellow, without changing the final white. This means that once the proportions of the three fundamental stimuli that will match each one of the spectrum colours has been determined, then light of any colour, however complicated its spectral distribution, can be fitted into the trichromatic reference system by substituting for each narrow wavelength band in turn, as if they were so many separate components in a mixed light, and adding up the three totals. This is the basis for all indirect colorimetry, and in particular the CIE system. This is named after the Commission Internationale d'Éclairage (International Commission of Lighting).

INTERNATIONAL CIE SYSTEM

Because the mixing of lights is additive in the strict sense, it is possible to transform results from one trichromatic system to another, as long as the matching stimuli used in one system have been measured in terms of those used in the other. It is even possible to use hypothetical reference stimuli that cannot exist as coloured lights, such as the X, Y and Z of the CIE system, where X, Y and Z are vectors in the three-dimensional colour-space lying wholly outside the region occupied by spectral colours and their mixtures.

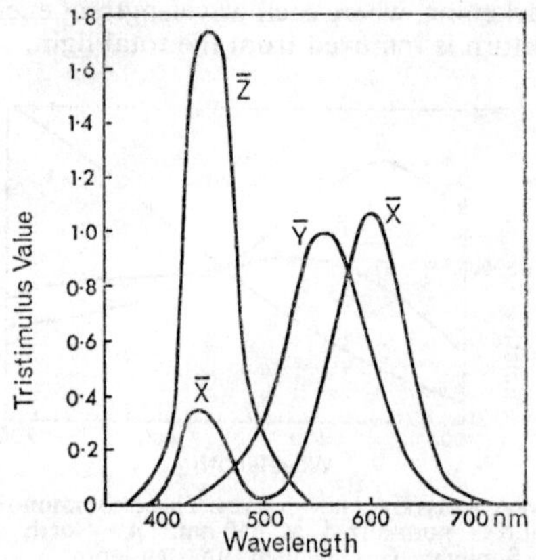

CIE TRISTIMULUS VALUES FOR SPECTRUM COLOURS. Colour matching functions of standard observer, with a 2° field of view. Y value, identical to curve for photopic human vision, expresses luminance of colour.

THREE-FIGURE SPECIFICATION. In the CIE system any colour is specified by a three-figure expression. This may give the X, Y and Z values derived by linear transformation of the R, G and B values of the matching stimuli, but more usually the quantity of light is separated from the quality of the light. In the three-coordinate colour-space, this means to separate the distance from the origin of the point representing the colour from the direction of that point.

The former measure, which specifies the luminance of the colour, is expressed by the Y value alone; for the colours of objects, it is usual to give the relative luminance compared to an ideal white for which Y = 100. It is possible to specify the luminance of a colour in this way because the spectral distribution of the function giving the Y value has been made identical to the curve of relative spectral luminous efficiency for photopic vision.

The luminance of a colour correlates in direction but not in magnitude with the perceived luminosity; that is to say, if luminance is increased, then luminosity also increases but not necessarily by the same factor. Various functions to convert luminance readings into luminosity have been suggested, but since the latter depends on subjective estimates and varies considerably with viewing conditions, no single formula has been found to be universally acceptable.

CHROMATICITY DIAGRAM. The quality of a colour, defined in the trichromatic system by the direction of the colour point from the origin, and called its chromaticity, can vary in two dimensions. This is shown on a chart plotting x against y, where $x = X(X+Y+Z)$ and $y = Y(X+Y+Z)$; this chart is called the chromaticity diagram.

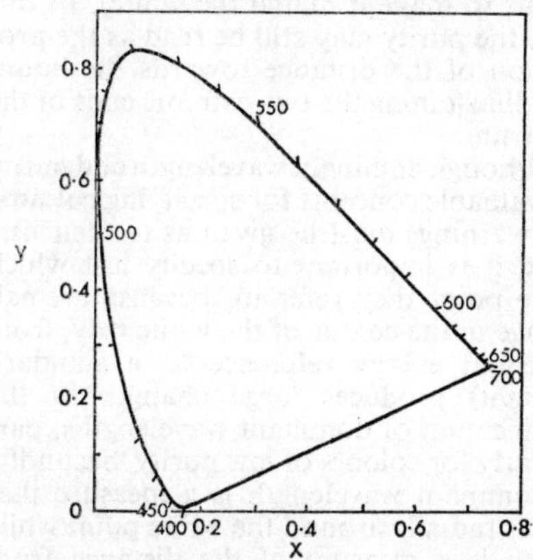

CIE x, y CHROMATICITY DIAGRAM. Curved boundary represents monochromatic spectrum colours. For red (600–700 nm), x is high, y fairly low; moving counterclockwise through yellow to green, y rises to maximum while x decreases, then for green to blue, both x and y decrease, finally towards violet (400 nm), x increases slightly.

The curved boundary on this diagram represents the monochromatic spectrum colours; for red, x is high and y rather low; then, passing up through yellow to green, x decreases while y increases to a maximum, after which, from green to blue, both x and y decrease until towards the violet, x increases again slightly.

The central point of the diagram ($x = 0{\cdot}333$, $y = 0{\cdot}333$) represents the colour of an equal energy source, a theoretically perfect white light. This must not be confused with a perfect (or ideal) white surface, which reflects diffusely all the light incident upon it. The colour of the white surface (as measured colorimetrically) depends upon the quality of the light illuminating it, just as much as that of any other surface.

As light with equal energy at all wavelengths is not readily available, the CIE has specified a few standard sources to represent typical illuminants (e.g., tungsten filament lamp, natural daylight) so that the comparison of surface colours can be simplified.

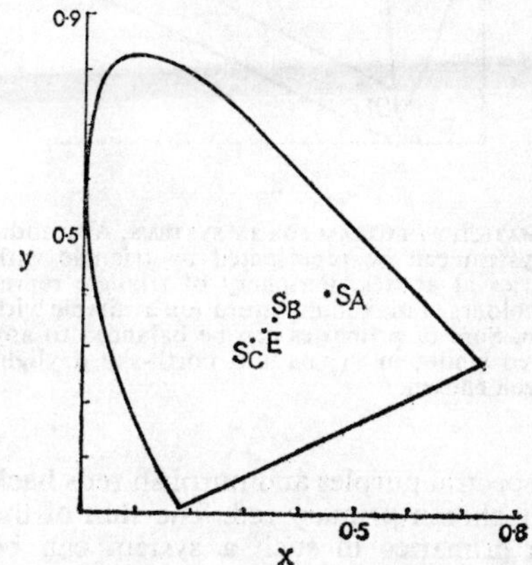

SOURCE LOCATION IN CIE CHROMATICITY DIAGRAM. E. Equal-energy white. Standard sources: S_A tungsten lamp; S_B tungsten lamp filtered to represent sunlight; S_C tungsten lamp filtered to represent daylight from overcast sky.

Given the point for white in a standard illuminant, this same point also refers to the whole series of neutral greys down to black, because the luminance dimension is not represented in the chromaticity diagram. Similarly, brown and olive colours plot in the same places as orange or yellow, and so on for other colours, changes in lightness or darkness alone making no difference to the chromaticity. To take account of this extra dimension, CIE specifications for surface colours are generally given in the form x, y, Y.

The CIE chromaticity diagram has various useful properties. Additive mixtures of two lights lie on the straight line joining the two colour points, so for any additive three-colour system (such as colour TV) the whole gamut of available colours lies within the triangle formed with one primary colour at each apex. The periphery of such a triangle represents the most saturated colours that can be given by the system, and can follow the spectrum very closely from red through yellow to green, though not so closely from green to blue, finally passing through the

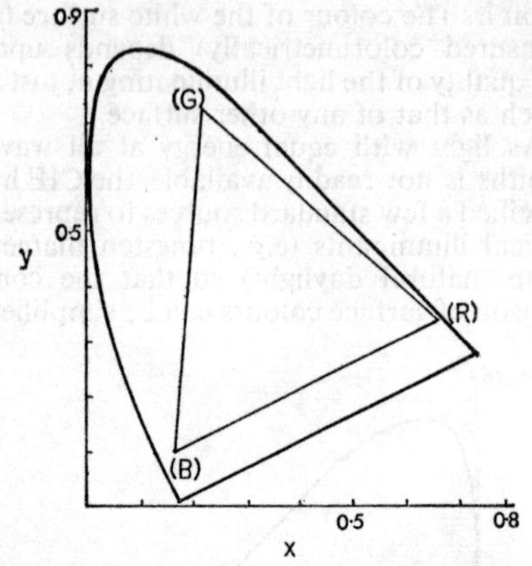

CHROMATICITY DIAGRAM FOR TV SYSTEMS. Any additive system can be represented by triangle with primaries at apexes. Periphery of triangle represents colours of maximum saturation available with system. Sum of primaries can be balanced to any required white; in TV, natural north-sky daylight has been chosen.

non-spectral purples and purplish reds back to the chosen primary red. The sum of the three primaries in such a system can be balanced to give whatever kind of white colour is desired; for colour TV this is chosen to be near natural north-sky daylight.

DOMINANT WAVELENGTH AND PURITY. Another useful way of specifying colours can be read from the chromaticity diagram; this depends on the fact that any colour can be matched by mixing white light and a suitable monochromatic light in the right proportions – more white light for pale and unsaturated colours, less for strong colours. A straight line from the white point through the colour point in question is produced until it cuts the spectrum locus: this gives the dominant wavelength of the colour. The distance from the white point to the colour point as a proportion of the whole distance to the spectrum locus gives the purity of the colour (where white has zero purity and spectrum colours a purity of 100 per cent). For purple colours, a complementary wavelength must be used, that is, the straight line must be produced on the other side of the white point to meet the spectrum locus (corresponding to the fact that the monochromatic light must be added to the sample colour to make it match the white). In this case, the purity may still be read as the proportion of the distance towards the boundary line joining the two extreme ends of the spectrum.

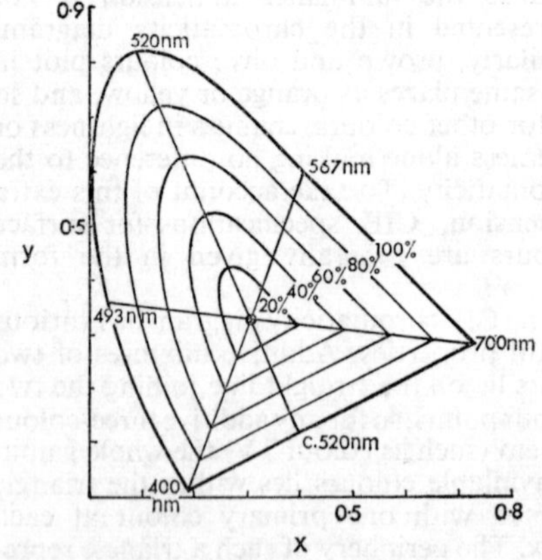

CHROMATICITY DIAGRAM AND DOMINANT WAVELENGTHS. Contours within diagram (along radial lines) represent series of constant dominant wavelength and different percentage purities from white (0 per cent) to spectrum colours (100 per cent) for source C.

Although dominant wavelength and purity are valuable concepts for specifying colours, two warnings must be given as to their use. First, it is important to specify just which white point they refer to, because a small change in the colour of the white (say, from an equal energy reference to a standard daylight) produces large changes in the specification of dominant wavelengths, particularly for colours of low purity. Secondly, as dominant wavelength is a measure that circles radially around the white point while purity is a measure of the distance from white, it is tempting to equate these two terms with the concepts of hue and saturation. In fact, there is only a broad agreement between the two sets of concepts.

If a set of actual colour patches is chosen to express series of constant hue and constant saturation (as in the Munsell system) and these patches are then measured colorimetrically and plotted in the chromaticity diagram, most constant hue series lie in curves spreading out from the white point, rather than straight lines, while the curves representing constant saturation series do not coincide with loci of constant purity. Moreover, for colours of different lightness, the points for the same hue and saturation do not always stay in the same place on the chromaticity diagram.

A final important discrepancy between perceived colours and colour points in the chromaticity diagram is in the magnitude of the spacing of colour differences. For in-

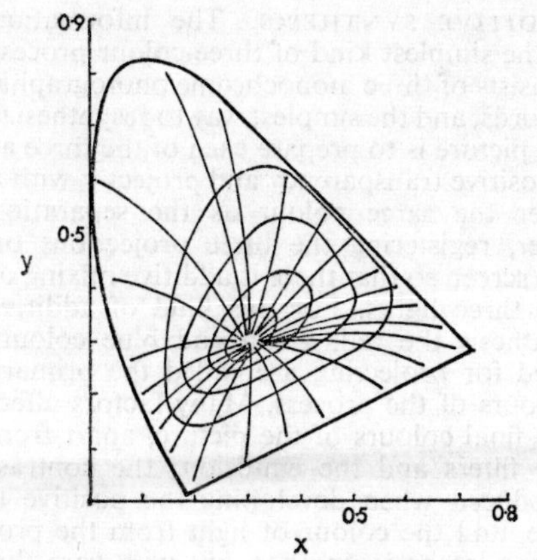

CHROMATICITY DIAGRAM AND MUNSELL COLOURS. Colours of constant hue (lines spreading from centre) and of constant chroma (rings) all at constant value.

stance, if a small area near the top of the diagram is chosen so that it covers a few green colours that can only just be told apart (by visual discrimination), then the same area at the bottom of the diagram near the violet end of the spectrum includes hundreds of discriminably different colours.

This non-uniformity of spacing is so striking that for some purposes where colour discrimination is of importance (e.g., for assessing the fidelity of colour films), a linear transformation of the chromaticity diagram is used which largely reduces the discrepancies. The new chart so produced is known as the CIE *u*, *v* diagram. But where it is of importance to assess colours in terms of the perceived variables of hue, saturation and lightness, it is probably better to convert the

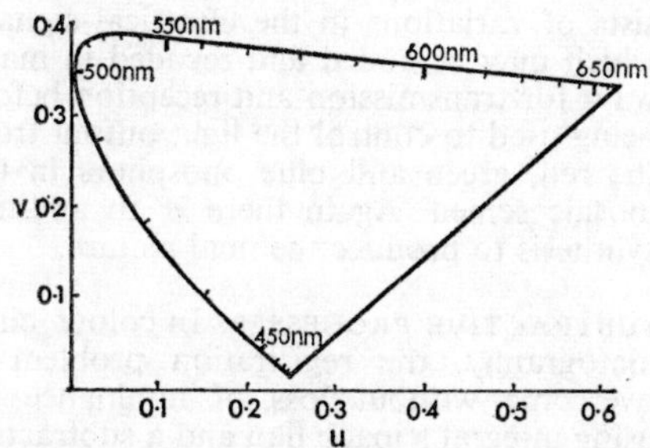

CIE *u*, *v* DIAGRAM. Chart produced by linear transformation of chromaticity diagram to make perceived colour discrimination more uniform in different regions of diagram.

colorimetric values into Munsell terms. In spite of many attempts, no acceptable formula has yet appeared that could be used to make such a conversion automatically, so the values have to be plotted on a graph and the Munsell notation read off. However, for many practical purposes this conversion is not required, and a colour can be specified adequately in terms of dominant wavelength, purity and luminance, or in *x*, *y*, and Y, or even directly in terms of the primaries, R, G, and B.

COLOUR MATCHING. The trichromatic system of colour measurement and specification is based entirely on experience with colour matching. As colour matching functions vary to some extent from one individual to another, a CIE standard observer based on average values for a number of real observers is used in colorimetry. The colorimetric specification of any sample of light (whatever set of terms is used) refers to the match, and not to the perceived appearance of the patch of light. For patches seen in isolation or in simple controlled viewing conditions, there is a close correlation between the colorimetric specification and the perceived colour, but this kind of specification does not take into account the context and surroundings of the colour patch, and these can affect the appearance. For this reason, it must be remembered that if a colour specification in CIE terms is converted into perceptual terms of hue, saturation and brightness (for instance, using the Munsell system), the result is true only for the correct conditions of viewing.

From the results of colour matching, the existence of three light receptors responding to different, but overlapping, bands of wavelengths has been deduced, and independent evidence for three such photopigments in the retina is forthcoming. This explains the trichromatic nature of colour matching and the additivity of colour mixing for small isolated patches of light. In order to explain the way the eye adapts to different illuminants and the effects of one colour on another, so as to be able to predict colour perceptions when given the colorimetric data for every part of a complex field of view, it would be necessary to know much more about the pattern of activity going on in the nerves of the eye and its connections with the brain.

COLOUR REPRODUCTION

The object of colour reproduction is to make a picture in which the colours perceived in its various parts are an acceptable representation of the colours perceived in the original scene. There is certainly no need for each colour to have the same spectral distribution as the original, even if it were possible to achieve this. Nor is it necessary to produce an exact colorimetric match for each colour, since the conditions of viewing the picture are rarely the same as for viewing the original scene. On the other hand, the basic trichromatic principles of vision underlie every successful system of colour reproduction.

TRICHROMATIC ANALYSIS. The process of reproduction falls into two parts, whatever medium is used. First, the original scene must be analysed to give tri-variant colour information for every part of it falling within the field of view required for the picture. Secondly, this information must be used to create suitable colours in each part of the picture in the chosen medium, whether cinema screen or television (prints to be viewed by reflected light need not be considered here, though the principles are similar). There may be many stages of processing the information before it is used to control the production of colours in the picture.

Although the theoretical basis of colour reproduction is largely understood, there is still scope for empirical knowledge and skill in getting the best results.

The trichromatic analysis of the original is made by using three kinds of light-sensitive receptors, each responding to about one-third of the spectrum but usually with some overlapping. The receptors may be photographic emulsions covering an image of the field of view or photoelectric cells scanning it. The field may be covered completely three times over or it may be divided up into a three-component mosaic pattern too small to be resolved by the eye at the usual viewing distance.

Although light-sensitive receptors may have differences in their spectral response characteristics, it is usual to control them with filters, each transmitting about one-third of the spectrum. These filters are red (for controlling the long-wave light), green (for the middle-wave) and deep blue (for the short-wave). Choice of the exact colours (or transmission curves) is usually the result of compromise between various conflicting requirements, but the names red, green and blue are sufficiently broad in meaning to cover every practical set of three-colour separation filters.

ADDITIVE SYNTHESIS. The information in the simplest kind of three-colour process consists of three monochrome photographic records, and the simplest way to resynthesize the picture is to prepare each of the three as a positive transparency and project it with a filter the same colour as the separation filter, registering the three projections on the screen so that there is additive mixing of the three lights. For this kind of additive synthesis the red, green and blue colours used for projecting are called the primary colours of the process. Many factors affect the final colours of the picture, apart from the filters and the emulsion; the contrast produced when developing the positive is one, and the colour of light from the projector lamps is another, but with care this method, first crudely demonstrated by J. C. Maxwell in 1861, remains one of the most faithful colour processes. The main drawback is the difficulty in registering the three components of the picture, particularly for cinematograph film.

One attempt to overcome the registration problem was to use a mosaic of tiny red, green and blue patches on a single film. The inherent disadvantage of this method for films is the relatively low luminance obtained; the brightest highlight has to be synthesized from red, green and blue areas side by side, and because each of these transmits only about one-third of the available light, at least two-thirds of the light in that area is unused, and in fact, wasted. Largely for this reason, mosaic colour films are obsolete, but for a medium in which the primary colours are produced by self-luminous phosphors rather than filtered light there is no such waste and the mosaic remains the most practical method.

In colour television, the original scene is analysed into three components in the electron camera, and the information consists of variations in the electrical signals, which may be coded and recoded in many ways for transmission and reception before being used to control the light output from the red, green and blue phosphors in the mosaic screen. Again there is an additive synthesis to produce the final picture.

SUBTRACTIVE PROCESSES. In colour cinematography, the registration problem is overcome without loss of luminance by using integral tripack film and a subtractive synthesis for the final picture. In the tripack, three layers of emulsion are sandwiched permanently together, and spectral separation is done with a combination of filtering and differential sensitization, so that each layer is again sensitive to about one-third of the spectrum. For the final synthesis, instead of adding varying amounts of light in each of the three spectral bands (long-wave, middle- and short-wave), the bands are controlled by absorbing light with transparent dyes in varying densities, one in each layer of the tripack.

This is called a subtractive process, because the final colours are achieved through controlled darkening or removal of light. Reference to the complementary or dark-strip spectrum shows that removing the long-wave part of the light gives cyan-blue, removing the middle-wave part gives magenta-red, and removing the short-wave part gives yellow. Dyes of these three colours can then be used to control the amount of light left in each of the three spectral bands. Where yellow and cyan overlap, only middle waves are left, making green; yellow and magenta give orange and red, while magenta and cyan give all the different shades of blue and purple. All three together in varying degrees give all the greyish, brownish and blackish colours, while to get white it is only necessary to remove all three dyes so as to leave the undimmed light from the projector lamp. This is why subtractive synthesis gives much more luminous results than additive mosaic films.

The intermediate processing of subtractive films may follow a number of courses. Reversal film with negative development, re-exposure and conversion of the silver residues to dye images, each in its separate layer, yields a single colour positive transparency on the original film. This is mostly used for amateur movies. When several prints of a film are wanted, negative colour film can be used, in which the negative images are directly converted to dye images, giving a negative transparency in complementary colours. Prints can then be made on to film of the same type, repeating the process and ending up with a positive transparency in the right colours. When the most brilliant colours are required, the information in the three layers of the film can be used to make separate gelatin images for dye transfer printing on to the final film. In general, much better dyes can be used with dye transfer processes, because their choice is not restricted by the chemical difficulties of colour development.

TWO-COLOUR PROCESSES. Before the widespread availability of integral tripack film, efforts were made from time to time to introduce two-colour processes, in which the final picture is synthesized from two primary colours. A film has two sides, and this makes a two-layer film pack much easier to process than a three-layer film, but the final picture must inevitably suffer severe loss of colour information. This loss can be mitigated to some extent through the use of dichroic dyes, for instance, one that is yellow when pale but becomes more orange or even red when dense, and one that is sky-blue when pale but purple-blue when dense. The reduction of colour may also be made less obvious by careful selection of the colour range in the original scene.

Inevitably the colour range of two-primary processes is two-dimensional, and the only way that this can be made to appear at all natural is to confine the illumination to low-temperature lamp or candle light which is so yellowish as to have virtually no short-wave component at all. The blue primary can then be dispensed with, and the picture synthesized from red and green only. But this restriction is too drastic for most feature films, and attempts to cover full daylight scenes with a two-colour process did little more than get a bad name for colour films in general. With improvements in three-colour processing techniques, there is now little hindrance to their use, and if they are still too expensive the better alternative is not two-colour but one-colour film, or plain black-and-white. This does not attempt to give a naturalistic impression of colours, but by reducing all colour values to a single scale of greys it raises no false hopes for colour reproduction and consequently no disappointment at bad colour rendering.

COLOUR MASKING. One of the limitations of subtractive processes, compared with additive, is the smaller range of saturation that can be achieved. The main reason for this is that the spectral distribution of suitable dyes is not sufficiently sharp. Ideally, each dye should absorb consistently across the band of the spectrum it is controlling (more or less strongly as required) and not at all in the other two bands, which should be controlled exclusively by the other two dyes respectively. But inevitably there is departure from this ideal pattern, and it is particularly difficult to get a good cyan dye because absorption bands always tend to be less sharp on the short-wave side. This means that if the cyan dye has sufficient absorption to control the long-wave end of the spectrum, it has rather a lot of unwanted absorption in the middle- and short-wave bands. As the density of the cyan dye varies to control the light in the long-wave band, so the density also varies in the other bands, upsetting the control of the other two dyes.

It is possible to correct this fault to some extent by masking the negatives of the short and middle bands (for the yellow and magenta dyes respectively) with a positive mask of a suitable density made from the long-wave negative. This does not eliminate the unwanted absorption, but it has the effect of compensating for variations fairly closely by allowing extra light through the other layers, resulting in cleaner and brighter colours. Similar masking can be done to compensate for deficiencies in the magenta dye; yellow dyes are usually good enough not to need this treatment.

The final reproduction, by whatever process, is never a perfect colorimetric match to the original, but improvements in technique are leading to pictures that are more and more acceptable as faithful renderings of the original perceived colours, and more and more able to give subjective pleasure.

R.W.Br.

See: *Colour cinematography; Colour principles; Colour systems; Visual principles.*

Books: *The Measurement of Colour*, by W. D. Wright (London), 1964; *The Reproduction of Colour*, by R. W. G. Hunt (London), 1967; *An Introduction to Colour*, by R. M. Evans (New York), 1948.

□COLOUR PRINCIPLES

Colour television is possible only because the effect on the human eye of any spectral distribution of light energy can be simulated by the addition of three monochromatic light radiations. These radiations must be situated at the long-wavelength end of the visible spectrum (red), in the centre (green) and at the short-wavelength end (blue). The mechanism of the human eye enables an addition of radiation at three discrete wavelengths to be observed as a particular colour.

BASIC SYSTEMS

ADDITIVE COLOUR MIXING. If three sources of light, red, green and blue, are projected simultaneously on to the same white screen, the brightness of light (assessed by an average observer, irrespective of colour) is measured in lumens. If the brightness of light reflected from the screen from the three sources when used independently is R, G and B lumens of red, green and blue light respectively, the brightness of the reflected light of the combined sources is:

$$L = R+G+B$$

where L is in lumens and has, in general, a particular hue.

When the sources are adjusted in brightness to give one lumen of white light so that L equals 1, then:

$$1 = r+g+b$$

where r, g and b are the quantities of red, green and blue in lumens required to give 1 lumen of white light.

There are three primary colours in the first example, so that there must be a portion of white light. If the amounts of red and green light are both greater than those required to be added to the blue light to produce white, then:

$$L = R'+G'+W$$

where R′, G′ and W represent the brightness in lumens of the red, green and white light comprising the total.

This shows that the light from any illuminated coloured area of a scene can be split into two components: the amount of white light present and the amount of pure coloured light present which must comprise one or two primaries.

Coloured light may therefore be defined by three parameters:

1. Luminance.
2. Saturation (i.e., the extent to which the light is coloured). Completely saturated light contains no white light and completely desaturated light is composed entirely of white light.
3. Hue, which is determined by the ratio of the primaries left after the deduction of the white light.

The representation of coloured light by the parameters luminance, saturation and hue, may seem more complicated than a simple statement of the luminance of the three primaries, but it is essential to the operation of colour television systems.

THREE-COLOUR SIMULTANEOUS SYSTEM. Since the light from any part of a scene can be synthesized by the addition of three primary colours, a colour television system may use three separate cameras to generate electrical signals corresponding to the quantities of red, green and blue light required. The cameras (actually three tubes in one camera assembly) produce electrical signals, corresponding to amounts of the required primaries, which are transmitted to three display tubes, each arranged to give light of the hue of one of the three primaries. The red, green and blue light thus obtained is superimposed by an optical system on a common screen and the original coloured scene reconstituted.

A three-channel, R, G, B system of this type is unsuitable for broadcast purposes because three separate transmission channels are needed. However, it is of interest because the three-tube camera arrangement is typical of that required in modern broadcast systems of colour television. The essential feature of the camera is that each tube is equipped with an appropriate optical filter.

The optical filters have characteristics which ensure that the brightness of the light reaching each camera tube corresponds to the relative brightness required for the appropriate primary in the display system. Thus the characteristics required for the optical filters in the cameras cannot be calculated without a knowledge of the precise hues of the display primaries. Finally, such a system could use the modern three-gun shadow-mask display tube which is, in effect, three tubes with red, green and blue phosphors superimposed.

SEQUENTIAL SYSTEMS. The simplest practical method of achieving colour television using only a single transmission channel was demonstrated as early as 1928 by J. L. Baird, using sequential colour analysis and synthesis.

More recently, a workable system was developed in America. A sequential system comprises a rotating disc of red, green and blue analysis filters placed in front of a single camera to intercept, field by field, the light reflected from the scene to be televised. The received picture is displayed on a monochrome tube and the light is fed through a similar disc comprising the corresponding colour synthesis filters. Thus, providing that the rotation of the two discs is synchronized, a coloured version of the original scene results.

While light from the scene is intercepted by one of the analysis filters, the camera output signal describes the variations of the quantity of that primary required for the synthesis. During the same period, the image reproduced by the receiving tube is presented to the viewer via the colour synthesis filter of the same primary so that the eye receives the appropriate proportion of the primary required in the coloured reproduction.

The three versions of the scene corresponding to the three primaries thus occur sequentially and the persistence of human vision causes the successive images to combine to produce the required result.

A disadvantage of the system, among others, is that the cycle must be completed in about 1/25 sec. if flicker is not to occur. This means that three times the bandwidth is required as compared with that necessary for a normal monochrome transmission. The system cannot therefore be regarded as a practical solution to the problem of broadcast colour television.

COMPATIBLE SYSTEMS

The above systems are impracticable because they do not produce a compatible signal, i.e., one that gives a satisfactory black-and-white picture when received on a conventional monochrome receiver and can also be radiated within the bandwidth occupied by a normal monochrome transmission.

As a colour picture source (camera, telecine, etc.) produces three electrical signals corresponding to the red, green and blue primaries, the transmitted signal, to be compatible, must contain a component for use by the monochrome receiver describing the brightness of the scene, together with a further component describing colour.

MIXED HIGHS. The basic, three-colour simultaneous colour system uses three camera tubes, sensitive to red, green and blue light respectively, providing three colour-separation signals which are conveyed to the three guns of a three-colour display tube. The signal from the red camera tube produces a picture consisting of the red component of the scene and the green and blue camera tubes produce signals which similarly cause the green and blue components to be displayed. The composite display then consists of a substantially faithful reproduction of the original scene. However, in order to provide a sharp picture, each of the three component signals must have the full bandwidth, i.e., the bandwidth required for a black-and-white transmission; thus such a basic system is too extravagant in bandwidth. Further, none of the three component signals is fully satisfactory for a black-and-white receiver. Two

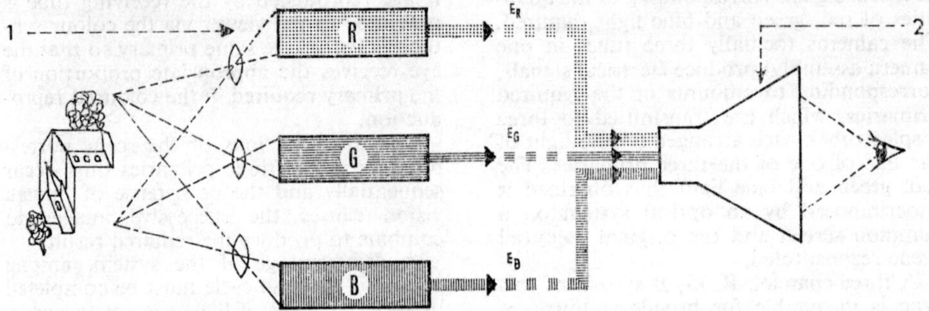

BASIC THREE-COLOUR SIMULTANEOUS SYSTEM. Three camera tubes, sensitive to red, green and blue light respectively, transmit separate signals to receiver. 1. Camera tubes. 2. Three-gun display tube.

properties of the human eye may, however, be exploited to permit the three signals to be coded in such a manner as to fulfil the requirements of compatibility.

The first of these properties is that the eye does not separate individual colours in finely detailed patterns: it is aware only of brightness changes. As the coarseness of the patterns is increased, orange and cyan become distinguishable and then, as the pattern grows even coarser, the eye begins to appreciate other colours.

The second of these properties is that the human eye is more sensitive to disturbance of brightness than of colour. Thus, if a colour television signal is subject to interference, it is important to ensure that the interference has less effect upon the instantaneous brightness of the displayed picture than on the instantaneous hue or saturation.

In modern colour television systems, advantage is taken of the first of these properties by employing what is termed the mixed-highs principle. As the human eye can perceive only brightness differences in fine detail, it is necessary to transmit only one wideband signal which describes the instantaneous brightness of the scene. As the observer can perceive colour in only the coarser patterns of the picture, a signal (or signals) representing colour information can be transmitted in reduced bandwidth. The second property of the eye may be exploited if the brightness signal is made appreciably greater in amplitude than the colour signal or signals.

One way in which the three colour-separation signals may be rearranged so as to take advantage of these factors is as follows.

The three colour-separation signals are combined in a precise manner to produce three other independent signals. The process is one of linear transformation and is performed by a device conveniently termed a matrix (M_1). The three new signals, E_Y, (E_R-E_Y) and (E_B-E_Y), are derived as follows:

$$E_Y = lE_R+mE_G+nE_B$$
$$(E_R-E_Y) = (1-l)E_R-mE_G-nE_B$$
$$(E_B-E_Y) = -lE_R-mE_G+(1-n)E_B$$

Considering first the luminance signal E_Y, the coefficients l, m and n are the relative luminances of the three primary colours used in the colour display tube and correspond respectively to r, g and b previously mentioned. Thus

$$l+m+n=1$$

and one lumen of white light is produced by l lumens of red, m lumens of green and n lumens of blue light. Assuming that white

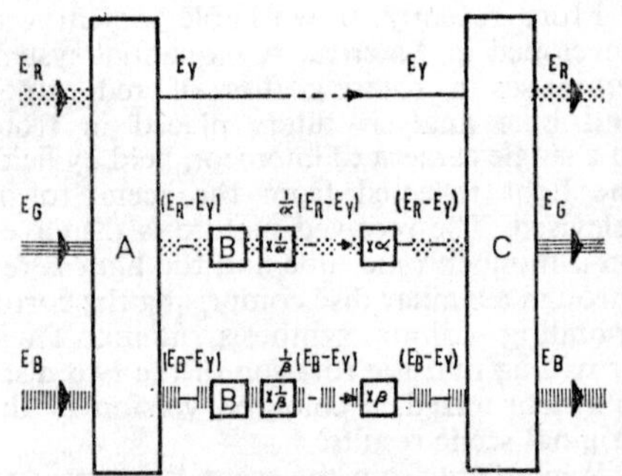

CONSTANT LUMINANCE MIXED-HIGHS SYSTEM. Wideband brightness signal provides picture sharpness; reduced bandwidth for colour signals since eye only responds to coarser details in colour. A. Matrix M_1. B. Low-pass filter. C. Matrix M_2.

light is displayed by the three-gun tube when

$$E_R = E_G = E_B = 1$$

then the luminance of any area of the displayed picture (whether coloured or not) is

$$Y = lE_R + mE_G + nE_B = E_Y$$

The E_Y signal carries the information concerning the instantaneous brightness (or luminance) and is termed the luminance signal. Thus, this signal is suitable for use by a conventional black-and-white display. For greys and white

$$E_R = E_G = E_B$$

and $(E_R - E_Y)$ and $(E_B - E_Y)$ are both zero. Thus, the colour-difference signals carry information about colour only. The mixed-highs principle permits them to be limited in bandwidth.

CONSTANT LUMINANCE PRINCIPLE. At the colour receiver, the luminance and colour-difference signals may be applied to a matrix (M_2) which is the inverse of that used at the transmitter, thus

$$E_R = E_Y + (E_R - E_Y)$$
$$E_B = E_Y + (E_B - E_Y)$$

and E_G may be derived from E_Y and $(E_G - E_Y)$, which is obtained from:

$$(E_G - E_Y) = -\left[\frac{l}{m}(E_R - E_Y) + \frac{n}{m}(E_B - E_Y)\right]$$

The effect of adding interference or noise to the colour-difference signals is as follows.

The red colour-separation signal fed to the display is

$$E_Y + (E_R - E_Y + \Delta_1) = E_R + \Delta_1$$

The blue colour-separation signal is

$$E_Y + (E_B - E_Y + \Delta_2) = E_B + \Delta_2$$

where Δ_1 and Δ_2 represent independent noise voltages, and the green colour-separation signal is

$$E_Y - \left[\frac{l}{m}(E_R - E_Y + \Delta_1) + \frac{n}{m}(E_B - E_Y + \Delta_2)\right]$$
$$= E_G - \frac{l}{m}\Delta_1 - \frac{n}{m}\Delta_2$$

Therefore, the displayed luminance is

$$Y = l(E_R + \Delta_1) + m(E_G - \frac{l}{m}\Delta_1 - \frac{n}{m}\Delta_2) + n(E_B + \Delta_2)$$
$$= lE_R + mE_G + nE_B + l\Delta_1 - l\Delta_1 - n\Delta_2 + n\Delta_2$$
$$= E_Y$$

Hence the interference or noise accompanying the colour-difference signals does not contribute to the displayed luminance. This principle, by which the primary signals are coded, is known as the constant-luminance principle.

As they are less susceptible to interference than the luminance signal, the colour-difference signals may be attenuated somewhat before transmission and restored in level by amplification in the receiver.

As yet, no values for l, m and n have been given. The three display-tube primary colours have been specified in terms of co-ordinates on the CIE diagram.

CHROMATICITIES OF DISPLAY TUBE PRIMARY COLOURS

	x	y	Y (*relative*)
Red	0·67	0·33	0·30
Green	0·21	0·71	0·59
Blue	0·14	0·08	0·11

The standard white is CIE Illuminant C (average daylight).

CHROMATICITY OF ILLUMINANT C

	x	y	Y (*relative*)
Illum. C	0·310	0·316	1·00

The luminosities of the red, green and blue primaries, when producing a white corresponding to Illuminant C, are in the ratio 0·3 : 0·59 : 0·11.

Hence,

$$l = 0{\cdot}3$$
$$m = 0{\cdot}59$$
$$\text{and } n = 0{\cdot}11$$
$$E_Y = 0{\cdot}3E_R + 0{\cdot}59E_G + 0{\cdot}11E_B$$

To obtain correct colour reproduction using these display-tube primaries, it is necessary to use correct colour-analysis characteristics at the camera. When ideal analysis characteristics are added in the ratio of 30 per cent red, 59 per cent green and 11 per cent blue, the resultant curve has the same shape as the luminosity curve (sensitivity of the eye versus wavelength). This ensures that the luminance signal conveys information about brightness only.

The following table gives the values of E_R, E_G, E_B, E_Y, $(E_R - E_Y)$ and $(E_R - E_Y)$ for various colours each having maximum brightness.

THE EFFECT OF GAMMA. Hitherto, it has been assumed that all the processes described are linear. In practice, however, the

SIGNALS FOR VARIOUS COLOURS

Colour	E_R	E_G	E_B	E_Y	(E_R-E_Y)	(E_B-E_Y)
White	1·0	1·0	1·0	1·0	0	0
Red	1·0	0	0	0·3	0·70	—0·30
Green	0	1·0	0	0·59	—0·59	—0·59
Blue	0	0	1·0	0·11	—0·11	0·89
Yellow	1·0	1·0	0	0·89	0·11	—0·89
Magenta	1·0	0	1·0	0·41	0·59	0·59
Cyan	0	1·0	1·0	0·70	—0·70	0·30

law connecting scanning-beam current and video drive voltage in display cathode ray tubes is non-linear, and the non-linearity cannot, in practice, be corrected in the receiver. The brightness of a displayed picture point, B_D, is related to the corresponding video drive voltage by

$$B_D = E_S^{\gamma}$$

where γ has a value of approximately 2·2 in three-gun colour tubes and approximately 2·5 in conventional black-and-white tubes.

In order to ensure correct reproduction, it is necessary that a compensating characteristic be applied to the signal before transmission. Thus a signal from a camera tube must be corrected to the form

$$E_S = B_S^{1/\gamma}$$

where B_S is the scene brightness of the point under consideration.

Thus, in the basic three-colour system, the colour separation signals require gamma correction and now become $E_R^{1/\gamma}$, $E_G^{1/\gamma}$ and $E_B^{1/\gamma}$. For convenience these will be written E'_R, E'_G and E'_B.

When the colour separation signals are combined, the luminance signal becomes

$$E'_Y = 0{\cdot}3E'_R+0{\cdot}59E'_G+0{\cdot}11E'_B$$

and the colour-difference signals are

$$(E'_R-E'_Y) \text{ and } (E'_B-E'_Y)$$

These changes affect the colour and compatible-monochrome pictures. In the colour receiver, the three colour-separation signals may be derived as follows:

$$E'_R = (E'_R-E'_Y)+E'_Y$$
$$E'_B = (E'_B-E'_Y)+E'_Y$$
$$\text{and } E'_G = E'_Y-0{\cdot}51(E'_R-E'_Y) -0{\cdot}19(E'_B-E'_Y)$$

When the luminance signal E'_Y is applied to a conventional black-and-white tube, the displayed luminance is

$$Y' = (0{\cdot}3E'_R+0{\cdot}59E'_G+0{\cdot}11E'_B)^{\gamma}$$

Correct reproduction is obtained when luminance is related to E_R, E_G and E_B by the equation

$$Y = 0{\cdot}3E_R+0{\cdot}59E_G+0{\cdot}11E_B$$

In general $Y' \neq Y$, but for grey and white areas of the scene

$$E_R = E_G = E_B$$

$$\text{i.e.,}\quad Y' = [E'_R(0{\cdot}3+0{\cdot}59+0{\cdot}11)]^{\gamma} = E_R = 0{\cdot}3E_R+0{\cdot}59E_G+0{\cdot}11E_B = Y$$

For, say, a saturated red of maximum brightness

$$E_R = 1{\cdot}0 \text{ and } E_G = E_B = 0$$

$$\text{Therefore,}\quad Y' = (0{\cdot}3E'_R)^{\gamma} = 0{\cdot}3^{\gamma}$$

However, the correct luminance is

$$Y = 0{\cdot}3$$

For a black-and-white tube $\gamma = 2{\cdot}5$ (say)

$$\frac{Y'}{Y} = 0{\cdot}3^{1{\cdot}5} = 0{\cdot}163$$

A black-and-white display is thus somewhat distorted in that saturated colours in the scene, particularly red and blue, are reproduced as greys that are too dark.

In the colour display, the luminances of large areas are correct (colour-separation signals are formed correctly at the receiver). However, contributions to the displayed luminance are made by both the luminance and colour-difference signals (except for greys and white) and in saturated coloured areas (particularly red and blue) the major contribution to luminance is made by the colour-difference signals which are restricted in bandwidth. Thus definition is lost in such areas and noise in the colour-difference channels can contribute to display luminance. A. V. L.

See: *Colour principles; Colour systems.*

Books: *Principles of Colour TV*, ed. K. McIlwain and C. E. Dean (New York), 1956; *Colour TV*, by P. S. Carnt and G. B. Townsend (London), 1961; *Constant Luminance Principle in NTSC Colour TV*, by W. F. Bailey (*Proc. IRE*, Jan. 1954).

⊕COLOUR PRINTERS (ADDITIVE)

The term additive derives from the fact that the exposing light in the printer is made up of separate red, green, and blue portions of the spectrum which are separately controlled as to intensity and are added together to make the composite light by which the print is exposed.

In printing colour positive from colour negative films, the printing exposure is adjusted to produce the desired density and colour balance in the positive, and to compensate for differences in negative density and colour balance from one scene to the next. This is especially important because it is usual to make release prints from the original negative for maximum colour fidelity.

NARROW-BAND ILLUMINANT

Colour information is contained in the colour negative as densities of dye in each of the three layers of the monopack emulsion. For maximum fidelity of colour in the print, the colour information contained in each layer of the negative must be printed to the corresponding layer of the print material without allowing the unwanted side densities of the negative dyes to affect the exposure in other layers. This desire to isolate the printing effects of the negative dyes on the positive layers leads to the use of a printing illuminant divided into three separated, and relatively narrow, spectral bands. The precise bands used depend on the specifics of the negative dyes and the positive sensitivities, but are typically located in the red, green, and blue portions of the spectrum.

Controlled intensities of these three bands may be added together to create a composite exposure at the printing gate.

ILLUMINANT ADJUSTMENTS

Motion picture printing is usually done at high speed. Typically, in contact printing, the films run at speeds of 180 ft. per minute or greater, and even in optical printing, the speeds have recently been increased to nearly 200 ft. per minute on the 35mm side of a 35 to 16mm reduction printer. It is necessary to change the illumination level and the balance frequently, and the printer must be designed so that the alteration may be made at each change of scene in the negative. The variation from one scene to the next may be quite large, and if the exposure level from one scene is allowed to carry over into the next, the error in density and colour balance will produce a short length of incorrectly exposed film and a visible and unwanted flash on the screen. The change should be made as nearly as possible within the frame line. In practice, it is impossible to make it in less than the height of the printing aperture on a contact printer, which works out to require actuation of the change mechanism in about 5 m.sec (milliseconds). In a step optical printer, the shutter-blanking time is about 7 m.sec on the fastest printers.

LIGHT CONTROL SYSTEMS

These two requirements, selective control of narrow spectral bands, and quick reaction time, have dictated the design of the light control system for colour printers. In most of the commercial designs, the light from an incandescent lamp is split into three channels by filters or dichroic reflectors. An attenuator in each channel controls the light flux in the channel.

By 1963, the commercial equipment most widely used had concentrated on two systems.

The Debrie Matipo Colour Printer uses a masking technique in which a short length of high contrast black-and-white film is exposed and developed to bear a pattern of black bars with clear spaces between. The

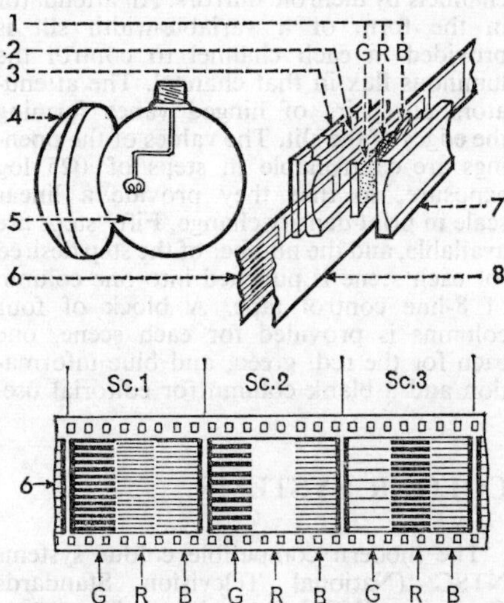

DEBRIE MATIPO ADDITIVE PRINTER. Simplified optical system. 1. Tricolour filters and beam splitter. 2. Beam condenser. 3. Heat absorbent glass. 4. Reflector. 5. Light source. 6. Control film band. 7. Beam collector. 8. Filter strip.

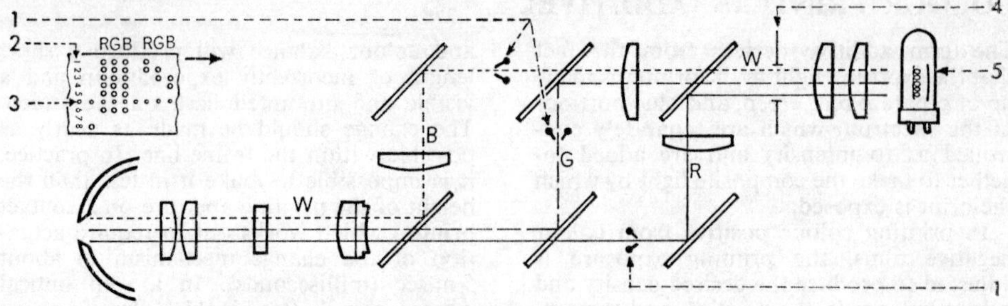

BELL & HOWELL ADDITIVE PRINTER. White light, W, from light source, 5, controlled by overall fader, 4, is split into red, green and blue beams, R, G, B, at surfaces of dichroic mirrors. Light valves, 1, in path of beams control their separate printing intensity in accordance with control strip (*top left*). 2. Control channels. 3. Transport perforations.

per cent clear area determines the average transmission of the film. Three strips of these bars are arranged parallel to the length of the film, one for each of the three colour channels. The light is divided into the three channels, each one passing through the appropriate control area, so that each is independently controlled. There is one control length of film for each scene in the negative, and the control film is advanced by an intermittent movement at the time of a scene change.

The Bell & Howell additive system is now used on both continuous contact and optical printers. The light from a 1,000 watt projection-type lamp is divided into three channels by dichroic mirrors. An attenuator in the form of a variable-width slit is provided in each channel to control the luminous flux in that channel. The attenuators are pairs of hinged vanes forming the edges of the slit. The values of the openings are controllable in steps of ·025 log exposure, so that they provide a linear scale in print density change. Fifty steps are available, and the number of the step desired for each scene is punched into one column of 8-line control tape. A block of four columns is provided for each scene, one each for the red, green, and blue information and a blank column for editorial use. The group of four columns is read simultaneously and the tape moves forward one group at a time.

MEMORY DEVICE

The attenuator is provided with an electromechanical memory so that as one scene is being printed, the slit width is maintained at the proper value while the tape is read and the new information for the next scene is stored in the memory. At the scene-change signal from the film, the vanes are driven quickly and directly from their positions for the scene just printed to the positions for the new scene. This vane movement takes place in about 5 m.sec, and without appreciable overshoot.

Since a production printer will make upward of a million scene changes in a year of normal use, these light valves must be extremely rugged.

In addition to the fifty steps of ·025 log exposure, which give a range of exposure of about 6 to 1, there are 14 additional steps which can be manually set to adjust for variations from one emulsion batch to another and for day-to-day variations in processing. M.G.T.

See: *Laboratory organization; Printers; Sensitometry and densitometry; Sensitometry of colour films.*

□COLOUR SYSTEMS

The modern compatible colour systems NTSC (National Television Standards Committee (US)), PAL (Phase Alternation Line) and SECAM (Séquentiel Couleur à Mémoire) are all based on the use of a luminance signal having a bandwidth equal to that required by the corresponding monochrome system, together with two colour-difference signals providing information about colour. If the luminance signal occupies the whole of the available bandwidth, then the colour-difference signals must be located within the luminance-signal band and each system must be so devised that all three signals interfere with one another to the minimum extent. Such an arrangement is known as band sharing.

In all three systems, the two colour-difference signals are effectively combined so as to form a single signal which is known as the chrominance signal; in fact, the systems differ mainly only in the ways in which they form their chrominance signals.

As compatibility is a prime requirement of a colour system, a complete colour signal must be of a form that is acceptable to receivers designed for the corresponding monochrome system, which may already be in the hands of the public when the colour system is introduced. The colour signal must, therefore, contain all the conventional synchronizing and blanking pulses, and the chrominance signal must be of a form such as not to interfere with the operation of the monochrome receivers.

To satisfy the foregoing requirements, the chrominance signal is transmitted to the colour receiver as the modulation of a suitable subcarrier whose frequency lies within the luminance band. However, the presence of the subcarrier results in an interfering pattern on the screen of the monochrome receiver and, in all three systems, measures are taken to minimize the visibility of this interference. These measures include the use of as high a subcarrier frequency as possible, thus ensuring that the interfering pattern has a fine structure. Further measures include (for NTSC and PAL) the use of specially chosen frequencies bearing particular relationships to the scanning frequencies; in the case of SECAM, the polarity of the subcarrier is reversed during certain lines and fields.

CHROMINANCE SIGNAL FORMATION

AMPLITUDE MODULATION. NTSC and PAL utilize amplitude modulation of the subcarrier to form the chrominance signal. In both cases, however, the suppressed-carrier method is used rather than the more conventional form. This ensures that the chrominance signal amplitude is directly dependent upon the amplitudes of the colour-difference signals; it is zero for all picture areas lying on the grey scale and is high only for picture areas of bright saturated colour. A further advantage of suppressed-carrier a.m. is that, for a given chrominance signal amplitude, it is more resistant to noise than is conventional a.m.

Normal forms of amplitude modulation (both suppressed carrier and conventional) are used for the transmission of only one modulating signal. In order to satisfy the requirement that two independent colour-difference signals be transmitted, NTSC and PAL utilize what is termed quadrature suppressed-carrier a.m.

QUADRATURE AMPLITUDE MODULATION. Quadrature a.m. may be regarded as the separate modulation of two subcarriers, of precisely the same frequency, which bear a quadrature phase relationship one with the other. Since the peaks of a sine wave coincide with the zero-crossings of a cosine wave having the same frequency, and vice versa, the two waves may be separately modulated, in amplitude, by two signals; the two modulated waves may then be added together to form a single resultant wave which is modulated in both amplitude and phase. The separate modulating signals may be derived from the resultant wave by examining its amplitude, first at the instants corresponding to the peaks of the sine wave and, secondly, at those corresponding to the peaks of the cosine wave. In NTSC and PAL coders, the quadrature modulation of the subcarrier by the two colour-difference signals results in the chrominance signal which, in turn, is added to the luminance and synchronizing signals to form the composite colour signal.

In an NTSC or PAL colour receiver, the chrominance and luminance signals are first separated by means of filters and the two colour-difference signals are then derived from the chrominance signal by means of two demodulating waves, representing the sine and cosine components, which are obtained from a local subcarrier oscillator in the receiver; this oscillator is maintained in synchronism with the master subcarrier source at the coder by means of a special synchronizing signal consisting of a short burst of subcarrier located in each post-synchronizing line-blanking interval (except for a few lines during the field-blanking interval). The two colour-difference signals and the luminance signal are then used to form the colour-separation signals required by the display.

PHASE ERROR CORRECTION. If, at the colour receiver, the phase relationship between the demodulating waves and the chrominance signal is incorrect, say due to phase distortion of the chrominance signal during transmission or to distortion of the burst, the colour-difference signals obtained may be erroneous; instead of the two original colour-difference signals, two mixtures of them are obtained.

The NTSC system does not incorporate any measures to prevent this form of distortion causing errors in the colour picture. However, in PAL, one of the two colour-difference signals used to modulate the subcarrier at the coder is reversed in polarity between alternate lines; a special form of burst is also used which indicates to the colour receiver whether the colour-difference signal of a particular line has been reversed or not. This periodic polarity reversal of one colour-difference signal enables the PAL system to minimize the effects of distortion.

In a simple form of PAL receiver, the polarity reversals of one colour-difference signal are compensated by a further reversal effected by a simple switch. In such circumstances, the effects of chrominance-signal phase distortion cause colour errors of opposite sign during successive lines, as displayed on the colour-receiver screen. Provided that the errors are not large, the eye averages them and the correct colour is seen. However, a PAL receiver may include an ultrasonic delay which permits the chrominance signal to be delayed by one line period; this arrangement allows the chrominance signal of one line to be combined with that from the previous line. In this way, an average chrominance signal may be derived electrically and large phase errors substantially corrected; in addition, the action of the delay enables the two colour-difference signals to be separated, one from the other, more easily.

FREQUENCY MODULATION. The SECAM system differs rather radically from NTSC and PAL. It does not attempt to transmit both colour-difference signals along with the luminance signal, but sequentially so that one colour-difference signal is transmitted during one line and the other during the succeeding line. Further, instead of amplitude-modulating the subcarrier, each colour-difference signal deviates the subcarrier frequency by an amount which is proportional to the colour-difference-signal amplitude; when the colour-difference signal is zero (grey scale) the subcarrier frequency has its normal value. Thus the SECAM chrominance signal consists of a frequency-modulated subcarrier whose amplitude is not zero in grey-scale picture areas. Various forms of pre-emphasis are used to modify the colour-difference signals prior to modulation and to vary the amplitude of the chrominance signal so as to increase its resistance to noise. As with NTSC and PAL, the SECAM chrominance signal is combined with the luminance signal to form the complete colour signal. As the colour-difference signals are transmitted sequentially, a special signal must be included in the complete colour signal indicating the colour-difference signal sequence.

At the SECAM receiver, the chrominance and luminance signals are again separated by suitable filters. However, as only one colour-difference signal is transmitted during any one line, the SECAM receiver must contain a one-line delay which enables the chrominance signal from one line to be available while the chrominance signal corresponding to the other colour-difference signal is received during the succeeding line. In this way, both colour-difference signals may be derived simultaneously, although only one of them is directly associated with the luminance signal being received at any one time.

The chrominance signals, obtained directly and via the one-line delay, are then passed to an electronic double-pole, double-throw switch which is actuated so as to result in two further chrominance signals that correspond to two continuous and separate colour-difference signals.

After amplitude limiting, the two colour-difference signals are recovered by frequency discriminators; suitable de-emphasis of the signals is also carried out. Finally, the recovered and corrected colour-difference signals are combined, as in NTSC and PAL, with the luminance signal to form the colour-separation signals.

COMPARISON OF SYSTEMS

The NTSC system has long been established in the USA and has also been adopted by Canada and Japan. Thus, it is principally in Europe that comparisons between the three systems have been made in order to determine the best national choice. As applied to the 625-line, 50-field standard, they have been intensely studied and tested by European broadcasting and industrial organizations under the auspices of a special Ad Hoc Group on Colour Television, set up by the European Broadcasting Union; the Ad Hoc Group have issued a report which summarizes the work carried out and the conclusions reached. The following are some of the points made in the report.

COMPATIBILITY. In general, NTSC is somewhat more compatible than PAL and SECAM. Little difference of compatibility was found between the latter two systems.

FUNDAMENTAL QUALITY OF THE COLOUR PICTURE. There is little significant difference between NTSC and PAL in this respect and both systems are slightly better than SECAM with regard to certain unwanted effects at horizontal boundaries between areas of different colour.

EFFECTS OF DISTORTIONS AND SIGNAL ERRORS. Of the three systems, SECAM is the most tolerant of chrominance-signal phase errors, although PAL with a delay-line receiver is about twice or three times as tolerant as NTSC (and PAL with a simple receiver).

PAL is more tolerant than NTSC with regard to unwanted attenuation of the chrominance-signal upper sideband.

In the simultaneous presence of certain distortions (amplitude non-linearity with loss of response at subcarrier frequency) and random noise, SECAM is not as good as NTSC and PAL.

RECEIVERS. The NTSC receiver is generally considered as the cheapest. An NTSC or simple PAL receiver requires more colour controls than a PAL receiver with a delay-line or a SECAM receiver.

The NTSC system is the most suited to the use of a single-gun display tube; SECAM has less flexibility in this respect than either NTSC or PAL.

RECEPTION UNDER ADVERSE CONDITIONS. With regard to the effects of co-channel interference, NTSC and SECAM are both slightly better than PAL (with a delay-line receiver) when the interference has zero offset; when the offset is about 250 Hz, NTSC is better than SECAM and slightly better than PAL (with a delay-line receiver).

The three systems are equally affected by weak echoes, as caused by multipath reception. When the echoes cause serious picture impairment, SECAM and PAL (with a delay-line receiver) are slightly better than NTSC.

In the presence of random noise, NTSC is the same as PAL, and SECAM is almost as good as NTSC for signal-to-noise ratios giving rise to a picture quality better than "rather poor". A.V.L.

See: *Colour principles* (TV); *NTSC colour system; PAL colour system; SECAM colour system.*

⊕**Colour Temperature.** Measurement of the
□colour quality of a light source, being the temperature on the Kelvin scale at which a black body must be operated to give a colour matching that of the source in question. In this definition, a black body is a temperature radiator whose radiant flux in all parts of the spectrum is the maximum obtainable from any temperature radiator at the same temperature. It is called a black body because it absorbs all the radiant energy that falls upon it. In colour cinemato-

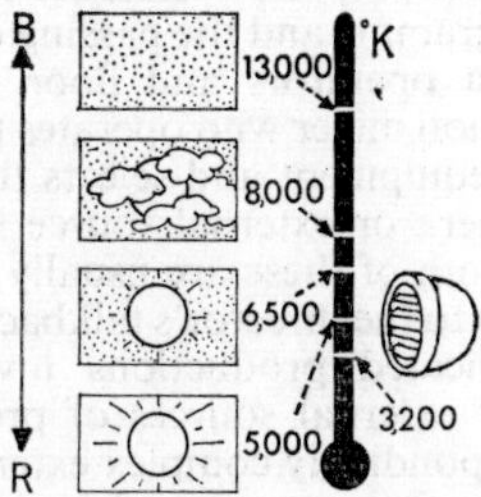

graphy and TV, to avoid colour distortion, the colour temperature of the lighting of the scene must be matched to that of the system which is to record and reproduce it.

See: *Exposure control; Transfer characteristics.*

⊕**Colour Triangle.** Method of representing
□colours as points within a triangle to permit the calculation and specification of a number of aspects of colorimetry. The three corners of the triangle represent three colour

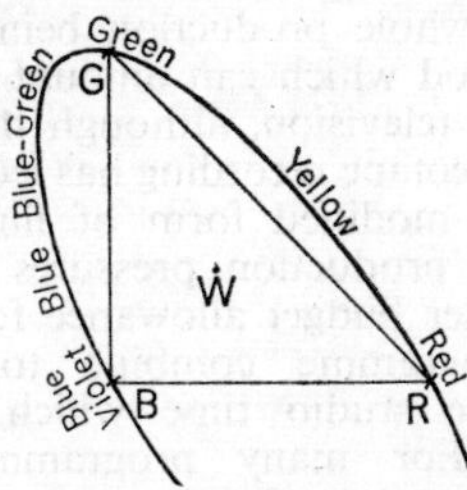

stimuli of red, green and blue lights, whose wavelengths are defined, and any colour which can be produced by mixing these three stimuli in different proportions can be represented by a point within the triangle. The position at which a mixture of the three gives the effect of white is known as the white point. Pure spectral colours fall outside the bounds of the triangle since they cannot be absolutely matched by mixtures of three stimuli.

See: *Colour principles; Telecine (colour).*

⊕**Coma.** Lens defect associated with images
□away from the optical axis. A non-axial point object does not give rise to a point image even in the absence of spherical aberration, and the best image which can be produced consists of a small patch of light with a tail to one side in the shape of a comet.

See: *Aberration.*

⊕**Combined Dupe.** Duplicate negative film in which picture and sound track images are printed on the same strip of film in their correct synchronization for printing. Also known as composite dupe.

See: *Printers; Laboratory organization.*

□**Combining Filter.** Passive device for feeding the outputs of two transmitters to a common aerial system.

□**Combining Unit.** Circuit designed to combine sound and vision outputs to enable them to use a common aerial array. The requirements of the combining unit are that a minimum loss should occur within the unit, that a minimum signal should be fed back to the other transmitter, and that the impedances of the inputs and output should correctly match those of the transmitters and aerial, in order to achieve a maximum transference of power.

It is usual to make up the combining unit from lengths of co-axial feeder joined together in various configurations. Addition of stubs, usually one-quarter wavelength long at certain parts of the assembly, can cause in-phase or out-of-phase reflections, and can give the effect of an open or short circuit to different frequencies. The aerial feeder and the associated transmitter outputs are connected to appropriate points on the assembly, and the two signals combine with a minimum loss of power to feed the aerial array.

See: *Transmitters and aerials.*

⊕**Commag.** Composite (combined) magnetic.
□Magnetic sound track striped on to a picture-carrying film.

See: *Comopt; Sepmag; Sepopt.*

□COMMUNICATIONS

The communication requirements of a television studio are unique and arise almost entirely from the "continuous take" method of operation originally imposed by the tradition and necessity of instantaneous transmission.

In film production, takes are usually short, the whole production being spread over a period which can amount to many months. In television, although the prevalence of videotape recording has led in some cases to a modified form of film studio techniques, production pressures and the much smaller budget allowance for a particular programme combine to restrict severely the studio time which can be allocated. For many programmes, this amounts to only a few hours, and it is a most unusual television programme which gets the use of a studio floor for more than three days.

The restriction on time available for rehearsal and the ephemeral nature of the television picture make it almost axiomatic that the programme director must spend most of his time off the studio floor, and hence out of physical contact with the actors, watching and directing the rehearsal on television picture monitors. It follows from this that a vital requirement of any television studio is a reliable and instantaneous system of communication between the director and his actors and the various members of his crew. Further, once the programme is on air, there can be no direct oral contact between the director and the actors, since this would be picked up by the microphones and be transmitted as part of the programme.

DIRECTOR'S TALKBACK

The personnel in the production control room usually consist of the director's programme assistant, responsible for the timing of the programme and the passing of cues to the camera operators and floor manager, and the vision mixer who operates the vision switching equipment and selects the appropriate camera or external source for transmission. Both of these are usually provided with access to the director's talkback system. On complicated productions involving a number of external sources of programme and correspondingly complex external communications, it is sometimes found expedient to provide an assistant director to deal with this part of the operation.

In general, the director's microphone is connected permanently in circuit, although a microphone-muting switch is provided so

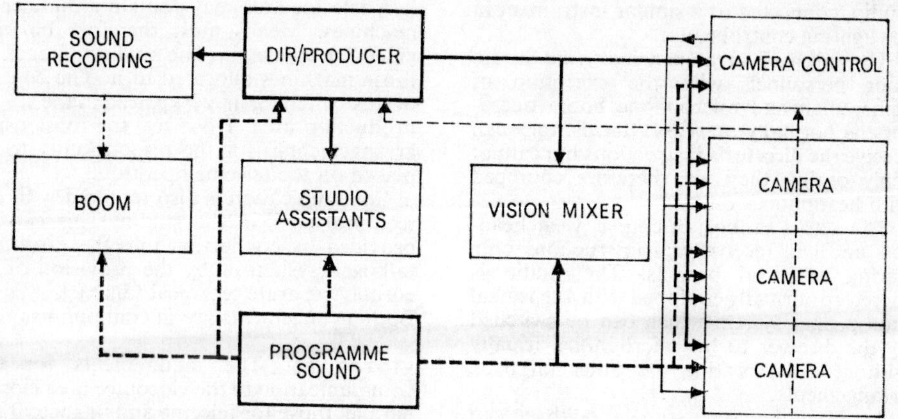

TALKBACK ARRANGEMENTS. Simplified block diagram. (*Heavy solid line*): director/producer talkback. (*Light solid line*): camera reverse talkback. (*Heavy dashed line*): programme sound. (*Light dashed line*): camera talkback (mixed to director/producer). (*Dotted line*): sound talkback.

that he can at will disconnect himself from the talkback system.

COMMUNICATION CHANNELS. The director must be able to communicate within the studio, with actors, cameramen, stage hands and the stage manager.

Within the studio control rooms, he needs to pass instructions to the sound controller or sound balancer, the lighting director, the camera control operators, and the sound boom operators. Each of these is interested in all instructions by the director, since any of them may directly or indirectly affect his own particular operation.

STUDIO COMMUNICATIONS

A television studio normally contains five distinct areas. The action takes place on the studio floor or stage. In the production control room, the programme director views the output of the cameras on preview monitors and listens to the sound on a loudspeaker.

The output of the various microphones and other sources of sound is mixed to form a homogeneous whole in the sound control room. The (vision) camera control room provides for adjustments to the electronics of the cameras to suit the particular scene. Adjustments to lighting levels and switching of lighting set-ups are made in the lighting control room.

All these areas need to have intercommunication facilities. There is therefore a requirement for other communication facilities which can be used independently of the director.

CHANNELS. The facilities provided vary in detail with different television organizations, but the basic arrangements are similar.

Camera talkback is primarily a two-way system linking each cameraman with his control operator in the vision control room. The outputs of the cameramen's microphones are mixed in a subsidiary circuit to form mixed camera talkback and fed into the director's talkback receiving system; thus the cameramen can also communicate freely with the director. Again, it is common practice to make this talkback available in all the control areas.

Sound talkback provides a two-way communication between sound control and the microphone boom operators on the studio floor.

Lighting talkback provides the lighting director with a means of communicating instructions to the studio electricians. The facilities vary greatly from studio to studio and depend on its physical size and the complexity of the lighting arrangements.

In general, since most lighting rigging is done when the studio is not in use for rehearsal, it suffices to give the lighting director means of switching his microphone to a loudspeaker network in the studio. In the larger studios, two-way VHF radio communication is also used. A transmitter of about 2 to 5 watts is adequate and the handsets have a power of a few hundred milliwatts.

For emergency use, it is common practice to provide a two-way circuit using ordinary hand telephone sets fitted with lamp cabling disposed at strategic points around the studio, connected to a similar instrument in the lighting control room.

The talkback arrangements to all studio floor personnel, with the exception of cameramen and microphone boom operators, is normally one-way, that is, they can receive the director's instructions but cannot reply orally; they are therefore equipped with headphones only.

The actors cannot, of course, wear headsets and can receive oral instructions only during rehearsal periods. The studio is therefore normally equipped with a rehearsal loudspeaker system which can be switched by the director to his microphone, usually without disconnecting the other talkback arrangements.

Again, in the larger studios, talkback can be supplemented with radio links to the floor managers.

EXTERNAL COMMUNICATIONS

In addition to the areas within the studio, communication is usually required with external areas, such as telecine (film scanners), videotape recording, presentation, master control, outside broadcasts on temporary locations and discussion programmes.

TELECINE. The basic need for a talkback system to a telecine area is to ensure the correct timing of filmed inserts in a live production. To this end, the director or his assistant needs to be able to talk directly to the operator of the machine assigned to his particular use.

In general, during transmission, the talkback is one-way, i.e., the telecine operator receives continuously the output of the director's microphone, and acknowledges the receipt of his cues by means of a buzzer and light.

The use of film in a production may be of relatively short duration and often only at the beginning and end of the programme. As a result, the telecine operator may be required to listen for long periods to talkback which does not in any way concern him. To reduce the temptation to the operator to reduce the talkback level and thereby risk missing his cues, it is common practice to insert a switchable attenuator, controlled by a non-locking key on the director's talkback panel. By this means, a reduced talkback level is sent to telecine except when the key is operated. The operator is thus assured that his connections to the studio have not been interrupted and those parts of the talkback directly concerning him are underlined.

A telecine area may contain a number of machines. Means must therefore be provided for connecting the studio talkback to those machines allocated to it. The control of this switching may be placed in the studio production area. However, the most usual arrangement is for the responsibility to be placed on the telecine operators.

Communication is also required with the technical staff in the studio. This can be provided by connection into the director's talkback system or by the provision of an entirely separate technical talkback system. Both arrangements are in common use.

VIDEOTAPE. The requirements for the communication to the videotape area closely parallel those for telecine and in general the arrangements are very similar, despite the use of videotape for recording as well as for playback.

PRESENTATION. Presentation's prime function is to weld the individual items in a transmission period into a homogeneous and smoothly-running programme and to accommodate the gaps in programming due to breakdowns, however caused. This department is therefore vitally concerned with programme-timing. A schedule of programmes containing only prerecorded items such as film and videotape can be accurately pretimed and a fixed and rigid switching sequence adhered to. With live programmes, no such accuracy of timing can be relied upon, and under- and over-runs, possibly only a matter of seconds, must be expected.

A rigid time schedule, with the implication that programmes may be cut, is not generally considered acceptable and it is normal to vest in presentation the overall control of programme-starting times, although difficulties arise when more than one centre is concerned in the transmission of a particular programme.

There is therefore a need for a separate and permanently connected two-way communication system between presentation and all programme-producing areas. In many cases, this need is met, for areas local to presentation, by one of the many commercially-available key-operated intercom systems, presentation being in all cases the master station.

Where programmes are interchanged between different studio centres, telephone circuits are usually rented, either on a permanent basis or for the programme duration, between the centres concerned. These can either terminate in a standard telephone instrument or on a local switchboard. The most usual arrangement is to provide a local switchboard connected to the local telephone exchange by permanently rented circuits, the connection between the centres being completed by trunk circuits rented for each programme period.

MASTER CONTROL. This is the technical parallel to presentation and is responsible for the technical aspect of the transmission and for the checking of the connections between studio centres. The technical communication facilities required are thus very similar to those provided for presentation and are usually supplied in the same manner.

OUTSIDE BROADCASTS. These are covered by mobile equipment operated in the vehicle or dismounted into a temporary studio, depending upon the requirements of the particular programme. Connection with the controlling studio centre can be by either line or radio link or a combination of both. It is usually possible to arrange for temporary cables to be installed between the programme site and the local telephone exchange and thence to the studio centre by permanent telephone circuits. This is the most common arrangement for the control of music connections. In the case of a simple programming, i.e., one concerned only with the output from one location, the connection is made directly to the presentation and master control areas. Two control circuits are normally provided, one for programme use, the other for technical control.

This simple system is not suitable for use on large multiple outside broadcasts involving a number of separate pick-up points. These occasions are mostly spontaneous and full rehearsal is not practicable. It is usual in this type of programme to use a studio as a central control point which also allows linking items to be introduced.

The need therefore arises for two-way communication between the central director and the director at each of the remote points.

One method is to provide a feed of the central director's talkback circuit to each site, and also a two-way telephone circuit to a telephone switchboard at the central studio. There are disadvantages to this system, such as the possibility of the switchboard becoming jammed and the slowness of operation, which can be an important factor in an unrehearsed programme of this type.

A refinement is to provide central director's talkback to each remote point, terminating on a loudspeaker. The return circuits are then all commoned and appear as a combined input to a loudspeaker in the central studio control room. The speaking arrangements at the remote point are such that the operation of the microphone disconnects the talkback loudspeaker, thus avoiding howl round (acoustic feedback).

The pick-up on the director's microphone of the output of his loudspeaker is sufficient to keep all other pick-up points aware of the conversation taking place.

This system works very well in practice, although care has to be taken in the commoning of the return circuits to avoid undue degradation of the signal-to-noise ratio.

DISCUSSION PROGRAMMES. In many programmes of the discussion type, a requirement arises for two or more people in different studios to hold a conversation by means of a microphone-loudspeaker connection. To avoid howl round, the microphone-loudspeaker loop has to be broken at one of the programme points. Various methods have been tried, notably voice-operated amplifiers and hybrid cancelling circuits, but the most usual and simplest arrangement is to provide the appropriate speakers with a deaf-aid-type earphone.

DESIGN CONSIDERATIONS

Talkback communications are generally tailored to fit the particular studio arrangements. There are, however, some considerations which affect all types of installation. Many of the recipients of instructions over the talkback system have no return circuit and are therefore unable to query any misunderstood instructions. It follows that a prime requirement is a high level of intelligibility and freedom from noise.

To this end it is becoming common practice to use broadcast quality microphones and, particularly in the case of the director's circuit, to fit slide back compressors, which have the effect of reducing unwanted pick-up and making the output level more independent of the speaker's position and the loudness of his voice.

It is also essential that the received levels at all output points can be individually adjusted so that each can be set to requirements without affecting the whole system.

All outputs should be sufficiently isolated, to prevent a mishap on any circuit affecting the whole system. Equipment should be rugged, simple and reliable. L.H. & E.C.V.

See: *Master control; Outside broadcast units; Studio complex* (TV).

⊕**Comopt.** Composite (combined) optical. □Optical sound track printed on to a picture-carrying film.

□**Compatibility.** Term applied to a colour TV transmission, indicating that it is receivable as a black-and-white picture on existing monochrome receivers, and can be transmitted within the bandwidth allocated to monochrome stations. All broadcast colour TV transmission is compatible, but closed-circuit TV is free of these limitations, though practical factors such as use of standard equipment often ensure that it is compatible.

See: *Colour principles* (TV).

⊕**Compatible Composition.** Composition of □the action photographed in a motion picture frame in such a way that no important part of the action is lost if the image is projected at a different aspect ratio. For example, in the photography of non-anamorphic (flat) wide-screen pictures, it is recommended that the composition of the action should be acceptable when shown at any aspect ratio between 1·65 : 1 and 1·85 : 1. Similarly, in anamorphic photography at an aspect ratio of 2·35 : 1, if essential action is composed to fall within the central 75 per cent of the image width, compatible unsqueezed prints for normal projection can be made without difficulty.

If composition must be compatible with television, width restriction is even more severe, because the TV aspect ratio is 1·33 : 1. If this requirement cannot be met, margins above and below the picture must be left blank on the receiver tube (kinescope).

See: *Wide screen processes.*

⊕**Complementary Colours.** Colours which □result from subtracting in turn the three

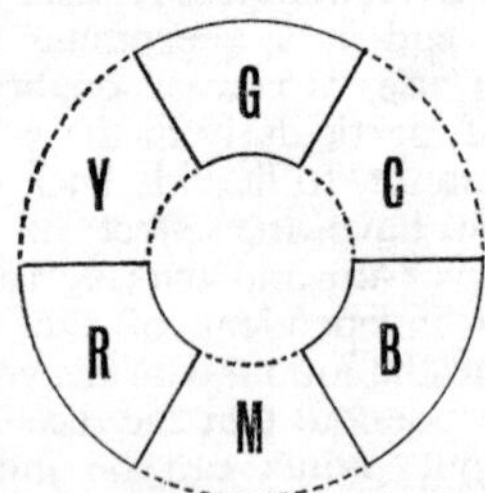

primary colours from the visible spectrum: minus-green (magenta), minus-red (blue-green or cyan) and minus-blue (yellow).

□**Composite.** Control signals, performing separate functions, added together to form a single waveform. Composite sync, for example, is a synchronizing signal which contains information to synchronize both line and field scan functions in the display device. Composite video is a video waveform comprising a picture signal for the brightness control of the display device, to create the displayed image, and composite sync to control the scanning functions within the display device.

Composite blanking occurs during the flyback of the line and field scans, when the picture signal must be suppressed. To ensure that the picture is fully suppressed before the scan flyback is initiated by the sync pulse, line- and field-blanking commences 1·5 μsec. before the sync signal. The back edges of the blanking signals are timed so as to allow the scanning circuits enough time to recommence their forward scan before the picture signal begins again.

Field drive is a pulse which provides a trigger to the picture-generating equipment which is coincident with the onset of the field sync period in the comp sync waveform.

Burst gating pulse, in a colour television system, is a short pulse provided during the post-sync period, that is, between the end of the sync pulse and the end of blanking. This gating pulse is used to time the start and finish of the colour burst in this post-sync period.

See: *Synchronizing pulse generators.*

□**Composite Picture.** The complete video signal, composed of video signals from more than one source, keyed or switched at predetermined time intervals.

⊕**Composite Print.** Film print combining picture and sound track, the two usually staggered in proper projector relationship for the gauge of film used.

□**Composite Signal.** Signal having both picture and synchronizing information.

□**Composite Sync Generator.** Supplies the signal required to synchronize the scanning circuits in the display devices that are reproducing the picture signal.

The short duration pulses are used to synchronize the line scan, and the long duration ones to synchronize the field scan processes. Equalizing pulses are employed in the 525 and 625 line systems, before and after the field sync period. These serve to stabilize the sync separator circuits in the display devices, to ensure more accurate field scan synchronization. During the equalizing and field sync periods, pulses occur at twice the line repetition frequency. Those pulses that occur half-way through the line scan period are ignored by the display line scan circuits.

Alternate field sync periods occur half-way through a line scan period. The effect of this is to cause the line scan, during alternate fields, to occur mid-way between the trace position of the previous field scan. This forms a 2 : 1 interlace and serves to reduce flicker on the displayed picture.

See: *Synchronizing pulse generators.*

□**Composite Video Signal.** Complete video waveform plus the synchronizing pulses.

⊕**Composite Master.** Fine-grain positive print in which picture and sound track images are recorded in printing synchronization and from which a combined dupe negative can be directly prepared.

⊕□**Compression (Sound).** Function of a device which transfers a signal from its input to its output and at the same time reduces the span of amplitudes of the signal. Compression is used in recording original sources such as large orchestras, in which the volume range is bigger than the recording system can accommodate, as well as in making transfers from a wide-range medium such as magnetic tape to a narrow-range medium such as 16mm optical sound track.

See: *Narrow gauge production techniques; Sound recording.*

⊕□**Compressor.** Device used to reduce the volume range of sound without producing noticeable distortion.

The waveform of sound signals usually has a marked transient character, with peak excursions far exceeding the average level. A sound mixer may wish to increase the relative loudness of a signal, but cannot do so by increasing the level, because of consequent overloading at peak signal. If the gain of the audio system is made dependent on the peak signal level, however, it is possible for him to obtain the desired effect without overload distortion.

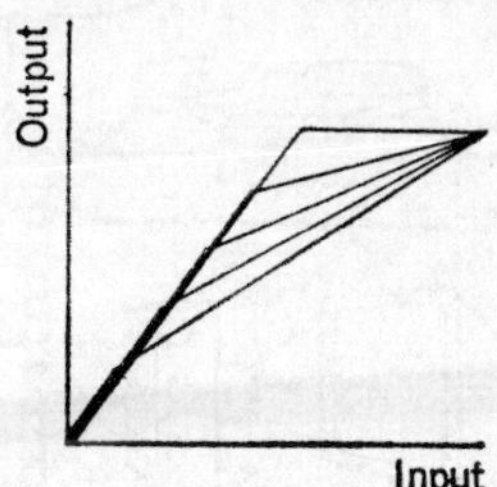

At very low levels, the output follows changes in input level faithfully, but as the input level is raised the amplifier progressively inserts an increasing amount of attenuation. The characteristic normally shown in graphs is a steady state characteristic in that it indicates the amount of attenuation inserted on an input signal of constant level. The instantaneous characteristic, however, is complex because the amplifier has finite response times.

The attack time is the time taken for the amplifier to react to a sudden increase in signal level. The release time indicates the rate at which the gain of the control amplifier rises when the input signal falls in level.

There are two distinct variables in the steady state characteristic. One is the position of the knee, and the other is the slope of the characteristic above the knee.

All four of these variables, attack time, release time, knee position and slope, can be varied according to the taste of the sound mixer and the effect desired.

See: *Sound recording; Transients.*

⊕COMPUTER ANIMATION

Animation consists essentially of line drawings, still or moving, occupying a fixed two-dimensional film area, and sometimes differentiated by tone or colour. Regarded in this way, animation is an obvious candidate for the computer. Lines of any degree of complexity may be represented by equations, while areas enclosed by lines may equally be specified for treatment in colour or tones of grey. However, it was only in the mid-1960s that computer techniques for animation began to be seriously developed.

PROGRAMMED CAMERA INSTRUCTIONS

The simplest arrangement is a combination of film and TV techniques. Programmed instructions are fed into a computer by any convenient method, e.g., punched cards. The computer converts these instructions into a set of frame-by-frame orders to a film camera to advance, and to a cathode ray tube beam to write the picture on its face. These orders are stored on magnetic tape, releasing the computer for its next task. The tape then actuates the camera and tube, much as tapes are programmed to

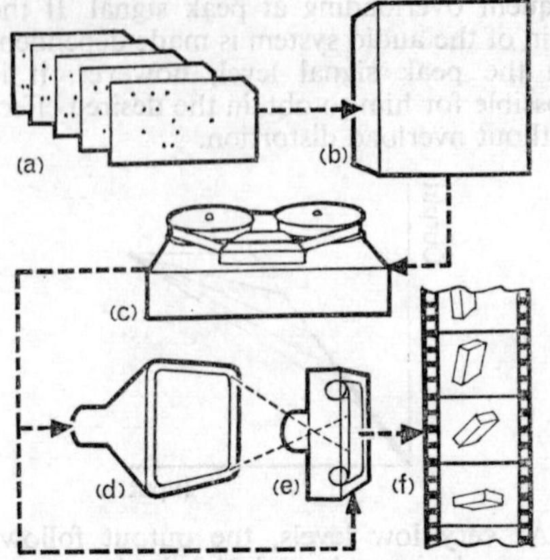

SCHEME FOR COMPUTER ANIMATION. Instructions to form moving image on cathode ray tube, together with information on animation camera dope sheets, are punched on cards (a). These are fed to computer (b), which combines them into programme recorded on tape (c). This tape later actuates a cathode ray tube display (d) and camera (e) to realize complete animated film (f).

actuate machine tools in milling and other operations.

Computer animation programmes share the advantages of other computer programmes. The repeated cycles of action so common in all instructional animation need to be written only once, and the computer then returns to them on a single instruction. Symbols may be used to give generality to a programme, and special values inserted later to produce different film sequences of the same genus. Sub-programmes may be developed and stored in a library for subsequent use, and whole programmes be linked together to produce longer films.

COMPUTER LANGUAGES

If an accepted mathematical computer language such as Fortran be employed, a scientist may use it to project his ideas directly on film, in the same way as he otherwise does with symbols and words on paper. But there is a concealed danger. Just as scientific writing is sometimes unclear, due not to confused thinking but to unfamiliarity with the art of words, so scientific animation may neglect the art of animation.

This should encourage partnership between scientist and animator, and for the first time the scientist can directly sketch out on the screen the very complex ideas he may wish to illustrate. It is then for the animator to show how best they can be tailored to a medium which is not merely expositional, but also interpretative.

BEFLIX

While early and predominantly scientific films were written in Fortran, and therefore required mathematical knowledge, others have been produced with the aid of BEFLIX, a non-mathematical language devised especially for this purpose and containing instructions such as PAINT, ZOOM and DISOLV.

The cost of Fortran films is relatively low, since only points forming lines have to be computed (there is no area shading or colouring), and straight lines are defined merely by their terminal points. BEFLIX, on the other hand, computes every point on the image surface, and so can produce tonal shading, but is more expensive.

This raises the question of resolution, since every determinate point in the picture requires a place in the computer memory. The tube-camera device employed by Bell Telephone for its early work had a 1024×1024 grid, comprising about 10^6 elements; but, as explained, by no means all of these were used at any one time. The BEFLIX grid had an effective resolution of 184×252, or about $4{\cdot}5 \times 10^4$. These figures are likely to be greatly exceeded in the near future, though the use of television techniques sets a practical limit of about $2{\cdot}5 \times 10^6$ information elements unless special equipment is developed.

Colour animation can be produced directly by a colour tube, or indirectly (and with better quality) by generating three black-and-white separations, which are combined on a colour printer to make a single colour print.

USE OF ANALOGUE COMPUTERS

It is in the exposition of complex mathematical and scientific ideas that computer animation is likely to show the greatest gain over classical man-drawn techniques. But a beginning has been made in the field of computer-drawn entertainment cartoons. A Japanese electronics firm has created a popular character whose body, eyes, mouth and legs consist of seven closed curves. Values in the equations of these curves are changed more or less quickly according to a programme, with the result that the character alters his expression and moves about the screen.

These cartoons are produced by an analogue computer, but it is difficult for the controlling artist to predict the effect of mathematical changes and so create the effect he desires. The method is said to be cheap, but the scope appears rather limited.

The same firm has developed a more flexible system, in which an artist draws two cells representing the terminal positions of a cartoon sequence, and indicates the essentials of the movements which are to link the first with the last. A computer is then employed to generate the intervening frames by interpolation, a method which requires the storing of much more data than the first, with a consequent increase of cost.

These are merely the first steps in what may prove a fruitful field of development, especially in instructional scientific animation. R.J.S.

See: *Animation techniques.*

Condenser. System of lenses or reflectors designed to collect light from a source over a large solid angle and used to increase the illumination obtained. When forming part of a projection system, a condenser is normally positioned so that the light source

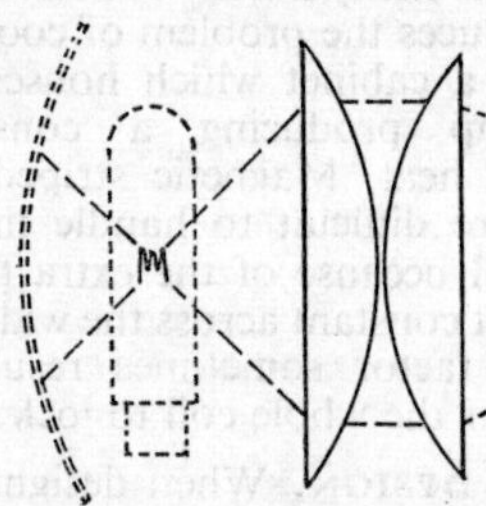

is imaged within the projection lens. However, the maximum diaphragm opening of the projector lens is not effectively employed unless the image of the source is larger than this opening.

Console. Desk or panel often of large size and elaborate construction used for the centralized control of picture monitoring and sound recording and dubbing operations. Consoles are of many types, and may

contain picture or sound monitors, or both, together with controls for cutting or fading from one source to another. They may provide for variable equalization, for checking the signal level from moment to moment, and for talkback and other communications facilities.

They may also contain controls for electronic special effects (in TV), and echo and reverberation devices (in sound recording).

See: *Master control; Sound recording.*

Contact Printing. Simplest and fastest method of film printing; negative and positive films are rolled together past an exposing

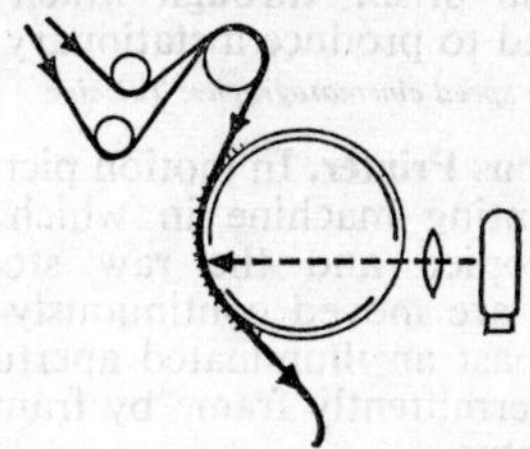

aperture (usually a slit), the illumination of which is automatically adjusted to control the printing exposure. Contrasted with optical or projection printing.

See: *Printers; Print geometry.*

Continuity. Correct follow-on of detail and movement between successive shots. Also, administrative member of film unit, almost always a woman. She is required to record the dialogue as spoken in each shot and briefly to describe the actions of the artistes. These notes are typed daily on to continuity sheets, primarily for the use of the film editor. She must memorize and/or record all relevant details of dress and make-up of artistes throughout the shooting of the film so as to be able to describe exactly what the appearance of every artiste should be at any point in the story. She must note the position of furniture and properties on each set up, particularly when, in the course of the artiste's action or to assist camera or sound technicians, anything is moved from its original position. She instructs clappers what number to record on each shot and notes the time each one takes to shoot.

In television, most of these duties are carried out by the production secretary.

⊕ **Continuous Motion Prism.** Device used in many high-speed cameras and some film viewers which, without intermittent movement, renders the image of each frame stationary for a short period, so that there

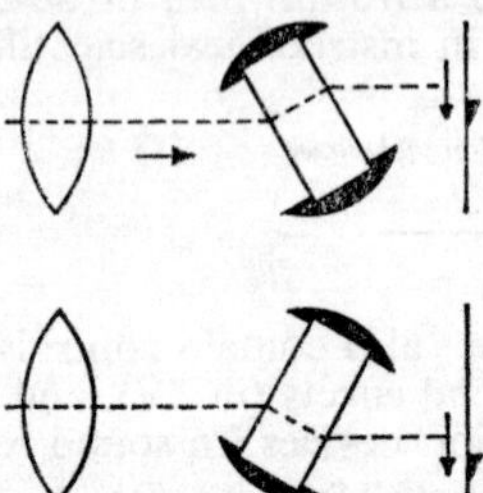

is an intermittent succession of images presented to the eye or the film. The effect is usually achieved by means of a rotating polygonal prism, through which the light rays bend to produce a stationary image.

See: *High speed cinematography; Telecine.*

⊕ **Continuous Printer.** In motion picture work, film printing machine in which the film being copied and the raw stock being exposed are moved continuously and uniformly past an illuminated aperture rather than intermittently frame by frame as in a step printer.

See: *Printers.*

⊕ **Continuous Projection.** Continuous film running with automatic rewind during projection. The film is wound in the same way as if it were on a spool, and is pulled from the centre and fed on to the outside of the coil after projection. The beginning and end are spliced together with the result that the film can be projected continuously time after time. This method is employed when cassettes are used.

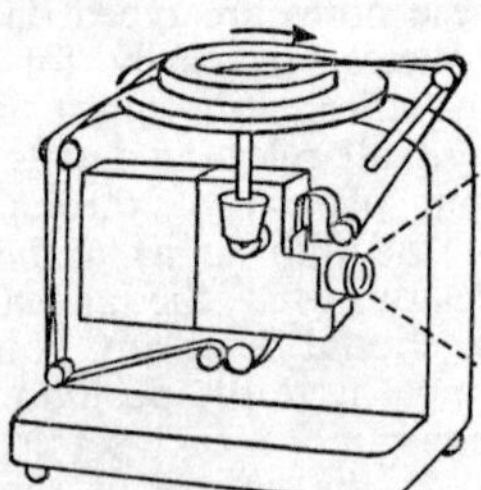

CONTINUOUS PROJECTOR. Film is drawn out of inside of coil and, after passing through mechanism, is rewound on outside of coil.

PROBLEMS. The continuous projection of film raises problems which are not present, or are unimportant, when a film is projected normally, rewound and projected again. As the linear speed of the film must be constant throughout its length, there must be relative movement between each coil and its adjacent coils because the diameter of the coils becomes less towards the centre. If this relative movement is retarded because there is friction between adjacent coils, a very great load builds up and the whole arrangement ceases to function. It is imperative, therefore, for the film to be lubricated on both sides and kept clean and free from loose dirt or dust. A good lubricant is carnauba wax which has a high melting point, but care must be taken to avoid too heavy an application. Usually immersion of the film in a 1 per cent solution, using a suitable solvent, is satisfactory. Even when lubricated, or treated in some other way, a film becomes sticky if it is allowed to get hot. This introduces the problem of cooling and ventilating a cabinet which houses a projector lamp producing a considerable amount of heat. Magnetic striped film is clearly more difficult to handle in a continuous coil because of the extra thickness which is not constant across the width of the film. This factor sometimes results in a tendency for the whole coil to lock solid.

CASSETTE DESIGN. When designing continuous loop attachments or cassettes, it is usual to make the centre as large as possible and so reduce to a minimum the change in diameter from the outside to the centre. Except with short lengths of film, it is usual for the coil to rest on a plate which is free to rotate. Pulling the film from the centre causes the coil and the plate to rotate, and because the outer turn is trying to travel faster than the inner turn, the film is taken up on the outside of the coil. The take-up tension of film is directly attributable to the friction between turns, so, with a short film having fewer turns, the coil must be wound more tightly than for a longer film.

With a carefully designed unit, it is possible to handle up to 400 ft. of 16mm film without driving the plate, but for any greater lengths the plate is driven instead of being rotated by the film being pulled from it. Usually lengths of 16mm film exceeding about 200 ft. are driven, and up to 1,200 ft. can be handled quite satisfactorily. In the case of 8mm film, lengths of up to 400 ft. with magnetic stripe can be loaded into undriven cassettes and function satisfactorily. This is a considerable advantage as 400 ft. of 8mm film provides a running time of 20 minutes at sound speed. H.S.H.

See: *Cassette loading for projectors; Daylight projection unit.*

⊕**Contrast.** Relation between the light and ☐dark tonal areas of a scene either in actuality or in their reproduction. The contrast in a scene depends both on the reflectivity of the objects in it and the manner in which it is illuminated, while in the reproduction these factors may be considerably modified by the characteristics of the intermediate process, which may be photographic, electronic or both in combination.

In the photographic sense, contrast refers to some of those factors of the sensitive emulsions and their processing which determine the conversion of various intensities of exposing light into corresponding areas of density on the film.

See: *Exposure control; Sensitometry and densitometry; Tone reproduction; Transfer characteristics.*

☐**Contrast Control.** Component of a television receiver or monitor governing the amplitude of the signal applied to the cathode ray tube. It is normally used together with the brightness control to give the required tone to the lighter parts of the picture.

⊕**Contrast (Lighting).** Range of lighting in-☐tensities in a scene, which must be related not only to aesthetic needs, but to the limitations of the medium, whether film or TV, black-and-white or colour. In the studio, the lighting contrast of a scene is principally established by the ratio of key light (from the main lighting source) to filler light (building up shadow illumination). A ratio of 4 : 1 is often quoted as a safe maximum for colour film, but this can be greatly exceeded if unusual effects are required.

See: *Exposure control.*

⊕**Control Band.** Strip used to control the intensity and colour of light for each scene when printing an assembled negative on a motion picture printer. A widely used type

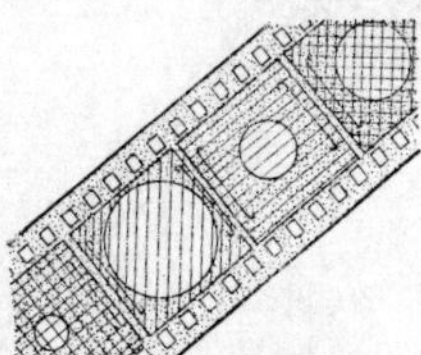

CONTROL BAND. Larger and smaller holes directly control amount of light from printing lamp reaching film. Colour filters are shown stapled over holes.

consists of a black paper strip 35mm wide in which holes of varying diameter are punched for each scene; the diameter of the hole itself directly controls the amount of light from the printing lamp and the strip is moved along one step each time a scene change on the negative passes. Colour filters may be attached to the strip by clips or staples to alter the colour of printing light at the same time.

In another form used on modern printers a control band in the form of a narrow paper strip carries a coded pattern of holes for each scene; these are read by a sensor and converted into electrical signals to alter the voltage of the printing lamp or to operate light valves in the optical system. This indirect system allows very high speeds of operation in comparison with the control band with filters inserted directly in the light beam.

See: *Colour printers (additive); Printers.*

⊕**Control Operator.** Film laboratory technician responsible for routine testing and control of one or more sections of film processing. One group of control operators is concerned with chemical control, the testing of chemical supplies, the analysis of solutions and the maintenance of replenishment rates and formulae to produce consistent results, while another group deals primarily with sensitometric work. This includes the photographic testing and rating of film stocks, the setting of developing times and the testing and correction of printers to provide uniform exposure levels. Another important aspect of control work is determining the required printing conditions and development characteristics in the production of master positives, dupe negatives and colour intermediates.

See: *Laboratory organization.*

⊕**Control Track.** Auxiliary sound track used to manipulate the volume of the main track or tracks, or to bring additional loudspeakers into action. Control tracks are often used in stereophonic sound installations.

☐**Convergence Circuits.** Circuits provided for a multi-gun colour tube to align the electron beams accurately with their corresponding phosphor spots across the complete picture. In this condition the electron beams intersect at a point so that the images corresponding to the three primary colours are correctly superimposed.

See: *Shadow mask tube.*

⊕**Conversion Filter.** Colour filter used on a camera to correct the colour temperature of the exposure light sources to match the characteristics of the film in use. Film balanced for exposure by artificial light sources requires an orange-red filter when used in sunlight or with high-intensity arcs; similarly, daylight film must be used with a blue filter when exposed with tungsten light sources. Allowance must also be made for the filter transmission factor in assessing the exposure required in these circumstances. Also called photometric filters.

See: *Colour temperature.*

⊕**Cookie.** Flag, or adjustably mounted opaque surface, perforated with a pattern of branches, flowers, etc., and set up in the studio to cast a shadow on an otherwise uninterestingly uniform surface. Cookies may also be translucent like scrims.

⊕**Copy Lens.** Lens which forms an image of the negative frame at the projection head on the positive stock in the camera head of an optical printer. Unlike normal camera lenses, copy lenses must be specially designed for short working distances and for more or less equal object and image separations.

⊕**Core.** Centre, usually made of plastic but occasionally of metal, upon which film raw stock is wound. To avoid abrasions caused by irregular winding, negative is invariably stored in rolls on cores instead of being

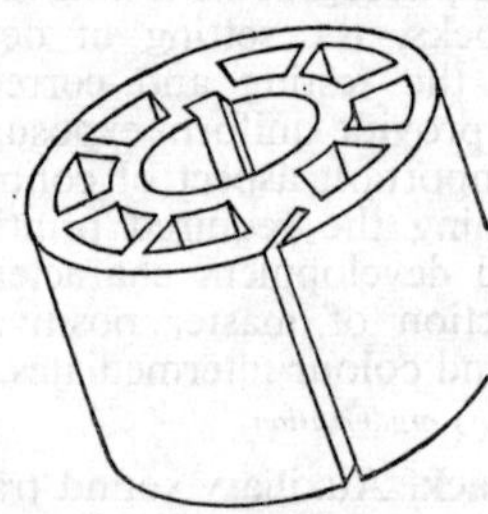

wound on reels. Cores are also used in the cutting room for winding up short lengths of positive film.

□**Corner Insert.** Picture inset in the corner of the main picture, formed by arresting at the appropriate moment simultaneous horizontal and vertical wipes. Thus, a vertical wipe from the bottom arrested at two-thirds the height of the picture and a horizontal wipe from the left arrested at two-thirds the width of the picture would leave a picture in the top right-hand corner occupying one-ninth of the image area.

See: *Special effects* (TV).

⊕**Coupler.** In colour film processing, a chemical used to produce a coloured dye image by combination with the oxidized developer by-products formed in the vicinity of the developed silver image. Couplers may form part of the developer solution, in which case each of the three emulsions of an integral tripack film must be developed separately with the required developer/coupler, but it is more usual for different couplers to be incorporated in the three emulsion layers, so that one developer solution can be used to form all three colour images at the same time.

Couplers incorporated in the emulsion may be used as colloidal dispersions or in combination with synthetic resins or as minute droplets of an oily nature to reduce any tendency to move within the gelatin emulsion layers during processing. Couplers which are themselves coloured may be employed so that the unchanged coupler in one layer can provide a masking image to help compensate for the unwanted absorption of the colour image produced.

See: *Chemistry of the photographic process; Colour cinematography; Integral tripack.*

⊕**CP Lamp.** Colour photography tungsten □lamp of 3400°K colour temperature balanced to Kodachrome film, but also suitable for exposing Eastman Colour, Ektachrome and other commonly used colour film emulsions.

⊕**Crab Dolly.** Type of movable camera plat-□form with all wheels steerable, so that it can move angularly or crabwise, as well as straight ahead and through the arc of a circle. Invaluable for shooting in confined spaces.

⊕**Crane.** Large camera trolley carrying a □boom, on the end of which is mounted the camera, usually with accommodation for one or more camera operators. The boom

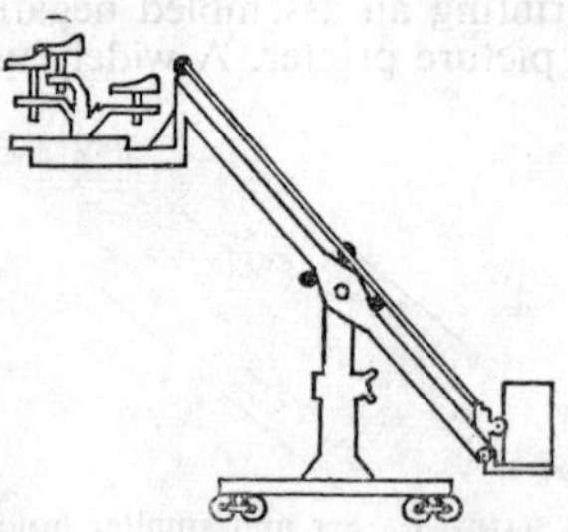

can be raised or lowered during operation, either hydraulically or by manual controls. Camera cranes often allow for an elevation of 20 ft. or more above ground-level, and mobility is achieved by means of powered wheels on the trolley.

□**Crane Truck.** A power-driven mount for the television camera. The term crane truck has become a generic one to describe any form of camera pedestal which is tracked by a person other than the cameraman. However, it is more properly used to describe a hand-propelled, four-wheeled truck incorporating a jib, also hand-operated, upon which is mounted the camera platform and cameraman's seat. The jib itself can also be rotated independently by hand wheel.

The truck is normally operated by two men other than the camera operator, one to propel the truck and the other to control rotation and elevation of the jib. The seat can be elevated and made to follow panning movements by the cameraman. Steering is controlled by a hand wheel, which operates the rear wheels. Cable guards are provided, and can be adjusted in height to prevent cables from being trapped beneath the wheels.

This type of unit has been largely replaced by motorized camera dollies, but has some limited applications where arrangements for powering a motorized truck are not easily made, and occasionally for outside broadcast use, where more facilities than are provided by lightweight dollies are required.

⊕□**Crawling Title.** Film or TV title which rolls upwards across the screen at a slow, steady speed. Sometimes used for long lists of names or explanatory introductions. Also called a roller title or creeping title.

⊕□**Credit Title.** Title which lists those creatively responsible for the film or TV programme to which it is attached. Credit titles may be placed at the start or finish, and when at the start are now often preceded by an introductory sequence.

□**Crispening.** Process of electrically sharpening the edges of a television image. The original picture is differentiated to give the rate of change of density moment by moment, horizontally. The resulting image has lines at all previous black-and-white boundaries. This differential signal is now added back to the original image with the result that the increased emphasis of the lines sharpens the boundaries of the picture.

Crispening in a vertical direction is more difficult since it involves storing a line of the image in order to compare it with the next one. Where necessary, this can be accomplished by a delay line.

Image orthicon camera tubes produce a crispening effect without external circuitry but crispening is often applied to Plumbicon tubes both in black-and-white and colour and to certain standards-converted pictures.

See: *Aperture correction; Camera* (*colour*); *Camera* (TV).

□**Crookes, William, 1832–1919.** British chemist and physicist, born in London. In 1878 produced the earliest form of cathode ray tube. This had evolved from his study of the creation of a vacuum within a sealed glass vessel by means of electricity. Crookes recognized the presence in his tube of particles of matter which he sensed were different from ordinary molecules. Roentgen, in 1895, enclosed a Crookes tube in a darkened box and found that mysterious rays were emitted which rendered fluorescent materials outside luminous. He called these X-rays. J. J. Thompson confirmed Crookes' theory about electrified particles and defined the nature of the cathode rays as being composed of electrons. Crookes also conducted original investigations into wireless telegraphy and wrote prophetic articles about it which inspired others (e.g., *Some Possibilities in Electricity* in the *Fortnightly Review*, 1892). He was knighted (1897), awarded the Order of Merit (1910) and was President of the Royal Society (1913–15).

⊕□**Cropping.** Restriction of the projected or transmitted image area of a film by the use of masks or apertures smaller than the image area available. An aperture of smaller height than the standard Academy may be used to show a picture of wide-screen aspect ratio. Similarly, cropping of the action at the sides of the picture may take place when a wide-screen print is used for television presentation.

See: *Film dimensions and physical characteristics.*

⊕**Cross Cutting.** Art of intercutting two independent sequences of action so as to reveal their relationship to one another.

⊕□**Cross Fade.** To fade up one scene as another is faded down. The equivalent of a lap dissolve in film editing. May be applied to vision, sound or lighting.

See: *Master control.*

⊕□**Cross Modulation.** Unwanted effect which can occur when two or more signals which are present in an amplifier modulate one another, giving additional outputs which are the sum and difference between the signals.

In particular, a type of distortion peculiar to variable-area optical film recording. When the modulated light beam from the recording galvanometer is reflected on to the film emulsion, the image of each wave is not confined to the area exactly under the beam. Because of the crystalline structure of the emulsion, the light scatters and the image spreads slightly. This raises the transmission of the negative, and produces lower-frequency components corresponding to the envelopes of groups of waves. The result is a very unpleasant type of distortion especially noticeable on sibilants, which sound as if two pieces of sandpaper were being scraped together.

See: *Intermodulation; Sound mixer; Sound recording.*

⊕**Cross-Modulation Test.** Method for determining the optimum printing characteristics for variable width sound track negative films; cross-modulation is a measure of the harmonic distortion introduced during processing dependent on the negative characteristics, the condition of printing and the characteristics of the positive stock used and its own processing. From any given negative and printing system there is only one optimum print density which shows a minimum of sound distortion, which must therefore be set as the print standard.

Basically, the method consists of making a series of prints at various densities from a representative test negative on which a high frequency signal, say, 6000 Hz, has been recorded alone and also modulated by a lower frequency of, say, 400 Hz. When the average transmission of each of the series is assessed on a reproducer, the print which shows the minimum content of the lower frequency is that in which minimum distortion occurs and is therefore the optimum density print.

See: *Cross-modulation; Distortion; Sound recording.*

⊕□**Cross-Over Frequency.** Almost all professional loudspeaker installations employ two or more speakers (or sets of speakers) to cover the audio band. Thus the low, middle and high frequencies may be assigned to separate speakers. The cross-over frequency is the central frequency at which the amplifier output is transferred from one speaker to the next, but there is a band of frequencies above and below the cross-over to which both speakers contribute an audio output.

See: *Sound reproduction in the cinema.*

⊕□**Crosstalk.** Signal from one channel detected in another as unwanted interference owing to inadequate isolation of the channels. If the channels in question are both sound channels, the effect is described as crosstalk, and if they are both vision channels it is referred to as crossview. In the latter case, if the interfering signal happens to be synchronous with the wanted signal, it can appear on the screen as a ghost image.

Crosstalk can also occur between a vision and a sound channel as is not uncommon in certain television receivers. Sound breakthrough into vision can be seen as a patterning on the screen; vision breakthrough into sound is usually audible as a background buzz.

The separation of two channels is usually specified by stating that a fully modulated signal on one channel produces a signal of $-x$ db on the other channel.

See: *Sound recording; Transmission networks; Vision mixer.*

□**Crossview.** Reception of an unwanted picture on a vision circuit analogous to crosstalk on sound. Usually present to some extent on switchers where a number of circuits occupy a small volume. Depending on the system, crossview becomes visible at about 30 dB difference in level between the wanted and unwanted signals.

See: *Transmission networks; Vision mixer.*

□**Crush.** Tendency to reduce excessively the tonal range of some part of the picture signal, e.g., white crush, black crush.

□**Crystal.** Generic term for devices based on piezo-electricity.

□**Crystal Control.** Technique of generating various electrical frequencies from one stable piezo-electric crystal source. The timing of the scanning functions in a television system may be derived from a crystal timing oscillator, whose inherent stability results in very stable scanning. This is vital if the television signal is recorded on tape recorders with poor servo stability, or if it is a colour signal which has to be encoded with a subcarrier which is frequency-related to the line scan rate.

A disadvantage of crystal control is that the picture frequency is no longer locked to the a.c. power line. Display devices with poor power supply filtering may then exhibit a moving hum bar. Improvement in receiver design, however, is such that crystal control is becoming general practice.

See: *Synchronizing pulse generators; Videotape recording (colour).*

□**Crystal Lock.** Operation of a television synchronizing pulse system from a stable crystal frequency reference rather than from mains frequency reference.

□**Crystal Mixer.** Superheterodyne mixer commonly used in microwave receivers, employing the non-linear properties of a crystal diode to improve the efficiency of frequency changing (mixing) as compared with a valve or most transistors.

See: *Microwave links.*

☐**Cue.** Signal to take some action, normally pre-arranged and anticipated. In television, the cue can take the form of a hand signal from the floor manager, a light signal, a spoken cue over the talkback system, a cue dot superimposed on the picture or a telephone call.

☐**Cue Card Holder.** Simple mechanical device attached to the camera for holding cue cards. The cue cards are an aid to the

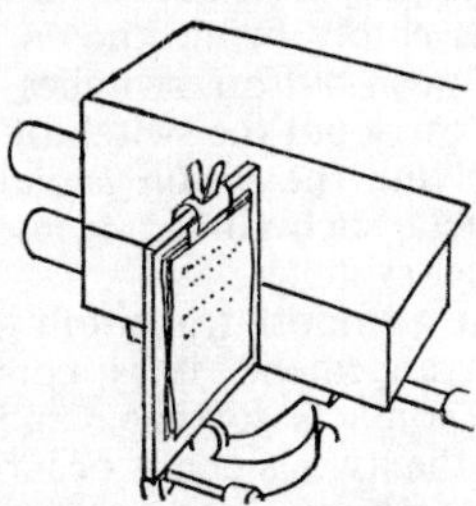

camera operator and contain information about the various shots he will be taking during a programme.

☐**Cue Circuit.** Auxiliary circuit provided on a vision mixer, switching system, microphone fader or the like to give a warning that a particular channel is selected. The term can also be applied to a telephone circuit over which cues can be given from a control point – for example to an outside broadcast (field pick-up) unit to warn it when to come on the air. A warning light is not always a cue light – an announcer may have a red light to warn when his microphone is live, and a green light to cue him when to speak.

In American terminology, a cue light is called a tally light.

See: *Master control; Vision mixer.*

☐**Cueing Device.** Method of presenting the script for an artist to read while appearing before the cameras. The cueing or prompting device may be attached to the camera itself and generally comprises a paper roll with the script typed upon it in large letters carried past a viewing area at a rate controlled by an operator.

⊕☐**Cue Marks.** Small marks appearing in the corner of a picture image to indicate the approach of the end of a programme section. In motion picture practice, cue marks are printed in the upper right-hand corner of four frames at two specified

positions near the end of the reel to warn the projectionist when to start the motor of the second projector and when to change over to that machine. In British and American usage, both these marks are in the form of circular dots, but in some European countries the change-over cue mark may be square.

The exact location of the marked frames has differed slightly in various countries but there is now fairly general agreement to provide a spacing of 168 frames, equivalent to 7 seconds, between the two sets and to put the change-over cue 24 frames (1 second) from the end of the picture image.

In television, cue marks are electronically keyed into the picture to provide similar warning signals and these usually take the form of a small square of black and white vertical bars generated by an unlocked oscillator. In British commercial television practice, the cue mark is displayed as a stand-by signal one minute before the end of the sequence and is removed exactly five seconds from the end. Five seconds is the standard run-up time for telecine channels and therefore the disappearance of the cue mark is the "roll telecine" signal. Television cue marks in other parts of the picture may also be used to convey special information, for example, emergency recall.

See: *Master control; Release print make-up.*

⊕☐**Curvature of Field.** Lens defect in which the sharpest image points do not lie in a flat plane but on a curved surface; usually becomes greater as the distance from the axis increases. Well-designed lenses can reduce this defect over the field required.

See: *Aberration.*

⊕☐**Cut.** In film, the instantaneous transition from one shot to the immediately following one which results from splicing the two shots together. In TV, the same effect is achieved by instantaneous transfer from one elec-

tronic picture signal to another. The alternatives to cuts are mixes or dissolves, and wipes.
See: *Editor; Master control.*

⊕**Cut Back.** Cut which brings back a shot of an actor or scene already established.

□**Cut Bank.** Piece of vision switching equipment used by a producer in a studio. It consists of a series of push-buttons for switching instantaneously from one source to another.
See: *Vision mixer.*

□**Cut in Blanking.** An instantaneous transition between two vision signals which occurs during the field suppression period so that no disturbance to picture information occurs during the visible lines.
See: *Vision mixer.*

⊕□**Cut-Off Frequency.** Frequency at which the response of an electronic system starts to fall away sharply, usually specified in terms of the decibel drop over a stated frequency range.

⊕**Cutout.** Small cutout figures, usually jointed, sometimes used in making animated films. By means of invisibly marked calibrations,

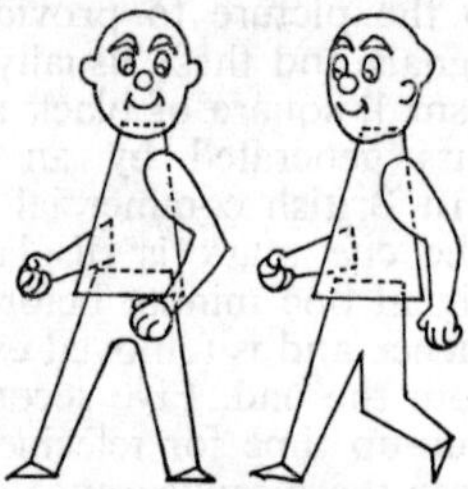

the cutouts are moved into successive positions, so that, when photographed a frame at a time, they give the illusion on the screen of continuous movement.
See: *Animation techniques.*

⊕**Cutter.** Film studio technician responsible for the editing, cutting, synchronization and assembly of rush prints into complete sequences in accordance with the director's intentions, and eventually the completion of the final cutting copies of the production. In laboratory work, the cutter is the operator who carries out the identification, matching and cutting of the original negative of picture and sound to correspond with the cutting copies supplied by the studio.
See: *Editor.*

⊕□**Cutting.** Synonymous with editing. Denotes the process whereby the raw material of a film or videotape recording is assembled into a coherent and compelling whole. The shaping process goes through a number of well-defined stages, from an assembly to a rough cut and thence to a fine cut. The sound cutter's task, often equally creative, is to build up a set of dubbing tracks which complement the effect of the picture.
See: *Editor.*

□**Cutting.** Effecting a transition between two signals by switching rapidly between them. To provide the facility of cutting between any of the inputs, it is common to allocate a push-button switch to each and to arrange the number of required buttons in a row or bank, the assembly being known as a cut bank. The push-button switches may be arranged to carry out the switching directly, i.e., by applying the vision signals to the actual contacts, such an arrangement being called a direct switcher.

To achieve a smooth transition with such a simple arrangement, it is common to arrange the contacts so that a momentary overlap of the two signals occurs at the instant of switching. Such an arrangement is called a lapping switch, while the inverse, an arrangement causing a momentary absence of signal during switching, is called a gapping switch.

A more common application of the cut bank is to control a switch matrix placed some distance away. This permits more flexible and economic signal routing and also provides the possibility of a faster transition during cutting. In this respect, if the transition is timed to occur during the field-suppression period and is fast enough to occur wholly within this period, then its smoothness and adequacy is maximum. Such an action is called an interfield cut or cut in blanking.
See: *Master control; Vision mixer.*

⊕**Cutting Copy.** Film editor's working reels of film, assembled from the original material and cut to a greater or lesser degree of fineness. The cutting copy is spliced together, and may contain build-up. It may also be marked up for optical effects, or for negative cutting. Also called a workprint.
See: *Editor.*

⊕□**Cyclorama.** Sky cloth; generally a large cloth stretched smoothly round the studio and forming, when suitably lit, a background for scenery seen against the sky. Also known as horizon or panorama cloth.
See: *Art director; Designer.*

⊕**Dailies.** First positive prints from the laboratory of the previous day's shooting. Also called rushes or rush prints.

See: *Laboratory organization.*

□**Dark Current.** Current in a photoelectric or photoconductive device when the device is supplied with its normal operating voltages and there is no light on the target.

See: *Camera* (TV).

⊕**Dark End.** All those parts of a film developing machine which, if the film path is exposed, must be kept in darkness or a very subdued light. In modern all-enclosed developing machines, dark and light ends have less meaning, but film must still be loaded on them in darkroom conditions.

See: *Processing.*

⊕**Daylight Loading Spool.** Spool with solid opaque flanges which fit tightly against the edges of the film wound on it; unexposed raw stock for use in the camera may therefore be safely handled in daylight while wound on such a spool since any fogging which results is limited to the outer layers only.

Daylight loading spools are very widely used in 16mm practice and occasionally for short lengths in 35mm work.

See: *Camera* (FILM); *Narrow gauge film and equipment; Narrow gauge production techniques.*

⊕**Daylight Projection Unit.** When film is projected in undarkened surroundings, it was until recently essential to use rear projection methods. An observer sees a picture on a front projection screen because it reflects the light thrown on to it by the projector. In daylight conditions, the light from the projector is reflected in exactly the same way as if the room were darkened, but the daylight is also reflected and, as it is stronger, the projected picture is hardly visible.

To avoid competition between the projected light and daylight, a translucent screen is still normally employed which allows light to pass through it but reflects very little. If the projector is in a cabinet and the only light passing through the screen is from the projector, a picture can be seen in daylight. The daylight passes through the screen in the opposite direction, and if the inside of the cabinet is painted flat black, the daylight is absorbed and not reflected back again through the screen.

However, new developments in screens of very high-gain reflectivity are now making front projection possible in lighted surroundings.

VIEWING ANGLE. Screens vary considerably, particularly as to the angle through which the picture is visible. Generally the finer the matting of the screen the narrower the angle, but the angle of the projection lens also has an effect on the viewing angle, i.e., a wide-angle lens contributes towards a wider viewing angle. As the projector produces only a limited amount of light, it must concentrate its output to compete with daylight. Lenticular screens have been made to

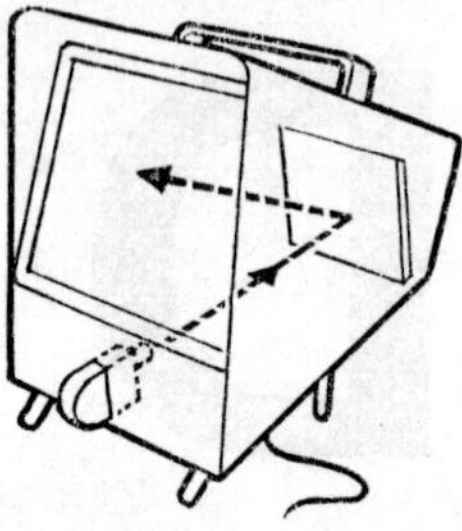

LOOP-CASSETE PROJECTOR. Film will run 4 mins. and is permanently sealed in plastic cassette needing no threading.

concentrate the light within limited viewing angles, both horizontally and vertically, but such screens are expensive and tend to have an obvious grain effect. The human eye automatically adapts itself to varying conditions of brightness, so the eyes of an observer standing in bright surroundings are less sensitive. This results in the same projected picture appearing to be less bright when viewed in daylight than in darkened surroundings. For these reasons, rear projected pictures to be viewed in daylight normally have a viewing angle within 30° on either side and are also fairly small.

Rear projection systems must use at least one mirror to reverse the image; in most cases three mirrors are used to reduce the size of the cabinet to a minimum. Mirrors must be front surfaced to avoid double images and absolutely flat to avoid distortion.

HOT SPOT. An inherent problem is the hot spot which appears in the centre of the picture. If the eye of the viewer, the screen and the lens are in the same plane, or the equivalent when mirrors are used, the viewer may see the lens of the projector in the centre of the screen if he focuses his eyes to the appropriate point behind the screen. The hot spot is the result of the increased amount of light passing straight through the centre of the lens, and its effect can be made less objectionable if the viewer is positioned so that the hot spot is near the bottom of the picture, where it is less noticeable. Other ways of reducing the hot spot are to employ a less translucent screen with consequent loss of brightness, or to reduce the amount of light passing along the axis of the optical system. This latter method also reduces the total amount of light and is difficult to achieve, but is employed when it is imperative to avoid a hot spot.

See: *Cassette loading for projectors; Continuous projection; Rear projection.*

□**D.C. Restoration.** Re-insertion, by clamping or other means, of the mean d.c. level of a television signal to the a.c. signal.

See: *Receiver.*

⊕□**Decibel.** Unit which expresses ratios of powers, voltages and currents. The scale is logarithmic, and in the case of power,

$$dB = 10 \log_{10}\frac{P_2}{P_1}$$

where P_2 and P_1 are the powers to be compared.

For voltages and currents respectively,

$$dB = 20 \log_{10}\frac{V_2}{V_1} \text{ and } 20 \log_{10}\frac{I_2}{I_1}.$$

The convenience of the decibel unit stems from the fact that the human ear responds equally to equal ratios of intensity change, more or less independently of the absolute level; so also does the eye with changes of brightness.

□**Decoder.** Device which separates the colour components from a standard television signal in a colour TV receiver.

See: *Receiver.*

⊕**Definition.** Sharpness of an image formed by an optical system or recorded on film, often measured in terms of its resolution as the number of lines or elements per millimeter which can be differentiated.

See: *Lens performance.*

□**Deflection Coils.** Coils external to the cathode ray tube in television systems with electro-magnetic scanning.

Wound in two distinct sections, one each for line and frame, often on a single former, they deflect the electron beam to scan the screen of the cathode ray tube.

□**Deflection Yoke.** Complete assembly of horizontal and vertical deflection coils in their special mounting.

⊕□**De Forest, Lee, 1873–1961.** American scientist and wireless pioneer, born in Iowa, son of a Congregational minister. At an early age, became interested in the subject of electromagnetic waves. In 1893 entered Sheffield Scientific School, Yale University, graduating in 1896, the year in which

Marconi arrived in England. Joined the Western Electric Company in Chicago in 1899, where he developed a new type of wireless detector, a self-restoring coherer, which he called a responder. In 1902 he founded the De Forest Wireless Company which strongly challenged the American Marconi Company. Two years after the invention of the thermionic diode by Ambrose Fleming in 1904, De Forest produced the audion, a three-electrode vacuum tube which resembled Fleming's diode but had one important difference – a small wire grid between the filament and the anode. The three-electrode tube, or triode, was able not only to rectify, like the diode, but also to act as a relay and amplifier. Its ability to amplify, however, was not fully realized until 1912 when De Forest began using his audions in cascade. By the end of his career De Forest had some 300 patents to his credit. In 1912, with some of his colleagues, he founded the American Institute of Radio Engineers.

Degradation. In photography and television, deterioration of the image from the original scene to its immediate reproduction, or from the latter to some more removed image arrived at by transmission or duplication. The degradation may be in terms of contrast, resolving power, or any other image characteristic, but the term is most frequently applied to the reproduction of colour, when hues are incorrectly rendered as a result of desaturation or by unwanted spectral absorptions.

See: *Image formation; Lens performance.*

Delay Cable. Special coaxial cable designed to considerably delay the passage of signals. Commonly formed of an inner coil wound over a magnetic powder core in the form of a flexible rod. Various delay times are available, and may reach tens of microseconds per foot. Sometimes used to obtain the necessary time constants in sync generators.

See: *Station timing.*

Delay Line. Transmission cable or equivalent path built up of elements introducing a delay in the passage of a signal. Used in colour receivers such as SECAM or PAL, also in vertical aperture correctors.

See: PAL *colour system;* SECAM *colour system; Standards conversion.*

Delrama. Trade name of an anamorphic lens system for both cameras and projectors which made use of curved reflecting surfaces to obtain the required distortion rather than cylindrical lenses or prisms.

See: *Wide screen processes.*

Demeny, Georges, 1850–1917. French pioneer of cinematography. Through an interest in physical education, joined the research establishment of Professor Marey in 1882, helping him to develop the chronophotographic apparatus used to record movement. Became interested in the study of lip movements in speech, and in 1891 developed the Phonoscope apparatus, in which pictures of lip movements, recorded by the chronophotographic apparatus of Marey, could be viewed in motion by a modification of the Phenakistiscope principle. The apparatus was patented in 1892, in which year he broke with Marey over its commercial exploitation. In 1893 he patented a camera design derived from Marey's latest apparatus, and in 1894 developed an improved model using a beater intermittent movement which he patented in that year. This principle formed the basis of the Chronophotographe camera-projector, using 60mm perforated film, constructed and marketed in collaboration with Leon Gaumont in 1896. A 35mm version was subsequently developed.

Densitometry. Measurement of the density of photographic images, used for the control of camera exposure, printing and processing operations.

Density. Factor which indicates the extent to which light is stopped by any portion of the image. If the intensity of light falling on a part of the image is I_0 and the intensity after passing through the image is I_1, the ratio I_0/I_1 is called the opacity, and the logarithm of this is known as the density, D.

Thus $D = \log I_0/I_1$.

The inverse ratio I_1/I_0, which is always less than 1, represents the transmission of

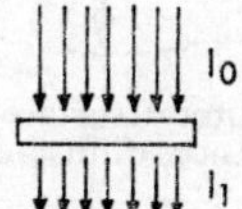

D	3·0	2·4	1·8	1·2	0·6	0·0	D
log I_1	0·0	0·6	1·2	1·8	2·4	3·0	
I_1	1	4	16	63	250	1000	

DENSITY. If a beam of intensity I_0 passes through an absorbing medium and is reduced to I_1, the opacity is I_0/I_1 and the density is $\log I_0/I_1$. For a value of I_1=1,000 a range of 1,000 to 1 for I_0 corresponds to densities from 0 to 3·0.

the image, but is usually multiplied by 100 and stated as a percentage transmission. Being stated on a logarithmic scale, a density of 0·30 represents a transmission of half or 50 per cent of the incident light, while a density of 1·00 represents a transmission of only one-tenth, or 10 per cent. The density of the maximum black area in a positive print of a motion picture film may exceed 3·50.

In colour photography, the same terminology applies, but density values are determined for particular spectral wavelength bands, usually three groups representing the red, green and blue regions.

Density is measured by the use of a densitometer, which compares the intensity of the light falling on the sample with that passing through it or which matches the transmitted intensity with that passing through a standard of previously determined density.

The value obtained depends on the form of illumination to which the sample is presented.

The density of a transparent photographic image may be measured by either specular or diffuse transmission and the values obtained may differ as a result of light scattered by the grains forming the image.

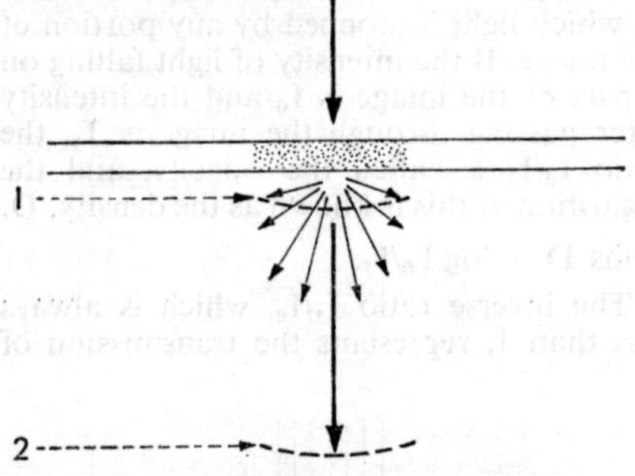

LIGHT SCATTER. Photocell surface catches more light at 1 than at 2. Hence, diffuse density lower than specular density.

Incident light is scattered by the grains forming the photographic image: the apparent transmission is therefore greater if measured by a photocell surface directly beneath the film which collects all the scattered light than from a surface farther away which only receives the directly specular transmission.

The diffuse density is therefore lower than the specular density. L. B. H.

See: *Exposure control; Sensitometry and densitometry; Sensitometry of colour films; Tone reproduction.*

⊕□**Depth of Field.** Permissible range of distances of the object plane of an optical

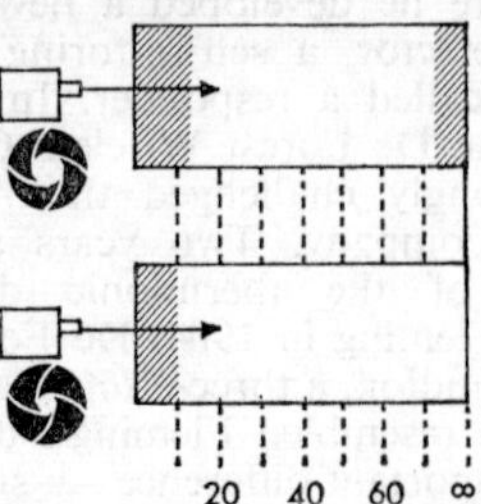

system, such that the circle of light resulting from a point object and produced at the image plane is still acceptable as an image of the point object.

Depth of field depends on the focal length and aperture of the lens and on the distance at which it is focused.

See: *Optical principles.*

⊕□**Depth of Focus.** In an optical system, the permissible range of distances of the image plane for a stationary object plane, over

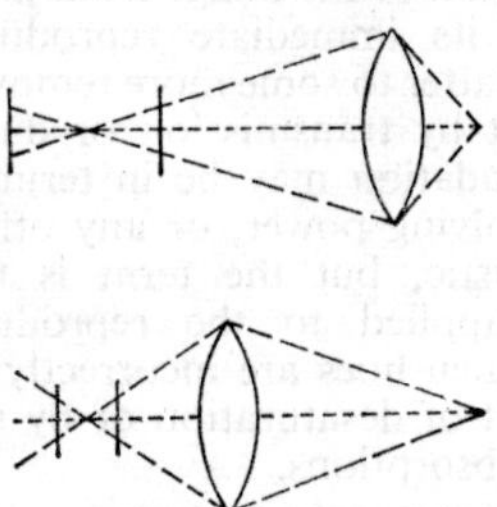

which the circle of light resulting from a point object is still acceptable as an image of the point object.

See: *Optical principles.*

⊕□**Desaturation.** Reduction of colour saturation often as a result of distortion or defect in the film or TV transmission system.

⊕**Desaturation Printing.** Style of colour reproduction less vivid than that normally obtained by standard processing. Desaturation techniques are used in which each of the colour component images is dulled by the addition of a proportion of the other components, giving the effect of grey added to it.

The most usual method is to prepare a dupe colour negative from separation masters printed from the original negative through wide-cut filters.

See: *Colour printers (additive).*

☐DESIGNER

The designer is concerned with the whole complex operation of bringing television pictures to the screen. His function is not solely to provide scenery for actors to move against or for the cameras to shoot. Nor is his work an isolated contribution to the production, beginning and ending with the provision of scenery.

The point of a production is its content. This is based on a writer's or producer's ideas, realized in pictures through a combination of speech, music and movement. With this conception the focal point becomes moving people.

FUNCTION

The scenery, therefore, though the designer's primary concern, is not his only responsibility. He is involved in other aspects of the production, about which he should have informed views. Since these are technical – for the main part lighting, camera movement, sound – each contributing to the process which interprets the script, the designer is a member of a team. This presupposes a basic knowledge of the whole of the team's work. His main function is not only creative on a pure design level but also that of a highly skilled technician. Except for the director (or producer) he can influence, more than any, the whole visual impact and point of the production.

But in spite of the need for technical knowledge, and the fact that a designer is a member of a team and cannot work in isolation, he must remain an individual. Amid the pressures and sheer complexity of television methods, it is easy for him to lose his individuality and for others to compel him to do so.

The artist working on his own is answerable only to himself. He can make what statement he wishes; he can change his mind: the result of his work is his statement only. The television designer's work is to interpret and support in moving pictures the needs of a script, but in doing this he should have his own point of view. His talent and training equip him to speak in more interesting visual terms than the writer or director. He must underline, develop, and extend their original concept, not just design to another's preconceived idea. Yet in establishing from the outset his own visual style for the whole production, he must design with the script and not against it; indulgence of his personal preferences for their own sake undermines his authority.

Thus his function is not confined to producing static backgrounds and pure visual style. He is involved in the whole practical and technical course of the production. His scenery not only backs the actors, it is an integral part of the camera movement. How and where the actors and cameras move depends on the designer.

THE PRODUCTION TEAM

The television production team consists of writer, director, designer and lighting director, senior cameraman, sound balancer and artistes. These are the creative people. When all is achieved in the studio, their work is in the hands of the engineers who can make or mar what the creative people produce visually.

The designer must be broadly conversant with the technical methods and problems facing each person in the team. He unavoidably influences their work and its quality; conversely, without them his contribution just does not exist. On average, the working time of a team on any single complex major production is rarely more than one month and is often less. Therefore, close co-operation and a conviction in ideas, combined with speed in decision, is essential. This does not happen without technical efficiency.

LIGHTING DIRECTOR. One aspect of the team's work which is particularly vital to the designer is lighting. Its importance cannot be too strongly emphasized. The whole purpose of the medium, moving pictures, hinges on this element, electronically and creatively. In his own field the lighting director is also a designer: he does not merely illuminate a scene. Whereas this is obvious to those who work in the film industry, it is not always clear in television.

The lighting director's problems are increased by the fact that the camera tube can reproduce only a limited tonal range. He has to ensure, too, that the lighting is suitable as cameras move from long shot to close-up or as the direction of the shot and/or the lens or even the camera is changed. Moreover, shooting is continuous on anything from one to four cameras on any one set and may be directed from any angle. Over all this is the sound boom. To get good pictures in these circumstances is a tall order.

Whereas, scenically, the designer leads the way in establishing the shape, width, depth and height of the scene, his creative statement must be accompanied by a certainty that the scene can be well-lit. The movement within the set in relation to furniture, how far the actors will be or should be from the background, where the boom must be, the direction of the key light, the angle of light on the actors, provision for good back-lighting, and camera space, are some of the physical aspects he must consider when setting his scale and proportion.

In demanding his own conception he must be technically correct. If in doing so he makes things impossible for the lighting director, he must alter his concept; there is no point in demanding something which will eventually fail through no fault but his own. Each must guide the other, building, extending, learning to take successful risks, playing safe as little as possible, forcing their other technical colleagues to do things they may initially disapprove of – but knowing when to step back. Allowing for frequent sub-standard home reception, they must make it possible in each of their contributions to get sparkle and brilliance into their pictures whether they be dark or light. It is not always a poor home receiver which is responsible for a dull, flat picture, but incorrect tones in settings and costumes in relation to the actor's face which, in turn, compels poor lighting.

All settings are produced in colour, as are the costumes. This is entirely for the benefit of the actors. The designer must therefore have an exact knowledge of the equivalent grey scale. This is comparatively easy where scene painting is concerned; the colours available are normally limited in number and hue. Moreover, they have been tested in front of cameras and are set out side by side on a chart with the equivalent grey scale and its tonal number. The basic tones are arbitrarily limited to seven between black and white.

The colour, and thus the tone, is selected by the designer with the lighting director's agreement. This choice is governed entirely by the face tone in relation to its physical separation from the background along with the general key and level of the lighting. The average background is tone 3 or 4. This does not prevent scenes being lighter or darker, but if they are, the designer must know with absolute sureness what he is doing and why.

Lighting is also affected by furniture, pictures, ornaments and curtains. The designer needs a free hand, but in enjoying this he must by instinct backed by knowledge introduce only those things which are tonally acceptable and undistracting behind the face. He must avoid materials with a pronounced sheen, large areas of glass, or highly polished furniture which cannot, if necessary, be dulled down. While observing these general rules, however, he must not avoid sparkle in favour of safety, even though he has to test his colleagues to exasperation. If he fails he must admit it and try again. Only through such trials can exciting things happen. But there is a very real problem in controlling colour and equivalent tone while providing stylistically correct dressing.

It is an even greater problem with wallpapers. The majority of those available are quite unsuitable; if they are not white then they are tone 1 or 2; or if they are tonally correct the colour is too saturated. It could be generally stated that where paper and paint are concerned, the average background tone is lower than films and depressing compared with everyday life. There is, therefore, all the more reason why a designer should not be influenced by the studio look but by what appears on a monitor screen.

COSTUME DESIGNER. The costume designer's job is equally exacting and often far more difficult. His work being mainly a matter of hiring existing costumes, he can have less choice or control and little chance of changing things in the studio. He must have a knowledge of colour and know the equivalent grey scale over the range of hues, what light does to various materials, and what the camera tube does to certain colours (especially saturated ones). All this must be known in relation to the actor's face. Although he is naturally in close liaison with the scenery designer, he needs to be in constant touch with the lighting director, too. The fundamental fact remains that there are the camera, the actor's face, the costume, the background – in that order. Face, costume and background have to be lit to harmonize colours and forms into controlled tones. There will be no successful solution unless each, though working individually, works in concert adjusting his particular statement in favour of the whole.

GRAPHIC DESIGNER. Graphic design is a comparatively new term in television; ten years ago people talked of captions, i.e., titles. These were rarely more than simple lettering on a light or dark screen, on some occasions superimposed over film or a still photograph. They are still in use and they often provide the most suitable solution.

Here the main advance is in the extended use of high-quality typography, suited to the proportions of the screen and more aptly pointing the style and mood of the production. Company or house styles have come into existence. The titles have become more of an integral part of the programme, as important as the scenery and costumes, to which they should be directly related. They set the style of the programme, sell it to the public at the beginning and close it in harmony at the finish. Furthermore, graphics have increasingly become part of the actual content of programmes. A greater use of film animation, photographic processes and illustrative art-work in titles, combined with high-grade typography, has long since closed the era of signwriting, bringing to television a new specialist, the graphic designer.

The complexity of his job is less onerous than that of the other two designers in that he is less involved with the production team. Working primarily on his own, in two dimensions, he is confined by the discipline evident in all good television design – simplicity and clarity, the tonal restrictions imposed by camera tubes, and the inferior picture reception on many home receivers.

DESIGN METHODS

PRELIMINARY DISCUSSIONS. The director and designer between them are responsible for presenting the pictures imagined by the author. The designer can rarely work with the author while he is writing the script so there is always the possibility of a lack of sympathy between them.

Some scripts contain detailed explanations of the writer's purpose. Directors can be equally specific in stating their practical requirements, but they should give a good designer a free hand in the early stages, knowing that his inherent talent is likely to provide better visual ideas. The designer and director will discuss the purpose of the script and its style, the environment, atmosphere, scale and character of the play, and perhaps some fundamental practical issues affecting the author's intention. The director may at this juncture indicate how he might like to conceive the camera work, stating the type of equipment which would be necessary. The budget must be assessed, and certain artistes may be discussed with reference to their personal and physical attributes.

In this way the author's known purpose and the director's broad intentions become clear. The designer, having secured in his own mind a visual picture of each scene, can put forward his views, accompanied by quick rough sketches. From this first discussion it may be found that the author has been too ambitious. Practical and economic reasons may compel cuts in the number of sets, or rewriting may be necessary to assist the studio operation. This can be initiated by the designer as well as the director, but it is the designer who has to make it all work. A similar discussion takes place with the costume designer, centring largely on character and the actors it is hoped to engage.

STUDIO PLAN. The economics of television in time and money dictate a way of working foreign to films and theatre; sketches normally come later, not at the outset. Therefore, this first discussion leads at once to considering the physical and technical layout of the studio plan. In all the stages through which he goes from the first rough layout to the first preliminary plan, the designer carries in his mind a picture of each scene.

This process begins with mapping out diagrammatically on a studio plan the approximate area and juxtaposition of each set, based on a shooting sequence ordered by the script, with a plan layout indicated by the director. This rough suggests camera shots, placing of furniture, and actors' movements. Extending this process through further discussions, with constant reference to the script, and an initial talk with the lighting director, the designer reaches agreement with the director. It is then clear that the general technical scheme works and he starts designing, but still in plan form.

In exact ¼-in. scale, on tracing paper, he draws the final shape and size of each set, paying attention to all that has been said about lighting, and movement of actors in relation to sound, establishing his pure design proportions at the same time. Most important of all, he draws on the plan of each set all the basic camera shots which will be used and the camera itself, as well as the boom position in relation to lighting. He may go further and suggest extended or alternative camera shots and draw these on, even to the extent of placing actors. Most certainly he will place essential set-dressing. In doing this, he is not only making it practical for the director to take each possible basic shot, establishing where the actors could move or sit, but he is sure of what will be seen in the background and is also proving that there is room for the technical equipment to operate.

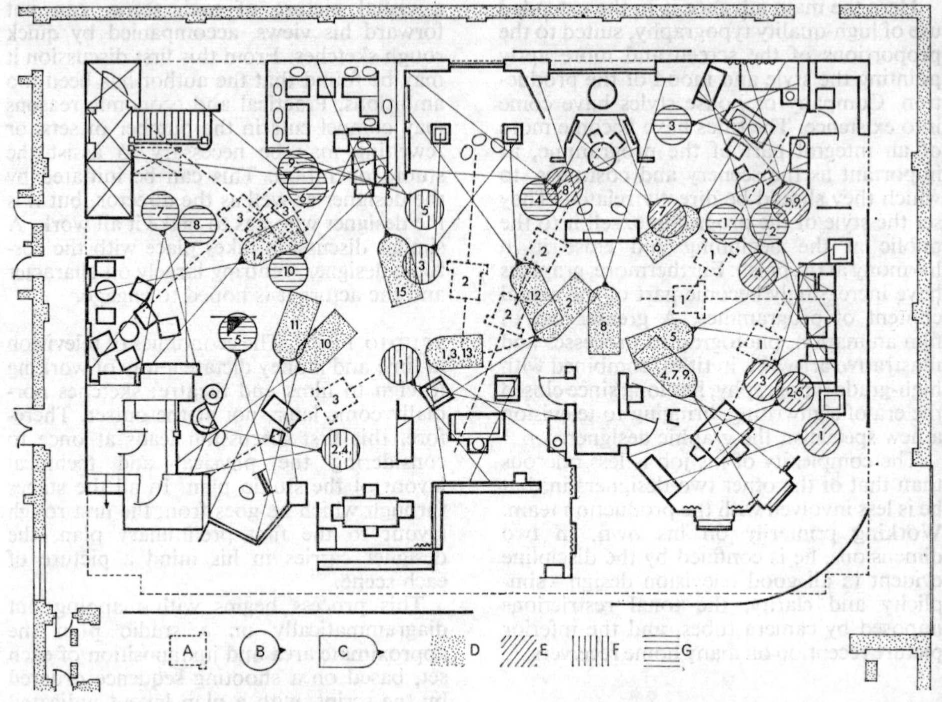

STUDIO LAYOUT FOR TYPICAL PRODUCTION WITH CAMERA AND BOOM MOVEMENTS. A. Tracking and crabbing microphone boom. B. Pedestal camera mount. C. Tracking and crabbing camera crane. D, E, F, G. Cameras, with successive positions of each shown by serial numbering, rays in front of camera denoting angle of field. Camera movements can thus be traced in relation to sets. Microphone booms, which can be extended over the set, swung and pulled back as required, need less movement of base than cameras.

The plans on tracing paper are placed over a studio layout. By manipulation of position, adjustment of angle, mirroring, cutting down or extending, the scenery for a multi-set show of up to twelve or more sets can be fitted into an area of, say, 90×60 ft. – and look good. Most important, it can be seen whether there is room for the equipment to move from one set to another in continuous transmission. This becomes the preliminary plan.

Further discussions with the whole team decide what other adjustments are necessary to lead to final agreement. There is no picture yet except in the designer's mind.

From this point, the designer on his own (but in constant touch with lighting) completes the final stages in planning and design with elevations, constructional drawings, consideration of painting requirements, properties and general set-dressing, with sketches and models where possible – in fact everything leading to the studio set-up and the time when camera rehearsal starts.

COSTUMES. The costume designer is faced with similar and exacting problems. Time and money are not on his side, preventing in most instances the designing and making of clothes. There are, however, sources of hire for any kind of costume from early periods to a limited range of present-day haute couture. Although costumes are designed specially when circumstances demand and permit it, his contribution is usually made only by selection. Using historical knowledge with his ability as a designer to conceive a picture of the whole production, he visits the many costumiers who are in business specifically for the entertainment world. He also buys direct from couturiers or large stores. He must not only possess and use this technical knowledge, he must have a deep understanding of an actor's technique. He must be a diplomat, flattering or firm. He must work swiftly with an unerring instinct for what is correct, knowing, too, how a hired costume can be altered to fit an actress, and indeed whether it is possible.

Although it is often unwise to generalize, it may be right to say that the first consideration is silhouette, confining contrasting detail to a minimum. Character is obviously a leading element. But in considering this, the actor's own personality as well as the persons he is portraying must be taken into account. Although it is a delicate matter persuading him, and certainly an actress, not to insist on something unsuitable for the part, the designer must consider his or her proper ego so that they can be persuaded to be at ease in what is right. The consultations, the research, the arranging and supervision of fittings, visits to rehearsals, the pure administration, and above all the endless travelling about from one couturier to another, finishing with the days in the studio, all within a space of rarely more than one or two weeks, calls for a special ability and a tough character.

GRAPHICS. Graphics play a significant part in the production. They are mainly concerned with titles. Whether these are still or animated on film (on occasion by hand), typography plays a prominent part. A sound knowledge is necessary of which typefaces are suitable in transmission. The bolder, less delicate ones are the better. The limitations of the medium eliminate some faces – sometimes those most suitable in style. In choosing his lettering, the designer must therefore balance the demands of style with technical limitations.

Titles are prepared by hot press, by hand, or by a photographic process. The latter is a method of photographing typefaces mounted on blocks and stove-enamelled white. Using a small point face, the lettering can be enlarged in the normal way to whatever size is required. It is a quick and extremely efficient method. Furthermore, the typography can be incorporated with art-work or other photographic backgrounds, the tone of the latter being carefully controlled to suit the demands of camera tubes without diminishing the power of the lettering.

Pure illustration and painting frequently form the basis of titles, again part of the programme content. Maps, charts, election results, racing and football scores – there are many other channels for the graphic designer's work using the methods described. Like the two other designers, he consults with the director and refers to the scenery designer to establish his style. The timing of the opening music, the moment when the titles come up, and over what – film, photographs, live scene, etc. – guide him in what he decides in letter selection and spacing, screen space, and tonal background. The limitations of the 4 to 3 aspect ratio of the screen can be an irritant in design layout and proportion. He must come to terms with this.

DESIGN PHILOSOPHY

REALISM. Realism may also be called representationalism, or naturalism – the latter misguidedly. Whichever expression is used, it means in practical terms an impression of reality or a semblance of it.

To illustrate the needs, think of a Victorian room, familiar in endless photographs. To reproduce it in exact architectural size with decorative detail in furnishings, ornaments and pictures, adding the wallpaper, and finally placing actors before it would reproduce life as it was. Place a camera in front of the scene and the result would be confusion.

The equivalent impression can be achieved by sweeping nearly half the content away. The guiding reason for this is clarity. The need for clarity, for practical reasons as well as those of discipline in design, is so that the viewer looks at the actor and not the scenery, though remaining conscious of the latter. By placing in the line of each camera shot a few selected furnishings, set before sparingly decorated and carefully proportioned architectural surfaces, the actor has a chance of surviving. It is not a question of what scenery the camera might see, it is that which it does see, and what it looks like behind a face.

This discipline is necessary however complex or simple the true life scene might be. The process of elimination can be carried to the limit where the scenery becomes absolute in terms of simplified realism, even to the point when it consists only of one or two significant details in the plainest architectural surrounding, and yet appears correct.

Realism can be distorted from actual life in that, while using real symbols, it can underline character rather than mirror a purely documentary view. It can be used still further to a point where fantasy takes over and is yet related to everyday experience. There are many variants and all require a special skill on the part of the designer.

The overriding problem is the size of the screen and the detail it can contain, the tonal performance of the camera tube, and what these limitations do to the focal point of the picture – the actor. The temptations are great: to design for himself and the effect on the studio floor; to produce life as it is or was with a conviction which reassures the actor. But unless on the screen this person stands clear and unconfused with what is set behind him, both will fail.

In all good design a process of elimination is necessary. In television it must be even more drastic. Including everything may well lead to seeing nothing.

STYLIZED AND ABSTRACT DESIGN. Stylization is a derived form of realism; this means purely decorative, painted scenery with a realistic connotation, seen mainly in variety, musicals and ballet, and sometimes opera.

Abstract design has no definite derivation; it is pure invention, original, evocative of mood through shapes or decoration, and is two- or three-dimensional. The spread of colour television should stimulate abstract design.

In both these styles the designer has a real chance to design, to be original, to display his talent as a pure painter. His contribution to the content of the programme, his influence on the style of that content, not only the enhancement of it, becomes more marked. He has greater opportunities to work in space, enlarging as it were the size of the small screen by using scenery up to even 30 ft. in height. In ballet, particularly, and indeed in any form of dancing on television, space, depth in movement and, above all, height are essential. Equally, such opportunities can present themselves in all aspects of variety and musical programmes.

The designer who is fortunate enough to have a creative painter's talent and the opportunity to work on such programmes in these styles needs to be exceptionally skilful technically. The discipline of simplicity and clarity of statement is now accepted, the designer's creative vision is presumed, and he is usually strong enough to persuade the director towards his own conclusions on visual interpretation. Nevertheless, all these are empty achievements if the designer cannot plan the exact camera movement. He must compel a certain basic camera operation. He must not only know what shots need to be taken, but just how this is to be achieved. He must plan it himself and draw it out, certain in his own mind that the lenses and angles he is suggesting are correct for the movement, and the height of the shots is practical and suitable for the artistes. If he is sure of the technicalities, he will know how foreground movement will relate to background scene and just what will be seen and how. He can then fulfil his prime function, that of the artist or painter, and be sure that the forms, proportions, scale and general picture composition he has in mind will actually appear on the screen.

These two styles are especially difficult for the television designer. Often designers are not the best technicians; it is perhaps unreasonable to expect them to be so. It is akin to expecting a pure abstract painter of talent to be automatically capable of transforming his two-dimensional work into three dimensions through architectural scale drawings. Yet this is in fact almost the case in television. The pressures and unavoidable speed of the medium compel the designer to acquire this technical facility. There is no time for extended discussion, experiment, or continual changes in studio layouts and camera plans before a final design is realized.

STUDIO BACKGROUNDS. These are the backgrounds used in talks, discussions, science programmes, panel games, many music programmes, news, sport, etc. Though related to display and exhibition work in the industrial design field, this kind of background design has no particular derivation. It is a style which supports in mood and through practical appliances a background which is in harmony with what is going on. It reflects through the use of furniture, fabrics and decorative surfaces new design standards of the day and, by the use of graphics, supplies visual aids to the programme in question. In general terms, it produces a non-representational set which is stimulating, relaxing or controversial, a mirror of the atmosphere in the programme content.

This expression of design requires skills similar to those of the abstract designer: an appreciation of what is happening visually in everyday life, publicly and domestically; a wide interest in all the arts, political awareness and an interest in people. To anyone who is asked to design for, say, a science or music programme without having any knowledge of, or sympathy for the subject, the task can be very taxing. Such work frequently calls for a specialist in this field. But, whoever he is, in expressing his own point of view he is as bound by the disciplines and limitations of the medium as any other type of designer.

Although there is a sensible tendency for the sake of standards and truth to use the designer as a specialist in one field or another, it is not impossible for him to be required to work in two or even more. Nevertheless, whichever field he works in, he must bear in mind various important points. The screen is small and must give an impression of space. The actor is the focal point. Clarity and simplicity are the essential disciplines of good design. Good pictures start with a camera, an actor and a background, and all three have to be related in terms of movement. Having achieved a broad practical solution to other people's problems, the designer can with confidence attend to his prime function of proportion and detail on a pure design level, assured that his own statement will be the one which appears on the screen in his own style. M.Y.

See: *Art director; Studio complex* (TV).

Book: *Technique of Television Production*, by G. Millerson (London), 1968.

⊕**Developer.** Laboratory operator engaged in running developing machines for film processing; developers are usually divided into negative and positive processing machine operators, the former being concerned with handling valuable original material, often in total darkness under critical conditions on fairly complicated but comparatively slow running machines. Positive developers, on the other hand, can operate with reasonably bright safelights, but their machines often run at much higher speeds. Developers are also sometimes divided into dark-side and light-side operators depending whether they are required to work in darkrooms or not.

See: *Laboratory organization.*

⊕**Developing.** Chemical process of converting the latent image of an exposed photographic material into a visible form. Normally consists of transforming the exposed silver salts of the emulsion into black metallic silver but may also be associated with the formation of colour images by dye coupling.

More generally, developing means the whole series of processing steps required to transfer the exposed film into a stable and permanent image of the character required and in this sense covers not only the actual developing of the latent image but also the removal of the unused or unwanted silver compounds (fixing), the removal of other chemical residues by washing and the removal of water by drying and conditioning.

See: *Chemistry of the photographic process; Image formation; Processing.*

⊕**Developing Agent.** In photographic processing, the active chemical which effects the reduction of the exposed silver halide in the emulsion to metallic silver. In the course of this reaction, the developing agent becomes oxidized and in a number of colour processes these oxidation products are used to form dye images by colour coupling around the developed silver grains.

In black-and-white motion picture processing, the most widely used developing agents are metol, hydroquinone and Phenidone, but in colour film processing the active agents are usually proprietary chemicals specifically chosen for their colour coupling characteristics.

See: *Chemistry of the photographic process; Image formation; Processing.*

⊕**Diabolo.** Pulley or roller freely suspended in a loop of moving film in a continuous processing machine and capable of rising or falling to compensate for changes of the film length as a result of processing. The axis of the roller being free to take up its optimum

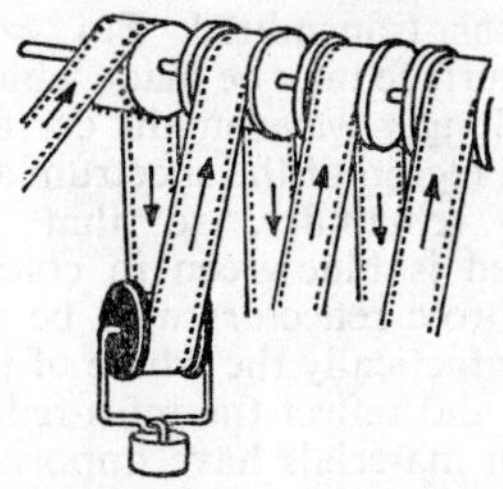

position, a diabolo permits the strand of film to be transported from one fixed roller to another with the minimum of strain. Also known as a floating pulley or roller.

See: *Processing.*

⊕**Dialogue Track.** Sound track or tracks on which dialogue has been edited for dubbing. Different types of sound (dialogue, music, effects) are usually segregated in this way for convenience of handling, and each type of sound may be allocated to several tracks to improve dialogue overlaps, and to allow different frequency correction to be applied to them.

See: *Editor; Sound recording.*

⊕**Diaphragm, Lens.** Adjustable opening formed □by thin overlapping plates usually placed between the elements of the camera lens to alter the amount of light reaching the film

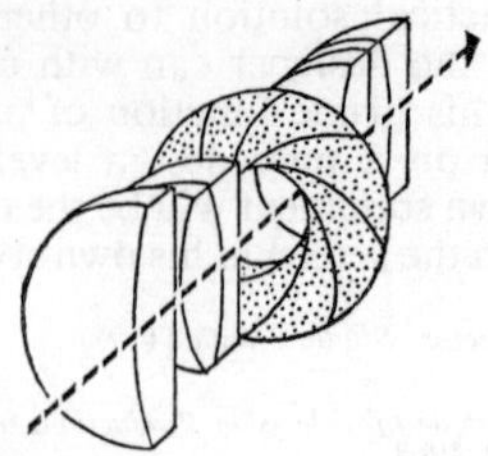

or other image-forming surface. Also called an iris.

See: *Lens types.*

□**Diascope.** Test card displaying device for use on a camera, or as a means of showing one or two slides. Consists of a low intensity light source operated at low voltage, usually via a transformer from the camera outlet.

The light passes through the transparent caption or test card and is focused on to the photocathode by means of a simple optical system. Not suitable for broadcast purposes; now rarely used.

⊕□**Dichroic Reflector.** Selectively reflecting surface from which light of certain wavelengths is reflected and other incident wavelengths transmitted. For example, a dichroic surface may be made which reflects only the longer wavelengths corresponding to the red region of the spectrum and transmits the remainder, so that the light transmitted is blue-green in colour. Similarly, dichroic reflectors may be chosen to transmit practically the whole of the visible spectrum and reflect the infra-red and heat rays: such materials have important application in projection optical systems where heat falling on the film must be minimized without undue loss of light.

A dichroically reflective surface usually consists of a series of extremely thin transparent layers deposited on the surface of a support such as glass, the refractive index of the deposited material and the thickness of the layers being such that in the selected wavelength region interference takes place between the rays reflected from the various surfaces.

The effect of a dichroic reflector is governed by the angle of incidence of the light reaching its surface, and the thickness of the interference layers must be chosen to conform to this.

Dichroic reflectors have particular application in colour processes involving beam-splitting optical systems since they permit very efficient subdivision of the available light into the required spectral regions.

In the television camera, dichroic filters may be used to analyse the subject in terms of the separate primary colours. Red and blue reflecting dichroics, in conjunction with plain front-surfaced mirrors, direct the red and green components to their respective pick-up tubes. The green tube takes the remaining image after its passage through both dichroics.

At the receiver, red and blue dichroics may again be used, but this time to reconstitute the image. The process is the same as at the transmitting end, but in reverse.

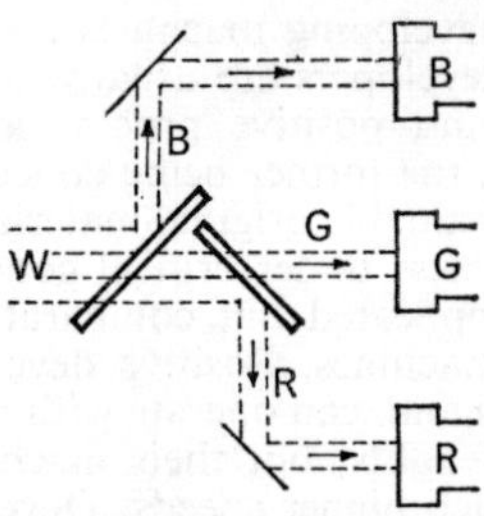

DICHROIC REFLECTOR. Two reflectors, selectively blue- and red-reflecting, split light, W, into three components for TV camera colour tubes.

It is important that the chromaticity of the coloured images that are combined in the receiver are equal to the system standards, if accuracy of reproduction is to be achieved. Likewise, it is essential that the transmitting analysis filters accurately analyse the image in terms of the system standards. To achieve these desired chromaticities, trimming filters are added to the optical system. It should be noted, however, that the spectral characteristics of the transmitting and receiving optical systems are not equal. This is because of the different natures of the analysing and reproducing functions.

See: *Camera (colour); Colour printers (additive); Telecine (colour).*

⊕**Dickson, William Kennedy Laurie, 1860–1935.** British inventor, born in Brittany of English and Scots parents. Wrote to Thomas Edison in 1879 expressing his interest in electricity and asking for a job. Despite Edison's refusal, Dickson with his mother and two sisters crossed the Atlantic in that year and settled in Virginia. Moving to New York in 1883, Dickson approached Edison again and this time was employed by him. He worked first on electric light and power developments, then carried out important work in ore milling, becoming the head of the research department working in this field at Edison's West Orange Laboratory.

Having done much of the official photography for Edison as part of his work, he was given the job of developing motion picture apparatus in 1888.

Dickson's efforts were largely responsible for the successful construction of the Kinetograph camera and Kinetoscope viewer in 1891. He left the Edison company in April 1895 and for a time worked with Latham, developing the Panopticon in the same year. After joining Marvin, Koopman and Casler, he worked on the American Mutoscope and Biograph Company's apparatus, and in 1897 returned to England as European supplier of Biograph subjects, including Queen Victoria's Diamond Jubilee, 1897, Pope Leo XIII, 1898, and the Boer War, 1899–1900. His contact with the Biograph Company ended around 1911.

□**Differential Gain.** In a colour video channel, the voltage gain for a small chrominance subcarrier signal at a given luminance signal level expressed as a percentage difference relative to the gain at blanking or some other specified level. Can cause changes of hue with changes of level.

See: *Transmission networks; Videotape recording (colour).*

□**Differential Phase.** In a colour video channel, the phase shift of a small chrominance subcarrier signal at a given luminance signal level, relative to the phase shift at blanking level. Can cause changes of hue with changes of level.

See: *Transmission networks; Videotape recording (colour).*

⊕**Differential Re-Exposure.** In some forms of reversal film processing, the emulsion is re-exposed to produce the final image; where the intensity of this re-exposure is controlled by the density of the first developed image, variations of original camera exposure can be compensated and the procedure is termed differential re-exposure.

See: *Processing.*

⊕**Diffraction.** Optical phenomenon, resulting □from the wave structure of light, occurring at the edges of illuminated objects. If light from a small source falls on an object with a sharp outline, the shadow produced is not absolutely sharp-edged but when examined by a microscope may show a series of alternate light and dark bands at its boundary. These result from interference between the wave components at the edge of the object which causes cancellation or combination at different positions according to their wavelength. If the illuminating light is monochromatic, containing only a restricted range of wavelengths, these bands or diffraction fringes appear light and dark, but with white light, containing many wavelengths, interference occurs at different positions for different wavelengths and the diffraction fringes appear coloured.

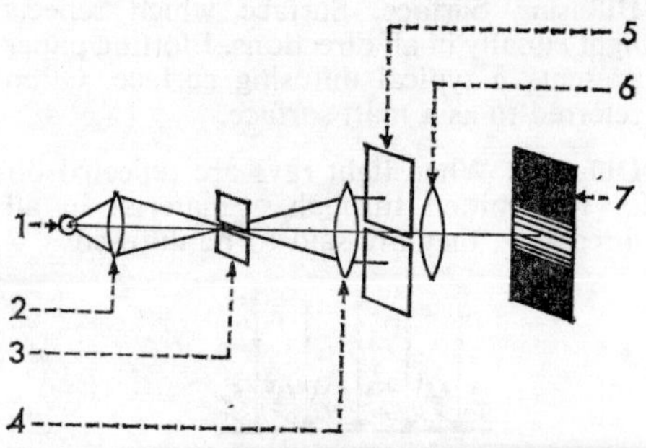

DIFFRACTION. Monochromatic lamp, 1, focused by lens, 2, on narrow aperture, 3, which acts as a source of small dimensions. Light passes through collimating lenses, 4 and 6, between which is a thin slit, 5, and is focused on the screen. Diffraction at edges of slit, 5, causes image on screen, 7, to be bordered by light and dark bars, the diffraction fringes.

This characteristic may be used to separate the component wavelengths of a beam of light into a spectrum by the use of a diffraction grating, which consists of a transparent plate ruled with a very large number of parallel straight lines, often 20,000–30,000 to the inch. When illuminated by a small source such as a bright slit, the diffraction fringe patterns from these lines combine in a regular fashion to produce a series of spectra on each side of the central slit image. Gratings can also be ruled on reflecting surfaces which may be in the form of concave mirrors to allow the spectra images to be formed without the use of additional lens systems.

Diffraction phenomena have important effects in the formation of optical images in photographic systems, particularly where small diaphragm stops are employed, and diffraction characteristics are made use of in certain processes such as Schlieren photography and in thermoplastic image recording. L.B.H.

See: *Light; Schlieren photography; Thermoplastic recording.*

⊕**Diffuse Density.** Density of a transparent photographic image measured in such a way that the total light transmitted through the sample is measured, no matter in what direction it may have been scattered by the image grains.

See: *Callier effect.*

⊕**Diffusing Surface.** Surface which reflects ☐light equally in all directions. Blotting paper presents a typical diffusing surface. Often referred to as a matt surface.

⊕**Diffusion.** When light rays are reflected off ☐or transmitted through a material in all directions, they are said to be diffused.

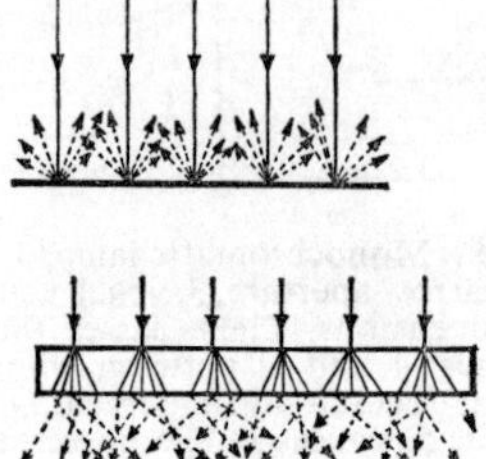

Diffusers are often placed in front of studio lamps to soften the light. Diffusion discs are sometimes used with camera lenses to produce a soft-focus effect, and may be manually controllable to provide different degrees of softness.

⊕**Dimension-150.** A system of wide-screen cinema presentation using optically corrected 70mm prints projected on to a deeply-curved screen.

See: *Wide screen processes.*

⊕**Dimmer.** Device for smoothly regulating the ☐current in a lamp so that its light output can be varied from extinction to full brightness. May employ a variable resistance, a tapped transformer, a variable choke, a magnetic amplifier or a controlled diode.

See: *Lighting equipment.*

⊕**DIN Print.** Form of 16mm motion picture print in which the picture image is laterally reversed with respect to the sound track

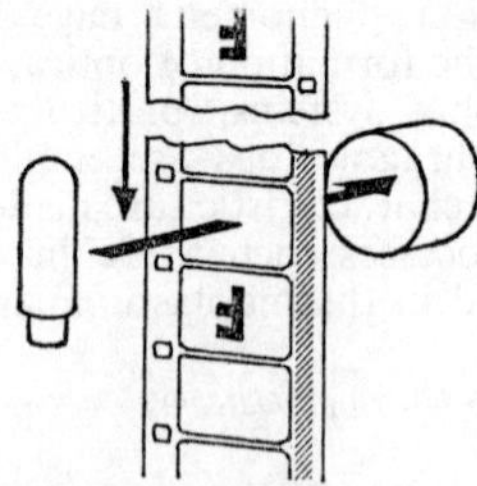

position in comparison with regular American and British standard practice. 16mm sound projectors requiring this form of print were formerly widely used in Germany and other European countries.

See: *Print Geometry* (16*mm*).

⊕**DIN Speed Rating.** System of specifying the sensitivity or speed of a photographic material; a logarithmic series is used in which an increase of rating by 3° DIN represents twice the speed.

⊕**DIN Standards.** Standards having international recognition in photography and motion picture practice, established by the Deutschen Normenausschuss in Germany.

See: *Standards.*

⊕**Dioptre.** Unit of power of a single lens, the ☐reciprocal of its focal length in metres, the sign + or — being used to indicate converging or diverging respectively. A 10+ dioptre lens thus indicates a converging lens of focal length 1/10 of a metre, or 10 cm.

⊕**Dioptre Lens.** Simple supplementary lens used as an attachment in front of the main camera lens to allow the photography of extreme close-ups. It is a positive convergent lens whose focal length is specified in dioptres and when used in front of a conventional lens focused at infinity it allows the photography of a close object situated at its own focal length distance. Thus a +2 dioptre lens, with a focal length of ½ metre, allows the photography of a close object ½ metre away, or 20 in.

Split-field dioptre lenses are similar supplementary lenses covering only part of the field of view of the main lens and thus allowing close-up and distant objects in different parts of the field to be equally sharp on the negative. Careful composition, lighting and focusing of the distant object are essential to the success of these effects.

All supplementary lenses are liable to upset the optical corrections of the main lens and it is therefore general practice to stop down as much as possible while using them.

See: *Cinemacrography.*

☐**Diplexer.** Specialized piece of equipment for joining power radio feeds. A diplexer sometimes denotes a unit for combining two feeds of the same frequency (joining of parallel transmitters), sometimes for splitting a feed to two separate feeders (to feed two aerial arrays).

A diplexer is also used in television receiver aerial systems, either for joining two aerials together to a common feeder (e.g., Band I and Band III) or for splitting the output of an aerial into separate bands for separate receiver inputs. This form of diplexer usually consists of two inductors and two capacitors in a simple bridge circuit, the inductors being diagonally opposite each other. The combined feed is taken across both the LC circuits, one single feed across one inductor and the other single feed across the capacitor of the other circuit, the junction of these two components being a common connection for all three feeds.

The L/C ratio is adjusted so that in one frequency band the major part of the signal is developed across the inductor, and in the other the signal is developed across the capacitor. This method gives very little attenuation to the two signals at their correct outputs, and a fairly high attenuation at the other output.

See: *Transmitters and aerials.*

□**Dipole Aerial.** A basic aerial arrangement in which a feeder is connected to the centre of two equal conductors or to the end of a single one arranged with a reflecting plate to give the equivalent of the first arrangement. The field from such an aerial is doughnut-shaped. The impedance varies not only with the length of the conductors but with their shape, and various types have been devised to give bandpass properties. Names such as turnstile have been applied to special shapes, with desired properties. The dipole unit can also be folded and it can be used with other conductors to modify the shape of the radiated or received field.

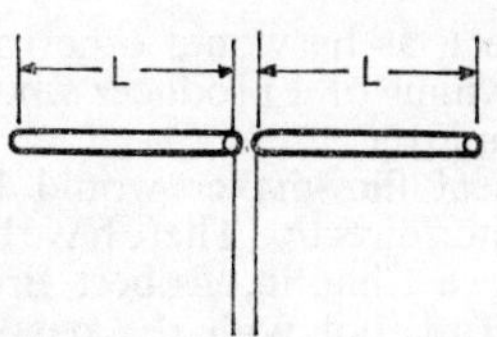

DIPOLE. Length, 2L, bears definite relationship to wavelength of transmission: 2L = ½ wavelength × 0·95, or L (ft.) = 234/frequency (MHz).

See: *Transmitters and aerials; Turnstile aerial.*

□**Directional.** General term applied to optical, ⊕electronic and acoustic devices such as screens, aerials (antennas), microphones and loudspeakers, denoting a limited angle of reflection, radiation or acceptance.

⊕**Directional Effects.** Processing fault caused by insufficient agitation in a film developing machine, leading to local exhaustion of the developer in the immediate neighbourhood of the film emulsion, especially in heavily exposed (dense) areas. The movement of the film through the developer then brings a new area of exposed emulsion through the pocket of exhausted solution, and this area is less fully developed than it should be. This gives rise to vertical tails and streamers. The remedy is more and better agitation.

See: *Agitation; Processing.*

⊕DIRECTOR

Film-making is essentially a co-operative job if only because of its technical complexity. But when a critic or a filmgoer speaks of a film as being made by a particular individual, he is usually referring to the director. The accuracy of this description depends on the relationship between three men: the producer, the writer, and the director. Sometimes the producer chooses the subject of his films and casts the director just as he casts the stars. Sometimes the producer and director are the same person. But in most cases the producer and his director form a two-man team, the producer being responsible for the economic, and the director for the creative aspects of the film.

PREPARATORY WORK

The director is usually the man who first has the idea of making the subject. It may be a book or a play, an episode of history or a story in today's paper; it may even be an original idea out of his own head. But, whatever the source, it is usually he who first decides that he wants to make a film of it, and in this sense, it is his film.

When a director decides to act as his own producer, he usually employs an expert associate producer to look after the economics of the film, but because he is producer as well as director, all creative decisions are his responsibility. Among these are the choice of writer, the approval of the script, the choice of cast, cameraman and other technicians, the supervision and approval of each stage of editing, the choice of composer, and the final dubbing or combining of the various sound tracks. In such a case, it is much easier for him to impress his signature on the picture than if he has to pay attention and respect, as he would otherwise, to the creative whims of a producer who may also be his employer.

The ideal film-maker would be a producer/writer/director. There have been some, for instance Chaplin, Robert Bresson and Jacques Tati, but with the passing of the years they seem to become fewer as films become more complicated and more expensive.

FINDING THE SUBJECT. The director is now nearly always a freelance. The days when studios kept directors under contract are gone, though a few production companies still include film directors on their boards. As a freelance, he earns nothing except when engaged on a film; he is therefore always on the lookout for a good subject. Experience has taught him that finding the right subject is half the battle, and that the best script-writing and technical work will never turn a poor subject into a good picture. So he reads a good deal, and there are a few ideas at the back of his mind which he has been trying for some time to translate into film terms but which have not yet come to fruition. One day he may meet a novelist who has recently published a book dealing with one of his pet ideas. Conversation might reveal that they have much common interest in the subject, although their approach differs considerably.

The director is usually no tycoon, so he does not try to produce the film himself. He finds and interests a producer whose views and tastes are broadly in sympathy with his own. To help persuade the producer to undertake the film, he gets an outline of his idea down on paper. He can do this himself after his discussion with the novelist, or he can persuade the novelist to provide it, perhaps on a speculative basis without payment, or perhaps by offering him a modest or nominal sum in return for an option to buy the story if or when the production is set up. If the producer likes the outline he will agree to try to find backers for the project. Meanwhile he and the director discuss who is to star in the film, for if they can tell prospective backers that one or two big box-office names are interested in playing in it, their chance of getting the necessary finance is greatly increased.

The producer may eventually succeed in obtaining financial backing for the production, including of course the purchase of the book, but almost certainly the sum of money available is smaller than the director had hoped. He has to develop the script with economies in mind which will not harm the subject. Ingenuity is called for, rather than sheer lavishness and, if properly used, can enhance rather than diminish the film. Choice between studio work and location work is considered early by the director; it is governed not only by his personal taste but by the requirement of the story, and it is considered early because the choice may well affect the construction of the script, on which the writer will by then have started work.

SCRIPTING. If the novelist is himself a script-writer he may well be employed to work on his own script. More probably he lacks such experience and a professional script-writer is used. Whichever course is adopted, the director collaborates closely with the writer. Together they decide what parts of the book must be kept at all costs, what can be desirably or safely sacrificed, and what must somehow be included though in a different position or form.

While the script progresses, the director and producer discuss the cameraman, the production designer or art director, the camera operator, the sound recordist, and other key members of the unit. They ponder over the choice of a studio suitable to their needs, or whether they should shoot the whole film on location, finding suitable interiors as well as exteriors, and thus dispense altogether with studio shooting. The needs of the subject dictate the director's choice; economic considerations influence the producer.

The director also continues with his casting plans, though some artistes will not definitely commit themselves until they read the completed script. Having agreed on the division of locale between exteriors and interiors, he looks for suitable locations – a "recce" (location hunt) – taking with him the cameraman if he is available. He may also take the art director, who has to design settings of the requisite character and may also have to reproduce sections of the exterior of certain buildings in the studio. If the production manager has been appointed by this time, he too can join the location hunt. He assesses the available accommodation, its cost and suitability, and he plans the movement of the unit and equipment, including the daily despatch of rushes to the laboratory.

The director returns to base, to finalize the script and to obtain the producer's approval to the changes which have been made. Just as the quality of the subject is important, so is the script equally vital. The script is the blueprint, essential to each department of the unit, and to none more than the director. But he must not regard it as sacred and inflexible, for by doing so he misses opportunities of making his film a true cinematic creation expressed in visual terms; he risks turning out a prosaic record of the written, or only of the spoken word. The result may be a well-photographed play, but it will not be a real film if it does not use the medium to the full, and this is almost impossible unless the director is flexible enough to modify and enrich his script during shooting. Nevertheless, without the script to guide him, he risks making a film which lacks a firm story line and consistent development. This is not to confuse story line with plot. Excessive plot can be so obtrusive that all character-drawing is lost. Most film-lovers feel that a balance between form and content is necessary for a film to be satisfying.

CASTING. Once the money has been obtained, the script approved, and the location and studio work determined, the director can complete his casting. He is usually helped by a casting director, who suggests actors, checks on their availability, and undertakes the necessary negotiations. But the final choice remains with the director.

In addition to screen tests to determine actors' suitability, make-up and lighting tests are needed for the principals once they have been cast. Dresses must be designed, wardrobe requirements worked out, and special make-up or wigs decided upon and ordered. The director has a guiding hand in all these activities, because his interpretation of the script has an important bearing on all of them.

SETS. The art director or production designer brings sketches and plans of studio sets. The director studies these closely to make sure they meet his needs. Sometimes he may ask for an extension to a set. Often he is able to suggest economies. If he fails to plan his own work in advance, such economies cannot be made, and wasteful overbuilding may result.

As in so many other aspects of film making, thorough preparation is essential if waste is to be avoided. The art department furnishes the director with copies of plans of all the main sets, with the help of which he can map out the work. He can decide the general direction in which each scene will be shot, and the principal positions and movements of both the camera and the actors. Such an exercise is most valuable: it compels the director, if he has not already done so, to imagine his work in visual terms, in which he can discuss with his cameraman not only the mood but the mechanics of each scene. The more he can do this during the planning stage, the more smoothly the work can be expected to go when it reaches the studio floor.

MAKING THE FILM

The shooting schedule, worked out by the production office in conjunction with the director, is like a jigsaw puzzle. Location work, studio work, time for building sets as well as for shooting on them, the availability of artistes, the time needed for make-up, all must be considered and squeezed in with a realistic estimate of the time required by the director. His opinion is vital. He may find that the production office has allowed more time than he needs for a particular scene. On the other hand, a scene may take six times as long as scheduled.

It is almost inevitable that, on the day scheduled for shooting to start, preparations are still incomplete. Despite every effort by the various departments and members of the unit, some important detail is not yet ready, or a supplier has failed to deliver something on the day stipulated. Flexibility is called for; the director must have prepared not only his first day's work but have a fairly clear idea of the second, so that the necessary switching around will not find him at a loss.

VISUALS AND STORYBOARD. Many directors include in their preparations what are sometimes called visuals. Having worked out his area of shooting and the movements of the cast in it as well as the movements of the camera, the lens he will use, and the number of angles he will need, he either makes sketches of each set-up himself or gets somebody from the art department to work them up from his own rough drawings. These can then be shown to all the key members of the unit at the start of the day's work and referred to at any time during shooting. It is quicker, and more explicit, to explain the requirements of a camera angle by even rough illustrations than by written or spoken words only. The camera crew's business is pictures and they are used to working in pictures.

The storyboard, so much used by advertising agencies, is the more sophisticated descendant of the visual. Used properly, it can be similarly valuable, but it should not become a rigid imposition.

DEALING WITH ACTORS. When the shooting starts, the director is handling actors and actresses. The characters who existed so clearly in his and the writer's minds have to be re-created by his cast, some of whom are ideal, and some a compromise, while others have potentialities which he must try to bring out. It is most important that he should win his actors' confidence: he discusses with them the characters they are going to play before shooting starts, preferably some days before they are first needed, so that they can learn their lines and walk on to the studio floor already knowing much of what the director wants of them.

Some directors believe that the actor should know as little as possible about the part, should be prevented from reading more of the script than those pages on which the character he is playing appears, and should in effect remain clay in the hands of the director. No doubt some actors respond to this treatment, but others are gratified at being taken more into the director's confidence and work all the harder to try to give him what he wants.

Nevertheless, the director's authority must be paramount. A film is shot out of continuity, unlike a play which can be rehearsed straight through from beginning to end. The script might include, for example, three scenes laid in the same setting, one near the beginning of the story, one in the middle, and one at the end. Because of limitations of studio space, and also because it is quicker, all three scenes must be shot consecutively. In this way, the unit moves on to the set only once, the general lighting of the set is done once only and, as soon as the work on it is satisfactorily completed, the set can be pulled down to make way for another to be erected in its place.

This shooting out of continuity, however, poses problems for the actor. The first scene completed, he is called upon to move straight to the scene in the middle of the film, and when that is finished, to the one near the end of it. Even if he is very experienced he is unlikely to be able to make the transitions without the director's guidance.

The director must keep the whole story in perspective and assess the mood, rhythm and pace which each separate scene needs to fit the jigsaw, the pieces of which he is now making.

It is almost impossible for the actor, concentrating on the scene in hand, to maintain such a perspective; for this reason if for no other the director is the arbiter and final authority on the studio floor.

To retain flexibility, however, the director must keep an open mind. Actors can make excellent suggestions for the enhancement of their parts, just as members of the unit can make excellent suggestions for the improvement of a scene. Unless the director is prepared to consider their suggestions, even though to accept them may imply some modification to his script and his shooting-plan, he may miss chances to improve the result. But he must be quick-witted enough to grasp all the implications of a suggestion, lest, although superficially attractive, it contain some flaw which will harm the characters or be inconsistent with some later development in the story.

SHOOTING SEQUENCE. Normally, exterior location work is undertaken before the interiors. Of course, weather conditions may dictate otherwise; with a film starting in winter it would be wiser to schedule the exterior work at the end rather than at the beginning of shooting. The advantage of starting with location work is that the building of sets in the studio can proceed simultaneously, together with other preparation, and that since the uncertainty of weather usually makes the location the more chancy part of shooting the film, it is as well to undertake it first; once the absolutely essential exterior shots have been covered, it may be possible, should the weather become very bad, to return to the studio and recreate chosen sections of the exteriors indoors.

The director can help greatly by being flexible in his shooting plan. He can consider all the work carefully with his cameraman with a view to agreeing on which shots must be done in direct sunshine, which would be preferable in sunshine but can at a pinch be done in indirect sun or light cloud, and those, if any, which will positively benefit by being shot in dull or even bad weather. Then he should prepare for each eventuality, so that with the help of his assistant director he can amend the order of shooting from day to day, and even from hour to hour, to suit the changing skies.

Even so, it is desirable to be able to shoot some interiors as well when on location; it is often possible to hire a small hall, or out-building, or shed, in which small sets can be built and shot, thus providing what is called weather cover for conditions so bad that all outside work is impossible. Anything is better than having the unit sitting around; it not only wastes money but lowers morale disastrously, and the drive and tempo of the work are lost.

ASSISTANT'S DUTIES. The assistant director is the director's right-hand man at all stages of shooting, but his work is primarily administrative rather than creative. He is responsible for the call, which ensures that the right artistes are in the right place at the right time. At the end of each day's work he consults the director about the call for the following morning; it will include not only members of the cast but the number, type, and dress of any crowd artistes who may be needed, as well as any special technical effects. He has to see that all the director's requirements for the day's shooting are provided, and that the work itself goes as quickly and smoothly as possible by maintaining good discipline on the studio floor, thus giving the director the opportunity to do his best work. His creative contribution is mainly concerned with crowd scenes, which he normally plans and rehearses himself, the director making any changes he may need only after his assistant has presented the action to him for comment. But despite this limitation he is in a good position to make suggestions for improvements to any scenes, and the director is well advised to listen to him. Good relations between them are of great importance to the efficient running of the unit, so that everybody knows that the assistant's requests are made with the director's voice.

WORKING WITH OTHERS

EDITING. While shooting progresses, another key member of the unit is at work – the film editor. Here again, the more the director takes the editor into his confidence the better. When seeing the rushes he tells him which takes of a scene he prefers; later, he goes through the whole sequence with the editor, script in hand, telling him how he intends each angle to be used. Otherwise, the editor might be faced with a mass of material which could be assembled in many ways. Unless the director has a clear vision of the effect he wishes to produce on the screen, the editor will be trying to solve a puzzle without the essential clue to its design. Once again preparation is essential. If the director has prepared his shooting properly, he also has a fairly clear idea of how he wants the sequence cut; which angle he intended for which lines of dialogue, which reaction shots he wants used at which point to illustrate a character's emotion, and so on; and he is able to describe how the emendations or enrichments to the script which he added during shooting can be incorporated in the editing of the sequence.

At various times during the shooting of the film, the editor shows sections which he has assembled to the producer and to the director. So the assembly of the film can proceed only a few days behind shooting, and when shooting comes to an end, a showing of most of the film can take place comparatively shortly afterwards, although some ingredients, such as model shots and optical work, will be missing. This assembly, often called a rough-cut, is in no way an attempt to present a finished picture, but rather an exposition of everything which has been shot, deliberately cut long rather than short. It is easier to judge the film's potential in this way, and from a practical point of view it is easier for the editor to reduce the length of a shot than extend it.

After running the rough-cut, the director discusses alterations with his editor; very likely the producer is also present.

When the alterations have been agreed on, the best course for the director is to go on a short holiday while the editor is carrying out the revisions. He will return with a fresher mind, able to recapture his original vision of the film, and to see how well or ill his work has measured up to it; some scenes are successful, some fail, others can be improved by altering their tempo or emphasis. It will all be easier to assess after even a few days' complete change.

The actual editing, as opposed to assembling or cutting, is a fascinating process of trial and error; what proved right in the director's last film will almost certainly be wrong in his new one. It has been well said that the only rule in film editing is that there are no rules. Above all, the director must detach himself sufficiently from his work not to regard any of it as sacred. The scene which took so long to shoot, on which he and the cameraman lavished so much thought and care, must, if necessary, be flung out without regret. The only yardstick is whether the scene is worth its place or not. Does it advance the story? Does it affect what has gone before or what follows? Does it illuminate character? It may be very entertaining in its own right, but once it has been removed the better progression of the action will show it to have been an irrelevance, so it must go and no tears be shed for it.

There are pitfalls: the more often the director and editor see the material the harder it is to assess its worth. This is especially true of comedy. Many a good gag has been thrown away because it no longer seemed funny in the projection room. First impressions are vital, and must be remembered. Showing the film or parts of it to colleagues may help, but mainly to confirm the director's own opinion. He must keep faith with his intentions.

SCORING. One other important contributor has yet to be called in – the composer of music for the film, though, of course, in the case of a musical, he has, in effect, been present throughout. The choice of composer is usually the director's, with the approval of his producer. As with the rest of the unit, sympathy with the subject of the film is most important, but selecting a composer may still be regarded as one of the inevitable gambles of film-making. Certainly, it is possible to evaluate the composer's previous work and thus his apparent suitability; to discuss the script with him at length and show him the result on the screen; to agree with him on the sequences which will be assisted by a musical score and on those which do not need its help; to hear suggestions for the musical treatment played on the piano. But it is not really until the day of the recording session itself, when the director hears the composer's work performed by the orchestra for the first time, that he can know if his gamble has come off or not.

It is then too late for any but very minor changes in the score, but if what he hears fulfils his hopes, then the addition of music can be one of the most rewarding moments in film-making.

DUBBING. After the music recording comes the dubbing, the process in which dialogue, music and sound effects are recorded on to the final, single sound track. At this stage, too, the director's influence is exercised, both in the choice of sound effects and in their use. In dubbing, much of the effect for which he has planned and striven can be finally made clear. But equally it may be clouded and weakened, even misunderstood and lost altogether in the technical process. At this last stage, as in the preceding stages, preparation is the key: the director should have made up his mind already on the balance which he needs to achieve his aim. The balance between music and sound effects in any one scene, or the choice between one and the other; the balance between dialogue and music, and that between dialogue and sound effects; the places where silence is more effective than any of the three.

For the director, preparation really means making up his mind on what he wants to say and how he wants to say it. Yet his film must look spontaneous, not contrived. A good deal is heard about spontaneous film-making but really it is only the appearance of spontaneity that is important, or even possible.

Jean Renoir once introduced one of his films by saying that when he made a film he threw away technique.

The young, enthusiastic cinéastes in the audience became very excited. Here was one of the world's acknowledged masters of film-making declaring publicly that technique was unimportant. Just what they had always said themselves! Slavish adherence to technique was what was wrong with films.

However, Renoir had not finished: "But before you can throw away technique", he continued, "you must *learn* technique. For if you have not learnt it, then how can you throw it away?" C.F.

See: *Director* (TV)*; Producer*.

□DIRECTOR

The TV director is the creative head of an artistic and technical team. He interprets an idea in visual terms. Whether it be entertainment, instruction or enlightenment, he must be clear about his purpose, for his contribution is the link between conception and realization. To achieve this, he brings together the talents of writer, composer, artistes, designer, camera and sound crew, lighting, wardrobe, make-up and engineering staff into a team whose objective is to carry out his own conception, for in the final analysis it is the director who selects what is seen on the screen. A good director is not one who sets himself up as supreme arbiter but one who can stimulate and wed the abilities of his production team into an honest presentation of the original idea.

To communicate this idea effectively, he must be familiar not only with the scope but with the limitations of his medium, for in television the director is obliged to work faster than in films and the theatre, and make an immediate impact on an audience distracted by the diversions of the home and the interruptions of commercial breaks.

BASIC QUALITIES

The director must know the grammar of television production, handed down from the motion picture industry but adapted to fit the restrictions of television. These restrictions also have their advantages, for the intimacy between performer and home viewer has made television unequalled as a means of mass communication.

RELATIONSHIP WITH PRODUCER. In film, TV and theatre the roles of producer and director have often been interchanged, but the title of producer lost some of its artistic connotation when Hollywood introduced legal and financial executives to control studio budgets.

Today in TV the director works under the administrative control of the producer, who is the executive in charge of organization and as such may well originate the idea or format or initiate the first written script. His creative contribution in the planning stages may be much or little, but certainly he will take charge of all finances.

All television must start with an idea. Whether or not it comes from the producer, it is the director's responsibility to broaden this outline and develop it into a detailed format. To achieve this he will require finance and must know how to budget the finances allocated by the producer. The producer must surround himself with the best available creative and technical team, and subsequently promote and publicize his project. A knowledge of contractual limitations, royalties and union agreements is essential. If he does not have a technical background, he should at least understand the limitations of his equipment, and be able to assess its availability. It is therefore essential that the producer choose a director with whom he can work in close association.

Since the director is responsible for the creative planning, scheduling and rehearsing of programmes, he must be able to refer administrative problems back to his producer. Only then can he have artistic freedom over his material.

In series productions, whether drama, light entertainment, outside broadcasts (known in the US as remotes) or documentaries, the producer's responsibility is greater. Here his duty is to establish an overall continuity from the individual styles of his team of directors.

The roles of producer and director are sometimes combined. In this instance the producer/director is answerable to an executive producer, the head of a station or the vice-president of a large network.

VETTING MATERIAL. All material, scripted or otherwise, should be vetted by the director to ensure that it is suitable for mass viewing. Apart from his own sense of good taste there are prescribed channels through which he will work. In the British Broadcasting Corporation (BBC) he is in constant communication with his group head or head of department, and they will together define the necessary limitations.

The British commercial production companies are governed by a more rigid set of rules laid down by the Independent Television Authority (ITA). Their charter clearly defines what may or may not be included in commercial programming. The United States has a similar organization, the National Association of Broadcasters (NAB), who govern the television code of that country. This Association keeps a watchful eye on the correct interpretation of the code to ensure that not only is the programme material not objectionable, but also that advertising within programmes is truthful and of adequate quality.

SHAPING THE SCRIPT. On the scripted show the producer either chooses the freelance director best suited to the particular style of production or schedules a staff director.

Once he has received the script commissioned by the producer, the director confers with the writer and script editor, and together they modify or rewrite the story so that it may be fully realized in the director's individual terms.

Passages of dialogue can be eliminated if they are better expressed in visual and sound terms only. The construction may be altered for dramatic effect and dialogue rewritten to suit an individual artiste under discussion.

PRODUCTION PLANNING

During this stage the director keeps in touch with the casting director or artistes' agents to check on the availability of talent, but he does not book it until the producer has allocated his budget allowance for the cast. By this time the director should have a breakdown budget for all departments and knows the limits of his expenditure. Some of these individual amounts may be subject to negotiation between director and producer. For example, should the director decide to film or videotape certain sequences instead of building studio sets, it may be possible to transfer money from the design allocation to that permitted for filming.

PRODUCTION TEAM. During these early days the director's production team begins to grow. Apart from his production secretary, he has a production assistant with him throughout. The titles used for television personnel vary from station to station and country to country. In the BBC the production assistant remains with the director during the planning stages and keeps in touch with all servicing departments to ensure that they understand the requirements of the production. He acts as assistant director on pre-filming, and on camera rehearsal days is in charge of studio floor management. Under him and also attached to the director's office is the assistant floor manager whose main responsibility is to acquaint the property department with script requirements, mark out the set for outside rehearsals and prompt any artiste who may forget the text.

Independent television in Britain uses the term production assistant to refer to a combination secretary and personal assistant, while the floor manager covers most of the duties of the BBC production assistant. In some companies the floor manager is attached to the programme from its conception, while elsewhere he is available on camera studio days only. Within various departments of the BBC also, floor managers may take charge of floor operations in the studio. In some US stations the term floor manager and studio manager are interchangeable.

The number of staff and the complexity of their respective jobs vary greatly according to the size of the television company.

SET DESIGN. Having clearly explained his intent to his staff, the director then meets his designer. They discuss the style of production, the atmosphere to be projected, whether the settings are to be realistic, authentic or stylized. Size, colour, materials and location of the sets within the studio floor space are decided. Several meetings may take place according to the time allowed for preparation and the interflow of ideas between designer and director. The designer submits sketches and models, and draws up a scale floor plan (usually ¼ in. to 1 ft.) showing the disposition of sets and props on the studio floor. This constitutes the rough plan to be discussed subsequently with lighting, camera and sound crew.

SHOOTING PATTERN. With this rough floor plan the director must work out his shooting pattern. For after all, television is an axial rather than a peep-hole medium. The theatre has its proscenium arch, apron stage or theatre-in-the-round where the audience has a static position and can only peep in on the stage from a single viewpoint. The intimacy of television permits the viewer to participate more immediately in the lives of the people on the screen. Their viewpoint is the actors' viewpoint and seldom can they sit back and watch objectively from a distance.

To achieve this axial movement the director must have sets designed to allow maximum freedom of access for his cameras and sound equipment without restricting the artistes' movements. But it is essential to focus visual attention, so, regardless of the subject matter, the director composes his pictures in shapes and patterns. These must be deliberately selected for the appropriate effect. Balance is essential, and unbalanced compositions, though often stimulating, may distract from his main purpose.

All this has to be considered to make the design fit the production.

As a television programme is basically a continuous show, the various movements of artistes and equipment enforce restrictions not only on the designer but also on the lighting supervisor. In the majority of television studios lighting equipment is suspended from a permanent lighting grid, thus leaving the floor space free for the movement of cameras, booms and all other technical equipment. Lighting is pre-set to accommodate the varied positions of artistes throughout the programme, unlike film studios where the action is discontinuous and each shot can be rehearsed and lit separately.

FINAL PLANNING. Casting has been continuing during this time and when artistes are contracted further discussions take place with wardrobe and make-up. The designer and lighting supervisor keep in touch with these departments on problems of colour and texture.

Eventually a pre-production meeting of heads of all departments available is held.

Compromises will then be made in design, lighting, sound coverage and camera planning to obtain the best overall coverage. A ceiling piece which may well have improved a particular camera shot may create lighting problems; particular lighting patterns may obstruct clear sound coverage; an extension to one part of the set may be required or an artiste's costume may provide difficulties for engineering personnel to balance picture quality.

Special camera, sound and electronic effects are now discussed, together with the director's technical requirements, for example, the number and types of cameras and microphones to be used, whether front or back projection is required, number of film or videotape inserts, and if so how many channels are needed.

All these and many more problems arise and must be settled, for after this stage there is little chance for further modification.

The director is by now thoroughly familiar with his script, and has probably worked out a rough blocking of the camera and sound dispositions in the studio, their major positions throughout the show and the difficulties of movement from set to set without tangling camera and boom cables.

The programme may also need to be edited, so the number of edits and method of videotape editing are decided.

Most of the engineering requirements have now been noted by the technical supervisor, who will implement them and check their use on the studio day.

In the US a technical director represents the various camera, sound and engineering departments and advises the creative director how best to obtain his visualization. This technical director remains with the director throughout rehearsals and concentrates on technical aspects of the production so that the director may concentrate more fully on the performance of artistes. In Britain, the director remains both artistic and technical director, but can always seek specific advice from his technical supervisor or the individual members of his crew.

OUTSIDE REHEARSALS

REHEARSAL SCRIPT. Having checked that all artistes are booked and all scripts amended, the director settles down to preparing a rehearsal script.

On this script is noted the basic blocking and the actors' moves, the location of camera and sound equipment, rough diagrams of shot composition and any difficult adjustments; also whether props or scenery have to be moved during the action to cope with camera and artiste movement; cueing positions and sound effects from the floor, such as the door or phone bell ringing. The main lighting effects should also be noted but are more often left until after the technical run-through. This then is the basis of the director's camera script.

REHEARSAL LOCATIONS. Outside rehearsals are then called and usually take place in church halls or empty and uncomfortable rooms, for very few television studios can afford to convert their valuable space into rehearsal rooms.

The assistant stage manager or assistant floor manager then marks up the rehearsal room. Marking tape is placed on the floor to represent walls, doors and windows, with chairs and tables standing in for the multitudinous and sometimes exotic properties.

READ-THROUGH. The artistes then foregather for a read-through of the script, after which the director discusses the programme's intent and individual characterizations.

Rehearsals start with artistes being shown the pattern of their movement within the sets. The director/artiste relationship is one of close personal involvement. The director has already decided the style and the mood which he wishes to engender. He looks at the project as a whole, a complete unity, whereas an artiste views it from the particular, bringing his specialized talent to his own part.

PACE AND TIMING. At rehearsals the director is conscious of the technical aspects of his production but must never let his artiste be aware of this, for to confine him to the straightjacket of a television lens may well limit the scope of his performance. Timing plays an essential part in the rehearsal schedule. Each artiste builds his performance at his own pace and the director must be aware of this and gauge the best means of helping him to reach his peak on performance day. His methods will vary from coaxing to outright coercion.

Apart from the pace of the show actual playing time must be considered. Every programme has a fixed time slot and for successful network presentation these time slots must be strictly observed.

TECHNICAL RUN-THROUGH. Despite the lack of sets and scenery in an outside rehearsal room, the artiste is expected to know his moves and reproduce them in relation to the floor markings which represent the studio design. Only when these moves are established can there be a successful technical run-through.

This run-through takes place a couple of days before studio camera rehearsals. It is attended by the lighting man, technical supervisor or technical director, designer, senior cameraman and sound mixer. By this time the director should have completed his camera script and had copies issued to the technical team. This complete script indicates every word spoken in the show together with detailed video and audio instructions. The technical crew can then follow the rehearsal and visualize the director's intentions, making notes of artistes' moves so that the set may be pre-lit for the studio rehearsal.

FILM OR VTR. Many television productions require more visual scope than the studio designer can recreate. It may, therefore, be necessary to record scenes on location for subsequent inclusion in the studio presentation.

This recording precedes the technical run-through. On a series programme, when a team of artistes are used throughout several shows, filming or videotaping can be planned and done some considerable time in advance. At other times the availability of talent may well limit this recording session to part of the period reserved for outside rehearsals.

The choice between film or videotape for these inclusions is usually settled by the demands of time.

Film takes longer to process, print and edit, whereas VTR can be recorded and played back immediately on location. With videotape, the director can select his takes on the spot and then needs only to edit them. The director may prefer videotape so that he can use a multi-camera technique and cut from shot to shot, thus eliminating editing time altogether. Film, however, still has greater flexibility for sound and editing. Whichever method is chosen, the end result has to be fed into the studio production on recording day.

CAMERA REHEARSALS

STUDIO REQUIREMENTS. The television studio is divided into two main areas: stage and control room.

Stage space varies in size according to the complexity of the production. An area of 40 by 50 ft. is adequate for local programming and less ambitious projects, whereas some of the larger network shows call for stages three times this size.

The three requirements for a television studio are area, height and adequate soundproofing. As the bulk of lighting is hung from the ceiling grid, height is essential. Some of the major US studios allow 40 ft., but many small stations get by with a 12 ft. ceiling. In this working area the designers' sets are assembled for the artistes to reproduce their performance before microphones and cameras.

TECHNICAL STAFF. Cameramen, sound crew and floor operational staff wear headsets to remain in constant communication with the director. He is in the control room, the centre through which the production is channelled and co-ordinated.

In this control room is located the engineering staff who adjust picture quality; the lighting supervisor with his control board; camera monitors for the director to see and select every camera shot; the mixing panel with which the vision mixer or technical director cuts, fades or dissolves from one picture source to another as indicated by the director; and the audio section with its turntables, tape recorders and special effects equipment.

BLOCKING THE SHOW. From this room the director begins to block his show. It is only now that his team get a full visualization of his intent. Cameramen have their shot cards but this only tells them their part in the overall plan. These individual camera shots have now to be integrated with sound, lighting, props, graphics, film or videotape inserts and music.

During the blocking, each shot is taken step by step while artistes familiarize themselves in the set with the mechanics rehearsed away from the studio.

It is essential that the director gives his instructions clearly and firmly so that no time is wasted over misunderstandings. The floor manager relays instructions to artistes and other personnel not in direct contact with the director, and he is responsible for the smooth running of the studio floor.

In the control room the director is flanked by his production secretary and vision mixer. His secretary is in charge of timing the show as well as previewing and precalling all shots. The vision mixer works from the camera script on which all editing points are noted. The director, however, will have to point out exactly when a cut is needed, or the tempo required for dissolves and fades. At this stage he will also have to relay all cues to the floor manager as well as the sound, camera and engineering staff.

STOP-AND-START REHEARSAL. When the blocking is complete, a "stop-and-start" rehearsal takes place to check that all crew members have remembered and can carry out their tasks. Small errors are ignored, and the rehearsal is stopped only when serious difficulties arise.

Difficult passages are again rehearsed before a run-through is attempted. Further notes are then given before the rehearsal proper. Inserts may be checked separately so that the vision mixer or switcher is clear when to change source from studio to film or VTR. During these stages lighting is adjusted.

DRESS REHEARSAL. Finally, a dress rehearsal, before which the director has had to explain patiently to every member of his team exactly what is required of him, while being ever conscious of the pressure of time.

TRANSMISSION OR RECORDING

Directing the transmission or recording is the most important part of the day and the culmination of weeks of preparation and rehearsal. Last-minute checks are carried out with the floor manager on the system of cueing to make sure that everything is ready, with cameras, sound, lighting and vision mixer to see that they clearly know their duties, and that film, videotape and title slides are in readiness.

The director's typical count-down procedure may sound like this:

"30 seconds to go, stand by studio, see your first shots cameras, 15 seconds, roll VTR, 10 seconds stand by film, 5 seconds roll film, stand by grams and studio, (on air) fade up grams and film, telecine slide stand by, super slide, lose super, telecine change slide, super slide, lose super, fade music, cue and mix to camera 2, camera 4 tighten your 2 shot, camera 1 steady, cut to 1, coming to 4 and cut."

Despite all these precautions, mistakes still occur during the show and the director is the only person who can counteract them by a split-second assessment and a rapid chain of commands to change that immediate part of the programme without marring the whole. E.F.P.

See: *Director* (FILM); *Producer.*

Books: *Techniques of TV Production*, by R. Bretz (New York), 1962; *TV Production Handbook*, by H. Zettl (Calif.), 1961; *Technique of TV Production*, by G. Millerson (London), 1967.

⊕**Direct Positive.** Positive print prepared directly from the original negative film without the use of intermediate stages; also describes the positive image obtained by the reversal processing of the original film exposed in the camera.

□**Direct Switcher.** Vision switcher in which the push buttons operate switches carrying the vision signal as opposed to sending a switching signal to a relay or matrix.

See: *Vision mixer.*

⊕**Discharge Lamp.** Light source of the arc type, containing a gas under high pressure through which the electric discharge takes place. Used in printers and projectors.

⊕**Discontinuous Spectrum.** Spectrum consisting of relatively few, well-separated lines as opposed to the broad, continuous emission band of most light sources. Feature of sodium and mercury vapour lamps, increasing their usefulness in some applications (e.g., film printers) and reducing it in others (e.g., studio lighting).

□**Discriminator.** Circuit which converts frequency excursions into amplitude excursions of the signal. Various circuits have been devised to eliminate from the final signal any amplitude excursions which were in the original mainly frequency-deviated signal.

See: *Receiver.*

⊕□**Dispersion.** Separation of light of mixed wavelengths into its constituents by refraction at the surface of a medium whose refractive index varies with wavelength. The dispersive power of a transparent medium such as glass is a measure of the extent to which wavelengths are so separated, and may be specified as:

$$\frac{\mu_B - \mu_R}{\mu_G - 1}$$

where μ_B, μ_G and μ_R are the refractive indices of the material for blue, green and red light respectively.

See: *Light.*

⊕□**Dissolve.** Result of the optical overlapping of two lengths of film, and formerly called a lap dissolve. As the first scene is progressively faded out, the second is faded in, so that the effect on the screen is of the first scene merging imperceptibly into the second, and taking its place. Also called mixes.

The term is equally applicable to television, where the progressive fading up of a signal from one picture source is accompanied by the fading down of another signal from a second source.

⊕**Distortion.** Lens defect in which the shape of □the image is incorrectly rendered, usually as a result of the magnification at the edges of the field differing from that in the centre.

□**Divcon.** Device for displaying printed messages on a television screen from information received from its computer-type store, a tape reader, or a keyboard provided. Developed in the US by RCA.

The apparatus consists of an operating console, a rack assembly which includes the computer-type store, and a converter for transforming teletype data to the Divcon format.

The console is a pedestal desk comprising the keyboard, tape punch, tape reader and controls.

The Divcon main frame is capable of producing:

Display Channel 1: 16 lines of 32 characters each.

Display Channel 2: 9 lines of 16 double-sized characters and one line of 15 double-sized characters for a "Datacaster".

Display Channel 3: 9 lines of 16 double-sized characters.

The input to the store can be as fast as 2,000 characters per second. As the apparatus is based on computer techniques it can readily store in digital form data messages composed of coded character information such as teletype or a computer language. The system provides means of translating this information into a form suitable for superimposing or otherwise displaying on standard television equipment. It is a convenient method of superimposing printed information over a picture and is widely used for election and sporting results, news headlines, tracking advertisements, foreign language translations, etc. The access to the store is random and hence corrections and alterations can be made without having to replace a whole block of data.

See: *Special events.*

⊕**D Log E Curve.** Characteristic curve of a photographic material showing the relation between the exposure and resultant density of the image.

⊕**D-Mount.** Standard form of interchangeable screw-threaded lens mount, generally used for 8mm motion picture cameras; the major diameter of the mounting screw is 0·625 in. with 32 threads to the inch and the focal plane is 0·484 in. behind the seating face.

See: *Narrow-gauge film and equipment.*

⊕**Dolly.** Wheeled mounting device for film □and television cameras. It allows the camera to be tracked forward and back by an operator other than the cameraman, while the cameraman controls the height of himself and the camera. A range of about 3–9 ft. is satisfactory. The cameraman can also pan and tilt the camera in the usual way. An additional desirable feature is that the unit can move sideways without forward or reverse movement, thus allowing an extended panning shot (crabbing). The dolly adds extensively to the flexibility of operation within a studio. It can be hydraulically powered, but must be completely silent in operation. It enables the cameraman to concentrate on his shot and maintain accurate focus, without having to bother about tracking. Thus long, smooth, accurate tracks can be obtained, and in addition there is a height advantage over the more usual pedestal.

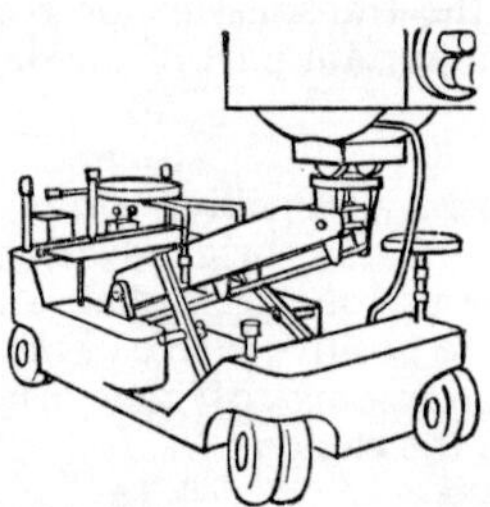

The most sophisticated dollies have four separate wheel units, two of which are used to drive the truck, and the other two, diametrically opposed to the driving wheels, for deceleration. The cameraman is able to rotate the camera independently of the mounting. The hydraulic drive moves the dolly at very slow speeds, but full torque is always available to bring the truck up to top speed, which is in the region of 12 ft. per second. Full braking is available at all times, even with motors at maximum power, and numerous safety precautions are incorporated to prevent misuse and accidents.

LIGHTWEIGHT OB CAMERA DOLLY. Camera dolly designed for outside broadcast (field pick-up) work. It is able to track reasonably well over rough ground, and is easily transportable to the OB site. A typical dolly can be tracked along a passage 54 in. wide, and turned 90° through a doorway 30 in. wide without reversing. Pneumatic and solid tyres can be quickly interchanged. The whole chassis is composed of steel tubes mounted with clamp brackets. A seat is provided for the cameraman. The unit can be manoeuvred by means of a push bar. A camera head mounting platform is provided for the pan and tilt head and the camera. Lens heights of from 5 ft. to over 6 ft. are obtainable by means of telescopic support legs.

LIGHTWEIGHT HEAD. Pan and tilt head constructed of lightweight alloys for limited weight cameras. Designed for outside broadcast units and for other specialized needs.

MOTORIZED PAN AND TILT HEAD. A power-driven remotely controlled device for pointing the camera where required. It is often used to control simple, repetitive operations. The complex variety of shots taken by a television cameraman during the course of, say, a play, would preclude the use of such a unit, but for continuity studio purposes, or for repetitive monitoring with an industrial camera channel, the equipment has much to commend it.

The panning and tilting are carried out by servo motors. The centre of tilt coincides approximately with the centre of gravity of the camera. Normally the panning movement is limited to about 45°, centred anywhere within a radius of 300°. The tilt can be controlled up to about 60° radius, and the speed of pan and tilt can be set to a predetermined rate. It is usual to rack-mount the servo-amplifiers separately.

See: *Camera* (FILM); *Outside broadcast units; Studio complex.*

⊕**Donisthorpe, Wordsworth, 1848–1914.** English barrister who spent over twenty years in the development of experimental motion picture apparatus. His first patent, filed in 1876, described a camera in which photographic plates could be exposed in rapid succession and the resulting pictures printed on a continuous strip, to be viewed intermittently, so producing an animated picture. In 1878 he suggested a combination of this Kinesigraph apparatus with the newly invented Edison phonograph, to be used to record and present dramatic productions. With W. C. Crofts, he patented an ingenious camera-projection apparatus for motion pictures in 1889. Surviving celluloid film exposed in this apparatus in 1890 shows images of remarkable quality. No sponsor could be found and it was never exploited.

⊕**Dope Sheet.** Analysis of film material prepared for library classification.

□**Doppler Shift.** A general phenomenon, of importance in satellite communications. Denotes an apparent change of frequency of a wave train caused by relative motion between the wave source and an observer. If a source emitting waves approaches the observer, the waves are crowded together and their frequency thereby increased. Similarly, if the source recedes, the waves are stretched out and their frequency is decreased. An example of the Doppler shift is in the change of note of a train whistle as the locomotive approaches and passes an observer.

See: *Satellite communications.*

⊕**Double-8mm Film.** Strip of film 16mm in width, half the width being exposed during the first run through the camera and the

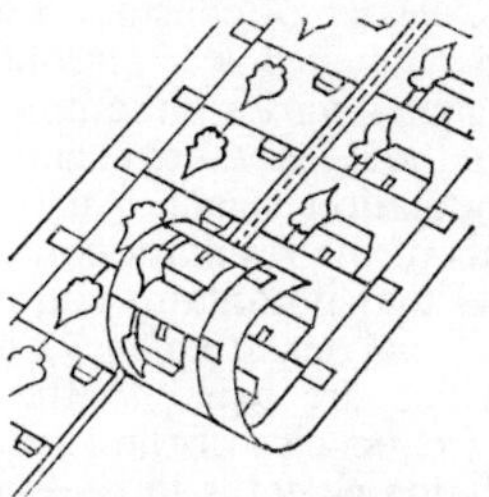

other half during a second passage. This double-8 film is slit after processing to provide the standard 8mm strip used in amateur cinematography.

See: *Film dimensions and characteristics.*

⊕**Double Exposure.** Combination of two images on the same strip of film by running it twice through the camera or printer.

⊕**Double Frame.** Image size occupying the space of eight perforations along the length of 35mm motion picture film, rather than

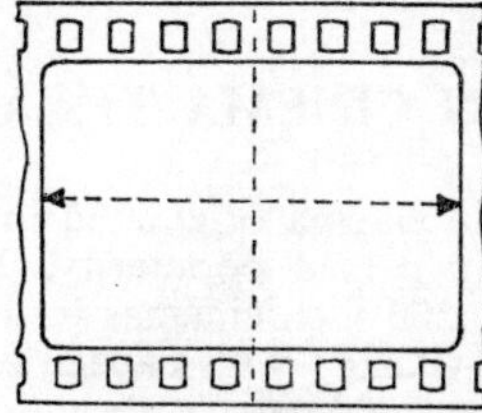

the standard four-perforation frame; the film runs through the camera horizontally rather than vertically. Double frame processes, such as Vistavision and Technirama, have been used to obtain a larger area of negative on 35mm film for wide-screen presentation.

See: *Wide screen processes.*

⊕**Double-Head Projection.** Projection of film picture and its accompanying sound on separate but synchronized lengths of film. Requires mechanically or electrically coupled picture and sound heads.

⊕**Double Perforated Stock.** Motion picture film with perforation holes along both edges. All normal 35mm and 70mm is perforated in this way, but 16mm stock may be double perforated for camera use and for making silent prints, or single perforated along one edge only for prints with sound tracks.

□**Double Sideband.** Transmission comprising the carrier frequency as well as both sidebands resulting from modulation of the carrier.

If an r.f. carrier of constant amplitude is modulated by a lower frequency, the resultant signal varies in amplitude. The envelope of the signal takes exactly the form of the modulation signal. Analysis of the components of the resultant complex waveform shows that it contains components at the original carrier frequency, plus components which are equal to the sum and difference frequencies of the two signals. If the modulating signal is in itself a complex waveform, then the highest frequency produced equals the carrier frequency plus the highest modulating frequency, and the lowest frequency equals carrier frequency minus the highest modulating frequency.

The signals above the carrier frequency are known as upper sidebands, and the signals below carrier are known as lower sidebands. The whole is called a double sideband transmission. The bandwidth necessary to accommodate the signal is twice that of the modulating signal.

See: *Transmission and modulation.*

⊕**Double 16 Millimetre Film.** Positive film perforated with 16mm sprocket holes so that, after printing and processing, it can be

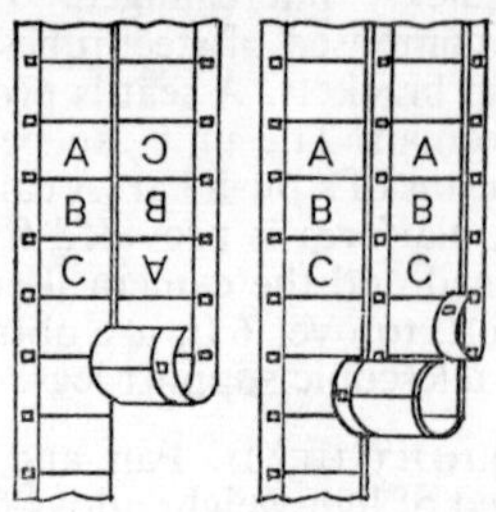

slit to provide two 16mm copies; its use is limited to the manufacture of large batches of prints at the laboratory.

See: *Film dimensions and physical characteristics.*

⊕**Double-System Sound Recording.** When film picture and sound are being taken simultaneously, it is usual for the sound to be recorded separately, as a rule on magnetic film or tape. Double-system recording usually requires special means for synchronizing camera and recorder.

See: *Sound recording* (FILM).

□**Downlead Impedance.** Characteristic impedance of the coaxial or twin transmission line between aerial and receiver.

□**Downwards Conversion.** Standards conversion to a lower (e.g., 625 to 405) line standard.

⊕**Drier.** Laboratory operator who handles the final stages of a developing machine for film processing, where the film passes from the washing tanks into the drying room for eventual winding up. With the general use of enclosed drying cabinets on processing machines, the subdivision of developer operators as driers has ceased to have much significance.

⊕DRIVE-IN CINEMA THEATRE

The drive-in cinema originated in America and quickly gained popularity. There are now over 4,500 such cinemas in the United States and Canada with capacities ranging from 200 to 2,800 cars, with an average of about 750. As a broad guide, the capacity of a drive-in cinema giving one performance per day is about 5 per cent of the total number of vehicles (other than lorries and heavy commercial vehicles) registered in the locality.

The choice of location must be made with care. It must not be more than five or six miles from an area which supports a population having a high proportion of cars, but not in a low-lying area where mist or fog are prevalent; it should be easily accessible from a main or good secondary road;

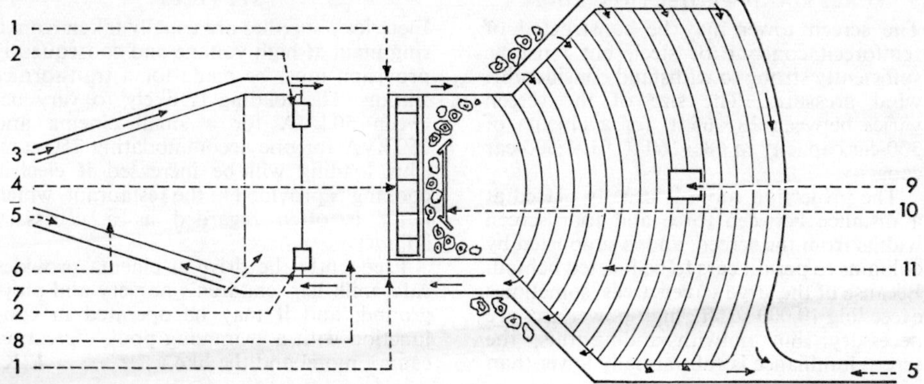

TYPICAL DRIVE-IN CINEMA FOR 650 CARS. 1. Control gates. 2. Pay boxes. 3. Main entrance. 4. Administration offices, restaurant, toilets, etc. 5. Exits. 6. Petrol and service area. 7. Waiting area. 8. Parking area. 9. Projection suite. 10. Screen. 11. Car bays.

transport and other authorities will impose restrictions upon positions of entrances and exits; and water, drainage and electricity supplies must be available.

LAYOUT

The general layout of the drive-in is substantially similar to the indoor theatre; that is to say, the viewing area is fan-shaped, enclosing a segment having an included angle of about 80°. The area should be oriented so that the light from the setting sun falls neither within the field of vision of the spectators looking at the screen nor directly on the screen surface, which would degrade the contrast of the picture. It is desirable that the perimeter of the area be well-wooded; if not, trees should be planted behind and on each side of the screen tower, and bushes below the screen. Besides affording protection against light and wind, this natural surround provides the décor.

A natural declivity of the ground towards the screen is desirable but a ramp for each car station is generally necessary.

DRAINAGE

The whole area must be adequately drained, and both car stations and roadways must be properly made up for the vehicular traffic by macadam or compounds of bitumen, small stones, sand or cement, etc. The subsequent wear due to tractional resistance is extremely small but that due to internal attrition is large, accounting for about 85 per cent of the maintenance costs. Deterioration due to weather accounts for 5 per cent, except where extremes of temperature do considerable damage. Adequate drainage must be provided not only for the interception of surface water that would otherwise flow across the area or would flood it, but also for the interception of seepage water.

On the other hand, effects of drought and high temperature can readily injure the area, particularly if the surface is of a clayey nature or of quartzite, sandstone or flints. Apart from the nuisance of dust, such surfaces will break down when rain occurs unless they have been treated with suitable binding media.

CAR BAYS

Car circulation requires to be carefully planned: it is usually on a one-way system with roadways behind each row of car bays sufficiently wide to allow cars to enter the bays without filling and backing. Each bay should be about 9 ft. wide to permit car doors to be opened sufficiently to allow the passage of a person.

At each pair of car bays is a post to carry two watertight speakers: frequently car heaters are also provided. The electrical conductors and cables for these facilities are laid underground.

In areas that are reasonably free from electro-magnetic disturbances, the sound from the film is transmitted to wires arranged around the periphery of the audience area. The patrons receive the sound via small solid-state amplifiers with speakers similar to transistor radio receivers. Elimination of wiring at each car position simplifies design and increases reliability. The result is that the installation and maintenance of such induction systems generally costs less than the provision of pairs of wires to each car bay post.

SCREEN AND PROJECTION ROOM

The screen tower may be constructed of reinforced concrete or steel, but must be sufficiently strong to withstand considerable wind pressures. The size of the screen varies between 28 × 65 ft. for a cinema of 300-car capacity to 68 × 160 ft. for 1,500-car capacity.

The projection room is usually placed at a distance between three and four screen widths from the screen, and is supported by columns to permit cars to be located behind. Because of the large screen areas, sometimes exceeding 10,000 sq. ft., high-power arcs are necessary; but notwithstanding this, the screen luminance is substantially lower than in indoor cinemas.

SERVICES

Electricity supplies are usually taken from a ring main at high voltage and consequently provision must be made for a transformer station. The loading is likely to vary between 50 kVA for a small cinema and 250 kVA for one accommodating 750 cars. This loading will be increased if electric cooking is provided in the restaurant, which latter is often regarded as a necessary adjunct.

Frequently the drive-in cinema provides café, milk bar, children's nursery and playground, and it may be operated in conjunction with a swimming pool, recreation centre, motel and the like. L.K.

See: *Cinema theatre; Projector; Screens.*

□**Drive Unit.** The unit which drives the large output stages in a high-powered television transmitter, normally of 5 kW upwards and operating in the VHF and UHF bands of 40 MHz to 850 MHz. Transmitters must be of a very high order of stability, and so are usually crystal-controlled. To obtain high power at these very high frequencies, the transmitter is designed in three sections: the drive unit, r.f. amplifier and output, and modulator.

The drive unit normally consists of a crystal-controlled oscillator followed by a number of frequency multipliers, since crystals cannot be manufactured to operate at the very high frequency required. The resultant carrier frequency may be modulated before feeding into a small power amplifier capable of delivering about 10 watts. This output is then used to drive the larger power amplifiers of the transmitter.

See: *Transmitters and aerials.*

⊕□**Dropout.** In magnetic recording, sudden momentary losses of signal caused by dust particles or other major imperfections of the tape.

⊕□**Drum (Sound).** Freely-rotating cylinder, often combined with a flywheel and supported by ball or roller bearings, round which optical or magnetic sound film is wrapped in a recorder or reproducer to give it the steadiest possible movement as it passes the record or reproduce heads.

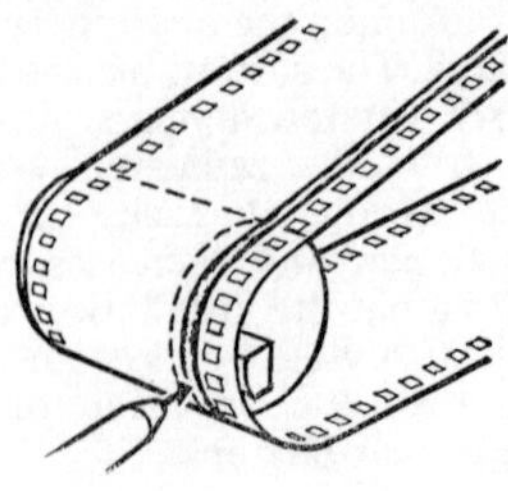

⊕**Drum Developing.** Process sometimes used in the field for processing camera tests. Short lengths of film may be developed by winding them on a drum which is immersed in turn in the developing solutions.

⊕**Drying.** Elimination of excess moisture accumulated within and on the surface of film during processing. In a continuous developing machine, surface moisture is taken off with a squeegee, and the film then passes into drying cabinets (dry boxes) where warm dry air is circulated.

See: *Processing.*

□**Dual Standard Receiver.** Receiver capable of receiving TV pictures on more than one line standard, e.g., 405/625 lines in Britain.

⊕**Dubbed Version.** Foreign-language version of a film or TV recording in which the original dialogue has been replaced by new dialogue carefully synchronized to the actors' lip movements.

See: *Foreign release.*

⊕□**Dubber.** Sound reproducer of the highest possible quality on which sound tracks (normally magnetic) are run in synchronism with other sound tracks on similar machines during the process of dubbing or mixing.

⊕□**Dubbing.** Combination, by mixing, of several sound tracks into a single track. In all film production, and in most types of TV production which are not transmitted live or recorded continuously, there is an editing process in which picture and sound materials are assembled from different sources. The sound materials (dialogue, music, sound effects, etc.) are edited on to a set of synchronized sound tracks, and it is the combination of these on to a single final sound track which is known as dubbing or mixing.

See: *Sound mixer; Sound recording.*

⊕**Dubray-Howell Perforation.** Form of sprocket hole for 35mm motion picture film, having the dimensions of the Bell-Howell perforation but with rectangular rather than rounded ends. Now obsolete.

⊕**Ducos du Hauron, Louis, 1837–1920.** French scientist, interested in photography from 1859. In 1864 he proposed a multiple-lens camera-projector apparatus for taking and presenting a series of pictures of moving objects, but the apparatus was never constructed. Patented a process for three-colour photography (1868). His book, *Les Couleurs en Photographie* (1869), described not only an additive system but also outlined most of the subtractive methods that have since been employed, although none of the processes would have been practical at the time. In 1891 he devised an anaglyph system for stereoscopic viewing. Died in poverty.

⊕**Dufaycolor.** Obsolete additive colour film process, in which the film was coated with a very large number of minute red, blue and green filters or screens, the proportion of the three colours being such that there was no resulting colour when the film was viewed at a distance. The filters were arranged in a regular grid called a reseau, so that duplicates could be made by exact registration. Film was threaded in the camera with its base and filter grid towards the light, so that the light reflected from each point in the scene was filtered according to its blueness, greenness and redness before it reached the emulsion. Thus the image was formed in silver opacities corresponding to the original colours, and it was only necessary to reverse these into transmittances to recreate the original colours when the film was viewed by transmitted light. When the film was projected, the transmission of each minute group of three-coloured filters through the emulsion over it corresponded to the brightness of the original light reflected from the corresponding point in the scene. However, the mesh of the reseau, the light loss in the screen, and the dilution of colour resulting from overlapping filter transmissions tended to make the system inefficient in spite of its many attractions, and it was abandoned.

See: *Colour cinematography.*

⊕**Dunning Process.** Early method for the combination of separately photographed foreground and background action in black-and-white special effects cinematography.

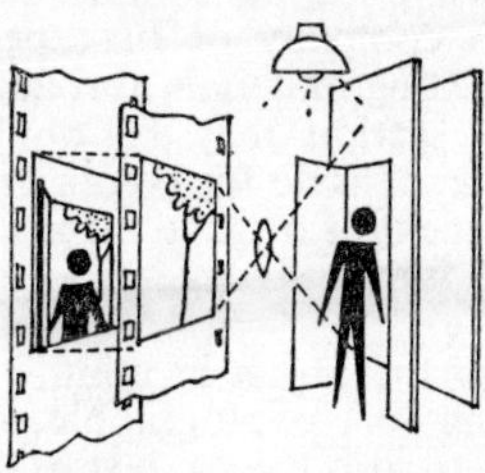

The foreground action was lighted with yellow light only in front of a uniform strongly illuminated blue backing. Panchromatic negative film was used in the camera as the rear component of a bipack in which the front film was a positive yellow dye image of the background scene. This yellow dye image was exposed on the negative by the blue light from the backing areas but the yellow light from the foreground passed through it and recorded an image of the foreground action at the same time.

See: *Special effects* (FILM).

⊕**Dupe Negative.** Short for duplicate negative, a reproduction of the original picture or sound track film negative. Dupe negatives may be prepared either to obtain characteristics not present in the original image, as in special effects work, or to protect and extend the production availability of the assembled original negative, as when dupe negatives are prepared for release printing simultaneously at different laboratories.

See: *Duplicating processes.*

⊕**Duplex Sound Track.** Type of symmetrical variable-area optical sound track on film,

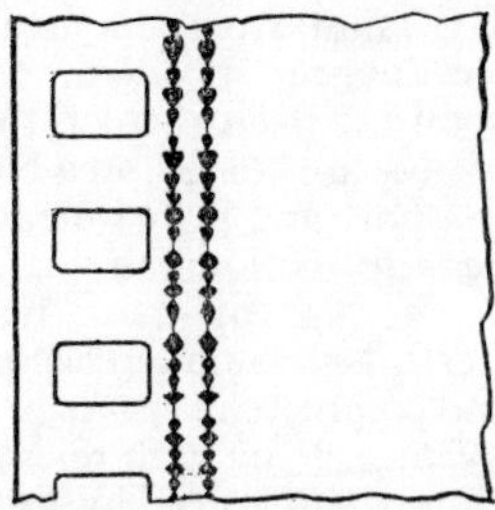

recognizable by the fact that in the negative the central axis and surrounding modulated area are opaque; in the equally common bilateral type of track they are transparent.

⊕DUPLICATING PROCESSES (BLACK-AND-WHITE)

Most motion picture productions include a certain amount of duplicated footage, if only in the opticals and stock inserts, while quite frequently the whole of a feature may be printed from a duplicate negative, either on the grounds of a policy adopted by the production company or of expediency in sending printing materials abroad. Less frequently, a section may be duplicated in order to compensate for errors in exposure or development of original material so as to match up with the scenes into which it is to be cut.

Whatever the object in mind, the aim in the making of photographic duplicates (dupes) is to produce a result that is indistinguishable from a print taken from the original negative. This is rarely achieved in practice, since there is some inevitable fall-off in quality, revealed as a loss in sparkle or definition in the dupe print, as a result of internal light scatter, emulsion graininess and so on. This loss can be kept to within reasonably narrow limits through the use of fine-grain stocks and by suitable control of exposure and development. Even so, it cannot be overcome altogether and its effect is magnified with each successive step in the duplicating process. Apart from this loss in definition, however, it is a comparatively straightforward matter to reproduce the photographic tonal quality of the original negative reasonably accurately in the dupe, provided that certain simple conditions are observed.

Two separate operations are involved in the making of photographic dupes: first, the preparation of the master positive from the original negative, and secondly, the preparation of the dupe negative from the master positive.

MASTER POSITIVE

A variety of film stocks is available for making the master positive, the choice depending on the use to which the latter is to be put. Some are coated on a bluish-grey-coloured support and prints made on them are referred to colloquially as blues or lavenders. The use of these materials is generally restricted to newsreels or to documentary and commercial shorts, where compatibility of processing with release positive film is an important consideration.

Others are coated on a clear support and have an extremely fine grain, so much so that the colour of the developed image is distinctly warm or reddish; it is for this reason that prints made on these particular films are referred to colloquially as red masters. It is regular practice for one or more fine-grain duplicating positive prints to be made from feature film negatives before bulk printing of projection prints is undertaken, and the red master is sometimes referred to also as the insurance print in the sense that it provides a means of supplying a replacement should the original negative be either damaged or lost.

These films are extremely slow, are sensitive to blue light only, and carry a yellow dye in the emulsion layer to limit the spread of the image through the depth of the layer and so increase the resolution obtainable.

It is general practice for duplicating positive films of 35mm format to be supplied with standard negative perforations. As an aid to recognition, the edge-marking carries the letter D before the footage number.

In making the master positive, the exposure through the negative is so chosen that the minimum density to be found on the developed print is not less than about 0·5 above the base. This is to ensure that all densities of the print lie above the toe of the characteristic curve of the duplicating positive material, a condition necessary to ensure clear highlights in the final, or projection print. Development of the master positive is to a gamma of about 1·3, for which purpose it will usually be found necessary to process it in the negative machine.

It is a mistake to regard it as an ordinary positive film and develop it as such in the positive machine. If this is done it is quite impossible to obtain a satisfactory result in the final print, which will exhibit the well-known duped appearance all too familiar to those who have tried this expedient.

While the master positive is generally darker than an ordinary print, it has a much lower range of tones, giving it an overall soft appearance.

DUPE NEGATIVE

Materials suitable for making the dupe negative include both panchromatic and blue-sensitive films coated on to grey-base support. Dupe negatives made on them are sometimes referred to colloquially as grey-backs. The speed of these films is very much less than that of camera films and grain size is correspondingly finer.

In making the dupe negative, the exposure through the master positive is so chosen that the minimum density on the developed negative is not less than about 0·4 above the base, i.e., above the toe in the characteristic curve of the material. As in the case of the master positive, this condition is necessary to ensure a satisfactory result in the final print. If the minimum density is allowed to fall below this value, the print shows a loss of shadow detail as well as insufficiently-dense blacks generally.

Development of the dupe negative is to a gamma of between 0·60 and 0·65, for which purpose the normal picture negative developer may be found suitable if appropriate machine conditions can be obtained; if not, a less active, or more dilute, developer may be required.

The dupe negative generally looks darker than the original negative as a result of its special characteristics. There may, however, be exceptions to this rule, as in optical printing work such as lap-dissolves, etc., where, if the whole of a scene is not to be duped, it is desirable to make a close match with the original negative density to avoid a noticeable jump in the resulting print.

In the duplication of a feature, or multiple-scene production, a very desirable aim is to finish up with a one-light duplicate

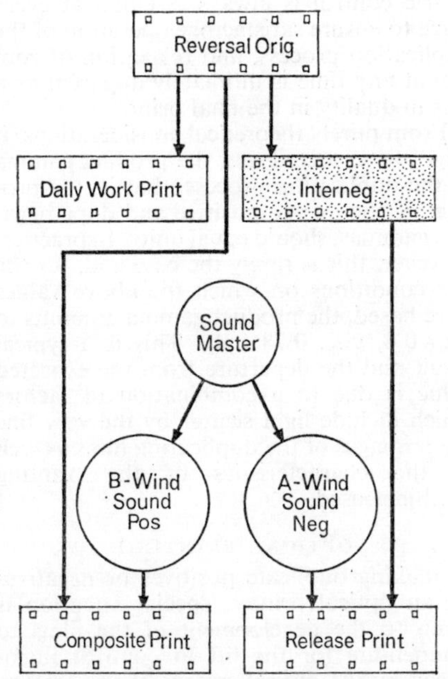

16MM BLACK-AND-WHITE REVERSAL ORIGINALS. Derivation of prints by direct contact printing and by contact printing from an internegative.

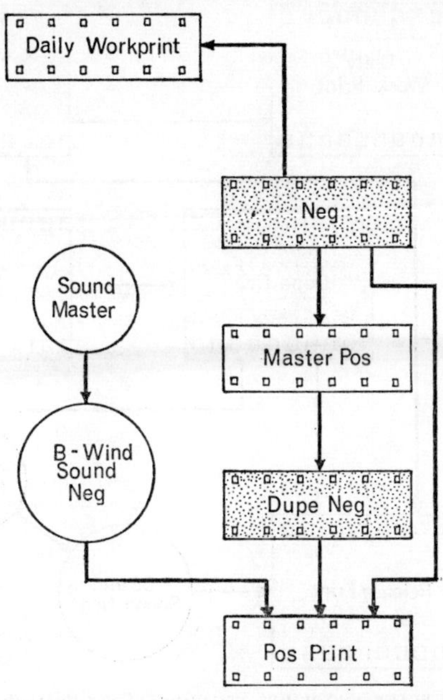

16MM BLACK-AND-WHITE NEGATIVE. Derivation of prints by direct contact printing and by contact printing from a dupe negative.

negative, that is to say, a negative which can be printed with no printer light changes throughout its length. To achieve this ideal condition, the grading of the master positive is accurately assessed in the first place and it is sometimes necessary to introduce further minor corrections when printing the duplicate negative.

OPTIMUM CONDITIONS

Strictly speaking, the particular values of master positive and duplicate negative gamma quoted above refer to the conditions of test under which they were determined. It is therefore to be recommended that each laboratory should establish for itself the optimum conditions that will give the most satisfactory result, since equipment variations, especially of printing machines, may sometimes affect the result. Such a test – it need be carried out only once to establish the procedure to be adopted – is best performed with a suitable test negative (a studio close-up is satisfactory) to which is joined a specially prepared step-tablet on 35mm film having some 15 full-frame steps with a step increment of 0·15 in density; i.e., a density range of 0 to 2·10.

If such a step-tablet is not available, a processed negative control strip can be used.

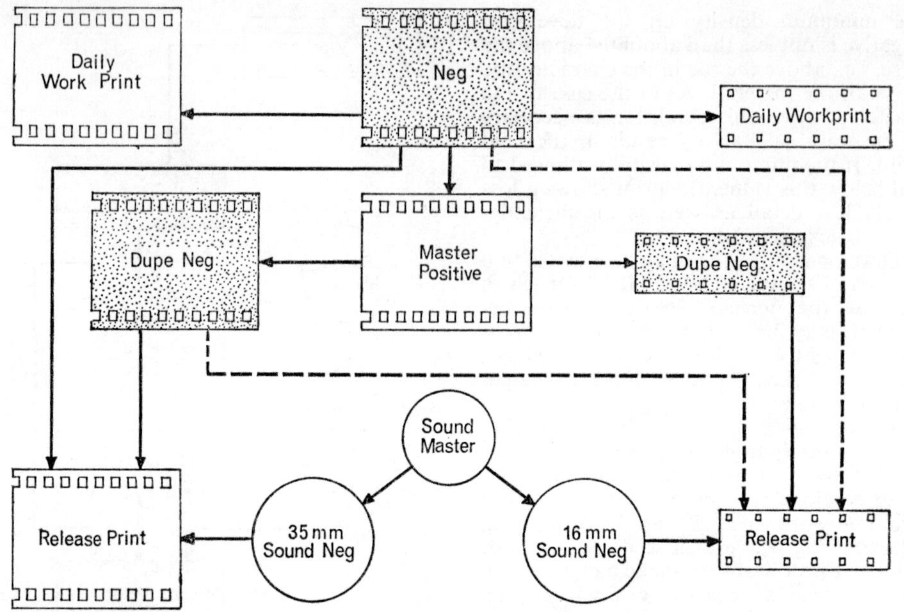

35MM BLACK-AND-WHITE NEGATIVE. (*Solid line*): derivation of 35mm prints by direct contact printing and from a dupe negative. (*Dashed line*): derivation of 16mm prints by reduction printing from the original negative and via a 16mm dupe negative.

In either case, the steps are marked which correspond most nearly to the average densities of important highlight and shadow areas of the picture negative.

Three prints are made from this composite test negative on duplicating positive stock and developed, together with appropriate sensitometer strips, to gamma values of 1·2, 1·3 and 1·4 respectively. From the resulting three master positives, three dupe negatives are prepared from each and developed to gamma values of 0·5, 0·6 and 0·7, respectively. Prints are then made from all nine negatives resulting from this test and compared with the best print that can be made from the original negative. This can be done visually with the prints from the camera subject, and densitometrically from a study of the readings made of the step-tablet prints.

In doing this, interest is centred on the part of the characteristic curve that lies between the two marked steps originally selected to represent average highlight and shadow areas of the picture negative. Density values which fall outside this range can safely be ignored for the purpose of this test.

The particular combination of master positive and duplicate negative gamma that is found to give the most acceptable result is adopted as standard practice. Sensitometric control is always essential at every stage to ensure satisfactory operation of the duplication process, and relaxation of control at any time is ultimately apparent as a loss in quality in the final print.

From purely theoretical considerations, it might be expected that the product gamma of the duplication process, that is, the product of the master positive and dupe negative gammas, should equal unity. In practice, however, this is rarely the case and, for the test conditions on which the above values were based, the product gamma amounts to $1{\cdot}3 \times 0{\cdot}6$, viz., 0·78 only. This is a typical result and the departure from the expected value is due to a combination of factors which include light scatter by the very fine grain images of the duplicating films as well as the characteristics of the printing machines used.

OPTICAL PRINTING

In making duplicate positives or negatives on an optical printer, special attention is given to the development of the films to compensate for the lift in contrast introduced by the optical system of the printer. This lift is due to the fact that a proportion of the light transmitted via the printing medium, be it negative or positive, is lost by scattering and does not reach the film being exposed.

Where the two films are held together in close contact, as in an ordinary printing machine, this scattered light is not lost and contributes effectively towards the formation of the image. The difference between the two techniques of printing is such as to give rise to different values of density even when exposures are made to an equal extent. With contact printing, diffuse densities are operative, whereas with optical printing, projection densities apply; the ratio between the two is known as the Callier coefficient, and its precise value depends on a number of factors such as the type of emulsion used, its grain size and so on. Projection densities are always higher than their corresponding diffuse densities.

In practice, the increased contrast observed in optical printing must be compensated through sensitometric control by appropriately restricted development. The same holds true of similar systems of printing where the two materials are not held in contact, such as in reduction printing and in making blow-ups.

The extent of the restriction necessary can only satisfactorily be determined by trial and error.

REVERSAL DUPLICATES

In printing 16mm reversal films, it is sometimes convenient to make copies by direct reversal, so obviating the need to provide an intermediate duplicate negative. For this purpose, slow, fine-grain films sensitive to blue light only are available. Development must be ascertained by experiment, but a low value of gamma (about 1·2) should be aimed at to avoid an over-contrasty result.

Where an appreciable number of copies is required, it is more usual to provide an intermediate duplicate negative, since this protects the original reversal film from damage that may be sustained in repeated runs through a printing machine. Its degree of development must be found by experiment, but will almost certainly be considerably less than that to which it is normally developed. R.H.B.

See: *Laboratory organization; Printers; Processing; Sensitometry and densitometry; Tone reproduction.*

Books: *Theory of the Photographic Process*, ed. by C. E. K. Mees and T. H. James, 3rd Ed. (New York), 1966; *Control Techniques in Film Processing, SMPTE Jnl*, (New York), 1960; *Motion Picture and Television Film*, by D. J. Corbett (London), 1968.

Duplitized. Coated on both sides. Films consisting of a base coated with emulsion on both sides were commonly used for obtaining two-colour prints by the dye-toning process.

Dye Coupling. Coloured dye images may be obtained by adding groups such as amines and phenols to certain developers, the image being formed by the coupling or condensation of the oxidized developing agent and the amine or phenol. The couplers may either be incorporated in the emulsion or added during processing.

See: *Colour cinematography.*

Dye Toning. Production of a coloured photographic image by converting the black-and-white image into a dye image proportional to the silver originally present.

Dye Transfer. Colour printing process in which a dyed image, usually formed by absorption of dye into a relief film, is transferred on to a blank film.

See: *Imbibition printing.*

Dynalens. Trade name of an American device for stabilizing the image of a film or television camera subjected to unavoidable vibration. No attempt is made to isolate the camera from the source of vibration. Instead, a correction is introduced into the optical path of the camera lens by altering the refraction of a variable element mounted in front of the lens. In essence, this element consists of two glass plates, normally parallel but flexibly connected to form a cell containing a transparent fluid. Gyroscopic sensing systems detect the vibration, and their outputs are applied through servo amplifiers to the glass plates to produce a compensating deflection. Both horizontal and vertical vibrations can be corrected, or at least mitigated, by this device.

See: *Stabilization (image).*

□**Dynamic Focusing.** Action of evenly focusing a raster over the whole of the face of a cathode ray tube. It involves corrections at the edges of the picture owing to the longer path from the deflection field to the edges of the tube.

See: *Camera* (TV); *Telerecording.*

Dynamic Frame. Cinema presentation in which the size and aspect ratio of the image shown on the screen is varied throughout the reel in accordance with the pictorial composition and dramatic requirements; 35mm film printed with a variable opaque mask is used.

See: *Wide screen processes.*

□**Dynode Effect.** A form of distortion produced by image orthicon tubes.

In an image orthicon, the beam, after scanning the target, is amplified in an electron multiplier section. The multiplier often consists of five dynode sections and a

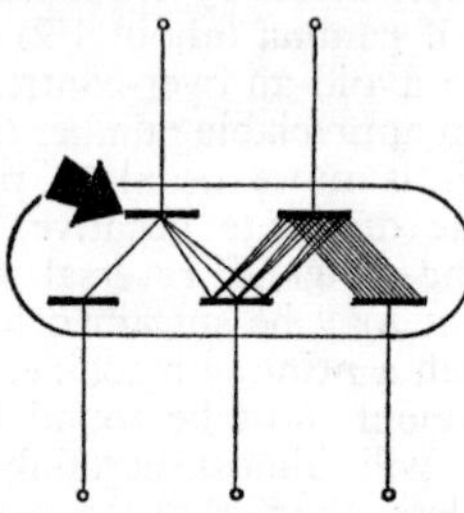

collector, and the return beam is directed on the first dynode under the influence of the persuader voltage. Occasionally imperfections in the dynode surface cause a small ghostly highlight with a spreading tail to appear on the picture.

The visual effect is very similar to that of lens flare, and hence has been described as dynode flare. Slight alteration of beam focus voltage or multifocus voltage on the tube will usually remove the effect, which is now not nearly so common as it was, owing to improved treatment of dynode surfaces.

See: *Camera* (TV).

□**Dynamic Range.** In sound recording, the ⊕range, measured in decibels, between the highest peak of the loudest sound to be recorded, and the softest whisper or quietest musical passage which must be audible. This dynamic range very often exceeds the capacity of the recording medium, and might prove disagreeable to the ultimate audience if it could be recorded and reproduced. The remedy lies in compression and peak limiting, but much can be done by skilful mixing to make the necessary reduction of dynamic range less noticeable.

See: *Sound recording.*

Eastman, George, 1854–1932. American inventor, born in Waterville, New York, USA. Moved with his family to Rochester, New York, in 1860, where his father ran the Eastman Commercial College. His father's death two years later brought hard times, and Eastman left school at 14, working first as an office messenger and later as a junior clerk. At 24 he became interested in photography, but was dissatisfied with the cumbersome equipment and awkward operation of the wet collodion process. Learning from British photographic journals of the new gelatin dry-plate process, he began making his own materials. In 1880 he leased part of a building in Rochester and started the manufacture and sale of gelatin dry plates.

Searching for substitutes for the bulky and heavy glass plate, Eastman produced paper roll films in 1884, devising in the next year a method for stripping the processed emulsion from the opaque paper base for printing. He decided to try to popularize photography by simplifying the process, and designed a new camera – the Number 1 Kodak. A box camera, it contained a roll of stripping paper film, enough for 100 pictures.

After exposure, the film was returned – in the camera – to Rochester for processing and printing, the camera being returned loaded with fresh film. Marketed with the slogan, "You press the button, we do the rest", it was an immediate success.

In August 1889, Eastman introduced a transparent celluloid roll film for the camera, which was used by Edison and other early workers for their motion picture experiments.

Eastman was a noted philanthropist who gave in his lifetime $20 million to the Massachusetts Institute of Technology, founded dental clinics in America and Europe, endowed schools of medicine and music, theatres and symphony orchestras. He pioneered employee benefit schemes and introduced profit-sharing, sickness and life insurance and pension schemes during the first quarter of the century when they were virtually unknown in industry. He died by his own hand in 1932 in the belief that his life's work was complete.

Eastman Color. Integral tripack colour film using negative/positive processes introduced for 35mm professional motion picture production in the United States in 1952. The negative material is characterized by automatic colour masking formed during development, the advantages of which have led to its very widespread use internationally.

See: *Integral tripack.*

⊕**Eberhard Effect.** Edge effect in processed film images which takes the form of accentuated density differences at the boundary between exceptionally heavy and light areas.

See: *Edge effect.*

⊕**Echo Effects.** Echo and reverberation effects □are often added during sound dubbing. Formerly this was achieved by the use of echo chambers, but variable delay lines are now used instead.

□**Echo Waveform Corrector.** Corrector for linear phase and amplitude distortions on a television signal, resulting from multi-path signals or echoes. Operates by mixing with the input signal other samples of the signal, advanced or delayed in time relative to the main signal and variable in amplitude and polarity.

See: *Transmission networks.*

□**Edged Captions.** Plain black-and-white captions can be made more striking by feeding the signal obtained from the scanned caption along an unterminated delay-line and adding the resultant delayed, differentiated waveform to the original signal. This gives an edged, almost three-dimensional effect.

See: *Special effects* (TV).

⊕**Edge Effect.** In photographic processing, local depletion of the developing agents in the solution at the boundary of a heavily exposed area of image can lead to the density of the edge of adjacent low exposure areas being less completely developed than would otherwise have been the case. In extreme cases an intensely heavy area may be surrounded by a light outline, sometimes known as the Mackie Line.

In the same way, the edge of a heavy area may receive enhanced development by the local diffusion of more active developer from an adjacent low density area. This accounts for the Eberhard effect, in which the inner boundary of a dark area is of greater density than the centre. In combination, the two effects result in increased local contrast at the edge, which in moderation improves apparent image sharpness, but when too extensive may be most objectionable.

Edge effects may also result from differential hardening of the developed image areas, especially in colour processes, and the resultant minute relief image can give rise to noticeable edge density differences on optical printing or projection in some circumstances.

See: *Eberhard effect; Processing.*

⊕**Edge Fogging.** Density caused by light penetrating the edges of a roll of film, perhaps loosely wound. Warped or insecurely fitted magazine covers can admit light and cause edge fogging. If fogging does not penetrate beyond the perforations into the picture area, edge-fogged negative may print satisfactorily.

⊕**Edge Number.** One of a series of numbers, combined with key lettering, printed at intervals of 1 ft. along the edge of most types of negative and duplicating film stock.

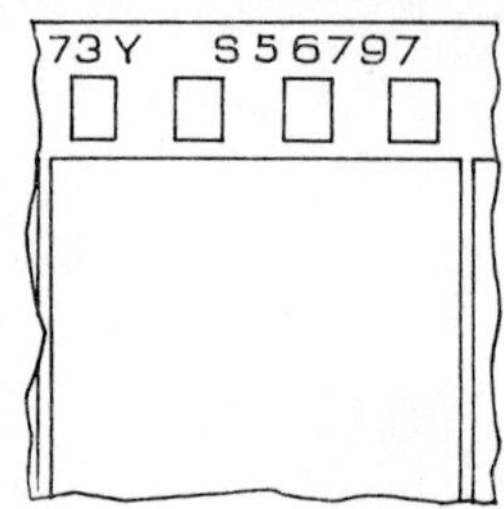

These numbers, incorporated in the film during manufacture, print through to positive stock not so marked. Also known as footage number and negative number.

See: *Code numbers.*

⊕**Edison, Thomas Alva, 1847–1931.** American inventor, son of a timber dealer. Had little formal education, but was taught the elements of reading, writing and arithmetic by his mother, who encouraged his interest in mechanical things. At the age of 10, had already begun to experiment and at 11 years constructed a telegraph outfit. His first job, at 12 years old, was as a newspaper and candy seller on the Port Huron to Detroit train, where he was encouraged by the trainman to continue his experiments in a corner of the baggage car.

In the summer of 1862, Edison became a telegraph operator and continued his experimental work over the next few years while working as an itinerant telegrapher. His experience led to an interest in multiplex telegraphy, and after an unfortunate practical joke which led to his dismissal, he decided to devote all his time to invention. By 1871 he had set up a manufacturing plant in Newark, New Jersey, with a staff of mechanics to make stock-market printers. From this workshop came a number of new devices, including a practical quadruplex telegraphic system.

In 1876 Edison made his greatest innovation – a research laboratory in Menlo Park, New Jersey, designed as a factory for inventions, employing a research team to investigate any and every interesting line of research which might lead to practical inventions. Between 1877 and 1887 this laboratory developed the carbon microphone which made Alexander Graham Bell's telephone a practical system of communication, the phonograph, the first commercially practical electric lighting system using incandescent lamps, the first full-scale electric locomotive in America, and many other devices. Edison opened a larger laboratory at West Orange in 1887, where his team led by W. K. L. Dickson developed the Kinetograph camera and Kinetoscope viewer in 1891. The Kinetoscope, publicly shown in 1893, was the first motion picture device to be commercially successful, and, although only a peep-show viewer, it stimulated other inventors, notably the Lumière brothers and R. W. Paul, to develop cameras and projection equipment, leading to the introduction of cinematography proper. Among Edison's important later inventions was the nickel-iron storage battery.

Edison had a distrust of theoretical science and mathematics, but this led him to tackle and solve practical problems which his contemporaries thought impossible. His research team were fed constantly with ideas from his fertile imagination, and although not all of these ideas were original, he and his assistants were able to convert them into commercially practical processes where others had failed. His combination into a commercially practical form of the principles of photographic analysis of movement, as developed by Muybridge and Marey, and the synthesis of movement, in such devices as the Zoetrope, was the most important single event leading to the introduction of cinematography.

Of almost equal importance was his insistence on the need for precisely positioning succeeding film images, which led him to introduce the 35mm ($1\frac{3}{8}$ in.) film standard, with its four perforations on each side of each picture, which has remained standard to this day. B.W.C.

⊕□**Editing.** Process whereby the raw material of a film or videotape recording (the individual shots) is assembled into a coherent and compelling whole. The shaping process goes through a number of well-defined stages, progressing from an assembly to a rough cut and thence to a fine cut. The sound editor's task, often equally creative, is to build up a set of dubbing tracks which complement and strengthen the effect of the picture. Also called cutting.

See: *Editor*.

⊕**Editing Machine.** Viewing machine designed to run short lengths of film forwards and backwards, as required in the process of cutting or editing. Often called a Moviola.

See: *Editor*.

⊕**Editola.** Trade name for a type of 35mm editing machine characterized by its compact form and use of an optical compensator prism to provide viewing without intermittent motion of the film.

⊕EDITOR

Film editing is the process of selecting shots, arranging and modifying them, in order to clarify and refine their form and content. It is concerned with the construction of a sequence of images (and sounds) of flowing continuity carried out by selecting the particular shot, trimming or expanding it to a certain length, and determining the order in which the shots will appear and the kind of transition between them. Often called cutting, film editing may be described as the art of composing with scissors. When his work is complete, the editor has cut down his material to between a quarter and a tenth of its original length.

The editorial process is an integral part of film production, and should be anticipated in writing the script or scenario.

FUNCTIONS

Editing is a stage-by-stage affair. In the course of a film's production, it begins as soon as the scenes have been recorded on film, the film has been processed in the laboratory, and the editor's copy of the camera original has been received. This work material (dailies, rushes or footage) consists of several takes, or repeat shots from the same camera position of each scene, the ones known to be unsatisfactory having for the sake of economy been eliminated at the laboratory stage and not printed.

The first stage of editing is the examining and cataloguing of picture and sound track footage. Dialogue and narration, recorded on separate strips of magnetic tape, are prepared by the picture editor. Music and effects tracks are usually dealt with by sound editors. The best takes are selected and joined in the order called for by the script to make a rough assembly. Subsequent refinements, in conjunction with the tracks, lead to a rough cut, then to a fine cut of the finished workprint, which shows the final continuity of the scenes, and the type of transition between scenes in terms of optical effects (fades, dissolves, wipes, etc.).

The manual operations of editing call for a high degree of precision so that the finished workprint and the several sound tracks are in perfect synchronization, that is, that corresponding sound and picture occur at exactly the same moment. There are various synchronous relations of track and picture, e.g., the simultaneous relation of speech sounds to lip movements, called lip-sync, or the proper interval between an off-screen sound and a player's reaction to it. Tracks and picture are said to be in sync when the desired relation has been established.

Film editing is a strange mixture of technique and art. It lacks the satisfying directness of fine arts and literary creation. Refractory, it yields its rewards only to those who accept and master its discipline. The editor's creative function comprises rewriting stories with films already shot, removing flaws, and sharpening performances. He watches his material closely for new suggestions or new viewpoints immanent in his scenes. Following clues to new approaches within the film itself, he may at times be able to chart a fresh course for the development of sequences, or even an entire film.

He can contribute values which nobody envisages while the film is being written or produced. The editor's work contributes decisively to the co-operative effort which is at the basis of any film making.

Editing is learned by experience. Studying films and film theory, analysing the work of other editors, deepens knowledge and skill. A sense of craftsmanship, professional integrity and commitment to quality are essential. Yet editing is based, to a great extent, on intangibles such as taste, a sense of timing, a feeling for rhythm, poetic and musical values, an ability to improvise, a deep affinity to and a good memory for things visual – and endless patience.

Pictures have their inner rhyme and reason. It is the editor's task to trace them. Editing can give a film its third dimension, an exciting dimension, always waiting to be explored and conquered.

EQUIPMENT

The editor's work room, the cutting room, is provided with mechanical equipment designed to secure and maintain the precision required at all stages of editing.

MOVIOLA. As a rule, the editor examines his material on the moviola, a brand name now used generically. In its simplest form, it has a picture-viewing head, projecting the image toward the viewer onto a small ground-glass screen, and a sound-playing head for either magnetically or optically recorded sound. Further sound and picture heads can be added. The two heads are connected by a drive shaft. Engaged, the moviola preserves the established synchronism; disengaged, it permits change in the relative alignment of picture and sound. Moviolas are available in both 35mm and 16mm gauges, and in combinations.

SYNCHRONIZER. For the manual handling of film, the editor uses a synchronizer. This is a device for maintaining the synchronous relation between picture and track on the cutting table. It consists of two or four large sprockets fixed to a common spindle, which hold the separate picture and sound reels in precise relationship while they are wound to and fro by the rewinds at either end of the table. Picture takes can be shortened or lengthened without disturbing the sound

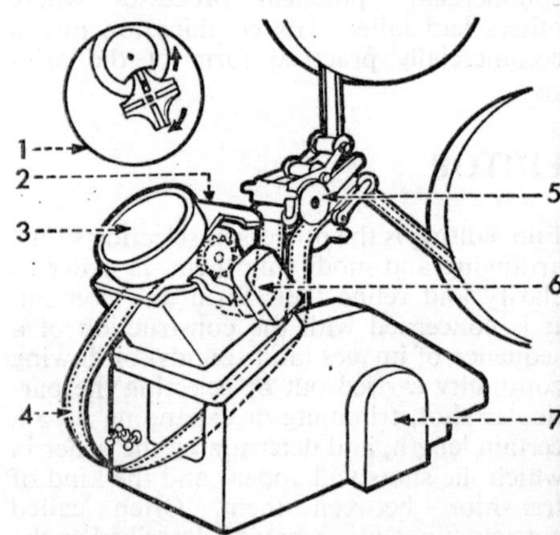

MOVIOLA MECHANISM. Basic Moviola picture head used (often without feed and take-up spools) for picture editing. Film is placed under hinged aperture, 2, and examined through magnifier, 3. It is fed to and from aperture by continuous sprocket, 5, with loops at 4 and 6, between which is intermittent sprocket driven by Maltese cross, inset at 1. Reversible motor, 7, actuates mechanism. No shutter is used.

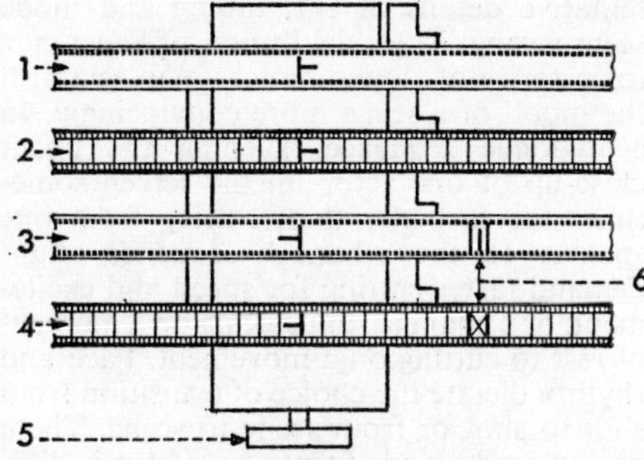

SYNCHRONIZING RUSHES. Outgoing first shot to be replaced by incoming second shot. Four films are laid on peripheries of four large sprocket wheels mounted on common spindle turned by handwheel, 5. 3. Outgoing magnetic sound. 4. Outgoing picture with synchronizing marks, 6. 1, 2. Incoming sound and picture.

and by the same token, words, sentences, syllables or objectionable noises can be removed from the tape without interfering with the continuity of the picture.

In a typical work session the editor plays picture and sound on the moviola, stopping the machine often to make grease-pencil marks on the picture or track, denoting changes to be made. Next, he takes the

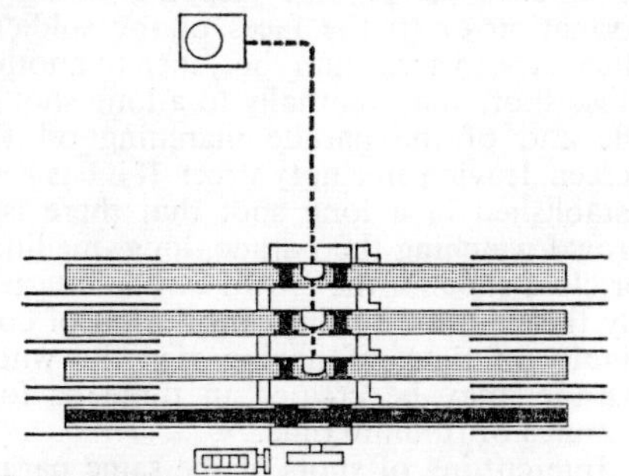

USE OF SYNCHRONIZER TO LAY SOUND TRACKS. Footage counter (*bottom*) reads cumulative footage and frames. Nearest synchronizer sprocket carries picture track, remaining sprockets sound tracks built up piece by piece to match picture. Three magnetic heads cover three tracks, and may be switched through amplifier to loudspeaker. Pulling film through or rocking handwheel enables invisible magnetic modulations to be identified and marked.

material to the cutting bench and pulls it through the synchronizer, possibly in assembly with a viewer or sound reader, in order to make the indicated changes. Then he may go back to the moviola for examination of the new version.

A footage counter calibrated in feet and frames, incorporated into both the moviola and the synchronizer, makes an accurate device for measuring the length of film, which can easily be converted to running time, in minutes and seconds.

SPLICER AND VIEWERS. The splicer most commonly used in the cutting room is equipped with a guillotine for cutting film, a roll of tape for joining two ends of film in a butt joint, and pins to hold the film securely in place during both operations.

The cutting bench is a work table equipped with a pair of rewind spindles, one at each end, used to wind film in either direction on reels or flanges. The rewinds may be hand- or motor-driven.

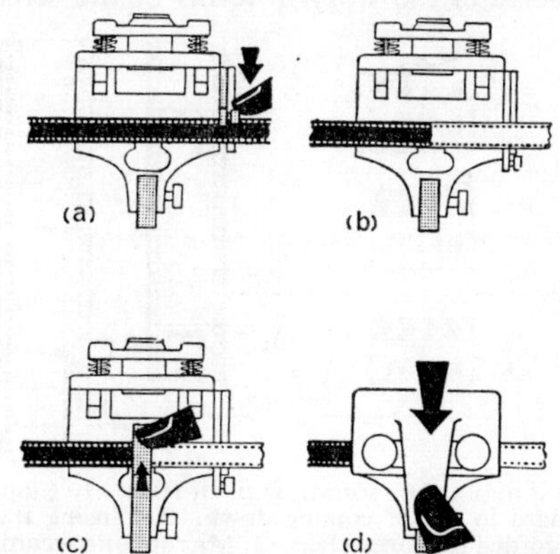

TAPE SPLICER OPERATION. (a) Left-hand film is brought to cutting edge and cut by knife. (b) Left and right lengths of film are abutted at centre of splicer. (c) Adhesive tape is pulled out and placed across intended joint. (d) Splicer clamp is brought down to make tape adhere, cut it at film edges, and punch registering perforations.

Small portable picture viewers and sound readers, through which the film is pulled by means of the rewinds, complete the list of basic mechanical apparatus used by the editor.

A projector for occasional viewings on a larger screen can be a welcome addition.

TECHNIQUE

The very first motion pictures were of events taking place before a stationary camera and recorded in a single shot, from as wide an angle as possible in order to give the viewpoint of the human eye. In moving closer to objects, it became evident that the viewing angle of the camera is narrower than that of the human eye, and that filmic representation of an event required a much larger number of shots than any direct viewing of the event would do. So early film makers started to tell a story by joining several shots together. Trial and error yielded the basic concepts of filmic continuity of time, space, action and idea, the principles of editing. Long before the introduction of sound, the work of pioneers such as Porter, Griffith, Pudovkin and Eisenstein elaborated and perfected these principles.

INHERITED CONVENTIONS. The editor of today inherits this immense system of conventions and techniques. It is always in a state of flux, some practices wearing out from overuse or misuse and fresh ones being added. Applying this set of conventions as the vocabulary of film continuity, the editor re-creates the story in terms of the screen, presents it moment by moment, and so tells the story in film. In joining together the strips of film which have been selected from the amorphous bulk of footage, the editor synthesizes a new reality from the isolated pieces in the individual shots. He creates whatever mood best serves the purpose of the film at any moment and shows what is, at a certain point, the most telling aspect of the story.

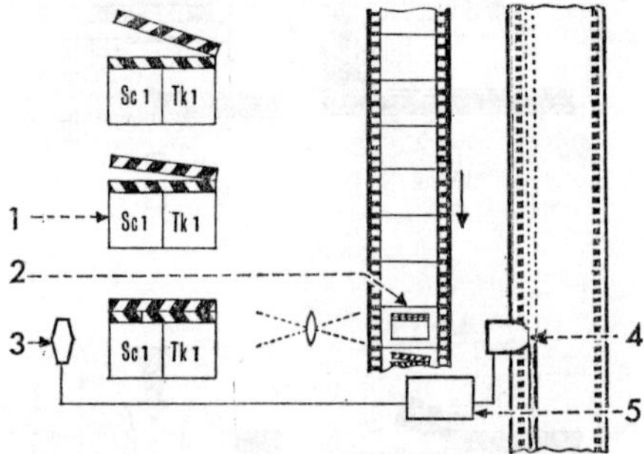

SYNCHRONIZING SOUND AND PICTURE. 1. Clapper board in act of coming down. 2. Closing frame recorded by camera lens. 3. Microphone recording sound of clappers closing via amplifier, 5, and magnetic head, 4. Cutting sync is established when sound and picture frames shown are aligned parallel.

The editor's first assembly, in which he may begin to shape some individual sequences, is very rough and probably two or three times as long as the specified length of the finished film. This version is carefully analysed reel by reel; further cutting brings it closer to the required form. During this succession of versions, the editor must constantly refresh his understanding of the film's intentions. He must judge the material, not as it is, but as it can be made to be.

As the editor approaches a fine cut, he sees that, viewed in context, many of the tentative details of cut, timing and mode seem wrong. The substitution of short cuts for a series of slow dissolves may establish the mood of a scene more convincingly. In a dialogue sequence, keeping the uncut close-up of one actor on the screen sometimes serves better than cutting from one speaker to the other. A sequence might demand faster cutting for speed and excitement, or a change from cutting on a moment of rest to cutting on a movement. Pace and rhythm dictate the choice of transition from shot to shot or from scene to scene. There are no ready-made formulae, and it is difficult to establish rules. Indeed, one of the few accepted facts is that a rule may be broken to achieve a compensating advantage.

THE LANGUAGE OF EDITING. A few examples illustrate the language in which editing communicates. Suppose it is desired to show a military parade. In almost every case, it is necessary to shorten the real time of the parade, say 20 minutes, to a suitable filmic time.

If shortening is the sole purpose, it can be accomplished by showing the parade beginning in a long shot, cutting to a medium shot of the soldiers marching, then perhaps to a close-up of feet stepping along in formation or to the faces of the soldiers, then back to a medium shot, then to another close shot, and eventually to a long shot of the end of the parade marching off the screen, leaving an empty street. If it has been established in a long shot that there is a crowd watching the parade, long, medium, or close shots of the crowd can be inserted. By this kind of cutting, an illusion of continuity of time, of having seen the whole parade, may be created in three or four minutes of running time.

Intercutting of shots of the same parade with shots of a parade of soldiers in a different kind of uniform marching in the opposite screen direction, using progressively shorter shots to create mounting tension, then cutting to a battle scene, might indicate the beginning of a war.

Shots of the parade, perhaps emphasizing the neat uniforms and brightly-polished accoutrements of the soldiers, and the jubilant crowd, intercut with shots of soldiers dressed in rags trudging wearily along a road, could suggest that the war is over and contrast victor and vanquished.

Juxtaposing shots of the parade with shots of a flock of sheep or a group of toy soldiers moving in the same screen direction and in a similar disposition on the screen, or with small children playing at war, might be used to make various implicit or explicit comments on these soldiers or on soldiers in general.

THE EDITOR'S ROLE. Ideally, on a high-budget feature film, editing may be integrated into the early stages of production. The editor is present on the set. He attends the daily screening of the previous day's shooting and discusses with the director the choice of takes and ideas for the editing of a sequence. There may be a second editor preparing the material in the cutting room, to whom he relays the instructions of the director; and there may be an assistant doing the routine organizational work, such as classifying and cataloguing the footage, syncing up picture and sound by means of the slates, and checking the camera and sound reports against the picture and track.

At the other extreme, perhaps on a low-budget instructional film, the editor is often first called in when all the material is in the can, and his only guide is a narration script; or he may be given a huge mass of material from which to create a film of his own. It often happens that a skilled editor makes a whole film from footage gathered from various sources, none of which was originally shot for the purpose. Here the editor's job assumes directorial proportions. He must elicit a convincing continuity and meaning from this medley of hitherto disconnected scenes and chisel congruity out of the incongruous. In either case the overall editing process follows the same pattern.

BUDGETED TIME. The time needed to edit a film cannot be easily calculated. The time allocated to the average feature film, however, is usually a minimum of thirteen weeks, with a safety period of two to four additional weeks allowed for in the budget.

Documentary half-hour films forming a typical television series are scheduled at an average of eight weeks for the completion of each film up to and including the dubbing session, with the understanding that during the last three weeks the editor starts on his next film of the series.

Throughout all the stages of his work, the editor is in close communication with some of the other specialists working on the film. His expertise and his intimate knowledge of the film can be borrowed to advantage by his colleagues; knowledge of their special procedures enables him to anticipate and evaluate the ways in which their contributions can be brought to bear on his own work. The editor works closest with the director, who is the chief instrument of unity in the film.

DIRECTOR'S METHODS. Directorial principles are difficult to systematize. However, practical experience shows that there are two main types of personality with whom the editor has to work. One is the sure-footed, precise director who in effect camera-cuts his scenes. The other, although knowing what he is about, lacks absolute vision, cannot fully and immediately articulate his concept and overshoots. By trial and error, he gropes for precise editing on the moviola.

Few directors today edit their own film, and not all are as capable of directing the editor as of directing the actors. Still, they like to place their personal stamp on the screen image.

A strong editor can offer invaluable help to those actor directors with mainly a background of live TV, stage, and radio experience. But even directors who started their careers in the cutting room admit that because they are emotionally involved with the work on the sound stage, their judgment is coloured. As a rule, they find it helpful to sit down with a capable and sympathetic editor and listen to his critical and detached views.

PRELIMINARY EDITING. An experienced editor has no difficulty in assembling a sequence the way it was filmed by the director. He probably tries first to create a sort of free assembly of his material, starting with whatever sequence can be put together after the first few days of shooting. Much depends on the preference of director and producer: some may want to see a sequence cut, and others be content to see their material once a week, assembled roughly. The trouble with perfecting single sequences at this stage is that when viewed later on in the context of the entire film, many details and cuts, as well as timing and mood, may seem wrong, and the scene may have to be re-edited from scratch. In any case, the process of assembling scene after scene continues until shortly after the shooting period, when between 15 and 18 reels, or 2½ to 3 hours' screening, have been put together. All the scenes, and most of the individual shots, are overlong.

Everything is very rough. And the film has to be brought down to the 100 or 150 minutes' running time usually allotted to feature presentations in the theatre.

CREATING THE FILM. Director and editor start analysing the film all over again. Reel after reel undergoes careful scrutiny. A week or more is at times spent over 10 minutes of film. Decisions arrived at weeks ago may be questioned. The director is not always aware of the strange chemistry contained in his images; and in searching for the principles which guided the director, the editor may discover unspoken directorial intentions in the images themselves. Each scene must be treated on its own merit.

The temptation to use material because it is there, or because it is beautiful, is great. But editing is a highly sensitive instrument for the control of timing. This applies to the spacing of events within the whole story as well as to the timing of the individual cuts. By considering the rhythm of the action, the highlights of the performances and photography, and the flow of the story, the editor can create scenes which seem perfectly natural and effortless. Bearing in mind the unity of the film, he can build, scene by scene, a lucid continuity. And so from the early stages of the rough cut, the editor gradually compresses the film into the essential footage of the fine cut where every waste frame has been eliminated from each single shot.

As his fine cut nears completion, the editor, together with director and/or producer, decides what kind of optical effects, dissolves, fades and wipes are to be incorporated into the film. He assists and supervises the optical effects specialists in the preparation of the layouts necessary for their work. An awareness and knowledge of their technical procedures enables him to evaluate the contribution they can make to the film.

COMPOSER. The editor may be assigned to work with the composer. Editors and film composers rely substantially on the same elements to create a completely integrated impression on an audience. Since the basis of film art is movement, both attempt to co-operate closely in accomplishing appropriate effects. This means that each of them often has pertinent and mutually stimulating suggestions to make to the other, regarding tempo, dynamics and mood. Total combination of picture and sound track is their main concern so that image and music work together, either in harmony with or in counterpoint to each other.

When the budget does not allow for an original score, a music editor, or scorer, and a sound effects editor cut and arrange pre-recorded library music. Here again the editor works within the obvious limitations towards the same goal.

MIXER. A close partnership exists between the editor and the mixer in the sound studio. In blending together, say, three voice tracks, four effects tracks and two music tracks to make one complete sound track, the mixer must match the feeling of the track to that of the picture at every moment. He relies almost entirely on the editor to convey to him the intent of the picture, as a whole and stage by stage, both before and during rehearsals. The editor has prepared a cue or log sheet for each track. These sheets designate where on each track a sound is present, what the sound is, how it is to be brought in and taken out (e.g., for a certain music cue, fade in and, at the end of the cue, cross over to another track), what special acoustical treatment is needed by a certain sound, etc. The final product of the mixing session is a re-recorded sound track which combines all the separate sound tracks produced in the editing process.

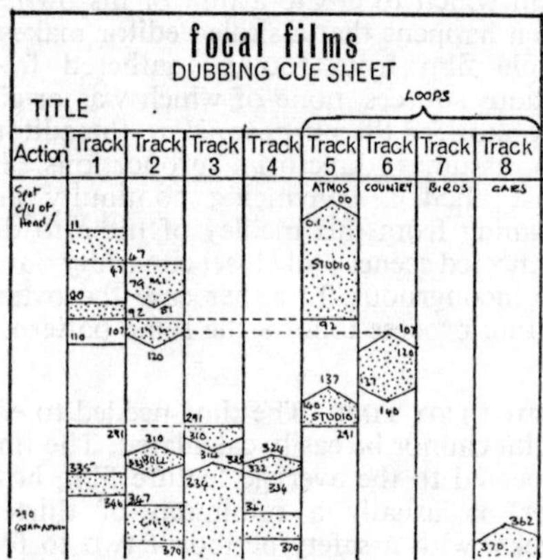

DUBBING CUE SHEET. Simple documentary film with four continuous tracks and four loops. Numbers show footages from start, which appear on counter below screen and/or on console. Tapering heads and tails to sound sections indicate fades. Other transitions are direct sound cuts. A horizontal format is often adopted. Feature film cue sheets are similar but much more complicated.

Mixing or dubbing is a continuation of editing. Some editors know from experience how things are opening up when the picture and all the tracks are being presented together for the first time. Conscious of the element of improvization inherent in their work, they may take along to the dubbing studio what amounts to a small truckload of

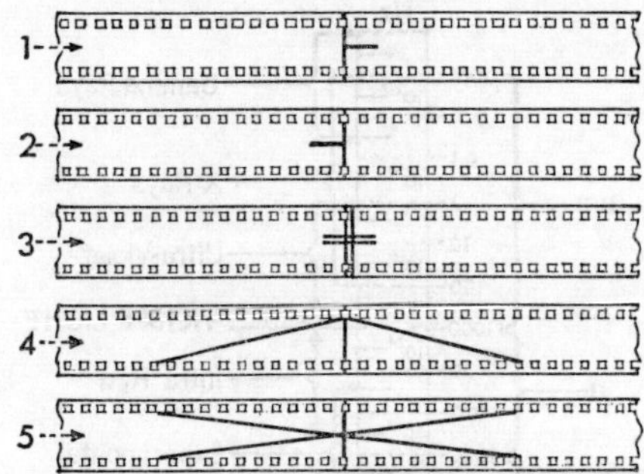

WORKPRINT MARKS FOR NEGATIVE MATCHING. 1. Cut. Lose right-hand portion. 2. Cut. Lose left-hand portion. 3. Ignore cut. 4. Dissolve. 5. Fade out (*left*) followed by fade in (*right*). These marks are not universally recognized, and many variations are found.

out-takes and trims of picture and track for every eventuality. Sometimes there are up to a dozen tracks to be mixed, calling for several pre-mixes. After such a quasi-rehearsal, the film goes back to the cutting-room for final corrections. Altogether, going through a mixing session is like the experience of the composer who listens to a full orchestra playing his score for the first time. Complicated and subtle effects which are to be incorporated into a final track cannot be mechanically prescheduled. A good editor operates with metronome sensitivity and precision, shaping nuances in film and track to a fine point where a few frames added or cut may give the desired effect.

After the dubbing session the workprint is turned over to the negative matcher and the editor is no longer physically involved in the final technical steps preceding the delivery of the married copy. The matcher handles the production's most precious asset in terms of capital investment, the film negative, which he cuts and splices to match the finished workprint. The workprint has been carefully marked and annotated with instructions for the matcher, comparable to a manuscript going to the printer. Once the negative has been cut, it cannot be restored.

FINAL TOUCHES. When the negative and the optical sound track have been completed, they are sent to the laboratory for the making of an answer print, the first trial print combining sound and picture. In conference with the timer, the laboratory specialist who controls the density of each shot in the print, the editor can contribute his knowledge of the photographic qualities and the intent of the various shots.

When this print has been viewed and, after any necessary alterations, approved, the editor's task is complete. P.V.F. & C.W.

See: *Director; Sound mixer; Sound recording* (FILM); *Special effects* (FILM); *Splicing; Spools, cores and cans; Videotape editing.*

Books: *Editors and Editing*, ed. by G. Gross (New York), 1962; *Technique of Film Editing*, by K. Reisz (London), 1956; *Film and its Techniques*, by R. Spottiswoode (Berkeley), 1953; *Technique of Editing 16mm Films*, by J. Burder (London), 1968.

□**Edit Pulse.** Magnetic pulse recorded on the control track of a transverse-scan videotape machine to mark the boundary between one picture and the next. Corresponds to the frame line in film.

See: *Videotape editing.*

□**Effective Radiated Power.** Product of the power fed into a transmitter aerial and the aerial gain.

□**Effects Bank.** That part of a vision switching matrix used in conjunction with special effects equipment.

⊕**Effects Filter.** Filter used in front of the camera lens at the time of photography to modify the character of the image on the film.

Examples are graded colour filters to control the rendering of sky areas, diffusion discs and fog filters.

See: *Cameraman* (FILM).

⊕□**Effects Track.** Sound track carrying sound effects (gunfire, footsteps, doors closing, etc.) not normally recorded when a film is being shot, but separately recorded or taken from library sources. Effects are assembled in proper synchronization to form edited effects tracks, and these are combined in the dubbing process with dialogue and music tracks to produce the final sound track.

See: *Editor; Sound recording.*

□**EHT Rectifier.** Extra high tension rectifier supplying anode voltage to a picture tube.

□**Eidophor.** Device for projecting a TV image in which the electron beam builds up an invisible image which acts in the manner of a slide, in that it can be illuminated by a powerful, conventional light source such as an arc lamp, to throw an image on to a large screen. Colour Eidophors employ three of these monochrome projectors equipped with additive colour filters.

See: *Large screen television.*

⊕**Eight Millimetre Film.** Smallest of the narrow-gauge films used for motion picture photography; originally introduced in 1932 purely for amateur use, but in the early 1960s increasingly for educational purposes.

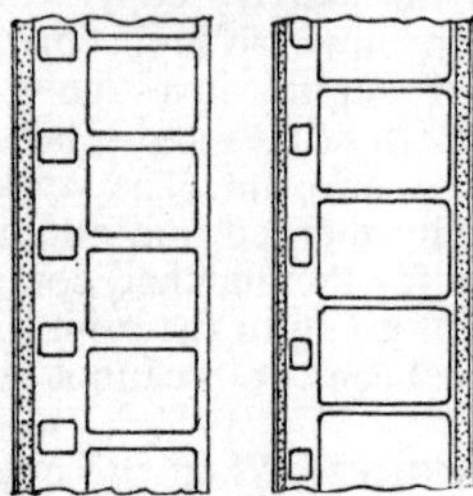

Since 1966, undergoing replacement by Super-8 film (*right*). Standard 8mm (*left*) as projected has a single row of perforation holes down one edge of a film 8mm wide and has 80 frames in a length of one foot. Now often referred to as Double-8 because it is supplied on 16mm film which is slit after processing.

See: *Film dimensions and physical characteristics; Narrow gauge film and equipment; Narrow gauge production techniques; Super 8mm film.*

⊕**Ektachrome.** Integral tripack reversal colour film process, normally used for motion picture work in the 16mm form. Ektachrome originals are not intended for direct projection and viewing but as a source material from which prints may be made, either directly on to another reversal stock or by way of a colour dupe negative on to colour positive film.

See: *Colour cinematography; Integral tripack.*

⊕**Electromagnetic Spectrum.** Spectrum which □embraces the continuous range of electrical and magnetic radiation, from gamma rays of 10^{-13}m wavelength through X-rays, ultra-violet, visible and infra-red radiations to television and radio waves up to 10^{9}m.

From UV frequencies downwards, a large part of the spectrum is used for communications. Photography covers the region from the near UV (about 200nm) through the visible spectrum into the infra-red (IR), ending at about 1,350nm. Laser beams bracket the visible spectrum and reach downwards into the IR and heat radiation region, also used in thermography. Microwave frequencies overlap those generated by lasers.

Downwards from the microwave region, with radar and inter-city TV links, the spectrum is used for TV and radio until, beyond LW radio, each channel absorbs too large a

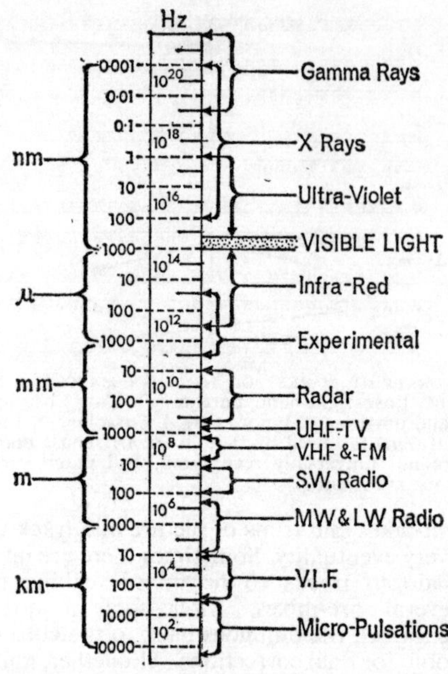

part of the band. Very low frequencies are used for direction finding. Below these, with wavelengths of thousands of kilometres, are natural radiations known as micro-pulsations.

⊕**Electron-Beam Film Scanning.** In a normal □flying-spot film scanner (telecine), an electron beam in a high-resolution cathode ray tube produces on its phosphor a fine spot of light which is scanned over the raster. This light is focused by a lens on to the film, which acts as a density modulator, and the modulated light falls on a photomultiplier, which generates and then amplifies a sequential signal.

In the electron-beam film scanner now undergoing development, a specially prepared film acts as its own source of light. The normal light-sensitive emulsion is coated with scintillators, colourless materials which emit 5 to 10 photons for every primary electron which strikes them. After printing and development of the photographic image in the usual way, the film is scanned by an electron beam impinging directly on it at the aperture in a vacuum. The light generated at the emulsion surface by the scintillators is amplified directly by a photomultiplier as before, no lens system being required. Light losses in the lenses and at the phosphor surface of the normal film scanner are thus overcome.

See: *Electronic video recording.*

⊕□**Electron-Beam Recording.** Photographic film, normally exposed by light in the visible and near-visible regions of the spectrum, is also sensitive to electron beams and obeys the reciprocity law under accelerations of approximately 18–20 kV. This principle forms the basis of telerecorders now under development, in which the film passes into a vacuum chamber where it is directly exposed by an electron beam modulated by the video signal.

Though this method has problems of its own (e.g., producing and maintaining a vacuum round the film, and neutralizing the charge which the beam builds up on the film surface), it eliminates some of the major difficulties of conventional telerecording, in which the film photographs a light image formed on the faceplate of a display tube. Among these are non-uniformities of the phosphor, producing shading and mottle, and halation in the thick glass faceplate.

See: *Electronic video recording; Thermoplastic recording.*

□**Electron Gun.** Unit comprising heater, cathode and control grid assembly of a cathode ray tube or television camera pick-up tube in combination with an accelerating electrode. Called an electron gun because it projects a narrow stream of electrons at a screen or target, under the control of the mean grid-cathode potential and the positive d.c. potential applied to the accelerating electrode (anode).

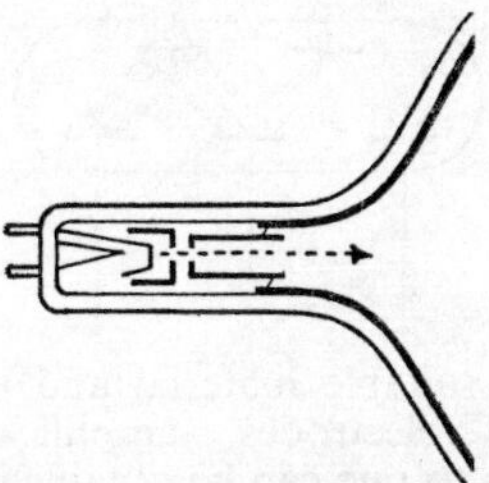

ELECTRON GUN. Arrow shows path of electron beam.

In the case of the picture cathode ray tube (kinescope), the electron beam so formed is modulated in intensity corresponding to the brightness variations of the transmitted picture by applying the picture signal between cathode and grid of the electron gun. After being suitably scanned, this beam strikes the phosphor on the tube screen, thereby producing the light and shade of the picture image. However, the transfer function relating light output to applied signal voltage is not linear, and this has to be compensated before transmission.

Shadow mask display tubes as used in colour television receivers contain three electron guns, one for each of the synthesis primaries red, green and blue.

⊕□**Electronicam.** System of programme production using combined film and television cameras, first employed in the US in 1955, and later developed in Germany under the name Electronic-Cam.

In both forms, a single lens system in each camera records an image on a 35 or 16mm film, black-and-white or colour, and on the faceplate of a TV camera tube. The latter feeds a viewfinder for the cameraman as well as a programme monitor, mounted alongside monitors for two or more other cameras which are filming the scene from different angles.

In the American system, now obsolete, a programme director, watching the monitors, electronically cut or dissolved at will between these sources to provide a live telecast, while the film in the cameras was later processed and edited in the normal way with the help of a videotape recording of the telecast. In this way, a high quality film version, with improved editing and shot selection, was available for subsequent transmission.

In the current German version, Electronic-Cam, based on the Arriflex camera, the complication of simultaneous transmission has been abandoned, and the electronic facilities are concentrated on providing the best possible conditions for simultaneous film recording, in which the cameras are controlled by a director who watches monitor screens and issues instructions to pan, zoom and dolly the camera as in a TV studio.

The bulky image orthicons of the American system have been replaced, first by vidicons, then Plumbicons, and the definition on the cameraman's monitor is good enough to enable him to follow focus and adjust the lens diaphragm setting if necessary without the help of an assistant.

The director, who can also start and stop individual cameras to save film, indicates his cutting choice with the help of different frequency tones, one allotted to each camera, which are recorded on a separate cueing track. Thus the editor, following this track, is told precisely where a cut is suggested from camera 3, say, to camera 1.

The object of the system is to speed up production by centralized control of multiple cameras in real time, thus saving on rehearsals and repeated takes at the expense of a possibly larger consumption of film, often a relatively small item in the production budget. R.J.S.

See: *Add-a-vision, Gemini; System 35.*

□**Electronic Converter.** Picture standards converter in which the line and/or field conversion is carried out entirely by electronic means. No picture is reproduced and rephotographed.

□**Electronic Editing.** Putting together of a videotape programme by electronic means, i.e., without cutting and joining the tape. The latest devices insert the signal correctly in step so that the resultant signal when the tape is replayed is free from interruptions which might cause the picture to roll or break up.

See: *Videotape editing.*

□**Electronic Spotlight.** Device for locally intensifying the brightness of a TV picture to give a spotlight effect by purely electronic means. A pulse is derived which has line and field components of the appropriate TV signal, and the signal thus obtained can be used as an electronic pointer or spotlight to bring the viewer's attention to parts of the picture, e.g., race cards on outside broadcasts, or schools broadcasts.

See: *Special effects* (TV).

□**Electronic Switch.** Device used to produce electronic special effects. In its simplest form consists of two identical non-linear amplifiers, fed with two different video signals, only one amplifier being allowed to conduct at any one time. The length of the conduction period of each amplifier determines the ratio of the two signals, and thus the amount of picture which appears on the screen from each of two picture sources.

See: *Special effects* (TV).

⊕**Electronic Video Recording.** Trade name for □a form of video presentation developed in the US by CBS. Picture origination is orthodox, i.e., by film camera or TV camera and VTR. The picture is then transferred to a special non-perforated 8·75mm film (i.e., 35mm × ¼) by an electron-beam recorder producing a line-free image independent of the line and frequency standard of a television original. The picture is reproduced in the home or school by a mechanism containing a flying-spot scanner, generating a signal which is fed into the aerial (antenna) socket of a normal TV set.

The 8·75mm film travels at 5 ips (cf. 16mm film = 7·2 ips), and is loaded in cassettes which are pushed into the player mechanism and require no threading. The film carries two picture tracks side by side, giving a total screen time of 60 min., and a magnetic sound track.

The EVR system, only applicable to black-and-white in its original form, has a colour capability. One of the picture tracks can carry a chrominance signal coded for NTSC, PAL or SECAM as applicable, which, when combined with the luminance signal derived from the picture record, actuates the appropriate circuits of the colour receiver into which it is fed. This halves the screen time of a cassette.

See: *Electron-beam film scanning.*

□**Electron Image.** Image formed on the target of a camera tube as a result of emission of electrons from the photo-cathode. It consists of a pattern of varying electrical charges which faithfully follow the original variations in light of the optical image on the photo-cathode. This electrical pattern, after processing, is eventually transferred to the receiver, which converts it back to an optical image on the cathode ray tube.

See: *Camera* (TV).

□**Electron Multiplier.** Device which amplifies a small electric current by subsequent bombardment of suitable electrodes and increased secondary emission at each stage.

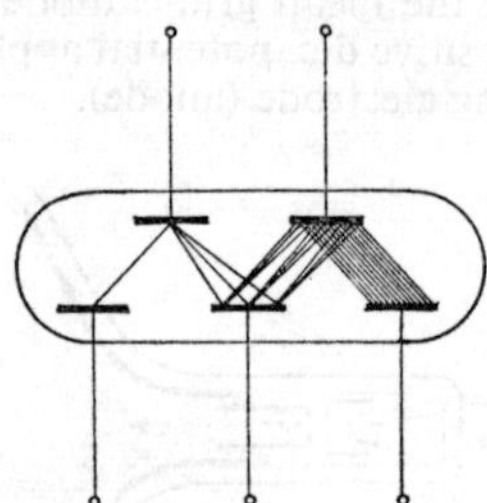

By using suitable material and up to ten successive electrodes, amplifications of thousands to one can be obtained.

See: *Camera* (TV); *Telecine.*

□**Electrostatic Focusing Electrode.** Electrode which forms part of an electron gun. Variation of its potential alters the focusing of the electron beam. Commonly used in receiver cathode ray tubes.

⊕**Elevator.** In motion picture film processing, an assembly of fixed and movable film rollers which can be arranged to accommodate a variable length of film, either to provide variable processing times on developing machines or to act as a reservoir to allow the end of the film to be temporarily stopped without stopping the whole machine.

See: *Processing.*

☐**Elster, Johann Philippe Julius, 1854–1920.** German physicist who carried out (1882–1889) with Hans Geitel a series of investigations of phenomena connected with the emission of electricity from hot bodies, which laid the basis for the development of the vacuum tube. Elster and Geitel experimented with glass bulbs, either exhausted or filled with various gases, containing an electrically heated wire and a metal plate, and observed that "electrified particles" were thrown off from the glowing wire in every direction.

☐**Emergency Bypass.** Simple and above all reliable switching system which, under emergency conditions, routes an input to the output without invoking any of the apparatus which it bypasses.

See: *Communications.*

☐**Emitron.** Form of iconoscope tube developed in Britain and named after the manufacturers, Electrical and Musical Industries. Now obsolete. Also called a super iconoscope.

⊕**Emulsion.** Gelatin layers coated on the base or support of a film and containing the sensitive materials in which the photographic image is formed.

See: *Film manufacture.*

☐**Encoder.** Apparatus used for transforming original colour signals, or colour and luminance signals, into the form required for transmission. This usually involves translating the colour information into some form of modulation of a subcarrier. The form used varies with the colour system. Other necessary information may also be inserted in the signal at the same time. In the US, often called a colorplexer.

See: *Colorplexer.*

☐**Energy Recovery Diode.** Device by which the unwanted part of the waveform of the signal used to drive scanning coils is employed to provide extra voltage for a receiver. In valve receivers this enables the mains voltage to be added to the recovered voltage without using a transformer to step up to the required higher voltage.

See: *Receiver.*

☐**Envelope Delay.** Rate of change of phase response with frequency. If the phase shift of a signal after its passage through a circuit is plotted for various frequencies, the envelope delay or group delay is the slope of the curve at a particular frequency.

⊕☐**Equal Energy White.** Light composed of equal energy of red, green and blue of the correct wavelengths. White light may still appear white over a fairly wide mixture of its components.

⊕☐**Equalization.** Technique of improving sound signal-to-noise ratio. As an audio signal progresses through an audio system, it very often suffers attenuation distortion. This variation of gain with frequency can occur over any portion of the audio spectrum. It is most likely to occur with electro-mechanical components, e.g., gramophone pick-ups, loudspeakers and microphones.

There are two ways to overcome this distortion. One is to anticipate the distortion which occurs further down the chain, and provide equal and opposite gain variation before the distortion occurs. If this is applied correctly, the overall gain frequency response is constant. This is called pre-equalization.

A second approach consists in inserting the equal and opposite gain correction after the distortion has occurred. This is called post-equalization.

A disadvantage of pre-equalization is that it is possible to overload the equipment prior to the distortion when correcting for a large loss of response. A disadvantage of post-equalization is that if the distortion is causing a large loss of response, the signal may approach the noise level of the system. Subsequent correction of the response may then raise the noise to an objectionable degree.

If it is practicable to apply pre-equalization by reference to the output of the signal chain, or if it is possible to predict accurately the distortion that will occur, pre-equalization is normally applied as far as the risk of overloading allows. Any residual distortion is remedied by post-equalization.

See: *Sound recording.*

⊕☐**Equalizer.** Network whose insertion loss varies in some desired manner with change of frequency. Equalizers are inserted into recording and reproducing sound channels to compensate for losses or to introduce changes into the overall frequency response.

See: *Sound recording.*

☐**Equalizing Pulses.** In the TV waveform, two short series of pulses having twice the frequency and half the width of line-synchronizing pulses, thereby preserving the same d.c. level, inserted just before and just after the chain of broad pulses.

Their function is to ensure that immediately prior to the beginning of the chain of broad pulses, which initiate field flyback, the condition of the field synchronizing circuits in receiving equipment is precisely the same at the end of odd fields as it is at the end of even fields. In the absence of equalizing pulses, the field sync integrator has twice as much time to recover from the last line sync pulse at the end of even fields as it has at the end of odd fields, due to the latter ending at the mid-point of a line instead of at the end of a line. Consequently, field flyback could commence too early after odd fields with the result that interlacing could be impaired.

Equalizing pulses are a feature of both the European 625-line and the American 525-line television systems.

See: *Scanning; Television principles.*

⊕**Erasing.** Process of eliminating previous □modulations from magnetic recording tape or film. May be effected by passing a suitable current through a special erase head on the tape recorder, or by placing the roll of magnetic tape or film in a strong magnetic field.

⊕**ES Cap.** Edison screw cap introduced in the □1880s and still used for domestic lamps in many countries. Also manufactured in various sizes for studio lamps of non-prefocus type, i.e., in which the position of the filament in relation to the reflector is immaterial.

⊕**Examiner.** Operator in motion picture processing laboratory or distributor's exchange responsible for the inspection of the physical condition of release copies on a rewind bench after printing and projection. Dirt and damage in the form of scratches and rubs, torn or broken perforations, distorted edges and weak splices must be noted and cleaned or repaired where possible.

The condition of the copy and the position of defects must be reported so that the source of damage during subsequent usage can be identified.

See: *Laboratory organization.*

⊕**Exciter Lamp.** In optical sound projection, a small lamp whose beam is focused on the sound track, which thus acts as a light modulator. The modulated light beam falls on a photocell cathode whose electrical output is amplified and converted by loudspeakers into audible sound.

⊕**Exposure.** Process of subjecting a photo- □sensitive surface to light. In photography, exposure results in a latent image on a photographic emulsion, and according to the reciprocity law, exposure is determined by the product of time and intensity of illumination. In television, exposure produces an electronic image which is scanned and removed as a signal.

See: *Exposure control; Tone reproduction.*

⊕**Exposure Index.** Number based on emulsion speed and latitude, exposure meter characteristics and techniques, and expected conditions of development, which enables the user of a film emulsion to determine the correct exposure under different light conditions estimated by an exposure meter or from tables.

See: *Exposure control.*

⊕EXPOSURE CONTROL

In any photographic reproduction process, control of the exposure of the sensitive materials used is of prime importance, both initially in the camera and subsequently in printing, if satisfactory tonal rendering is to be obtained. This is because of the limited tonal range inherent in photographic emulsions. When colour film is used, these requirements become even more critical, because incorrect exposure then gives rise to distorted colour reproduction as well. In cinematography, uniformity of exposure from scene to scene is of vital importance to avoid an inconsistent rendering of tones and colours.

BASIC PRINCIPLES

CHARACTERISTIC CURVE. Any photographic emulsion may be regarded as showing a definite relation between the intensity of exposure given and the resultant density of the image produced after processing, and this relation is represented by the characteristic curve of the material. For a given time of exposure, which is normally fixed in cinematography by the number of frames exposed per second, the intensity scale of this curve may be regarded as indicating various amounts of light reaching the film.

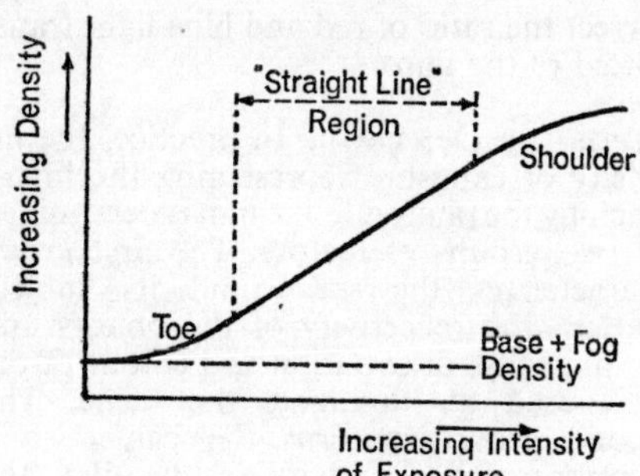

CHARACTERISTIC CURVE OF NEGATIVE FILM. Typical negative characteristic shows well-rounded toe and shoulder regions, between which lies very slightly curved portion commonly known as "straight-line" region. Base and fog densities set limit to maximum transmission of negative.

The range of intensities to which the material will respond is limited. At very low intensities no image is formed, while at very much higher levels a maximum image density is reached which cannot be increased no matter how much the intensity increases. Between these two extremes there is a region over which the density of the image varies substantially in accordance with the intensity. For good tonal reproduction all the images of important details of the scene being photographed should be found within this range, which is termed the straight-line portion of the characteristic curve.

The region where the image densities approach the minimum is called the toe of the curve, and where they approach the maximum, the shoulder.

Some very dark subject tones and shadows are bound to give only low image densities on the negative film, and are therefore within the toe region. When too much of the subject matter is recorded in this region, the result is underexposure of the scene as a whole.

Similarly, the image of very bright objects in the scene may approach the maximum density; overexposure means that too many of the light tones are recorded by densities in the shoulder region and are thus inadequately differentiated.

SPEED RATINGS. The sensitivity of a photographic material under normal conditions of use and processing is specified by its speed rating, sometimes called exposure rating or index. The subject is a complex one and there have been numerous attempts to establish a scale of speed ratings of practical value and international acceptance.

Essentially all rating systems are based on the determination of the minimum exposure required (in absolute units, such as metre-candela-seconds) to produce a density a specified amount above the basic fog level when the material has been processed under standard conditions. This density level is chosen to represent the point on the characteristic curve at which the toe begins to merge into the straight-line portion and thus corresponds to the minimum density of the image which can be expected to record tonal gradation satisfactorily.

Differences in speed between various materials may be specified in two ways. On the arithmetic scale, the speed figures are proportional to the reciprocal of the minimum exposures required. Thus a material with a rating of 80 has twice the speed of one with a rating of 40 and requires only half the exposure. On the logarithmic scale it is the log of the minimum exposure which determines the speed figure of a material; to provide convenient numbers the logs are usually multiplied by 10, so that a twofold difference in speed is indicated by an increment of 3° in the speed rating.

Both methods have been used in internationally-accepted speed rating systems, but in the field of professional motion picture photography the ASA Exposure Index on an arithmetic increment is the most widely adopted; there is an identical British Specification also.

The index figure for a black-and-white negative stock depends on the type of illumination used, so that it is usual to quote two speed values, one for daylight and the other for artificial or tungsten light. Modern fine-grain black-and-white negatives for studio use are rated on the ASA scale at 250 daylight, 200 tungsten, while high-speed stocks, with increased graininess, may be rated 500 daylight, 400 tungsten. Colour negative stocks are rated for tungsten only and an ASA index of 100 has been achieved.

The method of assessing the speed of reversal films is slightly different in detail, since the minimum exposure becomes the darkest part of the image, but exposure index ratings are given on the same internationally accepted scales. Black-and-white reversal stocks in general show the same range of speeds as black-and-white negative materials, but colour reversal stocks are made specifically for daylight or for tungsten lighting. Different daylight types may have ASA ratings from 50 to 200, while stocks for tungsten may be as fast as 160 ASA with normal processing. By so-called forced development, effective speeds as high as

800 ASA may be obtained at the cost of increased image grain.

LATITUDE. Closely associated with the speed rating of a film is its latitude, which may be defined as the range of exposure intensity over which it will satisfactorily reproduce tonal gradation; this is effectively represented on the characteristic curve of the material by the length of the mainly straight-line portion lying between the marked inflections of the toe and shoulder. The longer this portion the greater the range of intensities which can be rendered and the greater the latitude. It also means that if the brightness range of the subject matter is not too great, errors in exposure setting can still yield acceptable results. Modern black-and-white stocks have a considerable latitude, but that of colour materials is much more limited, especially in reversal films.

COLOUR TEMPERATURE. When colour materials are exposed, the colour quality of the light as well as its intensity is important. Colour films are manufactured to suit specific types of illumination, usually either daylight or tungsten-filament lamps, which have colour temperatures of about 5000°K and 3200°K respectively. The higher the colour temperature of the source, the greater proportion of blue light it contains, so if a film designed for use with daylight is exposed by artificial light, the layer of the colour emulsion recording the blue component is underexposed, even though the red layer is correctly exposed.

It is therefore essential to ensure that the light source is correct for the film, or to correct its character by the use of suitable colour filters on either light sources or camera. It is also important not to mix sources of light of substantially varying colour temperature in illuminating a scene, as differences greater than 200°K are noticeable in the resultant print.

Colour temperature meters are sometimes employed to check light sources: these measure the relative proportions of red and blue light in the illumination and show a direct reading in degrees Kelvin. They can also be used to indicate the corrective effect of a particular colour filter.

Colour negative films for professional motion picture photography are normally balanced for use with incandescent tungsten lamps of colour temperature 3200°K. When this type of negative is used with daylight, it is usual to mount an orange-red filter (e.g., Wratten No. 85) on the camera lens to correct the ratio of red and blue light transmitted to the film.

EXPOSURE CONTROL. In practice, the intensity of exposure representing the image reaching the photographic film is determined by two groups of factors. The first group characterizes the scene and its subject matter – the reflectivity of the objects and the intensity, distribution and colour of the light used to illuminate the scene. The second group concerns the camera and comprises the sensitivity of the film, the length of time of the exposure and the amount of light effectively transmitted through the camera lens.

In motion picture work, a number of the latter factors are standardized; for example, all professional cinematography uses cameras whose normal speed is 24 pictures per second, giving a basic exposure time of 1/50 sec. for a shutter opening of 175°.

Some cameras have shutters adjustable down to a narrow aperture, and shutter closing is occasionally resorted to as a means of exposure control in order to work at a desired lens aperture under fixed lighting conditions.

The sensitivity of the negative stock used is carefully controlled by the manufacturer within close limits and is processed by the laboratory under standard conditions unless specially requested; in addition, it is often the practice to use a particular batch of stock for uniformity throughout a production, especially in colour photography.

Because of the limitation of these items, the basic problem of routine exposure control in motion picture photography is concentrated on the factors of illuminating the scene and determining the lens aperture to be used, and even here in studio work it is frequently the practice to control primarily the set lighting to allow the camera to be used with a specific lens diaphragm setting.

LIGHTING

BRIGHTNESS RANGE. The reflectance of typical objects such as a piece of white cloth and a piece of dark cloth may differ by a factor of as much as 10 to 1, and in combination with strong directional lighting, such as bright sunlight, may give a range of subject brightness of several hundred to one. But the range which can be satisfactorily reproduced by any photographic process is far less than this, and steps must be taken by the control of the illumination of the scene to ensure that the brightness range of im-

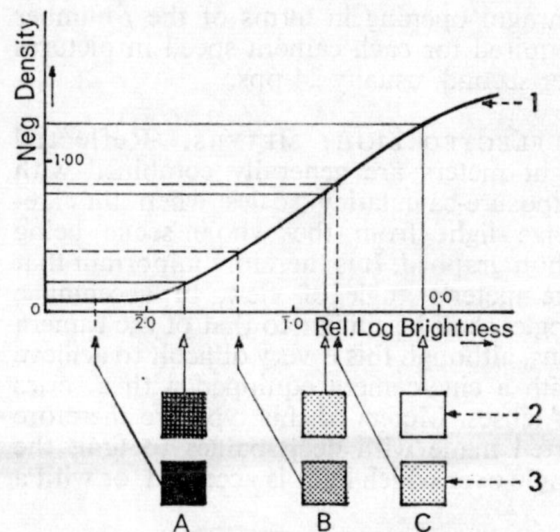

TONE REPRODUCTION SCHEME. Scene brightnesses reflected through characteristic curve (extending to parts of toe and shoulder regions) to produce typical span of negative densities. 1. Negative characteristic. 2. Blocks representing fully-lighted areas of scene. 3. Blocks representing shadow areas for 4 : 1 lighting contrast. A. Black velvet, 2% reflectivity. B. Mid-grey, 20% reflectivity. C. White card, 90% reflectivity.

portant subject matter is held within the scope of the process. Where the light sources are completely at the command of the lighting cameraman, as in the studio, this is done by their distribution so as to provide both a main directional light, known as key light, and generalized illumination, known as filler light, which falls evenly over the whole subject area and lightens the shadows produced by the main source.

LIGHTING CONTRAST. The lighting contrast is the ratio between the light value of the key plus filler and the filler alone.

In black-and-white cinematography, it is usual to employ a lighting contrast of the order of 4 : 1, although this may be exceeded for night effect sequences or other dramatic requirements. When using colour materials, it is general practice to use a lower lighting contrast, not exceeding 3 : 1, while in colour films shot for use on television an even lower ratio of 2 : 1 is recommended.

In exterior photography using natural light, the lighting contrast of each shot must be studied carefully and limited to the desired ratio by the most suitable means available. Shooting under a veiled but bright sky can give a suitably low lighting contrast, but with strong sunlight the brightness of highlight areas may have to be reduced by shading, and the harshness of shadows reduced by reflectors or supplementary artificial lighting. Conversely, under extremely dull and overcast conditions the natural light may be sufficient for the filler light only and the key must be provided by artificial sources. When artificial light sources are used in this way, it is essential that their colour temperature be matched to the natural daylight available.

For the efficient control of lighting contrast, the most important tool in the hands of the lighting cameraman is a light meter, which is often combined with adjustable scales to act as an exposure meter.

LIGHT METER TYPES. Light meters are basically of two types, one of which measures the intensity of light falling on the subject to be photographed (incident light meters) while the other type measures the light reflected from the scene towards the camera viewpoint and thus takes account both of the illumination and the reflectivity of the subject.

Reflected light meters may measure the light reflected from the whole of the scene viewed by the camera or may have a narrow acceptance angle so that they can be used to measure the brightness of comparatively small areas of the subject (spot brightness meters).

The overall scene brightness method is widespread in amateur cinematography and forms the basis of the automatic exposure control systems built into many amateur cameras, but in professional motion picture work the need for the accurate balancing of illumination from several sources normally leads to the use of incident light readings.

INCIDENT LIGHT METERS. Incident light meters consist fundamentally of a photocell receiving the illumination and a sensitive meter to measure the current produced by the cell; the meter may be calibrated directly to indicate the intensity of light falling on the cell surface, usually in foot-candles (fc), or it may in addition be provided with adjustable scales which show directly the lens aperture setting required for each time of exposure once the sensitivity (speed rating) of the particular film stock has been set.

Direct light-intensity readings are of particular value where the cameraman wishes to adjust conditions to provide a carefully controlled lighting contrast, and special forms of meter are available to indicate relative brightness ratios.

The photocell of an incident light meter is often mounted separately on a paddle at the

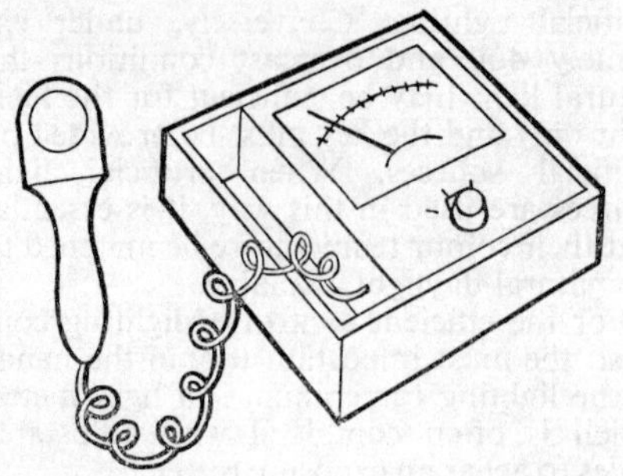

PADDLE-TYPE LIGHT METER. Photocell is mounted separately on end of flexible lead and held wherever required to measure incident light. This type of meter is frequently used for accurate measurement of screen illumination.

end of a lead connecting it to the meter, so that it may be conveniently used to explore the illumination falling on different parts of the subject, and in this form the receptive surface of the cell is usually flat. For studio work with artificial light sources, a scale reading up to 600 or 800 fc is normally sufficient, but for exterior photography in sunlight, values up to 10,000 or 12,000 fc may have to be measured, and the scale must be capable of extension to such levels.

Incident meters designed for direct indication of exposure may have conical or hemispherical translucent light collectors

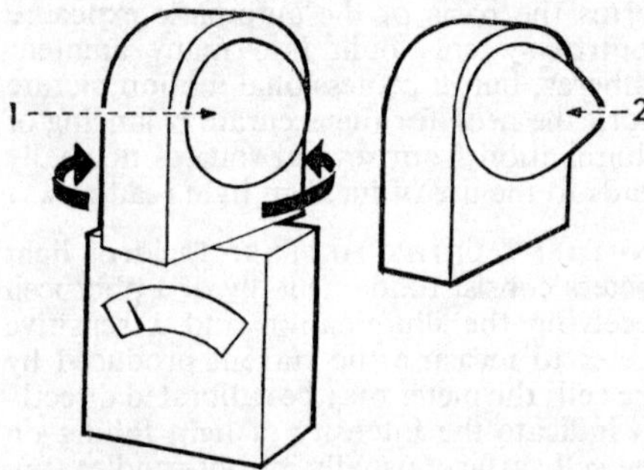

INCIDENT LIGHT METERS. Diffusing hemisphere, 1, over photocell is turned towards main source of light, but integrates light falling on it from many directions. In some designs the hemisphere is replaced by a diffusing cone, 2.

fitted over their cells so as to integrate the light falling on the subject from all directions; they are used close to the position of the subject being photographed with the cone or dome pointed towards the camera lens. The adjustable film sensitivity scale is calibrated in one of the internationally-recognized speed rating scales, such as ASA, and the meter reading shows the lens diaphragm opening in terms of the *f*-number required for each camera speed in pictures per second, usually 24 pps.

REFLECTED LIGHT METERS. Reflected light meters are generally combined with exposure-calculating scales when they receive light from the whole scene being photographed. It is therefore important that the meter's angle of view, or acceptance angle, shall be similar to that of the camera lens, although this is very difficult to achieve with a cine-camera equipped with a series of lenses. Meters of this type are therefore often made with deep baffles to limit the angle over which light is accepted, or with a

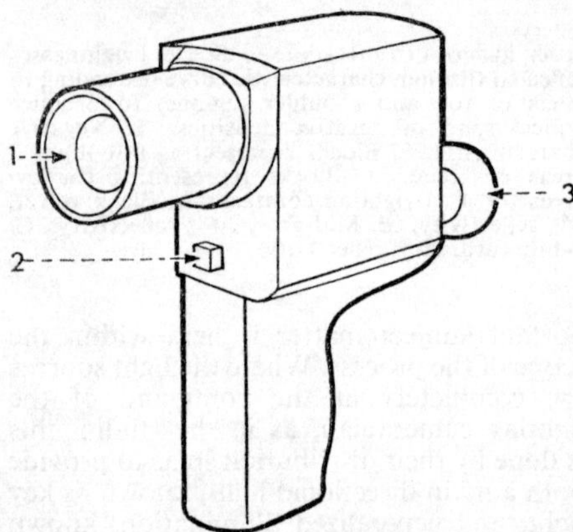

SPOT EXPOSURE METER. Reflected-light type of meter with narrow acceptance angle for reading brightness of small areas at a distance. Lens, 1, of instrument is directed at subject by viewing through eyepiece, 3. Button, 2, is pressed to take reading. Operating batteries are contained in handle.

lens or a lenticular surface which produces an image of the scene on the photocell surface. The logical extension of this principle is to use the camera lens itself to provide the illumination of the cell, and this is done by a reflex system in some advanced models of amateur cine cameras. Reflected light measurements are the basis of all automatic exposure control systems built into the camera, which adjust the effective diaphragm of the camera lens without attention from the user once the speed index of the film has been set.

Although the measurement of overall scene brightness is an extremely simple way of using an exposure meter, it can often give unsatisfactory results because of the character of the scene; for example, where a face is seen against the background of a bright sky, the meter reading is unduly dominated by the bright sky area and the face in the foreground areas will be underexposed in comparison. The value of the system is therefore limited to subjects of average brightness distribution or to the less critical requirements of amateur photography.

Professional use of reflected light readings is generally restricted to spot brightness meters with acceptance angles of 2° or less which can explore the brightness of small

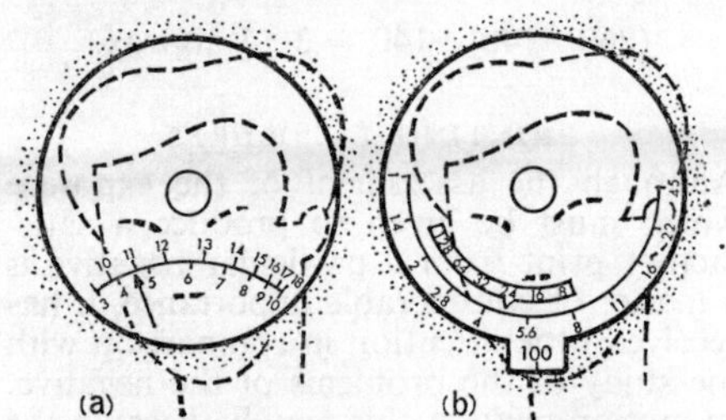

USE OF SPOT EXPOSURE METER. Centre of field of view shows point being measured, subtending 1° or 2° at meter. Bright illuminated scale (actually seen as white against face) is calibrated in foot-lamberts (a) or as lens aperture required at any camera speed for given film speed rating (b).

light and dark areas of the scene. In what is termed the binary method, a meter is used which reads the spot brightness of the main foreground subject, while at the same time the total brightness of the whole area is also measured. The two values are automatically combined to show the required exposure.

EXPOSURE CONTROLS

TIME. Of all the exposure factors which must be controlled by the motion picture cameraman, the length of time of exposure presents the least problem. The running speed of professional film cameras when sound is to be recorded or added must be very exactly maintained to ensure synchronization of the sound record. A normal speed of 24 pps is universal in motion picture cinematography, which is equivalent to an exposure time for each image of 1/50 sec. at a normal shutter opening of 175°. Even when high-speed cameras are used for slow motion effects and other forms of trick photography, it is usual to run them at simple multiples of this rate, so that the increase in lighting level required can be easily calculated. Problems occasionally arise when the printing level of a processed negative is assessed from a section at the beginning or end of a scene when the camera is not running at full speed, but these sections are usually obvious on inspection.

In the photography of silent films, the camera speed is normally 16 or 18 pps, but may not be very precisely controlled in amateur equipment; this corresponds approximately to an exposure time of 1/35 sec.

LENS SETTINGS. With a given film sensitivity and the time of exposure fixed, the motion picture cameraman must finally control the exposure of the film by choosing the best combination of scene lighting and lens opening. The nominal light transmission of a lens is represented by its *f*-number, which may be adjusted by the opening or closing of an iris diaphragm forming part of the lens assembly. The light passed by the diaphragm varies inversely as the square of its *f*-number, so it is usual to provide lenses with a diaphragm setting scale calibrated in increments of $\sqrt{2}$, thus *f* (1·0), 1·4, 2·0, 2·8, 4·0, 5·6, 8, 11, 16, 32, representing decreasing light transmission with increasing *f*-number. These values are known as stops, and an alteration of diaphragm by one interval of the scale, or one stop, increases or decreases the light transmitted by a factor of 2. A setting of *f* 4 thus gives twice the exposure of a setting of *f* 5·6 but only half that of *f* 2·8.

T-STOPS. In actual practice, the *f*-number, which is calculated geometrically by dividing the focal length of the lens by the diameter of the diaphragm, does not accurately represent the true effective light transmission of a lens because of losses by absorption of the glass components and by reflection at their surfaces. For professional work, therefore, it has become the practice to calibrate lenses with their equivalent effective transmission numbers or T-stops; the T-number representing the transmission which would be given by the corresponding *f*-number of a perfect lens having no losses whatsoever. Since the actual losses within a particular lens result from many causes, there is no fixed relation between its geometrical *f*-number calibration and the effective T-numbers, and the latter values must be established by actual measurements of light arriving at the image. T-stop calibrations, therefore, provide a much more accurate basis for exposure correction by diaphragm adjustment, especially when matched results from several different lenses must be obtained.

In studio photography, where the cost of artificial lighting equipment and power is considerable, it is usual for economic reasons to employ as large a lens transmission (low stop number) as possible, but this is complicated by the fact that the depth of field of the lens becomes smaller as the stop number decreases and may not provide sharp images of all the required action. On the other hand, close shots may be required with the background intentionally unsharp, so that the lens diaphragm will be opened up to a low *f*-number and the lighting of the scene correspondingly reduced. Each camera set-up must therefore be studied on its merits, and the skill of the lighting cameraman shows itself in his success in meeting the artistic and dramatic demands of the scenes with his available lighting resources so as to produce consistently-exposed negatives.

FILTERS. In exterior photography by natural light, the emphasis is more on the adjustment of the lens diaphragm to suit the available illumination once the extremes of subject brightness have been controlled. Under very bright sunlight conditions it may be impossible or undesirable to reduce the lens diaphragm setting sufficiently, and use will then be made of neutral density filters of known transmission or filter factor.

When colour filters are used on the camera to provide unusual contrast or tonal effects, or to correct the available lighting to the type of film used, their effect on exposure must be allowed for, and all filter manufacturers specify the factor by which the exposure level on a particular type of film must be increased when a given filter is employed. The value of lighting correction filters is sometimes specified as an alternative speed index for a given type of film; for example, colour negative exposed to daylight through a Wratten 85 filter is given an ASA speed rating of 32 in comparison with its unfiltered rating of 50 to tungsten.

PRACTICAL CALCULATIONS. A useful formula for the calculation of exposure is:

$$L = \frac{25 \times f^2}{S \times t}$$

where L is the required lighting level, in foot-candles (key+filler);

f is the lens diaphragm setting, or stop;

S is the sensitivity of the film used, on the ASA Exposure Index scale;

and t is the time of exposure, normally 1/50 or ·02 sec. for professional cinematography.

As an example, the lighting level required for exposing a scene to be shot at a lens setting of *f* 4 on colour negative of ASA rating 50 at standard camera speed is

$$L = \frac{25 \times (4 \times 4)}{50 \times 1/50} = 400 \text{ fc.}$$

To provide a lighting contrast of 3 : 1, the filler light would have to be of the order of 120 to 150 fc, and the key light alone, measured without the effect of filler, 250 to 280 fc, so that

$$(260+140) : 140 = 3 : 1 \text{ approx.}$$

PRINTING EXPOSURES

Although the assessment of the exposure which must be given to produce a satisfactory print from a particular negative is a matter of considerable importance, it has received little attention in comparison with the study of the problems of the negative. In motion pictures this may be because the exposure of the positive is a subject for the specialized processing laboratory, so that an incorrectly exposed print can be replaced at relatively little expense of time and material, whereas to retake a wrongly exposed negative may involve enormous production costs. However, the latitude of the high-contrast positive stocks used to make prints is extremely small, and exposure increments must be used which are very much smaller than the half-stop subdivisions which a cameraman regards as close limits.

On any film-printing machine, the factors affecting exposure are, as before, the sensitivity of the stock, the intensity of the light and the time of exposure. Each type of stock has its own sensitivity rating but careful checks from one emulsion batch to another are necessary to compensate for small variations.

PRINTER POINTS. On most production machines, the speed of film movement and hence the time of exposure is set, so the normal method of adjustment for various negatives is by light intensity. Printers are therefore provided with various systems of modulating this intensity in accurately determined steps which can be very rapidly operated while a roll of varied scenes of negative passes through the machine. The intensity steps provided are normally of logarithmic increment and are known as printer points. For many years there was a generally accepted convention that one point represented a step of ·05 log intensity, so that an increase of 6 points doubled the exposure, but with the greater accuracy required for colour printing it is now usual to work in half-point steps of ·025 log intensity variations.

There is unfortunately no uniformity between laboratories in their printer point scales. For a long time, a range of 21 points was regarded as adequate and the printing level for a correctly exposed negative thus lay in the middle of this scale, about printer point values 10 to 12. Close collaboration between cameraman and laboratory for reporting exposure ratings or printing levels on a mutually understood basis is essential for consistent work; with care and co-operation, negatives on studio productions can be produced which require no greater variations of printing level than two or three points.

NEGATIVE ASSESSMENT. The methods used by laboratories for the initial determination of negative printing level are numerous, although most of them rely basically on one or more test prints, sometimes in the form called the Cinex strip, in which each frame is given a different exposure corresponding to a known step on the printer scale, from which the best value is estimated by visual inspection. Densitometric measurements of the negative image may be made, but these are difficult to interpret if the scene does not include objects of known reflectivity, such as white surfaces.

Integrated densitometry of the whole frame area has been attempted, following similar automatic methods used in still photography printing, but the wide variation of subject distribution and the importance of a satisfactory visual continuity from scene to scene in a film are not easy to cope with by this means. Visual examination of the negative by an experienced operator also plays an important part, although the assessment of colour negative is extremely difficult.

CLOSED-CIRCUIT TELEVISION. The most encouraging method yet adopted for assessing the printing requirement for colour negative is the use of closed-circuit colour television to present a positive picture by scanning a frame of the negative; if the amplification circuits of the system are adjustable in steps calibrated to correspond with the printing scale in use, it becomes possible to adjust the controls until the most satisfactory reproduction on the TV screen has been obtained, and read off the printing level required.

In dealing with printing exposures, reference must be made to the stringent requirements of colour intermediate stocks and similar derivatives from reversal colour originals. These stocks have negligible latitude and the straight-line portion of the characteristic curve is excessively limited. The utmost accuracy of printing exposure level is therefore essential for both intensity and colour to avoid serious distortion of tonal and colour rendering. L.B.H.

See: *Cameraman* (FILM); *Printers; Sensitometry and densitometry; Tone reproduction.*

Books: *Exposure Manual*, by J. F. Dunn (London), 1959; *Professional Cinematography*, by C. G. Clarke (New York), 1964; *Theory of the Photographic Process*, ed. by C. E. K. Mees and T. H. James, 3rd Ed. (New York), 1966.

⊕□**Exposure Meter.** Device for determining the light flux incident upon or reflected from a scene which is to be recorded by film or TV cameras, the corresponding instruments being known as incident-light meters and reflected-light meters. Exposure meters contain a light-sensitive surface in the form of a

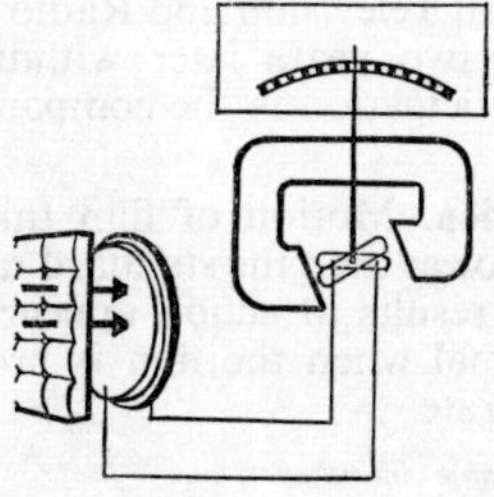

photovoltaic cell or a photoresistor. The cell generates an electrical current varying according to the light falling on it, while the electrical resistance of the photoresistor varies in the same circumstances. The resistor, therefore, is used in conjunction with a battery to produce a varying current. In each case the current produced drives a microammeter calibrated to give exposure recommendations.

In the reflected-light type of meter, the exposure meter is held near the camera position, and its acceptance angle is approximately the same as that of the camera lens. The incident-light type of meter is held close to the principal object being photographed, and has a 180° angle of acceptance, so that it receives all the light falling on the scene.

See: *Exposure control.*

□**Faceplate.** Front part of a camera tube on which light is focused to form the image. In a receiver cathode ray tube (kinescope), the front of the tube on which the picture is formed.

□**Face Tone.** Mean light level of a human face properly lit for television.

⊕**Fade.** Optical effect in which a shot (or TV □scene) gradually disappears into darkness (fade-out) or appears out of darkness (fade-in). Also decrease in strength of a received radio signal caused by increased attenuation, refraction of a beam or cancellation due to interference between direct and reflected signals.

⊕**Fader.** Control used in vision and sound □mixing for raising and lowering signal level and where necessary producing fade-ins and fade-outs. The fader is often a stepped variable resistance (potentiometer) across an input circuit.

See: *Sound recording; Vision mixer*.

□**Farnsworth, Philo Taylor, b. 1906.** American television pioneer, born in Beaver, Utah. One of the first to develop electronic scanning, and the inventor of the image dissector. Became interested in television by reading about it in magazines while still at school, and started studying photoelectricity and the cathode ray tube. Met an influential San Francisco businessman and through him received financial backing for his development work from a group of Californian bankers. He opened a laboratory in Los Angeles in 1926 and established the Crocker Research Laboratories in San Francisco. The following year he took out his first television patent covering a complete electronic system including a dissector tube. In 1929 the Farnsworth laboratories became Farnsworth Television Inc. of California. His system was at that date the only alternative electronic system to that invented by Zworykin.

For a time Farnsworth's activities were moved to Philadelphia while they were under Philco financing. In 1938, he became research director of the newly formed Farnsworth Television and Radio Corporation, but two years later withdrew from active participation in the company.

⊕**Fast Motion.** Motion of film through the camera slower than the standard rate, which therefore results in action appearing faster than normal when the film is projected at standard rate.

See: *Time-lapse cinematography*.

⊕**Feedback.** Technique of general application ☐to amplifiers, whereby some of the output is returned to the input. In this way, the gain and frequency response of the amplifier can be modified. As a rule, negative feedback decreases gain but improves linearity and stability; positive feedback gives rise to oscillation.

☐**Feeder.** Means by which radio-frequency energy is supplied from a transmitter to its aerial; in common parlance the term is also applied to the lead-in connecting a receiving aerial to its terminal equipment.

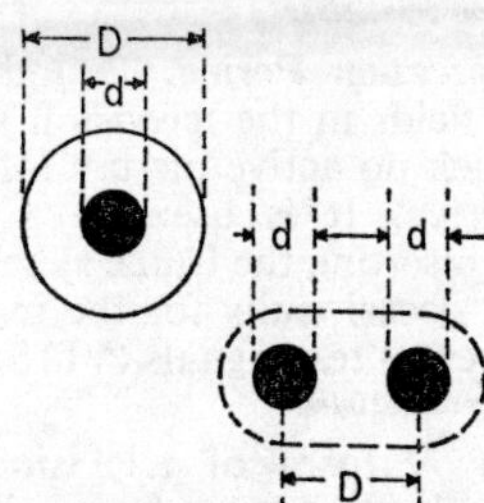

FEEDERS. (*Top*) Coaxial. D, d. Diameter of outer and inner conductors respectively. (*Bottom*) Parallel. D. Distance between centres of conductors. d. Diameter of conductors.

Two main types are in common use: coaxial and parallel.

COAXIAL FEEDER. In this, a central conductor is supported by means of a suitable insulating material so that its axis coincides with that of an outer tubular conductor which is then covered with an insulating sheath. Characteristic impedance Zo is:

$$Zo = \sqrt{\frac{R+j\omega L}{G+j\omega C}} \text{ ohms,}$$

where R = series resistance per unit length in ohms,
L = series inductance per unit length in henrys,
G = shunt conductance per unit length in mhos,
C = Shunt capacitance per unit length in farads,
ω = angular frequency ($2\pi f$).

At frequencies above a few hundred Hz, ωL and ωC become much larger than R and G respectively, and the expression becomes

$$Zo = \sqrt{L/C} \text{ ohms}$$

which is independent of frequency. The values of L and C are determined by the physical dimensions of the cable and the particular substance used to separate the conductors. Hence, the characteristic impedance may be alternatively expressed in terms of these as follows:

$$Zo = \frac{138 \log_{10} D/d}{\sqrt{k}} \text{ ohms}$$

where D = diameter of outer conductor,
d = diameter of inner conductor,
k = average dielectric constant of the insulating material.

Common values of Zo for currently manufactured cables are 50, 75 and 100 ohms.

Coaxial cables are essentially unbalanced, and typical attenuation figures vary, depending on size, from around 3 dB per 100 ft. at 100 MHz for solid polythene cables to about half this figure for cables having semi-airspaced insulation. The attenuation in dB is roughly proportional to the square root of the frequency.

PARALLEL WIRE. Consists of two similar conductors supported in such a way that the distance between them is constant throughout its length. Characteristic impedance is:

$$Zo = \frac{276 \log_{10} 2D/d}{\sqrt{k}} \text{ ohms}$$

where D = distance between centres of the conductors,
d = diameter of the conductors.

For low-power use it is manufactured in the form of a flat polythene ribbon about ½ in. wide in which the two conductors are immersed. The Zo is usually 300 ohms and its attenuation, again roughly proportional to the square root of the frequency, is typically 1·3 dB per 100 ft. at 100 MHz.

Each type of feeder must be terminated with a resistance equal to its characteristic impedance, otherwise reflections occur.

See: *Transmission and modulation; Transmission networks; Transmitters and aerials.*

⊕**Ferraniacolor.** Negative/positive integral tripack colour film process similar to Agfacolor and introduced for professional motion picture production in Italy in 1953.

See: *Colour cinematography.*

☐**Ferrite Isolator.** Device placed in a coaxial line or waveguide which uses ferrite in a permanent magnetic field to provide greater attenuation in one direction than another.

☐**Field.** In an interlaced scanning raster on a camera, telecine, monitor or receiver, each coverage of a raster is a field, and the two interlaced fields make a picture. (Other interlace ratios have been used, but 2 : 1 is universal in broadcasting.) The term frame, with its confusing ambiguity and reference to film, should be avoided.

See: *Television principles.*

□**Field Blanking.** Period between pictures which is blanked or suppressed to allow the scanning spot to return to the start of the picture. Also called field suppression.

The field-blanking period is much longer than the line-blanking period, and occupies 13–21 lines in the American 525-line system, and 18–22 lines in the CCIR 625-line system. It contains a group of field-sync pulses (broad pulses) flanked by groups of equalizing pulses, and the line-sync pulses continue throughout the field-blanking period.

See: *Television principles.*

□**Field Drive Signal.** Signal used to establish field sync in studio systems, for example, in non-composite working.

⊕□**Field Flattener.** Auxiliary lens which helps to correct for curvature of field, the lens aberration which causes a photographic image to assume a curved shape in contrast to the shape of photosensitive surfaces and projection screens, which are usually flat.

□**Field Identification Signal.** Signal (often called ident signal) incorporated in early forms of the PAL colour system, to identify alternate fields and so produce correct phasing of the R−Y component of the colour subcarrier which, in this system, is reversed on alternate lines of each field. In later developments of PAL, the same result is achieved more simply by swinging the phase of the colour burst through ±45°; there is no separate ident signal, and none is required in NTSC.

In the SECAM system, different components of the chrominance information are transmitted on succeeding lines, so that ident signals must be inserted in the waveform to ensure that the R−Y signal does not reach the B−Y output, and vice versa. This is achieved by sawtooth waveforms incorporated in five lines of the field-banking period of the transmitted waveform, the teeth being positive-going on the B−Y lines and negative-going on R−Y lines.

See: SECAM *colour system.*

⊕□**Field Lens.** Large-diameter lens used in optical relay systems to increase the light transfer from the primary-input lens system to the output. An aerial image is formed in the plane of the field lens by the input lens and this is re-imaged by the output system, generally on the photo-cathode of the pick-up tube. The use of a field lens reduces the size of other lenses needed and hence the cost of the unit.

See: *Camera (colour).*

□**Field Phasing.** Action of setting the phase of a picture frame with respect to a synchronizing source so that, on genlocking, a frame or picture roll does not occur.

See: *Picture locking techniques.*

□**Field Sawtooth.** Waveform which rises at a constant rate and then falls rapidly, at the frequency of field deflection. The simple signal is used for checking linearity.

□**Field Suppression.** Period between successive fields which is blanked or suppressed to allow the scanning spot to return to the start of the picture. Also called field blanking.

See: *Television principles.*

□**Field Suppression Period.** Period between successive fields in the television waveform during which no active picture information is transmitted. It is used as a reference period for inserting the frame synchronizing signal and occasionally for the insertion of vertical interval test signals (VITS).

See: *Television principles.*

□**Field Tilt.** A form of television picture distortion, also known as frame tilt. Normally, reference points on the syncs of the correct television waveform bear a constant relation to earth potential, but this relationship may be destroyed, e.g., by hum or when the signal is passed through an a.c. coupling. After the sync signals are removed from the waveform in a sync separator stage, the distortion of the video signal remains as a gradual brightness increase or decrease over the frame period, and the result on the waveform is similar to the addition of a frame-shading signal.

The introduction of a line-by-line clamping circuit removes such effects from the signal, provided that the signal has not been modulated by them. Usually the response of an amplifier is checked for frame tilt by applying a 50 Hz square wave to the input. When measured at the output, a typical response is residual tilt less than 5 per cent of the output level. Frame tilt is sometimes deliberately introduced in the form of frame shading to remove spurious shading signals from the output of a camera tube.

See: *Receiver.*

□**Field Time Base.** Circuits generating the sawtooth waveform required to deflect the electron beam vertically in a camera or receiver tube. Two complete fields interlace to produce a complete frame.

See: *Interlace.*

⊕□**Fill Light.** In studio lighting, light directed into the shadows to prevent excessive contrast. Also called filler and fill-in light.

FILM DIMENSIONS AND PHYSICAL CHARACTERISTICS

Cinematograph film consists fundamentally of a thin flexible transparent base material on which is coated a photo-sensitive layer having the characteristics required for recording the image. The base is usually a cellulose derivative, originally cellulose nitrate but now a less inflammable compound, while the sensitive layer is normally a photographic emulsion containing gelatin. Both the base and the emulsion contribute to the physical characteristics of film behaviour.

The dimensions of most of the generally accepted forms of motion picture film have been agreed internationally, and the data presented here are based on the publications of the standards organizations of the United States (USA, formerly ASA), Britain (BSI), France (AFNOR) and Germany (DIN), although not all these bodies have in fact issued a standard for every type of film subsequently described; this is particularly the case in film gauges other than 35mm. The following specifications are considered:

1. Raw stock: The width of the film and details of its perforation.
2. The picture image: The size and location of the picture area.
3. Sound: The size and location of optical and magnetic sound tracks on film.

HISTORY

In the early development period of the cinema, each inventor experimented with his own particular design of camera mechanism and film, but in 1889 collaboration in the United States between W.K.L. Dickson of Edison's laboratory, and George Eastman of the newly formed Eastman Kodak Company, produced a form of film which eventually became the professional standard still in use today.

INTRODUCTION OF 35mm. This film, $1\frac{3}{8}$ in. or 35mm in width, was perforated with rows of holes along each edge to permit the movement of the film by sprocket tooth drive. The width of the film between the perforation holes was 1 in., which was approximately the width of the exposed picture. Its height was almost $\frac{3}{4}$ in., corresponding to the length of film occupied by four perforation holes at $\frac{3}{16}$ in. spacing. A one-foot length of film thus provided 16 picture images, or frames, and took one second to run through a camera or projector at a rate of 16 pictures per second.

The technical and commercial success of Edison's equipment led to the widespread adoption of film of these dimensions, which was generally accepted internationally by the early years of the 20th century. Standardization of dimensions was initiated in 1917 by the Society of Motion Picture Engineers of America, and has been continued by the national and international standards organizations concerned. The introduction of sound in the late 1920s did not involve changes in the basic dimensions of the raw film stock, but only of the images recorded on it, and 35mm film continues to be the most widely used material for all professional entertainment motion-picture production and presentation.

SUB-STANDARD SIZES. The cost of 35mm material was too great for the amateur cinematographer, and numerous systems were introduced using narrower gauge materials, the so-called sub-standard sizes. Several of these, even as early as 1898 to 1900, used 35mm stock slit in half to $17\frac{1}{2}$ mm width or 35mm film with two rows of pictures side by side. Other proposals used film of 21, 22 or 28mm width, some being almost as large as the standard format. At the end of 1922, Pathé Cinema of France introduced a projector which used prints made on film 9·5mm in width, and in 1923 this was followed by a corresponding camera and reversal film for amateur photography. In the United States, Eastman Kodak had also been working on a similar programme and introduced the Cine-Kodak process, using 16mm film, in 1923.

Although both these systems were introduced about the same time, their subsequent history shows an enormous divergence. The 9·5mm film size was never accepted in the United States even in the amateur field and, although extensively used in Europe, it was never adopted by other than the amateur user. Unfortunately, 9·5mm film was often associated with cheap equipment of poor reliability, and it lost popularity in the face of later competition. The 16mm form, on the other hand, was found to be capable of such excellent results that its use soon extended far beyond the amateur field and, with the introduction of sound on 16mm film in 1939, it became established as an alternative professional medium, outstandingly well suited for presentation to smaller audiences in the fields of education and industrial publicity, as well as entertainment.

Film Dimensions and Physical Characteristics

SMALL-FORMAT SYSTEMS. With the improvement in the definition and grain of photographic emulsions, the possibility of a format even smaller than 16mm could be considered. In 1932, Eastman Kodak introduced a system for exposing two rows of pictures on 16mm film which was subsequently slit down the middle after processing, and projected as a strip 8mm wide. Each picture frame occupied one quarter of the area of the 16mm image and, although the size of the projected picture was restricted, the system provided a very adequate and relatively inexpensive method for amateur cinematography. It entirely displaced 16mm from this field, and replaced much 9·5mm usage. From 1961 onwards, 8 mm film, sometimes with sound recorded on an added magnetic stripe, began to be used for educational and industrial purposes with school classes and other small audience groups.

The most recent small format development has been the introduction of the Super-8 system by Eastman Kodak in 1965. Although still using film stock 8mm in width, a modification of the perforation hole size and arrangement allows a larger area to be used for the picture. The system allows improved sound-track reproduction and is expected eventually to replace the standard 8mm film in both amateur and commercial applications.

LARGE-FORMAT SYSTEMS. At the other end of the scale, film of dimensions wider than the standard 35mm had been suggested for the presentation of pictures on very large cinema screens, and at various times films of 42, 55, 65 or 70mm width have been employed. Of these, the only one which has achieved international use is the 70mm form of wide-screen presentation for which prints are usually produced from negative of 65mm width in the United States and Western Europe, and from 70mm negatives in Russia and Eastern Europe.

RAW STOCK DIMENSIONS

The basic dimensions of the various types of film on which motion picture images are exposed are specified by their width and the size and arrangement of the perforation holes. The main categories of stock width are nominally 8mm, 9·5mm, 16mm, 35mm and 70mm, and some of these may be found with different forms of perforation to meet particular requirements of use.

PERFORATION HOLES. The widely used professional 35mm film stock may be supplied with four different shapes of perforation holes. The earliest form of these, now referred to as the Bell & Howell type, has rounded ends which are arcs of a circle with straight parallel sides, or flats, intersecting these arcs at fairly sharp angles. This form of perforation allows accurate registration of the film in camera or printing mechanisms, but the acute corners were a source of weakness in prints which had to pass over the sprocket teeth of a projector a very large number of times, and the use of the B & H perforation hole is now limited to negative films for use in cameras,

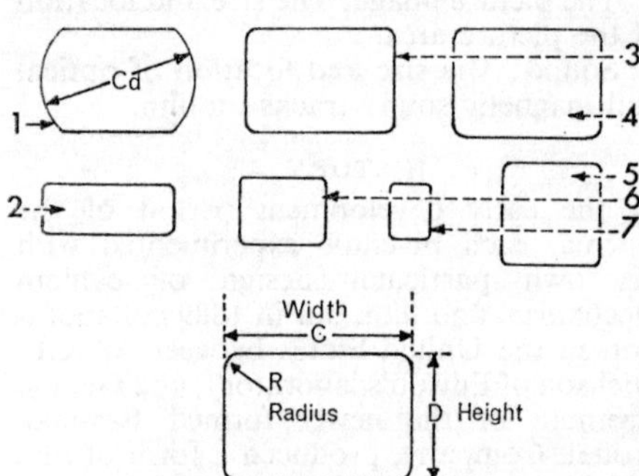

PERFORATION TYPES AND DIMENSIONS. 1. 35mm Bell-Howell (BH). 2. 9·5mm. 3. 35mm Dubray-Howell (DH). 4. 35mm positive Kodak Standard (KS). 5. 35mm Cinemascope (CS). 6. 16mm. 7. Super-8.

PERFORATION DIMENSIONS

Perforation Type	Inches C Width	Inches D Height	Inches R Radius	Millimetres C Width	Millimetres D Height	Millimetres R Radius
9·5	·0945	·0394	·006	2·400	1·000	0·15
16	·0720	·0500	·010	1·830	1·270	0·25
Super-8	·0360	·0450	·005	0·914	1·143	0·13
BH	(Cd. ·1100)	·0730	—	(Cd. 2·794)	1·850	—
DH	·1100	·0730	·020	2·800	1·850	0·50
KS	·1100	·0780	·020	2·800	1·980	0·50
CS	·0780	·0730	·013	1·980	1·850	0·33
Tolerances	±·0004	±·0004	±·001	±0·010	±0·010	±0·025

where it is standard, and to other stocks used for printing or projection in special effects work where accurate registration is essential.

In order to improve the life of release prints and avoid breakdown at the perforation angles, a rectangular hole with rounded corners was introduced for positive print stock in 1923; the height of this hole was slightly larger than the Bell and Howell hole, which helped to reduce interference by the sprocket teeth in the projector. This form has now become almost universal for 35mm release print material, and is referred to as the Kodak Standard, Eastman Standard, or simply Positive Standard perforation. At international conferences in 1934 and 1936, proposals were made to adopt this form of hole as the universal standard for all 35mm stocks, both negative and positive, but although this was accepted by Russia and a number of Eastern European countries, it was not favoured by the motion picture industry in the United States, which has continued to use different holes for negative and positive materials. Great Britain and the countries of Western Europe have conformed with the American practice. However, the Kodak Standard perforation has been accepted for both negative and positive materials of width greater than 35mm.

In some forms of printing, the difference in height between the B & H negative hole and the KS positive hole allowed relative movement and was therefore undesirable for image registration. An alternative rectangular hole with rounded corners having the same height as the Bell and Howell negative hole, was therefore introduced under the title Dubray-Howell perforation but, although very satisfactory for registration printing purposes, it has found only limited acceptance.

The fourth variety of 35mm perforation is the smaller rectangular hole with rounded corners introduced in 1953 as part of the Cinemascope wide-screen process in order to provide a larger film area for picture and track images; this form is used only for positive stock on which release prints with four magnetic sound-tracks are made.

Of the narrower film gauges, 16mm material is made with only one type of perforation hole, but is found in two forms, one with rows of holes down both edges of the film, and the other with only a single row down one edge to allow space for a sound track along the other edge.

Perforations of 9·5mm film take the form of wide rectangles with rounded corners and are located as a single row along the centre line of the film instead of down the edge as in all other types.

The original type of 8mm film was produced by slitting 16mm which had been perforated with regular 16mm holes down both edges at half the normal pitch intervals. This provided a material (Double-8) which could be processed on standard 16mm developing machines, and which was exposed by running through the camera twice, first to give pictures down one side of the film, and then in the opposite direction to give a second row of pictures down the other side. Although conveniently compatible with 16mm handling, the holes occupied an unduly large proportion of the film width, and the use of smaller perforations in the Super-8 format provides considerably more space for picture and track information. Super-8 film is normally exposed in the camera and processed as 8mm in width rather than as Double-8.

COMBINATION STOCKS. Although the main sizes of film used in camera and projector are those described above, several other forms of perforated film are produced to assist in the bulk manufacture of release prints in quantity. For example, because most professional laboratories are equipped with high-speed developing machines for 35mm film, a stock of this width, but perforated with two rows of 16mm holes, is made to allow two 16mm copies to be printed on the same piece of stock, which is then slit after processing to give two 16mm strips, the narrow outer margins being discarded. A Double-16 stock 32mm in width is also used for the bulk printing of 16mm copies, and has the advantage of needing only a single slit through the centre after processing.

Double-8 film is normally used for exposure in standard 8mm cameras, as well as for making copies, and double Super-8 may also be used in bulk printing. In addition, there are forms of 35mm stock which are perforated so as to be slit to yield four 8mm strips, or three 9·5mm strips, the unwanted additional width of the 35mm being provided with additional holes and used for film transport through processing machines.

SHORT PITCH PERFORATION. The printing of positive copies from an original negative is usually done on a continuous printer in which the two films are in contact on a rotating sprocket at the time of

Film Dimensions and Physical Characteristics

RAW STOCK DIMENSIONAL STANDARDS

Ref. No.	Type of Film	Perforations Type	Perforations Rows	Inches A Width	Inches B Pitch	Inches E* Margin	Inches F Inter-perf. Width	Millimetres A Width	Millimetres B Pitch	Millimetres E* Margin	Millimetres F Inter-perf. Width
	(i) Narrow Gauge Materials										
1	Standard 8 mm ...	16	1	0·314	0·1500	0·0355	—	7·98	3·810	0·90	—
2	Super-8	S-8	1	0·314	0·1667	0·020	—	7·98	4·234	0·510	—
3	9·5 mm	9·5	1	0·374	0·2977	(central perf.)	—	9·50	7·56	(central perf.)	—
4	Double-8 Standard ...	16	2	0·628	0·1500	0·0355	0·413	15·95	3·810	0·90	10·49
5	Double Super-8 ...	S-8	2	0·628	0·1667	0·020	0·516	15·95	4·234	0·510	13 11
6	16 mm, short pitch ...	16	1	0·628	0·2994	0·0355	—	15·95	7·605	0·90	—
7	16 mm	16	1	0·628	0·3000	0·0355	—	15·95	7·620	0·90	—
8	16 mm, perf. both sides, short pitch	16	2	0·628	0·2994	0·0355	0·413	15·95	7·605	0·90	10·49
9	16 mm, perf. both sides	16	2	0·628	0·3000	0·0355	0·413	15·95	7·620	0·90	10·49
10	Double-16, short pitch	16	2	1·256	0·2994	0·0355	1·041	31·90	7·605	0·90	26·44
11	Double-16	16	2	1·256	0·3000	0·0355	1·041	31·90	7·620	0·90	26·44
12	Double-16 on 35, short pitch	16	2	1·377	0·2994	0·096	1·041	34·975	7·605	2·44	26·44
13	Double-16 on 35 ...	16	2	1·377	0·3000	0·096	1·041	34·975	7·620	2·44	26·44
	(ii) Standard or Large Gauge Materials										
14	35 mm, short pitch, Negative	BH	2	1·377	0·1866	0·079	0·999	34·975	4·740	2·00	25·37
15	35 mm Negative ...	BH	2	1·377	0·1870	0·079	0·999	34·975	4·750	2·00	25·37
16	35 mm Cinemascope Positive	CS	2	1·377	0·1870	0·086	1·049	34·975	4·750	2·18	26·65
17	35 mm Positive ...	KS	2	1·377	0·1870	0·079	0·999	34·975	4·750	2·00	25·37
18	35 mm Dubray-Howell	DH	2	1·377	0·1870	0·079	0·999	34·975	4·750	2·00	25·37
19	65 mm, short pitch, Negative	KS	2	2·558	0·1866	0·117	2·104	64·97	4·740	2·97	53·44
20	65 mm Negative ...	KS	2	2·558	0·1870	0·117	2·104	64·97	4·750	2·97	53·44
21	70 mm Cine	KS	2	2·754	0·1870	0·215	2·104	69·95	4·750	5·46	53·44
	Tolerances for 1 to 18			±·001	±·0005	±·002	±·002	±·025	±·013	±·05	±·05
	except			(1) ±·002		(4, 5, 8, 9)	±·001	±·05		(4, 5, 8, 9)	±·03
	Tolerances for 19 to 21			±·002	±·0005	±·003	±·003	±·05	±·013	+·08	±·08
	(iii) Wide Film Recording Materials										
22	60 mm Recording ...	KS	2	2·328	0·1870	0·078	1·948	59·12	4·75	1·98	49·38
23	70 mm Recording ...	KS	2	2·756	0·1870	0·078	2·386	70·00	4·75	1·98	60·35
24	80 mm Recording ...	KS	2	3·150	0·1870	0·078	2·770	80·00	4·75	1·98	70·36
	Tolerances			+0—·01	±·002	±·008	±·002	0—·25	±·05	+·20	±·05

Maximum skewness, G, is ·001 for 4, 5, 8 to 18
·002 for 19, 20, 21
·004 for 22, 23, 24

G maximum is 0·03 for 4, 5, 8 to 18
0·05 for 19, 20, 21
0·10 for 22, 23, 24

A dimension, L, overall pitch for 100 perforations may be quoted:
L = 100 × B with tolerance ±·015 for 1 to 5, 14 to 21
±·030 for 6 to 13
±·060 for 22, 23, 24

L = 100 × B with tolerance: ±0·38 for 1 to 5, 14 to 21
±0·8 for 6 to 13
±1·5 for 22, 23, 24

* Where there is only one row of holes the value for E refers to the narrower margin.

exposure. The direction of printing light is more-or-less radial from the sprocket axis, the positive film being the outer layer and the negative film the inner layer on the sprocket. The surface speed of the two films is therefore not exactly the same and, to avoid slippage, the distance occupied by a given number of perforations on the negative should be slightly less than that of a similar number on the positive. For a sprocket of 1 ft. circumference and film thickness ·006 in., the difference in perforation pitch should be of the order of 0·3 per cent.

When film stocks were made using cellulose nitrate base, an appreciable degree of shrinkage took place during processing and storage, so that the processed negative was always shorter than the positive raw stock, even though both had originally been perforated to the same dimensions. However, since 1950 low-shrinkage safety base has been in use and negatives do not now normally shrink sufficiently to provide the necessary pitch difference. Film stocks to be used as negatives are therefore perforated to a pitch dimension slightly shorter than that of the corresponding positive material, the values for 35mm films being ·1866 in. (4·740mm) for negative compared with ·1870 in. (4·750mm) for positive. For 16mm films the values are ·2994 in. (7·605mm) and ·3000 in. (7·620mm) respectively.

RECORDING MATERIALS. Many photographic recording instruments use perforated film or paper as the sensitized material to be exposed, although the resultant images are examined frame by frame and not in continuous motion. Recording film of 16mm and 35mm width is perforated in the same form as positive film for motion picture use, but somewhat more

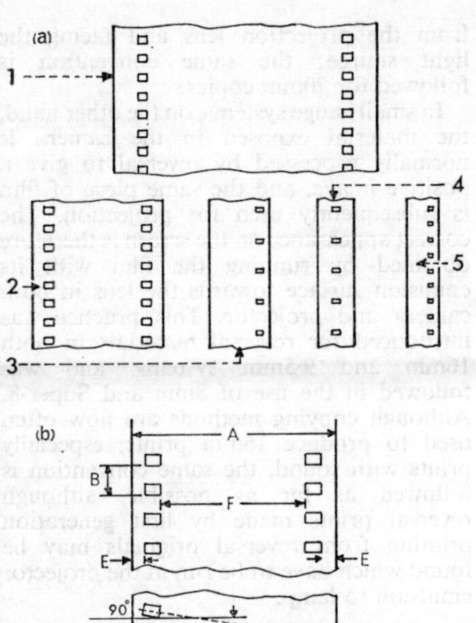

RELATIVE FILM SIZES AND RAW STOCK DIMENSIONS. (a) Relative sizes of various gauges of film (to same scale). 1. 70mm. 2. 35mm. 3. 16mm. 4. 9·5mm. 5. 8mm. (b) Basic measurements used in dimensioning of raw stock. A. Width. B. Perforation pitch. E. Margin. F. Inter-perforation width. G. Skewness of perforations.

generous dimensional tolerances are permitted. Recording materials are also made in 60, 70 and 80mm widths, but it should be noted that 70mm recording film has much narrower perforation margins than 70mm cine film, and cannot be handled on the same sprockets for printing or processing.

FILMING SPEED AND FILM POSITION

Early forms of cameras and projectors operated at speeds between 14 and 18 exposures a second, and a rate of 16 was eventually adopted as standard. The linear speed of 35mm film through the mechanism was therefore nominally 1 ft. per second or 60 ft. per minute.

FILM FOOT. The measurement of long lengths of film was thus conveniently made in units of 16 frames or one film-foot, and this has been continued as the unit of length in the industry. For many years the length of four perforations occupied by a frame on 35mm has been slightly less than .75 in., and therefore the actual length of a film-foot measures slightly less than the linear distance of twelve inches. The length of four standard positive perforations is actually 0·748 in. so that 16 frames measures 11·968 in. or 0·032 in. short of a true foot. This means that the length of a reel described as 1,000 (film) feet is almost 3 ft. short of the full linear distance.

SOUND SPEED. With the introduction of sound on film, the number of images per second was increased to 24 to provide greater length of film for a given time, and normal 35mm camera and projector speeds are now specified as 24 pps, $1\frac{1}{2}$ film-feet per second, or 90 film-feet per minute. This is referred to as sound speed, the 16 pps rate being termed silent speed.

The introduction of 16mm film followed the same conventions for speed, but since the frame height in this gauge is ·3000 in., the linear speeds of film are 24 ft. per minute for silent and 36 ft. for sound; in 16mm, 40 frames are exactly one foot in length.

SMALL FORMATS. Films of 8mm and 9·5 mm format were originally for amateur use only, and the general use of sound did not arise. The silent speed of 16 pictures per second was usual, although more recently international standards have favoured 18 pps. Where 8mm prints are made by reduction from 35mm and 16mm originals with sound, and are striped for magnetic sound tracks, they are projected at 24 pps. Standard 8mm film has 80 frames to the foot, so that linear running speeds are 12 ft. per minute nominal in silent pictures and 18 ft. per minute for sound.

Image rates proposed for Super-8 are 18 pps for silent and 24 pps for sound. The Super-8 frame is longer than Standard-8 and there are 72 frames to the foot, giving linear speeds of 15 ft. per minute silent, and 20 ft. per minute sound.

LARGE FORMATS. In the larger gauges of film used for wide-screen presentation, the speed of 24 pps has generally been maintained, although some of the earlier productions in the Todd-AO system were made at 30pps to reduce the effect of flicker at high levels of screen brightness. However, this raised difficulties when 35mm reduction copies were required for general release, and the 24pps speed is now universal. The frame of the 70mm print is five perforations high, so that the linear film speed is 25 per cent greater than for 35mm, giving the equivalent rate of $112\frac{1}{2}$ ft. per minute.

Since linear film speeds are based on the length of a frame interval, which is not an exact number of millimetres, film speeds expressed in the metric system give somewhat inconvenient figures, and there is no simple relation between metres and seconds as there is between film-feet and seconds.

METRICAL FILM SPEEDS

Film Speeds pps	Film Gauge	Metres/ Minute
24	70	34.20
24	35	27.36
24	16	10.97
16	16	7.31
18	Standard–8	4.11

In the United States, film for television runs at the same speed as in the cinema, 24 frames per second, but in Britain and some other countries where 50Hz electrical supplies are used, film for television is run at a speed of 25 frames per second.

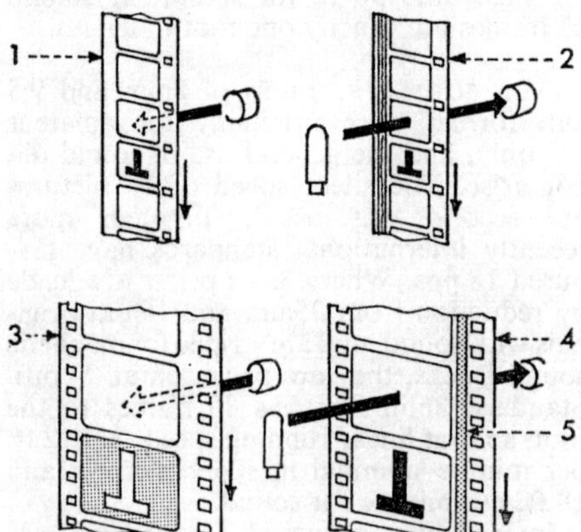

FILM USAGE DURING EXPOSURE AND PROJECTION. 1. 16mm camera: emulsion side of film faces lens. 2. 16mm projector: emulsion side again faces lens. (8mm and 9·5mm usage is similar.) 3. 35mm camera: emulsion side of film faces lens. 4. 35mm projector: emulsion side faces lamp. 5. Sound track. (65/70mm usage is similar.)

FILM POSITION. The difference between 35mm and 16mm, in the relation of the emulsion side of the film to the projection lens, arises from the characteristics of the materials originally used in these two gauges. In the camera, the original material must be exposed with its emulsion surface facing the light coming from the camera lens, and in 35mm usage the camera original is normally processed to give a negative image. From this negative the positive used in projection is made by contact printing, in which the emulsion surfaces of the two films are together. To produce the correct orientation of the image on the screen, the positive must therefore be projected with its emulsion surface away from the projection lens and facing the light source; the same convention is followed for 70mm copies.

In small gauge systems, on the other hand, the material exposed in the camera is normally processed by reversal to give a positive image, and the same piece of film is subsequently used for projection. The correct appearance on the screen is therefore obtained by running the film with its emulsion surface towards the lens in both camera and projector. This practice was introduced for reversal materials in both 16mm and 9·5mm systems, and was followed in the use of 8mm and Super-8. Although copying methods are now often used to produce 16mm prints, especially prints with sound, the same convention is followed as far as possible, although reversal prints made by first generation printing from reversal originals may be found which have to be run in the projector emulsion to lamp.

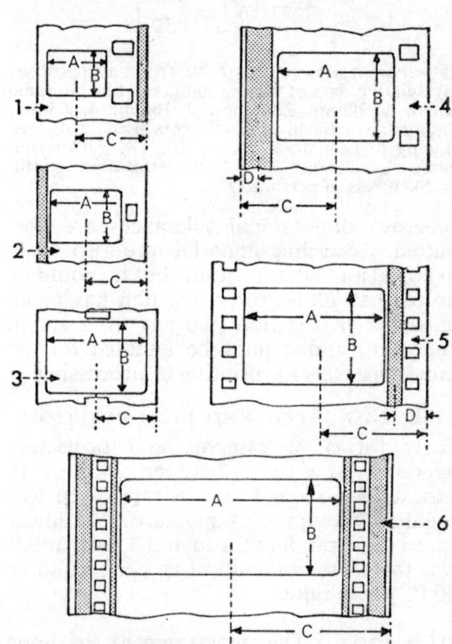

PROJECTED PICTURE AND TRACK DIMENSIONS. 1. Standard 8mm. 2. Super-8. 3. 9·5mm. 4. 16mm. 5. 35mm. 6. 70mm. Size of picture area exposed in camera and printed on film is slightly larger than projected image in both width and height. Films not drawn to uniform scale.

IMAGE SIZES AND AREAS

In its final presentation, the area of film used to produce the picture is determined by the aperture in the projector, but the

DIMENSIONS OF PROJECTED AREAS ON FILM

Type of Film	*Inches* A	B	C	*Millimetres* A	B	C
Standard 8 mm	·172	·129	·205	4·37	3·28	5·20
Super-8	·210	·157	·170	5·32	3·99	4·31
9·5 mm	·315	·242	·187	8·00	6·15	4·75
16 mm	·380	·284	·314	9·65	7·21	7·97
35 mm Movietone	·825	·600	·738	20·96	15·24	18·75
35 mm Wide Screen AR 1·85:1 ...	·825	·446	·738	20·96	11·33	18·75
35 mm Cinemascope AR 2·35:1 ...	·839	·715	·738	21·31	18·16	18·75
70 mm	1·913	·866	1·377	48·60	22·00	34·95
General Tolerances	—	±·002	—	—	±·05	—
Areas reproduced in Television from film:						
from 16 mm	·368	·276	·314	9·35	7·01	7·97
from 35 mm	·792	·594	·738	20·16	15·09	18·75

The projected area of the film image is specified by:
A Width of projector aperture.
B Height of projector aperture.
C Distance of vertical centre-line of picture from reference edge of film.

The projector aperture is usually made with rounded corners, of radius ·005 to ·010 in. (0·13 to 0·25 mm).

The dimensions of the image area *printed* on the film are usually larger than the projected area by ·020 to ·030 in. (0·5 to 0·8 mm) or more in both width and height.

The horizontal centre-line of the picture is centred mid-way between two perforations in standard 8, 9·5, 16 and 35 mm and on the centre of a perforation for Super-8 and 70 mm.

edge of the picture shown may also be defined by black masking borders on the screen itself. In order to avoid white edges being visible, the printed area on the film is made somewhat larger than the projector aperture both horizontally and vertically, and the size of the camera aperture used to expose the original negative is also larger than the projected image.

APERTURE DESIGNATIONS. In the period of the silent cinema, the camera aperture was approximately 24×18 millimetres (·945 in.×·709 in.) and the size of the projector area slightly smaller (23×17·25mm or ·906 in.×·679 in). This is sometimes referred to as the old Full Frame standard or Full Aperture.

With the introduction of sound, the width of the projected image was reduced to allow room for the sound track to be printed on the film, but the general proportions were retained by reducing the height to correspond. This Movietone projector aperture was fixed at 0·825 in.×0·600 in. (20·96×15·22mm) and the corresponding camera image at 0·858 in.×0·631 in. (22·05×16·03mm), which is referred to as the Reduced Aperture standard. These sizes continued in use until 1953, when the introduction of various wide-screen processes led to the larger Cinemascope frame for anamorphic methods of photography and projection, and to the smaller frame height of the so-called flat wide-screen methods. However, even though 35mm cinema projector apertures may be reduced in height, prints made from similarly masked camera negatives can cause difficulties when used in television, and the continued use of camera apertures 0·631 in. high is recommended, even if the picture action is composed for a smaller height.

In the narrow gauges the same practice is followed, the camera aperture being slightly larger than the projected image, but in general the smaller size of the film stock means that the difference between the two cannot provide as wide margins as in 35mm.

The position of the picture area in the

DIMENSIONS OF OPTICAL SOUND TRACKS ON FILM

Type of Film Track	*Inches* D	G	*Millimetres* D	G
Optical on 16 mm	·058	·071	1·47	1·80
Optical on 35 mm	·244	·084	6·20	2·13
General Tolerances	±·001		±·03	

Optical sound tracks on film are specified by:
D Distance of track centre-line from reference edge.
G Track width scanned by reproducer.

The *printed* width is normally ·020 to ·030 in. (0·5 to 0·8 mm) wider than the scanned area.

width of the film is normally referred to one edge of the film which is the guided edge in the projector mechanism. The picture image is centred on the centre line of the film only in 9·5mm, 16mm and 70mm prints, but not in 35mm sound copies or in 8mm and Super-8.

It is of interest to compare the relative areas of film used to provide the projected image in various gauges and formats and in the following list the standard Academy (or Movietone) Aperture is taken as 100; where the print is made from a different size of negative, the relative area of the corresponding negative image is given.

RELATIVE IMAGE AREAS

Gauge and Format	*Relative projected image area*
65 mm Negative and 70 mm. Positive ...	334·9
Technirama Negative AR 2·35:1 ...	220·2
Vistavision Negative AR 1·85:1 ...	194·9
35 mm Full Aperture Negative ...	124·2
35 mm Cinemascope AR 2·35:1 ...	121·2
35 mm Academy Standard AR 1·33:1	100·0
35 mm TV Transmitted Area	95·1
35 mm Wide-Screen AR 1·66:1 ...	83·0
35 mm Wide-Screen AR 1·85:1 ...	74·3
Techniscope Negative AR 2·35:1 ...	60·2
16 mm Standard	21·8
16 mm TV Transmitted Area	20·5
9·5 mm	15·4
Super-8	6·6
8 mm	4·4

FILM AREAS USED IN TELEVISION. The area of the frame used in television must be selected to allow for possible inaccuracies in the transmission scanning equipment and also for the much more considerable variations in the setting and adjustment of the domestic receiver in the home. It is therefore current practice to recognise reduced areas on the print for this purpose in both 35mm and 16mm.

In addition to the film area transmitted by the television chain, American practice specifies a safe action area within which all essential action must take place, and a smaller safe title area to avoid title information being lost at the edge of the domestic receiver screen. British standards, however, specify a single safe area of somewhat different proportions, which is intended to contain all essential information, both action and titles.

The selected areas are all centred on the same lines as in the corresponding cinema film practice but, because of the shape of the normal TV screen, the corners are more rounded.

In review rooms used for screening films produced for television use, it is recommended that the projector aperture should correspond to the dimensions of the safe action area.

SOUND TRACKS

The position of the photographic (optical) sound track on 35mm film lies inside the perforations at the side of the image which appears as the left-hand edge of the picture seen on the screen. In 16mm prints, the sound track occupies the opposite side adjacent to the right-hand edge of the picture on the screen. In both cases, the outer edge of the film nearest the sound track is made the guided edge in the projector.

INTERNATIONAL STANDARDS. For a number of years from 1934, a difference of interpretation of standard practice led to the issue of German standards (DIN) calling for 16mm prints with the track area on the opposite side, but subsequently international agreement was obtained for the convention described above. Sound track widths may be specified in mils, an abbreviation for milli-inches or thousandths of an inch. A 100-mil track is thus one which is 0·100in. in width.

The width of the track scanned by the

TELEVISION SAFE AREA SPECIFICATIONS

	Inches			*Millimetres*		
	A Width	*B Height*	*R Corner Radius*	*A Width*	*B Height*	*R Corner Radius*
35 mm Film						
Area transmitted	·792	·594	—	20·12	15·09	—
American safe action area ...	·713	·535	·143	18·10	13·59	3·63
American safe title area ...	·634	·475	·127	16·10	12·07	3·22
British safe area	·642	·523	·13	16·31	13·29	3·3
16 mm Film						
Area transmitted	·368	·276	—	9·35	7·01	—
American safe action area ...	·331	·248	·066	8·41	6·30	1·68
American safe title area ...	·294	·221	·059	7·47	5·61	1·50
British safe area	·297	·242	·060	7·54	6·15	1·5

projector sound head is narrower than the printed area, and it is, in fact, normal practice to make the area printed from the track negative overlap very slightly the area printed from the picture negative, so as to avoid a gap of clear film appearing between the two.

MAGNETIC TRACKS. The use of magnetic sound tracks on 35mm release prints was introduced as part of the Cinemascope system in 1953, four magnetic stripes being recorded and reproduced to provide stereophonic effects. To allow sufficient width for these, the release prints had to be made on stock perforated with narrow sprocket holes.

However, the use of prints with only magnetic tracks was limited to suitably-equipped cinema theatres and the combination Magopt print was introduced to allow more general distribution of such copies. The narrow perforations and the four magnetic stripes were retained, but an additional narrower than standard, could be scanned by the sound head of a regular 35mm projector.

The presentation of 70mm prints has always been associated with stereophonic sound, six magnetic records being used. Four of these occupy positions outside the perforations, two on each side, where they may be found either as separate magnetic stripes or more usually as two recording positions within a wider stripe along each edge of the film.

Magnetic sound stripes on 16mm prints have usually occupied the same position as the optical track, although magopt prints with half-width stripes have been prepared for special purposes.

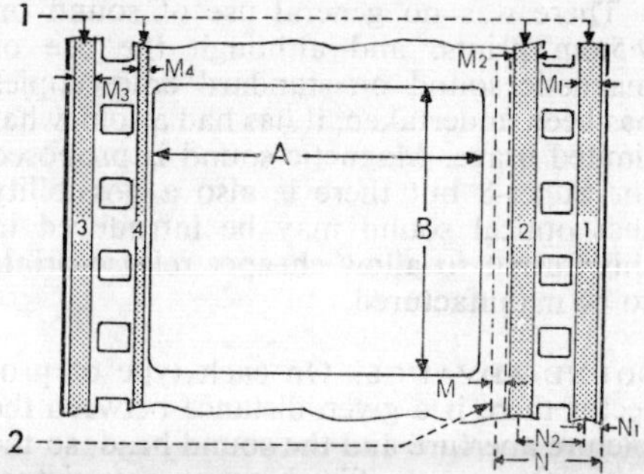

MAGOPT SOUND TRACKS ON 35MM FILM. 1. Magnetic tracks numbered as in Table. N_1, N_2, etc., distance from reference edge to centreline of tracks 1, 2, etc. M_1, M_2, etc., similar reference to track widths. For 70mm 6-track magnetic, track 1 is split into 5 and 1, track 2 keeps same position, track 4 becomes 3, track 3 is split into 6 and 4. 2. Optical track.

DIMENSIONS OF MAGNETIC SOUND TRACKS ON FILM

Type of Film		Inches: Width M	Inches: Distance from Edge N	Millimetres: Width M	Millimetres: Distance from Edge N
Standard 8 mm		·029	·015	·75	·39
Super-8*		·027	·297	·69	7·54
16 mm*		·100	·578	2·54	14·65
35 mm 4-Track Magopt					
Left Channel ...	(Tr. 1)	·063	·040	1·60	1·02
Centre Channel ...	(Tr. 2)	·063	·211	1·60	5·36
Surround Channel ...	(Tr. 4)	·038	1·188	·97	20·18
Right Channel...	(Tr. 3)	·063	1·388	1·60	33·99
Optical Track		·038	·262	·97	6·66
70 mm 6-Track Mag.†					
Left Channel ...	(No. 5)	·063	·052	1·60	1·32
Left Centre ...	(No. 1)	·063	·162	1·60	4·11
Centre Channel ...	(No. 2)	·063	·368	1·60	9·33
Right Centre ...	(No. 3)	·063	2·381	1·60	60·48
Right Channel...	(No. 6)	·063	2·586	1·60	65·99
Surround Channel	(No. 4)	·063	2·697	1·60	68·49
General Tolerances		±·002		±·05	

Magnetic tracks on film can be specified by the width of the magnetic stripe, M, and the distance of the centre-line of this stripe from the reference edge of the film, N. The magnetic recording and reproducing heads will normally be centred on the stripe centre-lines.

* An optional balancing stripe of nominal width ·020 in. (·50 mm.) may be applied to the opposite edge of the film outside the perforations but is not used for recording or reproduction.

† The pairs of stripes, Nos. 5 and 1 and 6 and 4, outside the perforation on both sides may each be replaced by a single stripe of width ·20 in. (5·04 mm.), on which the two tracks are recorded on their original centre-lines.

There was no general use of sound on 9·5mm prints, and although the use of magnetic sound on standard 8mm copies has been undertaken, it has had a somewhat limited usage. Magnetic sound is proposed for Super-8 but there is also a possibility that optical sound may be introduced in this format to allow cheaper release prints to be manufactured.

SOUND ADVANCE. On each type of projector there is a given distance between the picture aperture and the sound head, so the separation on the film between any picture frame and its corresponding sound must be accurately defined for synchronisation. This is usually stated as the appropriate number of frames rather than the actual measurement of distance, the accepted separations being:

35mm Optical sound : 21 frames ahead of picture.
35mm Magnetic sound : 28 frames behind the picture.
70mm Magnetic sound : 24 frames behind the picture.
16mm Optical sound : 26 frames ahead of picture.
16mm Magnetic sound : 28 frames ahead of picture.
8mm Magnetic sound : 56 frames ahead of picture.
Super-8 Magnetic sound : 18 frames ahead of picture.
Super-8 Optical sound : 22 frames ahead of picture.

Tolerances are $\pm \frac{1}{2}$ a frame for all except standard-8, where $\pm$ 1 frame is permitted.

PHYSICAL CHARACTERISTICS OF FILM

The dimensions of film can change greatly with varying conditions of temperature and humidity, its behaviour being complicated by the differing characteristics of the base and the emulsion. For many years, film base was essentially cellulose nitrate, but the highly inflammable nature of this material imposed many restrictions on its processing and use, and rendered it completely unsuitable for amateur use in the home where such precautions could not be enforced.

SAFETY BASE. Small gauge amateur cinematography therefore necessitated a safety (or non-flam) base and this was first provided by the use of cellulose acetate as the support material. The earlier types of acetate base were, however, more expensive and mechanically weaker than the available nitrate, so professional use of the latter for 35mm films continued despite the inherent fire risks; but further improvements were introduced by the use of cellulose acetate propionate or other mixed esters from 1937 onwards, and of high acetyl compounds such as cellulose tri-acetate, so that eventually base characteristics showing very considerable advantages over nitrate were obtained.

From 1951, safety base of very low flammability has been universal for professional as well as amateur purposes, and the data given below refers to the current types of tri-acetate in use since 1954.

COATING. In the course of manufacture, the base is often first coated with an intermediate layer known as the sub-strate to ensure optimum adhesion of the emulsion, while the opposite side may be coated with a thin gelatin layer or otherwise treated to reduce its tendency to curl during processing. Anti-halation backing may also be applied to this side and removed in the course of development.

For simple black-and-white materials, the emulsion coating may be only a single layer, but all modern colour film stocks are coated with several layers, the different emulsions being interspersed with other layers acting as colour filters or processing barriers.

Some early forms of colour film used two emulsion layers coated on opposite sides of the base (duplitized stock), but these are now obsolete.

DIMENSIONAL STABILITY. Dimensional changes in film may be temporary and reversible or permanent. The former can be a result of expansion or contraction due to temperature, humidity, tension, and wetting during processing. while permanent changes can be caused by the loss of solvent or plasticiser from the base, or by mechanical strain beyond the yield point of the material. As a result of its method of manufacture, film may show differential proportional changes in width and length, but the differences are usually small with the width changes slightly greater than the longitudinal. Because of the marked influence of temperature and humidity, all operations involving the critical maintenance of dimensions are carried out under controlled conditions, and measurements always made at specified levels, usually 70 per cent relative humidity and 70°F (21°C).

Film expands with heat, typical coefficients at normal operating temperatures being ·003 per cent to ·004 per cent per 1°F or ·002 per cent per 1°C. It also expands with increasing humidity, typical changes being ·006 per cent to ·009 per cent per 1 per cent change of relative humidity over the range 20 per cent to 70 per cent RH at room temperatures.

Film dimensions change very markedly on immersion in water or processing solutions. In sufficient time at 70°F, a gelatin emulsion may absorb two to three times its own weight of water and may increase in thickness by ten times or more. The base, on the other hand, absorbs only a few per cent of its weight and swells to only a limited extent. The combined effect is therefore much affected by the relative thicknesses of the base and its coating, and varies greatly from one type of stock to another. In modern positive stocks, the base thickness is of the order of ·0055 in. with the emulsion thickness about ·0005 in., and given sufficient immersion time such film can absorb water equal to some 15 per cent of its own total weight even when superficial liquid has been removed; dimensional expansion of the order of 0·4 per cent to 0·5 per cent can take place. At higher temperatures or in films having greater emulsion thickness, these values may be even larger.

MECHANICAL STRENGTH. In the wet condition, mechanical strength of the film is reduced and tension produces considerable elastic stretch which may become permanent under too great loading. Where film is processed without tension, the operations of wetting and drying usually result in only a small shrinkage of the order of 0·03 per cent, but tensions applied while passing through a continuous developing machine may offset this or even lead to elongation.

Once it has been processed, it is most desirable that the dimensions of the film, particularly negative, should change as little as possible with age, and modern materials are very stable. Permanent shrinkage is caused by the slow loss of solvent and plasticiser from the base, and is increased by storage at high temperatures and humidity; but if abnormal conditions are avoided, such shrinkage in negative materials does not exceed 0·1 per cent during the first six months rising to about 0·2 per cent after two years, after which very little further change takes place. The eventual shrinkage may be slightly greater for positive prints and for 16mm material, up to 0·3 per cent or 0·4 per cent. Nitrate film and earlier types of safety film may be found showing shrinkage greatly in excess of this, up to as much as 1·2 per cent in length and even more in width, and these require special handling during printing or projection. L.B.H.

See: *Aperture sizes; Compatibility; Film manufacture; Film storage; Narrow gauge film and equipment; Print geometry (16mm); Standards; Wide-screen processes.*

Book: *Eastman Motion Picture Film Reference Book*, 1968.

FILM MANUFACTURE

The techniques of manufacturing cine film stock are among the most complex in modern industry, and many of the most recent advances are confidential to the manufacturing organizations concerned. This makes it possible to give only a general outline of the manufacturing process.

Motion picture films consist essentially of two things: a light-sensitive emulsion layer and a suitable transparent flexible support. Since 1952, all 35mm and wider films supplied by the leading manufacturers have been on a high-acetyl cellulose acetate (or so-called tri-acetate) safety support; 16mm and narrower films are also manufactured either on such a support, or on acetate propionate.

THE BASE

The raw material of acetate film support is cotton. The long fibres of the cotton seed pods are used in the textile industry, and the short fibres, or linters, which are left, make an economical source of supply. The raw cotton is mechanically cleaned, degreased, bleached and washed and then dried. Next, the prepared cotton is treated with acetic acid and either acetic anhydride or sulphuric acid. The process is controlled to give a cellulose acetate with about 43 per cent acetyl content. This material is then dissolved in a mixture of two or more organic solvents to which is added a suitable plasticizer, and after filtration it is ready for conversion into film base.

DOPE COATING. The actual preparation of the base entails spreading the cellulose acetate dope in a uniform thin layer on a smooth surface, evaporating most of the solvents, peeling the film sheet off its temporary support, and completing the drying process.

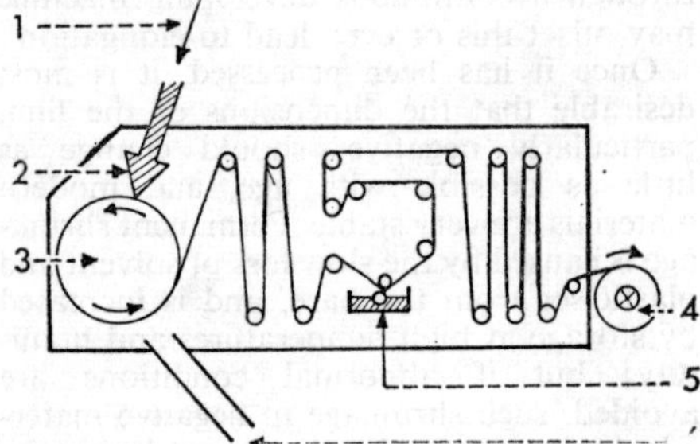

BASE-MAKING PLANT. Dope from storage, 1, is fed into hopper, 2, which spreads it on highly polished drum, 3. Air conditions are adjusted to make solvents evaporate, and base is peeled off drum and passed over series of heated drums to complete curing. Substratum is applied, 5, to ensure subsequent adhesion of emulsion, and base is wound up, 4, ready for coating. 6. Solvent recovery.

Dope coating is carried out either on an endless metal band or, more usually, on a drum.

In either case, the metal surface is a material such as silver plate or stainless steel and is very highly polished.

The dope solution is fed on to the surface by means of a suitable spreader, and the air conditions around the drum or band are adjusted to cause rapid evaporation of the dope solvents. The film base is peeled off the metal support when set, and passed over a number of heated drums to complete the curing. Great care is taken to ensure that no dust is present during this operation.

SUBBING. If photographic emulsions were coated directly on to this base, the adhesion would be very poor, and there would be a serious risk of the emulsion floating off the base during processing. A thin layer, called a sub-stratum, is therefore applied to the base before it is spooled up. The sub-stratum has a chemical nature intermediate between that of the base and that of the emulsion. It is usually prepared from a mixture of base solvents and gelatin dissolved in water. When a coloured support, such as the grey base used for negative film, is required, a suitable dye is mixed with the base dope and the completed roll of film base is then ready for the photographic emulsion to be applied.

THE EMULSION

The most vital stage in the manufacture of film is the preparation of the photographic emulsion. Practically all photographic emulsions consist of light-sensitive silver halides suspended in gelatin. The halide used in most motion picture films is silver bromo-iodide.

There are many reasons why gelatin is used as a vehicle to carry the silver halide. Its physical properties are almost ideal for the purpose. It dissolves easily in water at moderate temperatures, yet when cold it sets rapidly to a solid gel. It permits processing solutions to penetrate easily and rapidly to react with silver halide, and it prevents the halide crystals from clumping together or settling in a sediment.

Finally, gelatin contains minute quantities of complex organic compounds known as sensitizers, which have an important effect on the light sensitivity of the halide crystals.

RAW MATERIALS. In addition to gelatin, the main raw materials of emulsions are potassium bromide and iodide and silver

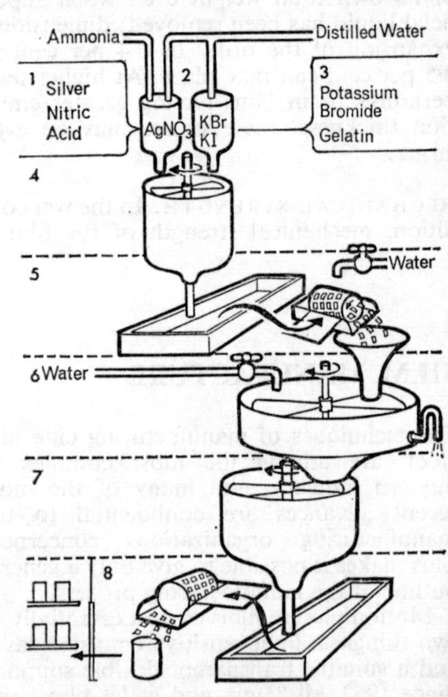

EMULSION MANUFACTURING PROCESS. Simplified schematic. Preparation of raw materials: 1, silver and nitric acid, 3, gelatin, potassium bromide and iodide, etc. 2. Mixing and metering. 4. Emulsification in emulsion kettle and first digestion stage (physical). 5. Setting and shredding. 6. Washing. 7. Second digestion stage (chemical). 8. Final stages to make emulsion ready for coating.

nitrate. These chemicals have to be of high purity, and many manufacturers prepare their own silver nitrate, for example, from silver bullion.

Silver nitrate is prepared from the bullion by treatment with nitric acid, and subsequent evaporation of the solution leaves silver nitrate crystals.

In production, emulsions are made in large silver bowls, fitted with heating and cooling jackets, and known as kettles.

A solution of a small quantity of suitable gelatin is prepared, and to this is added a mixture of potassium bromide and potassium iodide. This mixture is raised to a fixed temperature and stirred mechanically. A warm solution of silver nitrate is then allowed to flow into it at a controlled rate. Silver bromo-iodide, the light-sensitive part of the emulsion, forms.

DIGESTION. Some additional gelatin is added after this reaction and the solution is then maintained at a moderately high temperature for a time. During this digestion period, as it is known, the halide crystals grow in size and, chiefly as a result of this, the light sensitivity of the emulsion increases considerably.

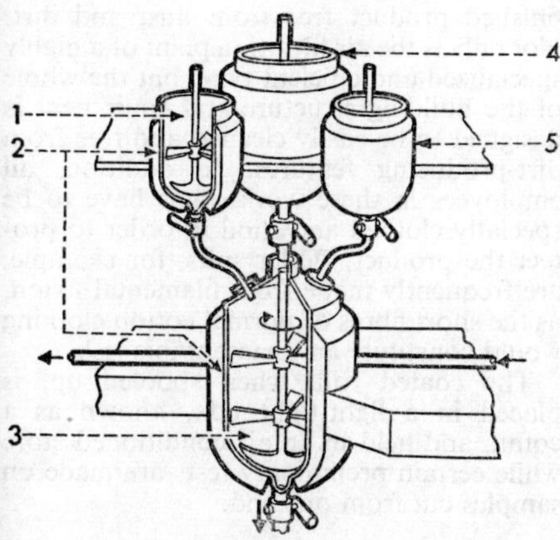

EMULSION KETTLE. Kettle, 3, with stirrers, is fed with solutions of silver nitrate, silver bromide, etc., from metering vessels, 1, 4, 5, water-jacketed, 2, to maintain constant temperature. Digestion follows emulsification in same vessel.

The emulsion is next cooled rapidly, set, broken up by a shredding machine, and then washed in running water.

During this wash, the water-soluble by-products of the emulsification, soluble halide and potassium nitrate, are removed. The emulsion shreds are then drained and re-melted, and a further gelatin addition made. Usually the emulsion is then subjected to a second period of digestion. Here again the light sensitivity of the emulsion is increased, but in this case not by crystal growth, but by modification of the characteristics of the silver halide crystals as a result of the action of the gelatin sensitizers referred to earlier. Hardening agents, spreading agents to promote even coating, and certain other chemicals are added to the emulsion, and it is then ready for coating on to the film base.

INDIVIDUAL VARIATIONS. This outline describes the emulsion making process in its simplest form. A large number of stages in the process are subject to adjustment by the

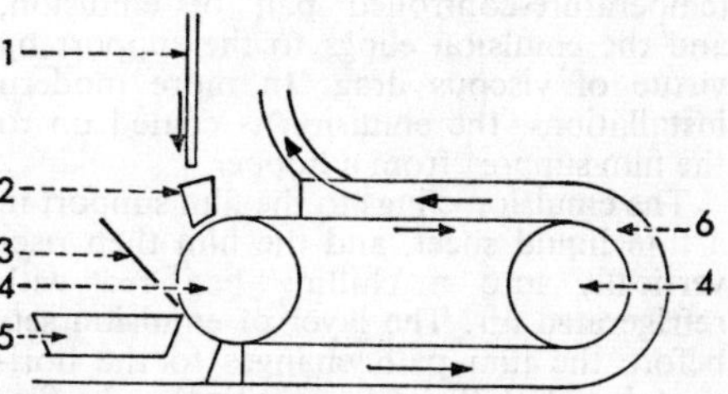

EMULSION SETTING AND SHREDDING MACHINE. Emulsion from kettle passes down pipe, 1, to applicator, 2, from which it flows on to wide steel band stretched over rotating drums, 4. Counter-flow of cool air, 6, helps emulsion to set, and gelled emulsion is then shredded by cutter knife, 3, and collected, 5, for transfer to washing process.

emulsion maker and it is by these means that he can produce the many different types of emulsion which are required for the present-day range of films.

In addition to control of the basic emulsion characteristics, the photographic properties of a film can also be adjusted by such methods as alteration of the total thickness of the emulsion, or even by the use of two emulsions coated one on top of another. The real problem arises in producing emulsions having characteristics which are extreme in one way or another and yet free from objectionable features such as high fog, for all the primary characteristics of speed, grain-size, contrast or gradation and fog are interdependent.

Emulsions made in the manner outlined are only blue-sensitive; panchromatic or other colour sensitivity is achieved by adding minute quantities of certain dyes to the emulsion.

Suitable dyes are absorbed by the silver

halide crystals and sensitize them to the wavelengths of light absorbed by the dye.

The emulsions for black-and-white films are made in the manner just described. Colour film emulsions are basically similar but colour films may carry three or more emulsions, with differing characteristics.

EMULSION COATING. On the emulsion-coating machine, the subbed film support is handled in rolls up to 2,500 ft. long and from 40 to 55 in. wide.

The emulsion is subjected to a further filtration and then fed by gravity to the coating machine.

The film support is drawn over a series of rollers to the emulsion-coating station. In the classical method of emulsion coating, the support is merely passed through a temperature-controlled pan of emulsion, and the emulsion clings to the support by virtue of viscous drag. In more modern installations, the emulsion is coated on to the film support from a hopper.

The emulsion clings to the film support in a thin liquid sheet, and the film then rises vertically into a chilling box fed with refrigerated air. The layer of emulsion sets before the film path changes to the horizontal, but chilled air is applied to the film for a further 20 seconds or so before it emerges into the drying alley.

In this section of the machine, which may be several hundred feet long, very efficiently filtered air, accurately controlled as to temperature and moisture content, is applied to the film to evaporate the surplus moisture.

The control of the drying air is of great importance, as variations in drying time, temperature or humidity can have great

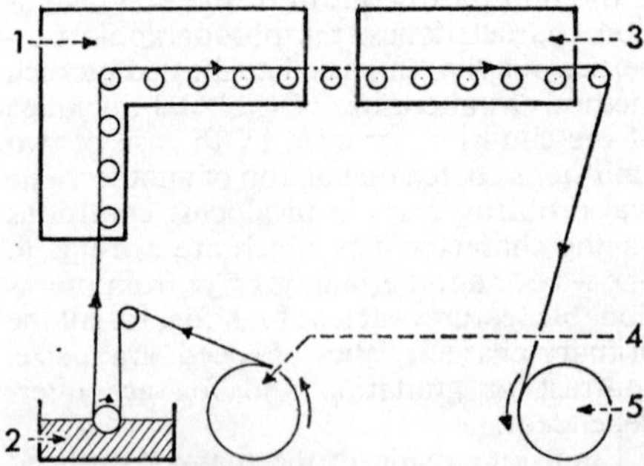

FILM-COATING MACHINE. Raw base, 4, is drawn through pan of emulsion, 2, which clings to it by viscous drag, then passes to chilled air box, 1, which sets emulsion layer before reaching drying track, 3. After further conditioning, film is reeled up, 5.

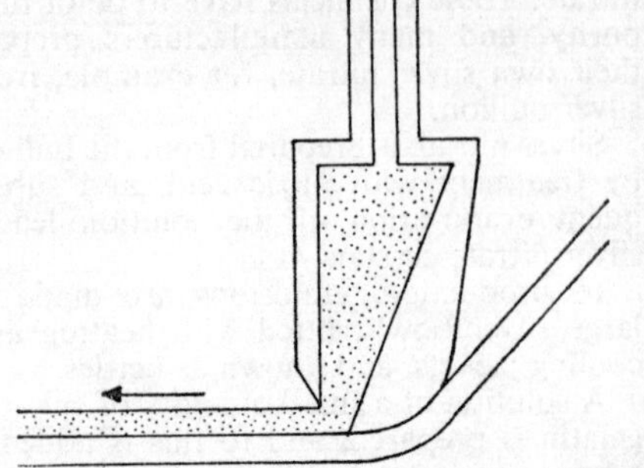

HOPPER COATING OF EMULSION. Modern replacement of emulsion pan, in which hopper deposits emulsion on base in thin uniform layer.

effects on the sensitometric characteristics of the emulsion.

After drying, the film is given a period of conditioning in which the final moisture content of the emulsion is adjusted by exposure to an atmosphere having a controlled moisture content, before the large roll is wound up on a spooling machine.

While the film is drying, and the emulsion still tacky, it can be likened to an enormous sticky fly-paper. For this reason, very special precautions have to be taken to keep the finished product free from dust and dirt. Not only is the air filtration plant of a highly specialized and efficient type, but the whole of the building structure and equipment is designed to be easily cleaned and free from dirt-producing features. In addition, all employees in these workrooms have to be specially clothed and shod in order to protect the product. Boiler suits, for example, are frequently made from filamental nylon, as the short fibres of normal cotton clothing would constitute an unacceptable risk.

The coated roll, when spooled up, is placed in a light-tight box, known as a coffin, and held in an air-conditioned store while certain preliminary tests are made on samples cut from one end.

FINISHING AND INSPECTION

After preliminary testing, the wide roll of film is slit into the appropriate widths. Inspection either by visual means, in safelighting, or by more refined methods, is frequently carried out during this stage, and the edges of the wide roll are rejected.

CUTTING AND PERFORATING

The slit film is usually cut into rolls of 1,000 ft. or such greater length as may be required. A single large roll will yield 30 to 40 strips of 35mm film.

The film is then perforated with great accuracy. The perforating punches are made from a special grade of steel, and each is ground and hand-lapped to an accuracy of 1/10,000 in. Powerful suction nozzles on the perforators ensure that any small particles of dust created in the punching operation are removed from the film before it is spooled up. Footage numbers, where applicable, and other data are automatically printed on to the film either during the slitting operation or during perforating.

These operations are again carried out in a carefully controlled atmosphere, and this ensures that when the film is finally wrapped and sealed into cans, the level of moisture of the air in the cans and in all the wrapping materials contributes to the satisfactory keeping qualities of the product.

Generally, manufacturers give useful information on the can label relative to the emulsion, wide roll, and the strip from which the film in the can was derived. Thus each product type has its own specific code number, followed by numbers indicating the emulsion mix used to make the coating in question, the number of the wide roll, and sometimes the number of the strip. Manufacturers differ in their coding procedure, and also the amount of information of this kind given on the can label, or on the film itself. Specific manufacturers should therefore be consulted should such information be required for their products.

INSPECTION. One of the most important jobs still remains to be done, that is, the careful testing of the product. This is carried out in three stages.

First, sensitometric tests determine the speed, contrast and fog levels of each batch produced.

Second, the examination of the parent rolls during the slitting operation is supplemented by examination of samples taken both across and along the parent rolls. These strips are exposed to a uniform light density and processed. Spots, static, kinks, streaks, uneven coating and all the other manifold defects which occasionally affect motion picture film can be readily detected and isolated by such tests. The perfection of the manufacturing process today is such that well over 99 per cent of film for sale is entirely free from defects.

As a final check, packed rolls, ready for sale, are taken from stock and subjected to practical tests closely reproducing trade usage. A statistical group is continually engaged analysing the results of these tests. These analyses are used to direct attention to any stage of the manufacturing process which could be modified to improve the quality of the product.

Testing is not complete when the film is sold. Samples of each coating are retained in the factory, and sensitometric tests are made on them at regular intervals in order to ensure that the keeping qualities of each batch are satisfactory. H.E.B.G.

See: *Chemistry of the photographic process; Image formation.*

Book: *Making and Coating Photographic Emulsions*, by V. L. Zelikman and S. M. Levi (London), 1964; *Photographic Emulsion Chemistry*, by G. F. Duffin (London), 1966; *Photographic Chemistry*, Vol. 1, by P. Glafkides (London), 1958.

⊕**Film Register.** Part of the film transport mechanism in a camera, telecine or projector which ensures that the frame to be exposed is in the correct position.

See: *Pin; Register; Registration.*

⊕**Film Slide.** Mounted photographic transparency, usually single 35mm frame for projection of still picture. Frame size is usually Leica format, 36×24mm, mounted in 2 in. square slide.

⊕FILM STORAGE

The conditions recommended for motion picture film storage depend on many factors: its chemical composition, the type of emulsion it carries, the length of time for which it is to be kept, whether it is exposed or unexposed, and so on.

Film consists of a transparent, flexible base or support, on which is laid an adhesive substratum on top of which the light-sensitive emulsion is coated.

The base is a wood cellulose derivative, and up to 1951 was of cellulose nitrate, a highly inflammable material preferred for 35mm film to the cellulose acetate employed in the narrower gauges because of its better physical characteristics when new, a requisite for professional production. By 1951, a cellulose triacetate base had been developed which met both safety and physical requirements, and within a year or two, nitrate base ceased to be manufactured.

Much old film of historic interest remains on nitrate base, because transfer to safety base, though highly desirable, is uneconomic. Proper storage and handling of nitrate film is therefore still important.

The light-sensitive emulsion, repository of the picture image, also presents storage problems, different according to whether it has been exposed or not. It consists of a gelatin binder containing silver salts and other chemicals, and the image, if black-and-white, is formed in finely divided metallic silver; if colour, usually in organic dyes.

Film storage may conveniently be considered under four main heads: short-term storage of unexposed film stock (raw stock); short-term storage of release prints; long-term storage of film originals and library material; and archival storage of nitrate and safety base film.

SHORT-TERM STORAGE

RAW STOCK. This may be negative material in the studio or print stocks in the laboratory. In both cases, it is important that separate batches of film should be kept apart.

Because of minor differences between different batches of camera stock, it is customary to order from the makers sufficient footage of the same batch number for the complete filming of a production – perhaps 100,000 ft. or more in the case of a major production, or 2–3,000 ft. for a 16mm industrial film.

Similarly, batches of print material may differ slightly, especially in the case of colour stocks. Each batch as it is received by the laboratory is tested, so that corrections may be applied to the printer conditions; thus it is again important that batches should be kept separate.

Because of the importance of maintaining a consistent degree of shrinkage in all laboratory films, cans should be opened and the film handled in air-conditioned surroundings.

Where the period of storage is not expected to exceed six months, storage conditions are not critical. Film is packed in taped cans and thus sealed against ingress of moisture, so humidity does not need to be controlled unless it is high enough to rust the cans and peel off the labels, as may happen under tropical conditions. Temperature, however, is important: in general, the lower it is kept, the less likelihood of eventual picture deterioration from fog or colour fading.

For short-term storage, it suffices if the temperature is maintained below 10°C (50°F) for colour films, with values 3–5°C higher for black-and-white.

As soon as the tapes are removed from cans of raw stock, different conditions obtain, for now the humidity of the surrounding air affects the film. Moisture will in fact condense on it if it has been taken quickly from a cold store, and untaped when at a temperature below the prevailing dew-point, which may be as high as 25°C (77°F) in temperate climates in summer, and higher in the tropics.

To prevent this damaging condensation, it is necessary to let the taped cans warm up to the dew-point temperature of their surroundings before opening them. This may take several hours.

Once out of the can, raw stock is also susceptible to damage from gases such as coal gas and exhaust fumes, as well as the vapours of many common solvents. In or out of its can, it is of course extremely sensitive to radioactivity, and all radioactive sources must be heavily lead-shielded if brought within 25–50 ft. of unexposed film.

RELEASE PRINTS. The storage facilities of most distributors of cinema films are based on the stringent requirements for nitrate film. Vaults are often built in the open, or on the roofs of buildings, and provided with sprinkler systems.

Now that all films are coated on safety base, the same degree of precaution is no longer necessary, and vaults might in fact be better situated in a gently warmed basement, where there would be less variation of temperature. Prints are stored in individual cans each containing not more than 2,000 ft. of film. Easy identification is a prime essential.

The commercial life of a film production is rarely more than a very few years, and the storage of acetate prints for such comparatively short periods presents few technical problems. When film has been placed in unsealed cans, it is extremes of relative humidity which tend to damage it more than extremes of temperature. So long as the RH remains between 25 and 60 per cent, film is not likely to suffer; though a temperature of 24°C (75°F) is recommended, a few weeks at 32–38°C (90–100°F) is not harmful.

LONG-TERM STORAGE

ORIGINALS AND LIBRARY MATERIAL. Every film studio and TV production organization has storage vaults for the valuable originals of its past productions and often many reels of library material. These do not justify the rigorous (and expensive) requirements of archival storage, because the time scale is measured by decades rather than centuries.

Most film studio vaults were originally built for the storage of nitrate films, and therefore meet the safety requirements for any nitrate material the library may still wish to store. For the storage of acetate film, the only serious fire hazard is from outside the vault, i.e., the spread of fire from neighbouring buildings which may destroy valuable film records. Strongly built nitrate film vaults are therefore an advantage even for the storage of safety film.

Temperature and humidity needs are much as for release prints, but the longer the period of storage the more desirable it is to maintain conditions near the optimum, and make provision for periodical rewinding and examination of older film. Where climatic conditions are severe, and especially where high relative humidities are experienced regularly, the cost of air-conditioning may well be justified to extend the life of important library material, especially if this is on nitrate base.

ARCHIVAL STORAGE

When film is to be preserved for the longest possible period, which with modern stable base material may well – like the related substance, paper – be measured in centuries, extreme precautions must be taken. It is, however, uneconomic to invest large sums in providing ideal ambient conditions when the film itself contains the seeds of its own destruction.

Tests should therefore be made on samples of each roll to ensure that residual hypo is at a sufficiently low value to prevent fading of the image. Colour films are not ideally suited to very long-term storage, since organic dyes are fugitive, and though much improved in this respect today, have not the lasting power of a well-fixed silver image. Moreover, the individual layers of a tripack film may fade at different rates, thus upsetting the colour balance.

Wherever possible, a set of black-and-white separation prints should be made of the three dye images in the original negative film. One print is made from the yellow original, the second from the cyan, and the third from the magenta. From this set, a colour print can subsequently be made by techniques which were in common use before the universal adoption of integral tripacks in the early '50s.

Since nitrate film, however well stored, deteriorates more rapidly than acetate under the same conditions, it is very desirable to transfer nitrate records to acetate so as to make it unnecessary to store them at all. But this is a counsel of perfection. Two printing stages are normally required if prints are ultimately needed from prints, or negatives from negatives, and the cost of this double operation may well exceed any practical budgetary allowance. Nitrate storage must then be properly provided for.

STORAGE OF NITRATE FILM. Nitrate film is subject to spontaneous decomposition, when it becomes rapidly more flammable and may eventually ignite spontaneously. It is therefore essential that nitrate films intended for long-term storage be rewound and inspected at regular intervals – say, once a year, or in tropical climates once every few months. The first onset of decomposition appears in the emulsion, which becomes tacky; rewinding in a dry atmosphere prevents the turns of film sticking together. Spotty amber discolouration is often noticeable.

In the next stage, a very characteristic unpleasant odour is given off, and the image begins to fade from the emulsion, the silver having become silver sulphide. If the content is of value, it is essential that the film should have been duplicated before this stage is reached. Next the base becomes sticky, and at this stage the film is completely useless. Finally the film disintegrates into a powdery substance which is explosive.

The rate at which nitrate film passes through these stages varies greatly, depending on factors such as the purity and stability of the original cellulose nitrate and the temperature and humidity of previous storage. Nitrate film is much more sensitive to high temperatures and humidities than acetate, and its rate of decomposition doubles for every rise of 5°C.

In the storage vaults, the following conditions should obtain:

1. Cans should be corrosion resistant, either of aluminium or stainless steel. In practice, where proper ambient conditions prevail, commercial cans are found to be satisfactory.

2. Cans should be placed flat on the shelves, and left untaped so as to allow the escape of gases. The rolls of film should be

placed between two perforated plastic discs.

3. Relative humidity should be 40–50 per cent. Temperature should be kept at or below 10°C (50°F). Lower humidities retard decomposition, but may increase brittleness.

A typical nitrate vault for archival storage holds 500 reels of film, and the vaults are arranged in blocks of 12 to 14. Precautions must be taken to prevent fire starting within the vault, or spreading to the vault from outside. An automatic sprinkler system is essential, together with a heavy-duty electrical installation and self-closing fireproof doors. If a fire starts within a vault, all the film in it is likely to be destroyed. Pressure vents are provided for the escape of gases which, in the event of fire, might explode and damage the vault, thus spreading the fire. Normal ventilation prevents the accumulation of gases due to decomposition.

STORAGE OF ACETATE FILM. With a very much slower rate of deterioration, the conditions for acetate archival storage are rather less critical. Corrosion-resistant cans are recommended, laid horizontally. Relative humidity of 40–50 per cent is the same as for nitrate, but the temperature may range from 15–27°C (60–80°F), so long as means are provided to prevent condensation during storage and when film is removed from the vault.

So long as combustible materials are kept out of the vaults, fire will not be expected to start within them; but their construction must still be fireproof to withstand damage from external fires. Sprinklers, however, are

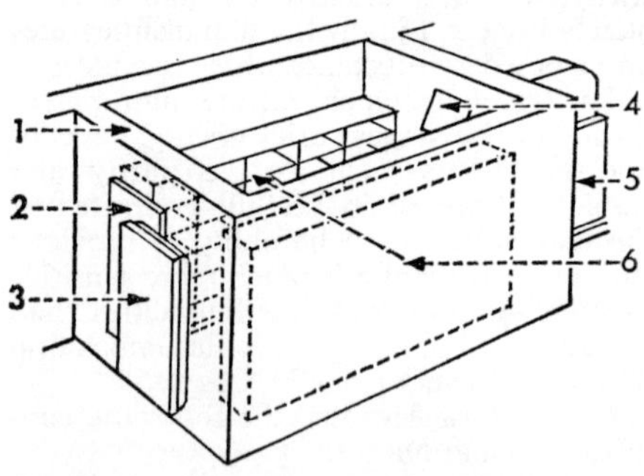

ARCHIVE STORAGE VAULT FOR ACETATE FILM. 1. Fire-resistant walls of 12 in. brick. 2. Sliding fire door. 3. Metal-clad airtight door. 4. Fire dampers to conform with local regulations. 5. Air-conditioning system. 6. Standard steel horizontal shelving, 12 in. deep for 1,000 ft. rolls, and 18 in. deep for 2,000 ft. rolls. Absence of sprinklers, with their risk of water damage to film, is conditional on nothing but safety film being stored in vault.

unnecessary and even undesirable, for serious water damage can result from their accidental release. If air conditioning is necessary to maintain proper temperatures and relative humidities, suitable fire dampers must be incorporated in the ducting to prevent the ingress of flame.

If colour negatives are to be stored for maximum life, certain additional precautions are recommended. Relative humidity should be reduced to 15–25 per cent, in the interests of long dye life, and even lower if the accompanying brittleness does not matter. This is the case if long conditioning periods can be allowed when the film is removed from storage and before it is handled.

Temperature also should be kept low, and a maximum of −20°C (−4°F) is the ideal. The high cost of refrigeration must be set against the alternative of storing three-colour separation prints at more normal temperatures.

FILM TESTS

The need for regular examination of nitrate film for stability of the base has already been emphasized. Both nitrate and acetate films are subject to fading of the image. Tests for these two causes of film loss must therefore be carried out.

FILM STABILITY TEST. The onset of decomposition on nitrate film can be predicted by its content of acid gas. The test is as follows: two $\frac{1}{4}$-in. punchings of film are placed in a test tube, and the tube sealed with a glass stopper round which is wrapped a piece of filter paper impregnated with an indicator dye, alizarin red, moistened with glycerine and water.

The tube is heated in an air bath to a temperature of 134°C. The time required for the bleaching of the lower edge of the paper is a measure of the stability of the base; any film which reacts in under 10 minutes is regarded as unstable, and the film must be duplicated immediately and the nitrate original destroyed. If the reaction time is from 10 to 30 minutes, the film is re-tested in six months; if from 30 to 60 minutes it is re-tested in a year; if over 60 minutes, it is re-tested in two years.

HYPO TEST. To prevent fading of the image, all films must be tested for residual hypo. The testing solution consists of:

Potassium bromide	2·5 gm.
Mercuric chloride	2·5 gm.
Water to	100 ml.

A film punching is cemented, base down, to a piece of leader film. A drop of this test solution is poured on it and left for 3 minutes.

The slightest trace of opalescence in the liquid indicates that the film should be re-washed.

TEST FOR NITRATE. BS.850, Definition of Cinematograph Safety Film, stipulates two requirements: it shall have a nitrogen content of not more than 0·36 per cent, a figure at which the proportion of toxic fumes on burning is inappreciable; and it shall be slow-burning.

There are laboratory methods for testing for nitrogen content. One is a burning test, in which a 21-in. strip of film – 35mm or 16mm – is placed on a device in which it is strained over a semicircular path. One end of the film is ignited; if the flame fails to reach a mark 18 in. from the ignited end, or if it takes two minutes to do so, the film is slow-burning.

A simpler and generally effective way of testing for nitrate is to cut off a frame of film, hold it vertically in a pair of pliers, and apply a match at the top. Nitrate film burns rapidly – safety film usually goes out, or burns very sluggishly.

A risk to be borne in mind is that nitrate and safety film may have been spliced together in a reel.

Safety film is marked in the spaces between the perforations with the letter S, or the letters of the word SAFETY. However, it is not safe merely to observe that a print carries this distinguishing mark, which may have been printed through from another film.

MAGNETIC RECORDINGS

All original sound recordings for films are now made on magnetic tape or magnetic-coated film. Many release prints have magnetic sound tracks, and much television material is recorded on videotape.

When film originals are preserved, the sound track must obviously accompany them. But it is generally agreed to be undesirable to keep videotape recordings. Equipment for reproducing them is costly, and furthermore, if recording standards should change such records would become obsolete, and reproducing equipment might be unavailable. Video recordings would therefore be copied on to photographic film.

STORAGE OF MAGNETIC MATERIALS. Magnetic sound tracks have proved to be far more permanent than had been expected when they were first introduced. They seem to be virtually immune to fading. But a serious difficulty is known as print-through.

When reels of magnetic film or tape are wound tightly, the sound modulations on one layer may print through the base to the adjacent layer, giving rise to spurious signals. The only way to avoid this fault is to rewind the material at frequent intervals.

Magnetic recordings are subject to pick-up from stray magnetic fields and static discharges. However, the regular degaussing of steel components, and general care in handling magnetic tracks, makes pick-up rare, and magnetic material may be stored on steel shelving. R.H.C. & R.J.S.

See: *Film dimensions and physical characteristics; Film manufacture; Library; Magnetic recording materials.*

Book: *Storage and Preservation of Motion Picture Film*, Eastman Kodak, 1967.

⊕**Film Strip.** Strip of film carrying disparate picture images intended to be separately projected in the manner of lantern slides. In

modern equipment, often synchronized with a separate sound accompaniment, usually on disc, which may carry subsonic pulses automatically moving one picture on to the next at the appropriate moment. In this way, the effect of film cutting, though of course with static shots, can be successfully achieved.

⊕□**Filter (Colour).** Transparent material having the characteristic of absorbing certain wavelengths of light in the visible spectrum and transmitting others more or less undiminished. A narrow-cut filter transmits a limited group of wavelengths only, absorbing all others, and thus gives a pure or saturated colour of light, while a wide-cut filter transmits a much more extensive spectral band and the resultant colour of light is less saturated.

⊕**Filter (Electrical).** In sound recording and □reproduction, a selective network which freely transmits frequencies within a determinate band, known as a pass-band, and substantially attenuates all other frequencies.

Filters may thus be designed to accept or exclude any range of frequencies, and their pass-bands may be brought very much nearer to the calculated requirements than is the case with optical filters.

Electrical filters are of many kinds: band-pass filters have a single transmission band, with more or less strong attenuation of frequencies above and below the pass-band. High-pass filters pass all frequencies above a stated cut-off point; low-pass filters pass all frequencies from zero (or some very low quantity) up to a stated cut-off frequency. Broadly-tuned filters have a wide pass-band, usually with gently sloping skirts; sharply-tuned filters have a narrow pass-band, usually with steeply-sloping skirts.

⊕**Filter Factor.** Numerical factor by which the □length of an exposure must be increased to compensate for the light absorption of an optical filter through which the exposure is made.

See: *Exposure control.*

□**Filterplexer.** Device incorporating vestigial sideband filter and vision and sound combining unit. All television transmissions use vestigial sideband transmission for the vision transmission in order to reduce the large bandwidth required. To achieve a vestigial sideband signal, a filter is necessary to remove the unwanted portions of the lower sideband. This can be done in the earlier stages of a transmitter, but more usually after the final output stages.

See: *Transmitters and aerials.*

⊕**Fine Cut.** Stage in the cutting of a film which marks a substantial advance on the early rough-cut and denotes that the film is nearing approval by the producer and director.

See: *Editor.*

⊕**Fine Grain.** In duplicating procedures, usually connotes fine grain master positive. A black-and-white positive print made direct from the original negative film on to a fine grain emulsion stock and used for the preparation of a black-and-white duplicate negative.

See: *Duplicating processes (B. and W.).*

⊕**First Assistant Director.** Film director's □assistant responsible for co-ordinating the work of the various technicians and craftsmen on the stage. Must be able to appreciate the amount of work involved and consequently time required by them to bring their preparations to the point of readiness between each shot. He must ensure that artistes are on the set ready for rehearsal or shooting when required. He maintains discipline and gives the instruction for film and sound cameras to start when the director is ready to shoot a scene. He has one or more assistants to help him on the stage and in the production office.

In television, the duties of unit manager and assistant director are combined in the grade of floor manager.

See: *Director* (FILM).

⊕**First Generation Dupe.** Duplicate negative prepared by printing directly from a master positive derived from the original negative film.

See: *Lens performance; Lens types.*

⊕**Fish-Eye Lens.** Lens in which retrofocus construction has been utilized to give maximum possible field of view. At an aperture of about f 8, the limiting field of

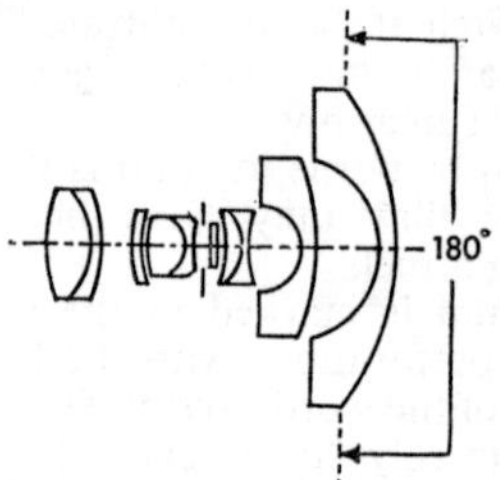

180° (i.e., a whole hemisphere) can be reached. The portrayal of so wide a field on a flat film surface is necessarily distorted, but fish-eye lenses usually contribute further distortions, some of which may in future be significantly reduced with the aid of parabolic surfaces.

⊕**Fishpole.** Light pole resembling a fishing rod □on which a microphone is slung, the whole being projected out over a scene when dialogue is being recorded. Fishpoles are often used in confined spaces where a boom would be inconvenient.

⊕**Fixing.** Process of removing unexposed silver halides from film after development in order to render the image stable under white light. The fixing solution converts the silver halide into soluble silver salts which can be removed by washing in water. Sodium thiosulphate is the most widely used agent.

See: *Chemistry of the photographic process.*

Flag. Small rectangle of wood or card mounted on a stand in the studio to keep

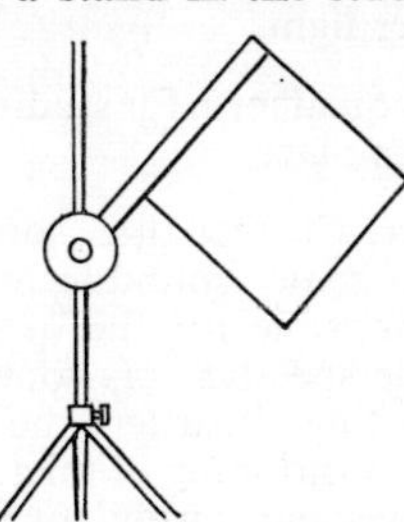

direct light off the camera lens or shade some part of the set.

Flange. Metal or Bakelite disc to provide smooth and uniform winding of film when mounted on a rewinder.

Flare. Undesired light arriving at the image plane in an optical system. This light may be uniformly distributed over the image, resulting in a reduction of image contrast, or it may be concentrated in specific areas of the image. Usually the result of reflections from surfaces within the optical system.

In a television tube, flare is light produced by excitation of the phosphor outside the area in nominal use and falling on that area.

See: *Coating of lenses; Phosphors for* TV *tubes.*

Flashback. Storytelling device in which the chronological unfolding of the plot is interrupted by a scene or sequence drawn from the past.

Flashover. Discharge caused by an excessively high voltage breaking down air or surface insulation. May be induced by surface moisture or a film of impurities. May lead to permanent breakdown of vacuum tubes and capacitors.

Flat. Low contrast rendering of restricted density range in a photographic reproduction. Flat lighting of a scene indicates the use of an illumination distribution providing a limited tonal range.

See: *Tone reproduction.*

Flat. Large flat mobile piece of plywood or other material usually of a standard size in a studio complex. Basic unit of film and television studio set building.

Flat Print. Motion picture copy suitable for projection with normal lens systems, in contrast with a squeezed print requiring anamorphic projection.

See: *Wide-screen processes.*

Fleming, John Ambrose, 1849–1945. English electrical physicist and engineer, born at Lancaster and educated at University College School, Gower Street, London. Graduating with a B.Sc. degree at the age of 21 years, he was for a brief period a schoolmaster, teaching science, before winning a scholarship for St. John's College, Cambridge, where he attended lectures by Clerk Maxwell. Was appointed Professor of Physics and Mathematics (1881) at University College, Nottingham, but in 1885 returned to London as Professor of Electrical Engineering at University College, a position he held until 1926.

Fleming was the first to focus cathode rays magnetically. In a note to *The Electrician* in January 1897, Fleming described an experiment in which he used a Crookes tube having a Maltese cross near one end and a coil traversed by current placed round the tube. He found that when a current was passed through the coil it deflected and focused the shadow of the cross on the end wall of the tube, depending on the amount of current passing through the coil.

He was scientific adviser to the Edison Electric Light Company and later to the Edison and Swan combination, and became scientific adviser to Marconi's Wireless Telegraph Company Limited in 1899. In this capacity he took a leading part in the design of the powerful transmitting equipment at Poldhu, Cornwall, erected by Marconi for his successful attempt to span the Atlantic by wireless in 1901. In 1904 he invented the thermionic valve.

Fleming was a prolific writer and the author of several textbooks. He trained many hundreds of engineers who later distinguished themselves in the radio industry, and was a notable lecturer, continuing to lecture until ninety years of age. He was a Fellow of the Royal Society and was knighted in 1929.

Flicker. Visual sensation produced by rapidly alternating periods of light and dark when their frequency is too slow to allow persistence of vision to give the impression of continuous illumination. The frequency at which the impression of flicker disappears increases with the intensity of the light period and is also affected by the conditions of viewing, such as the size of the area viewed and the accommodation of the eye to surrounding illumination.

In cinematography, the illusion of movement depends on the presentation of a series of bright images interspersed with dark periods, and the avoidance of flicker has been of fundamental importance. In the early days of the silent film, it was found that a black-out frequency corresponding to the intermittent movement of the film 16 times a second gave most objectionable flicker, even at the comparatively low screen brightness of that time. It was therefore necessary to introduce additional black-out periods even while the film was stationary in the projector gate and eventually a three bladed shutter in the projector giving three dark periods for each frame was adopted as standard. This produced a flicker frequency of 48 per second and was found generally acceptable.

With the introduction of sound film the frame rate was increased to 24 pictures per second and a two-bladed shutter could be used to give the same flicker frequency of 48.

This continues to be standard cinema practice, although the trend towards large screens of higher brightness has from time to time led to suggestions that the frequency should be increased still further. For example, the first Todd-AO productions were photographed for projection at 30 pps, giving a flicker shutter frequency of 60, but this innovation was not generally adopted. It has also been claimed that the use of illuminated surrounds to the screen in the cinema both enhances the picture and reduces the appearance of flicker, but such systems have not come into common use. With the current average practice for print density and screen brightness, a flicker frequency of 48 per second is acceptable.

In television, the use of interlaced scanning provides two fields per picture, and therefore with minimal complication doubles the picture repetition rate to 50 pps in Europe and 60 pps in the US.

See: *Visual principles.*

⊕**Float.** Unsteadiness of image as a result of irregular movement of film in the camera or projector gate at the time of exposure. Also denotes temporary removal of a set wall.

⊕**Flood.** Lamp which provides general set ☐illumination on the studio floor. Has only a small range of focusing adjustment, or none at all, and is used to light backings and provide filler light.

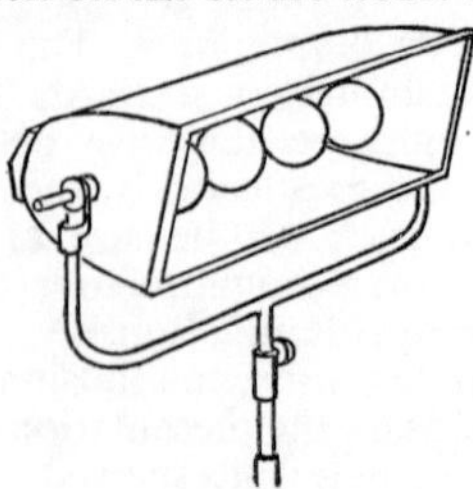

⊕**Floor.** Colloquial term for studio stage where filming takes place.

⊕**Floor Mixer.** Correct title, sound recordist ☐(mixer). The senior sound technician on the stage. Responsible for the quality, clarity, and the perspective of sound recorded during shooting. Instructs the floor sound staff on the positioning of microphones and of their movement during any scene.

There is no floor mixer in television, the function being carried out by the sound supervisor from the control room.

See: *Sound recording.*

⊕**Floor Plan.** Plan of a stage to enable the art ☐director and construction staff to decide where a set or sets are to be constructed.

In TV, plan of a studio with set or sets indicated on it for use by the director in planning camera positions and action.

⊕**Flop Over.** Scene in which the action is shown laterally reversed from left to right; this is usually done by making an optically printed dupe negative.

See: *Editor.*

⊕**Flutter (Picture).** Repetitive variation of image density or steadiness as a result of irregular exposure or location of the film in a camera, printer or projector.

⊕**Flutter (Sound).** Rapid cyclical deviation of ☐frequency usually caused by unsteadiness of motion of a film or magnetic tape either in a recorder or reproducer, or by deformation of a disc. The ear is very sensitive to even small amounts of flutter, which are particularly noticeable on held musical notes. Flutter on speech produces a kind of gargling effect.

This is a fault which cannot be corrected after it has occurred. Hence professional recording and reproducing equipments are designed for maximum motion steadiness, involving minimum friction and close bearing tolerances.

⊕**Flutter Bridge.** Equipment for determining ☐the flutter content of recording and reproducing equipment by null balance methods.

☐**Flyback.** At the completion of the scanning of one line in the camera or CRT, the beam has to be deflected rapidly to the beginning of the next line. This occurs during the line suppression period, and is known as the line flyback or retrace period.

Similarly, at the end of each field scan, the spot again has to fly back to the beginning of the next field scan, and this is known as the frame flyback or retrace period.
See: *Television principles.*

□**Flying Spot.** Film scanning by a cathode ray tube. The picture is scanned by a flying spot of light emitted from the tube, and focused on to the film. The amount of light passing through the film is modulated by the film content before striking a sensitive photomultiplier. The signal is subsequently amplified and processed.
See: *Telecine.*

□**Flywheel Circuit.** Tuned circuit of high Q (figure of merit) which maintains oscillations for a relatively long period analogous to the effect of a flywheel in a rotating machine.
See: *Receiver.*

⊕□***f*-Number.** Expression of the relative aperture of a lens at its various diaphragm openings, derived from the focal length divided by the effective diameter of the diaphragm. Standard *f*-numbers are 1·4, 2, 2·8, 4, 5·6, 8, 11, 16, 22, 32, etc., each being $\sqrt{2}$ times the previous number (with some rounding off for convenience).
See: *Exposure Control.*

⊕□**Focal Length.** In a thin lens, the distance from the centre of the lens to either principal focus. The equivalent focal length of a thick

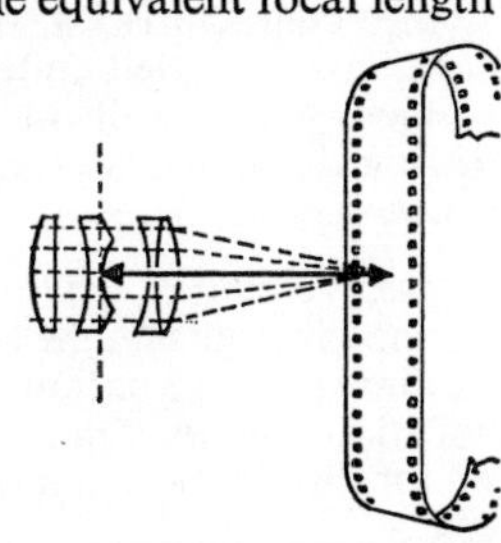

lens is the focal length of a thin lens of identical magnifying power. This is the constant of a lens upon which the size of the image depends.

⊕□**Focal Plane.** Plane through the principal focus of a lens, perpendicular to its optical axis.

⊕□**Focus.** Point to which converging rays of light converge, in which case the focus is real, or a point from which diverging rays of light appear to diverge, in which case the focus is virtual. Parallel rays of light, after passing through a lens system, converge to, or appear to diverge from, the principal focus. More generally, focus denotes the position in which an object must be situated so that the image produced by a lens is sharp and well defined; hence, an object is said to be in focus or out of focus.
See: *Optical principles.*

□**Focus Coil.** Coil concentric with the beam of a cathode ray or camera tube which produces a magnetic field shaped to bring the beam into a small spot at the target or screen.

□**Focus Modulation.** Variation of the focusing of a cathode ray beam as it is deflected, to compensate for the difference in distance from the scanning position to the point where the beam strikes the screen of the cathode ray tube or the target of a camera tube.
See: *Camera* (TV).

⊕**Focus Puller.** Operator on a motion picture camera who adjusts the optical focus, usually by turning a sleeve on the lens not conveniently placed (as in a TV camera) for operation by the cameraman himself.
See: *Camera* (FILM).

□**Focus Servo System.** System for remote adjustment of the focus of a television camera. A small handgrip controls a feed to the motor which adjusts the focus movement of the lens mount.

⊕**Fog.** In a photographic image, a detectable density which is not related to the exposed image itself. Fogging may be caused by exposure to light other than in the camera or printer, by exposure to X-rays or gamma radiation or by unsuitable storage or chemical processing conditions. The minimum density produced in a film material which has been processed without any exposure whatsoever is termed the fog level.
See: *Exposure control; Chemistry of the photographic process.*

⊕**Fog Filter.** Graduated and often movable type of diffusing filter placed in front of the camera lens to simulate the effect of fog.

⊕□**Foldback Loudspeaker.** High-quality loudspeaker system provided for reproducing music or effects on the studio floor for the benefit of the artistes' timing or reaction. To reduce feedback into the microphone to an absolute minimum, as a rule the foldback loudspeaker is directional and mounted in the dead area behind the microphone where its response is lowest.

⊕**Follow Focus.** Continuous change in camera ☐focusing necessitated by relative movement between the camera and its subject greater than can be accommodated by depth of field. In film, following focus is usually the function of a special technician, the focus puller.

⊕**Follow Focus Cameraman.** Second technician ☐on the camera team; responsible for manipulating the lens of the camera during shooting to ensure the image is in sharp or desired focus. He is also responsible for the security and cleanliness of the lenses and for day-to-day maintenance of the cine camera. Also called focus puller. In television these duties are carried out by the camera operator.

⊕**Follow Shot.** Shot in which the camera moves around following the action of the scene. Also called a dollying or trucking shot.

⊕**Foot.** Sixteen frames or 64 perforations on 35mm motion picture film, which in fact is very slightly less than 12 in. A foot of 16mm film contains 40 frames. The term screen-foot is often used for the length of film containing 16 frames in any gauge.
See: *Film dimensions and physical characteristics.*

⊕**Footage Counter.** Counter driven off a sprocket wheel which engages with perforations in the film as a means of measuring the footage which passes through any film machine. Footage counters are incorporated

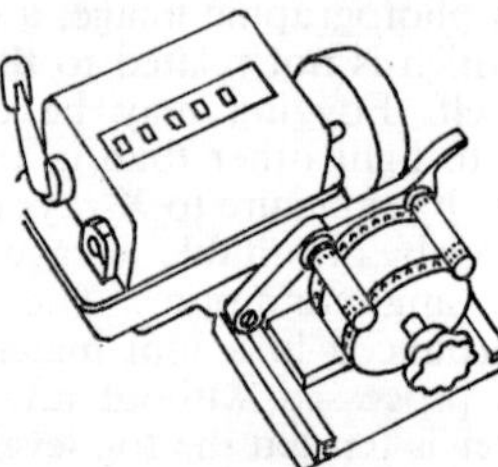

in cameras, printers, and many other film devices. Separate footage counters are used in cutting rooms. These consist of a sprocket over which the film is passed, coupled to a counting mechanism.

⊕**Footage Number.** One of a series of numbers, combined with key lettering, printed at intervals of 1 ft. along the edge of most types of negative and duplicating film stock. These numbers, incorporated in the film during manufacture, print through to positive stock not so marked. Also known as edge number and negative number.

⊕**Foot-Candle.** Unit of illumination. Intensity ☐of illumination at the surface of a sphere of 1 ft. radius, from a light source of one candle-power placed at the centre of the sphere. Equivalent to a luminous flux of one lumen per sq. ft.
See: *Light units.*

⊕**Foot-Lambert.** Unit of brightness or lumi-☐nance, defined as the brightness of a perfectly diffusing surface, when the total flux radiated is one lumen per sq. ft. of surface.
See: *Light units.*

⊕**Forced Development.** Although in processing motion picture film it is normal laboratory practice to provide standard developing conditions for specific photographic materials, it is sometimes necessary to improve the density of the image which can be obtained on an underexposed camera original by additional or forced development. This is usually obtained by increased time or temperature in the developing solution, or a combination of the two, and the increased density obtained is liable to be accompanied by increased image grain and contrast. The increased contrast is often advantageous in improving the tonal gradation of underexposed material but the increase of grain may become objectionable and often sets a practical limit to the extent to which development can be forced.

Within limits, black-and-white negative stocks and some colour reversal materials respond satisfactorily to forced development, but the characteristics of colour negative stocks are less well suited to this treatment which may result in distorted colour rendering if unduly extended.
See: *Laboratory organization.*

⊕**Foreign Release.** Version of a film specially prepared for showing outside the country of origin. Since film is an international medium of communication, versions must generally be produced for distribution in a number of countries.

There are two distinct approaches to the problem of foreign language release. Ideally, the foreign version of the film should be indistinguishable from a native production. The dialogue is dubbed into the native language, titles and inserts being translated. The details of a fully dubbed foreign release of this type must be planned from the original production script, and a number of proposals have been made to facilitate the international exchange and nomenclature of the appropriate material. There is also a British Standard recommendation for the preparation of export scripts to guide the producers of foreign versions of British productions. A duplicate picture negative is often supplied for printing in the foreign laboratory, but the translated dialogue is usually recorded in the foreign country and mixed with a music and effects track provided from the original production. Re-editing, including censor cuts, may be carried out on the dupe negative before foreign release printing.

Where the market is too small to warrant the cost of preparing a dubbed sound track, explanatory titles may be superimposed in the picture area in the appropriate language, since this method is practical for quite a small number of prints. The dialogue of the original version has first of all to be condensed to a minimum number of words and short phrases in the foreign language, and may then be reproduced on the print either photographically or by a mechanical etching method. In the former case, the words may be reproduced on a separate roll of film where each title is repeated on a sufficient number of frames to give time for reading, and used as an overlay when making prints from the picture negative. In another method, a single frame of film is prepared for each title and projected on an optical printer while printing the picture negative in contact, so as to appear on a predetermined number of positive frames at the appropriate positions. In both cases, the titles appear as white letters against the background of the scene.

The etching system has the advantage that it can be applied to finished prints, and in this method the words of the title are set up in type and etched or impressed into the emulsion of the print. Here again the effect is to remove the emulsion where the letter occurs, so that titles appear as white letters on the background scene.

Superimposed titles are normally positioned at the bottom of the picture frame and it is therefore desirable that the film should be projected on a screen of 4 : 3 aspect ratio.

It is, however, a major problem that at the present time many theatres are equipped only for wide-screen projection, which encroaches substantially on the area available for superimposed titles.

In some countries where it may be necessary to provide translated dialogue titles in several languages, they may be produced on a separate film which is projected below or at the side of the picture, and in some parts of the world it is the practice to engage a narrator to translate the original dialogue as the picture proceeds. L.B.H.

See: *Release print make-up; Sound recording* (FILM).

⊕**Format.** Image aspect ratio or size of motion picture film.

See: *Aperture sizes; Film dimensions and physical characteristics.*

⊕**Fourier Analysis.** Mathematical system of ☐analysis which expresses any waveform, however complex, as the resultant of a fundamental frequency and a given number of harmonics of calculable amplitude.

⊕**Foxhole.** Colloquial term for the small sprocket hole used on positive prints of Cinemascope films when copies with four magnetic sound tracks are required; named after the 20th Century Fox Company, who introduced the Cinemascope process. Also called a CS perforation.

See: *Film dimensions and physical characteristics.*

⊕**Frame.** Individual pictures on a strip of film. Frame lines are the horizontal bars between frames, which can sometimes be seen during projection. When frame lines are visible, the picture is said to be out of frame; a correctly framed picture is said to be in frame. Framing devices are incorporated in projectors and editing machines for moving the picture up or down until it is in frame.

⊕**Frame Counter.** Type of footage counter in which each foot of film is divided on a separate dial into the appropriate number of frames, depending on the gauge of film used.

⊕**Frame Line.** Black horizontal bars appearing

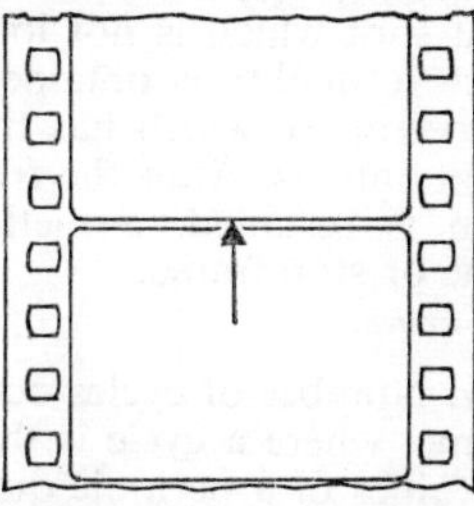

between successive picture images in positive film prints.

⊕**Frame Stretch.** Extending and slowing down the action of a scene by repeating each successive frame one or more times, usually by the preparation of an optically printed dupe negative. Films photographed at 16 pictures per second may thus be printed so as to project at the standard sound speed of

24 pps by repeating alternate frames, so that the sequence 1, 2, 3, 4, 5 . . . becomes 1, 2, 2, 3, 4, 4, 5 . . .

See: *Freeze frame.*

⊕**Framing.** In film projection, the act of putting the frame of film in register with the projector aperture. This done, the picture is said to be in frame; when the frame lines at top or bottom are visible, the picture is out of frame.

In moviolas and most narrow-gauge projectors, framing is effected by moving the aperture itself in relation to the picture. In the case of a projected picture, the projector itself must then be raised or lowered to reposition the picture on the screen.

Professional projectors have a device for moving the intermittent mechanism up or down, or altering its phase, so that the aperture, and thus the picture positioning on the screen, remains fixed. There is a limit to the distance it can be moved, and if this limit is reached with the picture still out of frame, the operator must return to the original position and continue framing in that direction, with unpleasant effects often seen on the cinema screen.

See: *Projector.*

□**Free-Space Propagation.** Manner of signal propagation when the transmitting and receiving aerials are completely isolated from all material objects.

See: *Transmitters and aerials.*

⊕□**Freeze Frame.** A single frame of a moving shot arrested for as long as required by an optical printing process (or in TV electronically) to concentrate attention on a phase of action, e.g., a football goal, or to lengthen a shot which is not long enough. Better picture quality is obtained by alternating two frames, which has the effect of reducing graininess. After the freeze frame, the action is resumed. Sometimes called hold frame or stop frame.

See: *Frame stretch.*

⊕□**Frequency.** Number of cycles occurring per unit of time, where a cycle is the complete series of values of a periodic quantity. The unit of frequency is the cycle per second (cps or c/s) now renamed the Hertz (Hz).

□**Frequency Changer.** Multi-electrode valve or equivalent transistor in a superheterodyne circuit which converts the input signal frequency (by heterodyning with a locally generated oscillation) to the intermediate frequency.

See: *Receiver.*

□**Frequency Divider.** Circuit for accurately dividing one frequency by a required integer to produce a lower frequency.

⊕□**Frequency-Response Characteristic.** Curve denoting the relative response of a system (or one or more of its components) to changes of frequency. The band of frequencies covered by the curve is usually the pass band of the system. In audio systems it is commonly the audio spectrum.

□**Frequency Translation.** In re-radiating or distributing a radio signal (e.g., by a relay station), it is usually convenient to change frequency in the device so that the amplified signal from the transmitter does not feed back into the receiver and so cause instability. Such devices are often called translators or frequency translators.

See: *Microwave links; Satellite communications.*

⊕□**Fresnel Lens.** Stepped-surface lens often used as a condenser lens in spot lamps, as for the same focal length it can be made

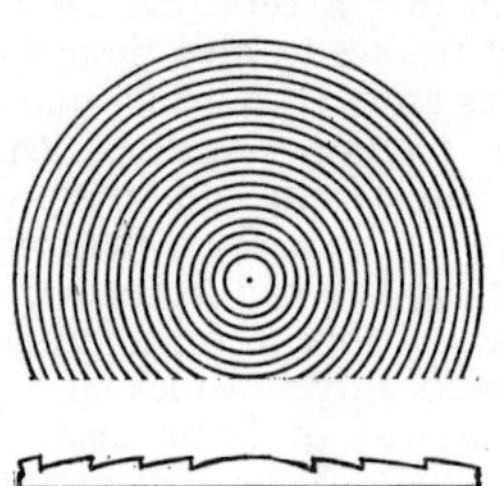

thinner and lighter than a normal lens, and is thus less likely to crack from the heat of the lamp.

See: *Lighting.*

□**Fresnel Zone.** The first Fresnel zone, important in signal transmission, is bounded by points where the distance from transmitter to receiver via the point is greater by one-half wavelength than the direct path.

See: *Transmitters and aerials.*

⊕⊕**Friction Head.** Type of panning and tilting tripod head which incorporates a sliding

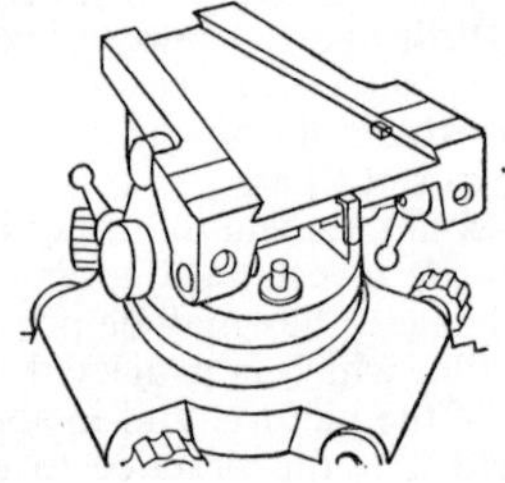

friction device to secure smoothness of camera movement. Also called a free head.

⊕**Friese Greene, William, 1855–1921.** Photographer who opened a series of studios in Bath, Bristol, Plymouth and London from 1874. Became interested in the possibilities of recording motion after meeting J. A. R. Rudge, a Bath mechanic who had devised a magic lantern apparatus with which an illusion of movement could be produced by the projection of a series of posed pictures. Between 1886 and 1889, demonstrated several Rudge lanterns to various photographic bodies.

With Mortimer Evans, a London civil engineer, Friese Greene constructed and patented a somewhat impractical camera for "taking photographs in rapid series", which from its construction was clearly intended to produce lantern slides for a Rudge-type lantern. This patent, claimed by some as the master patent for cinematography, made no mention of projection apparatus or perforated celluloid film, and proposed a picture frequency of only a few frames per second. In 1890, he demonstrated a two-lens camera built and patented by F. H. Varley, a London engineer, and in 1893 he patented a virtually identical design himself.

While not without ideas, Friese Greene had insufficient theoretical knowledge and technical skill to put them into effective practice.

⊕□**Fringing.** Effect caused by imperfect registration in the superimposition of two or more images. Fringing is particularly objectionable in colour synthesis, since the fringes will then be of different colours from adjacent parts of the images.

See: *Camera (colour); Imbibition printing; Telecine (colour).*

□**Front Porch.** Period of time (usually about 1·8 μsec.) preceding the line-sync pulse within the line-blanking interval in the standard TV waveform. Its purpose is to provide time for a high video signal amplitude (i.e., a white object) at the right-hand side of the picture to drop down to black level and thence to blanking level before the start of the line-sync pulse.

See: *Porch; Television principles.*

⊕**Front Projection.** Normal form of cinema projection where the image is thrown on to an opaque surface and viewed by reflection from the same side as the projector.

A specialized application of the term is the process used in special effects work for the combination photography of studio action with a projected background scene, which may be a still or motion picture image. The background scene is projected via a semi-reflective mirror on to a beaded screen of highly directional reflectivity,

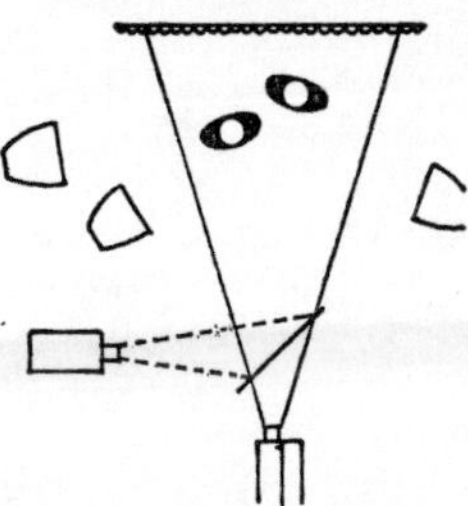

FRONT PROJECTION. Projector (*left*) and camera (*bottom*) are optically aligned so that foreground characters are shadowless when seen against projected image (*top*).

using a projector placed in a position equivalent to that of the camera.

The camera, with the same angle of view as the projector lens, photographs through the mirror both the normally lighted foreground action and the background reflected from the screen. Because the characteristic of the beaded screen is to reflect rays along their line of incidence, the system is efficient in its use of light and the background image is not degraded by light from the foreground falling on the screen.

See: *Screens; Special effects* (FILM).

⊕**Fujicolor.** A negative/positive integral tripack colour film process introduced for professional motion picture production in Japan in 1955. The negative stock embodies a type of colour masking.

Reversal colour stocks are also manufactured.

See: *Colour cinematography.*

⊕**Full Aperture (FA).** Aperture size of camera or projector used in 35mm motion picture work before sound-on-film processes were introduced, hence sometimes known as silent aperture. Still used for film strips and occasionally for motion picture photography where the maximum frame size on 35mm is required.

See: *Aperture sizes.*

⊕**Full Frame.** Maximum image size on 35mm film which can be obtained from a frame four perforations high and extending the full width of the film between the perforations. Corresponds to camera or printer exposure using the full aperture.

See: *Aperture sizes; Film dimensions and physical characteristics.*

⊕**Gaffer.** Senior electrician on a film unit. Responsible for ensuring that the lighting cameraman's requirements are satisfied. This includes provision of the precise types of lamps in the correct quantities, together with all ancillary equipment to be used with them. He must give instructions to the other electricians on the unit as to the positioning of lamps and their operation, including movements during shooting.

On location, responsibilities include the operation of the generators.

See: *Studio complex* (FILM).

⊕**Gain.** Amplification, or ratio of output to □input, of an electronic system or some section of it such as an amplifier. The term gain is used even when the amplification is less than unity.

⊕**Galvanometer.** Device employed in optical (photographic) recording to convert an electrical signal into the motion of a light beam. The signal is applied to a coil suspended in a magnetic field, causing it to rotate. To the coil is attached a small mirror, so that the varying signal is made to move a light beam which, after passage through a slit, forms a record on a moving film.

See: *Sound recording* (FILM).

⊕**Gamma.** Measure of the contrast of an □image reproduction process. In photographic sensitometry it is represented by the slope of the straight-line portion of the characteristic curve of the processed emulsion, where the relation between the density increment $\triangle D$ for a given log exposure increment, $\triangle \log E$, is constant:

$$\gamma = \triangle D / \triangle \log E$$

In television, a similar characteristic is given by the relation between the luminance increment on the receiver screen and the luminance increment in the original scene. Here, $\gamma = \triangle \log R / \triangle \log S$. In general, this relationship is not uniform over the whole tonal scale and the gamma value or contrast gradient at a given point is of importance.

For camera tubes and similar transmission devices, point gamma is defined as the instantaneous slope of a curve relating the logarithms of the incident light and the resultant output voltage, while for receivers and display devices, it is the instantaneous slope of a curve relating the logarithms of the input voltage and of the intensity of the resultant light output.

See: *Transfer characteristics.*

□**Gamma Correction.** Process of modifying the linearity of the amplitude response of a video amplifier in such a way that deficiencies in the gamma law of an associated light-sensitive or picture-display device are corrected. Unless the overall gamma of a system approaches unity, the grey scale suffers distortion. In monochrome television the eye is relatively tolerant of errors, but in colour television the greatest care is necessary to match gamma laws in the colour channels and produce an overall gamma close to unity.

See: *Colour principles* (TV); *Transfer characteristics.*

⊕**Garland.** Ring type of studio lighting unit incorporating a number of incandescent lamps, sometimes placed over a film camera.

⊕**Gasparcolor.** Obsolete colour print process originated in 1934 involving a bleach-out or dye destruction method in an integral tripack material.

⊕**Gate.** Aperture unit of cameras and projectors, so called because this unit often swings outward on hinges for threading and cleaning.

⊕**Gate Pressure.** Pressure applied to a film in the camera or projector gate to prevent it buckling or otherwise moving out of the focal plane.

□**Gating.** Operation of an electronic device acting as a gate to select a portion of a current signal for examination or storage, or to activate another circuit.

□**Gating Pulse.** Pulse designed to operate a system for a specified period to enable some other action to take place during the interval. Employed in colour TV transmitters and receivers.

⊕**Gaumont, Leon, 1863–1946.** French inventor and film producer. In 1895, became director of the Comptoir Général de Photographie, and with Georges Demeny in 1896 marketed the 60mm Chronophotograph film apparatus, the first of many professional and amateur motion picture machines to be produced by his organization.

In 1902 he devised a sound-on-disc sound film system, demonstrated in that year to the Société Française de Photographie; in an improved form, called the Chronophone, it was shown to the Académie des Sciences in 1910. From 1911 his *Films Parlants* were shown at his 5,000-seat Hippodrome theatre in Paris using mechanical amplification of the acoustic recordings to fill the huge auditorium. For his sound films, Gaumont developed the system of pre-recording the sound in the studio, and actors miming to it on location when the film was shot.

With Petersen and Poulsen, he developed the GPP sound system in 1928, using an optical sound track carried on a 35mm strip separate from the pictures. The first French sound film, *Eau de Nil*, was made by this method, but synchronization problems led to its abandonment.

Gaumont also pioneered colour processes; his three-colour additive system, Chronochrome, was patented in 1912 and publicly demonstrated in 1913.

⊕□**Gaussian Distribution.** A distribution of events which follows the curve or formula of the Gaussian distribution function. This is used to define a number of the properties of systems with random variations. In a number of occurrences with random variations superimposed, most results will occur at the mean value and fewer as the variations investigated get farther from the mean. The Gaussian distribution function gives the measure of this for true normal random variation such as is encountered naturally when a very large number of random causes affect the final result. Essentially it is a symmetrical exponential curve of the general form:

$$f(x) = Ke^{-ax^2}$$

which has a maximum value at $x = 0$ and decays exponentially as the value of $-x$ or $+x$ becomes larger, having a value of zero at $x = \pm\infty$.

□**Geitel, Hans Friedrich, 1855–1923.** German physicist and collaborator with Julius Elster in early investigations of the vacuum tube.

⊕**Gelatin.** Protein occurring naturally in animal hides and bones. When extracted and purified it forms the essential medium for the manufacture of photographic emulsions.

Hot solutions of gelatin set to a firm jelly at lower temperatures and this property permits the coating of thin layers of emulsion on film. Gelatin also absorbs substantial quantities of water, with resultant swelling, which allows the penetration of the chemical solutions used in photographic processing into the depth of the emulsion layer.

Also generic term for a screen of translucent material which diffuses and reduces the intensity of a studio lamp. Gelatins are often mounted on the lamp itself, but are sometimes on separate stands and are used to control the light falling on a dress or some other small part of the scene.

See: *Film manufacture.*

⊕**Gemini.** A type of combined film and tele-
□vision camera distinguished from most others of its kind by the fact that the

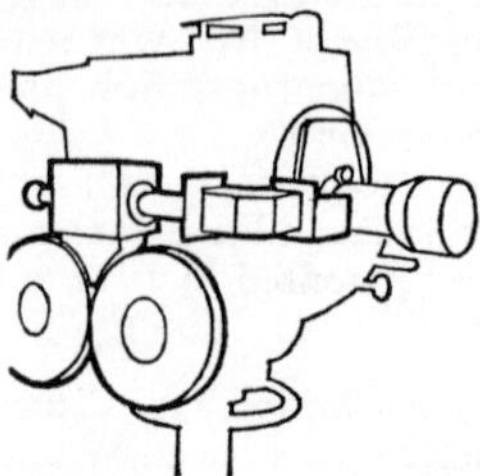

GEMINI. TV camera lens system incorporates semi-reflector and relay lens to provide optical image to inverted 16mm film camera.

primary unit is the TV camera. The Gemini consists of a zoom lens and beam splitter, the latter with an optical feedback to a 16mm Auricon camera, mounted upside down so that the large magazines do not impede the operator.

This unit can be attached to a standard orthicon or Plumbicon camera and provides a permanent film record, better in quality than a telerecording.

The black-and-white beam splitter directs 35 per cent of the incoming light to the film, and the alternative colour beamsplitter 70 per cent to compensate for the lower sensitivity of colour stocks.

Multicamera techniques (sometimes called system operation) may be employed with the Gemini, as in normal TV production. A small light marks the edge of the film in the camera selected by the director, as a guide in subsequent editing. As a rule, all the cameras run film all the time, but they may be started and stopped remotely to economize footage.

See: *Add-a-Vision; Electronicam; System-*35.

⊕**Generator.** Motor-driven source of elec-
□tricity, normally d.c., used to supplement existing power supplied when filming in the studio or on location. Arc lighting on location usually requires a generator, since normal power supplies are a.c.

⊕**Gevachrome.** Integral tripack reversal colour film process used for 16mm motion picture work. The camera original is of low contrast and is not intended for direct projection but for printing copies on to a corresponding reversal print stock.

See: *Colour cinematography.*

⊕**Gevacolor.** Negative/positive integral tripack colour film process introduced for professional motion picture purposes in 1953. The system has a close relation to Agfacolor, with which it is now unified.

See: *Colour cinematography.*

□**Ghost.** Vision signal received with a delay, as compared to the direct signal, usually caused by reflection from aircraft or prominent objects. The result is a fainter signal (ghost) displaced from the main signal.

In the VHF and UHF wavebands used for television transmission, the reflection of signals from tall buildings and other objects can present a serious problem at a receiver. This is due to the time difference between the arrival at a receiver aerial of the direct signal and the arrival of a reflection or echo.

Assuming that the direct signal is appreciably the stronger of the two, the synchronizing circuits of the receiver lock on to it and disregard the weaker signal. But the video section of the receiver is unable to ignore the picture information contained in the reflected signal, with the result that it appears on the screen as a ghost image. This

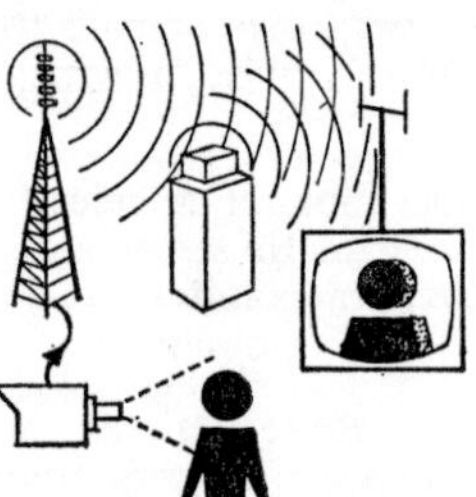

GHOST. Transmitted signal arrives by direct path and, delayed, by reflection from tall building.

is displaced to the right of the primary image by an amount corresponding to the difference in time taken by the direct and reflected signals in travelling from transmitter to receiver and may be anything up to the full picture width.

As an example, if the difference between the direct and reflected signal paths is one mile, then the two signals appear at the receiver separated in time by approximately 5 μsec. At the 405-line scanning speed this corresponds to a positional separation on the screen of a little more than 1/20 of the picture width, but on both 625 and 525 lines it is just under 1/10 of the picture width.

Ghost images resulting from reflections may be reduced or even eliminated by using either a more directional and properly orientated aerial at the receiver or a device known as an echo equalizer. C.W.B.R.

See: *Echo waveform corrector; Microwave links; Transmission networks; Transmitters and aerials.*

⊕**Glass Shot.** Trick shot in which part of the scene is a painting on glass which is photographed at the same time as the main action

seen through the clear portion of the glass; very spectacular scenes can thus be shown without the need for the whole of the set structure to be built in the studio.

See: *Art director; Special effects* (FILM).

⊕**G-Number.** The photometric aperture of a ☐lens opening may be specified as its G-number, rather than its relative geometric aperture or *f*-number. It is defined as the ratio of the field luminance to the illuminance produced in the focal plane of the lens, when measured under specified conditions. In general terms, the G-number of a lens is nominally equal to four times the square of its T-number, or to four times the square of its *f*-number divided by its transmittance.

See: *Exposure control;* f-*number; T-stop.*

⊕**Gobo.** Adjustable mask on a stand, used in the studio to shield the camera from direct light rays, or produce special lighting effects.

⊕**Gradation.** Accuracy with which a process ☐in a film or television chain renders a series of incremental steps of grey scale.

⊕**Grader.** Laboratory technician responsible for determining the colour and density at which a motion picture negative should be printed in order to produce a satisfactory positive image for projection. At the first stage, the grader is concerned with making rush prints from each scene photographed at the studio in the course of production; subsequently the grader must deal with the scenes assembled in sequence and in reels and ensure that the printing of each scene is such as to provide a consistent appearance throughout the whole production in accordance with the dramatic character intended by the director and lighting cameraman.

Experienced graders can become extremely skilful in assessing visually the printing levels required for black-and-white negative, but colour negative is very difficult to judge in this way and test prints or closed circuit colour TV analysers are often employed to assist the grader. Despite these aids, the final result of printing the large number of scenes which are assembled in a feature film depends greatly on the subjective judgment of the grader in viewing and correcting his first results, and this is a job in which personal skill and ability are of vital importance.

See: *Laboratory organization; Tone reproduction.*

⊕**Grading.** Process of balancing the density of successive shots in a completed film (and in a colour film the colour values also) in such a way as best to carry out the cameraman's original intentions in respect of lighting and atmosphere, and produce an overall harmonious effect. Grading is a subjective and difficult task, requiring a high degree of skill. In a black-and-white film the grader's decisions are made in terms of printer light settings for each shot, with occasional changes during the course of a shot. In a colour film, the grader must also specify the colour corrections, either as colour filters or as the required levels of the red, green and blue printing exposures.

See: *Sensitometry and densitometry; Sensitometry of colour film; Tone reproduction.*

⊕**Graduated Filter.** Non-uniform type of filter placed in front of a camera lens to give differential effects in various parts of the scene. An example is a sky filter which is yellow at the top shading off to colourless at

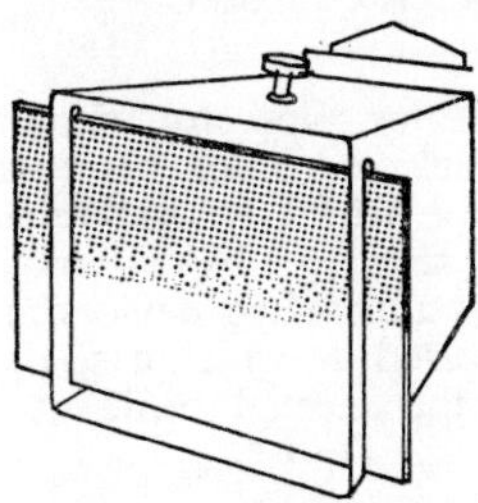

the bottom. It is designed to increase the density of a black-and-white rendering of blue skies without affecting the rest of the scene. The corresponding filter for colour has a neutral density at the top to retain the blue in an over-bright sky.

See: *Cameraman* (FILM).

⊕**Grain.** Basic constituent of the photographic image, consisting of metallic silver formed by exposure and processing of light-sensitive silver halide crystals in the emulsion. Individual grains are visible only under the microscope, but the clumping caused by the overlaying of grains at different depths in the emulsion can be observed when a mag-

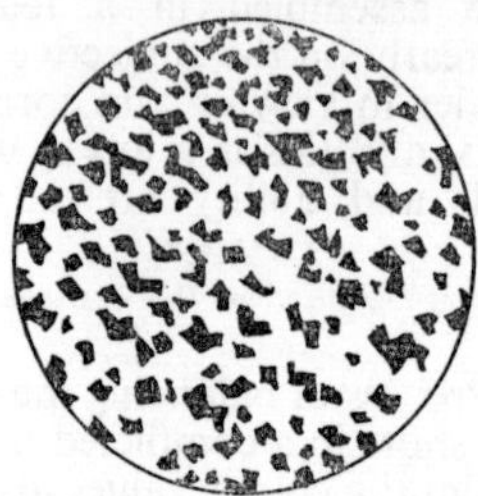

nified image is projected in the cinema, giving rise to the moving random pattern in some areas of the picture which is known as graininess.

See: *Image formation.*

⊕**Graininess.** Subjective factor of the photographic image affected not only by the structure of the image but also the viewing conditions involved. In viewing motion picture images on the screen, the persistence of vision of two or more frames greatly improves the apparent graininess of the picture, since the random clumping varies from frame to frame. When a single frame is viewed at comparable magnification, the graininess is much more objectionable.

Graininess is always most noticeable in the middle tones of the image, particularly in large uniform areas, and is rendered more visible by any factor increasing the overall contrast in reproduction.

See: *Image formation.*

□**Gram Operator.** Operator who produces at the signal selected fragments of music from disc or tape which he has rehearsed and probably selected beforehand. Electro-mechanical automated devices are now increasingly used, in which case the operator re-records the selected music into them.

See: *Studio complex* (TV).

⊕**Grandeur.** Early form of wide-screen cinema presentation using 70mm film.

⊕**Granularity.** Physical characteristic of the photographic image: granular structure of the silver deposits forming the image. A numerical assessment of granularity can be obtained by the use of a micro-densitometer scanning the image structure. In general, the most sensitive high-speed negative materials have larger halide grains and therefore show higher granularity, while positive stocks are finer in structure and have low granularity. In both types the degree of granularity can be significantly affected by processing conditions, particularly those affecting the contrast of the stock.

See: *Image formation.*

□**Graticule.** Transparency marked with limits or guide marks, placed over a signal trace on a cathode ray tube.

⊕**Green Print.** Release print which has been inadequately dried and therefore contains excess moisture. This causes swelling, with the result that a green print chatters in the film gate during projection and emulsion tends to peel off, causing abrasions and scratches. Lubrication improves the projection properties of green prints, but the real remedy lies in the laboratory. The term is used less than formerly.

See: *Laboratory organization.*

⊕**Greyback.** Colloquial term for duplicating negative films coated on a grey base support.

⊕**Grey Base.** Base in which a grey dye has been incorporated so as to reduce the danger of halation around the exposed image caused by light reflected into the emulsion layer from the base surfaces. Several types of motion picture film stocks, particularly black-and-white negative and duplicating materials, are coated on such a base.

The grey dye in the base is permanent and thus differs from other forms of anti-halation backing which are destroyed or removed in the course of processing.

⊕**Grey Scale.** In photographic sensitometry and processing control, a grey scale is a standard test object, consisting of a series of steps of decreasing reflectivity from white through grey to black. In some cases the scale may consist of a transparency with steps of varying transmission from clear to opaque and is then illuminated from behind for photography.

The steps of a grey scale are usually made to increase in density in a regular fashion by known amounts and the reproduction of the scale at various stages of a photographic process provides valuable information for exposure and development control. The uniform neutral reproduction of all tones of grey is also of fundamental importance in determining the colour balance in colour photographic processes.

See: *Sensitometry and densitometry; Sensitometry of colour materials.*

□**Grid Modulation.** Application of the modulating signal to the grid of the appropriate r.f. amplifier, producing modulation of the carrier frequency.

□**Grid Tube.** A colour display tube of relatively simple design and potentially high efficiency but which has not yet emerged from the laboratory stage of development.
See: *Colour display tubes.*

⊕**Grimoin-Sanson, Raoul, 1860–1942.** French inventor of early motion picture apparatus, notably the Cinéorama process of 1897, using ten projectors and a cylindrical screen, the films being produced by ten synchronized cameras recording a 360° view.

⊕**Grip.** Stage hand, but attached to the camera crew of a film unit. Responsibilities include laying tracks on which camera dolly or velocilator travels; movement of dolly or velocilator (and also of camera crane provided it is not electrically motivated); assisting camera crew in removal work connected with cameras or camera equipment to the requirements of the camera operator.
See: *Studio complex* (FILM).

⊕**Grivolas, C., d. 1938.** French industrialist who sponsored development of early motion picture processes, including one for stereoscopic projection by the anaglyph method. Became financial partner with the Pathé brothers in the development of the Compagnie Générale des Phonographes et Cinématographes.

□**Grounded-Grid Stage.** A class of valve circuits in which the control grid is grounded, or connected to the dead end of the input signal, and the live side of the input is connected to the cathode.

⊕**Ground Glass.** Piece of glass with a finely ground surface on which an image can be formed. Used in the viewfinders of film cameras, the image being often enlarged by a focusing magnifier. Ground film may also be placed in the camera aperture to form an image.
See: *Camera* (FILM).

⊕□**Ground Noise.** Residual system noise which in recording systems employing optical film or magnetic tape is principally due to the ultimate particle size of the recording medium. The term may also include residual noise generated by the amplifier channels.
See: *Noise.*

□**Ground Wave Transmission.** Radio transmissions effected by the wave travelling

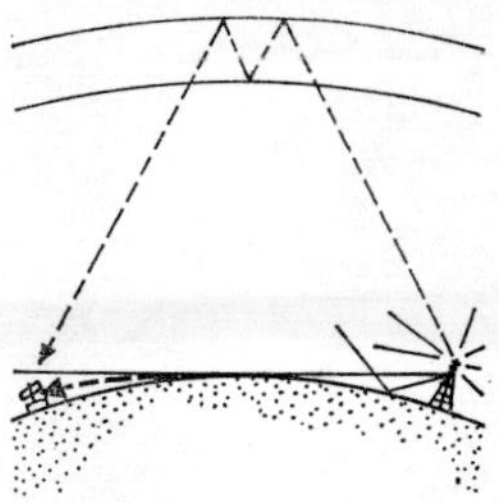

parallel to the earth's surface as opposed to being reflected from the ionosphere.
See: *Transmitters and aerials.*

□**Group Delay.** If the phase shift of a signal after its passage through a circuit is plotted for various frequencies, the group delay or envelope delay is the slope of the curve at a particular frequency. Rate of change of phase response with frequency. An important factor in the design of some colour circuitry.
See: *Transmission networks; Videotape recording* (*colour*).

⊕□**Guide Track.** Synchronized sound track recorded at the same time as its associated picture track, but not intended for reproduction to the final film or TV audience. It acts simply as a guide for recording the same dialogue under better acoustic conditions, usually in the studio, by the process of post-synchronization.
See: *Sound Recording* (FILM).

⊕□**Gun Mike.** Highly directional microphone which can be aimed like a gun at the source of sound.
See: *Microphones.*

⊕**Gyro Head.** Type of tripod head, nowadays little used, which incorporates a heavy flywheel driven at high speed by gearing from moving camera platform. Inertia of flywheel ensures steady movement of camera.
See: *Camera* (FILM).

⊕**Halation.** Unwanted exposure which may be ☐observed surrounding the image of a bright object recorded on a photographic film caused by reflection of light from the rear surface of the film base. To eliminate this effect, the base surface is often coated at manufacture with a jet black layer, known as the anti-halation backing, to absorb the light passing through the emulsion.

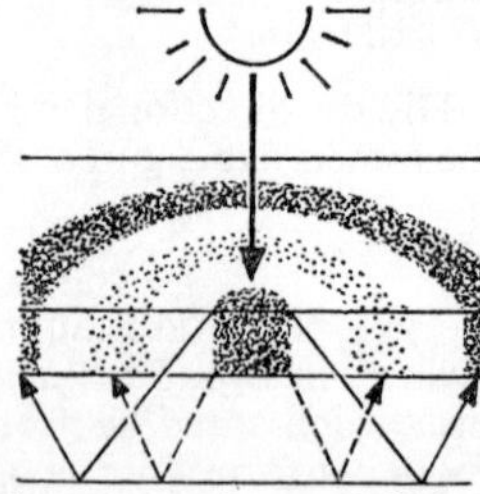

A similar scattering of light around a bright object may be met with in television wherever a phosphor is used and is worsened by glass mounts of conventional design which give multiple reflections.

See: *Exposure control; Image formation; Integral tripack.*

⊕**Halide.** Compound of one of the halogens, fluorine, chlorine, bromine or iodine, with another element. In photography the silver halides, as chloride, bromide and iodide, form the most important light-sensitive substances which are fundamental to all the emulsions used in motion picture work.

See: *Chemistry of the photographic process.*

☐**Hallwachs, Wilhelm, 1859–1922.** German physicist, born at Darmstadt and at one time Professor of Electro-technics and Physics at the Technische Hochschule at Dresden. Hertz's observation in 1887 that the behaviour of a spark gap was affected by illumination of the electrodes was investigated in the following year by Hallwachs, who found that a clean insulated zinc plate acquired a positive charge when directly illuminated by the light from a carbon arc. If a glass plate was placed between the arc and the plate, this effect disappeared, showing that it was the ultra-violet light which was responsible for the charge. A negatively charged plate lost its charges by illumination. The effect was found to persist if the metal was placed in a vacuum.

☐**Halo Effect.** Electrical effect which produces a similar appearance on the picture to halation on film.

⊕**H & D Curve.** Characteristic curve relating the exposure and the developed density of a photographic emulsion. Named after the

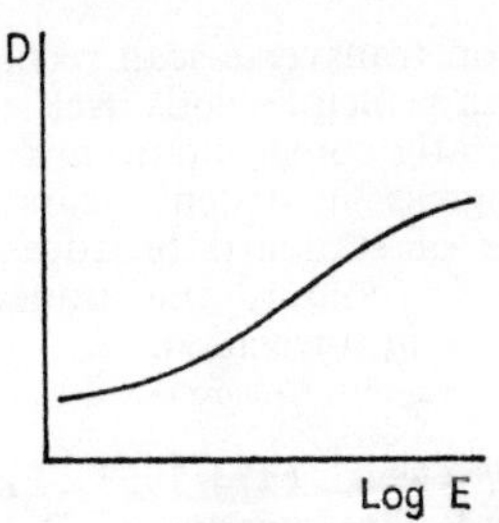

H & D CURVE. Logarithm of exposure plotted against density produces characteristic curve marked by relatively straight central section and pronounced shoulder and toe regions.

Swiss chemist, Ferdinand Hurter (1844–98) and the English chemist and photographer V. C. Driffield (1848–1915), whose joint researches established sensitometry on a scientific basis. They also devised photometers and densitometers, and worked on the theory of the latent image and of the development factor, now known as gamma.

See: *Characteristic curve; Sensitometry and densitometry.*

⊕**Hardener.** A chemical used to render the gelatin emulsion of a photographic film more resistant to physical damage and abrasion. Sodium sulphate may be used in processing solutions to reduce the swelling of the gelatin, and the fixing solution of hypo may contain potash alum to produce a harder emulsion after drying. Dilute formaldehyde (formalin) solutions may also be used for the same purpose. Hardeners may also be used at the beginning of a processing sequence to allow the use of developing solutions at higher temperatures.

See: *Chemistry of photographic process; Processing.*

□**Harmonic Filter.** Device for filtering out or removing unwanted harmonics or multiples of a signal which can be present at the output of a circuit.

□**Harmonic Generator.** Unit which produces harmonics or multiples of a fundamental frequency.

⊕**Head.** General term for the central mechanism of several film devices, such as a camera-printer and projector. It excludes magazines, light sources, lenses and other attachments and accessories.

⊕**Head, Camera.** Term sometimes used to denote the main camera body, comprising the intermittent movement, feed and take-up sprockets, and lens mounting, and their housing, but excluding external magazines and other accessories.

See: *Camera* (FILM).

□**Head Amplifier.** Stage or stages of video amplification, plus a stage of impedance transformation to match to the low impedance of the cable, located in the camera, as close to the tube base as possible. The amplifier must have a high signal-to-noise ratio, and to achieve this triode tubes have been used in preference to multielectrode tubes or transistors.

Owing to the camera tube load resistor and its associated shunt capacities, there is a loss in the higher frequencies, and to offset this, and to improve amplitude linearity, frequency-selective negative feedback is often employed. The low frequency response is maintained by a bass boost circuit, and the use of a cathode follower assists this and provides the low impedance output to the cable.

Camera head amplifiers (or head units) often include a polarity reversing facility (pos-neg switch), tube heater switch, and test signal input, and usually provide the connections to the tube base. Great attention is paid to screening and hum reduction, and it is desirable that the unit should be mounted on shock absorbers to prevent or reduce microphonic effects.

Improvements in transistors and solid-state devices generally are now bringing them into general use in the design of head amplifiers, and the small size of transistorized units with printed circuit boards makes them ideally suited to this purpose.

See: *Camera* (TV).

⊕**Head Leader.** Standard leader attached to the front of a projection print, containing synchronizing information which, in conjunction with the run-up time of the machine, determines how many seconds will elapse after the projector has been started before the first frame of the picture appears on the screen or is scanned in a telecine.

See: *Release print make-up.*

□**Headphones.** Earphones mounted on a headband as worn by personnel in a studio, so that they may hear sound signals which must not be picked up by the studio microphones. These signals may be a pre-listen facility being used by a sound mixer, or producers' talkback to the studio in general. Both earpieces in a pair of headphones may be fed with the same signal, or one may have a feed of the programme signal and the other talkback.

Since it is common to operate a number of headphone sets in parallel, they are usually of high impedance, say 10,000 ohms.

See: *Communications.*

⊕**Headset.** A pair of headphones with a □microphone attached to them. This gives the facility of reverse talkback, so that a sound boom operator can talk back to the sound mixer, or a television cameraman to his control operator. As light weight is essential to the user, and only telephone quality is required, the microphone is usually a carbon capsule.

See: *Communications; Sound recording.*

□**Head Unit.** In microwave transmission, the collection of apparatus housed in the head or unit fixed to the receiving or transmitting parabola or aerial.

See: *Microwave links.*

□**Helical Aerial.** Aerial assembly where the radiating element takes the form of a helix (a longitudinal spiral of constant diameter).

See: *Transmitters and aerials.*

□**Helical Scan.** Type of videotape recorder in which the tape is wrapped helically round a fixed drum of large diameter, while a rotating disc coaxial with the drum and mounted within it carries the recording head or heads. Head contact with the tape is through a circumferential slot in the drum. The helical wind of the tape imparts a slant to the head-scanning path, so that the whole width of the tape is used to carry video tracks. Problems arise, however, where the tape meets the drum and leaves it, and this gap must be kept as small as possible to reduce loss of recorded information.

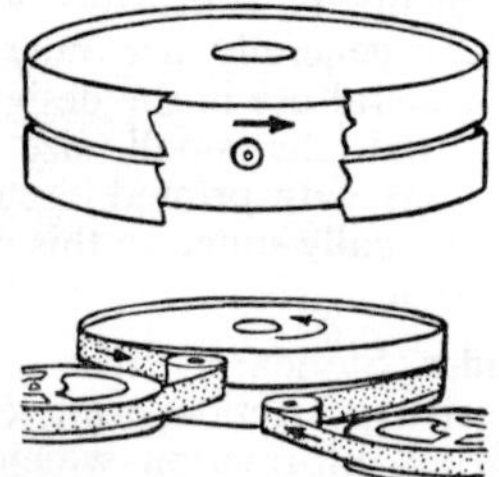

TYPICAL HELICAL-SCAN LAYOUT. (*Top*) Disc carrying one or more magnetic heads rotates rapidly in direction of arrow. It is enclosed in a fixed drum or shell with a circumferential slot through which the head tip projects as it rotates. (*Bottom*) Tape follows a helical path as it is pulled relatively slowly round this stationary drum by a capstan (not shown), so that the head sweeps across its full width at a small angle to the length of the tape. The feed and take-up spools are thus at different levels.

The tape velocity is relatively low (often 7½ips), and the disc velocity high (usually one revolution per TV field), so that the head-to-tape velocity is comparable to that achieved on transverse-scan recorders. The helical-scan principle lends itself to simpler and less costly construction, and has made great progress in recent years; but for recordings of standard broadcast quality, especially in colour, the transverse-scan recorder is still unrivalled.

See: *Transverse scan; Videotape recording.*

⊕**Hepworth, Cecil, 1874–1953.** English inventor and film producer. Began as a lanternist and lecturer; first patent granted when he was 21. After working as a travelling showman with motion pictures, he turned to film production; of his many early productions, one of the most famous is *Rescued by Rover* (1905), one of the earliest films to employ narrative techniques. Hepworth later devised and exploited the Vivaphone system for synchronizing discs played on an acoustic gramophone and projected film; it was in successful commercial use by 1910. Hepworth continued in successful film production until 1923.

⊕**Hertz.** Unit of frequency. One Hertz equals □one cycle per second. Larger units are the kilohertz, KHz (1,000 Hz), the megahertz, MHz (one million Hertz), and the gigahertz, GHz (one thousand million Hertz).

□**Hertz, Heinrich Rudolph, 1857–1894.** German inventor, born in Hamburg. First to confirm by actual experiment the predictions of Clerk Maxwell regarding the production of electromagnetic waves by electrical means.

In 1879 the Berlin Academy of Science offered a prize for research with the object of proving by direct experiment that electromagnetic forces required time for their propagation. The problem was complicated by the fact that there were no means then available of creating electrical oscillations of sufficiently high frequency. Hertz's attention was drawn to this challenge by Hermann L. F. von Helmholtz under whom he was studying in Berlin, but it was not until 1886 that Hertz decided to investigate the problem seriously. He then conducted a series of experiments in which he not only succeeded in generating these invisible waves but in detecting and demonstrating their presence in space at a distance of several metres from their point of origin. He also proved that a close relationship existed between these waves and those of light in that their velocity was identical.

Hertz published the results of his experiments between 1887 and 1889 in a series of papers addressed to the Berlin Academy of Sciences. Although he had succeeded in generating wireless waves (or Hertzian waves, as they came to be known) he did not immediately appreciate the possibilities of using the waves as a means of communication and left that for others.

One of Hertz's important discoveries made during the course of his experiments was that if ultra-violet light was directed on to the spark gap of his transmitter, the intensity of the spark increased and the potential required to produce it was much reduced. Hallwachs later investigated this effect and showed that it was photoelectric in character.

⊕**H.I. Arc.** High-intensity arc; arc in which the carbons contain the salts of various rare earths, making possible an increase of 4–8 times in the current passed per unit cross-section, with a corresponding increase of luminous intensity. Invented by Beck in Germany in 1910, the high-intensity carbon made it possible to raise progressively the light output of projection arc lamps which today reach about 40,000 lm.

See: *Lighting.*

□**High Frequency Carrier System.** System for conveying broadband signals over a cable using amplitude or frequency modulation of a sine wave carrier. Cable loss increases with frequency, but, if a carrier system is used, the ratio of the highest frequency to be transmitted to the lowest frequency can be very much less than in the original video signal, and so the amount of equalization to be applied is also less.

See: *Transmission and modulation.*

□**High Gain Aerial.** Aerial array which has a very high gain compared to a dipole, and thus produces a high effective radiated power for a given input.

See: *Transmitters and aerials.*

⊕**High Hat.** Small mount of fixed height which can be attached to the floor or to any place where it is desired to set the camera as low as possible.

⊕□**High Key.** Style of tonal rendering of a scene which emphasizes the middle and lighter tones at the expense of the darker ones. Best suited to subjects which convey a light and cheerful impression.

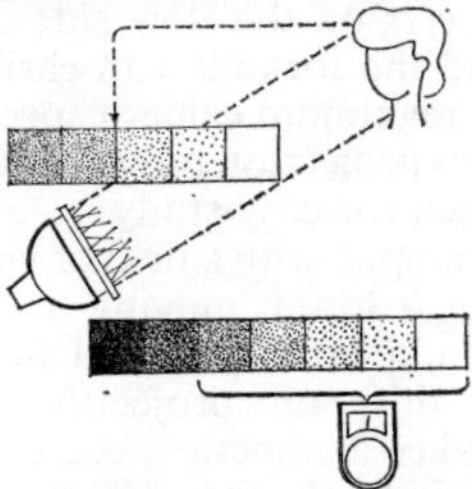

High key lighting avoids strong shadows and makes use of plentiful front light.

⊕□**Highlight.** Area of excessive brilliance in a scene, which exceeds the permissible range of lighting contrast.

⊕**Highlight Density.** Density or blackness of the film image in the highlight or white areas. Brightest part of the subject reproduced as the densest area in the negative and the lightest area in the print.

⊕□**High-Pass Filter.** Electrical circuit which passes only frequencies above a certain value, known as the cut-off frequency of the filter.

⊕HIGH-SPEED CINEMATOGRAPHY

Cine cameras have a wide range of taking rates, from time-lapse systems which may be as slow as even a few frames per hour up to the super high-speed cameras of scientific research which operate at a rate of many millions of frames per second. Whatever the taking rate, the projection rate is normally 16, 24 or 25 fps. The camera thus can change the apparent time scale of the subject being studied, turning it from a speed too slow or too fast to be followed into a normal understandable sequence.

BASIC PRINCIPLES

The test of an effective time change is that, on projection, the action should look ordinary and be easily followed. In still photography, a useful practical formula to decide the exposure time necessary to arrest action is

$$T = \frac{L}{500} \times V \text{ seconds}$$

in which L is the width of the field of view and V is the subject velocity or L is the frame width and V is the image velocity, keeping dimensions in common units.

IMAGE VELOCITY. In normal cine-photography and cine-projection an average frame exposure time is 1/50 sec. Putting this in the formula and rearranging it to find V in terms of L gives:

$$V = \frac{L.50}{500} = \frac{L}{10} \text{ per second}$$

This may be interpreted as saying that the image velocity should be such that the object crosses a full frame in not less than 10 sec. The result errs on the slow side but is near enough and the formula can easily be used to find the maximum camera speed to make the event on projection move slowly enough on the screen for easy study.

As an example, with a field of view of 3 ft., a shuttle in a loom, moving at 100 ft. per second, would cross the field in 3/100 sec. To make this, on projection, equal to 10 sec. implies a camera speed of 1000/3 or 333 fps. It is likely that 100 fps would do but it is certain that a normal cine-sequence would be quite inadequate.

EQUIPMENT

INTERMITTENT ACTION. Cine-cameras for normal speeds use intermittent action. They hold the film still during exposure and move it on one frame between exposures. Many standard cameras can operate at a variety of speeds but rarely at more than 64 or 100 fps. When pictures are taken just a few times faster than they will be projected, the effect is one of slow motion, but equally it is the beginning of high speed. The best cine-cameras are precision-engineered, mainly to give good registration. To reach higher taking rates, the precision is aimed at good film transport, particularly to alleviate stop-start stresses.

The limiting factor with intermittent action is the ability of the film to withstand these stresses, and the maximum rates are about 300 fps with 35mm film and 600 with 16mm. Beyond these rates the film breaks. Cameras of this kind are often referred to as photo-instruments and are associated with other cameras taking pictures on command, that is, the camera takes one picture each time it is pulsed. Both types are the backbone of photo-instrumentation in missile research.

CONTINUOUSLY-MOVING FILM. Cameras with higher taking rates are possible with moving film if that motion is continuous and not intermittent. Again, a very high engineering standard is demanded and film can successfully be driven at speeds approaching 200 miles per hour. For example, 16mm film can reach 250 ft. per second, which, at 40 frames per foot, means 10,000 full frames per second.

As the film moves continuously, it is in motion during the exposure, causing image blur. A way to avoid this is to eliminate relative motion of image and film, i.e., the image must be made to move with the film and at the same velocity. One solution is as follows.

Light passing through a parallel-sided block of glass is refracted, except for rays at normal incidence. A block of this type is placed between a camera lens and its image and rotated, thus deflecting the image. Over a small but adequate rotation angle, the image shift is approximately in a straight line and at a uniform velocity directly related to the rotational velocity of the block. If the block is not too thick, spectral dispersion is negligible.

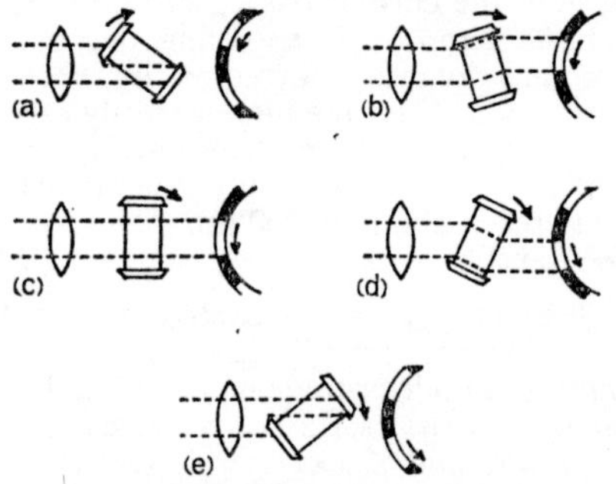

ROTATING PRISM COMPENSATION. (a) Top and bottom of prism act as shutter to keep image off film during inoperative part of prism revolution. (b) Prism transmits image to film on right, refraction holding image stationary on film in successive positions (c) and (d). Ends of prism again shutter film.

In theory this technique is full of approximations, but in practice their effects are sufficiently small to allow pictures of acceptable quality to be taken.

The block is usually called a prism, and has four or eight sides; it is driven by gears in synchronism with the sprocket wheel moving the film, and at high speeds the film take-up spool often has a separate drive. The presence of the block limits the choice of camera lenses and also the maximum usable aperture. Some designs have a relay lens system which allows a wider choice of objectives but again limits the

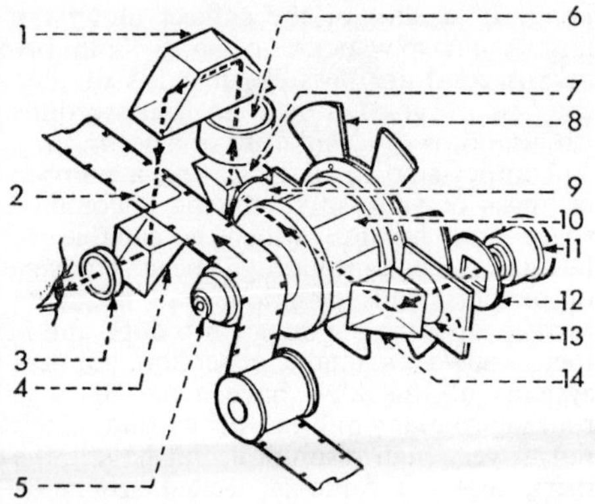

ROTATING PRISM SYSTEM WITH RELAY OPTICS. Lens, 11, backed by aperture mask, 12, forms image on film, 2, through relay system comprising first field lens, 13, first prism, 14, compensating prism, 10, second field lens, 9, second prism, 7, relay lens, 6, and penthouse prism, 1. Film is driven by sprocket, 5, and viewfinding is through eyepiece, 3, and finder prism, 4. This design enables multi-blade shutter, 8, to be combined with multi-face prism, 10.

maximum aperture. A recent design has two sprocket wheels which improve registration.

RIBBON FRAME AND STREAK CAMERAS. Maximum taking rate is determined by film strength and is 10,000 fps at full 16mm frame, double at half frame height, and double again at quarter frame height. This latter is often called ribbon frame and has limited use as it is only satisfactory for a subject with one dimension considerably longer than the other. If the direction of motion is predictable, any missile would make a satisfactory subject. The ultimate step in this reduction of frame height is to use a slit as in a streak camera, giving a race-camera-type record, but known in missile circles as ballistics-synchro. In this case, there is no need to use the rotating prism. Adding a slit near the film plane with the prism in use reduces the frame exposure time. The slit then operates as a focal plane shutter, but time distortion must be considered.

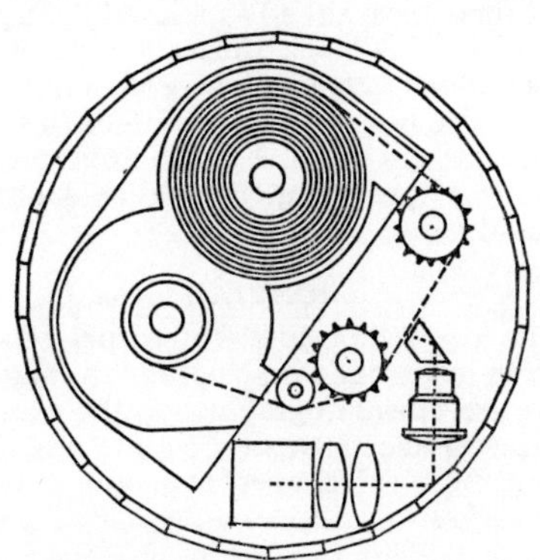

TYPICAL MIRROR-DRUM CAMERA. Image-to-film motion is compensated by small mirrors arranged on inside of rotating drum, within which camera mechanism is laid out. Image size and frame rate are altered by interchanging mirror drums: maximum speed, 40,000 fps, frame size, 4×4·5mm; or 2,000 fps, frame size, 18×22mm.

ROTATING MIRROR CAMERA. If the film is wrapped round a drum to give it support, its length is limited to the circumference of the drum, and in general this method gives insufficient speed gain to make it worth while.

The preferred mechanical solution for extremely high speeds with a limited number of exposures is the rotating mirror camera with optical compensation. The film is fixed

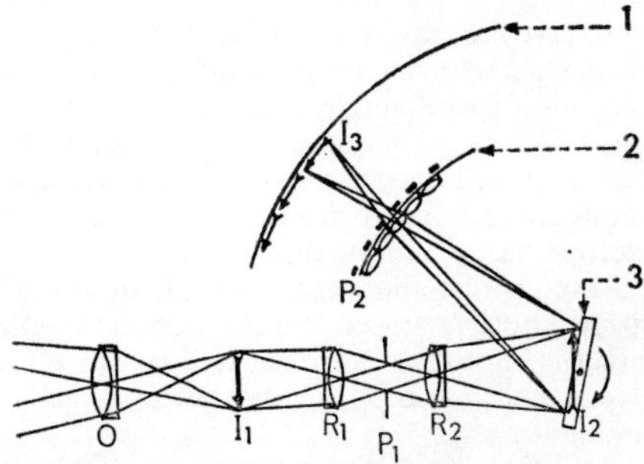

ROTATING MIRROR COMPENSATION. Objective lens, O, produces virtual image, I_1, transferred by relay lenses, R_1, R_2, to plane of rotating mirror, 3, at which real image, I_2, is formed. Arc of fixed lenses, 2, re-images object at I_3 on fixed band of film, 1, as mirror sweeps image through arc. P_1, P_2, stops Mirror rotates at upwards of 2,000 revs./second.

on an arc and is stationary during exposure. The image-forming beam is reflected to the film from a mirror which can rotate about an axis parallel to its surface. A light beam reflected in this way could, at some distance after reflection, have an apparent sweep velocity of any magnitude, even exceeding that of light itself. It would therefore appear that the largest cameras would be the fastest. This seeming paradox is limited by the nature of light in that the image detail possible is set by diffraction.

The light is moving along the film and so some image compensation is needed. If there is a real image formed at or very near the mirror surface, and if this is re-imaged on to the film by an arc of lenses, each of those lenses will produce a reasonably stationary picture on the film. Relative to each of these lenses there is no lateral image motion, only a motion of the light beam across the lens apertures. This allows the width of the lens apertures to be used as a control of exposure time of each frame.

A typical camera of this kind can produce 30 pictures of 20mm diameter at 2,000,000 pictures per second or up to 120 pictures of 8mm at 8,000,000 pictures per second.

All these cameras were developed either for explosive studies or for plasma physics research. In both cases the subjects are very bright and self-luminous. However, cameras of this kind have been successfully applied to non-luminous events by producing short-duration, high-intensity light sources, usually either high-voltage electronic flashes or explosive flash-bombs, and there has been some very fine colour work done in the explosive fields around 1,000,000 frames per second.

MULTI-CAMERA METHODS. There have been various systems in which a battery of very high speed single shot cameras make a short but rapid sequence of exposures, but unlike those previously mentioned, they produce too few frames to make a satisfactory motion picture sequence.

Such single-shot cameras include high-speed shutters based on the magneto-optic effect (Faraday) and electro-optic effect (Kerr cell) and the photoelectric effect (image tubes).

Additionally there is the Cranz-Schardin technique which combines an array of cameras with a similar array of spark light sources and an optical arrangement such that each camera uses only its own spark. This has been most satisfactorily employed in missile studies, e.g., bullets and shells.

RACE-TRACK, STREAK AND IMAGE DISSECTION. An interesting technique is that used in race-track cameras or photo-finish systems. A narrow sector of space (the finishing line) is continuously recorded on a moving film, and a horse or runner is photographed crossing the line virtually a slice at a time. If the film and subject motions are well chosen, the result is a strange single-shot picture in which one dimension only is actually space, the other dimension in the direction of film motion being more truly considered as time.

The same principle is used in streak cameras by high-speed photographers, for example, using a rotating mirror to move an image of a slice of the subject along the film. As image velocities of up to 40mm per microsecond are possible, instants of time can be resolved down to nanoseconds (thousandths of a millionth of a second).

Another step involves having a number of slices or slits, so producing a complex space–time mixture which is capable of being re-formed into a high-speed sequence of pictures. This technique can be taken further, dividing the image into dots, and is then known as image dissection. Optical systems of this kind have achieved rates running to many millions per second, albeit not at very high resolution, but having the merit that the cameras needed are much more economical to produce.

IMAGE CONVERTERS AND INTENSIFIERS. These are developed as super high-speed cameras, and models exist producing very short sequences at rates of many millions per second. However, they can have two other advantages for the cine-photographer or television engineer.

A simple image tube can have a photo-cathode and a phosphor such that an input of one radiation comes out as a different radiation: for example, the photo-cathode could respond to infra-red while its phosphor glows in the visual. This is an image converter and so can make any cine camera into, say, an infra-red cine camera.

More complex image tubes can have a number of stages with secondary emitter couplings such that from input to output there is a considerable electron gain and a resulting image intensification. These image intensifiers have already achieved light gains as high as ×10,000, thus allowing photography of subjects previously too faint to be filmed. The best demonstration of this with a cine camera is of cine photography by moonlight produced by Oude Delft in Holland.

LIGHTING

Up to a few thousand frames per second, lighting techniques are basically the same as in normal cinematography and the depth of field required can usually be quite restricted, permitting high lens apertures, often between *f* 4 and *f* 2.

LAMPS. At higher speeds, the extra lighting needed to counteract the shorter exposures could be difficult to produce except that the subject size is normally not large. It is rare that large objects have such high velocities in relation to their size that really high taking rates are needed. Modern improvements in tungsten lamps, e.g., the compact projector lamps with integral reflectors and the tungsten-halogen lamps, have proved quite valuable. The best single lamp for such duties is the pulsed xenon arc lamp, which can produce enough light to permit photography of a subject area of one foot square at 20,000 fps in black-and-white, and 10,000 in colour.

FLASHBULBS. A technique that has also been used for illuminating large areas is to arrange a battery of large photo-flash bulbs to be fired in a predetermined sequence. The bulbs can be arranged so that the illuminated area moves along with the action.

ELECTRONIC FLASH. At the ultra-high speeds, the events are mainly self-luminous, but when they are not, the usual auxiliary source is electronic flash. Considerable expertise exists in electronic flash circuit

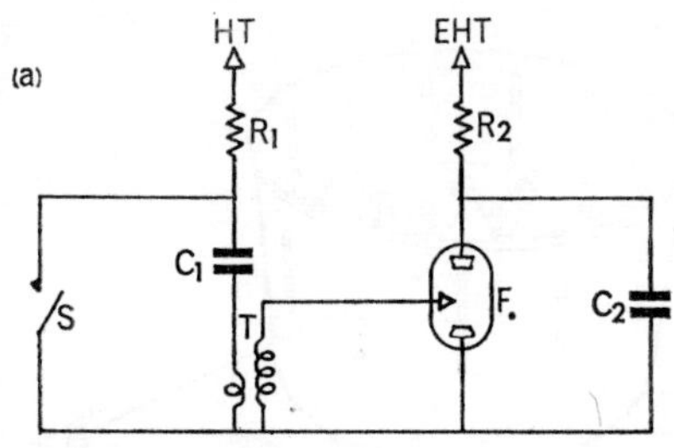

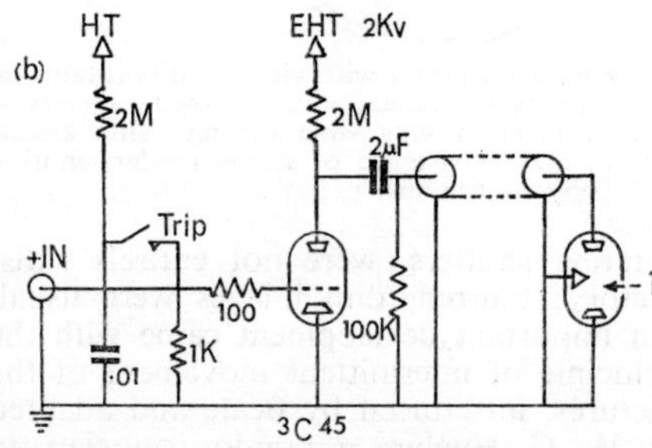

BASIC FLASH TUBE CIRCUITS. (a) Using trigger transformer, T; (b) using thyratron switch to permit use of much higher voltages than flash tube can withstand alone. Low-voltage tubes can be used to give exposures of a few microseconds in place of hundreds of microseconds. 1. Flash tube.

design to produce flashes essentially level in output for many different durations. The only other source used, especially for larger subjects, is the argon flash bomb, which utilizes the intense radiation produced as an explosively generated shock wave passes through argon in a transparent chamber.

SYNCHRONIZATION. At high speeds, the total camera-recording time is quite brief; even with cameras using 100 ft. of 16mm film, all this film can be used in less than a second. The cameras with stationary film are even more time-restricted, for example, 120 frames at 8,000,000 per second occupy only 15 μsec. It is therefore necessary to ensure that the event, camera and lighting all operate at the same time. These speeds are beyond human operation and so a variety of electrical and mechanical devices is used. Generally the camera produces some signal to initiate the event and, at the highest speeds, the events are generally amenable to electrical triggering. Alternatively, if the event must begin as soon or sooner than the camera, then delay circuits are arranged to start each at the correct instants to result in synchronism at the critical periods.

The framing rate is not always predictable and often varies during the run. So that the actual rate can be known, timing marks are made on the film by argon or neon lamps flashed on for very brief instants at repetition rates set by reliable crystal oscillators. These lamps can also be additionally pulsed to indicate the event start or at some known event instant.

In ultra-high-speed systems (mirror cameras), a light beam is reflected by the mirror to photo-multiplier pick-ups so that the mirror attitude, and hence the image position, are known. So long as the mirror is reasonably close to a preselected rotational speed (indicated by a tachometer) when the firing switch is closed, appropriate delays can be inserted between the photo-multiplier pick-ups and the event initiation.

When neither camera nor event can be synchronized to the other, two approaches are possible. One is to make numerous attempts, hoping that some will be successful – this, in desperation almost, has worked in nearly impossible situations. The other is to have a camera with what is called continuous access. This means that it always has an image being recorded somewhere in the system, as in a mirror camera with sufficient mirror faces so that there is always an optical path from objective to film. The task then is to ensure that light enters the camera only during the event period. As the light source is generally electrical or electronic, this is quite practical. G.H.L.

See: *Camera* (FILM); *Continuous motion prism; Time lapse cinematography.*

Books: *High-Speed Photography*, by R. F. Saxe (London), 1966; *Photographic Recording of High-Speed Processes*, by A. S. Dubovik, ed. G. H. Lunn (Oxford), 1968.

⊕HISTORY

Throughout recorded history, man has attempted to reproduce the world of movement. In its earliest form, the motion picture was a shadow of moving hands cast by firelight, perhaps upon a cave wall; in the East, shadow plays using silhouettes of manipulated jointed figures thrown on a screen by lamplight have been presented for thousands of years. The first step towards the modern motion picture came with the introduction of the magic lantern in the 17th century; by the early 19th century, lantern slides animated by gears, levers or wires were used to project simple moving images.

EARLY ANIMATION EXPERIMENTS

The phenomenon of persistence of vision, on which the cinema is based, although recognized for centuries, was first studied in the early 19th century; P. M. Roget, the English mathematician, in December 1824, demonstrated the principles and some of the effects of this optical illusion. Soon toys appeared using the effect: J. A. Paris, an English doctor, marketed the Thaumatrope in 1826 – a circular card carrying a picture on each side. When this was rapidly rotated on its axis, like a coin spinning on its edge, persistence of vision caused the two images to superimpose.

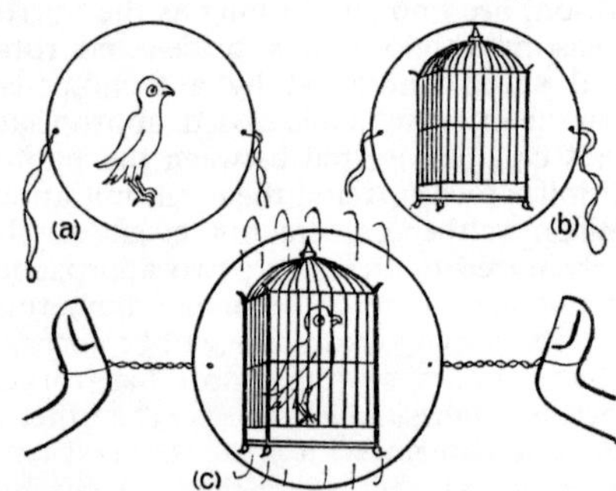

THE THAUMATROPE. Toy demonstrating persistence of vision, devised by J. A. Paris (1826) on basis of P. M. Roget's experiments (1824). (a) Bird on one side of card. (b) Cage on reverse. (c) Twirling of card reveals bird in cage.

From this demonstration that two pictures could appear to occupy the same place at the same time, developed the experiments of the Belgian scientist J. A. Plateau; in 1829 he laid down the theory of the phenomenon, and then introduced in 1832 the Phenakistiscope – the first instrument to animate a series of pictures through persistence of vision. A circular card, with slots round its circumference, carried between the slots a series of animated drawings. When the card was held in front of a mirror and rotated, the slots appeared to merge into one; the reflected pictures seen intermittently through the slots gave the impression of movement. An identical device was simultaneously introduced by Simon von Stampfer as the stroboscope.

A later development of this principle by W. G. Horner, the English mathematician, led to the introduction of the Zoetrope – a slotted cylinder in which a band of pictures was animated in a similar fashion. Molteni, the French optician, and others devised projection versions of this principle for the magic lantern, using transparent discs and rotating slotted shutters. The rapidly revolving pictures, intermittently viewed through

THE ZOETROPE. Drum with viewing slits rotates on vertical axis and contains card showing phases of action. Brief viewing time through slits arrests action, and persistence of vision renders motion seemingly continuous.

rotating shutters, were not entirely satisfactory; blurred, dim images were usual. An important development came with the principle of intermittent movement of the pictures, introduced by Beale and adapted by W. C. Hughes, a London optician, in 1884 in the form of the Choreutoscope, which in a magic lantern was used to project a series of stationary pictures in quick succession, each picture being changed for the next while covered by a moving shutter. The modern film is based upon this principle.

An alternative solution to the problem was devised by Emile Reynaud, the French showman, artist and inventor, in the Praxinoscope, in which a mirror drum was used to stabilize optically the images carried on a

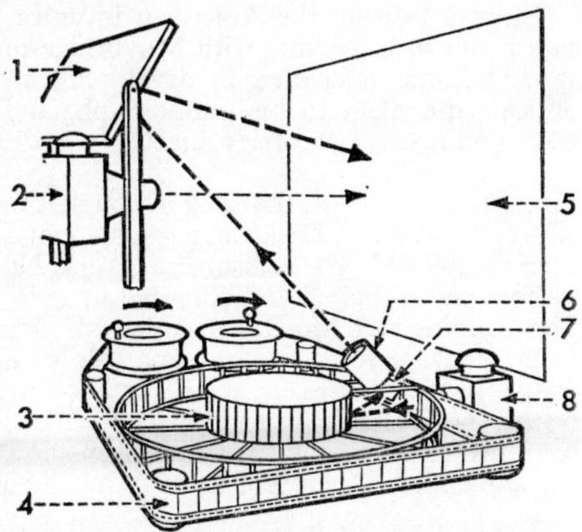

REYNAUD'S THÉÂTRE OPTIQUE (1892). Long band of film, 4, wound on spools and carrying animated images, is perforated to mesh with projections on large framing wheel spoked to mirror drum, 3. In operation, light from lantern, 8, passes through film, is reflected from facet of mirror drum, and again from fixed mirror, 7, and is focused by lens system, 6, through large mirror, 1, on to screen, 5, in front of audience. Fixed slide projector, 2, supplies backgrounds to animated action.

continuously rotating strip, a principle similar to that employed in modern animated viewers for film editing. A more elaborate version of this apparatus, using a long, flexible, perforated, transparent strip which carried hand-painted pictures, was used by Reynaud in the Théâtre Optique from 1892 to 1900. Reynaud was thus the first person to exploit commercially projected animated pictures – but hand-painted pictures were used, giving a result not unlike the modern cartoon film.

PHOTOGRAPHIC EXPERIMENTS. Although photography had been an accomplished fact since the 1830s, it was twenty years before the wet collodion plate made instantaneous exposures possible. As a result of the introduction of the gelatin dryplate, several experimenters, notably W. Donisthorpe, an English barrister, in 1876, proposed and patented methods of exposing plates in rapid succession to analyse movement, the results being viewed in a Zoetrope. In 1878 Donisthorpe suggested the combination of his apparatus with the recently invented phonograph of the American inventor, Thomas Edison, to record and reproduce dramatic productions.

Scientists began to employ chronophotography – the recording of photographs at regular intervals – to study movement. Pierre J. C. Janssen, a French astronomer, in 1874 constructed a photographic revolver with which the transit of Venus across the sun was recorded in a series of photographs around a circular plate.

Eadweard J. Muybridge, an English photographer working in California, pioneered the use of chronophotography in the study of human and animal movement,

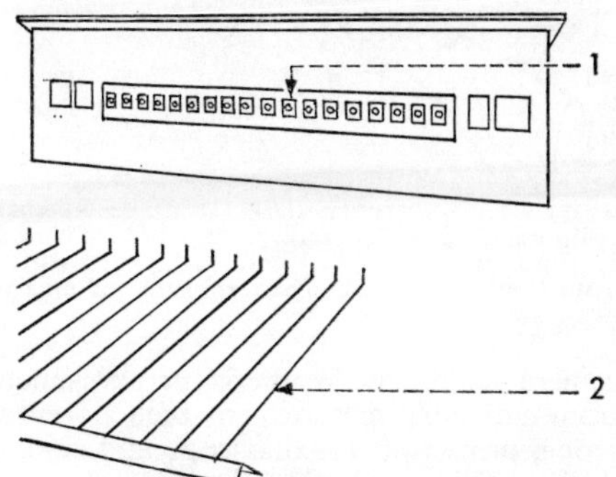

EARLY MUYBRIDGE EXPERIMENTS. First demonstration of the action of a running horse. About 1880, Muybridge was using a battery of some 24 still cameras, 1, triggered by a series of trip wires, 2, laid on the surface of a race track at Palo Alto, California, now part of the campus at Stanford University. In the early experiments, the camera shutters were released by breakage of the trip wires by the horse's hooves. This uncertain method was later replaced by a more sophisticated electrical contact system, using a rotating commutator wheel to release the camera shutters electrically at precise intervals.

using a battery of cameras exposing in rapid succession; between 1878 and 1890, with his published photographs and lectures using projected animated photographs, he made a major contribution both to the science of movement and to the development of the moving picture. A similar system of cameras was used by Ottomar Anschütz in Germany from 1885 onwards.

Muybridge's cumbersome battery of cameras was improved upon by Professor E. J. Marey, a French scientist investigating animal movements. He first devised a photographic gun in which twelve pictures could be recorded on a circular plate in about one second; by the mid-1880s, flexible paper roll films became available, and he devised in 1887 a camera, using this material, which for all practical purposes was the first motion picture camera as we know it.

COMMERCIAL APPLICATION

By the late 1880s, the two components of the motion picture process had been demonstrated: the recording of successive photographs of movement – as with Marey's

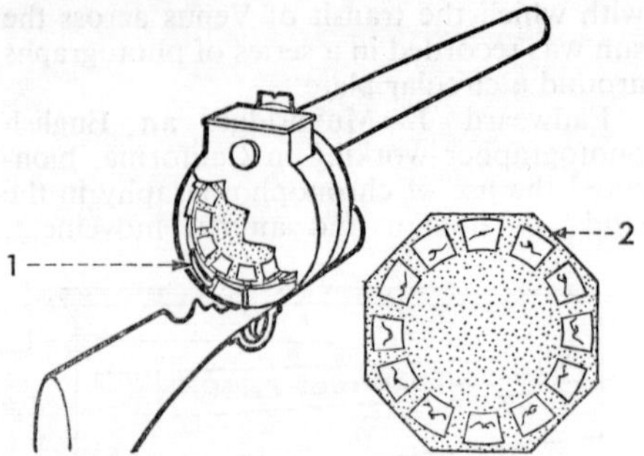

MAREY'S PHOTOGRAPHIC RIFLE (1882). 1. Ratchet, positioning areas of sensitized plate in turn behind aperture on release of trigger. 2. Sensitized glass plate showing 12 successive exposures of bird in flight.

camera – and the synthesis of movement from individual pictures – by the Choreutoscope, projection Praxinoscope and similar devices. These two principles had only to be combined in a commercially workable apparatus for cinematography to be an accomplished fact.

The work of Marey and Muybridge, together with the introduction of celluloid roll film in 1889 by the American inventor George Eastman, gave a stimulus to the work of a number of inventors in several countries. In Britain, L. A. A. Le Prince, a French chemist and artist (1888), W. Friese-Greene, a photographer (1889), W. Donisthorpe (1889) and F. H. Varley, an English civil engineer (1890), worked on and patented experimental apparatus, none of which was successfully demonstrated.

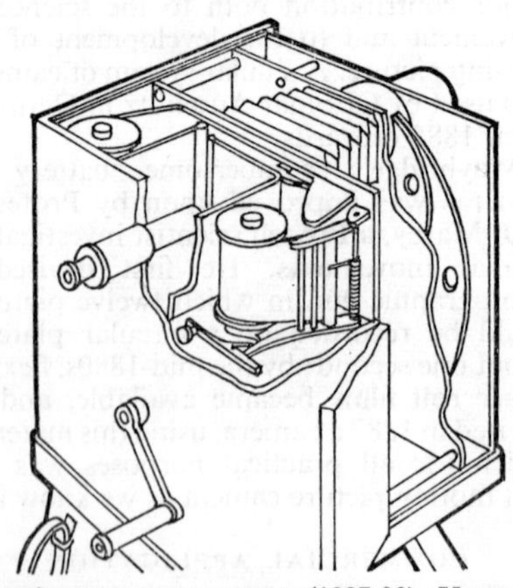

MAREY'S FIRST FILM CAMERA (1887-90). Unperforated film is drawn horizontally across aperture, guided by spring-loaded rollers. Very brief exposure through holes in contra-rotating shutters. View-finding through back of film.

Thomas Edison, the American inventor, as a result of a meeting with Muybridge in 1888, became interested in developing an optical equivalent to his phonograph, and instituted research on the project.

THE KINETOSCOPE. Largely through the efforts of W. K. L. Dickson, a young Scotsman employed by Edison, a workable camera was designed and patented in 1891. The machine used film 35mm wide with four rectangular perforations on each side of a frame $\frac{3}{4} \times 1$ in. in size, at the rate of 46 frames per second. The films so produced were presented in coin-operated peep-show viewers called Kinetoscopes, each machine

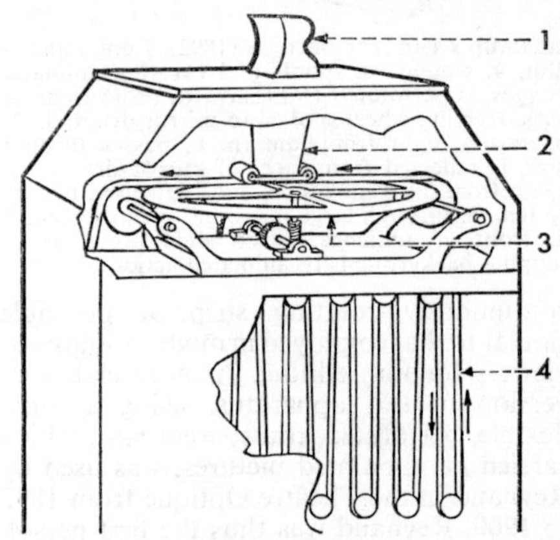

EDISON'S KINETOSCOPE (1893). Machine for viewing by a single person, looking down on to the film image through aperture, 1. A 56-ft. continuous band of film was pulled with a steady motion through a reservoir chamber, 4, and passed beneath a magnifier in the tube, 2. A revolving disc, 3, had a perforated rim which, in this model, was interposed between the film and the light source (not shown), thus very briefly illuminating each frame, so that it appeared to be stationary.

showing only to one person at a time. About 50 ft. of film circulated in an endless loop under a viewing lens, each frame briefly illuminated by a flash of light through a rotating shutter. Edison's Kinetoscope was shown publicly in 1893, and was the first commercially successful motion picture system.

THE CINÉMATOGRAPHE. The Kinetoscope's appearance and subsequent success stimulated inventors in other countries, notably the Lumière brothers, Auguste and Louis, French inventors who in 1895 designed, produced and demonstrated the Cinématographe – a combined camera, printer and projector with which they gave the first public demonstrations of projected motion pictures at the Grand Café in Paris on 28th December 1895. The apparatus, using a claw mechanism for the first time, was small, portable and easy to handle, and was in use all over Europe by the middle of 1896.

LUMIÈRE'S CINÉMATOGRAPHE (1897). The camera, when used as a projector, is set up in front of an arc lamp. After projection, the films, 17m in length, fall into a bin from which they are afterwards rewound. Spare films on shelf below mechanism.

THE KINETIC LANTERN. In Britain, photographer Birt Acres developed his Kinetic Lantern – a combined camera and projector – in 1895, and became the first Englishman successfully to demonstrate projected moving pictures, although his presentations were to private bodies. Robert Paul, a London instrument maker, designed and marketed his Animatographe cameras and projectors early in 1896 using a form of Maltese cross as intermittent mechanism. Workable apparatus was patented and demonstrated at this time by, among others, W. Latham, C. F. Jenkins and T. Armat in America, M. Skladanowsky in Germany and G. Demeny and H. Joly in France. The age of the motion picture had begun.

STANDARDIZATION. After the early years, in which a large number of film sizes and widths were in use, standardization was finally agreed by an international conference in 1909, which adopted the Edison 35mm width and perforation disposition. Except for minor modifications, it has remained unchanged until the present day.

There has been relatively little change in the basic principles of camera and projector design. The Maltese cross intermittent mechanism used by the English instrument maker R. W. Paul and others is still almost universally employed in cinema projectors; the claw mechanism used by the Cinématographe is found, in various forms, in most cameras. Until the coming of sound in the late 1920s, most cameras were hand-cranked, although clockwork, electric and even compressed-air drive had been utilized – the latter in the Aeroscope camera, 1912, devised by the Polish scientist, Kasimir de Proszynski. The need for constant speed with sound films brought with it universal use of electric or clockwork drive, and a standard running speed of 24 frames per second replaced the previous variety of speeds from 15 to 25 frames per second.

Camera lenses improved in quality and light-gathering capacity; variable focal length (zoom) lenses with a 4 : 1 ratio were first commercially used in the early 1930s. Wide-screen developments in recent years have required great increases in brightness of illumination, and the carbon arc lamp, universal since the earliest days, is beginning at last to be replaced by xenon enclosed arcs, most efficient and simple to operate.

SOUND

Attempts to accompany pictures with sound had been made by many of the pioneer inventors, notably Edison, some of whose Kinetoscopes in 1895 were fitted with synchronized phonographs. Several processes for electrically or mechanically synchronizing sound and pictures were patented before 1900, during which year short films of music hall turns were made to accompany records. By 1910 the English film producer C. Hepworth's Vivaphone system, a horn gramophone coupled to a film projector, had had some theatrical success. The Gaumont Chronophone system in 1911 used mechanically amplified, synchronized discs to fill a large auditorium with sound.

SOUND-ON-FILM. Mechanical methods of sound recording and synchronization were inherently unsatisfactory; sound-on-film recording was to provide the answer. Eugene Lauste, the French inventor, was among the first to achieve success in this field; by 1904

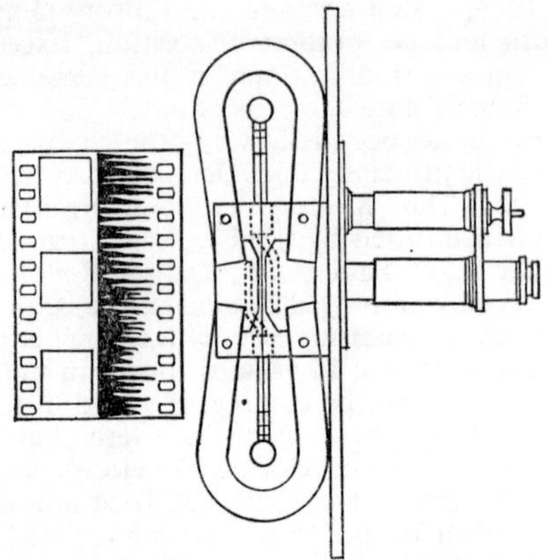

LAUSTE'S SOUND FILM SYSTEM (1912). Combined 35mm sound and picture, in which restricted picture area gave space for large sound modulations recorded by mirror galvanometer with built-in microscope alignment. Success was frustrated by lack of practical audio amplifiers.

he had recorded both picture and sound photographically on the same film and patented some methods in 1907. He devised sound systems in the years before World War I, but success had to await the development of means to amplify the faint signals generated by his reproducing equipment.

Lee de Forest invented the triode amplifying valve in 1907, and thus opened the way to commercial exploitation of electrically reproduced sound. The Tri-Ergon Company in Germany devised, in 1919, a glow-lamp light modulator for variable-density recording of sound, effectively demonstrated in Berlin in 1922; no film company was prepared to adopt it. With T. W. Case, another American inventor, Lee de Forest devised a modulator lamp for sound recording and with it demonstrated the Phonofilm system in 1923; Case went on with research, developing the Aeo-light modulator which was acquired by Fox in 1926, forming the basis of the Fox-Movietone system, while the American inventor, E. C. Wente, at Bell Telephone Laboratories, devised an electromagnetic light valve which was used in the Western Electric variable density system.

FIRST COMMERCIAL SOUND FILMS It was, however, the Western Electric sound-on-disc system which was first commercially exploited by Warner Brothers as the Vitaphone system, and used in 1926 for one and two reel shorts. Warner Brothers launched sound-on-disc features with *The Jazz Singer*, starring Al Jolson, on 23rd January 1927, while Fox-Case followed with a sound-on-film feature, *Seventh Heaven*, in May 1927. A sound boom began, with a battle during the next few years between disc and sound-on-film recording, won finally by the latter.

In 1928 the Radio Corporation of America (RCA) in collaboration with Westinghouse and the American General Electric Company (GEC) introduced a variable area recording system based on the Pallophotophone system devised by C. A. Hoxie in 1920, and based in turn on the early experiments of the French inventor, E. A. Lauste.

In Europe, Gaumont, with Petersen and Poulsen, introduced in 1928 the GPP system, which used a separate 35mm sound record film, but it was dropped owing to synchronization difficulties. In Germany the Tobis-Klangfilm organization was formed and entered into battle with RCA and Western Electric over the Tri-Ergon patents, a dispute settled amicably in 1930.

Both variable area and variable density optical sound tracks are still in use; greatly improved sound quality has resulted from technical developments such as the RCA system of push–pull variable area recording.

Although many experiments had been made with stereophonic sound recording to accompany films, the first commercially successful demonstration was in 1941 with a system designed for special presentations of Walt Disney's *Fantasia*; a separate, synchronized 35mm optical sound film carried a number of tracks operating separate loudspeakers in the auditorium. However, little development occurred in this medium until the introduction of magnetic recording.

MAGNETIC RECORDING. The earliest magnetic recording was of the kind demonstrated by Louis Blattner in 1929, with steel tape or wire on which sound was recorded in the form of magnetic variations. Little use was made of such methods by the film industry, until their refinement during and after World War II, when paper or plastic tape coated with magnetic iron oxide was introduced. This development was exploited by the film industry in two ways.

First, by coating four stripes of magnetic material on the film carrying the picture, very high quality stereophonic sound tracks could be recorded and replayed in the cinema with no synchronization problems; some current widescreen processes make use of these methods.

Secondly, magnetic recorders are used for virtually all original sound recording, replacing the previous cumbersome, delicate optical sound recorders. Not only does a greatly improved quality result, but a much increased versatility and flexibility is possible. The final sound track is either recorded on magnetically striped prints, as already described, or is transferred to photographic film used to produce conventional optical sound prints. A further development involves the magnetic striping of narrow-gauge films: 16mm, 8mm and Super-8 originals can thus have sound tracks recorded directly by the amateur or small-scale professional with great ease and low cost.

COLOUR

Colour, like sound, has been with the film from the start. Some of the earliest films shown in 1896 were hand-coloured – teams of girls, each allocated one colour, applied it with brushes to the film, frame by frame.

STENCIL METHODS. Hand-colouring became impractical with the increase both in the length of films and the number of prints required. Pathé Frères devised a method of preparing stencils by cutting, from each frame of a print of the film, areas corresponding to a particular colour, up to six stencil prints for as many colours being produced. Each stencil was run in turn through a staining machine, in register with an uncut print, to which colour was applied through the holes in the stencil.

The Pathécolor system, in use from about 1905 until the early '30s, made possible large numbers of elegantly coloured prints in a short time, once the stencils had been produced.

TONING AND TINTING. Less expensive, more general colour effects were achieved by chemical toning, in which the black-and-white image was converted to a coloured image, and tinting, in which an overall wash of colour was applied to the film. Sometimes both methods were combined for a picture. In such processes, the few colours were arbitrarily chosen; for more accurate rendering of colour, photographic methods had to be devised.

COLOUR PHOTOGRAPHY. Although the theory of colour photography had been introduced in 1855 and demonstrated in 1861 by James Clerk Maxwell, the Scottish physicist and mathematician, it was not successful in the cinema until the Kinemacolor process demonstrated in 1909 was launched in 1911, based on the English inventor G. A. Smith's two-colour additive process patented in 1906.

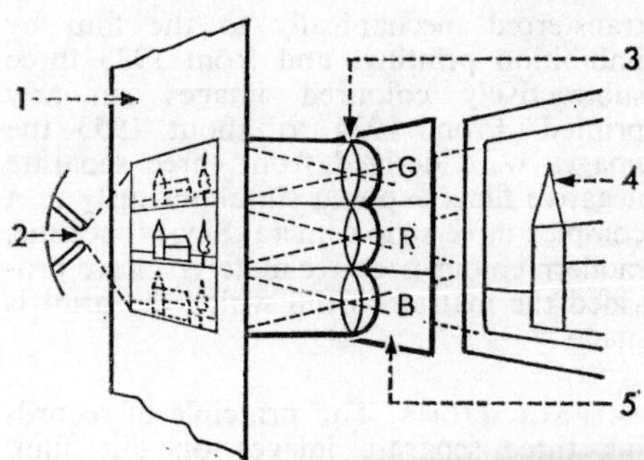

GAUMONT ADDITIVE COLOUR SYSTEM (1912). In the Chronochrome system, three frames on black-and-white film, 1, were effectively colour separation positives, and were projected simultaneously by illumination from arc source, 2, through triple lenses, 3, and red, green and blue filters, 5, to form composite colour image, 4, on screen.

It utilized persistence of vision to combine alternate black-and-white frames coloured by projection through a rotating red-and-green filter, at 32 frames per second, the black-and-white print being derived from a negative film exposed through a similar filter fitted on the camera. Despite limitations, it was commercially very successful.

Subsequently two-colour additive processes simultaneously mixing red and green images (e.g., Raycol, 1928–35) and three-colour processes mixing simultaneously red, green and blue images (e.g., Opticolor, 1934–39) were in limited commercial use. These systems involved much modified cameras and projectors; more convenient additive systems employed lenticular film (e.g., 16mm Kodacolor, 1928–35) or mosaic screens (e.g., Dufaycolour, 1931–47) enabling the film to be used in virtually any apparatus.

Since additive systems involve severe loss of projection illumination (up to 70 per cent being absorbed by the colour filters), they have been entirely superseded by subtractive colour processes in which colour images are formed directly on the film. The earliest forms of this system were two-colour methods; from Kodachrome (1915) and Prizmacolor (1919) to the most successful commercial process, Cinecolor (1932–52), all employed double-coated film carrying two images subtractively coloured.

The Technicolor process first appeared in two-colour form: from 1922 to 1928 using two positive films cemented back to back, from 1928 to 1933 using two coloured images transferred mechanically to the film by imbibition printing, and from 1933 three subtractively coloured images similarly printed. From 1934 to about 1953 the images were derived from three separate negative films exposed simultaneously in a complex three-strip camera. Since that time, modern colour negative materials have provided the matrices from which the print is made.

TRIPACK FILMS. The principle of recording three separate images on one film, although proposed in the last century by the French scientist Ducos du Hauron, was not turned into successful practice until the 1930s. Gasparcolor (1934) was a print process only, using film carrying two light-sensitive layers on one side and one on the other.

The first three-colour tripack film, Kodachrome, carrying three light-sensitive layers on one side of the film, was introduced in 1935 as a reversal film in 16mm and 8mm sizes. As well as appearing as a tripack 16mm reversal film in 1937, Agfacolor was introduced in Germany in 1941 as a negative-positive colour system, and taken up in a number of countries after World War II as the basis for several current colour processes.

A significant advance occurred with the introduction of the Eastman Color materials in 1951, using a system of integral colour masking giving improved colour rendering. These processes are extensively used throughout the world and, as well as resulting in a great increase in colour film production, have encouraged the development of a number of the wide-screen systems in current use.

WIDE-SCREEN

Many wide-screen systems have been devised in attempts to increase the realism of film presentation. There have been four methods:

MULTIPLE FILMS. The French inventor R. Grimoin-Sanson, in 1897, patented the Cinéorama process in which ten linked cameras recorded, and ten projectors reproduced, a 360° picture shown on the wall of a circular room. Although closed down after a first demonstration, due to fire hazard from the projector arc lamps, it was the forerunner of a number of similar processes.

A less ambitious, but more successful method was designed and demonstrated by Abel Gance in 1927. Three linked projectors filled a triple-size screen with a panorama taken by three synchronized cameras, or with a triptych of three different pictures.

The process was used by Gance for sequences of his film *Napoléon*.

The principle was once again revived in 1952 by the American designer Bert Waller as the Cinerama process, in which three cameras recorded a field of view three times wider and half as high again as the conventional camera, the images being combined on a large screen by three projectors, with a fourth film carrying synchronized stereophonic sound. The process has subsequently been changed to a single wide-film presentation.

The Circarama system (1955) of Walt Disney in the USA and the Russian Kinopanorama process (1959), known in Britain as Circlorama, have returned to the circular presentation of Grimoin-Sanson, the former using 11 projectors, the latter 11 pairs of projectors for a double-height picture. The films are shot by the same number of synchronized cameras, and accompanied by multi-track stereophonic sound.

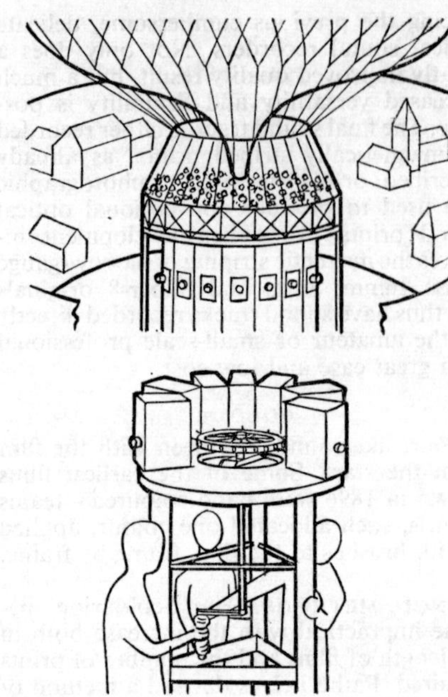

GRIMOIN-SANSON'S CINÉORAMA (1897). Ten coupled projectors (*above*) presented a circular image to spectators assembled in model of balloon nacelle overhead. Ten coupled cameras (*below*) were to be operated over Paris in nacelle of real balloon. First unsuccessful attempt at total cinema.

SQUEEZED PICTURES. A more convenient alternative to multiple films is the anamorphic lens, which compresses laterally a wider image on to standard film, with subsequent expansion of the distorted image on projection. Patented by the French scientist Henri Chrétien in 1927, the anamorphic lens system was first commercially demonstrated in the Fulvue process (1930) recording a double-width scene compressed into a standard frame. It appeared at a time of waning interest in wide-screen processes and was not then adopted. Chrétien demonstrated his system on a screen 33 × 297 ft. at the International Exhibition in Paris in 1937, using two anamorphic projectors combining two panoramic images.

The process was revived by 20th Century Fox as Cinemascope in 1953. Similar lenses are employed in a number of current systems.

MASKED IMAGES. An even simpler panoramic picture process involves masking the standard frame, with an original ratio of 1 : 1·33, so as to change the picture shape to longer formats, up to 1 : 2 in ratio. Shorter focal length projection lenses then enlarge this image on a wider than normal screen. A format of about 1 : 1·66 is now generally adopted.

Variable focal length projector lenses to increase the screen size during the running of the film, as in the Magnascope system, were used from 1926 for sequences in *Chang* and other films. Variation in the shape of the screen image has been tried from time to time. Fixed or variable masks were often employed to change the picture shape of silent films.

A more sophisticated process was devised by Glenn Alvey in 1955, with the dynamic frame system in which the picture format could be altered continuously to accompany changes in subject or dramatic content. Demonstrated in *The Door in the Wall*, the process was not taken up commercially.

Wide-screen processes, with their big increase in screen size, put a heavy premium on image quality. One answer to this problem is the employment of a negative larger than normal, printed by reduction to a standard format. Proposed in 1930, the principle was commercially exploited as VistaVision (1954–59) in which a horizontally running, double-frame 35mm negative film yielded standard reduction prints or contact prints for special large-scale representations. The Technirama process uses a a similarly disposed negative, but with slight squeezing of the image. The negative can then be printed to make fully squeezed or unsqueezed prints in any film size, from 16mm to 70mm.

An alternative system, Techniscope, introduced in 1964, employs a conventional 35mm negative film, on which unsqueezed images of half the normal height are recorded. Printed through an anamorphic lens system, conventional squeezed prints are produced from this wide format image.

WIDE FILMS. To fill a larger than normal screen, a wider film with bigger images can produce improved quality. In the early years of the cinema, wide films were the rule rather than the exception. G. Demeny (1896), the French pioneer of cinematography, used 60mm film, the Biograph (1897) 68mm film, while Lumière devised a 75mm film system in 1898, demonstrated on a 65-ft.-wide screen at the Paris Exposition Universelle of 1900.

After the standardization of film width in 1909, wide films did not reappear until 1929, when several new film sizes were demonstrated, including Magnafilm (56mm), Vitascope (65mm) and Fox Grandeur (70mm).

They were not widely adopted at the time, and as a trade journal commented: "This novelty, therefore, is to be tucked away, to be brought forward when the industry needs a fresh stimulant." This need arose in the early 1950s when, in addition to the previously described wide-screen methods, wide films made a comeback.

An early but unsuccessful example was Cinemascope 55 (1954), with a 55mm negative film to produce reduction prints on 35mm stock. The Todd-AO process (1955) employs a 65mm camera film, with prints on 70mm stock with stereophonic sound tracks. Several wide-screen systems use 70mm prints, which give large screen images of high quality.

STEREOSCOPY

STEREOSCOPIC FILMS. A further aid to realism involves methods of presenting pictures in relief. Two cameras (or a twin-lens camera, or a single-lens camera with a beam-splitting device) produce two images, recorded with a separation approximately the same as that of the human eyes. These two images must be presented by means which allow each eye to see only the picture appropriate to it, when they fuse to create an illusion of depth and solidity in the image on the screen.

ANAGLYPHS. The pictures may be projected in the anaglyph process, first proposed by A. d'Almeida in 1858 and demonstrated for motion pictures by C. Grivolas, the French industrialist, in 1897. The two images are coloured, usually red and blue-green, by dyeing or projection through filters. The audience views the superimposed pictures through filters complementary to those of the images, the red image being visible only through the blue-green filter, and vice versa. This process was commercially exploited in 1925, and later in 1935 as Audioscopics. The latter films were revived in 1950 as Metroscopics, but as the system cannot be used with colour films, it was quickly superseded by methods employing polarizing filters, which have no effect on colour rendering.

POLARIZING FILTERS. Such methods, although described in theory in the 19th century, were not commercially practicable until the introduction of inexpensive polarizing materials in 1932.

The use of Polaroid projection and viewing filters to ensure that each eye sees only the image appropriate to it was first demonstrated at the New York World's Fair in 1939.

Impressive colour stereoscopic films with stereophonic sound were demonstrated at the Telekinema at the Festival of Britain in 1951. The following year, a number of 3D feature films were produced in the United States; they enjoyed a brief popularity, but the problems of synchronization and registration of both images on the screen, the need to run two projectors simultaneously, and the complication for the audience of wearing viewing spectacles led to their abandonment in favour of wide-screen techniques.

PARALLAX STEREOGRAM. An alternative form of presentation removes the need for the audience to wear analysing spectacles. This is the parallax stereogram, in which the pictures are projected on a screen fitted with many vertical elements – wires, slats or embossed lenticles – so placed that each eye sees only the image intended for it. Suggested by A. Berthier in 1896, later developed by E. Estanave and the Americans F. E. and H. E. Ives, it has been used in a workable system by S. Ivanov, from 1941 onwards, in a Moscow cinema and later in other Russian cities. The system has commercial drawbacks, in that seating is critical and audience number are limited.

Further large-scale showing of stereoscopic films is unlikely until some major breakthrough simplifies the problems of viewing and projection.

AMATEUR CINEMATOGRAPHY

For economy and convenience, most amateur film processes have used film widths narrower than 35mm. One of the first systems designed for the amateur was the Birtac camera-projector of 1899; devised by Birt Acres, it used 17·5mm stock produced by splitting standard film in half. In the same year, the Biokam appeared, using 17·5mm film with a single perforation between each frame. The Gaumont Chrono-de-Poche camera employed 15mm film, while for projection only, the Pathé Kok apparatus marketed in 1912 used 28mm safety film prints reduced from 35mm productions.

16MM AND 9·5MM. Two amateur film systems were introduced in 1923: Eastman Kodak with a 16mm reversal safety film process and Pathé with the 9·5mm gauge, providing the opportunity not only to make home movies, but also to hire films for home cinema shows. The two systems ran side by side until World War II. Since then 9·5mm has virtually disappeared; 16mm, on the other hand, while declining as an amateur medium, is more and more used professionally in publicity, educational, industrial, propaganda and television film work.

8MM AND SUPER–8. To reduce further the expense of home movie-making, Eastman Kodak marketed the 8mm gauge in 1932. A 16mm film is run twice through a camera in which a series of pictures is exposed in turn down each half of the film width. After processing, the film is slit down the middle to produce an 8mm film for projection.

Since World War II, the number of amateur film-makers using this gauge has tremendously increased. In 1965, the Super-8 process was introduced, incorporating a number of improvements over regular 8mm, including a 50 per cent increase in picture area and provision for steadier, brighter pictures, together with greatly simplified loading and operation of both cameras and projectors. These developments, coupled with highly automated systems of exposure control, are certain to bring about even more widespread use of film-making by eliminating or reducing many of the former technical complications.

SOUND. Optical sound tracks were applied to 16mm professional production in 1930. The introduction of the magnetic striping process, first in 16mm in 1954, later for 8mm and now for Super-8 films has increased the amateur production of sound films, as good quality synchronized sound can be readily combined with the film.

Striking developments in the manufacture of 8mm film stock, particularly in colour, together with modern high-efficiency projector lamps and lenses, have made possible large high-quality screen images previously only practical with the much more expensive 16mm gauge; further development in this direction can be expected.

ANIMATION

The cartoon film, in a sense, antedates the cinema, in that the Phenakistoscope, Zoetrope and Praxinoscope, among others, all made use of animated drawn images. Among the earliest true cartoon films were those of Emil Cohl, the French cartoonist and photographer, working with Gaumont in 1907. His films, like those of his contemporaries, the American artists J. Stuart-Blackton and Windsor McCay, were very simple in style and execution. With at least 16 pictures to be drawn for each second of screen time, this was a necessity.

INTRODUCTION OF CELS. A technical advance came with the patenting by the American cartoonists Earl Hurd and J. R. Bray in 1914 of processes using celluloid sheets (cels) to portray only those parts of the subject that were to move; the background and other non-moving subjects were thus painted only once for each scene. This simplification of the process brought about an increase in the use of the cartoon.

DISNEY'S INFLUENCE. In 1917, Max Fleischer, creator of Popeye, made his first cartoon, and Felix, first of a long line of cartoon cats, was introduced by Pat Sullivan. Walt Disney entered the field with *Laugh-O-Grams* in 1921; the first *Silly Symphony* with sound appeared in 1929, and the first three-colour Technicolor cartoon, *Flowers and Trees*, in 1932, followed by a number of feature-length productions from 1937.

Among major technical developments pioneered by the Disney studios was the Multiplane animation camera with which complex animation effects of great realism could be produced.

In the late 1940s, the realistic image characteristics of many of the Disney feature cartoons gave way to a simplified style, sometimes approaching the abstract – a change pioneered by UPA and developed in many sponsored cartoons, particularly those designed for television. An important factor in this change was the development of pencils with which the creative animator could draw directly on to cels, reducing both the time and effort formerly involved in tracing from original designs.

DERIVATIONS. In addition to the cartoon, other forms of animation have been developed. Lotte Reiniger became famous for her delicate silhouette films; George Pal, Ladislas Starevich, a Russian of Polish parentage, and Ptushko developed the art of the puppet film, while Len Lye pioneered and Norman Maclaren developed the abstract film. Maclaren has refined the process of direct drawing of both picture and sound on 35mm strips, including some films with hand-drawn stereoscopic images and hand-drawn stereophonic sound tracks; he also exploited pixilation – animation of human subjects.

The high production costs of any animated film, especially the cartoon, have today led to its being confined almost exclusively to sponsored production or comedy cartoons for television, so that work done for the cinema has markedly declined.

SUMMARY

The technical history of film falls into three periods. The first (properly prehistory) covers about fifty years to the development in the 1880s of the basic essentials: camera, projector and flexible perforated strip of film. The second period, also of fifty years, perfects these elements, marks the rise of the silent film, and ends with the coming of sound in the late 1920s. The third period, the forty years to the present day, sees the popularizing of colour, the discovery of new and freer camera techniques, the penetration of film as a technical tool into more and more sciences, and the coming of the wide screen and multi-image picture.

Though yielding ground to videotape recording, film has remained an essential adjunct to television, and after years of audience decline, seems to have re-established itself as a continuing and complementary form of entertainment. B.W.C.

See: *Camera* (FILM)*; Chronology of principal inventions* (FILM)*; Cinema theatres; Colour cinematography; Wide screen processes.*

□HISTORY

Any history of television must begin with the earliest attempts at phototelegraphy, for the first television – a word which did not come into use until about 1900 – was confined to the transmission of still pictures. After this early experimental stage, phototelegraphy is not to be confused with television, but it is necessary to remember that they shared a common origin.

Perfected high-definition television as a public service did not become available until sound broadcasting was well established. Indeed, that seeing by radio should follow hearing by radio might be considered logical in terms of technological progress, and yet the possibility of vision at a distance was occupying the minds of men long before sound broadcasting, or even wireless telegraphy, became a reality.

ORIGINS

Television depends for its functioning on relationships between light and electricity. Therefore, to date the true origin of television, an appropriate time to select is the year 1839 in which Edmond Becquerel discovered the electrochemical effects of light.

Becquerel observed that when two plates of a metal were immersed in an electrolyte, an electric potential developed between them whenever a beam of light illuminated one of the plates. Becquerel did not suggest any practical use for this discovery, yet he had unwittingly hit upon one of the fundamental needs of any television system – the electric effect produced by light, or photo-electricity.

FIRST LIGHT-SENSITIVE CELLS. In 1873, the first photo-conductive effect of practical value was observed by a young telegraph operator, Joseph May, working at the Valentia Island (Eire) cable station. May was using some high resistances of a new type in the form of bars of selenium, when he noticed that his instruments behaved erratically whenever the sun shone on them. This effect was traced to variations in the electric resistance of the selenium bars according to the intensity of the light to which they were subjected.

While this revelation regarding selenium did not have any immediate practical result, it raised the possibility of making a simple light-sensitive cell with the minimum of apparatus. Soon afterwards, in 1877, an instrument appeared which enabled a beam of polarized light to be modulated by electrical means – the Kerr cell.

PICTURE TELEGRAPH. Meanwhile, in 1842, Alexander Bain, an English physicist, made the first proposal for a system of facsimile telegraph transmission over wires and this established some of the fundamental principles of modern practice. At the transmitter, letters were set up in metal type and reproduced at the receiver by a contact moving over chemically treated paper. Bain's method recognized the need for synchronous action in dividing up the picture into elements at the transmitter and receiver, and to achieve this he used electrically controlled swinging pendulums.

Another chemical telegraph was that of F. Bakewell in 1847. His system consisted of a transmitter and a receiver, each of which contained a revolving metal cylinder and a stylus tracing a spiral along its length. By the use of shellac on the transmitting cylinder, and by rotating the two cylinders exactly in step, the stylus at the receiving end could be made to trace on chemically-treated paper what was, in effect, a negative of the picture on the transmitting cylinder.

The Italian priest, Abbé Caselli, took out a British patent in 1861 for a successful process of phototelegraphy. A main component of his apparatus was a cylinder covered in tinfoil. He could send pictures, transmitting these in small portions, as well as handwritten messages, and for a time a commercial service was operated in France. However, it was not possible to stop regular telegraphic traffic during the long process of transmission so the image often appeared with dots and dashes superimposed.

One of the earliest schemes to draw a good deal of attention was that of M. Senlacq of Ardres, in Northern France, whose Telectrascope (1878) incorporated the important principle of scanning, though the mechanical details at this distance in time may seem a trifle obscure. Indeed, it is unlikely that Senlacq himself ever made his ideas work. In his first proposal, the picture to be transmitted was drawn by a piece of selenium moving across the surface of a screen of unpolished glass, while at the receiver the image was again drawn, this time by an ordinary pencil controlled by an electromagnet. As the selenium tracing point at the transmitter encountered light and shade in the picture, varying currents were passed to the receiver, where the pencil point was made to vary its pressure on the paper, thus reproducing the picture with the necessary light and shade.

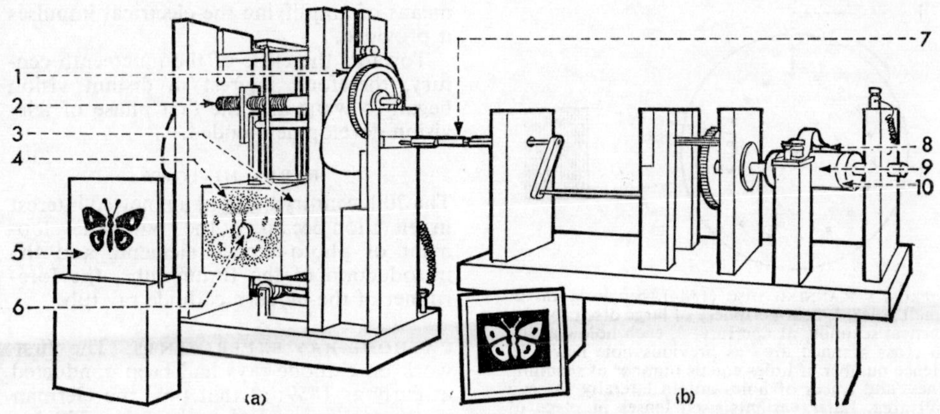

BIDWELL'S SCANNING PHOTOTELEGRAPH (1881). At transmitter (a), illuminated picture for transmission, 5, is set up in front of box, 3, containing pinhole aperture and selenium cell, 6. Shadowgraph image, 4, is scanned by traversing box through lead screw, 2, and raising and lowering it by action of cam, 1. At receiver (b), driven in model through coupling, 7, picture is reconstituted on sensitized paper, 9, wrapped on metallic drum, 10, through agency of platinum stylus, 8.

Senlacq may have realized that his original ideas were unworkable in practice because he published another proposal in 1880. His new transmitter was to consist of a large number of tiny selenium cells, each cell being connected to a corresponding section in the receiver screen in which was a platinum wire. The wire would glow as current passed through it. By a system of sliding switches, the selenium cells would be switched on one at a time and another switch, keeping in step, would switch on the corresponding platinum wire, the whole process being designed to operate very rapidly.

There were other schemes put forward about this time which, like those of Senlacq, were described in some detail, but were apparently never tested by actual experiment. These again were related to picture telegraphy rather than to television.

SCANNING MECHANISMS. The proposals made by the British pioneer Shelford Bidwell in 1881 were of some importance because he translated his ideas into practical experiments. Bidwell's machine was used successfully for transmitting silhouettes and it employed an effective scanning mechanism. A shadowgraph of the picture to be transmitted was projected on to the front of a small box with a pin-hole aperture and containing a selenium cell. The image was scanned by the motion of the box which was made to rise and fall in a vertical line and also move slowly from side to side. The receiver consisted of a platinum-covered brass cylindrical drum around which was wrapped paper soaked in potassium iodide. A platinum point traced out brown lines on the paper which varied in intensity according to the current through the selenium cell and thereby reproduced the shadowgraph picture.

The principle of scanning had now come to be recognized as vital to any picture transmission system, in order to avoid having multi-core connections between the transmitter and receiver. The scanning process meant dissecting the picture in various ways and then transmitting each little element in rapid succession and with the necessary synchronism between the transmitter and receiver. Provided the picture was scanned rapidly and frequently enough, the eye would receive the impression of a complete picture.

NIPKOW DISC. In 1884 Paul Nipkow, a German engineer, patented his scanning disc; the Nipkow disc came to be the best-known scanning system and the most widely used by later experimenters.

Nipkow's device exploited the basic principle on which television depends – that characteristic of the human eye known as persistence of vision. The disc, of large diameter, had near its periphery a series of holes arranged in the form of a spiral. These scanning discs were used at both the transmitter and the receiver, and were made to operate in synchronism. When the trans-

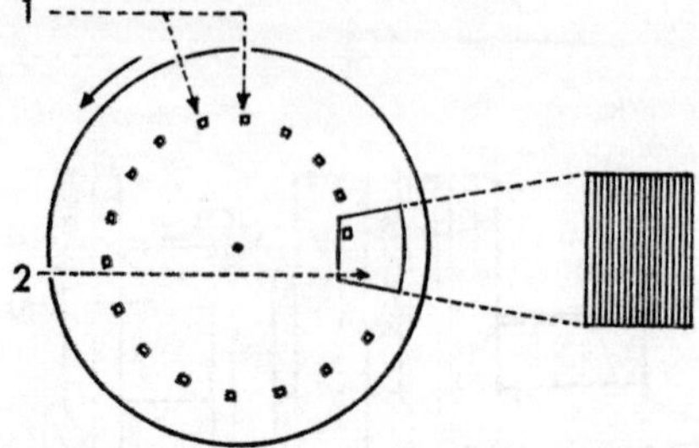

NIPKOW'S SCANNING DISC (1884). Single spiral of small holes, 1, near periphery of large disc provides vertical scanning at aperture, 2, each hole starting to cross scanned area as previous hole leaves it. Hence number of holes equals number of scanning lines, and traces of holes adjoin laterally to scan full area. Later variants used lenses in place of holes.

mitting disc was placed directly between a light source and an object and slowly rotated, the light shone through one hole at a time. After one complete turn, every element of the object had been illuminated by the narrow beam of light from the lamp. However, the area of image which could be scanned in this way was small in comparison with the size of the disc. It is not certain whether Nipkow ever put his ideas into practice. If he did, it can be assumed that his attempts to reconstruct the picture at the receiver were unsuccessful – not through any fault of the disc arrangement but because the currents passed by the selenium cell in his transmitter were too weak and no means of amplification were available.

A derivative from the Nipkow or aperture disc was the lens disc. In this, each aperture was replaced by a lens which was able to collect far more light from the source than could a plain aperture.

MIRROR DRUM. The next significant system of scanning to appear was that proposed in 1889 by Professor Lazric Weiller, also German. Instead of a disc with holes, he used a drum on which were mounted small mirrors angled in such a way that, as the drum revolved, the area of the image was scanned and each part of the picture reflected in turn on to a selenium cell.

Like Nipkow's disc, Weiller's mirror drum was to be employed later in the first television systems, but at the time they were conceived, neither the Nipkow nor the Weiller method, ingenious as they were, proved successful. There were two main reasons for this – selenium's slow reaction to light and the lack of any satisfactory means of amplifying the electrical impulses it provided.

Towards the close of the nineteenth century, therefore, interest in distant vision began to wane and the first phase of television development ended.

EARLY HISTORY

The 20th century brought a renewed interest in television because of the gradual development of photo-electric elements and the introduction of the Braun tube, the forerunner of the modern cathode ray tube.

CATHODE RAY EXPERIMENTS. The first work on cathode rays had been conducted as early as 1859; at that time the German mathematician and physicist Julius Plücker had given the name kathode ray (the "k" was soon dropped in favour of "c") to the discharge of electricity from the negative electrode or cathode in a vacuum tube when a high positive potential was applied to the anode, the electrode at the other end of the tube, thereby causing a fluorescent glow to appear on the glass of the tube.

The presence of these invisible rays emanating from the cathode was confirmed by the German physicist, J. W. Hittorf, and then by the British chemist and physicist, William Crookes, who observed that if another electrode was placed in the tube in front of the cathode – Crookes made the additional electrode in the form of a Maltese cross – it cast its shadow in the fluorescence on the wall of the tube.

The English physicist and engineer, Ambrose Fleming, inventor of the thermionic valve, later experimented with a Crookes tube and found that when he encircled the tube with wire and passed an electric current through it, the cathode rays could be deflected and, to some extent, focused.

BRAUN TUBE. It was in 1897 that Karl F. Braun, a German physicist, produced the cathode ray oscilloscope. This was similar to the Crookes tube but incorporated important modifications. Braun directed the cathode rays on to a diaphragm with an aperture and caused them to impinge on a mica screen at the end of the tube, coated with a fluorescent material. By applying an external magnetic field at the neck of the tube, he could deflect the rays so that the light spot where the rays hit the screen could be moved anywhere on the screen. Braun's interest, however, lay only in the rays themselves and not in using them for television.

A few years later, the German physicist A. R. B. Wehnelt contributed an important improvement to the Braun tube, namely the hot cathode which gave a much brighter spot of light on the screen but with lower voltages on the electrodes.

In 1887, Hertz had discovered that if ultra-violet light were directed on to a spark gap, a much lower voltage was required to produce the spark than when no light was applied. In 1905, Julius Elster and Hans Geitel in Germany developed a photo-electric cell based on the earlier work of their compatriots, H. R. Hertz and W. Hallwachs. This cell reacted faster to light variations than did the selenium cell.

RUSSIAN PROGRESS. In 1907, Boris Rosing in Russia conceived his own television system, using a cathode ray tube in the receiver and mechanical scanning in the transmitter with two mirror drums of an improved design. These reflected the light from the subject on to a selenium cell, but no means were available of amplifying the signals passed by the cell to the cathode ray tube. None the less, Rosing's system was a considerable advance towards modern television practice.

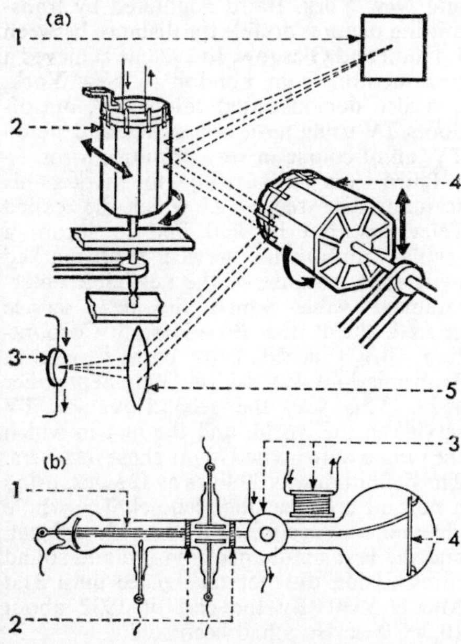

ROSING'S PATENT (1907). (a) Transmitter with mechanical scanning. 1. Picture or scene to be transmitted. 2. Mirror drum for horizontal scanning. 4. Mirror drum for vertical scanning. 5. Imaging lens. 3. Photocell. (b) All-electronic receiver. 1. Cathode ray beam. 2. Perforated discs, between which lie modulating plates fed by signal from photocell. 3. Scanning deflection coils, with sawtooth current obtained from rheostats on mirror drum shafts. 4. Fluorescent screen.

CAMPBELL SWINTON. In 1908 following Rosing's demonstration, but without any knowledge of it, A. A. Campbell Swinton, a British electrical engineer,

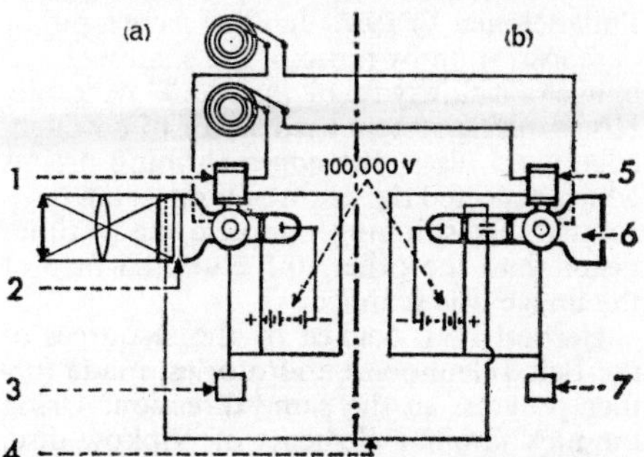

CAMPBELL SWINTON'S CATHODE RAY TV (1908–11). Version of 1911. (a) Transmitter. (b) Receiver. Image formed by lens in transmitter falls on mosaic screen of photo-electric elements, 2, in tube. Electron beam deflection is by generators, 1 and 5, at transmitter and receiver, with line connection, earthed at 3 and 7, to provide synchronization from common a.c. source. Picture signal is carried by line, 4, to deflection plates in receiver cathode ray tube, 6.

writing in *Nature*, proposed a complete television system using a cathode ray tube for both transmitter and receiver, and three years later presented his ideas in detailed form in a paper to the Röntgen Society. Campbell Swinton's ideas, which remained on paper, constituted a remarkable prophecy, since the modern television system is still fundamentally the same as that which he proposed.

LOW-DEFINITION TELEVISION

In 1907, the American wireless pioneer Lee de Forest, by inserting an additional electrode, the grid, into the thermionic valve (tube), made it possible to amplify signals electronically, and the following years saw notable improvements in the techniques of amplification. This revived interest in television, and in the 1920s practical results with mechanical scanning methods were at last achieved. Leading the field were Denes von Mihaly in Germany, Herbert E. Ives and C. F. Jenkins in America, and J. L. Baird in England.

In his first experiments, Mihaly used oscillating mirrors, but these produced no practical results. He then developed a scanning system which employed a small double-sided mirror rotating on the axis of a stationary drum fitted with a number of small mirrors on its interior surface.

SCANNING DISC EXPERIMENTS. Ives and Jenkins in America both developed methods of picture transmission, and Jenkins in 1923 successfully transmitted by radio a picture of President Harding from Washington to Philadelphia. In 1925, Jenkins gave a public demonstration by radio of the shadowgraph image of a slowly revolving model windmill. His scanning system used a pair of bevelled-edge glass discs, the edges forming prisms which deflected the beam of light as the discs rotated. By spinning one disc many times faster than the other, the entire surface of the image was scanned.

Herbert Ives, backed by the resources of the Bell Telephone Laboratories, made further progress in the same direction. Using the now familiar elements of Nipkow disc, photocell and glowlamp, he achieved a 50-line picture at a rate of 18 pps, and in 1927 transmitted full-face portraits in experimental picturephone between New York and Washington. Bell engineers also constructed a folded neon glowlamp covering an area of 5 sq. ft. to enable several viewers to see the picture at the same time.

During 1927 and 1928, Dr. E. F. W. Alexanderson in America, a consulting engineer working for General Electric (GE), was also experimenting with scanning disc equipment. During 1928, GE television broadcasts began from station WGY, Schenectady; they were half-hour transmissions three days a week, and appear to have been the first regular experimental television broadcasts in the world. The line standard was only 24.

During July of the same year, Jenkins began to transmit improved pictures from his experimental radio station W3XK near Washington, D.C., on a 48-line, 15 pps basis, but these were mostly scanned from film.

BAIRD'S CONTRIBUTION. John Logie Baird, a Scot, began his experiments at Hastings, England, in 1923, afterwards moving to London. His first equipment, using a Nipkow disc and selenium cell, was extremely crude, but by 1925 he was able to give a public television demonstration in a London department store. In 1926, at a demonstration to members of the Royal Institution, Baird achieved the transmission of moving images of the human face with

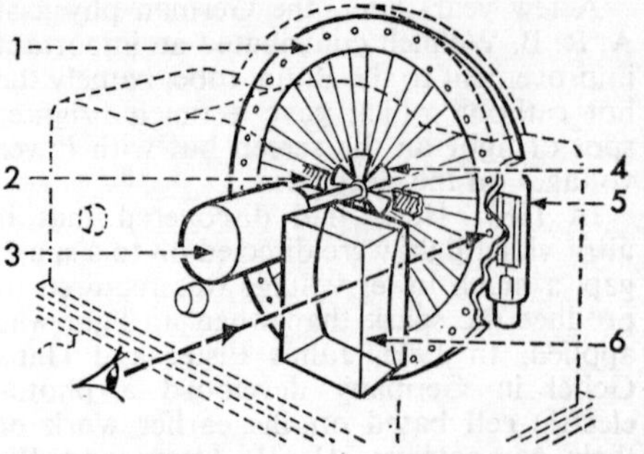

BAIRD'S TELEVISOR (1929). World's first commercially marketed receiver. Nipkow scanning disc, 1, is backed by neon tube, 5, modulated by video signal and observed through magnifier, 6. Disc is driven by motor, 3, and maintained at correct speed by multi-pole wheel, 4, on same shaft, turning between magnets, 2, fed by synchronizing current from receiver circuits.

some degree of light and shade. The picture definition was 30 lines at 5 pps, and the image area measured only $2 \times 1\frac{1}{2}$ in.

When, in 1927, the Bell pictures were sent 200 miles by telephone between Washington and New York, Baird countered by transmitting pictures double the distance between London and Glasgow. In 1928 he achieved a transmission from London to New York, and also demonstrated colour TV, out-of-doors TV using natural lighting, and stereo TV, all of course in very primitive form.

Baird was now ready to market his scanning-disc receivers, which he called Televisors, but he had first to secure a regular transmission service. With marked reluctance, because of the negligible entertainment value which the new service offered, the British Broadcasting Corporation (BBC) acceded to Baird's request. Transmissions started on 30th September 1929. This was the second regular TV service in the world, and the first in which the public were invited to purchase receivers. The definition was 30 lines at $12\frac{1}{2}$ pps, using a normal broadcasting channel. The whole channel was occupied by the video signal, and the first simultaneous vision and sound transmission did not take place until 31st March 1930. By the end of 1932 about 10,000 Televisors had been sold.

In the absence of an electronic camera, Baird used a flying-spot scanner, similar in basic principle to many modern telecine (film scanner) machines. A spotlight projector in conjunction with a revolving mirror drum caused an intense spot of light to scan the subject. Light reflected from the subject in an otherwise darkened studio was picked up by banks of photocells, and the output from these constituted the initial signal currents.

OTHER LOW-DEFINITION SYSTEMS. In Britain, the Gramophone Company developed a lens-drum system, and a receiver employing a Kerr cell to modulate the brightness of an arc lamp. Definition was 150 lines at 12½ pps. The Marconi Company gave two demonstrations in 1932 using Nipkow discs, and in one of these a picture was successfully transmitted by short-wave radio from Chelmsford to Sydney, Australia.

None of these experimental transmissions had appreciable entertainment value, and the mechanical systems on which they were based held no possibilities for the future. However, they encouraged public interest in television and thus laid the groundwork for the high-definition systems which required the resources of powerful corporations to bring them to fruition.

HIGH-DEFINITION TELEVISION

In this field, much of the pioneering was in America and under the impulsion of Vladimir Zworykin. As a young man, Zworykin had been a pupil of Rosing in Russia, and when he emigrated to the US he joined Westinghouse and began research into electronic television in 1920–21. His first iconoscope patent was filed in 1923, and in 1929 he demonstrated this earliest of electronic camera tubes at a meeting of the Institute of Radio Engineers in New York. In 1930 Zworykin moved across to RCA to head the Electronic Research Group at Camden, N.Y.

SYSTEMS APPROACH. The title of the Camden Group was significant, and one of its first achievements was the conscious recognition of the need for a systems approach to television – today an accepted concept, but then very much a novelty. This approach led to a careful study of the line and frame standards to be adopted, an investigation of practical synchronization between transmitter and receiver, and a decision to move up into the 40-80 MHz band to accommodate the wider frequency channels demanded by high-definition pictures. This presented severe problems of line-of-sight transmission, and gave rise to the first methodical field strength surveys.

STORAGE PRINCIPLE. By 1933 the iconoscope had been developed into a practical pick-up tube for television transmission. The important factor in Zworykin's approach was the idea of storage. He wanted to perfect an electronic camera which would accept light energy from the subject continuously and store it between successive scannings. In previous systems, the camera looked at each element of the subject for a very brief moment in each scan. By storing the light energy, it would be possible to achieve far stronger signals. This storage principle had been included in the proposals earlier put forward by Campbell Swinton, of whose work Zworykin was at this time unaware.

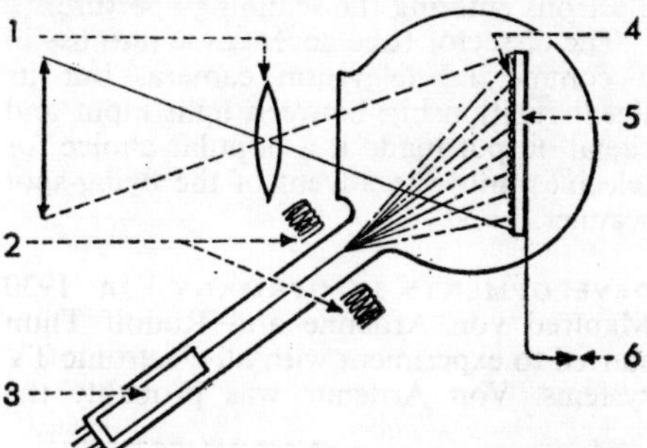

ZWORYKIN'S ICONOSCOPE CAMERA TUBE (1934). High-vacuum tube developed from gas-filled tubes of 1923–24. Object, imaged in lens system, 1, is focused on signal plate, 5, covered by thin mosaic of photo-sensitive globules, 4. Signal plate is discharged by collector electrode on wall of tube, leaving globules positively charged. Electron beam, 3, deflected by coils, 2, discharges globules and sends pulse of current to output circuit, 6. Between arrivals of beam, globules store light energy, thus greatly increasing sensitivity over dissector-type tube.

INTERLACED SCANNING. Between 1933 and 1935 important advances were made by the Camden Group. In their 1934 field tests, the number of scanning lines was increased from 240 to 343, the picture frequency remaining at 24 pps; an all-electronic synchronizing generator was developed and, most important of all, interlaced scanning was introduced to double the effective picture repetition rate and thus eliminate flicker. By this method, now universal, the picture is scanned twice, first along the even and then along the odd lines, each scan constituting one field. The two interlaced fields make up one complete picture or frame. Said to have been devised by two Britons in 1923, interlacing was independently invented and patented in 1933 by a member of the Camden Group, R. C. Ballard.

IMAGE DISSECTOR. Another electronic system developed in America was that of

Philo T. Farnsworth, who patented his camera tube, the image dissector, in 1927. Whereas Zworykin in the iconoscope had scanned the subject in successive stages, Farnsworth moved the whole electron image past a scanning aperture which dissected it into as many segments as the number of picture elements required in the transmitted picture. Since the image dissector did not employ the storage principle, it was extremely insensitive, though Farnsworth later remedied this by incorporating an electron multiplier, which made use of secondary emission to build up the small number of electrons entering the scanning aperture.

The dissector tube never came into use in a commercial television camera, but its linear relationship between light input and signal output made it a popular choice for telecines until the advent of the flying-spot scanner.

DEVELOPMENTS IN GERMANY. In 1930 Manfred von Ardenne and Rudolf Thun started to experiment with all-electronic TV systems. Von Ardenne was probably the first to devise a practical flying-spot scanner using a cathode ray tube. He built receivers on which pictures of reasonably good quality were demonstrated with 50 lines per frame.

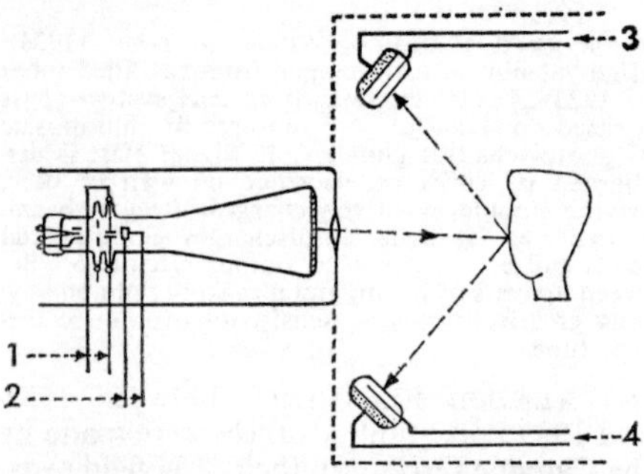

VON ARDENNE'S STUDIO SPOT SCANNER (1930). High-voltage cathode ray tube produces focused spot which scans subject in studio with frame scan, 1, and line scan, 2. Photocells in front of subject produce combined outputs, 3, 4, corresponding to point brightnesses. Contemporary phosphors had long decay times which rendered this proposal impracticable in 1930s.

DEVELOPMENTS IN BRITAIN. In 1931 a powerful industrial group, Electrical and Musical Industries (EMI) attacked the problems of electronic television. Like RCA, EMI from the first recognized the importance of a systems approach, and in the event gained a notable lead over their friendly rivals in the US by supplying the equipment for the world's first regular service of public high-definition television broadcasting. To do this, they allied themselves with an equally important group specializing in high-power transmitters and aerials.

EMI's development team was led by Isaac Shoenberg, a distinguished Russian scientist, and included several engineers seconded from the Columbia Graphophone Company in the US, notably the English pioneers A. D. Blumlein and P. W. Willans. More than a year was spent on perfecting a cathode ray receiving system fed by a mechanical film scanner which ultimately progressed up to 180 lines. Flicker problems led the team to adopt interlaced scanning, and by the end of 1932 they were ready to move on to the electric generation of the initial vision signals.

Co-operation with RCA gave them access to Zworykin's iconoscope, which at this stage suffered from spurious signals caused by secondary emission effects. Blumlein and his colleagues devised ingenious circuits to deal with these unwanted signals, and went on to reach the final solution to the problem, the stabilization of cathode potential. The resulting Emitron tube was in principle similar to the iconoscope, but its development must be regarded as quite separate.

In 1934 came the link-up between EMI and the Marconi Company, and the determination to press for a commercial system based on 405 lines and 25 pps, which to this day remains the standard for British VHF television. Baird's trial period was expiring, and the BBC, sole broadcasting authority in Britain, looked about for a more advanced system to replace it. Of those available, only the Marconi-EMI and the Baird high-definition systems met the minimum standards set by the investigating committee, namely 240 lines at 25 pps.

Baird proposed a flying-spot camera for close-ups, studio scenes and film scanning, and for larger scenes an intermediate-film process comprising a cine camera in which the exposed film was fed direct into a developing machine, processed, and scanned in negative form while it was still wet.

The committee recommended that both systems be employed alternately for a trial period of regular broadcasting, which started on 2nd November 1936 and continued until February 1937, when the Marconi-EMI system was adopted for the permanent service. Baird was bitterly disappointed; he had put all his faith in his mechanical systems and had laboured unceasingly at them with considerable success. Because of his enthusiasm and ability to arouse interest in what he was doing, he gave the industry a useful impetus and thus spurred others to undertake serious research. But his equipment had proved cumbersome and expensive, and though more sophisticated mechanical systems such as Scophony were belatedly developed, the future obviously belonged to electronics.

The BBC maintained a regular programme service from November 1936 to September 1939, when the London station was closed down at the outbreak of war, not to reopen until June 1946.

FURTHER PROGRESS IN USA. The Zworykin group, Farnsworth, A. B. DuMont and many others continued towards the realization of a high-definition commercial service, and by 1939 RCA were ready with a tested system employing 340 lines at 30 pps. Experimental transmissions from the Empire State Building were on 49·75 MHz for picture and 52 MHz for sound. Regular programmes were started by the National Broadcasting Company (NBC) on 30th April 1939, at the opening of the New York World's Fair, but in the following year the Federal Communications Commission (FCC) rescinded its go-ahead on the grounds that generally acceptable standards had not been established. The FCC then set up the National Television System Committee (NTSC), which recommended a 525-line standard at 30 pps. This standard came into force in July 1941, and has remained in effect from that time on. But five months later, American entry into World War II brought the close-down of commercial television.

MODERN DEVELOPMENTS

With the end of the war, the television broadcasting industry gathered new momentum, and the question of standards underwent review. Europe adopted a 625-line system, while France added an 819-line service. Britain remained on 405 lines, while the American 525-line standard was adopted in Canada and the Far East.

The establishment of new stations was hastened, and television coverage in Britain increased from its pre-war 25 per cent of the population to 46·5 per cent by 1949 and 98 per cent by 1957. In the US, where there were virtually no sets in use in 1946, the total had grown by 1952 to nearly 22 million, and by 1964 to 70 million.

CAMERA TUBES. When television services reopened in 1946, the camera tubes in use in Britain were the Emitron and the Super Emitron, or image iconoscope. In this tube, secondary emission of electrons from the mosaic was employed to increase the sensitivity to about ten times that of the iconoscope. The Super Emitron had been introduced in 1939, but the war had put a stop to its application. The similar image iconoscope was developed during the 1940s in the US.

RCA's orthicon tube – so called because the electron gun directly faced the mosaic instead of being set obliquely as in the iconoscope – represented a somewhat different approach, but was in turn soon replaced by the image orthicon, announced in 1946. This tube employs electron image

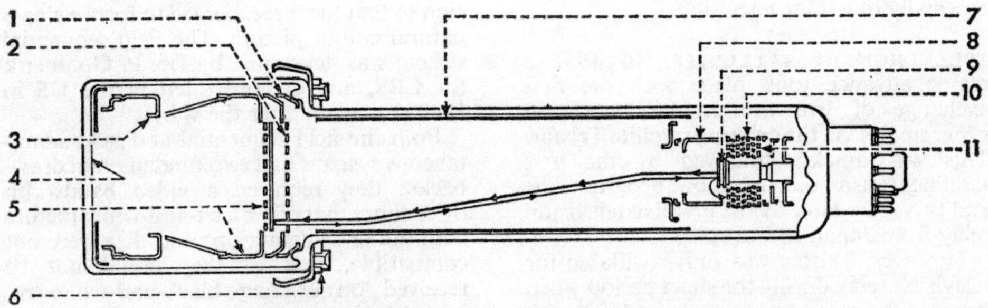

IMAGE ORTHICON CAMERA TUBE (1946). Image section consists of photo-cathode, 4, image accelerator, 3, target with wire mesh, 5, held in target cap and backed by field mesh, 2, also decelerator, 1, the whole held in position by shoulder base with locating bushes, 6. Central scanning section of tube comprises persuader, 8, and beam focus electrode, 7, with external scanning coils to deflect beam (*arrowed*). Return beam, after striking target, enters multiplier section of tube: 9, first dynode, 10, multiplier, 11, anode.

amplification (as in the image iconoscope), an efficient low-velocity beam (as in the orthicon), and an electron multiplier (as in later types of Farnsworth dissector). As a result, the image orthicon achieved a sensitivity almost 100 times that of any previous camera tube, together with outstanding freedom from spurious signals. It came into use in Britain in 1949 in its original 3-in. form, but in 1955 Marconi announced an improved 4½-in. version.

All the tubes so far mentioned are characterized by a photoemissive surface, but it is possible to make use of photoconductive properties to produce camera tubes with many times the output current of the image orthicon. In spite of the very early discovery of the photoconductive property of selenium, it was not until 1950 that the first commercial tube of this type, the vidicon, was announced by RCA. The high sensitivity of photoconductive materials made it possible to eliminate both the electron image amplifier and the image multiplier of the image orthicon.

Although the vidicon has a tendency to smear the picture where there is rapid movement, unless the illumination is very high, its compactness, cheapness and low operating cost make it highly suitable for many types of work, such as unattended closed-circuit television, interview studios and film pickup (telecine) equipment.

The next major advance came in 1964 with the announcement by Philips Gloeilampenfabrieken at Eindhoven of the Plumbicon tube. This is a photoconductive pick-up tube similar to the vidicon, but its photoconductor is of lead monoxide (PbO) instead of antimony sulphide or selenium. In the Plumbicon, the defective properties of the vidicon are greatly reduced, and it has already won an important place for itself, especially in colour television.

TELEVISION BY SATELLITE. In 1962 a major advance took place with the first exchange of live transatlantic television programmes via the orbiting satellite Telstar. This was quickly followed by the first satellite transmission of colour programmes, and two years later by the first live television relay from Japan to Europe.

However, Telstar was only available for television relay during the short period when it was in the correct position in its 157½-min. orbit between transmitting and receiving stations. To be commercially successful, a satellite had to be placed in a stationary orbit, and after several experimental launches, this was achieved by the American Early Bird in April 1965. It then became possible for the first time to transmit live television continuously in both directions between the United States and Europe, and by later satellites between the United States and Japan. Three weeks after the launching of Early Bird, Russia put its first communications satellite, Molniya I (Lightning), into orbit.

In 1966 the first live two-way television transmission took place between Britain and Australia, and in 1967 it became possible for the first time to girdle the whole earth in simultaneous transmission by combining the resources of American and Soviet satellites.

COLOUR TELEVISION. As long ago as 1909, A. C. and L. S. Andersen of Denmark proposed a colour television system consisting of a mechanical scanner, a selenium cell and a dispersive prism to split the subject into its colour components. A synchronous coloured disc in the receiver reconstituted the colours and displayed them on a screen. In the 1920s, both H. E. Ives in the US and Baird in England demonstrated crude colour systems.

In 1940 RCA disclosed an electronic and optical method of colour television using three orthicon cameras with a colour filter in each – red, blue and green. Three kinescopes were employed in the receiver, one with phosphors sensitive to red, the second to blue and the third to green. This became known as the field simultaneous system.

In the field sequential system, red, green and blue filters rotated in front of the camera tube in synchronism with the field scan and with corresponding discs at the receiver. The viewer thus saw a red, a green and a blue component field in rapid succession so that the three merged to form a single natural colour picture. The field sequential system was developed by Dr. P. Goldmark for CBS, and demonstrated in the US in 1940 and again after the war.

Both the field sequential and field simultaneous systems had two fundamental drawbacks: they required a video bandwidth three times that of a black-and-white picture with the same definition; and they were not compatible, that is, they could not be received on existing black-and-white receivers.

Nevertheless, by 1950 powerful electronic interests in the US, led by CBS, were pressing for the adoption of a non-compatible system; their opponents, RCA, took the

COLOUR TELEVISION SYSTEMS

Abbreviation	*Full Name*	*Country of Origin*	*Date of First Entry into Service*
NTSC	National Television System Committee	USA	1953
SECAM	Séquentiel Couleur À Mémoire ... ... (Sequence Colour and Memory)	France	1967
PAL	Phase Alternation Line	West Germany	1967

view that, with more than 10 million sets already in use, compatibility was essential. They therefore pushed to completion, with help from other sections of the industry, the first fully compatible colour system. In December 1953 this was endorsed by the NTSC, by which name it is now known.

COLOUR SYSTEMS. There are today three commercially workable and compatible colour transmission systems:

In the NTSC system, the two colour signals – the luminance component (brightness of picture) and the chrominance component (hue and saturation of colour) – are transmitted and received simultaneously. The NTSC receiver has separate controls for colour, hue and saturation.

The SECAM system was developed by the Compagnie Française de Télévision. In this system, the three primary colours, red, green and blue, are combined into two component signals which are then transmitted sequentially line by line. The reconstruction process at the receiver involves a delay process by which the first component is stored long enough for the second component to catch up. Colour control on the receiver is automatic.

The PAL system was developed by the Telefunken Company. It is similar to the NTSC system, two colour component signals being transmitted simultaneously but with the phase of the chrominance component reversed during alternate line periods. No hue control on the receiver is required, but a saturation control is usually provided.

All of these three colour systems use the shadow-mask picture tube, developed by RCA. The shadow mask is a tightly stretched thin metal sheet containing many thousands of regularly spaced circular apertures. It is positioned a fraction of an inch from the inside of the phosphor screen which is coated with an array of red, green and blue phosphor dots arranged in triangular units. Three electron guns generate three closely spaced beams which pass through the apertures in the shadow mask in such a way that the red beam falls only on red phosphor dots, the blue beam on blue phosphor dots, etc.

While the shadow mask tube is the only one in commercial use at present, various single-gun tubes based on different principles are being developed; two of these are the Philco-Apple index tube and the Mullard-Banana using mechanical vertical scanning.

SUMMARY

The prehistory of television spanned the eighty years from the first attempt at "seeing at a distance" in the 1840s to the development of electronic amplification in World War I. The next twenty years saw the struggle between low-definition mechanical TV systems of limited potential, and high-definition systems with their severe problems of wide channel bandwidths, camera tubes and general electronic design. By 1940, these difficulties had been substantially overcome, and after the interruption of World War II, television took shape in its present form. **R.W.B. & R.J.S.**

See: *Camera* (TV); *Camera (colour)*; *Chronology of principal inventions* (TV); *Colour systems.*

Books: *Invention and Innovation in the Radio Industry*, by W. R. Maclaurin (New York), 1949; *The History of Television*, by G. R. M. Garratt and A. H. Mumford (*Proc. IEE*, Vol. 99, Pt. 3A (London), 1952).

□**Hittorf, Johann Wilhelm, 1824–1914.** German physicist who did pioneer work on cathode rays.

⊕**Hold Frame.** A single frame of a moving
□shot arrested for as long as required by an optical printing process (or in TV electronically) to concentrate attention on a phase of action, e.g., a football goal, or to lengthen a shot which is not long enough. Better picture quality is obtained by alternating two frames, which has the effect of reducing graininess. After the hold frame, the action is resumed. Sometimes called freeze frame or stop frame.

See: *Time lapse cinematography.*

HOLOGRAPHY

Holography is a type of photography in which a pictorial image is neither formed nor recorded. Instead, light reflected from an object is recorded directly on an emulsion surface, without passing through a lens, so that all parts of the surface receive light from all visible parts of the object. No picture can be seen by looking at the developed emulsion directly. But, by a suitable reconstruction process, the object can be viewed in three dimensions, and movements of the observer's head, within the limits of the hologram's edge or frame, produce true effects of parallax. This is not the case with stereoscopic pictures, which are reconstructed from pictorial data in pairs of images taken from selected viewpoints.

PRINCIPLES

In holography, the complex array of wavefronts reflected from the object is intercepted by the plane of the emulsion, and there recorded. This again contrasts completely with ordinary photography, which merely records the amount of reflected energy, point by point. It was in its aspect of wavefront reconstruction that holography was invented by Gabor in 1947, and developed as a means of improving the resolution of the electron microscope.

INTERFERENCE PATTERNS. The holographic process depends on the ability of two light waves to cancel each other out (destructive interference) or to add to one another (constructive interference). Ordinary white light can produce interference patterns only in special conditions, for it consists of vast numbers of bursts of radiation, each differing in amplitude, frequency and phase from the others. The conditions necessary are established only when the two light beams originate from the same source and travel by slightly different paths before interference takes place. To achieve this, the light must be strictly monochromatic (i.e., of a single frequency), the wavefronts must be plane and uninterrupted, and the wave trains must advance continuously. Light of this kind is called coherent light.

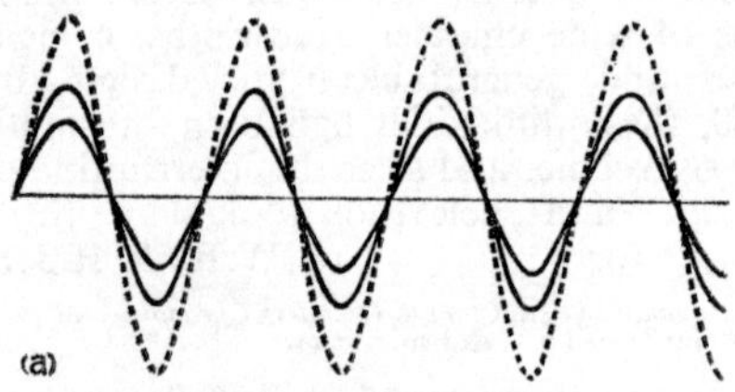

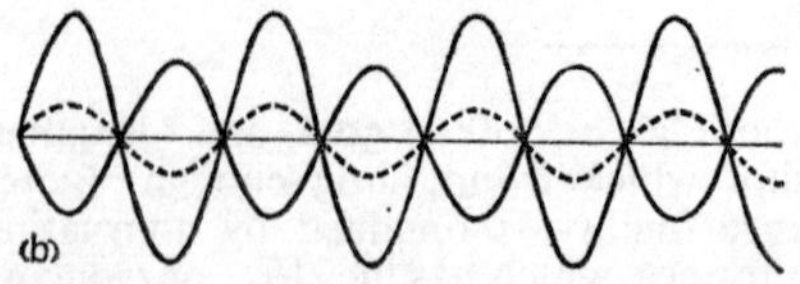

CONSTRUCTIVE AND DESTRUCTIVE INTERFERENCE. (a) Two in-phase wavetrains (*solid line*) add to form resultant wavetrain (*dashed line*). (b) Two wavetrains out of phase by 180° (*solid line*) subtract to form lower amplitude resultant (*dashed line*).

USE OF LASER. Holography was not practicable without a powerful source of coherent light, provided by the invention in 1960 of the laser. Lasers emit monochromatic light in plane, parallel, progressive wavefronts. To produce a hologram, light from a laser is split by a partially reflecting mirror. One of the beams so produced is directed on to the object, and reflected from it on to an emulsion surface. The other beam, called the reference beam, is directed straight on to this surface. Interference takes place between light from different parts of the object, producing a pattern on the emulsion which constitutes the hologram, and which, when viewed by

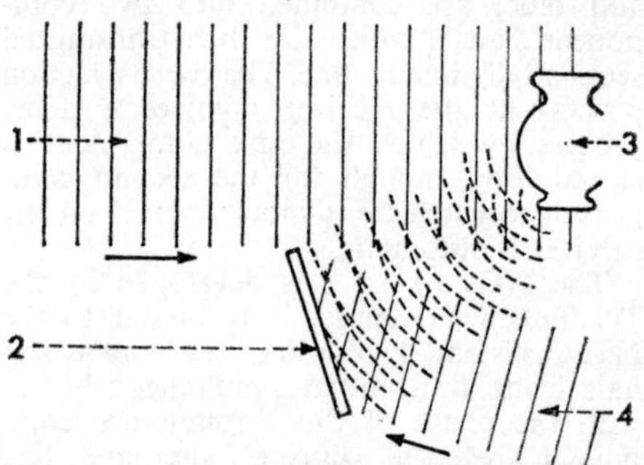

FORMATION OF HOLOGRAM. 1. Beam of coherent light from laser strikes object, 3. Complex system of wavefronts originating from object strikes the emulsion surface, 2. Second coherent light beam, 4, called reference beam, from same laser, is directed on to emulsion.

ordinary light, bears no resemblance to the original object. To interpret the hologram, i.e., to make the information it carries visible as an image, light from the laser is made to pass through the emulsion and a virtual or real image is produced. In this way, an image is formed which is in all ways (except in colour) indistinguishable from the object.

This last limitation can be overcome in two ways. Firstly, three colours of laser light may be used to make three superimposed holograms on a tripack colour emulsion. When this is projected with three similar laser beams, a reconstructed object in full colour is produced. Secondly, the Lippmann method of colour photography may be applied.

REFERENCE BEAM. The reason for using a reference beam is as follows. In all forms of photography, information is conveyed by means of the frequency, amplitude and phase of the light reflected from an object. In ordinary photography, the frequency carries the colour information. The phase information is lost, because only the amplitude of light contributes to the record. In holography, however, the phase as well as the amplitude is recorded in the interference pattern. Because this pattern consists of positive and negative light amplitudes, and because emulsions record only intensity, which is the square of the amplitude and therefore independent of sign, some way must be found of retaining this sign information.

The reference beam does this by adding a constant, uniform light field large enough to make all the light amplitudes positive. Thus, when these amplitudes are squared in the process of recording the intensity of light on the emulsion, no information is lost. It is because of this, and the fact that each part of the emulsion sees light from the whole of the object, that the hologram possesses its remarkable properties of three-dimensionality and true parallax. Holograms can in fact be damaged or broken up and still produce the entire picture when viewed.

OBJECT MOVEMENT. The above applies only to single holograms of stationary objects, and stationary here implies a severe limitation. Because the interference pattern is produced by the mixing of many wavefronts, the object must not move relative to the emulsion by more than a quarter of a wavelength of the light used, which in the middle of the visible spectrum is about a thousandth of a millimetre. Thus the strictest precautions are necessary to make a still hologram, especially as the exposure may take several minutes.

RECONSTRUCTION. In order to view a hologram, coherent light is directed on to it. This light is diffracted by the interference pattern in the emulsion and emerges to form both a real and a virtual image. The real image is on the opposite side of the emulsion from the light source, and the virtual image on the same side. It is this which is seen by the viewer.

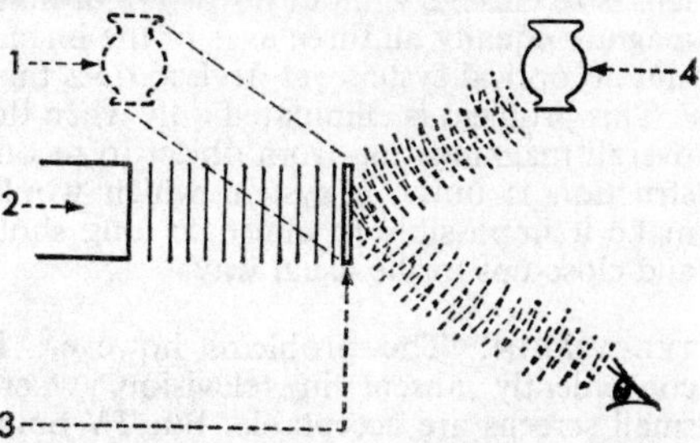

RECONSTRUCTION OF HOLOGRAM. Laser, 2, emits coherent light beam which strikes developed emulsion of hologram, 3. Light, diffracted by interference pattern on hologram, results in virtual image, 1, seen by eye of spectator, and real image, 4. Other methods of reconstruction are possible.

Variants of this reconstruction process have been developed for viewing other types of hologram, and for viewing normal holograms without using coherent light and thus without requiring a laser.

MOTION PICTURE APPLICATIONS

To make a motion picture hologram, with its portrayal of moving objects, object and emulsion must therefore be illuminated for only a very short time, for instance by lasers producing high-intensity pulses of a duration of the order of 20 nanoseconds, and of very high intensity. This also is important because, for holography to be possible at all, the whole scene must be illuminated with coherent laser light. At present this limits holography to indoor scenes.

Two methods have been suggested for recording multiple phases of action. First, a single emulsion can be used to record several different pictures simultaneously. Each picture can be reconstructed separately by altering the angle which the incident light makes with the surface of the emulsion when viewed.

The second method is to record successive exposures, as in ordinary cinematography. Both these approaches have been tried on a small scale, and they may eventually be used in combination.

SIZE LIMITS. Present-day holograms, however, measure only a few inches square, and thus only one or two people at a time can conveniently view them. Since holograms the size of a cinema screen are likely to remain impracticable, some projection process will be called for. This gives rise to great difficulties because, if the projection lens is to enlarge without distortion, it must magnify equally all three axes of the image, and no optical system yet devised does this.

This problem is eliminated only when the overall magnification from object to reconstruction is unity, a system which would make it impossible to alternate long shots and close-ups in the usual way.

TELEVISION. The problem, however, is conveniently absent in television, where small screens are acceptable. But TV holography presents other difficulties. The essence of television lies in the transmission of pictorial information over long distances. The amount of information which can be carried by either a cable or radio system is limited by the bandwidth available. Because it contains within itself all visible aspects of the object, a hologram requires very large bandwidths, for these aspects are represented by the complex interference patterns which have to be transmitted. In film, this problem is overcome by the use of emulsions with a resolution of the order of 2,000 lines per millimetre (about twenty times that of ordinary film). In TV, an equivalent high-resolution camera and its associated circuitry are needed, and these remain to be developed.

BANDWIDTH REQUIREMENTS. The requisite bandwidth depends on the fineness of the interference pattern, which varies with the method used for taking the hologram, but is greater the greater the angle of acceptance, i.e., the width of field. It has been shown that to transmit a small studio scene at 30 pps would require a bandwidth of 150,000 MHz, as against about 8 MHz for an ordinary television channel.

However, the resolution of the human eye is poor compared with that of the hologram, and a method might therefore be sought for shedding some of the superfluous information. Techniques of economizing bandwidth by selecting only essential information are now well developed, and it seems likely that the space requirements of a holographic TV system could be considerably reduced, though they would remain much greater than for ordinary TV systems, necessitating carrier frequencies far beyond the UHF or even the microwave bands. **S.J.P.S.**

See: *Laser; Stereoscopic cinematography.*

Books: *Bibliography on Holograms Jnl SMPTE*, **April 1966, Aug. 1966 and April 1967.**

□**Hook.** A recognizable distortion at the top of the television picture which bends vertical lines sideways, causing them to look rather like a hook. This effect is usually caused by poor recovery of the line frequency oscillator after the burst of field pulses. The extra half-line trigger pulses can upset the working of the line oscillator and it can take a few periods of oscillation to become stable again.

Hooking can also occur at the originating source of the signal owing to a similar fault, or from phase modulation of the television signal.

See: *Receiver.*

□**Hopping Patch.** In a telecine (film scanner), to produce an interlaced picture, the raster on the face of a flying spot tube can be displaced between alternate fields, so that the film frame may be scanned twice. Since the patch of light on the face of the tube changes position, it is called a hopping patch.

See: *Telecine.*

□**Horizontal Picture Shift.** Method of moving the position of the scanning raster on a tube in a horizontal direction.

⊕**Horner, William George, 1786–1837.** English mathematician who in 1834 described the Daedalum, the first form of Zoetrope.

□**Horn-Reflector Aerial.** Directional aerial which consists of a horn-shaped enlargement of the end of a waveguide. One wall of the horn turns inwards beyond the axis of the horn and reflects the waves through a large aperture at right-angles to the axis.

See: *Microwave links; Transmitters and aerials.*

Horse. Device used in film cutting rooms which consists of two parallel hinged arms mounted vertically on a bench and carrying a spindle half-way up. The arms are

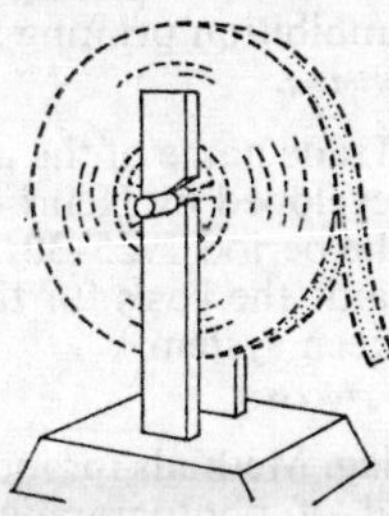

separated by a distance just greater than the width of the film to be mounted, and this film, wound in a roll, is supported on the horse spindle for unwinding or breaking down.
See: *Editor.*

Hot Spot. Area of excessive brilliance, either on a projection screen or in a scene which is being lighted for the film or TV camera.

Howell, Albert S., 1879–1951. American engineer and inventor. Held 65 patents, the first of which was for a 35mm projector, which led to the formation of the Bell & Howell Company. Among his important inventions were the Bell & Howell film perforator (1908) and a continuous motion picture printer (1911).

Howl Round. Natural frequency output or howl caused by a connection leading to a loop with greater than unity gain, i.e., positive feedback. Term originally applied to sound and also called acoustic feedback, but used for video also. When a camera sees its own picture in a monitor, part of the picture is likely to go above peak white and out of control, by the effect of positive feedback.
See: *Acoustics; Communications.*

Hue. Variable describing the similarity of a colour to one of the series ranging circularly through red, yellow, green, blue, purple and back to red. A colour either has a hue that can be fitted somewhere into this series or no hue at all (black, white and grey).
See: *Colour principles.*

Hughes, William C., 1844–1908. London optician and inventor. Devised a number of animated slide mechanisms for the magic lantern, including the Choreutoscope using an intermittent mechanism for picture transport, patented by Hughes in 1884.

Hum. Audio pick-up of mains frequency; by an extension of meaning video bars of mains frequency.

Hum Bar. Dark or light lines on the screen of a television receiver due to components at mains frequency (50 or 60 Hz) in the d.c. supply circuits. Usually produced by poor smoothing of the high-tension supply and is 50 (60) Hz in the case of half-wave rectification and 100 (120) Hz for full wave rectification. Another common cause is heater-to-cathode leakage in one of the vision circuit valves or the cathode ray tube (kinescope); this can add a component to the vision signal.

The fault manifests itself as a light and dark bar across the picture (two bars in the case of the doubled frequency). Normally the bars are stationary as the television signal is in synchronism with the supply frequency. Sometimes, however, the television signal is locked to a crystal oscillator, and differs slightly in frequency from the supply mains. In this case the bars slowly move up or down the picture.

The density of the bars compared to the picture depends on the severity of the hum, and can vary from a faint bar to one of similar amplitude to the television signal. In the latter case, the hum can cause malfunctioning of the synchronizing pulse separator, resulting in mis-triggering of the time bases.
See: *Receiver.*

Hum Balancer. Circuit used to feed hum into a system to cancel hum already present.

Hum Displacement. Cyclic positional shift in a CRT picture display, normally at mains frequency. When mains-locked, the picture suffers distortion of vertical lines; when unlocked, it acquires a disturbing wobble.
See: *Picture monitors; Receiver.*

Hurd, Earl, d. 1940. American cartoonist who, with J. R. Bray, developed and patented in 1914 the system of drawing cartoon images on celluloid sheets which thus carried only those portions of the scene which were to move. Static subjects and backgrounds could then be drawn on a separate sheet, combined with the "cels" for filming. This was a major step in the development of the cartoon film.

Hybrid. Device which allows two-way transmission of signals between point A and points B and C, but does not allow transmission between points B and C.

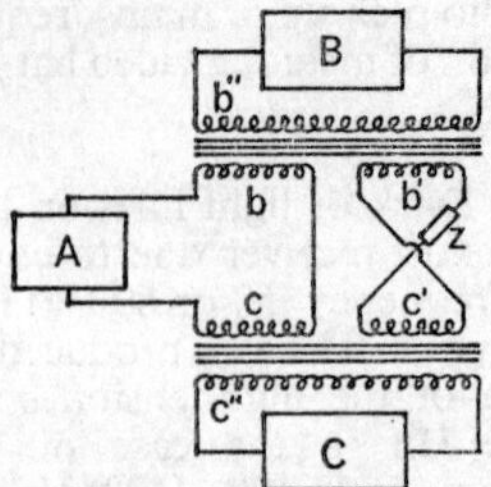

HYBRID. Signals from A follow path bb″ to B and path cc″ to C, with loss of 3 dB to each. Signals from B reach A by reverse route, but c′ produces opposing but equal flux to c, so no signal appears in c″, and C is isolated from B; equally B from C. Z adjusts balance; isolation of 70 dB is possible.

Hybrids are mainly used in two different types of circuit arrangement. If a sound source has to be split to feed two separate circuits, the hybrid prevents spurious signals, present on one circuit, from affecting the other. The second application is to a four-wire repeatered circuit, where the hybrid prevents the Go and Return channels from forming a closed loop when the ends of the system are restored to two-wire.

⊕**Hydrotype Process.** Photomechanical method of printing by dye transfer used to a limited extent in Russia to manufacture cinematograph release prints in colour. Basically an imbibition printing process.

See: *Imbibition printing.*

⊕**Hypergonar.** Trade name of the anamorphic lens system developed by Henri Chrétien in France over the period 1925–30, which subsequently became the basis for the Cinemascope wide-screen system.

See: *Wide screen processes.*

⊕**Hypersensitizing.** Methods for increasing the effective speed of photographic materials, especially negative stocks, after manufacture and before processing. Chemical solutions, the action of certain vapours and very brief uniform exposure to light have been employed with black-and-white films but are not well suited to colour stocks.

See: *Laboratory organization.*

⊕**Hypo.** Common term for sodium thiosulphate (hyposulphite) or its solution used as a fixing bath in photographic processes.

See: *Chemistry of the photographic process; Processing.*

☐**Iconoscope.** Early TV camera tube, invention of Vladimir Zworykin, and first to use the storage principle (build-up of charge throughout the interval between successive scans). Described in US Patent 2,141,059 (1923), the iconoscope employed a mosaic of insulated globules of photoelectric material on which the optical image was formed. Scanning was from behind this target, as in modern tubes, and the tube was gas-filled.

Later and more practical types of iconoscope were high-vacuum, and scanning was from the front side of the target. This tube underwent development from 1930–39, when it was superseded by the image iconoscope.

See: *Camera* (TV); *History* (TV).

☐**Identification Signal.** Signal placed on all important sound and picture links on major broadcasts, particularly international ones, to minimize errors and failures during the lining-up period. Picture identification is usually by test card with superimposed name of source, and sound by a repeating loop announcing the source in the appropriate language.

See: *Special events.*

☐**IF Amplifier.** Circuits of fixed frequency amplifying the intermediate frequency signal resulting from the heterodyne process. Because the frequency remains constant, these circuits can have better selectivity, stability and gain and be given a specific amplitude/frequency response.

See: *Receiver.*

☐**Image Diagonal.** Measurement taken across opposite corners of the optical image formed on the faceplate of a camera or receiver tube. Standard measurement of image size in Britain; in the US the tube face size is defined in square inches.

☐**Image Dissector.** Early TV camera tube of the non-storage type, invented by Philo T. Farnsworth and first described in US Patent 1,773,980 (1927). An optical image was formed on the photosensitive cathode, and the resulting photoelectrons were attracted to the anode by a high voltage. The electron image was scanned across an aperture in the target by deflection coils, and was thus dissected into elemental areas varying in signal strength with the original optical brightness.

Since this effect was instantaneous (i.e., non-storage), the sensitivity of the tube was very low, and Farnsworth later employed an electron multiplier to amplify the signal within the tube.

See: *History* (TV).

⊕IMAGE FORMATION

A real image, such as that to be recorded by a photographic material, consists of a two-dimensional distribution of light intensity, or, photometrically more correct, illuminance (I). This results in a distribution of exposures, I (x, y) × t, where x and y are the co-ordinates in the image plane.

Any layer which is to react to the exposure distribution by forming a picture must of necessity be particular in nature in the sense that a small area in the recording medium must respond to the incident light intensity by producing the correct tone value. There are, in principle, two ways in which this may happen.

In one, the light-sensitive units produce different amounts or areas of light-absorbing material, as in the half-tone processes.

The other is that used in photographic materials. These contain large numbers per unit area of tiny light-sensitive crystals, any one of which reacts to light (and subsequent development) by an all-or-nothing process: it either blackens completely or not at all, according to the quantity of light it has absorbed.

Thus the various tones between black and white are produced by local variations in population density of the developed blackened grains. It follows that all photographic pictures must be more or less grainy.

BASIC PRINCIPLES

THE EMULSION. The light-sensitive crystals in the photographic emulsion layer consist of silver halides: in plates and films, of silver bromide with a few per cent of iodide; in papers, of silver chloride or chlorobromide in all ratios. These crystals (often called grains), between a few tenths of a micron (μ) and a few microns in size, are suspended in gelatin; there are around 10^9 to 10^{10} per cm^2 of surface of a fast film. The coated dry emulsion layer is perhaps 10 μ in depth, so that the grains lie about 20 deep in the layer, overlapping to some extent but never in actual contact with one another.

An emulsion crystal 1 μ in size consists of about 10^9 pairs of silver and halide ions. Ions are atoms which have either lost or gained an electron; if silver bromide is formed from the elements, each neutral atom of silver loses an electron (becoming positively charged) to a bromine atom which gets a negative charge. The attraction between these positive and negative partners is the main force holding the crystal together.

LATENT IMAGE. Heavy exposure to light decomposes the crystals to their elements; silver is formed and halogen is released. One unit quantum of light can result in the production of one silver atom. This decomposition by light is described as print-out. Exposures many powers of 10 shorter than those required to cause visible print-out produce in the emulsion a condition known as the latent image. A grain in this condition is reduced to metallic silver in a suitable solution (a developer) much more quickly than an unexposed grain.

There can be no direct information on the exact nature of the latent-image condition, because there is no physical or chemical method by which it can be detected or studied except that of photographic development. The indirect evidence suggests that the latent-image condition is caused by a small speck of silver on the surface of the grain, where the developing solution can reach it. The specks may be very small; careful tests have shown that the absorption of no more than four quanta per grain, and hence the formation of four atoms of silver, is sufficient to produce a latent-image grain.

The light quanta may be absorbed by any of the 10^9 halide ions of a crystal, so that a migration process must occur for the resulting silver atoms to get together to form a single speck. The details of this migration process are the subject of latent-image theories. Since there is no definite agreement, it suffices to state that by considering the process of latent-image formation as the creation of a new solid phase – like water drops in a cloud created from water vapour – many of the features of the photographic process can be explained. Both physical faults of the crystal and chemical impurities play a part in ensuring that the light is utilized with utmost effect by forming a single silver speck on the grain surface.

Since but a few silver atoms are sufficient to cause the whole crystal to be rapidly turned into silver by the developer, the utilization of the latent image by development implies a very great increase in sensitivity compared with the print-out process: the amplification factor of development is around 10^8, depending, of course, on the grain size; this is the main reason for the high sensitivity of photographic materials. The high amplification factor is a unique characteristic of the conventional silver-halide system; most light-sensitive systems operate without any amplification, and of all those being investigated or used nowadays, only electrophotography, photopolymerization and the "free-radical" process possess an important amplification, but none has reached the sensitivity of silver-halide systems.

COLOUR SENSITIVITY. Silver halide crystals absorb only ultra-violet and blue radiation: they may be said to be colour-deficient. Photography by such blue-sensitive materials leads to completely false tone rendering, since the green, yellow and red parts of the spectrum are not recorded and appear black. By the addition of suitable dyes, which are adsorbed to the surfaces of the emulsion crystals, it has become possible to extend sensitivity to the whole of the visible spectrum and even to the infra-red.

An emulsion which is sensitive to blue and green light is described as orthochromatic. A panchromatic material is, in addition, sensitive to red light.

For a photographic material to render the colours of an object in the correct tones between black and white, its spectral sensitivity should be the same as that of the normal human eye, i.e. with a pronounced peak in the green. With the ordinary panchromatic material, sensitivity is too high in the blue and often also in the red; this can be corrected by using suitably coloured filters.

Often, however, correct tone rendering is not the aim, since it may cause two colours which appear different in the object because of their colour difference, but which are similar in brightness, to come out as identical shades of grey in the monochrome photograph. If their colour difference is to be brought out as difference in tone, contrast filters or a non-panchromatic material have to be used, i.e., the tone rendering is deliberately falsified.

NEGATIVE TO POSITIVE. Increasing exposure of an emulsion coating causes an increasing number of grains to reach the latent-image condition and in turn to be developed to black silver. This means that the tone scale of the object is reversed; light is reproduced as dark and vice versa. The original photograph is a negative.

The positive picture is then obtained by, in effect, rephotographing the negative, which thus acts as the original for this second photograph. Printing can be done by bringing the negative and the printing material together face to face and exposing the positive material through the negative (contact printing) or by projecting the negative on to the positive material (optical printing). Negative and positive thus work together to obtain a certain tone reproduction, and a right-sided picture.

EMULSION CHARACTERISTICS

CHARACTERISTIC CURVE. The transfer process from scene to photographic negative to positive can be considered quantitatively. The characteristic curve of a photographic material describes its response to the incident light, i.e., to the exposure, which is the product of the intensity (illuminance) and the time during which it acts. The concept of density is used as a measure of response. Density corresponds to the subjective impression of blackness and is defined as the logarithm of the ratio of the incident to the transmitted light (transmission density) or to the reflected light (reflection density). Since density is a logarithmic unit, for $D = 1$, 1/10 of the incident light is transmitted, for $D = 2$, 1/100, and so on.

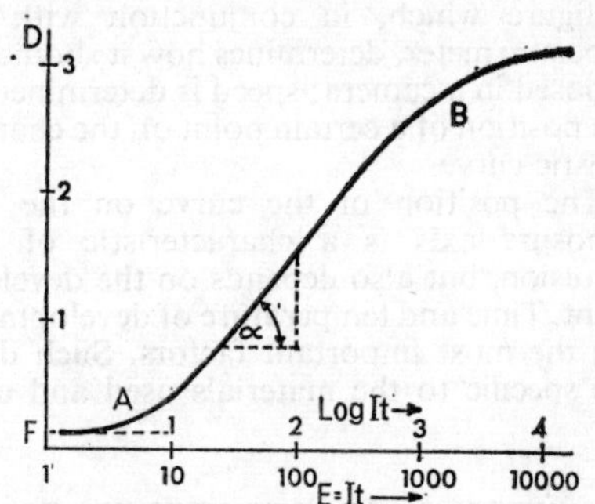

CHARACTERISTIC CURVE OF PHOTOGRAPHIC MATERIAL. Typical negative emulsion, for which density D is plotted against logarithm of exposure $E = I \times t$. Arithmetic exposure scale also shown. Gamma $(\gamma) = \tan(\alpha)$. A. Toe region of curve. F. Fog density. B. Shoulder region.

In the usual characteristic curve, density is plotted against the logarithm of exposure. The reason for using logarithmic units is purely convenience; a wide range of values can be accommodated in this way.

A certain minimum exposure is required before a negative material shows any response at all; below this exposure, a certain constant density is shown which is known as fog. With exposure increasing beyond the threshold value, response rises with increasing rapidity; this is the toe region which is followed by a region of constant slope, the tangent of which is called gamma, and in turn by a region of decreasing slope, the shoulder of the curve. With some materials, the region of maximum density is followed by one of negative slope, a phenomenon described as solarization, which is mainly of theoretical interest.

A photographic material is clearly of higher sensitivity the further to the left, i.e., towards lesser exposures, the characteristic

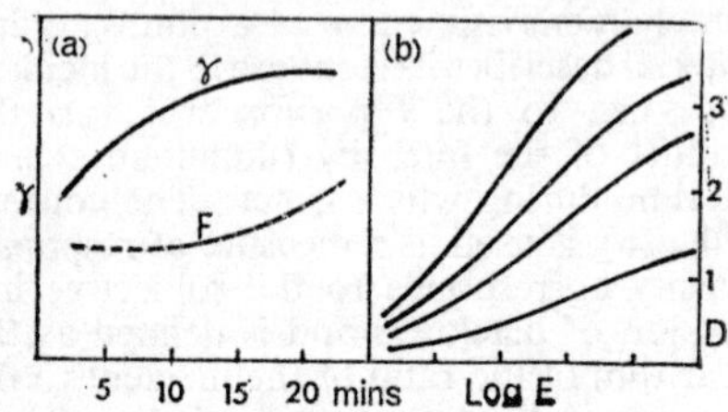

GAMMA AND DEVELOPING TIME. (a) Gamma-time curve for different times of development. F. Growth of fog against time of development. (b) Characteristic curves of same material.

curve is situated. The speed of a material is a figure which, in conjunction with an exposure meter, determines how it should be exposed in a camera; speed is determined as the position of a certain point on the characteristic curve.

The position of the curve on the log exposure axis is a characteristic of the emulsion, but also depends on the development. Time and temperature of development are the most important factors. Such data are specific to the materials used and also

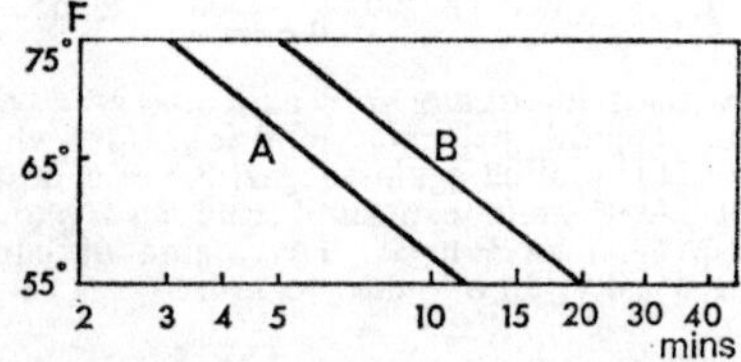

TIME AND TEMPERATURE. Relation between developing time and solution temperature for constant gamma values. A:γ=0·9. B:γ=1·1.

depend on the type of agitation used during development; they are normally supplied by the manufacturers but to be precise should be established under the working conditions of the user.

The characteristic curve of positive materials does not, in principle, differ from that of negative films; when the positives are transparencies there is no essential difference. The curves for positive papers differ from those of films in one basic aspect: their maximum densities are limited to values of the order of 1·8. This is because even the densest silver deposit reflects some of the incident light. The useful exposure range of a photographic paper is that range of exposures which leads to values of light and dark in which small tone differences are still discernible. The points at the end of the exposure range are those at which the characteristic curve still has a definite small slope of 0·1. Thus a paper cannot reproduce a tone range greater than 30 : 1. A transparency such as a positive film can reproduce a much wider range of the order of 100 : 1.

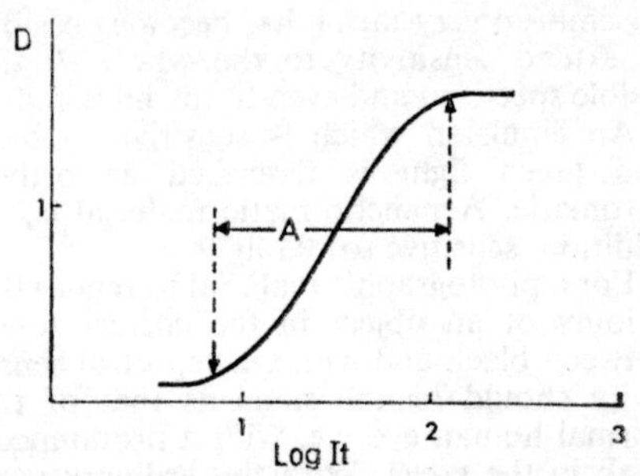

CHARACTERISTIC CURVE FOR PHOTOGRAPHIC PAPER. A. Useful exposure range, i.e. density range of a negative which could be printed on this paper. Much higher maximum densities are attainable with positive films, yielding tone ranges of 100 : 1, compared with 30 : 1 for paper.

TONE REPRODUCTION. The way in which the tones in the original scene are reproduced in the positive depends on the characteristic curves of the negative and positive materials used and on the exposures given in taking the photograph and then printing the positive from the negative.

In taking a photograph, the different areas of the film receive different exposures; the time of exposure is constant, but the intensity varies according to the brightness distribution in the original. Changing the shutter time and the lens stop varies the densities by which the different parts of a scene are recorded in the negative.

The simplest situation results if the camera exposure conditions are so chosen that the densities all lie on the straight-line part of the characteristic curve. If, in printing, only the straight-line part of the positive curve is utilized, the result will be a true reproduction of the tones of the original.

This simple approach is not possible in practice. In a paper print the straight-line

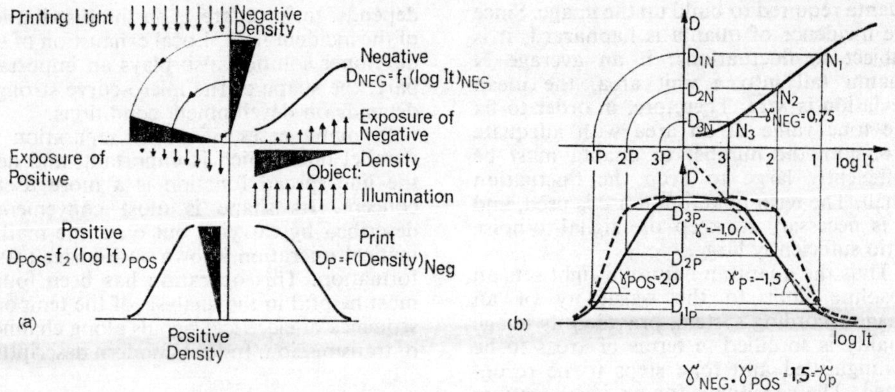

TONE REPRODUCTION IN THE PHOTOGRAPHIC NEGATIVE-POSITIVE SYSTEM. (a) Object, negative and positive represented by density wedges. Illuminated object produces negative exposure reflected through negative film characteristic to give negative density range. Printing light produces corresponding exposures on higher-gamma positive film, resulting in overall print reproduction in lower right quadrant. (b) Equivalent schematic for three sample tones in scene, 1, 2, 3, producing corresponding negative and positive densities. Overall reproduction (solid line, lower right quadrant) shows gamma = 1·5. Light dashed line shows idealized unity gamma curve for theoretically perfect reproduction. Heavy dashed line shows high-contrast reproduction with toe and shoulder compression.

portion is very short; moreover, to be acceptable, a paper print must contain all tones from black to white. A picture consisting of muddy greys looks unpleasant. Hence the curved parts of the characteristics have to be taken into account.

The system of tone reproduction can be most readily followed in the quadrant diagram. Representing the original scene are three sample tones 1, 2, 3, one a highlight, one a middle tone, one in the shadows. By proper choice of the camera exposures, these are recorded at 1N, 2N, 3N on the negative, giving densities D_{1N}, D_{2N}, D_{3N}. In printing, these negatives are uniformly illuminated; the three densities result in three exposures 1P, 2P, 3P to the positive material and, via the positive characteristic, in the densities D_{1P}, D_{2P}, D_{3P}. By means of the mirror quadrant these final densities can be compared with the original tones 1, 2, 3. For correct reproduction, they should coincide; here the three density values D_P may be shifted along the exposure axis by the same amount, which would merely correspond to altering the level of illumination of the print.

The curved shape of the positive characteristic is to some extent compensated for by utilizing the toe of the negative curve, resulting in reasonably good reproduction.

The transfer characteristic does not take into account all the factors entering the picture. Thus the wide range of tones occurring in natural scenes, for example, in shots against the light, is always compressed by the scatter of light in the lens and the camera. One per cent of light scattered uniformly over the negative material reduces an original tone range of 1,000 : 1 to 100 : 1. Scatter of light occurs with similar effect during optical printing.

DETAIL RENDERING. The foregoing has been concerned with the rendering of tones and colours in large areas of the original, both in negative and positive. The situation is altered considerably in the case of recording of fine detail. Here the granular nature of the photographic emulsion becomes important both during the taking of the photograph and in viewing it. Light is scattered in the emulsion from the bright parts of the optical image into the dark, and the resulting picture looks grainy and tends to lack fine detail. These facts make photographic materials very much inferior to the ideal receiver.

The ideal receiver is limited in its performance only by the quantum nature of light. In this theoretical model, an image is built up only in those areas on which the quanta fall. In an image, any small area should represent the luminance of the original; in other words, the tone of the original should be reproduced. The larger the number of tones to be reproduced, and the smaller the unit areas into which the original, and thus the image, may be divided, the larger is the total number of quanta required to build up the image. Since the incidence of quanta is haphazard, it is subject to fluctuations: if an average N quanta fall into a unit area, the mean deviation is $\sqrt{N}$. Therefore, in order to fix the tone value of an area with adequate precision, the number of quanta must be sufficiently large to keep the fluctuation small. The term quantum noise is used, and it is necessary to keep the signal-to-noise ratio sufficiently large.

Thus the quantum nature of light sets an absolute limit to the sensitivity of an image-recording system, provided its image quality is specified in terms of areas to be distinguished and tone steps to be recognized. The sensitivity for an ideal receiver with the performance of a stationary television image has been estimated to lie within 10^6 and 10^8 ASA units. By comparison, the sensitivity of a fast negative film is around 400 ASA.

For photographic emulsions, a detective quantum efficiency has been defined; this is the square of the ratios: (signal-to-noise ratio for emulsion) to (maximum signal-to-noise ratio possible), which latter is again governed by the quantum nature of light. Figures for this efficiency for modern emulsions are, if anything, below 1 per cent. This concept does not take the sideways scatter of light into account, but merely the grainy nature of the receiver and image.

LIGHT SCATTER. If a small point of light falls on a photographic material, light is scattered into the surrounding areas according to the point spread function of the material. This function depends both on the make-up of the emulsion and on the wavelength of light; thus with very fine-grain materials, light of short wavelength is scattered more strongly, thus leading to a wider spread than light of longer wavelengths. With more coarse-grain emulsions, where absorption is the more important factor, the situation may be reversed. The ability of an emulsion layer to yield clearly separate images of two neighbouring points of light, such as the images of two stars, will then depend in the first instance on their separation and on the shape of the spread function, secondly on the characteristic curve, which turns the resulting exposure distribution into a distribution of densities, and finally on the viewing conditions and the observer, i.e., on the criterion adopted. The characteristic curve applicable, however, is not the normal macro-curve discussed above but a micro-curve, the shape of which depends, unfortunately, on the distribution of the incident light. Local exhaustion of the developer solution also plays an important part; the shape of the micro-curve strongly depends on development conditions.

In many cases, it is the separation of parallel lines which is important, and here the line spread function is a more useful concept. Its shape is most conveniently described by carrying out on it the mathematical operation known as Fourier transformation. This operation has been found most helpful in the analysis of the temporal sequence of electrical signals along channels of transmission. Indeed, modern description

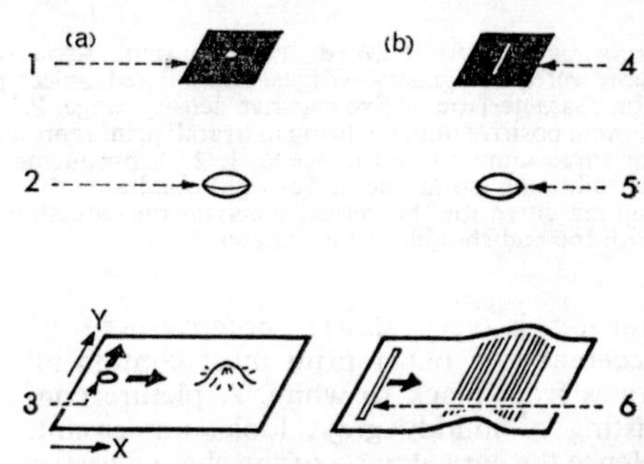

FORMATION AND SCANNING OF SPREAD FUNCTIONS. (a) Point spread function with pinhole, 3, for scanning. 1. Point source of light. 2. Lens. (b) Line spread function with slit, 6, for scanning. 4. Line source. 5. Lens.

of the optical characteristics of images, both those formed by optical systems and those recorded by image receivers, strictly follows the well-known electrical analogue with one simple difference: temporal sequences in electrical signals correspond to spatial arrays in the pictorial image. The case of the image receiver is a specially simple one in that the spread function is always symmetrical, whereas in optical and electrical systems unsymmetrical functions are possible.

The Fourier transformation of a function answers the following question: how may the function be imagined to be built up of sine waves of different frequency and amplitude? The result of this transformation of the spread function is the modulation transfer function of an emulsion, where the ratio $\frac{E_{max}-E_{min}}{E_{max}+E_{min}}$ is defined as modulation, E_{max} and E_{min} being the maximum and minimum values of a (spatial) sine wave of light intensity. The transfer function of a photographic material can also be determined more directly by exposing the material to a range of sine-wave distributions of intensity with different spatial frequencies and by determining, via the micro-characteristic curve, how the modulation changes with frequency. The resulting curve was formerly described as sine-wave response or contrast transfer function.

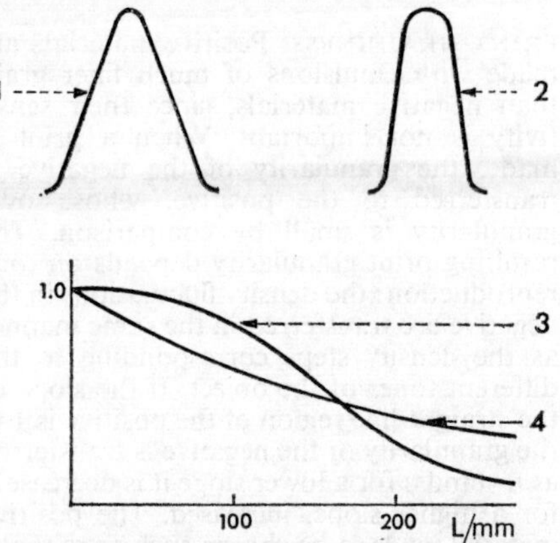

LINE SPREAD AND TRANSFER FUNCTIONS. Two photographic materials. 1. Has broad line spread function; corresponding transfer function, 3, is marked by high values at low frequencies, thus high edge sharpness. 2. Has narrow spread function; corresponding transfer function, 4, yields higher ultimate resolution.

PHOTOGRAPHIC MATERIALS AS LINEAR SYSTEMS. It is not self-evident that, by exposing a photographic material to a sinusoidal distribution of light, the resulting light distribution in the emulsion layer must also be sinusoidal with merely reduced modulation. The experimental fact that this occurs means that the material is a linear system in the sense of communication theory and makes it especially simple to treat mathematically. In particular, it means that a photographic material can be regarded as a member of a chain of optical elements. Thus, for example, a lens may, in any one part of its field, also be characterized by a transfer function. The total transfer function of lens plus film is then simply the product of the functions for lens and film at any one value of spatial frequency; in the same manner, haze between object and lens may be taken into account. The recognition of this situation by H. Frieser and later by Otto Schade paved the way towards modern methods of image evaluation.

The modulation transfer function, by definition, has the value 1·0 at zero spatial frequency and approaches zero at the limit of (line) resolution. For different systems its shape may vary widely, and the detail rendering of a system may be recognized from the shape of the function. A high value of the function for low frequencies and a short tail indicates high edge sharpness (acutance) and a poor limit of resolution; a more sloping shape with longer tail towards high frequencies means less sharp-looking edges, but a higher limit of resolution.

The modulation transfer curve of a photographic material denotes a performance which is strictly comparable to that of a frequency filter in electronic line transmission. A photographic material may be described as a low-pass filter for spatial frequencies, the transfer curve indicating the filtering action.

Some manufacturers now publish modulation transfer curves of their photographic materials. Since the concept is a purely optical one, these curves do not completely describe the behaviour of a photographic material, because they merely indicate the exposure distribution resulting from a spatial sine wave of incident light intensity. The resulting density distribution depends on development and is thus to some extent in the hands of the user. In particular, the use of developers giving a strong neighbourhood effect can sharpen the response of the emulsion, leading to more clearly defined edges. Unsharp masking methods can lead to the same result: a build-up of local density differences without an increase, or even with a decrease, in overall slope of the characteristic curve.

GRAIN AND GRANULARITY

THE GRAINY PICTURE. Any photographic picture looks grainy if enlarged sufficiently. The term graininess refers to the visual impression, and granularity to any objective measure which yields values that correlate with the graininess.

Graininess may be determined in two ways. One way is to prepare standard enlargements and determine the blending distance from the eyes of the observer at which the impression of graininess disappears. A second way is to prepare a range of enlargements of different degrees of magnification and visually compare the results with standards. Both methods suffer from subjective differences between observers and generally poor precision, so that the establishment of a recognized method of determining granularity has been a great step forward.

E. W. H. Selwyn has established that graininess and granularity arise from the purely random array of the developed silver grains in the image. The density in a small area of a uniformly exposed and processed piece of photographic material fluctuates, as

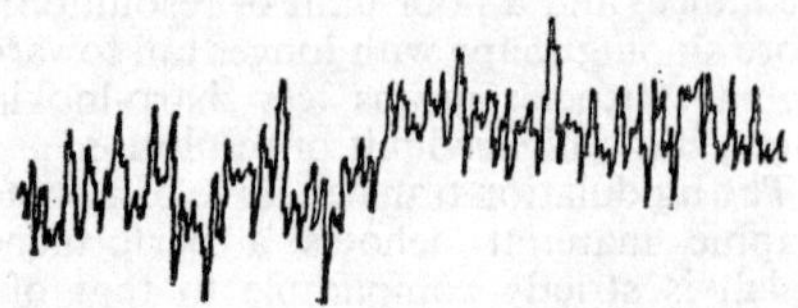

MICRODENSITOMETER TRACES. Two adjacent areas on a film with slightly different exposures and therefore slightly different mean density values. Chances of confusion are high.

is shown by any microdensitometer tracing; counting the number of times a certain deviation from the mean density value is reached and plotting this against density, gives a bell-shaped curve, the Gaussian distribution. The spread of this distribution, σ, is a measure of granularity; its actual value also depends on the size of the measuring aperture a, and is smaller the bigger that aperture. The Selwyn granularity G is defined as $G = \sigma \times \sqrt{a}$. For a density of 1·0 of a negative film and a scanning aperture of diameter 24 μ, σ has values of the order of 0·01 to 0·1, which means that 68 per cent of density values lie within, say, 0·9 and 1·0, and 1 in 1,000 outside 0·7 and 1·3.

Granularity is often said to be caused by clumps of silver grains. To the extent that the Selwyn formula applies, this is clearly incorrect, since a physical clump implies a deviation from randomness. In fact, genuine clumping is a rare incident and happens only with faulty emulsions or with infectious development such as with lith systems. Another case of genuine clumping is that of high-energy quanta, such as X-rays, or ionized particles, which each cause more than one grain to be made developable. The developed grains then tend to be close together and produce a higher granularity than on exposure to light, where more than one quantum is necessary to make a single grain developable.

SPEED/GRAIN RATIO. The practical sensitivity of a photographic material increases with grain size, since the grains require much the same number of quanta to be made developable, regardless of their size. If the average number of developed grains in a small area is N, the mean deviation is $\sqrt{N}$. Therefore the fluctuation relative to the total number of grains is larger the smaller their number, i.e., the larger they are, for the same average density. It follows that the faster a film, the larger tends to be its granularity; advances in emulsion technology consist in improving the speed/grain ratio; the last important improvement of this kind has been gold sensitization.

PRINT GRAININESS. Positive materials are made with emulsions of much finer grain than negative materials, since their sensitivity is not important. When a print is made, the granularity of the negative is transferred to the positive, whose own granularity is small by comparison. The resulting print granularity depends on tone reproduction: the density fluctuations in the negative are transferred in the same manner as the density steps corresponding to the different tones of the object. If the slope of the straight-line region of the positive is 1·0, the granularity of the negative is transferred as it stands; for a lower slope it is decreased, for a higher slope, increased. The positive material tends to be chosen such as to make the product of the gamma values of negative and positive, $\gamma_N \times \gamma_P =$ constant. The granularity of a high-gamma negative is therefore reduced in printing and vice versa. A useful measure of the granularity in the print can thus be obtained by defining as print graininess the ratio G/γ_N.

The graininess in a print, whether on positive film or on paper, is most pronounced in the middle tones: in the highlights, the density is too low to show much fluctuation, and in any case, the slope of the characteristic curve is here too low to reproduce the density fluctuations of the negative at their full value. The shadows are too dark for density variations to become noticeable. Since the middle tones in a print stem on average from a negative density of around 1·0, routine granularity measurements are usually made at that density.

Granularity as measured by transmission in a negative increases with density; very roughly $G \sim \sqrt[3]{D}$. It is advantageous, therefore, to keep camera exposures as short as possible: prints produced from overexposed negatives tend to look very grainy.

GRAIN AS NOISE. Graininess in a photograph is the exact spatial equivalent to the temporal noise in an electronic system, and both can be treated by the same mathematical methods indicated above in connection with the transfer function. Graininess tends to obscure small detail in a picture, and transfer function and granularity acting together set the limit of resolution of a photographic material. The way in which these factors collaborate is the subject of Shannon's information theory, originally developed for communication problems.

Because of granularity, in a small area, the number of tone steps which can be recognized with certainty is smaller the smaller the area, until, when it is small enough, it is sometimes possible to distinguish only two tones: black and white, or none at all.

The unit of information is the bit (short for binary digit, the unit also used in electronic digital computers). A bit of information in a photographic record is a recognizable tone step in a small area. if n is the number of areas (cells) into which the record may be divided, and m the number of recognizable steps in each cell, the total information capacity is $n \cdot \log_2 m$· It is clearly advantageous to make n as large as possible, even if that means that m is reduced, and since n and m both depend on the slope of the characteristic curve and on granularity, a curve plotting information capacity against density has an optimum. This lies at quite low densities, in the order of 0·1 above fog.

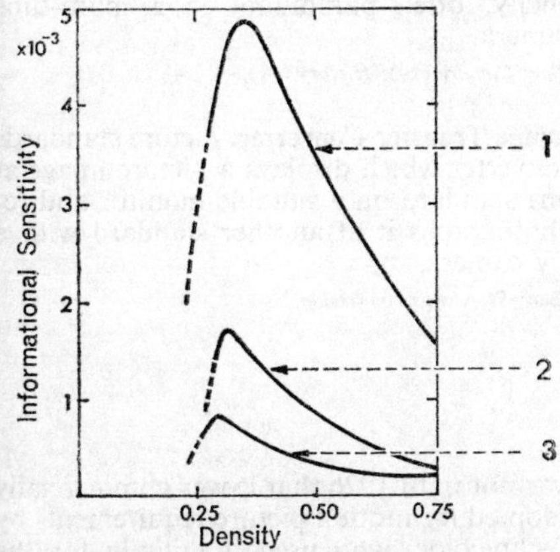

INFORMATIONAL SENSITIVITY AND DENSITY. 1, 2, 3. Three typical commercial films. Information capacity is highest at quite low density values.

A figure of merit is obtained by relating the exposure required with the number of bits per unit area that can be recorded. The detailed treatment of this situation is the subject of present-day research.

NOISE POWER SPECTRUM AND TRANSFER FUNCTION. Noise is defined more quantitatively than by the Selwyn granularity by the power which it contributes in the different frequencies. The result of Fourier analysis of a micro-densitometer recording is the noise power spectrum, which indicates the contribution that noise makes at different spatial frequencies. Frequencies higher than those at which the transfer function approaches zero are irrelevant; at frequencies lower than this critical one, the noise power indicates the degree of confusion granularity contributes to the recognition of lines of a spatial frequency which according to the transfer function should still be separated.

In both functions, factors affecting the recording of detail are plotted against spatial frequency. The transfer function indicates the ability to separate, strictly speaking, infinitely long lines of sinusoidal cross-section. The granular nature of the image is averaged out in the process of determining the transfer function. The noise power spectrum shows what happens if, instead of long lines, small areas in the picture are separated; the power of the signal which is indicated by the transfer function must be in a certain ratio to the noise power at that same frequency. The exact value of this signal-to-noise ratio necessarily depends on the criterion defining recognition, which is to some extent a matter of choice.

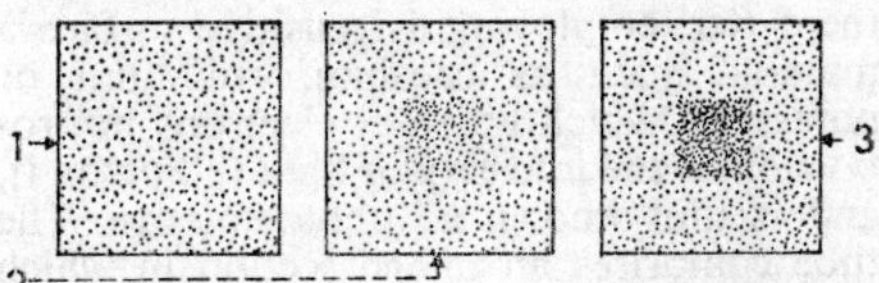

IMAGE CLARITY AND SIGNAL-TO-NOISE RATIO. SNR: 1, 4 : 1, 2, 19 : 1, 3, 46 : 1, showing that for a grainy image the ratio has to be quite high for clear recognition.

NOISE POWER SPECTRUM AND SELWYN GRANULARITY. At least for a fine grain material, the noise power spectrum of photographic granularity is essentially white: over the range it can be measured, which is limited by the transfer function of the microdensitometer, power is the same at all frequencies. The Selwyn granularity is essentially the integral over the noise power spectrum. Therefore, to compare the performance of different materials, it is irrelevant whether the information obtained from measurements of Selwyn granularity is used or the more detailed information from a noise power analysis.

The consideration of the noise power spectrum allows the printing process to be studied in more detail. A transfer function applies in transmitting the granularity of the negative to the positive; this transfer function may, e.g., be that of the enlarger lens. Even if the power spectrum of the negative material is white, i.e., if it has the same power at all frequencies, its higher frequencies are filtered out in the transfer process. Thus the noise power spectrum in the print is the product of that of the negative into the transfer function of the printing process for the lower frequency range, and simply that for the positive material for the higher frequencies. This does not, of course, mean that the positive contains high-frequency detail, since this is eliminated in the printing process along with the higher frequencies of the negative noise power spectrum; it merely means a more detailed description of the granularity pattern of the final picture. W.F.B.

See: *Chemistry of the photographic process; Exposure control; Lens performance; Lens types; Sensitometry and densitometry; Sensitometry of colour films; Tone reproduction.*

Books: *Chemistry of Photographic Mechanisms*, by K. S. Lyalikov (London), 1967; *Theory of the Photographic Process*, 3rd Ed., by C. E. K. Mees and T. H. James (New York), 1966; *Photographic Theory*, 2nd Ed., by T. H. James and F. C. Higgins (New York), 1960; *Introduction to Fourier Optics*, by J. W. Goodman (San Francisco), 1968.

□**Image Iconoscope.** Development of the iconoscope, having greater sensitivity, i.e., a greater electrical output for a given amount of light falling on the tube.

See: *Camera* (TV); *History* (TV).

□**Image Orientation.** For special effects purposes, the television image can be rotated about its centre. This may be done in two ways: by the use of rotating prisms on the front of the camera lens (this is the preferred method); or by rotating the deflector coils of the television camera.

□**Image Orthicon.** Camera tube introduced by RCA in 1945 and improved in 1949 and thereafter, employing as sensitive surface a material such as caesium, rubidium or potassium which liberates electrons in proportion to the intensity of light falling on it, and is thus known as photoemissive. The tube comprises an image section in which electrons emitted by the photo-cathode form an image on a thin glass target, an orthicon section in which the return scanning beam carries the signal, and a multiplier which amplifies the very weak output current. These elements are all arranged in a straight line, as the name of the tube implies.

In spite of its complication and rather temperamental behaviour, the image orthicon has for many years remained unrivalled as a studio camera tube, because of the crisp pictures it produces and because it has inherent compensation against overloading resulting from excessive incident light. Latterly, with the increasing emphasis on colour, it has given ground to the Plumbicon in the interests of compactness and simplicity, often paramount in a multi-tube camera.

See: *Camera* (TV); *History* (TV).

□**Image Transfer Converter.** Picture standards converter which displays a picture image at one standard on a suitable monitor and re-photographs it at another standard with a TV camera.

See: *Standards conversion.*

⊕IMBIBITION PRINTING

The term imbibition printing is used to describe a photo-mechanical reproduction process sometimes employed for the bulk manufacture of subtractive colour prints used in cinematography for general distribution. This process may be referred to as the dye transfer process, while in Russia the term hydrotype process is also used. Although the basic principles of the method have been recognized since their discovery by E. Edwards in 1875 and Leon Warnerke, a Russian engineer, in 1881, it was not until 1926 that it was commercially adopted for motion picture requirements by Technicolor, who used it initially for the production of two-colour subtractive prints, and subsequently from 1933 onwards for three-colour printing. In still photography, a similar process was later introduced by Kodak under the term wash-off relief.

PREPARING THE MATRIX

The first stage of any imbibition process is the preparation of a positive image, known as a matrix, capable of absorbing dye corresponding to the density of each part of the image required, and capable also of giving up that dye in the same proportion on transfer to another emulsion layer. The matrix therefore performs a function similar to that of the blocks or plates used for colour printing on paper and it must be capable of repeating the transfer operation over and over again in the manufacture of a large number of copies.

NATURE OF IMAGE. Although several different methods of preparing such matrices have been proposed, the only one which has been consistently employed has been that in

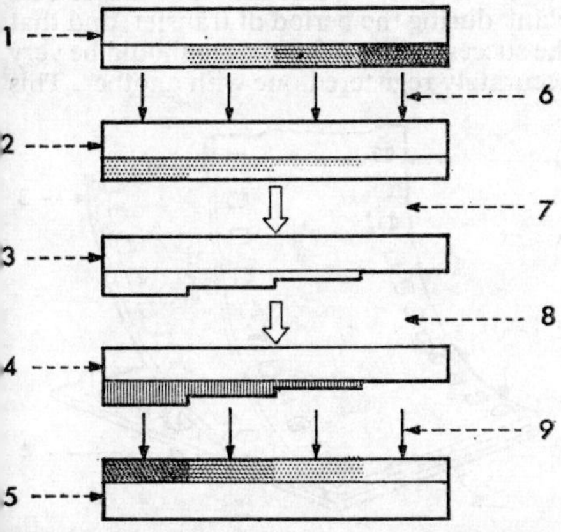

SEQUENCE FROM NEGATIVE TO DYE TRANSFER PRINT. 1. Negative (unexposed, left; fully exposed, right). 6. Optical printing stage. 2. Exposed matrix (reversed tonality). 7. Development. 3. Matrix relief image. 8. Dyeing. 4. Dyed matrix. 9. Transfer on to blank film. 5. Dye image positive print.

which the matrix takes the form of a relief image of varying thicknesses of hardened gelatin. To produce this, a suitable positive stock, usually coated with an unhardened photographic emulsion heavily loaded with absorbing dye or carbon, is printed from a negative by exposure through the base. The penetration of the light into the emulsion layer is determined by the transmission of the negative image at each point; where the negative is of minimum density, the exposure of the matrix emulsion in depth will be greatest, while where the negative density is greatest, there will be little or no exposure, and hence minimum penetration.

DEVELOPMENT. The exposed matrix is then developed in a tanning developer, which has the effect of hardening the gelatin in the immediate vicinity of the developed silver

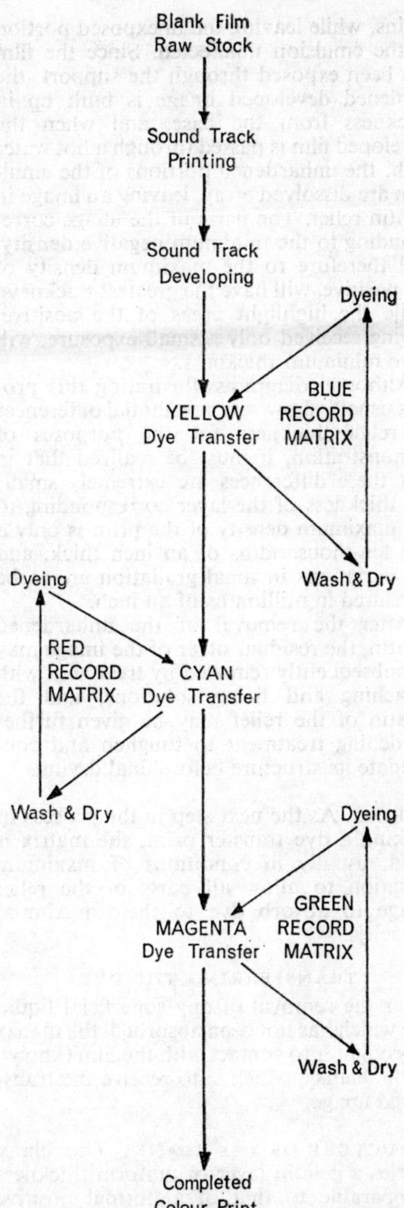

IMBIBITION PROCESSING SEQUENCE. Blank film is first printed with a B & W optical sound track which is then developed. The clear picture area next receives in succession the yellow, cyan and magenta dye images. Each is transferred from a separate matrix, corresponding to the blue, red and green records of the original negative, which has been made to absorb dye of the appropriate complementary colour. After the dye has been transferred to the blank, the matrix is washed and dried, and is then ready for re-use. A subtractive colour sound print is the result.

grains, while leaving the unexposed portion of the emulsion unaffected. Since the film has been exposed through the support, the hardened developed image is built up in thickness from the base, and when the developed film is passed through a hot water bath, the unhardened portions of the emulsion are dissolved away, leaving an image in gelatin relief. The parts of the image corresponding to the minimum negative density, and therefore to the maximum density of the positive, will have the greatest thickness, while the highlight areas of the positive, having received only a small exposure, will have minimum thickness.

Although diagrams illustrating this process usually show very substantial differences of relief thickness for the purposes of demonstration, it must be realized that in fact these differences are extremely small: the thickness of the layer corresponding to the maximum density of the print is only a few ten-thousandths of an inch thick, and the variations in tonal gradation are to be measured in millionths of an inch.

After the removal of the unhardened gelatin, the residual silver of the image may be subsequently removed by treatment with bleaching and fixing solutions, and the gelatin of the relief may be given further hardening treatment to toughen and consolidate its structure before final drying.

DYEING. As the next step in the process of making a dye transfer print, the matrix is dyed, usually in conditions of maximum agitation to allow all parts of the relief image to absorb dye to their maximum capacity.

TRANSFERRING THE DYE

After the removal of any superficial liquid dye which has not been absorbed, the matrix is brought into contact with the film (known as the blank), which is to receive the transferred image.

STRUCTURE OF THE BLANK. The blank carries a gelatin layer of uniform thickness comparable to that of a normal positive stock. This coating may be a black-and-white photographic emulsion to allow the printing of a silver sound track or a faint black-and-white picture image, if this is required to supplement the colour transfer.

The positive emulsion contains hardeners to suit the subsequent conditions of the transfer operation, and may contain mordanting agents substantive to the dye being used in order to improve dye retention and prevent unwanted sideways diffusion of the dye image.

REGISTRATION. To permit the transfer of dye from the matrix into the blank, the gelatin of the latter is already swollen by immersion in water, but the migration of the dye from one layer to the other occupies an appreciable time even at elevated temperatures, and it is necessary to provide means for keeping the two films in intimate contact for the whole of this period.

Where several successive transfers on to the same piece of blank are used to produce a three-colour picture, it is, of course, essential that there should be no relative movement whatsoever between the matrix and blank during the period of transfer, and that the successive transfer images should be very accurately registered one with another. This

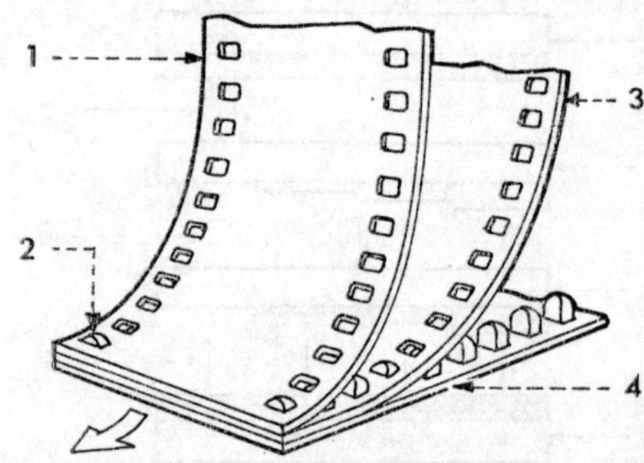

PIN BELT ENGAGING TRANSFER BLANK AND MATRIX. 1. Transfer blank (eventually becomes print), emulsion surface down. 3. Dyed matrix, emulsion surface up. 4. Flexible metal pin belt with pins engaging film perforations. 2. Matrix and blank emulsion surfaces maintained in contact for period of dye transfer.

presents no small problem in the manufacture of cinematograph film prints on continuous processing machines, and is solved by holding the matrix/blank pair in position for the time of transfer on an endless metal belt of film width, using rows of registration pins along each edge of the belt which engage accurately with the perforation holes of the two films. The dyed matrix and the blank are rolled down together on to the continuously moving belt, the pins of which hold them in contact and correct register, while the belt passes through a heated enclosure to accelerate the transfer of dye. On emerging from this enclosure, the two films are peeled away from the belt and then separated. The blank, which now carries a dye transfer image, is dried, while the matrix is washed and dried in preparation for re-use.

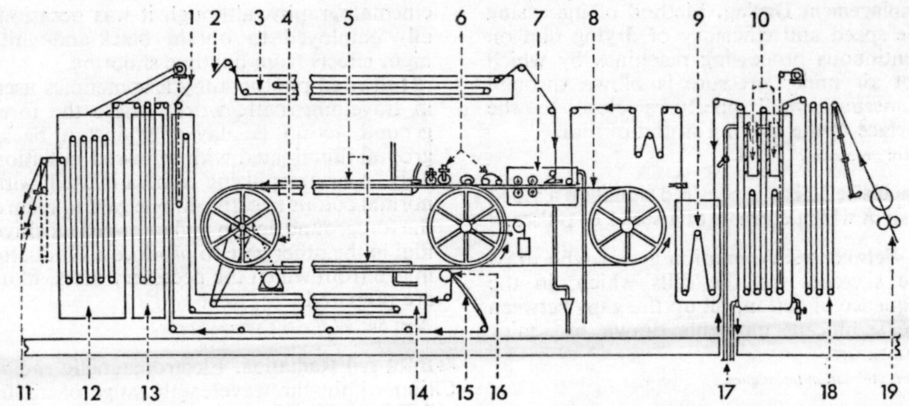

GENERAL ARRANGEMENT OF DYE TRANSFER MACHINE. Simplified schematic (from BP 307,659) of one bank of three-stage machine. 19. Matrix enters machine. 18. Feed elevator. 10. Dye tank with cascades. 9. Wash-back tank, where matrix is subjected to water jets to remove surplus dye. 7. Roll tank with jet of aerated water at meeting point of matrix and blank on pin belt, 8. 2. Blank from feed or previous machine bank. 3. Pre-wetting tank (water) to swell gelatin. 6. Rollers and seating belts to secure adhesion between matrix and blank, followed by suction and blow-off jets. 5. Enclosed hot table through which passes pin belt with matrix and blank to effect dye transfer (break indicates length much greater than shown). 4. Main drive by constant speed motor. 16. Stripper wheel. 14. Blank enters drybox to dry out before passing to take-up or next transfer stage, 1. 15. Matrix twists through 180° and passes to washing tank, 13, and drybox, 12. 11. Matrix reservoir and take-up. 17. Dye circulation feed and return pipes.

Where three-colour prints are made by this process, the blank film passes successively through three similar transfer paths receiving one of the subtractive primary colour images in yellow, cyan and magenta at each stage, to build up the final picture. Even with the control of image position obtained by the pins of the belt, the maintenance of colour registration requires the most careful control of the film dimensions by temperature and humidity at all stages, since misregistration by a few ten-thousandths of an inch can be objectionable at the large magnifications used to project the picture on to the screen of the cinema theatre. The use of low-shrinkage film base has undoubtedly assisted the solution of this problem. It is, of course, also essential to ensure the correct synchronization of the three matrices of the blank to the exact perforation during the whole of the process, in order to avoid the misregistration of the image of moving objects.

For many years it was possible to print matrices only from black-and-white separation negatives. These were as a rule produced in a special three-strip camera, in which separate strips of film were simultaneously exposed to the red, green and blue components of the scene, entailing the use of rather bulky equipment with triple magazines, two pull-down mechanisms, and registration of a high order of accuracy. In an animation camera, a filter wheel was used to expose three successive frames on each cel.

The introduction of integral tripack colour negative stocks led to the development of colour-sensitive matrix emulsions so that separation matrices could be printed directly from the colour negative through blue, red and green filters, and this is now the usual practice.

ADVANTAGES OF DYE TRANSFER

Despite its mechanical complexity, the use of a photo-mechanical process for the manufacture of large numbers of copies on a comparatively simple and inexpensive raw stock, offers a number of advantages, and the method has been extensively used by the Technicolor laboratories in the United States and Europe, and also to a limited extent independently in Russia.

Although the preparation of a set of matrices requires the use of high-precision optical printers with special film stock and processing, it is possible, once these have been prepared, to manufacture release prints in quantity at high speed with considerable consistency. L.B.H.

See: *Colour cinematography; Integral tripack; Technicolor.*

Book: *Colour Cinematography*, by A. Cornwell-Clyne, 3rd Ed., (London), 1951.

⊕**Impingement Drying.** Method of increasing the speed and efficiency of drying film on continuous processing machines by which hot air under pressure is blown through numerous small apertures close to the surface of the moving strand of film.

See: *Processing.*

□**Impulsive Noise.** Undesired signal on a video system which consists of a series of pulses.

⊕**In-Betweeners.** Animation artists who draw the sketches for the cells which, in the sequence of action, fill up the gaps between the significant moments drawn by senior animators.

See: *Animation techniques.*

⊕**Inching.** Act of moving a length of film forward or back by small amounts in a camera, projector, synchronizer, etc., to check proper threading or to establish synchronized relationships. In cameras and projectors, this is accomplished by a manual inching knob connected to the drive mechanism, often through a dog clutch.

⊕**Independent Frame.** System of film production techniques combining preconstructed sets on movable platforms with still and moving projection backgrounds. Used in the 1950s in an attempt to introduce greater efficiency into studio operations.

□**Index Tube.** A class of colour display tube (of which none at present is in commercial use) distinguished by having a single picture-writing electron gun and a referencing or index device designed to correct scanning errors and keep the beam on the appropriate red, green or blue phosphor as the receiver signal demands from moment to moment. For example, a thin line of phosphor can be kept under an electron spot if it is flanked by conducting areas which, by means of suitable circuits, are used to correct the beam tracking when excursions occur.

See: *Colour display tubes.*

⊕**Infra-red Drying.** Drying film after processing by infra-red sources or heat lamps rather than by hot air. Since both water and the silver of the photographic image are good absorbers of infra-red, the method can be very efficient in terms of energy used.

See: *Processing.*

⊕**Infra-red Photography.** Photography using the infra-red part of the spectrum. It requires the use of specially sensitized films, corrected lens focusing and special make-up of actors and has been little used in normal cinematography, although it was occasionally employed to obtain black-and-white night effects from daytime shooting.

Infra-red photography is sometimes used in travelling matte work, where the foreground action is played against a background illuminated with infra-red radiation only. A beam-splitting camera is used, with normal colour negative in one gate to record the foreground action and infra-red sensitive film in the other gate to produce a silhouette image from which the necessary matte films are prepared.

See: *Special effects cinematography.*

⊕□**Infra-red Radiation.** Electromagnetic radiation within the wavelength range of about 7,500 to 100,000 Angstroms, i.e., between the red visible light and UHF radio waves. Also known as heat radiation. Not appreciably scattered by fog or haze, and consequently photographs taken on infra-red sensitized plates may frequently disclose detail invisible to the naked eye. Photographic sensitivity, however, does not usually extend much beyond wavelengths of 13,000 Angstroms.

See: *Electromagnetic spectrum.*

⊕**Inkers.** Artists in an animation studio who trace the outlines of drawings prepared by animators on to cells.

See: *Animation techniques.*

□**Inlay.** In television special effects, method of combining two pictures to give a composite picture which contains full level components from selected parts of both original sources. A third source provides a signal which is used to key or gate out a portion of the background picture. The other source is then used to fill in the holes made in the background picture. The keying is carried out by an electronic switch.

See: *Special effects* (TV).

□**Inlay Mask.** When keying signals are electronically generated by means of a flying spot optical system, the shape and timing of the keying signal is determined by a mechanical cut-off, sometimes known as an inlay mask.

See: *Special effects* (TV).

⊕□**Insert.** Intercut shot of a static object. Inserts (books, letters, clock on mantlepiece, view from window, etc.) are usually shot separately from the main sequence of action in film, and are covered by a special insert camera in television.

See: *Editor.*

❑**Insertion.** In television special effects, the substitution of one picture for another over a part of the camera field. This can take two forms:
1. Inlay. A small portion of the output from one camera is inserted into that of another in the form of a window, both pictures being fully modulated;
2. Overlay. The subsidiary signal is inserted by keying itself into the main picture. The point at which keying occurs is dependent on the level of modulation in certain portions of the subsidiary picture.

See: *Special effects* (TV).

⊕**Inspection.** Periodic inspection of film to ensure that it is in proper physical condition, or to report on defects, is essential to accurate laboratory controls. Negative inspection is carried out by hand-winding on an inspection table, a metal table equipped with two rewinders, or a horse and a rewinder, between which is placed a light box, so that the operator can minutely inspect negative or other film for defects. Positive inspection in the laboratory is by projection, often at much higher than normal speed, with an inspector trained to detect faults under the special viewing conditions. Periodic inspection of narrow-gauge release prints is often by high-speed machines which can detect perforation damage and other faults.

See: ***Laboratory organization; Release print inspection and maintenance.***

❑**Instantaneous Characteristic.** Due to non-linearity of components, an amplifier usually has different properties according to the signal it is handling; e.g., the gain of a sound amplifier will probably not be the same for low-level speech as for a steady high-level sine wave. The instantaneous characteristic is the response to an occasional transient.

See: ***Sound recording; Steady state characteristic.***

⊕**Integral Screen.** Type of screen used in viewing stereoscopic (3-D) motion pictures in which the principles of Lippmann's integral photography or its derivatives are employed. More generally, any type of screen in which the separation of left- and right-eye images is a function of the screen itself, so that polarizing or anaglyphic glasses do not need to be worn.

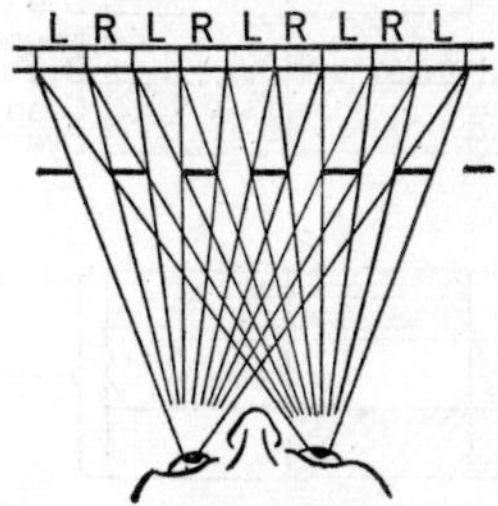

INTEGRAL SCREEN. **Screen slats or lenticulations block view of left-hand strips from right eye, and vice versa.**

See: ***Stereoscopic cinematography.***

⊕INTEGRAL TRIPACK SYSTEMS

One of the earliest suggestions for obtaining three differently coloured images from three superimposed differently sensitized emulsion layers came from Karl Schinzel.

COLOURING THE IMAGES

DYE DESTRUCTION. Schinzel proposed introducing an appropriately coloured dye into each emulsion layer at the time of coating, and then, following exposure and first development, destroying the negative silver images and the adjacent dyes in a bleaching solution. In this way, three superimposed positive colour images would be obtained. The colours would normally be cyan, magenta and yellow as these are complementary to the sensitivities of the three emulsion layers, red, green and blue.

Theoretically, this system of dye destruction can be used to form either negative or positive images, but in practice material of this kind suffers a serious disadvantage when considered as a taking film because its light sensitivity is so low as to be virtually useless for exposing in a camera. That dye-destruction materials can be used for printing was proved originally by the Gasparcolor process and has been shown more recently by the Cibachrome Print process.

INCORPORATED COUPLERS. Another way of separating the three images of an integral tripack was suggested by Rudolph Fischer in 1911. In each of the three colour layers at the time of coating, Fischer suggested incorporating a colourless substance which, by reaction with the oxidation products resulting from development, would form three differently coloured dye images, all of them associated with their respective silver images. But the silver images could subsequently be removed in a bleaching solution that would leave the dye images unaffected.

In due time, this system proved to be quite practicable and was first successfully operated by Agfa, both for negative films and printing materials.

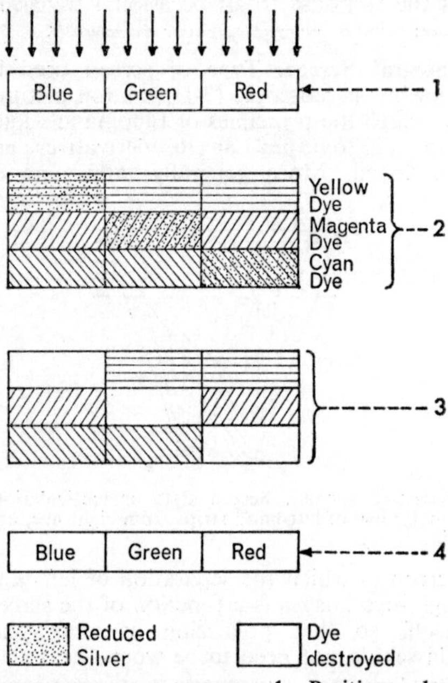

DYE DESTRUCTION PROCESS. 1. Positive colour transparency to be printed. 2. After exposure, printed material is processed in B & W developer to form negative silver images in three emulsion layers. Silver images are then removed in presence of suitable catalyst, along with dyes in contact with them, leaving composite positive dye image, 3. This is a reproduction, 4, of original, 1.

SEQUENTIAL DEVELOPMENT. There is another way of separately forming the three necessary colour images in an integral tripack. If, after developing the three separation records to form silver images, each of the three residual positive images is converted into appropriately coloured dye images by means of a sequence of separate processing stages, then a multi-coloured composite image results. Sequential colour development can be carried out in two different ways; one is by controlled penetration, and the other by selective re-exposure. Both these methods have been used for processing Kodachrome films, although the controlled penetration procedure was abandoned after a relatively short time in favour of selective re-exposure.

Because of the long sequence of steps

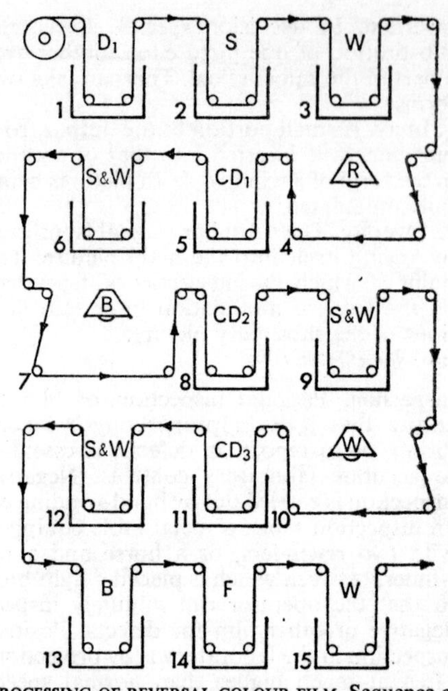

PROCESSING OF REVERSAL COLOUR FILM. Sequence of operations for film having no dye formers in original emulsion. Lengths of film, spliced into continuous rolls, are fed at constant speed through large number of tanks. Different times are provided by different lengths of loops within each bath. 1. First developer. 2. Stop bath. 3. Wash. 4. Exposure to red light. 5. Colour development to form cyan dye image. 6. Clearing (usually a stop bath and one or more washings). 7. Second re-exposure to blue light. 8. Development in yellow-dye forming solution. 9. Stop bath and wash. 10. Exposure to white light. 11. Magenta-forming development. 12. Clearing. 13. Silver bleaching. 14. Fixing. 15. Final wash.

involved in processing a Kodachrome type film, the processing machine and the control techniques are relatively complex. It is also very difficult, though not impossible, to apply the process to any form of material that cannot be run at uniform speed through a continuous processing machine.

Although it is possible to use the Kodachrome type of process to form a colour negative image, the method is, in fact, used commercially only to produce positives. The serious limitation of the system for colour negatives is that colour masking images cannot easily be obtained.

SUBSTANTIVE AND NON-SUBSTANTIVE. Over the years, those processes such as Agfacolor, Ektachrome, Kodacolor and the various Eastman Color motion picture films have come to be classified as substantive or coupler-containing films, whereas processes such as Kodachrome, where the couplers are in the developers, are generally said to be non-substantive films.

Although it is reasonably satisfactory to group materials such as Agfacolor and Ektachrome under the same general heading of coupler-containing processes, there are, in fact, basic differences between the methods used to introduce couplers into the emulsion layers of integral tripacks. Agfa's technique is to use organic colour-coupling compounds of a molecular structure that ensures that the couplers, while sufficiently soluble to allow their introduction into the emulsion before coating, become substantive or non-diffusing after coating. Ektachrome films, as well as the whole range of Kodacolor and Eastman Color motion picture films, incorporate colour couplers that are dispersed as very fine particles or globules each of which is protected by a membrane or envelope of a resin-like water-permeable but water-insoluble substance.

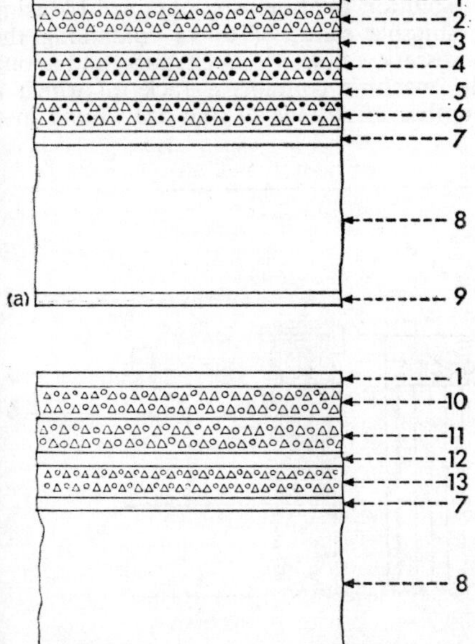

SCHEMATIC CROSS-SECTIONS OF EASTMAN COLOR FILM. (a) Negative film. 1. Gelatin supercoating. 2. Blue-sensitive emulsion with uncoloured yellow dye coupler. 3. Yellow filter layer. 4. Blue and green-sensitive emulsion with yellow-coloured magenta dye coupler. 5. Gelatin inter-layer. 6. Blue and red-sensitive emulsion containing reddish-orange coloured cyan dye coupler. 7. Substratum layer. 8. Film base. 9. Anti-halation layer. (b) Print film. 1. Gelatin supercoat protecting green-sensitive emulsion layer, 10, containing uncoloured magenta dye coupler. 11. Red-sensitive emulsion layer incorporating uncoloured cyan dye coupler. 12. Gelatin inter-layer. 13. Blue-sensitive emulsion layer incorporating uncoloured yellow dye coupler. Substratum, 7, binds all layers on to film base, 8. 14. Removable anti-halation backing.

Quite apart from the great advantage that an integral tripack material offers by allowing the use of normal exposing techniques and equipment, a substantive type colour material requires quite straightforward processing – to form either a negative or positive image by direct reversal or by printing. For these reasons, coupler-containing materials have come to be used more than any others for professional motion picture purposes.

COLOURED COUPLERS. It has been assumed so far that colourless coupling substances are introduced into the three emulsion layers of an integral tripack, but Eastman Color Negative film has two of the three couplers coloured at the time they are incorporated into their respective emulsions. Coloured couplers are used in the green- and red-sensitive emulsion layers and each is a colour which matches the unwanted absorption of the image dye it produces on colour development.

Coloured couplers and the masking images they produce are used in colour negative films because even the best available substantive dyes fall short of theoretical requirements in terms of their ability completely to absorb or transmit the light they should. For instance, the cyan dye is intended to absorb red light and at the same time transmit green and blue light completely. In practice, all cyan dyes absorb some of the green and blue light they should pass. Similarly, a magenta should absorb nothing but green light, which it should do perfectly; but in fact all magenta dyes absorb much green or red light.

Any undesirable or spurious absorption of one or more of the three dyes used in a colour system results in degraded reproduction – the extent of degradation depending upon the amounts by which the dyes fall short of theoretical requirements.

EASTMAN COLOR NEGATIVE. In Eastman Color Negative films, the spurious absorptions of two of the dyes, the cyan and the magenta, are compensated in an ingenious manner. To understand the mechanism, it will be useful to consider it layer by layer from the outer surface of the monopack down to the film base. The top layer of the film is a clear gelatin supercoating, merely intended to protect the emulsion layers from unnecessary physical damage during handling. The next layer is a blue-sensitive emulsion in which there is dispersed a colourless coupler capable of yielding a yellow image on development. The third layer is a yellow filter layer to prevent the passage of blue light to either of the two lower emulsion layers which, besides being green- and red-sensitive respectively, are also sensitive to blue light. The fourth layer is a green-sensitive emulsion in which there is dispersed a yellow coloured coupler capable of yielding a magenta dye when coupled with oxidized developer. Wherever magenta dye is generated, the original yellow colour of the coupler is discharged, but the unconverted coupler corresponding to the unexposed areas of the layer remains yellow and therefore provides a positive masking image that compensates precisely for the unwanted blue absorption of the magenta dye.

The next layer is a plain gelatin separating layer between the green- and the red-sensitive layers. The bottom emulsion layer is red-sensitive and contains a reddish-orange coloured coupler. During development, this coupler combines with oxidized developer in the area of the developed silver image to form an associated cyan dye image. Wherever the cyan dye is formed, the original orange-red colour of the coupler is discharged, but the residual unconverted orange-red dye forms a positive masking image to compensate for the spurious blue and green absorptions of the cyan dye. Finally, to complete the monopack there is a substratum to hold the red-sensitive emulsion on to the film support. On the opposite side of the film there is a soluble black anti-halation backing layer that is removed during processing.

COLOUR DEVELOPMENT

Motion picture film is almost always processed on machines through which the film is transported continuously at a constant speed. This means that the required sequence of solution treatments is obtained by using a suitable series of tanks containing the necessary solutions. Each tank unit along the machine contains a rack on which a number of strands of film are wound in a

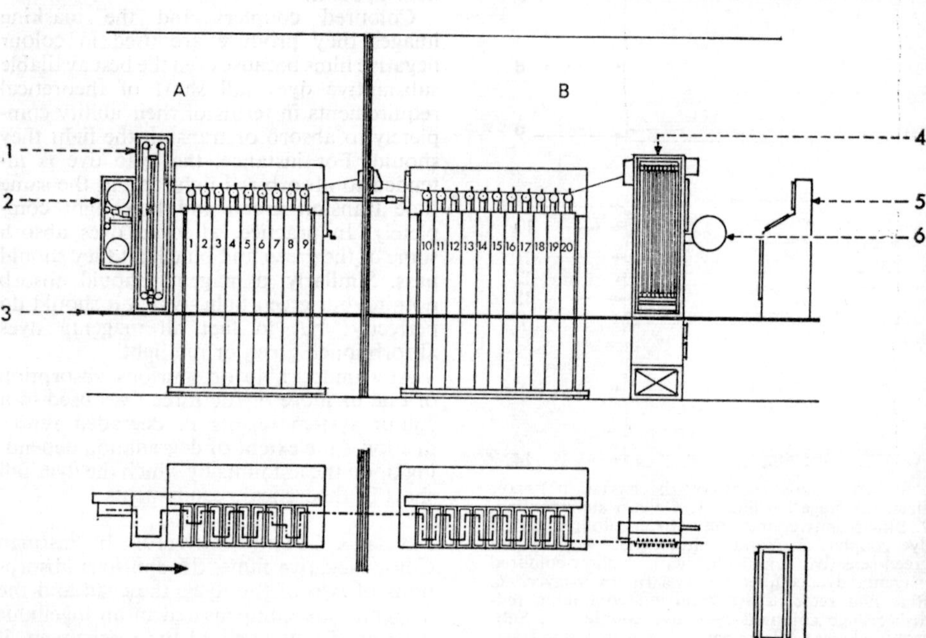

CONTINUOUS PROCESSING MACHINE SUITABLE FOR COLOUR FILM. 1. Film reservoir (elevator) from which exposed or blank leader film is fed into processing machine while new roll of film, 2, is being joined to preceding one. 3. Platform on which machine operators stand. Machine is divided into two parts: (A) operating in the dark, (B) in the light. Solution tanks numbered 1–20 may be used in single units or joined in groups to provide longer treatment times for some solutions. After following path of broken line, lower diagram, processed film passes into drying cabinet, 4, and is wound on take-up spindle, 6. Console, 5, incorporates all electrical switches and indicators.

helical path around sets of rollers at the top and bottom of the rack. In a typical machine, running at about 50 ft. a minute, each rack may hold 16 strands of film (eight strands on each side), so there may be a total of 100 ft. of film in each tank at any one time. If there are 20 or more rack and tank units in a machine, some 2,000 ft. of film are required to thread it.

MACHIN EFEED MECHANISM. Sometimes the film is driven through a processing machine by means of sprockets engaging in the perforations of the film. More often nowadays, friction or tendency drive machines are used. These do not use sprockets but depend upon the variations in the tension on the film at any point along the machine to control the speed at which the film is fed into each section. In other words, if the film strands in a particular rack become tight, the rate of removal of film from that tank is slightly reduced until sufficient film has entered the section to release the tension. This type of drive, although in a sense intermittent, is in fact so quickly self-compensating that for all practical purposes the film moves through the machine at a uniform speed.

Reacting solutions such as developer, bleach and fixer are pumped to the machine tanks through pipe lines that enable the solutions to be continuously circulated between the machine tanks and reservoirs in which heat exchange takes place for temperature adjustment. It is usual to control developer to within a quarter of a degree of Fahrenheit of the required temperature. Less critical solutions such as bleach and fixer can usually be allowed a latitude of plus or minus 2°F.

AGITATION. Because the agitation of processing solutions can have a significant effect on the rate and extent of the chemical reactions that take place, it is essential to control this variable in film processing. Movement of film through a developing solution at high speed does, of course, provide a form of agitation, but because it is a unidirectional movement of the film relative to the solution, it usually leads to undesirable effects such as directional streaks if it is not augmented by movement in some other direction.

Agitation by means of jets of compressed air, although very effective, is not acceptable in a developer, because oxidation of the developing agent would be excessive. It is usual, therefore, to introduce turbulation of the solution by means of submerged jets close to the emulsion surface of the film.

PROCESSING SEQUENCE. A typical processing sequence for a colour negative material usually commences with treatment in an alkaline pre-bath to soften the pigmented anti-halation layer coated on the back of the film. Removal of the softened layer often requires the use of soft rotating buffing wheels. Backing removal is usually carried out in a side tank through which water can be circulated continuously to ensure that the backing pigment is all washed away and none of it remains attached to either surface of the film.

A short spray wash follows and then the film enters the colour developing solution. Here the exposed silver halides are reduced to metallic silver and the couplers adjacent to the reduced silver react with the oxidation products of development to form coloured dye images *in situ* with the silver images. This reaction removes the original coloration of two of the three couplers wherever a negative silver/dye image has been formed. After removal of the remaining unexposed silver halide by a hypo fixing bath, the negative silver image itself is destroyed by bleaching followed by a second fix to leave only the coloured dye image. A final wash completes wet processing and the film then only has to be dried and possibly lacquered before it is ready for use.

TYPICAL COLOUR NEGATIVE PROCESSING SEQUENCE

Step	Minutes	Seconds	Temperature
Pre-bath		10	70°F
Spray Rinse		10–20	70°F
Colour Development	12		
Spray Rinse		10–20	70°F
First Fix	4		70°F
Wash	4		70°F
Bleach	8		70°F
Wash	4		70°F
Second Fix	4		70°F
Wash	7		70°F
Stabiliser	1		70°F
Spray Rinse		1–5	

The formation of the image in colour positive stock by colour coupling at the development stage is similar, and the processing sequence follows a series of steps resembling that of the negative. However, the need for a sound track on the film introduces an additional stage, since the infra-red absorption of the normally processed colour image is inadequate for a sound track to be used with the deep red sensitive cells standard in professional motion picture projector sound heads. It is therefore customary to redevelop the sound track image, after the regular bleaching stage but before fixing, by the application of a viscous stripe to that area only, so as to produce a silver image together with the colour track image. This silver image is confined to the upper layer, usually that carrying the magenta image, so as to obtain the sharpest possible sound record.

TYPICAL POSITIVE PRINT FILM PROCESSING SEQUENCE

Step	*Minutes*	*Seconds*	*Temperature*
Pre-bath		10	
Spray Rinse		10–20	70°F
Colour Development	14		
Spray Rinse		10–20	
First Fix	4		
Wash	4		
Bleach	8		
Wash	2		
Sound Track Development		10–20	
Wash		10–30	
Second Fix	4		
Wash	7		
Stabiliser		5–10	

The process for reversal colour film differs in that the first development is in a normal black-and-white developer in which the latent image from the original exposure is converted to silver; at this step no colour image is produced. The remaining undeveloped silver halide is then uniformly exposed, or in some cases chemically treated, so as to convert it to a latent image which can be developed at the next step in a colour coupling developer so as to produce coloured dye images corresponding to areas which were not originally exposed. The whole of the silver produced by both developing operations is then bleached and removed by fixation, so that the final result is a colour positive image on the film originally exposed in the camera.

In the most recent varieties of reversal processing, the secondary re-exposure is omitted as a separate stage, and a chemical agent having the same effect is incorporated in the colour developer. To permit shorter processing times, processing solutions at temperatures of 95–100°F rather than the standard 70°F are used, and the film is given a preliminary hardening bath to prevent the gelatin emulsion becoming unduly softened.

TYPICAL REVERSAL FILM PROCESSING SEQUENCE

Step	*Minutes*	*Seconds*	*Temperature*
Pre-hardening	2	35	95°F
Neutraliser		31	95°F
First Developer	3	20	98°F
Stop		31	95°F
Wash		31	98°F
Colour Developer	3	37	110°F
Stop		31	95°F
Wash	1	2	98°F
Bleach	1	33	95°F
Fix	1	33	95°F
Wash	1	2	98°F
Stabiliser		31	90°F

REPLENISHMENT. A source of variation that must be controlled in all forms of photographic processing results from the exhaustion of solutions that become either depleted or contaminated from continuous use. To offset such changes, replenishing solutions must be carefully formulated and introduced into working solutions at the exact rate to compensate for the decline in activity that takes place as the solutions are used.

When a satisfactory replenishment system has been established, it is possible to operate a processing machine for very long periods without discarding any of the solutions. Some used solution is, of course, removed from the working baths by the film itself and by the overflowing that results from the addition of replenisher to the system.

CONTROLS. Generally, results obtained by sensitometric measurement tend to be substantiated by chemical analysis, but occasionally the two cannot be reconciled and then there is little choice but to accept the sensitometric indications – particularly if they are supported by visual examination of current work.

Accurate knowledge of the concentrations of all the significant chemicals in all the working solutions of a process is the aim of the control chemist. He usually employs methods of analysis that have been worked out and recommended by the manufacturer of the film stock. The frequency of his analysis of any given solution depends upon its stability and relative importance in the system. Such variables as pH, developing agent, concentration of bromide, are checked frequently and results recorded in graphical form so that any trends can be more quickly detected.

Two further areas of control are involved when prints are made from colour negatives – the control of printing machines and of sound track quality.

Photometric measurements at the exposing gate coupled with test prints made from standard negatives normally provide all the data necessary to keep a printer under control. In fact it is sometimes possible to use a printer as a sensitometer on which control strips can be exposed for other purposes.

PRINTING COLOUR NEGATIVES

The only practical use for a photographic negative is for printing as it is valueless for direct viewing. This is why it is possible to incorporate colour masking images in a negative material and not in a reversal or a print material that will be viewed or projected.

There are two basic methods of printing colour negatives, one known as subtractive or white light printing and the other additive printing. The difference between the two methods lies in the type of printer used and in the quality of the results obtainable. To understand why the method of printing can cause differences between prints made from the same colour negative, it is necessary to consider the principles involved.

SUBTRACTIVE AND ADDITIVE PRINTERS. Ideally, each of the three component images comprising a colour negative should modulate only one of the three layers of the print material. In other words, the red record, cyan image, of a colour negative should, when printed, have no effect whatever on the red-sensitive cyan-forming layer of the print film, but should serve only to modulate the green-sensitive and the blue-sensitive layers. In practice, it is possible to approach these theoretical conditions more closely if the printing light is employed in the form of isolated red, green and blue rays, rather than as composite white light.

With a given light source, the amount of light available for printing additively is greatly reduced when it is passed through red, green and blue colour filters. It also tends to be rather more difficult mechanically to apply colour balance and density adjustments when using additive rather than white light printing.

In its simplest form, a subtractive motion picture colour printer is a white light printer in which colour corrections are effected by the introduction of low density colour filters into the printing beam. Changes in density can be made by varying the size of an aperture in the optical train. This type of printer can be made to operate at high speeds because there is no serious waste of light.

An additive printer combines red, green and blue light into a printing beam and the filters used to provide the coloured light are carefully chosen with respect to the spectral transmissions and sensitivities of the corresponding layers of the negative image and the positive film. Modern additive printers usually incorporate means of adjusting the individual contributing light beams so that both colour and density corrections can be effected at the same time. These adjustments have to be made very quickly between successive scenes in a film so that a minimum number of frames is involved during the change. If too long a time is taken, the change becomes obvious on the screen.

LIGHT VALVE OPERATION. Some modern colour printers operate at speeds as high as 180 ft. a minute, and to obtain maximum light efficiency and at the same time provide extremely rapid adjustment of the three printing beams, it is usual to employ some form of electromechanical light valve.

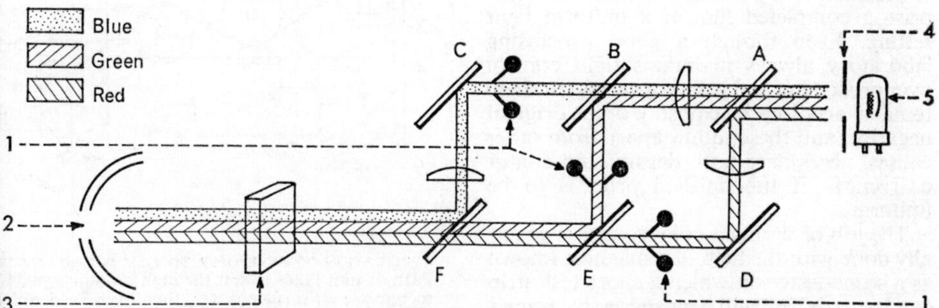

BELL AND HOWELL ADDITIVE PRINTER. White light from source, 5, is split into three beams by dichroic reflectors, A, B, C. A reflects red light and transmits both green and blue, B reflects green and transmits blue, while C reflects blue light only. A second set of dichroic filters, F, E, D, are used to recombine the three beams of light after they have passed through light valves, 1, which control the intensity of the separate beams in accordance with information from a punched tape control band. 2. Printing aperture. 3. Condenser system. 4. Overall light fader.

In a typical printer, light from a 1,000-watt incandescent lamp is split into three beams by means of an optical divider. A colour filter and a light valve is inserted into each of the beams before they are optically recombined and imaged on to the printing aperture. Because of the high intensity of the light they are transmitting and absorbing, the filters used in high speed colour printers are not dyed gelatin but are formed by vacuum deposition on glass so that they are not only extremely efficient but also resist deterioration.

Each separate light beam in the printer can be independently attenuated by its own light valve, and high speed scene-to-scene changes are initiated either by notches cut into the edge of the negatives or by means of a synchronized punched tape system.

In practice, while one scene is being printed with the light valve set to provide the required colour and intensity of light, a tape reader establishes the setting for the next scene and stores it. Then, when the end of the first scene is reached and a signal or cue is transmitted to the reader, the stored settings are brought into play within a few milliseconds. In fact a colour change can be made within a quarter of an inch of film with the printer running at nearly 200 ft. a minute.

GRADING. Even though the mechanical, optical, and electronic elements of a modern colour printer are highly refined and the operation of the machine is automatic, the information that must be fed into a printer before it can be used effectively still depends upon the skill and subjective judgment of a human grader or timer.

For a number of reasons, it is not possible to print the many different scenes that comprise a completed film at a uniform light setting. Even though a good processing laboratory always maintains tight control over processing and printing variables, there remain variations in exposure of the original negative, and these, quite apart from other causes, necessitate both density and colour corrections if the finished print is to be uniform.

The job of timing a colour print is generally done with the help of a machine known as a scene-tester on which a short test strip can be made that will bear successive frames of the same scene at slightly different colour balances. The frame that appears to yield the best result can then be selected and translated into terms of the required printer settings.

A rather more sophisticated method of grading now used by some laboratories is to employ what is in effect a closed circuit television system, so that a positive image on a colour tube can be obtained directly from a colour negative film. Adjustments to the colour of the image on the tube can then be made empirically until a satisfactory result has been obtained – after which the colour tube settings can be translated into printer light settings.

Although scene testers and image simulators can help a colour timer, his proficiency must depend very largely upon experience. Even when scenes have been graded and appear to be satisfactory in isolation, it is often found that further slight changes in colour balance are necessary when a number of scenes are assembled together in a complete print.

So far it has been assumed that release prints are made from original camera negatives, but in practice this rarely happens. Usually master positives and duplicate negatives are used for production printing, not only to safeguard the original negative but also to enable optical effects to be introduced into a film.

SOUND TRACK. In the days of black-and-white film production, it was a relatively simple thing to produce a print that combined picture and sound track. It is not

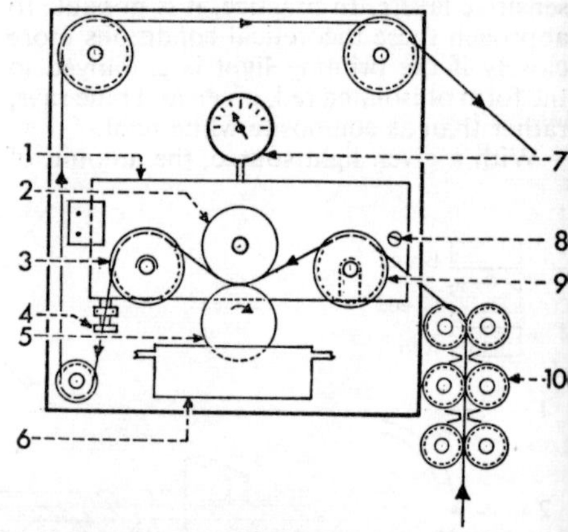

APPLICATION OF DEVELOPER TO FILM SOUND TRACK. Path of film takes it first through air squeegee, 10, to remove all surface water, then over fixed roller, 9, back-up roller, 2, and fixed roller, 3. 6. Reservoir from which applicator roller, 5, picks up supply of developer. Back plate, 1, carries film guide roller, 2, which can be adjusted precisely in relation to rotating applicator roller, 5. Adjustment is by pivoting back plate and roller about pivot, 8, by adjustment screw, 4, reading on dial indicator, 7.

quite so easy to provide a satisfactory sound track alongside a colour image. For one thing, the colour image is composed of dyes which, if used for the sound track, would not provide sufficient modulation to produce satisfactory sound quality. After early attempts to use different combinations of coloured tracks with differently sensitive photo cells, it was found that only a silver track is capable of producing an acceptable result.

The most usual way of forming a silver sound track on a colour film is to apply developer to the track area alone by means of a rotating applicator wheel. This is done at that stage of processing when the silver images that occupy the dye images have been reconverted to silver halides prior to being removed in a final fixing bath. At this point, if the silver halide in the sound track area is redeveloped, the resulting sound track is composed of silver and dye images and is not affected when the film enters the fixing solution.

CONTRAST CONTROL FOR TELEVISION. Colour film for colour television transmission must not contain images of too high a contrast, and the use of colour positive stock of lower gamma than that normally used for optical projection has been recommended in addition to flat lighting during original photography to reduce the tonal range of the subject matter. Since the contrast of colour positive materials cannot normally be altered by varying the developing process, as is possible in black-and-white, special stocks of inherently lower gamma have been produced, although this is bound to reduce the saturation of the colours which can be obtained. J.H.C.

See: *Chemistry of the photographic process; Colour cinematography; Colour printers (additive); Printers; Processing; Sensitometry of colour films.*

Books: *Reproduction of Colour*, by R. W. G. Hunt (London), 1967; *History of Colour Photography*, by J. S. Friedman (London), 1968; *Colour Cinematography*, by A. Cornwell-Clyne, 3rd Ed. (London), 1951; *Production of Motion Pictures in Colour using Eastman Color Films*, Eastman Kodak; *Manual of Processing Ektachrome Film*, Eastman Kodak.

Intensification. Chemical methods of increasing the density of a processed photographic image. Only of exceptional application in motion picture film processing.

Intensity Scale Exposure. Method of exposing step wedge tests of photographic materials in which the time of exposure is constant but the steps are varied in their intensity of light.

See: *Sensitometry and densitometry.*

Intercarrier Sound. System used in a TV receiver with f.m. sound, where there is a common i.f. amplifier for vision and sound. The f.m. sound passes through the envelope detector for vision and appears as a frequency modulated difference frequency equal to the sound and vision carrier spacing. This is subsequently limited and discriminated to detect the audio signal.

See: *Receiver.*

Intercut. Means of relating two independent sequences of action, or two single shots, in such a way as to give them a meaningful relationship. Similar to a cross cut.

See: *Editor.*

Interference Filter. Optical filter in which the wavelengths of the spectral band transmitted are determined by interference effects taking place between the thin transparent layers of different refractive index with which the filter is coated. Interference filters usually reflect those wavelengths which they do not transmit, and can be made to be effective in ultra-violet and infra-red wavelengths as well as in the visible spectrum.

See: *Camera (colour); Colour printers (additive); Telecine (colour).*

□**Interfield Cut.** Cutting or switching from one source to another during the vertical or field sync period. With random switching a good deal of picture disturbance is obtained, but in an interfield cut, the signal to cut is stored and a switch is made electrically or by relay only during the blanking interval. This gives no video disturbance, but if the syncs are too badly damaged and are not properly restored in the interfield cut, frame roll may be obtained on a flywheel sync receiver. Also called cut in blanking.

See: *Vision mixer.*

□**Interlace.** The technique of interposing two fields to give minimum flicker. A television raster is said to be interlaced if, as is common practice in broadcasting systems, the scanning lines comprising each complete frame of information are divided alternately into two consecutive fields in such a manner that the lines of one field interleave, or interlace, in position with those of the other. Consequently, although the horizontal velocity of the scanning spot both in the receiver and at the transmitter remains unaltered, the flicker rate of the reproduced picture, which would otherwise be unacceptable, is

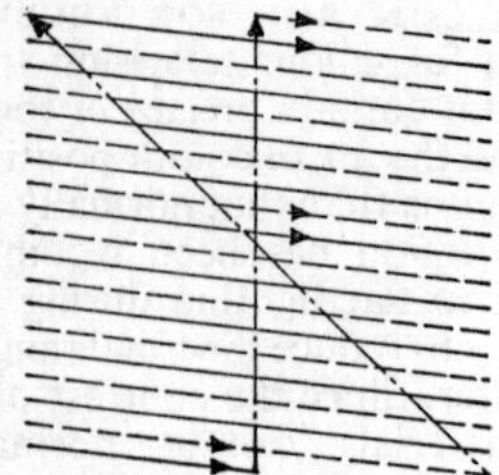
INTERLACE. Solid line represents first field, dotted line second or interlaced field.

increased in the ratio of 2 : 1 to a frequency which, owing to the persistence of human vision, is almost imperceptible.

To achieve the same result in any equivalent sequentially-scanned television system of the same resolution requires double the channel bandwidth.

Given that the intervals of blanking following odd and even fields are of equal duration, and that the number of lines contained within each complete frame is odd, then each field will necessarily consist of a whole number of lines plus half a line, and interlacing follows automatically.

See: *Scanning.*

⊕**Interlock.** System of interdependent motors □so controlled electrically that all turn at the same speed. In some types of interlock control, such as Selsyn, rotation by hand of one motor shaft turns the other motors correspondingly, so that there is angular identity between them, but with a margin of error for lead or lag.

The term is sometimes loosely applied to any system by which picture and sound tracks may be operated synchronously, even by mechanical coupling.

See: *Selsyn motor; Synchrostart.*

⊕**Intermediate.** In integral tripack colour processes, a tripack derivative from an original colour negative. May be either positive or negative in character. The colour master positive intermediate is sometimes referred to as an interpositive and the colour dupe negative derived from this may be called an interdupe. A tripack dupe negative derived by direct printing from a reversal colour original is termed an internegative.

See: *Integral tripack; Laboratory organization; Printers.*

□**Intermediate Film System.** Television transmission system, now obsolete, in which a film camera was used to record a studio or outside scene. The film was passed into a developing machine working at the same speed of 90 ft. a minute and the wet film was scanned, electronically reversed to a positive and then transmitted.

See: *History* (TV).

⊕**Intermittent Movement.** Stop-and-go movement of film in a camera or projector which enables each picture to be exposed or viewed at rest and then replaced by the next image. The term is also applied to the mechanism by which this effect is secured, usually a cam-actuated claw or a Maltese cross sprocket.

See: *Camera* (FILM); *Projector.*

⊕**Intermittent Printer.** Film printing machine in which the film being copied and the raw stock being exposed are moved frame by frame and held stationary at the time of exposure.

See: *Printers.*

⊕**Intermodulation.** Modulation of the com□ponents of a complex wave by each other, as a result of which waves are produced which have frequencies equal to the sums and differences of those of the components of the original complex wave. These sum and difference frequencies are not harmonically related to the basic tones, and they therefore produce a type of distortion very unpleasant to the ear. Intermodulation

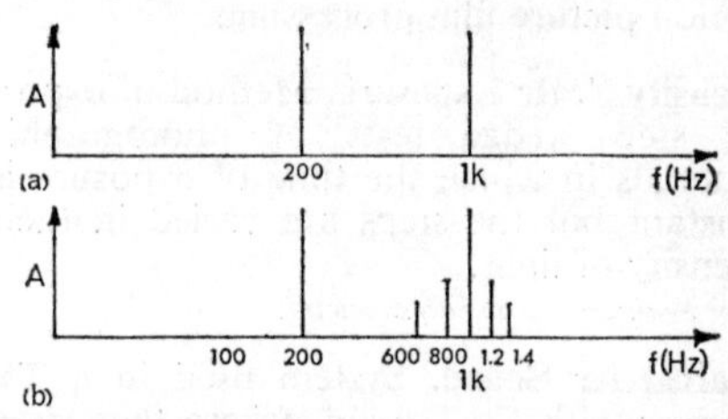

INTERMODULATION PRODUCTS. (a) Linear amplifier fed with a 200 Hz and a 1,000 Hz sine wave input produces an output containing only these frequencies. (b) The output of a non-linear amplifier contains in addition spurious sum-and-difference frequencies of (1,000+200) and (1,000−200) Hz, as well as (1,200+200) and (800−200) Hz, etc., higher products of lower amplitude.

is the result of passing a complex signal through a non-linear transmission system. The linearity of audio channels is often checked with the aid of a low-frequency high-amplitude signal modulated by a high-frequency low-amplitude signal.

See: *Sound distortion; Sound recording.*

⊕**Internegative.** Integral tripack colour duplicate negative derived directly from reversal colour master original camera film.

⊕**Interpositive.** Colour intermediate film with a positive image printed from a colour negative.

□**Inversion.** Inversion of a signal takes place when its polarity is reversed with respect to itself. An example of this occurs in an amplifying valve (tube) having a resistive load where the signal at the anode is an inversion of that at its grid.

This property of the valve is sometimes wrongly described as being a 180° phase shift. The two processes are quite different but yield the same result when dealing with symmetrical signals such as sine waves. With video signals and other non-symmetrical signals, the effects are not the same.

Inverting amplifiers are used in television to produce a positive picture from a negative, and vice versa. This is usually accomplished in a single valve or transistor stage with a split, balanced load. The signal appears in opposite phase in each load, and a switch selects the polarity required.

See: *Special effects* (TV).

□**Ion Burn.** Blemish produced on certain cathode ray oscilloscope tubes by ion bombardment of a small area in the centre of the scanned raster. Modern practice is to incorporate an ion trap near the electron gun of the tube, which effectively allows the ions to be removed from the beam prior to its being accelerated down the tube. The gun is angled, and the electron beam is corrected by a small magnet external to the tube, but as the magnet has little or no effect on the ions they do not reach the scanned area. Provided the gun points at some unimportant area, the ions can then do no harm.

See: *Receiver*.

□**Ionosphere.** Collectively, the various electrically conducting layers at differing distances above the earth, caused by various natural phenomena, e.g., X-rays and emissions from the sun. Of importance in all forms of radio transmission.

⊕□**Iris.** Means of controlling the amount of light which is allowed to pass through an optical system. Besides the use of neutral density or colour filters, this control can be exercised by means of a mechanical iris consisting of an opaque diaphragm having an almost circular aperture of variable diameter which is placed at right-angles to and concentric with the optical axis. It thus enables the effective aperture of the lens to be reduced, a process commonly referred to as stopping down. As a lens is progressively stopped down, the brightness of the image is reduced and the depth of field increased.

See: *Lens performance; Lens types*.

⊕□**Iris Wipe.** Printing effect providing a transition from one scene to another at the boundary of an enlarging or diminishing circle.

When using a television flying spot system, the generation of a circular wipe can be accomplished by means of a simple mechanical iris similar to that used in a camera lens. Electronic generation is usually by a combined line and field parabolic waveform switching signal.

See: *Special effects*.

⊕□**ISO Standards.** Standards published by the International Standardisation Organisation, the body co-ordinating the work of various national associations. They include a number of standards in the field of motion pictures and television.

See: *Standards*.

⊕**Ives, Frederick E., 1856–1937.** American printer who developed several processes for still colour photography and, with H. E. Ives, considerably improved the parallax stereogram method of stereoscopic photography.

□**Ives, Herbert Eugene, 1882–1953.** American physicist and specialist in electric telephotography, born in Philadelphia and graduated from the University of Pennsylvania in 1905. Son of Frederick Eugene Ives, noted for his work in photography and inventor of the first half-tone process. Herbert Ives's early work was concerned with photometry and illumination, followed by researches on vision and the phenomena of flicker. In 1924 he was working on a thin film cathode. From 1919 Ives worked with the Western Electric Company and Bell Telephone Laboratories and it was there in 1923 that he and his collaborators developed his method of transmitting photographs over telephone lines; this was used at the inauguration of President Coolidge in 1925 and then put into operation across the American continent.

By 1924 Ives was actively at work on a mechanical scanning television system and made some significant advances. At the Bell Telephone Laboratories in 1929, brilliantly tinted pictures about the size of a postage stamp were sent over wires from one end of a room to the other, using a three-colour three-channel method. Ives also demonstrated in 1928 the television of outdoor scenes by daylight and in 1930 a complete videotelephone system was set up between the Bell laboratories and 195 Broadway. It was maintained for over a year and was used by over 10,000 people.

⊕**Jacketed Lamp.** Tungsten halogen studio □lamp in which the filament tube (usually made of quartz) is enclosed in an outer envelope of glass, the space between being filled with inert gas. This prevents contact of atmospheric oxygen with the quartz-metal hermetic seals, a critical point in the design of such lamps, since oxidation of the seals may lead to fracturing and ingress of air.

Jacketing enables wattage to be raised and lamp life extended; it also reduces UV emission (with its sun-tanning effect on actors), and aids the design of lamps which fit existing light fittings (luminaires).

See: *Lighting.*

⊕**Janssen, Pierre Jules César, 1824–1907.** French astronomer of Norwegian birth, who in 1874 devised a photographic revolver to record the transit of Venus across the sun. It employed a Daguerreotype plate, rotated intermittently behind the lens by clockwork mechanism.

⊕**Jelly.** Diffuser placed in front of a studio □lamp to soften the light. Also a colour correction filter on a lamp.

⊕**Jenkins, Charles Francis, 1867–1934.** Amer-□ican television pioneer born of Welsh and French parents near Dayton, Ohio. Gave the first public demonstration of mechanical television in the US. At the age of 23 he took up inventing as a profession.

In 1892 Jenkins demonstrated his first moving pictures, and by 1894 had developed the Phantascope in which a continuously moving film could record or project a succession of stationary images through an optical stabilizing mechanism.

In 1913 he proposed "wireless moving-picture news" and, in 1916, founded and became first President of the Society of Motion Picture Engineers. He set up a research laboratory in 1921 to work on optics, radio, picture transmission and television.

In 1923 he sent pictures of President Harding by radio from Washington to Philadelphia, a distance of 130 miles; in the following year clear images of the signature of Herbert Hoover, then Secretary of Commerce, were sent from Washington to Boston, 450 miles.

Jenkins's demonstration of mechanical television came in 1925 when he used a revolving disc, the rim of which was lined with tiny lenses. In 1932 several Jenkins motor-driven scanners were demonstrated in New York, the receivers having large glass bull's-eye-type screens. The Jenkins Television Company was formed in 1929 to manufacture both transmitting and receiving apparatus. Programmes were announced and the company licensed a number of manufacturers under its patents to make sets, but before broadcasting was started Jenkins Television went into receivership.

Jenkins took out over 400 patents for inventions, including an airplane brake, altimeter, and a self-starter for automobiles.

⊕**Joiner.** Operator in a motion picture studio, laboratory or distributor's exchange primarily concerned with splicing film. In studio cutting rooms, a joiner may be employed to relieve the editor and cutter of the routine work, while in the laboratory, raw stock joiners, negative joiners and positive print joiners have their regular functions. At the exchanges, the joiner's work on release prints may include the attachment of titles, the removal of censor cuts, the repair of damaged copies and the replacement of leader sections.

See: *Editor; Laboratory organization.*

⊕**Joly, Henri, b. 1866.** French inventor. Invented and patented in 1895 an intermittent mechanism employing the beater principle. After a brief association with Charles Pathé, Joly collaborated with M. Normandin in the commercial production of film apparatus based upon his invention.

⊕**Jumbo Roll.** Large roll of film or coated □magnetic material before slitting.

⊕**Jump Cut.** Cut interrupting normal sequence of action. If a section is removed from a piece of film recording a continuous action, and the ends are joined together, moving objects will be seen to jump instantly into new positions. Jump cutting is occasionally used to produce whimsical or startling effects. Unwanted jump cuts may occur when shots are reassembled with wrongly calculated build-up, when the negative cutters have not been properly instructed. If a shot contains an unduly long static section sandwiched between two pieces of action, the static section (if indeed there is absolutely no movement in it) may often with advantage be invisibly jump cut.

See: *Editor.*

□**Jump Scan.** Method of film scanning which uses a displaced flying spot raster to scan a frame of film twice.

See: *Telecine.*

⊕**Kalmus, Herbert Thomas, 1881–1963.** American physicist. Formed a company for research in industrial engineering with Dr. D. F. Comstock in 1916. They developed an additive colour motion picture process, later discarded in favour of a two-colour subtractive system. This used separation negatives obtained with a beam-splitting camera, printed on separate positive films stuck back to back for projection. The Technicolor process so introduced was subsequently improved, first by the development of imbibition printing of the two subtractive images on to a single film, then by the introduction of three-colour printing and the three-strip camera.

The imbibition printing process devised by Kalmus and his associates in Technicolor Inc. has continued with great success to the present day.

⊕KALVAR PROCESS

The Kalvar photographic process is based upon the phenomenon of light-scattering, rather than upon that of light absorption as in conventional silver halide materials. Whereas incident light is absorbed by the silver grains within the developed silver halide film, it is reflected and refracted by the scattering centres within the developed Kalvar film.

BASIC PRINCIPLES

Kalvar motion picture films consist of a saran polymer coated on a base of transparent polyester. Within the thermoplastic resin, which is normally coated to a thickness of slightly less than 0·0005 in., there is uniformly dispersed a sensitizer consisting of an ultra-violet sensitive compound. On exposure to ultra-violet radiation, this photo-sensitive diazonium salt is decomposed, releasing nitrogen and other volatile products.

The internal pressures created by these decomposition products within the thermoplastic vehicle constitute a latent image of internal stresses. When heat is applied, the resin crystallites soften and the gaseous decomposition products expand. A re-orientation and ordered recrystallization of the polymer into microscopic vesicles or gas

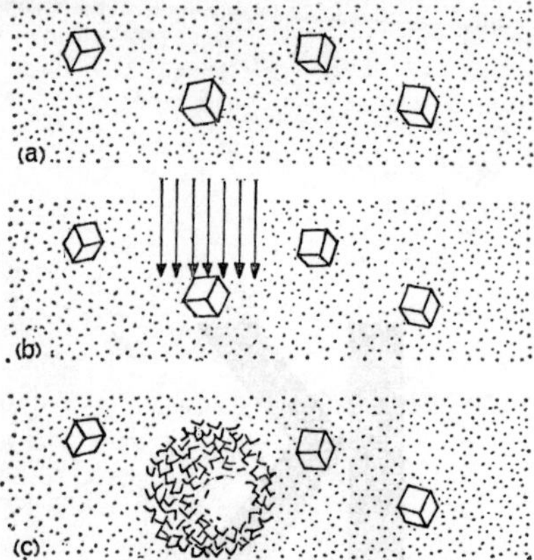

EXPOSURE OF KALVAR FILM. (a) Film before exposure. Within layer of thermoplastic resin, an ultra-violet sensitizer is uniformly dispersed. (b) Film after exposure. When exposed to ultra-violet radiation, 1, photosensitive salt is decomposed, releasing volatile products. (c) Film after development. Decomposition products constitute latent image in terms of internal stresses. Application of heat softens resinous crystallites and swells gaseous decomposition products to form vesicles or gas bubbles.

bubbles takes place. These vesicles, since they are of a different index of refraction from the surrounding medium, scatter light incident upon them and thus constitute the image.

The light-scattering vesicles vary in size from less than 0·5 microns to 2 microns in diameter and consist of cavities enclosed by a shell of more highly ordered crystallites than the surrounding medium. The vesicles are, as a result, highly resistant to environmental changes and mechanical stresses and provide an extremely stable image.

SENSITIVITY AND EXPOSURE. Kalvar film is a comparatively low speed material with primary photosensitivity in the near ultra-violet, peaking at 3850°A. The spectral response curve shows that the photosensitivity is not limited to a narrow peak but extends from below 3,500°A to above 4,300°A. The film is not photographically sensitive to ordinary levels of visible light for short periods of time and printing is normally carried out in undarkened rooms.

LATENT IMAGE STABILITY

The temperature of the film during exposure should not exceed 110°F. Temperatures above this result in a higher diffusion rate of the latent image-forming gas with subsequent reduction of the maximum density obtained. Experiment has shown it is best to develop the film within three minutes after exposure.

Since the latent image comprises a given amount of gaseous nitrogen, it has a definite decay time depending on the permeability of the emulsion's thermoplastic vehicle to nitrogen. The decay time can be adjusted by adding modifiers to the basic vehicle resin to increase or decrease its permeability. For reversal-processed materials, the latent image diffusion time is approximately 30 seconds.

DEVELOPMENT AND IMAGE STABILITY. Kalvar film is developed by heat. The processing is completely dry, requires no chemicals and is accomplished merely by applying sufficient heat to soften the plastic emulsion layer. Current film-development practice in roll film printer-processors employs a revolving drum coated with a high-temperature epoxy resin, with heat provided by a high wattage electric blanket laminated to the inside perimeter of the drum. A thermistor controls temperature to within $\pm 2°F$ of the desired setting.

FIXING AND IMAGE STABILITY. As with most photographic processes, Kalvar film calls for a fixing technique to provide image permanence. After exposure and development, the non-light struck areas of the film still contain undecomposed sensitizer. The fixing technique consists of exposing the film overall to ultra-violet light. Applying about four times the amount required for maximum exposure completely decomposes the residual sensitizer.

SENSITOMETRIC CHARACTERISTICS. The sensitometric units and standards currently used in silver halide photography do not apply directly to Kalvar photography. In any practical use, a photographic material is viewed or projected through an aperture of finite dimensions. With Kalvar, a substantial portion of the transmitted light is scattered outside the angle over which light is collected by the effective aperture. Correspondingly, the effective density of Kalvar film is strongly dependent upon the cone angle subtended by the light gathering element, whether it be the eye, a projector lens, or the integrating sphere of a densitometer.

These characteristics have been taken into consideration in the design and development of new measuring techniques for the photometric evaluation and process control of Kalvar photography, so as to provide measurements that will correlate with normal photographic experience in ultimate projection viewing.

APPLICATIONS. Special emulsions of the basic Kalvar process have been developed for photographic black-and-white duplication requirements, including roll microfilm, aperture cards, microfiche, aerial negatives, graphic arts sheet films, and motion pictures for screen and television. R.B.L.

See: *Image formation; Thermoplastic recording.*

Books: *Microfilm Technology, Engineering and Related Fields*, by C. E. Nelson (New York).

⊕**Kelley, William Van Doren, 1876–1934.** American inventor, born in Trenton, New Jersey. Started a showcard and display business with his brother, but after a commission from Biograph in 1910 for display devices for Biograph cinemas, became interested in motion picture processes, particularly in the use of colour.

In 1913 he formed the Panchromotion company, experimenting with an additive colour process based upon Kinemacolor but using an increased picture frequency. A development of this system, exploited by his company Prizma Inc., was used in the film *Our Navy* (1917). The additive process was not entirely satisfactory, and with Carroll H. Dunning and Wilson Saulsbury he opened the Kesdacolor laboratory, working on a two-colour subtractive process, used by J. Stuart Blackton of Vitagraph for *The Glorious Adventure* (1922).

In 1924 Kelley introduced the Kelleycolor two-colour subtractive process with imbibition printing. Some of the principles introduced in his early systems were used later in several very successful two-colour subtractive processes.

□**Kell Factor.** Early high-definition television systems were based on the supposition that, for a given bandwidth, the vertical and horizontal resolution of the picture at the receiver should be equal, since the eye has approximately the same acuity in both directions. This leads to the formulation of a simple relationship between the highest frequency to be transmitted, f_{max}, and the other relevant factors: $f_{max} = \frac{1}{2}Al^2f_p$ Hz, where A = aspect ratio, l = number of lines, and f_p = picture repetition frequency. Hence,

$$l = \sqrt{\frac{2f_{max}}{Af_p}}$$

However, viewing tests with a fixed bandwidth and a varying number of lines have shown that viewers prefer a greater number of lines than this relationship provides, i.e., they will accept a lower standard of horizontal resolution for the sake of a finer line structure. The Kell factor, K, allows for this, and its value is about 0·4 to 0·8. Af_p in the above expression then becomes KAf_p.

See: *Scanning.*

⊕**Kelvin Scale (Degrees K).** Scale of basic □temperature units, which are degrees Celsius (Centigrade) measured from absolute zero or minus 273°C.

Colour temperature is invariably measured on the Kelvin scale.

See: *Colour temperature.*

□**Kerr, Revd. John, 1824–1907.** Scottish physicist born at Ardrossan. Educated at a village school in Skye and subsequently at Glasgow University, he studied under William Thomson, afterwards Lord Kelvin. His name is associated with two great discoveries affecting the nature of light – the bi-refringence caused in glass and other insulators when placed in an intense electric field and the change produced in polarized light by reflection from the polished pole of an electromagnet. The latter, which caused considerable excitement when communicated to the British Association at its Glasgow meeting in 1876, was known as the Kerr effect.

The Kerr cell, the earliest device for modulating a beam of light electrically, still has applications today.

□**Keying Signal.** Signal which actuates an electronic switch in the production of special effects.

It can be generated electronically, or obtained from a video signal by passing the signal through a special effects amplifier.

See: *Special effects* (TV).

⊕**Key Light.** Lighting of the main object of □interest in a scene. Is usually established first by the lighting cameraman, who builds round it back lighting, cross lighting, effects lighting and filler lighting.

See: *Cameraman.*

⊕**Keystone.** Distorted shape of projected ☐image area when the optical axis of a projector is above (or less commonly below) the centre of the projection screen. The rectangular picture aperture appears on the screen as of inverted-keystone or keystone shape respectively. This can be partially corrected by suitably inclining the plane of the screen away from the vertical.

In large-screen projection television, the correction may be effected electronically by altering the lengths of the line scan.

See: *Cinema theatre; Large screen TV.*

⊕**Kinemacolor.** The first (1906) commercial cinematograph colour process; a filter wheel of red and blue-green sectors was rotated in front of the camera lens so that alternate frames of the negative were exposed to red and to blue-green. A black-and-white print was similarly projected through a rotating colour wheel to show an additive two-colour picture. Colour fringing of moving objects was, of course, unavoidable.

See: *History* (FILM).

☐**Kinescope.** American term for a film recording made from a picture tube. In Britain called a telerecording. Also, American term for the picture tube in a TV receiver.

⊕**Kinopanorama.** Russian system of multiple-film wide-screen presentation over 360° on the inner surface of a large cylindrical screen. In its most elaborate form, it used a total of 22 projectors, 11 covering the lower cylindrical area and 11 more covering a domed area above, but in general, live action material was restricted to the lower zone.

See: *Wide screen processes.*

⊕**Kino-Vario.** Method of multi-image film presentation developed in Eastern Germany. Three 35mm projectors give overlapping areas on the screen and the three films are printed with masks of various areas; the resultant combination of images can be continually varied for artistic and dramatic effects. Four-channel magnetic sound is used. Similar effects can be obtained by multiple printing on to a single film for normal use on one projector.

See: *Multi-image; Multiscreen.*

☐**Klystron.** Special type of valve used for producing oscillations at very high frequencies.

⊕**Knee.** In the graphical representation of a ☐tonal reproduction process, such as photography or television, a point or region of inflection on the characteristic curve, where the slope representing the rate of change alters, usually from a higher to a lower value.

⊕**Kodachrome.** One of the earliest of the integral tripack colour reversal processes, first introduced for 16mm amateur cinephotography in 1935; while still used for professional motion picture work in 16mm, it continues to have its widest use for the amateur as 8mm and Super-8.

The Kodachrome colour image is intended for projection on an original and is less suited for the preparation of copies by printing.

See: *Colour cinematography; Integral tripack.*

⊕**Kodacolor.** Originally a system of colour photography using a lenticular process employed for amateur cinephotography in 16mm about 1928 but subsequently obsolete. Kodacolor now refers to a colour negative/positive process used for still photography but not in motion picture work.

See: *Colour cinematography; Integral tripack.*

⊕**Kodak Perforation.** Name sometimes used for the standard form of sprocket hole in

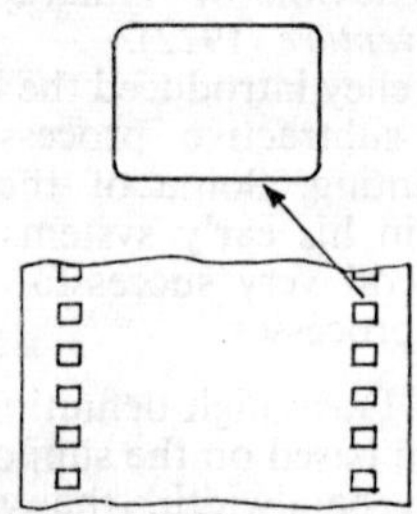

positive 35mm, 65mm and 70mm film; it is rectangular with rounded corners.

See: *Film dimensions and physical characteristics.*

☐**K-Rating System.** Method of stating the subjective effect of linear amplitude and phase distortions on a television signal. Sine-squared pulse and bar signals and field-frequency square wave signals are passed through the system. The output waveforms are compared with special oscilloscope graticules marked with limits for various K ratings, and the K rating of the system is stated as the largest value found, i.e., the worst rating.

See: *Transmission networks.*

LABORATORY ORGANIZATION

The function of the motion picture laboratory is to provide all the photographic processing and associated services required by studio operations during the production of a film and subsequently to manufacture the release prints for general distribution to cinema theatres. The laboratory organization reflects the differing demands of these two allied functions, services to the studios usually being more individually specialized, while release print operations call for the application of mass production factory methods for quantity output and quality control.

NEGATIVE DEVELOPING

If the costly work of motion picture production at the studio is to go forward without delay, it is essential for the production organization to view the result of each day's work as early as possible the following morning. The associated laboratory operations are therefore generally carried out during the night, with deliveries of exposed negative from the studio in the late evening and viewing of the resultant rush prints in the early hours of the next morning. They are then delivered to the editor of the production for synchronizing with sound records. For many years after the introduction of sound on film, the laboratory was concerned with the nightly development and printing of both picture and track negative; but since the universal adoption of magnetic film for original sound recording, regular rush printing is normally confined to picture material, of which the greater part is now colour.

DEVELOPING MACHINES. When the exposed negative is received at the laboratory, together with the corresponding camera reports or log sheets from the studio, it is usually given a laboratory identification number and developed as soon as possible. Modern negative developers are continuous-running machines designed to handle the negative with great uniformity and avoid undue tension on the strip of film, sprocketless tendency-drive machines being frequently employed. Running speeds in the range of 50 to 100 ft. per minute are usual for 35mm film, and the total processing time to pass from the feed end to the final take-up is of the order of one hour for modern colour negative stock. Black-and-white negative developing is a simpler operation requiring fewer different solutions, so that machine

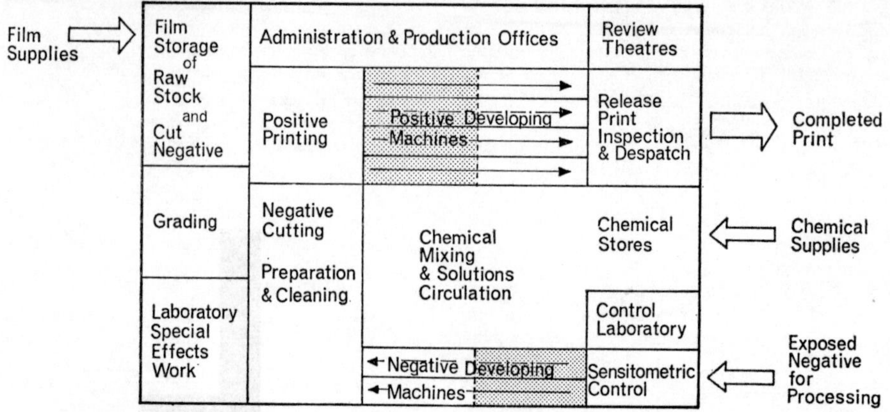

BLOCK DIAGRAM OF FILM LABORATORY. **Principal laboratory operations grouped to show departmental flow, but practical layouts seldom conform precisely to ideal arrangement. Auxiliary services not shown. Shaded areas indicate dark end of machines.**

speeds are generally greater and the total processing time considerably less.

Modern developing machines are equipped with accurate automatic control of solution temperatures to close limits and the chemical conditions of the various solutions are regularly checked by routine analysis; in addition, the overall performance of the developing process is tested at frequent intervals during the operating period by the sensitometric evaluation of test strips of film which have been given a standard exposure on a control sensitometer.

INSPECTION AND BREAKDOWN. Immediately a roll of negative has been developed, it is inspected for any physical defects which might have been caused either in the camera or during processing, and the negative is then prepared for printing.

The roll is first broken down into the individual scenes and those which have been indicated on the camera report as required for printing are joined together; if rush prints in colour are needed from only some of the scenes of a colour negative, while others are to be printed in black-and-white to save expense, the material is assembled in two separate rolls.

The scenes not specified for printing are spooled separately and stored for possible future requirements.

At this stage of breakdown and assembly, it is the practice of some laboratories to take short clippings of three or four frames which are identified with their scene and take number and used to make short test prints before printing the full length negative. After the negative scenes to be printed have been spliced together in rolls, it is usual to clean the film to remove any loose dirt which may have been picked up during handling.

RUSH PRINTING

After assembly the negative must be timed, that is, assessed for its required printing light levels. Despite careful control of exposure by the cameraman and of developing conditions in the laboratory, small variations of stock, exposure, subject matter and processing must be compensated.

TIMING METHODS. Black-and-white negative is often timed by visual inspection by experienced operators but a colour negative cannot be accurately assessed in this way for its colour printing requirements, so some form of test exposure method is frequently employed. Tests may be printed from sample clippings taken from the end of the scene, the results being viewed by the operator over a light box or on a slide projector to decide what corrections are to be made in printing the full scene.

A widely used alternative method is to print a short test section from a part of each scene in the assembled roll, using a special exposing machine which gives each frame of the test section a different printing level for both intensity and colour. The exposure given is accurately calibrated to correspond to known steps of the printing machines to be used. Each test strip may contain 12 or 16 frames of different printing levels, and after development on the positive developer the operator can usually interpolate accurately the final required level for each scene tested.

It is sometimes the practice to deliver these test strips to the cameraman at the studio with the laboratory report on the negative, particularly where only black-and-white rush prints of the full scene are being supplied for editing.

Because of the demands for the urgent delivery of rush prints, the time and labour occupied in printing and developing test sections before the complete scene can be printed often represents a serious delay in the course of the night's operations, which laboratories would be glad to eliminate. One method which is being adopted to avoid photographic tests is the use of closed-circuit colour television, with equipment to scan a selected frame of each scene of the negative and produce a positive picture on a colour television screen. If the amplification

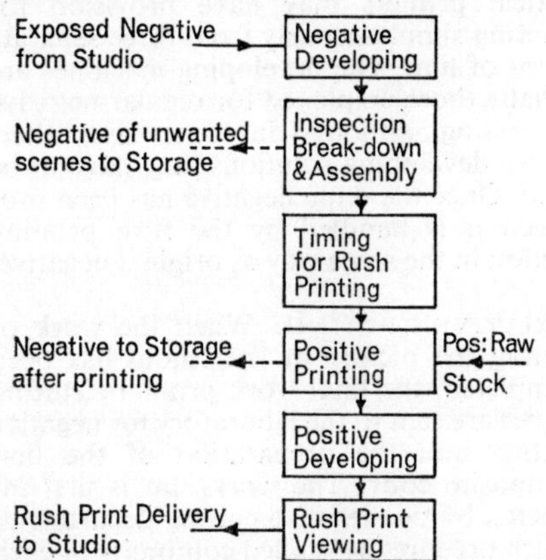

SEQUENCE OF RUSH PRINT OPERATIONS. Negative from studio usually arrives in evening; operational sequence must be effected by night shift for morning delivery. Breakdown and classification of stored negative takes care of print re-orders.

channels of this system are accurately calibrated in terms of the photographic printers being used, it is possible to assess the printing level required for each scene very rapidly and without the expense and delay involved in making photographic tests. Whatever the timing method employed, the required printer points for each scene are entered on the record sheet prepared for each roll of negative, and both the film and its data passed to the printing department for the next operation.

PRINTING MACHINES. In the majority of cases, the printing machines used for making rush prints are continuous contact printers similar to those for general release print production; but in the larger laboratories, for operational convenience, it is usual to organize rush-printing separately from bulk print manufacture.

Rush prints of picture negative are almost always printed without a sound track, because the corresponding sound exists at the studio as a magnetic recording. The type of rush print made is usually the exact equivalent of the negative format – 35mm wide-screen, 35mm anamorphic or 16mm – except in the case of productions photographed on large-area negatives such as 65mm film, where it is general practice to provide 35mm rush prints by anamorphic optical reduction.

According to the type of printers used, a control band corresponding to the timing data of the roll of negative must be prepared, either using colour filter packs for subtractive printing, or in the form of a punched paper tape to control the three light valves of an automatic additive printer. Modern colour printing machines are fast in operation, often with speeds of 100 to 180 ft. per minute, so that the actual time taken to print from a roll of assembled negative is one of the shortest of all the stages of laboratory work.

DEVELOPING MACHINES. After printing, the roll of negative is returned to storage in case reprint sections may subsequently be asked for by the editor, and the exposed positive film is passed to the developing section for processing. The developing machines used for rush prints are usually identical with those for handling release prints, and sprocket-driven machines running at speeds up to 250 or 300 ft. per minute are usual for black-and-white work.

For developing colour positive film, the longer processing times and the greater number of solutions usually restrict to somewhat slower rates the speed at which a machine of reasonable size can be operated; but even here, running speeds of 150 ft. per minute may be employed.

As in the negative developing processes, chemical analysis of the various solutions is carried out regularly, and sensitometric control test strips are processed on both printers and developers at frequent intervals.

INSPECTION. Before delivery to the studio, the developed rush prints must be viewed in a projection review theatre and a report on the negative exposure level of each scene prepared for the cameraman. If any defects are present on the negative, they must also be reported to the editor. At this stage, the closest co-operation between studio and laboratory is essential, and it is usual for the laboratory contact man who acts in liaison with the production team at the studio to view the appropriate rush prints himself or to deal with the viewers' reports at first hand. The contact man is responsible for ensuring that the rush prints supplied satisfy as far as possible the creative and artistic requirements of the director and cameraman; his advice and guidance to the timing operator and the viewer are of fundamental importance in ensuring satisfactory laboratory service.

RECORDS. Any outline of this stage of laboratory work would be incomplete without reference to the detailed records which must be meticulously kept: for every production handled, the identity of every scene and take must be recorded with the corresponding stock edge numbers which are printed along the margin of the negative. In the same way, the storage location of every scene and take, both printed and unprinted, must be known for ready access, while the rush print timing data, the position of negative defects and other information obtained from rush print viewing, must also be recorded. The design and use of efficient technical record systems is an important part of laboratory management.

PREPARATORY WORK

When the main period of production photography on a subject at the studio has been completed, the requirements for regular daily negative developing and rush printing disappear, although occasional reprints may be called for by the editor to replace damaged sections.

SPECIAL EFFECTS. The next stage of laboratory work is usually concerned with the preparation of the special effects scenes required for the completion of the picture. Here again close co-operation between laboratory and studio is essential, particularly where the effect arrived at necessitates the composite printing together of several pieces of negative which have been separately photographed by the studio's special effects department. Multiple printing of this type is sometimes done by the same department, sometimes by an independent specialist organization, but the laboratory is almost always called upon to process the master positive and dupe negative materials, while for the more straightforward effects such as fades, dissolves and titles on action backgrounds the work may be left entirely to a laboratory department working from the editor's instructions.

The laboratory section specializing in this work is concerned with the selection and assembly of the negative of the scenes specified, the printing and developing of the fine grain master positives or colour intermediates, the assembly of these and the subsequent printing and developing of the dupe negative containing the effect, which is then rush-printed and delivered to the editor of the production for his approval.

The printers used for this type of work are usually intermittent step-printers of specialized design provided with variable exposure shutters and often with optical printing functions to allow the size and position of the image to be altered. For some forms of special effects such as travelling mattes, optical printers may have provision for printing simultaneously from three separate strips of film. The developing machines are usually those employed for regular negative processing, although in some cases alternative developing solutions may have to be used. Once the dupe negative has been produced it is handled by the rush printing section in the same way as original negative.

NEGATIVE CUTTING. When the work of editing the picture at the studio has been completed, the final work prints or cutting copies are sent to the laboratory for negative cutting and the preparation of the first composite copy. The workprint is first inspected by the negative cutting department, which prepares a detailed continuity of each reel, identifying each scene and and take and its length by means of the stock numbers printed in the rush print from the negative. The negative of all the required scenes is then obtained from storage, where the rolls of negative after rush printing have previously been broken down into their individual scenes for ready access, and passed to the negative cutter concerned. Each negative scene is then cut to match the corresponding section of the workprint, the sections spliced together, head and tail leaders added, and a printer card prepared for each reel, showing the scene number, length and description, and leaving space for timing data to be added.

At this stage, the final dubbing of the

sound track at the studio has normally been completed and the transfer from magnetic recording to optical sound negative carried out. The exposed sound negatives for each reel of the subject are sent to the laboratory for developing and rush printing, and here again close co-operation between the studio sound department and the laboratory sensitometric control is called for. Sound track negatives are processed in high contrast black-and-white developing solutions, and cancellation or cross-modulation tests are usually processed and assessed by the same department to decide on the optimum developing and printing conditions for the negative concerned. After rush printing, track work prints for each reel are sent to the laboratory to allow the track negative to be made up in synchronization with the cut picture negative.

PROTECTIVE MASTERS. When the cutting and joining of a reel of picture negative is complete, it is usual to clean it, often with a solvent-cleaning machine with ultrasonic agitation, and prepare a protective master positive from which a duplicate negative could be made in the event of any serious damage to the original. For black-and-white pictures, this protection is normally a fine-grain positive or "lavender", but for colour negatives a colour interpositive is not regarded as being sufficiently permanent, and the recommended procedure is to prepare a set of three separation masters by printing three times through red, green and blue filters on to a suitable panchromatic fine-grain positive black-and-white stock. Protective master material, being intended only for use in an emergency, is usually stored quite separately from the corresponding original negative, sometimes outside the laboratory in accordance with insurance requirements.

TIMING. The picture and track negatives are now ready for printing, and the necessary timing data for each picture scene must be obtained; this may be derived from the records prepared at the time of rush printing or a new series of tests may be printed from the assembled reel on a scene tester. Where negative clippings of each scene have been taken prior to rush printing, they may be joined together in sequence corresponding to the cut reel so that a short test print can be made using only a few feet of positive stock. A closed circuit colour television analyser can provide a rapid method of determining scene-to-scene timing values. Whatever timing system may be employed, the corresponding control band must then be prepared for the type of printer in use, and the first composite print with picture and sound on the same strip of film printed and developed.

GRADING. No matter how much care is given to obtaining the first timing data, it is unusual for the first print from a cut reel of negative to appear perfectly consistent for colour and density for every scene, and some degree of correction for scene-to-scene evenness is generally necessary. These corrections are made on the basis of viewing under review theatre conditions by an experienced grader. At this stage, co-operation between the laboratory contact man, the grader and the editor or other representative of the studio production company can be most valuable. In many sequences the character of lighting may have been designed to emphasize the dramatic mood of the story, and it is essential that the grader understand and correctly interpret the director's or cameraman's intentions.

In making his corrections, the grader must not only ensure that each reel is uniform in appearance from scene to scene, but that successive reels of a feature film are also consistent so that there are no marked alterations of colour and density when a reel is changed in the theatre. All this calls for a considerable amount of viewing time, often on two screens with two prints projected side by side, as well as personal skill and experience on the part of the grader.

The first completely graded copy, known as the answer print or check print, is submitted to the production company for approval and may be used for previews or rehearsals prior to the first public showing. With the acceptance of the answer print, the stage of preparatory work by the laboratory may be regarded as complete, but it is usual to prepare a few specially selected copies under the supervision of the grader to provide the first show prints for the premiere performances, together with the studio show copy, the laboratory reference print and any other special requirements.

RELEASE PRINTING

After the approval of the answer print by the production company, the laboratory is usually working under the instruction of the distributing organization who specify their requirements for release copies. From here onwards the picture and sound track negatives are handled in the release print section where equipment is specialized for bulk production of copies.

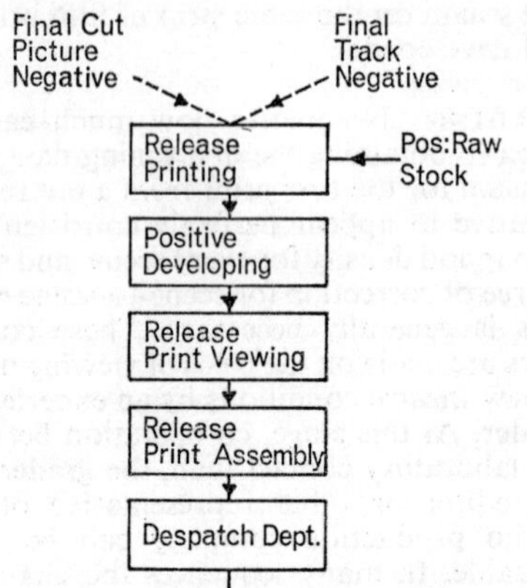

SEQUENCE OF RELEASE PRINT OPERATIONS. Since bulk print orders are seldom made direct from original negative, first stage usually entails duplicate printing to provide intermediate material and thus preserve original from risk of damage.

PRINTING MACHINES. Rotary contact printers on which the picture and track are printed at a single passage of the raw stock are usual; they may be run with the negatives joined in an endless loop to avoid loss of time in rethreading the printer for each copy. Other types of printers are reversible in direction of running, so that the negative need not be rewound, and repeated copies can be printed merely by feeding on a new roll of raw stock each time. In some laboratories a high-speed printer may be continuously fed with raw stock through a reservoir elevator and connected to a positive developing machine running at the same speed, so that time and labour are saved. Where a very large number of copies is required in a short time, as with newsreels, multi-head printers may be used in which several positive copies are produced at each pass of the roll of negative.

With all types of high-speed continuous printer, the operations are made as automatic as possible, and it is here that the advantages of modern methods of light-change control by punched tape programmes are of considerable importance. It is usual to include cleaning units, often in the form of brushes connected to vacuum lines, on the picture and track negative paths through the printer to avoid dust and dirt being picked up on the film. It is also general practice to put the negative through a wet cleaning unit at regular intervals during its printing usage.

INSPECTION AND DESPATCH. Release prints are edge-lubricated immediately after development and inspected, by projection on a small screen, to check that the quality of the copy is satisfactory and that it is free from dirt and physical defects, either on the copy itself or printed in from the negative. As in other sections, the work of the control department is vital in ensuring the consistent performance of picture and track printers and developers, and sensitometric tests and chemical analyses are made at regular intervals during all production periods, supplemented where necessary by cross-modulation tests on sound printing equipment.

After projection, prints are passed to the release assembly department with their corresponding record sheets or labels and are given a final inspection over a light-box. Where necessary, stock joins are cut out and individual reels may be joined up into double-reels for running in the cinema. At this stage also, censor titles and distribution trade marks are attached to the head-end of the first reel and the copies spooled for final delivery. The print of each part is then packed in a can labelled with its title, part number and copy number, and when all parts of a given copy are complete it is despatched to the distributor.

SPECIAL PROCESSES

REVERSAL MATERIAL. The processes outlined above are those which are normally employed in laboratories where professional motion picture film, colour or black-and-white, 35mm or 16mm, is being handled with negative-positive materials. However, in the 16mm field a number of reversal stocks are also used and involve some differences of procedure; reversal material calls for more steps in its processing, so that the developing machines have more solution tanks and the total time of processing may be longer.

In general, the same sequence of operations is followed for the preparation of reversal rush prints as for negative-positive materials, but the preparatory work for the answer print may include the printing of a full-length dupe negative from the reversal original if the final copies are to be made on regular positive stock. When this is done, fades and dissolves are incorporated in the dupe negative and no protective masters are necessary since the reversal original is not used for release printing.

On the other hand, if only a limited number of release prints is needed, it may be decided to make these as reversal prints from the reversal original; fades and dissolves are then included by the cutting of the original in A & B rolls and double-printing each copy on a printer equipped with a variable opening shutter or similar light-modulating device.

TELEVISION. Processing for television does not differ in essentials from regular theatrical motion picture work, and the same operations of rush printing and preparatory work are involved during the production period. However, in general, the number of copies required for television use is small and there is no stage corresponding to the mass production of prints for general cinema release. Laboratory work for television is therefore normally completed by the provision of a few selected copies after the approval of the answer print; these may be needed in both 35mm and 16mm formats.

The reproduction characteristics of the television system may call for prints to be made whose contrast and tonal range is lower than those of prints made for regular cinema projection; with black-and-white materials, the positive developing process can be modified to give lower gamma but this is generally impracticable when colour films are used and the need has been met by manufacturers producing a colour positive material of lower inherent contrast for processing under standard condition . L.B.H.

See: *Imbibition printing; Printers; Processing; Sensitometry and densitometry; Sensitometry of colour films.*

Books: *Control Techniques in Film Processing*, SMPTE (New York), 1960; *Photographic Processing Chemistry*, by L. F. A. Mason (London), 1966; *Motion Picture and Television Film Image Control and Processing Techniques*, by D. J. Corbett (London), 1967.

⊕**Lacquer.** Coating often applied to film to protect it against abrasion during use.

□**Lag.** Time constant of many phosphors and targets in television sufficient to cause smear and persistence on a moving object.

See: *Phosphors for TV tubes.*

⊕**Lambert.** Unit of brightness or luminance, □defined as the brightness of a perfectly diffusing surface, when the total flux radiated is one lumen per square centimetre.

See: *Light units.*

⊕**Lap Dissolve.** Method of mixing from one □picture to another. The first picture is gradually reduced in amplitude or brightness while the second picture is increased, until the second picture has replaced the first.

See: *Special effects.*

□LARGE SCREEN TELEVISION

Large-screen television denotes images bigger than those which can be formed on a directly viewed cathode ray tube.

These pictures may range in width from a few feet to over 30 ft. There are two methods available for producing them: the cathode ray tube projector and the Eidophor. In the past, an intermediate film process was sometimes used consisting of a telerecording system with rapid film development and a film projector. In the future, thermoplastic recording may become a competitor for producing large-screen pictures in monochrome or colour.

These projection systems may be used in theatres or in the open air for large audiences, or as back projection systems in television studios. The Eidophor is particularly suitable for the latter purpose because the illumination is continuous, from a lamp, and so no difficulty is caused by variation of the illumination at video frequencies.

Development of the monochrome cathode ray tube projector began in the 'thirties, and reached the stage to be described in the early fifties. It has to a great extent been superseded by the Eidophor system which was developed by the late Dr. F. Fischer of the Institute of Physics in Zürich and which is capable of greater brightness and has the advantage that it can be installed in a cinema projection room. Colour cathode ray tube projectors are still in use but are rapidly being superseded by Eidophor projectors giving a colour image.

CATHODE RAY TUBE PROJECTOR

The projection cathode ray tube is a special tube which can be operated at high voltage and beam current, and so produces a very bright picture which can be projected with a suitable optical system.

The Schmidt optical system was originally designed as a wide-angle astronomical camera, and it is used because of its wide aperture, an effective aperture of f0·85 being possible, and its relative cheapness. It consists of a spherical surface concave mirror together with an aspheric corrector plate, mounted so that its centre is at the centre of curvature of the mirror, to correct for the spherical aberration of the mirror. The cathode ray tube face has a spherical curvature to match the curvature of field of the optical system. As the corrector plate is a relatively weak lens, the chromatic aberration it introduces is small. Coma and astigmatism are also small.

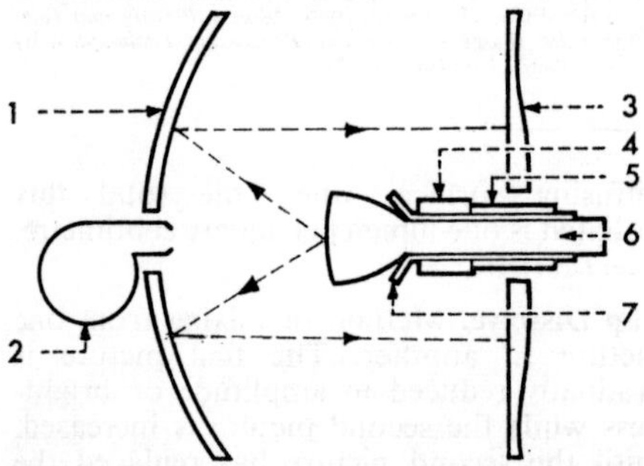

SCHMIDT OPTICAL SYSTEM. 1. Front-aluminized concave mirror with central aperture. 2. Blower for cooling tube face. 3. Corrector plate (curvature exaggerated). 4, deflection and, 5, focus coils of high-voltage cathode ray tube, 6, held in insulating glass sleeve, 7. Image is projected through corrector plate to screen on right (not shown).

The barrel distortion produced by the optical system partly cancels the pincushion distortion produced by the tube.

An area at the centre of the mirror, the same size as the tube face, is obscured because light from one part of the tube face could be reflected from this area and fall on another part of the tube face, decreasing the contrast of the projected picture.

Cathode ray tubes can be made sufficiently accurate for them to be changed without any adjustment to the optical system other than a check of the focus adjustment.

PERFORMANCE. One monochrome theatre television projector uses a 9¼ in. diameter magnetically focused and deflected triode tube with a 27 in. mirror and 18 in. or 23 in. corrector plate. The tube supply voltage is 50 kV. The average beam current can be 2 to 4 mA, with a peak current of 10 mA, from a tungsten filament.

Another example is a 7 in. electrostatically focused, magnetically deflected pentode tube with a 26 in. mirror and 22 in. corrector plate. The tube supply voltage is 80 kV. The average beam current is 1 to 2 mA, with a peak current of 6 mA from an oxide cathode.

Both of these systems are capable of producing pictures up to 18 ft. × 24 ft. in size, using an aluminium-sprayed front projection screen with a directional gain of two, and a highlight brightness of about 3 foot-candles.

SCREENS. Front projection screens are normally used and have the advantage that they can be perforated and have loudspeakers installed behind them. Beaded screens can have rather higher gains than two, but are difficult to construct in large sizes.

A normal back-projection screen tends to have a hot spot, that is, the picture is brighter as viewed along the line joining the viewer to the projector. Lens types of screen can have high gain, without a hot spot, but are very expensive.

EHT SUPPLY. The 80 kV supply required for the 7 in. tube uses a conventional mains transformer and rectifiers. The 18 kV electrostatic focusing supply needed is obtained by the use of a bleeder resistance across the 80 kV supply. When the beam current is high, the regulation of the supply will cause a drop in both the 80 kV and the 18 kV which will increase the picture size but will not affect the focus.

The 50 kV supply for the 9¼ in. tube is a radio frequency supply regulated to 1 per cent. The system uses a master oscillator, power amplifier and voltage tripler rectifier. Two rectifiers have about 17 kV peak radio frequency voltage on their cathodes and it is very difficult to make a transformer which is insulated for this voltage. For this reason generators are used which are driven by an a.c. motor with long, insulated shafts.

A large reservoir capacitor of 0·05 mfd is used. With a load current of 10 mA, the EHT would fall 200 volts per millisecond if the rectifier output were accidentally removed. The regulation of the tripler rectifier circuit without feedback cannot be very good because the size of coupling capacitors is limited by physical considerations. Hence a proportion of the output is compared with a stable d.c. supply, the difference being amplified and used to control the drive to the power amplifier. A large reservoir capacitor ensures that the EHT cannot drop severely before the feedback operates.

Both these supplies are potentially lethal, and the usual high-voltage precautions must be taken. In particular, a lethal charge can be stored by the reservoir capacitor.

In a conventional cathode ray tube, the EHT appears across the thickness of the glass inside the scanning coils. As it is difficult to make a glass neck which will stand the application of this voltage, a double thickness of glass is employed, either by the use of a double-walled tube or a separate insulating sleeve.

SPARKING. All cathode ray tubes are liable to spark internally from the anode to the low voltage electrode nearest to it. In the case of direct-viewed tubes with valve amplifiers, this generally causes no difficulty, but in the case of a tube fed with 50 kV through a 75 ohm cable, the prospective instantaneous fault power is 33 megawatts. Sparking tends to pit the electrodes and so make itself more prevalent. The peak current can be decreased by providing a series resistor at the tube end of the EHT cable. Protective spark gaps to earth are provided for the grid and cathode of the tube and a saturable series diode may be provided in the EHT supply to limit the peak current from it. This diode has a tungsten filament. It will pass 30 mA at an anode voltage of 670, and is insulated for 50 kV.

Because of this tendency to sparking, the beam current of the tube cannot be measured by a meter directly in the EHT circuit. In the case of an r.f. EHT supply, it can be measured proportionally from the HT current to the power amplifier stage. The mean brightness of the tube can be measured with a photocell, and this will give an indication of the average beam current.

X-RAYS. The X-radiation produced by these tubes is considerable. Lead-sprayed covers with lead glass windows are essential, and X-rays may leave the projector along the light beam. These X-rays, together with electron bombardment, also cause a brown (silicon) discoloration of the tube faceplate, which can be bleached almost completely by heating the tube with infra-red heaters. This procedure must be repeated about every twenty hours of tube life.

TUBE COOLING. As 100 watts or more may be dissipated in the tube face, the tube must be cooled by a blower. In the case of the 9¼ in. tube, another blower is used to cool the tube neck around the filament.

CIRCUITS. Vision amplifiers, a black-level clamp, the synchronizing signal separator and the scan generators are similar to those used in a high-quality picture monitor, the differences being that more scanning power and vision drive signal to the tube are required. Regulated HT supplies are used.

Cathode drive of the tube is preferable because slightly less drive is needed with a pentode electrode gun and because the vision output stage will be biased to run at low current when at black level. This results in a lower average anode current.

FOCUS MODULATION. In the 9¼ in. tubes, because of the high beam current, the beam is of large diameter as it passes through the deflection coils, and it converges to the faceplate at a relatively wide angle. The curvature of the tube face is set by the requirements of the Schmidt optical system, and so to obtain good overall electrical focus and sharpen the spot to the edges of the picture, modulation of the focus current is necessary.

Parabolic voltage waveforms at line and field frequency are generated by double integration of the appropriate synchronizing pulses. The line waveform is used to control the current to an auxiliary focus coil, while the field waveform is applied to the current regulator which feeds the main focus coil.

SPOT WOBBLE. As the current density to the screen in the highlights is very great, the screen phosphor saturates and also some gas may be released from it. These effects may be decreased by deflection of the spot vertically by means of a high frequency sine wave current at 10 MHz or more. This deflection is popularly called spot wobble. The current is fed into deflection coils of one turn on each side of the tube. This also has the effect of eliminating the line structure of the picture with little impairment of definition.

A getter may be flashed over the entire inside of the tube cone to absorb gas as soon as possible after it is emitted.

GAMMA CORRECTOR. Screen saturation decreases the brightness of the whitest objects in the picture. This can be corrected by providing a gamma corrector with increased gain for signals corresponding to white.

SCAN PROTECTION CIRCUITS. As a high average power is dissipated in the screen, rapidly acting circuits are needed to protect the tube against the results of scan failure.

If one scan fails, a line can be burnt on the screen, and if both fail a hole may be melted through the tube face, with subsequent collapse.

To reduce the risk of fire, as little equipment as possible is incorporated in the projector installed in the auditorium. However, the final vision amplifier and spot wobbler require short leads and so must be in the projector; likewise the blower and heater transformers. Most of the components can be contained in an airtight compartment so that in the event of failure no smoke can be released into the auditorium.

CATHODE RAY TUBE COLOUR PROJECTORS

Large screen compatible colour pictures can be produced by the use of three Schmidt projectors of the type just described.

It would be possible to combine the light from the three projectors using dichroic mirrors or partly reflecting mirrors. As large dichroic mirrors are very expensive, and too much light would be lost by the use of partly silvered mirrors, the projectors are normally placed side by side.

The cathode ray tubes are fed from the same high voltage supply and have common field and line scan generators, so that any variation of these affects the pictures equally and does not spoil the registration.

PICTURE DISTORTION. As the projector on the left, for example, is nearer to the left edge of the screen than to the right edge, the projected picture from it will be smaller at the left, both in the vertical and the horizontal direction. A typical proportion might be 5 per cent. The picture might also be about 10 per cent brighter at the left.

A horizontal correction can be carried out by the use of individual line linearity controls for the three tubes. The difference in brightness can be corrected by the use of shields to vignette part of the aperture of the projection systems.

The vertical correction is called keystone correction, and involves addition to the field scan current of a line sawtooth current, modulated with field sawtooth. This modulation must be carried out in a doubly balanced modulator so that no line sawtooth or field sawtooth is present in the waveform. One method of producing this waveform is to produce line frequency pulses amplitude-modulated with field frequency sawtooth, balancing out any constant line pulse and field sawtooth, and then to integrate this waveform to produce a sawtooth waveform.

SKEW CORRECTION. It is difficult to make scanning coils which produce field and line deflections exactly at right-angles. For this reason, skew correction may be applied.

A small amount of field frequency sawtooth current of controllable amplitude and reversible in phase may be used to deflect the beam in the line direction, to make the angle between line and field deflections identical for the three colours.

TUBE COLOURS. As the types of phosphor which will stand the bombardment of a projection tube are very few, the tube colours are blue, green and yellow, the yellow tube being used together with a red gelatin filter.

PERFORMANCE. One projector uses 4 in. tubes with a 30 kV supply, with 10½ in. mirrors in the Schmidt optical systems. The tubes are cooled by a blower. With a screen gain of two, it will produce a highlight brightness of 5 fc on an 8 ft. × 6 ft. picture.

Back-projection is not possible with this type of projector because the peak of brightness (hot spot) visible along the axis of projection with a back-projector screen will occur in a different direction for the three colours. This will result in severe colour shading.

THE EIDOPHOR

If an electron beam is played on a thin layer of an insulating oil, the surface of the liquid is deformed by electrostatic forces. If the focus of this beam is altered, the deformation is correspondingly altered. The beam can be deflected at line and field frequencies over the surface of the eidophor oil, and a pattern of deformations built up which can be projected by a special optical system.

The amount of light which can pass through the eidophor is not altered, but the light is refracted. For this reason the Schlieren optical system, which only passes refracted light, is used.

THE SCHLIEREN OPTICAL SYSTEM. The light source is a 2 kW xenon lamp, which is air-cooled. A cold light mirror reflects some of the heat back to the lamp while allowing the light to pass. The light is focused by a lens and is reflected from a grid of mirrors, which are arranged in parallel planes so that

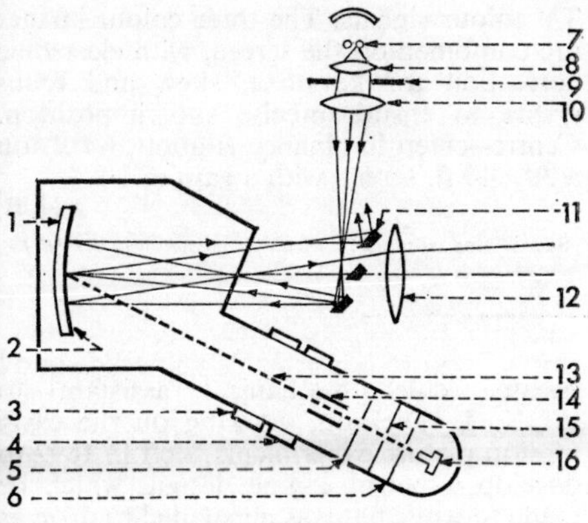

EIDOPHOR OPTICAL SYSTEM. Xenon arc source, 7, is imaged by condenser lens, 8, on to strip mirror, 11. Picture mask, 9, is imaged by lens, 10, on to concave mirror, 1, coated with Eidophor surface, 2, using Schlieren optics. When, in absence of modulation, Eidophor surface is undisturbed, light is blocked by mirror strips, 11, from reaching main projection lens, 12, which forms images on screen. Eidophor is deformed to produce image by electron beam, 13, from gun assembly of cathode, 16, grid, 6, anode, 15, aperture, 5, modulation electrode, 14, with deflection, 3, and focus coils, 4.

no light from the lamp can pass between them, on to the face of a concave mirror. On the face of this mirror is the eidophor oil film.

The grid of mirrors is in a plane passing through the centre of curvature of the convex mirror. The image of each strip mirror coincides in position with the mirror placed symmetrically on the other side of the centre of curvature. Thus, in the absence of

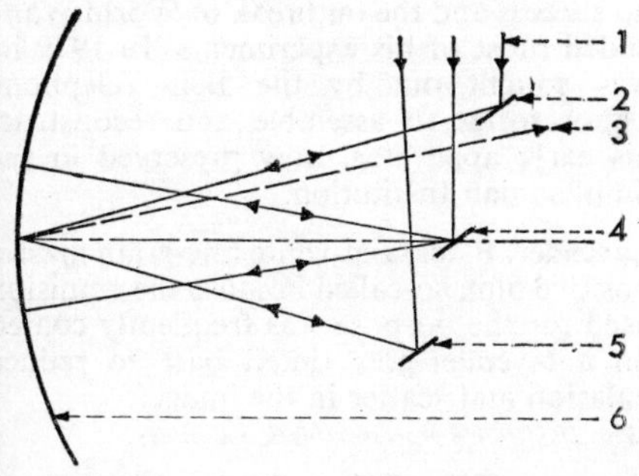

PRINCIPLE OF SCHLIEREN OPTICS. 1. Incident rays of light from projection source strike sections of strip mirror, 2, 4, 5, lying at radius of curvature of Eidophor-surfaced concave mirror, 6. Hence, rays reflected from mirror strips are always intercepted by strips. Extreme thinness of Eidophor surface leaves optics unchanged until point by point deformation causes deflection of path, with rays passing between strips, as at 3.

deformation of the oil film, light reflected from one strip mirror is reflected from the concave mirror and falls on another strip mirror. Thus no light passes between the mirrors to the projection lens. Deformation of the oil film causes refraction so that some of the light is able to pass the strip mirror and reach the projection lens. This focuses an image of the surface of the concave mirror on to the large screen.

ELECTRON GUN. The use of an electron beam necessitates a high vacuum, the system being continuously evacuated by an oil diffusion pump with a mechanical backing pump.

The electron beam passes a current of a few tens of microamperes from a supply of 15 kV. It is essential for focus modulation to be applied because variation of focus will vary the brightness of the picture. Picture modulation is applied to an electrostatic focusing electrode.

The life of the cathode is about thirty hours, and three interchangeable cathodes are provided.

EIDOPHOR. The type of oil used must have low vapour pressure, the correct surface tension, viscosity and electrical conductivity, the last of these deciding the time for which the image is retained. As all of these characteristics vary with temperature and the oil is subject to heat from the light source, it must be cooled. For this purpose it passes through a heat exchanger, then through a filter before being sprayed on to the mirror. Water flowing through the heat exchanger is cooled by a small refrigeration plant.

A blade is used to spread the liquid smoothly as the mirror rotates slowly, one revolution taking a few minutes.

The mirror, pump, oil reservoir, spray and scraper are in a cassette which can be closed so that it can be changed easily.

PERFORMANCE. This projector is capable of producing an illumination of 5 fc on an area of 400 sq. ft. With a screen gain of 1·8 the maximum picture size is 24 ft. × 32 ft., giving a highlight brightness of 3 fc.

One advantage of the Eidophor system is that it can operate with a long throw distance from the projector to the screen, and so can be installed in a cinema projection box.

THE COLOUR EIDOPHOR. A modified form of this projector can be used for field-sequential colour pictures, a colour wheel being inserted in the light path. The maximum picture size is reduced to about 16 ft × 21 ft.

There is now a triple-tube Eidophor projector, in which light from a 2·5 kW xenon source is split three ways by dichroic mirrors, and each is picture-modulated by an Eidophor fed by standard compatible TV colour signals. The three colour images are combined on the screen, with electronic correction for keystone, skew and focus errors to ensure precise superimposition. Centre-screen luminance is about 6 ft.L on a 30 × 40 ft. screen with a gain of 1·8.

J. H. B.

See: *Closed circuit TV; Medical cinematography and TV.*

□**Laser.** Acronym for Light Amplification by Stimulated Emission of Radiation. A maser which works in the visible and near-visible regions of the spectrum. Currently being developed for wide-band communications.

See: *Holography.*

⊕**Latensification.** Method of increasing the developable density of a photographic image by giving the exposed but unprocessed film a long exposure to an extremely low intensity light. Very occasionally employed on motion picture film negatives.

See: *Chemistry of the photographic process.*

⊕**Latent Image.** Invisible image existing on an exposed photographic film before development.

See: *Image formation.*

⊕**Laterna Magika.** Entertainment form combining live action on the stage with projected motion picture and still images. Developed in Czechoslovakia.

⊕**Latham, Major Woodville, d. 1911.** American inventor whose somewhat impractical development of the Edison Kinetoscope was, in 1894, greatly improved by Eugene Lauste, who developed the Eidoloscope projector embodying the loop-forming device invented by Latham, first shown in 1895.

⊕**Latitude.** Range of exposure of a photographic material over which the variation in density is substantially proportional to the variation in exposure intensity; it is directly related to the length of the straight line portion of the characteristic curve of the material and provides an indication of the exposure range over which satisfactory reproduction of a scene may be expected.

See: *Exposure control; Image formation; Sensitometry and densitometry.*

⊕**Lauste, Eugene Augustin, 1857–1935.** French inventor, born Montmartre, who, before the age of 23, had filed 53 devices in the French patent office. Joined the staff of Edison's West Orange Laboratory, 1887, where he became chief mechanical assistant to W. K. L. Dickson, working on the early motion picture experiments. Left in 1892 to develop a petrol engine design, which he made to work but was persuaded to drop as commercially impractical. In 1894 he worked with Major Woodville Latham, being engaged to carry out experimental work on Latham's ideas. Constructed the Eidoloscope projector, exploiting the Latham Loop, and demonstrated it in 1895. In the following year he joined the American Mutoscope and Biograph Company, as his former associate, Dickson, had done.

In 1900 an interest in sound recording, acquired while working for Edison, led him to develop ideas for sound film systems. By 1904 he had built his first sound-on-film recording device, applying on 11th August 1906 for a British patent, which was accepted on 10th August 1907. This was, in effect, a master patent for sound-on-film recording. A meeting in 1910 with Ernst Ruhmer led to collaboration between them until Ruhmer's death in 1913.

Although he had attempted to develop amplifying systems from 1912, Lauste had no success and the outbreak of World War I ended most of his experiments. In 1929 he was sought out by the Bell Telephone Laboratories to assemble and reconstruct his early apparatus, now preserved in the Smithsonian Institution.

⊕**Lavender.** Black-and-white fine-grain master positive film, so-called because the emulsion used for the purpose was frequently coated on a lavender-grey tinted base to reduce halation and scatter in the image.

See: *Duplicating processes (black and white).*

□**Lawrence Tube.** Colour display tube, named after the American physicist who first suggested the principle of operation; also known as chromatron, post-deflection focus and focus-mask tubes. It is an attempt to reduce the complexity of colour TV receivers, by removing the need for static and dynamic convergence.

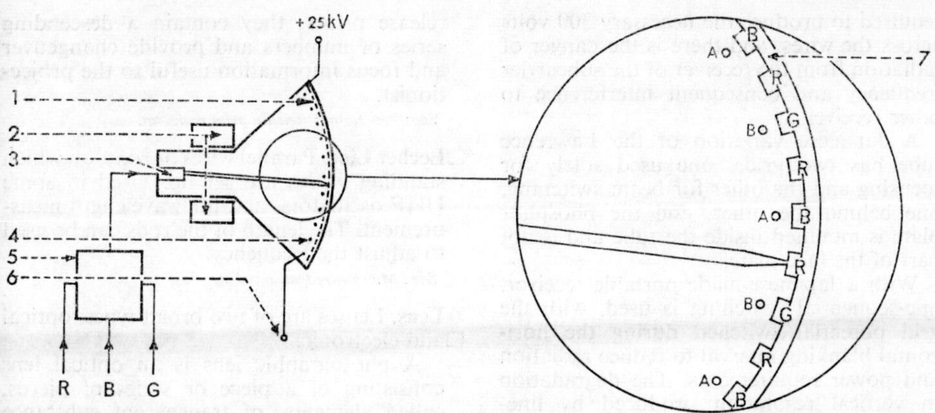

ʟAWRENCE TUBE (CHROMATRON). Electron gun, 3, emits single beam, deflected by scanning coils, 2. Gun is switched at colour sub-carrier frequency by electronic switch, 5, fed with R, G, B video information. Grid of vertical wires, 4, backs vertical phosphor strips, 1, 7, and has applied to it an a.c. switching and d.c. focusing voltage, 6. Enlargement (*right*) shows arrangement of phosphor strips with more frequent reds to compensate esser red sensitivity. All grid wires, A, are connected together, also grid wires, B, and switching potential s applied between them. Diagram shows a recent form of the tube.

Developed largely by the Paramount 'icture Corporation, its use in receivers to late is very limited, and it is more suited to small-screen portable receivers, where it can produce sharp, bright pictures.

It differs from the shadow-mask tube in he following ways:

. There is only one electron gun; versions aving three guns have been produced but n most versions the additional guns serve nly to reinforce the electron beam.

. There is no shadow mask; a wire grid, in oughly the same position relative to the hosphor takes its place.

. The phosphor, instead of being deposited n groups of three dots, as in the shadow-nask tube, forms vertical stripes on the face late (about 1,500 on a 19 in. rectangular ube).

The electron beam, attracted towards the inal anode, passes between the grid wires nd strikes the phosphor, the exact position epending on the potentials between the djacent grid wires. The size of the spot epends on the potential of the grid wires elative to the cathode and the final anode, ypical voltages being about 25 per cent of e final anode voltage, or 6 kV positive elative to the cathode.

In the absence of any switching voltage on e grid wires, the red phosphor strips are ctivated, and a completely red screen is roduced. Since red phosphors are usually ss efficient at turning beam current into ght, there are twice as many red stripes as ere are blue or green.

If a switching voltage, normally about 500 volts, is applied between adjacent wires the beam is deflected to one side after passing the grid, and strikes either the blue or green phosphor, giving a blue or green screen, depending on the polarity of the voltage.

Thus, provided that the geometry of the tube is correct, the colour of the screen can be controlled by adjusting the potential between the two sets of grid wires. This means that the convergence circuitry and hardware of the shadow-mask tube are not required and that the magnetization of the shadow-mask is no longer a problem, thus avoiding the need to de-gauss the tube. Also, the grid is about 85 per cent "transparent" as compared with the 25 per cent achieved by the shadow-mask tube.

In a colour TV receiver, means must be provided for keying the electron beam to the colour information to ensure synchronization between the sequentially-switched video drive voltage and the switching waveform on the grid wires. This is achieved by the use of an electronic switch between the source of demodulated colour information and the electron gun, and by a switching-voltage generator, working at the colour subcarrier frequency, driving the two sets of grid wires. The tube can, with correct phasing of the two switches, be regarded as having three guns, energized in turn, each gun activating the appropriately-coloured phosphor. Since the capacitance of the wires is about 1,000 pF, a considerable current is required to produce the necessary 500 volts across the wires, and there is the danger of radiation from the receiver of the subcarrier frequency and consequent interference to other receivers.

A Japanese variation of the Lawrence tube has two grids, one used solely for focusing and the other for beam switching, one behind the other, and the phosphor plate is mounted inside the tube and is not part of the face plate.

With a Japanese-made portable receiver, line-sequential switching is used, with the grid potential switched during the horizontal blanking interval to reduce radiation and power requirements. The degradation in vertical resolution produced by line-sequential switching is apparently acceptable on the 9 in. screen used in this receiver.

L.T.M.

See: *Display tubes.*

⊕**Leader.** Blank film consisting of a coated or uncoated base. Used for threading continuous developing machines so that undeveloped film can be drawn through them without stopping. Also used in large quantities in editing workprints and cutting dubbing sound tracks when it serves as blank film to fix the position of wanted sections of picture or sound.

Academy leaders (named after the Academy of Motion Picture Arts and Sciences) are leaders which are placed at the beginnings and ends of all reels of release prints; they contain a descending series of numbers and provide changeover and focus information useful to the projectionist.

See: *Processing; Release print make-up.*

□**Lecher Line.** Parallel wires or rods, on which standing waves are set up. Used in some UHF oscillators, and for wavelength measurement. The length of the rods can be used to adjust the frequency.

See: *Microwave transmission.*

⊕□**Lens.** Lenses are of two broad types, optical and electronic.

A photographic lens is an optical lens consisting of a piece or series of pieces, called elements, of transparent substance bounded by two curved surfaces (usually spherical), or by a curved surface and a plane. The lenses used in cinematography and in the television camera are converging types which form a real image of greater or less magnification, usually on the emulsion surface of a film, the photosensitive surface of a TV camera, or a projection screen.

In an electronic lens, light rays are replaced by rays of electrons and these are deflected by magnetic fields to form an image much as in an optical lens.

See: *Camera; Lens performance; Lens types.*

⊕□**Lens Hood.** Tubular or funnel-shaped device mounted on or in front of a camera lens to prevent stray light from entering the lens.

□LENS PERFORMANCE
⊕

One of the most important advances of the last few years is the development of a new way of assessing lens performance and relating it to the requirements of film and TV systems.

RESOLVING POWER

The earliest techniques of lens assessment were borrowed from the astronomers, who rated telescope objectives according to their ability to separate or resolve the images of double stars. This ability constituted the resolving power of the objective. The same general line of thinking was followed by the microscopists, who assessed the quality of microscope objectives by their ability to separate fine detail in such things as diatoms. The use of resolving power as a means of assessing the performance of photographic lenses was not adopted until many years later, and at the time there was good justification for it.

TARGETS. Once the notion of assessing the performance of photographic lenses by their resolving power had become established, it was necessary to develop reasonably standardized patterns for making this assessment. Such standardized patterns constitute resolving power targets. Without exception these targets comprise either groups of black stripes on a white background or white stripes on a black background (grey may be used in place of black for low contrast measurements, but for the present black and white are assumed). The width of a black or white stripe is equal to its separation from neighbouring stripes. Within any target or pattern there are a number of groups of stripes and spaces, with progressively decreasing widths of stripes from group to group.

These targets are set up in front of the lens and a photograph is made, or the target images are transmitted through a television

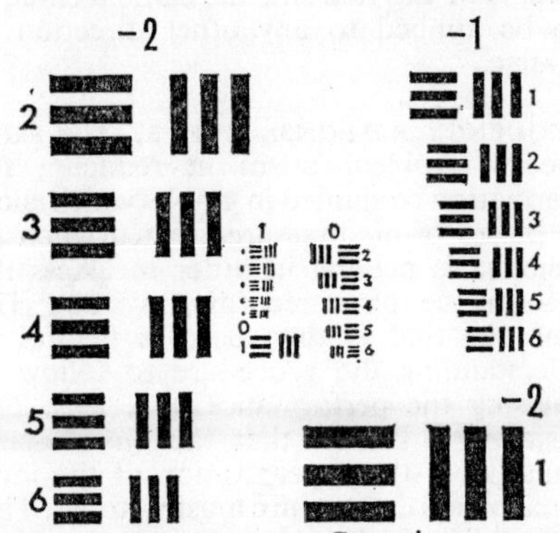

RESOLVING POWER TARGET. Comprises groups of black or white stripes, with spaces equal to the stripe width, and with a progressive diminution of width from group to group.

system or examined visually with a microscope. In any event, the resolving power of the lens is measured by starting from the coarsest groups and proceeding to the group in which it is just possible to detect that the stripe and space structure exists. If the total span, in such a group, of the width of a stripe plus its separation from a neighbour is equal to ·02mm, the lens has a resolving power of 50 lines per millimetre.

SPURIOUS RESOLUTION. It is often found that, on proceeding to groups having an even finer structure, the stripe and space pattern becomes apparent again. This is due to spurious resolution: the number of stripes is not what it should be and the lens cannot be said to be capable of resolving such fine groups, since it is providing inaccurate information.

The resolving power of a lens must be specified by stating the means by which it

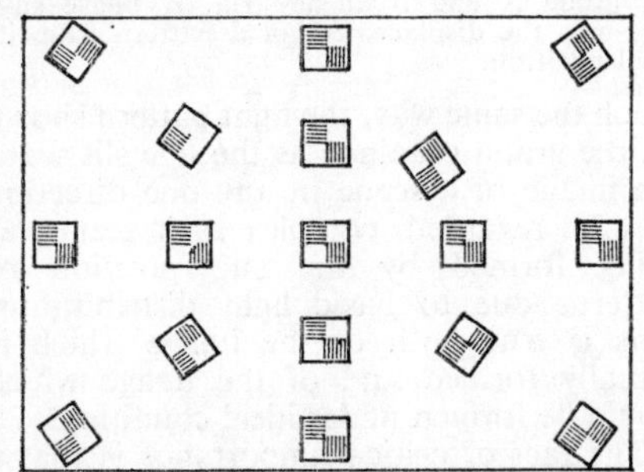

LENS TESTING. When testing the performance of a lens, resolving power targets are distributed over the field of view. One set of lines on each chart points towards centre of field to determine radial resolution; the other, at right angles, determines tangential resolution.

was evaluated, either photographically, visually, or through a TV channel. This is because each of these means of recording the information supplied by a lens imposes its own limitations upon the image. In film, there is graininess of the film itself and scatter in the emulsion; in TV, there are limitations due to the finite bandwidths of the transmission channels.

LIMITATIONS. The assessment in terms of resolving power served as a useful tool during the 1940s, largely because it was better than any other tool then available, but also because it reflected the information-detecting needs of military reconnaissance. Its limitations soon became apparent, however. For example, it was found that pictures which were aesthetically satisfying were not always provided by a lens with high resolving power. Moreover it was not possible, by using a resolving power alone, to analyse the interaction of all the elements in a complete process of picture presentation. Such a chain of elements, for example, might include a taking lens, a film or plate, an enlarging or projection lens, and finally the human eye. Alternatively the chain of elements might comprise a taking lens, an image orthicon, a transmitter and receiver, and a TV viewing screen. Each of these has its own limitations, which may be expressed crudely through a resolving power, and it is important to be able to combine them in a system to get the best possible final result. This cannot be done effectively by using resolving power alone.

Defining lens or film performance in terms of resolving power alone is like writing a biography of a man in terms of his life-span only, without regard to his activities within that span. Except for some unusual individuals, a reasonable life-span is a prerequisite to having a character that is worth writing about, but a long life does not automatically make a worthwhile subject. In the same way a reasonable resolving power is usually needed in order to ensure that a lens is worth considering, but by itself it does not guarantee a worthwhile lens.

FREQUENCY RESPONSE

The key to obtaining a more complete evaluation of either a photograph or a television lens lies in the concept of frequency response and in the associated ideas of optical transfer function (OTF) and modulation transfer function (MTF).

LIGHT DISTRIBUTION. To do this, the type of image actually produced by a lens must be considered. An image comprises a pattern of light and shade which is directly related to the light distribution in a scene in front of the lens. If a lens has an ideal performance, it is possible to predict what the light distribution will be in the image of a particular scene. But the faults in the performance of any lens do not, in fact, allow it to realize this ideal and the light distribution depends on the characteristics of the lens. The relation between the actual and ideal light distributions may be used as a means of assessing the lens performance. This comparison must be made at all points in the field covered by the lens.

To make this comparison at any point in the field, a small area surrounding the point is enlarged to a scale that makes it readily accessible for examination in detail. Across

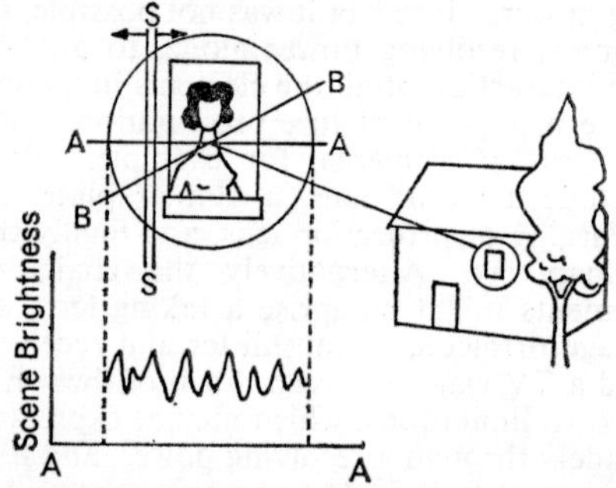

ANALYSIS OF PICTURE QUALITY. To analyse picture quality, a small area, indicated by circle in picture on right, is examined for brightness variation along directions such as AA and BB with the aid of a slit as shown at SS travelling along AA. (*Below*) Typical brightness curve, which may be analysed into a group of superimposed basic brightness patterns.

this enlarged area, a very fine slit is traversed, the direction of motion being at right-angles to the length of the slit. The light from the image which comes through the slit is recorded as a rather irregular graph drawn by a recording pen. The ordinate of the graph measures the total light from the image that is received from the entire length of the slit, as the slit is traversed perpendicularly to itself. Repetition of this procedure for all possible orientations of the slit then gives complete information about the distribution of light in the image. However, the information is in a form which is all but incomprehensible as it stands, without some means of processing it and reducing it to manageable proportions. Fortunately, such a means of reduction exists.

If a method is found to reduce the information contained in one direction of traverse of the fine slit, the same technique can be applied to any other direction of traverse.

FREQUENCY RESPONSE CURVE. The same type of problem arises in reducing the information contained in graphs which show the pulses of air pressure created when an orchestra is playing in order to assess the performance of a recording system. The technique that is then used is helpful in understanding the procedure to follow in analysing the performance of a lens. For music, it is known that the air pressure results from the superposition of the individual pulses due to pure musical tones. This is true of both the original performance of an orchestra and of the way in which the music is reproduced through a radio or record player.

The faithfulness of the reproduction can be described by giving a frequency response curve, which relates the intensity or loudness of any pure tone in the reproduction and in the original performance. In very

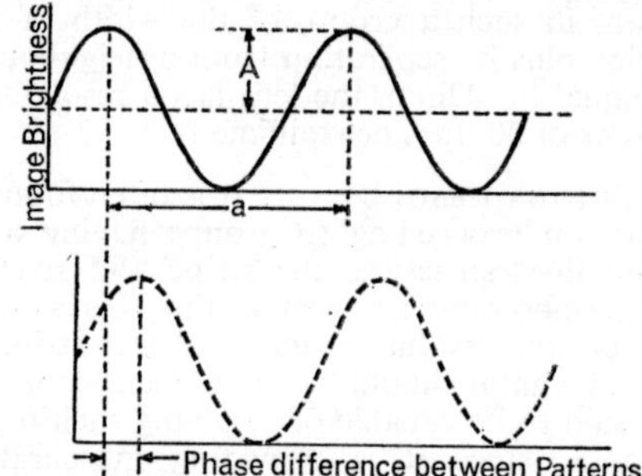

BASIC PATTERNS OF BRIGHTNESS CURVES. Any brightness curve can be built up by superimposing a set of basic patterns of this form, characterized by amplitude A and frequency 1/a. A phase shift measures the displacement of a pattern from its ideal position.

much the same way, the light pattern shown on the graph obtained as the fine slit scans the image of a scene in any one direction may be regarded, complex as it seems, as being formed by the superposition of patterns due to basic light distributions. This is true both of the image which is actually formed and of the image which would be formed under ideal conditions.

The fact of critical importance is that a basic (sinusoidal) pattern of light distribution in the scene in front of a lens, or what amounts to the same thing, a basic light distribution in the ideal image of a scene, always produces the same basic pattern of light distribution in the image plane of the lens, no matter what may be the state of aberrational correction of the lens, or even its state of focus. The departure from the ideal situation, in which the lens is free from aberrations and in perfectly sharp focus, is shown by a diminution in the brightness range in the actual image plane pattern, as compared with the brightness range to be expected in the ideal pattern. There may also be a slight shift of the pattern position.

MODULATION FACTOR. The general statements of the preceding paragraph can be quantified in the following way: suppose that the maximum brightness in the image of a basic (sinusoidal) light pattern is I_{max}, the minimum brightness is I_{min}, and the brightness range for the image is denoted by R_A, then

$$R_A = (I_{max} - I_{min})/(I_{max} + I_{min})$$

Suppose that the brightness range for the ideal situation, with an aberration-free lens, is denoted by R_I. Then a modulation factor M for this particular basic pattern is given

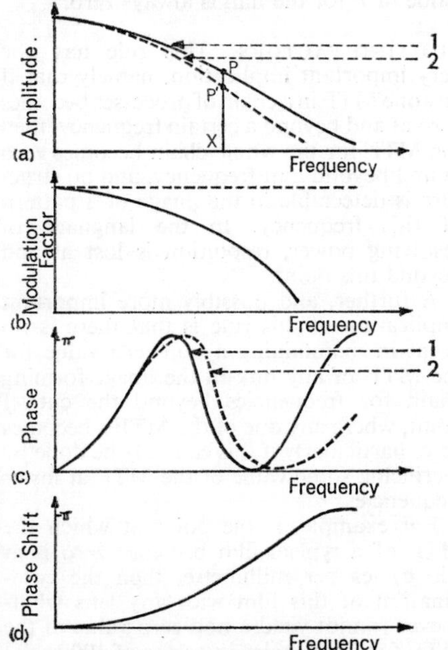

MODULATION TRANSFER FUNCTION. (a) 1. Curve representing distribution of amplitudes, if scene were reproduced under ideal conditions. 2. Curve representing distribution of amplitudes resulting from aberrations present. Modulation factor at a given frequency is P'X/PX. (b) Modulation factor curve or MTF. Dotted line shows ideal curve, unrealizable in practice. (c) and (d) Effects of phase shift similarly shown.

by the equation $M = R_A/R_I$, which defines the performance of the lens for this pattern.

To complete the description of performance, any shift of the image pattern relative to its ideal position must be stated. This is effected by means of a phase shift P, where P is measured in degrees and is given by the formula $P = 360 \times$ (pattern shift)/(pattern size). These two quantities, M and P, define completely how the lens reproduces this particular pattern.

It has already been pointed out that any arbitrary light distribution in the image plane, whether actual or ideal, may be built up by the superposition of basic light patterns. If it is known how each of these patterns will be reproduced by the lens, in other words if the values of M and P are known for each basic pattern, then it can be predicted how any light distribution in the image plane will differ from its ideal value.

OPTICAL TRANSFER FUNCTION. The complete set of values of M and P for all sizes of basic light pattern from coarse patterns to those of very fine pitch, measure the frequency response of the lens, and the statement of their values constitutes the optical transfer function (OTF) of the lens. The values of M alone, for the same range of patterns, constitute the modulation transfer function (MTF) of the lens. As a rule, the MTF is more important than the OTF, since the phase shift does not play a dominant role in determining picture quality.

In a typical representation of results, the value of the MTF is plotted in a graph against the frequency of a basic pattern in the image plane, e.g., if the size of a basic light pattern is ·02mm, the frequency of the pattern is 50 cycles per millimetre, and this is the term used to describe it.

Because the performance of a lens is not the same at all points in the field that it covers, it is necessary to present a number of graphs of the kind described above. Moreover, the frequency response or MTF is also dependent upon the direction in which the patterns are oriented, and thus must also be adequately shown in the presentation of results. As a rule, it is sufficient to show the results for the sagittal and tangential directions, in other words, for patterns which lie along and are perpendicular to a line joining the centre of the field to the point in the field at which the performance is to be evaluated.

COMBINATION OF OTF'S

Just as the performance of a lens may be fully described by giving its OTF for various points in the field that it covers, so can the performance of any other element in a film or TV system, which affects image quality, be described by an OTF.

For example, in recording an image on film, there are effects of graininess and scatter within the emulsion which degrade the quality of an image. These effects are symmetrical in form and are uniform throughout the area of the film. The phase shift is zero and the performance of the film is described by its MTF. It should be noted, however, that the MTF may depend to some extent on the level of exposure given to the film and on the type of developer.

In the TV application, there are limitations on the picture quality imposed by the finite bandwidths of the electrical circuits. These may also be expressed by means of an OTF. which is in fact identical with the frequency response and phase distortion curves of the electronics engineer.

CHAIN PROCESSES. It is quite simple to imagine more complex chains of processes which ultimately yield a useful picture, including in such a chain the formation of an image on the retina of an observer's eye. Each link in such a chain has its own OTF. The important fact is that the OTF for the complete chain of processes may be obtained in a simple way from the OTF's of the individual links.

The rule is that the value of M for the complete chain of processes is obtained, at any value of the frequency of a basic pattern, by multiplying together the individual values of M for that frequency, and the value of P for the complete chain is obtained by adding together the values of P for the individual links at that frequency.

This method of combination for all frequencies gives the OTF for the complete chain of image-forming processes. For example, in the simple case of image formation by an imperfect lens on a grainy emulsion, the OTF of the whole image-forming process under these conditions is obtained by multiplying corresponding values of the lens MTF and of the film MTF for the complete range of frequencies. The value of P for the complete process is equal to the value of P for the lens alone, since the value of P for the film is always zero.

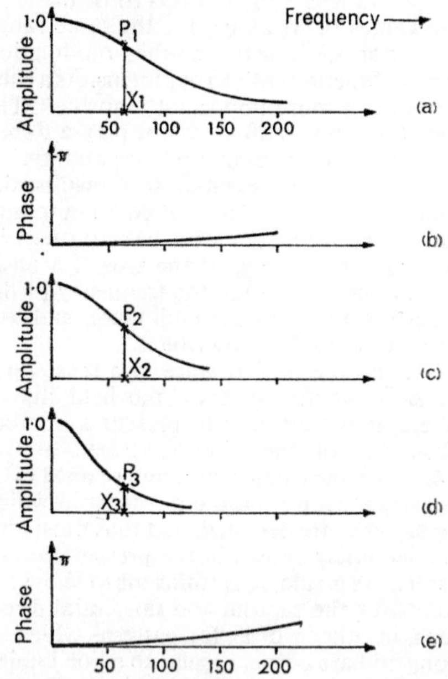

OTF FOR LENS-FILM COMBINATION. (a), (b) Characteristics of lens in terms of OTF curves for amplitude and phase respectively. (c) Characteristic of film in terms of MTF curve. (d), (e) Amplitude and phase characteristics of lens-film combination found by multiplication at each frequency, e.g. $P_3X_3=(P_1X_1)\times(P_2X_2)$.

LIMITING FACTORS. This rule has one very important implication, namely that if any one MTF in a chain of processes becomes zero at and beyond a certain frequency, then the MTF for the whole chain becomes zero at and beyond that frequency, and no structure is detectable in the image of a pattern of this frequency. In the language of resolving power, resolution is lost at and beyond this point.

A further, and possibly more important implication of this rule is that there is no point in maintaining a non-zero value for the MTF of any link in the image-forming chain for frequencies beyond the cut-off point, where any one of the MTF's becomes zero, particularly if this can only be done by sacrificing some value of the MTF at lower frequencies.

For example, if the point at which the MTF of a typical film becomes zero is at 100 cycles per millimetre, then the combination of this film with any lens whatsoever cannot yield a non-zero value of the MTF for frequencies in excess of 100 cycles per millimetre: the lens–film combination has a limiting resolving power of 100 lines per millimetre. This, of course, is scarcely a surprising result. Below this limit, however, the behaviour of the lens-film combination depends very strongly on the MTF of the lens. If it were possible to have a lens

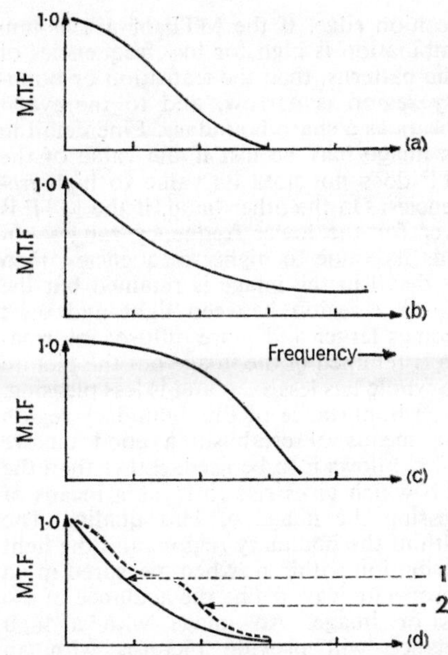

MTF FOR LENS-FILM COMBINATION. (a) MTF curve for film. (b) MTF curve for lens of higher resolving power. (c) MTF curve for lens of lower resolving power. (d) 1. Combination of (a) and (b). 2. Combination of (a) and (c). Lens (c) gives better picture quality.

with an MTF equal to unity up to 100 cycles per millimetre and falling off rapidly to zero beyond that point, then the maximum potentiality of the film would be realized, even though the lens by itself would not exhibit a resolving power much in excess of 100 lines per millimetre.

On the other hand, a lens whose MTF fell almost immediately to, say, 0·50 for frequencies below 100 cycles per millimetre, but maintained a non-zero value out to 200 cycles per millimetre, would by itself show a resolving power of 200 lines per millimetre, but the picture-producing quality associated with this particular lens-film combination would be inferior to that obtained with the 100 lines per millimetre lens cited above. This analysis brings out the fact that there is not 100 per cent correlation between resolving power and picture-taking ability. A high resolving power in a lens may turn out to be an empty value.

TV BANDWIDTH LIMITATIONS. In the case of TV systems, the limiting factor, in a very drastic way, is the finite bandwidth of the electrical circuits. An electron beam in the image orthicon, Plumbicon or vidicon scans the picture formed in the TV camera and, as it crosses light and dark areas, it gives rise to pulses of electric current. The number of such pulses which can be transmitted in one second is the bandwidth of the electrical circuits, and is primarily limited by the need to provide for a number of different television channels. The number of times per second that a complete picture must be transmitted in order to eliminate flicker, and to provide the illusion of smooth movement, is also quite well defined. From this it follows that the number of picture elements which may be contained in any picture is limited. Substituting the proper values for commercial television in the USA, where the bandwidth limitation is 4·2 MHz (million cycles per second), with a picture repetition rate of 30 per second, shows that the cut-off frequency of the television system corresponds to about 7 cycles per millimetre on the face of an image orthicon tube. The BBC standard of performance is a little higher and corresponds to a cut-off frequency of 8 cycles per millimetre. If a vidicon tube is used, with the same bandwidth and the same picture-repetition rate, then the corresponding cut-off frequencies are approximately 18 and 20 cycles per millimetre on the face of the vidicon. If closed-circuit television is used then it is theoretically possible to use greater bandwidths, with a better picture quality, and with cut-off points that have a higher frequency.

The cut-off frequency for film, even for colour film, may be taken as 80 cycles per millimetre or higher. In order to make even reasonable use of the film to record fine detail, it is desirable to maintain the MTF of a lens at non-zero values up to this limiting frequency. In some instances, this can only be done by sacrificing the value of the MTF in the lower frequency ranges: such lenses will not give a performance which is best adapted to television use. Nor can it be taken for granted that a lens which has been specifically designed for television use will necessarily give the best photographic performance, though there is a good chance that lenses for television and photographic use will be interchangeable.

APPLICATION OF OTF'S

The treatment of the image-forming process in photography and in television by means of OTF's, with particular reference to the MTF's of the lenses used, provides an extremely powerful tool for making a rational approach to an evaluation of these systems. To some extent it is an embarrassing tool in

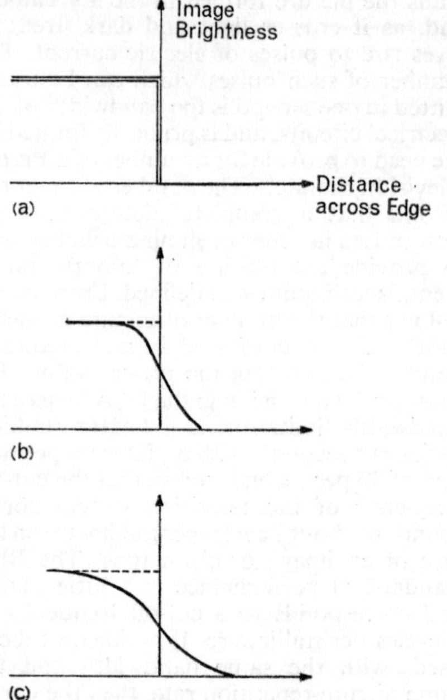

EDGE GRADIENT. (a) Acutance of a lens system is calculated by measuring the brightness across a boundary; (b) has the higher acutance but (c) has the higher resolving power.

that it provides so much valid information. While such a mass of data may be of great value to professionals in the field of optical engineering, it is very desirable that some way be found to summarize or epitomize it for widespread use. More research is needed to develop a means of extracting information from all the mass of data provided by OTF theory about the use of lenses in conjunction with photographic or television systems, that will at the same time be compact, meaningful and accurate.

ACUTANCE. One such approach rests upon the fact that a viewer is largely concerned with the way in which the boundary of an object is reproduced. Ideally the transition from light to dark across such a boundary should be abrupt. In actual fact a boundary region of gradual change replaces the sharp transition edge. If the MTF of a lens-film combination is high for low frequencies of basic patterns, then the transition or boundary region is narrow, and to the eye it appears as a sharp boundary. Fine detail in this image may be lost if the value of the MTF does not hold its value to high frequencies. On the other hand, if the MTF is lower for the lower frequency ranges but holds its value to higher frequencies, then fine detail in the image is retained but the boundary region between light and dark becomes larger and more diffuse: information is retained in the image but the picture as a whole has less snap and is less pleasing.

The importance of this boundary region as a means of establishing good picture quality allows it to be used, rather than the MTF which gives rise to it, as a means of assessing the image or lens quality. The width of the boundary region, and the light distribution within it, when measured in an appropriate way, define the acutance of the lens or image. An image with a high acutance will provide pictures with an overall snap and sense of sharpness even though fine detail may be lost.

With the advent of electronic computers, it has become a straightforward matter to calculate the OTF's for lenses as they are designed, and from this information to predict the overall OTF of a complex system, once the OTF's of other elements in the image-forming chain are known. It is possible, for example, to calculate the acutance of the complete system. Considerable progress has thus been made in recent years both in understanding the way in which individual stages in the production of an image may be combined, and in developing computer-based techniques to carry out the necessary calculations in a short time. There are, however, areas in which further research is still needed. These involve the relationship between the OTF values and aesthetic satisfaction, and ways in which OTF information may be accurately summarized. A.C.

See: *Camera; Image formation; Lens types.*

Books: *Fourier Methods in Optical Image Evaluation*, by E. H. Linfoot (London), 1966; *Photographic Optics*, by Arthur Cox (London), 1966; *A System of Optical Design*, by Arthur Cox (London), 1967.

⊕**Lens Turret.** Device on the front of a camera
□for holding four or five different lenses. Any of these may be brought into the taking position by rotation of the turret.

Lenses are often mounted divergently to prevent a wide-angle lens "seeing" a long-focus lens in the adjacent position.

See: *Camera.*

☐LENS TYPES

Some lenses may be used for either photographic (cine) work or for television. They fall quite naturally into two groups. In the first group are 16mm movie lenses and lenses for use in vidicon or Plumbicon cameras. In the second group are lenses for the standard miniature camera format (24 × 36mm) and lenses for use with image orthicon cameras.

This does not mean that the demands of photography and of television are identical. There are types of optical systems or lenses which are used for one application rather than the other, such as anamorphic systems in photography, or beam-splitter systems in colour television. What follows first considers those systems which may have a dual application in both photography and television, and subsequently the specialized systems for particular applications.

DUAL PURPOSE LENSES

COVERAGE. The grouping of lenses referred to above arises because of a reasonably close match between the sizes of the image areas to be covered in the respective uses.

The diagonal of the picture area of a vidicon camera is approximately 25 per cent larger than the diagonal to be covered on 16mm film. Whether this difference is significant depends on two factors. The first of these is the ability of the lens to cover the additional field of the vidicon without undue vignetting, which reduces the illumination in the corners of the picture area to undesirably low (even to zero) values.

The second factor is the extent to which the aberrational correction of the lens is maintained beyond the boundaries of the 16mm film format. This is evidenced by the image quality in the corners of the picture area, and satisfactory picture quality may not be maintained out to these areas.

In the case of non-zoom camera lenses having a focal length of the order of 1 in. and upwards, there is good reason to believe that the performance will be maintained to the limits of the vidicon field. There is not the same expectation for 16mm projection lenses. For focal lengths less than 1 in., the situation becomes more critical, from the points of view of both vignetting and aberrational correction, and each lens must be evaluated individually.

In the case of zoom lenses, the problems of optical design are rather more severe than for fixed focal length lenses, and the tendency on the part of lens designers is to design so that the lens will just cover the 16mm film area. This also leads to a reduction in size and weight. It is, therefore, not reasonable to expect a zoom lens designed for the 16mm cine area to cover a vidicon format. On the other hand, a lens designed for the vidicon format will certainly cover the 16mm format, but with a larger bulk and weight than is necessary.

TV TUBE COVER PLATE. A point which must be considered, in discussing whether a lens may be used both for photography and for television, is the effect of the cover plate on the vidicon or Plumbicon. The image surface in photographic use lies on the naked face of the film, but in a vidicon it lies behind glass that is ·094 in. thick. This thickness of glass introduces spherical aberration on axis, and coma and astigmatism off-axis, but except for some very-high-aperture or very-wide-angle lenses the effects are not important.

The picture area which has to be covered on the image orthicon is less than that of the standard 35mm format, and therefore many of the lenses developed for photography may be used for television. The thickness of the glass front plate of the image orthicon is ·135 in. and this introduces more spherical aberration than does the face plate of a vidicon. However, the electrical bandwidths of the vidicon and of the image orthicon in standard television use are identical, and this permits a greater latitude in image degradation on the image orthicon.

ZOOM LENSES. In recent years, many zoom lenses have been introduced for 35mm film use. While, in theory, such lenses may be used on image orthicons, it does not work out that way in practice. The zoom range of a 35mm lens tends to be restricted to about 3 : 1, in order to reduce weight and size. The application of zoom lenses in television, however, are such that the emphasis on

FILM AND TELEVISION FORMATS

Image	*Area (ins.)*
8 mm. Projection	0·129 × 0·172
8 mm. Camera	0·138 × 0·188
Super-8 Projection	0·158 × 0·211
Super-8 Camera	0·166 × 0·227
16 mm. Projection	0·284 × 0·380
16 mm. Camera	0·292 × 0·402
1 in. Vidicon	0·380 × 0·500
35 mm. (Leica)	0·964 × 1·429
Image Orthicon	0·960 × 1·280

weight and size is secondary to the needs of showing striking end results, and zoom ranges up to 10 : 1 and higher are current.

FREQUENCY RESPONSE. While it is possible to use some lenses interchangeably for both film and television, as discussed above, there is one very important factor which has to be taken into account in order to secure optimum results, namely that the frequency response characteristics needed for the two applications are not identical.

In the case of photographic use, an upper limit to the frequency response may be set by the granularity of the emulsion used. As a rough guide, the frequency response of the emulsion may be assumed to become zero at about 100 cycles per millimetre. To make full use of the capabilities of the emulsion, the frequency response of the lens should be maintained at non-zero levels to this frequency limit, even if some response is sacrificed at low frequency values. For television use, on the other hand, the frequency response limit is set by the bandwidth of the television channel. When this is translated into cycles per millimetre, the limiting value is about 8 cycles per millimetre for the image orthicon, and about 20 cycles per millimetre for the Plumbicon or vidicon. The response of the lens must be maintained as high as possible out to these particular frequencies, without reference to what happens at higher frequencies. In some cases, these two sets of demands are both quite well satisfied, particularly if the lens is stopped down, but in other cases only one or the other is satisfied.

LENS CONSTRUCTIONS

For all practical purposes photographic and television lenses, other than zoom lenses, may be grouped in the following categories: Petzval; Triplet; Symmetrical; Telephoto; Inverted telephoto or retrofocus; and Extreme angle (fish-eye).

PETZVAL LENSES. The Petzval type of lens in its original form comprised two quite widely separated positive achromatic doublets. A lens of this construction gives an excellent performance over a rather narrow angular field. The actual forms of the doublets have been subject to changes over the years, but the lens in its basic general construction still finds a widespread use for projection purposes. Its range of application has been increased by replacing the original doublets by doublets plus single lenses, and by adding field flatteners (negative lenses near the focal plane) to provide a wider angle of coverage. It still remains true, however, that the Petzval type of lens is used mainly for projection.

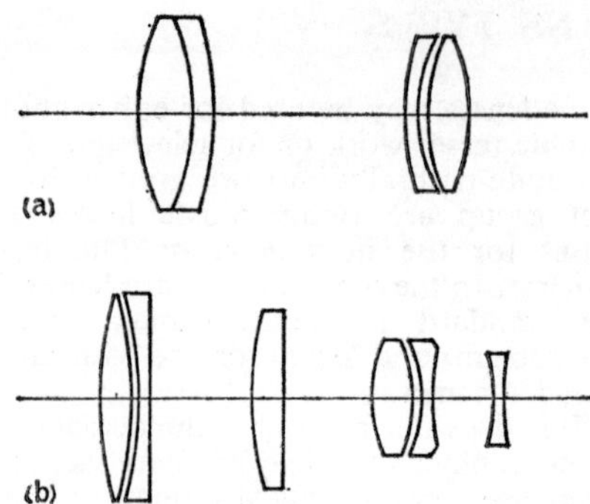

PETZVAL LENSES. (a) Basic type. (b) A more modern type, using a field flattener.

TRIPLETS. The triplet lens in its initial form comprised a biconcave lens of flint glass positioned between two outer positive lenses made of crown glass. In its original form it was difficult to achieve an aperture greater than $f\,2{\cdot}5$, but an angular field coverage up to 50–52° could be obtained.

In later developments of this type of lens, any of the single elements, or any combination of them, is replaced by a pair of lenses. Alternatively, the single elements may be replaced by cemented doublets or even triplets. Finally, both these lines of development may be combined to give a variety of modern high aperture lenses with good angular coverage. In some cases, evolution

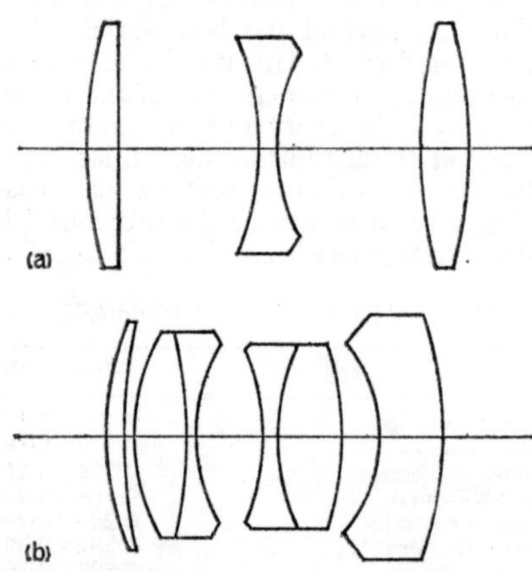

TRIPLET DERIVATIVES. (a) Original form of triplet lens. (b) Biogon, which may be regarded as a descendant of (a).

has been carried so far that it requires skill and a knowledge of the history of optics to recognize that a design has originated from a triplet or other form.

SYMMETRICAL LENSES. The earliest forms of symmetrical lenses comprised two identical sets of glass elements, facing each other across a central plane of symmetry. Advantage was taken of the fact that such a design

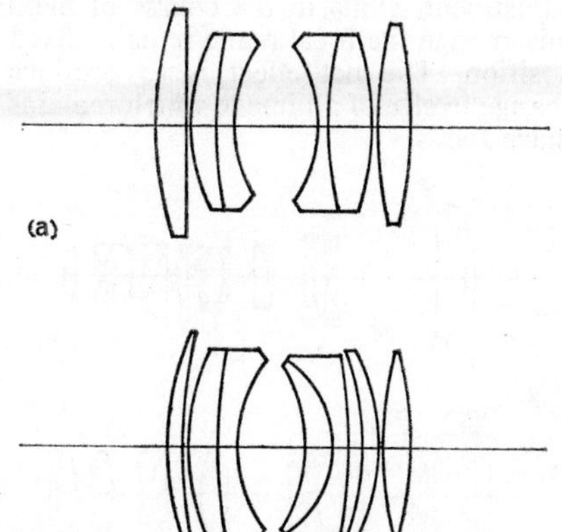

SYMMETRICAL WIDE-ANGLE LENSES. (a) Early version of the wide-angle sub-class of symmetrical lenses. (b) Modern elaboration of same basic approach.

tends to give an automatic correction of some aberrations, namely coma, distortion and lateral colour, and so leaves the designer free to concentrate on correcting the other aberrations. In later developments the exact

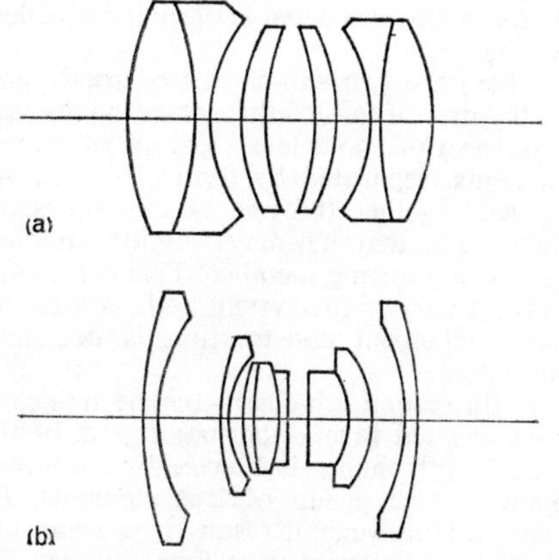

SYMMETRICAL HIGH-APERTURE LENSES. (a) Speed Panchro, starting-point for whole families of such lenses. (b) A more modern version, the Septon.

symmetry has been sacrificed, even though a high degree of similarity is retained between the two sets of elements.

There are two distinct sub-classes of this lens type. In the first sub-class the emphasis is laid on wide-angle coverage, up to 100° of total field or more, but the aperture is restricted to about $f4$ as a maximum. In this form of lens the best results are obtained if the outermost glass elements are negative lenses.

In the other sub-class, where the angular coverage is held to a maximum of about 50°, the basic form is variously known as the Biotar, Double Gauss or Speed Panchro construction. It comprises two outer single positive elements, between which are mounted two negative meniscus doublets. In later designs, the cemented doublets have been replaced by cemented triplets, and the single outer members have been replaced by pairs of positive lenses or by cemented doublets. Apertures up to $f1$ and higher have been realized with this lens form.

TELEPHOTO. The purpose of the telephoto form of construction is to provide a compact type of lens, in particular to provide a lens

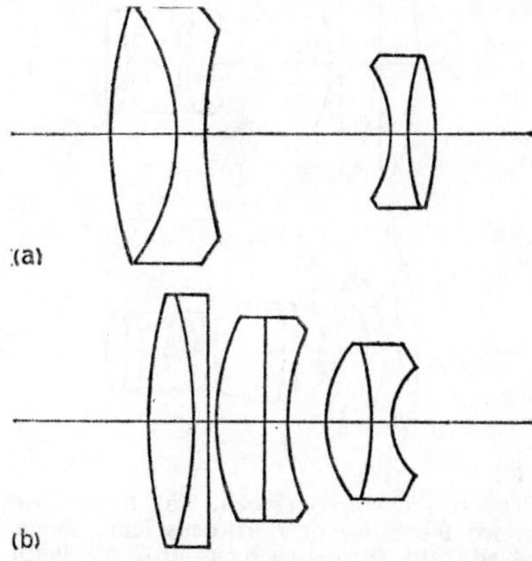

TELEPHOTO LENSES. (a) Early form of telephoto lens. (b) Form of higher aperture telephoto lenses.

with as short a distance as possible from the surface of the front glass of the lens to the focal plane. This distance, when divided by the focal length of the lens, gives the telephoto ratio. For lenses of the types described above, this ratio has a minimum value of about 1·2. In telephoto lenses it may be reduced to a practical minimum of about ·75 to ·80. A telephoto lens may, in fact, be defined as one in which the telephoto ratio is less than 1·0.

In order to secure this lower value of the telephoto ratio, the construction comprises a front group of positive lenses and a rear group of negative lenses. In earlier forms of telephoto lens, the aperture was restricted to about *f* 3·5, with a total field coverage of not more than 30°. In more recent developments, in lenses with fields restricted to about 12°, apertures up to *f* 1·8 have been realized.

INVERTED TELEPHOTO LENSES. The original aim of the inverted telephoto or retrofocus lens was to secure a larger back focal distance (i.e., distance from the last glass surface to the image plane) than may be obtained with any of the lens types so far described. This type of requirement is imposed, for example, when space must be provided for the pull-down mechanism in a cine camera or projector, or when a beam-splitting system is required for colour separation on film or in television applications. To achieve such a goal, this design form comprises a front negative group of lenses and a rear positive group of lenses.

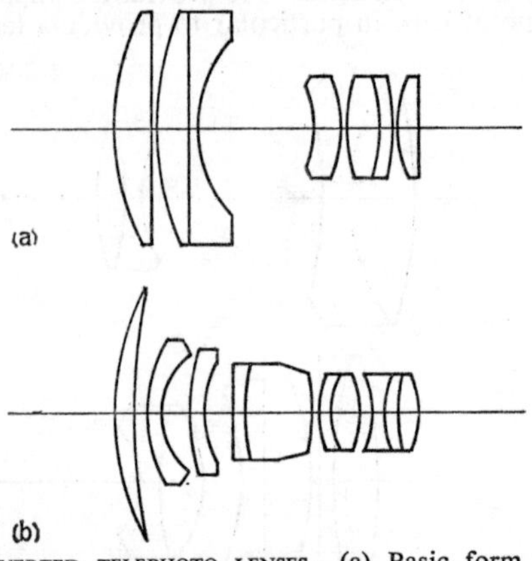

INVERTED TELEPHOTO LENSES. (a) Basic form of inverted telephoto or retrofocus lens, which has evolved into forms such as (b), a 20mm *f* 4 Flektogon.

In more recent times, other virtues have become apparent, including extended field coverage at high apertures, up to about *f* 1·4, in cine, television and still camera use.

FISH-EYE LENSES. This is a lens form in which the retrofocus construction has been pushed to an extreme to secure a maximum field of coverage. For example, with lenses having an aperture of about *f* 8, the whole hemisphere in front of the lens can be covered, and at an aperture of *f* 1 it is still possible to cover an angular field of 140°. Such lenses normally possess heavy distortion, although designs have been described in which the distortion is significantly reduced by using parabolic surfaces.

ZOOM LENSES. One of the most important developments of the past few years has been the widespread use of high performance zoom or varifocal lenses. In such a lens, the focal length may be changed by a simple adjustment, while in the course of making this change the focal plane remains fixed in position. The net effect is to produce a change in size of an image which remains in sharp focus.

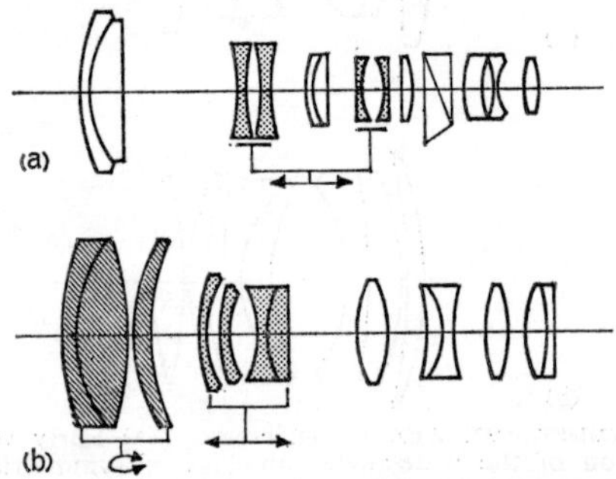

ZOOM LENS CONSTRUCTION. In an optically compensated zoom lens (e.g. (a) Pan-Cinor), groups of elements are moved in unison. In the mechanically compensated form (e.g. (b) Studio-Varotal), there is a relative displacement between groups of moving elements.

Zoom lenses of an earlier period were not remarkable for the quality of pictures which they produced, but the present generation of zoom lenses is capable of giving excellent results.

There are two sub-classes of zoom lens. In the first of these – the optical-compensation form of zoom lens – groups of optical elements, separated by fixed elements, are moved together through exactly the same distance, so that they may be rigidly attached to one supporting member. This is the only moving part of the system and, as a result, the mechanical construction is decidedly simplified.

In the second sub-class – the mechanically compensated form – the major part of the focal length change is effected by the movement of one group of lens elements. By itself, this movement results in a small but significant shift of the focal plane. To eliminate this shift, a second group of lens elements is moved by a cam through a carefully prescribed distance along the lens axis. Present-day techniques of glass manufacture and of element fabrication enable this cam-controlled movement to be accurately predicted, and to ensure that it will be the same from one batch of lenses to another.

In earlier types of the modern zoom lens, a range of focal length changes, from long to short, was obtained with a ratio of 3 : 1 or 4 : 1. In other words, the longest focal length that a lens would provide was three or four times the shortest. This ratio of focal length variation is still maintained on lower cost cine cameras, on still cameras (35mm and half-frame), and on 16mm cine cameras or vidicon cameras where size is important. Where some sacrifice of size and weight may be made, to secure rather striking visual effects, it is possible to have a lens in which the variation of focal length has a 10 : 1 or even 20 : 1 ratio. For image orthicon lenses this is achieved with an aperture of *f*4, while for vidicon lenses it is achieved at an aperture of *f*2.

Zoom lenses are sometimes designated as (f×z), where f is the minimum focal length and z the zoom ratio. Thus a (14×10) lens is a lens of 14 to 140mm focal length.

SINGLE PURPOSE LENSES

There are some types of lens or optical system which could theoretically be used for either cine work or for television, but which in practice are exclusive to one or the other. The most important of these types are anamorphic lenses in cine work, and colour separating systems for colour television.

ANAMORPHIC LENSES. With the normal type of lens, i.e., one which uses only elements with spherical surfaces (or, in the more general case, which has surfaces with rotational symmetry about the lens axis), the proportions of an object scene area recorded as an image are the same as those of the film area. In other words, if the film area used in a camera is a rectangle with sides in the ratio of 4 : 3, all of the scene which is recorded lies in an area with a 4×3 format.

To record a scene with a different format, the obvious step is to change the shape of the film area. This, however, may imply quite severe changes in equipment already in use, even to the extent of requiring its replacement by new equipment, and other means of achieving the same end result are worth looking for. An answer is provided by anamorphic systems.

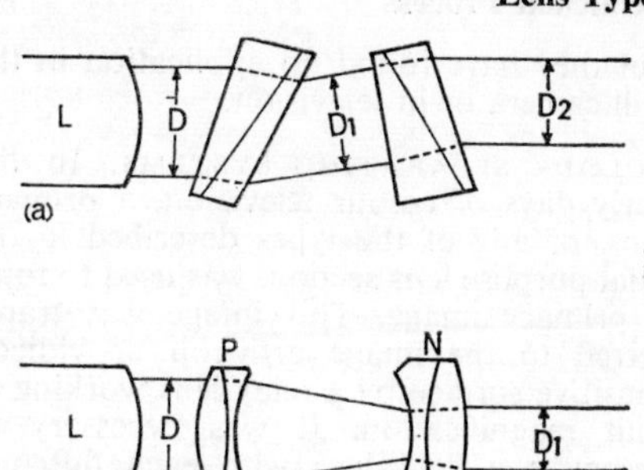

ANAMORPHIC SYSTEMS. (a) Prismatic form of anamorph depends on compression of beam of light from D to D_1 and D_1 to D_2. While an individual beam of light is compressed as it traverses a prism, the various beams of light are spread apart, but in one plane only. (b) The same result may be achieved by cylindrical lens systems comprising a positive element, P, widely spaced from a negative element, N, together acting as an afocal wide-angle lens in one plane, and having no effect on image spread in a plane at right-angles. Prism and lens anamorphs are achromatized, and are backed up by a standard taking or projection lens, L.

These are of two types. In the first, the desired compression or expansion of the image is produced by using optical elements with cylindrical surfaces. In the second, the same effect is obtained by the use of prisms and, moreover, the amount of image compression can be varied by changing the relative orientations of the prisms.

Anamorphic systems are used freely in 35mm film, and to a lesser extent in 16mm,

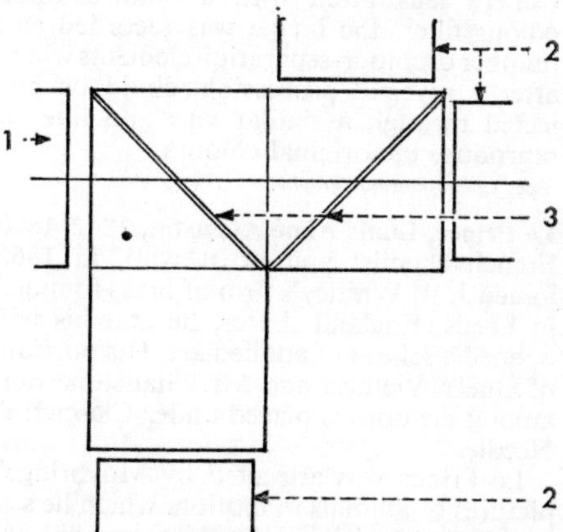

PRISMATIC BEAM SPLITTER. Prisms used in a colour TV camera behind the lens, 1, to provide images of three different colours on three separate vidicons or image orthicons, 2. In practice, the prism angles are less than 45° to reduce the angle sensitivity of the dichroic mirrors, 3.

but they have found no application in the still camera or in television.

COLOUR SEPARATING SYSTEMS. In the early days of colour television, a primary lens, of any of the types described in the dual-purpose lens section, was used to form a primary image. This image was transferred to the image orthicon or vidicon sensitive surface by a relay lens working at unit magnification. It was necessary to associate a field lens with each different primary lens, and this tended to degrade the picture quality: the same relay lens could be used with a variety of primary lenses.

Prisms with dichroic surfaces, i.e., surfaces which selectively reflect or transmit light of different colours, were disposed between the relay lens and the final image plane so that three images of different primary colours were formed on the faces of three image orthicons. The information contained in these three coloured pictures was used to generate a signal that could be transmitted for colour television.

More recent designs employ the same basic approach. Prisms with dichroic surfaces form three separate coloured images by means of dichroic reflectors on three (or in some cases four) pick-up tubes which may be image orthicons, Plumbicons, vidicons, or a combination of different types. The difference lies in the lens system; relay lenses are rapidly being phased out in favour of zoom lenses, thus greatly simplifying the optics and saving considerable weight.

The zoom lens unit can be detached from the camera for ease of transport, and alternative lenses with different focal ranges may be substituted. Zoom lenses for colour television cameras are specifically designed for use with their dichroically surfaced prisms, since these prisms introduce significant amounts of aberration. The designer must also provide an adequate back focal length to take care of the extended path through the prismatic optical system.

A.C.

See: *Camera; Image formation; Lens performance; Stabilization, image.*

Books: *Fourier Methods in Optical Image Evaluation,* by E. H. Linfoot (London), 1966; *Photographic Optics,* by Arthur Cox (London), 1966; *A System of Optical Design,* by Arthur Cox (London), 1967.

⊕**Lenticular Process.** Obsolete additive system of colour photography which made use of a black-and-white film stock, the base of which was embossed with a very large number of minute lenses or cylindrical ribs. When exposed through the base using a camera lens fitted with a banded three-colour filter, the image was recorded as a number of colour-separation elements which, after a reversal processing, could be projected through a similar lens and filter to reproduce the original colours.

See: *Colour cinematography.*

⊕**Le Prince, Louis Aimé Augustin, 1842–1890.** French chemist and artist who, in 1866, joined J. R. Whitley's firm of brass founders in Leeds, England. Later, he and his wife opened a school of applied art. His portraits of Queen Victoria and Mr. Gladstone were among the objects placed under Cleopatra's Needle.

Le Prince was attracted by Muybridge's pictures of animals in motion, which he saw in 1875; on visiting America in 1881, he began experiments on motion photography, and in November 1886 applied for a US patent for a 16-lens camera, granted 10th January 1888. A British patent, filed on the same date, included a more practical one-lens camera and projector, with which he took some pictures in 1888.

In 1890, Le Prince visited France with some improved designs and was seen in Paris on 16th September, boarding the train to Dijon. He was never seen again.

⊕**Leroy, Jean Acme, 1854–1932.** American inventor and photographer. Experimented unsuccessfully in the early 1890s on apparatus for taking and showing moving pictures. Is said to have acquired in 1893 a length of film exposed in Donisthorpe's apparatus. For this he designed a projector, which he modified to take perforated 35mm film after seeing the Edison Kinetoscope. With this machine, completed by February 1894, Leroy claimed to have shown Kinetoscope films to various private bodies. His evidence helped in the litigation against the Motion Pictures Patents Corporation, giving evidence of "prior art in the public domain".

□**Level-Dependent Gain.** Variation of gain of an amplifier with variation of input signal level.

□**Level-Dependent Phase.** Variation of phase shift through an amplifier with variation of input signal level.

⊕LIBRARY

The majority of film libraries exist to serve the needs of a parent organization – either a producing company, newsfilm agency, government department or a television authority. All are prepared to supply material to outside organizations subject to certain conditions. But different types of library exist, and definitions are difficult because many libraries carry on several functions and overlapping is frequent. There are factors, primarily concerned with ultimate intent, that serve to distinguish them, but basically the library techniques remain the same.

In the past, the main business of a production library was to exploit its collection, and had an interest in keeping its films only so long as they served this end; the preservation of films as an historical record was the primary responsibility of a national archive. Nevertheless, the film librarian, particularly when dealing with documentary and newsfilm material, made it essential to store and preserve films in the best possible manner.

That it is impossible to preserve and exploit a collection of films without spending money is becoming increasingly realized. For this work, expert staff are necessary. The value of librarians trained in the techniques of cataloguing and classification is gaining recognition and the modern film library has had to adopt many of the new methods developed in other types of library to keep pace with the demands of an omnivorous industry.

The need for a body to organize ideas and to act as a clearing house for information led to the creation in Britain of the ASLIB Film Libraries Group in 1955. This Group, which acts under the auspices of the Association of Special Libraries and Information Bureaux (ASLIB), was responsible for the publication of *The Directory of Film and TV Production Libraries*, which gives details of the principal film libraries in Britain. The need for standardization in film cataloguing led to the publication of *Film Cataloguing Rules* in 1963. This is a code intended for all types of film libraries. The Group has also organized courses for new entrants into the profession.

STAFF AND STAFFING

The size of the staff, and what proportion should be clerical, trained or qualified, technical or graduate, cannot be determined until policy decisions have been made as to the acquisition of material, its selection, methods of storage and preservation, the type of cataloguing and classification, and what auxiliary services such as transport, cleaning, messenger and postal arrangements are available in the parent organization. These decisions, plus a pre-ordained budget, influence the number and kind of staff to be appointed.

DUTIES. Normal staffs in larger film libraries may be as follows:

Chief Librarian (or Librarian);
Deputy Librarian (or Assistant Librarian, or Duty Librarian where shift work is involved);
Chief Cataloguer;
Cataloguers and information staff;
Acquisition and selection staff;
Sound Librarian;
Film examiners;
Vault staff;
Clerical staff and typists, including operators for document reproduction.

The three main groups into which the work can be divided are as follows:

1. Clerical and routine: keeping records, filing, shelving, typing, document reproduction.
2. Technical and professional: (a) film acquisition and selection, cataloguing, classification, indexing, handling enquiries; (b) film processing and duplication, ordering, preservation, storage and repair.
3. Administrative: framing and carrying out policy of the library; promotion and sales; contract work; reports and statistics; allocation of work, overtime, etc.; committee work; staff recruitment and training.

Not all libraries have, or need, so complex an organization, but some elements of it are essential even in libraries of moderate size.

QUALITIES. For film production librarians in particular, an interest in production methods and film processing is essential, together with a knowledge of cinematograph regulations and legal requirements. As well as an appreciation of current film library techniques, the librarian should have organizational ability and a clear understanding of the purpose of the library and its terms of reference. The duties associated with cataloguing, indexing, recording and filing all types of material call for tidiness, accuracy and method. Those who deal with users of the library should be willing and tactful, with a flair for obtaining information by looking in the right places.

SIZE. The libraries with the largest number of staff are those concerned with television and newsfilm. They normally work seven days a week and staff are required on a shift basis to ensure a fast service. One actual film library of this type has a staff of 22, allocated to administration, 2; film handling and supply, 13; and documentation and research, 7. The total stock is 25,000,000 ft. (35mm and 16mm) with an annual intake of 650,000 ft. (16mm). The annual output averages 250,000 ft. of 35mm and 500,000 ft. of 16mm.

An advantage of this particular library is that catalogues, film vaults and the laboratory are all immediately accessible and the laboratory is also working seven days (and seven nights) a week.

STORAGE AND PRESERVATION

The storage and preservation of film has been the subject of many reports. The most comprehensive is that of the International Federation of Film Archives, an English version of which is obtainable from the National Film Archive, London.

TEMPERATURE. The most favourable temperature recommended for nitrate film is 35°F and the maximum for acetate film 54°F. When stored with magnetic films and magnetic tapes, the optimum temperature recommended is 42°F.

Colour films in the form of coloured copies require to be stored differently according to the process used, but the producers of raw film do not seem to be in agreement concerning the storage properties of colour film. Kodak gives 0°F as the ideal temperature, while Agfa concludes that 59°F is acceptable. In practice, few film libraries are able to afford the installations necessary for low or constant temperatures. The National Film Archive in Britain keeps all stock at a constant temperature of approx. 55°F, copying their nitrate on to acetate when necessary. All stock must be examined at regular intervals for signs of deterioration.

METHODS. Negatives should be stored separately from positives. In some libraries they are kept in different coloured cans or with different coloured labels on the cans; in others they are merely distinguished by the prefix "N" in front of the location number. Each film as it is acquired should be numbered and placed in numerical order; this number should be scribed on the protective leader of each roll or reel.

Libraries run by feature and documentary producers generally file their films shot-by-shot from unused material. This material may be similar but additional to parts of a completed film; it may be material which has not been used in the film, or it may extend the coverage of a particular subject illustrated briefly in the film. This material is usually known as stock material. Newsfilm libraries, however, generally file their material in cut story form and keep surplus footage only when essential.

SELECTION. The high cost of film library operations forces most film libraries to adopt a rigid selection policy. Shots which are too short or of poor quality can be discarded immediately, although any unusual event or location may have to be kept in spite of being photographically inferior. What is retained in the library depends largely on its policy, but the principal aim is to provide for foreseeable demands and to ensure a balanced coverage. In most libraries, a holding period is maintained for intake before the final canning-up, in order to provide a reasonable time for decision before a film is finally discarded.

MICROFILM. The rate of intake of related material such as scripts, commentaries, shot-lists and cameramen's reports poses a serious space problem. This can be overcome by the use of unitized microfilm (i.e., microfilm frames mounted in apertures cut in data-carrying cards), a system combining the space-saving properties of microfilm with the advantages of rapid selection and filing associated with card indexes. The standard size of the card allows for punching and coding for rapid machine programming in the same manner as tabulating card systems; the use of tabulating machines is envisaged as a future development. The original aperture cards can be reproduced for daily access without disturbing the master file, and enlarged prints can also be forwarded to customers with the film. To avoid the cost of making unnecessary prints, microfilm readers are placed near viewing machines so that shot-lists and commentaries can be scanned as the film is viewed.

CATALOGUE AND INDEX

No verbal description can give a complete impression of a filmed sequence or be a substitute for viewing film. Nevertheless, in newsfilm and television libraries particularly, time is an important factor, and to work only from the film would prove impossible. Users of the library have therefore to rely on the catalogue entries and related information for their choice of material. Similarly, the cataloguers themselves have to work from secondary sources, otherwise the intake would pile up and never be adequately recorded.

Consistency of treatment and format of entries is important and accuracy essential. The cataloguing code published by the ASLIB Group enables a consistency of cataloguing practice to be achieved.

FORM OF ENTRY. The main entry in the catalogue should record, apart from the title and location number, details of production credits, country of origin, date of event or year of release, physical format, length in feet, running time and summary of contents. The detail required in a summary depends on the purpose of the library, but generally speaking it should describe accurately and objectively the content of the film and should be specific enough to serve as a basis for subject classification. Apart from subject entries, added entries may be made for all those persons and organizations recorded on the main entry.

Information that is liable to change, such as details of commercial or non-commercial distribution and the ownership of rights, may be recorded on a separate card or elsewhere. In some cases, availability for television may be necessary, with an indication for what country, countries or region the film has been cleared for broadcast or television. In certain cases, clearance may not be the same for image and sound track.

Newsfilms are arranged chronologically under the date of the event or of release. In many cases, the main entry takes the form of shot-lists, commentaries, contents sheets, release sheets or a combination of these. The added entries may be prepared from this material, or following a viewing of the film.

Picture materials other than complete films, usually referred to as stock shots, are entered directly under subject.

There are varying methods of recording film catalogue entries. Some libraries use loose-leaf forms, others books containing positive frame transparencies, but the majority use cards in vertical files. No matter how large the medium, however, the conclusion is often that there is not enough room to record all the details required. The ASLIB Group gave consideration to this problem and have drawn up a list of abbreviations required.

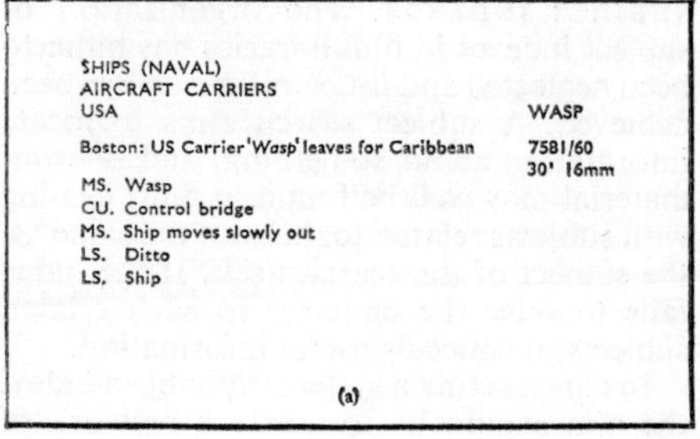

SHIPS (NAVAL)
AIRCRAFT CARRIERS
USA — WASP

Boston: US Carrier '*Wasp*' leaves for Caribbean — 7581/60
30' 16mm

MS. Wasp
CU. Control bridge
MS. Ship moves slowly out
LS. Ditto
LS. Ship

(a)

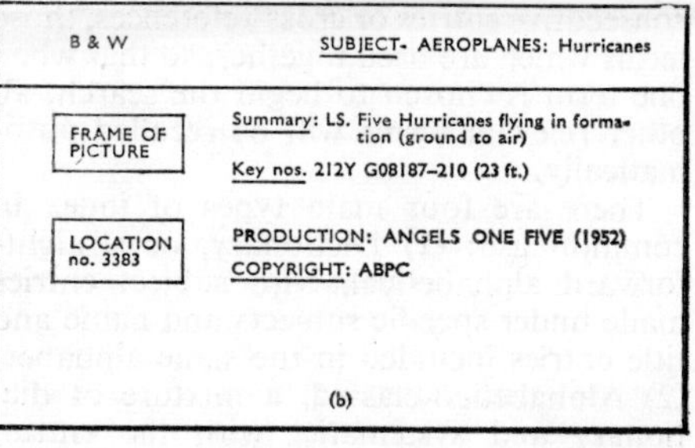

B & W — SUBJECT- AEROPLANES: Hurricanes

FRAME OF PICTURE — Summary: LS. Five Hurricanes flying in formation (ground to air)

Key nos. 212Y G08187–210 (23 ft.)

LOCATION no. 3383 — PRODUCTION: ANGELS ONE FIVE (1952)
COPYRIGHT: ABPC

(b)

TITLE: (8th February, 1912) PATHE [GAZETTE] — Location No. 8271 A(h)

BELFAST: THE CHURCHILL MEETING. Scenes and incidents of Mr. Churchill's visit to Belfast. A shot of mounted troops wearing capes (6–19); a contingent of the Irish Guards form fours and march off (41); policemen at a street corner search and check the identity of passers-by (50); a queue of people waiting to gain admittance to the hall (57). Arrival of Lord Pirrie and Mr. and Mrs. Winston Churchill. Lord and Lady Pirrie in close shot leaving their car (61–68); a similar shot of Mr. Churchill who is immediately followed by Mrs. Churchill (81).

35mm/st./pos. Scottish Film Council

SUBJECT REFS.
941.61[i]: 323.17 Belfast – Anti-Home Rule Meeting
PIRRIE, William James, 1st Viscount
PIRRIE, Margaret Montgomery, 1st Viscountess
CHURCHILL, Sir Winston Leonard Spencer
CHURCHILL, Lady Clementine

(c)

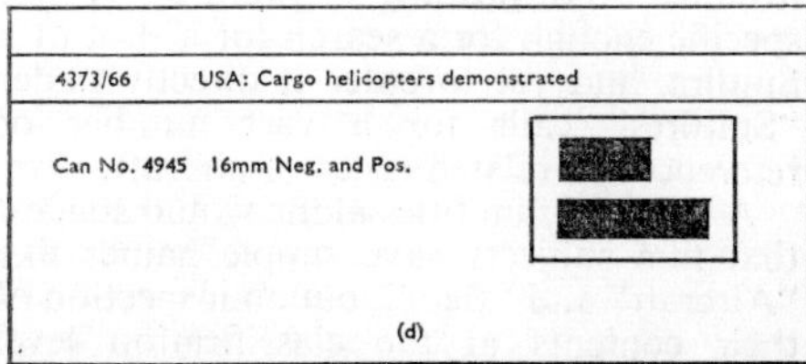

4373/66 — USA: Cargo helicopters demonstrated

Can No. 4945 16mm Neg. and Pos.

(d)

LIBRARY CARD LAYOUTS. (a) Subject entry for a news-film. Card gives brief details, with footage and film gauge, of a newsreel sequence. Further information is by reference to camera dope sheets accessibly filed. (b) Subject entry for a stock shot. Heading gives classification. Key numbers of negative enable original material to be quickly traced. Specimen frame provides general appearance of shot. (c) Main entry for historic newsfilm in National Film Archives. Absence of dope sheets necessitates detailed description. (d) Microfilm record card. Black area represents microfilm attached to card containing full dope sheet details and text of commentary.

SUBJECT INDEXES. The organization of subject indexes in film libraries has hitherto been neglected and little uniformity has been achieved. A subject search aims to locate information about something, and relevant material may well be found in films dealing with subjects related to, but not the same as, the subject of the search itself. If the index fails to refer the enquirer to such related subjects, it conceals useful information.

In constructing a systematic subject index, the aim should be to create a network of terms which brings together, by a series of consecutive entries or cross-references, those terms which are used together, so that when one term is chosen to begin the search, all other relevant terms will be recalled automatically.

There are four main types of index in common use: (1) Dictionary, or straightforward alphabetical, with subject entries made under specific subjects and name and title entries included in the same alphabet; (2) Alphabetico-classed, a mixture of dictionary and systematic, with the entries made under their appropriate classes, each class being subdivided alphabetically to accommodate the subjects; (3) Classified, based on a scheme of classification such as the Universal Decimal Classification; (4) Mechanized indexes, not based on manual searching, but on the use of machines of varying degrees of complexity.

DICTIONARY. All types have their problems. The dictionary catalogue calls for a multitude of cross-references and its principle of specific entry always raises the question "How specific is specific?". For instance, the heading "Aircraft" is not specific enough for a search for a shot of a Spitfire, and yet to enter it directly under "Spitfires" calls for a vast number of references to related types of aircraft.

A study of film titles alone would suggest that film subjects have simple names like "Aircraft" and "Bees", but an inspection of their contents at the classification level reveals that such titles are simple in appearance only. Each subject is in reality a complex aggregate of specific subjects, each of which is the main theme discussed from a particular aspect. Thus "Aircraft" can cover balloons, airships and aeroplanes, their manufacture and application to transport and warfare. But "Transport" and "Warfare" are terms of wide connotation and would require subdivision which calls for an alphabetico-classed catalogue. The compilers of a dictionary catalogue must therefore keep to the specific entry rule and make their entry for "Spitfire" under that term and refer from "Aeroplanes" and "Aircraft". The fact that the card following "Spitfires" may be "Spithead Review" has no relevance.

ALPHABETICO-CLASSED. The alphabetico-classed catalogue calls for an arrangement of subjects in classes and divisions with possible subdivisions. A decision has to be made as to the level at which the main headings and the subordinate terms are collected, and the type of subordinate terms to be subsumed and how they are to be linked with the main heading. Thus if "Transport" is selected as the main heading, the divisions may be "Air", "Land" and "Sea". The logical manner in which subdivision proceeds can be illustrated as follows: TRANSPORT (main heading), Air transport (division), Civil (subdivision), Aeroplanes (followed to any degree required).

Each group may be divided as minutely as the need occasions, until subjects incapable of further division are reached. The danger here is that the specific subject required may be lost in the welter of sub-headings. A decision would have to be made whether it would be more logical to choose the term "Air Transport" as the main heading and refer from "Transport" or to adopt "Aeroplanes", referring from "Transport", "Air Transport" and "Aircraft". Cross-references are essential to show users how the headings have been chosen, particularly where a minor term has been given prominence because of an immediate interest. The order within each subordinate heading should also be carefully determined – a chronological arrangement is usually preferred, although in some libraries the country of origin is the determining factor.

CLASSIFIED. If the hierarchy listed above is adopted, with main classes being divided into the minutest divisions, a classification scheme such as the Universal Decimal Classification is called for. This is a definite scheme in which numerals are provided which stand for subjects. These numerals, referred to as the notation, allow for continued division by decimals, and the result is a schedule suitable for classifying many subjects in considerable detail as follows:

629.12 Shipbuilding. Ships. Boats.
629.121 Primitive craft. Coracles, Rafts, etc.
629.122 Inland (fresh) water craft.
629.122.3 Barges.

A classified index such as the Universal Decimal Classification can fail to provide for particular subjects in the published schedules, and since its entries follow a sequence of notational symbols, a conversion table has first to be consulted; this usually takes the form of an alphabetical index.

MECHANICAL. Dissatisfaction with the alphabetical and classified indexes has led to the development of indexes based on punched cards. Punched card machines, like the classified index, need a method of translating notational terms into symbols that the machine can read. At the moment this is feasible only for a library covering a limited subject field. A good deal more theoretical study is needed before machines become a practical proposition for a film library covering the whole field of knowledge, but considerable progress has been made. When mechanical indexes are introduced, libraries using a notational system may have the advantage.

CHARGES AND CONTRACTS

Film libraries must charge for their material. When an application is made to a library for material, a contract has to be signed which specifies exactly what may or may not be done with the material, and what fees are payable. A search fee will almost certainly be charged whether or not a search is successful. In some cases, the search fee is not refunded even if the material is used. In others, it is credited if over a minimum footage is purchased.

Processing is normally chargeable at cost; if a duplicate negative is required, it will probably be necessary to make a duplicating positive first. This may be seen only as a charge on the invoice. Royalty charges are based on the footage actually used in the production, not the footage supplied, and vary according to the length, distribution and clearance required. D.G.

See: *Film storage; Library* (TV); *News programmes.*

Books: *Directory of Film and TV Production Libraries*, ASLIB (London); *Film Cataloguing Rules*, ASLIB (London), 1963.

□LIBRARY

The purpose of a television film library is to make its stock of various types of film material available for re-use. The organization and methods of the library are directed to achieve this end as efficiently as possible.

In addition to the normal difficulties experienced by film libraries, television film libraries encounter some special problems. One of the most significant is the volume of material that pours into their storage areas, requiring detailed documentation with little delay. The form of the material is likely to be varied, that is, 16mm or 35mm gauge film, optical or magnetic sound on combined or separate tracks, telerecordings or videotape recordings. The type of presentation ranges from a complete programme down to a new story, insert or stock shot.

The nature of television is such that only one or two copies of a production are required. There is consequently no incentive to provide insurance prints or duping material and quite frequently a programme or item has no matching cut negative. Moreover, the information conveyed tends to include technical and political developments immediately they occur, allowing little time for assimilation and association with related topics.

To overcome these difficulties, it is essential to establish the purpose of the library clearly, to organize the library into logical departments and to house and equip it to carry out its functions effectively. The purpose, scope and organization of a television film library varies according to the size and nature of its parent organization.

PURPOSE AND SCOPE

The main purpose of re-use can take several forms. One of the more obvious is the provision of complete programmes for repeat transmission. Another is the supply of stock shots from the library's resources. The building up of an archive of television material of permanent value also has re-use potential. It is generally considered necessary to provide special storage for master material in this category and to make duplicate material for production use.

The television film library is often responsible for acquiring stock material from other libraries for incorporation in programmes. The library does not normally purchase complete programmes, but it may well be concerned with their storage, cataloguing and checking for transmission. These show prints are held for the period of contract and returned when this has expired.

The library may also sell stock material to other organizations although, as with purchasing, it is not usually concerned with the sale of complete programmes. However, it must store, catalogue and maintain master material to provide copies for sales purposes.

SCOPE OF MATERIAL. The stock of a television film library can be divided into four forms.

Complete programmes and related material, which may be film or film recordings in either 16mm or 35mm gauge. It is usual for them to have separate magnetic tracks for transmission purposes. Related material can include film sequences, titles, optical tracks, music and effects tracks, commentary tracks, cutting copies, etc. The negatives may be cut or uncut.

Stock material, which can be acquired by the library at several stages of production. It is preferable to do this before editing, when fine grains and dupe negs can be made of suitable uncut footage. Trims and spares may be taken after editing and sometimes cut sequences from film inserts can be a useful form of stock. Libraries could improve their stock collections by well-planned special shooting but as a rule this is prohibited by cost.

News stories, which are usually 16mm with combined or separate magnetic sound tracks. They form a valuable source of material for re-use, not only for strictly news purposes. They provide a wider coverage than cinematograph newsreels, both in topics treated and frequency.

Videotape programmes and inserts, which are held by some libraries for short-term purposes such as repeat within a few months of original transmission. If the material is required for a longer period for archival or sales purposes, or requires substantial editing for re-use, it is usually transferred to film.

ORGANIZATION. All television film libraries store, document, select and examine their material. They also deal with enquiries relating to it. Depending on the size and type of library, these functions may be performed by one individual, may occupy several departments, or fall somewhere between the two extremes.

SELECTION. The stock of a television film library requires constant sifting and selection. This is usually effected by a policy agreed with production departments and administered by the film librarian, who is responsible for building up a collection of material related to the needs of the organization.

A systematic approach to retention and discarding is required for dealing with current material. It may be necessary to select material for archival presentation in conjunction with organizations outside the television service producing the programmes.

ENQUIRY SERVICE. This section deals with requests for film material from the library and arranges for its supply to users. The service can range from the provision of a programme for a viewing to a detailed research into subject material. Enquiry assistants should have a sound educational background and be available throughout transmission hours. The section should be provided with viewing equipment to enable them to select or assist in the selection of material to satisfy requests.

ACCESSIONS. Film is usually identified and shelved by numbers. These are allocated immediately the film enters the library and recorded with other details to provide information for cataloguing.

CATALOGUING AND CLASSIFICATION

Compared with other material, film is relatively inaccessible and easily damaged and its form prohibits arrangement on the shelf by subject. It is therefore essential to provide a detailed and accurate catalogue, using modern documentation processes and employing an up-to-date, flexible classification scheme.

FORM OF CATALOGUE. The form of the catalogue needs to be as flexible as possible. It must enable a large volume of entries to be added without difficulty and provide for expansion in terms of the number of entries and new approaches and innovations reflected in television output.

The form least suitable for accommodating these features is the printed catalogue which is out of date immediately and cannot systematically assimilate new additions. It is also expensive to produce. These factors outweigh the only significant advantage, which is the ease of duplication and distribution to library users.

The sheaf catalogue, consisting of looseleaf slips filed in binders, offers flexibility in terms of ability to add entries, but filing is not particularly fast as the binders have to be opened, the correct place found, and the binders closed again for each slip added to the catalogue. The slips also tend to be flimsy and do not stand up well to the wear usually suffered by a catalogue of this kind.

Some television libraries have adopted a variation of the sheaf catalogue, using very strong rigid binders containing sheets of thick paper rather larger than foolscap size ruled to accommodate six entries per page. These have been used for stock shot catalogues. On the whole, they do not seem ideally suited for this purpose as it is not possible to add entries in a flexible way without constantly retyping sheets of six entries. The binders also tend to be bulky and awkward to handle and make multiple entries rather difficult.

Some libraries have attempted to use visible indexes of various kinds. These have several drawbacks. Space for information on each entry is limited, duplication is difficult and the catalogue produced tends to be bulky and takes up a large area of useful space. Filing entries by means of narrow or interlocking slips tends to be time-consuming. These systems can have advantages if the information required on each entry is brief or if it is considered necessary to publish and circulate the library catalogue. Many visible index systems can be photocopied to facilitate distribution of information.

If it is not considered necessary or desirable to circulate copies of the catalogue, the most obvious choice seems to be a card catalogue. This has the advantages of ease of insertion of new entries, durability and ease of consultation, and is reasonably economical of space.

The choice of size of card depends on the information required in the entry. For most purposes, 6 in. × 4 in. is a suitable size, but for indexes consisting of a heading or classification number, 5 in. × 3 in. may be sufficient. Some types of catalogue, requiring space for more information, are better served by the 8 in. × 5 in. size.

It is not feasible to distribute a catalogue in card form except to provide duplicate copies which can be filed at recognized centres. Several identical entries are usually required for each item being catalogued, making copy typing wasteful and inefficient. The most satisfactory solution is to adopt a unit-of-entry system, i.e., making one basic entry, reproducing it and adding the necessary headings afterwards. Cards can be produced by stencil, spirit duplicating, dyeline, xerography or any of the usual processes of document reproduction.

COMPLETE PROGRAMMES. The most frequent approach to the cataloguing of complete programmes is by title. It is therefore usual to arrange the main entries in this order. The title entries contain details of all material held relating to the programme, film recording, videotape, film sequences, titles, and the various kinds of sound track. These items must be arranged to indicate clearly the accession number for each reel.

Other essential items include gauge, footage, duration, and transmission or recording date. Some libraries also incorporate a synopsis and credits in the entry though this may not be possible through lack of space. Moreover, this information can be obtained from other sources such as scripts, shot-lists and published details of programmes. Any copyright or contractual restrictions must appear on this entry.

It is also essential to be able to retrieve information by subject from complete programmes. It is not usually feasible to produce one descriptive entry for the programme which can be duplicated and filed under the headings or classification numbers required, as such a description would have to be detailed enough to show all aspects. This would result in a cumbersome entry and would not be helpful for retrieving sections of the programme. It is usually more satisfactory to produce analytical entries for sections of the programme, allocating headings or classification numbers as required.

NEWS FILM. The two most significant approaches to news are chronological and subject. The chronological aspect is most frequently catered for by a card or sheet setting out details of stories transmitted each day. They should include title of the story, gauge, footage, duration and copyright details.

Most news departments allocate an issue number for each day which provides a useful filing device.

The subject approach requires a description of the story visually, together with background information on the content of the sound track and the context or occasion of the story. The most convenient way of providing this information is to adopt a unit of entry giving the description together with gauge length, duration, form of sound, copyright and accession number.

The required number of cards can then be produced and headings or classification numbers added.

STOCK SHOTS. Stock shots are asked for only by subject. Unlike a complete programme, they have no title and no recognized order of issue as with news film.

The main body of the entry should consist of a detailed and accurate description of the shot. Many elements which consistently appear such as time, season, weather, etc., can be preprinted. Details of date, footage duration and accession number should also be included. This can form a unit entry, the required number of cards being then produced and the headings or classification numbers added.

Some libraries have a window corresponding to a frame of film punched out of the cards. Frames of positive are then fixed to the cards to show essential details of the shot. This can aid selection but in television libraries the operation is time-consuming and also adds to the bulk of the catalogue. It is quite feasible to manage without this if proper attention is given to producing an accurate description of the shot in the entry.

SUBJECT CATALOGUE. From the above description, it will be seen that it is necessary to make subject entries of some kind for complete programmes, news stories and stock shots. Re-use of film in television rarely follows the original purpose for which the material was shot. It is therefore helpful to have one subject catalogue for the whole library rather than have it divided by its form of presentation. The subject entries must be on the same format to make this possible, i.e., on the same size cards, using the same headings and the entries compiled on the same principles.

CATALOGUING RULES. To produce consistent and accurate entries, it is essential that the library should devise or adopt a cataloguing code of rules. The best results are usually obtained by taking a recognized code such as the *ASLIB Rules for Cataloguing Film*, adapting it if necessary to allow for variations in local practice. Codes designed for conventional libraries are also useful for such problems as forms of name, use of designations, etc. The most generally used of these is the Library Association's *Cataloguing Rules, Author and Title Entries*. Strict adherence to a set of filing rules is also important.

SUBJECT CLASSIFICATION. The range of programme material produced by most television organizations covers the whole field of knowledge. It is therefore necessary to choose a subject classification capable of accommodating this range and one which can be expanded to cope with new developments in all subjects and with advances in television technique. There are no classification schemes specially designed for film material or any other visual media. The choice is between adapting one of the classification schemes intended for printed material and devising a scheme for individual television libraries.

Some libraries have classified their material by the second method, using subject headings built up as the film collection grows. These systems tend to become inadequate as the size and complexity of the stock increases. The choice and allocation of headings is limited by the knowledge of an individual or group of individuals. Operation is also difficult for new staff who are unfamiliar with the headings. Related subjects are unhelpfully scattered by such an alphabetical arrangement.

It is, therefore, advisable to adopt one of the existing classification schemes, notwithstanding their disadvantage of being devised for printed material. The scheme most suitable for adaptation has proved to be the Universal Decimal Classification (UDC). This covers the whole field of knowledge, is published and kept up to date on an international basis and is the product of specialist advice and effort. The schedules are systematically arranged, with a numerical decimal notation. Its structure allows for maximum flexibility in the relationship between different subjects and the treatment of various aspects of subjects. It has tables of common auxiliaries for recurring factors such as language, time and place. It also has an alphabetical subject index.

The ASLIB Film Libraries Classification Committee has produced special tables to be used in conjunction with the scheme to enable such aspects as movement of the subject and movement of the camera to be included in the notation.

A good alphabetical subject index to the classified catalogue must be maintained to bring together aspects of subjects separated by the classification scheme and to guide users to the correct place of any subjects on which they are seeking material.

It is also important for the library to keep an authority file, that is, a record on cards defining the scope of each UDC number used. This helps to ensure consistency in the use of the classification.

Other advantages of using UDC are the resulting systematic arrangement according to a universally accepted scheme, ease of training staff who have worked in other libraries using the scheme and its suitability for use with mechanized information-retrieval methods which are developing in libraries and information services.

OTHER SOURCES OF INFORMATION. The catalogues need to be supplemented by information supplied by the production departments in the form of shot-lists, scripts, copyright and contractual agreements, publicity handouts, etc.

PREMISES AND STORAGE

Storage areas should be at ground-floor level where possible or well served with lifts and mechanical-handling assistance. The enquiries section should be easily accessible to users and provided with ample space for research and consultation of catalogues. All other servicing areas, such as cataloguing and film handling, should be conveniently related to those sections. The overall plan should be as flexible as possible to allow for expansion and possible changes in emphasis.

EQUIPMENT. Film libraries also require equipment for their special purposes. This comprises both editing and viewing machines. Library staff and users need viewing machines with double-speed facilities for cataloguing and selection of library material. A wide range of editing equipment is required for maintaining the library stock in good condition and exploiting it to the full. The equipment must cater for the gauges and varied forms and combinations of sound tracks used in television.

STORAGE. Film storage is usually considered in terms of film vaults designed to meet the needs and legal requirements of nitrate film holdings. Most television organizations have little of this type of material and require film stores which make the most economic use of the space available and allow easy access. The requirements seem best met by storing the film in long runs of open racking. This type of storage area accommodates 500 cans in 50 sq. ft. of floor space and is convenient to service.

Archival sections of the television film library require special storage with temperature and humidity control.

The racking should be specially designed to accommodate the varying sizes and gauges of cans used in television. For current storage, the cans are shelved on their circumference; for archival purposes, vertical stacking is considered more effective.

EXAMINATION. It is essential that the stock of the television film library should be kept in good condition. This necessitates a staff of experienced film technicians to carry out the checking, repair and replacement of film library holdings. This section also checks programmes for transmission.

It is necessary for reasonable financial provision to be made for maintenance purposes.

Ideally this section should handle all work on library master material, such as papering and reprinting sections, provision of special prints, preparation of safety tracks and insurance copies. It is impossible for library film to remain in satisfactory condition if production staff are allowed free access to the master material. A.H.

See: *Film storage; Library* (FILM); *News programmes.*

Books: *Film Cataloguing Rules*, ASLIB (London), 1963; *Cataloguing Rules, Author and Title Entries*, American Library Association (Chicago), 1964.

Library Shots. Shots too difficult and expensive to shoot for a particular production and so obtained from libraries. Great care must be taken not to use individual library shots so often that the audience recognizes their source and they lose reality.

See: *Library.*

Lift. Control associated with each picture-generating apparatus such as a camera, telecine machine, etc., whereby the operator can lift the picture signal bodily up or down in potential with respect to blanking level and so set the darker tones of the picture for optimum contrast in relation to black. This is achieved by altering the d.c. level of that portion of the generated video waveform containing the picture signal without affecting the blanking level, and is most conveniently carried out by modifying the d.c. conditions associated with the blanking pulse insertion circuit.

Because the lift control operates equally on all signal levels comprising the picture information, any adjustment of this control usually needs to be accompanied by an adjustment of picture amplitude to establish the highlights of the picture in relation to peak white level.

See: *Camera* (TV); *Telecine.*

⊕LIGHT

□

The study of light as a branch of physics is generally concerned with the nature and properties of those electro-magnetic radiations which can provide a signal to the eye, and which, when interpreted by the brain, produce the sensation of sight, that is to say, visible light. Photographic science is equally concerned with the same radiations, but its field also extends to similar radiations which, while they do not give rise to the sensation of visible light in the observer, do produce changes in photographic emulsions.

THE NATURE OF LIGHT

Both the intensity and the sensation of colour of visible radiations are dependent upon the temperature of the emitting source, and also upon the nature of the surface of the source.

EMISSION. When an electric fire is first switched on, the current flowing through the wire element causes it to warm up, emitting non-visible radiation, which is felt as heat. At this stage, there has been no visible change in the appearance of the element, but in a short time a further rise in the temperature of the wire causes the element to emit visible radiations which produce in the observer a sensation of redness. In the same way, if the current of an electric lamp is switched on, the temperature of the wire filament of the lamp rises to a level considerably higher than that of the element of the electric fire, and the radiation emitted by the lamp produces a visible sensation which is both more intense and whiter in colour than that from the fire. But even this radiation from the lamp is of relatively low intensity and yellow in colour in comparison with the visible radiation emitted by the sun, which is at a very much higher temperature than the filament of the lamp.

The sun, the electric lamp and the electric fire emit visible radiation. This radiation may fall on other objects and may be scattered so that some of the radiation enters the eye of an observer. These objects do not themselves emit radiation, but can be seen because the visible radiation from a self-luminous source, e.g., the sun, is scattered by the surfaces of the object.

WAVE MOTION. Most of the phenomena associated with light can be explained by the assumption that light is a wave motion, with some similarities to the waves produced on the surface of a pond when a stone is thrown into the water. There are, however, differences between light waves and water waves, one of which is that while the waves on water require a material medium for their propagation, the light waves do not, but can be propagated through a vacuum, as for example the light waves reaching the earth from the sun. Light waves are, of course, also propagated through transparent materials.

Two of the characteristic quantities associated with wave motion are wavelength and frequency.

Wavelength is defined as the distance between two corresponding points on successive waves, e.g., in the case of water waves

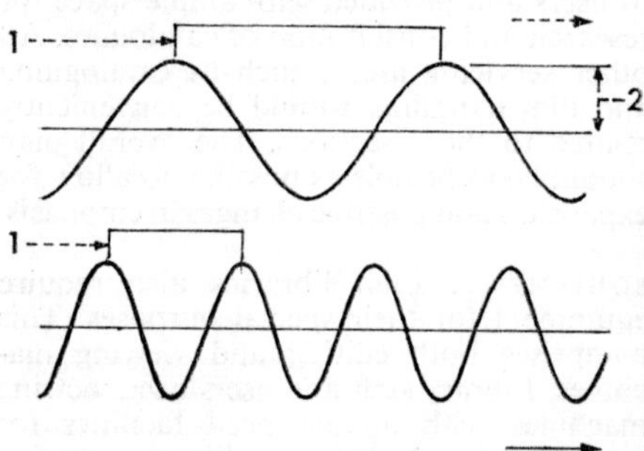

CHARACTERISTICS OF WAVE MOTION. 1. Wavelength. 2. Amplitude, or maximum displacement from mean value. Wavelength equals velocity divided by frequency, hence halving wavelength doubles frequency (*bottom*).

the distance between two successive crests or troughs. The frequency of the wave motion is the number of waves passing a fixed point in unit time.

The velocity of propagation of the wave motion = wave length × frequency, and in the case of light is 186,000 miles or 3×10^{10} centimetres per second in a vacuum. Its velocity varies in different media, the value in glass being approximately 2×10^{10} centimetres per second.

In the case of water waves, the individual particles of water do not move with the wave across the surface of the pond but oscillate up and down. This can be demonstrated by throwing a cork on to the surface of the water. Waves in which the oscillatory motion is at right-angles to the direction of propagation are known as transverse waves. The amplitude of the wave, that is, the distance from a crest to the undisturbed position, is indicative of the energy being transmitted by the wave.

Light waves are also transverse waves but differ from water waves in an important respect. The oscillations of water waves are in a single direction at right-angles to the direction of propagation, whereas light wave oscillations are in all directions at right-angles to the direction of propagation. Certain substances, however, such as Polaroid, transmit preferentially the transverse oscillations in a certain direction. The light is then said to be polarized. The reflection of light at certain angles from some surfaces, e.g., an expanse of water, causes the light to be partially polarized. The effect of an excessive light reflection from the surface of an expanse of water can be visibly reduced by the use of a suitably oriented sheet of Polaroid.

RECTILINEAR PROPAGATION. The waves produced on a pond travel outwards from the source, appearing as a series of concentric circles. Any radius to one of these circles can represent a direction of propagation of the wave and its energy. If there is an obstacle in the pond, the waves can be seen to enter the area beyond the obstacle, indicating that the wave can bend around an obstacle and that water waves are not confined to rectilinear, i.e., straight line, propagation.

Most phenomena, however, suggest that light waves travel in straight lines. This is in fact untrue, but is a result of the very small value of the wavelength of light, which is approximately 0·00005 cm. Under laboratory conditions, using extremely small objects, the diffraction of light can be demonstrated.

For most purposes it can be assumed that light does travel in straight lines, and the term "ray of light" can be used to indicate a direction of propagation.

INTERFERENCE. If two sets of waves of the same wavelength come together, the disturbance at any point is the sum of the disturbances which each would separately contribute in the absence of the other. It may happen that two sets of waves are such that their crests and troughs coincide, in which case the combined disturbance is large. If, however, the crest of one set of waves coincides with the trough of the other, there is no disturbance and no wave energy at that point.

When the crests of one set coincide with the crests of the other and the waves are in step they are said to be in phase. When they are out of step, the waves are said to be out of phase. The phase difference between two sets of waves is expressed as a circular angle. Two sets of waves of the same wavelength are in step if the phase difference is $0, 2\pi, 4\pi$, etc., and out of step when the phase difference is $\pi, 3\pi, 5\pi$, etc.

Phenomena which are produced by phase differences in two sets of waves of the same wavelength are known as interference phenomena.

The colours exhibited by thin films are due to interference of light waves.

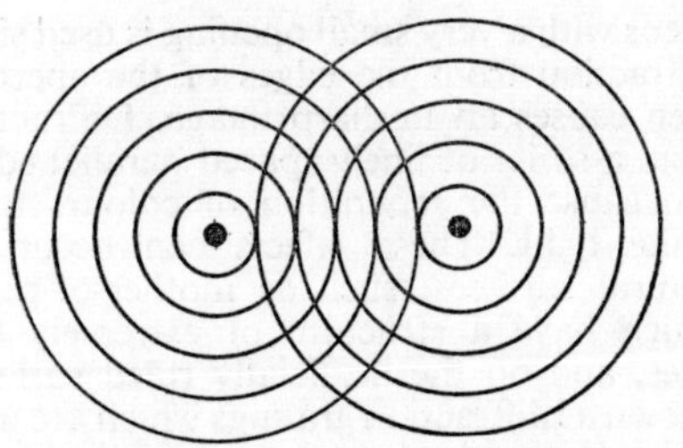

INTERFERENCE. Trains of waves radiating from two point sources interfere when they impinge on one another. Crests are represented by circles, troughs fall midway between. Coincidence of two crests or troughs produces constructive interference, i.e. maximum disturbance. Coincidence of crest and trough produces destructive interference, i.e. minimum disturbance.

DIFFRACTION. The effects which are produced by the ability of light waves to spread into the geometrical shadow produced by an object are known as diffraction phenomena. In photography diffraction can cause undesirable effects, particularly when

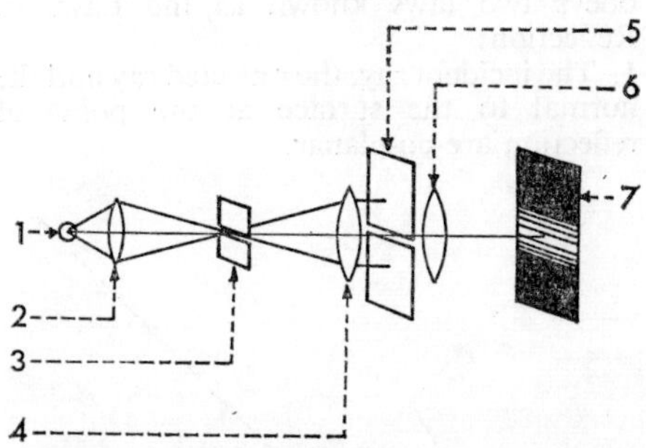

DIFFRACTION. Monochromatic lamp, 1, focused by lens, 2, on narrow aperture, 3, which acts as a source of small dimensions. Light passes through collimating lenses, 4 and 6, between which is a thin slit, 5, and is focused on the screen. Diffraction at edges of slit, 5, causes image on screen, 7, to be bordered by light and dark bars, the diffraction fringes.

a lens with a very small opening is used since diffraction from the edges of the aperture then causes an unsharp image. Diffraction from a series of finely spaced parallel edges can cause the separation of colours from white light. These effects can occur on natural surfaces, such as mother-of-pearl, which have a structure of extremely fine lines, and on the artificially ruled surfaces known as diffraction gratings which are used to analyse white light into its colour components as a spectrum.

REFLECTION AND REFRACTION

When light falls on the surface of an object, some of the light is scattered by the surface, some may be absorbed within the object while some may pass through the object. Reflection is concerned with the light scattered at surfaces. With a perfectly diffusing surface, light is scattered by the surface equally in all directions, but if a beam of light falls on a perfect plane mirror light is reflected in one direction only. With a perfect reflector all the light energy is contained in the reflected beam without loss by absorption.

The majority of surfaces are not perfect reflectors and a light beam striking an irregular surface is reflected in many directions by the irregularities.

REFLECTION. A plane wave front of light falling on a plane reflecting surface is reflected so that the resulting wave front is also plane. Any line perpendicular to the wave front is called a ray. A ray of light falling on a mirror is an incident ray, and after reflection is a reflected ray. Experiment shows that light reflected at a plane reflector obeys two laws known as the Laws of Reflection:

1. The incident ray, the reflected ray and the normal to the surface at the point of reflection are co-planar.

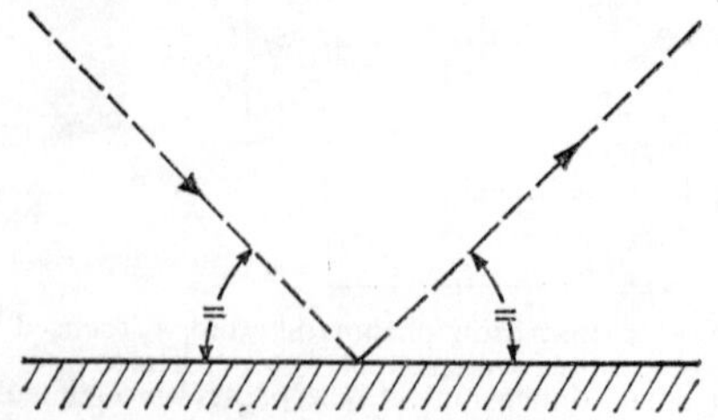

LAWS OF REFLECTION. Incident and reflected rays make equal angles to the normal at the point of incidence, and to a plane surface, and lie in the same plane.

2. The angle of incidence, i.e., the angle between the incident ray and the normal, is equal to the angle of reflection, i.e., the angle between the normal and the reflected ray.

When rays of light from an object fall on a plane mirror, an image of the object is seen beyond the mirror and is called a virtual image. (The image of a film projected by a lens on to a cinema screen, on the other hand, is called a real image.) A group of rays originating at a point on the object appear after reflection to have originated at a corresponding point on the virtual image.

The virtual image is located behind the mirror, and the same distance from the mirror as the object is in front of it. Another characteristic of the image in a plane mirror is that it is laterally reversed, e.g., a right-handed man when shaving sees his image in the mirror as a left-handed person.

REFRACTION. When light is incident on a plane surface of a slab of transparent material, most of the light passes through the surface into the material and emerges from the other side.

If the incident light is not normal to the surface, the direction of the ray is changed and it is said that the light has been refracted.

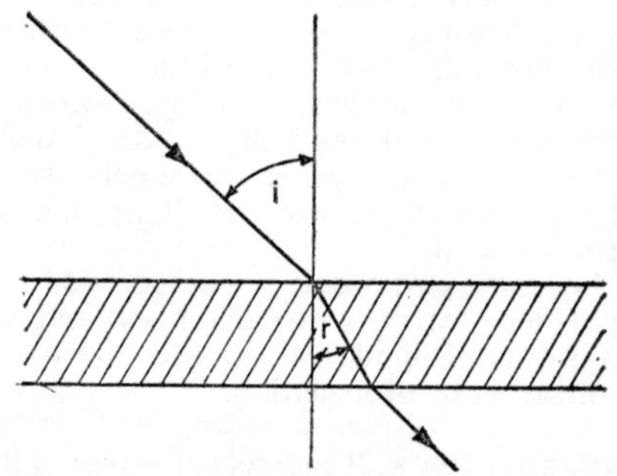

LAWS OF REFRACTION. When light strikes a boundary between two different transparent substances, the incident and refracted rays lie in the same plane, and the sine of the angle of incidence bears a constant ratio to the sine of the angle of refraction. If light is travelling from air (strictly, from a vacuum) into a denser medium, this constant is the refractive index of the denser medium.

The two laws governing refraction of light are:

1. The incident ray, the refracted ray and the normal to the surface at the point of refraction are co-planar.
2. The ratio of the sine of the angle of inci-

dence and the sine of the angle of refraction is a constant for the two substances concerned, and is known as the refractive index.

The angle of refraction is the angle between the refracted ray and the normal. The Greek letter μ is often used to express the ratio $\frac{\sin i}{\sin r}$.

When light passes from air to glass, the ray of light is refracted towards the normal as it enters the glass, i.e., the angle of refraction is less than the angle of incidence. When the light passes from glass to air, the ray is refracted away from the normal. The refractive index for light passing from air to glass s written as air μ glass, and that from glass o air as glass μ air.

The refractive index is also the ratio of he velocity of light in the two media, so that

$$\text{Air } \mu \text{ glass} = \frac{\text{velocity of light in air}}{\text{velocity of light in glass}}$$

$$\text{and glass } \mu \text{ air} = \frac{\text{velocity of light in glass}}{\text{velocity of light in air}}$$

$$\text{Air } \mu \text{ glass} = \frac{1}{\text{glass } \mu \text{ air}}$$

One of the effects of refraction is that a pond of clear water appears shallower than t really is, an important consideration in nderwater cinematography and television.

OTAL INTERNAL REFLECTION. When ight passes from a dense material like glass o a less dense material like air, it is refracted way from the normal. The angle of refraction is greater than the angle of incidence. As the angle of incidence is increased, a alue for this angle is reached where the mergent ray is along the air glass surface, e., the angle of refraction is 90°. The angle f incidence corresponding to this condition s called the critical angle. If the angle of ncidence is increased to a value larger than he critical angle, the ray of light does not merge from the glass but is reflected at the lass air surface. This type of reflection is nown as total internal reflection, and obeys he Laws of Reflection.

This phenomenon of total internal reflecon is employed in a number of optical nstruments, where it may be desirable to otate the path of a beam of light through 0° or 180°. The critical angle for glass is pproximately 42°, so that a ray of light ncident on a glass air surface at an angle of ncidence of 45° is totally internally reflected. right-angled glass prism can be used to otate the path of a beam of light through

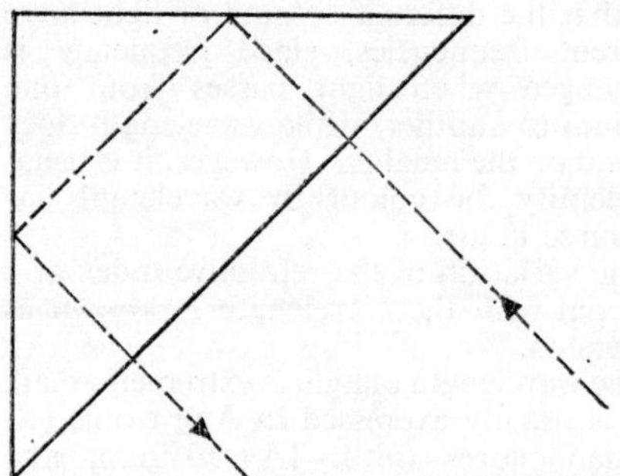

TOTAL INTERNAL REFLECTION. If light falls normally on the hypotenuse of a right-angled prism, it will be reflected internally from the other two sides, and will re-emerge parallel to its direction of incidence. This form of total internal reflector is called a Porro prism.

180°, by allowing the light to fall normally on the hypotenuse, and obtaining total internal reflection at the other two glass air surfaces.

DISPERSION

REFRACTIVE INDEX. The refractive index depends not only upon the two media, e.g., air-glass, but also on the colour of the light employed, the refractive index for blue light

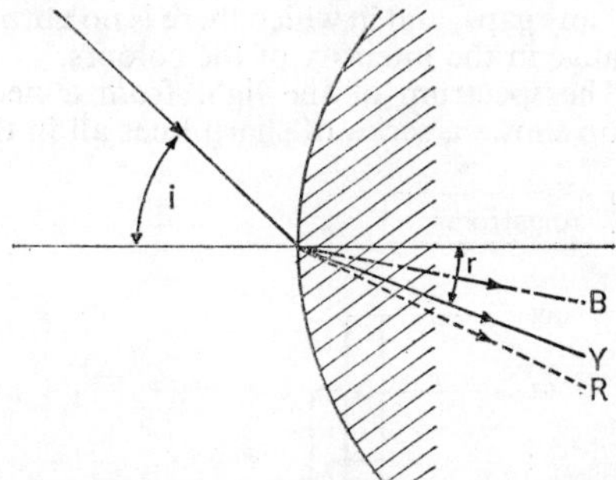

DISPERSION. White light entering a denser medium is not uniformly refracted: blue rays are refracted more than yellow, yellow more than red, so refractive index varies with colour of light, and is greatest for blue. Amount of dispersion shown greatly exaggerated.

being slightly greater than the refractive index for red light. A beam of white light, after passage through a glass prism, is spread out by the prism into a band which shows all the colours of the rainbow from violet to red. White light consists of a mixture of light of different colours, each having a different wavelength. It is more precise to say that the different colours of light have different frequencies, since frequency is unchanged when light passes from one medium to another, while wavelength does depend on the medium. However, it is usual to identify the colour by wavelength, as measured in air.

The variation of the refractive index of a material with the wavelength is known as dispersion.

The wavelength of light is extremely small, and is usually expressed in Angstroms (A) or nanometres (nm). $1A = 10^{-7}mm$, and $1nm = 10^{-6}mm$. Approximate values for the wavelengths of light are: blue 4,500A, green 5,000A and red 6,500A, or 450, 500 and 650nm.

THE SPECTRUM. The coloured band of light which is produced when a beam of white light passes through a glass prism is known as a spectrum and can be examined by a spectroscope. This instrument consists of a collimator, which produces a parallel beam of light from the source under examination, a suitable prism and a telescope for inspection of the beam of light after passage through the prism. The light from the sun, or from an incandescent filament lamp, produces a continuous spectrum, i.e., a spectrum in which all the colours are reproduced in a continuous sequence without any gaps, and in which there is no abrupt change in the intensity of the colours.

The spectrum of the light from a neon lamp shows a series of sharp lines all in the

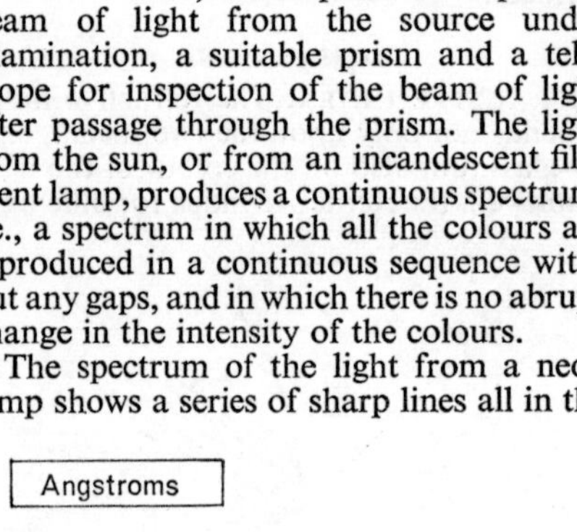

ELECTROMAGNETIC SPECTRUM. A. Gamma rays. B. X-rays. C. Ultra-violet radiations with near UV from 3,000–4,000A. D. Visible spectrum, 4,000–7,000A (400–700nm). E. Infra-red, far infra-red and heat rays, with fast infra-red photographic emulsions sensitive up to 9,000A, and slower special emulsions to 14,000A.

red region of the spectrum, and a mercury vapour lamp produces a spectrum consisting of a few sharp lines in the violet, green and yellow regions. Insertion of a green filter into the light of a filament lamp produces a spectrum in which the green part only is present.

With a suitable spectroscope and methods of examination, the spectrum can be shown to extend beyond the visible region, the region beyond the violet being known as the ultra-violet, and that beyond the red as the infra-red. The ultra-violet extends to wavelengths of a few hundred Angstroms while the infra-red extends to wavelengths as long as one millimetre. Both infra-red and ultraviolet radiations are capable of producing exposures on photographic emulsions, as also can X-rays and γ-rays, which are similar radiations belonging to a whole range of electromagnetic radiations extending from the long radio waves of 1,000 metres or more down to gamma rays of wavelengths of the order of 10^{-12} cm.

PHOTOMETRY: MEASUREMENT OF LIGHT INTENSITY

Since light is a form of energy, one way of measuring it is by exploring different parts of the spectrum by a heat-sensitive instrument such as a thermopile. This shows that the peak of the energy distribution is in the infra-red region, i.e., outside the visible spectrum. However, this type of measurement cannot be applied to the illumination falling on or reflected by a surface.

STANDARD LIGHT SOURCE. All systems of measurement of light intensity, and the definitions associated with them, are based on a visual comparison of the illumination from an unknown source with that from a standard source. In earlier times, the illumination from a candle of specified composition, dimensions, and rate of burning was used as the standard light source. The internationally accepted unit of light intensity is now the candela, defined as one sixtieth of the luminous intensity of a black body radiator maintained at the temperature of melting platinum.

The radiation emitted by a body is dependent upon the nature of its surface, a dull black surface being a better emitter than a highly reflecting white surface. The concept of an ideal radiator is a radiator capable of emitting all wavelengths in the visible, ultra-violet and infra-red parts of the spectrum. Such a body is also capable of absorbing all wavelengths and is called

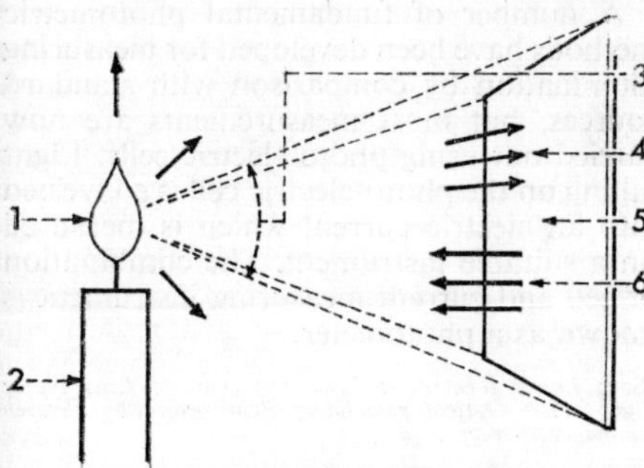

LIGHT UNITS. 1 Luminous intensity or candle-power, amount of light emitted in all directions by standard candle. 2. Light source of one candela. 3. Luminous flux; unit, the lumen, or light received by unit area at unit distance from source of one candela. Total emission of this source 4π (12.57) lumens. 4. Illumination, or luminous flux per unit area; unit, the lux (lumens per square metre) or lumens per square foot (formerly called foot candles). 5. Reflectivity, the proportion of incident light reflected by a surface. 6. Luminance, intensity of light reflected from surface; units, the nit and foot lambert. A perfect diffusely reflecting surface (reflectivity 100%) illuminated by 1 lumen per square foot reflects 1 foot lambert; illuminated by 1 lux, it reflects 1 nit.

perfectly black body. The perfect radiator is only approximately realized in practice. Stefan's Law states that the energy emitted per square centimetre per second from such a body is proportional to the fourth power of the absolute temperature. The distribution of the energy between the various parts of the spectrum varies with the temperature. A perfectly black body at a precise temperature emits a definite quantity of radiation, and can therefore be used as a standard.

Standard electric lamps can be calibrated in terms of the candela and other sources can be compared to these lamps.

The term candlepower is still used for the luminous intensity of a source of light, but it now refers to the candela.

A theoretical point source of light radiates equally in all directions. The quantity of light which crosses a surface depends upon the solid angle subtended by the surface at the point source. The quantity of light emitted in unit time from a uniform point source of one candle power in unit solid angle is defined as one lumen.

INTENSITY OF ILLUMINATION. The intensity of illumination of a surface is the quantity of light falling on a unit area of the surface in one second. The quantity of light crossing a surface in unit time is called luminous flux. Several different units are employed in the measurement of intensity of illumination, depending upon the unit of length employed, e.g., feet, centimetres or metres. The foot-candle is the intensity of illumination at the surface of a sphere of one foot radius from a source of one candle-power at the centre of the sphere. One foot-candle corresponds to a flux of one lumen per square foot. Other units with similar definitions are the centimetre-candle, often called the phot, and the metre-candle or lux.

THE INVERSE SQUARE LAW. If the source of illumination can be regarded as a point source, and if the illumination falls normally on the surface, the intensity of illumination at the surface is inversely proportional to the square of the distance between the source and the surface. In practice, calculations based upon the relationship are reasonably valid provided the physical dimensions of the source are small in comparison with the distance between the source and surface.

THE COSINE LAW. If the light from the source does not fall normally on the surface, but a ray of light makes an angle θ with the normal to the surface then the intensity of illumination at the surface is proportional to cosine θ.

If light from several sources illuminates the same surface, then the total intensity of illumination at the surface is the sum of the intensities of illumination from each separate source.

The intensity of illumination of a surface, distance ν from a source of illumination of I candle power, where the light makes an angle θ with the normal to the surface is given by $\frac{I \cos \theta}{\nu^2}$.

LUMINANCE. The appearance of a surface when viewed depends not only upon the intensity of the incident light but also upon the nature of the surface and the angle from which it is viewed.

Brightness (or luminance) is expressed as the equivalent candle power per unit effective area of the surface. If B is the brightness of a surface in a direction θ to the normal, and I is the effective candle power in this direction,

$B = \frac{I}{A \cos \theta}$, where A is the area of the surface.

If the surface is a perfectly diffusing surface, it appears equally bright in all directions from which it is viewed, i.e., brightness is independent of θ. But $B = \frac{I}{A \cos \theta}$, and since A is constant and B is independent of θ, I must be proportional to $\cos \theta$, i.e., the candle power of a perfectly diffusing surface in a direction making an angle θ with the normal is proportional to $\cos \theta$. This is known as Lambert's Law.

The lambert is the unit of measurement of brightness, and is defined as the brightness of a perfectly diffusing surface when the total flux radiated from a unit area of the surface is one lumen per sq. cm. If the area is one square foot, the unit is the foot-lambert.

A number of fundamental photometric methods have been developed for measuring illumination by comparison with standard sources, but most measurements are now carried out using photo-electric cells. Light falling on the photo-electric cell is converted into an electric current which is measured on a suitable instrument. The combination of cell and current measuring instrument is known as a photometer.

See: *Image formation; Lens performance; Lens types; Light units; Optical principles; Rear projection; Screen luminance; Screens.*

Books: *Fundamentals of Optics*, by F. A. Jenkins and H. E. White (New York), 1957; *Principles of Illumination*, by H. Cotton (New York), 1948; *Optics of the Electromagnetic Spectrum*, by C. L. Andrews (New York), 1961.

□**Light Application Bar.** During its television transmission, a film frame may be illuminated for only a part of each field scanning interval; whether or not this matters depends on the amount of storage, or memory, in the telecine pick-up tube. The ratio of illuminated to unilluminated time is called the light application ratio or time, and is often expressed in angular degrees, 360° representing a continuously lit frame.

If the film projection rate is not synchronized to the television picture frequency, and if there is insufficient storage in the telecine, it is possible for a horizontal black bar to float up or down the television picture. This light application bar is caused by the film frame not being illuminated during part of the active television field scanning intervals.

See: *Telecine.*

□**Light Application Time.** Time during which light is allowed to fall on a frame of film in a vidicon type of film scanner.

See: *Telecine.*

⊕**Light Box.** Box or flat section of an inspection table which contains a light source and is covered by a sheet of frosted glass. When film is held in front of a light box, the picture on it is easily visible.

⊕**Light-Change Board.** Board attached to a film printer, by the use of which the printer light settings determined by a grader can be preset, so that a whole reel of cut negative, with all its light changes, can be printed automatically.

See: *Printers.*

⊕**Light Change Points.** Exposure steps used in a motion picture film printer; it is usual to provide a logarithmic series of increments increasing by either ·05 log E or ·025 log E at each step.

See: *Printers.*

⊕**Light End.** Part of a film developing machine in which the film may be safely exposed to white light. Normally comprises the drying cabinets.

See: *Processing.*

⊕LIGHTING EQUIPMENT

Until recently there has been little in common between the methods and equipment used in film and TV studios.

Film lighting methods have been gradually developed from long-established techniques going back to the turn of the century. Lighting equipment in major film studios today reveals a very high percentage of old units dating back to the 'thirties but still giving useful service. Although for the most part bulky and cumbersome, they provide the lighting director with the means to produce either striking or delicate effects. It is true that this is at the expense of filmed minutes per day. But, in contrast with television, speed is often not the determining factor in feature film production.

A shooting period running into several months, using a single camera, with large numbers of set-ups and takes, gives great scope to the lighting director to evoke all that is in the scene by subtle lighting effects. Again, in contrast with television and the small picture, the giant screen of the cinema calls for a more elaborate and perfect lighting balance to bring out the details of big casts and the lavish settings in which they move.

When many hours and even days may be spent in lighting a single scene, the size and weight of the lighting units is no longer a matter of great importance.

INCANDESCENT LAMPS

These lamps, which greatly predominate in numbers used, fall into two classes, flood and spot. Floodlamps are fixed focus, or have only a limited focus range. Spotlamps have a wider focus range from the spot to the flood position. Both follow the same general design pattern: a well-ventilated housing containing the lamp holder, a polished aluminium or silver-backed reflector, and a Fresnel heat-resisting front lens with a honeycomb backing which minimizes filament striation and gives a smooth light output of constant colour.

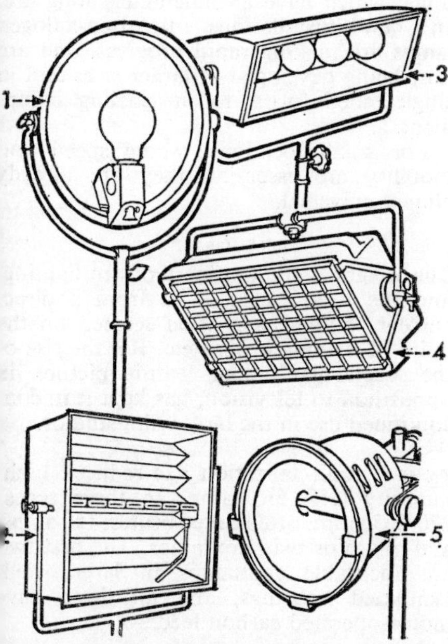

TUDIO LIGHTING UNITS (LUMINAIRES). 1. Skypan kW. Wide-angle, high-powered floodlamp, used or lighting backdrops, etc. 2. Large-source unit or shadowless floodlighting, powered by tungsten-alogen units in centre strip. 3. Quad, with four 50 or 500 W bulbs, for general purpose front or verhead lighting. 4. Tenlite, with ten bulbs up to 00 W each, and provision for rapid change of lter material. Similar units are designed for four kW tungsten-halogen lamps. 5. Tungsten-alogen 1 kW flood lamp, with bulb mounted ransversely.

FLOODLAMPS. Many different types of floodlamp are in common use, among them the following. *Broads* are lamps fixed horizontally, and longer than they are high. They are used to provide fill light. *Strip lights*, as the name implies, are long, narrow lamps containing several or many light sources for evenly lighting large areas of backdrop or cycloramas. A similar purpose is served by the 10 kW *skypan*, a unit made up by individual studios and consisting of an open bulb set in an aluminium reflector about 3 ft. square and of slightly dished shape, with a protecting wire cage in front of the bulb. It may be set high on tubular scaffolding or mounted on short legs as a floor unit.

Banks are units containing 2, 4, 6 or 12 bulbs within a single reflector, often individually switched. *Rifles* are for general floor illumination, covering an angle of about 60° by means of a single bulb set in a paraboloid housing. *Garlands* consist of a circle of reflectors around the camera housing up to eight photofloods.

SPOTLAMPS. Spotlamps are of more uniform design, and vary principally in the power of the light source they employ. Small 200 W lamps and larger lamps of 750 W and 2 kW are used for modelling of facial outline and for shadow removal. The 5 kW lamp has the widest application, and is used for front, back and cross-lighting, as well as for modelling. Still larger spotlamps are available in the 10 kW size.

COLOUR TEMPERATURE. Now that colour film is virtually universal, the colour temperature (CT) of lamp bulbs is of the greatest importance. Colour film is balanced to 3,200°K, and most studios use CT lamps in which the colour temperature is raised to this figure at the expense of useful lamp life, which is about 100 hours for the larger sizes. After this, the drop off in colour temperature may exceed 150°K and become unacceptable to the lighting director. Older bulbs may, however, continue to give useful service in less critical applications than spotlamps; for instance, in skypans used for lighting backings.

The larger bulbs contain a small quantity of fine loose tungsten granules. By swirling these around, much of the blackening inside

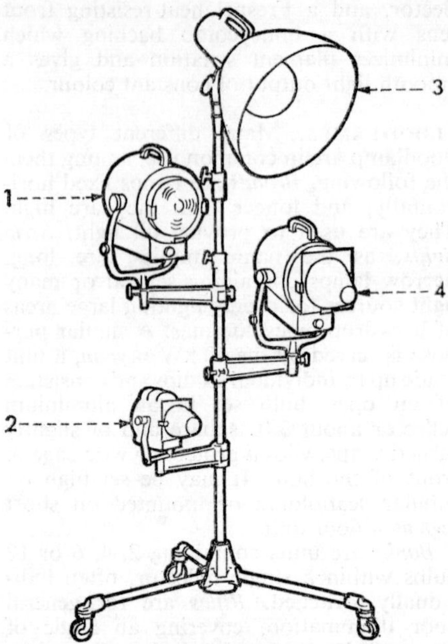

STUDIO LIGHTING UNITS (LUMINAIRES). 1. 2 kW spot with Fresnel lens; beam angle variable from about 10–45°. 4. Ditto, with conical snoot. 3. Scoop (500 or 1,000 W). 2. Miniature spot lamp (500–700 W) with snoot. Lamps mounted on telescopic stand with dismountable wheeled base.

the glass envelope can be removed, and the light output prevented from falling so rapidly. Eventually, however, the blackening becomes permanent, the bulb temperature rises, and blistering is likely to occur.

BASE TYPES. Bulb bases are mainly of the pre-focus or medium bipost type for 500 to 750 W sizes, while a larger bipost with hollow pins is used for sizes from 2 kW to 10 kW. Very good contact between the bipost pins and the holder is essential to prevent arcing, which may puncture the hollow pins and thus allow air to enter the bulb and so destroy it. The low ohmic resistance of cold tungsten causes a heavy surge of current to flow at the moment of switching on, and it is then that arcing is most likely to take place.

TUNGSTEN-HALOGEN LAMPS. The conventional tungsten lamps so far discussed are now undergoing replacement by the tungsten-halogen type. These are tungsten-filament lamps with a halogen (usually iodine or bromine) regenerative cycle. The bulb contains a small quantity of the halogen which quickly vaporizes at working temperature and combines with tungsten atoms evaporated from the filament and deposited on the bulb wall. A volatile tungsten halide is formed which diffuses back to the filament, where the high temperature decomposes it. This redeposits the tungsten, leaving the halogen free to continue its regenerative action.

For this effect to occur, it is necessary for the bulb wall temperature to exceed 250°C, but bromine functions with a rather wider range of wall temperatures than iodine, and eliminates the slight purplish tint characteristic of the iodine lamp when warming up.

The striking advantage of the tungsten-halogen lamp is its maintenance of virtually constant light output and colour temperature throughout its life. Early lamps of this type were exclusively of long, tubular construction, and double-ended. These called for specially designed housings, which are now available in many types. However, for economic reasons, the larger studios have been slow to replace their conventional units, which have an indefinitely long life. In new installations, tungsten-halogen lamps are making rapid progress, and are now being developed in larger sizes and in single-ended forms to fit existing equipment.

For small locations, where speed and mobility are essential, they are already almost universal.

ARC LAMPS

The weight and clumsiness of arc lighting and the need to power it from a direct current source, have often seemed on the point of making it obsolete. But the rise of the spectacular colour feature picture, in opposition to television, has kept it in continued use in the larger film studios.

TYPES. Rationalization has reduced high intensity (HI) arc lamps to three types 220/265 amps, often called brutes, 150 amps and duarc or twin-arc lamps. The first two are focusable spotlamps in large, well ventilated housings, and all three have motor-operated carbon feed.

CARBONS. Carbons used in this equipment are of two types, white flame and low colour temperature (LCT). On a colour production where white flame carbons are used, a filter is placed in front of the lamp to reduce the colour temperature from 6,000°K to 3,200°K. Only when arcs provide the sole light source is it possible to replace the lamp filters with a single camera filter.

FILTERS. LCT carbons burn at about 3,350°K, but for complete correction when incandescent and arc lighting are mixed, it is desirable to add a light yellow filter. Filters raised to a high temperature have a short life, and on large arc equipment filter frames extended well away from the housings are advisable.

RIGGING AND OPERATION

Lighting directors vary in their methods; some are outstanding in the large amount of light they call for, others in its economical use. Lighting foremen or gaffers stay fairly regularly with the same cameraman, and thus can anticipate their general requirements.

LAYOUT. Up to medium size sets, the layout is fairly consistent, varying only in rigging for black-and-white productions, colour, or day and night shooting. Generally speaking, set building is well advanced before the lighting cradles are positioned, and almost completed before lamps are rigged. An advance estimate of the lighting and loading requirements is obtained from the script, models and set drawings. Because of the varying height and shape of sets, very few of the spotting rails can remain in position for other sets that follow on. This means that re-rigging must take place for almost every change of set.

Stages vary in height from 30 to 40 ft. and the lighting cradles are suspended from the steel joists which run the full length of the stage, and are hung from the roof trusses. These joists are usually 5 ft. apart and contain numerous half-ton lifting tackles on movable skates. The cradles are raised from the floor into positions roughly conforming to the shape of the sets; they are then hung on chains at the various heights and deaded off, i.e., secured, to give a firm fixture. If there is any sway in the cradle, it causes a troublesome light movement when the operators are walking about.

CRADLE CONSTRUCTION. The cradles are normally made of wood, but a light duralumin rail about 10 ft. long and 3½ ft. wide is preferable. A safety hand rail varies in height according to the size of the lamps rigged, and a lower rail is fixed to the back section for extra safety. Three or four lighting brackets are permanently fixed on the front and back of the cradle, but it is general practice to fix the very heavy lamps on turtles (so called from their appearance) screwed directly to the floor.

Other forms of lamp spotting are achieved by building up tubular cradles, either suspended from above or raised from the ground. Indeed, most sets call for mobile towers capable of holding from one to three large lamps.

POWER GENERATION

VOLTAGE. Because arc lamps require a 115-volt, direct-current supply, it has always been film studio custom to maintain this low pressure for all lighting equipment, even in countries like Britain with a higher standard supply voltage. There are some incidental advantages: lamp filaments are more robust and so will withstand rougher treatment; low voltage is safer to handle when dealing with hundreds of pieces of mobile equipment; regulations are less stringent and no earthing is required. Drawbacks are

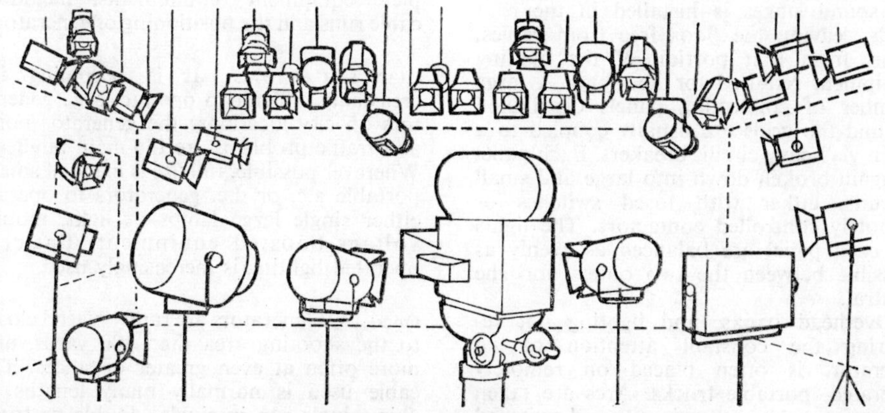

TUDIO LIGHTING LAYOUT. Conventional film studio lighting is based on lights rigged to sets and catwalks nd mounted on studio floor for prolonged single-camera shooting in restricted area.

heavier bus-bars, main cables and flexible leads to individual lamps, as well as more cumbersome switchgear.

LOAD. Where the main lighting load is taken from the commercial or outside grid supply, conversion plant is installed to give three-wire d.c. of 230 volts across the outers, 115 volts between an outer and neutral. This plant can comprise rotary converters, motor generators, silicon rectifiers or mercury arc rectifiers.

Because of the erratic and unpredictable load factor in film production, the power cost per unit rises rapidly as the maximum demand is increased. Most major studios therefore have a large standby diesel-driven generating plant which is brought into use to offset these peak conditions, especially during the winter months when the power demand is at its highest rate.

RIPPLE. A.c. ripple content, inherent in all d.c. supplies, must be kept as low as possible, since it is a constant source of trouble to the sound recording departments. Any conductor free to vibrate acts as a loud-speaker and emits audible notes at the frequency of the ripple. The flame of the carbon arc is such a conductor; it is free to vibrate and responds to a wide range of frequencies.

Most studio generators are designed with a special winding, screw-slotted armature slots, and a graded air gap on the pole faces to reduce ripple to a reasonable level. Some location generators require large capacitors to smooth out the ripple.

POWER DISTRIBUTION

The major portion of the supply distribution on sound stages is installed in the roof grids, leaving the floor free from cables, apart from that portion of the lighting equipment required for floor use. A large number of connecting panels distributed around the grids are usually coupled to a main via heavy circuit breakers. Each panel is again broken down into large and small circuits, either with fused switches or remotely controlled contactors. The lights on each panel are balanced as evenly as possible between the two outers and the neutral.

Overhead banks, and lighting not requiring the constant attention of an operator, is often placed on remotely operated, portable trucks. Arcs are taken direct from the heavy circuits and switched on and off by the operators, but three-wire, six-way spider boxes feeding the floor or lighting cradles for the incandescent equipment are controlled by the grid switchboard operator, or by remote contactors operated from the floor console.

Installed capacity on stages ranges from 60 to 100 watts per square foot of floor area, so that a stage of 18,000 sq. ft. requires about 1,500 kW.

LIGHTING ON LOCATION

Location lighting presents different problems according to whether the location is in a foreign country, or in a near or distant part of the home country.

ABROAD. Equipment for foreign locations is usually supplied by well-known hire firms or their representatives in the location country or a neighbouring territory. Except in the case of special gear, it is usually found more economical to work in this way and so avoid transport costs and customs problems. Most of the labour force is drawn from the studio, but this often has to be backed with local labour in the actual location area. With local locations, i.e., those within reasonable daily travelling distance from the studios, equipment is transported daily and is usually coupled at the location to a hired generator.

DISTANT HOME LOCATIONS. Units on location distant from the studio take sufficient equipment to meet all eventualities and include among the crew a maintenance man for repairs. Most locations of any size and duration are surveyed prior to the shooting date by the electrical gaffer allocated to the production. He collects all data for complete equipment requirements, including cable runs and the positioning of generators.

CITY LOCATIONS. It is becoming increasingly difficult to operate large generators on city locations; the generator noise and traffic problems are the main bugbear. Wherever possible, full use is made of small portable a.c. or d.c. generators to operate either single large lamps or light, mobile voltage-boosted equipment. Battery-operated lighting is increasingly used.

CABLES. Generators are rarely placed closer to the shooting area than 100 yards, and more often at even greater distances. The cable used is normally many lengths of ·2 in. single run in single, double or triple lines, according to load requirements.

Great use is now made of aluminium flex cables, especially in the ·2 size. These have a big advantage in weight saving, and also reduce freight charges on distant locations.

Quick portable connecting boxes are used to join up lengths; for crossing roads special flat cables, made in sizes up to ·5 sq. in., are often used to make it easy for traffic to pass over them.

ORGANIZATION. An experienced gaffer can be a great help to the sound recording engineer by positioning generators to suit prevailing winds and, if possible, by shielding them behind buildings. Cue signals back to the generator operator save a lot of running about and shouting; correct loading of vans is essential. A location unit often needs to change sites rapidly; foreknowledge of the next move and the equipment required helps greatly and time can be gained by intelligent loading. It is an advantage to have one van for cable only; when shooting at night, a small cable run out to supply operating lights from the generator is the first requirement for positioning the gear and the last for wrapping up when the shooting is finished.

DUAL-PURPOSE STAGES

The rise in the number of films required for television (and thus needing to be produced swiftly and economically) compared to orthodox films for cinema theatres, has made it imperative to modernize motion picture studios, and especially the lighting and rigging methods used in them. Sound stages must be quickly adaptable from shooting a major film production to shooting films made only for TV screens. This means remote control of equipment, with switching and dimming fittings permanently rigged on overhead lighting hoists, easily raised, lowered, positioned and plugged up.

CONVENTIONAL EQUIPMENT. It has already been explained that the wide gap between film and TV methods is dictated by an entirely different approach to shooting. As long as the demand for lavish film productions continues, the studios cannot slavishly follow TV practice, but must work out their own ideas. Thus with larger, higher and more complicated sets, and with lighting values often three or four times greater, the full use of pole-operated overhead equipment is ruled out, and spotting rails, heavily rigged for 5 and 10 kW incandescent and often for 150 and 225 amp arcs, must still be provided. Many more electricians are needed to be at the beck and call of the lighting director. He will demand delicate adjustment of the light output, and the fitting of many forms of adapter to the front of the lamp, such as cones for reducing the area of the beam, barndoors for adjusting the cut-off light, and dimmer shutters for lowering the light intensity without affecting the colour temperature.

TV-STYLE EQUIPMENT. A studio constructed to operate both the newer overhead hanging lights and conventional spot rails enjoys important advantages. Modern dual-purpose stages have done away with the old open space grids and their criss-cross of operating platforms; they employ a form of TV-style lighting grid comprising a series of suspended steel platforms arranged to form continuous clear track slots running longitudinally through the full length of the studio roof 30 to 35 ft. These clear tracks are at 2½ ft. centres, but infill panels between allow lamps to be spaced at every 5 in.; these latter tracks are restricted to a lateral movement of 6 ft. On these tracks, monopole telescopic hoists allow lamps to be drawn up tightly beneath the grid and positioned in almost any part of the roof area.

Transfer trolleys at the end allow the equipment to be moved rapidly to new positions or to adjoining stages. Cross transfer slots, if fitted in two or more positions, allow for speedier and more extended movement of the lamps.

ADVANTAGES. Because of the need to retain full facilities for hanging normal spotting rails, the grids are of heavier section than are used in TV studios. These new stages have already proved their worth in speeding up all rigging processes with added safety. For instance, monopoles may be used for the top lighting of all backings, for awkward corners where rigging becomes complicated, and across the narrow walls of composite sets, provided always they can be pole operated or reached from a spot rail.

However, 100 per cent pole operation is unsuitable for major films. In fact, the larger the production, the less the role they play. For small-budget films and for TV films with shooting schedules of ten days or so, their use is a tremendous asset. T.E-K.

See: *Cameraman* (FILM)*; Lighting equipment* (TV)*; Studio complex* (FILM).

Literature: *Evolution in Tungsten Lamps for Television and Film Lighting*, by C. N. Clark and T. F. Neubecker, *SMPTE Jnl*, April 1967.

□LIGHTING EQUIPMENT

Television lighting is based on incandescent lamps, most of which can be voltage-controlled to give the required amount of light. In general, the lanterns are mounted flexibly using a variety of bar or telescope techniques so that the position can very readily be adjusted. This is essential because, in spite of videotapes, retakes are not permitted in television to correct faults or unwanted effects. As in the theatre, once the show begins it goes through to the end. The theatre, however, is not concerned with varying viewpoints. Television cameras are, and the lighting must be capable of instant alteration to suit each shot. To achieve this, the following requirements are essential:

1. Pre-planning of lighting layout in detail by the lighting supervisor.
2. Facilities for quick and flexible rigging by the studio crew.
3. Quick setting and adjusting of the rigged lamps from the floor by electricians, initially to the plan and finally to the instructions of the lighting supervisor.
4. Control facilities to enable the lighting supervisor to preset and modify lighting at run-through and subsequently. Facilities to carry out the lighting changes required by the nature of the show by switching in and/or regulating the intensity of groups of lights. These facilities make written plots of the theatre type unnecessary – otherwise the show would be over before it was written down.

RIGGING PRACTICE

A modern television studio is equipped mechanically and electrically to provide an extremely rapid turnover of production. Often each 24 hours sees a different production set up overnight, lit, transmitted or taped, and struck. Certain productions may take double this time but they are a minority. In general, the optical systems of the lighting units themselves, spotlighting or softlighting, do not differ greatly from those in use in film studios; it is in the rigging and adjustment of these units that the difference occurs. To begin with, once rigged, all units can be directed for pan and tilt from the floor by attaching a pole. Spotlights can be focused in the same way. The suspensions themselves take the form either of short motorized bars hoisted up and down by cables or single-point telescope suspensions. The latter operate in conjunction with an all-over theatre-type grid and nowadays provision is invariably made for high-level loading and storage of lighting equipment. The grid also carries the socket outlets for the lighting circuits, and extra suspensions are provided for flying scenery.

Before the expansion of the public television services in the early 1950s, rigging in television studios was confined to methods inherited from film studios and could be described as building studio lighting specially around each setting required and then repeating the operation for other sets. Within a particular setting, the lighting for each camera shot was specially set up, though it was no longer in the hands of the cameraman as in film practice.

MOTORIZED BARS. Increased programme hours led the television companies to consider the economics of their lighting arrangements. There were marked differences in the approach to rigging. Some organizations settled early for motorized bars about 8 ft. long, each carrying feeds for four outlets of varying wattage and each acting in effect as a store for four or more lanterns. A centralized hoist control enabled any bars required for use to be lowered in quickly so that lanterns could be changed round overnight to a plan for a particular production before the scenery came in.

This form of suspension naturally provides far more outlets than are needed for a single production because they are distributed evenly over the studio wherever they

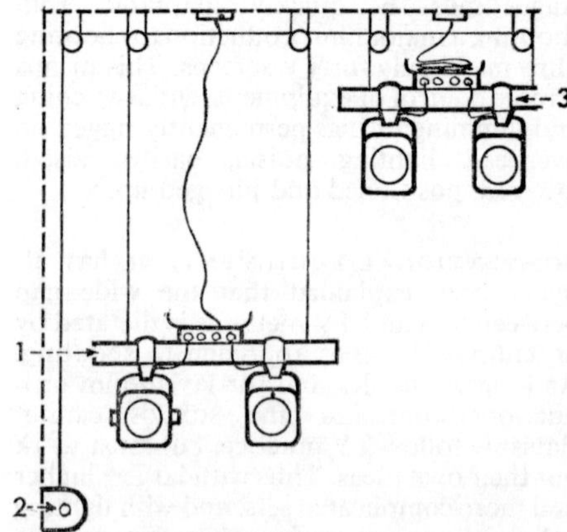

MOTORIZED BARS. Bars, 1, 3, are about 8 ft. long, and carry outlets for up to four lanterns mounted on them. Raising and lowering is by motorized hoist, 2, and lights are rapidly rigged overnight to correct angle and height for next day's production.

might be required – a saturation of outlets and lanterns for flexibility of use. It would be quite uneconomic to have all such outlets under full control, and patching is therefore necessary. This is the selection of circuits for control by a lesser number of channels, the effect of which is to discard outlets not in use for a particular production. For full flexibility this generally takes the form of a cord and jack system. To enable any cord to reach any channel on a large jack-field, built-in jumpers are provided, and there are ammeter test sockets to check doubtful loads. Where wattages of control channels vary, safety devices allow a 2 kW load to be plugged in to a 5 kW control channel but prevent a 5 kW load being connected to a 2 kW channel.

The existence of a surfeit of lanterns at fixed points all over the studio, only some of which are selected for control, produces need for a diagram, based on the studio layout, to show which outlets have been patched into the control and to some extent the pilot lamps also show a state of dim. This diagram is placed in the control room.

SLOTTED GRID. The most common alternative to motorized bars is the provision of an overall slotted grid with point suspension telescopes or monopoles to carry the lanterns. The first characteristic of this system is that the grid itself, on which the electrician walks, is in fact used for patching. It is here that the outlets are found and it is unnecessary to have more of these than are likely to be required for a production as lanterns are also plugged up to the grid. The telescopes slide along the slots in the grid and can be raised or lowered by means of a power operated tool from above. At first the lanterns were attached at floor level, but this wasted time, and the grid now has a loading bay at one end. Here retracted telescopes can be inserted into the chosen slot from a gallery, some six or seven feet below the grid, and lanterns can be taken from the adjacent store and attached to the suspension. The electrician then runs the telescope and its lantern to the required position in the slot, plugs up the lead, and sets the correct height of the unit.

There is usually a transverse bearer at the end of the grid slots over the loading bay so that telescopes with their lighting units attached can easily be moved from one slot to another. Another variant is transverse slots at regular intervals in the grid with a complex mechanical system to enable transfer to take place within the grid itself.

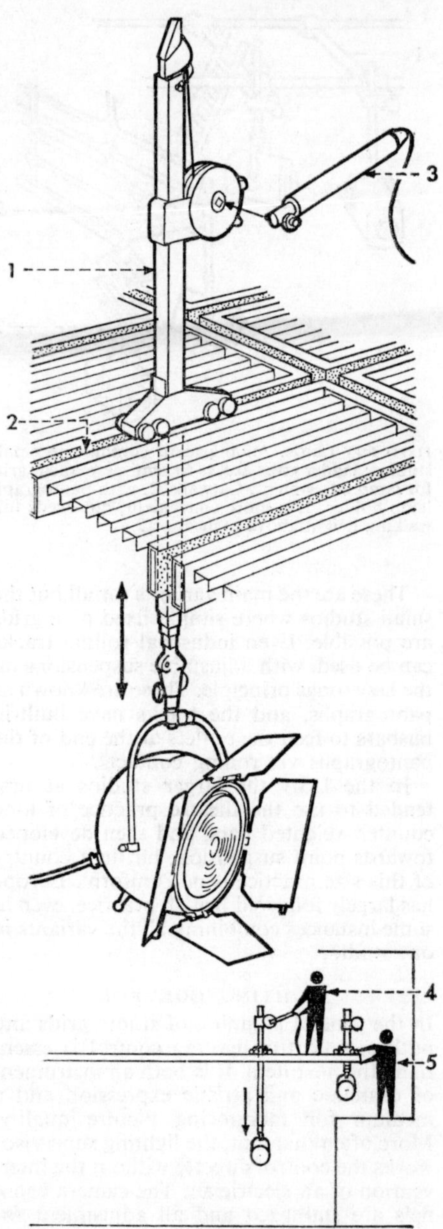

MONOPOLE UNIT. (*Top*) Monopole, 1, is mounted on slotted grid, 2, which allows rapid transfer from point to point. Power-operated tool, 3, enables light unit (luminaire) to be winched rapidly up and down on telescopic extensions. (*Bottom*) Operator, 4, stands on grid to move lamps across studio and raise and lower them. A second operator, 5, stands at a convenient lower level on the loading bay to transfer monopoles from slot to slot and change lamps.

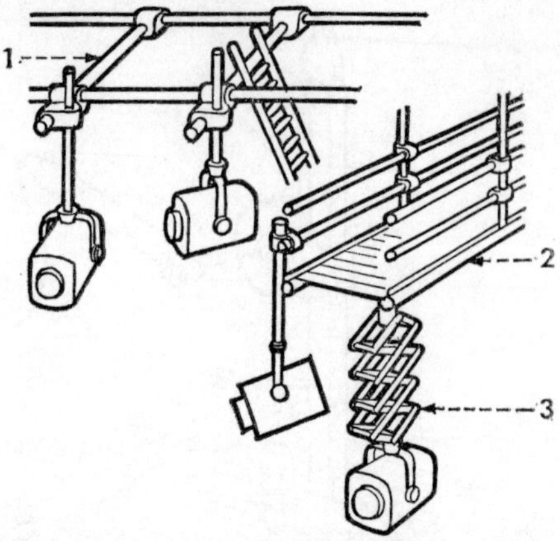

FIXED PIPE GRIDS. Arrangement common in small studios; ladder gives access to semi-permanent grid for lamp adjustment. Catwalk, 2, with pantograph lamp holder, 3, is more convenient, but needs full walking height above highest set.

These are the main variants for all but the small studios where simple fixed pipe grids are possible. Even industrial rolling tracks can be used, with adjustable suspensions on the lazy tongs principle. These are known as pantographs, and the tracks have built-in busbars to feed the outlets at the end of the pantographs via rolling contacts.

In the USA the larger studios at first tended to use the theatre practice of long counter-weighted bars, and then developed towards point suspension, but in a country of this size practice is not uniform. Europe has largely followed British practice, even in some instances combining all the variants in one studio.

LIGHTING CONTROL

In the general planning of studio grids and outlets, a central lighting control is essentially the next item. It is both an instrument of dramatic and artistic expression and a medium for monitoring picture quality. More often than not, the lighting supervisor works the control directly without the intervention of an electrician. The camera channels are stabilized and all adjustment for picture quality is made by the vision controller with remote iris control, and the lighting supervisor with his dimmer controls. This is the hands-off technique.

ONE-MAN OPERATION. It is now usual to aim at, but not always to achieve, a lighting control station for television which is capable of being operated by one man. Ideally this means that all controls should be within arm's reach of an operator seated with a good view of the television monitors. In smaller productions, one operator often handles all the lighting adjustments. Even in larger productions, he needs to do so when the lighting supervisor is on the floor of the studio.

SWITCHING TECHNIQUES. The main task is the centralized switching of all channels, and the control of dimmers. Most studios in Britain have a dimmer for every channel, so that the picture, as seen on the monitor, can be corrected for lighting balance without the delay of repatching the dimmers. This also permits all hanging spotlights to be of one size and yet to give reduced light when necessary. Dimmers are also used for stage-type lighting changes, particularly in variety or musical shows. Even in colour television, with its need for a constant colour temperature, dimming by up to 20 per cent of full voltage is practised here and in the United States.

Switching is principally concerned with the various scenes around the studio. If the lighting can be eliminated in scenes outside the range of the camera, less studio ventilation and mains capacity need be allowed. At a particular moment, the operator may be concerned with one lighting channel or a group of several channels, or the lighting of an entire scene area, or perhaps only with a particular effect on one area. To rely on dimmer levers for this selection is wasteful, nor can a dimmer lever be as quickly operated as an on-off switch. For simple control panels, this selection can be provided for by a three-position switch to form three groups. Thus the two preset levers to each lighting channel are mounted one over the other, and above these is a three-position switch. Each of these switches in its top position causes the presets to be mastered from the top pair of preset faders, in the bottom position from the bottom pair; and in the centre position from the centre pair. Thus presets can be reserved for lighting changes actually involving changes of intensity in respect of a number of dimmer channels and not wasted in mere switching of groups in or out.

The number of lighting changes in television varies greatly. Light entertainment requires many such changes. Drama needs very few except in the kind of mystery drama in which the characters are perpetually going in and out of rooms switching the lights on and off. All these are quick switching or crossfade cues, i.e., roughly half the channels are likely to be cut or faded out. As regards the balancing of levels for individual dimmer channels, experience seems to show that two presets (at the most three) are enough, provided there is a method of group selection. With all-electric dimmers, this may lead to the provision of more than two or three levers per dimmer in order to take care of the combined effect of the several scenes or areas of a scene.

MEMORY RELAY SYSTEMS. One feature common to all but the smallest controls is the memory action. This, without the use of supplementary selectors, enables channels to be selected for control and memorized for immediate recall as a group at the touch of a button. In multi-set productions, this means that groups of lanterns for each set can be invoked as required for the cameras, while for larger scale spectacular productions theatrical type changes can be brought easily under control. As many as 14, 20 or even 40 such groups are commonly used.

A popular form of lighting control for television used until recently was based on electro-mechanical dimmer banks with a common variable-speed motor drive to which each channel dimmer was connected by a servo using reversing clutches and a polarized relay. In these systems the basic principle of operation is to select a channel or channels and then, by master controls, apply movement to them. The inherent inertia of the system (i.e., no signal is needed to remain as last set) means that selection has to take place only for change and never to maintain the status quo. Selection is generally by organ-type stop keys or by reversing luminous pushes for each channel.

A memory relay system derived from organ combination piston practice was developed for stage lighting controls in 1934 and this, too, was used in television. In this system, stopkeys or pushes for the channels required are selected to form a group and then captured on to one of the memory pushes by use of a master presetter. At any time, that group can be recalled by using the appropriate memory push until a different selection is deliberately memorized. The memory pushes can be used in any order and can either add their contents to what has already been selected or replace it.

Allied to this channel selection, the control has two dimmer levers per channel which, working with the servo positioner circuit, enable precise dimmer positions to be selected at the control desk and achieved at the dimmer bank. The two presets thus offered have been found to be sufficient as, due to the select-for-change principle, they represent two positions ahead of what is in use.

Master controls offer the ability to move selected channels up or down to positions set on preset I or on preset II or to switch them on or off. Speed of movement is in the hands of the operator.

Servo-operated dimmers employ resistances or auto transformers. They give good flexibility of control and so have remained in use longer than might have been expected.

ALL-ELECTRIC DIMMERS. A more attractive dimmer in an electric circuit is a static device without mechanical moving parts. These are known as all-electric, and use saturable reactors, thyratrons and thyristors. Until the early 1960s, the simple saturable reactor was the cheapest form of remote control available. Later a transistorized amplifier with some degree of feedback was added to the control line and presetting was thereby made possible. Some improvement in load variation of the order of 4 to 1 was also effected, but to do this properly required the complication of magnetic amplifier forms of reactor. These, though used extensively in Europe and America, have not so far been adopted in Britain because the servo-operated transformer described is preferred.

The development of the thyristor pointed the way to a lower cost, lightweight, compact variable load unit which could act as a dimmer by passing the a.c. waveform wholly or in part, thus avoiding wasteful power loss. This method of dimming had an early run using thyratron valves but tended to be wasteful owing to filament heater losses, and unreliable because of ventilation sensitivity. But the modern thyristor also had disadvantages. The chopping of the waveform to get the dimming effect can lead to audible noise from certain lamp filaments and to the propagation of electrical interference. In consequence, a choke has to be added to delay the rise time of the wavefront. This choke forms an integral part of the dimmer module which still remains very compact. Modules are commonly collected together in racks of twenty.

For simple control, it is enough to provide duplicate or triplicate sets of dimmer levers with master faders on a panel so that two or three changes can be preset. Thus a three-preset control with variable load thyristors became cheaper than the earlier fixed-load saturable reactor with only one set of control levers. For more advanced control, it is necessary to have a direct equivalent of the memory group servo controls but in all-electric terms.

The first step was to do away with the stopkey or luminous push as a channel selector and integrate the functions by making each dimmer scale a rocking switch to perform this function. The rocking scale is internally illuminated in either white or red. When a dimmer channel lever is to be selected, its scale is touched and a reverser relay lights it in red. When the scale is touched again, the red light is extinguished. When selected, the particular channel or channels are connected to the red master dimmer of the preset concerned. If this master is raised, slowly or fast as the cue demands, the lights controlled by any selected channels are brought in proportionately until the master is full. The intensity levels are then transferred (automatically) to the corresponding white master dimmer, provided it is, as is normal, at full when the lever scales change to white. The red master is now free either to be returned to zero to collect further channels to bring in or to have selected channels transferred by touching the scales of some of the channels which are white but should now be taken out on the red master. The act of transfer to the white master can be imagined as parking of the channels concerned. It is true that parked channels are also subject to their master dimmer, which could be used for a cue, and are always subject to their individual control levers, but such channels are completely passive in respect of the memory action.

There are normally two presets, i.e., duplicate dimmer levers, and each has its own red and white master faders, and its own memory pushes. Roughly speaking, white display indicates lighting in use but not concerned with the present change, while red display indicates lighting in an active state of change. This distinction is very important as the operator has to concern himself mainly at the moment of change with levers displaying red. Those which are not illuminated at all can be ignored.

These systems are now relegated to the smaller studios, since full memory of channel-grouping and of the recorded levels of each channel at any moment (i.e, any cue) is available. The form of electronic memory may alter as technology improves, but in general the capability must be at least 250 channels, each recorded at any one of 32 discrete levels for 250 cues or "states of the lighting". Both recording and recalling are instant, i.e., with numerical references; but the playing of the cues in any order chosen is still the job of the operator, and at any time he can modify what he has recorded either before he brings it in or when it has arrived. Multiple pre-sets represented as channel levers are no longer required.

LIGHTING PRACTICE

IMAGE ORTHICON TUBES. General diffuse lighting from the direction of the camera gives a basic lighting level to suit the sensitivity of the camera tube. This is generally speaking 30–40 fc incident and obtained from scoops. These are simple spinnings with 1 kW silicon-sprayed lamps and are usually employed in pairs. An alternative soft light is the ten-light, rectangular in form and housing ten 200 watt reflector lamps with a diffuser and louvred front.

The picture resulting from this lighting would be flat and uninteresting, so modelling lighting must be added. For this, the general purpose lantern is the 2 kW spotlight with a 10 in. diameter Fresnel step lens plate giving a beam angle variable from 15° to 55° and fitted with adjustable barndoors to curtail unwanted spill.

The lantern is controlled for pan, tilt and focus from the floor by means of a pole with a bayonet or hook type of connection to operate the three facilities. The barndoors can be similarly adjusted.

For the modelling light there are three basic uses. First key lighting, which mimics the apparent source of light and gives a three-dimensional quality to the performer. Secondly, there is back lighting from behind the performer to lift him away from the background. Levels of intensity for key or back lighting are of the order of 60–80 fc incident.

Then there is set-lighting to give form to the scenery. This is to a lower level but must make the background look credible. With continual change of camera shooting angles, what is key for one shot may be back for another, and so on. This is one of the reasons for the importance of lighting control, as levels may have to be adjusted shot for shot as the function of the lantern changes. It is here again that the importance of detailed planning by the lighting supervisor is established.

Lastly there is the lighting of cycloramas or sky-cloths, which is usually effected by scoops or similar soft-light sources using a tubular lamp horizontally in a curved reflector. On to these sky-cloths are also projected different optical effects for variety of interest, chief among which is the ubiquitous moving cloud projection. Formal patterns of light are also required and are obtained from a number of profile spotlights which have an optical system designed to project the shape of an adjustable gate in the lantern which is then focused by the lens. Cut-out patterns can also be inserted here for further variation. Or again, a very old principle may be used – the Linnebach effect. Optically, this effect needs a compact source lamp in a housing with a blackened interior, no reflector or lens, but preferably some adjustable shutters on the front. These shutters must be capable of being angled as well as being pushed in and out, as their purpose is to confine the beam to the slide by means of which shadowgraph and silhouette effects can be produced.

VIDICON TUBES. For reasons of capital expense, vidicon cameras are often used in closed-circuit television for teaching and similar purposes. Vidicon cameras are quite sensitive but they have, at the lower levels of lighting, a pronounced lag or smear when performers or cameras move. This can be minimized only by increased lighting levels of not less than 200 fc incident. This level in the small studios likely to be used can be very uncomfortable. Recent study has suggested that for the smaller jobs in this field the soft-light should come from fluorescent tubes in the ceiling to give some 60 fc incident. These can be used for rehearsals, thus reducing the discomfort, and the modelling light can then be added for run-through and production. Such modelling light, assuming a ceiling or grid of some 12–15 ft., can be produced by a minimum of three 1 kW Fresnel spots with another in reserve for each area in use. Expansion of the production with more performers, set models, or several areas needs, of course, more lanterns.

SIMPLER TECHNIQUES. Emphasis in television lighting techniques has so far been on advance planning before arriving in the studio, but recently there has been a move to reduce this and certainly the amount of rehanging has been lessened by employing double-ended lighting units of 2 kW known as twisters, which function as either a softlight or a Fresnel spot. This principle has now been extended to a dual-purpose, dual-source lantern which can be used as softlight or Fresnel spot and in which the source can be either 2·5 kW or 5 kW at will. In the softlight end the sources are 4 × 1,250 W tungsten halogen lamps, while in the spot there are twin filaments each of 2·5 kW in the same lamp. Another idea worth consideration in simple studio setups is a more or less fixed lighting rig into the beam patterns of which the scenery and characters are moved rather than the lighting moved to them.

Since the lighting is in fact suspended aloft to keep the floor clear it may well be the harder item to move and adjust. This technique is also logical for interview or current affairs studios. **F.P.B.**

See: *Cameraman* (TV); *Lighting equipment* (FILM); *Power supplies; Studio complex* (TV).

Literature: ***Evolution in Tungsten Lamps for Television and Film Lighting***, by C. N. Clark and T. F. Neubecker (*SMPTE Jnl*, April 1967).

❑**Lighting Supervisor.** Controls and adjusts the lighting in a television studio. He assists the designers in drawing up the lighting plan for a production (i.e., a plan showing the type and position of the lamps) and oversees their setting by the studio electricians. He sets the brightness to an approximately correct value using a lightmeter and his monitor. The final adjustments of brightness are done during rehearsal. The modern trend is to set up television channels with a test signal and then alter the lighting so that the camera channel is producing a technically correct picture by the adjustment of light levels alone. Lighting supervisors must have a good knowledge of camera techniques and a general overall understanding of the technical operation and limitations of television apparatus. The responsibility is now largely that of the lighting supervisor to produce a satisfactory picture with the camera control operators doing much less than they used to. In some studios the lighting supervisor may control his dimmers and switches through an intermediary, the console operator.

See: *Studio complex* (TV).

⊕**Light Lock.** Opening to a dark room for photographic work protected by double doors, curtains or passages with baffles so that light is excluded while allowing free passage for staff and materials. Walls and

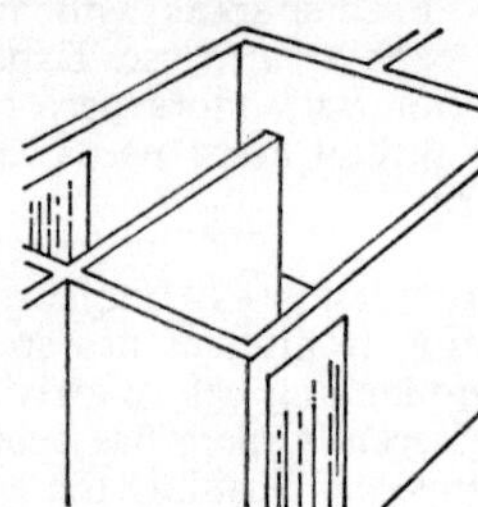

baffle partitions are matte black to prevent light being scattered into the work room.

□**Light-Sensitive Mosaic.** Sensitive surface on which light falls in a TV camera tube such as a vidicon or iconoscope. Consists of millions of caesium-coated silver globules coated on a sheet of mica, and thus insulated from one another. When light falls on this mosaic, electrons are emitted by the caesium, leaving a positively charged electrical image which can be detected by scanning with an electron beam to form a sequential TV signal.

Efforts are now being made to dispense with vacuum camera tubes of the type mentioned, and all their high-voltage circuitry, by means of another kind of light-sensitive mosaic: a matrix of minute photo-sensitive transistors which are read out by electronic switching. Such matrices are as yet in a primitive state, and cannot give the image resolution required by television.

See: *Camera* (TV).

□**Light-Transfer Characteristics.** Relationship between light input and voltage output.

See: *Transfer characteristics.*

⊕**Light Trap.** Mechanical device for excluding light while allowing the movement of film out of and into a magazine of a camera or printer. The film passes over rollers and felt- or velvet-covered lips which must be kept free from particles of dirt and emulsion to avoid abrasions and scratches.

Light traps are also used in the form of baffles to prevent light entering a photographic dark room at ventilation holes and fans.

⊕□**Light Units.** The following terms may be encountered where the quality or intensity of light has to be measured.

COLOUR. The units for the measurement of the wavelength of light, which determines its colour, are:

Angstrom unit, A, one-ten-millionth of a millimetre.

Millimicron, mμ, one-millionth of a millimetre.

Nanometre, nm, the modern term for the millimicron.

The range of the visible spectrum may therefore be stated as 4,000 to 7,000 A, or 400 to 700 mμ, or 400 to 700 nm, the last now being the preferred international term.

The colour of a near-white source of light can be specified as its colour temperature, measured in degrees Kelvin (°K).

POWER OF A LIGHT SOURCE. The power of a light source may be stated as its candle-power, but the international unit is now the candela (cd), which is almost identical with the standard candle.

The light emitted by a source (luminous flux) of power one candela which falls on one square unit of surface at one unit of distance from the source is termed the lumen (lm). A source of one candela emits in all directions a total of 4π lumens, or 12·56 lumens. The efficiency of an electric light source may be expressed as its output in lumens per watt of power supplied.

ILLUMINATION. The illumination of a surface may be regarded as the incident light coming from a source of given power at a given distance: the illumination from a source of one candela at a distance of one foot is known as a foot-candle (fc) or one lumen per square foot, and both terms are employed in the measurement of illumination for photography. The metric unit is one lumen per square metre (or metre-candle), termed the lux.

1 foot-candle = 1 lumen per square foot = 10·76 lux.

A unit for very high levels of illumination is the phot, which is one lumen per square centimetre, or 10,000 lux. The milliphot is very close to the foot-candle: 1 foot-candle = 1·076 milliphot.

BRIGHTNESS. The brightness or luminance of a source represents its light output per unit area: the brightness or luminance of a surface similarly represents the light intensity which it reflects per unit area. There are several units, of which the preferred international metric one is the nit, representing a brightness of one candela per square metre. There is also the foot-lambert, which is the luminance of a perfectly diffusing surface emitting or reflecting one lumen per square foot. Both these units are employed in measuring the brightness of cinema screens, and may be converted as:

1 nit = 0·292 foot-lamberts.

1 foot-lambert = 3·42 nits.

For brighter sources, one candela per square centimetre is termed one stilb (= 10,000 nits), and an ideal diffuser emitting one lumen per square metre has a luminance of one apostilb.

Since an ideal diffuser of one candela per unit area emits π lumens per unit area,

1 candela per square foot = 3·14 foot-lamberts, and

1 candela per square metre = 1 nit = 3·14 apostilbs.

A directional reflecting surface, such as a beaded cinema screen, may show a greater luminance from some directions than a perfectly diffusing surface: the ratio of actual luminance to that of an equivalently illuminated perfect diffuser is termed the luminance factor. However, a gain or luminance factor much greater than unity in one direction is always offset by factors less than unity when the screen is viewed from other directions. L. B. H.

See: *Light; Optics; Screen brightness.*

Light Valve. Mechanism used to control the intensity of light passing in a sound recording or film printing system. Consists of a narrow slit or gate of variable width which modulates a beam of light directed on to a moving photographic film. If the light valve is mounted at right-angles to the length of the film, a variable-density sound record results; if mounted parallel to the length of the film and suitably masked, a variable-area sound record is produced.

In the optical systems of film printers, the light valve may take the form of a pair or series of louvre shutters which can be opened and closed very rapidly in accurate increments to vary the area of the opening through which the beam of printing light passes.

See: *Sound.*

Lily. Board on the surface of which is a pattern of agreed colour shades and references white areas. These standard shades are photographed in colour on the beginning or end of a reel and provide a colour standard for the laboratories to work to.

Limiter (Sound). Device for preventing signals exceeding a preset limit. It is often vitally important that programme peaks should not exceed some particular value, especially when the programme is feeding a transmitter or recorder. It is usually impossible for a sound mixer to avoid having some programme peaks over the peak programme level. But to clip or square off the excess voltage excursions may well result in audible distortion. To cope with this situation, a limiting amplifier is generally used.

A limiter, as it is often called, behaves like a compressor with zero slope following the knee, this knee being positioned at peak programme level. Signals below peak programme pass through the limiter with no modification. Any signal exceeding this level causes the control circuits to insert attenuation with a very short attack time.

See: *Sound recording.*

□**Limiter (Vision).** Circuit which prevents the amplitude of a television signal rising above the level defined as peak-white. Such an excursion cannot be displayed, being whiter-than-white, and can overload amplifiers.

See: *Transfer characteristics.*

⊕□**Limiting Resolution.** Highest resolution which a system or device can produce under the limitations imposed by physical characteristics of lenses, emulsion grain, phosphor grain, photoactive surfaces, etc.

See: *Image formation.*

⊕□**Linearity.** Relationship between two quantities which can be graphically represented as a straight line.

□**Linearity Control.** Owing to inductance and capacity effects, a scanning coil is seldom truly linear in its deflection, and it is usual to modify the shape of the applied waveform to correct for this.

□**Line Blanking.** Suppression of signal in camera or receiver for a period sufficient to allow flyback of the scanning beam to the start of the next line. Amplifiers in the transmission chain are also blanked to prevent modulation of the transmitter during flyback periods. The line-blanking period in the American 525-line system is 10·8 μsec., and in the CCIR 625-line system 12 μsec. Within the line-blanking period is the line-sync pulse with its front and back porches.

□**Line Drive.** Horizontal timing pulse used to trigger circuits in the picture-generating equipment. A common position for the leading edge of this pulse is coincident with the leading edge of blanking, 1·5 μsec. before the leading edge of sync. The exception to this is when the leading edge of field-blanking occurs at the end of the odd field, in a half-line position.

See: *Synchronizing pulse generators.*

□**Line Drive Pulse.** Signal at line frequency distributed from the station pulse generator to allow triggering of the line scanning circuits at a time suitable to allow insertion of blanking, without loss of picture information.

□**Line Drive Signal.** Signal used to establish line sync in studio systems, for example, in non-composite working.

□**Line Period.** In the standard TV waveform, the duration of one line of information, comprising a line of picture signal and a line-blanking period, which itself contains the line-sync pulse with its front and back porches.

□**Line Phasing.** Adjustment of the line timing in a device such that two line signals occur simultaneously.

□**Line Sawtooth.** Waveform which rises at a constant rate and then falls rapidly, at the frequency of line deflection. The simple signal is used for test purposes for checking linearity.

See: *Television principles.*

□**Line Scanning.** Action of deflecting the electron beam in a horizontal direction to produce a line of picture in a TV camera, receiver, telecine, etc.

See: *Scanning.*

□**Line Strobe.** Special type of waveform monitor or oscilloscope for examining television signals which is capable of displaying the waveform of any selected single line or group of lines constituting the raster.

With the more common types of oscilloscope, the line display comprises a large number of line waveforms superimposed on one another. In the line strobe monitor, the time base is accurately triggered to scan at or near line speed once only per frame at a position which can be precisely adjusted to occur at the commencement of any given line.

The line strobe monitor is especially useful in examining the field-synchronizing waveforms or test signal waveforms sometimes inserted during the post-sync field-blanking period.

See: *Picture monitors; Synchronizing pulse generators.*

□**Line Structure.** Appearance of the picture or raster on a cathode ray tube caused by the method of building up the complete picture by a number of horizontal lines.

□**Line Suppression Period.** Period at black level between successive lines which is used as a reference period for inserting synchronizing signals, and also as a black level reference for line-by-line clamps.

See: *Television principles.*

□**Line Tilt.** A television picture distortion. Comparable with field tilt, line tilt is a gradual increase or decrease in the d.c. component over the course of the line waveform, owing to a.c. coupling or the addition of hum or other spurious low-frequency signals. Usually the amplitude is less than that caused by frame tilt, as there is less time for the line waveform to take up the new potential. The effect can be reduced by passing the signal through a keyed, or line-by-line clamping circuit, thus restoring the beginning of each line to the same potential with respect to earth. The visual effect is that of a gradual increase or decrease in brightness from left to right of the viewed television image. Line tilt is sometimes introduced deliberately in the form of line shading to counteract spurious signals from a camera tube, and a desirable specification is less than 2 per cent line tilt in any one line of the output signal.

See: *Receiver.*

□**Line Time Base.** Circuits generating the sawtooth waveform required to deflect the electron beam horizontally in a camera or receiver tube.

See: *Camera* (TV); *Receiver.*

□**Line-Up Time.** The time required for warming up and adjustment in a television studio; usually scheduled. For optimum performance from the complex equipment used in television studios, the usual practice has been to set aside a period of time, usually half an hour to one hour, just prior to transmission, called line-up time.

During this period, the equipment is adjusted to give its peak performance, i.e., lined up. Any faults that may have been present during the rehearsals are cleared and every effort is made to ensure that the technical performance of all the equipment is as high as possible. With the increased reliability and stability of modern equipment, there has been a tendency for line-up times to become shorter and in some cases to be dispensed with entirely.

See: *Studio complex.*

☐**Linkman.** Announcer who links various incoming broadcasts with suitable covering material.

⊕**Linnebach Effect.** A system for lighting by ☐the reflected light from coloured surfaces, named after Adolf Linnebach, a Dresden stage designer. He also devised a simple scene projection method using a large transparency and a point light source.

⊕**Lippmann Process.** Two photographic processes invented by the French physicist and Nobel Prize winner, Gabriel Lippmann (1845–1921), either being generally referred to as the Lippmann process.

The first, a colour process for still photography, is distinguished by being neither additive nor subtractive, in which a plate is exposed with its transparent panchromatic emulsion in contact with a mirror. Light passing through the plate is reflected back out of phase, setting up an interference pattern in the emulsion layer which becomes developable only where incident and reflected wave peaks coincide. Thus the distribution of the image is dependent on the wavelength of the exposing light at every point.

The developed negative is again backed by a mirror and is viewed by reflected light. At every picture point, light of all wavelengths, except that which gave rise to the image at the point, is absorbed. Thus the picture is seen in full and correct colour because the point pattern of wavelengths which can pass through the image and after reflection return to the viewer's eye is identical with that which went to form the original image.

In spite of its elegance, this process proved no more than a scientific curiosity because of difficulties of viewing and reproduction, and of producing emulsions both transparent and sufficiently sensitive. Many aspects of the process, however, foreshadow the modern hologram.

The second is a stereoscopic process of the integral type, a name invented by Lippmann to distinguish it from processes in which the separation of the binocular images is carried out at the viewer's eyes by anaglyphic or polarizing devices. Lippmann proposed that a celluloid sheet be embossed on both sides with dome-shaped projections in such a way as to form a large number of minute lenses, each imaging the object on the emulsion which coated the dome on the opposite side.

After development and reversal, the image could be viewed binocularly without optical aids. Since the taking conditions were exactly reproduced in the viewing, a stereoscopic effect would result. Moreover, some true movement parallax would be observed when the viewer shifted his head from side to side, as in a modern hologram.

The Lippmann stereoscopic process was the forerunner of many lenticular processes which, because of formidable optical and commercial problems, have had little commercial success. **R.J.S.**

See: *Stereocinematography.*

⊕**Lip Sync.** Precise correspondence between lip movements and speech sounds. Requires exact synchronization from frame to frame between picture and sound which, when the two are recorded on separate strips of film and tape running through different machines, raises engineering problems not encountered in the less critical field of sound effects.

The term is also used to designate simultaneous dialogue recording of voice and vision, in contrast to wild recording.

See: *Pulse sound; Sound recording.*

⊕**Liquid Gate.** In some forms of motion picture printing or projection, the film may be coated or enclosed with a liquid of suitable refractive index in the gate at the time of exposure in order to reduce the optical effects of scratches and abrasions on its surface.

See: *Laboratory organization; Printers.*

⊕**Liquid Head.** Type of tripod head in which ☐the bearing surface consists of a thin film of liquid, usually containing silicones, to give a degree of smoothness approaching that obtainable with a gyro head. Since the liquid head contains no gears, there cannot be any lost motion, the bane of the gyro head. Liquid heads are also much lighter.

See: *Camera* (FILM).

⊕**Living Screen.** Form of stage entertainment in which live actors perform in combination with projected images, both still and motion picture.

☐**Lobe.** Loops representing the points of equal strength radiation produced when the radiation pattern of an aerial is plotted graphically.

See: *Transmitters and aerials.*

⊕**Location.** Place, other than the studio or ☐studio lot of a film or TV production organization, where one of its units is shooting pictures.

☐**Locking.** Synchronization of a repeating waveform with a train of pulses or signals, applied to all signal sources to make them instantly interchangeable for master control purposes. A considerable exactness of synchronization is implied by the word lock. Where synchronization averages a correct figure, but the phase may be inexact, the expression spongy lock is used. Occasionally the word lock is qualified as hard lock, i.e., a very exact degree of synchronization.

See: *Picture locking techniques.*

⊕**Log Sheet.** Form, usually kept by a film camera assistant, which records all details of each shot and take, together with necessary instructions to the film laboratory. A copy of the log sheet is sent to the film cutting room, where it forms a master record of all scenes taken in the studio or on location. Also known as report sheet.

See: *Cameraman* (FILM); *Laboratory organization.*

⊕**Loop.** Slack section of film designed to provide necessary play when film is being fed from a continuously moving sprocket to an intermittently moving sprocket, thus avoiding the tearing which would otherwise take place. Originally called a Latham loop, after its inventor.

⊕**Loop (Sound and Picture).** Sound and picture loops are made from lengths of film joined head-to-tail to form an endless band, thus allowing continuous running through dubbers and projectors respectively. Sound loops are made up during editing and are used in dubbing to carry continuous sounds such as wind noise, waves breaking, bird song and the roar of traffic. Picture and sound loops together are employed in the post-synchronization process. Virgin loops are loops of magnetic track on which post-synchronized dialogue is recorded, the previous recording being automatically erased. Thus the latest attempt to match dialogue can always be held and reproduced over the monitor loudspeaker in synchronism with the picture. If satisfactory, it is transferred out to another track before erasure and repetition, if still further takes are needed.

See: *Sound mixer; Sound recording* (FILM).

⊕**Loop Elevator.** Series of fixed and movable rollers used for printing large numbers of release copies of a motion picture film. The negative is in the form of a continuous loop with the head and tail ends joined so that it can be run through a printer as many times as required without the loss of time caused by threading the film through the mechanism for each print. The elevator, or reservoir, can be adjusted to accommodate the length of the reel in use. Loop elevators may have capacities from 50 ft. to 200 ft. for advertising shorts and trailers but up to 2,000 ft. for reels of feature films.

See: *Laboratory organization; Printer.*

⊕**Lot.** Open area on which can be built sets for exterior scenes such as a street. The area is within the confines of the film company's site.

⊕**Loudspeaker.** Electro-acoustic transducer for ☐converting electrical impulses into sound waves. Most loudspeaker installations are designed to reproduce the whole audio spectrum, which is often broken down into two or three sections, each of which is assigned a loudspeaker with special characteristics and appropriate cross-over networks.

⊕**Low Key.** Style of tonal rendering of a scene ☐marked by predominance of dark tones with rich shadow detail, often designed to produce a dramatic or mysterious effect. Low-key lighting limits the important parts of the subject to a narrow range of dark tones while at the same time providing good modelling.

See: *Cameraman.*

⊕**Low Pass Filter.** Electrical circuit which ☐passes only frequencies below a certain frequency known as the cut-off frequency of the filter.

⊕**Low-Shrink Base.** Modern types of safety base showing comparatively small shrinkage. Nitrate film base and the earlier forms of acetate base showed marked shrinkage both during processing and in storage, which had to be allowed for in film mechanism design.

See: *Film dimensions and physical characteristics.*

☐**Low-Velocity Scanning Beam.** An electron beam decelerated to a very slow speed before scanning the target, employed in some types of camera tube, such as the image orthicon.

See: *Camera* (TV).

⊕**Lubrication.** Application of waxy materials to the perforation area of positive film prints to provide smooth running in the gate of a projector and avoid abrasion of the emulsion surface.

See: *Laboratory organization; Release print examination and maintenance.*

⊕**Lumen.** Unit of luminous flux. Quantity of ☐light emitted per second in unit solid angle, by a uniform point source of light of one candela intensity.

See: *Light units.*

⊕**Lumière, Louis, 1864–1948; and Auguste, 1862–1954.** French inventors whose connection with cinematography began in 1870 when their father set up a photographic business in Lyon. By 1882 they were manufacturing dryplates using a process devised by Louis, and by 1893 a flourishing business had developed.

The appearance of Edison's Kinetoscope in Paris in 1894 inspired Louis and Auguste Lumière to develop their own apparatus. Their chief mechanic, Moisson, constructed a prototype camera/projector, patented in February 1895, and demonstrated during a lecture at the Société d'Encouragement pour l'Industrie Nationale in March 1895. The film projected showed workers leaving the Lumière factory. At this meeting the Lumières met Jules Carpentier, a scientific instrument maker, who offered to manufacture the Cinématographe. More films were shown, and taken, at the Congress of Photographic Societies at Lyon in July 1895. On 28th December 1895 the first public demonstration, with payment for admission, of projected motion pictures was given at the Grand Café, Boulevard des Capuchines, Paris. In the next few months the Lumière apparatus was in use all over Europe.

Louis Lumière continued throughout his life to develop new apparatus and processes. His experiments in colour photography led to the introduction of the Lumière Autochrome plate, using a mosaic-type additive process, which enjoyed great commercial success from 1903 as a still process; it was never satisfactorily used for motion pictures. At the age of 72, he developed a practical stereoscopic film process.

⊕□**Luminaire.** US term for studio lamp or lighting unit.

See: *Lighting.*

⊕□**Luminance.** Luminous intensity of a surface in a given direction. Measure of brightness of the signal as opposed to hue.

□**Luminance Channel.** Part of the frequency spectrum of a colour transmission containing luminance information.

See: *Colour principles* (TV); *Colour systems.*

□**Luminance Signal.** Coloured lights of equal radiant energy may appear to the eye to be of unequal brightness because the sensitivity of the eye varies with the wavelength of the radiation observed. This variation of response has been measured in a large number of people, and an average response curve obtained, which has been adopted internationally as a standard luminance response. This curve indicates that the eye is most sensitive in the green region of the spectrum, and has a falling response towards the red and blue ends. When a coloured object is reproduced in monochrome, the grey scale response of the reproduced image corresponds to the luminance variation which the eye would perceive if the image was in colour.

In a monochrome system, therefore, the camera has a colour response equal to that of the standard luminance curve. In a colour television system, parts of the red, green and blue separation signals are added together in proportions equal to the relative luminance of the primaries. The resultant signal is called the luminance signal and is given the term E_Y. This signal is used to control the brightness of the image on both monochrome and colour receivers reproducing the signal.

See: *Colour principles* (TV); *Colour systems.*

⊕□**Luminous Flux.** Rate of flow of light, for which the unit is the lumen. Luminous flux refers to energy in the visible region of the spectrum in contrast to radiant flux which applies to energy in any part of the spectrum.

See: *Light; Light units.*

⊕□**Luminous Intensity.** Quantity of light emitted per second by a point source in a given direction into unit solid angle. The unit is the candela.

See: *Light; Light units.*

⊕□**Lux.** Metric unit of illumination. Intensity of illumination at the surface of a sphere of one metre radius, from a light source of one candela placed at the centre of the sphere. Equivalent to a luminous flux of one lumen per square metre.

See: *Light; Light units.*

□**Machine Leader.** Specially tough blank film, as a rule without emulsion coating, used to thread the film path on a continuous developing machine. When the machine is first started, the head of the film to be developed is joined to the end of the leader, which then pulls it through the machine. Similarly, when developing ceases for the day or shift, leader is attached to the tail of the last film. The machine is stopped when that film is processed and is then, full of leader, ready to start again.

See: *Processing.*

⊕**Macrocinematography.** Motion picture recording of small objects, but not so small as to require the use of a microscope.

See: *Cinemacrography; Cinemicrography.*

⊕**Magazine, Film.** Film containers forming part of picture cameras, sound cameras and projectors. Camera magazines are light-tight, the film entering and leaving them through light traps.

See: *Camera* (FILM); *Projector.*

⊕**Magnafilm.** Form of wide-screen presentation about 1930 which used a special film 56mm in width.

See: *History* (FILM).

⊕**Magnascope.** Early form of wide-screen cinema presentation using a projection lens of variable magnification.

See: *Wide-screen processes.*

□**Magnetic Deflection.** Deflection of an electron beam by means of a current flowing through appropriately placed coils producing a magnetic field.

□**Magnetic Focusing.** Constriction of an electron beam to a focus by a suitably placed and adjusted magnetic field.

⊕□**Magnetic Recording.** Recording by effecting magnetic variations in a ferromagnetic medium, usually a coated tape or film. As a rule, separate magnetic heads are employed for converting audio frequency electric variations into magnetic variations, reconverting these variations back into audio frequency signals, and erasing the magnetic record from the tape or film.

See: *Magnetic recording materials; Sound recording.*

⊕MAGNETIC RECORDING MATERIALS

□

The magnetic recording media used in the television and film industries are in the form of spools of tape, normally ¼ in. wide for audio recording, up to 2 in. wide for video recording, and in widths of 16mm, 17½mm or 35mm for sprocketed tape (magnetic film) for sound recording in synchronization with photographic film.

In the recording process, the tape passes across record heads so that, as the signal through the head is varied, a magnetic pattern is formed in the tape in conformity with the changes in the recording signal. The information stored in this way must be proof against accidental erasure or loss of strength in storage over long periods, yet allow deliberate erasure when necessary. Passage of the recorded tape over a replay head reproduces the original signal. The form of the tape is such that sections can be cut out and rearranged as required in the course of programme editing.

STRUCTURE AND PROPERTIES

These requirements are met by a tape comprising a layer of magnetic material applied to one side of a plastic carrier known as the base film, the latter providing the necessary strength for handling and editing. The materials used in the preparation of the magnetic layer provide as high a magnetic remanence as possible, i.e., retention of magnetization after removal of the field which has brought the material to saturation. They are also of coercive force high enough to resist erasure by accidental exposure to magnetic fields and to avoid excessive self-demagnetization of the recorded signal, but low enough to allow for deliberate erasure by running the tape over a.c. fields from an erasing head on the recorder. Coercivities in the range 200–350 Oersteds are normally used.

MAGNETIC MATERIAL. In order to obtain good resolution and high outputs which are stable in storage, present-day tapes use magnetic material in the form of finely divided particles of needle shape. These are less than one micron in length and behave as single magnetic domains. Their needle form allows them to be aligned during the processes of tape manufacture in the direction of tape-to-head movement, with consequent improvement in remanence in that direction. These particles are dispersed in as high a concentration as possible in a plastic resin binder which holds the magnetic layer together and secures it to the base film. The desired qualities of the plastic formulation are: good mechanical properties, storage life, lubrication (which minimizes wear of either tape or heads) and avoidance of the unwanted effects of static electricity which might be generated as the tape passes over heads and guides on the recorder.

TAPE SPEED. When the frequencies of the recording signal are high and/or the speed of passing the tape across the heads is low, the wavelength of the magnetic information stored on the tape becomes very short and the output from the tape is reduced. For the audio range, relative head-to-tape speeds of several inches per second (typically 7½ or 15 in./sec. for professional work – 1⅞ or 3¾ in./sec. for most domestic applications) are sufficient to ensure adequate performance over the frequency range required. At the much higher frequencies needed for video purposes, relative head-to-tape speeds some two or three orders of magnitude higher are necessary. This is normally achieved by using rapidly rotating heads which scan across the more slowly moving tape. Tapes for video use are made with their particles oriented approximately along the direction of the scan, i.e., across the tape.

NOISE. The signal-to-noise ratio which can be obtained from the overall record/store/replay system depends in part upon the recorder and its associated electronics, and in part upon the tape itself, which can give rise to noise effects in a number of ways. The statistical fluctuations of net magnetic moment of the particles passing the head in an erased tape produce a so-called basic noise which can be minimized by careful control of the particle size range of the magnetic material. Noise known as modulation noise occurs in the presence of a signal owing to inhomogeneity of the coating, or by asperities (roughnesses) on the tape surface which can vary the spacing of the tape from the heads. In an extreme form, owing to dust particles or other major imperfections, these asperities can cause sharp local losses of signal known as dropouts. Modulation noise can also result from minute irregularities in movement of tape across the head owing to frictional effects in the transport, giving rise to velocity changes and frequency modulation of the signal. Noise is also created by print-

BASE FILMS—RELATIVE PROPERTIES

	Cellulose Acetate	*Cellulose Triacetate*	*Polyvinyl Chloride*	*Polyester*
Strength	Low	Low	Average	High
Surface	Excellent	Excellent	Some indentation	Excellent
Stretch or elongation at break	Low	Low	Medium	High (Standard) Low*
Usable thickness	·0015 in. ·001 in.	·005 in. ·0015 in.	·0015 in. ·001 in. ·0005 in.	·0015 in. ·001 in. ·0005 in. ·0003 in.
Thickness uniformity ...	Excellent	Excellent	Medium	Excellent
Thermal properties	Average	Average	Average	High
Moisture resistance	Poor	Fair	Excellent	Excellent
Used for	Sound Tape Instrumentation Tape	Magnetic Film	Sound Tape	Sound Tape Video Tape Computer Tape Instrumentation Tape Magnetic Film

* In addition to the normal (standard) grades, which are orientated both longitudinally and crosswise, other types preferentially drawn in the longitudinal direction are used and are referred to generally as tensilized. This latter material has lower stretch properties.

through, or echo effect, where a strong signal recorded in one layer of a spool of tape records at lower levels on to an adjacent layer. This happens if there are particles whose coercivity is low enough under the conditions of storage of the tape to accept the recording.

BASE FILMS. Some early sound tapes were made using high quality paper, but these were soon discarded on account of rough surface and dimensional instability, and today tapes are mainly of the following types:

Cellulose Acetate (*CA*), produced from solvent solutions cast on metal bands, dried and then stripped off as clear film.

Cellulose Tri-Acetate (*CTA*), produced by the same method as cellulose acetate but with improved moisture resistance. Same material as non-inflammable cine film.

PROPERTIES OF BASE MATERIAL

Type	*Thickness (in.)*	**Tensile Strength per ¼ in. width (lb.)*	**Yield Point per ¼ in. width (lb.)*	**Elongation at break %*	**Elongation ½ lb. load %*	**Temperature Coefficient per 1° F.*	*Humidity Expansion Coefficient per 1% RH*	*Used for*
CA ...	·0015	5·6	4·5	25	·25	5×10⁻⁵	12×10⁻⁵	SP Sound Tape
CA ...	·001	3·5	3·1	30	·40	5×10⁻⁵	12×10⁻⁵	LP Sound Tape
PVC ...	·0015	10·0	4·75	50	·25	4×10⁻⁵	10×10⁻⁶	SP Sound Tape
PVC ...	·001	7·0	2·75	50	·34	4×10⁻⁵	10×10⁻⁶	LP Sound Tape
PVC ...	·0006	4·5	2·0	40	·40	4×10⁻⁵	10×10⁻⁶	DP Sound Tape
PE Normal...	·0015	10·0	6·0	100	·25	15×10⁻⁶	11×10⁻⁶	SP Sound Tape
PE Normal...	·001	7·0	4·0	100	·34	15×10⁻⁶	11×10⁻⁶	Computer LP Sound Tape Video
PE Normal...	·0005	3·50	2·0	150	·50	15×10⁻⁶	11×10⁻⁶	Computer DP Sound Tape†
PE Tensilized	·001	10·5	—	40	—	15×10⁻⁶	11×10⁻⁶	Computer
PE Tensilized	·0005	5·5	—	40	—	15×10⁻⁶	11×10⁻⁶	DP Sound Tape TP Sound Tape
PE Tensilized	·00033	3·1	—	40	—	15×10⁻⁶	11×10⁻⁶	QP Sound Tape
CTA ...	·0015	5·6	5·3	40	·25	2·8×10⁻⁵	6·4×10⁻⁵	SP Sound Tape Magnetic Film
CTA ...	·005	18·0	17·6	40	·06	2·8×10⁻⁵	6·4×10⁻⁵	Magnetic Film

NOTES: (*a*) SP = Standard Play
LP = Long Play
DP = Double Play
TP = Triple Play
QP = Quadruple Play

* All properties in M/D (machine direction) i.e. longitudinally

† Seldom used by leading tape manufacturers

Polyvinyl Chloride (*PVC*), prepared by hot rolling of the plastic at very high pressures and then stretching in a longitudinal direction.

Polyethylene Terephthalate (*Polyester or PE*), made by extruding the molten material and stretching in both the longitudinal and lateral directions.

Details of these tapes are given in the accompanying tables.

STORAGE

With the exception of tapes made from the cellulose plastics, which are more sensitive to moisture and become brittle with age, most tapes can be stored for very long periods provided they are wound at correct tensions and are kept reasonably clean.

However, most manufacturers recommend that, for long-term storage, temperatures of 60–80°F and a humidity range of 40–60 per cent are a necessary safeguard, as minute dimensional changes, while unimportant on individual lengths of tape, can cause deformation in the mass of tape wound on to reels even under normal tensions.

As print-through, where relevant, is affected by higher temperatures, it is always sensible to follow the maker's recommendations. After several months of storage, print-through can be much reduced if tapes are wound over once before replay, as a rapid reduction of printed signal takes place after separation of the layers.

The possibility of accidental erasure from stray magnetic fields was once considered to be a serious problem, but with normal commonsense precautions this is a factor which has seldom caused trouble. For better security, tapes may be stored in ordinary tin plate cans instead of cardboard cartons.

Storage data is available only for periods of up to twenty years or so, the age of the tape industry, but recordings on magnetic tape stored for such periods have been found satisfactory.

MANUFACTURE

All magnetic tapes, which consist of base film and magnetic oxide coating, are made by the same basic process, as follows:

1. Manufacture of magnetic iron oxide.
2. Manufacture of clear medium or binder.
3. Dispersion of magnetic powder in binder to produce oxide paint, usually known as dope.
4. Cleaning of base film.
5. Application of dope to base film to produce coated film.
6. Polishing of coated material to produce smooth finish.
7. Slitting of coated film to required width.
8. Perforating – magnetic film only.
9. Testing of slit tape.
10. Reeling on to spool ready for sale.
11. Final inspection and packing.

MAGNETIC OXIDE. Although some earlier tapes were made using cubic and rhombohedral shapes, almost all modern tapes contain synthetically produced gamma Fe_2O_3 in its acicular form, that is, needle-shaped crystals approximately 1 micron in length and 0·2 micron in width.

Typically, the starting material in the preparation of magnetic iron oxide is Goethite, a yellow hydrated iron oxide ($\alpha FeO.OH$). Although Goethite occurs in nature, it is always prepared synthetically for magnetic recording. This is done by oxidizing ferrous hydroxide which has been precipitated either partially or completely from a ferrous salt solution using an alkali, usually sodium hydroxide. The shape and size of the Goethite particles can be controlled by the concentration of solution and temperatures at the point of precipitation.

Goethite is dehydrated to Haematite (αFe_2O_3), which is non-magnetic, and this in turn is reduced to black magnetic ferroso-ferric oxide (Fe_3O_4). Usually both these

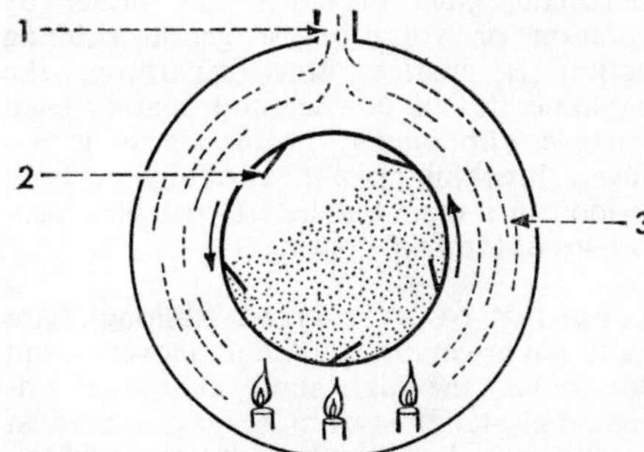

BATCH PROCESSING FURNACE. Processes magnetic iron oxide. 1. Flue. 2. Baffle plates. 3. Heat chamber.

reactions are carried out in one operation using a continuous or rotary batch type furnace at 300–450°C in a reducing atmosphere such as carbon monoxide or hydrogen.

The final gamma iron oxide (γFe_2O_3) is formed by slowly re-oxidizing the ferroso-ferric oxide with oxygen at an elevated temperature, but taking care that the exothermic reaction does not proceed above 400°C with reversion to the non-magnetic α form.

CLEAR MEDIUM OR BINDER. Conventional mixing systems are used to produce the binder. The two major requirements to be met are good adhesion to the base film and resistance to wear when run hundreds of times (for video) and thousands of times (for sound) across recording heads. Two main types in use are thermoplastics, usually based on one of the vinyl groups of resins, and thermo-setting, which may be based on iso-cyanate or phenolic systems. While the latter type tended to be favoured until quite recently, improved resins now available in the earlier group have become increasingly used owing to the greater ease of manufacture, coupled with the elimination of a tendency to become brittle under accelerated ageing conditions. To both types it is necessary to add compounds which assist the wetting of the oxides, as well as lubricants to eliminate excessive wear caused by friction as the tape passes across recording heads.

DISPERSION. While many methods of dispersion are used, they are all similar in principle. The oxide and binder are placed in a vessel containing small spheres of either stainless steel, aluminium oxide, steatite, porcelain, glass or silica and, either by rotation or vibration, a gentle rubbing action is created, thus separating the agglomerates of powder and coating each particle with binder. If this action is too severe, breaking-up of the carefully prepared oxide can impair the electro-magnetic properties of the final coating.

CLEANING OF BASE FILM. Although base materials are manufactured under very clean conditions, the high static charge of uncoated plastic film can cause occasional dust or slitting debris to adhere to the surface, and this, when wound into the roll, becomes stuck fast. It is therefore necessary to re-

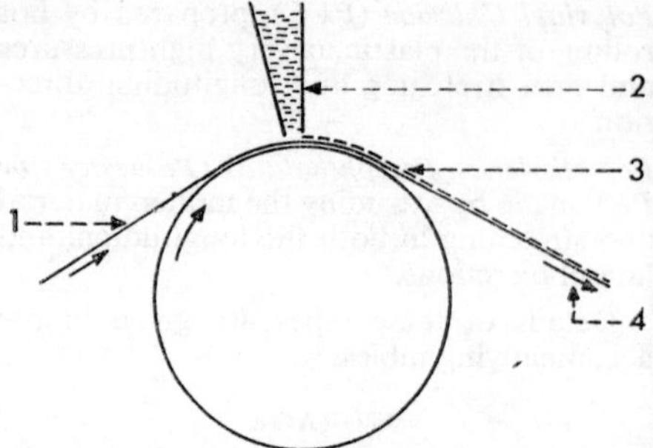

KNIFE COATING. Base, 1, is carried on roller beneath knife edge, 2, where coating, 3, is deposited. Coated tape then passes towards drying oven, 4.

clean film before coating, and this is done by passing it through liquids and cleaning by brushing, or by ultrasonic or other agitation. For information concerning the various types of base materials see table on page 432.

COATING. Factors which govern the choice of method of coating are:

1. Ability to produce extremely smooth surfaces.

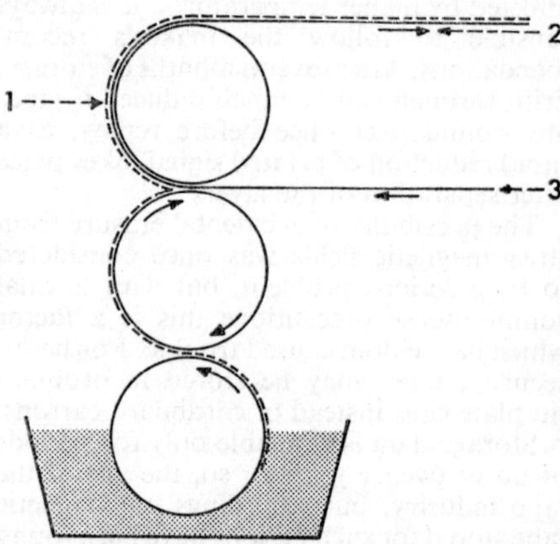

ROLLER COATING. Lowest roller picks up coating from tank, and transfers it to intermediate roller. This coats it on base entering at 3, and coated base, 1, passes towards drying oven, 2.

COATING METHODS

Method	*Advantage*	*Disadvantages*
Knife	Simple to operate. Easier to maintain dope in clean state.	Thickness uniformity more difficult as it is dependent upon accuracy of base film.
Reverse roll	Easy to operate.	Thickness uniformity dependent upon maintaining accuracy of many revolving components.
Gravure	Thickness uniformity good as constant amount of coating must result.	Difficult to change mean thickness without changing coating roll. Harder to keep dope free from impurities.

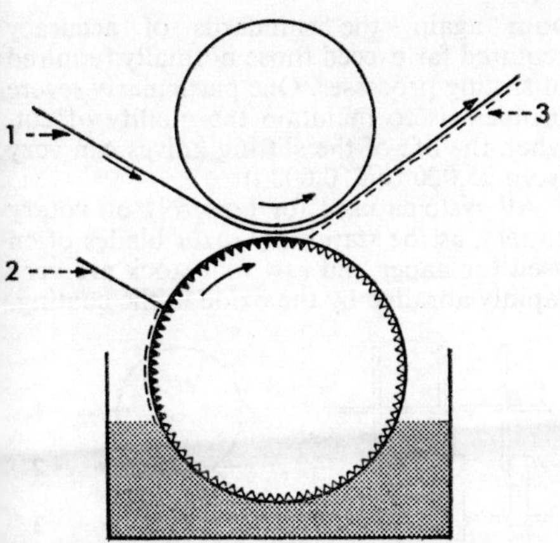

GRAVURE COATING. Indented gravure roller picks up coating material. After excess has been removed by doctor knife, 2, coating is deposited on base, 1, at contact with upper roller. Coated base then moves towards drying oven, 3.

2. Ability to obtain uniformity to the limits of ±0·00005 in., as a variation of ·0001 in. is equal to a sensitivity change of 1 dB when measured at 1 kHz at 15 ips.
3. Ability to deal with the transportation of base films down to ·0003 in. required for the thinnest types of audio tape.

The three main methods of coating in current use are summarized in the accompanying table.

During the wet stage of the coating, i.e., immediately following the point of application, the oxide particles must be magnetically orientated to obtain maximum output (sensitivity) at long-wave lengths. For example, a coating of acicular oxide particles in the unorientated state is 3 dB for 1 kHz at 15 ips lower than after orientation. To achieve this, the wet coating is passed through a uniform magnetic field to line up the magnetized particles.

This process makes it necessary to demagnetize the tape after coating, as some recording machines, either without erase heads or with lower strength erase systems, do not adequately remove the magnetism from the tape, and a d.c. signal remains.

To avoid foreign particles, either on the surface or within the coating, absolute cleanliness is vital. To meet the required standards, air supplied for drying is filtered down to 1 micron, and the coating material is refined by passing through filtration systems which only just permit the passage of the oxide particles.

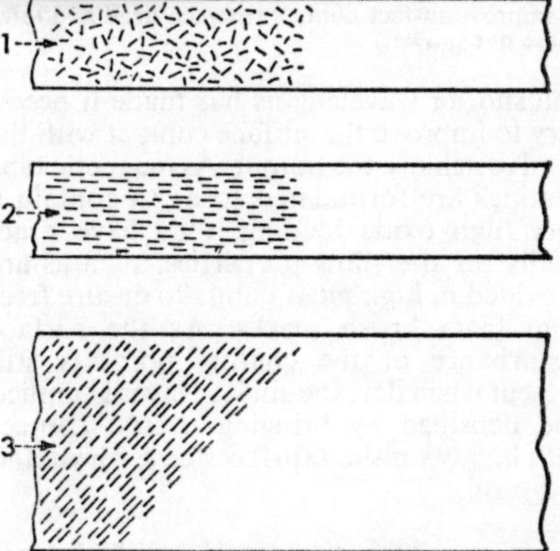

ORIENTATION OF MAGNETIC PARTICLES. 1. Unorientated particles immediately following point of application. 2. Particles orientated longitudinally for sound recording. 3. Particles orientated diagonally for videotape recording.

POLISHING. With the improvement in recording heads, the ability to record shorter

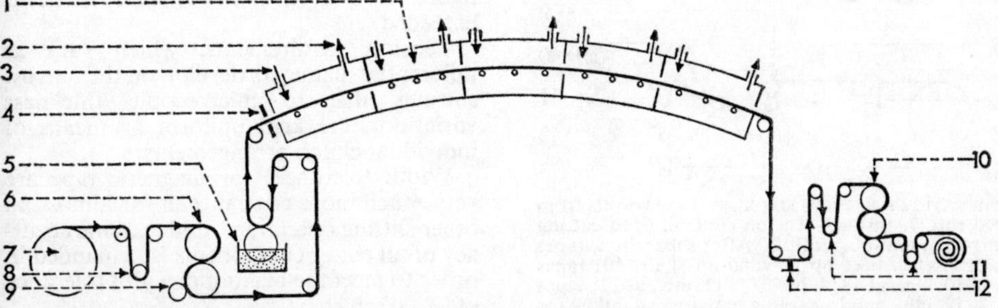

SCHEMATIC OF COATING MACHINE. Tape base from feed roll, 7, passes to tension rollers, 8, and preheat roll, 6, with nip drive, 9. Base takes up coating material, 5, and traverses magnetic field at 4 to orientate particles. Successive drying zones, 1, have air inlets, 3, and outlets, 2. Coated tape, after web guiding, 12, and tension control, 11, passes through polishing rolls, 10, before being spooled up.

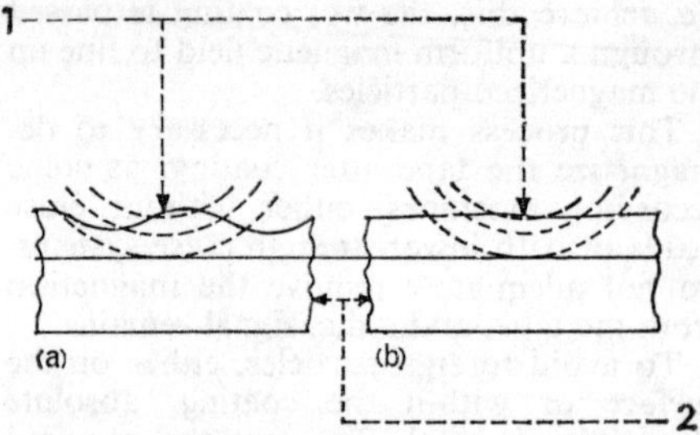

TAPE POLISHING. 1. Magnetic flux. 2. Oxide layer. (a) Unpolished tape. (b) Polished tape. Need to record ever higher frequencies has made it essential to improve surface contact between head and tape. (Base not shown.)

and shorter wavelengths has made it necessary to improve the surface contact with the head to achieve the output. As magnetic tape coatings are formulated so as to contain a very high oxide loading, they have practically no after-flow properties, such as are provided in high gloss paints to ensure freedom from brush marks. As the surface disturbance at the coating point is still present when dry, the surface layer is unified and densified by bringing it into contact with highly finished surfaces under heat and pressure.

SLITTING. The coated and polished tapes are prepared in what are known as jumbo rolls, which can be from 2,500 to 7,500 ft. in length and usually between 12 in. and 24 in. in width. The material is then slit to its final width, which varies from 0·246 in. for sound tape up to 2 in. in width for video, where,

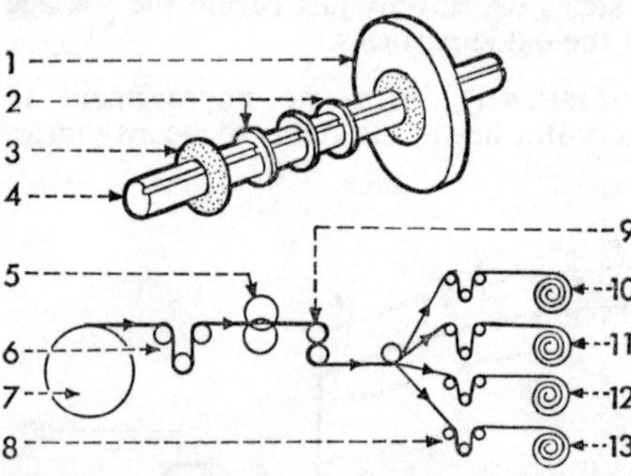

SCHEMATIC OF SLITTING MACHINE. Tape travels from feed roll, 7, through tension control, 6, to cutting knives, 5, and pull roll, 9. After separation, tapes 1, 5, 9, etc., take up at wind-up shaft, 10; tapes 2, 6, 10, etc., at 11; tapes 3, 7, 11, etc., at 12; tapes 4, 8, 12, etc., at 13. Reeling mandrels, 4, allow for minute tape thickness variations by clutch drive to each tape slitting, 1, through steel friction rings, 2, with paper washer between, to bakelite formers, 3, on which tape is wound.

once again, the standards of accuracy required far exceed those normally required of slitting processes. One particularly severe problem is to maintain the quality of cut, when the life of the slitting knives can vary from 25,000 to 250,000 ft.

All systems used for tape rely on rotary cutters, as the stationary razor blades often used for paper and raw film stock are very rapidly abraded by the oxide in the coating.

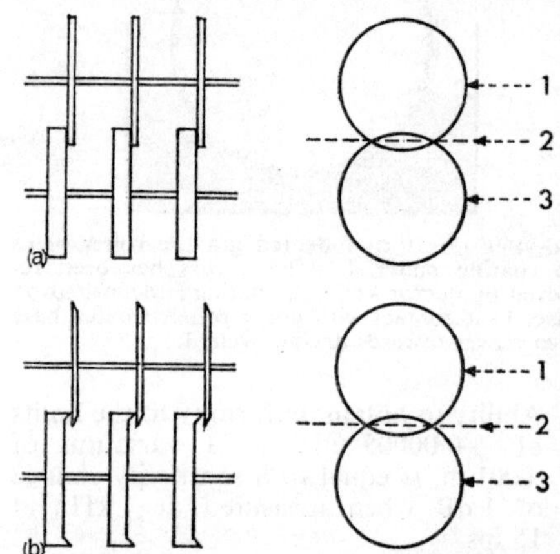

SLITTING SYSTEMS. (a) Disc cutting. (b) Shear cutting. 1. Top cutter. 2. Tape. 3. Bottom cutter. In shear cutting, the top cutter is spring loaded against the bottom cutter.

In setting up cutter assemblies, optimum adjustment of angles, pressures, penetration, speeds, etc., are required for each grade of tape to obtain the minimum edge burr and assure freedom from roughness. Continuous checking by microscope guarantees that the quality standards are being met.

Weave or curvature occurs unless the web is transported through the cutters without sideways movement, and special care has to be taken to control this so that a curvature figure of $\frac{1}{8}$ in. in 3 ft. of tape is never exceeded.

Tension requirements, which vary as differential diameters develop on the narrow slittings due to microscopic thickness variations, are kept uniform by means of individual clutch arrangements.

Width tolerances for magnetic tape are very much more critical than for almost all other slitting operations, and extreme accuracy of all cutter components is demanded in order to meet the maximum width tolerance of $\pm$·002 in.

PERFORATING. Magnetic film, which is used for obtaining exact synchronization of

REEL SIZES AND PLAYING TIMES—MAGNETIC FILM

Type	*Thickness (in.)*	*Spool*	*Length (ft.)*	*Overall Dimension (in.)*	*Playing Time (mins.) per track* 18 ips*	*Playing Time (mins.) per track* 7·2 ips†
35 mm	·00565	2 in. bobbin (50 mm)	1,000	10·4	11·1	
		4 in. bobbin (100 mm)	1,000	10·9		
17½ mm	·00565	2 in. bobbin (50 mm)	1,000	10·4	11·1	
		4 in. bobbin (100 mm)	1,000	10·9		
16 mm	·00565	2 in. bobbin (50 mm)	1,000	10·4		27·75
		4 in. bobbin (100 mm)	1,000	10·9		
		2 in. bobbin (50 mm)	1,200	11·35		33·3
		4 in. bobbin (100 mm)	1,200	11·85		
		2 in. bobbin (50 mm)	2,400	16·0		66·6
		4 in. bobbin (100 mm)	2,400	16·4		

* 35 mm film speed † 16 mm film speed

sound with photographic cine film, is manufactured on the same base material and is perforated on machines identical to those used for photographic stock. This operation is carried out after coating and slitting.

TEST PROCEDURE

The quality checks applied during the various processing stages are supplemented by end-product tests of the tape; these involve electrical tests on recording apparatus of very high quality as well as on commercially available recorders, and physical and mechanical testing.

AUDIO TAPES – BIAS. Amplitude recording for audio purposes is normally carried out by applying, together with the signal, an a.c. bias signal of frequency above the audio range. This has the effect of increasing the linearity of the recording characteristic with consequent increase of dynamic range. There is an optimum value of the intensity of the bias signal for any given head/tape combination in terms of the sensitivity, output for a given limit of distortion and modulation noise. This value is set by the coercivity of the material, the thickness of the coating, and the quality of the dispersion of the particles. It must be maintained substantially constant for any given tape type to ensure uniformity of performance and interchangeability. Control is exercised by determining the bias required for optimum output of the tape at long and short wavelengths for a given low level of distortion. Tolerances of ±5 per cent on the optimum bias of a master reference tape are applied. No absolute reference can be defined on such parameters as this, but most manufacturers have set their arbitrary reference tapes around sufficiently similar points for comparison of data to be possible.

MAGNETIC PROPERTIES. Coercivity and remanence measurements are made by magnetometry. By using a.c. as well as d.c. fields, curves may be plotted in the magnetometer giving magnetic performance data under anhysteretic conditions which provide more useful information about the value and behaviour of a tape specimen for recording than could be obtained from d.c. hysteresis loop measurement.

SIGNAL CHARACTERISTICS. The sensitivity at various frequencies, and the uniformity of these within a spool or from spool to spool, are measured using the bias levels appropriate to the tape. Sensitivity is again an arbitrary measurement against master reference tapes. Uniformity within a spool is held to ±¼ dB, dropouts being watched for and eliminated by a rejection of the tape when of frequent recurrence.

NOISE CHARACTERISTICS. Wide band noise measurements are made through internationally agreed aural weighting networks and referred as ratios to the output for 3 per cent distortion. Basic noise and modulation noise (typically 70 and 58 dB weighted rating at 15 ips) are separately reported and print-through checked after storage of a recorded tape for several days at high temperatures (110°F/45°C).

PHYSICAL CHECKS. Tapes are checked for wear life on recorders, and 100,000 passes or more of a tape loop round a commercial

REEL SIZES AND PLAYING TIMES—SOUND TAPE

Types	Spools			Playing Time (mins.) per track		
	Type	Dia. (in.)	Length (ft.)	15 ips.	7½ ips.	3¾ ips.
Standard Play (·002 in.) ...	European	11½	3,250	43	86	172
	N.A.B.	10½	2,400	32	64	128
	Cine	8¼	1,800	24	48	96
	Plastic	7	12,00	16	32	64
	Plastic	5¾	900	12	24	48
	Plastic	5	600	8	16	32
	Plastic	4	300	4	8	16
	Plastic	3	175	2	4	9
Long Play (·0015 in.) ...	N.A.B.	10½	3,600	48	96	192
	Cine	8¼	2,400	32	64	128
	Plastic	7	1,800	24	48	96
	Plastic	5¾	1,200	16	32	64
	Plastic	5	900	12	24	48
	Plastic	4	450	6	12	24
	Plastic	3	250	3	6	13
Double Play (·001 in.) ...	Plastic	7	2,400	32	64	128
	Plastic	5¾	1,800	24	48	96
	Plastic	5	1,200	16	32	64
	Plastic	4	600	8	16	32
	Plastic	3	400	5	10	21
Triple Play (·00075 in.) ...	Plastic	5¾	2,400	32	64	128
	Plastic	5	1,800	24	48	96
	Plastic	4	900	12	24	48
	Plastic	3	460	6	12	24
Quadruple Play (·0005 in.)	Plastic	3	600	8	16	32
	Plastic	3¼	800	11	22	43
	Plastic	4	1,200	16	32	64

transport are regarded as satisfactory. Tests are concurrently made to ensure that no significant abrasion of the head surface by the tape oxide is occurring. These tests are backed by laboratory checks of coating adhesion, cohesion, hardness, friction coefficients and tensile properties. Tapes are conditioned under adverse humidity and temperature and then examined to ensure absence of static electricity effect and satisfactory spooling on recorders at various tensions and winding speeds. The flatness, quality of slit edge, etc., are visually examined.

While all the tests are carried out on a statistical sampling basis, some are checked on each reel of tape, others on tapes representing each jumbo roll, while testing properties such as print-through need be related only to a batch of coated material from which several rolls have been made. In certain cases, tapes are checked throughout their entire length for such properties as sensitivity and uniformity which, combined with other statistically collected data, enables guarantees of performance to be given with the maximum degree of certainty. These tapes are called fully certified.

While it is important to measure the bias characteristics of batches of tape, all other electro-magnetic properties must be measured at a fixed bias which, once established for a given type of tape, determines the condition under which limits of interchangeability are checked. Without this safeguard, it could become necessary for users to line-up recording systems for each reel of tape.

Test standards present one of the greatest difficulties as, so far, no absolute reference has been agreed. This makes it necessary to define accurately the test methods, and, therefore, the machines on which results are obtained, and to refer back to a master reference tape; most manufacturers have managed to set their arbitrary levels around a sufficiently similar point to enable cross-references to be reliable and practical.

VIDEO TAPE. The audio track of the tape is subjected to the same tests as are applied to normal audio tapes. The video performance is the subject of tests of output, noise and dropouts. Intensive tests for dropout incident rate are made by playing back picture-free recordings (black level with syncs) and counting the rate at which errors in the form of isolated black or white dots due to signal loss at a dropout are occurring. The figure can be obtained automatically using counters, appropriate gating being applied to include dropouts of significant size only, but visual counts are usually carried out in addition to ensure that measured results relate in a practical way to the use of the material. A dropout rate of 10 per minute is regarded as acceptable.

The signal-to-noise behaviour under video conditions is also measured, again using subjective evaluation of the noise content seen on a monitor screen to back up the electronic measurement. The standards of signal-to-noise required for high-band monochrome and for colour recordings are somewhat higher than are necessary for low-band working, and the suitability of a tape for these applications can be assessed only by measurement under these more advanced conditions.

The wear life of a video tape is of paramount importance. Conditions of head-to-tape contact are much more severe than for audio tapes since the head tip has to be set to penetrate into the plane of the tape surface, and effects due to head clogging with consequent bands of noise or even loss of picture must be avoided. The working life of a tape under studio use in which head tip size and penetration have been carefully maintained at the levels recommended for video recording, can exceed 1,500 playings, but in other conditions this life may fall to a few hundred passes only. The design of the tape coating formula is crucial and must be correctly balanced to avoid head clogging.

COMPUTER TAPES

A computer tape is constructed and manufactured similarly to other magnetic tapes, but there are some differences in detail.

Almost without exception, polyester material in the thicknesses ·001 in. and ·0015 in. is used as base material, the thicker tape being the main requirement. This base is selected for its superior dimensional stability and strength, two factors of vital importance for this class of tape, and for ensuring a high standard of transport on the recorder.

Owing to the higher tape speeds of up to 150 ips normally used on computers, together with rapid reversals of the order of 1–2 milliseconds stop to start, computer tape coatings must possess special wear properties. At these higher speeds, lubricants must be built into the coatings to reduce high temperatures arising from friction, and the resins are selected to prevent degradation at high working temperatures. Adhesion of the coating to avoid minute flaking at the edges, and cohesion to avoid separation within the layers, must be of the highest order, otherwise particles could become detached and wound into the layers, causing dropouts.

FINAL STAGES

It remains only to spool the tape, give it a final inspection and package it.

REELING ON TO SPOOLS. While this is reasonably simple, it is most important to control the tension applied to the tape, as too slack a wind can cause the tape to drop off the centre line of the spool, whereas too tight a wind can result in tape distortion. Excessive tension nearly always results in curling of the plastic material toward the film, and this, if allowed to develop during winding, causes the tape to deform.

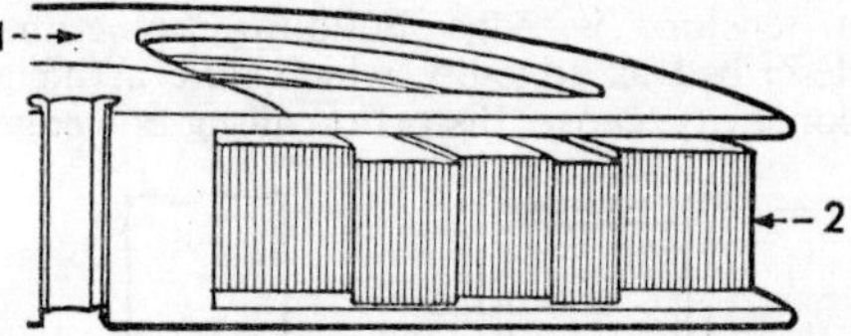

INADEQUATE WINDING TENSION. Tape, 2, winding on to spool, 1, tends to drop off centre line of spool when winding tension is insufficient.

If tape with the oxide coating innermost develops artificial concavity owing to too

REEL SIZES AND PLAYING TIMES—VIDEO TAPE

Tape Length (approx.) (ft.)	*Spool Size (dia.) (in.)*	*Playing Time* 60 Hz (mins.)	50 Hz (mins.)
400	8	5¼	4¾
800	8	10½	10½
1,200	8	16	15¼
2,400	12½	32	30½
2,600	12½	34½	33½
3,200	12½	42½	40¼
3,600	12½	48	36
4,800	12½	64	61¼
5,400	12½	72	69
7,200	14	96	92

high a tension at wind-up, and if it is left stored for any length of time, it will be found

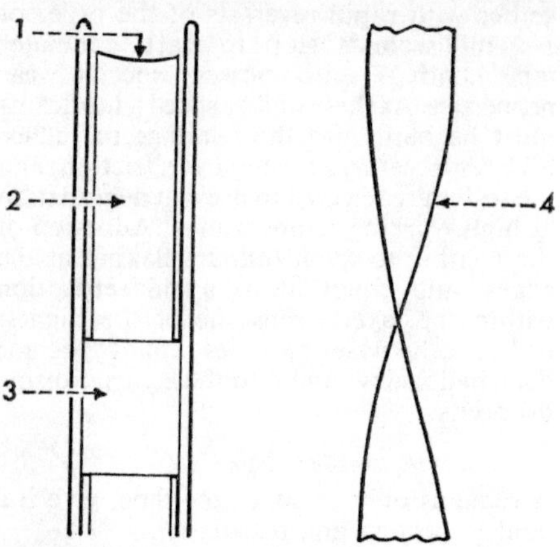

EXCESSIVE WINDING TENSION. Tape, 2, with oxide coating inwards, wound too tight on spool, 3, develops concavity, 1. If stored and then unwound, it tends to twist, as at 4.

that when unwound it twists. Another fault, either developed or accentuated by too high a tension, is edge build-up, sometimes described as edge-lip, which, like artificial concavity, causes distortion along one edge.

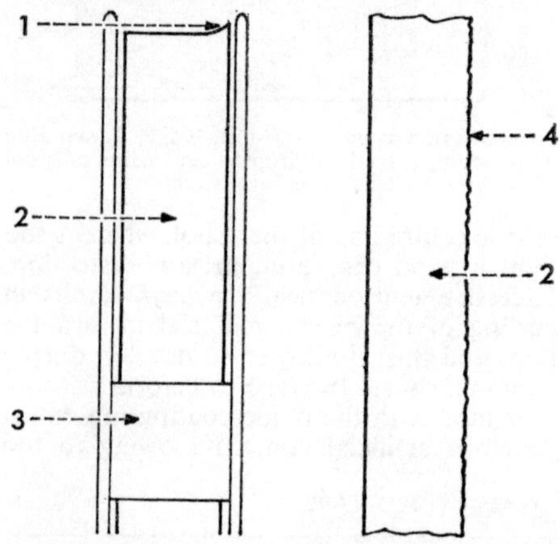

BURRED EDGE. Tape, 2, wound on spool, 3, has lip edge, 1, resulting from slitting burr or edge buffeting, and when unwound suffers from wrinkled edge, 4.

This fault can be due to the slitting burr, but can also be caused by edge buffeting against tape guides or spool flanges.

Modern recording machines are designed to provide the correct tension for their purpose and, if adequate maintenance is provided, no trouble arises. As too high a surface finish and varying coefficients of friction influence the tendency for tape layers to cling together or block, some tapes need less tension than others, but as a general guide tape tensions should be kept to the following figures:

TAPE TENSIONS

Base Thickness (in.)	*Tape Tension* (gm per $\frac{1}{4}$ in.)
0·0015	65–110
0·0010	40– 65
0·0005	30– 60

Although much has been said in favour of varying tension curves for winding tape, the most reliable and accurate method is to control at constant tension.

FINAL INSPECTION AND PACKING. The main requirement here is to make certain that sufficient time, usually not less than 12 hours, has elapsed for winding tensions to have taken effect.

FUTURE DEVELOPMENTS

Future developments in recording media can be expected to follow two trends: improvement of material and techniques along conventional lines, and the possible introduction of alternative systems based on other than magnetic coating of plastic tapes.

Technological improvement in tape manufacturing techniques and in the strength and uniformity of base films, particularly of polyester, are continually occurring, but major steps require a net gain in overall signal-to-noise ratio of the tape recording system, taking into account the basic causes of friction, asperity, and print-through tape noise. With the existing γFe_2O_3, variations in particle size distribution and dispersion can produce different compromises in tape design. For example, smaller particles give tapes of lower basic noise but more susceptible to print-through. The replacement of the γFe_2O_3 by alternative materials of higher remanence per unit volume in the same range of particle size will result in a gain of overall signal-to-noise ratio, and papers have already been published on the alternative use of other oxides, metal alloys and metal films, in which this result is achieved. Technological problems have to be overcome to achieve commercial availability of such materials, but a new tape based on chromium dioxide shows advantages for some forms of recording. J.W.

See: *Film dimensions and physical characteristics; Sound recording.*

Books: *Physics of Magnetic Recording—Selected Topics in Solid State Physics, Vol.* 11, by C. D. Mee (Amsterdam), 1964; *Magnetic Tape Recording*, by H. G. M. Spratt (London), 1958.

Mag-Opt. Cinematograph release prints in which both optical and magnetic sound tracks are provided so as to allow their use in theatres equipped with either system. In 35mm, the optical track occupies its normal position but is half the width of an optical-only track, and the magnetic tracks are the four provided for multi-channel sound reproduction.

Combined mag-opt tracks, each half the normal width, have been used to carry two different language versions on the same copy in 16mm.

Since the magnetic coating used is substantially transparent to infra-red radiation, it is possible to use mag-opt prints in which the two full-width tracks are superimposed rather than half-width side by side, but this procedure has not been widely adopted.
See: *Stereophonic sound processes.*

Main Title. Title which gives the name of a film or TV programme.

Maltese-Cross Mechanism. Intermittent pull-down device universal in 35mm film projectors, rarely used in 16mm projectors, based on the Geneva movement long em-

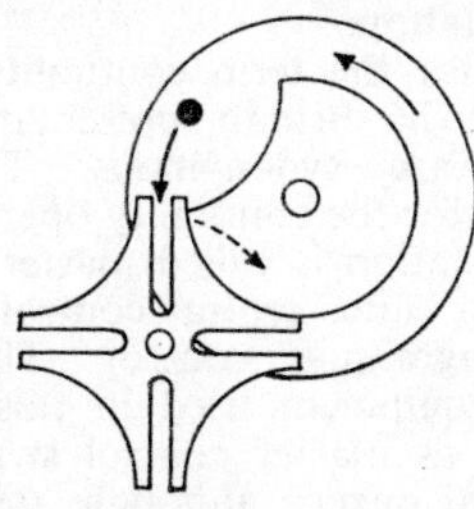

ployed in watchmaking to prevent overwinding. The device gets its name from the star-shaped wheel in the form of a Maltese cross which rotates intermittently.
See: *Projector.*

M and E Track. Music and effects track. When a film is dubbed and its component sound tracks are properly blended into a final track, it is often convenient to dub a separate track which contains all these elements except the dialogue. If a foreign version of the film is later made, it is then not necessary to have recourse to all the original tracks, but only to the M & E track.
See: *Sound mixer; Sound recording.*

Marey, Etienne Jules, 1830–1904. French inventor, graduated at Dijon in 1849. Went to Paris where he studied physiology at the Faculty of Medicine, and became a doctor in 1859. After a time in general practice, began experimental work in 1864, continued on his moving to the Collège de France in 1869, where, until 1880, he investigated the problems of animal locomotion. In 1881 an experimental station was set up at the Parc aux Princes where, inspired by the work of Muybridge, he began to use photography to record animal locomotion.

Marey designed the photographic gun, which recorded a series of exposures round the edge of an intermittently rotated glass plate, and used it to record the flight of birds. He also employed a fixed-plate camera in which a rotating shutter produced a series of stroboscopic exposures of moving subjects. With the introduction of flexible roll-films, first paper, later celluloid, he, with Georges Demeny, designed and operated in 1887 a camera in which a band of sensitive film was moved intermittently and exposed while stationary through a rotating disc shutter. This was the first working example of a motion picture camera, with which and subsequent improved versions, Marey made major contributions to knowledge of animal movement. He introduced slow-motion photography (as many as 700 frames per second being achieved in 1894) and cinemicrography as research methods.

□**Maser.** Acronym of Microwave Amplification by Stimulated Emission of Radiation. A low-noise microwave amplifier depending for its action on quantum mechanics.

⊕**Mask.** Form of light modulator sometimes consisting of a strip of film containing opaque areas designed to exclude or reduce the transmission of light. Masks are commonly used in trick photography and colour printing.

Also, a plate incorporated in a camera, projector or printer in which a window is formed to delimit the area of the picture.

Interchangeable masks may be provided for the aperture in a film projector to alter the height/width ratio of the picture thrown on the screen. This ensures that unwanted parts of the frame are not reflected from the screen surround.

This surround, usually of matte black material, is also called a mask.
See: *Printer; Projector; Screens; Special effects* (FILM).

⊕□**Masking (Colour).** Partial correction of unwanted characteristics in colour reproduction. In all practical colour film processes the component colour images are found to have spectral transmission and absorption characteristics which depart from those required in theory. Practical methods of masking include the preparation of additional strips of film with specially processed images to be used in combination with the original film, while in some integral tripack materials masking images are formed automatically in the course of processing. In TV colour reproduction systems masking may be introduced by electronic means.

See: *Colour cinematography; Integral tripack.*

□MASTER CONTROL

In any television station the master control forms a focal point where programmes originating from studios, outside broadcasts, network, film or videotape are combined into a complete, smooth presentation. It is here that station identification, commercial advertisements, public service films, trailers of future programmes, apology captions and other materials forming station breaks are inserted into the main programme. The responsibility of this area is also to route the complete programme to a single transmitter, or the required variants of this programme (e.g., with different language commentary or without commercial advertisements) to several transmitters or networks. Incoming recordings are also directed to the correct departments.

BASIC PRINCIPLES

The complexity of television broadcasting has increased over recent years due to the growth of networking, a wider variety of signal origination equipment, and longer hours of programme production per day. This has had a notable impact on the continuity or presentation operations of the master control area. Since the number of incoming sources is constantly increasing and presentation techniques are becoming more sophisticated, the master control switching must now be more exacting. Normally, during the main programme, the operator has very little to do because programmes come from a studio, film or videotape. His activities are concentrated during the few minutes of station breaks when he has to perform a complicated sequence of switching operations. The pattern of manipulations required to switch the correct vision and sound sources on-air at the right time, by means of a suitable transition, and to cue or start telecine or videotape machines, becomes so complicated that it taxes the operator's capabilities almost to the limit. It is not surprising that station breaks are sometimes called panic periods.

The continuity operations are carried out according to a daily schedule which, especially on commercial stations, is followed quite rigidly. The sequence of switching operations and the duration of each event are predetermined, and therefore the automation of master control is feasible and desirable. Automation can bring a number of advantages: elimination of human errors, relaxation of tension during station breaks and economy of manpower. Many stations have already automated or semi-automated the control of presentation and achieved smoother operation and saving in costs. Some other stations, however, find that manual control is best suited to their needs.

The terms used to define discrete parts of the master control area, their functions, and the switching equipment used in them, are by no means uniform and indeed vary from station to station.

In America the term continuity is universal, but in Britain presentation and continuity are synonymous. The area responsible for the continuity operations of the whole station is called master control, presentation suite, central control or continuity programme control. The main switching equipment used in this area is referred to as master control switcher or presentation mixer, although terms like programme switcher, continuity mixer, central control switcher are also used.

LAYOUT

It is difficult to describe a typical layout of master control because the make up of this area varies greatly from station to station depending on the needs that the station serves, the operational procedures adopted and the degree of automation applied. Most of the long established stations have grown gradually over the years, adding new facilities as the need arose. The expansion was very often influenced by the available accommodation or by the need to integrate new facilities with existing equipment.

A station's presentation activities are sometimes divided between several mutually supporting areas, such as presentation studio with its control room, master control room and network control room, and sometimes centralized in a single area. In general, the higher the degree of automation applied the more centralized are the presentation operations and the more distinct is the division between production and continuity.

PRESENTATION STUDIO. This is a small studio equipped with two or three vidicon or image orthicon cameras, which may be remotely controlled. The studio is used for programme introductions, interviews, weather forecasts and other minor programmes. The studio control room is equipped with conventional vision and sound mixers, camera remote control panels, talkback equipment, etc. Sometimes the vision mixer is adapted for married sound so that vision and sound are cut or faded together during simpler presentations, and the sound mixer's position is unmanned. The presentation studio supplements the facilities of the master control switcher and in case of emergency it can take over the switching of the station output.

In some stations, the presentation studio takes the form of a simple announcer studio without a control room. A fixed vidicon camera is mounted on top of a picture monitor in front of the announcer's desk. Green stand-by and red on-air lights are provided for cueing purposes. The main lighting is controlled by the announcer. The output of the studio is fed as a source to the master control switcher.

Another form of presentation technique is to combine the function of the announcer (in voice only) with presentation mixer operator. A small control room, called the continuity suite, is equipped with a combined sound and vision mixer, picture monitors, audio cartridge tape machines, disc reproducers (audio turntables) and talkback panels. An adjoining cubicle houses a remotely controlled caption scanner. From this suite, the announcer can produce simple station breaks by cutting and mixing between sources, selecting captions, providing announcements or operating cartridge tape equipment and disc reproducers.

The simplest form of the announcer studio is an announcer booth which is often a partitioned portion of the master control room. The booth can be used regularly for live announcements or it may be manned only during emergencies, all normal announcements coming from pre-recorded tapes.

MASTER CONTROL ROOM. When it is not supported by an elaborate presentation studio or a continuity suite, the master control room embraces all the presentation activities of the whole centre. In stations using entirely manual control, the room can be staffed by three persons: presentation or programme controller, vision mixer operator and sound engineer. The presentation controller co-ordinates the work of the other two, cues the pre-roll of telecine and videotape machines, follows and amends, as required, the programme schedule, and maintains communication with programme originating areas. The duties of the vision mixer operator and the sound engineer do not differ substantially from those of the corresponding studio control personnel.

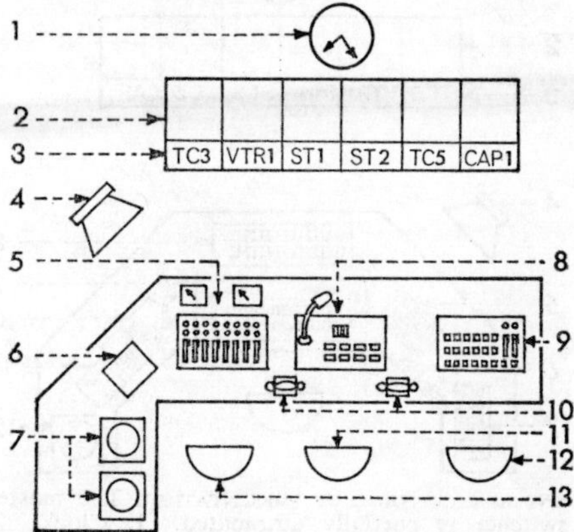

THREE-MAN CONTROL OF PRESENTATION. Vision operator, 12, and sound engineer, 13, are co-ordinated by presentation controller, 11. Sound engineer operates sound mixer, 5, cartridge tape machines, 6, and disc reproducers, 7. Vision operator works vision mixer, 9. Talkback equipment, 8, and telephones, 10, provide studio communication. Programme material is previewed on picture monitors, 2, with source indicators, 3. (TC, telecine; VTR, videotape recorder; ST, studio; CAP, caption scanner.) 1. Clock governing studio operations. 4. Quality loudspeaker.

The master control room is sometimes manned by a single presentation operator. To achieve this economy in manpower, the presentation mixer is at least partially automated and the whole control desk ergonomically designed.

The central position of the desk is occupied by a master switcher (presentation mixer) control panel. Sound and vision control are carried out from the same panel. If the switcher is automated, a display panel showing the state of memory is mounted above. A remote machine control panel permits starting and stopping telecine projectors and videotape recorders, changing captions, slides, etc. Cartridge tape machines and disc reproducers provide sound for vision sources which have no married sound. When live announcements are required, an announcer booth is used.

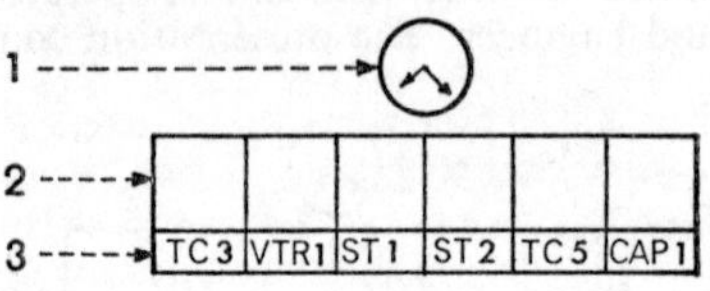

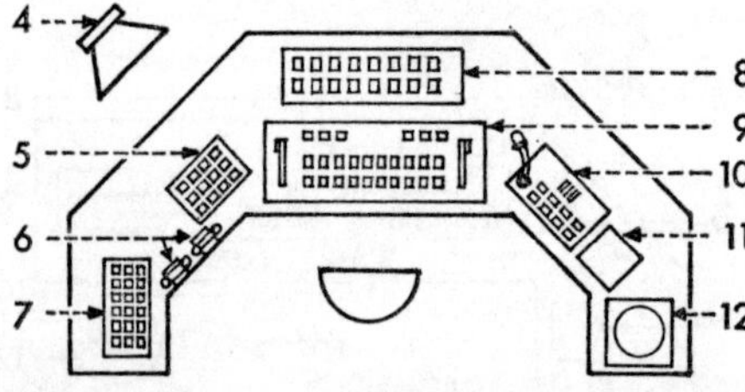

ONE-MAN CONTROL OF PRESENTATION. The master switcher is partially automated. 1. Clock. 2. Picture monitors. 3. Source indicators. 4. Quality loudspeaker. 5. Machine control panel. 6. Telephones. 7. Preview/emergency switcher. 8. Store display. 9. Presentation mixer. 10. Talkback equipment. 11. Cartridge tape machines. 12. Disc reproducer.

A separate preview switcher permits monitoring of any of the sources. In case of a failure of the main switcher, the preview panel can be used for on-air switching. Communication equipment includes a talkback system giving two-way communication between master control and signal origination areas such as studios, assigned telecine and videotape machines, etc., and several telephones used for internal and external calls or providing control lines to outside broadcast locations and to the main network control.

Behind the desk in front of the operator is a bank of monitors showing on-air, next event, off-air and one or several roving preview pictures. Each monitor has an indicator showing the source of the picture displayed on it. In a central position on the wall is a clock driven by the station master control pulses. A second clock, for emergency use, driven from mains may sometimes be provided. All the control panels have lights or indicators showing the state of the control, and buttons most commonly used are placed in the most accessible positions.

DUAL MASTER CONTROL. Some television centres have to feed simultaneously two separate transmitters carrying different identifications, continuity, and sometimes even different main programmes. In such a case, the two transmissions are normally controlled from two identical presentation mixers fed from the same sources. The outputs of these mixers are however combined in such a way that, when required, both transmissions can be fed from a single mixer. This facility is useful during certain periods of the day when both transmissions are identical. The second mixer can then be used for rehearsing more elaborate promotions or for prerecording other material.

NETWORK CONTROL. In all countries, television broadcasting tends to develop regionally, and each centre, besides feeding its local transmitter, has to make some contribution to a network. At any given time the programme supplied to a network may not be the same as the programme fed to the transmitter. In commercial television, the contribution to a network should be clean, i.e., without the advertisements which are fed to the local transmitter. A prerecorded play may be transmitted one evening and then sent to a network days later. To control the programmes to network a separate switcher may be provided.

A network control switcher is normally fed by the same sources as master control, with the addition of master control output. As a rule the facilities provided on the network switcher are limited: married switching of vision and sound is used and transitions between sources are by simple cutting. Sometimes fading to black is also included.

It is customary to combine the network switching with the general engineering supervisory functions such as quality checks and adjustments of all the incoming and outgoing signals, routing of communication circuits, sync pulse generator change-over, genlocking, etc.

SWITCHING EQUIPMENT

The switching equipment in master control differs from that used for sound and vision mixing in studios mainly because of the different scope of these two areas. In studios, programmes are produced from material containing a high proportion of the human element, which is not always completely predictable. Master control, on the other hand, has to combine ready-made programmes of predetermined length and sequence. Work in the studios is more creative, that in master control more mechanical, therefore tools used to perform the two duties must be different to achieve the best results.

In a studio production control room, the vision mixer operator works on cues from the director, who tells him when to switch, to which source, and what type of transfer to use. However well the play is rehearsed, there will be differences on the final night: actors may walk slower or talk faster; besides, the director may change his mind at the very last moment. Similarly, the work of sound mixer operator in the sound control room has a strong element of improvization to ensure that the final sound that leaves the studio has the correct dynamic range, balance, perspective and quality, and that it matches the picture at all times.

The work of master control is much more predetermined, because there are fewer variables: sequence of switching operations, duration of each event, and type of transfer to a new event are all laid down in a daily schedule. Of course, there are emergencies caused by an equipment failure or an unforeseen occurrence of national or international importance, but these are usually rare. Most of the sources fed to the master control have married sound and vision (e.g., studios, films, videotapes); therefore, separate switching of sound and vision from two control positions is not essential. Switching of separate sound sources such as announcer, audio tape, and disc reproducer with caption or slide vision sources can usually be carried out from a single control position.

SOURCES. The sources fed to a master control switcher can be divided into permanently connected sources and assigned sources which are connected only when required. The complement of permanent sources includes network, announcer's studio or booth, caption scanner, cartridge tape machines and disc reproducers. The caption scanner plays an important role in the continuity operations, because it supplies readily available pictures of station identification, clock, apology captions and other graphic material that can be quickly prepared.

Studios, remotes, telecine machines and videotape recorders form the second group of sources which are routed to the master control by special assignment switcher or by manual patching. At any given time in a large studio centre, two studios, two remotes, three or four telecines and three or four videotape recorders may be connected to the master control. The assignment of the sources permits a smaller-capacity switching equipment to be used and, in case of machine sources, facilitates their more efficient utilization. In smaller stations, studios and remotes may be connected permanently. The total number of vision sources varies between 12 and 25.

SWITCHER CONFIGURATION AND FACILITIES. The switcher configuration most commonly used for continuity purposes is of the next-channel type. Its main advantage is simplicity of operation and the ease with which it can be automated. Other configurations like A/B/cut or knob-a-channel with additional preselection of sources to each channel are used infrequently.

The next-channel or preset type switcher consists of two identical channels which are used alternately as transmission and preview/prelisten outputs. While one channel is on the air feeding a transmission line, the off-air channel is used to preset and preview the next event source (auto-preview). When the time comes, the preset source can be transferred to transmission and the channel which was previously on-air can now be used to preset the next event.

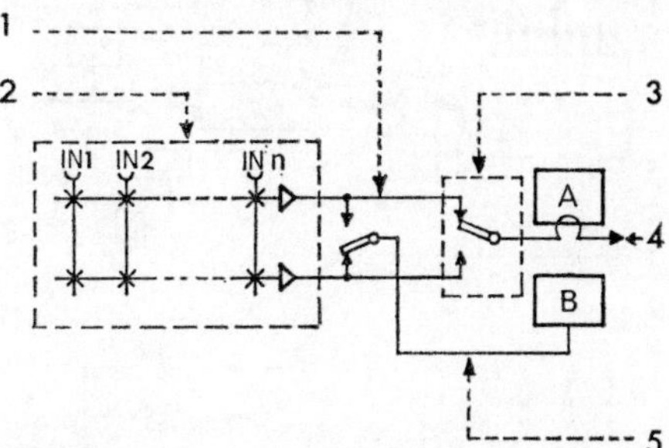

PRINCIPLE OF NEXT-CHANNEL CONFIGURATION. Two identical channels are used alternately as transmission and auto-preview/prelisten outputs. 1. Auto-preview changeover. 2. Switching matrix. 3. Transfer unit. 4. Transmission output looped via on-air monitor, A. 5. Next channel to auto-preview monitor, B.

The transfer to the next-event source can be by cut (cutting during field blanking interval), fade (fading the on-air source down, cutting to the new source at black and then fading up the new source), or mix (cross-mixing or lap-dissolving). On some presentation mixers, wipe (special effects) transfers are also used. The next-channel type of operation permits mixing and wiping between on-air and the auto-preview sources

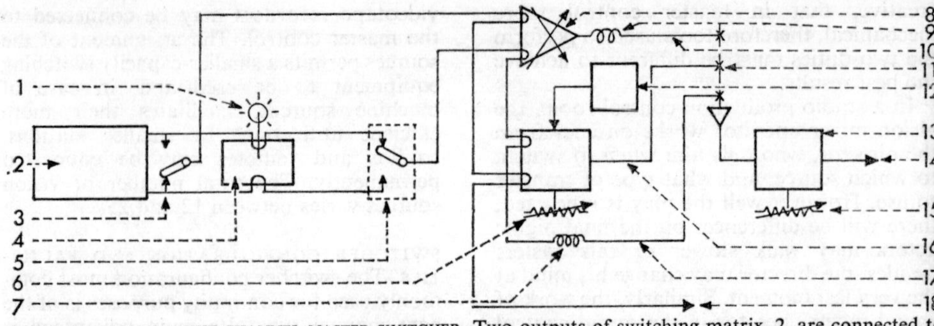

BLOCK DIAGRAM OF NEXT-CHANNEL MASTER SWITCHER. Two outputs of switching matrix, 2, are connected t auto-preview changeover, 3, and cut amplifier, 5. Output of auto-preview changeover, after looping throug mixing amplifier, 8, and electronic switch (with wipe control, 6, and fed by pattern generator, 12), is availabl as auto-preview output, 18. Output of cut amplifier, 5, after looping through mixing amplifier, 8, an electronic switch and passing through delay, 7, is connected as cut output, 17, to re-entry matrix, 11, whic is also fed by wipe output, 16, and through delay, 9, by mix output, 10. Transmission output, 14, is obtaine from processing amplifier, 13, with overall fade control, 15. Sync. comparator, 4, connected between tw outputs of matrix, lights pilot lamp, 1, to warn operator not to mix or wipe when next-event input is non synchronous.

only. A sync comparator may be connected between the two channels to warn the operator not to mix or wipe if the next-event source is non-synchronous with the on-air source. The warning is in the form of a red light, but sometimes the mix and wipe levers are also disabled.

The sound arrangements of the next-channel type switcher are similar to vision.

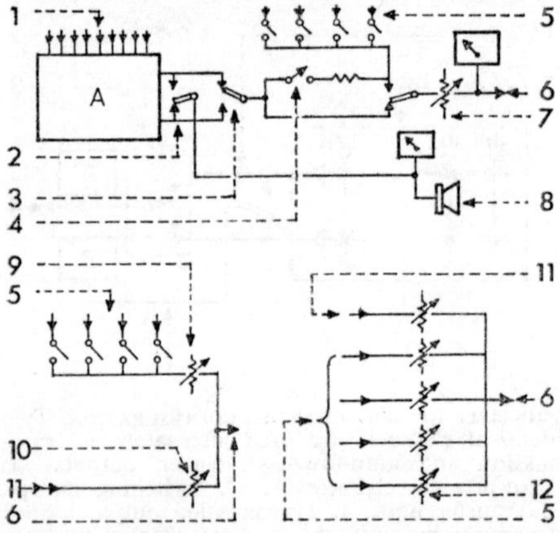

MASTER SWITCHER SEPARATE SOUND ARRANGEMENTS. (*Top*) Separate sound sources, 5, are switched to on-air by operating desired key. Main sound mix switch, 4, permits superimposition of any separate sound source over main married sound. (*Bottom, left*) Additional fader, 9, for separate sound sources. (*Bottom, right*) Each separate sound source has its own fader, 12. 1. Married sound sources to switching matrix, A. 2. Auto-pre-listen change-over. 3. Sound change-over. 4. Main sound mix switch. 5. Separate sound sources. 6. Transmission output. 7. Master sound fader. 8. Auto-pre-listen loudspeaker. 9. Separate sound fader. 10. Married sound fader. 11. Married sound input.

In a married (or audio-follow-video) opera tion, sound and vision are preselected on th same switching elements. In separat operation, sound is preselected in paralle with the vision on different channels. Durin cutting or fading sound follows vision during vision mix or wipe the sound i usually cut to the next source at the end o mix or wipe lever travel. Separate soun

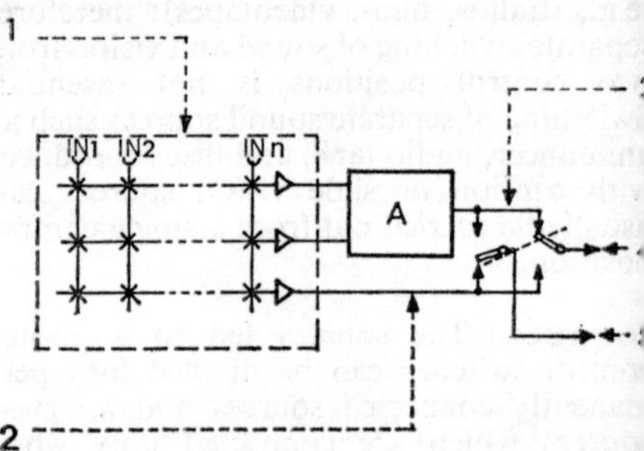

BYPASS/REHEARSAL FACILITY. During emergency th transmission is transferred to a roving previe output, thus bypassing most of electronic circuit A. During normal conditions, same facility can b used to free main mixer for rehearsals. 1. Switch ing matrix. 2. Roving preview output. 3. Bypa switch. 4. Transmission output. 5. Roving pr view or rehearsal output.

sources such as announcer, audio tapes an disc reproducers can be switched to on-a simply by operating a desired key. Mo elegant methods of employing an addition fader for separate sound sources or even fader per source are also used.

Besides the main outputs, the mast switcher normally has a number of auxilia outputs for roving preview/prelisten. In cas of a fault in the main transmission path, th output can be transferred to one of the preview outputs. In this condition most of the electronic circuits are by-passed and only simple cutting between sources is available. Further extension of this facility is possible: at the beginning of a longer programme the transmission can be switched to by-pass, and the main switcher with all the vision and sound facilities can be used for rehearsals.

SWITCHING MATRICES. It is convenient to visualize the basic switching network as a matrix in which the operation of a crosspoint switch connects the required source to the right destination. For master control use, each crosspoint may have a number of contacts to switch video, balanced sound, cues and sometimes machine start control. The matrices have a capacity of up to 25 inputs and at least three outputs (on-air, auto-preview and roving preview). Relays, uniselectors (rotary stepping switches) or semiconductors are used as basic switching elements.

Relay matrices employed in master control do not differ substantially from matrices used in studio mixers except that each crosspoint may be wired to switch balanced sound in addition to vision and cues. The control circuits necessary to give the next channel configuration are sometimes provided.

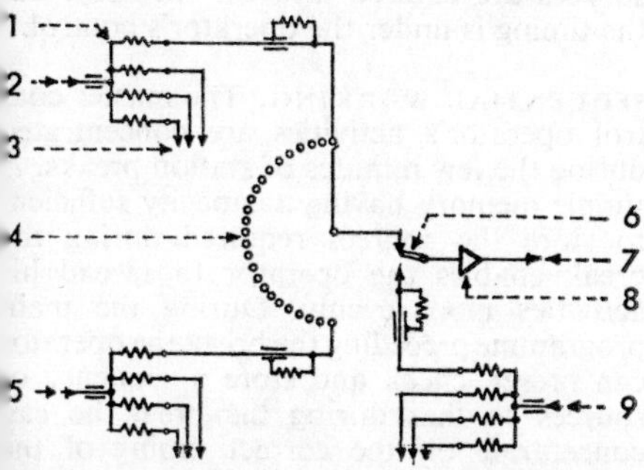

VISION CIRCUIT OF UNISELECTOR SWITCHING PANEL. Simplified version with only one output channel shown. First input, 2, has splitter pad, 1, which feeds uniselector, 4, and with outputs, 3, remaining three uniselectors. Identical circuits (not shown) up to 25th input, 5. Output goes to muting relay, 6, 12 dB amplifier, 8, to make up losses and becomes video output, 7. Sync. input, 9, provides muting signal to all four channels.

Uniselectors are often employed in larger switchers because of their high capacity and small volume. They have been used a very long time in automatic telephone exchanges, but their application in television switching is of comparatively recent origin. A typical uniselector has 25 positions on 8 levels, which is equivalent to 25 crosspoints having 8 contacts each. A common switching unit employs four uniselectors and routes any of 25 inputs to four independent outputs of sound, vision and cues. Two spare levels on each uniselector can be used to switch machine controls when required. All vision inputs are split four ways in resistive pads and each uniselector is fed via miniature coaxial cable terminated by 75 Ω. The loss of 12 dB is restored in high input impedance transistor amplifiers. The twisted pairs of the balanced sound inputs are simply looped over the levels of the four uniselectors. Two of the uniselector switcher outputs can be used in the next-channel configuration for the main transmission; the remaining two outputs can be used for previews.

Semiconductor matrices can be justified only for switching vision, and their use in master control is limited to cases where it is desired to control sound separately. Cues are obtained from auxiliary relays operated by the crosspoints themselves or directly by push-buttons. Sometimes silicon-controlled rectifier latching circuits are used to operate the crosspoints and light the cue lamps. The main advantage of semiconductor matrices – very fast switching speed giving instantaneous cut – is not utilized in master control switchers where the next source is preselected before being cut to on-air.

AUTOMATION OF MASTER SWITCHING

Automation is playing an increasing role in the operation of television stations, and master control is the area where its application brings most benefits. The principle of master control automation is simple: once the switching schedule is available, it has to be transferred into a suitable store or

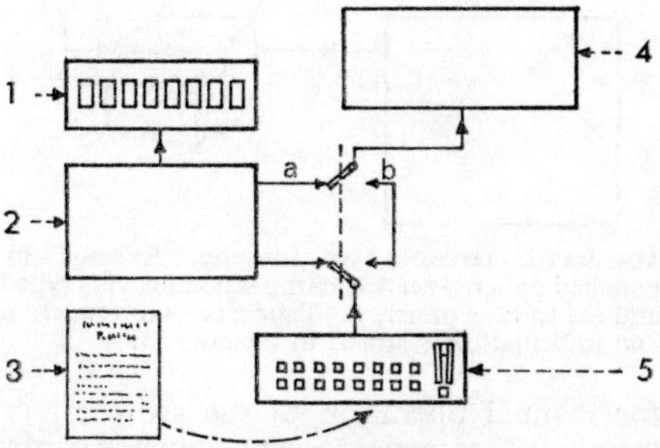

PRINCIPLE OF MASTER SWITCHER AUTOMATION. Memory, 2, with display, 1, is loaded manually from control panel, 5, according to switching schedule, 3. Master control switcher, 4, can be operated automatically (a) or manually (b).

memory which, in turn, will control the master switcher. The form of the switching schedule (programme log) is tailored to the type of automation used, but in general it contains the scheduled switching time, duration, video source, audio source and the description of each event.

MEMORY. The memory can be as simple as a number of multiposition switches or as complicated as a computer with magnetic drum storage. There are, of course, numerous alternatives between these two extremes: capacitors, relays, uniselectors, ferrite stores, beam switching tubes, punched paper tape or cards, magnetic tape, etc.

The choice of memory depends on the extent of automation desired. For a limited number of events, as for example the sources required for one station break, a relay or uniselector store is ideal. If, on the other hand, it is desired to put into store the operations required for a whole day's programme, a ferrite store magnetic tape or a computer is more suitable. Because of its large storage capacity, the computer, besides controlling the master switcher, is also used for general station applications such as billing, programme planning, etc.

A memory can be loaded either from a control panel, which very often is also used

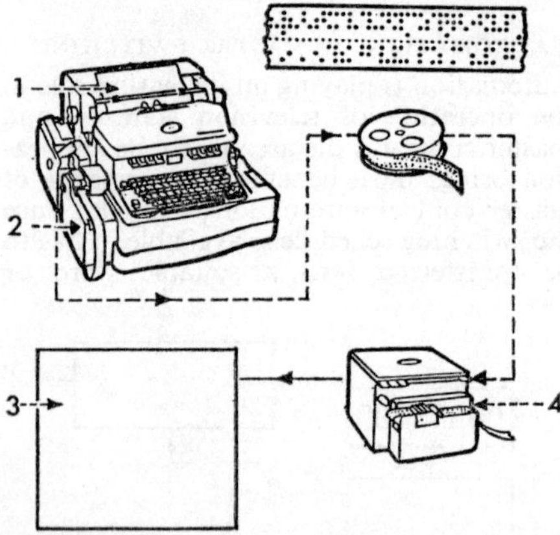

AUTOMATIC LOADING OF MEMORY. Effected by punched paper tape. Switching schedule, 1, is typed and fed to tape punch, 2. Tape passes to reader, 4, and information is stored in memory, 3.

for manual operation of the switcher, or from punched paper tape or punched cards. During the preparation of the daily schedule a special typewriter is used, and a punched paper tape containing the switching information for the whole day's programme is obtained as a by-product of typing. The tape is then loaded into a tape reader which, when energized, reads the switching information for a number of events into the memory, whether computer or uniselector store. A similar system is employed with punched cards. Punched paper tape or cards are rarely used to control the master switcher directly because they do not offer random access to stored information, and corrections are difficult to introduce.

A display panel showing a number of upcoming events stored in the memory forms an important part of any automatic switcher. On this panel, the operator can see the pattern of switching a few events ahead so that in an emergency he can take corrective action well in advance. The memory is normally backed up by a manual control panel which permits overriding of the automatic control.

The automation of the whole day's programme, which is controlled by a 24-hour clock, is termed real time working (true time or clock time). A simpler type of clock operation is cue clock (or elapsed time) working, where each station break is timed individually. In some stations, however, schedules are not sufficiently rigid to allow any form of clock operation. In this case, a sequential working is adopted in which sources are entered into the memory, but the timing is under the operator's control.

SEQUENTIAL WORKING. The master control operator's activities are concentrated during the few minutes of station breaks. A simple memory having a capacity sufficient to store the sources required during the break enables the operator to spread his activities more evenly. During the main programme preceding the break the operator can preset, check and store a sequence of sources so that during the break he can concentrate on the correct timing of the transfers to on-air.

This type of semi-automatic switcher can be operated in two modes: manual and automatic. In the manual mode the memory is inoperative and the switcher is controlled in a conventional way: a next-event married source is preselected by pressing one of the appropriate buttons (one button per source is provided), checked on an auto-preview monitor and prelisten loudspeaker and then taken to transmission by cut, fade or mix. In the case of a vision source which has no married sound, one of the separate sound keys is pressed and an announcer, tape or disc reproducer is switched to on-air.

When using automatic mode, the operator can enter into the store the sources required for a break. First the store has to be reset to the start position. With a small capacity store, a display panel usually shows the code of all sources entered into the memory, the actual position of the store being indicated by a light. By means of the appropriate preselect button, the first source is entered into the memory and the store steps on to the second event. In this way source after source can be entered sequentially into the store.

When the last event is entered, the store returns to the first position ready for the beginning of the break and the first source is automatically selected on auto-preview. At the beginning of the break, the operator presses the cut button or moves the fade or mix lever down and the first event is transferred to on-air; the store steps on and selects the second source to auto-preview.

Some memories have a facility of random access so that corrections of stored information are possible up to the moment when the source is transferred to on-air. Any unwanted next-event source can be by-passed by operation of a button (skip, discard next event).

In stations employing a large number of sound sources two separate stores are used, one for vision and another for sound. Similarly, the control panel is fitted with two rows of preselect buttons. The two stores can be loaded and corrected independently, but during the break they step in synchronism.

MACHINE CONTROL. During the station breaks, prerecorded programme sources are mainly used. Therefore, in addition to preselecting a machine as the next event the operator must make sure that it is started at the right time, watch it roll on the auto-preview monitor and then transfer it to on-air when it becomes stabilized. The machines assigned to master control are sometimes controlled from a panel mounting the buttons necessary for starting, showing, stopping, changing slides, etc.

Another solution integrates all buttons necessary for machine control with the main switcher control panel. The start controls of the machines are routed via the switcher, and any machine preselected as next event can be started by operation of a single button. The only limitation of this method is that a source should remain on-air for at least the time necessary to run-up the next-event machine. More complex systems are overcoming this limitation.

If multiplex telecine machines are employed, a separate preselect button per projector is used so that each projector is treated as a separate source. Any machine taken off the air can be stopped automatically.

In some stations, to simplify the operation and scheduling, the run-up (pre-roll) times of all the telecine and videotape machines are made equal. This can sometimes be achieved by adjustment of the machines; alternatively, a suitable length of leader can be added to films or tapes on faster machines. A delay circuit can be built into the switcher so that an automatic transfer to on-air takes place after a fixed delay following the operation of the start button. In this way, during the station breaks, the operator pushes the start button a few seconds before the transfer to a new source is required, and the remaining functions are performed automatically.

When all the material used during station breaks comes from prerecorded sources, the operation of master control is further simplified. The run-up times of all the machines are equalized, for example, to 5 seconds, and cartridge tape machines are used with caption scanners, slide projectors and other sources which have no married sound. Thus each machine is started 5 seconds before it is scheduled to be transferred to on-air. By inserting cues on films and tapes 5 seconds before the end of each spot, the next event machine preselected by the memory can be started automatically. The operator initiates the break by starting the first machine; all other operations, including the transfer back into the main programme, are automatic.

CUE-CLOCK OPERATION. The method of controlling switching operations by film and tape cues is simple and economical but the operator has no visual check on timings and only prerecorded sources can be used during the break. As an alternative, the automatic switching of the station break can be controlled by a cue clock on an elapsed time basis. A cue clock is a chain of transistor ring counters or of miniature uniselectors driven by one-phase pulses obtained from the station crystal clock. The memory in this case has a bigger capacity so that both the sources and the switching time of each event (in the case of machine sources the start time, which is a few seconds earlier than the actual on-air switching time) can be stored.

The memory display panel, in addition to source indicators, has corresponding time

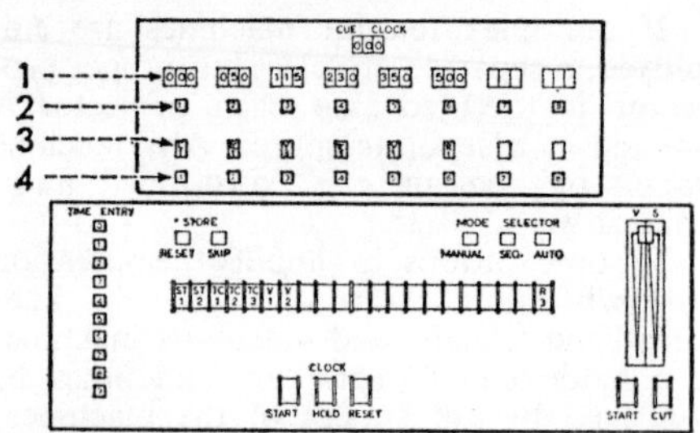

TIME-CONTROLLED MASTER SWITCHER. (*Bottom*) Control panel with conventional controls and 10 time entry buttons and three clock control buttons. (*Top*) Store display panel. 1. Time store indicators with, 2, random access buttons. 3. Source indicators with, 4, random access buttons.

store indicators for each event and a digital cue-clock display. With a three-digit cue clock the maximum duration of a break is 9 minutes 59 seconds. The 10 time entry buttons and cue-clock control buttons are mounted on the switcher panel together with other conventional controls.

Before the break, the operator can enter into the memory the sources and switching times required. To make a time entry, three buttons are operated sequentially. The time of the first event is 000. At the beginning of the break, the operator presses the clock start button, the cue clock starts counting, and the first event is initiated. The memory steps to the second event. When the cue-clock read-out coincides with the second-event time entry (e.g., 050), the second event is initiated. When it is desired to keep any event longer than the scheduled time the clock-hold button is operated just before the coincidence time for the next event. The cue clock stops and restarts only when the hold button is released.

In another system, the switching time is entered as actual duration time of each event. In this case, the cue clock, instead of being a count-up clock, operates as a count-down device. Ten seconds before the beginning of the break, the clock (set to 10 seconds) is started. When count-down reaches, for example, 5 seconds, the first-event machine is started. When the count reaches 0 the machine is transferred to on-air and the cue clock resets itself to the duration of the first event entered in the time store. The memory then steps to the second event. When the count again reaches 5 seconds the second-event machine is started, and so on. At the end of the break, the cue clock resets itself to 10 seconds and stops.

With a memory having a capacity sufficient for storing sources and duration times, the automatic transfers are by cutting only. Larger capacity memory or addition of a simpler transfer mode store permits more refined automatic presentation with fade, mix and wipe transfers.

The manual entry of information into the memory is sometimes substituted by an automatic read-in. During the preparation of the daily schedule, a punched paper tape or punched cards containing the switching information for all the breaks required during the day is prepared. The operator has to start the tape or card reader to load the memory for the first break of the day and then at the appropriate time to initiate it. At the end of this break, the tape reader is energized automatically and it reads the second break into the memory.

REAL-TIME OPERATION. The real-time control of automatic switching can usefully be employed only if the timing schedule is very strictly adhered to. If a play or a network programme over-runs or under-runs the scheduled time, the operator has to take action to delay or advance the beginning of the break. It is customary to assign real time only to the beginning of certain main programmes; the rest of the events are defined by their duration. The last event in a break is often used as a filler, its duration varied as required, so that its end coincides with the real time beginning of the main programme.

Today, when a larger and larger proportion of programmes is prerecorded, automation of the whole day's switching is coming into use. In emergencies, however, caused either by equipment fault or unforeseen incidents necessitating a sudden change of programme, there is no substitute for a human to take over control. H.M.

See: *Matrix; Special effects* (TV); *Vision mixer.*

□**Master Control Operator.** Operator of the controls which switch inputs from studios, videotape reproducers, telecines, etc., to one or more outputs. He not only has to switch the signals correctly, but must maintain levels of picture and sound, and also the quality of sync pulses, black/peak-white level, etc. In some set-ups, the switching function is taken away to a separate presentation control or is automated, leaving the master control operators with a supervisory function. These operators have roughly the same knowledge and skills as camera control operators, who do a very similar job.

See: *Master control; Presentation.*

Master Matching. In motion picture work, the assembly of the scenes of the original reversal master material, usually 16mm, to match the cutting copy or work-print. It corresponds to the stage of negative cutting where the original material is a camera negative.

See: *Laboratory organization; Narrow-gauge production techniques.*

Master Picture Monitor. A precision monitor placed at a key point in the control system and providing the operator with his main source of information. Generally the monitor can be switched to several points in the circuit to check the functioning of the apparatus. Both picture and waveform monitors are used in this role.

See: *Picture monitors.*

Master Positive. Specially prepared positive print made from the original negative film and used for the preparation of duplicate negatives rather than for projection. For black-and-white pictures and for sound tracks, the master positive is printed on a fine-grain emulsion of moderately low contrast, while for colour pictures the material used is a colour intermediate stock of low contrast and incorporating colour masking.

See: *Duplicating processes (B & W); Laboratory organization.*

Matrix. Strip of film carrying gelatin relief images used to transfer dye in the imbibition process of colour film printing.

See: *Imbibition printing; Technicolor.*

Matrix. A switching complex by which a number of electrical signal sources can be connected in any desired relationship to a number of outputs. Such matrices are employed in programme control units, and are coming into increasingly wide use with the need for ever more elaborate input-output switching. Additionally, the term matrix is applied to a network forming part of colour television coders and decoders whereby the three primary signals are correctly apportioned.

An output signal from each source is applied separately to the input side of the matrix which, for the sake of simplicity, may be thought of as a set of parallel wires.

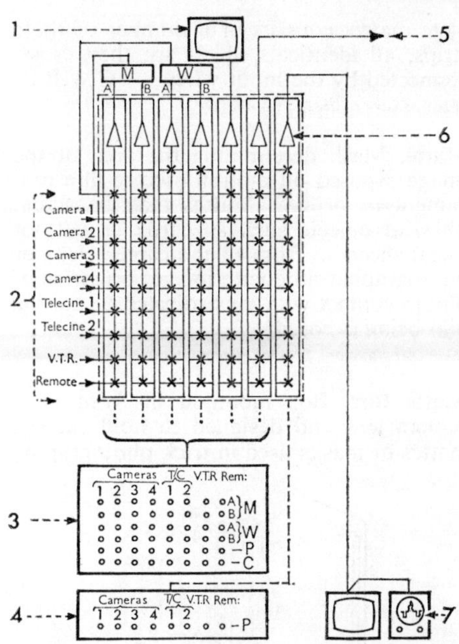

10×7 MATRIX. 1. Production preview monitor. 5. Studio output. M, W. Mixing and wipes amplifiers with A & B inputs. 6. Isolating amplifiers. 2. Inputs. 3. Production control panel with push-button operation for mixes (M), wipes (W), preview (P) and cut (C). 4. Engineering control panel. 7. Control operator's picture and waveform monitors. Rem. Remote (OB) source.

Similarly, the output side of the matrix can be regarded as another set of parallel wires crossed perpendicularly with the first set and arranged so that no two wires are physically in contact. Each output wire is associated with its own isolating amplifier and can be connected with one of the required output circuits. In a video switching matrix, outputs may be transmission, production preview, engineering preview or either of two pairs of circuits feeding special effects equipment.

At each point of the matrix where an output wire crosses an input wire, there exists a switching element which, when operated, enables them to be connected together. Switching elements used for this purpose may take the form of either a relay or a solid-state device containing diodes or transistors. These are usually physically grouped so that all the elements associated with a particular output wire are mounted as an individual sub-unit together with the appropriate isolating amplifier. If relay switching is used, these sub-units are commonly referred to as relay strips and the complete matrix consists of a number of these strips, all identical, which are then cross-connected by the input wires. C.W.B.R.

See: *Master control; Vision mixer*.

⊕**Matte.** Mask determining the area of the image exposed on motion picture film in a camera or printer. Mattes may be actual physical objects such as cards or cut-out metal sheets in front of the camera lens or photographic silhouette images on strips of film in contact with, or projected on to, the film being exposed.

See: *Art director; Special effects* (FILM).

⊕**Matte Box.** Box mounted in front of a camera lens and designed to hold camera mattes or masks used in trick photography,

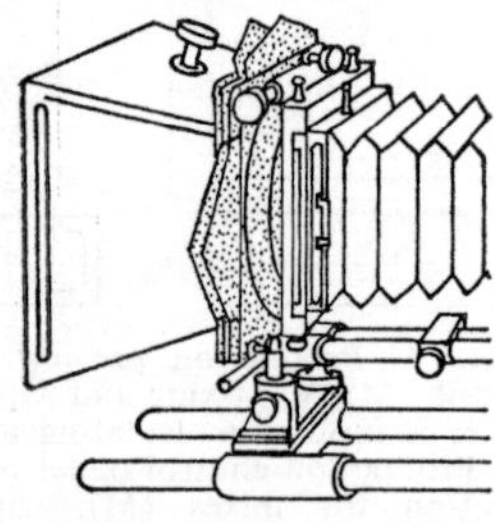

as well as filters. Usually combined with a sunshade (lens hood).

See: *Camera* (FILM).

⊕**Matte Loop.** Strip of film with clear horizontal bars separated by opaque areas which is used to expose the heavy density interframe lines on cinematograph release prints; for convenience, it is generally used as an endless loop in a contact printer.

See: *Printers*.

⊕**Matte Screen.** A diffusing projection screen having a reflection characteristic such that its brightness is substantially the same at all viewing angles. A matte screen is said to have a light gain of unity.

See: *Rear projection; Screens; Screen luminance*.

⊕**Maxwell, James Clerk, 1831–1879.** Scottish physicist and mathematician who made important contributions to the understanding of perception and the nature of colour. Demonstrated in 1861 the basis of the photographic synthesis of colour based upon the three primaries, red, green and blue.

⊕**McCay, J. Windsor, d. 1934.** American artist and cartoonist, pioneer of cartoon films. Early work included *Gertie, the Trained Dinosaur* (1909).

□**Mean Picture Level.** The mean (d.c.) level of the video signal.

□**Mechanical Scanner.** Any mechanical device, such as a mirror drum or Nipkow disc, which scans a scene point by point and line by line to produce an electrical signal.

⊕MEDICAL CINEMATOGRAPHY AND TELEVISION □

Although the technical aspects of medical cinematography are mainly concerned with the human subject, they are equally applicable to many aspects of biological photography, particularly to those involving experimental animals. To achieve success, not only must the correct technique be employed but there must also be an understanding of the subject being filmed.

EQUIPMENT

Most medical work is carried out with 16mm equipment, mainly for financial reasons but also because the equipment is particularly suitable for handling in confined spaces and where speed of manipulation is important. The bulkier 35mm equipment is occasionally used when still pictures from individual frames are required, such as the analysis of gait, and this size is still used by some workers, particularly in Germany, for cinemicrography. Eight-millimetre cameras have a limited use, partly as cheap recording equipment, and partly in conjunction with certain endoscopic apparatus where the available light may be sufficient to expose only a small image area.

As much medical work involves relatively small areas, reflex cameras are to be preferred because they avoid all difficulties arising from parallax. They become essential when filming inside a cavity such as the mouth for, even if a viewfinder can be sufficiently compensated for the parallax error, the angle of view would still be incorrect. A rotating turret with lenses from 15–100mm focal length has generally been used in the past but today the zoom lens is often chosen for greater flexibility combined with less bulk and weight. Unfortunately, most zoom lenses focus down only to 4 ft., so that they have to be used in conjunction with a suitable supplementary lens for close-up work.

In medicine, the timing and lengths of events may be difficult to estimate, so that 400 ft. magazines are desirable and, for the same reason, an electric motor is almost essential.

GENERAL CLINICAL RECORDING

The recording of clinical conditions in the studio in general calls for no particular techniques other than those normally associated with 16mm production.

Colour film is chiefly used and the majority of records are silent.

Synchronous sound films are confined to conditions where voice changes are associated with visual abnormalities such as some speech disorders and the endocrine disorder of myxoedema. When time relationships are of importance, a synchronous motor is used so that the film provides its own clock against which measurements can be made.

Spatial measurements may be required when recording the extent of limb movements and it is convenient to film these closely against a standard gridded background.

To ensure accuracy, the furthest possible camera position must be chosen with a suitably long focus lens so that the effect of parallax is minimized.

RECORDING GAITS. In both orthopaedic and neurological conditions, it may be necessary to record the patient's gait from both the lateral, antero-posterior and postero-anterior viewpoints. This would present little difficulty in a large film studio, but the confines of a hospital are a different matter.

When very much recording of this nature is to be done, a walking machine is sometimes used. This is similar to the horizontal portion of a moving staircase and is driven by an electric motor so that the speed can be varied to maintain the subject static in relation to the camera. True lateral and anterior views can be obtained without change in magnification and angle of view but unfortunately the effect of a moving floor is disturbing to the patient, who may already have a disordered gait, and such machines are generally confined to physiological studies on the normal subject.

In the absence of a walking machine, a studio at least 20 ft. × 30 ft. is needed and it is an advantage to have one end curved so that a greater area of unobstructed background may be used. Diffuse lighting minimizes background shadows and this should extend forward for 20 ft. to cover the antero-posterior view. For this latter view, a zoom lens allows the size of the subject to remain constant within the frame at all distances.

Slow motion taken at 64 fps is often done, and for research work a frame-by-frame analysis may be required.

When single frames of film are to be examined, it is an advantage to have a camera which allows the shutter to be closed to a quarter of its full opening (45°) so that any movement on each frame is more effectively frozen and greater accuracy of measurement achieved.

OPHTHALMIC CONDITIONS. Since the eye is a highly reflecting surface, a single light source is to be preferred, and this may need careful positioning to avoid the significant lesion being obscured. A small 100-watt spot lamp is sufficient to cover the 1½ in. field required for a single eye. When both eyes are to be filmed, the same principle holds good on a rather larger scale. Abnormal eye movements such as nystagmus are often filmed in slow motion and a frame-by-frame analysis may be needed for accurate study.

Infra-red emulsions are used for two purposes. First, in the study of the pupillary reactions to light, they allow the pupil to be filmed in the dark. The light stimulus need only be a torch bulb which will have no effect on the overall exposure. Secondly, infra-red may sometimes be capable of penetrating corneal opacities and revealing an underlying lesion.

SURGICAL OPERATIONS

The filming of all surgical operations imposes limitations on what might be the most desirable photographic techniques. First, the

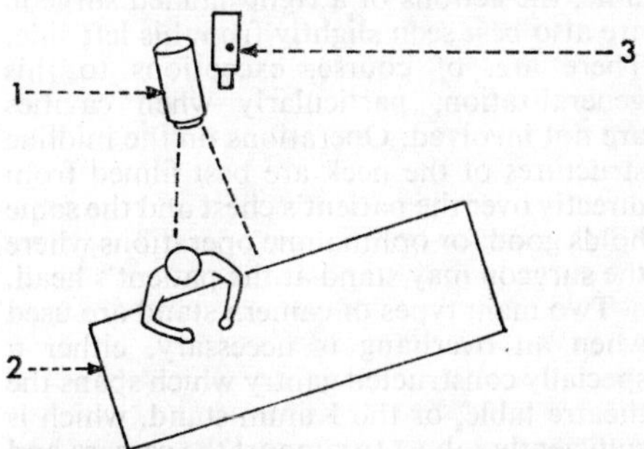

LIGHTING TECHNIQUES FOR SURGICAL OPERATIONS. 1. Spot lamp. 2. Operating table. 3. Camera. Best position for camera and single light source when filming operations involving cavities and small areas.

safety of the patient is paramount, and the camera team must be prepared to work with a minimum of equipment so as to avoid interference with normal operating theatre procedure. At the same time, they should naturally be familiar with the aseptic routine of a surgical operation. Secondly, the exact nature of the operation sequence to be filmed may well not turn out as planned so that the camera operator may have to cope with the demand for rapid changes in viewpoint and camera angles. Decisions have to be taken on the spot and a surgically knowledgeable camera operator can assist the editor by thinking ahead of possible difficulties, and minimize unfortunate jumps in the action by judicious changes in magnification and in viewpoint. This is all the more desirable because cutaway shots introduced purely to help cinematic smoothness become too obviously false in such technical films.

Limitations on lighting are imposed by the need to avoid overheating the patient, the avoidance of excessive highlights from moist surfaces and the fact that many operations involve filming into deep cavities. All these factors favour the adoption of a simple technique which can easily be adapted to changing circumstances. It is more important to obtain a picture of every stage than to strive after a perfectionist technique: no amount of editing will replace vital shots that have been missed.

CAMERA POSITION AND SUPPORT. The ideal camera position is from the point of view of the surgeon and, in practice, this involves working over his left shoulder. Most cameras are operated from the left side and this enables the instrument to be brought nearer the surgeon's viewpoint. At the same time, the actions of a right-handed surgeon are also best seen slightly from his left side. There are, of course, exceptions to this generalization, particularly when cavities are not involved. Operations on the midline structures of the neck are best filmed from directly over the patient's chest and the same holds good for ophthalmic operations where the surgeon may stand at the patient's head.

Two main types of camera stand are used when an overhang is necessary, either a specially constructed gantry which spans the theatre table, or the Kamm stand, which is sufficiently robust to support the camera and to provide a step on which the cameraman stands.

A zoom lens, when fitted with a supplementary for close-up work, is by far the

KAMM STAND. Allows a cine camera to be positioned centrally over a bed or operating table, and is robust enough to provide a stand for the operator.

most convenient objective. The zoom facility should be sparingly used, the main object being to provide a precise and quick method of framing the relevant action. An electric motor and 400 ft. magazines are, of course, invaluable.

LIGHTING. Fortunately, most surgical operations involve small areas rarely exceeding 15 in. in diameter, and these can be adequately illuminated by a 1,000-watt spot lamp.

A single light source can offer several advantages. When the depth of a cavity has to be filmed, the lamp can be focused into it, and over-illumination of the skin edges, which are both nearer to the lamp and lighter in tone, can be avoided.

With only one light source, highlights from moist surfaces are minimized. Shadows from a single axial lamp will fall behind their source and remain unobtrusive. This is important because, at a surgical operation, there are many pairs of hands close to the field of view but not visible to the camera and, if multiple lamps are employed, shadows of unknown origin continually pass across the screen.

Lastly, the single lamp permits a technique that ensures that the surgeon keeps his head and hands out of the line of sight of the camera. This is done by placing the spot lamp between the camera axis and the surgeon's head. As the other lights in the theatre are relatively dim, he thus obstructs his own view before that of the camera and is forced to play his actions to the camera.

In the operating theatre, a spot lamp is safer than other types on account of the mechanical protection offered by the housing and lens, while heat filters are readily fitted.

Lamps and particularly high-intensity sources operate at temperatures far in excess of the flash point of ether and other inflammable anaesthetics, and it is the responsibility of the photographer to point this out to the anaesthetist and ensure that no anaesthetic of this type is used.

CINE-ENDOSCOPY

Endoscopy can be said to start where cavity photography leaves off. A good definition is that cavity photography involves a hole in which the depth is less than its width, whereas true endoscopy is concerned with holes whose depths are greater than their width.

A great deal can be learned in medicine from the study of those regions which are accessible to an endoscope and these commonly include the respiratory tract from the larynx down to the bronchi, the oesophagus and stomach, the ear drum, the rectum and colon, the vagina and the bladder.

In addition to using the natural orifices, the surgeon may sometimes insert the endoscope through an artificial incision and this has been used to study the interior of the abdomen (peritoneoscopy) and recently even the ventricles of the brain. In all of these techniques, it is impossible for more than one person to see at once and, since the time available for viewing the subject is limited, cinephotography provides a permanent record for consultation and teaching.

The problem of cine-endoscopy consists of illuminating the subject to a high intensity, providing a satisfactory viewing system for the operation and being able to focus and operate a cine camera. All of these must be done simultaneously down a tube perhaps 40 cm long and 10mm in diameter.

OPEN TUBE ENDOSCOPY. Although many attempts had previously been made, the earliest practical system was evolved in the US during the 1940s and this is still the basis of much equipment today.

The system uses an open tube endoscope, all the optics being at the outer or proximal end. A beam splitter in front of the camera lens diverts a part of the light into a second matched lens which is coupled to that of the camera. The aerial image formed by the second lens is viewed through a telescope and used for framing and focusing. The development of the cine reflex camera has since enabled this viewing system to be simplified, as the image can be seen on the focusing screen. The light source is attached to the camera and is focused by a condenser lens on to a 45° mirror which surrounds the camera lens. The beam is focused into the proximal end of the endoscope, which is internally polished so that a large proportion of the light emerges at the distal end of the scope. This system has provided excellent colour pictures on 16mm film but is subject to two limitations.

The first is that the surgeon has not only to focus the camera himself but to manipu-

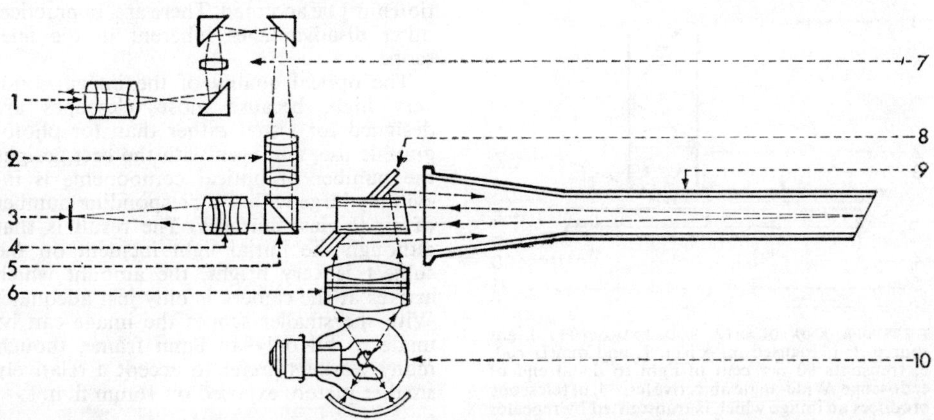

HOLLINGER-BRUBAKER SYSTEM FOR CINE-ENDOSCOPY. Illumination by lamp, 10, and condenser, 6, is focused down endoscope, 9, after reflection from 45° mirror, 8. Beam-splitting cube, 5, diverts 15 per cent of light reflected from object into viewfinding system composed of telescope objective, 2, reversing and erecting prisms, 7, and eyepiece objective, 1, which together enable aerial image to be viewed with correct orientation. Remaining 85 per cent of light is focused by camera lens, 4, on to film, 3. 2, 4, are 90mm lenses forming matched pair focused by single control. In practice, light source is mounted at one side, not below.

late the scope using a view of which only a narrow plane is in focus. Since most endoscopic procedures are potentially dangerous if not expertly carried out, any impairment of the surgeon's view may have serious results. This difficulty can be overcome by incorporating a beam-splitter and an optical system which allows the camera to be set at right-angles to the axis of the endoscope.

The surgeon in this way views down the axis of the scope and now has the equivalent of a naked eye view from the position that he is accustomed to, while a second operator focuses the camera.

The second limitation involves the size and length of scope that can be used. The larger bronchi, the oesophagus and the rectum are feasible but smaller orifices demand the use of an optical telescope which is not compatible with a proximal lighting system.

OPTICAL TELESCOPES. It is not possible to see satisfactorily through a very thin tube on account of the very narrow angle of view which results. For example, when looking into the bladder, only a few square millimetres of its wall are seen. An optical telescope overcomes this. At the distal end of the telescope there is a wide-angle objective lens which covers a useful area. The image plane is, of course, only a few millimetres behind the lens, but if a series of repeater lenses is placed up the tube, this image can be transferred to the proximal end of the scope

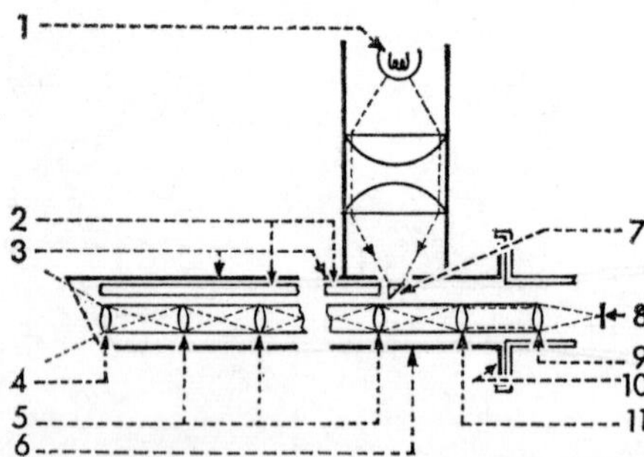

TELESCOPE AND QUARTZ ROD ENDOSCOPE. Light source, 1, is focused on prism, 7, and quartz rod, 2, transmits 80 per cent of light to distal end of endoscope. Wide-angle objective lens, 4, of telescope produces an image which is transmitted by repeater lenses, 5, and emerges from eyepiece lens, 11, as parallel beam which can be viewed by observer or by cine camera with film plane, 8, and objective lens, 9, focused at infinity. Camera can be hand-held, located against flange, 10. Telescope and quartz rod are both interchangeable, and are enclosed within endoscope sheath, 3, 6.

where it is brought to the focal plane of the eyepiece and emerges as a parallel beam. This beam can be brought to a focus either by the surgeon's own eye or by the lens of a camera.

The endo-telescope has been in use for direct observation for many years but its application to cinephotography has been impossible on account of the poor lighting associated with it. The light source must of necessity be alongside the wide-angle lens at the tip of the scope, and the minute size of lamp which could be accommodated has produced quite inadequate illumination. However, the quartz rod light guide introduced in 1952 enabled lighting intensities of a different order of magnitude to be achieved.

The light source can now be placed externally so that a high-intensity lamp can be used. This is focused on to the outer end of the quartz rod light guide which passes alongside the telescope. The light is transmitted by internal reflection down the guide and the efficiency is such that 80 per cent emerges at the distal end. This development has opened up a whole new range of cine-endoscopy and enabled routine records to be made of internal structures which previously had been but dimly seen.

The endoscopic telescope has a further advantage for cinematography in that the necessity for focusing the camera has been eliminated. The depth of field of the very short focus objective lens is such that the whole field remains sharp at all times, although the penalty of perspective distortion must be accepted. There are, in practice, other disadvantages inherent in the telescope.

The optical quality of the image is not very high, because most telescopes are designed for visual rather than for photographic use, and even with the best designs the number of optical components is inevitably large with a corresponding number of air-to-glass surfaces. The result is that although the initial light incident on the subject is very bright, the amount which arrives at the camera is only just adequate. With the smaller scopes the image can be made to fill only an 8mm frame, though many workers prefer to accept a relatively smaller picture exposed on 16mm film.

THE PROBLEM OF WEIGHT. It must be remembered that with all endoscopic procedures, the safety of the patient is of prime importance. The introduction of endoscopes demands considerable skill and delicacy of

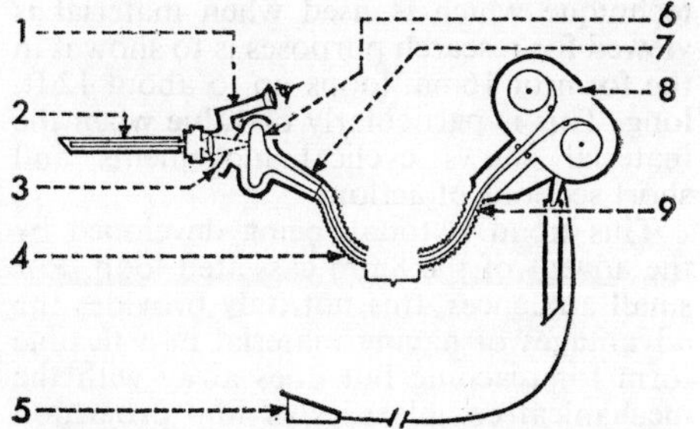

LIGHTWEIGHT CAMERA ATTACHED TO ENDOSCOPE. Heavy components, 8, consisting of 400 ft. film magazine and drive motor are supported independently on stand. Film, 9, and drive pass through flexible rubber tube, 4, to camera head, 7, which comprises film loops, 6, claw mechanism and mirror shutter, 3. This unit weighs only 28 oz. (794 gr.) and is attached to endoscope, 2. Operator views through eyepiece, 1, and controls camera with foot switch, 5.

touch on the part of the surgeon. When the endoscope is encumbered with extra light sources and a heavy camera (the smallest camera is heavy in comparison with an endoscope), the inertia of the equipment, even when counterweighted, makes the task very much more difficult, and good designs always incorporate a quick-release mechanism for the camera.

In one design only the bare essentials are retained in the camera head, which weighs a mere 28 oz. The heavy components of the camera, consisting of the 400 ft. magazine and motor, are located on a separate stand, the film and driving power being transmitted up a flexible tube to the camera head. This includes the gate, mirror shutter, viewing telescope, claw mechanisms and sprocket wheel.

The head is quite easily attached to the endoscope and is small enough to use as a handle for manipulation. A further advantage of the small size is that the distance between the endoscope and the examiner's eye is not significantly increased. This camera has great potential in addition to endoscopy for any procedure in which delicate manipulations are done, such as with the operating microscope.

INDIRECT LARYNGOSCOPY. **Indirect** laryngoscopy falls between cavity photography and true endoscopy. The technique is to hold a small mirror over the back of the tongue so that an image of the larynx, which lies some 2 in. below, can be seen. The difficulty arises from the fact that it is performed on the conscious patient and, as the laryngeal mirror is hand-held, there is no fixed optical axis on which the camera and light source can be aligned.

As a result studies on the phonating larynx have been confined to trained subjects who were prepared to co-operate by having a rigid mirror fixed over the back of their tongue. Using this principle with a 4 kW lamp, records have been obtained of laryngeal movement at 4,000 fps, showing the action of the vocal cords during vowel sounds of differing pitch and intensity.

Recently two methods of approach have achieved successful routine records on patients with pathological conditions of the larynx. One method uses the optical telescope system incorporating a right-angled prism to see over the back of the tongue. The apparatus is still fixed on an adjustable mount but the very-wide-angled view obtained makes accurate alignment far less critical, though the resulting picture suffers from perspective distortion.

The other method uses the beam-splitting attachment previously mentioned, allowing cameraman and laryngologist to work together.

The laryngologist is able to carry out his normal procedure of holding the patient's tongue with his left hand and holding the mirror in his right hand, thus maintaining flexibility of the patient's head. The camera assistant introduces the camera–beam-splitter–light source unit, which is suspended

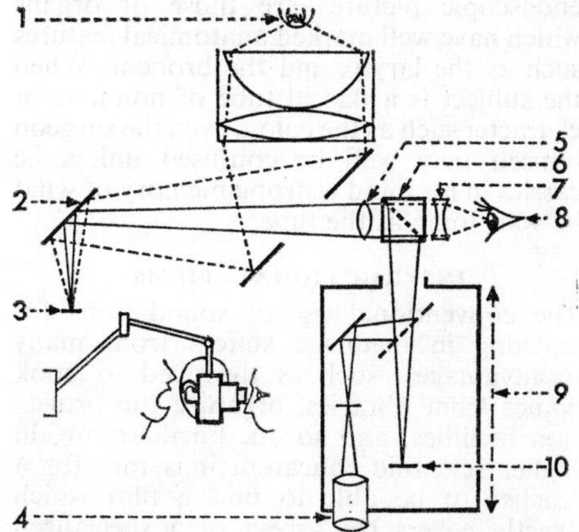

CAMERA FOR INDIRECT LARYNGOSCOPY. Light source, 1, illuminates mirror, 2, held over patient's tongue, focusing on plane of larynx, 3. Surgeon's eye, 8, looking through negative lens, 7, and film plane, 10, both see larynx through objective lens, 5, a partially silvered prism, 6, splitting the light 90 : 10 in favour of camera, 9. Camera operator uses focusing viewer, 4, and moves camera bodily to get sharp focus. (*Inset*) Whole unit is free floating between surgeon and patient on jointed support.

from an adjustable mount, between the patient and laryngologist, who now views through the beam-splitter. The free-floating mount allows the camera operator to follow all movements and focus the camera, but good teamwork is required for high quality results.

INTERPRETATION OF ENDOSCOPIC PICTURES. The projection and interpretation of the results of endoscopy suffer from a difficulty which does not occur with conventional cinematography – that of orientation.

When a surgeon inserts an endoscope into a tube such as the oesophagus or a hollow organ such as the bladder, he knows the position of its working end partly by his tactile sensations and partly by distance markers on the endoscope. If he rotates the instrument to the left and advances it a few inches he unconsciously incorporates this information in his interpretation of the view he sees. However, when the results of cine-endoscopy are projected the picture appears in a circle which remains centrally on the screen irrespective of where the endoscope is pointed. Within the circle the picture may move from side to side and objects may change their size, but as the audience lack any tactile sensation of having moved the instrument themselves it becomes very difficult to maintain a sense of orientation. It is for this reason that the most successful endoscopic pictures are those of organs which have well marked anatomical features such as the larynx and the bronchi. When the subject is a flaccid tube of nondescript character such as the colon, even the surgeon himself may well be confused unless he carries in his mind a strong memory of what he was doing at the time.

INSTRUCTIONAL FILMS

The conventional use of sound films for teaching in medicine suffers from many disadvantages, such as the need to book copies from libraries, organize the projection facilities, and so on. Furthermore, in higher scientific education it is rare for a teacher to be able to find a film which exactly covers the aspect of a specialized subject with the emphasis that he would choose.

A more flexible approach occurs when silent films are used in short sections each separated by a foot of black leader. This enables the film to be used as animated lantern slides with a live commentary that is specific to teacher and audience. Another technique which is used when material is viewed for research purposes is to show it in the form of 16mm loops up to about 12 ft. long. This is particularly of value when the material shows cyclical movements and short sections of action.

This trend is today being developed by the advent of the 8mm cassetted loop. For small audiences, this not only provides the advantages of having material in a flexible form for teaching but does away with the mechanical complexities of film projection and the need to use hire facilities. Subjects which are now available for sale in this form include many instructional techniques such as basic nursing procedures, bacteriological techniques, resuscitation techniques and instructions for patients. Factual information is represented by microscopic sequences of live material often in time-lapse, conditions seen on endoscopy, physiological experiments, and many privately owned loops on clinical conditions.

Cassetted film loops, being mechanically simple to operate, lend themselves to use by the student himself as opposed to by the teacher. The teaching laboratory which provides material for individual work by students can now add film in this form to the existing slides, tapes, specimens and microscopic sections that are available.

Film is not only being applied to teaching situations but is being introduced into examinations where the patient chosen for observation and diagnosis by examinees can be presented on film. This has the effect of presenting all examinees with identical information.

Finally film has a place in the treatment of hospital patients. In one form it is used as instructional loops for patients who need to learn techniques such as giving themselves injections or changing their ileostomy bags, while in the form of full-length sound films it is used in group therapy sessions for psychological disorders where the film is used to introduce new ideas and promote discussion.

CLOSED CIRCUIT TELEVISION

The use of closed circuit television (CCTV) in medical schools differs little from its use in other teaching fields in respect of the apparatus and techniques required. As a transmitting medium, it can be used to demonstrate patients to otherwise unwieldy numbers of students, which is of particular value in cineradiography and in psychiatry. Psychiatric interviews are difficult to conduct in the presence of students but a remote-controlled camera is unobtrusive, giving students an excellent view of the patient, while a teacher with the students can comment on the progress of the interview without disturbing the clinical situation.

MICROSCOPY. CCTV is especially useful when associated with the microscope. Few schools can afford a phase contrast instrument for all their students so that demonstrations of live tissue at high magnifications are otherwise impracticable.

Another situation occurs with the conventional microscope and a class each equipped with their own instrument and a similar stained microscope slide. The lecturer is handicapped, however, by uncertainty as to whether every member of the class has his microscope aligned on the part of the slide under discussion. But, by using a camera over his own microscope, he can demonstrate the desired field to every student. The fact that it is in monochrome is of no importance as the student will see the colour down his own instrument.

Adapting a television camera to the microscope is relatively simple, the main requirement being a means of support which allows the camera lens to be accurately aligned with the optical axis and close to the microscope eyepiece. The choice of camera lens is governed by the same factors as in cinemicroscopy. Most microscopes are designed for working vertically and some television manufacturers advise against their cameras being used in the vertical position for more than a short period, partly because dust may accumulate on the face of the tube, and partly because ventilation may be inadequate in this position. In this case, a right-angled prism will need to be used on the front of the camera lens.

PRESENTATION OF RADIOGRAPHS. In the presentation of radiographs to large numbers, CCTV has great potential. Lantern slides of X-ray films take a considerable time to prepare and yet they may be used but once. X-ray projectors are very expensive, are relatively inefficient with dense radiographs, and they cannot adjust the size of the image, so that if a chest film fills the screen, a dental film will appear like a postage stamp.

A television camera is, however, sufficiently flexible to adjust image size, contrast and density. In practice, it is important that the viewing screen used to illuminate the radiograph be fitted with adjustable masks so that extraneous background light does not affect reproduction quality. A 4 : 1 zoom lens on the camera will give a sufficient size range for most originals. Although automatic gain control will cope with most variations in density, it is convenient to have an overriding control so that unusual contrasts can be manually adjusted. Definition does not come up to the quality of a good lantern slide but the ability to zoom in on a small area does much to compensate.

TELEVISION OF SURGICAL OPERATIONS. In the operating theatre, CCTV has proved disappointing for teaching students. The lack of colour is a serious disadvantage and few surgeons can, while operating, produce the flow of commentary needed to clarify the picture. The few successful areas have been those in which the student normally sees little or nothing, such as surgery of the eye and middle ear. For the latter, a beam-splitting modification to a binocular microscope allows a television (or 16mm cine) camera to be attached and a greatly enlarged image can be seen on the screen. However, the small depth of field, lack of colour and technical difficulties associated with operations on the middle ear combine to give a picture of limited value.

In general surgery, the camera is often incorporated in the scialitic lamp, the controls being remotely operated. Though it is disappointing as a teaching medium, an unexpected use has emerged – that of keep-

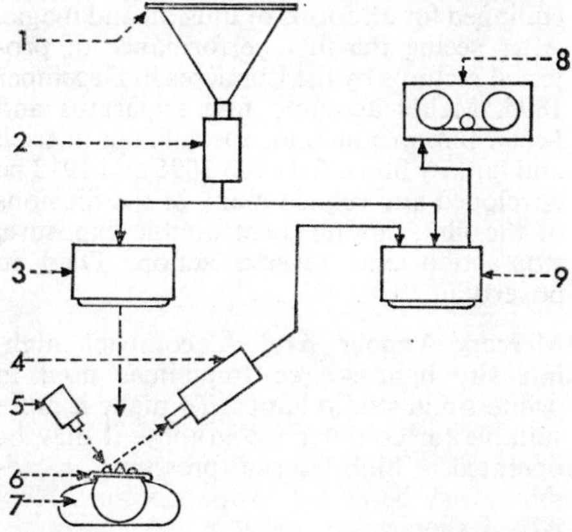

TELEVISION EYE MARKER. Set-up for obtaining records of scanning movements of eye. 7. Patient seen from above, positioned by headrest, 6. Spotlamp, 5, focused on eye reflects into vidicon camera, 4. Patient looks at monitor, 3, which derives its picture of test chart and other material, 1, on cine film via second vidicon camera, 2. Combined images of two cameras are displayed on second monitor, 9, watched by investigator, and recorded on videotape, 8.

ing those outside the theatre in touch with the progress of the operation. Modern techniques, particularly in neurosurgery, demand the use of much ancillary recording apparatus, and television has enabled the number of technicians actually present in the theatre to be kept to a minimum. Bulky apparatus can remain outside but the television screen keeps the operator in touch with the progress of the operation.

THE TELEVISION EYE MARKER. The television eye marker is an interesting application which would be almost impossible to achieve by any other method. The purpose of the technique is to study the manner in which the eye scans a picture or moving scene presented to it.

The subject sits with his head in an ophthalmic stand in front of a television screen on which the test pictures are shown. A small light is shone on to his eye at an angle and the reflex from it is picked up by a second camera. The image of this light reflex is then superimposed over the test image on a second monitor which is seen by the experimenter.

First a geometrical test chart is shown on the screen and the experimenter adjusts the reflex image electronically to coincide with points in the chart which he has asked the subject to look at. Thereafter the reflex image will continue to represent the point on the television screen at which the subject is looking.

When the eyes scan a scene they do so in a series of small rapid jerks, critical vision being confined to those aspects that the mind considers necessary for interpretation. This experimental set-up, therefore, is of great value to the psychologist studying perception. A child with reading difficulties can be seen to avoid looking at emotive words such as "mother"; car drivers when fatigued will tend to fixate more and more on the far point of a straight road, ignoring potentially useful information from each side; psychologically distasteful objects in a picture are skipped over and not noticed.

P.N.C.

See: *Cineradiography; Closed circuit systems.*

Books: *Research Films in Biology, Anthropology, Psychology and Medicine*, by A. R. Michaelis (New York), 1955; *Medical Photography*, by T. A. Longmore, P. Hansel and R. Ollerenshaw (London), 1962; *A System of Oph thalmic Illustration*, by P. Hansell (USA), 1957.

⊕**Méliès, Georges, 1861–1938.** French artist and illusionist. In 1888 bought the Théâtre Robert-Houdin in Paris, designed and equipped for all forms of illusion and magic. After seeing the first performance of projected pictures by the Lumières in December 1895, Méliès acquired film apparatus and began film production, specializing in trick and fantasy films. Between 1896 and 1913 he developed and refined many of the illusions of the film, among them double exposure, stop-action and reverse action. Died in poverty in 1938.

⊕**Mercury Vapour Arc.** A compact high-intensity light source sometimes used in printers and studio lamps. To make it more suitable for colour photography, it may be operated at high internal pressure, or cadmium may be added to the mercury. Now largely superseded.

High pressure mercury vapour arc lamps with metallic additions to improve their colour quality are also employed in some 35mm cine projectors, although the presence of strongly defined mercury bands in their spectrum raises problems in the reproduction of colour.

See: *Printers.*

□**Mesh Effect.** A type of blemish on the television picture. Camera tubes such as the image orthicon and vidicon which have a target mesh electrode can, under certain conditions of manufacturing misalignment, give an effect on the screen of the monitor as though the image were being viewed through a mesh or grill. This can usually be removed by a slight adjustment of the beam focus control, although occasionally a tube has a visible mesh pattern at the point of optimum beam focus.

Even if the effect is slight, dark scenes make such a camera tube unsuitable for broadcast use.

See: *Camera* (TV).

⊕**Messter, Oskar, 1866–1943.** German inventor and manufacturer. Inspired by the Kinetoscope in 1895, designed a projector, sold as the Kinematographe in June 1896. Subsequently designed and manufactured much cinematographic apparatus.

⊕**Microcinematography.** Motion picture recording on film by a combination of microscope and cine camera.

See: *Cinemicrography.*

MICROPHONES

Microphones are devices for converting the energy of sound waves into mechanical energy and then into electrical energy. A sound wave produces a difference of pressure between the faces of a diaphragm, which makes it vibrate. The vibration is used in conjunction with an electrical system to generate a voltage, the frequency and amplitude of which are determined by the frequency and amplitude of the sound wave.

BASIC TYPES

Moving coil and ribbon microphones work on the principle that, if a conductor moves in a magnetic field, a voltage is produced across its ends. In microphones, the magnetic field is provided by a fixed permanent magnet which has pieces of soft iron known as pole pieces attached to it so that an intense field is produced across a narrow gap.

MOVING COIL. In the moving coil microphone, the coil (the conductor) is held positioned in an annular gap by the diaphragm. When the diaphragm vibrates, the

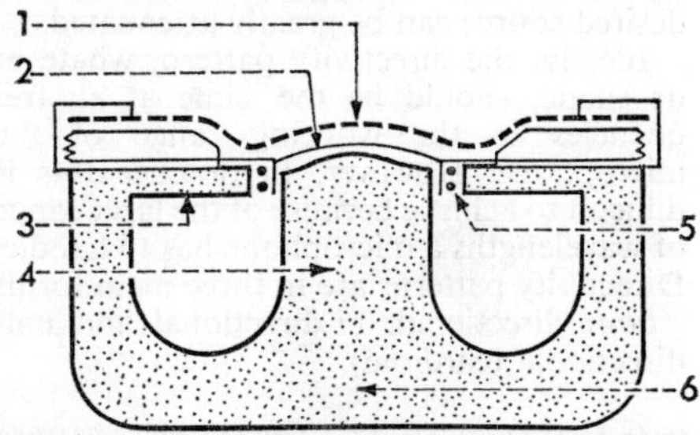

MOVING COIL MICROPHONE. 1. Perforated metal guard protecting, 2, diaphragm on extension of which the coil, 5, is wound. It lies in annular gap formed by central pole piece, 4, of permanent magnet, 6, and soft iron outer pole pieces, 3. Vibration of coil in magnetic field produces voltage.

coil moves in the magnetic field and a voltage is produced across its ends. Moving coil microphones are often referred to as dynamic microphones.

RIBBON. In the ribbon microphone, the diaphragm is also the conductor. It consists of a thin strip of metallic foil placed between two elongated pole pieces.

CRYSTAL. Crystal microphones work on a different principle which needs no magnetic field. If certain materials are mechanically deformed, e.g., bent or twisted, a difference of voltage is produced between the faces of the material. This is known as the piezoelectric effect, and it may be applied to microphones by using the forces produced by a sound wave to deform a thin layer of suitable material.

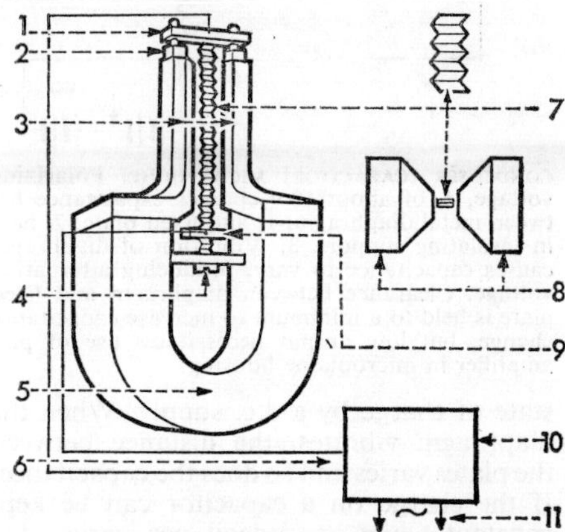

RIBBON MICROPHONE. Works on same principle as moving coil microphone, but ribbon acts as both conductor and diaphragm. Corrugated ribbon, 7, vibrates between pole pieces, 3, attached to permanent magnet, 5. Top contact, 1, separated from poles by insulator, 2, and bottom contact, 9, go to input, 6, of output transformer, 10, which raises impedance from less than one ohm. 8. Top view of pole pieces. 4. Ribbon tension adjustment. 11. Microphone output.

CONDENSER. In the condenser or electrostatic microphone, the diaphragm is one plate of a capacitor (condenser), the other plate being fixed. The capacitor is kept in a

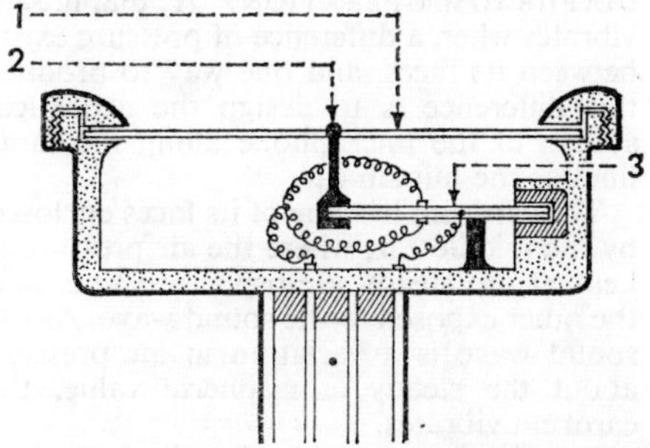

CRYSTAL MICROPHONE. 1. Diaphragm connected by link, 2, to crystal bimorph, 3, consisting of two thin layers of Rochelle salt, fixed together back to back, forming one side of the generator; exposed faces are connected to form the other side. Deformation of generator by movement of diaphragm sets up varying voltage differences between sides of generator. Other mechanical arrangements are possible.

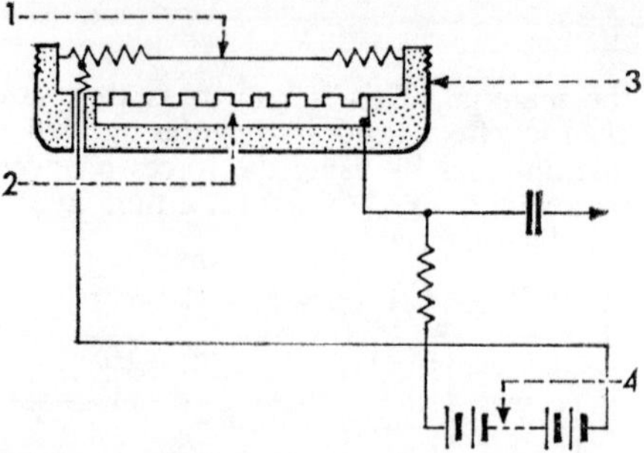

CONDENSER (CAPACITOR) MICROPHONE. Polarizing voltage, 4, of about 60 v charges capacitance between metal diaphragm, 1, and fixed plate, 2, held in insulating support, 3. Vibration of diaphragm causes capacitance to vary, producing alternating voltage. Clearance between diaphragm and fixed plate is held to a minimum to increase capacitance change, but low output necessitates use of pre-amplifier in microphone housing.

state of charge by a d.c. supply. When the diaphragm vibrates, the distance between the plates varies and so does the capacitance. If the charge on a capacitor can be kept constant while its capacitance varies, the voltage across the plates will vary as well. This type of microphone is in wide use today and gives excellent results though it has one practical disadvantage: the impedance of a direct conversion type of electrostatic microphone is extremely high. To overcome this, another form of the electrostatic principle is used in which the variations in capacitance modulate a radio-frequency oscillator. The r.f. is demodulated to give the audio-frequency output.

CHARACTERISTICS

DIAPHRAGM OPERATION. A diaphragm vibrates when a difference of pressure exists between its faces, and one way to produce this difference is to design the acoustical system of the microphone along the same lines as the human ear.

The eardrum has one of its faces enclosed by the middle ear, where the air pressure is kept at the steady atmospheric value, and the other exposed to the sound wave. As the sound wave is a variation in air pressure about the steady atmospheric value, the eardrum vibrates.

Exactly the same principle is applied to some microphones, and is called pressure operation. The inner face of the diaphragm is enclosed by an air chamber which has a pressure-equalizing tube functioning in the same way as the Eustachian tube in the ear. The outer face is open to the sound waves, and thus the diaphragm vibrates in sympathy with their frequency and amplitude.

Another way to produce the necessary difference of pressure between the faces of the diaphragm is to allow the sound wave access to both faces, but at different times. Instead of comparing the pressure variations of the sound waves with the steady atmosphere pressure, the pressures at two separated points in space are compared and their difference causes the diaphragm to vibrate. This is called pressure-gradient operation.

DIRECTIVITY PATTERNS. The graph of a microphone's response to sounds coming from different angles is called a polar diagram or directivity pattern. This is important as, in most cases, sounds reach the microphone not only direct but by reflected paths, as for instance from the walls of a room. The shape of its directivity pattern has a great bearing on how a microphone should be placed to obtain an acceptable balance between direct and reflected sound.

Another useful lesson from the polar diagram is that a microphone with known directional properties can be angled in such a way that the contribution from an undesired source can be greatly attenuated.

Ideally, the directivity pattern, whatever its shape, should be the same at all frequencies in the working range of the microphone. However, in practice this is difficult to achieve because of the large range of wavelengths a microphone has to handle. Directivity patterns are in three main forms – omni-directional, bi-directional and uni-directional (cardioid).

OMNI-DIRECTIONAL. The simplest pattern is a circle, indicating that the microphone is equally sensitive in all directions. This is the omni-directional pattern, characteristic of a pressure-operated microphone, which simply compares the sound wave pressure with that of the atmosphere, so that it does not matter from what direction the sound wave reaches the microphone. From the theory of pressure operation this would be true at all frequencies, but in practice the directivity pattern is dependent on frequency.

BI-DIRECTIONAL. The bi-directional pattern is often referred to as a figure-of-eight on account of its shape. It can be obtained by using a pressure-gradient device, such as a ribbon microphone.

CARDIOID. Cardioid is the term used to describe a one-sided or uni-directional pattern, which ideally should be heart-shaped. It provides the same degree of discrimination of direct against reflected sound as a figure-of-eight. The high ratio of front to back pick-up is very useful in practice.

The cardioid shape can be obtained with two microphone elements by combining the electrical outputs of a pressure-operated microphone and a figure-of-eight one. However, single-diaphragm cardioids of the moving coil, ribbon or condenser types have been developed and are now widely used. The pattern is derived by combining the driving forces on the diaphragm due to a pressure component and a pressure-gradient component. The combination is performed at the acoustic input, not on the electrical output as in the two-element type.

PATTERN VARIATION. In professional work it is often a great advantage to be able to vary the directivity pattern of a microphone, for instance to control the balance in a studio without having to move the performers or microphones.

One method is to use two cardioid units placed back-to-back so that their directions of maximum sensitivity are 180° apart. Condenser microphones using this technique are available in which the pattern is controlled by altering the voltage on one of the diaphragms relative to the common fixed plate. This causes the two units to combine in phase or out of phase, the pattern varying from omni-directional, through cardioid to figure-of-eight.

SENSITIVITY. The electrical output of a microphone for a given sound level should be as high as possible, so that an adequate signal-to-noise ratio is obtained at the start of the electrical chain from microphone to recorder or loudspeaker.

Since the microphone converts acoustical signals into electrical signals, the definition of sensitivity gives the electrical output for a given acoustic input.

IMPEDANCE. In practice, the impedance of microphones can range from a few ohms to several thousand ohms.

Moving coil and ribbon microphones are low-impedance generators, typical figures being about 30Ω (ohms) and 1Ω respectively.

With the ribbon, a transformer is normally built inside the case to step up the impedance, the turns ratio of the transformer controlling the output impedance. Some models have fixed impedances, typical values being 30Ω or 300Ω. With other types, a choice of impedance is available and this can be Low – 25Ω, Line – 600Ω or High – 50kΩ.

The advantage of having a low-impedance microphone is that the length of cable connecting it to the rest of the equipment can be as long as required, at least up to 100 yards. The disadvantage is that, with equipment having a high input impedance, a transformer must be used to match the microphone and its cable into the equipment. In addition, the sensitivity of low-impedance microphones is generally low.

A high-impedance microphone can be coupled straight into such equipment without a transformer, and the sensitivity is usually much higher. But the length of cable which can be used with a high-impedance microphone is severely limited. This is because the cable acts as a capacitance across the microphone output terminals. With a low output impedance, the effect of this shunting is negligible, but with a high output impedance, the shunting causes a serious drop in the signal passed on to the apparatus. The value of the cable capacitance depends on its length, and cables can be specified as having so many pF (picofarads) per foot. A value of 80 pF is a typical maximum shunting capacitance that can be tolerated before serious attenuation results. Depending on the type of cable, up to 20 ft. is about the maximum length that can be used with safety.

Moving coil microphones are available without transformers giving an output impedance of 30Ω. However, like the ribbon, there are many models available with a built-in transformer to give a choice of output impedances, and the foregoing remarks apply equally well to the moving coil types.

Crystal microphones have a very high output impedance. They consist essentially of two metallic skins separated by the crystal material, and so they behave rather like a capacitor – in practice a value of about 1,000 pF is representative. This means

MATCHING OF HIGH IMPEDANCE MICROPHONE

Input impedance of Equipment (Megohms)	*Frequency at which response is 3dB down on that at 1,000 Hz (Hz)*
1	200
2	100
5	40
10	20

that the impedance is not only high, but it is also frequency dependent since capacitive reactance is inversely proportional to frequency. With decreasing frequency, the internal impedance of the microphone goes up, and this can lead to a loss of bass frequencies unless the microphone is fed into a much larger impedance.

All high impedance, low capacity cables are susceptible to hum and interference pick-up and can generate noise signals if they are moved. It is therefore essential to take care with the cables of high impedance microphones.

Condenser capsules too have a high impedance, indeed much higher than a crystal type, the capacitance being about a tenth that of a crystal element. This means, in practice, that an amplifier is placed immediately adjacent to the capsule so that the leads are kept as short as possible. In fact, a professional condenser microphone normally consists of the capsule with its associated head amplifier. The performance of the microphone is the performance of the two together.

SPECIALIZED TYPES

PERSONAL MICROPHONES. It is sometimes convenient to place a microphone on the person, usually secured on the chest by a neck halter. This leaves a demonstrator's hands free and he can walk about while he talks. Chest microphones for each speaker are also often used for interviews or discussions.

"Lavalier" or "lanyard" are the usual terms to describe such microphones, and there are several types available. Almost all are moving coil units slung so that the diaphragm points upwards; the measured directivity pattern is omni-directional but this is very much modified by the close working position.

One difficulty of the chest microphone is that its position attenuates the vocal high frequencies. These frequencies, essential for good quality, are projected from the mouth over a much smaller area than are the low frequencies. Little or no radiation takes place in a downward direction. To correct the response, Lavalier microphones have a broad peak in the high frequency response between approximately 4,000 and 7,000 Hz. This peak is usually obtained by suitable internal design of the microphone but a current model uses an adjustable sleeve. When fully extended, the sleeve forms a cavity in front of the diaphragm which is broadly resonant over the required range. With the sleeve retracted, the frequency response is smoother and the microphone can then be used in the hand.

Noises produced by the user's breathing can cause accentuation of some of the low frequencies. Suitable correction is obtained by reducing the low frequency response of the microphone.

Clothing can also affect the reproduced quality if, for example, the microphone becomes covered by a close-weave tie. Such accidents must be avoided. The rustling noise caused by movement can often be restricted by clipping the microphone to the clothing.

NOISE-CANCELLING MICROPHONES. Widely used for commentaries from sports events, this ribbon microphone is placed very close to the mouth, the distance from mouth to ribbon being fixed by a bar against the upper lip. When struck by spherical waves from a near source, a strong bass frequency rise occurs. This is equalized out, with the result that plane waves from a distant source are heavily attenuated in the bass and middle frequencies, resulting in very good signal-to-noise discrimination.

HIGHLY DIRECTIONAL MICROPHONES. The microphone cannot always be placed near the source of sound, as, for example, in recording bird song, the crack of the ball on the cricket bat or the commands from the officer at a military parade. For these situations it is essential to have a microphone which picks up sound energy only over a small angle or which, in other words, has high directivity. This can be obtained by using interference effects, as is done with the wave type microphone.

The simplest and probably the oldest way of providing such interference is the parabolic reflector. The reflector picks up energy from a distant source on its axis as a plane wave which is reflected when it strikes the reflector surface. It is a property of a parabolic surface that, irrespective of what part of the reflector the wave strikes, each part of the wave arrives in the same phase at the focal point of the reflector. This is because the path length from an axial source to the focal point via the surface is the same, whichever part of the surface is used. Obviously, at other angles of approach these conditions do not apply, and maximum sound intensity at the focus is obtained only when the reflector is facing the source. Placing a microphone at the focus, facing into the reflector, produces a sound-collecting device of high directivity.

Another method is to use a series of tubes, of small diameter and graded length, fixed in front of the diaphragm of a pressure-operated microphone. When the tubes face the source, the contributions from each tube are arranged to arrive at the diaphragm in phase. For other angles of incidence, there is a phase difference between the various contributions, owing to the different path lengths each has traversed. There is thus maximum output when the tubes face the sound source.

A fundamental difficulty of the wave principle is that, at low frequencies, there is little or no directivity with practical sizes of microphone. No reflection takes place from an object unless the object is large compared with the wavelength. A practical size of reflector is 3 ft., which is effective down to about 800 Hz. Usually some form of bass cut filter is connected in the microphone's output circuit.

Another difficulty is that the reflector becomes very effective at high frequencies – short wavelengths – and there is extreme narrowing of the directivity pattern. This causes serious loss of high frequencies if the source or the microphone is moved. With some designs, this defect is corrected by introducing some defocusing – placing the microphone slightly away from the focal point and so broadening the high frequency directivity response. Other designs cover the outer part of the parabolic surface with absorbent material which is effective at high frequencies only. This reduces the effective size of the reflector and hence gives the same directivity at high and low frequencies.

With the tube type of microphone, the directivity at low frequencies depends on the maximum length – the larger the better because of the larger phase differences between the various contributions.

A modern version of the tubes idea is to allow the sound to enter a narrow slit running down a single tube coupled to a moving coil microphone. Standing wave effects are eliminated by a layer of fabric fixed over the slit.

Sound energy entering at the front end of the slit, however, suffers greater loss in travelling down the tube than energy entering further down the slit, causing amplitude differences between the contributions instead of only phase differences. This is counteracted by having more absorbent material over the slit near the diaphragm. High frequency loss in coupling the slit to the microphone is compensated for by having resonant cavities formed by small baffles fitted at 90° to the axis of the tube over the slit.

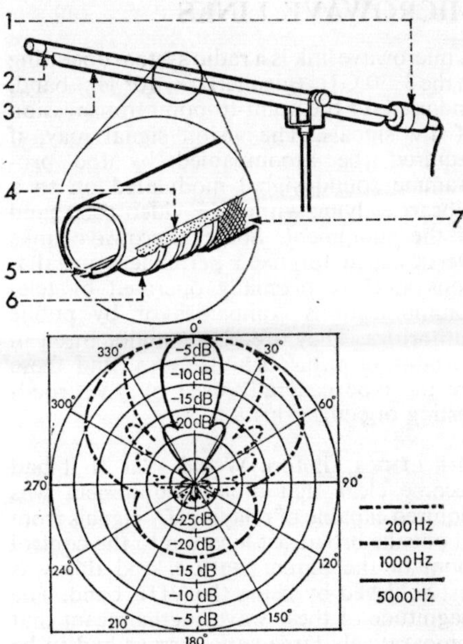

LINE MICROPHONE. Form of highly directional microphone employing narrow slit, 5, running along tube, 2, coupled to moving coil microphone, 1. Baffles, 6, form resonant cavities to increase high-frequency response. Absorbent material, 4, with gauze cover, 7, helps to eliminate standing wave effects. (*Below*) Polar diagram showing increase in directivity with increasing frequency.

Tube or slit microphones are sometimes called line or rifle microphones, and they have advantages over the 3 ft. parabolic reflector in weight and ease of handling. Smaller reflectors of 2 ft. diameter are available for work where high directivity is not required at low frequencies. The parabolic reflector is more sensitive, however, since there is a gain due to the larger amount of sound energy which is collected by the reflector.

With the line type, the maximum sound which strikes the diaphragm when the microphone is facing the source cannot be more than would have struck it in any case. In fact, there is some loss of sound energy when the tubes or slot are connected in front of the microphone unit. G.W.M.

See: *Sound; Sound mixer; Sound recording* (FILM)*; Sound recording systems.*

Books: *Microphones*, by A. E. Robertson (London), 1963; *Acoustic Techniques and Transducers*, by M. L. Gayford (London), 1963.

□MICROWAVE LINKS

A microwave link is a radio system operating in the 3–30 GHz (gigahertz = 10^9 Hz) band, and used for the point-to-point transmission of TV signals. The vision signal may, if required, be accompanied by the programme sound signal modulated on to a subcarrier lying within the video pass-band of the equipment. Some microwave links are designed for fixed permanent installations and are normally operated by telecommunications companies or by public authorities. They are also manufactured in portable or transportable form, and these are the type generally used by TV broadcasting organizations.

VHF LINKS. Before World War II it had become clear that mobile equipment was required capable of relaying TV signals from an outside broadcast location to the control room of the parent station, and this was first achieved by using the VHF band. The magnitude of these wavelengths meant that comparatively large aerial arrays had to be used to obtain enough aerial gain and, even so, considerable radio-frequency (r.f.) power was required, with the result that the VHF link equipment was big and ungainly. The microwave techniques developed during the war for radar and allied devices were, however, quickly adapted for TV purposes, and microwave links came into general use when TV broadcasting was resumed in 1946.

MICROWAVE ADVANTAGES. Microwave links are limited to line-of-sight conditions, but have a number of outstanding advantages over the VHF links previously used. At 7 mHz the wavelength is such that a 4 ft. diameter parabolic aerial reflector can achieve a beamwidth of 3 degrees (between 6 dB points), enabling a TV signal to be relayed over forty miles or more for a few watts.

Frequency modulation is used owing to the triangular nature of the noise spectrum so obtained (high-frequency noise proving less objectionable than low-frequency noise on a TV picture). This is made possible in the thermionic valve-operated equipments which are still in general use by a reflex Klystron valve in the output stage of the transmitter. Direct frequency modulation and fine tuning (and, if required, automatic frequency control) are attained by applying the appropriate video and d.c. control voltages to the reflector electrode, while coarse tuning is achieved by adjusting the physical size of the Klystron cavity. A typical mobile microwave link has an r.f. power output of between 100 mW and 3 watts, and these small aerial dimensions and low power requirements permit compact design.

In Britain, earlier equipments operated in the region of 4 GHz, but this band has now been re-allocated for other purposes, and 7 GHz and 11 GHz bands are now used. In general, the lower frequencies allow a wider beamwidth (for a given aerial reflector size) which makes the equipment less exacting with regard to aerial alignment and less susceptible to sway and vibration of the aerial platform. Heavy rainfall and small obstacles in the radio path do not have a serious effect. On the other hand, the higher frequencies give higher aerial gain (for a given reflector size) and correspondingly small r.f. plumbing, allowing lighter and smaller equipment design. Furthermore, the higher signal frequencies and narrower beamwidths make it possible to use a large number of TV channels simultaneously in the same band.

CURRENT EQUIPMENT. Microwave links in current use are typically as follows.

The transmitter and the receiver are generally similar in outward appearance, each consisting of aerial assembly and mounting, head unit, control unit, and interconnecting cable. The transmitter control unit, often in a suitcase type of container, is fed from the 50/60 Hz mains and includes power supplies, metering arrangements, video level controls, cable compensation circuitry and some video amplification. It is connected by up to 400 ft. of multicore cable to the transmitter head unit which holds the Klystron valve (appropriately cooled), modulator, tuning arrangements, and the r.f. waveguide plumbing. The head unit is clamped to the 4 ft. paraboloid and aerial feed, which is mounted on a collapsible tripod arranged to be adjustable in bearing and elevation. The receiver is similarly arranged, the head unit in this case containing receiver plumbing, Klystron local oscillator and crystal mixer, tuning controls, and intermediate frequency (i.f.) pre-amplifier, and the control unit containing the main i.f. amplifier, limiter/discriminator, AFC/AGC circuitry, video amplifier and level controls, power supplies, and metering arrangements. The multicore cable usually contains up to three

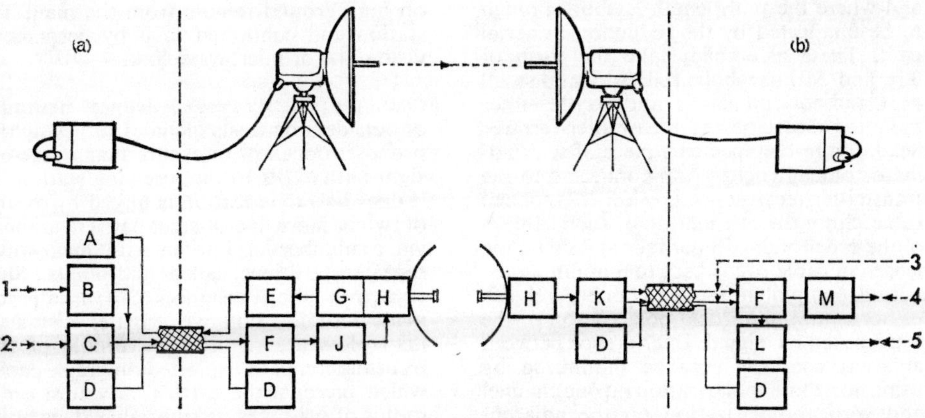

TYPICAL MICROWAVE RELAY: SIMPLIFIED BLOCK DIAGRAM. (a) Transmitter. (b) Receiver. In both, head equipment in proximity to parabolic reflector is linked by multi-way cable to equipment in portable container. A. Monitor output, fed by E, monitor amplifier. 1. Audio input to B, audio channel. 2. Video input to C, cathode follower stage. D. Intercom between transmitter and receiver to guide accurate positioning of reflectors for maximum signal. F. Video amplifier to J, d.c. restorer, and H, temperature-stabilized klystron associated with reference tuning cavity, G. From receiver klystron, H, signals feed to K, i.f. amplifier, limiter and discriminator. 3. Balanced push-pull signals to F, video amplifier, M, band suppression filter, and thus to 4, video output. Also to L, audio channel, and 5, audio output. Equipment is designed for standard colour transmissions and operation on 12-13 GHz band.

coaxials and about 20 twisted pairs, one of which is used to provide telephone communication between head and control unit positions. This is essential for aligning aerials, etc.

ANCILLARY EQUIPMENT. Most manufacturers also provide, as an optional extra, a means of multiplexing programme sound on to the vision channel. This is achieved by frequency-modulating the sound signal on to a subcarrier whose frequency is above the upper limit of the video signal but within the pass-band of the link equipment. The method of mixing is usually so arranged that failure of the sound circuitry does not also affect the vision signal. The two signals are separated by a filter at the receiving end and the sound subcarrier is then demodulated in the normal way.

Also provided as optional extras by some manufacturers are automatic frequency control of the transmitter, and a transmission monitoring device whereby the transmitted signal is sampled by a probe placed near the aerial and demodulated for display at the control position. This latter affords the only certain continuous check of the Klystron performance.

Microwave link equipment of the type described above may be installed in a two-ton truck, preferably in specially provided racking so that, while normally operated from the truck, it may, if required, be de-rigged and set up elsewhere. The vehicle serves as a mobile links control room and therefore also houses such items as radio-telephone equipment, test gear, automatic voltage regulator, telephones, video and audio patch panels, and heating, lighting and ventilation. It may thus be used as a transmitter, receiver or repeater station. During operation, the aerial assemblies may be placed on a platform on the truck roof or erected on a mobile tower vehicle or on the roof of a nearby building, as dictated by line-of-sight circumstances.

POWER REQUIREMENTS. Power consumption of a mobile microwave transmitter and receiver is of the order of 250–500 watts each, but the ancillary items need around 3 kW. A 3–4 kW diesel alternator is usually provided and may be towed by the control truck to sites where no mains supply is available. The radiotelephone is used for technical communication between the transmitting and receiving sites during initial setting up and aerial alignment and for liaison during actual transmission. It also enables an outside broadcast to be linked at very short notice without prior provision of telephone lines, since all essential features, i.e., vision, sound and control (cueing) circuits, are in radio form.

AERIAL ASSEMBLIES. In addition to the standard 4 ft. parabolic reflectors, smaller aerial assemblies are frequently used where space or other considerations so demand,

and where the path length is short enough to be unaffected by the reduction in aerial gain. These assemblies take the form of 3 ft. and 2 ft. parabolic reflectors and small horn and polyrod aerials, and may be either mounted direct on the transmitter-receiver head unit or clamped to a piece of scaffolding or other structure and connected to the transmitter/receiver head by low-loss coaxial cable. Since the attenuation of such cable is of the order of 0·4 dB per foot at 7 GHz, the length of cable run is kept to a minimum.

Link equipment frequently offers a choice of horizontal or vertical polarization of the transmitted r.f. signal. Interference between adjacent channels may be minimized by using horizontal polarization on one channel and vertical polarization on the adjacent channel.

A TV broadcasting company normally has one or more microwave receiving stations equipped for picking up signals relayed from outside broadcast locations. Depending upon topographical considerations, such a receiving station may be an integral part of the main TV station, or it may be situated on high ground remote from the main TV station and connected to it by permanent vision line or microwave link.

LOCATION SURVEYS. Advance planning of outside broadcasts includes surveying the proposed location to ensure that a line-of-sight path exists to the receiving station. If it does not, the location is linked by means of two or more line-of-sight paths in tandem via a number of intermediate microwave repeater positions called mid-points. Surveys over short distances can usually be made visually, but over greater distances map plots are required. A profile of the transmission path is plotted on graph paper which presents the earth's curvature on a radius of 4/3 times its true value. This gives an indication as to whether or not line-of-sight conditions are likely to exist, but in doubtful cases a test transmission is always arranged. T.A.K.B.

See: *Feeder; Transmitters and aerials.*

Books: *Foundations for Microwave Engineering*, by R. E. Collins (New York), 1966; *Microwave Engineering*, by A. F. Harvey (New York), 1963.

⊕**Mil.** An abbreviation for milli-inch, one-thousandth of an inch, which is the unit generally used in motion picture work to measure the width of a sound track or stripe. Verbal confusion is sometimes caused by the colloquial use of "mill" as short for millimetre, which is however always abbreviated to mm. or m/m. in the written form.

⊕**Miniature.** Scale model used to avoid the expense of the real thing. Usually applied to model sets constructed with false perspective and photographed from a carefully calculated viewpoint to combine with real action seen at the same time.

See: *Art director; Special effects* (FILM); *Studio complex* (FILM).

⊕**Mired.** Unit used for the measurement of □colour temperature (*mi*cro-*re*ciprocal-*d*egree), corresponding to the value in degrees Kelvin divided into one million.

See: *Colour temperature.*

□**Mirror Drum.** Scanning drum or cylinder, with a number of mirrors fastened to its periphery, each at a fixed angle relative to the adjacent mirrors, so that the image moves linearly with rotation of the drum. Used in some early forms of mechanical picture scanning.

See: *History* (TV).

⊕**Mirror Shutter.** Type of camera shutter in which the opaque blades take the form of mirrors. When the shutter is opened, light from the scene outside, after passing through the lens, falls on the film and produces an image; when the shutter is closed this light

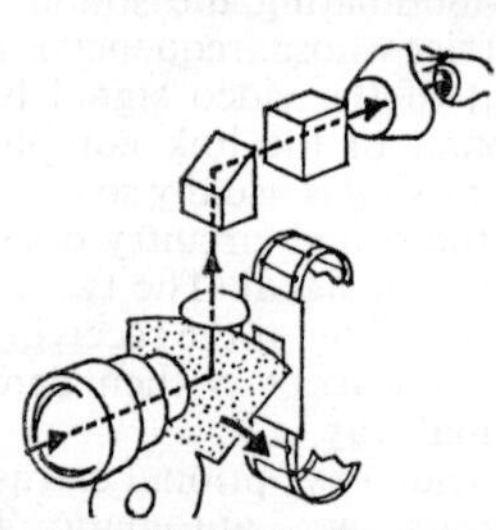

MIRROR SHUTTER. Shutter is shown in mid-position over aperture, obscuring film and reflecting lens image to ground glass above.

falls on the mirror and is deflected on to a ground glass, where it forms an image which can be seen by the cameraman. Thus cameraman and film see an identical image, and no problem of parallax arises. Also called a reflex shutter.

See: *Camera* (FILM).

⊕**Mitchell Mechanism.** Intermittent movement of very high quality, characterized by double

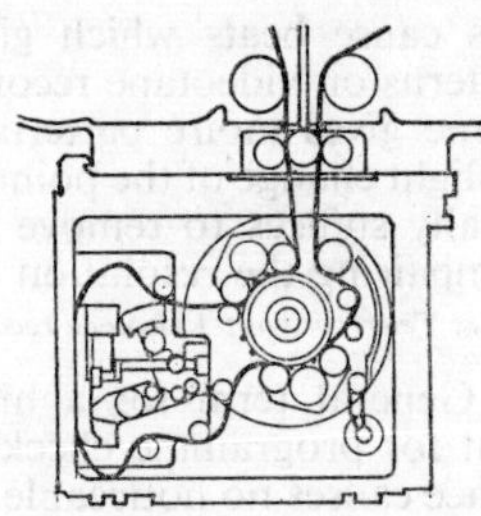

pull-down pins engaging perforations on both sides of the film, and by plunger-actuated register pins.

See: *Camera* (FILM).

Mix. In the picture sense, equivalent to dissolve. A sound mix means the combination or dubbing of several sound tracks into a single (mixed) track.

See: *Editor; Sound mixer.*

Mixed Syncs. A synchronizing signal consisting of line sync pulses and field sync pulses but containing no other information.

See: *Synchronizing pulse generators.*

Mixer. Circuits in a superheterodyne receiver in which the required signal and the local oscillator frequencies are mixed to produce the intermediate frequency (i.f.) signal.

See: *Receiver.*

Mixing. Action of combining two or more picture signals by adding arithmetically a controlled fraction of each.

A simple application is adding two signals to give a composite result known as superimposition. In such cases, precautions must be taken to ensure that the total permitted signal voltage excursion is not exceeded, and as each signal alone may reach this limit, the need arises for adding a controlled fraction of each.

This technique may be applied to achieve a smooth overlapping transition between two signals. Thus, assume the initial conditions are where signal 1 alone constitutes the output, i.e., the fraction of signal 2 is zero. An increase in this signal, accompanied by a corresponding decrease in signal 1, leads to the end condition where signal 2 alone constitutes the output. The transition between them has been achieved by a controlled superimposition more commonly called a cross-fade or lap dissolve.

See: *Master control; Special effects* (TV); *Vision mixer.*

Mixing (Sound). Process of combining the outputs of several microphones in original sound recording, or of several dubbers in re-recording. This highly skilled task is carried out by a sound mixer.

See: *Sound mixer.*

□**Mixing Equipment.** Equipment used to select one or a combination of signals in a controllable manner from a number of vision sources. The transition between signals can be by an instantaneous cut, a cross-mix, fade in or fade out. In more sophisticated equipment, additional means such as electronic effects are also available.

See: *Master control; Matrix; Special effects* (TV); *Vision mixer.*

Models. The design, preparation and photography of models and miniatures forms a most important part of special effects work in motion picture production, the scale of models ranging from the representation of the solar system on a table top to replicas approaching full size. Models may play their part as static scenes reproducing buildings and architectural structures impossible or too expensive to construct in reality or they may be used in action, whether in a realistic naval battle or train crash or in the representation of space travel or prehistoric monsters. Often models must be combined with studio production shots involving live actors by optical printing trick effects.

In all model photography, the factors of perspective, depth of field and speed of action are most carefully considered to ensure the appearance of reality, the problems involved becoming greater as the scale of the model is reduced. In general, model photography requires the use of short focal length lenses at close distances but in these circumstances the depth of field over which the object is sufficiently sharply in focus may be too small, so the level of lighting must be increased to allow the lens to be stopped down to give greater depth of field. Additionally, the action of models must be slowed down in photography to give the necessary time scale effect; this involves running the camera at a speed higher than normal, often by a factor of four times, so that the lighting level must be increased for this reason also to compensate for the reduced time of exposure.

Satisfactory model work therefore demands not only meticulous construction in detail and finish but also an accurate calculation of conditions of photography to give both optical and time-scale reality, and is recognized as a field where only the skill of the specialist can lead to success.

See: *Art director; Special effects* (FILM); *Studio complex* (FILM).

⊕**Modulation.** Process whereby the amplitude, ☐frequency or phase of a wave is varied as a function of the instantaneous value of another wave. The first wave is called the carrier wave, and is usually of a single frequency; its function is to carry the modulating waves, which normally comprise a band of frequencies.

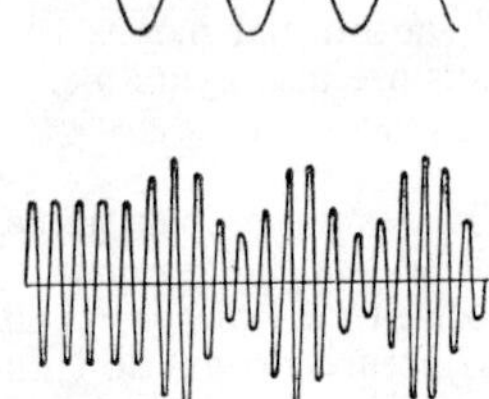

AMPLITUDE MODULATION. (*Top*) Modulating signal. (*Bottom*) Unmodulated carrier (*left*) and modulated section of carrier (*right*), showing low-frequency wave impressed on high-frequency carrier.

In amplitude modulation, the profile of the carrier as modulated by the train of modulating waves, is called the envelope. In frequency modulation, the amplitude of the carrier remains constant, but its frequency is varied from moment to moment by the modulating waves.

See: *Television principles; Transmission and modulation.*

☐**Moiré Patterning.** The line patterns produced by certain finer details of dots or lines through interference effects. When an image contains repetitive information (e.g., check pattern on a sports jacket, or a tiled roof), there is a possibility of an interference or beat pattern being set up between the image and the scanning lines used in the television system. On the monitor screen this looks as though the image, although itself stationary, contained varying degrees of high- or low-frequency patterning. The effect is sometimes felt to show that the system has a high resolving power, but it can be distracting to the viewer and, where possible, scenes or patterns of clothing which give this tendency are to be avoided. Certain telerecording systems, particularly those on 405-line systems, are prone to moiré patterning when re-scanned in television. The telecine scanning lines beat with the lines from the display tube recorded on the film. To counteract this, spot modulation to break down the line structure can be employed on recording, or the telecine scanning spot can be made astigmatic, or wobbled.

The harmonics of the f.m. carrier, particularly in lowband VTR recordings, sometimes cause beats which give similar moiré patterns on videotape recordings.

If a scene gives moiré patterning in the studio, a slight change of the point of optical focus usually suffices to remove the effect, without impairing the resolution too much.

See: *Telecine; Telerecording; Videotape recording.*

⊕**Monitor.** General term for a high-quality ☐instrument for programme checking whose performance causes no noticeable distortion of the material. Thus a sound monitor is a high-quality loudspeaker and amplifier, a picture monitor is a high-quality picture display. The use of either assumes that the conditions for viewing or listening are satisfactory.

A picture precision monitor is a display device in which linearity, geometry, grey-scale, brightness and amplitude/frequency response are all of a high order and hum and other defects are minimized. Such a device can be used to monitor a vision signal in the assurance that faults seen will be due to the signal and not to defects in the monitor.

See: *Picture monitors; Sound recording.*

⊕**Monitor (Sound).** Loudspeaker system of ☐very high quality used to check by ear the quality of sound being recorded in a studio. On the studio floor, monitor headphones may be used during live recording, but the ultimate check is by playback through the monitor loudspeaker system under ideal acoustic conditions.

See: *Sound mixer; Sound recording.*

⊕**Monitor Viewfinder.** Viewfinder external to ☐the camera and often to the blimp (if one is used) which enables the cameraman to watch the scene while the camera is turning. Since it is often mounted some distance from the lens axis, this type of viewfinder must be equipped with accurate compensation for parallax. To eliminate parallax at the source, a vidicon tube may be arranged to receive a picture off a reflex shutter, the amplified signal being fed to one or more remote TV monitor screens.

See: *Add-a-Vision; Gemini; System* 35.

☐**Monochrome Channel.** Channel in a colour television system capable of carrying a monochrome signal.

⊕**Monopack.** General term for an integral tripack colour film material. Specifically, a reversal film of this type similar to 16mm Kodachrome used for professional 35mm photography to a limited extent during the period 1940–50.

Monopole. Telescopic hanger, generally running on rollers, from which studio lights are suspended from a slotted grid. Height adjustment is by cable and winch.

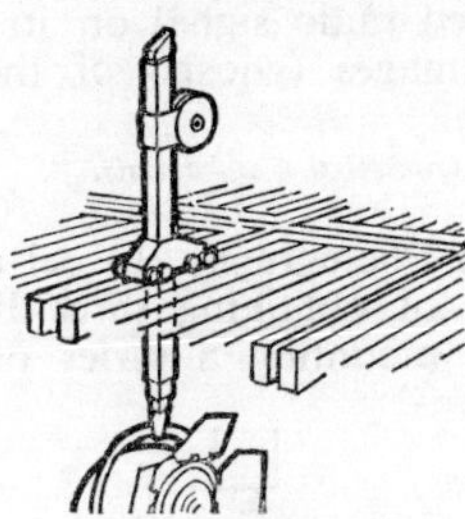

See: *Cameraman* (TV); *Lighting equipment* (TV).

Monoscope. Electron tube for generating a fixed television picture signal, derived from an image on an electrode inside the tube. Depends for its operation on the difference in secondary emission ratio between two materials forming the image pattern. In its usual form, has a life many times that of an image orthicon or vidicon tube.

Montage. In studio parlance, a rapidly cut film sequence, often containing many dissolves and superimpositions to produce a generalized visual effect. In a more specialized sense, montage means the type of film cutting originated by the Russian school soon after the Revolution, characterized by staccato transitions and violent alternation of apparently unrelated shots. Now obsolescent.

See: *Editor.*

Mop-Up Waveform Corrector. Amplitude and phase response corrector installed in a system to cater for slight inaccuracy of correction of individual items of equipment, the errors of each of which may be negligible.

Mordant. Chemical compound capable of forming an insoluble complex with suitable dyestuffs; mordants may be added to the emulsion of the blank film used for the dye transfer process to prevent unwanted diffusion of the colour image.

See: *Imbibition printing.*

Mosaic. A repeating pattern of very small colour filter elements, usually red, green and blue, through which a photographic image is exposed in some additive colour processes. In still photography, the mosaic was sometimes separate from the photographic plate, but in cinematography it was made integral with the film base.

See: *Colour cinematography.*

⊕**Movietone Frame.** Image size on 35mm film which became standard when sound-on-film processes were introduced and which allows space for the width of the sound track. Also called Academy Aperture or Reduced Aperture (RA).

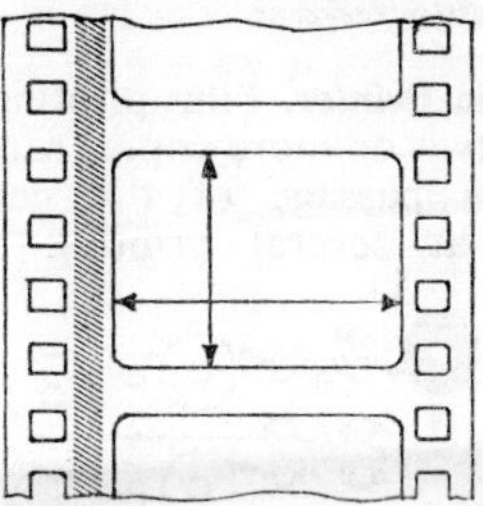

See: *Film dimensions and physical characteristics.*

⊕**Moviola.** In strict usage, the trade name of a particular kind of portable, motor-driven film viewing machine much used in editing. The term, however, is often applied generically to any type of editing machine. Moviola heads may carry a picture track only, or sound only, or may be coupled together to give synchronized operation. In this way, 35mm picture may be combined with 16mm sound, or several 16mm or 35mm sound heads may be coupled to a picture head, so as to reproduce a dubbing machine on a small scale. Most types of Moviola drive the picture film through an intermittent sprocket with loops above and below, but the shutter is usually omitted as the quality of the picture is not important. A few 35mm machines, and many 16mm ones, employ optical compensators so that the picture film can be pulled through them with a smooth continuous motion.

MOVIOLA. **Right-hand portion is basic picture viewer, shown with feed and take-up spools. To left is coupled sound reproducer.**

See: *Editor.*

□**Multiburst.** Video test signal having each line divided into a number of trains of specified frequency, the mean level being set in the region of mid-grey. A typical signal consists of bursts of 1, 2, 3, 4 and 5 MHz sine waves.

See: *Transmission networks.*

⊕**Multi-Head Printer.** Film printing machine in which two or more copies can be made from each passage of the negative by exposure at several printing heads on separate rolls of positive stock. Also refers to a printer in which exposure from two or more negatives is made on to one strip of positive at one passage of the stock.

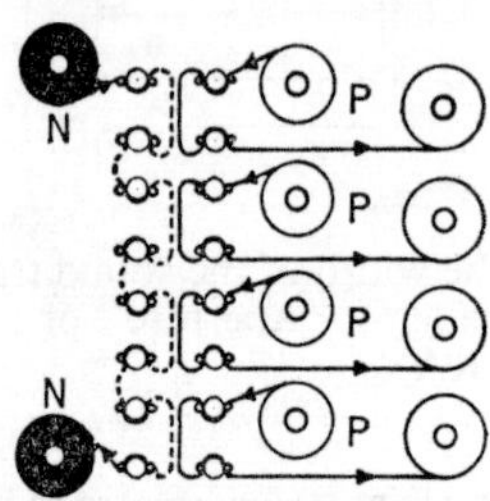

MULTI-HEAD PRINTER. Four printed copies, P, made simultaneously at a single passage of the negative, N. The negative film is exposed at successive apertures arranged between the feed and take-up spools. Much time is saved in bulk printing.

See: *Laboratory organization; Printers.*

⊕**Multi-Image.** Form of cinema presentation in which a multiplicity of images printed on a single film is projected by one projector on to one screen, thus contrasting with multiscreen. Pioneered by split-screen special effects, this technique has since been perfected with the aid of the 65/70mm film format, the increased image area providing adequate resolution for 10–15 sub-images of excellent quality, even when presented near the edges of the screen.

Multi-image films make use of expanding and diminishing pictures, and of panning effects whereby pictures complete in themselves can be made to traverse the screen from side to side or from top to bottom. In general, the screen may be used as a canvas on which a number of images can be moved and changed independently and with complete freedom. Multi-image films can normally be projected on standard 70mm equipment, or on 35mm equipment if reduction prints are available.

See: *Multiscreen; Wide-screen processes.*

□**Multi-Path Signals.** Signals reflected from the ionosphere back to earth by various combinations of paths. As a result, the received signal may arrive from a number of different directions at different time intervals. This can cause variations in strength of the received radio signal or, in television, multiple images (ghosts) of the received picture.

See: *Ghost; Transmission and aerials.*

⊕**Multiplane Photography.** In cel animation, a method of obtaining three-dimensional effects by mounting a series of cels one under another in different planes, the whole set being placed beneath the lens of the animation camera.

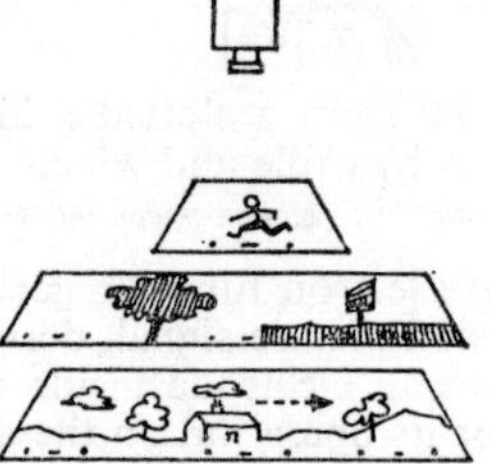

See: *Animation equipment; Animation techniques.*

□**Multiplexer.** An optical device for combining pictures from several sources (e.g., film and caption scanners) into a common signal-generating channel. This is generally achieved by moving mirrors, semi-silvered surfaces of prisms (ice blocks) or pellicles.

The tendency is towards single-channel outputs, since the multiplexer makes it impossible to preview the next item when the channel is already in use.

See: *Telecine.*

⊕**Multiscreen.** Generic term for a form of cinema presentation appearing under many commercial titles, in which three or more screens (not necessarily placed contiguously or even in the same plane) receive pictures from an equal number of projectors. First seen in 1900 as Cinéorama, later developed by Abel Gance for *Napoléon* (1925–27), but not perfected until the world exhibitions of the 1960s, when many limitations were removed by the use of larger film formats and better optical processes. Slide projectors may be used instead of films, as in the Czech Diapolyekran (112-screen mosaic). Multichannel sound, with from 2 to 12 channels, is usually employed to move aural attention with, or in counterpoint to, visual attention. More than six channels (the maximum for 70mm film) require at least some tracks to be transferred to a separate interlocked sound film.

Multiscreen techniques enable different images to be shown on all the screens, so that spatial cutting may be employed: that is, cutting in space between screen and screen, as well as in time on any one screen. The different images on the screens may also pursue their own continuity either in sympathy or in conflict with one another, or may come together into a single giant image covering all the screens, even if these are not precisely contiguous. In this case, the dark strips between the screens look rather like the leads in stained glass windows, interrupting but not fracturing the general effect. Some screens may be left blank from time to time in order to concentrate attention on the others.

Interlocked projectors are necessary for all multiscreen processes, and the complicated presentations are often highly automated to reduce the risk of breakdown. No standardization is yet in sight, and systems are often developed at great expense for a single application.

See: *Multi-image; Wide-screen processes.*

Multi-Standard. A piece of equipment capable of working on more than one television system of line and field scanning rates.

Multi-Track Sound. To provide stereophonic effects, two or more sound tracks may be accommodated on a motion picture release print, and reproduced through separate amplifier channels and loudspeaker systems. In the original Cinemascope process, provision was made for four magnetic tracks, two inside and two outside the perforations. With 70mm print systems, six magnetic tracks are used, the two outer stripes each being divided into two. In the original Cinerama process, the multiple sound tracks, of which there were seven, were all carried on a separate synchronized 35mm magnetic film.

See: *Sound recording* (FILM); *Sound reproduction in the cinema.*

Multivibrator. Oscillator having two stages each feeding into the input of the other. Depending on the circuit configuration it can be free-running or require to be driven by an external trigger signal.

See: *Synchronizing pulse generators.*

Music and Effects Track. Combination of all the components of a film sound track, except the dialogue, into a single sound track.

See: *M and E track.*

⊕**Mute.** Exclusively English term for silent film.

⊕**Mutilation of Prints.** Films may be made useless for projection by scoring them with one or more deep scratches along their length in a device called a mutilator. Prints from stock shot libraries for use in work-prints are often supplied in mutilated form to prevent unlicensed duplication or unauthorized use.

⊕**Muybridge, Eadweard James, 1830–1904.** English photographer, working for most of his life in the USA. With the introduction of photographic materials requiring only very brief exposure times, he became interested in the photography of moving horses and in 1873 began experiments sponsored by Governor Leland Stanford of California. By 1877 he had devised a system using a number of cameras, the shutters of which could be operated in rapid succession, triggered by trip wires struck by the moving horse. This system was soon replaced by electrical release of the shutters at accurately controlled intervals. With this multiple camera apparatus Muybridge made an extensive study of the locomotion of the horse and later of other animals and human beings. His photographs of the horse in motion were published in 1882, followed in 1887 by an 11-volume publication, *Animal Locomotion*, containing more than 20,000 pictures. He lectured all over the world, projecting moving pictures derived from his original photographs, using the Zoopraxiscope projector he had invented, a large-scale version of a projection Phenakistiscope.

Muybridge's work had considerable influence on the introduction of cinematography, stimulating many of his contemporaries, notably Professor Marey and Thomas Edison, whose subsequent work led to successful apparatus and the beginnings of cinematography proper.

Narrow-Band. Contrasting term to wide-band, used of TV transmission channels and multi-channel communication systems. Narrow-band, in contrast, is often applied to special forms of frequency modulation, of less than the normal f.m. bandwidth, for police, taxi and similar use. The term is also applied to channels which carry the audio band only.

NARROW-GAUGE FILM AND EQUIPMENT

Narrow-gauge is the term used to describe all film widths smaller than the professional 35mm standard width. The smaller films include 16mm, 9·5mm, Standard 8mm, and the more recent Super-8 and Single-8. For all these films the generic name sub-standard was first used, but is no longer favoured because it suggests inferiority.

All narrow-gauge films are, and always have been, coated on safety (slow-burning) base. Standard 35mm film, on the other hand, was coated until the early 'fifties on highly inflammable cellulose nitrate base. This hazard was one of the prime reasons for introducing film of lesser width on non-flam base for home use. Narrow-gauge film is cheaper to use than 35mm, and the equipment is less bulky.

Narrow-gauge films were originally intended for amateurs making home movies. Nowadays, however, 16mm especially is widely used for professional purposes, e.g., TV news, teaching, scientific records. Film libraries offer a huge choice of narrow-gauge films for entertainment, as well as instructional and publicity subjects. In some cases these films are reduction-printed from 35mm originals, but direct 16mm production has now become common.

GAUGES

HISTORY. The original narrow-gauge films were the first 17·5mm gauge (1899), and 28mm from Pathé in France (1912). Both were far ahead of the technical scope and economic needs of that time, and thus made relatively little impact on the world.

The real history of narrow-gauge films began when Pathé (France) introduced 9·5mm in 1923, followed six months later by Eastman Kodak with their 16mm film. Both these gauges offered reversal processing, in which the camera film is processed to give a positive image for projection; this saves printing a negative on to a separate film.

A decade later, Eastman Kodak introduced 8mm, with lower running costs and even more portable equipment. Meanwhile, Pathé in France aimed at the home cinema market by introducing printed films in a new 17·5mm gauge, destined to become a casualty of World War II.

The post-war period saw 16mm used more and more by professionals, while the amateur market moved towards 8mm, now technically much improved.

By the mid-1960s, when 8mm equipment sales had clearly passed their peak (particularly in America), Eastman Kodak introduced yet another gauge, Super-8, still 8mm wide but with smaller perforations, and making better use of the total film area. Fuji in Japan simultaneously introduced Single-8, identical to Super-8 in perforation and frame sizes but with a different camera cartridge loading system.

SIXTEEN MILLIMETRE

The 16mm gauge is the most important for professional narrow-gauge filming, because it has an adequate quality margin to give satisfactory duplicates or prints. Progress of 16mm has been greatly accelerated by its world-wide adoption for TV news and documentary filming. A wide variety of film-stocks, services and equipment is now available.

Though originally introduced for home movie making, 16mm is now, owing to its rather high running costs, almost exclusively used by professionals. So it is not surprising that equipment trends are more and more towards the professional. Some of the more serious amateurs still use 16mm, but mostly only for films where copies are needed, e.g., club productions, record films and competition entries.

SPOOLS. For camera use, 16mm film-stock is most commonly supplied in its original form, the 100-ft. daylight-loading spool. Here the film is wound between the closely fitting opaque cheeks of the spool, so that all but the end turns of film are protected from the light. Extra lengths of film (5 to 7½ ft.) at either end act as leader and trailer, to protect the underlying layers while the camera is being loaded and unloaded in ordinary lighting. Any light that may leak down between the spool cheeks and the edge of the film is safely absorbed by the perforation margins. The film is often on a grey base, which helps to absorb the light before it reaches the picture area. So long as the

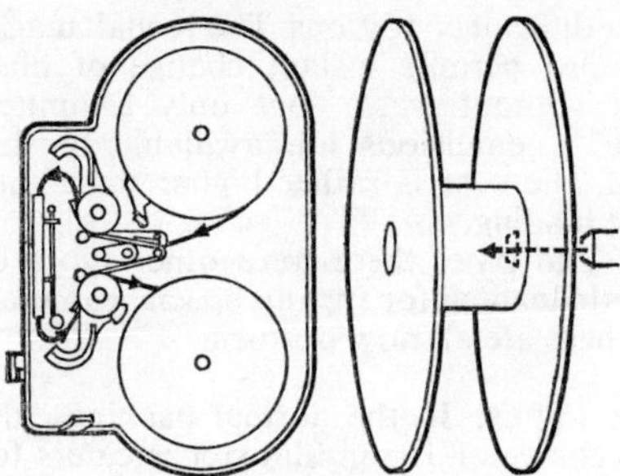

16MM DAYLIGHT LOADING SPOOLS. Film is wound between close-fitting flanges. Spools have square holes on inner sides, which engage with square drive shafts. Outer holes are round to prevent incorrect mounting in camera.

reel of film is not handled in unduly bright light, the image area remains unaffected.

For compactness, a few of the older cameras were made to take the smaller 50-ft. spool (which could also be used in cameras taking 100-ft. spools). But the smaller spool is rarely seen nowadays, and the 200-ft. spool is restricted to only one or two cameras.

MAGAZINES. For professional users, especially for synchronized-sound shooting, the trend is to cameras taking longer lengths of film, generally housed in a detachable magazine. Lengths of 200, 400, or up to 1,200 ft., wound simply on plastic cores, are usual. The film must be loaded into the magazine in the dark.

Still in use, mostly by amateurs, are small numbers of hand cameras taking Kodak's light-tight 50-ft. magazine. These cameras are very compact and simple and quick to reload. This can be useful when filming

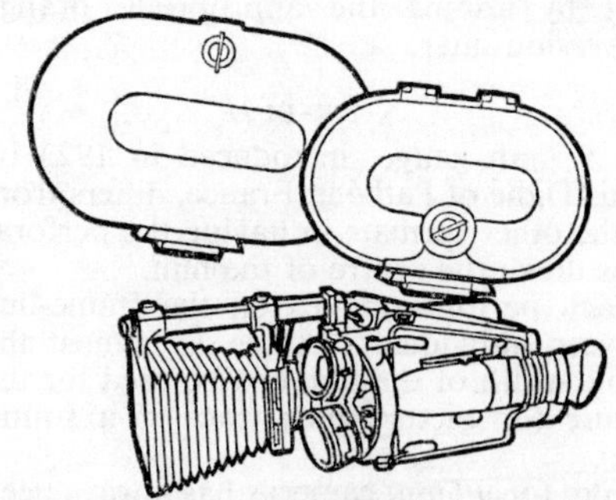

16MM CAMERA MAGAZINES. For professional use, separate magazines are more convenient than daylight-loading spools, and enable 200 ft. and 400 ft. magazines to be interchanged. 1,200 ft. magazines are considerably larger, but are sometimes used for interview shooting.

under difficult conditions. The Kodak magazine also permits instant change of film-stock without waste, but only a limited range of emulsions are available in this form. The cost is rather higher than with spool loading.

At one time, there were other types of cassette loading for various special cameras, but these are all now obsolete.

FILM TYPES. In the normal packings, the wide choice of 16mm film-stocks caters for almost every possible requirement. Black-and-white films are made in either negative or reversal types, and in different speeds from the slower fine-grain high definition films up to some of the fastest emulsions available.

Colour films fall into the same two groups: negative, and reversal. Of the colour reversal films, almost all professionals use one or other of the types made especially as a master for duplicating. The fastest type permits filming in low light levels that would have been impossible a few years ago. In most cases special forced processing can be used to give even greater sensitivity.

Amateur 16mm colour films are reversal types giving a positive image designed for projection; this is not optimum for duplicating mainly because the contrast is too high.

Colour films are matched for shooting with a specific light source, for example, either daylight or tungsten filament lamps. Amateur tungsten films are generally matched to overrun photoflood (about 3,400°K) lamps, while professional films are balanced to 3,200°K studio lamps which have a longer life but are slightly less efficient. Tungsten films can be used in daylight by adding the appropriate orange conversion filter.

NINE-FIVE

The 9·5mm gauge, introduced in 1923 by Louis Didié of Pathé in France, differs from all the other formats in having the perforations down the centre of the film.

Each perforation lies on the frame-line between individual pictures, so almost the whole width of the film can be used for the picture (or picture plus track if a sound film).

Film for 9·5mm cameras has always been available in light-tight chargers containing about 26 ft. of film, with a short loop of it protruding for insertion in the camera gate. A later type of film magazine is the 50-ft. Webo charger which contains the claw for film transport. The 9·5mm film is also available on 50-ft. and 100-ft. daylight-loading spools for some cameras, but because there is no perforation margin to absorb any light leaking down between the film and the spool cheeks, this type of load is not really satisfactory.

Nine-five has always remained strictly an amateur gauge, and it was never introduced into America. This, plus its French origin, meant that World War II dealt nine-five a death-blow. In fact, although many users still prefer this gauge, it is commercially obsolescent.

STANDARD 8MM

Introduced by Kodak in 1933 as a low-cost popular home movie gauge, 8mm is basically a design-variation on 16mm. Its progress was interrupted by World War II but it achieved great popularity in the 1950s and is still by far the most popular amateur gauge. Today, 8mm is in two forms – Standard-8 or Regular-8, and the new Super-8.

Standard 8mm film is initially 16mm wide and must be run through the camera twice. On its first pass through the camera, only half the film width is exposed. The film is then turned over and run through again to expose the other half. After processing, the laboratory slits the film into two 8mm-wide lengths, and joins them end-to-end ready for projection.

DOUBLE-RUN. The Standard 8mm camera film (sometimes known as Double-8) has perforations the same size and shape as on

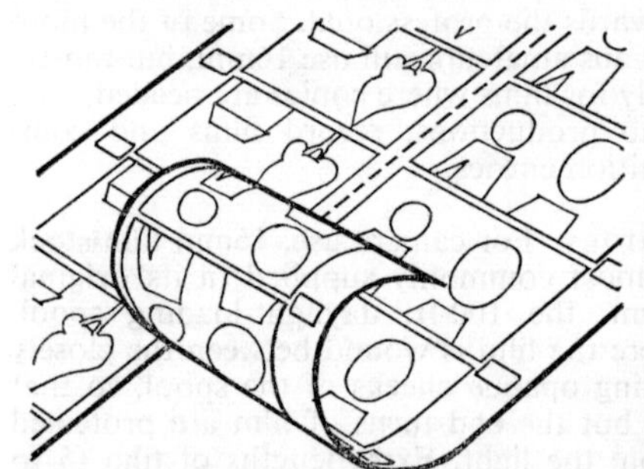

DOUBLE-8 FILM. Film of 16mm gauge, perforated with standard 16mm holes, but with twice the number in any length of film, providing a double run in a standard 8mm camera. After slitting, a double length of 8mm film is returned to the user.

16mm, but twice as many because the 8mm film contains twice as many frames in a given length as 16mm (80 frames per ft. on Standard 8mm). The double-run principle is used so that the film can be processed on the same machines as 16mm. This is of special importance with the complicated processing used for non-substantive colour films such as Kodachrome.

SINGLE-RUN. There have been several attempts to use single-run 8mm (that is, an 8mm wide camera film) for Standard-8. None achieved universal use, because such a film will not go through the standard processing machinery. Much more primitive processing procedures have to be used.

Of the single-run Standard 8mm film systems, the only significant one is Agfa's Movex cassette system, giving daylight loading of a 10-metre (about 33 ft.) length. The advantage of a cassette is that it eliminates the edge-fog that sometimes occurs near each end of a daylight-loading spool. Edge-fog would be especially objectionable on the unperforated edge of a single-run film, because it would extend directly into the picture.

SPOOLS. The commonest type of packing for Double-8 film is the 25-ft. daylight-loading spool. This is similar in principle to the 16mm spool, but smaller, and with different centre holes. An additional 4-ft. length of film is provided on each end of the 25-ft. length, making 33 ft. total. This allows for what is fogged in loading the camera on each of the two runs. Most of the extra leader length is generally cut off by the processing laboratory and a 50-ft. total length of 8mm wide film is returned on a spool ready for projection.

Besides the standard 25-ft. spool, Double-8 is also available in longer lengths, such as the 100-ft. spool for special cameras.

MAGAZINES. Double-8 is also available in Kodak-type 25-ft. magazine, totally enclosed, needing no film-threading, but still with the half-way turn-over. This type of magazine achieved some popularity in America, but never elsewhere because it cost more than standard spool loading.

The quick half-way turnover, possible with magazine loading, triggered some manufacturers in the early 1960s to revive an older idea of using a special plastic cartridge which could be pre-loaded by the user with ordinary 25f-t. spools of film. This could then be dropped into the camera, avoiding conventional threading both at the start and at the turn-over. This idea failed, mainly because there was a tendency for the end of the film to become detached from the spool at the half-way turn-round stage, with consequent jamming.

FILM TYPES. Nearly all 8mm filming is by amateurs, and they almost invariably use colour reversal film. Two types of 8mm film-stock are available, one balanced for daylight, the other for tungsten filament lamps of the photoflood type. Some black-and-white film-stocks are also made, including very fast emulsions.

The high quality of the camera original film cannot be matched by any sort of duplicate, no matter from what size original it is made. Careful laboratory work can produce an acceptable copy from a good 8mm original, while mass-produced prints reduced from 16mm and 35mm may be fairly good.

Standard 8mm has found quite wide use in teaching, in conjunction with the cartridge-loading Technicolor silent projector. In America, some 8mm magnetic sound films have been used for industrial sales and promotional purposes, again with a trend to cartridge-loading projectors.

SUPER-8

The completely new 8mm Super-8 format, introduced in 1965, is intended to replace Standard 8mm in both amateur and professional fields. The new format is properly called 8mm Type S, for it includes not only Eastman Kodak's Super-8, but also Fuji's Single-8. Both use the same sizes of

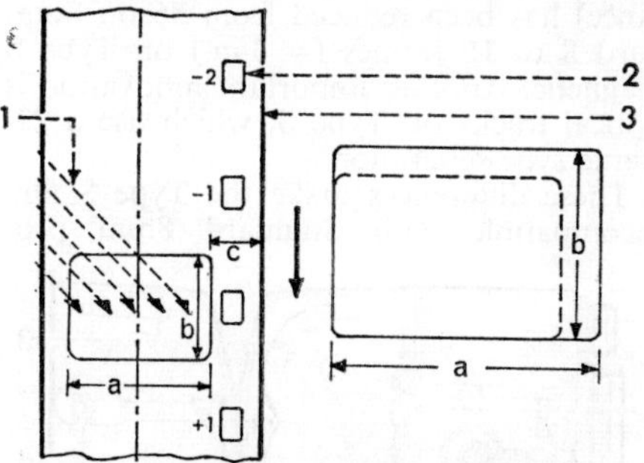

SUPER-8 CAMERA APERTURE IMAGE. 1. Direction of light striking emulsion surface of film. 2. Perforation used to position film at end of pull-down. 3. Edge used for guiding and dimensional reference.

a=0·245 in. nom.	6·22mm nom.
b=0·166 ,, ,,	4·22mm ,,
c=0·037–0·058 in. nom.	0·94–1·47mm nom.

(*Right*) Super-8 (solid line) and Standard 8mm (dotted); frame areas compared.

perforation, frame size, stripe width, etc.

The Type S format has technical and operating advantages over Standard 8mm, which has unnecessarily large perforations, wasting film area that could more usefully be employed for the picture. In the Type S version, the perforations have been made much narrower, allowing an increase in picture width. At the same time, the pitch (or distance between the centres of adjacent perforations) has been slightly increased, to one-sixth of an inch, or 0·1667 in. Thus the 4×3 proportions of the picture are kept the same as on the other gauges.

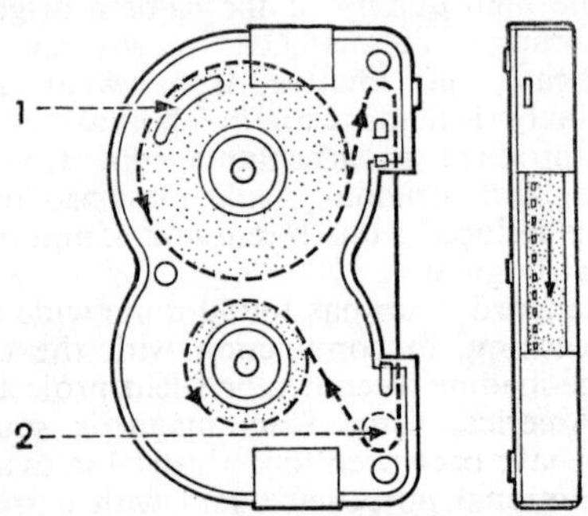

SINGLE-8 CARTRIDGE. Aperture and pressure plate are in camera, not part of cartridge. 1. Slot which engages pin on camera cover to adjust automatic exposure device to appropriate film speed. 2. Rubber roller permits film to be reverse-driven for special effects.

Another change on the Type S films is that the sound track has been moved to the unperforated edge, though the track width is the same as on Standard 8mm. The sync-separation (that is, picture-to-sound distance) has been reduced from 56 on Standard 8 to 18 frames (= 3 in.) on Type S magnetic. Another important innovation is optical tracks on Type S, which use a 22-frame sync-separation.

These differences make the Type S film incompatible with Standard 8mm projectors (nor is Standard 8 film compatible with Type S projectors).

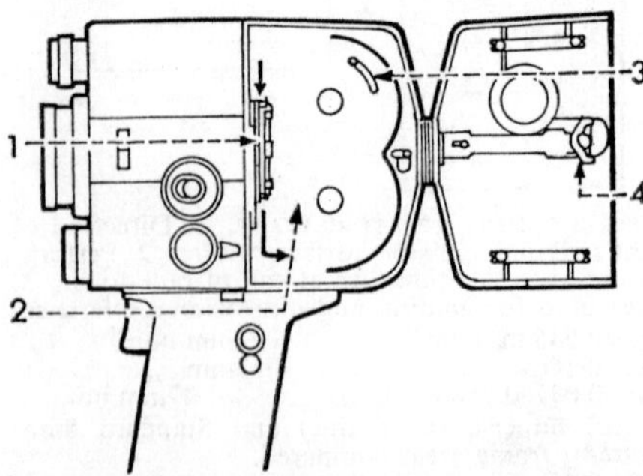

SINGLE-8 CAMERA. 1. Pressure plate. 2. Cartridge. 3. Film speed setting pin. 4. Devices for positioning film correctly and closing pressure plate when camera cover is fastened.

However, some manufacturers provide special dual-gauge projectors which can run both gauges with the minimum of fuss in changing from one to the other. Most of the progress in Super-8 projectors, however, has been towards simplicity in use. Cartridge loading and simplification by abolishing sprockets on many models have been important factors.

CARTRIDGES. Camera design underwent radical changes when Super-8 and Single-8 were introduced. Both these systems use single-run film (that is, 8mm wide) in a 50-ft. factory-loaded plastic cartridge. The cartridge slips into the camera in a moment. Notches on the cartridge automatically set the ASA speed and film type into the meter system of the camera.

The two rival systems, Kodak's Super-8 from America and Fuji's Single-8 from Japan, use completely different camera cartridges. So the user has to choose between one and the other.

The Super-8 cartridge has the feed and take-up rolls side by side. This gives a squat, but thick, cartridge shape. The Single-8 cartridge is tall and slim, with the rolls one above the other.

In some ways, the Fuji cartridge is the more versatile because it allows the film to be back-wound for special effects. Further, this cartridge should be capable of slightly better performance because the pressure plate is in the camera and not, as in the Kodak design, a part of the mass-produced cartridge.

To accommodate 50 ft. of film in the Fuji cartridge, which is surprisingly compact, a polyester thin-base film is employed for Single-8. This base is two-thirds the thickness of normal film but very strong mechanically. Polyester cannot be joined by normal film cement, so clear adhesive tape joins are generally used. However, special cements for polyester film are now being made.

FILM TYPES. Super-8 was launched on the market with just one type of film: colour reversal artificial light (Type A) material. All Super-8 cartridge cameras have a built-in conversion filter, so that the Type A film may be used in daylight. On many cameras when a movielight unit is fitted, it withdraws the filter automatically. Black-and-white reversal emulsions are also available in Super-8 cartridges in some countries where independent processing facilities are available.

Single-8 film started with only a daylight colour film, and in Japan also black-and-white films. Fuji have now added a Type A film, and the black-and-white films have been made available in certain countries, dependent on a factory-arranged processing service being in existence.

For scientific and professional users who wish to shoot on the 8mm Type S gauge, Super-8 is also available in double-run form on 100-ft. spools and 400-ft. rolls, in colour and black-and-white, for use in certain specialized cameras.

CAMERAS FOR 16MM

Only a few years ago, 16mm cameras could have been rigidly divided into two classes: moderately priced cameras for the amateur, and top-class engineered models for professional use. Now, however, the division is much less clear; 16mm cameras are no longer being made specifically for amateur use, but the best of the amateur cameras have been developed up to the point where they are mainly used by the smaller professional unit. Many 16mm cameras are used by amateurs for serious productions.

HAND CAMERAS. Cameras in this group, derived from amateur models, are portable and reasonably light in weight. Most are driven by a spring motor with governor control permitting a wide range of running speeds (from 8 to 64 frames per second, for example).

For professional use, a spring motor is generally a disadvantage, and those earlier camera designs still in production now accept an external electric motor drive. This eliminates the risk of the camera stopping in the middle of a shot. Newer camera designs dispense with the spring motor drive and use a built-in electric motor, often with an electronic-type speed governor working from a tiny reference generator on the end of the motor shaft.

Most hand cameras include a turret for rapid lens change, fitted generally with three lens positions of the C-mount fitting (1 in. diameter, 32 threads per inch, 0·690-in. flange-to-film distance or register). This is the internationally accepted standard lens fitting for 16mm hand cameras.

Some hand cameras include extra features such as a frame-counter, back-wind and variable shutter, allowing many effects to be produced in the camera. These features are generally of more use to the amateur than to the professional, who prefers to produce his effects at the printing stage.

VIEWFINDERS. All but the simplest 16mm cameras now have a true reflex viewfinder, showing the view through the taking lens itself. This gives parallax-free viewing, and in most models there is a ground-glass or a rangefinder device which permits visual focusing as well.

Even when the camera includes built-in reflex viewfinding, in some cases an external viewfinder may be provided. This proves useful under adverse conditions when the reflex image is too dim for convenient viewing.

Some reflex cameras use a beam-splitter in the lens, or between the back of the lens and the film gate. A small proportion of the light is reflected towards the viewfinder system, while the rest of the light goes straight through on to the film. The proportion diverted to the finder is lost to the film, so it must be kept small. Therefore the finder image tends to be short of light, especially if it includes a ground-glass for visual focusing.

Other 16mm cameras use a mirror-surfaced 45-degree shutter which diverts all the light into the viewfinder when the shutter is closed. When the camera is running, the

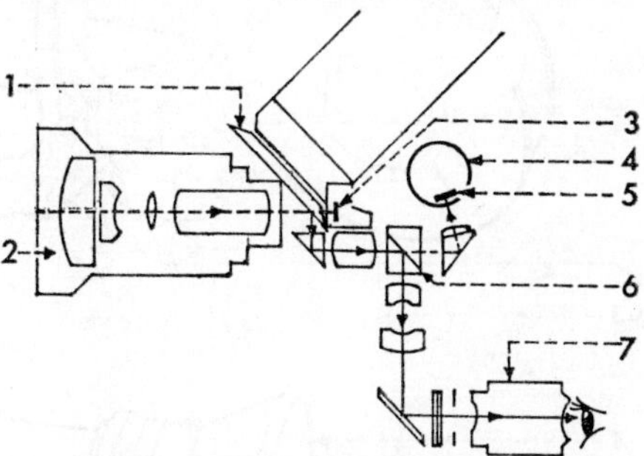

THROUGH-THE-LENS LIGHT METERING. Light through zoom lens, 2, forms image on film plane, 3, and is alternately reflected by mirror shutter, 1, when film is being pulled down. This reflected light is split at semi-reflecting prism, 6, a small part of it being bled off to a CdS photoresistor, 5, while the remainder forms an image viewed through a telescope eyepiece, 7, by the operator. A density wedge, 4, is adjusted to balance the system to film speeds from ASA 12 to 1600.

finder receives light only during the pull-down time. The rest of the time, the light passes through to the film and the view-finder does not receive any. As the pull-down time is about half the cycle, the finder image is brighter than with a beam-splitter design, but the mirror-shutter has the disadvantage that the image flickers when the camera is running.

METER SYSTEMS. One of the anomalies of 16mm cameras is that their design tends to be much more basic than the cameras designed for 8mm or Super-8. Nevertheless, some of the great technical developments in 8mm have reached 16mm cameras, and a number of the latter now include built-in CdS metering systems, usually with through-the-lens photocells. The metering systems on 16mm are often of the semi-automatic type, where the user has to adjust the lens iris until a pointer in the viewfinder matches a reference mark. This system gives the user more control of results than the fully-automatic mechanisms, though these, too, are now being built into 16mm cameras or lens units.

PROFESSIONAL CAMERAS. Those 16mm cameras designed expressly for the professional are not so limited in size and weight as models which started out as amateur designs, and can therefore be built more robustly to withstand the greater rigours and wear to which they will be subjected. Apart from the more solid construction, and more hand assembly because of the smaller numbers of cameras being made, the professional cameras are basically very similar to the hand cameras already described. Many of the more compact 16mm professional models are regularly used in the hand, and the gap between the two classes of camera has now narrowed almost to the point where the two types merge.

Professional cameras are invariably fitted with electric motor drive, and motor units are often interchangeable for mains or battery supplies. A tachometer is frequently fitted, so that the actual running speed can be set or checked. The motors can be of the wild (variable speed), governed d.c., synchronous a.c., or interlock type, depending on the application.

Lens mounts on many professional cameras are appreciably larger than the C-mount, giving better support to the lenses and greater freedom to the lens designer. A

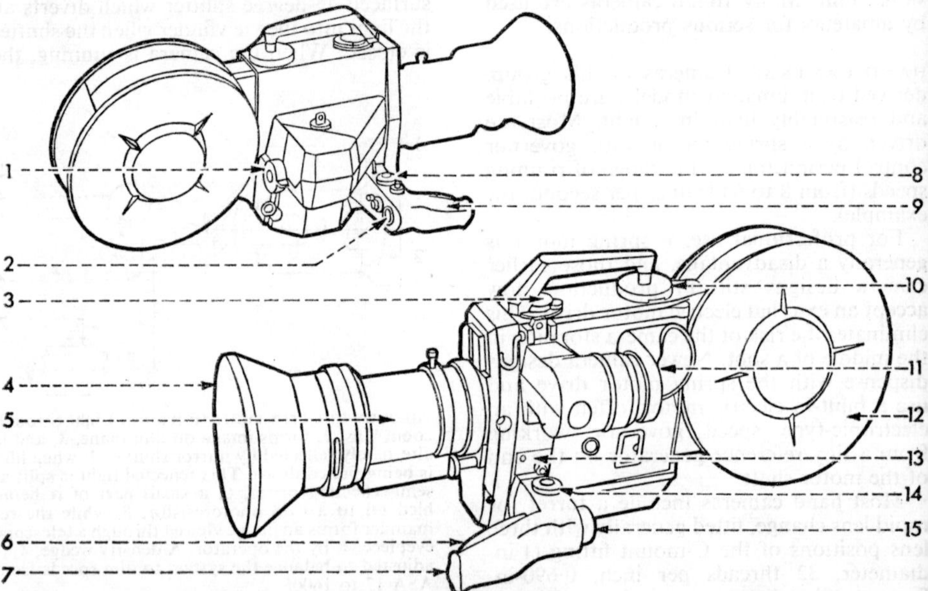

A PROFESSIONAL 16MM CAMERA FOR DOCUMENTARIES AND NEWSREELS. Camera is equipped with a single lens normally a zoom lens, 5, of 12–120mm focal length, with rubber sunshade, 4. Coaxial 400 ft. magazines, 12 with self-locking device, 10, are darkroom loaded, but will take daylight-loading spools. Film is self-threaded on attachment to camera with half-turn of loading knob, 1, for cutter, loop former and motor start. Hand grips, 7, face away from operator, who balances weight of camera and magazines on his shoulder. Grip incorporate power focusing servo control, 6, automatic diaphragm control with manual override switch, 14 speed-setting knob (16–50 fps), 15, control for servo motor speeds, 8, socket for power cable, 2, servo control for power zoom, 9. Automatic lens diaphragm setting is adjusted for film speed by control disc, 3. Reflex finder with eyepiece, 11, provides ×20 magnification, and enables an aerial as well as a ground glass image to be viewed. Signal lamp, 13, lights when end of film is reached, flashes when battery nears exhaustion. Commag recording facilities can be provided within camera. Internal three-phase motor ensures synchronous running at 24 and 25 fps for sepmag recording.

form of bayonet mount is normally employed to allow speedy substitution of lenses on the turret.

Film for professional cameras is normally contained in a magazine of 400-ft. capacity or greater mounted on the camera body and quickly interchangeable for reloading. A number of cameras that are basically 100-ft. spool-loading models can be converted to take an external magazine (fitted with its own take-up motor).

SOUND EQUIPMENT FOR 16MM

Almost all 16mm shooting is intended for eventual projection with a sound track. In some cases, the track is of the commentary-plus-music type, added after the film has been edited. More and more 16mm footage, however, is now being shot with synchronized sound.

There are two basic ways of shooting synchronized sound: single-system and double-system.

SINGLE-SYSTEM. In the single-system method, a specially designed camera is used which records the sound on the same film as the picture. The track is positioned along the edge of the film, taking the place of one row of perforations. Formerly the track was of the optical type, but is now almost invariably magnetic, requiring special pre-striped film stock.

The single-system sound camera is frequently used for television news filming, where speed of cutting is an important factor. Having the sound on the same film as the picture makes work simpler, despite the fact that the picture is separated from the sound by a distance of 28 frames (just over one second at normal sound projection speed).

The amplifier of a single-system camera is normally an external unit, arranged for operation by a second crew member, but capable of being operated by the cameraman if necessary. Models designed for one-man operation have been produced, but have not found much favour.

Older designs are basically intended for operation from the mains supply, with portable power units available for use in the field. Newer models tend to be fitted with d.c. motors, frequently using electronic speed stabilization, which allows a much lighter power supply to be used and reduces all-up weight.

The cameras are light enough to be mounted on a brace or chest-pod for greater mobility. Most cameras of this type are well enough silenced for shooting sound, even on interior locations, if care is taken in positioning the camera and microphone.

DOUBLE-SYSTEM. Double-system sound shooting involves recording the sound either on a separate film or on magnetic tape. The eventual production of the married print carrying both picture and sound comes later, after cutting and editing.

Sound on separate magnetic film is the most professional and direct method of shooting. This uses fully-coated magnetic 16mm perforated film. The camera and recorder can either be provided with special types of interlock motors, so that they run exactly in step including the run-up from standstill, or synchronous (note: *not* induction) motors may be used on camera and film-recorder; these only remain in step when up to speed and need a clapper mark each time the camera is started.

Much more common is for the sound to be recorded on tape, with one or other of the control-track synchronizing systems. These may be used with almost any constant-speed camera.

The camera must be fitted with either a cam-operated contactor or a pilot-tone a.c. generator. This provides either one pulse or one cycle (sometimes two) of control signal for every frame of film passing through the camera, and this is recorded on part of the width of the tape alongside the sound signal recording.

At the sound studio the tape is transferred on to fully-coated 16mm magnetic film, with the pulses kept in step with the film perforations. Editing takes place as for any double-system magnetic sound.

SILENCING. For critical sound shooting, especially interiors, some form of blimp or silencing is necessary to suppress the mechanical noise made by the camera. Best of all is a blimp that silences the camera to really professional standards. Usually the most important controls are brought out of the side of the blimp, so that focus, aperture, and sometimes zooming, can be adjusted during use. However, blimps are rather heavy and cumbersome and tend to restrict camera mobility.

In recent years, cameras have appeared with almost completely silent mechanisms, but nevertheless light enough to be hand-held. These cameras are being used more and more on candid filming, particularly for television interviews.

SYNCHRONIZATION. On some of the professional cameras, automatic slating facilities are fitted, providing a sync mark on the picture film at the same instant as a marker or bleep on the sound track. This does away with the need for the conventional clapper board to "mark" or slate each shot.

Professional 16mm filming with synchronized sound often needs great camera mobility. Then even the pulse-sync connecting cable from the camera to the recorder is too restricting. To do away with the cable link, the camera and recorder may be used quite independently by adopting crystal control.

With this system, the camera's drive motor is controlled by alternating current generated by an oscillator locked to a highly accurate quartz crystal. This holds the frequency constant within very narrow limits. A matched crystal in the tape recorder provides an identical signal which is recorded on the control track alongside the audio track. Some designs use an electrically maintained tuning fork in place of the crystal.

Slating can be by clapper in the ordinary way or automatically by radio link. A further advantage of the wireless control principle is that it may be used to synchronize several cameras at once to the one (or more) sound recorders, with each running independently.

EQUIPMENT FOR 9·5MM

With the decline in popularity of the nine-five gauge, almost no recent cameras have been produced for it. Though technically an excellent gauge, it is no longer commercially important. France is still the principal country making nine-five equipment, which is available in Britain but not in America.

EQUIPMENT FOR 8MM

Standard 8mm and Super-8 are both capable of giving excellent results, provided the camera-original film is projected. Copies from the original, whether contact printed or made by one-to-one optical printing, are normally only just good enough to be acceptable. So the 8mm gauges are used only when just the one copy of the film is needed, and when joins (due to editing) are acceptable. For almost all amateur filming, of course, extra copies are not required.

STANDARD 8MM CAMERAS. Up to the middle 1950s, 8mm cameras were basically similar to their counterparts in the larger gauges. For home movie making, there were a number of simple cameras with single-speed spring motor and a single lens, sometimes built-in, sometimes interchangeable to add versatility.

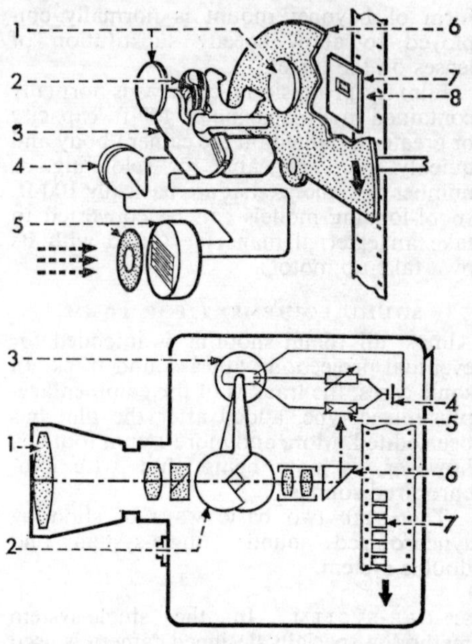

TWO AUTOMATIC DIAPHRAGM SYSTEMS. 1. Camera lens. 2. Diaphragm plates. 3. Moving coil instrument. 4. Battery. 5. CdS cell. 6. Shutter. 7. Film aperture. 8. Film. (*Top*) Conventional electric-eye system. Light falling on cell controls current flowing through moving coil instrument which works diaphragm plates to open or close aperture. Second diaphragm in front of cell adjusts sensitivity to compensate for different camera and film speeds. (*Bottom*) Self-compensating system. Meter cell measures light reflected from mirror shutter after passing through lens diaphragm. Moving coil is connected to cell through a bridge circuit, which causes diaphragm to open or close so as to ensure a standard light intensity reaching cell. Thus errors in diaphragm calibration are automatically cancelled. Camera and film speed settings are made by bridge circuit adjustment.

Viewfinders on the simple cameras were usually of the enclosed optical type, with a negative lens at the front and a positive viewing lens at the back. Sometimes the front lens had markings to show the field of view with other lenses. More elaborate models featured lens turrets for two or three lenses, multiple running speeds, and often improved viewfinders.

Two technical developments revolutionized the design of 8mm cameras: simple, low-cost fully-automatic exposure meter systems (the so-called electric eyes), and reasonably priced zoom lenses. These went into mass production towards the end of the 'fifties, and were soon to be found on all models except the very simplest and cheapest on the one hand and the most expensive and versatile on the other. These higher-priced cameras were intended for the more professional and scientific user, as well as the advanced amateur.

Early popular models had rather primitive coupled zoom viewfinders. But lens manufacturers soon produced zoom lenses with integral reflex through-the-lens viewfinding, and sometimes also with visual focusing facilities. These were quickly incorporated by camera manufacturers into their designs.

Automatic exposure control systems, in theory, can never be as accurate or as consistent as a properly used separate exposure meter. But the automatic systems work extremely well on most scenes encountered in practice. They are also much more convenient to use. They allow even beginners to obtain well-exposed films.

More elaborate cameras usually have provision for switching to manual control; this allows the user to compensate in scenes where the automatic mechanism is liable to be seriously misled.

Electric motor drive, used only rarely in the past, found greater application as small, powerful motors and improved batteries were developed. These removed the need for winding the spring after nearly every shot, or risking a stop in mid-scene.

SUPER-8 AND SINGLE-8 CAMERAS. All the cartridge cameras for Super-8 and Single-8 are designed to be simple to load and foolproof to use. Electric motor drive and fully-automatic metering systems have been adopted almost universally for these newer gauges, only the cheapest cameras having manual exposure setting. Most cameras adjust the meter system to suit the film-stock by a mechanism which engages speed coding notches on the film cartridge.

Super-8 cameras are primarily designed to take Type A (artificial light balanced) film-stock, and the necessary correction filter for daylight filming is built into every camera. The film cartridge carries a coding notch which indicates to the camera whether the contents are Type A, and the camera positions the filter accordingly.

On many models, this filter is removed automatically, and the necessary filter factor correction made, when a movielight unit with a suitable fitting is attached to the top of the camera for indoor filming. If one of the specially fitted movielight units is not available, a substitute key is provided for insertion in the movielight fitting. Other models use manual control for removing the filter when shooting by artificial light.

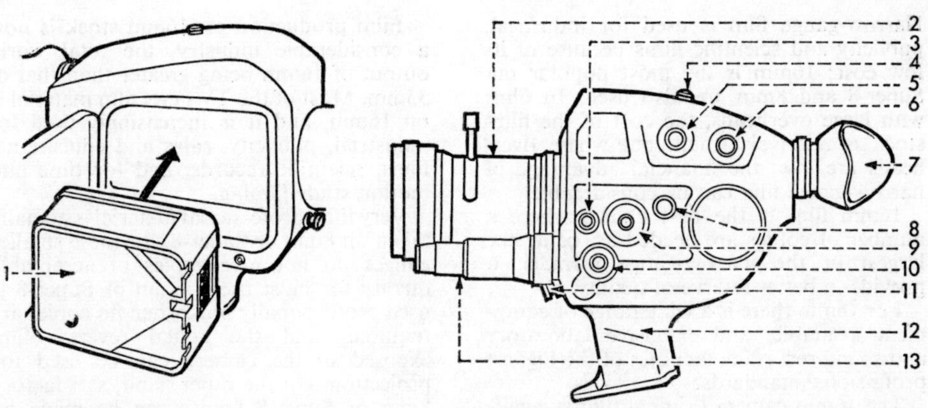

ADVANCED SUPER-8 CAMERA. 1. 50 ft. cartridge loaded into right-hand side of camera. 2. Interchangeable lens, with facility for 8–64mm zoom lens having iris diaphragm coupled to 13, micromotor under control of CdS cell metering light through lens. 5. Film speed setting control (8–400 ASA) coupled to automatic exposure system. 6. Frame rate setting (2–50 fps) interlocked with exposure system, so that camera speed can be altered while filming, and exposure held constant. 8. Footage counter. 9. Frame counter (0–100). 7. Eyepiece (×20 mag.) for reflex viewfinding system with, 3, retractable ground glass for alternative aerial-image viewing. 4. Variable shutter control for fading in and out. 10. Master switch with position for battery check. 11. Push-button stop-start. 12. Pistol grip with tripod socket. Drive motor and batteries are concealed in case.

The Fuji Single-8 system also makes provision for automatic operation of a conversion filter. But none of the early Single-8 cameras incorporated this feature. Therefore the cameras are generally used with either daylight or artificial light type film as appropriate. Conversion filters are available, in some cases with special means for correcting the meter system on particular cameras.

In the lower price class, cameras are very simple but nevertheless quite advanced technically. All the higher priced cameras are fitted with zoom lenses and reflex viewfinders. Some have provision for variable running speeds, but the range is generally rather restricted.

Super-8 cartridges do not permit back-winding, because the film in the feed side of the cartridge is loose on a central guide, and without any drive to it. It is therefore not possible to do superimpositions or dissolves with the Super-8 cartridge. However, the Fuji Single-8 cartridge does provide for reverse running or back-winding of the film.

SPECIALIZED CAMERAS FOR STANDARD AND SUPER-8. One enterprising manufacturer has produced a universal 8mm camera, with interchangeable magazine backs enabling any of the 8mm systems to be used: Double-8 reels, Super-8 cartridge, Single-8 cartridge, and Double Super-8 on 100-ft. reels.

Another manufacturer provides a Double Super-8 version of a camera also made for 16mm, and this can be used with 400-ft. magazines as well as 100-ft. daylight-loading spools.

One of the attractive features of the Super-8 and Single-8 film is that it provides for a relatively short separation between picture and sound track: 18 frames, or 3 in., for magnetic. This represents one second of running time at the normal 18 frames per second. So film from a single-system sound camera can be cut and edited with only a one-second overlap of the sound at each join. This is a more practical compromise than the 56-frame sync-separation on Standard 8mm. The Super-8 cartridge is not suitable for use in a single-system sound camera, nor is pre-striped film-stock at present available in such cartridges.

P.A.W.

See: *Camera* (FILM)*; Cassette loading for projectors; Continuous projection; Daylight projection unit; Editor; Film dimensions and physical characteristics; Narrow-gauge production techniques; Sound recording* (FILM)*; Splicing; Spools, cores and cans; Standards.*

Books: *Principles of Cinematography*, by L. J. Wheeler, 4th Ed. (London), 1969; *All-in-One Cine Book*, by P. Petzold (London), 1969.

●NARROW-GAUGE PRODUCTION TECHNIQUES

Narrow-gauge film is used for industrial, publicity and scientific films because of its low cost; 16mm is the most popular but Super-8 and 8mm are also used. In films with large overheads, the cost of the film-stock is relatively small, but where overheads are low, the financial advantage of narrow-gauge film can be considerable.

16mm film is the best choice where a number of copies are needed as, being the largest of the narrow-gauge formats, it provides a better reserve of quality.

For 16mm there is a wide range of equipment available, and extensive laboratory and sound recording facilities of the highest professional standards.

The 16mm camera original film is generally used for making 16mm copies, but it may also be used for making Super-8 or 8mm copies. It is even possible to use a 16mm original film for producing 35mm copies, e.g., for important news scoops, shots from expeditions, etc. Generally speaking, however, blowing-up to the larger size is not advisable.

Film production on 16mm stock is now a considerable industry, the total world output of 16mm being greater than that of 35mm. Most of the TV news film material is on 16mm, and it is increasingly used for industrial, publicity, sales and educational films, scientific records, and for time and motion study filming.

Very little professional material is actually taken on 8mm or Super-8, for these smaller gauges do not give copies of acceptable quality for most users. 8mm or Super-8 is used professionally only when no copies are required, and the actual reversal film exposed in the camera can be used for projection. On the other hand, satisfactory 8mm or Super-8 copies can be made by reduction-printing from a 16mm film.

EQUIPMENT AND FILM STOCK

The choice of camera equipment for narrow-gauge production depends on whether synchronized sound is wanted. If the script calls for much synchronized dialogue, the equipment must be much more elaborate than for filming a simple documentary where a commentary-plus-music track is added afterwards.

SOUND REQUIREMENTS. The inclusion of synchronized sound, then, requires more equipment, with all units designed or adapted to work together. The camera must either be blimped or an inherently quiet-running type. The recorder may be of the type taking 16mm fully coated magnetic film and synchronously driven with the camera. Sometimes this type of recorder is of relatively simple design, attached to the side of the camera and driven from it. This gives adequate quality for speech recording, but is not satisfactory for music.

Much 16mm sound filming is done with ¼ in. magnetic tape, with one or other of the pulse track systems to provide synchronization. The sound is recorded on one part of the tape, and the synchronizing pulse on another. The precise nature and location of the pulse depends on the system being used, but the basic principle is that the pulses recorded on the tape are in exact step with the frames on the film. Subsequently, these pulses are used to hold sync while the tape is re-recorded on to 16mm fully coated magnetic film.

With most sound recording systems, it is necessary to slate each shot, i.e., shoot a few frames of a clapper board with the scene and take numbers marked on it, at the same time speaking this information into the microphone, and then operating the clapper to provide a sync reference point on both sound and picture. Some cameras include a semi-automatic sync-marking device.

Much TV news filming is done with single-system cameras, which record the sound upon the same film as the picture. Nearly all such cameras record the track on pre-striped film, i.e., film with a magnetic stripe on the edge. The magnetic sound is always recorded 28 frames ahead of the picture. But clapper marking is still useful if the track is to be transferred to fully coated magnetic film for editing as double system.

In filming TV news with single-system sound, this material is generally edited as it is, and cuts are made in such a way as to hide the time-lag between each picture cut and its corresponding track cut 28 frames (just over one second) later. Sometimes a special re-recorder is used to transfer the sound from 28 frames in advance to level sync for cutting, and then to re-record it back again to the 28 frames advanced position for projection or transmission.

FILM STOCK. 16mm film stock is available in various packings to suit different cameras, e.g., 100-ft. spool, 400-ft. roll, etc. Moreover, the film is made either double perforated or single perforated, that is, with perforations down one edge only. Single-perforated film is known as A-winding or B-winding according to which edge the perforations are on. When the film roll is laid so that the film comes off it clockwise, A-wind has the perforations on the top side and B-wind has the perforations on the bottom edge.

Many cameras accept only double-perforated film; others accept either double or single perforated. For camera use, single-perforation film is always supplied B-winding; this brings the perforations on the correct side. There are two reasons for using single-perforated film. First, it allows the original to be striped for projection; this is of more interest to amateurs than to professionals, who want prints which can be made on single-perforated stock in any case. Second, when prints are to be made with optical tracks, double-perforated film tends to print traces of light leaks through the perforation holes and into the track area.

Single-perforated film is usually available pre-striped, that is, with a magnetic stripe already coated on it. The slight extra thickness of the stripe adds to the diameter of the roll for a given length. In the case of 100-ft. spools, the total length of film on the spool may be reduced by as much as 10 ft. so that the film does not overflow the diameter of the spool flanges.

Various emulsions are available on 16mm, and the choice will be dictated by whether the film is to be in black-and-white or colour, and by the type and intensity of the lighting. For filming in low light intensities, faster films are obviously desirable. On the other hand, high-speed films are too fast for filming in sunlight unless a filter (such as the appropriate neutral density) is used to cut down the light entering the lens.

The 16mm producer has also to choose whether to shoot on reversal or negative stock.

REVERSAL FILMS. The amateur uses reversal film for economy, as the film exposed in the camera can also be used in the projector. As there are no printing or copying stages, there is no loss of quality, so results can be outstandingly good, especially in colour.

The 35mm producer never uses reversal, but on 16mm reversal is widely used for the camera original, which acts as a master from which copies are printed.

The 16mm producer often selects reversal film for the following reasons:

1. Dust and dirt on a reversal master reproduce as black spots on a reversal copy, and are much less noticeable than the white spots which occur on a print from a negative. So reversal-reversal prints do not suffer from what is known as sparkle.
2. From a reversal master, it is possible to contact-print an internegative or an interpositive. This saves one stage when a duplicate master is required – and the saving of the extra stage is especially important for colour. This procedure also produces a final contact print with the standard emulsion position.
3. On black-and-white, the use of reversal stock gives an image with a finer grain.
4. On colour, the combination of available reversal master and reversal duplicating stocks gives noticeably better sharpness than the corresponding colour negative and colour print film.

Reversal master film, then, offers either straightforward duplicating to make an intermediate, and right-way-round prints from that, or sharpest possible colour prints on reversal printing stock. Negative film offers greater latitude in exposure. So with negative it is possible to apply greater compensation during printing, resulting in a print of even density from scene to scene. In the case of colour negative, considerably more colour correction can be applied than is possible when copying from a reversal original.

FILM HANDLING

FILMING. Shooting on narrow-gauge film is much the same as on 35mm except that the equipment is less bulky. The main difference is that the depth of field is greater. This is because the smaller the picture area on the film, the less is the focal length required to include the same picture angle. The shorter the focal length, the greater the depth of field.

Most 16mm cameramen use a zoom lens because it reduces the setting-up time for each shot. The exact framing desired can be achieved by adjusting the focal length on the zoom, thus avoiding the need to move the camera to and fro. Also, shots are more perfectly matched when taken on one zoom lens than with an assortment of prime lenses.

Besides the zoom, extreme wide-angle and extreme long-focus (telephoto) prime lenses are useful to cover shots where the range of the zoom does not extend far enough.

The relative compactness of narrow-gauge cameras often makes it tempting to dispense with a tripod. In some circumstances, e.g., for news filming, a tripod is a hindrance, but for anything else, it should always be used. Unintentional picture movement is irritating and distracting besides tending to reduce the apparent sharpness.

In much narrow-gauge production, the camera operator may also be the director. The director-cameraman, as he is called, needs a detailed understanding of film construction and film editing. For he must create, often without the aid of a firm script, a series of shots that can later be assembled coherently by the editor.

EDITING. The principles of film editing are the same for all gauges of film: assembly of the shots into the most effective storytelling order, cutting on action, setting the pace, and removing insignificant material.

Yet the mechanics of narrow-gauge editing are rather special. Equipment is mostly in the lower-budget bracket. The professional narrow-gauge producer edits almost invariably on 16mm, and, for synchronized sound, the editing is more involved because sound cuts have to be considered as well as picture cuts. Also, the sound is normally on a separate film which must be kept in frame-for-frame sync with the corresponding picture film. Much 16mm footage, however, is shot without sound, and this greatly simplifies editing.

Before beginning to cut the film at all, it is advisable to see it projected on the screen, several times if necessary. After that, the film may be wound through a small animated viewer, of the type which contains a rotating prism to give a static picture from the continuously moving film. Animated viewers of this type are normally placed on the bench between the rewind arms and operated by the motion of the film drawn through by hand.

For later stages of editing, especially for double-system picture and sound track together, one of the more elaborate motorized viewers, e.g., the Moviola, is generally used.

For anything other than urgent news material, editing is always done on a work print or cutting copy. The original master film – whether negative or reversal – should not be used for cutting as it is liable to be scratched. Most colour master films are susceptible to scratching.

SPLICING. There are two basic methods of joining films: cement splices and adhesive tape splices. In practice, picture films are almost always cement spliced. For amateurs using 8mm or Super-8, however, tape splicing is widely used for convenience, even though tape joins tend to show up more on the screen. For editing purposes, tape is useful for temporary joins, e.g., to try the effect of cutting from one scene to another.

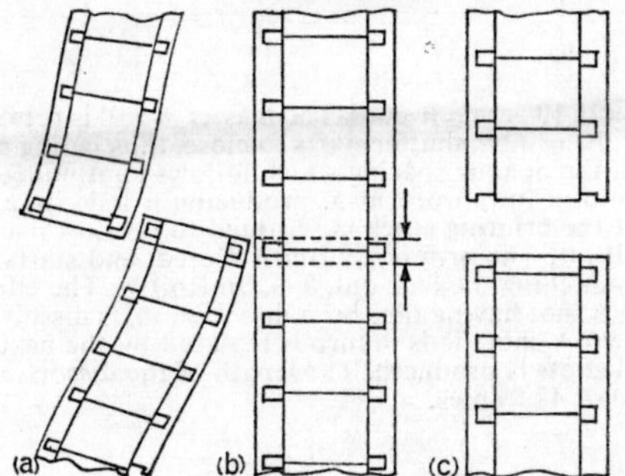

CEMENTED 16MM SPLICES. (a) To make a cement join, 16mm film is trimmed so that there is an overlap, the emulsion being scraped from the overlapping area on the one piece of film. Cement is applied and the two pressed firmly together so that they weld or fuse. (b) In a normal 16mm splice the overlap encroaches on two frames. (c) A neater join is obtained by using a splicer which trims one piece of film along the frame line. The overlap occurs along the bottom of one frame only and is much less noticeable.

Cement joiners for 16mm projection splices normally use a 0·1-in. wide join (sometimes $\frac{3}{32}$ or 0·093 in.), which overlaps the frame line both ways. That is, the edges of the join show on the top of one frame and the bottom of the next.

For joins in negative and reversal master material, a narrower join is preferable. The narrower overlap is strong enough for master film since film printers handle material more gently than the average projector.

The splice is rendered less visible not only by making it narrower ($\frac{1}{16}$ in. or 0·062 in., for example), but by cutting one of the ends exactly on the frame line, so that the join shows on one frame only.

The join is even less noticeable if the overlap is arranged so that it is along the bottom of the frame where the picture is usually darker.

The problem of joins showing on the screen is one which does not arise on 35mm sound film, where there is enough space between frames to permit a negative join to be made without overlapping on to the picture area. On 16mm, 8mm and Super-8 film there is no more than a thin line separating the frames. Thus with narrow-gauge films the join has to overlap at least one edge of the picture area.

Careful maintenance of cement splicers is essential for optimum efficiency. Further, on narrow-gauge film, it is important to ensure clean working, correct scraping, and to apply just the right amount of cement.

MASTER MATCHING. When the cutting copy or work print has been edited into its final form, and approved, the next operation is to get out the master film, be it negative or reversal positive, and to cut that exactly to match the cutting copy.

Footage numbers are provided along the edge of most professional film-stocks and print through on to the cutting copy. These edge numbers can be used to locate the scenes on the master film corresponding to those on the cutting copy.

Some film-stocks do not have edge numbers, but it is possible to have these printed in ink after processing. Some laboratories offer ink edge-numbering of the negative film or master film before printing, or the master film and freshly printed cutting copy may both be identically edge-numbered.

Many producers nowadays leave their original negative or master film with the laboratory, and return the cut work print to them for the master to be matched to it. The standard of cleanliness and film handling in laboratories is generally appreciably higher than in the producer's cutting room.

A & B ROLL ASSEMBLY. 16mm masters are often assembled by what is known as the A & B roll method. This makes it possible to print dissolves and superimpositions without making a special optically printed duplicate length of film. It is always preferable to avoid duplicating stages if possible, because of the inevitable loss of quality.

It has therefore become normal 16mm practice to make up the master into two rolls instead of one. Both rolls start from a sync mark frame on the leader so that they keep in step. The scenes up to and including the first dissolve are assembled in roll A, and followed by black leader film. The B roll consists of black leader up to the scene into which the roll A scene must dissolve. Thus, the two scenes which are to dissolve one into the other are overlapped by the length of the dissolve.

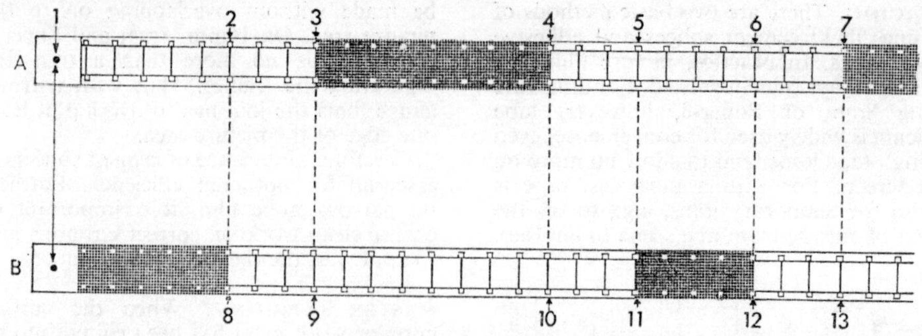

PRINTING 16MM A & B ROLLS. Each frame on diagram represents 10 actual frames. The master A roll is printed first, aligned with the printing stock on start mark, 1. At 2, the printer shutter starts to close, thus fading out the scene, and the shutter is fully closed 30 frames later at 3, the opaque spacing which follows continuing to blank the printer light. At 4, the shutter starts to open and is fully open at 5, producing a fade in, and shutter closing produces a fade out between 6 and 7. Next the printing stock is rewound to its start mark, and the matching start mark, 1, on the B roll is aligned with it. The printer shutter is closed, and starts to open at 8, opening fully by 9, and thus producing a fade in matching the fade out, 2–3, on Roll A. The effect of these superimposed opposite fades is a dissolve. The Roll B shot having thus been dissolved in, is dissolved out again by a B roll fade out, 10–11, which brings in the next A shot. This in turn is replaced by the next B shot with a fade in from 12 to 13. Thus a chain of dissolved shots is produced. The length of the dissolves is set by the length of each overlap, which is usually 30, 36, 40 or 48 frames.

Assembly of the two rolls continues frame-for-frame, using a synchronizer accommodating the A and B rolls as well as the cutting copy. Scenes are added in roll B until the next dissolve is reached, roll A being made up with black leader. The procedure is followed throughout both rolls.

In the laboratory, the A roll is put through the printer first, using a printer specially fitted with a means of fading in or out wherever desired. When the printer reaches the point at which the dissolve is to start, a fade-out is made. The printer shutter remains closed through the blank leader on that roll. When the second dissolve is reached, the fade shutter opens again, producing a fade-in, and remains open through the following scenes. The mechanics are slightly different for reversal master and for neg-pos printing, but the principle is the same.

When the A roll has been printed, the print stock is wound back to the start, and now threaded into the printer with the B-roll master. The start frame on the print stock is lined up with that on the B roll, just as it was on the A roll. When printing the B roll, the shutter remains closed until the beginning of the fade-in of the first dissolve. Thus the fade-in on the B roll is superimposed upon the fade-out on the A roll to give the dissolve. The rest of the scenes are printed from one roll or the other.

The same procedure produces superimpositions. Superimposed titles, preferably white-on-black, are inserted in the B roll in line with the scenes over which they are to be printed. The printer shutter is arranged to be open at that point on both the A roll to print the scenes and the B roll to print the titles. Thus the print shows the titles superimposed on the scenes.

When assembling A & B rolls, it is usually not necessary to use opaque leader between scenes, as the printer shutter is closed between each fade-out and the following fade-in.

CHECKERBOARD CUTTING. A & B roll printing was first introduced to make dissolves and superimpositions easier. Although it increases printing costs, it is better than having to make duped sections of the film for optical effects such as dissolves.

Now, however, there is an additional reason for A & B roll printing: it provides invisible splicing. This method is called checkerboard. It makes use of the fact that frame-line joins overlap one way only. The scenes are assembled into A and B rolls, e.g., the A roll carries scenes 1, 3, 5, 7, etc., while the B roll has scenes 2, 4, 6, 8, etc.

The length between each scene is filled with black leader film, and every join is made on a frame-line splicer with the direction of overlap always into the black leader film.

The direction of overlap must be reversed for the head and the tail of each shot. In practice, all the head joins are made on one roll when winding the roll through one way, and all the tail joins when winding through the other way. For single-perforation film, special 16mm splicers permit the splices to be made either way round.

The overlap area of every join is then only in the black leader film and does not show in the print. Care has to be taken that the black leader is completely opaque, because the printer light remains on throughout the printing of the roll (except of course where a fade or dissolve occurs). Transparent spots or semi-opaque parts of the black leader film print through and impair the shot printed from the other roll.

Fades and dissolves can be incorporated in rolls for checkerboard printing, dissolves being overlapped by the appropriate number of frames, as for non-checkerboard assembly. Superimpositions can be achieved by placing the title or scene to be superimposed in place of the black leader film at the appropriate point.

When using A & B roll printing, with or without checkerboard, it is possible to add a C roll, D roll, or any reasonable number of additional rolls for further uses.

For example, it is possible to superimpose more than two scenes upon the same picture. Secondly, alternate titles for a foreign language version can be added by using extra rolls of the master film. In this case, the titles of the original version are assembled into a C roll. Titles in the foreign language are assembled into an alternative C roll. Then the laboratory can be instructed to use the English C roll, French C roll, or whatever language is called for, with the appropriate sound track. This avoids the necessity for cutting titles in and out of the master whenever a different version is required. Diagrams, or any scenes containing wording that must be changed for the foreign versions, can also be placed in the appropriate C roll.

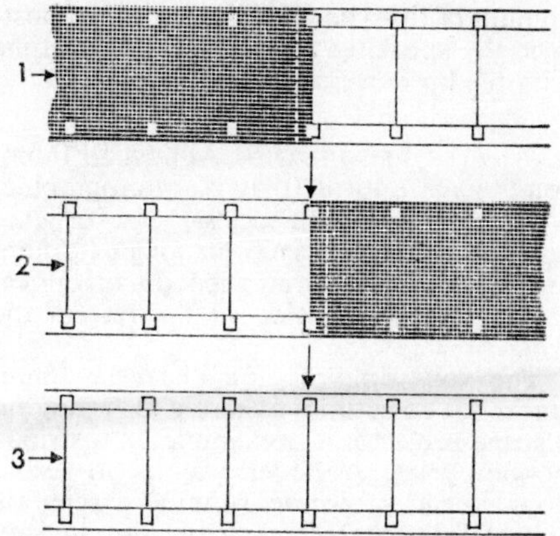

16MM ASSEMBLY SPLICING. 16mm joins overlap on to the frame area in normal assembly. 1. By using a frame-line joiner and splicing each shot to a length of black spacing, the overlap can be placed, not on another scene but on black spacing. 2. It is necessary to turn the joiner round when joining black spacing to film, i.e., at alternate splices. When the two rolls have been printed on to the copying stock, 3, the join is invisible.

SOUND TRACK

The methods to be used for preparing any sound track depend very much on its content and complexity. The range of possibilities is particularly great on 16mm, because productions may involve anything from a simple commentary-plus-music track to lip-sync and spot effects.

Apart from lip-sync dialogue and some effects recorded at the time of shooting, practically all 16mm tracks are made after the picture film has been edited. A small

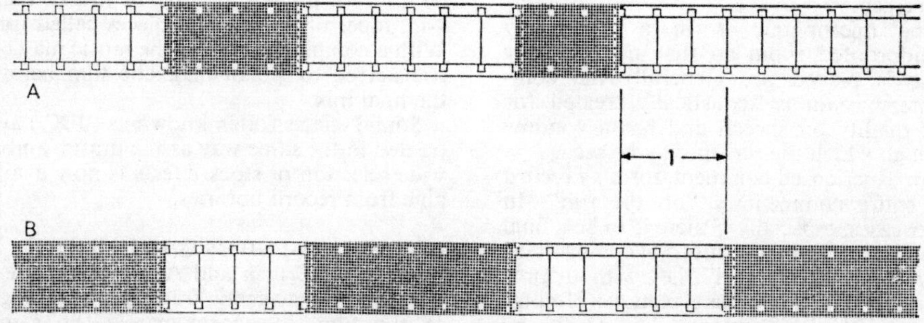

CHECKERBOARD ASSEMBLY OF 16MM MASTERS TO PRODUCE INVISIBLE JOINS. Each consecutive shot is placed in a different roll. Opaque black spacing is placed between alternate shots. At 1, however, the shots are overlapped to produce a dissolve just as in normal A and B roll printing. (Each frame on diagram represents 10 actual frames.)

amount of lip-sync may also be added post-sync, by speaking the words in exact time with the lip movements on the screen.

MAGNETIC RECORDING. Almost all 16mm prints have optical (that is, photographic) sound tracks. Nevertheless, the original recordings are all made on magnetic film, and then finally re-recorded on to optical track at the conclusion of the rest of the work.

The very simplest and cheapest 16mm tracks are sometimes made by recording on a stripe projector, for example on a striped no-join print. A better way is to use a double-head projector, running a separate roll of fully coated magnetic film in step with the edited picture film. Where a microphone is used to record the commentary on to the projector, it is essential to keep the microphone away from the machine noise.

The specialist sound recording studios use top-quality 16mm recorders and playback units, and interlock motors on every machine, including the projector. This permits all of them to be started and run in frame-for-frame sync. The final re-recording or dubbing session makes good use of this fact: it is normal practice to have several separate rolls of sound track all running through at the same time, with commentary, music, effects, lip-sync, and so on, the various rolls. The sounds from each are faded in to the master recorder at the appropriate points.

COMMENTARY. Sometimes the commentary is recorded as the commentator watches the projected picture: he begins speaking at each cue-point. Professional studios have a system of counter numbers, operated from the projector and shown under or alongside the screen.

The microphone is in a completely soundproofed room so that no projector noise is picked up. Generally the commentary room is acoustically treated for best quality on speech and has a window through which the screen may be seen.

An experienced commentator may record the entire commentary "on the run". In other cases, especially if there is to be a final dubbing session, the sections of commentary may be recorded "wild", i.e., without projecting the picture. These sections of commentary are later cut and laid to fit the picture, spaced out with blank magnetic film so that each section of speech comes in the right place.

MUSIC AND EFFECTS. A large proportion of narrow-gauge productions are documentaries, or industrial or educational films, so a specially composed music score is the exception rather than the rule. Where music is required, it is generally obtained from a recorded music library specializing in film work.

The library takes care of copyright problems, which can be quite formidable with ordinary commercial records.

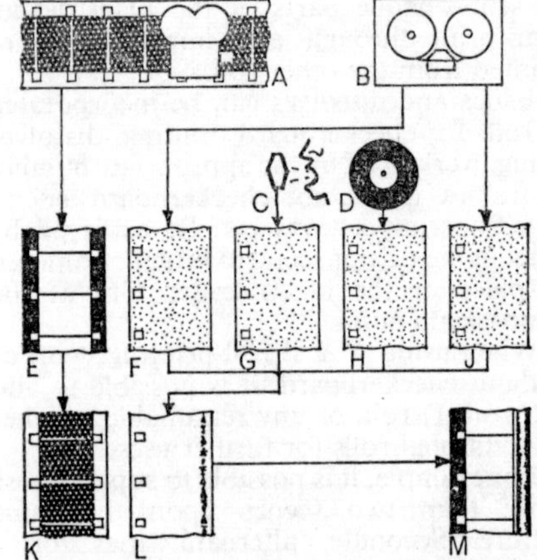

BUILDING THE SOUND TRACK. A. Picture shot silent. E. Workprint printed from it. B. Sound effects tape-recorded wild and edited on magnetic tracks, F, J. C. Commentary recorded on track, G. D. Additional effects or library music recorded on H. All sound tracks dubbed to magnetic master and thence to optical mixed track, L. This is synchronized with edited camera original, K, and printed to form composite sound and picture print, M.

For 16mm production, the library music is usually available on gramophone records or on ¼-in. tape. Records soon show signs of wear, so tape is preferred because of its quieter background. However, records may be used for the selection of the music, then ¼-in. tapes of the selected pieces called for. With a complicated track, the music may be transferred to 16mm magnetic film before the final mix.

Sound effects (often known as "FX") are treated in the same way as the music, and a wide selection of stock effects is now available from record libraries.

MAGNETIC SOUND TRACK In the past, it was normal British and American practice to record the magnetic track down the edge of the film, in approximately the same position as the sound track on an optical print, that is, on a 0·100 in. wide (100 thou., or in sound parlance 100-mil) track along the unperforated edge of the film. These tracks, made on fully-coated 16mm. magnetic film, were then standard with recordings made on 16mm. magnetic/optical projectors on magnetic striped film.

European practice, however, has been to locate the magnetic track approximately down the centre of the film. This is known as centre track, and is normally 0·200 in. (200-mil) wide. The centre-track position gives smoother contact with the magnetic heads, and the extra width gives improved quality.

Centre track is normally used for master recording on fully coated magnetic film and is becoming an international standard. Only tracks which must be played on a projector are made on 100-mil edge track.

JOINING MAGNETIC FILM. 16mm fully coated magnetic film is joined by applying adhesive tape to the back of the film as for 35mm magnetic joins. Tape joiners do the job speedily. The film ends are trimmed across, the two ends butted together and adhesive polyester-based tape is applied across the back of the join. The joiner then punches perforations in the tape to coincide with those in the film.

It is usual to cut the magnetic film diagonally, so that the join passes over the replay head more smoothly and with less likelihood of noise.

TRACK-LAYING. The process of editing 16mm magnetic sound tracks to synchronize with picture is basically the same as in 35mm production, though with less expensive equipment.

The usual equipment for 16mm track laying consists of an ordinary picture-synchronizer, with two or more coupled sprockets for running two or more lengths of film in step with one another. The picture film passes through one of the sprockets, and runs through an animated viewer; this provides a picture on a small screen. The magnetic tracks are run through the other channels of the synchronizer, each having a magnetic head. These pick up the sound which is then reproduced through an amplifier and loudspeaker.

A multi-sprocket synchronizer allows two or more sound tracks to be played back at the same time as the picture cutting copy is viewed. The small size of 16mm, compared to 35mm, enables a four-way synchronizer (picture and three tracks) to be used on a bench.

This device reproduces the sound track only well enough for identification, but it provides an excellent and speedy means of laying tracks in their correct relationship to the picture, and to each other. Subsequent stages of track-laying are usually done on a Moviola, which gives good enough sound quality for accurate editing of the track.

FINAL MIX. A complicated 16mm track is assembled in exactly the same way as a 35mm one: each component of the track is on a separate roll. Thus, there may be one roll of commentary, one of effects, one of lip-sync dialogue and one of music. Each begins from the same sync mark frame to keep in exact step with the picture.

For the final mixing, all the separate magnetic track rolls are threaded into playback units and the picture film threaded in the projector. All are then run through in frame-for-frame sync, and the dubbing mixer operator fades the sound from each machine in and out, as indicated in the dubbing script. Again it is usual to work on footage numbers, counting the start frame as 000. Measurements can be charted from the counter on the synchronizer sprocket.

Roll-back (sometimes called rock-'n-roll) is a system frequently used, in which the projector and sound dubber can be reversed in interlock and run forward again, thus eliminating the time-wasting process of removing reels, rewinding and re-threading.

TRANSFER TO OPTICAL. The final mix is usually made on fully coated magnetic film, because in the event of any error it can be wound back to the start and recorded upon again without any waste of stock.

When the final magnetic master has been approved, it is re-recorded on to an optical (photographic) track from which the release prints can be made.

Sometimes the final mix is made directly on to the optical recorder.

ORDERING OPTICAL TRACKS. Unlike 35mm film, the 16mm optical track may have to have the emulsion sometimes on one side, sometimes on the other; sometimes it must be a negative track, sometimes a positive. The correct choice depends on the type of picture master and the type of printing stock to be used.

When the sound print is to be contact-printed from a 16mm camera original, it must carry the emulsion on the side of the film facing the lamp and away from the screen. This is known as the non-standard emulsion position, because it is the opposite way round to a camera original which is the basis of all 16mm sound projector design. In practice, having the emulsion on the "wrong" side of the film means that the track is not in the plane of best focus of the sound optical system, so there will be a slight loss of high frequencies. However, the resulting difference is generally quite small. The optical sound track required for printing from the camera original must have the emulsion on the same side as the camera original. For convenience, this is known as a B-winding track, because it is the same way round as the camera film, which is also B-winding.

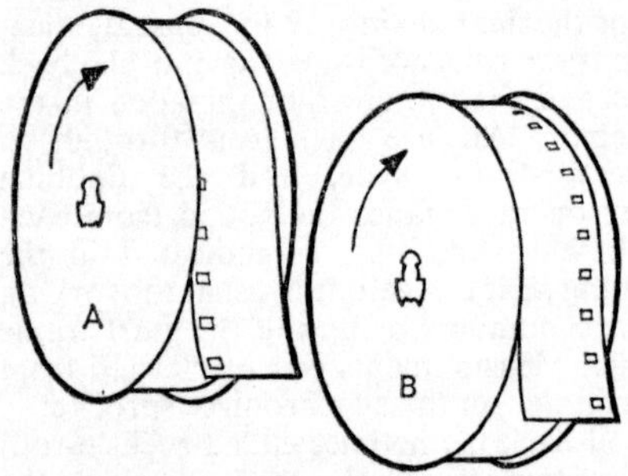

A & B WINDING. With the emulsion surface inwards and the film unwinding clockwise, the perforated edge is towards the operator for A winding, and away from him for B winding.

When the prints are to be contact-printed from a dupe, whether an internegative or an interpositive, these prints have standard SMPE emulsion position: the emulsion side of the print faces towards the screen. The optical track from which it is printed must have its emulsion on the same side as the dupe master, i.e., an A-winding track.

Thus a B-winding track is needed when the prints are to be made directly from the camera master, and an A-winding track when the prints are to be made from an interpositive or internegative.

Some printing processes require a negative track, others a positive track. The case of black-and-white is simpler than colour, where sound tracks are produced in unexpected ways from certain print stocks in the laboratory.

For black-and-white, a print to be made from a negative needs a negative sound track. A reversal print from a reversal master positive needs a positive sound track. Prints from an internegative need a negative sound track.

For colour, a print from a reversal master made on to the most common colour duplicating stock needs (unexpectedly) a negative sound track. The processing laboratory will give information on the needs of the printing stock they use. For prints from an internegative on to colour print film, a negative track is again required. (Of course, the winding will be different from a track for printing from the camera master.)

16MM RECORDING CHARACTERISTIC. 16mm film, being smaller than 35mm, passes the sound head at only 2/5 the speed of the latter, so the high frequencies tend to be less well reproduced.

It is usual when recording a 16mm optical track to adhere to a tonal characteristic of equalization which gives the most pleasing sound with normal projection. In practice, this involves emphasis of the upper frequencies that lie within the range of 16mm reproduction, say 4,500 to 6,000 Hz, and cutting off the highest frequencies which reproduce as background noise.

A magnetic master is given a rather similar characteristic, though the correction and the cut-off are slightly different.

Good sound tracks are especially important on 16mm, since it is rarely projected under the controlled acoustic conditions of a cinema. The great advantage of 16mm is portability, and its main purpose is to bring film into the church hall, factory or lecture room, where the acoustics may be very poor.

Yet, with care during the vital dubbing stages, the 16mm optical track can do an excellent job.

PRINTER START MARKS. Before the optical track is ready for printing, it must be marked up with a start mark on one frame of the leader. The picture film is fitted with a standard type of leader, and the track is unmodulated (or black on the print) during this run-in footage. The printer start or sync frame is normally well in advance of the printed start frame on the picture leader, to allow for threading the printer and for running up to speed.

In any sound print, the sound is not alongside the picture to which it relates, and the separation between the two depends on the gauge, and whether optical or magnetic.

For 16mm optical tracks, the sound is 26 frames in front of the picture.

Generally it is best to ask the laboratory to add the standardized leaders and to mark up the sync advance. Sometimes the pro-

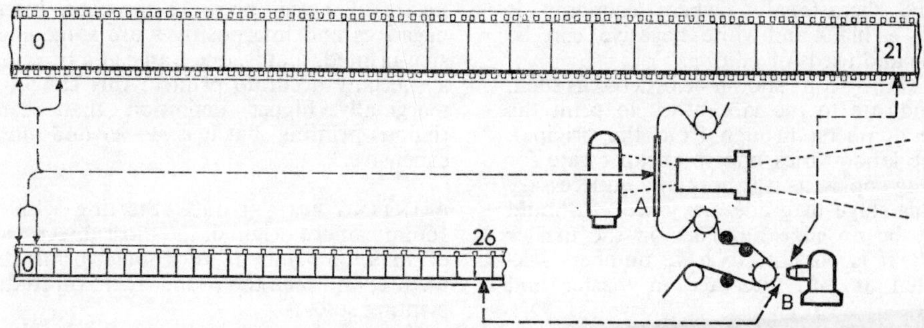

OPTICAL SOUND TRACKS. The picture and the sound that relates to it are not side by side on a print. In the case of optical sound tracks the sound is 21 frames ahead of picture on 35mm prints (*top*), 26 frames on 16mm prints. The reason for this separation of picture and sound is that at the instant that the picture A is being projected, the sound at B is being reproduced.

ducer does this, but he must make it clear to the laboratory whether the sound has been advanced or not.

If the picture and sound start marks are opposite one another (as they were on the synchronizer when editing) without the sound having been advanced, the start mark on both track and picture should be marked LEVEL SYNC. If the 26-frame separation has been allowed for, the track and picture should be marked PRINTER SYNC. To advance the track, the sound start mark frame should be moved 26 frames back, that is, farther in from the head of the reel.

Printer start marks are made by punching a hole in the middle of the film frame and marking the frame with an X in blooping ink or with a felt-tipped pen. Some laboratories like sync marks at the end of the reel as well as the front. This is because, for speed, they print the picture from the head of the roll; then, without wasting time rewinding, they print the sound from the foot of the roll.

When checking optical tracks before printing, it should be noted that a negative track normally sounds noisy and sibilant, because the characteristics of a good negative are not the same as those of a good print.

PROCESSING

THE 16MM LABORATORY. Most 16mm laboratories are specialists with the necessary costly equipment for proper servicing of this gauge. Where 16mm is handled in a large laboratory also handling 35mm, the narrow-gauge films are usually in a completely separate department. The ever-increasing proportion of colour has forced new standards of technical ability on the laboratories, and all phases of a 16mm laboratory's operation are now at an extremely high level.

Most laboratories offer additional services such as negative or master cutting to match the work print, provision of leaders, checkerboard joining, and so on. It is generally well worth the 16mm producer's time to make full use of these services.

PROCESSING CAMERA STOCK. In case there there is a need for any retakes, most laboratories offer an overnight service for processing the camera film and supplying rush prints or dailies.

In the case of negative film, it is usual to ask for a rush print to be run off immediately. Rush prints of colour negative can be black-and-white for economy. If there are many unwanted takes, the laboratory can cut these out before printing, provided they are properly slated with scene and take numbers.

With reversal film, producers often prefer to examine the processed film, when time permits, before ordering a cutting copy. In fact, the additional handling by the producer generally adds to the wear and tear and scratching of the master. The best way to minimize these is to examine it by rewinding it carefully over a light source. The better types of animated viewer can sometimes be used safely, but it is inadvisable to put master film through a projector even once. Waste footage or unusable scenes can be removed before the cutting copy is made, because the image on reversal master, unlike negative, can be seen clearly and the quality judged.

Cutting copies from reversal originals are usually made on reversal stock. Reversal colour film can be printed on to black-and-white reversal film; colour copying is rather costly. Occasionally, where economy is vital, a black-and-white negative can be made and used for editing.

A cutting copy should be ordered as such, to indicate to the laboratory to print the edge numbers through from the original. Some laboratories offer a cheaper rate for cutting copies, as it is normally unnecessary at this stage to grade every scene. Should there be no edge numbers on the master stock, it is possible to have numbers ink-printed at 1-ft. intervals on master and print.

Some types of black-and-white film can be processed either as reversal or negative, so when sending such stock to the laboratory it is important to state which is required.

Some film-stocks can be boosted in processing, giving a speed increase of generally one or two *f*-stops (an effective increase in the speed-rating of two or four times). The cameraman must decide at the time of shooting what degree of boosting he wants; exposures must be adjusted accordingly. This facility is particularly useful in emergencies when shooting in colour, if there is not enough light for normal exposure. Boosting increases grain-size, so should be avoided unless it is the only means of getting an acceptable picture. It is better to use a faster film, as the results are of higher technical quality and more predictable.

MAKING 16MM PRINTS. Most 16mm prints from a 16mm master are made by contact printing. The usual method is to draw the master film and print stock together continuously and smoothly past a light source of controlled intensity (and controlled colour, for colour printing), by means of a very accurate sprocket. Internegatives and interpositives are sometimes step-printed, that is, one frame at a time, on a specially accurate printer; this can give marginally higher definition than continuous printing, but it is slower and more expensive.

PRINTING PROCEDURE. Starting with a 16mm camera original, the most direct way of making prints is to contact-print the picture, and then the sound track, on to the printing stock.

If the camera original is a negative, it is printed on to positive stock. If the camera original is a reversal positive, it is printed on to a reversal duplicating film.

Printing directly from the camera original gives the best quality prints, but it gives no protection should the master film become seriously worn or damaged.

To protect the master, it is possible to make a duplicate master from which the prints can be made. Then the camera original can be stored in a safe place for future use should further dupes be required. Or the master can be used for printing, and the duplicate master put on one side as an insurance against damage to the camera original. The choice depends on the probable total number of prints wanted, and also on the costs of making prints by different processes and procedures.

If the camera original is a black-and-white negative, it is normal practice in both 35mm and 16mm to make a fine-grain master positive print, and from that to make a duplicate negative. Should this dupe neg, as it is called, become worn or damaged, another one can be made from the master positive. The same applies to the optical sound track, for which master positives and

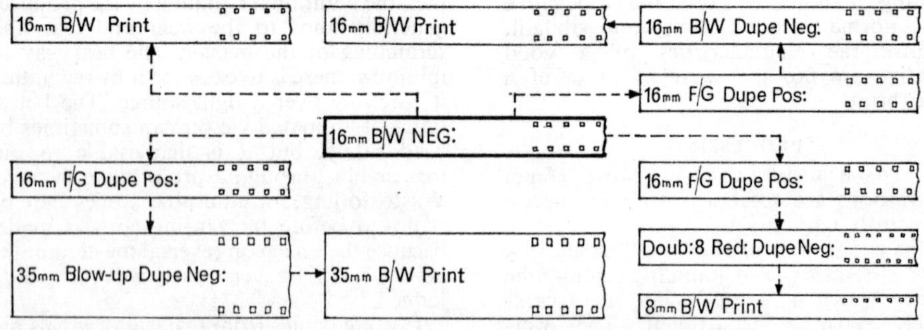

DERIVATIONS FROM 16MM B & W NEGATIVE. 16mm B & W print can be arrived at by direct contact printing, or through duping stages to provide protection for original negative. Blow-up or print-up can alternatively be derived from a 35mm dupe positive. The reduction dupe negative can furnish either Standard-8 or Super-8 prints, depending on perforation of stock used.

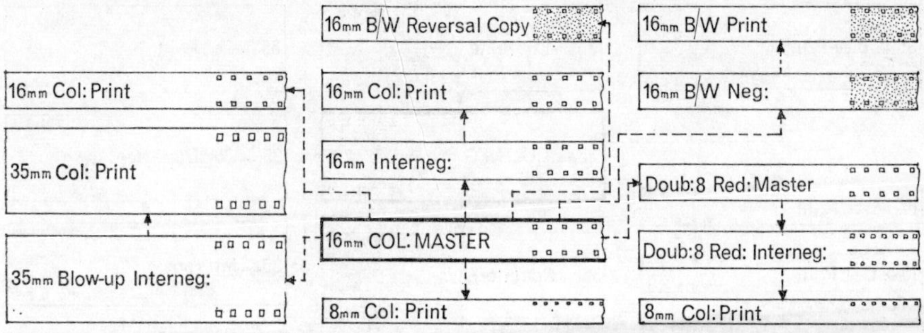

DERIVATIONS FROM 16mm COLOUR MASTER. Since colour master is a positive, contact colour prints are made by reversal, and those made through an internegative require one less stage than from a B & W negative. B & W derivations follow a similar course. 35mm blow-ups are preferably made at the internegative stage, as shown, rather than at the colour print stage. 8mm colour prints are preferably by reduction through a dupe positive and negative, and may be of either Standard-8 or Super-8 form. (B & W derivations shown dotted.)

dupe negatives are always on black-and-white stock, even when the final print is to be in colour. Alternatively, it is better to make new optical tracks by re-recording from the magnetic master.

In practice, the second way is more usual because, provided more than two or three prints are required, it is much cheaper. This is because positive prints cost less than reversal prints. On the other hand, an inter-positive may produce reversal copies that are of slightly higher quality. This is especially true in colour, principally because the colour duplicating stocks generally used have exceptionally good sharpness characteristics.

When 16mm masters are made up into A and B rolls, the prints made directly from them cost a little more than if the master was all on a single roll. This is because the laboratory first has to print the A roll, then the B roll, doubling the time involved.

When, however, the A and B rolls are printed on to a duplicate master, all the checkerboard assembly, dissolves, superimpositions, etc., are in the duplicate. Prints from this dupe are made by a normal single printing operation, and are therefore cheaper.

Before printing begins, each scene is carefully studied by a highly skilled operator known as the grader or timer. He decides how much light will be needed to give the correct density print from each particular scene.

In the case of colour film, he also takes into consideration the colour balance of each scene.

The first print may not show quite perfect grading, either for density or colour, but by the time the third or fourth print is made, the grading should be as even as it is possible to achieve.

No laboratory ever sends out prints which are seriously poor in grading, but usually some understanding is needed of the laboratory's problems in the first print, especially if wanted in a hurry, and if it is a colour print.

Some laboratories offer a slightly cheaper service for ungraded colour prints, that is, ones where the evenness of the camera original is such that it can all be printed on one colour balance. However, as more of the 16mm laboratories become equipped with sophisticated printers which readily give scene-to-scene colour correction, the trend is now towards colour corrected prints for everything other than first colour rushes.

Care must always be taken to see that the right type of 16mm sound track – that is, A-winding or B-winding, negative or positive – is provided to suit the master from which prints are being made. Where dupes are made from the picture film, it is common practice not to duplicate the sound track, but to order up another original optical track to be re-recorded from the magnetic master of the final mix.

Most laboratories offer a service for cleaning film before printing, and this is especially useful after the assembly of originals because they will have had a lot of handling when joining. It is well worth specifying that the film should be cleaned before first printing.

A proper cleaning operation, nowadays generally done by passing the film through a bath of special fluid with ultrasonic agitation, makes a great difference to the cleanness of the print, and even removes any wax crayon marks from the edited film.

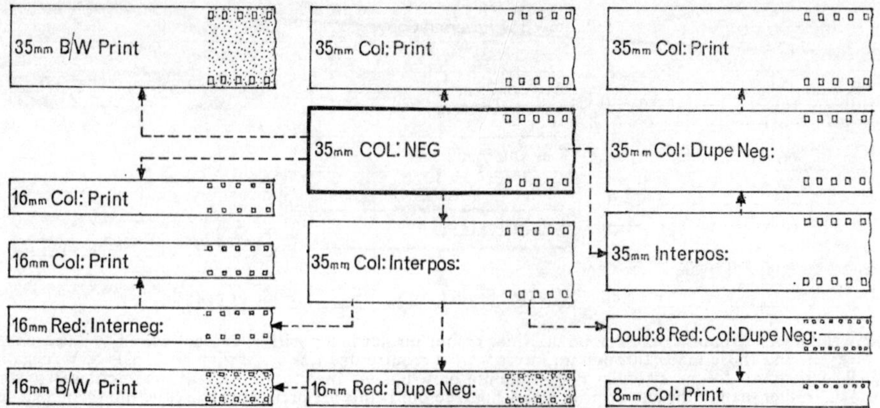

DERIVATIONS FROM 35MM COLOUR NEGATIVE. To preserve the original negative, materials are preferably derived from 35mm colour interpositive, though rush prints, both in colour and B & W, are made direct from negative. The colour interpositive thus becomes the master derivative from the original for bulk 35mm printing, and for reduction to 16mm and 8mm for release printing in those gauges.

REDUCTION PRINTING. By no means all 16mm prints are made from 16mm camera masters. Many films shot on 35mm are reduced to 16mm by making the prints on an optical reduction printer. On this machine, each frame of the 35mm film is projected, by a lens in the printer, on to the 16mm print stock. In other words, this printer consists basically of a 16mm camera looking into the gate of a 35mm projector head, with coupled drive to the two heads. Because every frame must be projected separately, the film must move intermittently, so this is a form of step-printer.

Reduction prints are always made to have standard emulsion position: the emulsion of the 16mm print faces the screen during projection, as would a camera original. So the sound optical system on the projector is accurately focused on the emulsion of the film, giving best high frequency reproduction.

Where a large number of 16mm prints must be made from a 35mm production, it is usual, and certainly more economical, to make a 16mm internegative by reduction printing, and to use this to make 16mm contact prints in the usual way.

The sound track has also to be reduced to 16mm, and it is possible to print this by means of a sound reduction printer – in this case a continuous printer and not a step-by-step type. Optical reduction printing of the track does not generally produce the best results, however, because the recording on the 35mm track has to be compressed into two-fifths of its original length, and the high frequencies suffer accordingly. In any case, the 35mm track was recorded with a tonal characteristic to suit cinema reproduction, and the reduction printing changes this characteristic and results in poor high-frequency reproduction with normal 16mm projection.

A far better method is to go back to the 35mm magnetic master track from which the negative track was made. From the magnetic master, a 16mm optical track can be re-recorded with the correct frequency response for optimum 16mm reproduction. This track can then be contact-printed on to the 16mm print stock after reduction printing of the picture.

TECHNICOLOR. The dye-transfer (imbibition) process is used by Technicolor for making large quantities of high quality 35mm, 16mm, 8mm and Super-8 prints.

Technicolor's imbibition process involves making a set of three colour separation matrices from the camera master film (if a negative, or from an internegative in the case of reversal master film). The master film is then put away, for all the prints are made from these matrices. Each of the three is soaked in a dye (cyan, magenta, or yellow, as the case may be) and each matrix carries the image in varying depths of gelatin. A heavier density soaks up more dye. Each dyed matrix is then brought into accurately registered contact with a blank film on which the sound track has already been processed, and the dye transfers from the matrix to the blank. When all three dyes have been transferred, the colour print is dried and wound up in a roll.

Because of the costly matrix-making procedure, the Technicolor imbibition process is not suitable for making small numbers of prints. However, the process is practicable for making more than about 40 prints. The advantage is that these large numbers of prints are made without having to wear out the master film. Once the matrices are made, the print-making is relatively less expensive than prints on multi-layer colour film.

Technicolor normally arrange all their own pre-printing work, for their transfer process is generally done on 35mm film, and most 16mm prints are made as two prints side-by-side on specially perforated 35mm film. Formerly, however, they made one print down the middle of differently perforated 35mm film, but this process is now obsolescent. Sound tracks, however, are produced differently. In either case, the prints are delivered as normal 16mm prints.

An important point should be borne in mind about sound tracks for 16mm Technicolor prints. The negative track should not be on 16mm stock, since double-rank prints need a double-rank sound negative. Technicolor always advise the producer on these technical aspects of using their bulk-printing service.

Technicolor also make 8mm reduction prints by the dye-transfer process, and this size is generally printed as quad-8, that is, four runs side-by-side across the width of the 35mm stock. When magnetic stripe sound is required, the film is striped and recorded in the quad-8 form before slitting.

A more recent addition to Technicolor printing services is Super-8; this is normally printed triple-run between the perforations of 35mm film with standard 35mm perforations which are used for registration during the transfer processes. Super-8 prints by Technicolor usually have an optical track, but a magnetic track is also available.

EIGHT-MILLIMETRE PRODUCTION

FILMING. When contemplating the production of films on 8mm or Super-8, it is well to remember that these smaller gauges do not have sufficient reserve of quality to permit satisfactory duplicating. They are really suited only to productions where no copies are required: where the camera original film is all that is needed.

For certain types of filming, therefore, the 8mm gauges may have to be considered. Cameras for these gauges are the smallest, the most portable, and the most highly sophisticated, while still being extremely easy to use. The 8mm gauges are ideal for amateurs, for whom they were designed.

Filming on 8mm or Super-8 is particularly easy because cameras generally incorporate every conceivable aid to good pictures: automatic exposure control, reflex viewfinding, zoom lenses, and so on. In the past, the most notable feature of 8mm cameras was the extreme depth of field of their lenses. A normal lens of about 12·5mm focus is in acceptably sharp focus all the way from about 3 ft. to infinity, at even a medium aperture of around *f* 5·6. Nowadays, however, the trend is towards longer and longer focal length zoom lenses, and the longer the focal length, the more accurate the user must be when setting the focus. Despite the provision of visual focusing and rangefinders on many cameras, the professional method of using a measuring tape to check the exact distance from film-plane to subject is worth while.

To make the most of the rather small 8mm (or Super-8) frame, it is advisable to use a higher than usual proportion of close and medium close scenes. In fact, the closer the better. Certainly it is best to avoid long shots which depend on fine detail for their effect. The situation is comparable with filming for television, where again fine detail cannot be adequately reproduced.

A fact to be remembered about 8mm (or Super-8) is that a camera original film is almost invariably of better quality than any form of dupe or reduction print, no matter how it is made.

Only projection film stocks are normally available on the 8mm gauges. Master film for duplicating is not obtainable because this gauge is generally considered too small to duplicate well. In particular, the loss of sharpness, and general increase in visibility of dirt, both cause disappointing results.

Of course, considerable quantities of amateur 8mm films are duplicated, generally by printing directly on to a colour reversal duplicating stock. When contact-printed, this gives a print with the emulsion on the "wrong" side of the film, so it cannot be cut into original camera material without the need for refocusing the projector lens each time. When the 8mm (or Super-8) original is copied by one-to-one optical printing, there is generally an additional loss of sharpness, or a serious increase in the visibility of dirt and dust on the original, or both.

EDITING. Film joins are made in the same way as with 16mm, using a splicer and film cement. Many amateurs find it easier to use special polyester adhesive tape joiners.

The splicers for 8mm and Super-8 range from the simplest splicing blocks to elaborate types which scrape or grind the two ends of the film to a wedge or chamfer shape; this gives a much smoother-running join than the simple overlap type.

Splicers for standard 8mm make the join across the frame line, just as on 16mm. But the usual practice on the better splicers is to make a frame-line cut (that is, to cut one of the ends across the frame line, and thus overlap only into one frame instead of two).

Super-8 film is designed with the perforation in the middle of the frame (instead of bisecting the frame line as on 16mm and standard 8mm). When the Super-8 film is joined across the frame line, the join is completely clear of the perforation. This gives easier scraping and a stronger join.

When joining 8mm or Super-8, great care is needed to make a clean join. The Fuji Single 8 film is on polyester base and cannot be joined by the usual film cements so is generally tape-spliced.

The very small size of the 8mm or Super-8 picture makes it imperative to use an animated viewer for editing. It is not possible to make out sufficient detail in individual frames with the naked eye. Some of the inexpensive animated viewers are rather rough on the film, and the professional user should be extremely critical when choosing editing equipment.

REDUCTION PRINTING. If 8mm or Super-8 prints are required, it is best not to shoot directly on the small gauge, but to shoot on 16mm and then have the prints made by reduction. Almost all commercial 8mm prints are made from masters shot on a wider gauge – often from 35mm.

Reduction prints made directly from the larger master tend to be expensive, firstly because reduction printing is carried out on step-printers which are slow compared to continuous types, and secondly because the intermittent movement of the film through the step-printer gate causes faster wear on the master. This is aggravated by the fact that optical printing shows up more dust and dirt on the master film than does contact printing. Nevertheless, where a small number of 8mm (or Super-8) copies are required, it is quite practical for them to be reduction-printed from the larger gauge master film.

When a larger number of prints is needed, the usual practice is to contact-print them from an intermediate dupe film. Much 8mm and Super-8 printing is on stock 16mm wide, using perforated Double-8 or Double-Super-8, as appropriate. It is then processed on the same machines as 16mm prints and slit after processing.

A Double-8 (or Double-Super-8) intermediate film may be an interpositive, for printing on to reversal duplicating stock, or it may be an internegative, for printing on to positive film. In the case of colour, the interpositive/reversal procedure gives the sharper prints because the film-stocks used are inherently a little sharper than their neg-pos equivalents. For commercial purposes, however, it is more usual to make the positive prints from an internegative, because this procedure is cheaper. With black-and-white, the dupe neg and positive printing stock are invariably used.

Some of the largest laboratories are equipped to produce Super-8 prints with magnetic sound by the quad-8 method: four rows of prints side-by-side on 35mm wide film-stock. The stock is pre-striped, and the magnetic track is re-recorded on all four tracks at the same time as the picture is printed by contact from the quad-8 internegative. After processing, the prints are slit up into the four 8mm wide runs plus a waste 3mm width. The internegative can be made from a 16mm or 35mm original. Multiple printing techniques result in greatly reduced costs, even though the equipment represents an enormous investment for the laboratory.

To get 8mm or Super-8 prints from larger originals, a choice must be made between quality and price. For quality, duplicating stages must be as few as possible. Thus, the best quality is obtained by reduction-printing directly from the larger original.

If intermediates are used, as is normal to protect the camera original as well as to minimize printing costs, the best quality is obtained by leaving the film in the large format until the last moment. Thus, a 16mm internegative reduction printed on to the 8mm film stock gives better quality, particularly in sharpness, than if the intermediate was itself in the 8mm gauge.

But because it is cheapest and most practical to contact-print the 8mm (or Super-8) prints, general practice is to make the reduction-printed double-run internegative and then to make the prints from that by contact.

Most laboratories can incorporate checkerboard cutting and superimpositions, but where the original is in A & B roll form, they are not equipped to make fades and dissolves on the reduction printer.

Where small numbers of 8mm copies are required, there is a choice of other methods. It may be that the camera original is too valuable (particularly in 35mm) to be used for making reduction prints, and yet high quality prints are required.

If the original is a 35mm or 16mm negative, it is possible to make a contact colour interpositive, or a black-and-white fine-grain master positive. This may be printed directly on to the 8mm stock on the 35-to-8mm reduction printer.

With a 16mm original, there is a wider choice, because of the availability of reversal duplicating stocks in the narrow gauges. An internegative can be made by contact, and reduction prints made directly from this on to positive print stock. Alternatively, an interpositive may be made and copied on to 8mm reversal duplicating stock by reduction printing. Each of these methods involves making the dupe in the larger gauge, and printing by reduction at the very last stage. This produces the best quality. At the same time the camera original is safeguarded as it is not used for printing.

8MM AND SUPER-8 SOUND PRINTS. Standard 8mm sound prints always have magnetic stripe tracks which have to be re-recorded in frame-for-frame sync with the picture film. This is done by sound studios specializing in this work. The prints may be on pre-striped film-stock, or the stripe may be added after processing. In the latter case, it is cheaper to stripe the unslit Double-8 stock. Some studios also record before slitting because they use modified 16mm recording units. For best results each stripe is recorded from the original magnetic master.

Super-8 prints may be either optical track or magnetic. In the case of optical track, total print cost is less than with striping and re-recording of a magnetic track.

Super-8 optical tracks are in many ways the more practical for all library prints, even though the results are not quite up to the standard of magnetic as regards low background noise level. The Super-8 optical negative (in some cases positive) track is normally in Double-Super-8 form, with two recordings on the same film. Some laboratories work side-by-side tracks, while others work down one side and up the other. Either way gives two prints at a time, and the film is slit into the two 8mm runs after processing, again on a 16mm machine.

Some laboratories also have facilities for optical reduction sound printing from a 16mm sound negative (or positive) on to the Double-Super-8 print stock.

Super-8 optical sound prints have the track 22 frames in advance of the picture, and not 26 frames as with 16mm. So if the Super-8 optical tracks are made by reduction printing from the 16mm negative (or positive), the printer sync mark position is different. With Super-8 magnetic prints, the separation is 18 frames, while on Standard 8mm magnetic it is 56 frames. W.H.B.

See: *Camera* (FILM)*; Editor; Film dimensions and physical characteristics; Narrow gauge film and equipment; Print geometry* (*16mm*)*; Sound recording* (FILM)*; Splicing; Spools, cores and cans; Standards.*

Books: *Technique of Documentary Film Production*, by W. H. Baddeley (London), 1963; *Technique of Editing 16mm Films*, by J. Burder (London), 1968.

National Reference Network. Network along which a reference signal is fed, to provide carriers for modulation and to be reinserted in suppressed-carrier signals.

Negative Assembly. The matching and joining of negatives, shot by shot, to match the work print of a film.

When the positive work print has been completed and approved, it is sent to the laboratory, where the first edge number of each shot is noted, so that the negatives of all the shots in the film (previously broken down and labelled from the camera rolls) can be collected in one place. They are then matched to the work print, one by one, using the edge numbers, and the negative is finally spliced in reels, cleaned and sent for grading.

⊕**Negative Perforation.** Standard form of sprocket hole used along the edges of 35mm negative and intermediate film-stocks, also known as the Bell and Howell perforation (BH). This perforation has two flat faces with the ends as circular arcs.

⊕**Neutral Density Filter.** Filter which has no effect on the spectral quality of the light it transmits, but has a transmittance less than unity. Used in film camera work for increasing the lens aperture at a given light level without altering colour values, and for reducing the effective contrast of a scene.

□NEWS PROGRAMMES

The production of television news programmes involves techniques not common to the rest of the television industry, although the equipment used for collection, editing and transmission of the available sources of picture material is basically the same. The studio and technical area devoted to news programmes demand particular attention to facilities for quick assembly of news items into a complete newscast programme.

PREPARATION OF NEWS MATERIAL

In news work, the most important factor is speed, but this does not imply a lowered standard. On the contrary, the hazards of this type of operation tend to make technicians quality-conscious, since they are permanently on guard against degradation due to unfavourable conditions.

PICTURE SOURCES. News in television, as opposed to sound radio, requires the picture to tell the story, and therefore, whenever possible, the story is illustrated. The sources of picture material are many, and may be events occurring almost anywhere in the world at the time of transmission, taped or filmed records of incidents a few minutes or a few hours old, or a combination of both. A constant influx of picture material is required which provides worldwide coverage of current affairs. This material may arrive in many forms, e.g., processed or unprocessed cine-negative or positive in either 16mm or 35mm gauge, videotape on any line standard, and live or recorded material transmitted by the British, European or US networks. In addition, there is a complete news service on teleprinter from news agencies.

The editorial newsroom is much like that of a national daily newspaper. Incoming items are sifted and assessed, and arrangements made for appropriate pictorial coverage, either by home-based film crews, by stringers (agents) in remote areas or, if the occasion warrants it, by an outside broadcast operation. Any event occurring in the network area serviced by a television station can be covered by a film unit attached to that particular centre. This film is then normally processed there, and transmitted as soon as available via the network to the central station for videotaping ready for the next newscast.

Further sources of newsfilm are syndicating services and members of foreign networks such as the EBU (European Broadcasting Union). In the latter case, a daily interchange of film material takes place, all items selected for use being line-standards converted and stored on tape. Still pictures are taken from the agencies, by direct collection or via a printer installed in the newsroom. Arrangements for shipping crews and equipment abroad with a minimum of delay are essential, and resident staff at airports facilitate matters considerably. Permanent or semi-permanent staff at selected remote places are frequently necessary, examples being the staffing of a Moscow office for daily recorded reports by telephone, and a resident reporter and camera crew in distant theatres of war.

COMMUNICATIONS. Any news service is completely dependent on its communications facilities, and access to them must be comprehensive. This entails a comparatively large number of local ends, i.e., permanent circuits to the various centres. These include vision, programme sound and control lines to the local network switching centre, and programme and control lines to the International Exchange. Additionally, circuits from the news agencies carry a round-the-clock teleprinter service, including wired pictures.

STANDARDS CONVERSION. A television news organization must be able to handle picture sources on all standards. For conversion between standards having the same field rate, the line store converter developed by the BBC is the best method. This operates as a totally electronic digital device, and picture degradation is negligible. For conversion between different field rates, the only practical method has, until recently, been the image converter, which utilizes a display tube operating on one standard and an electronic camera on the other. The BBC has, however, now put into service a highly sophisticated completely electronic device which converts the field rate from 60 to 50 fields, subsequently employing the line store method for line standard conversion. This technique eliminates the two main drawbacks of the earlier method, i.e., the 10 Hz flicker and the smearing on fast action.

For minimum cover in Eurovision and satellite two-way working, two of each type of converter are required. Ideally, optimum coverage should be provided by

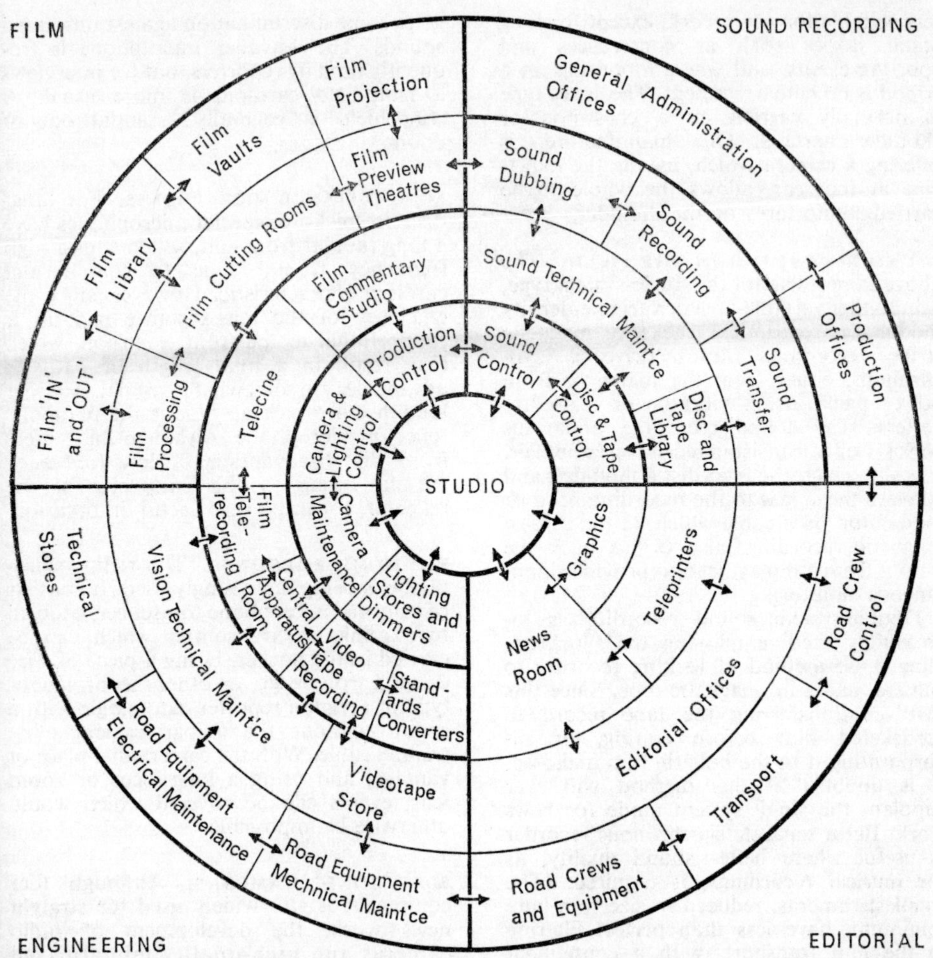

SCHEMATIC OF NEWS STUDIO COMPLEX. The studio itself, conceived as the centre of the production complex, divided into four quadrants: Film, Sound Recording, Engineering and Editorial. Concentric rings indicate degrees of importance of siting relative to studio. Inner ring comprises all areas which must have immediate access to studio. Next ring contains production facilities, and third and fourth rings show auxiliary services not directly required for transmission purposes. Arrows show principal interrelationships.

100 per cent duplication of equipment. Some coverage can be provided by recording the incoming standard, for later conversion, but this depends on availability of equipment.

EQUIPMENT

FILM CAMERAS. The cameras used in newswork range from a variety of light hand-held silent types to the somewhat heavier single- or double-system sound cameras. The main requirements for a silent camera are ruggedness, ease in maintenance and convenience in operation. A reflex viewfinder is often an operational advantage, but the benefit is lost if a strong filter is necessary, because the image then becomes difficult to see. Such a refinement as a variable shutter, which provides fades and dissolves, is not necessary for news. A self-contained battery drive to the motor is an advantage, and should provide sufficient power for about six 100 ft. rolls of film – a considerably greater footage than can be obtained from a clockwork drive.

The almost universally used sound camera is of American manufacture, and can be obtained in three main versions. The smallest, suitably modified, can accept 400 ft. rolls and, complete with zoom lens, weighs less than 20 lb. The larger types are less suitable for news work, except for long static stories such as conferences and sporting events, and where mounting on a tripod is no embarrassment. The small type is normally carried on a chest-pod or shoulder harness. One manufacturer is offering a camera which, having the magazine at the rear, allows the whole to be carried comfortably on the shoulder.

SOUND RECORDING IN THE FIELD. The above cameras are of the single-system type, using either variable-area or variable-density modulators for optical tracks, or magnetic stripe recording. Motor drive is synchronous, either from the mains or from power packs using vibrator or static inverters. The sound recording apparatus consists of a transistorized mixer/amplifier. It is slung over the recordist's shoulder, and conveys the signal to the recording head or modulator via a cable which, in the case of magnetic recording, also brings back the output from a replay head to provide simultaneous monitoring.

Double-system sound recording is increasingly used, employing the synchronizing pulse method of locking recorder to camera, and ¼ in. magnetic tape. Since this involves transferring the tape record to sprocketed film before editing, or incorporating it in the bulletin film make-up, it is doubtful if this method will ever supplant the single-system mode for news work. But a separate synchronous recorder is useful where better sound quality, as for musical recordings, is required. (The smallest cameras, reduced in size to a bare minimum, have less than perfect filtering in the film transport, with a consequent wow and flutter figure of 0·3 per cent average.)

Although up to now the transfer time has been a drawback, equipment is available for interlocking pulsed tape and projectors, reproducers, and videotape machines, and this may well lead to a new technique. One difficulty would still remain, however, the necessity for slating all takes, i.e., the simultaneous recording in sound and picture of a synchronized start mark.

MICROPHONES. Recording for news is generally carried out in conditions of high ambient noise levels, so suitable precautions must be taken. Except for microphones of the Lavalier type, which can be slung around the neck and to some extent concealed against the chest, a microphone with a cardioid polar diagram is preferable as it gives some discrimination against unwanted sounds. The Lavalier microphone is frequently used by reporters, but for interviews a hand-held cardioid is more usual. A windshield is generally essential out of doors.

LONG-DISTANCE MICROPHONES. For long-distance pick-up, special microphones have a long tubular front end, which adds a high frequency frontal lobe to the normal cardioid characteristic. However, since the efficiency of the long-distance pick-up is proportional to the front extension, which may result in a total length of 7 ft., its advantages are somewhat marginal. Another long-distance pick-up type is the parabolic reflector microphone. Although this suffers from the disadvantage of low frequency cut-off, owing to the finite size of the reflector, it can be very useful on occasion.

RADIO MICROPHONES. The radio microphone is being increasingly used. It consists of a normal microphone transducer, coupled to a miniature transmitter which can be carried in the pocket. Using a piece of wire down a trouser leg, or other inconspicuous type of aerial, a reporter can mingle with a crowd without the embarrassment of a trailing cable. With the camera in a place of vantage, and using a long-focus or zoom lens, events can be covered which would otherwise be impossible.

ELECTRONIC CAMERAS. Although film equipment is still widely used for straight news work, the development of radio cameras and high-quality miniaturized video recorders makes it practicable to use light and mobile vehicles which can get quickly to the scene. Custom-built units are available in the US and are now being designed in Britain.

The camera must be capable of being hand-held. Weight precludes a conventional viewfinder, but there is no reason why a zoom lens with reflex finder should not be used with a Plumbicon or one of the newer improved vidicon tubes. Where a camera cable to the vehicle is unacceptable, the radio camera, with its disadvantages of transmitter pack weight and currently inferior quality, has to be used.

Camera cables normally carry several circuits, but suggestions have already been put forward in the US for a new approach to miniaturization of mobile equipment. The ever-decreasing size of electronic components will eventually result in pulse generators so small as to allow each camera to generate its own pulses, and all cameras to be locked together. This immediately reduces the number of coaxial conductors in the cable, as only one coaxial is needed to convey the outgoing composite signal to the vehicle, the incoming locking pulses being carrier-borne on a frequency above the video band. Audio and d.c. circuits can be accommodated on one pair of wires carrying tally and intercom with d.c. isolation.

As the video unit would be used only where circumstances permitted, a number of advantages appear. A link or links back to base would give the ability to inject live material into the newscast. A considerable improvement in picture quality could be expected.

One common trouble with most newsfilm is that the exposure and contrast range of the negative are seldom optimum. If a negative having a contrast range of more than 7 : 1 (or, in the case of a positive print, more than 50 : 1) is transmitted through a telecine chain, the grey scale suffers, crushing either whites or blacks. On the other hand, an acceptable print for theatre projection may have a contrast of 100 : 1.

Excess contrast ranges in newsfilm are quite common. The advantage of the video camera is that the electronically processed picture is available on a monitor at the time of exposure, and adjustments may be made in conjunction with the waveform monitor to give the desired result. Correct framing in the action field and sharp focus are also more easily obtained in this way.

ARTIFICIAL LIGHTING. Standard studio-type spots and floods are used for static occasions such as a permanently equipped interview room, or when time is available to complete a large installation. For fast working, however, small portable units are brought into service. These are battery operated, an example being one which uses 30 volts of nickel cadmium cells which can be charged by connecting in-built chargers to the a.c. mains. The illuminator in this unit may be a photoflood bulb or a quartz iodine lamp. Between these extremes is the mains-connected photoflood in its own reflector. This can be mounted on a lightweight stand, or supplied with clips, and can be set up reasonably quickly.

HANDLING FILM MATERIAL

FILM PROCESSING. Newsfilm is almost invariably of 16mm gauge, and in most cases arrives in the form of unprocessed negative. This is developed on a machine similar to that used in the trade laboratory. In smaller news units, where the volume of work is less, a simpler semi-professional type of plant is installed.

To save time, the negative is used for transmission and syndication prints are made afterwards. Scrupulous cleanliness is therefore essential, and the negative is put through an ultrasonic cleaner before printing.

In theory, picture quality can often be improved by varying the development time to compensate for known deficiencies in the negative, but this is, in fact, seldom practicable.

FILM EDITING. Editing the negative for direct use is always hazardous, especially in conditions of speed and stress. There is seldom time for cleaning between editing and transmission and, owing to the necessary phase reversal on transmission, dirt shows as white specks, which are much more noticeable than the black specks of dirt on positive. The utmost care in handling is therefore necessary, as is efficient air filtration in the ventilation system.

ASSEMBLY OF FILM MATERIAL. It is convenient to assemble film stories in two alternating rolls, thus facilitating run-down to the start point of each story. If, however, during rehearsal or transmission, the running order is changed, thereby placing two consecutive stories adjacent on one roll, sufficient time must be given during the news reader's lead-in to allow the relevant story to be run down to the start point. Late stories which cannot be included in the two main rolls may be assembled for a third telecine machine.

The practice of including negative and positive in one roll is undesirable, but often unavoidable. In any case, where a single story contains a mixture of negative and positive, it is split over two rolls, since the necessary reversal of phase often cannot, in practice, be manually switched with sufficient precision. A theoretical disadvantage of mixing the two is that the correct focus setting for one does not hold good for the other, since the plane of the emulsion in the gate is displaced by the thickness of the base. Since refocusing is not practicable, the focus is split, and the loss is virtually unnoticeable.

It frequently happens that the two adjacent stories on a roll have different sound track arrangements, e.g., commag on one and sepmag on the other. This can be dealt with in two ways. In one, a mechanically locked sound head can carry the sepmag track which is located in spacing film so that its position corresponds with that of the picture. The second method, which is preferable since it occupies less editorial time, is to supply the sepmag sound as a single roll which is laced on a separate sound reproducer, this reproducer then being electrically interlocked with the projector prior to cueing the story. In this method, the editor has only to top and tail the sound, and to mark up the start position. In the first method he has to run all the film through the synchronizer in order to ensure correct synchronization throughout the roll.

FILM COMMENTARIES. Commentary on film stories can be either from the newscaster himself or from a separate commentator housed in a small sound studio. Using different voices not only gives variety but allows a choice to suit the story, e.g., male or female.

Scripts are written as film becomes available, normally from early afternoon onwards. Since the deadline can literally be the end of the bulletin, the changing news situation can result in many revisions involving re-writes, often while the news bulletin is actually being transmitted.

FILM TRANSMISSION. Ideally, the whole news bulletin should be rehearsed. This is seldom possible but every effort should be made to rehearse the newsreader, because his is usually the only part of the bulletin not previously timed.

The rehearsal of film is equally important, but for another reason. Unlike a graded print, which is reasonably consistent in light transmission qualities, a newsfilm may vary from almost clear acetate to a dense overexposed picture. Even with one rehearsal, it is difficult to establish optimum gain and lift setting; with no rehearsal the result may be unacceptable. Automatic light control in the telecine, however, can go a long way to easing this problem.

In news material, with contrast values often at variance with the optimum, the production of a reasonable grey scale is frequently a compromise. Development gamma is rigidly controlled, and if the telecine gamma correction circuits are capable of adjusting the overall gamma in accordance with that of the development process, that is about all that can be done towards obtaining consistent picture quality.

It is because of this inevitable variation in newsfilm quality that the vidicon is usually chosen for transmission purposes. The flying-spot scanner is intolerant of high density, although with good film it is capable of better performance. Furthermore, the vidicon offers a simple means of caption scanning.

Still frames on 35mm film can be made without difficulty and laced on the 35mm projector in a normal mixed gauge cross-fire assembly. The difficulty of using standard caption scanners on widely different sizes of material necessitates a different approach, and the 35mm still frame is a very useful adjunct to the usual method.

VIDEOTAPE RECORDING

Many years ago, before videotape was available, all contributions from the network had to be injected into the bulletin at the time of transmission. In commercial television, this introduced networking complications, since the number of circuits might be insufficient to allow the operation to be effected without precisely controlled cue switches. Late changes in the bulletin could call for a revision in the order of injects, and last-minute alterations in the switching schedule were always dangerous.

The use of tape has removed this hazard. Items may be transmitted via the network at convenient times during the afternoon and, in the event of failure, can always be repeated. Long story items, originating from OBs (remotes) in distant places, are often recorded at 7½ ips, which is more economical in tape usage than the normal 15 ips, and are subsequently edited to suitable length.

REPLAY. The use of tape in the bulletin has led to a new operational technique, whereby the tape machine figuratively takes its place alongside telecine as an operational tool, and is cued in a similar way. In this respect, it is becoming common practice when time permits to transfer to tape complicated film sequences, involving several telecine machines, so as to provide one composite picture for transmission. This eliminates accidents and simplifies the telecine operation during a busy period. In order to attain maximum reliability, with the shortest possible cue-in times, it is necessary to use frame and line locking devices, and also the dropout compensator, which, in one form, effectively removes dropouts by the substitution of the preceding line. This reduces considerably the number of synchronizing disturbances resulting from the dropout gaps in the record.

EDITING. From a news point of view, editing is most important. Long unrehearsed stories such as conferences and sports events have generally to be recorded entire and selected high spots abstracted for use. Speed is essential, especially when the end of the recording approaches transmission time. Since more than one projected programme may be using material from the same story, requiring different editing arrangements, recordings are frequently made in duplex, or even triplex, thus making available several masters.

Editing in news work is an assembly operation. Though wasteful of tape, the simplest method is to cut the master physically, and splice the sections together. Selected sections are usually dubbed on to a second short tape, and then properly spliced. This is uneconomic, as a spliced tape has limited use, and a better method is that of electronic splicing.

SPLICING. Two variations of electronic splicing are available. In the simpler form, a control track has to be pre-recorded, after which the items can be selected and recorded in sequence, correct sequential sync pulses having already been laid. A more elegant way involves the recording of the control track as editing proceeds.

The choice between physical or electronic splicing depends on the story. Sometimes it is quicker to make a physical splice. Again, it may be found that the longitudinal separation of sound and picture on the tape means that a good edit cannot be made without either sound or picture suffering. In this case an electronic splice is preferable, since in this way, sound and picture can be edited simultaneously within a frame or so. Where cutting of tape is necessary, an electronic device is used, whereby the actual editing pulse on the control track is displayed against a reference on a cathode ray tube, correct registration ensuring that the proper cutting point is located under the knife.

If the machine has a cue track, it is invariably used by the programme editor to indicate selected heads and tails, and to record any additional editing instructions which may be appropriate. The cue track shows the engineer where to make the edit. If it is known that time will not permit editing, two masters – or dubbings – can be used on two replay machines, and run in alternation. This method, however, is unsatisfactory as it offers too many chances of error.

SOUND RECORDING

STATIC SOUND RECORDING FACILITIES. A well-equipped sound recording, transfer and dubbing section is very important for the preparation of news and current affairs programmes. Essential requirements are the ability to record on both gauges, magnetically and optically, to be able to transfer a recording from any one to any other medium, and the facility for carrying out a dubbing using a minimum of at least eight sound sources.

While it is true that dubbing for a news bulletin proper is not normally required (transfer from commag to sepmag is all that may be needed), the material handled in other types of programme, such as topical magazines, calls for quite extensive dubbing, frequently entailing a combination of both gauges. Equipment should therefore include several 16mm and 35mm magnetic recorders, optical recorders for both gauges, magnetic and optical reproducers for both gauges, switched and continuously variable-speed $\frac{1}{4}$-in. tape recorders, including pulsed tape transfer facilities for all known types of pulse, a disc recording lathe, a dubbing desk with full single channel equalization, a commentary recording theatre or booth, and a number of preview theatres, equipped for single or double systems, both gauges, and having the ability to reproduce edge or centre 16mm magnetic track.

SOUND EFFECTS. An extensive sound effects library is important. While natural sound is recorded whenever possible – carrying the hallmark of authenticity – there are inevitably occasions where this is not practicable and library effects have to be used. Crews are supplied with miniature tape recorders and encouraged to use them. In this way, the library is constantly expanding. Even on a synchronized sound story, cutaways may have been shot on silent cameras and effects may well be required. The choice of effects or music is a very debatable matter, but it is usually acknowledged that one or the other is necessary. Except where dramatically essential, complete silence accompanying pictorial action is disturbing to the viewer, causing a break in continuity.

All master material is generally stored on $7\frac{1}{2}$ in. per sec. $\frac{1}{4}$-in. tape, but a number of effects are obtained on commercial discs.

In news transmission, where sound effects have to be mixed in at great speed and with a minimum of rehearsal, the choice of medium (tape or disc) depends on a number of factors. The simplest method is the acetate disc running at 78 rpm. With a special type of groove locator, selection can be made from any desired groove.

Discs are convenient, and for short news programmes not exceeding 15 minutes or so, the required discs can be racked adjacent to the turntables, ready to hand. Variations in pitch are obtained by using a variable-speed turntable. The crash-start turntable which allows the disc and pick-up to drop on to the platen, already turning at speed, is also useful for a start from an accurately located point.

Acetate-coated discs, when cut on a properly designed lathe, and with an amplifier having sufficient driving power, have an output level comparable with that of a commercial pressing. The acetate disc is free from noise but wears away quickly.

A second method is the use of a tape cassette device. A large number of effects can be held in readiness for use, access being by push-button selection, remotely-controlled if desired. A variation of this method is an apparatus which allows multiple effects to be selected via a piano keyboard and a number of group and track controls. This is a highly efficient instrument, offering a choice of more than a thousand effects, which, by use of an instant start facility, can be accurately synchronized to picture.

THE NEWS STUDIO

STUDIO EQUIPMENT. The equipping and operation of a news studio depends on the overall usage. A studio transmitting news bulletins only does not need equipping as comprehensively as a general purpose studio to transmit current affairs programmes, discussions, etc. For news only, a remotely controlled vidicon camera, fixed lighting, and a comprehensive caption, still and map scanner may be adequate. For more ambitious programmes, it is advisable to equip with image orthicon cameras, and a much more flexible lighting installation. In both types of studio, a prompting device for the news-reader is essential.

Caption scanners take different forms. A display equipment using a vidicon camera may be housed remotely from the studio. In this, the physical sizes of the displayed materials have to be kept within certain limits, but it is possible to perform superimpositions so that the picture signal entering the mixer already has the required combination. Alternatively, where the studio is a general purpose one, the captions, stills, etc., may be mounted on easels and scanned by normal studio image orthicon cameras. This method is completely flexible in dimensions and formats, both in respect of the original, and the amount to be shown. Superimpositions are made with two cameras using mixed viewfinders. The caption cameras, equipped with a full set of lenses and mounted on studio camera dollies, can be used for other purposes such as interviews and discussions.

PRODUCTION CONTROL ROOMS. Control room facilities are generally standard, but have a few additional features necessitated by the particular method of operation. Close co-ordination between newsreader and control room results from feeding a deaf-aid with director's talk-back and any programme material, or a mixture of both. This is essential in view of the constantly changing situation in news transmission. The talk-back installation is more than usually comprehensive, allowing full two way facilities between director and all operating points over a wide area. W.H.O.S.

See: *Camera; Editing; Laboratory organization; Microphones; Sound recording; Special events; Telecine; Video tape editing; Videotape recording.*

⊕**Newsreel.** Short film, usually not exceeding ten minutes in length, presenting current events with commentary and forming part of the programme of an entertainment cinema. Because of their topical content, newsreels are produced very rapidly and a new edition is usually issued once a week and sometimes more frequently for special occasions. Since their content is only of immediate interest for a limited period, a large number of copies of each newsreel is required to allow simultaneous distribution to as many theatres as possible. However, the widespread availability of television news in the home has reduced the importance of the regular newsreel in the cinema to vanishing point.

⊕**Newsreel Base.** Thinner form of base at one time used for economy as the support in black-and-white films for newsreel release prints.

Newton's Rings. Regular patterns of light and dark rings, often of distorted shape, caused by optical interference when two surfaces are separated by a very small distance and illuminated. When the light is white, the rings are coloured. The effect is sometimes observed in the step printing of motion picture film if the contact between negative and positive is imperfect.

See: *Light; Optical principles.*

Night Effects. Techniques of photography and lighting used to obtain the appearance of night in exterior scenes. In black-and-white work, the effect is generally obtained by contrasty lighting, assisted on occasion by the use of orange or red contrast filters, with a degree of under-exposure of the negative. In colour photography, it is usual to adopt the convention of an overall blue effect with accentuation of the yellow of artificial light sources, such as lamps and the windows of houses, but the satisfactory rendering of face tones and colours is often a problem. Polarizing filters may help to reduce the brightness of a blue sky. Considerable colour correction is often necessary when making prints from day-for-night colour negatives.

See: *Cameraman* (FILM).

Nine-point-Five Millimetre Film. One of the earliest of the narrow-gauge films used for amateur cinematography, first introduced in France in 1922–23; although popular in Europe for many years, it tended to be replaced by the more popular 8mm gauge

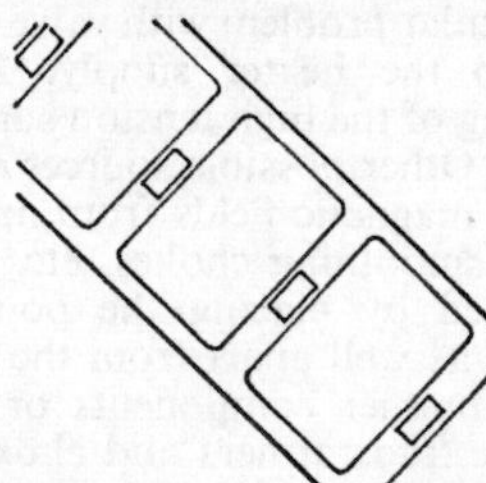

and is now obsolescent. This type of film is 9½ millimetres in width with a single row of perforations down the centre, having 40 frames to the foot.

See: *Film dimensions and physical characteristics.*

Nipkow Disc. Early mechanical scanning device for TV transmission and reception. Periphery of disc is perforated with one or more spirals of holes, arranged so that when one hole has passed across the scanned area, the next one is just ready to start on an

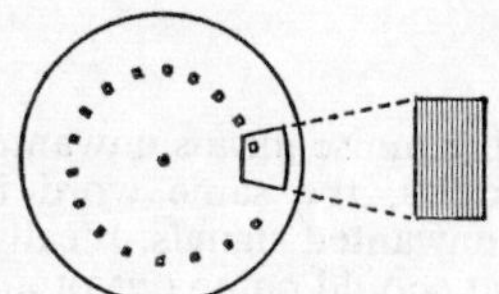

adjacent path. The difference of radius between first and last holes of the spiral thus determines the width or height of the picture, according to whether scanning is horizontal or vertical.

Nipkow, Paul, 1860–1940. German engineer born at Lauenberg and educated there and at Neustadt, where he became interested in the telescope and also the telephone. He was one of the foremost contributors to early TV development by reason of his invention of the scanning disc, the basic scanner used in the first television systems. He conceived the idea of a perforated spiral distributing disc and took out a patent for it in January 1884. As at this time electromagnetic waves had not yet been demonstrated and there was no wireless, Nipkow's ideas were directed to sending pictures over telephone wires.

Nit. Metric unit of luminance or brightness. Equal to one candela per square metre.

See: *Light units.*

Nitrate Base. Cellulose nitrate support used in the manufacture of most professional 35mm film up to 1951. The fire hazards in the handling of this material were considerable and it has now been replaced by various types of low-inflammability safety base.

See: *Film dimensions and physical characteristics; Film storage.*

Nitrogen-Burst Agitation. Method of agitating a developer solution by releasing periodic bursts of gas at the bottom of the tank to form groups of large bubbles which displace the liquid around the film surface. Nitrogen is used rather than air to avoid oxidation of the developer.

See: *Laboratory organization; Processing.*

Nodal Head. A specialized type of tripod head in which the camera pivots precisely around the nodal point of its lens, thus ensuring that its viewpoint is not displaced up or down, or from side to side. In some special effects processes this is essential to avoid optical distortions.

See: *Special effects* (FILM).

□NOISE

In acoustics, noise means unwanted sounds; in electronics, the same word is used to describe unwanted signals. Ideally, the only signal that should come out of an amplifier is an amplified replica of the input, but in practice the output also contains unwanted signals which give the well-known hissing, sometimes spluttering continuous background which is referred to as noise. With the advent of television, the use of this acoustical term has been widened to include visual effects and describes the random graininess seen as a continuous background to the picture.

Noise was once a serious problem in high-fidelity recording systems, but modern practice has led to it being reduced to a satisfactory level. Indeed, improvement in recording standards has been as much due to the reduction of noise levels as to any increase in frequency response.

TYPES OF NOISE

There are several types of noise, due to different causes: thermal or Johnson, shot, partition, flicker, excess, microphony, hum and induced noise.

THERMAL NOISE. Random fluctuations of electrons in conductors produce thermal noise. A metal is a conductor because it has free electrons which move when a voltage is applied. But without voltage there is still a random movement of electrons due to thermal agitation, and at any instant more electrons may be moving in one direction than in the other, giving rise to a noise voltage across the ends of the conductor. This voltage fluctuates from instant to instant depending on the predominating motion of the electrons. Thermal noise thus sets a limit to the lowest voltage which can be satisfactorily amplified.

SHOT NOISE. Both valves (tubes) and transistors are subject to shot noise. In the former, the electron emission from the cathode is not smooth, and the irregular variations in the number of electrons reaching the anode produce noise. In transistors, shot noise is due to the irregular flow of the current carriers.

PARTITION NOISE. Where the current through a valve or transistor is shared between two or more electrodes, partition noise arises. With a pentode, for example, there are small fluctuations in the number of electrons reaching the anode and screen grid because of the random sharing or partition. This produces noise which makes the performance of the pentode inferior in this respect to that of the triode.

FLICKER EFFECT. In valves, flicker effect is caused by variations in emission over small cathode areas.

EXCESS NOISE. Produced when a direct current passes through carbon resistors, and due to random fluctuations in contact resistance between the granules of the resistor material. Carbon microphones produce the same noise and the effect is also found in transistors and crystal diodes.

MICROPHONY. When valve electrodes vibrate mechanically, they produce variations in the anode current at audio frequency which are amplified with the input signal. This is microphony and is particularly troublesome in high-gain amplifiers, especially with the first stage, and it is important to choose valves with low microphony, to mount them on a spring base and then to cover them with cases lined with sound-absorbing material. The reason for the last precaution is that sound waves can often be a cause of microphony. Transistors are immune to this type of noise.

HUM. Interference originating from the power supply causes a characteristic hum. It is a particular problem with valve equipment owing to the heater supply. Inadequate smoothing of the high tension supply is also a source. Other possible sources of hum are the stray magnetic fields from mains transformers, smoothing chokes, etc. These can be avoided by placing the power supply components well apart from the remainder of the amplifier components or by orientating the transformers and chokes so that their magnetic fields do not affect other components, or by screening.

INDUCED NOISE. Noise can be induced by some form of coupling between the noise source and the equipment. This coupling can be by a magnetic field or by an electrostatic field, i.e., capacitance must exist. Careful placing of components, especially of those parts of the equipment where the signal level is very low and therefore prone to interference, together with magnetic and electrostatic screening, mitigate this type of noise.

MEASUREMENT

SIGNAL-TO-NOISE RATIO. The ratio of the required signal to the inevitable noise which is present is obviously important in determining the performance of technical equipment. It is essential to keep this ratio as high as possible at all points along the chain of equipment. The overall performance of the whole is determined by the weakest link, certain parts by their nature being more noisy than others. The signal-to-noise ratio of the input to such parts must therefore be sufficiently high.

NOISE FACTOR. Where the signal is weak, e.g., a microphone output, care must be taken to amplify the signal as soon as possible to a satisfactory level. Thermal noise of the source sets a limit to the lowest signal which can be handled and therefore the amplifier should add as little noise as possible. To assess this characteristic of an amplifier the noise factor is used; this is defined as the ratio of the total noise output power to the thermal noise output power. It is usually expressed in decibels:

$$\text{Noise factor} = 10 \log_{10} \frac{\text{total noise power}}{\text{thermal noise power}} \text{ dB}$$

Assuming that the load resistance is the same, this reduces to:

$$20 \log_{10} \frac{\text{total noise voltage}}{\text{thermal noise voltage}} \text{ dB}$$

To illustrate the use of the noise factor: if the amplifier adds no noise, the only noise is that due to the thermal noise of the source resistance.

$$\text{Thus the noise factor} = 20 \log_{10} 1{\cdot}0 \text{ dB} = 0 \text{ dB}$$

A practical figure is 3 dB.

More generally the noise factor N is given by:

$$N = \frac{V^2_{is}\, V^2_{on}}{V^2_{in}\, V^2_{os}}$$

where V_{is} is the input signal, V_{in} the noise being fed in by the source, V_{os} the output signal, and V_{on} the output noise.

WEIGHTED MEASUREMENT. In the measurement of noise in audio frequency work, it is sometimes important to bear in mind that the aural sensitivity of the ear is not the same at all frequencies and that a noise signal giving a certain objective reading on a measuring set might not give the same subjective effect when monitored on a loudspeaker.

In practice, this problem is solved by inserting a filter in the measuring set. The frequency response of this filter ensures that the readings obtained indicate the aural nuisance value. The filter is called an aural sensitivity network. Noise measurements obtained using such a filter are said to be weighted.

NOISE IN TELEVISION

There are several types of noise of particular importance in television. The effect of the noise depends on its amplitude relative to the signal, and it is usual to express the relationship in terms of the ratio of peak signal to rms noise.

RANDOM NOISE. The most common appearance of random noise is as a uniform structural moving dot pattern over the whole image, so that the image acquires a pronounced graininess. Such noise may be produced by thermal agitation or shot effects. Random noise can take other forms: synchronous, where the noise or hum pattern bears a definite relationship to mains frequency; impulse, due to faulty connections, which cause flashes of varying amplitude on the picture. It is also usual to group externally generated noise, for example, r.f. patterning and car ignition interference, under the broad heading of random noise.

WHITE NOISE. White noise is random noise whose rms. amplitude over an incremental bandwidth Δf remains independent of frequency, where $\Delta f/f$ is constant.

TRIANGULAR NOISE. In this case, if the incremental bandwidth Δf is taken at different frequencies, noise energy which increases in proportion to the increase in frequency of the considered bandwidth is termed triangular noise. This form of noise is usually less objectionable visually, due to its greater high frequency content, and gives a smaller grain structure over the image. G.W.M.

See: *Sound distortion; Sound mixer; Sound recording systems.*

Books: *Principles of Television Engineering*, Vol. 2, by R. C. Whitehead (London), 1965; *Audio Cyclopedia*, by H. M. Tremaine (New York), 1959.

⊕□**Noise Limiter.** Circuit for reducing background noise as, for example, by lowering the gain of an amplifier in the absence of modulation.

⊕**Noise Reduction.** Technique applied to optically recorded sound tracks, which reduces ground noise during passages of low modulation when this noise is most objectionable. Noise reduction is achieved

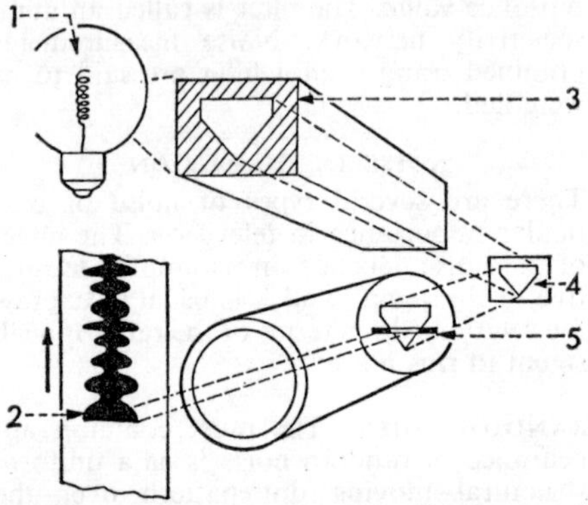

NOISE REDUCTION. Beam from recording lamp, 1, is shaped by mask, 3, to wedge form and imaged on galvanometer mirror, 4. Mirror movement under modulation varies width of beam at slit, 5, and thus alters exposed area as film moves past slit image, 2. Noise reduction biases mirror upwards so that only tip of light wedge covers slit, thus reducing exposure to minimum. Other systems use shutters.

by reducing the transmission of the positive print (i.e., increasing its density) during low-level passages. The envelope of the train of waves being recorded is rectified and applied to the galvanometer mirror or to shutters in the recording beam which mask off just enough of the track to allow passage of the modulated light beam from moment to moment.

See: *Sound recording* (FILM).

Non-Additive Mixing (NAM). Combination of two or more video signals in such a way that at any instant the largest signal alone is passed to transmission, the smaller signals being suppressed. Used for superimpositions and cross-fading in some vision mixers, ir gives effects similar to overlay. Thus, in a composite picture formed from two cameras, a relatively bright subject from one camera can appear solid, without darker detail from the other showing through. NAM was devised by RCA.

□**Non-Composite Signal.** A picture signal lacking its accompanying (composite) synchronizing signals, but having blanking informa tion. When television pictures are generate and processed within the technical area o a studio, it is not always necessary for th picture signal to be accompanied by syn chronizing signals. Where a number of pic ture sources have to be handled within studio area, an economy can be made b providing comp sync signals only at the fina processing point.

This non-composite mode of operatio has the advantage that sync distributio requirements are reduced to a minimum with resultant installation economy. Alsc when the comp sync is finally added to th picture waveform, no stripping of existin sync is required. This process of syn stripping always gives rise to some degree c picture distortion, and the fewer times it i performed, the better is the resultant pictu quality.

On the other hand, preview monitoring c the various picture sources within the studi complex demands either comp sync feeds t the monitors, or comp video signals from th picture sources. Owing to the number c monitors involved, the composite vide mode of operation is usually adopted.

See: *Synchronizing pulse generators.*

⊕**Non-Flam Film.** Film in which the base use is of low inflammability and has a slo burning rate as specified by various nation and international standard test method Non-flam film base is now usually cellulo triacetate or a similar polyester. Also calle safety film.

See: *Film dimensions and physical characteristics.*

□**Non-Linear Amplifier.** Amplifier in whic the output is not proportional to the inp signal, e.g., an amplifier with a logarithm gain, such as is used for altering the gamn of a television signal.

See: *Transfer characteristics.*

□**Normal-Reverse Switch.** Switch which r verses the horizontal and vertical scannir currents through the deflection coils in a T camera. This gives a lateral or vertical i version of the image, sometimes called f on a special effect.

See: *Special effects* (TV).

⊕**Notch.** Shallow cut in the edge of a sectic of negative film which actuates the ligh change board in certain types of film printe thus printing the section at the requir density.

See: *Printers.*

NTSC COLOUR SYSTEM

The NTSC system, named after the US National Television System Committee which devised it in the early 1950s, was the first colour TV broadcasting system in the world to be nationally standardized and successfully exploited commercially. Colour elevision in the US, after a relatively slow start ascribed in part to the high cost and complexity of receivers, had by the late 1960s made such headway that the majority of programmes were radiated in colour. By hat time, Japan and Canada had adopted he NTSC system.

BASIC PRINCIPLES

OMPATIBILITY. The requirements conronting the NTSC at the start of their work were difficult to meet. Colour transmissions had to be compatible, that is, receivable in black-and-white on sets already in use. They must occupy the normal bandwidth, in order to conserve the same number of hannels as before. The colour information, when picked up on a black-and-white set, hould create the minimum visible interference with the picture. Colour information needed to be so coded as to be liable to the east possible disturbance from co-channel nterference, or from attenuation or phase istortion either in landlines or on the air etween radiation and reception.

OLOUR SIGNALS. The basic principle of ne NTSC system (as indeed of all compatble colour TV systems) is to employ a uminance signal occupying the full bandwidth and describing the brightness of each oint in the scene, and a chrominance signal olely concerned with the colour information. he luminance signal alone actuates a blacknd-white receiver. The chrominance signal formed by combining two colour difference signals and, in the colour receiver, is plit into its two component parts; these, in urn, are combined with the luminance gnal in such a way as to give separate nformation about the required quantities f the three primary colours. The luminance nd chrominance signals occupy the same andwidth (called band sharing) and are rranged so as to interfere with each other s little as possible.

Because band sharing involves waveform nalysis and makes use of different systems f modulation, a mathematical approach is navoidable. The usual nomenclature is mployed, in which E'_Y denotes the luminance signal, and the two colour difference signals are $(E'_R - E'_Y)$ and $(E'_B - E'_Y)$, where the subscripts R and B denote red and blue. While E'_Y occupies the full video bandwidth, each colour difference signal takes up only about one-fifth to one-third of that bandwidth.

COMPOSITE SIGNAL. To combine these three signals into a composite signal occupying only the luminance-signal bandwidth, the NTSC system utilizes, for the colour-difference signals, the suppressed-carrier amplitude modulation of, in effect, two subcarriers having the same frequency but a quadrature relationship.

Ordinary amplitude modulation may be expressed as:

$$f(t) = [1+f(t)] \,.\, \cos \omega_s t$$

where f(t) represents the modulating wave (with maximum and minimum values of +1 and −1) and $\frac{\omega_s}{2\pi}$ is the carrier frequency.

Suppressed-carrier amplitude modulation by a function $f_1(t)$ is then

$$S_1(t) = f_1(t) \,.\, \cos \omega_s t$$

Another suppressed carrier amplitude modulation signal having the same carrier frequency but a quadrature relationship is

$$S_2(t) = f_2(t) \,.\, \sin \omega_s t$$

Adding these two signals together:

$$S_1(t)+S_2(t) = f_1(t) \,.\, \cos \omega_s t + f_2(t) \,.\, \sin \omega_s t$$

Multiplying this sum first by $\cos \omega_s t$ gives

$$[S_1(t)+S_2(t)] \,.\, \cos \omega_s t = \tfrac{1}{2} f_1(t)[1+\cos 2\omega_s t] + \tfrac{1}{2} \,.\, f_2(t) \,.\, \sin 2\omega_s t$$

If the twice-carrier-frequency components are removed by filtering,

$$[S_1(t)+S_2(t)] \,.\, \cos \omega_s t = \tfrac{1}{2} \,.\, f_1(t)$$

Similarly,

$$[S_1(t)+S_2(t)] \,.\, \sin \omega_s t = \tfrac{1}{2} \,.\, f_2(t)$$

and, in the more general case of multiplication by $\cos(\omega_s t + \theta)$,

$$[S_1(t)+S_2(t)] \,.\, \cos(\omega_s t+\theta) = \tfrac{1}{2} f_1(t) \cos\theta - \tfrac{1}{2} f_2(t) \sin\theta$$

The two modulating signals, $f_1(t)$ and $f_2(t)$, may thus be obtained, separately or added in any desired ratio, by multiplicative demodulation using a suitably phased subcarrier. In the NTSC system:

$$f_1(t) = k_1(E'_R - E'_Y) \quad \text{and}$$
$$f_2(t) = k_1 k_2(E'_B - E'_Y)$$

The composite colour signal (excepting synchronizing signals) is

$$E'_M = E'_Y + S_1(t) + S_2(t)$$
$$= E'_Y + k_1\left[(E'_R - E'_Y) \,.\, \cos \omega_s t + k_2(E'_B - E'_Y) \,.\, \sin \omega_s t\right]$$

where $k_1 = \frac{1}{1{\cdot}14}$ and $k_2 = \frac{1}{1{\cdot}78}$.

The values of k_1 and k_2 are so chosen that the excursions of the composite signal do not extend too far into the region normally occupied by synchronizing signals nor too far beyond white signal level. The colour-

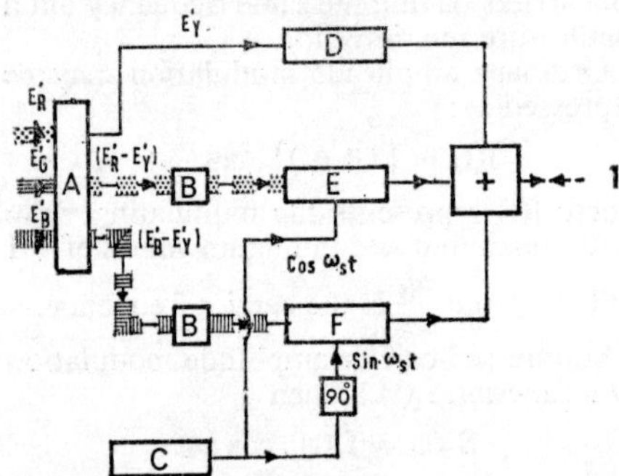

NTSC CODER. Simplified block diagram of transmitter circuits. A. Matrix M_1. B. Low-pass filter. C. Subcarrier generator. D. E'_Y delay. E. $(E'_R-E'_Y)$ balanced modulator. F. $(E'_B-E'_Y)$ balanced modulator. 1. Composite output signal.

difference signals are limited in bandwidth and fed to balanced modulators where they modulate (in a suppressed-carrier manner) the two quadrature-related subcarrier components. The resulting signals are added to the luminance signal (which has been delayed in order to compensate for the band-limiting and modulation processes) to form the composite colour signal.

The addition of the two subcarrier components

$$k_1(E'_R - E'_Y) \,.\, \cos \omega_s t \quad \text{and}$$

$$k_1 k_2(E'_B - E'_Y) \,.\, \sin \omega_s t$$

results in the chrominance signal. This consists of a subcarrier signal whose phase and amplitude are functions of the amplitudes of the colour-difference signals. The chrominance signal E'_C can be represented by a vector, where

$$E'_C = k_1\left[(E'_R - E'_Y)\cos \omega_s t + k_2(E'_B - E'_Y)\sin \omega_s t\right]$$

The amplitude of E'_C is the vector sum of the amplitudes of the two colour-difference components. Further, the phase ϕ of E'_C is determined by the ratio of their amplitudes.

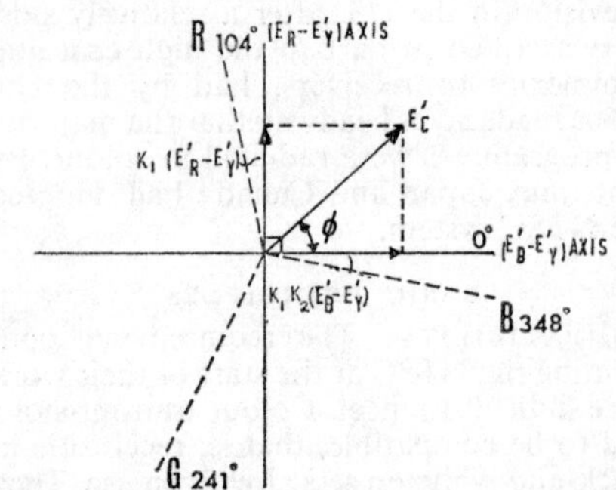

CHROMINANCE SIGNAL VECTOR DIAGRAM. E'_C a vector sum of the two colour difference components $(E'_B-E'_Y)$ and $(E'_R-E'_Y)$. Phase angle, ϕ, is determined by ratio of amplitudes of these components. R. Red. G. Green. B. Blue.

If now a subcarrier used for demodulation is represented by a further vector, its projections on the $(E'_R-E'_Y)$ and $(E'_B-E'_Y)$ axes give the respective proportions of these signals in the output of the demodulator.

$$E'_C = k_1\left[(E'_R - E'_Y)^2 + k_2{}^2(E'_B - E'_Y)^2\right]^{\frac{1}{2}}.$$

$$\angle \tan^{-1}.\left\{k_2\frac{(E'_R - E'_Y)}{(E'_B - E'_Y)}\right\}.$$

As the amplitude of both colour-difference signals depends upon the degree to which the colour is diluted by white (i.e., desaturated), the amplitude of E'_C is related to colour saturation. Further, as a given hue of colour results from colour-separation signals having amplitudes in a unique ratio, the same hue is described by a unique ratio of colour-difference signals. Thus the phase of E'_C is closely related to hue.

CHOICE OF SUBCARRIER FREQUENCY. As the chrominance signal is added to the luminance signal E'_Y to form the composite colour signal, it should interfere with the luminance signal as little as possible. As both colour-difference signals are zero for all points on the grey scale, the chrominance signal vanishes in all black, grey and white picture areas. In the coloured areas of the picture, however, the chrominance signal

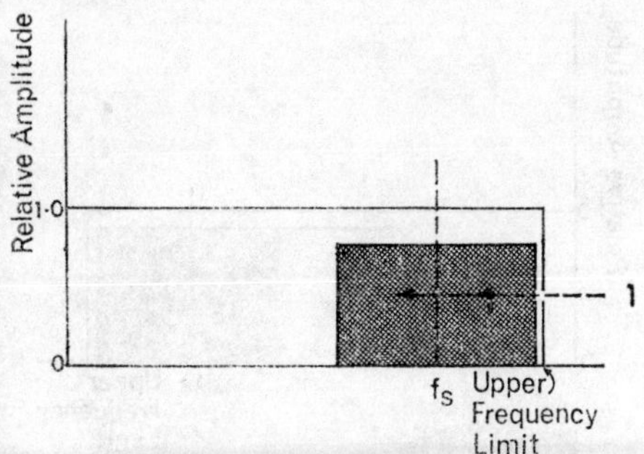

COMPOSITE SIGNAL SPECTRUM. Subcarrier frequency, f_S, placed high in video band, but with sufficient channel space above to allow for upper sideband of chrominance signal. 1. Chrominance signal sidebands.

present and appears as a pattern on the screen of the black-and-white receiver. Raising the subcarrier frequency reduces its visibility. However, to avoid cross-talk between the two colour-difference signals at least one of them is transmitted in a double-sideband manner. Hence, the subcarrier frequency is chosen so that there is sufficient space in the signal spectrum for the upper sideband of the chrominance signal.

The visibility of the chrominance-signal pattern on the black-and-white display and the interference between the luminance and chrominance signals are minimized by utilizing the "frequency interleaving" principle.

This is based upon the fact that the spectrum of a normal television signal, such as a luminance signal, consists of a series of harmonics of line frequency; associated with each line harmonic there is a set of symmetrical sidebands corresponding to field frequency and harmonics of field frequency. If the subcarrier frequency is chosen to be

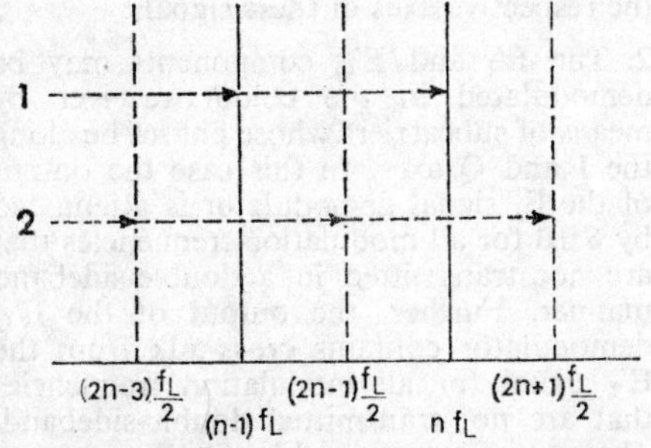

FREQUENCY INTERLEAVING. Interleaving of luminance and chrominance spectrum components in simplified form. Components corresponding to 1, luminance signal and, 2, chrominance signal.

an odd multiple of half the line frequency, its position in the spectrum lies midway between two line harmonics and mutual interference is reduced to a minimum. The principal sidebands of the subcarrier are spaced symmetrically from it by multiples of line frequency so that they, in turn, lie between the line harmonics of the luminance signal; the secondary sidebands of each principal subcarrier sideband (corresponding to field frequency and its harmonics) interleave the corresponding sidebands of the luminance-signal line harmonics.

The effect of using a subcarrier frequency which is an odd multiple of half the line frequency minimizes the visibility of the chrominance signal at the screen of the black-and-white receiver. The alternations of brightness are in antiphase on successive

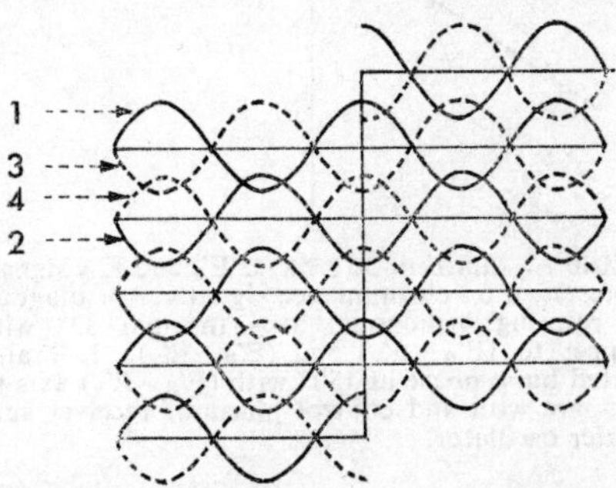

RASTER WITH SUBCARRIER AT $2\frac{1}{2}\times$ LINE FREQUENCY. Five-line raster shows alternations of brightness in antiphase on successive lines of same field and scans of same line. 1. Field one. 2. Field two. 3. Field three. 4. Field four.

lines of the same field and are also in antiphase on successive scans of the same line. The pattern is repetitive in a period of four fields and appears as fine dots which, owing to stroboscopic action, appear to "crawl" in several directions.

E'_I AND E'_Q SIGNALS. The two colours which are first distinguished one from the other, as the coarseness of colour patterns is increased, are orange and cyan. Hence, it would be reasonable to design the system so that, given an ideal receiver, the highest colour resolution occurred for these two colours. Further, in choosing as high a subcarrier frequency as possible it is necessary to ensure that at least one colour-difference signal is transmitted in a double-sideband manner.

In the NTSC system these considerations

have been taken into account.

The two colour-difference signals actually used in forming the chrominance signal are known as the E'_I signal and the E'_Q signal. These two signals can be formed either by direct matrix operation from E'_R, E'_G and E'_B, or can be derived from $(E'_R-E'_Y)$ and $(E'_B-E'_Y)$. They can be represented on the

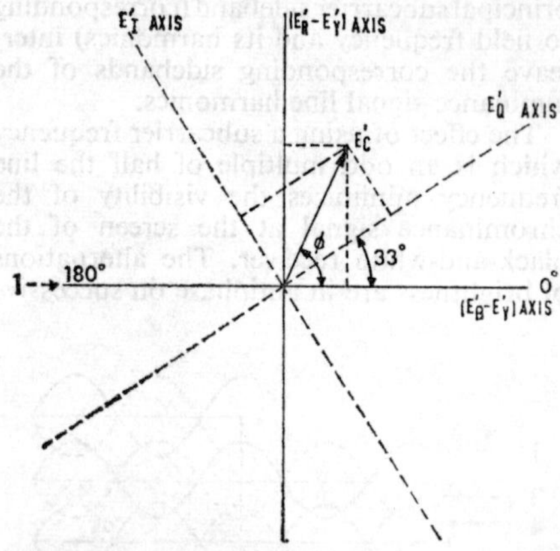

E'I AND E'Q CHROMINANCE AXES. E'_I and E'_Q signals represented on chrominance signal vector diagram by rotating orthogonal axes through 33° with respect to $(E'_B-E'_Y)$ and $(E'_R-E'_Y)$. 1. Transmitted burst phase at 180° with $(E'_B-E'_Y)$ axis to compare with and control phase of receiver subcarrier oscillator.

chrominance phase diagram by two orthogonal axes rotated with respect to the $(E'_R-E'_Y)$ and $E'_B-E'_Y)$ axes by 33°. The E'_I axis then describes the colours orange and cyan, and the E'_Q axis describes the colours green and magenta. The matrix operations deriving E'_I and E'_Q also incorporate the coefficients k_1 and k_2 so that the chrominance signal E'_C can now be written

$$E'_C = E'_I \cdot \cos(\omega_s t+33°) + E'_Q \cdot \sin(\omega_s t+33°)$$

where

$$E'_I = k_1(E'_R-E'_Y)\cos 33° - k_1k_2(E'_B-E'_Y)\sin 33°$$

and

$$E'_Q = k_1(E'_R-E'_Y)\sin 33° + k_1k_2(E'_B-E'_Y)\cos 33°.$$

In forming the chrominance signal, the E'_Q component modulates a suitable subcarrier component in a double-sideband manner,

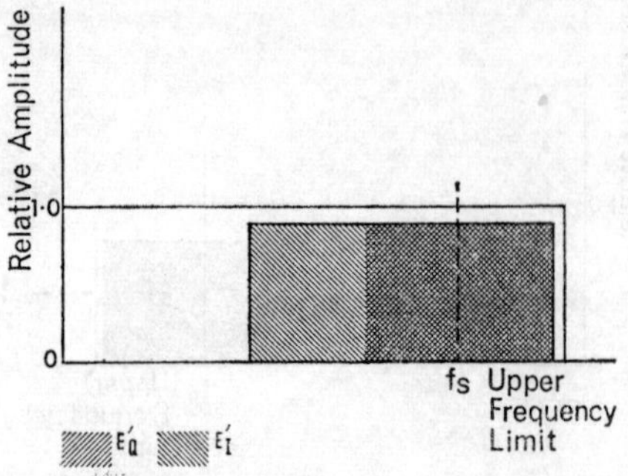

COMPOSITE SIGNAL SPECTRUM USING E'I AND E'Q. E'_Q colour-difference signal is a symmetrical double-sideband signal just reaching to upper video-frequency limit. E'_I is asymmetric double sideband, with wider lower sideband to carry orange-cyan information down to medium-resolution detail.

each sideband having a width equal to the spacing between subcarrier frequency and the upper video-frequency limit. The E'_I component modulates the quadrature subcarrier component in an asymmetric manner; the upper sideband is equal in bandwidth to the upper E'_Q sideband but its lower sideband is wider. This arrangement permits the subcarrier frequency to be as high as possible and also permits the lower sideband of the E'_I component to carry medium-resolution information about orange-cyan detail.

A composite signal using the E'_I and E'_Q chrominance signals may be used in either of two ways by the colour receiver decoder:

1. The decoder may accept only that portion of the chrominance signal spectrum which is double-sideband. This then enables the colour-difference signals $(E'_R-E'_Y)$ and $(E'_B-E'_Y)$ to be demodulated directly by means of subcarriers whose phases lie along the respective axes of these signals.

2. The E'_I and E'_Q components may be demodulated in the colour receiver by means of subcarriers whose phases lie along the I and Q axes. In this case the output of the E_I signal demodulator is attenuated by 6 dB for all modulation frequencies that are not transmitted in a double-sideband manner. Further, the output of the E'_Q demodulator contains cross-talk from the E'_I signal for all modulation frequencies that are not transmitted double-sideband. These must be removed by filtering.

COLOUR SYNCHRONIZING. The NTSC decoding process in the colour receiver requires the provision of the two subcarriers of precisely the same frequency but having a quadrature relationship. These c.w. subcarriers are required in order to demodulate the chrominance signal so as to provide the separate modulating components, for example ($E'_R-E'_Y$) and ($E'_B-E'_Y$). It is, therefore, necessary to provide a synchronizing signal in the transmitted waveform which controls, very precisely, an oscillator in the receiver; this oscillator then provides the two demodulating subcarriers. The synchronizing signal consists of a burst of subcarrier which is transmitted during the back porch of the composite signal. In the proposed standards for a 525-line NTSC colour system for use in the USA, the burst consists of eight to eleven cycles of subcarrier.

In a typical NTSC decoder, the phase of the burst is compared at the beginning of every line with the phase of the receiver subcarrier oscillator and any phase error is then corrected by a suitable reactance control operating on the oscillator. The transmitted burst in the composite video waveform has a phase of 180° with respect to the ($E'_B-E'_Y$) axis.

NTSC DECODING. The essential principles of NTSC decoding have already been outlined. a high-quality NTSC decoder utilizes the E'_I and E'_Q chrominance signals and can make use of the orange/cyan axis for medium-resolution colour detail. The composite video signal, from the detector in the

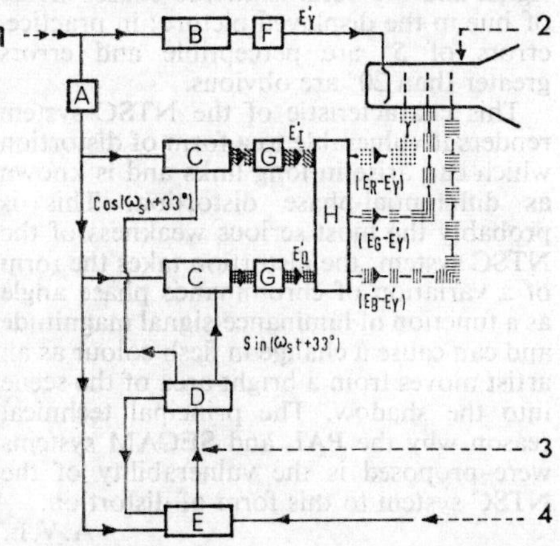

NTSC DECODER USING E'_I AND E'_Q. A. Band-pass filter. B. Band-stop filter. C. Synchronous demodulator. D. Subcarrier oscillator. E. Subcarrier locking circuits. F. E'_Y delay. G. Low-pass filter. H. Matrix. 1. Composite video from detector. 2. Three-gun display tube. 3. Control signal. 4. Pulses from line-deflection circuit.

radio frequency part of the receiver, is divided between two paths.

In the first path, a band-stop filter attenuates frequencies close to subcarrier frequency (to remove dots in large areas of the displayed picture) and the signal is then delayed to compensate for delays in the chrominance circuits.

The signal is now fed to the cathode of the three-gun display tube as the luminance signal E'_Y.

The second part of the output from the radio frequency detector is fed to a band-pass filter which accepts the whole of the chrominance-signal band but rejects that part of the luminance-signal spectrum falling outside the chrominance band. The output of the band-pass filter is fed to two synchronous detectors which are similar to balanced modulators.

The first of these synchronous detector is supplied with continuous subcarrier having a phase lying along the E'_I axis which results in a synchronous-detector output consisting of the E'_I signal; this is passed to a suitable low-pass filter and correction for the 6 dB attenuation of upper frequencies may be applied. The E'_I signal is then fed to the matrix.

In a similar manner, the second synchronous detector is supplied with continuous subcarrier having a phase lying along the E'_Q axis, which results in an output of E'_Q, which is then correspondingly passed through a low-pass filter to the matrix. In the matrix the E'_I and E'_Q signals are combined in various proportions to produce ($E'_R-E'_Y$), ($E'_G-E'_Y$) and ($E'_B-E'_Y$), which are then passed to the three grids of the three-gun display tube; the resulting grid-cathode potentials are then E'_R, E'_G and E'_B respectively. The two supplies of continuous subcarrier are provided by an oscillator which, as already mentioned, is controlled by the burst signal in the composite video waveform. The burst is separated from all other signal components by means of a keying-pulse derived from the line-deflection circuit of the receiver and is compared in phase with an output from the oscillator and a control voltage used to maintain phase lock of the oscillator. Such a decoder, however, is usually found only in a professional monitoring receiver; at present most NTSC colour receivers effectively decode using the ($E'_R-E'_Y$) and ($E'_B-E'_Y$) axes and do not take advantage of the potentially higher colour resolution along the orange/cyan axis. In such receivers the continuous subcarrier feeds are phased to lie along the $(E'_R - E'_Y)$ and $(E'_B - E'_Y)$ axes.

In some forms of receiver decoder, three synchronous detectors are used with demodulating subcarriers lying along the $(E'_R - E'_Y)$, $(E'_G - E'_Y)$ and $(E'_B - E'_Y)$ axes. Other variants are possible.

PRACTICAL APPLICATION

As used with the US 525-line, 60-field system, the NTSC system can be summarized as follows:

$$E'_Y = 0{\cdot}30E'_R + 0{\cdot}59E'_G + 0{\cdot}11E'_B$$

where $\gamma = 2{\cdot}2$.

$$E'_I = 0{\cdot}60E'_R - 0{\cdot}28E'_G - 0{\cdot}32E'_B$$
$$= 0{\cdot}74(E'_R - E'_Y) - 0{\cdot}27(E'_B - E'_Y)$$

$$E'_Q = 0{\cdot}21E'_R - 0{\cdot}52E'_G + 0{\cdot}31E'_B$$
$$= 0{\cdot}48(E'_R - E'_Y) + 0{\cdot}41(E'_B - E'_Y)$$

The luminance signal carries information describing brightness, while the colour-difference signals describe colouring. The luminance signal E'_Y is transmitted with the full system bandwidth of 4·2 MHz while E'_I and E'_Q are bandwidth-limited to approximately 1·5 MHz and 0·5 MHz respectively.

E'_I and E'_Q are then used to quadrature modulate a subcarrier of frequency $\frac{\omega_s}{2\pi} =$ 3·579545 MHz. This frequency is an odd multiple of half the line frequency in order to cause the minimum interference to the picture when the signal is displayed by a black-and-white receiver.

The composite signal (excluding sync pulses and burst) is:

$$E'_M = E'_Y + \left\{ E'_I \cos(\omega_s t + 33°) + E'_Q \sin(\omega_s t + 33°) \right\}$$

and, for colour-difference frequencies below 500 kHz, may be expressed as:

$$E'_M = E'_Y + \frac{1}{1{\cdot}14}\left\{ (E'_R - E'_Y)\cos\omega_s t + \frac{1}{1{\cdot}78}(E'_B - E'_Y)\sin\omega_s t \right\}.$$

In another form:

$$E'_M = E'_Y + \left\{ E'^2_I + E'^2_Q \right\}^{\frac{1}{2}}.$$
$$\angle \tan^{-1}.\ \frac{E'_I}{E'_Q} + 33°.$$

This is the sum of a luminance signal E'_Y and a chrominance signal whose amplitude is $\{E'^2_I + E'^2_Q\}^{\frac{1}{2}}$ and whose phase angle, with reference to the $(E'_B - E'_Y)$ axis is

$$\phi = \tan^{-1}\frac{E'_I}{E'_Q} + 33°$$

The amplitude of the chrominance signal represents saturation (depth of colour) and the angle ϕ describes hue.

In the colour receiver, either the colour-difference signals E'_I and E'_Q, or $(E'_R - E'_Y)$ and $(E'_B - E'_Y)$, are extracted by synchronous detection and are combined with E'_Y to form E'_R, E'_G and E'_B.

The synchronous detectors require a supply of continuous subcarrier. This is generated by an oscillator locked in frequency and phase to the subcarrier generator at the signal source by means of a burst of approximately 10 cycles of subcarrier located within the back-porch interval of line-blanking.

ADVANTAGES AND DISADVANTAGES. The NTSC system has many good features. These include compatibility, good utilization of the available bandwidth, the fact that no difficult problems are posed in receiver design, good signal-to-noise performance, etc. However, the fact that the hue of a picture area is described by chrominance phase angle means that any distortion in which the relative phase between the chrominance signal and the burst is altered causes errors of hue in the displayed picture; in practice, errors of 5° are perceptible and errors greater than 20° are obvious.

This characteristic of the NTSC system renders it vulnerable to a form of distortion which can arise in long links and is known as differential-phase distortion. This is probably the most serious weakness of the NTSC system; the distortion takes the form of a variation of chrominance phase angle as a function of luminance-signal magnitude and can cause a change in flesh colour as an artist moves from a bright area of the scene into the shadow. The principal technical reason why the PAL and SECAM systems were proposed is the vulnerability of the NTSC system to this form of distortion.

A.V.L.

See: *Colour principles* (TV); *Colour systems; PAL colour system; SECAM colour system; Sequential system.*

Books: *Principles of Colour Television*, ed. by K. McIlwain and C. E. Dean (New York), 1956; *Colour Television*, by P. S. Carnt and G. B. Townsend (London), 1961; *Constant Luminance Principle in NTSC Colour Television*, by W. F. Bailey (*Proc. IRE*), Jan. 1954.

Numbering Machine. Machine for printing edge numbers at regular intervals on raw stock or processed film.

Numerical Aperture. Measure of the resolving power of a microscope, defined as $\mu \sin \alpha$, where μ is the refractive index of the medium in which the object under examination is placed (air, oil, etc.), and α is the angle made by the optical axis with the most oblique ray entering the instrument. Resolving power increases with numerical aperture (NA).

See: *Cinemicrography.*

Off-Microphone. Cues or dialogue spoken from a direction in which the microphone is not pointing, and therefore recorded with less than the proper sound quality. Deleted during editing.

See: *Director* (FILM); *Studio complex* (FILM).

One-Light Dupe. Duplicate picture negative film in which the exposure and processing variations of the assembled scenes in the original negative have been corrected by grading so that the resultant dupe can be printed at a single printer light level without scene-to-scene alterations.

See: *Laboratory organization; Printers.*

One-Light Print. Print made with a single printer light setting, and thus ungraded. To enable sound and picture cutting to be carried on simultaneously, it is often desirable to have one or more duplicate workprints. These may be obtained quickly and cheaply if they are printed from the original workprint at a single printer light setting, picture quality being of little importance.

See: *Editor; Laboratory organization; Printers.*

Opaque. Term contrasting with transparency, such as a lantern slide, to denote a printed picture on opaque paper.

Optical Effects. Special effects prepared by the use of an optical printer, such as fades, dissolves or mixes, wipes, hold-frame, flop-over, etc. In particular, the most commonly used effects, fades and dissolves, are often termed simply opticals.

See: *Special effects* (FILM).

Optical Intermittent. Device which optically arrests the image on a continuously moving length of film, so that it can be seen by projection or direct viewing as if the film had been physically brought to rest by a

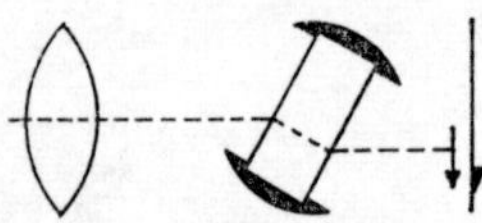

OPTICAL INTERMITTENT. Two-sided prism (more often polygonal in practice) deflects image downwards by refraction at same rate as film movement, until opaque prism ends cut-off rays from lens.

true intermittent movement. Optical intermittents are of the prism or mirror-wheel type, and are most commonly used in telecine machines and in film viewers for editing.

See: *Telecine; Telecine* (*colour*).

OPTICAL PRINCIPLES

Optics may be described as that branch of physics concerned with the behaviour of light, in particular with the geometrical characteristics of the paths followed by rays of light under given conditions, e.g., a reflection from a polished surface or passage through a transparent material such as glass.

When a ray of light passes from air into glass it is refracted or bent towards the normal (a line at right-angles to the surface at the point of entry) of the air–glass surface, providing the incident ray is not itself normal to the surface. When a ray of light passes from glass to air, it is refracted away from the normal to the surface. If the glass

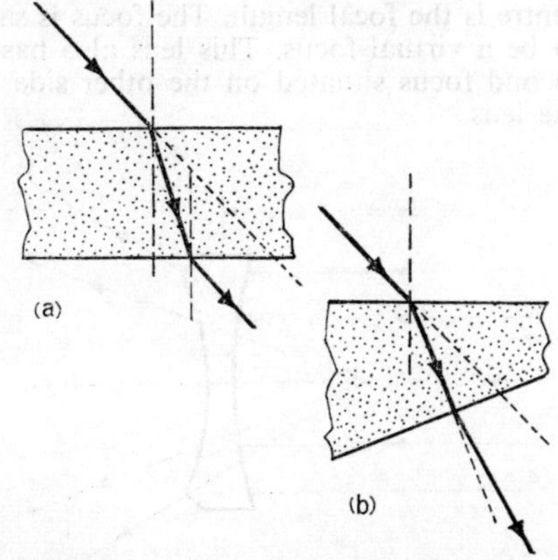

REFRACTION OF LIGHT. (a) Ray of light passing from air into glass is refracted towards the normal on entry, and away from the normal on re-emergence into air. (b) If the two glass surfaces are not parallel, the emergent ray is deviated from the incident direction.

is bounded by parallel plane surfaces, the emergent ray is parallel to the incident ray, but displaced from the incident ray by a distance depending upon the distance between the two glass surfaces, the refractive index of the glass (its power to bend light), and the angle of incidence of the incident ray.

If the plane glass surfaces are not parallel, they can be regarded as being part of the surfaces of a glass prism. The emergent ray of light is no longer parallel to the incident ray.

The angle between the incident ray and the emergent ray is called the angle of deviation, and its value depends upon the angle between the two glass surfaces, the angle of incidence of the incident ray and the angle of emergence, i.e., the angle between the emergent ray and the normal to the surface.

If the angle between the two glass surfaces is small, and the angle of incidence of the ray with the first surface is also small, the angle of deviation between incident and emergent rays is independent of the angle of incidence. If D is the angle of deviation, μ is the refractive index from air to glass, and A the angle between the two glass surfaces, then $D = A(\mu - 1)$.

Thus for any ray of light which is almost normal to the first surface, the angle of deviation depends only upon the refractive index and the angle between the glass surfaces, and increases as the angle between the glass surface increases.

THIN LENSES

A piece of transparent material bounded by two spherical surfaces is called a lens, and if the distance between the two surfaces is so small that it can be neglected it is called a thin lens.

DEVIATION. A line through the centres of curvature of the surfaces is the optical axis of the lens. Where the optical axis cuts the two spherical surfaces, these surfaces are parallel, and a ray of light passing through this part of the lens is undeviated. The ray does, however, emerge from the lens with a slight displacement, depending upon the thickness of the lens at this point. The point within the lens at which such a ray crosses the optical axis is known as the optical centre of the lens.

Moving out from the centre of the lens towards the rim, the angle between the surfaces, while remaining small, increases, and consequently the deviation, which depends only upon this angle and the refractive index of the material, also increases. Providing the radii of curvature of the surfaces are large in comparison to the diameter of the lens, it can be shown mathematically that the deviation produced at a point in the lens is proportional to the distance of that point from the centre of the lens.

Rays of light passing through the lens are deviated, either towards or away from the optical axis depending upon the spherical surfaces. If the lens is such that the centre is thicker than the rim, rays of light are deviated towards the optical axis and the lens is called a converging lens. If it is thicker at the rim than the centre, rays are deviated away from the optical axis and the lens is said to be a diverging lens.

LENS SHAPES. The following are some of the more common shapes of simple lenses: Bi-convex, in which the two surfaces are convex-to-air.

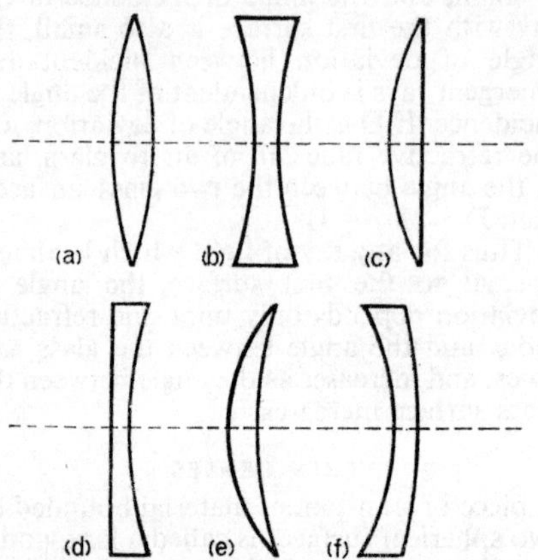

LENS SHAPES. (a) Bi-convex. (b) Bi-concave. (c) Plano-convex. (d) Plano-concave. (e) Convex meniscus. (f) Concave meniscus.

Bi-concave, in which the two surfaces are concave-to-air.

Plano-convex and plano-concave, which each have one plane surface associated with either a convex-to-air or concave-to-air surface.

Convex meniscus and concave meniscus, each having one surface convex-to-air and the other surface concave-to-air. The convex meniscus is thicker at the centre than the rim, while the concave meniscus is thicker at the rim.

PRINCIPAL FOCUS AND FOCAL LENGTH. A converging lens causes a narrow beam of parallel rays of light, close to, and parallel to, the optical axis, to converge and pass through a point on the optical axis on the far side of the lens. This point is called the principal focus of the lens, and the distance between the principal focus and the optical centre of the lens is known as the focal length. A screen held at the position of the principal focus shows a small concentrated spot of light. The focus is said to be a real focus. A plane through the focus normal to the optical axis is known as the focal plane.

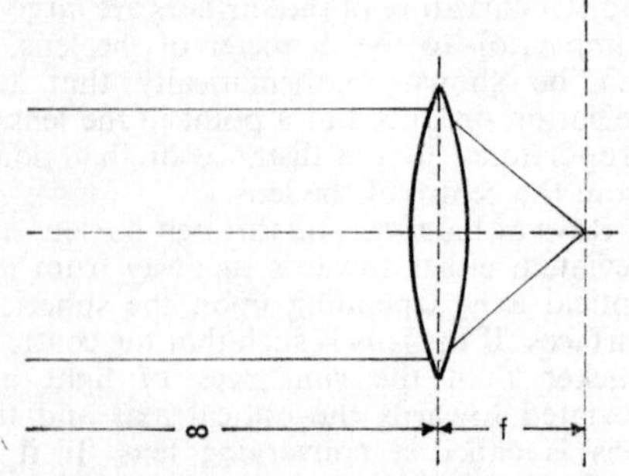

FOCUS OF CONVERGING LENS. Object at infinity forms an image in the plane of the principal focus of the lens, distant *f* from its optical centre.

There are, of course, two principal foci, one on each side of the lens.

In the case of a diverging lens, a similar beam of parallel rays diverges on passage through the lens, as though the rays had originated at a point on the optical axis on the same side of the lens as the incident beam. This point is called the principal focus, and its distance from the optical centre is the focal length. The focus is said to be a virtual focus. This lens also has a second focus situated on the other side of the lens.

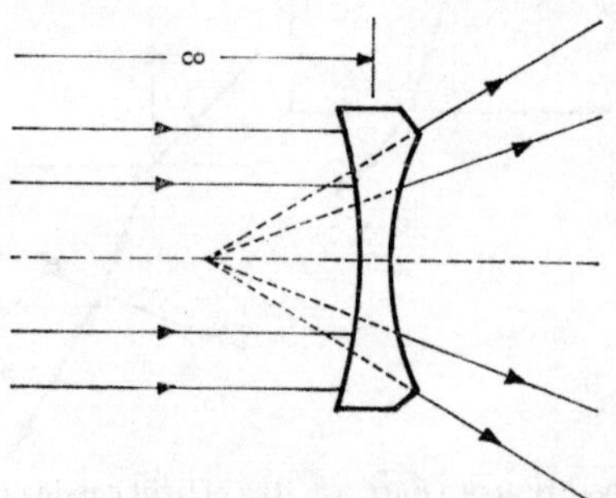

FOCUS OF DIVERGING LENS. Diverging lens bends parallel rays of light so that they diverge and appear to come from a point of focus in front of the lens. This point is the principal focus, distant *f* from the optical centre of the lens. This is called a virtual focus, and the focal length is negative. Diverging lenses are often called negative lenses and are used in photography only in special circumstances.

If r and s are the two radii of curvature, and the lens with a refractive index of μ is in air, simple mathematics can be used to show that the focal length f is given by:

$$\frac{1}{f} = \left(\mu - 1\right)\left(\frac{1}{r} + \frac{1}{s}\right)$$

It is, of course, necessary to use a sign convention in dealing with the radii of curvature of the two surfaces, which may be convex- or concave-to-air. One convention is to call a surface convex-to-air positive, and one which is concave-to-air negative. The focal length of a convex lens is positive, while that of a concave lens is negative.

CONJUGATE FOCI. Conjugate foci, or conjugate points, are defined as pairs of points, such that, if an object is placed at either point, its image is formed by the lens at the other point. The image may be real, in which case rays of light from the object, after passing through the lens, pass through the image point; or it may be virtual, in which case the rays of light from the object, after passing through the lens, appear to originate at the image.

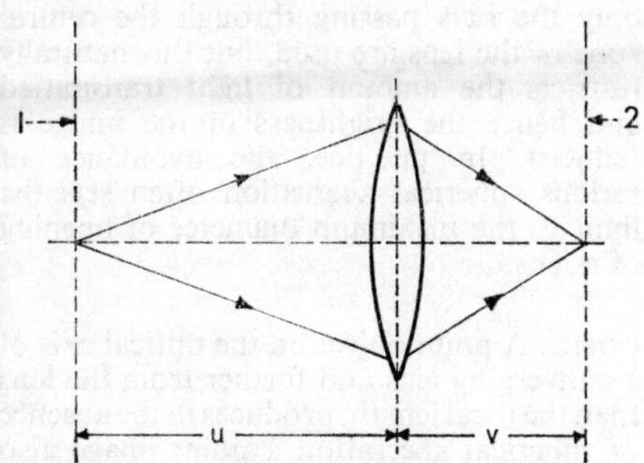

OBJECT AND IMAGE DISTANCES FOR THIN LENS. 1. Object plane. 2. Image plane. Then (1/u)+(1/v)=(1/f).

A mathematical formula can be derived relating the object distance from the lens, denoted by u, and the image distance from the lens, denoted by v, with the focal length of the lens.

$$\frac{1}{u}+\frac{1}{v}=\frac{1}{f}$$

In making calculations based upon this formula, it is necessary to observe a sign convention. One such convention uses a positive sign for distances measured from the lens to real objects and real images, and a negative sign for distances measured from the lens to virtual objects and virtual images.

The position and size of the image of a small object placed perpendicular to the optical axis can be found by graphical methods. On a scale representation of the object and lens, two convenient rays can be traced from the tip of the object to give the location of the tip of the image. The two rays are:

1. A ray from the tip of the object parallel to the optical axis is refracted by the lens to pass through the appropriate focus of the lens, or in the case of a diverging lens, the ray is refracted as though it originated at the focus.
2. A ray from the tip of the object which passes through the optical centre of the lens is undeviated.

The intersection of these two rays gives the location of the tip of the image.

LATERAL MAGNIFICATION. The relationship between the linear size of the image and the linear size of the object is known as the linear lateral magnification, denoted by m:

$$m = \frac{\text{linear size of image}}{\text{linear size of object}}$$

$$= \frac{\text{distance of image from lens}}{\text{distance of object from lens}}$$

$$= \frac{v}{u}$$

COMBINATION OF TWO THIN LENSES. The combined focal length of two thin lenses in contact can be determined by applying the lens formula separately to the two lenses. The image produced by the first lens becomes the object for the second lens, but if the image is real for the first lens, it becomes the virtual object for the second lens, and if the image due to the first lens is virtual, it becomes a real object for the second lens.

If f_1 and f_2 are the focal lengths of the two lenses, and f the focal length of the combination,

$$\frac{1}{f}=\frac{1}{f_1}+\frac{1}{f_2}$$

If the two thin lenses are separated by a distance x, the formula becomes

$$\frac{1}{f}=\frac{1}{f_1}+\frac{1}{f_2}-\frac{x}{f_1 f_2}$$

POWER OF A LENS. The power of a lens is defined as the reciprocal of the focal length. The unit of power is the dioptre, and a lens of focal length of 1 metre has a power of 1 dioptre.

$$\text{power in dioptres} = \frac{1}{\text{focal length in metres}}$$

$$= \frac{100}{\text{focal length in cm.}}$$

A convex lens has a positive power, while a concave lens has a negative power.

DEFECTS OF LENSES

SPHERICAL ABERRATION. Rays of light parallel to and close to the optical axis of a converging lens are deviated to pass through

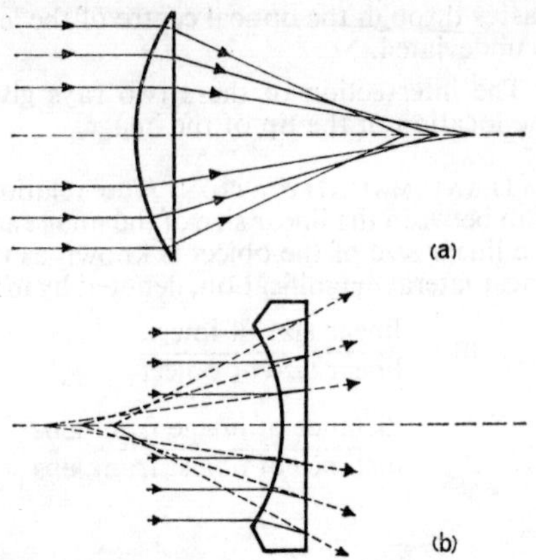

SPHERICAL ABERRATION. (a) With a simple converging lens, rays farthest from lens axis converge more strongly and come to a focus nearer the lens than rays close to the lens axis. Image is never fully sharp. (b) With a simple diverging lens, marginal rays appear to come from points nearer the lens.

the focus. Rays of light, still parallel to the optical axis but further away from the axis, do not pass through the focus but intersect on the optical axis at a point slightly nearer to the lens than the true focus. This point of intersection moves further in towards the lens as the radius of the zone of rays increases, reaching a limiting position for the marginal rays, i.e., those rays which pass through the edge of the lens. The distance between the true focus and this limiting position is called the axial spherical aberration.

A clear image of a small object, such as a pinhole, cannot be produced if all the possible incident rays are allowed to fall on the lens. The movement of a screen along the axis in the region of the image shows that, even in the best position, the image of the pinhole has an appreciable size. The smallest patch of light which can be obtained from the pinhole is called the circle of least confusion.

The correction of spherical aberration is of importance in the design of photographic lenses. It can be completely eliminated by modifying the surfaces of the lens so that they are no longer spherical, but the practical manufacture of such aspheric surfaces presents very considerable technical problems. In addition, since this modification can only be calculated for one particular object distance, such a lens will not be completely free from spherical aberration at other distances.

The effect of spherical aberration in a single lens can be minimized by correct selection of the radii of curvature of the two surfaces, but the general solution in photographic objectives is to correct the spherical aberration of a positive lens component by combining it with a negative lens in which the spherical aberration is equal in amount but opposite in direction. Such a combination can also be designed to correct other defects such as chromatic aberration.

Spherical aberration is also reduced if only the rays passing through the central zone of the lens are used, but this naturally restricts the amount of light transmitted and hence the brightness of the image is reduced. In practice, the avoidance of serious spherical aberration often sets the limit to the maximum diameter of opening of a lens.

COMA. A point object on the optical axis of a converging lens and further from the lens than the focal length, produces in the absence of spherical aberration a point image also on the optical axis. If the object is not on the optical axis, a point image is not produced.

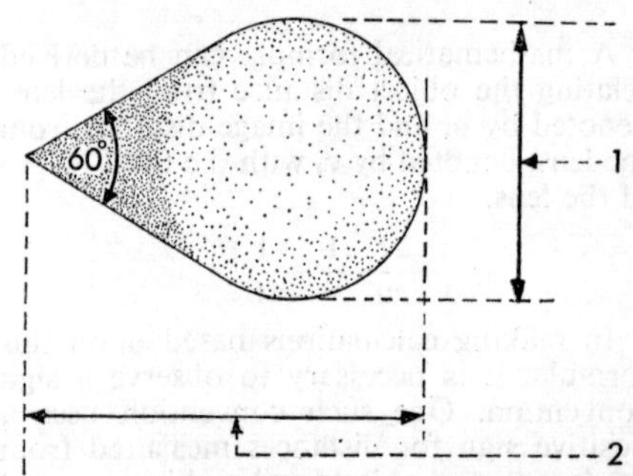

COMA. An uncorrected lens tends to reproduce off-axis image points as unsymmetrical patches flaring away to one side. A first order coma patch is still regular in shape, the length, 2, being, in practice, three times the width, 1. Higher order coma patches are more complicated and may have multiple tails.

Rays of light passing through a very small zone at the optical centre of the lens produce a point image in the image plane. If the rays of light are confined to a small annulus of the lens surrounding the central zone, the image produced is still in the same image plane, but is now nearer the optical axis, and is no longer a point but a circle of small radius. As the lens annulus increases in radius, the image moves in towards the optical axis and the circle of light constituting the image gets larger. The image produced by all rays of light passing through the lens is confined by the envelope of all the circular images, and appears as a patch of light similar to a comet with the maximum intensity of light at the point.

The defect can be reduced by the use of a stop at the lens restricting the rays to the central region, but only at the expense of the illumination of the image.

ASTIGMATISM. Even if a stop is used at the lens to reduce spherical aberration and coma, a point object an appreciable distance from the optical axis does not give rise to a point image. If the image of a point object is allowed to fall on a screen, and the screen is moved along the optical axis from a position near the lens, the image appears first as a straight line, and changes into an ellipse, a circle, a series of ellipses and finally another straight line at right-angles to the first. At no position of the screen is a point image produced. Both straight line images are closer to the lens than an axial image produced from a corresponding axial object at the same distance from the lens as the non-axial object.

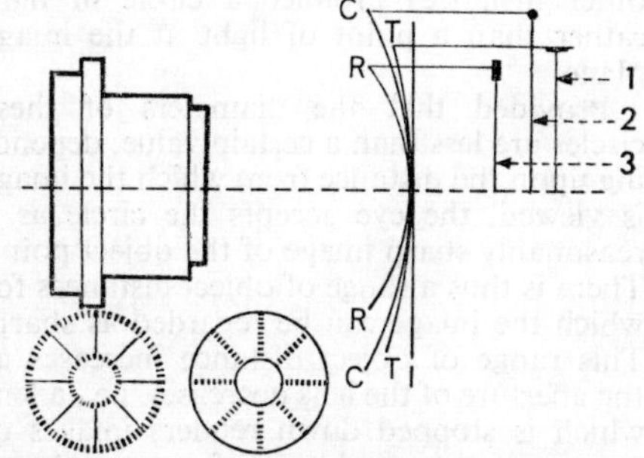

ASTIGMATISM. (*Top*) With astigmatism present in a lens, transverse lines, 1, are focused on a surface, T, circular discs, 2, on a surface, C, and radial lines, 3, on a surface, R, instead of correctly in image plane, I. (*Bottom*) An object containing both transverse and radial lines can never show both sharp at the same time.

If the object is in the form of a spoked wheel, with centre on the optical axis, the rim and the spokes cannot be focused in the same plane, and the outer edge of the spokes do not focus in the same plane as the centre of the wheel.

CURVATURE OF FIELD. The simple lens formula was derived on the assumption that the aperture of the lens and the size of the

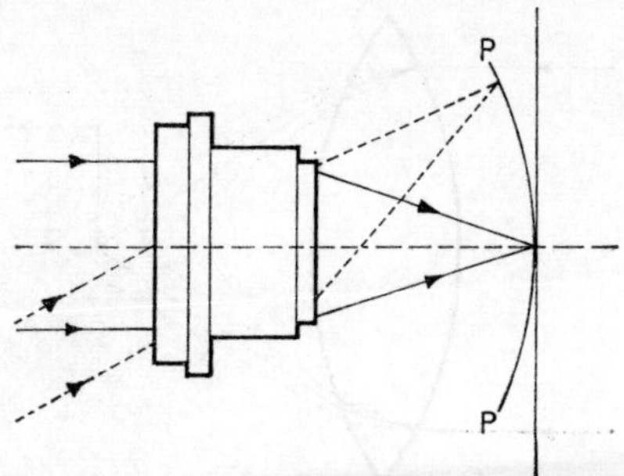

CURVATURE OF FIELD. When a lens shows curvature of field, all image points lie on a curved surface, P, instead of a flat plane.

object were small in comparison with the distance of the object from the lens. If these conditions are not met, it is impossible to produce a sharp image of all parts of the object on a flat screen. The edge of the object is brought to a focus closer to the lens than the centre of the object.

DISTORTION Distortion occurs when the magnification of the system varies at different parts of the image. This variation of magnification may be produced by a stop,

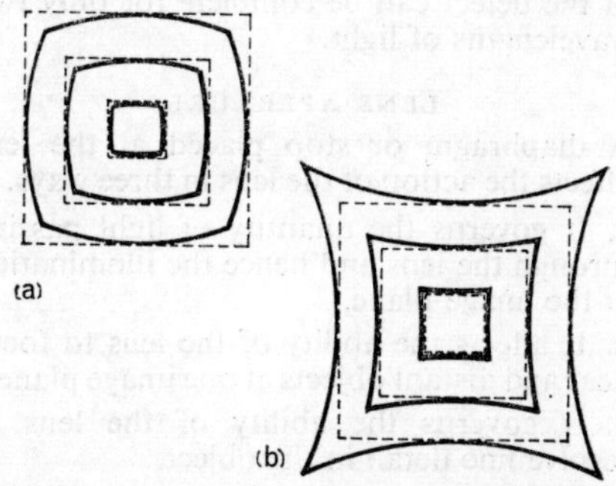

BARREL AND PINCUSHION DISTORTIONS. Distortions resulting from image magnification varying with distance of image point from axis. (a) Image points reproduced nearer to the axis than they should be cause barrel distortion. (b) Image points farther from the axis than they should be cause pincushion or hour-glass distortion. Dotted lines indicate shape of original object.

either in front of or beyond the lens. Barrel distortion is produced if the stop is placed between the object and lens, while pincushion distortion is produced when the stop is beyond the lens.

CHROMATIC ABERRATION. Chromatic aberration is due to dispersion of light by the material of the lens. A narrow axial

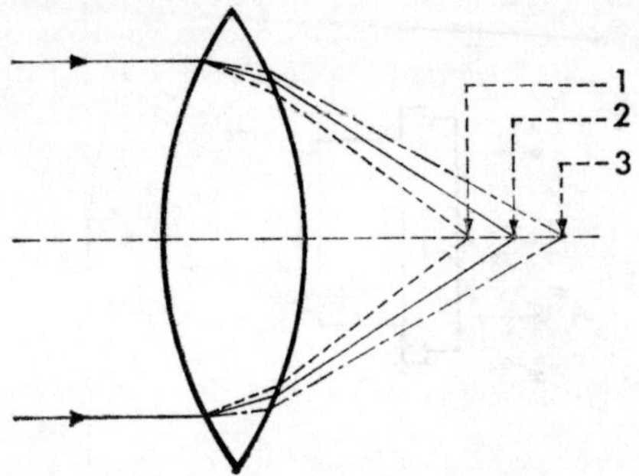

CHROMATIC ABERRATION. A simple lens refracts blue rays, 1, more strongly than green rays, 2, or red rays, 3. The latter therefore come to a focus farther behind the lens, and the image shows reddish-bluish colour fringes.

beam of white light is not brought to a single point focus by the lens. The violet component of the light is focused closer to the lens than the red component, while the green component is focused at an intermediate position. This effect produces coloured fringes at the edges of the image of an object.

The defect can be partially corrected by using a combination of two thin lenses made from different types of glass. The elimination of the defect can be complete for only two wavelengths of light.

LENS APERTURE

A diaphragm or stop placed at the lens affects the action of the lens in three ways.

1. It governs the quantity of light passing through the lens and hence the illumination at the image plane.

2. It affects the ability of the lens to focus near and distant objects at one image plane.

3. It governs the ability of the lens to resolve fine detail in the object.

A small diaphragm also reduces many of the lens defects mentioned above.

TRANSMISSION OF LIGHT. The quantity of light transmitted by the lens is proportional to the area of the lens diaphragm. Where object distances are large in comparison to the focal length, the area of the image is inversely proportional to the square of the focal length. The brightness of the image, which determines the photographic exposure required, is thus dependent upon the area of the stop and the focal length of the lens; in fact, it is proportional to

$$\frac{(\text{diameter of lens})^2}{(\text{focal length})^2}$$

The diameters of lens stops are selected so that the exposure required for one stop is twice that required for the next larger stop. The larger the diameter of the lens and the smaller the focal length the brighter the image and consequently the lower the photographic exposure required. Lens stops are made so that one stop requires twice the exposure of the next higher stop; and hence the ratio of the diameters of two adjacent stops is $\sqrt{2}$.

The standard *f*-numbers given to lens apertures produce the series, 1·4, 2, 2·8, 4, 5·6, 8, 11, 16, 22 and 32, each number being $\sqrt{2}$, i.e., 1·4, times the preceding number. This simple relationship neglects many factors which affect the transmission of the lens and the brightness of the image, e.g., the variations of absorption of light by different glass, and the light reflected at glass-air surfaces.

Relative apertures are therefore better based on actual measured transmissions of light.

DEPTH OF FIELD. Theoretically, only objects at a certain distance from the lens produce sharp images at a particular image plane. The pencils of light from objects at other distances produce a circle of light rather than a point of light at the image plane.

Provided that the diameters of these circles are less than a certain value, depending upon the distance from which the image is viewed, the eye accepts the circle as a reasonably sharp image of the object point. There is thus a range of object distances for which the image can be regarded as sharp. This range of object distance increases as the aperture of the lens decreases, i.e., a lens which is stopped down renders images of near and distant objects of greater sharpness than does a lens of a large aperture.

RESOLUTION. The number of separate, distinct lines which can be resolved per millimetre of image is theoretically dependent upon the diameter of the lens aperture; the greater the lens aperture, the greater the resolution. Lens defects, however, have a greater effect at large apertures so that even if a high resolution is obtained, the line images are unsharp.

See: *Image formation; Lens performance; Lens types; Light.*

Books: *Modern Optical Engineering*, by W. Smith (New York), 1967; *Photographic Optics*, by A. Cox (London), 1966; *Applied Optics and Optical Engineering*, ed. by R. Kinglake (New York), 1966.

⊕**Optical Printing.** Method of film printing in which an image of the film being printed is formed by means of an optical copy lens system on to the raw stock being exposed, rather than by contact printing where the two films are held together in contact at the time of exposure. Optical printing is imperative where the size or shape of the required print differs from that of the original, as in reduction, enlargement or anamorphic copying, and is also common in special effects work.

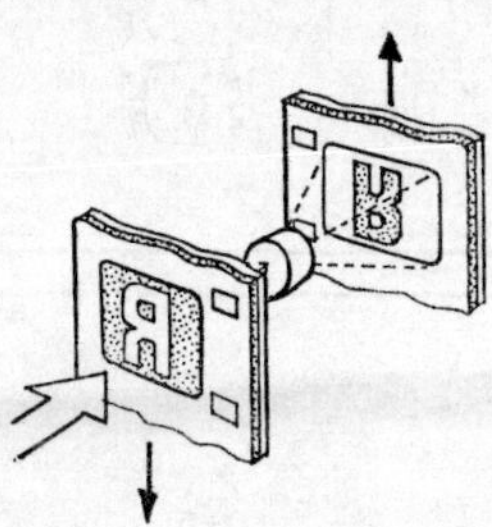

PRINCIPLE OF OPTICAL PRINTER. Separation of printing from printed film makes possible enlargement, displacement, motion reversal, etc. Top-to-bottom and side-to-side reversal of image geometry result from interposition of copy lens.

See: *Printers; Special effects* (FILM).

⊕**Optical Sound.** Method of sound recording (now little used for original recording, but still almost universal for 35mm and 16mm release prints) in which a narrow, modulated beam of light impinges on a light-sensitive film which is moved past it at constant speed. The light modulations are recorded either by the variable-area or the variable-density method.

See: *Sound recording* (FILM); *Variable area recording; Variable density recording.*

⊕**Optical Sound Head.** Device for converting optical sound recordings on film to electrical signals for reproduction.

See: *Projector.*

⊕**Order of Magnitude.** One order of magnitude ☐is a multiplying or dividing factor of 10, two orders 100, three orders 1,000, and so on.

⊕**Original.** Film exposed in a camera and thus of better picture quality than any subsequent derivative. As it is sometimes a negative and sometimes a reversal positive, original is a convenient generic term for it.

See: *Laboratory organization; Narrow-gauge production techniques; Printers.*

⊕**Orthochromatic.** Photographic emulsions whose sensitivity is effectively limited to the blue and green regions of the visible spectrum and excludes the orange and red.

⊕**Orwocolor.** A negative/positive integral tri-pack colour film process for which the stocks are manufactured at Wolfen in East Germany; the system is generally similar to Agfacolor and the negative does not include masking. A colour duplicating film is also made for reversal processing to obtain a dupe negative direct from the original negative.

☐OUTSIDE BROADCAST UNITS

Television transmissions originate from two sources: studio and remote. Remote here means all areas distant from fixed facilities, and therefore requiring temporary equipment to be moved in and set up. Despite the fact that studio broadcasts (including film recordings) bulk very large in all TV transmission, the work of remote outside broadcast (OB) units is extremely important.

OB transmissions provide the sense of immediacy, of bringing to the audience actual events when they happen, which has been the foundation of television's success. Some of these fall into the category of newsreel, and include outdoor sports and political happenings; others are events of note, such as a public concert in an auditorium, which cannot be transferred to the studio. Formerly they would have been recorded on film by well-established production techniques, and these techniques are still used by TV production groups.

But the delays of film processing and editing have led to the development of instantaneous methods such as live relay and videotape recording, which require a different kind of organization. Outside broadcasts call for a high degree of mobility and close-knit communication. They form an integral part of all television groups of any complexity, and need to be planned as a piece of logistics, with a base, a staff, vehicles and equipment of their own.

The OB section is housed in a building containing a garage, vehicle and technical workshops and stores, and administrative and planning offices. The vehicles and engineering staff work from this base, which is situated as centrally as possible in the area of operations, has a good approach to main traffic routes, and is reasonably near to the studio centre and main headquarters. The

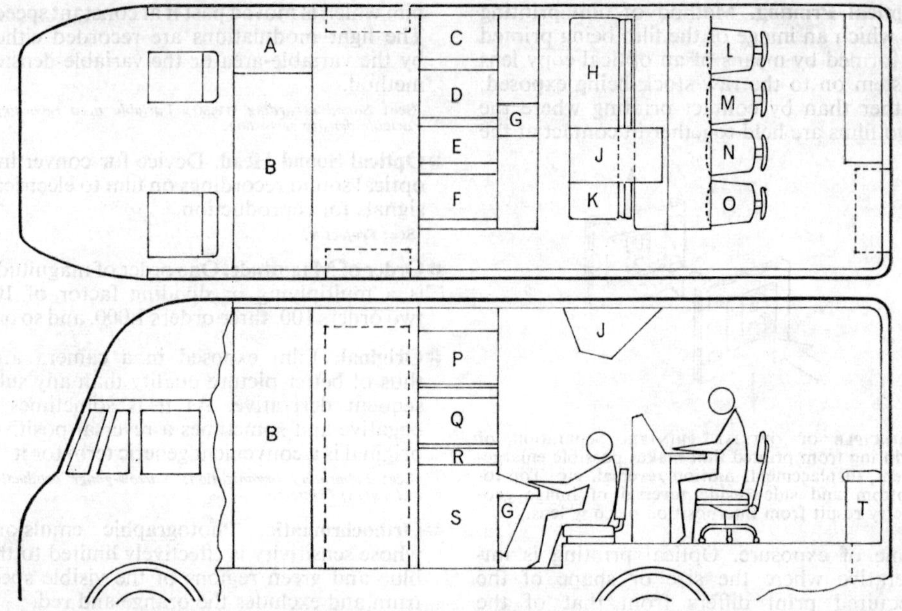

TYPICAL MOBILE CONTROL ROOM. Based on a general-purpose non-articulated vehicle. A. Bench top. B. equipment and storage. Doors on both sides of vehicle open into next compartment. C. Producer's picture monitor. D. Line picture monitor. E. Radio check receiver. F. Engineer's picture monitor. G. Oscilloscope. H. Vision control engineer. J. Loudspeaker (above). K. 2nd vision control engineer. L. Production assistant. M. Producer. N. Sound engineer. O. Engineering manager. P. Picture monitor. Q. Camera 1 picture monitor. R. Waveform monitor. S. Camera control panel. T. Stowage. Separate rigging, lighting and camera tenders carry bulky equipment to site.

production staff offices are not always attached to the base, but should be within a few minutes' journey for planning meetings and conferences. If the base is on the same site as the studio centre, some sharing of technical workshops and administrative services is quite practicable. The layout of the base can take numerous forms, depending on the land or buildings available and the scale of organization.

EQUIPMENT

The amount and nature of the equipment used at an OB site depends on the type of programme, time of day, producer's requirements and engineering needs. The general requirements are cameras and microphones suitably placed and mounted, and the OB vehicle, or mobile control room (MCR), as the centre for production and technical control of the programme. Also needed are vision, sound and communication circuits to the studio centre or a mobile recording equipment if the programme is being recorded at the site for transmission later. Local communications and commentary facilities, power supplies, and cables and rigging equipment are necessary. Auxiliary transport vehicles and lighting equipment, if the programme is indoors or during the hours of darkness, may be additionally called for.

OB VEHICLE (MOBILE CONTROL ROOM). The OB vehicle, in common with all items of television equipment, has been continuously developed by various organizations to suit their own requirements. Consequently, there are many designs, but they all have common features.

A typical general-purpose vehicle accommodates the camera control units, sound control, vision mixing unit, monitors, communications, production and technical staff. Power generators, radio links, videotape equipment, camera mountings, microphone booms, cables, etc., all vary according to need, and special fittings for them are not normally provided.

The overall size of the vehicle is governed by parking problems, manoeuvrability, and the limitations of access at sites. An articulated vehicle has the advantage of a smaller turning circle, a tractor which can be detached while the trailer is parked and used for other work, and shorter length when the tractor is detached. However, to meet local site conditions, it is often necessary to be able to drive away at short notice in case of fire or other emergencies, so that the tractor must often be retained. Here, a rigid four-wheeled vehicle is at an advantage because of its greater compactness.

The layout of the general purpose OB vehicle is a compromise and does not suit all circumstances, but it has been used successfully for many years and is founded on the following principles.

The producer has the overall responsibility of the broadcast and is centrally placed to give directions and to have a good view of all pictures. He is at the centre point of all programme communications and, in many cases, operates the vision mixer controls.

The production assistant notes all extra details of production requirements (e.g., camera directions) before or during rehearsal, and at the time of the programme gives preliminary directions to the crew and maintains liaison with the studio centre on programme timings, cues, etc.

The sound engineer must at every instant provide programme sound to match the pictures selected. Hence he must hear all crew directions and be able to anticipate the producer's intentions. The sound engineer must be placed to see all pictures and be able to react instantly to the producer's hand movement.

The engineering manager, or engineer-in-charge, is placed so that he can oversee all technical operations, and give guidance to the producer. The sound and communications jackfield is placed between him and the sound engineer and each can use the engineering communications.

The vision control engineers are seated in front of, and slightly below, the producer's team. Thus they have some separation from the activity behind them while they are still in sufficiently close contact. Modern equipment with stable performance, set up correctly, should not require more than two engineers for controlling four cameras, but uneven illumination, varying contrast due to passing clouds, haze or mist can make the task difficult.

A lighting supervisor and a make-up assistant are sometimes required, but no place is formally allotted to them. A separate one-way communication jack on the engineering manager's desk enables the lighting engineer to direct his electrician concerning lighting changes on the scene.

A single assembly of picture monitors is used for vision and sound control, lighting and production purposes. This makes for economy in space, weight, power consumption and cost, and minimizes the troubles due to variations in the settings of the individual monitor controls. One monitor is tied to each camera channel, two monitors are available for production and engineering preview and one for output monitoring or radio check. A special monitor line-up test signal can be switched to any or all monitors for setting-up and picture matching.

The general equipment area contains all unattended equipment or that which can be adjusted during rehearsal and left. The first category includes camera power supplies, sync generators and test signal generators, and the second facilities for remote cameras, radio microphone receivers and talkback transmitters. The area also includes mains distribution and fuses, and bench top and spares cupboard for minor repairs.

The technical equipment in the vehicle should all be demountable and capable of being set up in any other place in any desired arrangement. Typical examples are theatre broadcasts in busy towns where parking permission can be obtained only from late at night to early morning, important ceremonies requiring a closely coordinated coverage by, say, ten cameras, or a large sports meeting lasting perhaps a week with many extra facilities.

In these cases, the equipment is set up on temporary racking in a vacant room, or in a sectional hut erected for the occasion. One or two complete camera channels can be taken from the vehicle and set some distance along a processional route. During elections, it is advantageous to place single cameras each in different towns.

Alternative conceptions and layouts exist. For example, in sponsored broadcasting, the technical crew concentrates on providing the technical services, while the programme director deals with the programme content and performers. Where there are such clear divisions of interest, it is customary to partition the sections of the vehicle. A large vehicle may be divided into four separate areas for sound control, production control, video equipment, maintenance and air conditioning, and be 35 ft. in length.

COMMENTARY VEHICLES. For domestic broadcasting, the OB vehicle carries communication circuits between the producer and either one or two commentators. The commentator is seated so that he can watch the event and he has a picture monitor to enable him to relate his commentary to the picture being transmitted. He has a small communication unit with headphones for receiving producer's talkback, programme sound or radio check sound, a cue light which shows when his microphone is faded up, a telephone handset for speaking directly with the producer, and a press button for operating a light on the sound control desk to indicate when he wishes to be faded in or out. No special vehicle is required for this service.

When a programme is being relayed to foreign countries, the same picture signal is transmitted everywhere but each receiving country likes to have its own commentator, and special equipment has been designed for the purpose.

Foreign commentators are also seated to watch the event and, to minimize mutual disturbance, each is usually in a booth. Each has a picture monitor and a small box called a commentator's control unit. The general facilities of the control unit have been standardized and accepted by member countries of the European Broadcasting Union. In the centre of the panel is a key which, when moved to the right, connects the commentary microphone into circuit, when moved to the left connects to an interview microphone, and has a central off position. Two lamp indicators show which microphone is in circuit. Two headsets are provided, either of which can be switched by the user as a telephone to the foreign control line, or to the assistant in the foreign commentary vehicle, or to receive clean effects or programme sound. "Clean effects" comprises the complete sound programme but without any commentary. "Programme sound" is normally the home programme with commentary but, when specially required, a suitable guide commentary can be substituted.

RIGGING AND CAMERA TENDER. The amount of loose equipment taken to an OB site varies according to the practice of the TV organization and the nature of the programme, and can be considerable. Unless the operation is on the simplest scale, it is impractical to carry much of it in the OB vehicle. If it were packed with cameras, lenses, tripods, microphones, commentators' monitors, etc., installation at the site could not start until the vehicle arrived; it would then have to be unpacked in an order which might not suit the arrangements for installation, and equipment would be left in the open unprotected. The expensive technical equipment would be idle until the installation was completed, and the same applies in reverse order after the programme.

It is preferable, therefore, to have a separate rigging tender which carries cables, ropes, ladders, scaffolding and camera platforms, all loaded independently of the OB vehicle to a schedule planned earlier. In this way, all cables are laid in place ready for coupling when it arrives, and technical testing and setting-up commence without delay. At the end of the programme, the OB vehicle can drive away as soon as it is uncoupled, and the removal of cables can follow.

It is also desirable to have enough cable and tackle for the next site to be rigged while the programme is taking place at the present site. This allows more efficient utilization of equipment, improves the staff work pattern and reduces overtime.

A camera tender is also a useful provision. Image orthicon cameras are bulky and heavy; OB lenses can be large and numerous; camera tripods and panning heads are awkward; commentators' monitors do not like rough treatment; sundry delicate devices such as caption displays and prompters are often required. The camera tender can be fitted to carry all these safely and accessibly. This tender may also provide a repair bench separate from the busy activity in the OB vehicle, and if some special journey is necessary, the camera tender can be used.

LINES, RADIO LINKS AND VIDEOTAPE. These requirements are much affected by local conditions and by the relative needs for live and recorded material. Within cities, radio links have restricted use because of buildings obstructing the radio path and causing reflections. It is more reliable and convenient if the telephone authorities can provide satisfactory cable circuits for communications, programme sound and vision between the site and the studio centre. The TV organization must satisfy itself on the quality of the circuits by appropriate tests such as pulse and bar, frequency response, linear distortion and noise tests.

A mobile videotape recorder is invaluable for broadcasting in the evening a sporting event, such as a football match, which took place during the day. It can also be used for long events such as athletics meetings or political conferences from which highlights are to be selected for a shortened programme in the evening, and for built programmes which may be compiled from recordings made in separate sequences.

Radio link and VTR equipment, though fast becoming lighter and more compact, is still bulky and requires specialized operators. It is thus advantageous to equip separate vehicles for them. Such equipment need only then be sent to site when needed, and when more than one OB vehicle is provided, a smaller amount of radio and VTR equipment can be employed under scheduling arrangements.

POWER VAN AND AERIAL TOWERS. These units use mains power supplies whenever possible because they offer greater reliability, less cost, minimum parking requirements and the absence of engine noise and exhaust fumes. At country sites such as golf courses and motor race circuits, however, a power van of about 25 kVA capacity is required to supply the OB vehicle and mobile VTR or local radio link.

For an intermediate radio link relay point, a trailer generator is provided having a rating of 6–8 kVA, which is of convenient size for supplying one or two cameras on an isolated site or in a river craft.

An aerial tower vehicle is often employed at the OB site to raise the transmitting aerial above local obstructions such as trees, buildings and vehicles. A telescopic triangular section lattice tower gives best stability and it is raised through a system of steel cables, and a hydraulic ram powered from the road engine. For travelling, the mast is hinged down to near horizontal. The vehicle is provided with swivelling stay arms and stabilizing jacks, and the aerials and cables are set up with the mast in the vertical position at minimum height. The mast is then extended, guy ropes are secured, and final aerial bearing adjustments are made from the ground by operating an electrically powered aerial-turning unit on the mast head.

Intermediate radio link relay sites are placed on high ground chosen from map studies and surveys, and structures such as water towers are often used to support the aerials and give clearance above animals and moving vehicles which could affect the radio path.

Typical receiving sites are the roof of the studio centre or telephone exchange, or on the television transmitter mast at a subsidiary level. Duplicate video circuits are often provided from the studio to the transmitter to give a reverse OB circuit when needed, or protection for the main programme circuit.

LIGHTING. Where it is possible to contract with a local undertaking for the supply, erection and operation of OB lighting requirements, it may well be economic to do so, since the broadcasting requirements alone may be too variable or infrequent to support these facilities efficiently. On the other hand, at a smaller studio centre, not too intensively worked staff and equipment might be afforded for the alternate role. The lighting electricians work under the instruction of the lighting director, and it is very important, especially in places to which the public has access, to ensure safety by keeping all cables and equipment out of reach and to see that all suspended fittings are secure and have additional safety chains.

When the broadcasting organization provides its own OB lighting equipment, a large van is also necessary, fitted with shelves and bins on either side for the various luminaires (lamp units), spare bulbs, diffusers, fittings, ropes and suspension tackles, cables and junction boxes, switch trucks and sundries. A lighting contractor may also have available a generating set for lighting purposes.

SAFETY. Attention must be given to instituting and maintaining adequate safety measures to protect the public, the performers, the employees and the equipment. Cameras are often sited on scaffolding towers erected for the occasion and, for any but the smallest structure, a competent contractor is employed to construct the tower and accept primary responsibility for its soundness. The risk of falling objects from elevated camera platforms is a serious hazard. The area beneath is roped off, the camera with its lenses doubly secured, the platform has kicking boards and rails, and safety nets are fixed outside and beneath the platform.

EQUIPMENT EVOLUTION

The use of semiconductors has made welcome reductions in the size, power consumption and weight of circuits. In some equipment, such as camera channels, this has permitted more elaborate and sophisticated circuits; in vision mixers and sound mixers, it has allowed more channels to be built into a given desk size; it has improved devices such as radio cameras and microphones, and servo-controlled lenses have all become much more practical and reliable.

MOBILE CAMERAS. The roving eye was developed primarily to cover moving events. Now in various forms it comprises either a medium-sized van or large car which provides the programme output while travelling and it carries one or two cameras, sound facilities and communications, radio links, a power supply and the necessary cameramen, engineers and production staff. A radio link-receiving vehicle is required as a base station and this is usually close to an OB vehicle which, with its own static cameras, provides overall coverage and control of the programme.

A roving eye can be used at horse race meetings where it can be stationed in turn at the various starting points. For those racecourses which have a hard road parallel with a significant part of the course, and especially where there are jumps, the small roving eye can be used to good effect, driving 100 yards or so ahead of the field with its camera facing rearwards. The vehicle should have good acceleration, a smooth ride at up to 40 mph, and be small and unobtrusive in order not to interfere with the race or the spectator's view of it.

A radio camera is essentially a portable hand-held camera, with a back-pack containing video signal processing, synchronizing and deflection circuits, a radio transmitter and batteries, all of which can be carried by the cameraman wearing a harness. A vehicle is required for the base radio link equipment and to transport the radio camera and battery chargers and to provide minor repair facilities. The radio camera evolved from the 1-in. vidicon tube and miniature circuit techniques. It has various uses for obtaining close-up pictures or interviews in crowded places where the camera can be brought to the scene and taken away without the obstruction of a camera dolly and trailing cables. It can be used in a car or on a motor cycle as a simple roving eye, and even in a small aircraft or helicopter.

Its main disadvantage is the limited sensitivity of the vidicon tube: because of its mobility, rapid changes of illumination are presented to the camera which may be outside the range which it can handle satisfactorily. Some radio cameras are fitted with circuits which provide automatic correction of signal level over a 20 : 1 variation in illumination. Another radio camera employs a standard 3-in. image orthicon camera tube which gives the expected improvement of sensitivity, but the weight and size of the camera is appreciably greater so that it must be carried on the shoulder. The cameraman cannot reach the pre-set controls on his back pack and hence an engineer needs to be in attendance with him to make any necessary last-minute adjustment, assist with communications, and guide the cameraman when, peering in the viewfinder, his view of nearby people is restricted.

ELECTRONIC CONTROLS. The improved circuit stability of the camera channels and the removal of unwanted cross-coupling effects enable most of the electronic controls to be pre-set at line-up time. The channel operator then concentrates on obtaining the best exposure for the area of interest on the picture. At outdoor events, the illumination and scene contrast vary with the time of day, the weather, and where the area of interest is placed. For example, in misty conditions at a racecourse grandstand position, the contrast is low at the starting point and increases towards the finish. The aim here is to produce the clearest and most recognizable picture throughout the race. The operator achieves this by choice of gamma correction circuit, control of exposure and black level, and with due regard for noise and definition. According to circumstances, he also uses suitable colour and/or neutral filters mounted on a separate turret between the lens and the photocathode to obtain the most satisfactory picture effect from the equipment.

LENSES. Lenses are traditionally mounted on turrets, but some new designs of camera provide only for zoom lenses. These lenses are now extensively used and they are very valuable devices, but at wide angles they have a greater minimum object distance, a greater length and a smaller maximum aperture than corresponding fixed focal length lenses. Though camera cost, weight and mechanical complexity are reduced, there is a compensating tendency to add electrical servo-systems for the control of zoom and focus.

The benefits of servo-control derive mainly from smoother action. If a rate-following system is used for zoom control, a very smooth and virtually imperceptible slow zoom action can be achieved. An example of this is a twist-grip control by which the direction and speed of zoom depend on the direction and degree of twist applied to the control, which has a spring return. It is inherent in a zoom lens that the direct focus control operates coarsely at the narrow-angle end of the zoom travel and sharply towards the wide-angle end. This effect can be much reduced by applying a variable correction to the focus servo-system derived from information from the zoom servo-system on the position of the zoom travel.

It can be said that the OB cameraman uses the zoom control more as an operational on-air control than does the studio cameraman, and the focus control correspondingly less.

This results in some departures in the preferred form and position of the controls in the two cases. An accessory shot-box unit can be provided giving a number of pre-set zoom angles which can be set during rehearsal and selected by push buttons under the control of the cameraman, offering the effect of a set of turret lenses.

A problem with any design of camera is the need for quietness of operation, since whirrs and clicks distract either the viewing audience, the audience at the event who may have paid for their admission, or the performers.

Following the evolution of microwave techniques for radar, similar radio links are in wide use for television. With a parabolic dish aerial system at the transmitter and receiver, a high aerial gain is obtained such that the input power need be of the order of only 1 watt for point-to-point links up to 40 miles on a virtual line-of-sight path. These frequencies are less suitable than the lower frequencies for links from moving radio cameras or roving eyes, owing to increased multipath propagation effects and increased absorption from nearby fixed or moving objects.

New designs are becoming available which make use of solid state devices to the complete exclusion of thermionic tubes. While these make for increased portability and operation from batteries, the stability and reliability are also greatly improved and, once they are installed and set up, they require very little attendance.

MOBILE VIDEOTAPE RECORDING. The equipments in general use are studio-type four-head machines with their ancillaries mounted in medium-sized vans. Similar machines which can utilize the same tapes are in use in studio centres and hence no editing or replay difficulties arise, and the normal high standard of recording can be achieved. A single machine is usually sufficient, provided that a replay can be made on site to ensure that the recording has been successful. Two machines are, however, preferable.

In the newer recording machines which aim at smaller size and lower power consumption, the single head helical scan or slant track machines are making good progress.

While their present recorded quality limits them to backing up a four-head recording machine at a studio centre, they are approaching the point where they can be employed in OBs. But different makes of helical scan machines employ different recording standards, and the joining and editing used with transverse recorded tapes is not possible owing to the extended recording track; electronic editing must be employed with transfer to another tape. The tendency is towards further improvement and standardization. A.E.C.

See: *Camera* (TV); *Lighting equipment* (TV); *News programmes; Power supplies; Studio complex* (TV); *Videotape recording.*

Out-Take. Discarded take in the assembly stage of editing a film.

Over-Exposure. Exposure of a sensitive photographic material in a camera or printer in such a way that the intensity of light and/or the time of exposure are too great for satisfactory reproduction of the tonal range of the subject. With negative and reversal materials, over-exposure leads to the loss of detail in the lighter and highlight areas of the scene, while an over-exposed print from a normal negative appears much too dark.

In sensitometric terms, over-exposure means that too much of the tonal range of the subject falls in the upper region of the characteristic curve, and with negative materials the shoulder portion of this curve as it approaches the maximum density reduces tonal differentiation.

With colour materials, over-exposure can give rise to distorted colour rendering, particularly of white surfaces and other bright objects.

See: *Exposure control; Image formation; Sensitometry and densitometry.*

□**Overlay.** Technique of superimposing television pictures. Overlay is a similar technique to inlay but, in this case, the keying signal is generated from the fill picture. If the object (or person) to be overlaid on to another background is carefully lit against a black background, a large difference in signal level is obtained between the wanted and unwanted information. This signal can be suitably amplified and clipped in such a manner that the signal consists of a number of square waves, zero level for unwanted signal and peak level for wanted signal.

In colour television, by taking the output of one of the three chrominance channels in a colour camera, a signal can be obtained for overlay, for use in electronic special effects. The value of the signal as a key can be enhanced by the use of coloured filters in the luminaires or lighting units.

See: *Special effects* (TV).

⊕**Overloading.** Impression on a transmission □system of a signal of greater amplitude than it is designed to carry. The consequent deformation or distortion of the signal produces effects which are more or less objectionable to the eye in a video system and to the ear in an audio system.

See: *Picture monitors; Sound recording.*

□**Overscan.** The scanning of a greater area of the camera target than is normally used, effected by increasing the scanning currents, and hence the magnetic fields.

See: *Camera* (TV).

□**Overshoot.** Waveform distortion appearing on a reproduced picture as a narrow white outline at the extreme right-hand edge of dark areas, and conversely a thin dark outline at the right-hand boundary of light areas.

A positive pulse reproduced by a video amplifier having a level amplitude response and linear phase response has a square-topped form. If, however, the phase response at or just beyond the upper end of the passband departs from strict linearity in an upwards direction, i.e., higher frequencies are delayed with respect to lower frequencies, the front edge of the pulse rises too far, forming a pip, and the trailing edge

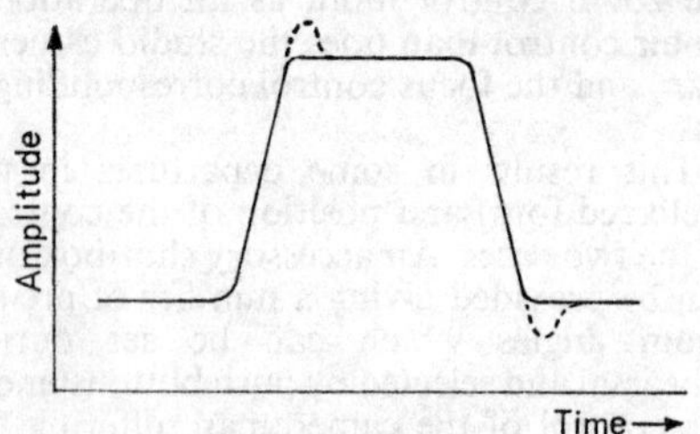

OVERSHOOT. Correctly formed pulse shown as solid line. When overshoot is present, leading edge of pulse (*left*) rises too far and forms a pip (*dotted*), while trailing edge (*right*) falls too far, causing a dimple.

falls too far, causing a dimple. Overshoot is then said to occur, and can result from introducing into an amplifier certain correction circuits for the purpose of extending the level portion of its passband.

See: *Undershoot.*

⊕**Oxidation.** Chemical reaction increasing the oxygen content of a chemical compound. Many developing agents used in photographic processing readily react with the oxygen of the air, and are so rendered useless for their purpose, while the resultant compounds may have adverse effects on the desired developing reaction. It is therefore essential to minimize contact between the developer solution and the air in mixing, storage and circulation; where agitation of the solution is provided by the gas-burst method an oxygen-free gas such as nitrogen must be used.

See: *Processing.*

Painted Matte Shot. Trick effect in which a part of the image which has been photographed in reality is combined with a painted scene to complete the effect. It differs from a glass shot in that the two components are usually combined by double exposure at different times, the portion of the image required for the painting having been kept unexposed by use of a matte at the time of shooting the live action so that it can be exposed later to the painted scene.

See: *Art director; Special effects* (FILM).

PAL COLOUR SYSTEM

PAL (Phase Alternation Line) is a modification of the NTSC system first disclosed in the Federal Republic of Germany in 1963 by Dr. W. Bruch of Telefunken, and mitigates a weakness of that system. In an NTSC transmission, the hue of a picture area is described by chrominance phase angle, so that phase distortions result in errors of hue in the receiver. Errors of 5° are perceptible; and those greater than 20° are obvious. Since phase distortions are common in landline and other long transmission links, a method of eliminating hue changes is very desirable.

BASIC PRINCIPLES

Briefly, in the PAL system, one of the two colour-difference signals is reversed in polarity during alternate lines. In a simple type of PAL receiver, chrominance signal phase distortion is made to produce colour errors of opposite sign during successive lines and, if the errors are not too large, the eye averages them out and sees the correct colour. In a more sophisticated type of receiver, a delay line is employed to hold the chrominance signal for one line so that it may be combined electrically with the signal for the next line. In this way, an average signal is derived which cancels out much larger phase errors.

CORRECTING PHASE DISTORTION. In the standard nomenclature, E'_Y designates the luminance signal, and the two chrominance signals are $(E'_R-E'_Y)$ and $(E'_B-E'_Y)$, the subscripts R and B denoting red and blue respectively.

The main modifications to the NTSC system consist of reversing the phase of the $(E'_R-E'_Y)$ chrominance component on

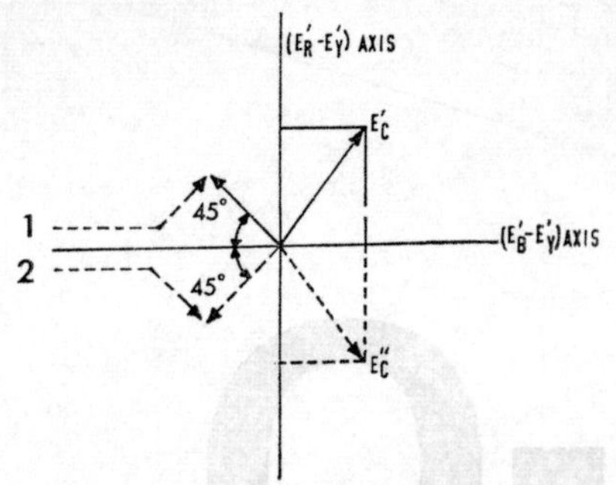

PHASE ALTERNATION OF $(E'_R - E'_Y)$ COMPONENT. This component is reversed in phase on successive lines, and the colour burst phase lags and leads the NTSC burst phase by 45° on successive lines. 1. Burst for E'_C. 2. Burst for E''_C.

successive lines and transmitting bursts which alternately lag and lead the NTSC burst phase by 45° on successive lines.

Thus, during one line, the chrominance phase is identical to that obtained with the NTSC system; the burst phase, however, lags by 45°. Let E'_C represent the chrominance vector during this line. During the succeeding line the $(E'_R - E'_Y)$ chrominance component is reversed, the $(E'_B - E'_Y)$ component remaining unaltered and the chrominance vector is now E''_C; the burst phase, during this line, leads by 45°. By means of a further process at the decoder, in which the $(E'_R - E'_Y)$ component of E''_C is again reversed, the vector E''_C is, ideally, made to coincide with E'_C. However, should a phase error between chrominance and burst be introduced in the path from coder to the decoder, the hue error introduced during one line is followed by an equal and opposite hue error on the succeeding line.

In the diagram, OA is the chrominance vector during one line (i.e., E'_C); this vector has an $(E'_R - E_Y')$ component OL and an $(E'_B - E'_Y)$ component OX. OB is the chrominance vector during the succeeding line (i.e., E''_C); OM is the $(E'_R - E'_Y)$ component and OX is again the $(E'_B - E'_Y)$ component. It is obvious that reversing the $(E'_R - E'_Y)$ component of OB in the decoder results in the vector OA (since OM=OL).

When distortion causes a chrominance relative phase shift θ, the vector during the first line is OC and during the second line is OD. The $(E'_R - E'_Y)$ component of OC is ON and that of OD is OP; the $(E'_R - E'_Y)$ components are OY and OZ respectively. Reversing the $(E'_R - E'_Y)$ component of OD at the receiver gives the component OR; together with the $(E'_B - E'_Y)$ component OZ, this results in the vector OE. Thus the hue distortion on succeeding lines is in equal, but opposite, directions. Therefore, if means are provided in the decoder for obtaining the average of the chrominance phases on successive lines, the effect of chrominance phase errors is substantially zero.

Although the NTSC system utilizes the E'_I and E'_Q colour difference signals to form the coded signal, the PAL signal is formed using $(E'_R - E'_Y)$ and $(E'_B - E'_Y)$. The chrominance signals corresponding to $(E'_R - E'_Y)$ and $(E'_B - E'_Y)$ are weighted in

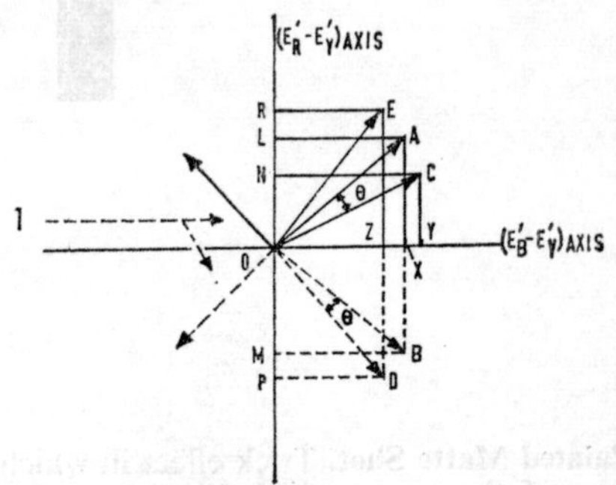

EFFECT OF CHROMINANCE PHASE ERROR. Phase error between chrominance and burst results in hue error during one line being cancelled by opposite error in succeeding line. 1. Bursts.

amplitude by the coefficients k_1 and k_2, as in NTSC. This is permissible for PAL, because cross-talk from one colour difference signal to the other during one line is exactly opposed by the corresponding cross-talk during the succeeding line. Thus an averaging process in the decoder can eliminate the effect of such cross-talk and, in principle, both colour difference signals may be transmitted using asymmetric sideband modulation of the subcarrier and both may describe medium-fine colour detail.

CODING AND DECODING. The PAL coder is very similar to the NTSC form. However, it includes a switch that reverses the phase of the subcarrier fed to the $(E'_R - E'_Y)$ balanced modulator and arrangements for adding to the inputs of both balanced modulators, burst-keying pulses of equal amplitude; thus the $(E'_R - E'_Y)$ chrominance component is reversed in phase during successive lines and the burst phase alternately lags and leads the phase of $-(E'_B - E'_Y)$ by 45°.

A decoder for the PAL system can have two main forms. The first is very similar to

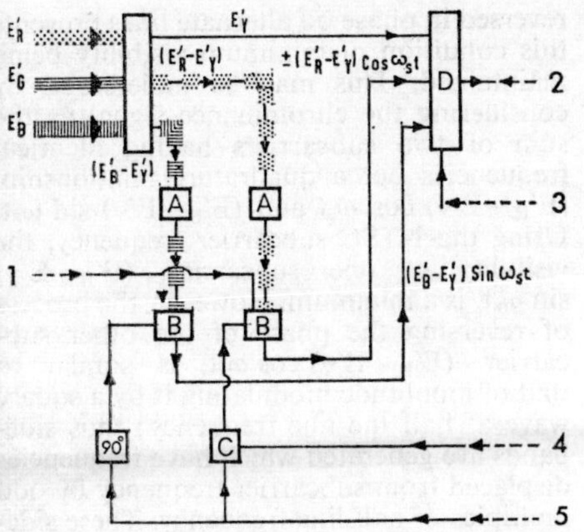

PAL CODER. A. Low-pass filters. B. Balanced modulators. C. Phase-reversing switch. D. Adder. 1. Burst-keying pulses. 2. Composite colour signal. 3. Sync. 4. Line pulses. 5. Subcarrier. The phase reversing switch, C, constitutes the principal difference from an NTSC coder.

an NTSC decoder. However, it includes a phase-reversing switch, in the circuit supplying subcarrier to the $(E'_R-E'_Y)$ synchronous detector, which corrects for the phase reversals in the received $(E'_R-E'_Y)$ chrominance component.

As in an NTSC decoder, a source of demodulating subcarrier is provided by a burst-locked oscillator. However, although the burst phase alternately lags and leads the $-(E'_B-E'_Y)$ axis by 45°, the speed of response of the oscillator control servo is

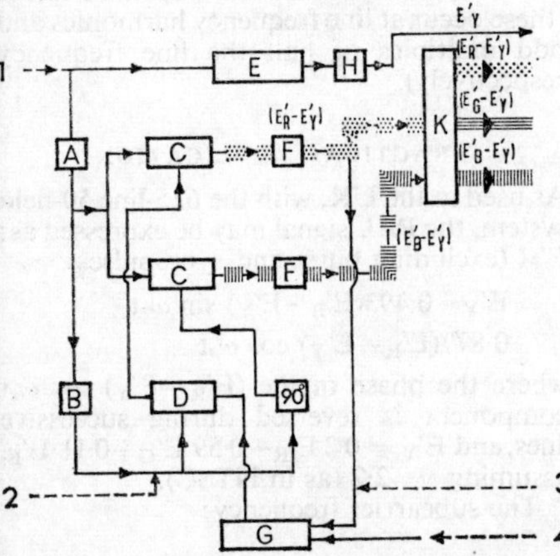

SIMPLE PAL DECODER. A. Band-pass filter. B. Burst-locked oscillator. C. Synchronous detector. D. Phase-reversing switch. E. Band-attenuating filter. F. Low-pass filter. G. Switch-operating circuit. H. Delay. 1. Composite signal. 2. Subcarrier. 3. Burst pulses. 4. Line pulses.

such as to adjust the oscillator phase according to the average burst phase, i.e., the $-(E'_B-E'_Y)$ axis. Thus, during the burst, the output of the $(E'_B-E'_Y)$ detector consists of identical negative pulses, once per line. If the phase-reversing switch is operating correctly, the output from the $(E'_R-E'_Y)$ detector during the burst consists of identical positive pulses.

The operation of the decoder phase-reversing switch is primarily carried out by means of line pulses but any mis-operation

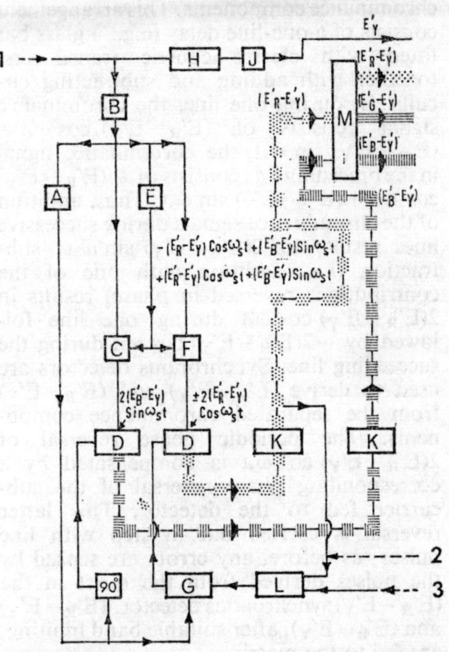

DELAY-LINE PAL DECODER. A. Burst-locked oscillator. B. Band-pass filter. C. Adder. D. Synchronous detector. E. One-line delay. F. Subtractor. G. Phase-reversing switch. H. Band-attenuating filter. J. Delay. K. Low-pass filter. L. Switch-operating circuit. M. Matrix. 1. Composite signal. 2. Burst pulses. 3. Line pulses.

of the switch is sensed by the polarity of the pulses derived from the burst in the $(E'_R-E'_Y)$ synchronous detector. This form of decoder is relatively simple. In the presence of a chrominance phase shift the colour picture displayed may show horizontal stripes in which alternate lines of a field are of different colour; when viewed at a distance the combination of the two colours approximates to the intended colour.

The second form of decoder performs electrically this process of averaging the positive and negative colour errors that may occur on successive lines. It incorporates a one-line delay which is also used to separate the $(E'_R-E'_Y)$ chrominance component from the $(E'_B-E'_Y)$ chrominance component.

The luminance signal is derived, as before, from the composite colour signal by means of a band-attenuating filter. The complete chrominance signal, after extraction by a band-pass filter, passes to the burst-locked oscillator and to an arrangement for separating the $(E'_R-E'_Y)$ and $(E'_B-E'_Y)$ chrominance components. This arrangement consists of a one-line delay (e.g., a glass bar fitted with electro-acoustic transducers) together with adding and subtracting circuits. If, during one line, the chrominance signal consists of $(E'_R-E'_Y)\cos\omega_s t+(E'_B-E'_Y)\sin\omega_s t$, the chrominance signal in the preceding line consists of $-(E'_R-E'_Y)\cos\omega_s t+(E'_B-E'_Y)\sin\omega_s t$. Thus, addition of the chrominance signals during successive lines results in $2(E'_B-E'_Y)\sin\omega_s t$; subtraction (i.e., addition with one of the contributions reversed in phase) results in $2(E'_R-E'_Y)\cos\omega_s t$ during one line followed by $-2(E'_R-E'_Y)\cos\omega_s t$ during the succeeding line. Synchronous detectors are used to derive $(E'_R-E'_Y)$ and $(E'_B-E'_Y)$ from the separated chrominance components. The periodic phase reversal of $2(E'_R-E'_Y)\cos\omega_s t$ is compensated by a corresponding phase reversal of the subcarrier fed to the detector. This latter reversal is carried out in step with line pulses; as before, any errors are sensed by the pulses derived from the burst in the $(E'_R-E'_Y)$ synchronous detector. $(E'_R-E'_Y)$ and $(E'_B-E'_Y)$, after suitable band limiting, are fed to the matrix.

During any one line, the $(E'_R-E'_Y)$ and $(E'_B-E'_Y)$ chrominance signals are the average of the $(E'_R-E'_Y)$ and $(E'_B-E'_Y)$ chrominance signals transmitted during successive lines. This approximately corrects for the errors arising from chrominance phase shift; however, for large phase errors, the average chrominance signals so obtained are reduced in amplitude and some desaturation of the picture results.

SUBCARRIER FREQUENCY. In the NTSC system, the subcarrier frequency is an odd multiple of half the line frequency. This results in minimum subcarrier visibility on the screen of the black-and-white receiver. However, if the same subcarrier frequency is used for the PAL system, the fact that the $(E'_R-E'_Y)$ chrominance component is reversed in phase on alternate lines prevents this condition of minimum visibility being maintained. This may be understood by considering the chrominance signal as the sum of two subcarriers having identical frequencies but a quadrature relationship, $(E'_R-E'_Y)\cos\omega_s t$ and $(E'_B-E'_Y)\sin\omega_s t$. Using the NTSC subcarrier frequency, the visibility of one subcarrier, $(E'_B-E'_Y)\sin\omega_s t$, is a minimum. However, the process of reversing the phase of the other subcarrier, $(E'_R-E'_Y)\cos\omega_s t$, is similar to that of amplitude modulating it by a square wave at half the line frequency; thus sidebands are generated which have frequencies displaced from subcarrier frequency by odd multiples of half line-frequency. These sidebands have frequencies that are harmonics of line frequency and, therefore, have maximum visibility. Thus the use of the NTSC subcarrier frequency results in a variation in the subcarrier visibility dependent upon the hue of the colour; the hue determines the relative contributions of $(E'_R-E'_Y)$ and $(E'_B-E'_Y)$.

As a consequence, the PAL system employs a subcarrier frequency that is displaced in frequency from the nearest harmonic of line frequency by about one-quarter of the line frequency; in the case of the 625-line system, the displacement is one-quarter the line frequency minus half field frequency. It can be shown that the patterns resulting from $(E'_R-E'_Y)\cos\omega_s t$ and $(E'_B-E'_Y)\sin\omega_s t$, using this frequency, are equally visible and that their visibilities lie about midway between the maximum and minimum values (these occur at line frequency harmonics and odd multiples of half the line frequency respectively).

PRACTICAL APPLICATION

As used in the UK, with the 625-line 50-field system, the PAL signal may be expressed as: E'_M (excluding burst and sync pulses) =

$$E'_Y+0{\cdot}493(E'_B-E'_Y)\sin\omega_s t\pm 0{\cdot}877(E'_R-E'_Y)\cos\omega_s t$$

where the phase of the $(E'_R-E'_Y)\cos\omega_s t$ component is reversed during successive lines, and $E'_Y=0{\cdot}3\,E'_R+0{\cdot}59\,E'_G+0{\cdot}11\,E'_B$, assuming $\gamma=2{\cdot}2$ (as in NTSC).

The subcarrier frequency:

$$f_s=\frac{\omega_s}{2\pi}=(284-\tfrac{1}{4})\,.\,f_{LINE}+\tfrac{1}{2}\,.\,f_{FIELD}=4{\cdot}43361875\text{ Mc/s.}$$

The bandwidth of E'_Y is 5·5 Mc/s and the bandwidths of $(E'_R - E'_Y)$ and $(E'_B - E'_Y)$ lie within the range 1·0–1·5 Mc/s.

The burst signal consists of 10 cycles of subcarrier. A.V.L.

See: *Colour principles; Colour principles* (TV); *Colour systems; NTSC colour system; SECAM colour system; Sequential system.*

Literature: *The PAL Colour Television System* (*Radio & Electronic Engineer*, Mar. 1967); *PAL—Selected Papers* (*Telefunken-Zeitung*, Mar. 1965 and June 1966).

Pan. Pivotal movement of the camera in a horizontal plane. Sometimes the term is used

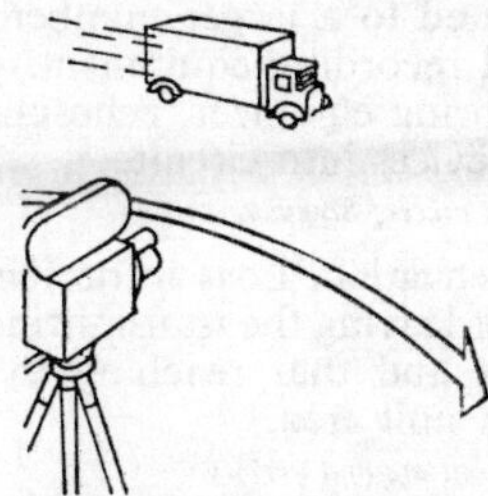

generally to describe movements of the camera in any plane.

See: *Camera; Cameraman.*

Panavision. Trade name of a series of wide-screen processes. Panavision uses 35mm film with anamorphic lens systems on camera and projector; the picture images have a lateral compression of 2 : 1. In Super-Panavision, the original photography is on 65mm negative without distortion but 35mm prints are made by anamorphic reduction. Ultra-Panavision uses 65mm negative anamorphically photographed with a 1·25 : 1 compression.

See: *Wide screen processes.*

Panchromatic. Photographic emulsions having an effective sensitivity over the whole of the visible spectrum from blue to red.

Pan Pot. During the dubbing of stereophonic sound tracks, it may be necessary to move a single (i.e., non-stereophonic) sound source across a set of loudspeakers, say from left to right of screen. The resemblance to camera panning led to the use of the term pan pot for the potentiometer by which this distribution of sound is effected.

See: *Sound recording* (FILM); *Sound reproduction in the cinema; Stereophonic sound processes.*

Pantograph. Device for suspending a studio light so that it may be moved up and down without effort while maintaining the horizontal and vertical axes unchanged. Usually made on the lazy-tongs principle using springs as counterweights.

See: *Lighting equipment* (TV).

⊕**Papering.** Inserting markers in a roll of film to indicate to the laboratory the start and end of sections which are to be printed or

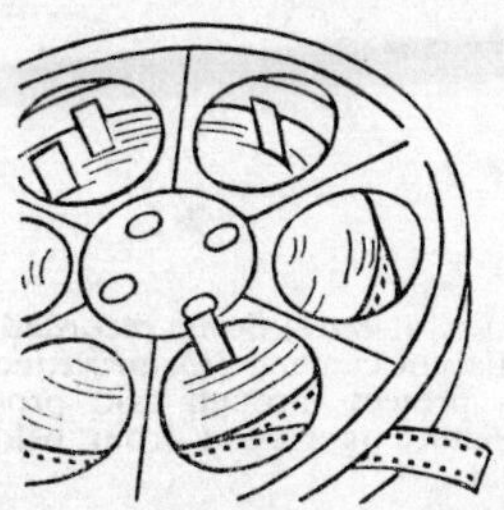

duplicated. Papers held by clips are no longer used because of the risk of damage to the film.

□**Parabolic Aerial.** Simple aerial or waveguide horn placed at the focus of a metallic parabolic reflector, directing energy at it. The reflector produces a beam in a similar way to the reflector of an electric torch.

See: *Transmitters and aerials.*

⊕**Parallax.** The apparent displacement of an observed object due to a change in the position of the observer. Binocular parallax results from the fact that the eyes are set about 2½ in. apart, so that the observer, even when motionless, has in fact two positions from which objects can be viewed. Provided an object is not too distant, the angle which it subtends at the eyes helps to determine its distance from the observer.

When the observer moves sideways, nearer objects appear to pass across more distant ones, and their relative speed is another way of estimating their respective distances. This is often called motion parallax, and is effective with a single-eyed observer, and thus with a motion picture or television camera.

Parallax is an important problem in those types of film camera viewfinder which are displaced sideways (less often vertically) from the lens-to-film axis. The difference of viewpoint then requires angular correction which is greater the shorter the object distance, and at very short distances is accurate only for a single object plane.

See: *Camera* (FILM); *Visual principles.*

⊕**Parallax Panoramagram.** A form of parallax stereogram in which multiple left and right images replace the single images of the simpler system. As a result, sideways movements in the observer's position bring fresh aspects of the subject into view, giving a more realistic appearance than a simple two-view stereogram, in which distortion results from change in the observer's position because movement parallax in the scene is non-existent.

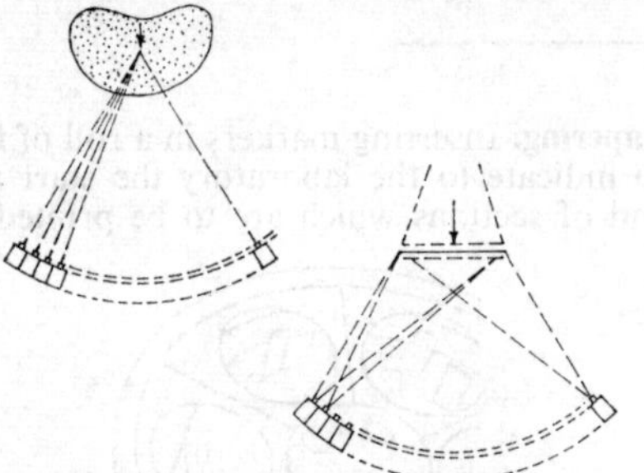

IVES' PATENT. (*Above*) Scene recorded by some 50 closely adjacent cameras; (*below*) equal number of projectors project through grid producing strip images. Picture is viewed from behind through similar grid.

See: *Stereoscopic cinematography.*

⊕**Parallax Stereogram.** A type of stereogram which may be viewed without optical devices, patented by the American inventor, Frederic Ives (1856–1937) in 1902. The camera imaged two views of the subject, taken from positions 2½ in. apart, on to a plate. In front of it was interposed a vertical grid of alternate opaque and transparent bands. The result was that the strips of the left- and right-hand views of the subject were interleaved, and when the plate was developed and a print made, it could be viewed through a similar grid to give a three-dimensional effect.

This simple parallax stereogram, limited to a single viewer at a fixed distance, was greatly elaborated in the 1940s and '50s, notably in Russia, in the hope of better adapting it to large cinema audiences.

See: *Stereoscopic cinematography.*

⊕**Paris, John Ayrton, 1785–1856.** English doctor who described and sold the Thaumatrope in 1826, the first toy employing persistence of vision for its effect.

⊕**Patch.** Small piece of cardboard, celluloid or metal with projecting teeth on either side to engage in film perforations, and of standard film widths. Used as a means of temporarily assembling film before splicing, and also for papering. A patch is also a film or tape overlay to join two lengths without loss of picture or sound modulations.

See: *Laboratory organization.*

⊕□**Patching.** Selection by plug and socket arrangement as in a cord type of telephone switchboard. In a lighting installation, the cords correspond to the dimmer circuits while the sockets each connect to an outlet, so enabling a smaller number of dimmers to be allocated to a larger number of outlets. In sound recording equipment, patching is used to bring equalizer, echo chamber and similar devices into circuit.

See: *Sound mixer; Sound recording.*

□**Path Attenuation.** Loss in decibels between the power leaving the transmitting aerial per unit area and that reaching the receiving aerial per unit area.

See: *Transmitters and aerials.*

⊕**Pathé, Charles, 1863–1957.** French industrialist. Began as a seller of phonographs, becoming an agent for films for the Edison Kinetoscope in 1895. After sponsoring the work of H. Joly in that year, he broke with the inventor and, with his brothers, founded the Société Pathé Frères. The first development of the new company was based upon the Lumière Cinématographe – Pathé producing separate camera and projector mechanisms.

In 1896 M. Grivolas, an industrialist, took over the financial responsibility of the company, establishing the Compagnie Générale des Phonographes et Cinématographes, with Charles's brother Emile responsible for the phonograph development. The organization Pathé set up became one of the largest in the world of the cinema, adding film production and filmstock manufacture to its activities. Among important processes it sponsored were the Pathécolor stencil colouring process and the 9·5mm amateur film system.

□**Pattern Generator.** A generator producing waveforms of simple geometric shape, e.g. sawtooth or square wave. These waveforms or patterns may be used to test TV systems and to trigger electronic special effects.

Test signals usually include a line sawtooth waveform for testing signal amplitude linearity, megacycle bars for measuring the mid- and high-frequency responses, and a 50 Hz square wave for checking low frequency performance. Additionally, the equipment produces a waveform which when displayed on the screen of a picture tube, produces a horizontal and vertical grating pattern to assist in correctly setting-up the scanning linearities of a receiver/ monitor. Each of these waveforms, and several others, may be inserted at will into the signal path.

Trigger signals are used to operate an electronic switch for the purpose of combining two picture sources into one composite whole. In this application, the pattern generator normally consists of line

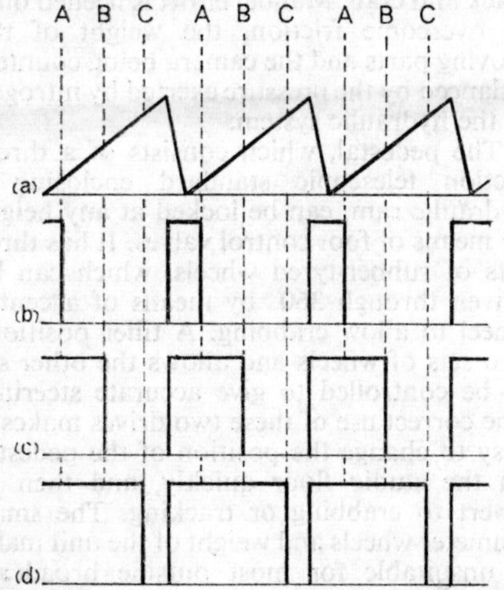

PULSE TRIGGERING. (a) Line sawtooth. (b) Pulse width when triggered at A. (c) Pulse width when triggered at B. (d) Pulse width when triggered at C. Resulting pulses of varying width provide keying signal which changes position from left to right of picture as required by the wipe.

and field sawtooth and triangular waveform generators. Suitable addition of these waveforms is then used to operate a trigger circuit which generates rectangular pulses. The width of the output pulse depends on the part of the sawtooth where the triggering occurs. Varying the triggering by the line sawtooth results in pulses which vary in width and which give a keying signal which changes position from left to right of the picture.

The normal patterns produced by these methods consist of horizontal and vertical wipes, diagonal wipes, rectangular, diamond-shaped or circular wipes.

If the oscillators are triggered at multiples of line and field frequencies, a multiple wipe results, such as a venetian-blind effect. Further effects can be obtained by modulation of the signals with sine waves to produce deckle-edged patterns.

It is also possible to make geometric wipes by means of an inlay table. W.J.C.

See: *Picture monitors; Special effects* (TV).

⊕**Paul, Robert William, 1869–1943.** English instrument maker and engineer; opened a business in Hatton Garden, London, in 1891. Was approached in 1894 with a request to make copies of Edison's Kinetoscope, not patented in Great Britain. As a result of this work, Paul began to devise apparatus of his own.

Collaborating with Birt Acres, he produced a working camera, with which Acres, in 1895, recorded several events of some importance, including the Derby and the opening of the Kiel Canal.

Acres left Paul after a clash of personalities, and Paul continued on his own, designing the Theatregraph projector, which was working by February 1896 and patented on 2nd March. The machine, renamed the Animatographe, was shown on 20th February at Finsbury Technical College, in March at Olympia, and from 25th March on it was installed at the Alhambra Theatre, Leicester Square, London, and used regularly, Paul being paid £11 a show. He marketed the cameras and projectors, over 100 projectors being made and sold, many overseas.

Paul also entered film production work, building, in 1897, an open-air stage at Muswell Hill for pioneering films using trick effects and models. He constructed a studio for film production in 1899, exploiting experiments in trick effects, slow motion and animation. He was a founder member of the Kinematograph Manufacturers Association which, among other things, in 1909 agreed to standardize on the 35mm film dimension. He closed down the motion picture side of his business in 1910 to concentrate on instrument making. His works were acquired in 1920 by the Cambridge Instrument Company.

□**Peak Brightness.** Brightness of a small portion of a cathode ray tube phosphor when excited by a peak white signal.

See: *Phosphors for TV tubes.*

⊕□**Peak Programme Meter.** Device for metering approximately the peak values of a sound signal. Used particularly in Britain as an alternative to the VU meter. When feeding a sound signal to a broadcast transmitter, it is vitally important that programme peaks are not allowed to exceed a certain maximum level.

Clippers within the transmitter ensure that damage is prevented to this equipment but, if these operate, it follows that distortion of the sound waveform will result. A normal sound signal contains many peaks which far exceed the average signal level. If the signal level is reduced to a point where none of these peaks reach clipping level, the transmitter works very inefficiently. In practice, therefore, programme peaks of very short duration are allowed to be clipped, raising the transmitter efficiency to a reasonable level while keeping the distortion low enough to be inaudible.

This line of thought led to the development of the peak programme meter (PPM), which integrates transients in the signal as follows:

1. A pulse of 4 msec duration gives rise to a meter indication which is 80 per cent of the indication which would occur if the peak were repetitive, as in steady state tone.
2. The time taken for the meter indication to fall 24 dB, upon the removal of the signal, is 3 seconds.

The meter is calibrated either in seven steps from 1–7 or in dB relative to 1 mW (dBm) in 600 Ω. When this level is applied to the PPM, it indicates 4 or 0, depending on the scale-marking used. Three calibration marks extend, in a linear fashion, at 4 dB intervals up and down from this reference indication, resulting in minimum and maximum indications of −12 and +12 dBm respectively.

The PPM may be modified by shunting the meter movement with a large capacitor to produce a slugged response. This results in a very slow movement of the meter with changes in programme level. This mode of operation is used for monitoring continuity of programme rather than for the accurate control of programme level. A switch is often provided to introduce the slugged facility when desired. L.S.

See: *Sound recording.*

⊕□**Peak-to-Peak.** Total measurement of a waveform from its most negative to its most positive point.

□**Peak White.** Potential at a point in the signal path which the brightest parts of the transmitted scene are allowed to attain. Other tonal values lie somewhere between peak white level and black level, which represent the two extreme values between which the picture signal can vary.

At those points where the d.c. component is present, the peak white level has a fixed value in the same way as black level is fixed, but at other points where the d.c. component is absent the potential corresponding to peak white is variable, depending on the mean value of the picture signal.

See: *Television principles; Transmission and modulation.*

□**Pedestal.** A column type of camera mount.

Early camera pedestals employed a system of springs and counterbalance weights to support the camera, allowing for ease of movement. Difficulty in obtaining smooth operation, however, resulted in the introduction of the hydraulic pedestal, which is the most common form of camera mounting in use in television broadcasting today. It provides a strong, stable mounting, easy to track and crab. Manual effort is needed only to overcome friction, the weight of the moving parts and the camera being counterbalanced by the pressure exerted by nitrogen in the hydraulic system.

The pedestal, which consists of a three-section telescopic standard enclosing a hydraulic ram, can be locked at any height by means of foot control valves. It has three sets of rubber-tyred wheels, which can be driven through 360° by means of a centre wheel to allow crabbing. A tiller positions two sets of wheels and allows the other set to be controlled to give accurate steering. The correct use of these two drives makes it easy to change the position of the pedestal on the studio floor quickly, and then to revert to crabbing or tracking. The small diameter wheels and weight of the unit make it unsuitable for most outside broadcast purposes.

See: *Camera* (TV); *Designer; Studio complex* (TV).

□**Pedestal.** In the transmitted television waveform, an artificial increase in height of the black level.

For many areas, but especially in Britain, the pedestal was established as a safety

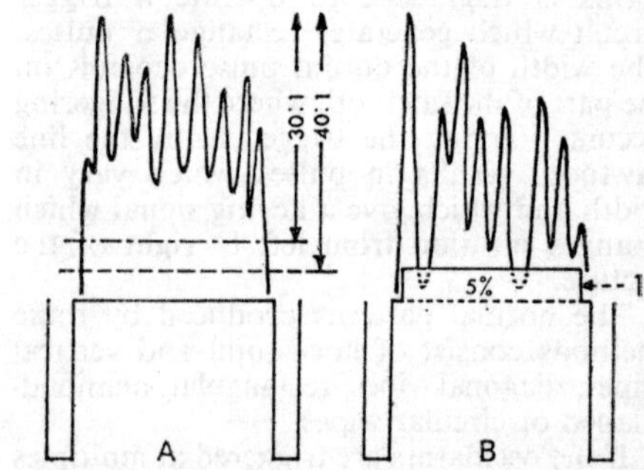

PEDESTAL. Artificial increase in height of black level, to impose limit on contrast range. 1. Pedestal. A. Video signal with indication of 30 : 1 and 40 : 1 contrast ranges typical of limits in colour and black-and-white respectively. Excursions to black level would come down to the solid line, an impossible contrast range. B. Video signal with 5% pedestal allowing no excursions below this level and imposing an artificial limit to the contrast range.

margin in the black region of the picture to avoid unacceptable blacking out in shadow areas. Today a fixed base level (which produces black crush) is no longer used. The procedure is to fix acceptable lighting contrast, usually 30 : 1 for colour and 40 : 1 for black and white, together with a working gamma usually about 0·5. This sets a black level about 11 and 16 per cent (black-and-white and colour) above true black and the signal is only allowed to fall below these levels in areas of exceptionally deep shadow.

See: *Television principles; Transmission and modulation.*

Pellicle. Extremely thin semi-reflecting film which reduces to vanishing point the double reflections present when a thicker support is used. A collodion film stretched on a steel frame on to which an aluminium reflecting layer is evaporated is a typical form used in telecine multiplexer systems.

See: *Telecine.*

Penthouse Head. Magnetic sound reproducer head assembly built above the main picture head of a 35mm cinematograph projector to play films with four magnetic stripes for stereophonic presentation, and often installed as a later addition to standard optical sound machines. Since the film reaches the picture gate of the projector after it has passed through the magnetic pick-up point, the synchronization of magnetic sound must be recorded 28 frames behind the corresponding picture.

See: *Projector; Sound reproduction in the cinema.*

Pentode Electron Gun. Cathode ray tube electron gun which has a cathode, control grid and three successive anodes. Employed in some receiver picture tubes.

See: *Receiver.*

Perforations. Holes along one or both edges of strips of motion picture film, used for its location and movement through mechanisms.

See: *Film dimensions and physical characteristics.*

Periodic Noise. Repetitive undesired signal in an audio or video network.

See: *Noise; Transmission networks.*

Perry, John, 1850–1920. British scientist, born in Ulster. With William Edward Ayrton he put forward one of the earliest proposals for "seeing by electricity". Was a professor of Engineering in Japan (1875–79) where he met Ayrton, with whom he collaborated in many researches and inventions. He was Professor of Mathematics and Mechanics at the Royal College of Science, South Kensington (1896–1914).

⊕**Persistence of Vision.** The action of the eye's
□light receptors does not cease instantaneously when the light stimulus is removed but rapidly diminishes over a brief period of time. If therefore a short light stimulus is repeated sufficiently rapidly, the eye is given the sensation of continuous illumination. It is this phenomenon which permits the illusion of movement to be provided in cinematography, successive stationary pictures being presented at a sufficient frequency to allow the eye to accept them as a continuous presentation.

See: *Visual principles.*

□**Persistent Phosphor.** Phosphor in which the image persists after excitation and whose decay law is such that a usable or viewable image remains for television purposes over the intervals commonly encountered.

See: *Phosphors for TV tubes.*

□**Phantom.** If balanced-pair circuits are carried on two pairs of wires, an extra phantom circuit can be provided by connecting between centre taps on transformers feeding these pairs. If the transformers and circuits are perfectly balanced, no crosstalk occurs between the three circuits.

See: *Hybrid.*

□**Phase Comparator.** Circuit which detects a difference of phase between two signals of the same nominal frequency, as in automatic tuning circuits.

□**Phaseless Boost.** Simple high frequency boost circuits can produce phase distortions in the signal. More complex circuits, however, are free from this defect and are known as phaseless boost circuits.

See: *Videotape recording (colour).*

□**Phase Lock.** Two devices are phase locked when they are synchronized not only in speed but also in time. Used in synchronizing generator circuitry.

See: *Picture locking techniques; Synchronizing pulse generators.*

□**Phase-Locked Loop.** Feedback system in which the phase of the signal is the most important factor. Employed in picture locking and videotape recorder (VTR) circuits.

See: *Synchronizing pulse generators; Videotape recording.*

□**Phosphor Grain.** Granular structure of the phosphor of a cathode ray tube, visible as a disturbance of the resolution of a picture.

□**Phosphor Saturation.** State where further excitation of a phosphor produces no increase of light output.

□**Phosphor Screen.** Coating of chemical substances deposited on the face of a cathode ray tube. When bombarded by electrons, the substances emit light.

□**Phosphors for Television Tubes.** The phosphor screen in a cathode ray tube converts the electron beam energy into radiant energy of various wavelengths. In television, the intention is to convert as much of it as possible into visible light of appropriate colours. The property involved, cathodoluminescence, is one aspect of phosphorescence. Light can be excited from phosphors by a wide range of means, e.g., bombardment by photons, X-rays, UV radiation or atomic particles, as well as by electron beams. The basic method is to raise atoms to a number of levels corresponding to higher energy so that they release light quanta of the characteristic type as they revert to lower levels.

Phosphors emit in several colours and the persistence of the emission also varies widely. In television, a short persistence is desirable. However, some persistence helps to reduce flicker of bright portions of the tube face and may be tolerated, if resolution does not suffer excessively through the afterglow. Most substances used as phosphors are crystalline and the crystals are often doped with small quantities of impurities known as activators. An early example was the hexagonal form of zinc sulphide (crystallized at 1,150°C) doped with copper or silver. In this case, the zinc sulphide does not exhibit significant fluorescent properties without the activator. Mixtures of different crystals giving separate colours may be used to get some desired overall effect, e.g., doped zinc sulphide (blue) and zinc cadmium sulphide (yellow) can give a fair approximation to white when used in the correct proportions.

In the selection and preparation of the phosphor material the following points have to be considered: purity and degree of doping; colour and brightness; susceptibility to ion burns and electron burns; ease of coating; and life of finished phosphor.

The material may be deposited on the glass surface of the television tube by spraying, settling, slurrying or evaporation at high temperature.

Spraying produces a dry material, and slurrying needs air drying. Settling was very popular at one time with black-and-white tubes and consists of a balance between settling particles and the chemical formation of a gel to hold them well enough to decant off surplus liquid without damage. The coating can then be dried. Evaporation has recently been used, mainly in the production of transparent phosphors. The luminescent efficiency of this last technique, however, is not high. A binder has to be incorporated in the crystalline mixture to ensure adherence to the glass, except in the evaporation technique, which needs a suitable substrate to which the crystals adhere. The finished coatings are inspected for uniform brightness, efficiency and faults such as pin holes. It is often possible to improve the performance of a phosphor by backing the material with a conducting layer such as aluminium, stannic oxide or cupric iodide. These prevent charges building up and, in the case of aluminium, almost double the brightness by reflection of the emitted light from the tube face. Modern black-and-white phosphors can work comfortably at 60 ft./lamberts and are often used up to 100 with a reasonable life. At beam currents of milliamps, however, damage is easily done to the phosphor by local overheating. At a steady EHT voltage, the relationship between voltage on the control grid and brightness approximates to a square law.

In colour tubes, obtaining a good red phosphor has always been difficult, but recently the use of rare earth elements such as yttrium vanadate in the phosphors has almost doubled the brightness.

The Radio Manufacturers' Association (of America) has at various times published characteristics of well-established phosphors under various RMA numbers – P1 upwards

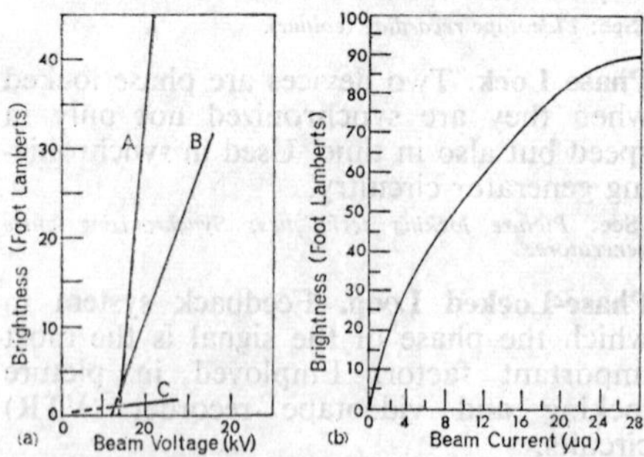

PHOSPHOR SCREEN BRIGHTNESS. A. Aluminium coated screen. B. Cupric iodide coated. C. Uncoated. Brightness of a transparent zinc sulphide (ZnS) screen as (a) function of beam voltage, (b) function of beam current.

These cover not only television use but radar and other techniques. They may be found in most specialized works of reference. E.C.V.

See: *Receiver; Shadow mask tube.*

Phot. Metric unit of illumination equivalent to one lumen per square centimetre.

See: *Light units.*

Photocell. Device for converting varying intensities of light into corresponding variations of electrical energy. Photocells make use of one of three effects: photoconductive, or change of resistance under the action of light; photoemissive, or emission of electrons from metallic surfaces; and photovoltaic, or the production of a potential difference across the boundary between two substances in close contact (the barrier-layer cell much used in exposure meters but now often replaced by the photoresistor).

See: *Exposure control.*

Photoconducting Tubes. Common but not wholly accurate description of such tubes as the vidicon and its variants whose resistance effectively varies as light falls on the target.

See: *Camera* (TV).

Photodiode. Semiconductor whose junction is light-sensitive, and can thus be used as a photosensitive device.

Photoflood. Type of incandescent tungsten lamp bulb in which excess voltage (boosting) is applied to the filament, greatly increasing the light output and shortening the life. Colour temperature is also raised. Photofloods are available with built-in reflectors.

See: *Lighting.*

Photometer. Form of precision light meter for the accurate determination of brightness.

Photomultiplier. Photoelectric device which uses a primary photocathode and whose photoelectric emission is amplified by successive secondary emission generation in a series of suitably arranged and charged targets in series. The whole assembly is compactly arranged in one envelope. Gains of several thousand can be obtained with stability and linearity.

A photomultiplier is incorporated in the image orthicon tube. Another application is in flying-spot telecine channels where the low light level produced by the scanning spot requires a high gain in the photoelectric tube to produce a reasonable signal level and acceptable signal-to-noise ratio.

See: *Camera* (TV); *Dynode effect; Electron multiplier; Telecine.*

⊕**Photopic Response Curve.** Graphical repre-
□sentation of the sensitivity of the average human eye to light of different wavelengths or colours.

□**Phototelegraphy.** Transmission of still photographs over a distance by wire or radio. A narrow transmission band is usual, with a time of transmission of the order of a minute or more.

□**Pick-Up Tube.** TV camera tube which converts an optical image into an electrical signal. Typical pick-up tubes are image orthicons, vidicons and Plumbicons.

See: *Camera* (TV); *Telecine; Telerecording.*

□**Picture Element.** Smallest item of information which can be resolved by a television system. It is a function of the number of lines in each picture, and the band width of the video signal.

See: *Television principles; Transmission and modulation.*

□**Picture Inversion.** Reversal of the signal polarity giving a negative picture.

□**Picture Line-Up Generating Equipment.** Equipment which generates a waveform specially designed for rapid and accurate adjustment of the operational controls of a picture monitor. Correct adjustment of the brightness control is essential if the black parts of the picture are to appear as black. The ambient light falling on the face of the picture tube has an influence on the adjustment necessary to achieve black. Incorrect adjustment causes detail in the shadows to be lost if the brightness control is set too low, or failure of any blacks in the picture to appear as black if the control is set too high.

The principle involved in the PLUGE waveform which enables this adjustment to be made is to provide two adjacent vertical bars with a very slight difference in brightness level between each other and the background. The background is arranged to be at pedestal level while one of the bars is 3 per cent above, and the other bar 3 per cent below, the background level.

Correct adjustment of the brightness control results in the −3 per cent bar blending into the background, leaving the +3 per cent bar clearly visible. If neither or both of the bars are visible, then the brightness control is incorrectly set. Additional pulses are associated with the vertical bars to provide a peak white signal. This is to ensure that the tube is passing the normal beam current for an average picture and enables the highlight brightness to be measured as well as the contrast control to be adjusted.

One version of the PLUGE waveform has two white bars at the extreme edges of the picture to enable the aspect ratio to be adjusted as well as checking the brightness and contrast control settings.

Another version of PLUGE includes a broad vertical stripe divided into two regions of definite amplitude. These two regions are arranged to produce peak white and a grey of 50 per cent peak white brightness, thus serving as a check on the contrast range of the picture tube.

The maximum benefit from the use of the PLUGE waveform as a means of setting up picture monitors can be achieved only if the controls are subsequently left untouched and if the monitor itself can be relied upon to maintain black level over a period of several hours. R.C.H.

See: ***Picture monitors.***

□PICTURE LOCKING TECHNIQUES

Picture locking techniques serve to synchronize different and often remote television picture sources (studios, outside broadcasts (OBs), videotape recorders, etc.), so that transitions between them by way of mixes and special effects may be made without disturbance to the received picture.

These techniques for locking together the output pulses from two or more synchronizing pulse generators are often described as genlock, a contraction of generator lock. This term also covers slavelock, in which one or more subordinate stations (slaves) are locked to a central station (the master).

To apply the techniques of mixing and special effects in programme production, it is necessary for all the picture signals to have their picture components occurring over the same time interval for any given line. This is automatically ensured if the associated synchronizing pulse trains are locked together and correctly phased. The function of genlock operations is to achieve this in circumstances where signals would otherwise be unlocked and therefore unsuitable for mixing or special effects.

BASIC PRINCIPLES

Understanding of genlocking methods requires a close examination of what is meant by picture locking. Two or more synchronizing pulse trains are said to be locked when the pulses have precisely the same line and field frequencies, i.e., the same number of line or field pulses occurring in a given time. However, pulses may have the same frequency but yet be displaced in time so that those in each train do not occur at the same instant.

A constant time difference between the pulse trains is termed a phase or timing difference, two signals being said to be correctly timed, or in phase, when the synchronizing pulses occur at the same instant. Since the pulses in two different trains are almost never completely identical in shape or duration, they can only be strictly in phase at one particular point on each pulse. Because the leading edge of the line synchronizing pulse is always used as a time reference, it is a point on this edge (the half-amplitude point) which is normally implied in discussion of line synchronizing pulse phasing.

The field synchronizing pulses of two trains are in phase when the groups of broad pulses occur over the same time interval.

In general, the term synchronization is applied to cover both locking and phasing of two or more synchronizing pulse trains so as to bring them into precise time coincidence.

Both frequency and phase aspects of a waveform are fundamentally measurements in terms of time, although phase is often expressed as a fraction of the period of one cycle of the waveform. The angular units are degrees (1 cycle equals 360 degrees) or radians (1 cycle equals 6·28 radians).

Angular phase differences may be converted into time differences by the following relationship:

$$t = \frac{\phi}{2\pi f}$$

where t is the time displacement in seconds between two waveforms of the same frequency, ϕ is the angular phase displacement in radians, and f is the frequency in Hz.

This relationship is useful because the practice has been established in television of quoting phase differences directly in time, usually in nano-, micro- or milliseconds as appropriate.

STATION SYNCHRONIZATION

Genlock usage is closely tied to station synchronization and to the operational techniques used in programme production

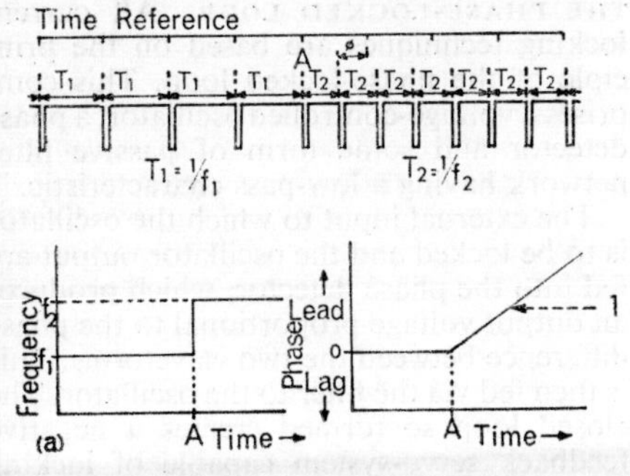

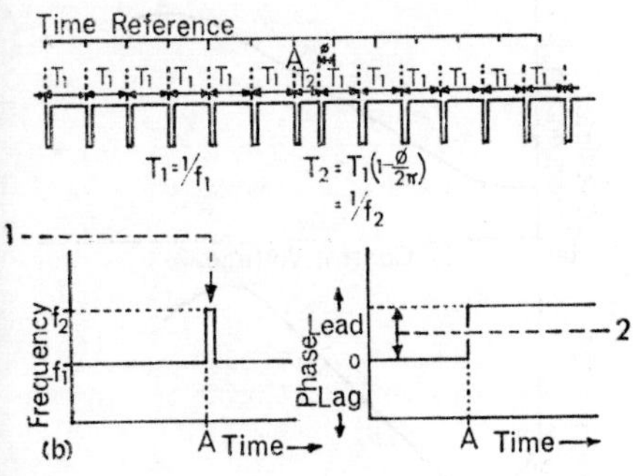

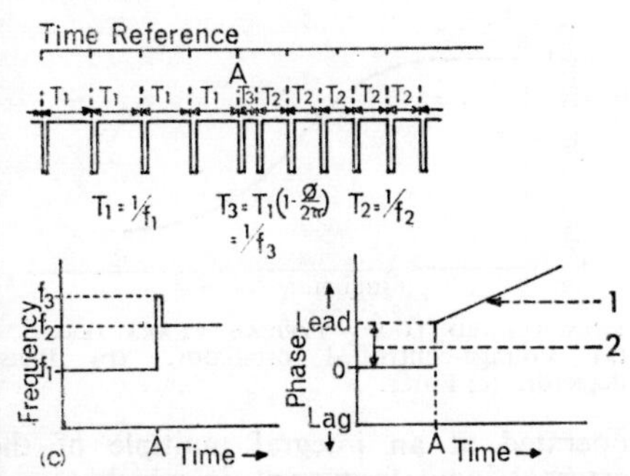

FREQUENCY AND PHASE CHANGES DURING TRAIN OF PULSES. Time reference marks are of frequency f_1; change occurs at point A.

(a) Change in frequency.

1. Slope $= d\phi/dt = 2\pi(f_2 - f_1)$.

(b) Change in phase.

1. $f_2 = \dfrac{f_1}{1-(\phi/2\pi)}$. 2. $\phi = 2\pi\left(1 - \dfrac{f_1}{f_2}\right)$.

(c) Change in frequency and phase.

1. Slope $= d\phi/dt = 2\pi(f_2 - f_1)$. 2. $\phi = 2\pi\left(1 - \dfrac{f_1}{f_3}\right)$

CENTRALIZED PULSE GENERATION. The ideal television signal should have the line and field frequencies constant and be free from sudden changes in phase of either the line- or field-synchronizing pulses. This ideal can be closely approached with a single station by employing centralized generation of all the synchronizing pulses used within the station. The various video signals are all routed to a central or master control point; the different time delays of the signal paths are equalized to bring all pulse trains into coincidence. Video signals can then be mixed or combined by special effects techniques without disturbing the frequency or waveform. This type of station operation also makes possible non-composite video mixing and processing, the synchronizing pulses being added to the video component at the output from master control. When composite working is used, mutilated synchronizing pulses can easily be removed, and replaced by those direct from the synchronizing pulse generator to ensure the best possible transmitted waveform.

OUTSIDE BROADCASTS AND NETWORKS. Many programmes contain material originating at some point outside the station. Common instances are outside broadcasts and network programmes. Each external source must be equipped with a synchronizing pulse generator to feed the other equipment and this is not locked to that at the station unless some form of generator locking system is used. Unlocked operation will prohibit the use of mixing and special effects between station-originated and external signals.

Large television studios have found it necessary to break away from the principle of a single central synchronizing pulse generator for the whole station. Many of them now use considerable numbers of generators, each one feeding certain areas within the station. To permit full integration of station facilities it is necessary to be able to lock all or a selection of the generators together.

LOCKING REQUIREMENTS AND PRINCIPLES

TECHNICAL PERFORMANCE. A synchronizing pulse generator locked to another waveform must still provide pulses conforming to the appropriate television standard. Limits are normally set for the deviation of line frequency and for the rate of change of line frequency, and it is particularly important to ensure that locking systems keep within these limits during the locking period. Special precautions are taken to ensure that system standards are met in the event of failure of the signal to which the generator is locked. To guarantee correct operation of station-mixing equipment, it may be necessary to specify stringent limits on the line phase stability against time, and against changes in frequency. Transient response is important in many applications, in which case oscillatory characteristics should be well damped.

Video recording with the magnetic videotape recorder has affected genlock techniques. Difficulties can arise during both recording and playback. During recording, the machine must be capable of following any change which may occur in line or field frequencies if defects in the recorded signal are to be avoided. The electro-mechanical parts of all machines in use have relatively slow response to changes and demand that frequency and phase deviations be kept to very small amounts if high standards of recording are to be maintained. During playback, it is desirable to have the reproduced signal locked to the station synchronizing pulses. With telecine machines, this is readily accomplished because the scanning is electronic. Magnetic tape reproducers employ electro-mechanical scanning and are more difficult to lock to the station pulses. Early magnetic machines could only be locked to field pulses and the output signals had to be treated as external sources, but continuous development has resulted in machines now being available which can be locked fully to the station synchronizing pulses. Even these modern machines are still limited in their ability to follow rapid changes in frequency and respond better to crystal-controlled drives.

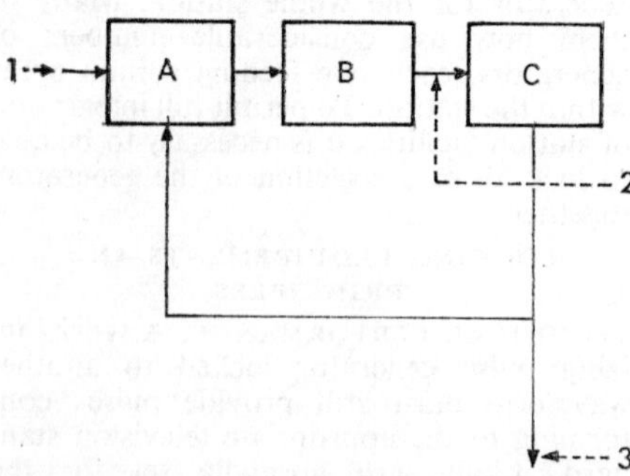

PHASE-LOCKED LOOP: BASIC BLOCK DIAGRAM. A. Phase detector. B. Filter. C. Variable frequency oscillator. 1. Input signal (frequency=f_{in}). 2. Control voltage. 3. Output signal (frequency=f_{out}).

THE PHASE-LOCKED LOOP. All current locking techniques are based on the principle of the phase-locked loop. This comprises a voltage-controlled oscillator, a phase detector and some form of passive filter network having a low-pass characteristic.

The external input to which the oscillator is to be locked and the oscillator output are fed into the phase detector, which produces an output voltage proportional to the phase difference between the two waveforms. This is then fed via the filter to the oscillator. The closed loop so formed creates a negative feedback servo-system capable of locking the oscillator frequency to that of the external input. The oscillator may be operated at an integral multiple of the external input frequency, in which case a divider is interposed between the oscillator output and the phase detector.

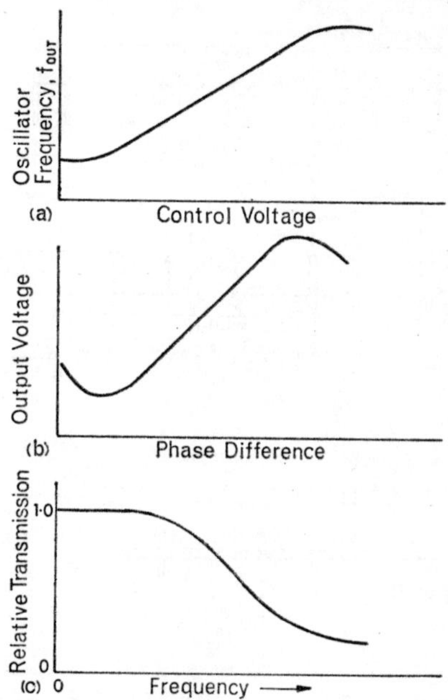

PHASE-LOCKED LOOP: TYPICAL CHARACTERISTICS. (a) Voltage-controlled oscillator. (b) Phase detector. (c) Filter.

The oscillator in television applications operates at twice the line frequency, except in colour systems where it may operate at the subcarrier frequency. Circuits commonly used are the multivibrator and the LC types, particular attention being paid to stability. Control of the frequency can be obtained by varying the potential applied to the timing circuits of the multivibrator, or by using a variable reactance circuit in the LC type.

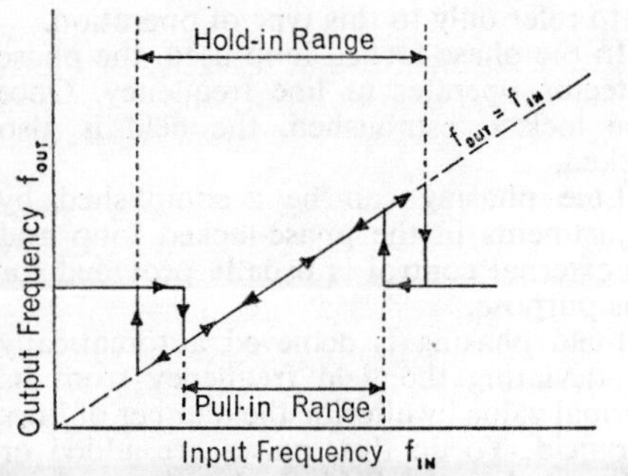

PHASE-LOCKED LOOP: TYPICAL LOCKING PERFORMANCE. When locked to input signal, $f_{out}=f_{in}$. Outside a certain range, oscillator is unlocked, and $f_{out} \neq f_{in}$. Due to hysteresis, range of lock often depends on direction in which output signal is changing.

The phase detector may operate at field, line or subcarrier frequencies, depending upon the type of locking technique being used. Many different circuits have been devised, each one requiring certain waveforms. The most popular circuits in use at the present time are the balanced diode bridge, which is fed with a pulse and a sawtooth waveform, and the coincidence circuit, which is fed with pulse waveforms only.

The filter has an important effect on the performance and, by suitable design, transient and noise characteristics can be improved.

Technical terms have been established by usage for the most important aspects of performance:
Hold-in Range is the frequency band over which the oscillator remains locked to the input frequency.
Pull-in Range is the frequency band over which the oscillator will come into lock with the input frequency when initially unlocked.

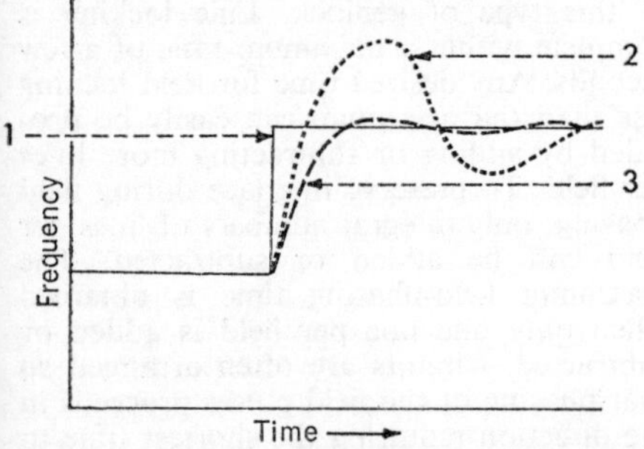

PHASE-LOCKED LOOP: TRANSIENT RESPONSE. Typical variations of output frequency in response to step of input frequency, 1, for loops with, 2, simple RC low-pass filter network and, 3, RC lead-lag filter network.

It is also termed Capture Range.
Lock-in Time is the time taken for lock to be established in going from unlocked to locked states.

POWER LINE LOCK

It has long been established practice in television to lock the field frequency to that of the public electricity supply where this is a.c. This procedure is known in Britain as mains lock and in the USA as line lock. It has the important advantage of minimizing the effects of stray power frequency hum which are evident as horizontal bands of light and shade and as distortion of the verticals on the received picture.

When the field frequency differs from that of the power line, small amounts of hum can easily be detected by their movement through the picture. Locking the field frequency to the power line frequency fixes the effects in relation to the picture and makes even moderate amounts of hum difficult to detect. Where power is distributed through a national grid system and the frequency is the same throughout the nation, all stations and OB units locked to the power line operate with the same average line and field frequencies.

The phase difference between signals locked to different parts of the power supply network is not constant because of local disturbances on the power line, and it is not possible to achieve good enough phase stability to permit the use of mixing and special effects between such signals. Transfer from station to OB signal, for example, must be by direct cut.

FIELD PHASING. To bring all field pulses at a given point, such as station master control, into coincidence to avoid field timebase roll-over on receivers when cutting takes place, field phasing methods can be used. If this is not done, the interposition of a fade to black during the cut is desirable to minimize the visibility of any roll-over caused by an excessively long or short field period.

EQUIPMENT. Circuits to permit locking the field pulses to the power line frequency are included in the majority of synchronizing pulse generators, so special equipment is not normally required for this purpose. Desirable performance characteristics are: the ability to remain locked over the likely range of power line frequency (typically ±5 per cent); the absence of severe surges during lock-in; a well-damped transient response when subjected to sudden changes in phase or frequency. The phase detector in the phase-locked loop operates at field frequency, being fed with a waveform from the field circuits within the generator and with a waveform derived from the power line. A variable-phase shifting network can be included in either of these waveforms for manual field phasing, and often forms part of the synchronizing pulse generator.

Provision is sometimes made for the input waveform to the phase detector to be supplied from an external source as an alternative to that from the power line feed. This feature is useful for OB operations when powered from generating sets, the external lock input being obtained from the station by a low power feed.

The phase stability between two trains of pulses from two generators locked to the same power line is seldom better than a few microseconds. Power line locking is gradually falling into disuse as television networks are extended and colour transmissions introduced, and it is likely to be abandoned completely by many countries in the near future.

GENLOCK

If video signals are to be mixed or used with special effects, it is essential that all line and field pulses are locked together. Moreover, phase differences must be small enough to avoid noticeable picture shift during mixing, etc. Techniques designed to accomplish this are referred to as genlock and include a variety of methods, each one suiting a particular application. The permissible phase error when expressed as a time displacement varies from a few nanoseconds for colour operation to a few hundred nanoseconds for monochrome operation.

QUICK GENLOCK. For the most accurate locking of both line- and field-synchronizing pulses, it is necessary to transmit from one generator to the other a signal containing both line- and field-synchronizing components. The simplest way of doing this is by using the composite video signal, so that the station is locked to the incoming signal.

Despite the fundamentally simple nature of this approach, it was not until the late 1940s that commercial equipment for this purpose became available, the main stimulus coming from the growth of networking in the USA. This technique quickly became established under the name of genlock and has enjoyed widespread popularity. The term genlock is still sometimes restricted so as to refer only to this type of operation.

In the phase-locked loop used, the phase detector operates at line frequency. Once line lock is established, the field is also locked.

Line phasing can be accomplished by adjustments in the phase-locked loop and an external control is usually provided for this purpose.

Field phasing is achieved automatically by deviating the field frequency from its normal value by altering the number of lines per field. To do this, pulses are added or subtracted in the field-divider chain of the synchronizing generator. Throughout field phasing, the line frequency remains locked to that of the input signal. When field phasing is correct, the normal number of lines per field is restored; complete locking of both line and field pulses then persists.

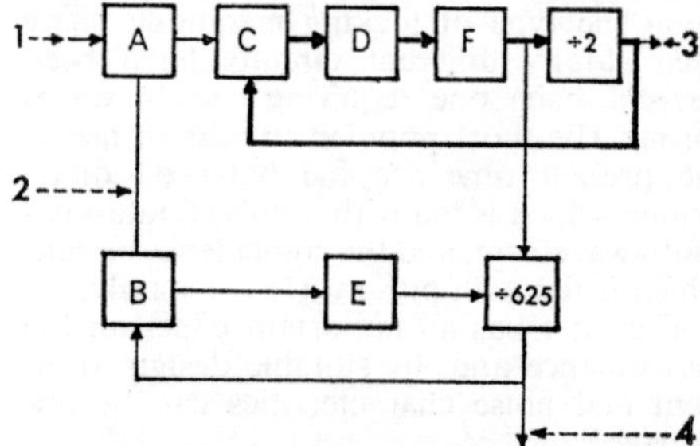

QUICK GENLOCK BLOCK DIAGRAM (625 LINES). 1. Composite video to A, sync separator. Thence, line sync to C, line phase detector, D, filter, F, twice line oscillator. 3. Line pulses. 2. Field sync to B, field phase detector, thence to E, pulse add-or-subtract generator. 4. Field pulses. (*Heavy line*) Line frequency locking obtained by phase-locked loop. Field phasing obtained by miscounting in field divider chain; miscount pulses are fed in from E, controlled by B.

Rapidity of lock is an important property of this type of genlock. Line locking is complete within a maximum time of a few seconds. Any desired time for field locking less than the maximum can easily be provided by adding or subtracting more lines per field. To preserve interlace during field phasing, only integral numbers of lines per field can be added or subtracted. The maximum field-phasing time is obtained when only one line per field is added or subtracted. Circuits are often arranged so that phasing of the field pulses proceeds in the direction requiring the shortest time to achieve correct phasing.

Equipment for quick genlocking is invariably designed to complement a particular synchronizing pulse generator, and interchangeability between different makes of generator and genlock is not often possible. This is because of the need to connect into the field-divider chain of the synchronizing pulse generator for field phasing.

Remote control panels are sometimes provided, enabling any one of a number of video signals to be selected for locking. Arrangements are usually made in the design for control of the generator to return to a crystal oscillator or for the generator to lock to the power line frequency in the event of failure of the video signal. A short time delay circuit is incorporated to guard against momentary interruptions of the video signal.

The rate at which field phasing takes place can often be selected by a switch. Rates corresponding to deviations of one to three lines per field are commonly available. A position giving an instantaneous field phasing is occasionally provided.

The line-phasing control can sometimes be located at a remote point and typically covers a range of $\pm 5\,\mu$ sec, about the in-phase position. This facility can be most valuable in compensating for variations in time delay due to the signal traversing different paths within the station.

Special attention is paid to minimizing the effect of noise on the video signal which could otherwise cause random phase errors (jitter). Precautions are also taken to ensure satisfactory performance in the presence of power line hum which is often evident on the incoming video signal.

To reduce the visibility of the inevitable disturbance during line locking it may be arranged to occur during the field-blanking period.

SLOW, OR CONTROLLED RATE, GENLOCK. The object of the slow genlock is to accomplish the change in line and field frequencies during the locking process in such a way that the effect on recording machines, especially magnetic types, is negligible. One method of doing this is to operate the phase detector in the phase-locked loop at frame frequency during the locking process so that locking of the field is obtained by varying the line frequency. The long time constants which can then be used for the filter in the phase-locked loop result in considerable reduction in the rate at which the line frequency can change. Any attempt to increase the time constants by a significant degree while operating the phase detector at line frequency, as in the quick genlock, would result in a quite unacceptable reduction in the pull-in range. The phase-locked

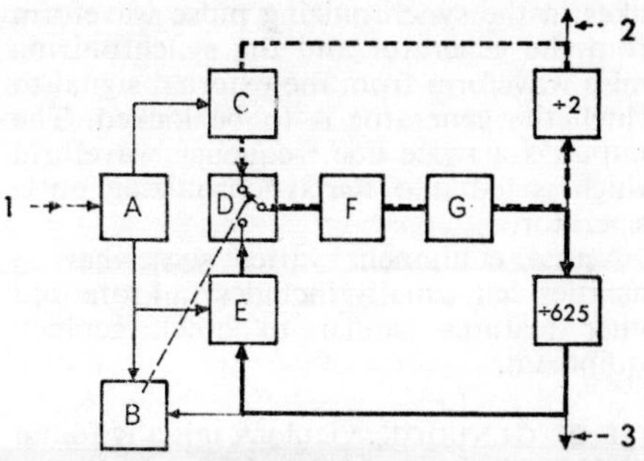

SLOW GENLOCK BLOCK DIAGRAM (625 LINES). 1. Composite video to A, sync separator feeding C, line phase detector and E, field phase detector. D. Electronic switch. B. Field phase sensor. F. Filter. G. Twice line oscillator. 2. Line pulses. 3. Field pulses. When field is unlocked, B operates D, connecting field phase-locked loop (*heavy line*). When field is locked and phase correct, B operates D to connect line phase-locked loop (*dotted line*), and line locking and phasing take place.

loop during locking resembles that used for power line locking, and similar circuits can be used.

It can be shown that the rate of change of line frequency is inversely proportional to the lock-in time, and lock-in times of about 30 seconds have been found to produce a suitable rate of change of line frequency for present-day requirements.

To overcome the phase jitter characteristic of field or frame frequency locking, an additional phase-locked loop operating at the line frequency is arranged to take over control of the twice line frequency oscillator when field locking and phasing has been achieved. The complete slow genlock thus includes two phase-locked loops, one operating at frame frequency and the other at line frequency, in conjunction with an arrangement to transfer control from one to the other.

Because the divider chain in the synchronizing pulse generator is not involved in the locking process, there will always be a constant number of lines per field, and this has led to the term constant line number being applied to this technique. Another term used is driftlock, because the variation in line frequency during the locking period resembles the drifting tendency of an uncontrolled oscillator.

The slow genlock can be made suitable for operation in conjunction with any synchronizing generator capable of accepting an external driving input at twice line frequency. In this form, the slow genlock takes in the synchronizing pulse waveform from the generator and the synchronizing pulse waveform from the external signal to which the generator is to be locked. The output is a twice line frequency waveform which is fed into the synchronizing pulse generator.

Actual equipment varies somewhat in facilities but usually includes fail-safe and other features similar to quick genlock equipment.

USE OF GENLOCK. Genlock initially found use in outside broadcast (OB) operations, as it still does today, although slow genlock is tending to displace quick genlock in some circumstances, owing to videotape recorder considerations.

When using genlock, the station accepts an incoming signal a short time before it is to be transmitted. At this time, another programme originating within the station is being transmitted. Some time during the period up to the time of transmission of the OB or network programme, the station-synchronizing generator is switched to genlock to the incoming signal. The synchronizing pulses and, therefore, the transmitted signal change in both line and field frequency, eventually coming into lock with the OB or network signal. When this has occurred, the transfer to the incoming programme can be made in any desired manner: by cut, mix or special effects. Moreover, during the transmission of the incoming signal, any additional programme material from within the station can be included by mix or special effects. However, signals from other OBs or networks cannot be so treated unless they happen to be locked to the first.

Disturbance of the transmitted signal created during genlocking is usually quite undetectable on a normal receiver.

SLAVELOCK

In national television networks all programmes may be routed through a central master station, in which case it is advantageous to have all the signals from outside slave stations correctly synchronized at the master point. Various systems have been evolved in recent years to achieve this, differing mainly in the way that the slave is controlled from the master and in the degree of automatic operation possible. To control the slave it is necessary to transmit some form of signal from the master. This may take either of two basic forms: reference signals or control signals.

REFERENCE SIGNALS. These are steady a.c. waveforms, usually sinusoidal, which are transmitted from master to slave continuously and to which each slave synchronizing pulse generator is locked. Suitable signals may be transmitted over land line or by radio waves. Telephone lines, because of their widespread availability, are attractive but the maximum frequency is limited to around 5 kHz or less. The use of a radio wave channel can easily overcome this but restrictions may be imposed by the lack of suitable channels and by the need to have aerials at master and slave points.

The ideal choice of frequency for the reference signal would be twice the line frequency, but almost as good results can be obtained by some small sub-multiple of this. The lower limit of reference frequency is set by the steady increase in jitter of the pulses as the frequency is lowered. Frequencies down to 1 kHz have been used successfully with 625- and 525-line systems.

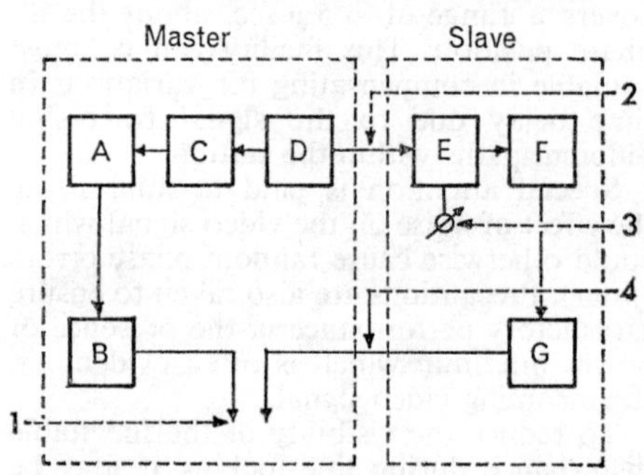

REFERENCE-SIGNAL SLAVELOCK BLOCK DIAGRAM. Master and slave stations. A. Sync pulse generator. C. Locking unit. D. Reference oscillator. 2. Reference signal fed to E, slave locking unit with, 3, manual field and line phase control. F. Sync pulse generator. B, G. Cameras. 4. Video plus sync. 1. To mixer.

A very convenient reference signal occasionally available at the slave point is the television signal radiated by the master station, in which case lock is easily obtained by the use of either quick or slow genlock units. This technique suffers from the disadvantage that it is no longer possible to lock to the master when the programme from the slave itself is being broadcast, because then the slave would be effectively locking to itself and frequency stability would be completely lost.

The main limitation to the reference signal technique is the necessity to carry out manual line- and field-phasing at the slave point to correct the phase of the signal reaching the master.

CONTROL SIGNALS. These are essentially d.c. signals appearing in the phase-locked loop between the phase detector and the voltage-controlled oscillator. The phase detector is situated at the master and the voltage-controlled oscillator at the slave. The loop is completed by the return video path from slave to master, the synchronizing signal being conveyed composite with the video signal.

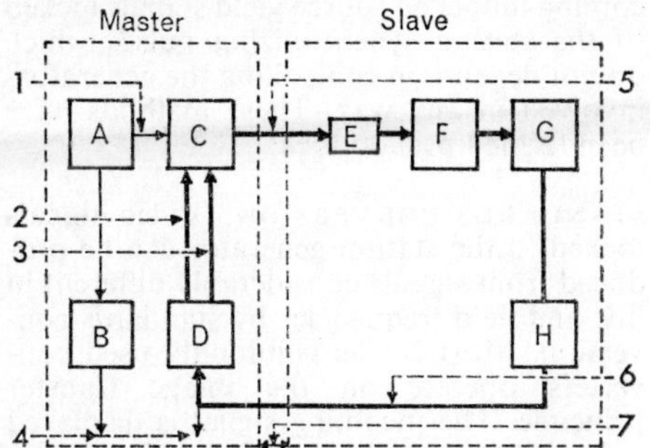

CONTROL-SIGNAL SLAVELOCK BLOCK DIAGRAM. Master and slave stations. A. Sync pulse generator with, 1, line and field feeds to C, line and field phase detectors. 5. Control voltage to E, noise filter, F, voltage-controlled oscillator, and G, sync pulse generator. B, H. Cameras. D. Sync separator fed with, 6, video plus sync. 2, 3. Line and field to phase detectors. 4. Outputs to mixer. (*Heavy line*) Phase-locked loop effecting automatic line and field phasing at master station. 7. Time delay approx. 1 ms. per 120 miles of each path.

The potential advantage of this method is that correct phasing of the slave signal is automatically achieved at the master. Full realization of this is hindered by the difficulty of transmitting the d.c. control signal without introducing spurious signals and noise which would directly effect the oscillator at the slave point, which in turn drives the slave synchronizing pulse generator. Modulation techniques could be used to overcome this problem to some degree. Serious difficulties are also introduced by the time delay between master and slave points which appears directly in the phase-locked loop circuit.

CONTROLLED REFERENCE SIGNALS. Reference signals as previously defined can be controlled in frequency and phase from the master. This technique offers considerable advantages. The signal conveniently takes the form of a sine wave within the audio band to facilitate transmission over telephone lines. The frequency chosen is a small integral sub-multiple of the twice line frequency, and the slave synchronizing pulse generator is locked to this.

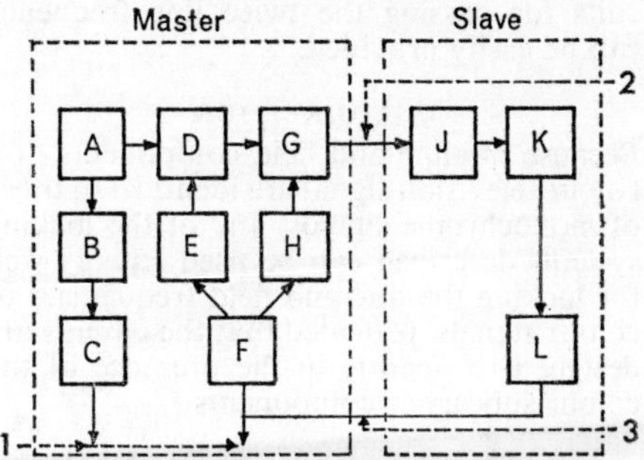

CONTROLLED-REFERENCE-SIGNAL SLAVELOCK. Master and slave stations. A. Reference oscillator controls D, frequency shifter, and G, phase shifter, sending, 2, controlled reference signal to J, locking unit, and K, sync generator. A also controls B, sync pulse generator. C, L. Cameras. 3. Video plus sync feed to F, sync separator, and thence to E, field phase detector, and H, line phase detector. 1. Outputs to mixer. Thus automatic line and field phasing are carried out at master station.

Control of the frequency and phase of the slave video signal is carried out by varying the reference signal sent out from the master station. This may be done manually by a continuously rotating phase shifter operating over the full 360° of each cycle, as in NBC's system named Audlock, or by slightly offsetting the reference frequency until field pulse coincidence is obtained, the frequency then being returned to normal.

Line-phasing is controlled by a phase shifter of limited range. Automatic line- and field-phasing can be done by using line and field phase detectors at the master point to control the reference signal. This type of system was first used by the Japanese television network NHK.

GENLOCK EQUIPMENT. Slow genlock equipment can be used for slave operation. For this application one slow genlock is required at the master for each slave station. Synchronizing pulses from the master generator and from the slave, after being separated from the video signal, are fed into the equipment. The twice line frequency output from the slow genlock is fed to the slave where it is used to drive the synchronizing pulse generator.

This technique is of limited application for remote slaves because the twice line frequency is well above the capabilities of telephone lines. It is particularly suited for applications within large stations where individual studios have to be locked to master control, because the necessary circuits for passing the twice line frequency can be easily provided.

COLOUR LOCK

Because the line and field components of a colour television signal are identical to those of monochrome signals, any of the locking systems described can be used equally well for locking the line and field frequencies of colour signals, provided that the circuits are designed to operate in the presence of the colour subcarrier components.

COLOUR GENLOCK. Locking a subcarrier oscillator to the colour burst by means of a phase-locked loop operating at subcarrier frequency provides colour genlock, or colour subcarrier lock. Line and field pulses are locked by a normal type of genlock suitably designed to work in the presence of the colour subcarrier signal.

The characteristics of the subcarrier phase-locked loop are chosen to attain a high degree of phase stability, especially for the NTSC colour system. The oscillator may be crystal-controlled to promote stability. Control of the frequency is then by means of a variable capacitance diode, or other variable reactance element.

The tight limits usually specified for the subcarrier frequency in television standards permit a narrow pull-in range to be used in the phase-locked loop and a high degree of noise immunity to be obtained.

The use of a separate phase-locked loop for subcarrier generation, quite independent of the line- and field-locking circuits, has the important practical advantage of permitting easy extension of monochrome genlock equipment to colour operation.

Colour genlock can also be obtained by locking the subcarrier oscillator to the incoming subcarrier and then deriving line and field pulses by dividing down. Line and field phasing is accomplished by miscounting field and line until phasing is achieved. This method is an extension of the quick genlock technique used in monochrome operation.

Alternatively, field, line and subcarrier can be locked successively by means of field-, line- and subcarrier-phase detectors forming separate phase-locked loops with the subcarrier oscillator. The divider chains for line and field frequencies maintain fixed-division ratios. This method is an extension of the slow genlock technique used in monochrome operation.

Colour genlock practice is well established for NTSC. Techniques for PAL tend to follow NTSC practice. In SECAM, the subcarrier is frequency modulated and the need for locking does not arise. In both PAL and SECAM, the chrominance phase alternation necessitates additional locking arrangements to ensure correct colour relationships between remote and local signals.

PSEUDO LOCK

Methods are available for making an incoming unlocked source yield signals locked to the station synchronizing pulses which do not depend on controlling the generators involved in any way. These methods have been termed pseudo lock.

STANDARDS CONVERSION. Video signals locked to the station generator can be produced from signals considerably different in line and field frequencies by standards conversion. Most of the commonly used converters operate on the image transfer principle. The incoming signal is displayed as a picture and viewed by a camera operating on the desired standard.

A more recent development is the all-electronic converter which uses a sampling technique employing high-speed electronic switches, gaining more stable operation with only a slight loss in picture quality.

The use of standards conversion offers an effective way of ensuring complete synchronization of all video signals within a television station regardless of their origin and avoids the need to control outside sources in any way. Its practical application as an alternative to genlock is restricted by cost and complexity of equipment and by degradation in picture quality.

CONTROLLED DELAY LOCK. When the incoming signal has the same field frequency as the station, it is possible to bring the remote signal into lock by introducing a controlled delay, so as to delay all the pulses in the incoming signal sufficiently to bring them into coincidence with those in the following frame of the station pulses. Any practical means of accomplishing this must preserve the full video bandwidth and the delay must cover periods of up to one frame. Moreover, the delay must be capable of rapid electronic control to enable automatic phasing to be carried out. If the practical difficulties involved could be overcome, this technique would be an attractive alternative to the present genlock systems.

HIGH STABILITY UNLOCKED OPERATION. The BBC has developed a system to provide approximate coincidence of the field pulses from a number of sources. Very precise high-stability oscillators are used to drive the synchronizing pulse generators, the stability aimed at being in the order of 2 parts in 10^8 per day over a temperature range of 0° to 45°C. For the 625-line system, this order of stability limits the change in field phase between any two generators to a maximum of about one line per half hour, if the oscillators are set initially to exactly the same frequency. For normal programme use, this change is small enough to permit direct cuts to be used between sources.

The fundamental frequency of the oscillator used is that of the colour subcarrier, to cater for colour as well as monochrome operation. Phasing of the field pulses is achieved by offsetting the countdown ratio from colour subcarrier to line frequency. A frequency offset of about ·02 per cent is used, which gives a time of about 3½ minutes for advance or retardation of the field pulse through a full frame.

A recent refinement of this technique using digital control techniques permits synchronization of remote sources to a master station, as in slavelock systems.

H.D.K.

See: *Master control; Station timing; Synchronizing pulse generators; Vision mixer.*

Books: *Phaselock Techniques*, by F. M. Gardner (New York), 1966; *Television Engineering Handbook*, by D. G. Fink (New York), 1957.

☐**Picture Matching.** Adjustment of the appropriate controls to produce similar ranges of contrast and brightness in a number of pictures, as, for example, from different cameras.

See: *Camera (TV); Picture monitors.*

☐**Picture Modulation.** Video signal applied to a display device to vary the brightness of the spot. General term for the modulation which converts a blank, unmodulated raster into a television picture.

See: *Television principles; Transmission and modulation.*

☐PICTURE MONITORS

The function of a picture monitor is the same as that of a television receiver in as much as it converts electrical signals into an intelligible picture but, whereas the link between a receiver and the picture source is by means of a television transmitter, a picture monitor gets its signals by means of a cable direct from the picture source. The associated sound signal, which is also transmitted to a receiver, is not supplied to a picture monitor, as separate equipment is used to monitor the sound signals.

Precision picture monitors are the means by which the technical quality of pictures is assessed. They are used in the engineering sphere of the television industry and must be of the highest technical performance and designed to precise specifications.

Production picture monitors are used to provide pictures in the control room for the producer and his staff, and have many other applications in the studio complex where the accent is on picture composition and programme content. They must produce bright pictures of high quality but need not conform to the same specification as that for the precision monitor.

Combined picture and waveform monitors are for the technical assessment and adjustment of the generated picture signals. One cathode ray tube (kinescope) in the monitor displays the picture and acts as a precision monitor while another cathode ray tube displays the waveform of the electrical signal forming the picture. The display can be switched to show the various waveforms necessary in order to make measurements and adjustments to the picture signal.

CONSTRUCTION

TUBE SIZE. The size of picture tube required in a picture monitor depends on the function the monitor has to perform. A 14 in. tube is normally adequate for the technical assessment of picture quality as the operator sits close to it. Similarly, a commentator's monitor should be small, light and easily transportable. Audience participation monitors, on the other hand, should enable the maximum number of viewers to see the transmitted picture and are as large as possible, often 24 in. in diagonal and suspended from the ceiling.

Monitors for the production control room vary in size depending upon the complexity and size of the control room. Large control rooms may have as many as 20 monitors in a bank covering one wall; in these circumstances the production staff must be back far enough to see all the monitors, which must be of adequate size to enable the picture content to be viewed without strain. Tubes of 17 or 19 in. diagonal are often used.

MONITOR SIZE. The size of a picture monitor depends on the dimensions of the picture tube fitted. In many instances the chassis carrying the circuits is common to a range of picture monitors and the external case dimensions increase with the tube dimensions. The frontal area of the monitor is seldom much larger than the area of the picture tube although some additional space is required for the essential controls which are usually immediately below the picture or in a narrow strip to one side.

A further facility provided on many monitors is prominently displayed on the front in the form of a cueing and designation indicator. The designation indicator consists of an illuminated panel into which may be fitted a designation strip which indicates the source of the picture signal or the function of the monitor, for example, CAM 2, T/C 3, PREVIEW, TX, etc. The cueing facility indicates which of the pictures displayed on a number of picture monitors is the one selected for transmission and may take the form of an illuminated panel or lamp or in some instances a picture frame round the perimeter of the picture area which glows red when that particular picture is on transmission.

MULTI-STANDARD MONITORS. Manufacturers can broaden their markets by building monitors which may be easily converted from one standard to another. The biggest changes are concerned with the scanning circuits which have to work at a different frequency. In standards conversion equipment, and in some studios, there is need to change picture monitors rapidly from one standard to another with the minimum of adjustment. This can be achieved by a switch on the monitor, but if many monitors are involved in the changeover, together with other equipment, it can be an onerous task.

One solution is to have remote system changing facilities, in which case one master switch can change all monitors at once. A more sophisticated method is one in which each monitor has a built-in system-sensing circuit which automatically operates relays in the relevant circuits and depends entirely upon the frequency of the incoming synchronizing pulses.

SEPARATE SYNC FACILITIES. If the picture monitor is to be used with non-composite video inputs, separate sync facilities are essential. The monitor should be able to accept the conventional negative polarity signals with syncs from 0·5 V to 6 V amplitude and provide reliable synchronization of the scanning circuits. The scanning circuits normally derive their synchronizing information from the composite video signal. In the case of non-composite signals, however, the input to the sync separator is switched from the video chain to a separate sync input socket. The switch is sited in an accessible position and in some cases remote selection of separate syncs may be provided.

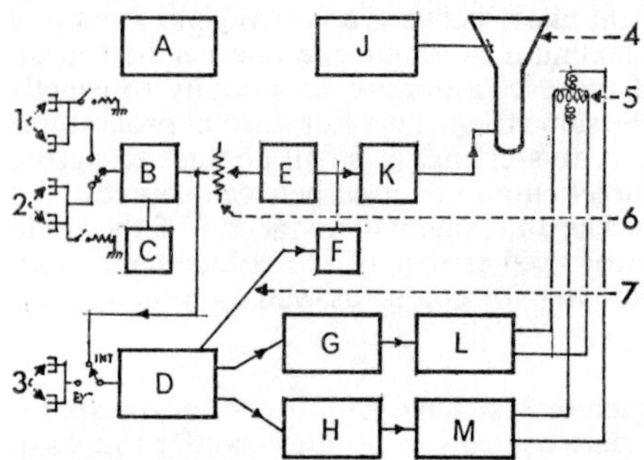

PICTURE MONITOR BLOCK DIAGRAM. Shows general resemblance to later stages of domestic receiver. 1. Video Input One. 2. Video Input Two B. H.igh input Z video amplifier with cable length compensator, C, and contrast control, 6, feeds E, video amplifier, and K, video output stage, to picture tube, 4. 3. Sync input with external-internal switch to D, sync pulse separator, and clamp pulse generator sending clamp pulses, 7, to black level clamp, F. From D, line and field pulse generators, G, H, with output stages, L, M, energize picture tube deflection coils, 5. A. Stabilized power supply. J. Stabilized EHT supply.

DUAL INPUTS AND BRIDGING FACILITIES. Picture monitors are often fitted with two pairs of signal input sockets. Each pair form an input which may be terminated by 75 Ω for monitors at the end of a cable or alternatively be unterminated and form a high impedance input which may be looped through the monitor and enable the signal to be monitored without affecting the characteristics of the line.

The facility for switching between two differing input signals may be desirable for several reasons. It is better when comparing pictures from different sources to view them on the same monitor tube because there are inevitable variations in the colour of a white phosphor. These minor differences make

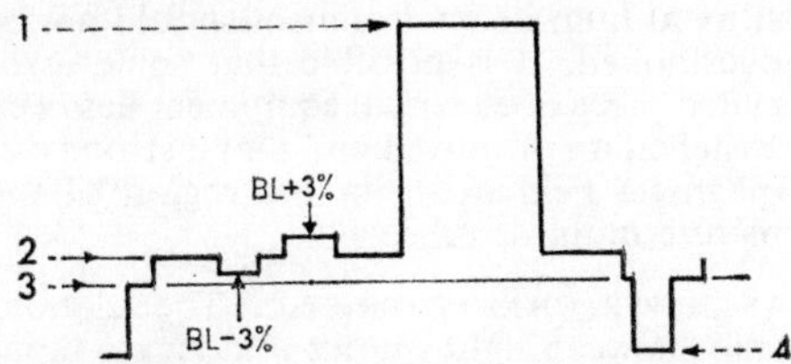

PLUGE WAVEFORM FOR LINING UP MONITORS. Picture Line-Up Generating Equipment establishes correct settings of contrast and brightness controls. On waveform, peak white, 1, black level, 2, suppression level, 3, sync tips, 4. Contrast control is adjusted to obtain desired highlight brightness from peak white bar, 1. Brightness control is adjusted under normal lighting conditions so that −3 per cent pulse cannot be seen and +3 per cent pulse can be seen. If neither or both bars are visible, control is incorrectly adjusted.

comparison of tonal gradations difficult when the pictures on two monitors are compared. Another use for the second input is when a test signal device such as PLUGE is permanently connected to the monitor for rapid checking of the control settings.

RASTER LINEARITY AND GEOMETRY. To ensure that distortion of picture shapes is avoided, the linearity and geometry of the raster on precision monitors must be of a high standard. Picture distortion of large circular objects can be particularly objectionable if the linearity of the scanning generators is not held to close limits.

A typical specification for linearity is between 1 and 2 per cent. This implies that the actual position of the spot on the raster is not more than 1 to 2 per cent away from its proper position and is expressed as a percentage of the distance between a reference point and the proper position of the spot. Some relaxation of the specification may be allowed in the corner areas of the raster because of the difficulty of maintaining good geometry over the curved areas of the screen. A technical assessment monitor usually has its raster dimension rather less than the whole tube face to ensure that none of the picture is lost in the corners and to give the best overall geometry.

CABLE LENGTH COMPENSATION. Where a picture monitor receives its signal over a long length of video cable, cable compensation is desirable. Coaxial video cables have a loss which increases with length and with increase in frequency and may be as much as 6 dB at 5 MHz over a length of 1,000 ft. In order to avoid the loss of definition that would inevitably result if this frequency response were not corrected, a calibrated correction circuit can be incorporated in the video amplifier of the monitor. This takes the form of an amplifier which has an amplitude/frequency response the inverse of the cable characteristic and the degree of boost may be varied to compensate for the length of cable in use.

PERFORMANCE

In general a picture monitor should be able to display a picture of good contrast with an input 6 dB below standard level of 1 volt in European countries and 1·4 volts in American countries. The video amplifier must be of more than adequate bandwidth for the system in use and free from any phase distortion which will give rings or overshoots on transients.

A typical h.f. response is ±0·25 dB from 10 KHz to 7 MHz and −3 dB at 10 MHz. This response is adequate for the 625-line system and more than adequate for the 405-line system.

The choice of picture tube, with its deflection and focusing arrangements, must be of a standard that enables a finely focused raster to be displayed with freedom from deflection defocusing or other aberrations of the beam.

PICTURE BRIGHTNESS. The actual numerical value of picture brightness depends on the use to which the monitor will be put.

The majority of monitors are used in television studio control rooms, etc., where the ambient lighting is normally very low, in which case brightness is not of prime importance. A commentator, on the other hand, may be doing an outside broadcast and attempting to view his monitor in broad daylight. In this case brightness is a vital factor even though special anti-reflective measures are taken such as a neutral density filter over the tube face.

The black parts of the picture cannot be blacker than the level of brightness produced by incident light on the tube face. Consequently the range of contrast between black and peak white will be considerably reduced unless the brightness of the peak whites can be increased accordingly.

A typical figure for the highlight brightness of a monitor is between 50 and 100 ft. lamberts or 170 to 340 nits. (1 nit = 0·292 ft. lamberts.)

FLICKER. The television picture is built up in sequence from a single fluorescent spot which covers the entire field 50 or 60 times a second in interlace, relying upon the persistence of vision of the viewer's eye to retain the whole image. This inevitably produces a flickering image which can be annoying in very bright pictures. The brighter the picture, the more conscious the eye becomes of the flicker frequency.

Raising the repetition rate significantly reduces the apparent flicker at a given brightness, and for this reason, monitors working on American standards at 60 Hz can operate with brighter pictures for a given amount of flicker.

RESIDUAL HUM. Non-synchronous video signals are frequently displayed on picture monitors, which implies that the field scanning rate and the mains supply frequency may differ slightly. In these circumstances any residual hum in the d.c. supplies could cause amplitude modulation of the deflection circuits, making the picture appear to breathe at the difference frequency between the mains supply and the field frequency. Amplitude modulation of the video amplifier supplies causes horizontal bars of brightness modulation to move up or down the picture. A further source of trouble is the presence of magnetic fields from mains transformers or smoothing chokes. Such a field can deflect the scanning beam and cause the picture to wobble. The specification normally allows for a maximum of one line displacement in the field direction and a fraction of a picture element in the line direction, both displacements should be gradual rather than sudden. It is possible to screen the tube by means of suitably shaped pieces of mumetal but this can be cumbersome on large tubes. It is usual, therefore, to screen the transformers and chokes and to position them in such a way that they have least effect upon the beam. Although a transformerless monitor might seem desirable because of the stray magnetic field problem, this is not acceptable with most valve monitors because they are often in metal cases and one pole of the mains supply would be connected to the case. Some monitors using the series heater technique have been designed but they have an isolation transformer mounted in the case. Transistorized monitors almost certainly need a transformer as their power supply voltages are much lower and cannot rectify the mains voltage directly.

In cases where a monitor which has been properly designed and meets specification suddenly gives trouble in respect of positional hum its environment should not be overlooked. It is possible that some associated piece of electrical equipment has been switched on or moved and that a strong a.c. magnetic field exists in the region of the picture monitor tube.

TRANSFER CHARACTERISTIC. The relationship between light output and signal input of a cathode ray tube is known as its transfer characteristic. This characteristic is not a linear one and is largely due to the fact that although the brightness of the fluorescence of the phosphor on the screen is related to the quantity of electrons striking the phosphor and their velocity, the beam current is related to the control grid voltage by a power law in the same way as the anode current of a triode is related to its grid voltage.

This non-linearity of relationship is measured in terms of gamma, defined as the slope of the curve produced when the

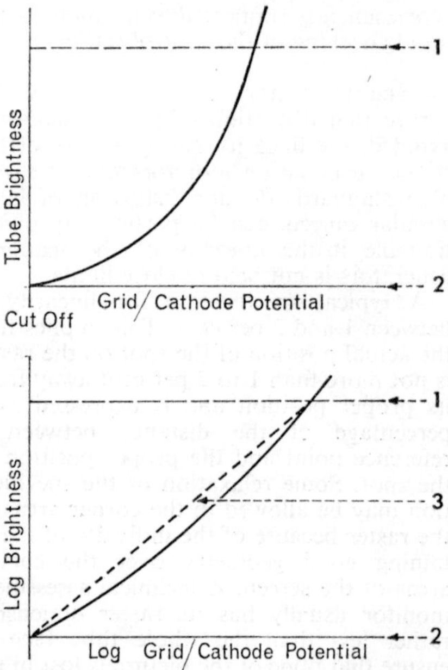

TRANSFER CHARACTERISTICS OF PICTURE TUBE. 1. Peak white. 2. Black. (*Top*) Relationship between tube brightness and signal input is nonlinear and would result in distortion of contrast ratio if left uncorrected. Distortion is measured in terms of gamma and is determined by re-plotting curve on logarithmic axes and measuring slope. (*Bottom*) Log curve is substantially linear, enabling average slope, 3, to be drawn and gamma calculated. Generally accepted figure for picture tube gamma is 2·5, and resulting distortion is corrected by making gamma of rest of chain equal to reciprocal of tube gamma, to give overall gamma of approximately unity.

logarithm of light output is plotted against the logarithm of input voltage. The average value of gamma for a picture tube is generally accepted as 2·5.

Since the object of a television system is to produce a picture on a number of cathode ray tubes with faithful reproduction of the tonal gradation of the original scene, the overall gamma of the system should be approximately unity. As the picture tube has a gamma of 2·5, the rest of the signal processing equipment must have an overall gamma of 1/2·5 to achieve unity gamma for the complete system. The responsibility for producing the gamma correction circuit rests with the picture generating equipment and not with the picture reproducing equipment; in this way, a single gamma correction circuit can compensate for all the cathode ray tubes in picture monitors and domestic television receivers. This implies that all the circuits in the video chain of a picture monitor have unity gamma as the incoming video signal will have been previously corrected to allow for the gamma distortion of the tube itself.

STABILITY

The stability of picture size and focus on the picture tube depends on the stability of the high voltage supplies to the electrodes of the picture tube. For this reason good quality picture monitors have a more complex EHT generation system than the average domestic television receiver. It is important that the regulation of the EHT supply should be equivalent to an internal source impedance of 1 MΩ or less for normal beam currents and that the voltage should be independent of picture width.

Whereas the domestic television receiver uses the flyback EHT principle derived from the line scan output stage, it is usual to find in picture monitors that a separate generator is provided using the flyback EHT or the ringing choke principle and includes a regulator circuit to ensure a constant voltage output regardless of width setting.

It is highly desirable that such a separate EHT system should be locked to the line scan frequency in use and not be free running or running at a multiple or sub-multiple frequency. Although the utmost care may be taken in the design of such a system in a monitor it is extremely difficult to eliminate the last traces of beat patterns from the raster.

BLACK LEVEL. The black level of a video signal applied to a picture tube must remain constant with changes of picture content. This implies that the d.c. component of the signal must be maintained or reinserted before application to the tube. A similar loss of d.c. component can be produced by the use of mean level automatic gain control although this technique is seldom used in picture monitors. In this case the gain of the amplifier is varied automatically by deriving a control signal from the mean of the video signal. Signals containing a predominance of black will have a low mean level, thereby increasing the gain of the amplifier excessively and causing the blacks to appear as mid-greys.

In the conventional video amplifier the d.c. component is maintained by a black level clamp operating on the back porch of the video signal. An efficient clamp has the advantage that it can substantially reduce the amount of undesirable distortions such as l.f. tilt or hum that may have got on to the signal during its generation or distribution.

Some picture monitors have an alternative method of maintaining the d.c. component in the form of a d.c. restorer. In this case the restorer works on the tips of the synchronizing pulses and consequently the black level depends upon the sync/picture ratio of the video signal and alters when the contrast control is adjusted, unless a compensation circuit is included. In the case of a non-composite signal the d.c. restorer works on the blanking period and the compensation circuit must be switched out.

The clamp or restorer circuit is usually included near the output of the video amplifier chain to ensure that any subsequent d.c. drift is kept to a minimum. Clamping of the signal at the input to the picture tube presents a problem inasmuch as the signal amplitude may be of the order of 60 volts, necessitating the generation of high amplitude clamp pulses.

The setting of the brightness control on a monitor is an important operation if the black level of the video signal is to produce black on the tube face. A misadjustment will result in loss of detail in the dark greys or failure to reach black at all. The setting up of monitors in this respect can be aided by the use of a special test signal known as PLUGE.

Streaking on smaller amplitude signals can be the result of a poor clamp or a poor medium frequency response in the amplifier. An ideal signal with which to test a monitor for streaking is a window pattern. This pattern consists of a white square on a dark background. If the monitor exhibits streaking then the dark background to the right of the white square will be slightly different from the background above, below and to the left of it.

ADJUSTMENTS. Where the monitor is left unattended or is placed out of reach, stability of adjustments and of performance are of prime importance. Most essential are stability of black level, obviating readjustment of the brightness control, stability of gain, obviating readjustment of the contrast control and stability of synchronization. Stability of picture size is of slightly less importance.

The required degree of stability is normally achieved by means of negative feedback in the amplifiers and by stabilization of all the important d.c. supplies including the potentials of the picture tube electrodes. A well-designed monitor should be able to accommodate a ±5 per cent variation of mains supply voltage without any readjustment of the controls. This requirement results in a more complex power supply circuit than that normally found in domestic television receivers.

PICTURE WAVEFORM MONITORS

It is convenient to consider here the waveform display unit usually combined with the picture monitor. The role of a waveform monitor is to act as a partner to the picture monitor in the technical assessment of

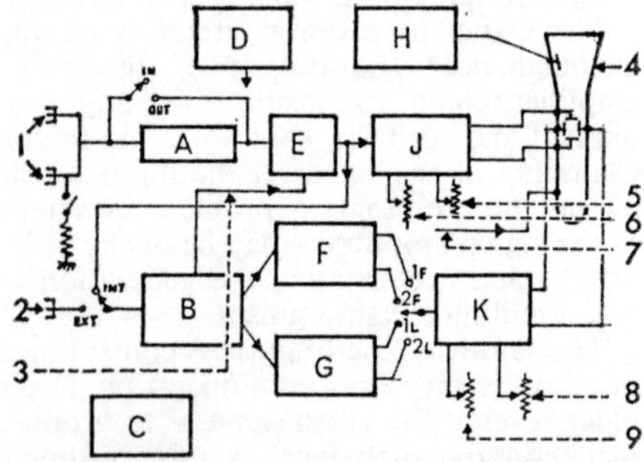

WAVEFORM MONITOR BLOCK DIAGRAM. 1. Bridged video input to A, switchable roll-off filter, and E, black level clamp. J. Balanced wideband Y-amplifier with gain control, 6, and vertical position control, 5, actuating vertical deflection plates of CRT, 4. 2. Sync input with external-internal switch to B, sync pulse separator and clamp pulse generator, sending clamp pulses, 3, to E. One-field and two-field sawtooth generator, F, and one-line and two-line sawtooth generator, G, feed via switch to K, balanced X-amplifier, with width control, 9, and horizontal position control, 8, actuating horizontal deflection plates of CRT. 7. Flyback suppression. C. Stabilized power supply. D. Y-amplifier calibrator.

television video signals. Fundamentally it is an oscilloscope specially designed to display the video signal at line and field sweep rates.

VIDEO SIGNAL AMPLITUDES. If a video waveform generated by any video source is to conform to the agreed standards, the amplitudes of the signal corresponding to peak white must be maintained at the correct level. Similarly, the black parts of the signal must fall at the agreed level. A predetermined amplitude of synchronizing information, also displayed on the waveform monitor, will make up a video signal of the correct sync-to-picture ratio.

None of the above measurements can be made with a picture monitor so an engineer or operator must use a waveform monitor in order to adjust and control the amplitudes of the video signal.

If the signal amplitude is the only parameter to be monitored, it is possible to compress the X-axis into a thermometer-like display on a narrow tube. This enables a number of such monitors to be mounted in a smaller space. Normally, however, other measurements are also made on waveform monitors and these measurements of time along the X-axis require a display of about four inches in width produced by a sweep of high linearity.

It is customary to provide a display which may be switched to show one or two fields of the waveform to enable field shading and field blanking measurements to be made; and similarly a display of one line for line shading and two lines for line blanking and front porch measurements. Some waveform monitors may also have a fast sweep that can be triggered at the start of each field scan to enable the first few lines of signal containing the field synchronizing information to be observed with greater accuracy.

PERFORMANCE. The trace provided by a picture waveform monitor must be bright and well focused and, if accurate measurements other than waveform amplitudes are to be made, the tube diameter should be at least five inches. Accurate amplitude measurements imply that an easily read calibrated graticule should be provided and that the Y-axis amplifiers must have stable gain together with a built-in means of amplitude calibration. Stability of gain in the amplifier is of little use unless the tube electrode potentials are also stabilized to ensure a constant deflection sensitivity.

Amplitude measurements of high frequency components in the signal cannot be

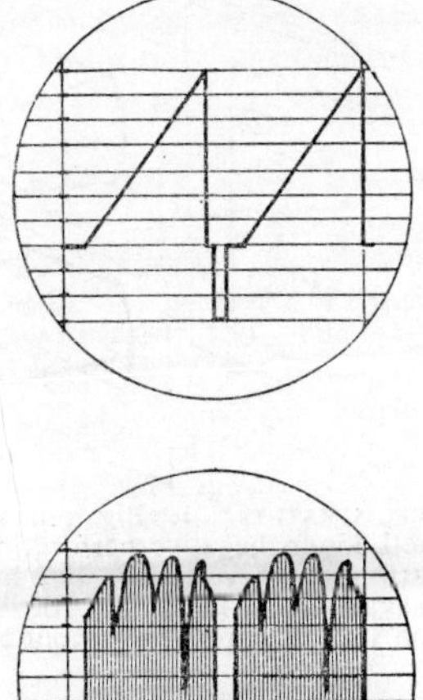

TYPICAL WAVEFORM MONITOR DISPLAYS. (*Top*) Two lines of composite test sawtooth signal aligned against calibrated graticule, enabling signal amplitude and sync/picture ratio to be adjusted, and sync pulse and blanking period observed. (*Bottom*) typical picture signal showing two fields. Operator can observe amplitude variations of all picture lines displayed side by side, together with intervening field suppression and field sync pulses.

made unless the bandwidth of the Y-amplifier is greater than the highest frequency contained in the video signal. A bandwidth of 3 MHz ±0·1 dB is adequate for the 405-line system and 5 MHz for the 625-line system.

For some operational purposes the high frequency components of the signal, including random noise, cause a fuzziness of the trace which detracts from the clarity of the display. In this case, the input to the Y-amplifier may be passed through a low pass filter which has a slow roll-off characteristic of a predetermined law to exclude the h.f. components of the signal, including the random noise, and thus remove the fuzziness from the trace. It should be possible to switch this filter in and out by means of an easily accessible switch on the monitor.

It is not always necessary to make actual time measurements along the X-axis of the display but the linearity of the sweep generator and of the X-amplifier is most important if a linear signal such as a line sawtooth is to appear as a linear sawtooth on the display.

Such a sawtooth test signal is frequently used to set up a video channel and to check that no non-linearity exists.

Similarly, gamma correction which is checked by deliberately distorting a sawtooth in a predetermined manner cannot be set up correctly if any distortion exists in the waveform monitor.

The waveform monitor can be called upon to display composite or non-composite signals and must be capable of deriving its own synchronizing signals from the video waveform in the case of composite signals and also to accept an input of mixed synchronizing pulses in the case of non-composite video signals. An internal/external sync switch should be readily accessible on the monitor.

In order that the display may be positioned accurately behind the calibrated graticule, X- and Y-shift controls should be provided. The d.c. component of the signal must be retained in the Y-amplifier; this is usually achieved by a black level clamp circuit and prevents the black level of the display from wandering up and down as the changing picture content varies the mean level of the signal.

It is a useful facility to be able to switch out the black level clamp because an efficient circuit will remove the hum from the display and the operator could be unaware that the video signal contains serious hum and l.f. distortion which, although not affecting the waveform monitor, could affect other processing equipment in the chain. R.C.H.

See: *Master control; Picture line-up generating equipment; Studio complex; Synchronizing pulse generators.*

Books: *Television Engineering Principles and Practice*, Vol. 4, by S. W. Amos and D. C. Birkinshaw (London), 1958; *Television*, by F. Kerhof and W. Werner (London), 1952.

Picture-Sync Ratio. Ratio of the voltage amplitude of picture information to that of synchronizing pulse in the television waveform. If the total signal amplitude is taken as 1 volt, the video information is usually 0·7 volt and the sync pulse 0·3 volt. With a very weak signal, such a ratio allows the picture to be just recognizable when the synchronizing signals are of sufficient amplitude to allow the receiver just to synchronize. Picture-sync ratio may be checked on a waveform monitor on which a sawtooth signal is aligned against a graticule.

See: *Television principles.*

□**Picture Tube.** Display tube in a TV receiver. Often simply called a cathode ray tube. In the US called a kinescope.

⊕□**Piezo-Electricity.** Phenomenon in which deformation of a non-conducting crystal produces an electric potential, or conversely by which an electric potential deforms the crystal. The first effect is put to use in microphones, pick-ups and other pressure transducers, and the second effect in headphones and loudspeakers. These are generically spoken of as crystal devices.

Certain crystals are cut so as to vibrate mechanically at a fixed rate, and by making use of both the above effects, these crystals can be caused to resonate in an electronic circuit to produce a stable and accurate frequency. Such crystal-controlled oscillators are common in many forms of radio-frequency transmitters and receivers.

See: *Microphones; Transmitters and aerials.*

⊕**Pilot.** Colour test strip accompanying black-and-white rushes from colour originals, which shows the cameraman how a scene would look if printed with different emphasis on the three colour components. Pilots provide the same guidance as Cinex tests in black-and-white filming.

See: *Laboratory organization; Printers.*

□**Pilot.** Signal which is transmitted continuously through a system simultaneously with the required signal. As it is a constant signal, it can be used for a continuous check on the characteristics of the system.

See: *Outside broadcast units.*

⊕□**Pilot Tone.** Widely used method in field sound recording of recording a control track in addition to the normal sound on a magnetic recorder. The control track can then be used to synchronize a studio recorder or a film projector to the sound reproducer.

See: *Pulse sync.*

⊕**Pin.** Register pin employed in many film cameras, and in some specialized printers and projectors, to ensure constant placement of the film frame by frame with reference to the picture aperture, provided that the perforations with which the register pin engages are themselves accurately positioned.

Two types of pin may be used: a big pin which fills the perforation snugly on all sides, and a little pin on the opposite side of the film which fits tightly in a vertical direction and controls the very slight rotation of the film about the axis of the big pin. Since the little pin has a side-to-side clearance, it allows for small changes in the width of the film without producing buckling or distortion.

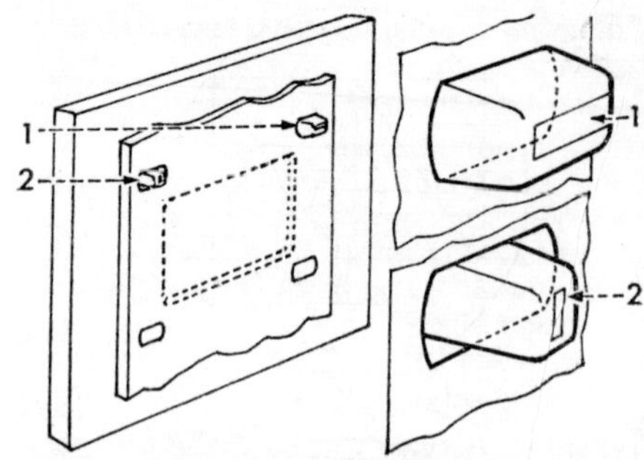

PILOT PIN REGISTRATION. 1. Big pin fully fits Bell & Howell 35mm negative perforation on all sides. 2. Little pin fits vertically, but horizontal clearance (exaggerated in diagram) allows minute expansion and shrinkage of film without causing it to buckle.

Tight-fitting pins cannot be used in studio film cameras because of the noise they produce as they move in and out of the perforations, but they may be necessary in optical printers and process projectors.

In some camera designs, a small pin provides vertical location, while a spring acting along one side of the film presses it against a machined edge on the other side which positions it horizontally. This may be called edge-and-point guiding.

See: *Animation equipment; Camera* (FILM)*; Printers; Special effects* (FILM).

⊕**Pin Belt.** Flexible metal belt the width of 35mm film and provided with rows of perforation pins along each edge which engage the perforation holes of long strips of film to hold them in contact and exact registration during a cinematograph film printing process such as imbibition.

See: *Imbibition printing.*

⊕□**Pincushion Distortion.** Defect in a lens or cathode ray tube deflection system which causes a square to appear pincushion-shaped, i.e., stretched out at the corners.

See: *Optical principles.*

⊕**Pitch.** Distance between two successive perforation holes along the length of a strip

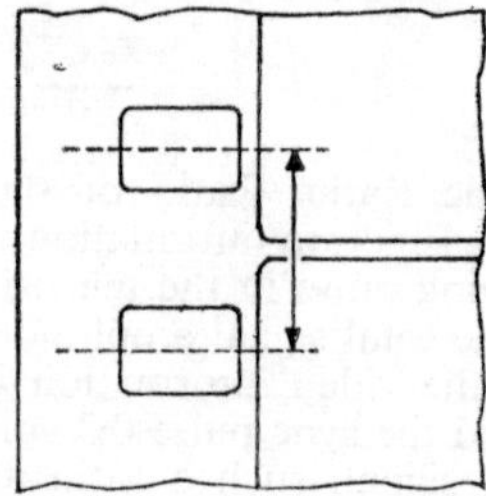

of film. Short-pitch negative film is used for contact printing.

See: *Film dimensions and physical characteristics.*

Pitch (Sound). Subjective quality of a sound which determines its position in a musical scale. The units of pitch and frequency are the same (Hertz, or cycles per second), but pitch and frequency do not necessarily correspond since one is subjective and the other objective.

See: *Sound; Sound recording.*

Pixillation. A form of film animation technique employing rapid cutting between still shots to give an appearance of movement. Moving shots may also be accelerated and made jerky by eliminating every second, third or fourth frame by skip printing on an optical printer, but in true pixillation the still shots are held just long enough to be appreciated as static.

See: *Animation techniques.*

Plasticizer. In the manufacture of film base, chemicals added in small quantities to the cellulose derivatives to ensure flexibility and avoid brittleness and cracking during subsequent use.

Plate (Background). Still or motion picture scenes photographed as backgrounds to be combined with foreground action. The most frequent use of plates is in the form of positive prints for back-projection shots, but front projection is also used.

See: *Special effects* (FILM).

Plateau, Joseph Antoine Ferdinand, 1801–1883. Belgian scientist who discovered in 1828 the principle of the synthesis of movement, using persistence of vision. In 1833, described the Phenakistiscope – the first device producing an apparently moving picture from a series of drawings viewed intermittently – a discovery made simultaneously by Stampfer. Plateau proposed in 1849 the use of a series of stereoscopic pairs of photographs instead of drawings in the Phenakistiscope; this was carried out later by, among others, Wheatstone and Coleman Sellers.

Playback. In general, the immediate reproduction of a recording. Also a technique in which music and song are pre-recorded under ideal acoustic conditions and then played back in a studio on loudspeakers, to which the performers mime and play in precise synchronism. The performers are recorded on a guide track which is subsequently synchronized to the original track, which then takes its place.

By means of playback, singers' voices can be substituted, facial contortions eliminated, and the picture recorded under conditions far from ideal for sound.

See: *Sound recording* (FILM); *Sound mixer.*

□**Plücker, Julius, 1801–1868.** German mathematician and physicist, born at Elberfeld. Studied at the universities of Bonn, Heidelberg and Berlin and in 1823 went to Paris, where he came under the great school of French geometers. Later taught mathematics at Halle and Bonn. Plücker made important contributions in the field of analytical geometry, introducing new methods and ideas, which included the formulation of six equations known as the Plücker equations. After his appointment as Professor of Physics at Bonn, he turned his attention to researches in magnetism, crystallography, spectrum analysis and the discharge tube. In 1859 he gave the name cathode rays to the rays which he had noticed emanating from the cathode of a vacuum tube. In 1865 Plücker turned to the study of line geometry, which he can be said to have founded.

□**Plumbicon.** A photoconductive camera tube derived from the vidicon, but employing a lead oxide target. Its principal advantages are a low dark current, producing a more stable black level, greater sensitivity, and a shorter time lag which reduces smearing at low light levels. Early examples suffered from a very restricted response in the red region of the spectrum, but this has latterly been improved.

Plumbicons are finding increasing application in colour cameras, and generally as a replacement for both vidicons and image orthicons. Plumbicon is a trademark of N.V. Philips Gloelampenfabrieken of Holland, by whom the tube was developed in the early 1960s.

See: *Camera* (TV).

□**Point Gamma.** For camera tubes and similar devices, the instantaneous slope of a curve relating the logarithms of the incident light and the resultant output voltage; for receivers and display devices, the instantaneous slope of a curve relating the logarithms of the input voltage and of the intensity of the resultant light output.

See: *Transfer characteristics.*

⊕**Pola Screens.** Polarizing filters designed for camera mounting. Rotation relative to the lens cuts down glare and unwanted reflections from glass surfaces, etc.

See: *Cameraman* (FILM).

⊕□**Polar Diagram.** A type of diagram, usually in circular or semi-circular form, in which distance from the centre represents the magnitude of one variable, and angle with respect to a zero axis the other. Frequently employed to show loudspeaker and microphone response at different frequencies and in different directions; also to show change of effective power radiated from aerials (antennae).

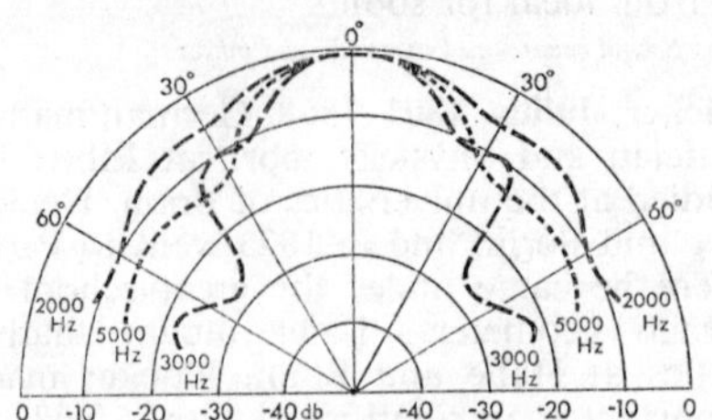

TYPICAL POLAR DIAGRAM. Loudspeaker response showing sound distribution in three frequency bands.

See: *Microphones; Screen luminance.*

⊕□**Polarization.** Modification of transverse wave motion so that vibration occurs in a restricted plane or series of planes rather than in all planes possible at right-angles to the direction of propagation. If the vibrations are limited to a single plane only, the wave motion is described as plane-polarized, while if the direction of polarization at any point rotates systematically around the axis of transmission the wave motion is in circular polarization.

Plane polarization of the electro-magnetic waves of light occurs on reflection from polished surfaces or by transmission through transparent crystalline substances which have the property of absorbing the vibrations in all except a single plane. Such materials are known as polarizers or polarizing filters and may be natural crystals, such as tourmaline and calcite, or synthetic, such as Polaroid. Pairs of polarizing filters with their planes of maximum transmission at right-angles are used in stereoscopic projection and viewing systems to allow the separation of the two images for right and left eyes when projected on the same screen.

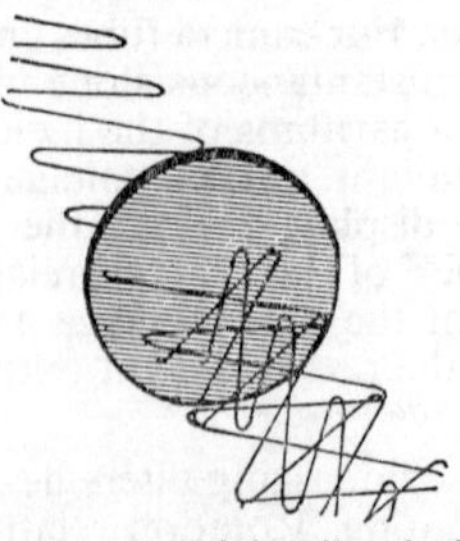

PLANE POLARIZATION. Light vibrating in all planes at right-angles to direction of motion falls on polarizing filter and emerges vibrating in a single plane.

In photography, polarizing filters may be used for the suppression of reflections from glossy surfaces or from water, for the reduction of the polarized light from a blue sky to make it appear darker and for the recording of scientific phenomena in which polarization changes take place in the material being photographed, as in stress analysis.

Pairs of polarizers may also be used for the control of the intensity of a beam of light by their relative rotation.

Polarization of the electro-magnetic waves used for radio and television transmission can also take place, the signals emitted by an aerial system being vertically or horizontally polarized according to its design. Circular polarization of radio waves is of importance in satellite communications by reducing the accuracy with which the dipole receiving antenna must be aligned with the transmitter when one is on the satellite and the other on earth.

See: *Stereoscopc cinematography.*

⊕**Polyecran.** Entertainment form developed in Czechoslovakia in which multiple still and motion picture images are presented with multi-channel sound effects. The component pictures are not intended to build up a single realistic representation on the screen but rather to provide contrast and counterpoint of the images for dramatic effect.

See: *Wide screen processes.*

□**Polygon Machine.** Telecine machine which uses a polygonal or multi-sided prism to arrest the image motion of a continuously moving film for purposes of scanning. The prism is sometimes called an optical compensator.

See: *Optical intermittent; Telecine.*

□**Polyrod Aerial.** Technique involving spaced rods of dielectric material as radiators or receiving aerials sometimes employed instead of parabolic or horn reflectors at microwave frequencies.

See: *Microwave links; Transmitter and aerials.*

□**Porch.** The brief interval which occurs between the commencement of line blanking and the leading edge of line sync and the somewhat longer interval separating the

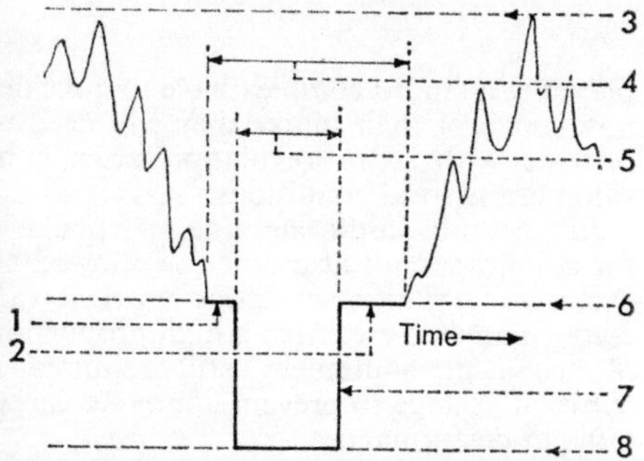

ELEMENTS OF LINE SIGNAL WAVEFORM. 1. Front porch (i.e., pre-sync line-blanking period). 2. Back porch (i.e., post-sync line-blanking period). 3. White level. 4. Line-blanking period. 5. Line sync period. 6. Black level. 7. Line sync pulse. 8. Sync level. Term porch derives from appearance of the waveform on either side of the line sync pulse.

trailing edge of line sync and the end of line blanking. These two very important features of a television signal waveform are correctly described as the pre-sync line blanking period and the post-sync line blanking period respectively. More succinctly, they are commonly known as the front porch and the back porch.

The purpose of the front porch is to allow a sufficient interval of time at the end of each line scan prior to the commencement of line sync, during which the signal is free to return to black level from whatever tonal value corresponds with the extreme end of the line; this latter may of course be any level in the range from black to peak white. Without such provision, receiver line flyback would occur later after those parts of a picture having white or near white at the right-hand edge than after areas of darkness. This would give rise to a reproduced image in which all vertical edges were ragged or not, according to whether the right-hand edge was predominantly light or dark.

The back porch is simply what is left of line blanking after the front porch and the line sync pulse have been accounted for. However, it also serves as a convenient part of the waveform for certain ancillary functions, notably d.c. restoration and, in the case of certain systems of colour television, it serves to contain a reference burst of colour subcarrier. C.W.B.R.

See: *Television principles.*

Positional Control Servo. Method of remotely controlling a mechanism, such as the focus ring of a lens. The mechanism follows the exact position of the control, movement taking place at a constant speed.

See: *Camera* (TV).

⊕**Positive Drive.** Method for the movement of motion picture film through continuous processing machines in which the drive is provided by a series of rotating sprocket drums having teeth which engage with the perforation holes along the edge of the strip of film. Positive drive systems can provide very high speeds of film transport and are mechanically straightforward but, because of the danger of distortion or damage to the sprocket holes of the film, they are usually employed only for machines processing positive prints.

See: *Processing.*

⊕**Positive Perforation.** Standard form of sprocket hole used along the edges of positive 35mm film stock, having the form of a rectangle with rounded corners. Also standard for 65mm negative and 70mm positive materials and sometimes also known as the Kodak Standard (KS) perforation. Other important types of 35mm perforation are the negative perforation (used in applications where accuracy of film register is essential), and the Cinemascope perforation, which increases the width available for picture and sound track on release prints only.

See: *Film dimensions and physical characteristics.*

□**Pos-Neg Switch.** Switch on a camera to change the picture from positive to negative or on a telecine machine (film scanner) to enable it to produce a positive picture from a negative film.

See: *Camera* (TV)*; Special effects* (TV)*; Telecine.*

⊕**Post-Synchronization.** Synchronization of sound to a previously recorded picture. Normally, a guide track is recorded with the picture. During editing, the guide track is broken down into loops and these are played on headphones to the actors who are watching the synchronized action loops on the screen. They are thus enabled to re-speak the dialogue with precisely matching lip movements. The advantages of post-synchronization are similar to those of playback: to make filming possible under very bad acoustic conditions, to substitute actors' voices where necessary, and to change dialogue from one language to another (usually called dubbing).

See: *Sound mixer; Sound recording* (FILM).

□**Power Separation Filters.** Filters which separate power supplies from signals to enable both to be carried on the same conductors, thus eliminating the need for special power conductors to distant repeaters.

□POWER SUPPLIES

Electrical power is required for production lighting, driving the television equipment, air-conditioning and ventilation, supplying auxiliary services, and lighting workshops, office areas, etc.

The biggest proportion of the load is for production lighting. Typically, in a studio, the equipment load is about 10 per cent of the lighting load, and air-conditioning of the control areas another 10 per cent of the total in a temperate climate. A small studio complex may use a few hundred kVA; a large one with a high proportion of colour production will approach 10,000 kVA at maximum demand.

In television studios, most of the lighting load is a.c.-supplied incandescent lamps, ranging up to units of 10 kVA. These are supplemented to some small extent by d.c. arcs driven from a supply in turn derived by rectification from the main a.c. intake, but the d.c. requirement is usually no more than 5–10 per cent of the total load, and confined to large, prestige productions.

INTAKE ARRANGEMENTS

It is not usual for television studios to generate their own power, which they purchase from the local electricity generating authority. It is brought into the studio area from public substations at high voltage, and reduced to the studio voltages by means of step-down transformers. The power companies' metering and fusing arrangements are installed here also.

REGULATION. Public supplies are usually given legal limits of permissible variation in voltage at given test points (not at the user's terminals as might be expected). If the local authority has been consulted early enough in the planning of the studio and the supply area is such that reasonable increases in load can still be provided, a rather more stable supply will be obtained at the ends of the high voltage feeders to the television station than the average industrial user gets.

The actual variation obtained from one studio complex to another can range from a few per cent daily to perhaps 20 per cent under conditions of national emergency, and may then even include a drop in frequency. As a public service, a television studio may hope to be treated as a priority user and to keep its power supply, under conditions of emergency, as long as hospitals, government installations, airports and other priority users. Accordingly, the engineers planning a studio complex have to consider how much of their intake they will need to provide with extra regulation after considering the local conditions.

In general, studio lighting, particularly for colour, cannot at present be allowed an excessive variation of either intensity or colour temperature. Also a high proportion of technical equipment still requires a constant voltage to prevent annoying variations in performance.

It is usual to find most technical areas and their equipment fully regulated. Some lighting for production is also supplied from separate regulators, but this is unusual in large studios, because of the high cost of such installations. Other methods are used to minimize the variation. Cable losses are reduced as much as possible and if all circuits are able to be dimmed this gives a fair measure of adjustment.

REGULATORS. Most regulators in common use are of the type which assess the incoming voltage and then buck or boost the line voltage from a series auto-transformer; they are often of the Variac toroidally-wound type, having close-spaced taps and a motor-driven wiper. The range of power-handling capacity is usually 1 kVA to 30 kVA. The most common arrangement is to place a suitably sized regulator to service a technical area from a nearby position.

The speed of operation of such devices has been increased over the years to about 25 per cent of the nominal line voltage per second. This is sufficient for ordinary load variations in mains voltage but is not able to cope with power surges such as flashovers or lightning strikes. The operating speed of the motor would have to be increased some two orders of magnitude to follow these surges. Modification to the terminal studio equipment by generally increasing the time constants of vulnerable circuits is a better solution to this problem.

Typical flashovers in industrial areas are due to a combination of ice and grime. In a complicated feed system, the magnitude of a surge can vary considerably over a few miles.

INSTALLATION. Starting from scratch, there is no absolute necessity for breaking up the intake into separate units, although this is often done for ease of servicing. In a studio complex to which new studios are added subsequently, additional smaller power intake areas, separate from the original ones, are often provided.

If it is desired to upgrade a power installation without removing existing transformers, it is safe and practical to parallel transformers which are identical in design and rating. Units in the range 500/1,000 kVA can be easily handled by local contractors, as their weight and size is within the scope of commonly available hoists and vehicles. The following points also need to be considered:

1. An alternative intake of high-voltage power is desirable. Two feeders from supply areas, a national or a local supply; or any other alternative with a changeover switch is a great help should one feeder be damaged or put out of use.
2. The cables from the low-voltage windings should be kept as short and as heavy as possible, to get the best regulation. Cable costs will be substantial.
3. Adequate ventilation must be provided to get rid of the waste heat from transformer losses. Partly open construction for mounting the transformers is popular.
4. The transformer must be accessible without too great difficulty to allow oil changing (if of the oil-filled type), examination of the core, and replacement in case of damage.
5. Oil-filled transformers are erected in areas with oil-retaining walls, or pebble-filled sumps capable of holding all the contained oil. The additional fire risks from using oil are countered by fire reducing installations such as carbon dioxide and a fire alarm system.

In general, fire insurance companies favour air-cooled transformer installations and charge a lower premium to encourage them.

STANDBY POWER. It is seldom economic to provide standby power for a whole studio complex. In favoured cases, such as for example where a national grid supply is backed up by alternative local priority use of a steam generation plant, it is not worth while having locally generated standby power at all. It is usual for those with a rather vulnerable supply to provide some standby power to run an emergency programme service.

This may well consist of a power supply to central control areas, videotape areas, some telecine machines, PTT terminal equipment and emergency lighting. Thus there may be standby generator sets, usually diesel driven, of some 200/300 kVA. The particular power feeds must be segregated and arranged with a changeover switch or relay. The operation of starting and changeover can be made automatic on failure of the usual mains supply.

It is seldom that anything bigger than a very small studio is included in the emergency arrangements; accordingly, emergency programme facilities have to be provided. Instantaneous changeover is not desirable; it should last some 10 to 15 seconds to let technical equipment clear itself of voltages. Fast re-connection often blows local fuses, because of surging.

DISTRIBUTION

The equipment used for power distribution is standard, and comprises normal switchgear, busbar chambers, cables, meters, overload trips, etc. Distribution from the transformers begins with cables from the secondary windings. These cables feed busbar chambers with copper busbars via suitable circuit breakers or switch fuses.

If more than one transformer is used, it is advisable to have either a switch or readily accessible links between separate busbars fed by each transformer. Transformers effectively paralleled in this manner should be identical in design and load rating. Cables from the main busbar chambers take the load to secondary distribution points, where further busbar chambers and switch fuses may be arranged. It is convenient for single-phase distribution to start here.

EARTHING. The earthing (grounding) arrangements in a television studio comprise the power supply with its nominal neutral, and two separate earths. One of these, sometimes called the machine earth, is connected to the casings of all power-using equipment; it often consists of a good water-pipe picked up near the power distribution point. The other, called a technical earth, is linked to the outer conductors of the unbalanced video cable complex, and often consists of copper earth rods. Despite the accepted practice of keeping the two earths separate and radiating outwards, they inevitably, in a large studio, come together at a number of points.

This can have unpleasant results, for shields and outer conductors may be found to be carrying power currents. Coaxial fittings are often the culprits, if they unintentionally form a junction between the machine and technical earths. Standardization of power plugs and other connections is by no means universal, and earth currents can often be traced to movement of equipment which has been wrongly connected internally.

Sometimes earth currents appear on long lines from remote sources, and these may be suppressed by the insertion of large coaxial chokes until the electricity generating authority has tracked them down and got rid of them.

STUDIO PRACTICE. British and European practice generally is to use three-phase a.c. supplies at about 400 volts between phases, 230/240 volts phase to neutral. Although these voltages are a little hazardous, the currents for the usual run of 2 kW, 5 kW and 10 kW lamps are conveniently low.

In the United States, the 110/117 volt mains are preferred. There is an increasing tendency to have one lamp-feed per dimmer, and to incorporate silicon-controlled rectifiers in each circuit. Black-and-white productions are usually supplied with some 25/50 watts per square foot of studio area. Colour productions require about three times as much. Luminaires are notoriously variable in their efficiency. A choice of efficient luminaires brings the load down to the lower wattage figure.

As the feed to the lamps requires a power distribution point, a patch panel, if used, silicon-controlled rectifier (SCR) dimmers and leads to the lamp, a considerable amount of cable is needed with a consequent risk of poor regulation. Generally, planning engineers keep runs as short as possible, make the cable as heavy as possible, and hope by these means to get by, if the mains are unregulated. Possibly, in the future, SCR dimmers will include a feedback arrangement which will keep the light constant despite mains variation.

SAFETY. It is vital in a studio installation to comply with all the local wiring and power regulations. Often it is an advantage to use an even heavier gauge of cable than the authorities demand in order to get good regulation. Switch fuses or overload devices must be of good quality and be properly set up. All this is normal good practice, but in one or two areas difficulties appear.

It is not convenient in a large studio, say of 4,000 sq. ft. and upwards, to have all the production lighting in one phase. Large studios, to get a balanced load, may need three-phase supplies. Hence there is a danger of high voltage or at least a lethal voltage, appearing between adjacent areas. The danger can be minimized by segregating the areas, and colour-coding plugs and boxes appropriately for each phase. Under these conditions, extra attention must be paid to good maintenance and earthing. Technical and workshop areas should be kept to a single phase. If any equipment is of the live chassis variety, workshop areas should have an isolating transformer for supplying test gear and soldering irons. Power plugs must be inspected when new gear is introduced to make sure that earths and neutrals have not been confused. Safety is improved if all technical equipment using transformers includes a heavy foil shield, properly earthed, between the mains winding and the rest.

PICK-UP – ELECTRICAL AND ACOUSTICAL. Pick-up in a properly installed studio is never electrostatic. Some magnetic pick-up from transformers and cables into audio gear is not unknown. Of late, SCR dimmers have added their quota. These dimmers have a very sharply rising waveform even with appropriate shaping circuits, and the interference range has been increased over that of normal power transformers. Radio frequency pick-up may also be traced to these devices.

It is difficult to give minimum distances as these vary so much with both the SCR design and the audio equipment used. As a rule of thumb, if the distance is less than 100 ft., measurements should be made on typical pieces of equipment. Lamps and their leads carrying the current waveforms produced by SCR dimmers have been known to induce hum into neighbouring microphones and their cables more readily than when carrying normal a.c.

Acoustic pick-up from vibration may result from transformers or busbars used for distribution. The frequencies are usually low and difficult to keep out of building structure. One solution is to mount them on separate concrete slabs floated on resilient material. The non-sinusoidal waveform of SCR dimmers also produces disturbance.

Busbars in trunking for distributing high power are not recommended. Owing to the spacing and the resonant properties of the bars, it is inevitable that an acoustic resonance will be produced at some load. Filling the chambers with glass wool is a poor solution, as, after a time, it packs down and ceases to attenuate the noise. Cables are preferable even if they are more costly, as the acoustic hum from a cable is negligible.

Butyl rubber insulated cables are suitable and may well be preferred to paper insulated cables as the ends do not need any special sealing.

Where the cable is armoured, the steel wires of the armour should not be allowed to go to earth other than at the regulation earthing point, to avoid earth loops. Insulated outer covers can be obtained.

POWER FOR OB AND LINKS. Wherever possible, power supplies for an outside broadcast should be obtained from a public supply. It is usual for television companies to arrange and own their own connection points at places where they often have to go for broadcasts. Examples of this are airports, to cover the arrivals of celebrities, homes of Presidents or Prime Ministers, etc. When a broadcast is planned in advance, it is often possible to arrange with the public authority supplying the power to provide the service. Alternatively, some private individual or corporation may be able to help.

Other provisions must be made for emergency broadcasts or broadcasts from remote areas.

A whole range of portable power equipment is available, attuned in its various forms to the job it has to do. A microwave link can operate from a small generator set of 1 or 2 kVA. Such units are usually diesel powered and towed behind a truck, and are normally not provided with special silencing. If enough power is needed to drive two vehicles with up to eight camera channels and a surplus for emergency lighting, the power requirements may reach 35 kVA. It is then common practice to silence the vehicles very thoroughly both to suit the needs of the broadcast and avoid annoyance to the neighbourhood. This necessitates thorough acoustic treatment and special silencers for the exhaust gases. It is prudent to have a single-phase output for this type of work, but three-phase outputs are not unknown. The latest models are fitted with good regulators.

Generators may be designed for towing or they may be incorporated in self-propelled vehicles. Despite automatic braking arrangements, towed generators sometimes turn over when the towing vehicle has to swerve. In general, they are a danger in the hands of a careless driver. Self-propelled vehicles, although more cumbersome and more expensive, are usually a wiser choice. A number of these vehicles, studio-based, are required to service outside broadcast (OB) units, two mobile generators, as a rule, to four OB units. D.W. & E.C.V.

See: *Lighting equipment* (TV); *News programmes; Outside broadcast units; Studio complex* (TV).

Pre-Amp. Amplifier fed from a source of very low output such as a photocell, capacitor microphone or video camera, which boosts a signal before further transmission in order to reduce capacity and other losses with consequent impairment of the signal-to-noise ratio. For this reason, pre-amplifiers are mounted very close to the signal source, and often in the same enclosure.

See: *Camera* (TV); *Microphones; Noise.*

Pre-Emphasis. Technique of raising the recording level of the higher audio frequencies to compensate for subsequent attenuation of these frequencies, thus safeguarding the signal-to-noise ratio.

In the vast majority of sound signals, the contribution to peak signal by frequency components above 1,000 Hz decreases with increasing frequency. If the audio signal in the transmitter/modulator is passed through a capacitance resistance network with a time constant of 50 μsec, the response at 10 KHz can be lifted by 10 dB relative to 1 KHz. This is not enough to cause the high frequency peaks of the signal to exceed those of the low frequency, so no overload problems arise. This h.f. boost process is called pre-emphasis.

At the receiver, the audio signal is passed through an inverse network, also of 50 μsec. This is called de-emphasis and ensures that the overall audio gain frequency response is constant. The advantage of this process is that the receiver de-emphasis provides attenuation of the noise between the transmitter modulator and the receiver detector. An improvement is thereby achieved in the random noise signal separation.

See: *Sound; Sound recording.*

□**Pre-Listen.** Facility which enables a sound operator to listen to a sound source before that source is connected to the main output of a sound mixer. Useful in locating the required section from a disc or tape recording which has to be inserted quickly and accurately into a programme.

See: *Master control; Outside broadcast units; Vision mixer*

⊕**Première.** First public showing of a film production; for an important feature film it takes place at a major cinema theatre with considerable publicity.

⊕□**Pre-Mix.** In the dubbing or re-recording process, an inconveniently large number of sound tracks may need to be combined. Pre-mixing is a technique for selecting groups of these tracks and mixing them; the smaller number of pre-mixed tracks is then combined into the final sound track.

See: *Sound mixer; Sound recording* (FILM).

⊕**Pre-Scoring.** Technique of recording a song or music track before the picture which is to accompany and synchronize with it. Singers and orchestra are then post-synchronized by playback.

See: *Sound mixer; Sound recording* (FILM).

□**Presentation.** Technique of presenting a programme in continuity. When a television broadcasting station goes on the air, the separate programmes transmitted from that station are linked together by a process called presentation. The presentation of a station's programmes gives that station an identity or personality which can be recognized by the general public. Since each station wants as large an audience as possible, this station personality is as attractive as the administrators can make it.

Studio facilities are provided to help them produce a programme that will hold the audience. This may include having station announcers, in vision, introducing programmes, weather forecasting services and local news, serving as a buffer between programmes that may differ substantially in character.

Presentation also performs the vital role of covering breakdowns. This may necessitate anything from a short apology for a break to the provision of stand-by programme material if the break is prolonged.

See: *Master control; Vision mixer.*

□**Presentation Mixer.** Sound and vision equipment used to present programmes in continuity. Since the main function of presentation is to link programmes together, the presentation mixer differs from studio mixers in that it must have sound control facilities as well as vision. The number of programme sources that may be selected may well be greater than the studio mixer's capacity, and some of these sources are also likely to be non-synchronous.

Sound and vision control have to be separate so that sound and vision from different sources may be added together. It must also be possible to superimpose vision and sound on any programme that has been selected for transmission. To achieve this last requirement, and in order to change between non-synchronous sources without incurring a field sync disturbance, provision must be made to phase lock the local synchronizing generator to the sync signals from any of the programme sources.

Three separate programme routes are often provided through the mixer. One of these is a straight in/out route from the source to local output. A second route takes the signal through a synchronous mixer where local signals may be added. The third is a straight in/out path to a separate output which may be feeding a network.

See: *Master control; Vision mixer.*

⊕**Preservatives.** Substances which, when applied to the surface of film, tend to lengthen its life by protecting it from scratching, or by helping it to retain its plasticizer so as to prevent it from becoming dry and brittle.

See: *Release print examination and maintenance.*

⊕**Pressure Plate.** In a camera, projector or optical printer, a plate which presses on the back of the film in order to keep the emulsion surface in the focal plane of the lens. Camera pressure plates often have indented surfaces, or are fitted with a bank of small rollers, to retard emulsion build-up and hence reduce scratching.

See: *Camera* (FILM).

□**Pretuned Receiver.** Receiver adjusted for reception at one particular carrier frequency.

⊕**Preview.** Presentation of a motion picture to an invited or otherwise selectively limited audience before showing to the general public.

□**Preview.** To view a vision signal on a monitor before selecting it for transmission or similar application.

Also applied to the equipment used for this facility – preview monitor, preview bank (on a vision mixer).

See: *Master control; Vision mixer.*

⊕□**Primary Colours.** In colour reproduction processes, those essential colour components from which all other available hues can be produced by mixture. In three-colour additive processes the primaries are red, green and blue light, while when using dyes

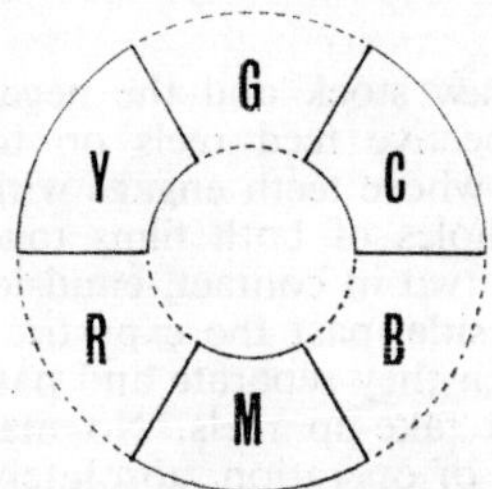

ADDITIVE AND SUBTRACTIVE PRIMARIES. Dotted sectors show additive primaries, red, green and blue. Opposite each in a solid sector is its corresponding subtractive primary, cyan, magenta and yellow.

or pigments in subtractive processes, the primaries are red-purple (magenta), blue-green (cyan) and yellow.

See: *Colour principles.*

Principal Focus. Point to which parallel rays converge after passing through a positive lens, or from which they appear to diverge after passing through a negative lens.

See: *Optical principles.*

Principal Plane. Plane through a principal point of a lens or lens system perpendicular to the optical axis.

See: *Optical principles.*

Principal Points. In a thick lens or a combination of lenses, the two nodal points along the axis which are optically equivalent to the centre of a thin lens. When the entrant and emergent rays make equal

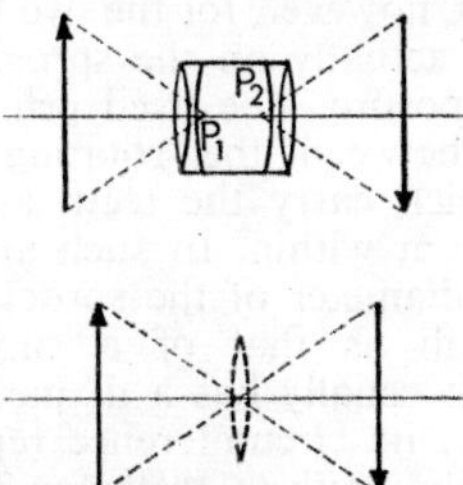

PRINCIPAL POINTS. (*Top*) P_1, P_2 are optically equivalent to centre of thin lens (*bottom*).

angles with the axis, the entering ray meets the axis at the first principal point, and the emerging ray appears to come from the second. Optical formulae for thin lenses may be used for thick lenses or combination systems if object, image and focal distances are measured from the principal points.

See: *Optical principles.*

Printer. Operator in a motion picture laboratory who prints cinematograph film. The term covers an extremely wide range of functions from the operator of a simple continuous contact printing machine for the bulk production of black-and-white copies to the technician in charge of a complex optical printing machine for special effects work. Release printing machines usually operate at high speed in dark rooms with safelights and their light change mechanisms must be automatically controlled from a punched card or tape programme so that the operator can concentrate on the feed and take-off of the rolls of picture and sound negative and positive stock. In the use of multi-head printers for bulk production, the negatives are often run as loops and the operator is principally concerned with handling the positive rolls.

The work of the optical printer, on the other hand, although carried out in the light, involves very exact setting up of an elaborate machine on which alterations may have to be made frame by frame as the work proceeds; while automatic and pre-set systems for lens movement and shutter opening are often employed, each job in special effects work is likely to be different and the operator must have a wide knowledge of the technical capabilities of his equipment.

See: *Laboratory organization; Printers.*

⊕**Printer Light.** Varying intensity of illumination provided by printers to compensate for differences in the exposure and thus the density of negatives, so that release prints and intermediate materials of uniform density can be produced. The illumination is arranged to vary in steps, called printer lights, determined by the grader and set up on light-change boards or by perforated control strips.

See: *Exposure control; Printer point; Printers.*

⊕**Printer Point.** Exposure step used in a motion picture film printing machine, also known as the light change point. A printer is usually equipped with exposure steps in logarithmic increments, and it is very general for a change of one printer point to produce an exposure change of ·05 log E. Half-points, corresponding to ·025 log E changes, are also frequently used. On any printing machine, the total range of exposure steps, usually 20 or 25 points, is called the printer scale.

The overall light level is set so that a correctly exposed negative prints at a mid-scale value around 10 to 12.

See: *Exposure control; Printers; Sensitometry and densitometry.*

PRINTERS

Equipment for printing motion picture film consists basically of means for moving exposed negative and unexposed positive film, emulsion facing emulsion, past an illuminated aperture, where the intensity of light can be accurately varied to give the desired printing exposure.

It is convenient to divide printers into two types according to how they move the film: continuous, where both negative and positive are moved at uniform speed, or intermittent, where the two strips are moved one frame at a time and held stationary for the period of exposure.

A further division is used to indicate the way in which the positive stock is exposed, either in contact with the negative strip, or by optical projection, in which an image of the negative is formed on the positive by a copying lens. Each type has its special field of application in motion picture laboratory work, continuous contact printers being normally employed for the rapid manufacture of large quantities of release copies, while intermittent printers, both optical and contact, are mainly used in preparatory work and in making special effects. A continuous optical printer is a rarity and has been employed only for the reduction printing of sound tracks.

CONTINUOUS PRINTERS

In its simplest form, a continuous contact printer provides film paths which bring the positive raw stock and the negative from their respective feed reels on to a drive sprocket, whose teeth engage with the perforation holes of both films together and move the two in contact, emulsion side to emulsion side, past the exposure aperture, after which they separate and pass to their individual take-up reels. No matter what the speed of operation, absolutely uniform continuous movement is essential to avoid irregular exposure, and the contact between the two must be consistently intimate during the whole exposure period. In some printers the driving sprocket is slightly displaced from the printing aperture, an arrangement sometimes termed the stationary gate. It is more usual, however, for the two films to be in contact actually on the sprocket at the time of exposure. The fixed printing aperture lies between the rotating sprocket flanges which carry the teeth and is illuminated from within. In such an arrangement, the diameter of the sprocket cannot be as small as that of a simple drive sprocket. It usually has a diameter of just under 4 in., its circumference representing a foot of film, with 64 teeth for 35mm and 40 teeth for 16mm.

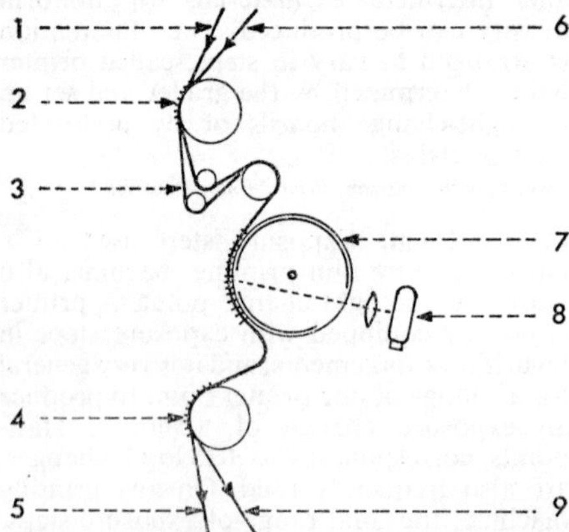

CONTINUOUS CONTACT PRINTER: FILM PATH. 1. Film from unexposed positive reel meets 6, film from image-bearing negative reel on 2, feed sprocket, and passes over tension rolls, 3, to main sprocket, 7. Here, positive is exposed in contact with negative by lamp and optical system, 8. 4. Take-up sprocket. 5. To exposed positive take-up reel. 9. To negative take-up reel.

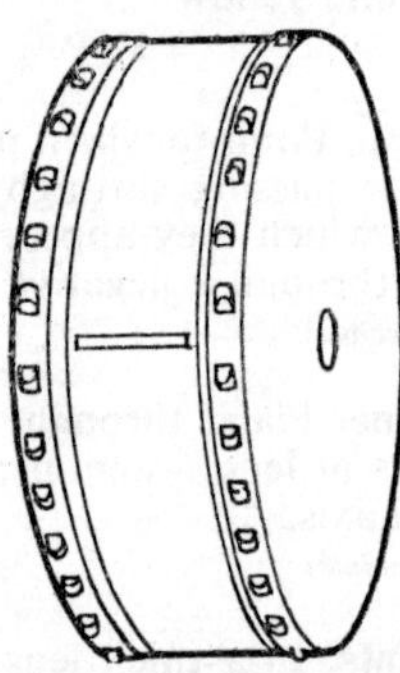

CONTINUOUS CONTACT PRINTER: SPROCKET DRUM. Exposure aperture is formed in fixed shell between rotating sprocket discs.

ELIMINATING FILM SLIP. Where two identical films are driven by the same sprocket, the outer layer moves on a circumference of slightly larger radius than the inner and some slip between the two is bound to occur. Relative movement of the two during exposure causes loss of definition of the printed images that can be serious, especially with sound tracks whose ability to reproduce high frequencies is dependent on a high standard of definition.

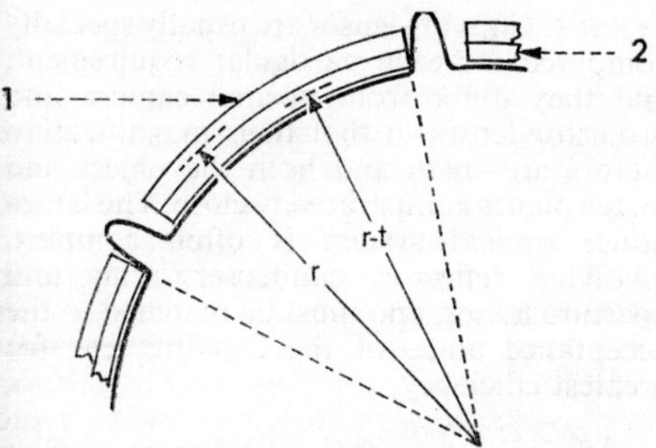

PITCH DIFFERENCE BETWEEN FILMS ON SPROCKET. To avoid slip and consequent loss of image definition when contact printing, perforation pitch of inner (negative) film, 2, must be less than pitch of outer (positive) film, 1, in the proportion $(r-t)/r$, where t=film thickness and r=average radius of outer layer.

The time during which slip may occur is made as short as possible by restricting the dimension of the aperture to a minimum in the direction of the film movement, but the greatest improvement is obtained by the use of films whose length for a given number of sprocket holes in the positive (outer) stock is slightly different from that of the same number of holes in the negative. For two layers of film of normal thickness wrapped on a cylinder of 4 in. diameter, it can be calculated that the length over a given angle is approximately 0·25 per cent greater for the outer layer than for the inner, and if this relation can be provided, slip from this cause is substantially eliminated.

HOLE AND SPROCKET DIMENSION. It has therefore become the practice to manufacture negative raw stock with an intentionally short perforation pitch to provide the desired difference from the positive. The standard hole-to-hole dimension for 35mm positive film is 0·1870 in. and the corresponding figure for negative is 0·1866 in. Similarly on 16mm, the positive raw stock pitch is 0·3000 in. and that for negative 0·2994 in.

The form of the sprocket tooth profile is carefully designed to provide smooth transfer of drive from one hole to the next on each of the two films and, in addition, lightly weighted rollers are provided in each path to give constant tension conditions.

OPERATION. Continuous contact printers can be made to run at high speeds, up to several hundred feet a minute, and are therefore well suited for the bulk production of large numbers of release prints. Black-and-white release prints are usually made from a specially combined dupe negative providing both picture and sound negative images on a single strip. When prints have to be made from separate picture and track negative, as is always the case in making colour prints, the printer has two exposing positions, or heads, so that both picture and sound images may be printed at a single pass of the positive raw stock.

To avoid the labour of rewinding the rolls of negative between each printing, simple adjustment of the aperture is provided so that the area exposed corresponds to the image position for running film either head first or tail first. For even greater efficiency in machine use, the negative is formed into large endless loops and the positive raw stock is fed in through a reservoir system. Then the printer can be run continuously without stopping to change reels and can be coupled to a continuous developing machine running at the same speed to provide a unified high-output processing channel for mass production. Continuous printing operations are usually carried out in a completely dark room with only safe-light illumination.

Continuous optical printers have been designed to permit reduction copies of sound track negatives to be made, such as 16mm prints from 35mm original negatives, but they have found only very limited application.

INTERMITTENT PRINTERS

In the intermittent or step printer, the two films, negative and positive, are held stationary for the period of exposure of each frame, after which the exposing light is cut off by a shutter while the two films are moved a distance of one frame so as to bring the next image into position for printing. The operation is therefore very similar to that of the camera and the intermittent mechanisms and associated shutters show a great likeness.

CONTACT TYPES. In a typical intermittent contact printer, the negative film and the unexposed positive raw stock are brought together, emulsion to emulsion, on to a continuously running feed sprocket which pulls the film off the reels. At the printing aperture, or gate, the two films are moved together by the reciprocating action of a shuttle mechanism during the period that the rotating shutter blade obscures the printing light, and are then held stationary in the gate by the pressure of the spring-loaded back plate while the shutter opens to give the exposure. In precision step printers, the two films are also very exactly positioned in the aperture by closely fitting register pins which engage in chosen perforation holes of the two films in the gate. When the shutter has closed after the exposure of each image, the shuttle moves the film the distance of one frame and the operation is repeated. The two films then pass over a second continuously moving sprocket to their separate take-up reels. Loops are formed in the paths of each film, both before and after the gate, to absorb film from and to the sprockets during the stationary period of the intermittent.

OPTICAL TYPES. In an intermittent optical printer, the same sequence of operations is followed, but the negative and positive films follow separate paths through separate but synchronized intermittent mechanisms, and the copying lens produces an image of

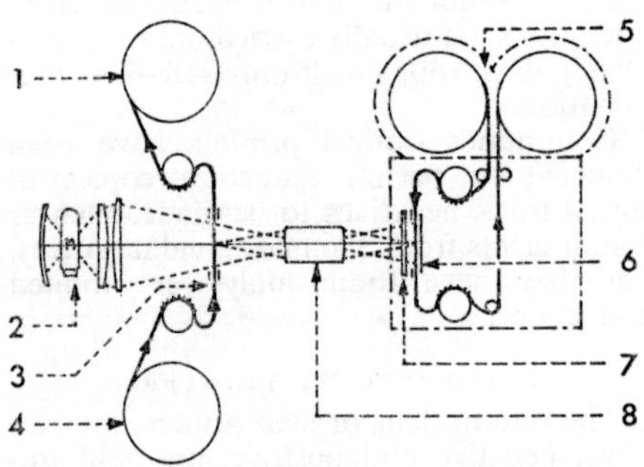

OPTICAL INTERMITTENT PRINTER: SCHEMATIC LAYOUT. Printing negative is fed upwards from reel, 4, through intermittent projector head, 3, to take-up reel, 1. Lamp and condenser, 2, form image of stationary frame in projector aperture through copying lens, 8, on to stationary frame of positive (printed) film in camera head, 6. During film movement, light is obscured by shutter, 7. Unexposed positive moves downwards from feed side of magazine, 5, and is wound on take-up side after exposure. Positive stock path may be completely enclosed, so that printer can be operated in lighted room.

the negative on the surface of the positive stock during the stationary period of exposure while the shutter is open. The direction of motion of the positive stock must, of course, be in the opposite sense to that of the negative to allow for the optical inversion of the image through the copy lens. The whole optical system must be carefully designed to provide uniform illumination of the printing aperture and a high standard of projected image quality at the positive plane.

LENSES. Copying lenses are usually specially computed for each particular requirement, and they differ from normal camera and projector lenses in that their magnification factors are small and both the object and image planes comparatively close. The lamphouse optical system is often complex, involving reflector, condenser, relay and aperture lenses, and must be matched to the acceptance angle of the copying lens for greatest efficiency.

WET PRINTING. Optical printing suffers from the disadvantage that it produces a print of higher contrast than normal contact printing, especially in black-and-white work, as a result of the Callier effect when using

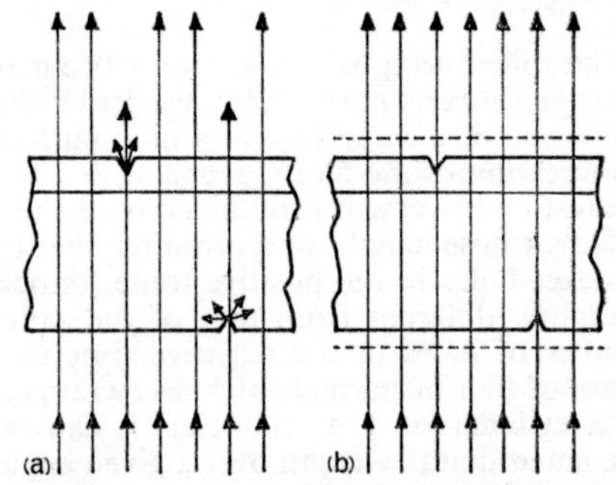

PRINCIPLE OF WET PRINTING. (a) Light transmitted by film is scattered and partially lost by surface scratches. (b) If film surface is coated with liquid of correct refractive index (dotted lines), light passes through it unimpaired.

specular illumination. The specular light also increases the apparent graininess of the print and emphasizes the presence of dirt, scratches and physical abrasions on the surfaces of the negative film.

The objectionable appearance of surface defects can be greatly reduced by coating the negative with a lacquer of similar refractive index to the film base or emulsion or by the use of a suitable liquid to provide a temporary surface layer at the time of exposure. Various volatile organic solvents are suitable and the film can be dipped into the liquid on its path to the aperture, so that a smooth layer is formed on each face by surface tension, or it can be contained in a cell of liquid between glass plates forming the aperture enclosure. After the exposure cycle, the liquid is removed from the surfaces of the film by air squeegees and the film is dried. The complete operation is known as wet printing, and by its use very severe scratches on both base and emulsion sides of the film can be made almost imperceptible.

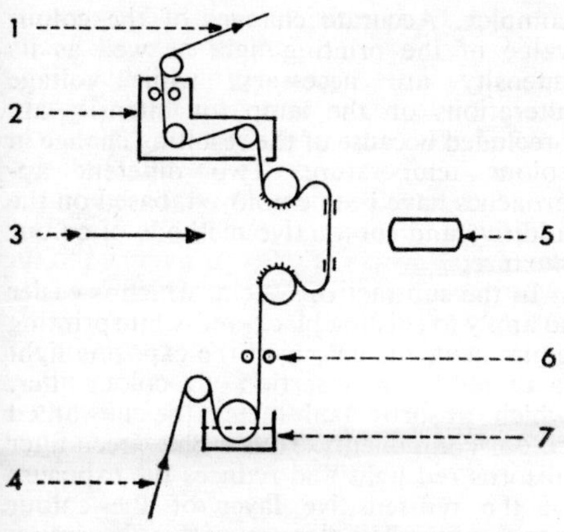

WET PRINTING: SCHEMATIC OF NEGATIVE FILM PATH. 4. Negative from feed reel passes through dip tank, 7, excess liquid being removed by air doctor jets, 6. Light, 3, from lamp house projects image of liquid-coated negative frame through copy lens, 5, and liquid is removed from film in drybox, 2, before passing to take-up reel at 1.

provided that the photographic image on the negative has not been disturbed.

OPERATION. Intermittent printers are slow in operation in comparison with continuous printers, and speeds of 10 to 20 fps (frames per second) are usually the highest rates which can be obtained on precision movements. They are therefore not well suited for printing the picture image on release copies (the sound track must always be printed continuously) but they play an important part in the film laboratory in making preparatory intermediates and for special effects work.

Prints in which the exact location of the image is important, such as back-projection plates, are always made on a step printer using registration pins, and for the same reason separation masters from colour negative must be printed on a step printer.

Where an alteration of frame size or image geometry is required, as in reduction, enlargement or anamorphic printing, an optical intermittent printer is essential, and one of its most frequent uses is for the preparation of reduction dupe negatives in 16mm format from 35mm original material.

In the field of special effects, the facilities offered by the independent operation of the negative and positive movements provide the basis of many cinematograph tricks such as stop frame, reverse action, skip frame for speed-up and stretch frame for slowed-down action. Movement of the lens and the film plane can provide zoom effects and position the image wherever it is required, while the accuracy of registration possible with an intermittent movement allows the combination of multiple exposures from several pieces of film. In the making of travelling matte shots, two strips of film can be used together in the projector head and imaged on to a third in contact with the raw stock in the camera head.

EXPOSURE CONTROL

In all film printing operations means must be provided to alter the intensity of the exposure to correct for variations of negative image density from one scene to another; for printing colour materials, the colour of the exposing light also has to be altered to compensate for differences of negative or to provide the colour effect required. Quite large changes of exposure may be required between successive scenes in an assembled negative, and the required exposure alterations must be made extremely rapidly, the ideal of course being an instantaneous change.

LIGHT INTENSITY CHANGES. For many years, most film-printing machines used incandescent tungsten lamps as their light sources, the intensity of which could be adjusted by variation of the voltage applied. This provides a practical means of exposure alteration but, except at low printing speeds, the thermal inertia of the tungsten filament involves too slow a response of light output for acceptable scene-to-scene changes. Consequently, voltage setting is now only used in continuous printing for overall adjustment of printer intensity to correct for positive stock sensitivity or similar factors which do not require alteration during the printing of a reel of negative.

Other methods of light modulation have been devised together with semi- and fully-automatic means of actuation to cope with the rapidity and the number of changes required in the course of printing a reel of film.

In continuous printers for black-and-white film, light intensity changes are usually effected by alteration of an area through which the light passes. Thus, a variation of the actual height of the slit at which the film is exposed effectively changes the time of exposure on film moving at a constant speed, but this can be unsatisfactory for sound track printing. Alternatively, a movable mask allows more or less of a diffused secondary source of light to illuminate the film, or a diaphragm in the illumination optical system is varied in size.

The diaphragm system may take the form of circular holes of varying diameter punched in a strip of opaque material and moved into position in the light beam by a rapid advance mechanism actuated at selected points along the length of the negative as it runs through the printer. A typical material for this control strip is a heavy black paper 35mm wide and perforated with holes along each edge similar to motion picture film. The diaphragm holes, which may be up to 20mm in diameter, are punched at intervals of five perforations according to the number of light changes required, and the whole strip provides a conveniently repeatable programme of the printing light values for a particular roll of negative.

PRINTER SCALE. All light change systems require a series of exposure increments of known value so that in the process of grading the negative for printing the effect of selected values may be estimated. In general photographic practice, equal increments of print density are produced by logarithmic increments of exposure. So the steps, or light points, in a film printer are arranged to give logarithmic increments of light intensity, and it was for many years a regular practice to provide a series of 21 points increasing in steps of 0·05 log exposure intensity. An increase of 6 printer points would thus double the intensity of the exposing light, while a decrease of 6 points would halve it. The range of steps from 1 to 21 is termed the printer scale, and the overall level of light would be set so that a correctly exposed negative would print at a mid-scale value of about 10 to 12. Grossly under- or over-exposed negative which is too light or too heavy to give satisfactory results from any of the 21 printing levels available is termed off-scale.

For some purposes, the step of 0·05 log E is considered too large for satisfactory grading and a smaller interval is employed. In some printing equipment, this has led to the adoption of an 0·04 log E point value, of which 8 are required to give double the exposure. More commonly, however, the number of intervals on the original scale is doubled by the introduction of half-point steps of 0·025 log E value, retaining the original full point numbers from 1 to 21 by ½'s.

COLOUR PRACTICE. With the widespread introduction of colour, the problems of printing exposure control have become more complex. Accurate changes of the colour value of the printing light as well as its intensity are necessary, while voltage alterations on the lamp for intensity are precluded because of the resulting change in colour temperature. Two different approaches have been employed, based on the additive and subtractive methods of colour mixture.

In the subtractive system, which is easier to apply to existing black-and-white printing equipment, the colour of the exposing light is altered by the insertion of a colour filter, which absorbs (subtracts) the unwanted colour components. Thus a blue-green filter absorbs red light and reduces the exposure of the red-sensitive layer of the colour positive film. The measure of this absorption can be given as the density of the filter to red light, and the printing effect of any filter can be thus assessed by colour densitometry.

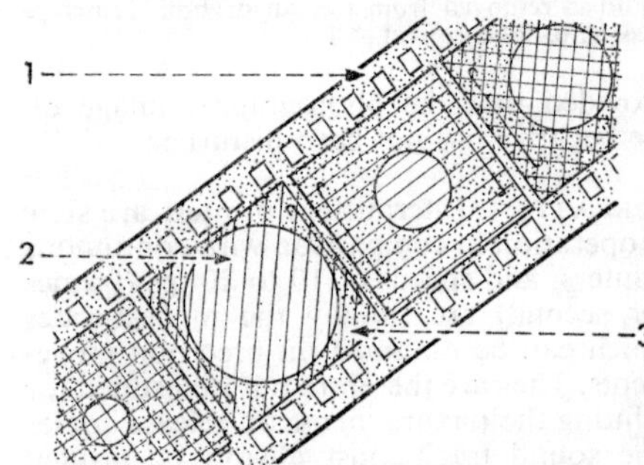

COLOUR PRINTING CONTROL BAND. Typical control band consists of opaque paper strip, 1, in which holes of various sizes, 2, are punched at regular intervals to control intensity of printing light; variations of colour are made by thin gelatin filters, 3, attached to strip by clips or staples at each hole. Perforations along edges are used to move strip in printer by one step each time a change in printing level is required.

In practice, laborious density measurements are avoided by the use of standard colour filters, usually of thin gelatin, which are manufactured with close tolerances to specific colour density values. These colour correction filters are made in the three primary colours, red, green and blue, and the three secondaries, cyan (blue-green), magenta (red-purple) and yellow, with specific density values in the series ·025, ·05, ·10, ·20, ·40, referring to densities to their complementary colours. Thus a filter specified as ·10 green (or briefly 10G) has a density of 0·10 to both red and blue light and negligible density to green light.

By the combination of a number of these standard gelatin filters, a pack of exactly the required absorption can be put together and used in conjunction with other methods of light intensity modulation. In one simple system, the filter packs may be attached to the control band strip by clips or staples, so that for each punched diaphragm hole there is also a selected group of colour filters. Another system uses a clear carrier strip to which the filters are attached by clips, neutral density filters of known density being used to control the printing intensity for each scene together with the required colour filters.

The additive system of colour printing relies on mixing carefully controlled intensities of red, green and blue light so as to provide the correct balance in the exposure. The optical arrangements for such a process are naturally more complex than those of a subtractive system. Either three separate lamps must be provided and individually controlled, or the light from a single lamp must be divided into red, green and blue portions which are then separately modulated and recombined before reaching the film being exposed.

Systems were developed in which the light from a lamp was used to illuminate three colour filters, the areas of which could be separately adjusted by the use of shutters or grids, but all these were inefficient in their transmission values and could not provide sufficient intensity for high-speed printing machines.

Modern additive colour printing machines have optical systems based on the use of dichroic filters which permit the red, green and blue components of the light from a single source to be efficiently subdivided by reflection and transmission at the filter surfaces. Modulating light valves giving accurately controlled quantities of light are situated in each individual colour beam and the three components are brought together, again by the use of dichroic reflectors, into the lens system which illuminates the printing aperture.

PRINTER LIGHT SOURCES. The most frequently used light source in motion picture printers continues to be the incandescent tungsten filament lamp, usually a high-efficiency projection type. To ensure long life and stability, it is desirable to under-run this type of lamp by operating at a lower voltage than its manufactured rating, but requirements of operating speed and stock sensitivity may make this impossible. Mercury vapour arc lamps are valuable in sound track printers for black-and-white film where UV printing is wanted, and high-pressure xenon arcs may come to be used for printing colour film if more powerful sources are needed for high-speed running.

VARIABLE SHUTTERS. In addition to changing the print exposure at each scene of negative, it may also be necessary to change the printing level smoothly over a series of frames. This is especially important in special effects printers where dupe negatives with fades or dissolves must be prepared. To provide a fade, for example, the exposure of the dupe must be gradually reduced over a specified number of frames from its full intensity down to zero, so that the image becomes fainter and fainter. Similarly, a dissolve or mix between two scenes is made by double exposing the film, so that over the required length the image of the first scene becomes fainter and fainter while, at the same time, the image of the second scene gradually increases from nothing up to its full density.

A step printer with a variable-opening shutter is commonly used to provide these effects. In most intermittent printer mechanisms, the period for which the film is stationary while being exposed is almost exactly equal to the time for which it is being moved to the next position, so that for a rotating disc shutter the maximum opening is 175° out of 360°. If the shutter is provided with a second disc which can be moved relatively to the first on the same axis, the opening can be reduced to any angle between 175° and 0°, which has the effect of reducing the exposure time. Actuation of the shutter opening by a cam driven by the main printer drive therefore reduces the image exposure from full to zero or increases it back again as the film runs through the printer. Selective gearing between the drive and the cam determine the number of frames printed during the complete closing or opening cycle.

In practice, it is usual to employ different cam profiles to produce different shutter opening sequences for fade effects and for dissolve effects to provide the most satisfactory visual impression. For a dissolve, in which the two successive exposures must be added, the shutter is opened or closed by uniform increments but, for a fade, a smoother transition to or from zero exposure is given if the shutter openings form a logarithmic series. It is usual to provide gearing which will provide the full shutter opening range over a number of frames which may be selected automatically, the standard lengths being 16, 24, 32, 48 and 64 frames.

Although the use of variable-opening rotational shutters is limited to intermittent printers, a similar feature is sometimes provided on continuous contact printers, which allows dissolve effects to be made by double printing from negative assembled in A & B rolls. In this form, a continuous change of light intensity must be operated, and this may be done either by the opening and closing of an iris diaphragm, or by a pair of blades moving to widen or narrow the width of the light source beam, depending on the particular optical system employed. In other forms, a graduated neutral density filter can be moved across the beam.

LIGHT CHANGE ACTUATION. When printing a reel of assembled negative scenes, a large number of changes in printing exposure may be required, often as many as 200, each of which must take place exactly as the join between the scenes reaches the printing aperture. The light change mechanism must therefore be actuated at an exact frame position along the length of the negative, and it was often the practice to notch the edge at the required points by the removal of a shallow sliver of film. The passage of this notch or nick was detected by a lightly loaded roller bearing on the film edge, which operated a micro-switch as the notch passed by to make the required change in printer light intensity.

This procedure is still widely used but the physical weakening of the edge of the film is a source of damage and, once made, the nicks are difficult to alter in position if the negative is recut or rearranged. Several alternative methods have therefore been developed in which the negative edge is not disturbed: for example, cues in the form of small metal tags or thin strips of conducting metal foil may be attached to the edge of the film at the required places to allow an electric impulse to be sent by bridging a gap between contacts as they pass through. Metal foil or spots of metallic paint may also be sensed by an electronic proximity detector which need not actually be in contact with the film at all, and a similar device can also be operated by spots of magnetic paint.

In simpler types of black-and-white printer, the required printing level is pre-set for each scene by the operator as a manual adjustment, the pre-set mechanism being released by the impulse from the notch or detector on the film at the required instant. With the greater elaboration required for colour printing, however, and the higher operating speeds of printing machines, the whole programme of scene-to-scene printing levels for a complete reel is now prepared in advance, either as a control band with colour filters or in the form of a punched tape coded for the required settings of the controls of an additive lamphouse. This punched tape control may incorporate the coding required to actuate the length and function type of a variable shutter for A & B printing. In some forms, it may even include punched coding to indicate the footage position at which a light change or shutter function is to operate, and when this is the case the negative film itself need not be nicked or cued in any other way. L.B.H.

See: *Colour printers (additive); Exposure control; Laboratory organization; Sensitometry and densitometry.*

Books: *Motion Picture and TV Film Image Control and Processing Techniques*, by D. J. Corbett (London), 1967; *Principles of Cinematography*, by L. J. Wheeler (London), 1964.

Printer Scale. Range of light intensity steps normally available on a film printer; the scale usually consists of 21 or 25 steps differing by exposure increments of ·05 log E, known as points, although some machines provide the same range in half-point steps of ·025 log E.

See: *Exposure control; Printers.*

PRINT GEOMETRY (16mm)

Cinematograph sound prints on 16mm film are made on stock which has a single row of perforations along one edge, and the soundtrack printed along the opposite edge outside the picture area. While there is now general agreement on the position of this track area in relation to the picture image, two forms of print may be encountered which differ in their thread-up position in the projector: emulsion to light, or celluloid to light, depending on the original material and the method of printing. Considerable inconvenience may arise when it is necessary to use the two types of print intercut in the same reel.

STANDARD AND CONTACT. Unfortunately there is not yet any universally recognized

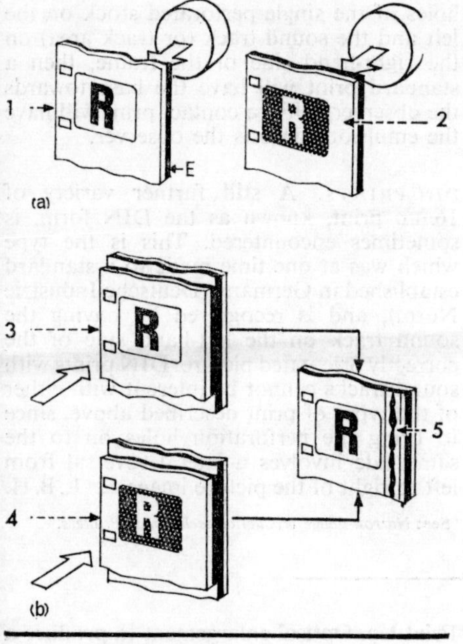

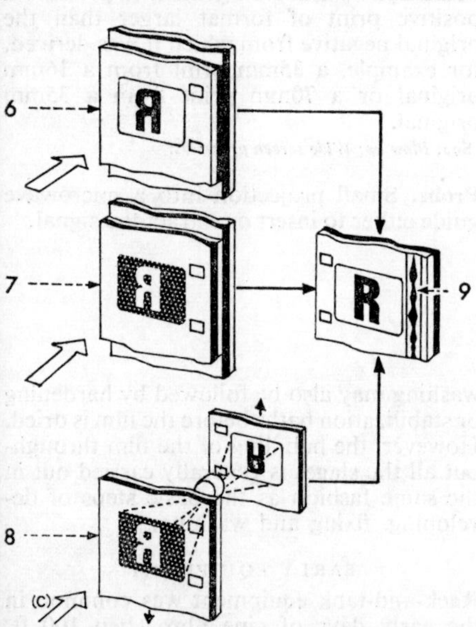

16MM PRINT GEOMETRY. (a) Both an original reversal master, 1, and an original negative, 2, are exposed in the camera with emulsion surface, E, towards lens. The image geometry of 1 is SMPE standard. (b) Normal contact printing, 3, from reversal master and, 4, from original negative, produces a print, 5, with contact image geometry different from SMPE standard. (c) Printing from contact intermediate, 6, gives a standard print, 9, but to obtain this geometry direct from the camera original, it is necessary either to print base-to-emulsion, 7, or preferably optically, 8.

nomenclature for the two types, which may be described as follows:

1. The form of print specified as standard in British and American Standard publications was originated by the Society of Motion Picture Engineers (now the Society of Motion Picture and Television Engineers), and was designed to have the same characteristics as an original reversal film exposed in the camera. An image on this type of print is seen the correct way round when viewed through the base, and if the copy carries a sound track, this is seen at the right-hand side of the correctly orientated picture. Such prints must be used in the projector with the base towards the lamp and the emulsion surface towards the lens.

2. The other type of print is that produced by making a normal contact print direct from a camera original, either negative or reversal. Since the normal printing operation is done with the emulsion surfaces of the original and the print stock face to face, the image on the resultant copy is seen the correct way round when the print is viewed from the emulsion side of the film, and in projection it is this side which must be towards the light source with the base towards the lens. As in the previous form, the sound track is seen at the right-hand side of the correctly orientated picture.

CONVERTING CONTACT PRINTS. It is important to recognize that a normal 16mm direct contact print from a 16mm camera original does not have the projection characteristics specified as the SMPE standard. However, it is often the practice to make release prints of an original reversal subject by means of an intermediate, either a dupe negative or an intermediate master, and such copies have the geometrical characteristics of a reversal original, and thus correspond to the SMPE standard.

To make prints of this form direct from an original 16mm negative, it is necessary to contact print with the original turned over, so that its base is next to the emulsion of the print stock. Note that it is necessary for the negative original to be turned over rather than the positive stock, because it is impossible to expose colour print stock through the anti-halation backing, which is not removed until the film is processed. Unfortunately, this form of printing with the emulsion surfaces separated results in a loss of definition of the image unless an extremely specular light source is used. This loss can be avoided by using an optical printer with the negative and stock in the correct relation, but this is often much less convenient.

In making 16mm prints by reduction from 35mm originals, which are usually negatives, an optical stage is always introduced, whether the copies are made by direct reduction printing from the original or by means of an intermediate reduction dupe. It is therefore possible to arrange for either form of geometry of the 16mm print to be obtained through the use of this optical reduction step, but unless specially requested, most laboratories make 16mm prints from 35mm originals in the form described as Standard.

RECOGNITION PRINCIPLES. The two forms of print can be easily distinguished by inspection. If the film is held so that the picture appears the correct way round, as it would be seen on the screen with the upper part of the image at the top, the perforation holes of the single perforated stock on the left and the sound track (or track area) on the right-hand side of the frame, then a standard print will have the base towards the observer, while a contact print will have the emulsion towards the observer.

DIN PRINTS. A still further variety of 16mm print, known as the DIN form, is sometimes encountered. This is the type which was at one time made to a standard established in Germany (Deutsche Industrie Norm), and is recognized by having the sound track on the left-hand side of the correctly orientated picture. DIN prints with sound tracks cannot be intercut with either of the types of print described above, since to bring the perforation holes on to the same side involves a lateral reversal from left to right of the picture image. L.B.H.

See: *Narrow-gauge production techniques; Printers.*

⊕**Print-Through.** Fault in magnetic recording □which occurs when, in a tightly wound roll of magnetic tape or film, a signal from one layer impresses itself inductively on the adjacent layer and so creates an undesirable background. Print-through is accentuated by high storage temperatures. It can be much reduced by winding through after storage and before replay.

See: *Magnetic recording materials.*

⊕**Print-Up.** Optical enlargement to produce a positive print of format larger than the original negative from which it was derived, for example, a 35mm print from a 16mm original or a 70mm print from a 35mm original.

See: *Blow up; Wide screen processes.*

□**Probe.** Small projection into a microwave guide either to insert or extract the signal.

⊕PROCESSING

The basic operations for processing motion-picture film are the same as for any other film: develop, rinse or stop, fix, wash and dry. The main difference is that cine film must be handled in much greater lengths and to a uniform standard, while colour film processing calls for some additional stages.

Modern integral tripack colour films normally require bleaching and fixing to remove the developed silver component of the image, while in reversal colour processing the developing operation usually takes place in two stages, the second of which provides the formation of the colour image by coupling. All these basic operations may be preceded by treatment of the film to remove anti-halation backing or to provide hardening for the gelatin, and the final washing may also be followed by hardening or stabilization baths before the film is dried. However, the handling of the film throughout all the stages is generally carried out in the same fashion as the basic steps of developing, fixing and washing.

EARLY EQUIPMENT

Rack-and-tank equipment was common in the early days of cine film when 100 ft. lengths of film were the general rule. It comprised a number of wooden frames which closely resembled one half of an old-fashioned domestic clothes horse, several wooden tanks, usually made of teak, big enough to permit complete immersion of one or more frames, a yoke to enable rotation of a frame, and a large-diameter drum constructed by spanning two disks of wood with slats, as in a water wheel. In use, the exposed film was wound round a frame and the ends were secured either by elastic or by a spring. The frame was then immersed in each of the solution tanks in turn for an appropriate time, until the processing was complete. Drying was accomplished by mounting the frame, pivoted about its centre, in the yoke, which allowed it to rotate. One end of the film was detached from the frame and secured to the drum, which was turned by a handle and the film transferred and wound up spiral-fashion with the celluloid down, thereby exposing the emulsion surface to the air for drying.

This type of equipment, though crude, provided a great deal of flexibility, and with enough tanks to contain the various solutions, every type of operation – negative, positive, and reversal processing as well as reducing, intensifying and tinting processes – could be carried out with minimum expense.

CONTINUOUS PROCESSING MACHINES

As the demands and standards of the industry increased, the need for more sophisticated methods of processing led to the development of continuous processing machines. In these, the film is caused to pass in a continuous length through tanks containing the necessary solutions and then on through the drying stages without traction being interrupted. The implications of such a requirement are that there should be means of attaching and removing rolls of film while the machine is in motion, means to control the flow and chemical activity of the solutions and adequate speed regulation to control and maintain development time.

The design and layout of a machine based on these requirements is almost infinitely variable, and in the early days most machines were specially designed to fit into a regular space within a building. They were large, and frequently extended through three rooms, the first being a darkroom in which development and fixing took place, the second a washroom (normally lit), and lastly a drying room. The ancillary equipment, including air compressors, temperature control gear, and chemical replenishment tanks, together with suitable pumps, was housed in yet another area.

Today, with the increasing use of 16mm film, particularly in television, many machines are manufactured as complete freestanding daylight-loading units, and only need connecting to suitable power, water and drainage supplies to be put into operation.

DESIGN CONSIDERATIONS. In a practical machine containing different liquids in separate tanks, only two basic film paths are possible. In the first, the film is guided by rollers into a series of long vertical loops in an elongated flat spiral formation, so that it effectively travels laterally from one side of a tank to the other. In the second, the film is guided into long vertical loops

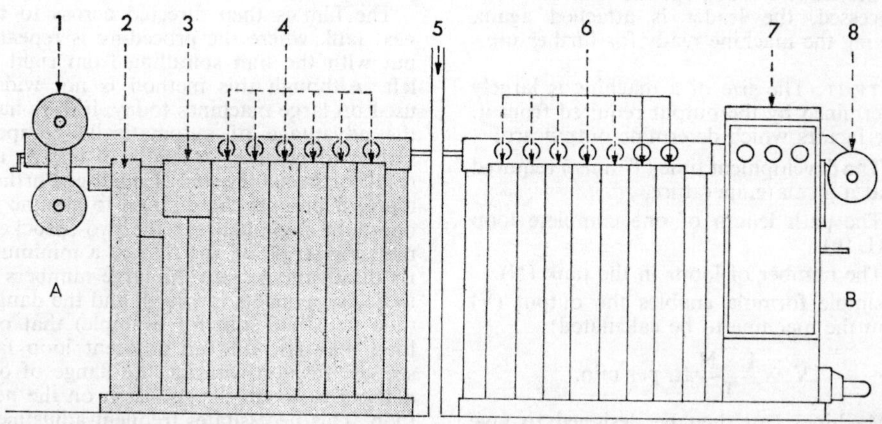

TYPICAL DARK AND LIGHT END COLOUR DEVELOPING MACHINE. A. Dark end. B. Light end. 1. Film feed, alternating between upper and lower spools. 2. Film reservoir. 3. Pre-bath treatment. 4. Six processing tanks. 5. Light wall through which film passes on its way to 6, seven additional processing tanks. 7. Drying cabinet. 8. Take-up spool. Capacity of small machine having one tank for each processing solution is about 600–900 ft. per hour, depending on time in first developer.

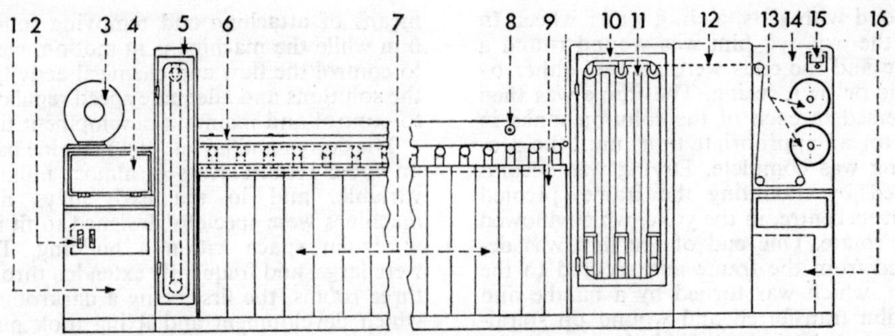

TYPICAL DAYLIGHT DEVELOPING MACHINE. **1. Solution pump and valve housing. 2. Flowmeters. 3. Enclosed film magazine. 4. Stapling compartment for attachment of new roll. 5. Film reservoir. 6. Daylight cover for first section of 7, processing tanks; open tanks are used for those after fixing stage. 8. Contact thermometer. 9. Final wash tank. 10. Drying cabinet. 11. Ducting for air jets. 12. Humidity control. 13. Film take-up. 14. Take-up housing. 15. Automatic temperature control. 16. Blower and main drive housing.**

in an undulating formation and travels longitudinally. The second method is not favoured because it means that alternate rollers will contact the emulsion side of the film. It is, however, sometimes used for convenience, and on these occasions the film is twisted to bring the base back into contact with the roller.

MACHINE LEADER. In a continuous machine, films are joined end to end one after another, and the film is suitably guided and driven by rollers on its path through each of the solutions. But first the complete film path must be laced or threaded, and for this a special tough film known as machine leader is used. The film to be developed is then joined to the end of it and is effectively pulled through the machine by the leader. When the last roll has been processed, the leader is attached again, leaving the machine ready for further use.

OUTPUT. The size of a machine is largely determined by the output required from it. The factors which determine output are:

1. The development time (T mins.) required at a given temperature,
2. The path length of one complete loop (L ft.),
3. The number of loops in the tank (N).

A simple formula enables the output (V) from the machine to be calculated:

$$V = \frac{L \times N}{T} \text{ ft. per min.}$$

Machines can then be designed to give the required path length or machine speed. In practice, very deep tanks are usually avoided because they are both cumbersome and difficult to house and clean.

FILM TRANSPORT. Film is normally driven and guided along the chosen path by rollers, and there are two principal ways of doing this. With positive drive, the film is propelled by its sprocket holes and with tendency drive, by friction rollers. Often a combination of both methods is used, loosely known as semi-tendency drive.

When positive drive is used, a driven shaft carrying the required number of sprockets is positioned horizontally across a tank. A loop of film is produced between the first and second sprockets and a large flanged weighted roller named a diabolo is placed in the loop. Since the film is engaged with the teeth on two adjacent sprockets which are fixed on the shaft, the loop remains of constant size, and successive loops spiralling from left to right are thus formed.

The film is then directed across to the next tank where the procedure is repeated but with the film spiralling from right to left. Although this method is not widely used on large machines today, it does have the advantage of automatically compensating for changes in length of film as the result of stretching and shrinking. Furthermore, it permits the diabolo to assume an optimum angle between the two sprockets, reducing stress on the film to a minimum. Its disadvantages are the large numbers of loose components involved, and the danger (due to a bad join for example) that one loop may rise and an adjacent loop fall, see-saw fashion, causing the flange of one diabolo to touch the emulsion on the next loop. This necessitates frequent adjustment to the level of the diabolos.

The most usual form of sprocket drive, however, is one in which the top driven shaft carries one sprocket only which is pinned to the shaft, the other positions on the shaft being occupied by free rollers which are assisted in their rotation by the driven shaft. The lower rollers, which form the spiral of film, are idlers, all mounted on a single floating shaft which is free to rise and fall by small amounts in accordance with changes of dimension of the film.

Positive sprocket drive is also used on the type of machine in which the film travels longitudinally, the length of the film between two successive sprockets being disposed as a series of loops, the rollers at the bottom of the loop usually being free to rise and fall to compensate for changes in film length during processing. Machines of this type are of straightforward design, but may occupy a considerable length and are

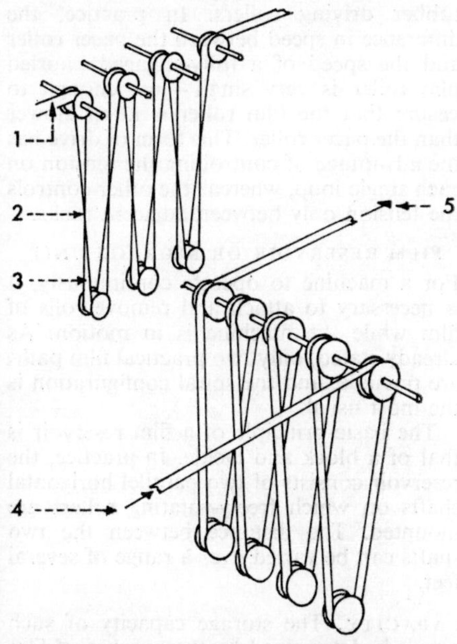

FILM PATHS IN PROCESSING MACHINES. (*Top*) Undulating or in-line film path. 1. Incoming film with emulsion outwards. 2. Twist in film to ensure that emulsion is again outwards at 3 for passage over bottom free roller. (*Bottom*) Spiral or transverse film path. 4. Film from preceding rack. 5. Film to next rack. Separate bottom rollers are diabolos, angling themselves to give straightest film path. Free rollers on single shaft are commoner.

usually, therefore, limited to comparatively simple developing operations.

In the tendency drive method, which is sprocketless, two shafts are positioned one at the top and one at the bottom of each tank. The appropriate number of rollers are mounted on the driven shaft so that they rotate when the shaft rotates. On the other shaft the rollers turn freely. None of the rollers carry sprockets.

In a typical construction, the two shafts are held apart by two vertical side members of channel section, in which the top shaft is carried in fixed bearings and the bottom shaft in bearing blocks which are free to move up and down guided by the channel sections forming the sides. These are known as racks and each tank is equipped with one or more of them. The drive is transmitted to the top shaft and rollers by friction, which enables the speed to vary between one rack and the next to accommodate stretching or shrinking of the film. Many ways exist to effect this differential drive but only two will be considered here.

DIFFERENTIAL DRIVE METHODS. To describe the orthodox principle of differential drive, it is necessary to consider the behaviour of only two complete loops of film, the first one being the last on its shaft in tank No. 1 and the second being the first on its shaft in the adjacent tank No. 2. The primary drive is rotated positively by gears or chains and is transmitted to the top film roller in tank No. 2 by a smooth belt, which imparts drive only when tensioned by a roller which is connected to the sliding bearing block assembly in the preceding tank No. 1. In operation, if the drive to the top roller in tank No. 2 is too fast, the bottom roller in tank No. 1 will rise, carrying the connecting rod and roller upwards, thereby reducing the tension on the smooth belt until the speed is reduced. This in turn permits the bottom roller in tank No. 1 to

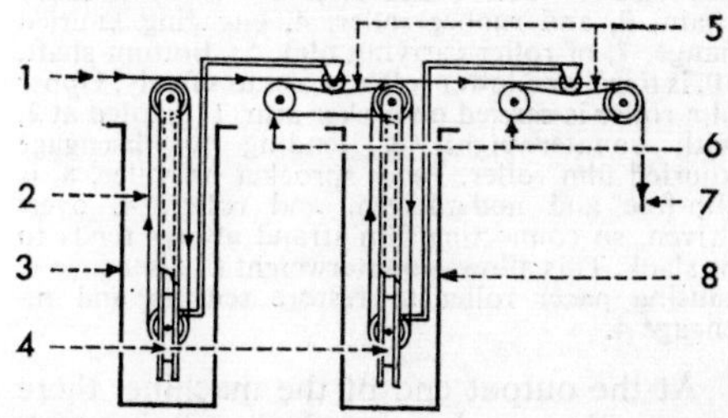

TENDENCY DRIVE: SCHEMATIC LAYOUT. 1. Film from preceding rack. 2. Last loop on its shaft in Tank One, 3, passing across to form first loop on its shaft in Tank Two, 8. 4. Sliding bearing block carrying bottom shaft and free rollers. 5. Endless smooth belt with tensioning roller connected to 4 of preceding rack. 6. Primary drive wheel, overdriven. 7. Film to next rack. If drive in Tank Two is too fast, rack in Tank One will rise, thereby reducing pressure of tensioning roller and allowing a measure of belt slip which causes rack in Tank One to fall.

descend and tension the belt. A state of equilibrium is then rapidly reached where the rate of slip on the friction drive equals the rate of feed from the preceding tank.

Thus the speed of drive to each film rack is automatically controlled and maintained by the position of the bottom shaft in the rack preceding it.

In a variation of the orthodox principle, the tension on each individual film loop is automatically controlled. A fixed spindle is mounted across the tank, on which are a row of rocker arms pivoted about their centre. Each arm carries a film roller at one end and a counterweight at the other. The flanges are knurled around their periphery and can be brought to bear on a plain rubber drive roller immediately beneath and parallel to the row of film rollers. The lower rollers are free and are mounted on a shaft with its centre fixed close to the bottom of the tank.

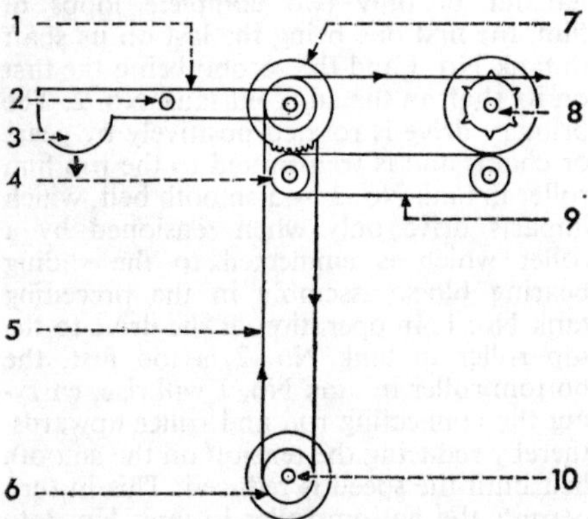

MODIFIED TENDENCY DRIVE. Drive is by primary chain, 9, and rubber roller, 4, engaging knurled flange, 7, of roller carrying film, 5. Bottom shaft, 10, is fixed, and lower roller, 6, rotates freely. Upper film roller is carried on rocker arm, 1, pivoted at 2, with counterweight, 3, tending to disengage knurled film roller. Pacer sprocket or roller, 8, is slip-free and underdriven, and roller, 4, overdriven, so connecting film strand always tends to be slack. This allows counterweight to disengage 4, causing pacer roller to restore tension and re-engage 4.

At the output end of the machine, there is a single sprocket, or pinch roller, which effectively provides a slip-free drive and is called a pacer roller or come-along roller. The relationship in speed between the maximum drive possible from the rubber rollers across the tanks and the pacer roller is fixed, so that the output of the pacer roller is always effectively the slower. When the machine is at rest, the counterweights on each rocker arm tend to disengage the knurled film rollers from the rubber driving roller. This is contrived in the following manner.

When the machine is started, the pacer roller rotates and increases the tension on the first film loop, thereby pulling the knurled film roller into contact with the rubber driving roller. This imparts a drive which in turn increases the tension on the next loop, and so on throughout the length of the machine.

In the reverse reaction, when a single film roller overdrives it will tend to produce a slack loop of film which allows the counterweight on the rocker arm to disengage the drive until equilibrium is restored.

Thus, throughout the length of the machine, the rocker arms are constantly oscillating as they engage and disengage the rubber driving rollers. In practice, the difference in speed between the pacer roller and the speed of a fully-engaged knurled film roller is very small – just enough to ensure that the film roller is never slower than the pacer roller. This form of drive has the advantage of controlling the tension on each single loop, whereas the other controls the tension only between adjacent racks.

FILM RESERVOIR OR STORAGE UNIT

For a machine to operate continuously, it is necessary to attach and remove rolls of film while the machine is in motion. As already stated, only two practical film paths are possible, and the spiral configuration is the most usual.

The basic principle of a film reservoir is that of a block and tackle. In practice, the reservoir consists of two parallel horizontal shafts on which freely-rotating rollers are mounted. The distance between the two shafts can be varied over a range of several feet.

CAPACITY. The storage capacity of such a unit is determined by the number of film loops available and the effective lengths of loops. Consider one loop of film formed between two top rollers and a single bottom roller hanging free. If one end of the film is anchored and the other end is pulled x ft., the bottom roller will rise $\frac{x}{2}$ ft. If 10 loops are formed, and the single free bottom roller is replaced by a guided shaft carrying 10 rollers, it follows that if 20 ft. of film is pulled out, the bottom shaft will approach the fixed top shaft by $\frac{20}{2\times 10} = \frac{20}{20} = 1$ ft. If the shafts are initially separated by

7 ft., at least 5 ft. of effective travel is available, permitting 100 ft. of film to be pulled out, while the shaft separation is reduced by 5 ft. The time for this to occur is known as the storage time of the elevator and depends on the film speed through the machine. It must be long enough for a magazine change and a film splice, plus a healthy safety margin. One minute is usually adequate.

RESERVOIR DESIGN. In a practical machine, the foregoing factors are used to determine the number of loops and the height of the reservoir cabinet. It is preferable to use a tall cabinet with a few loops rather than a short cabinet with many loops because the accumulated friction from the large numbers of rollers required for many loops imparts considerable stress to the film.

The method by which one shaft is guided parallel to others is a matter of convenience; in some machines, the top shaft has a fixed axis with the bottom one rising towards it, while others have the opposite arrangement.

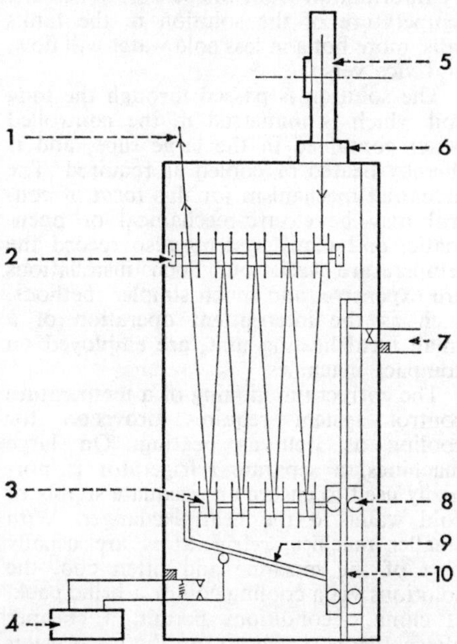

FILM STORAGE UNIT (RESERVOIR): SCHEMATIC LAYOUT. Film passes from magazine, 5, through attaching boot, 6, to fixed top shaft, 2, with free rollers. Bottom shaft, 3, is mounted on frame, 9, carrying rollers, 8, which guide its upward movement along fixed vertical bar, 10. At 1, film passes to first processing tank. If film entering reservoir is stopped for splicing in new film roll, film continues to move out at 1, and rollers, 8, climb on bar, signalling to operator through contacts and warning device, 4, that reservoir is in action. If re-loading is not completed by the time reservoir reaches top, switch, 7, brings machine to emergency halt.

In one widely used method, the bottom assembly moves upwards guided by rollers which bear on a single vertical rectangular bar, this effectively maintaining the lower shaft parallel to the fixed axis of the top shaft. A microswitch is located at the lower end of the bar to actuate a warning device whenever the bottom assembly rises, and another microswitch is located near the top limit of travel to switch off the machine drive motor when all the stored film has been used.

MAGAZINES. In a daylight-operated machine, exposed film is loaded into light-tight magazines which are then fixed to the input end of the storage unit via a light-tight device (often called a boot) which also traps the end of the film when a magazine is removed. The magazines must be loaded in a darkroom, and in one method, the exposed film is positively anchored and wound on to a core which is secured to a backplate. The backplate is then placed inside the magazine, and the end of the film is fed out through a light trap, so that a foot or so is available for joining. After the lid is secured, the remaining operations may be carried out in daylight.

FILM JOINING. Two basic means are used to join film for passage through a processing machine: an overlapped join and a butt join. Either can be achieved with appropriate metal staples or special adhesive tapes. The main requirements are that the joins should be made in the minimum of time, and that they must be reliable. There are proprietary devices to align the film accurately while the join is made and, where appropriate, align the perforation holes as well.

OPERATION OF RESERVOIR. While the film from a magazine is passing freely into the reservoir, the bottom assembly maintains a position close to the base of the reservoir cabinet. In this position, the two shafts are spaced at their greatest distance and the reservoir is fully charged. When the roll of film comes to an end, the tail remains positively anchored to the core secured to the backplate in the magazine. Since the machine is still running, the lower part of the storage unit must rise to maintain the feed to the first processing tank, and a bell or buzzer warns the operator of what is happening. He disengages the magazine from the boot, which automatically retains the end of the roll, breaks the film, replaces the magazine with a full one, joins the film and, on fully engaging the magazine in the boot, automatically releases the entrapped film, permitting it to flow freely into the reservoir again. To prevent the lower assembly from falling rapidly to the bottom of the cabinet, its movement must be restrained, and an air dashpot or damper is often used as a brake.

PROCESS CONTROL

SPEED CONTROL. For satisfactory operation of a processing machine, which must handle several different types of film, the development time must be variable. This can be achieved either by altering the path length or, more usually, by altering the speed. For large machines, it is common to use variable speed d.c. motors, having high torque at low speeds, which are fed from the regulated d.c. supply usually found in commercial laboratories. For compact machines such a supply is not always available and these normally operate from a.c. mains, necessitating a.c. variable-speed motors. These, however, are prone to speed variations caused by small changes in input voltage, resulting from alterations in the current as the heaters are switched off and on. In addition the mechanical loads on the machine can vary. A very satisfactory alternative is to use a mechanical speed variator driven by a brushless induction motor.

SOLUTION DISTRIBUTION AND CONTROL. To maintain a satisfactory and consistent standard of processing, the agitation, temperature and chemical activity of the developer must be kept within certain limits. In a practical system, agitation and temperature control are interdependent, as a large volume of developer cannot be easily maintained at a given temperature unless it is kept thoroughly mixed.

In large machines, solutions are mixed in separate tanks and passed to a main reservoir where the temperature is held at a nominal value. From these they are pumped as required through a heat exchanger to the processing tanks, from which they flow back to their respective reservoirs.

In smaller compact machines, the solutions are mixed and put directly into the processing tanks, and the pumps and temperature control are part of the machine itself.

TEMPERATURE CONTROL. In all but the very simplest machines, some attempt is made to maintain the temperature within known limits. For the developer in black-and-white processing, it is desirable to maintain a control of ±0·5°F, but many machines operate satisfactorily with a variation of ±1·0°F. Other solutions are more tolerant and ±5·0°F is usually satisfactory. In the processing of colour film, the developing operation is more critical and a tolerance of ±0·3°F is recommended, with ±2°F for the other solutions.

Many control systems exist, but in principle they all comprise a temperature sensor placed in or near to the solution tank actuating a unit which either heats or cools the solution as required.

In a large machine, the solutions are passed through a heat exchanger. This consists of a long coil of pipe housed inside a large tube connected to both a hot and cold water supplier differentially controlled by information from the sensor. When the temperature of the solution in the tanks falls, more hot and less cold water will flow, and vice versa.

The solution is passed through the long coil which is immersed in the controlled water contained in the large tube, and is thereby heated or cooled as required. The actuating mechanism for this form of control may be electro-mechanical or pneumatic, and many systems also record the temperature variation. Such installations are expensive, and much simpler methods, such as the intermittent operation of a small local heating unit, are employed on compact machines.

The correct functioning of a temperature control system requires provision for cooling as well as heating. On large machines, a separate refrigerator is normally used to ensure an adequate supply of cold water to the heat exchanger. With smaller machies, refrigerators are usually part of the machine and often cool the solutions via a cooling coil or a brine pack. If climatic conditions permit, a cheaper alternative is simply to use the wash water for this purpose.

RECIRCULATION AND AGITATION

Consistent and even development of motion picture film requires fresh developer continuously supplied to the emulsion surface. Failure to do this results in uneven development, streaks and directional effects, particularly where areas of high and low density are adjacent. The basic requirement, therefore, is to produce sufficient turbulence at the surface of the film to wash away exhausted developer and replace it by fresh developer. The movement of the film itself provides some of this agitation and the faster the movement the more effective it is. In most machines, film movement alone is insufficient and supplementary agitation must be provided. This can be by recirculation, submerged sprays, cascade or inert gas bursts.

RECIRCULATION. The commonest method of additional agitation is recirculation. It is simple, relatively inexpensive and easily applied to compact machines. Large installations make use of two variations.

In the first, a large centrifugal pump takes developer from the main reservoir and pumps it through the heat exchanger into the processing tank. The developer then overflows via a float valve (to prevent air being drawn in) back to the reservoir.

In the second method, one pump injects fresh developer from the reservoir, while another, with a much slower flow rate, operates a closed circuit, taking developer from near the top of the solution tank and pumping it back near the bottom. This enables smaller-diameter pipe lines to connect the overflow between the developer tank and the reservoir while still maintaining a high flow rate through the tank.

Smaller machines, with no reservoir, use one pump operating on the closed circuit principle. The degree of agitation should be at least the minimum above which any further increase produces no change in the density, gamma, and uniformity of the developed image.

SUBMERGED SPRAYS. This system differs from the recirculation method in that the output from the pump is directed through small submerged jets on to the emulsion surface of the film. To be effective, the jets must be positioned close to the film and give an even flow of developer across the whole width of the film. The velocity of the emerging jet often makes it necessary to provide back-up rollers behind the film to prevent it being driven away by the jets of developer. The output of the pump is used to greatest effect because all the flow is on to the film and not into the tank. However, the jets add to the cost and may take up too much space for a compact machine.

CASCADE. The cascade method of agitation is confined to specially designed machines. There are no tanks as such, the film being enclosed in splashproof cabinets. Thus the loops can be made longer than usual, often up to 10 ft. long. The bottom of the cabinet consists of a shallow trough rather like that used for a shower bath. The solutions are pumped at a high rate through large orifices on to the strands of film as they pass over the top rollers, and cascade down the film into the trough below, and so back to the reservoir. This method is not practicable on compact machines but is used successfully in some very large and fast machines for processing positive film.

INERT GAS BURSTS. In colour processing machines, the agitation is often provided by bursts of nitrogen. The system is arranged to release bursts of bubbles at regular intervals from the bottom of the tank. As the bubbles rise, they agitate the solution.

SPRAY DEVELOPING. Processing machines have been designed in which the film is not submerged in the solutions but passes through a series of cabinets where the liquids are directly applied in the form of a fine spray. A high rate of replacement of the used solution at the surface of the film is claimed in combination with chemical economies, but the dangers of developer oxidation in the presence of air require special precautions, including filling the cabinet with an inert gas such as nitrogen.

REPLENISHMENT

Whenever a chemical reaction takes place, active constituents are consumed and by-products are released into the solution. From a practical point of view there are two alternative methods of coping with this gradual change in the composition of the solution. The first is to use the solution to a point when its performance is no longer acceptable and then to replace it completely. This method, called batch processing is generally adopted with compact machines. The second is to keep the solution under constant analysis and attempt to maintain its activity within limits of acceptability. This requires a chemist with elaborate metering and measuring equipment and is therefore economical only where large volumes of solution are involved, as in a big laboratory.

The successful operation of a full replenishment system involves the analysing and preparation of the replenisher and the proper control of its addition to the main solution. It does not suffice to check the solution once a week and add several gallons of replenisher in one lot. Proper control is achieved by constantly metering the developer through a flowmeter so that chemical activity is maintained, and bleeding off an equal amount to ensure that the by-products are kept at a certain level. Initially this involves much testing and checking but eventually it is possible to arrive at a suitable flow rate for each material, and it is not unusual for big laboratories to work continuously for many months without replacing the bulk solution.

In some instances, the control is taken to the point of automatically reducing the flow of replenisher when machine leader is passing through the solution, as, apart from the carry-over from the previous tank, this material causes no chemical reaction because it has no emulsion.

In practice, the replenisher is usually introduced into the return pipe from the processing tank, so that it is well-mixed with the bulk solution. It is also necessary for the flow rate between the reservoir and the processing tank to be sufficiently high to avoid local exhaustion and excessive build-up of by-products in the actual processing tank.

VISCOUS PROCESSING

A form of processing which employs extremely small volumes of solution and eliminates both agitation and replenishment makes use of thin coatings of viscous liquids containing the necessary chemical constituents. A uniform layer of the viscous material is applied to the emulsion surface of the film, and allowed to remain for the required processing time before being removed by a vigorous spray jet of water. The reaction is usually carried out in enclosed cabinets in an atmosphere of air saturated with water vapour, and can be at elevated temperatures for rapid processing.

Both developing and fixing of black-and-white film can be done in this way, the final washing being carried out by water sprays.

Because the actual volume of solution required is that which is applied directly to the film, viscous processing machines can be very compact and can employ pre-mixed packaged solutions, so that they are particularly well-suited to the requirements of the small user who does not require an elaborate installation of large output. So far, however, the viscous processing of colour materials has not proved practicable.

AIR KNIVES AND SQUEEGEES

When film passes from one solution to another, it carries on its surface a significant quantity of the liquid contained in the previous tank, which contaminates or, in the case of water, dilutes the solution in the next tank. To minimize this carry-over, devices known as air knives and squeegees are fitted between tanks to remove the surplus liquid.

AIR KNIVES. This name is usually applied to a unit comprising two chambers, one positioned either side of the film, and arranged with narrow slits spanning the width of the film and inclined against the direction of its travel. Air at several pounds pressure is fed through these slits so that a very high velocity wall of air pushes the liquid back along both sides of the film as it travels through the unit. Suitable rollers are positioned to ensure that film is guided centrally through these jets so that joins do not foul them.

The design of air knives is a compromise between cheapness of manufacture and efficiency. The crudest design is two short tubes of small diameter (say ½ in.) which have a flat machined down their length which breaks into the bore. A stainless steel plate is screwed to the flat and is adjusted to leave a slit which is a few thousandths of an inch wide. One end of the tubes is blocked up and air is fed in at the other end.

More efficient designs use chambers which are shaped to utilize the principle of a

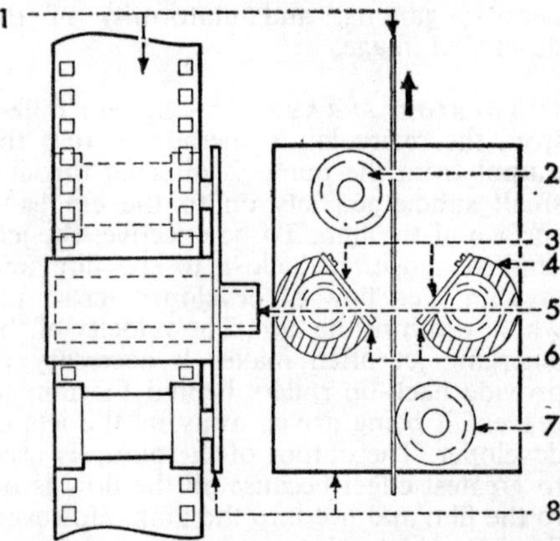

PRINCIPLE OF AIR KNIFE. 1. Film with guide rollers 2 and 7. 3. Stainless steel plates secured to flatted tubes, 4, to form air nozzles. 5. Air inlet to nozzles, with outlets, 6, forming jets against direction of film travel. 8. Mounting plate.

venturi to increase the air velocity, without an increase in the air consumption. A machine employing several air knives will require a very large air compressor to supply it because each unit may well use 3–5 cu. ft. air per minute depending on the pressure.

The high velocity airflow atomizes the liquid it is blowing off and these droplets can find their way into other solutions. In an enclosed daylight-operated machine, provision must be made to vent or extract this contaminated air.

SQUEEGEES. A squeegee is really a mechanical wiper, like a windscreen wiper blade. In this application, the blade is kept stationary and the film moves. Squeegees are commonly used on compact machines because they need no air supply, but they are likely to make scratches, and unless strict cleaning and adjusting routines are observed they can be troublesome.

A variation on the wiping squeegee is supplied by the suction squeegee, which is extensively used by some commercial laboratories. The unit comprises a suction box through which the film passes and in which a depression is maintained by the close proximity to the film of rubber-lipped blades. The blades are arranged in pairs on opposite sides of the suction box and are adjusted to leave a gap of only 0·006 in. so that they effectively serve as a suction seal. The required depression in the system is maintained by a vacuum pump, and separators or traps are included to prevent the sucked-off liquid entering the pump.

This system is more economical in terms of power than the compressor required for the air knife, but demands greater-diameter pipe lines for the low pressure airflow. It also has the big advantage of not spraying particles of solution into the free air, from where they might contaminate other solutions.

DRYING

When the film has passed through all the wet processes, the surplus water is removed by either an air knife or a squeegee. If these units are operating efficiently, both sides of the film are free from droplets of water, but the emulsion side, which can be likened to a micro sponge, still contains a significant amount of water entrapped in the swollen gelatin layers. This water should be evaporated, or dried out, so that the film is left with a uniform moisture content, or normalized with the surrounding air. This operation is usually carried out in a drying cabinet, in which several loops of film are exposed to a warm air flow long enough to effect the required amount of drying.

DESIGN CONSIDERATIONS. The most important consideration in the design of a drying system is the heat-versus-velocity combination. It is clear that warm air will dry film more rapidly than cold air and that a high velocity airflow will produce quicker drying than a gentle breeze.

In practice, the best solution is to use a lot of air with the minimum of heat and, on large installations where space is not at a premium, this method is generally adopted.

The very simplest design comprises a cabinet, or in some cases a room, which effectively encloses the banks or racks of film. This is fitted with a heater and a large fan fitted with a filter. The air is drawn through the filter, passes over the heater and is directed into the cabinet so that it blows from the dry end towards the wet end.

Although this method is widely used, it is not very efficient because a fairly large proportion of the available air never touches the film.

HIGH TEMPERATURE & HIGH VELOCITY. With the increasing use of compact machines, there is less footage of film, and therefore less time available for drying. To achieve the required result in the time available, there are two alternatives: to increase either the temperature or the effective velocity of the air.

A significant increase in temperature leads to a softening of the gelatin and a loss of the plasticizer in the base. This causes both shrinkage and brittleness. Furthermore, since the heat is usually provided by electric elements, this method is expensive in power.

The most elegant solution is to direct the air so that every bit of it is made to impinge on the film. This is achieved by the use of what is known as a plenum, which is a hollow rectangular chamber shaped to fit between adjacent banks of film. Air is fed into this chamber and emerges from both sides through rows of slits positioned to direct the air straight on to the film.

Thus the film path is effectively followed by rows of air jets positioned close to and coincident with it, so that all the emerging air impinges on the film. In this way film can be dried quickly with a short path length.

APPEARANCE OF DRYING FILM. Film is made up of two parts, the base which is usually cellulose tri-acetate, and the emulsion which consists of silver halide grains suspended in gelatin. When the film is wetted, the sponge-like gelatin absorbs more water than the base and in consequence swells more.

This action causes the emulsion side to become relatively longer than the base and the film curls with the emulsion on the outside.

As the film dries, the gelatin contracts, first causing the film to flatten and then to curl the other way, so that the emulsion side is concave.

This change of curl makes it easy to observe the pattern of drying and to determine the exact point in the cabinet where the film is dry. Ideally the film should appear dry in approximately half of the total drying time so that it can normalize itself during the remaining time. D.J.C.

See: *Chemicals; Chemistry of the photographic process; Laboratory organization; Sensitometry and densitometry; Sensitometry of colour films; Viscous processing.*

Book: *Motion Picture and Television Film Image Control and Processing Techniques*, by D. J. Corbett (London), 1967.

⊕**Process Shot.** Photographing action against a projected background or against a uniformly illuminated coloured backing for subsequent combination with a background scene by a travelling matte process.

See: *Special effects* (FILM).

⊕PRODUCER

The film producer's function begins with the conception of a film in its minutest form and continues long after its release. He has to merge hundreds of real and abstract creative financial elements into a practical course of action.

That merger begins with a basic literary property which will eventually evolve into a screenplay. This material can be taken from almost any literary form – a novel, a play, a short story, an original scenario or an original idea.

CASTING

CHOOSING A SCREENWRITER. Once the producer has selected his basic literary material, he needs a screenwriter. Often producers make the grave error of trying to go into scenario form with the novelist. This can be a mistake, because the novel form is very different from the screenplay form. A screenplay is only a stepping stone to telling the story with visual images as opposed to word pictures. And, while a novelist can draw word pictures, a film director or cameraman often cannot translate these words into meaningful images.

The producer discusses with his screenwriter the overall story and how to translate it into screen play form. It is important that a producer stay involved in the writing process to ensure that the basic concept of the new story form is not lost. He should be aware of a step outline, the headlines, the form the story will take, how it will develop, which characters will be retained, which characters will be dropped and which will be changed.

Certain screenwriters are virtually stars and deserve to be treated as such. But a producer makes a grave error if he allows creative control out of his hands. He should not be dictatorial about what he wishes to do, but by the same token if he is going to function properly as a producer, he must play an active part in the creative process.

CHOOSING A DIRECTOR. Once the screenplay is in good form – the so-called first draft screenplay – the next step in casting is the director. Since the director obviously will have very definite ideas in relation to the screenplay and will want changes, the final screenplay will be written with the advice and comment of the director as well as producer and writers.

The producer usually chooses a director in the same manner in which he casts an artist. He can cast to type or against type. Obviously, for a very light comedy, he would not choose a director whose past record has been one of pictures filled with heavy, stark realism. Some directors have no capacity for fantasy. Others are marvellous comedy directors but cannot do heavy drama. Choosing a director, then, is basically a matter of experience and a working knowledge of that director's method. Additionally, the producer should know if a director is a fast or slow worker. There are directors who cannot physically or emotionally do more than a limited number of set-ups in one day.

Various economic factors enter into picture-making here. With a limited budget, a producer needs a director who can do more than a few set-ups in a working day. Again, numerous retakes and excessive coverage are an expensive, impossible drain on a low-budget production.

CHOOSING TECHNICIANS AND CAST. Total casting of a picture is crucial. All elements must be cast with equal care – cameraman, sound engineer, editor, all of the key technicians involved in making the film.

As well as technical abilities, the human element must be considered. Cameramen, for example, are of differing mood and temperament. The producer must be able to judge whether the cameraman will get on well with the director, because this rapport between the director, his assistant directors, and the lighting cameraman is a key thing once the film goes on the floor. Here all elements must mesh.

Casting the actors involves the same process for the producer as casting a writer or director. Common sense dictates that the artist be right for the part. This does not imply that stereotyping or type-casting is a necessity.

MAKING THE FILM

When photography starts, the producer's function is to insure that the director and crew keep to the shooting schedule, because the old cliché, "time is money", is nowhere more important than in film-making with its extremely high daily operating expenses.

Among the multitudinous elements that a producer must also concern himself with are day-to-day administrative matters such as insurance, the personal and personality problems of the cast and crew, weather cover sets and the relationship and rapport of the cast with the director and the director with his crew.

EDITING. The completion of photography is still only the beginning for the working producer, because, while he has been supervising the making of the movie to this point, he must now concern himself with his own ideas as to how the film should be edited and cut. Many bad pictures have been turned into good pictures by imaginative and creative cutting and editing. The producer is also concerned at this time with the scoring of the picture, the sound mixing, the sound effects, opticals and titles.

TECHNICAL KNOWLEDGE. Throughout shooting and post-production, it is important that the producer have a good working knowledge of the technical side of the scene. He must know what will happen when his director, or his cameraman, says "switch over to an 18" or "switch over to a 25" (mm lens). He must know what is being discussed and what the effect of the lens change will be. Even though he may never process a foot of film, he should be able to appreciate the importance of colour correction when striking release prints. He must understand the difference between the mobility and portability of one camera as opposed to another. He must be aware of why a lighting cameraman asks for so many lights on a certain set and just how much wattage a location generator will pour out to the demands of the lighting cameraman, and how many brutes, bashers and pups he wants to light a set and what effect he is striving for.

He must have a knowledge of sets, what they are, what they cost and why. Labour relations are very important: he must understand the union and guilds, the demands that are made and the reasons for them. He needs to have a working rapport with the shop stewards as well as with the department heads of the studio.

MERCHANDISING. When actual production is complete and the picture in the can, the producer still has work to do.

Advertising and merchandising of the picture now become a prime consideration. He must work with the distributor to decide the best manner in which to exploit his product, and how to sell the picture to the film-going public. The business of merchandising a film continues long after its initial theatrical run. Re-runs, other subsequent appearances and television showings, including the special marketing and exploitation of films for television, must be dealt with.

FINANCE

Thus far, this discussion has been limited to the creative and technical roles of the producer. His role as the financial director is of equal importance.

In the early days of motion pictures, the large studios had staff producers – men who were paid a weekly salary, given assignments and who remained under contract to one studio. Today, the trend is toward the independent producer.

An independent producer can finance a film in many ways. Among them, he can put up his own money, borrow money from some private source or work through and with a large distributor.

The most common practice is the last, with the producer putting his own money into a literary property, developing the property to a certain point and then making an arrangement with a distributor for financing.

As an independent agent, the producer is in fact a partner. His income can be derived from a salary called a producer's fee, usually some percentage of the total budget of the picture, and he can own a percentage of the picture.

To describe and document fully the total role of the motion picture producer would take volumes rather than pages. Suffice it to say that each of the phases mentioned above is no more than a bare outline, for the modern motion picture producer's role covers every aspect of film making – creative, financial and technical.

Some involve themselves more than others. There are, in fact, two distinct types of producers – the packager who merely assembles the elements, takes his profits and moves along to another venture, and the working producer who treats each project as a continually unfinished labour of love. The package arrangement is simple and neat while the working producer invests difficult months, sometimes years, in a single picture.

Neither exists, perhaps, in the pure form, for the packager may become creatively involved against his will, and the working producer must have regard to budgets and schedules. But these are broadly distinguishable types. There is a place for both of them in the film industry. Each must choose his own style. K.H.

See: *Director; Producer* (TV).

□PRODUCER

A television producer usually exercises overall control of a project consisting of more than one programme. It may be a drama anthology made up of individual plays, a dramatic series with continuing characters, a group of documentaries with or without a common theme, an entertainment series or a long-running current affairs programme. To put these programmes on the air, the producer works with a team of directors, each of whom is responsible for the transmission or recording of one or more programmes in the series. The director is operationally in charge of the studio, staff, actors, performers, cameras, and all the technical apparatus. He works under the supervision of the producer, who retains editorial and artistic control and gives the programme series style, continuity and unity.

BASIC DUTIES

Whether the producer is himself responsible for conceiving the programme series depends on a number of factors. They include the practice obtaining in the organization for which he works and the circumstances in which he is employed.

He may be a free-lance brought in to produce a series of programmes which he has conceived. He may – again as a free-lance – be invited, because of some special skill, to look after a series conceived by someone else. If he is a staff producer, he may have thought up the idea for the series, or he may simply be allotted to it, because his services are available and he is well equipped for the job.

CHOOSING DIRECTORS. The producer's freedom to choose his own team of directors varies. In general, television organizations prefer to use directors who are already on their payrolls. Free-lance directors are an extra charge on the budget. So although a producer always attempts to handpick his team, he may find that he has to employ X or Y, who are by no means ideal for the kind of programme he has in mind, simply because the organization cannot afford to let directors on the payroll sit idle.

This problem would be solved if all but a very small quota of directors were on short term contracts, or contracted to direct a small number of specified programmes. But such a policy might well lead to the casualization of the industry.

BUDGETING. One of the producer's first duties on taking charge of a series is to prepare a budget estimate. This may be his own estimate of what the total cost of the series will be, expressed as a global figure, which he then tries to persuade his employers to allocate to the project. It may, on the other hand, be his attempt to achieve the best results within the limits of the sum his employers have told him is available.

The budget estimate deals with direct costs, which are variously defined by different organizations. They include such items as the hire of actors and musicians, copyright fees, author's fees, costumes, the cost of filming (stock and printing), transport, travel, accommodation for cast and technicians, subsistence money, sets (scenery to be constructed in the studio), and the hire of back projection or other special equipment.

There are also indirect costs, which can include such details as a proportion of the producer's own salary and the cost of employing the technical crews – cameramen, outside broadcast crews, film editors – according to formulae worked out by the cost accountants.

The producer's responsibilities are more than financial, however. He must also be aware of the effects of his technical decisions on the organization. Thus, if he decides to use a great deal of film, he must bear in mind that he will require a lot of editing time and a large number of cutting rooms and film editors. If he exceeds the resources of his own organization, he will need to hire outside facilities. Alternatively, he may force some other producer to hire outside facilities. Either solution will be unpopular, since it is the aim of most organizations to work within the limits of their own resources. From the producer's point of view it is also preferable to do so, since outside hire charges become direct costs on his budget and mean that he has to make economies in other directions – in the quality of the actors he can employ, or the kind of scriptwriter he can hope to hire. The ability to budget accurately and to induce his employers to agree to as large a budget as possible are among the first requirements of a producer.

SPECIALIZED KNOWLEDGE

The producer generally has to have some specialist knowledge outside the field of television.

DRAMA. A drama producer is expected to have a wide knowledge of the theatre and to keep himself abreast of developments in it. His expertise must include the ability to assess the talents and suitability for television of actors and actresses. He must know which writers working for the stage may be useful for television. He will want to see the work of stage directors and stage designers.

It is true that for advice on casting and on the choice of authors the producer can turn to experts within his own organization – the head of casting or the head of scripts – but unless he has some personal knowledge of these fields he will find it difficult to choose between the alternatives they offer him.

In the case of a drama anthology or a dramatic series he has to know what kind of script editor he wishes to collaborate with and who the best people are in that field. Alternatively he may decide to be his own script editor.

LIGHT ENTERTAINMENT. The light entertainment producer must be familiar with trends and fashions in entertainment, in pop music and in humour. He must know what artistes are available in all branches of show business – dancers, comedians, musicians and singers – and should have first hand knowledge of emerging talent at home and abroad.

CURRENT AFFAIRS. The producer of current affairs programmes is expected to have detailed and up-to-date knowledge of developments in home and foreign affairs, together with a feeling for the flow of the news, for trends in politics, and for the good story. He knows who the experts are on any subject of public interest and debate, and how to lay hands on them so that they appear in his programme and not that of the opposition. He has good contacts with the political parties at all levels from Cabinet members to backbenchers. He has to deal with politicians up to Ministerial rank in circumstances which are often extremely delicate. He must be seen to be at once expert and politically neutral, scrupulously fair to both sides.

SPORT. In sports programmes, the producer has his own problems, which can sometimes be almost as delicate as those posed by questions of political balance. Sport, like politics, is emotionally charged. Sports personalities are often highly extroverted. Promoters are quick to imagine they detect partiality. The public is well informed on the subject and feels strongly about it. Any suspicion of unfairness or ignorance will lead to vehement protests from sportsmen, promoters, and public. The producer must know how to handle them all.

HUMAN QUALITIES

TACT. Whatever area of television programming the producer works in – whether it be children's programmes, scientific, artistic or religious programmes, he has to deal with experts, who expect to be understood and to be treated with due respect and tact. One of the producer's most valuable contributions to a programme is to act as a middleman between the expert and the audience. This is particularly so in the case of scientific or medical programmes, where the experts find it difficult to believe that everyone does not understand their particular technical jargon or start off from their assumptions.

The producer is in these cases a kind of filter. He must be able to understand the expert and know the point he is trying to make. He must persuade the expert to break down complicated thought processes and to translate technical terms into something near ordinary speech. In short, the producer has to be for the expert an intelligent lay audience. Over and above this he must think constantly of visual methods of driving home the argument of the programme, of making it good television.

Tact might be defined as one of the supreme qualities of the producer, for he is constantly involved in difficult situations – between writers and script-editors, between directors and actors, with technicians, with his superiors, with the press, with agents. Tact, however, is useless unless combined with firmness and a clear idea of what he wishes to achieve, of what he believes must be fought for, of his minimal demands as the creator of the programme series.

JUDGEMENT. Some of his most difficult decisions have to be taken when he is called upon to act as censor or to give a ruling about taste. The point at issue may concern a scene in a play, a shot in a documentary, a gesture, a camera-angle, or a word or phrase in an interview or commentary. He may find that the onus for decisions of this kind is placed firmly on his shoulders. He may have no written code to guide him and must therefore use his own judgement.

In organizations such as the British Broadcasting Corporation (BBC), he is guided by an ingrained code, based on his own personal attitude, which is conditioned by his beliefs and upbringing; on a general code of accepted practice, unwritten but familiar to everyone working on programmes; and on proper sensitivity to the world around him. This is both a very liberal code and at times a very difficult one to apply.

If he finds the problem too complex or feels that it raises issues beyond his competence, he can refer it to his departmental head, who may be an executive producer with a number of producers working under him.

LEGAL CONTROLS

In commercial television in Britain, the producer must work within the framework of the Television Act, which is applied and interpreted by the officers of the Independent Television Authority. He is naturally free to refer any difficult question to his superiors; but a different element is introduced into the situation by the existence in the Authority and in the individual companies of programme clearance officers. These officers are responsible for seeing that programmes comply with the terms of the Act, which lays down that "nothing (may be) included in the programmes which offends against good taste or decency or is likely to encourage or incite to crime or to lead to disorder or to be offensive to public feeling".

The officers of the ITA have the right to see programmes at any stage and to discuss them in detail. They may object to a programme in whole or in part. It is, however, open to the producer to argue his case. If he is persuasive he may carry the day; alternatively a compromise may be agreed on. A really good producer normally contrives to see that matters do not go so far – or if they do, knows precisely why he wishes to insist on screening the programme in a debatable form.

Whichever type of organization the producer works for, decision of this kind are among the most difficult he has to make. In the case of recorded programmes or films, they can be taken with a certain amount of deliberation. With current affairs programmes they frequently have to be taken in the studio during the run through or even when the programme is already on the air.

ORGANIZATION

In general, current affairs programmes present a producer with some of his most complicated problems, not only editorially but also in the technical and organizational fields.

STATE OCCASIONS. Programmes like the coverage of a general election, a state funeral, an American Presidential or a British Royal occasion can be of immense complication. On a purely organizational level, the producer has to deploy his team of reporters and commentators, who may be scattered throughout the country or even spaced out across the globe. He must know how to get them to their posts, what facilities they will require when they arrive – camera crews, studio and recording facilities – and how to get their material back to base. He needs to be familiar with the television organizations in the areas or countries from which his material is to originate, to know what facilities he may hope to receive from these organizations, and, above all, what they are going to cost him. If he knows the right members of the foreign organizations personally, his task is greatly lightened.

NATURE OF MATERIAL. The material the producer requires from sources outside the studio may be live, i.e., provided directly by electronic cameras, on film, recorded on videotape for deferred transmission, or in the form of a filmed telerecording. It may even be stills. The producer may wish the pictures to be accompanied by natural sound only, so that commentary can be added in the studio. Alternatively, he may need the voice of the man on the spot.

His material can reach him by various routes. Film and tape can be flown in. Live pictures can be brought to the studio by coaxial cable, by cable supplemented by visual links (microwave), or by satellite. The producer knows how much his lines are going to cost him, how long it will take to set up the circuits, how many visual links are required, and who is to provide the gear.

INTERNATIONAL NETWORKS. In the case of programmes brought over the Eurovision coaxial network, which in turn links up with the Intertel network of Eastern Europe, he knows how to set up the necessary circuits. In the case of satellites, he is aware of their availability and has good contacts with the ground stations. Besides the coaxial cable for picture transmission, he books music lines for sound and control lines for communication with his commentators and the operational engineers. If the signal, or some of the signals, are to be originated by cameras with different line standards from his own, he must remember to book conversion machines.

EQUIPMENT. Outside the studios he has chosen his camera sites, which may have to be the subject of negotiation with civic authorities, the police, firms or individuals. Inside the studio he decides, with the engineers, what technical equipment is required to meet his editorial needs. Together they decide on the number of cameras and microphones. The cameras, in the case of an elaborate political programme, are required not only to photograph speakers, announcers, politicians, and studio activity generally, but to scan captions, stills, teleprinter tape, and results cards. The speakers, in turn, require monitors so that they can follow what is going on; so, too, do the staff on the studio floor. Internal communications by means of headphones or earplugs have to be set up between the control box and the floor. Other important matters of a non-technical nature include arranging that those taking part in the programme, which may be very prolonged, have food, drink and somewhere to rest.

PROCEDURE AND QUALITIES. During the broadcast, the producer is in the control room with the director, watching the flow of the programme, keeping an eye on the material coming in to the studio as it is displayed on the array of monitors, and taking editorial decisions as to its use. Once the operation is under way he cannot hope to interfere much. He is like a commander whose troops are now engaged and have been committed; but he must be ready to take major decisions and to sustain the morale of the complex team which by its joint effort is making the programme possible.

The ideal producer is a paragon of men. He is conversant with the technicalities of television and able to talk to the engineers in their own language. He is an expert in some field outside television, has tact and judgement and is painstaking and immensely attentive to detail. He is a good programme accountant and can inspire and lead others. Ideal producers are, in the nature of things, rare. A good producer is one of the most valuable assets of any television organization. S.H.

See: *Director; Producer* (FILM).

Books: *Factual Television*, by N. Swallow (London), 1966; *Techniques of Television Production*, by R. Bretz (New York), 1966.

□**Programme Switching Centre (PSC).** Centre at which programme items are switched in order to provide a complete programme.

See: *Outside broadcast units; Transmission networks.*

⊕**Projectionist.** Operator in motion picture studio, processing laboratory or cinema theatre who runs the film projection machines. In studio and laboratory review rooms, the projectionist is required to handle separate picture and sound track prints, the latter being both optical and magnetic records; his range of equipment is therefore varied and he is responsible for maintaining standard conditions of screen brightness and sound quality so that the results of production work can be correctly assessed. Although in the cinema theatre the projectionist is always handling married prints, he is the final link in the chain of operations which brings the motion picture to the public. His skill is responsible for ensuring an unbroken and faultless period of entertainment by presenting a clear, bright and sharp picture on the screen accompanied by good quality sound at a suitable intensity.

See: *Automation in cinema theatres; Projector; Cinema theatre.*

⊕**Projection Sync.** Synchronization of picture and sound tracks which takes account of the physical separation or stagger between the picture aperture and the sound head in projectors. This separation varies with the gauge of film and with the type of sound track, magnetic or optical. Projection sync is contrasted with recording sync, also called cutting or parallel sync.

See: *Advance (sound and picture).*

□**Projection Tube Television.** System of projecting a television image from the end of a high-voltage cathode ray tube or tubes on to a screen.

See: *Large-screen TV.*

⊕PROJECTOR

Motion picture projectors in current use cover a range of film sizes from 8mm to 70mm, and applications ranging from amateur showings through education, selling, television broadcasting, and data analysis, to the professional theatre. These different uses have generated an equal diversity of projector types, whose common function is to synthesize the illusion of motion out of a time series of still photographs.

BASIC REQUIREMENTS

FRAME FREQUENCY. The need for the synthesis of motion, and the psycho-physical response characteristics of human perception which make it possible, set some of the limits of performance which must be met by acceptable equipment. The human eye and mind have a limit at about 48 Hz (cycles per second) above which, at useful light levels, an intermittent light is perceived without flicker. This sets the minimum frequency at which the light beam must be interrupted by the shutter.

Depending on the amount of movement between one picture and the next, there is a frame frequency at which discrete pictures blend into smooth motion. Experience shows this to be at about 16 fps (frames per second).

If the positional movement from one picture to the next is greater than a fraction of one per cent of the frame height, the picture appears noticeably unsteady on the screen. There is an upper limit to the distance the film can move along the lens axis before the image loses its crispness of focus.

These characteristics of the perceptive process set rather stringent limits on the mechanical performance of the projector, and to a great extent dictate the structure and the precision of the film moving parts.

FILM TRANSPORT. Motion picture films are usually stored on some sort of spool. It is necessary to unreel the film from one spool, pass it through the picture mechanism in an intermittent motion, through a sound mechanism if there is one, in a smooth motion, and wind it on to a take-up spool. The intermittent motion at the picture gate, and especially the smooth motion at the sound gate, must be isolated from the usually irregular pull of the spooling drive. This means that, in all but the simplest equipment, there is a sprocket or similar drive which unreels the film from the feed spool and supplies it to the intermittent feed mechanism at the picture gate, and there is a similar sprocket which holds back against the tension from the take-up spool.

PROJECTOR PARTS. The basic elements, then, of a motion picture projector are:

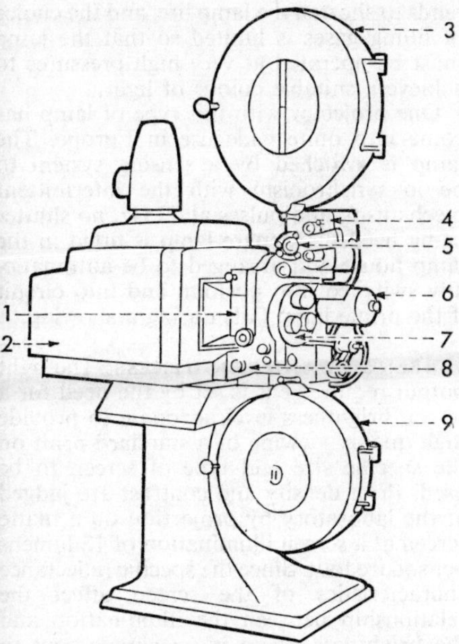

LAYOUT OF MODERN 35MM PROJECTOR. 3. Upper spool box, capacity 6,000 ft. of 35mm film. 4. 35/70mm penthouse magnetic head, with plug-in head clusters for four-track and six-track films. 5. Dual-gauge feed sprocket, with 35mm sprockets fitting between and below level of 70mm sprockets (other sprockets similar). 6. Drive motor. 7. Three-lens turret, replaceable by single anamorphic lenses. 1. Light beam passage from lamp house, 2, with water cooling ducts to remove heat from interchangeable 35mm and 70mm gates. 8. Optical sound head. 9. Lower spool box.

A spool on which the film is stored; an unspooling drive; a device for providing intermittent motion to the film at the picture gate; a picture gate and aperture; a device for generating uniform motion at the sound gate; a sound scanning and amplifying system; a hold-back arrangement against the spooling drive; the spooling drive; a lamphouse and lamp to provide the projection illumination; a projection lens.

It is in the form, size, and arrangement of these basic elements that various projectors differ, the specifics being to a large extent determined by the film size and the use to which the projector is to be put.

DESIGN PRINCIPLES

INTERMITTENT MOVEMENTS. The major point of difference between projectors for various services lies in the design of the components which produce the intermittent film motion at the picture projection gate.

In 35mm and 70mm projectors for the professional theatre, the Maltese cross, or Geneva movement is almost universally used to drive an intermittently-rotating sprocket. This advances the film rapidly to bring the next frame into the projection aperture, and then dwells stationary for as large as possible a part of the cycle.

Typically the pulldown time is 25 per cent of the total time for each frame. The projection rate is 24 fps, and the flicker frequency must be at least 48 Hz, so that it is necessary to have a two-blade shutter, resulting in a shutter transmission of about 50 per cent. For a Geneva movement to produce a steady picture, the cross-shaped member which is intermittently driven must be manufactured with extreme accuracy, and the driving member must be fitted very closely to the cross.

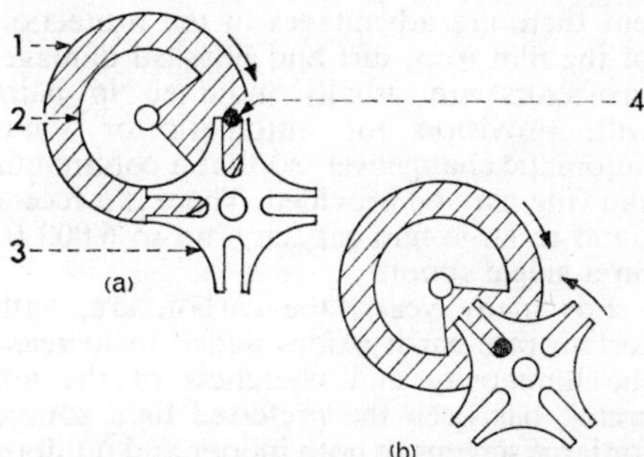

MALTESE CROSS INTERMITTENT. (a) Driving member, 1, rotates continuously. Convex face, 2, positions adjacent concave face of driven member, 3, to which intermittent sprocket is attached, and locks it in place. Driving pin, 4, is just entering slot of driven member. (b) Pin turns driven member which swings into cut-out area of guiding face. When pin leaves slot, driven member is again locked in position. Very accurate machining and fitting of all engaging faces are needed.

Projectors for 70mm film are usually dual-purpose machines. Two sets of sprocket teeth on each shaft, with the 35mm toothed flanges between the 70mm flanges, provide the drive, so that to change from one film width to the other, it is necessary only to adjust the film guides and change the aperture structure.

Projectors used for backgrounds for process photography during the shooting of professional motion picture films must meet especially strict requirements for picture steadiness. These projectors are equipped with film movement mechanisms derived from the shuttle advance and pilot pin registration used in professional cameras. In these movements, the film is advanced by a pin which engages a perforation, and is located by separate pilot pins which engage additional perforations. These pilot pins are full-fitting and are very precisely located, being either stationary or, if movable, very closely guided. These projectors are synchronized with the cameras so that their shutters are open at the same time as the camera shutters.

TYPES OF LAMP. For professional theatre operation, projectors are equipped with arc lamps, either carbon, or more recently xenon, operating at power levels up to 11,000 watts, and capable of providing a first-class level of illumination on very large screens. Most professional theatre projectors are designed with the film path completely enclosed from supply spool to take-up. To some extent this is a holdover from the days of inflammable nitrate film, but there are advantages in the protection of the film from dirt and physical damage. Projectors are usually mounted in pairs with provision for automatic or semi-automatic changeover, so that a continuous showing may be provided. There is a recent trend to large film capacity, up to 6,000 ft. on a single spool.

For many years, the carbon arc, with certain rare earth oxides added to increase the luminosity and whiteness of the arc crater, has been the preferred light source for large screens in both indoor and outdoor theatres. Lamps using carbons 11·5mm in diameter and operating at currents up to 180 amperes have provided acceptable illumination levels on large screens in major theatres and enormous screens in drive-in theatres. Light outputs may be as high as 40,000 lumens.

In the past decade, the xenon arc lamp has come into wide use, especially in Europe, with lamps currently being offered up to 6,500 watts, fully competitive with carbon except in the largest sizes, and having the advantage of a burning-life of more than 1,000 hours. In the xenon lamp, the arc is struck between tungsten electrodes in an atmosphere of xenon gas at a pressure of several atmospheres.

The advantages of the enclosed arc lamp can be further extended by the use of a pulsed power supply which turns the lamp on only during the time when the film is not moving. This makes possible a substantial reduction in average power drawn from the mains for the same effective illumination on the screen.

There are difficulties in the design of such a lamp, since the pulsating current tends to shorten the lamp life, and the choice of filling gases is limited so that the lamp must be operated at very high pressures to achieve a suitable colour of light.

One projector with this type of lamp has come into quite wide use in Europe. The lamp is switched by a sensing system to be in synchronism with the intermittent mechanism, and pulses at 72 Hz, no shutter being needed. A spare lamp is fitted in the lamp house and arranged to be automatically switched into position and into circuit if the prime lamp fails during a showing.

SCREEN SIZE AND BRIGHTNESS. The light output requirement is set by the need for a screen brightness level adequate to provide high quality viewing of a standard print on the specific size and type of screen to be used. Print density and contrast are judged at the laboratory by projection on a matte screen at a screen illumination of 12 lumens per square foot. Since the specific reflectance characteristics of the screen affect the relationship between the illumination and the brightness, there is no simple way to compute the maximum screen size from the total illumination available.

A matte screen operates as a diffuse reflector, appearing to have the same brightness from any viewing angle. If a screen returns light in a specific direction at a higher level than would a diffuse surface, it is said to have a gain in that direction. It is possible, by shaping the surface of a screen into lenticles backed by a reflective coating, or by using a layer of glass beads, to concentrate a very large portion of the illumination in the direction of a normal audience space. Gains of two or three are common, and some beaded screens have gains as high as four, giving a bright picture to an audience seated within a relatively narrow angle.

In the professional theatre, there is a considerable use of anamorphic prints and projection lenses. The Cinemascope process employs 35mm film with a 2 to 1 squeeze, and the later Cinerama releases use a 70mm film with a somewhat lesser squeeze ratio. The de-anamorphosing projection lenses may be constructed with either cylindrical optics or prisms. The prismatic de-anamorphosers can be adjusted over a range of squeeze ratios by altering the obliquity of the prism settings.

16MM EQUIPMENT. The smaller-format 16mm equipment is used for some theatrical showings, and for nearly all audio-visual

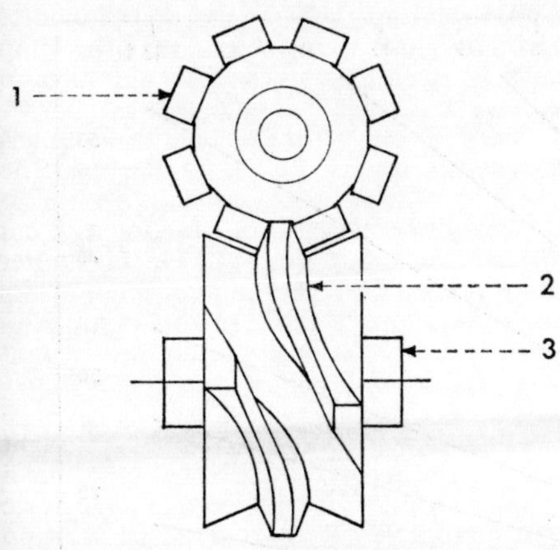

DRUNKEN SCREW INTERMITTENT. Driving member, 3, carries a screw thread, 2, which engages with driven member, 1. Thread profile is such that, for the period of dwell (film at rest in gate), the pitch is zero, and no movement is imparted to 1. For the drive period, the thread profile gives a quick sideways thrust to 1, moving it round to next space between teeth, with which the thread overlap engages.

instruction. The upper range of audio-visual projectors overlaps in performance with the lower range of professional equipment.

Many 16mm projectors designed for theatrical use have an intermittent sprocket drive for the film at the projection aperture, with structures similar to those already described for 35mm. The drunken screw mechanism is more popular for 16mm than for 35mm. Screen illumination is generally at a premium in 16mm, and the drunken screw is more easily tailored to give a fast pull-down and a correspondingly high shutter transmittance. Some designs have used the Geneva movement with some form of angular accelerator to give non-uniform angular velocity to the crank pin so that the pull-down time is shortened.

By far the largest use for 16mm films is as a medium of instruction in schools, industry, and the armed forces. Equipment for this service is designed for relatively easy portability, durability, simplicity of operation and low noise level, so that it can be operated without a sound-absorbing booth. Typical projectors weigh about 30 lb., have a film spool capacity of 2,000 ft., and a light output of 500 to 600 lumens from a 1,000 watt incandescent lamp. A new enclosed tungsten arc lamp gives light outputs up to 2,000 lumens in these 16mm projectors.

Audio-visual projectors universally employ a shuttle type of intermittent feed to advance the film at the picture gate. A reciprocating member, or an arm pivoted at one end and permitted to oscillate, carries one or more teeth designed to engage the film perforations. This member is driven by a rotating cam which moves the teeth in a rectangular path so that they advance to engage the perforations, move parallel with the film path to advance the film one frame, retract, and return for the next stroke. Such a movement has fewer precision parts than an intermittent sprocket and, because it locates each frame by the same elements which move the film down, makes unnecessary the ultra-precise angular indexing required by the star wheel of the Geneva movement or the drunken screw.

The desire for simplicity of operation in projectors for the school and industrial fields has led to the development of automatic threading arrangements to lace the film through the sprockets, the picture gate, and the sound system.

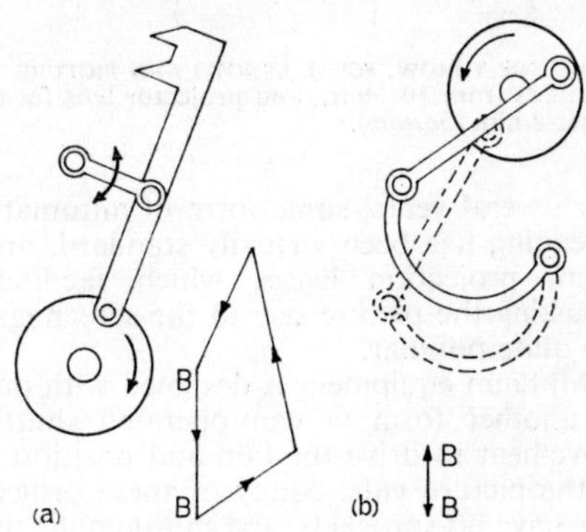

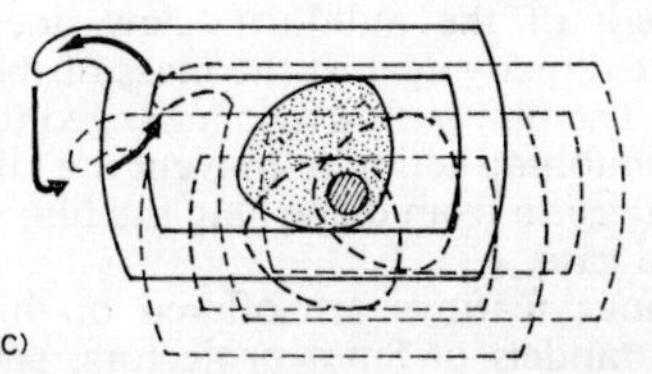

NARROW-GAUGE PROJECTOR INTERMITTENTS. (a) Claw movement with double claw. Arrowed figure traces path of claw. (b) Dog or beater movement. Curved member pushes down film loop, hence film movement BB is only half that of (a). (c) Shuttle movement. Cam on driving spindle operates in cage to give straight pull-down path.

8MM EQUIPMENT. The very small-format 8mm equipment has been designed primarily for the home movie-maker, with a premium on low cost and simple operation.

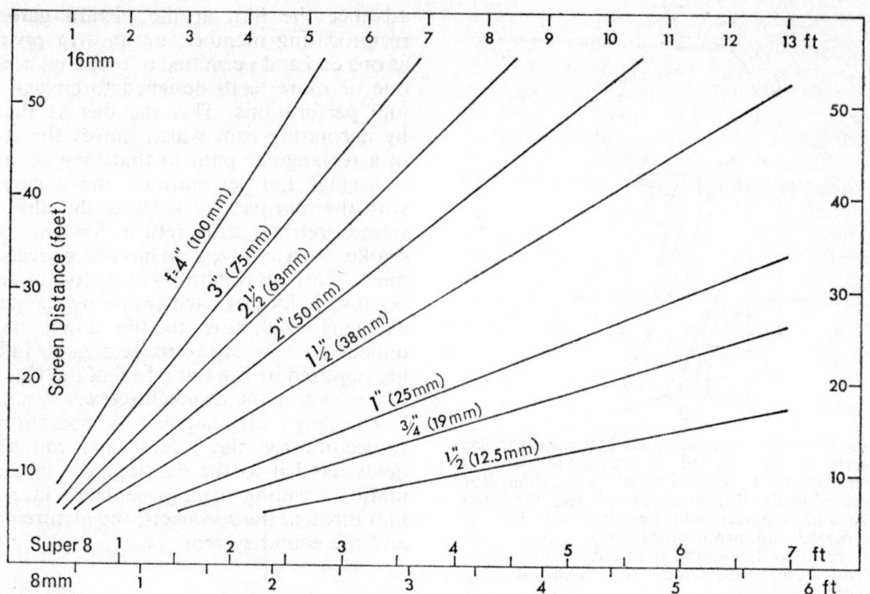

PROJECTOR THROW, FOCAL LENGTH AND PICTURE WIDTH. Picture width on screen may be read off for screen distances from 10–50 ft., and projector lens focal lengths from ½ in. to 4 in. for 16mm (*top*) and 8mm and Super-8 film (*bottom*).

For several years, some form of automatic threading has been virtually standard, and zoom projection lenses which facilitate adjusting the picture size to the screen size are quite popular.

All 8mm equipment is designed with one or another form of cam-operated shuttle movement to drive the film and position it at the picture gate. Many of these projectors have no sprockets; the shuttle pulls the film directly from the feed spool through a resilient member so that the high-acceleration jerk of the pulldown claw does not operate directly against the mass of the film spool. The film path is also arranged to give some snubbing action to prevent the take-up spool tension from disturbing the film at the picture gate.

Various features are offered on higher-priced models of 8mm projectors, such as provision for several different frame rates, so that film photographed at one rate can be projected at a different rate to give slow-motion or speed-up effects. Incandescent lamps are universally used. Newer designs are equipped with lamps having internal reflectors which image a compact filament into the optical system without the need for condensing lenses.

From about 1962 these small format films have found increasing application in education, industrial instruction, sales and some forms of commercial entertainment. There are 8mm projectors which use film in enclosed cartridges to be plugged in by the user. Some are equipped for sound recorded on a magnetic stripe on the film. Similar projectors using the larger picture area provided by the Super 8mm format are expected to become widely used for educational and industrial purposes.

Super 8mm copies can be made with either photographic or magnetic sound tracks and projectors of both types are commercially available, either for cartridge loading or for use with film on reels. Photographic sound is permanent, cannot be erased and provides the cheaper method of release printing copies. Magnetic sound is of higher quality and is not easily damaged by scratching the film. The greater cost of making magnetic prints may be overcome by the use of pre-striped film stock and highly efficient printing equipment.

SOUND SYSTEMS

Sound systems must read the sound information from the moving film and provide the signal to drive the appropriate speakers.

FILTERING WOW AND FLUTTER. All sound systems are designed to operate with film moving at constant velocity in both recording and playback. Film moves with a highly non-uniform velocity at the picture gate, and even though there is usually a continuously-rotating sprocket between the picture gate and the sound-scanning point, intermittent velocity components are introduced by the sprocket teeth themselves, and it is necessary to provide means for filtering out these irregularities. This filtering is usually accomplished by a combination of compliance in the form of spring-loaded rollers, resistance in the form of damping, and inertia in a flywheel which controls a drum over which the film is pulled taut. Constant speed motion is achieved with the film under the control of the drum. A good sound system will have a small fraction of one per cent of combined flutter at the perforation frequency and wow at lower frequencies.

PHOTOGRAPHIC SOUND. Photographic sound records exist as variations in density or width in an area adjacent to the picture and disposed along the length of the film. When these variations in light transmission are caused to modulate the illumination falling on a photocell, electrical variations result which can be amplified to drive loudspeakers. In 35mm practice the sound track is usually illuminated to a quite high level in a band of moderate width, and images the moving track on to a slit over the photocell. If the image is moderately enlarged, and the slit suitably narrow, the width scanned along the length of the film is small, and high frequencies can be resolved. In 35mm, it is quite usual to reproduce frequencies up to 10,000 Hz.

The scanning of 16mm sound tracks usually images a straight coiled lamp filament on to the film through a cylindrical optical system, producing a line less than ·001 in. in dimension along the film. The film modulates this light and the transmitted illumination falls directly on to the photocell.

MAGNETIC SOUND. Magnetic sound records are used in all film sizes, and the variations in magnetization recorded in the track cause variations in the current in a coil on the head structure as the magnetic record passes across a narrow gap in the pole piece structure.

There are many different variations in arrangement of magnetic sound records on film in professional equipment. Up to four magnetic records on 35mm sound film in the Cinemascope system produce stereophonic sound through appropriate speaker arrangements, and in 70mm, up to ten magnetic tracks are used, either on the picture film or on separate sound record film run on a separate reproducer in synchronism with the projector. M.G.T.

See: *Cassette loading for projectors; Cinema theatre; Flicker; Framing; Narrow-gauge film and equipment; Pulsed lamp; Rear projection; Screens; Screen luminance; Sound reproduction in the cinema.*

Book: *Picture Presentation Manual*, British Kinematograph Sound & Television Society (London), 1961.

Prompter. Device which displays on a cathode ray tube or on a paper roll the words of a newsreader or actor in rough synchronization with the action. An operator is usually needed to control the speed.

See: *News programmes.*

Props. Properties. Items of furniture (tables, chairs, beds, etc.) and smaller items such as books, telephones, lamps, used to dress the set.

Alternatively, property man. The man responsible for handling properties.

Prop. Plot. Detailed list of properties required for any particular programme showing the dates on which they will be wanted.

See: *Studio complex.*

Proszynski, Kasimir de. Polish scientist who devised and introduced the Aeroscope camera in 1912. The apparatus was powered by a compressed air motor, fed by cylinders of air compressed by a pump before use. One charge was sufficient for 600 ft. of film. Designed for hand holding, the mechanism incorporated a gyroscope for stabilizing the camera. It was much used by Cherry Kearton, the naturalist and photographer, in the production of animal films.

⊕**Protective Master.** Master positive print prepared from an assembled original negative and held for the production of a duplicate negative in the event of damage to the original. In black-and-white practice, the protective master is a fine-grain positive, but for colour production the protective masters may be either a colour intermediate positive, or a set of three separation positives printed on black-and-white panchromatic stock through red, green and blue filters. Since the long term stability of colour images cannot be guaranteed, permanent or archival protection is best provided by black-and-white separation masters.

See: *Duplicating processes* (*B & W*); *Laboratory organization; Library* (FILM); *Narrow-gauge production techniques.*

⊕**Pulldown.** The action of moving the film from one frame to the next in a camera or projector. The pulldown period is normally obscured from the lens by a shutter, and is followed by a stationary period used for exposure or projection.

See: *Camera* (FILM); *Telerecording.*

□**Pulse and Bar.** Short for sin² pulse and bar, a test signal consisting of a short duration pulse of sin² shape followed by a bar of the same peak amplitude and having sin² shape leading and trailing edges. The sin² signal has the lowest proportion of harmonics and hence a bandpass circuit will have less effect on the amplitude of such signals than it would have on other waveforms.

See: *Picture monitors.*

⊕□**Pulsed Lamp.** Type of high-intensity, short duration discharge tube which has had marked success in the 35mm film projection field and is also used in some vidicon telecines. Normal arc and incandescent sources require the use of a projector shutter to obscure the film while it is moved on from one frame to the next. Indeed, to eliminate flicker, it is necessary to interrupt the light beam at least once again when the film is stationary. The use of a pulsed lamp, which is of the high-pressure gas discharge type, eliminates the consequent loss of light, since it flashes only when light is required and therefore needs no shutter.

See: *Projector.*

□**Pulse Modulation.** A sine wave carrier can be switched on and off by a modulator, forming a pulse. If a train of pulses is transmitted, the pulses can be varied in position, repetition frequency, duration or amplitude, or groups of pulses can be sent carrying a binary code by their presence or absence. These variations can be used to carry information.

See: *Satellite communications.*

□**Pulse Phasing.** Preparatory to locking two or more picture sources so that they can be intercut without relative movement, a train of pulses may be phased by means of a variable delay so that the action of locking takes place with minimum disturbance. The two waveforms are usually displayed simultaneously during adjustment. The action implies that the required phase relationship is known.

See: *Station timing; Synchronizing pulse generators.*

□**Pulse Sound.** Alternative to separate sound carrier used in satellite communication to conserve bandwidth. Takes the form of a bipolar pulse placed in the back porch after the colour burst, and thus part of the colour signal itself. The pulse is position-modulated at intervals of about $\pm 1{\cdot}5\ \mu s$. At line frequencies of 15 kHz a pass band of 7 kHz can be obtained with signal-to-noise ratios approaching 30 dB., giving acceptable sound quality.

See: *Pulse modulation; Satellite communications.*

⊕□**Pulse Sync.** Method of synchronization between separate picture and sound recording cameras in which a precise and continuous one-to-one correspondence in time between the two records is not attempted. In fact, with unperforated magnetic tape in the recorder, this would be impossible: tape stretch and capstan slip cannot be eliminated.

To achieve perfect synchronism, a series of regular pulses is generated in the picture camera and recorded on the tape in one of various ways which do not interfere with the normal sound record. These pulses form what may be called electronic sprocket holes. Later, in the studio, the original tape is re-recorded to perforated film running through a recorder driven by a special motor kept in step by the pulses, so that the final sound record maintains precise synchronism with the original picture track. In this way, perfect lip sync can be achieved, provided that the speed errors have been kept to a practicable minimum.

Pulse sync systems require a cable to transmit the pulses from camera to tape recorder, and in highly mobile camera work this may prove inconvenient. Equipment has therefore been developed to transmit the pulses by radio link from the camera, sometimes in conjunction with the signal from a radio microphone. Alternatively, crystal sync may be employed. A piezo-electric crystal (or electrically maintained tuning fork) controls the camera speed; a corresponding device at the recorder generates identically timed pulses which are laid down on the tape. Since, as explained above, these pulses cannot control tape speed, a synchronizing transfer is still required.

Finally, devices are available for eliminating the old slate or clapper board which establishes initial synchronism between

sound and picture. In the wired system, a fogging light operated by a press-button on the camera simultaneously feeds an audio signal to the programme track on the tape. In the radio system, the audio signal can be fed through the radio microphone channel. In the crystal sync system, a clock recording time from a determinate start (and not merely a series of identical pulses) can be caused to mark both picture film and magnetic tape at the beginning of each shot.
See: *Narrow-gauge film and equipment.*

Pup. Versatile focusing spot lamp which uses a 500/750 watt bulb.

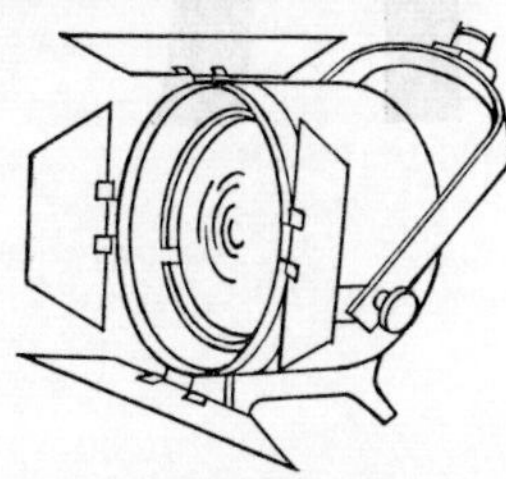

May be fitted with barndoors, as adjustable shades, or with a conical snoot.

Push-Pull Amplifier. Amplifier with two identical signal paths, so connected that they are in opposite phase to produce additive outputs of the wanted products, and cancellation of some unwanted distortion products.
See: *Transmitters and aerials.*

Push-Pull Sound Track. Optically recorded sound track divided into two equal parts, the two halves being exposed to light modulated in opposite phase, so that across the track the transmission is always constant. Played through an ordinary reproducer, the output is zero; but through a special reproducer, the normal benefits of push-pull (reduced distortion, etc.) are realized.
See: *Sound recording* (FILM).

Quad. Cable consisting of four insulated conductors twisted together in one lay in an overall single envelope. The conductors can be twisted in two pairs but it is very difficult to produce a smooth compact cable with this arrangement. The addition of a foil or woven shield produces a cable usually referred to as a squad (shielded quad).

Quad-8. 35mm film perforated with five rows of sprocket holes so that after printing and processing it can be slit to give four

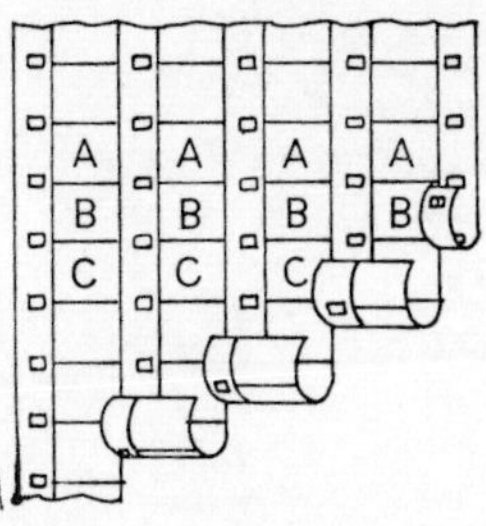

strips of 8mm or Super-8 print, the outer margin with one row of holes being discarded. The use of this stock is limited to the laboratory where large numbers of copies are manufactured.
See: *Double 8mm film; Film dimensions and characteristics.*

□**Quadrant Aerial.** Form of transmitting aerial, which is made up of two wings forming a quadrant. The angle between the two wings of the quadrant can be varied in order to produce the desired radiation pattern.
See: *Transmitters and aerials.*

□**Quadruplex Recorder.** Recorder which makes a record in four sections, such as that produced on videotape by a four-head recording device.

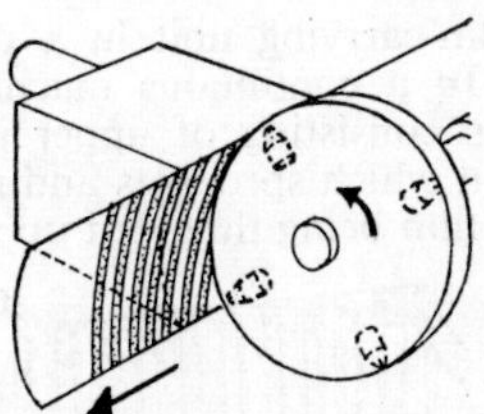

QUADRUPLEX VIDEOTAPE RECORDER. Tape is curved by vacuum chamber behind, so as to fit periphery of rapidly-revolving drum carrying four magnetic heads which trace successive transverse tracks across tape width.

The main purpose of quadruplex recording is to increase the effective head-to-tape speed over the figure obtainable with simple longitudinal recording or with most types of helical scan. At a tape speed of only 15 ips, a head-to-tape speed of some 1,500 ips is commonly achieved.
See: *Helical scan; Videotape recording.*

□**Qwart.** (Trade name from "quart in a pint pot".) Compact lightweight spotlight/softlight used in TV studios, particularly for saturation lighting. Combines a 2½/5 kW spotlight and a 2½/5 kW softlight in approximately a 2 ft. cube housing.
See: *Twister.*

R

⊕**Rack.** Film-carrying unit in a developing machine. In a continuous machine, racks are frames consisting of upper and lower spindles on which sprockets and rollers are mounted, film being threaded up and down

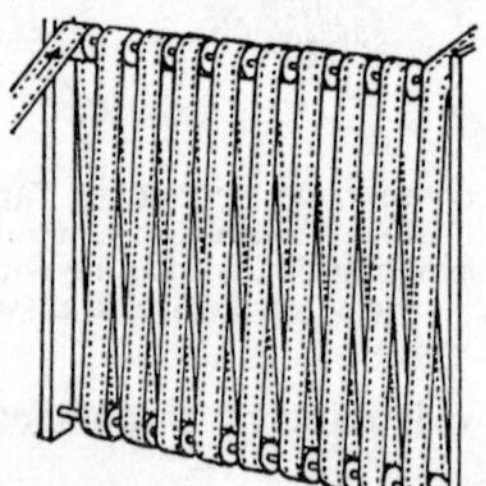

between them in long spirals. The racks are immersed in tanks containing processing solutions. Movable racks are called elevators.

See: *Processing.*

⊕**Rack-and-Tank Development.** Obsolete non-continuous system of film development in which a batch of film is wound on a rack and dipped in turn in tanks containing processing solutions.

⊕**Rackover.** Device employed on some high-quality film cameras to provide parallax-free viewfinding. Consists of a precision side-to-side slide carrying the camera body and viewfinder, while the lens remains fixed to the camera base on the tripod. Thus, from the ordinary taking position in which the film lies behind the lens, the camera is racked over by a handle to the viewing position in which the viewfinder takes the place of the film. Viewing and filming therefore cannot occur at the same time, and an outside or monitor viewfinder has to be provided.

With improvements in reflex viewfinding, which do not have this drawback, the rack-over is obsolescent.

See: *Camera* (FILM).

□**Radiating Element.** Elements of an aerial to which the feeder is connected in order to radiate the electro-magnetic waves. The remaining elements are used to modify the radiation pattern of the aerial.

See: *Transmission networks; Transmitters and aerials.*

□**Radiation Resistance.** Resistive component of an aerial impedance. This value, when multiplied by the square of the aerial current at the feed point, gives the radiated power.

See: *Transmission networks; Transmitters and aerials.*

□**Radio Frequency.** Term used to indicate the complete range of frequencies which may be used for the transmission of information by electro-magnetic waves.

Radio-Relay. Transmission system including relay stations where the signal is received and retransmitted by radio to compensate for losses.

See: *Microwave links; Transmission networks.*

Radio Television Camera. Small television camera, usually a vidicon or Plumbicon, which transmits the picture by radio link to the associated fixed equipment.

See: *News programmes; Outside broadcast units; Special events.*

Random Noise. Non-repetitive undesired signal on an audio or video network.

See: *Noise.*

Raster. In stereoscopic (3-D) projection, a grid of vertical (sometimes fan-shaped) alternately opaque and transparent slats which serve to separate the left- and right-eye images at the screen, so that the audience can dispense with polarizing or anaglyphic glasses. Rasters may also be formed of minute embossed lenses or lenticles.

See: *Stereoscopic cinematography.*

Raster. The pattern formed by the scanning spot of a television system. It is rectangular in outline and consists of a number of evenly spaced, horizontal, straight lines traced in a zig-zag path from left to right and top to bottom of the image in a manner very similar to the movement of the eye when reading.

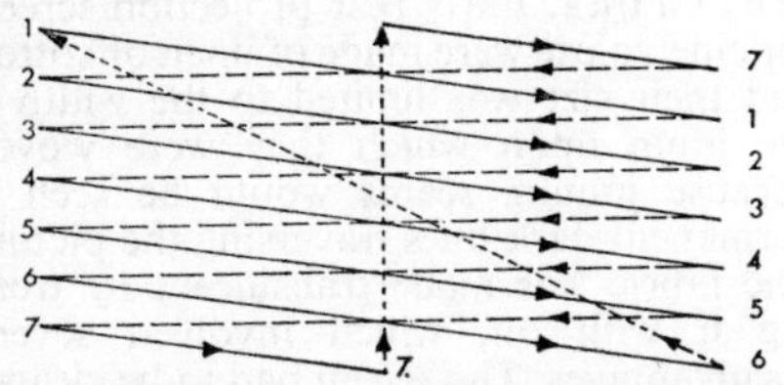

INTERLACED/RASTER. Simplified 7-line TV raster to show principle of interlace and flyback. Line flyback is shown in dashed lines, with end of line 1 flying back to start of line 3. At midpoint of line 7, field flyback occurs (dotted line), line 7 is completed and lines 2, 4 and 6 interlace. Blanking ensures the suppression of the scanning spot during flyback.

The spot is constrained to form this pattern in camera and display tubes alike by passing the electron beam through two electromagnetic scanning fields simultaneously acting at right angles. One of these is responsible for the horizontal or line deflection while the other produces the slower vertical or field deflection. Electrostatic deflection is today seldom used.

In the studio the camera evaluates in fine detail the variations of light and shade (and where appropriate, colour) that occur as the scanning beam traces a raster across an image of the scene to be transmitted. If the reconstituted information at the receiver is to occupy the correct relative position on the screen of the display tube, the scanned rasters at transmitter and receiver must be maintained precisely in step with one another. This is achieved by the inclusion of two sets of synchronizing pulses as an integral part of the transmitted video waveform. These pulses are used to synchronize respectively the line- and field-scanning circuits in the receiver with those at the transmitter.

Hence, a raster controlled in this way is said to be locked or synchronized, whereas in the absence of synchronization it is free-running. In addition, the design of scanning circuits must ensure that the forward trace in both horizontal and vertical directions takes place at constant velocity. Any departure from this results in positional distortion known as non-linearity.

The return of the scanning spot to the left hand side of the raster at the completion of each line is known as line flyback and the time allowed for this to take place, the line-blanking interval, is about 20 per cent of the full line period. Similarly field flyback from the bottom to the top of the raster occurs at the end of every field and must be complete within the permitted field-blanking interval of about 7 per cent of the total field period.

The type of raster so far described is referred to as sequential. Broadcasting systems use an interlaced raster.

In North America and Japan a 525-line, 30 frames per second, system is standard. The European standard (including Britain) is 625 lines, 25 frames per second, but Britain also has a 405/25 system, now obsolescent.

However, in all cases, the number of active lines present in the raster which actually contain picture information is less than the nominal figure by those lost during the field-blanking period.

The width of the active part of a correctly-adjusted synchronized raster in relation to its height, i.e., its aspect ratio, is in the proportion 4 : 3. C.W.B.R.

See: *Interlace; Scanning; Television principles; Transmission and modulation.*

□**Raster Shading.** Non-uniformity of brightness across the raster when a fixed input signal is applied. Differences from side-to-side are called horizontal shading and from

top to bottom, vertical shading. Difference between the centre and the corners is called parabolic shading.
See: *Camera* (TV).

⊕**Raw Stock.** Motion picture film before exposure or processing.
See: *Film manufacture*.

□**Reactance Tube Oscillator.** Oscillator frequency-controlled by a thermionic valve operated in such a way that it presents a variable reactance, i.e., a variable inductive or capacitative impedance.

⊕**Reader.** Portable device for rapidly reading picture film or magnetic soundtrack. In its picture form, a reader usually consists of a simple rotating prism type of head, com-

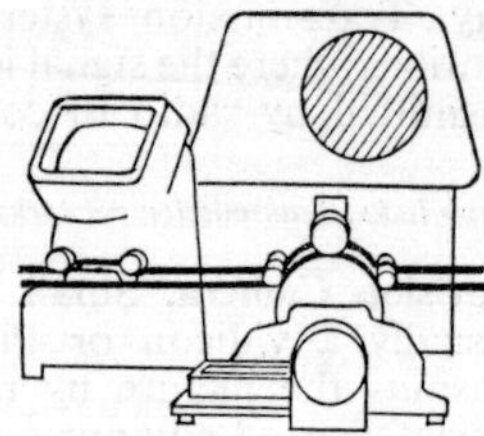

bined with sprockets and a magnifier for enlarging the picture. As a rule, film is pulled through an optical reader by means of a rewinder. The sound reader consists of a magnetic head and pre-amplifier, with film guides and sprockets for passing the film over the head. Sound readers are often combined with synchronizers.
See: *Editor; Narrow-gauge film and equipment*.

⊕REAR PROJECTION
□

Motion pictures can be projected on to the back of a translucent screen instead of on to the front of a reflecting screen. The quality of the picture is considerably improved. A wider contrast range can be handled and colours are much brighter. Highlights have an almost scintillating quality.

The improvement is due both to the inherent superiority of transmitted-light viewing systems and to the lack of interference with the image-forming beam. In normal front-projection systems, the light beam carrying the photographic image passes through the vitiated, sometimes smokeladen atmosphere of the auditorium, where a proportion of the light energy is scattered. The beam itself becomes visible because of the dust particles, smoke, etc., in its path; the light reflected from the screen and reaching the observer's eyes similarly passes through the light-scattering atmosphere. Each observer in an auditorium views the screen through, and at an oblique angle to, the light beam. Light scatter impairs the quality of the image and of the colour; it reduces both contrast and resolution.

Nevertheless, the enhanced picture quality of rear projection systems can be attained only if several difficult requirements are fulfilled. The first concerns the screen.

SCREENS

The function of any projection screen is to diffuse and re-radiate the image-carrying light beam incident on it so that the image is visible to the observer's eyes. In the case of a rear projection screen a part of the incident light is reflected, a part is absorbed, and the balance is transmitted. Both reflection and absorption reduce picture contrast and image brightness. The transmitted portion which provides the visual image is subjected to diffusion and scatter.

STRUCTURE. Early rear projection screens for cinema use were made of linen or cotton, and their size was limited to the width of the loom upon which they were woven, because joining seams would be seen as permanent dark lines traversing the picture. The fabric was made translucent by treating it with oil, which involved several disadvantages. The screen had to be cleaned and re-oiled at frequent intervals. There was little or no control over the absorption and diffusion characteristics of such screens.

The position is very different today. Screens are now made of plastic material which can be produced in widths of some 18 ft. without a join. In some cases a solution of ethyl-cellulose, or similar clear material, is sprayed on to a flat back board until a requisite thickness is built up. When the material has set it is peeled off its back board. When set, both sides of the material have a matt surface similar to that of ground glass. Such a screen has a high transmission factor but poor diffusion. This can be improved by spraying the material on to a back board having a suitably textured surface. A screen of this type is not depolarizing and can be used for stereoscopic projection using polarized light.

When a lower transmission value and higher diffusion are required, as is usual with larger cinema screens, very fine opaque or semi-opaque particles may be mixed with the ethyl-cellulose. In the case of small screens used with audio-visual equipment, the substrate is of transparent perspex or clear glass. This is sprayed on one side with a vehicle consisting of resins and solvents and containing very small particles of materials such as titanium dioxide, zinc oxide etc. Spraying in this manner permits a reasonably close control of the absorption-transmission factors of a screen.

LUMINANCE. The difficulty with rear projection screens, as with directional screens for front projection, is that it is not possible to provide a uniform luminance over the entire screen surface, nothwithstanding the fact that the incident light may be uniformly distributed. If three pencil rays of light of equal intensity fall on a perfectly diffusing screen, one in the centre, one near the left edge, and one near the right, three observers spaced across the screen see the three spots at equal brightness. If, however, the screen is a perfect reflector, e.g., a mirror, then the three pencil rays of light will be reflected at the same angle at which they are incident upon the screen. The centre observer then sees only the reflection of the centre pencil of light; the two outside observers, if stationed well beyond the outside edges of the screen, see no light at all.

The reflectivity characteristics of a directional screen are somewhere between that of the perfect diffuser and the perfect reflector; and the so-called gain of the directional screen is the ratio of the luminance of the reflected ray on its axis of maximum reflection and the luminance of the perfect diffuser.

The characteristics of a rear projection screen are similar. The three observers see a point of light near the edge of the screen at widely different luminances, depending on the angle at which the incident light strikes the screen, and the transmission and diffusion factors of the screen itself. It has been arbitrarily suggested that the screen characteristics and the optical system of the projectors should be such that the range of distribution of screen luminance for any one observer should not exceed a ratio of 1 : 2.

Thus the higher the gain of the screen, the smaller is the angle at which the 1 : 2 ratio is reached. For an average commercial screen having a unity gain, the angle at which half this value is reached is about 40°: with a gain of 6, the angle is reduced to about 15°. If a distribution range of 1 : 3 is adopted – which, perhaps is the acceptable maximum – the viewing angles are increased to 55° and 20°.

LENTICULAR SURFACES. To improve light diffusion and maintain a satisfactory light distribution, plastic screens have been made by spraying the ethyl-cellulose on a back board having small, regularly disposed indentations which produce a screen having a lenticulated surface. Each element on the viewing surface of the screen tends to spread the light rays more nearly to simulate a directional screen of very low gain. Although, by this means, the seating area may be widened, the overall luminance of the screen picture is reduced, and the contrast ratio of the image is impaired because of the greater light scatter and internal reflections.

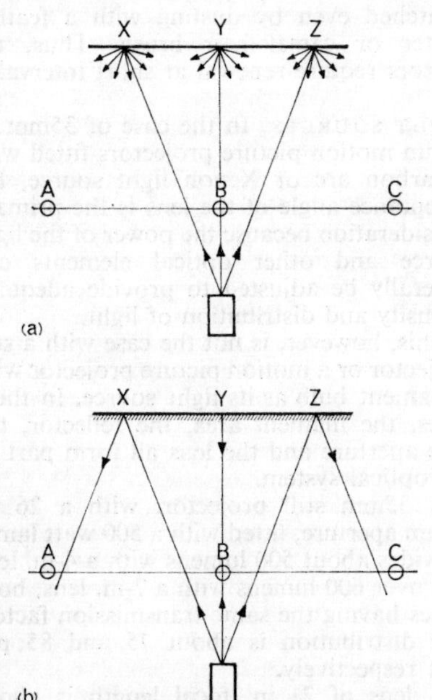

EFFECT OF DIFFUSING AND REFLECTING SCREENS. (a) Matte white screen acts as perfect diffuser. Consequently, spectators at A, B ,C, all see small screen areas X, Y, Z, at equal brightnesses. (b) When matte screen is replaced by mirror, spectator B sees the illuminated image at Y, but not at X or Z. A and C see no illumination at X, Y or Z.

LENS AND LIGHT SOURCE

WIDE-ANGLE LENSES. One substantial objection to the use of rear projection systems in cinemas was the large space required backstage between the screen and projectors, which necessarily had to be unobstructed for the passage of the light beam. To reduce the length of the throw as much as possible, it was the general practice to use lenses of very short focal length – focal lengths of 1⅜ in. and 1½ in. for 35mm film were not uncommon. This involves two major disadvantages: a lack of acceptable distribution of incident light across the screen and a severe restriction of the auditorium area over which the screen picture could be satisfactorily viewed. In many cases, the luminance at the edges of the screen was less than 10 per cent of that at the centre, and the area of high luminance at the centre of the screen was small, giving rise to a hot-spot. The area of the hot-spot is increased and its peak luminance is reduced as the focal length of the projection lens is increased and the gain of the screen is reduced.

FOLDED LIGHT BEAM. With 35mm motion picture projectors, lenses of 3½ in. to 4 in.

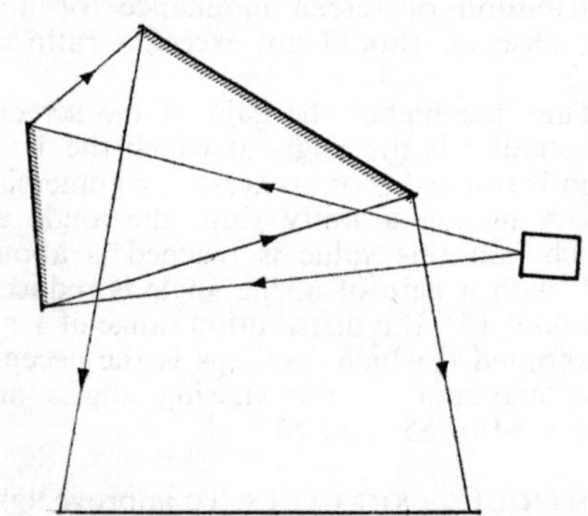

REAR PROJECTION WITH FOLDED LIGHT BEAMS. Two front-surfaced mirrors may be used to fold light beam and thus greatly reduce back-of-screen depth. Image reversal by projecting through non-standard side of film is required to produce correct left-to-right geometry when viewing from front of screen.

focal length, having acceptance angles of 14° and 12° are satisfactory, and can provide a luminance distribution that complies with international recommendations, but a 3½-in. lens requires a substantial distance between screen and projector – 4·21 ft. for every 12 in. width of screen for non-anamorphic systems.

Although the length of the light beam cannot be reduced for a given size of picture, the backstage distance may be shortened by folding the light beam by introducing mirrors in its path. This device also has disadvantages. At each reflection there is a loss of at least 4½ per cent of light energy due to absorption and scatter, and very much more if there is dust on the exposed surfaces. If the mirrors are rear-silvered, the projected light must twice pass through the glass thickness, which causes diffraction and impairment of the image.

In some cases, dependent upon incident and reflection angles or upon the thickness of dust upon the mirror surfaces, two or more images can be created. In falling upon the screen, these can cause ghosting similar to that in television reception, when transmitted signals are reflected by large objects and arrive after and out of phase with the direct signal.

To reduce light loss and scatter, the mirrors frequently have the silver coating on their front surface, but to avoid rapid oxidation the surface must be protected by a transparent lacquer coating. The surface, however, is very fragile and becomes scratched even by dusting with a feather duster or camel hair brush. Thus, the mirrors require renewal at short intervals.

LIGHT SOURCES. In the case of 35mm or 16mm motion picture projectors fitted with a carbon arc or Xenon light source, the acceptance angle of the lens is the primary consideration because the power of the light source and other optical elements can generally be adjusted to provide adequate intensity and distribution of light.

This, however, is not the case with a still projector or a motion picture projector with a filament bulb as its light source. In these cases, the filament area, the reflector, the gate aperture and the lens all form part of the optical system.

A 35mm still projector, with a 26 × 36mm aperture, fitted with a 500 watt lamp, provides about 500 lumens with a 4-in. lens and over 600 lumens with a 7-in. lens, both lenses having the same transmission factor. The distribution is about 75 and 85 per cent respectively.

A lens of 2¼ in. focal length is about optimum for the average 16mm motion picture projector, giving about 10 per cent more light than a 2-in. lens of the same transmission factor. An increase to 4-in. focal length reduces the light output by about 50 per cent.

With an 8mm motion picture projector, the optimum focal length of lens is about 1 in. The light output is about 30 per cent greater than a 3-in. lens, and the light distribution is about 10 per cent better.

APPLICATIONS

FILM AND TELEVISION. Rear projection systems have found favour in film and television studios for backgrounds. Adequate backstage space is usually available, projection lenses of appropriate focal lengths can be used, and screen and camera can be oriented easily to achieve optimum results. The transmission factor of the screen and the intensity of the projector arc can be arranged to suit the luminance levels required for the particular sequence. Stage lighting must be carefully arranged to prevent stray light falling upon the screen, as this would seriously impair the image quality.

VISUAL AIDS. Rear projection systems are also popular with manufacturers and users of visual aids for educational and similar purposes, because the system lends itself to being encased in a compact, integral unit. Almost invariably, the light beam is folded by a single mirror.

The picture image is created on a screen composed of a diffusing layer coated on to a rigid, transparent substrate. In these small systems, light scatter must be reduced to the minimum if the maximum of contrast is to be maintained to avoid loss of information. Because of the need for robustness and rigidity, the screen cannot be a thin membrane, but must have appreciable thickness, which increases light scatter.

One method of reducing this effect is to tint the screen a neutral grey which partially absorbs light, so that internally reflected light which necessarily follows a longer path is proportionately absorbed more than the direct light. The tinted screen is also helpful in reducing reflections of stray light, a cause of serious reduction in the quality of the projected picture, owing to further light scatter and reduction of contrast.

CONTRAST RANGE. In a cinema where the lighting arrangements are carefully controlled and the stray light falling upon the screen seldom exceeds one per cent of the screen luminance, a contrast ratio of about 120 : 1 is possible. In normal conditions of ambient lighting, under which audio-visual equipment is used, the stray light falling on the screen lies between 5 and 8 per cent of the centre screen luminance. Thus, the contrast range of the photographic image on the film cannot be reproduced correctly, and the details in the darker tones are lost. In many cases, screens are protected against stray light falling upon them by hoods with matt black inside surfaces.

Similar difficulties arise from the rear projection system which is common to nearly all television receivers. Here the image is created by an electronic beam impinging upon the phosphor coating on the inner surface of the front of the cathode ray tube which forms the screen. If a screen picture is viewed in complete darkness and in room conditions that permit no light reflections from walls, ceiling or floor, the darkest blacks in the picture of an average scene are raised to 1/50 to 1/60 of the peak highlights. This is due to light scatter and to internal reflections in the face plate. If sufficient stray light is now allowed to fall upon the tube to simulate normal domestic viewing conditions, the contrast falls still further, to 20 : 1 and possibly 15 : 1.

This condition explains, in part, why a normal cinema film – and particularly a colour film – intended for showing at a contrast range of about 100 : 1 reproduces badly on a television screen. The result of transmitting such a film results not merely in the loss of shadow detail; the entire picture becomes distorted in its tone relationships, and thus in films for television transmission, it is advantageous to reduce the slope of the highlights and lighter greys by continuing beyond the toe of the characteristic of the film emulsion so that the darker parts of the image are of steeper relative slope.

This is particularly necessary with colour films because compression of contrasts in the highlights (both density and colour) is less objectionable than in shadows. Additionally to this correction, some broadcasting authorities provide for compensation to the three colour (electrical) separation signals. L. K.

See: *Cinema theatre; Projector; Screen luminance; Screens.*

⊕**Rear Projection Unit.** A self-contained projector, sometimes with cassette loading, in which the optical path between lens and screen is usually folded by the use of mirrors to reduce size to a minimum. A translucent screen gives an image similar to that of a portable TV set. These units are often employed as sales and audio-visual aids.

See: *Cassette loading for projectors; Daylight projection unit; Rear projection.*

□RECEIVER

The television receiver is the final link in a chain which starts at the studio, where the television camera turns the scene to be transmitted into an electrical signal which is superimposed on a carrier and radiated from the television transmitter.

The television receiving aerial picks up this signal which is fed into the television receiver. The signal, so carefully encoded from the scene at the studio, is amplified until it can be decoded and returned again to a pattern of light and shade.

The television signal contains information on the brightness of any point on the picture (and in the case of colour, the shade of colour and its strength) and the receiver has to use this information to reassemble the original scene. The cathode ray picture tube (kinescope) is scanned by a beam of electrons so that the entire face is covered in a uniform way many times per second. Those parts of the scene which are dark at the studio do not contribute any signal, and hence the corresponding parts of the picture tube are also dark. Conversely, the brightest areas of the scene contribute maximum electrical signal and hence the picture tube is bright over these areas.

OPERATION

It is the task of the receiver to reconstitute the scene without removing essential information or adding false information and without altering the relative positions of the various parts of the picture, in just the same way that the sound component must be reproduced clearly and without distortion.

RECEIVER CONSTRUCTION. The various parts of a monochrome receiver are:

1. The tuner, to which the aerial is connected, and which enables any channel to be selected, amplified, changed in frequency and fed to:
2. The vision and sound amplifier which amplifies the signal so that it is large enough to be fed to:
3. The vision rectifier, which removes the wanted signal from the carrier on which it travelled from the transmitter. The sound signal is then amplified, rectified and fed to the loudspeaker via a sound output stage. The rectified vision signal passes to:
4. The video stage, which amplifies the wanted signal (now very like the output signal from the camera) and applies it to:
5. The picture tube, which turns the variations in electrical signal into variations in the brightness of its phosphorescent screen. The spot is moved systematically over the face of the picture tube by:
6. The time-bases or scanning generators, which provide current to the deflection coils, or yoke, fitted to the picture tube. By synchronizing the scanning generators to those in the camera in the studio, the pattern of light and shade on the face of the picture tube follows, more or less exactly, the scene at the studio. This synchronization is ensured by:
7. The synchronization separators, which, taking pulses inserted into the encoded signal at the studio, separates them from the picture information and feeds them to the time bases.

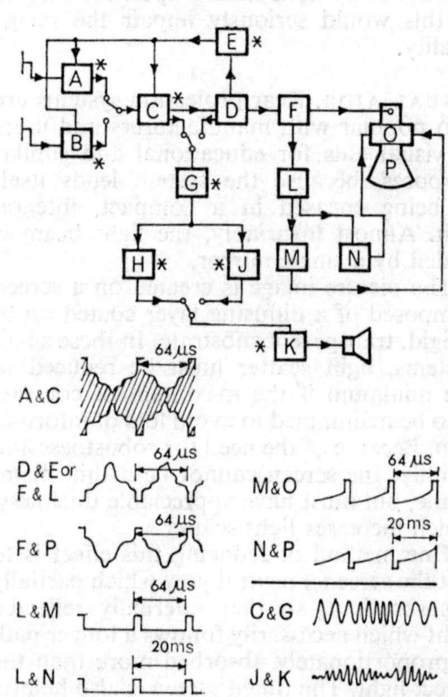

TYPICAL DUAL-STANDARD RECEIVER: BLOCK DIAGRAM. (* Transistor stages.) A, UHF tuner, and B, VHF tuner, receive signal from separate aerials, amplify it and feed it through selector switch to C, vision i.f. amplifier, and D, vision detector. Sound signal, which may be f.m. or a.m., is fed via i.f. amplifier, G, and a.m. or f.m. detector, H, J, to K, audio amplifier, and loudspeaker. Vision detector, D, feeds AGC detector, E, to provide automatic gain control to r.f. and i.f. stages. Demodulated signal from D, after amplification in video amplifier, F, goes to modulating electrode of picture tube (kinescope), P. Field and line sync pulses are separated in sync separator, L, and fed to line time base, M, and field time base, N, which drive deflection coils, O. (*Bottom*) Waveforms through receiver, as marked.

MONOCHROME. The path followed by a signal entering a monochrome television receiver is as follows:

From the aerial, the signal travels through the tuner, where its frequency is changed from the incoming channel (which is typically in the range 41·5 MHz to over 800 MHz) to the intermediate frequency.

This is so chosen that interference to and from the receiver is unlikely to occur. The two components of the signal, sound and vision, are detected and amplified separately, the audio voltage feeding the loudspeaker and the vision being applied, via the video stage, to the picture tube and also to the synchronizing separator. This strips the synchronizing pulses from the picture and applies them to the time bases which generate the sawtooth currents required to feed the picture tube deflection coils. Automatic gain control (AGC) is incorporated, which ensures that the contrast of the picture, once adjusted by the user, remains constant despite variations of input signal level. AGC is also used on the sound channel to keep the sound level constant.

The picture tube screen is scanned, for the first field, starting from the top left-hand corner and ending at the bottom centre. In the case of the British 405-line system, there are 187½ lines on the screen in one field. The second field, which, together with the first, constitutes a complete picture, also has 187½ lines, and starts at the top centre of the picture, ending at the bottom right-hand corner. The two remaining groups of 15 lines each carry no video information and are used to contain the intelligence required to ensure correct synchronization of the received scene with that transmitted.

COLOUR. In a colour receiver, other circuits are used to extract the information on the colour (hue) and colour-strength (saturation) of each point on the picture, from the transmitted signal. Thus the picture tube, in addition to reproducing light and shade, can also reproduce colour, and the relative proportions of three primary colours, usually red, green and blue, are controlled to produce the shade and colour strength of the various parts of the studio scene.

The colour video signal incorporates, in addition to the brightness modulation, a suppressed-carrier signal which contains the colour information in a manner which varies from system to system. This information is extracted at the video amplifier stage and decoded, that is to say, broken down

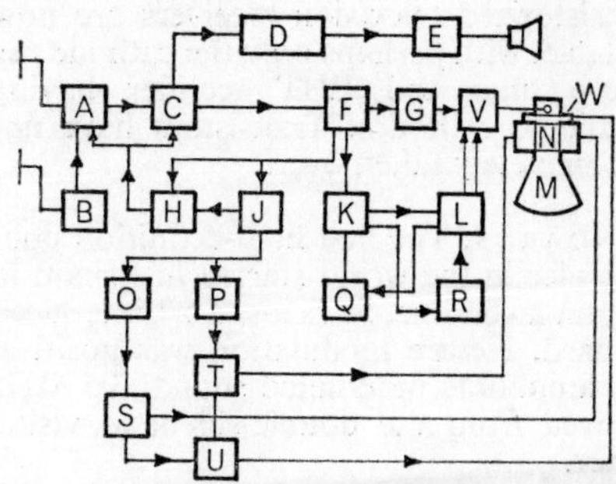

NTSC 525-LINE COLOUR RECEIVER: BLOCK DIAGRAM. A, B. VHF and UHF tuners. C. Video i.f. amplifier branched to audio i.f. amplifier, D, and output stage, E; video signal to video amplifier, F, thence luminance signal to video output stage, G, and matrix (decoder), V. F also feeds AGC circuit, H, which provides gain control for r.f. and i.f. stages, A, C, and sync separator, J, which in turn drives O, P, field and line oscillators, followed by field and line output stages, S and T. These provide current for deflection coils, N, of shadow mask tube, M, and also drive convergence circuits, U, supplying current to tube convergence coils, W. Chrominance circuits comprise K, colour i.f. stages, L, colour demodulator, Q, colour killer, and R, colour sync oscillator and burst gate. From demodulators, L, colour difference signals are fed to matrix, V, where, in combination with luminance signal from G, the three colour signals are formed which modulate the three beams in tube, M.

and combined with the brightness signal to produce three colour signals, corresponding to the red, green and blue components of the original scene.

These signals are applied to a three-colour picture tube, and produce a replica of the original scene. All current systems make use of the fact that the acuity of the eye to colour is small enough to render a high colour bandwidth unnecessary; this saves spectrum space.

MODERN DEVELOPMENTS. The receivers used in the earliest low definition public television experiments in Britain and the US operated in the medium-wave broadcast band. Increases in picture definition, and hence in the video bandwidth required satisfactorily to produce a picture, forced the carrier frequency out of the broadcast band and into the VHF region. Lack of sufficient channels in the VHF bands has now necessitated the use of UHF frequencies in both Europe and North America. Even higher frequencies are also in use in the US for educational broadcasting.

Development in vacuum tubes and, more recently, solid state devices has resulted in a reduction in size and fault rate in receivers.

Transistorized television receivers are now available, with perhaps only the cathode ray picture tube and EHT rectifier having thermionic cathodes. Transistors have not yet completely taken over.

STANDARDS. The first high-definition public service in the world started in Britain in 1936 on a 405-line, 50 fields, 2 : 1 interlaced standard. Picture modulation was positive, with amplitude modulated sound, 3·5 MHz removed from the double-sideband vision carrier.

This standard held the field in Britain until 1964, when it was joined by a UHF system, using the so-called Eastern European standards, which is steadily being extended. The standards for this system are 625 lines, 50 fields, 2 : 1 interlaced, negative picture modulation, vestigial sidebands, and frequency modulated sound with a sound/vision spacing of 6 MHz.

In North America and Japan, a 525-line system is in use, with a field frequency of 60 fps interlaced 2 : 1. Channel width is 6 MHz and sound/vision spacing is 4·5 MHz. Picture modulation is negative and sound modulation is f.m.

There are other European standards, including three variants of the 625-line system, differing in modulation type and direction, and an 819-line system. Thus receivers for use in several European countries, as well as in Britain, have to be suitable for operation on more than one standard. Only the British case, where the receiver has to be switched between 405-line standards on VHF and 625-line standards on UHF will be considered in detail.

THE BRITISH DUAL-STANDARD RECEIVER

It is convenient to divide the television receiver into the sections which are generally made in physically separate units, consider these sections in a general way, and then describe a typical hybrid receiver.

TUNERS. In Britain the standard TV aerial and downlead impedance, both on VHF and UHF, is 75 ohms, unbalanced. The tuners in the receiver, therefore, are designed to terminate a cable of this impedance.

In order to obtain the best possible noise figure, an exact 75 ohm match is not attempted, a compromise figure, rather higher than this, being usual. Separate tuners for VHF and UHF are a feature of present-day receivers, but integrated tuners are in development and limited use.

Tuners are now usually transistorized, with a turret on VHF and continuously tuned resonant lines on UHF.

I.F. AMPLIFIERS, DETECTORS, VIDEO AMPLIFIERS, AUDIO AMPLIFIERS. Again, transistors are in use, and a printed circuit construction is employed. Intermediate frequencies have been standardized at the following figures by agreement between the various manufacturers: for 405 lines, vision 35·65 MHz, sound 38·15 MHz; for 625 lines, vision 39·5 MHz, sound 6 MHz (intercarrier). Detection, both of sound and vision, is performed by semi-conductor diodes.

Transistors are not at present available for use at the high drive voltages required by the picture tube (typically 120 volts composite video) and valves are used in the video stages in non-portable receivers.

Audio amplifiers are increasingly transistorized and typically deliver a watt or two to a moving-coil loudspeaker, usually elliptical, for reasons of cabinet space.

TIME BASES. As in the case of the video amplifier, valves are preferred to transistors because of their greater tolerance of high voltage and power dissipation. High efficiency line output transformers, using an energy recovery diode to boost the HT supply to the line output valve are required to scan the wide-angle picture tubes.

Stabilization of both field and line circuits is usually employed to prevent valve aging and changes in mains supply voltage from affecting the size or proportions of the picture. In order to avoid false triggering of the line time base due to noise on the synchronizing pulses, flywheel circuits are almost invariably used on dual-standard receivers.

PICTURE TUBES. Monochrome receivers employ 19 in.- or 23 in.-diagonal rectangular picture tubes with a 110° deflection angle. Light outputs of over 100 ft. lamberts are obtainable with an EHT supply of 18 kV and the spot size is small enough for 625-line pictures.

POWER SUPPLIES. The all-valve receiver commonly uses a.c./d.c. techniques in which all the valve heaters are connected in series and with a resistor between the valve chain and the mains supply. Since the sum of the heater voltages of the valves is equal almost to the lowest input voltage found in Britain (200 volts), the series resistor is quite small, and other mains supply voltages are catered for by adding other resistors in series. A current limiting thermistor is often included in the chain to avoid damage to the valves when switching the receiver on.

Hybrid receivers, in which transistors are used as well as valves, employ a transformer to supply the heaters and the low voltage (typically 25 volts) transistor supply. They are thus only suitable for use on a.c. supplies. Silicon diode rectifiers are almost universal, having superseded the thermionic diode and selenium types, by virtue of their reliability, low voltage drop and small size. Their only disadvantage is that they require protection against short-term voltage transients, either from inside the receiver (e.g., flashover in the picture tube from the final anode to, say, the second anode) or from the mains supply. This protection can be given by the use of a few high-voltage capacitors and a series resistor, the former used for decoupling the diode, and the latter, in series with the diode, to limit the maximum current through the diode.

STANDARDS SWITCHING. Switching between standards on a receiver designed for Britain as distinct from a four-standard set used, say, in Belgium, is simplified by the fact that VHF signals carry only 405-line programmes, and UHF only those on 625 lines. A standards-change switch has to make the following alterations in going from 405 to 625; changes from VHF to UHF tuner; change from 405 to 625 IF; invert detector diode; change from 10,125 Hz line scanning frequency to 15,625 Hz.

It is now usual to gang the standards-change switch to the VHF turret tuner, which then has a position marked "625". This automatically brings the UHF tuner into use and makes the other necessary changes. Automatic standards recognition is in development, in which detection circuits respond to change of line frequency in the video signal and switch the time bases and detector polarity accordingly.

These circuits will be of value when integrated VHF/UHF tuners become general.

A PRACTICAL RECEIVER

A typical hybrid dual standard receiver for use in Britain can be considered as a series of block parts:

UHF TUNER. This consists of two transistors in the common-base configuration, the first acting as an r.f. amplifier and the second as a self-oscillating mixer (convertor).

Resonant-line r.f. circuits are used, and tuning is accomplished by a four-section variable capacitor controlled either by a rotating knob or by a push-button channel selector mechanism. The tuner is coupled to the i.f. amplifier by a filter which has rejection troughs at 31·5 MHz (adjacent vision), 33·5 MHz (sound) and 41·5 MHz (Channel 1 sound).

VHF TUNER. Three transistors are used, the first as a common-emitter r.f. amplifier, and the second as a separate oscillator. The r.f. and oscillator signals are fed to the common-base output stage.

Interposed between the output stage and the i.f. amplifier input are traps tuned to 39·65 MHz (adjacent vision), 33·5 MHz (adjacent sound) and 41·5 MHz (Channel 1 sound).

VISION I.F. AMPLIFIER. Three transistors are used in this block, all in the common-emitter configuration. On 405-line transmissions, where the sound is amplitude modulated, the first two i.f. stages are common to both sound and vision. The last i.f. amplifier has a sound rejection trap in its base which also serves as a sound take-off. Further trapping at 38·15 MHz (sound i.f.) and 41·5 MHz (Channel 1 sound) is also provided in the collector circuit of this transistor.

On 625 lines, the sound and vision are passed together to the vision detector.

VISION AND AGC DETECTOR. This is a simple diode detector which has a switch in its output to provide the correct video polarity on each standard. There is no d.c. component in its output. Trapping at 3·5 MHz on 405 lines and 33·5 MHz on 625 lines is incorporated.

The AGC detector provides an output to feed the AGC amplifier and also yields the 6 MHz sound i.f. signal, which is extracted from a 6 MHz trap circuit in series with the diode load.

AGC AMPLIFIER. The input to this circuit is provided by the AGC detector, which feeds in a voltage equal to the mean level of the vision signal.

The contrast control adjusts the d.c. level on the input signal and hence the video drive at which AGC is produced.

The one transistor used in this circuit provides at its collector a low-impedance source of direct voltage which is used to control the r.f. amplifier in each tuner and the vision i.f. amplifier. Threshold is used on the tuners, to ensure that they are controlled only after the input signal has risen to such a level that the overall noise factor is not spoiled.

After this point, control of the tuner r.f. amplifiers is necessary in order to avoid overloading and cross-modulation.

VIDEO AMPLIFIER. A pentode valve is employed in this circuit, which is a.c.-coupled, to avoid bias switching between standards.

Series-shunt inductance compensation is used in the anode circuit to give the required bandwidth.

SYNCHRONIZING PULSE SEPARATOR. A valve is employed in conjunction with semi-conductor diodes. The pentode valve is operated as a combined anode-bend and leaky grid detector, and provides two outputs, free of vision modulation and clipped to remove noise. These are applied one to each time base.

LINE TIME BASE. This block uses a mixture of valves and semi-conductor diodes, the oscillator being a pentode in a sine-wave circuit, driving the output pentode. The frequency is controlled by a reactance valve which is in turn controlled by a phase-sensitive double-diode circuit which compares the phase of a reference voltage from the line output transformer (flyback) with the line synchronizing pulse. Switching from 10,125 Hz to 15,625 Hz is achieved by adjusting the tuning of the oscillator coil.

The line output valve is a pentode operating above-the-knee in a stabilized circuit. In this circuit, a voltage derived from the output voltage is detected and applied as bias to the output valve, thus stabilizing the output level against changes in valve slope drive, and HT supply voltage (B+).

An efficiency diode (thermionic) is used to boost the HT supply to the output valve and a thermionic diode is used to provide the EHT supply required by the picture tube. This diode is heated by a single turn winding on the line output transformer and has applied to its anode a voltage at line frequency of about 18 kV, derived from an overwind on the line output transformer. This is rectified and smoothed before being applied to the picture tube.

The line output transformer is also used to match the line output valve to the line deflection coils.

FIELD TIME BASE. The field oscillator is the triode section of a multiple valve and is used as a blocking oscillator which obtains its HT supply from the boosted HT on the line time base. Integrated twice-line pulses are applied to its cathode and provide a source of synchronizing pulses which ensure interlace (i.e., correct vertical positioning of the two fields). The field output stage, fed from the charging capacitor of the blocking oscillator, uses overall frequency selective feedback to obtain the correct waveform shaping. An iron-cored transformer is used in the anode circuit to match the valve to the scanning coils.

DEFLECTION COILS. These are wound on a ferrite ring and usually consist of a pair of shaped coils for the line (or horizontal) deflection and a toroid for field (or vertical) scanning.

PICTURE TUBE. This is either a 19 in. diagonal or 23 in. diagonal cathode ray tube, often with an integral protective face plate, or some other built-in implosion guard. The video stage applies the composite video signal to its cathode and a brightness control, which consists of a potentiometer across the HT supply, adjusts its grid voltage.

The focusing electrode is connected to a suitable (usually fixed) voltage and the second anode is connected to the boost rail. A field blanking pulse is applied to the grid from the field output transformer, and consists of a negative pulse lasting for about six or eight lines during the field suppression interval. This reduces the visibility of field flyback (retrace) lines.

Line suppression is also sometimes applied to reduce the visibility of rope caused by overshoot on the back edge of the line synchronizing pulse which would otherwise be visible during the line suppression period.

SOUND I.F. AMPLIFIER. This has two transistor stages, each of which has two transformers in its collector. One of each pair is tuned to 38·15 MHz and the other to 6 MHz. Input to the first stage is taken either from the 6 MHz take-off coil in the vision AGC detector (625 lines) or from the sound trap in the vision i.f. amplifier (405 lines). The transformers in the second stage feed the sound detectors.

SOUND DETECTORS. On 405 lines, a simple diode detector is used, followed by a diode rate-of-rise noise limiter. This limiter blocks the audio channel on noise peaks and thus improves the signal-to-noise ratio. The audio signal is passed to a switch which selects either the 405- or 625-line output for feeding to the audio amplifier.

On 625 lines, a ratio detector is used to demodulate the sound signal, and its output is fed to the selector switch.

AUDIO AMPLIFIER. This uses four transsistors, and is transformerless. The first two stages are common emitter amplifiers and drive a complementary pair of output transistors in the single-ended-push-pull configuration. The loudspeaker is driven directly from the junction of the two output transistors. In order to reduce the variation of sound level on 405-line signals, a direct voltage is derived from the emitter of the first amplifying transistor and applied to the first transistor in the sound i.f. amplifier. This is not necessary on 625 lines.

POWER SUPPLY. A hybrid receiver requires three sources of power, one for valve heaters, one for valve HT supply and the third for transistor HT supply.

This is most conveniently obtained by an auto transformer which has taps to allow for different declared mains supply voltages. The valve heater supply, typically 100 volts at 0·3 amps, is obtained from a tap on the transformer; the valve HT supply, via a silicon diode rectifier, from the mains line output; and the transistor supply, via another silicon diode, from a further tap on the transformer (this can sometimes be made to correspond with a 20 volt mains voltage tap).

Filtering and surge protection circuits are required, the former consisting of resistance-capacity filters on the low voltage supplies and inductance-capacity filters on the valve HT. A series resistor and shunt capacitor are required on the HT rectifier to prevent damage due to mains transient surge.

MISCELLANEOUS DESIGN POINTS

Points which have to be considered during the design of a receiver, but which do not appear on the circuit diagram are many, and include those given below.

RADIATION. Radiation of electric, magnetic and X-ray fields can occur from television receivers, and the various national standards organizations have their limits and recommended practices for these properties. These standards cover radiation of signals due to time bases, i.f. and video amplifiers and oscillators, both as direct radiation from the receiver and its associated cables, and as injection of unwanted voltages into the mains supply network. X-radiation, for example, is controlled in Britain by BSS 415, which also lays down limits for other safety aspects, notably the potential of exposed metal parts, temperatures within the receiver and the access to the live parts of the receiver by a layman.

Temperature effects, which may result in drift of tuning, change in contrast, or change in picture size or geometry during the period of use of the receiver have to be considered, and exhaustive tests are made at the prototype stages.

Particularly with hybrid valve/transistor receivers, warm-up presents a problem, since some sections of the receiver will start to work immediately on switch-on, whereas others will take a minute or so to operate. Paralysis by overload of automatic gain control circuits (lock-out) has to be avoided and often quite elaborate circuitry has to be employed to avoid this.

SPECIAL RECEIVERS. VHF wired television in which signals are transmitted over wire instead of over the air, are common in the US and Britain and require receivers which are slightly different from normal receivers. In the US, the only difference is that the UHF tuner is not required, since the UHF signals are converted to VHF before being applied to the cable. In Britain, modification is usually necessary for reception of 625-line programmes on VHF since British receivers automatically switch the UHF tuner into use when 625 lines is selected.

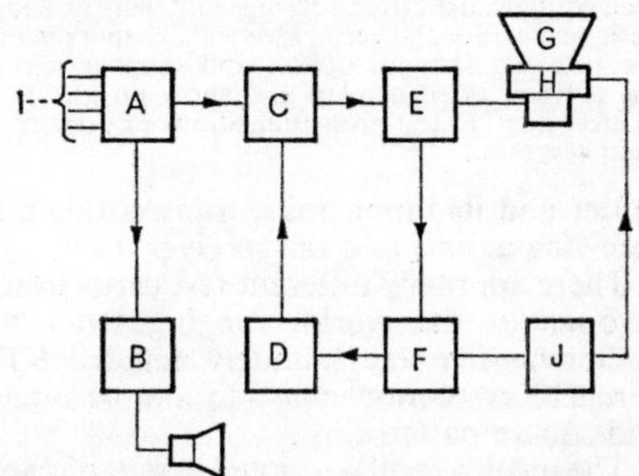

WIRED TV: HF TERMINAL UNIT. Incoming signals, 1, are separated in A, the audio signal (at audio frequency) feeding to transformer and volume control, B, and loudspeaker. Vision-modulated h.f. signal goes to video amplifier and detector, C, and output stage, E, modulating picture tube, G. Output stage, E, also supplies sync separator, F, driving time bases, J, and deflection coils, H. An AGC amplifier, D, maintains constant signal level from C.

HF wired television systems, used only in Britain, employ special receivers, which are much simpler than normal television receivers, being designed for an input in the range 3 MHz to 12 MHz at a level of several millivolts. Thus no tuner is required and such receivers are usually not superhets. Video amplifying and scanning circuits are very little different from those of normal aerial sets, but the sound is often fed into the receiver at audio frequency and at such a level that no amplification is required. Thus the loudspeaker can be driven from the input, via a transformer and volume control.

USE OF TEST CARDS

Test cards are geometrical patterns transmitted outside normal programme hours by broadcasting authorities and intended to check the performance of, and show deficiencies in, studio equipment, the transmitter and its input links, transmitting and receiving aerials and the receiver itself.

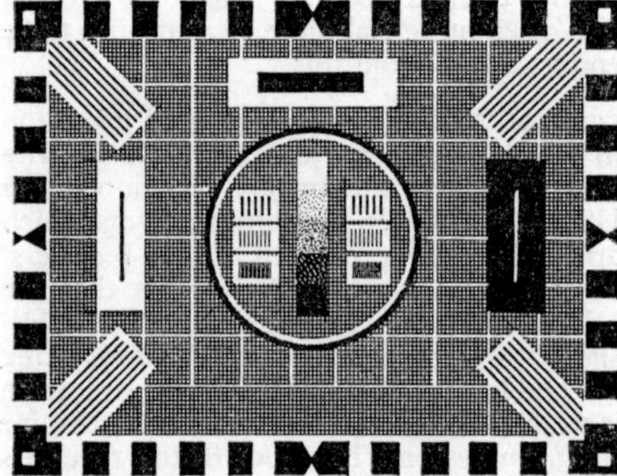

TYPICAL TEST CARD. (British Test Card D.) Castellated border establishes 3 : 4 aspect ratio. Central area contains definition gratings and density wedge with peak white and zero video (black) spots in end sections (not shown). Circles and squares help to check linearity of display and show up ghosting. "Letter-box" at top gives indications of poor transient response.

There are many different test cards in use throughout the world, ranging from the comprehensive and minutely detailed RTF (French) resolution chart to simple circle-and-square patterns.

The most recently adopted standard card, Test Card D, was evolved jointly by the BBC, ITA and BREMA to whom the copyright belongs. (There is a 625-line version, Test Card E, which differs only in respect of the definition gratings.) This card contains patterns designed to test geometrical and modulation linearity, frequency and transient response and direction of scanning.

Assuming that all the equipment up to the receiver input is operating correctly, many points of receiver performance can be checked.

ASPECT RATIO. Assuming that the aperture of the picture tube is of the correct 3 : 4 ratio, the castellated border should just fill the screen, with the crossover in the hourglass pattern at the centre of each of the edges. In fact, earlier CRTs were made with a ratio of 4 : 5 and most receivers still use this ratio, with the result that the side castellations are not visible.

LINEARITY. The circle in the middle of the card should not be distorted and should appear in the centre of the screen. The white lines which divide the card into squares should be equidistant, and the black-and-white gratings at the picture corners should be at 45° to the horizontal. Some distortion of these gratings is normal in commercial receivers, usually taking the form of curling of the outside ends of the gratings. Excessive cramping on the left-hand side is usually indicative of low emission of the line (horizontal) output valve, and the appearance of a vertical kink or fold-over about two-thirds of the way along the line scan usually shows that the efficiency diode is not working correctly.

Cramping of the bottom of the picture, with a consequent black-line between the bottom of the card and the bottom edge of the screen usually results from a faulty field output valve or leakage in its grid coupling capacitor.

SCANNING DIRECTION. There is not only the D (or E) near the bottom of the card, but also transmitter identification below the D to assist in ensuring that the scanning coils are correctly connected and positioned.

BANDWIDTH AND TRANSIENT RESPONSE. Restriction of bandwidth is seen by low definition in the definition gratings for loss of high frequencies and detail over the picture and streaking for poor low frequency response.

The "letter box" at the top of the card not only shows streaking but gives some clue to its cause if only the black (inner) or white (outer) rectangle is seen to streak, i.e. produce an effect on the next area to its right. The three central horizontal lines will, if there is low-frequency distortion, run through the circle and the rectangles at each side of the circle. Such distortion is often due to insufficiently high value or leaky coupling capacitors in the video circuit, failure of d.c. clamping diodes, or high heater-cathode leakage in the picture tube.

Poor transient response appears as distortion of transitions from black to white or white to black. Distortion may appear on the thin pulse bars situated one at each side of the circle as a white "echo" before or after the black bar or a black echo before or after the white bar. This shows that high frequency phase shift is taking place, usually associated with enhancement of contrast in the definition grating.

A smear, often seen after the gamma wedge or letter box, shows a lower frequency phase shift and is often associated with a reduction of contrast in the definition bars.

Either of these effects can be caused by faulty alignment of the i.f. tuned circuits or, in the case of a correctly aligned receiver, faulty positioning of the picture carrier on the Nyquist flank.

Ghosts, or delayed images, may also appear and usually show most clearly on the diagonal rulings or the white grid pattern.

GAMMA. Faulty modulation linearity shows as a failure to resolve the five steps of the wedge as approximately equal changes in contrast, together with disappearance of the peak-white spot in the centre of the white step and the zero-video (black) spot in the black step. This fault can be due to incorrect control setting, inadequate EHT regulation or low-emission valves in either the last vision i.f. amplifier or video amplifier stages.

TIME-BASE SYNCHRONIZATION. The black and white castellations at the sides of the card are intended to check the pulling-on-whites performance of the receiver. Poor transient response results in the video voltage failing to fall to black level from white, before the line synchronizing pulse appears, with the resultant late start of the next line. "Cogging" of vertical lines in sympathy with the castellations therefore appears. With the almost universal use of flywheel time bases, this effect is now much less often encountered but is almost invariably due to poor middle-frequency response.

Ghosts which are delayed by several microseconds can also cause cogging since the delayed picture information appears during the sync pulse period and may coincide with the leading edge of the line sync pulse.

COLOUR RECEIVER SYSTEMS

A television receiver designed to work on any of the three colour systems now in regular service, and using the shadow-mask tube, includes all the basic circuitry of a monochrome receiver, with additional sections which extract and process the colour information.

NTSC SYSTEM. The NTSC colour television signal conveys information on the colour components of a picture by modulation of a suppressed subcarrier. This modulation, in principle, consists of two colour-difference signals in quadrature, so that the phase of the resultant signal is related to the instantaneous hue of the picture, and its amplitude is proportional to the instantaneous saturation of the colour.

In order for the instantaneous phase of the colour (or chroma) signal to be established at the receiver, a frequency and phase reference must be provided, and this is the first respect in which the NTSC colour television system differs from monochrome.

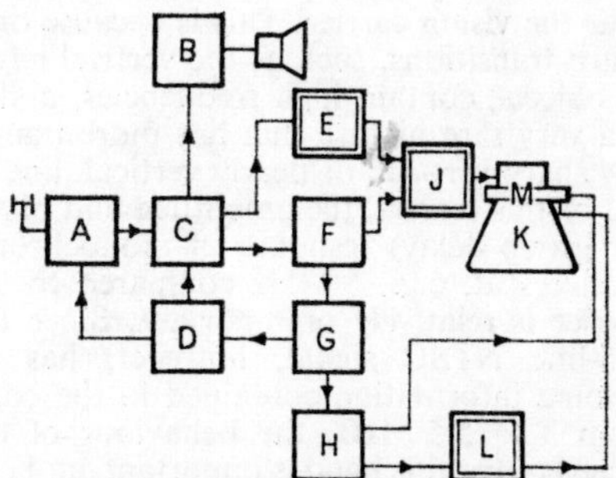

COLOUR RECEIVER: SIMPLIFIED BLOCK DIAGRAM. Tuner, A, feeds picture i.f. and detector, C, whence sound signal branches to audio i.f., detector and output stage, B. Luminance video circuits, F, feed luminance signal to matrix (colour decoder), J, which is also supplied with colour-difference signals from chrominance circuits, E. Matrix J derives separate R, G and B signals for modulating the three electron beams in shadow mask tube, K. Sync separator, G, provides necessary output to AGC stage, D, and drives line and field time bases, H, for tube deflection coils, M. Convergence circuits, L, maintain alignment of electron beams on colour triads forming phosphor screen. Double-lined blocks are peculiar to colour receiver.

A colour burst consisting of a few cycles of the colour subcarrier (approximately 4·43 MHz for a 625-line system), is transmitted on each line between the end of the synchronizing pulse and the start of the picture information (i.e., the back porch).

This burst is used to lock, in frequency and phase, an oscillator in the receiver, the oscillator output driving a pair of synchronous detectors. Thus an output dependent on both amplitude and phase can be derived from each detector, and the colour-difference signals regained, combined with the luminance information and applied to the shadow-mask tube.

Colour receivers have to be compatible, that is, display an acceptable monochrome picture when colour is not being transmitted. This necessitates a circuit known as the colour killer, which is capable of detecting the presence of the colour burst, and, in its absence, preventing the chroma circuits from operating and producing chromatic effects on the monochrome picture.

DISTORTION. The average energy content of a typical 625-line monochrome picture reduces as the modulation frequency increases, so that the average energy at, e.g., 5 MHz is only a few per cent of the energy near the vision carrier. This is because only sharp transitions, such as the vertical edges of objects, contain high frequencies, and it is a very rare picture that has more than a few sharp vertical, or nearly vertical, lines.

For this reason, the amplitude and phase (or group delay) response of monochrome receivers at, e.g., 5 MHz compared to the carrier is relatively unimportant. Since the 625-line NTSC signal, however, has its chroma information contained in the band from 3 to 5·5 MHz, the behaviour of the receiver over this band is important, and it is essential that a colour receiver does not appreciably attenuate or delay these signals. The effect of attenuating the chroma information is to reduce the saturation of the coloured areas, i.e., to make them paler, and the effect of phase error is to alter their hue (colour). Also, the amplitude and phase response must not change with voltage input level (these effects are known as level-dependent gain and level-dependent phase respectively), or the hue and saturation of coloured areas will alter with signal level. Thus the design of tuners, i.f. amplifiers and video circuits is more difficult than for monochrome receivers, and particular care has to be taken to avoid phase distortion in the chroma band, caused, for instance, by sound traps.

Cross-modulation between sound and chroma signals, in the common i.f. stages and the tuner, can also produce low frequency chroma patterning, whereas the same receiver does not show sound on vision cross-modulation effects on monochrome transmissions.

SECAM SYSTEM. The greatest weakness of the NTSC system is its poor tolerance of phase distortion, which can occur not only in the receiver, but also in the studio equipment, transmission links, and the transmitter. At least two systems have been devised in attempts to improve its performance in this respect. SECAM, a French system, used in Eastern Europe and France, transmits two frequency modulated colour difference signals and uses each one alternately for the duration of one line as the chrominance information. A delay line at the receiver holds one line until the next arrives so that all the chrominance information is simultaneously available. A SECAM receiver is more expensive than one designed for NTSC or for the other NTSC variant PAL.

PAL SYSTEM. This is basically the same as the NTSC system, but uses an ingenious artifice to reduce the effects of phase distortion. One of the two colour difference signals which modulate the suppressed colour subcarrier is changed in phase on alternate lines while the other is left unaffected. By the use of a delay line and an electronic switch at the receiver, signals on alternate lines are added and the colour difference signals extracted, thus restoring the original phase relationship. Any phase shift added by imperfections in the transmission path between the encoder at the studio and the detector in the receiver is the same on each of the two successive lines, and will therefore cancel out. Thus the hue of the picture is unaffected. The amplitude, and hence the saturation, is affected but this is subjectively less annoying than a change of hue.

A PAL receiver therefore needs a delay line and an electronic switch which are not required by an NTSC receiver. If, however, the eye of the viewer can be relied upon to average successive lines, the delay line can be dispensed with. A rather coarse horizontal line structure results and thus the picture on a receiver without a delay line (known as simple PAL, is considerably inferior to that on a receiver with a delay line (de luxe PAL).

PRACTICAL COLOUR RECEIVER

A colour receiver common to NTSC, SECAM and PAL contains sections which are additional to the circuitry of a monochrome receiver. In addition, the performance of the parts this receiver has in common with a monochrome receiver is modified because of the presence of colour information.

TUNER. This generally follows conventional monochrome practice, but with particular attention to good amplitude and group delay response, and with lower cross-modulation than is necessary for a monochrome receiver.

PICTURE I.F. AND DETECTOR. This is conventional, but the sound signal is extracted before detection and extra sound rejection circuits are added between the sound take-off point and the detector to avoid visible sound/chroma beat effects. As in the tuner, good amplitude and phase response is essential.

SOUND I.F. DETECTOR AND AUDIO. This is conventional except for above-average a.m. rejection to avoid chroma buzz.

LUMINANCE VIDEO CIRCUITS. A short delay-line is incorporated in this section to compensate for the delay suffered by the chrominance signal in its longer journey through the receiver. The capacity load of the video output stage is greater than that of an equivalent monochrome stage since the three shadow-mask tube cathodes in parallel have to be fed. Equalizing circuits are provided in the anode circuit to ensure correct balance of drive to the three tube cathodes.

SYNC SEPARATOR. This is conventional, requiring only particular attention to avoidance of distortion of the line and field blanking interval signals.

CHROMINANCE CIRCUITS. These contain all the special circuitry for extracting and processing the colour information encoded on the composite signal into a form suitable for applying to the shadow-mask tube.

NTSC CHROMINANCE CIRCUITS. The chrominance i.f. signal is extracted from the luminance channel, amplified and demodulated. The chrominance video signal is then applied to the synchronous detection circuits which are fed with two quadrature-related inputs at the colour subcarrier frequency from the colour sync oscillator. This is locked in phase and frequency to the colour burst via the burst gate. The I and Q signals are then decoded in the synchronous detectors and applied, after amplification, to the matrix. The colour killer operates in the absence of a colour burst to paralyse the chrominance circuits, in order to prevent undesirable chroma effects on monochrome transmissions.

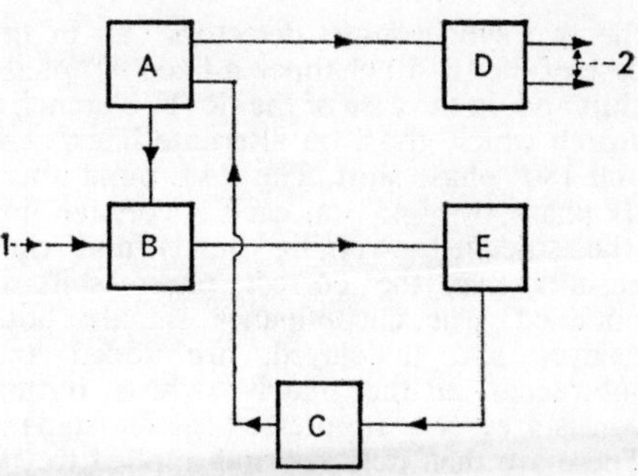

NTSC CHROMA CIRCUITS: BLOCK DIAGRAM. Chrominance i.f. signal is extracted and fed to chrominance i.f. amplifier and detector, A. Chrominance video signal is then applied to synchronous demodulators, D. 1. Colour burst (i.e., a few cycles of subcarrier transmitted on each line) is gated at B, and used to lock colour sync oscillator, E, in correct frequency and phase. Oscillator output is applied through colour killer, C (which paralyses chroma circuits in absence of colour burst, i.e., on monochrome) to control frequency and phase of demodulators, D. Their output consists of the two colour-difference signals needed by the colour decoder, together with the luminance signal.

PAL CHROMINANCE CIRCUITS. The chrominance circuits for a PAL receiver are more complex. As in the NTSC case, two synchronous detectors are used to decode the colour information, but there is a delay line interposed between the chroma input and the detectors. The subcarrier oscillator is synchronized by the colour burst and drives

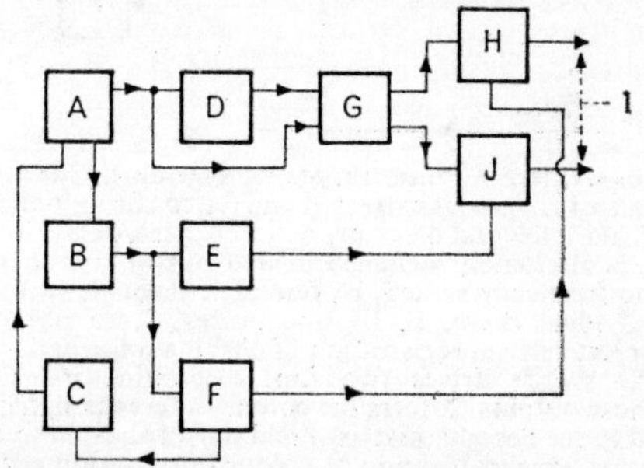

DE LUXE PAL CHROMA CIRCUITS: BLOCK DIAGRAM. Chrominance i.f. signal is extracted and fed to chrominance i.f. amplifier and detector, A. A 64 μsec. delay line, D, produces a one-line delay which averages positive and negative colour phase errors and separates the colour-difference signals in combination with a matrix, G. The matrix output goes to a synchronous detector, H, for the (R-Y) signal, and another, J, for the (B-Y) signal. Subcarrier oscillator, B, synchronized by the colour burst, drives J via a fixed 90° phase shift, E, and drives H via a switch, F, which on alternate lines gives zero and 180° phase shift. F also operates colour killer, C. Colour-difference signals, 1, are finally combined with the luminance signal in an output matrix, as in NTSC.

the two synchronous detectors via, in the case of the (B–Y) channel, a fixed 90° phase shift and, in the case of the (R–Y) channel, a switch which gives, on alternate lines, zero and 180° phase shift. The PAL burst alters its phase by ±45° on each successive line (the so-called swinging burst) and thus ensures that the correct phase shift is obtained. The chrominance signals, both delayed and undelayed, are added and subtracted in the matrix, whose output consists of two suppressed-carrier signals. These are then detected and applied to the output matrix as in the case of the NTSC receiver. Also, as in the NTSC case, the colour killer operates by silencing the chroma circuits on monochrome transmissions.

SECAM CHROMINANCE CIRCUITS. A SECAM receiver, unlike the NTSC or PAL types, needs no synchronous detectors, since the colour difference signals are sequential frequency modulated carriers.

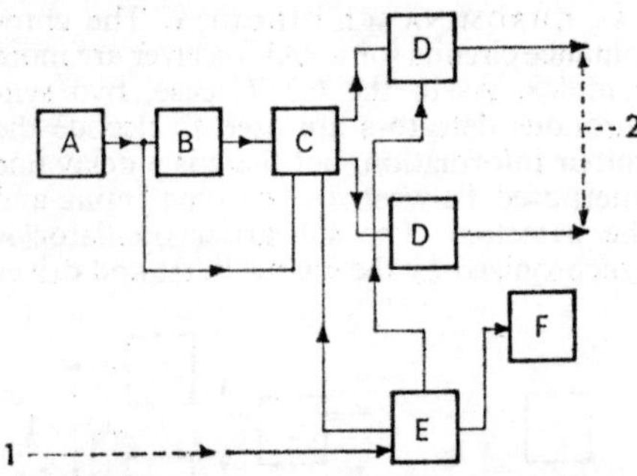

SECAM CHROMA CIRCUITS: BLOCK DIAGRAM. Chrominance i.f. signal is extracted and fed to chrominance i.f. amplifier and detector, A. A 64 μsec. delay line, B, is alternately switched in and out of circuit by line frequency switch, C, operated through switch and ident drive, E, by line pulses, 1, to obtain correct vertical registration of chroma information. The switch drives two f.m. discriminators, D, whose outputs, 2, form the colour-difference signals fed to the decoder matrix. Field pulses also fed to E ensure synchronization of colour information with the transmitted signal. F. Colour killer.

The delay line, as in PAL, is alternately switched in and out of circuit to obtain correct vertical registration of the chroma information. Two frequency discriminators are driven from the switch and their outputs form the colour difference signals.

Identification of the colour difference signal sequence in use on any given field is provided by an identification signal (often called an ident) radiated during the field-blanking period, which is different for successive fields. This signal is used to control the phase of the drive to the line-frequency switch to ensure synchronization of the colour information.

The colour killer uses the absence of the ident signal to suppress the chroma channel during monochrome transmissions.

MATRIX. This decodes the colour difference signals and, by mixing them with the luminance signal, restores the original R, G and B signals obtained from the camera at the studio.

The relative amplitudes of the three colour components are also adjusted in the matrix to obtain the correct colour balance. Unless this is obtained, correct hue (particularly noticeable on flesh tones) and an acceptable white on monochrome transmissions is not produced.

TIME BASES. These only differ from the conventional monochrome circuits in that (a) the line (or horizontal) time base, which also provides the EHT supply to the picture tube, is of higher power, since a regulated supply of 25 kV is required and hence the deflection power is also greater. In addition, the large neck diameter of the shadow-mask tube calls for more deflection power; (b) convergence signals, necessary to ensure dynamic convergence of the three beams at the face of the shadow-mask tube are derived from the time bases.

DEFLECTION COIL YOKE. This is of larger size than a normal monochrome type, since the neck of the picture tube is larger to accommodate the three electron guns, and also includes convergence coils to which the dynamic convergence signals referred to above are applied.

AUTOMATIC GAIN CONTROL. AGC circuits in colour receivers are very similar to those in good quality monochrome sets, and do not disturb the colour burst (or the ident signal in SECAM) or allow the burst or ident signal to disturb line clamping level.

CONVERGENCE CIRCUITS. Since the three guns in a shadow-mask tube cannot all occupy a position on the mechanical centre-line of the tube, there are differences in the geometry of the three rasters due to the different angles at which the electron beams strike the face-plate, and these differences are reduced to avoid registration errors.

A distorted sawtooth is derived from each time base and applied to the convergence coils and the amplitude and waveform of the convergence signals adjusted during manufacture and on installation. In addition to this dynamic correction, static convergence is controlled by adjustable permanent magnets placed round the neck of the picture tube.

MANUFACTURE

Television receivers are mass produced; the tooling required for all the parts which go to make up a complete receiver can cost a vast amount of money, and thus the receiver has to be made in large numbers to reduce the proportionate expenditure on tools and special test gear. A mass-produced receiver must be as easy to make as possible, must be able to use, as far as possible, commonly available parts and must not require individual alignment or testing procedures, which can be very time-consuming.

In a highly competitive market, a very small difference in the cost of a receiver may make a tremendous difference to its potential market. Special features such as remote control of channel selection, volume, brightness, and mains supply switching, and automatic contrast control with variation in ambient lighting, are available in the more sophisticated markets, but elsewhere the demand is for a well-built, straightforward, conventionally styled receiver.

The cost of the individual parts of a television receiver, i.e., cabinet, picture tube, chassis, loudspeaker, tuner, must be carefully balanced for the market envisaged in order to compete, and this need has led to many ingenious variations on a theme. The same chassis may be used for a dozen different models, from a 19 in. minimum-cost table model to a 23 in. luxury cabineted floor-standing receiver, perhaps with sound radio facilities.

COLOUR. The NTSC system is such that no real reduction in cost is possible by omitting some desirable, but not essential circuit, but its derivative, PAL, offers this facility.

By omitting the delay line, an acceptable, albeit somewhat flickery, picture is obtainable and this is thought by some protagonists of PAL to be an advantage of the system.

The cost of the shadow mask tube, by mass-production and refinement of processing techniques, has been greatly reduced in the past ten years, but still forms a considerable proportion of the cost of a colour set.

Other tubes, such as the Chromatron (or Lawrence) tube are in small-scale production in Japan, but it remains to be seen if they lead to a cheaper NTSC receiver.

L. T. M.

See: *Colour principles; Colour systems; Large screen television; Shadow mask tube.*

Books: *Principles of Television Engineering*, Vols. 1 and 2, by R. C. Whitehead (London) 1965; *Principles of Television Reception*, by W. Wharton and D. Howorth (London) 1967; *Television Engineering*, by S. W. Amos and D. C. Birkinshaw (London) 1957.

Reciprocity Law. Law of photographic exposure which states that if the light intensity multiplied by the time of exposure remains constant the effect on a photographic emulsion will remain the same. Increases of intensity can therefore be compensated by reciprocally proportionate changes of time and vice versa. In practice, the behaviour of a sensitive material may depart from this rule when the exposure times become extremely short or extremely long and this is termed reciprocity failure.

See: *Exposure control; Sensitometry and densitometry.*

Recirculation. Replacement of the solutions used in the tanks of continuous film processing machines to avoid chemical depletion. Generally done by controlled circulation to and from the machine tank from a large volume to which chemical replenisher solutions are supplied and where the temperature of the solution is controlled. In addition, recirculation of the liquid in the machine tank itself is often provided to ensure uniformity over the whole immersed film path and to obtain a degree of agitation at the surface of the film.

See: *Laboratory organization; Processing; Regeneration of solutions.*

⊕**Reconditioning.** Treatment of positive film prints with the object of removing oil, scratches and abrasions from the surface. While the exact methods are trade secrets, they are based on solvent cleaning followed by wax applications and polishing, and sometimes localized treatment of the emulsion to cause it to close over the scratch.

See: *Laboratory organization; Release print examination and maintenance.*

⊕**Recording System.** Complete channel consisting of microphone(s), mixing console, equalizers, compressors, monitoring and other devices, feeding into a magnetic or optical tape or film recorder.

See: *Sound recording.*

⊕**Recordist.** Senior sound technician on the stage. Responsible for the quality, clarity and perspective of sound recorded during shooting. Instructs the floor sound staff on the positioning of microphones and of their movement during any scene. Also called floor mixer.

See: *Sound mixer; Sound recording.*

⊕**Red Master.** Colloquial term for certain duplicating positive films of very fine grain coated on a clear support, so that the developed image is warm or reddish in colour.

See: *Duplicating processes (black-and-white).*

⊕**Reduced Aperture (RA).** Size of picture image established in 1929 when the development of sound-on-picture techniques necessitated room for a sound track. The corresponding aperture size in 35mm cameras and projectors became known as the reduced aperture in contrast to the original full aperture (FA). It is also referred to as the Academy Aperture or Movietone Frame.

See: *Film dimensions and physical characteristics.*

⊕**Reduction Printing.** In motion picture work, the printing of a copy on to a gauge of film smaller than that of the original, as in making 16mm or 8mm prints from 35mm negatives.

See: *Laboratory organization; Print geometry (16mm).*

⊕**Reel (of Film).** In general, the unit in which cinematograph film is handled, either as a cut assembled negative or as positive prints. At one time, the length of a reel was not greater than 1,000 ft. of 35mm film or

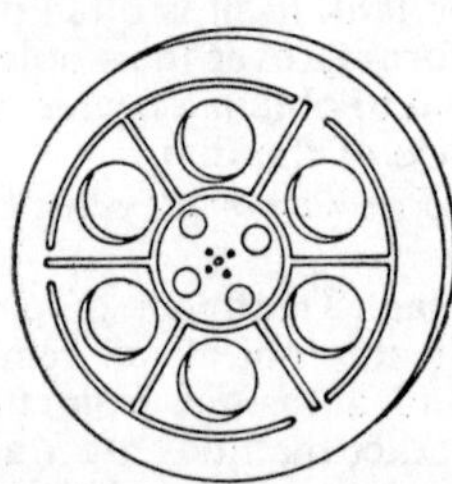

400 ft. of 16mm, corresponding to about ten minutes of projection time, and this is still known as a reel of film. It is current practice to use reels up to 2,000 ft. of 35mm negative film. The term is also used for the spool on which the reel of film is handled.

See: *Film dimensions and physical characteristics; Projector.*

□**Reference Circuit.** Circuit along which a signal is sent which is used as a carrier to be reinserted in suppressed-carrier signals.

See: *Colour systems.*

⊕**Reference Print.** Release print which has been approved by the producer and director of a film; often kept as a reference print to settle arguments about the quality of subsequent prints.

See: *Laboratory organization; Release print make-up.*

□**Reflective Network.** Network which provides an incorrect termination of a line and thereby causes a reflection.

⊕**Reflector.** Reflecting surface, frequently silvered, used to reflect light where it is needed. Projector and studio lamps use reflectors so as not to waste the light output from the back of the lamp. For location shooting, reflectors are often used to redirect sunlight either on to actors' faces or on to some shadowed part of the scene. Hard reflectors have a mirror-like quality and often produce highlights and hot spots. Soft reflectors have a pebbled or semi-matte surface, or employ diffuser net, to produce a less intense but more even illumination.

See: *Cameraman* (FILM).

⊕□**Reflector Lamp.** Type of incandescent projection lamp which contains a built-in rear reflector. By incorporating the reflector in

the bulb, it is possible to fix it in the optimum position and ensure that its surface does not suffer from dirt or contamination.

See: *Cold mirror reflector; Lighting equipment.*

⊕**Reflex Shutter.** Camera shutter which plays a dual role: to obscure the film during pull-down, and to form an image by reflection

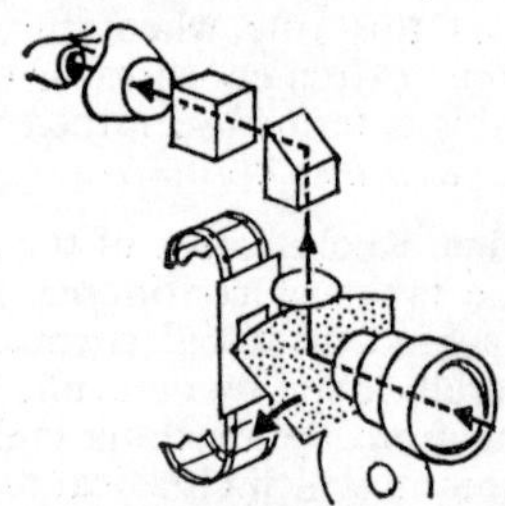

REFLEX SHUTTER. Shutter is shown in mid-position over aperture, obscuring film and reflecting lens image to ground glass above.

on a ground glass, which is relayed to the cameraman's eye. In a common form, the rotating shutter is silvered on its front face so that during pulldown it reflects the image into the viewfinder, the film and the cameraman thus seeing the picture alternately. The principal advantage of the reflex shutter is that it eliminates parallax viewfinder errors at source, and thus avoids the need for accurately setting up the finder for each shot.

See: *Camera* (FILM).

Refraction. Change of direction, or bending of a ray of light at the boundary surface where it passes obliquely from one transparent medium to another.

See: *Light; Optical principles.*

Refractive Index. Signified by n or μ. Ratio of the sine of the angle of incidence to the sine of the angle of refraction when a ray of light passes from a vacuum into a transparent medium. A very close approximation is obtained if the vacuum is replaced by air.

See: *Light; Optical principles.*

Refraction, Laws of. When a ray of light passes from one medium to another: (1) The incident ray, the refracted ray, and the normal at the point of incidence to the surface separating the media, are co-planar. (2) The ratio of the sine of the angle of incidence to the sine of the angle of refraction is a constant for the two media. The latter law is known as Snell's Law.

See: *Light; Optical principles.*

Regeneration of Solutions. Replacement of chemicals used up in the course of processing. Specifically, restoration of the required characteristics by the removal or alteration of unwanted components, for example, the regeneration of fixing baths by the electrolytic removal of silver.

See: *Processing; Recirculation.*

Register Pin. Positioning device used in certain high-precision film cameras, printers and projectors, in which a pin or pins enters one or more perforations when the film is stationary in the aperture so that each frame is locked in precisely the same position as the others. Precision of registration thus depends on precision of perforation. Also called a pilot pin.

See: *Camera* (FILM); *Pilot pin.*

Registration. The act of placing an image in an identical position to that of a simultaneous or successive image.

In a film camera, it is essential that each time the film comes to rest, its position in relation to the fixed frame of the camera is identical; otherwise, on projection, relative movement is visible in the form of picture unsteadiness. The same registration problems occur in intermittent-type printers and in projectors, especially those used in the studio for back projection plates.

In any colour printing process where the three colour images are produced in succession on the printed film, as, for example, in the imbibition process, a high degree of accuracy of register is necessary if colour fringing is not to be perceptible.

In colour television, precise registration is required both in the camera and the receiver: in the camera, because the incoming light is split up, usually into three spectral components (sometimes with an additional luminance channel), and the contribution of each component must be correctly related to each picture element; in the receiver, because the normal shadow-mask tube contains three electron guns, each of which must make its proper colour contribution to the correct picture elements.

See: *Camera* (FILM); *Camera (colour); Shadow mask tube; Special effects* (FILM); *Telecine (colour).*

Relay Lens. Lens used in the lamphouse system of an optical printer or projector in cinematography to produce an image of the condenser in the plane of the film being exposed.

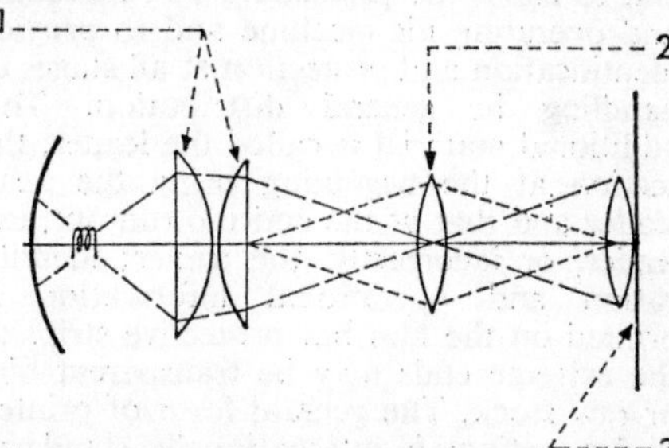

RELAY LENS. Relay lens, 2, placed at focus of condenser system, 1, forms image through aperture lens on film plane, 3, to provide uniform illumination for optical printing.

The term is also used to describe an optical system usually involving collimating lenses which can be introduced to lengthen the optical path in an image-forming system for camera or viewfinder without a proportionate change of image magnification.

See: *Camera (colour); Printers.*

Release. A film which has been distributed for public exhibition is said to be on release.

⊕**Release Print Examination and Maintenance.** Ideally, a print should be examined after every booking, since a projectionist is entitled to receive prints in good condition ready for showing. Narrow-gauge films are sometimes handled by inexperienced operators, imposing an extra responsibility on film libraries. In cinema theatre distribution, programmes are usually booked solidly for weeks at a time, and as prints are too costly for the distributor to have more copies made than are absolutely necessary, films must move by direct cross-over from cinema to cinema, and prints can be examined only at irregular intervals.

In the libraries handling narrow-gauge copies, inspection is carried out whenever a copy is returned, and automatic inspection machines are often used which run a film at high-speed with sensors to detect bad splices or torn perforations. Magnetic tracks may be checked for erasure or defacement by the detection of a low frequency signal of say 30 Hz at 20 dB below peak signal level superimposed at the time of recording. In the examining machine this signal can be detected even at high running speeds, and if it is absent the machine is automatically stopped to allow the track to be investigated.

In the course of inspection it may be necessary for the film to be cleaned, and suitable solvents, such as trichlorethylene, are used, sometimes with ultrasonic agitation. The film is cleaned by means of rotating brushes and polished on buffs before rewinding. When necessary, a print may be given special treatment to reduce or eliminate scratches by the use of lacquer or suitable solvents which soften the film surface before it is passed over a polished roll.

At the inspection stage, it may be necessary to cut out damaged sections or even to have such sections reprinted, particularly at the beginning and end of the copy where most of the handling damage takes place.

See: *Cinch marks; Inspection; Laboratory organization; Library; Lubrication; Release print make-up; Scratches.*

⊕RELEASE PRINT MAKE-UP

In addition to the essential picture and sound in a completed cinematograph release print, it is the normal practice to provide additional lengths of film at the beginning and end to assist the projectionist in threading and operating his machine and to provide identification and protection at all stages of handling in general distribution. This additional material is called the leader, the section at the beginning being the head leader and that at the end the run-out, tail leader, or incorrectly, the trailer; identification and operational information is printed on the film but protective strips at the extreme ends may be transparent film or raw stock. The general form of printed leader sections is internationally standardized to a certain extent although the differences of usage in motion picture and television practice provide a complication.

LEADERS

ACADEMY HEAD LEADER. For many years, the most widely used leader was that originated by the American Academy of Motion Picture Arts and Sciences, which was substantially adopted as an American standard (Z 22–55) in 1947 and in almost identical form as a British Standard (BS 1492) in 1948. This form of leader, sometimes with minor variations, had already become internationally recognized, a corresponding French standard having appeared in 1945, and it continues in regular cinematograph practice despite the revision introduced from 1965 onwards for the television user.

The word HEAD usually appears near the beginning, and the short identification section shows the subject title and the reel number printed both in large letters lengthwise along the film and on individual frames; similar information is provided in the sound track area, including the language version. The synchronizing section is normally opaque with the exception of a series of numbered frames, the first of which is marked PICTURE START at a distance of 192 frames (12 ft. in 35mm) ahead of the first scene in the reel. In the Academy leader these numbered frames occur at 16-frame intervals (originally one second at silent film projection speed) and run "11", "10", "9" etc., down to "3" after which all frames are opaque up to the beginning of the picture.

At a distance of 20 frames ahead of PICTURE START, a white line marked with a diamond indicates the corresponding sound start position for 35mm prints with optical sound tracks, and a similar diamond is

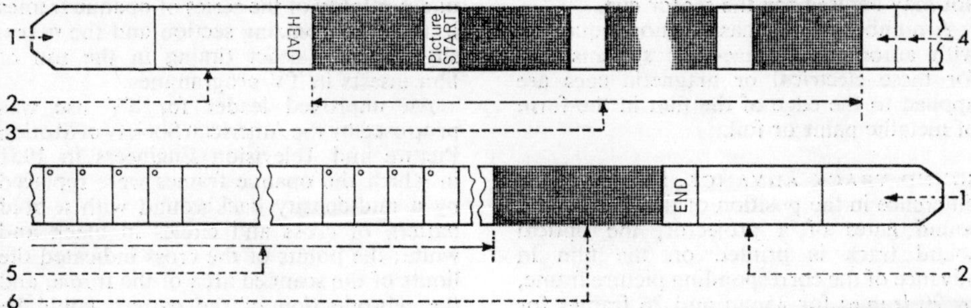

ESSENTIAL FEATURES OF RELEASE PRINT LEADERS. Leaders are designed to provide identifying and synchronizing information for the projectionist. They also protect the film itself from damage, which is most likely to occur at the beginning and end of a reel. (*Top*) Head leader. 1. Outermost section serves to protect the leader proper. 2. Identification of film, reel number, etc. 3. Synchronizing (sync, section) with numbered frames and sound head symbols. 4. First frames of picture. (*Bottom*) Tail leader. 5. Final picture section with cue marks to warn operator that a change-over is coming up. 6. Opaque run-out to prevent light reaching screen if change-over is slightly late. 1. Protective material. 2. Identification section. All types of release print leader contain these essential items of information.

repeated at the same distance in advance of all the other numbered frames. The projectionist can therefore check that his machine is correctly threaded-up by noting that a diamond marked frame is at the sound head when a numbered frame is in the picture gate.

Since 16mm prints are often made by direct reduction from a 35mm negative complete with Academy Leader, it has been found useful to include a frame indicating the position of the 16mm sound start 26 frames in advance of the picture start; this is identified by a line with a white circle and the number "16mm", but is not repeated for the other numbered frames.

RUN-OUT AND TAIL LEADER. After the last frame of the picture in the reel, a run-out section of 48 opaque frames is included, followed by an identification section showing the subject title, reel number and the words End of Reel, End, Foot or Tail. As at the head end, this information is printed both on individual frames and in large letters lengthwise along the film.

PROTECTIVE MATERIAL. In addition to the printed sections, a length of some 6 to 8 ft. of raw stock or clear film is attached at the very beginning and end of the reel to take the wear and tear resulting from projector thread-up, rewinding and transit. Damage to this part gradually shortens it and it is replaced when less than 6 ft. remains.

CHANGE-OVER CUES. An entertainment feature film may consist of four or more separate reels of film, and to provide continuous presentation it is essential to change from one projector to another without apparent interruption.

Before the change-over is made, the incoming projector must be fully up to speed. To warn the operator when to start his machine, a mark in the form of a circular dot is printed in the top right-hand corner of the frame of the picture eight seconds running time before the end of the reel; this is the motor cue. At a position one second from the end of the picture a second mark, known as the change-over cue, appears in the same position on the frame. When this is seen on the screen, it is the signal to switch over to the second projector and show its picture and sound. From experience with his own machines, the projectionist knows the time taken for a projector to get up to speed and therefore the number on the head leader which must be in the gate when the motor is switched on to give a smooth transition at the moment of the change-over cue.

Each of the two cue marks is printed on four consecutive frames at the positions specified, usually from holes punched in the picture negative, sometimes with a serrated die.

The size and position of the cue marks must allow them to be clearly seen by the projectionist in all conditions, including anamorphic and masked wide-screen presentation, and the dimensions are usually given in the appropriate standards. American and British practice is to use circular dots for both sets of cues but in France and some other European countries a square dot may be used for the motor cue.

A number of cinemas are now equipped with automatic change-over systems and for these electrical or magnetic cues are applied to the edge of the film in the form of metallic paint or foil.

SOUND TRACK ADVANCE. Because of the difference in the position of the picture and sound gates of a projector, the optical sound track is printed on the film in advance of the corresponding picture frame, by 20 frames for 35mm and 26 frames for 16mm.

This means that the sound for the first 20 frames of a reel is actually printed alongside the final series of black frames at the end of the head leader. Where important dialogue is taking place at a reel change-over it is sometimes the practice to record a short portion of the incoming reel at the end of the preceding reel to avoid loss in case of a bad operational change-over, but such exacting changes are avoided by the editor as far as possible in determining the reel assembly.

TELEVISION USAGE

LEADERS FOR TELEVISION FILM. As motion picture prints became used more and more for television transmission, certain difficulties arose in the use of the original Academy Leader, in particular the undesirability of the series of opaque frames in the synchronizing section and the necessity for very exact timing in the use of film inserts in TV programmes.

An improved leader for TV use was proposed by the American Society of Motion Picture and Television Engineers in 1951 in which the opaque frames were replaced by a mid-density background with a bold pattern of cross and circles in black and white; the points of the cross indicated the limits of the scanned area of the frame, and the reproduction of the pattern could be used to check transfer conditions and alignment before transmission began. Modified sound threading marks were introduced, and the alteration of the synchronization numbers to show 1 second intervals at 24 frames rather than the 16-frame interval of the Academy Leader was considered. But the 16-frame interval was retained in the hope that the new leader would be accepted by theatrical users as well as television. The numbers were shown on three consecutive frames at each interval.

This form of leader was not found acceptable for general cinema use but it was widely adopted for black-and-white television in the United States and to some extent in Britain and other European countries. Local differences of operating practice between organisations even in one country were still, however, considered

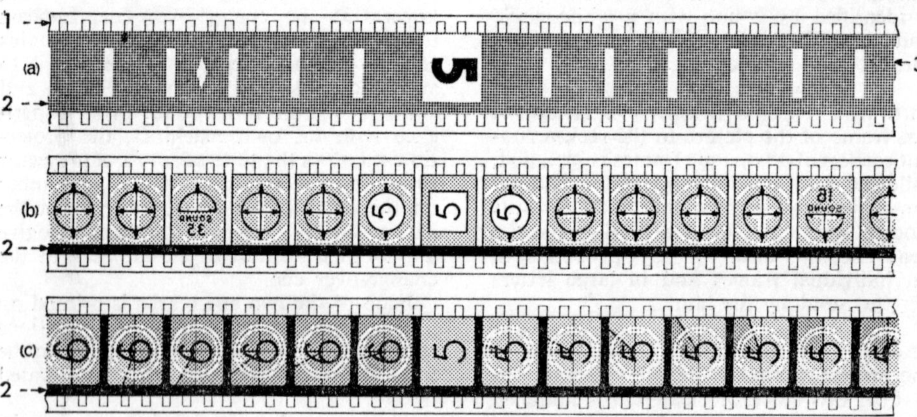

HEAD LEADERS: EXAMPLES OF SYNCHRONIZING SECTIONS. 1. Head end. 2. Sound track area. 3. Tail end. (a) Academy leader. (b) Television leader (SMPTE, 1951). (c) SMPTE Universal leader (1965). Academy leader has synchronizing numbers reversed top-to-bottom in comparison with the other two, so that they appear inverted on the screen. Television leader (b) has 35mm and 16mm sound start marks flanking each number and relating to a subsequent number located in the picture head. Arrow points on the mid-density frames indicate limits of the television scanned area of the frame. The Universal leader (c) shows a rotating radial line to indicate the passage of each second at standard projection speed as an aid to the accurate running-up of telecines (film scanners), telerecorders, etc. The synchronizing numbers are at 24-frame intervals (one second of time at US standards) instead of the 16-frame intervals of the Academy leader, which corresponded to one second at the old silent film speed.

necessary and many variations of either the original Academy form or the 1951 SMPTE version were used.

REVISED FORMS. Following a long period of discussion, the SMPTE proposed in 1965 a new Universal leader again intended to be acceptable for both theatrical and television practice.

Synchronizing numbers at intervals of 24 frames (1 second of time) were introduced and the count-down numbers thus became 8 to 2 instead of 11 to 3, and appeared on all frames. In the interests of the television user, the alternation of clear and opaque frames was changed to a series of middle densities with a continuously moving wedge pattern (animated clock) to denote the passage of each second and so allow very exact timing in the use of telecine equipment. The 35mm and 16mm optical sound start marks were slightly modified, and frames indicating the start frames for 35 and 70mm magnetic sound records were added.

The identification sections at the beginning and end were slightly lengthened but the same information is included with the addition of "Type of Sound", "Aspect Ratio", etc. It was recognized that European telecine equipment runs at 25 frames per second but the difference of timing was not considered important.

The SMPTE proposal was accepted as the American Standard in 1966 and is now used throughout for television films in the United States, although up to 1968 it has by no means become universally accepted for the theatrical distribution of motion pictures, for which the Academy Leader is still widely used.

A revision of the British Standard leader was also discussed during the same period and a new proposal parallels the American

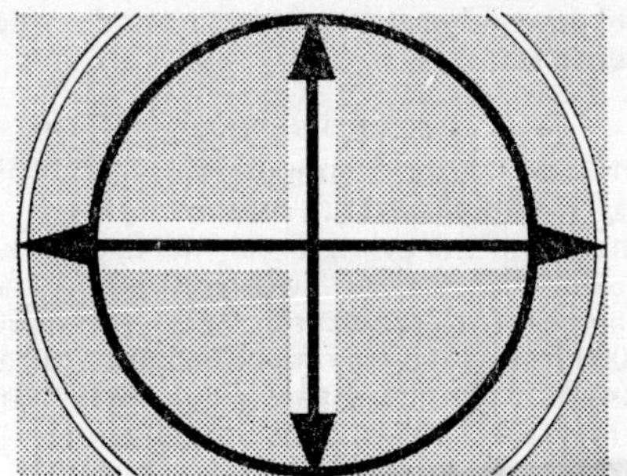

TELEVISION LEADER: MAIN BODY PATTERN. Points of cross correspond to limits of scanned area. Reproduction of pattern helps to check transfer conditions before start of transmission.

usage in having synchronizing numbers at intervals of one second, but the opaque frames are retained and the rotating sector is not included. Additional sound head start marks are provided as well as frames indicating the proportions and centring of different aspect ratio prints to allow the operator to check his projector aperture mask.

With the exception of the frames indicating aspect ratios, the proposed British Standard is very similar to a draft proposal of the International Standards Organization (ISO) in 1967; here again the identification sections give the title and reel number, the aspect ratio of type of presentation, the type of sound and the language of the version with a clear indication of the beginning and end of the reel in the appropriate language. The synchronizing section has numerals at 24 frames intervals from 7 to 2, and sound head marks are included for 35mm and 16mm optical tracks and for 35mm magnetic tracks.

In all these revisions, positions are allocated for the addition of control frames

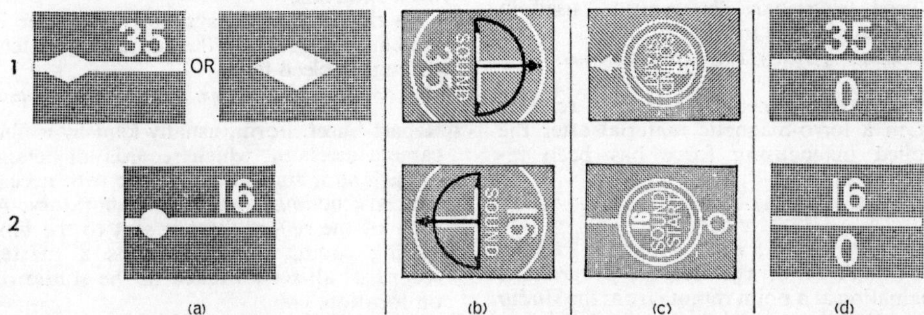

SOUND HEAD SYMBOLS USED IN SYNC LEADERS. 1. 35mm optical sound. 2. 16mm optical sound. (a) Academy leader. (b) Television (SMPTE, 1951). (c) SMPTE Universal (1965). (d) ISO Proposed (1967).

and other laboratory processing and sensitometric test data.

REEL LENGTH

For many years, 35mm cinematograph film was manufactured in rolls of 1,000 ft. in length and the printed reel was usually not greater than 950 ft. Two reels were joined together for projection but, for safety, 2,000 ft. was the maximum length of nitrate base film permitted as a single unit. Non-flam base film is now manufactured in 2,000-ft. lengths and 35mm negative is usually cut to give a reel length up to this as a maximum. A number of projectors are now capable of taking rolls up to 6,000 ft. in length, and reels can be joined together to allow a feature film to be shown with only a single change-over; however, the size and weight of such very large reels present handling difficulties in transport and the (nominal) 2,000 ft. reel is normal for 35mm general distribution.

Sixteen-millimetre prints, being smaller in weight and size, are assembled in reels up to 1,600 ft., and sometimes even 2,400 ft. The 1,600-ft. reel is convenient for handling feature film prints made by reduction since it is equivalent to two 2,000-ft. reels of 35mm; in television work the 2,000-ft. 16mm reel allowing an hour's programme to be handled as a single unit is widely used.

Seventy-millimetre prints present special problems of size and weight because of the greater width of the film and the increased thickness resulting from the applied magnetic stripe; in addition, because of the larger frame height, the playing time of a 2,000-ft. reel of 35mm requires 2,400 ft. in 70mm. 70mm prints are normally distributed on strong metal spools of nominal 3,000 ft. capacity with a separate transit case for each roll. **L.B.H.**

See: *Automation in the cinema; Cinema theatre; Projector; Release print examination and maintenance.*

Literature: *Motion Picture Presentation Manual*, British Kinematograph Society, 1962; *Recommended Practices*, SMPTE.

□**Release Time.** Operating time under defined conditions, of a piece of apparatus such as a sound limiter, for the gain to change from one value to another when the input signal is suddenly decreased in level; contrasted with attack time. It is usual to have much longer times for the release time. 0·1 second is not unusual, and for special purposes such as running a number of limiters in series on very long telephone lines, 4 or 5 seconds is recommended.

See: *Sound recording.*

⊕**Relief Process.** Method of processing a photographic image in such a way that the light and shadow are represented by varying thicknesses of hardened emulsion. In motion picture techniques, relief image processing is used to prepare the matrices used in imbibition or dye transfer printing.

See: *Imbibition printing; Integral tripack systems.*

⊕□**Remanence.** Magnetic flux density remaining in a ferro-magnetic material after the applied magnetising force has been removed.

See: *Magnetic recording materials; Videotape recording.*

□**Remote Pickup.** US equivalent term for outside broadcast. Live or recorded programme originating at a point remote from the studio and therefore requiring special pickup equipment and often the use of landlines or microwave links to transmit the video and audio programme signals to the home station.

See: *Microwave links; Outside broadcast units; Special events.*

□**Repeater.** Remote amplifier in a communications satellite, microwave chain, landline, etc., which receives signals, amplifies them, if necessary modifies them, and then retransmits them.

See: *Microwave links; Satellite communications; Transmission networks.*

⊕**Replenishment.** The making good of solution losses in a continuous developing machine caused by consumption and running to waste. In the batch system (used in small laboratories) the replenisher is added at fixed intervals.

In the continuous system, replenisher is fed continuously into the circulating system through a bleed tank.

See: *Processing; Recirculation; Regeneration of solutions.*

⊕**Report Sheet.** Form, usually kept by a film camera assistant, which records all details of each shot and take, together with necessary instructions to the film laboratory. A copy of the report sheet is sent to the film cutting room, where it forms a master record of all scenes taken in the studio or on location.

Also known as log sheet.

See: *Cameraman* (FILM); *Editor.*

Reprints. Additional prints for editing made from individual scenes of negative after the first rush prints have been supplied.

See: *Rush print.*

Re-Recording. Combination, by mixing, of several sound tracks into a single track. In all film production, and in most types of TV production which are not transmitted live or recorded continuously, there is an editing process in which picture and sound materials are assembled from different sources. The sound materials (dialogue, music, sound effects, etc.) are edited on to a set of synchronized sound tracks, and it is the combination of these on to a single track which is known as dubbing or mixing.

See: *Editor; Sound mixer; Sound recording.*

Reseau. Mosaic of minute colour filter elements forming a pattern through which a sensitive photographic emulsion may be exposed, in certain forms of colour film processes, particularly the Dufay system.

See: *Colour cinematography; Dufaycolor.*

Reservoir. Storage section in continuous film processing machines allowing the end of the film to be temporarily stopped without stopping the whole machine. Normally consists of two assemblies of film rollers, one fixed and one movable, so that the length of the film path between them can be increased or shortened. At the feed end

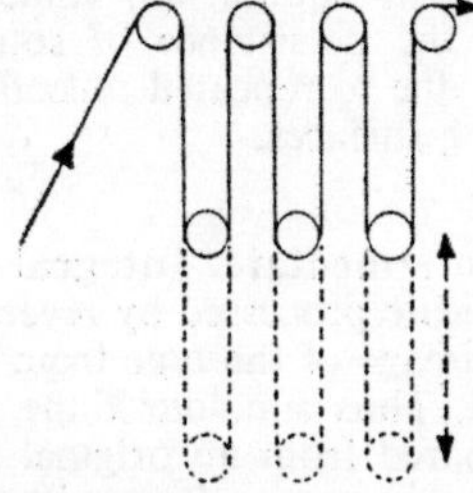

RESERVOIR. Normal film path in solid line; reserve capacity shown dotted.

of the machine, the normal running position of the two is widely separated, the film distributed among them providing a reservoir from which the machine can continue to draw when the feed has been stopped to join on a new roll. During this brief period, the distance between the movable and fixed rollers shortens as film is drawn from the reservoir and when the join has been made film must be fed in faster than the machine is taking it to restore the reservoir content.

At the take-off end of the machine, a similar roller arrangement normally operates with a minimum film content which is temporarily filled from the output of the machine when a take-off reel is stopped and changed.

The reservoir assembly is sometimes called an elevator, from the rising and falling of the movable rollers.

See: *Processing.*

Resolution. Ability of an optical system or photographic process to reproduce fine detail in the image.

In any optical system, a point object is reproduced as an image area of finite size and there is thus a limit to the proximity of two points on the object which can be recognized as two separate images without confusion. The resolving power of a system may thus be defined as the angular separation of two objects whose images can just be differentiated.

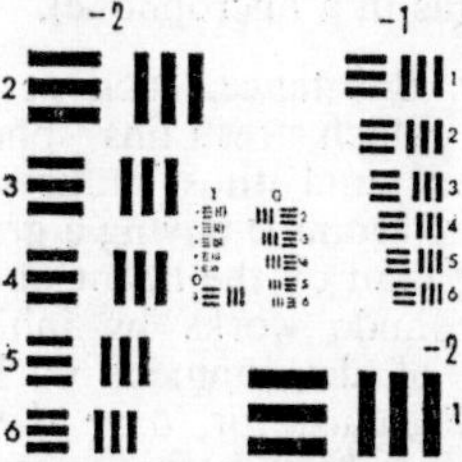

TYPICAL TEST TARGET. Ratio of line width between neighbouring pairs is 1 : 1.122, so that after six steps ratio of line width is doubled or halved. ($1{\cdot}122^6=2$.)

The resolution of a photographic material depends on its grain size and contrast, as well as the scatter of light within the emulsion layer, all of which may be affected by the colour of the exposing light.

Resolution is often stated as the number of line elements per millimetre which can be defined in the image, but actual figures vary considerably with the contrast of the test object, and the measurement of acutance is now regarded as of greater importance.

In TV a common method of expression is as "lines per picture height", to define horizontal resolution relative to the vertical resolution set by the picture standard.

See: *Acutance; Definition; Image formation; Lens performance.*

Resonant Aerial. An aerial tuned to the frequency it is desired to transmit or receive.

See: *Transmitters and aerials.*

Resonant Frequency. Frequency at which a tuned circuit (or other system) responds with maximum or minimum impedance, according to configuration, when driven by a sine-wave e.m.f. of constant amplitude.

⊕**Response Curve.** Curve connecting any two □characteristics of a transmission system, or of some part of such a system. Often the response at the output to a change in the

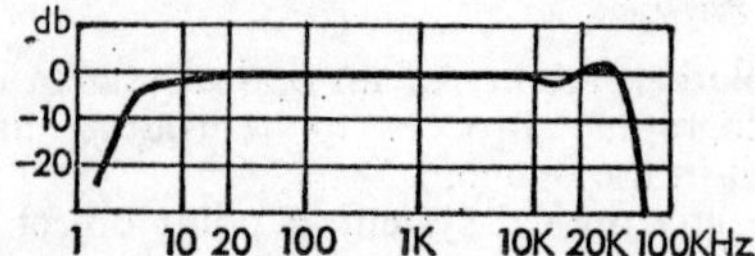

RESPONSE CURVE. Typical response curve of audio system or component, virtually linear from 20 Hz to 20 kHz, but with sharp drops outside extremities of audio spectrum.

input is measured in decibels, which may be converted to voltage at a stated impedance. The other variable may be frequency (as in an amplifier), or angle of incidence of sound (as in a microphone).

⊕**Restrainer.** Substance, such as potassium bromide, which restrains photographic development and thus reduces chemical fogging, the bromide having a greater effect on the fog than on the latent photographic image. Bromide works by inhibiting the formation of development nuclei on the unexposed grains; for, once the nuclei of reduced silver have been formed on these grains, their development as fog proceeds at the same rate as that of the exposed grains.

See: *Chemistry of the photographic process; Processing.*

⊕**Re-take.** To shoot again in picture and/or □sound a scene previously shot, but which has subsequently been found to be unsatisfactory. Also, the rephotographed and recorded scene itself.

See: *Editor.*

⊕**Reticulation.** Breaking up of film emulsion, and thus of any photographic image recorded on it, into wrinkles or fissures resembling the grain of leather. Usually caused by processing or washing at too high temperatures. The process is irreversible.

See: *Processing.*

⊕**Retrofocus Lens.** Lens formed of a front □divergent component widely separated from a rear convergent component, thus producing a long back focus, or distance from the rear surface of the lens to its focal plane. This construction, essential to many modern film and TV cameras, provides space behind the lens for a reflex shutter or a beam-splitter. Wide-angle retrofocus lenses also ensure more even illumination to the corners of the field (i.e., less vignetting) than lenses of normal contruction. Also called inverted telephoto lens.

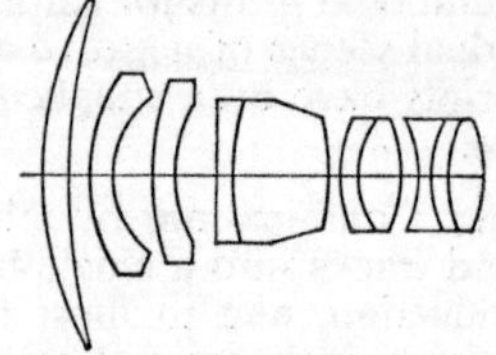

TYPICAL INVERTED TELEPHOTO LENS. The front negative group of lenses is backed by a positive rear group. This is a 20mm *f* 4 Flektogon.

See: *Camera* (TV)*; Lens types.*

□**Return Scanning Beam.** In certain types of camera tube, e.g., the image orthicon, the beam of electrons which is used to scan the electron image is reversed after discharging the target. After reversal, it returns back towards the cathode and is collected. This return beam is known as the return scanning beam. Since it is modulated by the light image on the target, it forms the signal output of the tube; in the orthicon, it is first amplified within the tube by a photomultiplier.

See: *Camera* (TV)*; Image orthicon.*

⊕**Reverberation Time.** Time taken for the □intensity of a reverberating sound to fall 60 dB below its equilibrium value. Reverberation is the persistence of sound at a given point due to repeated reflection from neighbouring surfaces.

See: *Acoustics.*

⊕**Reversal Intermediate.** Integral tripack colour duplicate processed by reversal so as to yield an image of the type from which it was printed. Thus a colour dupe negative can be prepared from an original negative in one stage, in comparison with the two stages required when going by way of a master positive to the dupe.

Reversal dupe negatives must be printed on an optical printer so as to reproduce the geometry of the original for eventual intercutting, but losses of colour quality are reduced by the elimination of one stage of duplication.

See: *Laboratory organization.*

⊕**Reversal Process.** Film process designed to produce a positive instead of a negative image after development. The latent image is developed to a silver image by primary development, then destroyed by a chemical bleach, and the remaining sensitive emulsion exposed by a flash exposure or darkened by chemical treatment. The film then enters a second developing bath before fixing and washing. Emulsions designed for reversal processing are characterized by fineness of grain, but their exposure latitude tends to be lower than normal.

See: *Chemistry of the photographic process; Integral tripack; Laboratory organization.*

⊕**Reverse Action.** Effect obtained by presenting the frames of a film in the reverse order to that in which they were photographed, so that action takes place backwards.

See: *Special effects* (FILM).

⊕□**Reverse Angle.** Shot taken from the opposite point of view to the preceding one; often, therefore, a reaction shot in a dialogue sequence.

See: *Editor.*

⊕**Rewind.** Geared rewinding device on which a reel or flange may be mounted and rotated rapidly by hand or electric motor. Motor rewinds are used in projection rooms and libraries where large reels of positive

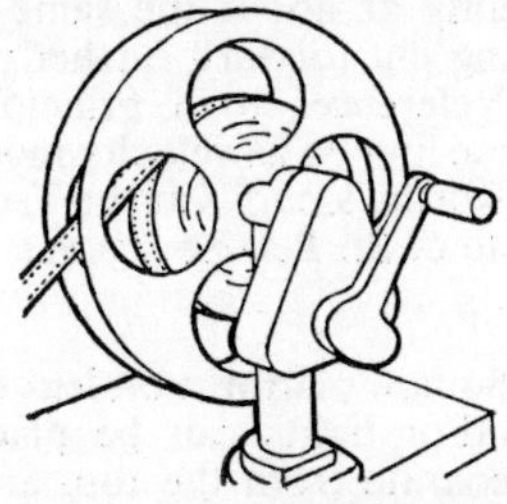

film have to be rewound rapidly. Negative rewinds are provided with a low gearing-up ratio to discourage over-rapid rewinding which may damage negative film. Positive rewinds are geared much faster since damage to positive film has less serious consequences.

See: *Editor; Spools, cores and cans.*

⊕**Reynaud, Emile, 1844–1918.** French showman, artist and inventor who devised a system using a mirror drum for optically stabilizing a series of images carried on a continuously moving band. It was a great improvement over the Zoetrope and avoided the mechanical complications of intermittent movements. His system was patented as the Praxinoscope in 1877, and the following year he brought out a table-top Praxinoscope Theatre and a projection Praxinoscope. In 1888 he patented a projector using a long flexible band of pictures, painted by hand and perforated for easy transport and accurate registration in the machine. The mirror-drum principle was used to project these drawings in action on a screen.

The Théâtre Optique, based on this apparatus, opened with regular public performances at the Musée Grevin in Paris in 1892. However, in 1900 Reynaud, suffering from the competition of the cinema and in disgust at the lack of discrimination of its audiences, threw his apparatus and films into the Seine.

□**RF Amplifier.** Amplifier operating at radio frequency; in a superheterodyne receiver, before the mixer/frequency changer. Normally tuned but can be aperiodic (untuned).

See: *Receiver; Transmitters and aerials.*

⊕□**Rifle Mike.** Highly directional microphone that can be aimed like a rifle at the source of sound. Also called a gun mike.

See: *Microphones; Sound recording* (TV).

⊕**Rifle Spot.** Type of studio spot lamp in which the light is concentrated by a para-

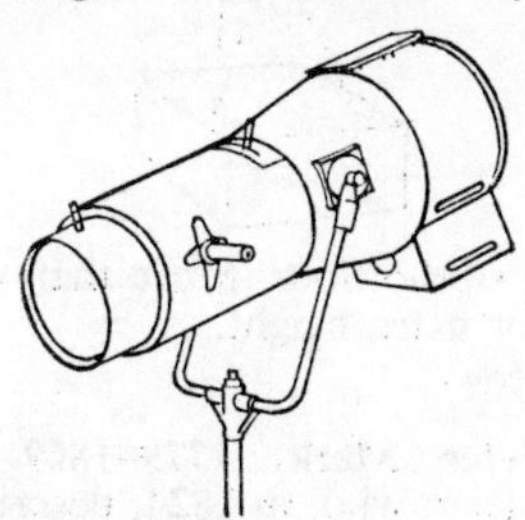

bolic mirror of fixed focus to produce a narrow beam at a distance.

See: *Lighting; Studio complex.*

⊕□**Rigging.** Placing studio lights in their preliminary positions, before the accurate adjustment called for when the action and camera movement have been established. In TV practice, virtually all lighting is rigged from above, but adjusted from the floor, and is independent of the sets. In film practice, lighting is rigged round individual sets, either mounted above them or placed on the studio floor.

See: *Lighting, Studio complex.*

□**Rigging Tender.** Vehicle carrying material for rigging a television show, e.g., material for camera rostra, lighting supports, etc.

See: *Outside broadcast units; Special events.*

□**Ringing.** The repetition of a sharp edge of a part of a picture as one or more signals of lower amplitude. This is a common defect in wide band video amplifiers, usually caused either by phase distortion in the amplifier or insufficient bandwidth compared to the frequencies which it is amplifying.

On the waveform of the video signal, it manifests itself as a number of small overshoots, both positive and negative, following a transient. When viewed on a television screen, it shows up as a number of faint vertical lines following a sharp change in picture level. It can also be easily seen on a test chart as faint images which are repeated after the blocks of frequency bars, normally located in the circle, towards the centre of the chart.

See: *Picture monitors.*

⊕**Riser.** Low platform for raising a prop, an □actor, or the cameraman, a few inches

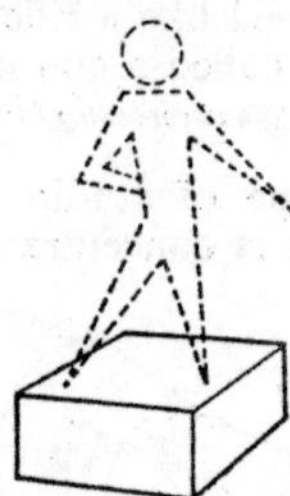

above the studio floor. More than one may be used for extra height.

See: *Art director.*

⊕**Roget, Peter Mark, 1779–1869.** British mathematician who, in 1824, described and explained the phenomenon of persistence of vision before the Royal Society in London. His was the first scientific investigation into the principle upon which the motion picture is based. He also compiled the famous Thesaurus which bears his name.

⊕**Roll.** Length of cinematograph film, usually wound on a core; negative at the stage of exposure and rush printing is usually identified as a production roll number. When the negative has been finally cut and assembled it is called a reel.

See: *Spools, cores and cans.*

⊕**Roller Title.** Film or TV title which rolls □upwards across the screen at a slow steady speed. Sometimes used for long lists of names or explanatory introductions. Also called a crawling or creeping title.

⊕**Roll Off.** Electrical output variation in □which the response falls off gradually, i.e., is rolled off rather than cut off sharply at a particular frequency.

See: *Sound recording* (TV).

□**Rollover.** Roll of a television picture in a receiver, such as is obtained when a field pulse is widely incorrect in phase and the local oscillator is set at a very different frequency from that of the incoming picture.

See: *Master control; Synchronizing pulse generator; Vision mixer.*

□**Rosing, Boris, 1869–1933.** Russian scientist who was the first to suggest a television system incorporating a cathode ray tube. In Russia he is regarded as the inventor of television. Rosing's system, for which he took out a British patent in 1907, used a mirror drum mechanical scanner for transmitting and a Braun tube as a receiver.

Only faint images could be received with this method owing to the crude nature of the photo-cells and cathode ray tubes then available and the lack of any suitable means of amplification; nevertheless, Rosing's proposals constituted a definite step forward from the impracticable schemes of the nineteenth century, and are comparable with those of Campbell Swinton made independently at about the same time. In 1911 Rosing put forward further proposals including references to a principle which later became known as velocity modulation.

One of Rosing's pupils at the Technological Institute of St. Petersburg was Vladimir Zworykin.

⊕**Rostrum.** Square platform on legs on which □camera and/or lights can be placed. The legs are separate from the top, are hinged and fold together so that they can be carried easily and occupy minimum space. The legs are made in varying heights from 1 ft. to 8 ft. or more.

See: *Art director.*

⊕**Rostrum Photography.** Photography employing a horizontal board on which cels

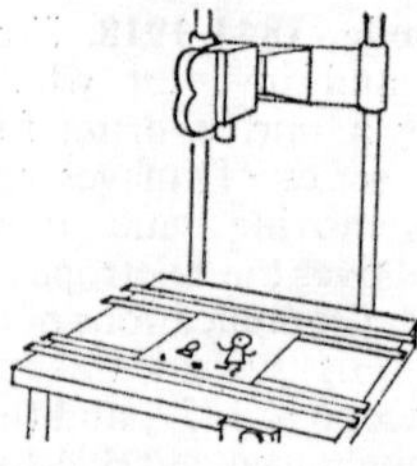

and other animation materials are mounted under an animation camera. The camera can be slid up and down on columns to alter the field of view and produce zoom effects.

See: *Animation equipment; Animation techniques.*

⊕**Rotary Printer.** Motion picture film printing machine in which the negative and positive move continuously by the action of a rotating sprocket drum on which they are brought in contact at the time of exposure.
See: *Laboratory organization; Printers.*

□**Rotary Stepping Switch.** Telephone switch selector which rotates from some impulses and steps at right angles from others. Used as part of line selection equipment in master control or elsewhere.
See: *Master control; Vision mixer.*

□**Rotating Disc Scanner.** Nipkow disc; an opaque disc perforated with a single spiral of small holes.
See: *History* (TV)*; Nipkow disc.*

□**Rotating Mirror Scanner.** Mirror drum with each mirror set at a slightly different angle so that with suitable optics a scanning action is obtained.
See: *History* (TV).

⊕**Rough Cut.** Stage in the editing of a film, between assembly and fine cut, when the first coherent outline of plot or theme becomes apparent.
See: *Editor; Narrow gauge production techniques.*

□**Routine Test Method.** Method of checking the amplitude-frequency and phase-frequency responses of a vision circuit by measurements of the output waveform when a sine-squared pulse is applied, the half-amplitude duration of which is 2T=1/fc, where fc is the designed maximum frequency of the circuits, and when a field frequency square wave is applied.
See: *Picture line-up generating equipment; Picture monitors; Transmission networks.*

□**Roving Preview.** Facility of viewing at will and without selecting for transmission, any input or output of a video-switching system. The equipment usually carries a row of push buttons to select the required circuit and a picture and/or waveform monitor to preview the signal. If a roving preview signal can be preselected for subsequent transmission, it is called an auto-preview.
See: *Master control; Vision mixer.*

⊕**Rudge, John A. R., 1837–1903.** British inventor and mechanic. Invented several devices for the magic lantern, designed to simulate movement on projection. Several of his lanterns were demonstrated by Friese Greene between 1886 and 1890.

⊕**Rush Print.** First positive print from the laboratory of the previous day's shooting. Hence also called daily.
See: *Editor; Laboratory organization.*

S

Safe Area. In the transmission of motion picture film by television, allowance must be made for the loss of the edges of the picture area as a result of inaccurate adjustment of the system, including the receiver in the home. It is therefore desirable to ensure that essential action and titles should be contained within an area somewhat smaller than the total area scanned and this is called the safe area.

Dimensions are specified separately for 35mm and 16mm film. The centreline of film and television pictures is the same, but because of the shape of the TV receiver screen, more rounded corners are called for.

American recommended practice recognises separate safe action and safe title areas, the former having 90 per cent and the latter 80 per cent of the height and width of the total area transmitted, but the British Standard specifies a single safe area for both.

When review rooms are being set up to screen films for television use, it is advisable to mask the projector aperture so that it corresponds to the reduced dimensions of the safe action area.

See: *Film dimensions and physical characteristics.*

Safe Light. Light used in film processing rooms and confined to a part of the spectrum to which the emulsion is insensitive. A blue-sensitive emulsion can be developed in the presence of quite bright yellow/orange light. Panchromatic emulsions require a much fainter green light. Colour emulsions must be developed in total darkness.

Safety Film. Film in which the base used is of low inflammability and has a slow burning rate, usually cellulose triacetate or a similar polyester.

Introduced before World War I for home movies, safety base was more expensive and mechanically weaker than nitrate base, and so failed to gain acceptance in professional 35mm film production. It was only from 1937 onwards that the use of cellulose acetate propionate and other mixed esters produced characteristics superior to those of inflammable nitrate base, and led to the complete replacement of nitrate by safety base in the early 1950s.

See: *Film dimensions and physical characteristics; Film manufacture; Film storage.*

Sample Print. Rush print showing the general character of the scene or reel but not embodying the final grading corrections.

See: *Laboratory organization.*

SATELLITE COMMUNICATIONS

Within a few years, communication by way of earth satellites has progressed from the passive balloon satellite Echo I to highly sophisticated active satellites. During this time, various satellites were launched and became operational on an experimental basis, each one more ambitious in design than its predecessor and providing improved reliability and technical capability.

It is, however, important when considering a satellite for use as a television point-to-point communication link, to study a number of fundamental questions, including the type of orbit in which the satellite should travel, the basic design of the satellite, the radio frequencies to be used in operation and how such a satellite link will fit into the pattern of terrestrial television systems and networks.

BASIC PRINCIPLES

ORBITS. Satellite orbits can fall into many varying configurations, but basically three main types can be considered: circular equatorial, circular polar, and elliptic orbits which are inclined at an angle to the equatorial plane.

Types of orbit range in height between 1,000 miles and 22,300 miles, the latter being the geo-stationary orbit height, and within these orbit ranges the periods vary between two hours and twenty-four hours respectively.

As most centres of world population lie within the latitudes 60°N to 60°S, the equatorial orbit is generally considered more favourable than the polar orbit for coverage of the earth. The advantage claimed for elliptic orbits is that the satellite moves relatively slowly at and around the point of apogee (its highest point), thus giving an extended mutual visibility period over large areas of the earth, whereas at perigee (the lowest point), the satellite is moving fast with respect to the earth over unpopulated areas. The satellite can also be injected into orbit at this relatively low height.

For all inclinations other than approximately 64° to the equator, the apogee of the orbit precesses around the orbit plane, this precession being due to the effects of perturbations resulting from the oblateness (variation from spherical shape) of the earth, and the effect of the moon and the sun. The earth's effect is mainly due to the inequality in its gravitational field and the non-uniformity in the magnetic fields surrounding the earth.

Circular orbits in the equatorial plane, on the other hand, have an advantage in that relatively simple attitude-control systems can overcome these perturbation effects and so produce a preferred type of orbit. The higher the altitude of orbit chosen the greater the area of the earth which is covered from each satellite; thus fewer satellites are required for a world-wide service. However, the higher the orbit the larger or more complicated the rocket launcher needed to place the satellite at the right altitude, although the rocket size can be reduced by the use of the parking orbit procedure employed with the Syncom and Early Bird satellites.

The further disadvantage of the lower altitude orbit is the risk of damage to solar cells and other solid-state devices, owing to the intense proton radiation within the inner Van Allen Belt which is present at heights between 1,000 and 3,000 miles in the equatorial plane, according to solar activity.

The geo-stationary orbit therefore appears to have many advantages, particularly its stationary effect which eliminates ground aerial tracking, its near freedom from Doppler frequency shift, and the smaller number of satellites required for world-wide coverage. It does have disadvantages, however, due to the signal attenuation and the delay in the signal transmission path, some 250 to 300 msec each way, but although this is a handicap in telephony, it is relatively unimportant for the transmission of television programmes.

TYPES OF REPEATER. Basically, two types of repeater can be considered for use with a communication satellite system: passive satellite and active satellite. The former operates on the principle that a reflecting surface returns a portion of the transmitted energy to the receiver but, owing to loss resulting from the energy being scattered and the long path lengths involved before reception, extremely high-power transmitters and super-sensitive receivers are required.

As a result of these factors, the use of high-altitude orbits presents operational difficulties and so necessitates the use of lower altitude systems, with the attendant disadvantages. However, as an offsetting factor, the passive system, not requiring any electronics at the satellite, is much less likely to suffer failure, and all maintenance and repairs necessary can be carried out at the ground installation.

In addition, as a result of improved techniques including new electronic advances, modifications can be embodied into the system without the necessity for replacing satellites.

The active repeater, on the other hand, contains electronic equipment which receives the signal from the ground transmitter and, after the amplification process and carrier frequency change, re-transmits the signal to the ground receiver. This method enables the respective receivers to deal with larger signal strengths than in the passive case, so reducing the effect of energy dispersion, resulting in greatly improved signal-to-noise ratios. This, in turn, means that higher-altitude orbits can be considered, resulting in larger terrestrial coverage with a smaller number of satellites and with less complicated tracking equipment associated with the ground installations. Of course, the active satellite when placed into orbit cannot at the present time be modified to take account of new and improved electronic techniques as they become available.

Both active and passive satellites have been launched into orbit and many more in both categories will be launched in the future. One successful passive satellite was the project Echo, which was launched into an orbit height of approximately 1,000 nautical miles. Examples of the active repeater type are Telstar, Syncom and Intelsat I and II.

Experience shows that the active satellite fulfils many more of the requirements for a communications system than the passive satellite and has, therefore, been much more widely adopted.

A decision to use active repeaters means that emphasis is placed upon the choice of the carrier modulation method to be used for the transmission of the information signals. This must take account not merely of telecommunications factors, but also the available electrical power in the satellite, weight, equipment reliability in space, and the most efficient use of the available bandwidth.

The choice of the carrier frequency employed between the ground and satellite is generally considered to be restricted within the limits of 100 and 10,000 MHz. Below 100 MHz, atmospheric and man-made noise affect the received signal, and above 10,000 MHz, the signal is affected by attentuation due to absorption from water vapour, oxygen and rain. The preferred frequencies for repeater service are in the 4,000 and 6,000 MHz bands, although difficulty does arise in their use as they are already assigned to terrestrial line-of-sight systems. Studies have been undertaken to determine whether frequency-sharing between terrestrial and satellite systems is possible, and sharing has now been agreed internationally.

MODULATION METHODS. Basically, three types of modulation system are under study for permanent systems and these include wide-band f.m., single sideband a.m., and systems employing pulse code modulation. The wideband f.m. system is at present the most widely used. It has many advantages, not the least being that it is well-tried and available; it is also capable of providing good signal-to-noise ratios by full exploitation of the frequency bandwidth.

A single sideband a.m. method has the advantage of simplicity and a maximum possible economy in bandwidth, but it requires greater transmitter power than f.m., important in view of the limited power supply capability of satellites.

The pulse code modulation method, however, is essentially robust, provides considerable inherent protection against interference, and requires lower transmitter power. So far, the system has not been developed to the same extent as conventional f.m., but in future it is expected to be a serious contender.

POWER SUPPLIES. The problem of providing electrical power under orbital conditions can be approached in two ways. One is to take stored or latent power up with the satellite, and the other is to equip the satellite with the means of generating power from external sources during the period of its operation. The latter is by far the most common method in current use, but some experimental forms of stored power are now being evaluated in space vehicles. The principal sources of power available in the generating type are solar, nuclear and chemical energy, the main methods of converting this power to electrical energy being photo-voltaic, thermo-electric and thermionic.

The use of nuclear power plants in particular avoids the need for storage batteries and allows a greater freedom in the design of the satellite. Such systems are now in active development and may well be in use within the next few years.

It appears that for satellite repeaters,

solar energy, in conjunction with silicon photo-voltaic cells and lightweight storage batteries such as nickel cadmium cells, will be the primary power source for some time to come.

The nickel cadmium cell, with deep discharging, can provide a power-to-weight ratio of about six to ten watt-hours per pound. However, this reduces the battery life, and with limited depth of discharge, the power-to-weight ratio is some three watt-hours per pound. To prevent evaporation of the electrolyte in space hermetically-sealed batteries must be used, and to avoid gassing, the potential at the terminals must be regulated so that it does not exceed 1·46 volts during the re-charging cycle.

SOLAR CELLS. The solar cell itself intercepts radiation and from it produces electrical power which can be arranged to operate the satellite equipment directly, or to charge the lightweight storage battery, so that the latter can provide the power for operating the equipment when the solar cells are hidden from the sunlight. Light cells have been used as power supplies for many years, but until quite recently selenium was the active material, and this has a very low efficiency of about 0·06 per cent. With the advent of semi-conductor techniques, newer materials can be applied to these light energy transfer requirements.

Present day solar cells almost exclusively employ silicon. They consist of an N-type wafer doped with phosphorus, antimony or arsenic on to which is deposited a thin P-type silicon layer doped principally with indium. To increase the efficiency and at the same time reduce the cell resistance, contact is made to the external electrical conductors by arranging the connection in a grid form rather than a point contact. The open circuit voltage so produced from the full sunlight is approximately 0·45 volts, decreasing rapidly when the received solar radiation falls below some 250 watts per square metre.

The efficiency of the silicon solar cell is rather low, typical values ranging between 8 and 14 per cent, but recently cells making use of gallium arsenide have been investigated and it is expected that efficiencies up to 15 per cent may be achieved.

Since solar cells have a low transfer efficiency, the remaining energy received from the sun is dissipated in heat and this produces further reduction in cell efficiency. Additionally, some of the solar radiation received on the surface of the cell is reflected. As the reflectivity of clean silicon at the appropriate wavelength is about 34 per cent, this reflection is a very serious loss, but the application of an anti-reflection coating to the face of the cell helps to reduce its effect.

An additional improvement results from the ability to maintain the solar cell structure as near as possible at right angles to the direction of the sun and, to achieve this, orientation equipment is necessary. In the more conventional designs, where the cell array is arranged around the periphery of a spherical satellite, power output is reduced by a factor of four on the comparable sun-seeking array.

HAZARDS OF SPACE. The solar cell is particularly prone to damage in space; even if this does not render it inoperative, it can reduce the already low efficiency. One hazard is collision with micro-meteorites and concentrations of free electrons in the outer Van Allen Belt. The cell can be protected by a glass cover, a thickness of 0·07 in. being sufficient to reduce this form of bombardment very considerably. The glass also increases infra-red emittance, so assisting in the reduction of heating within the cells.

Another hazard is collision with high-energy protons in the inner Van Allen Belt, the latter occurring between 1,000 and 6,500 miles out from the earth. While the total flux of protons within this belt has a predicted energy of some 40 m.e.v. (million electron volts), there is some evidence of a small amount of energy exceeding 100 m.e.v. It is impracticable to design a glass cover which will give any protection against the high energy protons within the inner belt, as glass will arrest only protons below 10 m.e.v.

A further serious hazard results from the intense solar cosmic rays which occur every few weeks, and the high energy end of the spectrum from these flares cannot be stopped even by several inches of lead shielding.

ORBITING SATELLITES

Studies in the design and operation of communication satellites have been conducted in a number of countries, and between 1962 and 1965 several satellite repeaters were launched and used on an experimental basis. The most significant of these satellites are as follows:

TELSTAR. Two satellites in this series were launched, Telstar I and Telstar II, the former having been put into orbit in 1962. Telstar II varied in design characteristics and orbit only slightly from its predecessor.

The design of Telstar I was arranged to provide only a small capacity but one in which the relayed signals in general conformed with the specification parameters of a fully-operational system. The satellite contained a single repeater and the power delivered to the transmitting aerial was approximately 2 watts, the repeater maximum gain being about 105 dB.

All of the satellite transmissions were controlled from the ground by the use of a communication system on a frequency of approximately 120 MHz, the same transmitter being used to operate telemetry equipment in the satellite. The satellite repeater was capable of relaying one television channel, or the equivalent telephone channels.

The microwave transmitter of the repeater used a travelling wave tube in the output stage. Including the telemetry command beacon functions, Telstar incorporated some 1,300 diodes and 1,000 transistors.

The orbit of the satellite was an ellipse inclined approximately 45° to the equator, with an apogee of 3,000 nautical miles and a perigee of approximately 500 nautical miles. In this way, the satellite passed in and out of the high-intensity region of the Van Allen radiation belt, so providing a great deal of information in connection with the research part of the experiment. With this orbit, the satellite had a period around the earth of approximately 2 hrs. 40 min., giving mutual visibility between the United States and Europe for periods up to approximately ½ hour for individual passes, with a total of approximately 1½ hours per day.

RELAY. Two satellites in this series were launched and operated from space and the first of the series was placed into an elliptic orbit with an inclination of approximately 50°. It had an apogee of about 3,000 miles and a perigee of about 1,000 miles. The satellite, weighing approximately 150 lb., contained two complete repeaters each capable of handling a television signal or 300 one-way telephone channels or 12 two-way telephone channels. The satellite was powered by solar cells and a nickel cadmium battery, although only one repeater could be operated at a time. This satellite, like Telstar, embodied equipment for studying environmental conditions, particularly flux and energy of protons and electrons in space. Such information and other operational data of the system was passed to the ground in the form of telemetry information over a VHF transmitter.

GEO-STATIONARY SATELLITES

SYNCOM. Three satellites in this series were launched, but the first, Syncom I, was destroyed when a rocket stage exploded on reaching its final orbit height. However, Syncom II was launched successfully and was subsequently used in a number of demonstrations. This satellite and Syncom III were placed in an orbit at a height of 22,300 statute miles above the surface of the earth, but Syncom II was not, in fact, in the geo-stationary orbit and was inclined some 33° to the equator.

The method of launching Syncom into stationary orbit is of some interest. Five and a half hours after launch, the satellite reached the synchronous altitude equivalent to the radius of a circular synchronous orbit and, at this time, a spherical solid-propellant apogee rocket was fired by a timer command to boost the satellite from the transfer ellipse into a circular orbit.

Following the apogee motor burn-out, the satellite was tracked, using the telemetry transmitter as a beacon, for about an hour to determine the orbital parameters, including the orbital velocity. If the satellite orbit drifted westward towards the desired longitude at the rate of approximately 5–10° per day, no velocity changes were needed. But if it drifted eastward, velocity had to be applied by ground command to cause the orbit to move west. Initial correction was made with a hydrogen peroxide axial velocity jet which supplied continuous thrust parallel to the spin axis in the same direction as the apogee motor.

By this method it was possible to place Syncom III into a true geo-stationary orbit from an elliptic orbit, using a rocket assembly which would otherwise have been incapable of putting the satellite into its final orbit.

The Syncom series of satellites was largely experimental, but was designed with a capacity for two-way voice channel repeaters.

The communication system with complete redundancy employed a frequency translation active repeater receiving signals on 7,361 MHz and translating them to 1,814 MHz for re-transmission back to the ground terminals. The satellites employed dual lightweight travelling wave tubes each with a nominal 2 watt output.

Syncom III was satisfactorily launched

in 1964 and placed in a geo-stationary orbit above the Pacific Ocean. From this point it was used experimentally for telephony across this ocean and, with modifications to some associated earth stations, for the Tokyo Olympic Games, 1964. It transmitted pictures from Tokyo to the West Coast of the United States.

EARLY BIRD. This satellite, now known as Intelsat I, was the first of a series of satellites to operate as geo-stationary repeaters. Early Bird was commissioned by the Communication Satellite Corporation (Comsat) and differs from the Syncom series in that the transponder has three times as much output power from the travelling wave tube transmitter; it also has a narrower-beam higher-gain aerial. These modifications give Early Bird an increase in effective radiated power of approximately six times that of Syncom III.

The solar cell panel area has been increased by 50 per cent on previous satellites, and its power capability from 29 to 45 watts. Early Bird frequencies are in conformity with the new allocations made by the International Telecommunications Union and receives signals on the link from ground to satellite on a frequency of approximately 6,000 MHz, and transmits 4 watts from satellite to ground on a frequency of 4,000 MHz. It can transmit television programmes or provide up to 240 two-way telephone-quality circuits or the equivalent in telegraph traffic. The receiving aerial pattern is symmetrical about the satellite's spin axis, and signals can be received by two receivers operating continuously. Each receiver has a flat bandwidth of 25 MHz and consists of a mixer local oscillator, intermediate frequency amplifier and a limiter amplifier.

The output from the receiver limiters mixed with the reference signal provides a frequency-translated low level output with a nominal carrier frequency of some 4,000 MHz. The outputs are connected to a single hybrid network which in turn is connected to the two travelling wave tube transmitters. Either travelling wave tube may be selected by command for use with both receivers, but the transmitters are interlocked so that only one may be used at any one time. The transmitter output is 4 watts, and each receiver with either transmitter is designed to operate at all times except when the satellite is in eclipse, whether the batteries are operating or not. The satellite goes into eclipse during the autumn and vernal equinoxes for about 70 minutes maximum each time; in addition, it is in partial eclipse for 20 days before and after each equinox.

The plan for launching Early Bird was similar to that used in placing Syncom III mid-way over the Pacific Ocean. The first and second stages of the thrust-augmented Delta rocket carried the third stage to an altitude of about 825 miles at a point above the equator and near the Greenwich Meridian.

Following the burn-out of the second stage, the third stage with its payload was aligned for a third stage firing; after coasting for some twenty minutes, the satellite was positioned over the equator at which time small rockets were fired to impart a spin to the third stage and the satellite combination. The third stage rocket was then fired, increasing the velocity of the satellite and reducing its orbital inclination by about 50 per cent. Shortly after the burn-out of the third stage, the satellite separated from its rocketry and began a five-hour coast to the apogee of an elliptical transfer orbit.

The first apogee position was above the Indian Ocean and out of sight of its ground/earth command station, so the satellite was allowed to continue another complete transfer orbit until it was in sight of the earth station. After this it was allowed to swing through two more orbits to a position where, some 39 hours after launch, it again came into full view of the control earth station. At the point of its fourth apogee, the solid-propellant motor was commanded to fire and the satellite was kicked into a circular, near-synchronous orbit above the Atlantic Ocean.

Corrective manoeuvres using the hydrogen peroxide control jets then brought the satellite to the desired 27·5°W longitude and reoriented the satellite attitude to a point where its aerial beam was across the North Atlantic area covering the eastern part of North America and Western Europe.

Since achieving its correct position in space, the satellite has been used for a large number of television programme exchanges and for many telephone and telegraph experiments.

CURRENT COMMUNICATIONS SATELLITES. Under arrangements now in force, communications satellites in the western world are owned by the international consortium Intelsat, in which many nations have an interest.

They are placed in orbit by government agencies, and are managed by the Communication Satellite Corporation (Comsat) on behalf of their owners.

At the present time (1968) Early Bird (renamed Intelsat I) has been operating successfully for more than two years. During this period the European terminals – Goonhilly (Britain), Pleumeur Bodou (France), Raisting (W. Germany) – have been extended to include Fucino (Italy). A Canadian station operates at Mill Village (Nova Scotia) to share the load with Andover (Maine, US).

LANI BIRD – HS 303A. Belongs to the second generation of Comsat-controlled satellites named Intelsat II. Lani Bird was launched successfully in 1966 over the Pacific Ocean, and works with earth stations at Brewster Flat (Washington, US), Paumalu (Hawaii) and Ibaraki (Japan).

Technically, Intelsat II satellites have about the same power as Early Bird, an expected life of three years compared to Early Bird's nominal 18 months, and better access arrangements. Because of the use of limiters, Early Bird can handle two-way transmissions between only two stations at a time. Intelsat II satellites have no limiting on the transponders, and thus allow a considerable degree of multiple access for general traffic. In television use, Intelsat II can be regarded as having two separate channels, either of which may be used for one-way television.

INTELSAT III. It is proposed to launch three of this new type of satellite in 1969 over the Pacific, Atlantic and Indian Oceans, all in geo-stationary orbits. The maximum capacity will be 1,200 telephone circuits, the life expectation is 5 years, and it is hoped that the effective radiated power (ERP) will be at least 100 watts.

On a technical level, the general circuitry will include two unlimited transponders with a bandwidth of 230 MHz. Input frequencies will be in the 5925/6425 MHz band, and output frequencies in the 3700/4200 MHz band.

To provide a complete global system, it is expected that at least six such satellites will be required.

The USSR has at least one communications satellite in successful operation, the Molniya I (Lightning) launched shortly after Early Bird.

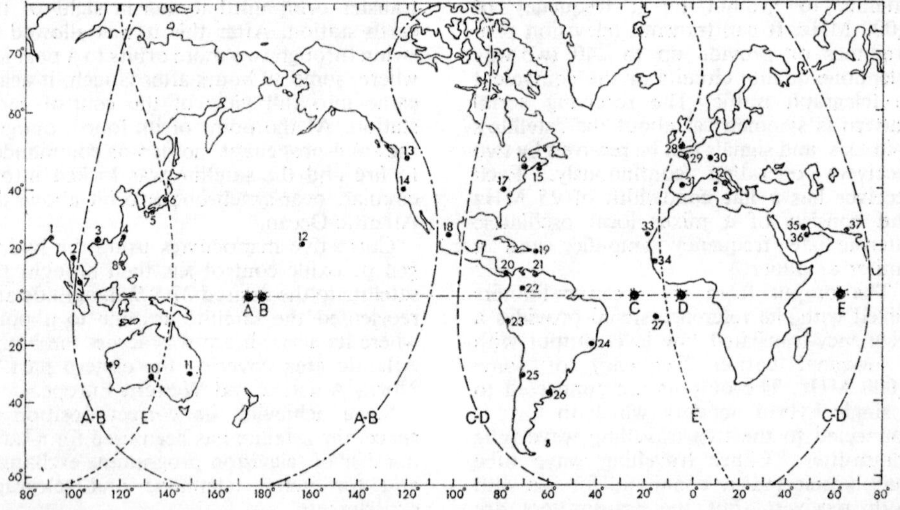

COMSAT SATELLITES AND ASSOCIATED EARTH STATIONS. A, B. Pacific Intelsat (II). C, D. Atlantic Intelsat (I, II). E. Indian Ocean Intelsat (III). Stations in service by end 1969: 1. Chittagong, Pak. 2. Sriracha, Thailand. 3. Hong Kong. 4. Chin Shan Li, Taiwan. 5. Yamaguchi, Japan. 6. Ibaraki. 7. Kuanton, Malaysia. 8. Tanay, Philippines. 9. Carnarvon, Aus. 10. Ceduna. 11. Moree. 12. Paumalu, Hawaii. 13. Brewster Flat, Wash. 14. Jamesburg, Calif. 15. Mill Village, N.S. 16. Andover, Me. 17. Etam, W. Va. 18. Tulancingo, Mex. 19. Cayey, P.R. 20. Utive, Pan. 21. Venezuela. 22. Colombia. 23. Peru. 24. Tangua, Brazil. 25. Melinilla, Chile. 26. Balcarce, Arg. 27. Ascension Is. 28. Goonhilly, Eng. 29. Pleumeur-Bodou, France. 30. Raisting, Ger. 31. Buitrago, Spain. 32. Fucino, Italy. 33. Las Palmas, Can. Is. 34. Dakar, Sen. 35. Kuwait. 36. Bahrain. 37. Karachi.

EARTH STATIONS

AERIAL SYSTEM. On the ground, transmitters and receivers are not faced with the same problems which have been shown to affect the satellite design, although the ground station equipment has problems of its own. In view of the extremely weak signal which is finally to be received, the transmitter and receiver aerials must be designed with very high gains and extremely narrow beamwidth capabilities, and thus tend to be large. For instance, at 4 GHz an 85 ft. diameter aerial is necessary to give a gain of 58 dB, and a 30 ft. diameter aerial for a 49 dB gain.

As a very weak signal is finally received, the importance of minimizing noise throughout the system is of paramount importance. Sky noise approximates 20–50°K within the usual carrier frequency band, provided the aerial is elevated sufficiently above the horizon. Elevation of 5° is usually regarded (gigahertz) and for a 50 per cent efficiency, as a minimum, though in particular geographical locations lower elevations may produce acceptable signals.

TRANSMITTER AND RECEIVER EQUIPMENT. The ground transmitter introduces no great design difficulties when operating in the microwave region. Normal transmitter techniques can be adopted, using power outputs ranging from 1–10 kW depending upon the requirements of orbit height, etc.

The ground receiver, on the other hand, is very much more complicated, and in its earliest stages it must be capable of detecting and amplifying an extremely small signal. This means that these stages must introduce the minimum inherent noise and, at the present time, this can only be achieved by the use of parametric amplifiers or masers. The simplest form of maser amplifier employs only one port cavity together with a circulator and as a result has a narrow bandwidth capability and can suffer from gain instability.

More recently masers employing travelling wave tube techniques have been introduced and provide greater stability and much increased bandwidth capability. The maser amplifier, however, is complicated by the additional components required to stimulate it into operation and in addition, it needs to be operated at a temperature below 4·2°K. That calls for the use of a liquid helium bath, which introduces a number of structural difficulties. Additionally, the maser installation needs to be mounted in close proximity to the focus of the aerial system.

The parametric amplifier does not have these limitations. It is, however, likely to introduce a small penalty in that it produces slightly greater inherent noise, although it is likely that a helium-cooled parametric amplifier in the near future will have a noise temperature of some 10°K over a bandwidth of 500 MHz.

The foregoing devices are designed to keep inherent noise at the lowest possible level and this is also one of the considerations when choosing the carrier modulation method. Wide-deviation frequency modulation is usually a first choice in order to obtain a good receiver signal-to-noise ratio, but of course at the expense of bandwidth. Unfortunately, the wider the bandwidth producing the improvement in the signal-to-noise ratio the greater is the threshold level presented by the frequency discriminator circuit of the f.m. demodulator, with resultant lower sensitivity.

SPECIAL CIRCUITS. Many ingenious circuits have been designed to offset this disadvantage, including an f.m. negative feed-back demodulator and also a dynamic tracking variable bandwidth demodulator. In the first type, the incoming i.f. signal resulting from the wide-band deviation has its deviation reduced by an f.m. negative feed-back loop, and feed-back gain is adjusted so that the resultant i.f. signal can be passed by a narrow-band filter without perceptible distortion.

The result is an improvement in the threshold point, while at the same time the wide-deviation frequency modulation signal at the input of the receiver still preserves the signal-to-noise advantage.

The dynamic-tracking variable-bandwidth demodulator, on the other hand, is so designed that a narrow-band filter tracks the instantaneous frequency of the received wide-deviation f.m. signal and, at the same time, the bandwidth of this filter varies in sympathy with the strength of the incoming signal.

In this way, once again, the advantage of wide-deviation frequency modulation of the carrier is retained while preserving reasonable threshold level at the discriminator.

A conclusion cannot yet be drawn to determine which method has the greater merit, and for some time experiments will continue using these and other approaches to solve the many problems of receiving

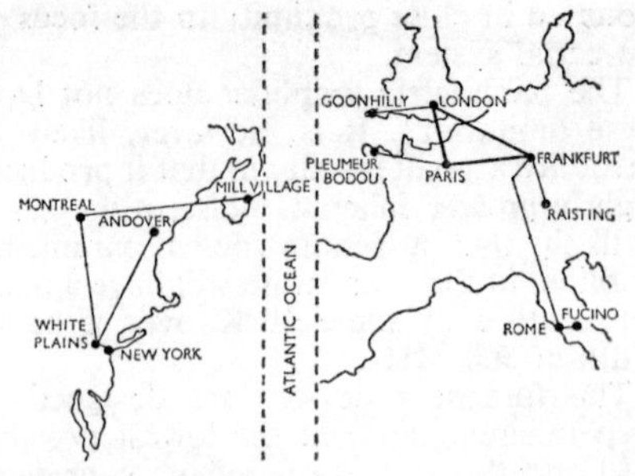

EARTH STATIONS (NORTH ATLANTIC AREA). Interconnections between principal earth stations and television broadcasting centres.

wideband but weak signals from communication satellites.

TERRESTRIAL CONNECTING NETWORKS
In a completely world-wide television network, the satellite point-to-point links need to be coupled to terrestrial networks allowing the whole to operate as a completely-integrated system. This imposes conditions in that the transfer from one type of link to the other must be achieved in a compatible fashion without appreciable attenuation or distortion. The present terrestrial networks operate in conformity with international regulations covering the various transmission parameters, and these conditions apply also to the satellite links.

Many large national networks operate at the present time including a vast network across the United States and Canada, and another in Europe, both East and West, which forms part of the Eurovision and Intervision network. Additional national networks will be established in various parts of the world, and the main purpose of the satellite links will be to connect these national television networks. By this method a complete international network is available, whereby transmissions can originate in remote corners of the earth and be communicated to local transmitters, using a complexity of networks made up of terrestrial radio, cable links and satellite relay connections. L. F. M.

NEW DEVELOPMENTS. Beyond the Intelsat-III series of satellites, planned for launch in 1969, an Intelsat-IV series is envisaged, with a capacity of 6,000 two-way voice circuits, or 12 two-way TV channels.

First in space for obvious reasons have been the international satellites capable of trans-oceanic multi-channel transmission, but there are several national areas sufficiently large to warrant separate satellite systems. Here the USSR has taken the lead, no doubt because its population, scattered over a vast land mass, has never justified the landline and microwave complex developed in the US. At present, nine Molniya-I satellites in 12-hour elliptical orbits, provide round-the-clock coverage of the whole country, each satellite traversing Russian territory for 8–10 hours in every 24-hour period. Some 20 ground stations in the so-called Orbit network receive the satellite transmissions originating from Moscow, and relay them to local TV stations. Newspaper texts, weather maps and sound radio programmes are also sent by satellite, as well as telegraphy and telephony.

The problem in Canada is not dissimilar to that of the USSR, but here a geo-stationary satellite system is planned, with special emphasis on the needs of the northland, much of it now out of reach of networked television programmes. A prototype ARCOM (Arctic Communications) satellite earth station has already been built.

More controversial are the proposals for domestic satellite operation in the US, already provided with an elaborate ground network. However, the growth of communications is now so rapid that a number of rival schemes have been put forward, all based on stationary satellites. The Comsat proposal for a pilot programme calls for two satellites, each with a capacity for 12 colour TV channels, or 21,600 trunk message channels, or 9,600 multi-point message channels, or various combinations of these. Other plans have been advanced by foundations and private corporations, all for achievement in the early 1970s.

The satellites, two or more in number, would be sited in equatorial orbit south of Mexico and on a central longitude of the US. Earth stations would be distributed on the East and West coasts, as well as in the Chicago area.

In Europe a scheme is being considered for knitting together the 21 countries of the European Broadcasting Union (EBU), which includes N. Africa, by a stationary satellite provisionally called Eurafrica, with a narrow-beam aerial to serve all Europe, Iceland and the Middle East, and a wide-beam aerial to span the whole African continent. This plan is expected to be realized in the mid-1970's R. J. S.

See: *Microwave links; News programmes; Special events.*

Saturable Reactor. Variable inductance (choke) in which the reactance of the load circuit is varied by altering the magnetic flux in the core by means of d.c. flowing in an auxiliary winding. Control of the low-current d.c. can thus be made to alter a much greater a.c. load current, e.g., in studio lighting circuits.

See: *Lighting equipment; Power supplies.*

Saturation. In colour reproduction, the spectral purity of a colour; saturated colours represent a limited group of wavelengths only.

See: *Colour principles.*

Saturation Lighting. The provision of a surplus number of luminaires regularly spaced above a studio to avoid delay in rigging them to the required positions. Generally each luminaire is movable over a limited travel and is of a combined soft and spotlight form.

See: *Lighting equipment* (TV).

Sawtooth Deflection. Voltage waveform in the form of a sawtooth, i.e., quick, linear rise and slower linear fall or vice versa.

See: *Scanning; Television principles.*

Scan Burns. Blemishes caused by overloading phosphors or other scanned surfaces with excess current, or by repeated use. Constant scanning by an electron beam gives rise to a loss in sensitivity over the scanned area. When a camera tube or a telecine flying-spot scan tube has been in use for some hours the loss in sensitivity can become apparent on the reproduced picture. If the scanned limits are adjusted so that a new part of the target or phosphor becomes part of the scanned area, the output can be higher over this area than over the area previously scanned. Normally, this causes no difficulty, but in circumstances where the limits on the tube have to be adjusted (e.g. for Cinemascope on telecine, or occasionally for zoom lenses on studio cameras), great care has to be taken to underscan the patch to obtain even sensitivity. Manufacturers are experimenting with target materials which do not age, and camera manufacturers have introduced the image orbiting facility, which continually shifts the scan patch at a slow rate.

See: *Phosphors for TV tubes.*

□**Scan Generator.** Generator of line or field frequency sawtooth voltage or current waveform, used to deflect a cathode ray beam in both cameras and receivers.

See: *Synchronizing pulse generators.*

□**Scan Linearity.** Accuracy of position, both horizontally and vertically, of any point in a reproduced image.

See: *Camera (colour).*

⊕□**Scanned Print.** Unsqueezed print made from an original anamorphic motion picture negative in such a way that the essential action is retained even though it does not fall in the centre of the composition in the frame. This is particularly important in copies of wide-screen subjects made for television transmission since only about half of the original aspect ratio of 2·35 : 1 can be reproduced.

See: *Safe area; Scanning printer; Wide screen processes.*

SCANNING

A television picture is produced by a small spot of light which covers the face of a picture display tube in a systematic manner at a rate sufficiently fast to avoid flicker and to give the effect of a simultaneous illumination. The instantaneous brightness of the spot is controlled at the receiver so that the light and shade of the picture being transmitted are reproduced on the face of the display tube.

The scanning pattern used in all contemporary television systems is one of slightly slanting horizontal lines. The spot starts at the top left-hand corner of the screen and moves steadily to the right and slightly down. On reaching the right-hand edge of the picture, the spot quickly returns to the left-hand edge, just below its original starting place, changes direction again and traces the second line, just below the first. This continues until the bottom of the picture is reached, at which point the spot is rapidly returned to the top left-hand corner and retraces the pattern again. A picture repetition rate of at least 50 Hz is required to avoid the effects of flicker at high brightness levels.

PRACTICAL SYSTEMS. All entertainment television systems currently in use employ interlace to obtain freedom from flicker while having a 25 Hz or 30 Hz picture repetition rate. In this system the picture is first scanned by half the number of active lines and is then re-scanned by the other

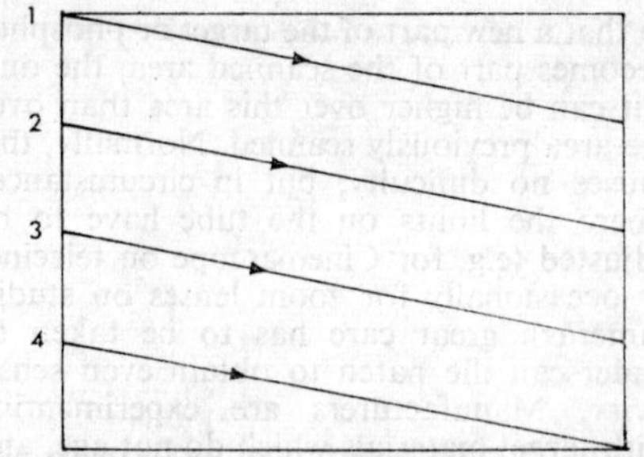

SEQUENTIAL SCANNING. Simplified to show field scanned by only four successive lines. The scanning beam is suppressed at the end of each line, and re-traces its path to the beginning of the next line to start the next scan. The whole process must be repeated at least 50 times per second to avoid flicker.

half, with the second set of lines sandwiched (or interlaced) between the first.

One half-scan of the screen is known as a field and the complete scan (two fields) is known as a picture.

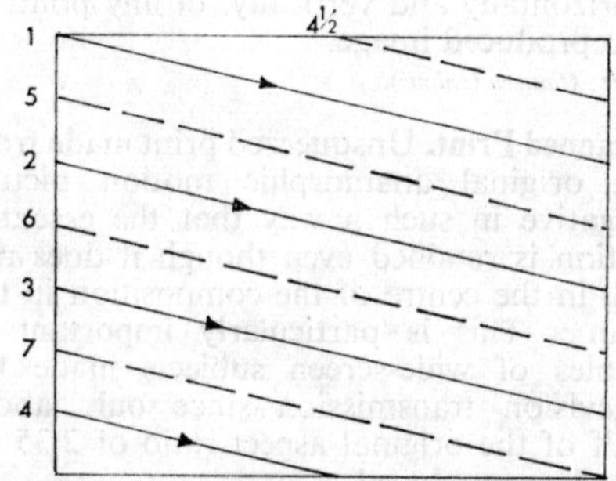

INTERLACED SCANNING. Simplified 7-line scan. First scan (*solid line*) ends at midpoint of bottom of screen. Suppressed beam then flies back to midpoint of top of screen and second scan (*dotted line*) starts scanning spaces between previous lines, ending at bottom right-hand corner. The eye sees 50 fields each second, though picture scan is only at rate of 25 per second (60 and 30 in N. American standard).

Since some time must be allowed for the spot to return across the screen from right to left and from bottom to top (the flyback or retrace time), picture information cannot be transmitted during this time and thus some information is lost. In a practical system, the scanning spot is blacked-out during these periods, known as blanking, and information is added, below black-level, to the transmitted waveform, to enable synchronism to be maintained between the transmitted and received pictures.

RECEIVERS. Two forces are required to move the spot, one vertical and the other horizontal. In a television receiver, these forces are provided by the vertical and horizontal time-bases which drive the deflection yoke.

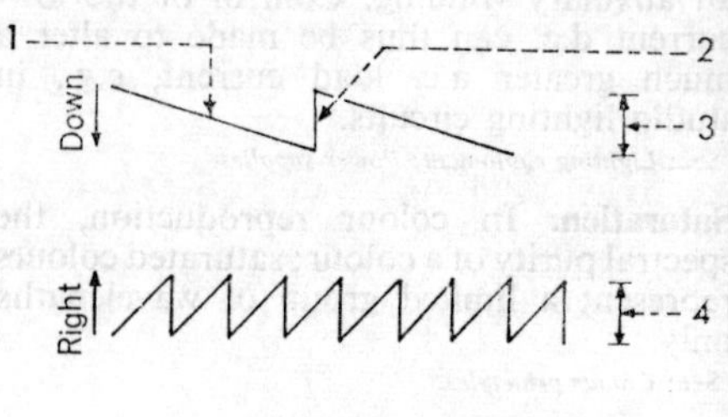

SCANNING WAVEFORMS. 1. Scanning stroke. 2. Flyback or retrace. 3. Height of picture. 4. Width of picture. These sawtooth waveforms are generated in timebase circuits which produce the basic unmodulated raster (line and field pattern) on the receiver tube.

In the very early days of electronic television, cathode-ray picture tubes employed electrostatic deflection, in which deflector plates inside the picture-tube controlled the position of the spot, and it was only necessary to apply a linear sawtooth waveform to the plates in order to obtain a linear movement of the spot over the screen. Now, however, magnetic deflection is used, and the spot is deflected by variation of the magnetic fields through which the beam of electrons has to pass before reaching the screen.

In one simplified arrangement the beam

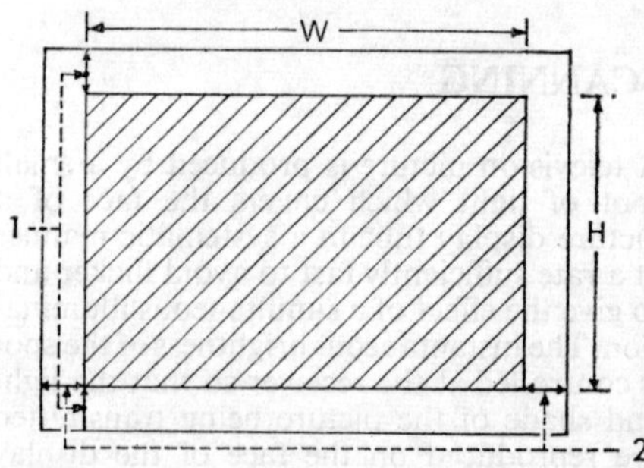

BLANKING PERIODS AND ASPECT RATIO. Blanking periods are those in which scanning is suppressed for purposes of line and field flyback. These periods are made use of to supply the receiver with line and field synchronizing information so that it is kept in step with the camera scanning beam at the transmitter. 1. Vertical blanking for field flyback and, 2, horizontal blanking for line flyback. Active line duration, W, and field duration, H, set aspect ratio, W : H, in all standard systems 4 : 3.

of electrons travelling from the cathode to the screen is subjected to the field from a pair of coils placed one each side of the neck of the tube.

By the laws of magnetism, the beam is deflected in the same way that a loudspeaker cone is deflected when a current is passed through the voice coil.

In a practical television receiver, two sets of coils are required, disposed round the neck of the tube, one pair providing the vertical component and the other the horizontal component of the scanning pattern.

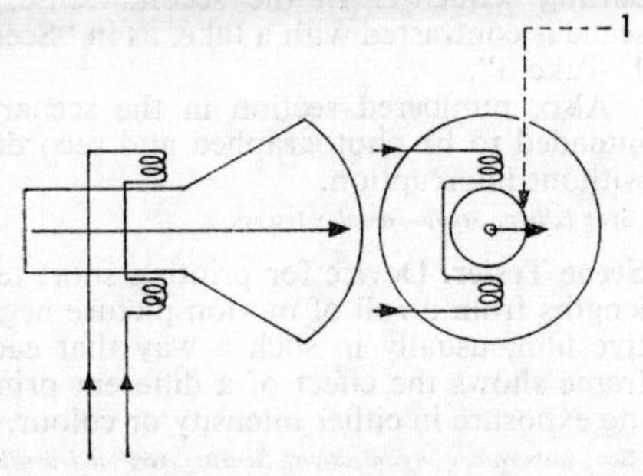

MAGNETIC DEFLECTION IN PICTURE TUBES. (*Left*) Section of tube showing scanning beam between horizontal deflection coils supplied with sawtooth waveform. (*Right*) View from screen end of tube. 1. Spot undergoing horizontal deflection. Second set of coils at right-angles to first provides vertical deflection.

CAMERAS. At the studio, scanning of the scene being transmitted is achieved by means similar to those at the receiver. The camera tube target, on which an optical image is focused, is scanned by a beam of electrons, the change in target potential being related to the instantaneous brightness of the picture.

SYSTEM PARAMETERS

		USA 525/60	CCIR 625/50
No. of active lines*	...	493	587
Line frequency (Hz)	...	15,750	15,625
Field frequency (Hz)	...	60	50
Picture frequency (Hz)	...	30	25
Active line time* (μ sec)	...	53	51·5
Aspect ratio	...	4:3	4:3

*Nominal values.

SCANNING WAVEFORMS. To obtain a reasonably shallow cabinet, modern cathode ray picture tubes have a large horizontal deflection angle, typically 110° for a monochrome tube and, for aesthetic reasons, have almost flat faces. To obtain a linear displacement of the spot on the screen, therefore, the scanning movement of the beam is a complicated one, and as the spot moves away from the centre of the screen it travels faster for a given deflection angle. Because of this, a linear sawtooth deflection current does not give good picture linearity, and a correction, known as "S" correction, must be added to the waveform.

In addition, since it is the current through the coil which determines the deflection, the voltage across the coil is not necessarily of the same waveform.

If the coil is a perfect inductance, with no losses, a square wave voltage produces a sawtooth current. Any resistive losses in the circuit, however, add a sawtooth component to the required waveform. In the case of the field coils, core losses in the yoke and the field output transformer also modify the required voltage waveform.

Transistor receivers have different coil

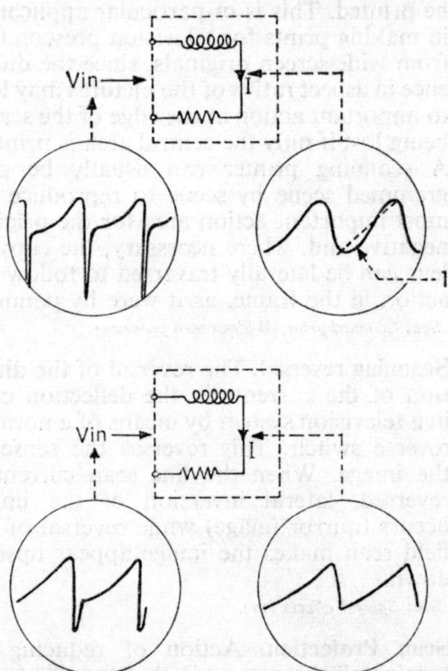

SCANNING WAVEFORMS. (*Top*) Horizontal deflection. Short picture tubes with wide and flat screens demand a more complex waveform than the simple sawtooth to give linear spot displacement per unit time, needed for undistorted reproduction. Hence, output sawtooth must be modified by "S" correction, 1. Deflection is determined by current, I, in deflection coils; because of losses and inductance, input voltage, V_{in}, does not correspond to output waveform. (*Bottom*) Corresponding vertical deflection waveforms.

inductances and resistances and, hence, different waveforms from those found in receivers using thermionic valves.

COLOUR. In a colour receiver, accuracy of scanning is of first importance, and, owing to the large diameter neck with three beams to be deflected, design of the deflection coils is not an easy task. All that has been said above, however, applies equally to colour and monochrome receivers. L.T.M.

See: *Television principles; Transmission and modulation.*

□**Scanning Coil.** Coil which produces a magnetic field at right angles to the axis of the neck of a cathode ray tube. This field deflects the beam to form the television scanned raster.

See: *Raster; Scanning; Television principles.*

□**Scanning Coil Assembly.** Deflection coil system of a cathode ray tube (kinescope) generally shaped to fit closely round the neck of the tube.

⊕□**Scanning Printer.** An optical printer in which the copying lens is continuously adjustable in the horizontal sense so that selected areas of the original negative may be printed. This is of particular application in making prints for television presentation from wide-screen originals, since the difference in aspect ratios of the pictures may lead to important action at the edge of the screen being lost if only the central area is printed. A scanning printer can usually be programmed scene by scene to reproduce the most important action area for the original negative and, where necessary, the copying lens can be laterally traversed to follow the action in the frame, as it were by panning.

See: *Scanned print; Wide-screen processes.*

□**Scanning reversal.** The reversal of the direction of the currents in the deflection coils in a television system by means of a normal-reverse switch. This reverses the sense of the image. When the line scan current is reversed, lateral inversion of the image occurs (mirror image) while reversal of the field scan makes the image appear upside-down.

See: *Special effects* (TV).

□**Scan Protection.** Action of reducing or turning off the current in the scanning beam of a receiver or camera tube in the event of failure of the scanning fields. Necessary to prevent burning of the target resulting from a stationary spot.

See: *Camera* (TV); *Ion burn; Receiver; Scan burns.*

⊕□**Scene.** Loosely-used term, which sometimes denotes the setting for a series of shots, and sometimes an individual camera set-up. In the first sense, a master scene means a continuous shot of a complete series of actions, afterwards to be covered by intercut closer shots to build up a comprehensive cutting sequence. In the second sense, a scene is contrasted with a take, as in "Scene 1, Take 3".

Also, numbered section in the scenario intended to be photographed and recorded without interruption.

See: *Editor; Studio complex* (FILM).

⊕**Scene Tester.** Device for printing short test lengths from a roll of motion picture negative film, usually in such a way that each frame shows the effect of a different printing exposure in either intensity or colour.

See: *Laboratory organization; Sensitometry and densitometry.*

⊕□**Schlieren Photography.** Method of photography making use of diffraction effects which allows small density differences in otherwise transparent gases and liquids to

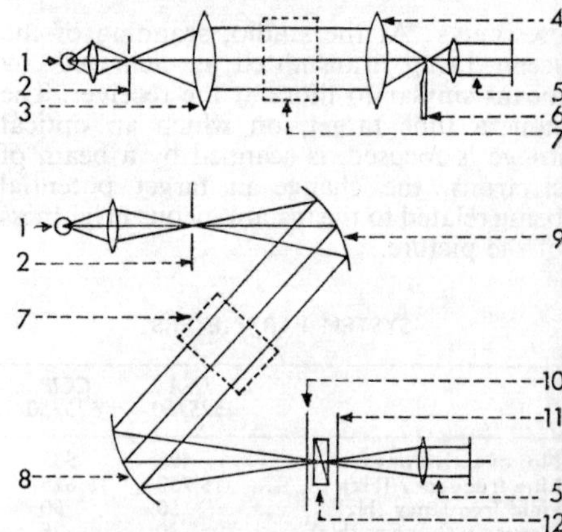

SCHLIEREN OPTICAL SYSTEMS. In both systems, rays from small high-intensity light source, 1, are focused on aperture, 2, and collimated to pass through test material, 7, finally traversing camera lens, 5. (*Top*) Lenses, 3, 4, are used for collimation, and knife edge, 6, is placed at rear focal point of 4. Diffraction reveals changes in refractive index of 7. (*Bottom*) Mirrors, 8, 9, used for collimating light rays. To record phase changes in light across test sample, knife edge is replaced by polarizing filters, 10, 11, with Wollaston prism, 12, between.

be recorded; it is of particular value in the study of aerodynamic problems in wind tunnels, in the investigation of fluid flows, and in ballistics for the examination of shock waves for high velocity projectiles.

In essence, variations of density within a liquid or gas, caused by either compression or heat differences, affect light passing through the medium both by refraction and by change of phase, and both characteristics may be recorded in different systems of Schlieren photography (Schlieren=streaks (German)).

In most Schlieren systems, the material to be examined, which may be a section of a wind tunnel or a liquid flowing in a transparent container, is placed in a beam of parallel light between two collimating lenses or concave mirrors illuminated from a small but intense source. This beam is focused on to the film of a camera and, for the recording of refraction effects in the material, a knife edge is located at the rear focal point of the collimating system. Variations of refractive index in the test material cause diversions of the parallel light passing through and, as a result of diffraction effects at the knife edge, these variations are recorded on the film as significant changes of local image brightness. The knife edge may be replaced by a constant deviation diffraction grating, and variations of refractive index in different parts of the sample are then made visible as Schlieren effects in colour which can be recorded by the use of colour film in the camera.

Where it is desired to record phase changes in light passing through parts of the test sample, the knife edge of the simple Schlieren system is replaced by a pair of polarising filters with a Wollaston prism between; this causes interference between rays passing through different zones of the sample which are recorded as light intensity differences on the camera film.

Schlieren projection systems are employed in the Eidophor and in the reproduction of thermoplastic recordings. L. B. H.

See: *Large screen television; Thermoplastic recording.*

Schmidt Optics. An optical system incorporating mirrors and an aspherical lens to give a wide field free of aberration combined with a large effective aperture.

See: *Large screen television.*

Schoenberg, Isaac, 1880–1963. Distinguished Russian scientist who became the chief architect of the British television system. Born at Pinsk and studied at the Technological Institute at Kiev. Afterwards, became Chief Engineer of the Russian Wireless Telegraph and Telephone Company and helped to install some of the earliest wireless stations in Russia. Came to England in 1914 and took British nationality in 1919. He worked first of all for the Marconi Company as a consultant, later becoming joint general manager. In 1928 he joined the Columbia Graphophone Company as general manager and then in 1931, when that company merged with the Gramophone Company to become Electric & Musical Industries Ltd., he became the new company's director of research. Shoenberg led the team of scientists which developed the Emitron, a camera tube based on Zworykin's iconoscope, and also the 405-line system adopted by the British Broadcasting Corporation in 1936 for the world's first public service of high definition television. He was awarded the Faraday Medal (1954) for his outstanding contribution to television development, and was knighted (1962).

⊕**Schüfftan Process.** Method for combining live action with models or paintings at a single photographic operation. A large silvered mirror is mounted in front of the camera at 45° and reflects the model or painting into the camera's field of view; portions of the reflective surface are then carefully scratched away so that the live action part of the scene can be seen through the glass of the mirror in the appropriate position with respect to the reflected model and the two components shot simultaneously. Named after its inventor, Eugen Schüfftan.

See: *Art director; Special effects* (FILM).

⊕□**Scoop.** Floodlight used in film and television studios and shaped like a grocer's scoop; a simple diffusing reflector and general service lamp, usually of 500 or 1,000 watts rating.

See: *Lighting equipment.*

⊕**Scope.** Popular abbreviation for wide-screen systems involving the projection of anamorphic images with a lateral expansion factor of 2 : 1; derived from the trade name Cinemascope and similar forms.

See: *Wide screen processes.*

□**Scophony System.** An early mirror-drum projection system used in television based on Schlieren optics, in which an acoustic wave generated from the video signal caused changes in refractive index in a liquid. The technique fell into disuse for many years but has recently been revived.

⊕**Scratches.** Scored lines which penetrate the emulsion surface of film or seriously indent its base; distinguished from abrasions by their greater severity. Scratches usually run vertically on the screen, and are caused by the gouging effect of hard particles in cameras, developing machines, printers and projectors. Negative scratches are the most serious type of scratch, because they damage negative or original material. They can often be identified when viewing a print by the fact that they are lighter than any part of the image. Positive scratches are usually of secondary importance, and can often be identified because they are black, having collected particles of dirt during projection and rewinding. If not too serious, scratches can be rendered invisible or at least less visible by reconditioning.

See: *Lacquering; Mutilation of prints; Reconditioning; Release print examination and maintenance.*

⊕**Scratch Print.** Motion picture print made from an assembled reel of negative without the scene-by-scene corrections of printing made in grading and used only as an indication of the action of the film, for sound track preparation, etc. Also called slash print.

See: *Editor; Laboratory organization; One-light print.*

⊕**Scratch Removal.** Treatment of positive release prints which have become scratched in the course of use in cinema theatres. Suitable solvents and polishing rollers or lacquers are used. In emergency negative film may be similarly treated.

See: *Lacquer; Reconditioning.*

⊕**SCR Dimmer.** Silicon controlled rectifier □used for lighting control.

⊕**Screen Brightness.** Luminance of the screen in a cinema theatre illuminated by the projector running under normal conditions but without film in the aperture. It is affected not only by the light output of the projector and its lens, but also the material and condition of the screen itself and the size of the image thrown. Although a degree of standardization has taken place, screen brightness is allowed considerable tolerance and there is lack of international uniformity.

See: *Screen luminance.*

⊕SCREEN LUMINANCE
□

Motion picture films and transparencies must be processed to provide densities and contrasts appropriate to the conditions under which they will subsequently be projected and viewed. The two most important factors affecting these conditions are the amount of light available in the projection system, and the reflective characteristics of the screen.

For the projection of 35mm motion picture films, powerful light sources are available, but too bright a screen picture shows graininess and objectionable flicker.

NATIONAL STANDARDS

Hitherto various national units for the measurement of luminance have been adopted by different countries. At the International Standards Conference, 1967, it was agreed to adopt the unit candela per square metre. In the following the US and UK references have been converted to this unit.

TEST CONDITIONS. In the majority of specifications screen luminance has to be measured with a photometer having an acceptance angle of $1\frac{1}{2}°$ or 2° and having the spectral sensitivity of a Standard Observer, as defined by the International Commission on Illumination, 1924. The photometer should have an accuracy of ± 10 per cent.

All national specifications and international recommendations require that the screen luminance shall be measured when the projector is operating under normal conditions at a speed of 24 frames per second with the lens focused on the film plane, but with no film in the gate. In the case of the British Draft Specification for 70mm projection, to reduce the heat intensity and so prevent damage to the projection lens, it is recommended that a film having a uniform transmission density of 0.3 shall be projected while taking measurements, or a glass filter of the same density be placed either behind or in front of the film gate. A film or filter of 0.3 density has a transmission factor of 50 per cent and thus the luminance measurements must be multiplied by two. The presence of the film or filter may affect the accuracy of the reading to the extent of ± 5 per cent, but this error is negligible.

US EXPERIMENTS. Experiments carried out in the US led to the following conclusions:

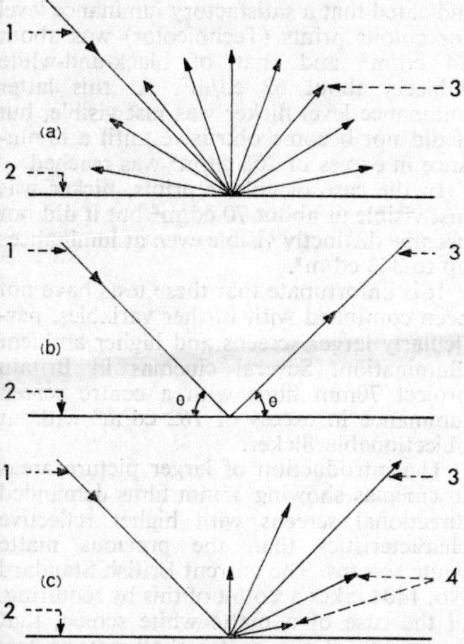

TYPES OF REFLECTION. (a) Pencil ray of light, 1, falling on matte white screen, 2, is reflected almost uniformly in all directions, 3. (b) When pencil ray, 1, falls on mirror, 2, it is reflected to 3, the angle of incidence, θ, being equal to the angle of reflection. (c) When pencil ray, 1, falls on semi-specular (directional) screen, 2, it is reflected in a way intermediate between (a) and (b). Angle of ray of principal reflection, 3, equals angle of incidence, but subsidiary rays, 4, are of lesser intensity. If lengths of lines of reflection are proportional to intensity of reflected rays, ends of lines may be joined to form polar curve of screen reflection characteristics.

As screen luminance is increased, the density of the print must be increased and the contrast scale lowered to maintain optimum quality.

The American standard for theatre screen luminance gives picture quality superior to that obtained at lower screen luminance. This current standard (PH.22.124.1961), now under revision, calls for a luminance at the centre of the screen of $55\ ^{+14}_{-21}$ cd/m² measured from a position at the longitudinal centre line of the auditorium and two-thirds distant from the screen to the rearmost row of seats. The luminance at a distance of 5 per cent of the screen width from the side edges of the screen and at its horizontal axis, must be between 65 and 85 per cent of the centre luminance.

Increasing the screen luminance from 34 to approximately 86 cd/m² improves quality in most cases, provided proper adjustment is made in the density and contrast of the print.

CANDELAS PER SQUARE METRE (NITS) AND FOOT LAMBERTS: REPRESENTATIVE VALUES

cd/m²	ft/L	cd/m²	ft/L
1	0·292	55	16·1
4	1·17	60	17·5
10	2·92	62	18·1
14	4·09	70	20·4
19	5·55	80	23·4
20	5·84	82	23·9
21	6·13	86	25·1
25	7·30	100	29·2
27	7·88	155	45·3
30	8·76	162	47·3
34	9·93	206	60·2
40	11·68	343	100·2

Quality within 10 per cent of the best can be obtained over a wide range of screen luminance levels, beginning at 34 and extending to about 155 cd/m².

These experiments did not give a definite explanation for the quality decrease at higher screen luminances. Visual discomfort or other subjective effects due to viewing pictures of high luminance in a dark surround may be important factors.

Further experiments are needed to determine the cause of the quality decrease and to measure quality as a function of screen luminance under various conditions of ambient illumination and picture surround.

BRITISH TESTS. Similar tests carried out by the British Standards Institution (BSI)

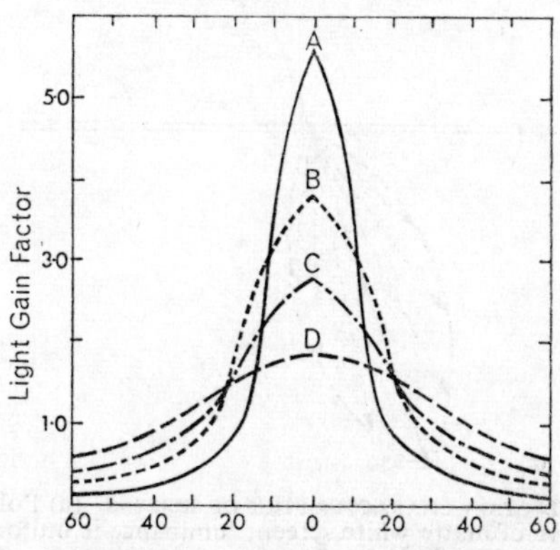

GAIN FACTORS FOR TYPICAL DIRECTIONAL SCREENS. A. Gain of 5·5 limits use of screen to long, narrow auditoria, since at viewing angle of only 18°, luminance falls to unity. D. Gain of 1·8. Screen has a much improved distribution and may be used in shorter and wider auditoria. B, C. Screens of intermediate gain.

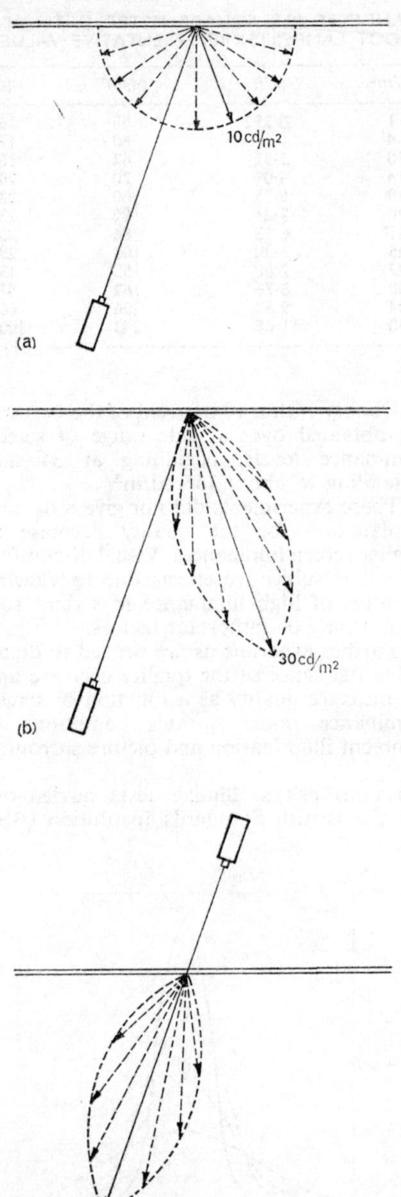

REFLECTION CHARACTERISTICS OF SCREENS. (a) Polar curve of matte white screen; luminance is uniform when viewed from any forward direction. (b) Polar curve of directional screen; gain factor is defined as ratio of luminance at angle of principal reflection to luminance of matte white screen. (c) Translucent rear projection screens have characteristics similar to directional screens; gain is controlled by the transmission factor of the screen material and opacity of substance sprayed on its surface. Gain factor of (b) and (c) compared with (a) is 3.

indicated that a satisfactory luminance level for colour prints (Technicolor) was about 34 cd/m² and that of black-and-white subjects about 62 cd/m². At this latter luminance level flicker was just visible, but it did not become obtrusive until a luminance in excess of 206 cd/m² was reached.

In the case of colour prints, flicker was just visible at about 70 cd/m² but it did not become distinctly visible even at luminances up to 343 cd/m².

It is unfortunate that these tests have not been continued with further variables, particularly larger screens and higher ambient illumination. Several cinemas in Britain project 70mm films with a centre screen luminance in excess of 162 cd/m² without objectionable flicker.

The introduction of larger picture areas in cinemas showing 35mm films demanded directional screens with higher reflective characteristics than the previous matte white screens. The current British Standard No. 1404 takes account of this by requiring, in the case of a matte white screen, that luminance at the centre shall not be less than 27 nor more than 55 cd/m², this luminance to be measured from any seat in the auditorium. In the case of directional screens, however, the luminance at the centre of the screen shall be not less than 27 and not more than 69 cd/m² when measured from any seat within a defined area in the body of the auditorium.

EDGE LUMINANCE. When tests and experiments were carried out for the preparation of the standard, it was found that a uniform luminance over the entire area of the screen was not desirable. It gave an impression of flatness, but when there is a diminution of brightness towards the edges, the picture appeared to be brighter and have more sparkle. It is a curious fact that a screen picture having a uniform luminance of 55 cd/m² appears less bright than one having a centre luminance of 55 cd/m² and an edge luminance of 41 cd/m². Thus the British Standard requires that the luminance at each side is measured on the horizontal axis should lie between 60 per cent and 85 per cent and preferably be 70 per cent of the centre luminance; where anamorphic projection is used, the edge luminance should be between 50 per cent and 75 per cent of the centre luminance.

This standard cannot be achieved throughout the auditorium where directional screens are installed; it is only within a restricted area of the seating that eminently satisfac-

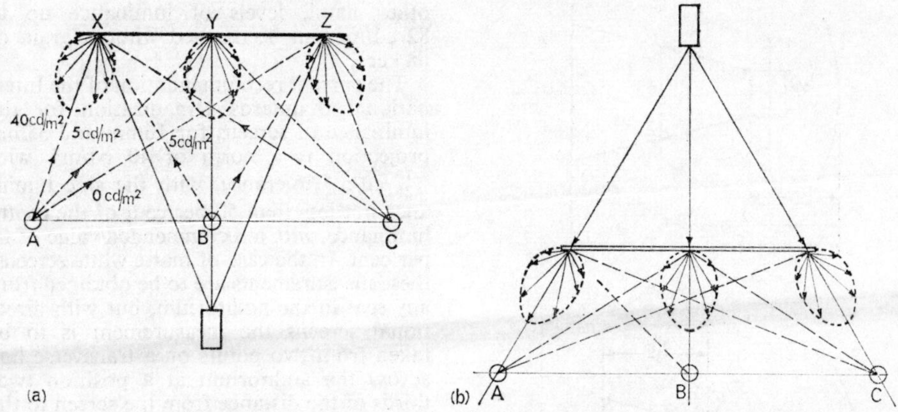

LIGHT DISTRIBUTION PROBLEMS. With semi-specular screens it is hard to obtain satisfactory light distribution. (a) Three equal-intensity rays of light fall on screen at X, Y, Z. Observer A sees the ray at X at, say, 40 cd/m², the ray at Y at, say, 5 cd/m², and no reflection from far side of screen at Z. C sees the reverse. B sees Y at 40 cd/m² and X and Z at 5 cd/m². (b) Same conditions arise with translucent rear projection screen. Light distribution improves with increasing focal length of projection lens, as cone of rays becomes narrower.

tory luminance conditions are obtainable and it is only in a few seats that equality of luminance from each edge of the screen can be achieved. It is in an attempt to overcome these deficiencies that directional screens are curved.

In the case of such screens, it is required that the luminance at the side of the screen measured from two points in the auditorium two-thirds of the distance from the screen to the rear seats and at a distance apart equal to the width of the screen shall be between 60 and 85 per cent (preferably 70 per cent) of the luminance at the centre of the screen for non-anamorphic projection and between 50 and 75 per cent for anamorphic projection. The specification also makes a provision for screen luminance at balcony level.

RUSSIAN AND FRENCH PRACTICE. Russia adopts a standard for centre of screen luminance of 40^{+20}_{-10} cd/m², with the following distribution:

USSR SCREEN LUMINANCE

Highest luminance at centre of screen cd/m²		Lowest luminance at edges of screen cd/m²	
		Aspect ratio up to 1·85:1	Aspect ratio exceeding 1·85:1
Upper limit	60	39	30
Normal	40	26	20
Lower limit	30	19·5	15

The current French specification gives a centre of screen luminance of $55\ ^{+10}_{-25}$ cd/m² for both matt white and directional screens, with a distribution of 65 per cent for normal projection and 50 per cent for anamorphic systems.

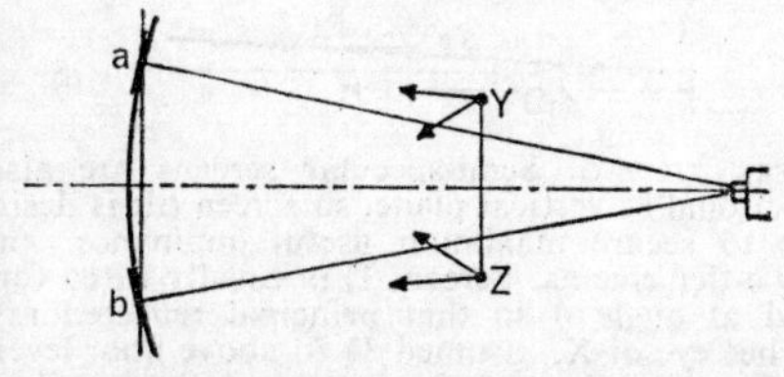

EFFECT OF SCREEN CURVATURE. Curvature of directional screen improves light distribution. Observers Y and Z see tangential screen elements *a* and *b* at equal brightness when illuminated by projector.

DIRECTIONAL SCREENS. The theory underlying the British Standard for luminance of directional screens, is based upon the fact that it is possible to turn two elements of the screen surface at the opposite edges tangentially to a curvature so that they will appear equally bright to two observers placed equidistant from the longitudinal centreline of the auditorium. The distance between these observers was determined from the results obtained from surveys of a large number of cinemas as one-third the length of the auditorium from the screen to the rearmost row of seats. Their longitudinal

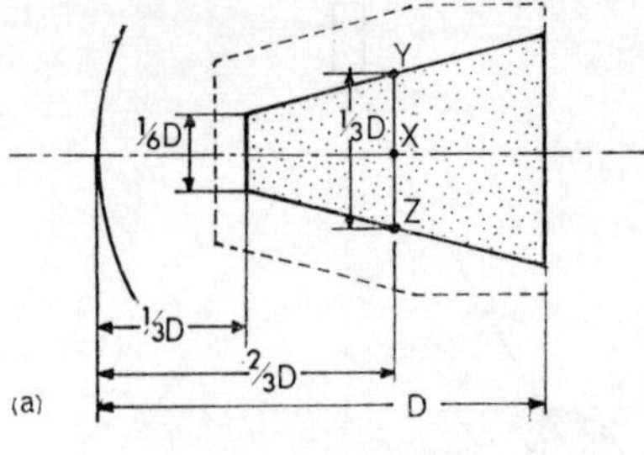

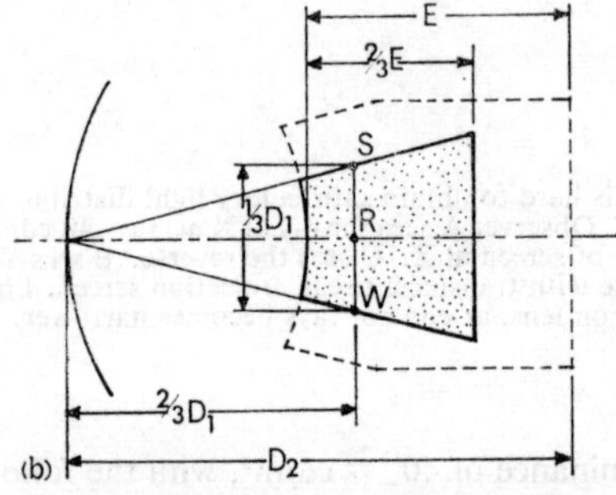

STANDARD FOR DIRECTIONAL SCREENS. Provisions of British Standard 1404. (a) Stalls. D. Distance from screen to rearmost stalls. Observers Y, Z at ⅔ D from screen and ⅓ D apart should see edges of screen at equal brightness. Observer X should see screen edges at 60–85 per cent (preferably 70 per cent) of centre screen luminance for normal projection, and at 50–75 per cent for anamorphic projection. (b) Balconies. D_2. Distance from screen to rearmost row of balcony. D_1 equals D or D_2, whichever is greater. Observers S and W should see screen edges at equal brightness. R's observations should be same as X's.

position is two-thirds the length of the auditorium from the screen.

The British standard has not received international recognition, but a proposed international standard requires the screen luminance for both 16mm and 35mm to lie between 30 cd/m² and 60 cd/m². For matt screens, the measurement is made from any seat in the auditorium; for directional screens from a point on the horizontal centre line at a distance equal to two thirds of the distance from the screen to the rearmost row of seats.

STANDARDS FOR 16MM. British Standard 2954 gives recommendations for screen luminance for the projection of 16mm films. It recommends a screen luminance not less than 27 and not more than 55 cd/m² with a lateral distribution of not less than 60 per cent. It is pointed out in the recommendations that a luminance of 21 cd/m² provides an acceptable picture and, on the other hand, levels of luminance up to 82 cd/m² can be reached without grain or flicker.

The current recommendation of the International Standards Organization for the luminance of screens for 16mm and 35mm projection is a norm of 40 cd/m², with $^{-10}_{+25}$ cd/m² tolerance, with the side luminance not less than 50 per cent of the centre luminance, with a recommended value of 65 per cent. In the case of matte white screens, these measurements are to be obtained from any seat in the auditorium, but with directional screens the measurement is to be taken from two points on a transverse line across the auditorium at a position two-thirds of the distance from the screen to the back row of seats, these two points being located at a distance apart equal to the width of the screen, one on each side of the centre line of the auditorium.

PROPOSALS FOR 70MM. An international proposal will probably go forward for the adoption of the foregoing recommendation for the projection of 70mm films on matt white screens. The suggestion emanates from Russia, where a number of large cinemas have recently been constructed, primarily for the projection of 70mm films. The screens installed are so large and the auditoria so wide that directional screens cannot be used advantageously.

This proposal has received some support from the US. One view is that all motion

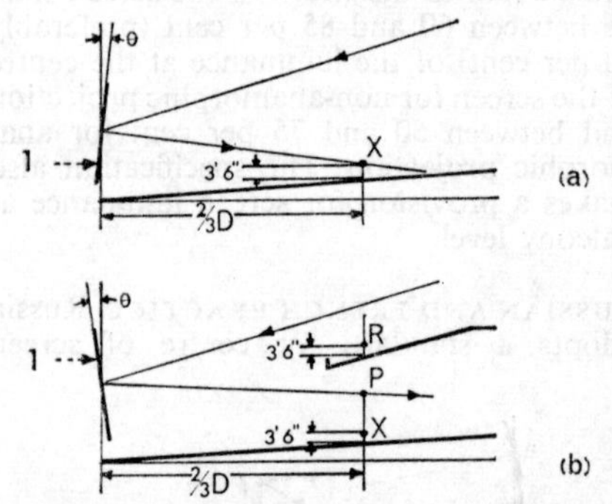

SCREEN TILTING. Semi-specular screens are also directional in vertical plane, so screen tilt is desirable to secure maximum useful luminance. (a) Single-tier cinema. Screen, 1, is usually tilted forward at angle θ so that principal reflected ray reaches eye of X, assumed 3½ ft. above floor level. (b) Two-tier cinema. Screen, 1, is usually tilted backwards at angle θ so that principal reflected ray passes through P, situated at mid-height between observers X and R. D. Distance from screen to rearmost row of stalls.

picture films should be made for the same luminance objectives, because this makes for convenience in processing laboratories' operations, and it is likely that the US Standard Specification for 70mm films will be similar to PH.22–124.1961 for 35mm and 16mm projection. The French proposals follow a similar line of thought. It is felt, however, that the philosophy is unfortunate; convenient laboratory operations should not be put first, but rather the presentation of screen pictures of the highest quality attainable irrespective of the size of film.

The British approach follows this line. All 70mm projectors installed in cinemas are capable also of showing 35mm films but they have a common lamp house optics and light source. However, the aperture of the film gate for 70mm projection has an area some four times larger than the aperture for 35mm film at 1·65 : 1 ratio. Thus about four times as much light passes through the 70mm aperture. The screens normally used in cinemas for 70mm projection are not four times larger than 35mm screens, and screen luminance is in practice greater when 70mm films are projected.

The British Standard now in the course of publication will require a luminance level at the centre of the screen for 70mm projection of 80 $^{+20}_{-19}$ cd/m².

The requirements for light distribution will be similar to those for 16mm and 35mm.

GERMAN 8MM PROPOSAL. There are no standards for screen luminance for 8mm projection, and the general view is that the diverse conditions under which these films are projected would require such broad limits as to make the standard of negligible value.

The German Standards Organization has, however, published a draft specification for the projection of 8mm and 16mm films which proposes a centre screen luminance of 16 to 47 cd/m² and a distribution to the edges in the ratio of 2 : 1.

REVIEW ROOMS

A review room is usually situated at laboratories and is used for appraisal of the photographic quality of a positive film before release.

BRITISH STANDARD. The British Standard 2964 for review rooms requires that the luminance at the centre of the screen for the projection of 35mm films measured from any seat in the auditorium shall be 37·5 ±3·5 cd/m² and, when a non-anamorphic optical system is being used, the luminance of each side of the screen measured on the horizontal axis shall lie between 0·6 and 0·75 times the measured luminance at the centre, and shall be as near as practicable to 0·7.

For 16mm projection, the luminance at the centre of the screen should be 34·3 ±3·5 cd/m² with a similar distribution.

The BSI has issued an amendment to the Specification for Screen Luminance for the Projection of 16mm films in Review Rooms, which amendment requires the same luminance levels for 8mm as for 16mm alm.

AMERICAN SPECIFICATION. The current American Specification PH.22/100/1967 requires a luminance at the centre of the screen of 55 ±7 cd/m² measured from a seat area within 15° on each side of the centre of the screen and at a distance within the limits of 2–4 picture heights from the screen, the distribution of the luminance to be 80 per cent ±10 per cent. The Specification also requires that the maximum luminance at any point on the screen measured from the viewing area shall not exceed 62 cd/m², and that the stray light on the screen shall not exceed 0·4 per cent of the centre screen luminance.

No definite decisions have been made internationally on screen luminance levels for review rooms, but in a recent draft proposal a range of 40–50 cd/m² with a distribution of 75 to 80 per cent was advocated. This proposal was for 35 and 70mm films of all types and for all screen sizes. At the International Conference in 1967, there was a body of opinion that the level for 35mm projection should be 40 $^{+15}_{-0}$ cd/m² with a luminance uniformity as for cinemas. There was some opposition to these levels for the projection of 70mm film.

There is no international recommendation or proposal for the projection of 16mm films in review rooms.

There are no national standard specifications or international proposals or recommendations for screen luminances for rear-projection systems, but compliance with the same requirements as for directional screens is to be expected.

TELEVISION

There are no published standards of screen luminance for television screens. With modern cathode ray tubes, the high-light luminance is about 206 cd/m² (black-and-white) and 103 cd/m² (colour), but the quality of a reproduced picture is largely conditioned by the physical and technical limits of the transmission system. This is particularly so for motion picture films, and thus, in national and international discussions, the tendency has been to seek agreement on the density and contrast of films and transparencies for television rather than to standardize screen luminances.

STRAY LIGHT

Stray light falling upon the surface of the viewing screen impairs the quality of all projected pictures; it debases the contrast of the picture and saturation of the image colours. The causes of stray light may be classified as:

1. Diffusion of the projector light beam from the projector lens, the glazing of the projection room ports, the presence of dust and smoke in the auditorium light beams, and, with rear projection and television systems, from the screens upon which the picture is presented.
2. Reflection of light from the screen on to the ceiling and walls of the auditorium and on the faces of the audience, re-reflected on to the surface of the screen.
3. Extraneous light emitted from the various luminous sources in an auditorium necessary for public control and safety, or from the ambient lighting usually present when audio-visual and television pictures are viewed. Light from these sources is invariably also reflected from walls and ceilings on to the screen.

It has been recommended that the total value of stray light falling upon a cinema screen should not exceed 1 per cent of the centre-screen luminance. This is seldom or never achieved: frequently the management lighting of the auditorium produces 1·25 to 1·75 per cent of stray light. The ambient lighting under which television and audio-visual equipment is viewed is difficult to assess. In the case of television reception, domestic viewing conditions vary between wide limits. The extraneous light falling upon the face of the screen is sometimes as high as 6 per cent excluding specular reflections from light sources such as table lamps which are often present. Similar conditions also apply to the viewing of audio-visual equipment, particularly in classrooms when screens are often viewed under indirect daylight conditions. It is difficult with such equipment to obtain a screen luminance in excess of 55 cd/m^2 and the screen should be so placed and so protected that extraneous light reaching it does not exceed 5 per cent of this figure.

Similar equipment used for advertising purposes also suffers from a high level of ambient lighting. To reduce degradation of the picture, it is a common practice to provide a hood lined with matt black material around the perimeter of the screen to protect it from stray light. L.K.

See: *Cinema theatre; Continuous projection; Daylight projection unit; Drive-in cinema theatre; Rear projection; Screens.*

⊕SCREENS

A screen is essentially a plane surface upon which the light rays from the projector impinge and which, either by the nature of its surface or its translucency, causes the projected picture to become visible to the spectator. Thus it is upon the screen in the cinema that the final result of all the work and endeavour of producers, artists, technicians and others engaged in making films is presented to the public. It is remarkable that only within recent years has the manufacture and surfacing of screens received scientific and technical consideration.

CONSTRUCTION

The majority of screens are of the reflective type: that is to say, the light rays fall upon its surface and are reflected back to the eyes of the spectator. Such screens are opaque and generally made of plastic material, such as polyvinyl chloride. To provide large screens, it is necessary to join several sheets of the material together. The seams are invariably arranged vertically and are heat-welded. The process calls for considerable skill because it is most important that the joins are not readily visible to the observer. Eyelet holes are provided around the perimeter of the screen to permit it to be laced to the surrounding steel or stout wooden frame. It is, of course, necessary that the screen be tightly and uniformly stretched to give a perfect plane surface free from corrugations.

PERFORATION SIZE AND SPACING. When the loudspeakers of the sound reproduction system are placed, as is usual, behind the screen, the entire area of the screen is perforated. The presence of the screen in front of the speakers attenuates the acoustic power which they deliver, and the attenuation increases at higher frequencies. An American specification requires that the attenuation shall not exceed 2½ dB. and 4 dB. at 6 KHz and 12 KHz respectively compared with the power level at 1 KHz. This is difficult to attain unless the perforations are increased in size or are more closely spaced, neither of which can be recommended because the effective area of reflection surface is thereby reduced, with a consequent reduction of picture brightness and definition.

A leading British manufacturer of screens offers two types of perforations, one having holes of ·0625 in. diameter, pitched $\frac{5}{16}$ in. horizontally and $\frac{7}{32}$ in. vertically, with intermediate, staggered rows similarly pitched; and the other having ·0468 in. diameter holes, pitched $\frac{3}{8}$ in. and $\frac{3}{16}$ in. respectively. The former attenuates sound by about 6 dB., and the latter about 8 dB. at 10 KHz.

SURFACE TREATMENT. The most important characteristic of reflective screens is their front surface. In its simplest form it is matt white, obtained by spraying uniformly with titanium, zinc or other oxide suspended in a volatile spirit or oil which does not readily oxidize and turn yellow. A matt white screen is not particularly efficient for projection purposes; the light falling upon it appears almost equally bright at whatever angle the front surface is viewed. In other words, such a screen reflects as much light towards the ceiling of the auditorium as it does to the seating area.

A matt white screen surface is known as a diffuse reflector. An improvement in effectiveness can be achieved by providing a different kind of surface to obtain a semi-specular characteristic. Such a surface lies between the extremes of a diffuse reflector and a specular one, the latter being, say, a silvered mirror which reflects light from its surface only at an angle equal to that upon which the light is incident to it.

A limited improvement is achieved by scattering very small pieces of thin glass on the wet surface, to which they adhere when the coating pigment dries. Such screens are known as glass-beaded. Although the reflective characteristic cannot easily be controlled in manufacture, it is invariably highly specular.

The intensity of a light ray measured at the angle of maximum reflection is about five times greater than that from a matt white surface. Such a screen is said to have a gain factor of 5. It can be used satisfactorily, however, only in long, narrow auditoria where the seating is confined around the longitudinal centreline. A person sitting at an extreme side seat would see the screen picture at his near edge at full brilliance; the picture at the centre would be barely discernible; and no picture would be seen at the far edge. Such a bad distribution would be unacceptable.

Another method of making a semi-specular reflective surface, and one which is now very widely used, consists of spraying the surface with very fine metallic particles, usually aluminium. The particles are suspended in a plastic vehicle which provides protection against the early oxidation of the metal. By varying the thickness of the coating and the size of the metallic particles, the reflective characteristics of the surface can be more closely controlled, and gain factors of between 1·5 and 6 can be obtained.

It must not be thought that by increasing the gain factor the quantity of light reflected is increased. This is not so; the total quantity of light reflected by these various screens is substantially the same.

Several attempts have been made to emboss a lenticular pattern on the screen surface, thus to form a large number of elements to reflect light downwards and sideways over the seated area. The cost and difficulty of making such screens have not justified their small advantages.

TRANSLUCENT SCREENS

Translucent screens are necessary where a rear projection system is used but, although common at one time, they are very seldom installed in cinemas today. The screen is of transparent plastic material, which may be naturally frosted, or the translucent effect may be obtained by spraying the surface with very fine particles of semi-opaque material. This latter method has the advantage of offering a control over the light transmission. The translucent screen can be used satisfactorily only in a narrow auditorium, the lateral distribution of light being unsatisfactory.

SCREEN MAINTENANCE

It is unfortunate that the surfaces of all screens are vulnerable, and deterioration begins immediately they are installed. Normal oxidization causes a loss of reflectivity and a yellowing of the surface: uneven discolouring is usually caused by rising air currents which carry and deposit on the screen dust particles, but the principal cause of deterioration is due to tar deposits from smoking.

This latter can be reduced by a properly designed ventilating system which draws all vitiated air away from the screen and towards the rear of the premises.

It is possible to re-spray in situ most matt white screens and some types having aluminized surfaces, but the number of times that they can be resprayed is limited. Tar stains cannot be killed and will quickly penetrate new coatings.

Respraying invariably reduces the size of the perforations and thereby attenuates the sound transmission.

In general it is advisable to renew screens when they have deteriorated, but the frequency of renewal is wholly dependent upon the quality of the local atmosphere and the efficiency of the ventilation system in the cinema. L.K.

See: *Cinema theatre; Continuous projection; Daylight projection unit; Drive-in cinema theatre; Rear projection; Screen luminance.*

□**Screen Saturation.** Limitation of the brightness of a cathode ray tube fluorescent screen by the rate at which energy from the electron beam can be transformed into light. Usually accompanied by burning.

See: *Phosphors for TV tubes; Scan burns; Scan protection.*

⊕**Scrim.** Type of flag made of translucent material, whose object is partly to cut off, partly to diffuse the light source near which it is placed in studio shooting.

See: *Cameraman* (FILM)*; Lighting equipment* (FILM).

⊕**Seating.** The arrangement of seating in a cinema theatre auditorium is of utmost importance in providing satisfactory motion picture presentation; viewing positions too close to or too far away from the screen can cause considerable strain to numbers of the audience, while positions very much to one side give a distorted picture of low brightness. These problems are accentuated by the use of wide-screen presentation techniques with very large, deeply-curved screens. It is generally recognized that it is undesirable for any seats to be closer to the screen than a distance of twice the screen height or further away than four times the screen width, but economic demands may override these considerations.

See: *Cinema theatre; Screen luminance; Screens.*

□SECAM COLOUR SYSTEM

SECAM (Séquentiel Couleur à Mémoire) was devised in France by Henri de France who published his first proposals in 1958. Later development has now much improved its performance, and it has been adopted by the USSR and the Soviet bloc, as well as by France and some Middle Eastern countries. It represents a more radical departure from the original NTSC system than does PAL, for it employs a fundamentally different way of conveying colour-difference signals from the coder in the transmitter to the decoder in the receiver.

BASIC PRINCIPLES

In NTSC, colour definition is the same as black-and-white definition in a vertical direction, but is reduced horizontally by the adoption of the mixed-highs principle, according to which fine detail need only be transmitted in monochrome. In SECAM, on the other hand, colour definition is reduced in a vertical direction also, for each of the two colour difference signals is transmitted during alternate lines only. In the usual nomenclature, if the $(E'_R - E'_Y)$ signal, where the subscripts denote red and luminance respectively, is supplied for line n, the succeeding line $(n+1)$ receives the $(E'_B - E'_Y)$ signal, and so on. The $(E'_R - E'_Y)$ information for line $(n+1)$ is also required by the receiver and is obtained by repetition of that of line n; it must therefore be stored for the requisite period by a delay device.

Since in SECAM the colour information is transmitted sequentially, the two colour-difference signals can have the same bandwidth; and since frequency modulation of the subcarrier conveys colour-difference signals to the receiver, the hue of the reproduced colour is not determined by phase. This helps to overcome a recognized weakness of the NTSC system, wherein phase distortions resulting from landline transmission produce unpleasant hue alterations.

In more detail, and continuing the nomenclature adopted for the basic NTSC system, the operation of SECAM is as follows:

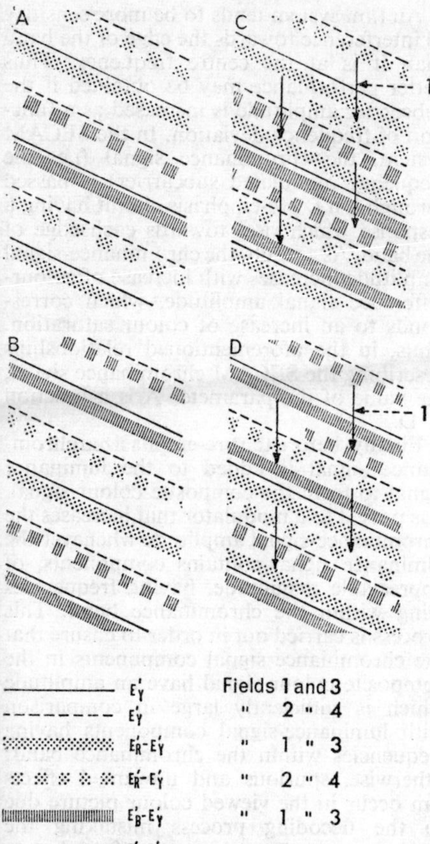

FOUR FIELDS OF SECAM 5-LINE PICTURE. A. Transmitted fields 1 and 2. B. Transmitted fields 3 and 4. C. Displayed fields 1 and 2. D. Displayed fields 3 and 4. Arrows, 1, show downward displacement of colour-difference information.

SEQUENTIAL COLOUR-DIFFERENCE SIGNALS. The fact that each colour-difference signal is transmitted on alternate lines has three consequences. First, in order to recover E'_R, E'_G and E'_B correctly in the receiver, it is necessary that both colour-difference signals and the luminance signal are available at the receiver decoder throughout every line period.

This means that some form of store must be provided in the receiver in order to enable one colour-difference signal to be stored while the other is being received. The store consists of a delay device (e.g. a glass bar fitted with electro-acoustic transducers, as used in PAL) which has a delay of one line period.

Thus, when one colour-difference signal, say $(E'_R-E'_Y)$, occurs at the input to the receiver delay device, the output of the delay device consists of the other colour-difference signal, $(E'_B-E'_Y)$, which was transmitted during the previous line period. By means of simple switching arrangements, it is possible to obtain $(E'_R-E'_Y)$ and $(E'_B-E'_Y)$ simultaneously, although some of the information has been displaced downwards in the picture by the spacing of one line in a field.

Secondly, the vertical information carried by the colour-difference signals is less than that carried by the luminance signal. This occurs because samples of information, taken along a vertical strip of picture, are spaced in each field by one line pitch for the luminance signal and twice the line pitch for a colour-difference signal. When the E'_R, E'_G and E'_B signals are reconstituted at the receiver, this, together with the fact that some of the colour-difference information is shifted downwards by one line spacing, leads to loss of vertical colour resolution and flicker effects on near-horizontal edges.

Thirdly, sequential transmission permits the two colour-difference signals to have the same bandwidth; the use of E'_I and E'_Q signals having different bandwidths offers no advantage.

USE OF FREQUENCY MODULATION. Since frequency modulation of the subcarrier is used to convey colour-difference signals to the receiver, the hue of the reproduced colour is not determined by phase. However, when the colour-difference signals are zero (i.e. the grey scale) the subcarrier amplitude is not zero; its frequency is merely undeviated.

The deviation of the subcarrier frequency is determined by the instantaneous value of the colour-difference signal applied to the frequency modulator.

The SECAM chrominance signal may be represented by:

$$E'_C = A\cos(\omega_s + D.\Delta\omega_s)t$$

where A is the amplitude of the f.m. subcarrier.

D is the modulating function causing deviation of the subcarrier frequency and has two values:

i.e. $D_R = K_R.(E'_R-E'_Y)$ for those lines during which the $(E'_R-E'_Y)$ signal is transmitted

and $D_B = K_B.(E'_B-E'_Y)$ for those lines during which the $(E'_B-E'_Y)$ signal is transmitted.

Further, $\omega_s = 2\pi f_s$

and $\Delta\omega_s = 2\pi\Delta f_s$ and is the deviation of the subcarrier frequency resulting when D has unit amplitude.

As is normal practice in f.m. systems, the colour-difference signal is pre-emphasized before being fed to the frequency modulator.

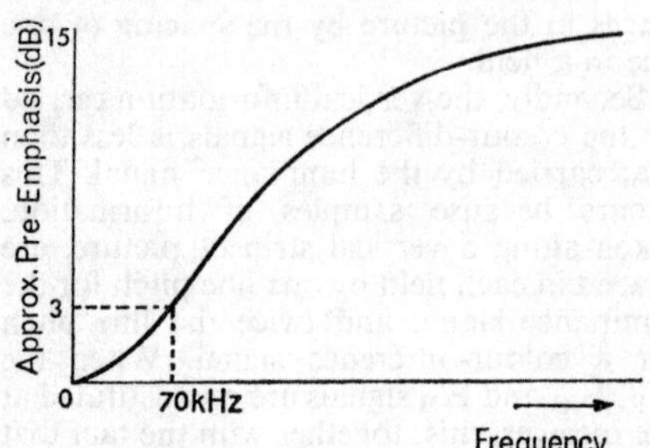

PRE-EMPHASIS OF COLOUR-DIFFERENCE SIGNALS. Typical pre-emphasis characteristic applied to signal before being fed to frequency modulator.

The subcarrier is present throughout the active portion of each line period and may be visible on the black-and-white receiver even in those areas of the colour picture which are black, grey or white. Because of this, the amplitude of the subcarrier is kept as low as possible to allow for the problem of reception in fringe areas. The visibility of the pattern produced on the screen of the black-and-white receiver is further reduced by reversing the phase of the frequency modulated subcarrier during every third successive line and every alternate field. This process exploits the frequency interleaving principle somewhat less effectively than in the case of the NTSC system.

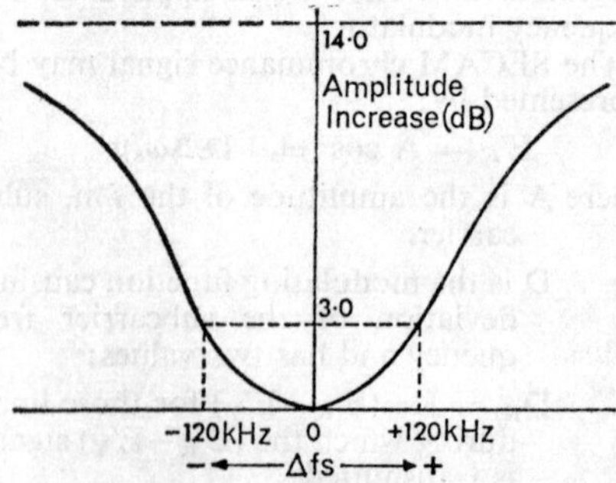

H.F. PRE-EMPHASIS OF F.M. SUBCARRIER. Subcarrier amplitude increased with increasing frequency deviation to reduce characteristic f.m. sensitivity to interference towards edge of band.

An f.m. system tends to be more sensitive to interference towards the edge of the band than it is at the centre frequency. Thus better performance may be obtained if the subcarrier amplitude is increased as a function of frequency deviation. In the SECAM system, the chrominance signal (i.e. the frequency modulated subcarrier) is passed through a h.f. pre-emphasis circuit having a response which rises towards each edge of the band. As a result, the chrominance-signal amplitude increases with increase of colour-difference signal amplitude, which corresponds to an increase of colour saturation. Thus, in the aforementioned relationships describing the SECAM chrominance signal, the value of the parameter A is a function of D.

Finally, before the pre-emphasized chrominance signal is added to the luminance signal to form the composite colour signal, it is passed to a modulator that increases the chrominance-signal amplitude whenever the luminance signal contains components, of appreciable magnitude, having frequencies lying within the chrominance band. This process is carried out in order to ensure that the chrominance signal components in the composite colour signal have an amplitude which is sufficiently large in comparison with luminance-signal components having frequencies within the chrominance band; otherwise, spurious and unwanted effects can occur in the viewed colour picture due to the decoding process mistaking the luminance-signal components for chrominance.

CODING AND DECODING. In a typical SECAM coder, the colour-difference signals $(E'_R - E'_Y)$ and $(E'_B - E'_Y)$ from the matrix, together with colour synchronizing signals consisting of sawtooth colour-difference signals occurring during the field-blanking interval, are fed to an electronic switch driven by line-frequency pulses. This switch connects them alternately to the input of the low-pass filter (typical bandwidth 1·0 MHz). The output of the filter is then fed, via the pre-emphasis circuit, to the subcarrier frequency modulator. After modulation, the phase of the modulated subcarrier is then reversed during every third line and every alternate field. H.f. pre-emphasis is applied and the chrominance signal is then amplitude modulated by a control signal consisting of the output of a rectifier; the rectifier is fed from the luminance signal via a band-pass filter selecting frequency components in the region of the subcarrier. Finally, the

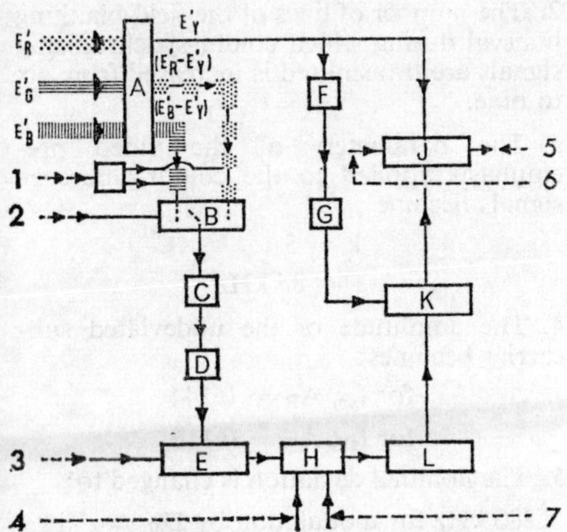

SECAM CODER. A. Matrix. B. Electronic switch. C. Low-pass filter. D. Video pre-emphasis. E. Frequency modulator. F. Band-pass filter. G. Rectifier. H. Phase reverser. J. Adder. K. Amplitude modulator. L. H.f. pre-emphasis. 1. Colour sync. 2. Line pulses. 3. Subcarrier. 4. Line pulses. 5. Composite signal. 6. Sync. 7. Field pulses.

processed chrominance signal is added to the luminance and sync signals to form the composite colour signal.

In a typical SECAM decoder, the composite signal input is fed to two signal paths. In the first, a band-attenuating filter attenuates the chrominance-signal components and provides a luminance signal for the matrix. At the input to the second path, a band-pass filter selects the chrominance components of the composite signal; the filter characteristic is the inverse of the h.f. pre-emphasis characteristic of the coder.

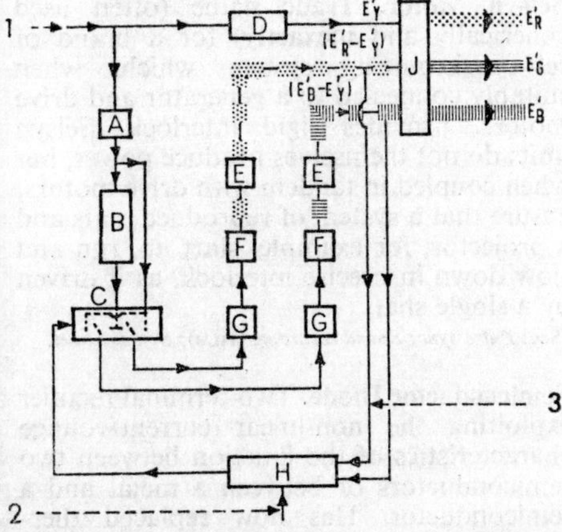

SECAM DECODER. A. Band-pass filter. B. One-line delay. C. Electronic switch. D. Band-attenuating filter. E. De-emphasis. F. Discriminator. G. Limiter. H. Switch-operating circuit. J. Matrix. 1. Composite signal. 2. Line pulses. 3. Colour sync.

The chrominance signal is now fed directly, and via the one-line delay device, to an electronic double-pole, double-throw switch operated by line pulses; any mis-operation of the switch due, say, to interference with the line pulses, is corrected at the end of each field by the colour-synchronizing pulses derived from the colour-difference signal detectors. Modulated subcarrier signals corresponding to $(E'_R - E'_Y)$ and $(E'_B - E'_Y)$ are thus made separately available and are fed to limiters and discriminators. The $(E'_R - E'_Y)$ colour-difference signals obtained are subjected to de-emphasis (corresponding to the video pre-emphasis applied to the colour-difference signals in the coder) and are fed to the matrix where they are combined with E'_Y to form E'_R, E'_G and E'_B.

PRACTICAL APPLICATION

As used in France with the 625-line 50 field system, the SECAM luminance signal can be expressed as

$$E'_Y = 0{\cdot}3\, E'_R + 0{\cdot}59\, E'_G + 0{\cdot}11\, E'_B$$

$$\text{assuming } \gamma = 2{\cdot}2.$$

The SECAM composite signal is expressed as

$$E'_M = E'_Y + A \cos(\omega_s + D.\Delta\omega_s)t$$

where $D_R = -1{\cdot}9\,(E'_R - E'_Y)$

and $D_B = 1{\cdot}5\,(E'_B - E'_Y)$.

Δf_s is the subcarrier frequency deviation produced by unit amplitude of pre-emphasized colour-difference signal and is equal to 230 kHz. (Unit amplitude of colour-difference signal is defined in terms of a scale in which, for white, E'_R, E'_G and E'_B have unit amplitude.)

The bandwidth of each colour-difference signal is approximately 1·4 MHz (after pre-emphasis). The subcarrier frequency (undeviated) is equal to 4·4375 MHz (284 times line frequency). The value of A for undeviated subcarrier (i.e. colour-difference signal zero) is 0·1. The subcarrier is interrupted at the end of each line-blanking interval and recommences before the start of each active line period.

The pre-emphasis characteristics are defined by

$$g_{v.f.} = \frac{1 + j\, f/f_1}{1 + j\, f/kf_1} \qquad k = 5{\cdot}6 \qquad f_1 = 70 \text{ kHz}$$

$$g_{r.f.} = \frac{1+j\,16F}{1+j\,1{\cdot}26F} \qquad F = \frac{f}{f_0} - \frac{f_0}{f} \qquad f_0 = f_s$$

The amplitude modulation of the chrominance signal which is related to the amplitude of luminance signal components $(E'_Y)_{CH}$ within the chrominance band is defined by

Gain 0 dB for $(E'_Y)_{CH} \leq 0{\cdot}2$

6 dB for $(E'_Y)_{CH} = 0{\cdot}4$

The colour-synchronizing signal consists of subcarrier signals corresponding to trapezoidal colour-difference signals transmitted during six lines of each field-blanking interval.

MODIFICATIONS. The specification outlined above may be modified in order to permit the optimization of certain parameters.

These possible modifications include:

1. The use of different undeviated subcarrier frequencies during the lines in which $(E'_R - E'_Y)$ and $(E'_B - E'_Y)$ are transmitted. During the lines carrying $(E'_R - E'_Y)$, $f_{sr} = 4{\cdot}40625$ MHz, while during the lines carrying $(E'_B - E'_Y)$, $f_{sb} = 4{\cdot}25000$ MHz.

2. The number of lines of the field-blanking interval during which colour-synchronizing signals are transmitted is increased from six to nine.

3. The parameters of the video pre-emphasis applied to the colour-difference signals become:

$$k = 3$$
$$f = 85 \text{ kHz}$$

4. The amplitude of the undeviated subcarrier becomes:

for f_{sr}, $A_R = 0{\cdot}144$

for f_{sb}, $A_B = 0{\cdot}116$

5. The nominal deviation is changed to:

280 kHz for modulation by $D_R = \pm 1$

230 kHz for modulation by $D_B = \pm 1$

6. The centre frequency of the r.f. pre-emphasis characteristic is 4·286 MHz.

A.V.L.

See: *Colour principles* (TV); *Colour systems; NTSC colour system; PAL colour system; Sequential system.*

Literature: Reports of the EBU Ad-Hoc Group on Colour Television. Oct. 1963 and Feb. 1965. Amendments Feb. 1966; CCIR Doc. XI/164 (France and USSR) (Oslo), 1966.

□**Secondary Emission.** The property of electrode materials to emit electrons when they are hit by other electrons. The striking electrons are called primary electrons and the emitted electrons (usually more numerous) are called secondary electrons. The ratio of secondary to primary electrons is called the secondary emission coefficient, and is a function of the material and the operating conditions.

See: *Camera* (TV); *Photomultiplier; Telecine.*

⊕**Second Generation Dupe.** Duplicate negative which has been derived from a preceding duplicate rather than from an original negative film. Although such second generation dupes are sometimes necessary to provide copies of special effects and trick work, the loss of image quality, especially in colour materials, as a result of the additional printing steps, leads to inferior results in the final presentation and their use should be avoided wherever possible.

See: *Laboratory organization; Printers.*

□**Selenicon.** Specialized camera pick-up tube of vidicon type using selenium as the active material. Used as luminance tube in some colour telecines.

see: *Telecine (colour); Vidicon.*

⊕**Sellers, Coleman, 1827–1903.** American engineer who designed and patented in 1861 the Kinematoscope, using stereoscopic pairs of posed photographs of machinery viewed in a modified Zoetrope.

⊕**Selsyn Motor.** Trade name (often used generically and inexactly) for a brand of self-synchronizing system which, when suitably connected to a generator and drive motors, provides rigid interlock. Selsyn units do not themselves produce power, but when coupled in tandem with drive motors, ensure that a system of reproduce units and a projector, for example, start up, run and slow down in precise interlock, as if driven by a single shaft.

See: *Pulse sync; Sound recording* (FILM); *Synchrostart.*

⊕□**Semiconductor Diode.** Two-terminal rectifier exploiting the non-linear current/voltage characteristics of the junction between two semiconductors or between a metal and a semiconductor. Has now replaced thermionic diodes in video and audio circuits.

⊕□**Semireflector.** A semireflecting surface is one which partly reflects and partly transmits the incident light, the proportions either being equal or having some predetermined ratio. The simple type of semireflector is spectrally neutral, so that in the case of a 50/50 division, the reflected and transmitted rays each have half the energy content of the incoming ray.

See: *Camera (colour).*

Sensitizer. Chemical added during manufacture to increase the sensitivity to light of photographic emulsions which are naturally sensitive only to the shorter wavelengths corresponding to the blue and ultra-violet portion of the spectrum. Sensitizers, which are usually dyes, allow this natural sensitivity to be extended to the green, red and even the infra-red region.

See: *Film manufacture; Image formation.*

Sensitometer. Instrument used in photographic processing to provide a standard modulated exposure on to a sensitive photographic material. After development, the series of densities resulting may be measured and used to assess the photographic characteristics of the material or development process.

In motion picture work, the short length of film so exposed is known as a test strip or sensitometer strip and the series of densities are referred to as a step wedge. It is usual to provide the exposure series in logarithmic increments, usually of 0·15 of log intensity, so that the resultant densities are also of uniform increment as far as possible. For cinematograph film materials, it is normal practice to make use of a standard short exposure time and vary the intensity of the light at each step by a series of accurately determined neutral density filters.

See: *Sensitometry and densitometry.*

SENSITOMETRY AND DENSITOMETRY

Sensitometry is the scientific measurement of the effect of light on photographic materials. Since the result of the exposure to light is not apparent until the material has been processed, it depends not only on the type and amount of exposure but also on the processing conditions. Sensitometry is therefore primarily concerned with the effect produced by variations in the conditions of exposure and processing, and also with the methods of interpreting the results quantitatively so that they can be applied to practical situations.

Densitometry is concerned with measuring the light transmission of a processed photographic image. The result could be expressed as a percentage or fraction of light transmitted by a particular sample under defined conditions but, in practice, it is found more convenient to work in terms of density, which is defined as the logarithm of the reciprocal of the fraction of light transmitted. For example, a piece of clear film might transmit 92 per cent of any light falling on it and would have a density of log 1/0·92 = 0·04, while a piece of exposed film might transmit only 1 per cent of the incident light and would have a density of log 1/0·01 = 2·0. Density may be measured directly on a densitometer.

SENSITOMETRY

APPLICATIONS. One important application is in the control of film processing conditions. For this purpose, sensitometric test strips are exposed and processed at regular intervals and, by a study of the results of these, any changes in the process can be detected before they become large enough to be noticeable on practical tests. Sensitometric equipment and methods may also be used to measure the effect of deliberate changes in exposing or processing conditions or to study the differences between types or samples of film.

A sensitometrically exposed test strip usually consists of a number of equal areas, or steps, successive steps receiving increasing amounts of exposure. The whole strip should cover the range of useful exposures of the sample, from an exposure too small to have any effect on the film to one sufficient to produce a density greater than would be encountered in practice. This is conveniently achieved by exposing the film sample in contact with a sensitometric exposing wedge, which has steps of known density value.

After processing, the density corresponding to each exposure level is measured on a densitometer, and the complete result for these particular exposing and processing conditions may be shown as a graph on which each density is plotted against the logarithm of the exposure producing it.

These points may then be joined by a smooth curve – the H. & D. or characteristic curve for these conditions. Similar curves for slightly different conditions, for example with different development times,

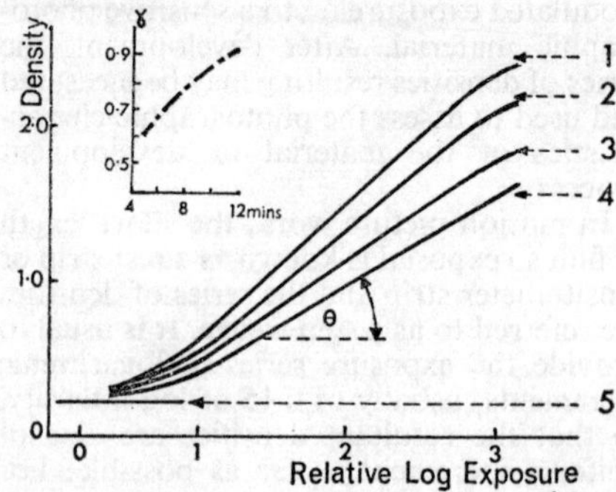

DEVELOPMENT SERIES ON NEGATIVE MATERIAL. Material developed to, 1, 12 mins., 2, 9 mins., 3, 6½ mins., 4, 5 mins., showing resultant characteristic curves. Gamma (γ) = tan θ. 5. Base density. (*Inset*) Gamma plotted against development time.

may be drawn on the same sheet and the differences between them can then be readily interpreted in terms of practical effects, such as speed, contrast or fog.

RECIPROCITY LAW FAILURE. The amount of exposure received by a film depends on the product of the intensity and the exposure time, so that the two are in a reciprocal relationship.

For most practical purposes, equal effective exposures always produce the same final density whether they result from a relatively high intensity exposure for a short time or from a low intensity exposure for a proportionately longer time. However, with exposure times shorter than 1/1,000 sec. or longer than 1/10 sec., some loss in speed may become apparent. This effect is known as reciprocity law failure and it becomes particularly noticeable in the case of colour films where, as well as an overall speed loss,

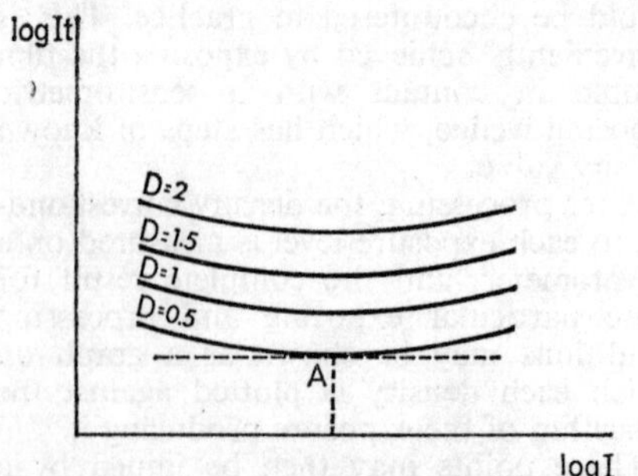

RECIPROCITY FAILURE. Log exposure (intensity $I\times$time t) plotted against log I. If density, D, depended solely on product It, curves would be straight lines, as shown dotted. In practice, curves show optimum intensity, A.

differences between the responses of the three layers may cause a change in colour balance.

EXPOSING CONDITIONS. The instrument used to give a carefully controlled exposure to the film sample is called a sensitometer. It consists essentially of a light source, a shutter and an exposing wedge, with provision for inserting filters.

In order to obtain results relevant to the practical use of the film, it is important to keep the sensitometric exposing conditions similar to those used in practice, particularly as regards the colour and time of exposure. For example, even with black-and-white film, the colour temperature of the light used for exposure has a significant effect on the results obtained. A tungsten filament lamp is normally used as the light source in sensitometers, and this is suitable for exposing print materials directly, although a heat-absorbing filter may be included to match that used in a printer. In the case of colour print materials intended for use with masked negative films, it is also necessary to include a filter with a transmission equivalent to that of the unexposed base of the negative material.

At colour temperatures above 3,000°K, the life and stability of a tungsten lamp decrease steadily, so that to expose correctly materials intended for studio or daylight illumination a colour temperature converting filter is used.

The British Standard for daylight exposure of films defines the constitution of a pair of liquid filters which together convert the light of a tungsten lamp to a close equivalent of daylight.

However, liquid filters are not generally convenient for routine use and blue glass or dyed gelatin filters are more often used in practice.

All filters tend to change their transmission to some extent under the influence of light and heat, and sensitometric exposures are therefore made after allowing any filters used to reach a steady temperature. These effects are mainly reversible but prolonged exposure to strong light can produce permanent fading or, in some cases, darkening of the filter.

CALIBRATED LAMPS. Sensitometer lamps are specially selected projector lamps which have been calibrated in terms of candle power and colour temperature by comparison with the physical standards maintained at a national standard organization, in

Britain the National Physical Laboratory. This is an indirect process involving several stages of intermediate substandards, and measurements are made under laboratory conditions on an optical bench.

A lamp may be calibrated and used at several different colour temperatures, but it is impractical to use one at colour temperatures above 3,000°K because of the decreased life and stability which result. All lamps tend to lose intensity gradually while in use, particularly when they are new, and for this reason new lamps are aged at a lower voltage before calibration. If long-term drifts are to be avoided, it is essential to check or replace the lamps at regular intervals.

However, the lower the colour temperature at which these lamps are used, the less rapid is the loss in power due to ageing. Since the intensity of the lamp is also reduced, it is not usual to calibrate lamps below 2,360°K, at which level a life of several hundred hours between calibrations may be expected.

The intensity of a lamp depends very much on its supply voltage, and some form of stabilized supply is essential. A change of only one volt in the supply of a 110 volt sensitometer lamp changes its intensity by over 4 per cent. For accurate work a d.c. supply is preferred, partly because a potentiometer method can then be used to control the voltage, although a reliable voltmeter with sufficiently open scale can be quite satisfactory with a.c. or d.c. supply. The voltage drop in the leads carrying current to the lamp cannot be ignored, so that the voltage of the lamp should be measured by using a second pair of leads taken to the lamp socket.

NEUTRAL DENSITY FILTERS. Often the lamp position can be adjusted, so that in spite of lamp changes, or the effect of ageing, the same intensity level can be maintained. However, when a considerable reduction in intensity is necessary, neutral density filters are used. Glass or gelatin filters are available, but these are never perfectly neutral and for precise work on colour materials a filter consisting of alloy sputtered on to glass is preferable.

EXPOSING WEDGE. The step tablet through which the sample is exposed is most commonly made photographically. For convenience, each step should differ from the next by the same density increment, but inaccuracies can be allowed for when plotting the results.

Each step must be appreciably wider than the reading head on the densitometer to be used, so that the edges of the step can be avoided, as well as any blemishes that may be present.

Step wedges can be obtained with a wide range of step widths and density increments. The density increment should be chosen to suit the material being tested. For high contrast materials, a smaller increment is preferable, but for the same number of steps, a wedge of greater density increment covers a wider exposure scale and this may be necessary for negative materials. An increment of 0·15 giving an exposure scale of 3·0 over 21 steps is a convenient compromise for most purposes.

On many sensitometers, different exposing wedges can readily be interchanged. The exposure scale of a wedge can effectively be extended by making two exposures, one through a neutral density filter, so that two overlapping scales are obtained.

Cast carbon step tablets are also available which are more neutral and less diffusing than photographic step tablets.

SENSITOMETER SHUTTERS. Time scale sensitometers used to be common, in which the intensity of exposure was kept constant and the different step areas were given different exposure times by means of a specially shaped sector wheel or drum. This avoided the difficulty of finding a neutral wedge but, due to reciprocity law failure, the results obtained did not quite agree with those from intensity scale exposures, which corresponded better with almost all practical

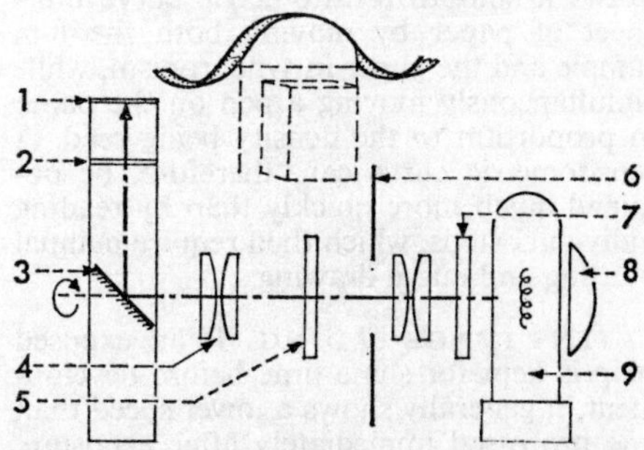

HERRNFELD INTENSITY-SCALE SENSITOMETER. Optical path. Lamp, 9, with mirror, 8, and heat-absorbing filter, 7, focuses filament image on objective lens, 4, through swinging shutter, 6, to give exposure period, and spectrum filters, 5. Objective lens focuses image on film plane, 1, backed by density wedge, through rotating mirror, 3, so that light beam evenly scans film plane, located as an arc round optical axis (see insert). 2. Adjustable slit (aperture plate).

conditions. Consequently time scale sensitometers have now largely been replaced by intensity scale instruments.

The shutter on many sensitometers is controlled by a synchronous motor so that the exposure time is dependent on the frequency of the a.c. mains supply. This is closely controlled and is perfectly reliable in practice. Simpler sensitometers may have a falling plate or pendulum shutter in which an aperture is uncovered briefly by the fall or swing of the shutter plate. A good quality camera shutter may be satisfactory in a sensitometer, particularly if the speed setting is kept unchanged. The actual exposure time and its consistency may be checked with an electronic timer.

CONTINUOUS WEDGE SENSITOMETRY. A continuous exposing wedge, in which the density increases smoothly from one end to the other, may be used in place of the step wedge. This type of exposure requires a more elaborate form of densitometer for interpretation, called a recording or curve-

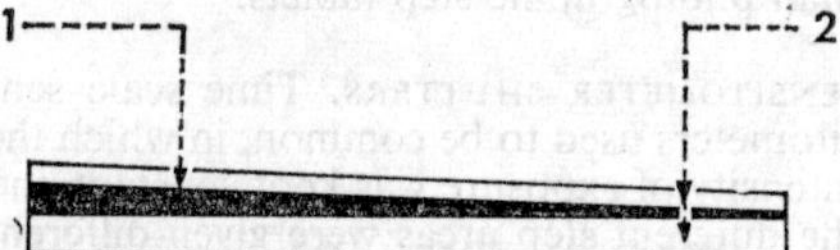

GOLDBERG (CONTINUOUS) WEDGE. Gelatin solution, 1, containing neutral colouring material, is coated between two glass plates, 2, set at an angle. Density increases with thickness of gelatin, thus progressively reducing light intensity of exposure towards thicker end of wedge. Continuous and step wedges are used in intensity-scale sensitometers.

tracing densitometer. This automatically draws a smooth sensitometric curve on a sheet of paper by moving both the film sample and the paper in synchronism, while simultaneously moving a pen on the paper in proportion to the density being read. A sensitometric curve can, therefore, be obtained much more quickly than by reading individual steps, which then require manual plotting and curve drawing.

LATENT IMAGE FADING. If an exposed strip is kept for some time before development, it generally shows a lower speed than one processed immediately after exposure. The curve shape may also be different. This effect, known as latent image fading, is not usually large enough to affect practical results but cannot be ignored in sensitometric work.

COMMERCIAL SENSITOMETERS. One well-known type of sensitometer has a shutter

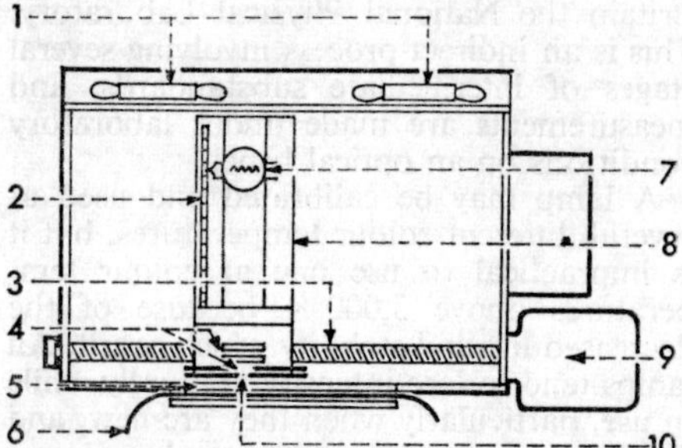

KODAK TYPE 6 HIGH INTENSITY SENSITOMETER. Lamp, 7, on carriage, 8, is moved at constant speed by lead screw, 3, driven by synchronous motor, 9, past step wedge, 5, against which is film sample, 6. Intensity adjustment is by moving lamp, 7, down centimetre scale, 2. Exposure is through filters, 4, and slit, 10. 1. Cooling fans.

consisting of two concentric drums which are rotated continuously by a synchronous motor. The inner drum has a series of longitudinal slots of different widths corresponding to different exposure times. The outer drum can be adjusted to leave only the required slot open. The film is held emulsion down on the exposing wedge by a pressure plate and, when the exposure is initiated, a simple flap shutter on the lamp house opens for one revolution of the drums, allowing light from the 500-watt standard lamp to pass to a large mirror which reflects it vertically upwards to the film. Exposure times between 1/100 and 1 sec. are available.

The intensity of illumination is scarcely sufficient to expose some of the slower print materials even with exposure times of 1 sec., which would in any case be suspect because of reciprocity law failure effects. If the lamp is brought closer to the film, it leads to unevenness along the length of the wedge. This difficulty is overcome in a different model by having the lamp and shutter slot move together along the length of the wedge. In this way, the 500-watt lamp can be as little as 12 cm. from the film. A three-phase synchronous motor turns a lead screw on which the lamp unit is mounted, and the direction of travel is reversed at each end by means of relays actuated by microswitches. Exposure times of 1/10, 1/25 and 1/50 sec. are interchanged by rotating a wheel on the lamp unit, while a similar wheel can contain four filters which are equally easily exchanged.

A completely different type of sensitometer is available in which the light source is a xenon flash tube, similar to that commonly used for still photography. The high voltage used is stabilized so that the amount of light emitted at each flash is very consistent. In this case, no shutter is required, and another advantage is that the colour of the light is similar to that of daylight so that no filter is required for daylight-balanced materials. Switches select exposure times of 1/100, 1/1,000 or 1/10,000 sec. Another model is available with a second flash tube to provide additional exposure times of 1/100,000 and 1/500,000. This model finds particular application in the study of reciprocity failure characteristics.

PROCESSING. When placed on the film processing machine, the sensitometric test usually gets the same treatment as all other work, being spliced between other samples or leader. Notice should be taken of the direction of passage through the machine as this can affect the results by streamer effect from heavily exposed areas which tend to exhaust the developer. The end of the strip having least exposure should, therefore, be to the front.

When it is necessary to carry out sensitometric tests on solutions independently, the conditions which must be carefully controlled are the time, temperature and agitation, particularly in the developer. Time and temperature normally present no great difficulty, although good thermostatic control is usually necessary.

Various methods of providing adequate and consistent agitation have been recommended. A camel hair brush may be used gently and uniformly on the submerged film surface throughout development. More elaborate methods involve the use of nitrogen gas bubbling through the solution or a thin paddle moving to and fro at a uniform rate close to the emulsion surface. The method used in the British Standard for Photographic Speed Measurement requires a vacuum flask which is rotated in a particular manner.

The conditions for fixing, washing and drying should also be controlled, but they are much less critical than those for development.

DENSITOMETRY

The simplest form of interpretation of processed strips is their examination on an illuminated viewer. For example, seen beside a reference strip, each step of the sample may be perceptibly darker or lighter than the corresponding step of the standard, indicating an overall speed difference. If each step of the sample were similar to the next higher step of the standard, the speed difference between them would be equal to the wedge increment. In the case of colour materials, slight differences in colour balance are quite easily detected if the wedges are nearly neutral. It may be useful to prepare a set of strips exposed through different colour correcting filters, so that by comparison against these, the extent of any colour bias may be judged. However, for all serious work, a densitometer is essential.

TYPES OF DENSITOMETER. Early densitometers were usually visual, and were operated by matching the brightness of two adjacent areas seen in an eyepiece. These were slow and tiring to use and have been almost entirely replaced by photoelectric densitometers in which a photocell is used to measure the light transmitted by the sample. This may be done either directly,

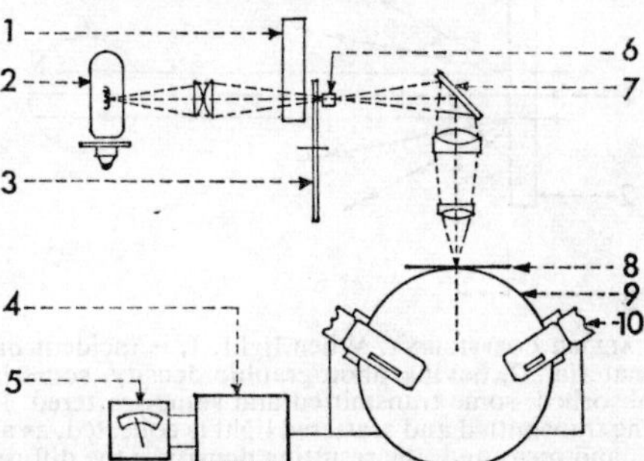

WESTREX INTEGRATING-SPHERE DENSITOMETER. Optical path. Light from lamp, 2, is focused through filter disc, 1, on to glass diffusing block, 6, which eliminates coil pattern of filament. Interrupter wheel, 3, chops light to frequency of 375 or 450 Hz. Exit face of diffuser is focused through mirror, 7, and objective lens system on to film sample, 8. Light passing through film undergoes multiple reflections at surface of integrating sphere, 9, and is collected by two photocells, 10, one red- and one blue-sensitive. Combined output is amplified, 4, and diffuse density directly read on meter, 5.

using the photocell with a suitable electrical circuit so that a meter reading in terms of density units is produced; or by a null method. In this case, the light passing through the sample also passes through a movable measuring wedge (the density of which increases along its length) and is then compared by the photocell against a reference beam from the same light source. When the intensity of these two beams is the same, the position of the measuring wedge represents the density of the sample. This method has the advantage that neither the light source nor the photocell need be particularly stable.

SPECULAR AND DIFFUSE DENSITY. When light falls on a glass or gelatin filter, some of the light is absorbed and some transmitted. Hardly any is scattered in a direction different from that of the incident light. However, the silver image of a photographic material scatters, or diffuses, a significant proportion of incident light, so that its density depends on the optics in which it is being used. If it is being printed by contact, virtually all the scattered light still reaches the print material and its density is relatively low. Most densitometers are designed to measure this value of density, which is called diffuse density. Precise conditions for this measurement are defined in British Standard 1384.

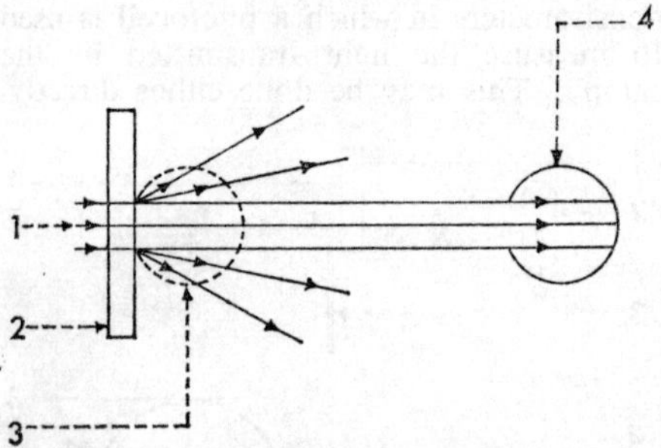

CALLIER COEFFICIENT. When light, 1, is incident on material, 2, having photographic density, some is absorbed, some transmitted and some scattered. If the transmitted and scattered light is collected, as at 3, and measured, the resulting density is the diffuse density, D_d. If only the transmitted light is collected, as at 4, and measured, the density is specular, D_s. Since less light is collected at 4 than at 3, specular density is higher than diffuse. D_s/D_d is the Callier quotient or coefficient.

If the film is being projected, much of the scattered light does not reach the lens, so that the apparent density, called the specular density, is higher. The exact value depends on the optics of the system and in particular on the condenser used and the aperture of the projection lens.

The ratio between the specular and diffuse densities of a sample is known as the Callier coefficient, or Q factor. It is fairly consistent for any type of emulsion.

The term printing density is used to define the value of density obtained when printing a negative material under practical conditions.

SPECTRAL SENSITIVITY. For reading the density of black-and-white film, the effective sensitivity of the photocell should ideally match that of the human eye for print film, while it should be mainly blue-sensitive if it is to read printing densities on negative materials. Since the silver deposit is more or less neutral, this does not make much difference in practice unless the emulsion is coated on a coloured base.

For reading colour film densities, it is of course essential to use a set of tricolour (red, green and blue) filters so that the effective densities of each dye layer may be measured independently. In fact, because each dye has some unwanted absorption, it is not possible to obtain completely independent density readings in this way. For example, the density of a sample read through a green filter is mainly due to the amount of magenta dye present, but also depends to a lesser extent on the amounts of yellow and cyan dyes present. The densities read in this way are called integral densities, and they are adequate for normal purposes. It is possible by more elaborate methods, or by calculation from the integral densities, to determine truly independent values for the individual dye layers, and these are called the analytical densities.

DIFFERENCES BETWEEN DENSITOMETERS. When two densitometers give different readings on the same sample, this is often caused either by a difference in optics, i.e., one is more specular than the other, or by a difference in colour sensitivity due to the colour filters or photocells used. In particular, the infra-red sensitivity of some photocells can cause difficulty with colour samples and this should be avoided by using an efficient heat-absorbing or reflecting filter.

COMMERCIAL DENSITOMETERS. One well-known type uses a barrier-layer photocell, the output of which feeds a sensitive ammeter calibrated in density units. Neutral range filters are withdrawn from the light beam to read densities up to 3·0. A colour model is also available which operates on a similar principle but uses a galvanometer to give the necessary increase in sensitivity.

Another type uses a photo-emissive vacuum cell in conjunction with a d.c. valve amplifier to read black-and-white or colour densities up to 3·0. The range switch operates electrically.

A third type is available in several different models which can read black-and-white or colour densities on a meter covering the range 0–4 in density. The use of a sensitive photo-multiplier enables narrow-cut tricolour filters to be used.

PLOTTING AND INTERPRETATION. For some control purposes it is sufficient to record on a control chart the densities of several steps of the wedge. Any gradual change of level will soon become evident and remedial action can be taken. If a sudden change occurs, a more complete study of the results may be required, and the first step is to plot a graph of the density of each step of the wedge against the logarithm of the exposure it has received. Although the absolute exposure for each step can be calculated, it is often sufficient to know the relative values and these are obtained directly from the densities of the corresponding steps of the exposing wedge.

The smooth curve drawn to pass through all the plotted points is called the characteristic curve for the film. From the shape and position of this curve, values may be derived which represent practical aspects of the film's performance under these conditions. The most important of these are speed, contrast and fog. Fog, or minimum density, is simply the density of an area which has received no exposure at all.

Speed and contrast may each be measured in a number of ways and it is important to select a method which is relevant to the practical conditions of use. One suitable method for negative materials is to find the point on the curve with a density 0·1 units greater than the fog density. The relative exposure at this point is a measure of the speed of the material. Contrast may be measured as the slope of the line joining this speed point to another point which has received 1·3 log units more exposure.

A common contrast measurement is gamma (γ), which is the slope of the straight line portion of the curve. This is not always easy to measure accurately since not all materials have a clearly defined straight line portion to the curve. Similar types of measurement are made for print and reversal materials, but they are based on the most important parts of the curves.

It is often useful to make a series of tests with increasing development times. When the results for each test have been calculated, curves can be plotted of speed, contrast and fog against time of developmeent, and from these the correct time for a required contrast, or for the best combination of values can be determined.

Results for colour film samples are drawn as three curves, one for each densitometer reading filter. The colour balance of the film is shown by the differences in density between the curves.

CONTROL FILM. The film used as a control for processing conditions should be particularly selected for uniformity. A sufficiently large quantity is then stored in sealed containers at a low temperature (deep freeze if possible) to minimize any changes due to keeping. Smaller quantities, say sufficient for a week's work, are then withdrawn as required. The containers must be allowed several hours to reach room temperature before being opened. After exposure, the film should be processed without delay to avoid the effect of latent image fading. When it is necessary to change to a new batch of control film, strips of the new and old films should be run together for a short while to maintain continuity. M.C.F.

See: *Colour printers (additive); Exposure control; Laboratory organization; Printers; Sensitometry of colour films; Tone reproduction.*

Books: *Sensitometry*, by L. Lobel and M. Dubois (London), 1967; *Theory of the Photographic Process*, by C. E. K. Mees and T. H. James, 3rd Ed. (New York), 1966; *Fundamentals of Photographic Theory*, by T. H. James and F. C. Higgins (New York), 1960; *General Sensitometry*, by Y. N. Gorokhovskii and T. M. Levenberg (London), 1965.

SENSITOMETRY OF COLOUR FILMS

The basic purpose of the sensitometry of colour films exactly parallels that of black-and-white: to assess the photographic characteristics of materials and their processing and to apply such data to control operations in which they are used, in the camera, in a printing system or through a developing process. Similarly, the fundamental steps are comparable: the controlled exposure of a photographic material on a sensitometer to produce a known series of latent image steps and subsequent examination of the processed film by means of a densitometer to establish the relation between exposure and image density. However, the nature of colour film images raises special problems and the theory of photographic colorimetry is complex, although practical operations can be reduced to a routine by the use of suitable instruments.

COLOUR SENSITOMETERS

The character of the sensitometric exposure must be as similar as possible to the actual use of the film. This means that short exposure times must be used to avoid problems of reciprocity law failure and a time of 1/50 sec., comparable to the exposure time of a single frame in a motion picture camera, is usual. The spectral quality of the exposing light should be that for which the film was designed and for motion picture colour negative and positive stocks this normally means a tungsten lamp of 3,200°K colour temperature. Some reversal films are intended for daylight exposure so that for these the sensitometer lamp requires correction by a standard colour filter to convert it to the spectral equivalent of daylight.

Since a fixed exposure time is used, the intensity variations on test strip must be obtained by use of an accurate step wedge consisting of a number of density steps of known increment; steps of 0·15 density, corresponding to a factor of $2\sqrt{}$ exposure, are usual. In some instruments where the resultant image is to be measured on an automatic recording densitometer, a wedge with a continuous uniform gradient of density rather than steps is preferred. It is a fundamental requirement that the wedge used shall be strictly neutral throughout, that is, it should produce the same change of intensity at each step for all wavelengths to which the film is exposed.

In some cases tests may require each of the red, green and blue sensitive layers of a colour film to be exposed separately, and for such work the sensitometer must be used with colour filters having narrow spectral band transmissions matching the sensitivities of the film.

COLOUR DENSITOMETERS

When an exposed sensitometer test strip has been processed, the colour densities of each step must be measured so that the characteristic curve for each of the three colours may be constructed. Photo-electric densitometers are universal for colour work, the most usual form being that in which the reduction of the intensity of a beam of light passing through the test film to fall on a photo-cell is amplified to give a direct reading on a meter calibrated in density units. Colour filters are inserted in the path of the light before it reaches the film so that effective densities in the red, green and blue regions of the spectrum can be read.

The light source is normally a tungsten lamp operating at a colour temperature of about 3,000°K for maximum life and, since the photo-cells normally used have appreciable sensitivity in the near infra-red region, it is essential to fit heat-absorbing and infra-red rejecting filters permanently in position. In addition, the sensitivity of the cell differs substantially with different wavelengths and it is most desirable that the gain factor of the amplifier should be automatically adjusted as each different filter is inserted so that the instrument gives equal accuracy for each colour.

The selection of the three colour filters used for densitometry is important and it is now usual to match these to the purpose for which the film is to be used. Thus, in measuring a colour negative, it is necessary to assess its effective densities when used to make prints on colour positive stock. The filters are therefore selected in combination with the cell characteristics to match the spectral peak sensitivities of the three layers of the colour positive film. Such filters are termed printing density filters and one widely used set is refered to as Status M filters.

When measuring colour positive print materials, on the other hand, the effective density required is that visible in the print; the filters and photocell must therefore have a response corresponding to that of the mechanism of vision, as specified for a standard observer. In this case the instrument uses colorimetric filters, a recognized set being known as Status A filters.

CHARACTERISTIC CURVES

By plotting the density of the film for each colour against the logarithm of the exposure for each step, the photographic characteristic curves for the material can be prepared, and from these important information can be obtained. For example, the gamma values, or slopes of characteristic curves,

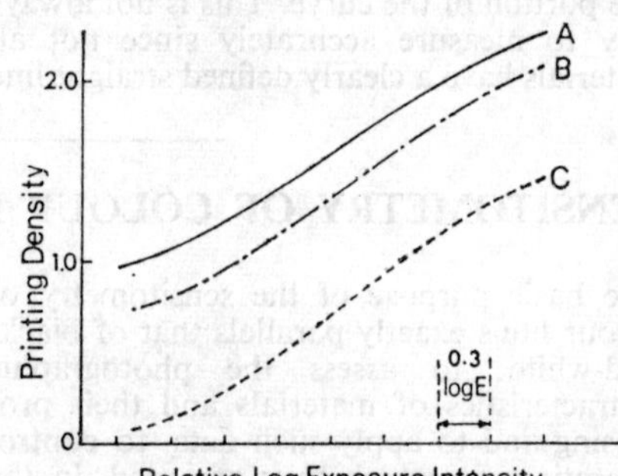

CURVES OF MASKED INTEGRAL TRIPACK NEGATIVE. Characteristic curves of typical material. Densities to blue (A), green (B) and red (C) light are plotted against logarithm of exposure. High minimum values to blue and green result from masking system which gives film an orange-yellow appearance in the unexposed areas.

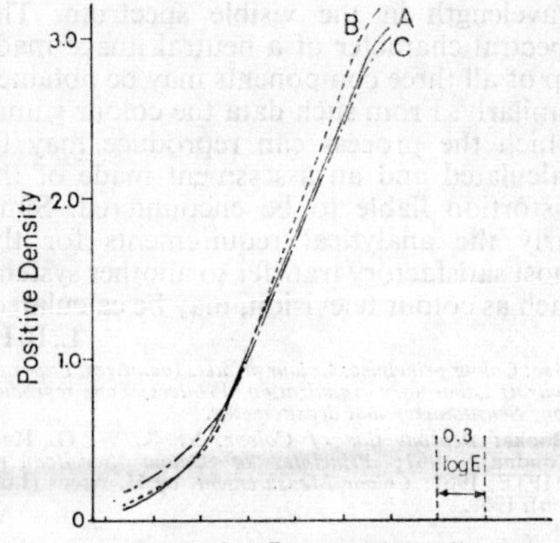

CURVES OF COLOUR POSITIVE FILM. Exposed so as to produce nominally neutral image in mid-tones. A. Yellow image read by blue light. For ideal reproduction the curves for each colour should be parallel. In this example, the higher contrast of the cyan image, B, read by red light, means that the shadow areas tend to be increasingly blue-green. Flatter toe of magenta image, C, read by green light, indicates that highlights and light tones will appear too pink.

may be compared for each colour record. Correct colour reproduction requires substantially equal gamma for the three colours for both the negative and positive stocks, since otherwise objectionable differences of

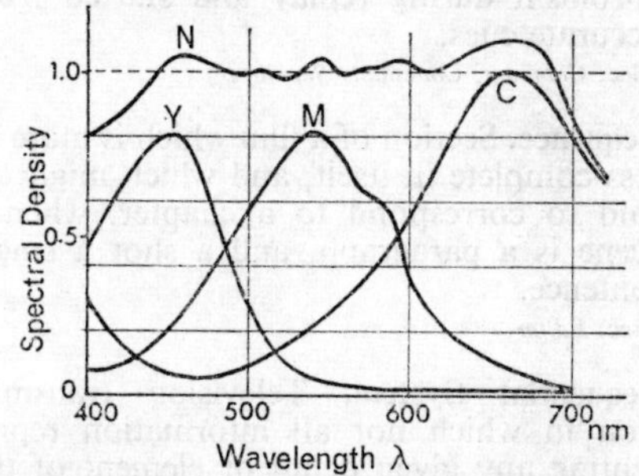

COLOUR POSITIVE FILM: SPECTRAL DENSITY CURVES. Curves for the three components, the yellow, cyan and magenta images. Together, these three images add to give a neutral grey with spectral curve, N.

colour rendering from highlight to shadow will be observed. The actual gamma values, as well as their relation in the three images, must also be maintained to the correct figure by processing control so that the resultant picture does not appear of too high or too low a contrast for its purpose. Detailed examination of the curves for two different film stocks allows their relative photographic speeds to be compared, so that their exposure requirements can be estimated both in intensity and in colour correction. When sensitometer test strips on a standard material are used to check the characteristics of a developing process, the exact curve shape of the three colour records is of great importance in indicating the chemical changes which may occur and which require correction.

Colour densitometry is also of value in determining the printing requirements of a particular negative image and is essential in making derivatives by way of colour intermediates. Densitometry of colour positive prints can also show whether their tonal range is satisfactory for this purpose, for example in colour telecine use, and if a standard object such as a neutral gray scale is included, the accuracy of colour balance in the reproduction can be determined.

AUTOMATIC PLOTTING

When colour densitometry is used for the routine control of photographic processes, the labour of reading and plotting large numbers of test strips is considerable and automatic recording densitometers offer great advantage. In these the sensitometer exposure is in the form of a continuously varying density wedge so that a smooth curve can be plotted by the instrument. In operation, the test strip is accurately located on a moving carriage which is traversed across the densitometer light beam while at the same time the pen of the recorder chart moves horizontally across the paper. The signals from the amplifier corresponding to the density values read are translated into vertical movements of the pen, so that as the test film is moved a curve is drawn showing the variation of density along the length of the strip. If the sensitometer exposure gives a suitable logarithmic rate of change of exposure along the strip, the curve plotted is the photographic characteristic curve.

In colour sensitometry the strip must be scanned three times with red, green and blue light and each density series plotted; this is normally done automatically and, at the same time, the colour of the recording pen is changed so that the three records can be easily distinguished.

SPECIAL REQUIREMENTS

While the comparatively straightforward methods of integral densitometry described above are normally sufficient for photographic process control, there are sometimes requirements for a more detailed analysis of colour images, as for example when the characteristics of a colour telecine system must be matched to a particular type of film. This entails measuring the spectral densities of a colour image, that is, the variation of density with the wavelength of light.

In modern colour films the three component images are the subtractive colours, yellow, cyan and magenta, but the wavelengths at which peak absorbtion occurs can differ significantly from one type of stock to another. Even more important, the unwanted absorption bands of each primary dye can vary greatly. The detailed specification of a colour process therefore demands the spectrophotometric curves for each dye showing its transmission density at each wavelength in the visible spectrum. The spectral character of a neutral image made up of all three components may be obtained similarly. From such data the colour gamut which the process can reproduce may be calculated and an assessment made of the distortion liable to be encountered. Similarly the analytical requirements for the most satisfactory transfer to another system, such as colour television, may be calculated.

L. B. H.

See: *Colour principles; Colour printers (additive); Exposure control; Laboratory organization; Printers; Tone reproduction; Sensitometry and densitometry.*

Books: *Reproduction of Colour*, by R. W. G. Hunt (London), 1967; *Principles of Colour Sensitometry*, SMPTE, 1967; *Colour Measurement*, by H. Arens (London), 1968.

□**Separate Mesh Vidicon.** Improved form of vidicon which contains an extra electrode in the form of a wire mesh. The result is to improve the resolution of the tube.

See: *Camera (TV); Vidicon.*

⊕**Separation Master.** Colour films are sometimes duplicated by preparing three separate strips to record the red, green and blue components of the original negative by printing three times on to panchromatic black-and-white stock through the corresponding colour filters. From these three separation masters, a colour intermediate negative or inter-dupe may be prepared by printing three times through the same colour filters on to a suitable integral tripack colour film stock.

See: *Laboratory organization; Printers; Separation negatives.*

⊕**Separation Negatives.** Negatives in which the component colours of the original have been separately recorded as black-and-white images, either on individual strips of film or as successive frames on a single strip. In two-colour processes, the separation images correspond to the blue-green and red-orange components, but in the more general three-colour systems they record the red, green and blue.

The original Technicolor process of colour cinematography used separation negatives simultaneously exposed in a special camera, and separation negatives are still sometimes made from original colour reversal films and from colour videotape for protection or duplicating purposes.

See: *Film storage; Separation master; Technicolor.*

⊕□**Sepmag.** Abbreviation for separate magnetic sound track, where the pictures are on one film while the corresponding sound is carried as a magnetic track or tracks on a second length of film, which can be of a different gauge. The two films must be run in synchronism during replay and started from an accurately located pair of cues.

See: *Commag; Commopt; Sepopt.*

⊕□**Sepopt.** An abbreviation for separate optical sound track, where the pictures are on one film while the corresponding sound is carried as an optical track on a second length of film, which can be of a different gauge. The two films must be run in synchronism during replay and started from accurate cues.

See: *Commag; Commopt; Sepmag.*

⊕**Sequence.** Section of a film which is more or less complete in itself, and which might be said to correspond to a chapter, when a scene is a paragraph, and a shot a single sentence.

See: *Editor.*

□**Sequential System.** Television transmission in which not all information representing any given detail or element of the picture image is transmitted at the same time. The term is particularly applicable to certain methods of colour television.

An underlying principle of colour television systems in general is that each element of the reproduced picture is constructed from three separate signal contributions representing respectively the red, green and blue components of colour present in the corresponding part of the original scene. In the NTSC, PAL and SECAM systems of colour broadcasting, all three colour components are analyzed simultaneously, and for this reason they are referred to as simultaneous systems. However, less refined systems exist in which the separation of the colour image into its three primary components takes place sequentially. For instance, in the field sequential system, filter-wheels having equal red, blue and green sectors are placed in front of cameras and receivers, and made to rotate in synchronism at such a speed that each colour sector is in position in front of its appropriate tube for exactly a field period. Other systems have been devised in which the colour sequence occurs line by line or even as a continuous series of red, blue and green dots. These are called respectively line sequential and dot sequential systems. The term also applies to a plain, uninterlaced raster.

See: *Colour principles; Colour systems.*

□**Sequential Scanning.** System whereby all the elements in one picture are scanned by one field scan, and all those in the next picture by the next field scan, i.e., the fields follow sequentially.

See: *Scanning.*

◑□**Set-Up.** Originally the position of the camera at the beginning of each scene. Now loosely used to refer to camera, microphones and artistes at the commencement of a shot or scene.

◑**Seventeen-point-Five Millimetre Film.** Form of narrow-gauge film produced by slitting standard 35mm film down the centre to produce two strips 17·5mm wide, each with one row of perforations down the edge. Although first introduced in the early period of motion picture history, it was never widely used as a system of picture presentation, but optical and magnetic sound recordings on this width have had some general application on grounds of economy.

See: *Film dimensions and physical characteristics.*

◑**Seventy Millimetre Film.** Largest gauge of film, used for wide-screen cinematography and for a number of photographic instrumental recording systems. In motion picture usage, the material has two rows of perforations with wide margins to allow the positioning of magnetic stripes along each edge, but 70mm instrumental recording film has much narrower margins; the motion picture frame used is 5 perforations high and there are just under 13 frames to the foot of 70mm film.

In the cinema industry in the United States and Western Europe, 70mm film is used as a print material only, copies being made from 65mm negatives or by enlargement from 35mm and usually provided with six multi-channel stereophonic sound records on magnetic stripes. In Russia and

Eastern Europe, however, 70mm width film is also used as a camera negative material.

See: *Film dimensions and physical characteristics; Wide-screen processes.*

□**Shading Correction.** Method of correcting the shading present in the output from a camera tube, by addition of correction waveforms which are in opposition to the tube shading.

See: *Camera* (TV); *Shading error.*

□**Shading Error.** Error produced by a camera tube when it does not faithfully reproduce a scene, but gives the effect of variations in light, horizontally or vertically, producing a shading effect on the picture. This is usually caused by some error in manufacture giving substantial change in sensitivity in one direction.

See: *Camera* (TV); *Shading correction.*

□**Shadow Mask Tube.** A display tube incorporated at present in virtually all colour TV receivers. It produces an additive colour synthesis through the use of three electron guns bombarding a screen formed by a mosaic of phosphor dots arranged in triads of red, green and blue. Interposed between the guns and the screen is the shadow mask, a perforated metal screen containing as many holes as there are triads, and set about half an inch from the back of the tube face-plate on which the phosphor dots are coated.

The three colour guns are set so that their beams converge and intersect at the plane of the mask. After passing through a hole in the mask, the red beam strikes only the red-emitting dot in the triad behind it, the green beam the green dot and the blue beam the blue dot. If the red beam alone is scanning the plate, its electrons hit only the red dots, and the mask shadows the green and

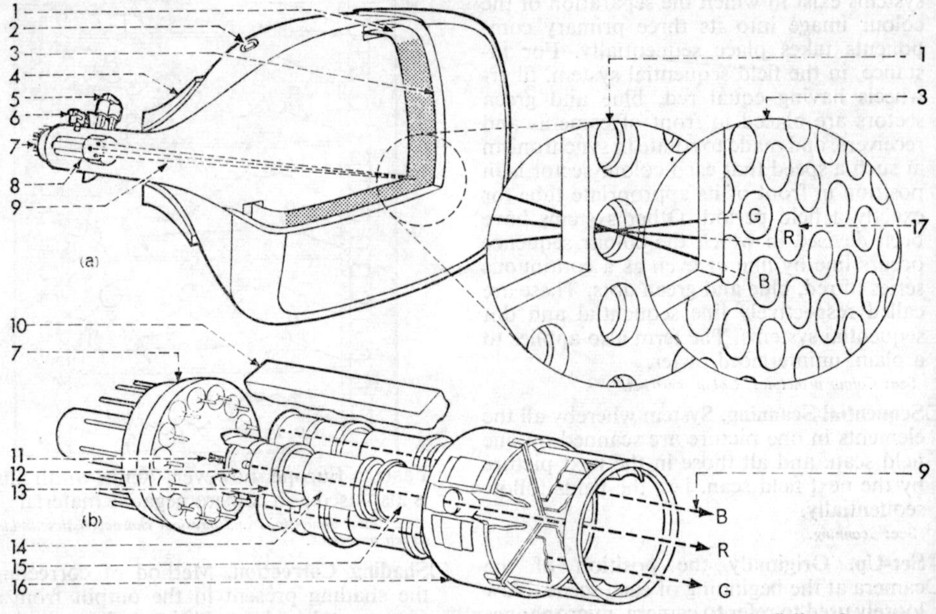

SHADOW MASK TUBE. (a) General layout. (b) Exploded view of electron gun assembly, 8. Salient features are the triple electron beams, 9, for red, green and blue, which together scan the phosphor screen, 3, through a shadow mask, 1, set just behind it. The beams are so converged that, through any hole in the mask, each "sees" only the appropriate colour phosphor in a triad of small spots, 17, of which there are as many as there are holes in the mask. All the remaining tube features are designed to ensure the proper working of this complex electron optics system. On base, 7, are mounted the three guns, held in place by support beads, 10. Each comprises heater, 11, cathode, 12, first anode, 13, focus anode, 14, third anode, 15. Convergence assembly, 16, contains three pairs of plates which act as extensions of external radial convergence magnets, 5. These provide separate radial beam adjustment, and lateral adjustment of blue beam only is effected by magnet, 6. Deflection yoke, 4, comprises conventional line and field scanning coils, and final anode button, 2, connects 25 kV supply to internal graphite coating of tube. Phosphor dot colour screen, 3, carries upwards of one million precisely positioned dots.

blue dots. Hence its name. So also with the green and blue beams.

In a 25 in. tube there are about 330,000 holes in the shadow mask, and consequently a total of nearly a million colour dots, all of which must be precisely located to register with the holes through which they are bombarded. The holes represent about 25 per cent of the total mask area, so that only one-quarter of the electrons from each gun reach the phosphor surface. This loss of efficiency is partly made good by increasing the EHT to about 25 kV, but the screen brightness of a colour receiver is still much less than that of a monochrome receiver.

With a total mean beam current of 1 mA, the mask absorbs ¾ mA, which at 25 kV means that it must dissipate about 20 W of heat. This poses problems of expansion and thus of electron-optical alignment.

SHADOW MASK. This is made of steel, which in a typical tube is 0·006 in. thick and mounted 0·6 in. from the phosphor screen. The holes are spaced 0·028 in. apart on a 25 in. tube, and 0·023 in. on a 19 in. tube. The shadow mask is mounted in such a way that heat expansion moves it slightly along the axis of the electron beams, so that its colour register is not impaired.

COLOUR SCREEN. Because of the extreme accuracy of registration required, each colour screen is manufactured in conjunction with the shadow mask which is to be combined with it. The tube face-plate, on the inner surface of which the screen is formed, is first coated with a continuous layer of green-emitting phosphors, basically cadmium zinc sulphide, formed into a slurry with polyvinyl alcohol and ammonium dichromate.

The shadow mask is then fitted into the faceplate, and the two are mounted on a device which contains an ultra-violet lamp whose rays are directed through a quartz rod to form a small source. The holes in the shadow mask act like a pinhole camera, so that the UV rays, following precisely the path of the electrons from the green gun, form spots on the faceplate. (Optical correction devices are used to simulate electron deflection effects towards the edges of the faceplate.)

The UV rays, passing through all the holes at once, thus illuminate only the areas which are to form the green dots in each triad. The effect of the UV is to polymerize the polyvinyl alcohol (made sensitive to it by the ammonium dichromate) and so to harden or fix it. It only remains, therefore, to remove the shadow mask and dissolve away all the unfixed phosphor, and repeat the process with the blue phosphor (like the green, basically cadmium sulphide) and the red (basically yttrium orthovanadate, and the most difficult to produce satisfactorily). In each case the UV lamp is displaced so that its ray pattern corresponds to that of the corresponding gun.

Finally the screen surface is aluminized to provide a return path for the beam currents and increase the effective screen brightness by reflection.

ELECTRON GUNS. Three separate guns are formed into a common assembly in the neck of the tube. Each emits its own stream of electrons, which is modulated by the picture signal for a particular colour and focused and accelerated by successive anodes. These anodes act as electrostatic lenses, concentrating the electrons into a narrow circular beam. The guns are terminated by a convergence device which forms an extension of the radial convergence magnets outside the tube.

EXTERNAL COMPONENTS. These serve several functions: static deflections for aligning the beams when setting up the tube; dynamic deflections for correcting the scan at wide angles; and the usual deflection coil assembly in the form of a yoke to provide line and field scans as in a monochrome tube, the three beams being moved as one to form the 625-line (or 525-line) raster.

Next to the deflection yoke, and on the neck of the tube, are mounted the radial convergence magnets above the convergence assembly within the tube. These three magnets move the three beams individually towards or away from the central axis, and thus serve to align them in register on the face of the shadow mask. Static convergence holds only in the central area, so special correction currents are supplied from the receiver to maintain convergence on the outer parts of the tube face.

Next in order is the purity ring magnet, a permanent magnet whose adjustment slightly bends all the beams together. Only when the three beams precisely cover the triad of phosphor spots are the purities of the three primary colours at a maximum.

Finally, nearest to the tube base is the blue gun lateral convergence magnet. A permanent magnet enables slight sideways correction to be applied to the blue beam only, or, on later receivers, to all three beams. Sometimes dynamic correction circuits are included.

OTHER FEATURES. This greatly simplified description has necessarily omitted many important features of the shadow mask tube. The final anode at 25 kV comprises the third anode on the three guns and a conductive coating in the flared section of the tube. This coating, with a similar grounded coating on the outside, forms a smoothing capacitor for the EHT supply, the tube glass acting as dielectric. Glass, phosphors and some metals when subjected to an electron beam at 25 kV emit potentially dangerous X-rays, but the tube face is made of heavy glass to reduce emission to a safe level, while further precautions are taken to prevent radiation above and below the receiver cabinet.

The shadow mask tube is heavy (about 38 lb.), complicated and expensive, but it is at present the only television display tube capable of giving reliable high-quality colour pictures. Its additive synthesis makes possible, at least theoretically, a rendering of colour values superior to the subtractive reproduction of film, which depends on the imperfect absorption characteristics of organic dyes. R.J.S.

See: *Colour display tube; Lawrence tube; Receiver.*

□**Shaping Network.** Linear network used, in circuitry such as synchronizing generators, to form pulses and edges of square waves to the shape required.

See: *Synchronizing pulse generators; Transmission networks.*

⊕**Short.** Short film, whose limiting length is differently defined in different countries for quota and subsidy purposes. The limit is usually 30 to 45 minutes. Sometimes called a short subject.

⊕**Shot.** Basic division of a film into elements within which spatial and temporal continuity is preserved. Shots are characterized according to camera angle, distance between camera and subject, and subject matter, but in each case the kind of picture of which they form part must be borne in mind: a close shot from an aeroplane might be taken 50 ft. away, and a long shot of a beetle 5 in. away from its subject.

Extreme Long Shot (ELS): An overall shot which depicts a very large area. Often filmed from a high vantage point to set a scene.

Long Shot (LS): An establishing shot which contains all the actors in a scene; a shot in which the object of principal interest appears distant from the camera.

Medium Shot (MS): Shows a scene at normal viewing distance, and cuts actors usually at the waistline. Also called a mid-shot.

Medium Long Shot (MLS): Intermediate in distance between a medium shot and a long shot.

Close-Up, Close Shot (CU, CS): Gives a closer view of its subject than a medium shot. A medium close-up (MCU) frames actors from head to chest, and an extreme close-up (ECU) the actor's head only.

Establishing Shot: A long shot which establishes the whereabouts of a scene, or shows a group of actors who take part in a scene.

Dolly Shot: A shot in which the camera moves bodily from one part of the set or scene to another on a dolly or boom. Also called a trucking or tracking shot. The term crane shot is similarly used.

Moving Shot: A shot from a normally moving vehicle such as car or aeroplane.

Pan Shot: A shot in which the camera pans or sweeps across the scene.

Reaction Shot: A shot which shows the effect of an actor's words on his hearers. Usually a close-up and shot silent.

Two-Shot: A medium shot in which two people appear.

See: *Cameraman* (FILM); *Director* (FILM); *Editor*.

❑**Shot Box.** Control unit for remotely operating a camera. The action of certain mechanical devices used in television can be controlled by servo-mechanisms which may include preset controls and switches. The unit which controls the servo-mechanisms and the limits between which they operate is called a shot box.

A shot box (or focus unit) can operate a zoom lens so that it may be set at any one position within its range, equivalent at that position to the focal length of a fixed focus lens. A typical arrangement gives a combination of seven settings equivalent to seven fixed focus lenses with the additional facility of movement if required. The latter can be made automatic and the speed of the zoom can be controlled.

Coupled to a motorized pan and tilt head, a shot box can be set up to give certain fixed panning and tilting operations.

Such servo-mechanisms, or combinations of them, can be of particular value in television broadcasting in presentation or continuity studios, where it is uneconomic to have a cameraman available all the time, or in industrial television where it is desired to monitor a sequence of operations, either regularly or at random. The operation of the unit can be accomplished electrically by using control potentiometers to unbalance an a.c. bridge, the output of which is fed to an amplifier, and this current is used to control the driving motors. Movement ceases when the balance position is restored. Provision can also be made for the control potentiometers to be preset, and the sequence of shots delayed, until a manual/automatic switch is operated.

See: *Camera* (TV).

⊕**Shot Listing.** Listing of shots in completed films or out-takes often made by film libraries for classification purposes.

See: *Library* (FILM).

⊕❑**Shot Noise.** Form of electrical noise or random unwanted signals present on a wanted signal. Often generated by thermionic valves.

See: *Noise*.

⊕**Shoulder.** In the characteristic curve of the tonal reproduction of a photographic process, the region of upper densities approaching the maximum possible from the material. In this region, increasing exposure produces smaller proportionate increments of density until a point is reached where no further increment is obtained; this is described as the levelling-off of the shoulder.

See: *Characteristic curve; Exposure control; Image formation; Sensitometry and densitometry; Tone reproduction.*

⊕**Show Copy.** Positive print of a motion picture film which has been specially selected for use at an important presentation, such as a première.

⊕**Shrinkage.** Reduction in both length and width which motion picture film may show as a result of processing, drying and storage. A typical temperature coefficient is 0·002 per cent for 1°C, while for humidity changes the coefficient is about 0·006–0·009 per cent for a 1 per cent change in relative humidity.

See: *Film dimensions and physical characteristics; Film storage; Telecine.*

⊕**Shutter.** Device for blanking off the film during its intermittent movement, so as to avoid blurring. Camera shutters vary in opening from about 170° to 250°, and often

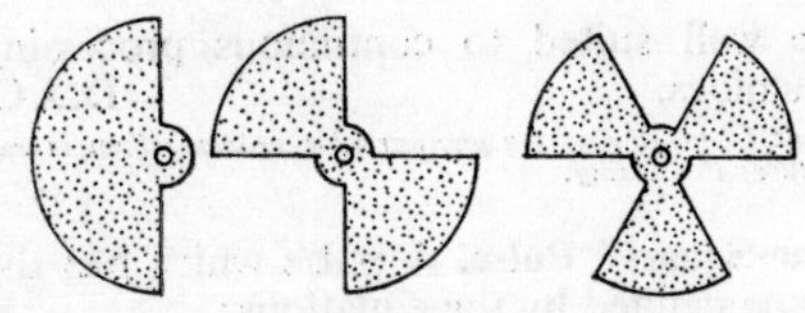

SHUTTER TYPES. (a) Single-bladed, of kind used in non-reflex cameras. (b) Two-bladed; one blade obscures film during projector pulldown, the other gives an extra interruption when film is stationary to produce a 48 Hz flicker frequency at sound speed. (c) Three-bladed; two interruptions of stationary frame to give 48 Hz flicker at silent (16 fps) speed.

take the form of mirrors for viewfinding purposes. To reduce flicker, projector shutters are so designed that they interrupt each image of the film on the screen more than once. In high-speed cameras, mechanical shutters are often replaced by electro-optical devices.

See: *Camera* (FILM); *Projector.*

Shuttle Pin. Part of intermittent movement in a camera or printer. Moves the film from one position to the next by engaging the perforation hole of the film, traversing it the required distance and withdrawing to leave the film stationary during the period of exposure while the shuttle returns to its original position in preparation for the next cycle.

See: *Pin; Registration.*

Signal. Form or variation with time of a wave train whereby information is conveyed in a transmission system.

Signal Plate. Electrode of a camera tube from which the output signal is taken.

See: *Camera* (TV).

Signal-to-Noise Ratio. Ratio of the amplitudes of the wanted signal to the unwanted noise in an electrical transmission system. Since the signal-to-noise ratio imposes a maximum limit on the effective bandwidth of the system, and hence on the amount of information it can transmit, it is a factor of prime importance in design.

See: *Sound and sound terms.*

Signal-to-Weighted-Noise Ratio. A measure of the significance of noise on an audio or video signal. The audibility of noise on sound or the visibility of noise on a television picture depends on the frequency band in which the noise falls. Weighting networks can be constructed such that when they are connected to the output of a system in the absence of signal, and root-mean-square meters are connected to them, the readings of the meters indicate the subjective amplitude of the noise. In the case of audio signals, the signal with which this is compared to find the signal-to-weighted-noise ratio is the normal root-mean-square output signal of the system; in the case of video signals, the signal with which it is compared is the peak-to-peak amplitude of the video output signal, usually excluding the synchronizing signals.

See: *Acoustics; Sound and sound terms.*

Silent Frame. Aperture size of camera or projector mask used before the introduction of sound-on-film techniques; also known as full aperture.

See: *Academy aperture; Film dimensions and physical characteristics; Movietone frame.*

Silent Speed. Number of frames per second taken by a camera for use without synchronized sound accompaniment. The normal rate is 16 or 18 fps, the slowest and consequently the most economical speed which renders motion with comparatively little flicker. However, in the days of the silent film, camera speed varied between about 12 and 20 fps.

Silver Recovery. When a latent image is developed, only the exposed silver halides are converted to metallic silver to form the visible image. In the fixing process, the unexposed silver halide is converted into a soluble compound and removed from the film.

The amount of silver thus removed varies according to the type of film, there being more silver in negative film than in positive, but a convenient approximation gives 1 oz. of silver released into the fix from each 1,000 ft. of 35mm film. Since large installations process millions of feet of film, the recovery of this silver is a worthwhile operation.

There are three practical methods of recovering silver from a fix bath: chemical precipitation, metallic replacement and electrolysis.

CHEMICAL PRECIPITATION. In this method the soluble silver compound in the fix is converted to an insoluble compound by adding a suitable precipitating agent. Many such chemicals exist, but the most common is sodium sulphide. Unfortunately, when sodium sulphide is added to fix, very unpleasant hydrogen sulphide fumes are produced. However, the reaction also produces silver sulphide which is precipitated and can be refined after settling, washing and drying.

METALLIC REPLACEMENT. If a penny is placed in a well-used fixing solution, the copper surface appears to become plated with silver, though in fact the copper is being replaced by it.

In practice, it is more common to use powdered zinc for this purpose. The zinc is introduced into the discarded fix and effectively changes place with the silver, which is precipitated and forms a sludge. A further variation of this method uses steel wool, and here the silver atoms change place with iron atoms so that the steel wool becomes transformed to silver wool. The efficiency of this method depends on presenting to the silver-laden solution the largest possible surface area of the substitute metal.

ELECTROLYTIC METHOD. If two electrodes are immersed in an aqueous solution of silver salts and a current is caused to flow, silver is deposited on the cathode or negatively charged electrode. This method is clean and free from unpleasant odours, it produces almost pure silver and, if properly controlled, can be introduced into the normal fixing solution to keep down the build-up of silver compounds, thereby considerably extending its useful life. This is therefore the most favoured way of recovering silver from a fixing solution.

In practice, certain requirements must be fulfilled to achieve satisfactory results. The biggest problem is to prevent the formation of sulphide caused by the electrolytic decomposition of the hypo. Provided that there is a reasonable amount of silver present in the fix and that the current density is not too high, the deposition of the silver occurs before the sulphide is formed.

There are two basic approaches to this problem. The first is to use electrodes with a very large area combined with a very low potential; the second is to use smaller electrodes, a higher current density and efficient agitation to prevent local desilvering of pockets of fix which would give rise to the formation of sulphide.

Ideally, some form of automatic current control would raise the current when the silver content was high and lower it as it decreased. At the moment this is not practical, and in most large installations samples are analysed at regular intervals and the current is adjusted so that the silver content is maintained just above the level at which sulphide formation occurs.

Several manufacturers produce units for recovering silver in this way which are well suited to continuous processing machines. D.J.C.

See: *Chemistry of the photographic process; Film manufacture; Processing.*

□**Sine-Squared Pulse.** A pulse which has the shape defined by the equations:

$$V = 0 \text{ for } t<0$$
$$V = \sin^2 t \text{ for } 0<t<\pi$$
$$V = 0 \text{ for } t>\pi$$

See: *Picture monitors.*

□**Single Echo.** Echo occurs on video signals in an analogous way to sound signals, and a single echo is a sample of the original signal which arrives later than the original does, owing to a reflection. It may be reversed in phase.

See: *Ghost.*

⊕**Single Eight Millimetre Film.** Narrow gauge film when used in the camera and processed as a single strip 8mm in width is sometimes referred to as Single-8 to differentiate it from the more general use of photography on the two halves of a 16mm wide strip (Double-8) which is only slit to 8mm width after processing. Single-8 film is now normally used with a frame size and perforations of the Super-8 standard.

See: *Film dimensions and physical characteristics; Narrow-gauge film and equipment.*

□**Single Sideband.** The technique of reducing the signal bandwidth by entirely suppressing the frequencies produced by modulation on one side of the carrier, and leaving information to reconstitute the signal.

Normal sound broadcasting uses the full band of frequencies generated by amplitude modulation of an r.f. carrier. These sidebands extend equally on both sides of the central carrier frequency and large channel widths are required if the modulating frequencies extend over a large number of octaves. To reduce this channel bandwidth all frequencies either above or below the carrier frequency can be removed. When this is done, the resultant carrier and sideband signal is known as a single sideband transmission.

The shape of the modulation envelope is similar to that of normal double-sideband modulation, though it is of lower amplitude, and a linear detector gives only slight amounts of distortion. When low frequencies are present in the modulating signal, sidebands very close to the carrier frequency are produced. It is impractical to remove one sideband only without affecting these lower frequency sidebands, and distortion is introduced into the lower frequency components.

See: *Transmission and modulation; Transmission networks; Transmitters and aerials.*

⊕**Single-System Sound Recording.** Method of □sound recording in which the sound is originally recorded at the same time and on the same strip of film as the picture image. The recording system is usually magnetic, but may be optical. The method has the advantage over double-system recording of simplicity and speed, but lacks flexibility in editing because sound and picture are in fixed relationship to one another.

See: *Double-system sound recording; Narrow-gauge film and equipment; Narrow-gauge production techniques; News programmes.*

⊕**Sixteen-Millimetre Film.** Most widely used narrow-gauge film for motion picture work, originally introduced in 1923 for the amateur but now regarded as a professional medium dominating the non-theatrical field of commercial, educational and industrial

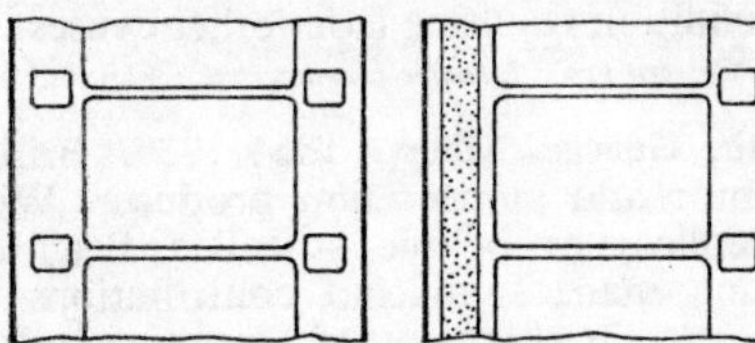

use, and also widely employed in television. This material is 16mm in width and has 40 frames to the foot; stock for use in the camera and for making silent prints is usually perforated with a row of holes along both edges, but for sound work one edge only is perforated so that the other can be used for the sound track, which may be optical or magnetic.

See: *Film dimensions and physical characteristics; Narrow-gauge film and equipment.*

⊕**Sixty-Five Millimetre Film.** Gauge of film used for the photography of certain wide-screen systems of motion picture presentation, particularly those using 70mm width positive prints. This type of material is 65mm in width, having rows of perforation holes along both edges and using a frame five perforations in height of which there are just under 13 frames in the length of one foot of film. Used only as a camera negative and as an intermediate stock for duplicating and never for release prints for cinema projection.

See: *Film dimensions and physical characteristics; Wide screen processes.*

□**Skew Correction.** Correction of a skew parallelogram-shaped television scan to a true rectangle. Skew error may be due to off-centre placing of a large-screen television projector.

See: *Large screen television.*

⊕**Skid.** A structure of three rods, in the form □of an equilateral triangle, with a rubber-tyred wheel at each corner. It is used to make a fixed tripod reasonably mobile on outside broadcasts by locking the legs of the tripod to each corner of the skid. The wheel mountings can be rotated through 360°, and in some versions one wheel mounting can be locked in an angular position to enable a straight track to be accomplished. Provision is also made to lower small jacks at each corner to prevent unwanted movement when the device requires to be locked.

See: *Cameraman; Pedestal.*

⊕**Skip Printing.** Form of optical printing in special effects cinematography by which only frames of the negative selected at regular intervals are printed on to the positive, usually to give the appearance of speeded-up action. Thus, printing only every third frame of the negative produces the effect of the action taking place three times as fast as it actually did.

Skip printing was also used in certain colour systems where colour separation images were recorded in successive groups of frames of the negative.

See: *Freeze frame; Hold frame; Special effects* (FILM).

⊕**Skladanowsky, Max, 1863–1939.** German showman who designed and patented the Bioskop apparatus in November 1895, and in the same month gave the first public presentation of projected motion pictures in Germany. His apparatus employed two linked projectors operating at eight frames per second, a dissolving-view mechanism being used to project frames alternately from each machine.

⊕**Sky Filter.** Graduated light filter used in black-and-white cinematography to darken the blue sky area without affecting the rest of the scene.

See: *Camera* (FILM); *Cameraman* (FILM).

⊕**Sky Pan.** Lighting unit (luminaire) often □made up by the individual studio and consisting of a 5 or 10 kW open bulb set in a slightly dished aluminium reflector, with a protecting cage in front of the bulb. May be mounted on tubular scaffolding, or on short legs as a floor unit.

See: *Lighting equipment; Tungsten halogen lamp.*

⊕**Slash Dupe.** Duplicate negative, usually in black-and-white only, printed directly from a cutting copy, work print or other normal positive print rather than from a specially prepared master positive. Generally used to provide a print of the picture action for editing or dubbing purposes where excellent photographic image quality is not essential. Normal release print stocks are used rather than the more expensive fine-grain duplicating materials.

See: *Duplicating processes (black-and-white); Editor.*

⊕**Slate.** Board placed in front of a film camera at the beginning or end of each take of each scene, identifying the scenes and takes, and giving the name of the film, the director and the cameraman.

□**Slave.** When two units are coupled together in such a way that one unit is controlled by the other, the controlled unit is called the slave, and the controlling unit the master.

See: *Picture locking techniques; Synchronizing pulse generators.*

□**Slavelock.** British term to describe a number of remote television stations running synchronously with a main station by means of radiated signals from the main station, the main station being the master and the remote stations the slaves.

See: *Picture locking techniques; Synchronizing pulse generators.*

□**Slaving.** The operation of making a local synchronizing pulse generator operate in lock with an incoming video signal in both line and field.

See: *Picture locking techniques; Synchronizing pulse generators.*

⊕**Slit.** In optical sound reproduction, a narrow slit past which the sound track moves at constant speed and on which light from the exciter lamp is focused. Since a sound wave cannot be resolved, and thus cannot be reproduced, if it is narrower than the slit, the slit width sets a limit to the high frequency response of the system in conjunction with the film speed. There is a close analogy between slit width in an optical reproducer and gap width in a magnetic reproducer.

Variable-width slits are also used in optical sound recording, and to vary exposure in film printers.

See: *Sound recording* (FILM); *Sound recording systems*

⊕**Slitting Burr.** Edge or lip caused by poor slitting of large rolls of film or magnetic material to smaller sizes.

See: *Film manufacture.*

□**Slot Aerial.** Instead of a piece of metal in air being used to radiate or receive a signal, it is possible to use a slot in a larger area of metal, e.g., in a cylinder. This slot can be tuned to resonate, and becomes a slot aerial.

See: *Dipole aerial; Horn reflector aerial; Transmitters and aerials; Turnstile aerial.*

⊕**Slow Motion.** In film terms, motion of the film in the camera faster than the standard rate, which therefore results in action appearing slower than normal when the film is projected at the standard rate.

See: *Camera* (FILM); *High-speed cinematography.*

⊕**Slug.** Piece of leader inserted in a workprint to replace damaged or missing footage. Also called build-up.

□**Smear.** Effect of lag in a photoconductive camera tube such as a vidicon, causing bright edges to smear when the object is moving or the camera is panned in conditions of insufficient light.

A similar effect is caused by poor low-frequency response, which may give a well-defined streak after a sharp boundary. Smearing is also loosely used to mean lack of definition resulting from other causes.

See: *Camera* (TV); *Transfer characteristic; Vidicon.*

⊕**Smith, George Albert, 1864–1959.** British inventor and pioneer film producer. With his colleagues of the so-called Brighton School, made significant contributions to the early development of film story-telling in the first few years of the cinema. Invented and patented in 1906 a two-colour additive process for colour motion pictures, which, with the financial backing of Charles Urban, was commercially exploited as Kinemacolor – the first successful colour motion picture process.

□**Smith, Willoughby, 1828–1891.** British engineer born at Great Yarmouth, who spent most of his life in telegraphic signalling and cable work. At the age of 20 years he entered the service of the Gutta Percha Company, London, and started experimenting on covering iron or copper wire with gutta percha for telegraphic or other electric purpose. From 1849–50 he was engaged on laying the Dover-to-Calais cable and later in the Mediterranean laying various cables. In 1865 he accompanied the *Great Eastern* when laying the Ireland-to-Newfoundland cable. On return from the Mediterranean he became manager of the Gutta Percha works, later formed into the Telegraph Construction and Maintenance Company Ltd.

Willoughby Smith was the first to draw attention to the photo-electric properties of selenium, the element used in all the early picture transmission and television systems.

This he did in letters to the Society of Telegraph Engineers in 1873 and 1876 after he had carried out investigations into certain observations, made by his chief assistant, Joseph May, at the Valentia Island (Eire) cable station, of the behaviour of some high resistances in the form of bars of selenium.

⊕**SMT Acutance.** Abbreviation for System Modulation Transfer Acutance, which is a factor proposed to represent the subjective impression of the sharpness of a complete picture presentation system by a single numerical quantity calculated objectively. It is based on derivatives of the modulation transfer functions for each stage of the process, and for a professional motion picture system the stages considered include the camera, the negative film, the printer, the positive film, the projector, the screen and the final observer.

The proposed basis for numerical rating of a system gives a figure of 100 for a process which is perfect in its sharpness reproduction at all stages. Ratings of above 90 are good, from 80 to 89 fair, and 70 to 79 only passably acceptable. In general, a difference of two to three units represents a difference in sharpness which can be definitely appreciated in a side-by-side comparison under standard conditions.

The convenience of the SMT acutance factor is that it allows the visual effect of altering the parameters of any stage or of introducing additional steps in the process to be calculated in advance and their significance assessed.

See: *Acutance; Image formation.*

⊕**Sodium Lamp Process.** Method of photographing the components required for the preparation of a travelling matte shot. The foreground action is lighted normally against a uniform yellow background illuminated with sodium vapour lamps having an extremely narrow spectral output. For photography, a beam-splitting camera is used with a special prism reflecting only the sodium light on to a black-and-white negative in one gate, while in the second gate the foreground action is recorded normally on colour negative. The black-and-white film recording the sodium light background provides a silhouette image from which the necessary mattes can be prepared.

See: *Special effects* (FILM).

□**Solar Cell.** A device which converts light into electrical energy, often used to power electric circuitry in satellites.

See: *Satellite communications.*

⊕□**Solid State Circuit.** Electrical circuit employing solid state or semi-conductor elements rather than thermionic valves (tubes). The most commonly used solid state element is the transistor.

⊕□SOUND

The source of any sound is a vibrating body. Examples are the vocal cords by which we speak, the string of a violin, the air column in a clarinet and the cone of a loudspeaker.

The vibrating body causes pressure variations in the form of wave motions in the air surrounding it. This can affect the ears to produce the sensation of hearing or be picked up by a microphone which converts the pressure variations into voltage variations.

BASIC PRINCIPLES

There are thus three essentials for the realization of sound – a vibrating body, a medium through which the sound wave can travel, and some form of receiver. Like the vibrating body, the medium can be in one of several forms.

In addition to air, metal and wood can also transmit sound.

WAVE MOTION. A medium can transmit waves because it possesses elasticity. Air is a gas and all gases have elasticity. If a layer of air has its volume changed, it returns to its original volume when the force causing the change is removed.

When a body vibrates, the layers of air immediately surrounding it are alternately compressed and rarefied. Owing to elasticity, these layers of air in turn affect further layers so that travelling outwards through the air from the vibrating body is a wave of alternate compressions and rarefactions superimposed on the normal steady atmospheric pressure.

Such a motion, in which the medium vibrates along the direction in which the wave is travelling, is called a longitudinal wave. In another type, the medium vibrates at right-angles to the direction in which the wave is travelling. The motion is then said to

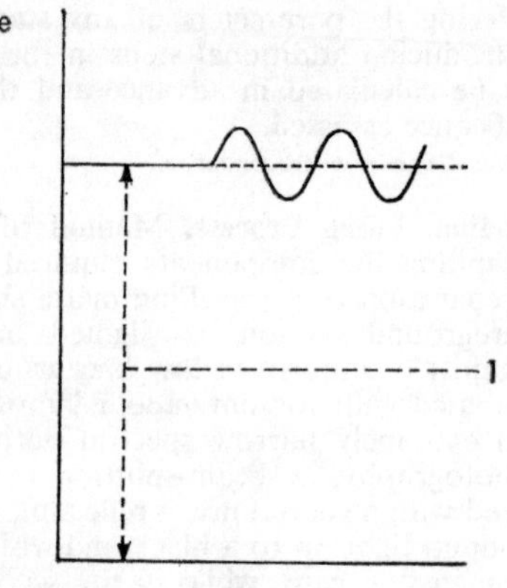

ACTION OF VIBRATING BODY ON SURROUNDING AIR. 1. Steady atmospheric pressure on which are superimposed air pressure variations caused by vibrating body. This pressure variation is very small, and is here magnified to make it easily visible.

produce a transverse wave. These waves occur, for example, on vibrating strings and on the surface of water.

There are two important points to be noted about all types of wave motion: first, energy is propagated and, secondly, there is no displacement of the medium as a whole.

The first point is obvious, as energy must be available to cause vibrations in the receiver. The second point implies that the particles of the medium simply vibrate about their normal position of rest.

So that the longitudinal motion of a sound wave can be shown in a simple sketch, a convention used is to show forward movements at varying heights above, and backward movements at varying depths below a reference line which corresponds to the position of rest. By joining the points so plotted, a normal transverse type of wave drawing is obtained.

SOUND TERMS

AMPLITUDE. Amplitude is the maximum value of a wave motion. For example, the maximum distance a particle moves from its rest position is the amplitude of displacement.

WAVELENGTH. When particles of a transmitting medium are displaced in the same direction by the same distance, they are said to be in phase. The minimum distance between two particles in phase is termed the wavelength (λ).

PERIODIC TIME. The interval of time between two points of equal displacement or of equal pressure is called the periodic time (T).

FREQUENCY. The number of complete changes of cycles, for example from one compression to the next compression, which pass a given point in one second is called the frequency (f). Frequency is measured in cycles per second (cps), which is now commonly expressed as hertz (Hz).

Frequency and periodic time are related as follows:

$$\text{frequency} = \frac{1}{\text{periodic time}} \text{ or } f = \frac{1}{T}.$$

VELOCITY. The velocity of a sound wave is related to the wavelength and frequency of the wave as follows:

$$\text{Velocity} = \frac{\text{distance travelled}}{\text{time}} = \frac{\lambda}{T} = \lambda \times \frac{1}{T} = \lambda \times f.$$

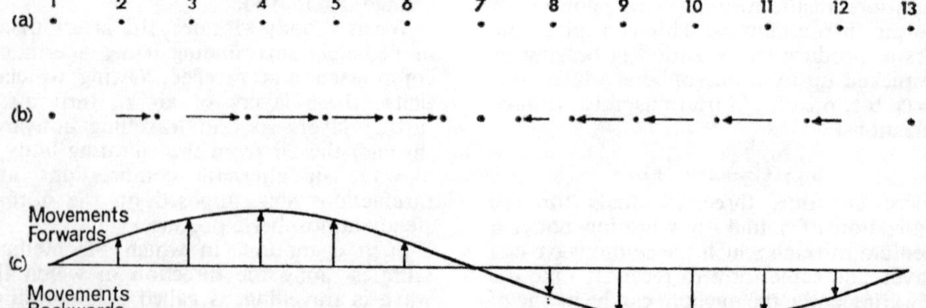

MOVEMENT OF AIR PARTICLES. (a) Particles represented by dots 1 to 13, assumed evenly spaced. (b) Relative positions of particles altered as shown by arrows under influence of sound waves radiated by vibrating body. (c) Displacements redrawn to give more usual wave form.

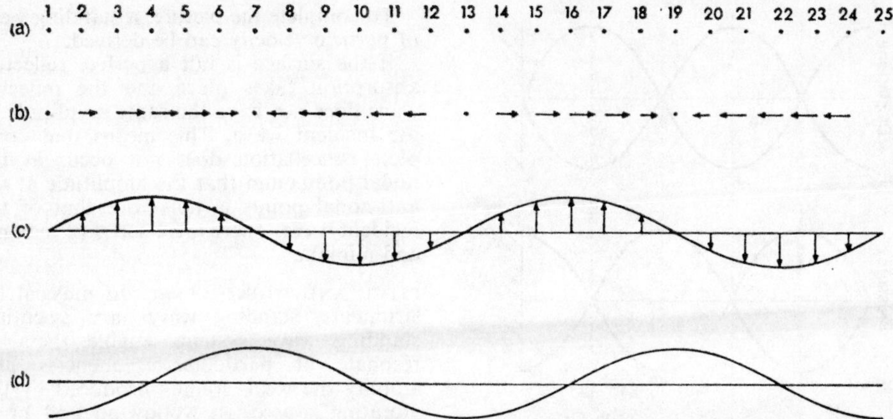

PARTICLE DISPLACEMENTS AND PRESSURE WAVES. (a) Particles at rest and evenly spaced out. (b) Particle displacement by action of vibrating body. (c) Displacements redrawn to show waveform. (d) Because of displacements, pressure at different points of space varies in a wave-like manner similar to displacement wave but shifted along axis. Zeros are at points of maximum displacement; maxima are at points of zero displacement.

The velocity of sound waves in air depends on temperature and humidity. For normal conditions, the velocity is about 1,130 feet per second or 344 metres per second.

It is interesting to calculate the wavelengths of two frequencies, one low and the other high, which lie within the audio range. At 50 Hz the wavelength is 22 ft. 7 in. and at 15,000 Hz the wavelength is 0·9 in. This large range of wavelengths creates many difficult problems in acoustics and acoustical transducers.

PITCH. The subjective quality of a note which enables it to be placed on the musical scale is known as pitch.

Pitch and frequency are obviously closely related. Frequency depends on the rate of vibration of the source; if this rate is altered, the frequency changes and so does the pitch.

Although pitch is mainly determined by frequency, there are circumstances in which the pitch of a note can be altered by changing its intensity while keeping the frequency constant.

INTENSITY. The intensity of a sound is the rate at which energy is transferred through the medium. It is measured in watts per sq. cm. The intensity of a sound wave is proportional to the square of the sound wave pressure.

DECIBELS. The ear can handle a very large range of intensities, something of the order of 1,000 billion to 1 ($10^{12}:1$). This is possible because the ear has a logarithmic response. Suppose that a sound wave has an intensity of x watts per sq. cm. If this value is increased to 10x and then again to 100x, the ear interprets these two changes as being equal changes in loudness, since the ratios $\frac{10x}{x}$ and $\frac{100x}{10x}$ are equal.

The logarithm of the intensity ratio to the base 10 is called the bel, i.e.,

$$\text{number of bels} = \log_{10} \text{intensity ratio.}$$

In practice, the bel is too large, and a smaller unit, the decibel, is used,

$$\text{number of decibels} = 10 \log_{10} \text{intensity ratio.}$$

In many cases, measurements are made of pressure instead of intensity, and since intensity is proportional to the square of the pressure it follows that,

$$\text{number of decibels} = 20 \log_{10} \text{pressure ratio.}$$

STANDING WAVES

The waves considered up to now have been waves which travel freely in space and meet no object which disturbs the uniform distribution of sound energy. But in practice there are many cases where waves do strike objects which reflect the sound energy, and when a wave is reflected the interference between the energy of the incident wave and that of the reflected wave sets up waves which have special characteristics of importance in acoustics. These are called standing or stationary waves.

ORIGIN. Suppose a sound wave strikes a rigid surface and, for simplicity, assume that

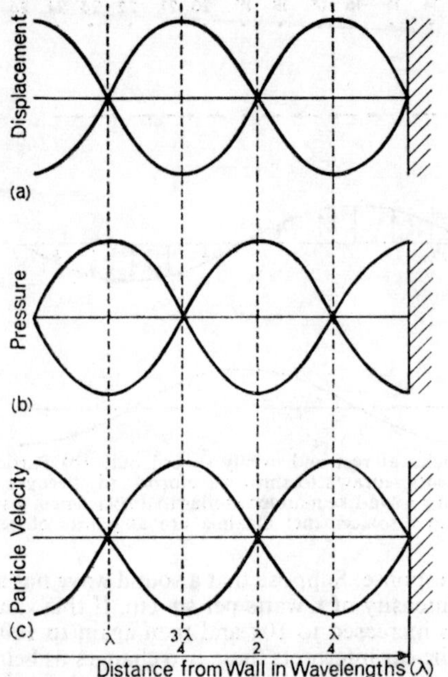

REFLECTION OF SOUND WAVE AT WALL. Standing wave patterns of sound wave reflected back from wall along original path. (a) Displacement wave with nodal (zero) points at wall, $\lambda/2$ and λ from wall. Anti-nodal points are $\lambda/4$ and $3\lambda/4$ away from wall. (b) Pressure and (c) particle velocity patterns deduced from (a).

the wave is reflected back along its original path with no loss.

Consider displacement first: there can be no displacement at the surface; at any instant of time, therefore, the phase of the reflected wave must be such as to cancel any displacement due to the incident wave. The effect of this requirement is that at certain points in space the two waves are in phase and at others they are out of phase. Where they are in phase, the amplitude of displacement is twice that of the incident wave and these points are termed anti-nodes. Where they are out of phase the displacement is zero, and these points are termed nodes. Anti-nodes are half a wavelength apart, and so are nodes.

Since these nodal and anti-nodal points are fixed in space and never move, the wave is called a standing or stationary wave.

Next consider pressure: there must be a pressure anti-node at the reflecting surface as there is a displacement node at that point. The pressure standing wave is similar to the displacement wave but shifted in space by a quarter of a wavelength.

To complete the picture, a standing wave of particle velocity can be derived.

If the surface is not a perfect reflector, absorption takes place and the reflected wave does not have the same amplitude as the incident wave. This means that complete cancellation does not occur at the nodal points and that the amplitude at the anti-nodal points is not twice that of the incident wave. There are a series of maxima and minima.

PITCH AND SOUND LEVEL. In musical instruments, standing waves are essential; standing wave systems enable them to resonate at particular frequencies and amplify the weak sounds produced by the vibrating source. In woodwind and brass instruments, standing waves are set up by reflections in the associated air column. The pitch depends on the length of the air column and, by a series of holes or valves, the player can alter the length and hence the pitch.

In room acoustics, the presence of standing waves is a serious fault. They can be detected by listening, at various positions, to a source of sound. In some places the sound level is found to be high, at others low. The variation in level is not due to the fall in intensity observed when moving away from the source; on the contrary, moving towards the source can reduce the level. The variation is due to the standing waves produced by reflections from the walls, ceiling and floor. The positions of maximum and minimum sound vary with frequency since the frequency of the standing wave is determined by the distance between the reflecting surfaces.

MICROPHONE POSITION. In sound recording and broadcasting, the presence of standing waves makes the positioning of the microphone difficult and therefore steps must be taken to control their production.

AUDIBILITY

AURAL SENSITIVITY. When a tuning fork is sounded, the waves it produces are immediately audible. Gradually the loudness gets less and less until the tuning fork cannot be heard. Yet a light touch with the fingers shows that the fork is still vibrating. The fork must still be causing the air to vibrate, but since the sound waves are inaudible there must be a minimum level of intensity necessary to stimulate the sensation of hearing.

Starting at the threshold of hearing, increasing the intensity increases the loudness. This process continues until listening gets uncomfortable and eventually painful. There is thus a level which separates sounds heard properly by the hearing system and those which are felt or are uncomfortable.

With a subjective characteristic such as hearing, these two levels are not absolutely fixed, but vary from person to person. To get a representative figure, many people are tested and curves are averaged to give the general picture.

When the threshold of hearing is examined, using pure tones, i.e., notes of a single frequency, it is found that the required minimum intensity of sound differs widely over the audible frequency range.

The ear is most sensitive in the region of 1,000–5,000 Hz. For frequencies above this range, the sensitivity falls until, at some very high frequency, increase of the sound intensity does not produce any sensation of hearing. Below 1,000 Hz, increased intensity is required to hear the note until finally a low frequency is reached which cannot be heard irrespective of its intensity.

The maximum audible frequency range is usually quoted as approximately 20–20,000 Hz; for adults a range of about 30 Hz to 17 kHz would be more representative.

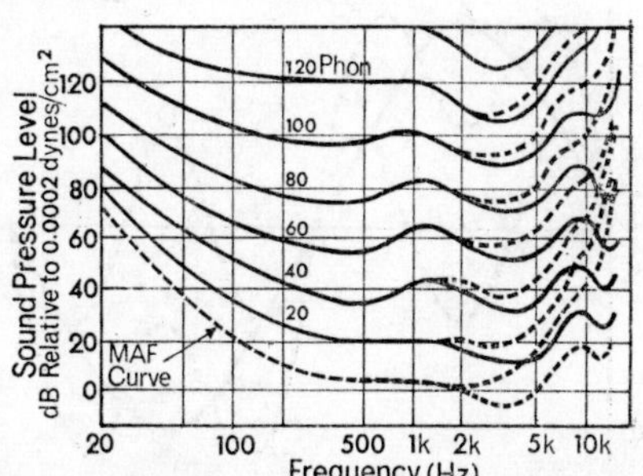

EQUAL LOUDNESS CONTOURS. Because the ear is not uniformly sensitive throughout the audible spectrum, loudness depends on frequency as well as intensity. Equal loudness curves represent constant level in phons. (*Solid lines*) Subjects aged 20. (*Dotted lines*) Reduced aural sensitivity to high frequencies in subjects aged 60. MAF Curve: minimum audible field, also with variation for subjects aged 20 and 60.

Increasing age leads to a marked loss of sensitivity at the high frequencies. The low frequency sensitivity also falls, but the decrease is not so noticeable. Experiments show that the loss of high frequency sensitivity with age for men is greater than for women.

This is balanced by the fact that women show greater loss at low frequencies.

LOUDNESS. Loudness is defined as the magnitude of the auditory sensation which a sound produces. Since the threshold-of-hearing intensity varies over the frequency range, the loudness of a sound does not depend only on its intensity, but also on its frequency. Thus the decibel cannot properly be used in loudness measurement since it is a ratio of two intensities. In assessing loudness, subjective techniques are used, the listener judging the effect of changing the intensity of the frequency. This leads to a unit of loudness level called the phon.

In using the phon, the technique is to compare the loudness of a reference note with that of the given note. The intensity of the reference note can be adjusted until it sounds as loud as the given note. The amount by which the reference note has been moved away from its threshold-of-hearing intensity value gives the loudness level of the given note in phons.

It is necessary to specify, not only the frequency of the reference note, but also its threshold-of-hearing intensity value. The reference note used is a pure tone of 1,000 Hz and the reference intensity is 10^{-16} watts per sq. cm., which corresponds to a pressure of 0·0002 dynes per sq. cm.

In normal listening conditions, an observer is readily able to judge when the loudness of the 1,000 Hz note is the same as the given note. To obtain this parity, the 1,000 Hz note intensity has been raised by a certain number of decibels above the reference intensity value. The number of decibels gives the equivalent loudness – or loudness level – of the given note in phons.

TYPES OF WAVEFORM

TONE QUALITY. Much of the pleasure in listening to musical instruments is due to the fact that each has its own quality or colour of tone. The word timbre is often used for this characteristic by which the various instruments are distinguished.

An oscilloscope shows that the sound waves from a tuning fork have a smooth sinusoidal waveform. A clarinet making the same note as the fork sounds very different, and the oscilloscope shows that the waveform is quite different too. Instead of the smooth shape produced by the fork, the waveform is such that, although the cycles repeat in the normal way, the shape of each cycle is irregular. The sound of the fork is said to be pure while that of the clarinet is called complex.

The note of the fork consists of only one frequency whereas the clarinet sound is a mixture of several frequencies. This difference can be demonstrated by means of a spectrum analyser, an instrument which separates the frequencies present in a sound and displays them as spaced vertical lines. The height of each vertical line gives the amplitude of that particular frequency.

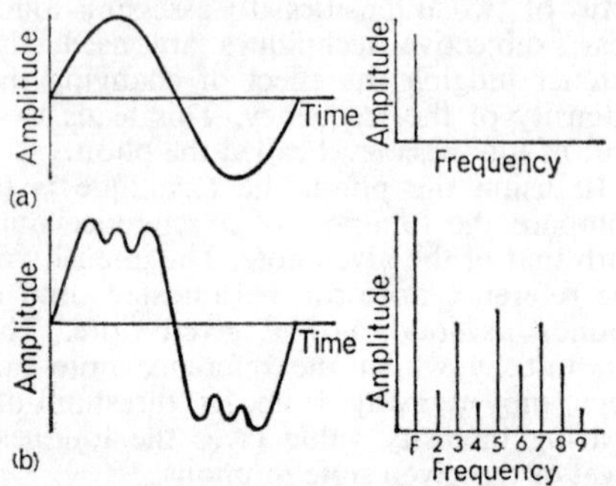

CHARACTERISTIC WAVEFORMS. **(a) Pure note, e.g., from a tuning fork. Spectrum diagram (*right*) shows single line at frequency of note. (b) More complex waveform of note from a clarinet. Spectrum gives amplitudes of fundamental and harmonics up to the ninth.**

The fork shows only one line, at the frequency marked on it, but the clarinet produces several frequencies simultaneously and thus a series of lines. The lowest frequency of the series is the same as that of the fork; this is called the fundamental. The other frequencies are the overtones of the fundamental. The spectrum analyser shows that the overtones do not all have the same amplitude, some being stronger than others. It is the presence of overtones and their relative strengths which gives the clarinet its characteristic tone.

HARMONICS. Most musical instruments produce complex waves, each type varying in its overtone content and thus having its own particular timbre. The addition of a fundamental and overtones produces a complex wave.

The overtones in some cases are exact multiples of the fundamental and are then termed harmonics. String, wood-wind and brass instruments all produce a series of harmonics in this way. In some other instruments, for example the xylophone and tubular bells, the overtones are not exact harmonics of the fundamental.

Another feature of the complex sounds from many instruments is that the time duration of the overtones varies. At the instant that the note is produced, many overtones are generated but the higher ones tend to be weak and die out quickly. The lower frequency overtones are generally much stronger and persist for a longer time. The fact that the higher overtones last for only a short time produces a tone quality for the initial part of the sound which differs from that of the remaining or continuing part. Attack is a word sometimes used to describe this effect.

Since the higher overtones making up the attack tone pass quickly, they are said to be transient in nature or simply transients. In acoustics there are many examples of transients occurring, i.e., rapid changes of amplitude with time, and since their presence is an essential for the musical timbre, any recording and reproducing system must be able to handle them with fidelity.

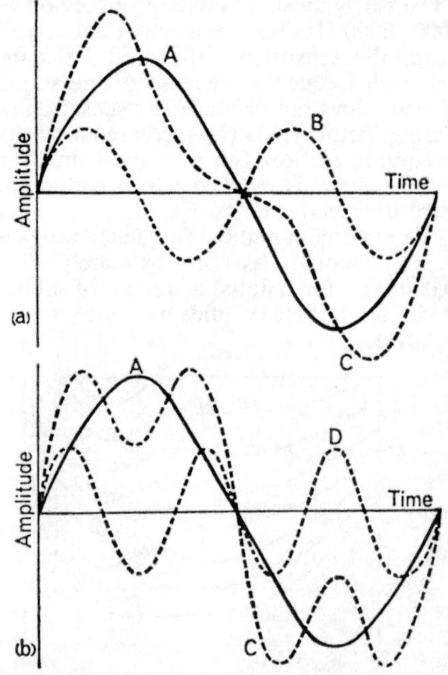

DERIVATION OF COMPLEX WAVES. A. Fundamental. B. Second harmonic. C. Resulting complex wave. D. Third harmonic. (a) Addition of fundamental and second harmonic to give complex wave. Shape of resultant depends on relative amplitude and phase of the two components. The ear ignores phase changes, so that alteration of phase relationship of components, though producing changes in waveform, makes no difference to tone quality. (b) When third harmonic is added to fundamental, resulting complex wave has shape quite different from (a).

SPEECH. The voice has a characteristic pitch. A male voice has a lower pitch than a female one and in turn the female voice is of a lower pitch than a child's. The average pitch for a man is about 130 Hz; that for a woman is about twice this. The actual pitch of any individual voice varies about these average values.

The fluctuating air stream leaving the mouth has a complex waveform, consisting of a fundamental note along with its harmonics. The frequency of this fundamental determines the pitch of the voice.

Natural male speech has an overall frequency range of 100–8,000 Hz; for women, the range is something like 200–10,000 Hz. Within these ranges, there are some frequencies which are more important than others to the understanding of what is being said.

Most of the energy in speech is contained in the low frequencies but they contribute very little to the understanding or intelligibility of speech. When only the low frequencies up to 500 Hz are heard, the speech sounds muffled and woolly and it is impossible to make sense out of it.

It is the high frequencies which contain the intelligence. Speech which has a severely restricted bass is still easily understood although it lacks power.

MUSIC. Music has a much wider frequency and intensity range than speech. From a frequency point of view, a high-fidelity reproduction system for music should be capable of handling a range from 30–15,000 Hz. The maximum intensity range of music directly heard is of the order of 70 dB. In practice, much music does not exploit this full range.

The frequency range just quoted is, of course, for the entire range of sounds produced by the musical instruments which are normally used. The range can be broken down, first of all into the individual ranges

ENERGY LEVELS OF MUSICAL SOUNDS

Origin of Sound	*Energy (watts)*
Orchestra of 75 performers, at loudest	70·0
Bass drum at loudest	25·0
Pipe organ at loudest	13·0
Trombone at loudest...	6·0
Piano at loudest	0·4
Trumpet at loudest	0·3
Orchestra of 75 performers, at average	0·09
Piccolo at loudest	0·08
Clarinet at loudest	0·05
Human voice:	
bass singing *ff*	0·03
alto singing *pp*	0·001
average speaking voice	0·000024
Violin at softest used in a concert ...	0·0000038

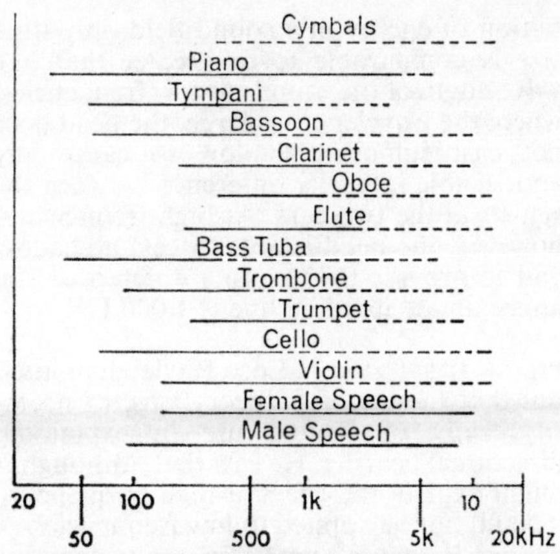

FREQUENCY RANGE. Male and female speech contrasted with a selection of musical instruments. (*Solid lines*) Fundamentals. (*Dotted lines*) Harmonics.

of the various instruments and then, for a particular instrument, into the range of the fundamental notes and overtones it produces.

DIRECTIONAL HEARING

The problem of how the direction of the source of a sound is located by the ears has exercised the minds of investigators for many years, and the subject has received increased attention with the development of stereophonic recording and reproduction.

The faculty of directional hearing is a complex one depending on many factors, but a start can be made by noting that the acoustic impression of the environment is conveyed to the brain by two channels which, of course, begin with the ears. Since they are spaced apart, the two ears are affected slightly differently. This difference provides acoustic clues which can be interpreted by the brain, using its built-in store of acoustical experience, and allows it to ascribe a direction to the source.

INTENSITY THEORY. An early attractive theory was that the signals at the two ears differed in intensity, the difference being produced by the shadowing section of the head. The listener locates the source as being in the direction of the ear receiving the greater intensity. An extension of this idea was that, where the person was unsure of the source's location, he turned his head until the sound and the ears were in the same line. This would then make the intensity difference a maximum and hence allow accurate location.

However, an object can affect the distribution of energy in a sound field only if its size is comparable to or greater than the wavelength of the sound. At low frequencies, where the wavelength is large, the head does not cast sufficient shadow to cause any appreciable intensity difference between the signals at the two ears. At high frequencies, however, the head is able to cast a shadow and so provide the intensity difference. The approximate dividing line is 1,000 Hz.

PHASE DIFFERENCE. Lord Rayleigh pointed out that the intensity theory was, of necessity, only of limited use in explaining directional hearing. He saw that, although it might explain the effect at high frequencies, it could not be applied to low frequencies.

Lord Rayleigh's next step was to examine the possibility that phase difference provided the brain with the necessary directional clues. His experiments led him to discover that there is sufficient phase difference at low frequencies to cause the signals to differ by an appreciable amount. He then concluded that both intensity and phase differences are utilized, depending on whether the frequency is low or high.

TIME DIFFERENCE. Later work during the 1930s showed, however, that the effect of a phase difference could be regarded in the same way as that produced by a time-of-arrival difference and that the latter was more likely to be directly noticed by the listener. Further work showed that this time difference would also be effective at high frequencies and that the intensity difference theory was suspect because the differences actually experienced seemed insufficient to give the known accuracy of location.

More recent research has tended to confirm the theory that it is time-of-arrival differences which enable the ears to locate the direction of a source. It has also been able to confirm that the ability to locate a source with accuracy depends on the nature of the source. If it is producing pure tones, the location is uncertain, but if the source produces complex waves, more precise location is possible. Sources producing a very wide range of frequencies, such as noises or clicks, can be located very accurately.

STEREOPHONY. A monophonic system, i.e., using one channel, cannot recreate the same acoustic impression a listener would receive had he been in the same position as the microphone. Many factors contribute to this discrepancy, but a considerable improvement in the naturalness of the reproduced sound can be obtained if a stereophonic system, i.e., one with more than one channel, is used. Listening to a stereophonic system gives a sense of spaciousness which the monophonic system lacks, and is due to the listener receiving more acoustic clues so that he is able to hear a better sound picture.

Stereophonic systems require more than one microphone; the outputs differ, depending on the distance and/or the angle of incidence of the sound waves to each microphone, and so when reproduced on spaced loudspeakers they produce a sound stage between the loudspeakers. The simplest arrangement is for two channels to be used with two electrically separate microphones feeding separate amplifiers which in turn feed spaced loudspeakers.

Monophonic systems have the advantage of simplicity and low cost, and can give very acceptable results, but produce, especially with music, a hole-in-the-wall effect.

Stereophonic systems, requiring even in the simplest case a doubling-up of equipment, microphones, amplifiers, loudspeakers etc., are more expensive, more complicated and require considerable attention to the matching of the two channels, but give more realistic reproduction. G.W.M.

See: *Acoustics; Sound distortion; Sound mixer; Sound recording; Sound recording systems; Sound reproduction in the cinema.*

Book: *Acoustics*, by G. W. Mackenzie (London), 1964.

□**Sound Balancer.** Operator who controls the level and quality of television sound and puts the sound programme together. Also known as sound mixer. He may work alone or with assistants who operate equipment producing music and effects in the same locality.

Cassette tape machines and other automatic aids reduce the sound balancer's need for assistants. As television is effectively live, all effects and music are added wherever possible during the production. Owing to the continuous action, the sound balancer is usually faced with a larger number of microphone inputs for selection compared with his opposite number in films. As well as operating the equipment correctly, he watches the quality of the product and helps in any problems on the studio floor.

See: *Master control; Sound recording.*

Sound Console. Sound controlling and adjusting apparatus housing all the controls required by a sound mixer to make up a complete programme.

It must be possible to adjust the signal level from every sound source individually. Frequency response shaping controls are usually provided on each of the sources. It may also be possible to switch sources into channels which feed subsidiary functions, such as a public address system.

Sources may be switched, with others, into separate group circuits. These group circuits may then be selected for insertion into the main output with a single level control for the whole group.

Facilities such as echo amplifiers may be available, and individual sources may be routed via these amplifiers. Control of the echo level is provided and often the echo duration time as well.

Operational controls of any limiter or compressor are provided, together with means of assessing the operating levels within these devices. An output signal level meter is used to check that overload levels are not exceeded.

Ease and convenience of operation are paramount in the design of a good sound console. The complexity depends on the type of programme material to be produced.

See: *Master control; Sound mixer.*

SOUND DISTORTION

The waveform of sounds leaving a loudspeaker should be a faithful replica of the original sound waves picked up by the microphone. Unfortunately, in sound recording and broadcasting this ideal is not possible.

Any change in the waveform which occurs in a recording or transmission system is referred to as distortion. The most important forms of distribution are: attenuation, phase, non-linear (harmonic and intermodulation), and amplitude.

ATTENUATION DISTORTION

It is essential to preserve the amplitude relationships of the component parts of the input signal. In passing through the recording or transmission system, the signal suffers some loss and this is overcome by amplification. Any variation in the loss or amplification with variation in frequency is known as attenuation distortion. Providing it is not excessive, attenuation distortion can be corrected by equalization. A common example of this type of distortion is loss of high frequencies owing to the effects of capacitance; by passing the signal through an equalizer which attenuates the low frequencies but not the high frequencies, the balance between them is restored.

PHASE DISTORTION

This occurs when the phase relationships of the component parts of a complex wave are altered. The waveform of a complex wave depends not only upon the amplitude of its different component frequencies but also their relative phase. If the time of transmission through the system varies with frequency, this relationship alters and the waveform changes.

Fortunately the ear cannot detect distortions of this kind unless they are excessive and this seldom occurs in practice.

NON-LINEAR DISTORTION

Harmonic. If a single frequency is applied to a system, the output should contain only

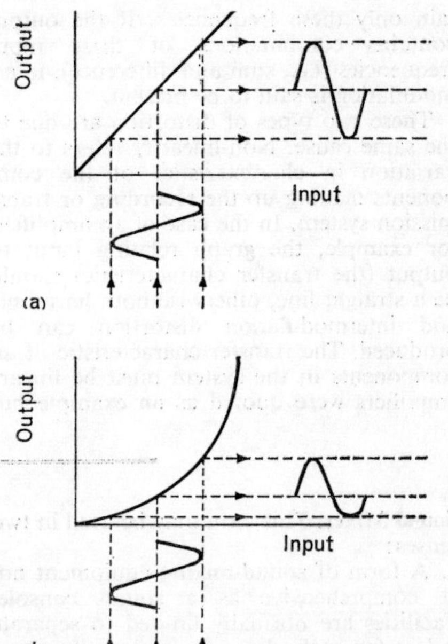

CURVATURE OF TRANSFER CHARACTERISTIC. (a) Signal applied to linear characteristic is transmitted unchanged; no distortion introduced. (b) Curved characteristic produces distortion of waveform, more marked the greater the curvature.

that single frequency. If the output contains harmonics of the input frequency, harmonic distortion is present.

Intermodulation. If two frequencies are applied to a system, the output should con-

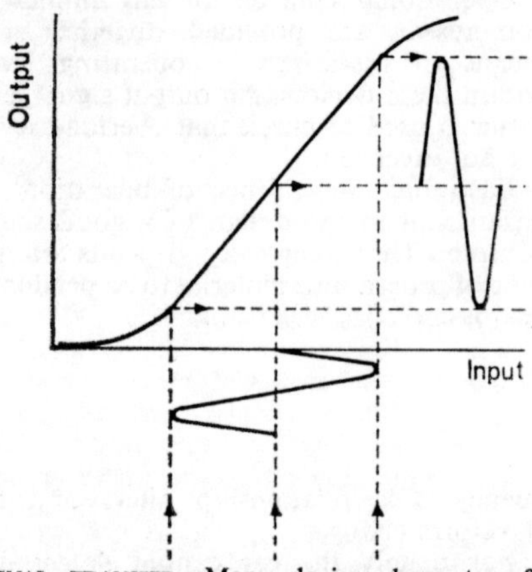

SIGNAL TRANSFER. Many devices have transfer characteristics which are partly linear and partly curved. To avoid distortion, the input signal must be confined to the linear part. Signal shown would distort if applied to any other part of characteristic. Larger amplitude signal would cause distortion of peaks, which would encroach on curved parts of characteristic.

tain only these frequencies. If the output contains combinations of these input frequencies (i.e., sum and difference), intermodulation is said to be present.

These two types of distortion are due to the same cause. Non-linearity refers to the variation in characteristics of the components making up the recording or transmission system. In the case of an amplifier, for example, the graph relating input to output (the transfer characteristic) should be a straight line, otherwise both harmonic and intermodulation distortion can be produced. The transfer characteristic of all components in the system must be linear; amplifiers were quoted as an example but the same requirement applies to transformers, recording and reproducing heads, loudspeakers, microphones, etc. In practice, many transfer characteristics are linear only over their working range but curved outside this; the problem is then to ensure that the input signal is applied to the straight part of the characteristic and that its amplitude does not cause it to run into the curved parts.

Non-linear distortion cannot be corrected, but of the two types, intermodulation is more objectionable to the ear as the combination frequencies are not usually in simple mathematical relationships to the input frequencies and thus sound discordant and unpleasant.

AMPLITUDE DISTORTION

This is defined as the variation of gain, or loss, of a system with the amplitude of the input. It is a form of non-linear distortion but the non-linearity applies only to changes in amplitude; the response to a steady amplitude is linear.

Amplitude distortion is introduced deliberately by compressors and limiters in sound recording and broadcasting since the systems in use can handle only a range of intensities between 25 and 40 dB, whereas the range in real life can be as wide as 60 to 70 dB. If no compression was applied, the weak sounds would be lost in the noise of the system and the loud sounds would produce non-linear distortion due to overloading.

This compression is carried out manually in the control cubicles attached to the studios and is carefully applied so as not only to satisfy the technical requirements of the system but also to preserve, for instance, the light and shade of music. G. W. M.

See: *Sound and sound terms; Sound mixer; Sound recording; Sound recording systems; Sound reproduction in the cinema.*

Books: *Telecommunications*, by W. Fraser (London) 1957; *Principles of Electronics*, by M. R. Gavin and J. E. Houldin (London) 1959.

⊕**Sound Mixer.** This term may be used in two □senses:

1. A form of sound-mixing equipment not as comprehensive as a sound console. Facilities are normally limited to separate faders for each channel, an overall output level control and a signal level meter.

2. The control operator who mixes various sound sources to produce a composite programme. Depending on the type and complexity of the final programme material, the process of mixing sound sources may be a simple routine operation, or may call for a highly skilled sound engineer with a strong aesthetic sense for sound quality. Sound mixers fall into several categories: floor mixers, who do original dialogue recording in the studio; music mixers, who specialize in the fine points of music recording; and dubbing mixers, who mix final sound tracks.

⊕SOUND MIXER

The recording and compilation of film sound tracks present both technical and aesthetic problems for the sound mixer, mainly associated with the satisfactory recording and reproduction of speech and music. Although the mixer is assumed to have a certain amount of technical knowledge, he does not necessarily have to possess a degree in electronics or engineering. Mixing is a task in which practical experience is of equal value to technical training, since much of the day-to-day recording routine consists of dealing with non-technical situations.

The sound mixer is responsible for the sound quality obtained for the duration of a film production, and his problems include choosing the right equipment, controlling the acoustic characteristics of the studio or set, recording under adverse conditions out-of-doors, and ensuring that the director's requirements are met. In addition, he may be called upon to record music, and carry out the final re-recording or dubbing process to produce a composite sound track. Then there is a transfer process to make a photographic sound negative.

The sound mixer must also possess an extensive knowledge of the various types of microphones, and be capable of assessing their exact performance from previous experience. The technique of microphone placement is also learnt mainly from experience, although a certain amount of basic information is available from textbooks and manufacturers' data.

Sound reproduction in a large auditorium is not the prime concern of the sound mixer, but he should be sufficiently interested to hear the results of his efforts in an average cinema, preferably with an audience.

MIXER CATEGORIES

It has been said that most people can be classified as being either ear-minded or eye-minded, according to which medium influences them most. Naturally the ear-minded person will make a far better sound mixer, since he has acute hearing and is able to concentrate on sounds in general. But he must also understand average requirements, so that he can produce recordings which are pleasing to a large percentage of people.

There are three categories of sound mixer employed in film production, each one a specialist in his or her particular field. These are the production or floor mixer, the music mixer, and the re-recording or dubbing mixer. While the last two categories are sometimes interchangeable, floor mixers do not always have the experience to carry out all three tasks.

FLOOR MIXER. The sound man attached to a production during filming is called a production or floor mixer. He is charged with the task of obtaining the best possible dialogue recording on the studio floor and under location conditions. Pre-production meetings are usually held so that the mixer can meet the director under whom he will be working, and become acquainted with the remainder of the film unit before shooting starts.

A floor mixer must possess a good knowledge of microphone techniques, and be familiar with all types of magnetic tape and film recorders and their use under varying conditions. He should also keep abreast of any new developments in this field. An understanding of acoustics is desirable, so that the appropriate action can be taken when necessary to correct the apparent liveness or deadness of a studio set.

MUSIC MIXER. Only a large studio can support a specialist music mixer; small studios do not have enough work to keep him fully employed. In fact small studios seldom have a scoring stage of their own and commission their music recording elsewhere. A music mixer is expected to be familiar with all types of orchestra, such as jazz, dance, and symphony, as well as choirs, soloists and a large variety of solo instruments. If he is to gain the confidence of the composer, the music mixer should be capable of reading and understanding a musical score, although he need not be an accomplished musician himself.

A good working knowledge of studio acoustics is essential, so that the varying conditions required by different types of music can be dealt with quickly and efficiently. Experience of microphone placement is also a vital qualification, since an expensive orchestral session is not the occasion to conduct acoustic experiments.

DUBBING MIXER. All film studios today employ at least two dubbing mixers. It is their task to mix all the original sound tracks and library material together after editing, and re-record a final composite track. This involves operating a comprehensive mixing console installed in a small review theatre, effecting control of sound quality as well as obtaining an aesthetic balance throughout the film.

Sound balance is an individual thing, and different treatment may be given to the same film by different mixers. In fact a film can be made or ruined by the re-recording process, and it is for this reason that the presence of the director or producer is welcomed. Indeed, they usually insist on expressing their opinions at all stages of the re-recording process.

FLOOR SHOOTING

Most sound films are recorded in studios the walls of which are covered with sound-absorbing material. Sets consisting of standard-sized plywood flats are erected with an open top for lighting, and an open front to allow access for camera, microphone boom, and additional lighting equipment. This open type of set has acoustics which correspond more or less to those of the studio, so that the character of the sound reaching the microphone is fairly dead – even in a large studio. Sets with four walls and a roof are sometimes used, and then the sound quality has a coloration due to reflections between the set walls. Large canvas backings are often erected which completely cover the studio walls, thereby substantially nullifying the acoustic treatment. The result is a delayed sound reflection giving a disturbing echo.

The floor mixer sites his console only a few feet away from the set, and judges sound quality on a pair of moving-coil headphones. If the sound appears to be too live, he may decide to erect a felt baffle covering part of the set not being photographed, or perhaps over part of the open end of the set. Less attention need be paid to set acoustics when

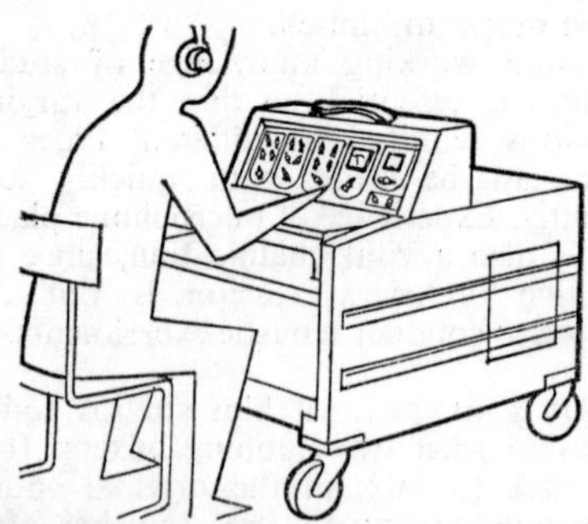

FLOOR MIXER. Operates close to set with simple wheeled console, and monitors with aid of high-quality headphones. His position on studio floor enables him to make rapid adjustments to microphones and localized acoustics.

a uni-directional or cardioid microphone is used, because reflections from set walls are not then so apparent.

SET NOISE. Film sets are notorious for introducing unwanted noise, since there may be as many as 50 or 60 persons present in the studio. Apart from the movements of the camera and crew, clicks from studio lamps can be heard with astonishing clarity. Other minor disturbances such as off-stage noises, electric fans used for wind effects, or heavy rain on the studio roof, can also spoil a scene. All these noises must be listened for, and reported to the director if they occur during a printed or selected take.

It is also the sound mixer's duty to warn the director if any dialogue is unintelligible, and also if one artiste overlaps another. The reason for avoiding overlaps is one of editing, for the inter-cutting of two scenes is difficult when the dialogue on the sound track is not clean. But since the mixer has no jurisdiction over the artistes' performance, the director may insist on overlaps to speed up the scene.

MICROPHONE PLACEMENT. Before deciding exactly where the microphone boom should be placed for a particular scene, the mixer and boom operator watch the first rehearsals. The operator eventually positions the boom alongside the camera, usually on the side opposite the main key light. He should do this before the set has been finally lit by the cameraman, so that any microphone shadows cast on the artistes or the walls of the set can be dealt with. Experienced cameramen can leave an area devoid of light beams about 8 ft. from the studio floor, so that the microphone is free to move in a shadowless zone.

The boom operator ensures that he can safely cover all movements of the artistes, while keeping to the microphone position required by the mixer. Where this is impractical, such as when dialogue is spoken near to the back wall of a set, a second microphone is employed attached to a fish-pole and suspended from the lighting gantry or set wall. If any lighting difficulties arise, a compromise microphone position is agreed upon by the mixer.

SOUND PERSPECTIVE. Binaural hearing gives a sense of direction and perspective. The distance of a sound is experienced by the ratio of direct to indirect sound waves. Out-of-doors there is very little indirect sound, and distance is judged by volume and quality alone. A close-up on the screen needs a close-up voice and a long shot a long shot voice. Since a microphone is a monaural device, the microphone-to-subject distance is varied to obtain perspective and thus maintain realism.

A good starting point is to place the microphone at about one-half the camera-to-subject distance, always providing that a normal angle lens is used such as a 35mm or 40mm on 35mm cameras, and a 16mm or 25mm on 16mm cameras. The mixer finds it useful to look through the camera viewfinder and see the exact perspective of each scene. The microphone should be positioned above the heads of the artistes, since this is the region of maximum intelligibility, and angled downwards at approximately 45°. It should never be placed closer than 2 ft. 6 in., although the usual working distance is much greater than this.

Close and distant artistes playing a scene together may sometimes be recorded with a single microphone, unless there are reasons which prevent this, such as camera movement, shadow problems, or one artiste facing in a difficult direction. In cases where two or more microphones are employed, it is safest to switch off those not in use since the effect of multiple microphones is increased reverberation. To cover more than one artiste at a time when they are facing each other, it is necessary to flatten the angle of the microphone so that it is almost parallel to the studio floor. This can often be done remotely by the boom operator. If excessive sibilance is encountered from any artiste, turning the microphone slightly sideways brings about an improvement.

VOLUME LEVEL. The mixer must contain all volume levels within the dynamic range of the recording system, although this may mean no more than raising a whisper or lowering a shout. Apart from his headphone monitor, the mixer has on his console a decibel meter calibrated from −10 to +2 dB. The average dialogue level is kept to approximately 12 dB below 100 per cent modulation, thus allowing an adequate margin for signal peaks. Since the meter indicates only the rms value of the signal, its sensitivity is increased by approximately 6 dB, so that better use can be made of the scale. This means that average dialogue shows −6 dB on the meter. The headphone monitor volume is adjusted in relation to the meter, so that a comfortable listening level is obtained.

In addition to setting the average volume level for speech, a certain amount of balancing between artistes is sometimes required. Experienced professionals try to match each other as nearly as possible, but a watch must be kept for low-level dialogue, especially from women and children. Untrained speakers with a nervous disposition can also prove difficult, and a request for them to speak up results in a far better recording than merely increasing the fader setting.

Whispered speech can cause concern, since it hardly registers on the dB meter and can only be assessed on headphones. A whisper increases the high-frequency content of the voice, and therefore the mixer must be on the lookout for sibilance. Camera noise can also be noticed if the fader is turned very much above its normal setting, and all other noises are also magnified.

Shouts are best dealt with by raising the position of the microphone or taking it further away, so that less direct and more indirect sound combine to produce a less peaky waveform. At the same time, the fader may have to be reduced to prevent excessive overload. Slight overload is sometimes unavoidable, and is perfectly acceptable since shouted voices are rough-sounding aurally.

Other loud sounds such as gun shots can occur during a normal dialogue scene. Even though blank cartridges are used, the steep wavefront of each shot must be kept within the limits of the system. A reduction of from 10 to 20 dB below the fader setting for dialogue may be necessary. Sometimes these floor-recorded gun shots are left in the final sound track, but it is quite usual to cut them out during editing and replace them with better ones.

LOCATION RECORDING. Recording out-of-doors away from the studio means changing over to portable tape equipment, but the principle of headphone monitoring is retained, also the dB meter. Dynamic microphones with a cardioid characteristic are employed in suitable windshields, so that any sudden air movements will not affect the microphone diaphragm and cause rumble.

In general, dialogue appears weaker out-of-doors due to the absence of any reflective surfaces, and the microphone must be kept fairly close. The mixer will be hard pressed to enforce his overhead position, since the angle of the sun may cause microphone shadows on the artistes. The only safe alternative is to place the microphone underneath the camera's field of view, directed vertically upwards. This gives a satisfactory dialogue recording, although footsteps may predominate, especially on gravel.

MUSIC RECORDING

The scoring of a film is primarily arranged between the composer, the director and the film editor. They decide exactly what kind of music is required, and at what footages within each reel the music sections will start and stop. The music mixer does not enter into any of these preliminary discussions, and is concerned only with the actual orchestral balancing and recording.

Music is recorded in a scoring stage where the walls are lined with a variety of sound-reflecting and sound-absorbing surfaces. The reverberation time is considerably higher than in a studio used for dialogue recording, and is usually in the vicinity of two seconds.

ORCHESTRAL COMPOSITION. Orchestras for film music vary considerably in size, and their composition is seldom identical for two pictures. Large symphony orchestras are employed for films of epic proportions, while smaller orchestras of 30 or 40 players are used for normal pictures.

Before the actual recording session, the orchestral composition is given to the music mixer who makes a sketch showing the position he requires on the scoring stage for each musician. He knows from previous mixing experience what type of music to expect, symphonic, dance, jazz, trad. or choral, and sets out the orchestra accordingly. The brass instruments will be on a rostrum at the back, then the woodwind section, and the strings at the front. This is a typical concert arrangement, with the first violinist sitting near to the conductor. Other arrangements are made for special purposes, such as separate brass, woodwind and percussion sections for split track recording, or for augmented dance orchestras with large rhythm sections.

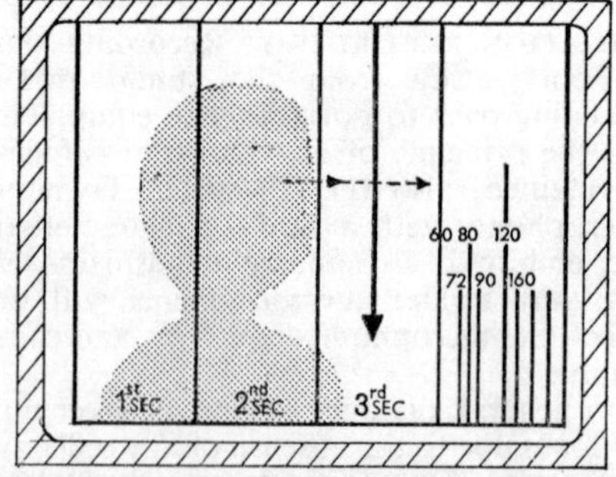

MUSIC-TIMING FILM. Synchronously projected on top of picture film for which music is being recorded. Conductor watches pointer moving towards instant of starting music section at end of fourth second. This final second is marked in metronome timing so that he can begin his beat at precisely correct moment whatever the music tempo.

MICROPHONE PLACEMENT. There are no rule-of-thumb methods for microphone placement although there are many schools of thought. Some of the finest orchestral music has been recorded with only a single condenser microphone, suspended about 8 ft. or 10 ft. above the conductor's head. This demands what is called an internal balance within the orchestra, the conductor controlling the dynamics of each player as at a concert.

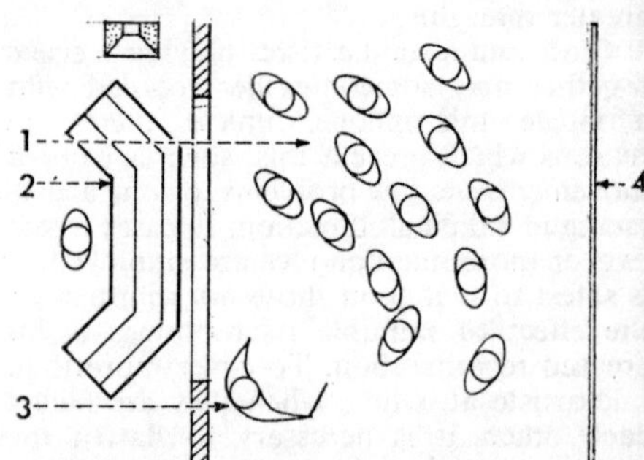

STUDIO ARRANGEMENT FOR MUSIC RECORDING. Conductor, 3, faces performers, 1, and screen, 4. Recording console, 2, is separated from studio by double-glass partition, so that mixer can watch action but hear only sound through recording system from theatre-type loudspeaker.

But film music is not concert music, and many solo passages from various instruments must be close to a microphone to achieve a correct balance. Therefore the mixer will blend the output from numerous microphones to obtain the required balance in the monitoring room, under the guidance of the composer.

It is usual to have three or four main microphones on a symphony orchestra for the string, brass, woodwind and percussion sections. This allows for some measure of control in mixing without interrupting the seating of the orchestra or their playing. Additional solo microphones are placed as required, allowing the solo instruments to predominate. Stereophonic music may require fewer microphones, since the separation between instruments is more pronounced on two or more tracks and an easier balance obtained.

LISTENING LEVELS. It has been established that sound mixers have a preferred listening level considerably higher than the average person. Only musicians approach the same level, since they are accustomed to hearing high intensities of sound from certain musical instruments. Since the ear does not have the same frequency response at all sound levels, the loudness at which loudspeaker monitors are used is quite important. At high levels, high and low frequencies appear relatively louder than middle frequencies, while at low levels the reverse is true. This means that the whole balance of sound is altered by the monitor volume setting.

It is fair to assume that the correct listening level is the one which matches the intensity of the original sound. For example, dialogue requires a lower monitor setting than does a large orchestra. While this is true to some extent, there are other factors to be taken into account. Dialogue is always reproduced at a much greater sound intensity in a cinema than in real life, but music is heard at almost its natural intensity. So the monitor volume is kept fairly high, especially in relation to the small size of the monitoring room, and it is the decision of the mixer to choose a setting which he finds most suitable for his needs.

The mixer finds it difficult to balance loud music at low monitor settings, although if a particular music section is required as a background for dialogue it is a mistake to monitor too high. The dynamic range of the recording system must also be taken into account for it is not nearly so accommodating as the human ear to low-level sounds.

RE-RECORDING

Re-recording, also known as dubbing, is executed in a specially equipped projection theatre set aside for the purpose. This is the process where the film begins to take on its final shape, and the first time that the director has the opportunity of viewing his work with a comprehensive sound track. The theatre contains a large mixing console for dealing with some 10 to 15 tracks, and has a variety of filters, equalizers and other frequency correction networks. At least two mixers are employed to operate the console, the senior mixer handling dialogue and music while the other looks after all the sound effects.

The mixers themselves are not responsible for physically laying the sound tracks in their correct place within each reel of

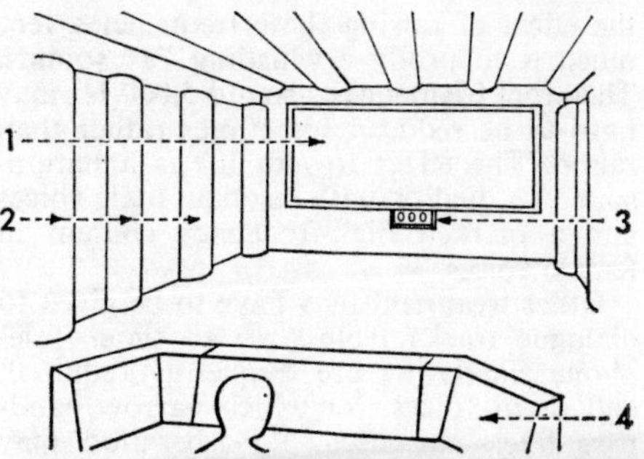

TYPICAL DUBBING THEATRE LAYOUT. Large screen, 1, simulates theatre conditions. Wall surfaces, 2, as well as ceiling, are faced with poly-cylindrical diffusers or other treatment to give desired acoustic characteristics. Footage counter, 3, tells mixer at console, 4, precise point reached by film being dubbed.

film, and are concerned only with re-recording material supplied to them by the editing department. The mixers' role is an extremely important one in picture-making, for their judgement alone determines the quality of the end-product from any particular studio. The same monitoring conditions apply to re-recording as during a music session, and most mixers prefer to have the loudspeaker monitor control a point or two higher than normal listening level.

DIALOGUE CORRECTIONS. The first task of the re-recording or dubbing mixer is to produce a smooth dialogue track, removing any irregularities in volume and frequency content. The dialogue tracks themselves have been prepared from material recorded over a period of from two to six months or more, by different mixers using different channels and different microphones. Although a standard frequency response may have been adhered to on the recording channels, there are bound to be quality changes introduced by set acoustics, microphone positions, and even artistes' voices varying in delivery and voice effort.

When correctly balanced, dialogue should have a good presence and intelligibility, with neither too much bass nor too much treble. For example, if male voices sound boomy due to set resonance or a poor microphone position, the response between 100 and 200 Hz (hertz = cycles per second) must be attenuated. With female voices, these figures become 200 and 300 Hz. Intelligibility can be improved by raising frequencies between 500 and 4,000 Hz, and in particular the region of 2,000 to 3,000 Hz. Above 4,000 Hz is the range of natural speech sibilants, and the effect of raising these frequencies very much is to produce whistling "s" sounds. Therefore frequencies around 5,000 Hz may have to be reduced sometimes rather than raised. The effect to aim for is a natural sounding quality, with resonant male voices and a marked high-frequency content in female voices.

Other treatment may have to be given to dialogue tracks from time to time. Telephone simulators are frequently required, and radio voices, for which narrow band-pass filters are used. Reverberation may have to be added in cathedrals, prison cells, or underground caves. Then there are numerous loudspeaker effects to be manufactured, combining band-pass filters with reverberation, and staggering tracks to produce an echo. Also used are background suppressors which automatically lower the background level between dialogue, especially on exterior recordings. Band-stop filters serve to remove narrow band interference from arc lamps, camera motors, or any high frequency sound.

MUSIC BALANCE. Music tracks will have been balanced and recorded under ideal acoustic conditions so, providing the frequency responses of the record, transfer and reproduction systems are identical and within working tolerances, no additional frequency compensation should be necessary during re-recording. But monitoring conditions are bound to vary between studios, and slight equalization is sometimes desirable to make music tracks sound correctly balanced to the ear.

For example, the bass end of an orchestral recording should be prominent, but not excessive or boomy, and this may vary according to the volume at which the music track is played. The violin quality should always appear clean and bright, but without any harshness to give an unnatural timbre. Woodwind and brass instruments should be heard with equal clarity if the recording is good, especially the trumpet. The piano is the most difficult instrument to judge, and in most cases is left unequalized.

The problem of balancing music against dialogue is one of personal taste and judgement, and it is seldom that any two people have the same idea of what constitutes a correct balance. The mixer tries to achieve what is correct in his opinion, and makes adjustments to the balance if requested by the director. But a music track seldom plays through a scene at one volume level, and it requires constant attention to achieve a satisfactory balance with all other tracks.

Generally speaking, music should never be played so low that it becomes lost in the background noise of an optical sound track, and never so high that signal peaks go into overload and cause harmonic distortion. Once music has been over-recorded in this way, the original quality can never be restored.

SOUND EFFECTS. The greatest number of tracks contain sound effects, and these are controlled almost exclusively by the second mixer. Apart from spot effects which synchronize with the action, and have to be assessed for quality and volume, there are a number of background tracks which require skilful blending. This often represents more work than can be handled with the equipment available, and some of the effects tracks are combined into a pre-mix.

Sequences involving car chases, fights and montages are pre-mixed whenever possible, since a large number of reproducing machines may be in use for only a short while during a reel. If all the dialogue tracks have already been pre-mixed, as is often the case, it is helpful for this pre-mix to be used as a guide during the balancing of the effects. Certain background sounds of a continuous nature, such as traffic, chatter, surf, birds and crickets, are often supplied to the mixers in loop form. It is always wise to pre-mix these and dispense with the originals to prevent wear and tear.

The first run-through of a reel with all tracks and pre-mixes determines whether any alterations or additions have to be made. If not, the reel is rehearsed several times, each run being nearer a correct balance than the previous one. Finally a take is made on to 35mm (or 16mm) magnetic film. A reference tone is recorded during the running-up period, and sometimes a brief announcement of the reel and take number. The track must be left clear for a sync pip lasting one frame immediately before the reel commences, so that any subsequent transfers on magnetic or optical can be accurately synchronized. The master dub is recorded across 35mm film on three separate tracks containing dialogue, music and effects. This gives flexibility when transferring, and permits the dialogue to be excluded when making combined music and effects tracks for foreign versions.

OPTICAL TRANSFERS. The mixer is responsible for transferring the master dubbed track of each reel on to an optical recording camera, which in theory means a simple run through without any quality or volume changes. Since the master is on three tracks, alterations in balance can still be achieved if necessary. New material can also be added during transfer, although this is not a desirable practice.

The reels of exposed sound negative are processed overnight, and the prints assessed the following morning in the re-recording theatre or other projection room. Assuming that there are no processing faults, and that the replay equipment is correctly aligned, there should be no difference in quality between the original magnetic dub and the photographic print. The superiority of the magnetic track will become apparent only when listening to a monitoring system with a flat frequency characteristic, and not a monitor which contains the necessary h.f. loss for optical replay conditions.

Since optical sound rushes are of no value after they have been checked, except as rolls of spacing for the cutting rooms, it is sometimes the practice to print only the first 100 ft. of each reel as an economy measure. The reason for having rush prints at all is to check the negative processing, and the resulting sound quality.

CHECKING ANSWER PRINTS

At the completion of a film, the sound mixer attends a private screening of an answer print. This is the first combined print from the sound and picture negatives, and is usually run in a preview theatre or occasionally a large cinema. This is a real check on the quality, for listening conditions in the studio have been confined to small theatres.

To make an accurate assessment in an unfamiliar cinema, it is prudent to run a reel of some other picture for comparison of general sound quality, as well as the required fader setting in the projection room. Otherwise criticisms of quality and volume may be directed at the sound mixer, when in fact the fault may be with the projection equipment or the cinema auditorium.

SOUND QUALITY. The first point that will probably be noticed is a lack of high frequencies as compared to a studio viewing room, especially with optical sound. This is due to the fact that the viewer is usually much further away from the loudspeaker system, although the decor and projection equipment can be contributing factors. Even the air in the auditorium acts as a high-frequency absorber, especially when the atmosphere is dry. Because of this, large cinemas usually have walls and ceilings offering less absorption to high frequencies, whereas converted theatres seldom have this feature.

Sound quality may vary at different positions within the auditorium. If the cinema is empty and has a balcony, the best listening position is usually just underneath. The balcony and front stalls may suffer from first order reflections, giving an unpleasant delayed echo. In fact, excessive reverberation is often the cause of poor intelligibility, and the film sound track is not always to blame.

Fortunately, modern cinemas are good from an acoustical point of view, and they also have an extension fader in the auditorium so that volume adjustments can be made by the sound mixer.

All stereo installations have a ganged fader for the back-stage loudspeakers, and a separate fader for those surrounding the auditorium.

The quality of any film sound track appears to vary when heard in different cinemas, owing to the coloration given by the auditorium and the particular projection installation. Only the skill of the sound mixer can produce a final track which will be satisfactory in a large number of cinemas.

J.B.A.

See: *Acoustics; Sound and sound terms; Sound distortion; Sound recording; Sound recording systems; Sound reproduction in the cinema.*

Books: *Acoustics*, by G. W. Mackenzie (London), 1964; *Microphones*, by A. E. Robertson (London), 1963; *High Quality Sound Production and Reproduction* by H. Burrell Hadden (London) 1962.

□**Sound on Vision.** A fault condition (due to mistuning or maladjustment) in which the sound signal appears as a cross-modulation of picture information, or causes break-up of the synchronization. At least 40 dB of sound signal rejection is required if sound-on-vision is to be avoided. In receiver design, therefore, one or more trap circuits is provided before the video i.f. amplifier to produce maximum attenuation at the adjacent sound channel frequency and zero attenuation at the video carrier frequency.

See: *Receiver.*

□**Sound Pre-Listen.** Arrangement which allows sound to be checked by the operator before it is broadcast or recorded.

See: *Master control.*

⊕SOUND RECORDING

The organization behind the smooth day-to-day running of a studio recording department is fairly complex. It must be reliable as well as efficient, since time is all-important in film production and a five-minute delay over a sound problem can cost a considerable sum of money. Furthermore, the output from the department must be consistent in quality, regardless of the difficulties involved in attaining that end.

STUDIO STAFF

RECORDING SUPERVISOR. The recording department in a major studio is staffed by many technicians, and the task of co-ordinating their various activities is the responsibility of the recording supervisor. He should be a good administrator, capable of placating irate producers, and a first-class technician, capable of solving all staffing and equipment problems. In addition he is responsible for all bookings made by various productions for services offered within the department, such as post-sync, effects recording, scoring, and re-recording or dubbing.

TRANSMISSION ENGINEER. The personal assistant to the recording supervisor is usually a qualified transmission engineer, who is responsible for the proper functioning of the departmental installations, and for the equipment requirements of all the productions working in the studio. He is continually reviewing modern trends in techniques and equipment, so that he can plan future installations and modifications to existing ones. He has a number of maintenance engineers, each allocated to a particular production, who work in close contact with their own sound mixer and attend to his requirements from hour to hour.

SOUND CREWS. Each sound mixer on a production has his microphone operator and sound camera operator. It is essential that they all understand each other's method of working, so that use of the talkback system can be cut to a minimum. The sound camera operator, although not necessarily a skilled engineer, should be familiar with all the basic types of recording equipment used in film studios, so that he can operate each one of them without instruction or supervision.

Sound crews are also employed within the department on post-sync and effects, scoring and dubbing. The mixers on these particular jobs seldom carry out all the recording for a production, except in a small studio. They are specialists in their particular field, and tackle a large number of pictures in the course of a year.

RECORDING SYSTEMS

There are four basic systems of sound recording used in film production today: magnetic film, magnetic tape, optical or photographic film and lacquer discs. Each system is used for a particular purpose, and they have all been devised from many years of practical experience.

MAGNETIC FILM. All prime recordings inside a studio are made on a standard 35mm, 17·5mm or 16mm magnetic recorder. The film perforations then act as a reference for synchronization to the relative picture material during editing, and ensure that sound and picture records remain correctly synchronized when passing through a magnetic reproducer or double-headed projector with sepmag facilities. The standard track width on 35mm is 200 mil (0·200 in.) located inside (and adjacent to) the perforations. There is room for three such tracks across the width of the film.

The standard track widths on 16mm using single perforated stock are 100 mil adjacent to the unperforated edge, 200 mil adjacent to the unperforated edge, and 200 mil located in the centre of the film.

The frequency response of a magnetic channel is maintained flat from 50 to 12,000 Hz ($\pm\frac{1}{2}$ dB), so that second, third, fourth or even fifth generation transfers can be made without any major alteration to the frequency content. It is also essential that strict control of azimuth is maintained to facilitate world-wide exchange of recorded material.

The standard reference film to which all 35mm magnetic recorders are aligned is the RCFM, available from the Society of Motion Picture and Television Engineers. It is an original recording of three 200 mil tracks on fully coated film, containing a series of spot frequencies from 50 to 12,000 Hz. Each film is individually calibrated and correction factors are provided. For 16mm recorders and reproducers, a similar film is available, type M16MF.

MAGNETIC TAPE. Tape equipment is far less bulky than that required for magnetic film, yet the sound quality can be maintained at a high standard. The overall frequency response is identical to that of film, and the recording characteristics usually adhere to the CCIR curve of 35 μsec at 15 ips or 70 μsec at 7½ ips.

In the absence of perforations, a camera pulse track must be recorded in addition to the audio signal, so that a camera speed reference is available when the selected material is transferred to 16mm or 35mm magnetic film.

OPTICAL FILM. Until 1949, all studio and location recordings had to be made on optical film using relatively bulky equipment. Since magnetic sound is still not universally used in cinema projection rooms, a photographic track is required for making release prints. Two types of track are in current use, variable density and variable area. The variable area type consists of a constant density track varying in width according to the modulation envelope. Variable density implies a constant-width track varying in density according to the modulation envelope.

Theoretically, area variation has the advantage over density variation since it is possible to obtain at least 6 or 8 dB more in volume. The well-defined outline between the exposed and unexposed portions of the track gives a sharper reproduction of transients, and the lower grain noise from the opaque portion permits a more faithful rendering of bass frequencies.

The advantage of the variable density track is that it replays satisfactorily under adverse conditions, such as a badly aligned projection system or when the film is well-worn. Each type of track reacts to overload in an entirely different manner. The signal peaks are completely chopped off on the variable-area type, causing undesirable distortion and a lack of clarity. A variable density track enters the overload point more gradually, and on certain sounds, such as explosions, as much as 6 dB can be accepted without noticeable distortion.

LACQUER DISCS. Modern acetate lacquer discs are occasionally used for playback facilities of previously recorded material. This applies particularly to musical sequences, where the sound has been pre-recorded in one complete take. The artistes mime their performance while the camera is set up in perhaps a dozen different positions, shooting only a small section at a time. These discs are usually cut with 96 grooves to the inch, and at 78 rpm. This makes accurate groove location a fairly simple process.

Quarter-inch tape playback is now employed to a large extent within the film industry. The tape contains, in addition to the audio track, a pulse reference track for synchronizing the playback machine to the camera motor supply system.

EQUIPMENT

Film recorders are made by a few specialist manufacturers. Portable tape recorders with pulse track facilities are numerous, as are microphones and other ancillary equipment.

MICROPHONES. Uni-directional cardioid microphones are preferable for dialogue

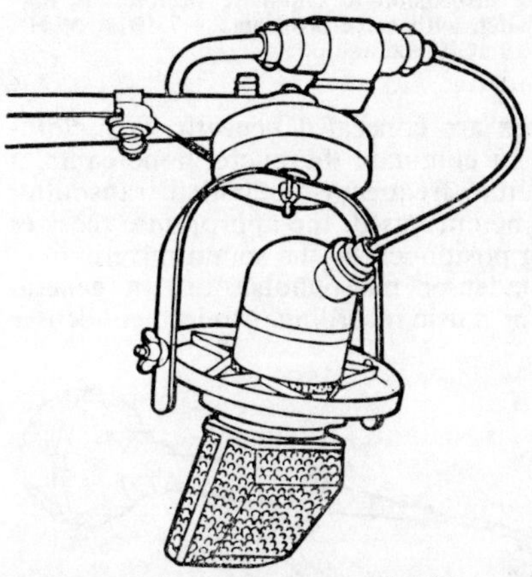

CARDIOID RIBBON MICROPHONE. Flexibly suspended inside ring, adjustable by wing nuts for downward angle. (*Above*) Cord drive from operator's end of microphone boom, providing rotation to follow artistes' movements on sound stage.

recording since they can be operated further away than omni-directional types and still obtain a satisfactory sound pick-up. Miniature condenser microphones are rapidly gaining favour. Their small weight makes them ideal for use on the end of a fishpole, and they can be supplied with pocket-sized power units.

Ultra-directional or line microphones are sometimes used when a closer sound pick-up is required than can be obtained with an ordinary cardioid microphone. The improvement is usually worthwhile, but for optimum sound quality this type of microphone must be accurately aimed at the required sound source.

Where it is impossible to position a microphone near the artistes, chest micro-

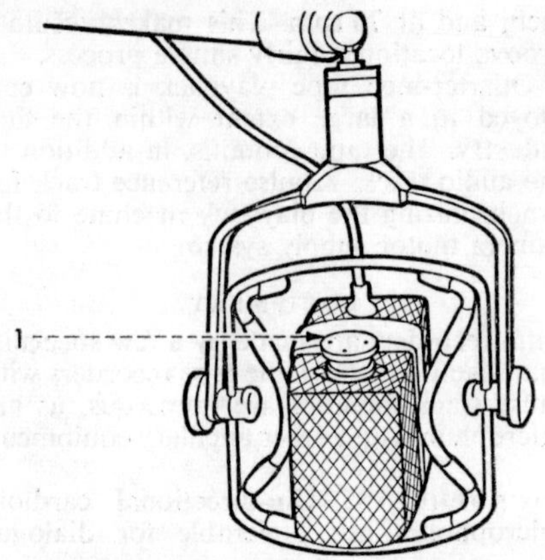

CARDIOID DYNAMIC MICROPHONE. Fitted with soft rubber suspension to eliminate rumble. 1. Bass cut switch with three positions, −7 dB at 50 Hz, −12 dB at 50 Hz, and off.

phones are concealed beneath their clothing. To eliminate the microphone cable, a miniature frequency-modulated transmitter is sometimes used, the appropriate receiver being positioned by the sound mixer.

Condenser microphones are in general use for music recording. Modern condenser

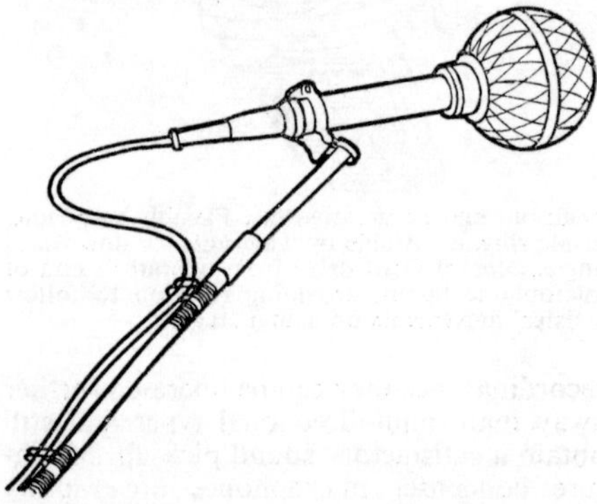

CARDIOID CONDENSER MICROPHONE. Equipped with spherical windshield for outdoor use, and mounted on a hand-held fishpole.

capsules are extremely sensitive, and have a flat frequency response from 50 to 18,000 Hz. Owing to a special double diaphragm construction, their directional characteristic can be adjusted to suit studio conditions. Bi-directional ribbon microphones are also popular.

BOOMS. A typical microphone boom provides an adjustable platform which enables the operator to stand 4 or 5 ft. above the ground, and the boom arm is capable of extending to about 15 ft. The microphone cradle rotates through 360° without transmitting any rumble, and a tilt control is sometimes fitted to permit the angle of the microphone to be remotely adjusted.

The wheelbase is less than 3 ft. so that the boom will pass through a standard doorway, yet is capable of extension so that the platform will support the weight of the operator without overturning.

MIXING CONSOLES. These vary in size from a small 2- or 4-way unit for dialogue recording, with volume indicator and monitoring facilities, to a 10- or 20-way console with tone compensators, filters, and reverberation controls for use during scoring or re-recording. The smaller units are often mounted on a trolley for moving around the film set, together with spare microphones and other accessories.

HEADPHONES. Moving-coil headphones are in common use and have a good frequency response. They have close fitting pads to exclude all direct sound from the ear, and are accurately matched to the impedance of the monitoring outlet.

PORTABLE RECORDERS. Many recorders with pulse track facilities are now available. They are transistorized and operate in the conventional manner, except in the method of recording the pulse. Some types use one half-track for audio and one half-track for pulse. Others employ a narrow pulse track, leaving about 80 per cent of the tape width for audio. Another device is to use the entire tape width for audio, and superimpose a narrow push-pull pulse track down the centre. The audio replay head ignores this track, and crosstalk is negligible.

The actual pulse may be derived from the a.c. supply to the camera motor, or from the output of a special tone generator attached to the camera mechanism. Some cameras are merely fitted with a pair of contacts which open and close once per frame, and these are used with an oscillator to give a 24 Hz pulse. Re-recording facilities for handling all types of pulse track are available at the studio.

RECORD AMPLIFIERS. Although portable equipment is transistorized, studio installations are mostly valve-operated. Limiter and compressor power amplifiers are in constant use to contain unexpected signal peaks within the dynamic range of the system, particularly with optical recording.

Limiter amplifiers reduce gain by a ratio of 10 to 1 as the signal volume reaches 100 per cent modulation. Compressor amplifiers, on the other hand, reduce gain at a predetermined level below 100 per cent, and at variable ratios such as 20 dB into 10 dB and 8 dB into 4 dB, giving 2 to 1 compression at different percentage modulation. All these amplifiers have two main characteristics: the attack time, the interval taken to reduce gain (usually 1 or 2 milliseconds), and the release time, the interval required to restore gain to normal (usually variable in steps from 0·1 to 1 second).

A small line compressor is also used, containing a light-sensitive cell and transistor amplifier connected to a miniature lamp. Loud signals cause the lamp to light, which increases the resistance of the cell, thereby reducing the level of the audio signal. The attack time is slower than that of a valve amplifier, but can be improved with special filament bulbs.

MAGNETIC FILM RECORDERS. A conventional film recorder consists of single or double drive sprockets, mechanical filter rollers, and two dynamically balanced flywheels and sound drums. Separate record and replay heads are mounted adjacent to and between the drums, but erase heads are seldom fitted. Magnetic film on plastic cores is loaded on to plates or split spools, and the entire recorder is driven by an interlock or synchronous motor.

One of the most disturbing features of film mechanisms is a tendency to 96 Hz flutter. When 35mm film is running at the standard rate of 90 ft. per minute, 96 perforations per second pass the recorder head. However, the resulting flutter has largely been eliminated by heavy twin flywheels and a spring roller filter damped by an oil-filled dashpot. Another frequency at which flutter is very perceptible is 3,000 Hz, owing to the high sensitivity of the ear in this region.

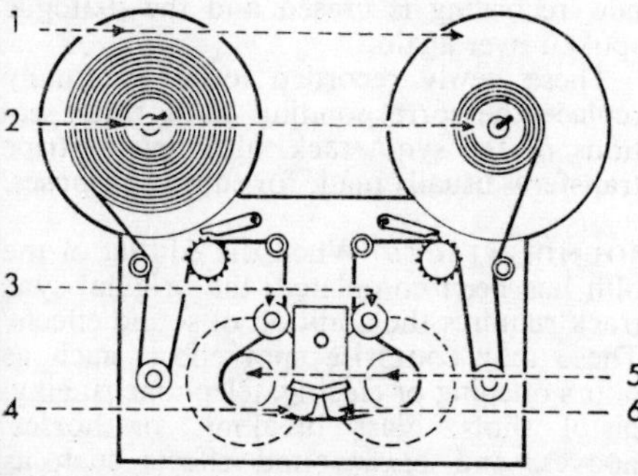

FILM PATH OF TYPICAL MAGNETIC RECORDER. 1. Metal flanges. 2. Film cores. 3. Film guide rollers. 4. Record head. 5. Balanced sound drums with flywheels attached to damp irregularities in film motion. 6. Monitor head.

DISC RECORDERS. At least one disc recorder is to be found in all studio recording departments. It consists of a heavy platter to support the acetate disc, a lathe bed for traversing the cutter, and a drive motor compatible with the studio motor system. Discs are recorded with the usual constant amplitude characteristic, similar to the RIAA standard. In Britain, the turntable speed is usually 77·9 rpm, but the American speed of 78·26 is also employed, and care must be taken to avoid confusion between the two speeds.

LOUDSPEAKER SYSTEMS. Cinema-type loudspeaker systems are used for monitoring, capable of handling from 20 to 80 watts according to the size of the room. Separate high-frequency and low-frequency units are fed from a frequency dividing network, the crossover frequency being approximately 500 Hz with a 12 dB per octave attenuation on either side of this point. The h.f. unit is fitted with a cellular horn, or slanting vanes, to provide an even distribution of high frequencies.

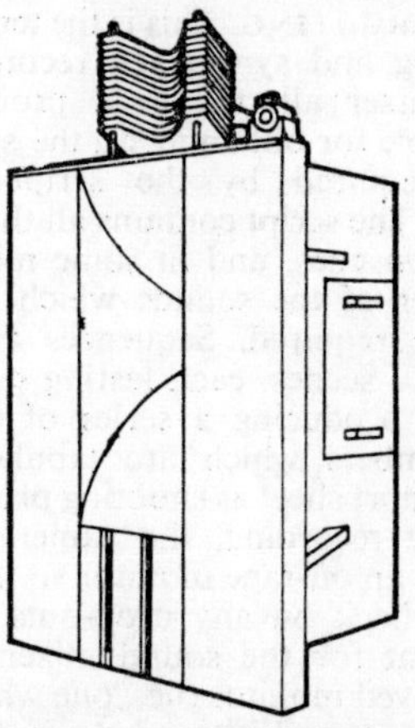

CINEMA LOUDSPEAKER SYSTEM. Typical arrangement with separate low- and high-frequency units. Low-frequency unit below has flared semi-exponential horn. High-frequency unit is fitted with acoustic vanes for improved sound distribution in large auditoria.

STUDIO RECORDING PRACTICES

A film sound track is made up from many individual sections, recorded at various times by different mixers, and often from several studios. The track begins with the actual sound recorded at the time of shooting, in synchronization with the picture camera, and is universally referred to as the sync track. This is edited, together with the picture material, and notes taken of any poor-quality sound which needs replacing.

Fresh recordings of dialogue are subsequently post-synchronized in a small projection room set aside for this purpose under the supervision of the film's director. Some additional material also has to be recorded, such as music and sound effects. The entire film is later re-recorded to obtain a composite magnetic track, followed by a transfer process to produce a photographic sound track negative. A combined music and effects track is another normal requirement, for use when dubbing the film into a foreign language.

FLOOR SHOOTING. This is the term applied to filming and sync track recording. The sound mixer allotted to a production is responsible for obtaining all the sync sound tracks required by the script and the director. The script contains all the dialogue and action cues, and in some measure the remainder of the sounds which will eventually be required. Sequences are broken down into scenes, each lasting only 1 or 2 minutes, producing a series of scene and take numbers which are tabulated on a sound report sheet as shooting progresses.

During recording, the camera operator listens to an off-tape monitor so that he can keep a check on any drop-outs. It is not convenient for the sound mixer to use a tape-delayed monitor (i.e., one which reproduces the already-recorded sound from a separate head), since he is sited close to the film set and reacting to visual cues.

At the end of each day, the recorded material is delivered to a sound transfer bay, where the selected takes are re-recorded on to 35mm magnetic stripe. All rolls of original sound have a reference tone of either 400 or 1,000 Hz on the front, usually recorded at a level of 10 dB below 100 per cent modulation. These master rolls are handled as carefully as photographic negatives, and stored in a vault. The stripe transfers are known as rushes, and are collected by the editing staff for synchronization to the picture.

LOCATION SHOOTING. Although the sound crew use small portable recorders, they may also be responsible for supplying power to the picture camera motors. This may mean looking after a three-phase rotary converter and battery supply, or a much smaller transistor inverter. In either case, additional equipment and cables are required.

Filming out-of-doors has the usual hazards of sound interference from noisy backgrounds, and the close proximity of an arc lamp generator can often frustrate attempts to obtain clean dialogue recordings. Such scenes are always announced as a guide track only, and marked accordingly on the picture number board. The sound is used as a guide when replacing the dialogue with a clean recording under studio conditions.

POST-SYNCHRONIZATION. Post-synchronized dialogue is recorded in a small projection room specially equipped for handling short film loops. The scenes to be synchronized are broken down into single sentences, and loops are made of picture, guide track, and an unrecorded or virgin magnetic track. Short sections of the picture and sound can then be examined as the loops are projected.

The sound mixer places the microphone in an optimum position to obtain the correct sound perspective for each loop, and adjusts the acoustic conditions of the room as he thinks fit. The artiste copies his or her original dialogue for performance and synchronization, while listening to the guide track on headphones. A fresh recording is made, and immediately played back over the loudspeaker system so that it can be compared with the original sound. If the sync or performance is not acceptable, the new recording is erased and the dialogue spoken over again.

These newly recorded loops eventually replace the corresponding guide track sections of the sync track, although a stripe transfer is usually made for editing purposes.

SOUND EFFECTS. When the editing of the film has been completed, the original sync track requires the addition of sound effects. These may comprise spot effects such as doors opening or closing, telephone ringing, pistol shots, glass breaking, or horses' hooves, and background effects such as traffic, birds, waterfall, or airport noise to replace the ambient sound behind post-synchronized dialogue.

A large number of these can be obtained

already recorded from a sound effects library, but many of the spot effects have to be tailormade to fit the film. Loops are made up as for post sync sessions, and the effects are manufactured and performed in synchronization with the picture.

SCORING. Film music is recorded in a specially contructed studio with well-defined acoustics, known as the scoring stage. It is designed to have a relatively high reverberation time, usually 2 or 3 seconds, and its size depends on the type of orchestra it is intended to use. Orchestras of 100 musicians require a cubic capacity of 300,000 cu. ft., but an average-sized orchestra of 30 to 40 musicians only requires about 100,000 cu. ft. Small pop groups and solo instruments can be recorded in almost any small studio with less reverberation time, controlled reverberation being added as required from an echo chamber or reverberation plate.

Since all music recording (except pre-scoring) will probably be carried out to picture, i.e., with the conductor facing the screen and watching the film run through, a full-sized screen and projection room are available. There is a large monitoring room and sometimes a separate vocal room, both having double glass windows facing the scoring stage and screen. The monitoring room has a reverberation time of from $\frac{1}{3}$ to $\frac{1}{2}$ second, achieved by the application of sound-absorbing surfaces to parts of the walls and ceiling.

The music mixer has at his disposal a console containing multiple microphone inputs and pre-amplifiers, linear faders, and mono or stereo switching on 1, 2 or 3 tracks across 35mm magnetic film. Each microphone channel has its own frequency compensators, together with individual echo mixture controls for adding reverberation when required. Monitoring loudspeakers are of the type installed in small cinemas, consisting of 12 in. or 15 in. cone l.f. units, crossover networks, and diaphragm type h.f. units fitted with cellular horns or acoustic lenses for directivity.

Music is recorded in sections, each one being identified with its particular reel of film by a number, 1M1, 1M2, 2M1, 2M2, etc. Start marks are made on both sound and picture films, and the sound camera operator also makes a start mark on the magnetic recorder. These start marks are located 15 ft. to 18 ft. before the actual scored section begins, so that the projector and recorder can run up to speed and settle down before the first note of music is played. A mechanical counter showing minutes and seconds assists the conductor in keeping to his music cues throughout a section, and the dialogue from the sync track is available to him on headphones if required.

Each selected take is immediately played back for checking, and is heard in the monitoring room together with the dialogue track. At the end of each day, the master music rolls are transferred to provide magnetic music tracks for the editing department.

RE-RECORDING OR DUBBING. This process is carried out in a medium-sized projection theatre whose acoustic conditions resemble those in an average-sized cinema as far as possible. The auditorium contains a large console operated by two or more sound mixers. The console is fitted with a number of rotary or linear type faders, volume indicator meters with a dB scale, and an assortment of filters and equalizers. These include high and low pass filters for attenuating frequencies below or above a selected value, hi-mid-lo equalizers for frequency correction, telephone simulators, and tunable band stop filters for removing objectionable sounds over a narrow band-width. Reverberation facilities are also available.

The projection room is equipped with a row of 35mm or 16mm magnetic reproducers, capable of being interlocked to the projector and sound recording camera. For every reel of picture there may be a dozen or more reels of sound, such as two dialogue tracks, two music tracks, and eight effects tracks. These are threaded on the reproducers with start marks opposite the magnetic heads, and the sound mixers given cue sheets to guide them through the reel. During every run, footages on the cue sheet are related to an illuminated footage counter below the screen, which provides an exact indication of the position within the reel.

The output from the console is split to cover three tracks on 35mm film: dialogue, music, and effects. This enables any slight correction in balance to be achieved during the next transfer stage, and allows the dialogue to be easily removed to make a foreign version. Crosstalk on the three-track reproducer should be of the order of −60 dB, otherwise there will be an audible leak of dialogue on to the adjacent music track.

The signals from the three separate channels are combined for monitoring pur-

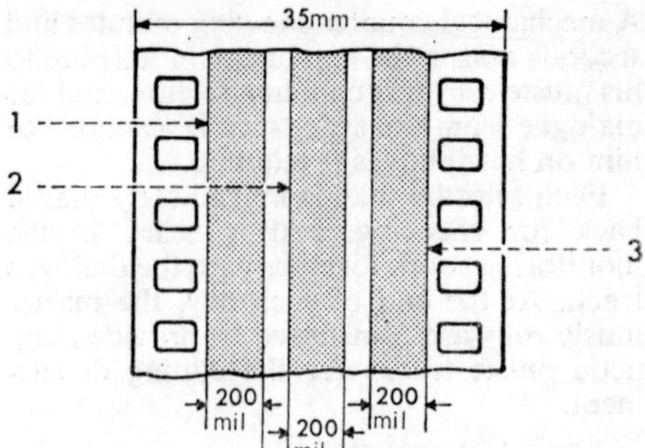

MAGNETIC TRACK LOCATION ON 35MM FILM. Three 200-mil (0·2 in.) tracks symmetrically positioned across 35mm fully-coated magnetic film. 1. Dialogue track. 2. Music track. 3. Sound effects track.

poses, and the result is balanced over a single loudspeaker system which has a standard cinema frequency characteristic (−5 dB at 4,000 Hz and −10 dB at 8,000 Hz).

ROCK'N ROLL. Also called roll-back, this is a motor control system very popular among technicians and producers. It enables the entire equipment to be stopped in interlock at any place within a reel, and then run back without losing synchronism as and when required. This permits the close examination of a film section to achieve a better sound balance, since the operator can run the section several times with all tracks without having to unthread the projector and sound reproducers. The installation also includes the magnetic film recorder, so that the dubbing mixer can insert a new section into a previous recording by operating a record/check switch at the appropriate time.

An extension of the roll-back system is incorporated in a dubbing machine of recent design, which is assembled from add-on units of similar appearance. They consist of horizontal film transport mechanisms of the kind preferred in continental Europe, in which film is loaded on flat plates or flanges. One unit is a magnetic or optical recorder, others are sepmag reproducers, and the last is a picture head with optical compensator in which the "stationary" frame is scanned by a vidicon tube. The signal output is processed and fed to a monitor screen in a small mixing theatre which does not need to adjoin the dubbing room. The film transport mechanisms are geared together on to a common driving shaft, and incorporate fast forward and reverse motion, there being no intermittent movements to limit film speed. 35mm and 16mm drive sprocket assemblies are interchangeable.

The only drawback to this system is the small picture size, though this could be overcome by projection with an Eidophor.

OPTICAL TRANSFERS

Final master-dubbed tracks are usually contained on 16mm or 35mm magnetic film, and occasionally ¼ in. tape carrying a reference pulse track. A photographic sound negative is now required, which can be combined with the edited picture negative to produce a composite or married print. The transfer to optical track is carried out by the sound mixer and his crew, using a special sound recording camera. This consists of a light-tight film chamber containing sprocket drive, sound drum, stabilizing rollers, exposure lamp, modulating device and optical system. A standard 1,000 ft. film magazine is secured on top of the chamber, and the camera is driven by either a three-phase synchronous or interlock motor.

Two types of modulator are employed: a mirror galvanometer which produces a variable-area track, or a light valve which produces either a variable-area or variable-density track, according to the way in which it is installed in the camera.

GALVANOMETER. In a standard RCA sound camera, light from the exposure lamp

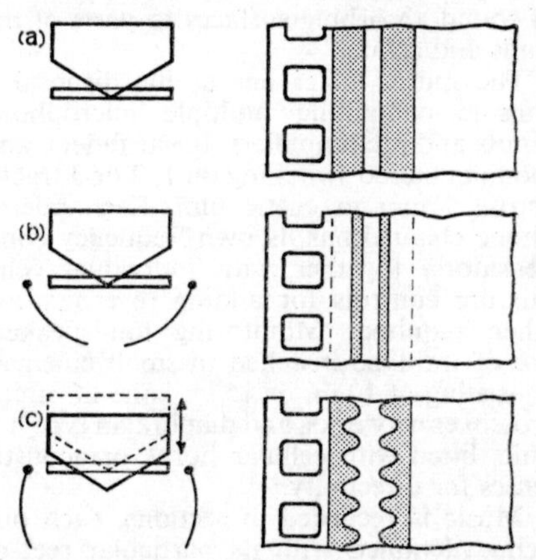

NOISE REDUCTION. (a) No audio signal, no noise reduction. (b) No audio signal, noise reduction shutters closed. (c) Full audio modulation, shutters swung clear.

is passed through a V-shaped mask and convex lens on to a mirror galvanometer. The shape of the mask is then reflected through a narrow horizontal slit, the image of which is focused on to the film through an ob-

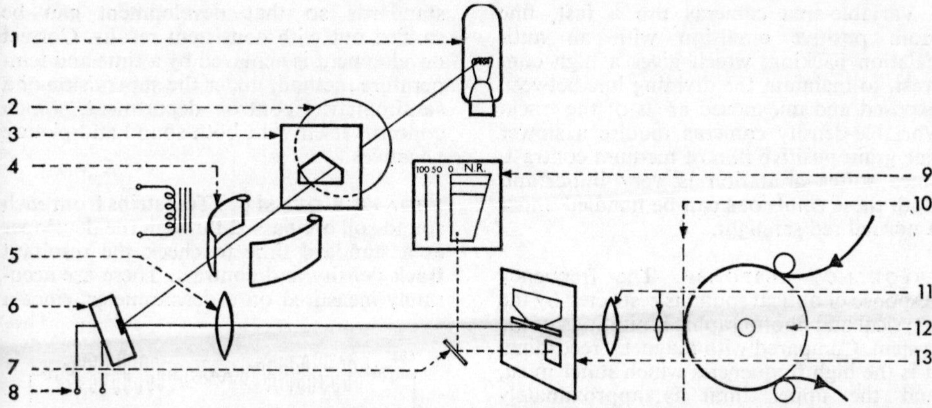

VARIABLE-AREA RECORDING: OPTICAL SYSTEM. 1. Exposure lamp. 2. Condenser lens. 3. Beam-shaping mask. 4. Noise-reduction shutters. 6. Galvanometer and mirror with audio input, 8. 5. Objective lens. 7. Mirror for monitor card, 9. 13. Plate with exposure slit and reflective area for monitoring. 11. Objective lens system. 12. Heavy sound drum to damp irregularities of movement of film wrapped round it in passage between feed and take-up sprocket. 10. Film plane at point of exposure.

jective lens system. Audio signals are fed to the galvanometer windings, causing the mask image to move vertically in front of the slit. This results in a varying width of light falling on the film, producing a photographic trace of the audio signal waveform.

To improve the signal-to-noise ratio of the resulting print, a pair of shutters are fitted which close in on the area outside the mask image when there is no audio signal. These shutters are biased from a ground noise reduction amplifier, and lift out of the way at the beginning of an audio signal.

LIGHT VALVE. Light valves are employed as modulators in Westrex sound cameras, and are virtually string galvanometers. They consist of two or more metal ribbons, clamped vertically under tension in a strong magnetic field. Audio signals cause displacement of these ribbons, and the incorporation of a horizontal slit in the optical system focuses a variable width trace of the audio waveform on to the film. Rotating the light valve through 90° and dispensing with the slit causes the varying light intensity passing through the horizontal ribbons to produce a variable overall exposure, and therefore a variable density-track.

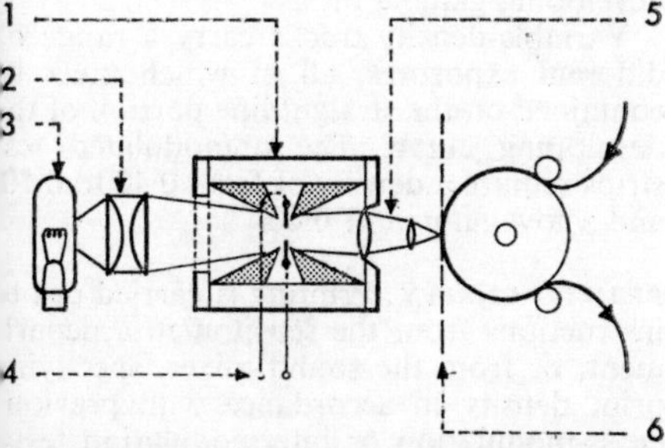

VARIABLE-DENSITY RECORDING: OPTICAL SYSTEM. Lamp, 3, and condenser, 2, form light beam modulated by light valve, 1. 4. Audio input to valve ribbons. 5. Objective lens system forms image on film plane, 6.

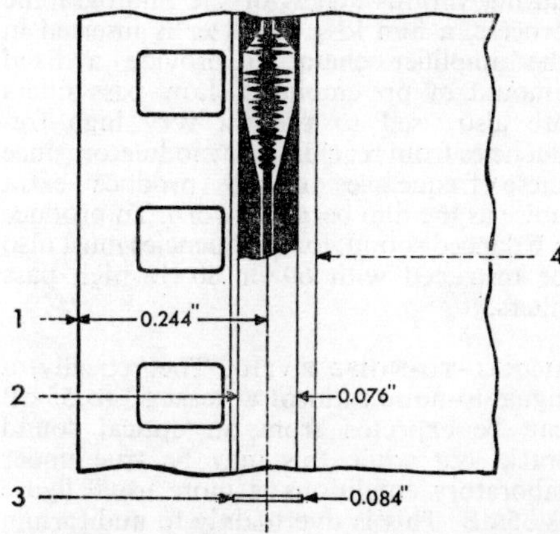

LOCATION OF VARIABLE-AREA SOUND TRACK. 1. Film edge to track centre. 2. Width of track image. 3. Width of projector scanning slit. 4. Limit of picture area.

Noise reduction is achieved by means of a bias voltage applied to the ribbons, which reduces the exposure of a density negative and the width of an area negative. The bias is automatically removed when an audio signal is received by the ground noise reduction amplifier.

Variable-area cameras use a fast, fine grain positive emulsion with an anti-halation backing, which gives a high contrast, to maintain the dividing line between exposed and unexposed areas of the track. Variable-density cameras require a slower fine grain positive film of medium contrast, since softer gradation is very important. Both these emulsions can be handled under a normal red safelight.

FREQUENCY RESPONSE. The frequency response of optical sound is restricted by the physical and photographic limitations of the system. Compared with magnetic recording, it is the high frequencies which suffer most, and the upper limit is approximately 6,000 Hz for 16mm and 8,000 Hz for 35mm. This is determined by the size of the slit through which the exposure is made, and the degree of resolution obtainable in the film emulsion. Theoretically a slit of 0·002 in. enables frequencies up to 9,000 Hz to be recorded. But the image formed on the developed film is always greater than the calculated value, and it is current practice to use a much smaller slit of 0·0005 in. The slit itself is of more generous dimensions, the correct-sized image being formed by the objective lens.

Since high frequencies may become lost during various stages of the photographic process, a film loss equalizer is inserted in the amplifier chain to provide a fixed amount of pre-emphasis. Low pass filters are also used to prevent very high frequencies from reaching the modulator, since these frequencies merely produce extra noise as the film becomes worn. To produce a balanced sound, low frequencies must also be restricted with 60 or 80 Hz high pass filters.

SIGNAL-TO-NOISE RATIO. Theoretically, a signal-to-noise ratio of at least 50 to 55 dB can be expected from an optical sound track, but while this may be true under laboratory conditions, a more usual figure is 35 dB. This is due mainly to auditorium noise masking low level signals, coupled with variations in transfer level of the magnetic track, and processing.

ASSESSMENT OF OPTICAL TRACKS

Once the optical sound negative has been exposed in the recording camera, the material leaves the care of the sound mixer and is collected by the processing laboratory. Chemists are responsible for maintaining all processing solutions at the correct standards so that development can be carried out with consistent results. Correct development is achieved by a time and temperature method, under the supervision of a sensitometric control department, which concerns itself with both sound and picture negatives.

NEGATIVE DENSITY. Test strips from each sound roll are passed through the developer at a standard time to check the resultant track density and contrast. These are accurately measured on a densitometer, since a purely visual inspection would be meaningless. The negative density required for variable area tracks is usually 2·8, and the test strips should come very close to this figure with correct exposure in the sound camera. Any adjustments required are carefully calculated, and the final developing time for the correct density is given to the machine operator. The correct contrast is obtained when the solutions provide a developing gamma of 3.

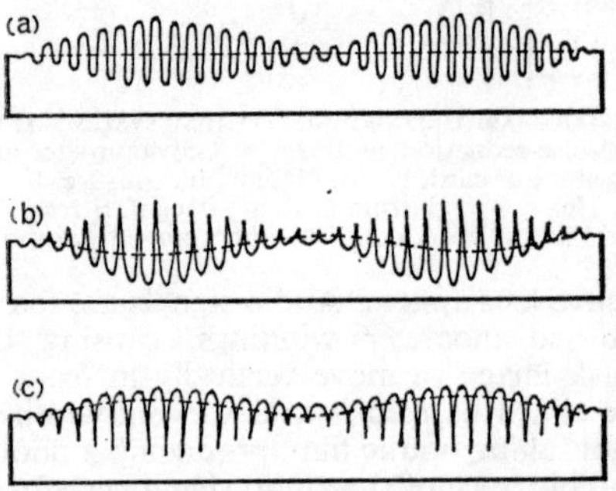

EFFECT OF EXPOSURE ON NEGATIVE IMAGE. Variable-area sound track. (a) Correct exposure, good sine waveform. (b) Under-exposure, distorted waveform. (c) Over-exposure, waveform filled in by halation.

Variable-density tracks carry a range of different exposures, all of which must be contained on the straight line portion of the developing curve. The unmodulated test strips require a density of from 0·40 to 0·50 and a low gamma of 0·40.

PRINT DENSITY. Printing is carried out to instructions from the sensitometric department, or from the sound mixer, specifying print density in accordance with previous cross-modulation or intermodulation tests. Each negative density requires a corresponding print density if distortion is to be avoided. The print density for area tracks is usually just over half the negative density although a similar value suffices for both negative and prints of a density track.

CROSS MODULATION. To find out the optimum exposure and processing conditions for both negative and positive, a series of cross-modulation tests is carried out by the sound mixer, preferably with each new batch of stock. Photographic emulsions tend to diffuse images to a small degree, depending on the grain structure and speed of the emulsion, as well as development. Image spread must be controlled to preserve a good high frequency response and to maintain a clean reproduction of transients. The effect of a large amount of spread in area tracks is a clearly audible distortion of sibilants, particularly the letter "s". These become grossly exaggerated, and sometimes split, owing to failure to obtain the exact geometrical image of the exposed waveform. Incorrect exposure in the camera can have a similar effect, with a distorted waveform giving a rough sound quality.

The sound negative is exposed at various illumination levels while recording a combined audio signal of 400 and 6,000 Hz at 80 per cent modulation. These tests are carefully processed, and a family of prints made at different densities from each negative density. All prints are then played on an optical reproducing machine, to which is connected a cross-modulation analyser. This comprises a volume indicator and selective filters for eliminating either the high- or low-frequency content.

The optimum print density is the one which produces the highest output from the high frequency, and the lowest output from the low frequency. Incorrect negative densities show up as a 400 Hz modulation, owing to changes in the average transmission of light through the film. If optimum conditions are met only at one end of the range of prints, the negative density is adjusted accordingly. Once standards have been set, they are adhered to, for that particular batch of emulsion. Fortunately, slight errors in the sound negative can be off-set in the printing.

INTERMODULATION. Incorrect processing causes harmonic distortion in variable-density tracks, and intermodulation tests are used to determine and control image spread. A combined audio signal of 60 and 4,000 Hz is recorded at near full modulation, and a range of negatives and prints obtained. The prints are replayed into an intermodulation analyser, and the percentage intermodulation of the 4,000 Hz tone by the 60 Hz tone is read directly on a meter.

It is possible to achieve good workable results without any elaborate signal generators and analysers by making test recordings of sibilant speech with several negative exposures.

A listening test carried out on the resulting batch of prints determines the most suitable processing conditions.

MULTI-TRACK SYSTEMS

Certain films using the Cinemascope format, and all 70mm productions, require a different treatment in re-recording to produce multi-track or stereophonic sound. The effect is actually wide-screen sound, emanating from several loudspeakers, and is not true stereo.

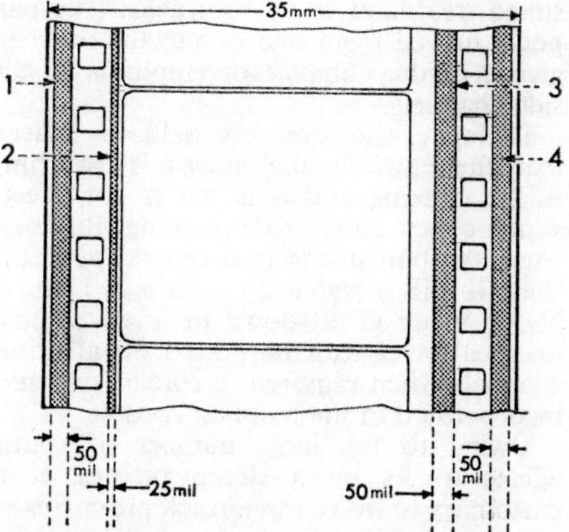

MAGNETIC TRACK LOCATION: CINEMASCOPE PRINT. Film as viewed from lens side in projector. 1. Right-hand track. 2. Auditorium or control track. 3. Centre track. 4. Left-hand track.

During re-recording, use is made of a standard multi-track magnetic recorder, with separate amplifiers and monitor loudspeaker systems. With the exception of music, which is recorded stereophonically, all original sound tracks are recorded mono. The reason for this is convenience of operation, since dialogue and sound effects are only marginally improved by being recorded stereophonically.

RECORDING REQUIREMENTS. The main recording channels are checked for gain and frequency response at least once per day, and all the loudspeaker systems balanced and phased with a 100 Hz tone. This is best checked on an output meter by combining the monitors, since standing waves in the re-recording theatre can make an aural test misleading.

The three-track reproducers are checked for frequency response, using the RCFM test film, and the three-way faders on the mixing console adjusted to give identical outputs to each channel. To eliminate phase shift, the three-track head clusters on recorder and reproducers must have their gaps correctly aligned, since second and third generation transfers will be made when re-recording effects and music material.

Since the final prints delivered to cinemas carry a magnetic stripe track with a frequency response of 50 to 12,000 Hz, high and low pass filters are not required.

PAN POTS. These are triple-ganged rotary faders which divide the output from a mono source into three channels, giving a pseudo-stereo treatment to mono tracks. The pan pot is moved from one side to the other to give a gradual change of emphasis to the side channels.

Dialogue, however, is seldom moved from the centre channel, since it is disturbing in the cinema, and it is the sound effects which benefit most from panning. But even when the pan pot is in a central position, there is still a signal on each side channel about 10 or 12 dB down in level. So that identical levels can be placed on all three channels when required, a spread switch is incorporated in the panning console.

Owing to the large number of sound effects tracks on a stereo picture, it is customary to make three-track pre-mixes so

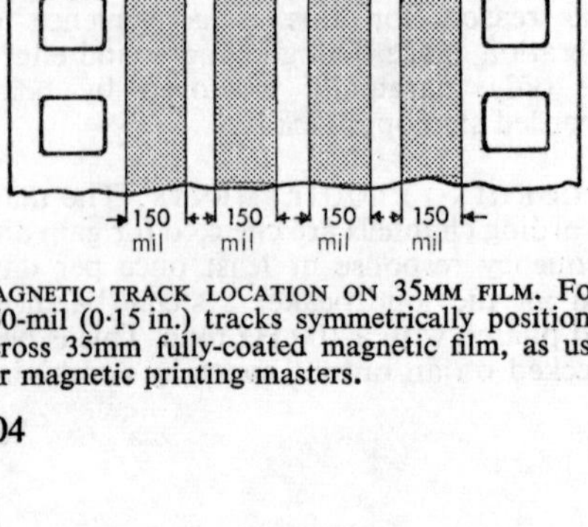

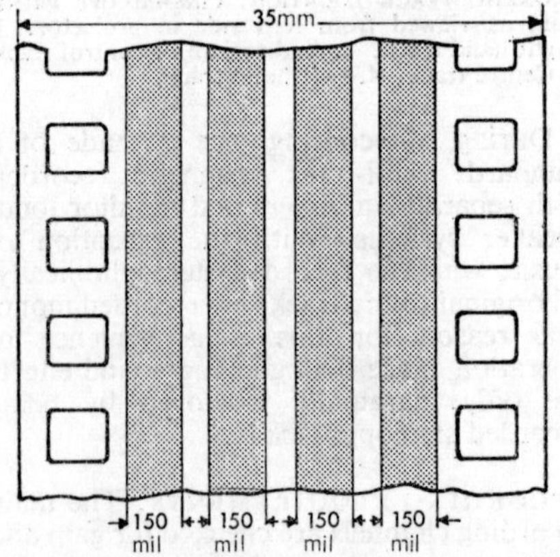

MAGNETIC TRACK LOCATION ON 35MM FILM. Four 150-mil (0·15 in.) tracks symmetrically positioned across 35mm fully-coated magnetic film, as used for magnetic printing masters.

that panning movements can be accurately controlled before recording the final track.

PRINTING MASTERS. The final dubbed track is retained by the studio for safety, and a transfer made to produce a three-track printing master. This is required for re-recording on to the striped copies of the picture.

If sound is to appear from the auditorium surround loudspeakers, a four-track printing master is required, which means installing a four-track head cluster in the recorder.

Material for the fourth track is prepared separately, and recorded on the printing master during transfer.

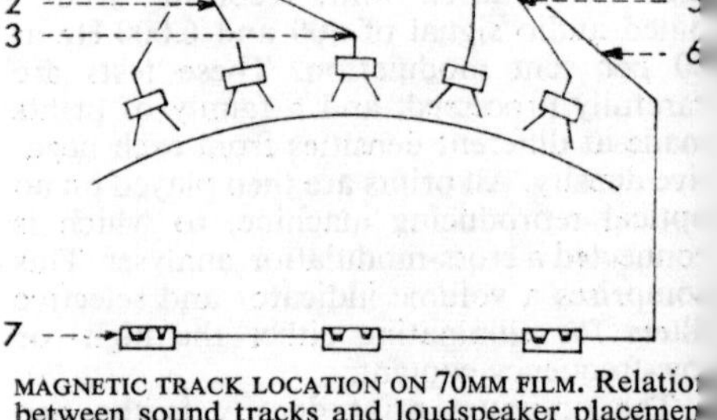

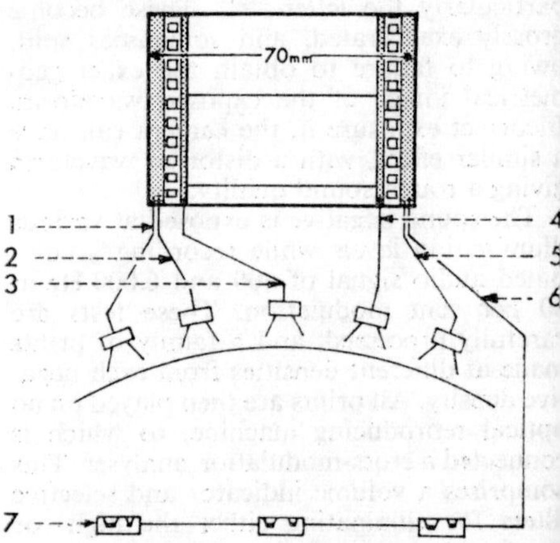

MAGNETIC TRACK LOCATION ON 70MM FILM. Relation between sound tracks and loudspeaker placement in auditorium. 1. Outside left track. 2. Inside left. 3. Centre. 4. Inside right. 5. Outside right. 6. Auditorium track. 7. Auditorium surround speakers.

For Cinemascope films, a 12 kHz control tone is superimposed on the fourth track, its purpose being to open up a gating amplifier in the cinema reproducing system just ahead of each audio signal. This is necessary because of the poor signal-to-noise ratio of the narrow fourth track.

Owing to the extremely wide screens used with 70mm films, five loudspeaker systems are used backstage in addition to the auditorium loudspeakers. The six tracks are obtained from either a six-track or four-track printing master. A six-track printing master must be made on a six-channel recording installation, using a six-track head cluster in a 35mm magnetic recorder. In the case of a four-track master, inside left and inside right tracks are made artificially with splitter networks. No 12 kHz

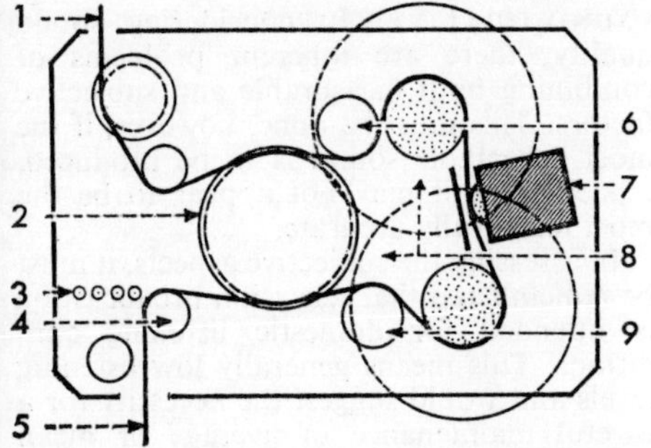

PENTHOUSE MAGNETIC HEAD FILM PATH. Penthouse head is mounted between top spoolbox and picture head, and has no driven sprocket. 1. Film feed from top spoolbox. 2. Free-running sprocket with guide flanges. 6, 9. Tensioned filter rollers. 7. Magnetic reproduce head. 8. Twin balanced flywheels and sound drums. 4. Sync-adjusting roller with alternative fixings, 3, to alter film path length. 5. Film pulled by feed sprocket in picture head below.

gating tone is necessary, since the stripe carrying the sixth track is wider and has a satisfactory signal-to-noise ratio.

MONAURAL MAGNETIC

It is anticipated that a monaural magnetic track will eventually replace optical sound in cinemas, resulting in improved sound quality. Various proposals for an industry standard have been put forward in Britain and Germany, although the US does not appear to be immediately concerned. It is felt in some quarters that there is room for improvement in optical sound standards.

The British proposal is to standardize on the square fox-hole perforations, so that existing Cinemascope installations can be used for reproduction with a replacement head cluster to accommodate a wider stripe. This would mean the existence of two perforation standards, and a complete alteration to laboratory film handling equipment. Cinemas without Cinemascope installations would have to conform to this standard, and trial conversions have already been made in a number of cases.

The German proposal is to retain the original 35mm perforations, for which the majority of film handling equipment all over the world is designed, and position a magnetic stripe over the area now occupied by the optical track. This system would cause less upheaval within the film industry, and prove more compatible when films were used for television purposes. In either case, the sound reproducer would be of the penthouse type above the picture gate, and the amplifier frequency response would be flat from 40 to 10,000 Hz with no h.f. loss in the loudspeaker system. J.B.A.

See: *Acoustics; Sound and sound terms; Sound distortion; Sound mixer; Sound recording* (TV); *Sound recording systems; Sound reproduction in the cinema.*

Books: *Elements of Sound Recording*, by J. G. Frayne and H. Wolfe (London and New York), 1949; *Audio Encyclopedia*, by H. M. Tremaine (London and New York), 1959.

□SOUND RECORDING

Television sound began as an amalgamation of many techniques from sound broadcasting and film. After a period of slow development, it has emerged as a fully creative and integral part of television. It is now realized that it differs considerably in artistic presentation from either sound broadcasting or film.

AIMS AND METHODS

Television sound is developing many unique features but the most important is the establishment of its own individual character and style. Basically, the main difference from film sound is the fact that the production is made either in one piece or in a few major pieces, and not, or rarely, by short takes as in film productions. Accordingly, this is a fast-moving operation that needs far more microphone inputs. Stops for retakes may grudgingly be conceded where a recording on videotape is occurring, but there are obviously no retakes on a live programme.

Microphones must be set up in all the scenes and sets in the television studio and the sound mixer must have a very good idea of the required setting. Seldom is there much time to switch, readjust or experiment.

In general, sound is complementary to the television picture, adding a further dimension to communication, but there are occasions when it can become the dominant force. The process of integrating all sections of a television production usually results in compromises all round, and so absolute standards are unobtainable. Nevertheless, the striving towards higher standards continues. To help in this, much new sound equipment is designed specifically for television, and should aid the closer integration of sound and vision.

SUPPLEMENTING THE PICTURE. Television is required to inform, entertain and educate. To help to carry this out it needs the effective sound communication of language, atmosphere and images. The physical method of this communication is by the control of electronics and the manipulation of acoustics, producing audio stimulations which have become conditioned to natural acoustics. This stimulation leads to psychological reactions which are also highly conditioned.

In its basic form, sound can complement and be easily identifiable with the visual images in both perspective and acoustical realism. The viewer is able to accept this situation without question and becomes relaxed and at ease. However, if this is reversed by producing sound which is alien to the picture, there is a disturbance of realism and this can produce stimulation and emphasis.

Many sound symbols and clichés develop and become subject to fashion, and the use of music to establish a mood is now one of the most common. If the establishment of a convention has accustomed the viewer to accepting a non-realistic situation, it will most probably produce a predictable reaction, for example of thoughts within the mind when hearing dialogue from a very close microphone.

From a picture, a series of symbols may suggest a wider environment. By the addition of background sounds, this environment can be filled out and further extended. For example, over a picture of a notice board directing "To the River", the addition of river tug-boats and crane noises over a background of lapping water considerably widens the overall image.

A sound montage can quickly evoke a rapid and overlapping series of reactions. This is because the ear is very effective in stimulating the imagination and in developing associations, whereas the eye tends to take what it sees literally.

Much auditory stimulation may be subconscious but it is nevertheless extremely powerful.

Considerable care and critical editing are necessary in avoiding over-complex sound tracks which demand excessive concentration, and any fatigue of this nature can be prevented by the interchange of visual and aural demands. By using the full range of sounds, there is little that cannot be communicated, but it requires a creative mind and technical skills to explore the possibilities fully.

SOUND QUALITY. In considering sound quality, there are inherent problems of combining both measurable and subjective factors. This must be done, however, if the most acceptable sound is to be produced, especially as it may not appear to be the most technically accurate.

In assessing the subjective aspects, it must be remembered that television broadcasting is intended for domestic listening conditions. This means generally low listening levels and would suggest the necessity for a careful maintenance of average or mean volume. However, due to the continuous nature of television, a reasonable range of sound levels and tonal quality is essential to avoid a flat and oppressive wall of sound, as this can produce auditory fatigue. The viewer must be allowed to divert and rest his hearing concentration in the same way as he is able to look away from the screen to rest his eyes.

Within the range of frequencies which produce the tonal structure of sound, the viewer generally prefers the sound system which he normally listens to, and has come to accept. The sound system of the average television receiver needs to be improved in quality if the tastes of the viewer are to be raised and satisfied, a factor which could well be developed technically and commercially. A balanced and undistorted frequency range is the most acceptable, high frequency distortion being particularly objectionable. A medium frequency range or band width of 80–8,000 Hz (hertz = cycles per second) is acceptable, though an extended range is available and is, in fact, used for sound balancing in the studio.

All of these considerations are of importance, but the overriding criterion of television sound quality is that it should be appropriate to the programme. Clear speech is most important to a news or discussion programme, while a pleasant blending of sound becomes a prime factor in a musical programme.

NETWORK METHODS. The sound network within television consists of permanent centres equipped with studios, videotape recorders and telecine machines. These feed their programmes into a master control switching centre, which also has a continuity studio for announcements. In addition there are mobile control rooms which cover outside broadcasts. They feed their programmes either to a master control via sound land lines or radio links, or direct to a mobile videotape recorder which has travelled with them. The master control feeds the sound output to its own transmitter and also to other master controls for different transmitters within the network.

The basic framework of a programme is built up from the sound signals from various microphones: music and sound effects usually come from tape or disc machines, film sound comes from the telecines, and videotape recording (VTR) machines replay any sections of a programme previously recorded. All of these are integrated smoothly and at the correct time via a sound mixing console to form the sound track of the programme. During this balancing process, echo can be added to the signals or they may be frequency corrected or equalized so as to produce a specific sound quality.

Programmes can be recorded on a video tape recorder in short sections which are later edited together. This process often results in uneven sound transitions and so it is necessary to dub or re-record the sound track to smooth it out and possibly to add extra music and sound effects.

There are two principal methods of recording and storing sound for television: magnetic tape and film. With the development of videotape recording it has become possible to store, alter and erase recordings at will. As a result, this is now the principal method used for programmes originating from television cameras.

Some re-recorded programmes begin as a VTR recording and are then transferred to film by a process called telerecording. Either 35mm or 16mm film is used with conventional film sound recording processes. If sound originates on film, especially on an optical track, considerable care is required to ensure that it is integrated smoothly with the standard television sound quality in the re-recorded material.

The dubbing or re-recording of sound on VTR is a very flexible process owing to the ease of magnetic recording and the high degree of synchronization possible with

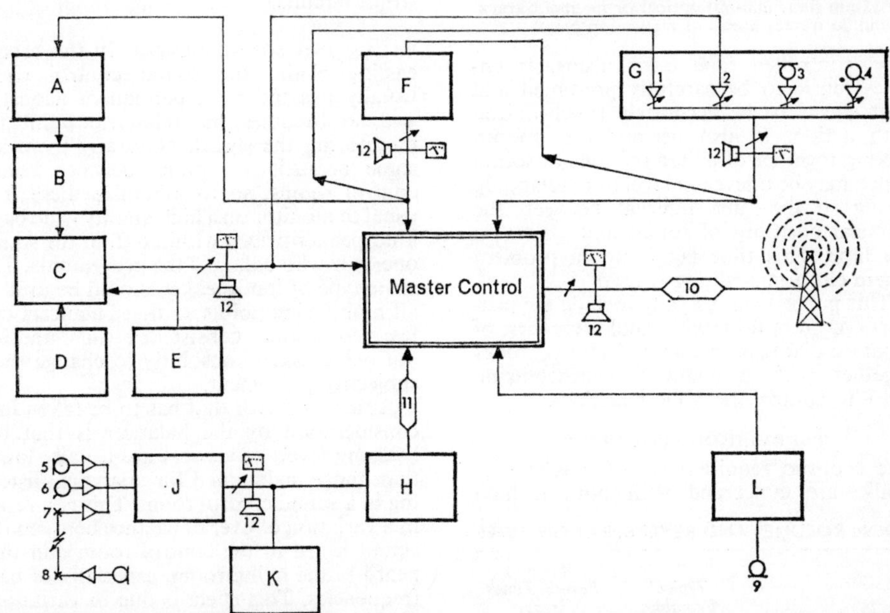

TELEVISION SOUND NETWORK. Block diagram of main programme sources feeding sound signals to Master Control for transmission. Studio microphones, 3, tape and gramophone machines, 4, film sound, 2, and VTR inserts, 1, are combined in a Sound Mixing Console, G. This mixed output goes direct to Master Control for live transmission, or to a VTR centre, F, for subsequent transmission via Master Control. An Outside Broadcast Unit, H, feeds Master Control via landlines or radio link, 11. Mobile Control Room, J, mixes together microphones, 5, tape output, 6, and radio microphone, 7-8. This output is fed direct to Master Control or to a Mobile VTR, K, for subsequent transmission. By telerecording (kinescoping), A, the output of a VTR machine can be transferred to film, B, for transmission via telecine (film scanner), C. Feature films, D, or topical film from mobile film unit, E, are also transmitted through a telecine, either direct or via a control room. In Master Control, with its Continuity Suite, L, and announcer's microphone, 9, all sources are linked and fed via landlines or radio links, 10, to the transmitter. At all important points, 12, sound is monitored on a peak programme meter (PPM) and high-quality loudspeaker.

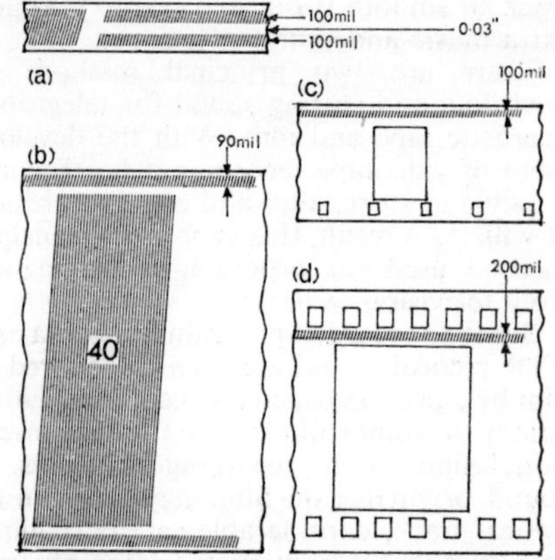

SOUND RECORDING METHODS FOR TELEVISION. (a) Quarter-inch sound tape: full-width track or two 100-mil tracks with 30-mil separation for stereo. No picture synchronization. (b) Two-inch videotape: 90-mil sound track above picture area, audio cue track below. 40 tracks to one picture; sound 9 in. (0·6 sec.) ahead of vision signal. (c) 16mm film: 100-mil optical or magnetic track. Sound 26 frames ahead of picture when combined on print. (d) 35mm film: 200-mil optical or magnetic track. Sound 20 frames ahead of picture on print.

servo-controlled tape mechanisms. It enables music to be carefully pre-timed and spot effects to be accurately synchronized with action. It also permits the smooth linking together of edited sections of sound which may be uneven in level or overlapping owing to the time interval between the physical positions of sound and vision on the tape. This time lag is not a problem where electronic editing is used.

This process can usefully include the production, on multi-track sound recorders, of separate dialogue, music and effects tracks together with a common synchronizing pulse to control the VTR machines.

TELEVISION ACOUSTICS

The acoustic requirements of a television studio are concerned with both a high degree of sound insulation and the acoustic performance of the enclosed volume. The insulation required from the outside is around 65 dB, with 45 dB being adequate between control rooms and studio. This is achieved by heavy or partitioned walls, insulated air locks on doorways and double glazing on windows.

The acoustic performance is indicated by the studio reverberation time and is determined by the absorption material cladding the walls, though the reflective nature of the floor and of the roof structure must also be taken into account.

The acoustics of a studio and those of a domestic living room are, of course, vastly different, and each can be subject to further changes due to studio set designs or domestic furnishings, so care is required when making assumptions about acoustic conditions by purely subjective assessment. It should also be remembered that the limits that are acceptable to acoustic analysis are much wider than those in associated electronic circuits.

ROOM VOLUMES AND REVERBERATION TIMES

	Typical Examples		*Optimum Reverb. Times (secs)*	
	Volume (cu. ft)	*Reverb. Time (secs)*	*Speech*	*Music*
Television Studio	400,000	0·9	1·0	1·8
Concert Hall	100,000	1·5	0·85	1·4
Television Band Room	13,000	0·75	0·75	0·9
TV Sound Control Room	3,000	0·6	0·7	0·7
Domestic Living Room	2,000	0·6	0·65	0·65

BALANCING SOUND LEVELS. In the broadcasting chain, the sound control room usually has the most permanent acoustics and so becomes the reference point for monitoring the signal. Naturally there are some acoustic variations between sound control rooms, so to minimize these it is usual to monitor on a high quality wide band loudspeaker placed within 6 ft. of the sound operator who balances the programme. The same type of loudspeaker should be used at all monitoring points so that balancers can learn to produce consistent results, and are not being asked constantly to change their subjective judgement.

A further factor that has to be taken into consideration by the balancer is that the listening levels of viewers are usually lower than those maintained for analytical listening in a sound control room. This can result in a variation of overall balance between the sound heard in the control room and that heard in the living room, especially of bass frequencies. This effect is due to variations in subjective loudness at different intensities of sound, as is shown by equal loudness contours.

The ear has an inherent weakness in judging absolute levels of sound and remembering them over a period of time, so it is necessary to have an agreed standard of measurement. This usually takes the form of a peak programme meter (PPM), which indicates peak voltages, the pointer having a fast rise time but a slow decay time. Over the measurable scale, it shows some 22 dB, which is the television dynamic range. The upper limit of this is maximum modulation of the sound transmitter. A standard reference point or zero dB is indicated by 4 on the PPM, which is equivalent to 40 per cent of maximum transmitter modulation, while the lower limit can lead to masking by noise and a serious loss of intelligibility.

The dynamic range of speech is some 30 dB and of music more than 50 dB, so the balancer manually compresses these into a smaller range by riding the volume control or fader. Average limits of PPM indications are specified for different types of programmes but the relationship between the loud and quiet passages depends upon his judgement. There are automatic electrical devices which control levels of sound by reducing gain with increasing level over the upper part of the transfer path. These are called compressors or limiters and need judicious use to avoid audible changes in sound quality.

The ear is very sensitive to changes in level and so it is important that sound transitions are carried out smoothly. This is especially so if any audible degree of automatic compression has been used. The energy in the wave envelope of compressed sound gives an apparent effect of being louder, though this does not show on a PPM. The ear is the best judge of this form of dynamic balancing which is easy to hear but very difficult to measure.

OUTSIDE BROADCAST (OB) SOUND. Programme demands are infinitely variable in outside broadcasts (remote pick-ups) and may range from a brass band in a town hall to a golf tournament. The acoustics of the location are usually unknown and poor, so special microphones and radio-microphones are used to cover long distances.

All the equipment should be basically simple and easily portable as it may be necessary to remove the sound mixing equipment from the mobile control room to a more suitable location. Some OB units have the minimum of equipment in a small van or car with one or two channels and are completely mobile with their own VTR machines and power supplies. At the other extreme, there are very large units which are as comprehensively equipped as a studio control room.

With programmes for live transmission, the sound is fed to the master control either by a radio link-up, radiating from one high point to another, or down a cable or land line. The frequency response of sound lines is subject to variations, and equalizers are introduced at various points to obtain a flat frequency response.

EQUIPMENT

The range of equipment used in television sound operation is varied and complex and is largely composed of commercially produced units, available in many instances as standard items, such as microphones and loudspeakers. Other equipment, such as mixing consoles and communication systems, is made to an individual specification and generally built on a modular basis from standard, or specialized, units.

There is a vital relationship between the equipment available and the operational techniques used. This can give rise to considerable variations in the choice and layout of equipment, operational emphasis and even the background and training of personnel. All of these factors, once they have become established, are difficult and expensive to change.

Equipment that is used on the studio floor needs to be robust in construction yet also neat and practical. It must be readily interchangeable, so standard connectors are important. In the sound control room, as many units as possible should be modular in construction and interchangeable with spares for maintenance work.

MICROPHONES. The main properties of a microphone are its directional characteristics, overall quality of sound, and any unique features that form part of its design for special applications.

Microphones for television can be classified into three groups, and in the first group are the simple and basic microphones which can be used on boom arms, on stands or over sets. They include capacitor, ribbon and moving coil types which are often used in music balances and they usually form the basic microphone complement.

In the second group are the microphones of specialized design. Examples are those for holding close to the mouth in noisy environments, and others which hang round the neck and have a falling bass response to reduce the effect of chest resonance. Then there are the so-called gun microphones which are extremely directional over a limited frequency range, and certain hand microphones for interviewing.

In the final group are the hybrid microphones which are particularly useful. Some

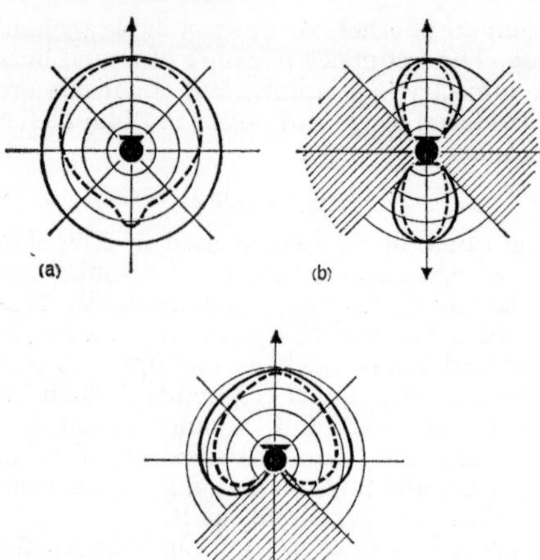

MICROPHONE POLAR RESPONSE CURVES. Polar diagrams in which 100 per cent response corresponds to curve reaching outermost circle, 0 per cent to curve falling in on microphone. (*Solid line*) Lower frequencies. (*Dotted line*) Higher frequencies. (a) Omnidirectional pattern; slightly directional for higher frequencies. (b) Figure-of-eight. Sides of microphone are dead, response is symmetrical front and back, higher frequencies more directional than lower. (c) Cardioid; heart-shaped pattern gives dead area at back, often convenient for speech recording, and for rejecting noise behind microphone.

of these are designed for use on booms but can be used as stand microphones; others are designed for use in the hand but are just as effective clipped to a floor or desk stand; while others can have their pick-up response varied from omni-directional through figure-of-eight to cardioid (heart-shaped) by remote control.

In television it is necessary to concentrate accurately upon the wanted sounds and to reject sounds such as extraneous studio noise during a drama, or other adjacent instruments in a band balance. This can best be achieved by the careful choice of microphones together with efficient positioning, and is the first step in producing the highest quality of sound.

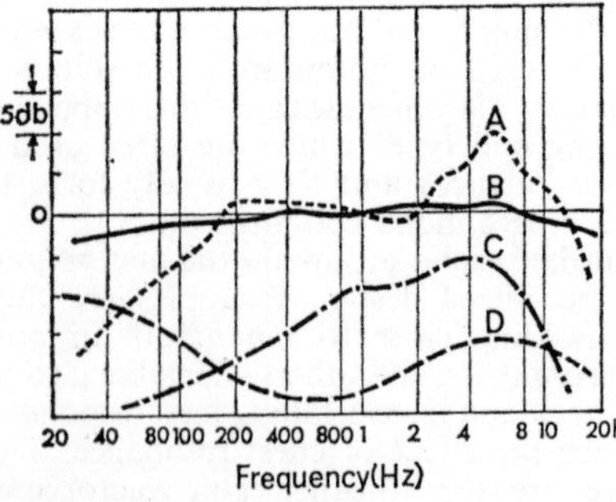

MICROPHONE FREQUENCY RESPONSE CURVES. A. Moving coil neck microphone, cardioid pattern at 0°. B. Condenser (capacitor) stand microphone, cardioid pattern at 0°. D. Same at 180° (back response). C. Ribbon noise-cancelling lip microphone; response to random distant sounds.

SOUND MIXING CONSOLES. As the complexity of television sound grows, so does the need for its comprehensive control. This is reflected in the design of mixing consoles, into which the programme circuits are fed for balance and quality control.

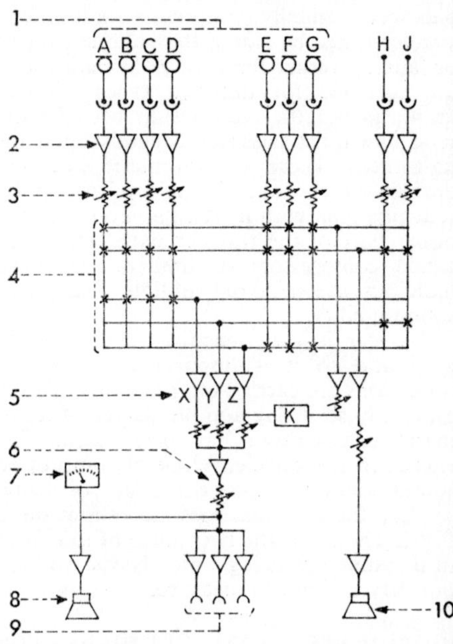

BLOCK DIAGRAM OF NINE-CHANNEL MIXING CONSOLE. 1. Studio floor microphones. A. Flute. B. Piano. C. Bass. D. Drums. E, F. Boom microphones. G. Hand microphone. H. VTR audio signal. J. Telecine audio signal. 2. Amplifiers and equalizers. 3. Faders. 4. Group selection network. 5. Groups X, Y, Z outputs. K. Echo unit. 6. Main output. 7. Peak programme meter (PPM). 8. Monitor loudspeaker. 9. Outputs. 10. Public address (PA) foldback to studio.

The evolution of sound mixing consoles has been from simple control panels with valve amplifiers in detached racks to large composite desks with modular valve amplifiers within the desk frame. Now the tendency is to integrate a transistor amplifier into a fader module and to include these on a large mixing panel. The future trend may be towards small standard transistor mixer units of 12 to 15 channels capable of being linked together as required.

A typical large studio sound console is likely to include 40 low-level microphone channels divided into four groups; high-level inputs for disc, tape, VTR, telecine, etc.; pre-set level adjustment on each channel; before-fader listen for checking inputs prior to mixing into the programme; echo mixture controls, regulating the amounts of echo on each channel; two separate echo circuits with individual controls over reverberation times; switching circuits controlled from vision mixer for synchronous vision and sound cutting; monitor points on important circuits, feeding the loudspeaker or programme meters; studio foldback circuits to relay selected sound to studio floor loudspeakers; public address circuits to relay selected sound to studio audience loudspeakers.

Foldback circuits are used in television studios to feed sound back to the floor for guide or mood control, but level must be limited so as not to disturb the recording or spoil the quality by coloration of the live sound.

The technical specification of a sound mixing console provides for an overall gain of 100 dB with low noise and low distortion. There should be little cross-talk between circuits, and the schematic design must ensure technical stability when ancillary equipment is integrated via the patch panel. This panel contains all circuit junctions of importance for increasing the flexibility of the desk, for example by the introduction of equalizers or compressors, and also for the easy checking of circuits.

TAPE AND DISC EQUIPMENT. The basic requirements of the tape and disc machines are maximum reliability and simplicity in operation in addition to high technical standards of recording and reproduction. These standards primarily are a flat frequency response of 40–15,000 Hz, a noise level better than 55 dB below maximum level and a speed accuracy better than 0·1 per cent. This last figure should be reached in less than one second to avoid frequency wow as the machine gains speed.

Tape machines run at 7½ and 15 ips with appropriate equalization, and have three heads, simultaneously to erase, record and reproduce so that checks can be made during recording; 15 ips is perhaps a luxury, but the quality is better and the length of tape involved not excessive. Fast forward and rewind spooling is by a variable-speed control, and the hand movement of the reels allows the exact location of a specific point on the tape. Marking the tape for editing should be done at a prescribed distance from the head, and the distance indicated on the editing block. The 45° editing cut is usually made with a razor blade, though some machines now have cutting by automatic scissors.

Disc replay machines run at 78, 45 and 33⅓ rpm, with equalization curves and high frequency cut filters available. Two isolated outputs are required, one for programmes and the other for cueing, by pre-listening on headphones.

Tape and disc have individual advantages for certain operations. Tape is the more flexible, whereas discs are quickly cued and repeatable, but only the availability of both ensures the maximum efficiency in what can be a complex operation.

EQUALIZERS AND FILTERS. The basic electrical circuits of the sound system are carefully designed to maintain a flat frequency response. However, there are occasions when it is desirable to change this by a small amount to produce a more balanced sound or by a large amount for a special effect.

Microphone equalizers usually increase and decrease by 10 dB: the bass from below 100 Hz and the treble from above 8,000 Hz. They may also have middle lift, or presence, at 3,000 Hz for speech equalization.

Frequency band equalization is also of considerable value. The equipment can produce a variety of responses by using variable gain frequency selective circuits spaced across the audio range.

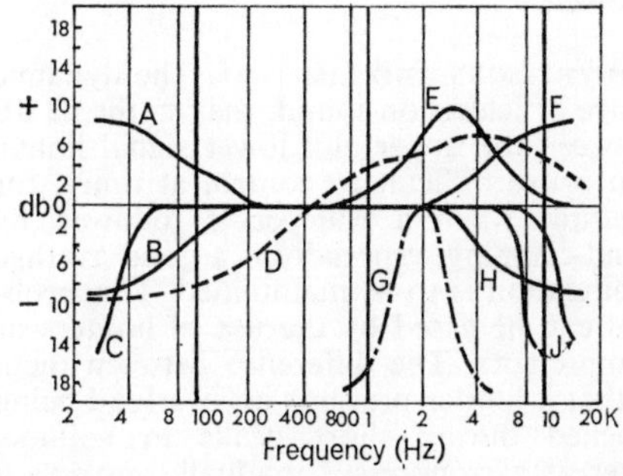

FREQUENCY RESPONSE OF FILTERS AND EQUALIZERS. A. Bass boost. B. Bass roll-off. C. Rumble filter. D. Frequency band equalizer. E. Presence boost. F. Top lift. G. Telephone distort filter. H. Treble roll-off. J. Needle scratch filters.

More extreme cut filters usually have slopes of 8 to 12 dB per octave and spot frequencies in the 40 to 200 Hz region, and in the 6,000 to 12,000 Hz region give bass and treble cut. Special filters are generally made to a fixed specification and examples are those for lip microphones or telephone distortion effects.

ECHO AND REVERBERATION UNITS. Echo can be described as the audible effect of a simple time delay, whereas reverberation is a complex and imprecise decay of successive reflections. Both are of value in creating acoustical atmosphere though reverberation is the more used.

Echo is produced by using the spacing between the record and replay heads of a tape machine as the time delay. A number of replay heads spaced apart can give a flutter effect, but echo is usually used only in electronic and popular music, as it is clearly synthetic in character.

Reverberation is the more natural sound and the highly reflective process can be achieved by using a hard-walled room, a suspended steel plate or a spring under tension. In each of these, a transducer injects the signal into a confined and reflective medium which has low absorption characteristics.

The most popular unit is the reverberation plate, its decay time being variable up to 4 seconds by the proximity of an absorbent panel, though equalization may be required for different settings to produce the best effect. These devices take the sound wave through what is effectively a low velocity path (coiled spring, loose plate, etc.), which is relatively undamped. Given an input drive and an output pick-up, various echo/reverberation effects can be evolved.

COMPRESSORS AND LIMITERS. The dynamic range of television sound, that is, the 22 dB between the upper and lower signal limits, can prove difficult to control at times, for example when a whisper is followed by hand-clapping, especially if a good average modulation is to be maintained. This problem can be eased by the use of limiters or compressors. The difference between them is that a limiter prevents an overload being reached during short peaks in volume, whereas a compressor gradually restricts a wide volume range into a narrower one.

The limiter can be adjusted with its operating point set at 2 dB below the overload point, and a 10 to 1 compression ratio. In order to reach the limit over this last 2 dB the input would therefore have to be increased by 20 dB, that is, 10 into 1 dB and so 20 into 2 dB.

It is possible with a compressor to control automatically a wide volume range into a manageable signal, so avoiding the variable quality that can sometimes be heard with excessive manual control. However, care must be taken to avoid unintentionally altering the basic sound quality, and also the subjective volume, by over-compression.

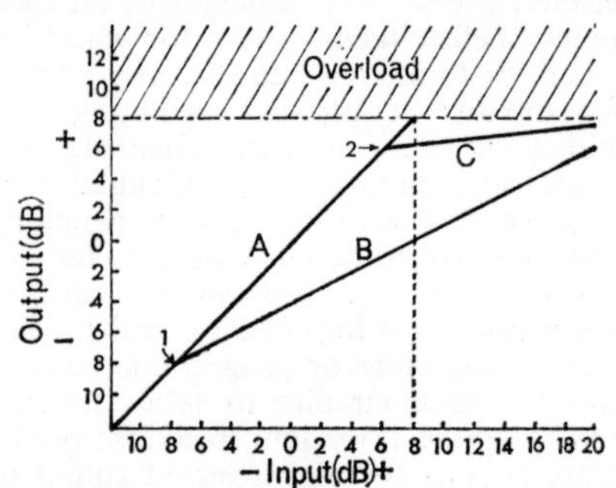

COMPRESSOR AND LIMITER CHARACTERISTICS. Graph shows input-output relation of linear system, A, which at +8 dB reaches maximum undistorted level and enters overload region. To cope with further increases of input, two approaches are possible. Using an early breakpoint, 1, the compressor applies a uniform compression ratio, B (here 2 : 1) to subsequent input increases. Using a later breakpoint, 2, the limiter applies a much more drastic compression ratio, C (here 10 : 1), so that virtually no increase of output occurs after the limiter has come into action.

COMMUNICATIONS AND TALKBACK. As most television is produced by a continuous process, there is the need for quick communications between everyone concerned. A typical discussion during rehearsals may best illustrate this.

The director points out that the microphone is casting a shadow on an artiste and the sound balancer asks the boom operator if he can change his position. The boom operator replies that the shadow is caused by the unusual angle of the light and so the balancer talks with the lighting supervisor. He, in turn, discusses the problem with the director, who decides to change the angle of the shot, and tells the cameraman.

All of this discussion is carried on by the simple operation of talkback switches, but the complete communication system is extremely complex, including some radio talkback systems, and it may contain upwards of 20 separate amplified circuits. Each control room area is in communication

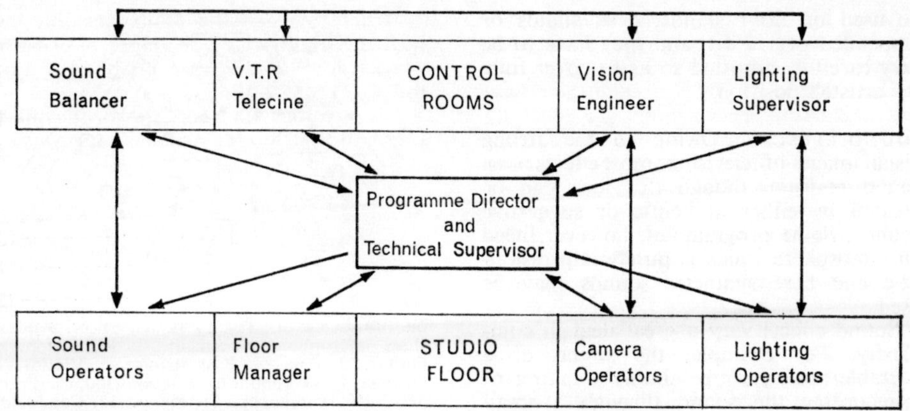

TALKBACK COMMUNICATION SYSTEM OF TELEVISION STUDIO. The programme director and technical supervisor are at the heart of a complex internal communications network linking them to all main production departments and operators, and linking these with one another.

with all others and also with the relevant floor personnel, who in turn need to be able to reply.

TECHNIQUE

The dictionary defines technique as "a mode of artistic execution, mechanical skill in art". The television programme can be looked upon as a composite art form to which a suitable technique of sound must be applied, but in developing an artistic technique it must not be forgotten that economic realism, technical possibilities and a careful attention to detail can play a very important part in achieving the final success of a programme.

THE SOUND CONTROL ROOM. This is the nerve centre of all sound operations, being well insulated from surrounding areas and similar in size to a domestic living room. There are usually three main groups of equipment, comprising the sound control console operated by the balancer, the tape and grams console with its operator, and equipment racks which contain talk back amplifiers and switches, distribution amplifiers, loudspeaker amplifiers and other ancillary equipment.

The sound balancer during rehearsals is in charge of studio and control room activities and ensures that at each stage the sound crew is in position, the correct music and effects are available on cue, and all microphones are set and adjusted. During transmission or recording he controls the sound operations by talkback and balances the various sound sources one against the other. He must be ready to react quickly to any problems that develop and to take remedial measures. Any deviations in level or quality need to be rectified before the viewer has time to be conscious of them, and a good layout of loudspeaker and vision monitors helps in ensuring a keen atmosphere for the concentration which is essential to a sound control room.

THE STUDIO FLOOR. The sound operations on the studio floor are primarily concerned with the placing of microphones, whether they are moving or stationary. In addition, there is much ancillary equipment that requires skilled attention, such as the positioning of fold-back loudspeakers.

The most universal and mobile method of microphone placing is by a microphone boom, which is a telescopic arm reaching out some 18 ft. from a high movable platform. The boom operator is able to look over the cameras, to follow the movement of the artiste and to position the microphone for the best pick-up, checking this on his headphones. He develops a visual imagination of what the camera lens can see and so puts the microphone on the outer edge of the picture, taking up the viewpoint of the camera that is being transmitted at the time. This is to maintain the correct perspective of sound to pictures.

The lighting casts shadows of the microphone towards the set, so side-lighting is used to keep the shadow out of camera range. This further indicates the skill that a boom operator needs to pick up the best quality of sound with the minimum of interference with other operations.

A number of stationary microphones may

be used on floor stands, desk stands or suspended over a set, and they have to be very carefully adjusted so as to cover fully the artiste's position.

SOUND EFFECTS. Owing to the strong visual images of television, most effects must sound realistic, though this idea can be created by either authentic or suggestive sounds. Some programmes, however, need an atmosphere which is purely impressionistic and here synthetic sounds may be used.

Sound effects vary a great deal in complexity. For example, the sound of a portable radio may be simply a matter of re-recording the sound through a small speaker mounted in a cigar box, whereas a gun shot effect may require an elaborate electronic device.

When recording effects, it is essential to get a clear and distinct sound quality, especially with short duration or spot effects, and the reduction of background noise by filtering can help. The effect may then be played into a quiet scene without a prelude of noises. If any continuous effects are used, they should be added separately, and carefully controlled to avoid monotony. The addition of occasional spot effects into a dialogue gap or during action reminds the viewer of the location. For example, in a railway carriage scene the continuous drumming of the wheels can be broken by the occasional train whistle and screeching brakes.

Continuous background effects are best provided on endless magnetic tape loops. These can be run continuously and contained within a moulded cassette for easy handling. Spot effects may also be loaded into a loop cassette and a location pulse recorded on a second track; the machine can then be programmed to cue itself automatically at the beginning of each spot effect.

Libraries of sound effects are kept on either tape or disc and require a comprehensive catalogue of the details of sounds, perspective, duration, etc.

MUSIC. There are many ways in which music is used. It can be pre-recorded or live, with musicians and singers in or out of vision, or they can be in the same studio or even remote from one another with the orchestra being relayed to the singer over a loudspeaker.

There are problems associated with all these methods and as a result of this many differing techniques have evolved.

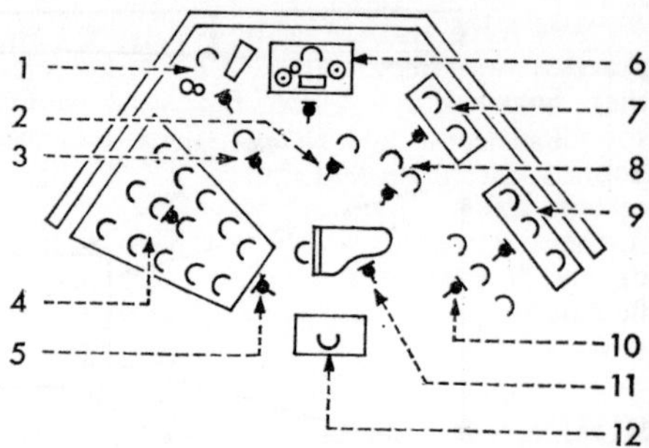

ORCHESTRA LAYOUT FOR TV TRANSMISSION. Symbols show microphone placement. 1. Percussion. 2. Guitar. 3. Bass. 4. Bass strings. 5. Violins. 6. Drums. 7. Saxophones. 8. Woodwind. 9. Trumpets. 10. Trombones. 11. Piano. 12. Conductor.

The type and number of microphones varies considerably. Usually only one is used for combinations of instruments possessing an inherent and internal balance, for example, symphony orchestras and chamber groups which play as an ensemble.

With popular and light musical groups this internal balance is not so strong and each instrument or group of instruments has an individual microphone. These are then balanced in the sound control room to produce the specific type of sound quality required.

If the musicians are not included in the picture, the sound balancer can arrange them and place his microphones so as to give the best results. However, if the musicians are to be seen, the director usually wishes to dispose them in such a way as to produce an interesting picture, and this can create sound problems.

The high quality of sound that can be achieved in studios specially designed for radio or disc recording has led the viewer to expect the same quality when watching television. If, however, this standard is to be achieved, the problems of separation of instruments, artistes' movements and studio acoustics must be minimized by careful planning and attention to detail. Small and intimate orchestral combinations are usually the most suitable for television, and musical arrangements need to be specifically made, with the balance in favour of clarity, especially when there is a vocalist.

Those who grapple with the sound problems in today's television studios may be surprised to learn how old are the arts of acoustics and studio design. In *The New Atlantis* (1624), Francis Bacon wrote:

"Wee have also sound-houses, where we practice and demonstrate all sounds and their generation . . . wee represent small sounds as Great and Deepe; likewise Great Sounds, extenuate and sharpe. . . . We have certain helps which set to the Eare doe further the hearing greatly. . . . We have also diverse Strange and artificial Ecchos, reflecting the Voice many times and as it were Tossing it. . . . We have also meanes to convey sounds in Trunks and Pipes, in strange Lines and Distances." J.E.T.

See: *Acoustics; Sound and sound terms; Sound distortion; Sound mixer; Sound recording* (FILM)*; Sound recording systems; Sound reproduction in the cinema.*

Books: *Technique of the Sound Studio*, by A. Nisbett (London), 1962; *Television Techniques*, by H. Bettinger and S. Cornberg (New York), 1957; *Sound and Television Broadcasting*, by K. R. Sturley (London), 1961.

SOUND RECORDING SYSTEMS

There are three principal forms of sound recording and reproduction in wide use at the present time: mechanical, photographic and magnetic.

MECHANICAL

Usually referred to as disc recording, the mechanical method records sound as a groove on a flat disc, the groove being of constant depth with lateral variations. The groove is cut by a cutter head which is basically an electric motor. An example is the moving-coil type in which a sapphire-tipped cutter is attached to a coil placed in a magnetic field. Sound signals cause the coil to vibrate, the amplitude and frequency of the coil vibrations being mechanical replicas of the amplitude and frequency of the original sound wave. Reproduction is effected by using the side-to-side movements of a needle or stylus running in the groove to drive some form of electrical generator.

The essentials of this system were conceived by Emil Berliner in Germany in the late 19th century, and were an improvement on the hill-and-dale, or up-and-down, recording on cylinders invented by American Thomas A. Edison in the 1870s. The advantages of using flat discs rather than cylinders, for storage and copying, were such that eventually the disc method came to be universally used.

DISC PRODUCTION. In modern practice, however, there is little direct recording on to disc. Records are usually produced in several stages. First the artiste's performance is recorded on magnetic tape. This is then re-recorded (dubbed) on a cellulose lacquer-coated metal blank from which a master with ridges instead of grooves is obtained. The master is used to produce a mother disc which has grooves exactly the same as the original recording. From the mother a stamper is prepared which, like the master, has ridges instead of grooves. The stamper produces the pressings or records which are sold to the public. The use of a stamper avoids wear on the master.

Modern fine-groove records are made of vinyl and are now usually run at 33⅓ or 45 revolutions per minute. The old coarse-groove records run at 78 rpm. It is important to use the appropriate stylus for the type of recording. The introduction of fine groove disc recording has not only increased the playing time by a factor of five, but has also led to a considerable improvement in quality.

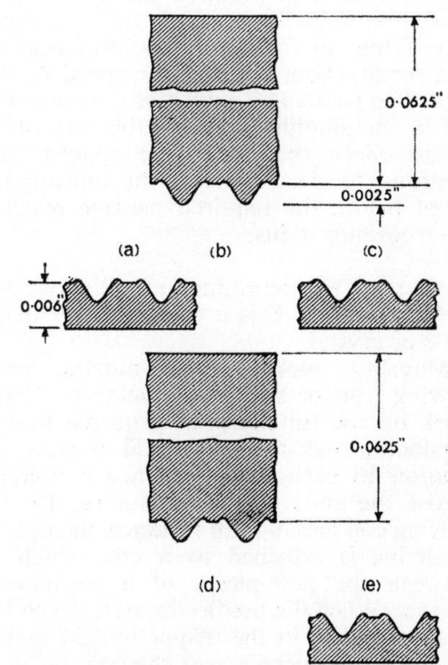

STAGES IN DISC MANUFACTURE. (a) Original direct disc coating. (b) Master, made by electroplating original disc. (c) Mother, formed as a shell on (b) in electrolytic bath. (d) Stamper or matrix, similarly produced. (e) Mass-produced record, made from (d). At each stage there is a reversal from ridges to grooves, and vice versa.

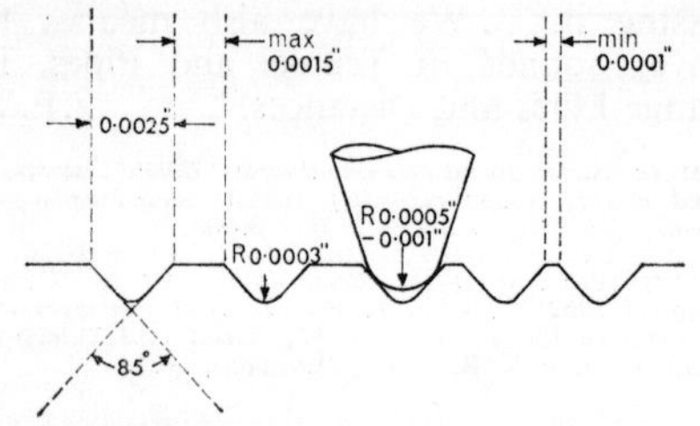

DIMENSIONS OF MICROGROOVE RECORDINGS. Groove width, approx. 0·0025 in. Wall width, max. 0·0015 in., min. 0·0001 in. Radius of groove bottom, less than 0·0003 in. Radius of stylus, 0·0005–0·0001 in.

VARIABLE GROOVE SPACING. Pre-recording on tape makes possible variable-groove spacing, so that when the amplitude of the signal increases, the radial speed of the recording head carrying the cutting stylus can be increased. This widens the separation between the grooves and so avoids a large-amplitude swing affecting the neighbouring grooves. Conversely, on quiet passages, the speed and groove spacing can be reduced, allowing more grooves to the inch, leading to a saving of record space.

This technique requires the use of two spaced reproducing heads on the tape deck. The output of the first, after amplification and rectification, controls the speed of the recording head. The output of the second is fed after amplification to the recording stylus. Since the heads are spaced, the recording head can be set to the appropriate speed before the required passage reaches the recording stylus.

PICK-UPS. For reproduction, a lightweight pick-up is used. This is usually either magnetic or crystal.

Magnetic pick-ups are moving coil, moving iron or variable reluctance. These work on the fundamental principle that a conductor and magnetic field moving in relation to each other produce a voltage across the ends of the conductor. In the moving coil pick-up, for instance, the needle or stylus is attached to a coil which is between the pole-pieces of a permanent magnet. When the needle vibrates, the coil is made to move in the magnetic field and a voltage is produced across the coil.

Crystal pick-ups use the piezo-electric effect, where a voltage difference can be produced between the faces of certain materials by twisting or bending the material. Rochelle salt is commonly used but some pick-ups employ ceramics such as barium titanate. The crystal is made up of two thin layers called a bimorph, of which one end is clamped while the other is attached to the needle. As the needle vibrates, the bimorph is twisted and a voltage is produced between its surfaces. Metal coatings provide the necessary electrical connections.

The needle or stylus used in reproducing pick-ups is usually sapphire-tipped. Diamond-tipped needles are more expensive, but have a longer life.

STEREOPHONY. Stereophonic records are now common. They provide the necessary two channels by using the 45/45 system. The groove carries information on the left side which is practically independent of that carried on the right side. The information is recorded as hill-and-dale variations on the groove walls. Reproduction is achieved by having a single needle pick-up with two independent generators so arranged that the two voltages are independently controlled by the groove walls.

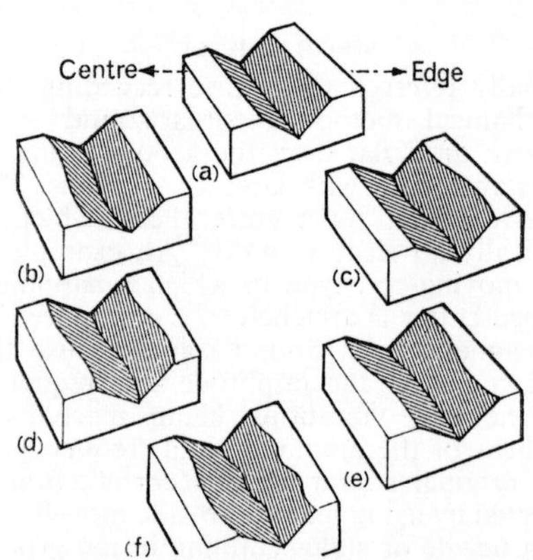

MODULATION OF 45/45 STEREO RECORD. (a) No modulation. (b) Only left-hand channel modulated. (c) Only right-hand channel modulated. (d) Channels modulated in phase with equal amplitude and frequency. (e) Channels modulated out of phase with equal amplitude and frequency. (f) Low frequency recorded on left-hand channel, higher frequency on right-hand channel.

PHOTOGRAPHIC

The principle of this method is to expose a film to illumination controlled by the sound signals to be recorded. The intensity and frequency variations of the transmission of the film are made to vary in sympathy with the amplitude and frequency of the sound waves. After processing, the sound record is reproduced by passing the film between a lamp of constant intensity and a photo-electric cell. The output from the photo-cell depends on how much light falls on it and this, in turn, is controlled by the sound track on the film.

There are two principal methods of photographic recording: variable-density and variable-area.

In variable-density recording, the intensity of the recording lamp is kept constant and the sound signals control the time of exposure. In variable-area recording, the intensity and time of exposure are kept constant but the width of the exposed area is varied by the sound signals. Both methods can give recordings of good quality.

Special film stocks are available to suit each method. Variable-density recording requires a long scale of low contrast, whereas variable-area requires a high contrast film to ensure that the exposed areas develop to a high density and the unexposed areas remain clear. With the latter the transition from exposed to unexposed areas must be sharp.

In one method of recording, a light valve is employed. This consists of a pair of ribbons spaced slightly apart and placed in a magnetic field. The electrical signals derived from the sound waves are passed through the ribbons which are thus caused to move.

In the variable-density method, the plane of the ribbons is at right-angles to the motion of the film, so that the image of the valve aperture formed on the film is transverse to the motion of the film. The instantaneous width of this image depends on the instantaneous spacing of the ribbon. Therefore the exposure, depending as it does on the width of the image, is controlled by the sound signals applied to the ribbons.

With variable-area recording, the movements of the ribbons are enlarged and focused on to the film so that a track of varying width is produced.

MAGNETIC

Magnetic recording gives better fidelity than either mechanical or photographic recording, and is now virtually universal for original recording.

PRINCIPLES. Essentially, the method consists of applying the sound signals to a recording head which produces a varying magnetic field through which is passed a tape coated with magnetic material. The disposition of the individual magnets of the coating is affected by the varying magnetic field so that the original programme is stored as a magnetic pattern on the tape.

Reproduction is effected by passing the tape over a head consisting of a core surrounded by a coil of wire. The magnetic pattern on the tape causes changes of magnetic flux through the core, providing a voltage across the coil.

The tape used has a plastic base, most commonly cellulose acetate, polyvinyl chloride or polyester. The coating is of magnetic ferric oxide. The tape width is ¼ in., the base thickness lies between 0·0006 in. and 0·0016 in. and the coating between 0·0004 in. and 0·0005 in. The whole width can be used to record one programme and this is termed single track. To economize in tape, two separate tracks (twin track) with stacked recording and reproducing heads are used. CCIR standards with twin track specify that the space between tracks should be at least 0·03 in. On some machines, multi-track recording with three or four tracks is possible. CCIR do not specify multi-track standards, but one quoted by one manufacturer is track width 0·045 in. with 0·02 in. spacing.

The comparative ease with which twin track recording can be provided makes magnetic recording eminently suitable for stereophonic recordings.

Perforated magnetic films are available in various widths, 35mm, 17·5mm and 16mm, for use in magnetic recorders and reproducers which are separated from the film

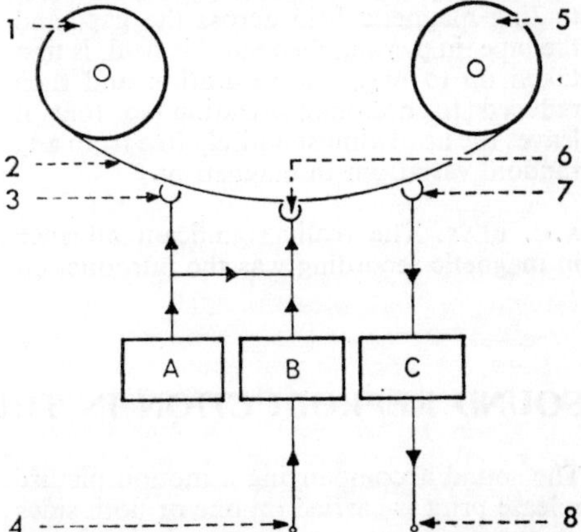

ESSENTIALS OF PROFESSIONAL TAPE RECORDER. Tape, 2, is drawn from feed spool, 1, to take-up spool, 5, past erase head, 3, recording head, 6, and reproducing head, 7. Signals, 4, to be recorded are fed to recording head via record amplifier, B, along with a.c. bias derived from oscillator, A, which also feeds erase head. Signals from reproducing head are fed via amplifier, C, to output, 8.

recorder and reproducer. In sepmag work, typical track widths, on both 35mm and 16mm, are 0·2 in. and 0·1 in., with special provision, if required, for 0·15 in.

TAPE SPEED. The tape transport system must draw the tape at constant speed past the heads; four speeds are now standardized at 15, 7½, 3¾ or 1⅞ ips on professional and domestic machines using the ¼ in. tape.

Most machines provide a choice of two speeds, some three, and some only one.

The choice of speed is controlled, in the main, by two factors: frequency response and playing time. The slower speeds give longer playing time but at the expense of high frequency response. Therefore the slower speeds (3¾ and 1⅞) are confined to speech while the higher speeds are desirable for music.

SOUND HEADS. Professional tape recorders use three heads, for record, erase and replay. Domestic machines usually have only two heads, one being used for both recording and replay. As the requirements for both functions are not exactly the same, this leads to a compromise at the expense of quality but the cost of the machine is kept down.

Before the tape is passed over the recording head, it must be demagnetized. This is the function of the erase head. High frequency currents (75 kHz is a typical figure), derived from an oscillator, are fed through a coil wound on a round core which has a small gap. The currents produce a fluctuating magnetic field across the gap, and the tape, in passing through this field, is first taken up to magnetic saturation and then reduced to non-magnetization so that it leaves the head almost entirely free from any random variations in magnetism.

A.C. BIAS. The really significant advance in magnetic recording was the introduction of a.c. bias, which has been the means of producing tape recordings of high quality.

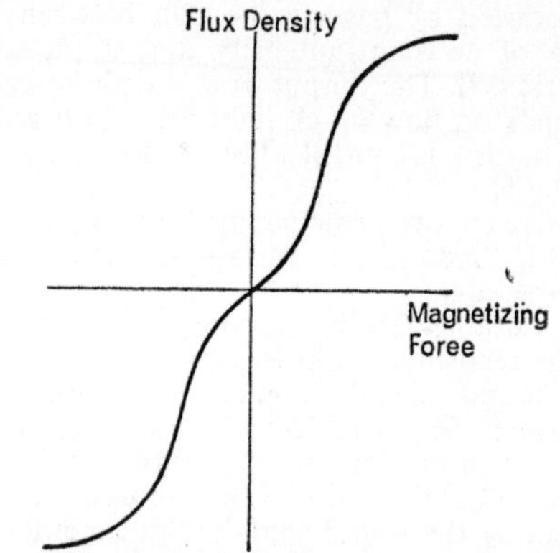

MAGNETIC TRANSFER CHARACTERISTIC. In absence of a.c. bias signal, curve relating magnetizing force to flux density is non-linear in region of origin. Application of proper bias to recording head corrects this curvature.

In changing the signal from a mechanical disturbance in the air to a magnetic pattern on the tape, the changes in amplitude and frequency on the tape must be faithful copies of the changes in the sound wave. Unfortunately, the magnetic transfer characteristic which relates the magnetic field to the current passing through the coil of the recording head is not linear. By applying to the recording head a supersonic signal as well as the programme to be recorded, the effect of the non-linearity of the transfer characteristic is neutralized and the recording is free from distortion. It is convenient to use the same oscillator for the bias current and the erase current. G.W.M.

See: *Magnetic recording materials; Sound and sound terms; Sound distortion; Sound mixer; Sound recording; Sound reproduction in the cinema.*

Books: *Disc Recording and Reproduction*, by P. J. Guy (London), 1964; *Tape Recording and Reproduction*, by A. A. McWilliams (London), 1964.

⊕SOUND REPRODUCTION IN THE CINEMA

The sound accompanying a motion picture release print is carried on one or both sides of the film either as an optical (photographic) recording or, less frequently, as a magnetic recording.

OPTICAL RECORDING

Photographic recording of the sound track has long been developed to a high degree of technical perfection and still gives excellent service in the film industry. However, the wide popularity of high-fidelity reproduction has drawn attention to the limited resolution obtainable by photographic methods. Light reflected from the recording galvanometer and focused on the sound track during the recording process is scattered in the sensitive emulsion layer, thus producing spurious exposures in areas of the track immediately adjacent to the recorded waveform.

When low audio frequencies are being recorded, this is not significant, but recorded waveforms at frequencies of about 5 kHz and above may suffer from valley fogging caused by the sideways spread of the exposure in the emulsion. This produces "shushing" noises, i.e., low frequency components occurring at the frequency of the envelope of the speech waves. Relatively high levels of harmonic distortion result.

In addition, the volume range between the maximum permissible signal that can be recorded and the inherent noise due to film grain and surface contamination is limited. Sophisticated techniques of ground noise elimination to ensure that the maximum area of track is exposed make it possible to achieve a signal-to-noise ratio in the region of 45 dB, but this standard is not maintained in the average film shown to the public.

MAGNETIC RECORDING

There are fewer limitations on magnetic recording. At the standard film speed of 18 ips, there would be little difficulty in recording signals having frequencies as high as 20 kHz if this was advantageous. Amplitude distortions are low and the signal-to-noise ratio achieved on prints released to the theatre is in the region of 45–50 dB. As the loudspeakers used in cinemas rarely have any significant output above 10–12 kHz, it is more usual to allow the recorded signal to fall away above this frequency.

Almost all original recording in the studio or on location is now on magnetic tape or film because of improved technical performance and operational convenience. Except for prestige theatres, the magnetic process has not achieved the same preeminent position in the cinema owing largely to the capital cost of adding magnetic track replay facilities to film projectors originally designed for films having photographic sound tracks.

MULTI-TRACK SYSTEMS

Though a single sound track is normally used when projecting standard 35mm films, many of the special formats such as Cinemascope, Todd-AO and Panavision employ up to six separate sound tracks for three or five separate speaker systems behind the screen, plus a track carrying sound to be reproduced through loudspeakers in the auditorium. An additional track may be used to carry signals for switching loudspeakers at appropriate points in the programme, but these control signals may be recorded at a supersonic or sub-audible frequency on one of the speech tracks.

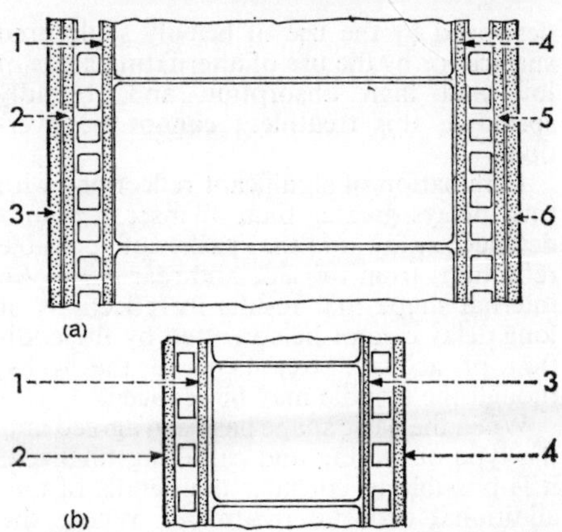

MULTI-TRACK LOCATION ON 70MM AND 35MM FILM. (a) 70mm. 1. Feeds centre screen loudspeaker. 2. Inside left. 3. Outside left. 4. Inside right. 5. Outside right. 6. Auditorium surround loudspeakers. 2, 3 and 5, 6 are in practice recorded on single magnetic stripes. (b) 35mm. Cinemascope. 1. Centre channel. 2. Left channel. 3. Surround channel. 4. Right channel.

Though multiple photographic tracks have been produced, all the current multi-track systems employ magnetic recording because of the superior sound quality obtainable. From many points of view, it is regrettable that magnetic track reproducer systems are largely confined to the major cinemas.

CINEMA ACOUSTICS

The acoustical performance of an enclosure such as a cinema theatre is largely determined by its shape. However, this tends to be dictated more by the site available, seating accommodation required, the local bye-laws and other statutory conditions than by the requirements of the acoustician. Where the acoustical needs can be allowed to influence the internal contours, the side walls should be designed so as not to produce coherent or reinforcing reflections in the seating area. It is equally important to ensure that reflections do not arrive at any group of seats more than about 40 msec behind the direct sound from the stage loudspeakers. Coherent or semi-coherent reflections of long delay greatly reduce the intelligibility of speech, though they do not have the same disastrous consequences on the reproduction of music. Coherence is destroyed by the use of heavily sculptured surfaces or by the use of alternating areas of low and high absorption and, broadly speaking, this treatment cannot be overdone.

Elimination of significant reflections with time delays greater than 40 msec demands detailed study of the paths of possible reflections from the side and rear walls. An internal shape that results in reflections of long delay cannot be corrected by the addition of acoustic treatment to the walls, though the trouble may be reduced.

When the basic shape has been agreed and the type of seating and carpeting finalized, it is possible to consider the details of any additional acoustic treatment. Where the floor-to-ceiling height at the back of the hall exceeds about 8 ft., it is advisable to include an efficient sound absorbent above the 6 ft. level. This is essential when the height of the rear wall exceeds 10–12ft., the need being quite independent of the reverberation time.

Large empty spaces behind the screen are to be avoided. Where the screen and speaker system are more than 8–10 ft. in front of the back wall of the stage, it is advisable to include a thick sound-absorbent blanket between speakers and rear wall in addition to the sound absorbent felt used as a backing to the screen. Again, the requirement is independent of the reverberation time of the auditorium.

When the hall shape approximates to the optimum for the particular site and the standard of soft furnishing is that currently accepted in cinemas, there is usually little need for additional acoustic treatment unless the volume per seat exceeds about 130 cu. ft. Target values of reverberation time have been accepted by the film industry

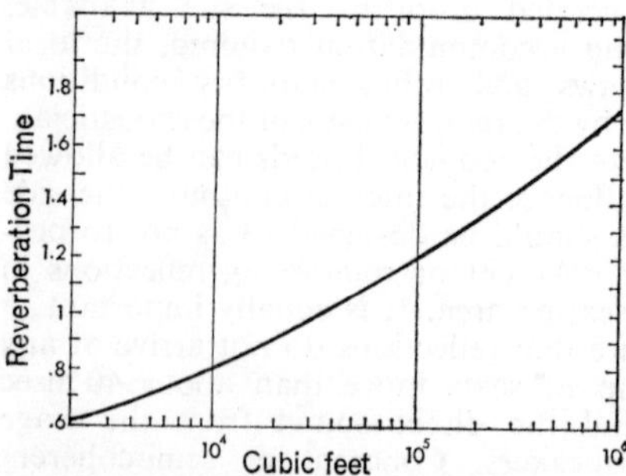

AUDITORIUM VOLUME AND REVERBERATION TIME. Optimum reverberation time (measured with no audience present) plotted against auditorium volume. Frequency: 500 Hz. Single-track optical sound.

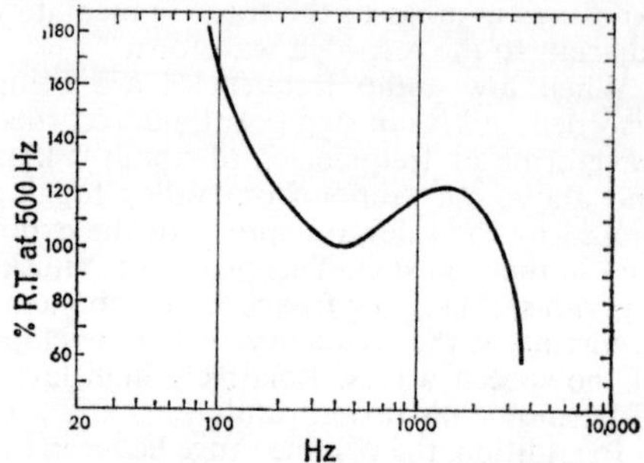

FREQUENCY AND REVERBERATION TIME. Suggested optimum for monophonic sound reproduction. Some authorities dispute need for increased reverberation time around 1,500 Hz, though this has been accepted by the motion-picture industry.

and relate hall volume and reverberation time at a typical centre frequency of 500 Hz and for a monophonic sound reproducer system. When a stereo system is to be installed, it is current practice to design for reverberation times about 10 per cent lower than the target values.

Achieving these values of reverberation time at a frequency of 500 Hz is not so important as ensuring that an acceptable reverberation time/frequency relation is obtained, and a relationship has been adopted by the cinema industry when a stereo sound system is to be installed. An increased reverberation time in the region of 1,500 Hz is called for, but the need for this has been disputed.

In arriving at a satisfactory acoustic treatment, it is essential to make an accurate assessment of the amount of absorption contributed by the building structure itself. At frequencies below 500 Hz, this type of absorption is often the major part of the total absorption and its neglect may have a significant effect on the shape of the reverberation time curve below 300 Hz.

LOUDSPEAKERS

The design and positioning of the stage loudspeaker system is vitally important for good acoustic performance. The sound radiated from the loudspeakers should, as far as possible, be confined to the seating area and not allowed to strike the side or rear walls of the auditorium. This ensures a high ratio of direct to reverberant sound, essential if speech is to have the maximum intelligibility. A speaker system having a polar diagram that varies little over the audio frequency range is necessary, a result usually secured by employing a two-unit system. Frequencies below about 500 Hz are radiated by a short horn or resonant enclosure driven by two 12 in. to 18 in. speaker units, the higher frequencies being radiated by a multi-cellular horn employing two driver units.

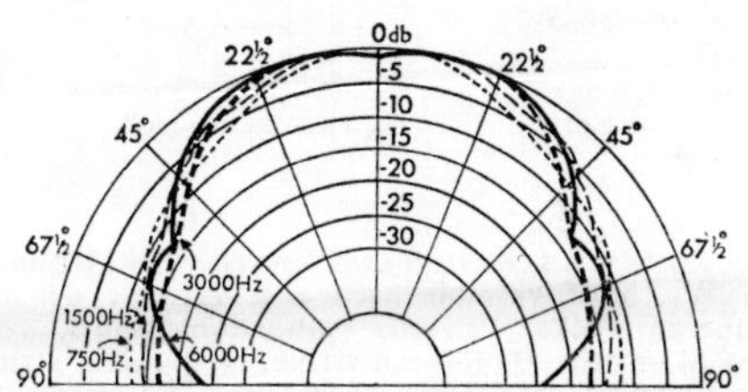

MULTICELLULAR LOUDSPEAKER: POLAR DIAGRAM. Horizontal response of 9-cell horn achieving well-maintained high-frequency sound distribution over wide angles. At 67½° (near extreme ends of front stalls), drop at 3,000 Hz is 15 dB, and at 6,000 Hz slightly less.

When a monophonic single channel system is being installed, the picture is most easily associated with its accompanying sound when the speaker system is mounted on the centre line and at two-thirds screen height. The high frequency speaker should be tight up against the back of the sound porous screen and enclosed by a thick sound-absorbent blanket.

Cinemascope requires three identical speaker systems, and 70mm wide-film systems five identical assemblies. These are nominally equally spaced across the screen but it is essential that each of the speaker systems be able to "see" the whole of the auditorium seating. This is often difficult when the screen is well away from the proscenium arch. In these circumstances, the sound beam from the high frequency horn tends to be obstructed by the proscenium. This is a valid argument for abolishing the customary arched enclosure of the stage, an inheritance from the days of the legitimate theatre.

STEREOPHONIC SOUND

All early films employed a single sound track operating a single channel sound reproducer system. However, the advent of Cinemascope, the first wide-screen system, made it advisable to produce a wider sound image to match the increased picture width.

When a monophonic single channel system is employed, the addition of further speaker systems in parallel with the centre stage speaker achieves little, for it produces an impression that all the sounds are emitted by the nearest speaker system. This is an undesirable illusion for the occupants of the side seating blocks, particularly when the action is taking place on the remote side of the screen.

A stereophonic sound system avoids this effect for, when properly designed and installed, the sounds appear from the position of the screen image even when this is located between the loudspeakers. Such a system has the further advantage that the sounds from a large source such as an orchestra appear to fill the whole of the stage and not merely the space occupied by the loudspeakers. An accurate reproduction of these spatial characteristics of the original sound adds a smooth character to reproduced music that allows its volume and frequency range to be raised, so that it more closely approximates to the original source of sound.

A good stereophonic illusion is secured only by the use of several completely separate channels between the original microphone and the final loudspeakers. Three are used in Cinemascope and its variants, five in the 70mm wide-screen systems.

Today, good examples of the five-channel stereophonic systems provide the best sound quality commercially available in any medium.

In spite of the limitations of a single-channel reproducer, the majority of Cinemascope installations are now operated with all three speaker systems reproducing the signal from a single monophonic track. A three- or four-track film is more costly to produce than a single-track film and it would appear that the advantages of the improved sound quality that result from a stereophonic sound system are insufficient to increase the cinema audience and thus justify the increased cost of production. The appearance of many pseudo-stereophonic recordings has done nothing to increase the public appreciation of true stereophonic reproduction.

Though multiple photographic tracks are technically feasible, the problems of recording and replaying a multiplicity of narrow tracks are simplified by the use of magnetic recording. These can be readily accommodated on wide-gauge 70mm film.

DRIVE-IN CINEMA

In areas south of about latitude 51°, twilight conditions exist for sufficient time every evening to make it possible to produce an adequately bright picture on an open-air screen. North of this line, the necessary brightness is difficult to achieve for a reasonable fraction of the evening hours.

Drive-in cinemas using very large picture screens viewed from the comfort of a private car have acquired considerable popularity, particularly in America and South Africa. Screen brightness values below 2 ft.-candles seem acceptable in the circumstances, though 10–30 ft.-candles are achieved in an indoor cinema of the normal type. The sound accompanying the picture is reproduced through a small loudspeaker normally stored on a light post adjacent to the parked car but clipped on to the inside of the car window when in use. Neither sound or picture quality approaches that achieved in an indoor cinema but the other advantages have sufficient appeal to the public to make drive-in cinemas economically viable.

NARROW-GAUGE SOUND REPRODUCTION

There are many situations where a small picture is adequate: advertising, educational use, home movies, shipboard shows, etc. For these purposes, 16mm or even 8mm film is satisfactory and indeed advantageous. The projectors are easily portable, provide a total screen illumination of 200–2,000 lumens (for 16mm films) and have a sound quality that is adequate for the purpose. Magnetic recording is more widely used in these narrow-gauge films than with 35mm because the slow speed of the film emphasizes the resolution limitation of photographic recording. At the standard 16mm projection speed of 7·2 ips, the frequency response that can be achieved using magnetic recording is superior to that obtained from a photographic track on 35mm film. Space limitation makes it impossible to include adequate mechanical filtering in the film drive and as a result wow and flutter are generally more obvious and set a limit to the performance obtainable.

Many 16mm and 8mm film projectors are equipped to record sound, allowing the same picture film to be used with tracks

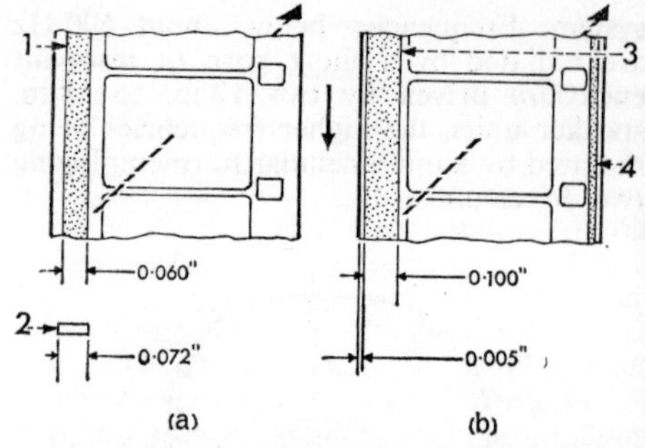

SOUND TRACK LOCATION ON 16MM FILM. (a) Optical. 1. Width of variable area and variable density squeeze tracks. 2. Area scanned by reproducer. (b) Magnetic. 3. 100-mil stripe. 4. Balance stripe to equalize thickness across film. Direction of travel through projector: downwards. Arrows through frames show light beam for standard front projection.

recorded in the language of the country in which the film is being distributed.

The use of magnetic recording allows an acceptable performance to be secured from a sound track on 8mm film moving at only 3½ ips. Recent improvements in the film format, with narrower sprocket holes (Super-8) and better film drive, have allowed the use of a narrow magnetic stripe and the achievement of wow and flutter figures in the region of ·25 per cent.

The achievement of good intelligibility in halls having poor acoustics requires the use of highly directional loudspeakers, and it is unfortunate that these are all relatively large. Both 16mm and 8mm film have their main applications where portability is essential and thus large loudspeakers are rarely acceptable. In consequence, the performance usually obtained from 16mm sound film is rarely as good as is possible, the limit being set by the size of the loudspeakers that can be transported. J.M.

See: *Acoustics; Cinema theatre; Drive-in cinema theatre; Intermodulation; Magnetic recording materials; Sound and sound terms; Sound distortion; Sound mixer; Sound recording; Sound recording systems.*

Books: *Film and Its Techniques*, by R. J. Spottiswoode (Berkeley), 1951; *High Quality Sound Reproduction*, by J. Moir (London), 1961; *Sound Reproduction*, by G. A. Briggs (London), 1949.

⊕**Sound Speed.** Film speed standardized at 24 frames per second at the time of introduction of the sound film, and universal today for all gauges of film when synchronized with sound tracks. Contrasted with silent speed. The term is now extended to 25 fps filming for 50-field TV systems.

See: *Silent speed.*

⊕**Sovcolor.** Negative/positive integral tripack colour film process for professional motion picture photography based on Agfacolor and developed in Russia from about 1950. This followed the Russian takeover of the Agfa plant at Wolfen during World War II. The process is in current use.

See: *Colour cinematography.*

SPECIAL EFFECTS

Very often in the course of professional film production the need arises for certain kinds of scene which are too costly, too difficult, too time-consuming, too dangerous or simply impossible to achieve with conventional photographic techniques. These scenes may call for relatively simple and inexpensive effects, as when optical transitions such as fades, wipes and dissolves are used to link different sequences. Or they may be much more demanding and costly, as when a city is destroyed by fire, when a full-size, non-existent building must be shown as part of a live-action scene, or when actors on the sound stage must be shown performing in locales which are hundreds of miles distant.

AVAILABLE TECHNIQUES

The solution of these and many other problems calls for the application of a set of non-routine photographic techniques known variously as special cinematographic effects, visual effects, optical effects or process cinematography. These effects can be described as involving either (1) in-the-camera techniques, in which all of the components of the final scene are photographed on the original negative, (2) laboratory processes, in which duplication of the original negative through one or more generations is necessary before the final effect is produced, or (3) combinations of the preceding two. The techniques which are presently available can be classified in the following fashion:

1. In-the-camera techniques:
   (*a*) Basic effects, such as changes in object speed, position or direction; image distortions and degradations; optical transitions; superimpositions; and day-for-night photography.
   (*b*) Image replacement, including split-screen (matte) shots; glass shots; and mirror shots.
   (*c*) Miniatures.
2. Laboratory processes: bipack contact matte printing; optical printing; aerial-image printing; and travelling mattes.

SPECIAL EFFECTS TECHNIQUES—THEIR COST AND OPERATIONAL CHARACTERISTICS

Type of process	*Complexity of operation*	*Relative cost for purchase, lease or construction of equipment*	*Relative speed of sound-stage operations*	*Relative cost to have the service performed by a contractor*	*Experience required for execution*	*Relative length of time required for execution, including all steps*	*Amount of sound-stage space tied up, beyond normal*
In-the-Camera							
Basic effects	Little	Nil	Fairly rapid	Low, if available	Little	Minutes	None
Image replacements							
Split-screen	Moderate	Nil	Slow	Low, if available	Moderate	Minutes or hours	None
Glass shots							
Painted mattes	Great	Low	Very slow	Quite high	Great	Hours	Slight
Photo mattes	Moderate	Low	Slow	High	Moderate	Minutes or hours	Slight
Cloud plates	Little	Low	Rapid	Low, if available	Little	Minutes	Slight
Mirror shots	Moderate to great	Low	Slow	High, if available	Moderate to great	Hours	Moderate
Miniatures	Great	Low to high	Extremely slow	Moderate to very high	Great	Days, weeks or months	Moderate to great
Laboratory Processes							
Contact matte printing							
Painted mattes	Great	High	Fairly rapid	High	Great	Days or weeks	None
Photo mattes	Moderate	High	Fairly rapid	High	Great	Days or weeks	None
Optical printing	Moderate	Extremely high	Not required	Low to medium	Moderate	Hours or days	None
Travelling mattes	Great	Medium to very high	Fairly rapid	Medium to high	Great	Days or weeks	None
Aerial-image printing	Moderate	Extremely high	Fairly rapid	Low to medium	Moderate	Days	None
Background Projection							
Rear projection	Great	Extremely high	Slow	Medium to high	Great	Days for preparation; hours on stage	Great
Front projection	Moderate	Low	Slow	Moderate, if available	Moderate	Days for preparation; hours on stage	Moderate

3. Combination techniques: rear projection and front protection.

Often, identical results can be achieved by using entirely different techniques. The selection of the right process for a particular assignment depends upon various factors, including image quality, flexibility, cost and scope of the production set-up. In-the-camera techniques produce superior image quality because all the effects are produced on the original negative. Compared to laboratory processes, however, they are relatively inflexible because they do not allow for correction or alteration of the image components after the original exposures have been made. In-the-camera effects require less equipment but involve more on-stage time for their execution than do laboratory processes, and so may run up labour costs considerably. Comparative cost factors and operational characteristics of the different techniques are summarized in the preceding chart.

By means of special effects cinematography, film producers are able to endow their pictures with production values which the budget could not otherwise sustain. They realize enormous savings in time and in set-construction costs and, even more important, their films display the stylistic touches and polish which have come to be associated with the theatrical product.

The procedures of the special effects cinematographer are as infinitely varied in their application as the kinds of production problem which can arise, for each effects assignment is a new one and is different in its peculiarities from every other one that has been done before. It is this variety of problems and solutions which makes the field so interesting; it is the same variety which also makes the work of the special effects cinematographer so complicated. There are few rules, if any, and mistakes are common. The tools of the art can range from simple, inexpensive devices which can be held in the hand to extremely costly machines weighing a ton or more. The length of time spent on an effects shot can range from a few minutes to several weeks. In the end, only familiarity with the tools and techniques available will provide the right solution for a particular problem, and only experience will provide consistently professional results.

IN-THE-CAMERA TECHNIQUES

BASIC EFFECTS. Basic camera effects are ordinarily performed with conventional production equipment. Some of the techniques involved are feasible for professional work; others are not.

Image distortions and degradations are easily achieved by interposing diffusion, fog, ripple and contrast filters between camera lens and subject. Mirrors and prisms can be used in similar fashion to warp and fragment the image.

Day-for-night effects are nearly always performed in the camera, a sunlit scene being photographed so as to produce the illusion of evening.

The speed, position and direction of moving objects can be altered through relatively simple camera adjustments. Objects can be made to move in reverse by running the film backwards in the camera. The speed of moving objects can be reduced by over-cranking the camera at higher-than-normal frame rates. Conversely, if the camera is undercranked, objects on the screen will move at accelerated rates of speed. If the camera is operated a frame at a time and the actors or other moving objects in the set are moved appropriately, then a novel "pixillation" effect is produced, whereby the performers appear, disappear and move instantly from one part of the frame to another.

Time-lapse transitions (fades, dissolves and wipes) are rarely performed in the camera, but in the laboratory, by means of optical, aerial-image or bipack contact matte printing.

IMAGE REPLACEMENT. Effects cinematography is often called on to replace selected portions of a scene with an entirely different kind of visual detail. The aim, in such cases, is usually the pictorial enhancement of the image or the saving of set-construction costs.

For example, the director may wish to show his actors in front of an elaborate hotel, or entering and leaving a department store. Quite often, however, he discovers that the particular kind of building and architecture which he desires does not exist or, if it does, cannot be conveniently photographed. When this happens, the structure is built on the sound stage where complete architectural, photographic and sound-recording control is possible. However, it would be a waste of time and money to build the entire structure, because the actors perform only in front of the bottom part of the set. So only the first floor is built on the sound stage and the remaining floors are added, artificially, through special effects cinematography.

The effects specialist may also be called

upon to place clouds in a barren sky, to add a ceiling to an elaborate ballroom set, to remove unattractive background detail in an exterior scene or to re-position an improperly situated tree or mountain-top. All of these problems involve image replacement – the addition of missing pictorial detail or the substitution of more attractive detail than appears in a real scene. Image replacement can be executed in the camera or in the laboratory. Among the in-the-camera techniques are split-screen (matte) shots, glass shots and mirror shots.

SPLIT-SCREEN (MATTE) SHOTS. Split-screen or matte shots are used for a variety of purposes. Sometimes dissimilar subject matter is shown stylistically in different sections of the frame. For example, two individuals in different locations can be shown talking to one another by telephone, one on the left side of the frame, the other on the right.

In other cases, the screen image is not literally split, nor is the replacement of images so obvious. Instead, a section of the original scene is realistically replaced with appropriate visual detail. For example, background landscape detail can be inserted into the window area of an interior set, or human figures placed on the balconies of distant buildings. This kind of replacement is called a matte shot. Properly executed, its mechanics are not supposed to be apparent to the audience.

All in-the-camera split-screen or matte shots require two or more exposures of the original negative. During first exposure, an opaque fibre or metal plate (a matte) is placed either in the external matte box or within the intermittent movement of the camera. This matte obscures selected portions of the frame. During second exposure, a counter-matte, whose outline matches that of the first exactly, is appropriately positioned so that it obscures those parts of the frame which were photographed during the first run, while allowing exposure of the remaining sections. Naturally, the camera is moved from one location to another between exposures so as to provide just the right combination of image components.

In-the-camera, split-screen and matte shots offer maximum image quality; however, the technique is awkward, time-consuming and imprecise. For these reasons, the effect is usually achieved in the laboratory.

GLASS SHOTS. The camera is aligned to photograph the live-action scene which is to be altered. A large sheet of clear glass is rigidly mounted a few feet in front of the camera. An artist paints appropriate representational images upon portions of the glass. These images obscure and replace certain visual components of the real scene. By viewing the emerging composite through the camera's eyepiece, the artist blends the real and artificial visual elements together, developing appropriate perspective, density, texture, brightness and colour relationships as he goes along. When completed, the complementary visual components from the

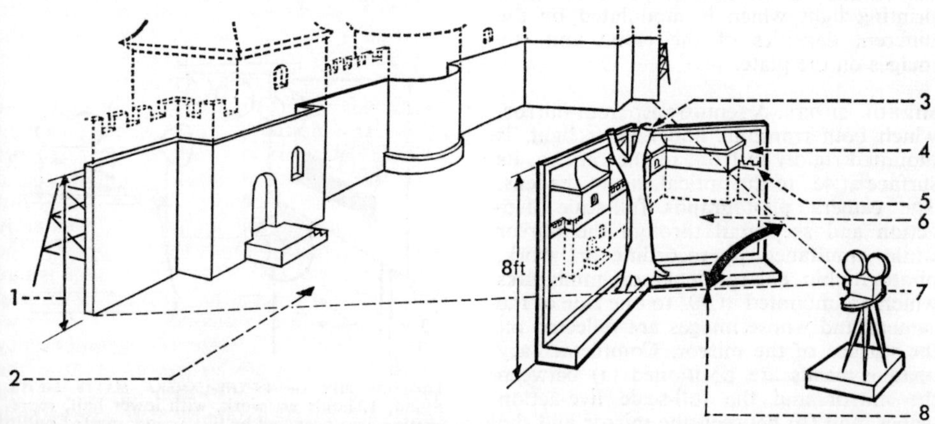

GLASS SHOT. Two large glass paintings are used to add architectural detail to a one-storey set. 1. Height limit of set construction. 2. Area within which action can take place. 3. Trunk of tree hiding junction between two glass panels. 4. Painting on glass. 5. Matte line dividing painting and clear glass, 6, which corresponds to action area. Camera is set on nodal head, 7, which pans about nodal point of lens system to maintain true perspective over arc, 8.

real scene and from the painted images on the glass are photographed simultaneously with one pass of the film through the camera.

Properly executed, this process produces a convincing composite. It requires little equipment or materials and offers maximum image quality, because duplication of the original negative is not required. Its success depends upon the availability of a skilled and experienced artist, however, with a flair for ultra-representational painting. This is a kind of skill which many otherwise fine artists find difficult to master. Alternatively, still photo enlargements can be prepared, sections removed with a knife or scissors, and the segmented enlargement used in front of the camera. Finally, three-dimensional miniatures (so-called hanging miniatures) can be mounted in front of the camera so as to merge realistically with background detail.

Another type of glass shot is used to add attractive clouds to an otherwise barren sky. Glass-base photographic diapositives are mounted in front of the camera. The upper section of the plate shows cloud detail; the lower section is clear. The diapositive is positioned so that the division between the cloud scene and the clear section is placed along the horizon. The live-action scene is photographed without modification, through the clear lower section of the plate. The sky area of the scene, which occupies the upper portion of the field-of-view, is photographed through the positive cloud section of the transparency. The barren sky of the real scene serves as a printing light which is modulated by the different densities of the cloud and sky images on the plate.

MIRROR SHOTS. A semi-transparent mirror, which both transmits and reflects light, is mounted rigidly in front of the camera, its surface at 45° to the optical axis of the lens. The camera photographs full-scale live-action and set detail through the mirror while simultaneously recording art work, photographic enlargements or miniatures which are mounted at 90° to one side of the camera and whose images are reflected off the surface of the mirror. Complementary opaque masks are positioned (a) between the mirror and the full-scale live-action scenes, and (b) between the mirror and the reflected art work or miniatures. The masks are made of heavy paper or card, out of which the complementary areas of the composite are cut with scissors or a knife. The

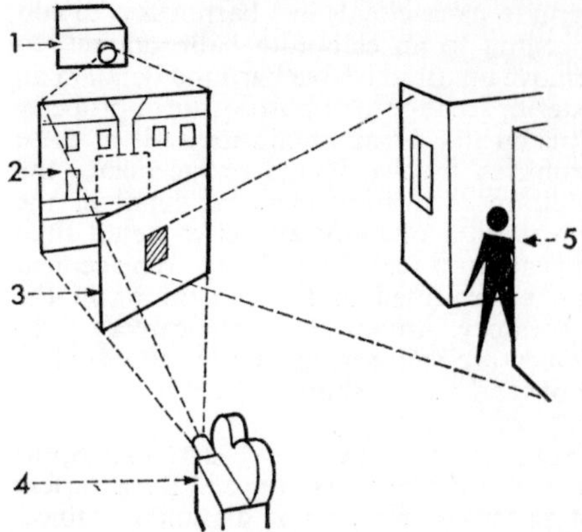

SCHUFFTAN SHOT. Camera, 4, picks up full-scale live-action component, 5, reflected off silvered area of mirror, 3. Simultaneously, the camera photographs through clear area of mirror either art work, miniatures or (as shown) a rear-projected image, 2, from projector, 1.

mask which stands between the mirror and the full-scale set holds back all the visual detail except that area in which the actors perform. The complementary mask, on the other hand, obscures from the reflection of the art work only the area into which the live action is being inserted.

An alternative but much more intricate technique calls for the use of a fully reflective, front-silvered mirror positioned at 45° to the camera's optical axis. Selected areas of the mirror's silvering are removed, allowing for the photography of live action

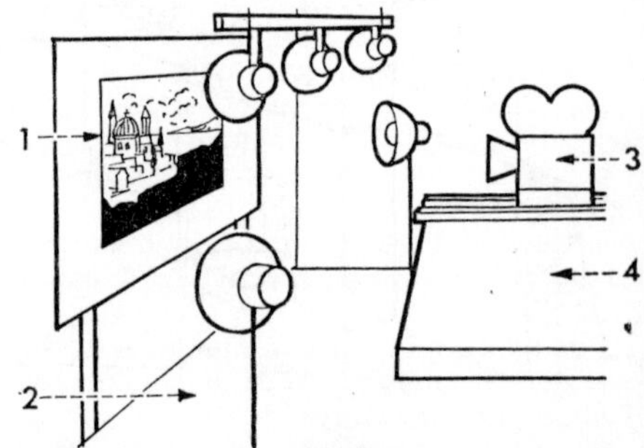

PHOTOGRAPHY OF IN-THE-CAMERA MATTE SHOTS. Stand, 1, holds art work, with lower half, representing area occupied by live action, matted out in black. Cement floor, 2, rigidly supports camera, 3, mounted on solid base, 4, surmounted by lathe bed in which movement is completely free from shake or play. Absence of relative movement between camera and art work is essential for success.

components through the desilvered portions of the mirror and art work which is reflected off the surface of the mirror.

Mirror shots (sometimes called Schüfftan shots, after the inventor who popularized the process) offer effective composites. Maximum image quality results because the effect is produced in the camera. The director and his crew are able to view the complete composite at the time that the shot is lined up, and the finished film is returned from the laboratory the following day with the production dailies. The process is time-consuming, however, and requires considerable experience for its proper execution.

Mirrors are also used in special effects cinematography to distort, fragment or multiply images, to simulate water reflections along the lower portion of the frame, to superimpose titles, art work or phantom figures over live action detail, and to produce inexpensive wipe transitions.

MINIATURES. Miniatures are representational scale models which are built, operated and photographed so as to appear genuine in character and full-scale in size. They are generally employed to simulate scenes which would be too expensive or dangerous to photograph in full scale. The burning of a city, the manoeuvring of a battleship at sea during a severe storm, or the explosion and collapse of a large building can all be simulated by the use of miniatures.

Miniatures can be photographed complete in themselves, as background detail viewed through windows or doors, or as image replacement components of a composite. In all three applications, miniatures are more versatile and believable than painted backgrounds or matte shots.

Miniatures offer a three-dimensional quality which the finest artist can only approximate in a two-dimensional representation. They can be lighted and photographed from a variety of angles and the lighting can be changed from shot to shot, whereas a matte painting has its highlights, its shadows and its perspective permanently fixed.

Finally, certain types of miniature can be operated in motion, thus enhancing the illusion of their reality.

Of all the tools of the special effects cinematographer, miniatures are probably the most expensive to use. They are difficult to build, to operate and to photograph, and even the best miniature work often appears unconvincing on the screen. Generally, believability is increased if the image of the miniature is combined with full-scale live action components by means of matte or optical printing techniques.

Mobile miniatures are particularly difficult to work with owing to time-scale differences which exist between the model and the full-size original. If a moving miniature is to appear realistic, its speed must be decreased in proportion to its reduced scale, because all linear dimensions appear to be magnified as the square of the magnification of time. In the case of gravity-fed components, this reduction is achieved by overcranking the camera an appropriate amount. The formula employed here is

$$\sqrt{\frac{D}{d}} = f$$

where D is the distance or dimension in feet for the real object, d is the distance of dimension in feet for the miniature (this fraction being simply the reciprocal of the scale of the model) and f is the factor by which the camera's operating speed is increased. Computations for typical miniature scales are as follows:

CAMERA-SPEED ADJUSTMENT FOR MINIATURES

Miniature Scale	*Camera-Speed Increase Factor*	*Adjusted Operating Speed for Camera (fps)*
1/2	1·4	33
1/4	2·0	48
1/8	2·8	67
1/10	3·2	77
1/12	3·5	84
1/16	4·0	96
1/20	4·5	108
1/24	4·9	117
1/36	6·0	144
1/48	7·0	168
1/64	8·0	192
1/100	10·0	240

For the sake of both convenience and believability, miniatures are built to as large a scale as the budget permits. Sometimes they are immense. For example, the model battleship used in the production of *Sink the Bismarck* was 30 ft. long, and was operated in a back-lot tank which measured 200 × 400 ft.

Where the budget permits, the scale usually ranges from 1 : 4 (3 in./ft.) to 1 : 12 (1 in./ft.). Surprisingly, it is often less expensive to build a large miniature than a small one. The cost of materials for a large miniature is relatively slight compared to the labour which goes into the detail work on a small one. If the miniature is sufficiently large, it can be built entirely in one scale. A single-scale miniature takes its own perspective and can be photographed from a variety of angles and positions. If, on the other hand, the miniature must be built to a relatively small scale, it may be necessary to force perspective by mixing the scales of the different components and distorting the shape of those pieces which are nearest to the camera, thus artificially increasing the illusion of depth.

Miniatures which burn, explode or collapse involve special problems. When they are burned, the type of fire desired determines the construction employed. For complete destruction, the entire set is built of soft pine; for a selective conflagration, those sections which are supposed to burn are built of pine, while those which must remain are made of plaster.

A mixture of paraffin (kerosene) and petrol can be sprayed over selected portions of the set to shape the fire. To simulate an oily, petrol tank fire, strips of paving or roofing tar are added to the inside walls of the miniature pieces. Where miniatures are expected to collapse, or are involved in impact action of any sort, the strength of the construction materials is reduced in exact proportion to the reduction in scale. If a miniature set, built to 1 : 10 scale, shows boulders in an avalanche crashing down upon a building, then the wood employed to construct the model building must be 1/10 the strength of that normally used in such a structure. Alternatively, if stronger woods are used, selected pieces must be deeply scored from within so that they will fracture easily when struck by miniaturized objects.

LABORATORY PROCESSES

All of the previously described in-the-camera effects, and many more, can also be achieved in the laboratory. The techniques require expensive and complicated equipment, but they involve much less sound-stage time and provide unlimited opportunity for experiment.

BIPACK CONTACT MATTE PRINTING. This versatile technique allows for the most complicated kinds of image replacement. The scene which is to be altered is first photographed normally (e.g., actors in front of a hotel set, of which only the first floor has been built). The negative of this shot is printed to a fine-grain master positive on a step printer. The master positive is then threaded into a process camera (a combination camera and step-printer) together with a roll of fine-grain duplicating raw

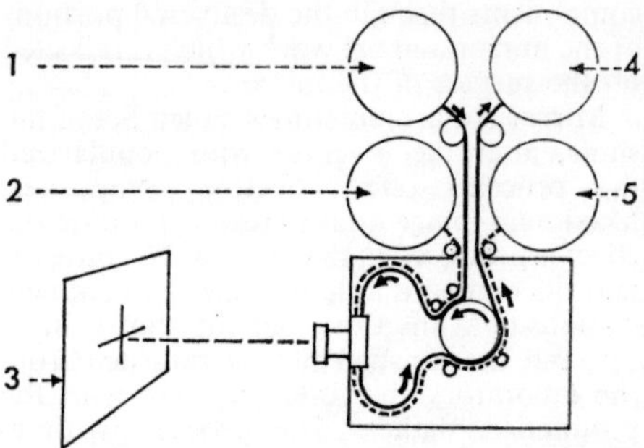

PROCESS CAMERA. Essentials only, as loaded for bipack printing. 1. Unexposed duplicating negative raw stock, wound emulsion in, so that at aperture it prints emulsion-to-emulsion with master positive film in chamber, 2, wound emulsion out. 4, 5. Corresponding take-up chambers. When matte board, 3, facing camera, is white and uniformly illuminated, this set-up functions as a step printer, the films moving forward intermittently frame by frame. If chosen areas of matte board are blackened, there is no exposure and selective printing takes place.

stock negative. The positive and negative are threaded in bipack – that is, with the emulsions of the two strips in contact with one another.

A white matte board is set up on a rigidly mounted easel in front of the process camera and the lens is focused on the board. So long as the entire white matte board is evenly and adequately illuminated, the

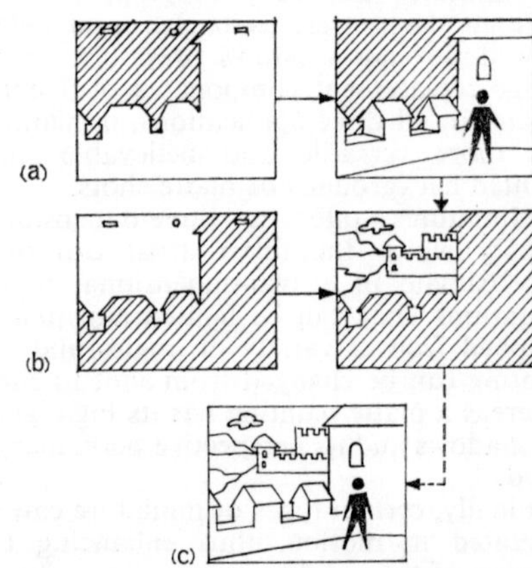

MATTE AND COUNTER-MATTE: COMPOSITE IMAGE. (a) Matte with positioning holes to provide peg-bar registration, outlined so as to mask unwanted detail in dark area (*right*). (b) Counter-matte of opposite tonality, similarly registered, and outlined to permit printing-in of wanted detail from another locale in transparent area (*right*). (c) Composite image combines live foreground action with derived scene in background.

camera functions as a step printer, printing a dupe negative from the entire frame area of the master positive. If, however, selected areas of the matte board are blackened with ink (matted out), then those areas, as imaged by the camera's lens, do not provide light to print sections of the master positive on to the dupe negative. For image replacement purposes, the matted-out area should correspond to those parts of the original scene which are unattractive or deficient, and which need replacement.

Following the first printing with the blackened matte board, a counter-matte is prepared which matches the contour of the blackened area exactly, and art work which complements the live action component is painted on to the white area of the counter-matte. The dupe negative raw stock is rewound in the camera and the master positive is removed. The raw stock is run through the camera again, this time by itself. During this second exposure, the art work is printed into place so as to blend with the previously printed live action scene. Thus a composite is made of two or more visual components.

Any combination of art work, still photographs, miniatures and live action scenes can be achieved with this technique, providing that actors and moving objects do not move outside of their own matted area. Bipack contact matte printing can also be used to add titles to live action backgrounds, to produce optical transitions and to print certain types of travelling matte effects.

OPTICAL PRINTING. Optical printing involves the re-photographing of one strip of film (a fine-grain master positive) on to another strip (duplicating negative raw stock), one frame at a time. The master positive travels through a printer head, the lamp house of which illuminates each frame in turn. The duplicating negative raw stock passes through a process camera which faces the printer head. By varying (a) the distance, position and movements of the process camera with respect to the printer head, and (b) the direction and speed of the two strips of film with respect to each other, the film maker can produce a tremendous range of visual effects and space-time distortions:

Optical Transitions – Fades, dissolves, wipes and push-offs.

Superimpositions – Successive printings of two or more strips of film.

Image Size Changes – Enlargements or reductions of the master positive image; zooms in and out; rotary movements of the image around its own centre.

Reverse Printing – Contrary movements of the positive and negative strips of film during printing; back-and-forth printing to extend too-short scenes.

Skip-Frame Printing – Elimination of selected frames during printing so as to increase the apparent speed of movement on the screen.

Multiple Printing – Increasing the number of times that selected frames are printed, thus slowing down or stretching screen movement.

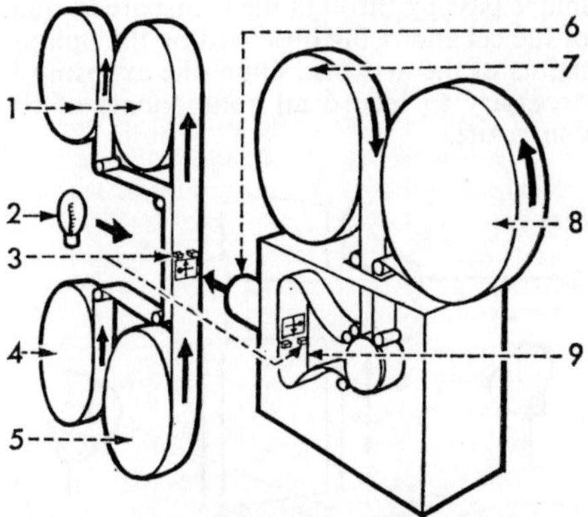

OPTICAL PRINTER LAYOUT. Process camera (*right*) faces printer head (*left*) through copying lens, 6, often in separate support to give adjustable magnification. Because of image reversal in lens, movement of film in printer is normally upwards from feed chambers, 4, 5, to take-up chambers, 1. Big registration pin, 3, is thus on opposite sides in printer and camera, where 9 is guided edge of film. Light source, 2, provides exposure. Camera feed chamber, 7, and take-up chamber, 8, are in orthodox positions. Camera is shown copying bipack master positives in printer for a travelling-matte shot.

Travelling-Matte Printing – Composite shots of moving actors or objects (photographed on the sound stage) with backgrounds which are photographed elsewhere.

Flip-Shots – Reversal of a shot's screen direction; upside-down printing.

Miscellaneous Distortions – Fog, diffusion and texture effects; warping, rippling, fragmentation and multiplication of images.

Optical printing can also be used for salvage purposes, whereby scratches are removed, exposure errors corrected, horizons re-aligned and frame lines re-positioned.

AERIAL-IMAGE PRINTING. Aerial-image printing combines all the best features and capabilities of both optical printing and bipack contact matte printing.

Aerial-image printing equipment is ordinarily built into an animation stand. A master positive of the background is optically projected from below the stand and is focused as an invisible aerial image into the plane of the stand's table top. This image, in turn, is re-focused by the lens of the camera above. Artwork painted on animation cels and placed on top of the stand's table is photographed simultaneously with the aerial image. With such equipment, matte paintings and titles are easily composited with live action backgrounds, the projected image passing through the transparent areas of the cel and being obscured by the opaque paints of the artwork. Only one exposure is necessary to record all components of the composite.

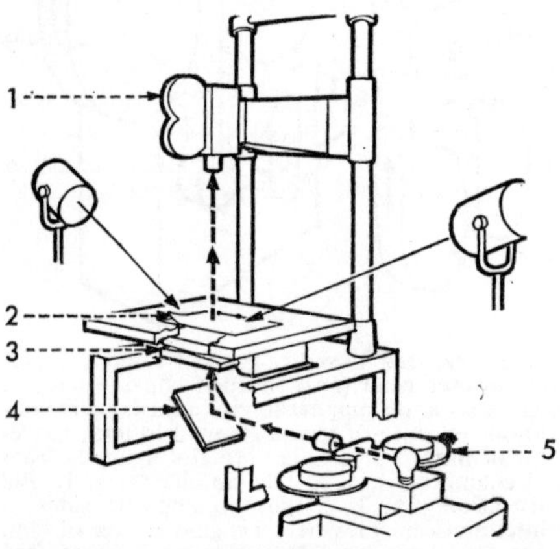

AERIAL-IMAGE ANIMATION PHOTOGRAPHY. Camera, 1, and motion picture projector, 5, are driven in synchronism. Projector produces aerial image in plane of cel, 2, through 45° mirror, 4, and large condenser lenses, 3. Optical system of camera picks up aerial image and transfers it to raw stock, thus enabling animated action on cels to be combined with live action on projected film.

Alternatively, some aerial-image printers are designed to operate horizontally, like an optical printer. In such designs, the artwork is imaged into the plane of the master positive, and two passes through the camera are necessary to fit the replacement image and the live action scene together.

TRAVELLING MATTES. Travelling matte cinematography involves a complicated body of technique by which the image of an actor photographed on the sound stage can be combined with a background scene photographed elsewhere.

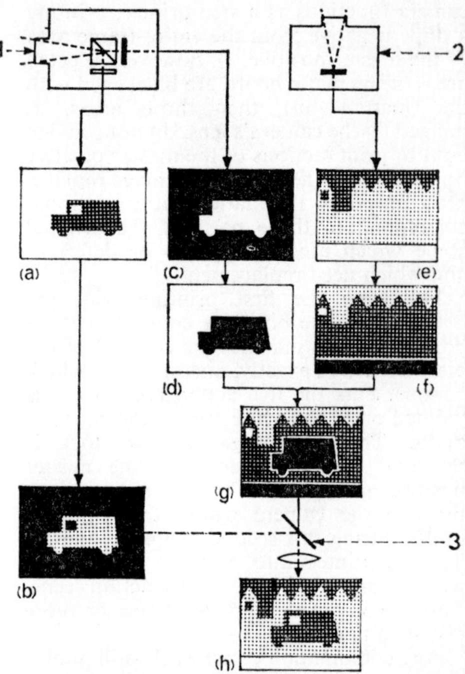

MULTI-FILM TRAVELLING MATTE SYSTEM. 1. Beam-splitting camera producing foreground action negative (a) from which positive (b) is printed. Camera also produces high-contrast matte negative (c) with no internal detail, usually by a colour-selection process. From this a positive (d) is printed. A normal production camera, 2, photographs background negative (e) in another locale, and from this is printed a positive (f). Bipack printing of (d) and (f) produces a bipack positive (g), and this is composited in a beam-combining process, 3, to produce the final result (h).

The process is infinitely versatile and offers many advantages to the professional film maker. Since the actor is photographed on-stage, ideal acoustics and isolation are available for sound recording. The actor and the background scene are photographed at entirely different times and in whatever order is convenient. Consequently there are great savings in location costs, since the crew and cast need not leave the studio. Finally, the process allows for certain types of scene which would be too dangerous to photograph conventionally, or which contradict the laws of nature. For example, individuals can be made to fly through the air, falling rubble and masonry can be made to engulf performers, and actors can perform in multiple roles with themselves, within the same shot.

The process is an intricate one which requires the preparation of a strip of film which bears an opaque silver silhouette, the outline of which exactly matches that of the actor's image on the production negative. This opaque mask, which is referred to as the male matte, can be produced in a variety of ways. Some systems employ a single strip of colour negative raw stock in a conventional camera photographing the actor who is posed in front of a deep blue backing. Later, in the laboratory, various printing operations are conducted, in the course of which different colour values are separated out of the original scene and recorded on separate strips of black-and-white duplicating stock. Ultimately, a male travelling matte is produced: a strip of film with an opaque silhouette against a clear field, the configuration of which silhouette exactly matches the outline of the actor in the original negative.

Another system which produces travelling mattes requires the use of a camera in which the raw-stock negative and a special travelling matte raw stock are run through separate intermittent movements simultaneously, and in frame-to-frame synchronism. By means of a beam-splitting prism, identically proportioned images are delivered to and recorded on each of the two films. One of the films records the scene as it is seen by the eye. The other, however, records only the set backing which is mounted behind the actors, thereby producing a female matte: an opaque frame area within which a transparent silhouette is positioned.

Multi-film systems such as these employ different wavelengths of light to expose the picture negative and the separate travelling-matte film. At present, either sodium-vapour, infra-red or ultra-violet radiation is used in the area of the set backing, and each of the three basic multi-film systems takes its individual name from the light source employed.

Once male and/or female travelling mattes have been produced, the final composite printing of foreground and background images is effected in the optical printer. Assuming, for example, that the shot of the actor on the sound stage must be combined with a background scene shot in a remote jungle, the master positive of the background scene is first inserted into the printer head of the optical printer in contact with the male matte, which is opaque in the area of the actor and clear in the surround. The printer head and process camera are set

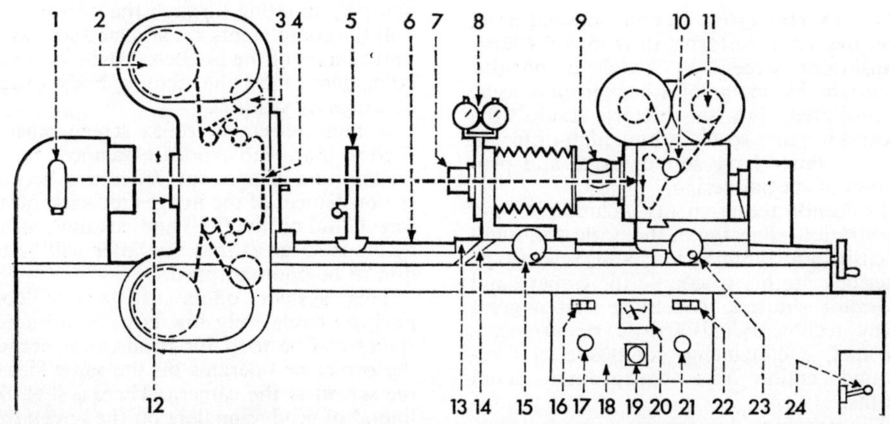

ELEMENTS OF OPTICAL PRINTER. Basic parts are projector, 2, through which film moves intermittently, and synchronized camera, 10, which photographs frame in projector aperture on to film wound up in chamber, 11. Projection lamp, 1, focuses beam of light on film in aperture, 4. Movable masks may also be placed in the beam, 5, and traversed or rotated with micrometer accuracy. Matte rolls, 3, may also be run in bipack. Camera lens, 7, supported on rigid base, 13, may be moved off centre by vertical and horizontal vernier adjustment, read on gauges, 8. Change of magnification is achieved by moving lens and/or camera by handwheels, 15, 23, or by motor drive actuated by lever, 24. Lens focusing is automatic by arm, 14, acting on cam between lathe-bed rails, 6, on which movable units, 5, 7, 10, travel with extreme precision. Reflex viewing aperture, 9, is for exact setting-up before shooting. Electrical controls are operated from panel, 18; light output of lamp is read on exposure meter, 12, and is set by rheostat, 21, and voltage is read on meter, 20. Printer functions are selected by knob, 17, which links or separates printer and projector, reverses motion, introduces skip-frame or double-frame printing, etc. Control, 19, enables driving motor to be speeded up to wind film forward and back to predetermined frame, with counter for projector, 16, and camera, 22.

running, and the background detail which appears around the actor's matte is printed on the dupe negative during first exposure.

The dupe negative is rewound to start position and the master positive of the background is now replaced with that of the actor's performance on the sound stage. The matte used during first exposure is replaced by its female counterpart, which is clear in the area of the actor's figure and opaque in the surround. A second exposure is now made, in which the actor's image is printed on the dupe negative, thus jig-sawing the two components together in the composite. Assuming many things – that the matte and counter-matte match one another properly, that neither matte has bled excessively during its preparation, that registration-pin positions have been kept consistent in all stages of the operation, that exposure and contrast of the two images have been properly balanced, and so forth – then a convincing composite results.

COMBINATION TECHNIQUES

As with the travelling matte process, back-projection techniques allow the combination of the image of an actor, photographed on the sound stage, with a background scene which was previously photographed elsewhere.

REAR PROJECTION. In conventional practice, the actor performs in front of a large translucent screen on which a positive photographic image of a background scene is projected. For satisfactory results, the process requires specially built high-diffusion screens, fairly large sound stages and projectors of special design.

Projection requirements include pilot-pin registration, silencing of the projector, high-intensity arc illumination, and Selsyn-type interlock motors to couple the camera and projector shutters. Assuming that a great many technical and artistic requirements are met, a convincing composite of foreground action and background detail results.

The system makes for great convenience, since the director, cameraman, art director and performers can all see the effect of the composite at the time of photography. The equipment is expensive, however, and the process involves considerable on-stage set-up time.

FRONT PROJECTION. Alternatively, the background image can be projected from the front of the screen. This is accomplished by means of a semi-transparent mirror or pellicle placed in front of the camera lens at a 45° angle to the camera's optical axis. A still or motion picture projector is positioned at 90° to one side of the camera. The background image is projected on to the surface of the mirror, from which it is reflected on to the screen. The camera photographs the screen's image and the actor by shooting through the mirror.

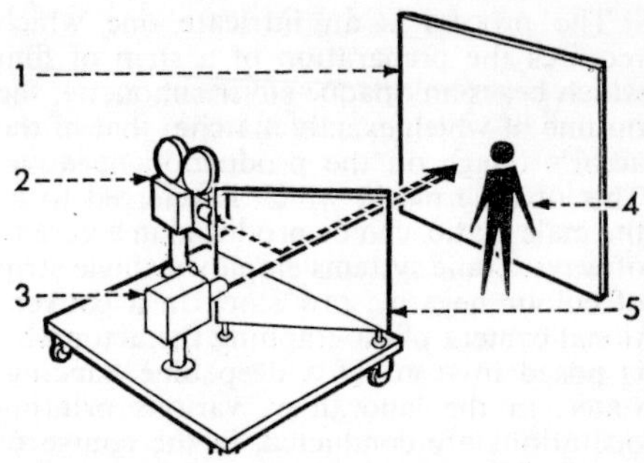

FRONT-PROJECTION PROCESS. Background image is projected by projector, 2, on highly reflective screen, 1, in front of which action, 4, takes place. Camera, 3, shoots through pellicle or semi-reflecting mirror, 5, which also turns projector beam through right angles. Principle is that shadows are eliminated because the reflected background image is projected directly along the camera's optical axis, so that the actor's shadow on the screen is precisely masked by his body. Hence, camera panning and movement are normally not practicable.

If the components are adjusted so that the optical axes of the projector and camera are coincident, then the actor's body exactly obscures its shadow.

A high-reflectance reflex screen must be used so that, with exposure balanced for the background image, the difference between the brilliances of the images reflected off the screen and off the flesh and costumes of the actors is so great that the latter will be too dim to be photographed.

This system offers satisfactory composites at relatively low cost. Small sound stages can be used for production because the projector operates on the same side of the screen as the camera. There is less likelihood of producing flare off the screen than in rear projection, and sharper background images can theoretically be expected. Unlike rear projection, however, once the optical components of the system have been aligned, the camera cannot normally be panned, tilted or dollied during the shot.

R.F.

See: *Art director; Blue screen process; Sodium lamp process; Special effects* (TV).

Book: *The Technique of Special Effects Cinematography*, by R. Fielding (London), 1969.

SPECIAL EFFECTS

The term special effects refers particularly to those most commonly used effects which are based on the technique of electronic picture insertion. This technique enables parts of the pictures from two television picture sources to be combined in various ways to form a composite picture for transmission. Applications include pictorial inserts, split-screens, distinctive captions and electronic wipes, used increasingly as an alternative to conventional cutting or dissolving between scenes. In addition to electronic picture insertion, some useful effects can be obtained by changing the nature of the picture produced by the camera itself.

In television centres, each studio of any size, and the presentation or master-control area, are equipped with the basic special effects facilities: an electronic switch and a pattern generator. Remote controls for these units are usually integrated into the design of the vision mixer control panel. Where special inlay cameras or scanners are installed in studio centres, the facility is normally shared between a number of studios.

ELECTRONIC PICTURE INSERTION

Electronic picture insertion is made possible by the sequential nature of the television scanning process, together with the normal practice of scan synchronization in cameras or other sources which contribute to a production. At any instant, therefore, the video signals from each camera relate to the same point in the standard picture format but, of course, carry different picture information, according to the picture detail being viewed by each camera. Thus if the video signal from one camera is switched to transmission during selected intervals of the complete scan in place of the video signal from another camera, a composite picture results, part of one picture being inserted into the other over an area determined by the timing of the switching operations.

This process is carried out in the electronic switch unit, in which the action of switching between the two sets of picture information is controlled by the amplitude of a keying signal, which may itself be taken from a camera.

Switching occurs whenever this signal exceeds a certain critical amplitude – the switching level.

While insertion is a convenient general term, there is nothing to prevent the contribution from either picture source occupying the greater part of the composite picture area. It is not possible, however, by electronic means, to insert the whole of one picture into a smaller area of another. A further limitation, using standard equipment, is that a part of the picture from one source can be inserted only in the corresponding part of another. The limitations are not serious in practice since the technique imposes no general restriction on camera movement or choice of shot.

TYPES OF INSERTION EFFECT. Insertion effects can be classified as inlay (also called keyed-insertion or static matte) or as overlay (also called self-keyed insertion or moving matte). In inlay the shape and size of the inserted area is independently determined, whereas in overlay it is made automatically to conform to the outline of a subject viewed by a camera. The effects are thus equivalent respectively to the stationary-matte and travelling-matte processes employed in the film industry. Electronic wipes, used as transitions between scenes, are produced by an extension of the inlay process, the area of the picture occupied by the new scene being progressively increased until it completely replaces the preceding one.

INLAY PROCESS

In the inlay process, the keying signal for the electronic switch is obtained from a camera or other picture source, or from a special effects pattern generator which pro-

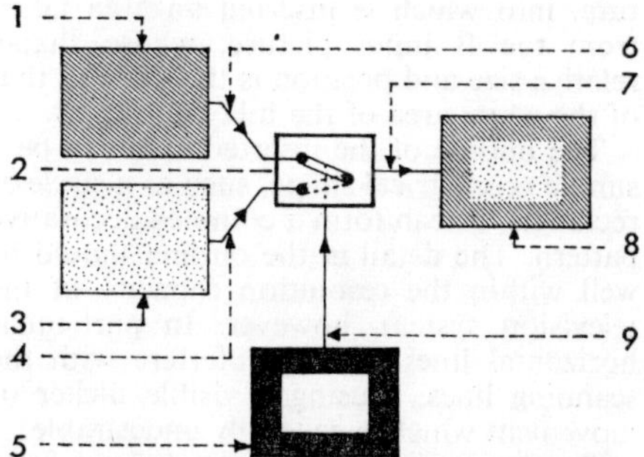

INLAY TECHNIQUE. Inlay or keyed insertion enables part of one TV picture to be inserted into another over an area whose shape and size is determined by a keying signal obtained from a camera viewing an inlay mask, or from a special effects pattern generator. 1. Picture A. 3. Picture B. 6, 4. Video signals A and B appearing at inputs of electronic switch, 2. Selected pattern or inlay mask, 5, generates keying signal, 9, which actuates switch. Resulting video output, 7, makes a composite picture, 8, in which a section of picture B is inserted into picture A.

vides for many purposes an equivalent electronically synthesized signal. Where a camera is used, it is directed on to an inlay mask, which can be in the form of a card having, on a black background, a white area corresponding in shape and relative size to the area to be inlaid. While the black background is being scanned, the keying signal does not reach the switching level, and video signal A is passed to the output of the electronic switch. During the periods in which the white area is scanned, however, the increased video level causes the electronic switch to operate, and video signal B replaces A at the output. The resulting

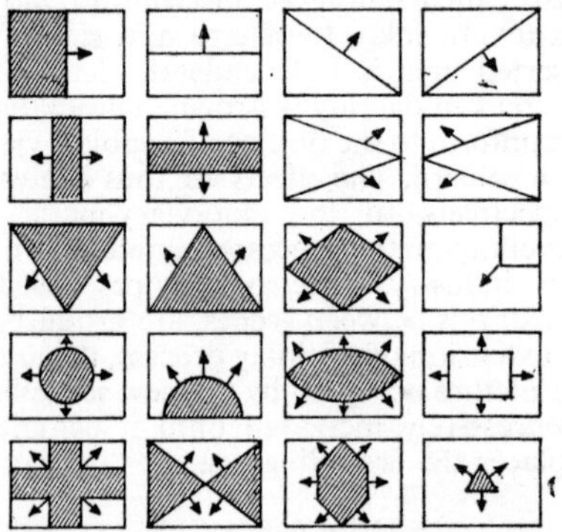

INLAY OR WIPE PATTERNS. Typical patterns available from special effects pattern generator; can be used for inlay effects or wipes. Arrows indicate pattern motion.

composite picture is thus the A input picture, into which is inserted an area taken from the B input picture, whose shape, relative size and position is the same as that of the white area of the inlay mask.

The outline of the inserted area can be a simple geometrical shape, such as a circle or rectangle, or can form a complex decorative pattern. The detail in the outline should be well within the resolution capacity of the television system, however. In particular, horizontal lines tend to interfere with the scanning lines, causing a visible flicker or movement which is generally undesirable.

The keying signal may be derived from any convenient source that can be presented with a mask in a suitable form. Slide transparencies can be used in a television slide scanner, and motion or animation can be imparted to the insert by the use of specially prepared film run through a telecine machine. The general requirements of the picture source are that it should have good resolution and be reasonably free of noise

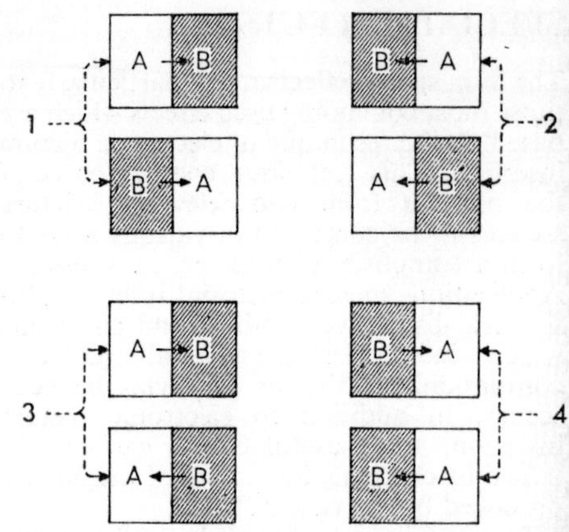

WIPE MODES. Special effects pattern generators usually have controls for direction or mode of the wipe, and for manual/automatic operation. 1. Auto. 2. Auto reverse. 3. Normal. 4. Normal reverse. Similar variations apply to other patterns.

and shading, so that a clean signal with sharp transitions is presented to the electronic switch.

INLAID CAPTIONS. The inlay process enables distinctive captions to be produced from plain black or white characters. The caption video signal, from a caption scanner or other source, is used as the keying signal for the electronic switch, and the video

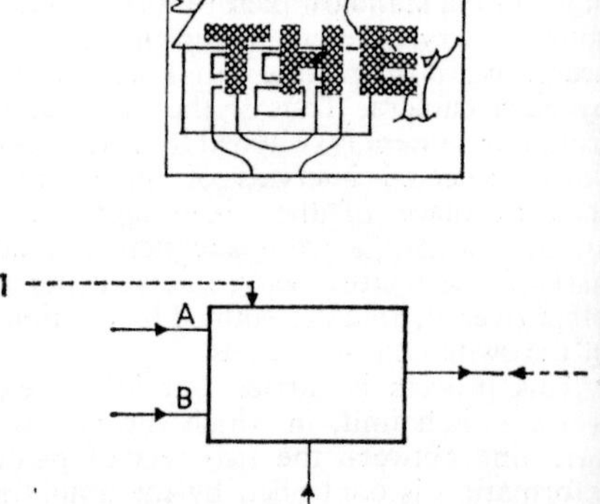

INLAID CAPTION TECHNIQUE. Lettering filled in with distinctive patterns or textures can be produced by inlay from simple black-and-white characters. Input A to electronic switch, 1, is the picture background. Input B is the video signal from a camera directed to a suitable patterned or textured surface. Keying input, 2, is the caption video with the letter shapes, and video output, 3, is the composite picture above.

signal into which the captions are to be inserted is applied to the A input. The characters in the caption are then filled in with a pattern or texture by directing a camera, connected to the B input, to a suitable surface or material.

ELECTRONIC SPOTLIGHT. The inlay technique enables an area of a picture to be raised in brightness relative to the level of the remainder of the picture, either functionally to draw attention to a particular subject, or to simulate a spotlight. For this effect the video signal from the picture source is taken to the B input of the electronic switch and also, via an attenuator, to the A input. A suitable keying signal is applied, defining the area to be treated. The attenuator is adjusted to lower the general brightness level of the picture sufficiently to achieve the desired result. Only the area defined by the keying signal will be at normal level. To simulate a spotlight, the keying signal can be taken from an inlay camera, and a mask with a circular hole manipulated so as to follow the artist, in the manner of an actual spotlight. Alternatively a pattern generator may be employed, with an additional positioning equipment.

APPLICATIONS OF INLAY. Inlay enables split-screen effects to be readily obtained. Familiar as a device for showing the parties engaged in a telephone conversation, the split-screen can serve to bring together subjects taken by cameras separated by distance. Split-screen and other insertion effects can be effectively introduced and terminated by a wipe of the same pattern. Functional inserts can show a clock, a commentator's face, or a scoreboard in sports events and panel games. Keyhole, binocular and similar inserts can be produced with the aid of a suitable mask in an inlay camera, and where a black background is required, a second live camera is not necessary.

Inlay can, on occasion, save studio space or set costs. Live artistes, taken by one camera, can be placed into a set in the form of a small model or photograph taken by another camera, or into a background from slides or film. The outline of the inserted area can be disguised by making it fit a doorway, window or similar feature. The set in which the artistes are actually situated can be reduced to a minimum, since nothing will be seen outside the insertion area; the artiste's freedom of movement is limited in consequence. In fact, one method of engineering a disappearing trick is to have the subject move out of the inlaid area. Another method is to mix or cut from the composite picture to that containing the background only.

OVERLAY PROCESS

The inserted picture area in the overlay process is made automatically to conform to the outline of a subject placed before a camera by using the video signal from the camera, both as the keying signal for the electronic switch and as one of the picture video inputs. The process is therefore similar to inlay, except that the subject itself, in effect, functions as the inlay mask. The subject to be inserted is thus free to move, and in any position within the picture format it is inserted in the picture from the second camera, which will generally be used to provide the background of the composite picture. This corresponds to the travelling-matte process in film.

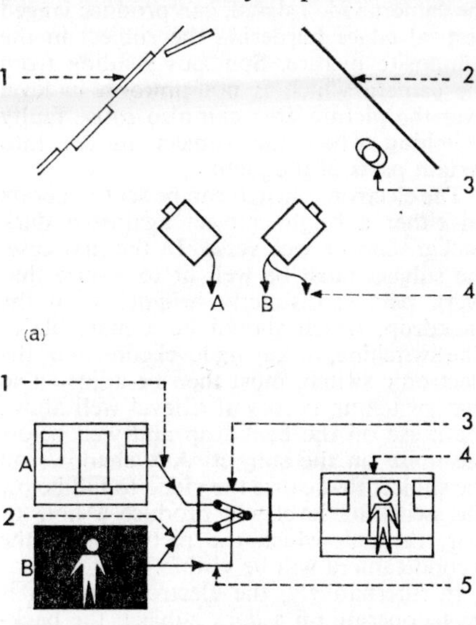

OVERLAY TECHNIQUE. Overlay or self-keyed insertion enables a subject placed against a contrasting backdrop to be inserted in a picture from another camera, the inserted area automatically conforming to the subject outline. (a) Scene. Set, 1, is the background and produces video output A. Lighted subject, 3, not necessarily in same locale, moves against black backdrop, 2, to produce video output B and keying signal, 4. (b) Video signal, 1, of background picture A is applied to electronic switch, 3; likewise video signal, 2, of foreground figure B, together with keying signal, 5. Output consists of composite picture, 4, in which the subject can move freely against the independently photographed background.

Since the action of the electronic switch depends on the level of the keying signal, the subject should be placed against a strongly contrasting backdrop and have a sharp outline. This will ensure that, as the camera scans across the outline, the camera video level will change a sufficient amount and rapidly enough for a clean switching action to be obtained. Otherwise noise, inevitably present to some degree superimposed on the camera video signal, can produce jagged vertical edges bordering the subject in the composite picture. Spurious shading from the camera which is not uniform in level over the picture area can also cause faulty switching when the subject moves into certain parts of the picture.

The electronic switch can be set to operate on either a bright subject against a dark background or vice versa. In the first case, the subject must be well lit to ensure that every part is distinctly brighter than the backdrop, which should be a matt black. The switching, or keying level control of the electronic switch, must then be adjusted so that switching occurs at a level well above the noise on the backdrop and well below the noise on the subject. Any shadows on the subject that cause the video to fall below the switching level will produce a hole or tear, through which the picture from the second camera will be visible.

If, alternatively, the electronic switch is set to operate on a dark subject, the backdrop must be bright, without shadows, and the subject must contain no excessively light areas. While this may be less difficult in some cases than avoiding shadows on a bright subject, the darker subject will be less prominent in the final composite picture.

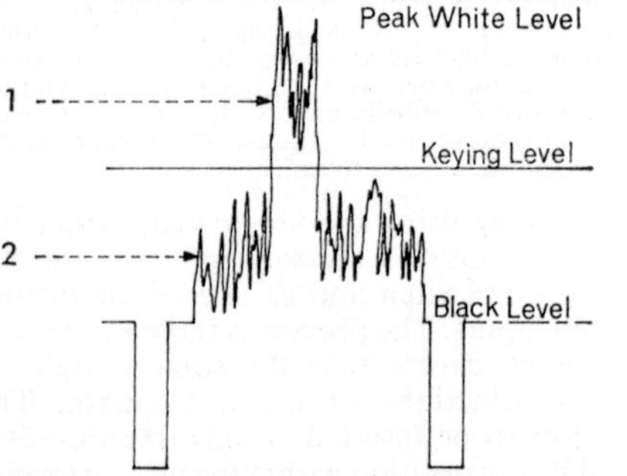

OVERLAY KEYING LEVEL. Keying, or switching, level is adjusted to be lower than the level of the darkest parts of subject, 1, but higher than lightest parts of backdrop, 2. Diagram of one picture line shows how margin is reduced by superimposed noise or spurious shading.

In practice, it is difficult to achieve an entirely natural appearance in overlay effects since the outline caused by the switching action is always evident to some degree. It can, however, be made less obvious by, for example, inserting into detailed rather than plain backgrounds.

COLOUR SEPARATION OVERLAY. To ease the problems involved in normal or brightness-separation overlay, colour separation can be employed to differentiate between the subject and the backdrop. The subject is lit

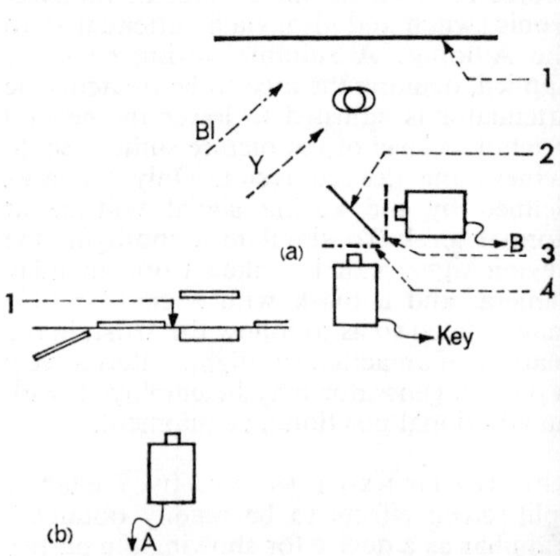

COLOUR SEPARATION OVERLAY. Property of colour aids in separating subject from backdrop. (a) 1. Blue backdrop with subject lighted in yellow. Video output B is derived from camera with yellow filter, 3, facing semi-silvered mirror, 2. Keying signal camera sees backdrop through blue filter, 4, yellow-lighted subject producing a "hole" and thus greatly improving signal contrast. (b) Video output A is derived directly from set, 1.

with yellow light and a blue backdrop is used, lit with blue light. By means of a semi-silvered mirror, two cameras are directed on the subject and their images precisely aligned. One camera is fitted with a blue filter, and its video output is used as the keying signal. The other camera has a yellow filter, and provides the picture video input for the electronic switch. Because little yellow light passes the blue filter, the keying-signal camera sees the subject as dark against a bright backdrop. Similarly the video camera sees the subject normally lit against a dark backdrop. A third camera provides the background picture. The complexity and inconvenience of the process prevents it being widely used.

In colour television studios, the essentia

equipment for colour separation overlay is already present. Using a bright, saturated blue backdrop, the keying signal can be taken from the blue channel of the colour camera, the subject being suitably lit with white light. While blue is suitable for human subjects, the other primaries may be better for certain other subjects, or mixtures of two primaries, the keying signal then being derived from a correctly related mixture of video signals from the two corresponding channels of the camera.

EDGED CAPTIONS. Captions can be inserted by the overlay process having a contrasting, electronically produced border on the vertical edges of each character. Apart from giving a solid, three-dimensional appearance to the characters, captions of this kind have the practical advantage that they show clearly against both light and dark backgrounds. While the captions may be either white with black borders or black with white borders, the former tend to be more effective.

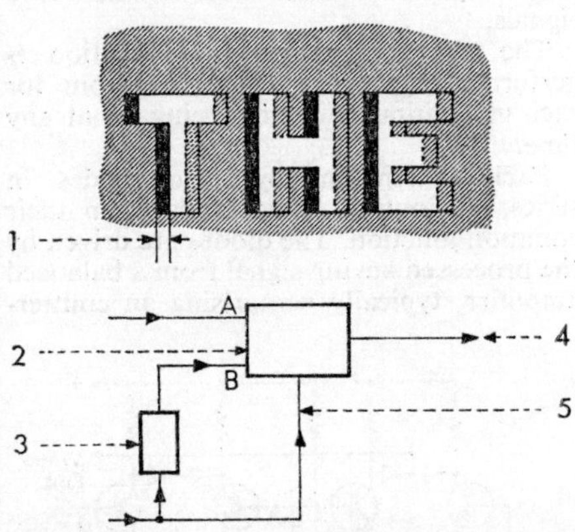

EDGED CAPTIONS. Captions which show up well against both black and white backgrounds can be obtained by overlay by delaying, 3, arrival of caption video B at electronic switch, 2, in comparison with keying signal, 5. Video output, 4, shows black edging due to relative delay, 1. Dotted part of each character is clipped off, but this can be avoided by increasing duration of keying pulses formed in electronic switch.

The effect is achieved by changing the normal timing of the video signal applied to the electronic switch relative to the timing of the switching action itself. Normally the internal and external delays are adjusted so that, in the overlay process, the step in the B video signal representing the edge of the subject to be inserted coincides exactly in time with the changeover of the switch. By delaying the B video signal slightly, the switch operates before the step arrives and reveals the caption background as a border on the left of each character. A disadvantage of this simple method is that the right-hand edge of each character is clipped off by an amount equal to the width of the border. This can, however, be compensated by modifying the electronic switch to increase the width of the keying pulse derived from the keying signal. A further increase in the pulse width gives a border also on the right-hand edges of the characters.

APPLICATIONS OF THE OVERLAY PROCESS. As with inlay, overlay enables a subject taken by one camera to be placed in a scene viewed by another camera or on slides or film, but without the restriction on movement of the subject. Overlay tends to be used principally for trick effects or unnatural situations, although it can be used, like inlay, for space or set economy. Human subjects can be transformed into dwarfs or giants, and can be seen to change size by tracking the camera or by the use of a zoom lens. As with inlay, appearances or disappearances can be effected by cross-fading between the background scene and the composite picture.

Action on the set, or camera movement, can control insertion. If, for example, the electronic switch is set to insert a second camera picture on black, and an area of black is concealed behind a door, then opening the door will expose the black area and the inserted picture will be seen through the doorway. A camera can be panned across to a black or white area with a similar effect. This principle will also produce a wipe in which one picture appears to push the other off the screen.

INSERTION EFFECTS IN COLOUR

The same techniques can be employed in colour as for black-and-white television. Pattern generators are, in any case, suitable for colour, and most modern electronic switches are designed for colour signals of the NTSC type. For some types of electronic switch, and generally for colour systems in which the subcarrier is frequency-modulated, it is necessary to take keying signals before the addition of the subcarrier since otherwise faulty operation of the switch will result.

EQUIPMENT

The principal items of equipment required for producing inlay and overlay effects are

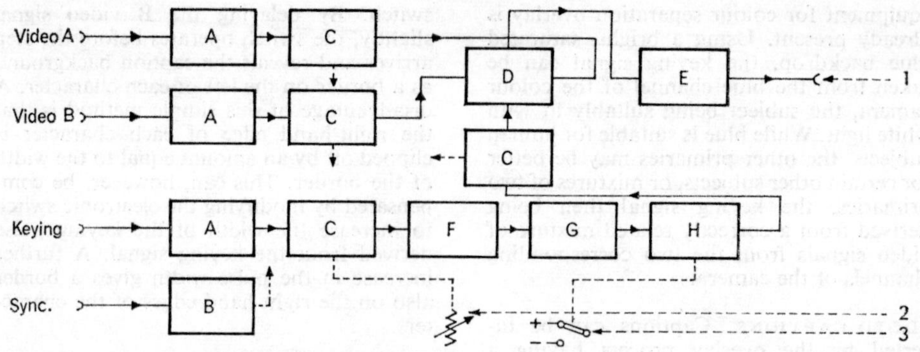

ELECTRONIC SWITCH: BLOCK DIAGRAM. Video inputs A and B are amplified, A, undergo black level clamping, C, and are applied to switching stage, D. Keying signal input, after amplification and black level clamping, is fed to Schmitt trigger, F, shaper, G, inverting amplifier, H, and black level clamp, C, before operating switching stage, D. Switched output is fed to output amplifier, E, to produce video output, 1. Remote controls include keying level, 2, and picture polarity, 3. B. Clamp pulse generator fed by sync pulses.

an electronic switch, a pattern generator and, where necessary, an inlay camera.

THE ELECTRONIC SWITCH. Also called the special effects amplifier, the electronic switch is an essential item of equipment for all insertion effects, being the unit in which the video signals from the contributing sources are combined. In function, it is equivalent to a simple two-position changeover switch, but must be capable of operating in less than the time taken to scan the smallest picture element, approximately 0·1 μsec (microsecond).

The switch has two inputs, usually designated A and B, to which the contributing video signals are fed, and an output at which, at any instant, one only of the input signals appears.

The operation of the switch is controlled by the video level of a special keying video signal, fed to a third input. Input A is switched to the output when the keying signal level corresponds to black and input B when it corresponds to white. The exact level at which the changeover occurs is adjustable, and the sense of operation can usually be reversed if desired.

The source from which the keying signal is obtained depends on the type of effect required.

ELECTRONIC SWITCH CIRCUIT ARRANGEMENT. In a typical electronic switch, two identical video amplifiers pass the signals from inputs A and B to the switching circuit, each including a black level clamp to ensure that the relative d.c. components are reproduced correctly at the switch output. Semiconductor diodes serve as the actual switch elements, the diodes used having in particular a very low capacitance when nonconducting to minimize crosstalk of high-frequency components between the video signals.

The changeover switching function is performed by two switch circuits, one for each video input, one only being on at any time.

Each switch employs two diodes in series, the output being taken from their common junction. The diodes are driven by the processed keying signal from a balanced amplifier, typically comprising an emitter-

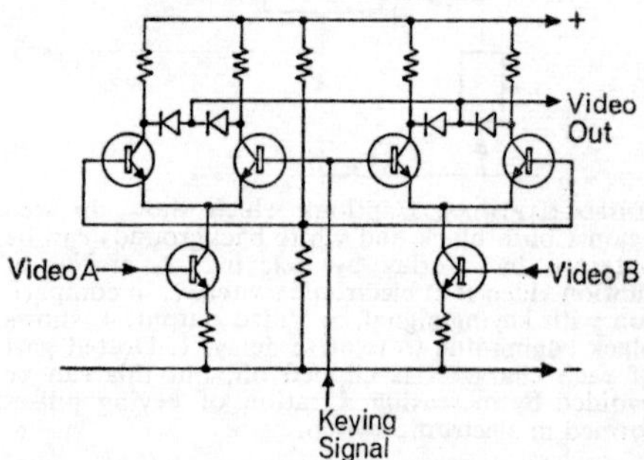

ELECTRONIC SWITCH. Simplified diagram of widely-used type of switching circuit.

coupled transistor pair, to the emitter circuit of which the video signal is applied via a third transistor collector. The final video stage is an output amplifier.

The keying video signal is amplified and applied to a clamp circuit, d.c. connected to a trigger stage. Variation of the clamped potential provides keying level adjustment. The rectangular wave is improved in shape in succeeding stages to provide a constant-amplitude keying waveform with rapid transitions for the switching stage.

Very careful design, especially of the switching stage, is necessary in order to avoid transients being visible in the final picture at the instant of switching.

A critical test is to apply the same video signal to both inputs; the insert outline should not be visible under normal viewing conditions.

THE PATTERN GENERATOR. The pattern generator forms part of the standard special effects facilities with which most studios are equipped. It is installed usually as part of the vision-mixing or control facilities, primarily to provide electronic wipes, but is also used for inlay effects. It produces electronically synthesized keying signals for the electronic switch corresponding to a range of insert or wipe patterns based on simple geometrical shapes.

One hundred or more patterns are usually provided, of which about ten to twenty can be preselected to be immediately available by means of push buttons at the vision control desk. The patterns include simple vertical, horizontal and diagonal straight-line divisions of the screen, a central circle, a central and corner rectangle, and so on. Each pattern can be used statically, for inserts, or can have motion imparted to form a wipe.

The wipe action is controlled by a lever, generally fitted alongside those used for mixing and fading. As the lever is moved down from one extremity to the other, the screen area occupied by the A picture gradually decreases, being replaced by the B picture which finally occupies the whole screen.

The form of the area change will depend on the pattern selected. In the case of the straight-line patterns, the boundary moves across the screen at right-angles to the direction of the line.

The enclosed patterns, such as the circle, produce iris wipes, the inserted area increasing in size from zero to beyond the point at which the boundary moves outside the limits of the screen.

Switches are usually provided for selection of the wiping mode, either normal, auto, normal-reverse or auto-reverse. Taking the case of a simple horizontal wipe, in the normal mode, moving the wipe lever down from the A position to the B position produces a wipe moving from left to right, the B picture replacing the A. Returning the lever to the A position produces a wipe from right to left, restoring picture A. In the auto mode, however, the wipe direction is always left to right. In the two reverse modes the direction of wipe is reversed in each case. Equivalent variations apply to other patterns.

For inlay effects, a suitable pattern is selected and the wipe lever is adjusted so that an insert of the required size is obtained. Separate adjustment of vertical and hori-

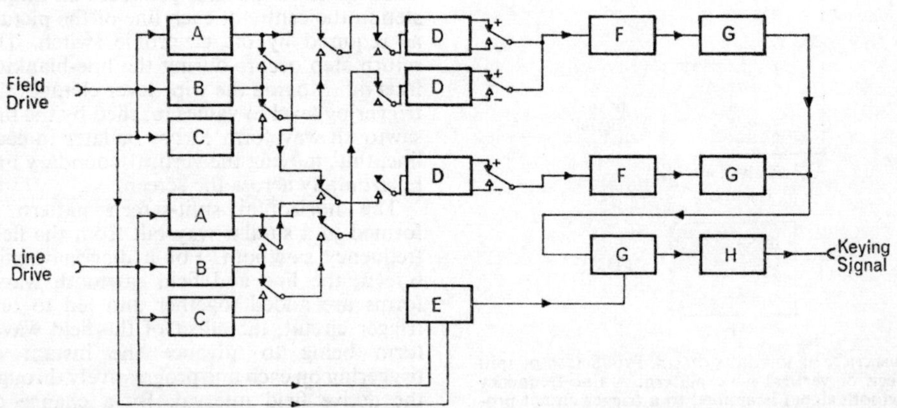

PATTERN GENERATOR: BLOCK DIAGRAM. A. Sawtooth generators. B. Triangle generators. C. Parabola generators. D. Inverter amplifiers for polarity reversal. E. Blanking generator. F. D.c. restorers. G. Trigger circuits. H. Output amplifier.

zontal components of patterns is often provided. This may be in the form of a separate knob for control of aspect ratio, or twin wipe levers may be employed which must normally be moved together, but can be separated for independent control of vertical and horizontal pattern edges. On standard equipment, the enclosed circular, rectangular and similar patterns occupy a central position in the screen. If they are to be moved, a positioner unit and control are added, although most practical requirements are met by the standard type of pattern generator. Important subjects usually demand a central insert, and for subsidiary subjects that must not encroach too far on the main centre of interest, the corner inserts available on standard equipment are suitable.

PATTERN GENERATOR CIRCUIT ARRANGEMENT. Most pattern generators follow the same general principle, although there may be differences in the number and types of pattern available. Typically, two sets of waveforms are formed from the system line and field drive pulses fed to the unit, one set repetitive at line frequency and the other at field frequency. Each set consists of a sawtooth, triangular and parabolic waveform, and inverting amplifiers provide each waveform in either positive or negative polarity. By means of relays, remotely controlled from the control panel pattern selection buttons, the waveforms are selected in groups and fed after d.c. restoration to one or two bistable trigger circuits whose outputs are combined. The wipe lever controls the threshold, or triggering level, of the bistable circuits. Rectangular waveforms generated by the trigger circuits are improved in shape by further stages and passed to the output.

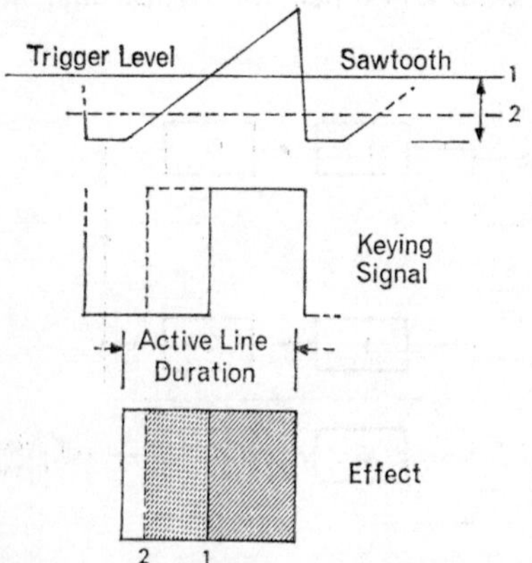

FORMATION OF KEYING SIGNALS. Production of split screen or vertical wipe pattern. A line frequency sawtooth signal is applied to a trigger circuit producing a pulse of duration corresponding to time when triggering level is exceeded. This level is changed by moving the wipe lever, which moves the dividing line on the screen (e.g. from 1 to 2).

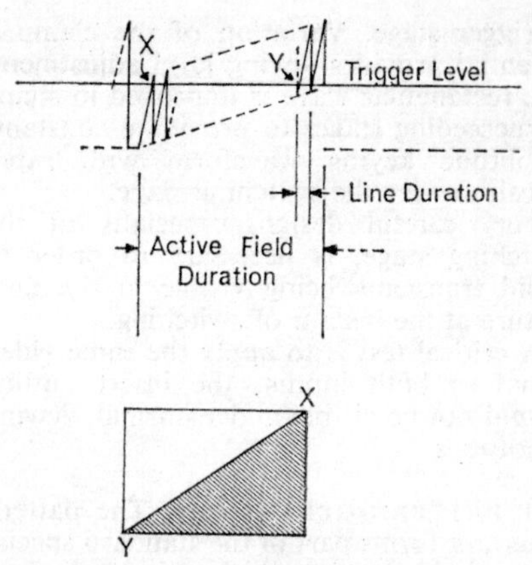

FORMATION OF KEYING SIGNALS. Production of diagonal pattern by adding a line and a field sawtooth waveform. Effect of field sawtooth is to advance (as shown) or retard instant of triggering on each line through duration of field.

For the vertical split-screen pattern, only the line frequency sawtooth waveform is required, and this is fed to one trigger circuit, the other not being needed for this pattern. The wipe lever, at the mid position, sets the triggering level at the value reached by the sawtooth waveform at the point halfway through each active line period. The trigger circuit thus produces an output step in the centre of each line of the picture as required by the electronic switch. The return step occurs during the line-blanking interval. Moving the wipe lever changes the triggering level to values reached by the line sawtooth waveform earlier or later in each line, thus moving the vertical boundary line horizontally across the screen.

The horizontal split-screen pattern is formed in a similar way but from the field frequency sawtooth. For a diagonal split-screen, the line and field sawtooth waveforms are added together and fed to one trigger circuit, the effect of the field waveform being to advance the instant of triggering on each line progressively through the active field interval. By a change of polarity of one waveform, either picture diagonal is obtained.

By the same principle, the line and field

triangular waveforms give vertical and horizontal bar patterns. The second trigger circuit is employed for patterns in which vertical and horizontal timings must be independently determined. This is the case for the central rectangle, for which the line and field frequency triangular waveforms are fed to separate trigger circuits. Where separate controls are provided for independent adjustment of vertical and horizontal components of certain patterns, each varies the triggering level of one of the trigger circuits. For other patterns, the triggering levels are usually varied together by the single wipe lever. The parabolic waveforms are required for the circle, and other patterns containing parts of circles. It is generally possible to feed in additional waveforms from an external source to extend the range of patterns. Often these waveforms are repetitive at multiples of the line and field frequencies.

Since a very small variation in the instant of triggering is readily noticeable in the final picture, a very high order of accuracy and stability is necessary in the waveform amplitudes and shapes and in the trigger circuit threshold level.

In standard pattern generators, provision is not made for control of position of the circle, etc. This is possible with additional equipment, which operates by introducing adjustable delays in both the line and field pulses fed to the pattern generator, but inevitably worsens the stability of pattern shape and position.

THE INLAY CAMERA. Since the full facilities of a studio camera are not required for scanning inlay masks, a special camera, using a vidicon tube or a flying-spot scanner, is sometimes employed. A convenient arrangement is to have the camera mounted vertically below a horizontal glass plate on which inlay masks can be rested, with a source of illumination above. A flying spot scanner can be similarly arranged, with the cathode ray tube below and the photo-tube above. Elaborate inlay scanners have been constructed having the facility of changing the effective size of the inlaid area. This is done with a suitable optical system by varying the distance between the camera, or cathode ray tube and the mask, or by employing a zoom lens. Remote control of this facility, and of mask positioning, has also been employed. Variation of effective mask area and position can be achieved by adjustment of the vertical and horizontal scanning amplitudes and shift currents in the case of the flying spot scanner. However, with a vidicon camera, the properties of the tube impose some limitations on the rate at which such adjustments can be made.

Inlay masks can be cut out of card, or similar opaque materials giving a clean edge, or can be painted or photographed on transparent materials. Modelling clay is useful where irregularly shaped masks are to be made, for example to match pictorial outlines in a scene. The inlay scanner can be used to produce electronic wipes that cannot be obtained from a pattern generator. An opaque card, for example, is drawn gradually across the initially fully illuminated working area until it is completely covered, so that the B picture progressively replaces the A picture. The wipe can proceed in any desired direction, and the card cut to a variety of shapes. Iris wipes require either an actual iris mechanism or a fixed mask whose effective size is changed by relative movement or zoom lens. Complex iris and other wipes, not available from the pattern generator, may be obtained with greater scope and convenience from film. Where the inlay scanner is to be used extensively for wipes, it is desirable for the action to be motorized for a steady and uniform motion, and to be remotely controllable.

Unusual wipe effects can be obtained with the aid of opaque liquids or flowing dry materials such as sand as the masking media for the inlay scanner.

VISION MIXER. Preferably, selection of the video signals for wipes and inserts is by means of separate rows of buttons for the A, B and keying signals. To reduce cost, the A and B rows can be shared with the mix facility, with the limitation that the mix and wipe facilities cannot be used at the same time. The keying signal can similarly share the preview row. Apart from any such practical limitation, the composite special effects video signal from the electronic switch can be fed into the mixer for cutting, mixing or fading in the same way as any other video signal.

Particular attention needs to be paid to the equalization of delays in the case of special effects equipment, since in the formation of keying pulses within the electronic switch, a greater delay typically occurs than is normally encountered in system units.

Provision must be made in special effects installations for feeding the special effects signal to camera viewfinders as an aid to the cameramen in aligning their pictures for inlay and overlay insertions. Studio cameras normally incorporate means to accept an external signal for this purpose.

OTHER ELECTRONIC EFFECTS

Throughout the history of television, many methods have been explored experimentally for modifying the video signal to achieve unusual effects. These include abnormal modes of operation of camera pick-up tubes, giving a distorted grey-scale or peculiar edge effects, or passing the video signal through frequency-selective networks to emphasize certain components. This latter technique is occasionally employed to achieve a relief effect, the higher video frequencies being raised in level relative to the lower frequencies. However, apart from electronic picture insertion, only a few other effects or auxiliary techniques are in general operational use.

THE RIPPLE EFFECT. The ripple effect, in which the picture dissolves into ripples as though reflected in disturbed water, is easily produced electronically. It is done by displacing the timing of the camera line scan sinusoidally. In some cameras, the effect can be produced by simply adding a suitable sine-wave signal to the line drive pulses fed to the camera. While this may upset the operation of some cameras, manufacturers can generally advise on the correct method of injection of the sine-wave signal. A suitable frequency for the signal is typically in the range 150 to 500 Hz. When the frequency is exactly equal to a multiple of the field frequency, the ripple pattern is stationary. Motion of the pattern downward occurs when the frequency is increased, upward when it decreases. The greater the departure from the nearest multiple, the faster is the motion. The ripple usually accompanies a fade or mix, and the effect is often enhanced by camera defocusing. The ripple can be smoothly introduced or removed by means of a fader on the sine-wave signal source.

POLARITY REVERSAL. The normal sense of variation of the video signal from a camera or scanner is for the signal to increase positively with increasing light received by the camera. If the electrical polarity of the signal is reversed, so that it increases negatively with increasing light, the resulting picture has the appearance of a photographic negative, and cameras and other television picture sources are generally fitted with a switch for polarity reversal. This facility is required in film cameras or scanners to enable negative film to be transmitted, and in studio cameras it can be used to produce a negative picture as a special effect.

For greater flexibility, and in order to have control of negative pictures at the vision mixer, a separate negative picture amplifier can be installed. A source from which a negative picture is required can be selected at the mixer for processing by the negative picture amplifier, and the resulting negative signal returned to the mixer where it can be cut or mixed with other sources. This permits the negative picture to be cross-faded, for example with the positive picture from which it is derived, creating an unusual effect.

In a negative picture video signal, only the picture excursions during the active part of each line must be inverted. The blanking and synchronizing components must always retain normal polarity. For this reason, the polarity reversal stage in cameras precedes the blanking mixer and the point at which synchronizing pulses are added. In negative picture amplifiers, the inverted signal is suitably re-blanked.

SCANNING REVERSAL. Most television camera channels are provided with a pair of switches for reversing the connections to the pick-up tube deflector coils, and hence reversing the direction of the two scans. Operating the field (vertical) scan-reversal switch has the effect of inverting the picture and, similarly, operating the line (horizontal) scan-reversal switch has the effect of reversing the picture left to right. With both switches operated, the combined effect is the same as would be seen if the camera, or the scene, were turned upside down. The orthodox application for scan-reversal is in restoring the normal image orientation when it is necessary to take camera shots via mirrors. Effects that would otherwise necessitate the use of mirrors can often be achieved with greater operational convenience by using scan reversal.

CUE DOTS. In networks of television stations, cue dots are added electronically to the picture at the originating station to give advance warning of the ends of programmes, or segments of programmes, to operators of other stations taking the programme from the network. Such a warning enables locally-originated material to follow without loss of time, for, if the following item is on film, the projector must be started 5 seconds in advance, the time nominally taken for the machine to reach correct speed. Difficulty is otherwise experienced, particularly in the case of programmes whose ends or breaks cannot be accurately predetermined.

The cue dot takes the form of a small rectangle, filled in with moving black and white bars for better visibility against the picture, and placed in the top right-hand corner of the picture. In this position the cue dot will not normally be seen on domestic receivers, being hidden by the corner radius of the cathode ray tube mask.

The cue dot is typically switched in one minute before the end of the programme or before a break in the programme and is switched out exactly 5 seconds before the end. Film projectors are started on the disappearance of the dot. The cue dot signal is usually superimposed on the outgoing video signal from the originating studio, under control of a production assistant or the station master control operator.

Cue dots placed in the top left-hand corner of the picture are sometimes employed in video tape recording. Where a recorded programme is required to have film inserts added during actual transmission, the cue dot is added during recording so as to give a start cue for the film projector operator.

A method of insertion for cue dots has been developed which avoids the need to place additional electronic circuits in the transmission chain. When the cue dot is switched on, a blanking signal is added to the standard system blanking pulses distributed to each camera to produce a black area for the dot to occupy. This area is then filled with a moving pattern of bars by

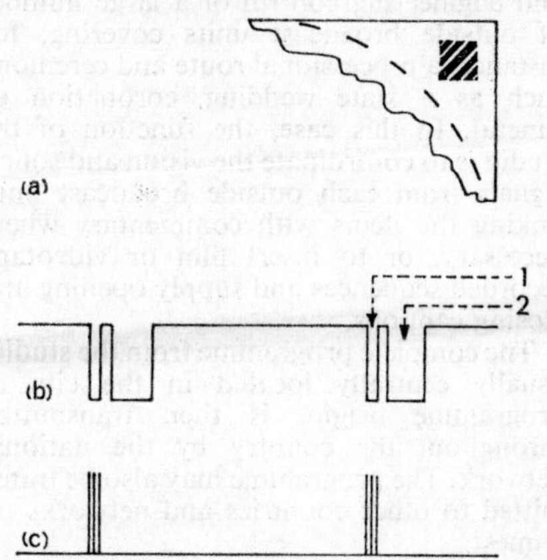

PRODUCTION OF CUE DOTS. (a) Cue dot superimposed on outgoing picture by originating studio to give warning of end of programme section. In domestic receivers, dots are normally hidden by curved mask over tube face, as shown dotted. (b) 1. Special cue dot blanking pulse added to normal line blanking pulse, 2. (c) Cue dot signal consists of burst of 1 MHz rectangular waveform superimposed on outgoing video signal, producing black and white bars. Burst can be given motion or flicker by suitable frequency adjustment.

superimposing an oscillation at approximately 1 MHz on the video signal at the studio output, from a source of high impedance. A.N.H.

See: *Master control; Special effects* (FILM); *Vision mixer.*

Book: *NAB Engineering Handbook*, Ed. A. Prose Walker, 5th Edn. (New York), 1960.

□**Special Effects Amplifier.** Provides a suitable form of signal for keying or switching purposes in the production of electronic special effects. The construction and circuitry can take various forms, but in general the amplifier alters the contrast of the signal to make changes in contrast more apparent, allowing switching signals to be derived from changes in grey scale in the picture signal.

□SPECIAL EVENTS

Television programmes covering special events need many facilities additional to those required by the direction and control of cameras and sound sources within the studio. Special events fall under two broad headings: national broadcasts and international broadcasts.

In both cases, three items are of prime importance to ensure the smooth coordination of all the programme sources used in such complex broadcasts: the switching and previewing system employed; the communications system linking all participants; and adequate time for personnel to rehearse with both the production and technical facilities.

NATIONAL BROADCASTS

A studio may be used for the production and engineering control of a large number of outside broadcast units covering, for instance, a processional route and ceremony such as a state wedding, coronation or funeral. In this case, the function of the studio is to co-ordinate the vision and sound signals from each outside broadcast unit, linking the items with commentary where necessary, or to insert film or videotape recorded sequences and supply opening and closing captions.

The complete programme from the studio, usually centrally located in the city of programme origin, is then transmitted throughout the country by the national network. The programme may also be transmitted to other countries and networks on request.

STUDIO FACILITIES. Most television studios contain switching and mixing equipment for some ten vision and twenty sound inputs. The normal complement of four cameras together with telecine, video and audio tape recording and playback machines, plus the floor microphones required for the particular production, comprise the facilities for a typical programme.

A studio required to control a variety of outside broadcast sources in addition to all the normal internal facilities, clearly needs considerable supplementary switching equipment. In practice, for such occasions, the studio vision and sound mixers must be capable of selecting the signals from at least 12 outside broadcast sources in any sequence, to be mixed with studio, film, caption or videotape material. In all, 20 vision and 30 sound sources presented to the studio mixers would not be unusual.

OUTSIDE BROADCAST FACILITIES. The television coverage of a procession through a city centre followed by a ceremony such as a wedding, may involve some 40 cameras to do full justice to what is usually a magnificent occasion. The normal complement of cameras in an outside broadcast vehicle is four, or sometimes five, and it follows that to deploy, say, 36 cameras on the route and ceremony coverage alone requires at least eight outside broadcast units.

It is unlikely that so large a number can be found by one broadcasting organization in one city and, in practice, the company responsible for mounting the programme usually requests the assistance of such units from other regions or associated companies. These units and crews, together with local ones, move into their carefully planned and prepared positions some days before the broadcast.

Where a site requires a heavy concentration of cameras, e.g., a cathedral or abbey, several outside broadcast units may be involved. Because of the large size of an outside broadcast vehicle, some 30 ft. long by 8 ft. wide and 12 ft. high approximately, and weighing many tons, a collection of several units with their cable-tenders presents a real parking problem in a city centre. For this reason, parking permission may be withheld and alternative arrangements have to be made.

TEMPORARY CONTROL ROOMS. Wherever possible, a room of suitable size in the building in which the broadcast is taking place is converted into a temporary control room into which the dismantled control equipment from the outside broadcast units is reassembled and cabled to the cameras in and around the building. If a suitable room inside the building cannot be made available, space on adjacent ground can sometimes be found for the erection of a temporary hut in which to build the control room.

Although outside broadcast equipment is heavy, it is designed for removal in units for reassembly elsewhere. The time required for this operation depends entirely on the facilities involved and thus on the amount of equipment to be dismantled, reinstalled and tested, and also on the manpower available. A control room designed to handle, say, 12 cameras and 40 sound sources, with commentary mixing and cueing facilities, standby equipment and emergency power supplies, takes several days to assemble and test before it can be handed over to the production personnel for rehearsal.

OUTSIDE BROADCAST UNITS IN TANDEM. The number of vision and sound circuits required to connect each outside broadcast location to the central control studio is considerable and, when multiplied by the number of sites involved, becomes unwieldy and the operation expensive. Some economy of circuits may be achieved by making one outside broadcast unit the slave of another, i.e., working in tandem, although this method is normally used only when the two units are not more than about half a mile apart. The complete sound-and-vision output of the slave unit is routed via line or micro-wave circuits to one input of the sound and vision mixers of the master outside broadcast unit, which can, on cue from the central studio director, switch the slave unit through to the control studio.

ROUTE COVERAGE. At early planning meetings, the format and timing of the programme are fully discussed and, in the case of a ceremonial procession, for example, each outside broadcast unit director knows how much of the route the cameras under his control are required to cover. Camera and sound pick-up positions are agreed, together with the type of lenses and angles required to give the best visual effect, each director being responsible for his section of the route. However, the overall control and timing of the production is the responsibility of the studio director, and for this purpose a foolproof and comprehensive communications system is required, embracing all contributing locations.

PRODUCTION CONTROL. To enable the outside broadcast unit directors to speak with the studio director, a two-way production control telephone system is set up from each site, terminating on a manned switchboard in the studio. Incoming calls are routed to the control room where they are answered by the director or his assistant.

TALKBACK CONTROL. For normal productions, the director uses a talkback system to communicate with the studio floor crews and personnel in the control rooms associated with his production. The basic system consists of a microphone on the production desk and a suitably amplified distribution system. This carries the director's instructions to headphones worn by operators on the studio floor and to loudspeakers in the various control rooms within the studio centre.

However, for major broadcasts, during which the studio director has to co-ordinate programme material from outside sources, an extension to the normal system is required. Two systems are in current use.

EXTERNAL OR OMNIBUS TALKBACK. In this version, an output of the director's studio talkback, via a press-to-talk key, is connected to a network of sound circuits throughout the area or country concerned. These circuits terminate on loudspeakers in the outside broadcast units, control rooms and switching centres involved and allow the personnel concerned with the programme contributions to hear, simultaneously, the studio director's instructions.

A variation of this system enables the studio director to speak to individual locations by selecting separate keys in the studio.

A disadvantage of the omnibus talkback system is the inability of the recipient to reply immediately to the studio director's message or instructions. The production telephone system described earlier is adequate for most purposes, but an urgent message could be held up while a less important call is being made to the studio director.

INSTANT TALK-IN. This system combines the omnibus talkback system with facilities for immediate reply from each remote location to the studio director.

A special four-wire combining instrument is needed at each distant location, and consists of a loud-speaking telephone carrying omnibus talkback, in conjunction with a standard telephone handset.

Under normal working conditions, omnibus talkback is heard from the loudspeaker, but is transferred to the earpiece when the handset is lifted, while the handset microphone is made live to a loudspeaker in the studio director's control room.

Special networks are required to combine the lines from each site to the studio and great care must be exercised in the control of both signal and signal-to-noise levels before the combined network is presented to the studio loudspeaker.

ENGINEERING CONTROL. Engineering control from the studio to all external contributing sources is by means of a telephone system similar to that used for production control. Incoming calls are routed to the sound and vision control rooms where technical queries and fault conditions are dealt with by the appropriate engineers.

When scores of lines are involved, the time allowed for the establishing and checking of sound, vision and control circuits prior to rehearsal has to be strictly scheduled in order to avoid severe congestion in the control rooms. Some hours are usually allocated for this purpose, priority being given to the most distant and complex sites and to those required for transmission first.

EXCHANGE TELEPHONE SYSTEM. In addition to the above systems, a standard exchange telephone is installed in each outside broadcast unit as a safeguard against a breakdown in the special communications systems. This telephone is also used during the cable rig period, before the main equipment and lines are installed in a temporary control room.

REVERSE PROGRAMME SOUND. For rehearsal purposes, an additional sound distribution network is set up carrying complete programme sound from the studio to a loudspeaker in all the participating outside broadcast units. If commentators are employed along the route, it is essential that they should be able to hear each other to pick up handovers and cues, etc.

During transmission, each commentator may listen in his headphones to radiated programme sound received by the off-air check receiver carried for that purpose. On rehearsal day, this is not possible, so complete programme sound on the reverse circuit from the control studio replaces radiated sound.

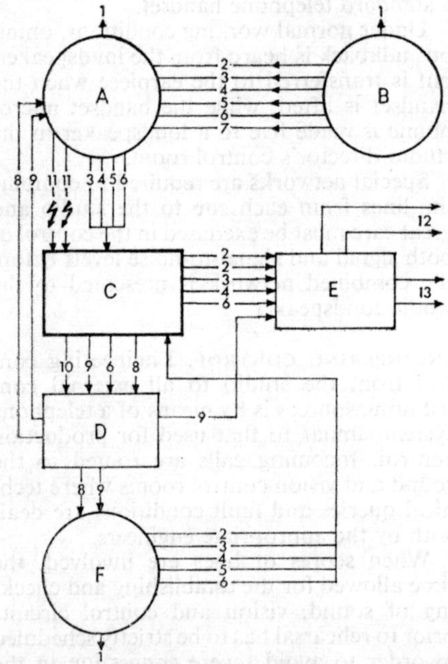

OUTSIDE BROADCAST SITE CONNECTIONS. For clarity, only three OB (remote pickup) sites, A, B, F, are shown. All have direct links, 1, to exchange telephone line. B is slaved to A. A, B, F, feed into Control Studio, C, output of which goes to Network Switching Centre, E, and thus to vision and sound feeds to national and overseas networks, 12, and satellites, 13. Distribution centre, D, provides reverse sound and omnibus talkback. Principal vision, sound and communications circuits are: 2, vision (duplicated), 3, main music (or effects), 4, reserve music, 5, production control, 6, engineering control, 7, local talkback, 8, reverse sound, 9, omnibus talkback, 10, omnibus talkback reserve, 11, microwave vision links.

COMMENTARY FACILITIES. Experience has shown that the commentary coverage of a processional route is best handled by one commentator, usually located in the central studio, with preview and transmission picture monitoring facilities. The commentator wears headphones and receives cues and guidance from the studio director who, of course, has preview facilities of the entire outside broadcast coverage. This arrangement avoids continual handovers from commentator to commentator along the route and the needless repetition of similar descriptive phrases.

CLEAN EFFECTS MIXING. If commentators are used along the route, it is essential that mixed background sound effects, excluding commentary, are fed to the controlling studio at all times during the broadcast. If, for example, a camera high on the roof of a building is taking a long shot through a telescopic lens of a procession, the sound effects must match the picture. For this reason, it is often preferable to take sound from the outside broadcast unit whose microphones are closer to the procession than the unit which is providing the picture. The studio sound balancer is then able to cross-fade or mix the sound effects sources from any outside broadcast site, thus shadowing the vision switcher and following the visual coverage of the procession.

The output of the clean effects mixer then forms a background to the commentary which is added in a second or complete programme mixer under the control of a second sound balancer.

FOREIGN COMMENTARY MIXING. If foreign language commentators are covering the event, it is usual to provide several outputs of mixed clean effects which are made available to as many separate sound mixers as are required, one for each commentator. To these separate mixers each commentator's microphone output is added to become a complete programme feed to the country concerned. These foreign commentators are often housed in the central studio area but they naturally have no part in the domestic programme arrangements.

Monitors displaying the transmitted picture are provided in soundproof cubicles from which the commentators speak. In order to assist them to identify personalities who may not be well-known to them, it is usual to provide them with headphones carrying a feed of the domestic commentary for guide purposes and general information.

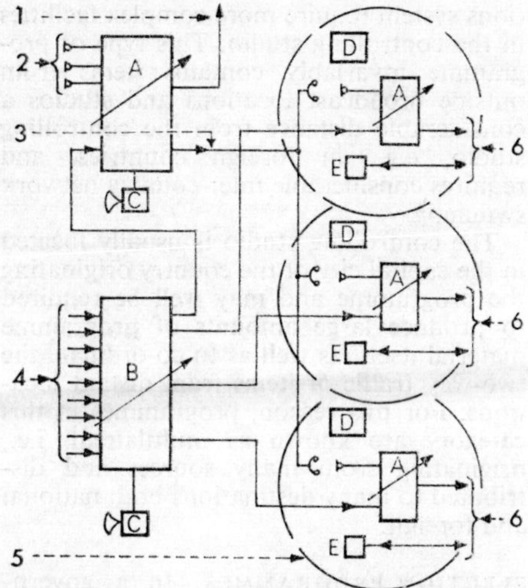

DOMESTIC AND FOREIGN COMMENTARY FACILITIES. Simplified diagram showing only three cubicles, 5. Inputs, 4, from OB sites go to clean effects mixer, B, and thence to commentary cubicles. Each contains a picture monitor, D, a headphone feed, 3, a complete foreign programme mixer, A, a cueing telephone, E, and connections, 6, to the foreign country. Domestic programmes are prepared with the aid of commentary microphones, 2, and a feed of clean effects in mixer, A, with output, 1, to national networks. Separate monitor loudspeakers, C, are provided for effects only and complete programme mixing areas.

Where foreign commentators are located at the site of the broadcast, separate sound mixing facilities, as described above, must be installed in or near to the main control room and cubicles erected at vantage points from which the commentators may view the proceedings.

This naturally complicates the lines requirements from the site as each foreign commentator requires a separate sound circuit to his own country plus a control telephone line to his broadcasting organization for cueing, etc. The vision signal, being international, is normally a common feed to all participating countries and organizations.

LINE TESTING. It follows that broadcasts of such complexity impose a heavy load upon the lines engineers and equipment in the telephone exchanges and repeater stations through which the circuits are routed. To assist the engineers in these control rooms to identify, test and hand over the scores of circuits involved prior to the broadcast, it is customary to transmit from each outside broadcast location an identification signal on both the sound and vision circuits.

IDENT SIGNALS. Identification signals are fed to line as soon as the outside broadcast unit is in position and the lines are available. The vision identification consists of a caption stating the location and the name of the organization servicing that site, and contains reference black-and-white video signal information.

The sound identification is provided by a continuous tape loop on a cassette-type tape replay machine giving verbal information corresponding to that of vision, interspersed with a reference tone at the accepted standard level.

The control studio also identifies its own outgoing sound and vision circuits in a similar fashion.

These signals are of great assistance to the engineers in the lines termination room of the broadcasting authority through which all circuits pass on their way to and from the controlling studio.

SLAVING (PULSE GENERATOR LOCKING). In order to ensure smooth vision transitions from studio items to outside broadcast sources or pre-recorded videotape inserts, it is customary to slave the studio cameras to the remote source. Without slaving, it is not possible to mix vision sources whose synchronizing pulses are derived from different generators. As each outside broadcast unit has a separate synchronizing pulse generator to drive its own cameras, it follows that each such unit or pre-recorded videotape insert is remote and cannot be mixed with a studio picture unless slaved.

Slaving is achieved by making the synchronizing pulses from the studio pulse generator exactly coincident in time and phase relationship with those of the pulses derived from each remote source just before that source is taken for transmission.

In order to minimize the disturbance caused to the transmitted picture when either slaving or deslaving, or when cutting between remote sources when slaving is not possible, certain precautions have to be taken. Each remote vision source is requested to phase the field-synchronizing component of the generated waveform to be coincident with that of the station synchronizing pulse generator. This assumes that the common reference to which all synchronizing pulse generators are locked is the electricity supply frequency (50 Hz in the UK) and does not apply to broadcasting organizations using pulse generators whose reference is derived from a crystal-controlled oscillator.

A technician is usually employed in the production control room to switch the remote vision source to the studio slaving apparatus a few seconds before that source is taken for transmission.

STANDBY FACILITIES. In important broadcasts every possible precaution against technical failure is taken. Vital links in the transmission chain are duplicated or alternative by-pass facilities are made available. The heart of the switching system consists of the studio sound and vision mixers, and emergency switch banks, giving reduced but adequate facilities to maintain transmission, may be brought into service in the event of a breakdown of the main mixers.

There are heavy power requirements for the central studio and control room area, which includes telecine, videotape recording, the lines termination room and master control, and a supply failure would be disastrous. Standby diesel electric generators are arranged to take over automatically and can be on full load within seconds, surge-proof fuses being fitted to all equipment to ensure continuous service during the change-over.

If the supply failure is local to the building, a permanently available reference frequency from the main city generating station ensures that the synchronizing pulse generators stay locked to the frequency of the national electricity supply grid.

In the case of a multi-outside broadcast programme, outside broadcast units from many organizations throughout the country may well take part. Their equipment will undoubtedly be of various types and manufacture and the servicing problem over many days, often hundreds of miles from base, can be critical. The co-ordinating company in this case would arrange for a supply of basic equipment to be available at a central location for maintenance on an exchange part basis.

In the event of street closures during the day of a state procession, maintenance vans equipped with radio telephones and in contact with the central studio are situated at vantage points along the route so that a request for help to the central co-ordinating studio from any outside broadcast unit may be quickly answered.

INTERNATIONAL BROADCASTS

Broadcasts such as the coverage of a government election, the report on a national budget, or a programme designed to inaugurate a television satellite communications system require more complex facilities in the controlling studio. This type of programme invariably contains items from outside broadcast locations and studios a considerable distance from the controlling studio, e.g., in foreign countries, and requires considerable inter-country network switching.

The controlling studio is usually located in the capital city of the country originating the programme and may well be required to produce large amounts of programme material itself, as well as to co-ordinate the two-way traffic of items from distant locations. For this reason, programmes in this category are known as multilateral, i.e., originating from many sources and distributed to many destinations both national and foreign.

ELECTION PROGRAMMES. In a government election, the control studio is required to fulfil two main functions: to act as an originating news centre, and to co-ordinate the network switching system used for channelling programme contributions from distant parts of the country and from overseas.

The essence of this type of programme is the speed with which polling results from constituencies can be presented on-air. As it is not possible to rehearse either the programme content or the sequence of events, a communications system which links all the participants is particularly vital. The studio may well be on transmission for more than eighteen hours, either continuously or with only a short break, and this alone represents a severe strain on personnel and equipment.

Typically, election results start to come in late in the evening of polling day and the broadcast usually starts with a general introduction of the various experts and commentators who discuss and interpret the results as they come in. It is also usual to have at an early stage a preview of the sites covered by outside broadcast units, and to interview members of the public for views and comment.

In order to accommodate results from several hundred constituencies throughout the country, no matter how rapidly they are declared, the presentation system must be flexible and fast. A method must be found, for instance, to interrupt an interview in order to switch quickly and cleanly to an outside broadcast site perhaps several hundred miles distant in order to witness the moment of the declaration.

By the early hours of the morning of the following day, the outcome is usually evident and, with only results from a few distant constituencies to come, arrangements are made for interviews with the leading politicians of the major parties. Foreign reaction is sampled by taking live interviews with personalities in studios in overseas countries.

These studio facilities and their associated inter-country circuits are often required at very short notice. However, most broadcasting organizations are sympathetic to urgent requests of this nature on a reciprocal help basis.

NETWORK REQUIREMENTS. Although the main political party headquarters in the capital city is usually covered by outside broadcast units, together with at least one public square or meeting place where crowd reaction scenes and interviews can take place, it is usual to have outside broadcast units to cover as many key constituencies or marginal seats as possible. Areas where strongly controversial issues are raised, or main industrial centres, may bring the total to 25 locations to be covered by outside broadcast units. It is likely that only 12 or so in and around the capital city can be fed directly to the control studio by line or microwave circuits. The remainder must be routed via the television trunk circuits and the network switching centres administered by the telephone company for the broadcasting organizations involved.

SWITCHING CENTRES. To ensure the greatest possible utilization of the inter-city television trunk circuits, it is necessary to set up regional programme switching centres at strategic points. Into these centres are fed the sound and vision signals from several outside broadcast sources to be switched, on cue, to one or more of the main line circuits to the controlling studio.

Programme switching centres are usually set up in studio premises and staffed by production personnel. Where possible, the normal circuits to and from the studio and the telephone company's terminal are used, augmented by microwave or temporary circuits as required.

During an election broadcast, contributions may be required in random sequence from any location throughout the country. The demand for pictures and sound is made with little warning and certainly without time for rehearsal. It is imperative, therefore, that the switching operation is as speedy and as foolproof as possible and that the personnel responsible for it are kept in close and continuous contact with the studio director. For this purpose the omnibus talkback system is extended not only to all outside broadcast units and studios supplying programme contributions, but simultaneously to the switching centres where it terminates on loudspeakers beside the switching operators.

Special switching systems are required to ensure that sound and vision from each source are switched simultaneously by the operation of a single button to the selected outgoing lines. This is important when many sources are being continuously and rapidly switched and ensures that sound from one outside broadcast unit and vision from another does not become inadvertently selected in moments of high drama. Picture preview and sound prelisten facilities are provided to monitor continuously all incoming signals and the output of the selector switch banks.

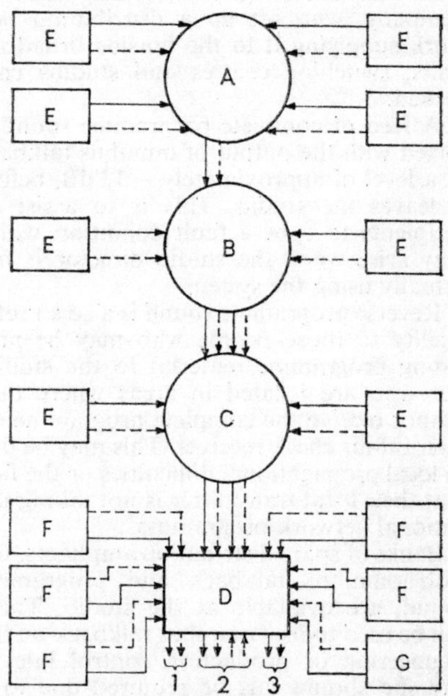

MAJOR OUTSIDE BROADCAST: PROGRAMME CIRCUITS. Greatly simplified. (*Solid lines*) Vision circuits; (*dotted lines*) Music/effects circuits. A, B, regional city programme switching centres (PSC). C. Capital or central city PSC. D. Central controlling studio. E. OB unit or studio. F. Local OB. G. Foreign studio. 1. VTR recording facilities at airport for delayed transmission to overseas countries. 2. VTR recording and replay channels. 3. Direct transmission to national and overseas networks.

Much of the vision and sound distribution network can be provided by the national television distribution system operated by the telephone company for the country's broadcasting authorities. However, broadcasts on this scale, particularly if more than one television channel is mounting a similar and simultaneous programme, impose a heavy strain on programme distribution equipment and lines. Auxiliary circuits and microwave radio links are required at short notice, often involving considerable distances, possibly over difficult terrain. The overall circuit requirements are tremendous and the success of such an exercise depends largely on the degree of co-operation existing between the telephone company concerned and the broadcasting authority.

OMNIBUS TALKBACK. As omnibus talkback is essential to this type of broadcast, special arrangements are made to ensure that vital cues are not lost in the event of a failure in the system. Two feeds of omnibus talkback from different amplifiers are fed from the controlling studio to the telephone company, who set up a distribution network supplying it to the outside broadcast units, switching centres and studios concerned.

A feed of complete programme sound is mixed with the output of omnibus talkback at a level of approximately -12 dB, before it leaves the studio. This is to assist all recipients to spot a fault condition which may arise when the studio director is not actually using the system.

Reverse programme sound is also a useful facility to those people who may be providing programme material to the studio, but who are located in areas where they cannot receive the complete programme on their off-air check receiver. This may be due to local propagational difficulties or the fact that their local transmitter is not taking the national network programme.

Banks of spare distribution amplifiers, fed with omnibus talkback and programme sound, are available at the studio. These can be used to feed omnibus talkback on the engineering or production control line to any site should this be required due to a local talkback line failure to that site.

In the event of a main trunk circuit failure carrying omnibus talkback, standby circuits are available to be switched into service.

It is important that all personnel concerned with switching operations fully understand the mechanics of the system and the nomenclature of the circuits in use. Only the control studio director has the complete picture of the state of the network and only he can decide which source he wants switched, and on to which trunk circuit, according to the priority of the material he is being asked to accept.

STUDIO FACILITIES. For the speedy presentation of results, essential for an election type of programme, the studio is converted into a practical newsroom manned by a staff of editors, sub-editors, managers and people skilled in the rapid processing of information.

Its main function is the immediate display of results as they become available together with live pictures of the actual declaration from those sites covered by outside broadcast units. Comment upon the course of events and trends indicated, statistical analysis and comparison with previous occasions, is conducted by a panel of experts in another part of the studio.

Where constituencies are not covered by outside broadcast units, and live picture coverage is therefore not available, up to 100 specially engaged news reporters may telephone the result to the studio direct. For this purpose, 20 manned telephone positions with free line switching facilities and having a common number are installed on the floor of the studio. On receipt of a call, the operator passes the information immediately to the newsroom staff where it is printed on a caption card by a special typewriter for display before the studio cameras.

The current state of the parties in terms of numbers of seats gained or lost is usually either inlaid permanently into the transmitted picture, or superimposed at regular intervals. The electronic signal for this is derived from a camera fitted with a pick-up tube having a non-image-sticking characteristic, viewing a display device capable of being continuously altered.

The on-air presentation of the newsroom in action, the display and discussion of results as they come in and the general continuity of the programme, is the responsibility of the studio director and the linkman in the newsroom with whom he works closely.

The function of the linkman is to provide visual continuity between the various activities of the newsroom and to introduce, on cue from the studio director, items from outside broadcast and distant studio locations. For this purpose he is given an independent system of private talkback from the director, which he hears on a miniature earpiece. In order to talk to the linkman, the director must operate a different key and use a separate microphone from those used for both normal and omnibus talkback. This degree of privacy is required so that the linkman shall not be confused by the almost continuous stream of instructions issued by the director to the studio crews, outside broadcast units and programme switching centre operators.

On occasions when the director wishes to cue simultaneously a distant studio or outside broadcast commentator, and also allow the linkman to hear his instructions, he can talk to both by using an override switch which connects omnibus talkback to the linkman's talkback.

REVERSE PROGRAMME SOUND. During both rehearsal and transmission, the output of the control studio's complete programme sound mixer is fed by line to all outside broadcast sites and regional studios who are contributing programme inserts.

The distant end of the circuit terminates on a loudspeaker beside the director, but if required a parallel feed may be extended to an earpiece worn by the commentator or interviewer. By this means, a distant commentator may hold a two-way conversation with the studio linkman, or engage in a conference discussion around the network with other commentators whose replies, via the control studio, he may hear in his earpiece.

When this facility is used, there is a very real danger of howl-round (acoustic feedback) occurring owing to the complete programme sound in the commentator's earpiece being picked up by the microphone into which he is speaking.

Loudspeakers are sometimes used on the reverse programme sound circuit to allow an audience or a panel in a studio to hear questions from a distant commentator or interviewer, and here the risk of howl-round is naturally greater. Although in practice the safe working limits for loudspeaker amplifiers used under these conditions are predetermined before transmission, should a howl-round develop, the quickest way of stopping it is for the control studio to cut the feed of reverse programme sound to network. Transmission is not directly affected, as this facility is a monitoring or second order communication system, but it would naturally destroy a question-and-answer session if the questioned could not hear the questioner.

For this reason, it is customary to attenuate the feed to network from the control studio by predetermined steps before cutting it altogether. This anticipates that the acoustic loop will be broken and stability restored, while a working, although lower, level of sound power from the loudspeakers is still maintained.

STUDIO DISPLAY FACILITIES. As a background to the linkman, a large screen television projector or a bank of picture monitors may be used to indicate the next source or sources to be taken for transmission, or to display the distant picture during two-way conversations and interviews.

Whichever method is used, special switching arrangements must be made to allow any of perhaps 25 sources to be fed to the display system. A duplicated set of small standby monitors, having the same vision feeds as the main display, is built into the desks of the linkman and interview panel. Large screen monitors are suspended at vantage points throughout the studio, showing the transmitted pictures for all to see.

Certain precautions must be taken when a television camera views a monitor displaying a picture whose line standard is the same as that of the camera. Owing to an effect known as photo-pulse, a heavy black and white horizontal bar appears in the camera output, its vertical position varying with the movement of the camera viewing it.

Its severity varies proportionately with the brightness of the displayed picture, and thus with the distance at which it is viewed. It is caused by the combined effect of phosphor decay time of the picture tube and the beat of the scanning rasters of the cathode ray tube and the camera pick-up tube.

Cathode ray tubes having a longer persistence phosphor than normal are used in monitors designed for this purpose. Tinted implosion screens in the monitors enhance the contrast of the displayed picture by attenuating the reflected studio lighting, which can seriously degrade the result unless great care is taken.

DATA PROCESSING At least five teleprinters, with an additional five fed in parallel for standby, are connected to the national press information service to provide up-to-the-minute information for inclusion into the programme. For immediate display, special arrangements are made for a camera to view the selected message as it is being typed, thus avoiding the delay caused by retyping in larger characters as in the case of caption cards for constituency results.

In recent years, increasing use has been made of electronic computers to forecast the result based upon the analysis of early trends. The computer must be programmed with extensive and detailed information based upon previous occasions and continuously fed with up-to-the-minute data as it becomes available. For this reason, it is often installed in the studio, modern solid-state advances now making this a practical proposition.

RECORDING FACILITIES. In a fast-moving programme such as an election, results are often declared in quick succession and remote vision sources have to be switched rapidly to the studio and transmitted without delay. Where a result in a constituency is about to be declared while another result is already on-air, the studio director may decide to route the second result to a recording channel. Because of the speed at which results may be declared, it is possible that several recordings may accumulate in this way. The studio director must then decide the sequence for transmission according to priority and arrange for playback from the recording channel of the results which he has in hand.

Again, a flexible input switching system is required to route any one of several vision and sound sources to the input of the selected videotape recording channel. The switching system is controlled from the production control room, and as soon as the machine finishes recording the item, it is wound back ready to transmit without delay to the studio.

In the case of a government election, considerable interest in the result is displayed by foreign countries who usually send representatives of their broadcasting organizations to cover the event.

Recorded highlights of the broadcast are made available for transmission direct via satellite, or the appropriate network, to the countries concerned. Where satellite facilities are not available and time zonal differences are suitable, aircraft may be used to fly the recording to its destination. Where flights are east to west, e.g., Europe to USA, advantage is gained by flying with the sun.

In order to gain time when tapes are flown long distances, a feed of the complete programme may be sent via microwave radio link to a mobile videotape recording channel located at an international airport. In this way, a recording can be airborne within minutes of completion, carried by either scheduled or specially chartered aircraft.

When standards conversion is required, and time does not allow a separate recording through an intermediate film process, arrangements are usually made to pass the signal through the appropriate apparatus inserted in its normal transmission path.

The equipment for this, using either pure electronics or electron-optical processes, may be sited at the transmitting or receiving earth station in the case of a satellite communications system, or at a convenient intermediate point in the television network of the countries involved in the interchange.

From all of the foregoing it will be apparent that television studios involved in the production of special events spread their influence over great areas, sometimes intercontinental, invariably international. They encompass wide-ranging and highly diverse problems which add spice to the daily diet of broadcasting fare. M.M.

See: *Communications; News programmes; Master control; Outside broadcast units; Satellite communications; Station timing.*

⊕□**Spectrum.** Effect obtained when electromagnetic radiations are resolved into their constituent wavelengths or frequencies. One example of a spectrum is the coloured bands produced when white light passes through a prism. The coloured bands, in order of increasing wavelength, are described as violet, indigo, blue, green, yellow, orange and red, but normally there are no clearly defined limits between them.

A continuous spectrum is one in which all wavelengths are represented in a continuous sequence without gaps, and in which there is no abrupt change in the intensity of energy in the sequence.

A line spectrum is one in which only certain wavelengths or lines appear. When white light is passed through a medium capable of absorbing particular wavelengths or bands of wavelengths, the spectrum of the transmitted light is known as an absorption spectrum.

Wavelengths which have no visible effect may well form part of the spectrum and are important in many aspects of photography. The region of wavelengths shorter than

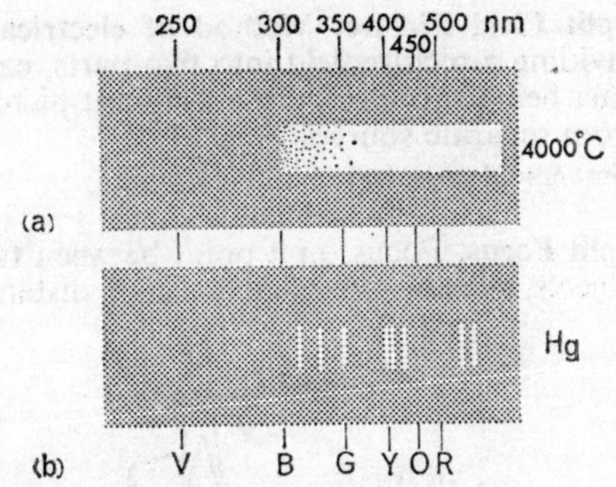

SPECTRA. (a) Typical continuous emission spectrum of a solid: positive pole of a carbon arc at 4,000°C. (b) Typical line spectrum: mercury, from an arc enclosed in glass, absorbing wavelengths below 310 nm.

visible blue light is called the ultra-violet, and that of wavelengths beyond the red region is termed infra-red.

See: *Electromagnetic spectrum; Light; Optical principles.*

Specular Density. Density of a transparent photographic image measured in such a way as to neglect that portion of transmitted light which is widely scattered by the grain of the image structure.

See: *Callier effect; Density; Diffuse density.*

Specular Reflection. Directional reflection of light at a smooth reflecting surface.

See: *Light; Rear projection; Screens.*

Speech Channel. Channel suitable for telephony, with a frequency range of 250 to 3,400 Hz maximum. Also called a telephony channel.

See: *Transmission networks.*

Speed. In motion picture practice, the linear rate of film movement through a camera, projector or processing machine, stated as pictures per second (pps) or feet per minute (fpm). Photographically, the sensitivity of a film being exposed, the chemical activity of a developer solution or the light-transmitting power of a lens.

The speed of a lens is usually expressed by an *f*-number, which is based on purely geometrical considerations and does not take account of the light absorbed, reflected or uselessly scattered during its passage through the lens. The effective transmission of a lens may be measured by a T-number, which takes account of these losses.

See: *Exposure control; f-number; T-stop.*

Spherical Aberration. Failure of a lens with spherical surfaces to bring the rays from a point to a point focus. The rays which pass through the marginal or peripheral part of the lens are, in fact, brought to a shorter focus than those which go through the central part of the lens. Spherical aberration can be corrected by combining two or more elements of different dispersion, or by

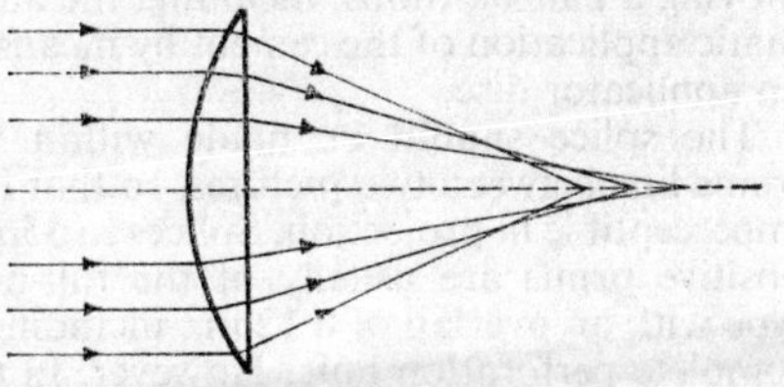

SPHERICAL ABERRATION. Rays farthest from lens axis converge more strongly and come to a focus nearer the lens than rays close to the lens axis. Image is never fully sharp.

fundamental design changes to make use of non-spherical refracting surfaces.

See: *Aberration; Optical principles.*

⊕**Spider Box.** Colloquial term for a junction □box into which two, four or more lamps may be quickly plugged for studio or location use.

See: *Lighting equipment.*

⊕**Splicing.** Satisfactory splices are necessary for every aspect of film making and showing. In narrow-gauge films, because their smaller dimensions leave less area for splicing, the problem is equally important. The basis of a splice is a weld between the base of the two films, so that the emulsion and substratum must be completely removed from one film end.

Perfect registration of the perforations and alignment of the film edges is essential, and for this reason the once universal hand-splicing methods have been replaced by various types of splicing press, even in the projection room.

One of the most popular types of splicer for 35mm grips each film in a hinged clamp within which it is registered by pins which fit perforations. Between the two hinged clamps is a fixed knife, the left-hand film being cut by bringing down the right-hand clamp, and the right-hand film by the left-hand clamp. The emulsion and substratum are removed from one end by a guided scraper. The right-hand clamp is then lifted and cement applied to the left-hand film. As the right-hand clamp is brought down with the end of the film projecting, pressure is applied to the splice by a spring finger. The blocks of this type of splicer may be electrically heated, ensuring a firmer weld and quicker drying. Other types of splicer perform these operations more or less automatically, and in the splicers widely used in studios and laboratories, the film clamps are operated by foot pedals. Another model widely used in laboratories is controlled by moving a pair of knobs, including the automatic application of the cement by means of an applicator disc.

The splice should be made within the frame line between two pictures, so that it is imperceptible in projection. Splices in 35mm positive prints are usually of the full-hole type with an overlap of 0·12 in., including a complete perforation hole. However, in the case of negative films a narrower overlap between the perforations is preferred, and a width of ·050 in., or even less, must be used. In narrow-gauge films the absence of the interframe area between pictures on 16mm and 8mm means that it is not possible to make an invisible splice, and in the assembly of original material for printing, it is usual to adopt checkerboard cutting in A and B rolls, so that no splice is seen at each scene change on the resultant release print. Most narrow gauge splicers are similar in principle to the 35mm splicer described, but a diagonal splice is sometimes preferred since it gives a larger overlap and travels more smoothly over smaller diameter sprockets.

An alternative method of splicing is to cut both films to form a butt joint without overlap, and to make the splice with special adhesive transparent tape perforated to match the film perforations. Splicers are available to do this semi-automatically using standard tape which is perforated on the

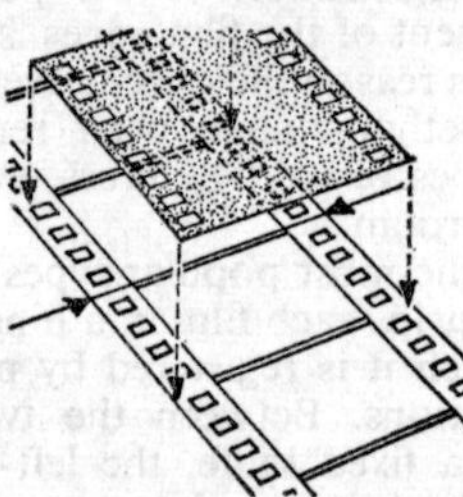

TAPE SPLICE. The ends of the two films are butted together, and pre-perforated adhesive tape is registered by the splicer on top of them.

machine at the time of making the join. This method is common in cutting rooms, and is particularly favoured for 16mm films because it avoids the overlap being projected. It is also essential in using certain film bases, such as Cronar, for which no completely satisfactory cement is yet available. L. B. H.

See: *Editor; Narrow-gauge production techniques.*

□**Split Field Picture.** Method of electrically dividing a picture field into two parts, each part being supplied with a different picture from separate sources.

See: *Special effects* (TV).

⊕**Split Focus.** Focus on a point between two □objects, widely separated in their distance

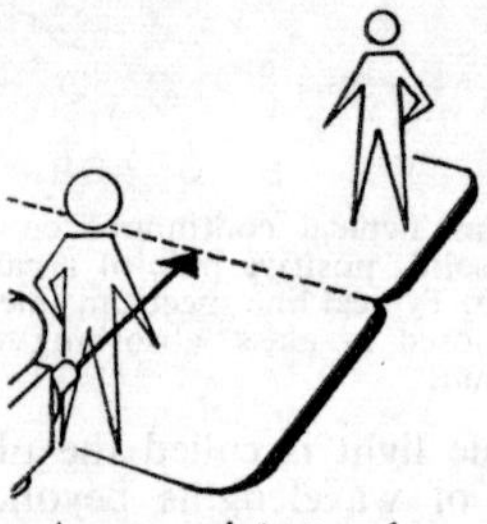

SPLIT FOCUS. Arrow points to plane of sharpest focus.

from the camera, so that both are included within the depth of field with approximately equal definition.

⊕**Split Reel.** Type of reel with one detachable side. Film on cores can be mounted on the

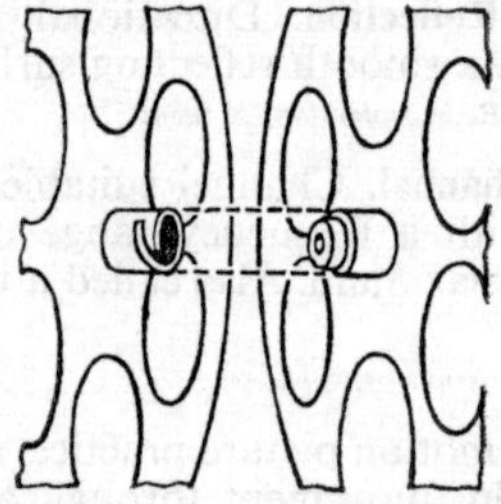

reel and secured by replacement and locking of the flange.

See: *Spools, cores and cans.*

⊕**Split Screen.** Effect shot where two separate □image areas are exposed or printed on each frame, usually with the boundary between them made as unnoticeable as possible. In television, the same effect is produced by combining the output of two cameras.

See: *Special effects.*

⊕**Spools, Cores and Cans.** Film is handled and transported either as tightly wound rolls on plastic cores (bobbins) or on spools with side-plates (flanges) to protect the edges. Spools, or reels as they are often termed, are almost always used for 16mm material but, to save weight, 35mm prints are generally wound on cores for shipment and transferred to spools only for inspection and projection. Except for daylight loading camera spools, raw stock is always supplied by the manufacturer wound on cores, which are therefore widely used in laboratory processing operations.

Cores are normally made the exact width of the film – 16mm, 32mm, 35mm, etc., as the case may be – and the most widely used size has a diameter of 50mm (nominal 2 in.) designed to fit on a 25mm (1 in.) shaft which has a driving key engaging with a keyway on the core. Cores with their own keys engaging with a slot in the shaft were at one time generally used for negative stock but these are now confined to certain types of camera magazines. With the tendency to use longer rolls of film and wider stocks such as 70mm, a larger diameter core of 3 in. or 75mm is recommended, since its use helps to reduce the difference in tension on the film between the beginning and end of a roll; cores of 73mm, 80mm and 100mm diameter also exist.

Spools for 16mm film handling are made in a range of standard sizes to take from 50 to 2,400 ft. of film; 800 ft. and 1,600 ft. spools are the most popular for motion picture subjects while 2,000 ft. is widely used in television. The 2,400 ft. spools allow the average feature film to be shown with only one break for rethreading if two projectors are not available. Sixteen millimetre spools may have either metal or plastic flanges and are all made to mount on a $\frac{5}{16}$ in. spindle (nominal 8mm) which is of square cross-section with a driving key on one flange and circular on the other. Similar flanges are used for 8mm and Super-8mm spools, which are normally made in sizes from 50 to 800 ft. capacity. In order to keep film tension on projection as low as possible the ratio of the spool centre diameter to flange diameter must be as high as possible; for 16mm reels of 1,600 to 2,400 ft. capacity an inner diameter of approximately $4\frac{3}{4}$ in. or 120mm is recommended.

Thirty-five millimetre reels are available in fewer sizes, usually 1,000, 2,000 and 3,000 ft. capacity, corresponding to 300, 600 and 900 metres, of which the 2,000 ft. size is by far the most widely used. Spindles on projectors vary, so spools with different hole sizes are required; the British Standard specifies a $\frac{3}{8}$ in. hole but the American standard for a $\frac{5}{16}$ in. spindle with driving key is popular, and projectors of European manufacture with 10mm diameter spindles are also found. Some projectors take special reels of up to 6,000 ft. capacity.

Seventy millimetre projection practices also vary but the greater weight of the reel of film always requires a substantial spindle, European practice providing 10mm spindles and American standards $\frac{1}{2}$ in.; in both cases a strong driving pin is provided on the projector to engage with a corresponding hole in the flange of the spool. Spools of nominal 3,000 ft. or 4,000 ft. capacity are generally employed.

For general storage and transport, rolls of film are packed in circular metal cans, which for 16mm and 8mm material contain the film on its spool; 35mm film is not normally transported on spools in Britain and Europe and the printed roll of film on a 2 in. core is usually packed in a can similar to that supplied by the manufacturer of the raw stock for either 1,000 ft. or 2,000 ft., the latter being usual for feature length films. In the United States, special spools in transit cases are more often used for general distribution. L.B.H.

See: *Film dimensions and physical characteristics; Film storage; Release print examination and maintenance; Release print make-up.*

⊕□**Spot.** Focusable lamp, the basic element in studio lighting, the beam of which can be narrowed to a spot to provide a key light.

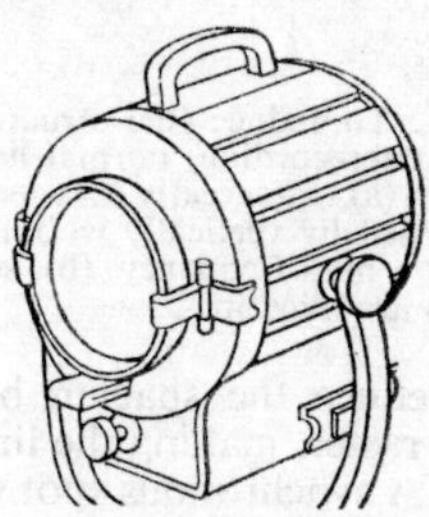

Since the focus is adjustable, the beam can also be widened to avoid over-intense lighting of a small area.

See: *Lighting equipment.*

⊕□**Spot Brightness Meter.** Photometric instrument used in the control of exposure and lighting contrast by the cameraman to measure the reflected light from small areas of the scene being photographed. Unlike regular exposure meters, spot brightness meters measure the light received over a narrow acceptance angle and can be accurately sighted to select a particular area of highlight or shadow in the field of view from the camera position. Measurement of the brightest area of the scene can help to determine the correct exposure within the latitude of the film stock being used. Com-

parison of highlight and shadow areas enables lighting conditions to be adjusted to produce the required contrast range.
See: *Exposure control.*

⊕**Spotting.** Location of individual words or sounds on a sound track by means of a Moviola or sound reader. As magnetic modulations are invisible, this calls for a good deal of dexterity. The term is also used for marking up dialogue footages on a sound track for foreign language captions.
See: *Foreign release.*

□**Spot Wobble.** Method of minutely moving the electron beam of a cathode ray tube sinusoidally in a vertical direction at a very high speed while it is tracing out a horizontal line. The effect is to thicken the line

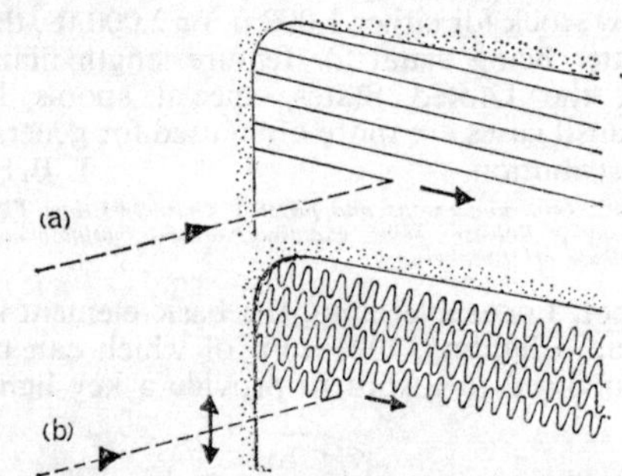

SPOT WOBBLE. To reduce line structure visibility, especially in telerecording, normal horizontal line scan, shown at (a) with greatly exaggerated spacing, may be modified by vertically wobbling the scanning beam at a high frequency, (b), so that spaces between lines are filled in.

and thus reduce the spacing between the lines of the raster, making the line structure less visible. A synchronous spot wobble at a rate that is an exact multiple of the line scanning frequency has been used in telecine and telerecording techniques but little improvement has been obtained to outweigh the extra complication.
See: *Telerecording; Standards conversion.*

⊕**Spray Processing.** Form of continuous film processing in which the film passes through the necessary solutions applied in cabinets in the form of a spray of fine droplets rather than through liquids in deep tanks.
See: *Laboratory organization; Processing.*

⊕**Sprocket.** Film driving wheel carrying regularly spaced teeth of the correct pitch and separation to engage with film perforations and to transport the film through various types of mechanism.
See: *Camera* (FILM)*; Printers; Projector.*

⊕**Sprocket Hole.** Perforations along one or both edges of motion picture film for its location and movement through transport mechanisms, which frequently use toothed sprocket wheels for this purpose.
See: *Film dimensions and physical characteristics.*

□**Spur Band.** Band of frequencies in a multichannel system which is separated out to feed a spur off the main network.
See: *Transmission networks.*

□**Spurious Shading.** An unwanted signal which gives the effect of brightness changes over the line or frame period of the television picture. The term is usually reserved for the description of signals generated in the camera tube, although it is sometimes used to describe spurious signals due to poor l.f. response of amplifiers.
See: *Shading correction; Shading error.*

□**Spurious Signal.** Unrequired signal obtained fortuitously by some imperfect property of the device employed.

□**Square Wave.** Wave of rectangular form often used where the leading or trailing edge is intended to be the reference point.

⊕**Squeegee.** Roller or wiper blade, usually of rubber or flexible plastic, used in contact with the surface of the film passing through

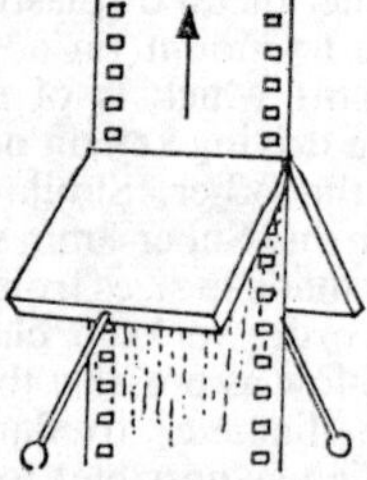

SQUEEGEE. Rubber or plastic blades shed moisture downwards as film passes upwards to roller and thence to next tank.

a continuous processing machine to reduce the amount of liquid carried over from one solution tank to the next. Because of the danger of scratches from the blade, the same effect is often performed by a jet of compressed air, known as an air-squeegee or air-knife.
See: *Processing.*

⊕**Squeeze.** Colloquial term in motion picture work for the horizontal compression of the image in anamorphic systems of wide-screen presentation. In most varieties of these processes the picture on the film is squeezed laterally, this distortion being corrected on projection by the use of a lens system giving greater magnification in a horizontal than in a vertical sense.

See: *Wide screen processes.*

Stabamp. Abbreviation of stabilizing amplifier – a video amplifier designed to remove transient disturbances and bring a television signal into a standard condition suitable for transmission or recording. Applications are at the termination of a cable or link circuit and following vision-switching equipment. Controls, manual or automatic, typically include picture amplitude, sync pulse amplitude, peak white clipper, clamping and set-up.

See: *Vision mixer.*

Stability. General term used to denote the steadiness of camera and projector mechanisms in reproducing scenes without unwanted vertical, horizontal or fore-and-aft movement. Film shrinking gives rise to such unwanted movement and its correction is a vital element in telecine (film scanner) design.

See: *Camera* (FILM)*; Pin; Registration.*

Stability (Sound). To eliminate undesirable frequency fluctuations, professional sound recorders and reproducers must provide a highly stabilized motion to the recording medium, whether tape or film. Non-perforated tape can be driven with the requisite steadiness by comparatively simple means: a well-balanced flywheel and capstan round which the tape is wrapped. Perforated film must be driven by sprockets, and this causes an irregular motion, called sprocket-hole modulation, whose frequency depends on the pitch of the perforations and the speed of the film. With 35mm film running at the standard speed of 90 ft./min., the frequency is 96 Hz. Elimination of this spurious frequency requires more elaborate sound stabilization, and oil damping and magnetic drag are often used in addition to a heavy and well-balanced sound drum and flywheel.

See: *Sound recording; Sound recording systems.*

Stabilization. Rendering a processed photographic image as permanent as possible and eliminating the effect of chemical residues in the emulsion; applies particularly to the processing of colour film, where a stabilizing bath often forms the last operation before drying.

See: *Chemistry of the photographic process; Processing.*

Stabilization (Image). One of the problems which is particularly troublesome with long focal length lenses, both in film and television, is the movement of the image due to vibration of the camera.

Many methods have been proposed to reduce this vibration, particularly in television. The most straightforward is to stabilize the whole camera with gyroscopes, or even to gyro-stabilize the enclosure in which the camera is mounted.

To obtain lower cost and reduce weight, methods of stabilization have been developed which rely on the control of an optical element. The most successful of these, currently on the market, is probably the Dynascience system.

In this system, the displacement of the camera is sensed by gyroscopes, but no attempt is made to control or eliminate it. Instead, a signal is fed to solenoids which move plane parallel plates of glass. Liquid is trapped between these plates and forms a liquid wedge. As the plates are tilted relative to one another, the angle of the liquid wedge is changed, and the deviation of light rays traversing the liquid undergoes a corresponding change. By matching the vibration displacement (angular only) of the camera with the wedge angle of the liquid prism, the movement of the image on the face of the vidicon or image orthicon is greatly reduced or even eliminated entirely.

The Dynascience system requires a good high frequency response from a servo system, and other methods of stabilization have therefore been proposed which do not require such a response. Among these is an

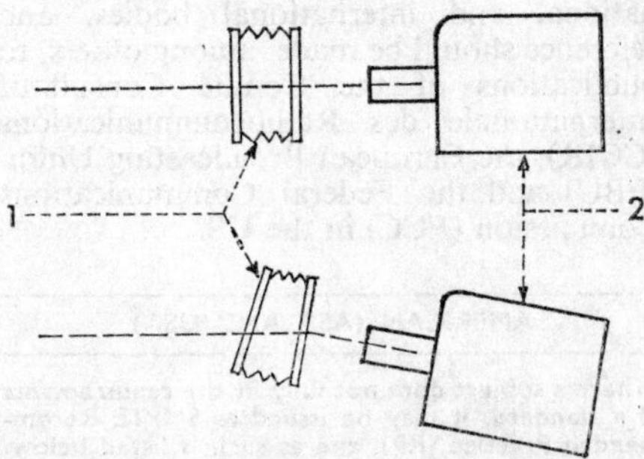

PRINCIPLE OF DYNALENS SYSTEM. 1. Liquid prism placed in front of lens, rigidly mounted to film or TV camera, 2. (*Top*) Plates of liquid prism are parallel; no refraction takes place. (*Bottom*) Camera tilted upwards, so that scene would move downwards in aperture. Front plate of liquid prism is tilted by servo motor system so that rays of light from scene enter camera lens at same angle relative to optical axis as before.

extension of the variable-angle wedge used in the rangefinders of early cameras. Two glass elements are used, one rigidly attached to the camera, the other stabilized by means of a gyroscope. There is a small gap between the elements, and the surfaces facing one another across this gap have a common centre. The gyro-stabilized member pivots about this common centre.

Purely electronic means of image stabilization have also been proposed in which the camera movement is sensed, and a signal then generated which modifies the deflection of the scanning beam in the image orthicon. So far, this system has been applied only to black-and-white television, but there seems to be no basic reason why it should not be applied to colour. A.C.

See: *Lens types.*

⊕ □ **Stage.** Studio areas in which sets are built and filming carried on. Except for some special effects and insert shooting, all stages are soundproofed for recording, and are often called sound stages. Each studio complex contains up to a dozen of these, the number depending on its production throughput.

In television, these areas are usually called studios.

See: *Studio complex* (FILM).

⊕ **Stampfer, Simon von (1792–1864).** Professor of Geometry at the Institute of Vienna. Discovered the principle of the synthesis of movement by persistence of vision simultaneously with, but independently of, Plateau, introducing his Stroboscope in 1833.

⊕ **Standard Eight Millimetre Film.** Term used for the original form of 8mm narrow gauge film in contrast to the later Super-8. The standard-8 frame is somewhat smaller, with 80 frames in the length of one foot of film, compared with 72 frames to the foot in Super-8.

See: ***Film dimensions and physical characteristics; Narrow-gauge film and equipment.***

⊕ □ STANDARDS

This tabulation lists the principal standards and, where relevant, recommended practices recognized by national and international standards bodies in the fields of film and television. The countries comprised are America, Britain, France and Germany, together with the International Organization for Standardization (ISO).

In the field of television, standards relating to transmission are set by different national and international bodies, and reference should be made, among others, to publications of the Comité Consultatif Internationale des Radiocommunications (CCIR), the European Broadcasting Union (EBU) and the Federal Communications Commission (FCC) in the US.

AMERICAN (ASA AND USA)

Where a subject does not fully fit the requirements of a standard, it may be issued as SMPTE Recommended Practice (RP), and as such is listed below.

* Signifies Under Committee Review.

† Proposed Standard or Recommended Practice.

‡ Draft Standard.

FILM DIMENSIONS

Each title contains a designation for the perforation shape (BH, DH, KS, CS), or the number of rows of perforations (1R, 2R, etc.), depending on which is the significant factor, and the perforation pitch without decimal point. The figures in brackets refer to positions of perforations. All Standard Nos. are preceded by "PH22" unless other letters are shown.

8mm perf. Super-8, 1R–1667	149–1967
16mm perf. 8mm, 2R–1500	17–1965
16mm perf. Super-8:	
2R–1664 (1–3)...	151–1967
2R–1664 (1–4)...	168†
2R–1667 (1–3)...	150–1967
2R–1667 (1–4)...	167†
16mm 1R–2994...	109–1965
16mm 1R–3000...	12–1964
16mm 2R–2994...	110–1965
16mm 2R–3000...	5–1964
32mm 2R–2994...	141–1965
32mm 2R–3000...	71–1965
32mm 4R–2994...	142–1965
32mm 4R–3000...	72–1965
35mm perf. 8mm, 5R–1500	RP 28–1968
35mm perf. Super-8:	
2R–1664 (1–5)...	169†
5R–1667	165†
35mm perf. 16mm:	
3R–2994 (1–3–0)	171†
3R–3000 (1–3–0)	170†
35mm perf. 32mm:	
2R–2994	73–1966
2R–3000	138–1964
35mm BH–1866	93–1964
35mm BH–1870	34–1964
35mm CS–1870...	102–1964
35mm DH–1870	1–1964
35mm KS–1866...	139–1964
35mm KS–1870...	36–1964
65mm KS–1866...	45–1965
65mm KS–1870...	118–1967
70mm perf. 65mm KS–1870	119–1967

8MM FILM
Film length, 25 ft. spool capacity ... 143–1965
Film usage
Camera 21–1964
Projector 22–1964
Image areas
Camera 19–1964
Projector 20–1968
Lamps, projector
Base-up type 84–1964
Base-down type 85–1964
Lens focus scales 74–1965
Lens mounts 76–1960
Reels, projection ... ... 23‡
Sound, magnetic
Stripe 88–1963
Reproducing characteristic 134–1963
Sound record 135–1962*
Splices
Laboratory type 77–1965
Projection type 24–1965
Spools, camera
25 ft. 107–1964
100 ft. 173–1968
Test films
Magnetic sound
Azimuth 129–1962
400 Hz signal level 130–1962
Flutter 128–1962
Multifrequency 131–1962
Photographic, registration ... ... RP 19–1965

SUPER 8MM FILM
Camera cartridge
Aperture, pressure pad, film position 159.2–1968
Cartridge, cartridge–camera fit ... 159.1–1968
Film length, camera run (50 ft. capacity) 159.5–1968
Notches ... ... 166†
Pressure pad flatness, aperture profile 159.3–1968
Take-up core drive 159.4–1968
Film usage
Camera 156–1968
Projector 155–1967
Image areas
Camera 157–1967
Projector 154–1968
Reels ... ... 160†
Sound, magnetic
Stripe, 1R 161–1968
Sound record ... ... 164†
Splices
Cemented, for projection ... ... 172.1‡
Tape, for projection ... ... 172.2‡

16MM FILM
Cores for raw stock film 38–1964
Edge numbering 83–1965
Film usage
Camera, 2R 9–1965
Camera, 1R 15–1964
Projector, 2R 10–1964
Projector, 1R 16–1965
Film winding, A & B ... ... 75‡
Image areas
Camera 7–1964
Printer
Contact pos. from neg. and reversal 48–1965
16 to Super-8 (neg./pos. and reversal) 153–1967
16 to 35mm enlargement ratio ... 92–1953*
Projector 8–1968
TV review room 148–1967
Lens focus scales 74–1965
Lens mounts 76–1960
Lamps, projector
Base-up type 84–1964
Base-down type 85–1964

Reels 11–1966
Screen brightness, review room ... 100–1967
Sound
Magnetic
30-mil stripe 101–1963
50-mil magoptical stripe 127–1962*
100-mil stripe 87–1966
200-mil stripe 97–1964
Perforated 8mm 136–1963
Perforated Super-8 (1–4) 162–1968
Perforated Super-8 (1–3) ... ... 176‡
Picture-sound separation 112–1958*
Location of synchronizing signal ... RP 25–1968
Photographic 41–1957
Spindles
Camera ... ... RP 24–1967
Projector 50–1960*
Projector ... ... RP 34–1968
Splices
Laboratory 77–1965
Projection 24–1965
Spools, camera, daylight-loading
50–400 ft. 174–1968
Sprockets ... ... RP 1–1950
Test films
Magnetic sound
400 Hz signal level 132–1963
Azimuth alignment 114–1959*
Flutter 113–1966
Multi-azimuth 126–1961
Multifrequency 140–1965
Photographic sound
400 Hz signal level 45–1962*
3,000 Hz flutter 43–1961*
5,000 Hz, 7,000 Hz sound focusing... 42–1962*
Buzz track 57–1963
Multifrequency 44–1963
Registration ... ... RP 20–1965
Sound projector ... ... RP 18–1964
Scanning beam, laboratory type (corrected) 80–1950*
Theatre test reel 79–1950*
Theatre test reel ... ... RP 35–1968
Test methods, sound distortion
Cross modulation, variable area ... 52–1960
Intermodulation, variable density ... 51–1961

35MM FILM
Cores for raw stock film 37–1963
Film usage
Camera 2–1961
Projector 3–1961
Projector (anamorphic) 103–1966
Image areas
Camera 59–1966
Printer
35 to 16mm (16mm pos. prints) ... 46–1946*
35 to 16mm (16mm dupe neg.) ... 47–1946
Release picture sound continuous contact 111–1965
Projector 58–1968
Aspect ratio 2·35:1 106–1965
Lens and mounts for motion picture projectors 28–1967
Reels 4–1965
70/35mm 147–1966
Screen brightness
Indoor theatres 124–1961*
Review rooms... 133–1963
Sound
Magnetic
4 150 mil records 108–1958
4 records, release prints 137–1963
35/17½mm 1 or 3 200 mil records... 86–1962*
Perforated Super-8 stripe, 5R ... 163–1968
Photographic 40–1967
Double width push-pull
Normal centreline... 69–1960

Offset centreline 70–1960
Spindles, rewind ... RP 21–1966
Sprockets 35–1962*
Test films
Magnetic
Azimuth Alignment... ... ... 99–1955*
Flutter 98–1963
Photographic
1,000 Hz balancing 67–1960*
7,000 Hz sound focusing ... ... 61‡
9,000 Hz sound focusing 62–1960*
Buzz track 68–1962
Scanning beam, uniformity... ... 65‡
Theatre test reel 60–1959*
Subjective test reel ... ... RP 33–1968

70MM FILM
Image areas, projector... 152–1968
Reels, 70/35mm 147–1966
Splices, reinforcement... ... ... RP 23–1967

MISCELLANEOUS, FILM
Density measurements
Calibration of densitometers ... RP 15–1964
Spectral diffuse 117–1960
Transmission 27–1960*
Graph paper ... ... RP 22–1966
Lamps, projector, 4-pin, pre-focus ... 175–1968
Lens
Aperture calibration 90–1964
Lens mounts, high-speed motion picture cameras ... ... RP 3–1957
Nomenclature, film
Sections 1–4 ... ... 56–1961*
Sections 5–7 ... ... 56a–1964*
Photometric performance, incandescent lighting units ... ... RP 4–1958
Release prints (universal leader) ... 55–1966
Reversal colour film speed ... ... 146†
Safety film 31–1967
Screen brightness (drive-in theatres) RP 12–1962
Sensitometric Strips ... ... RP 14–1964
Synchronization, sound-picture ... RP 25†
Unsteadiness, high-speed, camera ... RP 17–1964

TELEVISION
Alignment test pattern ... RP 27.1–1968
Density and contrast range, films and slides ... ... RP 7–1962*
Image areas
16mm film 96–1963
35mm film 95–1963
Safe action and title area ... ... RP 8–1968
16mm projector, monochrome film chains full storage basis 91–1955
Slides and opaques 94–1954*
Slides and transparencies for TV ... 144–1965

VIDEO MAGNETIC TAPE RECORDING
Labels ... ... RP 26–1968
Leader
Monochrome ... ... C 98.2–1963
Colour ... ... C 98.9–1967
Modulation practices ... ... RP 6–1967
Patch splices ... ... RP 5–1964
Records, characteristics of audio ... C 98.3–1963
Record dimensions, video, audio and tracking control ... ... C 98.6–1965
Record, tracking control ... ...RP 16†–1968
Reels ... ... C 98.5–1965
Speed ... ... C 88.4–1963
Tape dimensions ... ... C 98.1–1963
Tape vacuum guide ... ... RP 11–1968
Test tapes
Multifrequency
15 ips ... ... C 98.8†
7·5 ips ... ... C 98.11†
Primary audio level
15 ips ... ... C 98.7†
7·5 ips ... ... C 98.10†
Signal specifications for alignment tape ... ... RP 10–1962*
For use with RP 6 (practice LBM)
15 ips ... ... RP 29–1968
7½ ips ... ... RP 30–1968
(practice LBC) 15 ips ... ... RP 31–1968

D.P.B.

BRITISH (BS)

All numbers are preceded by BS except those marked CP.

FILM DIMENSIONS
35mm film ... ... 677 (Pt. 1)–1958
16mm film ... ... 677 (Pt. 2)–1958
8mm film ... ... 677 (Pt. 3)–1958
65 and 70mm film ... ... 677 (Pt. 4)–1965
Magnetic sound film 2981–1958

8MM FILM
Claw-to-gate distance, spool loading projectors 3747–1964
Projectors, light output 930–1962
Spools for projectors 2013–1960
Spool containers 2835–1957
Sprocket dimensions
Perforated 8mm film 4157–1967
8mm (type S) film 4328–1968

16MM FILM
Lenses for projectors 3389–1961
Projectors
Sound and picture, portable, performance of... 3675–1963
Light output 930–1962
Release prints 1585–1949
Screen luminance 2954–1958
Spools
2,000 ft. 2014–1960
2,400 ft. 3801–1964
Containers for 2835–1957
Test film for projectors 1488–1948

35MM FILM
Aspect ratio (film) 2784–1956
Lenses for projectors 1590–1949
Release prints 1492–1948
Remote control panels for projection, layout of 3933–1966
Screen luminance 1404–1961
Spools, 2,000 ft. 1587–1949
Sprockets
Broad tooth 1967–1965
Universal 3946–1965
Test film for projectors 1985–1953

MISCELLANEOUS
Acoustics
Reproduction safety requirements ... 415–1967
Measurement: preferred frequencies 3593–1963
Glossary of terms 661–1955
Musical pitch 880–1950
Artificial ear for calibrating earphones 2042–1953
Audiofrequency transformers for cine equipment 1793–1952
Audiometers, pure tone 2980–1958
Band-pass filters, octave 2475–1964
Carbons for projection arcs, diameters 1964–1953
Cells, photoelectric, for sound film apparatus 586–1953
Clips, film, bite of 1379–1947

Cores for motion picture and magnetic film 4356–1968
Exit signs for cinemas 2560–1954
Self-luminous 4218–1967
Exposure index of photographic negative material 1380–1962/3
Exposure meters, photoelectric ... 1383–1966
Film
Density and contrast range (black-and-white) 3115–1959
For television, picture areas... ... 2962–1958
Safety definition 850–1955
Strips and film slides 1917–1968
Strips, containers and notes for ... 2698–1960
Filter factors of photographic material 1437–1948
Lamps
Exciter 1015–1961
Expendable flash bulbs 2833–1968
Spotlight, studio 1075–1961
Lenses, camera
Colour transmission of 3824–1964
Dimensions, attachments 1618–1961
Markings, definition 1019–1963
Resolving power 1613–1961
Lighting fittings for TV productions... 4015–1966
Lighting for cinemas, maintained CP 1007–1955
Loudspeakers
Cone diaphragm 1927–1953
Performance of 2498–1954
Mercury arc rectifier equipment ... 1698–1950
pH scale... 1647–1961
Projection equipment, installation CP 412–1953
Screen luminance, laboratories and review rooms 2964–1958
Scripts, export, motion picture ... 4007–1966
Sound recording
Frequency characteristics 3154–1959
Frequency variation 1988–1953
Magnetic
on tape 1568–1960
6 track, on 70mm prints 4298–1968
Spotlights, studio 2063–1963
Television, measuring performance of receivers 3549–1963
Test films for azimuth alignments of sound heads 2829–1957
Transformers, audiofrequency, cinematograph 1793–1952

W.J.R.

FRENCH (AFNOR)

FILM DIMENSIONS
Double 8mm ... ... S 24–301 1966
16mm with 2 rows of perforations ... ... S 24–102 1966
16mm with 1 row of perforations ... ... S 24–103 1966
Double 16 ... ... S 24–106 1951
35mm stock ... ... S 24–002 1966
35mm perforated 2×16mm with 3 rows of holes ... S 24–109 1967
35mm perforated 3×9·5mm ... S 24–202 1960
35mm perforated 4×8mm ... S 24–303 1967
65mm ... ... S 24–401 1966
70mm ... ... S 24–501 1966

8MM FILM
Camera aperture for Double-8 S 24–302 1960
Projector
Aperture ... ... S 26–301 1960
Spools ... ... S 26–302 1960

9·5MM FILM
Camera aperture ... ... S 25–203 1960
Optical sound reduced from 35mm ... ... S 25–204 1960
Printing dimensions
Silent film ... ... S 25–201 1960
With optical sound ... ... S 25–202 1960
Projector aperture
Silent film ... ... S 26–201 1960
Optical sound ... ... S 26–202 1960

16MM FILM
Lamps
Exciter for sound heads ... S 26–006 1951
Projector
Cap up ... ... S 26–009 1959
Cap down ... ... S 26–010 1958
Negative aperture ... ... S 24–104 1966
Picture aperture and sound head position ... ... S 26–101 1948
Photocells for sound heads ... S 26–007 1951
Projector
Lamps ... ... S 26–102 1948
Lenses ... ... S 26–105 1955
Optical characteristics ... S 28–003 1951
Spools ... ... S 26–104 1958
Screen
Characteristics ... ... S 27–007 1955
Dimensions ... ... S 27–005 1955
Sound
Magnetic tracks on release prints ... ... S 25–104 1967
Track dimensions ... ... S 24–105 1948
Splices
Positive ... ... S 25–103 1960
Negative ... ... S 25–108 1960
Spools for stock ... ... S 24–107 1960
TV title areas ... ... S 24–009 1955

35MM FILM
Auditoriums
Acoustics ... ... S 27–002 1945
Dimensions ... ... S 27–001 1960
Cores
300 metre (1,000 ft.) rolls ... S 24–006 1955
120 metre (400 ft.) rolls ... S 24–003 1955
Intermediate ... ... S 27–011 1960
Editing picture and sound ... S 24–007 1951
Leaders and trailers ... ... S 25–003 1945
Lamps
Exciter, for sound heads ... S 26–006 1951
Projector
Cap up ... ... S 26–009 1959
Cap down ... ... S 26–010 1958
Lens mount, camera ... ... S 24–012 1958
Negative aperture ... ... S 24–004 1966
Photocells for sound heads ... S 26–007 1951
Projector
Aperture and sound head position ... ... S 26–001 1945
Lenses ... ... S 26–005 1951
Optical characteristics... ... S 28–003 1951
Rewind cores ... ... S 26–003 1948
Spool box ... ... S 26–004 1948
Projection room port openings S 27–004 1948
Release print assembly ... S 24–008 1960
Safety film specifications ... S24–001 1960
Screen
Characteristics ... ... S 27–007 1955
Dimensions ... ... S 27–005 1955
Luminance ... ... S 27–003 1948
Sound
Optical track ... ... S 24–005 1945
Magnetic track dimensions ... S 24–011 1958
Spools for 600 metre (2,000 ft.) S 26–002 1948
Sprockets
Intermittent ... ... S 24–010 1955
Projection ... ... S 26–008 1956
Wide screen framing ... ... S 24–009 1967
TV title areas ... ... S 24–009 1955

MISCELLANEOUS

Auditorium dimensions ...	S 27–001	1960
Defects on cine film – definition and symbols	S 20–004	1967
Light strength for picture image printing	S 25–006	1955
Lamps		
3,200°K, characteristics ...	S 27–008	1958
Arc, mirror dimensions ...	S 28–010	1960
Lenses, photometric calibration	S 28–002	1951
Panoramic size prints with stereo sound	S 25–007	1958
Test films for		
Continuous sound spectrum frequencies	S 28–015	1967
Fixed sound frequencies ...	S 28–016	1967
Gate position	S 28–014	1967
Image stability (steadiness) ...	S 28–011	1967
Uniformity of (projector) gate illumination	S 28–013	1967
Tests for		
Arc lamp mirror	S 28–008	1960
Camera image stability (steadiness)	S 28–012	1967

L.B.H.

GERMAN (DIN)

8MM FILM

Copies with magnetic track ...	15881	1963
Film usage, camera and projector	15852–2	1965
Projector		
Lenses	15744	1963
Lenses for Super-8	15844	1967
Spools	15821	1961
Raw stock	15851	1965
Screen luminance	15671	1966
Spools for Double-8	15822	1960
Sprockets	15823	1952
Test films	15806–1	1965

16MM FILM

Daylight loading spools ...	15632	1967
Film usage, camera and projector	15602	1960
Projector lens	15744	1962
Raw stock	15601	1967
Screen luminance	15671	1966
Sound		
Track position and scanning...	15603	1959
Magnetic reference film ...	15638	1958
Magnetic track		
Dimensions and recording characteristics ...	15655	1961
Single-perforated stock ...	15681	1958
Double-perforated stock ...	15682	1961
Spools and cans	15621	1961
Sprockets	15625	1961
Test films (series of tests) ...	15606	1963

35MM FILM

Camera and projector dimensions for		
1·275:1 format	15502	1959
1·66:1 format	15545	1965
Anamorphic format	15546	1963
Camera usage for single film stereophotography	15543	1961
Cans and spools	15521	1960
Cores	15531	1959
Cores	15599	1943
Leaders and trailers	15598	1965
Projector		
Optical characteristics and dimensions of lenses ...	15741	1965
Designations	15579	1967
Raw stock		
Magnetic coated film (17·5 and 35mm) with single and double track records ...	15552	1961
Safety film ...	15551	1962
Screen luminance for indoor theatres	15531	1965
Sound		
Magnetic sound head... ...	15568	1965
Optical sound head	15566	1962
Prints with 4 or 6 magnetic tracks	15554	1961
Prints with 4 magnetic and 1 optical track	15555	1967
Prints with 1 magnetic track	15572	1963
Track position and scanning slit position	15503	1959
Sprockets		
Feed	15520	1951
Intermittent	15522	1951
Narrow tooth	15530	1959
Test film		
Magnetic (17·5 and 35mm) ...	15522	1951
Series of tests for sound ...	15506	1959
		1967
Test methods for frequency response	15572	1966

65MM AND 70MM FILM

Prints (70mm) with 6 magnetic tracks	15702	1965
Projector lenses: optical characteristics and dimensions ...	15741	1965
Raw stock	15701	1964
Sprockets (70mm)	15730	1965

MISCELLANEOUS

Air conditions: storage and processing of cine film ...	15556	1958
Film counters (measurement of cine film) defects – definition and symbols	15581	1960
Lighting fittings	15560	1967
Projector		
Arc carbons	15742	1942
Measurements with incandescent light	15748	1965
Measurements with arc or discharge lamp	15749	1966
Speeds of exposure and projection of cine film	15577	1967
Synchronizing film and tape ...	15575	1964
Winding direction and emulsion position	15576	1963

L.B.H.

INTERNATIONAL (ISO)

FILM DIMENSIONS

Double 8mm	R 486	1966
16mm cine film perforated on one and two edges	R 69	1958
35mm	R 491	1966

8MM FILM

Emulsion position for silent film		
In camera	R 28	1956
In projector	R 29	1956
Image area, camera and projector	R 74	1958

16MM FILM

Emulsion position		
Silent film in camera	R 25	1956
Silent film in projector ...	R 26	1956
Sound film in camera... ...	R 27	1958

Image area		
Camera	R 466	1965
Projector	R 359	1963
Sound		
Optical...	R 71	1958
Magnetic strip		
Film perforated on both edges	R 163	1960
Film perforated on one edge	R 490	1966
35MM FILM		
Aspect ratio: maximum for projector apertures for non-anamorphic cine film ...	R 358	1963
Emulsion and sound recording positions		
In cameras	R 23	1956
In projectors	R 24	1956
Image area, camera and projector	R 73	1958
Sound		
Optical...	R 70	1958
Sound records and scanning area of 35mm double width push-pull sound prints ...	R 72	1958
Location of magnetic recording heads for		
3 sound records on 35mm ...	R 162	1960
1 sound record on 17·5mm ...	R 162	1960
4 sound records on 35mm ...	R 360	1963
MISCELLANEOUS		
Acoustics – preferred frequencies	R 266	1962
Safety film	R 543	1967
Standard tuning frequency ...	R 16	1955

W.J.R.

See: *AFNOR Standards; ASA Standards; DIN Standards; USA Standards Institute.*

Literature: *SMPTE Jnl.*, Publications of Federal Communications Commission (FCC); British Standards Institute; Association Francaise de Normalisation; International Standardization Organization.

STANDARDS CONVERSION

Throughout the years, there has been steady improvement in television equipment so that increasingly high standards of performance have become possible. As each country introduced a television service, it was naturally anxious to start on the best standard available. There are, therefore, a number of different television standards in use throughout the world.

In addition, the power frequencies employed are in some cases 50 Hz and in others 60 Hz. This has led to the development of television standards which differ in bandwidth, line rate, and field frequency. In Britain, the 405-line 50-field per second television system is being replaced by the more advanced 625-line 50-field system and the bandwidth has been increased from 3 MHz to 5·5 MHz. On the other hand, in North America, a 525-line 60-field system is well established.

The situation is even further confused by the introduction of colour television. The first commercial colour television system (NTSC), now standardized throughout North America, presented certain transmission problems. In an attempt to minimize these problems, alternative colour systems have been devised in Europe.

There are, therefore, many television standards in use throughout the world, and if programmes are to be exchanged between countries, broadcasting authorities must be able to change the incoming signal from one television standard to another. The problem is particularly severe in Britain where two systems are in use.

Standards conversion was introduced in the earliest days of television but, even now, leads to an inevitable loss in picture quality. Until recently, the loss has been severe and standards conversion has been employed only when unavoidable. A more complete understanding of the process has led to the development of better standards converters and it is now possible to produce a monochrome picture with only minor defects, quite acceptable for most normal purposes. The converters described here are designed specifically for monochrome work. Colour presents more difficult problems, and their solution is briefly alluded to.

BASIC PRINCIPLES

Standards converters to change television signals from one standard to another use both electrical and optical techniques.

When considering optical problems, the performance of a system is often quoted as patterns per millimetre. This spatial frequency clearly has an equivalent in MHz in the signal representing a television line. The response to a signal can be expressed as a time function and, in particular, the response of an electrical system to an indefinitely narrow impulse is of particular interest. This approach has much in common with the spread function of the optical world. The optical designer has long dealt with a picture in two dimensions and to complete the analogy both the horizontal and vertical resolution of a television picture must also be considered. Any vertical strip of the picture can be said to have a frequency response in a very similar manner to that which is often quoted in the horizontal direction so that both the electrical and optical systems can be said to have characteristic frequency responses and apertures in two dimensions, horizontal and vertical.

The simplest type of standards converter consists basically of a TV camera operating on one standard looking at a picture displayed on a tube on the other standard.

More sophisticated converters employ computer-like techniques in which information is sampled line by line and interpolated to produce the new standard. Standards converters of all types possess two fundamental properties: information storage and interpolation, i.e., averaging the stored information.

INFORMATION STORAGE. The converter must store the picture signal as it arrives and keep it until required for the reconstruction of the output picture. Converters which change the line rate must therefore have a storage capacity of at least one line and devices capable of field rate conversion must store at least one field.

Information can easily be stored in the afterglow of a cathode ray tube phosphor or in the target of a camera either as a semiconductor lag or as capacity storage, etc. An analogue signal can alternatively be stored in specially fabricated capacitor or inductor memories.

Magnetic recording or ferrite core stores may also be used but the storage mechanism then differs from the others in that it can only deal with digital information, i.e., an on-or-off or two-state signal. The other forms of information storage are analogue in type and can handle a television signal without special processing. The digital method of storage however has the advantage that in theory it can be made to introduce no errors; it thus reduces picture degradation to a minimum.

INTERPOLATION. The television picture is a modulated raster whose structure is dependent on the line standard employed. Any new television standard will in general have a raster which lies in the spaces between the old raster and coincides only rarely at the crossover points of input and output rasters. The correct signal on the output raster must be estimated from information on adjacent lines or, in mathematical terms, interpolated between these lines.

This process is similar to the technique of drawing a smooth curve through a series of points on a graph. Information is given about a number of discrete points (or lines) in the picture and the most likely value of the picture information which lies somewhere between them has to be estimated.

In a somewhat different way, field-rate conversion requires position interpolation to maintain the illusion of steady movement. There are therefore two types of interpolation: between lines and between fields (or pictures).

Interpolation between lines can be considered as a careful modification of the vertical resolution. On the one hand, a picture with the best possible resolution is required; on the other hand, the presence of the different line structures must not cause unwanted spurious signals and beat waveforms in the output. In the simplest cases, vertical resolution is controlled only by de-focusing. Much better results can, however, be obtained if the system response in a vertical direction is carefully tailored to an optimum. In the early camera equipment, spot wobble was used to modify the vertical aperture; in later apparatus, elaborate precautions are taken to ensure that the vertical response is as good as possible.

OPTICAL CONVERTERS

Any device which will display a complete picture without relying on the storage capabilities of the human eye can be used in conjunction with a conventional camera to produce a standards converted picture. The simplest converter principle is thus that of film recording. The vertical resolution on the displayed picture is modified by spot wobble in such a way as to remove the line structure and this picture is then recorded on film. Storage for an indefinite time is available and standards conversion to different line or field standards is relatively straightforward. Telecine machines or film scanners have been made to operate on any

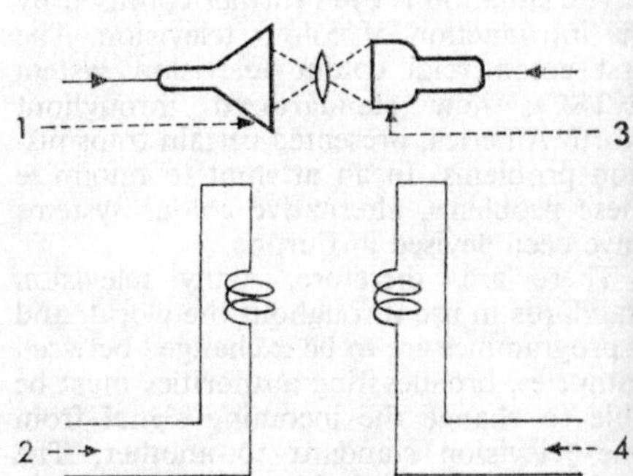

CONVERSION BY OPTICAL IMAGE TRANSFER. 1. Cathode ray tube (kinescope) display with, 2, scanning at input standard. 3. Camera with, 4, scanning at output standard.

line standard on both 50 and 60 Hz field rates.

The facility of easy conversion to any standard is one of the reasons for the continued use of film in television.

The earliest converters employed techniques similar to film recording. A display, suitably modified with spot wobble, was 'photographed' not by film but by another television camera.

For many purposes, however, the storage time available in the afterglow of the cathode ray tube phosphor was insufficient for the process of standards conversion and a camera tube was required which also had an inherent storage capability. Some of the earliest experiments were based on the cathode potential stabilized (CPS) tube and converters still in use today employ this pick-up tube. Other converters of the same type use the image orthicon and vidicon camera, and the choice between them is largely governed by the amount of storage available.

CPS CAMERA TUBE. The CPS camera is now no longer used except for conversion which involves changes in the field rate. Here a very long storage time is required and the storage capacity of the CPS tube must be assisted by a long afterglow phosphor. Unfortunately, the storage of pictures in the afterglow of the phosphor is a relatively difficult process because the signal is continuously decaying and the loss must be corrected externally. The situation is further complicated when changing field rates, because it may be necessary to scan portions of a picture more than once. In order to compensate for this variation, the signal is modulated in amplitude, either by a pre-set device or by some form of automatic amplitude control employing a reference white signal in the display. The control of the vertical aperture and the storage process is relatively elementary, and even under optimum conditions there is a noticeable loss of resolution and some movement blur in pictures derived from converters of this type.

VIDICON CAMERA TUBE. Vidicon converters have the advantage that the lag of the vidicon target can be made sufficiently long to permit standards conversion between different line and field rates. The limitations described earlier are, however, still present and although the equipment has many uses there is an observable loss in resolution and an increase in the movement blur.

IMAGE ORTHICON CAMERA TUBE. Converters based on the image orthicon camera can also change both the line and field rates of a television signal, but this converter has the same limitations as the other camera tube devices and differs only in the characteristics of the tube employed and its storage capabilities.

The tube is operated normally in the linear mode (i.e., unity gamma output). Image orthicon tubes are available with different characteristics depending on the mesh spacing within the tube. The close-spaced mesh gives a better signal-to-noise ratio but more movement blur and the wider-spaced tube has a shorter storage (or lag) characteristic, sufficient for line rate conversion but with a worse signal-to-noise ratio.

MAGNETIC TAPE CONVERTERS

Magnetic recording forms an alternative store of picture information which can be used as a fundamental part of a standards converter. A number of proposals have been made to use video tape recorders for this purpose, particularly in Japan. Both 625- and 525-line television programmes are used in Japan, and television standards converters are of great interest. The video tape recorder can also be adapted to provide elementary interpolation between lines and fields. Two examples will be quoted: one to achieve line standards conversion; the other field rate conversion.

LINE RATE CONVERTER. The line rate converter is particularly interesting in that it operates on a picture basis, i.e., it interpolates between adjacent lines on the complete picture and is not confined to adjacent lines on one field. Moreover, it is capable of handling an NTSC colour signal. The recorder employs seven record heads mounted on a revolving drum. Each scans the two-inch magnetic tape transversely, recording a track containing six television lines. The machine is therefore similar to the quadruplex recorder with four heads except that the head drum diameter is increased so that seven heads can be acommodated instead of four. The replay drum also has seven heads and rotates at the same speed, but the drum diameter is smaller so that only about five lines are replayed during the time taken to record six lines of the original signal. An external electrical delay of one line is used to provide a second signal and the appropriate interpolation made between the two signals to conceal defects due to omitting information from the unrequired television lines. In order to interpolate between adjacent lines in the picture, it is necessary to add a field delay. In the 525-line system, the interpolation would then be carried out between adjacent lines of a complete picture numbered, for example, lines 1 and 262.

This standards converter is unusual in that it has been designed to deal with a colour picture. The colour information is supplied as a conventional coded signal and it remains in coded form up to this point in the converter. The process of changing the line rate has, however, disturbed the colour coding and it is necessary to decode the signal into its fundamental red, green and blue components and subsequently re-encode it on the required colour standard.

FIELD RATE CONVERTER. The second converter is designed to change the field rate of the television signal and uses a helical scan tape recorder to store the information. Conversion from 50 to 60 Hz television is achieved by first recording the incoming signal field by field, i.e., 50 fields per second. There are two heads on the drum or head disc mounted side by side recording identical tracks.

On replay the speed of the tape remains unchanged but the head drum is now driven by a 60 Hz signal and the machine scans the tape at a slightly different angle because of the increased head velocity. A wider replay head is used, however, to ensure that there is always one recorded track beneath the replay head.

Unfortunately this system modifies the duration of the vertical interval and gives an incorrect line rate. Subsequently, conventional line standards conversion is employed to correct the signal. Achieving field conversion initially by means of the magnetic recording does, however, ease the design of this optical converter.

LINE STORE CONVERTER

Basically, this standards converter consists of a means of writing the incoming video signal into a suitable store and then reading out the stored information at a rate appropriate to the output scanning standard. The form which the converter takes is determined primarily by the storage medium employed, and homogeneous surfaces such as phosphors, magnetic tape and electrostatic storage, have hitherto been used to store a continuous flow of analogue information. However, the energy level of information stored by such media is relatively low, and the resulting signal-to-noise ratio of the recovered signal is marginal by broadcasting standards.

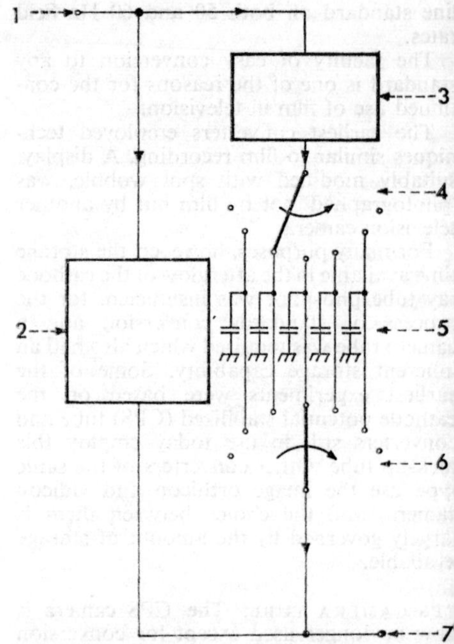

PRINCIPLE OF LINE STORE CONVERTER. 1. Signal input to, 2, sync converter and, 3, interpolator, leading to the electronic input switch, 4, rotating at input line frequency and sampling about 600 picture elements per line. These are distributed into capacitor store, 5, from which they are recovered by output switch, 6, rotating at output line frequency to give converted signal output, 7.

Discrete storage elements such as magnetic cores, capacitors, and inductors are able to store much larger quantities of energy but, being discrete, require individual means of access and hence much more complex circuitry.

Modern technology, however, now has made it feasible to utilize several hundred electronic circuits in a single piece of apparatus with a high degree of reliability and compact design, and it is now possible to use discrete storage elements to provide the capacity required by a line-store converter.

FUNDAMENTAL COMPONENTS. The line store type of converter depends upon the repeated application of sampling techniques. There are three principal components. First is an electronic input switch which cuts the television signal into about 600 picture elements per line and directs them to the second component, a capacitor store.

Once each line, the switch deposits one picture element in each capacitor, and once each line of the output standard a second switch collects the appropriate information from each capacitor of the store. Leaving the second switch there is, therefore, a train of picture elements assembled to form the required output signal. The first switch rotates at input line frequency and the second switch at output line frequency.

In practice, slight modification to the signal is required to give a satisfactory picture by an interpolation process; this can be external to the store or built within the capacitor store itself.

The synchronizing waveform is not converted and a television waveform generator is included to provide the correct synchronizing waveform.

This type of converter has a performance which can be precisely optimized; its detailed design highlights the fundamental problems which many standards converters have to overcome.

NUMBER OF STORES. When the signal is stored as discrete picture elements, consideration of fundamental sampling techniques gives the number of stores required.

Assume f_o is the system band width and f_h is the horizontal line rate. A line-store converter must provide a sufficient number of stores to accommodate the information corresponding to one line of the video signal and the theoretical minimum number of stores will be twice that number of line harmonics which lie within the video bandwidth or $2f_o/f_h$. Alternatively, this may be considered as a statement that the transmission of signals of bandwidth f_o by a train of samples requires a rate of at least $2f_o$ samples per second. The 405-line television system has, for example, a bandwidth of 3×10^6 Hz and a line rate of $10{\cdot}125\times10^3$ lines per second. Standards conversion using this sampling technique will, therefore, require

$$\frac{2\times3\times10^6}{10{\cdot}125\times10^3} = 594 \text{ samples per line.}$$

Assuming that that portion of the line period occupied by blanking needs no storage capacity, this number may be reduced to

$$\frac{594\times82}{100} = 485.$$

However, these simple calculations assume the use of perfect low-pass filters before and after the sampling process, and practical filters have a limited rate of cut-off necessitating an increase in the number of samples per second for a given nominal cut-off frequency. Naturally, the minimum sampling frequency consistent with an acceptable level of spurious signal is used in order to keep the cost of the store assembly to a minimum.

The low-pass filter is one which has its first stopband zero at a frequency of about 1·25 times the nominal cut-off frequency and the minimum attenuation throughout the stopband is greater than 30 dB. The filter has negligible group delay distortion and is representative of good modern practice. Even so, a sampling frequency of 10·5 MHz is the practical minimum. This frequency corresponds to 545 stores per active line period or 1·17 times the value for ideal low-pass filters. In practice, a more convenient number such as 576 is chosen.

INTERPOLATION TECHNIQUES. To analyze interpolation in a line-store converter, it is best to consider the signal which is handled by one of the store-elements. The input to the store-element capacitor consists of a train of samples of approximately picture-element duration and recurring at the line-frequency of the input signal. The samples are amplitude modulated by the

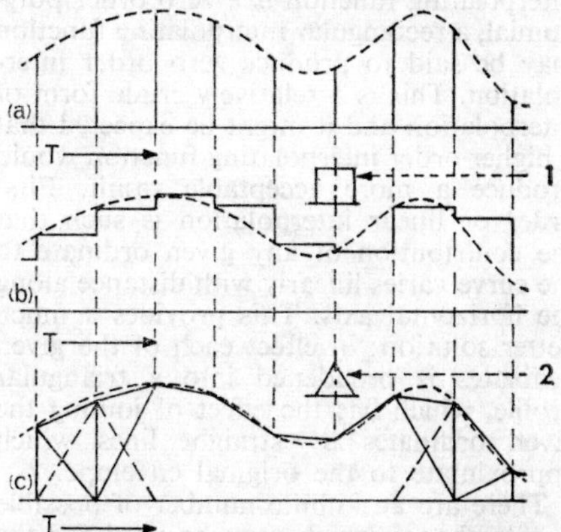

INTERPOLATION BETWEEN ADJACENT LINES. (a) Curve to be matched. Ordinates show sampling intervals. (b) Holding characteristics of simple capacitor store may be regarded as rectangular scanning aperture, 1, giving rather poor approximation to original curve. (c) Linear interpolation effectively broadens each of the given ordinates into a triangular profile, equivalent to a triangular scanning aperture, 2, providing much better approximation to envelope of original curve.

video signal and describe a vertical strip of picture one element wide. This signal will be a close approximation to the variations of brightness down the corresponding vertical strip of the original scene.

Suppose that standards conversion is required to a new standard having, for example, a lower line frequency. The input picture elements are supplied to the store at one frequency and the output picture elements are demanded as a series of pulses at some slower rate. As these output pulses seldom coincide with input samples, some means of extending the time for which the latter exist is essential. A simple store holds the value of the input pulse until a new value takes its place. It is now possible to recover the video signal stored in the array of storage elements by re-sampling the store contents with pulses of output line frequency and hence to reconstitute a video signal suited to the output standard.

The rectangular holding characteristic of this store may be examined by several analytic techniques. It may be considered as an interpolating aperture or as a filter acting in the vertical axis of the picture and having a frequency response which is the transform of the interpolating aperture. This simple store may be said to have an interpolating function generated by a rectangular aperture such that the value of the output is that of the nearest ordinate.

As the polynomial corresponding to this interpolating function is a zero order polynomial, a rectangular interpolating function may be said to produce zero order interpolation. This is a relatively crude form of interpolation and it might be expected that a higher order interpolating function would produce a more acceptable result. First order or linear interpolation is such that the contribution of any given ordinate to the curve varies linearly with distance along the horizontal axis. This provides a much better solution; in effect each of the given ordinates is broadened into a triangular profile, which has the effect of joining the given ordinates by straight lines which approximate to the original envelope.

There are an infinite number of possible profile shapes which may be used for the interpolation process but they must clearly satisfy the condition that if all the given ordinates have a constant value the resulting interpolated function must also have a constant value. It is also desirable that the major contributions should come from adjacent samples since, if a large number of consecutive samples are called upon to contribute simultaneously, more complex instrumentation will be required.

By using spectra obtained by a Fourier transform of the interpolating functions, it is possible to consider the interpolation process in terms of a vertical frequency response. It can be shown that a vertical aperture of general shape $\frac{\sin x}{x}$ will pass all the information in the band f_o and stop all spurious signals beyond this limit. In practice, the ideal interpolation characteristic can be considered as some convenient approximation to this shape.

SWITCHING TECHNIQUES. The switches in the converter perform the dual function of sampling the signal and directing these samples to the appropriate part of the line store.

The electrical size of each capacitor within the store is determined by the performance of the semi-conductor switches, and the specification of these switches is, in turn, a function of the switching system used in the converter.

In all converters of this type, the write and read circuits have much in common and a high speed diode switch is a basic

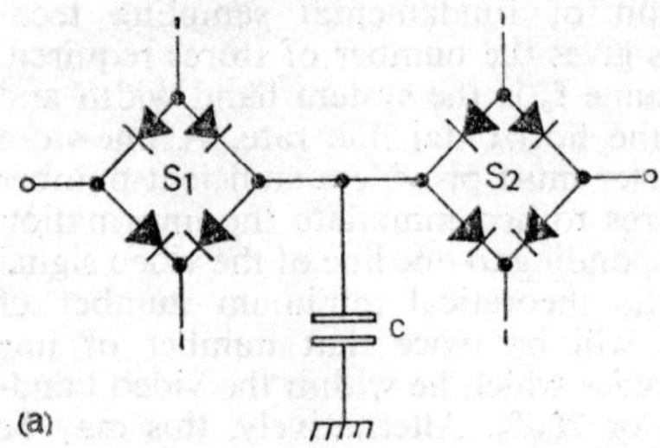

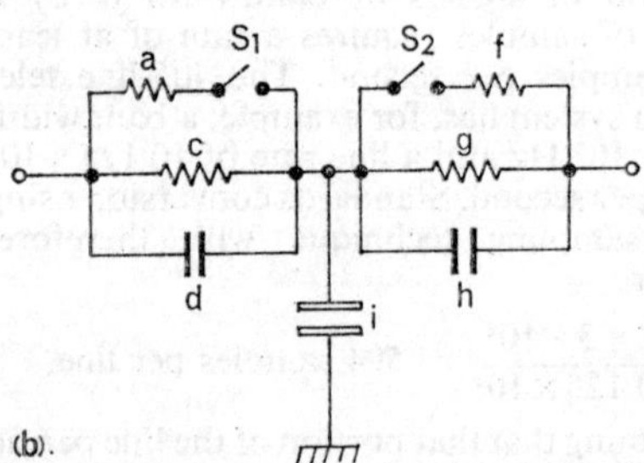

BASIC SWITCH AND CAPACITOR STORE. (a) S_1, S_2 Two switch circuits, with planar silicon diodes arranged to form a four-diode bridge. C. Storage capacitor. (b) Equivalent practical circuit with S_1, S_2, representing diode switches in (a). i. Storage capacitor, typically 500 pF, with charge time of 20 nsec.

component in any design. This switch must be capable of rapidly charging or discharging the store capacitor with extreme precision and the charge and leakage time constants of the practical switch must be approximately 20 nsec. and 20 msec. respectively if the charge/discharge process is to be completed with sufficient accuracy. In addition, the cross talk due to stray capacity must be kept to an acceptably low figure and the storage capacitor must be, say, greater than a hundred times the stray capacity.

Planar silicon diodes arranged to form a four-diode bridge can meet this switch specification quite easily and C_o, the storage capacity, is typically 500 pF.

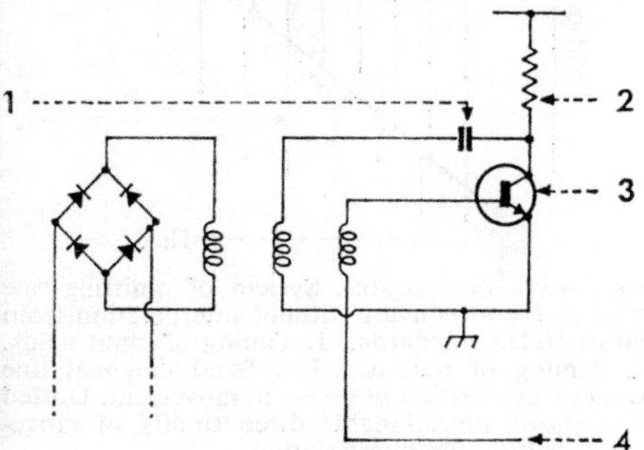

TRANSISTOR DRIVE TO DIODE SWITCH. Capacitor, 1, charges via resistor, 2, to supply potential. When trigger pulse, 4, arrives from shift register, transformer coupling causes transistor, 3, to fire as a blocking oscillator, discharging capacitor and causing large current pulse to flow through diode bridge.

The four-diode switches are driven by transistor pulse generators and the design is identical for both writing and reading switches. The two switches fundamental to the line store converter consist in practice of 576 similar pairs of switches. The moment of switch closure is controlled by a shift register and each switch conducts for a period of about 50 nsec.

In a typical transistor switching pulse generator, the capacitor charges via a resistor to the supply potential. When a trigger pulse from the shift register arrives, the transformer coupling causes the transistor to fire as a blocking oscillator, discharging the capacitor and causing a large current pulse to flow through the diode bridge, making it conduct.

WRITING AND READING TECHNIQUES. Many of the technical problems in the design of a converter of this type can be solved in more than one way, and there are at least two designs which use the same principles but achieve their result by different means.

As already explained, the charging of the storage capacitor involves the transfer of a considerable amount of signal energy, and low impedance switching circuits are necessary. Any one stored picture element is never required in two parts of the picture without modification, and new information must be extracted from the store to construct each line in the output signal.

Destruction of the unwanted portion of the old information in the store is therefore necessary in order to avoid degrading the converted picture. This operation can be achieved either by the writing or the reading circuits. If the writing circuits are low impedance, the potential across the storage capacitor will be raised to the appropriate value independent of any residual signal in the store. This is a case of a destructive writing process.

The reading circuits may be either high or low impedance, but if a low impedance reading circuit is employed this can also be arranged to empty the storage capacitor and so destroy any information remaining in the store. A destructive reading system has the advantage that it passes a considerable amount of energy to the output circuits and there is negligible loss of signal power. No amplifiers are required at any stage within the matrix of, say, 600 switch and store combinations.

On the other hand, the converter with destructive reading processes must write before reading and the converter must avoid reading the same store twice without recharging the storage capacitor in the intervening period. This is of particular importance when converting upwards (say 405 to 625 lines).

If non-destructive reading is employed, the performance requirements of the individual writing switches is considerably eased because there is no modification of the stored charge in areas of constant brightness, and some spurious signals can be avoided by using this technique.

Non-destructive reading by high impedance circuits does, however, mean that the output power is small and this can introduce problems of its own. This leads to the development of two-stage switching where connection is made to the storage capacitors by using multiple switches. It is like a telephone dialling technique where two numbers have to be dialled to obtain connection to the appropriate store.

REALIZATION OF A VERTICAL FILTER

Earlier discussion has shown that in addition to time redistribution (storage) to make the incoming picture information fit the new line length, it is also necessary to modify this information. Although the processes of time redistribution and interpolation have been considered separately, in the final instrument the two functions may well be interdependent.

NARROW BANDWIDTH INTERPOLATION. Interpolation has been described as the necessary operation of separating the wanted and unwanted information, and these components of the signal on any store-element can be separated by a simple low pass filter. Interpolation can be achieved by inserting an array of some 600 low-pass filters, one to each signal path. In this case the rectangular holding characteristic of the earlier store has been modified to a more satisfactory shape as defined by the filter parameters. The simple low-pass filter is formed by splitting the storage capacity into two parts and placing an isolating inductor between the two halves of the storage capacitor.

WIDE BAND INTERPOLATION. The large number of filters involved suggests that an alternative design could be used which would permit the filter to be inserted in the signal path at a point where only one would be required. Further consideration of this problem reveals that the filter must be relatively complex and, in order to affect the vertical resolution, it must have a comb-shaped frequency response with zeros at 15 kHz intervals.

This can be achieved in practice by combining the output and input signals of a wide band line delay in suitable proportions. Interpolation has been considered as the process of estimating the magnitude of a required picture element given information from adjacent lines. Since the relative position of the output information with respect to the incoming raster is known, a suitable mixture of the two pieces of information can be made to achieve the required interpolation. The precise mixture required is determined by the aperture to be simulated. If the proportions of the mixture are controlled, for example, by modulators, they in turn must be controlled by a waveform constructed from the required aperture shape. Using a line delay interpolator, it is therefore possible to carry out the processes of interpolation before (or after) putting the picture element into the capacitor store.

FIELD STORE CONVERTERS

Converters which can change the field rate by electronic means need a storage capacity of at least one field of the television signal. They must carry out the normal processes of interpolation between lines and also achieve some form of interpolation between fields.

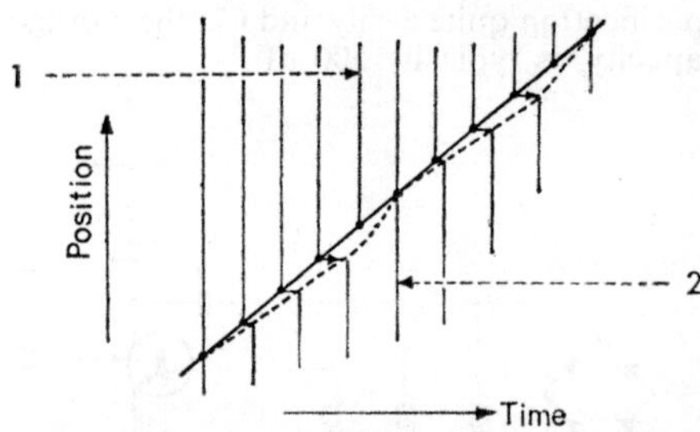

FIELD-RATE CONVERSION. System of omitting one field in six to convert without interpolation from 60 to 50 Hz standards. 1. Timing of input fields. 2. Timing of output fields. Solid diagonal line shows true position of object in movement. Dotted line shows objectionable discontinuity of movement, calling for interpolation.

For example, conversion from 60 to 50 Hz standards, by the simple process of omitting one field in six, introduces a discontinuity into steady movement which is objectionable. Interpolation between fields is necessary to achieve a satisfactory portrayal of motion.

A simple converter based on this principle has been contructed by the BBC and although it has some limitations, it nevertheless will provide a service. Unfortunately the frequencies chosen for use in Europe and North America are 50·0 fields per second and 59·94 fields per second and are not, therefore, in the simple ratio of 5 : 6. The simple converter outlined above can deal with these signals only if a non-standard output is accepted. Alternatively, a video tape recorder can be used as an intermediate step in the process and small corrections in speed or field rate of the signals can be easily made.

In the meantime, both the BBC and NHK (Japan) are developing more advanced converters. The BBC machine will be free from the limitation of the earlier machine and will give standard pictures even when converting between signals not related in frequency by the exact ratio 5 : 6. Such a converter will be free from any limitations of this sort and will eventually be invaluable in converting between SECAM, PAL and NTSC signals.

COLOUR CONVERTERS

A converter which makes no change to the line and field rate is known as a transcoder. However, conversion to or from the NTSC colour system will be inevitably associated with a television standards converter because of the difference in line and field rates between NTSC and PAL-SECAM countries.

All the earlier converters were designed to deal with a monochrome signal and conversion of a fully coded colour signal presents a major problem. Attempts have been made to convert the decoded RGB signals by optical means but considerable degradation of the picture quality occurs. More successful attempts in the future may well involve converting the luminance and colour difference signals separately. In this case, the colour difference signals have to be put on a special sub-carrier. The line store converter is theoretically capable of handling the coded signal although conversion in colour will inevitably involve changing the colour standards as well. The interpolation process will probably have to be modified to handle the subcarrier but, by one means or another, conversion in colour is undoubtedly possible using this type of converter. Colour-capable field store converters of the all-electronic type have now been shown to be practical but complicated devices.

FUTURE DEVELOPMENTS

The line store converter can give an excellent result but is fundamentally incapable of changing the field frequency. Even when operating on nominal 50 Hz, television systems conversion is sometimes necessary between a system locked to power frequencies and systems locked to an exact 50 Hz. Optical transfer converters are therefore still necessary for special occasions. Vertical aperture correctors can improve the performance but, because the stored energy is so much less, the signal-to-noise ratio inevitably limits the performance. Further developments in standards conversion depend upon the ability to store (or delay) more and more information. When these delays have been developed to the point where their performance does not unduly degrade the television signal and their price is economical, there will be a number of other applications. It is possible for electronic converters to operate between two similar standards where the field rates are not equal but are nominally 50 Hz.

In 1968, the BBC used the latest standards converter to supply Europe with pictures from the Olympic Games in Mexico. This converter was free from the limitations of the earlier machine and operated, in colour, between signals only nominally related 5:6. This converter in principle provides a solution to the problem of converting between signals of approximately equal field rates.

In the future there will be line store converters using picture interpolation, i.e., interpolation between lines in a complete picture rather than a single field. Field rate conversion by electronic means is possible but a compromise will have to be made between movement blur and the erratic movement caused by insufficient interpolation between pictures. This interpolation will inevitably require the use of further delay or storage units.

Since the output signal can never contain more information than the input signal, conversion in an upwards direction will necessarily produce an inferior picture. The term upwards direction is intended to imply conversion from 525 lines to 625 lines; 50 fields to 60 fields or SECAM to NTSC, in fact any system where the output standard requires more information than is available on the input standard.

Future developments will probably include coding the picture into digital form and using computer techniques to achieve a variety of new signal processes including standards conversion and similar techniques.

P.R.

See: *Image formation; News programmes; Special events Television principles.*

Literature: *Electronic Standards Conversion for Transatlantic Colour TV*, by E. R. Rout and R. E. Davies (*SMPTE Jnl.*), Jan. 1968.

Starevich, Ladislas, 1892–1965. Of Polish parents, born and educated in Russia. Pioneer of the puppet film, his earliest products in Russia involved animated models showing the life of insects. From 1913–1919 he produced "spectacle" films by conventional methods, but returned to animated films when, from 1921 onwards, in France, he produced a number of puppet films of considerable quality.

⊕**Start Mark.** Mark incorporated in a standard film leader to determine the exact footage between it and the first frame of the picture. When one or more sound tracks accompany a roll of picture film, the start marks on all of them establish the correct synchronism.

See: *Release print make-up.*

⊕**Static Marks.** Marks caused by discharge of static electricity through careless handling of rolls of undeveloped film, particularly in a dry atmosphere.

Such static markings often show as unwanted exposures, after processing, in a characteristic branched form.

See: *Cold climate cinematography; Film storage.*

⊕**Stationary Gate Printer.** Motion picture printer in which the film being exposed is moved continuously across a fixed aperture or slit by a separate driving sprocket.

See: *Printers.*

□STATION TIMING

A synchronizing pulse generator is used by television stations to lock together different picture sources both at field and line frequencies. This makes it possible for cuts and mixes between these sources to be carried out without disturbance of synchronization at the receiving end, thus avoiding picture roll-over and other undesirable effects.

But there are larger delays, possibly adding up to several microseconds, which occur in vision equipment and cables, and which cannot be harmonized by a sync pulse generator. In order to provide a signal which is exactly to specification, those signals which would otherwise arrive too early at the outgoing transmission line must be delayed. This is the function of station timing.

The picture as viewed on the home receiver should not move when a cut or mix is carried out, nor should definition deteriorate seriously while mixing, for instance, from a picture to the same picture with an inlaid caption. In colour television, there must be no temporary or permanent change of hue during cuts or mixes.

TIMING OF SIGNAL SOURCES

TIMING OF TELECINE. As the outputs of telecines may be used as sources in studios, they must be timed to occur earlier than any studio. If vision tape recording machines are fitted with Intersync or Pixlok, they must be timed in a similar way.

The telecines must have their feeds of line drive, mixed blanking and mixed synchronizing pulses delayed so that each machine produces an output signal with standard timings and the signals from all machines occur at exactly the same time.

TIMING OF A STUDIO. The camera output signals must have timings correctly to specification and must arrive at the inputs to the vision mixer at the same time as telecine signals.

To achieve this, the correct amount of time delay must be incorporated in the feeds of line drive, mixed blanking and mixed synchronizing signals to the camera channels, and in the vision lines from the telecine room. A variable delay adjustment to allow for the length of camera cable is normally incorporated in the camera channel.

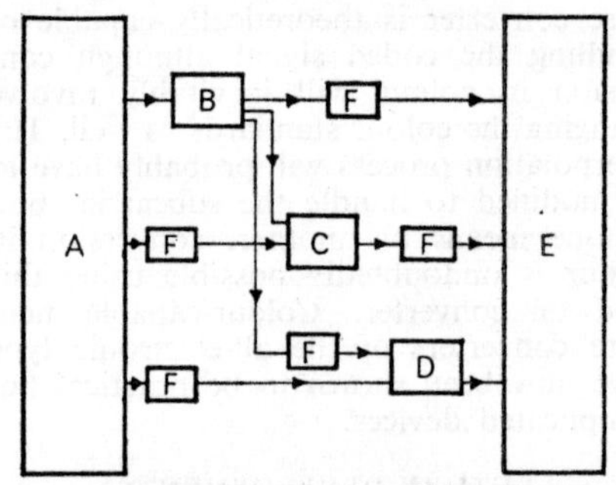

FUNCTIONS OF STATION TIMING. Object is to ensure that timing pulses from sync pulse generator, A, arrive at master control, E, at exactly the same time, no matter by what route they travel. Programme sources shown are telecine, B, and two studios, C, D. Telecine feeds master control direct, or through C or D. Hence delays, F, of different magnitude must be inserted in all lines to equalize pulse arrival times to that of the shortest route, from studio D.

In order to match the timing of channels on the cut bank of a vision mixer to the mix banks and special effects banks, delays may be inserted in the feeds of the channels to the cut bank. These can be adjusted by checking the output of the vision mixer with an oscilloscope or picture monitor triggered by a source of constant timing, and cutting between a channel on the cut bank and the same channel through the mix bank. This should be particularly accurately adjusted, because if a sequence of cuts is to be followed by a mix, such a cut will be carried out without any change of picture. Any timing error will show as a picture shift.

If the output of the vision mixer has new blanking and synchronizing signals added to it in a line clamp amplifier or stabilizing amplifier, the feeds of these pulses must be timed. If the blanking is wrongly timed, when the black level of the signal coming out of the stabilizing amplifier is varied, a step will be visible on the waveform at the beginning or end of line-blanking. Timing of the synchronizing signals can be checked by switching the stabilizing amplifier from local to remote synchronizing, when, if the timing is incorrect, the length of the presyne suppression, which is easy to check, will change.

Accurate timing adjustment must be empirical, because the pulses have finite rise times and they are normally clipped in camera channels and stabilizing amplifiers before use, at a level which might fall anywhere on the waveform. This can lead to variations of timing from those calculated of 100 nsec. or more. An accuracy of adjustment of timing of 50 nsec. is a reasonable standard to aim at.

OPERATION OF STUDIOS TOGETHER There are two main methods of timing a television studio centre where studios may have to be used in connection with each other.

The first method employs a central apparatus room containing the vision mixer units of all the studios. In this case, any camera channel can be fed as a source into any studio vision mixer by short links which introduce no timing errors. This reduces the station timing problem to that of a single studio.

For the other method, the signals into the vision mixer of one studio must be later than the output of the other studio. If it were then thought necessary to use the second studio as a synchronous source in the first studio, most of the station timings would have to be changed. Two studios are not frequently used together, however, and the combinations of studios to be used in this way can generally be foreseen.

To adjust station timings using this method, the path of vision signals through the station which has the most delay must be found. This will probably be a telecine fed into one studio, with the output of the studio fed as a source into another studio and the output of the second studio fed through the master control to the transmission line.

Pulse feeds to both these studios must be delayed to make the telecine signals coincident with camera timings in the studios. Delay must then be inserted in the telecine feed direct to the second studio to make this timing correct, and in the feed of telecine direct to master control, and in the feed of the first studio direct to master control.

Timing adjustments must be carried out in the correct order and, when an error is present, careful thought must be given to the way of correcting it. Adjustment of the wrong delay may correct one error while making another timing wrong which was previously correct.

COMPOSITE AND NON-COMPOSITE OPERATION. Signal sources can provide composite or non-composite signals, the composite signals containing synchronizing signals as well as picture signals. Each of these systems has advantages and disadvantages.

When operating with non-composite signals, mixing, fading and inlay of two or more monochrome signals can be carried out simply without any consideration of synchronizing signals. Black level clamping can be carried out using the synchronizing signal as clamp pulses, and then the correct amount of synchronizing signal can be added before the signal is fed to the outgoing transmission line. Timed synchronizing signals need be provided only at this point.

However, preview monitors must have separate feeds of synchronizing signals and, if remote composite signals are to be fed to the same monitor, the synchronizing source must also be switched.

When operating with composite signals, after mixing, fading or inlaying monochrome signals, black level clamping must be carried out in the post-sync suppression period and then the synchronizing signals may be clipped off and replaced with new ones.

With NTSC and PAL colour signals, it is more convenient to operate with composite signals as otherwise the timing problems of preview monitors are almost insoluble.

COLOUR STUDIO TIMING. Studio timing for the NTSC and PAL colour systems is similar to that for monochrome except that the vision mixer circuits must be arranged in such a way as to keep the delay through the circuits identical to within a nanosecond or two, whichever path through the mixer is used.

Station Timing

The phase of the subcarrier drive to each colour coder is adjustable to make the burst phase of each source into the mixer identical to within a degree. If this is not done, when a cut or mix is carried out there will be a temporary change of hue, as the locked subcarrier oscillator in the receiver cannot respond to rapid changes of burst phase. For this reason, it is essential to eliminate any change in delay when a mixing amplifier is brought into use, as its delay might correspond to 30° or more of subcarrier phase shift.

Studio timing for the SECAM system is similar to that for monochrome.

DELAY CABLES AND NETWORKS

Although as much as possible of the timing is carried out by delaying the input pulses to equipment, some delays in vision lines will be needed. The high frequency loss of lines or delay networks through which pulses are fed is generally negligible because clipping circuits which decrease the rise time of pulses are used in the equipment which they feed, but loss in a vision line is important.

Ordinary coaxial cable has a loss which, measured in decibels, rises approximately proportionately to the square root of frequency. Its loss at frequencies around 1 MHz is considerable when delays of the order of 1 μsec are involved. To correct accurately for this loss, a complicated network is needed.

A typical double-screened cable has a loss of 0·3 dB per 100 ft. at 1 MHz and 1·0 dB per 100 ft. at 10 MHz. As 660 ft. is required to give a delay of 1 μsec., the loss will be about 2·0 dB at 1 MHz and 4·8 dB at 6 MHz.

As great lengths of cable will be required, they will be large, heavy and expensive.

DELAY CABLE. Special delay cables are available in which the delay is increased by the use of a spiral inner conductor. This also increases the characteristic impedance.

A typical example has a characteristic impedance of 1,100 Ω, a delay of 1 μsec. for 55 cm length and a loss for 1 μsec. of about 1·7 dB at low frequencies, 2·5 dB at 2 MHz and 5 dB at 4 MHz.

To connect such a cable to 75 Ω lines, isolating amplifiers must be used, the frequency characteristic of which can be adjusted to correct for the high-frequency loss of the cable.

DELAY NETWORKS. The lumped constant delay network consists of sections of m-derived low pass filter, m being 1·27. It is found that this value of m gives the flattest curve of delay against frequency. This calls for mutual inductance coupling.

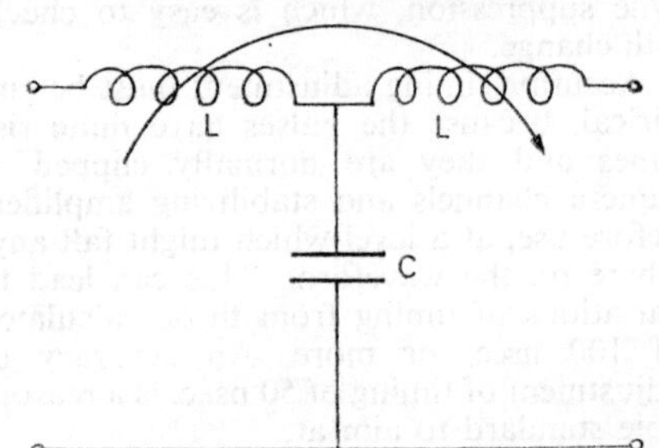

DELAY NETWORK. One section of m-derived low-pass filter with mutual inductance coupling of coils, L. The flattest possible curve of delay against frequency is desirable.

One such network has a characteristic impedance of 75 Ω, a delay of 0·025 μsec. per section and cutoff frequency of about 20 MHz, and a loss per μsec. of 0·25 dB at 1 MHz, 1·5 dB at 5 MHz and 3 dB at 10 NHz.

When delay networks are inserted in vision lines, a passive equalizer is constructed to correct for the high frequency loss of the delay circuit together with its associated cabling. An amplifier must be used to restore the signal to standard level. J.H.B.

See: *Camera* (TV); *Master control; Special events; Synchronizing pulse generators.*

Book: *Pulse, Digital and Switching Waveforms*, by J. Millman and H. Taub (New York) 1965.

□**Steady State Characteristic.** Due to nonlinearity of components, an amplifier usually has different properties according to the signal it is handling; e.g., the gain of a sound amplifier will probably not be the same for low level speech as for a steady high level sine wave. The steady state characteristic is the response to a defined continuous signal whereas the instantaneous characteristic is the response to an occasional transient.

See: *Transients.*

⊕**Step Printer.** Motion picture film printing machine in which the negative and positive are moved intermittently a frame at a time and held stationary for the period of exposure. A step printer may expose the two films in contact at a single gate or the negative may be optically imaged on to the positive by means of a copy lens.

See: *Optical printing; Printers; Special effects* (FILM).

⊕**Step Wedge.** Form of test on a sample of □photographic material in which a series of exposures increasing in known steps is given; the increments are usually in a logarithmic progression.

After processing, these yield a series of areas of increasing density which may be measured to allow the characteristic curve to be drawn.

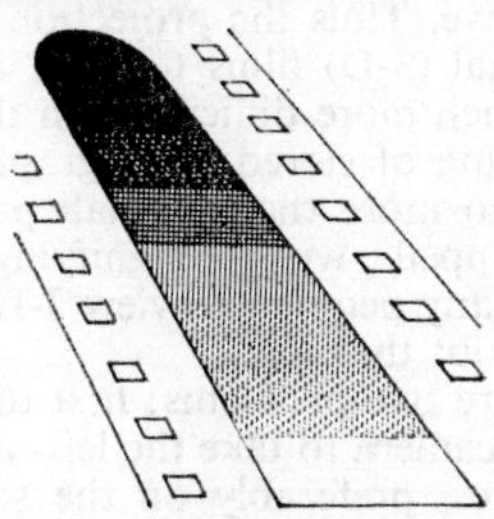

STEP WEDGE. The wedge comprises accurately determined density increments on a film sample, these being chosen to suit the film material under test. For high contrast materials, a smaller increment is preferable. Cast carbon step tablets are also used.

In film processing laboratories, step wedges are used to determine the characteristic of film stocks and the performance of printers and developers.

In television, they are generally designed to facilitate the setting up of a camera or telecine channel to a given gamma law or degree of black-and-white stretch, the densities being chosen so that, when properly adjusted, the output signal shows equal increments between steps. Frequently, two identical grey scale step wedges are arranged in opposite senses to compensate for line shading errors.

See: *Exposure control; Sensitometry and densitometry.*

⊕**Stereophonic Sound Processes.** Wide-screen film processes used for presentations in large-scale motion picture theatres often make use of stereophonic systems for enhancing the realism of the accompanying sound. When the screen width subtends a large angle at the eyes and ears of the audience, there is some justification for making sound and action emanate from the same part of the screen.

In full stereophonic systems, four or five sound tracks may separately feed an equal number of loudspeaker clusters behind the screen, while two further tracks feed speakers at the sides of the auditorium. Additional speakers can be cued in to provide sound coming from behind the audience. Simpler stereophonic systems use only three or four sound tracks.

The justification of stereophonic sound in the cinema does not really lie in reproducing the normal auditory sensations of the spectator. Binaural hearing provides little information about the distance of a sound source, and often only vague indication of its direction, because of multiple reflections from room walls as well as from objects in the open air. It therefore supplies less useful information than binocular vision, and since the loss of the latter is seldom felt in the cinema, the former is even easier to dispense with.

Stereophonic recording is valuable in orchestral reproduction, but this type of material is seldom encountered in film and television. Its justification must therefore be found elsewhere: in matching sound to extreme picture width, as already stated; in affording reproduction of exceptionally high fidelity and dynamic range; and in producing unexpected and sensational effects, as of sounds which come from behind the audience and therefore seem to envelop it in the action.

See: *Binaural recording and reproduction; Sound reproduction in the cinema.*

⊕STEREOSCOPIC CINEMATOGRAPHY

Stereo-cinematography provides moving pictures which present to the audience a measurable dimension of depth, thus recreating a solid world resembling that seen by binocular (two-eyed) human beings.

BASIC PRINCIPLES

All films and still pictures give some impression of depth, for the one-eyed observer has many clues to the depth dimension, among them perspective, aerial perspective (near objects sharp, distant ones fuzzy), masking of a farther object by a nearer, relative movement causing parallax changes, texture gradients with variation of distance (e.g., road and wall surfaces). These non-binocular factors, singly or in varied combinations, account in large measure for our sense of a three-dimensional world, and all are reproducible in normal films, taken with a single-lensed camera. They create, however, only an impression of depth, since the pictorial image is demonstrably a flat one. The stereoscopic film, when properly perceived by a two-eyed spectator, produces an image whose extent in depth can be physically measured, though it has no physical existence and occupies a position in space which is different for every spectator in the cinema.

BINOCULAR VISION. The binocular placement of objects in space depends on the rangefinding principle. The two eyes, set some $2\frac{1}{2}$ in. apart in the human head, see an object from slightly different positions, and thus produce images on the retina which are distinguishably different. As the object recedes from the spectator, the angle which any point on it subtends at his eyes becomes smaller, until it can no longer be discriminated. This distance is known as the limit

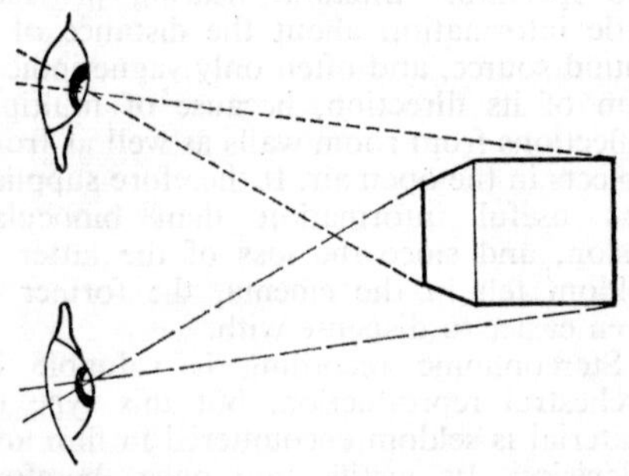

DIFFERENCE BETWEEN VIEWS SEEN BY TWO EYES. Solid object seen in plan presents different aspects to the left and right eye of the spectator. Since the two retinal images are different, the distinction between them must be maintained throughout the stereoscopic transmission system.

of binocular vision, and if the minimum discriminable angle is one minute of arc, it is about 700 ft. This is a normal figure, but trained observers can distinguish down to 20 seconds of arc or even less. Within this range, binocular vision certainly contributes to our sense of the solidity and proper placing of objects, though by no means so much as was formerly supposed.

METHODS. Two completely different systems exist for recording three-dimensional space and later reproducing it at will: holography and stereoscopy. Holography works by storing multiple wave-front impressions on film, and releasing them under suitable conditions to re-create the object in space. At the present time, holography is under active experimental development and its application to film and television has yet to be demonstrated, though it has immense possibilities, for it requires no viewing aids or special screens to project images in three dimensions.

Stereoscopy, on the other hand, necessitates two lens-and-film systems recording separate left- and right-eye images. This separation must be carried right through to the eyes of the ultimate audience in the cinema. Two pictures are projected in overlapped superimposition on the screen, and for every spectator, no matter where he sits, the left-eye picture must alone be visible to his left eye, and the right-eye picture to his right eye. Thus the projection of three-dimensional (3-D) films to large audiences is very much more difficult than the individual viewing of stereo photographs, which requires no more than a small partitioned box equipped with magnifying lenses. Indeed, many people can view 3-D pictures even without that aid.

There are two problems: first to devise a workable camera to take the left- and right-eye pictures, preferably on the same strip of film; second, to project these pictures (again, preferably, from one strip of film) in such a way that image selection is properly carried out. Both these problems are hard to solve, that of projection especially so.

PROJECTION

There are only two places where image separation can be achieved: at the spectators' eyes and at the screen. If at the eyes, there must be as many separators as spectators; if at the screen, one alone will suffice. This gives screen separation a strong inbuilt advantage; but technical problems have long delayed its development.

SEPARATION AT EYES. The earliest proposal for separation at the eyes (invented by D'Almeida in 1858) was to furnish each spectator with a pair of spectacles provided with a rotating or oscillating shutter which alternately obscured the left and right eyes. The spectacles were to be synchronized with a projector which successively projected the left- and right-eye pictures. Noise, complication and electrical hazard put this scheme and its further developments out of court. Its principal rivals, the anaglyphic and polarized light systems, both have the advantage of eliminating all moving parts and making the image selection by purely optical means.

Ducos du Hauron (1891) was the first to suggest the anaglyph method of separating stereo images by printing them in complementary colours and viewing them through reversed complementary-colour filters. In the application of anaglyphs to the cinema, the two images are printed on a strip of film in two complementary colours whose absorption curves overlap as little as possible. They are then viewed on the screen by spectators through filters of the same complementary colours, made up into a pair of spectacles. Thus, if the left-eye image on the film is formed in blue-green, and the right in red, the left eye of each spectator sees the blue-green image through a red filter which absorbs the red image and thus renders it invisible; whereas the blue-green filter over the right eye enables the red image on the screen to be seen, but blocks out the blue-green one.

This system has the merit of cheapness, because the two single-coloured images can easily be formed on a two-layered film. Gelatin viewing spectacles are also inexpensive.

Among the drawbacks of anaglyphic separation are: (a) it can only be applied to black-and-white films, because the colour property is already used up in effecting image separation; (b) good black-and-white quality cannot be achieved, only a muddy and indecisive tone; (c) using the best available dyes, there is still considerable spill-over from one eye to the other (lack of extinction), causing unpleasant ghosting of the stereo image; (d) the fact that each eye sees an image in a different colour causes what is known as retinal rivalry, which in many people produces more or less severe headache after a few minutes, and in some an acute feeling of nausea.

Polarized projection eliminates all the above drawbacks, except that extinction of the unwanted image, though much improved, is by no means perfect. The principle of polarized projection, patented by Anderton in England in 1891, did not become commercially practical until 45 years later, when E. H. Land in the US invented Polaroid, a plastic polarizing sheet material which was comparatively cheap. Normal light is passed through, say, the left-eye image on film, and polarized in a vertical plane. If the resulting polarized image is projected on to a metallic screen which reflects without depolarizing it, it may be seen by a spectator equipped with a vertically-orientated polarizer over his left eye. If he rotates this polarizer, the image gradually darkens until at the horizontal it is virtually extinct.

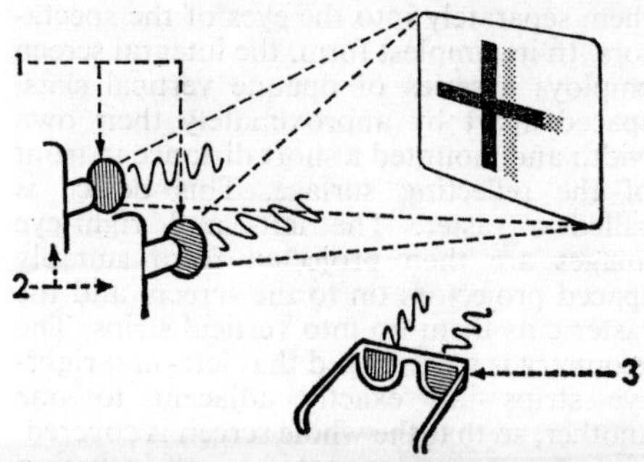

POLARIZED PROJECTION. Projectors, 2, projecting left- and right-eye views, are respectively fitted with vertically and horizontally oriented polarizing filters, 1. The resulting plane-polarized light beams form left- and right-eye images on a screen coated with a surface (usually metallized) which does not depolarize light on reflection. Hence, when viewed through filters, 3, polarized at the same angles as 1, the two images are seen exclusively by the correct eyes.

If, then, the right-eye image is projected through a horizontally-polarized filter, it is seen through a similarly oriented filter placed over the audience's right eyes, while their vertically polarized left-eye filters enable them to see the left-eye image only. Instead of using vertical and horizontal Polaroids, it is customary to adopt a V-orientation, set at 45° to the horizontal in either direction.

The simple system here described entails the use of film images set side by side and separately polarized. A highly ingenious development by Land and Mahler, the Vectograph, made it possible to superimpose the two images which were self-polarized, i.e., created in terms of polarization instead of in dyes or silver halide. The application of the Vectograph to motion picture film, in colour as well as black-and-white, was undertaken by the Polaroid Corporation during the 1950s, but never materialized commercially.

SEPARATION AT SCREEN. The wearing of any type of viewing spectacles is something of a nuisance, and much inventive activity has been devoted to finding a satisfactory means of separating the stereo images at the screen. Apart from the early work of Ives in America and of Gabor in England in the 1940s, virtually all this research and experiment has been carried out in the Soviet Union.

The integral screen, as it is called, is a device for forming the two sets of stereo images and at the same time channelling them separately into the eyes of the spectators. In its simplest form, the integral screen employs a series of opaque vertical slats, spaced apart by approximately their own width and mounted a short distance in front of the reflecting surface. This device is called a raster. The left- and right-eye images are then projected from suitably spaced projectors on to the screen, and the raster cuts them up into vertical strips. The geometry is so arranged that left- and right-eye strips fall exactly adjacent to one another, so that the whole screen is covered. If, now, a spectator seats himself in such a position that, to his left eye, the raster bars hide all the right-eye strips and the raster spaces admit the left-eye strips, while for his right eye the opposite is the case, he will achieve a perfect image separation.

This simple system has numerous drawbacks. If the raster bars are wide, they mask appreciable areas of the image and picture quality is correspondingly poor. If they are narrow (e.g., in the form of fine wires), their exact parallelism over the height of a large screen is impossible to achieve. In any case the luminous efficiency of the screen is low. And, worst of all, only one transverse row of spectators can enjoy perfect image separation, and these must hold their heads absolutely motionless. Sideways movement of the head results in a gradual mixing of the images until a pseudoscopic position is attained, when the left- and right-eye images are reversed.

To remove some of these crippling limitations, a better screen had to be devised. The first step was to replace the linear raster with a radial raster, in which the bars

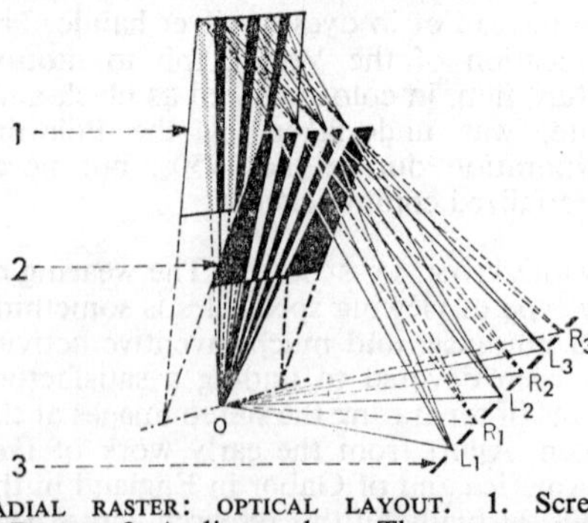

RADIAL RASTER: OPTICAL LAYOUT. 1. Screen. 2. Raster. 3. Audience plane. These are arranged so that prolongations of their planes intersect along a single axis, on which the point, O, is the centre of convergence for radial lines of viewers. L_1R_1, L_2R_2, L_3R_3 are representative pairs of eyes, and are also left and right projection centres for the stereo images.

radiate out fanwise from an imaginary centre situated below the bottom of the screen. By this means it was possible to provide eye separation for spectators in rows of seats which radiated fanwise from the front of the auditorium. However, since the left- and right-eye viewing zones themselves increased in separation from the front towards the back, there was still only one viewing distance which precisely fitted the normal interocular. A further improvement, to overcome the poor luminous efficiency of the solid raster, was to substitute thin lath-like lens sections for the solid slats of the original raster screen. This device has been termed a lenticular raster stereoscreen.

Next, to reduce the effect of sideways head movements producing an increasing fuzziness as spillover occurs from the left- into the right-eye zones, and vice versa, Russian inventors have proposed to project multiple stereopairs taken by cameras having two or more pairs of lenses, so that head movements yield a round-the-corner view of objects exactly as in real life. This, of course, still further complicates both the taking and projection of stereo pictures.

None of these devices has been constructed or demonstrated successfully in Western countries, and it is doubtful whether any of them would provide pictures of a quality sufficient to attract paying audiences at a time when straightforward large-screen projection has attained exceptionally high standards.

THE CAMERA

The problem of devising a 3-D camera to take moving pictures for projection on to very large screens is of an altogether different order of severity from that of a stereo camera for individual viewing. Except for very close subjects, the latter merely consists of a camera body adapted to take two lenses approximately eye-distance apart, and fitted with synchronized focusing, diaphragms and shutters. With some simple provisos about the focal lengths of these lenses and those of the stereo viewer, a perfectly orthostereoscopic picture can be secured—i.e., one whose apparent space geometry is identical with that of the recorded scene. Anyone who has used such a camera and viewer with colour film will have remarked on the uncanny verisimilitude of the picture.

LENS SEPARATION. The need for a picture which can be projected on a very large screen upsets this simple arrangement. When the observer is looking at very distant objects (usually called infinity objects), the optical axes of his eyes are parallel. To reproduce this condition in the cinema, the left and right screen points representing an infinity object must be eye-distance (2½ in.) apart. (Wider separation of these points would entail the audience squinting outwards, which is painful.) Assume that, if the left- and right-eye films were registered by their perforations, these points would coincide.

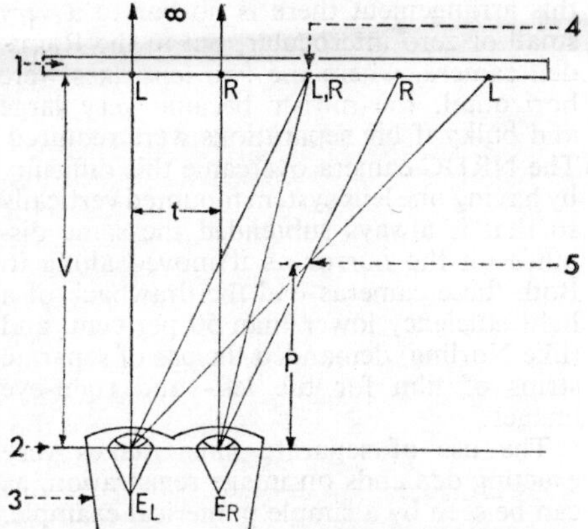

CONSTRUCTION OF IMAGE POINT IN SPACE. E_L, E_R are spectator's eyes in plane, 2, spaced t apart, wearing image separators, 3, and viewing left- and right-eye images projected on screen, 1. If L and R image points have separation, t, on screen, image is seen at infinity (i.e. very distant). (If more than t, eyes are forced to squint outwards and suffer eye-strain.) If L, R coincide, 4, image point is seen on plane of screen, as in a flat film. If L, R are crossed over on screen by distance $-t$, image point, 5, appears half-way from spectator to screen, no matter what his viewing distance, V. Concept of nearness factor, V/P, enables image to be situated in space for all spectators, and in this case is two.

Assume also that the magnification from 35mm film to screen is 500, which corresponds to a screen about 35 ft. wide. If, then, an object 12 ft. from the camera is to be imaged on the screen at the same time as the distant object—a commonplace in "flat" films—it can be shown by simple geometry (assuming 2-in. camera lenses and separation of 2½ in.) that the corresponding points on the screen will be separated by nearly 16 in. These points will be crossed over, i.e., the spectator's left eye will be looking at the right point and vice versa. This causes the object represented to come forward into the cinema space, and a separation of six times the eye spacing will make it reach out five-sixths of the distance from screen to spectator. If he is sitting 60 ft. from the screen, his eyes will be converging on a point only 10 ft. away, but they must remain focused on the screen at 60 ft. in order to see the picture.

So wide a disparity between the planes of focus and convergence would put an impossible strain on many people's eyesight, and could not be tolerated. The remedy is to reduce the separation of the camera lenses, which reduces the parallax, or difference of position of recorded points on the left and right images, when referred to a datum point such as the edge of a perforation. All professional stereo film cameras must therefore make provision for reducing lens separation well below the interocular distance of 2½ in. Many enable the operator to increase the separation above interocular to produce exaggerated or hyperstereoscopic effects when distant objects alone occupy the field.

Only the amateur, projecting on to small screens, can afford to dispense with these adjustments and make do with a simple beamsplitter attachment which fits to the front of his normal camera lens. However, this produces a frame of awkward proportions and very small size. For professional 3-D it is essential to be able to make use of existing screens and to record on image areas at least of standard frame size, since stereo films, with their emphasis on perfect realism, demand exceptionally high image quality.

PRACTICAL CAMERAS. The first 3-D camera to satisfy professional requirements was that of J. A. Norling, employed to make the first 3-D film presented by polarized projection to large audiences (Chrysler exhibit,

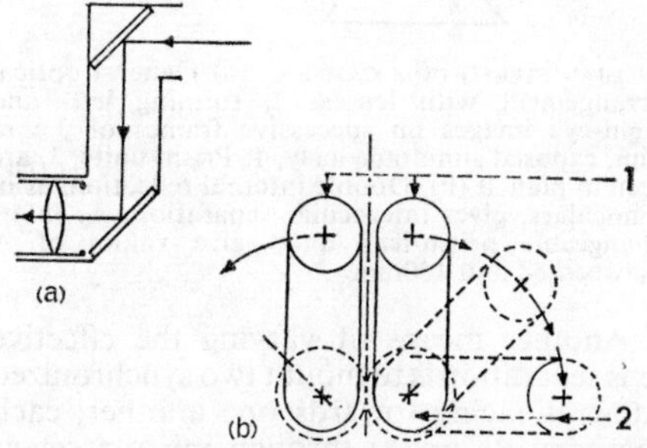

VARIABLE INTEROCULAR: NORLING CAMERA. (a) Periscope tube showing path of rays to camera lens. (b) Front view of camera. 1. Minimum interocular distance. 2. Half maximum interocular, when periscope tubes are swung down into horizontal position. Periscope tube swung down to give maximum interocular separation when other tube, geared to it, is in a similar position.

New York World's Fair, 1939). It used two 35mm films with centrelines only 1½ in. apart, and the lenses operated on the swivelling-periscope system, through prismatic optical tubes, whereby they could be swung closer or farther apart. With 50mm $f\,3{\cdot}5$ lenses, the interocular could be varied from 1½ to 4½ in., and with 80mm lenses 1½ to 7 in.

A comparable camera, but using only a single 35mm film, has been employed in the Soviet Union. This camera, the PSK-5, produces a stereopair on full-size frames arranged one above the other, the pull-down operating over a distance of eight perforations instead of the usual four. The lenses are centred on the frames, but the prismatic tubes are angled so that their viewpoints are at the same level. Five fixed interocular distances are obtained by substituting different prismatic tubes. Both these designs severely restrict the relative aperture of the lenses used.

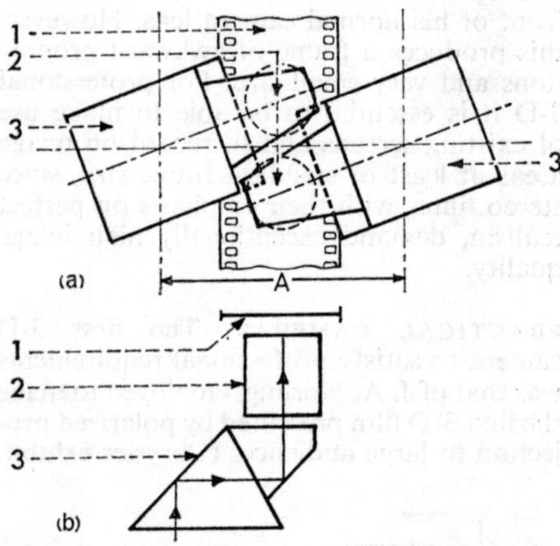

RUSSIAN STEREO CINE CAMERA. (a) General optical arrangement with lenses, 2, forming left- and right-eye images on successive frames of 35mm film, exposed simultaneously, 1. Prism units, 3, are seen in plan at (b). Double internal reflection, as in binoculars, gives interocular separation, A. Interchangeable prism/lens units give values of A between 38 and 130mm.

Another means of varying the effective lens separation is to mount two synchronized cameras facing towards one another, each receiving its image through mirrors set at 45° to its optical axis. If the cameras are made to slide, as on a lathe bed, they may be accurately moved towards and away from one another by a single long lead screw with opposite threads. This design entails the use of separate films.

With these designs it is clear that nothing approaching zero interocular can be achieved, and indeed if modern high-aperture lenses of large diameter can be fitted to them at all, the minimum interocular may be not much less than 2½ in.

SEMIREFLECTOR METHODS. Therefore, in another class of cameras, specially suitable for close-up and scientific work, the two individual units are mounted at right angles to one another, and a semi-reflecting front-surfaced mirror is set between them. With this arrangement there is no bar to a very small or zero interocular, but in the Ramsdell camera, where the two lens axes were horizontal, the mirror became very large and bulky if big separations were required. The NRDC camera overcame this difficulty by having one lens system mounted vertically so that it always subtended the same distance on the mirror as it moved along it. Both these cameras had the drawback of a light efficiency lower than 50 per cent, and (like Norling) demanded the use of separate strips of film for the left- and right-eye images.

The use of separate films makes very exacting demands on image registration, as can be seen by a simple numerical example. Infinity objects, as already explained, are represented on the screen by points 2½ in. apart. Objects seen in the screen plane have zero separation. Thus the parallax

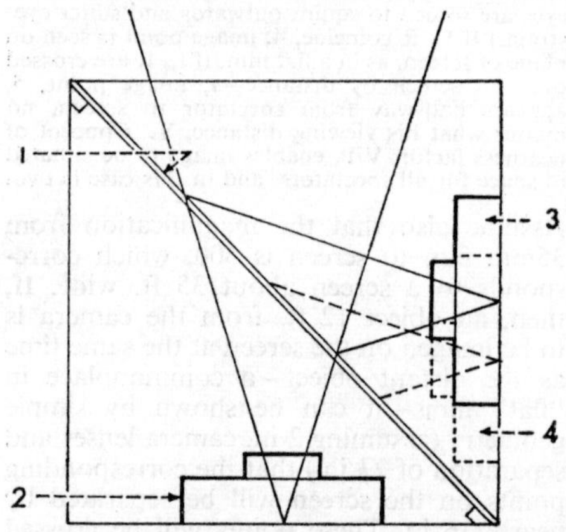

VARIABLE INTEROCULAR: RAMSDELL CAMERA. Two camera units, 2, and 3, 4, are mounted with axes at right-angles, and are separated by a semi-reflecting mirror or pellicle, 1. In this way, interocular separation can be reduced to zero (camera position, 3) where both cameras have same viewpoint. Alternative position, 4, gives a right-eye viewpoint to the camera relative to camera, 2. Continuous variation is possible.

range of all objects except those which are intended to come forward into the cinema space, are only 2½ in. at the screen. Assuming, as before, a film-to-screen magnification of 500, this corresponds to a parallax of only 0.005 in. on the film. And if this screen-to-infinity distance is divided into, say, ten separable planes, each corresponds to a parallax of 0·0005 in., or half one-thousandth of an inch. Two separate film intermittent movements, pulling down and positioning separate strips of film, can scarcely attain this order of accuracy.

CONVERGENCE CONTROL. Apart from the ordinary lens functions of focus and diaphragm setting, which must be coupled in all save the single-lens beamsplitter type of stereo camera, there is one other important stereo adjustment, the control of convergence. If the two camera lens systems point straight ahead (zero convergence), it is possible by suitable angling of the projectors to arrange for infinity points to be the correct 2½ in. apart on the screen. Once this is established, the projectors of course remain fixed. All closer points then take up their proper places in space nearer to the audience.

But, as shown in the above calculation of nearness, this may cause some objects to be imaged much too close for comfort to the audience's eyes. One solution is to reduce the lens separation, but this has disadvantages which are described later. Another is to "push" the scene away by converging the camera lenses. Carried too far, this will cause the infinity points to exceed a separation of 2½ in. on the screen, but if the range of object distances is not too great, the method is effective; the most distant object is simply set at "infinity" by suitable adjustment of the lens convergence.

Angling the lens axes to achieve this would be mechanically unworkable. In the Norling camera, and in almost all subsequent stereo cameras, the same object is achieved by sliding the lenses a short distance towards one another relative to the film and in a plane parallel with it. This decentring produces in effect an inclination of the optical axes. Care has to be taken not to exceed the covering power of the lenses by using more of their outer edges than the designer intended.

The professional stereo camera, therefore, with its separate films, intermittent movements and stereoscopic adjustments, all requiring to be coupled to work as one, is a highly complicated and expensive instrument and must be built with the maximum attainable precision. Even so, owing to limitations which have in part been outlined above, it cannot place before the ultimate audience an accurately recorded picture of three-dimensional space.

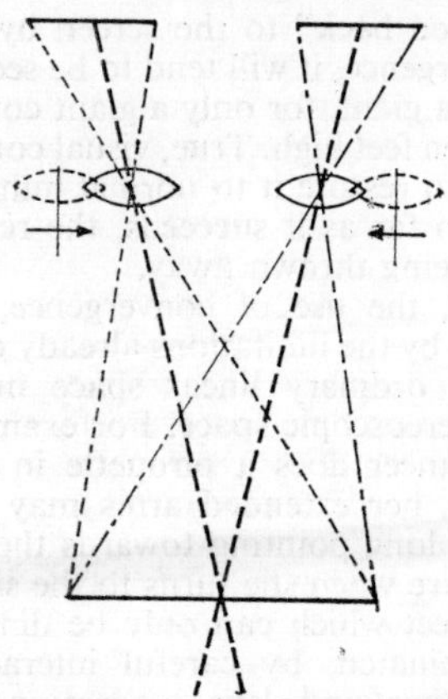

LENS CONVERGENCE. Instead of physically converging the two camera optical systems and films, it is possible, by displacing the lenses inwards, to converge their optical axes, e.g., to the point shown. The same principle is used in animation cameras to make off-centre zoom movements.

STEREO SPACE AND PSYCHOLOGICAL SPACE

The ordinary flat film has the supreme merit of not reproducing real space at all. The third dimension is left entirely to the audience's imagination, and by a universally recognized convention of vision, flat films (like flat photographs and representational paintings) do not resemble wallpaper but are accepted as portraying the living, solid world.

AUDIENCE CREDIBILITY. The better the reproduction of a film, the more clearly is depth rendered. Moreover, the spectator is prepared to make adjustments which are difficult in film 3-D. At the local cinema, a close-up of his favourite film star's face and lips will seem to him palpitatingly close and real; but if this is followed by a long shot of the prairie across which she is being pursued, he will not be a whit disturbed that everything is now small and far away.

Compare this with 3-D. If the heroine's head is to appear in large close-up, it ought to be made to come close to the spectator, even at the cost of eyestrain; if, to improve continuity with the following long shot, it is "pushed back" to the screen by the use of convergence, it will tend to be seen as the head of a giant, for only a giant could have a head ten feet high. True, visual convention will try to restore it to normal human size; but in so far as it succeeds, the realism of 3-D is being thrown away.

Again, the use of convergence, though imposed by the limitations already outlined, converts ordinary linear space into non-linear stereoscopic space. For example, if a ballet dancer does a pirouette in medium close-up, her extended arms may be four times as long pointing towards the camera as they are when she turns to the side. This is an effect which can only be diminished, not eliminated, by careful interaction of convergence and lens separation. For if lens separation is reduced enough to eliminate the need for excess convergence, the tendency to "giantism" greatly increases, and if this is not noticed, the scene simply appears flat.

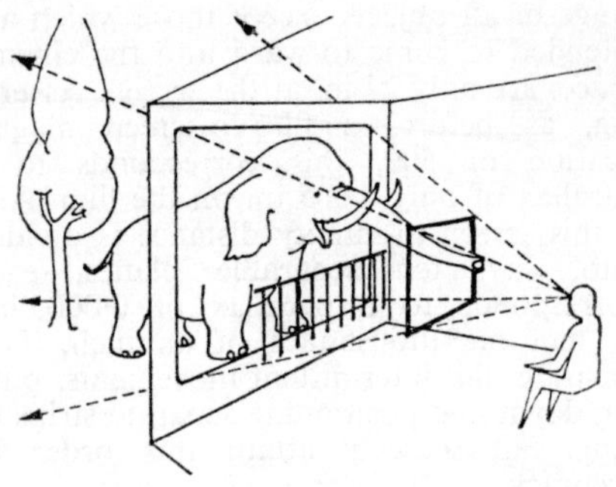

LIMITATIONS OF STEREOSCOPIC SPACE. In absence of stereo window formed between spectator and screen, railings in front of elephant would not be seen in their proper position, since they would seem to be standing in cinema auditorium. Only cantilevered objects like elephant's trunk would truly stand out. Stereo window increases usable volume in front of screen, but at the expense of making the screen seem much smaller, i.e. the size of the window.

FRONT-OF-SCREEN LIMITATIONS. Worst of all, the attempt to use the cinema space for the creation of images in front of the screen runs foul of two difficulties. If ground-supported objects (such as trees and people walking) are simply brought forward, they do not seem to occupy audience space, for the mind boggles at the supposition that they are standing on air. Instead, they merely seem bent back behind the screen. If, on the other hand, a stereo window is formed forward of the screen (which is technically easy to accomplish), this window is seen close to the spectator, say halfway to the screen proper, and looks as if it were the screen. The impression, however, is of a very small, almost television-size screen, and this is a direct negation of all that the cinema at present stands for, with its emphasis on the panoramic screen and on spectacles of gigantic size.

The three-dimensional film does in fact work best (apart from the animated film) for the sensations it was originally designed to project: animals like the giraffe lunging into the audience, the swung cricket bat, the ball hurtling at the spectator's face. To accomplish even these limited effects satisfactorily, technical progress of an improbably high order has to be made in the design of a combined-image projection film such as the Vectograph, or (if spectacles are to be eliminated) of integral screens vastly superior to those realized with all the skill and ingenuity of Soviet designers.

Even if all this were to be done, it is improbable that the resulting three-dimensional picture would provide a sufficiently increased realism or entertainment value to warrant its extra complexity. 3-D films may play a minor role in the scientific world as they continue to do in that of the still photograph, where they can offer perfect realism for a single viewer with relatively simple equipment. Only the further development of holography will show whether an extra measurable dimension justifies itself commercially in terms of added entertainment value. R.J.S.

See: *History* (FILM); *Holography; Visual principles.*

Books: *Theory of Stereoscopic Transmission*, by R. and N. Spottiswoode (Berkeley), 1953; *Stereoscopy*, by N. A. Valyus (London), 1966.

☐**Sticking.** A form of undesirable image retention. If a camera tube with a thin glass target (such as an image orthicon) is exposed to a sharply-focused image for several minutes, the image is retained by the target and takes a finite time to decay. Subjects of high contrast, e.g., black-and-white captions and test cards, quickly cause sticking, and the length of time that the image takes to decay depends on the age of the tube and the operating temperature of the target. It is longest with a tube which has completed several hundred hours, and with a target temperature which is lower than that recommended.

The process involved is not fully understood, but it appears that chemical changes of the target materials occur through electrolysis by the charging currents involved in signal production at the target. The retained image can be made to decay quickly by exposing the tube to a defocused peak white of high brightness (about 25 ft. lamberts), but this in its turn tends to increase the noise level in the image section of the tube for a period.

Sticking should not be confused with lag, which is a very much more short-term form of image-retention. Manufacturers are experimenting with target materials which do not stick, and modern broadcast cameras employing image orthicons are fitted with image-orbiting to prevent the focused image remaining stationary on the target for a long period.

See: *Camera* (TV).

⊕**Stilb.** One candela per square centimetre or □10,000 nits. An ideal diffuser emitting one lumen per square metre has a luminance of one apostilb.

See: *Light units; Screen luminance.*

⊕**Stock.** Motion picture film material, usually before exposure and processing. Also called raw stock.

⊕**Stop Bath.** Solution used in photographic processing to halt the chemical action of the previous stage; a typical example is the use of an acid stop bath following an alkaline developer.

See: *Chemistry of the photographic process; Processing.*

⊕**Stop Frame.** Technique for arresting the motion of a film shot, essentially by printing successive frames from a single frame of action, i.e., using the selected frame as if it were a still picture. If this frame is part of a short series of stationary frames, it is better to print in a cycle A-B-A-B or A-B-C-A-B-C so as to reduce the graininess of the stop motion shot, since grain is randomly distributed. In practice, the cycle most commonly used is A-B-C-B-A-B-C, so that the negative film in the printer head can always be moved to an adjacent frame. Stop frame is sometimes called hold frame or freeze frame.

See: *Special effects* (FILM).

□**Stop Key.** Plunger switch used in lighting consoles originally designed for organ stops.

⊕**Stop, Lens.** Fixed aperture designed to limit □the amount of light passing through a lens. The term is also applied to any specific setting of a movable lens diaphragm, and thus designates the effective speed of a lens which is working at other than its full aperture.

See: *Lens types; Optical principles.*

⊕**Stop Motion.** Device by which the motion of a camera, printer or projector may be stopped automatically after the advance of each frame, and motion resumed, often for another single frame, only by pushing a button or under the control of a timer.

Stop motion is used frequently in macro- and micro-cinematography to speed up very slow movements, e.g., growth of plants. It also forms the basis of an animation technique in which models or cut-outs are moved between the exposure of individual frames. In fact, stop motion is an essential part of all animation cameras.

See: *Time lapse cinematography.*

□**Storage Characteristic.** Manner in which a device such as a light-sensitive mosaic or target holds its charge for a period of time.

See: *Camera* (TV); *Telecine; Telerecording.*

□**Storage Switching.** Device which stores switching information until it is required. When a large number of programme sources have to be switched in and out of a transmission chain, it is often convenient to use a storage type switcher.

In this device, a source is preselected by relays, switching matrices or motor-driven switch banks. Sound as well as vision may (or may not) be selected at the same time, depending upon programme requirements. When needed for transmission, the preselected source is immediately available for selection. Without this preselection sequence, an appreciable delay may occur while the switching mechanism hunts for the required source.

The final switch to the transmission output may be either manual or automatic. Programmed switchers may operate from time signals to start programmes on some predetermined time schedule. A disadvantage of this entirely automatic process, however, is that if a programme under- or over-runs by a few seconds, the overall programme continuity will appear very crude and uneven. Manual operation is therefore usually the preferred method.

See: *Master control.*

□**Store Element.** Element having some storage property, electrical, magnetic, acoustic, etc., used in a computer or comparable device such as an electronic converter.

See: *Standards conversion.*

Storyboard. Sequence of still pictures, usually drawn but sometimes photographed, which outlines the story of a sequence or an entire film by highlighting its key points. Storyboards are invariably used in the early stages in making animation films, and are common in TV and film commercials.

See: *Animation techniques.*

Streaking. Defect seen on a picture when distortion of the low frequency components of the television signal occurs. A streak can appear on the right hand side of fairly large areas, and is usually black after white areas and white after dark areas.

See: *Receiver.*

Strip. Part of the original wide roll in which motion picture film is manufactured and coated. The strip number or slitting number gives the identification of the particular piece of film in its parent roll.

See: *Film manufacture.*

Striping. Applying to a length of photographic film a narrow stripe of magnetic material on which a sound track may be subsequently recorded.

See: *Magnetic recording materials; Sound recording systems.*

Stroboscopic Effects (Strobing). Any fast panning motion over static objects, especially if viewed at high screen brightness levels in the cinema, gives rise to broken and jerky edge effects. These are known as stroboscopic effects, or strobing. Ever since the birth of the cinema, spoked wheels revolving backwards have been a familiar sight on the screen. Similarly, if the camera moves past or pans across railings, the vertical bars are seen to skip and break up

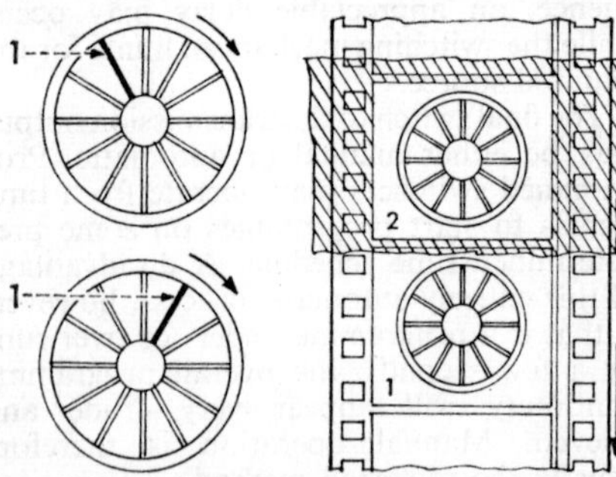

STROBING. First frame to be exposed is the lower one, with spoke, 1, in position shown. By the time the upper frame is in the aperture (shaded), spoke 1 has shifted to another position. If all spokes are alike, the wheel appears stationary. If not, or if spokes have moved less than one complete phase, wheel appears to turn backward. If more than one complete phase, it appears to turn forward, often at incorrect speed.

in a most disturbing fashion. The origin of this effect is extremely complicated, involving both physical and psychological factors.

Basically, the smooth continuous movement of real objects is recorded on film in a series of static exposures, usually 24 or 25 to the second. Of the twenty-fifth of a second thus available, about half is devoted to the exposure. During the other half, the film is obscured from the lens by the shutter while it is pulled down, and a new picture area (frame) positioned for the next exposure. Thus if a piece of action, e.g., one turn of a wheel, occupies one second, only half a second of action is recorded, and this is split into twenty-five successive phases (at 25 fps).

If, when filming a rotating spoked wheel, the spoke pattern is in a certain position during the exposure of frame 1, and happens to be in exactly the same position (1, 2 or more spokes later) at the exposure of frame 2, the wheel appears stationary; if the spoke pattern is slightly retarded at each successive frame, the wheel appears to be turning slowly backwards, and if it is advanced, slowly forwards.

Much the same process occurs with the railings, but in all cases other factors are involved. Some are physical. The normal frame exposure (1/48 or 1/50 sec.) creates appreciable blurring on fast-moving objects. Reducing the shutter angle (if this is possible) lessens the blurring, but also decreases the proportion of the total event actually recorded on film. Thus shutter angle affects strobing. Some factors are psychological, or psycho-physical. Thus strobing due to fast panning over fixed objects appears worse, the higher the screen brightness, and shots which are intolerable when projected in 35mm studio conditions may be quite acceptable with 16mm projection.

Slower panning provides a remedy for these undesirable effects, as does an increase in shooting speed, i.e., in number of frames per second. R.J.S.

See: *Camera* (FILM); *Visual principles.*

Stuart-Blackton, J., 1875–1911. American artist and cartoonist who pioneered animated films. Among his early productions were *The Magic Fountain Pen* (1907) and *Humorous Phases and Funny Faces* (1906).

In *The Haunted Hotel* (1906) he originated the one-turn one-picture technique in which objects are shifted a short space between successive frames, and thus appear to move by themselves. This mystifying trick gave US film makers a useful lead over their French rivals.

⊕STUDIO COMPLEX

The studio can be defined as the workshop for cinematography and sound recording—two operations usually referred to as shooting. In film terms, the studio is called a stage. The word studios, however, is loosely used to include all the buildings on a film company's site, whatever their function, for many services must be provided by technicians and craftsmen before and after shooting.

Each service requires an area in which to carry out its function, and it is these areas and services, together with the studio itself, which comprise the studio complex.

The services in the complex are those concerned with:

1. The preparation of settings—the sets—in which shooting takes place: art direction, construction, scenic painting, set dressing, property and electrical departments.
2. Artistes: casting, dress design, wardrobe and make-up.
3. Shooting, assistant direction, continuity, lighting, camera, sound and still photography.
4. Optical effects, sound effects, model work and other specialized services.
5. Finishing the film: editing, music, re-recording, dubbing and negative cutting.

SET PREPARATION

In terms of area, the services concerned with the preparation of settings make the greatest demand. Planning of this part of the complex is arranged so that work passes from the art director's design through the various stages of construction – machine shop, carpenters, joiners, plaster and/or paint shops – in the order in which work will be carried out until finally the set can be erected on the stage.

Supplies of timber, plaster, paint, nails, screws, door and window furniture – in fact anything required by the craftsmen in the manufacture of the sets – should be easily available but subject to proper control. These requirements demand stores situated near at hand and adjacent to a road on which heavy vehicles can travel when delivering bulk supplies.

DRESSING. Once the set is erected, it is furnished with whatever is appropriate to the type of scene required by the story. This is referred to as dressing the set, and furniture is usually hired as needed. When it is received from the hirer, it is necessary to check it and store it under security arrangements until it is used to dress the set. After use, it is returned to the furniture store for checking and held there until it is returned to the hirer.

Inevitably, a number of smaller items are required before the set is fully dressed: telephones, both modern and ancient, lamps, ornaments, books, bric-a-brac and, in fact, any of the small items that make a room look lived in. These are the hand properties and their scope is such as to require a further store – the property store.

LIGHTING. Artificial light is normally available on each separate stage, the number and type of lamps depending on the policy of the studio-owning company and on the requirements of the lighting cameraman and kind of film stock being used. Colour photography, for example, needs much more illumination than black-and-white.

Reserves of lamp housings, lamps, cables, junction boxes, dimmers, and all kinds of electrical spares and maintenance items are held in an electrical store, and the size and number of these items is such as to need a large building.

The studio site containing many stages may need small electrical stores to service groups of two or three stages each. In such cases, the main electrical store may be remotely situated but its proximity to a main road is essential.

STORAGE. After shooting is complete, the set is struck, i.e., taken to pieces, and removed from the stage. Much of it can be used again. These usable pieces – flats, doors, windows, fireplaces, staircases – must be catalogued, and details of each piece made available to art directors working at the studios. Storage of these pieces requires space and lay-out which makes identification simple, and removal and re-storage easy.

ARTISTES' SERVICES

CASTING AND DRESS DESIGN. The function of the casting department is to advise on the selection and contracting of artistes. An office in which to keep records of artistes and their agents is all that is needed. The dress designer, similarly, needs only office accommodation. Both casting office and dress design can easily be located outside the main studio complex area.

WARDROBE. The wardrobe department receives costumes, dresses, suits and uniforms,

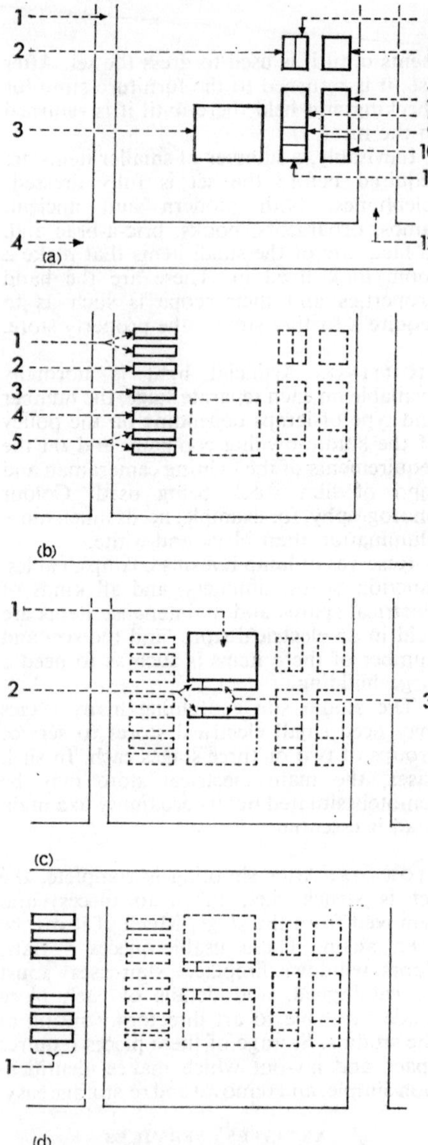

BREAKDOWN OF TYPICAL STUDIO COMPLEX. Access system: 4, public highway, 12, studio road for heavy traffic, 1, studio road for light traffic. (a) Services concerned with preparation of settings. 2. Prop store. 5. Electrical store. 6. Furniture store. 3. Sound stage I (studio). 7. Set store. 8. Assembly, plasterers, painters. 9. Carpenters, joiners. 10. Timber store. 11. Machine shop. (b) Services concerned with artistes. 1. Dressing rooms. 2. Wardrobe. 3. Unit offices. 4. Make-up. 5. Dressing rooms. (c) Services concerned with shooting. 1. Sound stage II. 2. Dark rooms. 3. Camera equipment. (d) Services concerned with finishing the film. 1. Editing suites.

purchased or hired, and issues them to the artistes for whom they are intended. The wardrobe department also holds a stock of such clothes as maids' dresses, police uniforms, legal gowns, clergymen's cassocks, and so on, required by the artistes who appear in front of the cameras to add atmosphere to the scene.

It is the responsibility of the wardrobe staff to store and generally look after the costumes. They may be required to alter costumes according to the demands of the story. The torn dress of the heroine or the bedraggled clothing of the beggar have the appearance of reality as a result of the wardrobe staff's art.

The wardrobe itself needs to be sited so that vehicles can reach it easily, but at the same time near artistes' dressing rooms and the stage itself.

MAKE-UP. Make-up salons have sufficient places, both in males' and females' rooms, for several artistes to be made up at the same time. The basic purpose of applying make-up is to assist the photographer by covering pigment in the skin to avoid shadows which would otherwise appear. But it is much more. The make-up staff are artists able to produce character and age in a face as a result of applying their make-up.

They are expert, too, in making wigs, beards and moustaches, applying them to head or face so that, even in the largest close-ups, artificiality is not discernible. All their work must be completed each day before the artiste can start work.

During the day, it may be necessary to apply a completely new make-up, for instance to make the artiste appear younger or older, and it can happen that shooting is held up until this is completed. So it is convenient and sensible to locate the make-up rooms near to the stage and to dressing rooms.

Make-up is often a long process and artistes have to report very early in the morning.

Comfortable chairs and a clean, inviting atmosphere are most desirable so that the artiste may relax before the tensions of the day's work begin.

SHOOTING

THE STAGE. The stage itself is the centre of the complex. It is soundproof, properly heated in winter, cool in summer, efficiently ventilated, equipped with an adequate electricity supply and wired for sound. It is

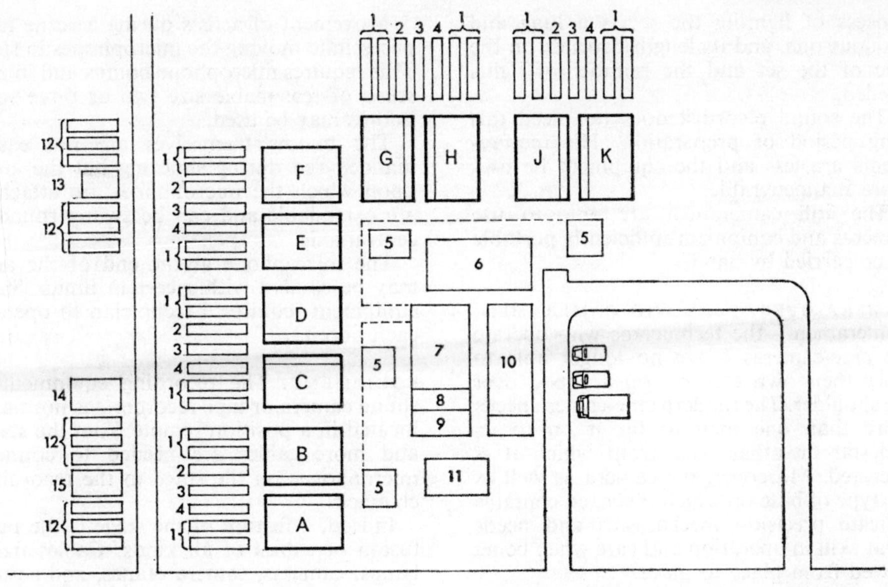

MULTIPLE STUDIO COMPLEX. A-K. Sound stages with auxiliary services appropriately grouped. 1. Dressing rooms. 2. Wardrobe. 3. Unit offices. 4. Make-up. 5. Props and electrical. 6. Main electrical store. 7. Assembly, plasterers, painters. 8. Carpenters, joiners. 9. Machine shop. 10. Set store. 11. Furniture store. 12. Editing suites. 13. Dubbing theatre. 14. Viewing theatre. 15. Recording theatre.

large enough to accommodate sets and high enough to allow lamps for photographic purposes to be hung from above, yet electricians need to be able to reach any lamp easily. The floor must be absolutely level so that cameras can move freely in any direction without vibration. It must be possible to fix sets firmly to the floor.

The occupancy of the stage falls inevitably into three phases. The first phase is the period of set construction during which the set is erected, dressed, lit and made ready for use.

The second phase covers the shooting of the scenes.

The third phase is the reverse of the first. Lamps, furniture and props are removed, the set is struck and the floor cleared for erection of the next set. So the cycle is repeated.

During phases one and three, shooting cannot take place, yet the economics of film-making necessitate continuous shooting once it has commenced. The answer lies in the provision of several stages. A group of two or three may then form the centre of a particular studio complex. While shooting takes place on one stage, striking and set construction can take place on others and all phases of work may continue uninterruptedly.

The services using the stage during actual shooting are:

1. Administrative staff – assistant direction and continuity.
2. Creative technicians – lighting cameramen, sound recordists and still cameramen.
3. Technicians – camera operators, cameramen, sound camera operators and boom operators.

The producer, director and production manager are not included in the complex.

ADMINISTRATIVE STAFF. The administrative staff require office accommodation but little else. Assistant directors act as the channels through which the director's needs are made known. They are responsible also for the organization of the day-to-day shooting on the stage.

Continuity is responsible for keeping a full and detailed record of what happens while artistes are performing each scene.

CREATIVE TECHNICIANS. The lighting cameraman is the man who decides what kind of lights, how many of each kind and in what positions they are to be fixed. The process of lighting the set is a long and arduous one, and its length depends on the size of the set and the number of lights needed.

The sound recordist does not need this long period of preparation. His requirements are less and the equipment he uses more manoeuvrable.

The still cameramen are able to use cameras and equipment sufficiently portable to be carried by hand.

CAMERA OPERATORS AND CAMERAMEN. Cameramen – the technicians who operate the cine-cameras – are no longer able to carry their own cameras on a tripod over the shoulder. The modern cine-camera needs more than one man to lift it on to its pedestal or other base from which it is operated. Moreover, the camera as well as the type of base on which it is fixed contains delicate precision mechanisms and needs great skill in operation and care when being moved from place to place.

The camera operator is able to tilt the camera up or down on its pedestal head or move it from side to side – to pan – on the same head. All these movements must be made with complete smoothness and at any speed from dead slow to very fast.

Movements, however, which entail the camera moving away from, or nearer to, the object or person to be photographed, use one of the types of camera cranes made for the purpose. Such cranes are able to move camera and crew in any direction. The larger cranes may raise them to the height of the set or higher to enable a bird's eye view of the scene to be obtained. All these movements must be possible during the action of any scene.

With a large crane, a small team of crane operators is required, some responsible for movement forward or backward, others responsible for operating the crane arm in a sideways or up and down direction.

Most of this equipment, including the camera itself, is operated electrically. The cables required can themselves become problems when manoeuvrability across the stage floor is necessary. Further technicians are therefore employed to handle them.

SOUND CAMERA AND BOOM OPERATORS. To record dialogue and the sounds which it is intended should be recorded, microphones are placed, on the directions of the sound recordist, in positions where they do not appear in the picture but pick up only what is to be heard.

Movement of artists during a scene may necessitate moving the microphones in step. This requires microphone booms and in any scene of reasonable size two or three such booms may be used.

The booms themselves are not easily manoeuvred during shooting but the arms upon which the microphones are attached are extendable and can be swung round a central axis.

The microphone at the end of the arm may be rotated within certain limits. Such equipment requires a technician to operate each boom.

EQUIPMENT. The recording equipment – sound camera or tape recorder – is normally located in a position remote from the stage and more cables are needed to connect microphones on the stage to the recording channels.

Indeed, a feature of the stage is the profusion of cables of all kinds. Cables from lamps, cameras, camera cranes, and sound cables, hang from set to wall and often appear to litter the floor in utter confusion.

Movement of the camera or the positioning of microphones often necessitates removal of some piece of furniture or part of the set. Specialist craftsmen must be present to carry out such work as required and with the least possible delay.

Cameras, camera cranes, microphones, booms and recording equipment are extremely expensive. They must be constantly cared for and maintained and available at a moment's notice. The studio-owning company therefore arranges for them to be hired to any producer renting facilities and at other times maintains them at the central base.

During the shooting of a film, it is desirable to have a room in which the most valuable equipment can be housed in safety when not in use and where darkroom facilities are also available for camera magazines to be loaded and unloaded. Such accommodation should be a part of the stage building.

All the technicians required during actual shooting of the film are part of the production or film unit. It is not unusual for most of them to work on a free-lance basis rather than to be permanent members of the staff of the studio-owning company.

SPECIALIST SERVICES

MODEL WORK. The essential make-believe of film-making has created a demand for specialists who, for reasons either economic or practical, are called upon to create the illusion of reality within the studio.

SPECIAL EFFECTS. Photographing and recording an intimate scene in a car or railway carriage driving through the streets of a busy city or roaring through the countryside, for example, would be impossible in reality.

The special effects department is called in to deal with this type of scene, which uses, often enough, what is known as rear projection. For this process, the camera crew first go to the location through which car or train is to be seen to travel, and photograph the scene from a moving car or train, taking care to exclude any part of the vehicle.

Then, on the stage, the print of this moving shot is projected on to a translucent screen through which the picture can be seen.

On the side opposite to the projector, the interior of car or railway carriage is built. While artistes play their scene inside the set, the moving shot is projected on to the screen behind, and what is seen through the windows is photographed at the same time.

SOUND EFFECTS. The sound experts are required to produce and record sound which in reality would not be possible or over which the required degree of control could not be exercised.

A scene in a mountain hut with a shrieking wind outside has no sound of storm while it is being photographed and recorded in a set in the stage. It is, in fact, desirable to eliminate from the recording of any scene all such loud and continuous background noises. These are produced by the experts in the special effects department and recorded separately. They are married to the recorded dialogue of the scene later by the re-recording staff who are part of the services employed in the finishing of the film.

The illusion of reality can also be obtained by the use of models. The re-enactment of the Battle of Trafalgar, with full size ships at sea, would be quite impossible. The cost alone would be prohibitive. The model department provides the answer. Here experts build model ships correct in every detail.

The tiny sails react to wind currents and miniature guns fire. In a tank erected in a stage where waves are created artificially and against a background of artificial sky and clouds, which can be made to move if need be, the battle is re-enacted. The ships are set in motion, change direction as required, fire their cannons, hitting some ships and sinking others, and the photographed result is of the battle as seen from a distance.

This kind of model re-enactment calls for specialist camera work too.

These models and special effects service technicians are the wizards of the film-making world and often produce the required effects with commonplace materials. They are meticulous craftsmen and technicians and, though the area allotted to them is small, they need to use any or every service and occupy a stage for as long as it is necessary to build, overcome and shoot their more complicated problems.

FINISHING SERVICES

EDITING. At the end of each day's shooting, the exposed film is rushed to laboratories, developed and printed. These prints – the rushes – are returned to the editing department with all speed.

The sound record will already be there. It is made separately on magnetic tape and requires no development.

The editing suite requires an editing room, assembly room and a store for film – the film vault. The function of the editing department is to receive all rushes, including sound, and select from them sections of film – cuts – which when joined together tell the story smoothly and succinctly and meet the intentions of producer and director.

Proper selection requires technical knowledge, skill, creative ability and patience. The editing suite is preferably located, therefore, away from the hurly-burly and tensions of the stage.

Each day, producer, director and principal members of the production unit see the rushes in a theatre to satisfy themselves that no shot needs to be retaken. The first editing service, therefore, is to prepare the rushes for projection in a theatre. The recorded sound has to match the picture print exactly – be in synchronization – and this work is carried out by the film assembler.

Each separate shot covering the whole day's shooting must be checked. Where the length of sound track does not tally exactly with that of the picture print, the assembler must shorten either tape or film (if the camera or recorder has continued to run after acting ceased) or insert blank track of the required length if a scene has been shot without sound until an exact match is achieved.

The assembler uses a variety of mechanical aids – joiners, rewinders, viewers, sound heads, speakers and measuring machines. Because sound is recorded on a variety of bases he has available a range of equipment sufficient to deal with any of the present methods of sound recording.

The shape of the picture, too, may vary even though a 35mm negative stock is used in each case. Attachments to the viewer in the form of either a mask or lens should be provided or be available to suit each such variation in shape. If 65mm picture film is being used, the whole range of picture editing equipment is duplicated to deal with this wider gauge film.

Each shot is carefully numbered according to its position in the story on both picture and sound track when the scene is shot in the stage. This enables the assembler to join them in sequence. The reels of film so assembled can then be viewed by director and editor who decide which, if any, to discard. When unwanted shots have been removed, the editor takes over and editing begins.

He considers thousands of cuts before he and the director are satisfied that the correct selection and relation of cut to cut has been made. When this has been done, the edited film is made up into appropriate reels, with the cuts joined. Picture and sound are still separate.

This is, however, not the end of the work in the complex. The track contains only dialogue recorded in the stage, and the varying conditions under which this has been done produce unevenness in the recordings which has to be rectified.

There is, as yet, no music unless the film is a musical. Now that the length of the film is known and its shape established, these matters may be dealt with.

MUSIC. This may be an integral part of the film. In the musical, artistes sing or dance, and music is essential to their performance.

But the stage is not the ideal place in which to record music. Sets, lamps, furniture and the size and construction of the stage itself produce sound recording problems. The considerably increased range of sound frequencies produced by an orchestra over those produced by the human voice make these problems so serious that a special recording theatre is necessary.

A theatre is required in which films may be projected and in which an orchestra and solo artistes may perform, a recording room where sound can be recorded and a control room.

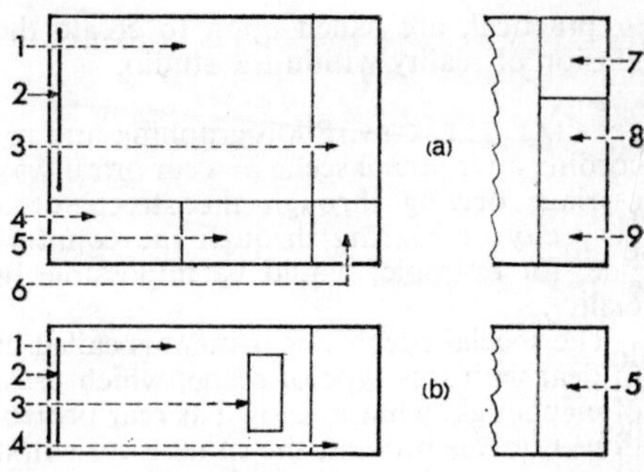

RECORDING AND DUBBING FACILITIES. (a) (*Left*) Recording theatre. 1. Orchestra. 2. Screen. 3. Control room. 4. Soloists. 5. Store for musical instruments. 6. Microphone store. (*Right*) Over. 7. Recording room. 8. Projection room. 9. Rewind room. (b) (*Left*) Dubbing theatre. 1. Theatre. 2. Screen. 3. Console. 4. Recording room. (*Right*) Over. 5. Projection room.

This theatre is constructed with special acoustic properties. Sound baffles hinged to the walls assist the recordist in controlling the liveliness of sounds. It must be large enough to seat a full orchestra, wired for sound and provided with microphone booms and/or mike stands to enable correct positioning of microphones.

An auxiliary room is desirable, attached to the theatre and separated from it by a glass partition or window through which solo artistes may watch the conductor while their performance is recorded separately from that of the orchestra. Such a room allows sounds produced by soloists and orchestra to be properly balanced.

The recording room should be provided with the full range of recording equipment to match whatever type is being used in the stage.

The projection room requires not only projectors but all the necessary ancillary equipment to accommodate the varying dimensions of picture shape and the whole range of sound reproducing apparatus. The theatre can then be used for purposes other than recording, the viewing of rushes for example.

For recording purposes, projector and recording machine must be linked so that they may run together at exactly the same speeds. This ensures that the recorded sound and picture will be in synchronization (in sync).

The control room is the nerve centre of the whole operation. It is here that sounds from a number of sources, from microphones or reproducing machines, are received and amplified, filtered or otherwise treated to achieve the proper balance of sound between them. All are then mixed together and forwarded as a combined sound to the recording room.

All the rooms – theatre, recording room, projection room and control room – must be linked by a communication system, the centre of which is the control room.

Where music is an integral part of the film, it is, whenever possible, recorded before the scenes of which it is a part are shot in the stage. In the recording theatre, soloists and/or orchestra sing and play the music to be used so that it can be recorded in the best possible conditions.

When the scene of which it is a part is to be shot, the recorded music is played back in the stage and artists sing or mime to the recorded voice or dance to the music they hear while they are being photographed.

Where music is being used as a background, it has to be played to the projected film so that conductor and technicians can see the picture as they carry out their work. In this case, projector and recording machine are linked and run at the same speed to ensure that recorded sound will be in sync with the picture.

The sound recordist needs to be more than a good sound technician. He must understand music too. He may find that sound on one particular channel is inadequate and so discusses with the conductor the balance between various sections of the orchestra. This may lead to certain instrumentalists being moved to a position in the recording theatre where a separate microphone can pick up their music more particularly.

He must be able to arrive at his decisions with speed. It is much too expensive to keep a large orchestra waiting while frequent adjustments to, or re-positioning of, microphones are being made.

But it is round the sound recordist that the whole operation revolves. Through his communication system he knows when projectionist, sound cameraman and performers are ready and gives the signal to start each operation.

RE-RECORDING. Literally, this means recording a sound track again using the original recording for the purpose. There are many reasons for doing this. The original shooting is recorded in conditions which vary so much from day to day that uniformity of standards is nearly impossible. This can be put right by re-recording.

Sound which is required to provide background to a scene may already have been recorded on a sound system different from the one the production team is using. So it is re-recorded, and the recording theatre can be used for the purpose.

A large selection of sounds and background noises exist in the sound library but these are usually much longer than are needed in the film so the required length is re-recorded from the original. A great many individual re-recordings of this nature are required to provide a realistic sound background to the film, each one timed to the exact length needed.

The lengths of re-recorded sound are then delivered to the editor to join and make into sets of reels to match the picture reels. The re-recorded sound of the original dialogue makes one set. The music track makes a second. When no background music is wanted on a part of the film, silent track is added for the required length. Lengths of re-recorded background sound matched exactly to the picture make a third set.

There are many occasions when a number of background sounds are needed at the same time. Each one forms part of a separate reel so it is possible to end up with as many as 10 or 12 different sets of reels of re-recorded sound including the re-recorded dialogue and music. All these are re-recorded yet again but now all together. This complicated operation is referred to as dubbing.

Where many films are being made at the same time, the re-recording theatre is used so much that it is desirable to have a separate dubbing theatre. This theatre is in principle the same as the re-recording theatre. It is smaller as no space is needed for orchestra, soloists, or microphones. What is needed is sufficient room for a screen, projection and sound reproducing apparatus, re-recording machines and the control room.

The projector is capable of dealing with different picture formats, and the range of sound reproducing machines cater for all types of track.

The control room needs enough channels on the control panel to accept the number of reels of separate sound and there is a communication system so that the sound recordist can speak to the staff in the projection room and the sound camera operator in the recording room.

All the pieces of equipment are interlocked so that they start together, reach the correct speed together and stay at this constant speed through each operation.

The sound recordist uses his skill to mix the sounds coming in on each channel to achieve a proper balance throughout the whole film. This requires great sensitivity, concentration and dexterity. It is a slow process. Every reel of film must be rewound and threaded again into its respective machine before a further rehearsal or recording can be made.

In this dubbing phase, final re-recorded sound is usually on multi-track magnetic film, but is re-recorded to single optical track for standard theatre release.

MANAGEMENT

The services in the studio complex are provided for the director making the film; his production manager plans the use of them.

PRODUCTION MANAGER. Production management starts with the scenario, and it is upon this that the size and scale of the production of the film is calculated.

The first duty of the production manager is to break down the script. All scenes in which action takes place in the same set are tabulated. Against each scene are listed the characters who are to appear in it and a note of any special dress, costume, hand-prop or make-up.

Film Title: LOOKING GLASS PEOPLE
Set: LOUNGE

Sc.Nº	Character	DressNº	Time	Remarks
4	David Sue Judith	1 1 2	Night	
7 8 9 10	Alex Sue John David Mary	3 2 3 1 1	Morn:	Practical Fire Drinks on Tray
71 72 73 74 75 76 77	Sue Judith	7 3	Late After:	Chessboard

SECTION OF BREAKDOWN. Correlates characters, scene numbers, time of action and props.

When the scenario has been completely broken down in this way the production manager calculates the length of time each scene will take to rehearse and shoot, and the shooting schedule is drawn up. It shows the ideal order for the sets and the length of time required to shoot all scenes in each.

LOOKING GLASS PEOPLE
January

Set		1 M	2 Tu	3 W	4 Th	5 F	6 Sat	7 S	8 M	9 Tu	10 W
Lounge		X	X	X							
Art Gallery					X	X	X		X		
Study										X	X
Character	**Artiste**										
Alex		X	X	X		X			X		X
David		X			X	X	X		X		
Frances		X	X	X							
Sue		X								X	X
Judith					X	X	X				
John		X							X		
Mary					X		X				
Extras					40	12	12				

SECTION OF SHOOTING SCHEDULE. Indicates set and character (artiste) requirements for specific dates, together with numbers of extras needed for crowd scenes.

The shooting schedule is not finalized until the views of the services in the complex have been considered and accommodated as far as is practicable. The casting director may require a different order of sets because a principal artist cannot be available on his first scheduled day. The construction manager may require more time between sets to allow his staff to complete their work.

STUDIO MANAGER. The practicability of the production services fulfilling their functions efficiently and on time depends, however, upon the studio manager, who is responsible for administration. His concern is to keep every member of the staff fully occupied and use all facilities to the maximum.

Although the services are inter-related, the time taken by each to carry out its individual function differs considerably.

Plans of all the sets, details of furniture, property, dress and make-up are sent to each department concerned and each assesses the time they need to make or obtain what is required.

With all this information available, the studio manager allots the individual stage to be used for each set. He can assist a production unit greatly by allotting two adjacent stages for their use. Dressing rooms, wardrobe, make-up and offices as well as camera, sound store and editing suite can then become their permanent base during the making of the film.

When all departmental heads are satisfied they can carry out their functions as required and the studio manager has decided on the stage or stages to be used, the shooting schedule is approved and work begins.

PLANNING FOR CONTINUITY. This planning of services and stages is a constant process, for the aim of a studio-owning company is to continue to let studio space to new production units as shooting on a film finishes. The illness of an artiste calls for an immediate change of plans to ensure continuity of shooting.

More films being made at one time require more services and stages. It would be extravagantly wasteful to repeat the complex layout for each film unit, so some of the services – particularly those concerned with preparation of settings – are made larger and are sited more centrally. Mechanical trucks are then used to transport heavy pieces of equipment, partly constructed sets and furniture with speed and ease.

Facilities for services concerned with artistes and with the shooting of the film are not centralized but repeated. The studio manager allocates the particular suite of dressing rooms with wardrobe accommodation and make-up facilities – the unit suite – related to the stage or stages he has decided are the best ones to fulfil the demands of the shooting schedule.

The editing suites associated with unit suites are then allocated to the finishing services, but at least twice as many editing suites as unit suites are needed.

As new production units come in to take the place of those going out, the complex can be organized to provide for an even flow of work in all departments and the stages can be fully occupied. V.A.P.

See: *Art director; Camera* (FILM); *Cameraman* (FILM); *Director* (FILM); *Editor; Lighting equipment* (FILM); *Producer* (FILM); *Sound mixer; Sound recording* (FILM); *Studio complex* (TV).

STUDIO COMPLEX

A major television network generally has one or more large studio centres, where the bulk of the studio productions are originated, and a number of smaller regional or local stations which rely on the network for most of their programme hours. A studio centre or studio complex is, in effect, a television programme factory, and houses on one site the facilities required to prepare scripts, scenery, furnishings and costumes as well as the studios and their immediate technical equipment.

The studio equipment represents a large capital investment with relatively high depreciation costs, and the organization of a studio centre must use the plant to the best advantage. The production costs are even higher, and the aim must be to get each programme on to the floor as quickly as possible, to avoid delays during rehearsals and transmission, and then to get the studio clear for another production. Unlike a film studio, the television studio is used only for those operations which cannot be done elsewhere.

MAIN DEPARTMENTS

Three main elements come together when a production goes on to the studio floor:

1. The cast, who have studied scripts, been rehearsed, wear costumes and make-up.
2. The scenery, which has been designed, constructed, painted, erected in the studio and dressed with furniture and properties, and which will be seen in association with the graphic work for captions and titles.
3. The technical facilities for lighting, ventilation, controlling cameras, balancing sound and recording the whole programme or transmitting it immediately.

To co-ordinate these elements an administration is required to ensure the efficient working and financial support of the operation.

In addition, the same site may house news and film units and outside broadcast (mobile) equipment, and possibly even the transmitter itself. It is rare, however, for a single site to be large or well placed enough for access and aerial systems to permit it to accommodate all these departments, except in the case of a fairly small station.

The larger the studio complex, the bigger are these departments, and the traffic of scenery, properties, equipment, cast and personnel becomes very heavy. The planning of studio centres is complicated by the need to keep the layout as compact as possible and to achieve the minimum cross-traffic, yet allow for easy expansion or development of the site.

PROGRAMME DEPARTMENT AND CAST. Long before the cast arrive at the studio, the scripts are written, edited and duplicated for distribution. The casting department, in consultation with the producer and director, arrange auditions, and copyright and other contractual matters are settled by the legal experts. Then work starts in the rehearsal

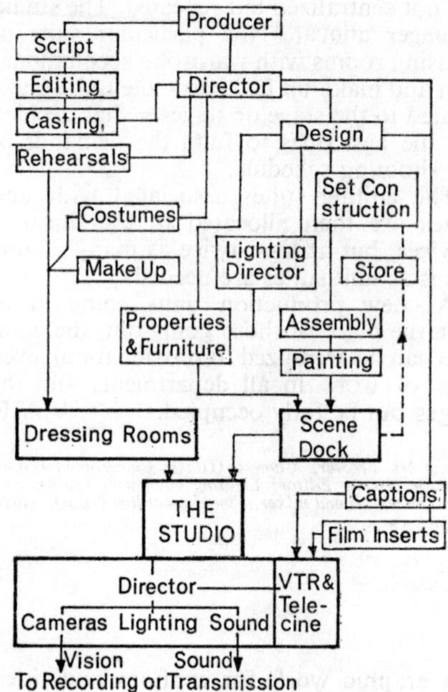

STUDIO INTERRELATIONSHIPS. Simplified diagram of relations between departments and personnel in the production and recording or transmission of a drama programme.

rooms, which must be large enough for the floor plans of each scene to be marked out full size. Scenes can overlap, and generally an area half that of the studio is ample. Furnishings and other bulky properties are required during rehearsals, so access for these must be provided. Telephones, toilets and canteen need all to be fairly accessible.

In the final stages of the production, dressing rooms will be needed, and these can be large enough for two or three artistes, or for a dozen or so, or may be used by a single star. All must have storage for costumes, well-lit mirrors, and ample showers and toilets. Easy passage to the studios is desirable, and the provision of a green room (artistes' foyer) and quick-change rooms near the studio help to meet this need.

The wardrobe department needs storage for costumes (particularly where television serials involve the repeated use of the same costumes) and workshops with fitting rooms where these can be made or altered. Cleaning and pressing equipment is in constant demand.

Make-up rooms are usually very close to the studio and have large mirrors lit by normal tungsten lamps, or colour-matching lamps for colour television.

SCENERY. The set designer works closely with the director and has detailed the scenery by the time the cast are in rehearsal. Drawings are passed to the construction shop where the sets are made up from stock items and specially built details. Ideally, only about 20 per cent of any set is newly built and large storage areas are necessary to hold the stock of flats, backcloths, rostra and other items.

Sets are built up in large sections on a flat floor so that they can be quickly taken down and re-erected in the studio. Large doorways are necessary to studios, stores and construction shops and the passages joining them must be wide and high. Flat trollies are commonly used for transport and a 16-ft. (5 metre) headroom is common.

Flats are constructed of timber frames covered with thin hardboard or canvas, similar to stage scenery, but plastics are becoming more common to add texture and depth. Plaster, as used in film studios, is not greatly used. Backcloths are painted in whatever is the local theatrical tradition – on the wall in Britain, flat on the floor in most Continental European countries. Many studio centres have elaborate sliding frameworks to hold flats and cloths being painted, the artist remaining on a raised platform while the scenery is moved up and down.

The scene stores occupy a large volume since height is essential, and for easy access they should be on the same level as the studios and workshops. Properties occupy as much space as can be allowed, although in metropolitan areas with a theatrical tradition, television studios can usually hire anything they need. Storage, good indexing and ingenious craftsmen are essentials of the prop department. Foliage is usually real, because television productions are seldom on the studio floor for more than two days.

Where outside contractors are employed, good access for large scenery and furniture vans must be provided, involving covered loading bays of ample height and unrestricted passages to studios and scene docks.

TECHNICAL FACILITIES. The equipment and techniques of television are still developing rapidly and take different forms in different countries. A German studio specializing in the telerecording of large-scale operas bears little resemblance to an American network outlet which combines commercial radio with television. The general principles remain valid, but the scale varies and the details of equipment depend on when it was installed.

For each studio (or pair of studios), a suite must be provided from which the director can see all the picture sources and control cast and crew. For small studios, a single room can contain all these facilities but in larger studios, with larger crews, division into separate functions is advisable.

The exact division of responsibilities varies with different organizations, but typically the people involved in the production are: Director; Production Assistant (PA), effectively secretary to the director and primarily responsible for timing the show; Vision Mixer (unless the director is doing this himself); Technical Supervisor (Technical Director in the US) who is the resident crew chief and responsible technically; These, together with Scene Designer, Executive Director (Producer) and other specialist advisers, sit in the main production control room; Camera Control operators (Racks Operators); Sound Balancers and assistants; Lighting Supervisor and assistants.

It is becoming common practice to put together in one room all the people responsible for picture quality – camera control operators, lighting control, and even make-up and costume staff. The combination of increased stability of equipment and improved control of lighting has made possible a rationalization of picture-matching and control techniques, and a simplification of the control panels.

The sound console, tape and disc equipment, etc., is best housed on its own away from other talkback and where the high loudspeaker levels found necessary by sound balancers do not disturb others.

The studio control suite will therefore have production, vision, and sound control rooms, with the addition sometimes of a viewing room (sponsor's booth) where both studio and control rooms can be seen, an announcer's booth (common in the US where commercials are injected into the programme itself) and sometimes a telecine, slide or caption channel for the exclusive use of the studio.

TELECINE. To enable facilities to be shared by several studios and by transmission control, it is common to house telecine, slide (diapositive) and caption machines in a central area. Because of the dust hazard, videotape recording (VTR) machines are generally isolated or mounted in pairs apart from other equipment.

CENTRAL APPARATUS ROOM. The pulse generators, test signal generators and other common signal-handling equipment are housed in a central apparatus room, usually in rows of 19 in. racking. With the increased stability of camera equipment and vision mixers, helped by transistorized, thin film and integrated circuits, it is possible to put more equipment in this central area and have only operational controls in the studio control suites.

PRESENTATION. All the equipment so far described is associated with the studio production. The VTR machine has enabled studio productions to be scheduled quite independently of the transmission timetable.

Thus artistes and studio staff can work more or less regular hours and the studio loading can be planned for optimum use of crews and plant.

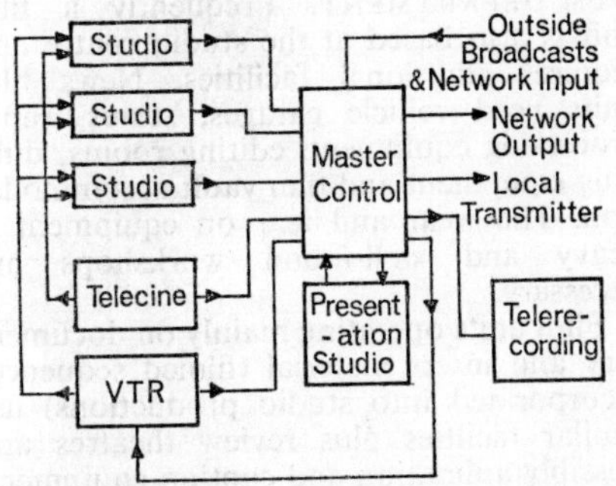

PROGRAMME ROUTING IN THREE-STUDIO CENTRE. Simplified diagram showing relation between picture input sources, master control, and programme outputs.

To control the transmission, maintain good continuity, arrange network switching, cue commercials, etc., a further control suite is necessary. Frequently the engineering functions of transmission control are combined with the general station engineering monitoring and control into Master Control. Presentation is then carried out in a control room, generally with an associated announcer's booth. This area of operations varies widely with the locality and size of the station, and between commercial and state broadcasting organizations.

The larger studio centre is usually as centrally placed as land availability permits, and this is unlikely to be a suitable site for the very tall aerial system that a large transmitter requires. Smaller studio centres are, however, often co-sited with their transmitters, and radio frequency interference with low-power equipment can prove very troublesome. This is particularly so when low-frequency sound transmitters are involved. There is, of course, an economy of monitoring equipment and staff when the transmitter is controlled from the studio master control, but this does not necessarily imply co-siting.

Cable or microwave links to transmitters and incoming and outgoing connections to other parts of the network are required. This may well necessitate a tower or mast with passive reflectors to get sufficient height and is a factor in the siting of studio centres. The control and monitoring of these links requires a great deal of equipment.

Similarly a large number of telephone, programme and control circuits are needed to terminate at the studio centre and the associated equipment is usually bulky.

FILM DEPARTMENT. Frequently a film unit is also based at the studio centre and requires additional facilities. News film units need vehicle garages, stores, rapid processing equipment, editing rooms, dubbing equipment and film vault accommodation. The wear and tear on equipment is heavy and well-found workshops are necessary.

Film units operating mainly on documentary and insert material (filmed sequences incorporated into studio productions) use similar facilties plus review theatres and possibly animation and caption equipment. Frequently, however, these activities are contracted out to specialist companies.

Quite apart from these uses of film, the commercial station needs a large library of advertising films, assembly rooms and an organization for handling the continuous flow of new films. All studio centres have to store a large amount of film for normal telecine use, and vaults are required to store this in conditions of fairly constant temperature and humidity, VTR tape poses a similar storage problem and vaults are frequently combined, and can occupy a surprising amount of space.

OTHER TECHNICAL AREAS. For the day-to-day operation and maintenance of a studio centre, additional technical areas are necessary. These include mechanical workshops where heavy plant such as camera pedestals and cranes can be repaired; electronic maintenance; technical stores; and a lighting workshop, best located close to the power grid; and a camera tube test rig where the quantity of tubes justifies it. Where telerecording (kinescope recording) is undertaken, darkrooms are necessary and usually a small processing plant for test strips, unless a larger installation is available on the site.

OUTSIDE BROADCAST (MOBILE) UNITS. Where the location of the studio centre and the space permits, the OB units are generally housed on the same site. The vehicles – scanners, tenders, microwave link units, utilities and aerial/camera towers – all take up valuable ground-floor area. In addition there are stores of cable, scaffold tubes, tarpaulins, platforms and other heavy equipment such as lighting generators and lights.

The advantage of common siting is mainly the convenience of administration and the ability to share maintenance and stores facilities. In a small centre with two or three vehicles only, the arguments for common siting are of course very strong.

VENTILATION AND AIR CONDITIONING. A very high proportion of the power drawn from the mains is dissipated as heat, particularly in studio lighting, and because of the continuous shooting and long working day of television productions, provision must be made for air circulation to reduce the temperature rise in the studio. The extent to which air cooling plant is necessary depends on the ambient temperature, but in tropical studio centres, air conditioning plant can account for half the building costs.

For control rooms, a flow of cool air directed towards the faces of operational staff improves working conditions and reduces the stresses considerably – enough air changes to remove cigarette smoke are essential. For studios in temperate climates about four changes of air per hour are adequate for image orthicon lighting. To avoid air noise being picked up by microphones, the velocities in ducts and outlets must be kept low, and expert guidance is essential in planning studio air-conditioning installations.

Equipment racks require ventilation and, if the ambient temperature is high, cooling is necessary. The stability of equipment is improved if the operating temperature is kept constant and for this reason alone some cooling may be justified. A mixture of valve and solid state equipment in the same racks needs careful attention to local temperature rise.

MAINS POWER SUPPLIES. A large studio centre draws a heavy electrical load and will almost certainly have its own sub-station with high-voltage feeders. Where possible these should come in duplicate from different parts of the supply network to improve reliability. It is seldom possible to have adequate standby generation plant for more than emergency services. Unlike film studios, television does not use d.c. supplies for lighting in great quantity, although most studios have a small amount of 110 volt d.c. supplies for arc lighting.

PLANNING

SITING AND LAYOUT. Whatever the size of the project, the site must be chosen with regard to its relationship to links or transmitter, its access by public transport and road, and the availability of electrical power and other services. A quiet site is desirable, although aircraft noise is so common in domestic surroundings that the occasional breakthrough into transmissions is rarely noticed.

A flat site is not by any means essential and the skilful use of levels can help to isolate the different forms of traffic. Ample car parking is necessary, and any layout must be capable of expansion with the minimum of disturbance to production. Television buildings, rather like airports, are invariably in the process of alteration or expansion if the service they provide is successful. Again like airports, once opened they cannot easily be withdrawn from service for very long, and so the possibility of expansion must be constantly considered in planning.

Although some well-known centres have been built with studios placed one above the other, this must always be a more expensive form of construction and justified only where the site is restricted. A vertical layout means that large lifts (elevators) must be used to bring scenery to the studios, and vertical air ducts occupy valuable area. Expansion is particularly difficult and noise interference between areas is always a hazard. The RAI-TV studios in Rome provide a very well-planned example of this vertical arrangement, justified in this case by the limited but valuable site.

By contrast, the BBC Television Centre in London has its studios arranged on one level in a circular layout with an extension making the form of a question mark. This results from the shape of the site and is very compact. The circular arrangement of rectangular studios leaves a series of wedge-shaped areas between studios which are rather difficult to use economically and any

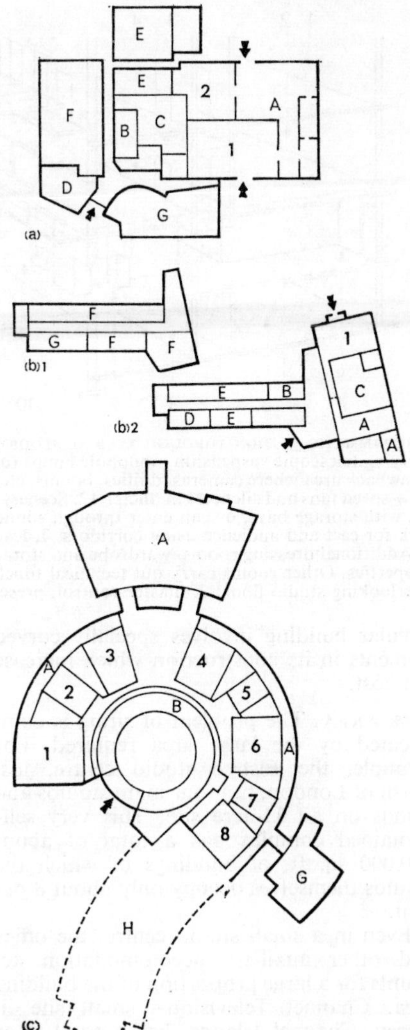

CONTRASTING STUDIO LAYOUTS. A. Scenery and properties. B. Dressing rooms, etc. C. Technical areas. D. Canteen. E. Film department. F. Offices. G. Radio/news department. H. Extension area. Numbers refer to studios. (a) Station KOGO, San Diego, California. One-level layout, except for ventilation plant and viewing rooms. Studio 1 is 3,200 sq. ft., Studio 2 is 2,000 sq. ft. Site allows for unlimited expansion. (b) Channel TV, Jersey, Channel Is. Single-storey studio block, joined to two-storey office block, (b) 1 over (b) 2. Studio 1 is 1,000 sq. ft. Overall shape dictated by site and by need to shield studio from traffic noise. (c) BBC TV Centre, London. Typical floor plan from complex multi-storey layout. Building is a tight fit in a lozenge-shaped site, and groups eight studios (with space for five more) round a circular core. Technical areas extend under central light well; offices occupy upper floors of hollow tower. Studio 1, for scale, is 11,000 sq. ft. in area.

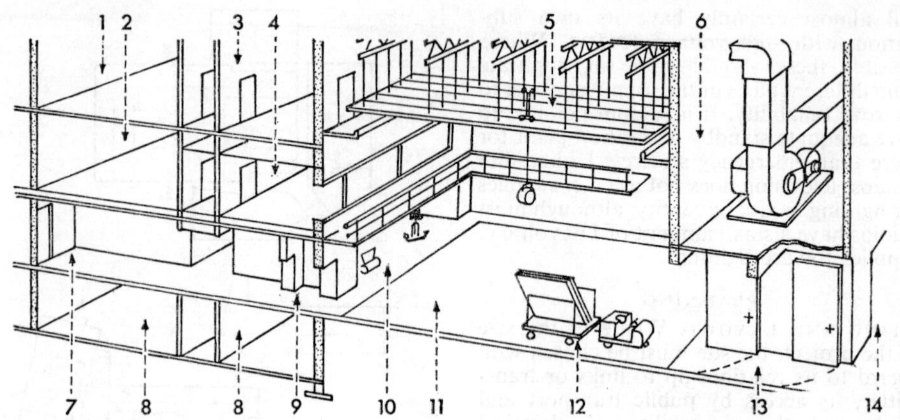

DIAGRAMMATIC SECTION THROUGH TYPICAL STUDIO CENTRE. Studio floor, 11, with overhead lighting grid, 5, carrying telescopic suspension monopole lamps (only one shown). 3. Lighting dimmers and lamp store. 10. Drawback area where cameras, dollies, booms, etc., are stored while studio is reset. 6. Ventilation plant with slow-speed fans and silencers in ducts. 12. Scenery tractor and trolley brings flats from main scenery corridor, 14, with storage bays, or can enter through sound lock, 13, large enough for car or small truck. 9. Sound lock for cast and audience using corridors, 7, leading to make-up, quick-change rooms and artistes' foyer. 8. Additional dressing rooms, wardrobe and storage areas. 1. Rehearsal rooms with access for furniture and properties. Other rooms carry out technical functions: 4, studio control suite with double-glazed windows overlooking studio floor; 2, master control, presentation, telecine and VTR.

circular building involves specially curved elements in its construction which increase the cost.

SITE AREA. The problem of siting is complicated by the large area required. For example, the Elstree studio centre, just north of London, has four main studios and stands on an 11-acre site. This very self-contained complex has a total of about 400,000 sq. ft. of buildings of which the studios themselves occupy only about 8 per cent.

Even in a small studio centre, the office and other auxiliary accommodation accounts for a large proportion of the building area. Channel Television's small site in Jersey, Channel Islands, has a total floor space of about 11,000 sq. ft. with the studio of 1,000 sq. ft. and announcer's booth of 90 sq. ft. accounting for about 10 per cent of this.

SIZE. The VTR machine has made very large studios less valuable, because productions can be built up scene by scene if very big sets are essential, but studios of 8–10,000 sq. ft. are commonly required for drama and light entertainment. About 4–5,000 sq. ft. is a good general-purpose studio area, while about 2–2,500 sq. ft. is as small as practicable for productions involving more than one set.

Presentation and news studios can be very much smaller than this although, to avoid frequent changes of lighting rig and set, they are often made large enough to allow four or five sets to be left standing. This practice is very common in the smaller stations in the US where a two-camera studio of 4,000 sq. ft. is not considered unreasonable. Semi-permanent sets are arranged around such a studio which requires very little labour, while the large volume makes air conditioning much easier.

Announcers' booths are often too small, causing acoustic problems and making ventilation difficult to achieve without noise being picked up by the microphones. A spare is always available; to have only one microphone is to invite breakdown at a critical moment.

PROPORTIONS. A square plan allows little scope for set layout in the studio (just as square rooms are difficult to furnish) and acoustic considerations suggest that studios should have proportions such as three units for height, four units for width and five units for length. Circular and other non-rectangular shapes are often suggested, and the clear sweep of cyclorama that such a shape produces is certainly attractive. The design of lighting suspension would be complicated, however, and the grouping of studios would leave odd shapes to fill between them.

The height depends upon the type of lighting suspension system, as the minimum

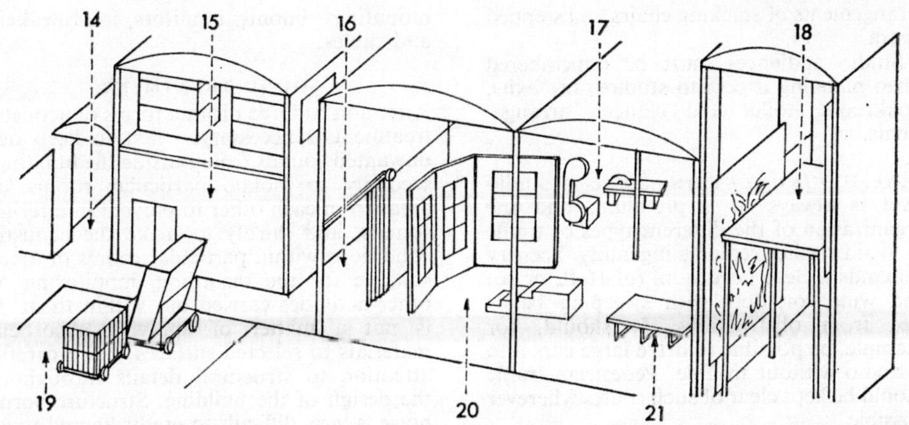

DIAGRAMMATIC SECTION THROUGH TYPICAL STUDIO CENTRE. **14. Main scenery corridor with storage bays, leading to 15, property storage area and workshops. Scenery arrives from pre-assembly area, 20, where complete sections of sets are put together on flat floor, and painting completed ready for removal to studio through tall doors, 16, to pass loaded scenery trailers. Flats are made or modified on wide benches, 21, to suit current productions. Adjacent machine shop, 17, for new set building has timber storage nearby. Scenic artists work on raised platform, 18, raising or lowering flats and backcloths on guides so that the section they are working on is within reach. New sets, with trollies of furniture and properties, are stored in scene dock area, 19, adjacent to studio, and kept till they are needed. Old sets and dressings are cleared from the studio into a similar area, and are then permanently stored or modified for re-use.**

height is set by the space required for lights above scenery. An 8 ft. flat (2·4 metres) can be regarded as minimum backing for a standing person – but 10 ft. or 12 ft. (3·0 to 3·6 metres) is much more desirable. The cyclorama must be as high as possible, and 30 ft. to the grid and 20 ft. cyclorama height are common for a studio of 4–6,000 sq. ft.

FLOORING. The movement of cameras and microphone booms over the studio floor demands a flatness of the order of 1 part in 1,500 to avoid shake on long shots. The surface must be quiet under studio traffic, long-wearing and easily repaired if accidentally damaged. It should be of a neutral colour, unharmed by being repeatedly painted and scrubbed, resistant to oils, water, paint solvents and impact. The joints must be flat and unobtrusive and as impervious as the remainder of the surface.

This is an almost impossible specification to meet, and there are many opinions as to the most satisfactory material. Nowadays some very promising plastic materials are being produced, and rubber has been used with success for years in the US. Industrial flooring compounds such as magnesium oxychloride have been tried, but the corrosive effect of the magnesium chloride would seem to be a great disadvantage. An experimental floor of polyvinyl acetate laid over hard asphalt gave good service but broke up when oil leaked over the surface from a hydraulic camera pedestal.

In Britain, the traditional floor covering is a heavy grade of linoleum laid on hard asphalt in turn laid on carefully levelled concrete screed. The jute backing is removed from the lino before it is laid to prevent water spreading from the joints by the wick effect of the fibres. The lino is allowed to spread for as long as possible before sticking down and all joints are rolled well firm. The asphalt is provided as a plastic layer to even out any irregular drying of the screed, which is carefully laid and may have a final trim with terrazzo grinders to make it flat. It is altogether a long and expensive business and is invariably done by specialist contractors.

AUDIENCE SEATING. Many light entertainment programmes demand an audience for full effect, and special television theatres with fixed, raked audience seating and theatrical stages are generally used by the larger organizations. In addition, smaller audiences are often required in the studio and must be seated so that the layout can be readily varied. The audience becomes, in effect, part of the setting.

Methods used vary from ingenious folding and extending units capable of being run on wheels from the studio, through lightweight bleachers of the type used for temporary seating at outdoor shows, to arrangements of stacking chairs and stepped rostra.

Studio audiences must be remembered when planning access to studios, fire exits, cloakroom, toilet and canteen arrangements.

TRAFFIC CIRCULATION. Space at studio level is always at a premium, and the organization of the different types of traffic to avoid clashes requires ingenuity. Scenery demands a clear headroom (of 16 ft. or so) and wide corridors with sweeping bends free from obstructions. It should, for example, be possible to drive large cars into a studio without trouble. Pedestrian traffic should be kept clear of such routes wherever possible.

The cast need rapid access between studios and dressing rooms, make-up, wardrobe, greenroom, toilets and canteen. They may be wearing cumbersome costumes or so little as to demand freedom from draughts. They also require easily available telephone call-boxes.

Production and technical staff must have quick access to the studio floor, to control rooms and to central areas. Cable routes must be kept short and ideally every studio should be the same distance by cable from the central area. The technique of central switching matrices considerably simplifies signal routing, but pulse timing problems remain.

SERVICE SUPPLIES. Apart from vision, sound and lighting equipment, productions frequently demand additional services. For example, hot and cold water are required for practical sinks and baths, which also need drainage. Rain in considerable quantities may be called for and will be collected in waterproof sheeting and drained or pumped away. Often small swimming pools and fountains are needed, and the flat floor of a television studio does not allow tanks below floor level as in a film studio.

Gas can generally be supplied from portable bottles but utility electrical supplies for cookers, etc., are often in demand. Camera cranes designed for film studio use often require d.c. power and, if no arc supplies are available, a local rectifier is necessary. Compressed air is often useful for paint spraying and portable tools.

During the setting period, it is as well to have cameras and other equipment out of harm, and a draw back area is frequently provided. This does not need to be full height and can be used for spare camera mountings, booms, monitors, loudspeakers and cables.

SOUND TREATMENT

Attention to three distinct forms of acoustic treatment is necessary – first to keep out unwanted sounds from outside the building, secondly to isolate particular rooms or areas from each other to prevent interfering sounds, and thirdly to make the acoustic conditions within particular rooms or areas suitable for the recording, monitoring or other activities carried out within them. It is not a matter of applying absorbent materials to selected surfaces, but a careful attention to structural details throughout the design of the building. Structure-borne noise is very difficult to eradicate and must be prevented by providing resilient joints at suitable points in the building.

External noise can often be minimized by careful siting of unimportant areas as buffers between quiet areas and the noise source. Planting shrubs and making earth embankments, for example, reduces traffic noise considerably. Ground vibration caused by trains or very heavy traffic is extremely difficult to combat and may put a site out of consideration.

Only weight will keep down low-frequency noise, and so consultation between architect and acoustic expert is essential in the early stages of planning. There is a danger of over-engineering acoustic isolation and an appreciation of the sort of attenuations that are really necessary will often save a great deal of cost. The methods of recording and presence of background noise in television studios impose acoustic conditions less stringent than in film studios. A house is seldom as quiet as a cinema auditorium.

STUDIO ACOUSTICS. The presence of scenery has a considerable effect on the reverberation time of a studio and it is the working condition that is important. The floor must be so hard as to provide little low-frequency absorption, the walls are partly or wholly obscured by scenery, so a large proportion of the low-frequency treatment has to be in the roof space. The aim is to provide an approximately uniform reverberation time over the transmitted audio range, free of discontinuities which would cause coloration of the sounds.

Television studios are not made as dead as film studios, nor yet as live as concert halls, and depend to some extent on the type of use. The reverberation time can be related to the total volume of a studio.

Control rooms present rather difficult problems due to the large areas of glass and the banks of picture monitors. Carpets provide valuable absorption and reduce traffic noises. Monitoring and sound control rooms demand particular care and must be given adequate volume if their acoustics are to be satisfactory.

WINDOWS AND SOUND LOCKS. Studios are entered through pairs of sound-proof doors enclosing an area known as a sound lock. A large sound lock volume again makes a high attentuation easier to achieve and relatively lightweight, sound-proof doors are effective if they lead into a large and highly absorbent sound lock. Doors settle and possibly twist slightly with use, so it is important that sealing arrangements are made adjustable or very resilient.

Where adjacent control rooms are listening to the same programme sound on loudspeakers, there is no need for great isolation, and single-glazed windows and partitions giving some 30 dB attentuation are adequate. Between control rooms and the studio, rather better attentuation is necessary, 45 dB or more, so that double-glazing and heavier partitions are required. Where the attenuation must be of the order of 50 dB, it becomes necessary to use isolated walls and very heavy glazing, with window areas reduced as far as possible. Such treatment applies to announcers' booths, quality check rooms and some speech-recording studios used for dubbing. P.R.B.

See: *Camera* (TV); *Cameraman* (TV); *Communications; Designer; Director* (TV); *Lighting equipment* (TV); *Master control; Power supplies; Producer* (TV); *Station timing; Synchronizing pulse generators; Telecine; Videotape recording; Vision mixer.*

Books and Literature: *Planning TV Stations* (RCA), 1969; BBC Monographs 13 and 14; CIE Report E-3.1.9.2 (Washington), 1967; *Television* (*US*), June 1964.

□**Studio Floor Manager.** The controller of crowds, visitors, unwanted technicians and others who may hold up studio work. He is also usually responsible for seeing that the items of technical equipment scheduled for use on the studio floor have actually arrived. Modern studios give him a radio or radio loop communication to the director which can be used while action is proceeding without causing interference.

See: *Studio complex* (TV).

□**Subcarrier.** A signal included in the spectrum of a television signal such that it gives a minimum of interference with other information. It is used to convey some essential discrete part of the total signal, e.g., colour information or sound. In colour television, the subcarrier is a constant frequency which is used as a reference for the colour information and other signals in the total transmission.

The subcarrier signal is modulated by the picture colour information and added to the luminance signal to form the encoded signal for transmission to the receiver.

The frequency of the subcarrier is placed at the high-frequency end of the video spectrum to reduce the visibility of interference patterns. To avoid interference between the colour and luminance signals, the frequency of the subcarrier is made an odd multiple of half the line scan frequency. An effect of this is to cause the subcarrier which appears on the displayed picture to occur in opposite phase during succeeding frames. When viewed by the eye, it results in a great reduction in the visibility of this signal.

See: *Colour principles* (TV); *Colour systems.*

□**Subcarrier Drive.** Sine wave signal at subcarrier reference frequency.

□**Subcarrier Oscillator.** Oscillator which generates the colour subcarrier frequency which is subsequently modulated by colour signals.

See: *Colour principles* (TV); *Colour systems.*

□**Subcarrier Sideband.** The sideband(s) of the subcarrier in a colour transmission.

□**Subrefraction.** A less than normal refraction of a radio beam, by the air through which it passes, caused by temperature gradients in the air.

See: *Transmitters and aerials; Tropospheric propagation.*

⊕**Substandard Film.** Obsolete term for smaller gauges of film, 16mm, 9·5mm, 8mm and Super-8, in contrast to the 35mm standard gauge.

See: *Narrow-gauge film and equipment.*

⊕**Substantive Process.** Colour photographic process in which the colour forming material, or coupler, is incorporated with the emulsion layer at manufacture, rather than provided from the developing solution during processing.

See: *Colour cinematography; Integral tripack systems.*

⊕**Subtitle.** Title superimposed on a film or ☐TV image for the pupose of translating foreign dialogue.

See: *Foreign release.*

⊕**Subtractive Colour Processes.** Methods of colour film reproduction in which transparent dyes are used to remove (subtract) the unwanted portion of the white light spectrum. The subtractive primary colours used are normally respectively red-absorbing (minus-red = cyan), green-absorbing (minus green = magenta), and blue absorbing (minus-blue = yellow). All current colour film processes are basically subtractive.

See: *Colour cinematography; Integral tripack systems.*

⊕**Subtractive Printing.** Process for the printing of colour positive film in which the colour of the exposing light is modulated by interposing colour filters which absorb that part of the white light spectrum which is not required.

See: *Colour printing (additive); Printers.*

⊕**Successive Frame Process.** Method of motion picture colour photography in which three colour separation negative images were recorded on one strip of film by photographing each frame three times successively through blue, red and green filters. The resultant negative was subsequently optically printed by the use of a skip-frame mechanism. The process was naturally restricted to the photography of animated cartoon and puppet subjects in which the movement from frame to frame could be controlled and was rendered obsolete by the introduction of single-strip integral tripack colour negative materials.

⊕**Sulphide Track.** Sound track on colour film. In many colour processes, the silver image produced in the film must be removed to leave the dye image, which is, however, unsuitable for use as a sound track. In some reversal systems, the silver halide of the emulsion may be converted to silver sulphide to provide a sound track image which is not destroyed by the silver bleach stage.

See: *Integral tripack systems.*

⊕**Sun Gun.** Trade name for a compact hand-☐held lamp, using a tungsten-halogen bulb, which is normally worked off batteries to increase mobility, and is also provided with a charger. A limited degree of focusing is possible.

See: *Lighting equipment; Tungsten halogen lamp.*

⊕**Super Eight Millimetre Film.** Most recent form of narrow gauge film for cinematography, introduced in 1965 to overcome some of the limitations of the small frame size of the original standard 8mm. Super-8

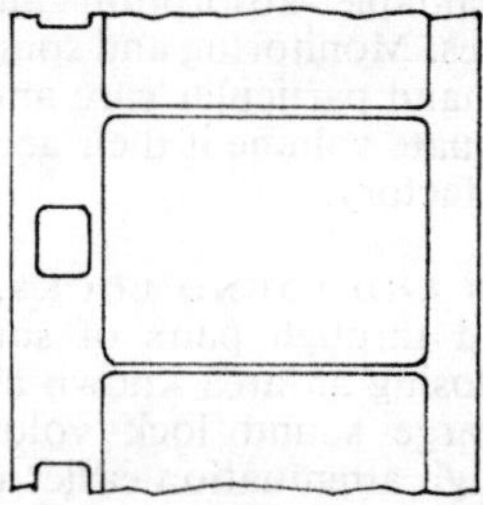

uses a frame size approximately 50 per cent greater in area, having a single row of smaller perforations along one edge and 72 frames in a film length of one foot. As well as being used for amateur photography, Super-8 copies will be made with optical and magnetic sound tracks and are expected to have extensive use in the educational and industrial fields, and even in television.

See: *Film dimensions and physical characteristics; Narrow-gauge film and equipment; Single 8mm film; Standard 8mm film.*

☐**Super Emitron.** British term for the super iconoscope.

☐**Supergroup.** When a large number of telephone channels are to be carried on a broad-band circuit, a group of channels are fed as amplitude modulation on close-spaced low carrier frequencies. A number of these groups shifted in frequency so that their band limits are adjacent form a supergroup.

☐**Superheterodyne.** Receiver using the principle of modulating the desired signal frequency by another generated locally and displaced by a fixed amount. Sum and difference frequencies result; the difference signal, the intermediate frequency, is then amplified and detected. Has the advantage that most of the amplification can be carried out at a fixed frequency, and so with better selectivity, stability and gain.

See: *Receiver.*

⊕**Superimpose.** To photograph or print one ☐image on top of another in such a way that they keep a constant relation to one another. Two shots may be superimposed in this way, as in dream and recollection effects, or a title may be superimposed on a shot. In TV, superimpositions are effected electronically.

See: *Special effects.*

⊕**Superimposed Titles.** Lettering superimposed on pictorial background shots in main and credit titles, or for the translation of foreign dialogue (captions). Captions are usually photographed on a continuous band of

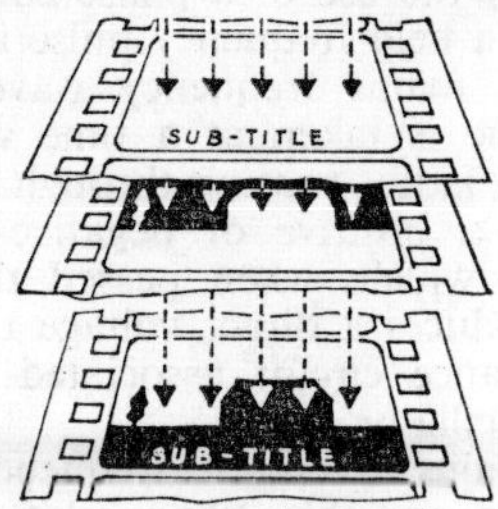

transparent film, which is overlaid on the picture negative and printed simultaneously to make a foreign language release print.

See: *Foreign release.*

⊕**Superscope.** System of producing film copies for wide-screen cinema presentation by anamorphically printing from normally photographed 35mm negative, and then projecting through an anamorphic lens.

See: *Wide-screen processes.*

□**Suppressed Carrier.** Technique of transmitting information with one sideband and no carrier, or vestigial alternative sideband. Requires synchronous detection technique.

See: *Colour principles* (TV); *Synchronous detector; Vestigial sideband.*

⊕**Swish Pan.** Type of panning shot in which the camera is swung very rapidly on its vertical axis, resulting in a blurred sensation when the image is viewed, which is unlike that produced by a corresponding movement of the eyes. Sometimes used as a means of transition between scenes.

See: *Camera* (FILM); *Cameraman* (FILM).

⊕**Synchronization.** In motion picture film handling the establishment of the picture and sound track records in their correct relative position. At the stages of editing and general preparatory work where picture and track are handled as two separate strips, it is usual to mark the material so that a point on the sound track is exactly level with the corresponding frame of the picture, and this is referred to as editing sync, cutting sync or level sync.

However, when the picture and sound tracks are combined on a single strip of film to provide a married print, the point on the sound track must be displaced along the length of the film by a standard distance from the corresponding picture image because of the separated positions of the picture and sound gates in the projector. Picture and track negatives prepared to produce this separation in the print are said to be in printing sync.

See: *Editor; Film dimensions and physical characteristics.*

⊕**Synchronizer.** Device used in the cutting room for maintaining synchronism between two or more lengths of film. It consists of two or more sprockets rigidly mounted on

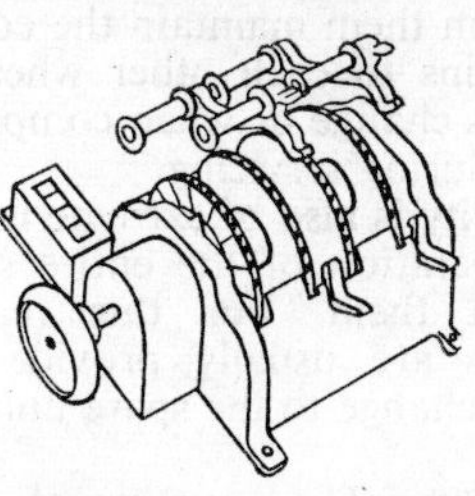

a revolving shaft. The films are placed on the sprockets and accurately positioned by their perforations, so that they can be moved along by rewinds while maintaining a proper synchronous relationship. Synchronizers are often provided with magnetic sound readers.

□SYNCHRONIZING PULSE GENERATORS

These devices provide electrical drive signals to picture-generating sources, and synchronizing signals for addition to the picture signals. Provision can be made to phase-lock the picture frequency to the a.c. power line or, alternatively, to derive the drive pulses from a stable frequency source such as a crystal. Another facility that can be provided is to phase-lock the generator output to some external television signal from, for example, another studio.

MONOCHROME

Monochrome generators normally provide four outputs: line drive, field drive, mixed blanking and mixed synchronizing pulses. In addition to these, generators for the NTSC and PAL systems of colour television must provide a colour subcarrier, the frequency of which is in exact frequency relationship to the line and field frequencies. SECAM generators must produce a field identification signal.

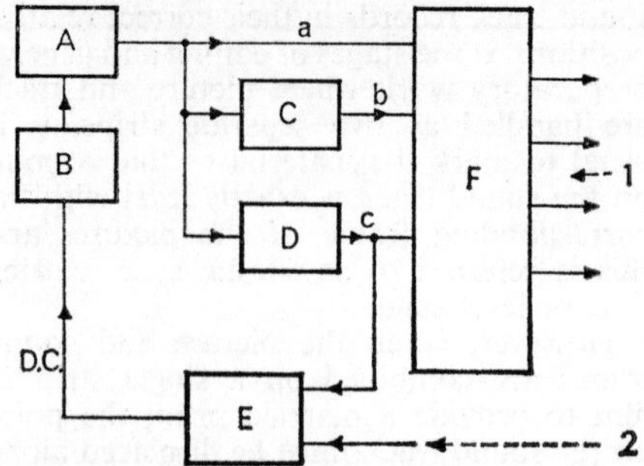

MONOCHROME SYNC PULSE GENERATOR. A. Master oscillator producing, a, twice line frequency. C. Divider by 2 produces line pulses, b. D. Divider by line standard (405, 525, 625 or 819) produces field pulses, c. Phase comparator, E, compares frequency with a.c. supply frequency, 2, and controls oscillator through variable reactance circuit, B. Line and field pulses are fed to a pulse shaper, F, and emerge as output pulses, 1.

Modern synchronizing pulse generators are very complicated pieces of equipment. They are designed so that they need no variable adjustments and so that the output pulses from them maintain the correct time relationships to each other when ambient conditions change or when components are changed during servicing.

Reliability is also of extreme importance, as the operation of the entire station depends on them. For this reason, two generators are usually provided, with a switch to change to the spare unit.

INTERLACE. The system of interlaced scanning universally used requires that the line scan frequency shall be an odd integral multiple of half the field frequency. This relationship is obtained by using a source of twice line frequency and dividing its frequency by two to produce line frequency, and by 405, 525, 625 or 819 according to the standard in use, to produce field frequency.

MASTER OSCILLATOR AND MAINS LOCK. The twice line frequency master oscillator must have an output free from jitter from one cycle to the next, and free from phase modulation. If there is no requirement for the field frequency to be equal to the mains supply frequency, a crystal oscillator can be used. However, if, for some requirement, field frequency has to be equal to the mains supply frequency, a normal tuned oscillator is used together with a variable reactance circuit to vary its frequency. This mains lock is sometimes carried out because the effect of hum displacement and modulation on domestic receivers is far less visible if it is stationary on the picture. The field frequency can be locked to the mains frequency by the use of a phase comparator, in which a field frequency pulse is used to sample a mains frequency waveform. If the sample is taken at a time when this waveform is not passing through a zero of voltages, a positive or negative pulse is produced which, when passed through a filter, produces a direct voltage to control the reactance circuit associated with the master oscillator.

The mains reference frequency is fed through a variable phase shift network because, for some telecines or telerecording machines, it may be necessary to adjust the phase of the pulses with respect to the mains phase.

If the mains supply frequency waveform applied to the phase comparator is a sine wave, the loop gain of the mains lock system varies as the supply frequency varies. To avoid this, a sawtooth or trapezium waveform is used. A trapezium waveform produces a larger output from the phase comparator for the same phase shift.

MAINS LOCK FILTER. If phase modulation of the master oscillator is present, signals corresponding to a vertical line on the object being televised are not exactly equally spaced in time. If the picture is then displayed on a receiver which has flywheel synchronization and hence a stable line timebase, phase modulation shows up as curving of the line. To avoid this, a smoothing circuit of long time constant must be used in the control voltage line to the reactance circuit, in the absence of crystal drive.

MAIN DIVIDER. The number of lines in a television picture has been chosen to have small factors, i.e., 405 = 5×3×3×3×3;

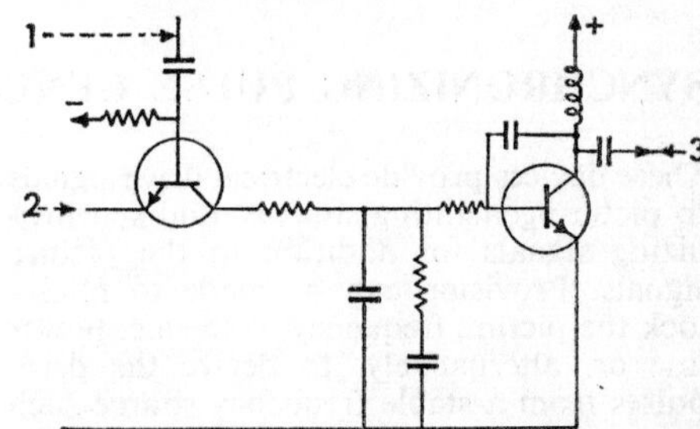

PHASE COMPARATOR. Comparator is followed by filter and reactance circuits, as shown. 1. Field frequency pulse. 2. Supply frequency sawtooth. 3. To oscillator tuned circuit.

525 = 7×5×5×3; 625 = 5×5×5×5; 819 = 13×7×3×3.

With the exception of 13, it is easy to make stable circuits to divide frequencies by these factors. Multivibrators, blocking oscillators and step counters have been used for this purpose. However, modern synchronizing pulse generators generally employ binary counters which are, in fact, logic circuits.

LOGIC CIRCUITS. These are circuits in which diodes, transistors or valves are used purely as switches. Because of this their performance is nearly independent of valve or transistor characteristics.

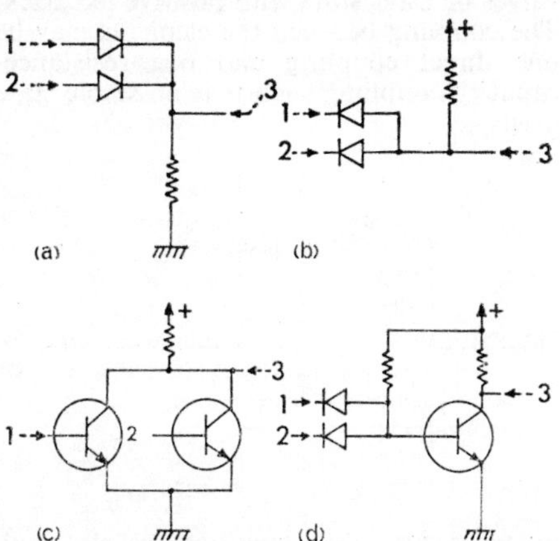

TYPICAL LOGIC CIRCUITS. 1, 2. Inputs. 3. Output. (a) OR circuit. (b) AND circuit. (c) NOR circuit. (d) NAND circuit.

Logic circuits of various types may be used in synchronizing pulse generators. If positive logic is used, a signal corresponds to a positive level and a 0 signal to a level near ground potential. The circuits are by universal agreement given names closely corresponding to their grammatical usage, OR, AND, NOR and NAND (i.e., NOT AND).

An OR gate is a circuit such that the output is positive if any one (or more) of the inputs is positive.

An AND gate is a circuit such that the output is only positive if all the inputs are positive.

A NOR gate is a circuit such that the output is near ground potential if any of the inputs are positive.

A NAND gate is a circuit such that the output is only near ground potential if all the inputs are positive.

A large number of different circuits can be used for these gates. The NOR and NAND gates considered here always contain a valve or transistor to provide a phase reversal.

BISTABLE CIRCUIT. This consists of two NOR gates with the output of each connected to one input of the other. If a positive signal

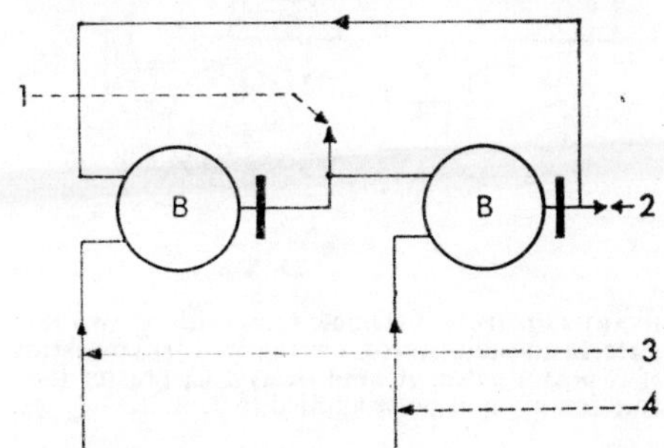

BISTABLE CIRCUIT. B. NOR gates. 1, 2. Outputs. 3. Set input. 4. Reset input.

is applied to one gate, it generates a negative signal at this gate output. This turns off the other gate, thereby holding the first gate in the conducting state when the input signal is removed. The circuit remains in this condition until a positive signal is applied to the second gate.

Opposite phases of the output of this circuit are available at the outputs of the two gate circuits. One stable state of the bistable circuit is called the set state and the other state is called the reset state.

BINARY COUNTER. This is made from the bistable circuit by the addition of a steering network which ensures that a pulse on a single input line is fed only to that gate which is non-conducting. This steering network consists of two AND gates and delays which are greater than the duration of the input pulse. When the input pulse is applied, it can only pass through one AND circuit to the delay circuit; after this delay, the bistable circuit changes state. Although the other AND circuit could then pass an input pulse, the input pulse has ended before this time. The delay and AND gate may be integral parts of the counter, and not obvious components in the circuit.

A binary counter produces one output pulse for two input pulses. If binary counters are to be used for division of frequency by a number which is not a power of two, some modification of the way in which the circuits are coupled must be used. One

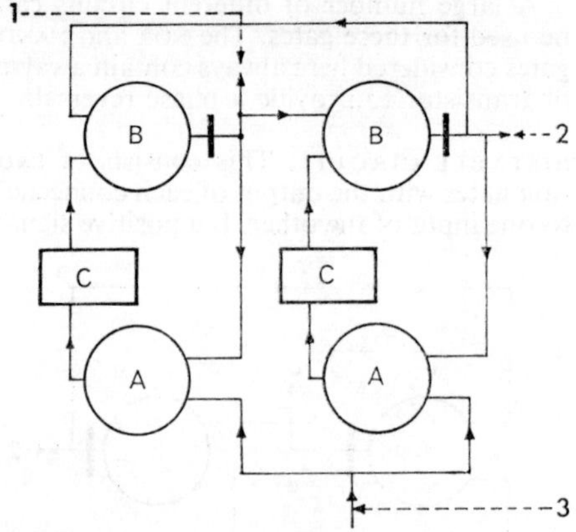

BINARY COUNTER. Bistable circuit with its two NOR gates, B, and addition of steering network consisting of two AND gates, A, and delays, C, greater than duration of input pulse applied to 3. 1, 2. Outputs.

method involves the use of feedback, and another involves the inhibition of an input pulse.

COUNTERS WITH FEEDBACK. Feedback signals can be applied from the output of the counter to earlier stages so that some of the possible stages are passed through without an input pulse occurring.

These signals must be delayed and fed through an OR gate to the binary counter input.

INHIBITION OF AN INPUT PULSE. This involves the use of a bistable circuit which is set by the output pulse of the counter and reset by an input pulse. An output of this bistable circuit is delayed and applied together with the input pulse and an AND gate. The output of the AND gate forms the input

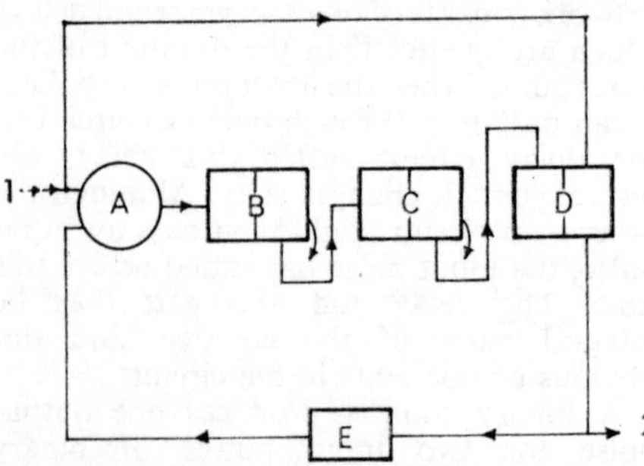

INHIBITED-PULSE BINARY COUNTER. Circuit to count 2^n+1, where binary counters, B, C, count 2^n. Thus counter may be used for counting 5's. Basic circuit requires addition of bistable circuit, D, delay, E, and AND gate, A. 1. Input. 2. Output.

to the binary counter system. When an output pulse occurs, the AND gate is turned on again. If the counters on their own count 2^n, this system will count 2^n+1. This has the advantage of serving to count in 5's, a submultiple of both 525- and 625-line standards. As the bistable circuit is reset by an input pulse, its output waveform has one edge with little delay from the input pulses.

The principles of feedback and inhibition of an input pulse can be combined to give a counter which divides the input frequency by any integral number with little delay from the input pulses.

MULTIVIBRATOR. This consists of two valves or transistors with positive feedback. The coupling between the elements may be one direct coupling and one resistance-capacity coupling, such that no stable state exists.

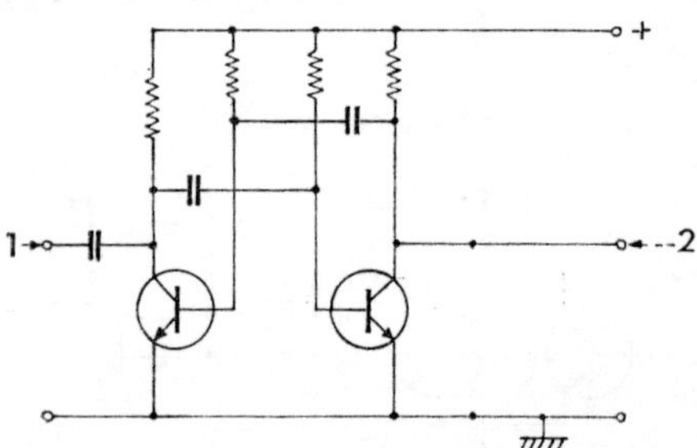

MULTIVIBRATOR. Two transistors coupled with positive feedback, and capable of acting as a relaxation oscillator or as a frequency divider. 1. Input. 2. Output.

Relaxation oscillation occurs at a frequency set by the time-constants of the couplings. This frequency is not very stable, and pulses fed into the circuit at a frequency which is slightly higher than a small multiple of the natural frequency initiate the change-over of the circuit from one quasi-stable state to the other, and so lock the multivibrator. The multivibrator is then acting as a frequency divider.

STEP COUNTER. The step counter uses two diodes and two capacitors. If a square wave of voltage is applied to this circuit, the series capacitor is charged on one half-cycle and connected in series with the output capacitor on the other half-cycle. The voltage on the output capacitor increases by steps, approaching exponentially the voltage of the applied square wave.

If a circuit is added which discharges the

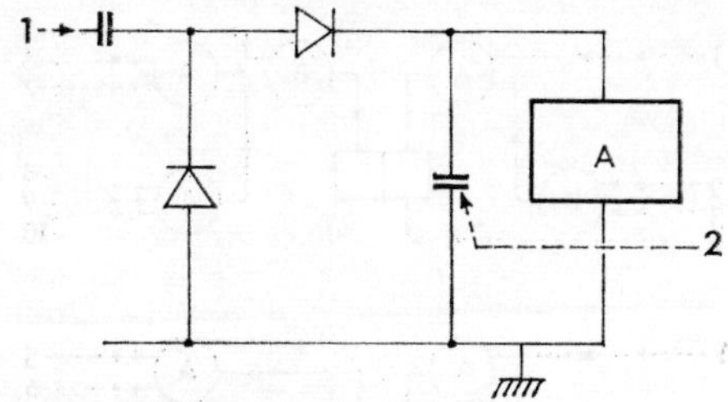

STEP COUNTER. Two diodes, fed by square wave input, 1, chaıge output capacitor, 2, by steps. Discharge circuit, A, discharges 2 when voltage on it exceeds a stated value, so that resulting circuit may be used as a frequency divider.

output capacitor when the voltage on it reaches a certain value, the resulting circuit acts as a frequency divider. A suitable device for discharging the capacitor is the blocking oscillator.

LINEAR STEP COUNTER. An improvement can be made to the step counter by returning the input diode to the output voltage of the counter, using a cathode follower for isolation, so providing steps of nearly equal voltage. A frequency divider using this circuit reliably counts to larger ratios than a divider using the simple circuit.

BLOCKING OSCILLATOR. This is a circuit using a single valve with positive feedback through a transformer. Oscillation is started by decreasing the bias to the point where the loop gain becomes unity. One half-cycle of the oscillation occurs during which the anode and grid currents rise. In the variety of this circuit being considered, either the anode current or grid current charges a capacitor which then prevents further oscillation.

FIELD FREQUENCY TIMING. Pulses of correct length for the equalizing pulse and field pulse intervals and for field-blanking can be generated by monostable circuits, by field frequency pulses, or by counters.

If a monostable circuit is used, the stability of its pulse length can be improved by feeding twice line frequency pulses into the circuit. At a time near to the end of the quasi-stable state, the circuit is more sensitive to these pulses, and one of them can turn the circuit back to the stable state at the exact time that it occurs.

Pulses can be derived from the main twice line frequency to field frequency divider chain at the correct times for the beginning and end of the field intervals. These can then be used to trigger a bistable circuit.

Another method uses a bistable circuit' gate circuit and counter. The field trigger sets the bistable circuit, which turns on the gate and enables twice line frequency pulses

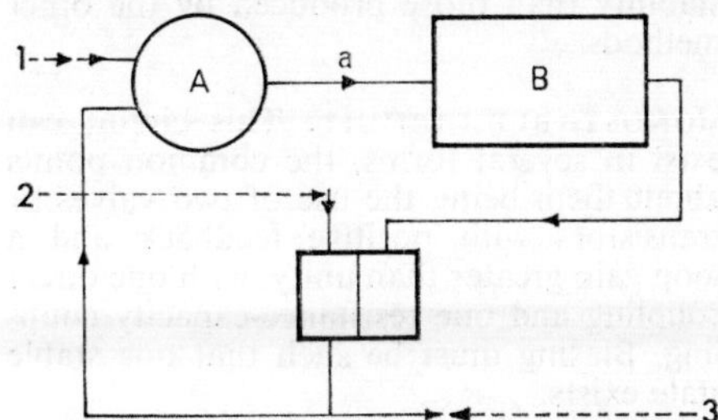

COUNTER FOR FIELD TIMING. Circuit counts up to number of pulses required, and then stops by resetting a bistable circuit and switching an AND gate. 1. Twice-line frequency pulse input. 2. Field trigger input. 3. Output. A. AND gate. B. Counter to *n* pulses. a. Line carrying *n* pulses.

to pass into the counter. As the counter reaches its full count and resets, it resets the bistable circuit, and so turns off the input pulses. The output of the bistable circuit can be used as an output wave-form, or output pulses can be taken from the counter chain.

LINE PULSE TIMING. The leading edges of line-synchronizing pulses, equalizing pulses and frame synchronizing pulses must be exactly spaced in time. If the pulses are to be used for automatic time compensation of a colour vision tape recorder, the accuracy of the timing should be a few nanoseconds. To ensure such accuracy, the leading edges must all be derived from a common source.

One method of doing this uses pseudo line-synchronizing pulses and pseudo equalizing pulses which start earlier than the final pulses required, applying them together with the broad pulses used for field synchronization to an OR gate. These pulses may be generated by monostable circuits. As the broad pulse waveform is positive up to the time when the required leading edges occur, its edges will form the leading edges of the composite waveform.

A second method generates pseudo line-synchronizing pulses, and pseudo field-synchronizing pulses which start later than the final pulses required and applies them together with equalizing pulses to an AND gate. The leading edges of the equalizing pulses then become the leading edges of the composite waveform.

A third method uses gated trigger pulses from the twice line frequency source to set a bistable circuit which is reset at the correct time by other delayed and gated twice line frequency trigger pulses. This method can generate pulse lengths of greater stability than those produced by the other methods.

MONOSTABLE CIRCUIT. This circuit can exist in several forms, the common points about them being the use of two valves or transistors with positive feedback and a loop gain greater than unity, with one direct coupling and one resistance-capacity coupling. Biasing must be such that one stable state exists.

In most of these circuits, a stable state exists in which one of the active elements A is conducting and the other B is not. An input trigger is fed into the circuit in such a way as to make element B conduct. This stops A conducting. The positive feedback keeps the circuit in this quasi-stable state until the charge on the coupling capacitor has changed sufficiently for the element A to start conducting again. When its gain is sufficient for the loop gain to rise to unity, a cumulative action occurs which returns the circuit to its original stable state. The duration of the quasi-stable state is principally set by the time constant of the resistance-capacity coupling.

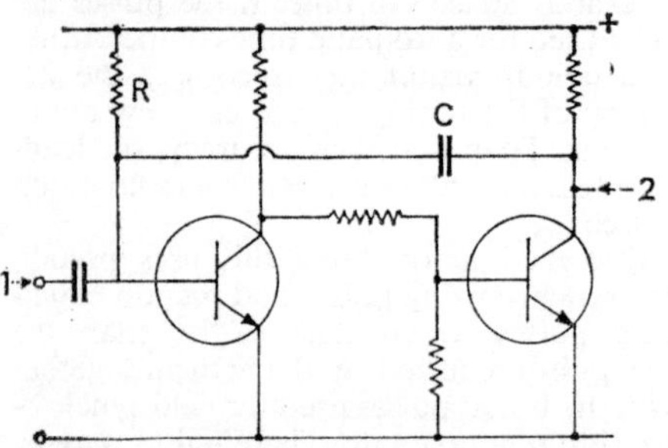

TYPICAL MONOSTABLE CIRCUIT. Two transistors with positive feedback and loop gain greater than unity, one direct and one capacitor coupled. 1. Input trigger. 2. Output pulse.

If PNP and NPN transistors are used, together, the stable state may be a state in which neither element is conducting or both elements are conducting.

FORMATION OF OUTPUT PULSES WITH BISTABLE CIRCUITS. To form the mixed synchronizing signal, the bistable is set by twice line frequency trigger pulses at a time corresponding to the leading edge of the pulse, alternate pulses being gated out

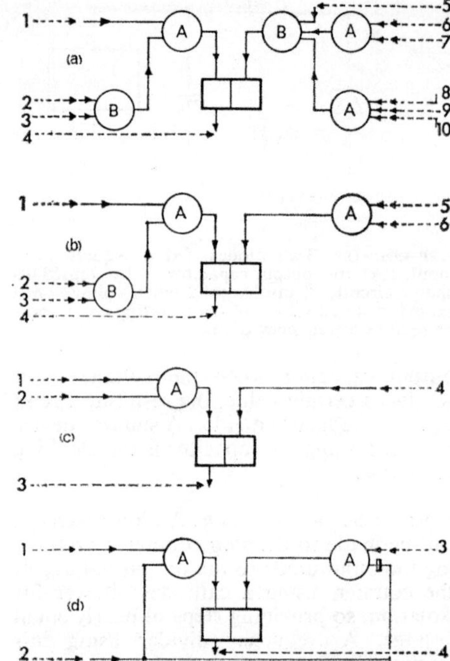

FORMATION OF OUTPUT PULSES. Using bistable circuits (625 lines). A. AND gate. B. OR gate. (a) Output, 4, mixed sync pulses. 1. Twice-line-frequency start. 2. Line frequency. 3. Field frequency, 7½ lines. 5. End of field pulses. 6. End of line pulses. 7. Negative field pulse, 2½ lines. 8. End of equalizing pulses. 9. Field pulse, 7½ lines. 10. Negative field pulse, 2½ lines. (b) Output, 4, mixing blanking pulses. 1. Start blanking. 2. Line frequency. 3. Field frequency, 7½ lines. 5. End blanking. 6. Negative field pulse, 25 lines. (c) Output, 3, line trigger pulses. 1. Start line trigger. 2. Line frequency. 4. End line trigger. (d) Output, 4, field trigger pulses. 1. Start sync. 2. Field pulses, 7½ lines. 3. End sync.

except during the equalizing pulse and field pulse intervals. It is reset by the first pulse to arrive. Trigger pulses for the end of equalizing pulses and the end-of-line synchronzing pulses are gated in at the required times, and pulses for the end of the field synchronizing pulse are always present.

Mixed blanking signals can be formed in another bistable circuit set by twice line frequency pulses at a time corresponding to the beginning of blanking, alternate pulses being gated out except for a few lines at the beginning of field-blanking. It is reset by twice line frequency trigger pulses which are gated out for the field-blanking period.

Line drive and field drive signals can also be formed simply in bistable circuits.

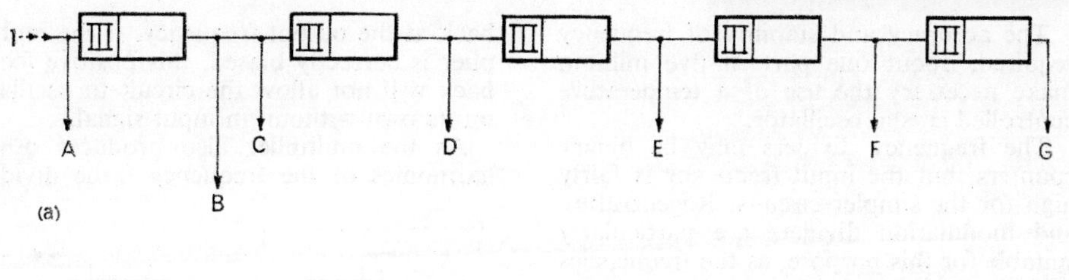

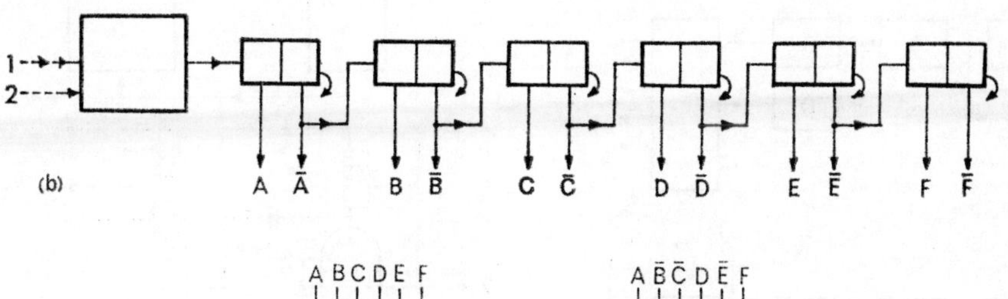

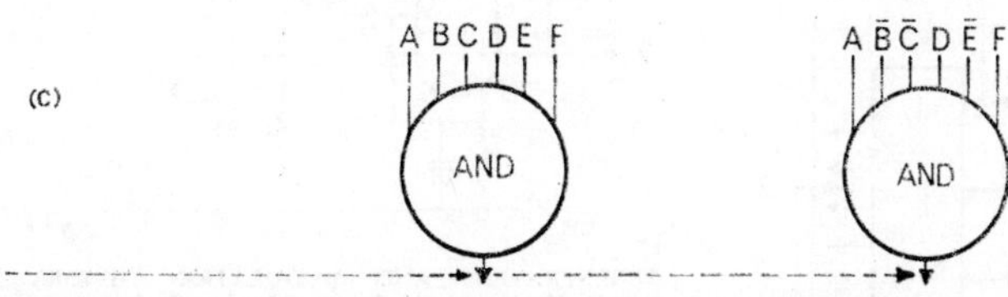

FORMATION OF TRIGGER PULSES. (a) By means of successive delay lines. 1. Twice-line-frequency pulses in. A. End of field sync pulses. B. Start of line trigger. C. Start of blanking. D. Start of line and field sync pulses, and equalizing pulses. E. End of equalizing pulses. F. End of line sync pulses. G. End of blanking. (b) By means of oscillator and counters. 1. Twice-line-frequency pulses in. 2. Keyed oscillator. Ā, B̄, etc. Not-A, not-B, etc. (c) Two of 64 possible trigger outputs, 1.

FORMATION OF TRIGGER PULSES. The trigger pulses required for driving these bistable circuits can be derived from a delay line or from an oscillator together with a counter circuit.

If an oscillator is used, its oscillation must start in the same phase on each line. This can be carried out either by having a master oscillator at this high frequency and dividing its frequency down to twice line frequency, or by having an oscillator which is stopped and restarted twice every line.

CLOCKED BINARY COUNTERS. If coupling between binary stages is carried out through an AND gate, the second input to the AND gate being the counter input pulses, all the changes of state of the counters occur precisely at the time of the input pulse. This is of advantage when the counters are being used with a high input frequency.

If the outputs of stages of this counter are fed into multiple-input AND gates, an output will only be produced when each of the binary stages is in the correct state. By choosing the correct phase of feed from each stage, any delay can be obtained in steps of one cycle of the oscillator frequency.

OUTPUT FILTERS. If the rise times of the generator output waveforms are very short, overshoots may be caused in the equipment. These may be obviated by the use of low pass filters in the output leads.

COLOUR

The main addition to a monochrome synchronizing pulse generator to convert it for the NTSC or PAL colour systems is a stable colour subcarrier oscillator with a divider to provide twice line frequency drive to the monochrome generator.

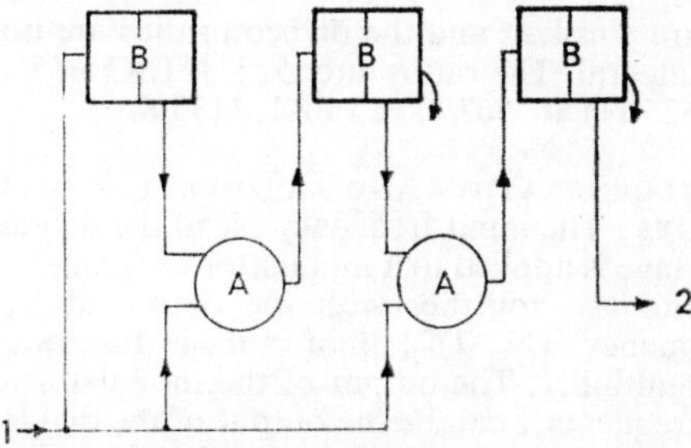

CLOCKED BINARY COUNTERS. 1. Input. 2. Output. A. AND gates. B. Binary counters.

The accuracy and stability of frequency required, about one part in five million, make necessary the use of a temperature controlled crystal oscillator.

The frequency dividers may be binary counters, but the input frequency is fairly high for the simpler circuits. Regeneration and modulation dividers are particularly suitable for this purpose, as the frequencies are constant and the division ratios are not integral. The ratios are 525 NTSC 455/4, 625 NTSC 567/4, 625 PAL 1135/8.

COLOUR SYNC PULSE GENERATORS. (a) NTSC. A. Crystal oscillator generating subcarrier, 1. B. Divider down to twice line frequency, *c*. C. Divider by 525 or 625 to give field pulses, *a*. D. Divider by two for line pulses, *b*. All pulses to pulse shaper, E, and output, 2. (b) PAL. A. Crystal oscillator. F. Frequency changer generating subcarrier frequency, 1. B. Divider down to twice line frequency, *c*. C. Divider by 625 to give field pulses, *a*. D. Divider by two for line pulses, *b*. All pulses to pulse shaper, E, and output, 2. Field pulses feed to divider by two, G, to produce field identification pulses, 3. *d*. 25 Hz sine wave.

REGENERATION AND MODULATION DIVIDER. The input frequency Nf to the divider stage is applied to a modulator or frequency changer, together with the output at frequency (N−1)f of a tuned frequency multiplier. The output of the modulator at frequency f can be the output of the divider and is also the input to the frequency multiplier. This divider will not start to operate unless there is some positive feedback at the output frequency. If the multiplier is correctly biased, this positive feedback will not allow the circuit to oscillate on its own without an input signal.

As the multiplier also produces other harmonics of the frequency f the divider can be used to divide the frequency by a non-integral rational number.

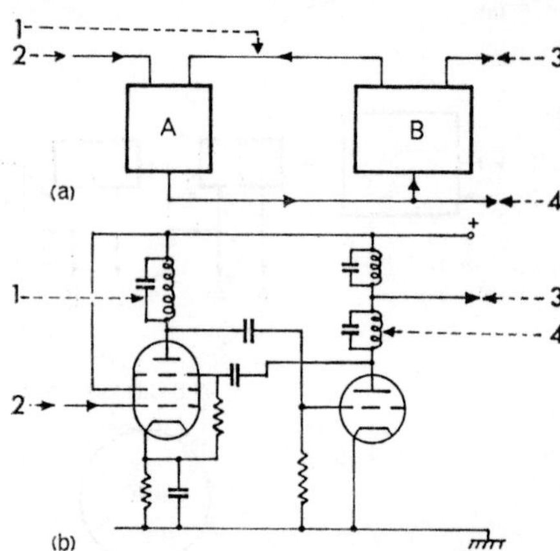

REGENERATION AND MODULATION DIVIDER. (a) Input, 2, to modulator, A, is at frequency Nf, together with input, 1, at frequency (N—1)f, from tuned frequency multiplier, B, of multiplying factor, M. 3. Frequency Mf, if required. 4. Frequency f. (b) Circuit details. 2. Frequency Nf applied to signal grid of pentode. 1. Anode (plate) circuit of pentode tuned to frequency f. 4. Anode circuit of triode tuned to frequency (N—1)f. 3. Output at frequency Mf.

The output signal will be accurately locked in to the input phase signal.

FIELD IDENTIFICATION SIGNAL. Various standards have been proposed for the PAL and SECAM systems. All of them involve a 25 Hz signal being fed from the synchronizing pulse generator to each colour coder.

In both these systems, the transmitted signal is different on alternate lines. The switching can be controlled by a binary counter dividing the line frequency by two. In the SECAM system for example, to prevent one camera coder transmitting a red chrominance signal while another camera coder transmits a blue chrominance signal at the same time, these binary counters must be set to the same phase by the 25 Hz signals.

Identification pulses during the field-blanking interval also need this 25 Hz signal for their generation. J.H.B.

See: *Camera* (TV); *Master control; Special events; Station timing.*

Book: *Pulse, Digital and Switching Waveforms*, by J. Millman and H. Taub (New York), 1965.

□**Synchronizing Pulse Regeneration.** Elimination of distortion of synchronizing pulses by clipping or with a bistable circuit. This is possible because the useful information is carried only in the leading and trailing edges of the pulses.

See: *Synchronizing pulse generators; Vision mixer.*

□**Synchronizing Pulses.** The parts of the television waveform that control the line and field repetition rates of the studio equipment and the receiver. In order that the scanning circuits in the receiver keep step with those at the transmitter, synchronizing pulses are included with the transmitted video signal. They comprise a system of constant amplitude, rectangular pulses operating negatively below picture black level. A line-synchronizing pulse occurs at the end of each line period and initiates line flyback in the receiver from its leading edge. Similarly, field synchronizing pulses, inserted at the end of each field period, initiate field flyback.

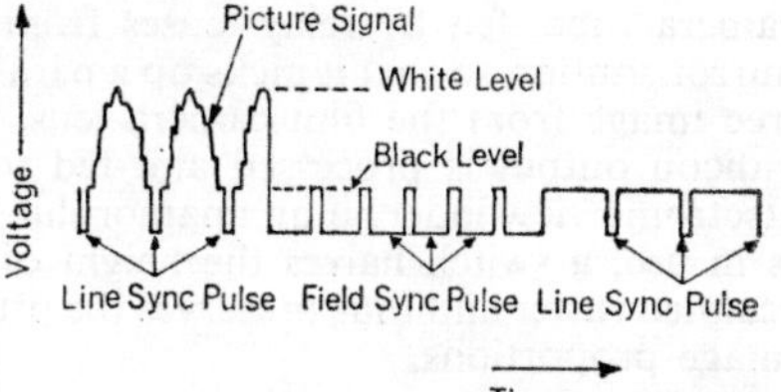

LINE SYNC AND FIELD SYNC PULSES. Schematic representation showing basic elements of sync pulses. The picture signal is shown as positive-going (i.e., upwards) with respect to black level, but may in practice be negative-going. The line sync pulses occupy a small but significant fraction (about 10 per cent) of the useful line period, and the field sync pulses occupy some 15–20 complete line periods. Line sync is preserved during field flyback by the use of broad pulses occurring at twice line frequency.

To enable the receiver to distinguish between line and field sync pulses their duration is made significantly different. Line sync pulses are relatively short and occupy about 10 per cent of a line period but field sync pulses have a duration of several complete line periods. However, it is important that line synchronism be maintained continually, even during field flyback, otherwise some lines at the top of the raster may be horizontally displaced from their correct position. For this reason, the field sync pulse is divided to form a chain of broad pulses occuring at twice line frequency. The leading edges of these broad pulses are arranged to occupy the same relative positions that would have been occupied by the leading edges of line sync pulses had these latter not been interrupted, and therefore serve in their absence to preserve line synchronism.

It might at first appear necessary only to divide the field sync pulse into line frequency components and not twice line frequency but in that event, since interlacing demands that alternate fields must finish at the end of a line and at the mid point of a line respectively, the pattern of broad pulses after odd fields would have to be different from that following even fields. By dividing at twice line frequency an identically similar field sync signal can be used in each case. It should be noted that receiver line scan circuits are only sensitive to every alternate leading edge as appropriate; the remainder which occur in intermediate positions are ignored.

See: *Television principles.*

□**Synchronizing Separator.** A circuit in a television receiver to separate picture and synchronizing information. The composite television signal contains both picture information and synchronizing information. To ensure accurate reproduction of the picture, the synchronizing pulses must be removed or separated from the composite whole.

The sync separator is basically a clipper circuit which removes all the information above a certain level, leaving the synchronizing pulses at its output. These pulses contain information for synchronizing both the line and frame oscillators, and it is therefore necessary to split the pulses into line pulses and field pulses.

The field information is contained in a block of television line signals, and is similar to the line sync pulse but much longer in time. To maintain line synchronism during this period, the field pulse is interrupted at half-line intervals to inject line pulses in the opposite polarity.

If the signal is differentiated by a short time constant circuit, a series of pulses at line frequency are obtained, containing extra half-line frequency pulses during the field pulse interval. If a long time constant is used, a large pulse is obtained from the field information with only slight line information present; the latter can be removed leaving a field pulse to trigger the field time-base oscillator.

The half-line pulses are necessary to maintain interlace.

With the system in use, field triggering occurs alternately half way along a line.

See: *Receiver; Synchronizing pulse generators; Vision mixer.*

□**Synchronous Detector.** When suppressed carrier transmission techniques are used, the signal information can be deciphered in the receiver only if the suppressed carrier can be restored in phase. The technique can be extended to two signals in quadrature as in the NTSC and PAL colour systems.

See: *Colour principles* (TV); *Suppressed carrier; Vestigial sideband.*

⊕**Synchronous Speed.** Identical speed of camera, sound reproducer and projector required to ensure that the film action keeps its proper tempo and the sound its proper pitch.

Varies according to the gauge of film used. Dialogue recording cannot begin until sound and picture machines are both up to sync speed.

See: *Sound recording* (FILM).

⊕□**Synchrostart.** Trade name of a device for ensuring the synchronous run-up of three-phase synchronous drive motors, e.g., those used for a projector and dubber system, which would otherwise reach synchronism at slightly different times on account of unequal load, etc., thus upsetting the precise relationship of sound and picture tracks.

The motors are fed through a Synchrostart generator which gradually increases the supply frequency to that of the line (50 or 60 Hz), thus enabling the drive motors to keep in step until they reach synchronous speed.

See: *Pulse sync; Selsyn motor.*

⊕**Sync Mark.** Synchronizing mark or marks used by film editors to fix the relationship between picture and sound tracks on which dialogue and sound effects may be separately recorded.

These marks, which are often made

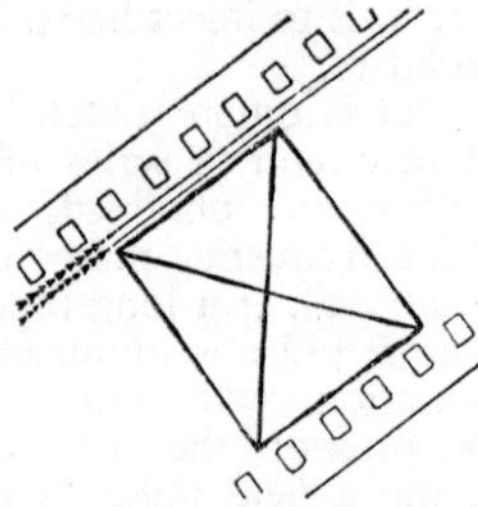

SYNC MARK. Temporary synchronizing mark (aligned on picture and sound tracks) made by sound editors when cutting and track-laying. They are often numbered to prevent alignment of wrong marks.

with wax pencils on the film surface, are essential to maintain synchronism during the fluid process of editing, when both the picture and a multiplicity of tracks are undergoing frequent change. They are easily erased and remade as the film is moved backwards and forwards through the synchronizer.

Frames referred to as sync marks are also printed in the head-end leaders of release prints to assist the projectionist in threading-up his machine with the film in the correct relation between the picture and sound gates.

See: *Editor; Release print make-up.*

□**System Blanking.** Combination of horizontal and vertical blanking which is added to a television signal to ensure that no picture information is present during the synchronizing and retrace period of a receiver.

See: *Blanking pulse; Television principles.*

⊕□**System 35.** A combined film and television camera system first demonstrated in 1965, and based by the manufacturers on the Mitchell Mark II 35mm reflex camera. To the sound blimp door is attached a vidicon camera tube, fed by relay lenses from the mirror shutter, so that it picks up a parallax-free image from the film camera lens. The vidicon output is processed and fed to an electronic viewfinder; if an anamorphic lens is in use, a switch halves the height of the scanned raster and thus preserves the proper image proportions.

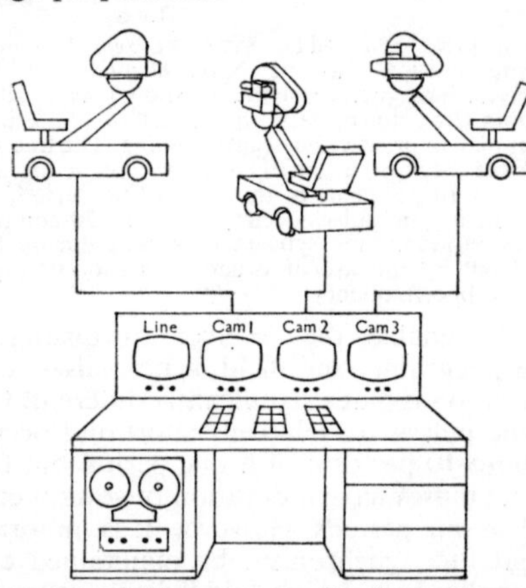

SYSTEM 35. Multi-camera use for direct film production. TV outputs are fed to a mixing console and switched to line as required. VTR previewer unit provides instant playback.

The vidicon output is also normally fed to separate monitors, so that the film's director and the director of photography can check action and lighting from any convenient point on or off the studio floor. A special videotape recorder, termed a previewer, can also be fed from the vidicon output to store all or part of the programme being recorded, and repeat it later at will. The previewer also has slow and stop-motion facilities.

System 35 can be used in both single and multi-camera configurations. Multi-camera film techniques resemble TV practice in that several cameras, operated simultaneously, feed their pictures to a group of monitor screens. The programme director, watching these, can cue the cameramen, start or stop the cameras, and insert reference marks to indicate subsequently to the editor which camera take he is to select.

See: *Add-A-Vision; Electronicam; Gemini.*

⊕**Tachometer.** Speed-measuring device, often mounted on a film camera, and graduated in frames per second, so that when used with a wild motor, the actual speed of the camera is visible to the operator.

See: *Camera* (FILM).

⊕**Tail Leader.** Leader attached to the end of a roll of film. On release prints, the tail leader protects the film itself from damage, and also identifies it by name. On rolls of negative, the tail leader also carries a start mark so that, with its accompanying sound, it may be printed backwards from the end.

See: *Release print make-up.*

⊕**Take.** Scene, or part of a scene, photographed without interruption, and repeated as often as necessary to perfect the action. Details are recorded on a slate.

⊕**Take-Up.** In motion picture film machinery, a reel or spool on to which film is wound after exposure or printing. In a camera, the take-up side is that half of the magazine into which the exposed film passes.

See: *Camera* (FILM)*; Projector.*

□**Talkback.** Form of communication, usually from a control to operators, and vice versa. When a production is in progress in a studio, it is important for the producer to be able to speak to the studio staff without disturbing the performer or making himself heard on the studio microphones. In a sound programme the artiste himself may also need to hear the producer through headphones.

In the production control room, a microphone is provided for the producer. This is connected to a distribution system with numerous output points around the studio. A member of the studio staff can wear a pair of headphones, which can be plugged into any one of these output points.

The floor manager has to be able to move rapidly from one part of the studio to another. To provide him with talkback without a trailing lead necessitates the use of a radio link. A careful choice of carrier frequency and aerial position is necessary if "dead spots" are to be avoided within the studio area. To avoid interference with other services, VHF frequencies are often used.

□**Tally Light.** Light in a TV studio which warns that a particular camera is in operation. In Britain called a cue light.

See: *Communications; Outside broadcast units; Special events; Studio complex* (TV).

⊕**Tank.** Container, usually rectangular in shape, for holding processing solutions in a

film developing machine. A few such machines carry their solutions in tubes; spray machines require neither tanks nor tubes at the developing stage.
See: *Processing.*

⊕**Tanning.** Process of hardening the gelatin associated with a photographic image, particularly used in development to produce relief images.
See: *Imbibition printing.*

⊕**Tape.** Generic term for any non-perforated ☐ribbon-like material coated with a magnetic or similar recording medium. Films are distinguished from tapes by having perforations. Tapes may be described as audio tapes or videotapes according to their intended use.
See: *Magnetic recording materials.*

⊕**Tape Splice.** Film splice made with an overlay of transparent adhesive tape by the butt

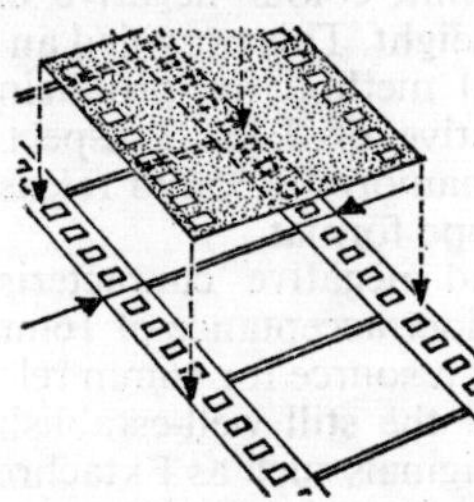

method. Pre-perforated tape may be used to line up with the film perforations, or perforations may be punched out by the splicer.
See: *Editor; Narrow-gauge production techniques.*

⊕**Target.** Very small flag (i.e., shade for keeping direct light off camera lens), circular in shape and some 3 to 9 in. in diameter.

☐**Target.** Part of a camera tube on which the electron image is formed.
See: *Camera* (TV)*; Image orthicon; Plumbicon; Vidicon.*

☐**Target Capacitance.** Capacitance possessed by individual elements of the target of an image orthicon to each other, and also to the target mesh. It is this capacitance which is charged by the emission from the photocathode to form an electrical image on the target.
See: *Camera* (TV)*; Image orthicon.*

☐**Target Layer.** Layer of material deposited on the glass target of a TV camera tube to give it its special electrical storage characteristics.
See: *Camera* (TV).

☐**Target Mesh.** Fine wire mesh screen, with approximately one million holes per square inch, placed in front of the target of an image orthicon camera tube to collect secondary electrons emitted from the target. The open area is as large as possible, usually being about 75 per cent.
See: *Camera* (TV)*; Image orthicon.*

☐**Target Voltage.** Voltage applied to the target of the camera tube to ensure correct functioning.

⊕TECHNICOLOR

Although the term Technicolor has sometimes been incorrectly used to mean colour motion picture films in general, it is in fact a trade mark referring to the products and processes of an organization which has specialized in this field since 1915, and which since 1926 has been particularly associated with the manufacture of colour prints for the professional cinema by the dye transfer or imbibition process.

The Technicolor process was initially a two-colour system; it subsequently employed a beam-splitting camera to provide three colour-separation negatives from which matrices were made for dye transfer. After the introduction in 1951 of integral tripack types of colour negative, special cameras were no longer necessary, and since that time most Technicolor dye transfer printing has been carried out using matrices prepared direct from original colour negatives.

WIDE-SCREEN APPLICATIONS

The wide-screen processes introduced in the early 1950s brought new problems but, as matrices were always printed optically, the dye transfer process permitted various different print formats to be derived from the same original negative without the preparation of an intermediate dupe negative. This facility was of particular importance when negative sizes greater than the standard 35mm, such as Vistavision or 65mm, came into use to provide pictures of improved quality on large screens. At this period, Technicolor themselves introduced a large-image, wide-screen anamorphic

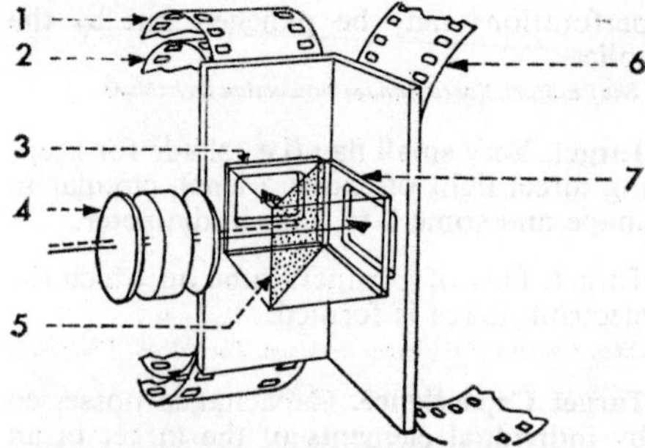

THREE-STRIP CAMERA. Original Technicolor camera, used from 1934-53, in which three records were produced on black-and-white films at two apertures. 4. Lens system. 5. Sputtered gold mirror dividing light between the two apertures. 1. Blue-sensitive film with blue-absorbing coating in bipack with, 2, panchromatic film receiving red image. 3. Magenta filter transmitting red and blue light. 7. Green filter transmitting green light only to panchromatic film, 6.

process under the title Technirama, which used a double-frame area of 35mm film with anamorphic optical systems both in the camera and at the reduction matrix printing stage.

SPECIAL EFFECTS

Matrix printing allows some special effects, such as fades and dissolves, to be printed from the original negative without the necessity of intermediate colour masters and duplicate negatives. At first, this was carried out by printing each matrix record twice from negative which had been assembled in A and B rolls with the use of a variable opening shutter during printing to provide the necessary superimposed exposures. However, matrix printing is always done on an optical printer and it is possible to run the matrix positive stock forwards and backwards independently of the movement of the negative. This allows fades and dissolves to be produced in the matrix when printing from negative assembled in a single roll by what is termed the auto-optical method.

The manufacture of narrow-gauge release prints by dye transfer has proved to be of increasing importance with the wider use of this medium in commerce and education, and the production of imbibition prints in 16mm has now been extended to 8mm and Super-8 formats.

LATER DEVELOPMENTS

Among the later developments of wide-screen film processes, Technicolor introduced the Techniscope system in 1962, again employing anamorphic optical printing at the stage of matrix preparation, but working from a 35mm colour negative of half the standard height. This provided an extremely economical method for producing an original negative of 2·35 : 1 aspect ratio, to yield an anamorphic 35mm release print of Cinemascope format.

Improved negative characteristics have brought wider acceptance of 16mm original negative as a source for 16mm release prints in place of the still well-established 16mm reversal originals such as Ektachrome, from which a colour intermediate negative is made as the source from which matrices are printed for dye transfer. L.B.H.

See: *Colour cinematography; Imbibition printing; Techniscope; Wide-screen processes.*

Technirama. System of film production for wide-screen cinema presentation using a double-size frame on 35mm negative photographed with an anamorphic compression factor of 1·5 : 1.

Special cameras similar to the Vistavision design were employed, in which the film moved horizontally eight perforation holes for each frame. In making 35mm release prints, optical reduction was employed with a further lateral compression factor of 1·33 : 1, so that the resultant print had a total compression of (1·5 × 1·33), i.e. 2 : 1, and thus was suitable for projection with Cinemascope equipment. Technirama made good use of available 35mm film area, and was popular from 1956–1966.

See: *Wide screen processes.*

Techniscope. System of film production for wide-screen cinema presentation using a half-size frame on 35mm negative photographed without distortion. Anamorphic enlargement is used in printing full-frame 35mm copies, which are projected with an anamorphic lens.

It was the improvement in grain and resolution of integral tripack films in the early 1960s which made large-area negative systems less necessary. Techniscope, introduced in 1963, made it possible to cut camera negative consumption to one half, using a camera with a simple modification to achieve two-perforation pull-down.

The reduced negative area imposes some limitation of the projected screen size.

See: *Wide screen processes.*

TELECINE

Frequent use is made of films as programme material for television broadcasting, and programmes which are intended only for television are nevertheless often produced on film. Film has the advantage of being a permanent record which is compatible with any line or field television standard – and colour film can be broadcast on any colour television system.

The relatively simple and portable nature of film cameras makes it commonly more convenient to film location shots for television drama than to record them on videotape: such shots are called film inserts. For the same reason, most television newsreel items are shot directly on film.

The machines which play back films on television are called telecine machines or film scanners. They fall into two broad categories:

One type of telecine machine consists essentially of a television pick-up camera looking straight into the lens of a normal intermittent-motion film projector. Vidicon television cameras are particularly suitable for this operation, and such film playback machines are then referred to as vidicon telecines.

A second type of television film machine is called a flying spot telecine. Normal television cameras are not used in flying spot telecines, nor does a projector lamp illuminate the film. Instead, the light which is shone through the film comes from a cathode ray tube, which is scanned with 405 or 625 lines, or whatever is appropriate to the broadcasting standards in use. This raster is not modulated at all and forms a very bright plain rectangle of light. A lens projects an image of the raster on to the film frame, and the light which goes through the film is collected by a photocell which converts it into a television signal.

Various specially designed types of film transport mechanisms are used with flying spot telecines. In some designs, the film moves continuously; in twin-lens machines, each lens produces one field of the two-field television picture and the film motion provides part of the vertical scanning movement; in polygon machines, a rotating optical system compensates the film movement to give a stationary image of each film frame. Intermittent film movements are also common in flying spot telecines; a very fast pulldown is required for the film during the television field-blanking interval so that the film is stationary during the active scanning period of the television field.

GENERAL PRINCIPLES

Telecine machines convert the optical images previously recorded on standard cinematograph film into electrical television signals suitable for broadcasting. Of the different sizes of film stock available, only standard 35mm and 16mm film are normally used for television transmission; 8mm film is used when either economy or news value necessitates it, but the resultant television picture quality is significantly below that produced by live TV cameras. Super-8 film, with its larger image area, may eventually increase the popularity of the 8mm gauge.

The television picture aspect ratio of 3 : 4 implies that only a part of the total picture area of wide-screen or anamorphic film frames can be transmitted. Continuous selection of the most appropriate part of the wide frame during transmission is technically feasible but normally leaves much to be desired aesthetically. Telecine machines provide facilities for converting both optical and magnetic sound tracks into electrical signals and often for running one or more separate magnetic sound reproducers in synchronism with the film in the telecine.

SCANNING STANDARDS. The number of scanning lines per picture used for any particular television broadcasting standard, e.g. 405 and 625 lines in Britain, raises only minor problems in the design of telecines. However, the two standard television picture

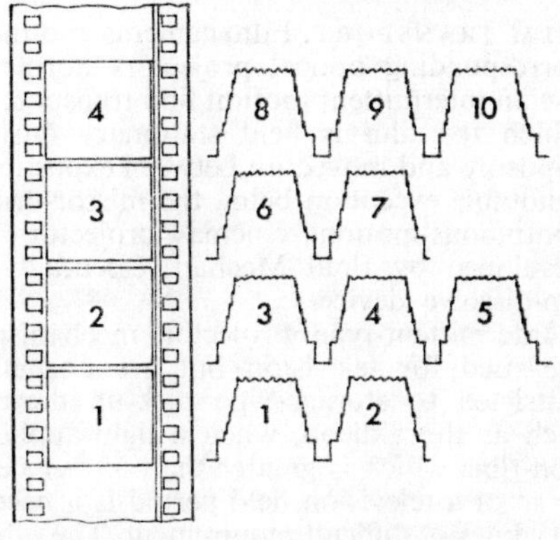

60-FIELD STANDARDS. Film frames must be scanned alternately with two and three television fields, so that 24 film frames are transmitted every second and occupy 60 fields of TV time (i.e., two film frames to five TV fields).

repetition rates of 25 per second (e.g., European television) and 30 per second (e.g., American television) impose different restrictions on the telecine design.

Standard 35 and 16mm film is shot at 24 fps (frames per second); it can be reproduced at the rate of one film frame to one television-picture-interval on 25 pps (picture-per-second) television standards, when the 4 per cent increase in playback speed is not normally noticeable. However, film shot specifically for television in European countries is nowadays usually exposed at 25 fps.

For film transmission on 30 pps television systems, either some form of storage is needed in the telecine – such as that offered by television camera pick-up tubes – and/or other means of translating the frame rate is required. For example, the odd film frames can be scanned with two television fields and the even film frames with three television fields. In such a case, each pair of film frames is transmitted in the time of five television fields.

FILM SHRINKAGE. Film is not completely stable in its physical properties, and changes in dimensions and resilience can continue to occur for months after processing, depending on the conditions of storage, the action of chemical hardeners in the emulsion, the number of passes through projectors, the heat in the projector gate, and so on. The film transport mechanism must be designed to accommodate film shrinkages of 2 per cent if loss of film register and damage to the film are to be avoided.

FILM TRANSPORT. Film cameras and their corresponding optical projectors normally use an intermittent-motion film transport in which the film is held stationary during exposure and moved on between exposures, a notable exception being the mirror-drum continuous-motion cinema projector as developed by Emil Mechau, essentially a lap-dissolve device.

Intermittent-type projector mechanisms are used for television but are normally restricted to storage-type pick-up devices such as the vidicon, when a light-application-time which is greater than 35 per cent or so of a television field period is a necessary but not difficult requirement. The other part of the cycle can be used for a leisurely pull-down. Incandescent light sources are mainly used with such projectors, although pulsed light sources are practical.

Intermittent-motion projectors can be used with non-storage type telecines but in this case the film pull-down must be accomplished during the television field-suppression period if all the active scanning lines of the television picture are to be utilized. Very high accelerations are therefore required, raising problems of picture stability immediately after the pull-down, and of damage to film and splices by the large forces used to produce the film acceleration and deceleration. These are especially severe in 35mm film, with its greater mass. Pneumatic devices have been used for such fast pull-downs, but acoustic silencing of the mechanism is difficult. Because of these problems, fast pull-down mechanisms have been commercially successful only for 16mm film telecines.

Continuous-motion film transports can be used in television and produce less wear of the film. The motion of the film image can be arrested optically, or the film movement can be used to provide part of the field scanning for the television image dissection. However, the necessity to scan each film frame twice to achieve the complete interlaced television raster means some complication in the arrangements to ensure that the two interlaced field rasters maintain the relative positional accuracy necessary for interlacing, as the film is steadily moving on. Thus a twin-lens optical system may be employed, one lens for each field, directed at different areas, or a single lens and a raster which alternates at field frequency between two positions, so that each patch is scanned twice. The latter arrangement is known as a jump-scan or hopping-patch telecine.

SYNCHRONIZATION. In general, intermittent-type mechanisms do not need accurate synchronization with the television field rate, the storage properties of the pick-up device reducing the noticeability of the application bar. For 30 pps operation, however, 2-3-2-3 field scanning techniques are necessary.

Continuous-motion mechanisms of the twin-lens or jump-scan types require sophisticated servo mechanisms to ensure phase lock between the film picture-rate and the television waveform.

Continuous-motion mechanisms which rely on optical compensation to arrest the film image do not need to be run in synchronism with the field pulses even when used in non-storage telecines, and such mechanisms are usually completely multi-standard.

Their main drawback is that of any lap-dissolve device – difficulty of producing constancy of picture brightness, maladjustments which may lead to flicker, and a noticeable increase in movement blur due to there being two images present which are not coincident during most of the cycle.

VIDICON TELECINES. Modern storage-type telecines use photo-conductive pick-up tubes of the vidicon type. These tubes are mounted in conventional television cameras, which look at standard intermittent-motion film projectors. The optical system usually incorporates a field lens and may also include the multiplexing of sources.

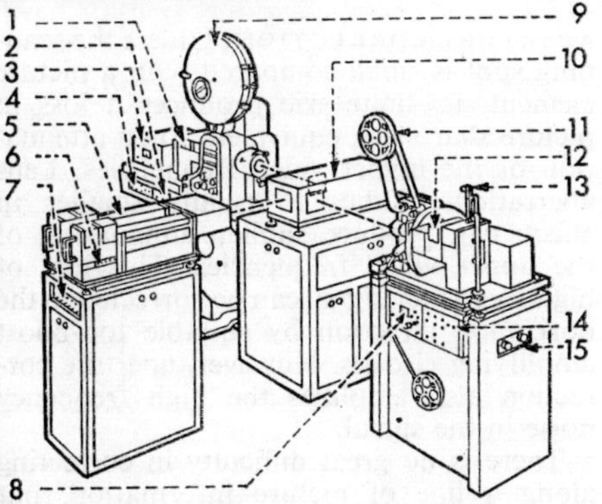

MULTIPLEXING PROJECTORS INTO CAMERA CHAIN. 1. 35mm projector, 12, 16mm projector and, 5, dual slide projector, multiplexed by 10, optical multiplexer containing vidicon camera. 35mm projector: 2, interlock control panel, 3, sound amplifier, 4, local control panel, 9, 3,000 ft. spoolbox. Slide projector: 6, local slide control panel, 7, mains switches and fuses. 16mm projector: 8, sound amplifier, 11, provision for 4,000 ft. spool, 13, pulley for closed loop operation, 14, local and interlock control panel, 15, mains switches.

FLYING SPOT TECHNIQUES. The first television pictures were produced on the flying spot principle, with the subject to be televised in a darkened room lit by the flying spot raster formed by rotating a disc of pin-holes in front of an incandescent source. Photo-electric cells picked up the light reflected from the subject; the photo-cells were placed in similar positions to the lamps used for scene illumination in film studios.

Modern telecine machines generate the flying spot by projecting an optical image of an unmodulated television raster from the face of a cathode ray tube on to the film frame area. A photo-multiplier tube on the other side of the film converts the received light into an electrical signal proportional to the light intensity.

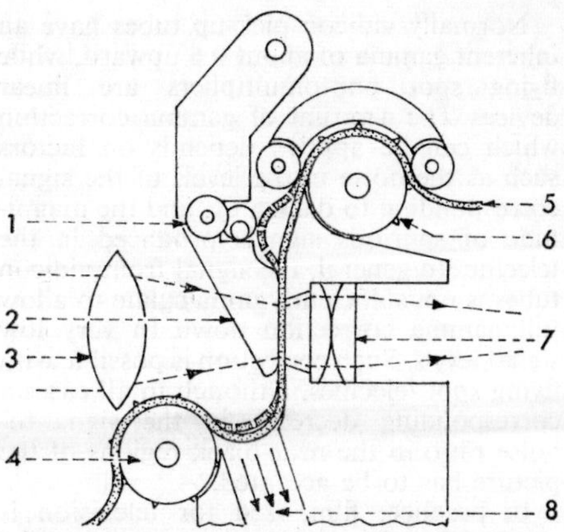

PNEUMATIC PULLDOWN. Raster of flying-spot scanning tube is imaged on 16mm film, 5, by lens, 3, and reaches colour receptors via condenser, 7. Sprocket, 4, feeds film continuously downwards but sprocket, 6, is driven by a double crank to obtain non-uniform angular velocity and an alternate 2-frame 3-frame dwell period. During this period, film is held by clamping gate, 2, but overruns above gate to take dotted path. When clamp is released by eccentric drive, 1, suction, 8, moves film from dotted line to solid line positions, through height of one frame. Clamping gate is closed and cycle restarts, taking 1·14 msec, i.e., time of blanking pulse.

TRANSFER CHARACTERISTICS. The input-output characteristics of the telecine machine, including the film transfer characteristic, are required to match inversely the transfer characteristic of the domestic television display tube. The most pleasing reproduction occurs when the picture-generating equipment has a transfer exponent of 0·4, although this results in an overall system gamma of slightly above unity.

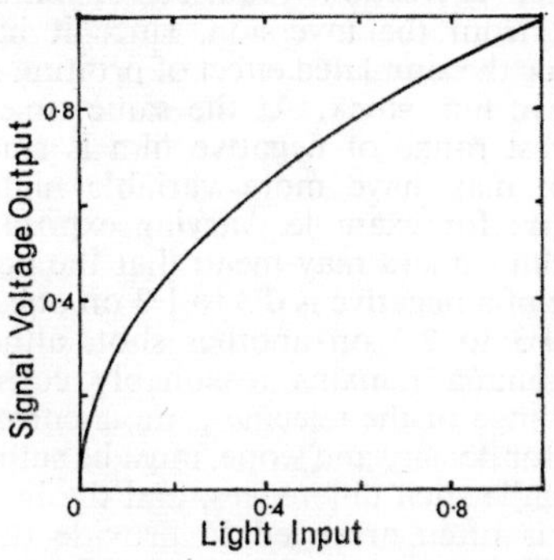

LIGHT INPUT AND SIGNAL VOLTAGE OUTPUT. To correct for input-output characteristic of viewer's display tube, picture generator must have non-linear input-output relationship. Power-law exponent is usually between 1/2·2 and 1/2·6 to compensate for tube gamma of about 2·5.

Normally vidicon pick-up tubes have an inherent gamma of about 0·6 upward, while flying spot photomultipliers are linear devices The a mount of gamma correction which can be applied depends on factors such as the noise in the levels of the signal corresponding to dark grey, and the magnitude of spurious signals produced in the telecine. In general, the signal from vidicon tubes is not sufficiently immaculate to allow full gamma correction down to very low signal levels. Such correction is possible with flying spot telecines, although in all cases a corresponding decrease in the signal-to-noise ratio in the near-black regions of the picture has to be accepted.

In practice, film used for television is processed as for general cinema release, the gamma of positive film being standardized at around 2·3, and negative film at about 0·65. Outdoor scenes can have a contrast range of 500 : 1 or more. The film contrast range may be 200 : 1 or so. Although monochrome display tubes viewed in dark conditions can exceed a contrast range of 100 : 1, TV receivers are normally used with a high level of ambient light on the tube face, and a consequent decrease in effective contrast range. Practical factors limit the transmitted signal-contrast range to about 50 : 1. Since most domestic monochrome receivers lose the d.c. component in the video signal, the spatial distribution of light and dark tones is important, to ensure that the average value remains in the mid-grey region.

NEGATIVE FILM. Negative film can be transmitted in telecine machines by electrical inversion of the signal. However, the gamma correction required is different, apart from the inversion, since it has to include the simulated effect of printing on to positive film stock. At the same time, the contrast range of negative film is smaller, but it may have more variable highlight density; for example, varying exposure in the film camera may mean that the density range of a negative is 0·3 to 1·3 on one shot, and 1·3 to 2·3 on another shot, although the gamma remains reasonably constant. The range of the telecine gamma-corrector, both for density and slope, must be sufficient to handle such differences, and the inverter used is often arranged to provide the required simulation of printing on positive film stock. Alternatively, it may be assumed that the absence of the film printing process compensates for the receiver's display tube characteristic, so that correction is needed only for the negative film contrast law.

After electronic signal-processing is completed, negative film is characterized by a noise level which increases from black to white in the reproduced picture.

Negative film requires careful handling since its low contrast results in an overemphasis of scratches and defects, while dirt and dust are reproduced as particularly objectionable white spots.

Since the emulsion on negative and normal positive stock is on opposite sides of the film base, the telecine must provide for focusing on either side of the base for sound and vision. This facility may also be required for some types of positive film prints.

APERTURE CORRECTION. Unless the scanning spot is small compared with a picture element, its finite size produces a loss in picture sharpness, equivalent to an attenuation of the higher video frequencies. Lens aberrations, although usually smaller in effect, can produce similar attenuation of the upper video frequencies. This loss of high-frequency signal can be corrected in the horizontal direction by suitable top-boost amplifying circuits. However, aperture correction also amplifies the high frequency noise in the signal.

There is no great difficulty in correcting along a line of picture information, but vertical aperture correction must involve remembering what the previous line looked like. Practically, it involves storage in delay lines and the comparison of elements in one line with the equivalent element in the previous line. This is involved and difficult and the technique has not come into common use.

Both the finite spot size and the usual residual lens aberrations produce errors in the transient response of the system which are symmetrical about the transient. Their correction, therefore, requires circuits which also have a symmetrical transient response at all levels of the correction; that is to say, amplitude correction is required which is free from phase distortion.

The film image itself has a loss of resolution for fine detail, i.e., an effective high frequency loss, particularly in 16mm film prints. Such loss of resolution may also be compensated for by increasing the telecine aperture correction; however, this increases the noise in the signal and emphasizes blemishes and grain in both the film and, in flying spot machines, the scanning-tube phosphor screen.

The amount of aperture correction used is therefore a compromise, and may be, typically, sufficient to produce 60 per cent modulation at 5 MHz for 625 lines.

SPOT WOBBLE. When a telerecording with the line structure of the original raster is played back in a telecine machine having a fairly good vertical resolution, objectionable patterns arise from the beat effect between the two line-scanning structures. It is customary partially to mask the line structure in the telerecording by putting vertical spot wobble on to the original picture display. However, such spot wobble necessarily deteriorates the resolution of, for example, fine bars inclined at 45° to the line scan.

The best practice is to apply just sufficient spot wobble to a telerecording to make the reproduction on vidicon telecines tolerable, and to apply further spot wobble to the flying spot machine, which has better vertical resolution.

Much work has been done on spot wobble devices other than sinusoidal. All have some pattern or structure defects. For a time, synchronous spot wobble, i.e., each line effectively in synchronism, was the favourite. Usually random sinusoidal spot wobble is employed.

VIDICON TELECINE

Most types of television camera pick-up tube have been used in telecine machines. Where film inserts in a studio production were obtained from a telecine, it was at one time the custom to employ the same type of pick-up tube in it as in the studio cameras

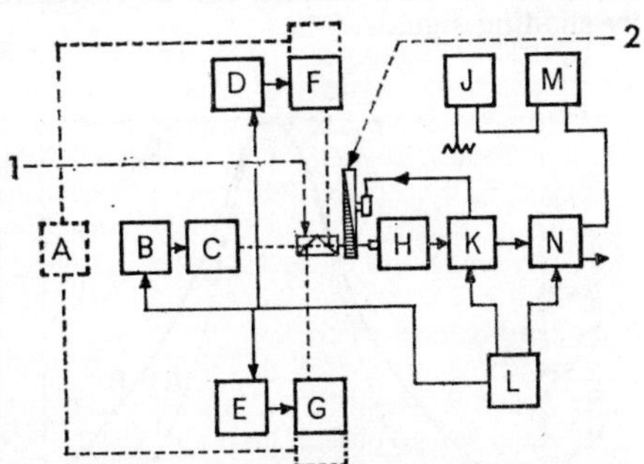

MULTIPLEXED VIDICON TELECINE: BLOCK DIAGRAM. Light beams from 35mm projector, F, 16mm projector, G, and slide projector, C, are multiplexed in prism block, 1, and after traversing filter wheel, 2, are focused on target of vidicon camera, H. (Dotted blocks on projectors are interlock motors wired to patch panel, A.) Vidicon output is fed to camera control unit, K, and gamma and light control, N. Auxiliary equipment comprises control panels, B, D, E, L, picture monitor, J, and waveform monitor, M.

in an attempt to match the picture characteristics of the two sources. However, the development of the vidicon tube, with its economic advantages of low initial cost and long life, has given it a virtual monopoly in telecines.

VIDICON TUBES. Special vidicon tubes have been developed for telecines, and although the normal 1 in. size tube is principally used, 1½ in. and 2 in. vidicon tubes also exist.

The vidicon tube uses a photoconductive target in which the optical image focused upon it produces changes of resistivity in

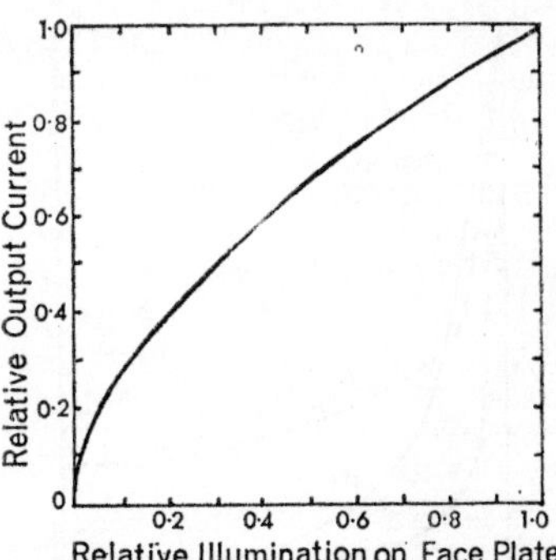

VIDICON OUTPUT-INPUT CHARACTERISTIC. Output current as function of light input. Typical dark current is 0·01 on relative output current scale.

the target layer. Campbell Swinton attempted to make a pick-up tube of this type in 1927, but it was not until 1938 that Miller and Strange showed the idea to be practicable, and it was only after World War II that vidicon tubes became generally available.

Amorphous antimony trisulphide is normally used for the target layer. It is evaporated in an atmosphere of one of the heavy rare gases on to a transparent conducting signal plate to form a spongy layer one or two microns thick. In operation, its front surface is stabilized at cathode potential, while the surface in contact with the signal plate is given a positive potential of up to 50 volts. Increasing this voltage increases the sensitivity of the target, but also increases the dark current. Because of the positive potential, a current flows through the target layer and the charge that reaches the exposed surface is removed by the scanning electron beam to produce a fluctuating signal current proportional to the illumination of the target.

GAMMA. The signal output is not linearly proportional to the light which falls upon the target between scans, but corresponds to a power law with an exponent of about 0·6; this effect is due to the fact that the recombination rate of positive and negative carriers in the target increases with the illumination.

LAG. If the target has a large capacitance per unit area, the potential rise of the surface will be small for a given charge on the target. Under these conditions, the acceptance of current from the scanning beam will be small, and many scans will be required to remove the full charge from the target.

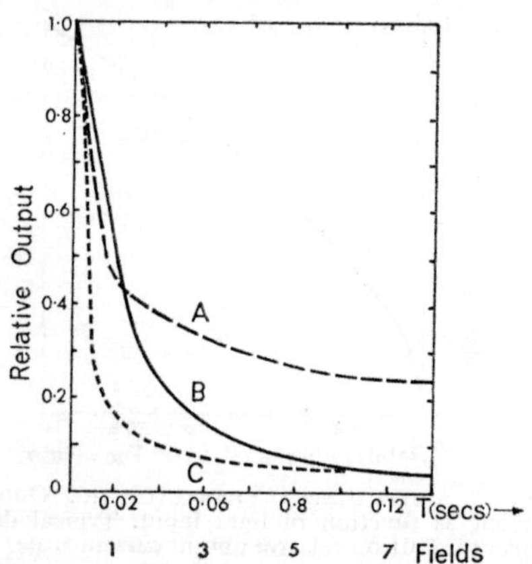

TYPICAL VIDICON LAG CURVES. Relative output after removal of peak white signal plotted against time (secs.) since signal was removed. Number of fields is given for 50 Hz standard. Curve A. Output current of vidicon after removal of peak white signal of 0·4 μA, with high target voltage of 50 v. Curve B. Same with 0·3 μA output and target voltage of 25 v. Curve C. Same with 0·4 μA output at 25 v.

On the other hand, the photoconductive lag is significant for target illuminations below 35 lumens per sq. ft., since the rate of decay of conductivity is faster if the image brightness is high. The lag effect increases with increasing potential across the target.

RESOLUTION. The vidicon target is essentially a high-resolution target despite its small area, and resolutions of 2,000 lines have been achieved, expressed as elements of picture height. The resolution is dependent on a number of factors, including the design of focus and deflector coils and the electric field conditions in the deceleration region close to the target. Separate-mesh vidicons, which enable a higher potential to be applied to the mesh than to the wall anode, give improved resolution. Typical wall-anode voltage is 280, and maximum corner resolution with minimum beam-landing errors occurs when the mesh is 150 volts above this wall-anode voltage. Higher values of both voltages improve the resolution further but call for larger scanning currents and reduce the sensitivity. Dynamic focusing by scanning-waveform modulation on the wall anode enables a further improvement in corner resolution to be obtained.

SIGNAL-TO-NOISE RATIO. The efficiency of the vidicon target in accepting the scanning electron beam appears to be high because of the spongy nature of its surface and its low secondary-emission coefficient. Peak white signal currents of 0·4 μA are practicable, and signal-to-noise ratios of 42 dB for 405 lines and 38 dB for 625 lines are typical. Since little gamma correction is required, the noise in the dark areas of the picture is not exaggerated in the subsequent signal processing. The noise spectrum after signal processing is not uniform over the video frequency range, being triangular in distribution.

SPURIOUS SIGNALS. The vidicon is free from optical burn-in and halo effects due to electron redistribution. There is no charging signal and the dark current is low but may vary from point to point on the target owing to varying thickness of the target or other non-uniformities in the layer. Such variations in dark current can be corrected by shading signals.

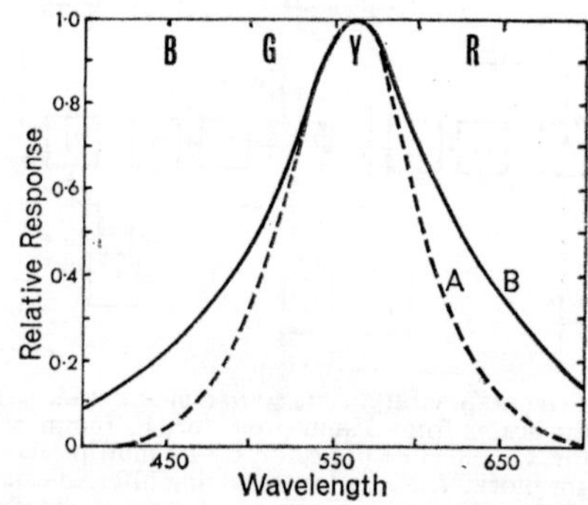

VIDICON SPECTRAL RESPONSE. A. Typical vidicon response curve. B. Photopic response curve of standard observer. Vidicon response curve can be broadened considerably and moved up to 100 nm towards blue end of spectrum for special purposes.

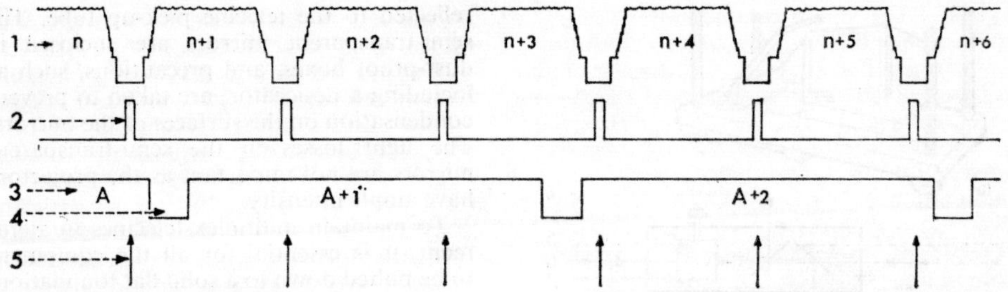

PULSED-LIGHT SYSTEM FOR 60-FIELD STANDARDS. 1. Series of TV fields. 2. Light pulses. 3. Film frames with, 4, pull-down periods. 5. Exposures of film. Pull-down is synchronized with field-blanking pulses in such a way that film frames are alternately exposed to two and three fields.

SPECTRAL RESPONSE. Ideally the combined response of the vidicon tube and the light source should approximate to the photopic response curve of the eye, but considerable departures from this shape are acceptable in practice, particularly if only monochrome film is to be used in the telecine. The typical vidicon tube has a response to red light which is only slightly down on its response to blue light. However, special targets with either enhanced ultra-violet, or enhanced infra-red, response are available.

ADVANTAGES OF VIDICONS. For telecine use, the vidicon makes a cheap and stable pick-up tube, and modern types have a reasonably uniform dark current across the target area. Since the sensitivity of the pick-up tube is not important in telecine applications, the vidicon can be operated under conditions which are particularly suitable for film transmission. Nevertheless, although vidicon telecines have been developed to a very high quality in the United States, in Europe flying spot machines have been the subject of greater development effort and give better results.

PROJECTORS. Standard intermittent-type film projectors can be used in vidicon telecines. In practice, the theatre-type projector is usually modified to a greater or lesser degree to suit the particular requirements of television. A typical modification is to provide the lamphouse with a remotely-controlled variable aperture which allows the intensity of the light source to be modified without change in colour temperature. Two-bladed shutter arrangements can be used on 25 pps systems, with the projector driven by a synchronous motor operating from the 50 Hz mains, since the application bar has a sufficiently low visibility.

For 30 pps systems, however, the 2-3-2-3 sequence has to be adopted for the relationship between frame projection and field-scanning frequencies; five-bladed shutters are then employed. Other arrangements have been used, such as transporting the film at a regular 24 fps and using a pulsed light source during the field-blanking period. The phase relationship between the pull-down and the field-blanking pulses must be such that one pull-down begins immediately after field pulse n, the next pull-down finishes just before field pulse (n+3), while the succeeding pull-down begins just after field pulse (n+5), and so on.

A 16mm optically-compensated projector has been used in the US with both flying spot and vidicon chains. The compensator and film drive are rigidly locked together and the positional compensation is independent of the speed of operation so that special effects calling for non-standard frame speed are practical. No shutters are required as an optical lap-dissolve is used to change from one frame to another. However, very precise mechanisms and absolute cleanliness of the optics are necessary if flicker is to be avoided.

In intermittent projectors, the light through the thin film is usually partially collimated, but diffused light effects a form of black expansion and white crushing which requires less gamma correction and hence improves the signal-to-noise in the blacks. At the same time, scratches and abrasions of the film become much less visible. However, effective resolution appears to suffer with diffused-light optics.

MULTIPLEX OPERATION. Since film projectors and television cameras usually have long-throw optical systems, there is little problem in designing optical arrangements which allow two or more film projectors to be used with the same pick-up tube and

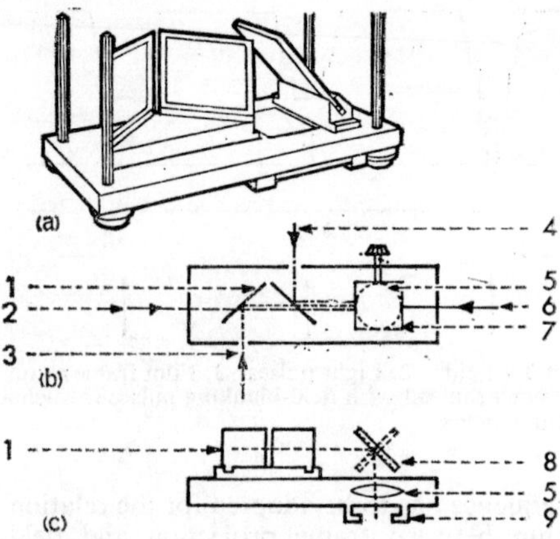

OPTICAL MULTIPLEXING LAYOUT. (a) Perspective view. (b) Plan view. (c) Side elevation. 1. Semi-reflecting pellicles. 2. Light path from slide projector. 3. Light path from 35mm projector. 4. Light path from 16mm projector. 5. Field lens. 6. Light path for testing. 7. Moving mirror. 8. Mirror in normal position (dotted, test transmission). 9. Test slide carriage.

electronic chain. Such arrangements are called multiplex designs.

A common arrangement is to provide one camera channel with a 35mm film projector, a 16mm film projector, and one or two slide projectors. Semi-transparent mirrors, normally of the tightly-stretched thin-film

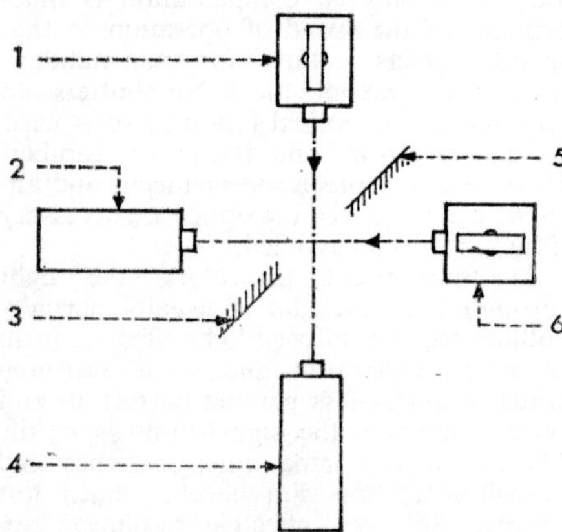

CROSS-FIRE VIDICON MULTIPLEXER. 2, 4. Vidicons. 1, 6. Projectors. By moving mirror, 3, into light path, image from projector, 6, is reflected into vidicon, 4. By moving mirror, 5, into light path, image from projector, 1, is reflected into vidicon, 2.

or pellicle type, are arranged so that the light from any of the projectors can be reflected to the telecine pick-up tube. The semi-transparent mirrors are enclosed in dust-proof boxes, and precautions, such as including a desiccator, are taken to prevent condensation on the surfaces of the mirrors. The light losses in the semi-transparent mirrors are not important as the projectors have ample intensity.

To maintain multiplex telecines in alignment, it is essential for all the equipment to be bolted down to a solid flat foundation. By international agreement, the height of the light path from the floor has been standardized at 4 ft.

One particular form of multiplexing is known as a cross-fire arrangement. Two telecines are arranged so that their optical paths cross. Each telecine is normally used as a separate installation. However, should the camera chain develop a fault, a front-surface plane mirror can be inserted into the light path to reflect the light from the operating projector on to the other camera.

The arrangements so far described are called optical multiplexing. Electronic multiplexing is possible, but is not normally used with vidicon telecines.

FLYING SPOT TELECINE

Normal television cameras are not required in flying spot telecines. Nor is a projector lamp used to illuminate the film. Instead, an image of an unmodulated television raster, appropriate to the transmission standards in use, is focused on to the film frame. On the other side of the film, a photocell measures the amount of light transmitted by the film and produces an electrical signal which is linearly related to the light intensity. The electrical signal is processed to correct for various effects, such as the finite size of the spot on the scanner tube producing the raster, the afterglow characteristic of the scanner tube phosphor, phase distortion in the amplifiers, and the transfer characteristic of the receiver cathode ray tube. At the same time, the black level of the signal is firmly established and all signals suppressed during the line- and field-blanking intervals.

In the early days of television, when mechanical scanning was the rule, film was the most reliable means of obtaining a television picture. The intermediate-film process (circa 1935), in which a film camera records the studio scene, and the film is rapidly processed and transmitted by a mechanical scanner, was for a short time the preferred method of picture generation.

A film scanner with a continuous-motion

mechanism, a double optical system, and a Farnsworth image dissector as the pick-up device, was developed just before World War II. After the War, work on this type of system was recommenced, but the flying spot principle was used. Mechanical methods of producing the flying spot raster are practicable, using, for example, rotating discs. The use of a laser light beam and rotating diamonds with suitable facets has recently been proposed. However, currently available flying spot telecines all use a cathode ray tube to generate the picture-dissecting raster.

SCANNER TUBE. The cathode ray tubes used for producing the flying spot raster have optically-flat screens and operate at high voltages (30 kV) and high beam currents (300 μA) to provide a very bright scan. Air is blown on to the tube face to cool it. Dynamic focusing eliminates any variation of current density over the raster with consequent variations in afterglow. Corrector magnets around the tube face ensure a raster free from pin-cushion distortion. The scan linearity is adjusted to correct for the flat screen. A screen with a very uniformly deposited phosphor is required, because any grain structure will be reproduced on the enlarged image displayed by receivers. At the same time, any burning of the raster on to the phosphor screen or the glass face plate will make it impracticable to shift the raster or alter its size – as, for example, in transmitting anamorphic images. In general, all tubes seem to burn with use to some extent.

The special response of the phosphor-emitted light should cover the visible spectrum with reasonable uniformity. The afterglow must be as short as possible, because

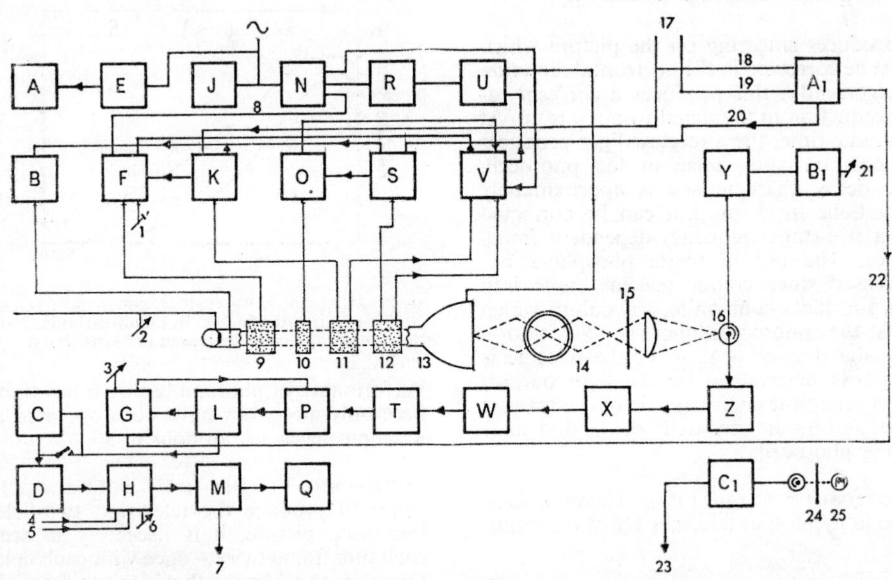

FLYING SPOT TELECINE: BLOCK DIAGRAM. Spot scanner CRT, 13, has focus coil, 9, focus modulation, 10, and field and line deflection coils, 11, 12. Polygon optical system, 14, arrests motion of film, 15, and modulated light is focused on photocell, 16. This is powered by h.t. (B) supply, B_1, with video gain control, 21, the cell being fed through network, Y. Cell output is direct to head amplifier, Z, thence to aperture correction, X, long afterglow correction, W, phase correction, T, video B amplifier, P, gamma correction, L, black level control, G, set by fader, 3, (clamp pulses, 22, supplied from sync converter, A_1), inverting amplifier, C, filter and phase correction, D, and blanking and sync mixing, H. Here, mixed syncs, 4, and mixed blanking pulses, 5, are fed in, and lift is controlled by fader, 6. Thence the signal passes to video distribution amplifier, M, and to output, 7. Waveform monitor, Q, provides a final check. Drive pulses enter system at 17, with line and sync pulses, 18, 19, to sync converter, A_1. Line drive, 20, goes to line scan, S, with scan protection circuits, V. Focus modulation, O, is controlled by S. Field scan, K, also protected by V, gets its drive pulses from 17, and in turn controls focus current stabilizer, F, with focus control, 1. Flyback suppression, B, is applied to grid of CRT, and beam current is set by rheostat, 2. Optical sound track, 24, modulates beam from exciter lamp, 25, and photocell output is stepped up by head amplifier, C_1, with output, 23. Remaining units constitute power supplies. Mains input goes to video power unit, E, with h.t. stabilizer, A. Head amplifier power is provided by J, and scanner power by N, with stabilizer, R. EHT generator, U, supplies final anode of CRT, 13.

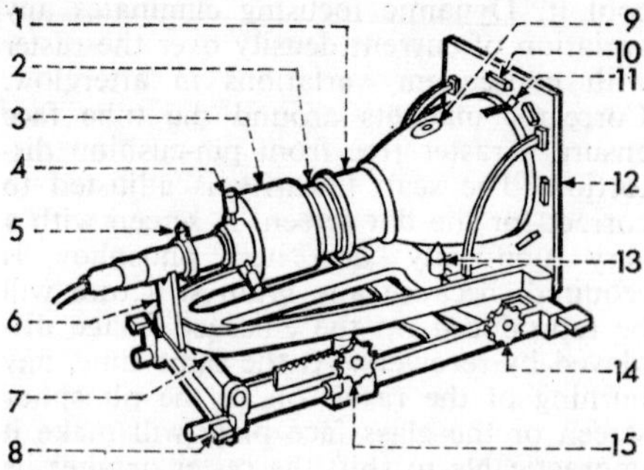

SCANNER TUBE FOR FLYING-SPOT TELECINE. Tube is supported on cradle, 7, by perspex supports, 13 (one not shown), and bridge, 6. Transverse adjustment is by handwheel, 14, and longitudinal adjustment by rack and wheel, 15. Carriage, 8, is attached to main housing. Mask, 12, is flexibly positioned on tube face by insulated springs, 11. Tube components: 1, deflection coils, 2, focus modulator coil, 3, focus coil assembly, 4, focus adjusters, 5, centring magnet, 9, anode button, 10, magnets for correcting pincushion distortion.

it produces smearing on the picture which must be corrected in the horizontal direction electronically: this produces a corresponding reduction in the signal-to-noise ratio. At the same time, the afterglow light produces unnecessary shot noise in the photocell. The decay characteristic is approximately hyperbolic in shape, but can be corrected by a five-stage frequency-dependent impedance. The use of white phosphors has increased since colour telecine came into use. For black-and-white, any colour which suits the photomultiplier is satisfactory. The signal-to-noise ratio of the telecine is primarily determined by the light output/short afterglow characteristic of the scanner tube, and by the sensitivity of the first stage of the photocell.

PHOTOMULTIPLIER TUBE. The photocells used in flying spot telecines are of the multi-

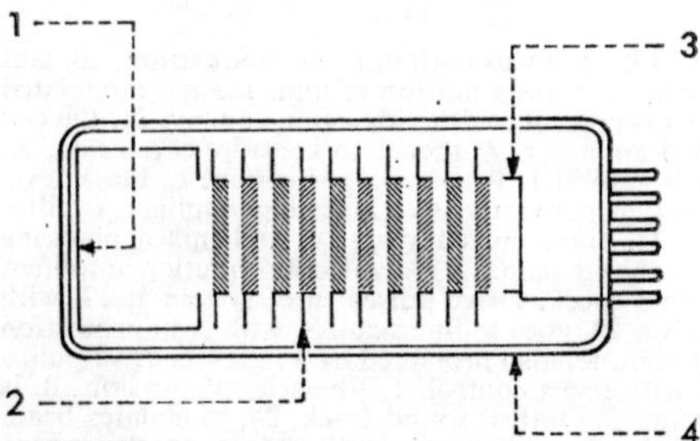

VENETIAN-BLIND PHOTOMULTIPLIER. 1. Semi-transparent photocathode. 2. Venetian-blind dynodes. 3. Collector. 4. Glass wall.

plier type in which the original photoelectrons give rise to secondary emission electrons, which give rise to more secondary-emission electrons, and so on. The number of secondary-emission stages may be 10 or more, with a maximum gain of 4 or 5 at each stage. Caesium antimony cathodes of the semi-transparent type are usually employed, although enhanced red sensitivity can be obtained with more complicated cathodes of the trialkali kind (Sb—K—Na Cs). The secondary-emission surfaces (commonly Cs—Sb) are called dynodes, and may have various geometrical shapes. The venetian-blind arrangement is used in telecine photomultipliers.

Peak-white output current is usually of the order of 100 μA but can be higher; there is a linear relationship to light input. The signal-to-noise ratio is proportional to the square root of the incident light intensity.

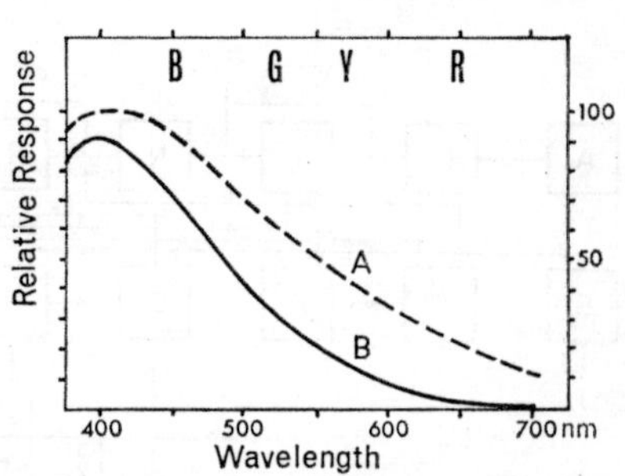

SPECTRAL RESPONSE OF TYPICAL PHOTOMULTIPLIERS. A. Tri-alkali cathode. B. Bi-alkali cathode. Note that cathode A gives increased red sensitivity.

Performance of photomultiplier tubes stabilizes only after they have been operated at low light-levels for an hour or so.

TWIN-LENS FLYING SPOT TELECINE. In order to produce the television two-field-interlaced picture, it is necessary to scan each film frame twice – once with each field. One way of achieving this is to use a double optical system, with a shutter to block the lens system which is not required for the particular field being scanned.

The film movement is regarded as a part of the required vertical scanning movement, and it accounts for just over half the field scan – the remainder is provided by the flying spot raster, which has rather less than half the normal height with respect to the line scan.

In order that both fields shall be registered to within a quarter of one line separation, which corresponds to about 0·0004 in. for

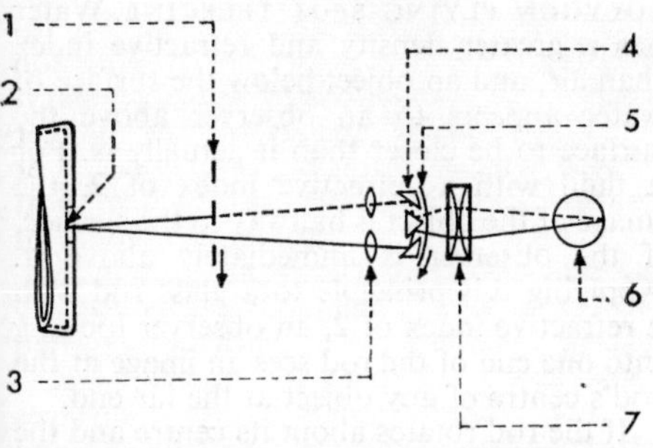

TWIN-LENS 35MM FLYING-SPOT TELECINE. Basic optical arrangement. 2. Scan on projection CRT. 1. Ten-bladed shutter shown at instant of time halfway through second scan. 3. Twin objective lenses with variable spacing to correct film shrinkage. 4. Combining mirrors. 5. Film moving continuously in curved plane. 7. Condenser system. 6. Photocell.

35mm film, great precision is required in the uniformity of the film movement and the geometry of the lens systems. In particular. the amount of shrinkage of the film affects the registration; since 35mm film moves 0·374 in. between successive field scans, a variation in film length of 0·1 per cent will produce a misregistration of a quarter of a line-spacing.

The effects of film shrinkage are corrected by altering the spacing between the two objective lenses. The shrinkage is measured continuously by a spring-loaded roller which lightly tensions a loop of film 83 perforations long. Variations in film length produce corresponding movements of the roller, and a damped mechanical coupling alters the lens spacing. Each of the optical systems

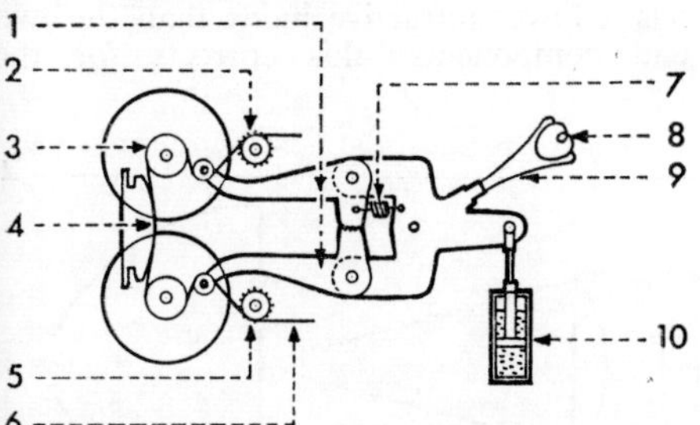

TWIN-LENS TELECINE: MECHANICAL FILTER. Method of smoothing out short-term variations in film velocity. 4. Curved gate with film, 6, passing over twin flywheel rollers, 3. Hold-back and feed sprockets, 2, 5, drive film continuously. Between sprockets and rollers are film tension arms, 1, carrying additional rollers which bear on film. These are tensioned by spring, 7, and damped by dashpot, 10. 8. Framing cam with filter springs, 9.

is folded to bring the two images of the raster to the correct positions on the film. The folding of the light rays is effected by prisms or plane mirrors arranged to produce images of the raster which are tilted slightly so that they fit tangentially on the cylindrical surface of the film. The film is curved along its length as it goes through the gate, so that it is given rigidity across its width.

A multi-bladed shutter is arranged to black out the objective which is not in use. At the moment of changeover from one field to another, the upper half of the picture shows through one optical system and the lower

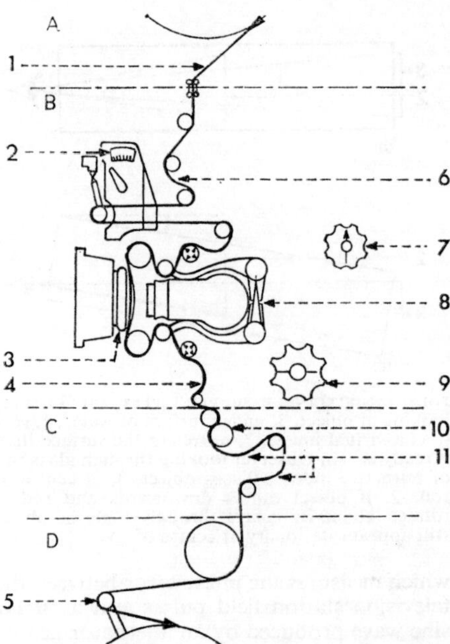

TWIN-LENS TELECINE: FILM TRANSPORT MECHANISM. A. Film feed section. 1. Emulsion side of film. B. Picture head. 2. Shrinkage indicator. 6. Free loop. 7. Framing knob. 8. Film tension indicator. 3. Film gate. 4. Free loop. 9. Manual control knob. C. Sound head. 10. Scanning drum. 11. Free loops. D. Film take-up. 5. Clutch operating arm controlling take-up tension.

half through the other lens. The precise changeover is carried out by the flying spot returning to the top half of the picture.

A condenser lens behind the film collects the transmitted light and focuses an image of the objective lens on to the photocell cathode.

Short-term stability of the film motion is achieved by flywheels and a mechanical filter which are symmetrical about the film gate so that film shrinkage variations do not affect framing. Below a few cycles a second, the film velocity is determined by the main motor drive. The speed of the motor can be controlled by an electronic servo system

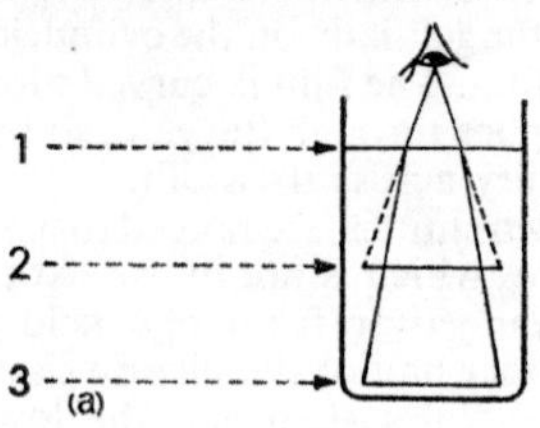

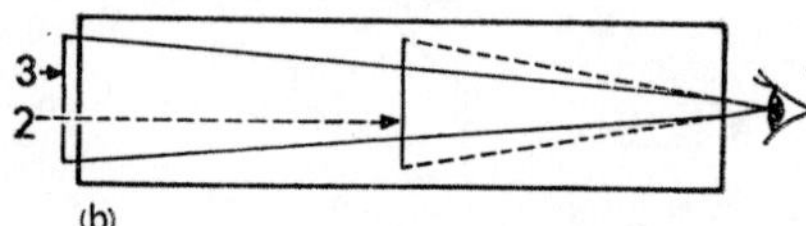

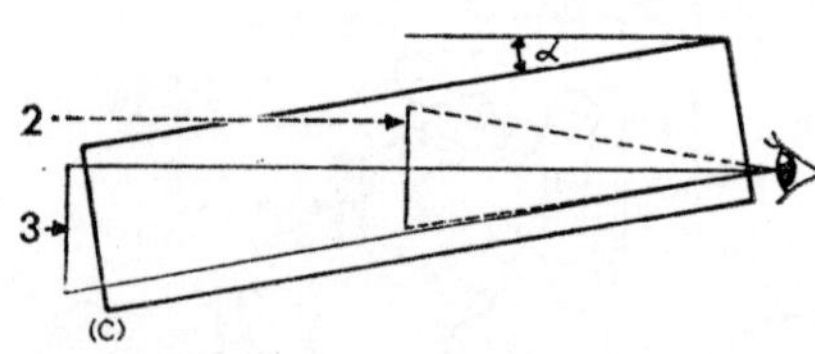

POLYGON TELECINE: BASIC PRINCIPLES. (a) Observer looking at object, 3, under surface of water, 1, sees it as a virtual image, 2, nearer to the surface than it really is. (b) Observer looking through glass rod of refractive index 2·0 sees object, 3, at centre of rod, 2. If object moves downwards and rod is rotated about its centre through angle α, object still appears stationary at centre of rod.

which measures the phase error between the television station field pulses and a 50 Hz sine wave produced by an alternator geared to the film traction mechanism.

POLYGON FLYING SPOT TELECINE. Water has a greater density and refractive index than air, and an object below the surface of water appears to an observer above the surface to be closer than it actually is. For a fluid with a refractive index of 2, the image of the object is halfway to the surface, if the observer is immediately above it. Applying this principle to a glass rod with a refractive index of 2, an observer looking into one end of the rod sees an image at the rod's centre of any object at the far end.

If the rod rotates about its centre and the object moves at the same speed, the observer will see a stationary image at the rod's centre.

Replacing the rod by a polygonal prism, and the object by a film frame, provides the essential elements for an optical compensation of the movement of film. Each pair of facets acts like the glass rod, and if the polygon is rotated with film wrapped around one side, a stationary image is produced at the centre of the polygon. As the edge between adjacent facets comes into view, a lap-dissolve between the two film frames takes place.

Conversely, if the system is arranged so that a virtual image of a scanning raster is produced at the centre of the polygon, the rotation of the polygon will keep the real image of the raster moving in register with the film.

In practice, some modifications to this simple device are necessary. In the first place, the polygon glass cannot have a refractive index of 2. The centre of the polygon is therefore removed and a three component core inserted into the polygonal cylinder. The centre component of the core has a lower refractive index than the two end components; this corrects for the

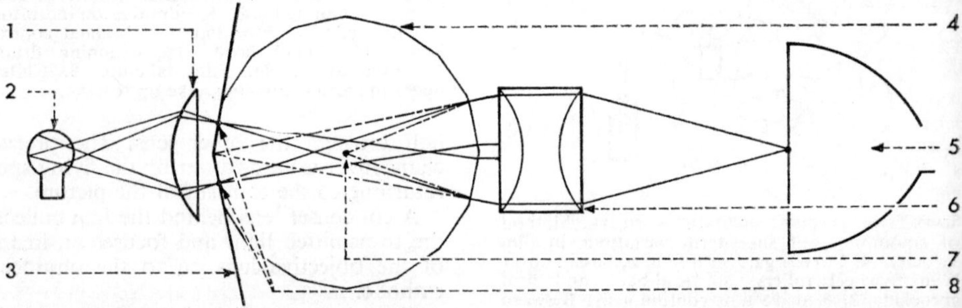

POLYGON TELECINE: PRACTICAL LAYOUT. Scanning CRT, 5, produces virtual image of raster, 7, at centre of rotating polygon, 4, through scanning lens, 6. Film, 3, is wrapped on polygon so that each frame lies on one facet. Real image of raster then appears at 8, and moves in unison with film frames. Raster light modulated by film density falls on condenser, 1, and is collected by photocell, 2.

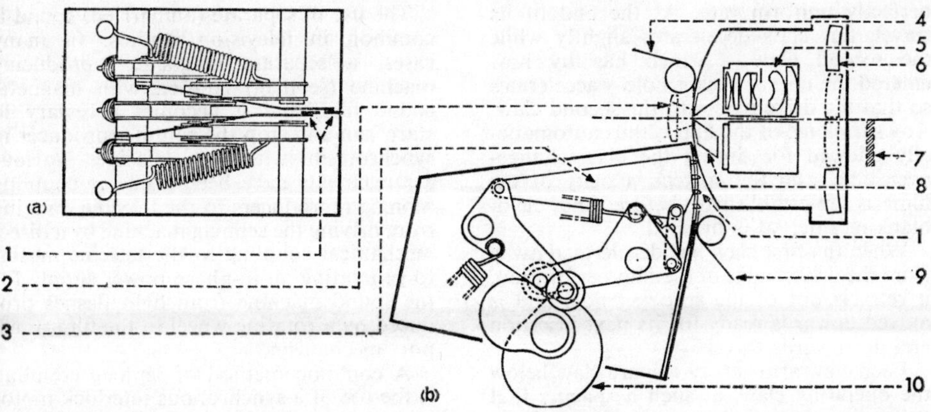

CLAW PULL-DOWN 16MM TELECINE. Two sets of double claws work alternately to give continuous film motion. (a) Detail of claws, 1. (b) General arrangement with 4, 9, film path through optical system. CRT (right, not shown) is obscured by shadow of shutter blade, 7. Image of two field rasters is formed through supplementary lenses, 6, and main objective lenses, 5, and focused by way of beam-folding rhombs, 8, on to film in curved gate. Modulated light is collected by condenser, 2, and focused on photocell (not shown). Cams, 10, operate claw plates, 3, each carrying a pair of claws, 1, which engage film and alternately pull it down to give continuous motion, the change-over period automatically compensating for film shrinkage.

inadequate refractive index of the polygon and ensures correct movement of the real image of the scanning raster. Careful shaping of the ends of the core produces an afocal air gap between the stationary core and the rotating polygon.

The image of the raster on the scanning tube is focused just beyond the facets of the polygon and then, by a second lens system, on to the moving film. Correction for the chromatic aberration of the polygon can be achieved in the lenses. The polygon moves the raster image at the same speed as the film, and a condenser behind the film collects the transmitted light and directs it on to a photo-multiplier cell.

To reduce the overall length of the system, mirrors are used to fold the optical system back on itself.

FAST PULL-DOWN FLYING SPOT TELECINE. A new type of 16mm film telecine uses a very fast pneumatic pull-down mechanism to transport the film. During the scanning period, the film is held stationary so that there is no problem about registration of the two field scans. When the picture scanning is completed, the film is accelerated by an air blast and decelerated very quickly so that it moves by one perforation spacing in less than 1·2 msec. Very careful design is required to ensure that the film tensions produced are not sufficient to damage the film or splices and that the film is quite stationary by the time the next picture scan begins. To reduce the stresses, the air pressure is applied across the full width of the film. Very fast run-up times are achievable, such as 0·3 second for stable picture and sound.

CLAW-TYPE TWIN-LENS FLYING SPOT TELECINE. For 16mm film, greater precision is required in a twin-lens film traction mechanism to achieve the same picture steadiness and field-to-field registration achieved in 35mm telecines. A recent development is the use of an alternate claw mechanism to produce continuous motion of the film, as distinct from the intermittent motion which is used in film cameras and projectors.

At the same time, larger-aperture twin-lens optics make the signal-to-noise ratio at 45 dB comparable with that on 35mm machines.

The claws engage the film sprocket holes in a position as close to the picture gate as is practicable – in fact, about one inch below the centre of the gate. Each of the two claw arms is actuated by a pair of cams; one cam controls the movement of the claw into and out of the film perforations, while the other cam controls the vertical movement of the claw.

The claw movements are phased so that each claw in turn enters a film sprocket hole without touching the film; the claw then accelerates gently to take up the film drive, and then pulls the film down at a perfectly uniform rate. At the end of its travel, this claw decelerates slightly while the second claw – which has by now entered the next sprocket hole – accelerates so that the drive passes to the second claw. Any shrinkage of the film is thus automatically allowed for during the claw change-over time, and the correct velocity of the film is re-established before the field-blanking interval is finished.

When the first claw has decelerated away from the driven edge of the film perforation, it retracts out of the sprocket hole and is moved upwards ready for its next insertion and downwards travel.

Each claw arm carries a spare claw below the operating claw, at such a spacing that the spare claw enters a sprocket hole in advance of the one engaged by the active claw; the spare claw makes contact with the driven edge of its film perforation only if the active claw encounters a damaged sprocket hole.

APPLICATIONS

REPRODUCING COLOUR FILM IN B & W. Vidicon tubes have a colour response which approximates to that of the eye, and the light source used in Vidicon telecines is similar to that used for optical projection. Thus, colour films in vidicon telecines give rise to a grey scale in the reproduction which is sufficiently orthochromatic – in the etymological sense of giving correct relative intensity to all colours – to be acceptable as a black-and-white reproduction of a colour original.

However, some flying spot machines have scanner tubes which use mainly blue and ultra-violet light to reduce afterglow effects. Such machines are not suitable for the reproduction of colour film.

SOUND ARRANGEMENTS. Telecines are required to reproduce sound as well as vision. On normal 35mm and 16mm film, the sound may be recorded on a single track by either optical or magnetic means. The telecine must therefore have facilities for reproducing either the comopt or commag sound.

The sound pick-off arrangements are similar to those in theatre-type projectors, irrespective of whether the film traction is by an intermittent or a continuous-motion mechanism. As usual, the film must move past the sound head at a constant velocity without cyclic or random variations in the audio range, and the optical sound head must have provision for focusing on either side of the film.

The use of separate (unmarried) sound is common in television studios. In many cases, a separate magnetic reproducing machine (sepmag) is used with magnetic sound film, and it becomes necessary to start, run and stop the sound reproducer in synchronism with the telecine. Various arrangements have been used for coupling sepmag reproducers to the telecine, ranging from driving the sepmag machine by a direct mechanical coupling to the telecine motor, to generating an in-phase power supply for the sound machine from light flashes produced by a rotating wheel in the film transport mechanism.

A common method of sepmag coupling is the use of a synchronous interlock motor system. A master three-phase motor of the wound rotor and stator type is mechanically coupled, sometimes through a mechanical filter, to the driving shaft of the main telecine motor. A similar three-phase slave motor on the sepmag reproducer is electrically connected to the master motor, so that rotor and stators of the slave are in parallel with the corresponding elements of the master motor. Applying single-phase power to a stator circuit achieves phase lock between the motors, preparatory to running-up in synchronism.

Pilot tone methods are also used, whereby a master oscillator distributes a reference frequency to which the motors of both machines pull into synchronism.

MATCHING FILM TO TV QUALITY. There are some aspects of film telecine picture quality which it is difficult to match to the type of picture produced by studio image orthicons. However, if adequate control of the contrast law is available in the telecine, the transfer characteristic of the combined film-telecine process can be matched to the image orthicon grey scale. Correct choice of source intensity and target voltage in the vidicon telecine will match the lag effects, while the colour responses of the two picture sources can be matched by the correct choice of colour-correction filters during the film shooting.

The cinema film-making industry is a prolific and natural source of programme material for television, but feature films produced for theatre release may have characteristics which are not ideal for television presentation. The lower contrast range of television viewing makes for poor reproduction of many high contrast scenes which look magnificent under optical projection in darkened cinemas. At the same time, the television format, and the relatively poor detail resolution, make the reproduction of long shots and of scenes with innumerable details rather unsatisfactory.

Feature films for theatre release are shot without regard to the mean level of the picture content, and the action may well continue for some time with the majority of the picture content near either black or peak white; the reproduction on mean-level automatic gain control (AGC) receivers is then unsatisfactory.

For films shot on wide-gauge stock it is usual to obtain 35mm prints for telecine operation; very few studios are equipped for 70mm film. By altering the ratio of line-scan amplitude to field-scan amplitude in telecines, it is possible to unsqueeze anamorphic prints but, since the viewer's receiver has an invariant height-to-width ratio, an operating difficulty arises. If all the picture width is transmitted, the full height of the viewer's screen is not used. If the scanning amplitudes are adjusted so that the picture fills the receiver screen vertically, picture information is lost at the left and right hand sides of the film frame. If the telecine has a manual control of horizontal picture shift, the most appropriate part of the film frame can be transmitted.

LIMITATIONS OF TELECINE PICTURES. The most obvious limitation of telecines is their inability to handle the full contrast range of which the film stock is capable. Signal-to-noise ratios are not yet high enough, while the phosphor-screen granularity of flying spot scanners is sometimes obtrusive. The resolution of 16mm film – particularly colour film – is inadequate for television. Movement effects in telecines can be objectionable, while film splices in continuous-motion machines can produce vertical bounce of the picture. Optically-compensated telecines with their numerous glass-to-air surfaces can produce significant halation effects, as can the flying spot scanner tube.

The amount of wow and flutter which is added to the sound track is usually small and is negligible with continuous-motion machines.

Filtering of the field- and line-scanning frequencies from the sound signal can produce some restriction of the audio-frequency range. G.B.T.

See: *Master control; Telecine (colour); Telerecording; Transfer characteristics.*

Books: *TV Film Engineering*, by R. J. Ross (New York), 1966; *Motion Picture and TV Film Image Control and Processing Techniques*, by D. J. Corbett (London), 1968.

TELECINE (COLOUR)

Telecines, or film scanners, are devices for turning motion picture images into television signals. They are of two main types: those in which a television camera looks into the lens of a film projector and scans the frame of film held stationary there by the pull-down mechanism; and those in which a flying spot from a cathode ray tube produces a scanning raster on the film, the latter thus acting as a modulator, so that a photocell on the far side generates a current which varies from point to point according to the density of the image.

Telecines for colour television use film transport mechanisms identical to those for monochrome television. Image dissection also takes place in a similar fashion: that is to say, by scanning the target of a vidicon pick-up camera tube, or by focusing a flying-spot raster on to the film frame.

After the light has passed through the film, it gives rise – in monochrome television – to an electrical signal which is proportional to the intensity of light passed by the film; this electrical signal is processed to correct for various effects such as the finite size of the scanning spot and the input-output characteristic of the receiver display tube.

In colour television, the light which passes through the film is analysed into its red, green, and blue components, and each of these components produces its own electrical signal. These three television waveforms are each separately processed to ensure their suitability for broadcasting. The output of the colour telecine is thus three voltage waveforms, E'_R, E'_G, E_B', each representing the intensity of one of the primary colour images. These are the signal voltages which are needed by the receiver display-tube to enable it to reproduce – in register – red, green and blue images which will be combined by the viewer's eye into a convincing representation of the original scene.

In practice, the colorimetric accuracy of the original colour negative film stock, of the positive colour film printing process, and of the telecine colour analysis, are all open to criticism – as is the reproduction

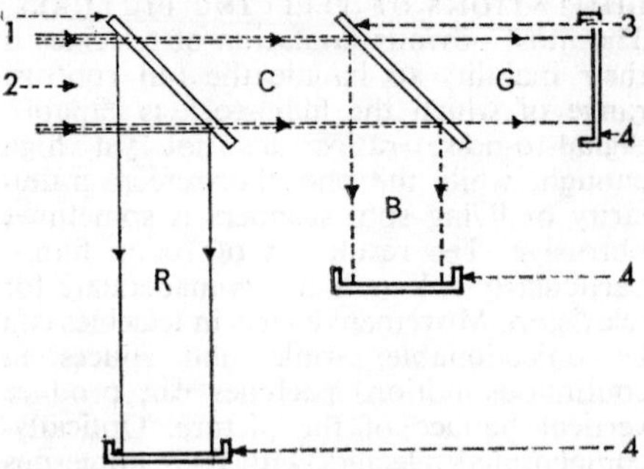

PRINCIPLE OF COLOUR SEPARATION. C. Cyan. R. Red. G. Green. B. Blue. Incoming coloured light beam from film, 2, strikes red-reflecting dichroic mirror, 1, redirecting red component at right-angles, and transmitting cyan (minus-red) beam to blue-reflecting dichroic mirror, 3, leaving green component for onward transmission. 4. Pick-up receptors for the separated colour components.

process in the viewer's receiver. When all the processes combine to produce a colour television picture, significant colour errors can occur; at the same time, the available television electronic techniques offer the possibility of correcting certain of these colour errors, including some due specifically to the photographic processes.

Nevertheless – despite the colour errors already mentioned – the convenience and availability of motion pictures have led to widespread use of colour film in broadcasting.

COLOUR ANALYSIS – GENERAL

In principle, the colour film has already analysed the scene to be transmitted into the requisite red, green and blue components and it is only necessary for the telecine to measure the relative densities of the cyan, magenta, and yellow film layers, ideally at spectral frequencies at which the ratios of the wanted dye response to the unwanted responses are maximum. But this approach comes up against an obstacle which has so far prevented its adoption: the need for securing the highest possible signal-to-noise ratio in colour television broadcasting. Noise level is therefore a crucial factor and calls for more detailed consideration.

SIGNAL NOISE. All signal sources produce random fluctuations, called noise, in the wanted signals. The frequency spectrum of the noise signals is normally much wider than that of the television signals, and noise components occur at all signal frequencies in the television band.

The unwanted noise signals are processed with the wanted television signals, and further noise signals may be inadvertently added during transmission and reception.

The television signals which are demodulated at the receiver may contain noise fluctuations which have come from any of the camera tubes or telecine photocells and these components will be added or subtracted during the signal matrixing processes, in the same way as the red, green and blue primary signals. During the signal decoding process at the receiver, these noise components will be decoded along with the red, green and blue signals.

However, noise which has been picked up in the propagation process, or at any time after the signal leaves the studio encoder, will also be treated as if it were a genuine television signal. In particular, noise in the high frequency part of the luminance band, say between 3 and 5·5 MHz, which will only produce small-area brightness fluctuations on a monochrome receiver, may produce low frequency (0→1·4 MHz) colour noise on a colour receiver. The receiver treats any signal around the sub-carrier frequency as a low frequency colour signal. For example, a signal at either 4·44 or 4·42 MHz (British Standards), i.e., 4·43±0·01 MHz, will be decoded as a 10 kHz colour signal. The general effect is to produce rather noticeable horizontal streaks of varying colour on the colour receiver. Precisely the same television signal may produce quite a clean monochrome picture on a black-and-white receiver. Such colour noise is called parc-noise and is analogous to the cross-colour effects produced by high frequency luminance signals.

In practice, therefore, telecines treat the light from colour film as if it had the same spectral energy distribution as an original scene. Colour vidicon telecines, in fact, can have the same type of colour analysis optics as live pick-up television cameras.

SEPARATION OF COLOUR IMAGES. Both vidicon and flying-spot telecines use the same principles to separate out the red, green and blue images. Partially-silvered mirrors could be used to split the light into three paths, with a red filter of dyed gelatin in one path, a green filter in the second path and a blue filter in the third path. A camera tube or photocell would then convert each light beam into an electrical signal.

Such arrangements make an inefficient use of the incoming light since the thin layer of silver on the mirrors reflects equally the light of all frequencies. The red filte then absorbs all the reflected green and blue light which, if it had not been reflected by the partially-silvered mirror, might have been used to form the green and blue signal components.

DICHROIC MIRRORS. By using interference effects, it is possible to make mirrors which reflect wavelengths at one end of the spectrum but transmit all other visible light (for example – a thin film of oil on water).

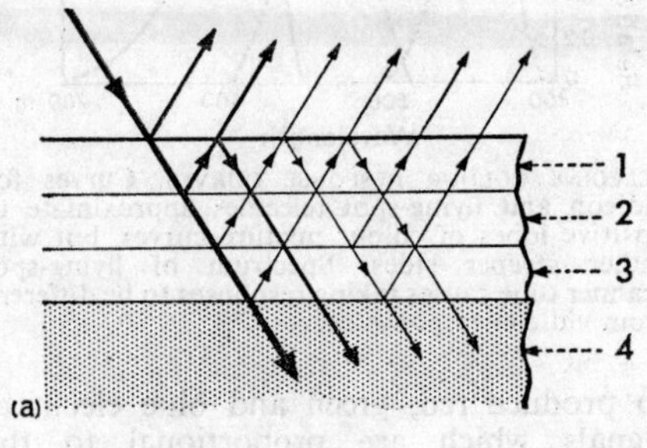

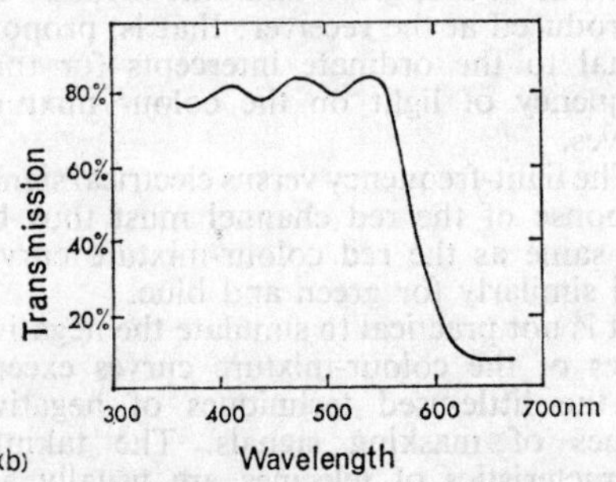

OPERATION OF DICHROIC REFLECTOR. (a) 1, 2, 3. Layers of alternating high and low refractive index. 4. Glass substrate. Layer thicknesses are so controlled that the principal rays emerge in phase for some wavelengths and out of phase for others. Rays which have been reflected several times are less important. (b) Transmission response of 7-layer dichroic transmitting blue and green light and reflecting red light.

Dichroic mirrors are plates of glass which have been coated with very thin transparent layers of alternately high and low refractive index (n) materials; for example, zinc sulphide (n=2·3) and magnesium fluoride (n=1·36). At each interface some of the incident light is reflected; the thicknesses of the layers are adjusted so that, for example, the reflected red rays cancel each other, while the blue reflected rays reinforce. There may be as many as ten or twenty layers. The changeover from reflection to transmission may occur in a band of wavelengths of about 40 millimicrons (mμ); the more layers there are, the sharper the transition that can be made.

The centre frequency of the changeover is determined by the thickness of each layer and the angle of incidence of the light. An increase in transition wavelength of 30 mμ may occur in changing from 45° to normal incidence; if the light is not collimated at the dichroic mirror, there may be a change of colour from one part of the raster to another.

The rear suface of the glass may produce displaced images of the more intense parts of the picture, but whether this is important or not depends on the design of the optical system as a whole.

ASTIGMATISM IN DICHROIC MIRRORS. The finite thickness of the glass on which the dichroic layers are deposited produces an asymmetry between the vertical and horizontal directions which may have to be corrected. Two similar sheets of glass,

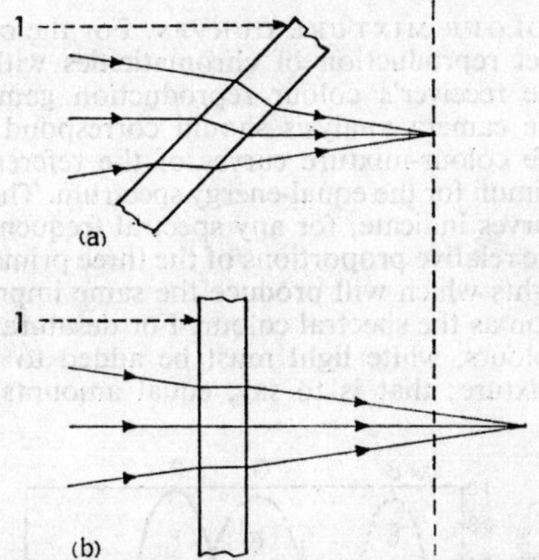

ASTIGMATISM PRODUCED BY MIRROR SUBSTRATE. Glass plate, 1, supporting dichroic surfaces has a different effect on a converging beam in the horizontal plane (a) and the vertical plane (b), thus producing an astigmatic focus difference between horizontal and vertical lines in the image.

without the dichroic layers, may be inserted in the light path at the same angle as the dichroic mirrors but rotated 90° about the axis of the light beam. Additional corrector plates may be needed to ensure that all rays pass through the same total thickness of glass.

REFERENCE COLOURS. The colour analysis is determined by the choice of primaries for the display tube, and the reference white adopted for the receiver. These have been standardized for most simultaneous colour systems at the values adopted by the American FCC for the NTSC system.

CIE CHROMATICITY CO-ORDINATES FOR THE FCC REFERENCE STIMULI

	x	y	u	v
Red	0·670	0·330	0·477	0·352
Green	0·210	0·710	0·076	0·384
Blue	0·140	0·080	0·152	0·130
White	0·310	0·316	0·201	0·307

Although the original shadow-mask display-tubes had phosphors which glowed with very nearly the correct chromaticities, contemporary display-tube screens depart considerably from the recommended standards, including the rather blue reference-white to which their makers recommend that they should be normalized. In spite of this, the transmission analysis is based on the assumed use at the receiver of the FCC primaries.

COLOUR MIXTURE CURVES. For the correct reproduction of chromaticities within the receiver's colour reproduction gamut, the camera analysis should correspond to the colour-mixture curves of the reference stimuli for the equal-energy spectrum. These curves indicate, for any spectral frequency, the relative proportions of the three primary lights which will produce the same impression as the spectral colour. For desaturated colours, white light must be added to the mixture; that is to say, equal amounts of

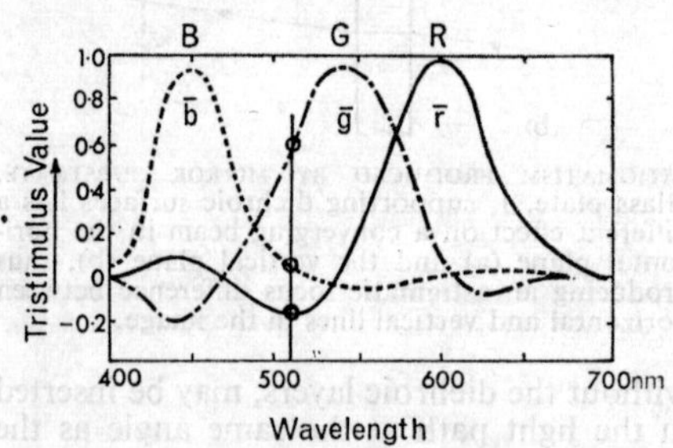

IDEAL COLOUR ANALYSIS CURVES. Based on use at receiver of FCC reference stimuli. Relative amounts of light required to match given spectral colour are found by drawing vertical ordinate through frequency of spectral colour, as shown at 510 nm. Intersection of ordinate with the three curves gives relative proportions of red, green and blue lights required. The occurrence of negative lobes should be noted, i.e., portions of curves falling below zero value.

all three primary lights must be added at the receiver.

The colour mixture curves are thus the ideal taking-response curves for the telecine colour analysis. For any spectral colour in the film image, the colour analyser is required

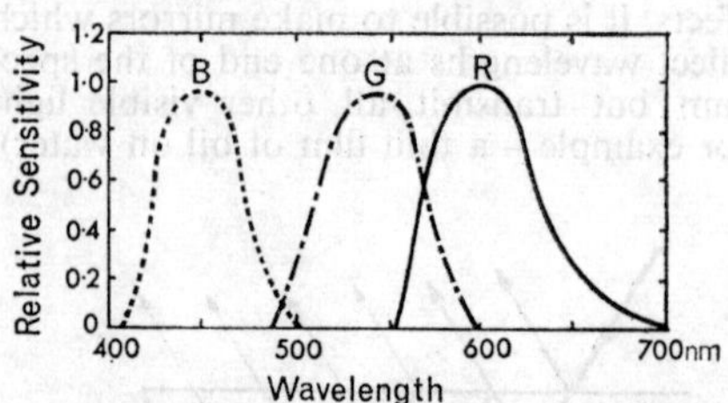

TELECINE COLOUR RESPONSE CURVES. Curves for vidicon and flying-spot telecines approximate to positive lobes of colour mixture curves, but with rather steeper sides. Spectrum of flying-spot scanner tube causes taking responses to be different from vidicon response curves.

to produce red, green and blue electrical signals which are proportional to the amounts of red, green and blue lights to be reproduced at the receiver: that is, proportional to the ordinate intercepts for that frequency of light on the colour mixture curves.

The light-frequency versus electrical signal response of the red channel must thus be the same as the red colour-mixture curve, and similarly for green and blue.

It is not practical to simulate the negative lobes of the colour-mixture curves except by the little-used techniques of negative values of masking signals. The taking-characteristics of telecines are usually arranged to approximate to the positive lobes of the colour mixture curves. Slightly reducing the width of the skirts of the curves gives some correction for the reproduction errors which are caused by omitting the negative lobes.

COLOUR ANALYSIS IN VIDICON TELECINES

THREE-TUBE TELECINES. The normal type of intermittent film projector is used to transport and illuminate the colour film. A single optical system throws an image of the stationary film frame into the three-tube colour camera. It is usually necessary to use relay optical systems for the image formation, so that there is room to incorporate the colour-splitting devices. Dichroic mirrors separate the incoming light into its component red, green and blue images, and front-silvered mirrors reflect the light into

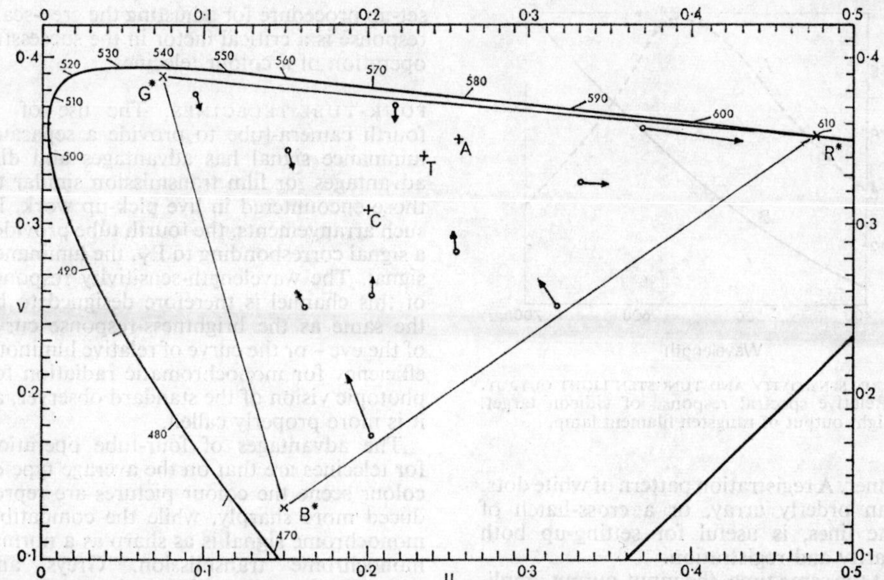

ERRORS RESULTING FROM TAKING POSITIVE LOBES ONLY. The original colour is represented by the tail of the arrow, and the reproduced colour is indicated by the position of the arrowhead. It will be seen that, in general, most of the chromaticities suffer some decrease in purity, but that there are significant dominant wavelength changes for some of the red chromaticities. The decrease in purity shown is roughly proportional to the original purity of the chromaticity which is being reproduced. In general, the shape of the positive lobes does not vary a great deal with a change in the reproducing reference stimuli; such changes mainly affect the optimum shape for the ideal negative lobes. A. Standard Illuminant A. C. Standard Illuminant C. T. Proposed Television Illuminant.

three camera tubes which for convenience are mounted alongside each other.

Red, green or blue light-filters are inserted into each light path to trim the wavelength-sensitivity response curve for each channel so that the overall response of the system, including light source, optical components, dichroics, correcting plates, trimming filters, and vidicon target sensitivity, corresponds

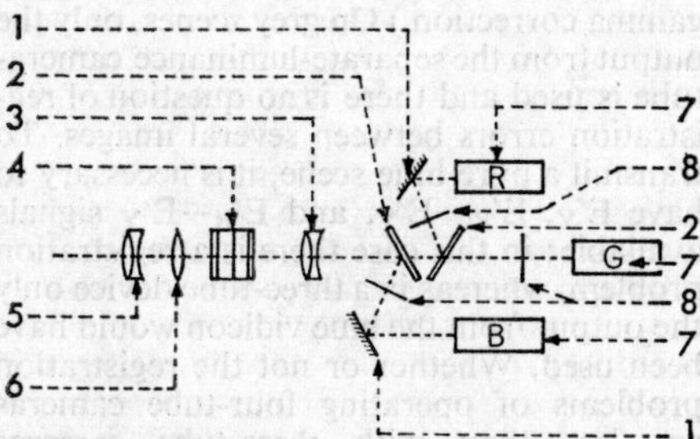

THREE-TUBE VIDICON TELECINE CAMERA. Schematic layout. 5. Objective lens. 6. Condenser lens. 4. Flat plates to correct mirror astigmatism. 3. Relay lens. 2. Blue-reflecting and red-reflecting dichroic mirrors. 8. Filters. 7. Vidicons. 1. Non-selective mirrors.

to the appropriate positive lobe of the colour-matching curves. Given the percentage transmission or response, at any one wavelength, for each of the contributing factors, then the overall response is obtained by multiplying the percentage responses together.

In all other respects than their wavelength-sensitivity response, the red, green and blue channels are required to have identical performance characteristics. In particular, it is important to ensure that the transmitted red, green and blue images are precisely in register. To this end, care is taken to ensure that the image and the dissecting scanning raster are identical on each vidicon target. This is a very difficult requirement to achieve, and some registration errors in the transmitted picture usually remain.

Static registration of at least one part of each colour image can be obtained by altering the amount of vertical and horizontal shift applied to two of the rasters relative to the third.

Registration of the other parts of the image is then a matter of adjusting the geometry of each scan in an appropriate

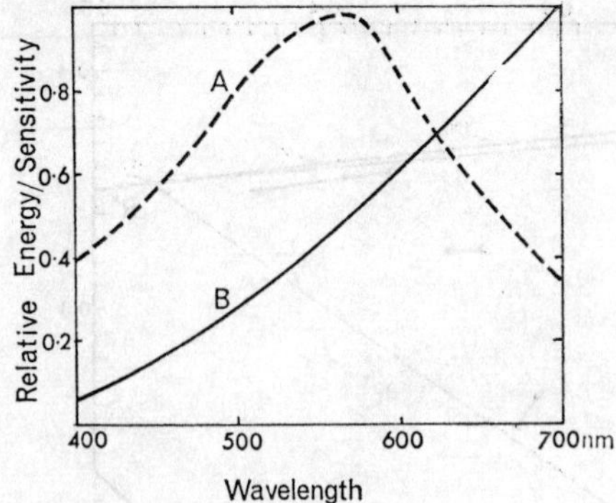

VIDICON SENSITIVITY AND TUNGSTEN LIGHT OUTPUT. A. Relative spectral response of vidicon target. B. Light output of tungsten filament lamp.

manner. A registration pattern of white dots, in an orderly array, or a cross-hatch of white lines, is useful for setting-up both linearity and registration.

At the same time, the input-output amplitude characteristic of each channel must be identical, otherwise a neutral grey-scale will be reproduced with varying amounts of unwanted colouring at different luminance levels.

To ensure the identical operation of each camera tube, neutral density filters may be inserted in two of the optical paths.

Differential amplifiers feeding the waveform oscilloscope enable an accurate assessment of the difference in output response of each channel to a grey-scale to be made. Lift (blanking level), gain and gamma correctors must then be adjusted to minimize these variations in input-output characteristic.

The eye is very sensitive to departures from neutrality of the grey-scale, and the

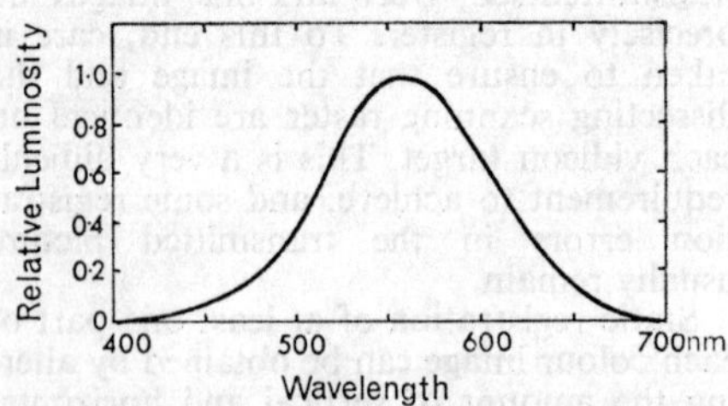

LUMINOSITY FUNCTION FOR PHOTOPIC VISION. Internationally agreed curve for a standard observer for vision in bright light. Represents relative subjective brightness of light from different parts of spectrum, assuming same energy in each section of equal bandwidth. Sometimes called the Vλ curve.

set-up procedure for adjusting the grey-scale response is a critical factor in the successful operation of a colour telecine.

FOUR-TUBE TELECINES. The use of a fourth camera-tube to provide a separate-luminance signal has advantages and disadvantages for film transmission similar to those encountered in live pick-up work. In such arrangements, the fourth tube provides a signal corresponding to E_Y, the luminance signal. The wavelength-sensitivity response of this channel is therefore designed to be the same as the brightness-response curve of the eye – or the curve of relative luminous efficiency for monochromatic radiation for photopic vision of the standard observer, as it is more properly called.

The advantages of four-tube operation for telecines are that on the average type of colour scene the colour pictures are reproduced more sharply, while the compatible monochrome signal is as sharp as a normal monochrome transmission. Greys and whites in the colour picture do not depend on the precise registration of the three colour tubes, all of which can be of relatively poor resolution since they are used only for the narrow-band chrominance signals.

The incoming light beam is split up into four paths, the separate-luminance channel and three other paths corresponding to the normal red, green and blue trichromatic channels. Some loss of sensitivity may be experienced compared with the three-tube arrangement, but adequate light-intensity is available from a standard projector.

REGISTRATION PROBLEMS. The red, green and blue channels are used to provide colour-difference signals only, i.e., $E'_R - E'_Y$, $E'_G - E'_Y$, and $E'_B - E'_Y$, (dashes signify gamma correction.) On grey scenes, only the output from the separate-luminance camera-tube is used and there is no question of registration errors between several images. To transmit a pure blue scene, it is necessary to have E'_Y, $E'_R - E'_Y$, and $E'_B - E'_Y$ signals available; in this case there is a registration problem, whereas in a three-tube device only the output from the blue vidicon would have been used. Whether or not the registration problems of operating four-tube cameras are less than with three-tube cameras depends on the type of colour picture which is normally transmitted, and to a first approximation it may be said that the four-tube arrangement becomes increasingly to be preferred as the colour saturation of the

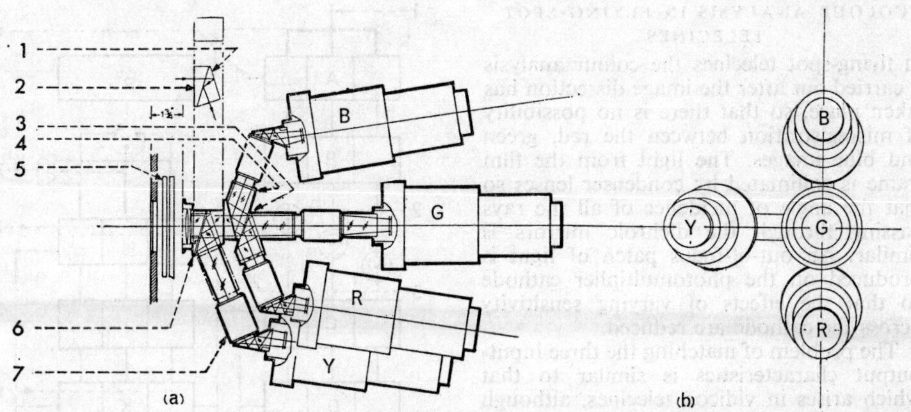

FOUR-TUBE VIDICON TELECINE CAMERA. The four-tube camera employs three colouring (chrominance) tubes, R, G and B, and in addition a fourth tube, Y, providing a signal corresponding to E_Y, the luminance signal. (a) General optical layout of camera. Light enters camera from left through objective lens (not shown). 5. Neutral density and colour filter wheels. 2. Luminance (Y) prism, with fully-reflecting surface, 1, shown displaced from its position on the optical axis for clarity. 6. YRGB dichroic-surfaced prism to give the light reflected to the Y tube the approximate shape of the luminosity function of the standard observer. 7. Relay lens in each optical path to the four vidicons to provide the necessary extension of length. 4. Colour trimming filters in each path. 3. Dichroic prisms to eliminate the astigmatism, ghost images, and colour shading of dichroic mirrors coated on glass plates. (b) Rear view of optical system, looking from vidicons towards lens.

picture decreases.

The range of colour saturation obtainable with contemporary film is limited, and the statistical distribution of colours in typical television scenes shows a marked preponderance of low-saturation areas. In the present state of the art, four-tube vidicon telecines show less registration error than three-tube telecines. The first effect of misregistration is not to show noticeable colour fringes but to reduce the apparent edge-sharpness (acutance) of objects in the picture.

SIGNAL MAKE-UP AND GAMMA-CORRECTION. When the signal from the separate-luminance tube has been gamma-corrected it becomes $E_Y^{1/\gamma}$; that is to say, it is not the compatible monochrome signal normally transmitted, which is

$$E'_Y = 0{\cdot}30E'_R + 0{\cdot}59E'_G + 0{\cdot}11E'_B.$$

Since

$$E_Y^{1/\gamma} = (0{\cdot}30E_R + 0{\cdot}59E_G + 0{\cdot}11E_B)^{1/\gamma}$$

the separate luminance signal is larger than the compatible monochrome signal E'_Y except for greys, when the two signals become identical.

Transmitting $E_Y^{1/\gamma}$ instead of E'_Y leads to noticeable colour errors in receivers which have been designed to operate with E'_Y.

These errors can be modified by also transmitting $E'_R - E_Y^{1/\gamma}$ instead of $E'_R - E'_Y$, and $E'_B - E_Y^{1/\gamma}$ instead of $E'_B - E'_Y$, when a non-linear circuit in the receiver can correct the value of the green-difference signal which is obtained.

Various other methods of reducing the colour-errors caused by separate-luminance operation and the subsequent transmission of $E_Y^{1/\gamma}$ have been proposed, but current practice is to use the red, green and blue camera channels to produce $E'_R - E'_Y$ and $E'_B - E'_Y$ without reference to the separate-luminance signal. The $E_Y^{1/\gamma}$ is corrected for its increased amplitude by a low-frequency correction signal derived from the colour tubes.

The large areas of the picture are then colorimetrically correct, while the fine detail still enjoys the increased resolution of the uncorrected E'_Y signal.

The various methods of formulating the transmitted signals give rise to different signal-to-noise ratios in the received picture, but the differences are small. If $E_Y^{1/\gamma}$ is transmitted it is, in principle, possible to make the colour receiver obey the constant luminance principle.

At the same time, the grey-scale reproduction on monochrome receivers is correct if the contribution from the rectified sub-carrier dots is avoided.

COLOUR ANALYSIS IN FLYING-SPOT TELECINES

In flying-spot telecines the colour analysis is carried out after the image dissection has taken place, so that there is no possibility of misregistration between the red, green and blue images. The light from the film frame is collimated by condenser lenses so that the angle of incidence of all the rays passing through the dichroic mirrors is similar. An out-of-focus patch of light is produced on the photomultiplier cathode so that the effects of varying sensitivity across the cathode are reduced.

The problem of matching the three input-output characteristics is similar to that which arises in vidicon telecines, although the linear nature of the signals from the photomultipliers makes the task somewhat easier. There is little advantage in separate-luminance operation; since the photomultiplier outputs are linear, E_Y can be formed, if desired, before gamma correction, by matrixing the red, green and blue output signals after they have been corrected for afterglow and aperture distortion.

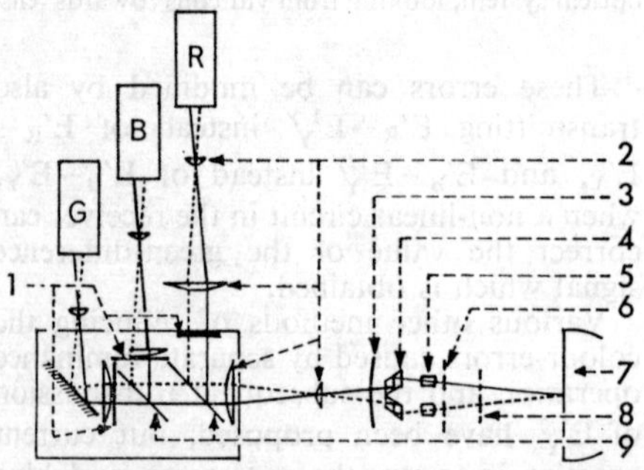

THREE-TUBE FLYING-SPOT COLOUR TELECINE. Optical layout. Scanner tube, 7, with light interrupted by shutter, 8. Shutter, 6, precedes object lenses, 5, and beam-folding rhombs, 4. Up to film path, 3, layout is identical with black-and-white telecine. Dichroics, 9, with condenser lenses, 2, and filters, 1, effect trichromatic analysis to three chrominance tubes, R. G. B.

PHOSPHOR COLOUR RESPONSE. Ideally, the phosphor should glow with an equal-energy white light and have the same short afterglow to all colours; the latter can be achieved by using a single-component phosphor screen but the colour response then leaves much to be desired. The light output is relatively poor at the red end of the spectrum, although it continues at a low level into the infra-red region. The discrepancy in red, green and blue signal amplitude on grey can be overcome by adjusting the relative gains in the photomultiplier cells.

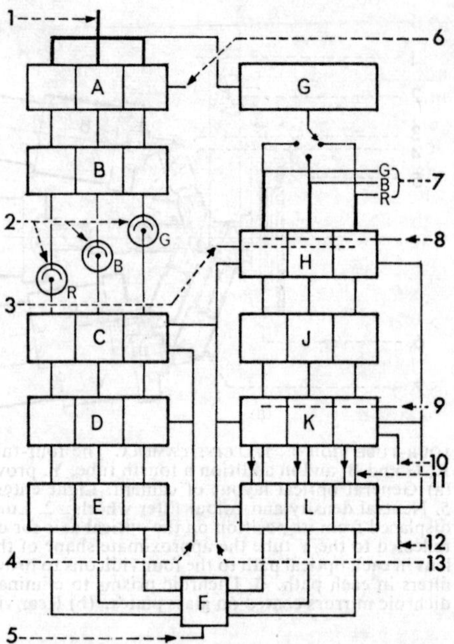

FLYING-SPOT COLOUR TELECINE: VIDEO CHANNEL. Block schematic: each function triplicated to provide a separate chain for the three primary colour signals. 1. Power supplies. A. Photocell HT (B) supply unit. B. Photocell network. 2. Photocells. C. Video head amplifier. D. Video pre-amplifier. E. Afterglow and phase corrector. 4, 6. 250 v. stabilized supply. F. Signal selector switch. 5. Test signal facility. 7. Composite video signal outputs checked by, G, monitors. H. Blanking and sync mixing amplifier with, 3, mixed blanking and, 8, mixed sync pulses externally supplied. J. Network and phase corrector. K. Gamma and black level control with clamp pulses, 9. 11. Black level clamp to L, video B amplifier. 10, 12, 13. Stabilized power supplies at +300 v., +250 v. and −100 v. respectively.

PHOTOMULTIPLIER COLOUR RESPONSE. The colour response of the photomultiplier is largely determined by the type of cathode employed. Normally a bi-alkali type is used for the green and blue cells and a tri-alkali type for the red photomultiplier.

COLOUR FILM REQUIREMENTS

LUMINANCE RANGE. Neither film nor television is yet able to reproduce either the full range brightness, or the complete gamut of chromaticities, which can occur in everyday scenes. Furthermore, the ability of television to reproduce what is recorded on the colour film is rather different from the ability of the film to record natural scenes.

The colour film may well record a luminance range of 150 : 1 or more. On the other hand, the ability of the television system to reproduce such a luminance range is largely determined by the performance of the shadow-mask tube and the conditions in which it is viewed. With the limited highlight brightness of contemporary display tubes, a contrast range of 50 : 1 for the large areas of the picture is probably a good average. The luminance ratio which can be obtained between small adjacent areas is very much less than the large area ratio, due to halation effects, excitation of phosphor grains by adjacent grains, and stray illumination. Again, the small-area ratio which can be obtained with television is less than can be recorded on colour film.

In photography it is now axiomatic that the most pleasing reproduction of a scene is obtained when the overall transfer characteristic is such that the brightness range of the reproduction process is used to the full, but only just to the full. That is to say, the reproduced brightness scale may be compressed or expanded with respect to the original scene, but the tone range of the receiver should be just fully utilized. Within limits, the telecine can use over-gamma correction of the television signals as a method of compressing the film contrast range to that of the overall television system.

CHROMATICITY GAMUT. Using a technique such as over-gamma correction to reduce the contrast range of the scene has the undesirable effect of reducing the saturation of the reproduced colour. The nonlinear correction has to be applied to all three signals and the amount of correction increases with the level of each tristimulus signal so that the largest signal suffers the greatest reduction.

However, limitations in the transmission characteristics of the available film dyes cause the reproduced purities to be significantly less than those in the original scene. At the same time, the colour analysis carried out by the telecine reduces the reproduced saturation still further. Desaturation of the reproduced colours with respect to the original scene is thus a hazard of using colour film in colour television; at the same time, there will be luminance errors and some hue errors.

In an attempt to overcome the effects of dye deficiencies, many colour photographic processes are designed to have average gamma values greater than unity so as to increase the saturation of the film record. Such high values of film print gamma make the matching of the film luminance range to the television reproduction range more difficult.

The gamut of chromaticities which the film will record is large, and in general can be matched by the available gamut of chromaticities at the television receiver. Whereas the subtractive process of the film can produce its maximum saturations only at low transmission ratios, because of unwanted absorptions in the dyes, the television receiver, with its additive colour system, produces its maximum saturations at relatively high luminance levels. For any given colour at a particular purity, the film process limits the maximum brightness at which it can be reproduced, while the television system limits the minimum brightness at which it can be displayed.

These reproduction limitations of the combined film-television process are insignificant for yellow colours but are noticeable in the reproduction of pure blues and highly saturated green colours.

FILM PROCESSING. In monochrome television, it is possible to adjust the exposure and the film processing to suit the particular requirements of television. Little can be done in this respect with colour film. Integral tripack materials present very complex problems in design, since the lower layers can only be exposed and processed through the outer layers: it is essential that the balance of the layers is maintained, so that the reproduction of greys remains neutral over the whole density range. In practice, it appears that commercial processing of colour film is possible only under standardized conditions and that it is impracticable to attempt to develop colour film to a lower gamma than normal.

Conversely, incorrectly-exposed film exhibits such marked errors in colour reproduction that it is rarely suitable for broadcasting. Colour telecines must thus accept the contrast law of the colour film as normally processed, but do not have to cater for wide variations in minimum density.

FILM SHOOTING TECHNIQUES. The easiest and most satisfactory control of luminance range and chromaticity gamut is that available when dressing and lighting a scene before film shooting begins. Fully illuminated white objects of any significance in the scene should not have a reflectance value greater than 0.7 and, except for special effects, it is useful if there is always a reference white with about this reflectance in the scene.

Black areas of any size, say greater than 1 per cent of the picture area, should not have any significant detail in them if their reflectance value is less than 0·04. At the same time, the ratio of key light plus fill light to the fill light alone should not be more than 2 : 1 in the key position. Backgrounds should be held down to an illumination of about half that of the foreground.

Such precautions can do much to ensure that the brightness range of the colour film print matches that of the television system, and that flesh tones are satisfactorily placed on this grey scale; the effect of a flatly lit scene can be avoided by the use of judicious backlighting and modelling. Since both the film processes and the television analysis desaturate chromaticities, it is advantageous to use rather more colourful drapes, costumes and sets than would be used for a theatre production. Make-up should slightly exaggerate the natural colouring, and should be designed to bring all characters rather closer together in their skin colouring than is conventional.

Television films are viewed under rather different ambient conditions from theatre release prints. It is usual for television receivers to be looked at in ambient light with many familiar objects in the field of view. Such conditions tend to stabilize the adaptation of the eye and make it more critical of variations in colour balance between individual shots and different reels. It is thus more important in shooting colour film for television to ensure that colour balance variations due to both lighting and processing are reduced to a minimum. One recommendation is to preface each shot with a grey-scale illuminated by the scene lighting, and each processing run with a piece of standard exposure density step wedge.

TELECINE ALIGNMENT. Before transmission, it is necessary to ensure that the telecine is in colour balance by adjusting the tracking of all three channels and altering the relative gains of the channels so that they produce the same signal output on neutral greys and whites. It is helpful in overcoming variations between films if the grey scale used for the telecine colour balance has been shot, printed and processed on the same stock and at the same time as the film which is to be transmitted.

MONOCHROME REPRODUCTION. For many years to come, it is anticipated that a substantial proportion of the audience viewing a colour film broadcast will be receiving the transmission in monochrome. All important areas of the scene must therefore be differentiated not only in colour but also in luminance. However, as is well known from many years of shooting colourful scenes on panchromatic black-and-white film stock, it is possible for materials which have good colour contrast to have little brightness separation.

Some experienced art directors can assess the relative luminance values in a scene by viewing it through a 100 : 1 neutral density filter; at low light levels, the eye – like the television system – desaturates colours. A closed-circuit monochrome television camera with a spectral response corresponding to the $V\lambda$ curve for photopic vision, is perhaps more precise.

COLOUR ERRORS IN COLOUR FILM REPRODUCTION

The colour television transmission is designed on the assumption that the viewer's eye-mind complex behaves in the same way as that of the so-called "standard observer". This is probably true in a statistical sense, although the standard observer data is based on a very small sample of the population. Individual viewers, however, may see the reproduced colour picture in different hues; this is particularly true of male viewers as the incidence of anomalous colour vision is far higher in men (8 per cent) than in women (0·2 per cent).

The television transmission also assumes that the receiver will use the FCC primary colours for synthesizing the colour picture, and that for neutral greys the receiver will reproduce Standard Illuminant D, corresponding to overcast north-sky daylight. It is further assumed that the viewer's eyes will be adapted to this rather cold daylight and the transmission is monitored by the producer on this basis, for pleasing rather than accurate colour reproduction.

Despite the subjective licence given to the producer, the engineering of the system is based essentially on producing as accurate a colour match to the original as is expedient. However, certain rather basic factors limit what is practicable in this respect.

SCENE ILLUMINATION. If he were present at the original scene, the viewer's eyes would have adapted to the prevailing colour balance and he would probably have been

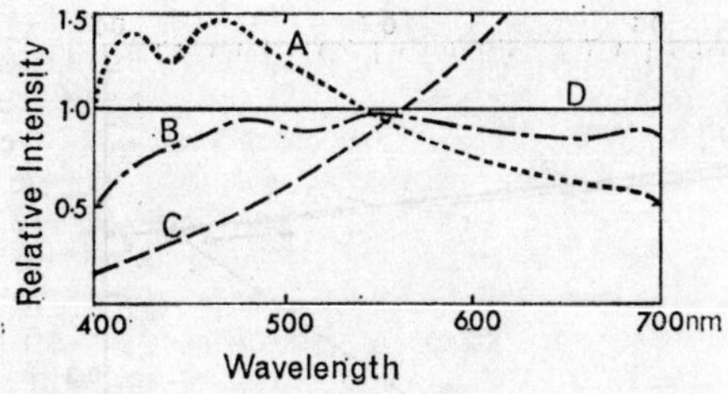

LIGHT SOURCES AND ENERGY DISTRIBUTIONS. A, North sky, B, sunlight, and C, tungsten lamp, have widely different spectral energy distribution with wavelength, but the eye adapts to the illumination of the scene it is observing. D. Equal energy white for comparison. All curves equalized at 550 nm.

unaware of the colour temperature of the scene illumination. The film records the scene in more absolute terms; if it is negative film then a substantial correction can be made to the colour balance when it is printed on positive film stock.

NEGATIVE FILM ANALYSIS ERRORS. As the dyes used to restrict the wavelength responses of the tripack layer are far from ideal and have unwanted responses outside the desired acceptance band, the negative film will not represent the original scene with colorimetric fidelity. Some correction for these unwanted responses may be obtained by the use of integral colour masks in the negative film itself.

The errors due to the unwanted lobes represent a contamination of the desired signal by the other signals, i.e., crosstalk

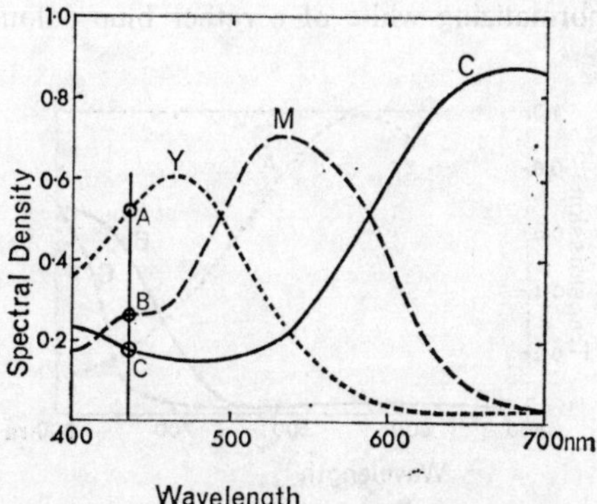

TYPICAL DYE RESPONSE CURVES FOR COLOUR FILM. Yellow, magenta and cyan responses, with ordinate at 430 nm exhibiting a response not only in the yellow dye layer (A), but in the magenta (B) and cyan (C) layers, because of overlapping skirts of curves.

between the three colour channels. In general, the film record represents colours with lower saturation and lower luminance than they had in the original scene, together with small hue changes.

FILM PRINTING ERRORS. These are similar in character to those of negative film, but integral colour masks cannot be used in the positive film stock because of the overall colour cast which they produce to visual inspection. On the other hand, complete control of the illumination is possible in the printing process, and corrections can be applied for overall colour casts in the negative film.

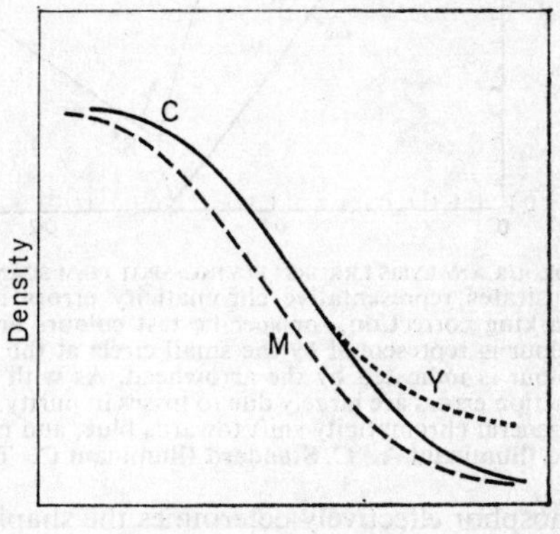

RESPONSE OF FILM DYE LAYERS. Characteristic curves of cyan, C, yellow, Y, and magenta, M, dye layers in colour film may not track or coincide precisely over the exposure range.

In the case of both negative and positive film, imperfections in the film stock and in the processing can cause relatively high or low density in one or two of the film layers, leading to an overall colour cast. It is also possible for the three layers to track in density in the mid-tones but to depart from colour balance in the highlights and/or in the shadows.

In television terms, one or more of the layers may be said to exhibit white crushing with respect to the other two layers, or black crushing.

TELECINE ANALYSIS ERRORS. In general, errors due to the colour analysis curves of the telecine are smaller than those due to the imperfect film dyes, but are additive to those errors. In flying-spot telecines, the restricted light-output of the scanner-tube

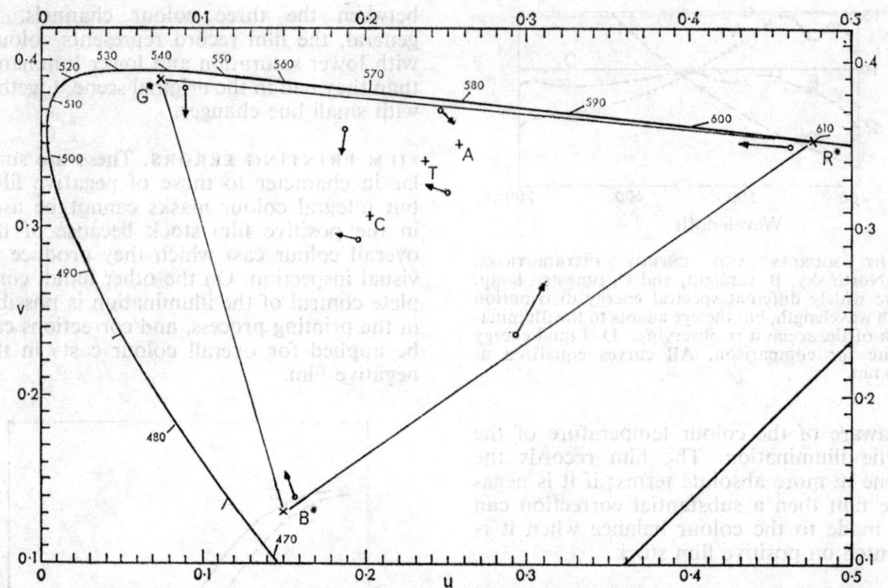

COLOUR ANALYSIS ERRORS: FLYING-SPOT FILM SCANNER. This 1960 Nearly Uniform Chromaticity scale diagram illustrates representative chromaticity errors in the reproduction of a flying-spot film scanner without masking correction, for specific test colours around the colour triangle. The chromaticity of the original colour is represented by the small circle at the tail of the arrow, while the chromaticity of the reproduced colour is indicated by the arrowhead. As with errors resulting from taking positive lobes only, the reproduction errors are largely due to losses in purity. At the same time, it should be noted that cyan colours show a general chromaticity shift towards blue, and magenta colours show a similar shift towards red. A. Standard Illuminant A. C. Standard Illuminant C. T. Proposed Television Illuminant.

phosphor effectively determines the shaping of the high-frequency side of the blue taking-curve and the low-frequency side of the red taking-curve, and the resulting curves are rather too narrow.

In both vidicon and flying-spot telecines, the response to radiation frequencies just outside the visible spectrum is small but not always negligible. Any combined response of the film and telecine to non-visible radiation will produce a difference between the appearance of the film on optical projection and its appearance on television.

For example, most colour film stocks are transparent to infrared rays of wavelength greater than 800 nm, even in the black areas. However, certain imbibition type prints have a significant transmission to wavelengths from 680 nm upwards. In such cases, the telecine will integrate the combined response of the overall system to wavelengths from the red response lobe upwards; the display tube will then reproduce this integrated response at the chromaticity of the red primary.

The effect will be to produce an undesired red flare signal which will be particularly noticeable in the shadow areas. In this case, a sharp cut filter which rejects all wavelengths above 670 nm will remove the effect.

RECEIVER SYNTHESIS ERRORS. Most manufacturers of colour tubes recommend that they should be set up to produce a normalizing white of a rather blue colour

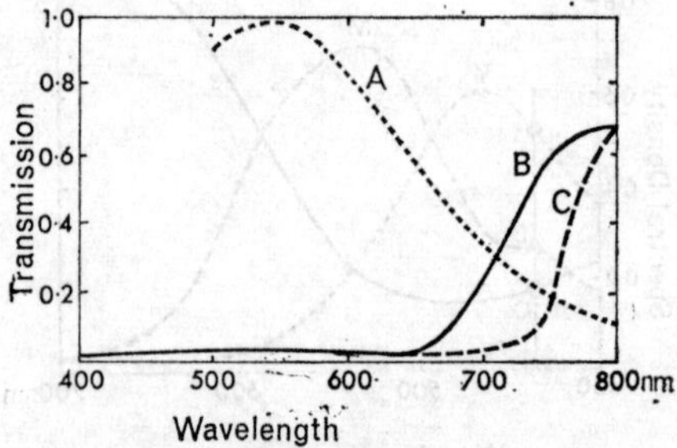

VIDICON INFRA-RED RESPONSE. Transmission through black areas of, B, imbibition print and, C, integral tripack print. Intersection with A, response curve of red-sensitive vidicon, indicates spurious response in infra-red region causing undesirable red flare, especially in shadow areas, unless sharp-cut filter is used.

(10,000°K) rather than Standard Illuminant D; their intention is to obtain a higher picture brightness, which is normally limited by the output of the red phosphor. On the other hand, there is evidence that viewers prefer the colour rendition which results when a warmer white is used by the receiver; a Television Illuminant has been proposed at the chromaticity u = 0·23, v = 0·33.

The transmission is gamma-corrected to an exponent of 2·2, whereas shadow mask tubes are usually operated under conditions where they have a gamma of around 2·8. This results in some hue errors, and an increase in saturation for colours with purities around 30 to 70 per cent.

Ambient light at the receiver reduces the luminance contrast range, desaturates most of the colours and, if the ambient light has a different chromaticity from the receiver normalizing white, adapts the viewer's eyes away from the receiver colour balance. The widespread use of all-sulphide phosphors and rare-earth activators has raised the brightness of television displays and reduced these effects. At the same time, since the new phosphors have chromaticities which are significantly different from the FCC reference stimuli, they have introduced other colour errors. As far as is practicable, the new phosphors obey the Ives-Abney-Yules compromise, and the chromaticity errors resulting are largely desaturation errors, together with luminance errors. Some improvement in colour accuracy would be obtained if the camera taking-responses were matched to the new phosphors, but the required changes in the positive lobes of the colour mixture curves are small; it is obviously impractical to change the taking-responses every time a new display-tube is marketed.

ELECTRONIC CORRECTION OF COLOUR ERRORS

Many of the colour errors which can occur in the complete film-telecine-receiver-viewer system fall into two distinct classes; a loss of colour balance at some or all luminance levels, and a loss of saturation in some or all chromaticities. Both these types of error

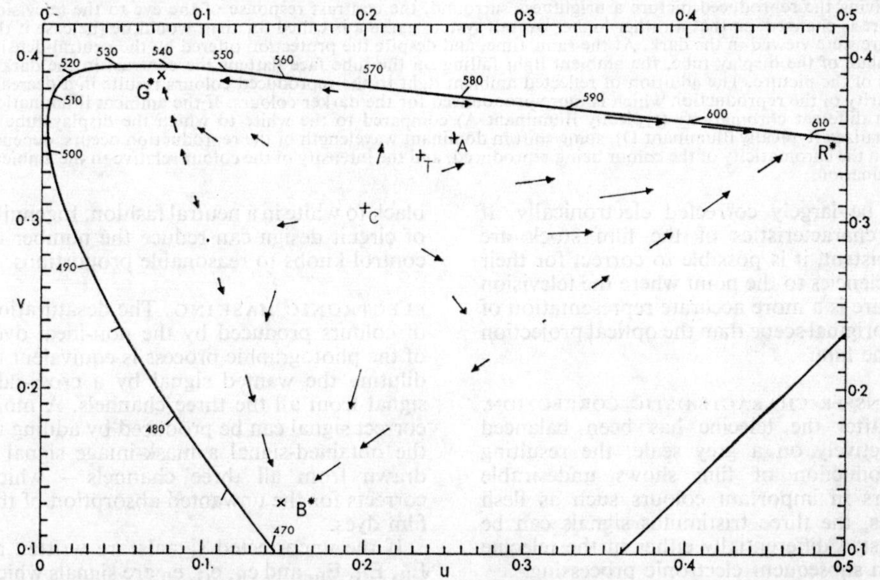

CHROMATICITY ERRORS ON DISPLAY TUBES. The signal formulation of colour television systems calls for the primary colour separation signals to be gamma-corrected to a value of =1/2·2. When these signals are reproduced on a colour display tube with a higher gamma than 2·2, significant colour errors result. The magnitude of the errors are indicated on the CIE Nearly Uniform Chromaticity scale diagram, where the chromaticity of the original colour is represented by the tail of the arrow, while the chromaticity of the reproduced colour is shown by the arrowhead. The errors shown have been calculated for a gamma of 2·8 for the display tube. It will be seen that the rise in the overall system gamma produces an increase in the reproduced purity of all those colours which fall in the mid-range of the purity scale. The general effect, therefore, of the overall gamma rise is to offset the decrease in reproduction purity caused by the analysis errors in the film itself and in the telecine.

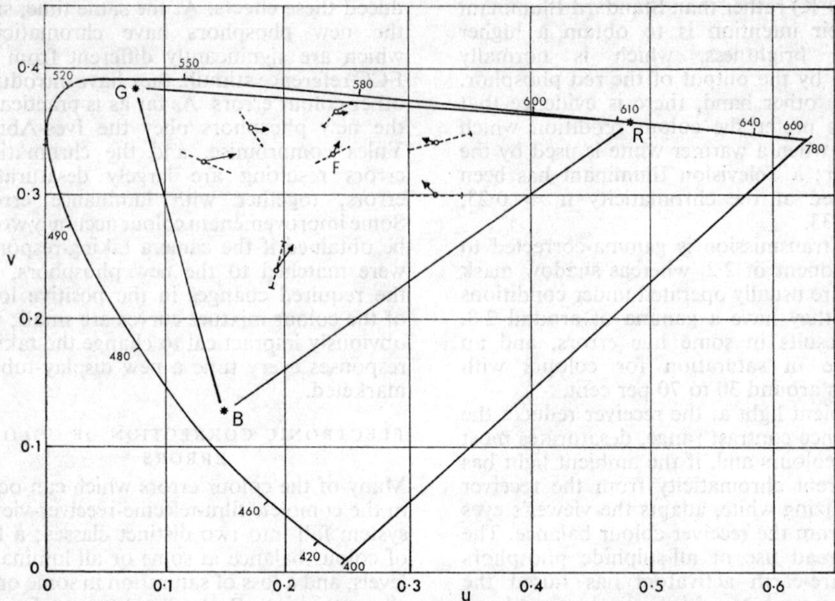

EFFECT OF AMBIENT LIGHT ON REPRODUCED CHROMATICITY. Television pictures are not normally viewed in the dark but in the presence of at least some ambient illumination. This ambient illumination has several effects. By giving the reproduced picture a brightness surround, the contrast response of the eye to the television picture is changed, so that a rather higher overall system gamma is called for than would be the case if the picture were viewed in the dark. At the same time, and despite the protection offered by the neutral-density faceplate of the display tube, the ambient light falling on the tube face flattens the contrast in the darker parts of the picture. The addition of reflected ambient light to the reproduced colours results in a decrease in purity of the reproduction which is more pronounced for the darker colours. If the ambient illumination has a different chromaticity (typically illuminant A) compared to the white to which the display tube is normalized (typically illuminant D), some shift in dominant wavelength of the reproduction occurs, depending on the chromaticity of the colour being reproduced, and the intensity of the colour relative to the ambient illumination.

can be largely corrected electronically. If the characteristics of the film stock are consistent, it is possible to correct for their deficiencies to the point where the television picture is a more accurate representation of the original scene than the optical projection of the film.

TRANSFER CHARACTERISTIC CORRECTION. If, after the telecine has been balanced objectively on a grey scale, the resulting reproduction of film shows undesirable errors in important colours such as flesh tones, the three tristimulus signals can be adjusted differentially either in the telecine or in subsequent electronic processing.

Lift and gain of any channels can be adjusted to make the three characteristic curves overlap in one part of their mid-range. Adjustment of the contrast law of those channels where the input-output characteristic shows the wrong shape, or black crushing, or white crushing, will then make all three characteristics track from black to white in a neutral fashion. Ingenuity of circuit design can reduce the number of control knobs to reasonable proportions.

ELECTRONIC MASKING. The desaturation of colours produced by the non-ideal dyes of the photographic process is equivalent to diluting the wanted signal by a cross-talk signal from all the three channels. A more correct signal can be produced by adding to the obtained-signal a mask-image signal – drawn from all three channels – which corrects for the unwanted absorption of the film dyes.

If the uncorrected signals are written as E_R, E_G, E_B, and e_R, e_G, e_B are signals which accurately represent the original scene, then

$$-\gamma \log e_R = \log E_R + [k_{11} \log E_R + k_{12} \log E_G + k_{13} \log E_B]$$

$$-\gamma \log e_G = \log E_G + [k_{21} \log E_R + k_{22} \log E_G + k_{23} \log E_B]$$

$$-\gamma \log e_B = \log E_B + [k_{31} \log E_R + k_{32} \log E_G + k_{33} \log E_B]$$

The logarithms are to base 10 and arise because the light transmission (T) through a film of density (D) is given by $D = -\log T$. The values of the constants k_{11} etc., will vary from film stock to film stock. A representative set of equations, omitting the gamma correction, might be

$$\log e_R = \log E_R + [0{\cdot}1 \log E_R - 0{\cdot}1 \log E_G]$$

$$\log e_G = \log E_G + [-0{\cdot}2 \log E_R + 0{\cdot}6 \log E_G - 0{\cdot}4 \log E_B]$$

$$\log e_B = \log E_B + [-0{\cdot}2 \log E_R - 0{\cdot}6 \log E_G + 0{\cdot}8 \log E_B]$$

LINEAR MATRIX. For small errors it is possible to approximate such equations to a linear matrix form which, however, will only hold over a small luminance range.

$$e_R = E_R + a_{12}(E_G - E_R) + a_{13}(E_B - E_R)$$

$$e_G = E_G + a_{21}(E_R - E_G) + a_{23}[E_B - E_G]$$

$$e_B = E_B + a_{31}(E_R - E_B) + a_{32}[E_G - E_B]$$

For unaltered transmission of the grey scale, the mask signals must vanish when $E_R = E_G = E_B$.

CORRECTION OF TAKING-CHARACTERISTICS. Such a linear matrix may also be used, with different coefficients, to correct for the wrong shape of taking-characteristic in the telecine colour analysis. Even if the optimum positive-only characteristics are already used in the telecine, a suitable linear matrix may halve the inevitable colour errors.

SIGNAL-TO-NOISE RATIOS. In general, the use of electronic masking tends to reduce the signal-to-noise ratio of the resultant colour picture, although the effect is not the same for all colours. In particular, it may improve the signal-to-noise ratio of the red channel, which normally has the lowest ratio of the three channels. The increased light available by using broad taking-characteristics tends to balance the deterioration in signal-to-noise produced by the correcting linear matrix.

It is feasible to restrict the bandwidth of some or all of the mask signals, and hence the amount of noise added to the main signals. The result is that large-area colours are more accurate while, at the same time, the deterioration in signal-to-noise ratio is minimized. G.B.T.

See: *Master control; Telecine; Telerecording; Transfer characteristics.*

Books: *Television Film Engineering*, by R. J. Ross (New York), 1966; *PAL Colour Television System*, by G. B. Townsend (IEE Monograph).

☐**Telecine Operator.** Operator who adjusts and tends the machines which produce the electronic picture for the many television programmes carried on film, and for film inserts into studio productions. He needs the know-how of projection – care and handling of films, etc. – and deals with the special problems of the various types of telecine.

The trend is towards remote control and starting of telecine machines but a telecine operator is still needed, e.g., on an old feature film, which needs considerable variation of the controls to get the best results.

See: *Telecine.*

☐**Telemetry.** Passing of information relating to measurements by radio or wire from a remote site, e.g., satellite to earth.

See: *Satellite communications.*

☐**Telephone Pair.** Pair of wires in a cable capable of carrying a telephone circuit.

☐**Telephony Channel.** Channel suitable for telephony, with a frequency range of 250 to 3,400 Hz maximum. Also called a speech channel.

See: *Transmission networks.*

⊕☐**Telephoto Lens.** Long-focus lens comprising a positive front component spaced away from a negative rear component, such that the total length from front surface to film plane is less than the focal length of the whole lens.

This design makes for a compact lens of short back focus. In the inverted telephoto lens, the back focus is greater than normal, so that a wide-angle (short focus) lens may be used in a camera where a prism or reflecting shutter has to be interposed between lens and film. The term telephoto is sometimes loosely used for a lens of any construction which is of greater than normal focal length.

See: *Camera; Cameraman; Lens performance; Lens types; Optical principles.*

TELERECORDING

The terms telerecording (in Britain) and kinescope recording (in North America) denote the record which is obtained on film of a television programme by photographing in a suitable camera the TV patterns appearing on a cathode ray tube. Sometimes the full term television film recording is used.

The main use of telerecording, in terms of film footage used per annum, is in broadcast television. A telerecording may be made during transmission and televised later to give a repeat broadcast: alternatively a telerecording may be made without simultaneous transmission and edited if necessary to provide a programme of selected material for transmission later.

Telerecording is also used for a number of non-broadcast purposes, particularly as a means of obtaining a record by closed-circuit television of such operations as the launching of a space rocket, the exploration of the sea bed and the examination of radioactive environments.

Recordings on magnetic tape (videotape) are now increasingly common, but will not for a long time replace telerecordings because equipment capable of projecting film optically and scanning it for TV reproduction is available world wide. By contrast, professional videotape reproducers are not yet universal, and must operate on the same line and field standards as the recorder unless a standards converter is used. These are now more common, but there remains a restriction on the interchange of videotape recordings which does not apply to telerecordings.

BASIC PRINCIPLES

The basic components of all telerecorders are:

1. An amplifier for the television signal to be recorded.
2. A cathode ray tube (kinescope) with appropriate scanning arrangements fed from the amplifier.
3. A lens forming part of a cine-camera viewing the cathode ray tube.
4. A film on which the lens forms an image of the cathode ray tube picture. These four components control the quality of the recorded image.

The aim in telerecording is, of course, to record all the picture information on the film; the ease with which this may be done depends upon the scanning system in use. With sequential scanning and a long field-blanking interval, it is sufficient to arrange for the pulldown of the film to take place during field blanking and to obscure the lens by a rotating shutter during this time. High quality telerecordings can be produced using such a scanning system designed to suit the recording process.

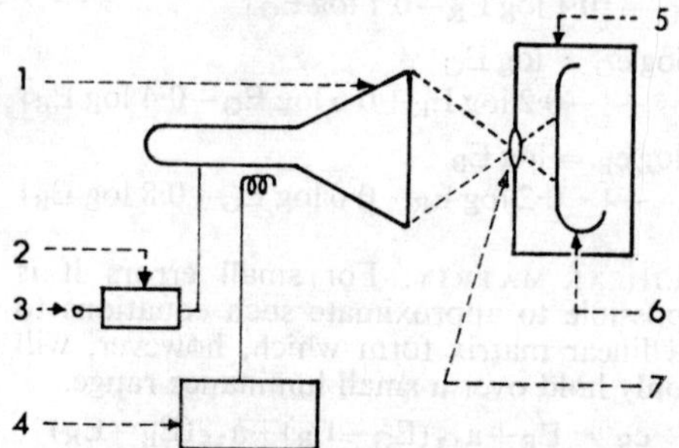

BASIC COMPONENTS OF A TELERECORDER. Cathode ray tube (CRT), 1, with scanning circuits, 4, is modulated by amplifier, 2, supplied with video input signal, 3. Objective lens, 7, transfers picture image on tube to film, 6, in what is essentially a cine camera, 5.

Broadcast television, however, uses interlaced scanning and a short field-blanking interval. The basic problem is the mechanical one of pulling down the film in a small percentage of the exposure cycle, and the various ways in which this can be dealt with have led to the development of a number of different types of telerecorder.

Moreover, the standard picture frame rate for film is 24 per second, whereas the television frame rate is 25 for the 405-line (British) and 625- and 819-line (Continental) standards and 30 for the 525-line (North American) standard. This is dealt with by accepting a frame rate of 25 per second for 405- and 625-line telerecordings and providing special shuttering and electronic arrangements for 525-line telerecorders such that a 24 fps film is produced from the 30 fps television signal.

Apart from the pulldown problem, other considerations are the reproduction of moving objects, the gradation, grain, noise and resolution of the picture and the synchronization of the camera with the television signal. Furthermore, the sound record may be recorded on the same film as the picture (optical or magnetic) or on a separate, synchronized magnetic film.

RECORDING METHODS

NORTH AMERICAN SYSTEMS. Production of a 24 fps film from a 30 fps television signal requires an indirect approach more easily understood after considering British and Continental systems.

BRITISH AND CONTINENTAL SYSTEMS. The following table gives the field blanking for these standards expressed in milliseconds (msec) and as an angle referred to the complete picture cycle of 360 degrees:

FIELD BLANKING PERIODS

System	Field Blanking (msec.)	(deg.)
405 lines	1·4	12·6
625 lines	1·2	10·8
819 lines	2·0	18·0

The normal time for pulldown and registration in cine cameras is around 180°, and the problem of speeding up the pull-down part of the cycle to a figure sufficiently near 10·8° to be acceptable is considerable. The acceleration and deceleration of the film must reach values of thousands of times the acceleration due to gravity. This has not been achieved for 35mm film but has been at least very nearly achieved for 16mm film for which the required acceleration is somewhat less.

Three methods have been developed to avoid this stringent requirement: the suppressed field system, the stored field system, and the partial stored field system, so that, with the fast pulldown system, four methods in all are available.

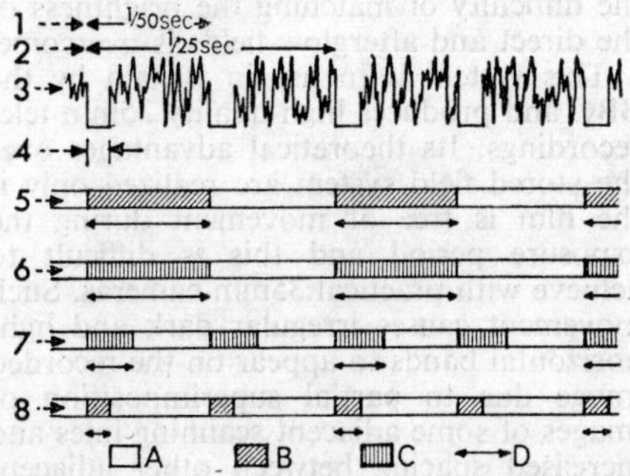

TELERECORDING METHODS FOR 50-FIELD SYSTEMS. British and Continental standards. 1. Field period. 2. Picture period. 3. Video waveform (sync pulses not shown). 4. Field blanking period. 5-8. Practical methods, with A, information recorded directly (shutter open), B, information lost (shutter closed), C, information recorded by CRT afterglow, D, film pull-down. 5. Suppressed-field method. 6. Stored-field method. 7. Partial stored-field method. 8. Fast pull-down method.

Another way of avoiding the quick pull-down difficulty is by using a continuous-motion rather than an intermittent film-traction mechanism, with a mirror drum to give a series of stationary images on the film on the principle used in some high-speed cameras. Though successful in telecine applications, this method was abandoned some years ago, at least for broadcasting purposes, mainly because of the difficulty of maintaining the correct alignment of the mirror drum.

THE SUPPRESSED FIELD SYSTEM. Historically the oldest system, it was first used by the British Broadcasting Corporation in 1947 for making telerecordings of important national events.

The principle is simple: a camera having a normal length pull-down and a 180° shutter is driven by a motor synchronized in speed and phase to the television picture so that alternate fields only are recorded, pull-down occurring during the unrecorded fields. The recorded picture contains only half the normal number of lines and the gaps between them are filled by the use of spot wobble – a rapid oscillation of the cathode ray tube spot in a vertical direction which has the effect of broadening the lines. It is essential to fill the gaps to avoid a moiré pattern on scanning the film with a telecine.

The number of active lines in the 405-line (British) television system is 377 so that suppressed field recordings on this system contain only 188½ lines. There is a reduction in signal-to-noise ratio of about 3dB. Such recordings look better than might be expected but the loss of half the information in the vertical direction is shown up rather badly by the stepped appearance of parts of the recorded picture which are near the horizontal. Suppressed field recordings are no longer used in British television when good quality is important.

The number of active lines in the 625-line (Continental) system is 603 to 607 so that suppressed field recordings on this system contain about 302½ lines. The extra number of lines as compared with 405-line recordings makes all the difference in practice and the vertical resolution is acceptable to the viewer watching his television set at home.

Telerecorders operating on the suppressed field principle have the advantage of simplicity and reliability, as the camera is operating under normal filming conditions and the electronic or optical complications of other methods are avoided.

Suppressed field telerecording is widely

used on the Continent on the 625-line and 819-line television systems.

THE STORED FIELD SYSTEM. To overcome the disadvantages of discarding half the picture information as in the suppressed field system, while using a camera with a normal film pull-down time, a cathode ray tube having a phosphor screen with a long afterglow may be used.

The obscuration of the shutter, which covers pull-down, occurs during one field period as with the suppressed field system, but during this time the field in question is being stored by the phosphor screen. During the next field, the shutter is open and this field is recorded directly on the film. At the same time, owing to the afterglow, the first field is recorded also. Both sets of information are held discretely by the interlacing of the two images.

Cathode ray tube phosphors have a light decay characteristic which is generally exponential in shape, and for the present purpose the light output should be down to about 15 per cent in 1/50 sec.

It is apparent that with a characteristic of this type the two interlaced fields would not be equal in brightness and hence in exposure. The exposure can be equalized, however, by providing a larger signal to the cathode ray tube during those fields which are to be recorded by afterglow. Furthermore, the amplification of the signal that is needed varies with the position on the cathode ray tube face, since the shorter the time between excitation of the phosphor and commencement of exposure, the smaller the loss in exposure – this applies both to the direct and afterglow fields.

In practice a suitably-shaped correction waveform, having a shape rather like an integrated sawtooth and a repetition frequency of 1/25 sec is applied to a variable-gain amplifier carrying the video signal to be recorded. Spot wobble is used to remove the residual line structure.

The range in brightness of the cathode ray tube spot which has to be provided to give uniform exposure on the film is about 3 : 1. The maximum permissible brightness of the spot is set either by phosphor saturation or by the increase in spot size which occurs when the beam current is increased. The variation in gain means that for most of the time the cathode ray tube is being used over a lower contrast range than for normal operation.

The variation in spot size with beam current makes a compromise setting of spot wobble amplitude necessary, since the scanning lines traced by the afterglow field are wider than those traced by the direct field. It is important not to let phosphor saturation occur at peak brightness as this prevents a correct match being made between the two fields.

Telerecording by the stored field system is in use in Europe and in Britain, where one of the independent companies has since 1955 employed equipment operating on this principle to produce good-quality 35mm telerecordings.

THE PARTIAL STORED FIELD SYSTEM. In this system, the pull-down time is reduced to about 5 msec, corresponding to 45°, and phosphor afterglow is used to record the part of the picture occurring during this period. For the 405-line standard, in which the field blanking is 1·4 msec, this means that about 3·6 msec or about 1/5 of the active part of a field is displayed while the shutter is closed.

The period of storage required is less than in the stored field system, and the phosphor used is such that the light output decays to about 8 per cent in 1/50 sec.

Exposure correction must again be provided, but since the shuttered period is less than the active field period, it is possible to shutter each field equally (instead of each pair of fields) and to apply a correction at field rate instead of frame rate. The correction is achieved by means of a graded neutral density filter attached to the face of the cathode ray tube, and has the advantage that all fields are modulated similarly and the difficulty of matching the brightness of the direct and afterglow fields is overcome.

This system is in use in Britain by the BBC and produces high-quality 35mm telerecordings. Its theoretical advantages over the stored field system are realized only if the film is free of movement during the exposure period and this is difficult to achieve with practical 35mm cameras. Such movement causes irregular dark and light horizontal bands to appear on the recorded image due to partial superimposition of images of some adjacent scanning lines and increased spacing between other adjacent scanning lines, resulting in alternately increased and reduced exposure.

The disadvantages of the system include the occasional appearance of a slight dark line near the top of the picture. This is caused by an inherent defect in the graded filter, which is made by grinding a thick piece of neutral density glass to approximately the ideal shape. Another drawback is that the driving motor must be very precisely phased to the television signal.

THE FAST PULLDOWN SYSTEM. It has not been possible to use this system with 35mm film but some years ago 16mm equipment was developed in Britain using a camera in which an accelerator mechanism speeds up the motion of the claw during pulldown.

The cathode ray tube has a phosphor with negligible afterglow and no exposure correction is necessary. The camera photographs two consecutive interlaced fields on each film frame, pulldown occurring substantially in alternate field-blanking periods.

To achieve reliable equipment in which the film is not damaged and the mechanism has a reasonable life of some thousands of hours, a pulldown time of not less than 2 msec has to be accepted. For the 405-line standard in which field-blanking is 1·4 msec this leads to a loss of 0·6 msec, or 6 lines, in alternate fields.

In practice, a shutter is used with an obstruction period of 2 msec on all fields, leading to an interlaced recording from which a total of 12 lines is missing. This is small enough to be ignored.

To move the film and stop it completely in 2 msec requires a high gate pressure and this tends to cause a pile-up of emulsion in the gate. Tests are necessary to establish the suitability of a particular film stock from this point of view.

Several 16mm telerecorders based on this camera are in use throughout the world and give satisfactory results with appropriate servicing.

Recently another fast pulldown camera of American design has become available for telerecording. This uses an interrupted blast of air to blow on the lower loop, such that when pulldown has occurred, one of the sprocket holes engages with a fixed pin. No intermittently moving parts engage with the film. Pulldown occurs well within the field-blanking period and the use of a fixed pin leads to a steady recorded picture. Adjustment is critical and the air noise of such devices is a problem.

NORTH AMERICAN STANDARDS. Telerecording on the 525-line, 60 field/sec standard, used principally in North America and Japan, poses a different problem. The television picture rate is 30 per second whereas the standard film frame rate is 24 per second, and to avoid overlapping information it is necessary to discard either (*a*) half a field in two and a half or (*b*) one in five.

The commonest method is known as picture join and requires a cathode ray tube with short afterglow and a camera with a pulldown time no greater than the duration of half the active field, i.e., 8·33 msec. Pulldown occurs every 1/24 sec, so this corresponds to 72°. It is not so difficult to design a camera to achieve this as to obtain pulldown in the field-blanking period on the British and Continental standards.

The sequence of events forming one complete recording cycle is as follows: the shutter opens, allowing fields 1 and 2 to be recorded in interlace on the film, forming film frame 1. Pulldown of the film then occurs and the shutter is closed during half the active period of field 3. Film frame 2 is exposed to the second half of field 3, the whole of field 4 and the first half of field 5 so that effectively two interlaced fields are

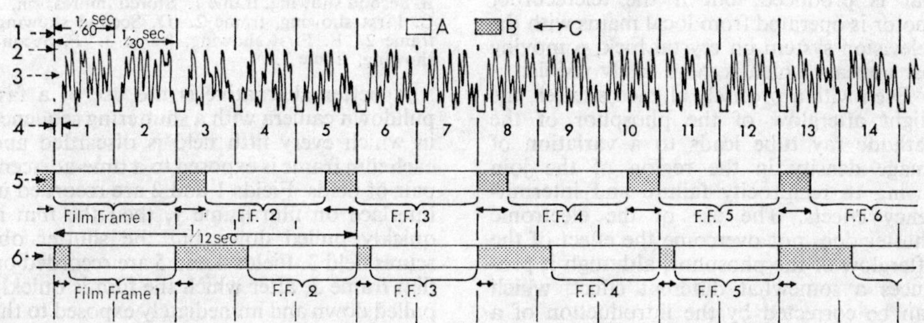

TELERECORDING METHODS FOR 60-FIELD SYSTEMS. (N. American standard.) 1. Field period. 2. Picture period. 3. Video waveform (sync pulses not shown). 4. Field blanking period. 5, 6. Practical methods, with A, information recorded directly (shutter open), B, information lost (shutter closed), C, film pull-down. 5. Picture-join method. 6. Fast pull-down method.

recorded on film frame 2 as well as on film frame 1. Pulldown again occurs, and the shutter is closed during the blanking plus the second half of the active period of field 5.

This completes the cycle of events, lasting 1/12 sec, in which five television fields are recorded on two film frames. The cycle repeats starting with fields 6 and 7 being recorded on film frame 3.

The picture join occurs on alternate film frames between pairs of half fields such as 3 and 5 and it is essential that such pairs of half fields should join without gap or overlap and that the composite field so formed should interlace with a field such as 4 occuring between the two half fields. There is then no loss of information beyond what must be discarded to obtain a 24 fps film from a 30 fps television signal.

In early telerecorders using the picture join principle, the edge of the camera shutter was relied upon to obscure the unwanted portions of the two fields. This is mechanically difficult to arrange with sufficient accuracy and some recent designs use an electronic shutter consisting of a circuit which counts the lines in each field and blanks the cathode ray tube beam so that, for example, in film frame 2 the contribution from field 3 starts at the beginning of the first line after half the field has been scanned and the contribution from field 5 finishes at the end of the line before this.

It is difficult to avoid slight visibility of the picture join in the telerecording. Any slight inaccuracy of the mechanical shutter, when one is used, gives underexposure or overexposure at the picture join and shows as a horizontal dark or light line. When the telerecorder motor drive system is locked to the television field frequency a stationary bar is produced, but if the telerecorder motor is operated from local mains with the television system on crystal lock, a moving bar appears which is more easily visible.

Even with a precisely aligned shutter, the slight afterglow of the phosphor of the cathode ray tube leads to a variation of image density in the region of the join owing to reciprocity failure and intermittency effects. The use of the electronic shutter does not overcome the effect of the afterglow of the phosphor, although it produces a somewhat different effect, which can be corrected by the introduction of a tilt waveform to the signal being recorded.

The effects of the picture join are not particularly objectionable to the layman, however, and telerecordings with picture joins have been an accepted part of television in the US for many years.

It is possible to avoid the picture join by recording only one television field on every film frame, instead of two, by having the shutter closed for a longer period. This gives a result like the suppressed field system used on the 625-line standard in that the number of scanning lines is halved, except that the vertical definition is less in the ratio of approximately 5 : 6.

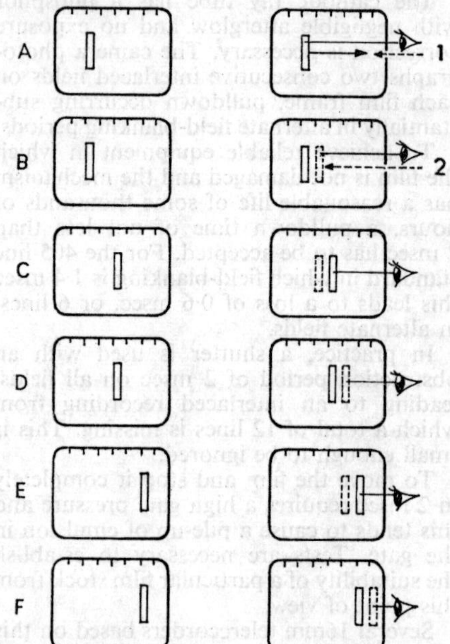

CINEMA REPRODUCTION OF A MOVING OBJECT. (*Left*) Screen images of uniformly-moving object, each frame interrupted once (i.e. making two appearances) by two-bladed shutter. (*Right*) Subjective doubling of images by spectator's eye. A. First showing, frame 1. Eye follows moving object, 1. B. Second showing, frame 1. Stored impression, 2. C. First showing, frame 2. D. Second showing, frame 2. E. First showing, frame 3. F. Second showing, frame 3.

A better alternative is the use of a fast pulldown camera with a shuttering sequence in which every fifth field is discarded and each film frame is exposed to a time-adjacent pair of fields. Fields 1 and 2 are recorded in interlace on film frame 1, then the film is quickly pulled down but the shutter obscures field 3. Fields 4 and 5 are recorded on film frame 2, after which the film is quickly pulled down and immediately exposed to the next pair of fields.

This scheme has been carried out with the British fast pulldown 16mm camera already described, using a different gearbox to give the required pattern of pulldown at three-field and two-field intervals and the different shuttering arrangement. Because pulldown takes 2 msec whereas field-blanking takes 1·14 to 1·33 msec, some lines are lost at top and/or bottom of the picture. The shuttering is arranged to retain the same number of lines in each field.

REPRODUCTION OF MOVING OBJECTS. Both television and the cinema simulate the movement of objects by a series of stationary images. The degree of perfection of the illusion depends on the particular system used, and in telerecording both film and television processes are involved.

In making a normal film, a moving object is reproduced as blurred in the direction of movement and, in the usual case of a shutter having an opening of 180°, the width of blur equals the width of the gap between successive images.

Normally, in projection, a shutter is used such that each frame is thrown on the screen twice, to reduce flicker to a negligible level. The eye follows the movement steadily at the average speed of movement and, owing to the storage characteristic of the eye, while one image is being viewed the previous one is still visible. The result of this is a subjective doubling of moving images. The final result as viewed on the cinema screen is a combination of this doubling and the blurring which occurs in the camera. These imperfections are not, however, of great practical importance and the reproduction of moving objects in the cinema has long been accepted as satisfactory.

In the case of telerecording, other effects occur and these effects are different according to the system used. A single image of the moving object is recorded with the suppressed field system, but multiple images are recorded with the fast pulldown system, the partial stored field system and the stored field system.

So much for the moving image as recorded by the telerecorder. But in order to transmit this image, it is necessary to scan it in a telecine, and here more complicated effects occur.

An ideal telecine would scan exactly in register with the lines on the film, each scanned field corresponding with the recorded field on the film, but in practice the output signal at any instant may be derived partly from one field and partly from

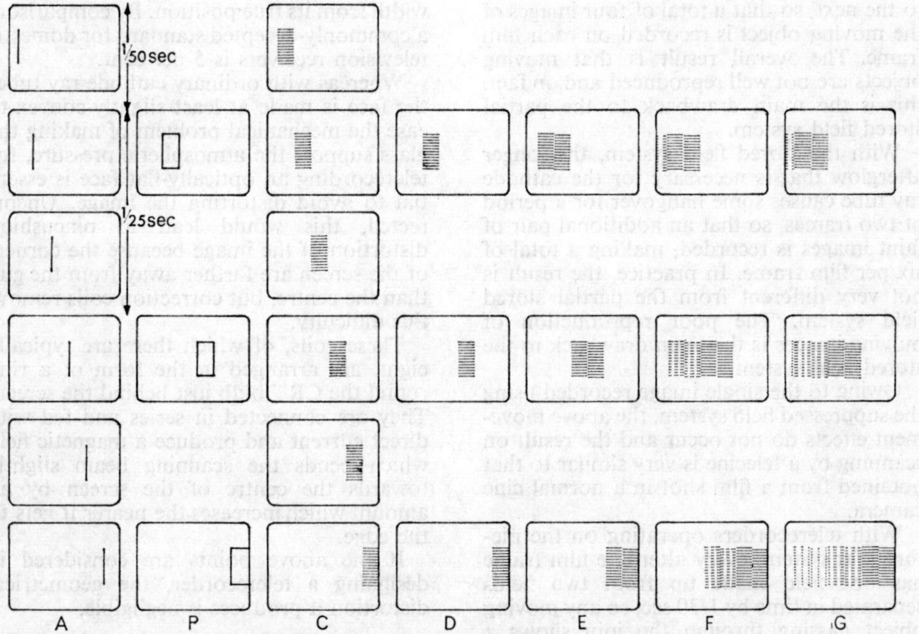

REPRODUCTION OF MOVING OBJECT. Cine camera and 50-field television and telerecording contrasted. A Actual movement of object in 1/50 sec. increments. B. Cine camera image at 25 pps. C. TV image. D. Telerecorded image, suppressed-field system. E. Telerecorded image, fast pull-down system. F. Telerecorded image, partial stored-field system. G. Telerecorded image, stored-field system.

another, owing to the impracticability of achieving precise registration. Because spot wobble is applied to the telerecorder, the images of stationary objects on the telerecording do not have a line structure and are correctly reproduced by the telecine, but the images of moving objects, which have every other line displaced in the direction of movement, are reproduced as an interference pattern between the recorded lines and the scanning lines of the telecine.

In Britain it is the standard practice to underscan the recorded film frame in the telecine by some 3 per cent, and under these conditions on the 405-line system the interference pattern takes the form of about 12 horizontal bands per picture height. Bands 1, 3, 5, etc. are derived from, say, the first of the two recorded fields and bands 2, 4, 6, etc. from the second.

This rather complicated effect causes a breaking up of images on movement which is tolerable for slow or very fast movement but can be irritating for certain rather quickly-moving objects.

The effects described above are evident with the quick pulldown system. With the partial stored film system they are modified by the afterglow of the CRT phosphor. This causes some hangover from one frame to the next, so that a total of four images of the moving object is recorded on each film frame. The overall result is that moving objects are not well reproduced and, in fact, this is the main drawback to the partial stored field system.

With the stored field system, the longer afterglow that is necessary for the cathode ray tube causes some hangover for a period of two frames, so that an additional pair of faint images is recorded, making a total of six per film frame. In practice, the result is not very different from the partial stored field system. The poor reproduction of moving images is the main drawback to the stored field system.

Owing to the single image recorded using the suppressed field system, the above movement effects do not occur and the result on scanning by a telecine is very similar to that obtained from a film shot in a normal cine camera.

With telerecorders operating on the picture join system, every alternate film frame has one field made up from two fields separated in time by 1/30 sec, so any moving object passing through the join shows a discontinuity. When the fast pulldown system is applied to the 525-line standard, this effect is avoided and movement is reproduced much as with this system on British and Continental standards. The difference is that every fifth field is discarded so there is a slight jump every 1/12 sec.

These effects on telerecorders using North American standards are not so serious as the multiple images caused by the CRT (kinescope) afterglow on telerecorders operating on British and Continental standards.

GEOMETRY OF THE IMAGE

The recorded image in all types of telerecorder should, of course, be free from distortions of shape, so that, for example, a circle viewed by the television camera remains a circle when recorded on the film.

The only two sources of imperfect geometry in the telerecorder itself are the cathode ray tube and the lens. Although some lenses give a little barrel or pincushion distortion, this is small in a lens used for a telerecorder, and the only geometrical distortion that need be considered is due to the CRT.

It is of primary importance to have horizontal and vertical scanning generators of good linearity and an accurately made scanning coil assembly. The displacement of any point of the image should be limited to not more than ½ per cent of the picture width from its true position. By comparison, a commonly-accepted standard for domestic television receivers is 5 per cent.

Whereas with ordinary cathode ray tubes the face is made at least slightly convex to ease the mechanical problem of making the glass support the atmospheric pressure, for telerecording an optically-flat face is essential to avoid distorting the image. Uncorrected, this would lead to pincushion distortion of the image because the corners of the screen are farther away from the gun than the centre, but correction coils remove this difficulty.

These coils, of which there are typically eight, are arranged in the form of a ring round the CRT bulb just behind the screen. They are connected in series and fed with direct current and produce a magnetic field which bends the scanning beam slightly towards the centre of the screen by an amount which increases the nearer it gets to the edge.

If the above points are considered in designing a telerecorder, the geometrical distortion it produces is negligible.

RESOLUTION

The aim in telerecording is to reproduce the television picture on the film with as little loss in resolution as possible.

The losses in resolution which occur are optical and electron-optical, the recording amplifier being designed to have a negligible variation of frequency response over the video frequency range. In the cathode ray tube the size of the spot, due to optical and electron-optical considerations, is not small enough to avoid loss of resolution.

The lens is subject to aberrations, and produces an image on the film which causes a further loss in resolution. To these losses may be added those which result from printing the telerecording and transmitting it in a telecine.

LIMITING RESOLUTION. In optical work, the term limiting resolution is widely used and, for a given lens or film, gives the spacing of a test pattern of black-and-white bars which is the smallest that can just be resolved. It is, however, widely realised that the subjective sharpness of the images produced by two lenses with the same limiting resolution may be very different due to the different shapes of the resolution curves plotted against the spacing of the test pattern.

In considering resolution in a telerecorder, the term limiting resolution is of little use, as the numerous sources of loss of resolution already mentioned have different-shaped resolution curves, and the final resolution obtained on the film is the result of these separate resolution curves acting in turn.

To make possible a unified approach to the problem of evaluating the different resolution curves involved, these curves will all be considered in electrical terms. For a television signal produced according to particular scanning standards, there is a maximum rate of transition from a black area to a white area and vice versa, both in the vertical direction (due to the finite number of lines), and in the horizontal direction (due to the scanning spot size). In telerecording this rate of transition must be reproduced as far as possible on the film.

To test the overall resolution of the telerecorder, a signal consisting of square pulses is often used, the appearance on the cathode ray tube being of thin vertical black-and-white bars of equal width. If the equipment were perfect, the spacing of the bars would have no effect on the difference in photographic density between the images of the white-and-black bars on the film. In practice, of course, the various sources of loss of resolution result in a decreased density range when the bars are closely spaced as compared with when they are widely spaced. On rescanning the film in a telecine, the resultant electrical signal has a greater amplitude for widely-spaced bars than for closely-spaced bars. By taking the ratio between the output signal amplitude from the telecine and the input signal amplitude to the telerecorder, for various numbers of black-and-white bars per picture width, the overall response curve for the telerecording process, including the telecine, is obtained.

RESOLUTION LOSS DUE TO THE CRT. The first requirement of a cathode ray tube to be used for telerecording is that its spot is small enough to permit adequate resolution with the number of scanning lines required by the scanning standard in use.

It is not sufficient to be able to separate visually all the lines on the screen, which is often all that is expected of a domestic television receiver, but the illumination of a point on any given scanning line must as far as possible be independent of the illumination of adjacent points. Unwanted light from one line falling on another or from one point falling on an adjacent point in the same line restricts the rate at which the brightness of the image can change in both the horizontal and vertical direction. This leads to a loss of contrast in the reproduction of the black-and-white bar test

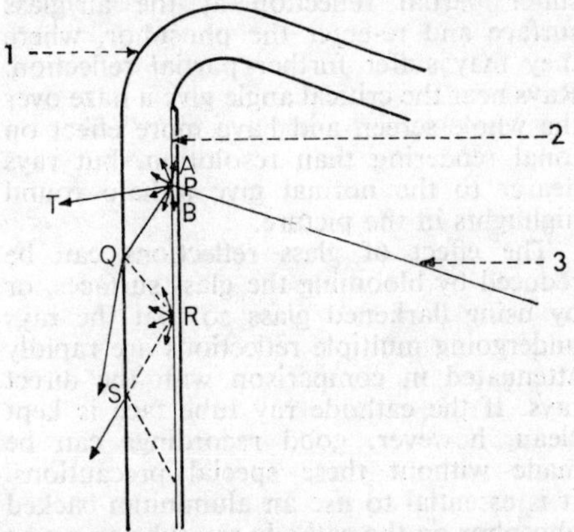

LIGHT SCATTER FROM POINT ON TUBE PHOSPHOR. 1. Glass of tube face and wall. 2. Phosphor layer. 3. Electron beam. PA, PB. Scatter due to reflection from one particle to another. PQRS. Multiple reflections within glass, with emission from front surface at Q and S. PT. Principal ray used by lens to form required image.

patterns, and contributes to the falling-off of the overall response curve with the frequency of the input signal.

The factors which cause spreading of any given image point to nearby points are as follows. First of all, the electron gun used in the cathode ray tube must be such that the electron beam is sufficiently fine. Apart from good gun design, this requires a high final anode voltage for the cathode ray tube (say 25 kV). Practical electron beams, however, do not have uniform electron density, and some electrons land on the cathode ray tube screen outside the main area of the beam. These of course cause the phosphor to glow in the areas between the scanning lines and lead to a loss of response.

Secondly, the light emitted from a given point on the phosphor tends to spread radially in the phosphor itself. This is due to reflections from the particles of phosphor: the effect is an increase in spot size.

Thirdly, reflections occur in the cathode ray tube glass face. Whereas some rays are used by the lens to make the image, others suffer partial reflection at the air/glass surface and re-enter the phosphor, where they may suffer further partial reflection. Rays near the critical angle give a haze over the whole screen and have more effect on tonal rendering than resolution, but rays nearer to the normal give a halo round highlights in the picture.

The effect of glass reflections can be reduced by blooming the glass surfaces, or by using darkened glass so that the rays undergoing multiple reflections are rapidly attenuated in comparison with the direct rays. If the cathode ray tube face is kept clean, however, good recordings can be made without these special precautions. It is essential to use an aluminium-backed phosphor on the cathode ray tube as, apart from increasing the light output by reflecting the light emitted towards the gun of the cathode ray tube, the aluminium backing attenuates considerably any light passing through it and prevents unwanted reflections from glass surfaces behind the screen.

An important source of image degradation is the granularity of the CRT phosphor. The usual way of depositing the phosphor screen is by settling of particles from a liquid, giving a granular surface. Special precautions must be taken with cathode ray tubes intended for telerecording to ensure that the size of the particles is small enough. If this is not done, the picture appears to have superimposed on it a fixed granular pattern.

A well-known effect which causes loss of resolution in the horizontal direction only is aperture distortion. Having a finite length in the direction of movement (i.e., in the direction of horizontal scanning) the spot takes a finite time to pass through its own length, and during this time any variation of intensity due to signal modulation will not be reproduced on the screen at full amplitude. The shape of the curve relating amplitude of response to signal frequency depends on the shape of the scanning spot, but for a circular spot with luminance falling-off from the middle as is the case with cathode ray tubes, the curve has approximately a cosine squared form.

Fortunately it is not difficult to correct for a loss of this shape, and a suitable aperture-correcting amplifier which gives increasing amplification to the higher video frequencies can be employed. Since aperture distortion is purely a reduction of the amplitude of the higher frequency components of the signal, and does not involve any phase shift of the components, an aperture-correcting amplifier having inherently correct phase is to be preferred.

In practice, having provided an aperture-correcting amplifier for a particular telerecorder, the amount of correction may be increased above that theoretically necessary to compensate for aperture distortion alone to correct also, at least partially, for many of the other losses of resolution which are now under discussion.

Before leaving the consideration of those aspects of the cathode ray tube which affect resolution, mention should be made of deflection defocusing. Given a cathode ray tube which is capable of producing a fine spot in the centre of the screen, it is essential that when the spot is deflected to form a picture, the diameter of the spot does not increase at the corners.

In the first place, the design of the deflection coils must be such that electron-optical aberrations do not increase the spot size; this is not difficult with a well designed deflection assembly. In the second place, since an optically-flat-faced cathode ray tube is used in the interests of best image geometry, the length of the electron path from the gun to the screen is greater in the corners and if this causes the spot to go out of focus there, focus modulation must be employed for optimum results.

The focus coil which focuses the electron beam on to the cathode ray tube screen is, of course, normally fed with a fixed current, but for focus modulation a current varying in time with the scanning is superimposed on the direct current. The waveform of this varying current is the sum of parabolas at horizontal and vertical scanning frequencies, the amplitude of these two components being adjusted to give uniform focus over the whole cathode ray tube screen.

In short, with an aluminium-backed screen, a very fine electron beam and a well-designed deflection assembly (with, if necessary, focus modulation), CRT resolution loss can be held to acceptable limits.

The measurement of this loss is most conveniently carried out with a film camera. In practice, the telerecording camera itself is used, and the effect of the lens and film will be considered next, before passing to the question of overall resolution.

RESOLUTION LOSS DUE TO THE LENS. A lens for a telerecorder must provide sufficient resolution for the scanning standard in use, and the resolution obtained depends to some extent on the operating conditions. The maximum aperture will usually be at least f2·8, although if the sensitivity of the film stock and other considerations allow, it will be operated at about f4. The magnification at which it is used depends on the width of cine film (35mm or 16mm) and the diameter of the CRT face. The magnification will be appreciably smaller than the range of magnifications for which ordinary camera lenses are designed and the resolution obtained from a normal camera lens operated at this small magnification will, in general, be worse. The dimensions of the equipment determine the focal length approximately, the nearest available focal length being chosen and the exact camera-to-cathode ray tube distance being chosen to suit the equipment.

The colour correction which is applied in designing a normal camera lens makes the lens suitable for all visible light and when a cathode ray tube having a phosphor colour near the middle of the spectrum is used, with panchromatic film in the camera, the normal camera lens is suitable in this respect. When the cathode ray tube colour is near one end of the visible spectrum, say blue, and blue-sensitive film is used in the camera, there may be an advantage in using a lens which is corrected only for blue light, as in designing the lens it may then be possible to reduce other aberrations.

It is normal in camera lenses to find that the resolution curve falls off gradually as the limiting resolution is approached. For television, however, not only for telerecorders but for cameras and telecines, the ideal lens resolution curve would follow the same shape as the frequency response of television video circuits, i.e., it would maintain full resolution for test pattern spacings up to a certain minimum, and then fall rapidly.

Although some consideration has been given to this aspect of lenses for television, and special lenses have been designed, telerecording can be achieved with a lens having a certain amount of droop or gradual fall-off in its resolution characteristic. This loss of resolution can be largely compensated in the horizontal direction by the aperture-correcting amplifier.

For checking the performance of a lens, the best method is to use it to photograph a resolution chart on high resolution film, and then to measure the image density ratio of black bar to white bar for a range of spacings of the bars.

RESOLUTION LOSS DUE TO THE FILM. To obtain a high enough resolution, a fine-grain film in the camera is essential. It is much easier to choose a film stock for a 35mm telerecorder than for a 16mm telerecorder, because the 35mm frame is four times larger. In choosing a film having as fine a grain as possible, attention must be paid to emulsion sensitivity, and where films having extremely fine-grain emulsions are used, their insensitivity requires the lens to be used at as large an aperture as possible consistent with good lens resolution and the film to be given high gamma development to obtain a sufficiently high density range.

Cine film has a resolution curve which falls away gradually to zero at a limiting resolution. Manufacturers quote the limiting resolution figure but this is no more than a guide to the loss of resolution to be expected at intermediate points on the curve.

The resolution curve of the film may be measured with a lens of known characteristics by exposing a suitable test chart through the lens, and then measuring the density of the images of the bars of the test chart.

RESOLUTION LOSS DUE TO PRINTING. When copies of the telerecording are required, prints must be made, and the printing operation involves a further loss of resolution. Contact printing is to be preferred because, in general, less resolution is lost than with optical printing. Optical printing is necessary in normal 16mm filming to obtain a standard SMPE print which has the emulsion towards the lens in the projector, but with 16mm telerecording the horizontal scanning on the cathode ray tube may be reversed so that contact printing yields a standard print.

Resolution losses due to printing are not serious when the film in the camera is of the 35mm gauge, but the definition of a 16mm original can be visibly worsened by printing. It is mostly for this reason that direct reversal film is widely used in 16mm Continental telerecorders, as the processed film from the camera can be used directly in the telecine.

OVERALL RESOLUTION LOSSES. The overall loss of resolution in a telerecording, due to the causes mentioned, should be judged under standard viewing conditions. In broadcasting, this means using a telecine and picture monitor. Resolution losses that occur in telecines are common to ordinary films and telerecordings, but with a normal film the resolution loss due to the telecine is not troublesome (at least for a good flying-spot 35mm machine), whereas with a telerecording the additional loss due to the telecine can be significant.

The overall result obtainable at the present time is usually quite acceptable, but continued attention must be paid to all the steps in the process to produce consistent results day after day. In sound and videotape recording, the losses and distortions with properly operated and maintained equipment are small, but even the best telerecordings show some loss of sharpness and the worst are easily distinguished by the layman.

GRADATION

In normal cinematography in which negative film in the camera is exposed to an illuminated scene and a positive print is obtained from the negative, the gradation of the print depends on the density/log exposure characteristics of the negative and of the print.

In telerecording, additional transfer characteristics are involved. These are the log output voltage/log scene luminance characteristic of the television camera, the log output luminance/log input voltage of the recording cathode ray tube, the log output voltage/print density of the telecine and the log output luminance/log input voltage of the viewing cathode ray tube of the television receiver. Additional or alternative film processes used in telerecording involve other film characteristics.

All these characteristics may be considered as having a central straight portion with a slope known as the gamma, and non-linear distortions at both ends. Knowing all the characteristics, it is not difficult to calculate the overall gamma of the process from the original scene to its reproduction on the cathode ray tube of the television receiver.

The transfer characteristics of the television camera and of the receiver CRT are such that for live television an acceptable result is obtained. The high gamma of the CRT is offset by the low gamma of the pick-up tube in the television camera or by correction circuits in the camera when necessary.

CINEMA FILMS AND TELERECORDINGS. A large part of television programmes consists of film, most of which have gradation suitable for projection in cinemas. Telecines must provide from such films a good picture on the home television receiver and must therefore have a low gamma to give an acceptable result. Vidicon telecines have an inherently low gamma due to the vidicon tube itself, whereas flying-spot telecines which are themselves linear have adjustable circuits for reducing the gamma.

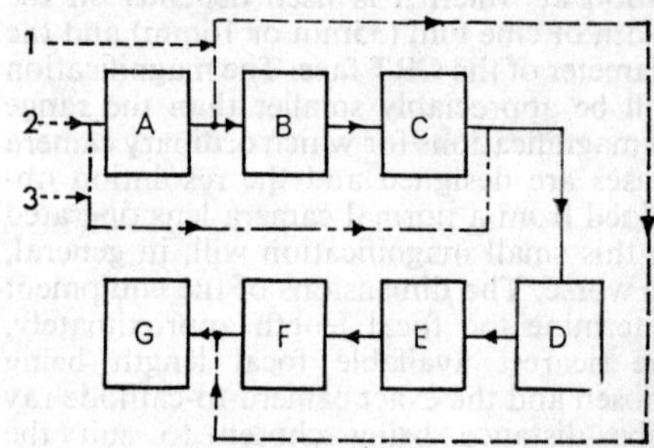

PROCESSES HAVING OTHER THAN UNITY GAMMA. Blocks A-G indicate such processes in telerecording chain following 2, original scene. A. TV camera. B. Recording amplifier. C. Recording CRT. D. Negative film. E. Positive film. F. Telecine (film scanner). G. TV receiver. Arrows, 1, indicate signal route for live television; arrows, 3, for televising a film.

It is not convenient to reset the telecine gamma when changing from normal films to telerecordings, so the aim is to make the overall gamma of telerecordings at least approximately equal to that of ordinary films. Furthermore, telerecordings are often shown on optical projectors for assessment purposes and a distorted tonal gradation would be a disadvantage, although it could be corrected in the telecine for television viewing.

It must, however, be said that the

contrast range of the average film intended for projection in the cinema is greater than can be really adequately reproduced by the television system (telecine plus home receiver) and when possible, films for telecine reproduction should have a maximum density in the blacks of not more than about 1·9. For best results, this rule is observed in telerecording.

Having now considered the transfer characteristics involved in live television and in the reproduction of a film by telecine, we can turn to the additional characteristics involved in telerecording. These are the characteristic of the recording CRT and the characteristic of the recording amplifier.

GAMMA CORRECTION. Recording cathode ray tubes, in common with those used for normal viewing, have a gamma considerably above unity, and this requires compensation. This is done in the recording amplifier, which feeds the recording cathode ray tube, by including circuits with a transfer characteristic having a gamma of less than unity. Corrections for the top and bottom ends of the characteristics are also carried out by appropriate circuits in this amplifier. The aim is to obtain a linear transfer of voltage from the amplifier input to the telecine output.

The CRT characteristic may depart from constant gamma at the ends. At the top of

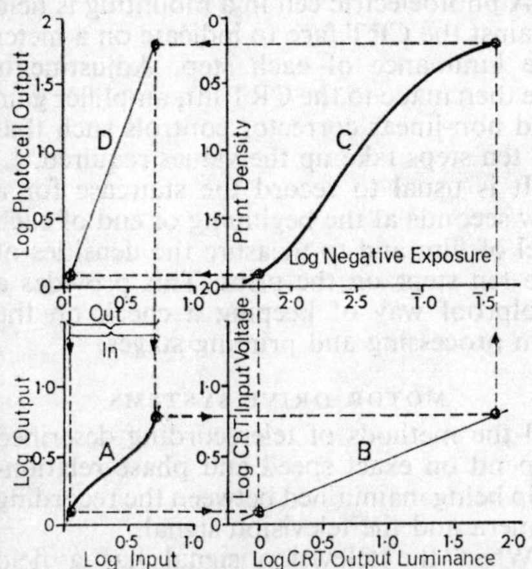

CORRECTION OF GRADATION IN TELERECORDING. Quadrant A shows how recording amplifier characteristic is chosen so that overall gamma of unity results. Ends of curve are so shaped that, after reflection by straight recording-CRT characteristic (Quadrant B), they compensate for distortion at ends of combined negative and positive film characteristic (Quadrant C). Telecine characteristic (Quadrant D) is that of a flying-spot type, with gamma-correction circuit and gamma of 0·4.

the characteristic (maximum luminance), it is possible to get phosphor saturation, such that an increase of log input voltage does not give the increase of log output luminance to be expected from the slope of the central part of the characteristic, but it is usual to avoid driving the CRT up to this point. It is also usual for the characteristic to be compressed at the bottom.

It is useful to consider typical values of the gammas involved. Assuming a telecine gamma of 0·4, a print of gamma 2·0, a negative of gamma 0·65 and a cathode ray tube of gamma 2·5, the required gamma correction in the recording amplifier must have a value of $\frac{1}{0{\cdot}4\times2{\cdot}0\times0{\cdot}65\times2{\cdot}5}$ or 0·77 for a linear voltage transfer to be obtained.

The additional correction which must be applied by the recording amplifier takes account of the toe of the negative emulsion and the toe of the print characteristic. Any reduction in slope of the CRT characteristic for the darker tones can also be allowed for.

HIGH GAMMA RECORDING. In the example given, the gammas of the negative and print have values typical for normal cinematography. However, a common practice is to develop the negative to a higher gamma of about 1·0 instead of 0·65 and to compensate for this by displaying a reduced-contrast picture on the CRT. This method of operation has certain advantages: the effective sensitivity of the film stock is increased so a slower, finer grain film may be used, or alternatively the lens aperture may be reduced, or the CRT beam current reduced – these changes all tend to give improved definition. Also, since the CRT luminance corresponding to black is well above the zero light level, the effect of flare from light into dark areas is reduced.

There are, however, disadvantages associated with operating the CRT at low contrast. Any spurious modifications to the signal such as phosphor grain, raster shading or blemishes on the tube face become exaggerated in the final picture. Also the non-standard development requires special treatment at the processing laboratory.

DIRECT POSITIVE RECORDING. Instead of producing a negative record in the camera, a positive image may be recorded directly if a negative picture is displayed on the CRT. There is no difficulty in providing a phase-reversing stage in the recording amplifier together with appropriate non-linear correction.

If only a single copy is required, this method has the advantage of saving printing costs and eliminating the loss of definition which occurs in printing. If several copies are required, it is necessary to print a dupe negative and make prints from this.

Many negative film stocks cannot be developed to a high enough gamma to give sufficient maximum density for the dark parts of the picture and if high-energy developers are used there is a danger of increasing the grain size unacceptably. A practical solution is to use the high-gamma technique with a very fine grain copying film stock.

REVERSAL RECORDING. It is common practice, particularly in Continental Europe, to use reversal film in telerecorders. The CRT display has a positive picture, with appropriate non-linear correction to suit the film. The grain tends to be smaller than with the usual negative-positive process and there is no loss of definition due to printing. Prints can be obtained, if required, either by making a dupe negative and printing from this or by making reversal prints from the original film. With the reversal process it is somewhat more difficult to control the processing to obtain the wanted gradation than with the negative-positive process.

CHOICE OF FILM EMULSION. The various methods of telerecording described have the same requirements for transfer characteristics.

However, the suppressed field, the fast pulldown and both the North American systems require a short-afterglow CRT which has a blue phosphor, whereas the stored field and partial stored field systems need a green or orange phosphor to obtain the necessary length of afterglow.

It is thus possible to use a film stock sensitive to blue only in the first case, but a panchromatic stock is needed in the second.

The sensitivity required is not quite the same for the different methods. Whereas the fast pulldown system uses all the light, the suppressed field system rejects half of it and therefore needs either a film twice as sensitive or a lens one stop larger. The stored field system and the partial stored field system both involve a reduction in light as compared with the fast pulldown system.

It can be seen that a film stock must be chosen suitable for the telerecorder in use. It is not, however, difficult in practice to select a film with acceptable characteristics.

CONTROL OF GRADATION. To obtain the wanted gradation, each step in the chain must be properly controlled. This means that the non-linear correction in the recording amplifier must be set up, the maximum and minimum values of CRT luminance, corresponding to white and black, must be set, and the development and printing of the films must be carried out as specified. Furthermore there must be negligible drift in any of the parameters concerned.

The electronic equipment must be free from drift and, in particular, the CRT heater and EHT voltages must be stabilized and the picture signal clamped at black level. The developing baths for negatives and positives are checked frequently by developing exposed sensitometer strips (using the CRT as a source of light) and measuring the resultant densities on a densitometer.

For setting up the non-linear corrector in the recording amplifier and the maximum and minimum luminance levels of the picture on the CRT, and for controlling gradation throughout the whole process, a staircase signal giving a number of discrete levels of luminance in the picture is used. This signal typically takes the form of a ten-step staircase at line repetition frequency, giving a presentation on the CRT of ten vertical strips in ascending luminance, the first representing picture black and the tenth picture white.

A photoelectric cell in a mounting is held against the CRT face to indicate on a meter the luminance of each step. Adjustments are then made to the CRT lift, amplifier gain and non-linear corrector controls such that all ten steps take up the values required.

It is usual to record the staircase for a few seconds at the beginning or end of each reel of film and to measure the densities of the ten steps on the print. This provides a foolproof way of keeping a check on the film processing and printing stages.

MOTOR DRIVE SYSTEMS

All the methods of telerecording described depend on exact speed and phase relationship being maintained between the recording camera and the television signal.

When the television signal has a field repetition rate locked to the 50 Hz mains supply, a mains-driven synchronous motor drive is sufficient. Some means must be provided for establishing the correct phase relationship, such as manual rotation of the motor stator through the required angle. This requires the stator to be mounted on suitable bearings, to have flexible connections and to be capable of being clamped in position. It is also necessary to provide an indication of the relative phase, and this may be done by superimposing on an oscilloscope presentation of the television waveform a pulse derived from a pickup device on the motor.

However, locking of the television signal to the mains supply is increasingly rare, and crystal lock is becoming universal. With crystal lock it is necessary for the telerecorder driving motor to be run in speed and phase coincidence with the television waveform. Several methods exist of driving the motor in this way.

One method is to pass field pulses derived from the television signal through a filter to give a sine wave, and then to amplify this to drive the motor. Means for setting the correct phase position, such as the oscilloscope presentation described above, must be available.

Another method uses a drive motor with a three-phase stator fed from the mains and a three-phase rotor fed from a special three-phase supply from an oscillator. The phase of the supply from the oscillator is determined by a circuit which compares the phase of the field pulse with the phase of the mains. The advantage of this method is that most of the power to drive the motor comes from the mains and only a relatively small amount is taken from the electronic circuit.

The accuracy with which the motor must be held in phase with the TV signal varies with the method of telerecording in use.

For the suppressed field and stored field systems, shuttering and pulldown must occur during the field suppression, so the tolerance is ± 7 lines about the midpoint of operation for the 405-line system and ± 6 lines for the 625-line system.

For the fast pulldown systems, both Continental and North American, the phasing is more critical, as some of the picture is lost at top and bottom, even for precisely correct phasing, due to the pulldown being slightly longer than field blanking. A phasing error does not move the picture but causes more to be lost at the top than the bottom, or vice versa. A desirable phasing tolerance is ± 2 lines.

The partial stored field system is very critical on phasing, as the graded filter operates correctly only when it is exactly positioned in relation to the picture and when the shutter opens at exactly the right time such that the contribution to the film exposure from the CRT afterglow is equalized over the picture. The phasing tolerance is about ± 1 line (405-line system) and a positional servo is required to maintain this.

The picture join system used on the 525-line system is different from the others in that a phasing error does not alter the number of lines recorded on each film frame. All that happens is that the position of the picture join moves vertically on the picture. A continuously-moving picture join is objectionable, so a motor drive system is necessary but need not be extremely accurate.

COLOUR TELERECORDING

The obvious approach to colour telerecording is to display a colour television picture and to record the image on colour film. However, the shadow-mask cathode ray tube as used in colour television receivers is not a suitable high-quality display device, because of the relatively coarse dot structure and the poor colour saturation obtainable with present colour negative films when exposed to the phosphors used in current shadow-mask tubes.

The trinoscope offers a better solution, with its three cathode ray tubes, one providing each primary colour, with dichroic mirrors to combine the three images into a single coloured image. It gives improved resolution and, with appropriate phosphors and colour filters, good saturation on colour negative film.

However, the problem of film transport remains. To obtain a good-quality picture, 35mm film is preferred, and there is at present no commercially proved 35mm quick pulldown camera which has proved itself for colour telerecording with 50 Hz standards. The use of long-persistence phosphors in a trinoscope to dispense with quick pulldown has also not been successful. In North America, the 525-line, 60-field scanning standard does not call for such a quick pulldown, and the use of a trinoscope with a semi-quick pulldown camera becomes practicable.

A radically different approach to colour telerecording is possible if the programme is first recorded on videotape as is usually the case today. This method is known as the colour separation method, or by its trade name of Vidtronics. A. M. S.

See: *Master control; Telecine; Telecine (colour); Vidtronics.*

Books and Literature: *Television Engineering Handbook*, by D. G. Fink (New York), 1957; BBC Engineering Monograph Series No. 1 (London).

□TELEVISION PRINCIPLES

Television has been defined as "The art of instantaneously producing at a distance a transient visible image of an actual or recorded scene by means of an electrical system of telecommunication."

Generally it is necessary first to produce an optical image of the scene, whether that scene is from a studio or from a cinematograph film. Each image can be visualized as comprising many small points or elements, and the detail in a scene depends on the amount of light reflected by the various elements. If all the elements reflect an equal amount of light, then the scene is blank and no intelligence is conveyed. If the elements reflect differing amounts of light then intelligence is conveyed.

The function of a television system is to transmit this intelligence to a distant point and there reconstitute an image of the original scene so that the distant observer receives substantially the same intelligence as if he were looking at the original scene.

The intelligence received from a scene in the form of light must therefore be converted into electrical impulses for transmission by cable and/or radio link. The camera pick-up tube is the equipment which, together with associated amplifiers, effects this conversion. In the case of certain types of telecine equipment (flying spot), a photoelectric cell is used, while with videotape recorders, magnetic heads convert the stored scene information into electrical impulses.

SCANNING METHODS

The problem with television is to transmit the variations of light and shade of each element to the distant end and there to reproduce these variations. Fortunately, because of persistence of vision, it is not necessary to reproduce all the elements simultaneously. The reproduction of the elements can follow in succession provided the whole scene is repeated quickly enough, in a similar way to that used in showing motion picture film where 24 frames are shown per second and an impression of continuous movement is given. To eliminate flicker in the cinema, however, it is necessary to interrupt each stationary picture at least once, so that the effective repetition rate is 48 fps.

It is therefore possible to obtain and use intelligence from each element successively to produce a complete image of the scene. This process is known as scanning.

PICTURE FREQUENCY. A finite time must elapse during the examination of all the elements of a scene. However, this time must be short enough for the persistence of vision of the human eye to overcome flicker and give the impression of natural movement.

Both these effects depend upon the number of complete pictures transmitted per second, known as the picture frequency, while flicker is dependent also on image brightness, the receiving screen phosphor and other factors. It is therefore difficult to determine a minimum value for the picture frequency but satisfactory results can be achieved with a value of 25 pictures per second, provided that the effective repetition rate can be doubled as in film.

SCANNING PATTERN. Technically, it is more convenient to use a scanning pattern in which the elements are arranged in horizontal rows. The scanning beam then moves along each row of elements from left to right returning rapidly to the left-hand side at the end of each row just as the eye reads a page of writing.

There is, of course, a vertical movement associated with the scanning process which is slow compared with the horizontal movement, the beam moving downwards through a distance equal to the vertical dimensions of an element during the time occupied by each horizontal excursion and return. When the beam reaches the right-hand bottom corner of the image it is returned to the top left-hand corner so that the whole process may be repeated.

There is one vertical sweep of the scanning beam for each transmitted picture in the method described, which is known as sequential scanning. Now provided that the

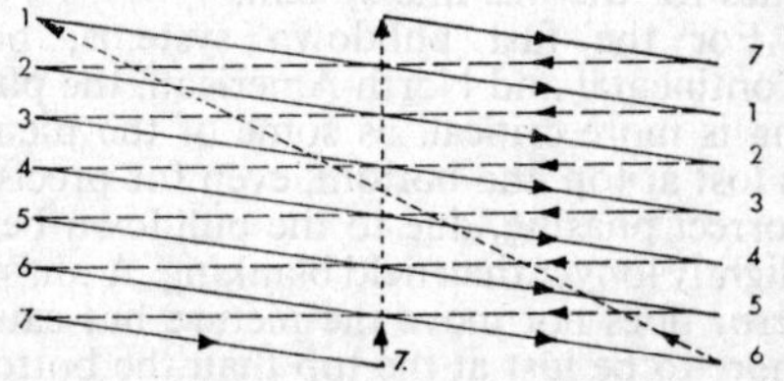

TWIN-INTERLACED SCANNING. Schematic 7-line raster to demonstrate course of scanning spot. Starting with line 1, spot flies back (dotted line) and scans line 3, then line 5. Line 7 is interrupted at mid-point, and field flyback returns spot to top of raster to complete line 7. Next the spot scans in succession lines 2, 4 and 6, from the end of which it returns to restart the cycle with line 1.

scanning beams of the transmitter and receiver are kept in step, there is no need for lines to be scanned in any particular order. There are certain advantages to be gained if, for instance, the odd numbered lines (1, 3, 5 etc.) are scanned first, followed by the even numbered lines (2, 4, 6 etc.). This type of scanning is called twin interlaced and, in this system, there are two vertical sweeps of the scanning beam for each complete picture.

The half pictures composed of odd or even numbered lines are called fields (or frames) and the number of fields transmitted per second is known as the field frequency. In a twin interlaced system, the field frequency is twice the picture frequency. Since it doubles the effective number of picture repetitions per second, this technique substantially reduces objectionable flicker.

All modern broadcast television systems use twin interlacing.

SPOT REGULATION

As each element of the original scene is scanned, the spot of light on the receiving tube must follow and be in a similar position on the line at the same instant. The intensity of the light spot must also vary in a similar manner with the amount of light present in the original element. In practice, the position of the spot at the receiver is regulated by synchronizing signals, while the brilliance of the spot is controlled by the picture signal.

SYNCHRONIZING SIGNALS. Two types of synchronizing signal are necessary.

The line synchronizing signal (line sync) is transmitted when the scanning beam at the transmitter reaches the end of each line and initiates the return of the spot at the receiver from right to left.

The other synchronizing signal is transmitted when the scanning beam at the transmitter reaches the bottom of the image. This is the field synchronizing signal (field sync) and it initiates the return of the spot to the top of the image at the receiver. The return is carried out at a higher speed than the scan and is termed flyback.

It is necessary to separate the two types of synchronizing signal at the receiver so they must have different distinguishing characteristics. Both types of synchronizing signal are inserted between the parts of the picture signal proper, which exists only during the scanning of lines.

PICTURE SIGNALS. To express the magnitude of picture signals, fixed reference levels are necessary. In any scene, the limits of brilliance of the elements are black elements (no light being reflected) and white elements (all light being reflected).

It is possible to fix magnitudes of electrical signal to correspond to these two limits. An area of no reflection has a magnitude of signal referred to as black level and similarly an area of maximum reflection has a signal referred to as peak white level.

White level may be positive or negative with respect to black level and the magnitude of the picture signal is usually expressed as a potential difference (p.d.) with respect to black level.

VARYING LEVEL POTENTIALS. The electrical signal containing the information of a scene is a varying one. It consists of both a varying a.c. component and a varying d.c. mean value, although attempts are made to keep this last value reasonably constant.

For example, consider two identical scenes, one more brightly lit than the other. They will produce two picture signals of similar waveform, one being nearer white level than the other. In order to reproduce these scenes correctly, it is necessary for the direct component, which depends on the average brightness of a scene, to be present at the cathode ray tube of the receiver. This is achieved by including in the television signal a reference level that is transmitted once per line. The instantaneous p.d. from this level depends on the light reflected from the element being scanned. Since no element can reflect less light than black the picture signal must lie entirely to one side of the black level.

When the d.c. component is lost, the picture varies about a mean value indicating zero volts. Black level is negative with respect to the mean value of the picture waveform having a value equal to the d.c. component. Furthermore, as the d.c. component varies, so will the black level, being less negative when the scanned area is dark than when it is bright. Therefore, at the points in a television system where the direct component is missing, the black and white level potentials will vary with average picture brightness.

WAVEFORM COMPONENTS

The complete television waveform is made up of the vision (or picture) signals and various synchronizing signals. The synchronizing signals are negative-going rectangular pulses whose datum line is made to coincide with the datum point (black level)

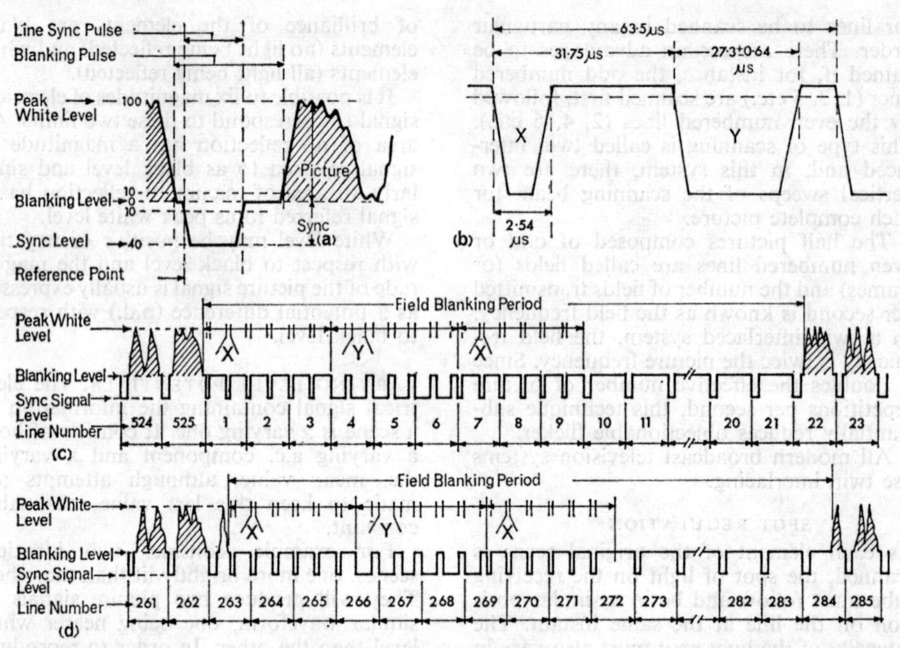

525-LINE 60-FIELD TELEVISION WAVEFORM. Vision modulation shown conventionally as positive-going. (a) Detail of line sync pulse and blanking pulse with signal levels. Line sync pulse (4·76 μs) is preceded on left by front porch (1·59 μs) and followed on right by back porch (4·76 μs). Length of blanking pulse, 11·11μs. (b) Detail of equalizing pulses, X, and broad pulses, Y. (c) Signal at end of even fields, showing six equalizing pulses, X. Six broad pulses, Y, and again six equalizing pulses, X, occupying nine lines in all. Total field blanking period is optionally between 13 and 21 lines. (d) Signal at end of odd fields with same notation. Time duration of pulses is between 10 per cent amplitude points; tolerances and times of rise of pulse edges are specified in standards. Blanking level is the same as black level.

of the positive-going picture signal. The polarity of the transmitted or radiated picture and sync pulses varies according to the system. In the British 405-line system, the sync pulses are negative-going with respect to the modulation of the signal while in the 525-line American and 625-line CCIR systems the sync pulses are positive-going with respect to the vision signal.

It is important that the shapes of the picture and sync signals remain unaltered during transmission and reception, the separation of the two signals being readily achieved by circuits responding to sense and polarity of the applied waveform.

Three significant levels in a television system are those corresponding to white, black and sync levels.

An important parameter in the television waveform is the ratio of picture to sync amplitude (picture/sync ratio). In Britain and the US, this ratio is arranged to be 7 : 3 or 7·25 : 2·5.

VISION FREQUENCY RANGE. The number of fields (or frames) and the number of lines scanned per second are two predominant frequencies in the vision signal. The former is known as the field (or frame) frequency and the latter the line frequency.

The line frequency (in Hertz, or cycles per second) is given by the product of L and P, where L is the total number of lines in the picture and P is the number of pictures per second. In the British 405-line system, L = 405 and P = 25; the line frequency is then 10,125 Hz. In the American system (FCC), L = 525 and P = 30 so that the line frequency is 15,750 Hz; while in the 625 CCIR system, L = 625 and P = 25, so that the line frequency is 15,625 Hz.

The vision (or video) signal contains the fundamental and harmonics of both field and line frequencies.

PICTURE DETAIL. In addition to this synchronizing information, there is the picture detail which consists of a pulsating direct signal. The upper frequency limit of the combined signal depends on three factors: (*a*) the picture frequency, (*b*) the number of lines in the picture, (*c*) the shape

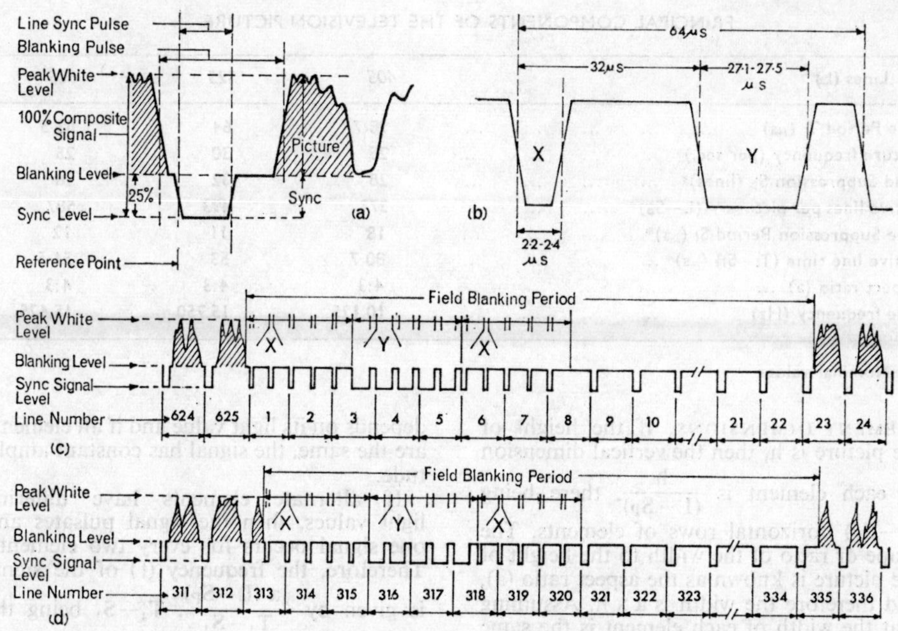

625-LINE 50-FIELD TELEVISION WAVEFORM. Vision modulation shown conventionally as positive-going. (a) Detail of line sync pulse and blanking pulse with signal levels. Sync level, shown as 25 per cent of composite signal amplitude, may be set as high as 30 per cent. Line sync pulse (4·7 µs) is preceded on left by front porch (1·55 µs) and followed on right by back porch (5·8 µs). Length of blanking pulse, 12·05 µs. (b) Detail of equalizing pulses, X, and broad pulses, Y. (c) Signal at end of even fields, with five equalizing pulses, X, five broad pulses, Y, and again five equalizing pulses, X, making $7\frac{1}{2}$ lines. Time duration of pulses is between half-amplitude points; tolerances and times of rise of pulse edges are specified in standards. Blanking level is the same as black level.

of the picture.

At the end of each line a line-sync pulse is introduced during the period of black level known as the line suppression period. This period has a duration (S_1) which is an appreciable fraction of one line (T_1) or 1/line frequency.

At the end of each field (frame), a field sync signal is introduced during a black level period called the field suppression period.

This period has a duration which is an appreciable fraction of the field period and has the effect of suppressing a number of lines, say SP.

Therefore, the number of active lines in the picture is (L—SP).

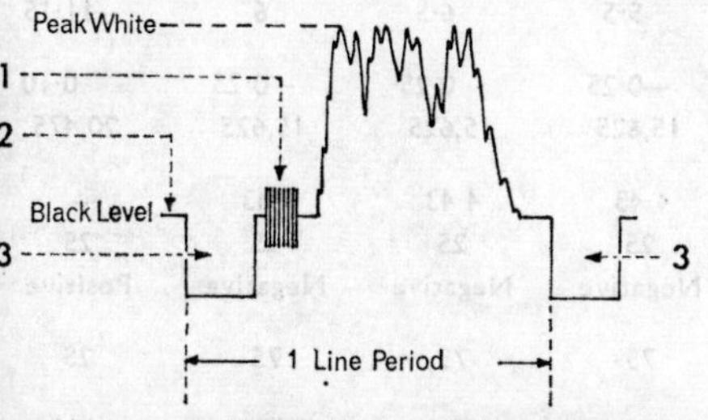

COLOUR BURST. Burst, 1, in relation to the line synchronizing pulse, 3. It occupies part of the post-sync line suppression period (back porch), following the line sync pulse and the pre-sync line suppression period (front porch), 2. The colour burst lasts for approximately 2 microseconds, i.e. about 8–10 cycles of unmodulated subcarrier.

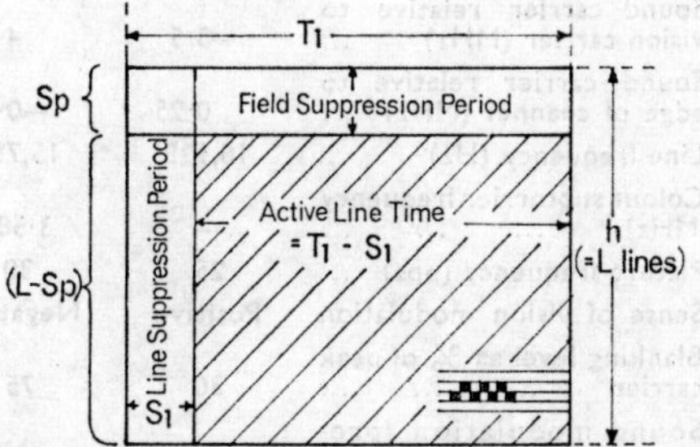

PRINCIPAL COMPONENTS OF TV PICTURE. (*Bottom, right*) Picture elements shown schematically. Number of vertical elements equals number of active lines=(L—Sp). Since elements have same width and height, aspect ratio, a, enables number of elements per line to be calculated as a(L—Sp).

PRINCIPAL COMPONENTS OF THE TELEVISION PICTURE

Lines (L)	*405*	*525*	*625*
Line Period T_l (μs)	98·7	64	63·5
Picture frequency (per sec.)	25	30	25
Field Suppression S_P (lines)*	28	32	38
Active lines per picture $=(L-S_P)$*	377	493	587
Line Suppression Period S_l (μs)*	18	11	12
Active line time (T_l-S_l) (μs)*	80·7	53	51·5
Aspect ratio (a)	4:3	4:3	4:3
Line frequency (Hz)	10,125	15,750	15,625

*Nominal values

ELEMENT DIMENSIONS. If the height of the picture is h, then the vertical dimension of each element is $\frac{h}{(L-S_P)}$, there being $(L-S_P)$ horizontal rows of elements. The shape or ratio of the width to the height of the picture is known as the aspect ratio (a), and therefore the width is a x h. Assuming that the width of each element is the same as its height, $\frac{h}{(L-S_P)}$, the number of elements in a line is $\frac{\text{picture width}}{\text{size of element}} = ah\frac{(L-S_P)}{h} = a\,(L-S_P)$.

The signal corresponding to each element depends on its light value and if all elements are the same, the signal has constant amplitude.

If alternate elements have differing light values, then the signal pulsates and one signal occurs for every two elements. Therefore, the frequency (f) of the signal is given by $\frac{\frac{1}{2}a\,(L-S_P)}{T_1-S_1}$, T_1-S_1 being the active line period.

In the 405-line British system, a = 4/3, L = 405, S_P = 28, $T_1 = 98\frac{2}{3}$ μsec (microseconds) and S_1 = 18 μsec. Substituting these values in the above formula gives f = 3·11 MHz and this is the upper limit at which amplifiers must perform for satisfactory reproduction. The vision band-

TELEVISION STANDARDS

Number of lines/ field frequency	1 *405/50*	2 *525/60*	3 *625/50*	4 *625/50*	5 *625/50*	6 *819/50*
Vision bandwidth (MHz) ...	3	4	5	6	5	10·4
Channel width (MHz) ...	5	6	7	8	8	14
Sound carrier relative to vision carrier (MHz) ...	−3·5	4·5	5·5	6·5	6	−11·15
Sound carrier relative to edge of channel (MHz) ...	0·25	−0·25	−0·25	−0·25	−0·25	0·10
Line frequency (Hz) ...	10,125	15,750	15,625	15,625	15,625	20,475
Colour subcarrier frequency (MHz)	—	3·58	4·43	4·43	4·43	—
Picture frequency (pps) ...	25	30	25	25	25	25
Sense of vision modulation	Positive	Negative	Negative	Negative	Negative	Positive
Blanking level as % of peak carrier	30	75	75	75	75	25
Sound modulation (pre-emphasis in μsec.)	AM	FM ±25 KHz (75)	FM ±50 KHz (50)	FM ±50 KHz (50)	FM ±50 KHz (50)	AM

1. UK only. 2. N. America, Japan. 3. Austria, Germany, Italy. 4. Poland, USSR. 5. UK & Europe (UHF) 6. France: an alternative 625/50 standard is also used.

widths for 525 (American), 625 (CCIR) and 819 (French) line systems are 4, 5 and 10·4 MHz respectively. The lower limit is about that of the field frequency, 50 Hz, but the d.c. component must be restored.

The conception of discrete elements is used only to determine the frequency of the signal and has no physical reality in the camera pick-up or reproducing tubes.

TELEVISION STANDARDS

Historically Britain established in 1936 the first high-definition television standards for regular transmissions. The system employed 405 lines and a field frequency of 50 Hz and continues to this day for British VHF television. Other countries began television transmissions later and set up other standards, but the basic television waveform is similar for all systems. The British 405-line system (which is to be gradually phased out) uses positive vision modulation and amplitude modulated sound with the sound carrier located below the vision carrier. Apart from the French 819-line system, all other systems employ negative vision modulation and frequency modulation for sound with the sound carrier above the vision carrier.

British transmissions in the UHF bands are on the CCIR (European) 625-line system with frequency modulation of sound having a frequency deviation of ± 50 KHz.

The US uses 525 lines with a 60 Hz field frequency and frequency modulation with a deviation of ± 25 KHz.

Television broadcasts are confined to Band I (approximately 40–80 MHz), Band III (approximately 170–220 MHz) and Bands IV and V (approximately 450–960 MHz). Channels within these bands are allocated by international agreement and the majority of countries adhere to the allocations, which are necessary to prevent or reduce interference, particularly in Europe, where many countries in close proximity need to use channels in all bands.

Restrictions on radiated power, aerial polarization, geographic separation of stations on the same frequency and other technical precautions are used to minimize mutual interference. J.D.T.

See: *Colour burst; Field blanking; Line blanking; Porch; Raster; Scanning; Sequential system; Standards; Time of fall, time of rise.*

Books: *Television Engineering Principles and Practice*, Vols. 1 and 2, by S. W. Amos, D. C. Birkinshaw and J. C. Bliss (London), 1953; *Radio*, Vol. 3, by J. D. Tucker and D. Wilkinson (London), 1965; *Final Acts of the European VHF/UHF Broadcasting Conference* (ITU, Geneva).

TELEVISION BANDS I AND III (VHF)

	British	*American*	*European*	*French*
Band I (MHz)	41–67·75	54–88	40–68	41–68
Channels	B1–B5	A2–A6	E1–E4	F1–F4
Band III (MHz)	175·75–215·75	174–216	174–223	162–215
Channels	B6–B13	A7–A13	E5–E11	F5–F12

TELEVISION BANDS IV AND V (UHF)

European—470–960 MHz	Channel width 8 MHz
American—450–885 MHz	Channel width 6 MHz

□**Televisor.** First commercially produced (1932) television receiver devised by John Logie Baird. It used a rotating Nipkow disc and glow tube.

See: *History* (TV).

□**Telojector.** Trade name of a slide projector commonly used with telecine equipment. It has two magazine discs each holding six or eight 2 in. by 2 in. slides and two lamp assemblies the outputs of which are multiplexed together into a common projection lens. Lamp filaments are switched on alternately, the unilluminated slide magazine then moving forward one step by release of an indexing stop restraining a stalled torque motor.

See: *Telecine.*

⊕**Tendency Drive.** Method, sometimes called sprocketless, for transporting motion picture film through continuous processing machines in which the film is moved by the friction of a large number of rotating rollers or discs, the speed of which is automatically controlled by the conditions of the film path. Tendency drive machines can provide uniform movement with only small tension on the strands of film and, since there are no sprocket teeth to engage and possibly damage the perforation holes, they are particularly employed for the processing of original negative material. Machines in which a number of tendency drive positions are combined with a limited number of driving sprockets are termed semi-tendency drive.

See: *Processing.*

⊕□**Tenlight.** Soft light used in film and television studios consisting of ten bulbs carried in a rectangular housing giving a very well diffused source of light.

See: *Lighting equipment.*

⊕**Test Film.** Standard film designed to test one or more characteristics of a projector. Test films are used to check picture steadiness and definition, sound frequency response, and optical alignment of the sound track.

See: *Standards; Projector.*

□**Thermistor.** Temperature-sensitive resistor in which the resistance decreases with operating temperature. Useful as a compensating, measuring or protective device.

⊕THERMOPLASTIC RECORDING □

Invention and development of thermoplastic recording (TPR) is an outgrowth of television projection investigations by Dr. William E. Glenn during the early 1950s at the General Electric Company's Research Laboratory in Schenectady, New York. The system is still at an experimental stage and many problems are yet to be overcome.

Essentially, it is a method of electron beam recording of information in picture or other form on a deformable plastic film. The signal is usually received from a television camera, although other types of scanning devices or symbol generators can be used as well. It is also possible to record various types of analogue signals and digital information. The system has several inherent characteristics which provide a group of potentially useful operational advantages. The principal features are a very high recording bandwidth, inherently to several hundred megahertz, instant development of the image, and high recording density. Colour is also readily recorded.

PRINCIPLES OF OPERATION

The recording medium uses a plastic film, the base of which is a polyester film. On top of the film base are deposited a thin transparent conductive coating and a layer of thermoplastic material. The thermoplastic softens when heated to a temperature in the range of 85°C to 100°C (depending on specific material used), and becomes solid again when recooled to temperatures below about 70°C to 80°C.

The thermoplastic coating on the tape is softened by heating with a resistance type heater. The tape is then passed under an electron gun which deposits negative charges on it. The beam of the electron gun is deflected to scan the tape and is modulated with the input signal to produce a pattern identical to the recorded subject.

In the softened state, the thermoplastic coating is deformed by the attraction of the negative charges to the conductive layer of the film. The deformed pattern is proportional to the charge pattern laid down by the electron beam, and takes the form of ripples on the film. The developing of the picture is essentially instantaneous. The recorded picture is fixed by cooling the thermoplastic tape to its original solid condition.

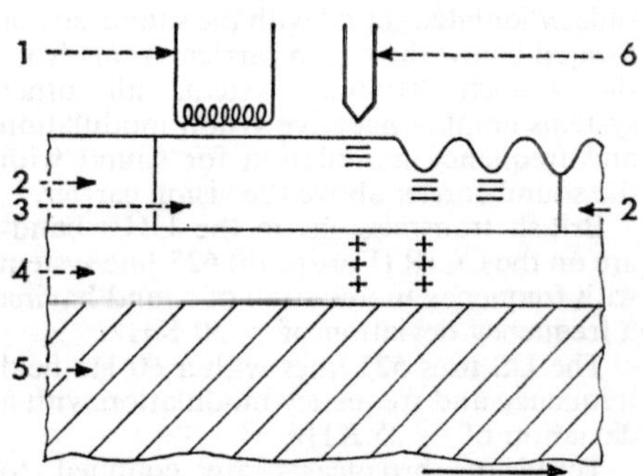

SCHEMATIC OF RECORDING METHOD. (*Top*) 1. Heater. 6. Electron beam. (*Bottom*) Sectional view of tape, moving left to right. Top layer is thermoplastic, solid at 2, before and after passage under heater; liquid at 3, in vicinity of heater and electron beam. 4. Conducting coating. 5. Film base.

Erasure of the information to enable the tape to be re-used can be accomplished by reheating the tape above the melting point of the thermoplastic material.

The tape reels, electron gun, tape transport and heater are all housed in a vacuum chamber, which can be quickly evacuated by vacuum pumps after changing the recording tape.

The projection or readout technique used in the TPR equipment makes use of a Schlieren optical system. If no picture (ripples) is present on the film, the light source is imaged by the condensing lens on the bar beyond the film, and no light gets through to the screen. Where ripples on the tape are present, the light is bent around the bar to the screen. The bar thus acts as a shutter, to permit projection of light on the screen only in places corresponding to ripples on the film. Refraction angles of 30° are typical, permitting use of simple optical design and wide tolerances.

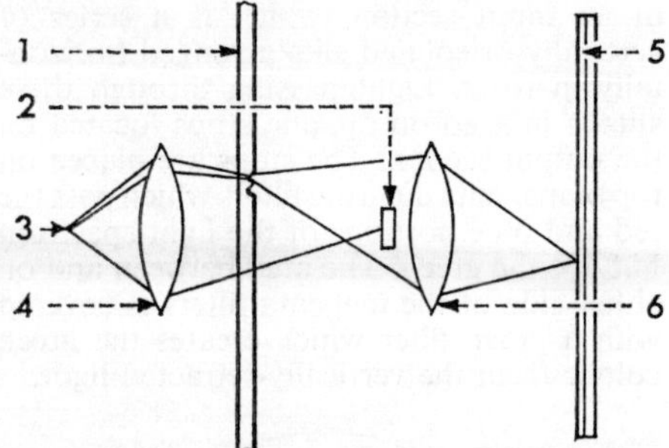

SCHEMATIC OF READ-OUT SYSTEM. Optical projection, with light source, 3, and condenser lens, 4. Thermoplastic film, 1, has deformations which refract light past Schlieren bar, 2, which with projection lens, 6, enables image to be formed on projection screen, 5.

Projection optics are included in the recorder and can be used to monitor recording quality, thereby permitting adjustments to be made. The picture is projected immediately after it is recorded, the delay being usually about one half second. The recorder, however, is not required for playback. Rather, the tape can be removed from the recorder and played back on a projector which can be a standard motion picture type, modified with Schlieren optics.

SYSTEM POTENTIAL AND LIMITATIONS
Theremoplastic recording has two inherent characteristics which theoretically permit recording bandwidths of 1,000 MHz or more. The first of these is that the recording tape is grainless so that resolution is not restricted by minimum grain size obtainable. The other factor is the great flexibility of design which can be achieved in the electron gun structure. The beam, operating in a vacuum, is easily deflected, having very low inertia. Beam current is usually in the range of 1–2 μA. Aperture size and shape can be adjusted, with resulting spot diameters on the tape of 5 μ or less. Consequently, horizontal resolution of 2,000 or more lines is within the capability of the electron gun, using 16mm tape. For wider tapes, 8,000 or more spot diameters can be recorded in the horizontal direction, the limitation being the maximum width for which acceptable focus and deflection can be provided by the electron gun.

The electron beam can be arranged to scan the tape in an interval of less than $\frac{1}{2}$ μsec, so that maximum recording bandwidth is primarily a function of practical limitations of the speed of vertical movement of the tape. Recording rates of 850 MHz have been achieved on laboratory equipment incorporating fast horizontal scan.

For recording of video signals from a television camera, there are, of course, source limitations on bandwidth, with a maximum short term prospect of perhaps 20 MHz. Design of the video circuits in the recorder also currently limits realization of maximum recording bandwidth for television signals to about 40 MHz.

The use of very small beam spot diameters and vertical scan spacing of 3,000 television lines per inch can produce recording densities of up to 5 television pictures per square inch of tape, using an 8mm tape width. For the 16mm width, recording density is approximately 20 pictures per square inch, allowing for space used by sprocket holes.

Contrast ratios of 300 : 1 have been observed using carefully adjusted optics. A more typical contrast ratio is of the order of 100 : 1. The grey scale can be corrected to be quite linear over the contrast range.

Brightness of the projected picture is satisfactory for screen sizes up to about 4 ft. × 3 ft. A xenon arc lamp can be used as a light source, a 500 watt lamp producing approximately 150 screen lumens. Resulting screen brightness is 12 ft. lamberts on a 4-ft. screen or 50 ft. lamberts on a 2-ft. screen. The use of screens with optical gain can, of course, increase brightness where wide viewing angles are not required.

Limitations in signal-to-noise ratio are most directly related to the extent of optical irregularities in the tape base and the conductive and thermoplastic layers. Meaningful measurements have not been made on noise level since additional development work on the recording tape is required from several standpoints. Expectations are that the level of dirt will be comparable to that in high-quality motion picture film. On the other hand, the thermoplastic tape is characterized by lack of grain, the principal source of noise in photographic film.

BEAM SCANNING AND READOUT METHODS.
The thermoplastic tape is scanned horizontally, while vertical tape motion can be accomplished by either of two methods. One technique employs intermittent tape motion, where the tape is pulled down at a rate of (usually) 24, 25, or 30 fps. Vertical

sweep is applied to the vertical deflection section of the electron gun so that all horizontal lines of each frame are scanned between each tape pulldown interval. The lines can be scanned sequentially, using modified television cameras, or a 2 : 1 interlaced signal can be accommodated by scanning the two fields for each frame. Additional design effort is being directed toward perfecting the quality of such interlaced signal recording.

Vertical scan can also be achieved by continuous motion of the tape at a constant rate. The lines of each frame are horizontally scanned sequentially as the tape is moved under the electron beam. If a 2 : 1 interlaced signal is to be received, the two fields are recorded separately, and optically interlaced on readout, either by projection of the fields at the field rate (twice the frame rate) or by simultaneous projection of the two fields at the frame rate.

VACUUM EQUIPMENT. The electron gun used in thermoplastic recording operates in a vacuum enclosure which must be maintained at a pressure of less than 10^{-3} torr. The enclosure also includes tape transport, windup and payoff reels, and most of the projection optics used to monitor the recording. The need periodically to open the enclosure to change tape requires careful design of compartment seals and a vacuum-pumping system which can quickly reduce enclosure pressure to the required level. Pump-down time is determined by the size of pump used and is typically about 2 minutes but can be less.

COLOUR RECORDING. Colour recording is accomplished by modulating a single electron beam simultaneously with three primary colours: red, green, and blue, received from a colour television camera.

The green signal is recorded in the same manner as black-and-white, by creating parallel grooves perpendicular to the tape length (horizontal in the recorded picture). Groove depth is varied (signal modulation) by vertically defocusing the beam where green picture points are not desired.

Red and blue are recorded as velocity-modulated diffraction gratings along the sweep as the beam moves across the tape in scanning each line. These changes in horizontal spacing of picture points are accomplished by modulating the beam at two different frequencies, one for red, one for blue, in the range of 10 to 20 MHz.

Intermediate colours are obtained by mixing the three primary colours in varying amounts. White light is obtained by the presence of all three signals in the proper ratios, while black occurs when all colour signals are zero.

The readout Schlieren system is composed of an input section, which is a series of vertically-orientated slits arranged horizontally in rows. Light passing through these slits is imaged on opaque stops located on the output section. The stops are placed on top of magenta dichroic filters which pass the red and blue portions of the light spectrum but not the green. The area between and on either side of the magenta filters is covered with a green filter which creates the green colour from the vertically-refracted light.

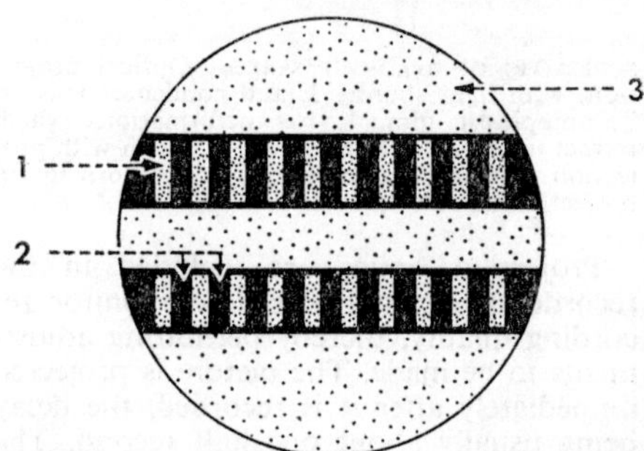

SCHLIEREN PROJECTION OF COLOUR RECORDINGS. Arrangement of stops and filters on Schlieren output section. 1. Magenta dichroic filter. 2. Output Schlieren light stops. 3. Green filter.

In forming red and blue, the individual light sources imaged on the output bars are deflected horizontally, forming a diffraction spectrum. The modulation frequencies are selected so that when a red output is desired, most of the red light goes through the open sections of the magenta filter and most of the blue light is stopped by the Schlieren output stops. In the same manner, blue colour signals cause most of the blue light to pass and stop most of the red. Of course, the slots are designed properly to accept the higher orders of diffraction as well as the first order. The green portions of the light spectrum are unable to pass through the magenta dichroic filter.

Green is formed by refracting the light vertically to pass through the green filters located above and below each of the light stops. This technique is essentially the same as used in the TPR monochrome Schlieren system.

In summary, red and blue are obtained by selection from horizontally-diffracted light containing the full light spectrum.

Green is obtained by vertical refraction of white light through green filters. Intermediate colours are obtained by mixing the three primaries.

MATERIALS AND EQUIPMENT

THERMOPLASTIC RECORDING TAPE. The recording system places requirements on tape characteristics which make tape design and manufacture one of the most critical aspects of thermoplastic recording. In evaluating various commercially available materials, deficiencies in one or more parameters are invariably found, and consequently new formulations have been developed which can provide the needed characteristics in combination.

APPLICATIONS. The principal near term use for thermoplastic recording is expected to be the recording of video signals received from a television camera. The equipment uses the transparent film described above in a 16mm width. The prototype floor-mounted vertically-oriented equipment is approximately 5 ft. in height and weighs 7–800 lb. Weight reduction will be effected in future designs.

Recording tape, on reels, provides a total recording time of 35 minutes at 30 fps. Use of larger reels and/or thinner tape base is expected to increase recording time to an hour or more.

Better portability has also been achieved by a recorder designed for 8mm tape. Including vacuum pumps and electronic circuits, this recorder weighs approximately 150 lb., and has a volume of less than 3 cu. ft. T.F.M.

See: *Electron beam film scanning; Electron beam recording; Electronic video recording; Large screen television; Schlieren photography.*

Literature: *Thermoplastic Recording: A Progress Report*, by W. E. Glenn (*Jnl. SMPTE*), 1965; *Thermoplastic Recording Systems*, by N. Kirk (*Jnl. SMPTE*), Aug. 1965.

Thirty-five Millimetre Film. Standard gauge of film used for all forms of professional motion picture work, introduced in the early pioneering days of the cinema about 1889 and still the most widely used material for entertainment purposes both in the cinema theatre and in television. This gauge is 35mm in width and has rows of perforation

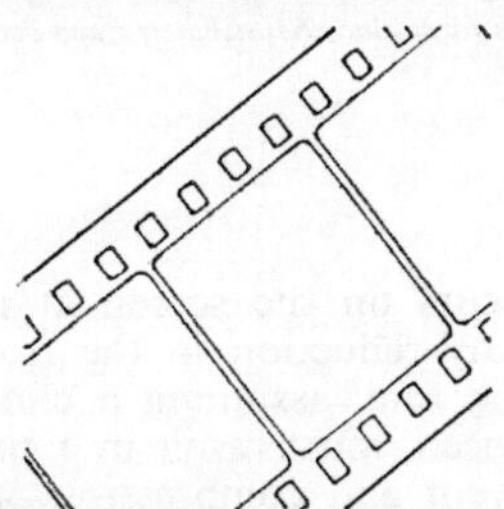

holes along both edges; each picture frame is four perforation holes high and there are sixteen frames in the length of one foot of film. Several types of perforation hole have been used in the industry from time to time, but current practice has now reduced these to three, for negatives, positives and Cinemascope prints, the latter having a smaller hole to allow space in the width of the film for four magnetic sound tracks.

See: *Film dimensions and physical characteristics.*

Thirty-two Millimetre Film. Motion picture film in strips 32mm in width suitably perforated so that after printing and processing it may be slit into two 16mm or four 8mm copies. The number of ranks or rows of perforation and their relative positions differ

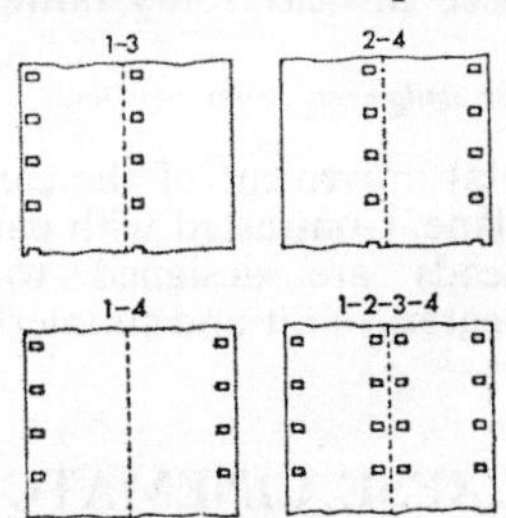

32MM FILM. Numbers designate ranks of perforations.

according to the type of printing and the requirement for sound or silent 16mm prints.

See: *Film dimensions and physical characteristics.*

Three-Colour Processes. Systems of colour reproduction in which the visible spectrum is divided into three sections, normally red, green and blue, for the purposes of recording and presentation.

See: *Colour cinematography.*

Three-Strip Camera. Motion picture camera for colour photography in which three separate strips of negative film recorded respectively the red, green and blue components of the original scene. The optical system normally made use of a single objective lens with a beam-splitting prism and colour filters dividing the light to the three negatives.

See: *Technicolor*

⊕**Threshold.** Minimum exposure of a photographic material which just produces a detectable density above the basic fog level after processing.
See: *Exposure control; Fog; Tone reproduction.*

⊕**Throw.** Distance from a projector lens to the screen on which it forms an image. There is a widespread misconception that the longer the throw, the more light is required to produce a screen picture of given size. The only light factor directly dependent on throw is atmospheric absorption and scatter caused by, e.g., cigarette smoke.
See: *Cinema theatre; Projector; Screens.*

⊕**Thyratron.** Gas-filled or mercury vapour
□triode thermionic valve used as an electronic switch for relatively heavy currents, as in studio lighting circuits.
See: *Lighting equipment; Power supplies.*

⊕**Thyristor.** Solid state rectifier in which the
□conducting / non-conducting characteristic can be rapidly switched by a potential applied to one terminal. The silicon controlled rectifier (SCR) is a thyristor. Frequently used in heavy-duty power rectifier circuits.
See: *Lighting equipment; Power supplies.*

⊕**Tilt.** Pivotal movement of the camera in a
□vertical plane. Contrasted with pan. Special tripod heads are designed to provide extreme degrees of tilt and counterbalancing

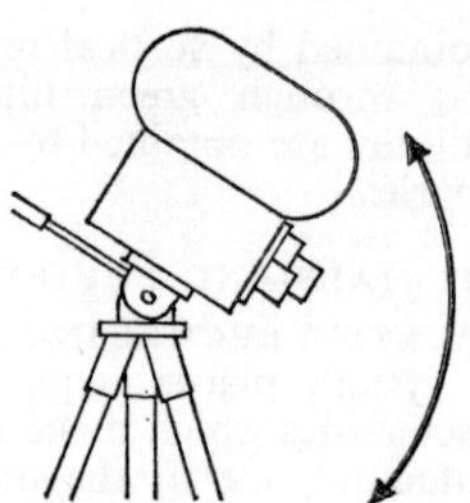

springs prevent the camera falling forwards or backwards.
See: *Camera; Cameraman.*

□**Time-Adjacent Fields.** Fields which follow one another in time but which do not necessarily contain similar information.
See: *Raster; Scanning; Television principles.*

□**Timebase.** Sawtooth oscillator producing a deflection of a cathode ray beam to display a waveform or to produce a TV raster.
See: *Raster; Scanning; Television principles.*

⊕**Time/Gamma Series.** With most photographic materials the contrast of the image as indicated by its gamma value, increases as the time of development is lengthened. To establish the processing characteristics of a developer, it is therefore usual to develop a series of standard step wedges with varying development times and establish the exact relation between those times and the gamma value obtained.
See: *Image formation; Sensitometry and densitometry.*

⊕TIME LAPSE CINEMATOGRAPHY

The basis of time lapse cinematography is compression of the time scale whereby movements which are so slow as to be imperceptible to the human observer are recorded on film for subsequent projection or analysis. The technique can be used in two ways. First, as a qualitative tool to produce filmic material for the visual analysis of events occuring over a long period of time and to create the illusion of growth and decay. Instances of such phenomena are plant growth and rain penetration of buildings. The second use is for quantitative recording. Film produced under these conditions is not necessarily meant for projection but for picture-by-picture analysis, e.g., the recording of traffic flow and meter readings.

When an event is recorded on film at 16 fps (or 24 fps) and projected at the same rate, the time magnification is unity and the event occurs on the screen at its natural speed. Any reduction in the frequency of the taking rate, assuming a constant projection speed, must result in a speeding up of the event and compression of the time scale. Time magnification is then less than unity and is given by the formula:

$M = F_t/F_p$

where M = time magnification
F_t = camera picture frequency in fps
F_p = projector picture frequency in fps

A taking rate of one frame per hour projected at 16 fps gives a time magnification of:

$$M = \frac{1/3600}{16} = \frac{1}{57,600}$$

The range of time lapse cinematography

includes pictures taken at a frequency of 8 fps down to 1 frame every 12 hours.

BASIC REQUIREMENTS

The apparatus required for time lapse cinematography can be divided into three components, viz. the motion picture camera, lighting and the programmer.

It is immaterial what make of motion picture camera is used provided it has facilities for taking single pictures. Similarly, the film gauge is not of paramount importance.

SINGLE-PICTURE CONTROL. If the camera is spring-driven, it must have a single picture button and some form of electro-magnetic tripping mechanism to operate it. A solenoid is usually employed to operate the single

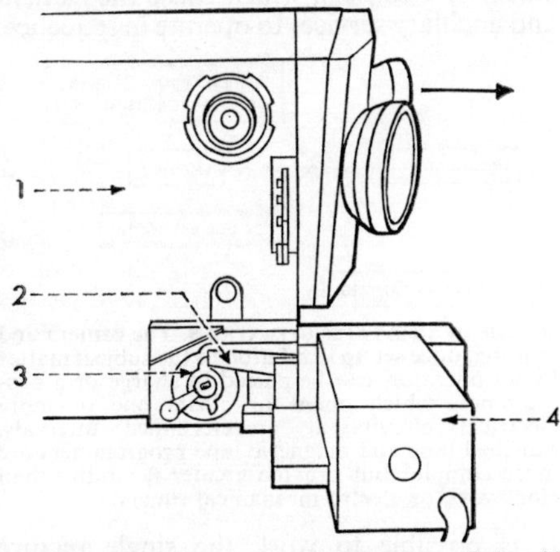

RELEASE FOR CINE-KODAK SPECIAL II. Solenoid electric release, 4, lifts movable arm, 2, to release cam, 3, which thereupon rotates one revolution, exposing a single frame. 1. Shutter operating arm.

picture button but this type of release can cause vibration. Motion picture cameras which are not spring-driven, however, usually have a single picture shaft which is ideal for time lapse work. This shaft must be turned through exactly 360° to advance and expose the film precisely one image at a time. The rotation of this shaft may be continuous or intermittent.

CONTINUOUS DRIVE. The simplest continuous drive involves reducing, by means of a gear box, the speed of rotation of an electric motor coupled to the direct drive shaft of the motion picture camera. Different gear ratios are used to provide a change of frequency which may extend from 24 fps to 2 fpm (frames per minute).

INTERMITTENT DRIVES. In an intermittent drive, there is a mechanical link between the continuously-rotating drive shaft and the camera shaft. In this way, it is possible to transform a continuous motion into an intermittent one. The standard method of doing this is to use some form of clutch, such as a dog clutch, a slipping plate clutch or a coil spring clutch.

DOG CLUTCH. The dog clutch assembly consists of two multi-toothed gear wheels,

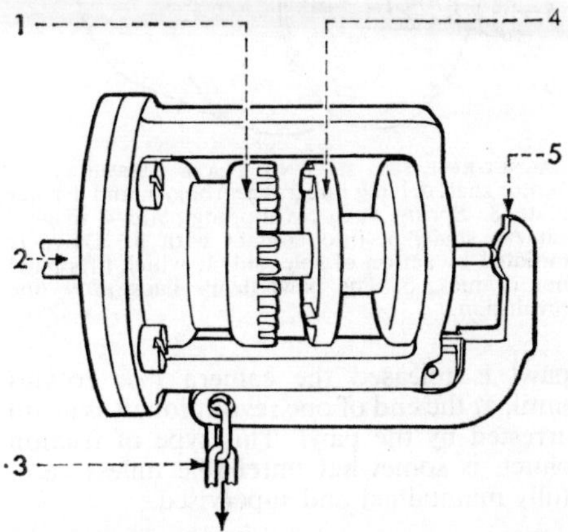

MULTI-TOOTH DOG CLUTCH MECHANISM. Link, 3, goes to control gear which determines rate of picture taking. Downpull on 3 pushes in 5, and engages dog-tooth wheel, 4, connected to camera shaft, 2, with multi-tooth wheel, 1, which takes its drive from a gearbox.

one driven by a continuously-rotating electric motor, the other attached to the camera drive shaft and normally at rest. These two wheels are held apart by an internal compression spring except when a picture is being taken. Then an impulse, either mechanical or electrical, causes the two wheels to mate and rotate. As soon as the whole assembly has made one complete revolution, the camera gear wheel moves away from the driving wheel and stops.

SLIPPING PLATE CLUTCH. There are many slipping plate clutches but basically they all consist of two rotating discs running in contact with each other. One is driven by an electric motor, the other connected to the camera shaft. To prevent this shaft from moving except when an exposure is required, a notch is formed in the disc attached to it into which a pawl can be dropped by an electro-magnetic release. Normally the two discs slip against each other, but when the

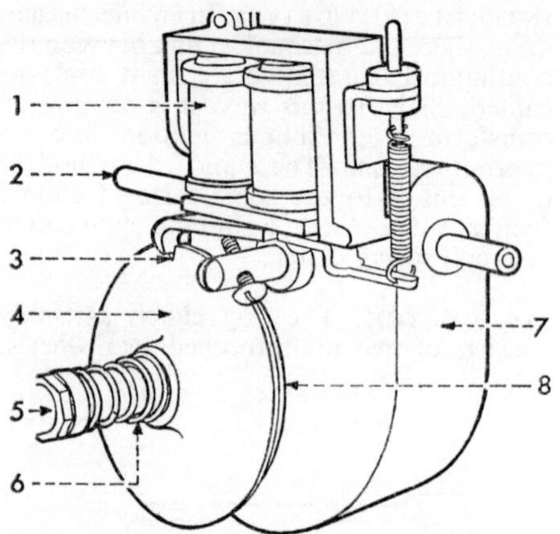

MAGNET-RELEASED SLIPPING PLATE ASSEMBLY. 2. Motor shaft driving integral gearbox, 7, and driving plate, 8. Spring, 6, forces slipping plate, 4, driving camera shaft, 5, into contact with 8. Drive is initiated by action of solenoid, 1, which lifts pawl out of nick, 3. The pawl drops back after one revolution.

pawl is released the camera disc rotates until, at the end of one revolution, it is again arrested by the pawl. This type of friction clutch is somewhat unreliable unless carefully maintained and supervised.

COIL SPRING CLUTCH. The coil spring clutch is a simple but more reliable method of joining two co-axial but independent shafts, so that a continuously-rotating electric motor can impart one revolution to the shaft of a motion picture camera. The amount of power capable of being transmitted by this device is limited only by the strength of the spring.

TIME-MARKERS. As time lapse filming deals with the recording of movement or change, however slow, it is essential that a time scale should be incorporated on the film if measurements are to be made from it. A daily, weekly or monthly calendar can be used to record the passage of time at very slow rates of filming. An integral time marker is sometimes built into the camera mechanism, either as a mechanical pointer or a neon lamp, the images of which are recorded photographically on to the film. When the event marker is outside the camera, an image of a clock face or stop watch is transferred optically on to the film. There are other parameters which may with advantage be recorded in the same picture as the image, e.g., humidity and temperature, but at present there is no inexpensive way of doing this.

THE PROGRAMMER

The object of the programmer is to provide, at pre-determined uniform intervals of time between exposures, impulses, either mechanical or electrical, which cause the camera and ancillary services to operate in sequence.

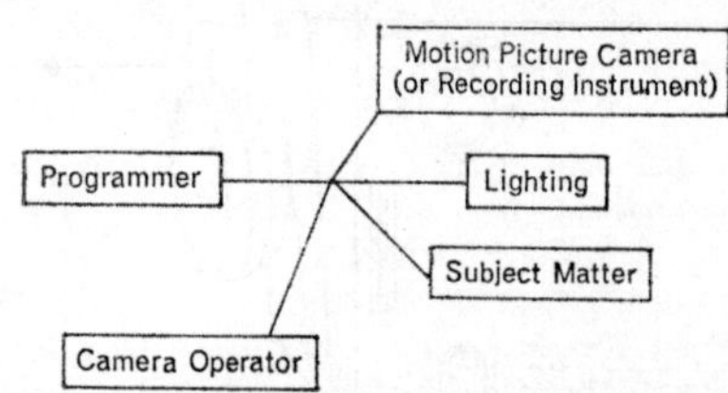

SCHEME OF TIME-LAPSE OPERATIONS. The camera and lighting, once set up in relation to the subject matter by an operator, can be placed in charge of a programmer, which opens or closes one or more electrical circuits at predetermined intervals. Punched tape and magnetic tape programmers are more complex but provide greater flexibility than clockwork or electro-mechanical timers.

It is possible to work the single picture button or shaft manually when there is a long interval between pictures, e.g., once every 12 hours; but operator fatigue and human fallibility make this method unreliable. Water has been used to give time signals either by means of a water wheel or by the alternate filling and emptying of two small buckets to which are attached a mercury switch and a pair of contacts.

EFFECT OF FREQUENCY ON CHOICE OF PROGRAMMER

Frequency of Event	*1 per day*	*1 per hour*	*1 per minute*	*1 per second*	*5 per second*
Manual	R	R	NR	X	X
Clockwork	R	R	R	NR	X
Electro-mechanical	R	R	R	NR	NR
Electric	X	X	R	R	R
Electronic	X	X	NR	R	R
Punched tape, Magnetic oxide, Punched card	X	X	NR	R	R

R=recommended NR=not recommended X=unsuitable

CLOCKWORK TIMERS. Clockwork motors, or large watch mechanisms, are cheap and highly reliable pieces of apparatus with which to generate a timing signal, preferably electric, which can trigger off a camera drive motor and other equipment. They run for 24 hours or longer and, provided the spring which produces the motive power is in good condition, last for years. Some modification is needed to the clock mechanism so that the second and minute hands can be made to act as contact breakers.

ELECTRO-MECHANICAL TIMERS. A synchronous electric motor acts as a master control which in turn governs the behaviour of the programmer. Using this method between each cycle of operations, all the apparatus is at rest except for the synchronous motor.

The rotation of the shaft of the synchronous motor closes a pair of contacts which actuate a second motor.

A second motor drives a spindle to which are fastened a series of cams. These actuate micro-switches associated with the electrical circuits needed to control the starting and stopping of the camera drive motor, the single picture clutch mechanism, the turning on and off of lights, or the removal of a shutter from in front of the light source, e.g., in microscopy.

One electro-mechanical timer makes use of a large wheel fastened to the shaft of an electric motor. Around the circumference of this wheel are a number of holes each capable of holding a solid metal pin. The arrangement and number of these pins is determined by the operator and each pin operates a micro-switch which controls one or more electrical circuits.

ELECTRONIC TIMERS. These depend for their operation on the charge and discharge of capacitors. The electrical qualities of these capacitors can be made to determine both the exposure times and the intervals between. In the most versatile electronic timer, the exposure and interval are variable and independent of each other. This is known as a variable mark space ratio. The electrical components must be carefully chosen and calibrated, special care being taken to ensure freedom from thermal drift and ageing.

PUNCHED TAPE AND MAGNETIC OXIDE TIMERS. Punched tape and magnetic oxide timers are other methods of controlling the cycle of operations in a time lapse sequence. One system uses punched tape in the form of a loop. This loop of instructions is passed over a sensing head from which impulses are sent to the camera and ancillary equipment. Magnetic oxide recording tape has also been used in a similar manner to the remote control of the projection of colour transparencies linked to a prerecorded commentary.

Another type of equipment uses punched cards arranged in such a way that up to eight pieces of electrically operated equipment can be controlled by one unit.

OTHER TYPES OF TIMER. Pneumatic timing devices have been designed, but it is difficult to manufacture a precision air leak that is both reproducible and reliable. One further class of timer is a self-actuating one, whereby the commencement of the phenomenon starts the filming programme. This technique is used only when the timing of the event is uncertain or variable, e.g., the deposition and erosion of sandbanks. This phenomenon is filmed when the tide is at a constant height but, as this varies with each tide, the time lapse sequence is initiated by the tide closing an electrical circuit at the appropriate level.

LIGHTING

Daylight is the simplest form of lighting to use but the most difficult to control. It is inconsistent both in intensity and duration, but may have to be used if the subject matter is large, e.g., a building in course of construction. The shadows cast by the sun vary according to the time of day in position, size and shape.

Filming may have to be confined to one period of the day to avoid the sun shining into the camera lens.

ELECTRONIC FLASH. A more controllable form of illumination is the electronic flash tube. This is a most flexible light source and can be obtained in a wide variety of shapes and sizes, and with ample electrical energy to illuminate the most ambitious project. The spectral energy distribution of flash tubes varies, depending on the rare gas mixture with which they are filled. The flash energy of the tube can be chosen to suit the particular lighting problem, while the duration of the flash can be either in the millisecond or microsecond range. The rate of flashing can also be controlled and depends on the size of the capacitors and associated circuitry. The synchronization of the flash discharge with the shutter open

RELATIONSHIP OF ILLUMINANT TO SUBJECT MATTER

Subject	Lighting							
					Xenon			
	Daylight	*Tungsten Filament*	*Electronic Flash*	*Zirconium D.C. Arc*	*D.C. Arc*	*A.C. Arc*	*Tungsten Halogen Compact Source*	*Mercury Vapour*
Exteriors memo-motion	R	NR	NR	X	X	X	X	X
Interiors memo-motion	NR	R	NR	X	X	X	X	X
Plant growth	NR	NR	R	X	X	X	X	X
Rain penetration of buildings and corrosion studies	NR	R	R	X	X	X	NR	X
Cloud formation and air pollution studies	R	X	X	X	X	X	X	X
Microscopy	X	R	R	NR	R	NR	R	R
Meter readings	NR	R	R	NR	NR	NR	NR	NR

R=recommended NR=not recommended X=unsuitable

position of the cine camera presents no difficulties whatever the speed of operation.

TUNGSTEN FILAMENT. Tungsten filament light sources are still, however, widely used in time lapse experiments. The heat generated by these lamps can be a nuisance, especially when dealing with biological materials. This can be minimized by placing heat filters in front of the lamps, if small enough, or the lamps can be switched off between exposures or run at a reduced voltage.

Alternatively, a separate shutter can be placed between the light source and subject. Tungsten lamps can be used for time lapse colour filming but the glass envelope darkens with age, causing a change in colour temperature. The newer tungsten halogen lamps are more suited for colour work as the glass (or quartz) envelope remains clear during the whole of its useful life and the colour temperature of the source is more suitable.

OTHER SOURCES. Mercury vapour lamps and compact source arcs, including the xenon arc and the tungsten iodide arc, can also be used. The tungsten iodide arc is a recent development and uses tungsten electrodes and metallic iodide fillings within a quartz envelope.

The arc is well stabilized and when run on a.c. seems to be more stable than the older d.c. zirconium arc.

Self-luminous subjects such as the heating and cooling of an iron girder under stress present an unusual problem, i.e., the difference in intensity of the phenomenon as it approaches and falls away from maximum brilliance. This can be controlled by the human observer or by using a cine camera equipped with an automatic iris.

APPLICATIONS

Time and motion studies depend to a large extent on photography for the production of evidence and although still photography is much used for this kind of work the cine camera can often be of great assistance in determining a flow pattern. Analysis of such a film strip, picture by picture, usually produces suggestions for a possible improvement in the work layout with a corresponding increase in efficiency. The correct siting and arrangement of car parks and children's playgrounds are instances. Movement of pedestrians on restricted crossings, and of cars in relation to them and to the utilization of road area, form a related application of time lapse, which makes possible an integration of time information into a comparatively short length of film.

Time lapse techniques can be used in studying air pollution from factory chimneys, in recording the growth of fruit trees or even the movement of tree roots, cell division as seen through a microscope, the movement of clouds and corrosion studies. As long as there is movement, the time lapse camera can record it and present it to the observer in a graphic and intelligible manner. K. M.

See: *Camera* (FILM); *Cinemacrography; Cinemicrography; Lighting Equipment* (FILM).

Books: *Macrophoto and Macrocine Methods*, by A. and I. Tölke (London), 1969; *Photography for the Scientist*, by C. E. Engel (New York), 1968.

Time of Fall, Time of Rise. Terms used in connection with pulse edges. Because synchronizing pulses initate flyback by means

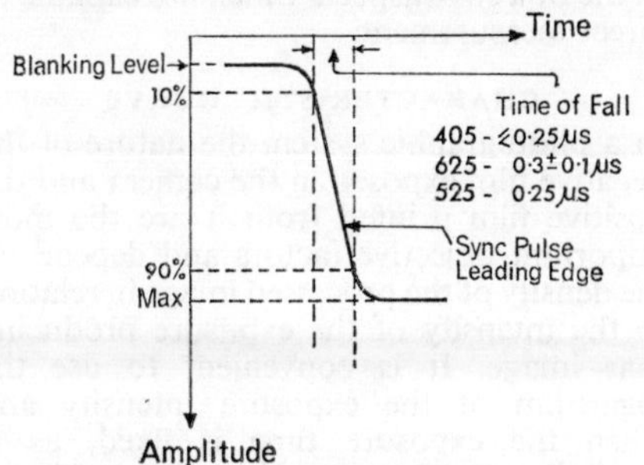

TIME OF FALL. Time of fall from 90 per cent to 10 per cent of maximum amplitude as specified for principal television systems.

of a trigger action associated with their leading edge, it follows that the precise instant at which the trigger is effective in line-scan circuits is governed by the steepness of the pulse itself. This is a measurable quantity usually expressed in terms of the time taken for the pulse to fall (or rise) between 10 and 90 per cent of its maximum amplitude and is known as the time of fall or time of rise as appropriate.

See: *Synchronizing pulse generators; Television principles.*

Time Scale Exposure. Method of exposing step wedge tests of photographic materials in which the intensity of light is constant but the time of exposure is varied for each step.

See: *Sensitometry and densitometry; Sensitometry of colour films.*

Timing. In addition to the obvious meaning of determining the exact time taken to project a scene or sequence of film, this term is sometimes used to mean the operation of grading.

See: *Exposure control; Grader; Grading.*

Tinted Base. Black-and-white prints using film stocks coated on coloured transparent support made for special sequences, for example, blue base for moonlight effects, red for fire, orange for candlelight, etc. Rarely used after 1940.

Title. Written material which appears on a film or TV screen, and is not part of the scene. Titles are of many sorts, e.g., credit, main, end, creeper, roller, sub-, superimposed.

Titra Titling. Method of making superimposed titles for foreign language translation by printing them directly on to a standard release print.

See: *Foreign release.*

Todd-AO. System of wide-screen cinema presentation using 65mm negative for photography and 70mm copies with stereophonic magnetic sound tracks for projection.

See: *Wide-screen processes.*

Toe. In the characteristic curve of the tonal reproduction of a photographic process, the region of lower densities approaching the

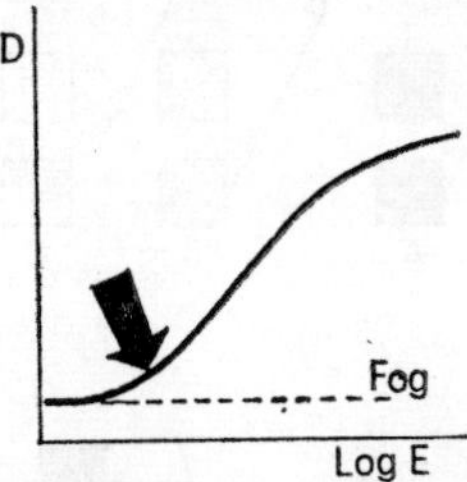

basic fog level of the material. In this region the relation between the log exposure and the density produced is not linear.

See: *Characteristic curve; Exposure control.*

Toned Track. Sound track for colour film. The dye images of colour film are not suitable for sound tracks because of their lack of absorption of the infra-red to which the sound reproducer photocell is sensitive. It is sometimes necessary to alter the sound track image by chemical methods, such as toning to Prussian blue.

See: *Integral tripack systems.*

TONE REPRODUCTION

In cinematography, tone reproduction deals with the relation between the brightness aspects of the original scene photographed and those of its eventual representation on the screen in a motion picture theatre; the photographic characteristics of the film material employed and the optical characteristics of the camera and projection lenses must be taken into consideration. In television, a similar relation between the original scene and its representation on the screen of a television receiver involving both optical and electronic factors is referred to as the transfer characteristic of the system.

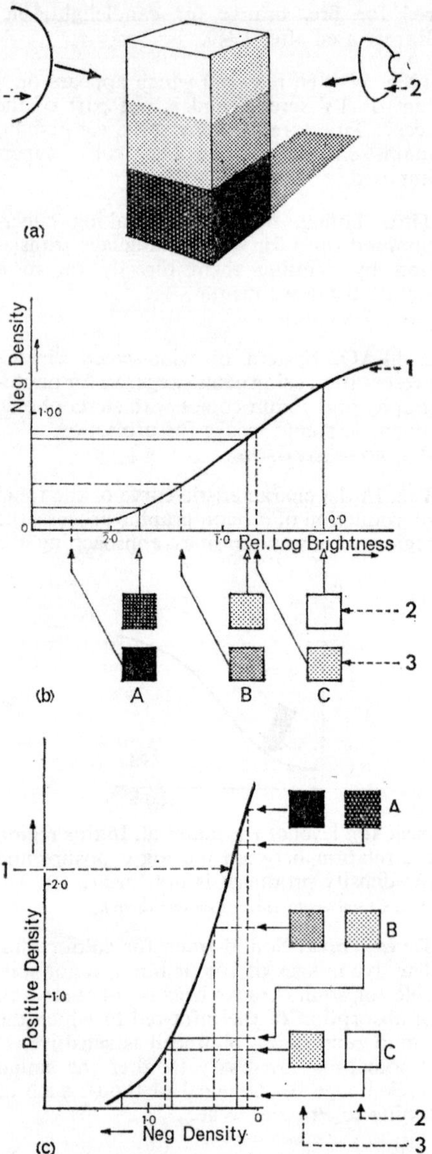

REPRODUCTION OF OBJECT. (a) Object with three reflectances corresponding to a white of 90 per cent reflectivity, a mid-grey of 20 per cent and a black of 2 per cent, as lit with lighting contrast of 4 : 1. 1. Key light. 2. Filler light. (b) Same object and lighting recorded on negative characteristic, 1, employing toe and shoulder regions. A, B, C, represent the three sections of the object, with high-lit areas, 2, and low-lit areas, 3. (c) Same object recorded as at (b) and printed to positive film with characteristic, 1, and same areas of 2 per cent, 20 per cent and 90 per cent reflectivity lighted at 4 : 1. Note low maximum density of negative and high maximum density of print.

In both cases there are subjective factors concerned with the visual processes of the observer under conditions of viewing as well as the objective aspects which are capable of direct measurement.

CHARACTERISTIC CURVE

In a photographic system the nature of the negative film exposed in the camera and the positive film printed from it are the most important objective factors and depend on the density of the processed image in relation to the intensity of the exposure producing that image. It is convenient to use the logarithm of the exposure intensity and when the exposure time is fixed, as is normally the case in motion picture photography, the densities of the image may be shown in relation to the brightness scale of the original. This relation is non-lineaı and when shown graphically is known as the characteristic curve of the photographic material.

BRIGHTNESS RANGE

In the photography of real scenes the brightness range is determined by both the reflectivity of the objects in view and the variation of their illumination between lighted and shadow areas, known as the lighting contrast. Object reflectance may range from 90 per cent of the incident light for a brilliant white card down to 2 per cent or less for black velvet. In exterior scenes with bright sunlight and deep shadows, the natural lighting contrast may be of the order of 100 : 1, but under studio conditions it is under the control of the lighting cameraman and the ratio does not usually exceed 4 : 1. Although in practice the effective brightness range of a natural exterior scene may occasionally reach a ratio of 1,000 : 1, this is very rare and a study of actual scenes has indicated an average value of 160 : 1, while in studio work it is normally appreciably lower.

FLARE AND LENS COATING

At the first step, the recording of the tones of the original scene is determined by the optical characteristics of the camera lens and the photographic characteristics of the film stock. Light is reflected and scattered from the surfaces of the glass components of the lens and its mounting and distributed over the image as flare; this has a much greater effect on the areas representing dark tones and shadows and therefore compresses the tonal scale. The amount depends on the details of the lens design and the character of the scene and is much reduced by the treatment of the glass surfaces with anti-reflective coatings, or blooming, now standard for modern lenses. Flare has been studied quantitatively, but usually for still cameras and outdoor scenes, and data for modern motion picture lenses under studio conditions is not readily available. It has been estimated as reducing the actual tonal range to half by raising the recorded density of extreme black and shadow areas, but the effect is probably less with modern lenses and studio lighting.

TONE CONTROL

Modern negative stocks are designed to yield a low contrast photographic image and to provide a linear relation between the logarithm of the exposure and the resultant density on the film over as wide a range of intensity as possible. The slope, or gamma, of this relation on the D log E curve is usually between 0·60 and 0·70 and the substantially straight-line portion can extend over a log E range of 2·3 or more. Despite this wide latitude, it is essential to ensure that all important shadow detail is reproduced on the negative by densities appreciably above the minimum fog level within the range of the minimum useful gradient in the toe area of the characteristic curve. This is accomplished by control of exposure at the time of photography and by the correct balance of key light and filler light in the studio.

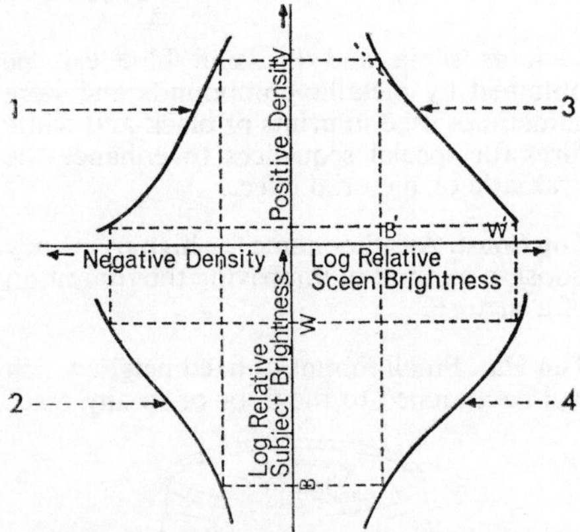

QUADRANT DIAGRAM. Objective characteristics of stages of film reproduction. 2. Camera and negative process characteristic (W=subject white, B=subject black) transfers to 1, printing and positive process characteristic, thence to 3, projection characteristic (flare and ambient lighting), and 4, overall objective tonal reproduction.

The positive process from negative to positive also introduces both optical and photographic characteristics, although it is usual to represent the combination of both these in the print-through curve which shows the relation between the density of the negative and the resultant print density for a particular process. Differences are introduced by the use of diffused or specular illumination, by optical or contact printing, and by variations of granularity of the negative image and its light-scattering effect. The effects of different methods of printing are much less marked with the dye images of colour negatives than with the silver grain images of black-and-white.

Positive film stock designed for motion picture projection shows a high contrast in its characteristic curve, with a low minimum level to allow good screen brightness for the white areas of the projected image. It might be thought that to reproduce the tones of the original scene the negative and positive processes together should provide a factor of unity, so that

$$\text{Gamma}_{neg} \times \text{Gamma}_{pos} = 1$$

but this is not the case in practice. Positive film prints can be obtained with a minimum density of 0·10 (80 per cent transmission) and a maximum of 2·8 (0·16 per cent transmission) or more, which thus give a range of transmitted light intensities of 500 : 1. However, in the cinema theatre, this is drastically reduced even under the best projection conditions by flare in the projection lens system and by ambient light in the auditorium. Information applicable to modern lenses and colour materials is lacking but the range which can be effectively presented on a cinema screen of open-gate brightness 16 ft. lamberts is probably not greater than 70 or 80 : 1. It varies not only with lens design and theatre conditions, but also with the average density and tone distribution of the scene; in all cases the effect of flare is small in the light areas and most marked in shadows and heavy densities. Considerable compression of the tonal range present in the print is therefore unavoidable in this region and it is usual to operate the positive process with a print-through gamma appreciably greater than 2. For colour materials the need for maintaining colour saturation calls for gamma values exceeding 3.

The shape of the positive stock photographic characteristic curve also introduces a tonal compression at the lighter end of the scale, and a compromise must be made between prints too light to show adequate highlight gradation and too heavy for satisfactory screen brightness. The whole sequence, as demonstrated graphically in a quadrant diagram, shows that the overall reproduction characteristic relating the original object-brightness scale to its representation on the cinema screen is in the form of a non-linear S-curve indicating increasing compression at both light and dark ends of the scale. Artistic skill in the studio and technical competence in photography and processing are required to ensure that as much as possible of the picture information falls within the satisfactorily presented tonal range.

TV REQUIREMENTS

In the chain of operations from the television camera to the domestic receiver a number of similar factors must be considered; photographic characteristics have their electronic equivalents in non-linear amplifiers, while final viewing conditions introduce marked tonal compression in the shadow areas. All these factors are included under the general heading of transfer characteristics but motion picture film for television transmission is a special case. It is generally accepted that both the limited tone range available on the receiver and the characteristics of the telecine system call for a lower contrast than is usual for motion picture theatrical practice. In photography, a lighting ratio not exceeding 2 : 1 is recommended and a positive of 20 per cent lower contrast is used, so that in the print the heaviest shadow areas required to show gradation do not exceed a density of 2·0 when the white area density is 0·30.

In both cinema and television presentation, the subjective factors affecting the sensation of tone-reproduction are of importance, since the viewer's eye is accommodated to low brightness levels. Under domestic TV viewing conditions, the surrounding illumination may have a brightness level of 3 or 4 ft. lamberts, while the brightest level on the TV screen is about 30 ft. lamberts; in the cinema theatre, the ambient lighting level is very low and the brightest area on the screen about 10 ft. lamberts. This may introduce a subjective compression of the tonal scale reproduced in addition to the objective factors previously considered. Studies have indicated that for satisfactory subjective tone reproduction in a darkened motion picture theatre, the objective reproduction process must have an appreciably steeper gradient than is required for either photographic transparencies or paper prints viewed in a normally lighted room. L. B. H.

See: *Exposure control; Screen luminance; Telecine; Telerecording; Transfer characteristics; Viewing television.*

Books: *Theory of the Photographic Process*, Ed. C. E. K. Mees and T. H. James, 2nd Ed. (New York), 1966; *Fundamentals of Photographic Theory*, by T. H. James and F. C. Higgins (New York), 1960; *Television Film Engineering*, by R. Ross (New York), 1967; *Motion Picture and TV Film*, by D. J. Corbett (London), 1968; *Basic Sensitometry*, by L. Lobel and M. Dubois (London), 1967.

□**Tone Source.** Electrical signal of dependable level used for alignment of programme level. The tone source provides a single frequency signal of 400, 800 or 1,000 Hz, usually at a level of 1 mW in 600 Ω. A stability of the order of 0·1 dB is desirable.

On sound mixing equipment it is often possible to switch the outgoing line from the mixer output to the tone source. This gives operators who are to receive the subsequent programme an indication of the programme level. In some cases, periods of continuous tone are interrupted by verbal announcements or morse signals indicating the programme source. This is done where a central switching centre has to handle a large number of sound sources, e.g., as in the case of the Eurovision network.

⊕**Toning.** Conversion of a black-and-white photographic image in silver to another colour by chemical methods. Certain colours such as sepia and Prussian blue can be obtained by metallic compounds and were sometimes used in prints of black-and-white films for special sequences to enhance the dramatic or pictorial effect.

□**Top Boost.** Another name for high frequency boost, a method of improving the resolution of a picture.

⊕□**Top Hat.** Small mount of fixed height which can be attached to the floor or to any place

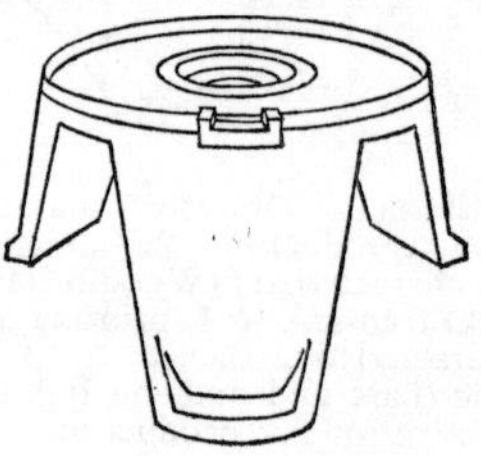

where it is desired to set the camera as low as possible.

□**Toroidal Coil.** A coil wound in the form of a toroidal helix, the torus being a doughnut-shaped surface. Commonly employed when a minimum of pick-up or radiation is required. It has a very low external field.

See: *Receiver.*

⊕**Track, Sound.** Generic term embracing tape and film, to describe a sound record of ribbon form, either separately or as a synchronized accompaniment to a film or TV recording.

See: *Editor; Sound recording.*

⊕□**Tracking.** Movement of the whole camera when making a shot. Sometimes called dollying. A tracking shot is a shot taken with the camera moving. Sometimes called a dolly shot.

⊕□**Track Laying.** Process of editing and synchronizing the many sound tracks (from 10 to 30) to accompany a film or videotape recording. Track layer is a term sometimes used for sound cutter, i.e., a person who specializes in editing sound tracks and preparing them for dubbing.

See: *Editor; Sound mixer.*

⊕□**Transducer.** A device which changes energy, motion or some manifestation of these from one form to another. Properly used with adjectives to qualify the noun, e.g., electro-mechanical transducer – a device which changes electrical energy into mechanical energy.

□TRANSFER CHARACTERISTICS

The overall requirement of a television system is the ability to produce light levels at the viewer's receiver which correspond to the levels in the original scene. This applies both to studio scenes and film transmissions. Ideally, it is expected that the relationship will be linear, i.e., changes in light levels in the original scene will be faithfully reproduced in the viewer's receiver. The actual characteristic which is obtained may be referred to as the transfer characteristic.

In practice, this requirement of linearity cannot be fulfilled except over limited ranges, and the problem is to maintain the linear response of the system as closely as possible, and over as great a contrast range as possible.

BASIC PRINCIPLES

The transfer characteristic of either the whole or part of an image-forming system is the relationship between changes in the signal level at the input and output. The signal levels may be of an electrical nature, as in the case of a video amplifier input and output; or levels of light intensity, as in the case of a camera tube coupled to an amplifier chain and producing a cathode ray tube picture. This relationship may be expressed graphically in linear or logarithmic co-ordinates.

GAMMA REQUIREMENTS. The use of logarithmic co-ordinates is of particular value where the characteristic is non-linear, and may approximate to a power law, i.e., of the form $y = x^n$, where **n** may be any positive value and need not be an integer. Quite a number of relationships have such a logarithmic law. If the power law is constant over the range of signal levels considered, the slope of the characteristic, plotted logarithmically, is known as the gamma of the system. This term is of photographic origin, from the slope of the density/log exposure curve for a photographic emulsion. Where the slope of the log/log plot is not constant, the term gamma must be qualified by reference to the range of signals over which it applies. Where the gamma is measured at a point on the curve, the term point gamma is often used. Since these properties of an electrical and a photographic system are similar, it is convenient to use the same terms.

Although the overall requirement is for a linear response, it is not, in practice, necessary for all individual parts of the system to be linear. Certain essential components of the system have inherent non-linearity. For the overall response to be linear, the overall gamma should be unity, and therefore the product of the gammas of the many components in the chain should be unity.

In general terms, the gamma of the receiver is decided by the characteristic of the cathode ray tube (kinescope), which can be considered to have an intrinsic gamma of just over 2, approximately a square law characteristic. This means that, to a first approximation, the rest of the system must have a complementary power law with a gamma of 0·5. As the only other non-linear element in the chain is the originating transducer (assuming a linear amplification chain), or combination of transducers, it follows that this original transducer must produce a 0·5 gamma signal. In practice, few devices have an intrinsic gamma of 0·5, so that non-linear amplifiers or gamma correction circuits are used. Fortunately, such devices as camera tubes and film scanning systems have gammas of unity or less, so the problem of correction is not a difficult one, and is well within the scope of commonly met electronic circuits. The basic solution, therefore, to the problem of obtaining an overall system gamma of unity does not appear from elementary considerations to be difficult.

LIMITING FACTORS. However, transducing devices will handle only a specific signal or light range, and this places a limiting factor on the whole system. For example, the receiving cathode ray tube has its range of brightness limited at the lower end by a flattening of its characteristic and by the presence of ambient light falling on the tube face when it is used in a lighted room. At the maximum end of the range, the brightness is restricted by screen saturation, limited beam current, and defocusing effects. In practice, this means that the cathode ray tube, under ideal operating conditions, can produce a contrast range not exceeding 100 : 1, which, in conditions of domestic viewing, may be reduced to 20 : 1 or even less. The camera tube similarly has electronic limitations which may reduce its effective contrast range to around 20 or 30 : 1. Similarly, even a good telecine scanning system will not reproduce films whose contrast range is excessive. The transfer characteristics of motion picture processes are similarly limited at each end of the range.

While the basic problem of transferring light values faithfully from the origin of the television system to the viewer's screen with a unity gamma is, in theory, capable of a simple solution, in practice it is difficult to maintain and set up such a system. An extreme case is the use of film telerecording in which the television image is directly photographed from the screen of a cathode ray tube, and then eventually transmitted through a telecine machine. In this chain there are a total of six inherently non-linear elements, each with its own transfer characteristic over a limited contrast range. The many optical systems present in any originating device provide a further limitation from the effects of light scatter. In addition, halation may act as a constant minimum light level over parts or the whole of a particular picture area.

MONOCHROME FILM REQUIREMENTS. The transmission of film over a television system means that, to obtain optimum reproduction in terms of the overall transfer characteristic, the contrast range and gamma of the motion picture process should be suitably matched to those of the replay telecine machine. This, however desirable, is not always possible as the television system often has to handle films made for cinema projection. Where films are made specifically for television transmission, it is possible to take the requirements of the television system into account, in particular those requirements which concern the contrast range and gamma of the film product. In general, the films produced for screening have a much higher contrast range than can be handled by the television chain.

For a normal release print to reproduce satisfactorily over the television system, the density range of the print that can be handled by the machine should include all the vital information and all essential composition. This is not always possible, and the film loses detail and composition by television reproduction. It is normal to expose the negative film to the scene, and then process this negative to a standard gamma of 0·65. This negative is then used to produce a positive print, the printer light setting being varied throughout this process to provide a visually matched series of pictures. The print film is then processed to a print-through gamma of 2·2 to 2·4, giving an overall gamma of the complete process greater than 1·3, but in practice close to 1·3 when allowance is made for averaging toe and shoulder distortion.

It is desirable that where films are made for television transmission, the processing conditions remain the same as those used for normal cinema release prints. It is generally neither possible nor desirable for commercial processing laboratories to vary their conditions to suit individual customers. Accordingly, all that may be done practically is to define the density ranges of the print to suit the reproducing machinery. This implies fixed negative densities and thus an appropriate lighting range to produce the whole chain. In practice, there is seldom the time or facilities to light a film essentially for television reproduction. The overall effect is rather flat, and, on projection and to the unpractised viewer, displeasing.

Various recommended practices and standards based on commercial processing have laid down that the only parameters which are to be controlled are maximum and minimum densities and face tones. This assumes that the numerous other variables have been satisfactorily controlled by the laboratory.

STANDARD FILM PROCESSING DENSITIES

Standard	*Minimum Density*	*Maximum Density*	*Face Tone Density*
US (SMPTE) RP7	0·3 (excluding highlights)	2·0	not less than 0·15 or more than 0·50 added to the minimum density
UK BS3115	0·4	2·0	0·70–0·90

CAMERA TUBE CHARACTERISTICS

The professional monitors used for viewing in a television studio approximate to a linear amplifier followed by a cathode ray tube, used under controlled conditions of low ambient lighting. The cameras, with their associated camera tubes and gamma correcting circuitry, produce their image on these monitors.

The assumption is then made that the picture will be similarly produced on the home receiver via the transmitter chain and this is probably valid for a home receiver of high quality. But since home receivers vary both in their properties and adjustment, corrections to the camera tube characteristic are made to match the professional monitor characteristic and to give an overall unity gamma for this chain.

IMAGE ORTHICON TUBES. Users are well aware that, beyond the published characteristics, less definable things happen with this type of camera tube. One reason why such an extremely expensive and complicated piece of equipment has persisted in the face of competition from simpler devices is its favourable picture quality, in addition to its simpler characteristics.

Apart from the usually measured properties of excellence of image reproduction, namely, resolution, sensitivity, lag, and signal-to-noise ratio, it has a less well-defined property of producing a black border which hardens the edge of black-to-white transitions, quite beneficially so far as the appearance of the image is concerned.

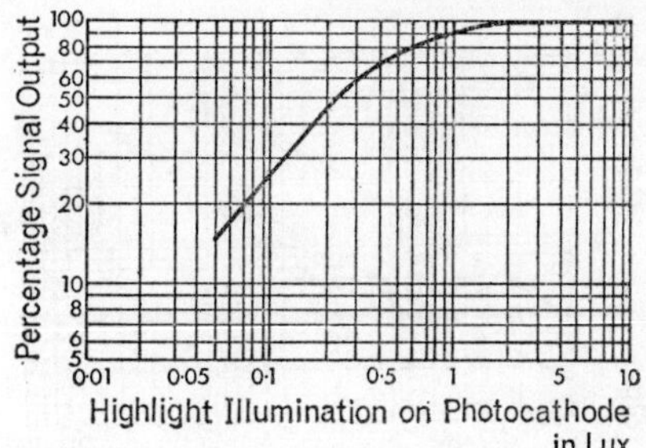

IMAGE ORTHICON TRANSFER CHARACTERISTIC. The well-defined knee at upper end of curve helps tube to handle scenes of very high contrast without overloading. Operating point is normally at knee in colour cameras and ½-2 stops above for black-and-white.

Up to a point, the human eye has similar properties and, therefore, finds this kind of picture more natural. The typical image orthicon picture is described as one which will travel, even over long lines. In conditions which tend to degrade the image, it stands up well. It is normally used with an attendant amplifier whose gamma may be altered continuously or in steps. Suitable cards of different opacities may be viewed by means of a camera, and the voltages equivalent to the card opacities displayed by strobing out an appropriate part of the picture and reading off the voltages. Adjustment may then be made to get the best results. A typical example shows an effective gamma of 0·80 over a range of 60 : 1. This would be reduced to an overall value of gamma 0·5/0·6 by a correcting amplifier, over a similar range.

VIDICON TUBES. Again, these tubes, as a class, have both well-defined and less well-defined characteristics, of which the latter are known but not often documented.

Apart from the main tube characteristic, vidicons as a class suffer from a black level shift, which is almost the opposite of the effect obtained in the image orthicon. With the vidicon, the area adjacent to a black-and-white transition shows a rising characteristic and hence small black areas surrounded by white lift above nominal black level. This is a fact which tends to add to the soft quality picture of the vidicon.

The effective gamma of the characteristic for large areas is typical of the class, namely 0·65/0·70. Curvature of the toe is controlled by the dark current which varies considerably between unselected specimens. The

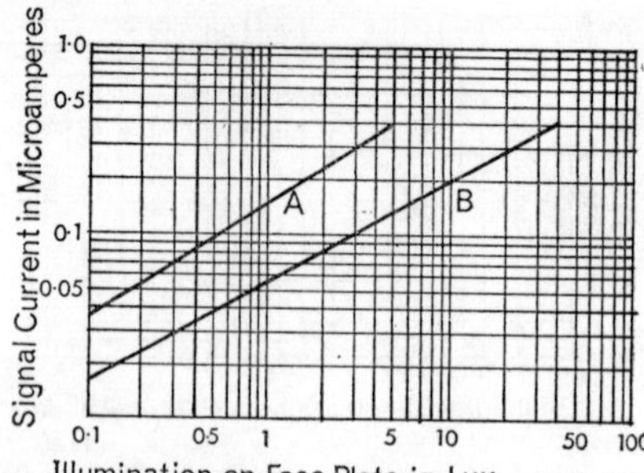

VIDICON TRANSFER CHARACTERISTIC. Typical tubes, with A, dark current=0·02 μA and B, dark current =0·004 μA, latter giving lower sensitivity but better black level stability. Gamma at 0·65-0·70 enables a tube to be used without correction on most pictures.

tubes will stand quite a high light level (witness their use in some telecine machines) but, in general, noise intervenes in the dark portions at about twice the amount of low level light which an image orthicon will handle.

The commonly met gamma of 0·65/0·70 is so close to the desired gamma of 0·50 that, taking into account the rather soft picture, it is seldom that any extra gamma control is applied to a camera using vidicon tubes. The slightly higher overall gamma helps to match with image orthicon quality, which may be used as a standard yardstick.

PLUMBICON TUBES. Although the Plumbicon tube is basically a photoconducting device, its properties are also controlled by the carefully contrived photodiode structure of the sensitive layer. This produces a typical saturated diode response which has the effect of keeping the dark current down to an extremely low value. As a result, the black level conformity is good.

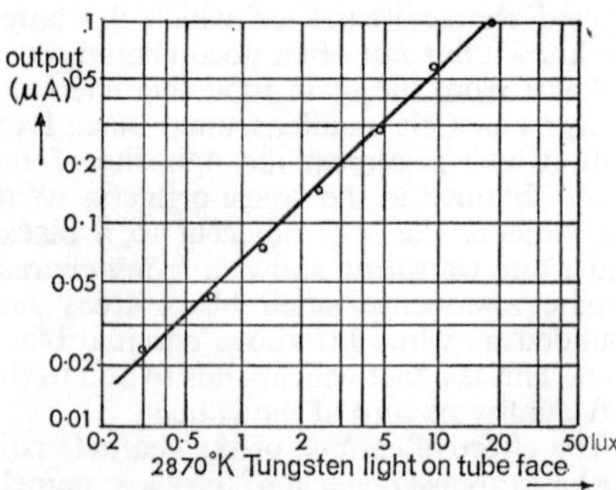

PLUMBICON TRANSFER CHARACTERISTIC. Like the vidicon, this tube has a linear characteristic, but with a gamma approaching unity and thus normally needing correction.

The light transfer characteristic has a gamma approaching unity. It is linear over an extremely wide range, and accordingly the sensitivity of the tube can be expressed in μA (microamperes) per lumen (300 μA has been quoted). Obviously, for black-and-white working, an attendant gamma-correcting device must be used.

The wide range of linearity makes the Plumbicon very suitable for use in colour cameras.

TELECINE

FLYING-SPOT TELECINE. In the flying-spot telecine a small image of a cathode ray tube spot is focused on to the film. The light which passes through the film is collected and handed on to a photonultiplier where it is turned into the equivalent electric current. The device would be absolutely linear if all the light were collected, but the optics give a specular integration and, accordingly, a value of gamma is introduced, which varies according to the image geometry. The overall gamma values normally range from 1·05 to 1·25.

It is usual to make the gamma correction amplifier variable over the range so that an overall gamma of 0·35 to 0·50 can be obtained. The extension downwards helps in the presentation of feature films devised for projection with a relatively contrasty and wide range image. A good modern flying-spot machine can cope with a contrast range expressed electrically as 44 dB, i.e., about 150 : 1. This is more than adequate for the average home receiver although some high contrast feature films may exceed this range.

For the showing of negative film as a positive image, a further correction of gamma −2 is required in the chain to simulate the photographic printing process. Negative originals are seldom used except for newsreel work, where the exact overall gamma is less important than the news content.

VIDICON TELECINE. Basically, a vidicon telecine is a vidicon camera looking at a film instead of a studio scene. Hence vidicon telecine starts off with the 0·65/0·70 gamma of the vidicon tube. The first examples of telecine produced using these tubes had no overall gamma correction, and the results were markedly inferior to those obtained from flying-spot machines.

The latest machines usually employ a gamma correction of 0·7, hence the overall gamma may be expected to be in the range 0·45/0·50. The signal-to-noise ratio, however, is inferior to that which may be obtained from flying-spot techniques, and seldom exceeds 35 dB. This, however, is usually adequate for the material with which the vidicon telecine has to cope.

TELECINES IN PRACTICE. There is a considerable spread in the transfer characteristic between different studios and often between individual machines in the same operating area. The reason is difficult to determine, but telecine machines have a long life and are expensive to replace, and techniques have varied over the years. Progressive changes of ideas on the shape of the gamma-correcting curves are the biggest variables.

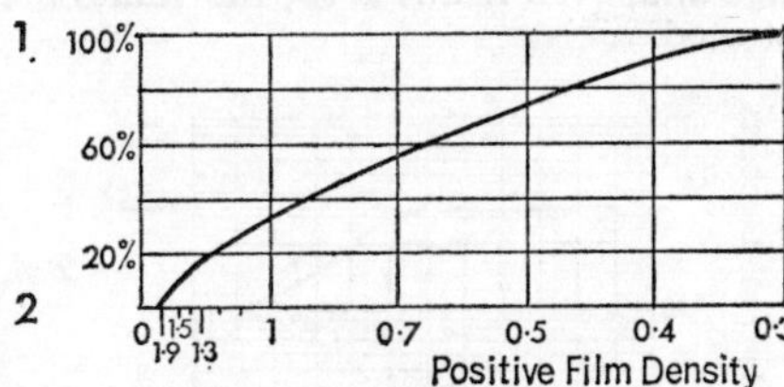

EBU STANDARD TELECINE TRANSFER CHARACTERISTIC. 0 per cent, 2, represents black level and 100 per cent, 1, represents white level on transmission, so that effective print density range is between 1·9 maximum and 0·3 minimum.

The mixture in Europe of vidicon and flying-spot machines has not helped. Vidicon tubes from different sources are also capable of a small gamma variation. Of late, a good deal of study has been made both of suitable measuring techniques and possible standardization. Exact measurements are complicated by side effects, and 18 possible sources of error have been noted.

The situation is complicated by the differences between three types of film: feature films, films made for television, and telerecordings (kinescopes), all of which show some properties of their own which may require different treatment on the telecine machine.

VIDEOTAPE

The videotape process is essentially a frequency deviation modulation and demodulation process and, basically, it is linear. For all practical purposes, the transfer chaıacteristic on large areas is unity.

TELERECORDING

The problem of getting the best images by photographing a picture displayed on a cathode ray tube is extremely involved. Owing to technical difficulties of timing, the image may have to be stored in a persistent phosphor, and may appear in almost any colour but red. Halation effects and multiple reflections interfere to various degrees. Although the desirable properties of the system can be broadly laid down, telerecording still tends to be an art as well as a science, in that limits and usages are strongly influenced by experience and the practical results attained.

One unorthodox technique is to reduce the contrast range of the display unit from 60 : 1 in the original scene to 10 : 1 on the display tube. The negative is then processed at about unity gamma. The combination gives a reasonable range of negative densities. This low contrast display technique is claimed to give less flare from whites into the black areas on the final film.

The difficulties encountered by laboratories in producing non-standard prints are such that the average commercial print of gamma 2·0/2·4 is accepted.

In general, the normal conception of gamma does not apply rigidly to these techniques and can be quoted merely as an indication for the control strips. Graphical presentation is often helpful in setting up the system.

MONITORS AND RECEIVERS

MONITORS. Scenes are composed in the television studio on professional monitors, which are usually extremely consistent and linear, and no attempt is made to alter the contrast. D.c. restoration is the rule unless for some special reason the user wishes to see the effect of an unrestored receiver on the picture. Peak brightness of at least 100 lumens can now be obtained. The transfer characteristic is thus basically that of the tube and is in the range of 2·0/2·2 for most black-and-white tubes.

It is arguable that the reference point of any signal fed to the monitor should be peak white and that the excursions downwards towards black level will, in fact, never reach it in a correctly adjusted system, because black level is minus infinity (log zero) on a logarithmic scale. Some stations try to work with their displayed signal this way but it is also arguable that ambient light of the same order of brightness requires a rather higher level on the tube face than the true theoretical one.

The more usual custom of letting the valleys intersect the black level to some small extent probably gives a more pleasing picture under conditions of ambient light by slightly distorting the transfer characteristics at low light level.

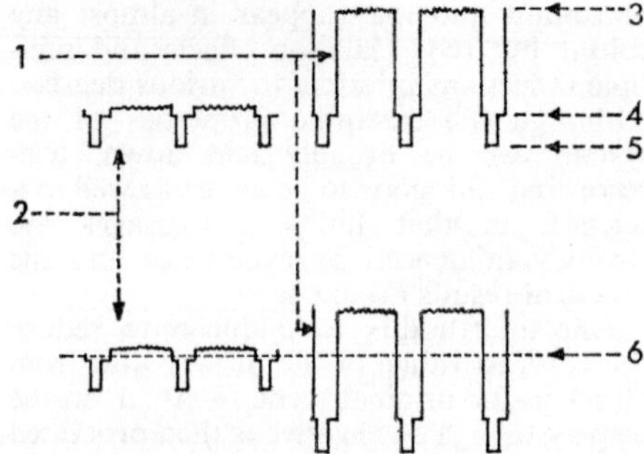

WHITE AND BLACK VIDEO INFORMATION. 1. Two lines of white information. 2. Two lines of black information. 3. Maximum white level. 4. Blanking level. 5. Sync level. (*Top*) With d.c. component present (d.c. restoration). (*Bottom*) Without d.c. component. Black information is raised to prevailing a.c. axis, 6, reducing brightness difference between white and black.

TELEVISION RECEIVERS. It is not unusual for the display unit of the television receiver to differ widely in its performance from that of a professional monitor.

To get a bright contrasty picture, the transfer characteristic of the receiver is often increased by the manufacturer's design of the driver stage, ana examples have been noted which have raised the picture to the power of 3·0 instead of 2·0. Manufacturers' associations have seldom been able to control these practices. Such receivers give a bright contrasty picture, but produce unpleasant effects on face tones, particularly when the face is against a bright background.

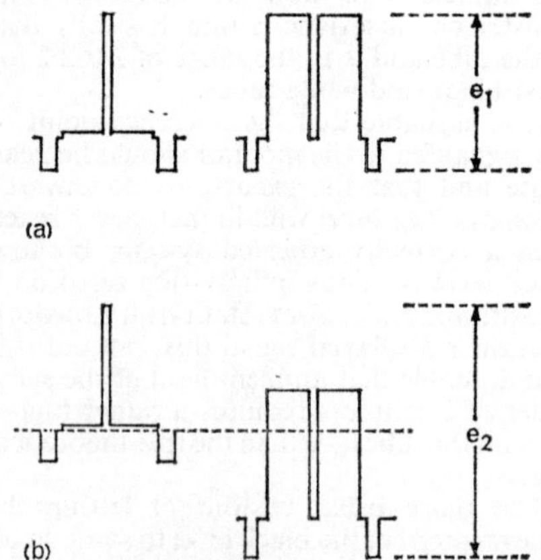

DYNAMIC RANGE REQUIREMENTS. (a) With d.c. component present, required range is e_1. (b) With d.c. component absent, full peak white may require enlargement of range to e_2.

The situation is further complicated by the widespread use of unrestored television receivers where the mean value of the signal sets the level. This is somewhat remedied by the fact that the more modern receivers do not work entirely on average signal but are also influenced by relatively small amounts of peak white and will set up a fairly accurate black level from as little as 10 per cent peak white distributed over the picture.

In extreme cases, an all black level is increased to the mean grey, and an all white level is brought down. Spikes of white on a mainly black picture are clipped or crushed unless the system has a greater than normal range and, even if this is so, the values are distorted.

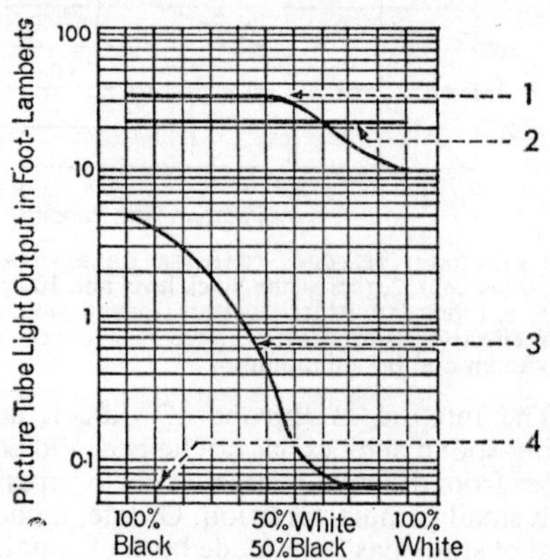

PICTURE TUBE LIGHT OUTPUT RANGE. With d.c. restoration: whites, 2; blacks, 4. Without d.c. restoration: whites, 1; blacks, 3. Input white and black signals held constant.

Whereas with d.c. restoration, as the balance of the picture is changed from all-black to all-white, the levels stay constant, the distortion of the transfer characteristic of an unrestored receiver varies over an infinite range according to the picture pattern. With a mainly black picture, the whites crush and the black level becomes grey. As the balance changes, the crushing continues but the addition of white to the picture drops the black level. At the point where the mean picture level is restored, the balance is nearly correct and crushing ceases. As the white content is further increased, peak white is forced down.

Thus the transfer characteristic is reduced in a not very predictable manner except when the picture happens to have an average value which suits the circuit design and sets the peak white and black level correctly.

Television stations attempt to balance their pictures and to avoid small patches of lit subject on a black background, but considerable variation still occurs. Unfortunately, there are some small advantages under adverse conditions for mean value (unrestored) receivers, which are also usually the cheapest. Engineers have been attempting to get d.c. restoration obligatory for many years, but with little success.

Mistaken settings of gain and brightness controls can obviously distort the transfer characteristic. The distortions necessary to overcome the effect of high ambient light are also likely to reduce the range of the contrast. For good viewing, not only should the receiver be of correct design, but it should be set up in the same ambient light as was used on the station monitor.

COLOUR

The display device, i.e., the colour tube, may be considered as three separate units or channels, for each of the separate primary colours. Despite the difference of colour, the gamma of each channel is the same and equal to that of a black-and-white display unit, namely 2·0/2·2. It is fortunate that the power law is so constant, otherwise a colour tube would give unexpected results when driven by a monochrome signal. That it will take a monochrome signal and give a reasonable response and an untinted image is evidence that the gamma of each colour channel is the same over the useful range.

Accordingly, in colour cameras, each colour channel is usually gamma-corrected before the individual colour signals are added in the matrix or colorplexer. Thus the figure of approximately 0·5 as the overall gamma of the camera chain as in black-and-white reproduction is again established.

To take an example, it is normal practice to couple a Plumbicon with a gamma corrector of, say, 0·4/0·6 and adjust to give the best response. The mid-position would give 0·5 with the unity gamma of the Plumbicon. It should be borne in mind, however, that this is the simple technique based on the customary red-green-blue colour addition of 0·30R+0·59G+0·11B. Other methods of processing the signal have been suggested, involving unity gamma at this point and subsequent alterations.

Colour in telecine is treated in the same general fashion. According to the technique used, flying spot or vidicon, the appropriate correction is made for each colour channel to the same extent as it is for black-and-white reproduction. D.P. & E.C.V.

See: *Camera* (TV); *Exposure control; Phosphors for TV tubes; Picture monitors; Receiver; Screen luminance; Transfer characteristic; Telecine; Telerecording; Viewing television.*

⊕□**Transfer Function.** Expression which denotes the fidelity to the original of an optical reproduction through a lens system, or of a combined optical and electronic transfer through a television system.

In modern methods for assessing the performance of optical systems, particularly in combination with photographic film, the concept of spatial frequency analysis in association with contrast reproduction has become fundamental.

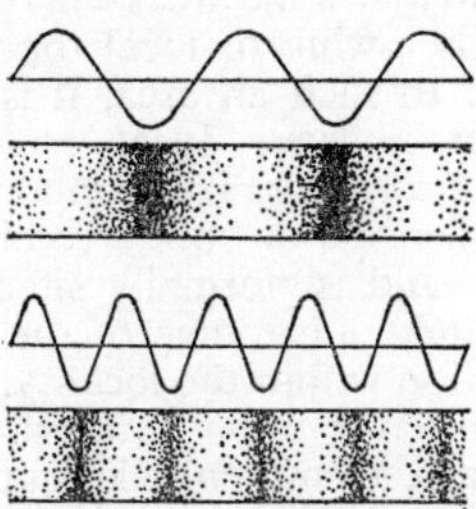

SPATIAL FREQUENCY. Sinusoidal wave formed in density variations can be expressed as a frequency in terms of number of repetitions per unit length.

Considering a test object to be reproduced consisting of a repeating pattern in which the brightness of light and shade varies sinusoidally, the spatial frequency indicates the number of times the pattern is repeated in a given distance, for example, with a frequency of 20 cycles per millimetre. The variation of brightness of the pattern from maximum to minimum, or contrast, is termed the amplitude.

With any particular reproduction system, the amplitude of the image recorded at any particular spatial frequency value may be lower than that of a perfect system: the ratio between the actual value and the ideal is termed the modulation factor, which can never be greater than unity. A curve showing the way in which the modulation factor varies with frequency is termed the frequency response of the system; the relation between the reproduced amplitude and the frequency is referred to as the modulation transfer function (MTF).

In a system involving lenses, the reproduced image may involve a change of phase, or displacement of maxima and minima relative to the original, especially at higher spatial frequency values. The full specification of a lens performance by its optical transfer function (OTF) therefore involves an indication of phase shift, although the image-forming performance of a photographic film involves no phase alteration and is therefore fully specified by its MTF.

The combination of the separate transfer functions of the steps in a reproduction process involving several stages allows the overall performance to be assessed and the influence of alterations in the various steps to be estimated so as to obtain optimum results. L.B.H.

See: *Image formation; Lens performance.*

⊕□**Transfer (Sound).** Duplicated copy of a sound track, normally produced by re-recording rather than a printing process. Distinguished as an optical transfer (i.e., to an optical track), or a magnetic transfer. 16mm transfers are re-recordings from a larger gauge, normally magnetic, original; they have special frequency characteristics and sometimes additional compression adapted to the limitations of 16mm projection.

See: *Sound recording.*

⊕□**Transients.** Transient waves are those which do not show significant periodic features. The simplest example is a step in a d.c. voltage. Wave trains in general may be classified as steady state, semi-transient, or transient, but each category merges into the next and there are no exact lines of demarcation.

Steady state transmissions such as a continuous tone are readily broken down into a series of frequency components bearing a harmonic relationship to each other, and the amplitude of each may be calculated. This is called Fourier Analysis.

Analysis of the Fourier components of a transient such as a step in a steady state d.c. signal is more involved. The analysis postulates an infinite series of frequencies whose amplitude over an element of frequency is predictable. For the case considered, the components can be represented by a smooth curve asymptotic to the x and y axes.

This simple case is modified as soon as an element of periodicity enters into the transient; for example, a separate pulse of reasonably short persistence, or a short train of such pulses.

The Fourier Analysis of a single square topped pulse is again an infinite series of frequencies but already a periodic property is noticeable. A plot of the amplitude shows maxima at harmonics of a fundamental frequency based on the pulse duration being half that frequency.

As the number of pulses involved increases, so does the effect noted above and ultimately the whole of the energy is concentrated in the harmonics when the pulse train becomes infinite.

Thus any transient signal has components ranging up to infinity. The original form of the transient may be seriously distorted or modified by a number of factors as it is impossible to transmit the components unchanged through a network under all conditions, such as: insufficient bandwidth, which eliminates certain frequencies; phase distortion; and undamped reactive components giving resonances which may select certain frequencies and allow them to persist.

To go back to the simplest case, if the step in the d.c. voltage has insufficient bandwidth a "ring" is obtained at the step such as may often be seen at the edge of a black/white transition in a television picture.

In television, the signal carrying the picture includes steady-state signals such as the synchronizing pulses and semi-transient components, e.g., the picture information. All the above distortions are important in television transmission of the picture.

Transients exist equally in sound, but are seldom treated with the same care as those occurring in picture. In speech recording of the best quality it is unwise to assume that a distortion of transients is immaterial to the quality of the final product. E.C.V.

See: *Sound distortion; Transmission networks.*

□**Translator.** A means of increasing the effective range of TV transmission.

When large-scale broadcast coverage is necessary on UHF and VHF bands, it is almost certain that some areas will receive very poor signals, owing to local topographical conditions. In such an area, it is common practice to improve local reception by means of a translator. This contains some of the basic elements of both a receiver and a transmitter and is normally sited on high ground to take advantage of the best possible reception within the locality.

The translator input stages are very similar to a conventional television receiver with an r.f. amplifier and mixer. The local oscillator is normally crystal-controlled to ensure good stability. The intermediate frequency produced by the mixer is amplified to a suitable level and fed to a second mixer. A second crystal-controlled local oscillator then mixes with the signal to produce a new r.f. signal which lies within the normal receiver band, but on a different frequency from the original signal. This r.f. signal is then suitably amplified, and a power output of 3 to 50 watts is usually produced and then re-radiated for local reception.

By siting in a good position, and by use of low-noise techniques, an r.f. signal of much higher strength is made available in the poor reception areas. The equipment is completely self-contained and requires no personnel for day-to-day operation.

⊕**Translucent Screen.** Semi-transparent screen used for the back-projection of motion pictures in which one surface of a transparent material is made sufficiently diffused to allow an image to be formed on it. The less the diffusion the brighter the picture but the more directional the light distribution, while a greater degree of diffusion allows the picture to be observed satisfactorily over a wider angle but at lower brightness.

See: *Rear projection; Screen luminance.*

□TRANSMISSION AND MODULATION

Telecommunication techniques are concerned with the transmission of information between fixed points or over large areas. The information to be transmitted varies considerably and may be separated into specialized categories among which are telephony, telegraphy, sound broadcasting, television, telemetry, etc.

All systems have common factors. Information is converted into electrical signals which are then varied with time by the process known as modulation. With some transmissions (e.g., sound broadcasting) the information is sent continuously, while in TV finite samples only are sent.

TRANSMISSION PRINCIPLES

High frequency electrical signals are used to radiate over long distances while the lower frequency electrical signal representing the information is made to modulate the high frequency signal. Another method is to send the electrical information by cable.

Thus, the electrical signals used vary from the very low frequencies encountered in the intelligence to be transmitted, to the ultra-high frequencies used for the carrier frequencies of modern communication systems.

THE ELECTROMAGNETIC SPECTRUM. Electromagnetic waves are of a similar nature to light waves and travel at the same speed. This speed (or velocity of propagation) is approximately 186,000 miles per second or 300,000,000 metres per second.

The connection between frequency and wavelength is given by the formula $v = N\lambda$ where:

v = velocity of propagation of electromagnetic waves (3×10^8 metres per second),

N = frequency in Hertz (cycles per second),

λ = wavelength in metres.

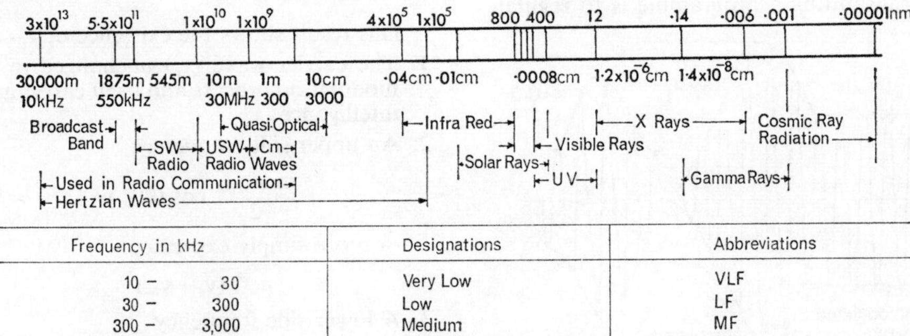

Frequency in kHz	Designations	Abbreviations
10 – 30	Very Low	VLF
30 – 300	Low	LF
300 – 3,000	Medium	MF
3,000 – 30,000	High	HF
30,000 – 300,000	Very High	VHF
300,000 – 3,000,000	Ultra High	UHF
3,000,000 – 30,000,000	Super High	SHF

ELECTRO-MAGNETIC FREQUENCY SPECTRUM. Upper side of spectrum is calibrated in nanometres (nm). The designations below are those approved by the FCC for communications use and generally adopted.

The electromagnetic spectrum has a range from about 3×10^8 to 3×10^{-12} metres so that it is sometimes more convenient to express frequency in kilohertz or megahertz.

Thus,

$$N\,(\text{kHz}) = \frac{300{,}000}{\lambda\,(\text{metres})}$$

$$\text{and } N\,(\text{MHz}) = \frac{300}{\lambda\,(\text{metres})}$$

The radio frequency portion of the electromagnetic spectrum extends over a range of wavelengths from 30,000 metres to 0·03 centimetres. This includes television broadcasting and radar.

CARRIER TRANSMISSION. The transmission of radio or television signals is accomplished by the use of a radio frequency many times the value of the highest intelligence signal it is desired to transmit. This radio frequency is known as the carrier frequency and the radiation it causes is termed the carrier wave. The carrier frequency for a particular station is chosen according to the distance over which it is desired to transmit the intelligence.

Each transmitting station is allocated a different carrier frequency; the same frequencies are allocated only to stations transmitting the same programme or to stations sufficiently separated geographically to prevent mutual interference. The radiated carrier wave induces a corresponding electromotive force (emf) in receiving aerials and is accepted only by receivers tuned to the appropriate frequency.

MODULATION TECHNIQUES

AMPLITUDE MODULATION. One method of transmitting a programme is to regulate the envelope or amplitude of the carrier wave in accordance with the audio and video frequencies from the studio. The carrier wave is then said to be amplitude modulated (a.m.). When the amplitude of the modulated envelope varies from maximum to zero value the carrier wave is said to be modulated to a depth of 100 per cent. This represents the maximum modulation depth possible without introducing distortion.

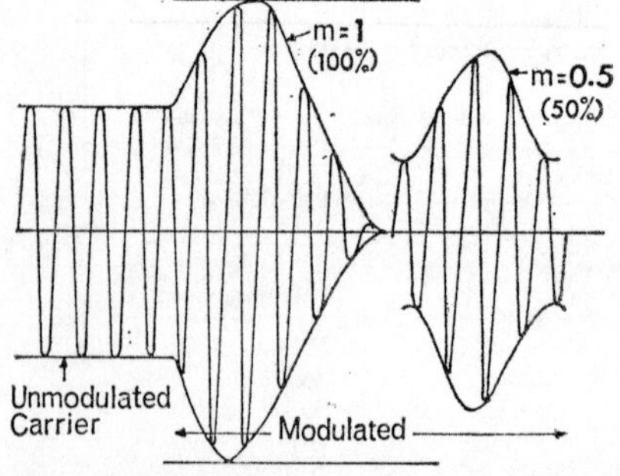

AMPLITUDE MODULATION. Envelope of modulated carrier wave follows outline of lower-frequency modulating wave. 100 per cent depth of modulation may be expressed as modulation factor (m) = 1. Similarly, at 50 per cent modulation, m = 0·5.

Another way of expressing this is to refer to the degree of modulation denoted by the symbol m, and in the above case m = 1.

$$m = \frac{\text{average envelope amplitude} - \text{minimum envelope amplitude}}{\text{average envelope amplitude}}$$

Similarly, a carrier wave modulated to a depth of 50 per cent has a value of m = 0·5, and so on.

Consider a single audio frequency:

$$v_{af} = m\,V_{max} \sin pt$$

and the carrier frequency

$$v_r = V_{max} \sin \omega t.$$

The process of amplitude modulation has the effect of multiplying these two instantaneously applied signals together. Thus the resultant modulated carrier is

$$v = V_{max} \sin \omega t\,(1 + m \sin pt)$$

where m = modulation depth described above.

Expanding,

$$v = V_{max} \sin \omega t + m\,V_{max} \sin wt \sin pt$$

$$\therefore\ v = V_{max} \sin \omega t + m\frac{V_{max}}{2} \cos (w-p)\,t - m\frac{V_{max}}{2} \cos (\omega + p)\,t.$$

This result shows the existence of

1. The carrier wave (v_r) unchanged in the modulation process and itself carrying no intelligence.
2. An upper side frequency

$$\frac{m\,V_{max}}{2} \cos (\omega + p)\,t$$

or more simply expressed

$$(v_r + v_{af})$$

3. A lower side frequency

$$\frac{m\,V_{max}}{2} \cos (\omega - p)\,t$$

or more simply expressed

$$(v_r - v_{af})$$

The difference between these two side frequencies is twice the modulating frequency ($p/2\pi$), and the amplitude of each of the side frequencies is m/2 times that of the carrier wave amplitude.

SIDEBANDS. When the modulating information is made up of several frequencies or a band of frequencies (as in practice), then ($v_r + v_{af}$) and ($v_r - v_{af}$) each comprise a band of frequencies known as the upper and lower sidebands respectively.

The extent of the sidebands is entirely dependent on the frequencies modulating the carrier wave and any restriction of the sidebands results in a corresponding loss of intelligence.

POWER. The power in the carrier is proportional to V^2_{max}. The power in the upper side frequency is proportional to

$$\frac{(m\,V_{max})^2}{2} = \frac{m^2\,V^2_{max}}{4}$$

The power in the lower side frequency is proportional to

$$\frac{(m\,V_{max})^2}{2} = \frac{m^2\,V^2_{max}}{4}$$

so that the increase of power in the sidebands when the carrier is modulated to a depth of 100 per cent by a sine wave is proportional to

$$\frac{V^2_{max}}{4} + \frac{V^2_{max}}{4} = \frac{V^2_{max}}{2}$$

therefore the total power radiated is proportional to

$$V^2_{max} + \frac{V^2_{max}}{2} = \text{carrier power } (1+\tfrac{1}{2})$$

In general, the effective amplitude of a sinusoidally modulated wave is proportional to $\sqrt{1+\frac{m^2}{2}}$ so that the total power radiated when the depth of modulation is denoted by m is proportional to

$$V^2_{max} + \frac{m^2 V^2_{max}}{2} = \text{carrier power} \times (1+\frac{m^2}{2})$$

For example, a 50 kW transmitter when modulated to a depth of 60 per cent delivers a total power of

$$50\,(1+\frac{0{\cdot}6^2}{2}) = 59 \text{ kW}$$

The same transmitter when modulated to a depth of 80 per cent delivers a total power of

$$50\,(1+\frac{0{\cdot}8^2}{2}) = 66 \text{ kW}$$

and at 100 per cent modulation it delivers a total power of

$$50\,(1+\frac{1^2}{2}) = 75 \text{ kW}$$

The percentage modulation in practice varies about the mean value (because of the varying intensities of speech and music), and to give a good signal-to-noise ratio it should be kept as high as possible.

INTERFERENCE. In practical broadcast or communication systems, there is always a certain amount of noise, arising from motor car ignition, switching clicks, atmospheric discharges, etc. The desirable features of any communication system are freedom from interference from unwanted signals and a high signal-to-noise ratio. Both these factors are influenced by the type of modulation employed.

When the transmitter and receiver are close to one another, there is little to choose between various systems. However, in providing a broadcast service, it is desirable to obtain the maximum possible coverage with the minimum number of transmitters.

Systems employing amplitude modulation suffer from noise consisting of sharp electrical impulses such as are produced by motor car ignition and electrical machinery – the interference known as static interference which includes so-called atmospherics.

The energy spectrum of atmospheric noise is fairly constant over the first few megahertz, falling off rapidly as the frequency increases. Very sharp pulses of electrical noise have a fairly uniform frequency distribution. In both cases, however, the energy distribution is uniform and the noise energy received is substantially proportional to the bandwidth of the receiver.

Random noise constitutes a variation of amplitude but has no orderly frequency variation. Therefore a signal having large changes in frequency should be capable of separation from random noise. This has led to the widespread adoption of frequency modulation.

FREQUENCY MODULATION. Frequency modulation (f.m.) employs a carrier of constant amplitude whose instantaneous frequency is varied at the modulation frequency. It is then possible to design a

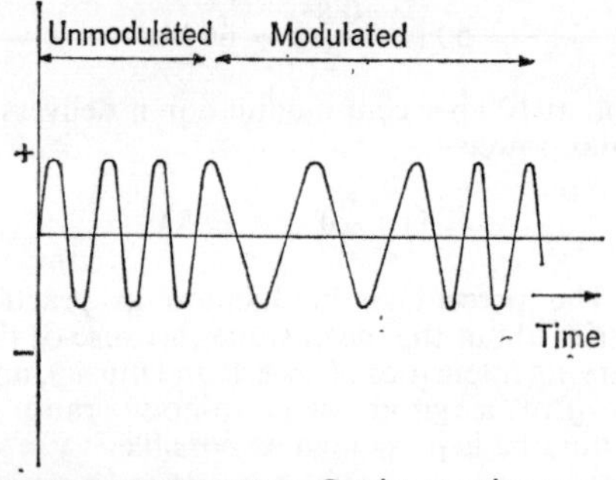

FREQUENCY MODULATION. Carrier remains at same amplitude, whether modulated or unmodulated, but its frequency varies from instant to instant at the modulating frequency.

receiver that is sensitive to frequency changes but not to amplitude changes. Hence the effects of impulsive interference are considerably reduced.

A frequency-modulated wave can be expanded into a trigonometric series and the spectrum it occupies can be determined. As with an amplitude-modulated wave, the frequency-modulated wave is composed of sine waves, but it contains not three terms but hundreds of significant components when the frequency excursions are large. For small deviation, the first order sidebands are similar to those obtained for amplitude modulation except for phase difference.

A signal of unit amplitude having an instantaneous angular frequency may be expressed, $\omega(t) = p+\Delta\omega \cos qt$ and will have an instantaneous voltage value of

$$e = \cos\left\{pt+\frac{(\Delta\omega)}{q}\sin qt\right\}$$

where p is the average carrier frequency, $\Delta\omega$ is the peak variation of this frequency, and q is the audio (or modulation) frequency rate of variation.

It is convenient to express the frequency deviation ratio, or modulation index as it is sometimes called, as

$$m_f = \frac{\Delta\omega}{q}$$

It is defined in words as the ratio of the peak deviation of the carrier frequency to the audio rate at which it occurs.

Thus, substituting m_f for $\Delta\omega/q$, the instantaneous voltage value

$$e = \cos(pt+m_f \sin qt)$$

Expanding gives

$$e = \cos pt \cos(m_f \sin qt) - \sin pt\,(m_f \sin qt)$$

The significance of the expressions $\cos(m_f \sin qt)$ and $\sin(m_f \sin qt)$ must now be sought. It is evident that they are periodic functions having a period of $2\pi/q$. Also every single-valued function $m_f \sin qt$ must repeat itself with the same period. It is possible, therefore, to represent the functions as a Fourier series and then solve the resultant power series by the use of Bessel functions. When this has been done, the instantaneous voltage value

$$\begin{aligned} e = {} & J_0(m_f)\cos pt + J_1(m_f)\cos(p+q)t \\ & -J_1(m_f)\cos(p-q)t \\ & +J_2(m_f)\cos(p+2q)t+J_2(m_f)\cos(p-2q)t \\ & +J_3(m_f)\cos(p+3q)t+J_3(m_f)\cos(p-3q)t \\ & +J_4(m_f)\cos(p+4q)t+J_4(m_f)\cos(p-4q)t+\ldots\ldots \end{aligned}$$

This equation shows that the spectrum consists of a carrier and an infinite number of sidebands whose amplitudes are various order Bessel functions of m_f.

From the above it is evident that the modulation index m_f is directly proportional to the maximum frequency swing of the carrier $\Delta\omega$ and inversely proportional to the modulation or audio frequency. In practice, $\Delta\omega$ for high fidelity broadcasting is fixed at values of ± 25, ± 50 or ± 75 kHz. So for an audio frequency of 15 kHz and $\Delta\omega$ of, say, ± 75 kHz, $m_f = 5$.

Substituting these values in the above formula, eight significant amplitudes are found on each side of the carrier frequency. Therefore a bandwidth of $8\times 15 = 120$ kHz on either side of the carrier frequency would be occupied. For 50 Hz modulation frequency the value of the modulation index becomes

$$\frac{75\text{ kHz}}{50\text{ Hz}} = 1{,}500.$$

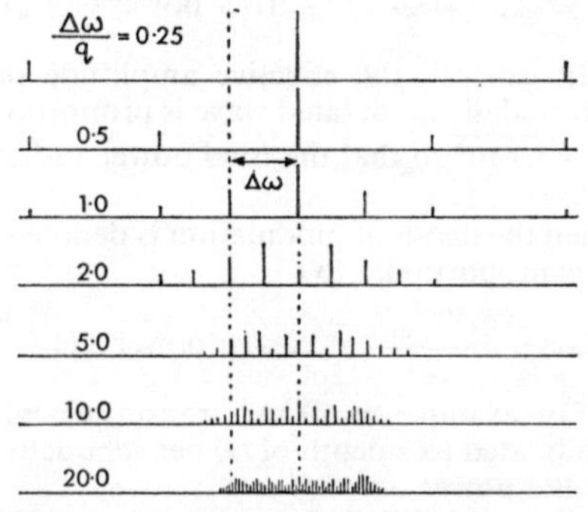

SPECTRUM OF FREQUENCY-MODULATED SIGNALS. Signals shown are sine-wave modulated in frequency with the peak deviation ($\Delta\omega$) held constant, and the audio modulated frequency (q) varied.

Examination of Bessel tables shows that an increase in the modulation index results in an increase in the number of significant side frequencies. This may result from increasing the deviation ratio and keeping the audio frequency constant, or from keeping the deviation ratio constant, as in practice, and varying the audio frequency from a high to a low value. So in the case of the 50 Hz modulating frequency, there will essentially be 1,500 side frequencies occupying a total bandwidth of $2\Delta\omega = 150$ kHz.

Summarizing, then, in frequency modulation the number of important side frequencies is larger for the low signal frequencies than for the higher modulation frequencies, even though $\Delta\omega$ remains constant for all modulation frequencies.

TRANSMISSION TECHNIQUES

Electromagnetic waves used for radio communication (including television transmission) reach receivers by different paths over various parts of the earth's surface, and the propagation path is determined by the frequency transmitted. The frequency used depends, among other things, on the distance at which the transmission is to be received.

There are two principal modes of transmission. The first is that known as ground wave transmission, which occurs over short distances when the electromagnetic waves travel directly between the transmitting and receiving aerials. The second, long-distance transmission, requires the wave to be reflected alternately once or more from the ionized layers of the upper atmosphere (D, E and F layers) and from the surface of the earth in order to reach the receiver.

GROUND WAVE OR DIRECT TRANSMISSION. This type of transmission is the usual one for long and medium wave broadcasting for the intended service area of the transmitter. The wavelengths used are from 2,000 to 150 metres. These transmissions tend to follow the curvature of the earth and are attenuated by the earth depending on the resistivity of the surface, there being less attenuation over sea than land. The wavelength also determines the loss due to the earth's surface, which increases as the wavelength decreases. This is noticeable, in practice, from the difference in service areas of stations at opposite ends of the medium waveband.

INDIRECT TRANSMISSION OF ELECTROMAGNETIC WAVES. Very low frequencies (below 30 kHz) have ground wave ranges of 1,000 miles or more and give world-wide reception by successive reflections from the D layer and earth's surface.

Low frequencies (30–300 kHz) behave in a similar way to very low frequency transmissions but have a ground wave of only hundreds of miles.

Medium frequencies (300 kHz–3 MHz) behave differently through the frequency range. As the frequency increases so the range of the ground wave decreases. The high frequency end of the medium range, therefore, is useful for communication purposes such as ship-to-shore, police, fire, ambulance services, etc., where ranges of no more than tens of miles are required.

High frequencies (3–30 MHz), where ground wave range is only a few tens of miles, are used for long-distance transmission, depending on multiple reflections for their successful operation.

Very high frequencies (30–300 MHz) are used for television transmission and are in the range where no reflection occurs from the ionosphere except under freak conditions. The range of these frequencies exceeds the visual range, but by an amount which becomes less with increase of frequency, since at high frequencies there is less refraction round the earth's surface. The received signal is the vector sum of the ground-reflected ray and the direct ray, the surface wave being greatly attenuated.

Reflections from objects such as bridges and metal-framed buildings are experienced at these frequencies. These reflections produce multi-path signals in the receiving aerial which can sometimes have values comparable to, or even greater than, the direct signal. This situation results in two or more images of the picture on the television screen and is sometimes termed ghosting. The difference in time between the direct or first image and the ghost image, seen on the screen, is a measure of the distance from the receiving aerial of the object causing the reflection.

Ultra-high frequencies (300–3,000 MHz) behave in a similar manner to VHF but are attenuated more quickly and are subject to more interference by reflections from objects. A line of sight path between transmitter and receiver is desirable, and care is necessary in built-up areas to achieve satisfactory UHF television reception.

TRANSMITTER POWERS. The effective radiated power (ERP) of a transmission is the product of the transmitter output power and antenna power gain. The ERPs used in television transmission depend on many factors, including the nature of the terrain, the population distribution in the area to be served, site available, whether the horizontal radiation pattern of the aerial is omnidirectional and so on. It is always necessary in practice to reach a compromise on the various factors, particularly as international agreements regarding powers and channels have to be observed.

As the frequency increases antenna power gains increase; typical antenna power gains in Band I are 4–5 times; in Band III 10–12 times, and in Bands IV and V gains of 20–50 times are realized.

Although the propagation losses of VHF and UHF are similar, the poorer signal-to-noise ratio of the UHF receiver requires a higher value of received signal than on the VHF band. This in turn means higher ERPs for UHF transmissions to achieve similar service areas to VHF and/or more transmitters to cover the same area.

Typical values of ERP for Band I television transmissions are 60–100 kW; for Band III 100–200 kW and in Bands IV and V ERPs of 500–1,000 kW are usual with filler or local stations of 10–20 kW.

The power of the associated sound channel is a quarter of the vision power for the British 405-line standard, a fifth for CCIR European standards and one-half the vision power for American FCC standards.

WIRED SYSTEMS. Wired systems are methods of distributing sound and/or television signals by means of cable. This method of distribution is called community television (CTV) or central aerial television (CATV). It is usual in Britain to refer to distribution in towns or large housing estates as a CTV system, while confining the use of the term CATV system to multiple outlets in one building.

There are two main methods of distributing television signals over cable systems:

1. High Frequency (HF) employing carrier frequencies around 8–12 MHz and using a pair of conductors for each programme source.
2. Very High Frequency (VHF), where the direct distribution of carrier frequencies of 40–230 MHz over coaxial cables is employed.

The HF method has the advantages of lower cable distribution losses, and simplified receivers can be used with the associated sound channel fed at audio frequency. The disadvantages of this method are that the choice of receiver is limited to special models produced for use on this type of system and

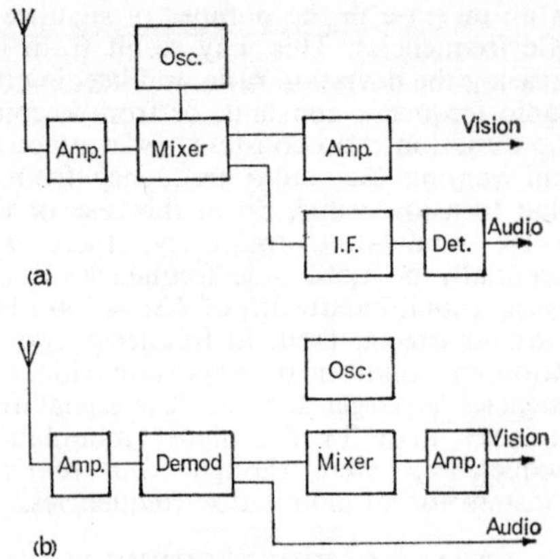

WIRED HF DISTRIBUTION SYSTEMS. Alternative high-frequency systems employing the 8–12 MHz band and using a pair of conductors for each programme channel. Cable losses are low, and the sound is fed to a simplified receiver at audio frequencies, but non-standard receivers offset some of the economic advantages of simplicity.

that the operating company has initially to decide how many pairs should be allowed in the multiway cable for all future requirements. Adding additional cable is very expensive.

The VHF method of distribution by coaxial cable has the advantage that the operating company may add programmes as required without any alteration to the network. The limiting factors are the bandwidth of the amplifiers used on the network (and these are easily arranged to be wide band, 40–230 MHz) and the adjacent channel selectivity of the television receivers used. These, incidentally, are standard domestic sets, and the subscriber can choose from a wider range than is available for HF subscribers. Extending the frequency range may well be expensive as the repeater amplifiers used may need respacing owing to loss alterations in the cable.

It is necessary for care to be taken in the selection of channels and signal levels in a VHF system but providing this is done excellent results are achieved.

The transmission of a UHF channel (Bands IV and V) is accomplished by converting it to a convenient spare VHF channel. The conversion is carried out at the sending end where other terminal equipment is located.

The terminal equipment is necessary for both systems and consists of aerial arrays to receive the signals to be distributed, receivers, channel converters, main sending amplifiers, test equipment, etc.

The installation of distribution network and amplifiers is carried out by fixing the trunk and spur cables under the eaves of buildings. In the case of VHF systems, the amplifiers are wall-mounted in metal cabinets, but HF systems use more widely spaced repeaters which, because of their larger size, are housed in kiosks. However, the increasing use of transistors is helping considerably to reduce the size of amplifiers used for these purposes. By feeding power down the coaxial cable, the necessity of providing metered power supplies along the routes is reduced.

In the case of new building projects, the modern trend is to have all cables underground as with other services, i.e., gas, electricity and telephone. However, in the US and Canada the trend is to use pole mounting for both the distribution cable and the amplifiers, and suitably equipped vehicles are needed to deal with maintenance problems. J.D.T.

See: *Television principles; Transmission networks; Transmitters and aerials.*

Books: *Radio*, Vol. 3, by J. D. Tucker and D. Wilkinson (London), 1965; *Radio Engineering*, by F. E. Terman (New York), 1937.

TRANSMISSION NETWORKS

Television transmission networks consisting of vision and programme (music quality) channels, together with associated control circuits, are used to interconnect: studios; switching centres and television transmitters of national broadcast services; buildings and sports arenas, etc., for national closed-circuit television transmissions; and two or more countries for the international exchange of broadcast programmes or for closed-circuit transmissions.

The vision and programme channels are tested and interconnected at switching centres set up in or near those cities where rapid switching of incoming channels to one or more outgoing channels is needed. The vision channels are designed and maintained to meet the requirements for the transmission of monochrome and/or colour signals to the television standards in use, i.e., 405-, 525-, 625- or 819-line definitions.

All vision channels are designed to transmit the highest nominal video bandwidth involved, e.g., 4·2 MHz for 525-line and 5, 5·5 or 6 MHz for 625-line systems, the value for the latter depending upon the particular 625-line television system required.

International transmissions require the use of point-to-point radio-relay or line systems (where practical) or communication satellites. For international transmissions between countries using different line standards, conversion equipment is necessary.

DESIGN AND PERFORMANCE

BROADCAST NETWORK DESIGN. Programme switching centres (PSCs) are set up by the broadcast organizations to control the contribution and distribution of programmes and to effect switches (by direct and remote control) on vision and programme channels.

Network switching centres (NSCs) in or near the main cities provide an interconnection point between the local channels and studios, programme switching centres and transmitters and the main channels to other cities and countries.

The NSCs are normally operated by the network organization. Switching at these centres, usually on a scheduled time basis, can result in reduced channel requirements.

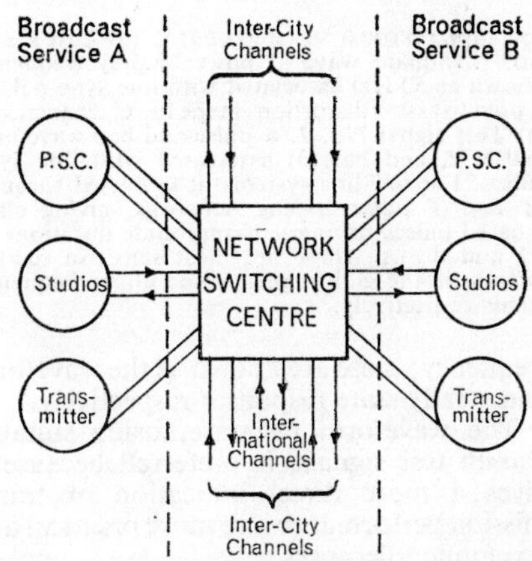

NETWORK SWITCHING CENTRE. Inter-city and local vision and programme channels terminate in these centres, which provide flexibility and control. Two network broadcast services (A and B) are shown, each with its programme switching centre (PSC), studios and transmitter.

Economies in equipment, standby plant and common services are secured, especially on the main channel routes, by integrating vision channels with other telecommunication services.

CLOSED-CIRCUIT NETWORK DESIGN. Closed-circuit networks do not usually involve inter-city channels.

The local channels are normally provided direct between the sites, with no intermediate flexibility points.

PERFORMANCE OF MONOCHROME NETWORKS. The transmission performance of a vision channel is substantially equivalent to that of a linear network and can be expressed as a function of either time or frequency; these are known as the waveform and steady-state responses respectively.

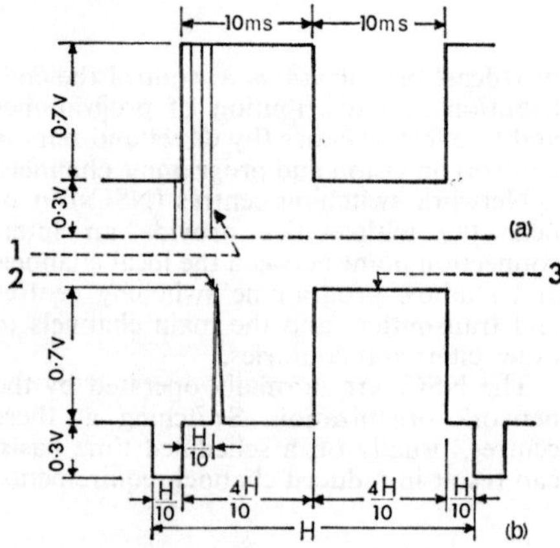

LINEAR WAVEFORM MEASUREMENTS. (a) Test signal No. 1. Square wave at power supply frequency (shown as 50 Hz) associated with line sync pulses, 1, used to assess distortions in the lower frequencies. (b) Test signal No. 2, a pulse-and-bar waveform (pulse, 2, and bar, 3) associated with line sync pulses. For 625-line systems, it is passed through 2T and T pulse-shaping networks, giving sine-squared pulses having half-amplitude durations of 0·2 and 0·1 μs. These are most sensitive to distortions in the mid-frequency and upper-frequency bands respectively.

The waveform response, using suitably chosen test signals, is preferred because it gives a more direct indication of transmission performance and more practical and economic tolerances.

The internationally accepted recommendations of the CCITT and CCIR for the transmission of monochrome vision signals form the basis of all vision channel performance specifications. In these recommendations, a hypothetical reference circuit is defined as one which has, between video terminal points, an overall length of 2,500 km (1,600 statute miles) and two intermediate video points.

The three equal-length video sections are lined up individually and then connected together without overall adjustment, correction, standards conversion, or synchronizing pulse regeneration of any kind.

Test waveforms for the various line systems are specified, together with detailed performance requirements applying at the video terminals.

From the concept of a hypothetical reference network and its performance requirements, it is possible to define a national reference network and its design performance.

In Britain the national reference network takes the form of a chain of four inter-city vision channels totalling 500 route miles, and eight local channels (two about 25 miles long and six under 6 miles long).

General requirements and target performance limits for main and local channels are specified. The general design requirements specify impedance at the video interconnection points, signal amplitude and d.c. component and insertion gain.

The standard level of a composite video signal at an interconnection point is:

$$1 \text{ volt peak-to-peak (p–p) in } 75\ \Omega$$
$$= +2{\cdot}2 \text{ dB relative to } 1 \text{ mW in } 75\Omega$$

The performance limits include random, periodic and impulsive noise, cross talk, non-linear distortion of the picture and synchronizing signals and linear waveform distortion.

THE K-RATING SYSTEM. When a television picture is received over two paths, one path having a slightly longer time for transmission than the other, a ghost image is seen. It is particularly noticeable following a sharp vertical edge. This picture distortion is shown on a waveform display as an echo after an amplitude change.

In the K-rating system for judging the quality of vision channels, the distortion caused by a single echo that is x per cent of the required received pulse amplitude and more than 1⅓ microseconds from the received pulse is given a rating of x per cent.

Other forms of picture impairment are compared with the ghost images by observers.

Types of waveform distortion that give the same degree of annoyance to the observers as the single echo distortion are given the same rating factor.

To combine the results into a suitable form to facilitate linear waveform distortion measurement, two test signals are standardized, and, after transmission over the channel under test, are displayed on an oscilloscope. For very accurate measurements, the displays are photographed and the rating factor is calculated from microscope measurement from the photographs. This is known as the acceptance-test method.

Where a quick, simple method is required, oscilloscope graticules showing rating limits are used. This is known as the routine-test method.

Maximum permissible rating factors are agreed for the national reference chain, main channels and local channels.

NON-LINEAR DISTORTION. A picture signal suffering from non-linear distortion produces inaccuracies in the reproduced picture's brightness gradation. The more important form of this distortion is known as line-time non-linearity, affecting the coarse detail of the picture.

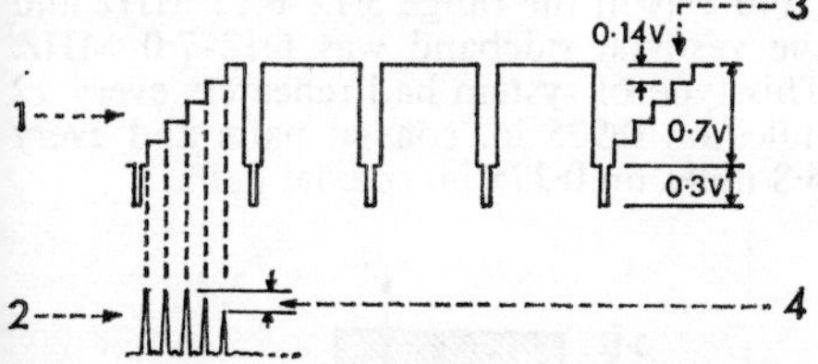

NON-LINEAR WAVEFORM MEASUREMENTS. 1. Transmitted waveform with five risers, 3, each of 0·14 v. Intervening three picture lines can also be switched to black or white level. 2. Received signal after differentiation. 4. Indication of degree of non-linearity.

To measure this distortion on a vision channel, a special five-riser staircase test signal is connected to the channel input; at the receiving end the signal is passed through a differentiating and shaping network. The five pulses produced are approximately sine-squared in shape and their amplitudes are directly proportional to the riser amplitudes and indicate the degree of non-linearity.

The three intervening lines of the test signal are set first at white level, then at black, and the non-linearity is measured for both conditions, the higher value being taken as the result.

PERFORMANCE OF COLOUR NETWORKS. A colour television channel can be considered as a combination of a luminance channel and a chrominance channel. The luminance channel can be specified and tested in the same way as the equivalent monochrome channel. The presence of the chrominance channel necessitates additional specification clauses and tests; the more important are luminance/chrominance cross-talk, luminance/chrominance gain (differential gain) and delay inequalities (differential phase) and the noise in the chrominance channel.

CABLE TRANSMISSION

CABLES. Several types of cable are used to provide local and main vision channels. In the past, especially outside Britain, screened, balanced pairs have been used to provide local channels. One of the latest forms of balanced-pair cable has six screened pairs, each consisting of two 40 lb. copper conductors, located by means of expanded polythene within a copper cylinder. The loss-frequency characteristic of a pair of this type follows approximately the law:

$$7{\cdot}7\sqrt{f} + 0{\cdot}3 \text{ dB per mile}$$
$$(f = \text{frequency in MHz})$$

Three sizes of coaxial pair have been used, i.e., pairs having outer conductors of 0·975, 0·375 and 0·174 in. diameter. The 0·975 in. has been in use on an inter-city route for many years; 0·375 in. pairs have been used for all other inter-city line systems. The 0·174 and 0·375 in. diameter coaxial pairs have been used extensively for local channels. The loss-frequency characteristic of a 0·375 in. diameter coaxial pair follows approximately the law:

$$3{\cdot}75\sqrt{f} + 0{\cdot}01f \text{ dB per mile}$$
$$(f = \text{frequency in MHz})$$

The dielectric of the three sizes of coaxial pairs used is effectively dry air, polythene discs or equivalent being used to maintain the spacing. The characteristic impedance of the coaxial pairs is 75 Ω. Coaxial cables are gas (dry air) pressurized to improve their reliability.

Video systems employing unbalanced video transmission on 0·174 and 0·375 in. coaxial pairs are used to provide local channels up to 25 miles long and also vision-tie circuits between line/radio terminals and network switching centres, i.e., part of a main channel.

These systems are capable of transmitting monochrome and colour signals on any line-standard with very low distortion; they provide lower cost, better performance channels than the alternative line systems of either video transmission on screened, balanced pair or vestigial-sideband carrier transmission on coaxial pair.

REPEATERS. One type of transistor amplifier used in video systems on coaxial pairs has a gain of 20–24 dB, which is flat with frequency. A repeater may consist of several of these amplifiers connected in cascade with various waveform-correcting networks. The repeater spacing required is dependent on the repeater performance, coaxial pair attenuation and type of signal to be transmitted. Typically, a 625-line colour channel provided on 0·375 in. coaxial pair will have a repeater spacing of about 3 miles.

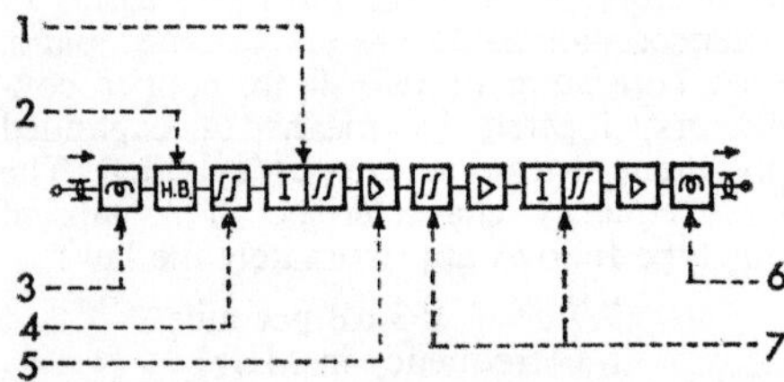

VIDEO REPEATER: BLOCK SCHEMATIC. Signal from coaxial cable is fed to 3, coaxial choke, 2, hum balancer, 4, line temperature compensating waveform corrector, 1, mop-up waveform corrector, 5, amplifier, 7, two waveform correctors, one specially designed for particular routes. After further amplification and second coaxial choke stage, 6, signal is ready for retransmission.

Several difficulties arise with video transmission. First there is a problem of power supplies for immediate repeaters. Local power supplies with reliable standby arrangements can be costly. Power feeding over a coaxial pair cannot be at mains frequency, because 50 or 60 Hz is within the video signal spectrum; d.c. power feeding introduces problems in the design of power-separation filters.

However, in situations where it is not economical or convenient to have intermediate video repeaters in surface buildings and operating from local power supplies, it is possible to have a few transistor repeaters in special watertight boxes in the ground and to power feed these repeaters over interstice or other pairs, bunched if necessary to reduce their voltage drop.

Secondly, there are problems with low-frequency interference (at frequencies where proximity effect is not present) and hum. Coaxial pairs usually have steel tape(s) over their outer conductors, but the shielding effect decreases below about 10 kHz and there is a risk of low-frequency interference. Induced voltages and differences of earth potential between the earths at equipped points give rise to longitudinal currents.

Low-loss video transformers having adequate waveform performances could solve these problems but they are difficult to design. The solution used is to isolate the outers of the coaxial pairs from earth and to insert coaxial chokes to reduce the effect of longitudinal currents, typically by 22 down to 10 dB for one choke. This measure, together with the use of special hum balancers and high send levels from repeaters, makes it possible to reduce low-frequency interference and hum levels sufficiently below signal levels.

HIGH FREQUENCY CARRIER SYSTEMS. These are particularly suited to inter-city routes because they can transmit either television or telephony or both; power feeding at mains frequency to a considerable number of intermediate repeater stations is possible. An early carrier system, suitable only for 405-line television, used a line frequency range of 3–7 MHz; the carrier frequency was at 6·12 MHz, the lower sideband was in the range 3·12–6·12 MHz and the vestigial sideband was 6·12–7·0 MHz. This type of system had repeaters every 12 miles on 0·975 in. coaxial pairs and every 4·8 miles on 0·375 in. coaxial pairs.

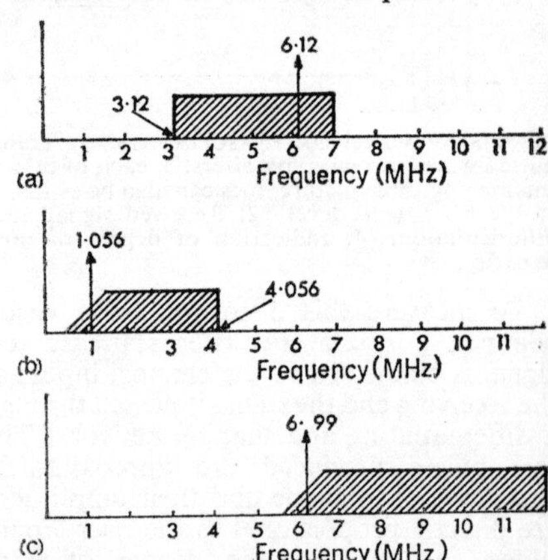

HIGH-FREQUENCY CARRIER SYSTEMS. Line frequency bands of (a) 3–7 MHz, and (b) 0·5–4 MHz systems. (c) Interim allocation of 12 MHz system suitable for a 625-line, 5·5 MHz, bandwidth TV channel.

A later system suitable for 405-line television or telephony (960 speech channels in the standard 16-supergroup band) used a line frequency of 0·5–4·0 MHz on 0·375 in. coaxial pairs, with repeaters every 6 miles

The 12 MHz line system now developed could be made suitable for the transmission of 625-line vision signals. This system, with three-mile repeater spacing on 0·375 in. coaxial pairs is normally for the transmission of 45 supergroups (2,700 speech channels), but alternatively the transmission of one 625-line television channel in each direction would be possible.

Residual mop-up waveform correction on high-frequency carrier systems can be carried out using echo waveform correctors with either transversal or reflective networks.

RADIO RELAY

Radio-relay systems, with line-of-sight paths, are extensively used on inter-city routes and in international networks for the transmission of telephone and telegraphy, including programme and control channels associated with television, and television, both monochrome and colour.

Internationally agreed recommendations such as those of the CCIR and CCITT, together with the international broadcasting unions such as EBU, for these systems relate to the 2,500 km hypothetical reference circuit, and include overall system performance, baseband and intermediate frequency characteristics, modulation of the radio-frequency carrier and r.f. characteristics in multi-channel systems.

The general overall performance requirements are similar to those for line systems, so that the two types of system give similar overall quality of transmission and can be joined together. However, there are special requirements for radio-relay systems to take into account their liability to fading. The amount of fading that can be tolerated has to be an economic compromise between system cost and the grade of service. From the international recommendations and a knowledge of fading characteristics, it is possible to specify the signal-to-weighted-noise-ratio for an individual main or local vision channel under free-space propagation conditions over every section of the radio system, and when there is a 30 dB fade on any section.

Stringent requirements for overall channel gain stability can be met under all but the most severe fading conditions by the use of limiters and automatic gain control in the receivers and repeaters.

MODULATION. Frequency modulation is used on modern radio-relay systems in preference to other forms of modulation because a better signal-to-noise ratio is

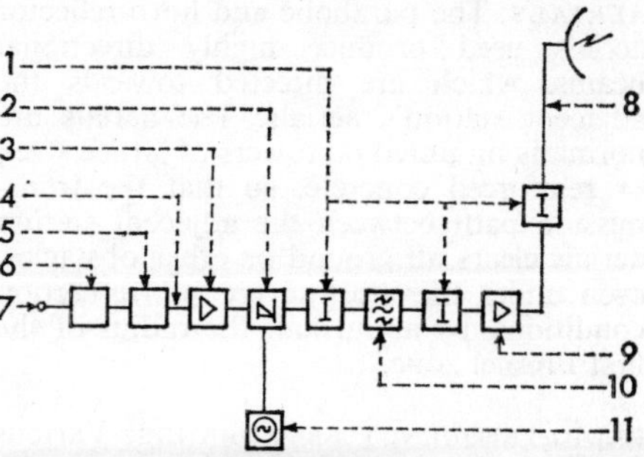

SINGLE BROADBAND CHANNEL MICROWAVE TV LINK. Block schematic of transmit terminal. Video, 7, is fed to pre-emphasis unit, 6, and frequency modulator, 5, whence i.f., 4, is amplified in i.f. stages, 3, and fed to mixer, 2, with microwave oscillator, 11. Microwave filter, 10, and travelling wave amplifier, 9, are separated from one another and from the rest of the system by ferrite isolators, 1, Which reduce the level of echo signals resulting from slight mismatching. A waveguide, 8, feeds the signal to a parabolic aerial.

obtained compared with a.m. under comparable conditions, the linearity in the i.f. and r.f. stages is not so critical as in a.m., limiting can be used to help achieve gain stability with fades, and the required bandwidth is less than on pulse modulation systems.

The peak-to-peak frequency deviation used on modern systems transmitting 625-line signals is 8 MHz. Pre-emphasis of the video signal modifies this deviation and is employed to reduce distortions, particularly differential gain and phase, and to facilitate the use of similar modulators for telephony and television.

The standard i.f. band on 2, 4 and 6 GHz radio-relay systems is centred on 70 MHz.

I.F. AND R.F. EQUIPMENT. The i.f. equipment includes modulators (baseband–i.f.), i.f. amplifiers and demodulators (i.f.–baseband). The i.f. amplifiers make good most of the transmission loss in a system and typically have a gain of 60–80 dB.

The r.f. equipment includes mixers (i.f.–r.f. and r.f.–i.f.) and r.f. amplifiers, usually of the travelling-wave type. Typical output power of a transmit travelling-wave amplifier is 5 W.

Waveguides are used to connect the r.f equipment to the aerials. Ferrite isolators having a non-reciprocal attenuation characteristic, e.g., a forward loss of 0·3 dB, and a reverse loss of 30 dB, are used to reduce the levels of echo signals arising from any mismatches in the microwave elements.

AERIALS. The parabolic and horn-reflector aerials used produce highly directional beams, which are directed towards the adjacent station's aerials. The aerials are normally mounted on towers of lattice steel, or reinforced concrete, so that the transmission path between the adjacent station aerials clears all ground or other obstacles, even under the most severe sub-refraction conditions, by about half the radius of the first Fresnel zone.

RADIO-FREQUENCY BROADBANDS. Various frequency bands have been internationally allocated to fixed radio services, and certain preferred bands within these broad allocations, having a total bandwidth of 3,200 MHz, have been recommended for international (and national) radio-relay systems.

PREFERRED RADIO RELAY BANDS

Band (MHz)	*Designation*
1,700– 1,900	2,000 MHz spur band
1,900– 2,300	2,000 MHz band
3,790– 4,200	4,000 MHz band
5,925– 6,425	lower 6,000 MHz band
6,425– 7,110	upper 6,000 MHz band
10,700–11,000	11,000 MHz band

The 2,000, 4,000 and lower 6,000 MHz bands are being fully exploited on inter-city routes and are channeled to give 20 (6+6+8) bothway r.f. broadband channels spaced 30 MHz apart. Each of these bothway r.f. broadband channels is suitable for 960 (1,800 in some cases) telephone circuits or a 625-line colour vision channel.

Multi-broadband channel radio-relay systems use branching filters to combine several r.f. channels for connection to one aerial waveguide feed. Four transmitters and four receivers, using r.f. broadband channels in a single-frequency band, can be combined for connection to the waveguide feed of a parabolic aerial.

The wider band horn-reflector aerials can be used, in conjunction with low-loss circular waveguides and polarization band-branching networks, to transmit or receive signals in the 4, 6 and 11 GHz bands.

Broadband protection channels contribute to giving radio-relay systems a high reliability; they are shared on the basis of one protection channel to several operational channels and are automatically switched, on an end-to-end basis, to replace any faulty or fading operational channel, i.e., when either the level of a pilot receiver over an operational channel drops or the noise in a narrow band associated with an operational channel worsens to a predetermined level.

REBROADCAST RECEIVERS. Specially developed receivers are used to connect low and medium-power vision (and sometimes associated sound) transmitters to a main broadcast network, in cases where a point-to-point cable or radio-relay system cannot be justified or where a reserve facility is required should a direct network connection fail.

The receivers are normally pre-tuned, and their directional receiving aerials have to be capable of receiving weak signals from the network in the presence of relatively strong signals radiated locally.

The automatic gain control system used in the receiver has to be capable of compensating for fades in the presence of interference signals.

OUTSIDE BROADCASTS

Temporary vision channels connecting outside broadcast sites to the permanent networks are provided, using one or more of the following methods: video transmission on spare telephone pairs, or on balanced and coaxial pairs; injections at intermediate repeater stations on permanent cable and radio-relay systems; on radio protection channels provided that the operational channels are given priority under fault conditions; and temporary radio-relay links.

The video transmissions on telephone, balanced and coaxial pairs are amplified, using specially designed video repeaters with a variable gain/frequency characteristic. The repeaters offer high rejection to interfering longitudinal signals and a gain of up to approximately 60 dB to the wanted transverse signals.

The latest design of repeater uses transistors and can be battery or mains operated.

The repeater spacing depends on the line-standard required and the cable-pair loss. For 625-line colour transmissions, the repeaters are spaced approximately one mile apart on 20 lb. telephone pairs and approximately four miles apart on 0·375 in. coaxial pairs.

Injections at intermediate repeater stations require the use of specially designed cable and radio equipments which accept the OB signal (1 v. p–p video) and provide a line-frequency signal, e.g., 0·5–4·0 MHz, and an i.f. signal (centred on 70 MHz) respectively. Temporary radio-relay links employ equipments in the 7 and 11 GHz bands. The latter have the advantage of using smaller parabolic aerials, but account has to be taken of possible increased path attenuation in the presence of fog, heavy rain and snow. Line-of-sight transmission paths are required, and the distance between repeater points is very dependent on the terrain and the achievable aerial heights; at best, the distance can be up to 20–30 miles at the high frequencies.

Temporary closed-circuit vision channels are provided, using techniques similar to those used for outside broadcasts.

SWITCHING AND MAINTENANCE

Network switching centres are usually equipped to switch rapidly a number of incoming vision and associated programme channels to a number of outgoing vision and associated programme channels. The switching operation may be manually or electrically actuated within the NSC, or remotely controlled from a programme switching centre.

MULTIPLE SWITCHING. On large switching equipments, it may be necessary to pre-set each connection pattern and to initiate a multiple switching operation under the control of a master switch, in turn controlled by an electronic clock.

On certain transmissions there may be a requirement for the switching operation to be carried out during the field-blanking interval. Several types of remote control are possible. The simplest uses a narrow-band circuit such as a teleprinter circuit or a telephony channel. Another method is to use pulses sent out over the vision channels during the field-blanking periods. Certain lines in this period on a 625-line signal are allocated to this function.

The design specification for network switching (and distribution) equipments has to be extremely stringent, especially on crosstalk, and has to take into account that several of these equipments may be in circuit in a network transmission. The vision-switching matrix may use either low-capacitance relays, or semiconductor diodes, or uniselectors, or a combination of these.

MAINTENANCE. A very high standard of reliability and performance is required on television networks, and special maintenance techniques have been developed. Preventive maintenance is performed outside programme hours when it is possible to schedule regular inspections and performance checks on transmission and power equipments, cables and aerials and to carry out measurements on overall channel responses.

During programme hours when most channels are in use, the NSCs are equipped to monitor channels in order quickly to identify, locate and clear any faults.

TEST SIGNALS. The use of test-line signals facilitates the localization of gross distortions during programme hours. In this technique, special test signals are inserted by the controlling PSC on allocated lines in the field-blanking period of the signal, and test signals can be examined on waveform oscilloscopes capable of line selection. This facility enables gross distortions occurring anywhere between the controlling PSC and the transmitter to be quickly localized during programme hours.

One form of test-line signal consists of a bar (period of white level), a sine-squared pulse and a five-riser staircase on one of the allocated lines. Automatic monitors have been developed to monitor this line and to operate an alarm should the linear and non-linear distortions (and also noise) exceed a predetermined value.

PROGRAMME CHANNELS

Programme (music quality) channels associated with vision channels in a television network and on outside broadcasts are provided, depending on the plant available, in the following ways: on deloaded audio pairs; on screened pairs; on phantoms, derived from pairs in carrier cables; on the same radio carrier as a television channel, e.g., by the use of an f.m. subcarrier; on high-frequency cable and radio-relay systems using special programme channel equipment. This translates the audio-frequency signal for inclusion in a basic group of 12 speech channels occupying the range 60 to 108 kHz. One version of the equipment accepts a music channel of 50 to 10,000 Hz and translates it into the band 86 to 95·95 kHz, displacing the three speech channels 4, 5 and 6.

Programme channels on satellite communication links and on point-to-point radio-relay and cable links can also be provided by making use of modulation of part of the synchronizing pulse. This is achieved by pulse modulation. J.B.S.

See: *Outside broadcast units; Picture line-up generating equipment; Satellite communications; Special events; Television principles; Transmission networks; Transmitters and aerials.*

Books: *Television Engineering*, IEE Conference Report Series No. 5 (London), 1962; IEE Conference Publication No. 46, *Parts* 1 *and* 2 *of the International Broadcasting Convention* (London), Sep. 1968.

TRANSMITTERS AND AERIALS

The purpose of a television transmitting installation is to generate vision and sound carrier signals, and to modulate these with the vision and sound programme signals originated in the studios. The composite signals are amplified and fed to the transmitting aerial, which is designed to radiate them throughout the service area. The signals are picked up by viewers' receivers, and in some cases by the receiving aerials of re-broadcast or translator stations which amplify the signal, change the frequency, and re-broadcast it from a transmitting aerial.

BASIC PRINCIPLES

EARLY TRANSMISSION STANDARDS. The first public television test transmissions with a 30-line system used two standard medium wave sound transmitters, one for the vision and one for the sound signals. It soon became clear that a much wider transmission spectrum would be necessary for worthwhile picture transmissions and this was not possible using medium wave transmitters. The transmitters themselves were incapable of sending a wide band signal and, in any case, such a transmission on medium waves would cause severe interference with other transmitters using nearby frequencies. For this reason, modern television transmitters use very high frequencies (VHF) and ultra-high frequencies (UHF).

The first British public service transmitters set up in the 1930s used a sound signal carrier frequency of 41·5 MHz and a picture signal carrier frequency of 45 MHz, in all, a band of 3·5 MHz. Both the picture and the sound transmitters were amplitude-modulated and, at the end of a series of tests, a 405-line picture signal was adopted. Since then, other line standards and different modes of modulation have been brought into use. Throughout the years, there have been many attempts to achieve international standardization of television transmissions but, although some progress has been made, several different standards are still in use.

TYPES OF TRANSMISSION AND FREQUENCIES. The signal may be transmitted from the aerial with either vertical or horizontal polarization. The frequency and type of transmission used affect the service area of the transmission. At one extreme, signals in VHF Band I diffract over hills and around large buildings relatively well, but tend to suffer interference from electrical equipment and motor vehicles. On the other hand, signals in UHF Bands IV and V, and to a lesser extent in VHF Band III, tend to travel along straight paths and leave shadow areas in valleys and behind tall buildings.

For this reason Band V transmitting aerials are usually supported as high as possible – up to 1,250 ft. and more.

TRANSMISSION STANDARDS

No. of lines/ field frequency	1 405/50	2 525/60	3 625/50	4 625/50	5 625/50	6 819/50
Line frequency (Hz)	10,125	15,750	15,625	15,625	15,625	20,475
Nominal video bandwidth (MHz) ...	3	4·2	5	6	5	10
Nominal r.f. bandwidth (MHz)	5	6	7	5·5	5·5	14
Sound carrier relative to vision carrier (MHz)	—3·5	4·5	5·5	6·5	6	—11·15
Sound carrier relative to nearest edge of channel (MHz)	0·25	—0·25	—0·25	—0·25	—0·25	0·02
Nominal vestigial sideband (MHz) ...	0·75	0·75	0·75	0·75	1·25	2
Polarity of vision modulation	Positive	Negative	Negative	Negative	Negative	Positive
Type of sound modulation	AM	FM	FM	FM	FM	FM
Ratio of radiated powers of vision and sound	4:1	5:1	5:1	5:1	5:1	4:1

1. U.K. only. 2. North America, Japan. 3. Austria, Germany, Italy. 4. Poland, U.S.S.R. 5. U.K. and Europe (UHF). 6. France: an alternative 625/50 standard is also used.

CHOICE OF SITE. In the selection of a site for a television transmitting installation, the following requirements must be satisfied.

The transmitting aerial must be situated so that it provides a good signal to the population to be served. Where possible, a high site in the centre of the area is chosen. Usually the aerial has to be supported on a mast or tower so that signals can reach over hills and other obstructions.

The site must be large enough to accommodate the mast or tower, as well as the building to house the transmitting equipment and auxiliary apparatus. A tall mast which is supported by wire rope stays or guys, requires a large, level site about twice as wide as the mast is tall. A self-supporting tower can be accommodated on a much smaller site. Tall masts and towers are a hazard to aircraft, and are not permitted near airports. They may also detract from local amenities, or conflict with area-planning requirements – important factors in densely populated areas.

Electricity supplies must be available for the equipment.

The cost of the programme link from the studios to the transmitter can be considerable and, where this is important, as in the case of a station intended to serve an isolated town, the transmitter is usually located close to the studio centre.

It is not easy to find a satisfactory site for the transmitting installation and this is one reason why several different television and sound services are often transmitted from the same site. Another advantage of this arrangement is that simpler and more satisfactory receiving aerials are possible. On the other hand, the areas covered by transmissions in the VHF and UHF bands are not the same, so that a good site for Band I (VHF) television transmissions may not be satisfactory for Band IV (UHF) transmissions.

In countries with several television services, the same channel may be allocated to a number of transmitters and the possibility of interference with other transmitters must be considered when choosing a site. Sometimes there are advantages in locating the transmitter towards one end of the service area and using a directional transmitting aerial to beam signals to cover the area.

TRANSMITTING POWER. The power of the transmitters used depends on the service area to be covered, the frequency band in use, the surrounding terrain, and the gain and height of the transmitting aerial. In fairly level country, good coverage can be obtained using a relatively small number of high-power transmitting stations, but in mountainous areas a much larger number of low-power transmitters is usual.

EFFECTIVE RADIATED POWER. The power radiated from a transmitting station is usually quoted in terms of the effective radiated power (ERP). The ERP is equal to the power of the transmitter that would be required to transmit the signal using only a simple dipole aerial. Large multi-element transmitting aerials are often used and give considerable gain when compared with a simple dipole. When such aerials are arranged to be directional, there can be a further increase in the effective gain in those directions. In Band III, for instance, an ERP of 200 kW is quite common using transmitters of 5 kW output power.

In calculating the ERP an allowance must be made for the power lost in the feeders from the transmitters to the aerials.

MASTS AND TOWERS. When a transmitting aerial has to serve a large area, it is usually supported at the top of a tall mast or tower. Steel masts of more than 1,000 ft. in height are now often used. Towers, which may be of steel or concrete, are not usually so tall, because their cost increases rapidly above about 500 ft. When a concrete tower is used, the transmitting equipment may be installed inside the tower above ground level, thus saving the cost of station buildings. High-gain aerials send out signals in the form of narrow horizontal beams and, because of this, the mast or tower must support the aerial rigidly since the signal at the receiving site would vary if the mast or tower shook in strong winds.

Large aerial assemblies at the top of a tall mast very much increase the total wind resistance of the structure, and the strength, weight, and cost of the mast are then considerably greater.

Steel masts are usually constructed in the form of vertical beams of triangular or square cross-section with solid round or angle steel legs at the corners and braced with a lattice of steel angle. Most are slender, with sides from 1 to 2 per cent of the height, and are supported by several sets of three or four stays or guys which are anchored at the ground to concrete blocks at a radius equal to about three-quarters of the height.

Powerful red lights to warn aircraft of the hazard are usually fitted to the outside of

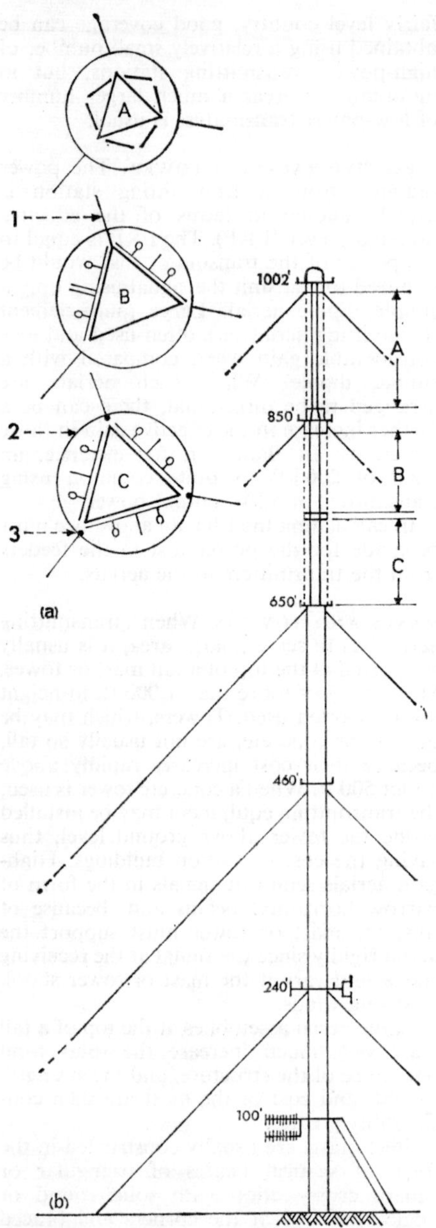

AERIAL ARRANGEMENTS ON TYPICAL TALL MAST. (a) Aerial arrays in plan. A. Band V array on 4 ft. 3 in. triangular section at top of mast for Programmes 1 and 3, with below it a Band V array for Programmes 2 and 4. B. Band III array (Programme 1). 1. Corner screen. C. Band III array (Programme 2). 2. 12 ft. diameter fibreglass cover. 3. Parasitic element. (b) General construction of mast. To 650 ft. level, 9 ft. circular section; above this, triangular section shown in (a). At 100 ft. and 240 ft. VHF communications aerials.

the mast at regular intervals, and in some countries regulations require the mast to be painted in contrasting red and white sections.

The mast or tower must be earthed for protection against lightning. Where the resistance of the earth is low, it is usually sufficient to connect buried copper or aluminium plates to each of the mast stays and to the mast base, or to each of the tower legs. On sites where the earth has a high resistance, much more elaborate precautions may be necessary, however, to ensure that lightning discharges are not conducted into the station building along the r.f. feeders or away from the site along service or programme cables. In extreme cases, it may be necessary to cover the whole site with a network of buried copper tapes to which the mast or tower, the feeders and the transmitting equipment itself are bonded at regular intervals.

TRANSMITTER STANDARDS. The task of a television transmitter installation is to generate the sound and vision carrier frequency signals and to modulate these with the sound and vision signals coming from the studio centre. The composite radio-frequency signals are then separately amplified to the required power, combined together, and fed to the transmitting aerials. The arrangement of the equipment depends on the carrier signal frequencies, the television standards in use, and particularly on the kind of modulation employed for the vision and sound carrier signals.

In Britain, two standards are currently in use. In Bands I and III (VHF), the vision carrier is amplitude-modulated with a positive vision signal, so that the greatest power is transmitted at times when the picture is white. The sound signal is also amplitude-modulated. In Bands IV and V (UHF), on the other hand, the vision carrier signal is amplitude-modulated with a negative vision signal, and the greatest power is transmitted during picture-synchronizing signals. The sound carrier signal is frequency-modulated. In the United States, all television transmitters use this system.

For either system, it is usual to employ separate transmitters for the sound and vision carrier signals, which are combined in a combining filter at the output of the transmitters. In the case of rebroadcast or translator installations, and low-power UHF transmitters, common radio-frequency amplifiers are possible, and are sometimes used.

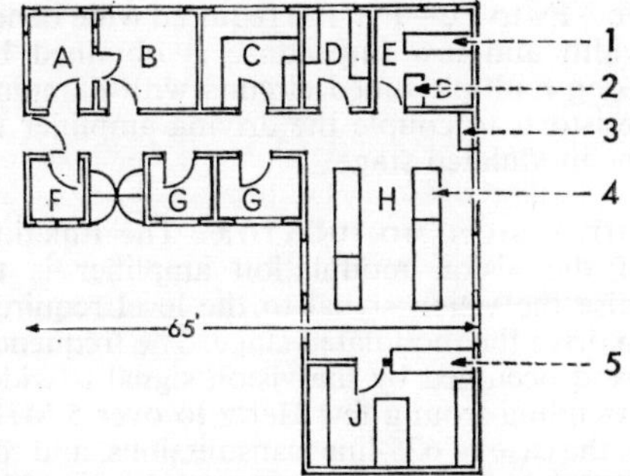

PLAN OF TYPICAL TRANSMITTING STATION. A. Kitchen. B. Staff room. C. Workshop. D. Film or slide scanner. E. Control room with, 1, programme input equipment and, 2, control desk. 3. Aerial switching facilities. F. Toilets. G. Offices. H. Transmitter hall with, 4, transmitters. 5. Mains switchboard. J. Spare parts store.

VHF VISION TRANSMITTERS

Most VHF vision transmitters are built to deliver an output power of between 500 W and 50 kW. Where larger powers are required, two transmitters may be operated in parallel. Designs vary principally in the power level at which the carrier signal is modulated and in the kind of r.f. circuits used but, in any case, it is usual to use tetrode or triode valves (tubes) in grounded

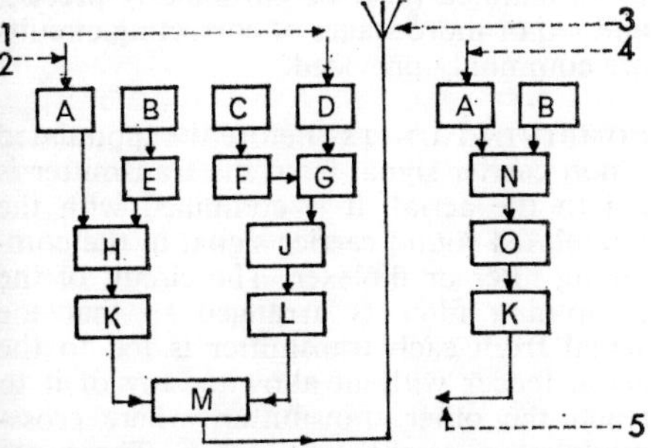

TYPICAL VISION AND SOUND TRANSMITTERS. Block diagram. (*Left*) 1. Vision signal input to vision transmitter, C, D, F, G, J, L. 2. Sound signal input to a.m. sound transmitter, A, B, E, H, K. Exciter or drive unit, C, is the master oscillator generating the carrier frequency. This is amplified in carrier signal amplifier, F. Vision signal, 1, is amplified in modulation amplifier, D, and used to modulate the carrier signal in the modulated stage, G. Modulated signal is fed to r.f. amplifiers, J, vestigial sideband and harmonic filters, L, and combining unit or diplexer, M, where vision and sound signals are combined for transmission from aerial, 3. Sound signal chain comprises: A, modulation amplifier, B, drive unit, E, carrier signal amplifier, H, modulator and frequency multiplier, K, harmonic filter. (*Right*) FM sound transmitter. Sound signal, 4, to A, modulation amplifier. B. Drive unit. N. Modulator and frequency multiplier and, O, r.f. signal amplifier. K, Harmonic filter. 5. Sound signal to combining unit.

grid stages as r.f. amplifiers. In recent designs, the low-power stages use solid-state circuits.

THE EXCITER OR DRIVE UNIT. To reduce the interference with other transmitters, a highly stable carrier frequency is required; this is generated in the exciter and usually a quartz crystal resonator contained in a temperature-controlled oven is used in the oscillator circuit to obtain a stability of one or two parts in a million or even better. In some solid-state designs, the whole oscillator circuit is contained in the oven.

Because quartz crystal resonators with a frequency as low as about 20 MHz have the best stability, the oscillator frequency is usually only 1/12 or 1/24 of the required carrier frequency. The oscillator stage must then be followed by the required number of frequency multiplying stages. Finally, a stage of amplification may be necessary to raise the output power to the few watts usually required to drive the r.f. amplifiers.

RADIO-FREQUENCY AMPLIFIERS. One or more stages of Class C r.f. amplification follow the exciter. These raise the power level to that required for the modulated stage. At present, grid modulation is invariably used in VHF vision transmitters. The power level at which the modulation is to be carried out is one of the most important decisions that the equipment designer must make. Any r.f. amplifying stages following the modulator must have a linear characteristic so that the modulated signal does not suffer distortion of the picture grey scale. The amplifier stages, therefore, have to operate in the Class A mode, which is comparatively inefficient.

The power dissipated in the valves must be removed, usually using air-blast cooling, but sometimes, in high-power transmitters, water or vapour cooling is used.

Except perhaps in the case of stages using wide-band distribution line circuits, each amplifier stage has two or more tuned circuits which must be adjusted to the correct frequency for satisfactory operation.

In short, high-power linear r.f. amplifier stages tend to be expensive and require careful adjustment for satisfactory operation. On the other hand, the size and complexity of the modulator increases rapidly as the required output signal level is increased. Further, when one or more stages of linear r.f. amplification follow the modulator, much of the required frequency-shaping of the output signal is carried out by the tuned circuits in the amplifiers, so simplifying the vestigial side-band filter, or even rendering one unnecessary.

Transmitters with 0, 1 or 2 r.f. amplifying stages are in use and prove satisfactory. The power gain per stage usually lies between 10 and 25.

RADIO-FREQUENCY TUNED CIRCUITS. Twin-line circuits, often called lecher lines, are common in balanced push-pull stages in which two valves are used in opposition. The circuit should be electrically symmetrical and, to achieve this, valves and tuned circuits are symmetrically arranged inside the cabinets or screening boxes. Coupling between stages is easy where the stages can be mounted close together, but otherwise presents problems to the designer.

In coaxial line circuits, the r.f. currents are confined to the inside of the outer, and the outside of the inner line. Most VHF and UHF valves have electrode connections intended to connect directly to such coaxial lines, permitting a neat, efficient and self-screening assembly. Provision must, however, be made to feed cooling air to the valve anode radiators, and to open the lines to change the valves.

The resonant frequency of the coaxial circuit is usually adjusted by moving a sliding disc, which shorts together the inner and the outer lines. Because the contacts between the disc and the lines carry the full r.f. circulating current, great care must be taken with their design and maintenance. For these reasons, coaxial line circuits tend to be more expensive to manufacture than twin-line circuits.

THE MODULATED STAGE. The design of the grid-modulated stage is probably the most critical part of the vision transmitter, and this is so for colour signals in particular. In a grid-modulated stage, a carrier signal Fc is mixed with the vision signal Fv to produce signals Fc, (Fc−Fv) and (Fc+Fv). The latter are called the sideband signals and carry the picture information. Usually part of one of the sideband signals (depending on the system in use) is later removed in the vestigial sideband filter.

Because grid current flows during part of the vision signal cycle, the r.f. driving stage must have a low output impedance so that the r.f. input voltages will not vary as the vision signal varies. Further, since the grid current has components varying at sideband frequency, this output impedance must be low throughout the vision frequency band Fc−Fv to Fc+Fv. The required wide bandwidth and low impedance is obtained by using multiple-tuned circuits with damping resistors, to couple the driving amplifier to the modulated stage.

THE VISION MODULATOR. The function of the vision modulation amplifier is to raise the vision signal to the level required to drive the modulated stage. The frequency band occupied by the vision signal is wide, extending from a few Hertz to over 5 MHz in the case of 625-line transmissions, and the modulated stage presents a considerable, largely capacitative load to the modulation amplifier. The output impedance of the modulated stage must, therefore, be low and shunt-regulated amplifiers or cathode follower stages are often used for this purpose.

The standard input level is one volt; the output voltage depends on the power level at which the modulation takes place and is generally in the range of 150 to 400.

In addition to amplifying the signal, the modulation amplifier must pre-distort it to compensate for amplitude and phase distortions in the modulated stage and in subsequent r.f. amplifying stages. The pre-distortion required where colour signals are to be handled must be particularly precise, and ten or more stages of correcting circuits are commonly provided.

COMBINING UNITS. Before the modulated vision carrier signal from the transmitter is fed to the aerial, it is combined with the modulated sound carrier signal in the combining filter or diplexer. The circuit of the combining filter is arranged so that the signal from each transmitter is fed to the aerial feeder without allowing any of it to reach the other transmitter, where cross-modulation might take place. There are various designs of combining filters, but the majority use a bridge circuit, which may also include frequency-selective filters. Often the harmonic filters, and any filters needed to attenuate the vestigial sideband, are incorporated in the combining filter unit.

POWER SUPPLIES. The overall efficiency of a conventional television transmitter using valve stages is about 25 per cent. Although a proportion of the total power is consumed as alternating current – for instance, in the air-blowers and valve filaments – the direct current required for valve anode supplies, etc., is considerable. Critical stages may need a supply stabilized against

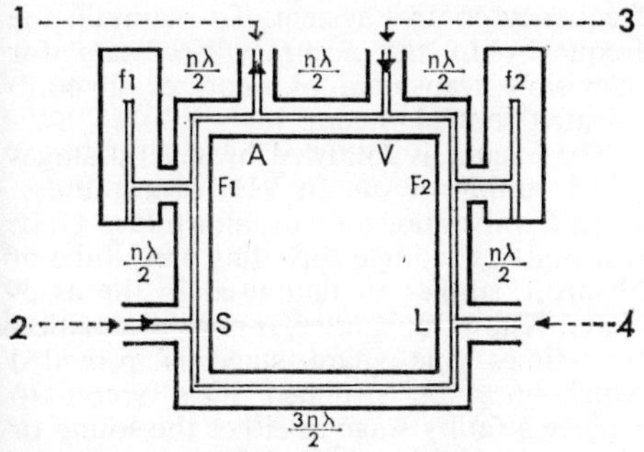

MAXWELL BRIDGE COMBINING UNIT (DIPLEXER). Comprises a ring of coaxial feeder, A, V, F_2, L, S, F_1, linked to frequency-selective filters, f_1 and f_2, which place a low impedance across the ring at video frequency. F_1A and F_2V are half-wavelength lines which transform this low impedance to a high impedance at A and V, so that vision power, entering the bridge at 3, flows directly to the aerial at 1. At sound signal frequency, the filters present a high impedance at F_1 and F_2. Sound power, 2, entering the ring at S, divides and flows in both directions. At L, connection to the load, 4, and at V, components arrive out of phase, and cancel, so that no power is fed into the load or back to the vision transmitter. At A the components arrive in phase and add, so that all the power flows into the aerial. Similarly, reflections back from aerial flow into load, 4.

any mains voltage and load current variations.

Recent designs of transmitter use silicon diodes in the rectifying circuits but, in older transmitters, gas-filled valve rectifiers are often connected in three-phase bridge circuits. In either case, meters to read the current taken by individual stages, and protective devices in the form of fuses and overload circuit breakers are required.

Generally, silicon diode rectifiers need circuits to protect them against voltage surges, but they are smaller, more efficient, and have a longer life than valve rectifiers.

MONITORING. A cathode ray tube waveform monitor is required to check the level of the radiated carrier-frequency signal at the transmitter output. It is usually arranged to display the envelope of the carrier signal, and is often built into the transmitter. In addition, monitor points are provided to feed a mobile oscilloscope which can be used to examine the waveforms at important points in the circuit. One of these is often a small probe in the final aerial feeder.

UHF VISION TRANSMITTERS

Because of the very much higher frequencies employed, UHF vision transmitters differ from VHF vision transmitters. First, at UHF, there is a choice of one of three families of valves for the r.f. amplification stages: triodes or tetrodes similar to those used at VHF or, in addition, travelling wave tubes, or klystrons. The power gain attainable with travelling wave tubes and klystrons is much larger than that possible with triodes and tetrodes but, on the other hand, they are more complicated and more expensive, and the overall electrical efficiency possible is much lower.

Although authorities differ on the best choice, successful transmitters have been built using each of these types of valves. The trend is to use klystrons for large and remotely controlled transmitters of 10 kW and greater output and travelling wave tubes for the smaller transmitters of 1 kW or less.

The UHF transmitter which uses triode valves differs from its VHF counterpart principally in that extra frequency multiplication stages are necessary between the crystal oscillator and the first r.f. amplifying stage and also in the tuned circuits in the power stages. At VHF, either balanced (lecher) line circuits or coaxial line circuits are possible, while at UHF pot or cavity resonators are necessary.

Since klystron and travelling wave r.f. amplifiers have a much larger stage gain than triode or tetrode amplifiers, a klystron amplifier may require 5 watts or less of r.f. drive to produce 25 kW r.f. output signal. The output required from the modulated stage is, therefore, quite small, and the use of absorption modulators becomes possible. The absorption modulator is connected between the output of the exciter and the input to the klystron and, in operation, absorbs or diverts part of the output power of the exciter and prevents it reaching the klystron. When a diode (absorption) modulator is used in conjunction with a circulator, an elegant and economical design of modulated stage results. As with grid modulation, pre-distortion of the vision signal in the modulation amplifier is necessary.

SOUND TRANSMITTERS

Two different types of sound transmission are in use for television. The most common uses a frequency-modulated (f.m.) signal, but amplitude-modulated (a.m.) signals are used in France and also in Britain (for the 405-line service). An f.m. signal is transmitted at constant level and the carrier frequency of the signal is varied fractionally in sympathy with the sound signal. An a.m. signal is transmitted at constant frequency, but its level is varied in sympathy with the sound signal. The transmitters required for these two types of transmission differ considerably.

AMPLITUDE-MODULATED SOUND TRANSMITTERS. These usually employ anode modulation of the r.f. output stage. Often the transmitter is constructed in a similar form to the associated vision transmitter and uses a similar exciter and similar Class C r.f. amplifying stages as far as the modulated stage, where the r.f. and amplified sound signals are mixed.

The sound signal power required is somewhat more than half the average r.f. power, and several amplification stages are needed in the sound modulation amplifier. To achieve good efficiency, the later stages are often operated in the Class B push-pull mode and the signal is fed to the modulated stage using a modulation transformer or inductor.

As in the case of the vision transmitter, power supply, protective metering and monitoring circuits are required to complete the transmitter.

FREQUENCY-MODULATED SOUND TRANSMITTER. Since, in the case of frequency modulation, the sound carrier signal varies in frequency, the exciter unit used is quite different from that used in vision transmitters and amplitude-modulated sound transmitters. There are two principal methods for frequency-modulating the carrier signal. For the first, a quartz crystal oscillator is used, as in the case of exciters in a.m. transmitters, but the circuit is so arranged that the application of the sound modulation signal varies the phase of the r.f. signal about its mean value. Several stages of frequency multiplication follow until this small variation has been increased to frequency deviations of ±50 kHz or 75 kHz.

The disadvantage of this arrangement is that the range of linear modulation is small, and a considerable number of harmonic generators may be necessary to achieve the required frequency deviation. It may then be necessary to subtract from the signal a second signal from a constant-frequency crystal oscillator to produce the carrier signal frequency required. Even then it is difficult or impossible to obtain a wide enough deviation (modulation range) to accommodate most types of stereo sound system.

In the second common system, the modulating voltage varies the frequency of a reactance tube oscillator. The mean frequency of this oscillator must, however, be held stable by means of an automatic frequency control system. To maintain the frequency to the accuracy necessary for television transmissions requires a complicated control system.

The exciter is followed by several stages of r.f. amplification. In VHF transmitters, Class C valve stages are usual, while in UHF transmitters a single travelling wave tube or klystron, similar to that used in the associated vision transmitter, is normal practice. Sometimes one spare stage is provided which may be switched into circuit to replace a faulty stage in either the sound or vision carrier signal amplifiers, and in the case of some low-power transmitters, a common output stage is used for both signals.

In order to avoid cross-modulation between the sound and the vision carrier signals, the sum of these must be kept well below the normal working level of the stage and, even then, the use of a common output stage presents problems for amplitude modulated sound signals, which are more sensitive to cross-modulation.

INPUT EQUIPMENT

TRANSMITTER INPUT EQUIPMENT. In most cases, the transmitting installation is some distance from the studio, and the signals reach the transmitter via programme links followed by various items of equipment referred to as programme input equipment. Normally, amplifying, switching and level-measuring equipment is provided at the transmitting station for both vision and sound signals.

REBROADCAST TRANSMITTERS. Rebroadcast transmitters receive the signal broadcast from a parent transmitter, amplify it and rebroadcast it on a different channel. The change to a new channel is necessary to avoid interference with the signals from the parent transmitter. Rebroadcast installations often use translator equipments in which the received signal is amplified, the frequency is changed, and the signal is again amplified and rebroadcast, without the actual sound and vision signals ever being produced.

Recent translator equipments employ entirely solid-state circuits to deliver from 5 to 50 watts of r.f. power. Where more power is required, valve (or, for UHF services, travelling wave or klystron) amplifiers are used.

TRANSMITTING INSTALLATION MAIN POWER SUPPLIES. The overall efficiency of a transmitting installation is usually rather low, lying between 5 and 25 per cent. The total power requirements of a high-power television station are therefore considerable. To protect the installation against failure of power, two separate sources of supply are often provided. The primary supply is usually obtained from the electricity supply authority at medium voltage (6 kV or 11 kV), and transformers and associated switchgear must be provided to convert this to the standard mains voltage. Where a second source of supply is provided, this may also come from the electricity supply authorities using a separate feeder or, alternatively, a diesel alternator installation may be provided. In either case, change-over switchgear, sometimes automatic in operation, is required.

The supply from the transformers or change-over switchgear is taken to one or more main switchboards from which it is distributed to the transmitters and auxiliary equipment.

Much of the technical equipment is sensitive to mains supply voltage variations, and automatic voltage regulators are often fitted in the feed from the main switchboard to keep the equipment supply voltage constant.

AERIALS

The effective radiated power of a transmitting installation depends on the transmitter power and on the aerial (antenna) gain. The most economic arrangement is usually a high gain aerial, and arrays of up to 100 ft. or more in height are in use. The complete aerial consists of one or more vertical arrays of radiating elements round a supporting structure.

The disposition of the elements and the method of feeding radio-frequency power to them must be arranged so that the required horizontal and vertical radiation patterns are obtained.

DIRECTIONAL CHARACTERISTICS. Until a few years ago, most aerials were designed for a uniform omni-directional radiation pattern, but more recently many directional aerials have been brought into use to serve irregular areas or because of the need to avoid interference with transmitters using the same frequency in other areas.

The signal field strength received at a point remote from the aerial array is the sum of the signals received from each element of the aerial. The vertical directional pattern (polar diagram) of a high-gain aerial is therefore determined by the contributions from a large number of separate radiating elements.

It is possible, for instance, as in the case of the superturnstile aerial, to arrange a ring of elements which, when they are fed with signals in phase and at the same power level, give a horizontal radiation pattern which consists of a narrow horizontal circular main lobe and a number of subsidiary lobes with sharp minima between them. In its simplest form, such an arrangement is unsatisfactory, especially if the aerial is supported on a high mast, because most of the power is radiated out to the horizon, while viewers situated near the aerial, in the direction of one of the minima, receive a poor signal.

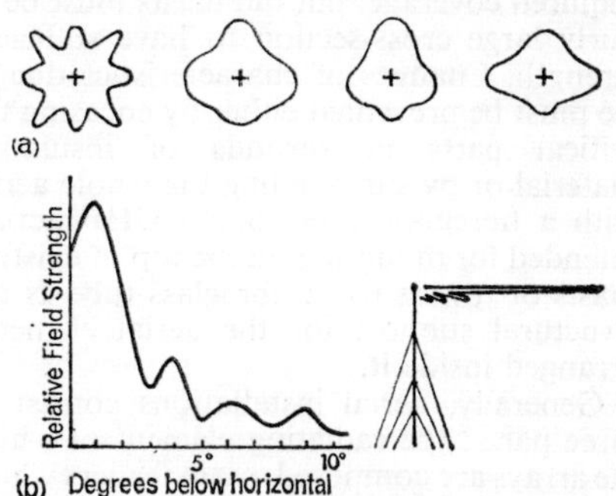

AERIAL SIGNAL DISTRIBUTION. (a) Horizontal distribution of different aerials shown by polar diagrams in which the field strength in any direction is measured by distance of point on curve from centre of plot marked by cross. (b) Two methods of displaying the vertical radiation pattern of the aerial. (*Left*) Graphically. (*Right*) Polar diagram showing relative strength of lower lobes.

To overcome these difficulties, the whole pattern is tilted downwards by about 1° so that the maximum of the main beam is directed at the edge of the service area. The beam tilt is usually achieved by altering the phase of the signals fed to the upper and lower parts of the aerial. It is also possible to mount the whole array with a downward tilt. The sharp minima in the pattern can also be reduced by varying the power or the phase of the signal current fed to the component parts of the array.

The superturnstile aerial illustrates how variations of the horizontal polar diagram can be achieved by merely varying the phase of the signals feeding the dipole elements of which it is composed. Similar variations occur if the position of the radiating elements is varied, so producing variations in phase of the signal at the receiving aerial.

Practical aerials often use a combination of phase, position, and signal amplitude to achieve the desired characteristics. When deciding on the exact arrangement to be adopted, the designer must consider other factors in addition to the radiation pattern, at carrier frequency itself. Among these are variations in input impedance and in the radiation pattern with frequency. Normally, the impedance must be kept within 5 per cent of the nominal value, and the radiation pattern within 3 dB throughout the band for which the aerial is intended.

The effect of the supporting structure and of icing may also be important. Design is easiest when a slender structure is used, allowing a small number of elements to be arranged round it close together to give the required coverage, but tall masts must be of fairly large cross-section to have sufficient strength. Changes of characteristics due to ice must be prevented either by encasing the critical parts in shrouds of insulating material or by surrounding the whole aerial with a fibreglass tube. Some UHF aerials intended for mounting on the top of existing masts or towers use a fibreglass tube as the structural support for the aerial elements arranged inside it.

Generally, aerial installations consist of three parts; the radiating element of which the arrays are composed, main feeders which conduct the radio-frequency power from the transmitters to the top of the mast, and distribution feeders which divide the power and feed it to the aerial elements.

RADIATING ELEMENTS. Transmitting aerials comprise many forms of radiating elements, but the majority fall into the following categories of resonant elements:

1. Dipole elements which are a half wavelength long. Variations include folded dipoles, batwing dipoles and slot aerials. Dipole elements are usually centre-fed, using coaxial feeders, although other arrangements are also used.
2. One-wavelength aerials, sometimes called full-wave dipoles, used in various ways and often end-fed from twin-line or coaxial feeders. Also other feeding arrangements.
3. Long resonant aerials. Various resonant aerials, two or more wavelengths long, e.g. zig-zag elements used in the mesney aerial. Such aerials are often fed with twin-line feeders.

TRAVELLING WAVE AERIALS. Many quite different aerials use the principle of the travelling wave or dissipative line in which the signal is gradually attenuated (owing to radiation resistance) as it travels from the input of the aerial, so that little, if any, remains to be reflected from the far end. Helical and some slot aerials work on this principle.

DISTRIBUTION FEEDERS. The distribution feeders transmit the power from the main feeders (which run up the mast or tower) to the aerial elements. Each element must receive the correct power at the correct phase angle. Matched, unbalanced (coaxial) or twin-line feeder lines are most often used for this purpose. The power from the main feeders is divided in one or more distribution or power-dividing networks which incorporate elements of selected characteristic impedance to divide the power between the various distribution feeders while at the same time maintaining the correct impedance at the output of the main feeders. When large arrays have to be fed, a branching arrangement is often adopted, the power being split into, say, 4, then 16, and finally 64 parts.

Some aerials, such as the quadrant aerial, use a resonant twin-line feeder system in the final stage of power distribution, while in some kinds of travelling-wave aerials the distribution feeder itself is, in effect, a continuous radiating element.

MAIN FEEDERS. Main feeders conduct the r.f. power from the transmitters to the aerial, and may be of considerable length. A design of feeder capable of carrying high power with very little loss is therefore necessary. The feeder is subject to weather and temperature variations, and the design of the feeder and supports must allow for this. Usually an air-spaced, or semi-air spaced, coaxial feeder is used, in which an inner conductor is supported by discs or a spiral of low-loss insulating material, at the centre of the outer tubular conductor.

Traditionally, feeders were made of brass or copper tube in sections of, say, 10 ft. in length, joined by spring plug joints in the inner conductor and bolted flange joints in the outer conductor. At intervals of 100 to 200 ft. a joint is provided to take up differences in expansion of the mast and the feeder. The joints between the feeder sections and the expansion joints are subject to wear and need regular maintenance.

FLEXIBLE FEEDERS. Recently, long continuous runs of flexible or semi-flexible feeder have been developed. One popular design of flexible feeder uses a corrugated copper or copper-plated steel tube for both inner and outer conductors. The inner conductor is supported inside the outer conductor by a multi-layer spiral of polyethylene tapes. The whole assembly is continuously manufactured from strip in a machine designed exclusively for the purpose.

To achieve the desired characteristic impedance throughout the length of the feeder, tight tolerances of all dimensions, but particularly of concentricity of the inner and outer conductors, are necessary. A variation of characteristic impedance of less than 2 per cent from the nominal value (50Ω or 60 Ω) is usually required. Semi-flexible feeders are manufactured according to a similar principle but use an extruded aluminium alloy outer conductor.

WAVEGUIDES AND SPECIAL PURPOSE FEEDERS. At ultra-high frequencies, the use of waveguide feeders is possible but, even in Band V, the dimensions of the waveguide are large and, on a mast, can add considerably to the wind resistance. Although waveguides have a large power-handling capacity, they are more expensive than coaxial feeders. The use of waveguides is therefore at present confined mostly to high-power installations.

New forms of low-cost waveguides, similar in external appearance to large flexible coaxial feeders, are now becoming available, and these may prove more attractive than the rigid rectangular waveguides of the past.

In addition to coaxial feeders and rectangular and circular waveguides, various other types of twin-line and surface wave feeders have been employed from time to time, especially where television signals have to be radiated from a mast used also as a medium wave radiator.

AERIAL MAINTENANCE ARRANGEMENTS. In order to provide a measure of protection against aerial and feeder breakdowns, it is not unusual, especially in Britain, to divide the aerial into an upper and lower half, and feed these from separate main feeders. In this way, it is possible for maintenance work to be undertaken on one half of the aerial or on one feeder while the other half is still in use.

In installations using two pairs of transmitters in parallel, it is then possible to connect each pair of transmitters to its own main feeder, but it is more usual to combine the output of the two pairs of transmitters by a diplexing network, and then to divide it in the desired ratio. In this way, unwanted variations of the aerial radiation characteristics, due to variations in the power and phase of the output of individual transmitters, are avoided. A.J.

See: *Television principles; Transmission and modulation; Transmitters and aerials.*

Book: *Radio and Television Engineers' Reference Book*, Eds. J. P. Hawker and W. E. Pannett, 4th Edn. (London), 1963.

⊕**Transparency.** Still picture used for projection as a background in the photography of process shots in cinematography.

□**Transponder.** A transmitter-receiver, often used in active satellites, which transmits signals automatically but only when interrogated. The term is sometimes wrongly used to mean a device such as a translator for the reception and re-transmission of a television signal.

See: *Satellite communications.*

□**Transverse Scan.** Type of videotape recorder in which the tape movement is along a straight line as in a domestic recorder, but in contrast to helical scan. The recording heads (of which there are four in the normal quadruplex type) are mounted on a small drum revolving at very high speed. The axis of the drum is parallel to the length of the tape, so that the heads sweep across it, the tape being curved or bowed to the exact radius of the drum by application of vacuum. Owing to the forward motion of the tape, the video tracks are slightly slanted, and the heads are so arranged that there is some overlap of information at the edges. This edge information is subsequently erased and replaced by longitudinal tracks recording sound and often cue information.

See: *Helical scan; Videotape recording; Videotape recording (colour).*

□**Trap Circuit.** Circuit presenting a very high (or very low) impedance to an unwanted frequency and rejecting (or absorbing) it.

See: *Communications; Sound recording* (TV).

□**Trap Valve Amplifier.** Multi-output sound-isolating amplifier providing a means of obtaining a number of separate outputs from a common input signal.

The term trap valve indicates the prime function, i.e., to isolate an output from unwanted signals which might otherwise be fed into the amplifier from the other output loads. It also serves as a buffer between the output loads and the common input signal source.

The separate outputs may be derived from separate valve (tube) stages with a common input. When this is done, very high output isolation is possible. Alternatively, a single amplifier can be designed with a low output impedance. Separate series resistors then feed the separate outputs and provide the required output impedance. In addition, the series resistors also provide attenuation to unwanted signals fed back into the amplifier. The amplifier stage output impedance must be kept very low to provide this attenuation to unwanted signals, and maintain the required degree of output isolation.

See: *Communications; Sound recording* (TV).

⊕**Travel Ghost.** Vertically overlapping projected image of bright objects produced when the projector shutter is incorrectly synchronized with the intermittent, so that the film is starting to move, or has not quite ceased moving, when a frame is projected on the screen. An analogous effect is produced when camera shutters are out of synchronism.

See: *Shutter; Stroboscopic effects.*

⊕**Travelling Matte Process.** Combination of moving action photographed in the studio with background scenes separately recorded. This allows actors to be shown against scenes from distant locations or in settings which it would be impossibly expensive to build in the studio. In all travelling matte systems the foreground is photographed in such a way that it not only records the action but also allows the preparation on strips of film of a pair of complementary silhouettes, one of which shows foreground objects as opaque areas on a clear background, while the other shows them as clear areas against an opaque background. When printing the dupe negative to combine the components, the matte with the opaque foreground is used to reserve the appropriate area when printing from the background master and the complementary matte is used when printing from the foreground master.

See: *Special effects* (FILM).

□**Travelling Wave Amplifier.** Microwave amplifier using a travelling wave tube. The tube has a wire helix, along the axis of which travels an electron beam. When a microwave signal is fed in at the cathode end of the tube, its speed of travel along the tube is delayed by the helix so as to be about the same as that of the electrons in the beam. In

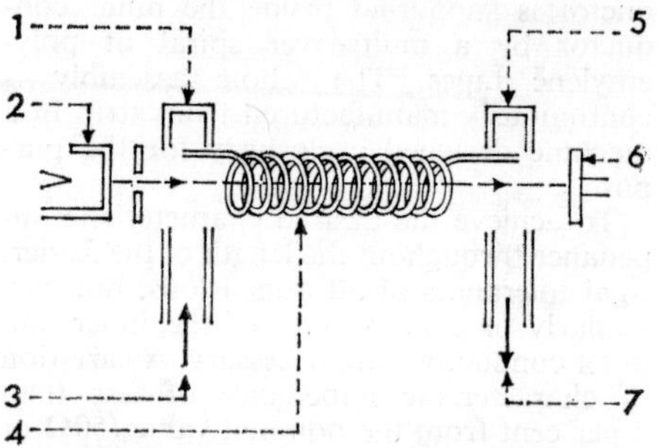

TRAVELLING WAVE TUBE. Electron gun, shown schematically with cathode, 2, focuses electron beam through helix, 4. Microwave input energy, 3, joins electron beam in input transducer, 1, and its speed is slowed by magnetic field in helix so that it draws energy from the electron beam. Beam circuit is completed by collector, 6, and amplified output, 7, is drawn off from output transducer, 5.

these circumstances, the signal can draw energy from the beam and so an amplified signal can be taken from the anode end of the tube. In association with a special aerial, a broad bandwidth can be secured by travelling wave amplification.

See: *Microwave links; Transmitters and aerials.*

⊕**Triacetate Base.** Film base of low inflammability and slow burning characteristics consisting essentially of cellulose triacetate, used for the manufacture of safety film.

See: *Film dimensions and physical characteristics; Film storage.*

□**Triad.** Minute group of three colour-sensitive phosphors repeated many thousands of times to form the colour-reproducing surface of a shadow mask television display tube.

See: *Receiver; Shadow mask tube.*

⊕**Trial Print.** First composite print from an assembled reel of picture negative with its corresponding sound track.

See: *Laboratory organization.*

⊕**Triangle.** Triangular or three-pointed star-shaped device which receives the three legs

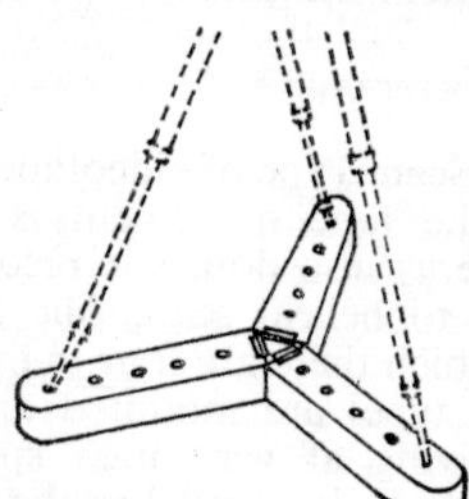

of a tripod to prevent them slipping apart. Can be folded for easy carrying.

⊕□**Trichromatic Analysis.** An analysis based on the three-colour (red, green, blue) theory. First, the original scene is analysed to give trivariant colour information for every part of it falling within the field of view required for the picture. This information is used to create suitable colours in each part of the picture in the chosen medium, whether cinema screen or television.

The trichromatic analysis of the original scene is made by using three kinds of light-sensitive receptors, each responding to about one-third of the spectrum, but usually with some overlapping. The receptors may be photographic emulsions covering an image of the field of view, or photoelectric cells scanning it. The field may be covered completely three times over or it may be divided up into a three-component mosaic pattern too small to be resolved by the eye at the usual viewing distance.

See: *Camera (colour); Colour principles; Telecine (colour); Trinoscope.*

□**Trichromatic Camera.** Camera using the three-colour principle (i.e., employing no separate luminance tube).

See: *Camera (colour).*

□**Trigger Pulse.** Pulse which initiates the operation of a circuit or device which then carries out its specific function and awaits another trigger pulse before repeating the process. Used, for example, in circuits producing a single line scan.

See: *Receiver; Sync pulse generator.*

⊕**Trim.** Sections of shots which are left over when the wanted parts of the shots have been incorporated in the workprint of a film. Trims must be carefully classified and put away as, in the progress of working from a rough cut towards a fine cut, they are often needed for incorporation in the film. Also called an out.

See: *Editor.*

□**Trinoscope.** Colour receiver display tube arrangement in which three separate monochrome projection picture tubes (kinescopes) are fed respectively with R, G and B output signals. R, G and B colour filters or mirrors in the path of the three tubes effect an additive combination of the three colour images in proper register. Trinoscopes have applications in colour telerecording, but are not used in domestic receivers, and for large-screen television have in recent years given ground to Eidophor projectors.

See: *Telerecording; Trichromatic analysis.*

⊕**Tripack.** Type of film having three layers of sensitive emulsion coated on a single support for the recording or reproduction of colour images.

See: *Colour cinematography; Integral tripack.*

⊕□**Tripod.** Three-legged stand for camera support, on which a rotating head is mounted for panning and tilting.

See: *Camera; Crane; Dolly; Velocilator.*

⊕TROPICAL CINEMATOGRAPHY

There are two distinct types of climate in the tropics, and each presents its own characteristic hazards to the cinematographer. In hot-and-wet areas, the unfavourable conditions make normal photographic manipulations more exacting. They frequently sap the energy and the enterprise of the traveller, and in addition they promote an insidious mould growth, which can appear both on equipment and on sensitive materials.

Hot-and-dry areas (such as deserts), on the other hand, are often invigorating, unless the temperature is excessive; untoward effects on both equipment and materials are found, but the major difficulty against which precautions must be taken is the presence of dust or sand. This ubiquitous hazard can ruin expensive equipment in a short time, and the photographer must always be on his guard against it.

EQUIPMENT

CAMERA CHOICE. A camera for use in the tropics should be chosen with great care. Reliability is the most important feature, because, in the absence of a competent mechanic, even a trivial fault could put the instrument out of action for many weeks. The camera should be made by a manufacturer of high repute and should be robust and generally free from awkward projections; gadgets and the more complex refinements are likely to find little use under tropical conditions.

Ease of handling, lightness in weight, and optical excellence are factors whose importance must be determined by the individual. The amateur film-maker living in Singapore clearly has very different demands from those of a nomadic desert explorer. It is of great value to have available lenses of various focal lengths, but personal choice

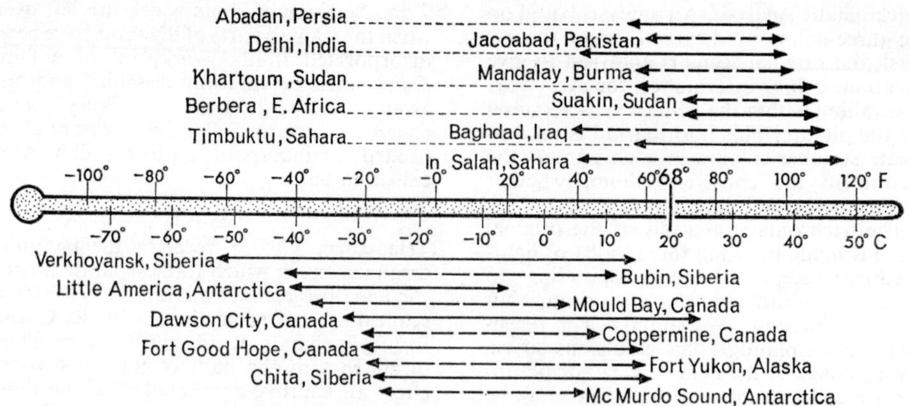

TEMPERATURE EXTREMES. Range of temperatures in typical hot and cold regions shows that difference from normal 20°C is much greater in cold than in hot regions.

will have to decide between the undoubted convenience of the zoom lens, and the somewhat superior definition likely to be given by separate lenses in a turret head.

The advantages of some measure of automation in exposure adjustment are real but, if serious work is being undertaken, the additional complexity of the mechanism (and its consequent liability to break down) must be considered. The high-contrast lighting conditions likely to be encountered can mislead an exposure meter; on balance, full automation is probably better avoided.

The film gauge is dictated by the ultimate purpose of the photographer, as well as by his experience; if intended for the cinema or television showing, 16mm or 35mm film must be used at 24 fps (to enable a sound track to be added later). Negative material (almost certainly colour stock) is then necessary. The less ambitious photographer can make do with the single copy obtained from reversal stock.

If it is impracticable to purchase a camera specially for tropical use, the existing instrument should receive a thorough professional overhaul before departure, to ensure that even minor faults are discovered and corrected in good time. Familiarity with the mode of use of all equipment must be assumed; activities in the tropics frequently impose an unusual physical strain on the operator and in such conditions the manipulation of a strange instrument becomes particularly difficult.

ACCESSORY CHOICE. The usual leather or plastic case is normally inadequate; additional precautions need to be taken if extensive travelling is involved, and the provision of strong wooden containers for all the photographic equipment is advisable. Such containers should be partitioned and padded to fit each item, so that individual components will not roll about in transit.

In dry desert regions, rigid precautions have to be taken to ensure that all the equipment is adequately protected from dust or sand, and individual plastic bags or boxes should be provided for each item. The camera demands particular attention; the use of two canvas or plastic bags over the instrument itself is recommended, and the carrying case should, in addition, have a closely-fitting canvas case to act as a final protection. Such elaborate precautions will prove their worth if the operator is unfortunate enough to be caught in a sandstorm.

In wet or humid regions, there is a tendency for leather or plastic materials to become covered with a mould growth when the humidity rises above 60 per cent RH, and since regular cleaning then becomes necessary, the storage boxes should be designed so that all areas are accessible. It is also wise to obtain a supply of silica gel, and to sew quantities of, perhaps, an ounce at a time into small canvas bags.

Silica gel is an excellent drying agent, which can reduce the humidity under which the equipment is stored so that mould formation is prevented. When the drying agent is saturated, its efficacy can easily be restored by warming it to about 300°F, for two hours.

The usual advice to use a tripod to keep the camera steady is even more important in the tropics, where excessive heat or physical exhaustion may cause the operator to give unwanted movement to a hand-held camera.

A lens hood, a photo-electric meter, and a UV filter (for colour work) will probably be regarded as essential. The normal UV filter found adequate in temperate zones should be replaced by one with a higher UV absorption.

A neutral density filter, to reduce the light intensity to a more convenient level, can often be of great value. The depth of field of cine lenses is so great that at the small apertures normally associated with intense tropical lighting, differential focusing techniques are impracticable, and backgrounds may be distractingly sharp.

EQUIPMENT MAINTENANCE. Before departure, all equipment should be checked and overhauled, professionally if possible. Since the conditions encountered are more severe than those met in more temperate regions, the precautions taken against them must be more rigorous, and regular inspection and maintenance of all equipment is essential. The frequency of such inspection will be dictated by the severity of the conditions prevailing, and the importance of ensuring satisfactory results, but a check at least once a fortnight is advisable.

This regular inspection should reveal any tendency towards mould growth on appropriate parts of the equipment; any such growth should be wiped off carefully with a clean cloth, preferably impregnated with a fungicide such as sodium pentachlorphenate. Particular attention should be paid to the use, and the renewal, of the silica gel drying agent already mentioned.

In dry areas, careful examination for the presence of dust or grit should be made regularly for, in spite of all precautions, the equipment will have to be exposed to some measure of hazard during actual use. A small vacuum cleaner is ideal for this type of inspection, but in its absence a small brush can be used to dislodge any foreign material. The type of brush which contains a rubber bulb, and thus acts as a miniature blower, can be of real value here.

Regular maintenance should naturally include an inspection of the mechanical state of all the equipment, and all the aspects which would receive attention in temperate zones should be watched with even greater care in the tropics. Minor faults may be detected and rectified before they have a chance to become important; the heavy expense and delay liable to be caused by the need to send the instrument away for repair should underline the wisdom of a regular personal overhaul of all equipment.

FILM STORAGE AND USE

The most serious single problem awaiting the cinematographer in the tropics is the liability of damage to his films by the heat, both before and after exposure. The first and most obvious precaution is to make sure that the films are as fresh as possible when purchased. Elaborate arrangements can be made with the manufacturers to have regular supplies flown out, if the importance of the work in hand justifies this, but most photographers will have to be content with the normal commercial channels. It is wise to select a dealer who has a rapid turnover, so that a given delivery will not lie on his shelves for many months. In addition, preference should naturally be given to the dealer who provides some form of cooled storage for his stock.

STORAGE. After purchase, every precaution should be taken to keep the films cool; if a refrigerator is available, it should be used – though there is no need to freeze the film, since temperatures around 40–50°F are quite suitable. In the absence of an orthodox refrigerator, cruder devices can be made effective; the evaporation of water, long employed in unglazed porcelain butter-coolers, can be used to keep a film store at a temperature substantially lower than that of the surroundings. A little ingenuity will enable a box (perhaps 1 cu. ft.) to be constructed; if the films are placed in this, and hessian soaked in water used to cover it, satisfactory cool storage can be attained, particularly if the box can be kept in a draught. Frequent renewal of the cooling water will be required, but there is no need to use precious drinking water, as brackish pond water will suffice.

In the absence of any such device, a substantial reduction in temperature can be obtained by burying the films some 12 in. in the ground. This expedient can be of great value in desert expeditions, and it is a wise precaution to adopt for even an overnight stay; the precise location of the buried films must be carefully marked, or they may be lost for ever.

After exposure, even greater care should be taken to ensure that the emulsions are kept as cool as possible, and that the delay before processing is brief. This warning is all the more important with colour films, since untoward conditions can lead to a colour bias, as well as to some decrease in speed. If necessary, arrangements should be made to fly the exposed material promptly to a processing station, and the operator should obtain a report on the technical quality of the first few reels, if only to give him confidence before he exposes the greater part of his stock. In addition to all the above precautions, there is the obvious need to keep all films (and indeed all equipment) out of the direct rays of the sun; it is very easy to forget this.

If processed materials are stored in hot, humid conditions, fungal decomposition of the gelatin occurs, and also some fading of the dyes in colour films. Fungal action may be retarded by keeping the films in tightly closed containers, in which a sachet of silica gel has been inserted; the resulting dry atmosphere discourages the growth of the destructive organisms. Each film should have its own separate container and supply of silica gel. The use of a fungicidal treatment, at the end of the processing sequence, is sometimes recommended, but it is not always satisfactory; cetyl tetrammonium bromide tends to soften the gelatin unduly, and zinc fluosilicate is highly poisonous – though both are effective in preventing fungal growth. Sodium pentachlorphenate may be the best choice.

There is no certain cure for the slow deterioration of the dyes in colour film; but its effect can be diminished by keeping the films as cool as possible; a thorough rewashing (preferably using a hypo elimination technique) ensures that action by residual processing chemicals is avoided.

If films are to be stored in tropical conditions for long periods, a regular routine inspection should be made, a representative sample being systematically chosen, examined and projected, say, every three months. The onset of deterioration may then be detected before much damage has been caused.

EXPOSURE. The high intensity of the light, and the fact that the sun may be almost vertically overhead, present real problems in tropical cinematography, especially if reversal films are being used. Subject contrast tends to be high, and dull-black uninteresting shadows are difficult to avoid. Black eye-sockets in portraiture can arise from such overhead lighting, and they are very unattractive. Only experience can reveal how best to overcome these disadvantages. Diffuse reflectors should be used whenever

TROPICAL PROCESSING

Temperature	*68 to 75°F 20 to 24°C*	*75 to 85°F 24 to 29°C*	*85 to 95°F 29 to 35°C*
Developer			
Formula	Kodak D.76	Kodak D.76	Kodak D.76
Sulphated	No	10%	10%
Time	10 to 8 min.	12 to 8½ min.	8½ to 6½ min.
Stopbath	3% acetic acid	3% acetic acid 5% sulphated	3% chrome alum 5% sulphated
Time	1 min.	1 min.	1 min.

A "sulphated" developer has had a quantity of sodium sulphate added, to suppress gelatin swelling; "10 per cent sulphated" means that 100 g. of anhydrous sodium sulphate have been dissolved, before the final volume is made up to 1 litre. If *hydrated* sodium sulphate (Epsom Salts) is used, 225 g. are required in place of 100 g.

The development times quoted are given as a range, corresponding approximately to the temperature range indicated; they are based on an assumed 10 min. at 68°F as standard. The reduction in activity through the addition of sodium sulphate results in extending the times for sulphated developers.

Fixer	Normal	Twice normal hardener	
Washing	15 mins. running water	15 mins. running water	6–8 2-min. rinses in static water

Fixer: Normal fixer with hardener should be used, and sulphating is not necessary. Above 80°F, it is advantageous to double the proportion of hardening agent normally used, to increase hardening efficiency. This doubling decreases the effective life of the solution, as the tendency to deposit sulphur on standing is greatly increased; with some formulae, the rate of deterioration is so high that doubling the hardener proves impracticable.

The fixing time must be determined by experiment; it should be between 2 and 3 times the clearing time under the conditions used.

Washing: The immersion time in water should be kept to a minimum. For exposure determination, negative permanence is not required, and these conditions may be found excessive; if permanence of the negatives is essential, a hypo elimination technique should be used.

Drying: Accelerated drying in a cabinet may prove hazardous, as the soft gelatin may be softened still further; a draught of air blown over the film will dry it readily, if the humidity is low, but the air employed must be carefully filtered from dust and insects. If the film is simply left to dry without any assistance, the time needed may prove unduly long, and again precautions against both dust and insects will have to be taken.

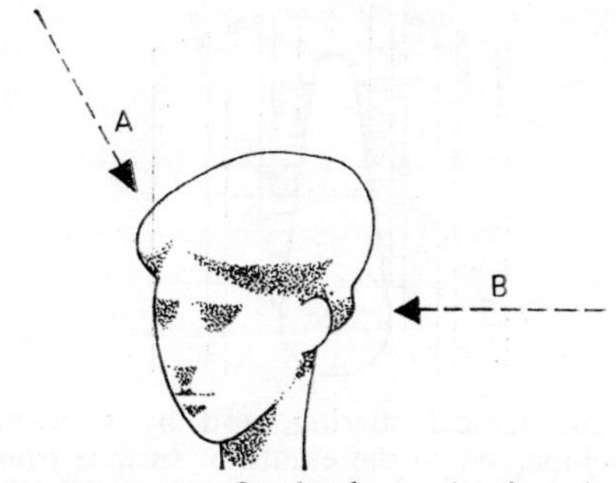

SHADOW EFFECTS. Overhead sun, A, gives rise to deep shadows in eye sockets and elsewhere, which can be lessened by reflected or auxiliary light directed as at B.

possible to provide fill-in light. Any diffuse shade which may be available should be employed, and it is well to avoid the middle of the day, when the sun is at its highest. The use of a reflector to scatter some light into the shadows can be a great boon; a white linen sheet makes a useful and easily portable reflector, and gives an adequately diffused beam. A more specular device, such as a polished metal sheet, would prove embarrassingly bright to the subject.

Skill in determining the exposure level can similarly be attained solely through experience; it is a common fault to overestimate the brilliance of the lighting, and thus to underexpose. The difficulties caused by excessive contrast should be studied carefully, and some experiments to determine the optimal exposure conditions will be amply rewarded later. For instance, a close-up portrait will demand more exposure than a full-length figure because of the need to reproduce more faithfully shadow details in the features of the first shot; it is probably well to avoid as far as possible the composite scenes, in which recognizable personnel and architectural or natural features are both major sources of interest.

PROCESSING. Bulk processing of cine films in tropical conditions is better avoided, but it is sometimes advisable to develop a few feet, if only to ensure that the exposure levels employed are correct. The modifications to the normal treatment referred to in the table should therefore be taken; they refer exclusively to black-and-white film, as colour processing in such conditions is too difficult to control and should be left alone. The sequence relates to the production of a negative; if reversal stock is being used, it is wise to avoid processing, and instead – purely for exposure determination purposes – to produce trial negatives only. If this unorthodox procedure is adopted, the operator should previously have performed a series of experiments, under temperate conditions, to correlate the characteristics of negatives obtained in this way with those of the correctly exposed and processed film.

Temperatures are normally at their lowest just before dawn, and processing is best postponed until then; the personal inconvenience is counterbalanced by the employment of relatively cool solutions, thereby reducing the hazards of tropical development. D.H.O.J.

See: *Cold climate cinematography.*

Books: *Photography on Expeditions*, by D. H. O. John (London), 1965; *Camera in the Tropics*, by G. C. Dodwell (London), 1960.

□**Tropospheric Propagation.** Propagation of TV signals by sky wave through the troposphere or weather-forming region approximately one-half to 10 miles above the earth's surface, instead of close to the surface by ground wave. At distances of more than 50 miles from a transmitter, the sky wave becomes of increasing importance because of the marked attenuation of the ground wave. Temperature and pressure variations in the troposphere cause the refractive index of the atmosphere to alter, so that high-frequency waves are reflected back to earth, causing fading and interference.

⊕**T-Stop.** System of lens calibration which makes allowance for the varying transmittance of different lenses. These stops are so defined that, if a lens has 100 per cent transmittance, its T-number is the same as its *f*-number.

See: *Exposure control.*

⊕**TTL.** Acronym for through-the-lens, denoting an exposure meter device which operates behind the camera lens and therefore receives light from the same area as does the film. Often coupled with the diaphragm setting lever in semi-automatic and fully automatic cameras, now common in 8mm and coming into use in 16mm, to provide rapid adjustment to changing light conditions. Sometimes used to denote through-the-lens viewfinding.

See: *Narrow gauge film and equipment.*

☐**Tuner.** RF and frequency changer (or mixer and oscillator) stages of a receiver, i.e., those circuits which are tuned to receive a particular signal.

See: *Receiver.*

⊕**Tungsten Halogen Lamp.** Development of ☐the normal tungsten filament lamp, in which a small quantity of one of the halogens, usually iodine or bromine, is introduced into the evacuated bulb.

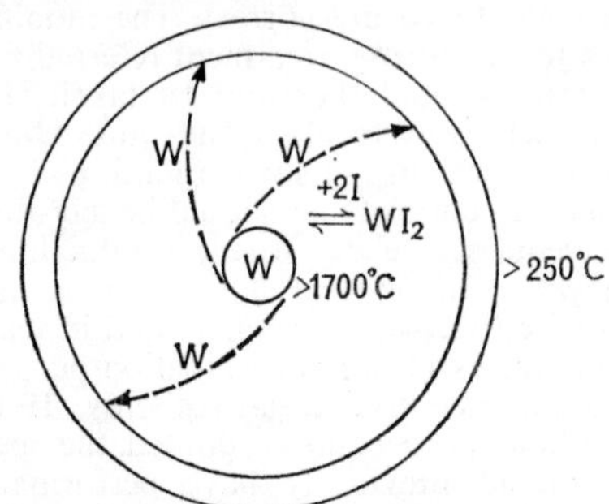

HALIDE CYCLE. Free halogen, here shown as iodine, combines with tungsten (W) atoms evaporated on bulb wall, forming gaseous tungsten iodide (WI_2). This redeposits tungsten on filament at centre, freeing iodine to continue cycling. Temperatures shown are minimum for operating cycle continuously.

In a normal tungsten lamp, tungsten atoms are evaporated from the filament, thinning it and reducing its life, and are deposited on the inner surface of the bulb, causing undesirable blackening. In the tungsten-halogen lamp, the halogen vapour combines with the tungsten atoms on the bulb wall (provided its temperature is 250°C or more) to form tungsten halide vapour, and this diffuses back to the filament where it decomposes at about 1,700°C, redepositing the tungsten, while the halogen atoms are set free to repeat the regenerative cycle. As a result, the filament may safely be worked at a higher temperature, giving greater luminous efficiency as well as longer effective life.

See: *Lighting equipment.*

⊕**Tungsten Lamp.** Incandescent lamp (as ☐distinguished from arc lamp) with a tungsten filament in an evacuated or gas-filled bulb.

⊕**Turbulation.** In continuous film processing, a means of carrying away oxidation products from the neighbourhood of the emulsion surface, where pockets of exhausted developer may give rise to unwanted streaks. Turbulation, also called agitation, is often effected by compressed air or nitrogen jets,

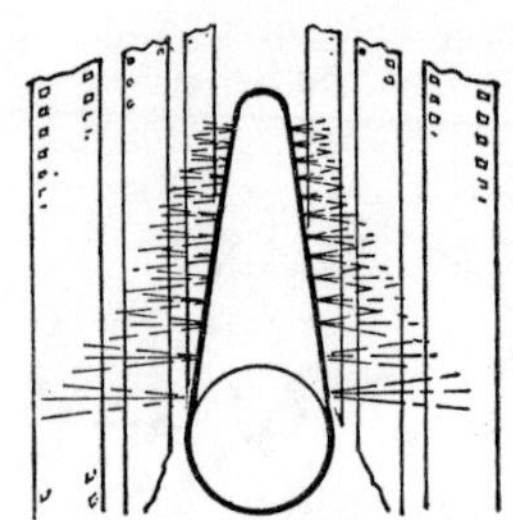

by mechanical stirring, and by spraying developer on to the emulsion surface from submerged jets.

See: *Nitrogen burst agitation; Processing; Spray processing.*

☐**Turnstile Aerial.** A particular form of television transmitting aerial.

The essential requirements of a television transmitter aerial (except in special locations) are that the radiation pattern should be as omnidirectional as possible in the horizontal plane, and very directional in the vertical plane. This is necessary if it is intended to give maximum coverage in all directions around the aerial array.

When a simple dipole is used in a horizontal plane, the pattern of the signal strength around the aerial, or the polar diagram as it is normally called, looks like a figure of eight. If a second dipole is introduced at right-angles to the original, but still in a horizontal plane, it is possible by suitable phasing of the two radiated signals to give a diagram which is approximately circular. The response will give the desired type of coverage in the horizontal plane. The two crossed dipoles are known as a turnstile aerial. Stacking of a number of turnstiles in a vertical direction increases the signal radiated horizontally and reduces that radiated in the vertical direction.

An alternative type of transmission aerial in common use is a slot aerial. The radiation pattern from a slot aerial is similar to that of a normal dipole. If two slot aerials are crossed in a similar manner to the dipoles, a similar change to the polar diagram results. The two slot aerials when crossed are known as a batswing or super turnstile aerial.

These aerials also are usually stacked in order to reduce vertical radiation and increase horizontal radiation.

See: *Transmitters and aerials.*

⊕**Turret (Lens).** Revolving mount attached to ☐the front of a camera carrying three or more lenses, enabling them to be swung into position in turn in front of the photographing aperture. Lens turrets may be actuated electrically from the back of a camera, and may be connected to remotely controlled

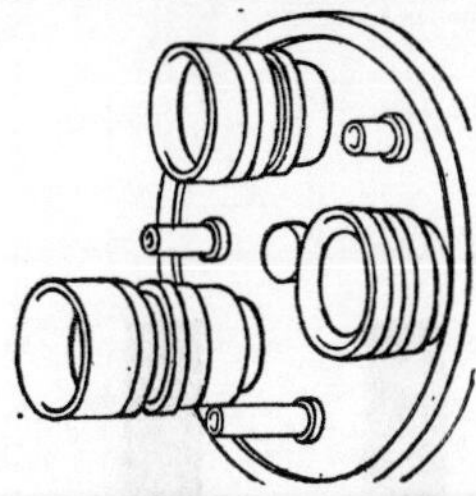

devices for focusing each lens and controlling the position of the iris diaphragm.
See: *Camera.*

□**Turret Tuner.** Tuner in which the coil assemblies are carried in a rotating turret bringing the required set into contact with the rest of the active circuit.
See: *Receiver.*

⊕**Turtle.** Turtle-shaped base on which studio lamps may be mounted to bring them close to the studio floor.

□**Twin-Interlaced Scanning.** Process by which adjacent lines in the scanned image belong to alternate fields, i.e., line 1, line 3, line 5, etc., form part of the odd field, and lines 2, 4 and 6 form part of the even field. Compared with a sequential system of the same picture frequency and number of lines, a twin-interlace system requires only half the transmitted bandwidth, but it also conveys only half the picture information. Its purpose is to double the effective picture rate and thus raise the flicker frequency from an unacceptable 25 or 30 Hz to an acceptable figure of 50 or 60 Hz.
See: *Interlace; Raster; Scanning; Television principles.*

□**Twin Lens Scanner.** In a telecine (film scanner), two separate lenses may be used to focus two separate images of the scanning tube on to separate places. This is done to achieve double scanning of a frame of film.
See: *Telecine.*

□**Twister.** Combined spotlight and soft light in the same housing and using the same bulb. Movement of the bulb beyond the normal focus travel away from the Fresnel lens causes an auxiliary reflector to direct the light on to a diffusing reflector to form a soft light. The object is to have only one type of lamp or luminaire in a television studio and thus considerably reduce rigging time.
See: *Lighting equipment* (TV); *Qwart.*

⊕**Two-Colour Processes.** Systems of colour reproduction in which the visible spectrum is divided into the blue-green and orange-red regions for recording and presentation. Although extensively used in early colour film processes, the inherent inability of two components to reproduce a satisfactory range of hues rendered all such systems obsolete when three-colour processes became readily available.
See: *Colour cinematography.*

□**Two-Field Picture.** One complete frame of a television picture is made up of two fields, normally called odd and even. These two fields are interlaced to form a complete picture.
See: *Scanning; Television principles.*

□**Two-Three-Two-Three Scanning.** Method of displaying a 24 fps film on a 60-field television system (the North American standard).

Successive frames of the film are exposed to two field scans and then three field scans respectively. Thus a second's worth of film contains 12 frames scanned twice, and 12 frames scanned three times, making 24 plus 36 or 60 fields.
See: *Telecine.*

U

Uchatius, Franz Freiherr von, 1814–1861. Austrian officer and inventor who in 1853 described two devices developed from the Phenakistiscope, for projecting moving pictures.

⊕**Ultra Semi-Scope.** Japanese system of wide-screen motion picture presentation using 35mm with a frame height of two perforations for both photography and projection. As the frame is only half the normal 35mm frame in height, negative and print costs are reduced, and since the image area has an aspect ratio of 2·35 : 1, no anamorphic optics are necessary. However, both cameras and projectors must be modified to handle the half-size frame, and the sound track must be specially recorded for reproduction at a speed of 45 ft. per minute.

See: *Wide-screen processes.*

⊕**Ultrasonic Cleaning.** Method of cleaning film release prints by passing them through a cleaning fluid which is subjected to agitation at ultrasonic frequencies. This effectively dislodges small particles of dirt embedded in the film.

See: *Release print examination and maintenance.*

□**Ultrasonic Delay Lines.** Lines of glass or steel a few inches long in which an acoustic wave can be delayed by the time interval of the line of a television signal. Used in SECAM and PAL colour TV systems and in vertical aperture correctors, drop-out compensators, etc., where information in one picture line must be delayed in order to compare it with the next line.

See: *Colour systems; Standards conversion; Videotape recording (colour).*

⊕□**Ultra-Violet Radiation.** That part of the electro-magnetic spectrum having wavelengths from approximately 4,000 Angstroms to a few hundred Angstroms. The longest ultra-violet waves have wavelengths just shorter than those of visible violet light. Radiation from the sun is rich in these wavelengths, and they are also produced by arc lamps and mercury vapour lamps. They affect a photographic emulsion.

⊕**Under-Exposure.** Exposure of a sensitive photographic material in a camera or printer in such a way that the intensity of light and/or the time of exposure are insufficient to ensure satisfactory rendering of the tonal range of the subject. With negative and reversal materials, under-exposure means inadequate rendering of shadow detail or even complete failure to record shadow gradation of the original scene. An under-exposed print from a normal negative, on the other hand, appears

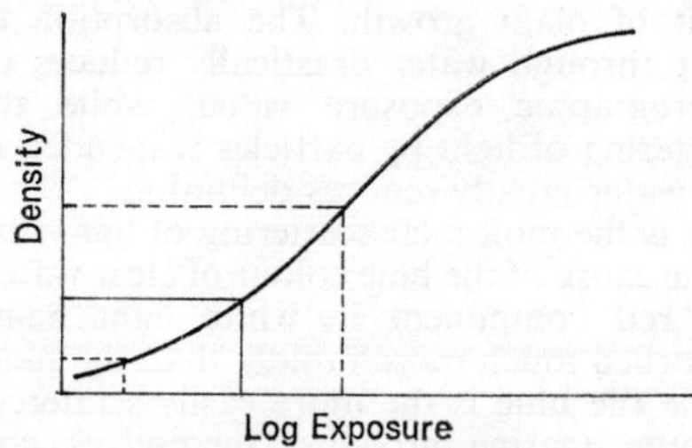

UNDER-EXPOSURE. Mean exposure is indicated by solid-line intercept. Shadow exposure (*dotted, left*) then falls well into the region, while highlight exposure (*dotted, right*) fails to make full use of "straight" portion of characteristic curve.

much too light overall and the highlight areas of the scene show little or no gradation.

In sensitometric terms, under-exposure means that too great a part of the tonal range of the subject falls in the toe region of the characteristic curve of the photographic material. In colour work, under-exposure can give rise to seriously distorted colour rendering as well as unsatisfactory tonal reproduction.

See: *Exposure control.*

❑**Undershoot.** A form of distortion of the television signal. It is possible to so modify

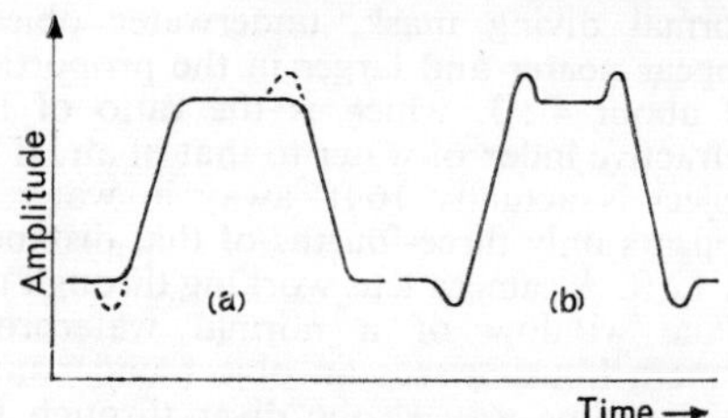

UNDERSHOOT. (a) Correctly formed pulse shown as solid line. When undershoot is present, leading edge of pulse (*left*) starts with a dimple (*dotted*), and trailing edge (*right*) with a pip. (b) When undershoot is introduced to correct for overshoot, a symmetrical waveform results.

the phase response of a video circuit that a positive input pulse waveform is reproduced with undershoot.

Phase equalizers designed to correct the pulse response of video amplifiers having overshoot not only have the property of reducing the amplitude of the overshoot but in addition introduce undershoot. When the applied equalization is optimum, the degree of undershoot introduced is equal in amplitude to the residual amount of overshoot which remains after correction. The reproduced pulse then has a symmetrical form.

⊕UNDERWATER CINEMATOGRAPHY

The term underwater photography is used to cover a multitude of ways and means of producing photographs of objects under water. The depth conditions alone can range from a picture taken in an aquarium to a shot from a bathyscaphe at the bottom of the Mariana Trench in the Pacific Ocean at a depth of 36,000 ft. – nearly seven miles down.

In film photography for entertainment purposes, every technique is acceptable as long as the final illusion is realized on the screen. Feature film cameramen are required to be artistic and imaginative, but photographically speaking they must be cheats and cunning illusionists. In most feature films which include undersea photography, studio tank shots are often intercut with the real thing, and in some cases complete undersea film sequences are shot under the controlled conditions of the studio tank. A completely truthful approach is confined to cameramen who undertake documentary films dealing with sea exploration, diving, salvage and marine research.

UNDERWATER CONDITIONS

EFFECT ON DISTANCE. When the human eye is used underwater, the rays of light come to focus behind the retina, as in farsightedness, and only blurred images can be seen. A diving mask enables the eyes to work in air, their natural element. Seen through a

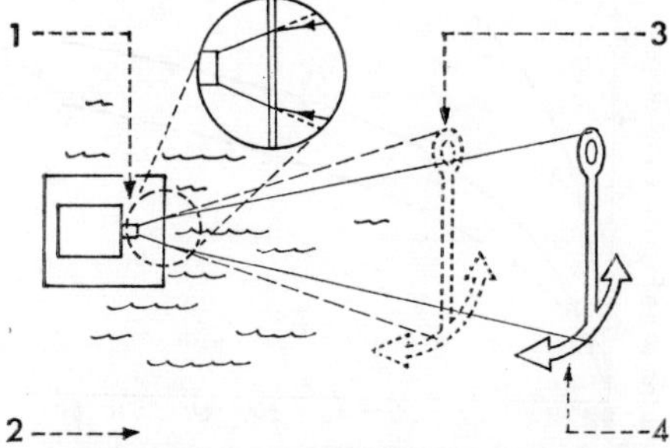

FOCUSING UNDER WATER. Refraction causes objects under water to appear nearer than they really are; focus must be set to three-quarters of real distance. 1. Camera surrounded by air in enclosure submerged in water, 2. Real object, 4, is seen at 3.

normal diving mask, underwater objects appear nearer and larger in the proportion of about 4 : 3, which is the ratio of the refractive index of water to that of air. If an object is actually 16 ft. away in water, it appears only three-fourths of that distance, or 12 ft. A camera lens working through the plane window of a normal waterproof camera housing sees an object underwater in the same way as the diver through his mask, and the camera focus must therefore be set on the apparent distance, or at three-quarters of the real measured distance.

Optical correction systems to maintain the full angular field of the lens can be incorporated in the camera housing, in which case correction lenses or dioptres replace the plane glass or perspex window in front of the lens. Mounted in a watertight manner with the front surface in contact with the water, they become part of the fish-eye lens combination. With this arrangement, the image is the same size as if the object were seen in air, and the focus has to be set to the real distance, while the depth of field obtained remains unchanged. However, these submarine lens attachments, superior though they may be in some ways, are not as yet in great demand as their expense, size and weight are all factors which militate against their use.

VISIBILITY. Visibility underwater, even under the best conditions, is rarely more than 100 ft., comparable to visibility through a heavy luminous mist on dry land. Light penetrating water is absorbed and scattered in its passage, the intensity rapidly decreasing with increased depth.

Light that can be detected by the human eye rarely penetrates to a depth much greater than 300 ft., which is just the lower limit of plant growth. The absorption of light through water drastically reduces its photographic exposure value, while the scattering of light by particles suspended in the water greatly reduces definition.

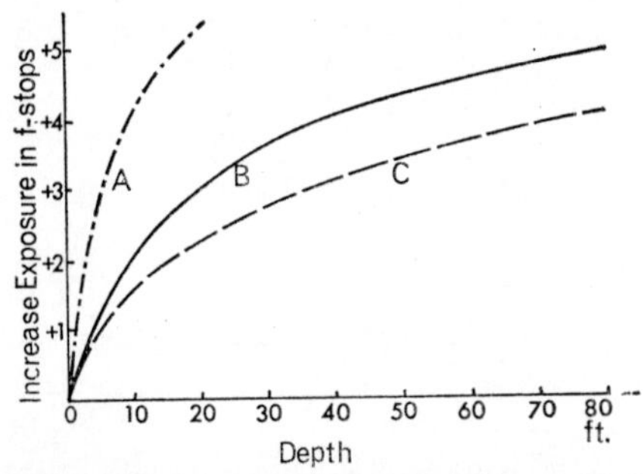

EXPOSURE UNDER WATER. Exposure using natural light increases with depth, very rapidly if the water is turbid, A, less so if it is clear, B, or exceptionally clear, C.

It is the molecular scattering of light that is the cause of the blue colour of clear water, the red component in white light being absorbed much more rapidly than the blue, while the blue is the more easily scattered. Minute marine growths, themselves coloured, can also act as filters to colour the transmitted light, and the reflection of the colour of the sky is another factor which changes the apparent colour under water. In combination with rapidly decreasing intensity of light, there is also a process of selective colour filtration with increased depth.

In black-and-white photography, this absorption has to be taken into account for exposure, and in colour photography it imposes serious restrictions. As little as 3 ft. below the surface, roughly half the red light has been filtered out, while at about 16 ft. practically all the red component has gone and with it about half of the orange. At over 30 ft. all the orange has gone and most of the yellow, while beyond 60 ft. the yellow has gone and most of the violet. At 100 ft. a blue-grey hue veils everything, and below 100 ft. it is a world of only black and grey without colour differentiation.

LIGHT AND COLOUR

To see and photograph all colours at all depths, there is no alternative to underwater lighting. Flash bulbs and electronic flash units may help in still photography, but in cine photography at depth a continuous source of light of some considerable power is needed. For very close work, self-contained battery-powered units are often used, but where large areas are to be filmed, underwater lamps suspended on long cables and powered from generators on the surface are usually employed. A wide range of equipment has been produced for this purpose.

FILTERS. Using a fast panchromatic black-and-white emulsion, filming with reasonably good contrast and clarity can in fact be undertaken without lights in clear water down as far as 200 ft., but in practice the maximum depth is often no more than 60 ft. At shallow depths, filters to increase tonal contrast between colours may be cautiously used, but the manufacturers' filter factors, which are given for normal daylight photography, can be used only as a basic indication underwater. The effective colour of underwater light must be taken into consideration, and the filter factor can change from a daylight value of ×3 to a factor of over ×10 under certain underwater conditions. It is often helpful to use the same filter over the light meter as is used in front of the camera lens to assist in calculating the exposure.

Filming in colour underwater without lights is very much more restricted than in black-and-white, and the overall predominance of blue-green must be accepted. The most widely used colour film in professional motion picture photography is now Eastman Colour negative; this colour-balances for a light temperature of 3,200°K, and the standard filter for photography in daylight is the Kodak Wratten No. 85. Although prints made from colour negative can be corrected for overall colour and density, it is often an advantage to correct the overall colour on the negative by the use of filters.

If underwater photography takes place at depths where all red light has been absorbed, there is no filter that can be used on the camera to restore this condition, but at shallower depths the colour correction red series filters (CCR) are often helpful in addition to the normal 85. Below 40 ft., the CCM series at times improves the chromatic rendering with Eastman Colour negative.

MAKE-UP. Make-up for underwater photography in colour is a specialized subject. The make-up must, of course, be waterproof, and because of the lack of red, and the softening of contrast underwater, make-up characteristics must be emphasized. A generous application of Pan-Cake with make-up retainer is used over the entire face and body, and then dry rouge is added to various parts of the body, forehead and skin, and blended into the foundation with a damp sponge. Lipstick is applied generously, and the make-up man emphasizes colour and contrast to the utmost to help the cameraman get a clear and effective picture.

EQUIPMENT

LENSES. As a general rule, wide-angle lenses are used as much as possible underwater to compensate for the apparent closeness of objects, and to allow there to be less water distance between the camera and the object being photographed.

CAMERA HOUSINGS. Divers are subjected to an additional pressure due to the weight of water above them, and for every 33 ft. of descent the pressure is increased by one atmosphere, or 14·7 lb. per sq. in. At 33 ft. the pressure corresponds to two atmospheres, or nearly 30 lb. per sq. in., and at 66 ft. three atmospheres, or nearly 45 lb. per sq. in. To remain waterproof, therefore, a camera housing must be pressure compensating, pressure resistant or internally pressurized, and all the controls must operate through watertight glands.

Small amateur cine cameras intended for use just below the surface may be enclosed in a simple plastic bag with a plain glass window, but for professional work elaborate housings, in some cases incorporating built-in aqualungs, are essential. The demand valves are built into the lowest part of the case and spring-loaded air vents are provided in the highest part of the lid so that a positive pressure is maintained with no possibility of water entering. Elaborations include camera cases that are operated by side handles with twist-grip focus controls, illuminated dials and diaphragm attachments. Whatever the weight of the complete housing in air, in water it has to be completely hydrostatic, i.e., balanced to equilibrium. It should also be as hydrodynamically shaped as possible, so that it can be swum for tracking shots at reasonable speeds in all directions. However, the most elaborate equipment is only a tool in the hands of the cameraman, who himself must be an experienced diver to get the best results in all circumstances. S.S.

See: *Camera* (FILM); *Tropical cinematography; Light; Optical principles.*

Book: *American Cinematographer Manual*, 2nd End., ASC (Hollywood), 1966.

□**Uniselector.** Telephone switch selector often used in television as part of line selection equipment in master control or elsewhere.

⊕**Unit.** Technicians and craftsmen appointed □to work as a team on the making of a film. The number varies according to the demands of the scenario.

⊕**Unmod. Track.** Unmodulated track, or sound track completely devoid of modulations. Used in editing a synchronized track to fill up the gaps between the modulated sections carrying dialogue, music or effects. Since magnetic heads are unaffected by optical density changes, photographic film of any kind whatever, provided it has the right gauge and perforations, may be used for unmodulated track.

See: *Editing; Sound recording* (FILM).

☐**Unrestored Receiver.** Television receiver with partial or no d.c. restoration of the exciting signal.

⊕**Unsqueezed Print.** Motion picture print for normal projection made from an anamorphically photographed negative by optical correction.

See: *Wide-screen processes.*

☐**Upwards Conversion.** Standards conversion to a higher (e.g., 405 to 625) line standard.

⊕**USA Standards Institute.** American organization responsible for establishing national standards, including those for film and television.

Formerly the American Standards Association (ASA).

See: *Standards.*

⊕**UV Printing.** Printing sound tracks on black-and-white film, using only the ultraviolet portion of the exposing light. These shorter wavelengths reduce the amount of scatter in the emulsion of the positive stock and provide a more clearly defined image.

See: *Printers; Intermodulation.*

□**Van Allen Belts.** Belts of high-energy radiation encircling the earth, important in satellite communications because they may damage electronic components.

⊕**Varamorph.** Film projector lens system in which the anamorphic lateral magnification factor can be varied by adjustment.
See: *Wide-screen processes.*

⊕**Variable-Area Recording.** Type of optical sound recording in which the sound track is divided laterally into opaque and trans-

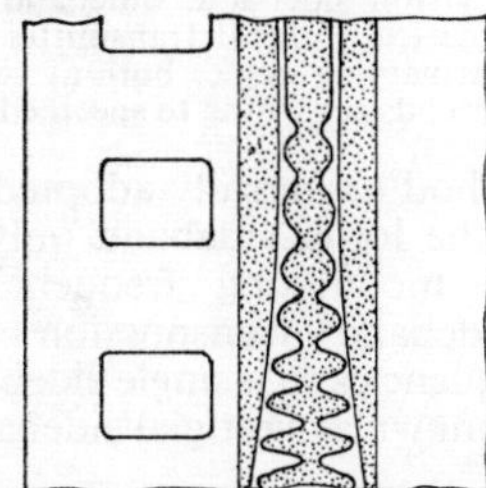

parent areas, the sharp demarcation line between them forming an oscillographic trace of the waveform of the recorded signal.
See: *Noise reduction; Sound recording* (FILM).

⊕**Variable-Density Recording.** Type of optical sound recording in which sound is recorded as a series of density graduations perpendicular to the edge of the sound track and extending across its full width. The distance between the gradations is determined by the

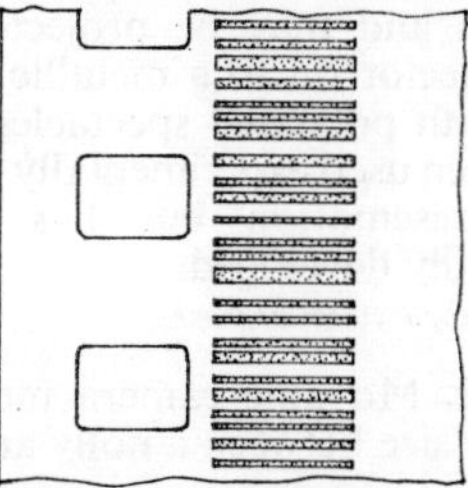

recorded frequency, while the amplitude of the signal governs their density. These gradations, which are recognizably separate from one another, are often called striations.
See: *Sound recording* (FILM).

⊕**Varley, Frederick H., 1842–1916.** London civil engineer, collaborator with W. Friese Greene. Designed and patented a stereoscopic camera/projector demonstrated in 1890 by Friese Greene, with whom he also patented a photometer in that year.

⊕**Vaultkeeper.** Operator in film studio or processing laboratory responsible for the storage of film and its associated reception, identification and movement. In the handling of both negatives and positive prints, a complete system of movement records and location cards is essential to ensure that every item called for by the production organization can be produced without delay from among the thousands of film cans which have to be stored. In laboratory work, vaultkeepers are also concerned with handling positive stock in large quantities from its delivery from the manufacturer up to its distribution to the printing departments.

⊕□**Vaults.** Storage places for the long-term and archival keeping of films and videotapes. Vaults are of fireproof construction, and are either provided with means of temperature control and humidification, or are so situated that these factors remain as constant as possible. For safety reasons, vaults are so constructed that each one contains no more than 1,000 to 2,000 cans of film or tape. When nitrate film has to be stored for archival purposes, much more stringent precautions have to be taken than for modern safety base.

See: *Film dimensions and physical characteristics; Film storage; Library.*

⊕**Vectograph.** Form of photographic image for stereoscopic viewing in which the two views corresponding to the right and left eyes are printed as polarized relief images on opposite surfaces of a transparent base. Both aspects are therefore contained on a single film and may be projected from a single projector on to a metallic screen for viewing with polarized spectacles. The system has been used experimentally for motion picture presentation but has not been commercially developed.

See: *Stereoscopic cinematography.*

⊕□**Velocilator.** Movable camera mount intermediate in size between a dolly and a crane.

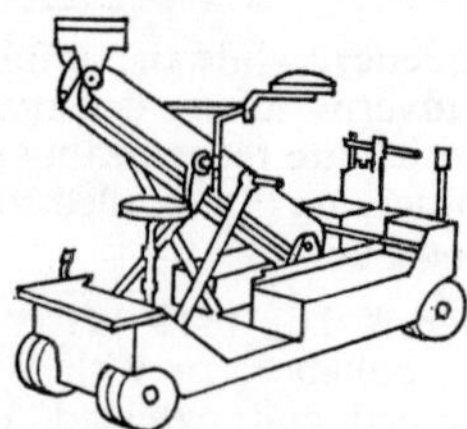

It carries a heavy camera to a height of 6 ft. but is not intended to be raised or lowered rapidly while the camera is running.

See: *Camera; Crane; Dolly.*

□**Velocity Control Servo.** Method of remotely controlling a mechanism, in which the velocity or speed of change and not the position is governed by the remote control.

□**Vertical Aperture Correction.** Correction in a vertical sense for the finite diameter of the scanning beam of the camera tube. Standard aperture correction corrects horizontally.

□**Vertical Bounce.** Picture instability in a vertical direction, giving the appearance of bouncing up and down.

□**Vestigial Sideband.** A preferred technique of shaping the transmitted signal to limit the bandwidth and produce a readily handled signal. The large bandwidth of television signals makes it very desirable to use some form of single sideband transmission to reduce the channel width below that necessary if both sidebands were to be used. The components of a television waveform extend from 50 Hz to nearly 6 MHz in 625-line transmissions. The lower frequency components preclude the use of single sideband transmission in its simplest form.

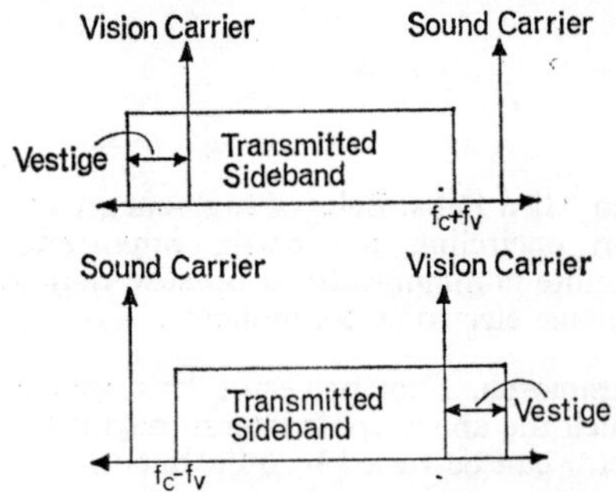

VESTIGIAL SIDEBAND. (*Top*) Upper-sideband transmission, with vestigial sideband on lower side of vision carrier. (*Bottom*) Lower-sideband transmission, with vestige above vision carrier. Sound carriers are respectively above and below the transmitted vision sideband. Sideband limits are shown schematically. Ideal transmitter characteristics slope at varying angles both at vestigial and sound carrier ends according to specified standards.

The method universally adopted removes a part of the lower sideband only, leaving the lower modulating frequencies as a double sideband transmission and the higher frequencies as a single sideband. This system is known as vestigial sideband transmission.

If the receiver had a normal bandpass response which covered the full range of transmission, the output after a linear detector would be twice as great for the lower frequencies as for the high. To overcome this effect, the receiver bandpass is adjusted. Over the sloping part of the response curve, the lower sideband contributes less to the output the further away it is from the carrier. Conversely, the upper sideband contributes more. As the carrier is at a level exactly half that of the flat portion of the curve, the combined output of the two sidebands is the same as that of the single sideband part of the transmission for the higher frequencies.

See: *Transmission and modulation; Transmitters and aerials.*

□**Video Repeater.** Amplifier which is inserted at a point in a video line to compensate for the attenuation in the line.

□VIDEOTAPE EDITING

As a picture recording medium, videotape now ranks equal with film, though its cost per unit time may be higher when equipment costs are included, and it is tied to a particular set of television line and frequency standards, thus making international diffusion more difficult than with film. The great convenience of videotape is its instant playback, coupled with the possibility of erasure and frequent re-use, some of the advantages which gave so striking a lead to magnetic sound recording over optical recording.

With videotape recording (VTR) thus in general demand, it became imperative to develop editing facilities as good as those which had long been available for film. There were certain inherent difficulties, not least that the magnetic video record is invisible, so that editing by eye, an important short cut in film, is in practice impossible. The studio machines on which videotape is recorded and reproduced cannot give a stationary image of one picture, and require special accessory equipment if they are to be modified to do so.

To overcome these drawbacks and give facility and speed, two radically different techniques have been developed, physical editing and electronic editing.

PHYSICAL EDITING

BASIC METHOD. In this, the earliest technique, videotape is cut and spliced together exactly as if it were film or magnetic sound track. The tape speed is $15\frac{5}{8}$ ips (15 ips in the US), so that one picture occupies $\frac{5}{8}$ in, of tape at the standard European 25 pps, and $\frac{1}{2}$ in. of tape at the American 30 pps. Film is viewed and marked on a standard transverse-scan videotape machine, to which a device may be added for making visible a stationary frame, a great help in determining the precise position for a cut. Stopping is not instantaneous, but occupies an accurately known time of about $\frac{1}{4}$ sec.

CUTTING SOUND. Sound is staggered by 14·8 frames, so that cutting of sound and picture together can only be accomplished in newsreel fashion by utilizing natural pauses in speech. For more sophisticated cutting techniques, the sound is transferred, often on to a simple $\frac{1}{4}$-in. tape recorder which will run synchronously for 30 seconds or so. For longer runs, pilot tone devices are used.

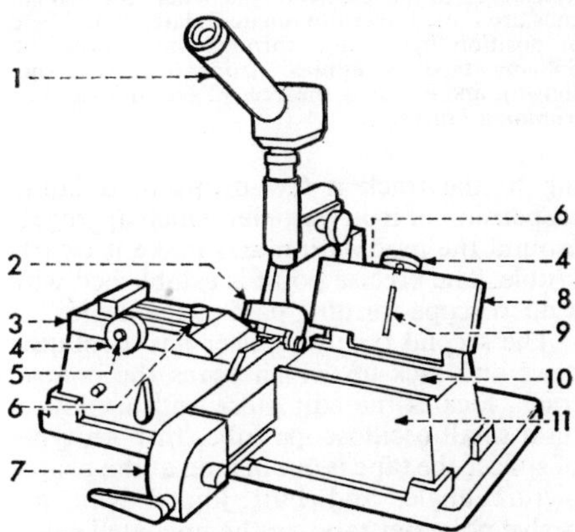

TAPE SPLICER WITH MICROSCOPE PULSE LOCATION. Tape splicer section is conventional and comprises a heavy base, 11, with left- and right-hand hinged plates, 3, 8, and lock knobs, 5. Tape is placed, coating upwards, in guide, 10. Edit pulse is identified by painting with powdered iron suspension and examining under microscope, 1, the pulse being precisely positioned under the cross-wires by adjusting wheels, 4, which rotate friction rollers, 9, and thus slide the tape slowly in the guides. When the pulse is located, the tape is locked with screws, 6, and guillotined with knife, 2. Splicing tape is dispensed to central area from magazine, 7, and applied to base side of tape before trimming edges.

SPLICERS. The method of cutting and splicing can best be understood in relation to the layout of the videotape with its three sound tracks, audio, cue and control. The control track carries an edit pulse which occurs once per picture and enables the pictures to be cut correctly in frame. This is the function of the tape splicer, of which there are two types. The first makes provision for revealing the edit pulse by apply-

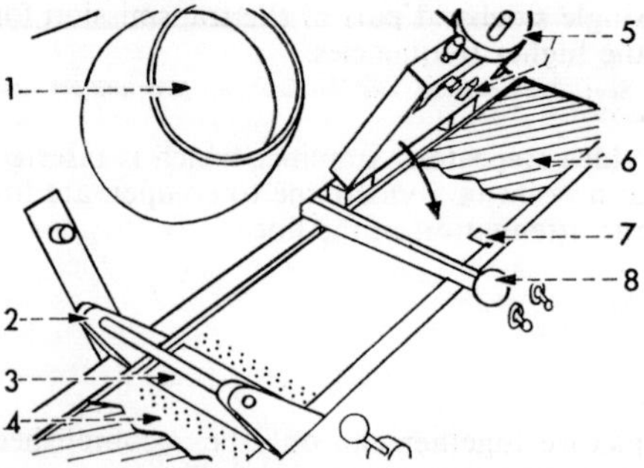

TAPE SPLICER WITH ELECTRONIC PULSE LOCATION. Tape is laid coating downwards on the bed of the instrument, 6, and clamped with the hinged plate over the rotating magnetic head, 7, precise positioning of the pulse being effected by turning handwheel and friction roller, 5, which slides tape slowly in the guide. The edit pulse is displayed on oscilloscope tube, 1. Tape is then guillotined, 8, and transferred to the left-hand side, where the two cut ends are butted over illuminated plate, 3, and held in position by suction through small holes, 4. Splicing tape is applied from dispenser (not shown), and excess is sheared off on both sides by trimming knives, 2.

ing to the track a few drops of a liquid suspension of iron particles which aggregate around the modulation and make it clearly visible. The precise point is established with a microscope forming part of the device.

The second type of splicer has a rotating head and pick-up which scans the control track, locates the edit pulse and displays it on a small oscilloscope tube. In both types of splicer the tape is cut in a jig at the proper picture angle, and butt joined with adhesive polyester tape on the uncoated side.

DISADVANTAGES. This technique of simple physical editing is still widely used in black-and-white and colour, and splice breaks are extremely rare. However, since the original tape is necessarily cut, it is usual to make protection copies in advance. Some television organizations are opposed to physical editing, not only because of loss of tape and extra wear on video heads, but because handling it in and around the splicer tends to produce slight creases and the working in of dust and other foreign matter, which can cause picture dropout in the neighbourhood of the splice. The consequent loss of quality can be reduced to acceptable limits by reducing handling to a minimum and by the use of electronic dropout compensators in the reproduce circuits of the VTR during transmission.

These drawbacks have been overcome by the newer techniques of electronic editing.

ELECTRONIC EDITING

BASIC METHOD. This process is akin to that of sound dubbing or re-recording, and consists of the transfer of a selected part of an unbroken tape record to a precise position in another unbroken tape record.

Suppose a live studio programme is being recorded. A sequence of action occurs which runs its course and ends. The tape is wound back and replayed. At the desired cutting point, normally some seconds before the VTR is stopped, a cue is inserted on the cue track by a control on the electronic editing unit. The videotape recorder is reversed from its stopping point to a point before the cue, a short overlap of action is arranged on the studio floor, and when the edit cue is reached, the electronic editor switches automatically to the studio output, erasing the remainder of the first record.

In this way a spliceless tape is gradually built up, always with the possibility of keying in inserts wherever necessary to improve the visual flow. These inserts need not originate on the studio floor, but can be keyed in and out by edit pulses from telecines (film scanners), caption scanners or other VTRs.

OUT-OF-STUDIO USE. The same arrangements may be used outside the studio with two or more VTRs as in a sound dubbing session; but because of the very high unit cost of the equipment, the number of VTRs is kept to a minimum. A switch setting on the electronic editor gives a rehearsal facility which enables a cut to be seen on the monitor without any actual transfer taking place; the edit cue simply shifts the video signal output from VTR 1 to VTR 2 or back again, and feeds it into the monitor. When the precise cutting point has been determined, another control makes the actual transfer, erasing any previous signal on the tape and recording picture only, sound only, or picture and sound.

Television production does not advance shot by shot, as does film, but moves in long sections in which the outputs of several cameras are combined in smooth continuity. Thus electronic editing is required only for joining sequences, or adding inserts, and is therefore different from film editing.

AUTOMATED EDITING. This makes it possible to envisage a degree of automation inconceivable in film, but already well on the way to achievement with videotape. The cue track can be used for recording time information, down to the individual picture, in digital form. The recorder can then be programmed to take a video signal from one source up to a predetermined time, then cut or dissolve to another source for a stated period, and then go on to a third. With rapid wind facilities forward and back, this saves valuable time in editing, for example, a political speech which is to be combined with library background shots on film.

TRANSFER TECHNIQUES. It has been assumed that the videotape recorder of machines used in electronic editing are of studio transversescan design, with output of full transmission quality.

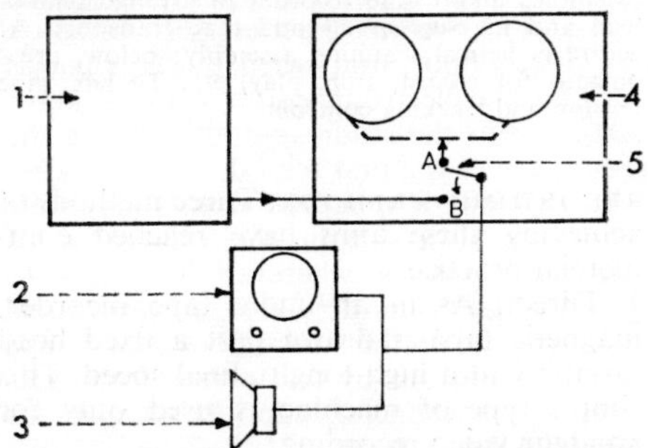

ELECTRONIC EDITING: REHEARSAL AND TRANSFER. Output of first VTR, 1, goes to switch, 5, on second VTR, 4. To rehearse, output of 4 passes to picture monitor, 2, and loudspeaker, 3, with switch in position A. When the cued edit point is reached, switch goes from A to B, thus transferring monitor to VTR, 1. If editing point is found satisfactory, machine is switched to transfer mode; when edit point is again reached, VTR, 4, starts recording, thus adding the new scene to previous scenes already on this tape. In this way, an edited reel is built up on VTR, 4, without any cutting of tape.

These machines are extremely expensive, however, and it is sometimes impossible to allocate them to editing. A convenient solution is to transfer the video signal to an editing copy on lower-quality helical-scan recorders, and to use these recorders for all the slow, detailed work of fixing cutting points for the various transitions. Once the digitally-timed cue marks have been established, the actual transfer, a comparatively rapid process, is effected on transverse-scan machines.

The same object can be achieved by a hybrid process. The original live programme is telerecorded on film as well as being recorded on videotape. The telerecording is then edited by ordinary film techniques, and any cine film inserts are cut in. Finally, the videotape recording is electronically edited to correspond with this telerecording. To produce satisfactory results, all telecines, telerecorders and VTRs must be locked to picture frequency.

COLOUR EDITING

With NTSC colour, where the waveform repeats itself every two fields (one interlaced picture), editing techniques are no different from those employed in monochrome, and a 30 Hz edit pulse is used (30 pps). With PAL and SECAM colour, a 4-field cycle is used, calling for a 12½ Hz edit pulse, instead of 25 Hz as in European black-and-white. Editing cannot, therefore, be quite so precise in its timing. A.P.W.M. & R.J.S.

See: *Editing; Videotape recording.*

Videotape Operator. Technician who operates and services the machine on which a large proportion of all studio output is recorded for editing and/or subsequent transmission. He may also help with editing and care of tapes. In Britain the functions of operation and maintenance are combined. Elsewhere they may be separated.

□VIDEOTAPE RECORDING

Videotape recording (VTR) is a direct alternative to film. Both employ cameras and recording media. But whereas, in a film camera, the optical image is focused directly on to the film, so that camera and film are part of one another, in VTR the camera and tape recording medium are separate. The camera is a television camera, black-and-white or colour as the case may be. The tape on which its image is recorded runs in a tape recorder remote from the camera.

VTR PRINCIPLE

The principle is the same as sound recording on tape. Whereas the audio tape recorder registers the sound spectrum converted into electrical impulses by a microphone and amplifier, the VTR registers the visual spectrum, converted into electrical impulses by the TV camera. The vastly greater range of frequencies the VTR must record, however, calls for complications in the method of recording far beyond what is needed for sound.

COMPARISON WITH OTHER METHODS. Since VTR is a method of recording images picked up by a television camera, it is in competition not only with direct filming, but with telerecording (kinescoping), i.e., the transfer of the TV image directly to film by means of a special camera. The two techniques have somewhat different fields of use, but there is a great deal of overlapping, and each has its advocates and exponents.

Direct filming and telerecording have the merit of producing a record on a universal medium, employable everywhere irrespective of local television transmission standards. On the other hand, they involve laboratory processes and consequent delays, and call into play photographic techniques not always readily available in an otherwise electronic environment.

Videotape recording can be instantly played back; its editing techniques are now highly advanced and do not involve cutting the tape; the tape itself can be erased and used many times.

On the whole, therefore, VTR is likely to make headway against film for studio re-recordings, although where permanence is important, or where recordings must be sent abroad for transmission, film still has many advantages. But the two techniques are becoming increasingly interchangeable, for transfer from videotape to film, in both black-and-white and colour, is now a commercial process.

RECORDER LAYOUT. The basic mechanical layout of a videotape recorder does not differ from that of an audio tape recorder, whether amateur or professional. It comprises a tape transport which draws magnetic tape at constant speed from a supply reel, across a set of magnetic head assemblies, and finally winds it up on a take-up reel.

But the recording of video signals makes two prime demands: the acceptance of a very wide frequency band ranging from near zero frequency (d.c.) to above 4 MHz (megacycles per second); and the achievement of an extremely high order of timing accuracy, so that in the record and reproduce processes there is exact synchronization of the position of the heads in relation to the tracks.

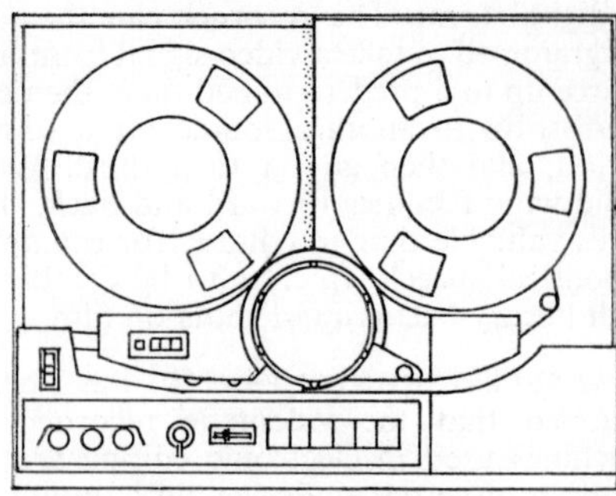

TYPICAL VIDEOTAPE RECORDER LAYOUT. Closely resembles audio tape recorder in arrangement of feed and take-up spools and tape transport. At centre is helical scanning assembly; below, press buttons for record, stop, play, etc. To left, tape tension and tracking controls.

RECORDING METHODS. Three methods of achieving these aims have reached commercial practice.

1. Direct. As in an audio tape recorder, magnetic tape is drawn past a fixed head assembly at a high longitudinal speed. This simple type of machine is used only for amateur video recording.

2. Helical scan. One or two rotating heads are mounted on a large diameter drum, which turns at a high speed. Around the drum the tape is wound helically, but it travels slower than the periphery of the drum. Because of the helical wind, the track pattern on the tape is at an angle to the edge of the tape. Thus, though the tape is transported at a relatively low speed, the head-to-tape speed is relatively high. This more sophisticated arrangement is employed in closed-circuit and educational VTRs.

3. Transverse scan. A small-diameter drum carrying four equally-spaced heads rotates at a very high speed on an axis parallel with the length of the tape, which is curved by a vacuum device to the shape of the drum. Thus, as the tape is drawn past the rotating assembly at a low speed, the heads in turn traverse it at a very high speed, and when one head passes away from the farther edge, the next head has already reached the nearer edge. Because of the longitudinal movement of the tape, there is a slight angle between tape and tracks.

Transverse scan recorders are employed universally in professional broadcasting organizations, where the demand is for the highest quality equipment. Only in the 1960s, when low-priced vidicon channels of good quality became available, were manufacturers able to apply the knowledge gained from these more advanced machines to the production of helical scan and direct recording machines of acceptable performance.

EARLY MODELS. The requirement for a tape recorder able to record and reproduce both picture and sound became apparent with the growth of the television broadcast industry after World War II. Large broadcasting organizations and private electronic manufacturing companies in the United States and Britain invested capital for the research necessary to produce such an instrument. In 1956 the Ampex Corporation, then a small Californian company which specialized in the manufacture of precision audio recorders, demonstrated to the Annual Convention of the National Association of Broadcasters in Chicago a transverse scan recorder with a 4·0 MHz video bandwidth suitable for recording and reproducing television signals. The machine had a continuous playing time of one hour and was accepted by broadcasters for use as a studio television recorder.

Two years later a large American equipment manufacturer (RCA) announced their transverse scan video recorder which was fully compatible with the Ampex machine, and thus the track patterns of these two machines became the standard of the broadcast industry for television tape recording.

MAGNETIC RECORDING

BASIC PRINCIPLES. When a magnetic material is placed in or near a magnetic field, the molecules of the material become oriented in accordance with the direction and density of that field. In magnetic tape recording, the material consists of microscopic iron oxide particles bonded to a plastic tape. On the core of a highly permeable material are wound a number of turns of wire to form the magnetizing device known as a head. During recording and reproduction, the gap in the core contacts the iron-oxide-coated tape, thus completing the magnetic circuit of the core through the tape surface.

If the current through the core winding is varied at a given rate and the tape is moved past the gap, a magnetic pattern, directly proportional to the coil current, is laid down on the tape as it leaves the gap. When the process is reversed, and the magnetized tape passes the gap with no coil current flowing, the magnetic circuit is completed again and a voltage, proportional to the rate of change of the magnetic flux at the gap, is induced in the coil.

If both the recording and reproducing processes are performed at the same constant speed, the voltage induced during reproduction will follow faithfully the current applied during recording. The wavelength of the signal recorded on the tape is a function of tape velocity and signal frequency as shown in the formula:

$$\lambda = V/f$$

where λ = tape wavelength, V = tape velocity and f = signal frequency.

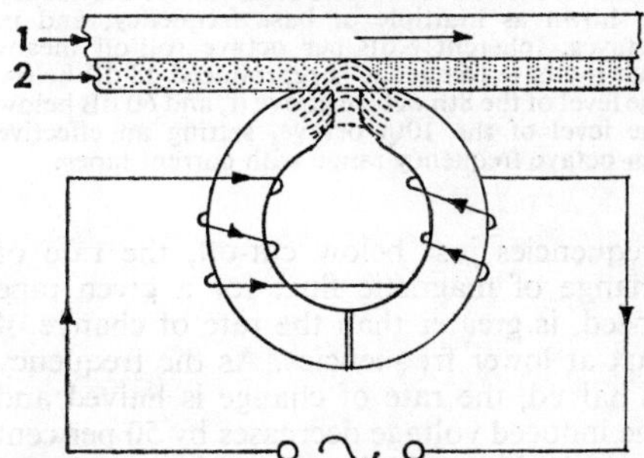

RING-TYPE RECORDING HEAD. Coil of wire to which audio or video signal is applied is wound on a ring core of magnetic material, broken by a very short gap, here greatly exaggerated for clarity. The tape with its coating, 2, on base, 1, contacts the head and thus closes the gap. Hence the magnetic field penetrates the coating, producing a magnetic pattern proportional to the coil current. Conversely, a recorded tape passed over the same head excites a voltage in the coil. If the tape speed remains the same, this voltage faithfully follows the waveform of the recording current.

From this formula it can be seen that, for a given speed, the wavelength becomes progressively shorter as the frequency is increased, and, conversely, as the tape speed is increased, the wavelength for any given frequency increases proportionally. The upper frequency limit of the recording head is determined only by the inductive reactance of the coil and its distributed capacitance. The frequency limitations of magnetic tape recording are thus due almost entirely to reproduce head limitations.

FREQUENCY LIMITATIONS. The voltage induced in the core coil during the reproduce operation is proportional to the rate of change of the magnetic flux at the gap. When the tape wavelength equals the width of the gap, the rate of change becomes zero, resulting in no induced voltage. Frequencies above this cut-off frequency cannot be reproduced dependably. In addition, at

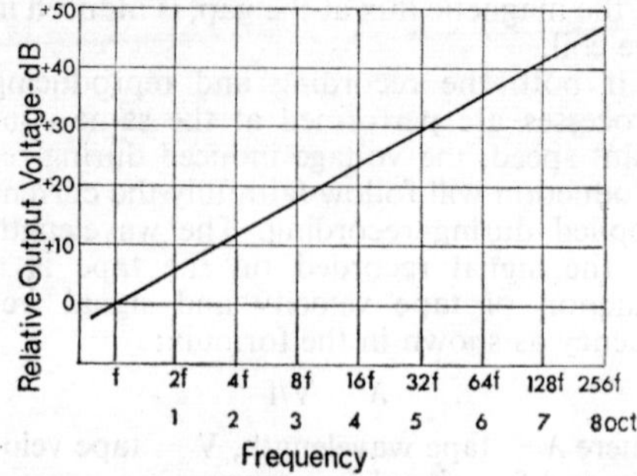

ROLL-OFF WITH DESCENDING FREQUENCY. Frequency is shown as multiple of base frequency, and in octaves. Inherent 6 dB per octave roll-off means that the base frequency is reproduced 48 dB below the level of the 8th octave above it, and 60 dB below the level of the 10th octave, setting an effective ten-octave frequency range with current tapes.

frequencies just below cut-off, the rate of change of magnetic flux, for a given tape speed, is greater than the rate of change of flux at lower frequencies. As the frequency is halved, the rate of change is halved and the induced voltage decreases by 50 per cent or 6 dB. If a recording is made at constant current, covering all the frequencies up to cut-off, for a given tape speed and a given reproduce head, the reproduce head output will indicate a characteristic 6 dB per octave roll-off with descending frequency. This 6 dB per octave roll-off is characteristic of all magnetic tape recording systems and limits to approximately 10 the number of octaves that can be practically reproduced. In a 10-octave recording, the signal-to-noise ratio for the first octave will always be 60 dB poorer than the signal-to-noise ratio for the tenth octave. Only radically improved tape with lower inherent surface noise and improved future pre-amplifier design can remove this 10-octave limitation.

Thus the bandwidth that can be reproduced on magnetic tape is limited to approximately 10 octaves by the nature of the reproducing system; and the maximum frequency that can be reproduced is determined by the reproduce gap width and tape speed. Because the minimum gap width capable of being made commercially is approximately 0·00005 in., the only way to extend the upper frequency limit is to increase the tape speed (tape-to-head velocity). For example: to reproduce a 1 MHz signal using a commonly encountered 0·0001 in. reproduce gap would require a tape speed of 100 in. per second (ips) or 30,000 ft. of tape for one hour's playing time; to reproduce a 4 MHz signal with the same 0·0001 in. gap would require a tape speed of 400 ips or 120,000 ft. of tape for one hour's playing time.

FREQUENCY MODULATION. A television signal extends from near d.c. to above 4 MHz, a bandwidth of almost 18 octaves. The practical bandwidth limit of magnetic tape recording is approximately 10 octaves. A means is needed to transform the near d.c. to 4 MHz signal into higher frequency sideband information. The video signal is used to frequency modulate a carrier whose resting frequency lies just outside the passband of the video signal. The polarity of the modulation is such that an increase in brightness level of the video signal causes an increase in carrier frequency of the modulator. Zero signal, which corresponds to the negative tip of the synchronizing pulse on the television waveform, corresponds to zero deviation of the carrier which, in a frequency spectrum drawing, is shown to be 3·4 MHz. The maximum carrier frequency corresponds to the peak white of the television signal and is 4·7 MHz. The total deviation in this system is 4·7—3·4 or 1·3 MHz.

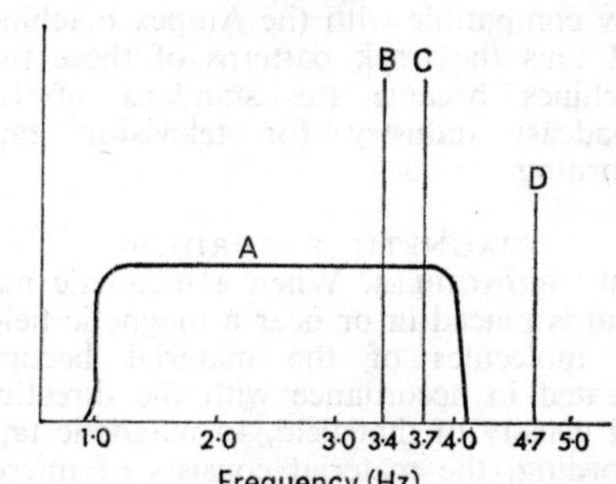

HELICAL-SCAN RECORDER: FREQUENCY SPECTRUM. A. Bandpass of lower sideband. B. Sync tip frequency. C. Black-level frequency. D. Peak white frequency.

This modulation process produces an upper and lower sideband. The upper sideband is attenuated by the limited passband of the head-to-tape process, and all the information is therefore contained in the one lower sideband. Because the band from 0·5 MHz to 7 MHz represents only slightly more than 4 octaves, these signals are easily recorded and reproduced.

DIRECT RECORDING

The earliest attempts to record wideband video signals on tape used the longitudinal or direct recording principle. In this system, the head-to-tape speed and the transport speed are the same, and to recover a 2 MHz signal using commercially available heads requires tape speeds of approximately 150–200 ips. Tape transport speeds of this order introduce a multitude of problems. Extremely thin tape must be used to achieve satisfactory playing times; the maximum playing time on a two-track machine using 10 in. reels is approximately 25 minutes. At such speeds, it is difficult to track the tape satisfactorily through the transport, and it is also difficult to obtain good head-to-tape contact; both are essential for high frequency recording. Many early attempts at video recording used this technique and it is still employed in home recorders. It is relatively simple and inexpensive.

To reduce the number of octaves of the television signal to be recorded, the low-frequency, field-rate video information is converted to a series of high-frequency pulses before recording. In the reproduce process these high-frequency pulses are used to regenerate new field sync and field blanking, which is superimposed on the reproduced television signal. Various manufacturers have experimented with this method, and one proposed system records two audio and two video tracks on a ¼ in. tape. The audio system uses conventional heads but, because of the high head-to-tape speed, it is necessary to frequency modulate the audio signal on a carrier so as to transfer the intelligence to a higher part of the frequency spectrum.

The highest possible video frequency that can be recovered by such a system is approximately 2 MHz, and at present its mechanical and electrical performance is considered unreliable for professional applications. To recover higher frequencies from tape, it is necessary to use a system in which the video head-to-tape speed and the transport tape speed are not the same.

HELICAL SCAN RECORDING

In this technique the magnetic tape is wound around a large diameter drum in the form of a helix. The drum rotates at a fixed speed and the video head or heads are mounted on the periphery of the drum. The head-to-tape speed in this configuration is dependent on the head drum speed, and the diameter of the drum. The head drum speed is usually chosen to be the same as the picture repetition rate of the television signal to be recorded. This enables the drum to be phased so that the video head parts from the tape along one edge and enters along the other edge during the field blanking period of the television signal, so that any distortion or disturbance caused by the momentary interruption to the signal from the tape occurs during the blanking time, and is not displayed on the reproduced picture.

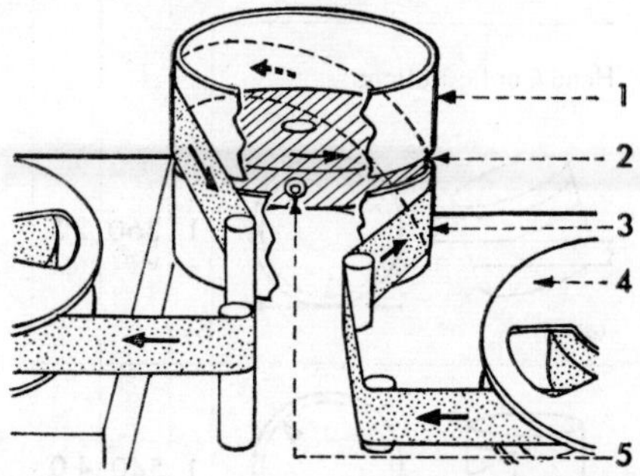

HELICAL SCANNING ASSEMBLY. A hollow drum, 1, consists of an upper and lower part, both non-rotating. Between the two is a slot, 2, through which protrudes a magnetic head, 5 (sometimes two heads), mounted on a revolving drum driven by a motor. After leaving the feed spool, 4, the tape, 3, is wound helically upwards around the fixed drum, and is pulled by a capstan drive which must overcome the wrapping friction; finally it is taken up on a spool at a higher level. The tape is pulled through comparatively slowly, and is scanned very rapidly by the head, 5, producing slantwise diagonal records the full width of the tape. While the basic principle remains as described, many mechanical configurations are possible.

PRACTICAL CONFIGURATIONS. Some of the principal configurations of helical scan recorders are best shown diagrammatically, with the tape path, the number of video heads used, the minimum angle of wrap, theta, and the minimum tape tension multiplication factor k where

$$k = \frac{T_{out}}{T_{in}} = e^{\mu\theta}$$

The first configuration uses only one head and theta (θ) = 360°. However, it has serious drawbacks if continuous recording is required; the edges of the tape are not available for recording of secondary signals such as audio or control track.

The second configuration is also a helical scan, using only one head. However, in a practical arrangement, continuous recording is not possible as the exit and entry guides must be of a finite size and there is a gap in the recording in this region. To reduce this to a practical minimum, fixed exit and entry guides are used, but this results in an angle of wrap of 540°. Even with such a layout the

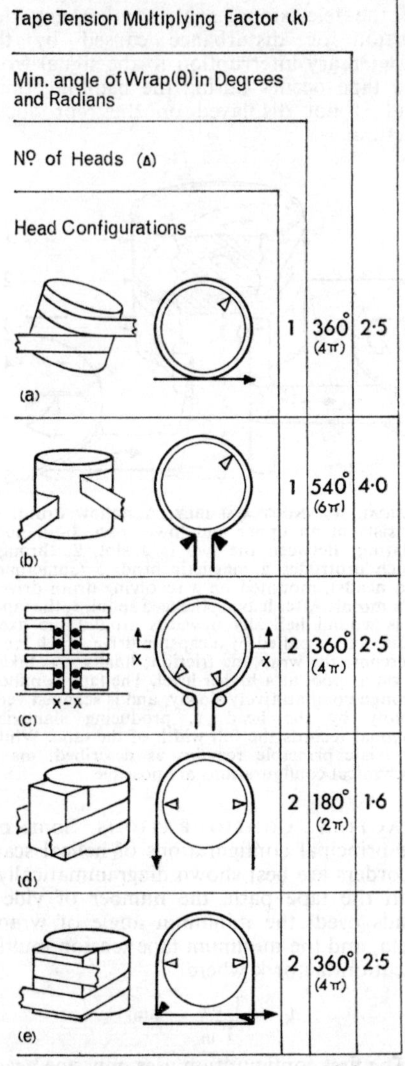

HELICAL-SCAN HEAD CONFIGURATIONS. Schematic. It is important to keep tape tension multiplying factor, k, as low as possible, other things being equal.

$$k = T_{out}/T_{in} = e^{\mu\theta},$$

where T = tape tension, and θ, angle of wrap, is in radians. (a) Full 360° wrap, no unscanned gap, but tape edges cannot be used for subsidiary tracks. (b) Fixed guides keep unscanned gap to a minimum, but wrap = 360°+90°+90° (for the two guides) = 540°. (c) Two heads at different levels (shown in section) overcome gap problem, and fixed guides are replaced by rollers to reduce effective wrap angle. (d) Simple two-head layout with reduced wrap angle. (e) Same layout but with fixed corner guides and crossed-over tape path, giving 180°+90°+90° = 360° wrap.

drop-out in the recordings will be approximately equal to the entire vertical blanking interval in a recorder of this class.

The third arrangement attempts to overcome some of the disadvantages of the previous one. To reduce friction, rotating exit and entry guides are used and the angle of wrap becomes 360°. This implies an even larger drop-out in the recorded signal, and to overcome this disadvantage a second head is mounted on a second drum and is used to write a separate signal during the time the drop-out would normally occur. This configuration requires the use of two video heads and introduces two head switching transients per television field.

The fourth configuration represents the simplest two-headed helical scan recorder. Two heads are used and as the angle of wrap around the scanning assembly is slightly greater than 180°, continuous recording can be achieved. The angle of wrap is reduced to a mere 180° and tape friction problems are held to a minimum. To ensure tracking around the helix, helical edge guides are required on the scanning assembly itself. These are expensive to manufacture and align, and interchangeability is extremely dependent upon the mechanical tolerance of the tape itself. This is an elastic medium and in a practical case is subject to variations in straightness and width. Furthermore, this design requires that auxiliary components, such as supply and take-up reels, auxiliary head stacks, capstan and pinch rollers, must all be mounted at compound angles to one another, complicating the manufacturing problems for the rest of the transport.

In the fifth layout two heads are used and continuous recording is possible. In order to accurately guide tape around the scanning assembly, fixed exit and entry guides are used and mounted at an angle to the scanning assembly. These can be very accurately positioned and designed so that long-term interchangeability becomes a practical reality. All auxiliary components are mounted in parallel planes and design of the entire transport and scanning assembly is simplified.

TAPE SPEED. When the highest frequency that the reproduce head must pass has been established and 0·00005 in. minimum practicable gap size accepted, the necessary tape speed can be calculated. A 7 mHz cut-off frequency requires a head-to-tape speed of approximately 600 ips. To achieve this, it is necessary to rotate the head at high speed past the tape and advance the tape at a moderate speed through the transport.

In one particular machine of this configuration, the actual head-to-tape speed used is 650 ips, corresponding to a cut-off frequency of 7 mHz for a ·00006 in. reproduce head gap width. The rotating head assembly consists of two heads on the periphery of a 7 in. diameter drum, each head spaced 180° apart; this is the fifth layout mentioned above. The tape is 2 in. wide and is wrapped cylindrically around the scanning assembly. The two heads are identical in construction and are fed identical f.m. signals simultaneously. Each head is 7·5 mil wide and lays down a 7·5 mil wide track on the tape. Due to the 3·7 ips forward

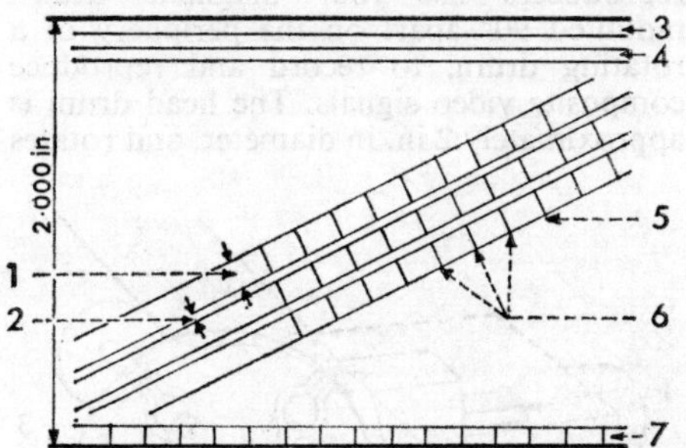

HELICAL-SCAN TRACK POSITIONING. Track positions on a typical recorder using 2 in. tape. 1. Video track width, 7·5 mils. 2. Guard band, 2·2 mils. 5. Video area, separated, 6, by line sync pulses 3, 4. Audio tracks. 7. Control track for timing playback.

tape motion, a 2·2 mil separation exists between tracks. Also owing to forward tape motion, the tracks are laid down at a 9° angle across the tape. Audio tracks are recorded on the upper edge in the conventional longitudinal manner. A control track is recorded at the lower edge of the tape, for timing control of playback.

TAPE TRANSPORT. Details of the transport layout of a helical recorder using this scanning assembly are shown diagrammatically. Tape on a supply reel is B-wind and tape on a take-up reel is A-wind. The main reason for this choice is that it allows the minimum length of unsupported tape between the reel and the scanning assembly, thus avoiding unwanted vibrations in the tape. Tape from the supply reel passes directly over the tape tension sensing arm, over the video erase head, around the entrance guide, up and around the scanning assembly, over the rotating video heads, around the exit guides, over an auxiliary

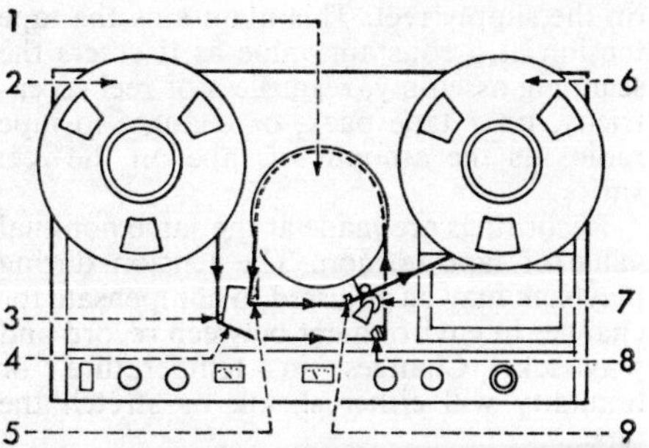

TYPICAL HELICAL-SCAN VIDEOTAPE RECORDER LAYOUT. Tape unwinds from feed reel, 2, and passes round tape tension arm, 3, over video erase head, 4, and entrance guide, 8, to helical scanning assembly, 1. After recording, tape leaves by exit guide, 5, passes over auxiliary head stack, 9, pulled by capstan and pinch roller, 7, and is wound up on take-up reel, 6.

head stack which contains the control track and audio heads, through the capstan pinch roller assembly which drives the tape at precise speed, and directly on to the take-up reel.

The most important aspect of this layout is that it enables all tape-guiding or tape-contacting elements to be mounted on, or positioned with reference to, the scanning assembly itself, which is a rugged, rigid unit made of only two D-shaped castings. In this way, all critical elements that guide or contact the tape are held rigidly with respect to one another.

TRACKING. The tracking of tape around the helix is controlled solely by the angular position of the exit and entry guides. Videotape is neither exactly straight nor of exactly uniform width and thus cannot be guided solely by its edges. Furthermore, the camber and width of the tape can change during its life due to temperature and humidity changes or stretching on the tape transport.

TENSION. An essential requirement in helical scan recorders is that the tension of the tape be carefully controlled at all times. The accuracy of the timing information of the video signal is dependent upon maintaining accurate and stable tape dimensions. A high performance tape tension servo is required to maintain a stable tape tension under all conditions. Tape from the supply reel passes over a tape tension sensing arm before it enters the scanning assembly. The arm is the input to a servo which controls the drag of a holdback band brake mounted on the supply reel. This maintains the tape tension at a constant value as it enters the scanning assembly, regardless of reel eccentricity, poor tape pack, or changes in tape radius as the amount of tape on the reel varies.

Recordings are made at the same nominal value of tape tension. The tension during playback may be adjusted to compensate for changes in environment between record and playback. Changes in temperature or humidity will either shrink or stretch the tape.

The tape transport configuration chosen may lend itself either to open loop or closed loop capstan drive. With proper design and special attention to the layout of the pinch rollers and accurate control of tolerances in the bearing supporting the capstan, the open loop has been found to be capable of results adequate for closed circuit television operation. Very small imperfections in the bearings or the capstan shaft result in appreciable changes in tape tension or tape velocity within the scanning assembly, giving rise to undesirable errors.

In this way, the tape tension servo not only maintains the tape tension constant, due to such gross changes as reel radius, but responds to all tape tension changes up to 10 Hz and, in effect, serves as a high pass filter for tape tension variations entering the scanning assembly.

TRACK GEOMETRY. With a tape speed of 3·7 ips, the geometry of the tracks laid down on the tape is such that the line synchronizing pulses are aligned on adjacent recorded tracks. This results in several very important advantages in the helical scan recorder.

As an entire field of the television signal is recorded with a single head in these recorders, playback, stop motion or slow motion is possible if line sync pulse alignment is used.

At any tape speed other than that at which the recording was made, the video heads must at some time or other cross from track to track and, at this point, signals will be read from both tracks at once. When the tape geometry is arranged to give line sync pulse alignment, the same information is picked up from the two adjoining fields and, of course, the sync timing is the same. Because of this, monitors will remain locked as the heads cross track. In this way, stop frame operation is possible without resort to complicated and unreliable distortion of the tape guiding. Slow motion is only possible at all when line sync pulses are aligned.

Furthermore, this feature ensures that any errors that may affect tape tracking will have a less serious effect on the picture stability, thereby considerably enhancing the overall performance of the recorder.

TRANSVERSE RECORDING

The transverse recorder, introduced in 1956, has been standardized by television organizations throughout the world for the recording and reproducing of broadcast television signals, both in black-and-white and colour.

HEAD ASSEMBLY. Transverse recorder/reproducers use four magnetic heads, mounted 90° apart on the periphery of a rotating drum, to record and reproduce composite video signals. The head drum is approximately 2 in. in diameter, and rotates

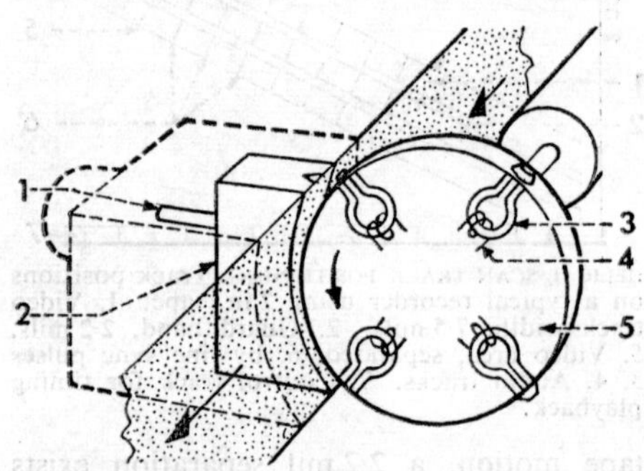

TRANSVERSE-SCAN RECORDING HEAD DRUM. Drum, 5, carries four heads, 3, with coils, 4, whose tips protrude and penetrate slightly into the magnetic coating of the tape. The tape is curved to the drum radius by the guide, 2, and is held in close contact with the guide by applying vacuum, 1.

at 250 revolutions per second. A head-to-tape velocity of approximately 1,500 ips is produced by the video heads as they move laterally across the surface of the 2 in. wide magnetic tape. Each video head records a lateral track 10 mils wide. The reel-to-reel tape speed of 15 ips provides a 5·6 mil separation between each recorded track. Centre-to-centre track spacing is 15·6 mils.

Using the 625-line, 25-frame European television standard, 16 to 17 horizontal lines of picture information are recorded as each head sweeps across the tape. The 625 lines which constitute one frame are contained in 40 successive parallel tracks occupying $\frac{5}{8}$ in. of tape length.

Conventional stationary heads record the

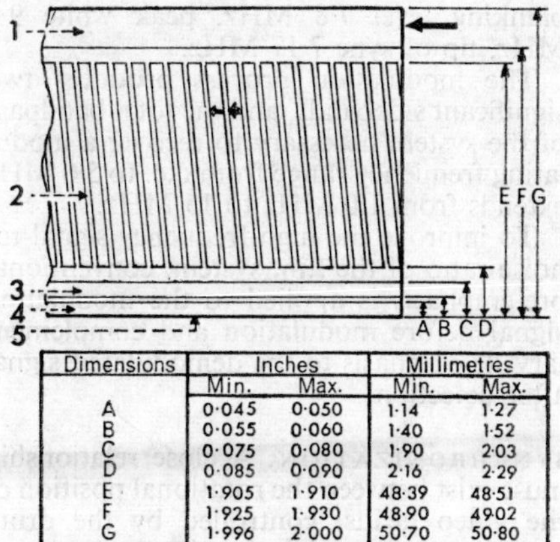

Dimensions	Inches		Millimetres	
	Min.	Max.	Min.	Max.
A	0·045	0·050	1·14	1·27
B	0·055	0·060	1·40	1·52
C	0·075	0·080	1·90	2·03
D	0·085	0·090	2·16	2·29
E	1·905	1·910	48·39	48·51
F	1·925	1·930	48·90	49·02
G	1·996	2·000	50·70	50·80

TRANSVERSE-SCAN TRACK POSITION AND DIMENSIONS. 1. Audio track. 2. Video tracks. 3. Cue track. 4. Control track. 5. Guided edge of tape.

audio, cueing and control tracks along longitudinal areas at the edges of the tape. When the recording process is complete, four synchronized tracks have been recorded on the magnetic tape: the video signal, recorded laterally across the tape, the audio sound track, recorded longitudinally along the top edge of the tape, and the control and cueing tracks, recorded longitudinally in parallel paths at the bottom edge of the tape.

Because the heads are mounted 90° apart on the drum, while the tape is curved to contact 120° of the drum periphery, there is an overlap, or duplication of video information that appears at the end of one track and again at the beginning of the following track.

A strip 95 mils wide on the bottom edge of the tape, including most of the redundant video information, is subsequently erased. A longitudinal audio track 75 mils in width, is recorded along the upper edge of the tape in the conventional manner. This provides a 20 mil guard band separating the audio and video information. Since the magnetic particles in the tape are so orientated as to favour the transverse video signal, they are at right-angles to the optimum for the longitudinal audio recording. But this is compensated by the high recording speed and wide track.

CONTROL TRACK AND ERASE HEADS. The control track is 50 mils wide, and is recorded longitudinally along the lower edge of the tape. Both the 250 Hz control signal, and frame pulse information, are recorded on the control track. The cue track, 20 mils in width, is recorded longitudinally between the control track and the video information, with 10 mils separation from each.

The audio and cue erase head assembly is located on the tape transport between the video and the audio head assemblies. It erases redundant video information from the longitudinal area of the tape allocated to the audio and cueing signals, just before these are recorded. No separate erasure is required or provided for the control track. The control track head is located beneath the slip ring sub-assembly on the video head assembly. In addition to the control track signals, this head records the 25 pulses per second frame pulse, necessary for accurate tape splicing.

MECHANICAL ARRANGEMENT. The video head assembly is a plug-in unit that includes the video head drum, slip rings and brushes, head drum motor, vacuum tape guide, magnetic tachometer, and the control track head. All video head assemblies are interchangeable from machine to machine of the same manufacturer. The video head tips are made from a hard material that is a composition of aluminium, iron and silicon. The head tips wear due to the high head-to-tape speed, and their life is approximately 250 hours.

A magnetic pick-up coil is mounted on the rear motor bracket. A small disc is fitted to the rear motor shaft. The disc is made of a nonferrous, and therefore nonmagnetic, material. Inserted in the disc are two or three small magnetic plugs. As the motor rotates, the plugs pass the pick-up coil and complete a magnetic circuit which produces a small electrical impulse at the coil output. This pulse is amplified and processed into a square-wave signal whose frequency and phase represent the true position of the head drum motor and video heads. This relationship is required for the reproduction of the recorded tracks.

The video heads are mounted coaxially on the shaft of a 50-watt hysteresis synchronous motor that operates from three-phase, 115-volt, 250 Hz power generated by a power driver. 50 Hz power supplies are assumed throughout. Corresponding figures for a 60 Hz supply are readily calculated.

The tape transport is similar to mechanisms found in other professional quality audio recorders. The tape is moved by the rotation of the capstan, from the supply reel, across the supply reel idler, and the stationary video erase head. It then passes the rotary video head drum, the control

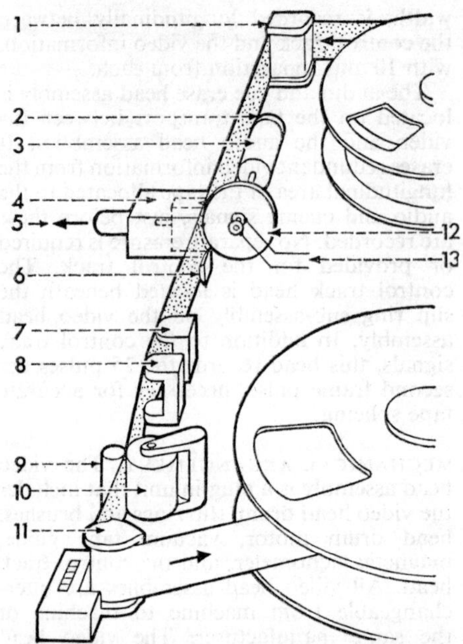

TRANSVERSE-SCAN TAPE TRANSPORT AND HEAD LAYOUT. Tape from feed reel passes over full-width erase master head, 1, and is curved at tape guide, 4, with the aid of suction vacuum, 5, to radius of recording head drum, 12. This is driven by drum motor, 3, having control signal generator, 2, and electrical connections, 13, shown purely schematically. Tape next passes control track head, 6, sound and cue track erase heads, 7, and sound and cue track recording heads, 8. Tape drive is by capstan, 9, with pressure roller, 10. Tape timer, 11, reads in minutes and seconds on adjacent counter.

track head, the audio and cueing heads, the capstan and its associated pressure idler, and over the tape timer assembly to the take-up reel. The vacuum tape guide accurately positions the tape as it passes the video head assembly for height, and for curvature that conforms exactly with the curvature of the video head drum surface. The uncoated rear surface of the tape is held in contact with the concave guide surface of the tape guide by means of a vacuum.

Video information is recorded on the tape in the form of a frequency-modulated carrier. During the modulation process the video signal is clamped to a d.c. level at the time of occurrence of the back porch, that period in the line blanking interval between the end of the line synchronizing pulse and the start of the active picture.

This enables definite carrier frequencies for the brightness levels of the various parts of the signal to be established. In the 625-line colour system, typical frequencies are: blanking level 7·8 MHz, peak white 9·3 MHz, tip of sync 7·17 MHz.

The modulation process produces two significant sidebands, and the total bandpass of the system necessary to recover a modulating frequency range from d.c. to 5·0 MHz extends from 1·0 MHz to 15 MHz.

To improve the high-frequency signal-to-noise ratio of the f.m. system, conventional pre-emphasis is applied to the modulating signal before modulation and complementary de-emphasis to the demodulated signal after detection.

SYNCHRONIZATION. A close relationship must exist between the rotational position of the video heads, controlled by the drum motor rotation, and the longitudinal position of the tape, controlled by the capstan motor rotation. The precise relationship of these two factors that existed during the recording process must be recreated and constantly maintained during the process of reproduction. The capstan servo system accomplishes this by locking the rotation of the capstan motor to that of the head drum motor.

During recording and reproduction, a control signal is generated by a magnetic tachometer mounted on the head drum motor shaft. The symmetrical square wave output of the tachometer processing electronics corresponds in frequency to the nominal 250 rps speed of the head drum. During the recording process, the 250 Hz output of the tachometer is electronically divided by four to 62·5 Hz, amplified, and used to drive the capstan synchronous motor. The 250 Hz tachometer output is also recorded longitudinally along the bottom edge of the tape, and is used as a control track during reproduction.

When the equipment is used for reproduction, the capstan servo system compares the signal from the reproduced control track with the signal from the tachometer. The capstan motor speed is controlled by the results of this comparison, increasing or decreasing to maintain the exact relationship between the angular head position and the longitudinal tape position that existed during the record mode. Two servo loops are used in the head drum motor servo system. One, the slow servo loop, provides the proper input frequency to the head drum synchronous motor. It eliminates fast phase motor correction requirements, and ensures the recording of field sync at the same physical location on the tape. The other, a fast servo loop, stabilizes the drum motor and suppresses its hunting tendency. The drum motor is normally locked to the field sync of the input video signal during the record mode. It can be locked to any of three sources during the reproduce process – external 50 Hz reference, 50 Hz mains or the 50 Hz component of studio complete sync.

Field sync is stripped from the composite video signal and compared in frequency and phase with an oscillator running at a nominal frequency of 250 Hz, five times the field frequency rate. The oscillator is phase-locked to the incoming field reference, and the oscillator output, after further processing, is used as the timing reference to control the speed and phase of the rotating video head drum.

TAPE STRETCH AND HEAD INTRUSION. A tape guide servo system is necessary to position the vacuum tape guide and control the contact pressure between the rotating video heads and the tape. A certain amount of head intrusion into the tape is necessary to ensure proper head-to-tape contact. This intrusion causes tape stretch, an effect that can be used to advantage when controlled.

Tape stretch produced by head intrusion is local in nature. Were it not for this phenomenon, head velocity would be reduced in proportion to the amount of head wear during normal head life, and would cause timing errors. As long as the degree of tip penetration does not exceed the elastic limit of the tape, the amount of tape swept in a given time will remain constant through the range of tip projection present during the life of the heads.

The position of the vacuum tape guide regulates contact pressure between the video heads and the tape, which in turn establishes the degree of tip penetration. Thus, as the guide approaches the video head, tip penetration is increased.

RECORDING. During the recording process, the full width of the 2 in. magnetic tape is erased by the video erase head before it makes contact with the rotating video head drum.

The incoming composite video signal is amplified, clamped and then used to frequency modulate a high-frequency oscillator. The oscillator may be of the multivibrator type operating at carrier frequency or a simpler higher-frequency oscillator or oscillators whose outputs are heterodyned to produce a resultant lower frequency corresponding to the carrier frequency. This carrier extends from 7·16 MHz to 9·3 MHz. The modulator output is amplified to a level that will supply sufficient current to the four recording heads to just saturate the magnetic particles on the tape. Amplification takes place in four individual wideband amplifiers located in a housing mounted to the tape transport beneath the video head assembly.

All amplifier outputs are fed through slip rings to their respective video heads. Each head receives the signal from its record amplifier, and records that signal during the time it is moving across the tape. Recent designs, however, eliminate slip rings by incorporating small coupling units which are effectively rotary h.f. transformers.

The audio recording circuits consist of a high-frequency bias and erase oscillator which drives an erase head that erases a longitudinal area along the top edge of the tape before the tape passes the audio record/reproduce head. Another erase head located near to the bottom edge of the tape erases a space for the cue track.

The audio signal is routed through an audio record amplifier. mixed with a high-frequency a.c. bias signal from the bias and erase oscillator, and supplied to the audio record/reproduce head. This head records the audio signal longitudinally on the erased area at the top edge of the tape.

Cue channel audio signals are amplified and mixed with high-frequency a.c. bias. and then supplied to the cue channel record/reproduce head. The cue channel is recorded longitudinally between the lowest edge of the remaining video information and the control track at the bottom edge of the tape.

REPRODUCTION. In the reproducing process. the magnetic flux recorded on the tape induces a voltage in the rotating video heads as they traverse the tape. This voltage is fed through slip rings or the rotary transformer to pre-amplifiers in the head channel assembly. Each head output is fed to a separate pre-amplifier.

The four pre-amplified outputs are then fed to an electronic switcher and sequentially gated. The switching from one head output to the next occurs during the front porch position of the line blanking period. The switched output is fed to the demodulator where the f.m. signal is subjected to limiting to remove unwanted amplitude modulation before it is demodulated to the original video form. The demodulated signal is then fed to the processor.

In the processor, the sync pulses, stripped from the video signal, are amplified and

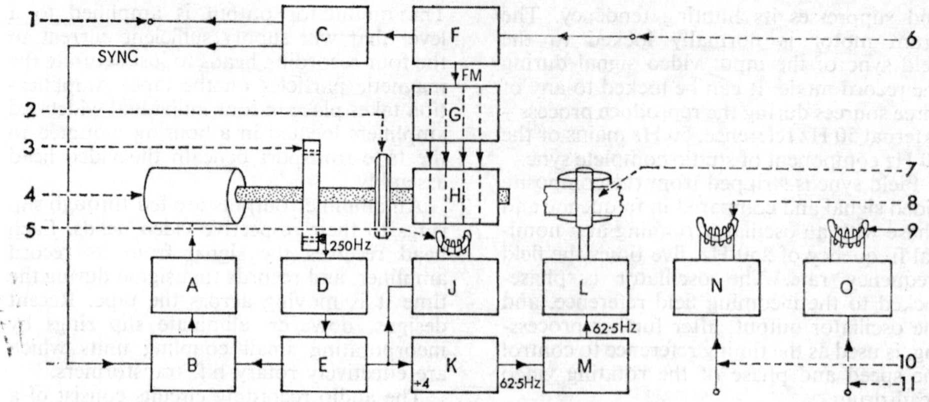

PROFESSIONAL TRANSVERSE-SCAN VIDEOTAPE RECORDER: BLOCK DIAGRAM OF RECORD MODE. **Shown for 50 Hz operation. Video head drum, 2, is mounted on the shaft of synchronous motor, 4, and carries rotary signal coupling assembly, H, and timing disc, 3. Composite video signal, 6, enters modulator, F, and is converted to f.m. signal. This signal is fed to four individual record drivers, G, and through the coupling assembly to the head drum. The r.f. signal is demodulated by C, and video output, 1, is used as a record monitoring signal. Synchronizing information from demodulated output, C, is used by the drum servo system, B, as a timing reference. This signal is compared with the drum phase signal derived from a tachometer pickup head, D. After amplification by E, the drive signal developed after comparison is amplified by power amplifier, A, which supplies power to motor, 4. The drum phase signal, E, is divided by four, K, amplified by power amplifier, M, and drives capstan motor, L, which pulls tape, 7, at constant speed through the transport. The 250 Hz drum tachometer signal is processed by record amplifier, J, and is recorded as a reference track for playback by the control track record/playback head, 5. Main audio signals, 10, are fed to audio record amplifier, N, and audio head assembly, 8. Auxiliary audio signals, 11, are fed to cue record amplifier, O, and to cue head assembly, 9.**

clipped to shorten the rise time and to remove any noise, and new line and field blanking pulses are generated. All pulses are then re-combined with the video portion of the signal, reforming the composite video signal.

Also during reproduction. the magnetic flux of the recorded audio track induces a voltage in the audio head as the tape passes the head. This voltage is routed to a reproduce amplifier, where it is amplified and equalized. The output level is monitored by a conventional VU meter connected across the outgoing 600 Ω line.

The control track head reproduces the recorded 250 Hz control track signal. This signal is amplified and used as one input to a phase discriminator circuit in the capstan signal generator. The second 250 Hz signal applied to this circuit is originated by the magnetic tachometer as it scans the rotation of the head drum. This signal is processed by the drum servo system unit, and its phase is compared with that of the first. If there is a phase difference, an error voltage is produced that controls the frequency of a Wien bridge oscillator running at a nominal 62·5 Hz.

The output of this oscillator is amplified by a power amplifier, which drives the synchronous capstan motor. The speed of capstan rotation is thus controlled by a servo system action to produce a constant phase relationship between head drum rotation and the longitudinal position of the tape. In this manner, the instantaneous relationships that existed during the recording process of the video heads and the video information recorded laterally on the tape, are re-created and automatically maintained during the reproduction process.

AUXILIARY EQUIPMENT

SERVO SYSTEM. During the recording of a television signal on magnetic tape, the head drum motor is locked to the field synchronizing pulses derived from the incoming video signal. The capstan motor is locked to the drum motor by the drum's tachometer signal. During reproduction, the servo system is normally locked to either external 50 Hz mains or field synchronizing pulses derived from the studio pulse generator. The playback field rate of the reproduced tape signal is therefore the same as the external reference. The tape replay field and line rate bear no precise relationship to other studio television equipment. Therefore the signal from the tape recorder must be treated as a remote source to the studio and cannot

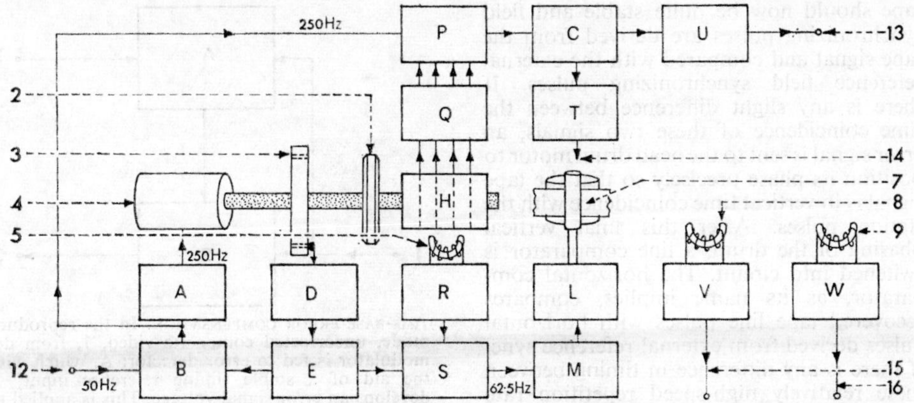

PROFESSIONAL TRANSVERSE-SCAN VIDEOTAPE RECORDER: BLOCK DIAGRAM OF REPLAY MODE. Tape, 7, is pulled by synchronous capstan motor, L, driven by power amplifier, M, from a signal whose frequency, nominally 62·5 Hz, is determined by capstan servo comparator, S, from a reference, E (the drum tachometer signal), and the tape control track signal picked up by head, 5, amplified and processed by R. The speed of drum motor, 4, is determined by external 50 Hz reference, 12, fed to the drum servo system, B, and compared with signal, E, derived from pickup head, D. The difference between these signals controls a 250 Hz signal which drives drum motor, 4, after amplification by A. The r.f. signal from the tape is picked up by video heads, 2, amplified by four pre-amplifiers, Q, the sequential output of which is selected by switcher, P, whose continuous output is demodulated, C. The resultant video signal is cleaned up by processor, U, which strips sync, regenerates blanking, and recombines sync to produce video output, 13. Main audio channel signals from the tape are picked up by playback amplifier, V, which feeds audio line, 15. Secondary audio channel signals are driven by head, 9, and amplifier, W, to line, 16. Automatic correction of tape guide errors is achieved by T, which compares the instantaneous timing of the reproduced video signal from C. Error signal produced, 14, controls position of the video head female guide.

be handled in the same manner as cameras and telecine equipment, whose scanning rates are locked to the local studio pulse generators.

These considerations limited the production flexibility of early video recorders, and it became necessary to develop a system whereby the head drum motor and capstan motor on the video recorder could be driven directly from station field synchronizing and line synchronizing pulses, so that the signals from the video recorder could be mixed and switched with other signal sources in the television studio. A new servo system was accordingly developed for the video recorder.

COMPARISON SIGNALS. When recording, 25 Hz frame pulses are derived from the incoming video. The pulse corresponds to the field synchronizing pulse after the half-line of video information on the television signal. These 25 Hz pulses are mixed with the 250 Hz control track pulses and recorded on the servo control track along the bottom edge of the tape. During record, the phase of the video head drum is precisely timed so that the third serration of the field synchronizing pulse is laid down in the physical centre of the 2 in. magnetic tape.

When reproducing, the capstan pinch roller engages against the capstan and starts to pull the tape through the transport. The control track head, which is always in contact with the tape, recovers the 25 Hz frame pulses; they are processed and fed to a circuit that compares their time relationship with 25 Hz pulses derived from the studio pulse generator. The difference signal from the comparator-type circuit is used as an error signal to control the frequency of the capstan oscillator which, in turn, controls the speed of the capstan motor.

This comparison continues until the frame pulses from the tape are in time coincidence with the frame pulses from the studio generator. After coincidence is established, the capstan servo system reverts to normal operation, i.e., the 250 Hz control track pulses are compared with the 250 Hz signal derived from the drum tachometer. During this initial lock-up time, the head drum motor is locked in the normal way by comparing its 250 Hz tachometer signal with the external 50 Hz field sync pulses derived from the external reference complete sync from the studio pulse generator.

VERTICAL AND HORIZONTAL CORRECTION, The television signal recovered from the tape should now be quite stable and field synchronizing pulses are derived from the tape signal and compared with the external reference field synchronizing pulses. If there is any slight difference between the time coincidence of these two signals, an error signal is sent to the head drum motor to position its phase precisely so that the tape signal is in vertical time coincidence with the station pulses. After this final vertical phasing of the drum, a line comparator is switched into circuit. The horizontal comparator, as its name implies, compares recovered tape line pulses with horizontal pulses derived from external reference sync. If there is any difference in timing between these relatively high-speed repetition rate pulses, an error signal is developed which is fed to an amplifier with a complementary frequency response and phase response to that of the video head drum; this pre-distorted signal is then used to apply correction directly to the head drum motor by phase modulating the sine wave signal used to drive the video head.

The vertical rate correction in this process ensures that the field synchronizing pulse leading edge recovered from tape is within ±10 μs in relation to the studio field synchronizing pulse. The line pulses recovered from tape are phased to be within ±0·2 μs in relation to the studio field pulses. To achieve this horizontal timing accuracy requires the servo to be capable of controlling the instantaneous position of the video head drum to within a tolerance of better than 0·04° positional error on playback relative to record.

ELIMINATING TIME-BASE ERRORS. During the recording and playback process, time-base errors may be introduced by head quadrature error, geometric distortion of the tape, mechanical load factors, or the passage of a splice, each of which will cause a relative displacement of visible picture elements during tape reproduction. An accessory has been developed which eliminates or greatly reduces the effect of these errors on the reproduced picture.

Under the 625/25 television standard, there are 625 horizontal lines of picture information in each completed frame.

Uncorrected composite video from the demodulator is fed to the input of the unit. One feed of the video signal goes to a sync separator, and the leading edges of sync are used to trigger a flywheel oscillator running at line rate. The AFC circuit in the oscillator has a relatively long time constant so that the oscillator frequency represents the average of the line repetition rate. The output from the oscillator is then compared with the uncorrected line sync pulses and, as a result of this comparison, an error signal voltage is developed. This voltage is fed in anti-phase to a voltage controlled delay line. The delay line consists of inductors and capacitors, the capacitative elements being variactors. A variactor is a reversed biased silicon diode whose capacitance changes with change of applied d.c. bias. The error signal is fed to all the variactors in turn. As the variactors change their capacity, so the delay of the delay line changes. The video signal is fed to one end of the delay line and is delayed in proportion to the error derived from the comparator. The output video signal is therefore completely time corrected and contains no instantaneous timing errors. This signal is then fed to the processor and to the output of the machine.

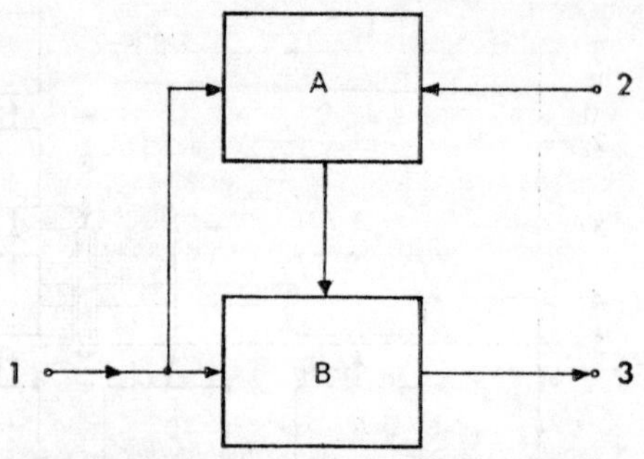

TIME-BASE ERROR COMPENSATION. In the reproduce mode, uncorrected composite video, 1, from demodulator is fed to error detector, A, which with the aid of a stable timing reference input, 2, develops an error signal voltage. This is applied to a voltage-controlled delay line, B, along with the signal, 1, so that resulting composite video output, 3, contains no instantaneous timing errors.

ELECTRONIC EDITOR. The electronic editor accessory provides the convenience of a single control, by which a master videotape can be assembled electronically from studio video signal sources or other tape segments with complete continuity of servo control signals, and without any physical cutting and splicing of the tape. The original tape segments are not disturbed in any way and may be re-used any number of times.

The electronic editor modifies and controls the switching logic of the videotape recorder. It co-ordinates the action of the field lock circuit in the servo system to maintain the correct phase relationship between the master tape signals and the incoming new signal. In addition, it automatically allows for the distance between the planes of the erase head gap, and the video record heads, by precisely synchronizing the successive applications of erase current and video record signal, to cause the first new frame of an insertion to follow precisely its immediate predecessor on the master tape, thus maintaining complete continuity of the video tracks on the tape comprising the television signal. E.F.M.

See: *Telerecording; Videotape editing; Videotape recording (colour).*

Books: *Physics of Magnetic Recording*, by C. D. Mee (Amsterdam), 1964; *Techniques of Magnetic Recording*, by J. Tall (New York), 1958; *Videotape Recording*, by J. Bernstein (New York), 1956.

□VIDEOTAPE RECORDING (COLOUR)

Recording colour television signals on magnetic tape produces additional problems to those already overcome in standard monochrome systems. There are three principal methods of producing colour signals: the NTSC system developed in 1953 (and in use in the US since that date) and the European variants PAL and SECAM.

These systems have much in common. The NTSC system has wide application and raises rather more difficult recording problems than do its variant forms. If the problems involved can be overcome for NTSC, only instrumentation changes are needed to record in any of the other systems.

BASIC PRINCIPLES

The additional difficulties of making a satisfactory colour recording are mechanical errors which give rise to instabilities in the timing of the replayed signal, and the electrical performance of the circuits in the signal system.

MECHANICAL ERRORS. The NTSC system uses a colour subcarrier to transmit the colour information, as well as the normal luminance signal of standard monochrome practice. The phase and amplitude, at any instant in time, of this subcarrier, give colour and saturation information; the phase is measured relative to a fixed oscillator in the coding equipment. In order to synchronise the decoding oscillator in the receiver, a burst of unmodulated subcarrier is transmitted at the start of every active line, and the design of the oscillator is such that there is very little drift in its phase over the length of the line while it is free-running awaiting the next synchronising burst. Any small timing errors in the coded subcarrier after the decoding oscillator has been locked change the phase relative to the fixed carrier, and so cause an incorrect colour signal at the output of the demodulator. Since it is possible to see a colour change corresponding to phase errors of less than 10 degrees, the stability of the machine must be at least as accurate as this for satisfactory reproduction of colour signals. For the European systems, the subcarrier frequency is about 4·4 MHz; 10 degrees is thus a timing error of about $6\frac{1}{2}$ nsec (nanoseconds) $=6{\cdot}5\times10^{-9}$ seconds.

The timing error can be interpreted as an error in the positioning of the playback head with respect to the tape; it is produced either by tape movement error or, in the case of moving heads, by head velocity errors. Thus the mechanical position of these heads must be fixed precisely, and the varying mechanical and frictional forces on the head eliminated as far as possible. To keep all forces and dimensions such that high timing accuracies can be achieved is extremely difficult; in fact, it is not possible to reach this standard by mechanical means only, so electronic time correction techniques have to be used in addition.

SIGNAL PROBLEMS. The problems which arise from the signal system are: arranging for the television signal to be changed into a form suitable for recording on to tape, and reconstituting this signal on replay without distortions. The video signal covers the range of frequencies up to 5·5 MHz; the highest practical frequencies that can be recorded on tape at present are in the region of 12 MHz. In comparison with conventional modulation methods – normal f.m. systems use signals of 20 kHz bandwidth and a carrier of 100 MHz – the carrier and the signal are very close together and it is difficult to achieve good linearity. The proximity of the two signals causes considerable problems in reproducing the correct colour carrier phase and amplitude with changing luminance level, showing on the screen as a change in colour where the brightness is changing. By using the highest possible carrier frequencies and taking great care with frequency and phase responses throughout the signal path, the errors can be minimized and an acceptable signal produced.

EQUIPMENT

A high head-to-tape speed is necessary to reach high frequencies. The easiest way to achieve this is to run the tape at high speed past a stationary head, but tape consumption is high and the mechanical wow and flutter of the transport causes poor stability.

LOW-COST MACHINES. The most promising low-cost machine is that using helical scan; the tape is wrapped around a drum which rotates and carries the video head with it. The tracks lie diagonally across the tape and head-to-tape speeds of 500 to 1,000 ips are possible. Typical figures for stability are 3 to 10 μsec; controlled stretching of the tape provides a further fine correction to about 200 nsec, too great a variation even for synchronous monochrome and greatly exceeding the requirements of a tightly locked colour signal. However, new stabilizing techniques are being developed to improve this figure by at least another order of magnitude, so that in domestic use there are prospects of a helical scan machine for colour.

In addition to the stabilizing problems, the head-to-tape speed is not sufficient for recording high frequency signals around 8 MHz without affecting the colour performance.

BROADCAST MACHINES. Broadcast machines are of the four-head type; the heads are mounted on a drum which rotates perpendicularly to the tape which is held around the head by a vacuum guide. They are known as quadruplex machines and have significant advantages over the helical scan design. With the drum at right-angles to the tape, linear tape fluctuations have only a minor effect on picture stability, causing the head to move off the centre of the recorded track. This is not significant for small variations as the frequency modulated signal from the tape passes through a limiter in the replay amplifiers. The head-to-tape speed is about 1,500 ips, allowing high frequencies to be recorded.

CORRECTION TECHNIQUES

For satisfactory operation in the NTSC system, the time stability of the replayed signal must be closely controlled, because any small time changes are reflected in phase changes in the colour subcarrier, and so, from the manner in which the colour information is decoded, in the reproduced colour.

EARLY CORRECTION TECHNIQUES. The first examples of videotape recorders were not nearly accurate enough for this stability, and a figure of 500 nsec for line-by-line errors was difficult to achieve. The early machines were almost all of the four-head variety, with the heads mounted on a drum rotating transversely across the tape; on playback, positional error signals were derived from the drum at a rate equal to its rotational velocity. Since this is only 250 Hz, this stability figure was not altogether unexpected.

With a colour television service in operation in the US it was desirable to be able to record the colour waveforms on to magnetic tape in the same way as was being done with monochrome signals. There were many proposals for methods of recording, ranging from frequency band splitting to driving the mechanical servo systems in the recorder directly from the colour subcarrier frequency, but these were not adopted finally, since it was thought desirable to aim for a system which was fully compatible with existing monochrome methods. The phase errors were accepted and special processes were developed in playback to compensate for the various colour errors. There were two systems in use, designed by the two major companies manufacturing videotape recorders.

HETERODYNING. The first (RCA) used a heterodyne technique whereby the incoming chroma information, separated from the luminance signal by simple filters, was multiplied within one channel by a subcarrier derived from the stable station subcarrier, and in another with a frequency derived from the tape burst. Two signals were finally produced with equal and opposite phase errors, and differing in frequency by 3·58 MHz (the US colour subcarrier frequency) so that on subtractive mixing a new subcarrier was generated at the correct frequency and the errors in the chroma information were cancelled out. This final matrixing left a stabilized colour signal whose standard was low but of broadcast quality.

Since the correct phase can be measured only at the time of the colour burst, there are still errors in the rest of the picture, and these become progressively worse the further in time they are from the correction point.

DECODING. The second system (Ampex) involves decoding the colour information on the tape replay output into the two colour signals I and Q, using as the decoding subcarrier a signal produced from the tape burst and thus containing all the phase errors at that time. The colour information is then encoded using this time the station subcarrier, which is at constant phase, to produce a corrected NTSC signal. However, with the correction occuring as before at line rate, the final performance was limited by the performance of the mechanical parts of the video recorder; any drift here shows as colour errors in the picture, particularly on the right-hand side, that is, further from the correction point. This is a fundamental limitation in any system which corrects once per line, and remaining errors are an indication of the stability of the machine. While marginally satisfactory on the 525-line system, these early techniques are not adequate for 625-line working.

MODERN TECHNIQUES. Modern VTR practice has improved considerably since the first generation machines. The servo stability has been gradually increased, both by improved mechanical design and by measuring the errors in position more frequently. The rotating headwheel is often provided with air bearings in preference to more conventional ball-race types, and this decreases the frictional resistance by more than one half and also makes the drag more constant. An added advantage is that the speed of response to a correction signal can be increased without loop instability.

In playback, the drum can be given correction signals corresponding to the time difference between station synchronizing pulses and pulses from the tape, so that a line-by-line correction can be applied to the drum itself. The natural errors remaining can be made as low as 100 nsec, which represents a considerable advance on the early machines although still not satisfactory for colour reproduction. With the machine locked to the station synchronization pulses, the average colour phase is correct, and the error changes relatively slowly along the line owing to the mechanical inertia of the rotating parts. The amount of this error is almost as low as can be achieved readily for such a system and further corrections are electronic.

The picture at this stage is not absolutely satisfactory for all the requirements of even monochrome working, because the tape cannot be regarded as fully synchronous; for example, a stability of 100 nsec precludes split screen operation and associated techniques. Stability is further improved by inserting a voltage controlled delay line into the video output of the machine, and making a line-by-line comparison between the tape and station synchronization pulses. Any error is translated into a d.c. signal which is amplified and applied to the delay line in the correct phase to change the line length so that the instantaneous timing error is corrected. The incoming signal passes through a fixed small delay before entering the variable line to enable the sync comparison to be completed by the time the picture signal arrives. This ensures that the error signal applied to the line corresponds to the correct television line. In between

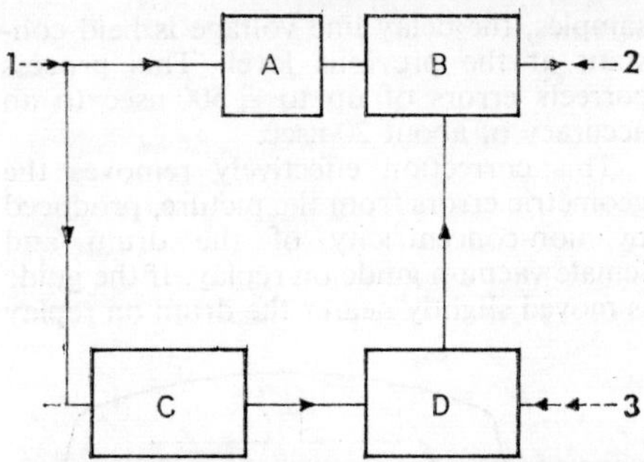

CONTROLLED DELAY LINE TIME CORRECTION. 1. Uncorrected video input. A. Fixed delay. B. Variable voltage-controlled delay line. C. Sync or burst separator. D. Error detector and amplifier. 3. Station syncs or subcarrier fed in to D. 2. Corrected video output.

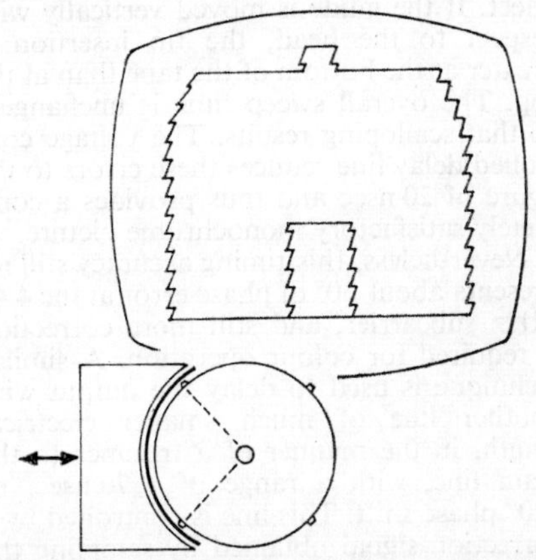

TIP INSERTION ERRORS. Incorrect horizontal alignment of female tape guide in relation to head drum produces venetian-blind type edges on vertical objects. There are about 20 of these discontinuities in the full screen height.

samples, the delay line voltage is held constant at the previous level. This process corrects errors of up to ±500 nsec to an accuracy of about 20 nsec.

This correction effectively removes the geometric errors from the picture, produced by non-concentricity of the drum and female vacuum guide on replay. If the guide is moved slightly nearer the drum on replay than it was on record, the penetration of the heads into the tape is greater; thus the time of replay of a video line changes, because the tape is stretched. This results in discontinuities in the time base which produces the characteristic venetian blind effect. If the guide is moved vertically with respect to the head, the tip insertion is greater at the bottom of the tape than at the top. The overall sweep time is unchanged, so that scalloping results. The voltage controlled delay line reduces these errors to the figure of 20 nsec and thus provides a completely satisfactory monochrome picture.

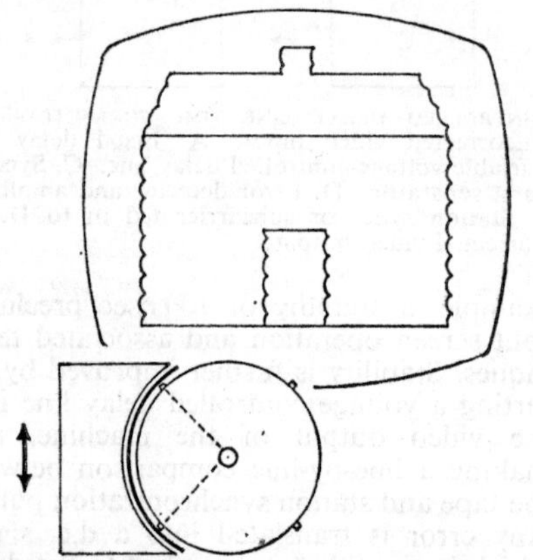

GUIDE HEIGHT ERRORS. Incorrect vertical alignment of female tape guide in relation to head drum produces scalloped edges on vertical objects, again with about 20 to the full screen height.

Nevertheless, this timing accuracy still represents about 30° of phase error at the 4·43 MHz subcarrier, and still more correction is required for colour operation. A similar technique is used to delay the output with another line of much smaller electrical length, in the manner of a trimmer to the main line, with a range of ±70 nsec, or 220° phase shift. This line is controlled by a correction signal obtained by sampling the tape colour burst signal with pulses produced from the station subcarrier; the final remaining error in the output signal is then ±6° at the start of a line. The visible faults increase from left to right across the picture, but with a modern machine these slow changes do not amount to more than 2° in addition to the error present at the beginning of the line.

TIP AND GUIDE POSITIONS. In playback, the mechanical parts of the head must be in the same position with respect to the tape as in the original recording. Both the vacuum guide height and the insertion of the tips of the video heads into the tape are particularly important. In monochrome, a tape reproduced with an incorrect non-standard guide height or tip insertion can be played back satisfactorily by adjusting the replay machine to the same positions used during the recording. In colour, the variations in tape to head pressure across the tape, produced by a non-concentric combination of tape, tips, and guide, produce differential distortions on playback which it is not possible to eliminate using the methods so far described. There are optimum positions for the guide and tips which must be used on both record and replay.

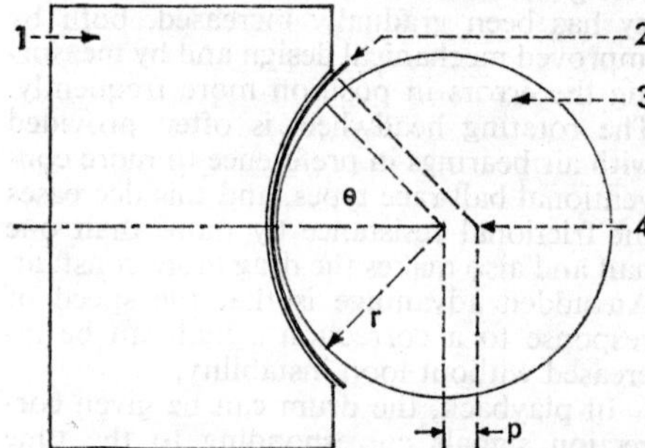

HEAD GEOMETRY MISALIGNMENT. Not to scale. 1. Vacuum guide. 2. Tape. 3. Head drum. 4. Centre of guide curvature displaced by p from centre of drum.

Assuming a correctly recorded tape, a wrong tip or guide position produces timing errors changing regularly across the band of some 16 lines scanned by each head, that is, across each sweep of the head over the tape, which although corrected at the start of each line still has a considerable rate of change of phase. Consequently there are noticeable errors by the end of the line. These vary in magnitude depending on the position of the head in its pass over the tape, so that the picture is broken up into marked bands with the hue changing across each band. The errors are not easily visible for small deviations from the ideal positions except on highly saturated pictures or test signals.

The error resulting from a very small misadjustment may be readily calculated. With each sweep taking 1/960 sec to perform, the maximum variation in the timing for a 0·0001 in. tip insertion error will be 0·14 μsec, or 220° phase error. A similar calculation may be performed for a guide height error; for an error in height of 0·0001 in. the timing is incorrect by 0·025 μsec, or 40° phase error.

Electronic correction can be applied to the start of the line to reduce this, but cannot correct the error that occurs by the end of that line, produced by the rate of change of phase; the tip insertion error produces an almost linear timing difference from line to line, so that the rate of change is almost constant. If it were exactly constant, the phase at the end of the line would be the same for all lines, while still being incorrect. Making allowances for small changes, the phase error varies from 3½° to 5° across the head track in the example mentioned above, so that there is a small differential change of about 1½° across the head band of some 16 television lines.

However, the rate of change of phase for a guide height error is not constant; there is zero rate of change in the centre of a head band, with positively or negatively increasing phase change on either side, and the error at the join of two head bands is about ten times that at the centre. For lines near a head switch, the phase changes by about 4° for a height error of 0·0001 in., while at the centre of a head sweep there is zero error. This 4° change is very objectionable and shows the sensitivity of the replayed signal to guide height errors.

VELOCITY ERRORS. While it is possible to play back satisfactorily a correctly recorded tape with only very minor errors, it is not possible to make copies from one tape to another as the minor errors which are marginally acceptable on the first master tape build up into errors which are not satisfactory. Furthermore, even though the machine can be perfectly set up at the start of a recording, small pieces of magnetic oxide build up on the guide surface and effectively alter both the tip insertion and guide height in a manner which can never be exactly reproduced in playback. Practically then, it is very difficult to produce good colour pictures even using these correction techniques. But the remaining errors are substantially constant on a head-to-head basis, and once the error has been measured for a given television line the same amount of correction can be added each time this line occurs to remove the last traces of errors.

For the correction of these velocity errors, the control signal to the coarse delay line is varied continuously (rather than once per line) in such a manner as to remove the majority of the hue errors. The control voltage is measured at the start of the line and also at the start of the next line; the difference between these two, which represents the velocity component of the change in error, is stored. All lines occuring in one revolution of the head are measured in this way, so that there are about 64 stores in the equipment. In the second and subsequent revolutions of the head this stored information is added to the normal correction voltage so that the delay line is varied continuously across a line and thus is able to correct for the differential components of the various errors. Any change in the amount of correction needed is also measured and the information is changed in the relevant store.

COMPATIBILITY. One of the major problems in colour recording is thus compatibility from one machine to another. Unless the playback rotating head is very similar to the recording head, there are errors. It is difficult to ensure that all four pole pieces on the drum have the same projection, and, if one differs from the remainder, the effect is equivalent to using an incorrect tip insertion. Typical heads selected at random may have differences between tips of up to 0·0001 in., with 0·00005 in. being common. This is the same order of error described above, so that, without the velocity compensation, interchanging of tapes is not altogether satisfactory.

ELECTRICAL PROBLEMS

MODULATING SYSTEMS. One of the most significant improvements of modern machines over the early versions has been in connection with the f.m. recording process. In the more usual communications systems, the carrier frequency and the modulating signal frequency are widely different, but in video recording the two are close together, so that the mathematical approximations usually made are not valid for determining the performance of the machine. The frequency spectrum consists of the carrier modulated by the signal frequencies producing an infinite number of sidebands both upwards and downwards from the carrier; in particular, those extending down to zero frequency pass through the origin and appear as if they were reflected about this point. They can then be near first-order sidebands and can cause interference with these depending on their relative amplitudes.

This can be shown diagrammatically for the f.m. system in use until recently for 625-line recording. The carrier frequency is 5·54 MHz, corresponding to the picture black level; sync tip is 5·0 MHz and peak white is 6·8 MHz. In an NTSC colour system, there are large amounts of 4·43 MHz colour carrier on the luminance signal and, with a peak white signal modulated with this frequency, lower sidebands appear at 2·37 MHz and −2·06 MHz; this latter appears reflected about the zero frequency axis to appear as 2·06 MHz, and beats with the first sideband producing a pattern (known as moiré pattern) at about 300 kHz.

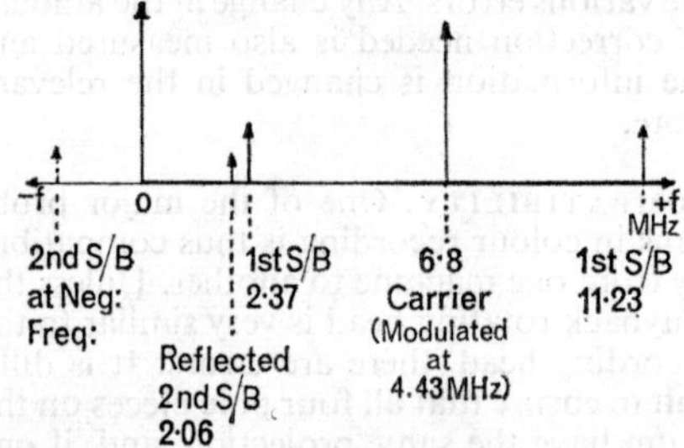

SPECTRUM FOR FM SYSTEM. Modulation at 4·43 MHz gives first sideband above carrier at 11·23 MHz, first sideband below at 2·37 MHz. Second sideband appears at −2·06 MHz (2·37+2·06=4·43) and is reflected to appear at +2·06 MHz. This is only about 300 kHz away from the first sideband.

Interferences are produced in the output as well from the sidebands of the harmonics of the signal beating with the carrier frequency and its sidebands; as the carrier frequency is increased, the spurious outputs increase in frequency; the second harmonic terms move upwards at twice the rate at which the carrier frequency is changed, and the third harmonic terms at three times this rate.

The magnitude of the output from each sideband is dependent on its distance from the carrier, and as the carrier is increased the spurious output changes, increasing with the increasing frequency. At the output of the video demodulator, there is a high quality linear phase low pass filter, with a pass band defined by the video system in use, for example 6 MHz in the 625-line system. Thus, as the carrier is increased, the spurious outputs pass outside the acceptance band of the filter, and so are removed. The filter has a sharp cut-off, so that at the instant of the frequency of the unwanted sidebands passing through the cut-off frequency, the unwanted output suddenly drops; the total level of the interfering signals is thus not changing smoothly but in discrete jumps from one shelf to another. The order of this shelf is defined as the order of the interfering sideband which produces the largest unwanted signal; the system described with the 5·54 MHz carrier is a second order shelf system.

The parameters chosen for the 625-line colour operation give a third-order shelf and use a carrier frequency of 7·8 MHz, with a deviation to 9·3 MHz at peak white. This system is known as high band operation by reference to the low band frequencies used before the introduction of this standard. The moiré patterning is about 32 dB below the wanted signal, with highly saturated colour signals at high luminance levels. Both the amplitude and frequency of the beat pattern change with carrier frequency, so that a test signal such as colour bars is useful to show these variations both in the electronics-to-electronics mode of the machine and in the tape replay. The new standard for the 525-line system uses a carrier of 7·9 MHz, a deviation to 10·0 MHz, and since this gives a fourth-order shelf (the video bandwidth is also less) has a moiré patterning some 38 dB below the wanted signal.

The deviation was determined from a knowledge of the maximum spurious outputs that can be tolerated – the carrier for the chosen shelf – and, following this, the deviation of the luminance signal must be

Television Standard	*Frequencies (MHz)* Sync Tip	Blanking	Peak White	*Signal Deviation (MHz)*	*Signal Noise Ratio (dB)*
405/525 Monochrome ...	4·28	5·0	6·8	1·8	48
525 Colour	5·5	5·76	6·5	0·74	39
625 EBU Low Band ...	5·0	5·5	6·8	1·3	44·5
625 High Band	7·16	7·8	9·3	1·5	43
525 High Band	7·06	7·9	10·0	2·1	46

translated into signal-to-noise ratios. Deviation cannot be increased beyond certain limits since this causes differential gain and phase distortions in the replay amplifiers. The deviations for high and low video frequencies are usually different to accommodate the conflicting requirements, with some 6–8 dB of boost applied to the higher frequencies during recording, with a corresponding cut applied during replay.

TRANSDUCER ERRORS. The playback system must have a comparable performance, and here the first component is the rotating head assembly. With the prime requirements of linearity and freedom from colour differential errors, any resonance in the heads over the pass band must be removed. The final frequency response, while not necessarily flat, must be such that there is no group delay distortion – this usually implies a flat or Gaussian-type response curve. Each head channel amplifier is equipped with various antiresonant networks to balance out the variable head peaks, so that it is a matter of minutes to correct errors after changing heads. For the higher carrier frequencies, it is important to keep the stray capacitances at the head as low as possible in order to maintain a good frequency response and a high output signal. To ensure this, the head preamplifiers are mounted on or at the rotating head assembly.

The r.f. signal must be obtained from the moving part of the head assembly, and older machines used conventional sliprings for this purpose. This gives rise to crosstalk between two adjacent heads which is particularly marked at colour frequencies. The effect on the picture is to produce a few lines of saturation banding at the transition between the heads, particularly between those with sliprings close together. This method of coupling to the heads is also prone to noise and the rings need repeated cleaning. Modern machines use screened rotating transformers which not only remove the crosstalk but are much more reliable.

Even with each head channel adjusted to give a signal free from differential distortions, the overall frequency response of the channel can vary slightly. With the NTSC system, the saturation of the picture is directly proportional to the level of the recovered subcarrier, and small variations in the channel equalizations cause marked head banding. There are also small variations caused by changing contact pressure of the heads or by tape thickness changes, so that the subcarrier tends to change in a random manner. The separate equalizers for each head channel can be adjusted by hand using a test signal, but this does not account for random changes.

In a colour programme, the combination of various tapes into one master is inevitable, and without a signal such as colour bars after every splice, it is difficult to control the equalizers accurately enough so that no saturation banding is visible on highly coloured areas. A change in equalizer settings is almost certainly required if the sections of tape spliced together were not recorded on the same machine. To avoid the necessity of making changes manually, there are devices which measure the replayed chroma burst level on a head-by-head basis, and automatically feed a correction signal to the respective channel equalizers to correct for any errors. This technique substantially removes saturation banding from the picture, and makes colour editing possible between different master recordings and from one machine to another.

PAL RECORDING. The PAL system, which originated in Germany, requires similar precision to make broadcast quality recordings. With the colour phase alternating line by line, any differential errors tend to be cancelled in the viewer's eye, since each line with its error is followed by one with an equal and opposite amount of distortion. PAL uses a delay line technique to measure these errors by comparing these two sequential lines, and by addition to remove them at the expense of halving the vertical resolution. Since nominally the same frequency of subcarrier as NTSC is used, the required stability of the replayed signal is governed by the ability of the decoding system to cancel out the errors. Phase errors of 25° are corrected but, as this represents a timing error of about 16 nsec, both the sync time delay line and the subcarrier phase correcting delay line are needed, and similarly the automatic subcarrier amplitude correcting device; in essence the same machine is required for the two systems, with the exception that the tolerances on playback are made slightly easier in the PAL system.

Certain new synchronizing techniques must be developed for playback in order to replay the signal locked exactly to the station reference pulses. The NTSC system, with the interlaced dot pattern (produced by the correct choice of subcarrier frequency) is such that the colour signal is picture-conscious; that is, for a constant non-moving picture, information is repeated every second television field.

The PAL system differs in that, as well as the picture component changing as in the NTSC system, the burst phase is switched line-by-line by ±45°; thus PAL information cycles on a four-field (two picture) basis.

For this reason, existing NTSC machines, which are purely picture conscious, must be modified to recognize the 7·5 kHz burst change frequency so that both picture and PAL phase may be correctly synchronized.

This technique is not entirely satisfactory, since, in the event of a poor splice or large drop-out in the tape which might interrupt the vertical lock of the picture, the whole four-field phasing process must be repeated, with consequent picture break-up and loss of continuity. This process could take from three to four seconds to achieve, so that a modified form of synchronization may have to be used if the tape has many points of sync loss. Simply locking to the line rate, and not to line and frame, produces a satisfactory colour picture and has a rapid recovery from discontinuities; however, truly synchronous working (for example, with other equipment for techniques such as split-field pictures) is not possible with line-only locking.

SECAM RECORDING. The French SECAM system uses frequency modulation of the colour subcarrier, which is again at about 4·43 MHz. Colour difference signals, such as (R−Y) (the red signal minus the luminance) and (B−Y), are produced in the coder to modulate the carrier sequentially line by line with first one colour difference signal and then the other. The decoder in the receiver must use a delay line to store the colour information from each line so that complete demodulation of Y, R and B (and hence G) can be achieved at all times. The vertical resolution is less than half that of NTSC, and also less than PAL but not by so large an amount.

Using the SECAM system, the problems of colour tape recording are undoubtedly easier, and the results achieved on older machines are much superior to the first NTSC recordings. Being a f.m. system, the process has a natural immunity to small amounts of phase change and considerable resistance to amplitude changes. However, for completely satisfactory operation on the 625-line standard, it is necessary to use the high band carrier frequencies for the f.m. recording.

With a standard recorder, operating in the EBU 625 monochrome standard, the differential phase errors from the tape are of the order of 80° at 4·43 MHz, produced mainly by group delay distortion in the f.m. path. At any one frequency, this appears as phase modulation and is detected by the SECAM decoder giving colour errors; in particular, transitions from white to black are followed by colour fringes. Large areas of constant colour suffer hue alterations as the luminance component changes. The magnitude of these effects depends on the exact parameter of the SECAM system chosen, but in all the proposed versions the errors are still too large for satisfactory broadcasts. These errors do not appear when the high band carrier frequencies are used for the same reasons described for the NTSC system earlier. To keep the phase errors low, the sync correcting delay line is used, and the fine correcting delay line produces a small but visible improvement.

The subcarrier level, which does not change significantly with hue and saturation changes in the picture, is about 200 mV, which is much less than the maximum of some 860 mV in the NTSC system. Thus the moiré patterning produced is small and, in practice, not visible. The compatibility of the monochrome picture is considerably degraded by the continual subcarrier pattern, which on the NTSC signal changes with the saturation and is readily noticeable only on areas of high saturation. In addition to its low level, the SECAM subcarrier is limited in the decoder, as in most f.m. processes, and this again tends to reduce the moiré since this is an amplitude variation. For similar reasons, there is no need for automatic saturation control devices.

TAPE LIFE. The average life of a colour tape is very similar to that of a standard monochrome tape, which is of the order of 75 to 100 playings. This long life depends very largely on the cleanliness of the video tape areas and, for this reason, they should be air conditioned at a slightly higher pressure than surrounding areas. D.P.R.

See: *Differential gain; Differential phase; Gaussian distribution; Group delay; Telerecording; Videotape editing; Videotape recording.*

Literature: *A Color Videotape Recorder*, by C.E. Anderson and J. Roizen, *SMPTE Jnl.*, Oct. 1959; *TV Tape Time Stability*, by A. Harris, *Wireless World*, Nov. 1962; *Techniques for Multiple Generation Colour Video Tapes - Today and Tomorrow*, by C. H. Coleman, *Royal Television Soc. Jnl.*, Vol. 11, No. 8; *Design of a High Band Videotape Recorder*: Four papers by M. O. Felix, C. H. Coleman, P. W. Jensen and J. Roizen, International Conference on Magnetic Recording, IEE (London), July 1964.

☐**Vidicon.** Small, compact camera tube introduced by RCA in 1952, working on the photo-conductive principle. The optical image is focused on to a transparent target of a material such as antimony trisulphide, whose resistance varies with the amount of light falling on it. The reverse side of the target is scanned by an electron beam, and the circuit is completed by a transparent signal plate in contact with the target. The signal plate is returned to earth through a load resistor across which the output is developed.

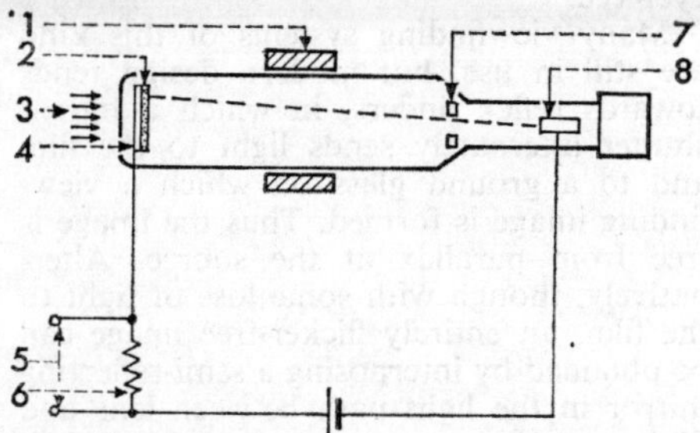

VIDICON. Light from camera lens, 3, is focused on target, 2, a photoconductive layer at cathode potential faced by a transparent signal plate, 4. The scanning beam (dotted) causes a current to flow whose amplitude varies from point to point of the target according to its conductivity. This current sets up a voltage across load resistor, 6, to produce video output, 5. The scanning beam is emitted by cathode, 8, focused by anode, 7, and deflected by scanning coils, 1. Only the basic elements of the tube are shown.

Vidicons are characterized by good sensitivity and resolution, rapid warm-up, but slow response to changes of light intensity at low overall light levels. This produces a characteristic smearing effect in which objects appear to leave traces as they move across the receiver screen.

The vidicon finds wide application in telecines, caption scanners and industrial and educational television.

See: *Camera* (TV)*; Image orthicon; Plumbicon.*

☐**Vidicon Channel.** A camera channel using a photoconductive tube of vidicon type. For broadcast use, modern vidicon channels produce television pictures of high quality, although not as good as those of the image orthicon. Their main disadvantage is that they suffer from lag, or short-term image retention. Vidicon channels are suitable in the studios for presentation and continuity, where there is only limited camera movement, and also for film scanning and some closed circuit purposes, where high incident light levels are available.

On outside broadcasts (remotes) vidicons are used where the size or weight of the camera is of importance, but they are not common in studios where the highest standards of picture quality are required. The camera tube characteristic, together with the small amount of black stretch which is usually incorporated, gives a satisfactory tonal range on the reproducing screen.

It is usual to provide the channel with a lens turret or zoom lens, a viewfinder and remote operation of the camera control unit. Modern developments include increasing the potential applied to the target mesh of the tube, which improves the electron optics, reducing spuring shading signals and improving resolution. Fully transistorized versions of the channel are available.

Cheap television equipment lacking studio quality is often adequate for industrial purposes.

Owing to its small size, convenient shape, and relative stability, the vidicon camera tube, which uses the principle of photoconductivity to produce its signals, has found an ideal use in industrial camera channels. It is suitable for mounting in difficult situations, requires little attention, and will handle a large range of lighting conditions satisfactorily. It is relatively cheap to produce, its small size makes possible the use of small aperture lenses, keeping costs low, and it needs no gamma correction for industrial use. Vidicon channels are not usually provided with a viewfinder. Many versions of the channel have been produced for different specific purposes in industry.

See: *Closed circuit systems; Outside broadcast units; Special events.*

⊕**Vidtronics.** Trade name for method of transferring from colour videotape recording to standard motion picture film for telecine use.

The colour videotape is replayed with the VTR output processed to provide the red instruction only, fed to the input of a black-and-white telerecorder. The tape is replayed again to provide the green information and a third time to provide the blue information. The three lengths of black-and-white negative film obtained are used as colour separation negatives which are then printed in register to produce 35mm and 16mm copies with the corresponding sound track.

See: *Telerecording.*

Viewer. Operator in motion picture processing laboratory responsible for the visual inspection of completed prints by projection in a review room. He must judge whether the final product of the release printing process is satisfactory for general quality of the picture image and sound, its density and colour, and its freedom from physical defects such as scratches and abrasions and dirt particles. He must report any suspected unsatisfactory processing conditions to the appropriate operating department.

See: *Laboratory organization.*

Viewer. Device for the rapid viewing of reels or short lengths of film; usually consists of

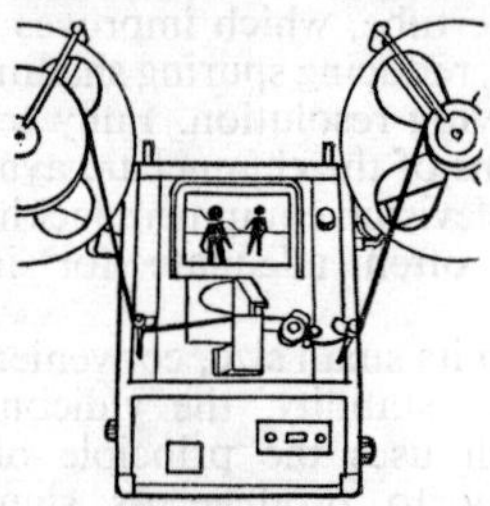

an optical compensator of the rotating-prism type, and hand rewinds for running the film forward or back at will.

See: *Editor library.*

Viewfinder. In motion picture practice, means of viewing, through the camera lens or otherwise, to compose the scene and watch it during filming. In early film cameras, the only method of through-the-lens viewfinding with the film in motion was by looking through the back of the film; with the advent of anti-halation film backings, this became impossible.

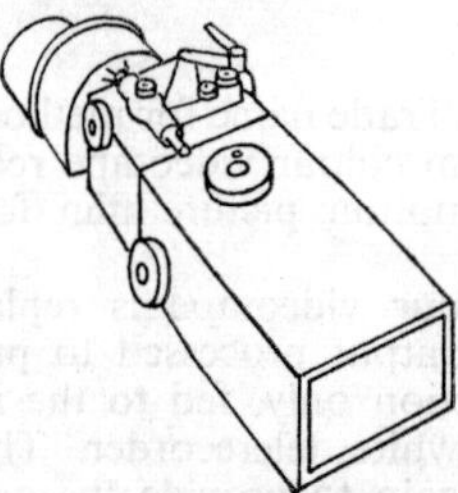

AUXILIARY FINDER. Large finder, with rectangular viewing aperture, mounted to outside of camera. Handwheels operate movable viewing mattes for different lens focal lengths and aspect ratios. Cam action sometimes provided to compensate for parallax errors on close-ups.

Other solutions were sought. With the film at rest, perfect viewing was provided by the rackover system, in which the film was displaced behind the lens, and its place taken by a viewing tube. But with the film running, parallax errors were unavoidable, and ingenious devices were designed to reduce the lateral (or vertical) shift of viewpoint to a minimum, and also to correct the consequent error by angular movement of the finder, sometimes automatically by cam connection with the focusing sleeve of the lens. This type of monitor finder often produced a large image which could be viewed with both eyes and did not require the operator to keep his head close to an eyepiece.

Many viewfinding systems of this kind are still in use, but modern design tends towards reflex finders, in which a mirror shutter alternately sends light to the film and to a ground glass on which a viewfinding image is formed. Thus the image is free from parallax at the source. Alternatively, though with some loss of light to the film, an entirely flicker-free image can be obtained by interposing a semi-reflecting mirror in the light path between lens and film. R.J.S.

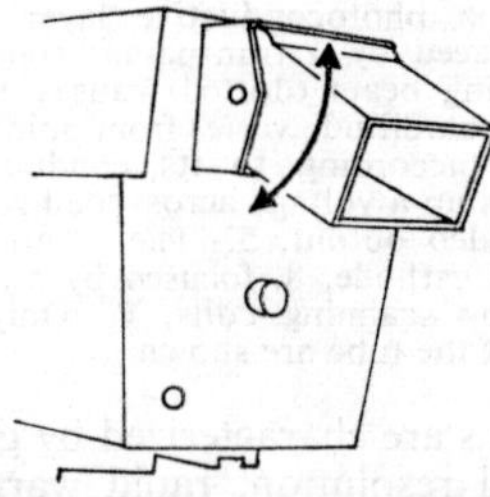

VIEWFINDER. Small CRT screen provides picture viewing facility for cameraman; finder pivots for convenience in making steeply angled shots.

In television practice, a wholly different device is used.

Basically, the modern viewfinder consists of a picture monitor, with a flat-faced black-and-white tube, usually mounted above the camera tube. It is fed with a television waveform from the camera control unit and its scanning circuits must be fed with suitable triggering pulses. Properties which are important in the performance of a viewfinder are resolution, to enable the cameraman to find the best focus quickly and accurately, linearity of both frame and line, to enable him to frame the shot accurately, and stability, to prevent him having continually to adjust the brightness, contrast, width and height controls of the viewfinder. Provision should be made to give indication in the viewfinder of the position of the lens turret. A hood prevents the high level of ambient light in the studio from falling on the viewfinder screen.

A slightly overscanned image allows a small margin of error in framing the shot. It is also usual for provision to be made for an external video signal to be fed into the viewfinder for test purposes, or for lining up captions or special effects (mixed viewfinder facility). The whole of the viewfinder should ideally be able to be tilted independently of the camera to allow for differing heights of cameramen and shots at maximum pedestal extension. J.W.S.

See: *Camera; Cameraman.*

□**Viewing Television.** Many factors influence the arrangement and adjustment of television sets to get the best results, and in the home it is usually much more difficult to get good viewing conditions than in a studio control room. Good operational practice demands that the brightness ratio of the composed picture should, wherever detail is required, be kept to a value which can be tolerably reproduced in good home viewing conditions.

If detail is allowed to range from peak white to black level, as sometimes occurs, the brightness ratio is infinite and a correctly set up receiver loses a lot of the low value details as a result of flooding by ambient room light. The linear grey scale of the receiver also becomes distorted at the lower values. If the picture is composed and held to a range of 40 : 1 or even 30 : 1 it has a good chance of survival, but higher ratios are acceptable for captions or pictures requiring a large black area for dramatic effect. With a range of 40 : 1, the lowest level transmitted is equivalent to a modulation of 16 per cent above suppression level.

In control rooms a figure of about 30 ft. lamberts peak brightness is commonly chosen, allowing for ambient light giving about 1 ft. lambert on the tube face.

Authorities vary over optimum viewing distance. The angle of viewing a television receiver is seldom as large as that of even the old-fashioned 4 by 3 ratio cinema screen. People also vary considerably in their awareness of the lines on the picture. A good average viewing distance is six times the height of the tube face.

To get the best home viewing it is desirable to cut down the ambient light on the tube face to as low a figure as possible, producing 3 ft. lamberts as the practical minimum. Black-and-white tubes having a maximum brightness of upwards of 60 ft. lamberts, there is little problem in getting a good tone scale reasonably undistorted by the room lighting. The lower maximum brightness of colour tubes, however, calls for the composition of the picture in the studio to a contrast range of only 30 : 1 for presentation in normal good home viewing conditions.

See: *Phosphors for TV tubes; Picture monitors; Receiver; Shadow mask tube; Transfer characteristics.*

⊕□**VI Meter.** Volume indicator; a device used for monitoring the amplitude of signals in a recording system to ensure that the recorded sound track is neither under-modulated nor over-modulated. Under-modulation in this sense means that the average programme level is too low, so that the full dynamic range of the medium is not employed, and the signal-to-noise ratio is less than it might be. Over-modulation means that signal peaks make too large and probably too frequent an incursion into the recording medium's non-linear region, where unpleasant distortion is set up.

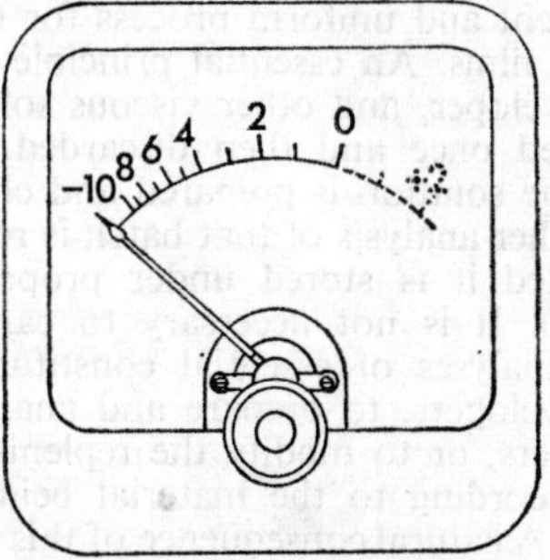

VI METER. Rms meter for monitoring level of studio sound recording. Average dialogue level is held to −6 dB to allow adequate margin for peaks.

The VI meter enables the sound mixer to avoid both under- and over-modulation. It may take the form of a meter, usually calibrated in decibels above and below the level of full modulation. It may consist of a row of neon lamps which light up in sequence as the modulation level is raised. Occasionally, an oscilloscope tube display is used as a VI meter.

A further distinction may be made in the kind of response designed into the VI meter. The neon and oscilloscope types normally give instantaneous response and therefore show peak readings, no matter how short the peak may be. But the ear readily accepts momentary over-modulation, and many VI meters are therefore designed with a damping factor which gives the needle a slower response.

See: *Sound distortion; Sound mixer; Sound recording.*

⊕VISCOUS PROCESSING

Conventionally, motion picture film is processed in deep tanks using bulk solutions of the chemicals (developer, fixer, etc.) required to complete the desired reaction. In recent years considerable effort has been expended in devising another way to process film with the object of attaining greater flexibility, equal or better reliability and a reduced need for chemical and sensitometric control of the process cycle. One such method is the application of the chemicals to the film in the form of viscous solutions. The viscous solution is laid on the emulsion surface, remains there for a predetermined time and is removed by a water spray. The next viscous solution is then applied, or the film is washed and dried.

AIMS AND ADVANTAGES

An important objective of viscous processing is to reduce the amount of chemical and sensitometric control required to achieve a consistent and uniform process for motion picture films. An essential principle is that the developer, and other viscous solutions, are used once and then discarded. Thus, once the solution is prepared and checked, no further analysis of that batch is required (provided it is stored under proper conditions). It is not necessary to carry out daily analyses of essential constituents of the developers, to prepare and analyse replenishers, or to modify the replenishment rate according to the material being processed. A natural consequence of this system of use-and-discard is a reduced risk of process drift, with the result that the amount and extent of sensitometric control can be minimized.

Practical experience has shown that viscous processing has several advantages. Since solution agitation is absent, problems of uneven developments do not occur, as they do sometimes with immersion processing. Additionally, each individual immersion-processing machine may give a different sensitometric result even with equivalent solution chemistry and process time, because it is very difficult to match precisely the degree and pattern of agitation from one machine to the next. Conversely, viscous solution application techniques can give precise and predictable results irrespective of the machine design. Processing machines for viscous application can be made to be extremely versatile. Thus one machine can readily handle several different kinds of film within one day with only brief periods required for adjustment of process times and change-over of solutions.

EQUIPMENT

Several processing machines have been built to apply one or more viscous solutions to the film. Each viscous application stage requires the provision of a solution supply reservoir, a positive displacement gear pump, a heat exchanger and a coating hopper. During the time the viscous solution remains on the film emulsion surface, the film should remain in a temperature- and humidity-controlled atmosphere. Finally, provision has to be made for the efficient removal of the viscous solution.

SOLUTION SUPPLY RESERVOIR. A typical design of solution supply reservoir provides for the viscous solution to be supplied in an air-tight polyethylene-lined container. A probe is inserted into the container to allow

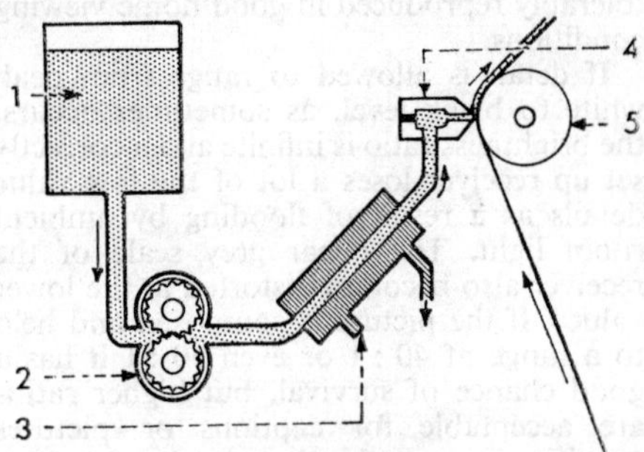

DEVELOPER FEED: SCHEMATIC LAYOUT. 1. Solution reservoir. 2. Gear-type metering pump. 3. Heat exchanger with tempered water flow. 4. Hopper for applying solution. 5. Backup roller.

the solution to be drawn off. Once the solution has been prepared and checked, it is important that it should be stored at a temperature which should not exceed about 70°F. In such conditions, the shelf life of the solution is about 18 months.

POSITIVE DISPLACEMENT METERING PUMP. Care should be taken in the selection of the solution pump as it is very important to ensure a precise and accurate supply of solution. Apart from this prime consideration, the materials used within the pump should be resistant to the chemicals used – particularly bleach solutions.

In designing the equipment, it is necessary to relate the pump speed to the speed of the film through the machine. Except in the case

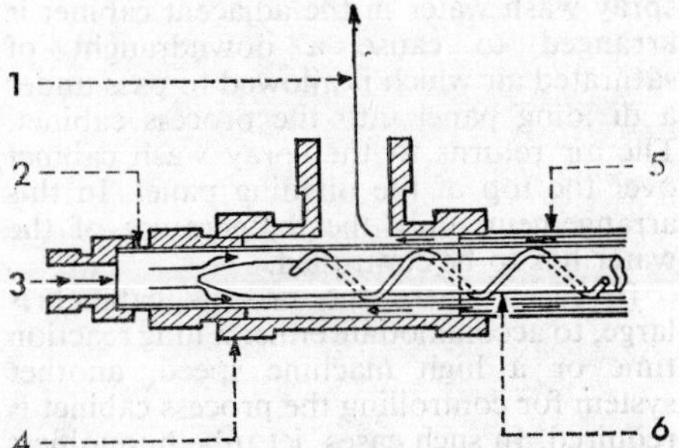

HEAT EXCHANGER. 1. Water flow through commercial heat exchanger tee, 4, extended by 1 in. O.D. tubing, 5. 2. ¾ in. O.D. tubing carrying developer flow, 3. 6. ⅝ in. O.D. rod wrapped with 0·018 in. wire constraining developer to flow in a helical path.

where the equipment is to be used invariably to process one particular film under known fixed conditions, it is desirable to be able to vary the pump speed and film speed in an easily controllable manner. One method of achieving this is to drive both film and pump (or pumps) from a single variable-speed motor-drive unit with a polyphase drive motor of good speed regulation. The unit drives a chain which in turn drives a group of jack shafts, one for each pump and one for the film pacer. The coating thickness is adjusted through the ratio of the jack-shaft gears, and the film speed is adjusted by the variable reduction on the drive unit.

HEAT EXCHANGER. Process times with viscous systems tend to be short and take place at elevated temperatures. The heat exchanger should be designed so that relatively small volumes of solution are heated, and be simple and cheap to construct. A typical design consists of a simple tube and shell with a solid rod through the centre. The rod has a stainless steel wire wound in a helix round the outside. The wire helix just fills the space between the tube and the rod, and forces the solution to flow in a helical path through the exchanger.

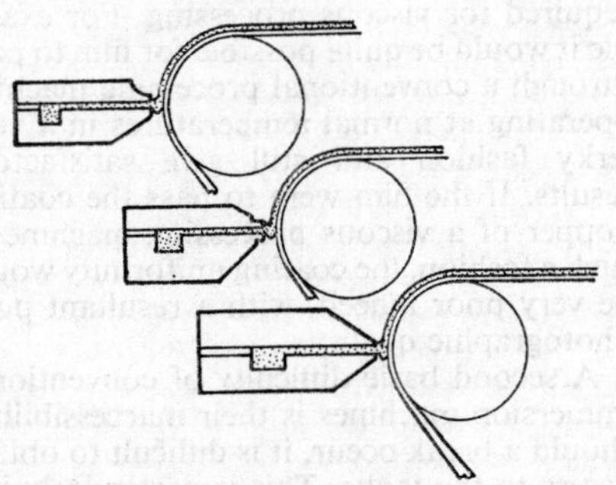

COATING HOPPERS. Three shapes of hopper lips; the straight bevelled shape (*bottom*) is usually found satisfactory. Distance between hopper lips and film has to be carefully controlled.

COATING HOPPER. The hopper design is of the fixed-orifice-extrusion type because the amount of material flowing through the hopper is determined by the metering pump and is independent of viscosity. Typical hoppers consist of a bottom-plate assembly with a distributor box milled in. This is to ensure uniform supply pressure across the width of the orifice. A shim between the top and bottom plates determines the depth of the orifice. The top plate is merely a cover and is designed to be easily removed for cleaning.

Several shapes of hopper lips have been tried. The straight bevelled shape has been found generally satisfactory, and usually has a depth slightly less than the thickness of the coating to be applied.

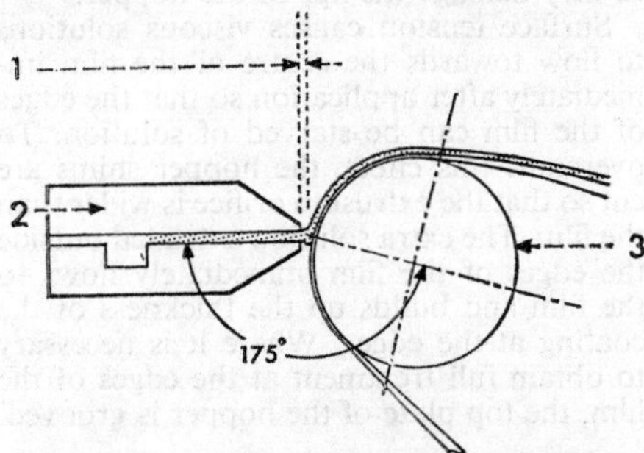

HOPPER AND BACK-UP ROLLER. Width of gap, 1, between lips of hopper, 2, and film is critical; it should be about the same as the thickness of the viscous coating. The back-up roller, 3, should be angled as shown for optimum coating of developer on to film.

The distance between the hopper lips and the film has to be carefully controlled. If the lips are too close, they will disturb the applied coating; if they are too far from the film, the head of solution is lost and air bubbles are formed under the coating. Generally the gap between the hopper lips and the film should be about the same as the thickness of the coating.

The angle between the centre line of the orifice and the outer diameter of the back-up roller is important. When a coating unit was set up so that this angle could be varied, it was found that if the coating hopper was moved down it was forced into the coated solution; if it was moved up there was a tendency for air bubbles to be drawn under

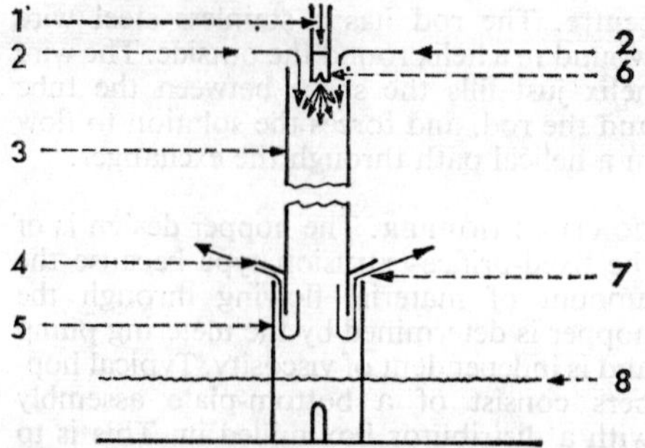

JET TUBE HUMIDIFIER. Tempered water, 1, enters humidifier tube, 3, through spray nozzle, 6, and draws with it an air flow, 2. Splash guard, 5, confines water flow, and deflector ring, 4, causes air driven upwards from water surface, 8, to be expelled, 7.

the coating. An angle of 175° was found to be optimum.

The coating hopper is usually spring-loaded so that splices can readily be accommodated. Metal splices should not be used as they damage the lips of the hopper.

Surface tension causes viscous solutions to flow towards the centre of the film immediately after application so that the edges of the film can be starved of solution. To overcome this effect, the hopper shims are cut so that the extrusion orifice is wider than the film. The extra solution extruded outside the edges of the film immediately flows to the film and builds up the thickness of the coating at the edges. Where it is necessary to obtain full treatment at the edges of the film, the top plate of the hopper is grooved.

PROCESS CABINET. After application of the viscous solution, the film has to be held in a controlled atmosphere for the required time. Generally the film is run over a series of helically wound racks in empty cabinets. These racks are conventional in design, but the transport spools should be deeply undercut to prevent transfer of viscous developer through the perforations on to the spools.

While it is not necessary for the cabinet to be watertight, it should be sufficiently airtight to retain the temperature- and humidity controlled atmosphere required for accurate and reliable processing, and should be insulated with foam sheeting or similar material. Since viscous development usually takes place in a saturated atmosphere, shields should be provided to prevent condensed water vapour from dropping on to the film.

Several methods have been used to control the temperature and humidity in the process cabinet. In one application, the spray wash water in the adjacent cabinet is arranged to cause a downdraught of saturated air which is allowed to pass under a dividing panel into the process cabinet. The air returns to the spray wash cabinet over the top of the dividing panel. In this arrangement, only the temperature of the water has to be controlled.

In cases where the process cabinet is large, to accommodate either a long reaction time or a high machine speed, another system for controlling the process cabinet is required. In such cases, jet tube humidifiers are used. In principle, water sprays located inside the process cabinet are used to saturate and control the temperature of the air. Provision is made to ensure that water droplets do not dilute the viscous solution on the film and cause spots of lighter density. The temperature of the water and, therefore, of the cabinet may be controlled manually with a water-temperature regulating valve, or automatically.

SOLUTION REMOVAL. In cases where the process time is short, it is particularly desirable to remove the solution quickly and efficiently. Flat spray jets operated at pressures between 15 and 35 psi, and impinging on the film at an angle of 45° against the direction of motion of the film, are effective in removing the viscous solution. With higher film speeds it may be necessary to use more than one jet.

PROCESSING MACHINE DESIGN

Existing processing machines may be converted for viscous development or machines may be specifically designed.

CONVERTED MACHINE. A basic difficulty of converted machines is that they do not usually have a uniform drive of the accuracy required for viscous processing. For example it would be quite possible for film to pass through a conventional processing machine operating at normal temperatures in a very jerky fashion and still give satisfactory results. If the film were to pass the coating hopper of a viscous processing machine in such a fashion, the coating uniformity would be very poor indeed, with a resultant poor photographic quality.

A second basic difficulty of conventional immersion machines is their inaccessibility; should a break occur, it is difficult to obtain access to the tanks. This is particularly important because film bearing a viscous solution is very slippery and difficult to handle.

It is preferable to build a machine

specially for viscous processing. Such a machine may be built on one level, with removable side panels for ready access to the development chamber and with a carefully designed film transport system to ensure uniform travel.

In a particular experimental design, the wet section consists of a single cabinet divided along its length by a partial separator to form two processing chambers each with a single rack. One rack is used entirely for running with viscous coatings on the film and the other entirely for spray wash. The film is given its first viscous coating and runs along the rack for the required time. It then transfers to the other chamber for the spray rinse. At the completion of this step, the film is transferred back to the first chamber for an application of a second viscous solution, and so on through the process.

The hoppers and transfer rollers are mounted in such a manner that they can be moved easily to any position along the racks. The machine is, therefore, very flexible in operation and easy to change from one process to another.

PROCESSING TECHNIQUES

BLACK-AND-WHITE FILMS. Experimentally, the whole range of black-and-white camera, duplicating and positive films has been processed viscously. A basic property of viscous processing is its ability to produce better definition than immersion or spray processes.

At the same time, the grain structure is more apparent, not because it is larger but because it is better defined. Sometimes the screen contrast seems to be higher on the viscous processed negative; this is due to the edge effects produced by viscous processing and is not an effect of sensitometric contrast differences.

In commercial practice, viscous processes have not so far been applied to the development of black-and-white films in places where well-established and stable immersion or spray processes already exist. However, a number of television stations and other organizations in the United States, Europe and Africa are using an ultra-fast viscous processor for certain black-and-white films. In all cases, the object has been to provide a processing service which is convenient, readily accessible and does not require a significant capital outlay in chemical mixing and control facilities.

COLOUR FILMS. The experimental processing of some colour films has also been carried out using viscous techniques. In particular Eastman Colour Print Film and Eastman Colour Internegative Film have received detailed study. The results, as with black-and-white films, show an improvement in definition and a slightly more apparent grain.

The results so far indicate that viscous development will give a sharper image than conventional immersion processes but further refinement of the process is still required. Apart from the advantages in technical control which viscous processing would seem to offer, such techniques may also give the processing engineer the opportunity of designing a processing machine which will readily handle a wide range of films of different types. B.J.D.

See: *Processing.*

☐**Vision Distribution Amplifier.** An amplifier which provides a multiplicity of equal outputs from a single input signal. A common arrangement comprises a variable gain video amplifier of high input impedance, thereby permitting bridging of the input signal, followed by an output stage having high current gain. Several outputs are derived by connecting simple resistive networks to this output stage, each providing a signal at the amplitude normally used for distribution (1 volt or 1·4 volts) and of impedance suitable for matching a coaxial cable, usually 75Ω. The variable gain stage is used to produce this output from an input signal of varying amplitude. It is common to provide attenuation as well as gain in order to accommodate input signals greater or smaller than the required output amplitude, the range being typically +6 dB to −8 dB with respect to the output.

Such amplifiers are used for distributing vision signals and pulses. Pulses usually demand a greater output amplitude (about 2–4 v.), and this calls for an output stage of greater capacity. Alternatively, the amplifier specification may be relaxed with respect to that for vision signals.

See: *Vision mixer.*

☐**Vision Drive Signal.** The video signal used to drive the grid (or cathode if the grid is held to a constant potential) in a cathode ray tube.

□VISION MIXER

In a television studio or outside broadcast unit, picture signals available from local sources, e.g., cameras, telecine machines, video tape machines, etc., and similar remote picture sources, are selected in turn in order to form the composite programme. The apparatus used to accomplish this selection is called the vision mixer.

Vision mixers vary considerably in size, facilities, design, complexity and price. So much so that they are often custom-made to suit the needs of a particular location. Standard vision mixers are, indeed, available but many are considerably modified before or after delivery in order to meet the requirements peculiar to the location in which they are to be installed.

The main difference between vision and sound mixers is the fact that a large number of vision sources can simultaneously be shown and viewed on separate monitors, whereas with sound this would cause confusion. Accordingly, vision previewing techniques are easier and a rather different approach has evolved.

BASIC PRINCIPLES

The simplest form of vision mixer in current use is the direct cutting A-B mixer. The unit consists of two separate rows or banks of interlocking latching push button switches. The output of each switch in a bank feeds a common busbar. Each busbar is in turn connected to some form of fading amplifier

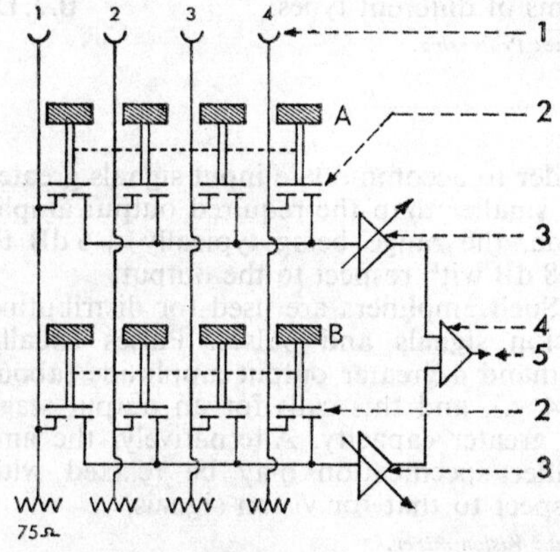

DIRECT CUTTING A-B MIXER. A. A bank. B. B bank. 1. Inputs fed to cut button switches, 2, and to 75-ohm terminating resistors. Switch outputs go through A and B busbars to fading amplifiers, 3, and the outputs are combined in combining amplifier, 4, with mixed output, 5.

which combines the outputs of both fading amplifiers. The input signals to the mixer are fed to the appropriate contacts of corresponding push buttons on each bank and to a 75 ohm terminating resistor. When a button is depressed, the input signal is routed through to the appropriate fading amplifier and thence via the combining amplifier to the mixer output.

FADING AMPLIFIERS. These may be either of valve or transistor design, and their purpose is to fade the vision signal up or down. They are fitted close to the vision switching matrix, and their gain is controlled by potentiometer faders mounted on the vision mixer control panel carrying d.c. potentials arranged so that when the fader is at zero the gain of the amplifier is also zero. The high impedance of the amplifier permits signals to be looped through its input to other units of the mixer after which they pick up their correct termination. Such units bridge the line with a minimum of interference. The outputs from two or more fading amplifiers may be combined into one output by a unit called a combining amplifier; the whole arrangement is then known as a mixing amplifier.

SYNC RE-INSERTION. The position of the faders dictates not only the picture signal amplitude but also the amplitude of the synchronizing pulses comprising part of the complete vision signal at the output of the mixer. Fading out both faders in a simple A-B mixer would, therefore, mean the complete removal of the synchronizing pulses. In these circumstances, equipment such as picture monitors connected directly to the mixer output and relying on the synchronizing pulses for their time-base locking, would run free. To avoid this, synchronizing pulses fed directly from the local synchronizing pulse generator are re-inserted into the vision signal by a line clamp amplifier or stabilizing amplifier immediately following the mixer.

GAP AND LAP SWITCHING. On simple A-B mixers the push button contacts are arranged to break before make. This is gap switching; on depressing a button, the button previously depressed releases and its switch contacts open before the contacts of the operated button close. This prevents the paralleling of the two input sources during the time of the cut, a necessity when one considers the mixer as a whole.

For instance, suppose input one is selected on both banks and bank A is on-air. It is now desired to mix to input two. Input two must be selected on the B bank prior to the mix. If the switch contacts on the B bank were to lap, i.e., make before break, input two would momentarily be paralleled with input one and a flash of the input two picture would be seen at the output of the mixer via the A bank.

Gap switching means, of course, a momentary break during each cutting operation undertaken on the on-air bank, and sync reinsertion is essential at a later stage if severe disturbances are to be avoided on the received pictures. However, despite sync reinsertion, gap switching means moments of black, when no picture is transmitted, which can be disturbing to viewers. This limitation is overcome in more sophisticated mixers.

CUE LIGHTS. As well as switching, fading or mixing the vision signal, the vision mixer is usually required to switch the camera-on-air cue lights, picture monitor indicator lights and perhaps indicator lamps on the mixer control panel itself. Vision mixers are usually operated in locations of low illumination, and indicators are necessary to show which button is actually depressed or which fader or bank is on-air.

On simple mixers, cue lights are usually operated by additional contacts on the push-button switches. These are generally wired in series with switch contacts on the faders, which are open when the fader is fully faded out and closed the instant the fader is moved towards the on position, remaining closed throughout its further travel. This allows cue lamps associated with an input on-air via one bank to be lit while inputs selected on the other bank not on-air remain out. When mixing takes place between the A and B banks, the indicators associated with both inputs selected light during the period of the mix.

MASTER FADER. This is located on the vision mixer control panel and operates on its own fading amplifier which forms part of a following line clamp or stabilizing amplifier. Such an arrangement permits pictures mixed by the A-B bank faders to be faded down to or up from black level, while still retaining the desired mixing ratio. This is often needed at the beginning and end of programmes where captions are superimposed on the opening or closing shots.

The master fader is located in that part of the circuit where it operates on the picture signal only and does not affect the amplitude of the synchronizing pulses; it can, therefore, be used to fade non-synchronous signals selected on the mixer.

PREVIEW SELECTORS. In some locations, the number of picture monitors that can be made available is limited, due to cost, space or even desirability from an operational point of view. In any event, it is frequently impossible to tie one picture monitor to each input of the mixer. In such circumstances one or more switchable preview monitors may be fed via additional rows of push button switches so that the operator may select any signal for display on the monitor. These preview selectors are arranged to switch with a gapping action, the momentary break in the signal during switching being of no great importance.

This arrangement works well enough in locations where all the preview selectors may be mounted within the vision mixer control panel, such as in outside broadcast units, or where all the operators can view common picture monitors. In studios, however, where individual preview selectors may be required in locations remote from the vision mixer, e.g., in the vision control room, lighting control room, sound control room or even on the studio floor, the use of direct switching cut buttons leads to considerable wiring difficulties.

RELAY SWITCHING MATRIX. A better solution to the problem of providing multiple switched outputs is to use a common switching matrix in which electro-magnetic relays controlled by the selector push buttons switch the vision signals. The relays may be operated remotely from the control panels, and one relay matrix may be built to serve all the individual selector panels situated in the various control rooms associated with a particular studio. Cue lights, etc., may be operated from additional contacts on the relays.

This method permits most of the complex video and cue light wiring to be confined to the relay matrix, thus reducing the cost of the mixer as a whole and the complexity of the control panels. During a production, the mixer control panel is in constant use, and it is often impossible to gain access to it even to rectify a simple fault. The switcher matrix on the other hand is generally rack-mounted in the vision control room where access is easier.

DO-NOT-MIX INDICATORS. It is not possible to mix non-synchronous signals together and still retain broadcastable pictures, because the synchronizing pulses and the picture signals of non-synchronous signals will almost certainly be out of time with one another. This is well known to all vision mixing operators but sometimes, during a production, mixing is attempted in error. As an added safeguard, a large indicator lamp incorporated in the control panel lights whenever a non-synchronous signal is selected.

The indicator may be operated in various ways, the simplest being from additional contacts on the push buttons or relays of nominated inputs to which non-synchronous signals are always fed. Depressing the buttons corresponding to any of these inputs operates the "do-not-mix" lamp. This simple method does not, however, meet all requirements. The studio, or outside broadcast vehicle as the case may be, might slave (genlock) to the incoming remote signal, thus making it synchronous. The "do-not-mix" indication would then be incorrect. Many circuits are in current use to extinguish the "do-not-mix" indicators in such circumstances, but perhaps the simplest and best is that using a device known as a sync comparator to operate the indicator in the first place.

SYNC COMPARATOR. The unit is incorporated, in the case of simple mixers, across the outputs of the A and B bank selectors. The two inputs (one from the A and one from the B bank) are fed to sync separators whose outputs are combined in such a way that if both inputs are synchronous, no output is obtained, but should the inputs be non-synchronous or even mis-timed one with the other, an output is produced. This output can be made to operate a suitable relay which in turn operates the "do-not-mix" light.

The sync comparator may also be used to operate the sync changeover relay in the following line clamp or stabilizing amplifier. This relay selects either the synchronizing pulses from the station sync-pulse generator or the synchronizing pulses which form part of the incoming video signal, depending on whether this is synchronous or non-synchronous. These pulses are later combined with the outgoing picture to reform the complete video signal. Should an incoming non-synchronous source be made synchronous by slaving and this input be selected on the mixer, the relay must remain in its normal position and feed syncs from the local sync pulse generator to the picture/sync combining circuit of the line clamp or stabilizing amplifier. If, for some reason, the slave lock is lost and the source reverts to its non-synchronous condition, the relay must operate to feed syncs derived from the incoming signal instead.

The use of sync comparators to control the "do-not-mix" indicator and the sync changeover relay, ensure that the operation of these devices is completely automatic.

MIXER TYPES

THE A-B CUT MIXER. In this arrangement, the outputs of the A and B banks are combined and returned to one input of a separate cut bank. The cut bank is arranged to lap-cut, thus eliminating the momentary period of black during cuts which is a feature of the simple A-B mixer.

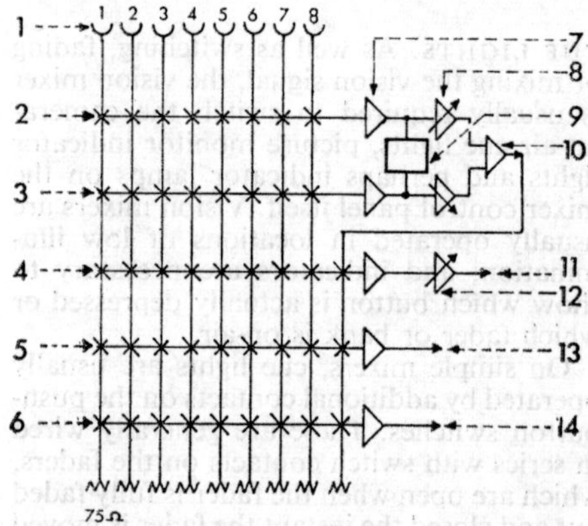

A-B CUT MIXER. 1. Inputs. 2. A bank. 3. B bank. 4. Cut bank. 5. Preview One, with output, 13. 6. Preview Two, with output, 14. Outputs of A and B banks are fed to isolating amplifiers, 7, with sync comparator, 9, and are individually faded, 8, and subsequently combined, 10, before being returned to an input of cut bank, 4. Output of this bank is through its own isolating amplifier, followed by master fade, 12, to mixer output, 11.

This mixer permits cutting along the cut bank from any input to any input, mixing between synchronous sources selected to the A and B banks and cutting to superimpositions which may be adjusted to the correct proportions and previewed before being taken on-air. Non-synchronous signals may also be faded by a control panel-mounted fader operating on the following line clamp or stabilizing amplifier.

THE A-B-C-D CUT MIXER. Both the A-B and the A-B cut mixers permit the super-

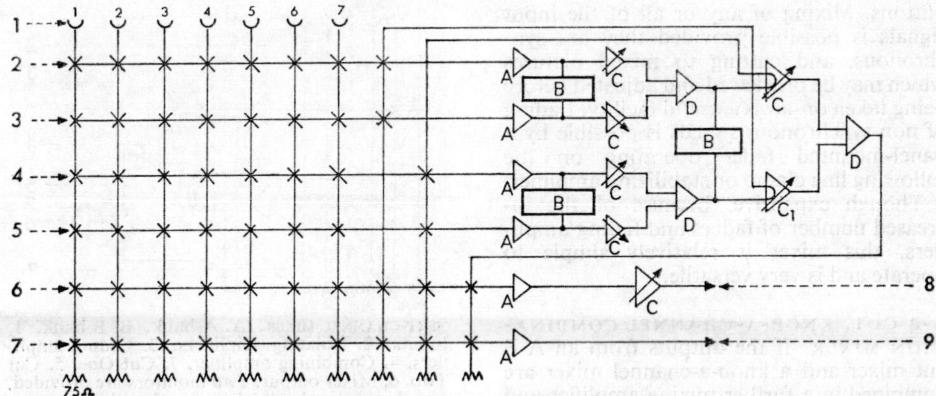

A-B-C-D CUT MIXER. 1. Inputs. 2. A bank. 3. B bank. 4. C bank. 5. D bank. 6. Cut bank. 7. Preview. 8. Output. 9. Preview. A. Isolating amplifiers. B. Sync comparators. C. Fading amplifiers. C_1. Master fade. D. Combining amplifiers. This mixer enables two sets of two sources to be mixed separately, and the resulting outputs to be themselves mixed.

imposition of two input sources only. However, in certain productions it is possible that three or four inputs may need to be mixed. A combination of these mixer types is then required, and the resulting apparatus is known as an A-B-C-D cut mixer. Five banks of selectors are employed; the first four feeding the A-B-C-D faders respectively, are arranged to gap, and the fifth, the cut bank, to lap. The output of the A-B combining amplifier is looped via an additional fading amplifier back to an input of the switcher matrix. The output of the C-D combining amplifier is similarly treated. The outputs of the two additional fading amplifiers are combined and again fed back to the switcher matrix (cut bank only).

This mixer permits cutting from any input to any input by operation of the cut bank push buttons and cutting or mixing sources at the output of either the A-B and C-D combining amplifiers, i.e., cutting or mixing from superimposition to superimposition. Non-synchronous signals may also be faded by a panel-mounted fader operating on the following line clamp or stabilizing amplifier. This type of mixer can be difficult to operate on fast and complicated productions and is not particularly popular with vision mixing operators.

KNOB-A-CHANNEL MIXER. As its name implies, this mixer utilizes a fader bank, with a fader knob on each input channel. The outputs of the fader amplifiers are combined and the signal returned to an additional input on the relay matrix. A cut bank is also incorporated and the output of the fader combining amplifier may thus be cut up on-air. The cut bank selectors must be arranged to lap.

The "knob-a-channel" configuration permits cutting from any input to any other input by operation of the cut bank push

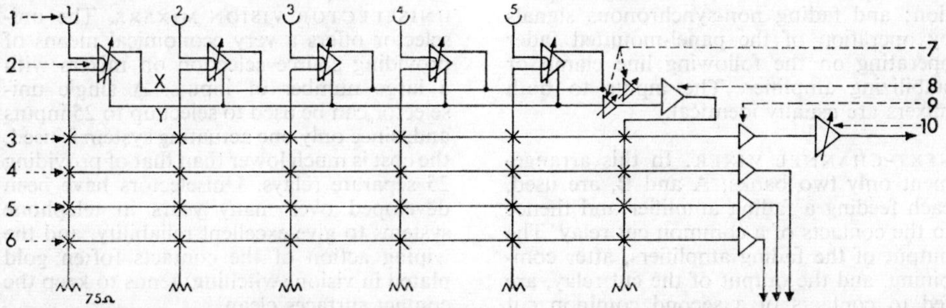

KNOB-A-CHANNEL MIXER. 1. Inputs. 2. Fading amplifiers, with a fader knob for each channel. 3. Cut bank. 4, 5, 6. Previews One, Two and Three. 7. X and Y bank fading amplifiers. 8. Combining amplifier. 9. Isolating amplifiers. 10. Master fade.

buttons. Mixing of any or all of the input signals is possible provided they are synchronous, and cutting to mixed pictures which may be previewed and adjusted before being taken on-air is a useful facility. Fading of non-synchronous signals is possible by a panel-mounted fader operating on the following line clamp or stabilizing amplifier.

Though expensive, because of the increased number of faders and fading amplifiers, this mixer is relatively simple to operate and is very versatile.

A-B CUT, KNOB-A-CHANNEL COMBINATION MIXER. If the outputs from an A-B cut mixer and a knob-a-channel mixer are combined in a further mixing amplifier and small switcher, cutting from any input to any input can be achieved by operation of the cut-bank push buttons on either mixer. The inputs to both mixers are usually identical.

Mixing of any or all of the input signals, provided they are synchronous, can be executed on the knob-a-channel mixer, while mixing of any two input signals, provided they are synchronous, can be accomplished on the A-B cut mixer. Cutting or mixing facilities from superimposition to superimposition (provided the signals are synchronous) and the fading of non-synchronous signals by a panel-mounted fader operating on the following line clamp or stabilizing amplifier, are also provided.

KNOB-A-CHANNEL, KNOB-A-CHANNEL COMBINATION MIXER. The outputs from two separate knob-a-channel mixers can be combined in a further mixing amplifier and small switcher to permit cutting from any input to any input by operation of the cut buttons on either mixer; mixing any or all of the inputs (provided they are synchronous) on both mixers; cutting or mixing from multi superimposition to multi superimposition; and fading non-synchronous signals by operation of the panel-mounted fader operating on the following line clamp or stabilizing amplifier. The inputs to both mixers are usually identical.

NEXT-CHANNEL MIXER. In this arrangement only two banks, A and B, are used, each feeding a fading amplifier and thence to the contacts of a common cut relay. The output of the fading amplifiers, after combining, and the output of the cut relay, are fed to contacts of a second common cut relay, the output of which represents the output of the mixer. The control panel

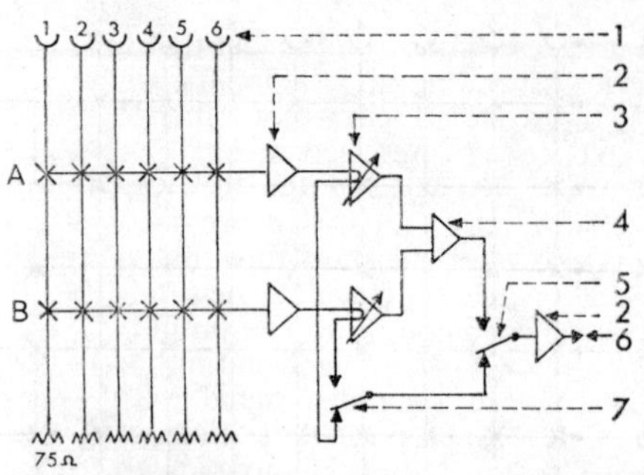

NEXT-CHANNEL MIXER. A. A bank. B. B bank. 1. Inputs. 2. Isolating amplifiers. 3. Fading amplifiers. 4. Combining amplifier. 7. Cut One. 5. Cut Two. 6. Mixer output. Two monitors are provided, one always showing bank on-air, the other bank not on-air.

utilizes a single row of push buttons and the circuit is arranged so that they operate source selectors of the bank not on-air.

Usually two picture monitors are provided, one always showing the on-air picture and the other the output of the bank not on-air, thus making preview automatic. A cut button switch operates the first common cut relay and allows the operator to cut between sources selected on the A and B banks. A fade button switch operates the second common cut relay and switches the output of the combining amplifier to the mixer output. A fade or mix is then made by fading down one channel fader to black and fading up the other. When this latter fader has reached the top of its travel, the first common cut relay operates followed by the second common cut relay, and the mixer output is restored to normal, i.e., to the output of the first common cut relay.

This type of mixer allows the operator to cut between sources, fade, mix and superimpose sources and to cut to mixed sources.

EQUIPMENT DESIGN

UNISELECTOR VISION MIXERS. The uniselector offers a very economical means of providing source selection on mixers with a large number of inputs. A single uniselector can be used to select up to 25 inputs and, since only one actuating system is used, the cost is much lower than that of providing 25 separate relays. Uniselectors have been developed over many years in telephone systems to give excellent reliability, and the wiping action of the contacts (often gold plated in vision switching) tends to keep the contact surfaces clean.

The big drawback to uniselectors is the time interval which must elapse during source reselection and the fact that this time varies depending on how far the wiper arm has to travel. Uniselectors are therefore only employed in preselect mixers where the source is first selected and then cut, mixed or faded up on-air, or to provide preview selection when the short time interval taken by the wiper to reach its new position is relatively unimportant. The output of each uniselector is wired via a muting relay; this replaces the outgoing signal with black level and syncs during the time when the wiper arm is in motion and avoids a quick succession of pictures appearing at the output, as the wiper arm passes over the contacts fed with the various vision input signals.

Uniselectors may be controlled in vision mixers either by touch button switches or by the latching, interlocking type. Touch buttons have to be held down until the wiper arm has reached its desired position, which can be indicated by a small lamp either beside or within the button and operated from additional contacts of the uniselector. Latching switches do not require to be manually held down until the selection has been completed and are thus more popular with operators.

As more than one switch bank or selector level is provided on each uniselector, the device may be used to select the appropriate sound, communication circuits and indicator lamps, coincident with the selection of the video signal.

MATRIX SIZE. Before embarking upon the design of any vision mixer, it is necessary to decide on the matrix size, i.e., the number of inputs and the number of outputs required. Too small a matrix will limit the facilities available; if it is too large it is uneconomical, physically cumbersome and has reduced frequency and crosstalk performance. In recent years, manufacturers have been disinclined to make a different size of matrix for each vision mixer ordered, and have concentrated instead on producing two or three standard matrix units of known performance. With these basic bricks, they are more easily able to supply the customer with any design of mixer he desires at competitive prices and acceptable delivery dates.

The following may be taken as typical of currently available equipment: 6 input 3 output; 12 input 3 output; 14 input 6 output. Should it be desired to provide for, say, 18 inputs with 3 outputs, two standard matrices may be used provided certain precautions are taken.

MIXER RELAYS. Earlier mixers used open type telephone relays and were unsatisfactory for switching video signals. They had inadequate mechanical stability, poor dust protection and poor contact action owing to the unsuitability of the contact material used. Frequent contact cleaning and adjustments were necessary, and relays became regarded as thoroughly unreliable for vision switching.

Plug-in relays sealed in cans are used in some more modern matrices but these suffer from an insufficient number of contacts. Failure to seal contaminants out of the can has also been a major problem. A great advantage of plug-in relays, however, is rapid replacement in the event of failure.

The latest relay matrix manufactured in Britain uses plug-in relays designed primarily for the computer industry and having a large number of change-over contacts. Each moving contact is a pair of independent gold plated wires, comb-operated.

FREQUENCY RESPONSE. The capacitance of each input busbar is quite high and, in addition there is a deliberate capacitance loading at each crosspoint, of the order of 100 pF, via the contacts of the relay when that relay is not energized. This loading is incorporated to reduce the inevitable variation in capacitance and thus its effect on the frequency response when varying numbers of banks are made to select any one input. The total input capacitance may be of the order of 700 pF. The reactance at 5 MHz of 700 pF is only 45Ω and this is too low to be placed directly across the 75Ω input coaxial cable. This difficulty is overcome by loading the input busbar with an inductance at each crosspoint, so that it looks like a 75Ω resistive load to the video source.

D.C. BLOCKING. Vision inputs from many sources may have d.c. components superimposed upon them. The value of this d.c. varies from source to source, and cutting from one source to another can produce unpleasant surges. This can be eliminated by blocking the d.c. with a very large capacitor placed between the incoming coaxial cable and the matrix. Another method is to provide an L-C bridged T input circuit at each input to the mixer; this has the advantage that it provides a return path for the superimposed d.c. signal, a requirement of some apparatus. Yet another method is to feed each input to the matrix via an isolating amplifier.

RELAY CONTROL. Whether the relays in the vision switching matrix lap or gap depends not only on the relays themselves, but also upon the controlling push button switches. If the push buttons lap, then, of course, the relays they directly control will also lap.

Latching type interlocking push button switches are generally mass produced and have wide mechanical tolerances. They usually gap or lap at random. Precision types in which the gap-lap action is predictable have been made, but their expense is such as to make their inclusion in vision mixers uneconomic.

To circumvent the problem, touch button switches are used instead; these are non-latching and non-interlocking, and their contacts make only so long as the button is held down by the finger against the action of a spring. The relays normally used with them are two coil devices each having an operating coil and a separate hold coil.

CROSSTALK. The breakthrough of an unwanted signal is termed crosstalk. In a relay matrix, it is most visible when all inputs but the wanted one are synchronous and the wanted signal is non-synchronous, and when all outputs take the wanted signal. In these conditions, the very high frequencies associated with the sharp edges of the sync and blanking pulses of the synchronous signal to some degree break through on to the wanted non-synchronous signal and may be seen moving about in the background of the picture.

The crosstalk increases as the number of inputs and outputs increases, i.e., as the size of the matrix or number of crosspoints increases. A crosstalk ratio of 44 dB at 5 MHz is adequate for most matrices.

In a uniselector matrix, crosstalk is a first order problem only from the bank contacts adjacent to the selected one and so it is independent of the number of inputs. The design of the uniselector is fortunate in that the capacitance between adjacent contacts is low, of the order of 0·4 pF giving a crosstalk isolation of some 60 dB at 5 MHz.

SEMICONDUCTOR SWITCHING. Instead of relays, uniselectors or direct cutting push-button switches, it is possible to use semi-conductor switching circuits at each cross-point in the vision matrix. The popularity of this form of switching has grown enormously in recent years due to the fact that solid state switches are free from the hazards of dust and contact contamination which plague mechanical systems. Extreme reliability in switching can thus be obtained. The main disadvantage is that each switch can control one circuit only, i.e., the vision circuit in vision matrices. The various cue lights, etc., must, therefore, be switched by an additional mechanical matrix or by contacts on the control panel push buttons, necessitating these buttons being of the latching type. Both methods considerably increase the cost of the mixer and, in the latter case, the complexity of the control panel.

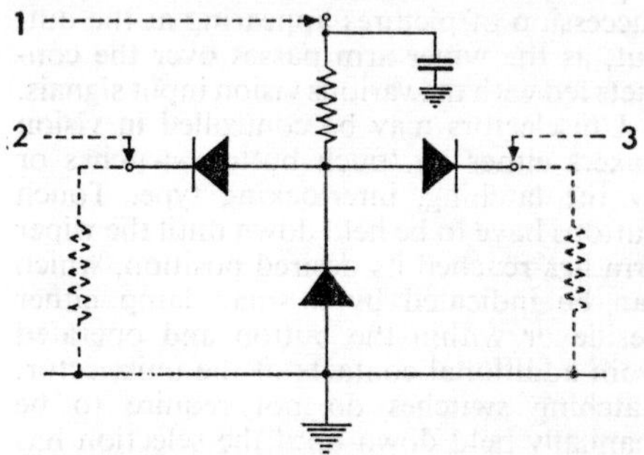

THREE-DIODE SEMICONDUCTOR SWITCH. Typical circuit making use of diode high forward/reverse impedance ratio. 1. Control, positive on, negative off. 2. Input. 3. Output.

There are a great many semi-conductor switching circuits in existence. They all rely on the excellent forward and reverse impedance ratios obtainable with semi-conductor diodes. The differences lie mainly in the choice of circuit arrangements, e.g., some use only two diodes while others use three.

CUT IN BLANKING. Diode switches may be caused to operate during the field-blanking interval and thus cuts from source to source are possible outside the active picture time with negligible disturbance on the final video output. Two control signals are necessary comprising a d.c. voltage with one usually derived from the field synchronizing pulses superimposed.

The operation is as follows: a d.c. control voltage of around 20 volts is applied through a high resistance, and the polarity of this is reversed by operation of the cut button. At the same time, the cut button selects large amplitude positive or negative pulses derived from the field synchronizing pulses. A capacitor is connected between the control point and earth. In the "on" position, a negative d.c. voltage is applied to the control point and the capacitor and, at the same time, selection of positive pulses from field syncs is made.

On arrival of the first of these positive pulses, the switch is turned on because the pulse voltage more than offsets the negative d.c. voltage, so making the control point and charge on the capacitor positive. The time constants are arranged so that the capacitor charge does not appreciably change during the time of one field and is recharged by each successive field pulse; thus the switch remains on.

On selecting another source by its push button, the button already made is released and this applies a positive d.c. voltage to its control point together with large amplitude negative pulses, the first of which reverses the polarity of the capacitor charge, thus cutting off the switch. Cut in blanking may also be incorporated in mechanical relay switchers by using two matrix outputs feeding two diode switches whose outputs are combined and controlled by the necessary control voltages. The relays then virtually preselect the signal, the final cut being accomplished by the diodes. Both relay banks and the diode switches are operated from a single row of push buttons.

VISION MIXER CONTROL PANELS. Owing to the variety of vision mixers in current use, the design of the control panels differs widely. No other piece of equipment used in television broadcasting differs so much in utilization and, therefore, the design variations are infinite.

Because of these variations, control panel layout is usually left to individual purchasers who have their own specific ideas and requirements and who instruct the manufacturers accordingly. For this reason, it is possible to deal only with the basic requirements of control panels which are invariably mounted in a horizontal position as part of the production control desk. The need to leave desk space of some 14 in. in front of the control panel for script purposes, and the fact that in Britain most vision mixing operators are female with limited reach, means that very compact units have to be manufactured.

Cut buttons are usually along the bottom of the panel nearest to the operator's hands. During a broadcast, the operator may be viewing the script or the picture monitors,

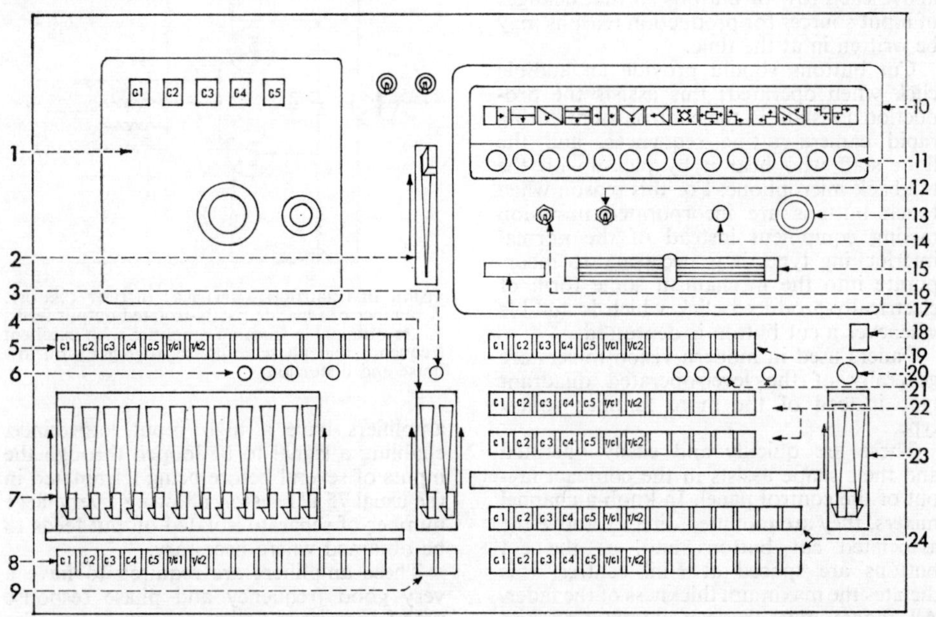

CONTROL PANEL LAYOUT OF KNOB-A-CHANNEL A-B CUT MIXER. 1. Control panel of "blinge" unit to produce ripple effects by adding a sine-wave signal to the line drive pulses fed to the camera. Selection of cameras 1-5 by buttons. 2. Master fader (line clamp). 3, 6. Do-not-mix warning lights. 4. X-Y group faders. 5. Preview buttons. 7. Knob-a-channel faders, X mixer. 8. Cut bank. 9. X-Y cut buttons. 10. Special effects facilities selected by switches, 11. 12. Wipe/inlay select switch. 13. Manual/auto control switches. 14. Cut select switch. 15. Special effects wipe control. 16. Normal/reverse switch. 17. On-air warning light. 18. Emergency cut button. 19. Emergency cut selectors. 20, 21. Do-not-mix warning lights. 22. Wipes A & B selector for special effects. 23. A-B faders, Y mixer. 24. A-B select for Y mixer.

and the buttons are often selected and depressed by touch. Accordingly, the interlocking latching type of button which remains down when depressed and has a definite click action is preferred by the operators. General opinion seems to confirm that the buttons should be spaced at ¾-in. centres with a separation of ⅜-in. which sets the maximum size of the button at ⅜-in. diameter.

Because many of the buttons are covered by the operator's fingers during a production, separate indicator lamps mounted on the control panel immediately above the cut buttons are preferred to indicator lamps which form an integral part of the cut button itself.

Where a row of buttons exceeds ten in number, it is desirable to break them up into short sections of four or five to assist the operator in finding the correct button quickly. On large mixers where many cut and preview buttons are used, it is common practice to colour code them for the same reason.

Identification strips of white plastic material are usually mounted immediately above each row of buttons so that changes of input sources for production reasons may be written in at the time.

Cut buttons should provide an audible click when operated; this assists the production assistant in the control room during rapid camera-cutting sequences, and the floor staff who hear it by pick-up on the talkback microphone. For this reason when touch buttons are incorporated in vision mixing equipment instead of the normal interlocking type, it is necessary to incorporate into the mechanism some form of electrically-produced click which is audible whenever a cut button is depressed.

Faders used in modern vision mixers are generally of the lever-operated quadrant type instead of the more familiar rotary type.

These are quickly and easily operated and their shape assists in the compact layout of the control panel. In knob-a-channel mixers, they are mounted directly above the associated cut buttons and, as the cut buttons are spaced at ¾-in. centres, this dictates the maximum thickness of the fader. All faders must be self-illuminating and generally give two indications: a white light operates as soon as the fader is moved toward the on position, and remains lit the fader is taken down again or taken on-air, in which latter case the indication changes to red.

ISOLATING AND VISION DISTRIBUTION AMPLIFIERS

ISOLATING AMPLIFIERS. The function of isolating amplifiers is to isolate the mixer matrix from other external apparatus. They may be of valve or transistor design and have a very high input impedance, a 75Ω output impedance and unity gain. They are required to have a very good frequency and phase response with low noise and distortion characteristics and good gain stability, because the video signal may have to go through several of them in series as it passes through the mixer. They are usually fitted very close to, or within, the case containing the switching matrix.

VISION DISTRIBUTION AMPLIFIERS. These may be of either valve or transistor design and their purpose is to provide several isolated outputs from one input signal. There are usually four or five outputs, each capable of delivering a signal amplitude of 1 volt peak-to-peak into a 75 Ω load. The

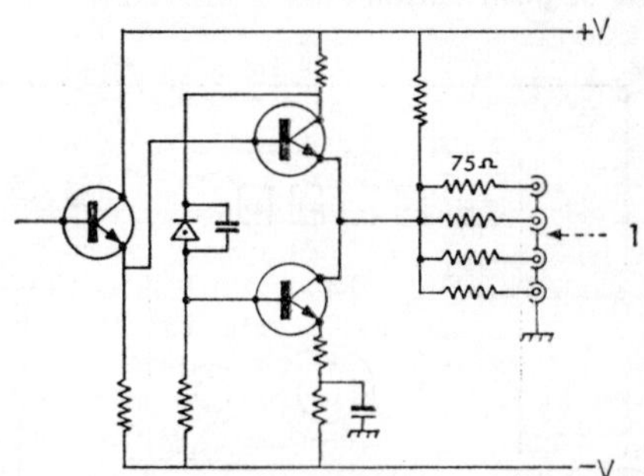

VISION DISTRIBUTION AMPLIFIER: OUTPUT CIRCUIT. A number of separate partly-isolated output feeds, 1, are obtainable from an amplifier with excellent frequency and phase response, together with low noise and distortion.

amplifiers have a high input impedance, enabling a signal to be looped through the inputs of several before being terminated in the usual 75 Ω resistor, allowing a very large number of separate isolated output feeds to be obtained where necessary.

These amplifiers are required to have a very good frequency and phase response with low noise and distortion characteristics and good gain stability, since the video signal will pass through many of them en route from source to transmitter. Their gain is usually variable over the range +6 to −6 dB, so that small variations of input amplitude may be corrected where they arise due to losses encountered in the interconnecting lines between the various control rooms, etc.

SIGNAL STABILIZATION AMPLIFIER
Television signals are frequently subjected to various distortions due to the imperfect low frequency and phase response of equipment, particularly that involved in switching or radio and cable links. In addition, interference such as hum pick-up from electric supply mains and high frequency noise is often impressed upon the video signal, causing impairment. If the waveform of the particular distortion is long, compared with the duration of one television line, it may be sampled during the line-blanking period and an error difference signal between the distorted and undistorted waveform derived. This error voltage is then used to reduce the offending distortion, and equipments using this technique are known as signal stabilization amplifiers or line clamp amplifiers.

CLAMPING CIRCUIT. The basis of either amplifier is a clamping circuit activated by specially-generated pulses which cause it to be operative during the back porch period and virtually disconnected from the video waveform at all other times.

The arrangement of a series capacitor clamp is that the input video waveform is fed to the control grid of a valve via a series capacitor with the usual grid leak omitted. The control grid is also returned to one of the null points of a switch in the form of a Wheatstone bridge with the other null point connected to a reference potential. The bridge itself is composed of four matched diodes. Positive-going clamp pulses from a multi-vibrator triggered by video synchronizing pulses are amplified and fed to one input of the bridge. These clamp pulses are also phase reversed and fed to the other side of the bridge. The pulse duration is arranged to be slightly shorter than the back porch period and is timed to occur immediately after the line synchronizing pulse.

VARYING GRID POTENTIAL. The clamping pulses cause the diodes forming the Wheatstone bridge to conduct, switching the reference potential to the grid of the input video valve previously mentioned and thus to one side of the series capacitor, the other side of the capacitor being at the potential of the signal back porch.

When the applied reference potential is equal to the back porch potential, the grid potential remains unchanged. When, however, the two potentials differ, the series capacitor charge changes and the grid takes up a new potential such as to oppose the original distorting voltage excursion. The grid remains at this potential during the subsequent line period and until the next clamp pulse period because of the absence of a grid leak.

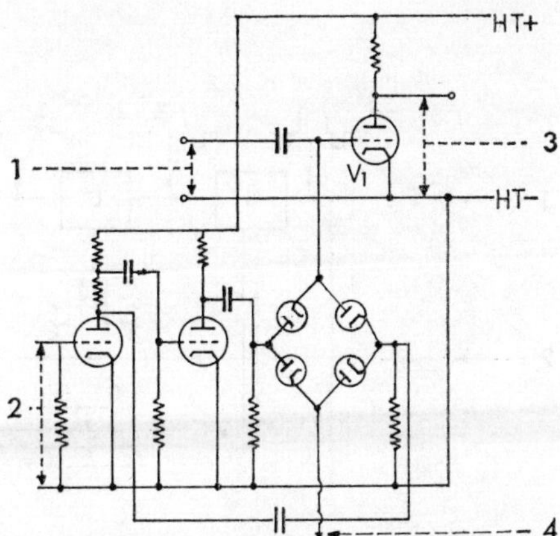

SERIES CAPACITOR D.C. CLAMP. Clamp circuit, operative only during the back porch period of the line sync pulse, and designed to reduce distortion arising in switching equipment, radio and cable links, etc. 1. Video input. 3. Clamped video output. 2. Clamping pulse input from separate source. 4. Reference potential applied to null point of Wheatstone bridge comprising four matched diodes. Opposite null point goes to grid of V_1 in lieu of normal grid leak.

The diodes forming the Wheatstone bridge circuit must be very similar in characteristics, otherwise the bridge will be unbalanced and clamp pulses will appear at the grid of the video input valve. In practice, it is impossible completely to match the characteristics of the diodes and inevitably a small proportion of the clamping pulses do get impressed upon the video signal. The effect is minimized, however, by arranging that the amplitude of the video signal at the clamping point is large compared with the amplitude of the unwanted clamping pulses.

SYNCHRONIZING PULSE SELECTION. The video signal on entry into the line clamp or stabilizing amplifier is split into a picture chain and a sync chain. In the sync chain, the picture component is removed by clipping, and the surviving sync pulses are fed to the contacts of a changeover relay. Sync pulses direct from the station sync pulse generator are also fed to the other side of the same relay. Operation of the relay selects one or other sync source. The chosen pulses, after cleaning and reshaping, are

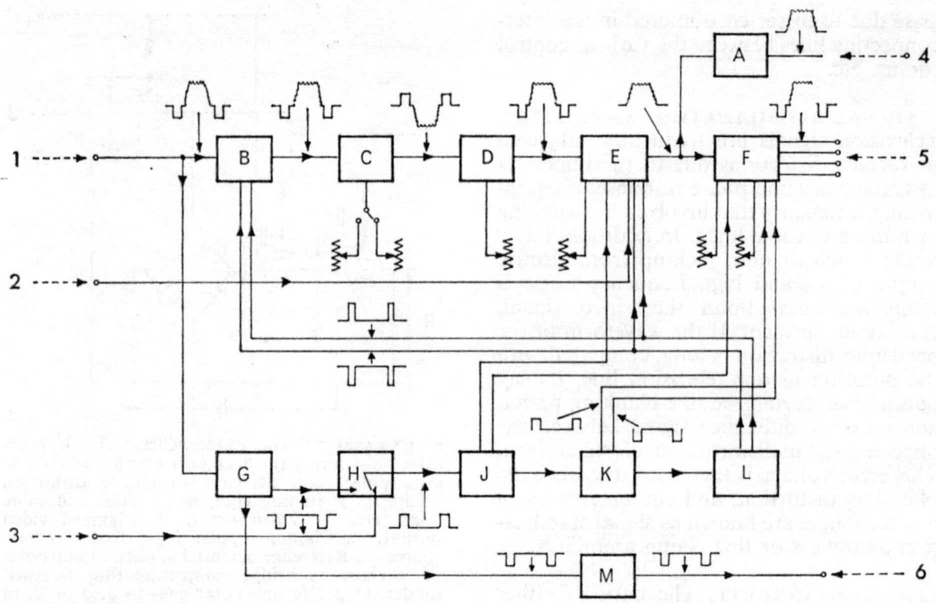

LINE CLAMP AMPLIFIER: BLOCK DIAGRAM. (Waveforms shown between blocks.) Video signal, 1, splits into picture chain, B-F and A, and sync chain, G-L and M. In sync separator, G, picture component is removed by clipping. Sync changeover switch, H, enables syncs to be selected from separator or from station sync input, 3. Selected sync is cleaned and reshaped in shaper, J, from which it is fed back to the video line. It also goes to phase splitter or mixing stage, K, and clamp pulse former, L, and from separator, G, a sync output, 6, is available through sync amplifier, M. In the picture chain, the picture amplifier clamp, B, is followed by fading amplifier, C, with local and remote controls, and picture gain amplifier, D. Picture separator, E, has a set-up control for establishing correct voltage separation between black level and line blanking. It is also fed with station blanking input, 2. Clamp pulses from L are also applied here. Non-composite signals, 4, are also derived through amplifier, A, and the picture signal is combined with sync pulses from the sync chain in sync mixer, F, incorporating adjustable white clipper to prevent peak overloading. Composite signal is then fed to video outputs, 5.

added to the picture component at a later stage in the apparatus.

This selection of synchronizing pulses is necessary in order to permit synchronous signals to be faded and mixed by the vision mixer. The position of the mixer faders dictates not only the amplitude of the picture component but also the amplitude of the synchronizing pulses comprising part of the complete vision signal. In this case, the relay position is such that station synchronizing pulses are re-inserted into the video waveform.

NON-SYNCHRONOUS SIGNALS. In the case of non-synchronous signals, the re-inserted pulses must obviously be derived from the original source, since the station pulses would be out of time, and the relay position is arranged accordingly. Non-synchronous signals cannot be faded prior to entry into the line clamp amplifier. This fading can be accomplished in the line clamp amplifier after the point at which the picture and syncs are split into their separate chains and before the picture and sync components are recombined. The sync changeover relay may be operated either by a manual switch, the contacts of nominated vision mixer push buttons, or relays to which non-synchronous signals are always fed, or by a sync comparator unit.

PULSE SHAPER. The output from the sync changeover relay is fed to a pulse shaper circuit which amplifies the synchronizing pulses and clips them top and bottom. This process removes any small amplitude irregularities at top and bottom of the synchronizing pulses which might occur due to the presence of high frequency noise. It also effectively improves the rise time of the pulses, which might have become impaired on passage through preceding equipment with a limited high frequency response. Outputs from this sync shaper drive the multi-vibrator which generates the clamping pulses and feeds a mixing stage for recombination with the picture component of the signal.

In the picture chain, the complete video waveform is amplified and then fed to a line clamp operated by clamping pulses derived from the clamp pulse generator. The output from this clamp circuit is fed to a fading amplifier whose gain is controlled by a local potentiometer or by a potentiometer fader mounted on the vision mixer control panel. Operation of either fader permits the complete video signal to be faded to or from black level, and after amplification the distorted syncs are removed from the video signal by clipping. Synchronizing pulses from the separate sync chain are, of course, available for recombination with the picture at a later stage.

SET-UP LEVEL. The picture set-up, i.e., the voltage separation between the picture black and the line-blanking is then adjusted. This may have become changed due to the signal passing through amplifiers with a non-linear gain characteristic.

The set-up level may be reduced by clipping at a higher point than the original blanking level, thus cleaning the porches of any small amplitude irregularities, e.g., impressed noise.

The set-up level may be increased by clipping the signal below the original blanking level and thus creating new porches at a correspondingly lower point on the waveform.

FINAL DISTRIBUTION. The picture signal and the synchronizing pulses from the sync chain are now combined in a further amplifier. Provision is usually made to enable the sync pulse amplitude to be adjusted in order to maintain the correct picture/sync ratio, and a peak white clipper prevents the picture whites from overloading any following aparatus.

The combined video signal is finally fed to a distribution amplifier incorporated in the apparatus which may provide as many as five outputs.

Outputs are provided, from the sync separator before shaping, for genlocking purposes, and from the picture chain after amplification and set-up correction but before sync re-insertion, for apparatus requiring non-composite picture signals.

F.B.B.

See: *Master control; Presentation; Presentation mixer; Special effects* (TV); *Special events.*

□**Vision Switcher.** Person who operates the keyboard which selects the picture from a chosen camera. In the US most directors do their own switching. In Britain, the action is usually delegated by word of mouth to the person who operates the keyboard. Also known as the vision mixer.

See: *Vision mixer.*

□**Vision Switching Matrix.** An array of M × N low capacitance relays or semiconductor switches, used to connect any one of M input vision lines to any one of N output lines.

See: *Matrix; Vision mixer.*

⊕**Vistavision.** System of film production for wide-screen cinema presentation using a double-sized frame on 35mm negative which is then printed by optical reduction. Now obsolete.

See: *Wide-screen processes.*

⊕VISUAL PRINCIPLES
□

During the 19th century, scientific psychology became established as a study in its own right. The Industrial Revolution and the age of mechanization had led to the provision of instruments with which men could make quantified observations to answer specific questions, and much of the work of the early physiologists and physicists was concerned with discovering how experiences are related to aspects of nervous structure and of the physical world. In the course of these early investigations, it was discovered that the senses could be fooled in a number of ways – the classical illusion figures are good examples of how the eye and brain can err in the interpretation of certain sensory information. So, too, the phenomenon of apparent movement – the perception of continuous movement by intermittent stimulation of the visual system – was elicited and used as the basis for the greatest of the media of mass entertainment, cinema and, latterly, television.

The detection of real movement which is so essential for survival is achieved in one of two ways by man. When the eye is stationary, the image of a moving object sweeps across the retina of the eye, successively stimulating the receptor cells. Velocity signals are thus generated and transmitted to the brain. If, however, the eye moves to follow, or track, a moving object the image on the retina remains relatively stationary, and the velocity information is given by the rotation of the eyes in the head. In the latter case the background to the moving object will move across the retina and produce velocity signals, but these are not essential since movement against an unstructured background can still be detected.

APPARENT MOVEMENT

Consider now apparent movement. If a series of still pictures are presented to the eye at a certain rate, the observer's percept, or 'picture', is one of continuous action. The perception of this action depends upon two basic characteristics of the visual system, persistence of vision and the phi-phenomenon.

PERSISTENCE OF VISION. This phenomenon was first described in 1824 by Peter Mark Roget. He noted that when a rotating spoked wheel was viewed through a narrow slit in a screen, for a certain speed of rotation and a certain illumination the spokes were seen as stationary. This effect is due to the inability of the eye to follow and signal rapid changes in brightness.

In the simple case of a steady light, of a duration of several seconds, the sensation of light develops as follows. Instead of beginning at the same time as the stimulus, the sensation starts after a delay, the latent period. It increases to a maximum then falls to a more or less constant level until the light is extinguished. When the stimulus is removed, the sensation remains for a time and then disappears progressively. The duration both of the latent period and of the sensation depends upon a number of factors, such as the subjective intensity of the light, its wavelength and the state of adaptation of the retina, and varies from 50 to 200 msec. The persistence of the sensation after the extinguishing of the stimulus can be demonstrated by means of an after-image, most easily produced by exposing the eye to a photographic flash. The positive after-image thus created imparts to the retina an inhibition which affects all subsequent visual processes of a like nature, but encourages processes of an opposite kind, so that a second stimulus falling on the same area gives rise to a negative after-image.

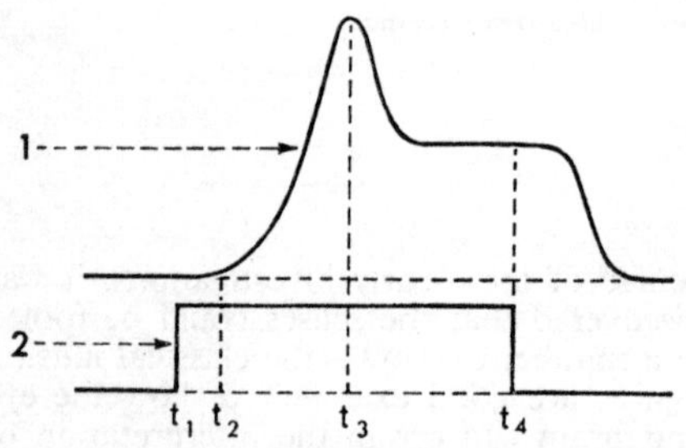

SENSATION FROM A CONSTANT STIMULUS. Light stimulus, 2, is switched on at t_1, remains at constant brightness, and is switched off at t_4. Sensation, 1, however, starts at t_2 after a delay (t_2-t_1), the latent period. It grows to a maximum at t_3, then falls to a more or less constant value till t_4, and then persists for a time until it finally dies away.

In the case of an intermittent light, the flash rate has to be such that the fusion of successive stimuli occurs when the sensation of the first stimulus lasts until the sensation of the next appears. Since the sensation is never shorter than 50 msec., continuous sensation can occur if the stimuli follow one another within this duration. The flash rate at which the light is seen as continuous is known as the critical flicker frequency and has a maximum at around 50 flashes per second, the actual value at any time depending upon the factors noted above. Use of a flash rate below the critical frequency produces flicker, a characteristic of early silent pictures.

PHI-PHENOMENON. The second characteristic underlying the perception of apparent movement is the so-called phi-phenomenon. This was studied experimentally and reported by Wertheimer in 1912. Wertheimer was one of the founders of the Gestalt school of psychology and was concerned with enunciating principles of "how things appear" to the observer. This school arose due to dissatisfaction with the ideas of the time that the eye and brain worked like a camera, mirroring the environment in the percept; and its followers established the existence of a number of important brain processes which modify the information from the external world.

Wertheimer studied the effects of presenting fixed short lines of light, separated in space, the second being presented some time after the first. If the interval between the two exposures is short (1/32 sec.) the two stimuli will appear as two and as simultaneous; and if the interval is relatively long (1/6 sec.) the lines are again seen as two, but successive. At some interval whose duration is between these two intervals, an appearance of movement is seen, the optimal value being around 1/16 sec.

MOVING PICTURE SYSTEMS

CINEMA. Before the explanations of the sensory aspects of the above phenomena had been clearly elucidated, apparatus had already been devised to create movement from still pictures. The earliest devices were no more than drums with vertical slits, static positions of objects being painted on the inner wall and viewed through the slits while the drum was rotated at an appropriate speed, thus achieving a crude effect of movement. Viewing was restricted to one person at a time, and it was not until the invention of celluloid film around 1890 that instruments were constructed to project a programme of reasonable duration. In the projector, the film is moved past the lens intermittently, at a rate of 24 fps (frames per second). A rate of 16 fps would be adequate for the perception of movement, and this is in fact the speed of projection of silent film, but the higher rate was adopted in the late 1920s to give sufficient speed to the sound track for recording the higher frequencies. At both these rates, flicker would be observed, and to avoid this a rotating shutter interrupts the projection of each frame, while stationary, once in the case of sound film and twice for silent film to achieve a projection rate of 48 screen appearances per second.

It is interesting to compare what the observer perceives and what the camera records. As a camera films a scene, say at 24 frames per second, each frame is exposed to the scene for about 1/48 second, and then obscured for an equal time during which the film is moved forward to the next frame. When the film is projected, the observer not only sees movement but has the illusion of seeing 100 per cent more than he actually does, to the extent that, in a film 1½ hours long, the observer really sees only 45 minutes of action.

TELEVISION. In television, the sensory specification is met in a very different way. The method of communication is electronic rather than photographic and involves the serial transmission of a considerable quantity of detail. The picture transmission rate is 25–30 per second (varying with the power frequency), each picture being built up as a series of light elements of varying brightness. This is, in fact, a sequential process, the phosphor on the inside of the television tube being activated by the electron beam which is deflected across the tube face from right to left. From close examination of the screen, this process can be perceived as a series of horizontal lines the height of the screen, each line being made up from a number of light elements.

With the original British transmission system of 405 lines, the picture aspect ratio being 4 : 3, the total number of elements in each complete picture is in excess of 2×10^5. Each pair of elements requires only one cycle of signal current to convey it, so that with a picture repetition rate of 25 per second, the highest signal frequency, and thus the vision bandwidth, is about 3 MHz. However, a 25 pps presentation would cause disagreeable flicker, and to double the picture taking and transmission rate would double the bandwidth required.

This problem is solved by the device of interlace. One complete field is scanned with the evenly-numbered lines, spaced apart to allow the odd-numbered lines to fall between them on the next scan. Each field occupies 1/50 sec, so that 50 fields are scanned per second, and the flicker frequency is raised to the unobjectionable figure of 50 Hz. However, the picture information transmitted still corresponds to 25 pps, and the bandwidth requirements are unchanged.

During the last twenty years, screen size has doubled from 9–12 in. to 21–23 in., and despite improvements in the design of cathode ray tubes and receiver circuits, picture quality at a constant viewing distance has deteriorated with increased visibility of the line structure. This is similar in effect to viewing a photographic enlargement from too short a distance, when the grain structure can be detected. The solution to this problem is to increase the number of lines; the US has a 525- and Europe a 625-line system. With this goes an increase of bandwidth to 4 and 5 MHz respectively. Even with these new standards, the TV picture is markedly inferior in information capacity to 35mm black-and-white film. With a frame size of 18 × 24mm and a practical limit to resolution of about 60 lines per mm, the information capacity per frame works out at about 15×10^5 elements, or 5 times the 3×10^5 elements achievable with current television standards.

High quality projection on large screens viewed at short distances demands this superior performance.

REALITY OF THE FILM AND TELEVISION PICTURE

There is considerably more involved in the process of representing the real world by cinema or television pictures than the mere creation of action within the scene. It is important to compare and contrast the percepts derived from direct vision of the real world, and those obtained from the representation of that world on a flat surface. It was in fact the Gestalt school of psychology which first demonstrated that the percept is not just the sum of the stimuli presented to the eye and transmitted to the brain. The basic fact that the perceptual system is prone to error in certain circumstances is proof that there cannot be an invariant relationship between the information coming into the system and the percept. Possibly the most famous example is the Necker Cube. If this is fixated for some

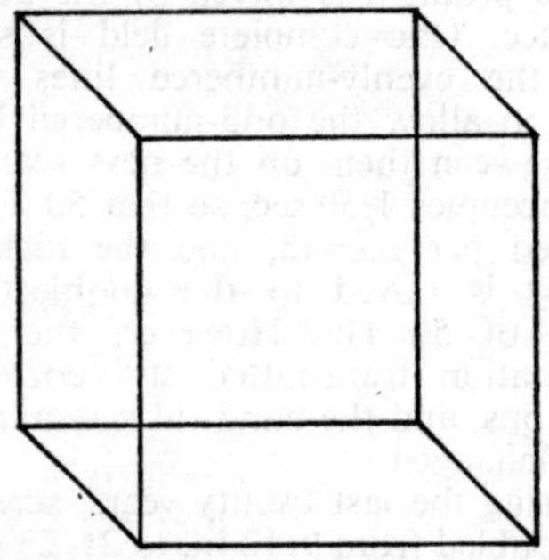

NECKER CUBE. The figure alternates in depth, the observer's visual system entertaining alternative hypotheses and never settling for one solution.

seconds, it will be seen that the cube reverses – the front face become the back face, and vice versa. This change takes place despite the fact that there is no change in the retinal image. Given that the brain has an active role in the generating of our percepts, what is its contribution? The most satisfactory way of answering this question is to look briefly at the characteristics of the entire visual system. The mechanisms can be divided roughly into two sections: sensory and perceptual.

SENSORY CHARACTERISTICS

In many respects the eye is similar to a camera, with the retina corresponding to the film. The image is focused on to the network of rods and cones, the receptor cells which transmit to the brain. It is now known that the cones and rods have different functions – the former function in daytime vision (photopic) and the latter in vision at night (scotopic). Whereas the cones are responsible for our perception of colour, the rods are exclusively a light-dark sense, nor do they contribute to the accurate perception of form.

The eye works over an enormous range of light intensities. ranging from 10^6 ft. lamberts to 10^{-6} ft. lamberts (i.e., 10^{12} : 1). The upper limit is set by the pain induced by the high luminance. The luminance of a white object in strong sunlight is 10^4 ft. lamberts, in room light 10 ft. lamberts and in moonlight 10^{-2} ft. lamberts. The cones respond down to 10^{-3} ft. lamberts and the rods from 10 to 10^{-6} ft. lamberts. That such a range can be tolerated is due to a process called adaptation; only small changes can be dealt with by the action of the pupil whose effective range is 16 : 1.

ADAPTATION. The process of adaptation is only now being understood, and is due to the bleaching action of light on the photochemicals in the retinal receptors. Light adaptation is fast, but dark adaptation is very slow and it takes more than thirty minutes for the eye to become completely dark-adapted. In addition to adaptation to light and dark, there is also chromatic adaptation. Artificial light looks less yellow as we adapt to it, but adaptation never restores all colours to their daytime appearance.

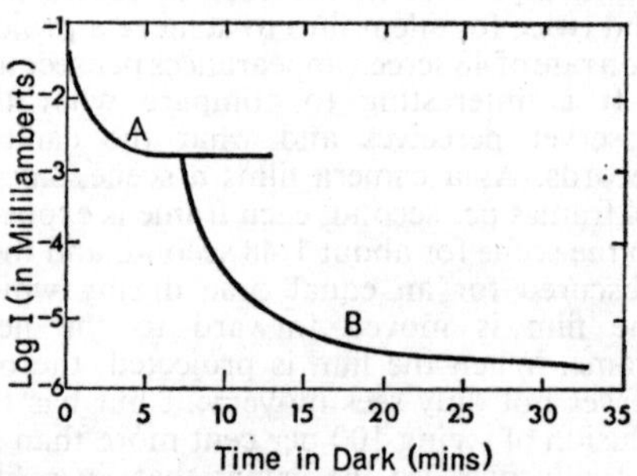

DARK ADAPTATION. Increase of sensitivity of the eyes in the dark. Curve A shows how the cones adapt, reaching a maximum sensitivity after about 5 min. Curve B shows that rod adaptation is slower, with marked increase up to 20 min., and giving considerably greater sensitivity.

In the cinema, the state of adaptation of the retina is determined by the mean illumination on the screen, and this is con-

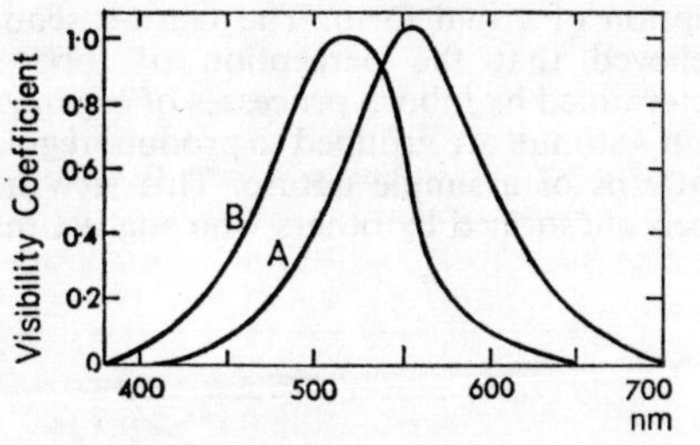

PHOTOPIC AND SCOTOPIC RESPONSE. Sensitivity curves of the light-adapted eye, A, (photopic response) and dark-adapted eye, B, (scotopic response). The shift along the spectrum from B to A as a result of light adaptation, when the cones take over from the rods, is known as the Purkinje shift.

siderably less than the daytime level. The situation is generally rather different when television is viewed. It is rare for the viewer to watch in a darkened room, and the adaptation of the eye is determined by the general level of the illumination.

TEMPORAL AND SPATIAL FACTORS. The response of the eye to a flash of light has already been mentioned. However, the duration of a stimulus is as important as its intensity. This is formalized as Bloch's Law, $I \times T = C$, and is linked to the Bunsen-Roscoe Law of Photochemical Equivalence, the keystone of all photochemical processes. Bunsen and Roscoe found that the amount of photochemical produced during a reaction was dependent upon the value of the intensity of illumination multiplied by the time of exposure – if a given amount is produced by a certain intensity over a given time, the same amount will be produced by half the intensity acting for double the time. Such a relation holds over a limited range for the density produced on photographic materials and is the basic principle of exposure. Bloch's Law holds provided that the duration of the stimulus is less than the retinal action time which has been shown to be in the range 50–200 msec.; otherwise intensity is the only determinant.

DIFFERENCES IN INTENSITY. The three perceptual aspects of colour are brightness, hue and saturation, but for reproductions in monochrome only the perception of brightness is involved. In the absence of colour, the observer is prepared to accept the reproduction of a multi-coloured scene in tones of grey, but will be critical of the exactness or inexactness with which the luminance differences in the original are reproduced in the picture viewed by him. It is these brightness differences which provide the observer with information which enables him to separate objects from the background.

The smallest difference in intensity which can be detected is directly proportional to the background intensity. This is known as Weber's Law, $\Delta I/I = C$. This is easy to demonstrate: take a second candle into a room which is lit by one candle and the increase in brightness by the addition of the second will be readily detectable. Now take the candle into a room lit by a 100 watt electric lamp and there will be no detectable difference. We can detect a change in intensity of about 1 per cent of the background illumination. The law holds over a wide range of intensities but breaks down at low intensities, the ratio having to be increased for detection of the difference.

The most familiar method for scaling sensation goes back to Fechner (1860), who thought that Weber's Law furnished the key to the measurement of mind. It is an indirect method and is based on the fact that the just noticeable difference of Weber's Law must get bigger as the scale of intensity is increased. Fechner found that the strength of the sensation varies directly as the logarithm of the stimulus, except at very high or very low values of the stimulus. In practice, departures from the law are to be expected, since the nature of the function depends upon the adaptation level of the eye, the duration of the stimulus, the area and portion of the retina stimulated, and the magnitude and distribution of the simultaneous stimulation of other retinal areas.

This final qualification introduces the phenomena of contrast. A factor which affects brightness is the intensity of the surrounding areas, a bright area being enhanced by a dark surround and the intensity of a colour being increased by viewing it when surrounded by its complementary colour. These effects are no doubt due in large part to the action of retinal mechanisms – though the brain probably has some role – which serve to make the outlines of objects more distinct than in the retinal image.

Contrast is concerned in the reproduction of luminances. In an outdoor scene, the range of luminance from the darkest to the lightest part of the subject (excluding the sun or any reflected images of it) is much less than might be thought. Measurements taken on a large number of typical scenes show a variation from 30 : 1 to 800 : 1 with a mean of about 160 : 1. With such a luminance range, there is no need for the eye to change its level of adaptation as the gaze wanders over the scene.

Since the range of luminance that can be recorded on film or by television is considerably less than the above maximum figure, (about 200 : 1 for film and 30–50 : 1 for television), little can be done to reproduce the appearance of the original with precise fidelity. However, viewing conditions are relevant, and the level of adaptation of the eye is especially important. As noted above, in the cinema the level is correctly determined by the mean level of illumination of the picture. For viewing television, the mean level is determined by the ambient light in the room. Increasing the adaptation level of the eye decreases its sensitivity and its ability to discriminate all the tones actually present on the screen, which further impoverishes the final percept compared with the real scene viewed by the television camera. To minimize these losses, television should be viewed in a darkened room – with the brightness adjusted to eliminate glare and its attendant uncomfortable effects.

VISUAL ACUITY. The resolving power of the eye should bear some relation to the spacing of the cones in the retina. However, this is a simplification and in fact visual acuity represents an interaction of many factors, ranging from the scattering within the eye, diffraction, tremor of the eye, illumination and retinal interaction. Acuity is highest in the fovea and drops markedly from the fovea out to the far periphery. For lines or dots the central fovea can resolve to about 30 sec. of arc, but a wire subtending only 0·5 sec. (about 1/40 of the diameter of a single cone) can be resolved thanks to the factors noted above.

Acuity is of importance both in relation to the grain size of the film, and the light element size in television. Considerable care has to be taken when enlarging a negative to keep graininess to a minimum, and this question takes on a more important role when the extent of the enlargement from the film to the cinema screen is considered.

PERCEPTUAL MECHANISMS

FORM PERCEPTION. The first step in the perceptual process is the disentangling of the things of the world from the stuff of the world – the separation of figure from ground. This distinction was first made by one of the Gestalt psychologists, who indicated the importance of contour formation to the perception of visual form. The Gestalt school believed that the perception of form is determined by inborn processes of organization – stimuli are grouped to produce regular patterns of a simple nature. This view has been questioned by others who suggest that many aspects of form perception depend upon learning processes in infancy. Whichever view is correct, there is no dispute over the importance of the process.

FIGURE AND GROUND. At some times a white vase is seen, at others a pair of faces in silhouette. Perception is not determined by parts of the stimulus; it is a dynamic interpretation by the brain of the sensory data.

DEPTH PERCEPTION. The fact that we have two eyes is important to our perception of depth. Whenever we fixate an object, the eyes pivot so that the images are projected on to the foveas. This pivoting of the eyes is known as convergence, and the amount that the eyes have to be converged is signalled to the brain to provide information about the object distance. The serious limitation to this method is that only the distance of one object at a time can be indicated – the object whose images are fused by the convergence angle. When many objects are concerned a different strategy is appropriate.

Since the eyes are separated, each receives a slightly different (disparate) image. In some way which is as yet not understood, the brain computes the distances of the different objects in the visual field from the information contained in the two disparate images.

These two very different mechanisms are in fact linked together – the angle of convergence adjusts the scale of the disparity system. Thus when the eyes focus a distant object, any disparity between the images is taken to represent a greater difference in depth than when the eyes are converged for near vision. Without this interlocking of the systems, distant objects would look closer together in depth than near objects of the same separation because the disparity produced by a given difference is greater the nearer the objects.

Binocular (stereo) vision is only one of the many ways in which we see depth and only functions for comparatively near objects, after which the difference between the images becomes so small that they are effectively identical. For distances greater than about 20 ft. we are in effect one-eyed, and make use of a number of other cues which are often classed as the monocular cues although for near vision they doubtless provide information complementary to that obtained from the binocular system. These include the relative size of familiar objects, relative clearness (objects appear less sharply outlined and more blue as the distance increases), interposition (the obliteration of part of the contour of an object due to the presence of another in front) and accommodation. These cues, in general, derive from the art of perspective painting, the invention of the Renaissance, and compelling depth can be obtained by reproducing photographs of a scene on a large screen in the cinema, since the observer seems to be surrounded by the picture and gets very important peripheral gradients of movement, texture, etc.

Finally, motion parallax can play an important role when binocular disparity fails, because the salient lines in the environment are horizontal, as on a flight of stairs or a landing field. A small horizontal movement of the head results in very different views of the object, but sequentially, and there appears to be relative motion within the visual field. By moving to the right, the objects move relatively to the left; but the angular displacement of distant objects is much less than that of near ones. Monocular depth perception enforced by blindness or partial blindness of one eye seems to result in dependence very largely on motion parallax.

THE PERCEPTUAL CONSTANCIES. It has been shown that the sensory stimulation from the cinema and television screen is very restricted in extent compared to the range of stimulation experienced in the real world. Nonetheless, the scenes represented on the screen appear to the observer to be remarkably lifelike. This rather surprising result is due to the existence and operation of perceptual mechanisms which serve to restrict the range of sizes, shapes and brightnesses of objects; since the visual system has built-in limiting mechanisms, the limitations of the media matter rather less than might be expected.

White paper still looks white in a dark room and coal still looks black in sunlight, despite the fact that the light reflected from the paper is less intense than that from the coal. This phenomenon is known as brightness constancy and is one of a class of phenomena, important because their operation contributes an invariance to the percept which is not present in the image at the eye. The others are size and shape constancy – the size of an object remains remarkably constant although the image on the retina is halved with each doubling of the distance, nor does the shape of an object conform to the laws of perspective. A tilted penny still looks round although the retinal image is elliptical.

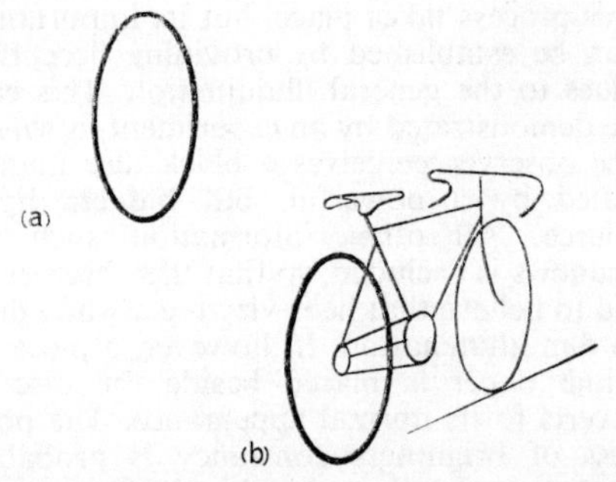

PERSPECTIVE AND INTERPRETATION. (a) The isolated shape is seen to be elliptical. (b) When combined with other shapes into a known form, the ellipse is immediately interpreted as the perspective view of a circular wheel.

The operation of these scaling mechanisms is never perfect – the size, shape or brightness being a compromise based on viewing conditions. This tendency is often called phenomenal regression to the real object.

The disentangling of the factors of importance in the constancy processes is as yet incomplete, but it appears that the operation of the mechanisms is extremely dependent upon the information coming into the visual system at the moment, rather than to learning or experience, although the attitude or set of the observer can influence his interpretation.

BRIGHTNESS CONSTANCY. The operation of the brightness constancy mechanism is

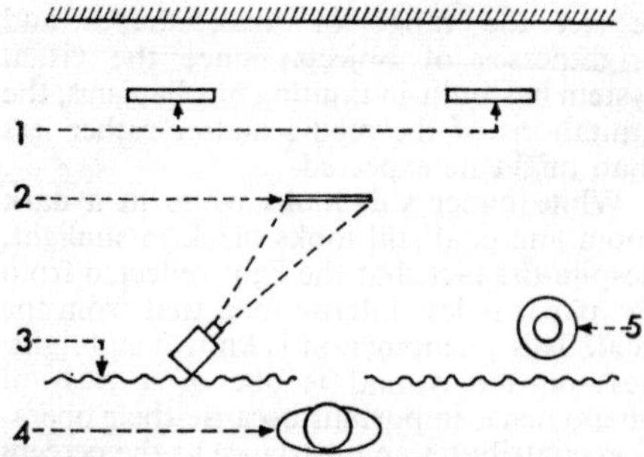

GELB'S EXPERIMENT. Observer, 4, views a black disc, 2, illuminated by a powerful lantern hidden from him by a screen, 3. A dim ceiling light, 5, lights up objects, 1. Due to brightness constancy scaling processes the disc, 2, is seen as white by the observer. When, however, a piece of white paper is placed on the disc it is seen to be black.

due to the estimating of the brightness of the object in relation to the general illumination of the background. It is not clear how this process takes place, but its importance can be established by providing deceptive clues to the general illumination. This can be demonstrated by an experiment in which the observer perceives a black disc illuminated by a powerful, but hidden, light source. All other information such as shadows is excluded, so that the observer is led to believe that he is viewing a white disc in dim illumination. If, however, a piece of white paper is placed beside the disc it reverts to its normal appearance. The process of brightness constancy is probably more complex than this, involving contrast effects at a peripheral and central level.

SIZE CONSTANCY. This term covers two rather different but related processes. In one, the distance may be estimated when there is knowledge of the size of the object, and in the other the size of an unknown object

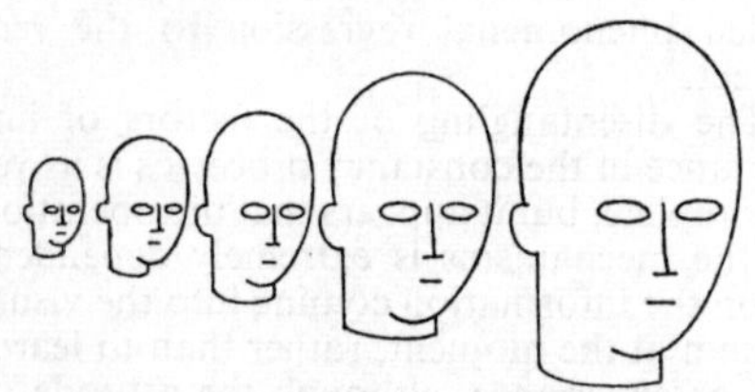

SIZE CONSTANCY. Change in size of observer's retinal image of a man's head as he approaches. Normally the head is seen as a constant size because the constancy scaling processes of the brain compensate this change by taking account of the distance change.

may be estimated on the basis of the distance and the angle subtended at the retina by its image. Whichever process is involved at any time, the basic consideration is that for constancy scaling to occur there must be depth or distance information available to the observer. If the normal cues to depth are eliminated by viewing an object through a narrow tube (reduction tube), the perceived size varies directly with that of the retinal image.

SHAPE CONSTANCY. This process was thought to depend upon the correct estimation of the distance of the relative parts of a foreshortened object. The presence of depth cues may contribute to the operation of the mechanism but shape constancy still holds in the absence of most spatial cues. The degree of constancy in these circumstances appears to be related to movement, the rate of change of the shape providing the necessary information.

THE TWO DIMENSIONS OF THE SCREEN

The perceptual mechanisms, which normally operate in a 3-dimensional world, must adapt to the flat cinema or television screen. When the third dimension is removed, depth has to be suggested by the perspective features of the scene. There is much more to the matter than the pure geometry of the situation, since factors such as light and shade are important to determine the orientation of objects in the scene. Perspective information alone is ambiguous, and the visual system depends on monocular distance cues for correct interpretation of the scene. With two-dimensional representation, binocular cues are not available. In fact the binocular mechanism can work against the perceiver of a two-dimensional scene, so perhaps we should normally view a motion picture with only one eye!

Since the binocular cues are the more important in our everyday perception, the conclusion that must be drawn in their absence is that the perspective of the scene must be modified in such a way that restriction in the operation of the various scaling mechanisms does not result in inappropriate percepts. The problem can, of course, be overcome when artificial studio sets are used as background for pictures or television productions, since the perspective can be modified to produce the correct depth features to the naked eye. Less can be done to overcome the problem when films are being shot on location, with the action taking place against a real background. There are occasions when effective use can be made of inappropriate size or depth, as Arnheim has pointed out. When projected, the camera's image of an approaching locomotive becomes larger and larger, until it dominates the whole visual field. In the real world, an observer would never gain such an impression, but because of its incongruity it can arouse his emotions. He can feel fear, awe or a sense of power – and this can contribute to the establishment of rapport between the observer and the action in the film. The aim must surely always be to involve the observer in the action to the maximum degree – he should lose the feeling of standing outside.

With the smaller size of the television screen, it is very much more difficult to stimulate the emotions visually. The edges of the television screen are always present and produce the effect of watching the action on the screen through a window – the observer is merely an onlooker, separated in space from the events on the screen. This effect is further enhanced by other factors such as the considerable difference between the size of the objects and people on the television screen and in the real world. People or objects of such a size in the real world could only be distant and remote from the observer. In general, television concentrates on close-up scenes to compensate for these factors, and the extent of the success of this approach is evident when one considers the diminution of effect which feature pictures suffer when they appear on the television screen.

As a rival to television, during the last few years the cinema has struggled to increase the realism of its presentations, and many experiments have been undertaken. Some have been concerned with increasing the size of the projected picture so that it dominates the vision of the observer completely by filling the whole visual field and removing the frame of reference of the edges of the screen. Technically this process is achieved in a variety of ways, from the use of special (anamorphic) lenses, which compress the picture on the film and then expand it when used on the projector, to the construction of special cameras and projectors which handle film twice the width of the normal 35mm.

Probably the ultimate method has been the development of Cinerama – the impression of being driven at speed along a narrow mountain road or being whisked up and down on a scenic railway is very effectively communicated to the observer. The brain is unable to accept that the whole of the visual field can change without movement on the part of the observer, since in the real world there would be systematic movement of the whole of the image on the retina only if the observer were walking, running or being transported. This is similar in principle to another well known movement effect. Most travellers have at some time noticed that when their train or bus is beside another stationary vehicle it is impossible to tell (without reference to some fixed object) which one starts to move first. Since movement is a relative process, the brain has to make the best bet on the information available and it can be very wrong.

Although increasing the width of the screen may confer greater realism to films shown in the cinema, it causes problems when these films are shown on television. The television screen aspect ratio (1·33 : 1) was determined by the aspect ratio of the early films. Change to an aspect ratio of say 2·35 : 1 means that either the extreme ends of the film have to be sacrificed, or the film has to be compressed, leaving an empty space at the top and bottom of the television screen. This compression reduces the size of the scenic content of the film, causes a loss of contrast and definition, and generally impoverishes the quality of the percept.

THE ROLE OF SOUND

An aspect of the reality of the cinema which is taken for granted is sound. There is little doubt that the addition of a sound track to the film changed the whole character of the cinema. After vision, hearing is the next most important sense since it also provides us with advance information of what is likely to happen – it is a second early warning system and is biologically useful to survival compared to the other senses such as touch which operate only in direct relation to external events. Its use rids many visual scenes of ambiguity and extends the reality of the action, since the attributes of the characters do not have to be made evident in their clothing or make up – no longer need the villain be dressed in black and the hero in white! J. A. M. H.

See: *Colour principles; Stereoscopic cinematography; Stroboscopic effects; Viewing television.*

Books: *Eye and Brain*, by R. L. Gregory (London), 1966; *Light, Colour and Vision*, by Y. Le Grand (London), 1957; *The Retina*, by S. L. Polyak (Chicago), 1941; *Development of an Art from Silent Films to the Age of Television*, by A. R. Fulton (Oklahoma), 1960.

⊕**Vitarama.** Form of multiple-film wide-screen presentation using eleven projectors showing a composite picture on the inner surface of a dome; a feature of the New York World Fair of 1939.

See: *Wide-screen processes.*

□**Voltage Doubler Rectifier.** System which enables a direct voltage of nearly twice the peak voltage of the applied alternating supply to be produced. It operates by charging capacitors in parallel and discharging them in series. Voltage triplers are also used. Employed in some large-screen TV projectors.

See: *Large-screen television.*

⊕**Volume Compression.** Function of a device □which transfers a signal from its input to its output and at the same time reduces the span of amplitudes of the signal. Compressors are much used in recording original sources such as large orchestras, in which the volume range is bigger than the recording system can accommodate. They are also used in making transfers from a wide-range medium such as magnetic tape to a narrow-range medium such as 16mm optical sound track.

The term compressor is usually applied to a device which compresses proportionately all the way up the volume range. Limiters, by contrast, transfer the signals linearly up to a predetermined point, after which they prevent the signal from rising any further. Hence they are often called peak limiters.

See: *Sound distortion; Sound recording.*

W-Z

□**Wall Anode.** Electrode placed inside the wall of a camera tube, along its axis, to improve tube performance.

⊕**Waller, Fred, 1886–1954.** American designer of multiple film presentation systems. Devised the first model of the Cinerama process in 1937 but failed to interest exhibitors at the New York World's Fair. From it, Waller developed a gunnery trainer which used five films projected on a hemispherical screen to simulate combat conditions. In 1946, Waller returned to the Cinerama process, which he demonstrated in 1948, and which was first commercially exploited in 1952.

⊕**Washing.** In a film developing machine, removal of the hypo and soluble by-products which remain in the film after fixing. Normally carried out in ordinary running water.
See: *Processing.*

⊕**Water Spots.** Defects on processed film caused by the presence of drops of water remaining on the emulsion surface of the film in the drying cabinet.
See: *Processing.*

□**Waveform Monitor.** Cathode ray oscilloscope specifically designed for displaying television signals. The X or horizontal axis represents time, and the Y or vertical axis represents the amplitude of the signal.
See: *Picture monitors.*

□**Waveform Oscilloscope.** Cathode ray oscilloscope with calibrated vertical amplification and timebase, used for making measurements of waveforms.
See: *Picture monitors.*

□**Waveguide.** Conducting tube along the inside of which radio waves can travel, reflected from the sides. A rectangular waveguide must have one dimension greater than half the wavelength of the waves.
See: *Microwave links; Transmission networks.*

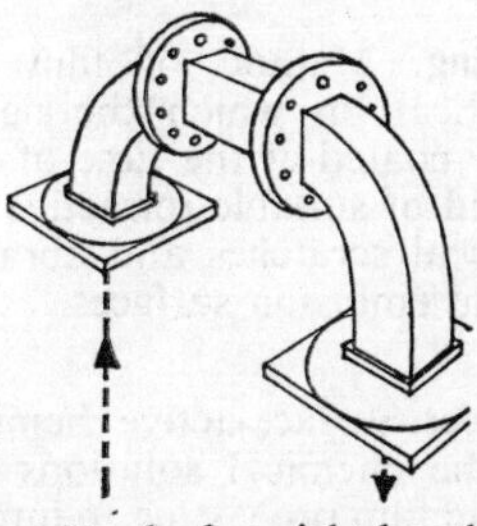

WAVEGUIDE. Standard straight lengths and curved shapes can be assembled in the form of plumbing.

⊕**Waxing.** Application of wax to the edge areas of film, to prevent intermittent dragging and consequent piling up of the emulsion during projection, which may occur especially if the film has been improperly dried.

See: *Laboratory organization; Release print examination and maintenance.*

⊕**Weave.** Unwanted side-to-side movement of film in camera or projector.

□**Wehnelt, Arthur Rudolph Berthold, 1871–1944.** German physicist who in 1905 added an important improvement to the Braun tube – the first cathode ray tube – by introducing a hot cathode. This took the form of a strip of platinum coated with oxides and heated electrically to red heat. The increased electron emission so obtained enabled a lower potential to be applied to the anode of the tube and gave a more brilliant fluorescent spot, although the life of the emitting surface was short.

□**Weiller, Lazare,** (Dates unknown). German professor who in 1889 originated the mirror drum scanning disc as an alternative to the earlier Nipkow disc. Instead of the latter's spiral of holes, the Weiller disc was in the form of a drum mounted with a number of tangential mirrors. The mirrors were so angled that, as the drum revolved, the area of the image was scanned in a series of lines and projected on to a selenium cell. The mirror drum was used substantially by later experimenters and played an important part in the development of mechanical television.

⊕**Weston Speed.** Obsolete arithmetical system of emulsion speed classification based on practical tests with reference to the photoelectric exposure meter of that name.

See: *Exposure control.*

⊕**Wet Printing.** Method of film printing, usually optical, in which the negative is temporarily coated at the time of exposure with a liquid of suitable refractive index to cover physical scratches and abrasions on the base and emulsion surfaces.

See: *Printers.*

⊕**Wetting Agent.** Surface-active chemical often added to the chemical solutions or wash water used in film processing to improve the uniformity of the liquid application and avoid the presence of drops which might cause irregular drying marks.

See: *Processing.*

⊕**Wheatstone, Sir Charles, 1802–1875.** English inventor who introduced the reflecting stereoscope in 1832. Subsequently devised, about 1860, a method for viewing stereoscopic pairs of posed photographs of stages of movement in a modified Zoetrope employing a form of intermittent movement.

⊕**Whip Pan.** Rapid panning movement of camera producing blurred image. Also known as swish pan or zip pan.

□**White Bar.** A number of lines of peak white used as a reference or test signal.

See: *Picture line-up generating equipment.*

□**White Crushing.** A form of peak distortion in the television image resulting from amplitude non-linearity, affecting the higher amplitude portions of the video signal. Since the signals corresponding to white are at the peak of the waveform, they are often the first part to be distorted. They usually distort to a greater degree than signals corresponding to other tonal values, with the possible exception of signals at or near black.

The non-linearity causes the video signal to be compressed, so that the tonal values no longer have the same range. The resulting compression on the television screen is a loss of detail in the whites in the picture. If the amplitude distortion is severe, the white compression becomes white clipping, showing as a complete loss of tonal detail in the white parts of the picture.

See: *Black crushing; Black stretch; Transfer characteristic.*

□**White Level.** TV signal level representing full peak excursion above or below the sync level, according to whether modulation is positive or negative. It represents a peak-white object.

See: *Television principles.*

□**Whiter-than-White.** An excursion of the television waveform signal above the normal peak (white) level.

Can be obtained in colour television waveforms.

See: *Transfer characteristics.*

⊕□**Wide-Angle Lens.** Relative term describing lenses of shorter focal length than normal, consequently giving lower than normal magnification and a wider field of view. A lens which is a normal lens on a given image area becomes a wide-angle lens if the image area is enlarged.

See: *Lens types.*

⊕**Widescope.** Early form of wide-screen presentation (in 1921) in which two pictures were shown side by side from two separate projectors.

See: *History* (FILM); *Wide-screen processes.*

⊕WIDE SCREEN PROCESSES

The term wide-screen is used to describe a variety of forms of film projection in which the aspect ratio (width-to-height proportion) of the picture shown is greater than the 4 : 3 format adopted as standard in the early days of the cinema. Since 1952 various types of so-called wide-screen systems have come to dominate entertainment cinema theatre usage both in the United States and in Europe, although the earlier format continues to be extensively used in the smaller film gauges, and in film for television. The presentation of these processes is often associated with cinema theatre screens of large dimensions, and it was not by accident that the appearance of wide-screen coincided with the brief boom in stereoscopic (3-D) movies. One object of the ultra-large screen was to stimulate peripheral vision and thus give an illusion of 3-D which the smaller screen had lacked.

However, the term wide-screen process should be regarded as referring to the proportions of the picture rather than to any actual size of projected image.

The many variations in general commercial use may be divided into three main groups:

1. Systems using anamorphic projection in which the image on the film is optically compressed (squeezed) in the horizontal sense and expanded to correct proportions by projection with lenses giving greater magnification laterally than vertically.
2. Systems using normal projection lenses in which the projector aperture is masked to give the required aspect ratio. The image on the film used in this way is not optically compressed and copies of this type are often referred to as flat prints. For both the foregoing groups, film stock of larger size than the standard 35mm width may be employed.
3. Multi-film systems where two, three or more projectors with separate prints are used simultaneously to build up the projected picture by a series of images, generally side by side.

EARLY METHODS

Mainly as a result of the commercial success of the cinematograph equipment produced by Edison, the most widely-used size of frame on 35mm film was approximately 1 in. wide × ¾ in. high or 24mm × 18mm. This format was in use in the early 1890s, and accepted as a general standard by 1909. When the use of optical sound on film became established in 1929, it was necessary to reduce the width of the frame to allow room for the optical sound-track, but the proportion of approximately 4 : 3 was retained by reducing the frame height. This format is often referred to as the Movietone frame, after the sound-on-film process which led to the alteration, and made use of a projected area on the film 0·825 in. × 0.600 in. (20·96mm × 15·25mm) with an aspect ratio of 1·37 : 1. This remained the international standard for commercial entertainment cinemas for some 25 years.

Experiments to provide larger and more spectacular presentation than could be obtained from normal 35mm film had, in fact, begun in the very early pioneering days of the cinema, usually by the use of several projectors side by side. Demonstrations of

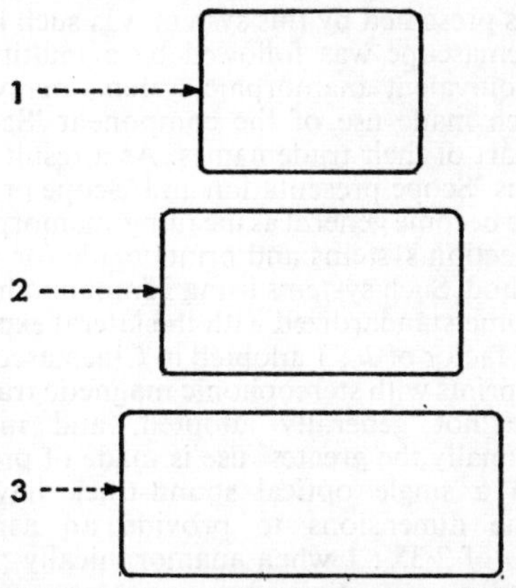

COMPARISON OF PROJECTED ASPECT RATIOS. Ratios of pictures as they appear on screen, height shown as equal for all. 1. Movietone (1·37 : 1), standard from 1928–52. 2. Wide screen (1·85 : 1). 3. Cinemascope (2·35 : 1).

Cinéorama in Paris in 1896 used ten projectors to cover a huge screen in a full circle, and Widescope (1921) using two films and the Triptych screen productions of Autant-Lara and Abel Gance (1925–27) were other examples. Systems using film widths greater than 35mm were also demonstrated from time to time, of which the Grandeur system using 70mm film, Magnafilm with 56mm film, and similar processes had a brief vogue in the period 1929–31.

The Magnascope process in use between 1926 and 1932 differed from other systems in that it made use of a special projection lens of variable magnification which could be operated to enlarge the picture on the screen during the most spectacularly dramatic sequences. Experiments using projection lenses with different magnifications in the vertical and horizontal dimensions (anamorphic projection), were also taking place about this time, demonstrations being made by Chrétien in Paris in 1927 and in the United States in 1930.

MODERN METHODS

CINEMASCOPE. Despite further demonstrations by Chrétien's system, including a display at the Paris Exhibition in 1937, it was not until 1953 that the process using anamorphic lenses was adopted for commercial cinema theatre usage under the trade name Cinemascope, initially with a format giving a projected picture with an aspect ratio of 2·55 : 1, and using four magnetic sound-tracks for stereophonic reproduction.

The commercial success of the feature films presented by this system was such that Cinemascope was followed by a multitude of equivalent anamorphic systems many of which made use of the component 'Scope as part of their trade names. As a result the terms 'Scope presentation and 'Scope prints have become general as meaning anamorphic projection systems and prints made for this method. Such systems using 35mm film have become standardized with the lateral expansion factor of 2 : 1 adopted in Cinemascope, but prints with stereophonic magnetic tracks were not generally adopted, and internationally the greatest use is made of prints with a single optical sound-track having frame dimensions to provide an aspect ratio of 2·35 : 1 when anamorphically projected.

As the anamorphic type of wide-screen presentation achieved recognition and financial success, there was naturally a desire to find alternative methods which should have a similar appearance to the audience in the cinema without the additional expense of special optical systems for photography and projection. This was done at first by cropping or masking the top and bottom of the standard Movietone frame so that a picture of general wide-screen proportions was projected. Where screens of greater width could be installed in the theatre, projection lenses of shorter focal length were used to give a larger picture than before to imitate the Cinemascope effect.

However, the arbitrary cutting of the top and bottom of the normal picture composition was clearly undesirable, and attempts were made by film production and exhibition interests to arrive at agreed standards for the composition, photography and projection of wide-screen pictures other than those using anamorphic systems. Unfortunately, no completely uniform international standard has become established, but a number of recommended practices are accepted for flat wide-screen prints with various aspect ratios ranging from 1·65 : 1 to 1·85 : 1.

LARGE-FORMAT FILMS. With the increasing use of very large screens of wide-screen proportions in the cinema, the quality of image obtainable on 35mm film tended to be considered inadequate for definition and grain. This led to a renewed interest in the use of film of larger dimension than 35mm, both for the negative in original photography and as the positive for projection. For this reason, the Vistavision system was introduced in 1954 reviving an earlier idea in the use of a film frame twice the normal 35mm size obtained by running the film horizontally in both camera and projector. No anamorphic optical systems were used with Vistavision and the recommended format for composition was 1·85 : 1.

Shortly afterwards the Todd-AO process was introduced, using 65mm width film in the camera and 70mm width prints for projection, with six magnetic sound-tracks for stereophonic presentation, and there were brief uses of a negative 55mm in width.

In due course, the difficulties caused in distribution and exhibition by the use of positive prints of many different dimensions was recognized, and horizontal double-frame projectors of the Vistavision type had only a very limited use. Commercial practice in the entertainment cinema has generally become limited to 70mm prints for the most important presentations in leading theatres in major cities, with 35mm prints, both anamorphic and flat, for general release. In many cases, however, the size of film and the proportions of the image on the original negative differ from those on the final print, and various forms of optical printing involving anamorphic reduction or enlargement are employed.

ANAMORPHIC OPTICS. Anamorphic projection systems using 35mm film have been standardized with a lateral magnification twice that of the vertical magnification, and the image on the positive print is correspondingly compressed. Thus, the image of a square object appears on the film as a rectangle twice as high as its width, while the image of a circle is an ellipse with its

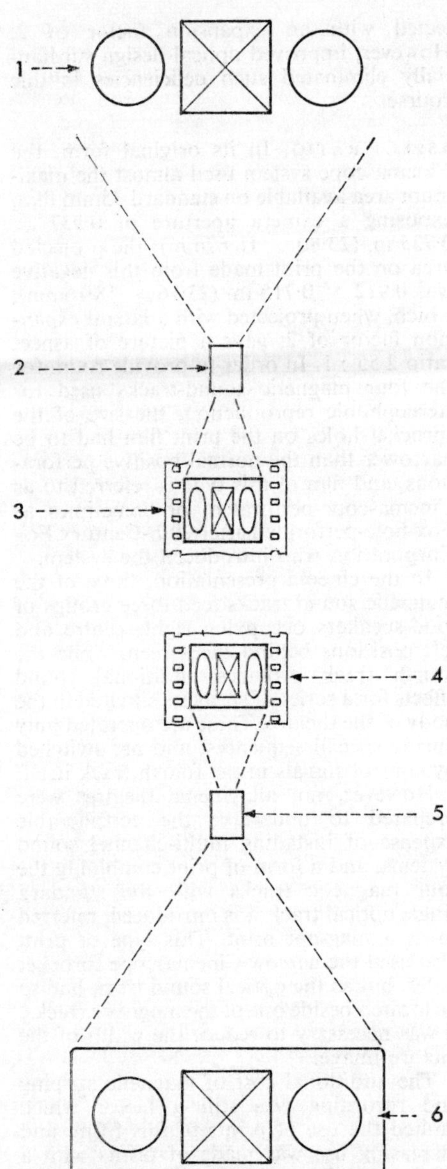

ANAMORPHIC PHOTOGRAPHY AND PROJECTION. 1. Original scene; circles and square show correct proportions. 2. Anamorphic camera lens applies a compression factor (normally 2 : 1) to the width of the scene, while leaving the height proportionally unaltered. Circles become ovals, and the square becomes a rectangle, as seen on negative, 3. After the processing and printing stages, the positive print, 4, is projected through an anamorphic lens, 5, with the same squeeze ratio as the camera lens. This forms an image on the cinema screen, 6, which is geometrically identical with the original scene, 1.

longer axis vertical and twice the length of the horizontal axis. Images of this nature can be obtained in various ways, the commonest method being by original photography using camera lenses with suitable anamorphic attachments.

The original Chrétien system described as the Hypergonar, and later adopted in Cinemascope, made use of cylindrical lenses in front of a normal camera lens system, and variants of this principle have been widely used. A concave cylindrical lens with its axis of curvature vertical acts as a wide-angle attachment in the horizontal dimension only, so that the combination has effectively a shorter focal length horizontally than vertically and the resultant image of any object is smaller in width but not in height than would have been the case without the cylindrical attachment. A similar concave cylindrical lens in front of the projection lens restores the image to its normal proportion on the screen.

The same effect can also be obtained by the use of a pair of wedge-shaped optical prisms which have the effect of compressing a parallel light beam in one plane only. The use of a pair of wedges tilted in opposite directions allows some of the unwanted optical aberrations of the system to be corrected. Since the anamorphic effect of this arrangement is determined by the angular position of the wedges in the beam, this arrangement was suitable for optical systems in which a variable degree of compression was required and formed the basis of the projection system known as the Varamorph. A third procedure, employed in the Delrama system, made use of curved reflecting surfaces in front of a normal lens, these being surface-silvered mirrors for projection and toroidally-formed faces of a pair of glass prisms for camera lens attachments.

In all cases, projection anamorph systems are designed and optically corrected for long working distances, but in camera systems for original photography the anamorphic attachments must be capable of accurate adjustment to match the focusing of the camera lens, and the requirements of optical design and mounting are very much more critical. Even though this was recognized, many of the earlier camera anamorphic systems suffered badly from loss of resolution, and variation in the compression factor at different object distances. This was particularly marked in close-ups, where the effective compression factor became less than 2 : 1, giving rise to a distorted appearance when the resultant prints were pro-

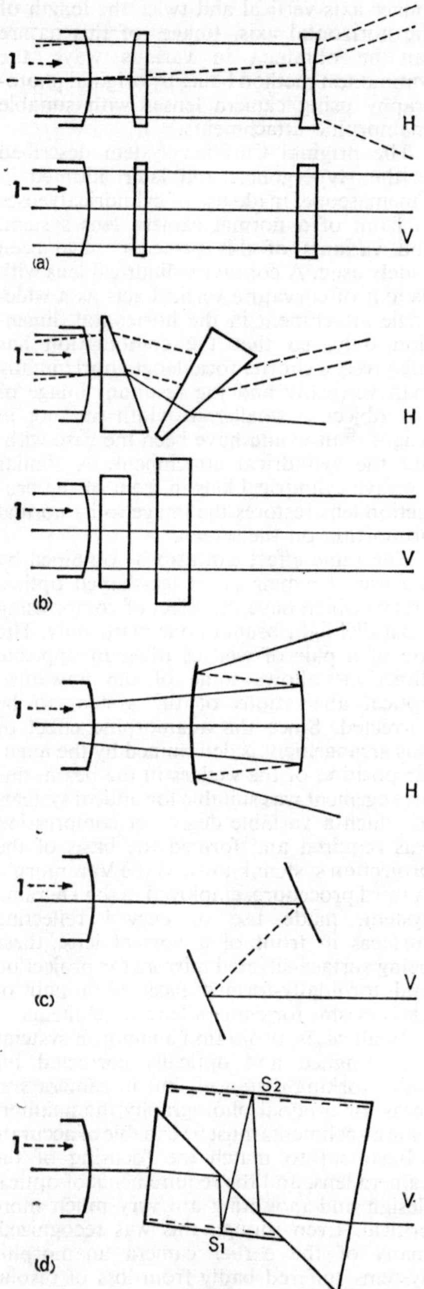

ANAMORPHIC LENS ATTACHMENTS. 1. Front component of normal lens. H. Horizontal plane. V. Vertical plane. (a) Cylindrical attachment. (b) Prismatic attachment. (c) Mirror Delrama. (d) Prism Delrama. Surfaces S_1 and S_2 are cylindrical reflecting surfaces giving divergence in the horizontal plane.

jected with an expansion factor of 2. However, improved optical design substantially eliminated such deficiencies in due course.

ASPECT RATIO. In its original form, the Cinemascope system used almost the maximum area available on standard 35mm film, exposing a camera aperture of 0·937 × 0·735 in. (23·80 × 18·67mm); the projected area on the print made from this negative was 0·912 × 0·715 in. (23·16 × 18·16mm), which, when projected with a lateral expansion factor of 2, gave a picture of aspect ratio 2·55 : 1. In order to provide room for the four magnetic sound-tracks used for stereophonic reproduction, the size of the sprocket holes on the print film had to be narrower than the normal positive perforations, and film of this type is referred to as Cinemascope-perforated, or sometimes as Fox-hole-perforated after 20th-Century Fox Corporation who introduced the system.

In the cinema presentation, three of the magnetic sound tracks feed three groups of loud-speakers occupying right, centre and left positions behind the screen, while the fourth track provides additional sound effects for a series of speakers situated in the body of the theatre. These are operated only during special sequences, and are switched by control signals in the fourth track itself.

However, not all cinema theatres were prepared to undertake the considerable expense of installing multi-channel sound systems, and a form of print combining the four magnetic tracks with the standard single optical track was introduced, referred to as a mag-opt print. This type of print also used the narrow Cinemascope sprocket holes, but as the optical sound track had to be located beside one of the magnetic tracks, it was necessary to reduce the width of the picture image.

The additional cost of magnetic striping and recording was still a factor which limited the use of prints of this form, and increasing use was made of prints with a single optical sound track only. These could be made on film having the normal standard positive perforation holes, and their use required no alteration of cinema equipment other than the provision of anamorphic projection lenses and the appropriate screen.

The projected area on such prints was 0·839 × 0·715 in. (21·3 × 18·16mm), giving a picture of aspect ratio 2·35 : 1. The extensive use of either mag-opt or optical anamorphic release prints greatly limited the use of the original Cinemascope format of 2·55 : 1, and practically all feature films photographed for anamorphic presentation on 35mm film are now composed for the aspect ratio of 2·35 : 1.

SUPERSCOPE SYSTEM. In addition to photography on 35mm film with 2 : 1 anamorphic compression lenses, a number of other methods were introduced either to provide improved photographic quality, or to reduce cost. In the Superscope system, normal camera lenses without anamorphs were used to give an image on 35mm negative extending the full width between perforations. In making prints from this negative, a horizontal compression and vertical enlargement was introduced by optical printing, so as to produce a copy suitable for anamorphic projection of the Cinemascope type. Prints of aspect ratios 2 : 1 and 2·35 : 1, were produced but the process was somewhat inefficient, since only a limited part of the available negative area was used.

OTHER SYSTEMS. 20th Century Fox introduced a system using large size film 55mm wide, the negative frame having four times the area of the 35mm Cinemascope image; 2 : 1 anamorphic compression was used on the camera lenses and 35mm release prints were made by optical reduction. The larger area of negative provided images of improved quality for grain and definition, but the cost of film was considerably increased, and the system was soon abandoned.

Another system using a larger area of negative was Technirama, in which 35mm negative was exposed to give an image twice the size of the normal frame positioned horizontally along the length of the film. Special cameras similar to the Vistavision design were employed, in which the film moved horizontally eight perforation holes for each frame, and the camera lens system had a lateral anamorphic compression factor of 1·5: 1. In making 35mm release prints, optical reduction was employed with the introduction of a further anamorphic lateral compression factor of 1·33 : 1, so that the resultant print had a total horizontal compression of (1·5 × 1·33) : 1, giving 2 : 1, and was thus suitable for projection with Cinemascope type equipment.

This process made good use of the available area of 35mm negative film and, although requiring special cameras, was widely used in Europe and elsewhere for some ten years from its introduction in 1956.

Ultra-Panavision is another anamorphic process using large-area film, in which 65mm-width negative is exposed with anamorphic camera lenses having a horizontal compression factor of 1·25. From such negatives, contact prints are made on 70mm-width film, and projected with the corresponding horizontal expansion of 1·25 : to give a picture on the screen having an aspect ratio 2·7 : 1.

These 70mm prints may be provided with six magnetic tracks for stereophonic reproduction over five loudspeaker systems behind the screen, together with additional effects over speakers in the auditorium. Copies for general release on 35mm film are made from the same negative by optical reduction printing introducing a further horizontal compression factor of 1·6 : 1, to give a print suitable for Cinemascope projection.

With the improvement in the grain and resolution of integral tri-pack colour film materials in the early 1960s, the need for large-area negative systems became less marked, and in 1963 the Techniscope system was introduced as an economical means of providing anamorphic release prints. This process uses standard 35mm film in cameras modified to pull down two perforations per frame instead of the standard four perforations. Normal camera lenses of short focal length are used to produce an image half the normal size with an actual aspect ratio on the film of 2·35 : 1. From this negative, 35mm copies are made by anamorphic optical printing, the vertical dimension being enlarged by a factor of two to produce a print of Cinemascope characteristics.

This process offers savings in the cost of negative since only half the normal footage is required for a given length of time, but the small size of the negative area used introduces some limitations in eventual projection on to very large screens.

FLAT PROCESSES. Where 35mm film is used, the so-called flat wide-screen picture is generally presented by projection with normal spherical lenses from a frame size 0·825 in. (20·95mm) in width, and a height varying from 0·498 in. (12·65mm) for an aspect ratio of 1·65 : 1 down to 0·447 in. (11·35mm) for an aspect ratio of 1·85 : 1. Masks are sometimes used in the camera to limit the dimensions of the frame exposed on the negative to a particular aspect ratio, but it is more general to recommend a larger exposed negative frame size and limit the height by the projector aperture mask. Occasional use is made of mag-opt tracks

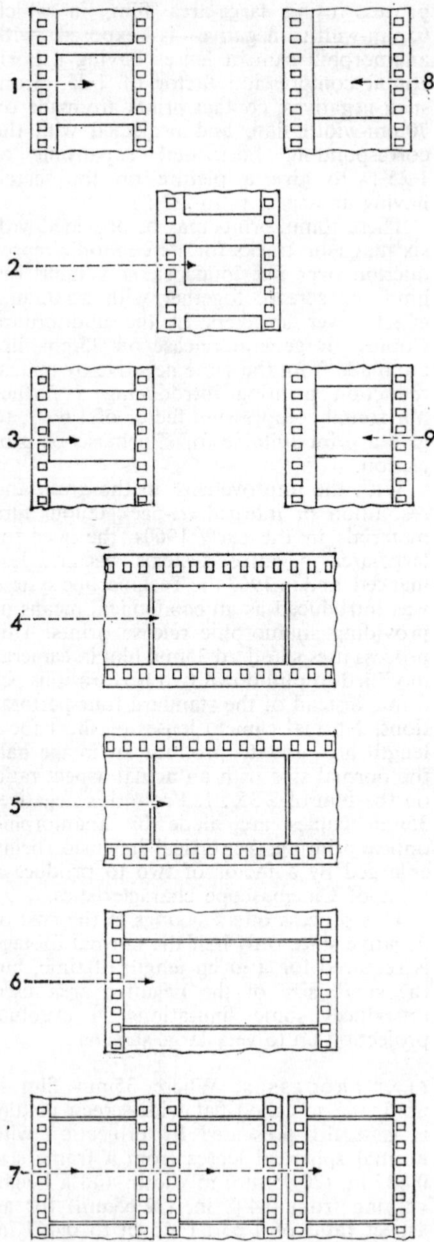

COMPARISON OF NEGATIVE FRAME AREAS. 1. Movietone (1·37 : 1). 2. Wide screen (1·85 : 1). 3. Superscope (2·35 : 1). 4. Vistavision (1·75 : 1). 5. Technirama (for 2·35 : 1). 6. 65mm Todd-AO, Super-Panavision, single-film Cinerama, etc. 7. Original three-film Cinerama and Cinemiracle (2·1 : 1 as seen from central position). 8. Cinemascope and equivalent processes (for 2·35 : 1). 9. Techniscope (2·35 : 1).

on flat wide-screen prints, but only for limited distribution requirements because of the additional expense.

In flat wide-screen systems on 35mm film, there is a good deal of wasted area above and below the composed image, and the small frame projected is sometimes found unsatisfactory for grain and definition when required for enlargement on to the very large screens of important first-run theatres; for such presentations, prints of larger area may be required and the use of 70mm-width film for this purpose has become frequent.

Having originally appeared as the Grandeur system in the silent cinema, prints of this type were reintroduced together with six-channel stereophonic sound as the Todd-AO process in 1955. The projected image size on the film is 1·913 × 0·866 in. (48·6 × 22·0mm), giving a picture of aspect ratio 2·2 : 1. The film carries four magnetic stripes on which are recorded six sound tracks for reproduction over five loudspeaker systems behind the screen and over a series of effects speakers in the auditorium.

In the Todd-AO process, the camera film is 65mm in width, as the extra width for the magnetic stripes is not required in the negative. The Super-Panavision system is similar but, in Russia, a negative width of 70mm, corresponding to the positive size, has been adopted as the standard system for wide film.

When necessary, prints for 70mm presentation can be made from originals of other sizes by optical printing; for example, the Technirama frame has almost the same height as the 70mm projected frame, so that only a rectification of the compression given by the Technirama camera lens is necessary to produce a 70mm copy. The required lateral expansion of 1·5 : 1 is introduced by anamorphic optical printing. Again, with the improved characteristics of negative materials and optical systems, 70mm prints are sometimes made from 35mm negatives of the Cinemascope type by optical enlargement including a 2 : 1 anamorphic factor in the horizontal dimension.

For general release purposes, 35mm flat prints may be required from other forms of original negative. Where original photography has used 65mm film, these can be made by straight optical reduction, but from original negative of the Cinemascope type, anamorphic printing is required. This involves correction of the compression originally introduced by the camera, and the flat print produced in this way is sometimes referred to as an unsqueezed copy.

Such prints are usually made of an aspect ratio of 1·85 : 1, and this naturally results in some of the composition at the extreme edges of the picture being lost. When prints of less wide aspect ratio, such as 1·65 : 1, are made by unsqueezing, this loss of composed subject matter may become quite marked, and can be very serious when prints of the 1·33 : 1 format are made from Cinemascope originals for use on television, or as 16mm copies, since it is possible that essential action may be obscured or completely lost. For these purposes, recourse may have to be made to the use of a scanning optical printer capable of traversing the negative area when making the unsqueezed print, so that the most appropriate section of the composition can be selected for each scene.

MULTIPLE FILM PROJECTION SYSTEMS

Although the use of two or more projectors showing pictures side by side on very large screens was demonstrated on several occasions during the early development of the cinema, these more complicated forms of presentation were naturally restricted to special occasions, such as international exhibitions or at special theatres built or modified for the purpose. The general adoption of multiple film methods of presentation was therefore outside the scope of the normal channels of cinema exhibition for dramatic entertainment, but the influence which these methods exerted on the eventual presentation and publicity of the wide-screen processes which came to be used for general distribution was quite considerable.

VITARAMA. In 1939, a multi-film system under the title Vitarama was presented at the New York World Fair. This used an assembly of eleven projectors to cover a vast curved screen in the form of the inner surface of a quarter sphere. Out of this system was developed an anti-aircraft gunnery trainer which came into use during World War II as the Waller Trainer; in this, five projectors were used to cover a large screen forming the inner surface of a dome on which were projected films of moving aircraft targets to provide aiming practice to trainees using gun sight units at the centre of the dome.

CINERAMA. After the war, development of a similar system for entertainment purposes continued and resulted in the introduction of Cinerama in 1952. Three projectors in separate projection rooms were used to throw three pictures on a deeply-curved screen. The images overlapped slightly, but soft-edged masking at the projectors was used to blur the joining of the pictures and blend the edges as imperceptibly as possible.

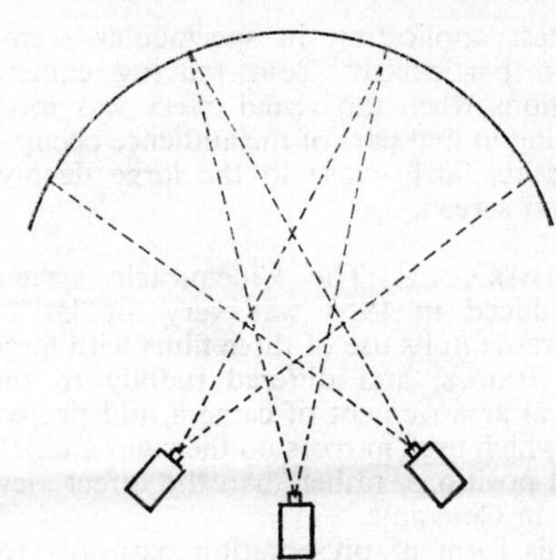

CINERAMA. In its original form, images of three separate films in three projectors were thrown on to a deeply curved screen. Audience seated between projectors and screen.

The frame size on each of the three 35mm films was larger than standard, being six perforations high and the full width available between perforation holes, the projected frame having dimensions 0·985 in. wide by 1·088 in. high (25·02×27·64mm). The three films together thus used nearly six times the area of a normal 35mm frame and allowed high standards of definition and brightness to be obtained on very large screen areas.

The picture films themselves carried no soundtracks, but eight magnetic records on a separate 35mm strip running in synchronism were reproduced over five loudspeaker systems behind the screen and eight more distributed at the sides and rear of the auditorium. Because of the deeply-curved screen, the aspect ratio of the complete display varied considerably according to the viewing position, but was generally stated as 2·1 : 1 when viewed from a position close to the central projector.

Photography of material for Cinerama was carried out by an assembly of three cameras mounted to correspond to the three axes of projection. The angle of the field of view was therefore effectively fixed, and problems of image distortion along the matching edges and in the corners of the frame imposed some limitations on the closeness of objects in the scene. The Cinerama process therefore found its greatest application in spectacular scenic shots, particularly from moving camera positions when the visual effect was most striking to that part of the audience occupying seats fairly close to the large deeply-curved screen.

CINEMIRACLE. The Cinemiracle system introduced in 1958 was very similar to Cinerama in its use of three films with large area frames, and differed mainly in the optical arrangement of camera and projector, which used mirrors on the right and left hand positions, rather than the direct view used in Cinerama.

This form of presentation naturally required very considerable modifications of the theatre in which it was shown, so that it was generally restricted to major cities large enough to provide an audience for the same programme over a period of several months. The limitations of the camera viewpoint and the expense of film and equipment prevented any widespread use for dramatic feature films, although one or two subjects were produced in this form, and in due course the greater convenience of 65mm negative for original photography and of 70mm prints for projection on very large screens led to the virtual abandonment of the three-film system; later productions under the Cinerama credit were photographed on a single strip of 65mm negative using a range of anamorphic camera lenses with a squeeze factor of 1·25 : 1. From these negatives, 70mm prints were made by optical printing at which small changes of image geometry were introduced to make the print more suitable for projection on a deeply-curved screen. Cinerama subjects produced in this form thus correspond very closely to the characteristics of 70mm prints described in the preceding section, and copies can be made for distribution and use in any theatre equipped with suitable 70mm projectors. 35mm copies for general distribution can also be made by anamorphic reduction printing to the Cinemascope format at a single operation, whereas the preparation of single-film copies from three-film Cinerama originals was a much more complex operation, involving multiple printing stages on an optical printer.

SCREENS. The use of a deeply-curved screen was an essential part of the Cinerama form of presentation using three projectors, and the large area of film projected allowed very large screen areas to be satisfactorily illuminated. The width of screen might be as great as 60 or 70 ft. measured directly from edge to edge, with a depth of curvature 20 ft. or more. It was therefore possible to distribute the seating in the auditorium in such a way that for a large number of spectators the screen subtended a wide angle and actual movement of the eye and head was necessary to follow action from one side of the screen to the other. Even when the main action was centred in the screen, the effect of picture area extending out to the limits of vision on right and left could be extremely spectacular since the audience was almost unaware of the screen edges.

This so-called wrap-around screen became a hall-mark of Cinerama publicity, which made use of such phrases as "audience participation" and even "stereoscopic presentation". Similar phrases were used when the Cinemascope process was introduced, and its first presentations were associated with curved screens which helped in some of the optical problems of anamorphic projection, although the curvature used was never extended to provide any wrap-around effect. Many of the same features were associated with the initial presentation of the 70mm wide-screen process in the Todd-AO version.

While the use of deep-screen curvature, often as great as one third of chord width, was necessary with Cinerama with its three specially-positioned projection rooms at the level of the centre of the screen, the use of excessive curvature in more normal cinema theatre projection conditions introduces very serious distortions to viewers over a considerable part of the auditorium. For example, the distortion of horizon lines when projecting downwards at a steep angle on to a markedly curved screen was often ludicrous. In the course of development and general adoption of wide-screen processes, much less stress has been laid on deep curvature, and in general the current practice for normal theatrical presentation is now to use screens which are only very slightly curved and which assist in maintaining a reasonably high standard of definition over the whole picture area from wide-angle projection lenses.

SPECIAL PRESENTATION SYSTEMS

FULL CIRCLE. The ultimate in wide-screen presentation is, of course, the full circle of 360° in which the audience is surrounded by picture in all directions, and there have been numerous systems aimed at providing this effect, usually as special displays at exhibitions. Probably the first of these was

Cinéorama, using ten projectors, at the Paris Exposition of 1896, and in general it has been found necessary to provide nine, ten or eleven projectors to provide the necessary angular coverage, although Theaterama at the New York World Fair of 1964 used only six 35mm projectors with anamorphic lenses.

Disney's Circarama system originally used eleven 16mm projectors, but subsequently employed nine 35mm projectors, while one form of the Russian system Kinopanorama used a total of twenty-two,

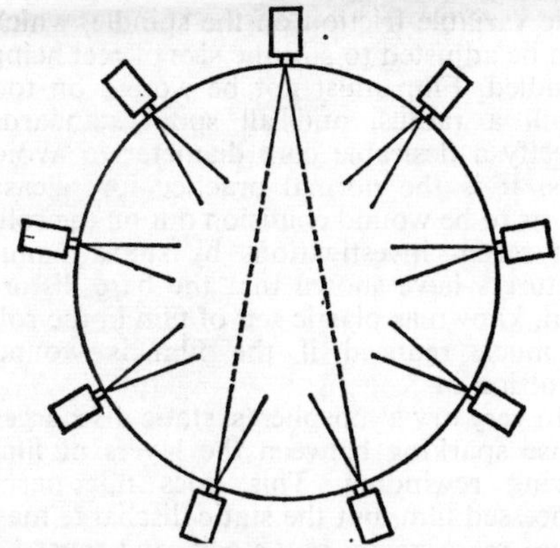

CIRCARAMA. Nine projectors arranged round the outer wall of a cylindrical screen, with lenses in narrow slots dividing screen into segments, are used to cover full 360 degrees. Audience is in central area of auditorium.

eleven of which completed the 360° of a cylindrical screen while eleven more were used to fill in part of a domed area above. However, this full presentation was only used with animated cartoon subject material, live-action photography being limited to the lower tier of eleven panels.

Presentation methods of this type make very stringent demands on projected image steadiness and uniformity of screen brightness and colour for satisfactory presentation and, unlike Cinerama, there is usually no attempt to blend the edges of the projected pictures imperceptibly one into the other. There are usually narrow dark strips between each image panel, giving the impression of a series of windows looking out over the scene, and these dark strips often provide the apertures through which the projected beam enters the circular enclosure.

Photography for full-circle presentation is carried out by means of a corresponding group of cameras rigidly mounted together and using lenses of fixed angle to cover the 360° field with very limited overlapping. The camera position has to be sufficiently elevated to allow an extensive view in all directions clear of foreground obstructions and, as a result, subject matter is usually limited to purely natural scenic material of the travelogue type, However, very spectacular effects can be produced in this way, particularly from mobile camera positions. Full-circle presentation is almost always accompanied by multi-channel sound effects provided from a separate magnetic record reproduced over a number of loudspeakers distributed all around the viewing area and sometimes above and below in addition.

TRIPTYCH. In contrast with the full-circle form of presentation, which is practically limited to the showing of pictorial scenes, other systems of multi-image presentation have been developed in which the object is dramatic and emotional effect rather than the reproduction of actuality. The triptych presentations of Abel Gance made use of three projected images side by side, but the scenes shown on the two outer panels were not regarded as an extension of the central scene but rather as providing images of dramatic contrast to supplement the theatrical effect of the central action.

This concept was taken further in the Polyecran presentation developed in Czechoslovakia in the late 1950s, in which nine motion-picture and five still-picture projectors were used to provide groups of images, both live-action and abstract, on an assembly of nine screens for dramatic and emotional effect. Similar specialized presentations have subsequently been shown at a number of international exhibitions.

As a further development, combinations of moving-film images, projected stills, and live actors in stage settings have been attempted on several occasions. One of the most effective of these was the Magic Lantern presentation developed in Czechoslovakia, first shown at the 1958 Brussels World Fair, and the development of similar concepts has continued, for example, the Living Screen system.

However, such forms of entertainment belong more properly to the creative art of the live theatre and have little, except some technical equipment, in common with the art and practice of the cinema as we know it. L.B.H.

See: *Film dimensions and physical characteristics; Screens.*

Book: *American Cinematographer Manual*, 2nd Edn., ASC (Hollywood), 1966.

⊕**Wild Motor.** Motor not running at synchronous speed. Usually refers to a camera motor used for silent shooting, and therefore not requiring to be synchronized.

⊕**Wild Track.** Sound track recorded otherwise than with a synchronized picture. In film usage, wild track normally carries sound effects, random dialogue and general background noise. Sometimes called non-sync.

See: *Editor*.

□**Willans, Peter, 1900–1943.** British television pioneer working in the EMI team of the 1930s headed by Isaac Shoenberg.

⊕**Winding.** Process of transferring film or □tape from one reel, spool, etc., to another. Winding, although the simplest of operations, is the cause of a large proportion of film damage. There are several possible reasons.

In the first place, the reel being rewound has considerable inertia, whereas an even tension must be maintained during winding if the turns of film are not to cinch and cause abrasions. A sudden jerk may even snap film or break a splice. On the other hand, if the film between the reels is allowed to run slack it may rub on the bench and be scratched.

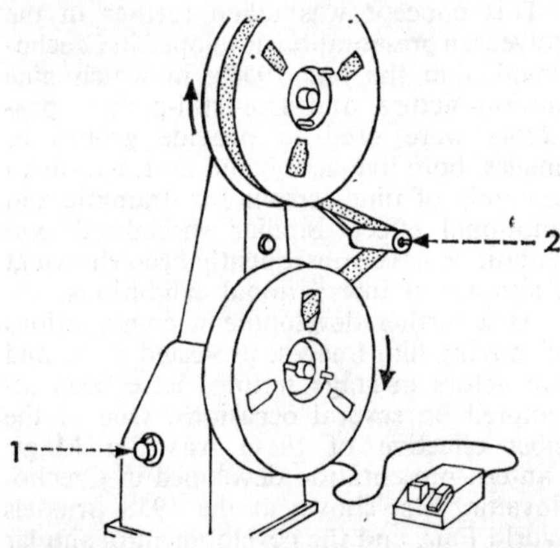

VIDEOTAPE POWER WINDER. Operated by foot switch at side; motor speed controlled by rheostat, 1. Sensing roller, 2, protects against a runaway supply reel by applying a brake and cutting off the motor. Constant take-up tension is maintained.

Motor-driven rewinders are increasingly used in the cinema and laboratories, but hand winders are still usual for inspection purposes. Each winder may carry the rolls of film horizontally or vertically, and some types are found with spindles set at an angle, which have the advantage that the spool need not be fastened by a catch on the spindle as in the vertical type. The spindles of the winders may carry either spools or plates for winding film on to cores. In the form known as a tight wind the film is wound on a standard plastic core under the pressure of a weighted roller providing a film guide which ensures a compact and neatly wound roll.

It is important that the two winder heads should be correctly aligned, and while some degree of tension is necessary to ensure close winding, it must not be excessive. Some of the better types of winders, particularly those used for negative, can provide variable friction on the spindle, which can be adjusted to suit the size of reel being handled. Film must not be wound on too small a radius, and all spool standards specify a desirable core diameter to avoid this. It is the normal practice for release prints to be wound emulsion out on the roll, but recent investigations by stock manufacturers have shown that the base distortion, known as plastic set, of film in the roll is much reduced if the film is wound emulsion in.

In very dry atmospheres, static discharges cause sparking between the layers of film during rewinding. This does not harm processed film, but the static discharge may cause marking on raw stock, and must be avoided by the maintenance of suitable humidity conditions in the film handling rooms. L. B. H.

See: *Laboratory organization; Release print examination and maintenance.*

⊕**Wipe.** Effect which provides the transition □from one scene to the next at the boundary of a line moving across the image until the second scene has entirely replaced the first. The boundary may be a sharply defined line, when the effect is called a hard-edge wipe, but is usually more or less diffused, or soft-edge wipe. Wipes are varied in form, ranging from a single straight edge moving

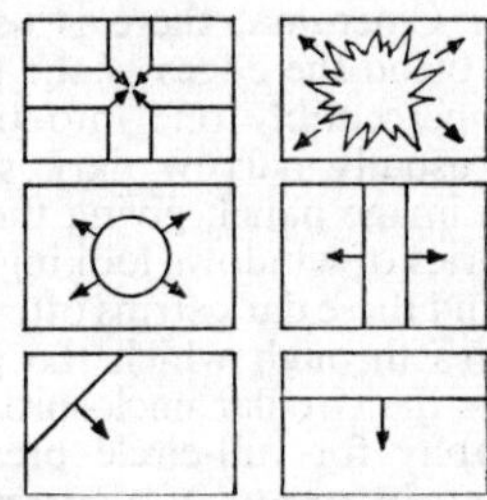

WIPE. Type and direction of wipe indicated by hard lines and arrows. In practice, wipes are often soft-edged and thus less perceptible.

vertically, horizontally or diagonally to multiple edges and geometric patterns moving simultaneously in different sectors of the frame. Wipes with expanding or contracting circular outlines are called iris wipes, while an expanding irregular form is termed an explosion or burst wipe.

Wipes of simple form may be made on an optical printer equipped with a mechanism which can move one or more blades across the field near the plane of the negative, the sharpness of the wipe edge being determined by the extent to which the blade edge is in focus. More elaborate wipes are made using animated matte films with the required patterns in silhouette.

Wipes may vary in length but are usually kept short for dramatic effect, and are often employed in advertising trailers and similar subjects where a startling visual impact is needed. At the beginning or end of a sequence a wipe, often an iris wipe, from or to black screen is sometimes used.

The same effect is achieved in television by replacing the output of one camera by another by means of an electronic switch.

See: *Special effects.*

⊕**Workprint.** Film editor's working reels of film, assembled from the original material and cut to a greater or lesser degree of fineness. The workprint is spliced together and may contain build-up. It may also be marked for optical effects or carry instructions for negative cutting. Also called a cutting copy.

See: *Editor; Laboratory organization; Narrow-gauge production techniques.*

⊕**Wow.** Colloquial term for a type of cyclical □frequency deviation in sound recording and reproduction which is properly termed flutter. Wow usually refers to a slow frequency deviation such as, for example, is caused by a once-per-turn slowing down of a phonograph turntable due to frictional drag or deformation of the disc. A steady tone shows up this defect most easily.

See: *Sound distortion; Sound recording.*

□**Writing Beam.** Electron beam in a cathode ray display tube (kinescope) which produces an image on the tube face. Term occasionally used to distinguish this beam from an index or pilot beam which, in certain colour receiver tubes, is used to guide the writing beam in producing the correct colour.

See: *Colour display tubes.*

⊕**Xenon Lamp.** Compact-source, high-pres□sure discharge lamp, containing traces of inert xenon gas, extensively used in medium power film projectors as a substitute for the carbon arc. It requires a rectifier and starting

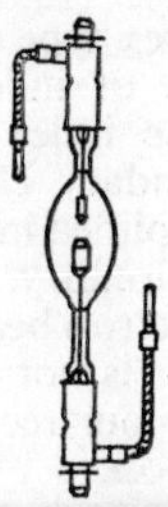

circuit, but needs no adjustment during use, has a high light output and reasonably long life, and provides good colour rendering.

See: *Projector.*

□**Yagi Aerial.** A specialized but simple television aerial with marked directional properties, named after its inventor. Yagi aerials are sometimes used for transmission purposes, but normally only for low power applications, as in a translator.

Yagi discovered that placing a parasitic element on a line with the aerial in its plane of polarization modified the normal polar diagram. If the parasitic element were tuned slightly lower in frequency, it reduced the sensitivity on its side of the aerial and increased it on the opposite side. Similarly, if the element was tuned to a higher frequency, it improved sensitivity on its own side and reduced it on the opposite side. It is thus

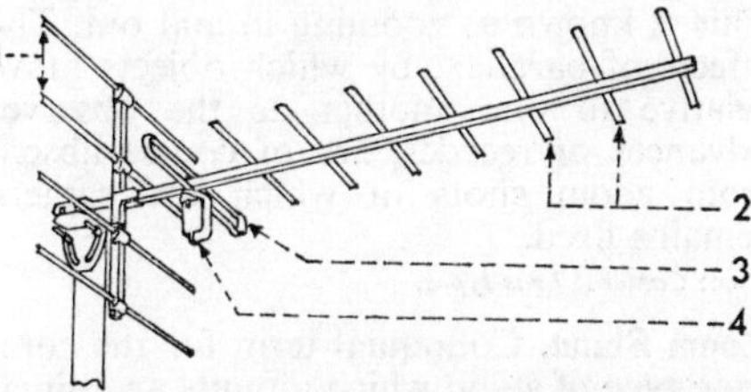

YAGI RECEIVING AERIAL. Tuned dipole element, 3, is made more directional and thus more sensitive to wanted signal by incorporation of directors, 2, and reflectors, 1. Balun unit, 4, is required to match dipole output to unbalanced coaxial downlead, when used.

possible to build up an assembly of elements in front of and behind a dipole, usually one reflector behind and a number of directors in front, and thus greatly increase the sensitivity in one direction.

See: *Transmitters and aerials.*

□**Zebra Tube.** A colour display tube, so called from the vertical red, green and blue-emitting stripes which make up the phosphor screen, which closely resembles the apple tube.

It is of the index type and, like the apple tube, makes use of index strips over the phosphors. These index strips, instead of giving out secondary electrons, emit UV light which is amplified in a photomultiplier. By ingenious circuitry, the need for a separate pilot electron beam is eliminated.

The zebra tube is not at present in commercial use in colour receivers.

See: *Colour display tubes.*

⊕**Zip Pan.** Rapid panning movement of camera producing blurred image. Also known as swish pan or whip pan.

⊕□**Zoom.** Real or apparent rapid motion of the camera towards its object. In actuality filming, usually accomplished with a variable-focus (zoom) lens, in animation usually by moving camera and object relatively to one another.

See: *Animation equipments; Animation techniques.*

⊕□**Zoom Lens.** Special lens which has a continuously variable focal length (and thus magnification) over a certain range, usually between 3 : 1 and 20 : 1. Once focused, the image remains in focus over the complete range of focal lengths. Similarly, the effective aperture remains constant, irrespective of focal length.

The use of variable magnification makes the camera seem to advance towards or recede from its subject at any desired speed. This is known as zooming in and out. The effects of parallax, by which objects move relative to one another as the observer advances or recedes, are of course absent from zoom shots in which the camera remains fixed.

See: *Camera; Lens types.*

⊕**Zoom Stand.** Colloquial term for the common type of stand which mounts an animation camera, in which the camera can move up or down on a vertical column so as to zoom towards or recede from the animation film mounted horizontally on a rostrum beneath. Zoom stands often incorporate automatic cam devices for following focus during the zoom movement.

See: *Animation equipment.*

□**Zworykin, Vladimir Kosma, b. 1889.** Russian inventor of the iconoscope and father of the modern electronic system of television. Born at Mourom in Russia and educated at the Gymnasium there and at the Technological Institute in Leningrad. There he gained his first contact with television, studying under Boris Rosing, who was developing a television system using a Braun cathode ray tube as a receiver and a double mirror wheel for transmitting.

Zworykin went to Paris in 1913 as a research student under Paul Langevin. During World War I he was in the Signal Corps of the Russian Army, but after the revolution he decided to emigrate to the United States, where he arrived in 1919. There he obtained employment as a research engineer with the Westinghouse Company. By this time he was already visualizing a method of transmitting, as well as receiving, television signals by electronic means.

His early attempts to work on television were thwarted by the emphasis placed on sound broadcasting, but by the end of 1923 he was able to apply for a patent on his first iconoscope, a charge-storage type of transmitting tube. The iconoscope – the electronic eye of the television camera – was finally produced in practical form in 1928 and given its first public demonstration at a meeting of the Institute of Radio Engineers, New York, the following year.

In 1929 Zworykin obtained the backing of the Radio Corporation of America for his system and became Director of Electronics Research for RCA at Camden, New Jersey. The iconoscope, after modification, became the basis of the Emitron camera developed in England by EMI. After World War II, Zworykin took a leading part in the development of colour television and the electron microscope.

Film & Television

A Basic Anatomy—by *Raymond Spottiswoode*

⊕ □

Optics 994
- Frequency response 994
- Lens design 994
- Dichroic filters 994

Modulation 995
- Picture elements 995
- Modulation techniques 996

Colour 996
- Colour variables 997
- Colour measurement 997
- CIE system 997
- Additive colour film 997
- Subtractive colour film 997

Visual Principles 998
- Reproducing motion 999
- Visual adaptations 999

Sound and Acoustics 1000
- Sound as wave motion 1000
- Frequency and Intensity 1000
- Magnetic recording 1001
- Optical recording 1001
- Acoustical problems 1001

⊕

ORIGINS AND MATERIALS

History 1002
- Flexible film 1002
- Talkies 1002
- Animation and colour 1003
- Wide-screen 1003

Materials and Storage 1003
- Film base 1003
- Emulsion structure 1004

Gauges and Processes 1004
- Non-standard gauges 1005
- Aspect ratios 1005

Narrow-gauge Processes 1006
- Eight-millimetre advantages 1006
- Equipment similarities 1006

The Photographic Image 1006
- Latent image 1007
- Spatial frequency 1007
- Thermoplastic recording 1008
- Kalvar process 1008

Colour Processes 1008
- Tripack principles 1008
- Practical considerations 1009
- Dye imperfections 1009
- Printing methods 1009

The Camera 1009
- Camera construction 1010
- Noise suppression 1010
- Lenses 1010
- Accessories 1011
- High-speed work 1011

CREATIVE TECHNICIANS

Producer and Director 1012
Lighting Cameraman 1012
Art Director 1013
Sound Mixer 1013
Editor 1013

THE STUDIO COMPLEX

Studios and Lighting 1014
Special Effects 1015
- Image replacement 1015
- Laboratory techniques 1016
- Optical printer 1016

The Laboratory 1016
- Developing agents 1017
- Printing processes 1017
- Imbibition techniques 1018

Laboratory Control 1019
- Characteristic curve 1019
- Toe and shoulder 1019
- Slope and gamma 1019

Sound Recording 1020
- Synchronized sound 1020
- Playback and post-sync 1020
- Multiple sound tracks 1020
- Dubbing equipment 1021
- Stereo recording 1021

SPECIAL TECHNIQUES

Animation 1022
- Cel process 1022
- Computer technique 1022
- Animation equipment 1022

Difficult Environments 1022
- Low temperatures 1023
- Tropical conditions 1023
- Underwater 1023

Film and TV in Medicine 1023
- Operations 1023
- Radiography 1023
- Macro and Micro 1023

The Film in 3 Dimensions 1024
Library 1025

PROJECTION AND THE CINEMA

Release Print Make-Up 1026
The Cinema Theatre 1026
- Projector features 1026
- Sound types 1027

Sound in the Cinema 1027
Hybrid Techniques 1028

Film & Television: A Basic Anatomy

FUNCTIONS AND ORIGINS

History 1029

BASIC PRINCIPLES

Scanning 1030
- Frequency and flicker 1030
- Raster lines and resolution 1031

Transmission Channels 1031
- Picture elements 1031
- Modulation methods 1031
- Bandwidth requirements 1032

The Television Signal 1032
- Blanking and synchronization 1032
- Line and field synchronization 1032

COLOUR PRINCIPLES

Bandwidth Problems 1033
The Colour Signal 1034
Reception on Monochrome Sets 1034
Colour Difference Signals 1035
- Encoding principles 1035
- Decoding problems 1035
- Suppressed carrier modulation 1036
- Quadrature modulation 1036
- Detection methods 1036
- Other considerations 1036

Nature of the TV Signal 1037

THE CAMERA

The TV Camera 1037
- Construction 1037

Tube Types 1038
- Photo-emissive tube 1038
- Image orthicon 1038
- Return beam modulation 1038
- Orthicon characteristics 1038
- Photo-conductive tube 1039
- Vidicon 1039
- Plumbicon 1039
- Auxiliary equipment 1039

Colour Cameras 1040
- Four-tube and three-tube 1040
- Beam splitter 1040

CREATIVE TECHNICIANS

Cameraman 1041
Producer 1041
Director 1042
Designer 1042

STUDIO OPERATION

Studio Complex 1043
- Continuous recording 1043
- Inserted material 1043
- Production control 1044

Master Control 1044
Picture Monitors 1045
Transfer Characteristics 1045
- Gamma principles 1045
- Gamma control 1046

Vision Mixers 1046
Electronic Special Effects 1046
- Special effects generator 1046
- Insertion effects 1048
- Inlay process 1048
- Overlay process 1048
- Other effects 1048

STUDIO SYNCHRONIZATION

Picture Locking Techniques 1049
- Synchronization problems 1049
- Types of lock 1049
- High Stability operation 1049

Synchronizing Pulse Generators 1050
Station Timing 1050

PICTURE RECORDING

Videotape Recorder 1051
- Frequency band requirements 1051
- Transverse scanning 1052
- Helical scanning 1052
- Servo circuits 1052
- Colour problems 1052
- Editing 1052

Telecine 1053
- Camera type 1053
- Flying spot type 1053
- Colour 1053

Tele-Recording 1054
- Pull-down requirements 1054
- Stored-field techniques 1054

OUTSIDE SOURCES

Outside Broadcast Units 1055
Special Events 1056

TRANSMISSION

Microwave Links 1057
Transmission Networks 1057
Satellite Communications 1058
Standards Converters 1059
- Line standard development 1059
- Converter techniques 1059
- Digital converters 1060

Transmitters and Aerials 1060
- Siting considerations 1060
- Modulation methods 1061
- Aerial design 1061

RECEPTION

Monochrome Receivers 1062
- Superhet principle 1062
- Tube design 1062
- Electron gun 1062
- Scanning generators 1062
- Scanning coils 1062

Colour Receivers 1064
- Shadow mask tube 1064
- NTSC system 1065
- PAL system 1065
- SECAM system 1065

Large Screen Television 1066
- Projection tube 1066
- Eidophor 1066

CLOSED CIRCUIT TELEVISION

Equipment 1067
Application 1067

FILM AND TELEVISION COMPARED

Audience Size 1067
Production 1068
Sound 1068

FUTURE TRENDS

Film 1069
Mixed Techniques 1069
Television 1069

Film & Television

A Basic Anatomy—by Raymond Spottiswoode

This survey of the subject matter of the Encyclopedia is interspersed, section by section, with lists of major and minor entries in the body of the work. Thus the survey presents an overall view of film and television technology in both theoretical and practical aspects; the lists help the reader to see at a glance where he is most likely to find the detailed information he wants.

Key entries appear first in capitals; these are not necessarily the longest articles, but are the most relevant to the subject under review, and may appear more than once in different contexts.

Film and television are the joint subject of this book, and have one fundamental element in common, that they take pictures of the world in movement, store them for a fraction of a second or indefinitely, and transfer them to a distant place for reproduction before an individual or a mass audience.

Because film is relatively old as a medium and television relatively new, the differences between the two have received more attention than their basic similarities. Film is thought of as a storage medium, TV as a method of instantaneous transmission. Film is photographic, TV electronic. Film seeks an audience of thousands under one roof, TV entertains in the living room.

In this volume, film and television are interlinked. Recognition of basic similarities helps towards an understanding of how it is that film and television have so often borrowed techniques from one another, or have grown in unexpected directions in mutual competition.

The limits of the survey are in general technical, and comprise the tools and processes by which images in picture and sound are formed, manipulated, transmitted and received; but both media have creative aspects which cannot be disentangled from the technical, and the roles of the producer, director, cameraman, editor, sound mixer and others, are therefore included.

To help the less technical reader, something may be said on a few of the basic concepts which underline both film and television.

OPTICS

Light and optical principles are a convenient starting point, for the outside world must be imaged on a light-sensitive surface, whether in a TV or a film camera, and that imaging can be accomplished only by a photographic lens. The camera lens is the first link in a long chain between the world seen and the world reproduced, and the fidelity of reproduction is of prime concern to the engineers of both media.

The standards and test methods they inherited were based on concepts carried over from astronomical photography: the criterion of definition was the ability to discriminate between points (e.g. stars) imaged very close together on the sensitive surface. From this was derived the concept of resolving power, tested by the familiar patterns of lines of varying width and spacing set in groups at different orientations about the field of the camera lens, and transmitted by many television stations.

Frequency response

During recent years, however, resolving power has increasingly yielded ground to concepts based on the reproduction of wave trains of high and low frequency by electronic and photographic systems. Instead of concentrating on the ability to separate closely-set parallel lines, researchers have paid more attention to the boundaries of objects, marked as these are by sudden changes of photographic density or screen brightness. It has been found that boundary sharpness, which depends on good high-frequency response in the system, correlates better with what the observer judges to be a crisp picture than does a high resolving power. Even more important, the frequency concept fits well with the scanning process of a television system, in which fine detail is represented by high frequencies and coarse detail by lower frequencies. The use of what are called transfer functions has made it possible to sum together the losses in the many stages of film and television transmission.

Lens design

The use of lenses in both media is not, of course, confined to cameras. Wherever an image is transferred from one piece of film to another (except by contact printing), or is projected on a screen, or a television image is recorded on film, or derived from it, or projected, a system of lenses is called into play. Some of these are calculated for optimum performance in an individual piece of equipment; others, like the modern zoom lens, must fulfil the functions formerly assigned to a wide range of individual lenses.

Today, lens design relies largely on the use of computers, but debate continues on whether the computer merely extends the skill of the traditional lens designer with his optical and mathematical expertise, or whether by means of standard programmes it can replace him altogether.

Dichroic filters

Among optical devices of recent development, none is more versatile in application than the dichroic filter or reflector, which separates portions of the colour spectrum by transmitting one part and reflecting the rest with the least possible loss by absorption. Dichroic filters do this by means of numerous extremely thin transparent layers deposited on a support such as glass. The refractive index of the material deposited, together with the thickness of the layer, is calculated so that interference takes place between rays reflected from the various surfaces, but only over a chosen range of wavelengths. Dichroic filters find many applications as beam-splitters in colour film and television.

COATING OF LENSES—LENS TYPES—LIGHT—OPTICAL PRINCIPLES—Aberration—Achromatic lens—Acutance—Anti-reflection coating—Aperture lens—Aperture, lens—Apostilb—Astigmatism—Barrel distortion—Bloom—Brightness—Chromatic aberration—Circle of least confusion—Collimation—Coma—Condenser—Copy lens—Curvature of field—Depth of field—Depth of focus—Diaphragm, lens—Diffraction—Dioptre—Discontinuous spectrum—Dispersion—Distortion—Electromagnetic spectrum—Field flattener—Field lens—Fish-eye lens—Flare—f-number—Focal length—Focal plane—Foot candle—Foot lambert—G-number—Infra-red radiation—Interference filter—Iris—Lambert—Lens—Light units—Lumen—Luminance—Luminous flux—Luminous intensity—Lux—Newton's Rings—Nit—Phot—Photocell—Pincushion distortion—Principal focus—Principal plane—Principal points—Refraction—Refractive index—Refraction, laws of—Relay lens—Retrofocus lens—Schmidt optics—Specular reflection—Speed—Stilb—Stop, lens—Telephoto lens—T-stop—Ultra-violet radiation—Wide-angle lens—Zoom lens

MODULATION

An image formed by a lens on black-and-white film becomes after development a pattern of densities which can be viewed by reflected light (as in a print) or by transmitted light (as in the transparency of the motion picture). In the transparency the light is modulated, point by point, by the opacity of the image: light is absorbed where the image is dense and more freely transmitted where it is more transparent. Thus the film acts as a light-modulator, and the modulations of light intensity perceived by the viewer should correspond as nearly as possible with those he would have perceived had he looked at the original scene. The monochrome picture responds more or less equally to all wavelengths of light; but a colour picture acts selectively towards different wavelengths, and therefore may be considered as a colour modulator as well as a density modulator.

Picture elements

In film, this concept of modulation is convenient; in television it is essential. To send a picture through a wire or across space it is necessary to break it down into individual, unitary elements which can be transmitted one at a time. Their position is determined by their place in a scanned structure, called a raster, resembling lines of type set on a page, but closer together so that there are no spaces between the lines. An electron beam scans the raster like a reader's eye on a page: from left to right of each line, then quickly back to the start of the next line, and at the bottom of the page, quickly to the top of the next page. In television, the information is arranged to decay by the time the foot of the picture is reached, so that the next picture can be written progressively in place of the first one.

The varying light intensity of the scene sets up an electrical wave train of varying magnitude (amplitude) in the camera. Since the picture is scanned at fixed speed, rapid changes of light intensity, corresponding to fine detail, produce rapid current changes corresponding to high frequencies. Slow changes of intensity, or coarse detail, produce slow current changes or low frequencies. Thus the scene is converted into a wave train of varying amplitude and frequency.

Modulation techniques

This wave train, after complex processing, is radiated from a transmitter. It is impossible to do this directly, however, for the range of frequencies is far too wide: from a few cycles per second up to several million cycles per second. This difficulty is overcome by modulating these video-frequency signals on to what is called a carrier wave of very much higher (radio) frequency than the highest modulating frequency. The video-frequency band is thus quite small compared with the lowest of these radio frequencies.

There are two principal kinds of modulation. The constant amplitude (i.e. height and depth) of the carrier can be impressed with the varying amplitude of the modulating signal; this is called amplitude modulation (a.m.). Alternatively, the constant frequency of the carrier can be impressed (or deviated) with the varying signal amplitude. In this case the carrier amplitude remains constant, but its frequency varies; hence the process is called frequency modulation (f.m.).

Both a.m. and f.m. are employed in television, and both have advantages and drawbacks. With amplitude modulation, there is a limit to the permissible amplitude of the modulating signal; it can be varied down to zero and up to the maximum output of the transmitter, which cannot exceed that of the final stage. With f.m. there is no such limit, since the frequency deviation can be made as large as desired; it is this which gives a high ratio of signal to receiver noise in f.m. radios. On the other hand, whereas an ordinary double-sideband a.m. transmission occupies only twice the bandwidth of the highest frequency to be transmitted, an f.m. transmission generates an infinite number of sidebands. In practice, the higher-order sidebands are suppressed, but the bandwidth required is still a large one.

Vision modulation on all public television systems is a.m., but the accompanying sound is almost invariably f.m. Videotape recorders are also frequency modulated.

LENS PERFORMANCE—SCANNING—TELEVISION PRINCIPLES—TRANSMISSION AND MODULATION—Band—Fourier analysis—Hertz—Lag—Modulation—Raster—Subcarrier—Vestigial sideband

COLOUR

The stuff of film and television is, therefore, a modulated carrier. In film, the carrier is a photographic image formed in silver halide or organic dyes, and supported on a transparent flexible base; it is modulated by density changes in black-and-white, and by density and spectral changes in colour film. In television, the radiated carrier is a wave train of high frequency on which the signal variations are impressed; brightness (luminance) variations convey the picture to all receivers, and colour (chrominance) variations may be separately transmitted on a subcarrier to convey the extra information needed to actuate colour receivers.

Since the world around us is a coloured one, as perceived by normal human eyes, it is natural to think of colour transmission as indeed the norm, and black-and-white as a special case. Historically the opposite is true, but this is because the basic photographic receptor – the silver-halide emulsion – produced an image in terms of simple opacity, i.e. light exclusion or blackness. The theory of colour was already beginning to be known, as it applied to the human eye; but the application of this theory to emulsions and dyes, though foreshadowed in the 1860s, took long to realize, and in television still longer.

Colour variables

The starting point is the world of coloured objects as seen by the human eye, and here three variables are met with: hue, which describes the similarity of a colour to one in a series ranging from red to yellow, green, blue, purple, and so back to red (black, white and grey thus have no hue); saturation, which defines the position of a colour in terms of its difference from grey, so that dull and pale colours have a low saturation, and vivid colours a high one; finally, brightness (now more usually called luminosity), which ranks a colour on the scale from darkness to lightness. These three factors serve to define any colour, and may conveniently be thought of as a 3-dimensional system: hue, varying with angle about a central axis, saturation or chroma increasing radially from this centre, and luminosity increasing vertically along the axis. Subjective colour systems such as the Munsell system make use of a thousand or more colour samples arranged in a colour solid of this kind.

Colour measurement

Objective colour measurement, though arriving at a similar result, starts from a different set of premises: that a wide range of colours can be matched by adding together red, green and blue lights in appropriate proportions. The intensities of each light required to produce an exact match measure objectively the colour of the sample. Furthermore, any component in a mixed light may be replaced by a mixture of lights which matches the component; in this way, light of the most complicated spectral distribution can be matched to the trichromatic primaries by substituting in turn for a narrow band of wavelengths.

Moreover, it is possible to alter the trichromatic primaries so long as they can be added in some proportion to produce white. Indeed, in a system whose object is to specify and match the widest variety of sample colours, the primaries themselves can be imaginary – that is, no actual lights can match them. This is not so surprising as it may seem at first: ideal or imaginary lenses, free of all aberrations, are used in optical analysis, weightless levers in mechanics, point sources of light in illumination, distortion-free components in electronics, though none of these actually exists.

CIE system

The internationally accepted CIE system is of this kind, and the amount of the three primaries required to match a unit amount of energy for each monochromatic light from one end of the visible spectrum to the other is represented by x, y and z curves, x corresponding approximately to red, y to green and z to blue. Furthermore, the primaries are so chosen that the y curve is identical to the luminosity curve of the eye of a standard observer – that is, the eye's relative response to light of equal intensity at different wavelengths.

The amounts of the primaries required to match a monochromatic colour are called tristimulus values, and when combined with the curve of a non-monochromatic colour are written X, Y and Z. Since Y corresponds to the response curve of the eye, it also represents the luminance of the colour under analysis.

Since the CIE system is basically a system of colour matching by addition of components, it has a direct application to television where mixed colours are broken down in this way in the camera, and subsequently recombined in the receiver by the addition of coloured primary lights in suitable proportions. Moreover, the compatibility essential to colour television is based on transmitting a separate luminance component, the Y signal, matched by CIE standards to the response of the viewer's eyes, plus chrominance components to provide the proper saturation and hue. A colour set makes use of all these signals; a monochrome set takes the Y signal only, to give a balanced black-and-white picture.

Additive colour film

Colour may also be reproduced additively by film. By the use of three-colour filters the three-colour components are recorded on separate black-and-white films, and the pictures are later recombined using white light similarly filtered. Problems of register and the complication of multiple films made this additive system impracticable for projection. Instead, inventors attacked the problem of producing minute three-colour mosaics on the film itself. These were extremely insensitive in the camera and wasteful of light in projection, and were in turn abandoned.

Subtractive colour film

In current subtractive colour films, three separate emulsion layers, red-, green- and blue-sensitive, are superimposed on the base, and each records approximately one-third of the spectrum, though with some unavoidable overlap. In development, corresponding negative silver images are produced in the three layers. If the film is of the reversal type, in which a positive colour image results, it now undergoes re-exposure and second development, as a result of which the black-and-white tonality is reversed and corresponding dye images are developed up in the layers in colours complementary to their original sensitivities. The silver is then bleached out.

If, for example, the film has been exposed to a green patch of light, only the middle layer responds. In a negative film, this would colour develop as magenta, or minus-green. But on reversal, positive images are formed in the other two layers, namely yellow (minus-blue) and cyan (minus-red). In projection, these layers act as selective stoppers, or subtractors, of the white light which passes through them. The yellow layer subtracts blue from the light beam, and the cyan layer subtracts red, leaving a green beam of light to portray the original green patch of colour.

The colour carrier, therefore, may be thought of as a spectral modulator: in additive processes, universal in television and common in colour film printing, it is an adder of primary lights; in subtractive processes, universal for the final stage of film reproduction, it is a subtracter of the primary colours.

COLOUR PRINCIPLES—INTEGRAL TRIPACK—Additive colour process—Beam splitting system—Chroma—Chromaticity coefficient—Chromaticity diagram—Colour balance—Colour temperature—Colour triangle—Complementary colours—Desaturation—Dichroic reflector—Equal energy white—Filter (colour)—Filter factor—Fringing—Hue—Kelvin Scale—Mired—Photopic response curve—Primary colours—Saturation—Spectrum—Three-colour processes—Trichromatic analysis

VISUAL PRINCIPLES

The picture thus presented on the film and television screen with all the resources of optical, electronic and colour techniques, bears on closer inspection surprisingly little resemblance to the original it is supposed to portray. If that original is a subject at rest, the differences stem only partly from technical deficiencies which have either not yet been eliminated or are inherent in the medium. To these are added the fact that a camera lens looking at a static scene observes it quite differently from the human eye. The camera either takes it in all at once, pans relatively slowly over it, or zooms in on it. The eye has a much more restless motion; its attention moves ceaselessly from point to point, scanning the subject but not with the steady uniformity of the TV camera's electron beam. It probes in depth also, with the result that it produces a mental synthesis of what the observer finds interesting, rejecting what he does not wish to see.

Reproducing motion

When the subject is in motion, another difference is added. The eye sees objects in more or less their true continuity of movement. The film camera chops up the scene into time-slices at a rate of no more than 24 or 25 per second. Because a shutter cuts off the light while the film is being moved down from one frame to the next, these slices represent only about one-half of what goes on in front of the lens. In a still subject this does not matter, since all the slices are the same; but in its rendering of motion, the film camera simply leaves out half of what is there.

The TV camera reproduces motion rather differently. Whereas the film camera sees all that it does see simultaneously, the TV scanning process ensures that what appears at the top left of the scene is examined before what appears at the bottom right. In practice, moreover, the lines of this scanning field do not adjoin but are spaced apart, thus allowing a second field to be scanned and interlaced with the first to form a single complete picture. In this way 50 or 60 fields are scanned each second to produce 25 or 30 pictures.

Visual adaptations

These unrealistic ways of looking at the world work only because of an act of subconscious co-operation by the eye and brain. The fundamental requirement – to accept phases of movement as continuous movement – is known as persistence of vision and was discovered in the 1820s. To reduce jarring and blurring of objects as they move across the field, large numbers of pictures per second are desirable, but this wastes film or, in television, bandwidth. The accepted figure for both is 24 to 30. But from another point of view this is insufficient; the phenomenon of flicker occurs when the picture is formed on a bright screen. In television, as has been said, the problem is overcome by interlace, which doubles the effective picture frequency; and in film by blanking out each picture at least once while it is stationary in the projector gate.

Though these are the basic visual adaptations, they are by no means the only ones. The eye works within an enormous range of light intensities, something like a million million (10^{12}) to one. This range cannot be compassed at any one time by action of the iris, but demands a process of slow adaptation to the prevailing light level. However, the perceived range of brightnesses on cinema and television screens is immensely lower – about 70 or 80 : 1 and 30 or 40 : 1 respectively between adjacent tones – and the viewer must adjust to this difference also.

Again, binocular vision enables the mind to form a fairly accurate estimate of the depth of a scene; but stereoscopic film and TV transmission are hardly practical, and existing systems portray the solid world as flat.

That these systems do in fact convey to the viewer a very lifelike and acceptable portrait of the real world is due to faculties of vision always being called into play but enhanced when an extra strain is put on them. The perception of form, distinguishing figure from background, the essential from the irrelevant; perceptual constancy, whereby objects tend to appear the same shape when viewed from any angle, the same size over a wide range of distances, the same colour when seen by orange evening light and at high noon; depth perception which, when binocular vision is lacking, replaces it with the aid of clues derived from motion, texture, masking and perspective; these and other factors work to cover the gaps and lessen the limitations in an imperfect system of picture and sound transmission.

Striking proof of this is found in a comparison between 35mm film and television. A well-projected 35mm picture contains at least twice as many information elements as the TV picture, and with 70mm film having about twice the image area, the ratio is about four times. Furthermore, the TV picture is marred by a line structure much more noticeable than the grain structure of film, both viewed at normal distances. These differences are fully noticeable when a critical eye is brought to bear on them; but so well does the mechanism of perception work to assimilate the image to the real scene it is known to represent that the television image is accepted by the average viewer as virtually the equivalent of the film image, when both are in colour or both in black-and-white.

VISUAL PRINCIPLES—Flicker—Persistence of vision—Photopic response curve —Stroboscopic effects (strobing)

SOUND AND ACOUSTICS

Film and television are dual media; picture is always accompanied by sound. The two normally remain separate until the final stages of transmission, when in film the sound track or tracks are printed on to the image-bearing carrier, the release print; and in television the sound carrier is radiated in proper frequency relationship to the still quite distinct picture carrier.

Though technically true, this wholly misstates the practical relationship between sound and picture techniques and the staffs responsible for them. Though sound is in a sense the subordinate member of the team, its role is an essential one, its practice complex and difficult, and its relationship with the picture very close indeed.

An understanding of basic sound terms, recording methods and acoustic principles is therefore vital.

Sound as wave motion

Sound is transmitted by alternate compressions and rarefactions of an elastic medium, usually air. It can be represented on paper as a transverse wave motion, and discussed in the same terms as electromagnetic waves such as the TV signal: amplitude, wavelength, frequency, velocity and so on. This is convenient because sound is frequently converted into electrical signals and back again in devices called transducers, among them microphones and loudspeakers; and also between one type of recording medium and another, as by gramophone (phonograph) pick-ups, magnetic recording heads, and so on.

Frequency and Intensity

Audible sound frequencies range from about 30 to 20,000 Hz (1 Hertz = 1 cycle per second), which is nine and a half octaves; but the range of sound intensities accepted by the ear is immensely higher, indeed very similar to the range of light intensities accepted by the eye, about 10^{12} to 1. It is therefore important to establish a unit of intensity which corresponds to the subjective measure of loudness, and this is achieved by use of a unit based on common logarithms or powers of 10, the bel and its tenth part, the decibel (dB). If sound intensities are raised in successive steps from 1 unit to 10, then to 100, 1,000 and so on (i.e. 10^0, 10^1, 10^2, 10^3), each will seem the same amount louder than its predecessor, not ten times as loud. The bel scale is based on these powers of 10.

The intensity range of normal sounds, from gunfire to the softest audible whisper on a quiet night, is about 100 dB, but this is much wider than can be compassed by a practical recording and reproducing system, and would be uncomfortably large in the cinema or the home. It is the first limitation which most concerns the sound technician. The upper limit is set by the modulating capacity of the system, called 100 per cent modulation, the lower limit by the inherent random noise caused by irregularities in the recording medium itself (film, tape, disc, etc.) and by electron disturbances in the amplifying and signal-processing equipment. The ratio of these two figures, the noise sources being lumped together, is called the signal-to-noise ratio (SNR). In commercial sound systems it seldom exceeds 70 dB, and may be much lower; it is a concept of vital importance, not only in sound recording but throughout television engineering. Noise, too, since it sets the lower limit of the SNR, is the subject of intensive study.

Magnetic recording

Of the many methods of sound recording available in film and television one is all but universal: magnetic recording. The medium is a fine suspension of minute needle-shaped particles of a ferro-magnetic material coated on to a flexible base either of plastic or motion picture film. These particles are magnetized by passing them through a magnetic field set up by the variations of the sound signal, and thus produce a pattern in magnetization which can be reproduced by similar means. The higher the tape speed past the magnetic gap of the record and reproduce heads, the higher the frequency that can be reproduced, since the wavelength is greater. A tape speed of 15 ips (inches per second) past the heads more than suffices for the audio spectrum; but for recording the picture signal very much higher head-to-tape speeds are needed, of the order of 1,500 to 2,000 ips.

Optical recording

The only important exceptions to the use of magnetic recording are release prints of films in cinemas not equipped for stereophonic sound, and most narrow-gauge release prints. Here the old method of optical recording continues. An image is formed on the unexposed film of a narrow slit, and this is exposed by a beam of light modulated by the audio signal. When the sound track is developed and printed, it in turn is used as modulator of a light beam, thus setting up light variations corresponding to the original signal which are converted into electrical variations and thence into sound.

Acoustical problems

These are the methods by which the engineer records and reproduces sound. The original pick-up of the sound calls into play the neighbouring science of acoustics. Because the film and television recordist works under exceptionally difficult conditions, often competing with ambient noise and usually forced to keep his microphone too far from the speaker or actor by the demands of the camera, he has to master problems set by standing waves, excessive reverberation and so forth. These are difficulties he shares with the larger world of sound recording which includes radio and the gramophone, but in film and television his problems are made more acute by this dual nature of the medium and by the great complexity of the final sound track, especially in the film studio.

ACOUSTICS—NOISE—SOUND—SOUND DISTORTION—SOUND RECORDING SYSTEMS—Acoustic backing—Acoustic feedback—Decibel—Feedback—Filter (electrical)—Frequency response characteristic—Ground noise—Hum—Hum balancer—Intermodulation—Limiter (sound)—Linearity—Low pass filter—Magnetic recording—Optical Sound—Periodic noise—Pitch (sound)—Polar diagram—Random noise—Remanance—Reverberation time—Roll off—Shot noise—Signal—Signal-to-noise ratio—Signal-to-weighted-noise ratio—Stability (sound)—Transients—Variable density recording—VI meter—Wow

Because the development of film preceded that of television by a good 50 years, its techniques are more settled and there is less scope for rapid advance. Differently regarded, film has for some time attained the goal of projecting motion pictures in excellent colour on to very large screens with extremely high fidelity to the original, and with accompanying sound tracks which can be made to give a full stereophonic effect to most parts of a big auditorium. Current developments are towards greater mobility and silence of camera equipment, faster and more flexible handling of now all-but-universal colour film, and improvement in the narrower film gauges and the equipment available for them.

ORIGINS & MATERIALS

HISTORY The motion picture has by now a history of respectable length, going back a century and a half to the first researches in persistence of vision, the characteristic of eye and mind which makes it possible to represent the continuity of the real world by a series of separate static pictures. If this were not so, and if an infinite number of pictures were needed, practical motion pictures and television would be an impossibility.

In the earliest experiments, often presented to the public in the form of toys (e.g. the Phenakistiscope, 1832), phases of movement were portrayed in separate drawings which could be viewed in quick succession for very brief periods so as to combine into an illusion of movement.

Flexible film Only with progress in photography, and especially the development of a fast emulsion, were instantaneous pictures made possible, so that at least 10 or 12 could be taken each second, just enough to give a flickery continuity. These emulsions were first coated on glass backings (e.g. Marey's photographic gun, 1882), too cumbersome for so-called successive photography to be practical. With the commercialization of celluloid in the late 1880s, it was at last possible to make a flexible continuous band of film.

Talkies Some few of the pioneers of the motion picture were men of science, but most of them were ingenious artisans and practical showmen. Their names cluster thickly in the last decades of the 19th century, before film techniques had evolved. By World War I, some of the basic camera, projector and printer mechanisms still current today were in commercial production, and sound recording was already possible. The "talkies" did not in fact appear until the economic crisis of the late '20s, for the film industry, lacking a coherent research programme, has often waited until almost too late to introduce an innovation.

Animation and colour

Animation techniques began to evolve about 1907, and colour processes in the camera arrived at the same time, having been preceded by hand-tinted films which had been a success since the earliest movies of 1896. But the technical problems of colour were very severe, and it was not until 1934 that colour films really became established, with a 3-colour process using three separate strips of black-and-white film in the camera. Modern integral tripack films, in which three colour-sensitive layers are superimposed on a single base, were not widely available until after World War II.

Wide-screen

The wide-screen revolution which started with Cinemascope in 1953 was accompanied by the general introduction of stereophonic sound (later restricted to the largest theatres), and led to the development of many new picture processes, some of which survived only a few years. The 1960s saw the growth, chiefly at international exhibitions and World Fairs, of multiple and encircling screens and images, in which the picture could be fragmented and combined at will in a way little resembling the cinema of the past.

Through these years of change, the basic material of film has undergone development to make it more stable and less inflammable, but its widths and perforation gauges have survived with surprisingly little alteration.

CHRONOLOGY—HISTORY—Acres, B.—Anschutz, O.—Armat, T.—Ayrton, W. E.—Bray, J. R.—Case, T. W.—Chrétien, H.—Cinéorama—Cohl, E.—De Forest, L.—Demeny, G.—Dickson, W. K. L.—Donisthorpe, W.—Ducos du Hauron, L.—Eastman, G.—Edison, T. A.—Friese-Greene, W.—Gaumont, L.—Grandeur—Grimoin-Sanson, R.—Grivolas, C.—Hepworth, C.—Horner, W. G.—Howell, A. S.—Hughes, W. C.—Hurd, Earl—Independent frame—Ives, F. E.—Janssen, P. J. C.—Jenkins, C. F.—Joly, H.—Kalmus, H. T.—Kelley, W. V. D.—Kinemacolor—Latham, Major, W.—Lauste, E. A.—Le Prince, L. A. A.—Leroy, J. A.—Lumière, L.—Magnafilm—Magnascope—Marey, E. J.—Maxwell, J. C.—McCay, J. W.—Méliès, G.—Messter, O.—Muybridge, E. J.—Paris, J. A.—Pathé, C.—Paul, R. W.—Plateau, J. A. F.—Proszynski, K. de—Reynaud, E.—Roget, P. J.—Rudge, J. A. R.—Sellers, C.—Skladanowsky, H.—Smith, G. A.—Stampfers, S.—Starevich, Ladislas—Stuart-Blackton, J.—Uchatius, F. F. von—Varley, F. H.—Waller, F.—Wheatstone, C.—Widescope

MATERIALS AND STORAGE

Film is a transparent cellulose substance, whose function is to act as a flexible support for the light-sensitive emulsion. Up to the early 1950s, the universal 35mm film base was a dangerously combustible material, cellulose nitrate, and this dictated stringent conditions for film storage which to some extent persist today. Since the changeover to safety base (until then confined to narrow-gauge films), all film has been less inflammable than paper, and this has greatly simplified its handling and transport.

Film base

Of more technical significance, modern film base material is less liable to unwanted swelling and shrinkage with the inevitable changes in temperature and humidity it undergoes while being processed and during storage. These relatively small changes of size are important because the film image is magnified up to 700 times when projected on a wide screen, an enlargement of about half a million times in area. Very small dimensional changes during processing may result in serious misregistration of the three colour images, and can also give rise to unpleasant jittering and weaving of the picture as a whole on the screen.

The raw material of acetate film base is cotton – the short fibres which remain when the long ones are separated out for textile production. The prepared cotton is treated with acetic acid and other chemicals to give a cellulose acetate which is in turn dissolved in organic solvents. This produces a dope which is thinly spread on a smooth surface, and the solvents evaporated. The film base thus formed is cured on heated drums, and a substratum applied to it which helps the emulsion layer to adhere.

Emulsion structure

Emulsions consist of light-sensitive silver halides suspended in gelatin, which is made from the bones and skin of animals. Gelatin not only allows processing solutions to penetrate it and react with the silver halide; it contains minute quantities of complex sensitizers which increase the light-response of the halides themselves. The other constituents of emulsion are silver nitrate, potassium bromide and iodide, and to these minute quantities of various dyes are added to extend the response of the emulsion from the basic blue region to which all silver halide emulsions are sensitive, to the rest of the visible spectrum. Colour emulsions are not basically different, though they may contain couplers to form the dye image; in tripack films, three silver halide emulsions are coated in succession on the same base, with appropriate interposed and underlying layers for filtering and other purposes.

The emulsion is prepared by a lengthy process of heating, cooling, shredding and washing, and is finally coated on the base under conditions of total cleanliness, since before final slitting, the wide strips of film with emulsion drying on them resemble enormous fly-papers.

FILM MANUFACTURE—FILM STORAGE—Acetate—Base—Blue—Edge number—Emulsion—Exposure index—Float—Footage number—Halide—Jumbo roll—Lavender—Low shrink base—Nitrate base—Orthochromatic—Panchromatic—Plasticiser—Preservatives—Raw stock—Safety film—Sensitizer—Shrinkage—Slitting burr—Sprocket—Sprocket hole—Stock—Strip—Tinted base—Triacetate base—Vaults

GAUGES AND PROCESSES

The classical gauge of film is 35mm, which has remained basically unaltered since first proposed (as 1⅜ in.) by Thomas Edison in 1889. Edison provided four perforations at each side of each picture area or frame, which are engaged by the teeth of driving wheels called sprockets, to move it through cameras, projectors and printers, and into which positioning pins may be inserted to hold each frame in register when the film is at rest. Sprocket drive and register pin positioning remain basic to all film technology, for with the very large magnifications already mentioned, image unsteadiness on the film must not exceed a few ten-thousandths of an inch. This is especially critical in processes such as back projection, in which a picture taken by one camera forms a background to a picture taken subsequently by another. If the slightest relative movement is perceptible, the illusion of looking at a real scene in the background is destroyed, as many cinema-goers will have noticed.

Non-standard gauges

The world standardization of 35mm film in 1909 did not discourage the introduction of many other gauges – the wider ones for extra large screen presentation, the narrower ones for use by the amateur in home movies and by the low-budget professional unit. Of these gauges few have survived: 70/65mm, 16 and Super-8, together with the now obsolescent 9·5mm and standard 8. 70 and 65mm employ the same perforation standard, the wider film providing space for extra sound tracks on the release print.

A few other gauges have been developed exclusively for laboratory use; it is often convenient to print several ranks of narrow-gauge pictures on one larger-gauge film, subsequently slitting them and removing any unnecessary edge perforations. The perforation has also undergone development and change, to provide maximum steadiness in the camera, maximum strength in the projector, convenience in the laboratory, or extra space for picture on a narrow-gauge film.

These changes in the basic material of film could have led to chaos and thrown away the great advantages of world standardization not as yet enjoyed by television. In the main, however, worldwide efforts to create acceptable standards have kept pace with the changes; the work of national bodies in the main film-producing countries is reflected at an international level by the International Standardization Organization (ISO).

Aspect ratios

Until the early '50s, the most stable characteristic of film was its proportion on the screen, four units wide to three units high, the same for all gauges and for television. The wide-screen revolution launched by Cinemascope upset all this, and new aspect ratios, all of them exceeding 4 : 3, came in. Since the limiting factor was usually the top of the proscenium arch from the back of a theatre with a balcony, screen height could seldom be increase; only a sideways stretch was possible.

To squeeze this wider picture on the old 4 : 3 film area, special cylindrical lenses and mirror systems were devised, thus giving rise to many new processes, some of only transient interest. Cameras with horizontal film movement produced enlarged pictures of the same 8-perforation format as 35mm still cameras, and 65mm film reappeared after a 25-year oblivion. Of these processes there remain today in common use only the original Cinemascope 2 : 1 horizontal enlargement of a 4-perforation 35mm frame (under various trade names), and similarly a 5-perforation frame on 65mm film. An economical process in which a 2-perforation 35mm frame is exposed unsqueezed in the camera and afterwards enlarged vertically and squeezed on to a 4-perforation Cinemascope frame is also current.

FILM DIMENSIONS AND PHYSICAL CHARACTERISTICS—STANDARDS—WIDE SCREEN PROCESSES—Academy aperture—Academy standards—AFNOR standards—Anamorphic processes—Aperture plate—Aperture sizes—ASA standards—Aspect ratio—Bell Howell perforation—Blow up—BSI standards—Cinemascope—Cinemascope perforation—Cinerama—Circarama—Compatible composition—Cropping—Delrama—Dimension-150—DIN speed rating—DIN standards—Double frame—Dubray-Howell perforation—Dynamic frame—Film register—Flat print—Foot—Format—Foxhole—Frame—Frame line—Full aperture—Full frame—Hypergonar—ISO standards—Kinopanorama—Kino-Vario—Kodak perforation—Laterna Magika—Living sereen—Mask—

Movietone frame—Multi-image—Multi-screen—Negative perforation—Nine-point-five millimetre film—Non-flam film—Panavision—Perforations—Pitch—Polyecran —Positive perforation—Quad-8—Reduced aperture—Reel (of film)—Safe area—Scanned print—Scanning printer—Scope—Seventeen-point-five millimetre film—Seventy millimetre film—Silent frame—Silent speed—Single eight millimetre film —Sixteen millimetre film—Sixty-five millimetre film—Squeeze—Standard eight—Super eight millimetre film—Superscope—Technirama—Techniscope—Thirty-five millimetre film—Thirty-two millimetre film—Ultrasemiscope—Unsqueezed print—USA Standards Institute—Varamorph—Vistavision—Vitarama—Weston speed

NARROW-GAUGE PROCESSES

The narrower film gauges have lent themselves to an increasing variety of uses. The amateur who formerly bought 16mm equipment has increasingly shifted to the more economical 8mm and now Super-8, a film of the same width in which narrower perforations and a longer pitch or distance between them have increased the picture area by nearly 50 per cent. However, the use of 16mm has itself increased, since it has become the favoured gauge of professionals making industrial, educational, scientific and travel films, and is widely used for TV filming, though film itself is under vigorous challenge from videotape recording.

Eight-millimetre advantages

To provide maximum simplicity for the less-skilled user, Super-8 cameras and projectors are today technically much more sophisticated than is common in 16mm or 35mm equipment, where expert operation can be counted on. Film cassettes notch into the camera to indicate emulsion speed and set a colour-correction filter when it is needed. Exposure adjustment is calculated automatically through the lens itself under varying light conditions, with allowance made for the film speed, filters (if any) and pictures per second. A single zoom lens normally replaces a fixed set of lenses, and the zoom movement is often motorized. The trend towards spontaneous camerawork under conditions of the most extreme difficulty is already extending these developments to 16mm and even 35mm cameras.

Equipment similarities

As with the camera, narrow-gauge equipment shows important differences in practice, but no departure from the principles of standard-gauge 35mm. Production processes also run parallel, but with a more restricted range of film emulsions and a general simplification of techniques. The smaller picture area means that the margin of quality is reduced, and scrupulous care is needed to keep image degradation to a minimum. This is one reason for the wider use of reversal materials in 16 and 8mm; producing a positive image from the camera film, and thereafter going from positive to positive, makes it possible to reduce the number of intermediate stages. But simple hard-and-fast rules cannot be laid down, and narrow-gauge production is very much a discipline of its own.

FILM DIMENSIONS AND PHYSICAL CHARACTERISTICS—NARROW GAUGE FILM AND EQUIPMENT—NARROW GAUGE PRODUCTION TECHNIQUES—PRINT GEOMETRY (16MM)—A & B printing—A & B winding—Checkerboard cutting—Double perforated stock—Double sixteen millimetre film—Eight millimetre film—Electronic video recording—Sixteen millimetre film—Splicing—Substandard film—Super eight millimetre film—TTL—Viewer

THE PHOTOGRAPHIC IMAGE

The material of film, the flexible image- and sound-bearing strip, with its appropriate gauge, perforation standard and aspect ratio, is the medium's carrier, but it is the image itself which is the heart of the photographic process. Though colour is today virtually universal, image formation, whether in colour film or black-and-white, is effected by silver halide emulsions whose origins go back to the early years of the 19th century.

Silver halide is present in the form of crystals, often called grains, of about a micron (a thousandth of a millimetre) in size. There are about a thousand million in each square centimetre of the emulsion surface, which is about 20 grains deep. Each crystal, moreover, consists of about a thousand million ions – atoms which have lost or gained an electron. Exposure to light decomposes these crystals, one quantum of light producing one silver atom, with release of the halogen. If every crystal required to be decomposed in this way, extremely high intensities of light would be needed for full exposure.

Latent image

The secret of photography is that once a very few atoms of silver – no more than four or five – have been produced by the same number of quanta of light, all the rest of the thousand million atoms in the grain are capable of being developed into silver. This exposed but undeveloped image is the latent image, whose investigation has proved so immensely difficult because there is no means of detecting it except by the process of development which radically alters it. The basic amplification factor – atoms developed to atoms exposed – is thus of the order of a hundred million, and this has enabled silver halide photography to stay ahead of its recent non-conventional rivals such as electrophotography, of which xerography is an example.

Grains having been developed from silver halide to black silver, the image is densest where the light falling on the emulsion has been brightest, and thus its tone rendering is the reverse of the correct one. It is a negative, and requires what is in effect a re-photographing to turn it into a positive. Nor is reversal film an exception to this rule, for here the negative silver image is developed and bleached out, and the residual silver halide image is re-exposed, and being of opposite tonality, develops to a positive; thus the negative and positive stages are combined in a single emulsion.

By its very nature the photographic image is a grainy image, and light is inevitably scattered from the brighter into the darker parts of the image, deteriorating the rendering of fine detail and producing an appearance of graininess. It is appropriate to use the term noise in this connection, and to speak of the signal-to-noise ratio of which the emulsion is capable.

Spatial frequency

As already indicated, the modern method of describing the definition or resolution of the image closely parallels the frequency-amplitude characteristic long familiar in electronics: the temporal sequence of signals corresponds to a spatial array in the image, and the term frequency can be applied to both. At a given frequency, loss of resolution results in a reduced level of modulation. What is known as the modulation transfer function (MTF) of either lens or film has, then, a value approaching unity at very low frequencies, but this reduces to zero at the limit of resolution. Between these values the curve may take many different shapes. It may continue for some distance at a high value, and then rapidly fall off (this shape lends itself well to limited bandwidth systems such as television); or it may start to decline at a fairly low frequency, but maintain this gentle slope to a very high frequency before reaching zero (this shape suiting high resolution applications).

Thus, in electronic terms, the lens-plus-emulsion system may be thought of as a low-pass filter, but one in which the response curve is susceptible to fairly wide variation. The MTF's of lens and film (now often available for both) can be multiplied together to give a combined product; but it must be remembered that the development process can produce special edge effects, not unlike those of crispening in electronics, which will modify the final result.

Thermoplastic recording

Non-conventional rivals to photography have been mentioned, and one such has reached the margins of film application. Thermoplastic recording is a method of electron-beam recording of information in pictures or symbols on a deformable plastic film. The information is written by scanning, not by simultaneous exposure as in photography, and the source is therefore a TV camera or other modulated scanning generator. The film surface is made deformable by heating it, and is exposed in vacuo to the electron writing beam. Cooling hardens the image, and read-out is by a simultaneous, not a scanning process, for the image impressed on the film as a picture built up of minute deformations can be projected by Schlieren optics, as in the Eidophor.

Kalvar process

Another unconventional but photographic process, the Kalvar, has film applications in black-and-white, but for printing only. Within a thermoplastic resin layer coated on the base, there is dispersed an ultraviolet-sensitive compound. Exposure decomposes this substance, releasing nitrogen and other volatile products which set up a latent image of internal stresses. This image is developed, or rendered visible, merely by the application of heat, which creates minute vesicles or gas bubbles. These have a different refractive index from the surrounding medium, and so form a projectable image in terms of light-scattering, rather than the light-absorption of conventional silver halides.

EXPOSURE CONTROL—IMAGE FORMATION—KALVAR PROCESS—LENS PERFORMANCE—NOISE—THERMOPLASTIC RECORDING—TONE REPRODUCTION—Acutance—Anti-halation coating—Apex system—ASA speed rating—Characteristic curve—Definition—Electronic video recording—Exposure—Fog—Grain—Graininess—Granularity—Halation—Limiting resolution—Reciprocity law—Resolution—Response curve—SMT acutance—Transfer function

COLOUR PROCESSES

In current practice, virtually all colour film processes are of the subtractive type in which complementary colours are formed in a pack of three superimposed colour layers called an integral tripack. Some of these are reversal processes, producing a camera positive and, if used for duplication and printing, going from positive to positive. The majority, however, are negative-positive, that is, producing a camera negative and alternating tonality at each stage. The broad principles of all are the same.

Tripack principles

In the camera film, the top layer records the blue region of the spectrum, and is an elementary blue-sensitive emulsion. Below it is a yellow filter layer which prevents blue light penetrating further and exposing the next two layers which are also blue-sensitive in addition to their specific sensitivity: usually green for the second and red for the bottom layer. These four layers in superimposition are less than a thousandth of an inch thick.

In principle, therefore, each layer records one-third of the spectrum; all the layers are developed in a black-and-white developer to produce corresponding spectrally-selective images in negative tonality, and in each layer the dyes are developed up at the same time in colours complementary to the response of the original layer; finally, the silver images and the yellow filter dye are bleached out, and the film is fixed and washed. From this negative in tonality, a positive can be printed which, on projection, yields the original colours by subtraction of lights as explained in an earlier section.

Practical considerations

In practice the operation is by no means so simple. The three emulsions must be carefully balanced for speed, contrast and other characteristics. Optical filters and emulsion layers cannot be produced with the relatively sharp cut-off of their electronic counterparts, so the yellow filter will let through some blue light to affect the green- and red-sensitive layers, and the three layers themselves will have overlapping sensitivities. Moreover, these sensitivities can only be balanced to one colour temperature of light; when the energy distribution alters, the emulsion layers respond differently and the colour rendering is thrown off balance. This is why, in the studio, all light sources must be closely matched to a single colour temperature; and why a radical change, such as from blue-biased daylight to yellow-biased tungsten light, requires a completely different colour emulsion or the use of a heavy filter which reduces photographic speed.

Dye imperfections

Furthermore, the dye images produced in each layer by the couplers incorporated in most colour films, do not have ideally sharp, non-overlapping responses: the magenta dye (minus-green) absorbs much blue light it should pass, and similarly the cyan dye (minus-red) absorbs some green and blue. In Eastman Color negative film this is ingeniously compensated by using couplers yielding a colour which acts as a filtering mask to eliminate the unwanted absorptions. Incidentally, it is these coloured couplers which give these negative films their peculiar orange cast.

Printing methods

Though all practical colour films are subtractive, additive processes are widely used in colour printers in which white light is divided into three narrow spectral bands by dichroic filters, and the coloured light beams are then separately varied in intensity to provide colour control before being recombined on the film.

COLOUR CINEMATOGRAPHY—IMBIBITION PRINTING—INTEGRAL TRIPACK SYSTEMS—TECHNICOLOR—Additive printing—Agfacolor—Anscochrome—Anscocolor—Automasking—Bipack—Bleach—Cinecolor—Coupler—Dufaycolor—Duplitized—Dye coupling—Dye toning—Dye transfer—Eastman color — Ektachrome — Ferraniacolor — Fujicolor — Gasparcolor — Gevachrome — Gevacolor—Intermediate—Internegative—Interpositive—Kodachrome—Kodacolor Lenticular process—Lippmann process—Masking (colour)—Monopack—Mosaic—Original—Orwocolor—Separation master—Separation negative—Substantive process—Subtractive colour processes—Successive frame process—Subtractive frame process—Tanning—Three-colour processes—Three-strip camera—Toned track—Tripack—Two-colour processes

THE CAMERA

All film is exposed in a camera, though cameras may differ so much in design that their common features are hard to recognize. Normally, of course, the lens is stationary and the film moves past it, but in some high-speed cameras the film is stationary and the lens moves. Normally, the film movement is arrested for each frame of film to be exposed, but in some cameras the image is optically deflected so that over a short distance it keeps pace with and exposes the continuously moving film.

Camera construction

However, for all but a well-defined group of special purposes, film cameras have much in common: a light-tight film magazine, from which the film is unwound as it is fed to the exposing aperture, and into which it is wound up again after exposure; an intermittent movement which arrests a frame of film behind this aperture, and then moves it on again after exposure, the stopping and moving on each taking about the same time; a shutter to obscure the film from the lens during movement to prevent blurred exposures; one or more feed sprockets to keep the film moving steadily through the camera, with loops above and below the aperture so that the intermittent motion does not tear it. All this is enclosed in a housing which carries a motor to drive the film. Finally, there is an optical system to form an image on the film, and means for the cameraman to view this image, or one as close to it as possible.

The most difficult requirement to meet is that of registration: the positioning to within a few ten-thousandths of an inch of each frame of film with respect to those which precede and follow it. In the more elaborate cameras this is effected by very accurately machined pilot or register pins which engage with one or more perforations when the film comes to rest in the picture gate, and disengage just before it is moved on again to the next frame.

There is still a great deal of individuality in the way in which these elements are designed and combined. In the studio camera, the intermittent movement is of such complexity and precision that, though no bigger than a matchbox, it may cost as much as an entire hand-held camera. This in turn is likely to have a movement of extreme simplicity, seemingly incapable of providing rock-steady pictures, yet in fact able to do so to all but the most exacting standards. In virtually all cameras of recent design, viewfinding is by a rotating (occasionally oscillating) shutter directing the lens image alternately on to the film and on to a ground glass which the operator sees through a magnifying system. Parallax errors are thus eliminated at source.

Noise suppression

Professional cameras are almost always synchronized with sound recorders, and must therefore be made as noiseless as possible. Hitherto, this has necessitated rigid enveloping covers called blimps, with hinged doors for changing magazines and threading film through the running gear and aperture, all necessarily heavy and clumsy. In recent years, 16mm self-blimped cameras have appeared, in which the principal cause of noise, the flapping loops and intermittently moving film, has been damped at source by careful control of claw design, film path and so on. Self-silencing is now being extended to 35mm models.

Lenses

Traditionally, cameras are fitted with lens turrets, revolving heads on which three or four lenses of different focal length are mounted for rapid substitution in front of the film. Turrets have always made sound blimping difficult, and the development of the zoom lens is rendering them unnecessary. For most purposes, the definition of a zoom lens, whose focal length is variable over a range of 4 : 1 or more, is, at all but the largest apertures, as good as that of the separate lenses it replaces. The modern tendency, as on television cameras, is therefore to scrap the turret in favour of a single zoom lens, which often has a detachable range extender. A zoom lens also simplifies the problem of taking exposure readings through the lens (TTL) with a cell mounted inside the camera; but this technique, all but universal on Super-8 cameras, has only recently reached 16mm, and has yet to be applied to 35mm. For the studio cameraman, manual exposure setting is essential and occupies a mere fraction of the time taken to set up lights; but in action photography the cameraman is now in competition with TV cameras, some equipped with automatic gain controls, and he will soon demand equivalent speed of lens diaphragm setting, whatever gauge of film he uses.

Accessories

The basic camera is of little use without a host of auxiliary devices and accessories: lens shades and filter holders, a footage counter and tachometer calibrated in frames per second, motors for various supply voltages, both a.c. and d.c., tripods and other mounts of different height, dollies, velocilators and cranes, the various sizes of wheeled vehicle necessary for tracking and rise-and-fall shots.

High-speed work

As described, the camera serves many uses from single-frame photography – the one-picture-at-a-time technique used in animation – to process photography, or rephotographing of a length of film with optical modifications, to normal camerawork at a considerably higher speed than the standard 18 or 24 fps (frames per second). Intermittent movements are reasonably steady up to a maximum speed of 300 fps (35mm) and 600 fps (16mm), but this is uncomfortably fast, and the severe acceleration risks ripping the perforations. For these and higher speeds to about 16,000 fps, a rotating prism or optical compensator carries the image along with the steadily moving film, eliminating relative displacement for just long enough to give an adequate exposure. This range of speeds in fact covers the vast majority of high-speed applications in industry, and rotating-prism cameras are manufactured in commercial quantities. For ballistic, missile and aerospace research, much higher speeds are often needed.

So long as the film moves, a powerful motor is required to drive it at high speeds, and at very high speeds so much time and film are consumed in acceleration that the system becomes too wasteful. It is therefore better to hold the film stationary and scan it with a rotating device. In the specialized framing camera, for instance, a mirror rotated at upwards of 25,000 revolutions per second images the object on an arc of film through a series of lenses, and can attain speeds up to 8 million images per second for recording periods of less than a millionth of a second.

CAMERA—COATING OF LENSES—HIGH-SPEED CINEMATOGRAPHY—LENS TYPES—STABILIZATION (IMAGE)—TIME LAPSE CINEMATOGRAPHY—Acceptance angle—Aperture lens—Blimp—Breathing—Buckle switch—Cam—Cartridge—Cassette—Changing bag—Charger—C mount—Claw—Continuous motion—Conversion filter—Daylight loading spool—D mount—Dynalens—Edge fogging—Effects filter—Exposure meter—Flutter (picture)—Fog filter—Friction head—Gate—Gate pressure—Graduated filter—Ground glass—Gyro head—Head, camera—High hat—Inching—Intermittent movement—Lens hood—Lens turret—Liquid head—Magazine, film—Matte box—Mirror shutter—Mitchell mechanism—Monitor viewfinder—Nodal head—Optical intermittent—Pin—Pola screens—Pressure plate—Pulldown—Rackover—Reflex shutter—Register pin—Registration—Shutter—Shuttle pin—Skid—Sky filter—Slow motion—Speed—Stability—Sequence—Synchronous speed—Tachometer—Take-up—Tophat—Triangle—Tripod—TTL—Turret, lens—Velocilator—Viewfinder—Weave—Wlid motor

CREATIVE TECHNICIANS

The cinema's complex production process is conveniently introduced through its primary creative workers, men who must use its tools and at the same time demand more than these can give, the prime stimulus to progress.

PRODUCER AND DIRECTOR

Even before a treatment has been written, the producer weighs up the possibilities of a novel, a play or an original idea, and plans how it can most advantageously be cast, financed and shot. Usually an independent promoter rather than a studio employee, he often invests money of his own and draws in the backing of a distributor. This gives him a large measure of control over the choice of actors and the appointment of the chief members of the production team. Taste and temperament determine how direct an influence he exerts from script-writing onwards to the film's completion. He may collaborate or delegate authority, but power over the production is ultimately his.

The execution of the film is the responsibility of the director, who co-operates on the script, handles the actors, visualizes every scene to be shot on the studio floor or on location, and supervises all the final stages of production. As with the producer, the directors' powers are very large, but film making is essentially a co-operative job, since he normally delegates many of them to the principal members of his staff. For the handling of the cast he has sole responsibility. He must win their confidence, however much or little he discusses the script with them. He may have to shoot out-of-continuity, and so must be able to keep the whole story in perspective, determining its mood and rhythm from start to end, even though the pieces of the jigsaw are not yet assembled.

DIRECTOR—PRODUCER—Breakdown—Close-up—Continuity—First assistant director—Scene—Sequence

LIGHTING CAMERAMAN

In feature production, the responsibility for creating the film in terms of camera movement, colour, light and shade, rests mainly with the lighting cameraman or director of photography, though contributions to all his decisions may come from the director, set designer and others. The actual operation of the camera is left to a separate camera operator, assisted by a focus puller and others. The visual style of the film is thus very much in the cameraman's hands, and his knowledge of colour emulsions and laboratory techniques is invaluable in resolving difficult problems.

Lighting is his chief weapon, and on an elaborate set the lighting alone may occupy a day or more, before the camera can begin to roll. It is this striving for technical, indeed aesthetic perfection, which is one of the distinguishing factors of the feature film; television production, if only because of the pressure of time, assigns the cameraman a quite different role.

In documentary films, with very much smaller units, the lighting cameraman is necessarily more of a jack-of-all-trades, and may himself operate the camera.

CAMERAMAN—Angle, camera—Camera operator—Crab dolly—Crane—Dolly —Exposure meter—Focus puller—Follow focus—Follow focus cameraman—Follow shot—Gaffer—Grip—High key—High light—Hot spot—Key light—Log sheet—Low key—Luminaire—Monopole—Neutral density filter—Night effects—Pan—Photoflood—Report sheet—Retake—Scrim—Shot—Split focus—Spot brightness meter—Swish pan—Take—Tilt—Tracking—Tracking shot—Whip pan —Zip pan—Zoom

ART DIRECTOR

The art director or set designer comes on to the production at script stage to translate the ideas of producer and director into detailed scale drawings which can be realized at the proper time, and to the estimated cost, on the studio floor. More and more, however, with the trend to realistic backgrounds, the designer finds himself responsible for searching out suitable locations and modifying them, their buildings and surroundings, to the needs of the director and his script.

Inside and outside, his task is also to advise on trick methods (special effects) by which the cost of settings can be reduced. He must therefore be a master of a wide variety of techniques for synthesizing reality out of a strange assortment of bits and pieces, though the task of carrying out these effects may be left to specialists.

ART DIRECTOR—Floor plan—Props—Prop plot

SOUND MIXER

Sound is by its nature a separate sphere of activity; and whereas the picture elements are far too diverse for the detailed management of one person, a recording supervisor is in charge of the whole sound department, and has under him one or more sound mixers responsible for the sound quality of an entire production from start to finish. Most other technicians are eye-minded, but the sound mixer is ear-minded, his attention concentrated on fine points of sound detail often inaudible to other members of the unit. As a rule, the sound man attached to the production during shooting, the floor mixer, hands over to a music mixer when the film is to be scored, and to a dubbing mixer when the many constituent sound tracks are combined into their final balance.

A mixer's problems include picking the right equipment for each job, controlling the acoustic characteristics of a set or an outdoor location, supervising the placing and movement of microphones, obtaining the best balance from a pop group or a large orchestra, and contributing to the clarity and perspective of the final sound track which is built up from a multitude of separate elements.

MICROPHONES—SOUND MIXER—Clapper board—Clappers—Console—Equalization—Equalizer—Fader—Fishpole—Floor mixer—Gun mike—Head set—Mixing (sound)—Panpot—Peak programme meter—Recordist—Sound mixer

EDITOR

The editor's role is often thought of as starting only when the shooting of the film is finished. In fact, an assembly editor starts to put the film in correct order after the first day of production; and when the chief editor takes over, his task is not merely to realize what was in the script writer's mind, and later in the director's mind, but what more can be achieved with lengths of film which, in the cutting room, acquire a new life and rhythm of their own. This is creative editing, and its tools are among the simplest in the armoury of film: a pair of scissors, a small viewing machine, a splicer and a synchronizer to relate sound to picture.

With these the editor builds up his picture gradually, refining it as he goes along and turning it from a rough cut into a fine cut. At this stage he works in close association with the director, who has his own ideas how he wishes it to look. After the fine cut has been approved by the director, producer and sometimes distributor, the editor is once again left to himself to assemble the complex dialogue, music and effects tracks. Finally, the director reappears to join with the editor and dubbing mixer to produce a final sound track which must enhance or counterpoint the picture at every stage, and at the same time achieve a convincing unity of its own.

EDITOR—Bin—Bloop—Blooping machine—Breakdown—Build-up—Buzz track—Can—Cement, film—Clip—Code numbers—Coding machine—Core—Cross cutting—Cut—Cut back—Cutter—Cutting—Cutting copy—Dialogue track—Editing—Editing machine—Editola—Effects track—Fine cut—Flange—Flashback—Footage counter—Frame counter—Insert—Horse—Intercut—Jump cut—Main title—Montage—Moviola—Numbering machine—Out-take—Reader—Reprints—Reverse angle—Rewind—Rough cut—Shot—Slug—Splicing—Split reel—Spools, cores, and cans—Spotting—Start mark—Synchronization—Synchronizer—Sync mark—Tape splice—Track, sound—Track laying—Trim—Unmod track—Work-print

THE STUDIO COMPLEX

Notwithstanding the trend to location shooting of studio pictures, the area in which technical operations are centred is necessarily large and complex. The overall decline of film production under the impact of television had the effect of closing down all but the biggest and most efficient units. Many of these have been further modernized to adapt them to the faster tempo of television serial production, but otherwise the pace of change in studio design has been relatively slow, because high standards had been achieved long before television became a threat.

STUDIOS AND LIGHTING

The studio floors, or sound stages, are the heart of the complex, but the largest area is that occupied by set preparation. The art director's designs pass to the machine shop, the carpenters, joiners, plaster and paint shops. Sets are dressed, and, after shooting, struck, i.e. taken to pieces, and the re-usable parts stored. The artistes' services occupy another part of the complex, with the casting department, costume design, wardrobe and make-up.

Studios require elaborate soundproofing against aircraft noise and the prevailing rumble of cities and highways; height to accommodate catwalks for lights; air conditioning to maintain constant comfortable temperature when summer heat is added to heavy lighting; powerful generating equipment to provide that lighting; and services for camera, sound and other technicians.

A special effects department provides back and front projection, models and all other devices needed to synthesize reality by trick processes. Cutting rooms adjoin the studio, and review projection rooms are available for rushes and production screenings. Often a processing laboratory has been built on the lot; if not, it will be within easy reach by car. As a rule there are special studios designed for recording small and large orchestras, and dubbing rooms with many-channelled consoles complete the final stages of film making.

It is in the lighting department that recent years have seen the greatest changes. Dual-purpose stages have been built to serve conventional film productions and rapid shooting of television serials. Heavy lighting units, both on the floor and rigged above the sets, are retained to serve the exacting needs of slow and careful film lighting. To these are added a large number of grid-mounted overhead lamps manoeuvred and set by long poles from the floor, which for television remains unencumbered by lamps, cables and junction boxes.

The old heavy incandescent floor lamps are gradually being replaced by tungsten-halogen lamps of greater brilliance, longer life and more constant colour temperature. In these ways the different needs of film and television are being reconciled.

STUDIO COMPLEX—LIGHTING EQUIPMENT—Action—Backlight—Baffle—Barndoor—Barney—Basher—Bell-Howell mechanism—Bipost lamp—Booster—Booster light—Box set—Brute—Cold mirror reflector—Contrast (*lighting*)—*Cookie—CP lamp—Crab dolly—Crane, camera—Cyclorama—Diffusion—Dimmer—Dolly—ES cap—Fill light—Flag—Flat—Flood—Floor—Fresnel lens—Garland—Gelatin—Generator—Gobo—Jacketed lamp—Jelly—Linnebach effect—Location—Lot—Pup—Reflector—Reflector lamp—Rifle spot—Rigging—Riser—Rostrum—Scoop—SCR dimmer—Set-up—Sky pan—Slate—Spider box—Spot—Stage—Sungun—Target—Tenlight—Thyratron—Thyristor—Tungsten halogen lamp—Tungsten lamp—Turtle—Unit*

SPECIAL EFFECTS

Not all scenes are best filmed directly by the conventional method of setting up a camera in front of them and shooting in the normal way. Departures from this norm are called special effects. Some are relatively simple, involving changes in direction of movement of objects, a faster or slower speed, image distortions and optical transitions such as fades, dissolves and superimpositions, and changing of day to night; these can be carried out in the camera, though this is not always the easiest or best way to achieve them.

Some scenes, however, are too large or costly or time-consuming to shoot in the form which the script calls for, and these require a more complicated approach. Miniatures and some forms of image replacement, including split-screen shots, glass and mirror shots, are also in-the-camera techniques. Laboratory processes involving the separation and subsequent recombination of different parts of the image are often cheaper and more satisfactory in the long run, for though the result cannot be seen at once in a viewfinder, as with the techniques last mentioned, a great deal of studio time is saved, and this is usually the costliest element of all.

Finally, there are combined techniques such as front and rear projection in which a second element, usually a background, is photographed, processed in the laboratory, and then projected in such a way as to combine with actors and foreground objects on the studio floor. Here again, the final effect can be seen in the camera viewfinder.

Image replacement

Image replacement is a name often given to all the techniques in which an element in the scene is deleted, and a preferred element substituted.

Often this is accomplished by a matte shot, which requires two or more exposures of the negative. If carried out in the camera, an opaque matte is placed in the matte-box or adjacent to the film plane, and this matte is shaped in outline to blot out the part of the scene which is not to be recorded, for instance the window area of an interior set. The film is exposed to the rest of the set, but remains unexposed where the matte blacks it out. The camera is then moved to a location where background action outside the window, and perhaps at a great distance, can be conveniently staged. A counter-matte obscures the previously exposed area and reveals the window area. A second exposure is made.

Laboratory techniques

Matte shots such as this are usually carried out in the laboratory. The unwanted window detail in the studio is not blacked out, and the negative is exposed in the ordinary way. It is then printed to a fine-grain positive and threaded into a process camera in contact with an unexposed fine-grain duplicating negative, the two films running through the camera together. The intermittent movement must be of a high-precision register pin type. The process camera is then focused on a white board, on which the area corresponding to the unwanted parts of the scene is blacked out. As the two films are exposed, frame by frame, by the light on the board, the black matted areas provide no light, so the corresponding negative areas remain unexposed. The negative film alone is then re-exposed with a counter-matte on the board containing art work (for example, the top of a set too large to build in the studio); or it may be re-exposed in contact with another film on which a scene has been recorded in the proper place.

Optical printer

A much-used device is the optical printer, a small precision projector facing a camera on a lathe bed on which fore-and-aft movements can be effected. The projector (sometimes called the printer head) carries the master positive to be modified; the camera the dupe negative which receives the image. One of these instruments may be run forwards, the other backwards; their speeds may be different; the frame centres may be changed; zooms may be produced; shots may be turned upside down or left-to-right.

Many other special effects techniques, notably the highly developed travelling-matte process, are too complex for description in a brief summary.

SPECIAL EFFECTS—Back projection—Barndoor wipe—Bipack—Blue-back shot —Blue-screen process—Colour difference process—Dissolve—Double exposure—Dunning process—Fade—Fast motion—Flop over—Frame Stretch—Freeze frame —Front projection—Glass shot—Hold frame—Infra-red photography—Iris wipe—Lap dissolve—Mask—Matte—Miniature—Mix—Models—Pin—Plate, background—Printed matte shot—Process shot—Reverse action—Schüfftan process—Skip printing—Slow motion—Sodium lamp process—Split screen—Stop frame—Superimpose—Transparency—Travelling matte process—Wipe

THE LABORATORY

During the course of production, film makes many passages through the processing laboratory, at once a highly automated chemical engineering plant and a complex organization for dealing with a multitude of orders, large and small. Here the film is developed, printed and duplicated or copied.

Developing agents

Development, fixing, washing and drying are lumped together under the term processing. They employ basic chemical techniques long standardized in photography. Considering that photographic development goes back more than a century, the number of practical developing agents is surprisingly few, and fewer still are employed in film processing. Colour development is a process in which a coloured insoluble dyestuff is formed in association with silver. There are certain developing agents whose oxidation products (resulting from reducing silver halide to silver) combine with colour couplers to form insoluble dyes. By altering the colour coupler, the same oxidized developing agent can give dyes of different hues, so that when the colour couplers are incorporated in the emulsion layers, as is the case with most professional colour films, only a single developing stage is needed, with consequent saving of time and simplification of equipment.

Processing machinery

Developing takes place on long, continuous machines, in which the film is often pulled by friction rather than by sprockets and perforations as it passes up and down through the solutions and is finally dried in a hot air cabinet. Colour film on developing machines running at up to 100 ft. per minute takes about an hour to pass from one end to the other. The developing solutions are circulated through heat exchangers to keep them at a constant temperature, and are continuously replenished to the proper strength.

Because of high production costs, it is essential to get rush prints back to the studio early in the day after the negative has been shot. Night shift work ensures early development, and the negative then goes to the printing department. Here it is timed or graded: in black-and-white, this means altering the printing light exposure shot by shot to bring the print to a uniform normal density and so compensate for errors in camera exposure; in colour, it means adjusting as well for off-balance caused by variations in lighting colour temperature and other departures from standard. Timing is a skilled operation requiring expert judgment. When timing colour film, the operator usually makes a series of test exposures in which both light intensity and colour are varied. This is rapidly developed on a fast positive machine, and the result guides him in his final choice of printer settings.

Printing processes

In the printing department, the majority of printers are of the continuous contact type; that is, the printing and printed films move continuously at a fairly high speed, and are guided together into close contact at an aperture where the exposing light is controlled in intensity according to the grader's settings. Intermittent (or step) printers, such as those already described under Special Effects, run much more slowly and their economic use is confined to precision work. The contact intermittent printer is often employed to make duplicate positives and negatives (in colour called interpositives and internegatives), while the optical intermittent printer, since it provides complete separation of printing and printed film, is useful for image modification, including reduction from a larger to a smaller film gauge.

The laboratory is equally concerned with all later stages of production: special effects, optical effects (fades, wipes, dissolves), matching of the final negative and sound track to the work print received from the editor, and bulk printing of copies for release to the exhibitors. At this stage a radically different colour process may appear. Many release prints are made from colour internegatives (to protect the original colour master) by printing to a subtractive colour positive of the kind already described.

Imbibition techniques

The alternative is imbibition printing, or dye transfer, which depends on the preparation of a matrix film carrying a positive image capable of absorbing dye in proportion to the density of the image, point by point, and giving it up in the same proportion on transfer to another emulsion layer. Thus the matrix fulfils the same function as the blocks used for colour printing on paper.

The matrix image is in the form of a relief in hardened gelatin, but the differences in thickness corresponding to differences of density are to be measured only in millionths of an inch. Three matrices are dyed with the three subtractive colour primaries, yellow, cyan and magenta, and are brought in turn into contact with the blank film which is to receive the final image. Contact must be maintained for some time to allow the dyes to migrate from matrix to blank, so the two are held together on a continuously moving flexible belt of film width, equipped with register pins which engage the perforations. Since the transfer operation must be carried out in three successive stages, registration of the images is extremely critical, and temperature and humidity, along with other factors, have to be maintained within the strictest limits.

The matrices are used repeatedly, but only the making of a large number of release prints can justify their necessarily high cost. For television printing few copies are needed and normal contact prints are made. The difference from film practice is confined to using printing materials of lower inherent contrast to suit the lower brightness range imposed by the television process.

CHEMISTRY OF THE PHOTOGRAPHIC PROCESS—COLOUR PRINTERS (ADDITIVE)—DUPLICATING PROCESSES—IMBIBITION PRINTING—LABORATORY ORGANIZATION—PRINTERS—PROCESSING—SILVER RECOVERY—TECHNICOLOR—VISCOUS PROCESSING—Additive printing—Agitation—Air knife—Answer print—Auto-optical—Backing removal—Blank—Booster (chemical)—Candela—Cascade—Checkerboard cutting—Check print—Chemical mixer—Chemicals—Cinex strip—Circulating system—Combined dupe—Composite print—Composite master—Contact printing—Continuous printer—Control band—Control operator—Cross modulation—Cross-modulation test—Dailies—Dark end—Developer—Developing—Developing agent—Diabolo—Desaturation printing—Differential re-exposure—DIN print—Directional effects—Direct positive—Drier—Drum developing—Dubber—Dubbing—Dupe negative—Eberhard effect—Edge effect—Elevator—Examiner—Fine grain—First generation dupe—Fixing—Forced development—Green print—Greyback—Hardener—Hydrotype process—Hypersensitizing—Hypo—Impingement drying—Infra-red drying—Intensification—Intermittent printer—Joiner—Lacquer—Latensification—Lavender—Light box—Light change boards—Light end—Light lock—Light trap—Lily—Liquid gate—Loop elevator—Lubrication—Machine leader—Master matching—Master positive—Matrix—Matte loop—Mercury arc—Mordant—Multihead printer—Negative assembly—Newsreel base—Nitrogen burst agitation—Notch—One-light dupe—One-light print—Optical effects—Optical printing—Oxidation—Pinbelt—Positive drive—Printer—Printer light—Printer point—Printer scale—Protective master—Rack—Rack-and-tank development—Recirculation—Redmaster—Reduction printing—Reference print—Regeneration of solutions—Relief process—Replenishment—Reservoir—Restrainer—Reticulation—Reversal intermediate—Roll—Rotary printer—Rush print—Safe light—Sample print—Scene tester—Scratch print—Second generation dupe—Slash dupe—Spray processing—

Squeegee—Stabilization—Stationary gate printer—Step printer—Stop bath—Tank—Tendency drive—Toning—Trial print—Turbulation—UV printing—Viewer—Washing—Water spots—Waxing—Wet printing—Wetting agents—Winding

LABORATORY CONTROL

There is another aspect to the laboratory, no less important than the sequence of developing and printing film. Extremely fine processing control is required because the image exposed in the camera can be altered in tonal range and contrast by the chemical conditions of the developing solutions, and by the developing time and the temperature of the bath.

Characteristic curve

The reason for this goes back to the formation of the photographic image. The transfer process, in which the input is a pattern of light intensities and the output a pattern of image densities, is mediated by what is called the characteristic curve of the emulsion. If this were a straight line, a straight scale of point intensities in the scene would appear in the developed negative as a straight scale of densities. The tilt of the curve would represent a compression or stretching of this linear scale. At the extremes, a flat curve would produce a zero change of density for an infinite change of light intensity, and a vertical curve an infinite change of density for an infinitesimally small change of intensity.

Toe and shoulder

Midway between these extremes is the slope at which equal light intensity changes produce equal density changes in the emulsion; but the real state of affairs is complicated by the fact that the characteristic curve, though reasonably straight in its middle section, slopes away to form a toe (low light intensities) and a shoulder (high light intensities), so that its shape is an elongated inclined *S*. Camera exposure determines where on this curve the range of subject light intensities lies, over-exposure shifting it up towards the shoulder and under-exposure down into the toe, close to the fog density which represents noise and mush. Development varies the shape of the curve and particularly its slope or degree of contrast; and since this must be standardized for optimum image rendering, it is also necessary to control very precisely the chemical and physical developing conditions which influence it.

The characteristic curve, or transfer characteristic of film is of special interest because in shape it resembles that of other transfer devices such as valves (tubes) and transistors, as well as television camera tubes and display tubes (kinescopes). Since photographic and television images are transferred onwards from film to film and one electronic device to another, it is often important to trace the tonal reproduction of an image through a long series of such processes in order that the inevitable compressions and distortions, especially in the toe and shoulder regions, may be made as unnoticeable as possible in the picture finally seen by the viewer.

Slope and gamma

The slope of the straight part of the transfer curve is called the gamma (γ) of the developed film or the electronic device; point gammas (tangents to the curves) are often derived in the toe and shoulder regions. Ideally, the overall gamma for the transfer process should be unity, but practical requirements are usually best met by a figure rather higher than this.

Whereas television gammas can be calculated from voltage readings, film gammas must be established indirectly by sensitometry and densitometry: a known series of accurate exposures is made on a test strip, and the corresponding series of densities is read off. From these figures the curve is drawn, and the necessary conclusions reached about processing conditions.

These techniques, here outlined in terms of a single emulsion, apply equally to colour film in which three superimposed silver halide emulsions are exposed simultaneously, and three curves may have to be plotted and read.

SENSITOMETRY AND DENSITOMETRY—SENSITOMETRY OF COLOUR FILMS—TONE REPRODUCTION—Callier effect—Characteristic curve—Contrast—Densitometry—Density—Diffuse density—D log E curve—Flat—Gamma—Gradation—Grader—Grading—Grey scale—H & D curve—Highlight density—Intensity scale exposure—Latitude—Over-exposure—Pilot—Point gamma—Sensitometer—Shoulder—Specular density—Step wedge—Threshold—Time/gamma series—Time-scale exposure—Timing—Toe—Under-exposure

SOUND RECORDING

Something has already been said about the processes of sound recording and the technicians who work them. Their equipment, though paralleled in other branches of the recording industry, has gained from much specialized development in film and television.

Synchronized sound

Film synchronization is an old problem. The camera and its associated, but physically quite distinct sound recorder, must turn as one, so that speech can be recorded without the least error in synchronism, a single frame (1/24 sec.) being just detectable. In feature film production, elaborate motor interlock systems with their own generators can be provided; for documentaries and TV newsreel units something much simpler and more compact is needed. Recorders with a normal capstan drive produce a slight tape slippage impossible to eliminate; the camera is therefore made to generate a pulse which is recorded on the tape along with the audio signal, and the tape is re-recorded through a machine whose speed is regulated by these pulses. This establishes exact synchronism between camera film and tape re-recorded to perforated magnetic film; the pulses may be thought of as magnetic perforations.

In some cameras this problem is eliminated at source by recording along the edge of the picture film which has been pre-striped magnetically. This is called single-system recording in contrast to the double-system of separate camera and recorder, with corresponding terms, commag and sepmag, for the combined and separate magnetic tracks. Optical photographic tracks are similarly described as comopt and sepopt.

Playback and post-sync

Other specialized sound techniques include playback, in which a singer and/or orchestra is recorded under ideal acoustic conditions, and then repeats the performance in front of a camera, keeping precise time with the help of a loudspeaker which plays back the recording. The opposite technique is post-synchronization (also called dubbing and post-syncing), in which an actor is first recorded on a rough sound track on location, where ambient noise may be very high. Afterwards, in the studio, he listens through headphones to short sections of this dialogue on loops and repeats the words with the proper intonation. When exact synchronism is achieved, the section is recorded. This technique is also convenient for dubbing films into foreign languages.

Multiple sound tracks

Feature films, and to a lesser extent documentaries, have very complex sound tracks, for nearly all sound effects are recorded after studio shooting is finished, so as to secure ideal acoustic conditions without wasting valuable studio time. When the film is ready for final dubbing (a term confusingly applied to two quite different techniques), it may have as many as 20 or 30 sound tracks, and to render these manageable by two or at the most three sound mixers, they must be pre-mixed in groups. Such is the excellence of magnetic recording that quality suffers little by third- or even fourth-generation transfers. (First generation is the term applied to an original sound or picture record. Its first transfer is a second-generation record, and so on. This useful concept is common in discussion of deterioration of quality in film duplication and sound re-recording. It is closely linked to problems of signal-to-noise ratio, since the signal deteriorates and the noise increases with each generation of transfer.)

Dubbing equipment

The dubbing console, in addition to a basic set of faders, one for each sound channel, is equipped with a large array of equalizers to alter the frequency response of the tracks, emphasize or cut out frequency bands, and produce special effects such as telephone voices; echo and reverberation can also be introduced. To help the dubbing mixers handle a complicated programme throughout a reel of film, the editors prepare a log sheet marked with the footage at which each sound track comes in and goes out, and where significant change of level is to occur; a footage counter on the console and below the screen mark the progress of the reel. Even with these aids, it may take as long as a day to re-record ten minutes of film to the complete satisfaction of producer, director, editor and dubbing mixer, whose views by no means always agree.

Stereo recording

Further complications are introduced if, as often now happens, music has been recorded stereophonically. A mono version of the mixed track is made for theatres only so equipped; in the stereo version, the monophonically recorded dialogue and sound effects are artificially given stereo separation by pan pots (panning potentiometers) which distribute the signal to the required loudspeaker channel.

The final sound track is a multi-track magnetic transfer for the larger theatres, and a single optical sound track for the rest. The sound department's responsibility ends at this point, for over the final reproduction in the cinema – or the classroom or factory in the case of narrow-gauge prints – they exercise no control.

SOUND RECORDING—Attenuator—Audio spectrum—Azimuth error—Bilateral sound track—Binaural recording and reproduction—Blimp—Boom—Capstan—Coercivity—Commag—Comopt—Compression (sound)—Compressor—Control track—Crosstalk—Cut-off frequency—Double-head projector—Double-system sound recording—Dropout—Drum, sound—Duplex sound track—Dynamic range—Echo effects—Equalization—Equalizer—Erasing—Exciter lamp—Flutter bridge—Flutter, sound—Foldback loudspeaker—Galvanometer—Guide track—High-pass filter—Howl round—Interlock—Intermodulation—Light valve—Lip sync—Magopt—M & E track—Mil—Mixing—Monitor—Multitrack sound—Narrow band—Noise limiter—Noise reduction—Off-microphone—Optical sound—Overloading—Piezo electricity—Pilot tone—Playback—Post-synchronization—Pre-amp—Pre-emphasis—Pre-mix—Pre-scoring—Print-through—Pulse—Pulse sync—Push-pull amplifier—Push-pull sound track—Recording system—Re-recording—Rifle mike—Selsyn motor—Sepmag—Sepopt—Single-system sound recording—Slit—Sound console—Stereophonic sound process—Striping—Synchronization—Synchrostart—Transfer, sound—Variable-area recording—Variable-density recording—VI meter—Volume compressor—Wow

SPECIAL TECHNIQUES

ANIMATION The cinema does not rely solely on the real world for its raw material: animation, the art of the cartoon, is a synthetic world fashioned from paintings on paper and celluloid, puppets and models, even patterns drawn by hand on the film itself.

Cel process The essential device is the cel, a rectangle of celluloid about 8×10 in. in size, positioned very accurately on register pins in front of a camera which normally points straight down on it. The cels depict phases of action previously painted on them, and are replaced after one, two or more frames of film have been exposed on each; the fewer the frames and the closer the phasing, the smoother the action. Several cels can be superimposed on the animation table, the underlying ones depicting backgrounds, or objects which move less frequently.

Cel animation, perfected in the era when full-length cartoons were commercially feasible, is an expensive technique, requiring a large staff of animators, in-betweeners, inkers and painters. In recent years, the technique of the cel has been ingeniously adapted to more fluid and flexible forms, widely applied to TV commercials, instructional films and short cartoons, and greatly aided by the liberation of popular art from the shackles of pure naturalism.

Computer technique The computer is beginning to play a part in animation as a rapidly programmed device; it can, for instance, instruct a television display tube to present a pattern by a kind of pointilliste technique, in which all the light elements are on or off. It can also be used to feed instructions to a modified animation camera of orthodox design, replacing the laborious work of the cameraman in traversing and replacing cels, zooming the camera, and so on.

Animation equipment The animation camera is a high-precision register pin instrument, mounted on a vertical stand which holds it in stable relationship to the art work beneath. The camera is supported on one or two columns, along which it can be slid up and down to provide zooming or tracking movements. Great rigidity of construction is necessary, since the camera film may have to be wound back to provide successive exposures between which there must be no noticeable movement.

The cels fixed to their peg tracks are movable with micrometer accuracy either north-south or east-west on a lead screw device called a compound, which also rotates to provide diagonal movements. Many other attachments are available, including projection from below the table of an aerial image which enables actuality to be combined with animation.

ANIMATION EQUIPMENT—ANIMATION TECHNIQUES—COMPUTER ANIMATION—Aerial image photography—Animascope—Cel—Cutout—In-betweeners—Inkers—Multiplane photography—Pantograph—Pixillation—Rostrum photography—Storyboard—Zoom stand

FILMING UNDER DIFFICULT CONDITIONS Orthodox film cameras and equipment are designed to work at normal room temperatures and within a reasonable span of outside conditions. When conditions become more severe, as in tropical climates, in the arctic, and under water, they continue to be usable, but require more or less modification if they are to continue to function properly.

Low temperatures

At very low temperatures, both equipment and cameraman function at low efficiency. The output of dry cells and storage batteries decreases, while at the same time heavier demands are made on them, for camera oils get thicker and mechanical clearances may be reduced. It is therefore wise to provide in advance for a camera overhaul and re-assembly with special free-flowing lubricants. The film itself is liable to become brittle, and perforations may break in the jerky stop-and-go movement of the intermittent mechanism. Means for keeping the equipment warm whenever possible are therefore very desirable, even though electrical devices must consume additional current.

Tropical conditions

In the tropics the problems are quite different. Almost any camera will work well, and the chief consideration is usually light weight to avoid fatigue in transport. Moulds and fungus tend to grow wherever the relative humidity is very high; advance tropicalization is often advisable; silica gel in the transport cases is a useful desiccant on location. Film suffers more than equipment; raw stock cans should be kept taped and therefore hermetically sealed until just before use, and refrigerated to about 5–10°C if possible. After exposure, these precautions are even more important, especially for colour film. Development should follow as soon as practicable.

During shooting, the prevailing overhead sunlight or bright glare tends to produce excessive contrast, and matte reflectors are desirable.

Underwater filming

Underwater sequences in feature films are usually shot in a shallow studio tank, where lighting and oxygen equipment can be operated conveniently and without hazard. Real underwater sequences are the prerogative of cameramen making exploration, diving and salvage films, as well as the keen amateur. Because the refractive index of water is about 1·33, or 4/3, objects underwater appear to be at only three-quarters of their real distance, and camera focus must be correspondingly altered. Correction optics can, however, be incorporated in the front of the special camera housing. Because water pressure increases by one atmosphere (14·7 lb./sq. in.) for every 33 ft. of descent, pressurizing is usually applied to the inside of the housing, and controls are brought out through watertight glands. The camera is balanced to equilibrium in water.

Underwater photography imposes severe restriction on colour, for 16 ft. below the surface nearly all red light has gone, and below 100 ft. there is virtually no colour at all. Self-contained battery powered lamps are useful only at very short range, and heavy lighting powered by cable from the surface is essential for longer distances.

COLD CLIMATE CINEMATOGRAPHY—TROPICAL CINEMATOGRAPHY —UNDERWATER CINEMATOGRAPHY

FILM AND TELEVISION IN MEDICINE

The recording of clinical conditions involves no special techniques, but many medical applications are concerned with very small areas, both inside and outside the body, and compact easily-handled equipment is essential. On grounds of cost, size and weight, 16mm is the preferred gauge, reflex viewfinders are essential, and zoom lenses enable the field of view to be quickly and accurately controlled.

Operations

Films of surgical operations are best shot from the surgeon's viewpoint, but an overhang may be necessary and a gantry may be used to span the operating table. Since excessive heat must be avoided, a single 1 kW spot lamp often suffices. Where cavities in the body are much deeper than they are wide, simple techniques of this kind no longer suffice, and an endoscope must be brought into play. In one type of device, all the optics are at the outer end, but in the endo-telescope a wide-angle lens is placed at the end of the tube with repeater lenses along its length. The light source is outside, but a quartz rod light-guide transfers fully 80 per cent of it to the farther end.

Closed-circuit television (CCTV) is used to present surgical operations to medical audiences, but colour is an essential element, and is a considerable addition to the equipment cost.

Radiography

Radiographs (X-ray films) can be very successfully presented by CCTV, and the flexibility of the TV camera makes it easy to adjust image size, contrast and density, all of which are likely to vary widely.

Cine-radiography, the taking of X-ray images combined with motion, necessitated the development of lenses of exceptional speed, and ratings of $f\,0{\cdot}7$ had been attained by the time that electronic image intensifiers became practical. These give a light amplification of more than 1,000 as compared with a standard fluorescent screen, and may be coupled to a cine camera. Alternatively, the image may be focused on a television pick-up tube, in which case an effective light amplification is obtained by the gain of the video amplifier.

Macro and micro

The twin techniques of cinemacrography and cinemicrography also have extensive medical and biological applications, though their fields are much wider. Cinemacrography is the filming of small objects otherwise than with the compound microscope. It makes use of supplementary lenses attached to normal camera lenses or of extension tubes. Additional extension gives a higher magnification, but at the expense of effective aperture, so that exposures are very much increased. The piping of light by a quartz rod or bundle of fibres is a convenient technique when lighting small objects.

In cinemicrography the film camera acts as recorder of the magnified image seen through the eyepiece of a compound microscope, and plays no part in its formation. The technical problems, therefore, are principally those which apply to microscopy itself, though it is essential to provide for a watching eyepiece through which the operator can monitor the specimen or event while the camera records it. This is achieved by a semi-silvered mirror in the light path, and a 10 per cent bleed to the operator's eye is usually sufficient.

CINEMACROGRAPHY—CINEMICROGRAPHY—CINERADIOGRAPHY—MEDICAL CINEMATOGRAPHY AND TELEVISION

THE FILM IN THREE DIMENSIONS

All the special techniques of filming so far discussed, together with those of mainstream cinema, continue to portray the solid visible world in the flat two dimensions of a painting. This seeming anomaly has vexed inventors for a century. For still photography, they have argued, the convenience of the simple print or transparency is perhaps unchallengeable; but with added spectacle and movement, realism and colour, the third dimension would surely be a powerful attraction.

The technical difficulties are formidable. The separation of left- and right-eye views must be accomplished, either at the screen with a single device, or at the spectators' eyes with as many devices as spectators. To eliminate the inconvenience of individual spectacles, inventors have devised special screens from which left- and right-eye views are automatically picked out by the left and right eyes of all members of the audience. However, the optical problems are so severe that they are now unlikely ever to be overcome to the extent of producing a large-screen picture of the highest cinema quality.

In recent years, interest in the traditional 3-D film has waned, and its place has been taken by a completely new technique still in its infancy. Holography does not make use of images projected on a flat screen, but creates images in space by reconstruction of interfering wave fronts. Hitherto it has depended on laser beams with exposures too long for motion picture photography, but these limitations are likely to lift as the technique advances.

HOLOGRAPHY—STEREOSCOPIC CINEMATOGRAPHY—Anaglyphic process—Parallax—Parallax panoramagram—Parallax stereogram—Polarization—Raster—Vectograph

LIBRARY

By no means all the cinema's material stems from current production: a high proportion of what is past is kept for posterity and incorporated in other films. Preservation of old footage requires special techniques, but by itself would prove useless if classification and filing failed to keep pace with this gathering flood of material.

It is desirable to identify and enter on the library's records not merely one but several leading characteristics of a shot or short sequence. An insignificant figure in the back row of a group photograph may eventually turn out to be a Churchill or a Hitler; but the single grouping "Politician" will successfully disguise this fact from subsequent searchers. Combining simple search and access with detailed classification is a difficult task.

The Universal Decimal Classification (UDC) has been widely adopted in film and television libraries. The UDC covers the whole field of knowledge, and is published and kept up to date on an international basis. Its use for both books and films helps to make library staff more interchangeable, and it is well adapted to automated methods of information retrieval, though these have so far made little headway.

A good cataloguing system and a highly trained staff are prerequisites of a film library, which is in many ways similar to its counterpart in television. Equally essential is adequate and convenient vault space, where film is preserved under controlled physical conditions, since extremes of temperature and humidity can permanently damage it. As a general rule, high values are dangerous, low values safe. The bigger the change in either factor when the film is removed from storage and brought into room surroundings, the longer it must be given to stabilize itself. Nitrate film is liable to decompose and combust spontaneously; all of it is now 20 years old or more, and conditions for storing and examining it are necessarily much more stringent than for safety base film.

FILM STORAGE—LIBRARY—Dope sheet—Mutilation of prints—Papering—Shot listing—Vaults

PROJECTION AND THE CINEMA THEATRE

RELEASE PRINT MAKE-UP

If a film has been produced by a special technique for research purposes, it may be projected very much as it comes from the camera. But the normal process is to cut and match the negative to the work print and synchronize the sound track in the laboratory, after which release prints are made in small or large numbers, in one or more gauges, and sometimes in different formats on the standard gauge.

The one additional piece of material required is the leader, which provides much useful information in standardized form: the title and reel number, a sync section to enable picture and sound starts to be threaded into their respective heads with the right separation, and a series of numbers to tell the operator how far he is from the first frame of the film to help him in making accurate change-overs.

The high cost of prints and the heavy wear they are subjected to in repeated running have brought into being a minor industry for their inspection and repair. High-speed machines detect bad splices, nicked perforations and defective sound tracks. Cleaning and reconditioning machines, capable of eliminating all but the worst scratches, postpone the day when replacement will become necessary. These techniques are as widely applied to 16mm as to the standard gauges.

RELEASE PRINT MAKE-UP—Abrasion—Advance (sound and picture)—Answer print—Balancing stripe—Caption—Censor title—Cinch marks—Crawling title—Credit title—Cue mark—Distribution title—Dubbed version—Foreign release—Head leader—Inspection—Inspection table—Joiner—Leader—Newsreel—Patch—Reconditioning—Release—Release print examination and maintenance—Roller title—Scratches—Scratch removal—Spools, cores and cans—Start mark—Sync mark—Tail leader—Ultrasonic cleaning—Waxing

THE CINEMA THEATRE

In recent years it could safely have been said that the cinema theatre had more past than future, at least in countries where television was a powerful rival. Growing up in the music-hall era, it fell a prey in the 1920s and 1930s to giantist architectural tendencies. Though the decline in audiences in the 1950s was partly offset by the introduction of wide-screen presentation, many thousands of cinemas closed. In this cold economic climate, few new ones were built, and modern ideas were given little scope. Only projection installations reflected the new techniques. More recently, both in Britain and the US, there are signs of a revival in cinema building.

Some countries have laid down ideal requirements relating to auditorium design, projector location, and areas where seating is and is not permitted, and a whole body of practice has grown up around safety, lighting and ventilation requirements.

Projector features

The essential elements in the cinema theatre are the projector and screen. The projector shares many features with the camera, from which it historically derived: film magazines, feed and take-up sprockets, an intermittent mechanism and shutter, and an optical system, all of very rugged construction. To them is added a light source, traditionally a high-intensity carbon arc, but now often a compact xenon arc which needs no adjustment when running. Some projectors pulse this arc to the movement of the film, thus eliminating the shutter.

Projectors in the larger theatres are often dual-standard, 35 and 70mm, with provision for multi-track stereo reproduction. Automated operation is widespread, with control of a long sequence of action, such as house lights, curtains, cassette tape music, and the functions of the projectors themselves, including all that is involved in changing over at the end of a reel to another projector. Change-overs, a frequent source of trouble, have been reduced in number by the adoption of large spools, capable of holding 30 to 60 minutes of film.

The same tendency to simplification appears in the narrow-gauge projector. Self-threading removes the chief bane of the inexpert operator. Cassette loading eliminates the bugbear of loose film and the risk of breakage, and is invaluable to the schoolmaster and salesman who do not need to run long programmes.

Screen types

It is on the screen that the image makes its final appearance after these Odyssean journeyings, and here too there are great variations in practice. Back projection, in which projector and light beam are on the opposite side of the screen from the audience, has many attractions, but optical and screen design problems have restricted it to two fields: process plate projection in the studio, and narrow-gauge compact projection, principally of films in cassettes. For standard front projection, there is a wide choice of screens. The matte screen (sheets and white paper are matte reflectors) provides uniform distribution of reflected light, but also wastes light on walls and ceiling.

The trend to very large theatre screens, and the need to economize light in narrow-gauge projectors, has encouraged the development of reflective screens of different properties. The object of all is to send a larger proportion of light back towards the projector, i.e., to the audience area; but its distribution is equally important, since viewers in marginal seats do not wish to see the nearer edge of the screen much brighter than the farther edge, nor do viewers in the middle expect a brilliant hot-spot in the centre of the picture. The screen has therefore to be tailored to the needs of the auditorium, hall, schoolroom or living room in which it is to be used.

AUTOMATION IN CINEMA THEATRES—CINEMA THEATRE—DRIVE-IN CINEMA THEATRE—NARROW GAUGE PROJECTION—PROJECTOR —REAR PROJECTION—SCREEN LUMINANCE—SCREENS—SOUND REPRODUCTION IN THE CINEMA—Ambient lighting—Arc lamp—Auditorium—Beaded screen—Beater mechanism—Brightness (screen)—Cartridge—Cassette—Cassette loading for projectors—Changeover—Continuous projection—Daylight projection unit—Diffusing surface—Discharge lamp—Exciter lamp—Film slide—Film strip—Framing—Front projection—Gate pressure—Gate—H.I. arc—Inching—Intermittent movement—Keystone—Loop—Maltese-cross mechanism—Matte screen—Optical intermittent—Penthouse head—Pressure plate Première—Preview—Projectionist—Projection sync—Pulldown—Pulsed lamp—Screen brightness—Seating—Short—Shutter—Shuttle pin—Subtitle—Superimposed titles—Take-up—Test film—Throw—Titra titling—Title—Translucent screen —Travel ghost—Weave—Xenon lamp

SOUND IN THE CINEMA

Sound is often the last element to be considered in theatre design. In the past, standards of reproduction in the cinema were governed by the limitations of optical recording, and even more by an obsolete code of practice. Today this no longer applies, and theatres equipped with multi-track stereophonic reproduction provide the best sound it is possible to hear commercially. Three-track original stereophonic recording of music effects a striking separation of groups of instruments; it also diminishes distortions which result from a single assembly of loudspeakers trying to reproduce a whole orchestra at once.

The picture, however, is nearly always the dominant element in film; with action on the screen to guide the eye, the audience pays little attention to the direction the sound comes from, and this has had an unfortunate effect on the commercial success of stereophony.

In the narrow-gauge field, the sound track often has to fight against difficult acoustic conditions, and a wide frequency and volume range is of less importance than maximum intelligibility of speech. Special sound track transfers are therefore made for 16mm reproduction, and separate bass and treble tone controls (if intelligently used) enable the projector to be adjusted to suit prevailing conditions.

SOUND REPRODUCTION IN THE CINEMA—Binaural recording and reproduction—Cross-over frequency—Loudspeaker—Multitrack sound—Stereophonic sound processes

HYBRID TECHNIQUES

Though film had established itself long before the arrival of television, and so tended to insulate itself against the sweep towards electronic techniques, these have now made some mark, and nowhere more than in the sphere of the camera itself. The speeding-up of studio production requires long sections of action to be filmed without a break; and since this action must be dissected into shots taken at various angles and distances, several cameras must operate at once, as they do in the television studio.

In the new hybrid techniques, film cameras are equipped with electronic viewfinders, which are compact pick-up tubes fed through a beam-splitter by the same image which is simultaneously recorded on film and seen through the camera shutter. This camera tube furnishes a large parallax-free picture to the operator, and another to a vision mixer, who monitors the outputs of the various cameras on screens and can cut or mix from one to another. The combined output is fed to a videotape recorder for instant transmission or subsequent playback. Finally, the camera film is available for more careful shot selection by subsequent editing under the guidance of the vision mixer's own choice, which he marks on the magnetically-striped picture film in the cameras by pulsing a signal as he cuts from one camera output to another.

These techniques, under development for a decade or more, still somewhat uneasily occupy the middle ground, while the purely electronic approach makes striking advances.

HYBRID TECHNIQUES—Add-A-Vision—Electronicam—Gemini—System 35

Film and television have for long been separated in popular thinking by an artificial barrier. The very name, film or motion picture, stresses the link with photography, a means for permanently recording an instantaneous image; whereas television is commonly thought of as an essentially impermanent medium, concerned solely with the transmission of an image to a distant place.

In fact, both film and television store and transmit pictures. The film emulsion, after a rapid exposure, carries a latent image which is developed into a record and may remain unaltered for many years. The TV camera tube also stores an optical image, using an electrode called the target to do so, but this image needs to remain in being for only a fraction of a second until it is discharged by a narrow pencil of electrons in the form of a beam which scans the target from side to side and top to bottom.

FUNCTIONS AND ORIGINS

In cinematography, the method of transmitting the image – that is, passing it on from stage to stage of production towards the ultimate spectator – is relatively slow and clumsy, for it requires the transfer of physical material, i.e., the film itself. In television this transmission takes place at approximately the speed of light, for the picture, once it has been scanned from the target, is in the form of an ordinary modulated electric signal of fundamentally the same kind as in sound broadcasting.

Thus the most basic difference between film and television is that in film the picture starts and remains simultaneous, all there at the same time; whereas in television it starts simultaneous, but is at once converted into sequential form and so becomes a high-frequency signal. It is put back into simultaneous form as a scanned image at the receiver in the viewer's home and in studio monitors used to check picture quality. It may also be converted into film form for more permanent storage as a telerecording, or into a magnetic image by videotape recording (VTR), which is closely akin to the magnetic tape recording of sound. This interchangeability of the image between the scanned picture, film and VTR is one of the great conveniences of television production.

HISTORY

As with the motion picture, so the pre-history of television goes back more than a century to the first efforts of inventors to send photographs by wire, the process now called phototelegraphy. This led to research into substances whose electrical properties were affected by light falling on them. There then remained the problem of converting a picture into a sequential signal in a wire, solved by the invention of the scanning process in which each picture element is in effect surveyed in an ordered sequence from left to right, and from top to bottom. Although the two basic television processes, electric image conversion and scanning, were thus clearly defined by the 1880s, means were lacking to effect them quickly enough to give a realistic portrayal of objects in motion, and it was only with the development of cathode-ray and vacuum tubes that this became practicable some 50 years ago.

For another two decades the advocates of mechanical scanning by discs and mirror drums held their ground, and it was these which were used to give a low-definition service on medium waves between the wars. By the outbreak of World War II, all-electronic high-definition systems were technically feasible, and in Britain a regular service was more than three years old. Some signal standards had already been established, and when the war ended the era of mass television began.

Colour soon followed in the US in compatible form (i.e., receivable in black-and-white on monochrome sets), and the 1950s also saw the rapid growth of networking by landline and radio links. In the 1960s, communications satellites in stationary orbits made possible uninterrupted TV transmission across oceans and continents.

CHRONOLOGY—HISTORY—Ayrton, W. E.—Bain, A.—Baird, J. L.—Becquerel, A. E.—Berzelius, J. J. B.—Bidwell, S.—Blumlein, A. D.—Braun, K. F.—Browne, C. O.—Campbell-Swinton, A. A.—Caselli, A. G.—Chemical telegraph—Crookes, W.—De Forest, L.—Elster, J. P. J.—Farnsworth, P. T.—Fleming, J. A.—Geitel, H.—Hallwach, W.—Hertz, H. R.—Hittorf, J. W.—Ives, H. E.—Jenkins, C. F.—Kerr, J.—Mechanical scanner—Mirror drum—Nipkow, P.—Perry, J.—Phototelegraphy—Plucker, J.—Rosing, B.—Rotating disc scanner—rotating mirror scanner—Schoenberg, I.—Scophony system—Smith, W.—Televisor—Wehnelt, A.—Weiller, L.—Williams, P.—Zworykin, V.

BASIC PRINCIPLES

SCANNING

Basic to television, though foreign to film, is the scanning process by which a picture is broken down into sequential form and later reconstituted on a monitor or receiver screen. Of the many possible types of scanning, one is universal and easily remembered because it resembles the reading of a book. Scanning is in lines which run from left to right, the electron beam returning rapidly by flyback from the end of one line to start the line below. When the picture is complete, the spot flies back to the top left-hand corner, and the process is repeated. The rectangular pattern formed by the beam is called a raster, and normally has an aspect ratio of 4 : 3, i.e., 4 units wide to 3 high, the same as the standard Academy film picture.

Frequency and flicker

The standard frequency of the picture scan is half the prevalent mains frequency, which is 50 Hz (cycles per second) in Europe and 60 Hz in the US, so that it is 25 and 30 pps (pictures per second) respectively. The number 25 is conveniently close to the standard film rate of 24 pps, but the North American standard raises special problems in the transmission of film. As with film, these rates are not high enough to avoid flicker, and means have to be found to increase them.

In television there is nothing to correspond to the projector shutter. So the odd-numbered lines of the raster are scanned first; then, the beam returning to the top, the even-numbered lines. Two interlocking pictures are thus produced in the same time as a single picture by simple sequential scanning, so 25 pictures are effectively replaced by 50, and 30 by 60. This greatly reduces flicker. The technique is called interlace; the individual scans are called fields, and the two fields form a picture. The word frame should not be applied to the scanning process to avoid confusion with film.

Raster lines and resolution

Since each line is only one picture element high, the number of lines in a picture gives a measure of the detail it can resolve vertically, and is 405 for the original (now obsolescent) British system, 525 for the North American and Japanese, and 625 for the CCIR (European) system which includes British UHF transmissions, all Eastern European countries, and some in the Middle East. The numbers 405, 525 and 625 are all odd numbers, because in this way the two fields can conveniently be made different, so that the receiver can easily recognize which is which. Combining the numbers of lines and fields, the North American system can be compactly described as 525/60 and the CCIR as 625/50.

The vertical resolution is set by the number of lines, but the horizontal resolution is limited in a rather less obvious way. It is determined, in conjunction with the line standard, by the total bandwidth or span of frequencies available, and this is governed by national and international allocation of channels.

RASTER—RASTER SHADING—SCANNING—TELEVISION PRINCIPLES —Flicker—Flyback—Interlace—Line scanning—Line structure—Sequential scanning—Time-adjacent fields—Time base—Twin interlaced scanning—Two-field picture

TRANSMISSION CHANNELS

To get an idea of the frequency bandwidth needed for television, it is convenient to take the American system with its 525 lines and arrive at a rough figure for the maximum video frequency which has to be transmitted. Since the line structure fixes the vertical resolution, it is logical to make the horizontal resolution neither greater nor less than this, which will be achieved when a chequerboard pattern of alternating black and white squares, 525 from top to bottom and extending across the raster can just be discriminated at normal viewing distance.

Picture elements

Since the raster is one-third wider than it is high, the horizontal elements must be one-third more than 525, i.e., 700. The total number of elements is then 525×700, or 367,500. This picture is transmitted 30 times per second, so the number of picture elements per second is 30×367,500, or just over 11 million.

Returning to the alternating black and white squares which are being scanned by the camera electron beam, it can be seen that when the beam just covers one white square its current is maximum, as it sweeps across the transition to black it decreases, and when it just covers a black square it falls to a minimum. Thus one complete alternating cycle of current corresponds to two squares, not one, so the frequency band needed to transmit 11 million picture elements is only 5½ million cycles per second, now called 5·5 MHz (megahertz). However, this is about 500 times the frequency band required for a sound radio transmission, so that if all the radio stations on the medium-wave band were abolished, there would still not be room for a single TV channel.

Modulation methods

In order to transmit this video signal, it is necessary to modulate or impress it on a carrier wave. By the normal method used in sound radio, this produces two sidebands, one equal to the carrier frequency (f_c) plus the maximum modulation frequency (f_m), i.e., (f_c+f_m), and the other equal to ($f_c - f_m$). This gives a bandwidth of $2f_m$, and in the example just given, the 5·5 MHz bandwidth becomes 11 MHz. Because many channels are needed for national TV coverage, it is extremely important to economize bandwidth, and this has led to the universal adoption of vestigial sideband transmission, in which one sideband is cut down to a width usually between 0·75 and 1·25 MHz, the other remaining at its full width.

Bandwidth requirements

In the US the channel width, allowing for vestigial bandwidth transmission and adequate station separation to prevent interference, is 6 MHz, and in Britain lies between 5 and 8 MHz. These very wide frequency bands can only be transmitted in the upper reaches of the radio spectrum, where there is sufficient room to accommodate many channels, and the practical limits today are between 40 and 890 MHz. The lower reaches of this band, from 40 to about 230 MHz, are called Very High Frequency (VHF), the middle section is allocated to other uses, and the upper parts of the band from 470 to 890 MHz are called Ultra High Frequency (UHF). Altogether, 82 channels are provided within these limits in the USA, and 58 in Britain.

TELEVISION PRINCIPLES—TRANSMISSION AND MODULATION—Band—Base band—Bothway broad band channel—Broad band—Channel—Electromagnetic spectrum—Kell factor—Single sideband

THE TELEVISION SIGNAL

Since the scanned signal must eventually be reconstituted in the receiver as a picture, it cannot be transmitted merely as a continuous video waveform like the audio waveform which a radio set converts back into sound. This is because the video signal at every moment represents a point in the camera raster, and must be reconstructed as an identical point on the receiver raster if it is to appear as a coherent picture; otherwise, as happens sometimes on faulty TV sets, a scrambled and unintelligible pattern appears on the screen. In addition, the scanning process itself must not damage the image structure in the camera.

Blanking and synchronization

In the TV camera or film scanner, a pattern of electric charges is built up on a target which is proportional to the point brightness of the scene. The scanning beam removes these charges as it traverses the target, thus producing the signal. But if the beam were at full strength during flyback, it would discharge the target at the wrong time, as also when it reaches the bottom of the target and flies back to begin again. Thus the beam must be suppressed during line flyback, and again during field flyback. This suppression is called blanking, and blanking pulses which interrupt the camera output after each line, and at the end of each field, are an important part of the television signal.

These blanking pulses waste potential picture space, but are used to provide another essential of a scanned picture: synchronization. If the original picture is to be faithfully reconstructed at the receiver, it is clearly essential that the electron beams at both ends of the chain sweep in unison, line by line and field by field, and therefore picture by picture. This is known as synchronization, and the synchronizing signals are commonly called syncs, which are of two kinds, line syncs and field syncs.

Line and field synchronization

In the TV studio is a master generator, the synchronizing generator, which produces line and field syncs, and these are interleaved at the correct intervals by an electronic switch, and applied to the already blanked signals. Sometimes it is more convenient in the studio to handle blanked signals from different sources without their accompanying syncs, and these are called non-composite in contrast to the composite signals in which picture and syncs are combined.

The line blanking periods are not quite filled by the line syncs. The short period before the line sync is called a front porch, and the longer period after it a back porch. Both porches play a role in the synchronizing process.

The field blanking periods are much longer than the line blanking periods, and indeed occupy some 13 to 22 lines, according to the system employed. These lines must be deducted to give the number of active picture lines. When the field sync pulses are being transmitted, the line sync must be maintained so as not to interrupt the action of the scanning generator.

Back porch—Blacker-than-black—Black level—Black level clamp—Blanking—Blanking pulse—Blanking signal—Broad pulses—Crispening—Dark current—Envelope delay—Equalizing pulses—Field blanking—Field drive signal—Field identification signal—Field sawtooth—Field suppression—Field suppression period—Field time base—Front porch—Group delay—Instantaneous characteristic—Line blanking—Line drive—Line drive pulse—Line drive signal—Line period—Line phasing—Line suppression period—Modulation overshoot—Peak-to-peak—Peak white—Pedestal—Picture element—Picture modulation—Picture-sync ratio—Porch—Sawtooth deflection—Signal—Undershoot—White crushing—White level—Whiter-than-white

COLOUR PRINCIPLES

All broadcast colour TV must be compatible: the signal must produce a good black-and-white picture when picked up on a monochrome receiver, and in its totality can occupy a bandwidth no greater than that long ago standardized for monochrome transmission.

BANDWIDTH PROBLEMS

At first sight it may seem impossible that a bandwidth only just adequate for black-and-white can be made to convey the much greater quantity of information needed to transmit an image in full colour. The first attempts at colour television therefore ignored compatibility and were based on well-tried photographic principles – the analysis of a scene into its red, green and blue (RGB) constituents by taking three complete pictures in terms of these three primary colours. Most photographic colour processes today are subtractive, but the simple additive analysis first made by Clerk Maxwell is best suited to television. In these early experiments, R, G and B analyzing filters mounted on a rotating disc in front of the camera tube produced signals corresponding to the redness, greenness and blueness of each element in the scene. These signals were transmitted to the receiver and fed to a monochrome display tube in front of which a similar colour disc revolved in precise synchronism with that at the transmitter. This disc synthesized the colours with the help of the viewer's persistence of vision, and so reproduced the scene before the camera.

However, what is possible in the two-dimensional medium of photography, where different images can be recorded on top of one another in separate emulsion layers, becomes impossible in the one-dimensional medium of the TV signal. To eliminate flicker, the three-colour cycle in this early sequential system had to be repeated at least 24 times per second, so that the bandwidth had to be multiplied by three, and the system was not only incompatible, but extremely wasteful of bandwidth.

THE COLOUR SIGNAL

The solution to the problem is thus seen to centre round the television signal itself; it was devised in the USA by the National Television System Committee (NTSC) and embodied in the world's first commercial colour service in 1953. Its principal variants, PAL (Phase Alternation Line) and SECAM (Séquentiel Couleur à Mémoire), basically resemble NTSC.

No attempt is made to transmit successively or simultaneously three separate coloured pictures. Instead, the signal is split into two parts, one which carries the brightness information about each point in the image, and one which carries the colour information. The brightness (luminance) signal is thus able to actuate a black-and-white receiver, giving it exactly the same information as it gets from a monochrome transmission; the colour (chrominance) signal must carry all the information needed to instruct the colour tube in a distant receiver, so that point by point and line by line it renders the closest possible approximation to the colour in the original scene. Because of the bandwidth limitations already mentioned ingenious methods of economizing the colour information transmitted have had to be devised.

RECEPTION ON MONOCHROME SETS

In circuit analysis, the basic signals are called E_Y, the luminance signal, and E_R, E_G and E_B, the three colour signals (E stands for electromotive force or voltage). In this simplified treatment the signals are referred to as Y, R, G and B. The luminance signal, Y, can be generated as a separate signal by using a fourth tube in the colour camera, but it can also be derived from the R, G and B signals in the following way.

To use any single colour signal by itself would be unsatisfactory because it would give the wrong tonal balance in a monochrome set. The human observer, standardized for international convenience, has a sensitivity or response to the three primary colours such that if their intensities (saturations) are ranged from 1 to 0, three-tenths of red added to about six-tenths of green and one-tenth of blue produces a faithful rendering of the scene in black-and-white. Thus the Y signal may be derived by using these proportions of the colour signals, and, stated accurately, the equation is:

$$Y = 0{\cdot}30R + 0{\cdot}59G + 0{\cdot}11B$$

If a pure saturated green is transmitted, $G = 1$ and $R = B = 0$, so $Y = 0{\cdot}59$. This is three-fifths up the saturation scale and so represents a fairly light grey on a black-and-white set. Applying the equation to a saturated blue ($B = 1$), $Y = 0{\cdot}11$ or a much darker grey. This shows that the normal eye is a great deal less sensitive to blue and somewhat less to red than it is to green. Here, then, is the method by which a satisfactory rendering of a colour scene can be produced on a monochrome set, and the Y signal, made up to this formula, must be transmitted in the same form as a monochrome signal in order to actuate this set in the normal way.

COLOUR DIFFERENCE SIGNALS

The next problem is to transmit the colour or chrominance information so as to produce in a colour set the requisite R, G and B signals. This is done by forming what are called colour-difference signals to be used in conjunction with the Y signal. Thus if a signal (R—Y) is added to Y, it is clear that R results. Similarly, (B—Y)+Y produces B. However, a third (G—Y) signal to give G is avoided by going back to the previous equation and writing:

$$0{\cdot}30(R-Y)+0{\cdot}59(G-Y)+0{\cdot}11(B-Y) = 0,$$

and $$-(G-Y) = \frac{0{\cdot}30}{0{\cdot}59}(R-Y)+\frac{0{\cdot}11}{0{\cdot}59}(B-Y)$$

Encoding principles

This shows that by adding together the proper proportions of (R—Y) and (B—Y), —(G—Y) can be obtained, the negative sign indicating that the polarity of the signal must be reversed to yield (G—Y). The assembly of the luminance and colour-difference signals into a waveform suitable for transmission is called encoding, and when the colour-difference signals have been separated or decoded in the colour receiver, they can be combined thus with the luminance signal, Y:

$$(R-Y)+Y = R;\ (G-Y)+Y = G;\ (B-Y)+Y = B,$$

so that for the colour receiver Y renders the point brightness of the scene correctly, while R, G and B add the necessary colour. For a black-and-white receiver, the Y signal gives all the information to produce a tonally-balanced monochrome picture. (In practical NTSC, a rather more complicated system is employed, in which the red and blue colour-difference signals are combined to form what are known as I and Q signals, in order to increase the amount of colour information transmitted.) In all practical systems, the encoded colour information is modulated on to a carrier which is transmitted within the bandwidth of the luminance signal. This is called the colour subcarrier.

Decoding problems

The problem now remains to transmit the two colour-difference signals in the standard frequency band along with the luminance signal, and to decode them correctly at the receiver. How this is done can only briefly be indicated here. The basic technique is called band sharing; a second carrier, or subcarrier, is transmitted at a frequency some way removed from the main carrier frequency but within the video bandwidth. On this subcarrier, the colour-difference signals are modulated. There are two difficulties: how to prevent the colour signals encroaching on the Y signal (so marring the picture built up in terms of brightness with unwanted chrominance information), and how to keep the colour-difference signals separate from one another during transmission so that they can be decoded without confusion in the receiver.

The first difficulty cannot be completely solved. The approach initiated by NTSC is to accept that colour differences cannot readily be detected by eye in picture areas of fine detail. These correspond to the higher frequencies, since the greater the image detail, the more rapidly the electron beam in the camera is modulated as it scans at fixed speed across the target. Two consequences follow. The bandwidth occupied by the colour difference signals can be greatly restricted as compared to the luminance signal, and so the frequency of the colour subcarrier can be set high in the transmitted frequency band at about 4·4 MHz. Some interference remains, and may be seen as a fine pattern of dots on the screen; these often arrange themselves to form a series of parallel lines, inclined to the vertical and particularly noticeable in moving images.

Suppressed carrier modulation

The two colour-difference signals are transmitted by amplitude modulation of the colour subcarrier but to minimize dot patterning due to the subcarrier, a suppressed-carrier technique is used. In addition the subcarrier frequency is very accurately maintained at a value carefully chosen to minimize patterning. The use of suppressed-carrier modulation has the advantage that the visibility of the dot patterns is dependent on the intensity of the colour: the patterns are most in evidence in areas of the picture where colours are saturated and are completely absent from areas where there is no colour, i.e., in white or grey areas.

Quadrature modulation

The second difficulty mentioned above – that of transmitting two independent colour-difference signals on a common subcarrier – is solved by the use of quadrature modulation. The subcarrier is generated at the transmitting end of the chain in a very accurate and stable source and is suppressed-carrier modulated by one of the colour-difference signals. A second subcarrier is obtained by phase-shifting the original carrier by 90° and this is suppressed-carrier modulated by the second colour-difference signal. The two modulated subcarrier signals so obtained are combined to produce a single signal (the chrominance signal) which is still, of course, at the subcarrier frequency. This form of modulation produces a resultant signal with an amplitude which represents the saturation of the colour and with a phase (relative to that of the original subcarrier) which represents the hue.

Detection methods

To detect a suppressed-carrier a.m. signal it is first necessary to recreate the carrier. Detection can then be achieved by reversing or switching the modulated signal at carrier frequency. To detect the chrominance signal, therefore, it is necessary to switch it using an oscillator operating at carrier frequency and in phase with the original subcarrier at the transmitting end; this yields the first colour-difference signal. To obtain the second, the chrominance signal is switched using an oscillator operating at carrier frequency, but at 90° to the original subcarrier. Provided double-sideband a.m. is used for both colour-difference signals, there is no mutual interference in the quadrature detector and each of its two outputs is free of interference from the other. The quadrature detector thus requires for its operation, in addition to the chrominance-signal input, a reference signal at carrier frequency and in phase with the source of subcarrier at the transmitting end. Such a reference signal is transmitted in colour television systems in the form of a few cycles of subcarrier frequency (known as the colour burst) following the line sync signal at the end of each line.

Other considerations

This greatly simplified account of the NTSC system ignores many important features of it concerned with colorimetry (colour measurement and interpretation) and with the processing of the colour signal, especially the need to correct the contrast factor (gamma) of the colour tubes. The differences between NTSC and its chief variants, PAL and SECAM, are best left to the discussion of the receiver.

NATURE OF THE TV SIGNAL

The signal, then, whether monochrome or colour, is the true material of television, the one-dimensional counterpart of the piece of film which carries all the picture information from the original camera facing a scene to the ultimate spectator facing a screen. On the way, the piece of film undergoes inspection, duplication, modification by optical effects, alteration of its contrast and colour balance in the laboratory. Similarly, the TV signal undergoes monitoring, copying by VTR and telerecording, modification by inlay and overlay, fading, dissolving and cutting, gamma correction, crispening, and a host of other processes. That the signal is invisible (and even when rendered visible in its linear form bears no seeming resemblance to the picture) is sometimes confusing to the newcomer from the photographic arts. To those trained in electronics this, of course, presents no difficulty.

COLOUR PRINCIPLES—COLOUR SYSTEMS—NTSC COLOUR SYSTEM—PAL COLOUR SYSTEM—SECAM COLOUR SYSTEM—Burst gating—Burst keying pulse—Burst-locked oscillator—Burst phase—Burst signal—Chroma—Chromaticity—Chromaticity coefficient—Chromaticity diagram—Chrominance—Chrominance channel—Chrominance vector—CIE system—Colour bar—Colour burst—Colorplexer—Colour triangle—Compatibility—Decoder—Degradation—Desaturation—Differential gain—Differential phase—Encoder—Fringing—Hue—Luminance channel—Luminance signal—Matrix—Monochrome channel—Reference circuit—Saturation—Sequential system—Subcarrier—Subcarrier drive—Subcarrier oscillator—Subcarrier sideband—Suppressed carrier—Synchronous detector—Trichromatic analysis

THE CAMERA

THE TV CAMERA

The TV camera is the most important signal-originating source. It not only transmits live scenes, but also converts film pictures into signals, often in a much modified form called flying-spot scanner.

Construction

Essentially, a TV camera consists of two parts, an optical section and an electronic section, the latter frequently subdivided. The optical section is essentially the same as the lens system of a film camera, since its purpose is to form an image on a light-sensitive surface. In fact there is a good deal of interchangeability between lenses for film and TV cameras, since the areas of film formats and TV camera targets are comparable. However, a lens which is ideal for a system of high bandwidth (i.e., capacity to transmit a large amount of pictorial detail) such as film, may not be best to use in television where the resolution is drastically limited by the available bandwidth.

In terms of practical design, the more mobile operation of the TV camera in the studio has put a premium on quick and accurate operation; wide-range zoom lenses with servo motors for focus and diaphragm setting controlled by the operator are fast becoming universal.

The electronic section of the TV camera centres on the tube which converts the optical image into an electronic signal. In broadest outline, this tube contains a light-sensitive surface whose electrical properties can be made to vary point by point by the light falling on it. The optical image is thus converted into an electrical image, and the remaining function of the tube is to scan this image with a pencil of electrons, and so discharge it to form a signal which can be amplified and processed. The scanning waveform, complete with blanking pulses and interlace, is generated outside the tube and camera, and fed to it by cable. When the light-sensitive surface has been discharged, the picture again builds up and is again discharged by the electron beam, this process continuing at field frequency, i.e., 50 or 60 times per second.

TUBE TYPES

TV camera tubes work on one or other of two principles: the photo-emissive and the photo-conductive.

Photo-emissive tube

Certain materials such as caesium, potassium and rubidium emit electrons when illuminated, their number being proportional to the intensity of the illumination. In the image-orthicon tube, the sole but still very important survivor of the photo-emissive group, light from the lens passes through the window of the tube and is focused to form an image on the photo-cathode, the back surface of which gives off electrons in the way just described. These are accelerated by voltages applied through coil windings outside the tube and fall on a target set behind and parallel with the photo-cathode. The target in turn emits a larger number of secondary electrons, which are collected by a fine mesh immediately in front of it. Once these negatively charged electrons have been removed from the target, the image remaining on it is built up of positive charges. These leak through the target to form an image on the reverse side, and the target must be very thin to prevent the charges spreading out laterally. This is called the image section of the image-orthicon tube.

Image orthicon

Return beam modulation

The rear side of the target with its image of positive charges is scanned by an electron beam from a gun, as in a cathode-ray tube, and all sections of the tube are aligned on a common axis. Hence the name orthicon (Greek orthos = straight), for in earlier tubes such as the iconoscope, the gun had to be mounted at an angle to the target plate. Electrons from the beam are deposited on the target to neutralize the positive charges, and where these charges are large (corresponding to white in the original scene), many electrons are subtracted from the beam so that it becomes weak, whereas where the charges are small (corresponding to black) the beam retains most of its original strength. Thus the return beam from the target is modulated by the picture brightness. Though the variation of current between white and black (i.e., the modulation depth) is a good 50 per cent, the current itself is extremely minute, usually less than a microampere; the return beam is therefore directed into an electron multiplier, which makes use of secondary emission repeated over several stages to build up the current until it can be handled by an ordinary amplifier.

Orthicon characteristics

Even this much simplified description of the image orthicon reveals it as a very complicated device, and in operation it is rather delicate and temperamental. However, when properly handled, it can produce pictures of the highest quality, free from smearing – the smudging effect when an object seems to leave part of itself behind as it moves across the screen, rather like the track of a snail – and with excellent edge contrast. The latter quality is especially valuable, since any camera scanning beam, when it traverses a vertical edge of high contrast between black and white, tends to degrade its sharpness and subjectively give the whole image a rather blurred and indefinite appearance. This is because the beam is of finite width, and thus produces a signal of gradually increasing or diminishing strength as it crosses the boundary, instead of the rapid rise or fall required. This is called aperture distortion, and can be corrected by electronic processing, but the image orthicon tends to produce a black border round a bright object which makes the picture appear crisp and of high contrast. It is also very sensitive, having a speed rating equivalent to ASA 500–1,000.

Photo-conductive tube

A very much simpler type of camera tube, the vidicon and its derivatives, is based on the photoconductive effect. Certain substances such as antimony trisulphide (Sb_2S_3) and amorphous selenium (Se) have the property of varying their electrical resistance according to the intensity of light falling on them. Their conductivity rises with the intensity of the light, but the relationship is not a straight-line one, and the condition of increased conductivity persists when the light intensity is suddenly reduced. It is this time lag which produces smearing, especially at low light levels, an inherent defect of the simpler types of vidicon.

Vidicon

The vidicon tube consists of a light-admitting window backed by a signal plate which is both transparent and conductive, and from which the picture signal is taken. This plate is in turn backed by a target on which the optical image is focused, having passed through the window and signal plate. The target is of photo-conductive material, and its rear face is scanned by the electron beam. Thus the whole tube is in one section, instead of the three sections comprised in the image orthicon.

The voltages on the various parts of the tube are so arranged that when the target is dark (as when the lens is capped) and its conductance very low, few of the electrons in the scanning beam are captured by the target; this is called a low dark current. If the lens cap is now removed, an image in terms of brightness is formed on the target. Point by point, and more or less according to the brightness, its conductivity increases. Electrons from the scanning beam are therefore able to pass through the target to the signal plate, more where it is bright and fewer where it is dark. These electrons return from the signal plate to the tube cathode via a load resistor, in which the varying current generates a varying voltage, which is the required picture signal. This is sufficiently large to be passed direct to an external amplifier.

The vidicon camera is light, compact and cheap, and has excellent sensitivity and resolution. It is widely used in portable camera channels, in telecines (film scanners) and in news and presentation studios, and is virtually universal for closed-circuit television.

Plumbicon

The advantages of the vidicon are retained, and its defects very largely overcome, by the Plumbicon tube (more generically called a lead-oxide vidicon) in which the photo-conductive layer is of lead monoxide (PbO). Its dark current is very low, giving a more stable black level, and it suffers much less from smearing at low light levels. In construction it differs little from the basic vidicon, and its compactness makes it very suitable for colour cameras. The lead-oxide vidicon may eventually oust the image orthicon from its present position as the standard studio camera tube.

Auxiliary equipment

The viewfinder in all TV cameras is fed from the signal output and thus produces a scanned picture free from parallax error. Each camera is connected to its Camera Control Unit (CCU) by a cable which, in the case of a four-tube colour camera, may comprise as many as 100 wires. The CCU contains auxiliary services, among them waveform monitoring, aperture correction, gamma correction, an adder for the blanking signals, and power supplies.

COLOUR CAMERAS

As will be clear from the description of the colour signal, the colour camera records separately and simultaneously the redness, greenness and blueness of the scene before it, and this information is then encoded and processed to produce the television signal. Separate tubes are used for the three-colour analysis, but all are fed with the same optical image through a single lens backed by a colour beam-splitter system.

Four-tube and three-tube

There are two principal types of colour camera. The four-tube camera, in addition to R, G and B tubes, has a separate tube which contributes the luminance signal, Y. In three-tube cameras, Y is derived from the three colour signals by taking balanced proportions of each as already explained. In the analysis of colour systems it was shown that a G signal can be derived from the R and B colour-difference signals with the help of the Y or luminance signal. A few 3-tube colour cameras have been designed for this RBY pattern. Finally, there are some 2-tube colour cameras which incorporate a luminance tube and a single chrominance tube. The chrominance tube may employ a fine-textured series of RGB stripes plus black keying stripes, the sequential output of which is fed to an NTSC encoder; or it may be a standard tube in front of which revolves a red and blue filter, so that one whole field is scanned through each filter. If, say, the R field is held in a magnetic store for the requisite one-sixtieth of a second, it can be fed, along with the B and Y signals to an NTSC encoder, and there processed into standard form. In spite of their compactness and simplicity, however, these 2-tube cameras have not yet been widely adopted.

In all colour cameras, the tubes may be image orthicons, Plumbicons or vidicons, and in 4-tube cameras vidicons may be used in the chrominance channels with orthicons or Plumbicons in the more critical luminance channel.

Beam splitter

In the optics of these cameras, an important role is played by the beam-splitter. In addition to the mirrors or prisms needed to direct the image into the three colour tubes, there is a set of dichroic filters which pass all light frequencies except those they are designed to reflect; unlike semi-reflecting mirrors, the amount of light they absorb is very small. The spectral passbands of these filters and the colour characteristics of the system as a whole, greatly complicate the design of a colour camera channel, which also includes the usual viewfinding monitor and a Camera Control Unit.

CAMERA—CAMERA (COLOUR)—CAMERAMAN—Add-A-Vision—Aperture correction—Beam current—Beam landing error—Beam splitting system—Camera cable—Camera control operator—Camera control unit—Camera operator—Collector—Crab dolly—Crane truck—Depth of field—Depth of focus—Diaphragm, lens—Dolly—Dynode effect—Electron gun—Electronicam—Electron image—Emitron—Exposure—Exposure meter—Face plate—Focus servo system—Follow focus—Gemini—Head amplifier—Iconoscope—Image dissector—Image iconoscope—Image orthicon—Light-sensitive mosaic—Line time base—Low velocity scanning beam—Mesh effect—Monitor viewfinder—Overscan—Pedestal—Photoconductive tubes—Photomultiplier—Pick-up tube—Plumbicon—Positional control servo—

Pre-amp—Radio television camera—Relay lens—Retrofocus lens—Return scanning beam—Scan burns—Scan generator—Scanning reversal—Scan protection—Secondary emission—Selenicon—Separate mesh vidicon—Shading correction—Shading error—Shot box—Signal plate—Smear—Split focus—Spurious shading—Spot brightness meter—Sticking—Storage characteristic—Streaking—System 35 *—Target—Target capacitance—Target layer—Target mesh—Target voltage—Trichromatic camera—Turret, lens—Velocilator—Velocity control servo—Vertical aperture correction—Vidicon—Vidicon channel—Viewfinder—Wall anode*

CREATIVE TECHNICIANS

With the signal established as the essential television material and generated in the camera, it is time to step back and look at the roles of some of those responsible for all television productions.

CAMERAMAN

Because TV studio production is essentially continuous, whereas film production is broken down into individual shots and takes, the role of the cameraman is different. The TV cameraman has the responsibility of carrying out complex movements of the camera with their attendant critical focus changes, knowing that if he makes a single mistake a long section of action may be spoiled – or, if the programme is live, a mistake will go out on the air.

On the other hand, since several cameramen of equal status may be operating on the floor at the same time, lighting is placed in the hands of a separate supervisor, whose job it is to set up in advance a long lighting sequence, balanced for different camera viewpoints, which can then be released into action by a more or less automated process. Exposure control, including alteration of lens iris setting, is also out of the cameraman's hands, so that he can concentrate on the difficult demands of composition in movement, and on getting everything to work out right throughout a sequence which in film would be broken down into many set-ups and would take many times as long to shoot.

PRODUCER

In charge as a rule of a series of programmes, the producer chooses the director and prepares budget estimates. He decides on the proportion of live shooting to film in a programme, bearing in mind time and cost of production, as well as the resources of the organization for which he works. If his speciality is drama, he must keep abreast of the theatre and film in all principal countries, assess acting talents and their adaptability to television, and study the work of stage directors and designers. If he is in light entertainment, he must know what artistes are available and what new talent is emerging; if in current affairs, his acquaintance must include a wide range of experts on all topics, and he needs a flair for handling politically delicate situations. Though there may be a higher court of appeal in matters of taste and censorship, he has to exercise a discriminating judgment on what may and may not go on the air.

On the level of organization, the producer may have to deploy an army of reporters and commentators right across the world; to keep them serviced, he must know what facilities they will need and how to get them, whether their material can be transmitted back via radio signal or landline, or must be stored on film or videotape and thus airfreighted home.

When a programme goes out live on the air, or, as is more common, is recorded on videotape, the producer is always available to the director of a programme for consultation and advice.

DIRECTOR As the creative head of both an artistic and a technical team, the director in Britain is expected to have overall control of the mechanics of production with a technical supervisor working under him. In the US, a technical director takes full responsibility for camera, sound, and engineering, and advises the director on how best to achieve his effects. These are only general tendencies, however, and no hard and fast divisions can be laid down, for practice varies widely from one television organization to another.

Working under the administrative control of the producer, the director starts with an idea or a rough script, which the writer works up into more complete form in close association with him. The director then assembles (or is allocated) a production team, an important member of which is the designer. Casting is discussed with the producer, and decisions are made as to how sets are to be laid out and built so that the production will fit into the allotted studio space. On the basis of the script and the floor plan, the director works out his shooting pattern, deciding how much is to be pre-filmed or pre-taped, and what his subsequent editing requirements will be.

First rehearsals take place outside the studio in rooms in which the studio floor plan is carefully marked out. Here the pace and timing of the production are determined. There follows a technical run-through in the studio, in which lighting is rehearsed for the first time, and levels and positions of light units preset, so that smooth transitions can be automatically effected, but over-ridden if there are last-minute changes in pace or dialogue. Next, the director calls for camera rehearsals, in which camera and microphone boom movements are practised through long stretches of action, and intercommunications are tested.

When all is satisfactory, the director moves forward to the dress rehearsal and the live (on-air) performance or recording. Here, when all the contributing factors in the production, human as well as mechanical, fully interact for the first time, the director must be capable of making instant decisions if anything goes wrong.

DESIGNER Producing scenery is not the only role of the designer; he must also grasp and help to realize the overall pictorial side of a programme. An integral part of this today concerns the production of titles – both those of series and individual programmes – which are often the work of a separate designer who specializes in graphics.

The set designer co-operates closely with the lighting director, since some shots may have to circle a full 360 degrees. The mobile microphone boom as well as three or more mobile cameras have to be considered, for the dynamics of design are just as important as the statics. Both are in the minds of designer, director, and producer as they map out available studio space in terms of the script order, so as to be able to shoot continuously through long stretches. Tracing paper plans are placed over the studio layout, and from this starting point, the designer produces elevations and structural drawings, as well as supervising painting, properties and set dressings. He must also collaborate with the costume designer on colour and style, so as to produce a unified and attractive visual whole.

CAMERAMAN—DESIGNER—DIRECTOR—PRODUCER—Boom operator—Camera control operator—Camera operator—Dubber—First assistant director—Lighting supervisor—Sound balancer—Sound mixer—Studio floor manager

STUDIO OPERATION

STUDIO COMPLEX

The production centre of a television organization is a massive assembly of equipment with high depreciation costs brought about by rapid technical progress. It is therefore planned for maximum throughput of programmes, and its design and layout vary greatly according to the size of the complex and the number of auxiliary services it must house and operate.

Basically, it has to make provisions for administration and for a wide range of technical facilities (sound and picture recording, vision mixing and switching, generation of pulses for synchronizing all picture sources, power generation for lighting, etc.), as well as all those activities concerned with the studios themselves, actors and their needs, and the design and erection of scenery. In addition, the complex may house news and film units, and outside broadcast (remote pick-up) equipment, but it will often not contain the transmitter itself. The geographical requirements of a centre housing some hundreds or thousands of technicians and artists frequently conflict with the lofty situation demanded for the transmitting aerial. Only the smaller studio centres are co-sited with their transmitters.

Continuous recording

The studios themselves differ from film studios in one essential respect; they are planned for continuous recording of long stretches of action, not single, often-repeated short takes. The studio floor has therefore to be kept as free as possible of cables; those that remain (mostly camera cables) have to be adroitly manoeuvred by hand. Lighting units, therefore, are suspended on telescopic posts from the ceiling, and can be moved this way and that on grids without electrical disconnection. Long sequences of variable lighting can be pre-set and programmed to come into action in the proper sequence.

Live transmission is now rare except for news and discussion programmes because, even in countries not divided into time zones, casts of actors are seldom available precisely at the required programme times. Recording is usually on videotape, but sometimes on film by telerecording (kinescoping) or by the hybrid methods already described. In spite of continuously running action, some subsequent editing is needed, and both film and tape editing rooms form part of the studio complex.

Inserted material

Not all of the programme originates on the studio floor; inserted material, often shot earlier on location, is fed from telecines (film scanners), videotape recorders and caption scanners. Sound has an important department of its own, and intercommunication between all the personnel on a programme – arranged so that accidental pick-up on the microphones cannot occur – is necessary to ensure the smooth interlinking of actors, cameras, lighting, microphones, telecines, and the rest of the closely knit organization.

Production control The end product of all this activity is the recorded programme itself, and it reaches videotape or film by way of the production control room, where the picture sources terminate in fading and switching equipment. This enables them to be kept on monitors, cut in or out, and faded or dissolved from one to another. Here, too, orders are given for the run-up of machines playing pre-recorded material, the timing of which has to be extremely exact. And here electronic special effects, ranging from simple wipes to elaborate combinations of foreground action and background scene separately shot, can be introduced as modifications before recording.

These are some of the steps in the production of a single programme, but the studio complex requires many programmes to constitute a single day's transmitted output, programmes which must themselves be dovetailed together into a smooth sequence, broken at proper intervals by station identification, weather and news reports, and in some cases commercials and sponsors' announcements. At the heart of this overall programming activity is Master Control.

ANIMATED CAPTIONS—LIGHTING EQUIPMENT—MAGNETIC RECORDING MATERIALS—MICROPHONES—POWER SUPPLIES—SOUND—SOUND DISTORTION—SOUND MIXER—SOUND RECORDING—SOUND RECORDING SYSTEMS—STUDIO COMPLEX—Attack time—Autotransformer—Caption roller—Caption stand—Chopping—Cold mirror reflector—Compressor—Crosstalk—Cue—Cue card holder—Cueing device—Diffusion—Dubbing—Dynamic range—Echo effects—Effects track—Equalization—Equalizer—Fishpole—Flood—Floor plan—Flutter (sound)—Flutter bridge—Foldback loudspeaker—Generator—Ground noise—Guide track—Gun mike—Headphones—Headset—High key—Howl round—Interlock—Intermodulation—Jacketed lamp—Key light—Limiter (sound)—Monitor (sound)—Pantograph—Peak programme meter—Phantom—Photoflood—Pre-emphasis—Premix—Pulse sync—Qwart—Release time—Re-recording—Reverberation time—Rifle mike—Rigging—Rostrum—Saturable reactor—Sepopt—Skid—Sky pan—Sound console—Spider box—Stop key—Talkback—Tally light—Telojector—Thyratron—Thyristor—Tone source—Transfer (sound)—Twister—VI meter—Volume compression

MASTER CONTROL

Here, too, scale has a determining effect on the nature and complexity of the equipment. In a very small independent station, an announcer (in sound only) may himself act as presentation mixer operator. His control room is equipped with a sound and vision mixer, picture monitors, cartridge tape machines, audio turntables and controls for a caption scanner, sufficient to produce simple station breaks by purely manual means.

In more elaborate studios, manpower may be greater, but a high degree of automation makes possible more complex continuity and presentation. The switching schedule between the many programmes available is organized into a programme log to aid the crew, and transferred to a memory, which may be as simple as a uniselector of the kind used in telephone exchanges, or as complex as a magnetic drum store.

Even if all programme sources are pre-recorded, completely automatic operation throughout the day is normally impossible. There may be technical faults at the station or at remote pick-up points, emergencies or crises may require personal announcements, and the automatic switching system must therefore be capable of being overridden by a human operator.

But the tendency today, especially in commercial stations, is to operate by a cue clock on an elapsed time basis, so that each event follows the previous one at a predetermined moment. Non-commercial national broadcasting organizations, independent of advertising and precisely timed station breaks, rely more on manual control.

COMMUNICATIONS—MASTER CONTROL—PRESENTATION—PRESENTATION MIXER—Cue circuit—Cue mark—Diascope—Divcon—Gram operator—Identification signal—Line-up time—Master control operator—Prompter—Rotary stepping switch—Sound prelisten

PICTURE MONITORS

An essential element in picture selection during programme recording or transmission is the monitor, a display unit superficially resembling a domestic TV set, but receiving the video waveform by line and therefore requiring no r.f. or demodulating circuits. Sound also is separately processed, and reproduced on high-fidelity loudspeakers. In other respects also, the picture monitor differs markedly from a standard receiver.

The linearity and geometry of the raster are very much better, and a picture element is no more than one to two per cent away from its theoretical position, except perhaps in the corner areas where curvature and large beam deflections make this order of accuracy excessively hard to achieve. Picture brightness is very carefully adjusted to suit ambient conditions; too brilliant a picture produces unavoidable flicker. The black level of the video signal is kept constant through fluctuations of average brightness in the picture content, whereas in many black-and-white receivers it moves up and down. This means that the d.c. component of the signal must either be maintained or reinserted before it is applied to the tube.

Complementing the picture monitor is a waveform monitor which is fundamentally an oscilloscope designed to display the video signal at line and field sweep rates. It enables a constant check to be kept on peak white and black levels, and on the amplitude of the synchronizing pulses.

PICTURE MONITORS—Bouncing—Clamp amplifier—Clamping diode—Clamp pulse—Colour bar—Graticule—Line sawtooth—Line strobe—Master picture monitor—Monitor—Monoscope—Multiburst—Pattern generator—Phosphors for TV tubes—Picture line-up generating equipment—Pulse and bar—Ringing—Routine test method—Scan linearity—Sine squared pulse—Vertical bounce—Waveform monitor—Waveform oscilloscope—White bar

TRANSFER CHARACTERISTICS

The monitors and their associated equipment help station engineers to maintain overall faithfulness of picture reproduction from the cameras onward towards the transmitter. Hitherto, for the sake of simplicity, it has been assumed that there were no problems in the overall picture transfer process – that, more precisely, the brightness of each object point on the receiver screen would be identical with that of the same point as recorded by the camera in the original scene. This convenient assumption is, however, very far from the truth.

Gamma principles

In the photographic process there is a curve, the characteristic curve, which relates the density of the image to the logarithm of the intensity of the light which exposes it. The slope of this curve, called gamma, is constant over its middle part, and here $\gamma = \Delta D/\Delta \log E$, where D is density and E exposure. Overall gamma, which measures the contrast and thus the tonal reproduction of the photographic process as a whole, is found by multiplying together the individual gammas of the different picture transfer stages. When this overall gamma is unity, equal changes of exposure of the original scene correspond to equal density changes in the final reproduction. In practice, it is not possible to confine exposure to the straight part of the characteristic curve, and results most pleasing to the eye may not correspond exactly with an overall unity gamma. Nevertheless, this is the point of departure for final corrections to the system.

Exactly the same considerations apply to the television process, except that image density is replaced by the log of the camera output voltage or of the display tube input voltage, as the case may be. Since the gamma of the display tube is fixed by its design, it may be used to govern the variable parts of the transmission chain. Its value is about 2·5 to 2·7, so that the rest of the system should have a gamma rather less than 0·4 to give an overall figure of unity. In practice, an overall figure of about 1·2 is found to give more pleasing results, the extra contrast helping to correct in monochrome for the loss of colour in everyday scenes.

Gamma control

Television signal processing makes it simple to alter gamma by the use of non-linear amplifiers, which amplify more or less than proportionally as the signal is raised. But this does not solve the problem completely, for the range of light intensities in an average scene far exceeds the range of the transmission system as a whole, which is even more cramped than that of film. Again, as in photography, part of the average scene is therefore imaged, not on the straight part of the characteristic curve, but on its curved upper and lower parts. It is then necessary to consider the compression or crushing of the dense blacks and lighter whites in the scene, and control their reproduction so that they benefit from the full capacity of the system without ever exceeding it.

Finally, attention must be paid to the transfer characteristics of each element in the train, cameras, amplifiers, telecines, VTRs, etc. in order to ensure a satisfactory picture at the receiver.

What has been said applies to colour as much as to monochrome reproduction, which is the easier to understand in television where the luminance information is a separate signal. But in a more detailed analysis than that given in the section on Colour Principles, account must be taken of the gammas of the colour-difference signals.

TRANSFER CHARACTERISTICS—Black crushing—Black stretch—Break point—Burn out—Crush—Flat—Gamma—Gamma correction—Knee—Limiter (vision)—Overshoot—Picture matching—Point gamma—Undershoot

VISION MIXERS

The processed and gamma-corrected signals have now to be combined by the programme controller, who, previewing his picture sources on monitor screens, cuts, fades or dissolves them to produce the recorded or transmitted programme, or introduces more drastic modifications with the aid of electronic special effects units.

Video mixing and switching are therefore of the greatest importance to picture presentation. In the control room, they are paralleled with audio mixing and switching, employing techniques basically inherited from film and radio and therefore requiring no detailed description. Essentially, the video mixing unit consists of a matrix or grid of connections, along one side of which are fed in the various inputs (cameras, telecines, VTRs and other remote picture sources), while the intersecting connections feed fading and combining amplifiers with outputs to preview and on-the-air. The matrix, therefore, is a device for connecting at will any input to any output. If this is effected by instant push-button action, a cut results; if by way of a fader from one source, a fade-in or fade-out; if by way of two faders worked in opposite directions, a lap dissolve; if two faders set at some positive value a superimposition.

This simple outline conceals a mass of intricate detail. For example, non-synchronous signals, i.e., those in which the sync pulses are not in step with one another, cannot be mixed to produce an acceptable signal. Warnings must be given to the controller of non-sync sources, and picture locking applied before they are brought in. Sync comparators are often used to provide this information automatically. Other problems concern the avoidance of crosstalk between channels, and the elimination of faults attendant on mechanical relays. These are now giving way to semiconductor devices (solid state switches) at each crosspoint in the matrix which rely on a high ratio of forward to reverse impedance. They are usually diode rectifiers.

In common with all television control equipment, vision mixers vary greatly in complexity, both in input-channel capacity and ability to premix sources before making the final output mix.

VISION MIXER—Bank—Crossview—Cut—Cut bank—Cut in blanking—Cutting—Direct switcher—Effects bank—Fader—Interfield cut—Mixing—Mixing equipment—Non-additive mixing—Prelisten—Preview—Roving preview—Stabamp—Storage switching—Trap valve amplifier—Uniselector—Vision distribution amplifier—Vision switcher—Vision switching matrix

ELECTRONIC SPECIAL EFFECTS

The normal scene transitions – cuts, fades and dissolves – can all be made within the video mixer, but wipes and other special effects are generated in separate equipment which can be added if required to these more basic facilities. Television has one fundamental advantage over film: it consists of a sequential signal which can be processed and modified in real time. Compare this with film wherein each frame requiring a special modification – even so simple a one as a fade – must be re-photographed in a slow, intermittently-moving optical printer; and if two such picture strips are required, as in a dissolve, they must be separately passed through the optical printer.

Special effects generator

The possibilities of video signal processing are seen to good advantage in the use of a special effects generator, a device of fairly recent introduction and still in rapid development.

Two cameras originate different picture signals, their outputs synchronized line by line and field by field. The output of one camera is replaced by that of the other halfway along each line for a number of successive fields. Two half-pictures are seen on the receiver screen, side by side, in the effect known as split-screen. If the transition takes place at the left-hand edge of the screen for all the lines composing the first fields, and then progressively across the picture towards the right-hand edge in succeeding fields over a period of a few seconds, a left-to-right vertical wipe results. Other wipe patterns are similarly produced.

Insertion effects

These are known as insertion effects because the output of one picture source is introduced or inserted into the output of another. Their operation depends on an electronic switch, essentially of the same simple two-position kind as a domestic light switch, but capable of operating in less than the scanning time of a single picture element, about 0·1 microsecond.

Insertion effects are of two main kinds, inlay and overlay. Inlay effects are those in which the insertion shape is predetermined, as in the examples just given; these correspond to stationary matte effects in film. Overlay effects are those in which a boundary in the picture itself keys the output of the second source; these correspond to travelling-matte effects in film.

Inlay process

In more detail, inlay effects are produced as follows. The keying signal required to work the electronic switch may be derived from a telecine (film scanner) in which a black-and-transparent mask in the form of the wipe pattern is projected and scanned as a length of film. This is, therefore, a strict electronic equivalent of special effects printing in film. A more sophisticated device, however, is a pattern generator which produces electronically the required sequence of keying signals, either statically or in the form of a wipe. Thus in the previous example, a split screen is simply a left-to-right vertical wipe arrested at midpoint. Usually, a hundred or more patterns are available in this way, and are controlled on the video mixing panel in the same fashion as simple dissolves and fades.

Overlay process

In the overlay process, in which camera signals produce their own keying, the subject in effect acts as the mask. This is achieved by making the video signal from one camera act both as a picture input and as a keying signal for the electronic switch. The subject to be isolated is thus free to move, and wherever it is at any moment, it triggers itself into the picture from the second camera, which normally provides the background for the composite picture.

It is important that the subject be placed against a sharply contrasting background, either lighter or darker than itself, so that keying is rapid and certain. Even so, the outline round the subject caused by the switching action tends to be visible on the screen. More complex methods using contrasting colours, with filters, help to reduce this effect and lend themselves naturally to colour TV cameras.

Other effects

Other picture modifications include ripple effects, achieved by sinusoidally displacing the camera line scan at a frequency of about 300 Hz; polarity reversal of the signal, which gives the effect of a negative image; and scanning reversal, which turns the picture upside down or left to right, and thus can be made to nullify the effect of mirrors inserted in the light path.

SPECIAL EFFECTS—Brightness separation—Corner insert—Dissolve—Edged captions—Electronic spotlight—Electronic switch—Fade—Hold frame—Image orientation—Inlay—Inlay mask—Insertion—Inversion—Iris wipe—Keying signal —Non-additive mixing—Normal-reverse switch—Overlay—Pattern generator—Picture inversion—Scanning reversal—Special effects amplifier—Split field picture —Split screen—Superimpose—Wipe

STUDIO SYNCHRONIZATION

PICTURE LOCKING TECHNIQUES

As already mentioned, precise synchronism must be established and maintained between picture sources if they are to be perfectly matched in transitions. A simple film analogy shows why. If a dissolve is being made in an optical printer and one shot is in frame while the other is one perforation out of register (in 35mm film), the resulting print will have a horizontal bar across it and the upper part of one picture will appear below the bar. Similar effects can be produced by loss of sync in television; usually the loss is only momentary, and results in a roll-over effect (incoming picture moving up or down the frame before stabilizing itself) which is sometimes seen on receivers.

Synchronizing problems

These faults are eliminated by picture-locking techniques, which are in any case needed for networking and other multi-station applications. It is essential that the line and field frequencies of all picture sources are identical, so that the same number of line and field pulses occur in a given time. But this is not sufficient, for sync pulses may have the same frequency but yet be displaced in time, so that the pulses in each train do not occur at the same instant. Nor is this all. Passage through signal-processing circuits, landlines and other transmission links distorts the sync pulses, so that if two or more were superimposed, their outlines would not coincide. Thus a single part of the pulse, the leading edge, is used as a time reference, and it is a point on this edge, the half-amplitude point, which precisely determines the phasing of the line-sync pulse. Since the time-period of a line is accurately laid down for each television standard, phase differences are usually expressed as time differences rather than as angles.

Types of lock

The actual method of locking is common to all techniques, and consists of an oscillator, a phase detector and a filter network. The external input to which the oscillator is to be locked, and the oscillator output, are fed into the phase detector which produces an output proportional to the phase difference between the two waveforms. This output goes via the filter to the oscillator, thus forming a negative feedback loop which locks the oscillator frequency to that of the external input. The types of picture lock used are many, and include mains lock (line lock) which uses the fixed frequency of the power line: genlock, which relies on a special synchronizing generator; and slavelock, in which programme sources are routed through a central master station which synchronizes the subsidiary or slave stations.

High Stability operation

Because of phase differences which occur over a national power line system, caused by varying load factors, etc., power line lock is not often used, though it has the great advantage of reducing picture interference caused by hum at 50 or 60 Hz and their multiples, as the case may be. Genlock, in its many forms, is the system of widest application, but a recent rival for special purposes is high-stability unlocked operation, using separate oscillators at the different sources to drive the respective sync pulse generators. These oscillators must remain accurate within two parts in a hundred million over a period of 24 hours, and when subjected to a wide range of temperatures.

PICTURE LOCKING TECHNIQUES—Gating—Gating impulse—Locking—Phase comparator—Phase lock—Phase locked loops—Roll over—Slave—Slavelock—Slaving

SYNCHRONIZING PULSE GENERATORS These devices, generally called sync generators, produce the basic stable frequencies to which, as already explained, all picture sources are locked. Interlaced scanning requires the line scan frequency to be an integral multiple of half the field frequency. The source, therefore, is of twice the line frequency, and this divided by two to give line frequency and then by 525 or 625, or whatever other standard is used, to give field frequency. To simplify frequency division, TV line standards are chosen to have easy factors: $525 = 7 \times 5 \times 5 \times 3$, and $625 = 5 \times 5 \times 5 \times 5$. Although multivibrators have been used in the past, logic circuits assembled from solid-state devices are now more common, since they can be made to provide a high order of stability. If the system is to be locked to power line frequency, a normal tuned oscillator is used to generate the master frequency; but if the field frequency is to be established independently, a crystal oscillator takes its place.

Frequency divider and gating circuits convert this master frequency into the line sync pulses, equalizing pulses, field sync pulses and other essential elements of the TV waveform, the skeleton or framework to hold the whole electronic system together at transmitter and receiver as a vehicle of the video signal. When colour is added to monochrome, the requirements are very severe; the colour subcarrier oscillator must have an accuracy and frequency stability of about two parts in a million.

SYNCHRONIZING PULSE GENERATORS—Composite signal—Composite sync generator—Composite video signal—Crystal control—Crystal lock—Delay cable—Frequency divider—Mixed syncs—Multivibrator—Piezo electricity—Reactance tube oscillator—Shaping network—Square wave—Synchronizing pulse regeneration—Synchronizing pulses—System blanking—Trigger pulse

STATION TIMING Although the sync pulse generator produces pulses in correct time relationship to one another, this relationship may be upset by delays sometimes amounting to several microseconds, which result from passage through equipment and in transmission lines. If these delays were equal, no problem would be created, but this is not the case in practice. Outputs of telecines may be used as sources in a studio, which in turn feeds the vision mixer, thus the telecine outputs must be timed to occur earlier than any studio output. Camera cables may be of any length from a few yards to a thousand feet or more, and so important is this source of variation that the camera control unit normally incorporates a variable delay. By the insertion of such delays, all the inputs to the vision mixer can be timed to coincide exactly with respect to their line and field sync pulses, so that they can be cut and mixed without any loss of stability and quality.

This is especially important in colour transmission, where changes of hue will result from mistimings of only a nanosecond (10^{-9} sec.) or so.

Signal sources can be operated with non-composite outputs, that is, without addition of sync pulses, and in this condition they can be mixed and faded with no regard to synchronizing problems until the sync pulses are finally added before transmission. The picture cannot, of course, be viewed in this form, and preview monitors must therefore be separately fed with sync signals.

STATION TIMING—Composite—Composite picture—Delay cable—Pulse phasing—Time of fall, time of rise

Attention has so far been directed only to the camera in the studio as a source of TV signals. Other origins are almost equally important, and it is convenient to classify these into recorded sources and live sources.

PICTURE RECORDING

There are at present only two practical methods of recording a moving image for television: film which makes a spatial record of the picture, and videotape which makes a scanned or sequential record of the TV signal waveform. To convert the film record into a TV signal, each picture in turn must be scanned in a device resembling a film projector; this is called a telecine or film scanner. The videotape recorder, like an audio tape recorder, is a single instrument which is used both to record and reproduce. Finally, it may be convenient to turn the TV signal into a film record, a process known as telerecording or kinescoping. This is because film is universally standardized, whereas the image on videotape conforms to standards which may differ from country to country both in field rate and number of lines.

VIDEOTAPE RECORDER

The magnetic recording of sound and picture entails the use of magnetic tapes and films as complex in their manufacture as photographic films. Manufacturing problems result from the need for a very high signal-to-noise ratio, and exceptional homogeneity in the particles of magnetic material, which are coated on a base of cellulose tri-acetate (safety film base) or of polyester or PVC. The particles are of magnetic iron oxide, and are usually needle-shaped, of no more than one micron in length and 0·2 microns in width. These particles are oriented in the direction of travel of the record and reproduce heads, longitudinal for audio and transverse for video recording, to give maximum signal output.

Frequency band requirements

The basic difficulty of videotape recording results from the wide frequency band of the video signal, from direct current to 5·5 MHz. This corresponds to an infinite range of octaves, and even with d.c. restoration which sets the lowest frequency at the field frequency, say 50 Hz, the range is nearly 17 octaves. This is to be compared with the 9 octaves required for a high-quality audio recording spanning 50–20,000 Hz, and cannot be compassed by a tape system, in which the output rises by 6 dB per octave at constant amplitude. For the videotape system, a range of 102 dB (17×6) would be required, and the practical maximum is only about 60 dB above noise level.

The problem of signal-to-noise ratio is solved by the carrier method which simultaneously raises the lowest and highest frequencies. If, for instance, the 50 Hz base frequency is raised to 0·5 MHz, a 7 MHz maximum frequency requires a span of less than 4 octaves.

The high-frequency recording requirement, however, is intensified by the carrier technique, which raises the maximum recording frequency. In practice, the maximum frequency which can be reproduced from magnetic tape depends on the tape speed and the width of the reproduce head gap (corresponding to the slit in an optical system). At its minimum value of 0·0001 in., the reproduction of a 5 MHz signal would require a tape speed of 1,000 inches per second (ips) or 300,000 feet per hour, a totally impracticable figure, if the record were to be made longitudinally as in conventional audio recording.

Transverse scanning

This problem is solved in most professional instruments by adopting transverse scanning to use the full 2-in. width of the tape which advances at the moderate speed of 15 ips. Four magnetic heads are equally spaced round a drum whose axis of rotation is parallel to the length of the tape, which is curved to conform to it. This drum rotates at 250 revolutions per second, producing a head-to-tape velocity of about 1,500 ips. On the CCIR 625-line standard, one interlaced picture is contained in 40 parallel tracks occupying only $\frac{5}{8}$ in. of tape length. In fact, frequencies up to 10 MHz can be successfully recorded and reproduced by transverse-scan VTRs. Audio cueing and control tracks are recorded longitudinally near the tape edges by conventional stationary heads.

Helical scanning

Exceptional engineering refinement is required in the construction, balance and drive of the rotating drum, and a less costly type of videotape recorder, known as helical scan, is often used for less critical applications than broadcast quality transmissions, such as educational and other closed-circuit recording. In this design the tape is wound helically round a drum and moved forward at a relatively slow speed, while a disc within the drum carrying record and reproduce heads rotates rapidly to give a high head-to-tape speed.

Servo circuits

By methods such as this, the basic problems of passband and wide frequency range are overcome, but there still exists the necessity of preserving during playback (the reproduce mode) the precise positioning of the video heads in relation to any point on the tape that existed previously in the record mode. Only in this way can the synchronizing information be correctly read off. The fact that the two types of drive in transverse-scan machines must be interlocked with extreme accuracy – a drum drive for the heads at 15,000 rpm and a capstan drive for the tape at 15 ips – indicates the complexity of the servo circuits required to achieve professional picture standards.

Colour problems

Colour VTR recording magnifies the severity of these problems, because in NTSC, time errors of the kind indicated result in phase changes which bring with them unwanted alterations of the reproduced colour. A phase shift of 10 degrees is detectable in this way, corresponding to a time displacement of only $6\frac{1}{2}$ ns (nanoseconds) with a 4·43 MHz subcarrier. In the earliest VTR's, positional errors of 500 ns were common. Ball bearings for the rotating head drum have been replaced by air bearings, decreasing the frictional resistance and increasing the speed of response to correction signals. Equally important, these signals now provide more complex and refined control of head position not merely at the start of each line but during the course of the line itself. Today, entirely satisfactory videotape recording and reproduction of colour signals is possible, and manufacturers of professional equipment have now standardized provision for colour and monochrome.

Editing

The usefulness of videotape would be much restricted if it could not be edited in the same fashion as film. Indeed, the simplest editing techniques resemble those of film in that they involve actual cutting and splicing of the tape. The invisible frame lines are identified by pulses on the longitudinal control track, which can be picked out by a magnetic scanning head or rendered visible by applying to them a few drops of a suspension of iron particles. A more advanced technique known as electronic editing makes use of the re-recording process to transfer the signal from one tape to another at a predetermined cutting point, so that on this second tape an edited picture can be built up shot by shot without ever having to cut the original tapes recorded from the camera. If these tapes are marked with timing pulses, a cutting schedule can now be programmed and read into a computer memory which will effect the actual transfers automatically.

MAGNETIC RECORDING MATERIALS—VIDEOTAPE EDITING—VIDEOTAPE RECORDING—VIDEOTAPE RECORDING (COLOUR)—Closed loop drive system—Coercivity—Electronic editing—Helical scan—Jumbo roll—Phaseless boost—Phase locked loop—Print-through—Quadruplex recorder—Transverse scan—Videotape operator

TELECINE

In spite of the growth of VTR, much programme material is still recorded on film, and means of converting the film image into a TV signal long ago reached a high level of development. With the universal film rate of 24 pps, film scanning at 25 pps on the European standard is simple, the slight speeding up of action and raising of sound pitch passing virtually unnoticed. Film for TV is normally shot at 25 pps instead of 24. The North American standard of 30 pps, however, requires more complicated transposition, since a speeding up of 25 per cent would be wholly unacceptable.

Camera type

The simplest type of telecine employs a vidicon tube which looks at the aperture of a standard film projector, back-illuminated from a small lamp house. Since intense brightness is easily achieved, the smearing effects of a vidicon are reduced to a minimum, and its high resolution and good signal-to-noise ratio are an advantage. For the 30 pps standard, it is arranged that the odd film frames are scanned with two fields and the even frames with three fields, so that two frames produce five fields and 24 frames 60 fields, which is the required relationship. Sometimes continuous-motion projectors with optical compensators are used to reduce film wear.

Flying spot type

The other principal type of telecine is the flying-spot scanner. A high-voltage cathode ray tube produces a very fine and intense spot which scans an unmodulated interlaced raster on the film frame in the projector. This frame is not otherwise illuminated; instead, a photomultiplier placed behind it receives light modulated by the density of the film, and from it generates an amplified signal proportional to this density from moment to moment. Thus, whereas the camera telecine sees the film frame, the flying-spot telecine scans it.

In some flying-spot telecines, an optical compensator is used with continuous film movement; in others, notably for 16mm film, a very fast pull-down advances the film by a frame during the field-blanking period, so that the two interlaced field scans are made on stationary film, and no registration problems can arise.

Both camera and flying-spot types of film scanner have advantages and drawbacks, but these are fairly evenly matched and neither type has ousted the other.

Colour

Telecines of the kinds described reproduce either colour or black-and-white films in monochrome. When a colour signal is required from a colour film, the basic mechanisms remain the same but dichroic filters as used in colour cameras separate the light (projector light or flying spot) after modulation by the film into its red, green and blue components. These are then gamma-corrected and processed electronically.

With three- and four-tube camera telecines, as with equivalent colour cameras, there is an image registration problem, since all the tubes, in spite of the interposition of prisms and filters, must have matching fields correct to a single picture element or so. This problem does not arise with flying-spot telecines, since the image dissection is carried out first by beam scanning, and only afterwards does the colour analysis take place.

Common to both types are the usual colorimetric problems found in colour cameras, but to these are added the imperfections in reproduction of the colour film itself. Electronic compensation can, in theory at least, actually improve on the film rendering.

TELECINE—TELECINE (COLOUR)—Automatic light control—Cropping—Flying spot—Horizontal picture shift—Jump scan—Light application bar—Light application time—Masking (colour)—Multiplexer—Pellicle—Photomultiplier—Polygon machine—Pos-neg switch—Safe area—Scan burns—Scanned print—Scanning printer—Spot wobble—Step wedge—Telecine operator—Twin lens scanner—Two-three-two-three scanning

TELE-RECORDING Known in the US as kinescope recording, this is a method, alternative to videotape, of recording a programme or indeed any televised event, by means of motion picture film capable of being reproduced anywhere in the world irrespective of TV standards.

Pull-down requirements In a telerecorder, a film camera photographs the TV image on a screen, but this basic simplicity masks some formidable difficulties. The television picture consists of two interlaced fields, and the full picture information is not available until both fields are complete, after which follows an extremely short field-blanking interval before the first field of the next picture begins. The problem, therefore, is to pull down the film in the camera in this interval which corresponds to no more than 11 degrees of shutter rotation in the 625-line system, in contrast with the 180 degrees for such action in a normal film camera.

With American standards, the problem is different, since 30 TV pictures (60 fields) have to be reduced to 24 film frames per second. Usually this is achieved by discarding half a field in every 24, and recording five fields on two frames. During the discarded half field, the film can be pulled down, and this interval corresponds to 72 degrees of shutter movement which greatly simplifies the pulldown problem.

Reverting to the 625 (and 405) line systems, it is found that the extremely short pull-down times required are unattainable with 35mm, but just or almost attainable with the smaller mass of 16mm film. Even so, the accelerative and decelerative forces on the film are many thousand times the force of gravity. Fast pull-down movements may be mechanical, with special accelerator devices, or by means of an interrupted flow of air which sucks on the lower loop and positions the film on a fixed pin when the downward movement is complete.

Stored-field techniques A quite different approach is to produce the picture on a tube with a comparatively long afterglow, so that, during the field obscured by the shutter (when pull-down takes place), the tube face is storing it, and during the next field both appear on the phosphor and are recorded. The stored field which gives its name to this technique is fainter than the following field, but its contribution to the signal can be equalized by amplification.

In the North American system already outlined, a different sort of difficulty stems from the discarded half field during which pull-down occurs. This necessitates a picture join in the middle of the frame area between two half fields, and very precise shuttering and drive mechanisms are employed to make this join invisible, or nearly so.

Colour telerecording, a late starter in Europe but commonplace in the US, may be accomplished most simply by displaying a colour TV picture on a shadow mask tube and recording it on colour film. This leads to very inferior colour saturation, and a better approach is to use a trinoscope, with separate R, G and B display tubes and dichroic mirrors to combine the three images. Alternatively, a videotape method may be used to separate the three colour records.

TELERECORDING—Electron beam film scanning—Electron beam recording—Electronic video recording—Moiré patterning—Spot wobble—Vidtronics

OUTSIDE SOURCES

Picture sources remote from the studio are today sufficiently specialized to warrant separate description. Often combining direct pick-up, videotape and film, they may also form part of an elaborate communications system to bring distant countries together for important events. Thus internal and international networks formed by cable, radio and satellite transmission may all be linked with mobile picture originating equipment.

OUTSIDE BROADCAST UNITS

These units, called remote pickups in the USA, are truck-mounted, often on an articulated vehicle, and contain a mobile control room which houses production staff, video control engineers and their equipment, including sync generators, together with one or more sound engineers and recorders. Picture and sound signals may be channelled direct to the station by land-line or by a radio link with its own mobile transmitter and aerial (antenna) tower. Alternatively, if the remote programme material is not urgently needed for transmission, it may be pre-recorded on a mobile VTR and brought back to base with the unit. Lighting equipment may be hired on site or brought out from the station in a special lighting tender which travels ahead of the OB vehicle and so can be set up before the production and engineering team arrive.

With the transistorization of all equipment, a great reduction in size and weight has been achieved in recent years, so much so that a complete colour camera channel can now be carried on the cameraman's back, with a cable feeding the signal to the camera control unit. The extra simplicity of black-and-white makes it possible to incorporate a small transmitter in the back pack, so that the cameraman is entirely free of cables. Automatic gain control circuits prevent tube overloading under sudden changes of lighting, so that the operator has only to aim, focus his zoom lens and shoot. Just as World War II perfected the highly mobile film camera with its reflex shutter, so today's unceasing battles and civil riots have helped to develop instant, on-the-spot television.

OUTSIDE BROADCAST UNITS—Linkman—Radio relay—Rigging tender

SPECIAL EVENTS

Newsreel units belonging to a TV station or network organization usually add to the facilities of the OB unit a complete film unit with compact 16mm cameras and rapid processing equipment. Editing facilities are needed for both film and VTR, and the choice between them tends to be governed by speed, which favours VTR, and international dissemination which favours film. Today, however, with the development of standards conversion equipment of high quality, this asset of film is dwindling, but for library storage, rapid access and subsequent re-use, it still has notable advantages.

The library is an essential auxiliary to all newsreel and special events units, and unexpected happenings create an urgent demand for background information, obituaries and the like. Most of this material is held in film form, and its cataloguing is on the same lines as for a film production library. Videotapes of complete programmes are also stored in the library.

Where highly complex events such as national elections are programmed for international transmission, the number of sources becomes so great that station video switching and mixing facilities often need to be extended. A total of twenty vision and thirty sound sources is not unusual. The presence of foreign-language commentators necessitates special sound mixing arrangements, whereby clean effects (i.e., mixed sound effects excluding commentary) are made available for final mixes, one mixer being allocated to each language. These commentators receive an earphone feed of the domestic commentary to enable them to identify important personalties whom they might not otherwise recognize.

In the organization of such programmes, the talkback arrangements, though entirely separate from the signal channels, are almost as important as the signal itself. Under the omnibus talkback system, the programme director can speak to all outside broadcast units, control rooms and switching centres. By an extension of this system known as instant talk-in, personnel in each remote location can silence their own loudspeakers and address the programme director.

Since smooth transitions must be made between remote picture sources controlled by their own sync pulse generators, it is usual to lock these sources to the studio pulse generator shortly before they come on the air.

LIBRARY—NEWS PROGRAMMES—SPECIAL EVENTS—Programme switching centre

TRANSMISSION

In the earliest days of television, an individual station generated its own programme and radiated it from a neighbouring transmitter. Programme origination and transmission were two quite separate things. Today, with the immense extension of networking and the growth of programmes themselves of worldwide extent, this distinction has become blurred. The components of many programmes are transmitted inwards to the studio before they themselves take to the air; other programmes, such as those of closed-circuit TV, may never be transmitted at all.

MICROWAVE LINKS

For the inward transmission of programme material to TV stations, point-to-point links at extremely high frequencies are common practice, enabling wide bandwidths to be used without interference. The microwave region of the spectrum extends usefully from about 3–30 GHz (1 gigahertz = 10^9 Hz), so that at the lower frontier of 3,000 MHz (3 GHz) the frequency is far above the UHF band extending to about 800 MHz. At these extremely high frequencies, transmission is purely line of sight, necessitating careful surveys before the path is established. But the advantages are manifold. The aerial, or dish, need be no larger than 3 or 4 ft. in diameter, the radiated power is a few watts down to milliwatts, and the equipment together with its power supplies can be packed in a small truck.

The 7–11 GHz band is most commonly used, the lower frequencies providing rather greater beam spread so that aerial alignment at the transmitter and receiver are less critical. Heavy rainfall and small obstacles in the path have less effect. On the other hand, the higher frequencies make for maximum economy in the size and weight of equipment.

For simplicity, sound is often multiplexed on to the video channel, that is to say, modulated on a subcarrier above the upper limit of the video signal but within the passband of the microwave system. Power consumption of the entire unit is likely to be no more than 250 to 500 watts. The only serious limitation is the line-of-sight requirement, and this may entail the use of a mobile, power-elevated tower to raise the transmitter dish clear of neighbouring obstructions.

MICROWAVE LINKS—Helical aerial—Parabolic aerial—Polyrod aerial—Probe—Slot aerial—Travelling wave amplifier

TRANSMISSION NETWORKS

Next in order, but overlapping with the microwave links, come the distribution systems by which stations and countries are linked to interchange programme material and complete programmes before the latter are radiated to the public. Today these networks are designed for colour transmission and must meet rigorous requirements laid down by international agreement.

A form of test known as K-rating has been devised to assess waveform distortion. It is based on two concepts: one, aperture distortion, already mentioned in connection with the camera; the other, ringing, which produces spurious edges preceding or following a vertical edge in a picture. These two effects – gradual change of brightness at what should be a sharp transition, and spurious edges – result from degradations in the waveform as it passes through the many electronic stages of its course from signal source to receiver. Some of these degradations arise from phase alterations in the signal which themselves are frequency-dependent; and it is these delays, called group delays, which at the higher frequencies create ringing.

Though picture impairment is ultimately a subjective judgment, standard test signals, including K-ratings, are extremely useful in deciding objectively whether a given transmission reaches an acceptable level of fidelity to the original.

Since a colour channel transmits both luminance and chrominance information, the tests just described are applied to the luminance signal. In addition, luminance/chrominance crosstalk (i.e., leakage of one channel into the other) is estimated, together with phase and gain differences between the two, which result in unwanted alterations of colour.

Cable networks are frequently used to link TV stations and consist either of screened (shielded) balanced pairs of conductors, or of coaxial pairs in which, as in a receiver aerial downlead, a central conductor is surrounded by insulating material or air, encircled by a braided shield. Air losses are less than those of solid insulator, and spaced-out polythene or other discs support the central conductor.

Very often a high-frequency carrier of about 12 MHz is used, on which up to a thousand narrow-band speech channels can be modulated in addition to a wide-band TV channel transmitted in each direction. Repeater stations to amplify the signals and improve the TV waveform are required every few miles, but these are small and automatic in operation.

Alternatively, microwave links may be employed of the type, and using the wave bands discussed in the last section, but with permanent equipments and fixed sites set on high ground.

Special network switching centres are situated in or near large towns to handle the very large amount of traffic which is routed in and out of TV stations along networks for programme origination and transmission.

TRANSMISSION NETWORKS—Acceptance test method—Balanced pair—Black bar—Cable compensation circuits—Double sideband—Echo waveform corrector—Ghost—High frequency carrier system—Hybrid—Ionosphere—K-rating system—National Reference Network—Pattern generator—Picture line-up generating equipment—Power separation filters—Pulse and bar—Quad—Reflective network—Repeater—Speech Channel—Spur band—Steady state characteristic—Supergroup—Telephone pair—Telephone channel—Translator—Video repeater

SATELLITE COMMUNICATIONS

Satellites for TV relay, a dream less than a decade ago, are today a commonplace. Historically, the first to be launched were the orbiting satellites, initiated by Telstar I (1962). The short period during which such a satellite is above the radio horizon for transmitter and receiver makes it commercially of little value, and it was not until the launching of the first successful synchronous satellite (SYNCOM II, 1963) that the era of satellite transmission really began. At an altitude of some 22,300 miles, the satellite is in a geo-stationary equatorial orbit, and thereafter retains an essentially fixed position relative to a point on the earth's surface.

Satellites are of two kinds; passive satellites merely reflect an incoming signal back to earth and thus require no power supplies, while active satellites receive and retransmit a signal, not necessarily on the same frequency. All commercial communications satellites are active, and are powered by solar cells; they operate in the 4–6 GHz region.

Earth stations are equipped with large aerials to give maximum gain so as to realize the highest signal-to-noise ratio from what is necessarily a very weak incoming signal.

Current commercial communications satellites (except those operated by the USSR for internal TV networking) are owned by an international organization, Intelsat, on which many nations are represented.

SATELLITE COMMUNICATIONS—Doppler shift—Frequency translation—Pulse modulation—Pulse sound—Solar cell—Telemetry—Transponder—Van Allen belt

STANDARDS CONVERTERS

Communications satellites symbolize the internationalizing of television, and this in turn highlights the lack of worldwide standards. In many branches of technology, inventions in their early stages show little promise, and each country advances along its own path. When success is finally achieved so much money has been invested that there are strong conservative reasons for maintaining the status quo. An example is alternating current power generation, which lies at the heart of the television problem. Before World War I, Europe tended to centre on 50 Hz, North America on 60 Hz, though many other frequencies were common. The Atlantic Ocean seemed to make agreement on a common standard unnecessary, and investment in generating equipment soon made it impossible.

Line standard development

When television developed in the '30s, it was natural to tie picture frequency to power line frequency for ease of synchronization, and thus the 50-field (25 pps) system took secure hold in Europe and the 60-field (30 pps) system in North America and Japan. Line standards have a more complicated history. The British 405-line transmissions provided the first public service in the world, and with a 3 MHz video bandwidth, difficult enough to achieve more than 30 years ago, gave excellent horizontal resolution of detail with a line structure which was not too obtrusive. The American standard, established a few years later in 1941, provided for a 4·5 MHz video bandwidth with 525 lines. After the War, the French, wishing to have the highest-definition service in the world, chose 10 MHz and 819 lines.

Today, two basic standards are becoming established. British 405-line and French 819-line will be phased out, leaving the North American 525-line, 60-field, and the CCIR (European) 625-line, 50-field systems. Between these there must be interconvertibility, both in monochrome and colour, while 405 and 819 lines (monochrome only) will for some time need to be converted.

Converter techniques

Converters fulfil two functions: they store the input picture signal until the output picture is reconstructed, and they interpolate between adjacent line and field information. The purpose of interpolation is to avoid the disagreeable effects of simple repetition of lines in going up, for instance, from 525 to 625, when 50 lines from each field would require repetition. This would create a jerkiness in the rendering of movement which can be palliated by a redistribution of the 525-line information to spread it evenly into its new 625-line form. Storage and interpolation, which is an essentially mathematical process, are functions of computers, and it is therefore to be expected that converters, both of line-changing and field-changing type, should assume two quite different computer forms, analogue and digital.

The analogue converter smoothly and continuously builds up the input picture and continuously extracts the output at its new line and/or field frequency by a medium which provides both storage and interpolation. The earliest and simplest converters used film for storage, a mere defocusing of the telerecording spot sufficing to wipe out the input line structure. Dropping or repeating fields enabled the field frequency to be altered. The film was then re-scanned at the output standard by a telecine.

In the second generation of converters, film was replaced by the display tube phosphor. A long decay time gave the necessary storage, and interpolation of a crude kind was effected by spot wobble, an up-and-down movement of the scanning beam at high frequency which effaced the visible line structure.

Digital converters

The latest converters have gone over to the digital principle, because analogue types with their blurring and decay times have a low energy capacity, and therefore a relatively poor signal-to-noise ratio which brings with it picture degradation. Digital converters work by disassembling the input picture into lines, and if necessary picture elements, storing them in a delay device and then distributing the information into the output picture in a way which minimizes discontinuities. Because of interlace, adjacent lines in a complete picture are separated in time by a whole field period, 20 ms in the 625/50 system, and delays of this order necessitate the use of fused quartz lines in which ultrasonic waves are caused to travel.

Additional problems are presented by the colour signal, in which a colour subcarrier is present at different frequencies in 525-line NTSC and 625-line PAL. Colour conversion, however, is now an accomplished reality.

STANDARDS CONVERSION—Downwards conversion—Electronic converter—Image transfer converter—Spot wobble—Store element—Upwards conversion

TRANSMITTERS AND AERIALS

The final links between the TV networks, national or commercial broadcasting authorities, and the ultimate viewers consist of transmitters operating within internationally defined bands in the VHF and UHF regions of the radio spectrum, as explained in an earlier section. Since only 82 wideband station channels are available in the US at present, and 58 in Britain, many low-power stations, geographically separated as far as possible, are required to work on the same channel in order to provide an adequate signal in pockets of difficult reception. When freak transmission conditions prevail, as often in summer, this may lead to co-channel interference with consequent break-up of the received picture.

The problems of main stations are different. They must provide as large a service as possible, and consequently their aerials are sited as high as possible, preferably near the centre of large conurbations, though this may be difficult in practice because of danger to aircraft approaching or leaving airports up to several miles away. Transmitting aerials tend, therefore, to be displaced towards more remote high ground, where they are sometimes open to amenity objections and are more difficult to service and repair in the event of breakdown.

Siting considerations

The service area of a station differs according to the band it occupies. In the lower bands (the VHF region), the signal tends to diffract satisfactorily round large buildings and small hills and similar obstructions, whereas in the upper UHF regions propagation is much more by line of sight. Hills and buildings tend to cast shadows and to reflect the signal, resulting in multi-path transmission accompanied by unpleasant ghost images at the receiver. Siting of UHF transmitters is therefore a critical and difficult process.

The power to be radiated by a transmitter depends on many factors such as frequency band, terrain, gain and directionality (called directivity) of the aerial. In flat country, a few powerful transmitters are normal, in mountainous country, many low-powered ones. The effective radiated power (ERP) of the transmitter is defined as the power required to transmit an equivalent signal by means of a simple dipole aerial. Since actual transmitting aerials are usually multi-element arrays of high gain and high directivity, the ERP may be forty times as much as the actual output power of the transmitter.

Modulation methods

The function of the transmitter installation is to generate the sound and vision carrier frequencies allocated to the station, and modulate these with the sound and vision signals in their final programme form coming from the studio centre. In Bands I and III (VHF), both Western and Eastern European countries employ amplitude modulation (a.m.) for the video signal and frequency modulation (f.m.) for sound, except in the British 405-line and French 625 and 819-line systems which have a.m. sound. These exceptions also have positive vision modulation, which means that the greatest power is transmitted on peak white. All the rest have negative vision modulation, with greatest power at the farthest remove from peak white, i.e., on the sync pulses. The US and Japan, like the majority of Western and Eastern European countries, employ a.m. for vision and f.m. for sound, and modulate their video signals negatively.

The accuracy of the carrier frequency to one or two parts in a hundred million is maintained by a crystal-controlled oscillator working at a much lower frequency and followed by frequency multipliers. This drive unit is in turn followed by one or more stages of radio-frequency (r.f.) amplification to raise the power level to that required for the modulated stage. After modulation, a part of one sideband is removed by the vestigial sideband filter, and the vision and sound signals are combined in a diplexing filter which prevents cross-modulation between the two signals.

In UHF transmitters the very much higher carrier frequency may make it advantageous to carry out the final stage of amplification by devices developed originally for radar, such as klystrons and travelling-wave tubes. These have much larger stage gains, but lower overall electrical efficiencies. In modern transmitting equipment, all the stages except the final power stages are normally solid state (i.e., transistorized).

Aerial design

The design of the transmitting aerial (antenna) is complex and critical. Its radiation will be polarized in a vertical or horizontal plane, and receiver aerials must be similarly oriented. This affords considerable discrimination against distant signals on the same channel which are oppositely polarized. Since the aerial is mounted as high as possible, it is usually given an effective downward tilt of a degree or so (often by altering the phase of the signals fed to the upper and lower parts of the array) so as to cover the area in the vicinity of the mast. Directional aerials are now common, especially if a sector of the radio horizon at the transmitter is water or uninhabited mountain.

TRANSMITTERS AND AERIALS—Beam tilt—Cavity resonator—Circular waveguide—Class A amplifier—Class B amplifier—Class C amplifier—Co-channel interference—Combining filter—Combining unit—Diplexer—Dipole aerial—Downlead impedance—Drive unit—Effective radiated power—Feeder—Ferrite isolator—Filterplexer—Free space propagation—Fresnel zone—Grid modulation—

Ground wave transmission—High gain aerial—Horn reflector aerial—Lecher line—Lobe—Multipath signals—Overloading—Path attenuation—Polar diagram—Polarization—Quadrant aerial—Radiating element—Radiation resistance—Resonant aerial—Single echo—Subrefraction—Travelling wave amplifier—Tropospheric propagation—Turnstile aerial—Waveguide—Yagi aerial

RECEPTION

The television signal has been nursed with immense care through all these multitudinous stages until it is radiated into space at the transmitter. Now, in its few remaining steps it often suffers such mutilations that the home viewer would be amazed if he could compare his picture with that on the studio monitor. To the hazards of transmission, such as multi-path signals reflected from hills and steel buildings, are added inevitable losses and distortions in receivers built to a competitive price as well as picture faults resulting from maladjustment and lack of maintenance. Finally, the TV set itself is usually viewed under the worst possible conditions, with strong ambient light striking the tube face and degrading the picture contrast. It is a tribute to the power of television as a medium and to the skill of the engineers responsible for programmes that it survives handicaps which would cripple its rival the motion picture.

MONOCHROME RECEIVERS

The receiver installation starts at the aerial. In theory, separate aerials should be provided for each channel, but where the signal strength is high a single aerial suffices to cover each VHF band, with two aerials for the UHF band if channels at both ends are receivable. Coaxial or flat twin-wire cables feed the signal to the receiver, and if this downlead is a long one, booster amplifiers may be needed at the mast head or the set. A high signal-to-noise ratio is essential, and the aerial must be carefully located and oriented to maximize the wanted signal and reject as far as possible ghost or interfering signals. Directional arrays are now available, especially for the UHF band, with high inherent gains.

Superhet principle

In a monochrome receiver, the signal is first r.f. amplified, then passes into a superheterodyne mixer. The superhet principle is a device for converting different incoming frequencies (in this case the different channel frequencies) to a fixed frequency called the intermediate frequency (i.f.). This is achieved by mixing the signal with the output of a tuned oscillator to give a constant beat (or difference) frequency with the incoming signal. The advantage of fixed frequency i.f. amplifiers is that they require no tuning and can easily be designed for high gain. When sound is frequency-modulated (vision being always a.m.) the video and audio signals may be amplified together at i.f. and then passed to separate detectors which demodulate them. The f.m. sound passes through circuits common to f.m. radio sets and is fed to the loudspeaker.

After demodulation, the video signal is amplified further so as to be able to drive the picture tube (kinescope), and a second output from the video amplifier is taken to a sync pulse separator which strips off the video signal and separates the field and line sync pulses.

Tube design

The picture tube, in spite of great research efforts to simplify it radically, is still a cathode-ray tube of basically conventional design, though greatly refined and with much larger beam deflection angles than formerly, so that a large screen can be scanned by a short beam in a compactly short tube. Modern scanning circuits and glass-blowing techniques have made deflection angles of 110 degrees possible. On the inside of the tube faceplate is deposited a layer of light-emitting phosphor, backed by an aluminized layer which is permeable by the electron beam but stops the larger ions which would otherwise damage the phosphor. It also increases picture brightness.

Since the atmospheric pressure on an evacuated 19 in. tube is about 1½ tons, there is a serious risk of implosion (followed by explosion of glass fragments) if the tube is fractured. This was formerly prevented by a sheet of armoured glass placed between tube and viewer, which had several drawbacks; it has now been replaced by a tough glass fibre skin bonded to the faceplate.

Electron gun

The electrons which bombard the phosphor screen are generated by a gun which is an extension of the principle of an ordinary valve (vacuum tube) but with a much higher anode (plate) voltage of about 10,000. The anode, set in the neck of the tube, has a central hole through which some of the electrons from the heated cathode pass, and these form the scanning beam. They are further projected towards the phosphor screen by the fact that the inner surface of the tube flare is coated with graphite and connected to the anode. The phosphor screen itself also takes up the anode voltage. To produce the smallest possible spot, focusing magnets ring the neck of the tube.

Scanning generators

It then remains to scan the picture raster and modulate the electron beam with the picture signal. To produce the raster, two scanning waveform generators (often called time-bases from their use in oscilloscopes) are needed, one for the line scan and a much slower one for the field scan. Both waveforms are of sawtooth shape; a linear sloping rise to a peak followed by an abrupt fall to zero representing flyback. These generators are triggered by the line and field sync pulses already separated in the sync pulse separator. This aspect of set design is a critical one, and much ingenuity has been expended on it.

Scanning coils

The waveforms thus generated and synchronized are applied to line and field scanning coils of curious and complex shape which encircle the neck of the picture tube near the point where it flares out. Thus when the tube is switched on, the focused beam scans the phosphor screen in a rectangular raster precisely synchronized with that generated at the transmitter. This beam is modulated by the output of the video amplifier which, as in normal radio practice, could be applied to the grid of the display tube but in fact is applied to the cathode, while the grid has a variable d.c. voltage fed to it to control picture brightness.

This in briefest outline is the method by which a sequential signal is synthesized at the receiver into a reproduction of the transmitted picture; all the main features of it apply equally to a colour receiver.

PHOSPHORS FOR TV TUBES—RECEIVER—VIEWING TELEVISION—VISUAL PRINCIPLES—Afterglow—AGC detector—Automatic frequency control—Brightness control—Chromatron tube—Cogging—Contrast control—Crosstalk—Deflection coil—Discriminator—Dual standard receiver—Dynamic focusing—EHT rectifier—Electrostatic focusing elcetrode—Energy recovery diode—Field tilt—Flashover—Flywheel circuit—Focus coil—Focus modulation—Frequency changer—Halation—Halo effect—Hook—Hum bar—Hum displacement—IF amplifier—Image diagonal—Intercarrier sound—Ion burn—Lag—Line tilt—Line time base—Magnetic deflection—Magnetic focusing—Mixer—Pentode electron gun—Phosphor grain—Phosphur saturation—Phosphor screen—Picture tube—RF amplifier—Scanning coil—Scanning coil assembly—Scan protection—Screen saturation—Sound on vision—Superheterodyne—Synchronizing separator—Toroidal coil—Tuner—Turret tuner—Unrestored receiver—Vision drive signal

COLOUR RECEIVERS

The principles of a compatible colour system were explained earlier, together with the means of sorting luminance from chrominance information, and thereafter at the receiver further separating the two colour-difference signals by synchronous demodulators.

Up to the video detector, therefore, colour receiver circuits are substantially the same as those for monochrome. The scanning circuits are also very similar, though they require a larger power output. It is in its separate luminance and chrominance channels, its decoder, convergence circuits and display tube that the colour receiver requires different components and circuitry.

Shadow mask tube

Almost all current colour receivers use the shadow-mask tube developed by RCA. This extremely complex device employs separate R, G and B electron guns mounted parallel and as close together as possible, which converge their beams on to a metal plate about $\frac{3}{4}$ in. behind the phosphor screen and parallel with it. In the 21 in. tube this plate is perforated with about a third of a million circular holes, and registered in front of each hole is a group of three phosphor dots, red, green and blue, called a triad, so that there are about a million separate dots in all. The shadow mask is so called because, looking along the red electron beam, it would only be possible to see through each hole a red dot, the other dots in the triad being obscured or shadowed by the edge of the hole; and so with the green and blue beams. However, though the electron beams are focused as sharply as possible, they cover several holes at once, and there is a large electron loss by absorption at the metal plate. Hence, the tube is only about 20 per cent efficient, and a very high anode voltage is required to energize phosphors which are themselves of low efficiency, since they are required to produce light of the correct R, G and B spectral response. In fact, the luminous efficiency of the whole system is tied down to the least sensitive phosphor, which is the red one.

The three electron beams, operating together, sweep a raster exactly as in the monochrome tube; and since each is modulated with its appropriate colour signal, it excites a dot (or rather a small group of dots) of the same colour. Thus the triads produce a three-colour additive synthesis as the three beams sweep over them, and a colour picture results. Though the method of operation is thus very simple, the shadow-mask tube is a triumph of design and manufacturing ingenuity, and its complex precision assembly makes it a costly unit to produce. Much research has been directed to the design of a single-gun colour tube of greater simplicity, but none has so far emerged as a satisfactory commercial proposition.

One of the difficulties of the three-gun tube is that the convergence does not maintain itself automatically as the beams sweep from side to side; and as the point of convergence alters so the three rasters get out of register.

Consequently, associated with the scanning circuits, there is need for the dynamic convergence circuits and coils mentioned above.

NTSC system

It remains to say something of the method of separating the luminance and chrominance channels and decoding them to drive the three guns in the shadow-mask tube. It is here that the differences between NTSC, PAL and SECAM receivers occur, reflecting corresponding differences of matrixing at the transmitter. In the most general terms it may be said that in NTSC receivers the gamma-corrected signal, after amplification and filtering out of the colour subcarrier, goes directly to the matrix, the network device in which signals are added and subtracted. The chrominance signal after amplification feeds the two synchronous demodulators which, as explained in the earlier section, separate the two colour-difference signals and pass them to the matrix. In the matrix, suitable addition and subtraction results in the appearance of gamma-corrected R, G and B signals which are then applied to the three guns of the display tube. This process is called decoding.

The burst of subcarrier frequency known as the colour burst which occupies the back porch of each line sync pulse is applied to a special oscillator, often of crystal-controlled type, and this very accurately maintains the phase of the two synchronous demodulators essential for correct separation of the colour-difference signals.

PAL system

The PAL receiver does not differ greatly from NTSC, but in its more advanced forms requires a delay line by which successive picture lines of alternating phase are compared and averaged. To do this it is necessary to store a line for about 64 microseconds, the time taken for a complete line scan in the CCIR system. This is difficult to do electrically, so the chrominance signal is applied to a crystal transducer resembling those used in crystal pick-ups, whose ultrasonic output is applied to a glass bar about 8 in. long. An output transducer similar to the input one reconverts the signal, which as well as being delayed, is attenuated by about 20 dB.

SECAM system

The SECAM transmission system, adopted by France and the USSR, though again basically similar, differs in important respects from NTSC and PAL, and this is necessarily reflected in receiver design. The two colour-difference signals are not transmitted simultaneously, along with the luminance signal, but sequentially, one on one line, the other on the succeeding line, this cycle then repeating. Furthermore, the colour subcarrier is frequency modulated; the frequency is altered (deviated) by an amount proportional to the amplitude of the colour-difference signal.

At the receiver a delay line is required, but its function is more fundamental than the PAL delay line. The chrominance signal from one line must be made available while the chrominance signal for the succeeding line is being received, in order that the two colour-difference signals may be derived simultaneously. One of these relates to the contemporary luminance signal, but the other relates to luminance information supplied on the previous line.

The colour-difference signals are recovered by frequency discriminators, no synchronous demodulators being required, since f.m. replaces suppressed-carrier a.m. The combination of the colour-difference and luminance signals in a matrix which yields R, G and B outputs to drive the display tube follows NTSC and PAL practice.

COLOUR DISPLAY TUBES—RECEIVER—SHADOW MASK TUBE—Apple tube—Banana tube—Colour killer—Convergence circuits—Decoder—Delay line—Dichroic reflector—Electron gun—Grid tube—Index tube—Lawrence tube—Matrix—Triad—Trinoscope—Ultrasonic delay lines—Writing beam—Zebra tube

LARGE SCREEN TELEVISION

In the earliest days of low-definition TV, it was the dream of inventors to put a televised picture on the large cinema screen. Forty years later this is no longer difficult to do, but, for commercial rather than technical reasons, it is seldom attempted. However, large-screen TV is extremely useful for providing live backgrounds in studio production, for closed-circuit projection to big audiences, and for occasional transmissions of major sporting events to motion picture theatres.

Projection tube

The normal receiver display tube cannot for obvious practical reasons be made in sizes as large as a cinema screen. The simplest solution, therefore, is to cause the tube phosphor to emit enough light to act as a scanning projection beam. With the aid of extremely high anode voltages and beam currents, together with very wide-aperture lens systems of the reflecting Schmidt type with aspherical surfaces, this can in fact be done, and with three such systems colour projection is possible. Direct projection tubes of this type are less popular than they were a decade ago. The wide-angle picture beam makes it necessary to mount the projector in the audience space inconveniently close to the screen; and in spite of all possible efforts to increase luminous efficiency, the screen brightness tends to be inadequate, especially on very large screens and in colour.

Eidophor

A more widely used device, the Eidophor, operates on an entirely different principle. A Swiss invention of great ingenuity, it remained under development for many years before it was commercialized. The root principle of the Eidophor is the complete separation of the light source from the picture generator. The light source is a xenon arc lamp of a type often used in cinema projectors. The beam from this lamp is modulated by a scanned picture built up by an electron beam in the usual way.

The heart of the Eidophor is the modulator. If an electron beam impinges on a thin layer of an insulating oil, the surface of the liquid is deformed from point to point by electrostatic charges. As the beam is deflected at line and field frequencies over the surface of the oil, a pattern of optical deformations is built up which corresponds to the point densities of the picture. This pattern in microscopic indentations cannot be directly projected like an image on film; but as the varying densities of the film modulate the light beam in a conventional projector, so the Eidophor image is made to modulate the light beam by a special optical system based on Schlieren principles. For colour projection, the beam from the arc lamp is split into R, G and B beams by reflection from dichroic mirrors, and these three separate colour beams are modulated by three standard Eidophors, each fed by its appropriate colour signal, and mounted in a single cabinet. Finally, the three picture-modulated light beams are recombined by accurate convergence on to the screen. Correction circuits for keystone, skew and focus errors are needed to ensure that all parts of the picture are in perfect register.

LARGE SCREEN TELEVISION—SCREENS—Eidophor—Keystone—Projection tube television—Schlieren photography—Skew correction—Voltage doubler rectifier

CLOSED CIRCUIT TELEVISION

This term covers all TV systems whose signals cannot be picked up on ordinary domestic receivers. Many of these systems contain microwave or other links which stretch the notion of a closed system, since access is then theoretically possible, but the term is a convenient one and its general meaning well accepted. In principle, closed-circuit systems need not adhere to accepted TV broadcast standards, and high resolution channels with wider passbands and higher numbers of lines have been developed for space and missile research. The majority of closed circuit systems at present conform to commercial standards if only because the complex electronic devices used in any TV channel would be prohibitively expensive if manufactured in small numbers.

EQUIPMENT

Closed-circuit equipment differs little from station equipment, except that as a rule it is simpler and more portable. Vidicon channels are all but universal, though Plumbicons are becoming increasingly popular. Cameras (e.g., for traffic control and store supervision) are often unmanned, and may be remotely controlled for pan and tilt as well as lens iris adjustment. Mixing and other programme switching facilities are either non-existent or very simple, but educational TV is a notable exception to this rule. However, with its interchangeable programme material, education is better considered as an offshoot of institutional TV than a form of closed circuit.

APPLICATION

Closed-circuit TV finds wide application in entertainment, especially in hotels, passenger ships and airliners. Programme material is either taped or originated in small studios, or both. As a warning system in power stations, the steel industry, on rocket sites and elsewhere, it eliminates the human being from points of danger, and has one inherently safe characteristic: if the picture goes off the screen, the cause may be an event at the camera site, and not merely a fault in the system.

As an information source, CCTV is vital in air traffic control centres at airports, in city traffic management, in document transmission, in stock exchange operations and the like. As a medium of instruction, it plays an important role in medicine (especially in the operating theatre), in training simulators for pilots, and in the dissemination of lecture and research material in education.

CLOSED CIRCUIT SYSTEMS—MEDICAL CINEMATOGRAPHY AND TELEVISION

FILM AND TELEVISION COMPARED

AUDIENCE SIZE

Though there are major areas of interaction between film and television, there are also areas where one or the other excels. In the presentation of large-screen colour pictures to audiences of a thousand or more, film has little competition. In particular, 35mm and 70mm film have an information storage capacity greatly exceeding that of broadcast TV systems, and the absence of line structure makes the large image very much pleasanter to view at close range. The development costs of a wideband 1,000-line or 2,000-line system with very much more powerful Eidophor projection would not be expected to yield a profit from the declining revenues of the motion picture theatre. Photographic film with its low cost and universal standardization should hold its own in this field for many years.

Equally, the position of television in home entertainment is unassailable and likely to become stronger. Home movies have enjoyed a long spell of popularity in spite of the drawbacks of separate projector and screen and the inconvenience of a darkened room. It is a strong temptation to use the standard TV receiver as a display mechanism for canned as well as off-the-air programmes, and devices for doing this by photographic means as well as by videotape, in each case with electronic (scanned) read-out, are today well on the way to commercialization.

PRODUCTION Film alone can be used for motion pictures to be shown in cinemas since the TV and film images are not interchangeable, and an image from which information has perforce been omitted can never be upgraded. In television production, film and videotape are strong competitors, with assets and drawbacks which have already been mentioned, but the trend is increasingly toward videotape with its advantages of instant playback and erasability. Electronic editing is becoming more convenient and efficient, and the perfecting of standards converters has gone far to overcome the handicap of lack of uniformity in the world's television standards, to which videotape recorders are necessarily tied.

On the other hand, where extreme mobility is important and instant transmission not required, 16mm equipment (and in future Super-8) has the advantage in size and weight over even the most miniaturized of portable TV camera channels. Rapid film processors working at higher than normal temperatures have greatly cut down the time for developing both colour and black-and-white film. If the short processing delay can be accepted, film can undoubtedly be edited more sensitively and conveniently than videotape, even when this is handled with the latest available equipment.

For studio productions there has been a tendency of uncertain momentum to combine film and television techniques. A reflex camera shutter alternates the image to film and to a TV monitor, and a remote programme director can start and stop several cameras, feeding a cut and mixed electronic output to videotape, while retaining a film record for more careful and precise editing later on.

SOUND Sound techniques are essentially common to both media. The prolonged multi-camera recordings of the TV studio put a premium on mobility and extreme precision of microphone movement, while crowd and interview recordings have encouraged the design of highly directional microphones. Film studios excel in meticulous microphone placement and in the editing and mixing of very elaborate multiple sound tracks. The development of studio lighting techniques in the two media has followed a parallel course.

In the transmission of TV sound, pulse code modulation (PCM) may now be used, as in satellite relays. The signal is sampled at twice the highest frequency to be transmitted, or more, and the amplitude at each point is represented by a binary code. This pulse modulation is multiplexed on to unused portions of the TV signal, such as the back porch.

FUTURE TRENDS

Predictions of trends in the two technologies must take many factors into account, among them the resources available for investment, the inherent developability of the underlying sciences, and the wayward variation of popular demand. Sheer individual inventiveness, lighting unexpectedly on the solution to a seemingly insoluble problem, is still a factor which cannot be ignored.

FILM

Film technology is based largely on organic chemistry and on optics and mechanical science. Progress in fundamental research on the photographic image has always been difficult and slow; for instance, the speed of commercial emulsions doubles in periods usually not less than ten years when account is taken of all the other characteristics in which no deterioration is permissible. Hence, slow, steady progress is to be expected here, with colour emulsions virtually eliminating black-and-white, and the processing laboratory coming to be regarded as a central plant for many types of image modification and transfer, including videotape to film.

With Super-8 firmly established, the standard 8mm gauge will follow 9·5mm into oblivion as low-priced consumer equipment is progressively scrapped, leaving 65/70, 35 and 16mm gauges for professional use. The changeover to Super-8 is encouraging the growth of multi-rank printing facilities, with magnetic and optical sound tracks competing for preference.

In the field of camera design, silent cameras can be expected to oust externally-silenced models, and the improvements already achieved in 16mm will be extended to 35mm. In the laboratory, colour film duplication will be simplified by new stocks designed for reversal processing so that dupe negatives can be prepared from original negatives in a single stage.

MIXED TECHNIQUES

The prospect in film is thus of substantial but relatively slow advance. In television, however, the pace is likely to be much faster, if only because electronics has behind it the impulsion of military and space research with their immense budgets. Film techniques should prove indirect beneficiaries, as more and more interfaces between film and television are developed. Film camera viewfinding using vidicons to give the cameraman a large parallax-free picture can also furnish pictures to remote monitors where they can be seen by the director, producer and lighting cameraman. Cineradiography has already gained from the electronic image intensifier, faster in its development than emulsions and high-aperture lenses. Techniques are now well advanced for exposing motion picture film directly by electron beams in vacuo, promising to improve and simplify telerecording (kinescoping). Flying-spot scanners of greatly reduced cost may make it possible in education, and even in the home, to transfer such an electronically recorded film for display to the input of a TV receiver. In this way, film may resist the increasing inroads of videotape recording.

TELEVISION

Spectacular as is the progress of electronics, the next decade may hold few complete surprises in television, where already the main goals have been attained and design is becoming stabilized. When equipment has reached a high state of refinement, the remaining problems are often extremely difficult to solve. Even the retirement of the comple andx already elderly image orthicon tube is not at all certain.

Among the nearer advances, PCM may be extended to providing telex, telepicture, or other information services as part of the television signal receivable in any home equipped with a small adjunct to the standard set. It is also likely that wired receivers will be fitted with devices for initiating signals which can be multiplexed back to a central station where, for example, answers to problems set on educational programmes are instantly processed by computer and individually replied to. Such two-way exchanges of information, and the coupling of TV sets to a variety of audio-visual aids, are among the striking gains which mass-produced micro-electronics will bring.

The faster spread of colour TV in countries less wealthy than the US is primarily an economic problem, pointing to replacement of the expensive shadow mask tube which alone today is able to give pictures of adequate quality. Simpler single-gun tubes, whether of the chromatron or beam-indexing type, require further large-scale investment if their problems are to be successfully surmounted after many years of indecisive development.

Looking farther ahead, it is possible that photo-sensitive solid state devices may become available for cameras, in which the picture signal is derived without a scanning beam by matrix-type connections to the third of a million sensitive picture elements required in monochrome, and an equal number of triads in colour. Were such a mosaic to be commercialized, it would almost certainly lend itself to making a flat display device, thus realizing the dream of a picture-frame tube.

The time-scale for such radical inventions cannot be predicted, but the next decade should see striking progress in signal processing, where purely electronic techniques have great possibilities. The principle of digitalizing information which is at the heart of the modern colour standards converter is applicable in other fields. It has the merit of noise immunity, guaranteeing a durable signal under difficult transmission conditions. Its convenience as a multiplexing device has already been mentioned in connection with PCM.

Signal processing of this and other kinds may be used not merely to arrest but to reverse the degradation of picture quality which occurs in the long television transmission chain with its cameras, videotape recorders, film scanners, special effects generators, microwave links, satellites and landlines.

For the highest quality of picture transmission and reproduction, especially in terms of information content, film will long remain unrivalled; but for flexibility and versatility, television has a decisive lead. In the final analysis, however, the two media are inseparably linked; each is impoverished if divorced from the other.

* * *

Index

Figures in bold face refer to page numbers of entry titles in Encyclopedia. Figures in *italic* refer to pages where illustrations occur. Key references only are given to subjects which appear repeatedly throughout the book, such as cathode-ray tube and 35mm film.

Left- and right-hand columns are referred to thus: 787/1 & 787/2.

A and B roll assembly, 487
A and B rolls, 1/1, 377/1, 495/1, 498/2, 576/1
A and B winding, **1**, *492*
A, B and C printing, 1/2, 116/1
A-B-C-D cut mixer, 958
A-B cut, knob-a-channel mixer, 960
A-B cut mixer, 958
Aberration, **2**, 50/1, 157/1, 274/1, 393/1, 396/2, 753/1, 820/2
Aberrational correction, 391/1
Abrasion, **2**, 128/1, 306/2, 309/1, 427/2, 428/2, 619/2, 978/1
Absorbers, 6/2
A and B printing, **1**, 53/1, 116/1
Absorption
(air), 8, *9*
audience, 5/2, table 7
coefficient, 4/2, *7*, *8*, table 7
light, 183/2
low frequency, 8/1
resonant, *8*
sound, 6, 720
spectrum, 752/2
studio, 8/2
units, 4/2
Academy
aperture, **2**, 71/2, 282/1, 471/2
leader, 2/1, 384/1, 622 ff
standards, **2**
Academy of Motion Picture Arts & Sciences (US), 2/1, 384/1
A.C. bias, 718
Acceptance angle, **2**, 268/2 ff, 606/2
Acceptance-test method, **2**, 895/1
Accession, library, 402
Acetate, **2**, 428/2
film, storage, *292*, 398/1
(*see also* Safety base)
Achromatic
colours, 181
doublets, 392/1
lens, **2**
Acids, 117, 119/1, 121/2
Acme peg, 23/1
Acoustic
backing, **2**
feedback, **2**
materials, 7 ff., 141/1
Acoustics, **3**
cinema, 141
room, 680/2
studio, 800
television, 708
Acres, B, **10**, 124/2, 319/1, 324/2, 539/2
Action, **10**
Action props, 45/2
Actors, dealing with, 232
Acuity (*see* Acutance)
Acutance, **10**, 347/2, 390
rating 11/1
SMT, 11/1
Adaptation, eye, 970
Add-A-Vision, **11**
Additive
colour synthesis, 188, 669/2
mixing, 184/2, 185/2, 190
photographic exposure, 42/1
printing, **11**, 195/1, *361*, 575/1
process, **11**, 172, 245/1, 396/1
Ad hoc group on colour television, 198/2
Administration, 233/1, 235/1
Advance (sound & picture), **12**
Advertising films, 38/1
Aeo-light, 114/1, 320/1
Aerial, 907
assemblies, microwave, 467
batswing, 916/2
dipole, 12/1, **229**, 908/1, 916/2
gain, **12**
horn-reflector, 338/2
mast, *902*
omnidirectional, 12/1
parabolic, 466/1
power gain, 892/1
radio relay, 898
receiver, 12/1, 228/2
signal distribution, *907*
slot, 916/2
standard, 12/1
system (for satellites), 639
towers, 529
transmitter, 12/1
turnstile, 916/2
Yagi, 989/2
Aerial image, 274/1, 455/2
photography, **12**, *730*

A

Aerial image
printing, 730
projector, *24*
Aeroscope camera, 125/1, 319/2, 599/1
AFNOR standards, **12**, 961/1 ff.
Afterglow, **12**, 765/1, 848
After-image, 968/2
AGC (*see* Automatic gain control)
AGC amplifier, 611
detector, **12**
Agfa Filmfabrik, 176/2, 177, 356/1, 477/1
Agfacolor, **12**, 125/2, 176, 304/2, 322/1, 356/2 ff., 398/1, 722/2
Agfa-Gevaert, 12/2
Agitation, **12**, 114/1, 118/1, 121/2, 122/2 229/2, 359, 507/2, 584, 916/1
Air absorption, 8, *9*
Air-glass interface, 165/1
Air knife, **12**, *586*, 756/2
Air-squeegee, 13/1
Air traffic control, 161
Alexanderson, E. F. W., 330/1
Alexandra Palace, 61/2
Alkali, 116, 118/2
Alvey, G., 323/1
Amateur cinematography, 276/1, 284/1, 324, 370/2, 474 ff., 485/2, 507/1
Ambient light, **13**, 63/1, 652/2, 843/1, 951/1
American Academy of Motion Picture Arts & Sciences, 622/1
American Broadcasting Company, 93/1
American Mutoscope & Biograph Company, 382/2
American Standards Association, 50/1, 922/2
Ampex Corp., 93/1, 929/1, 943/1
Amplifier, **13**
Amplitude
corrector, 471/1
distortion, 686
modulation, 197, *470*, 511/2, *888*
non-linearity, 978/2
sound, 678
Amplitude-frequency response, 2/2
Amplitude-modulated sound transmitters, 906
Anaglyph, 125, 245/1, 324/ 776/2
Anaglyphic process, **13**, 124/2
Analogue computers, 206
Analytical density, 664/2
Anamorphic
lens, 125/2, 217/1, 340/2, *395*, 980, *982*
photography and projection, *981*
printing, 803/1
process, **13**, 132/1, 204/1, 281/2, 756/2, 816
projection, 979/1 ff.
Anastigmat, 50/2
Andersen, A. C. and L. S., 127/1, 334/2
Angénieux, 94/2, 96/1, 129/2
Angle
beam tilt, 57/2
camera, **13**
of deviation, 519/1
of refraction, 409/1
of view, 71/2, *82*, 86, 89/1, 136/1
shooting, 45/1, 48/1
Angstrom, 410/1, 424/2
Angular field, 392 ff.
Animal Form, 34/1
Animal Locomotion, 473/2
Animascope, **13**
Animated
captions, **14**, 15, 16
films, 214/1, 241/1, 771/2, 784/2
photographs, 317/2
viewer 486/2, 498/1
Animating movement, *34*
Animation, 13/2, *15*, *16*, 166/2, 205/1, 354
cameras, 75
colour, 206/2
desk, *33*
early experiments, 316
equipment, **19**
history, 25, 325
photography, 730/1
production, *32*
stand, *20*
techniques, **25**, 561/1, 783/2
Animatographe, 125/1, 319/1, 539/2
Animator, 26/2, 30/1, 32/2, 33, 34/1
Animographe, 38
Announcer, studio, 443/1
Anschütz, O., **39**, 317/2
Anscochrome, **39**
Anscocolor, **39**
Answer print, **40**, 259/2, 375/2 ff., 693
Antenna (*see* Aerial)
Anti-halation coating, **40**, 284/2, 306/2, 308/1
Antinode, 7/1, 680/1
Anti-reflection coating, **40**
Aperture, 211/2
camera, 2/1, 41/1
correction, **40**, 83/1, 96/2, *98*, 820, 854/2
designations, 281
disc, 328/1
fixed, 783/1
full, 71/2, 673/2
lens, **40**, 42/1, 89/1, 132, 297/1, 305/1, 365/1, 392/2, 499/2, 524
mask (projector), 632/1
plate, **40**
printing, 570/2
projector, 41/1, 300/1
sizes, **41**, 301/2
unit, 303/1
Apex system, **42**
Apostilb, **42**, 425/1, 783/1
Apple Tube, **42**
Archival storage, 291 ff., 405/2, 600/1, 924/1
Arc lamps, **42**, 63/2, 414 ff.
Arctic communication (ARCOM), 640/2
Argon flash bomb, 315/1
Armat, T., **42**, 319/1
Armstrong, E. H., 127/1
Arriflex camera, 68/2, 261/2
threading path, *69*
Art director, **43**
Artificial lighting (news), 503
Artificial light (Type A) film, 483/2
ASA speed rating, **50**, 265/2, 270/1, 346/1
ASA standards, **50**, 758/1 ff.
ASLIB
film libraries classification committee, 404/2
film libraries group, 397/1
Rules for Cataloguing Film, 404/1
Aspect ratio, 41, **50**, *134*, 135/1, 204/1, 211/2, 249/2, 603/2, 979/1 ff., 982
automated control, 52
television, 614
Asperity, tape, 431/2
Aspheric corrector plate, **50**, 378/1
Assembly of news film, 503
Assistant director, 233, 236/1
Assistant floor manager, 236/1
Association Française de Normalisation, 12/2
Association of Special Libraries & Information Bureaux, 397/1
Astigmatism, **50**, 88/2, 93/2, 391/2, 522, *523*, *833*
Attack
time, **50**, 205/2
tone, 682/2
Attenuation, 205/2, 229/1, 263/2, 273/2, 294/1, 425/1, 685, 693/2, 925/2
Attenuator, **50**
Audibility, 680
Audience
seating (studio), 799
sight lines, 139
Audio
amplifier, 613
recording, 431
spectrum, 300/2
signal, 363/1
spectrum, **50**
tapes, 437
(*see also* Sound)
Audio-follow-video, 446/2
Audioscopics, 125/2, 324/1
Audio-visual
projectors, 597/1
screens, 605/1
units, 607/2
Auditorium, **50**, 135 ff., 652/2 ff., 693/2, *720*
Auditory fatigue, 706/2
Audlock, 551/2
Aural sensitivity, 680
network, 509/2
Auricon camera, 304/1
Autant-Lara, C., 123/1, 979/2
Automated
animation, 38/2
editing, videotape, 926
lighting, 52
master switching, 447 ff.
Autochrome plate, 429/1
Auto-masking, **50**, 121
Automatic
cel cycling, 22/2
changeover, 51/1
colour masking, 178, 251/2
compression, 709/1
diaphragm systems, *482*
dissolve 21/1
exposure 22/2, 72/1, 75/2, 267/2 ff.
exposure meter, 483/2 ff.
fade 21/1
field-phasing, 551/2

Automatic
  flare correction, 99/1
  focus, 22, 72/1
  frequency control, **51**, 467/1
  gain control, 155/2, 609/1, 613/2, 618
  inspection, 622/1
  light control, **51**
  line-phasing 551/2 ff.
  plotting (densitometry), 667
  projection, 144/2
  recording densitometer, 666/1, 667/2
  rewind, 208/1
  slating, 482/1
  standards recognition, 611/1
  temperature control, 84/2
  threading (projector), 597/2
  transfer, 450/2
  voltage regulator, **51**
  X-ray processor, 154/1
Automation in cinema theatres, **51**
Auto-optical, **53**, 816/2
Auto-preview, 445/2 ff.
Autotransformer, **53**
A-winding, 485/2, 495/2
Axial mode, 6
Ayrton, W. E., **53**, 126/2
Azimuth error, **53**

Back focal distance, 394/1
Background
  artist, 35
  blue, 61/1
  effects, 714/1
  mobile, 35/1, 36/2, 49/2
  projectors, 595/2
  scene, 301/2, 354, 561/1, 730/2 ff., 735/1 ff., 910/1
  sound, 692/2
  static, 28/1
Backing removal, **54**, 359/2
Backing view, 47/1
Back
  light, **54**, 422/2
  porch, **54**, 515/1
  projection, 48/2, 49/2, **54**, 377/1, 561/1, 732/1
    plates, 573/1
    screen, 887/2
Bacon, Francis, 714/2
Baffle, **55**
Bain, A., **55**, 126/1, 326/2
Baird, J. L., **55**, 191/1, 329/2, *330*, 332/2, 334/2, 865/1
Bakewell, F., 126/1, 326/2
Balance, colour, 182/1
Balanced modulators, 512/1
Balanced pair, **56**, 158/2, 895/2
Balancing sound levels, 708
Balancing stripe, **56**
Ballard, R. C., 331/2
Ballistics-synchro, 313/2
Banana tube, **56**
Band, **56**, 115/2, 228/2, 900/2
  limits, 56
  sharing, 197/1, 511/1
Bandpass, 924/2
Bandwidth, 6/2, 191/2, 193/1, 196, 392/1
  colour, 85
  holography, 338
  limitations, 389
  reduction of, 674/2
  transient response, 614
  videotape, 930/1
Bank, 56
  lighting, 413/2
Barndoor, **56**
Barn-door wipe, **56**
Barney, **56**
Baron, A., 125/1
Barrel distortion, **57**, 523
Base, **57**, 306/2
  distortion, 988/2
  film, 167/2, 284 ff.
  magnetic tape, 432
Base band, **57**
Base-making plant, *286*
Base Types (lamps), 413
Basher, **57**
Batch, **57**, 290/1
Batch processing furnace, magnetic oxide, *433*
Bath, **57**
Batteries, 73/2
Battery belt, *167*
  drive, 501/2
Bayonet mount, 71/1
BBC (*see* British Broadcasting Corporation)
Beaded screen, **57**, 301/2, 378/2
Beam current, **57**, 79/2
Beam landing errors, **57**
Beam scanning & readout methods (thermoplatic), 867
Beamsplitting
  camera, 174, 354/2
  prism, 89/1, 95/1, 731/1
  system, **57**, 85/1, 86/1, 88, 153/1, 226/2, 304/1, 394/1, 455/2, 457/2, 459/2, 779/2
  viewfinder, 479/2
Beam tilt, **57**
Beater mechanism, **57**, 217/2, 366/2
Beat pattern, 470/1
Becquerel, A. E., **58**, 126/1, 326/1
Bedford, A. V., 127/2
BEFLIX, 206
Bell, A. G., 253/1
Bell & Howell additive printer, *196*
  Company, 339/1
  mechanism, **58**
  perforation, **58**, 276/2, 499/2
  registration, 70/2
Bell Telephone Laboratories, 39/2, 206/2, 320/1, 330, 365/2, 382/2
Berglund, R., 126/2
Berliner, E., 715/1
Berthier, A., 324/1
Berthon, R., 172/2, 173/1
Berzelius, J. J. B., **58**
Bessel functions, 890/2
Bias, **58**
  audiotape, 437/1
Biconcave lens, 392/12
Bidwell, S., **58**, 126/2, *327*
Bilateral sound track, **58**, 245/2
Bin, **59**
Binary
  counter, 805, *806*, *809*
  digit, 349/1
  method, 269/1
Binaural recording and reproduction, **59**
Binocular vision, 776, 973/1
Biograph, 125/1, 323/2, 369/1
Biograph Company, 43/2, 227/1, 369/1
Biokam, 125/1, 324/2
Bioskop, 675/2
Biotar, 393/2
Bipack, **59**, 62/2, 128/2, 728
Bi-plane cineradiography, 154/2
Bipost lamp, **59**
Bi-refringence, 369/2
Birtac camera-projector, 10/1, 125/1, 324/2
Bistable circuit, *805*
Bit, 349/1
Black bar, **59**
Black body, 199/1, 410/2 ff.
Black crushing, **59**
Blackening (lamp bulbs), 413/2
Blacker-than-black, **59**
Black level, **59**, 81/1, 157/1, 426/2, 541/1, 557, 861/2, 884/2
Black level clamp, **59**
Black spotting, 61/2
Black stretch, **60**
Black and White
  derivations
    16mm negative, *247*, *494*
    35mm negative, *248*
    16mm reversal, *247*
  development, 122
Blackton, S., 166/2
Blank, **60**, 352, 430/1
Blanking, **60**
  period, *642/1*
  pulse, **60**
  shutter, **60**
  signal, **60**
  signals, mixed, 808/2
  stage, 84
Blattner, L., 125/2, 320/2
Bleach, **60**
Bleached whites, 64/1
Bleaching, 117
Bleep, 482/1
Blimp, **60**, 74/2, 481/2, 485/1
Bloch's law, 971/1
Blocking oscillator, 807
Blocking the show, 238
Bloom, **60**, 165/2, 877/1
Bloop, **60**
Blooping machine, **61**
Blow-off, 13/1
Blow-up, **61**, 249/1
Blue, **61**, 246/1
Blue-back shot, **61**
Blue screen process, 13/2, **61**, 179/1
Blumlein, A. D., **61**, 63/2, 127/2, 332/2
Blur, image, 990/1
Boom, **61**, 696, 713/2
  operator, **61**, 688/2
Boominess, 5/2
Boost
  h.f., 567/2
  phaseless, 83/2
  processing, 494/1
Booster, **62**
  chemical, **62**
  light, 42/2, **62**
Bootstrap circuit, **62**
Bothway broad band channel, **62**
Bouncing, **62**
Box set, **62**
Braun, K. F., **62**, 127/1, 328/2
Braun tube, 127/1, 328/ 630/2
Bray, J. R., **62**, 325/1, 339/2
Breakdown, **62**
  station, 568/1
Break point, **62**
Breathing, **62**

Bresson, R., 230/1
Brewster, P. D., **62**
Brewster, Sir D., **63**
Brightness, 62/1, **63,** 85/2, 181, 190/1 ff., 262/1, 267/2 ff., 274/2, 412/1, 424, 429/2, 875 ff., 971/1
  constancy (vision), 973
  control, **63,** 543/2
  curve patterns, *386*
  limitation (tube), 654/1
  picture (monitors), 555
  range, 266, 387/1, 876
  ratio, 951/1
  screen, 57/2, **63**
  separation, **63**
  unit, 298/2
  (*see also* Luminance)
Brighton school, 676/2
British Broadcasting Corporation, 55/2, 235/2, 236/1, 330/2, 332/2, 645/2, 770/2, 847/2, 848/2
  Television Centre, 797/1
British standard leader, 625/1
British Standards Institute, 63, 760/2
Brittleness, 292
Broad, 413/2
  band, **63**
  protection channels, 898/1
  pulses, 63, 264/1, 811/1
  signal, 62/1
Broadcast
  colour videotape machines, 942
  network design, 893
  studio (acoustics), 5/1 ff.
Browne, C. O., **63**
Bruch, W., 127/2
Brute, **64,** 414/2
BSI standards, **63,** 760/2
Buckle switch, **64,** 73/2
Budgeting, 44, 235/2, 257, 590
Build-up, **64,** 676/2
Bulk-printing, 497/1, 571/1
Bunsen-Roscoe law of photochemical equivalence, 971/1
Burn out, **64**
Burred edge, magnetic tape, *440*
Burst, 198/1
  gating, **64,** 204/2
  keying pulse, **64**
  phase, **64,** 534/1
  signal, **64,** 537/1
Burst-locked oscillator, **64**
Butt splice, 815/1
Buzz track, **64**
B-winding, 485/2, 495/2

Cable, 273
  camera, 76, 99
  capacitance, 463/2
  compensation circuits, **65,** 555
  correction, 83
  transmission, 895
Cables (lighting), 416
Calculations, exposure, 270
Callier effect, **65,** 249/1, 572/2, *664/1*
Cam, **65**
Camden group, 331
Cameflex, threading path, *69*
Camera
  accessories, 73
    animation, 22
  animation, 20
  aperture, 41/1
  cable, **100**
  caption, 506/2
  channel, 76
  choice, tropics, 911
  control operator, **100**
  control room, 201/1
  control unit, 76, **100**
  design, 92
  effects, animation, 36
  electronic, 502
  exposure chart, 35/2
  field of view, *36*
  film, **66,** 479
  hand, 479
  history, 66/2
  housings, underwater, 921
  indirect laryngoscopy, *457*
  inlay, 741
  level, 47/2
  Mitchell Mk. II, 812/2
  mobile, 529
  news, 501
  noise, 56/2
  operator, **112,** 788
  position and support, surgery, 454
  PSK-5, 780/1
  pulse track, 695/1
  radio, 502/2
  rehearsals, 238
  remotely controlled, 672/1
  script, 238/1
  self-blimped, *74*
  speed, 312/1
  talkback, 201/2
  techniques, animation, 35
  television, **76**
  television, colour, **85,** *92, 94, 95*
  3-D, 778
  tubes, history, 333
  types, colour, T96
Camera speed—adjustment for miniatures, 727
Camera speed measurer, 814/1
Camera tube
  blemish, 641/1
  characteristics, 881
Campbell Swinton, A. A., **112,** 127/1, *329,* 331/2, 630/2, 821/2
Can, **113**
Candela, **113,** 410/2 ff., 424/2 ff., 429/2, 507/2, 646/1
  conversion table, 647/2
Candlepower, 411/1
Cans, film, 754/2
  storage, 290 ff.
Capstan, **113**
  servo mechanism, 936/2
Caption, **113**
  animated, 14/1
  edged, 737
  inlaid, 734
  roller, **113**
  stand, **113**
Capture range, 547/2
Carbon arc, 42/2, 596/1
  automated, 52/1
Carbons, 414
Cardioid microphone, 462, 502/1, 689/2, *695, 696*
Carey, G., 126/2
Carpentier, J., 429/1
Carrier transmission, 888
Carrier wave, 242/1, 470/1
Cartoon, 25, 166/2, 325, 339/2
  animation, *29,* 38
  characters, *27*
  computer-drawn, 206/2
  scene elements, *34*
Cartridge, 69/2, **113,** 478 ff.
Cascade, 113, 123/1, 585
Caselli, A. G., **114,** 126/1, 326/2
Case, T. W., **114,** 320/1
Cassette, 69/2, **114,** 208
  loading for projectors, **114**
Casting, 231, 588, 785
Cataloguing, 397 ff., 402 ff.
Cathode
  drive, 379/2
  follower stage, **114**
  potential stabilized (CPS), 78, 765/1
Cathode ray tube (CRT), 126/2, 127, 159/1, 211/2, 261/1, 329
  early history, 328, 561/2, 630/2, 978/1
  oscilloscope, 62/1, 328/2, 971
  projector, 378
Cameraman
  animation, 35, 37
  film, **101,** 240/2, 788
  television, **106,** 240/2
Cathodoluminescence, 542/1
Cavity resonator, **115**
CBS, 262/1, 334/2
CCIR standards, 717/2, 758/1
CdS meter, 480/1
Cel, *23,* 62/2, **115,** 125, 325, 339/2
  animation, 28/1, 472/2
Celluloid, 66/2, **115**
  roll film, 251/2, 318/1
Cellulose acetate, 2/2
Celsius, 369/2
Cement,
  film, **115**
  splice, 478/1
Cemented
  doublets, 392/2 ff.
  triplets, 393/2
Censor title, **115**
Centigrade, 369/2
Centimetre candle, 411/2
Central aerial television (CATV), 892/1
Central apparatus room, 795
Central control 442/2
Chain processes, 388
Chang, 323/1
Change-over, **115**
Change-over cues, 623
Changing bag, **115**
Channel, 62/1, **115**
  communication, 201
  conversion, 159/2
  gain stability, **115**
  TV, table of, 115
Chaplin, C., 230/1
Characteristic curve
  film, **115,** *116,* 240/1, 264/2, *265,* 266/1, 302/2, 308/2, 343 ff., 370/2, 382/1, 659/2, 666, 672/2, 775/1, 875/2, 876 ff.
  television, 302/2, 775/1, 876 ff., (*see also* Transfer characteristics)
Characteristics of wave motion, *406*
Characterization, animation, *27,* 30/1, 38/2, 206/2
Charger, **116**

Charges and contracts, library, 401/2
Checkerboard cutting, 1/2, **116,** 488, 498/2, 754/1
  invisible joins, *489*
Check print, 116, 375/2
Chemical
  fogging, 628/1
  mixer, **116**
  precipitation, 673
  telegraph, **117,** 126/1, 326/2
Chemicals, **116**
Chemistry of the photographic process, **117**
Chopping, **123**
Choreutoscope, 124/2, 316/2, 318/1, 339/1
Chrétien, H., **123,** 125/2, 323/1, 340/2, 980 ff.
Chroma, **123,** 181
  circuits, block diagrams, *617 ff.*
Chromatic
  aberration, 2/2, **123,** 523, *524*
  adaptation, eye, 970/2
  colours, 181/1
Chromaticity, **123,** 226/2
  coefficient, 89, **123**
  diagram, **124,** *185 ff.*
  display tube primary colours, table, 193
  gamut, telecine, 839
  illuminant C, table, 193
Chromation tube, **124,** 382/2
Chrominance, **124**
  axes E'_I and E'_Q, *514*
  channel, **124**
  circuits, 617 ff.
  detector, 64/1
  information, 274/1
  phase, 534/1
  signal, 197, 511 ff., 655/2 ff.
  signal vector diagram, *512*
  vector, **124**
Chronochrome, 125/1, 303/2
Chronology of principal inventions
  film, **124**
  television, **126**
Chronophone, 125/1, 303/1, 319/2
Chronophotographe, 217/2, 303/1
Chronophotographic microscopy, 144/1
Chronophotography, 124/2, 317/1
Cibachrome print process, 355/2
CIE
  chromaticity co-ordinates for FCC reference stimuli, table, 834
  system, **128,** 184, 185
  tristimulus values for spectrum colours, *184*
  u, v diagram, *187*
Cinecardiography, 155/1
Cinch marks, **128**
Cinecolor, **128,** 321/2
Cine-endoscopy, 455
Cine-Kodak, 275/2
Cinema
  acoustics, 6/1, 719
  design, *140*
  films and telerecording, 856
  floor rake, *139*
  light beam angles, *139*
  lighting, 13/1
  theatre, **132,** *139, 140*
  visual principles, 969
Cinemacrography, **128**
  dark-ground illumination, *131*
  specialized equipment, *130*
  transmitted light illumination, *130*
Cinemascope, 55, 123/1, 126/2, **132,** 134, 136, 277/1, 281/2, 283/1, 299/2, 323, 340/2, 473/1, 596/2, 599, 641/2, 645/2, 703/2 ff., 719/1, 721, 816, 869/1, 980 ff.
  perforation, **132**
Cinématographe, 124/2, 318 ff., 429/1, 538/2
Cinemicrography, **144,** 441/2
Cinemiracle, 126/2, **152,** 986
Cinéorama, 125/1, **152,** 307/1, 322/1, 472/2, 979/2, 987/1
Cineradiography, **152**
Cinerama, 126/1, 133/2, 135, 137, 139/1, **156,** 322/2, 473/1, 596/2, 975/1, 977/1, *985*
Cinex strip, **156,** 271/1
Circarama, 126/2, **156,** 322/2, *987/1*
Circle animation, *16*
Circle of least confusion, **157,** 522/1
Circlorama, 322/2
Circuit module, 95/2
Circuitry (camera), 82, 99
Circular waveguide, **157**
Circulating system, **157**
Clamp amplifier, **157**
  pulse, **157**
Clamping, 216/2
  circuit (vision mixer), 965
  diode, **157**
Clapper board, **157,** 485/1
Clapper/loader, 102/1
Clappers, **157**
Class 'A' amplifier, **157**
Class 'B' amplifier, **157**
Class 'C' amplifier, **157**
Classification, library, 397/1, 402
Claw, **157,** 319/2
  pulldown, 16mm telecine, *829*
Clean effects, **158,** 528/1
  mixing, OB, 746
Cleaner, **158**
Cleaning film, 495/2, 918/1
Cleaning units, 376/1
Clinical recording, 453
Clip, **158**
Clippers, 84, **158**
Clocked binary counters, 809
Closed circuit systems, 80/1, 97/2, **158,** 271, 334/1, 362/2, 373/1, 375/1, 389/2, 458, 934
  air traffic control, *162*
Closed circuit systems
  network design, 894
Closed loop drive system, **164**
Close-up, 103, **164,** 228/2, 672/1
Cloud effects, 726/1
Clumping (grains), 348/1
C-mount, **164,** 479/1
Coating, **164**
  film, 284, 286
  tape, 432, *434, 435*
  viscous processing, 953
Coating of lenses, **164** ff.
Coaxial cable, 158/2, **166,** 217/1, 895/2 ff.
  feeder, 200/2, 273
  miniature, 447/2
Co-channel interference, **166**
Code numbers, **166,** 289/1
Coder, 197/2, **263**
  NTSC, 511 ff., *512, 515*
  PAL, 534 ff. *535*
  SECAM, 656 ff., *657*
Coding machine, **166**
Coercivity, **166,** 431/1 ff.
Cogging, **166**
Coherent light, 336/2 ff.
Cohl, Emile, 25/2, **166,** 325/1
Coil spring clutch, 872
Cold climate cinematography, **167**
  processing, table, 169
Cold light mirror, 380/2
Cold-mirror reflector, **170**
Collector, 77/2, **170**
Collimation, **170**
Collimator, 410/1
Colloidal silver, 40/1
Coloration (sound), 5/2, 6/2
Colorimetry, 89, 184, 199/2, 665/2
Colorplexer, **170,** 263/1
Colour
  analysis, 89/2 ff.
    telecine, 832 ff., *834 ff.*
  balance, 97/1, 168/2, **171,** 195, 306/2, 361/2 ff.
  bar, **171**
  burst, 54/2, 64, **171,** 274/1, 563/1, 615/2, *863*
  cinematography, **172**
  converters, 771
  correction, 97, 361/2, 855
  couplers, 120, 121, 177/2, 178, 357/1
  coupling developer, 177/1
  densitometers, 666
  development, 118/1, 120, 122, 358 ff., *579*
  difference process, **179**
  difference signal, 64/1, 85/2, 191 ff., 196/2 ff., 511 ff., 534/2 655 ff.
  display tube 42/1, 56/1, **179,** 209/2, 307/1, 354/1, 382/2, 669/2, 911/2, 989/2
  duplicate, 628/2
  editing, videotape, 927
  eidophor, 381
  errors, electronic correction, 843
  fidelity, 195/1
  film, 389/2
    density, 664/2
    storage, 398/1
    telecine, 838
    television, 179
    viscous processing, 955
  formers, 176/2
  genlock, 552
  history, 321
  information, 85/2, 195/1, 263/1
  insertion, electronic, 737
  intermediate positive, 599/2
  intermediate stock, 271/2
  killer, **180,** 617/1
  lock, 552
  matching, 187
  mixture curves, 834
  negative, 361/1, 178/1, 179/1, 195/1, 266/1
    storage, 292/2
  networks, performance, 895
  noise, 832/2
  positive film, 11/2, *667*
  principles

C

Colour
  film, **180**
  television, **190**
 printers, additive, **195**
 printing, control band, *574*
 receiver, 383/2, **609**, 615, 616, 842, 885
 recording, thermoplastic, 868
 resolution, 513/2, 655/2
 saturation, 512/2
 scanning, 644
 screen manufacture, 670/2
 sensitivity, emulsions, 343
 sensitometers, 665
 separation, 396, 496/2, *832*
  negative, 949/2
  overlay, 736
 sixteen mm derivations, *485*
 sound track, 802/1, 875/2
 standard, 425/1
 standards converter, 766/1
 stereoscopic films, 324/1
 thirty-five mm derivations, *495*
 studio timing, 773
 symbolism animation, 31
 synchronizing, 514, 656/2 ff.
 synthesis, 301/1
 systems, **196**, 335
 telecine, *838*
  characteristics, 668/1
 telerecording, 859
 television systems, **190, 196, 511, 533, 654**
  comparison of, 198
  history, 55/2, 334 ff., 396/1
  sequential, 668/2
 temperature 90/2, 150/1, **199**, 209/2, 210/1, 266, 369/2, 413, 414/2, 424/2, 543/1, 574/2, 660/2 ff.
  meter, 266/1
 test strip, 560/1
 triangle, **199**
 units, 424, 468/1
 wheel, 370/1
Coloured coupler, 50/2, 178, 357
Colourist, 35
Columbia Graphophone Company, 332/2, 645/2
Coma, 88/2, **200**, 391/2, 393/1, *522*
Comandon, J., 144
Combinations, animation, 35/2
Combination stocks, 277
Combined dupe, **200**, 571/2
Combining filter, **200**
 unit, **200**, 904
Commag, **200**, 503/2, 505/2
Commentary
 facilities, 746, *747*
 recording, 490
 vehicles, 527
Commentator, 527/2
Commission Internationale d'Eclairage, 184/1
Communications
 news, 500/2
 studio, **200**, 712, 814/1
Communications Satellite Corporation (Comsat), 637/1, 638/1
Communications satellites, current, 637
Community television (CTV), 892/1
Comopt, **204**
Compagnie Française de Télévision, 335/1
Compagnie Générale des Phonographes et Cinématographes, 307/1, 538/2
Comparison signals, videotape, 939
Compatibility, 191 ff., 197/1 ff., **204**, 335/1, 380/1, 511/1
 videotape, 945
Compatible composition, 102/1, **204**
Complementary colours, 181/1, **204**
Compliance, 8/2
Composite, **204**, 729/1 ff.
 colour signal, 197/2, 512/2 ff., 656/2 ff.
 copy, 374/2 ff.
 dupe, 200/1
 exposure, 195/1
 image, 356/1
 picture, **204**, 510/1, 733/1 ff.
 print, **204**, 700/2
 master, **205**
 signal, **204**, 511, *513 ff.*, 559/2, 773
 sync generator, **204**
 video signal, **205**, 548/1
Composition, camera, 101
Compound, 21/2
 animation, *23*, 24
 microscope, 145
 rotation, *24*
Compression
 anamorphic, 980/2 ff.
 ratio, 134/2
 sound, **205**, 250/2
 volume, 976/1
Compressor, **205**, 425/2, *712*
 amplifier, 697/1
 volume, *205*
Comptoir Général de Photographie, 303/1
Computer
 animation, 39, **205**, *206*
 languages, 206
 store, 240/1
 studio, 752/1
 tapes, 439
Comsat, 640/2
Comstock, D. F., 367/1
Concert hall acoustics, 6/1, 8/2, 9, table, 5/1
Condensation, 290/2
Condenser, **207**
Condenser microphone, 461 ff.
Conjugate foci, 521
Console, **207**
 operator, 423/2
Constant luminance principle, 193
 mixed-highs system, *192*
Construction methods, set, 46
Contact printer, 22/1, 376/1, 571
Contact printing, 195/1, **207**, 249/1, 343/2
 narrow gauge, 494/1, 497/2
 print geometry, 577/2
Continuity, 28, 33/1, **207**, 256, 257/1, 374/2, 442, 949/1
 planning, 793
 sketch, 48
 suite, 443/2
Continuous access, 315/2
Continuous drive, 871
Continuously-moving film cameras, 75/2, 312 ff.
Continuous motion prism, **208**
Continuous printer, **208**, 339/1, 570, 571/2
Continuous processing, 12/2, 354/1, 430/1, 563/2, 579
Continuous projection, **208**
Continuous wedge sensitometry, 662
Contouring, 97, 98/2
Contours-out-of-green, 95/1, 98/2
Contours-out-of-luminance, 96/2
Contrast, 65/2, 99/1, 105/1, 116/1, 118/2, 121/1, **209**, 248/2, 274/2, 295/1, 298/2, 302/2, 306, 405/1, 541/1
 control, 63/1, **209**, 363, 756/1
 lighting, **209**
 measurement, 665/1
 range, 503/1, 544/2, 607, 880/2
 reduction, 499/2
 transfer function, 347/1
 visual perception, 971/2
Control
 band, **209**, 373/2 ff.
 development, 360
 film, 665
 operator, **209**
 panels, vision mixer, 963
 room, 238/2
 signals, slavelock, 551
 strip, 247/2
 tape, 196/1
 track, 11/2, **209**, 560/1
 track and erase heads, video, 935, 938/1
Controls
 camera, 84, 111
 cinema, 143
 colour balance, 97/1
Convergence circuits, **209**, 618
 control, 781/1
 magnets, 671
Conversion filter, **209**
Converting contact prints, 16mm, 577/2
Cookie, **210**
Copy lens, **210**
Core, **210**, 754/2
Coner insert, **210**
Correction, colour, 209/2
 videotape, 942
Cosine law, 411
Costumes, 46/1, 220, 222
Counter footage, 73, 298/1
 step, *807*
 field timing, *807*
Coupler, **210**, 356/2 ff.
 competitive, 121
 incorporated, 355
CP lamp, **210**
CPS emitron, 78, 765
Crabbing, 240/2, 540/2
Crab dolly, **210**
Cradle construction, lighting, 415
Crane, **210**, 48/1
Crane truck, **211**
Cranz-Schardin technique, 314/1
Crawling title, **211**, 630/1
Creative editing, 258
Creative responsibility, 108
Creative technicians, 787
Credit title, **211**
Creepie-peepie, 112/2
Creeping title, 211/1, 630/1
Crispening, **211**
Crispness of focus, 594/2
Critical angle, 409/1
Crocker Research Laboratories, 272/2

Crofts, W. C., 241/1
C roll, 489/1
Cronar base, splicing, 754/1
Crookes, W., 62/1, 126/2, **211**, 328/2
  tube, 295/2, 328/2
Cropping, **211**
Cros, C., 175/1
Cross cutting, **211**, 363/1
Cross fade, **211**, 469/1, 510/1
Cross modulation, **211**, 616/1, 703
  picture, 693/1
  test, **212**
Cross-over frequency, **212**
Crosstalk, **212**, 513/1 ff., 541/2, 699/2, 947/1, 962
Crossview, **212**
Crown glass, 2/2
Crush, 60/1, **212**
Crystal, **212**, 560/1
  control, **212**, 482/1
  lock, **212**
  microphone, 461 ff.
  mixer, **212**
  oscillator, 549/1
  pick-ups, 716/1
  sync, 600/2, 601/1
CS perforation, 299/2
Cue, **213**, 443/2
  card holder, **213**
  circuit, **213**
  clock, 448/2 ff.
  dots, electronic, 742, *743*
  light, 814/2, 957
  marks, **213**
  sheet, 258/2, 699/2
  track, 505/1
Cueing
  device, **213**
  track, 261/2
Cue-in times (news), 504/2
Current affairs, producer, 591
Curvature of field, **213**, *523*
Curvature, tape, 436
Curved screen, 985/2
Cut, **213**, 445/2 ff.
Cut back, **214**
Cut bank, **214**, 956 ff.
Cut in blanking, **214**, 363/2, 962
Cut-off frequency, **214**, 389/2
Cutout, *15 ff.*, **214**
Cutter, **214**
Cutting
  copy, **214**, 493/2, 989/1
  film, **214**, 253 ff., 561/1
  raw stock, 288
  room, 254/2, 255, 339/1
  sound, videotape, 925
  spatial, 473/1
  sync, 594/2, 803/2
  techniques, animation, 17
  television, **214**
Cyan, 178, 189/1, 204/1, 355/1 ff.
  colour development, 177/1
  dye, 189/2 ff.
  dye former, 177/1
Cycles, frequency, 300/1, 310/2
Cyclorama, **214**, 413/2, 422/2

Daedalum, 338/2
Daguerre, L. J. M., 124/1
Dailies, **215**, 253/2, 493/2, 631/2
d'Almeida, J. C., 124/2, 324/1, 776/2
Damage from winding, 988/1
Dark adaptation, eye, 970/2
Dark current, 81/1, **215**
Dark end, **215**
Dark-ground illumination, *148*
Darkroom, 423/2, 424/1
Data processing, elections, 751
Day-for-night effects, 724/2
Daylight, 193/2
  developing machine, *580*
  exposure, 660/2
  loading spool, *69/2*, **215**, *475/1 ff.*
  projection unit, **215**
  speed, 265/2 ff.
D.C. blocking, vision mixer, 961
D.C. restoration, **216**, 884/2
Debrie Matipo colour printer, *195/2*
Decay time
  phosphor, 12/2, 541/2
  sound, 4/1, 8/2
Decibel, **216**, 679
  meter, 689/1
Decoder, **216**
  NTSC, 514/2 ff., *515*
  PAL, 534/2 ff., *535*
  SECAM, 657/1 ff., *657*
  videotape, colour, 943
Decomposition, film, 291/2, 292
Découpage, 102/2
De-emphasis, 198/2
Definition, **216**, 676/2
  sound, 5/1
Deflection
  circuit, 84
  coils, 77/1, **216**, 612
  defocusing, 854/2
  magnetic, 84
  yoke, **216**, 671/1
de Forest, Lee, 114/1, 125/1, 127/1, **216**, 320/1, 329/2
De Forest Wireless Company, 217/1
de France, H., 654/1
Degradation, **217**
Degrees K, 369/2
Dejoux, Jean, 38/2
Delay
  cable, **217**
  line, 83/2, 98, 198/2, 211/1, **217**, 252/1, 536/1, 651/1 ff., 820/2, 918/1, 940/2
  lock, controlled, 552
  networks, *774*
  signal, 772/1 ff.
  time, 5/1, 217/1
Delrama, **217**, 981/2
Delta-L system, 97
Demeny, G., 125/1, **217**, 303/1, 319/1, 323/2, 441/2
Demodulation, 197/2, 515/1
Densitometer, 309/1
  commercial types, *661*, *662*, 663 ff.
Density, 36/2, 115/2, 195/1, 211/1, **217**, 246/2, 247/1, 252, 264/2, 265, 297/2, 298/2, 302/2, 306/2, 308/2, 311/2, 342 ff., 361/2, 382/1, 659/2 ff., 876 ff., 923/2
  diffuse, 218/1, **227/2**, 249/1, 664/1
  gradation, *10*
  optimum, 212/1
  ring zones, 10/2
  sound track, 702/2 ff.
  specimen, 149
  specular, 218/1
Densitometry, **217**, 271/1, 663
Depth of field, 132, 149/2, **218**, 270/1, 365/1, 469/2, 486/1, 497/2, 524
Depth of focus, **218**
Depth perception, 972
Desaturation, 217/1, **218**
  printing, **218**
Design
  auditorium, 135
  cinema, 133
  talkback equipment, 203
Designer
  animation, 33
  television, **219**
  (*see also* Art director)
Detail, rendering of, 345
Deterioration
  film, 292/1
  image, 217/1
Deutsche Industrie Norm, 578/2
Deutschen Normenausschuss, 228/2
Developer, 118, **225**
  X-ray, 154/1
Developing, **225**, 602/1, 626/2
  agents, 116, 117 ff., **225**
  colour, 177
  instantaneous, 866/2
  machines, 371, 373
  process, test, 667/2
  time and temperature, *344*
Development, 118, 122, 306/2, 342/2 ff., 628/1
  colour, 120
  Kalvar, 368
  matrix, 351
  series, negative material, *660*
Diabolo, **225**
Diagrams, animated, 27, 30, 31, 37
Dialogue
  corrections, 691
  level, 689, 691/1
  recording, 563/2
  synchronization, 698/2 ff.
  track, **225**
  (*see also* Speech)
Diaphragm, lens, **225**, *226*, 268 ff., 365/1, 524/1
  light change, printer, 574/1
  microphone, 462
  setting scale, 269/2
Diapolyekran, 472/2
Diapositive, 726/1
Diascope, **226**
Dichroic
  coating, 170/1
  dyes, 189/2
  reflector, 85/1, *87*, *88*, 91/1, 94/2, 165/2, 195/2, 196/1, **226**, 380/1, 396/1, 575/1, *833*
  reflector problems, 88
Dickson, W. K. L., **226**, 253/1, 275/1, 318/2, 382/2
Didié, L., 476/1
Didier, L., 175/1
Differential
  drive, processing, 581
  gain, **227**
  phase, **227**
  phase distortion, 516/2
  re-exposure, **227**
  signal, 211/1
Diffraction, **227**, 407
  effects, 644/2
  grating, 227/2, 408/1, 645/1
  spectrum, 868/2

Diffuse density, 218/1, **227**, 249/1, 664/1
Diffuser, 366/1
Diffuse transmission, 218/1
Diffusing
  screen, 452/2
  surface, **228**
Diffusion, **228**
  controlled, 177/1
Digestion, emulsion, 287
Dimensional stability, 284
Dimension–150, 135/2, **228**
Dimensions
  film, 66/2, 284/1
  magnetic tracks, 283
  optical tracks, 281
  projected areas, 281
Dimmer, **228**, 420/2, 421
DIN
  print, **228**, 578
  speed rating, **228**
  standards, **228**
Diode switch transistor drive, *769*
Dioptre, 129/1, **228**, 521/2
  lens, **228**
Diplexer, **228**
Dipole, 12/1, **229**, 908/1, 916/2
Direct imaging, 94, 95
Directional, **229**
  effects, **229**
  microphones, 464
  screen, 55/1, *647*, 649
Directivity
  aerial, 12/1, 907 ff.
  microphone, 462
Director
  animation, 33
  film, **229**
  television, **234**
Director's methods, 257/2
Director's talkback, 200
Directory of Film and TV Production Libraries, 397/1
Direct
  positive, **239**
  positive telerecording, 857
  sound, *3*, 4/1
  switcher, 214/2, **239**
  videotape recording, 930
Disc
  manufacture, stages of, *715*
  production, 715
  recorders, 697
Discharge lamp, **239**
Discoloration, silicon, 379/1
Discontinuous spectrum, **239**
Discriminator, **239**
Discussion programmes, 203
Disney, W., 26/1, 27/1, 320/2, 322/2, 325
Dispersion, 2/2, **239**, *409*
Dissector tube, 332/1 ff.
Dissolve, **239**, 487/2, 575/2, 576/1, 816/1
  animation, 30/2
  automatic, 21/1
Distortion, **240**
  aperture, 854/2
  attenuation, 685
  colour, 198/1, 271/2, 298/2, 531/2, 919/1
  effects, 199, 729/2
  harmonic, 212/1, 703/1
  image, 62/2, 250/1, 370/1
  image, special effects, 724/2
  intermodulation, 364, 686/1
  lens, 294/2, 393/1, 523
  light, 217/1
  non-linear, 685, 895
  in NTSC, 616
  optical, 507/2
  peak, 978/2
  phase, 629/2, 685
  picture, 84/2, 274/2, 338/1, 380, 426/2
  signal, 60/1, 252/1, 370/2, 784/1
  sound, 53/2, 211/2, 212/1, 263/2, 364/2, 539/2, 989/1
  tape, 439/2
  waveform, 532/1, 894/2
Distribution, power, 565
Disturbance effects, sound, *6*
Divcon, **240**
Divergent lens mounting, 390/2
D log E curve, 116/1, **240**
D-mount, **240**
Documentary film, 105, 257/1
Document transmission, 162
Dog clutch, 871
Dolly, 110, 211/1, **240**
  shot, 672/1, 879/2
Dollying, 298/1
Dominant wavelength, 186
Donald Duck, 26/1
Donisthorpe, W., 124/2, **241**, 317/1 ff., 396/2
Do-not-mix indicators, 958
Door in the Wall, 323/1
Dope sheet, 30/1, **241**
Doppler shift, **241**
Dot sequential system, 93/1, 669/1
Double
  exposure, 21/1, **241**, 533/1
  frame, **241**
  frame exposure, 22/2
  Gauss lens construction, 393/2
  sideband, **242**, 513/1
Double-8mm film, **241**, 260/1, 277/2, *476*, 476/2 ff.
Double-head projection, **242**
Double-perforated stock, **242**, 485/2
Double-run, 476
Double-16mm film, **242**, 277/2
Double-system sound
  recording, **242**
  16mm, 481
Doublet, 2/2
Downlead impedance, **242**
Downwards conversion, **242**
Drama producer, 591
Dressing, set, 785
Dress rehearsal, 239
Drier, **242**
Driftlock, 549/2
Drive, camera, 73
Drive-in cinema theatre, **242**, *243*
  sound reproduction, 721
Drive unit, **244**
D roll, 489/1
Dropout, **244**, 431/2, 437/2, 438/2, 504/2
Drum developing, **244**
Drum (sound), **244**
Drying, **244**, 354/1, 587
  defect, 977/1
  marks, 123, 978/1
Dry processing, 368/2
Dual inputs, monitor, 554
Dual master control, 444
Dual-purpose
  lantern, 423/2
  lenses, 391
  projector, 595/2
  stages, 417
Dual-standard receiver, **244**, 610
Dubbed version, **244**
Dubber, **244**
Dubbing, 234, **244**, 258/2 ff., 298/2, 428/1, 441/1, 471/2, 490/1 ff., 505/2, 537/1, 563/2, 568/1, 627/1, 676/1, 691/1, 699
  cue sheet, *258*
  mixer, 687, 700/1
  videotape, 707/2
  theatre, 791/2, *691*
Dubray-Howell perforation, **245**, 277/1
Ducos du Hauron, L., 176/1, **245**, 322/1, 776/2
Dufaycolor, 125/2, 172/2 ff., **245**, 321/2
  reseau, *173*
Dufay, L., 173/2
Dufay system, 627/1
Du Mont, A. B., 333/1
Du Mont, T. H., 124/2
Dunning, C. H., 369/1
Dunning process, **245**
Dupe negative, 61/1, 200/1, **245**, 246, 294/2, 296/2, 299/2, 362/2 ff., 494/2, 658/1, 675/2
Duplex sound track, **245**
Duplicate master, 494/2
Duplicate positives, 248/2
Duplicating film, 61/1
Duplicating (narrow gauge), 497/2
Duplicating processes (black-and-white), **246**, 294/1
Duplitized, 128/2, **249**, 284/2
Durbar of Delhi, 172/1
Dye, 120, 121, 189/1, 195/1, 287/2
  anchoring, 176
  coupling, **249**
  destruction, 355, *356*
  image, 189/1, 249/1, 356/1 ff.
  response curves, *841*
  toning, **249**
  transfer, 174 ff., *175*, 189/1, **249**, 340/1, 350/1 ff., *351*, 471/1, 815
Dyeing, 352
Dynalens, **249**, *757*
Dynamic
  focusing, **249**
  Frame, 126/2, **249**, 323/1
  microphone, 461/1
  range, **250**, *884*
Dynascience, 757/2
Dynode, 826/2
  effect, **250**

Early Bird, 334/2, 637
Earphones, 111
Earthing, 565
Earth stations, 639
  N. Atlantic, *640*
Eastman Color, 126/1, 210/2, **251**, 322/1, 356/2 ff.
  internegative film, 955/2
  negative film, 178, *179/2*, *357*, 921/1
  positive film, 178/2, 179/2, 955/2
Eastman, G., 66/2, 124/2, **251**, 318/1
Eastman Kodak, 172/2, 173/1, 176/1, 177/2, 178, 275 ff., 324/2, 474/2 ff.
  reversal film, 179/2

Eastman standard perforation, 277/1
Eau de Nil, 303/2
Eberhard effect, **252**
Ebicon, 91/2
EBU (*see* European Broadcasting Union)
Echo, 4/1, 6/2, 304/2, 674/2, 688/1
  distortion, 894/2
  effect, **252**, 432/1
  electronic, 199/2
  equalizer, 304/2
  production, 692/1
  waveform corrector, **252**
  unit, 712
Echo I, 633/1
Edge
  build-up, tape, 440/1
  effect, **252**
  enhancement, 97
  fogging, **252**, 477/1
  gradient, *390*
  lip, tape, 440/1
  luminance, 648
  marking, 246/2
  number, 166/2, **252**, 298/1, 374, 487/2, 499/1, 517/1
  sharpness, 347/2
Edge-and-point guiding, 560/2
Edged captions, **252**
Edison and Swan, 295/2
Edison Company, 43/2
Edison Electric Light Company, 295/2
Edison, T., 43/1, 66/2, 124/2, 125/1, 226/2, 251/2, **252**, 275, 317/1, 318/2 ff., 382, 429/1, 473/2, 538/2
  kinetoscope, *381*
Editing, 166/2, 214, 233, **253**, 428/1, 471/2, 676/1, 789, 812/1, 921/2
  machine, **253**
  narrow-gauge, 486, 497
  news, 503, 505
  sync, 803/2
  videotape, 925 ff., 940
Editola, **253**
Editor, **253**, 375/2
  animation, 35
  electronic, videotape, 940
Edit pulse, **259**, 925/2
Educational
  films, 39/1, 276/1
  television, 163
Edwards, E., 350/1
Effective radiated power (ERP), **259**, 891/2, 901
Effects
  bank, **259**
  fitter, **259**
  track, **259**
EFS (Electronic Filming System), 11/2
EHT
  rectifier, **259**
  supply, 378
  supply, receiver, 612/1
Eidoloscope, 382
Eidophor, 159/1, 180/2, **259**, 380 ff., 645/1, 700/2
  optical system, *381*
Eight millimetre film, **260**, 276 ff., 324, 476 ff., 482 ff.
  production, 497
  projector, 597
  reduction prints, 497/1
  sound prints, 499
  (*see also* Single-8 *and* Standard-8)
Eight point seven five mm film, 262/1
Eisenstein, S. M., 256/1
Ektachrome, 179/2, 210/2, **260**, 356/2 ff., 816/2
Ektacolor, 178
Elapsed time working, 448/2
Election programmes, 748 ff.
Electrical and Musical Industries (EMI), 61/1, 63/2, 263/1, 332/1, 645/2, 988/1, 990/2
  2001 camera, 95/2
  (*see also* Emitron)
Electrical telescope, 126/2
Electrolytic fixing, 674
Electromagnetic
  spectrum, **260**, *410*, 887
  waves, indirect transmission, 891
Electron
  gun, **261**, 381, 671
  image, 79/1, **262**
  multiplier, 77, 79/1, 127, 250/1, **262**, 332/1
Electron-beam
  film scanning, **260**
  recording, **261**
Electronic
  controls, OB camera, 530
  converter, **262**
  editing, **262**, 708/1, 926, *927*
  field rate converter, 500/2
  flash, 315, 873
  image stabilization, 758/1
  spotlight, **262**
  sprocket holes, 600/2
  switch, **262**, *738*
  video recording, **262**
  viewfinder, 11/1
Electronic Filming System, 11/2
Electronicam, **261**
Electro-optic effect, 314/1
Electrophotography, 343/1
Electrostatic
  focusing electrode, **262**
  microphone, 461/2 ff.
Element dimensions, picture, 864
Elevator, **262**, 627/2
Elster, J., 127/1, **263**, 303/2, 329/1
Emergency bypass, **263**
Emergent ray, 519/1
Emission, light, 406
Emitron, 77/1, *78*, **263**, 332/2, 333/2, 645/2, 990/2
Emulsion, 10, 118/2, **263**, 284, 286, 303/2, 306/1, 308/2, 342 ff., 355/1 ff., 410/2, 659/1
  batch, 57/1
  choice of, telerecording, 858
  coating, *288*
  crystal, 342 f.
  manufacturing process, *286*
  pile-up, 978/1
  surface, 2/1
Encoder, 170/2, **263**
  (*see also* Coder)
Endoscope, 129
  pictures, interpretation, 458
Energy distribution, typical light sources, *90*
Energy recovery diode, **263**
Engineering manager, outside broadcast, 527/1
Enlargement, 61/1
Enquiry service, library, 402
Entertainment systems, 159
Envelope, 888
  delay, **263**, 307/2
Equal energy source, 185, 186/2
Equal energy white, 90/2, **263**
Equal loudness contours, *681*
Equalization, **263**
Equalizer, **263**
  and filters, 711
  phase, 919/2
Equalizing pulses, 204/2, **263**
Equilibrium intensity, 3/2, 4/2
Equipment, cameras, 102
  care of, 109
Erasing, **264**
Erasure, accidental, tape, 433/1
Errors, colour, 89/2, 97
  mechanical, colour VTR, 941
ES cap, **264**
Estanave, E., 324/1
Eurafrica, 640/2
European Broadcasting Union (EBU), 500/2, 528/1, 640/2
Eurovision, 500/2, 593, 640/1
Evans, Mortimer, 301/1
Event marker, 155/1
EVR (*see* Electronic video recording)
Examination, library, 405
Examinations, film for, 458/2
Examiner, **264**
Excess noise, 508
Exciter lamp, **264**
Exciter unit, VHF, 903
Exit requirements, cinema, 140
Exposing machine, 372/2
  (*see also Cinex strip*)
Exposing wedge, 661
Exposure, 2/1, 36/2, 115/2, 131, 149, 169/2, **264**, 302/2, 306/2, 308, 342 ff., 660/2 ff., 876 ff., 918/2
  control, **264**, 266, 269, 573
  determination, 132
  index, 265, **271**
  intensity, 382/1
  Kalvar, 368
  meter, 2/1, **271**, 755/2
  range, 382/1
  rating, 42/1, 265
  test, sound track, 703/1
  tropics, 914
Extension tubes, 129, 131/2
Exteriors, 103 ff.
External communications, 202
Eye
  and brain, 967 ff.
  function, 182, 184
  marker, TV, *459*
Eyepiece, watching, 145, 146/1
Eyestrain, 140/1
Eyring's formula, 4/2

Faceplate, **272**
Face tone, **272**
Facsimile telegraph transmission, 126/1, 326/2
Fade, 239/2, **272**, 445/2 ff., 488/1, 575/2, 816/1
  automatic, 21/1
Fader
  module, **272**, 710/2
  rotary, 704/1
Fading amplifiers, vision mixer, 956

F

Fading, in storage, 291/1, 292/2, 293/2
Fantasia, 26/1, 125/2, 320/2
Fantasmagorie, 166/2
Farnsworth, P. T., 77/1, 127/2, **272,** 332 ff., 341/2
  Television, Inc., 272
Fast motion, **272**
Fast pulldown
  telecine, 829
  telerecording, 849
FCC (*see* Federal Communications Commission)
Fechner's law, 971/2
Federal Communications Commission (FCC), 333/1, 758/1, 862/2
  specified colour primaries, *89*
Feedback, **273**
  acoustic, 2/2, 203/2, 339/1, 751/1
  loop, 2/2
Feeder, aerial, **273,** 908
Feed
  film, 68, 69
  radio, 228/2
Felix the Cat, 25/2, 26/1, 325/1
Ferraniacolor, 179/2, **273**
Ferrite isolator, **273**
Fibre optics, 156/2
Field, **273**
  blanking, 60/1, 204/2, **274**
    interval, 603/2
    periods, 847
    pulse, 612/2
  deflection, 603/1
  drive signal, **274**
  flyback, 264/1, 603/2
  frequency, 862/2
    timing, 807
  identification signal, **274,** 810
  locking, 548/2
  phasing, **274,** 547 ff.
  rate converter, 766, *770*
  sawtooth, **274**
  scanning, 642/1
  scan synchronization, 204/2
  store, 93/2
    converters, 770
  suppresion, **274,** 863/2
  synchronizing, 63/2, 264/1
    period, 205/1
    pulse, 544/2
    waveform, 426/1
  tilt, **274**
  time base, **274,** 612
Field
  coverage (lens), 394/1
  flattener, 40/2, **274,** 392/1
  lens, *87*, **274**
  masking, *36*
  of view, 13/2, *36*
Field pick-up (*see* Outside broadcast)
Field sequential system, 334/2, 669/1
Field simultaneous system, 334/2
Fifteen mm film, 324/2
Fifty-five mm film, 323/2
Figure and ground illusion, *972*
Fill light (filler light), 62/1, 209/1, 270/2, **274/2,** 296/2
Film Cataloguing Rules, 397/1
Film
  areas, TV, 282
  coating machine, *288*
  colour television, 363/2
  commentaries, news, 504
  department, studio, 796
  dimensions and physical characteristics, **275**
  foot, 279
  formats, table, 391
  joining, laboratory, 583
  laboratory operations, *372*
  length, 52
  manufacture, **285**
  register, **289**
  scanner (*see* Telecine)
  scanning, 917/2
  shooting techniques, telecine, 839
  sizes and raw stock dimensions, *279*
  slide, **289**
  slip in printing, 570
  position, 279, 280
  sound recording, *701*
  speed, metrical, table, 280
  stability test, 292
  storage, **289**
    reservoir, *583*
  strip, **293**
  television, 607/2, 632/1, 878/1, 880/2
  types, narrow gauge, 476 ff.
Filming
  narrow gauge, 486
  speed, 279
    for TV, table, 486
Films Parlants, 303/1
Filter, 72/1, 259/1, 270/1
  analysis, 191/1
  banded, 172/2, 396/7
  band-pass, 294/1, 515/2, 536/1
  band-stop, 515/2, 692/1
  colour, 132, 149, 165/2, 188/1, 209/2, **293,** 305/2, 321/2, 361/2, 415, 666/1
    correction, 96/1, 266/1, 366/1, 574/2
    synthesis, 191
  contrast, 507/1
  conversion, 484/1
  dichroic, 88/1, 868
    (*see also* Dichroic)
  electrical, **294**
  elements, 173, 471/1
  factor, 270/1, **294**
  heat-absorbing, 666/2
  high-pass, 294/1
  infra red rejecting, 666/2
  lamp, 414/2
  layer, 177/2
  low-pass, 294/1, 347/2, 515/2, 559/1, 702/1, 770/1
  narrow-cut, 293/2
  neutral density, 93/2, 96/1, 270/1, 661
  optical, 191, 363/1
  photometric, 210/1
  polarizing, 168/2, 324, 507/1, 561/2, 562/2
  printing density, 666/2
  sensitometry, 660/2
  separation, 188
  sky, 305/2
  status A, 666/2
  status M, 666/2
  stripe, 86/1, 93/1
  trimming, 91, 94/2, 226/2
  underwater, 920
  UV, 168/2
  vestigial sideband, 294/1
  Wratten, 266/1, 270/1
  wide-cut, 293/2
Filterplexer, **294**
Finance, producer, 589
Finder
  monitor, 950/2
  reflex, 950/2
  (*see also* Viewfinder)
Fine cut, 254/1, 256/1, 258/1, **294**
Fine grain, **294**
  master positive, 494/2, 599/2, 620/1
Fire insurance, 565/1
Fire precautions, 292/1
First assistant director, **294**
First generation dupe, **294**
Fischer, F., 377/2
Fischer, R., 176/2, 355/2
Fischinger, O., 26/1
Fish-eye lens, **294,** 394
Fishpole, **294**
Fixation, 118/1, 121
Fixer, 117/2, 122/1
  life, 121/2
Fixing, 117, **294,** 309/1, 673/2
  Kalvar, 368
Flag, 210/1, **295**
Flange, **295**
Flare, **295,** 876 ff.
  correction, 99
Flashback, **295**
Flash bulb, 315
  exposure, 629/1
Flashover, **295**
Flashpoint of ether, 455/1
Flash tube circuits, *315*
Flat, 46/2, **295**
  print, **295,** 979/1
  processes (wide-screen), 983
Fleischer, M., 25/2, 26/1, 325/1
Fleming, J. A., 125/1, 127/1, 217/1, **295,** 328/2
Flicker, 191/2, **295,** 332/2, 363/2, 389/2, 655/2, 917/1, 969/2
  critical frequency, 968/2
  effect, valves, 508
Flint glass, 2/2
Flip-shots, 729/2
Float, 45/2, **296**
Floating pulley, 225/2
Flood, **296**
Floodlamps, 413
Floor, **296**
  manager, 236/1, 238/2, 294/2
  mixer, **296,** 620/1, 687
  plan, 236/2, **296**
  shooting, 688, 698
Flooring, studio, 799
Flop over, **296**
Flowers and Trees, 325/1
Fluorescent
  materials, 182/2
  screen, 152 ff., *154*
  tubes, 423/1
Fluoroscopic viewing, 153/2
Flutter
  bridge, **296**
  picture, **296**
  sound, **296,** 697/1, 722/2, 989/1
Flyback, 60/1, **296,** 642/1, 861/1
Flying spot
  analyser, *85*
  camera, 127/2, 330/2 ff.
  caption scanner, 158/2
  EVR scanner, 262/1

Flying spot
film scanner (telecine), 260/2, **297,** 338/1, 365/2, 366/2, 543/1, 817/1 ff., *825, 826,* 882
inlay camera, 354/2, 741/1
Flywheel circuit, **297**
FM sound, 363/1
f-number, 42/1, 129/1, 131, 268/2 ff., **297,** 305, 753/1
Focal
length, 71/2, 72/1, 82/2, *86,* 129, 228/2, **297,** 394/2
plane, **297,** 394/2, 520/2
Focus, **297,** 497/2, *520*
coil, **297**
modulation, **297,** 379, 381/2, 854/2
fuller, 102/1, **297**
servo system, **297**
unit, 672/1
Focusing
beam, 262/2
lens, 21/2
microscopic, 72/2
raster, 249/2
Focus-mask tube, 382/2
Fog, **297,** 343/2, 665/1
chemical, 118/2, 122/1
filter, **297**
level, 265/2, 877/1
Fogging, 215/1
light, 601/1
mark, 61/1
Foldback
circuits, 711/1
loudspeaker, **297**
Folded light beam, rear projection, *606*
Follow focus, 22/2, **298**
cameraman, **298**
Follow shot, **298**
Foot, **298**
Footage
counter, 255/1, **298,** 299/2
number, 166/2, 252/2, 289/1, **298,** 487/2
Foot-candle, **298,** 411/2, 424/2
Foot-lambert, **298,** 412/1, 424/2 ff.
Forced development, 265/2, **298**
Forced perspective, *46*
Foreign
captions, 756/1
commentary mixing, 746
commentator, 528/1
release, **298,** 808/1
titles, 875/2
version, 441/1, 489/1
Format, 150, 275 ff., **299,** 323/1
Form perception, 972
Fortran, 206
Fourier
analysis, **299,** 349/2, 886/1
series, 890/2
transformation, 346/2
Four-track printing master, 704/2
Four-tube
cameras, 95
telecine colour analysis, 836
Fox, 20th Century-, 114/1
Fox-Case, 320/2
Fox Grandeur, 323/2
Fox Movietone, 125/2, 320/1
Foxhole, 132/2, **299,** 705/2
perforated film, 982/2
Frame, 273/2, **299,** 917/2
counter, **299**
frequency, projector, 594
line, **299**
rate, 75/1
roll, 363/2
shading, 274/2
size, 279
stretch, **299**
tilt, 274/2
Frame-for-frame sync, 490/1, 499/1
Framing, **300**
Free head tripod, 301/1
Free-radical process, 343/1
Free-space propagation, **300**
Freeze frame, **300,** 335/2, 783/1
Frequency, **300,** 406/2, 678
changer, **300**
changes, sync pulse, *545*
deviation, 546/1, 890/1
divider, **300,** 804
fluctuation, 757/1
interleaving, 513/1
limitations, videotape, 929
modulation, 198, 466/1, 655, 889, *890,* 897, 930
sound transmitters, 906
range, 56/2, 706/2
response characteristic, **300,** 385, 392
curve, 386, *711*
vision mixer, 961
translation, **300**
Fresnel
lens, *300,* 413/1
spot, 423/2
zone, *300*
Friction
cassette, 208/2
drive, 359/1
head, **300**
Friese-Greene, W., 66/2, **301,** 318/1, 631/2, 923/2
Frieser, H., 347/1
Fringing, **301**
Front porch, **301**
Front projection, 215/2, **301,** 732
process, *732*
Frosted surface, 39/1
Fujicolor, 179/2, **301**
Fuji Single-8, 477/2 ff.
Full aperture, 281/1, **301**
Full-circle presentation, 986
Full frame, 281/1, **301**
Fulvue, 123/1, 125/2, 323/1
Fundamental, 61, 682/1
Furnace monitor, vidicon, *161*

Gabor, D., 336/1, 777/2
Gaede molecular air pump, 127/1
Gaffer, **302**
Gain, **302**
control, remote, 84
gamma, 62/2
light, 57/2, 77/2
power, 12/1
variable, 12/2
screen, 653/2
Galvanometer, **302,** 700
Gamma, **302**
and development, *344*
colour film, 666/2
television, 193
correction, 62/2, 84, 99/2, **302,** 379, 820/1, 857
non-unity processes, *856*
photographic material, 116/1, 246/2 ff., 309/1, 343/2 ff., 870/2, 877
receiver, 615
signal, 510/2
telecine, 822
television, 879
telerecording, 856/2
Gamma rays, 260/1, 410/2
Gance, A., 125/2, 322/1, 472/2, 979/2, 987/2
Gap and lap switching (vision mixer), 956
Gapping switch, 214/2
Garland, **303,** 413/2
Gas-burst agitation, 532/2, 585
Gas discharge tubes, 183/1
Gasparcolor, 125/2, **303,** 322/1, 355/2
Gate, picture, 12/1, **303**
Gate pressure, **303**
Gating, **303**
Gating pulse, **303**
Gauges (narrow), 474
Gaumont, 166/2, 325/1
Chrono-de-Poche, 324/2
Chronophone, 125/1, 319/2
Chronochrome, 125/1, 303/2, *321*
Gaumont, L., 217/2, **303,** 320/2
Gaussian distribution, **303**
Geissler tube, 39/1
Geitel, H., 127/1, 263/1, **303,** 329/1
Gelatin, 118/1, 120/2, 121/2, 122/2, 263/1, 284/2 ff., 286 ff., **303,** 342, 351/1, 358/1
dry plate, 251/1, 317/1
emulsion, 309/1
relief, 174 ff., 352
Gelb's experiment, *974*
Gemini, **304**
General Electric Company, 95/1, 320/2, 330/1, 866/1
Generator, **304**
standby, 565/1
Geneva movement, 441/1, 595/2
Genlock, 548, 550
Geometry, image, telerecording, 852
Georges, H., 127/2
Geo-stationary satellites, 636
Gertie, the Trained Dinosaur, 452/2
Getter, 379/2
Gevachrome, **304**
Gevacolor, 179/2, **304**
Ghost, 88/2, 212/2, **304,** 472/2
Ghosting, 606/2, 891/2
Gibson and Schroeder, 98/2
Glass-beaded screen, 653/1
Glass shot, *47,* **305,** *725*
Glenn, W. E., 866/1
Glorious Adventure, 369/1
G-number, **305**
Gobo, **305**
Goldmark, P., 334/2
Goldovsky, E., 137/1
G.P.P. sound system, 303/2, 320/2
Gradation, **305**
telerecording, 856 ff., *857*
Grader, **305**
Grading, 106/1, **305,** 362, 375, 495/1, 875/1
print, 40/1
Graduated filter, **305**
Grain, **305,** *306,* 342
and granularity, 347

Grain
  as noise, 348
Graininess, 118/2, 300/1, **306**
Gram operator, **306**
Gramophone Company, 331/1
Grandeur presentation, 125/2, **306,** 979/2, 984/2
Granularity, **306,** 392/1
Grandeur presentation, 125/2, **306,** 979/2, 984/2
Granularity, **306,** 392/1
Graphic designer, 220
  technique, 31
Graphics, 223
Graphs, presentation of, 16/1 ff.
Graticule, **306**
Green print, **306**
Greyback, **306,** 246/2
Grey base, 40/1, 286/1, **306,** 475/1
Grey scale, 220, 303/1, 305/1, **306**
Grid, lighting, 418/1 ff.
Grid
  modulation, **306,** 903/2
  tube, **307**
Griffith, D. W., 256/1
Grimoin-Sanson, R., 125/1, **307,** *322*
Grip, **307**
Grivolas, C., 125/1, **307,** 324/1, 538/2
Groove locator, 506/1
Gross, A., 26/1
Ground glass, **307**
Ground noise, **307**
Ground wave transmission, **307,** 891
Grounded-grid stage, **307**
Group delay, 263/2, **307**
Guide track, **307,** 563/2, 698/2
Gun mike, **307,** 629/2
Gyro head, **307**

Halas and Batchelor, 26/1
Halation, 40/1, 306/2, **308**
Half-amplitude point, 544/2
Halide, **308**
  cycle, *916*
Hallwachs, W., **308,** 311/1, 329/1
Halo effect, **308**
H & D curve, 116/1, **308,** *309,* 659/2
Hanging miniatures, 726/1
Hardener, 117, 121/2, **309**
Hardening, 252/1, 351/1 ff.
Harmonic, 6/1, 682
  distortion, 685/2
  filter, **309**
  generator, **309**
Haunted Hotel, 784/2
Hazards of space, 635
Hazeltine Corporation, 127/2
Head, **309**
  amplifier, 83, **309**
  assembly, VTR, 934
  camera, **309**
  drum, VTR, 936/2
  leader, **309,** *624*
  tips, life of, 935/2
  unit, **310**
Headphones, **309,** 696
Headset, **310**
Head-to-tape contact, 937/1
Hearing
  directional, 683
  threshold, 681/1
Heat exchanger, viscous processing, 953
Heating, cinema, 142
Heat
  radiation, 354/2
  reduction, projectors, 226/7
Helical aerial, **310**
Helical scan, 159/2, **310,** 928/2, 931 ff., *932, 933*
Helmholtz resonator, 8/2
Hepworth, C., 125/1, **310,** 319/2
Herschel, J., 124/1
Hertz, 300/1, **310**
Hertz, H. R., 126/2, **310,** 329/1
Heterodyning, colour VTR, 942
HF
  carrier systems, **311,** *896*
  wire TV, 159/1, 614/1, 892
H.I. arc, **311**
High aperture lens, 392/2
High contrast, 81/1
High-definition
  scanning systems, 159/1
  television, 331, 369/1
High-fidelity, 50/2
High frequency
  carrier system, **311,** 896
  content, 40/2
  wired system, 892
High gain aerial, **311**
High gamma recording, telerecording, 857
High hat, **311**
High-intensity light source, 460/1
High key, 101/2, **311**
Highlight, 79/1, 122/1, 175/1, 188/2, 246/2, 250/2, **311**
  density, **311**
  exposure, **311**
High-pass filter, **311**
High-speed
  cameras, 75, 208/1
  cinematography, **311**
  printers, 571/1
High stability unlocked operation, 552
High temperature, drying, 587
Hill-and-dale recording, 716/2
History
  detail, 19, 25, 66, 132, 144, 152, 172 ff., 275, 350, 474, 578, 979
  film, **316**
  television, **326**
Hittorf, J. W., 62/1, 328/2, **335**
Hold frame, 300/1, **335,** 783/1
Hollinger-Brubaker cine-endoscopy, *455*
Hold-in range, 547/1
Hologram, 427
Holography, **336,** *337,* 776/1
Home movie making, 10/1, 474 ff.
Home movies, 324/2
Home videotape recorder, 931/1
Home viewing conditions, 951/1
Hook, **338**
Hoppin, H., 26/1
Hopping patch, **338**
  telecine, 818/2
Horizon cloth, 214/2
Horizontal compression, film, 756/2
  picture shift, **338**
  shading, 603/2
Horner, W. G., 316/2, **338**
Horn-reflector aerial, **338**
Horse, **339**
Hotels, CCTV in, 159
Hot spot, 216, **339,** 378/2, 606/1
How round, 2/2, 203/2, **339,** 751/1
Howell, A. S., 58/1, **339**
Hoxie, C. A., 320/2
Hubley, J., 27/1
Hue, 181, 186/2, 190/2, 227/1, **339,** 516/2, 655/2
  error, 534/1
Hughes, W. C., 316/2, **339**
Hum, 60/1, **339,** 508
  balancer, **339**
  bar, 212/2, **339**
  displacement, **339**
  residual, 556
Human figure, 32
Humidity
  effect on film, 284/2, 288, 290/2 ff.
  effect on tape, 433/1
Humorous Phases and Funny Faces, 784/2
Hurd, E., 62/2, 325/1, **339**
Hurter & Driffield, 116/1
Hybrid, **339,** *340*
Hydraulic pedestal, 540/2
Hydrotype process, **340,** 350/1
Hypergonar, **340,** 981/2
Hypersensitizing, **340**
Hypo, 121/2, 122, **340**
  test, 292
Hypothetical reference circuit, 894/2 ff.

Iconoscope, *77,* **127/2, 263/1,** 331/1, 333/2, **341,** 424/1, 645/2, 990/2
Ideal radiator, 410/2
Identification (ident) signal, 274/1, **431,** 618/1, **747**
IF amplifier, **341**
Illuminant
  CIE, 193/2
  cinemicrography, 148
  printer, 195
  time-lapse, 874
  visual adaptation, 182/1
Illumination, 90, 130, 146
  dark-ground, 131/1, 148
  Köhler's, 131/1, 146/2
  Lieberkühn, 147/2
  overstage, 147/1
  unit, 298/2, 424/2, 429/2
Image
  amplifier, 334/1
  area, 211/2
  blur, 312/2
  clarity and signal-to-noise ratio, *349*
  compensation, 313/2
  compression, 395/2
  contrast, 295/1
  converter, 314, 500/2
  developed, 117/2
  diagonal, **341**
  dissector, *77,* **127/2, 272/1,** 314/2, 331, **341,** 825/1
  dye, 178
  expansion, 395/2
  formation, **342,** 388/2
  grain, 298/2
  iconoscope, 77, 127/2, 333/2 ff., **350**
  intensifier, 152, 314
  multiplier, 334/1
  orbiting, 641/2, 783/1

Image
orientation, **350**
orthicon, *76*, *78*, *83*, 93 ff., 127/2, 158/2, 211/1, *333*, **350**, 389/2, 506/2
angle of view, 86/2
beam, 428/2, 628/2, 758/1
characteristics, *79*, *881*
cineradiography, 153/2
distortion, 250/1
history, 333/2 ff.
lens, 391/2 ff.
lighting, 422
mobile, 530/1
standards converter, 765
sticking, 782/1
temperature, 84/2
photographic, 10, 228/1, 305/2
point in space, 3-D, *779*
quality, 388/1
replacement, 724
residual, 177/1
retention, 782/1, 949/1
simulator, 362/2
size, 280, 301/2, 471/2, 729/1
spread, 703/1, 854/1
stability (Kalvar), 368
surface, 391/2
transfer converter, **350**
tubes, 314/2
velocity, 312
Imbibition
printing, **350**, 369/1
prints, 53/1
process, 174/1 ff., 340/2, *351*, 496/2
Impact, sound, 9/1
Impedance, microphone, 463
Impingement drying, **354**
Impulse noise, **354**
In-between, *29*, 30/1, **354**
Inching, **354**
Inching knob, 74
Incident light meter, 267/2, *268*, 271/1
Incident ray, 519/1
Independent frame, **354**
Independent producer, 590
Independent Television Authority (ITA), 235/2, 592/2
Index strip, 42/1, 990/1
Index tube, **354**
Induced noise, 508
Industrial films
animation, 38/1
graphic techniques, 30/2 ff.
Industry, closed circuit TV in, 160 ff.
Information
capacity, TV/film, 969/2
retrieval (*see* Library)
storage converters, 764
systems, 161
unit, 349/1
Informational sensitivity and density, *349*
Infra-red
absorption, colour film, 359/2
camera, 314/2
drying, **354**
filter, 363/2
photography, **354**, 731/2, 453/2
radiation, **354**, 260/1, 410/2
Inkers, **354**
Inlay, **354**, 355/1, *733*, *734*
mask, **354**
Input equipment, transmitter, 906
Input pulse, inhibition of, 806
Insert, **354**
Insertion, **355**
loss, 263/2
picture, 733
Inspection of film
factory, 289
laboratory, 264/2, 355, 372 ff., 950/1
Installation
power, 564
wired television, 893/1
Instantaneous characteristic, **355**
Instantaneous slope, 302/2, 561/2
Instant talk-in, 745
Institute of Radio Engineers, 217/1
Instructional films
animation, 25, 31/1
editing, 257/1
medical, 458
Instructional systems, closed circuit, 163
Instrumental recording film, 669/1
Insulation, sound, 8
Insurance, 375/1
Insurance print, 246/2
Integral
density, 664/2
screen, **355**, 777/2
tripack, 12/2, 39/2, 50/2, 120/2, 188/2, 251/2, 260/1, 273/2, 301/2, 303/1, 304/1, 353/2, **355**, 364, 370/2, 470/2, 525/2, 628/2, *666*, 668/1, 722/2, 815
Intelsat satellites, 637/1, 638/1, 638, 640/1
Intensification, **363**
Intensity
illumination, 411
light, 55/1, 343/2 ff.
scale exposure, **363**
sound, 679, 683/2
theory, 683
visual perception, 971
Intercarrier sound, **363**
Intercut, 256/2, 354/2, **363**
Interdupe, 364/1
Interference
colour television, 192/1 ff., 513/2
filter, **363**
light, 165, 212/1, 226/2, 227/1, *336*, *407*, 470/1, 507/1
narrow band, 692/1
transmission, 889, 896/1
Interfield cut, 214/2, **363**
Interiors, natural, 103
Interlace, 127, 205/1, 264/1, 273/2, 331, **363**, *364*, 603/2, 641/2, 804, 969/2
Interlock, **364**, 502/1, 658/2
Intermediate, **364**
film system, 332/2, **364**, 824/2
frequencies, 610/2
Intermittent movement, 65/2, *67*, 70, 157/2, 300/1, 312, **364**, 441/1, 468/2, 518/2, *595*, *597*, 673/1, 871
Bell-Howell, 58/1
early, 42/2, 316/2, 339/1, 366/2
telecine, 818/1
Intermittent printer, **364**, 374/2, 571, 575/2
Intermodulation, **364**, 703
International broadcasts, 748
International CIE system, 184
International Commission of Lighting, 184/1
International Commission on Illumination, 646/2
International Exchange, 500/2
International Federation of Film Archives, 398/1
International Radio Regulations, 56/2
International Standardisation Organisation, 365/2, 625/2
International Telecommunications Union, 637/1
Internegative, **364**, 496/1
Interpolation techniques (conversion), 764, *767*
Interpositive, **364**
Intersync, 772/1
Intertel, 593
Intervalometer, 75/1
Intervision, 640/1
In-the-camera (special effects), 724, *726*
Inverse Square law (light), 411
Inversion, **365**
Inverted telephoto lenses, *394*
Ion burn, **365**
Ionosphere, **365**
Ion trap, 365/1
Iris, 84/2, 226/1, **365**
wipe, **365**
Isolating amplifiers (vision mixer), 964
ISO standards, **365**
Isotropic radiator, 12/1
ITA, 235/2
Ivanov, S., 324/1
Ives, F. E., 125/2, 324/1, **365**, 538/1, 777/2
Ives, H. E., 125/2, 324/1, 329/2 ff., 334/2, **365**, 777/2
Ives' patent, *538*

Jacketed lamp, **366**
Jankers, Dr., 152/1
Janssen, P. J. C., 66/2, 124/2, 317/1, **366**
Jazz Singer, The, 125/2, 320/1
Jelly, 177/2, **366**
Jenkins, C. F., 42/2, 319/1, 329/2 ff., **366**
Television Co., 366/2
Jet
nitrogen, 12/2
air, 12/2, 359/1
Joiner, **366**
Joining magnetic film, 491
Joins (*see* Splices)
Joly, H., 319/1, **366**, 538/2
Joyeux Microbes, 166/2
Jumbo roll, **366**, 436/1, 438/1
Jump cut, **366**
Jump scan, **366**
telecine, 818/2

Kalmus, H. T., **367**
Kalvar process, **367**, *368*
Kamm stand, 454/2
Kearton, C., 599/2
Keller-Dorian process, 172/2, 173/1
Kelley, W., **369**
Kelleycolor, 369/1
Kell factor, **369**
Kelvin, Lord, 199/1, 266/1, 369/2
Kelvin scale (degrees K), **369**

Kerr
  cell, 314/1, 326/1, 331/1, 369/2
  effect, 369/2
Kerr, Rev. J., **369**
Kesdacolor, 369/1
Keyboard, sound effects, 506/1
Keyed insertion, 733/2
Keying
  pulse, 515/2
  signal, **369,** 733/2 ff., *740*
Key light, 209/1, 267/1, 270/2, **369,** 422/2, 755/2
Keystone, **370**
  correction, 380/1
Kinemacolor, 125/1, 172/1, 321/1, 369/1, **370,** 676/2
Kinematographe, 460/2
Kinematograph Manufacturers Association, 539/2
Kinematoscope, 658/2
Kinescope, 62/1, 261/1, 334/2, **370**
Kinescope recording (*see* Telerecording)
Kinesigraph, 124/2, 241/1
Kinetic lantern, 10/1, 124/2, 319
Kinetograph, 124/2, 227/1, 253/1
Kinetoscope, 42/2, 124/2, 125/1, 227/1, 253/1, 318 ff., 382/1, 396/2, 429/1, 460/2, 538/2, 539/2
Kinopanorama, 126/2, 156/2, 322/2, **370,** 987/1
Kino-Vario, **370**
Klystron, **370,** 466/1
Knee, 79, 205/2, **370**
Knife-edge exposure, *10*
Knob-a-channel mixer, 959
Kodachrome, 120/2, 125/1 ff., 176/1, *177*, 179, 210/2, 321/2 ff., 356/1 ff., **370,** 477/1
  processing, 177
Kodacolor, 125/2, 173/1, 177, 321/2, 356/2, **370**
Kodak
  No. 1 camera, 251/1
  standard perforation, 277/1, **370,** 563/2
  Type 6 sensitometer, *662/2*
Kodak, Eastman, 350/2, 398/1, 475/2 ff.
Köhler's illumination, 131/1, *146*
Koko the Clown, 25/2
K-rating system, **370,** 894

Laboratory
  control, 355/1
  organisation, **371**
  reference print, 375/2
Lacquer, **377**
Lacquer discs, 695
La Cucuracha, 174/2
Lag, 80 ff., **377,** 676/2, 822, 949/1
Lambert, **377**
Lambert's law, 412/1
Lamp
  arc, **42/2,** 63/2, 311/1
  calibrated, sensitometer, 660
  high-speed photography, 314
  life, 413/2
  power supply, 62/1, 150/1, 228/1 661/1
  projector, 596, 606
    automation, 521
  sodium and mercury vapour, 239/1
  studio, 57/1, 296/1, 300/2, 413, 418/1, 601/1, 755/2, 917/2
  tungsten halogen, 150/1, 366/1, 916/1
  Xenon, 989/1
Land, E. H., 777
Langevin, P., 990/2
Language of editing, 256
Lani Bird satellite, 638
Lanyard microphone, 464/1
Lap dissolve, 239/2, 247/1, **377,** 469/1, 818/2, 828/2
Lapping switch, 214/2
Large-format film systems, 276/1, 279, 980
Large-screen TV, 55/2, 180/2, 259/2, 370/1, **377,** 675/2
Laryngoscopy, indirect, 457
Laser, **382,** 336
Latensification, **382**
Latent image, 264/2, 342, 367/2, **382**
  fading, 662
  stability (Kalvar), 368
Lateral colour, 393/1
Lateral magnification, 521
Laterna Magika, **382**
Latham loop, 382/2, 428/1
Latham, Major W., 319/1, **382**
Latitude, 266, **382**
Laugh-O-Grams, 325/1
Lauste, E. A., 125/1, 319/2, 320/2, **382**
Lauste's sound film system, *320*
Lavalier microphone, 464/1, 502/1
Lavender, 246/1, **382,** 375/1
Lawrence tube, 180/2, **382,** *383*
Laws of reflection, *408*
Laws of refraction, *408*
Layout
  artist, 33
  lighting, 415
  master control, 442
Leader, 64/1, 309/2, **384,** *623*
  Academy, 2/1, 622
  television film, 624, *625*
Lecher line, **384**
Lee, F. M., 125/1
Legal controls, producing, 592
Leica format, 289/2
Le Mouvement, 144/1
Length of reel, 620/1
Lens, **384**
  acceptance angle, 2/1
  anamorphic, 13/2, 71/2, 123/1, 134/2, 323
  aperture, 40/2
  camera (film), 71
  camera (television), 76, 81, 110
  cinemascope projection, *138*
  collimation, 170/1, 621/2
  condenser, 130, 207/1
  constructions, 392
  convergence, *781*
  converging, 228/2, 519/2
  copy, 210/1, 572/2
  cylindrical, 134/2, 981/2
  defect, 213/2, 240/1, 521, 274/1
  deviation, 519
  dioptre, **228/2**
  disc, 328/1
  diverging, 519/2
  electronic, 384/2
  electrostatic, 671/1
  field, 274/1
  fish-eye, 294/2
  Fresnel, 300/2
  hood, 72/1, 168/2, **384**
  hypergonar, 123/1
  inverted telephoto, 628/2
  lenticular, 93/1
  long focus, 845/2
  macro, 129
  mirror, 152/2
  outside broadcast, 530
  performance, **384**
  power, 228/2, 521
  projector, 596/2, 923/1
  quality, 390/2
  relay, 86, 93
  separation, 778
  settings, 269
  shapes, *520*
  split-field, 228/2
  stops, 524/2
  studio, 96/1
  supplementary, 128, 228/2
  telephoto, **845/2**
  testing, *385*
  types, **391**
  underwater, 921
  Varotal III, 95/1
  wide-angle, 606, **978/2**
Lens-drum system, 331/1
Lenticular
  process, *172*, 321/2, **396,** 427/2
  raster stereoscreen, 778/2
  screen, 605, 653/2
Le Prince, L. A. A., 66/2, 318/1, **396**
Leroy, J. A., **396**
Les Couleurs en Photographie, 245/1
Level-dependent gain, **399**
Level-dependent phase, **396**
Level sync, 493, 803/2
Lever animation, *16*
Levy, T., 127/1
Library
  card layouts, *399*
  film, **397**
  material, storage, 291
  shots, **405**
  television, **401**
Library Association, 404/1
Lieberkühn illumination, *147*
Lift, **405**
Light, 182/2, 190/2, 263/2, **406**
  adaptation, eye, 970/2
  amplifier, 153
  application bar, **412**
  application time, **412**
  automation (projector), 52/1
  box, **412**
  collector, 268/1
  control, printer, 195, 376/1, 573
  distribution, lens, 385
  end, **412**
  energy distribution curves, *183 841*
  folded, projection beam, 606
  gain, 57/2, 314/2
  incident, 218/1, 343/2 ff., 408/1
  intensity, 149/2, 183/1, 217/2, 265/1 ff., 410/2
  leaks (perforation), 485/2
  lock, **423,** *424*
  loss, lens, 164/2
  meter, 267, 480
  modulator, 264/2, 701
  reflection (lens), 60/2, 165
  scatter, 65/1, 99/1, 212/1, *218*, 248/2, 308/1, 345/2, 346, 367 ff., *853*, 920/2

Light
  sensitivity (emulsion), 287/1
  source (*see* Lamp)
  transfer characteristics, **424**
  trap, **424**
  underwater, 920
  units, 113/1, 410/2, *411*, **424**
  valve, 361, **425,** 701
  (*see also* Illumination)
Light-change, printer, 573
  actuation, 576
  board, **412**, 510/2
  points, **412**
  punched-tape control, 376/1
Light entertainment (producer), 591
Lighting
  cinema, 141
  contrast, 209/1, 267, 311/2, 507/1
  control (television), 420
  control console, 11/2
  day-for-night, 104/2, 105/1
  director, 219
  equipment (film), **412**
  equipment (television), **418**
  exposure, 266
  exterior, 103 ff.
  grid, 417/2
  high key, 311/2
  high-speed cinematography, 314
  interior, 102 ff.
  key, 369/2
  low key, 428/2
  outside broadcast, 529/2
  studio, 785
  supervisor, **423**
  surgery, 454
  talkback, 201/2
  time-lapse, 873
Lightness, colour, 181/2
Lightning (*see* Molniya)
Light-sensitive cells, first, 326
Light-sensitive mosaic, 77, **424**
Light-splitting prism, 11/1
Lightweight camera attached to endoscope, *457*
Lily, **425**
Limiter (sound), **425**, *712*, 697/1
Limiter (vision), **425**
Limiting frequency, 389/2
Limiting resolution, **425**
Line
  animation, 15 ff.
  blanking, 54/2, 60/1, 204/2, **425,** 603/2
  clamp amplifier block diagram, *966*
  compressor, 697/1
  deflection, 603/1
  drive, **425**
  drive pulse, **426**
  drive signal, **426**
  frequency, 862/2
  locking, 548/2
  microphone, 465/2
  period, **426**
  phasing, **426,** 548/2
  rate converter, 765
  sawtooth, **426**
  sequential switching, 384/1, 669/1
  spectrum, 752/2
  spread function, 346/2, *347*
  store converter, 500/2, *766*
  strobe, **426**
  structure, **426,** 756/1
  suppression period, **426,** 612/2, 863/1
  -sync pulses, 63/2, 263/2, 544/2, 807, *811*
  tilt, **426**
  time base, **426,** 612
Linearity, **425**
Linearity control, 274/2 **425,** 614,
Linear matrix (telecine), 845
Line-of-sight transmission, 466/1 ff.
Line scanning, 296/2, **426,** 603/2, 875/1
Line shading, 426/2, 775/1
Line-up time, **426**
Linkman, **427,** 750 ff.
Linnebach, A., 427/1
Linnebach effect, 423/1, **427**
Lippmann, G., 427/1
Lippmann process, 337/1, 355/2, **427**
Lip sync, 254/1, **427,** 489/2, 600/2
Liquid gate, **427**
Liquid head, **427**
Listening levels (mixers), 691
Live action animation, 14/1
Living Screen, **427,** 987/2
Load, lighting, 416
Loading, film, 113/2, 114
Lobe, **427**
Location, 46/1, 238/1, **427**
  design, 47
  hunt, 230/2
  lighting, 416
  shooting, 698
  sound recording, 689
  surveys (outside broadcast), 468
Locking, **427,** 545, 547/2, 600/1
Locked sound head, 504/1
Logic circuits, *805*
Log sheet, 258/2, **428,** 626/2
London Television Station, 55/2
Long-distance microphones, 502
Long focal length Lens (vibration), 757/2
Long shot, 672/1
Loop, **428**
  cassette, 114/2, *216*
  8mm cassetted, 458/2
  elevator, **428**
  16mm, 458/2
  (sound and picture), **428**
Loss
  cable, 83, 100/1, 166/1
  light, 260/2, 269/2
  signal, 65/1, 200/1
Lot, **428**
Loudness, 681
Loudspeaker, **428**
  cinema, *697*, 720
  frequency, 212/1
  monitor, 691/1, 697/2
  multicellular, polar diagram, *721*
Low contrast colour materials, 179/2
Low-definition television, 329
Low-gamma stock, 363/2
Low key, 101/2, **428**
Low pass filter, **428**
Low-shrink base, 353/1, **428**
Low-velocity scanning beam, **428**
Lubrication, film, 208/2, **428**
Lubszynski, H., G., 127/2
Lumen, 42/1, 190/1, 192/2, 298/2, 411/1, 424/2, **428,** 429/2
Lumière, L. and A., 124/2, 125/1, 253/1, 318/2, 323/2, **429,** 460/1, 538/2
  Cinématographe, *319*
  intermittent movement, *67*
Luminaire, 366/1, *414*, **429**, 641/1, 675/2
Luminance, 63/1, 302/2, 411, **429**
  channel, **429**
  colour, 184/2 ff., 190/2
  constant, 85/2, 193
  range, 838, 971/2
  rear projection, 605
  signal, 97, 192/2 ff., 196 ff., **429,** 511 ff., 536/1, 655 ff., 836/2
  screens
    American, 646/2, 651
    British, 647/2 ff.
    French, 649
    German, 651
    Russian, 649
  tube, 95 ff., 658/1
  unit, 42/1, 298/2, 424/2
  video circuits, receiver, 617
Luminosity, 85/2, 181/2, 185/1, 193/2
  curve, 193/2
Luminous
  flux, 298/2, 411/1, 424/2, **429**
  intensity, 113/1, **429**
Lux, 411/2, 424/2, **429**
Lye, L., 26/1, 325/2

McCay, J. W., 325/1, **452**
McLaren, N., 26/1, 325/2
Machine control (station breaks), 449
Machine feed (processing), 359
Machine leader, **430,** 580
Macintyre, Dr., 152/1
Mackie Line, 252/1
Macrocinematography, **430**
Macro-curve, 346/1
Magazine, film, 22/1, *68*, 114/2, **430,** *475*, 583
Magenta, 204/1
  colour developer, 177/1
  dye, 178, 189/1, 355/1 ff.
Magic Fountain Pen 784/2
Magic Lantern presentation, 987/2
Magnafilm, 125/2, 323/2, **430,** 979/2
Magnascope, 125/2, 323/1, **430,** 979/2
Magnetic
  deflection, *643*
  dub, 692/1
  film, 481/2, 485/1, 694
    recorder, 697
    storage, 293, 398/1
  master, 495/1, 496/2
  particle orientation, *435*
  pick-up, 716/1
    coil, VTR, 935/2
  recording, 53/2, 320, **430,** 719, 929
    materials, **431**
    narrow gauge, 490, 722
  remanence, 431/1
  reproducer, 699/2
  sound, 284/1
    head, 541/1
    track, 283, 490, *700*, *703*, *704*, *719*
  tachometer, 936/2
  tape, 125/2, 126/1, 485/1, 502/1, 694

M

Magnetic
converters, 765
transfer, 886/1
characteristic, *718*
Magnetometry, 437/2
Magneto-optic effect, 314/1
Magnification, 129/1, 131, 149
measurement, 146
Magoo, Mister, 27/1
Mag-opt, *283*, **441**, 982/2
Mahler, J., 777/2
Mains lock filter, 804
Maintenance, aerial, 909
screen, 653
transmission networks, 899
tropics, 913
Main title, **441**
Make-up, 786, 921
Maltese cross mechanism, 43/2, 67/1, 295/2, 319/2, 328/2, 364/2, **441**, 595/2
Management lighting, 141/2
Management, studio, 792
M and E track, **441**
Manufacture, receiver, 619
Marconi, G., 62/2, 112/2, 217/1, 295/2
Marconi Company, 82/1, 217/1, 331/1 ff., 645/2
Marconi-EMI television system, 55/2, 61/2, 332/2
Marconi Mark VII tube, 95/2
Marconi Wireless Telegraph Company, 295/2
Marey, E. J., 66/2, 124/2, 144/1, 217/2, 253/2, 317/2, **441**, 473/2
film camera, *318*
Married print, 700/2, 803/2
Married sound, 443/1, 446/2
Martinez, 178/1
Maser, **441**, 639/1
Mask, 211/2, 250/1, 305/1, 323, **441**, 452/1
inlay (TV), 741
projector, 211/2, 441/2
special effects cinematography, 726, 731/1
Masking, 347/2
animation, *36*
colour, 50/2, 189, 195/2, 210/2, 357/2, **441**
automatic, 178/1
electronic, 844
Master
control, 203, **442**
operator, **450**
fader, vision mixer, 957
matching, **451**, 487
music rolls, 699/2
oscillator and mains lock, 804
picture monitor, **451**
positive, 246, 248/1, 362, **451**
reference tape, 437/2, 438/2
station, 550/1
switcher, *447*, *450*
Masts and towers, 901
Matching film to TV quality, 830
Matching of high impedance microphone, table, 463
Materials, acoustic, 7
Matrix (film), 174/2, 175/1, 178/1, 350, 352/1, **451**, 497/1
making, 175
printing, 816/1
Matrix (TV), 85/2, 89/2, 92/1, 97/2, 170/2, 192/2, 193/1, **451**, 514/1, 618
green, modification of, *91*
size, vision mixer, 961
Matte, 22/1, 72/1, **452**, *728*, 731/1 ff., 910/1
box, **452**, *71*
loop, **452**
screen, **452**, 653/1
shots, *49*, 725
Matt surface, 228/1
Maxwell bridge combining unit, *905*
Maxwell, J. Clerk, 124/2, 188/2, 295/2, 321/1, **452**
May, J., 126/2, 326/1, 677/1
Mean picture level, **452**
Mechanical
compensation, zoom, 394/2
index, library, 401/1
scanner, **452**
scanning, 55/2, 127/1, 365/2, 507/1, 630/2
sound recording, 715
television, 366/1
Mechau, E., 818/1
Medical cinematography and television, 163, **452**
Mees, C. E. K., 176/1
Méliès, G., **460**
Memory
device, 196
master control, **448**
relay systems, lighting, 421
Merchandising films, 589
Mercury vapour arc, 42/2, **460**, 874/1
Mesh
effect, **460**
spacing, 765/2
Messter, O., **460**
Metallic replacement, fixing, 674
Meter, photo-electric, 168/1
Metre-candle, 411/2
Metroscopics, 324/1
Mickey Mouse, 26/1
Microcinematography, **460**
Micro-curve, 346/1
Microdensitometer, 10, *11*, *348*
Microfilm storage, 398/2
Microgroove recording dimensions, *716*
Microphones, 61/2, 307/2, **461**, *462*, *465*, 695
news, 502
placement, 680, 688, 690
response curves, *710*
support, 294/2
television, 709
Microphony, 508
Microscope light trap, *145*
watching eyepiece, *145*
Microscopy, 144, 459
Micro-slating, *146*
Microwave
aerial, 562/2
amplifier, 910/1
links, **416**, *897*
relay, block diagram, *467*
Mil, **468**
Miller & Strange, 821/2
Millimicron, 424/2
Milliphot, 424/2
Miniature, **468**, 727
microphone, 696/1
tape recorder, 505/2
Miniaturized equipment, 502/2
Mired, **468**
Mirror drum, *313*, 316/2, 328, 329/1 ff., **468**, 630/2, 631/1, 645/2, 978/1
Mirror
camera, 315/2
effects, 726/2
grid, 381/1
projector, 216/1
reflex, 73/1
rocking, 75/2
rotating, 63/1, 75/1
scanning system, 127/1
semi-reflecting, 73/1
shots, 726
shutter, **468**
Misregistration, 99/2
Missile, 312 ff.
Mitchell
camera, 11
mechanism, *70*, **468**, *469*
Mix, 240/1, 445/2, **469**
Mixed highs, 85/2, 191 ff.
Mixed syncs, **469**
Mixer, **469**
floor, *688*
relays, 961
sound, 258
Mixing, **469**, 627/1, 696
equipment, **469**
frequency, 213/1
light, 184, 186/2
sound, 259/1, **469**
Mobile camera, 529
control room, *526*
designs, 30/1
generators, 567/2
links, control room, 467/2
recording unit, CCTV, *163*
transmitting equipment, 466/1
videotape recorder, 528/2, 531
Mock-ups, 49
Modelling lighting, 422/2 ff.
Models, 12/2, 48, **469**, 727, 788
Modelscope, 129/2
Modes, acoustics, *6*
Modulated stage (VHF), 904
Modulating
signal, 197/2
systems, colour videotape, 945
Modulation, **470**
envelope, 674/2
factor, 387, 885/2
index, 890
methods, 634
noise, 431/2
signals, 511/2
techniques, 888
transfer function, 11/1, 98/1, 346/2 ff., 385/2, *387*, *389*, 677/1, 885/2
Moiré patterning, **470**, 946/1
Molniya (Lighting), 334/2, 638/2, 640/2
Molteni lantern, 316/2
Monaural magnetic track, 705
Monitor, 444/1, **470**
characteristics, 883
filming, 155
picture, 11/2, 261/2
sound, **470**
viewfinder, 72/2, 74, **470**
Monitoring room, 699
Monitoring transmitters, 905
Monochrome
channel, **470**

Monochrome
receiver, 609
reproduction, 840
Monocular cues, 973/1
Monopack, 176, 195/1, **470**
Monopole, *419*, **471**
Monoscope, **471**
Monostable circuit, *808*
Montage, sound, 706/1
Mood, 256
Mop-up waveform corrector, **471**
Mordant, **471**
Mosaic, 173, 188, 341/1, **471**
colour films, 188/2, 189/1
screen, 172/2, 174/1, 188/2, 321/2
Motion Picture Patents Corporation, 43/2, 396/2
Motorized bars, *418*
Motors
camera, 479/1 ff.
clockwork, 167/1
drive, telerecording, 858
interdependent, 364/1
Mount
camera, 211/1, 240/2, 878/2, 924/1
free-floating, 458/1
lens, 164/2, 240/2, 916/2
Mounting, microscope and camera, 145
Movement, apparent, 968
Movex cassette, 477/1
Movielight, 483/2
Movietone, 114/1
aperture, 281/1, 282/1
frame, **471,** 979/2
Moving coil
matte, 733/2
microphone, 461 ff.
object, reproduction of, *850, 851*
picture systems, visual principles, 969
Moviola, 253/2, *254*, 300/1, **471,** 486/2, 491/2, 756/1
MTF (*see* Modulation transfer function)
Mullard banana tube, 335/2
Multi-aperture plate, 52/2
Multiburst, **471**
Multi-camera techniques, 11/2, 304/1, 314, 813/1
Multi-channel sound, 472/2, 987/2
Multicolor, 125/2
Multi-film systems, 731/2, 979/1
Multi-film travelling matte, *730*
Multi-head printer, 376/1, **472,** 569/2
Multi-image, 370/1, **472**
Multilateral programmes, 748/2
Multi-layer film, 175/1 ff., 178/1
Multi-path signals, **472**
Multiplane photography, 325/1, **472**
Multiple
camera system, 39/1
exposures, combination of 573/2
film presentation, 322, 370/1, 562/2, 976/1, 977/1, 985
printing, 498/2, 729/2
stereopairs, 778/2
switching, transmission, 899
techniques, animation, 37
Multiplexer, **472**
Multiplexing, 467/1, *819, 821, 824*
Multiplex operation, telecine, 823
Multiscreen, **472**
Multi-standard, **473**
monitors, 554
Multi-track sound, **473,** 703, 708/1, 719
Multivibrator, **473,** *806*
Munsell system, 186/2, 187/2
Music, 561/1, 683, 714, 790
composer, 234/1, 258
dynamic range, 709/1
editor, 258/1
hall, 133
mixer, 687, 699/1
recording, 5, 568/1, *690*
track, balance, 692
Musical instruments, sound, 682/1
Musical sounds, energy levels, 683
Music and effects track, 441/1, **473,** 490
Mute, **473**
Mutilation of prints, **473**
Mutoscope, 227/1
Muybridge, E. J., 124/2, 253/2, 317/2 ff., 396/1, 441/2, **473**

NA (*see* Numerical aperture)
NAB (*see* National Association of Broadcasters)
Nanometre, 410/1, 424/2
Napoléon, 322/2, 472/2
Narrow band, **474**
Narrow-band illuminant, 195
Narrow bandwidth interpolation, 770
Narrow gauge, 647/2, 669/1, 675/1
film, 1/1, 260/1, 275/2, 507/1
and equipment, **474**
production techniques, **484**
projection, 57/2
projector, 300/1
sound reproduction, 722
Narrow tube, monitor, 558/2
National Association of Broadcasters (NAB), 235/2
National broadcasting Company, 333/1
National Broadcasts, 744
National Film Archive, 398/1
National Film Board of Canada, 26/1
National Physical Laboratory, 661/1
National reference network, **499,** 894/2
National Television System Committee 333/1 ff., 511/1
(*see also* NTSC)
Necker Cube, 970/1
Negative
aperture, 40/2
assembly, **499**
assessment, 271
cutting, 374
density, 702
developing, 371
dye image, 178/1
film analysis errors, telecine, 841
film, telecine, 820
frame areas, wide screen, *984*
lenses, 393/2, 394/1
matcher, 259/1
material, 40/1
colour, 178/1, 189/1
meniscus doublets, 393/2
number, 252/2, 298/1
perforation, **499**
picture amplifier, 742/2
Negative-going sync pulses, 862/1
Negative-positive colour process, 174, 176/2
Neon glowlamp, 330/1
Network
control, 444
room, 443/1
international, 593
methods, sound, 706
monochrome, performance of, 894
national, 749
news, 504/2
switching centres (NSCs), *893/2* 899/1
Neurath, O., 27/2
Neutral
balance, 171/1
density filter, **499,** 913/1
graded, 848/2
grey-scale, telecine, 836/1
News
camera, 481
film, classification, 403
storage, 398/2, 399/1, 402/1
filming, TV, 485/1
material preparation, 500
programmes, **500**
studio complex, *501*
Newsreader, 504/1
Newsreel, **506**
base, **506**
Newsroom, editorial, 500/1
Newton's rings, **507**
Next-channel mixer, 960
switcher, 445/2 ff.
NHK, 93/1, 770/2
Niépce, J. N., 124/1
Night effects, **507**
Nine point five mm film, 125/2, 275/2 ff., 324, 476, 482, **507**
Nipkow, P., 126/2, 327/2, **507**
disc, 327, *328*, 330/1, 331/1, **507,** 631/1, 865/1
Nit, 424/2 ff., **507**
Nitrate
base, 428/2, **507**
film, 290 ff.
storage, 398/1
Nitrogen-burst agitation, **507**
Noctavision, 55/2
Nodal head, 49/1, **507**
point, 569/1
Node, 7/1, 680/1
Noise, 79/2, 307/1, 431, 437, **508,** 889/2
camera, 56/2, 60/2
electronic, 193
factor, 509
in television, 509
levels, permissible, 9
limiter, **510**
measurement, 509
power spectrum, 349
prevention, 9
reduction, **510,** *700*, 701/2
splice, 60/2
Noise-cancelling microphones, 464
Norling, J. A., 779/2
camera, 779/2, 781/1
Non-composite
signal, **510,** 559/2, 773
video mixing, 545/2

Non-flam film, 284, **510**
(*see also* Acetate *and* Safety film)
Non-linear
amplifier, **510**
distortion, 895
Non-linearity, 60/1
Non-perforated film, 262/1
Non-reverse switch, **510**
Non-sync sound track, 988/1
Non-synchronous signals, 958/1, 966
Normandin, M., 366/2
North American Standards (telerecording), 849
Notch, **510**
NRDC camera, 780/2
NTSC
colour system, 85/2, 97/1, 127/2, 196/1, 197 ff., 274/1, 333/1, **511**, 615
converter, 765
genlock, 552/1
recording, 947/2 ff.
Numbering machine, **517**
Numerical aperture, 149, **517**
N.V. Philips Gloeilampenfabrieken, 561/2
(*see also* Philips, N.V.)

Oastler College, 163/2
Object and image distances, thin lens, *521*
Objective, microscope, 129/1
Oblique
lighting, 147/2
mode resonance, 6
Off-air reception, 159/2
Off-centre zoom, *21*
Off-microphone, **518**
Omnibus talkback, 745, 750
One-light
dupe, 247/1, **518**
print, **518**
Opacity 217/2
Opaque, **518**
Open-tube endoscopy, 455
Ophthalmic conditions, 453
Optec, 129/2
Optical
compensator, 75/1, 313/2, 394/2, 562/2
converter, 764
divider, 362/1
effects, 1/1, 53/1, 272/1, **518**
enlargement, 578/2
film recording, 695
image transfer, *764*
intermittent, **518**
principles, **519**
printer, 40/2, 65, 374/2, *525*, 572, 644/1, 675/2, *729*, *731*
printing, 248, 343/2, 469/2, 497/2, **525**, 729
recording, 302/1, 718
reduction printer, 496/1
sound, 525, 676/1, 702/1
head, **525**
track, 53/2, 58/2, 325/1, 494/2, 510/1
telescope, endo-, 456 ff.
tracks, 491, 702
transfer function, 385/2, 387 ff., 886/1
transfers, 692, 700, 886/1
transitions, 729/1
Opticolor, 125/2, 321/2
Optics
camera, TV, 86
direct, 95
relay, 95
Orange-red coloured coupler, 178/2
Orbiting satellites, 635
Orbit network, 640/2
Orbits, satellite, 633
Orchestra
layout, TV, *714*
recording, 690, 699/1, 976/2
Order of magnitude, **525**
Organization, producing, 592
Orientation, magnetic particles, VTR, 935/1
Original, **525**
Orthicon, 78, 127/2, 333/2 ff.
Orthochromatic, 343/1, **525**
Orwocolor, **525**
OTF for lens-film combinations, *388*
(*see also* Optical transfer function)
Oude Delft, 314/2
Our Navy, 369/1
Output
filters, 809
pulses, bistable circuits, 808
Outside broadcast, 466/1, 626/1, 744, 949/2
camera dolly, 240
genlock, 550/1
power for, 567
recording, 504/2
site connections, *746*
sound, 709
units, **525**, 796
vehicle, 526
Outside broadcasts, 111, 203, 898
and networks, 545
Out-take, **531**
Out (trim), 911/2
Overexposure, 265/1, **531**
Overlay, 355/1, **531**, 733/2, 735, *736 ff.*
Overloading, **532**
Over-modulation, 951/2
Overscan, **532**
Overshoot, **532**, 630/1
Overstage illumination, *147*
Overtones, 682/1
Oxberry peg, 23/1
Oxidation, 116/2, 118/2, 119, 120, 122/1, 359, **532**

Pace and timing, 237
Pack shot, 128
Paddle-type light meter, *268*
Painted matte shot, *47*, **533**
PAL colour system, 171, 196/1 ff., 274/1, 335/1, **533**, *534*, 616
genlock, 552/2
videotape recording, 947
Pal, G., 325/2
Pallophotophone, 320/2
Pan, **537**
and tilt head, 672/1
motorized, 241
shot, 672/1
Pan pot, **537**, 704
Panavision, **537**, 719/1
Panchromatic (film), 125/1, 246/2, 287/2, 343/1, 427/1, **537**, 668/1
Panchromotion, 369/1
Pan-Matrix, 178/1
Panning, *36*, 300/2, 978/2
effect of, 676/2, 784/1
head, 110
shot, 240/2, 803/1
Panopticon, 227/1
Panorama cloth, 214/2
Panoramic (360°) projection, 125/1
Pantograph, *24*, 420/1, **537**
Paper
positive, 344
roll film, 251/1
tape, 432/1
Papering, **537**
Parabolic
aerial, **537**
reflector microphone, 464/2, 502/2
shading, 604/1
surfaces, 394/2
waveform, 365/2, 379/2
Parallax, 132, **537**, 781/1, 950/2
compensation, 470/2
motion, 973/1
panoramagram, **538**
stereogram, 125/2, 324, 365/2, **538**
Parallel sync, 594/2
wire feeder, 273
Parallelogram-shaped scan, 675/1
Parametric amplifier, 639/1
Paramount Picture Corporation, 383/1
Parc-noise, 832/2
Paris, J. A., 124/1, 316/1, **538**
Partially-reflecting mirrors, losses, *88*
Partial stored field system, telerecording, 848
Particle velocity, 7
Partition noise, 508
Pass-band, 294/1, 300/2
Patch, **538**
Patching, **538**
Patching, lighting, 419/1
Path attenuation, **538**
Pathé, C., 366/2, **538**
Cinema, 275/2
Frères, 307/1, 321/1, 324/2, 474/2 ff.
Kok, 324/2
projector, 125/1
Pathécolor, 321/1, 538/2
stencil process, 125/1
Pattern generator, **538**, 733 ff., *739*
Pattern variation, microphones, 463
Paul, R. W., 10/1, 43/2, 67/1, 125/1, 253/1, 319, **539**
Pay-TV, 160
Peak
brightness, **539**
excursion, 978/2
limiting, 250/2
performance, 426/2
programme level, 425/2
programme meter, **539**, 708/2
signal, 205/2
white, 425/2, 861/2, 884/2
Peak-to-peak, **540**
Pedestal, **540**
camera, 211/1
Peg
bar, *23*, *35*
floating, 23

Peg
  tracks, 22, *23*
Pellicle, **541**
Penthouse head, **541**
  film path, *705*
Pentode electron gun, **541**
Perceived colour, 181
Perception, 27/2
Perceptual
  constancy, vision, 973
  mechanisms, 972
Perforated paper, 55/2
Perforating
  film, 288
  tape, 436
Perforations, 67, 275 ff., 276, **541**
  size and spacing, screens, 652
  types and dimensions, *276*
Perforator, 339/1
Periodic noise, **541**
  time, 678
Peritoneoscopy, 455/1
Perry, J., 126/2, **541**
Persistence of vision, 124/1, 316/1, 321/2, 327/2, 364/1, **541**, 630/1, 758/2, 968
Persistent phosphor, **541**
Personal microphones, 464
Personnel
  animation, 32
  production, 200/2
Perspective
  and interpretation, *973*
  set building, *45*, 47
  sound, 688
Perutz film, 173/1
Petersen, A. C. G., 125/2, 303/2, 320/2
Petzval lenses, 392
pH, 117/2, 119, 120, 121/2, 122/1
Phantascope, 42/2, 366/1
Phantom, **541**
Phase Alternation Line (PAL), 196/1 ff., **533**
  field identification signal, 810/2
  sync pulse generator, 810
  videotape recording, 947/2 ff.
  (*see also* PAL colour system)
Phase
  comparator, **541**, 804/2
  correction, PAL, 533
  corrector, 471/1
  detector, 546/2
  deviation, 546/1
  difference, 544/2, 684
  error, 64/2, 199/1
    correction, 197
  lock, **541**
  locked loop, **541**, 546, *547*
  reference, 64/1, 171/2
  response, 263/1
  stability, 548/1
Phase-frequency response, 2/2
Phaseless boost, **541**
Phenakistiscope, 124, 217/2, 316/1, 325/1, 473/2, 561/1
Philco-Apple tube, 335/2
Philips, N. V., 80/1, 94/1, 127/2, 152/2, 334/1, 561/2
Phi-phenomenon, 968
Phon, 681/2
Phonofilm, 114/1, 125/1, 320/1
Phonograph, 124/2, 125/1, 253/1
Phonoscope, 217/2
Phosphor, 12/2, 188/2, 261/2, 308/1, 314/2, 541/2
  all sulphide, 843
  colour response, 838
  dots, 180/1, 669/2
  grain, **541**
  granularity, 854/1
  saturation, **542**
  screen, 335/2, **542**
    brightness, *542*
  strips, 42/1, 383
Phosphors for television tubes, **542**
Phot, 411/2, 424/2, **543**
Photion, 125/1
Photo-
  cathode, 76/2, 77, 78/1, 79, 153/1, 262/2, 314/2
  cell, 72/1, 114/1, 127/1, 188/1, 218/1, 267/2 ff., 329/1, 330/1 ff., **543**, 664/2
  conducting tubes, 80, 334/1, **543**
  diode, **543**
  electric
    densitometer, 666/1
    effect, 314/1
  electron, 77, 78/1, 79
  emission tubes, 77/1
  finish, horse race, 314/1
  flood, 62/1, **543**
  instrumentation, 312/1
  multiplier, 260/2, 315/2, **543**, 664/2, 862
    colour response, *826*, 838
  pulse, 751/2
  telegraphy, 58/2, 114/1, 126/1, 326/2, **543**
Photographic
  experiments, early, 317
  gun, 124/2, 441/2
  image, 306/1, **342**
  materials as linear systems, 347
  revolver, 66/2, 124/2, 366/1
  rifle, 66/2
  sound recording, 716
  speed measurement, 663/1
Photometry, 309/1, 410 ff., 412/2, **543**
Photophone, 125/2
Photopic
  and scotopic response, *971*
  response curve, **543**
  vision, luminosity curve, *836*
Photopolymerization, 343/1
Physical characteristics, film, 284
Pick-up
  electric and acoustic, 566
  tube, 91, **543**
Pick-ups, 716
Picture
  area, 280/2 ff.
  components, TV, *863*
  defect, 784/1
  degradation, 64/1
  detail, 862
  editor, 253/2
  element, 77, **543**
  frequency, 860
  gate, 12/1
  i.f. and detector, 617
  interpolation, 771/2
  inversion, **543**
  join, telerecording, 849/2 ff.
  line-up generating equipment (PLUGE), **543**, 555/1, 557/2
  locking techniques, **544**
  matching, **553**
  modulation, **553**, 610/1
  monitors, 451/1, 543/2, **553**, *554*, 950/2
  quality, 13/1, *386*
  scanning, 642/1
  shift, 772/2
  signals, 471/1, 861
  sources, 500
  sync ratio, **559**, 862/1
  telegraph, 326
  tube, receiver, **560**, 610, 612
    light output range, *884*
    transfer characteristics, *556*
Picturephone, 330/1
Piezo electricity 212/2, 461/2, **560**
Pigment layer, 40/1
Pilot, **560**
  beam, 42/1
  pin, 58/1, 70/2, *560*, 621/1
  tone, 481/2, **560**
Pin, 5, **560**
  belt, 17/1, *6352*, **560**
Pincushion distortion, *523*, **560**
Pinhole camera, 126/2
Pinocchio, 26/1
Pipe grids, lighting, *420*
Pitch
  perforation, 277 ff., **560**
    difference, *571*
  sound, **561**, 679, 680
Pixillation, 32/1, 325/2, **561**
Pixlok, 772/1
Plasticizer, **561**
Plastic cartridge, 477/1
Plate
  aperture, 40/2
  background, 48/2, **561**
Plateau, J. A., 124/1, 316/1, **561**, 758/2
Platen, 23
Playback, **561**
Playing times, tape (*see* Reel sizes)
Plotting and interpretation, densitometry, 664
Plücker, J., 126/1, 328/2, **561**
PLUGE, **543/2**, *555/1*, 557/2
Plumbicon, 11/1, 80/1, 81, 82/2, 83/2, 92/2 ff., 94 ff., 97/1, 98/2 ff., 100/1, 127/2, 158/2, 211/1, 261/2, 334/1, 391/2 ff., 396/2, 502/2, **561**, 603/1, 882, 885/2
Pneumatic pulldown, *819*, 829/1
Point gamma, 302/2, **561**
Point spread function, 346/1
Polar cinematography (*see* Cold climate)
Polar diagram, 462/2, **562**
Polarity
  reversal, 198/1, 742
  signal, 365/1
Polarization, 94/1, 96/1, 166/1, **562**
  signal, 468/1
Polarized
  light, 39/1, 369/2
  projection, 125, 126/1, 777/1
  relief images, 924/1
  spectacles, 924/1
Polaroid, 125/2, 407/1
  Corporation, 777/2
Pola screens, **561**
Polychromide, 125/1
Polyécran, **562**, 987/2
Polyester
  base, 367/1, 478/2
  tape joiner, 498/1

P

Polyfolium chromodialytique, 176/1
Polygon machine, telecine, **562,** 828
Polyrod aerial, **562**
Popeye, 26/1, 27/1, 325/1
Porch, **562**
  (*see also* Back porch *and* Front porch)
Porous
  absorbers, 6 ff.
  structure, 2/2
Portable recorders, 696
Porter, E. S., 256/1
Positional control servo, **563**
Positive
  development, 628/2
  displacement pump, viscous processing, 952
  drive, **563**, 580/2
  lens, 392/2, 393/2 ff.
  perforation, **563**
  print film processing sequence, 360
  standard perforation, 277/1
Positive-going sync pulses, 862/1
Pos-neg switch, 309/2, **563**
Post-deflection focus, 382/2
Post-equalization, 263/2
Post-synchronization, 428/1, **563,** 698
Post-sync line blanking, 563/1
Poulsen, A., 125/2, 303/2, 320/2
Power
  camera, location, 698/2
  carrier, 889
  distribution, 416
  gain, 12/1, 13/1
  generation, lighting, 415
  lock, 547
  requirements, mobile TX, 467/2
  separation filters, **563**
  stations, 160
  supplies, **564,** 613
    receiver, 610
    repeaters, 896/1
    satellite, 634
    studio, 796
    transmitters, 904, 906
  transmitting, 901
  van, 529
PPM, 540/1
  (*see also* Peak programme meter)
Praxinoscope, 124/2, 316/2, 318/1, 325/1, 629/1
Pre-amp, **567**
Precision
  monitor, 451/1
  step printer, 572/1
Pre-emphasis, 198/1, **567,** *656*, 657/2, 702/1
Pre-equalization, 263/2
Pre-listen, 445/2 ff., **567**
Première, **568**
Premises, library, 405
Pre-mix, 259/1, **568,** 692/2, 704/1
Pre-production, 43/2, 44, 45/1
Pre-recording, 561/1, 716/1
Pre-roll time, 449/2
Pre-scoring, **568**
Preselection, 783/2
Presentation, 202, 442, **568,** 795, 949/1
  control, *443, 444*
  mixer, 443/1 ff., **568,** 672/2
  studio, 443, 672/2
Preservation of film, 398
Preservatives, **568**
  developer, 116, 118/2, 121/2
Pre-set
  fades, 420/2
  switcher, 445/2
Pressure
  plate, 21, 70/2, **568**
  waves, sound, *679*
Pre-striped film, 485/2, 499/1
Pre-sync line blanking, 563/1
Pretuned receiver, **568**
Preview, 445/2, **568**
  monitor, 957/2
  monitoring, 510/2
  selector, vision mixer, 957
  switcher, 444/1
Previewer. VTR, 813/1
Primary colours, 11/2, 89, 185/2, 188/2, 190 ff., 192 ff., 226/2, **568,** *569*
Principal
  focus, 520, **569**
  plane, **569**
  points, **569**
Print
  density, 196/1, 702
  geometry (16mm), **576,** *577*
  graininess, 348
Printer
  card, 374/2
  continuous contact, *570*
  control band, **209**
  light, 518, **569**
    setting, 412/2
    sources, 575
  multi-head, **472/1**
  (operator), **569**
  optical intermittent, *572*
  point, 270, 373/1, **569**
  rotary, **631/1**
  scale, 569/2, 574, **576**
  setting, 362
  start marks, 492
  step, **774/2**
  sync, 493
Printers, **570**
Printing, 428/1
  A and B, 1/1, 488
  beam, 361/2
  characteristics, 212/1
  colour negatives, 361
  effect, 56/2
  errors, 841
  exposure, 195, 196/2, 270, 644/2
  illuminant, 195/1
  level, 373/1
  light, 209/2, 425/1
  machines, 373, 376
  procedure, 494
  sixteen mm. A and B, *488*
  sync, 803/2
  wet, 572 ff., *573*, **978/1**
Print-out, 342/2
Prints for TV, 377/2
Print-through, 293/2, 431/2, 433/1, 437/2, 438/1, **578**
  curve, transfer, 877/2
  gamma, 880/2
Print-up, **578**
Prism, 22/1, 75/1, 96/1, 174/2, 208/1, 312/2, 396/1, 409, 457/2, 459/1, 562/2
  beam splitter, 89/1, 94/1, 95/1, *96*, *395*
  optical, 981/2
  polygonal, 828/2
  rotating, 604/1
Prismatic tubes, 780/1
Prizmacolor, 125/1, 321/2, 369/1
Process
  cabinet, 954
  camera, *728*, 729/1
  control, 584
  photography, 595/2
  shot, **588**
Processing, 60/2, 113/2, 225, 229/2, 262/2, 298/2, 306/2, 309/2, **578,** 619/1, 621/1, 627/1, 756/1, 783/1, 978/1
  camera stock, 493
  cold climate, 170
  colour, 210/1, *358*
  continuous, 12/2, 54/1
  control, 659/1 ff.
  film for TV, 880/2
  Kalvar, 368/2
  machine, 359/1
    film paths, *581*
    viscous, 954
  narrow-gauge, 493
  news, 503
  sensitometric test, 663
  sequence, colour negative, 359
  temperature, 628/1
  time, 371/2
  tropical, 915
  viscous, 955
  X-ray, 154
Producer
  film, **588**
  outside broadcast, 527/1
  television, **590**
Producer/director, 229/2
Production
  assistant, 236/1
    OB, 527/1
  control room, 201/1
    news, 506
  manager, 792
  planning, 235
  secretary, 207/2, 238/2
  team, 219, 236
  techniques, animation, 28
Professional 16mm cameras, 480
Programme
  channels, transmission, 899
  circuits, OB, *749*
  continuity, 783/2
  controller, 443/2
  control units, 451/1
  department and cast, 793
  level, 878/1
  log, 448/1
  matrix, 52/1, 53
  routing, *795*
  switcher, 442/2
  switching centre (PSC), *594*, 893
  timing, 202/2
Programmed camera instructions, 205
Programmer, time-lapse, 872
Projected aspect ratios, comparison, *979*
Projected picture and track, dimensions, *280*
Projection
  density, 249/1
  lamp, 600/1
  optics, 867/1
  room, *142*, 244, 790/2
  stereo, 776

Projection
  sync, **594**
  tube television, **594**
Projectionist, **594**
Projector, 54/2, 67/2, 114, 215/2, **594**
  aperture, 2/1, 41/1
  double-headed, 490
  dual gauge, 478/2
  lamp, 315/1
  location, 138
  remote control, *143*, *144*
  telecine, 823
  television, large-screen, *159*
  throw, focal length, picture width, *598*
Prompter, 213/1, **599**
Properties, 45
Proportion, studio, 798
Prop plot, **599**
Props, **599**
Proszynski, K. de, 319/2, **599**
Protective
  leader material, 623
  master, 375, **599**
Pseudo
  equalizing pulses, 807/2
  field-sync pulses, 807/2
  line-sync pulses, 807/2
  lock, 552
  stereo, 704/1
Ptushko, A., 325/2
Publicity films, 25
Pudovkin, V. I., 256/1
Pulldown, 57/2, 69, 394/1, **600**, 847
  telecine, 818
Pull-in range, 547/1
Pulse
  and bar, **600**
  generation, centralized, 545
  generator, 503
  modulation, **600**
  phasing, **600**
  shaper, vision mixer, 966
  sound, **600**
  sync, **600**
  track, 485/1, 696/2, 700/2
  triggering, *539*
Pulsed
  lamp, 596/1, **600**
  light system, 60-field TV, *823*
  tape, 505/2
  track, 502/1
  Xenon arc lamp, 315/1
Pulses, 154/2
Punched tape, 362/1, 373/2, 376/1, 576/2
Puppet film, 771/1
Pup, **601**
Purity
  light, 186
  ring magnet, 671/2
Push-pull, **601**
  sound track, **601**

Quad, **601**
Quad-8, 497/1, **601**
Quadrant diagram, **601**, *857/1*, *877/1*
Quadrature
  amplitude modulation, 197
  subcarrier component, 514/2
Quadruplex recorder, **601**
Quality, 199
  colour picture, 199
  print, 620/2
Q factor, 664/1
Quantum
  light, 345/2 ff.
  noise, 346/1
Quartz, 903/2
  crystal resonator, 903/2
  rod light guide, 456/2
  tube arc, 42/2
Quick genlock, 548
Qwart, **601**

Race-track, streak & image dissection, 314
Rack, **602**
Rack-and-tank development, 359/1, **602**
Rackover, 22/1, **602**
Radial raster, optical layout, *778*
Radiant energy, 182
Radiating element, **602**
Radiating elements, 908
Radiation, receiver, 613
Radiation resistance, **602**
Radio
  camera, TV, 530/1, **603**
  microphone, 502
Radio Corporation of America, 320/2
  (*see also* RCA)
Radio frequency, **602**
  amplifiers, VHF, 903
  broadbands, 898
  detector, 515/2
  tuned circuits, VHF, 904
Radiograph, 152/1, 459
Radio link
  and videotape, 528
  receiving vehicle, 530/1
  relay site, 529/1
Radio Manufacturers Association (America), 542/2
Radio-relay, **603**, 897
  preferred bands, table, 898
Ramp function, 83/2
Ramp, studio, 47/2
Ramsdell 3-D camera, 780/2
Random noise, 509, **603**
Random phase errors, 549
Rank-Taylor-Hobson, 95/1, 96/1
Raster (stereoscopic), **603**, 778/1
Raster (TV), 166/2, 249/2, 273/2, 363/2, *513*, **603**, 644/1
  linearity and geometry, 555
  shading, **603**
Ratio, colour, 171/1
Raw materials, emulsion, 286
Raw stock, **604**
  dimensions, 276, 278
  storage, 290
Raycol, 125/2, 321/2
Rayleigh, Lord, 684/1
RCA, 79/2, 93/2, 95/1, 125/2, 127/2, 320/2, 331/1, 332 ff., 700/2, 929/1, 942/2, 949/1, 990/2
RCFM test film, 704/1
Reactance tube oscillator, **604**
Reader, **604**
Read-through, 237
Real image, 337, 342/1, 408/2
Realism, 223
Reality of film and TV picture, 970
Real-time operation, 448/2, 450
Rear projection 216/1, **604**, 732
Rebroadcast
  receivers, 898
  transmitters, 906
Receiver, 63/1, 199, **608**
  characteristic, 884
  colour, 194/1, 197, 615 ff.
  hybrid, 611/1
  microwave, 466/2
  monochrome, 197/1, *608*, 609/1 ff.
  unrestored, 884/2
  wired TV, *613*
Reception under adverse conditions, 199
Reciprocity
  failure, 619/1, 659, *660*, 663/1
  law, **619**
Recirculation, 584, **619**
Reconditioning, **619**
Reconstruction, holograms, 337
Recording
  amplifiers, 696
  and dubbing facilities, *790*
  channels, checking, 703/2
  facilities, elections, 752
  gaits, medical, 453
  materials, 278
  methods, videotape, 928
  room, 790/2 ff.
  supervisor, 694
  sync, 594/2
  system, **619**
    sound, 694
    video, 937
Recordist, **620**
Records, laboratory, 374
Rectilinear propagation, 407
Red light exposure, 177/1
Red master, 246/2, **620**
Reduced aperture (RA), 281/2, 471/2, **620**
Reducing agent, 117
Reduction, 118/2, 119/2, 120/1
  dupe negative, 573/1
  printing, 249/1, 496, 498, **620**
  prints, 16mm, 578/1
Reel, 754/2
  of film, **620**
  length, 626
  sizes and playing times
    magnetic film, 437
    sound tape, 438
    videotape, 439
Reeling on to spools, tape, 439
Re-exposure, selective, 177/1
Reference
  beam, 336/2
  circuit, **620**
  colours, telecine, 833
  frequency, 748/1
  note, 681/2
  print, **620**
  signal, 97/1, 550
  strip, 663/1
  tone, 692/2, 698/1
Reflected-light meter, 267/2, 271/1
Reflection
  density, 343/2
  in CRT, 854/1
  laws of, 408/1
  losses, lens, 164
  signal, 304/2
  sound, 3/2, 5/1, 720/1
Reflecting stereoscope, 978/1
Reflective network, **620**

Reflectivity, 266/2 ff.
screen, 605/2
Reflector, 207/1, **620**
lamp, **620**
Reflex shutter, 468/2, **620**
viewfinder, 72/2 ff., 479/2
Refracting stereoscope, 63/1
Refraction, 239/2, 408, 519/1, **621**
laws of, 408/2, **621**
Refractive index, 164/1, 239/2, 363/1, 368/1, 409, **621,** 645/1, 920/1
Regeneration and modulation divider, 810
Regeneration of solutions, **621**
Register pin, 21/1, 22/1, 70/2, 71/1, 560/1, **621**
(*see also* Pin, Pilot pin)
Registration, 21/1, *35*, 67/2, *70*, 71/1, 85/2, 86/1, 174/1, 188/2, 276/2 ff., 300/1, 301/1, 352, 560/2, **621**, 670/2, 827/1, 829/2
colour, 353/1
scan, 97/1
telecine, 835/2, 836
Regular 8 film, 476/2
Regulation, power, 564
Regulator, automatic voltage, 51/2
Rehearsals, 237 ff.
Reiniger, L., 325/2
Relay
control vision mixer, 962
lens, *87*, 396/1, **621**
matrices, 447/1, 957
satellite, 636
system, 170/2
Release, **621**
print, 306/2, 441/1
examination and maintenance, **622**
make-up, **622**
operations sequence, *376*
storage, 290
printing, 375
machines, 569/2
time, 205/2, **626**
Relief
effect, electronic, 772/1
film, 249/1
image, 351/1
process, 626
Remanence, 437/2, **626**
Remote
control, cinemas, 144
pickup, **626**
station, 676/1
Remotes (*see* Outside broadcasts)
Renoir, J., 234/2
Repeater, 896
station, 166/1, 467/2 ff.
types, 633
video, 898/2
Replenisher, 62/1, 119
Replenishment, 360, 585
Report sheet, 428/1
Reprints, **627**
Reproduction
colour film in B and W, 830
moving objects, telerecording, 851
object, tone scale, *876*, *877*
video, 937
Re-recorded programmes, 707/2
Re-recording, 568/1, **627**, 691, 699, 791
Rescued by Rover, 310/2
Research Studio of Radio-Diffusion-Française, 38/2
Reseau, 173/2, **627**
Reservoir, 262/2, 582, **627**
Resolution, 10, *11*, 81/2, 83, 97/2, 99/1, 206/2, 338/2, 349/1, 369/1, 524, 541/2, **627**
rating, 11/1, 216/2
telerecording, 852, 855
vidicon, 822
Resolving power, 10/1, 131, 149/2, 156, 384, *385*, 388/2, 389/1, 517/1
Resonance, 5/1, 6
Resonant
absorber, 8
aerial, **627**, 908/1
Response curve, **628**
Respray, screen, 654/1
Restrainer, 117, 118/2, 121/2, **628**
Re-take, **628**
Reticle, 21/2, 22/1
Reticulation, **628**
Retrace, 296/2, 642/1
Retrofocus lens, 394/1, **628**
Return scanning beam, **628**
Reverberation, 4/1, 5, 692/1
plate, 712/1
time, 4/2, 6/2, 8/2, **628,** 690/1, 699/1, 720
units, 712
Reversal
film, 39/2, 121/1, **176/2**, 189/1, 260/1, 271/2, 280/2, 304/1, 322/1, 324/2, 474/2, 485, 493/2, 577/2
processing, 122, 227/1, *356*, 360/1
speed, 265/2
intermediate, **628**
process, 376, **628**
recording, 858
Reverse
action, **629**
angle, **629**
printing, 729/2
programme sound, 746, 751
Reversal duplicates, 249
Review room
projection, 632/1
screen luminance, 651
Reversal-reversal print, 486/1
Rewind, 255/2, 293/2, **629**
Rewinding, 291
Reynaud, E., 124/2, 316/2 ff., *317*, **629**
Reynolds, R., 152/1
Rushes, 698/1
RF amplifier, **629**
Ribbon frame and streak cameras, 313
Ribbon microphone, 461 ff.
Richter, H., 26/1
Rifle mike, 465/2, **629**
Rifle spot, 413/2, **629**
Rigging, 415, 418, 528/1, **629**
tender, **629**
Ringing, **629**
Ring-type recording head, video, *929*
Ripple, 416
Ripple effect, electronic, 742
Riser, **630**
Rocket sites, 161
Rock'n roll, dubbing, 700
Rodda, S., 127/2
Roget, P. M., 316/1, **630,** 968/1
Roll, **630**
Roll-back, 491/2, 700/1
Roller
diabolo, 225/2
title, 211/1, **630**
Roll off, **630,** *930*
Rollover, **630**
Röntgen, W., 152/1, 211/2
Room
acoustics, 3/1, 4/1
resonances, 6
volumes and reverberation times, 708
Rose, A., 127/2
Rosing, B., 62/1, 127/1, 329, **630,** 990/2
patent, *329*
Rostrum, **630**
photography, **630**
table, 36/1
Rotating
disc
analyzer, 191/1
scanner, 126/2, **631**
mirror
camera, *312*, 313, 314/1
scanner, 126/2, **631**
prism viewer, 208/1
Rotary printer, **631**
Rotary stepping switch, **631**
Rough cut 233/2, 254/1, 258/1, **631**
Routine test method, **631,** 895/1
Roving preview, 444/1 ff., **631**
Rudge, J. A. R., 301/1, **631**
Ruhmer, E., 382/2
Run-out and tail leader, 623
Run-up time, 449/2
Rushes, 156/1, 233/1, 253/2
Rush prints, 214/1, 215/1, 371/1, *373*, 493/2, **631**, 632/2
Rush printing, 372 ff.
Russian
cinemas, 137/1
progress, history, 329
screen luminance, 649
stereo camera, *780*
Wireless Telegraph & Telephone Co., 645/2

Sabine's formula, 4/2
Safe
area, 282/2, **632**
light, **632**
Safety
film, 2/2, 125/1, 284, 290, 293/1, 510/2, **632,** 910/2
lighting, 13/1, 141/2
mark, unreliability, 293/1
outside broadcast, 529
power, 566
Sample print, **632**
Satellite communications, 160/1, 334, 500/2, 562/2, 600/2, **633**, *638*, 909/1, 923/1
Saturable reactor, **641**
Saturation, 181, 186/2, 189/2, 190/2, 516/2, **641**
lighting, **641**
Saulsbury, W., 369/1
Sawtooth, 62/2, 274, 380/1
deflection, **641**
Scale models, 727
Scan
burns, **641**

Scan
  conversion, 159/2
  generator, **641**
  linearity, **641**
Scanned print, **641**
Scanner
  motor-driven, 366/1
  tube, telecine, 825
Scanning, **641**, *642*, *643*
  aperture, 332/1
  beam, 76/2, 77/2, 78/1, 79/1, 80/2, 82/2
  coil, **644**
    assembly, **644**
  crystal timing, 212/2
  direction, 614
  failure, 644/1
  film, 297/1
  history, 272/1, 326/2, 327, 330
  mechanical, 329, 468/1
  methods, 860
  pattern, 641/1, 860
  printer, **644**, 985/1
  reversal, **644**, 742
  spot, 40/1, 83/1
  standards, telecine, 817
Scan protection, 84, 380, **644**
Scene, **644**, 814/1
  elements, *34*
  illumination, 840
  tester, 362/1, 375/1, **644**
Scenery, 794
Scene-to-scene printing levels, 576/2
Schade, O., 347/1
Scheele, C. W., 124/1
Schinzel, K., 355/1
Schlieren
  optical systems, 380, *381*, *644*, 866/2, 868/2
  photography, 227/2, **644**
Schmidt optics, 159/1, 378/1 ff., **645**
Schneider, W., 176/2
Schüfftan
  process, **645**
  shot, 48, *726*, 727/1
Schulze, J. H., 124/1
Scoop, 422/2, **645**
Scope, **645**
'Scope print, 980/1
Scophony system, 333/1, **645**
Scoring, 234, 699
S correction, 643/2
Scratches, 2/1, 70/2, 128/1, 306/2, 427/2, **646**, 978/1
Scratch print, **646**
Scratch removal, 619/2, **646**, 792/2
SCR dimmer, 566, **646**
Screen
  area, 134/1, 135/1, 136/1
  brightness, **646**
  curved, 136/2, *649*, 986
  cylindrical, 370/1
  diffusing and reflecting, *605*
  drive-in cinema, 244
  front projection, 215/2
  illumination, 16mm, 722/1
  lenticular, 216/1
  light distribution, *649*
  luminance, **646**
  rear projection, 604/2
  reflectance, 63/1
  reflection, types of, *647*, *648*
  saturation, **654**
  size and brightness, 596
  tilting, *650*
  translucent, 215/2, 216/1, 303/2
  view of, *139*
Screen-foot, 298/1
Screens, 57/2, 63/1, 378, **652**
Scrim, **654**
Script, 62/2, 230
  breakdown, 792/1
  shaping of, 235
Sczcepanik, J., 127/1
Seating, 136 ff., 140, **654**
SECAM colour system, 127/2, 171/2, 196/1, 197/1, 198, 274/1, 335/1, 616, **654**, *655*
  genlock, 552/2
  recording, 948
Secondary emission, 78/1, 79/1, 262/2, 332/1, 333/2, **658**
Second generation dupe, **658**
Security, 140, 161
SEC vidicon, 91/2
Selection, library, 398, 402
Selective re-exposure, 356/1
Selenicon, 91/2, **658**
Selenium cell, 327, 328, 329/1, 330/1, 676/2
Self-synchronizing system, 658/2
Sellers, C., 561/1, **658**
Selsyn motor, 364/1, **658**
Selwyn, E. W. H., 348/1
  granularity, 348/1, 349
Semi-automatic
  light meter, 480/1
  switcher, 448/2
  sync marking, 485/1
Semiconductor
  diode, **658**
  matrices, 447/2
  switching, vision mixer, 962
Semireflecting film surface, 541/1
Semireflector, **658**, 780/2
Semi-specular surface, 653
Semi-tendency drive, 865/2
Senlacq, M., 126/2, 326/2
Sensation from constant stimulus, *968*
Sensitivity
  film, 265/1 ff., 287, 344/1 ff.
  Kalvar, 368
  microphone, 463
  scale, 268/1
Sensitization, differential, 189/1
Sensitizer, 286/2, **659**
Sensitometer, 361/1, **659**, *661*, 662
Sensitometric characteristics, Kalvar, 368
Sensitometry, 302/2, 306/2, 309/1
  and densitometry, **659**
  of colour films, **665**
Sensory characteristics, visual, 970
Separate
  luminance camera, 91/2, 97/1, 98/2
  mesh vidicon, **668**
  sync, monitors, 554
Separation
  master, 375/1, **668**
  negatives, 63/1, 174, 175/1, **668**, 802/1
  of colour images, telecine, 832
  picture and sound, 594/2
  positive, 599/2
  prints, 291/1, 292/2
  signal, 170/2
Sepmag, 504/1, 505/2, **668**, 830/2
Sepopt, **668**
Sequence, 257/2, **668**
Sequential
  colour-difference signals, 655
  development, 356
  raster, 603/2
  scanning, **669**
  signal, 260/2
  system, 191, **668**
  working, master control, 448
Séquentiel Couleur à Mémoire, 196/1, 654/1
  (*see also* SECAM)
Series capacitor d.c. clamp, *965*
Service supplies, studio, 800
Servo
  controls, 11/2, 74, 82, 530/2
  motors, 75/2
  system, videotape, 938
Servo-controlled tape mechanisms, 708/1
Set, 46, 231
  acoustics, 688/1
  design, 45 ff., 236
  estimator, 44/2
  layout, *44*
  lighting, 422/2
  noise, 688
  preparation, 785
Setting up
  animation, 35
  camera, 102/2
Set-up, **669**
  level, vision mixer, 967
Seventeen-point-five mm film, 10/1, 275/2, 324/2, **669**
Seventh Heaven, 320/2
Seventy-five mm film, 125/1, 323/2
Seventy mm
  film, 382, 126/2, 135, 139/1, 156/1, 228/1, 276 ff., 306/1, 323/2, 472/1, **669**
  prints, 980/2 ff.
  process, 875/2
  projection, 595/2, 646/2, 721
  screen luminance, 650
Shading
  correction, 84 **669**
  error, 81/1, **669**
  signal, 78/1, 274/2
  waveform, **669**
Shadow board, 24
Shadow effect, tropical, *915*
Shadowgraph, 327/1, 330/1, 423/1
Shadow mask tube, 89/2, 127/2, 179/2, 191/1, 261/2, 335/1, 618/2, 621/2, **669**, *670*
Shape constancy, vision, 974
Shape distortion, *26*
Shaping network, **671**
Sharpening, 211/1
Sharpness, 10, *11*, 98/2, 216/2, 252/1
  subjective, 677/1
Shepherd, S., 175/1
Ships, closed circuit TV, 159, 161
Schoenberg, I., 332/2, **645**, 988/1
Shooting
  in colour, 104
  pattern, 236/2
  schedule, 46/2, 231/2, 792/1
  sequence, 232
Short, **671**
Short-pitch
  film, 561/1
  perforation, 277
Shot, **671**
  box, 531/1, **672**

S

Shot
  listing, **672**
  noise, 508, **672**
Shoulder, 116/1, 265/1 ff., 344/1, **672**
Show copy, 375/2, **672**
Shrinkage, film, 278, 285, **672**, 757/1, 818, 827
Shutter, 691, **672**, *673*, 701/1
  blanking, 60/1
  control, 73, 266/2
  electronic, 850/1
  louvred, 55/1
  mirror, 479/2
  phase, monitor filming, 155
  rotating, 776/2
  sensitometer, 661
  See also Reflex shutter
Shuttle pin, **673**
Shuttle-type gate, 21/1
Sibilance, 689,703/1
Sidebands, 294/1, 889, 924 ff., 904/2
Side-pip, 81/1
Siemens and Halske, 127/1, 173/1
Signal, **673**
  characteristics, tape, 427/2
  current, 81/1
  errors, effects of, 199
  make-up and gamma-correction, telecine, 837
  noise, 832
  plate, 78/1, 79/1, **673**
  problems, colour videotape, 941
  processing, 97
  stabilizing amplifier, vision mixer, 965
  waveform elements, *563*
Signals for various colours, table, 194
Signal-to-noise ratio, 79/2, 83/1, 263/2, 346/1, 349/2, 509, 567/1, **673**, 701/1, 702, 719/1, 765/2
  telecine, 822, 845
Silencing, 16mm camera, 481
Silent
  aperture, 301/2
  frame, **673**
  photography, 269/2
  prints, 242/1
  speed, 279/2, **673**
Silhouette
  diagrams, 27/2
  effect, 423/1
Silica gel, 912/2
Silly Symphonies, 26/1, 325/1
Silver, 177, 118/2, 120/2, 121/2, 122, 305/2 ff., 352/1, 359/2, 673/2
  halide, 117, 118/2, 119, 120/1, 121/2, 122, 286/2, 305/2, 308/1, 359/2, 342 ff., 673/2
  image, 177, 178/1, 355/2
  recovery, **673**
  sound track, 363/1
Simultaneous
  colour system, 191/2
  dialogue recording, 427/2
Sinding-Larsen, A., 127/1
Sine-squared pulse, 2/2, **674**
Sine wave, 346/2
Single
  echo, **674**
  sideband, **674**
Single-8 cameras, *478*, 483
Single-8 mm film, 478/2 ff., **674**
Single-perforated film, 485/2
Single-picture control, time lapse, 871
Single-purpose lenses, 395
Single-roll printing, 1/1
Single-run 8mm film, 477
Single-shot cameras, 314/1
  picture, 314/1
Single-system cameras, 485/1
  sound recording, 481, **675**
Sink the Bismarck, 727/2
Site, transmitter, 901
Siting and layout, studio, 797
Sixteen-lens camera, 396/1
Sixteen mm, 275 ff., **675**
  camera, 151/1
  cinemas, 137/2
  equipment, 475
  films, professional, 179, 475
  history, 125/2, 324
  Internegative, 498/2
  Laboratory, 493
  print recognition, 598
  prints, making, 494
  production, 485
  projector, 596
  recording characteristic, 492
  screen luminance, 650
Six-track printing master, 704/2
Sixty-eight mm film, 125/1, 323/2
Sixty-five mm film, 126/2, 276/1, 323/2, 472/1, **675**, 875/2, 980/2
Sixty mm film, 125/1, 323/2
Size
  constancy, vision, 974
  monitor, 554
  studio, 798
Skew correction, 380, **675**
Skid, **675**
Skin tones, 94/2
Skip
  printer, 174/1
  printing, **675**, 729/2
Skladanowsky, M., 319/1, **675**
Sky
  cloth, 214/2, 422/2
  filter, **675**
  pan, 413/2, **675**
  wave propagation, 915/1
Slash
  dupe, **675**
  print, 646/2
Slate, **676**, 814/1
Slating, 482/1, 485/1, 502/1, 814/1
Slave, **676**
Slavelock, 550, *551*, **676**
Slaving, **676**
  OB, 747
Slepian, J., 127/1
Slide projector, 865/1
Slip frame, 32/1
Slipping plate clutch, 871
Slit, **676**
Slitting
  burr, **676**
  film, 288/2
  tape, 436
Slope, 205/2, 302/2, 343/2
Slot aerial, **676**
Slow genlock, 549, 551/2
Slow motion, 312/1, 441/2, **676**
Slug, **676**
Small format, 276, 279
Smear, 377/1, **676**
Smearing, 80/2, 92, 949/1
Smith, G. A., 125/1, 171/2, 321/2, **676**
Smith, W., 126/2, **676**
SMT acutance, 11/1, **677**
Snell's Law, 621/1
Snow White, 26/1
Société Française de Photographie, 303/1
Society of Motion Picture & Television Engineers, 577/2, 624/2, 694/2
Society of Motion Picture Engineers, 275/2, 366/1
Society of Telegraph Engineers, 677/1
Société Pathé Frères, 538/2
Sodium lamp process, **677**
Soft focus, 228/1
Softlight, 423/2, 865/2, 917/2
Solar cell, 635, **677**
Solarization, 344/1
Solid state, 99
  circuit, **677**
  image intensifiers, 156/2
Solution
  distribution and control, 584
  supply reservoir, viscous processing, 952
Solvent
  developer, 118/1, 121/2
  silver, 121/2
Som Berthiot, 129/2
Sound, **677**
  absorption, 6
  acoustics, **3**
  advance, 284, 499/2
  arrangements, telecine, 830
  balancer, **684**, 713/1
  camera and boom operators, 788
  channel, 263/2
  compression, 205/1
  console, **685**
  control, 685/1
    room, 201/1, 708/2, 713, 790/2 ff.
  crews, 694
  cutter, 214/2, 879/2
  detectors, 612
  distortion, **685**
  editor, 254/1
  effects, 505, 692, 698, 714, 789
  energy, 3/1, 4/2
  engineer, 443/2, 527/1
  equipment, 16mm, 481
  films, history, 319 ff., 325, 382/2
  frequency range, *683*
  head, 12/1, 718
  i.f. amplifier, 612
  insulation, 8
  intensity, 3/2, *4*, 9/2
  limiter, 50/2, 626/1
  magnetic, 599 (*see also under* M)
  mixer, 205/1, 445/1, 567/2, 684/1, **686**, **687**, 691/1, 694/1 ff., 951/2
  mixing console, 707/1, 710
  negative, 375/1, 700/2
  on disc, 303/1, 320/1
  on vision, **693**
  persistence of, 628/2
  photographic, 599
  pre-listen, **693**
  quality, 4/1, 693, 706
  reader, 756/1
  recording, 56, 61/1, 114/1, 250/2

Sound, recording,
binaural, 59/1
film, **694**
in the field, 502
news, 505
optical, 923/1
systems, **715**
television, **705**
recordist, 296/2, 791/1 ff.
reduction printer, 496/1
reproduction in the cinema, **718**
rushes, 693/1
speed, 279, **722**
supervisor, 296/2
synchronizing, 560/1
talkback, 201/2
terms, 678
track, 64/2, 209/2, 225/2, 244/2, *255*, 258/2, 259/2, 282, 441/1, *490*, *493*
advance, 624
colour film, 362
development, 360/1, *362*
narrow gauge, 489
position, 16mm, 577/2, *722*
Technicolor prints, 497/1
UV, 922/2
variable-area, exposure, *702*
transfer, 698/1
transmitters, 905
travel, 9
treatment, studio, 800
waves, 3/2, *678*, *680*
Sovcolor, **722**
Sparking, 379
Spatial frequency, 347
analysis, 11/1, 885
Special effects
amplifier, 738/1, **743**
film, 12/1, 48, 54/2, 59/2, 245/2, 301/2, 374, 469/2, 472/1, 518/1, 575/2, 675/2, **723**, 789, 816
laboratory, 728
techniques, cost and operational characteristics, table, 723
television, 63/2, 262/1, 354/2, 355/1, 369/2, 510/2, 532/1, 538/2, 548/1, **733**
Special events, 160/1, **743**
Specialized cameras, 74
for standard and Super-8, 484
Specialized techniques, X-ray, 154
Special receivers, 613
Spectral
bands, 195
characteristics, 88/2, 89/1, 90, 226/2
distribution, 183/1, 184/1, 188/1, 189/2
purity, 641/1
reflectance curves, *183*
response, *81*, 188/1, 368/1, 823
sensitivity, *92*, 343/1, 664
separation, 188/2
transmittance, 183/2
Spectrophotometric curve, 668/1
Spectroscope, 410
Spectrum, 182 ff., 189/1, 227/2, 239/1, 410, **752**, 887
analyzer, 682/1
Specular
and diffuse density, 664
density, **753**
illumination, 572/2
reflection, **753**
Speech, 5/1, 6/1, 33/1, 61/2, 64/2, 682
channel, **753**, 845
dynamic range, 709/1
Speed, **753**
camera, 673/2
control, processing, 584
developing machines, 371/2
emulsion, 50/1, 169/2, 178/2, 179/2, 228/2, 340/2, 344/1, 665/1, 978/1
index, 270/1
lens, 40/2, 783/2
light, 887/2
Panchro, 392/2
printing, 195/1, 373/2
ratings, 265, 268/1
videotape, 932, 934/2
Speed/grain ratio, 348
Spherical
aberration, 391/2, 521, *522*, **753**
mirror, 378/1
Spider box, **753**
Spill light, 56/2
Splice, film, 1/2, 115/1, *487*, *489*, 815/1
Spliceless tape, 926/2
Splicer, 255
frame-line, 488/2
narrow gauge, 498/1
videotape, 925
Splicing, 366/2, 487 ff., 505, **753**
Split
field picture, 84/2, **754**
focus, 503/2, **754**
reel, **754**
screen, 472/1, 735/1 ff., **754**
Spools
camera, 475 ff.
cores and cans, **754**
daylight loading, **215**
Sport, producer, 591
Spot, **755**
brightness meter, 267/2 ff., **755**
exposure meter, *268*, *269*
regulation, 861
shape correction, 854/2
wobble, 61/2, 127/2, 379, **756**, 821, 848
Spotlamp, 59/2, 413, *755*
Spotlight, 422/2, 601/2, 917/2
electronic, 735
profile, 423/1
Spotting, **756**
Spray processing, 12/1, 359/2, 585, **756**
Spread function, 346/2
Sprocket, **756**
Sprocket hole, 58/1, 245/1, 283/1, 299/2, 370/2, 563/2, 571, **756**
modulation, 757/1
Sprocketed tape, 431/1
Sprocketless drive, 865/1
Spur band, **756**
Spurious
frequency, sound, 757/1
resolution, 385
shading, **756**
signals, 78/1, 80/1, 340/1, **756**
suppressing, 60/1
Square wave, **756**
Squeegee, **756**
See also Air knife
Squeeze, 13/2, **756**
Squeezed pictures, 323
Squash, animation, 26/2
Stabamp, **757**
Stability, **757**
monitors, 557
sound, **757**
Stabilization, **757**
image, 249/1, **757**
Staff
administrative, 787
library, 397
sound, 694
technical, 238
Stage, 46/2, **758**, 786
Staircase signal, 858/2
Stampfer, S. von, 561/1, **758**
Stand (animation), 20, *20*
Standard
8mm film, 260/1, 279/2, 284/1, 476 ff., 482, **758**, 866/1
cameras, 482
film processing densities for television, 881
Illuminant D, 840/2, 843/1
light source, 410
luminance response, 429/2
reference film (RCFM), 694/2
16mm print geometry, 577/2
Standardization, 319
Standards, **758**
Standards (TV), 610
converter, 262/1, 500, 552, 752/2, **763**
switching (receiver), 611
telecine, *817*
transmitter, 902
Standby
power, 748
signal, 213/2
Standing waves, 7/1, 679
Stanford, L., 473/2
Starevich, L., 325/2, **771**
Start mark, 699, **772**, 814/1
Static
marks, **772**, 988/2
matte, 733/2
Stationary gate printer, **772**
Station
break, 442 ff.
identity, 568
synchronization, 544
timing, **772**
Steady state characteristic, 205/2, **774**
Steel industry (closed circuit TV), 160
Stefan's Law, 411/1
Stencil methods, 321
Step
counter, 806
printer, 195/2, 573/1, **774**
printing, 494/2, 507/1
wedge, 247/2, 363/1, 659/2 ff., **775**, 875/1
Stephenson, W. S., 127/1
Stereo cineradiography, 155/1
Stereophonic
recording, 61/1, 283/1, 690/2, 703/2, *716*, 717/2
sound processes, 125/2, 126/1, 134, 209/2, 320/2, 322/2, 324/1, 473/1, 537/1, 721, **775**, 980 ff.
Stereophony, 684, 716
Stereoscopic
camera, 13/1, 63/1, 778 ff.
projector, early, 923/2
cinematography, **775**
films, 125/1, 323
pair, 978/1

Stereoscopic
pictures, 355/2
photography, 365/2
process, Lippmann, 427/1
projection, 125, 307/1, 562/2, 603/1, 605/1
space and psychological space, 781, *782*
television, 55/2
viewing, 924/1
window, 782/1
X-ray, 155/1
Stereoscopy, 323/2
Sticking, **782**
Stilb, 425/1, **783**
Still frames (use in news), 504/2
Stock, **783**
material, 398/2, 402/1
shots, 399/1, 473/2
Stop
bath, 123, **783**
frame, 300/1, 335/2, **783**, 934/1
key, **783**
lens, 269/2, **783**
motion, 22, 75, **783**
Stopping down, 365/1
Storage
cold climate, 170
film, **289**, 398, 405, 113/1
magnetic recordings, 293/2, 433, 578/1
negative, 210/1
set, 785
tropical, 913
vault, 924/1
Storage (TV) characteristic, 77/2, **783**, 815/2
principle, 331, 341/1
switching, **783**
Store, converter, *768*
Stored field system, telerecording, 848
Store element, 767/2, **783**
Storyboard, *28*, 30, 33, 48/1, 231, **784**
Stray light (on screens), 652
Streaking, **784**
Streaks (developer), 12/2, 359/1, 916/1
Streamer effect, 663/1
Strength, mechanical, film, 285
Striations, 923/2
Stringer, 500/1
Strip, **784**
light, 413/2
Stripe
projector, 490/1
transfer, 698/1
Stripper, 21
Stroboscope, 316/2, 758/2
Stroboscopic effects (strobing), 35/1, 441/2, **784**
Structure and properties (magnetic tape), 431
Stuart-Blackton, J., 325/1, 369/1, **784**
Stubs, aerial, 200/2
Studio
absorption, sound, 8/2
backgrounds, 224
capacity, *5*
communications, 201
complex
film, **785**
TV, **793**
facilities, elections, 750
floor, 713
manager, **801**
lamp, 63/2, 170/1
lighting, 641/1, *413*, *415*
manager, 236/1, 792
monitors, display, 751
news, 506
plan, 221, *222*, *786 ff.*, *794*, *797 ff.*
power supply, 566
recording practices, 698
reverberation, *5*, 708/2
show copy, 375/2
size, 238
station timing, 773
Style (camerawork), 101
Stylized and abstract design, 224
Subbing, 286
Subcarrier, 85/2, 197/1, 198/1, 274/1, **801**
components, 512/1
drive, **801**, 513/1
frequency, 512, 513/1, 536/1
oscillator, **801**
sideband, **801**
Subject indexes, library, 400, 404
Subjective
loudness, 708/2
noise, 673/2
sound, 561/1
Submerged sprays, 585
Subrefraction, **801**
Sub-standard, 275, 474 ff., **801**
(*see also* Narrow-gauge)
Substantive process, 356, **801**
Subtitle, **802**
Subtratum, 286/1
Subtractive
colour processes, 188, 321/2, **802**
dyes, 174/1
printer, 361
printing, 574/2, **802**
Successive frame process, **802**
Sullivan, P., 25/2, 325/1
Sulphide track, **802**
Sun gun, **802**
Sunshade, 72/1
Super 8mm film, 151/1, 260/1, 276 ff., 324, 477 ff., **802**
cameras, *477*, 483
prints, 497/1, 498/2
Super-Emitron, 77/2, 333/2, **802**
Supergroup, **802**
Superheterodyne, 127/1, **802**
Super iconoscope, 263/1, 802/2
Superimpose, **802**
Superimposed
images, 13/2
titles, 299/1, 488/1, **802**
Superimposition, 30/2, 240/1, 301/1, 469/1, 487/2, 498/2, 506/2, 510/1, 531/2, 568/2, 729/1, 776/2
Super-Panavision, 537/1, 984/2
Superscope, **803**, 983
Superturnstile aerial, 907/2
Suppressed
carrier, 197/1, 511/2, **803**
signal, 620/1
field, telerecording, 847
Surface
defects, film, reduction of, 572/2
reflections, reduction of, *165*
treatment, screens, 653
Surgical operations, 453
Swinging burst, 618
Swish pan, **803**, 978/2
Switch
electronic, 733/1 ff., 738
matrix, 214/2
Switcher configuration and facilities, 445
Switchgear, cinema, 143
Switching
centres, regional, 749
equipment, 444
matrices, 447
schedule, 448/1
techniques, converter, 768
lighting, 420
Switching-voltage generator, 383/2
Symbols, animation, 27/2, 30/2
Symmetrical lenses, 393
Sync
advance, 492/2
comparator, 446/1, 958
generator, camera, 74
leaders, 622 ff., *624*, *625*
levels, 862/1
mark, 166/2, 482/1, **812**
pip, 692/2
pulse separation, 61/2
reference point, 485/1
re-insertion, vision mixer, 956
separation, 478/1 ff.
track recording, 698/1
(*see also* Synchronizing)
Synchronism, 157/1, 772/1
Synchronization, 315, 600/2, **803**
by perforations, 694/2
camera, 482
picture and sound 254, *256*, 284/1, 427/2, 541/1, 563/2, 594/2, 601/1
rushes, *255*
signal, 427/2
telecine, 818
videotape, 936
Synchronized
sound and vision, animation, 31
sound track, 307/2, 485/1, 561/1
Synchronizer, 254, 488/1, 491/1, 504/1, **803**
Synchronizing
pulse
generator, 548/2 ff., **803**, *804*, *810*
regeneration, **811**
selection, 965
separator, 612
pulses, 603/2, 708/1, **811**, 875/1
separator, **811**
signal, 197/2, 510/1, 512/1 ff., 861
See also Sync
Synchronous
detector, **812**
speed, **812**
Synchrostart, **812**
Syncom, 636
System 35, **812**
System blanking, **812**
System modulation transfer acutance, 11/1, 677/1
Systems approach, 331

Table top animation, 22
Tachometer, 74, 480/2, **814**
Tachyscope, 39/1, 42/2
Take, **814**

Take-up, 22/1, 68, 69, **814**
Taking characteristics, 89
  correction of, telecine, 845
Tail leader, **814**
Talbot, W. H. Fox, 124/1
Talkback, *201*, 310/1, 444/1, 506/2, 712, *713*, 745, **814**
Tally light 213/1, **814**
Tangential mode resonances, 6/1
Tank, 13/1, **814**
Tanning, **815**
  development, 175, 351/1
Tape, **815**
  and disc equipment, 711
  cassette, 506/1
  certified, 438/1
  checks, physical, 437
  drive, 164/1
  guide servo system, 937/1
  life, colour video, 948
  manufacture, 433, *436*
  recorder, audio, *717*
  speed, 431, 718
  splice, *255*, 487/1, *754*, **815**
  splicer, video, *925*, *926*
  stretch and head intrusion, 937
  thermoplastic recording, 869
  transport, video, 933, 935/2
Target, 78/1, 79, 80/1, 262/2, 643, **815,** 949/1
  capacitance, **815**
  image retention, 782/1
  layer, **815**, 821/2
  lead oxide, 561/2
  lens, 384
  mesh, 79, 460/2, **815**
  voltage, 80/2, **815**
Tati, J., 230/1
Taylor, Taylor & Hobson, 152/2
Technamation, 39
Technical
  director, TV, 237/1
  facilities, TV studio, 794
  run-through, 237
Technicolor, 125, 174, 175/1, 176/1, 178/2, 321/2, 325/1, 350/2, 353/2, 367/2, 496, 648/2, 668/1, **815**
  cassette projector, 477/2
  three-strip camera, *174*, 175/1, 178/2, 353/1, 367/2, **869**
Technirama, 126/2, 241/2, 323/1, **816,** 983/1, 984/2
Techniscope, 126/2, 323/2, **816,** 983/2
Tecography, 152/1
Telecine, 51, 80/1, 159/2, 202, 260/2, 330/2, 332/1, 334/1, 338/1, 412/1, 504, 543/1, 562/2, 563/2, 757/1, **817,** 917/1, 949/2
  alignment, 840
  characteristics, 882
  colour, **841**
  gamma correction, 504/1
  operator, 202/2, **845**
  projector, 865/1
  standards converter, 764/2
  tube blemishes, 641/1
Telectrascope, 126/2, 326/2
Telefunken, 127/2, 335/1
Telekinema, 324/1
Telemetry, **845**
Telemicroscope, 151/2
Telephone pair, **845**
Telephony channel, 753/1, 802/2, **845**
Telephoto
  lens, 393, **845**
  ratio, 393/2
Teleprinter, 500/2
Telerecorder, 261/1, *846, 847, 849*
Telerecording, 470/1, **846**
  characteristics, 883
  colour, 859/2, 949/2
Telescope and quartz rod endoscope *456*
Television
  bands, table, 865
  channels, 115/2
  dynamic range, 709/1
  eye marker, 460
  formats, table, 391
  illuminant, 843/1
  lenses, *81*
  principles, **860**
  processing for, 377
  projector, 378/1
  screen luminance, 651
  standards, table, 864, 865
  surgical operations, 459
  viewing, 13/1, **951**
  visual principles, 969
  waveform, *862, 863*
Televisor, 55/2, 330/2, **865**
Telejector, **865**
Telstar, 334/1, 635
Temperature
  climatic extremes, **167, 911**
  control, 84, 584
  developing, 121/2, 122/2, 359/1 ff.
  effect on film, 284/2, 288, 290/2 ff.
    effect on receivers, 613/2
  Kelvin scale, 369/2
  library, 398
  tape storage, 433/1
Temporal and spatial factors, vision, 971
Tendency drive, 359/1, 581/2, *582*, **865**
Tenlight, 422/2, **865**
Tension
  take-up, 208/2
  tape, 439/2 ff., 933
Terrestrial connecting networks, 640
Test
  card, 226/1, *614*
  conditions, screen, 646
  exposure, 372/2
  for nitrate, 293
  for vision circuit, 631/1
  negative, 247/2
  procedure, tape, 437
  signal, 59/2, 274/2, 471/2, 538/2, 600/1, 978/2
    colour, 171/2
    transmission, 899
  strip, 373/2
    sound, 702/2
Thalofide cell, 114/1
Thaumatrope, 124/1, *316/1*, 538/1
Theaterama, 987/1
Theatre
  acoustics, 5/1, 6/1, 9/1
  sound, 790
  television, 160
  (*see also* Cinema)
Theatregraph, 539/2
Théatre Optique, 124/2, 317/1, 539/2, 629/2
Thermal noise, 508
Thermopile, 410/2
Thermoplastic
  recording, 227/2, 377/1, 645/1, **866**
  resin, Kalvar, 367/1
Thin lenses, 519
Thirty-five mm film, 275 ff., **869**
Thirty-two mm film, 277/2, **869**
Thomson, J. J., 211/2
Thompson, W., 369/2
Threading, 68, 69
Three-colour
  analysis, 911/1
  picture, 352/2 ff.
  print, 63/1
  printing, 174/2, 367/2
  processes, 188/2, 245/1, **869**
  simultaneous system, 190, *192*
  television, 365/2
Three-dimensional
  animation, 37
  effect, titles, 252/1
  films, 776/2
  (*see also* Stereoscopic)
Three-figure specification, colour, 184
Three-gun tube, 193/1, 194/1
  See also Shadow mask tube
Three-sixty degree projection, 307/1, 322/1, 370/1
Three-strip camera, *174*, 175/1, 178/2, 353/1, 367/2, **869**
Three-track printing master, 704/2
Three-tube
  cameras, 93
  telecines, colour analysis, 834
Threshold, **870**
Through-the-lens, *479*, 915/2
  See also TTL
Throw, 606/1, **870**
Thun, R., 332/1
Thyratron, **870**
Thyristor, **870**
Tilt, 300/2, **870**
Timbre, 681/2
Time
  base, 610, 615, 618, **870**, 940
  compression, 66/1, 75/1
  delay, sound, 5, 6/1
  entry, 450/1
  exposure, 269
  difference, hearing, 684
  magnification formula 870/2
  markers, time-lapse, 872
  range, cameras, 66
  scale exposure, **875**
Time-adjacent fields, **870**
Time/gamma series, **870**
Time-lapse cinematography, **870,** *871 ff.*
Time of fall, time of rise, **875**
Timer, laboratory, 495/1
Timers, time-lapse, 873
Timing, **875**
  animation, 32
  crystal control, 212/2
  high-speed cinematography, 315/2
  laboratory, 362, 372/2, 375
  station, 772
Tinted base, **875**
Tip and guide positions, colour videotape, 944
Title, 12/2, 23/1, 211/1, 223/1, 299/1 **875**
Titra titling, **875**
T-number, 269/2, 305/1, 753/1

Tobis-Klangfilm, 320/2
Todd-AO., 126/2, 279/2, 296/1, 323/2, 719/1, **875**, 980/2, 984/2
Todd, M., 135/1
Toe, 116/1, 265 ff., 343/2 ff., **875**
Tolerance, emulsion, 57/1
Tone
  control, 877
  quality, 681
  range, 116/1, 266/1, 345/2
  rendering, 343/1
  reproduction, 265/1, *267*, 344, *345*, **875**
  source, **878**
  steps, 349/1
Toned track, **875**
Toning, 321, **878**
Tools, animated captions, 15/1
Top boost, **878**
Top hat, **878**
Toroidal coil, **879**
Toshiba, 93/1
Total internal reflection, *409*
Tracer, 35
Track geometry, videotape, 934
Tracking, 22/1, *36*, 540/1, **879**
Track laying, 491, **879**
Track, sound, **879**
Traffic circulation, studio, 800
Traffic control, closed-circuit TV, 162
Training, art director, 43
Training simulators, 164
Transcoder, 771/1
Transducer, **879**
  errors, colour videotape, 947
Transfer
  characteristic, **879**, 881 ff.
    effect on sound, *685*, *686*
    monitors, 556
    photographic, 345/1
    telecine, *819*, 844, *883*
    telerecording, 856/1
  curve, 60/1
  function, 346/2, 349, *885*
  sound, **886**
  videotape, 927
Transients, **886**
  disturbances, removal, of, 757/1
  grey scale, 99/1
  sound, 682/2
  response, telecine, 820/2
Transistorized recorder, 696/2
Transistor receivers, 643/2
Transition
  scene, 988/2
  signal, 214/2
Translator, 300/2, 906/2, **886**
Translucent screen, 54/2, 653, **887**
Transmission
  characteristics, 628/1
  film, 196/1, 217/2, 343/2, 504/1
  light, 524
  number (lens), 269/2
Transmission (TV)
  and modulation, **887**
  networks, **893**
  point-to-point, 466/1
  standards, 900
  types and frequencies, 900
Transmittance, lens, 915/2
Transmitted-light viewing, 604/1
Transmitter
  microwave, 466/2
  powers, 891
Transmitters and aerials, **901**
Transmitters and receivers (for satellites), 639
Transmitting station plan, *903*
Transparency, 188/2, 189/1, 289/2, **909**
  in animation, 36/2
Transponder, **909**
Transport
  film,
    processing, 580
    projector, 594
    telecine, 818
  pins, 21/1
Transverse scan, **909**, 928/2, *934 ff.*
Trap circuit, **909**
Trap valve amplifier, **909**
Travel ghost, **910**
Travelling matte process, 48/2, 49/2, 61/1, 179/1, 354/2, 374/2, 573/2, 677/2, 729/2, 730, **910**
Travelling wave
  aerial, 908
  amplifier, **910**
  tube, *910*
Treatment, 28
Triacetate base, **910**
Triad, 180/1, **910**
Trial print, **910**
Triangle, **910**
Triangular noise, 509
Trichromatic
  analysis, 188, **911**
  camera, 91/2, **911**
  system, 184 ff.
Trick photography, 48 ff., 61/1, 305/1, 441/2
Tri-Ergon Company, 125/1, 320
Trigger pulse, *809*, **911**
Trim, **911**
Trinoscope, 859/2, **911**
Tripack (*see also* Integral tripack), 322, **911**
Triplets, *392*
Tripod, 168/2, 675/2, 910/2, **911**
  head, 307/2, 427/2, 507/2
Triptych, 322/2, 979/2, 987
Troland, L. T., 175/1
Tropical
  cinematography, **911**
  processing, 914
  storage, 290/2, 291/2
Tropospheric propagation, **915**
Trucking, 298/1
Trucolor, 126/1
T-stop, 269, **915**
TTL, **915**
Tube
  camera, 76
    cover plate, 391
  picture, 612/2
    monitor, 553 ff.
    projection, 379 ff.
Tuner, 610, **916**
Tungsten
  iodide arc, 874/1
  lamp, 183/1, 210/2, 413/2, 874, **916**
    halogen, 150/1, 315/1, 366/1, 414, 423/2, 802/1, **916**
  light, speed, 265/2 ff.
Tuning fork, 681/2
Turbulation, 12/2, 359/1, **916**
Turner, E. R., 125/1
Turnstile, aerial **916**
Turntable, 506/1
Turret (lens), *71*, **916**
Turret tuner, **917**
Turtle, **917**
Twentieth Century-Fox, 134, 173/1, 299/2, 323/1, 982/2 ff.
Twenty-eight mm film, 324/2
Twin-interlaced scanning, *860*, 861, **917**
Twin lens scanner, 818/1, 826, *827*, **917**
Twister, 423/2, **917**
Two-colour processes, 128/2, 172, 189, 321/2, **917**
Two-colour subtractive system, 174, 367/1
Two dimensions of screen (adaptation of eye), 974
Two-field picture, **917**
Two-three-two-three scanning, **917**
Two-tube cameras, 92
Type A film, 478/2 ff., 483/2
Type S film, 477/2 ff.

Uchatius, Franz, **918**
UHF
  bands, 56/2, 902
  tuner, 611
  vision transmitter, 905
Ultra-close-up attachment, 129
Ultra-high-speed systems, 315/2
Ultra-Panavision, 537/1, 983/1
Ultra Semi-Scope, **918**
Ultrasonic
  cleaning, 495/2, 503/2, **918**
  delay lines, 198/1, **918**
Ultra-violet
  light
    in colour-screen manufacture, 671/1
    Kalvar process, 367/2 ff.
    matte processes, 731/2
  radiation, 182/2, 260/1, 363/2, 410/2, **918**
Under-exposure, 265/1, **918**, *919*
Under-modulation, 951/2
Undershoot, **919**
Underwater cinematography, **919**, *920*
Uniform chromaticity, 89/2
Unique hues, 181/1
Uniselector, 447, **921**,
  vision mixer, 960
Unit, **921**
Universal decimal classification, 400/2, 404/2
Ungraded colour print, 495/2
Unmod. track, **921**
Unrestored receiver, **922**
Unsharp image, 408/1
Unsqueezed print, 641/2, **922**, 984/2
Unwanted
  reflections, 561/2
  signals, 508/1, 756/2
UPA, 26/1, 27/1, 325/2
Upwards conversion, **922**
Urban, C., 676/2
USA Standards Institute, 50/1, **922**
US Continental Army Command, 163/2
USSR, screen luminance, table, 650
UV
  emission, lamps, 366/1
  printing, **922**
U, V triangle, 90/2

Vacuum
cell, photo-emission, 664/2
equipment, thermoplastic, 868
tape guide, VTR, 936/1
tube, 263/1
Valley fogging, 719/1
Van Allen belts, 633/2, **923**
Varamorph, **923**, 981/2
Variable
angle wedge, 758/1
area recording, 58/2, 211/2, 245/2, 320/2, 425/1, 695/1 ff., 717/1, **923**
density recording, 320/1, 425/1, 695/1 ff., 717/1, **923**
focal length lenses, 82, 990/1
groove spacing, 716
magnification, 990/1
shutters, 21/1, 575
Variables of perceived colour, 181
Varifocal lenses, 394/2
Varley, F. H., 301/1, 318/1, **923**
Vaultkeeper, **923**
Vaults, 290/2 ff., **924**
Vectograph, 777/2, 782/2, **924**
Velocilator, **924**
Velocity
control servo, **924**
errors, colour VTR, 945
head-to-tape, VTR, 934
light, 406/2, 409/1
sound, 678
Venetian-blind photomultiplier, *826*
Ventilation
cinema, 142
studio, 9/2, 796
Vertical
and horizontal correction, VTR, 939
aperture correction, 767/1 ff., **924**
bounce, 924
illuminators, microscope, *148*
interval test signal (VITS), 274/2
scanning, 56/1
shading, 604/1
Vestigial sideband, **924**
VHF
bands, 56/2, 902
channels, 159/1
links, 466
transmitter, 903
tuner, 611
wired TV, 613/2, 892
Vibrating body, action on air, *678*
Vibration, camera correction of, 249/2, 757/2
Video
amplifier, 612
drive voltage, 194/7
recording, thermoplastic, 869/1
repeater, block schematic, *896*
signal amplitudes, monitors, 558
switching, 451/2
telephone system, 365/2
transmission, 896/1
(*see also* Vision)
Videotape
editing, **926**
operator, **927**
recorder, 11/2, 159/2, 310/1, 601/2, 909/1
history, 929
layout, 928
recording, 156, 431, **927**
colour, **941**
communications, 202
genlocking, 546/1
news, 504
storage, 293/2
tests, 438
transfer characteristics, 883
winder, *988*
Vidicon, *80*, 92, 95, 158/2, 159/2, 261/2, 334/1, 424/1, 502/2, 504/2, 506/1, 530/1, 603/1, 676/2, 700/1, 741/1, 812/2, *822*, *835 ff.*, *842*, **949**
channel, 423/1, **949**
characteristics, 881
converter, 765
lens, 391/2 ff.
resolution, 389/2
telecine, 600/1, 819, 821, 882
Vidtronics, 859/2, **949**
Viewer, 255, **950**
Viewfinder, *72*, *74*, 84, 111, 168/2, 307/1, **950**
electronic, 111/1
monitor, 74
parallax, 537/2
parallax-free, 602/1
sixteen mm, 479
Viewfinding
devices, 72
rackover, 72/2
reflex, 72/2, 73/1, 145/2
Viewing
animation, 2/2, 22/1
angle, 215
television, **951**
Vignetting, 86/1, 391/1, 628/2
VI meter. **951**
(*see also* PPM)
Virtual
focus, 297/2, 520/2
image, 337, 408/2
Viscous processing, 586, **952**, *953*
Visibility, underwater, 920
Vision
control engineer, 527/1
detector, 611
distribution amplifier, **955**, *964*
drive signal, **955**
frequency range, 862
i.f. amplifier, 611
mixer, 238/2, 446/2, 741, **956**, *958 ff.*, 967/1
control panel, 733 ff.
operator, 443/2, 445/1
modulator, 904
switcher, 239/1, **967**
switching matrix, **967**
transmitter, block diagram, *903*
Vistavision, 126/2, 241/2, 323/1, 815/2, 816/1, **967**, 980/2, 983/1
Visual
acuity, 972
adaptation, 181
aids, rear projection, 607
principles, **967**
recognition, 27
symbols, 14/1, 27
Visuals and storyboard, 231
Vitagraph, 369/1
Vitaphone, 125/2, 320/1
Vitarama, 135/2, **976**, 985
Vitascope, 43/1, 323/2
VITS, 274/2
Vittum, 177/2
Vivaphone, 125/1, 310/2, 319/2
Voltage, lighting, 415
Voltage doubler rectifier, **976**
Volume
compression, **976**
indicator, 951/2
level, 689
reduction, 205/1
sound, 209/2
von Ardenne, M., 332/1
spot scanner, *332*
von Helmholtz, H. L. F., 310 2
von Mihaly, D., 329/2
von Stampfer, S., 316/2

Wall Anode, **977**
Wall, E. J., 172/1, 175/1
Waller, Fred, 322/2, **977**
Walton, G. W., 127
Wardrobe, 785
Warner Bros., 320/1
Warnerke, L., 175/1, 350/1
Warning
signal, 213/2
systems, 160
Warming-up time, 426/2
Washing, 122 ff., **977**
Wash-off relief, 350/2
Water spots, **977**
Waveform
components, 861
measurements, *894 ff.*
monitor, 426/1, 451/1, *558 ff.*, **977**
oscilloscopes, **977**
response, 894/1 ff.
scanning, 643
sound, 681 ff., *682*
Waveguide, 909, **977**
Wavelength, 260/1
colour, 182 ff.
light, 239/2, 406/2, 410/1
sound, 678
Wave motion
light, *406*
sound, 677, *678*
Waxing, **977**
Wear, videotape, 439/1
Weave
camera, **978**
tape, 436
Weber's law, 971/2
Webo charger, 476/1
Wehnelt, A. R. B., 127/1, 329/1, **978**
Weighted measurement, noise, 509
Weiller, L., 126/2, 328/1, **978**
Wente, E. C., 125/1, 320/1
Wertheimer, 968/2
Western Electric Company, 217/1, 320, 365/2
Westinghouse Company, 127/1, 320/2, 331/2, 990/2
Weston speed, **978**
West Orange Laboratory, 382/1
Westrex sound cameras, 701/1
Wet collodion plate, 317/1
Wet printing, *572 ff.*, **978**
Wetting agents, 117/2, 123/2, **978**
Wheatstone, Sir C., 561/1, **978**
Wheel stop, *148*
Whip pan, **978**

White
  bar, **978**
  clipping, 978/2
  crushing, **978**
  level, **978**
  light, 185, 190, 192/2, 263/2, 409/2
    printing, 361
  noise, 509
  point, 186/2, 199/2
Whiter-than-white, **978**
Whitley, J. R., 396/1
Wide-angle lens, *393*, **978**
Wide band
  interpolation, 770
  video amplifier, 629/2
Widescope, **978**, 979/2
Wide screen
  aspect ratio, 50/1, 204/1, 211/2
  film dimensions, 276 ff.
  film for television, 641/2, 644/1, 831/1
  filming speed, 279/2
  colour, 815
  history, 322 ff.
  processes, 133 ff., 156/1, 228/1, 306/1, 430/2, 669/1, 675/1, 875/2, 918/1, 967/2, 976/1, 978/2, **979**
  sound, 320/2, 703/2 ff., 775/2
Wild
  motor, **988**
  track, **988**
Williams, P. W., 332/2, **988**
Wilson, G. R., 125/1
Winding, **988**
  film, 210/1
  tape, *439 ff.*
Windows and sound locks, studio, 801
Wipe, 210/1, 365/2, 445/2, 739 ff., **988**
Wired systems, *892*
Work-book, animation, 30/1
Workprint, 214/2, 254/1, *259*, 374/2, 499/1, 676/2, **989**
Wow, 722/2, **989**
Wow and flutter, projector, 598
Wratten filter, 266/1, 270/1
Writing and reading techniques, converter, 769
Writing beam, 42/1, **989**

Xenon arc, 42/2, 596/1, 874/1
  flash tube, 662/2
  lamp, **989**
X-rays
  danger of from tube, 671/2
  discovery, 211/2
  effect on grain, 348/1
  history, 152/1
  image intensifier and optics, *153*
  processor, 154/1
  projection tube, 379
  TV monitor image, filming, *155*
  wavelength, 260/1, 410/2
X, Y, Z colour values, 184/2

Yagi aerial, **989**
Yellow, 178, 189/1, 204/1, 355/1 ff.
  dye, 183/2, 190/2, 245/2, 246/2
  -forming developer, 177/2

Zebra tube, **990**
Zero signal, 59/2
Zip pan, 978/2, **990**
Zoetrope, 39/1, 253/2, *316*, 317/1, 325/1, 338/2, 629/1, 658/2, 978/2
Zoom, **990**
  devices, animation, 20
  lens, 11/2, 71 ff., 82, 95/2, 165/2, 168/1, 304/1, 319/2, 391 ff., *394*, 452/2 ff., 482/2 ff., 486/1, 497/2, 598/1, **990**
  stand, **990**
Zoopraxiscope, 473/2
Zworykin, V. K., 77/1, 127/2, 272/2, 331 ff., 341/1, 630/2, **990**
  iconoscope, *331*